American Society of Plant Biologists Centennial Compendium

In celebration of 100 years of ASPB

Published in 2025 by the American Society of Plant Biologists
15501 Monona Drive
Rockville, MD 20855-2768 USA
aspb.org

Plant Physiology
ISSN 0032-0889
EISSN 1532-2548
https://academic.oup.com/plphys

The Plant Cell
ISSN 1040-4651
EISSN 1532-298X
https://academic.oup.com/plcell

Table of Contents

Part 1: History of the American Society of Plant Biologists, 1924-2024

Part 2: Centennial Reviews from *Plant Physiology*

Table of Contents (continued)

Part 3: Centennial Reviews from *The Plant Cell*

Table of Contents (continued)

Part 1

History of the American Society of Plant Biologists 1924-2024

CONTRIBUTORS

Ivan Baxter ibaxter@danforthcenter.org

Judy Callis jcallis@ucdavis.edu

Kent Chapman Kent.Chapman@unt.edu

Maarten Chrispeels mchrispeels@ucsd.edu

Machi Dilworth mfukudilworth@gmail.com

Eva Farré farre@msu.edu

Elisabeth Gantt egantt@umd.edu

Alan M. Jones alanjones@bio.unc.edu

Brian Larkins Larkins@arizona.edu

Crispin Taylor ctaylor@aspb.org

PREFACE

Brian Larkins

As the American Society of Plant Biologists (ASPB) celebrates its Centennial in 2024, we thought it appropriate to describe important events that occurred since Jack Hanson's History of the American Society of Plant Physiologists was published in 1989. This period saw many changes in the Society including its membership, governance, publications, and, notably, its name: in 2000, Society members voted to re-name ASPP the American Society of Plant Biologists (ASPB). While our report is not as thoroughly documented as Hanson's, we have endeavored to describe notable changes that occurred in the Society's leadership, publications, and the scope of activities. We also explain how plant science research and its funding during this period were responsible for many of the changes that have shaped the current operation and vision of the Society.

Hanson's overview provides insight into decisions made by the founders and early leaders of the Society, how the Society was run, and how its founding journal, Plant Physiology, came into existence. But he decided not to include details describing how the science of plant physiology evolved over the first 75 years. He decided, "Only in a few cases did developments in plant physiology seem to impinge on the progress or functioning of the Society (e.g., advances in the science have had sweeping effects on the content of the journal [Plant Physiology], but little to do with the problems of publishing it, other than decisions on editorial appointments." This clearly has not been the case in the last 25 years. Indeed, the application of novel techniques in genetics, cell and molecular biology during this period not only revolutionized our understanding of plant physiology, but also impacted membership of the Society, particularly the number and involvement of women, and the number and diversity of its publications. Two new journals were created, and several books and textbooks were published. While originally largely an American society, ASPB has become even more international and impactful. It has also increased its role in providing leadership for funding plant research in the US and teaching plant science at the primary and secondary educational levels.

It was not possible for us to comprehensively describe the Society's operation during the past 25 years, and we apologize if we have overlooked people and events others believe should have been included. While the opinions and assessments described herein are our own, they have been vetted among ourselves and our colleagues.

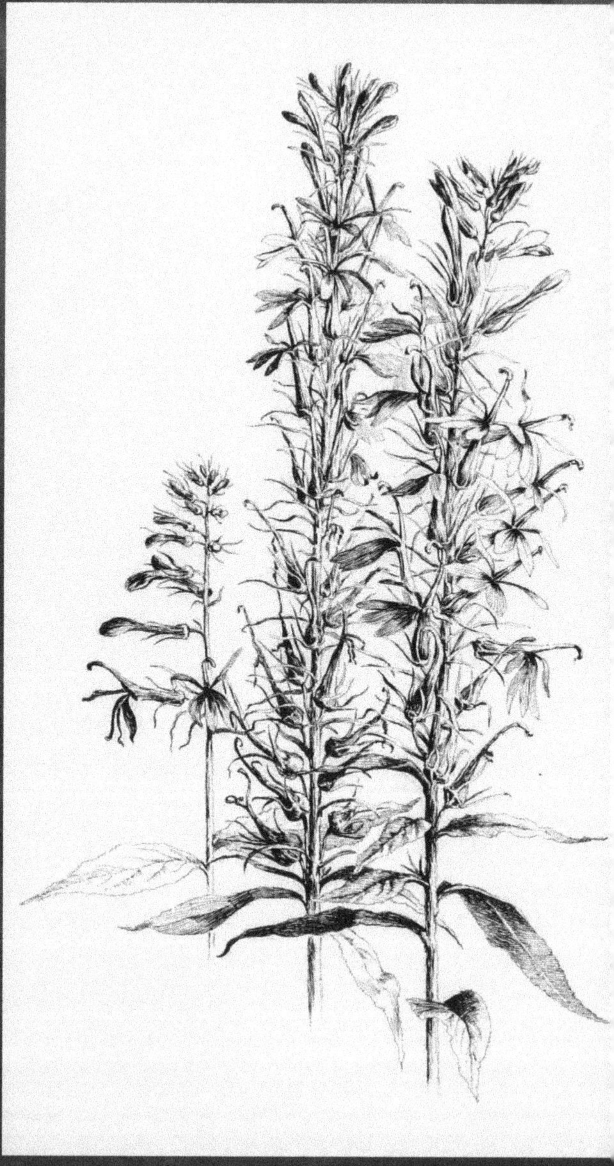

PART 1: ORIGIN OF THE AMERICAN SOCIETY OF PLANT PHYSIOLOGISTS

Elisabeth Gantt

In 1989, the American Society of Plant Physiologists (ASPP) published an account by Jack B. Hanson of the founding and history of the Society, entitled *History of the American Society of Plant Physiologists* (https://aspb.org/about/history). Hanson's synopsis of ASPP, which later became the American Society of Plant Biologists (ASPB), attempted to capture the essence of the early history of the Society; details of ASPP's first 75 years are thoroughly described in his book. Many dedicated plant biologists participated in the early efforts to form ASPP, including Dr. Charles A. Shull, initially at the University of Kentucky and later at the University of Chicago, who was its first President. Early supporters for the creation of a plant physiological society that would separate from the Botanical Society of America (BSA) included Drs. R.S. Harvey, University of Minnesota, B.E. Livingston, Johns Hopkins University, and many others, such as J.P. Rose and L.O. Regeimbal. The decision to separate from the BSA was not unanimous among plant physiologists, and it was plagued by a multitude of ideas and directions. Young plant physiologists were largely adherents of one or two extant scientific societies, particularly the BSA and/or the American Association for the Advancement of Science (AAAS).

Early moves toward creating a new society involved the conceptualization and writing of a Constitution and Bylaws. Despite differences of opinions, essential agreements were reached at least by the majority of members who favored independence from the BSA, and in 1925 the Society was recognized by AAAS. The first ASPP officers were elected in 1924. At that time ASPP had 104 members and was exploring the possibility of publishing its own journal. That required not only the enthusiasm to collect and evaluate manuscripts, but also the fiscal means to cover publication costs. The fiscal situation was eased by a generous financial contribution of $20,000 from Mrs. Charles A. Shull. Her benevolence gave ASPP time to organize a membership dues structure, decide the charges for a journal subscription, and attract library subscriptions for it. Publication of the journal, *Plant Physiology*, began in 1926, with Dr. Charles A. Shull as its first Editor-in-Chief.

The official founding of the Society was in 1924, with the following member-elected officers: Dr. C.A. Shull, President; R.P. Hibbard, Vice-President; and R.B. Harvey and W.A. Gardner, Secretary-Treasurers. At the time, the Constitution was still a work in progress. Revisions and improvements to the Constitution lasted several years, and it wasn't until 1929 that the official Constitution and Bylaws were adopted.

With determination and great dedication, ASPP survived and grew even through difficult times.

Basically, they provided for an annual meeting and election of officers for the Society and their terms of service (President and Vice-President for one year and Secretary-Treasurer for two years). It took several more years, until 1932, before the Society was officially incorporated in the District of Columbia. This status provided for such things as the legal retention of its name, its establishment as a non-profit organization (and associated tax-exempt status), and its ability to accumulate the costs for publications and provide awards. Further improvements and clarifications to the Constitution continued to be made over the ensuing years, with a major one in 1987. That revision reflected major and minor changes that the Society had been following, including the creation of an Executive Secretary-Treasurer. Prior to this time, these duties had been effectively performed at the home of William H. Klein, the Director of the Smithsonian Radiation Biology Laboratory, with outstanding help from his wife, Winifred Klein. Through the efforts of Patricia Cherry, a gift of land and building by the Gude Family, a member of whom served as the US Congressional Representative for Montgomery County, MD, provided the Society a permanent address and home in Rockville, MD, where the business of a large and effective organization could occur. Prior to receiving this valuable gift from the Gude Family, ASPP rented space at the campus of the Federation of American Societies for Experimental Biology for a short period.

The early annual meetings of ASPP were held either with AAAS or BSA. One hundred and four members attended the first meeting in 1924 (Table 2 in Hanson's book), but by 1930 attendance had increased four-fold to 424. For many years thereafter ASPP met jointly with other science-based societies, especially with the American Institute of Biological Sciences (AIBS). The last joint meeting with AIBS was held in New Orleans (1976). In later years, the Society met with similar biological societies, with input and/or coordination of AIBS. On a relatively regular four-year basis there was a joint meeting with the Canadian Society of Plant Physiology. ASPP meetings were regularly held on college campuses, which generally made them less expensive. But this also limited their size and access to certain conveniences. Consequently, the meetings were eventually moved to commercial establishments.

With determination and great dedication, ASPP survived and grew even through difficult times. One such period was during the economic depression of the mid 1930s and World War II, which ended in 1945. Subsequent improvement of the economy and renewed interest in plant physiological research led to an increase in membership from 424 in 1930 to 1,923 in 1963 (see Table 7 and Table 12 in Hanson's book).

Consistent with practices at the time, in its early years all the Society's officers were male. In 1973-1975, Mary Stiller was elected Secretary (Table 16, in Hanson's book). Afterward, there were no women in leadership until 1985, when Elisabeth Gantt was elected Secretary. Eventually, change came about partially because more women became plant physiologists. There were discussions about providing care for children during the annual meetings, especially if both parents were in attendance. There were also discussions for how women scientists could play productive roles in the Society, and these discussions were of interest to both men and women. They led to designating women as chairs of oral presentations and becoming elected officials of the Society.

The increased participation in plant physiology research following World War II is reflected in the growth of publications in *Plant Physiology*: there was an increase in yearly pages from 636 in 1930 to 779 in 1963 (Tables 7 and 12 in J.B. Hanson's book). Also, there were more subscribers due to increased membership in the Society. Especially noteworthy was an increase in meetings on special topics and the inclusion of international presentations. Beginning in the mid-1950s, federal agencies made more funds available for competitive research grants, and this trend continued, particularly in the 1970s and 1980s, with increases in national and international collaborations and a concomitant national and international increase in publications.

Over time, the Society became more internationally inclusive in practice, if not yet in name. *Plant Physiology* contained publications from an international clientele and included a wider geographical representation of individuals as Associate Editors. ASPP grew into an important, internationally effective organization that made many significant contributions to the field of plant science and training of people to improve the world's food supply and benefit its environment.

From its beginning, the Society honored outstanding members for their scientific contributions and extraordinary service with special awards. The first of these was the Charles R. Barnes Honorary Life Membership Award, which was first given to Charles A. Shull in 1925. Other awards were the Stephen Hales Prize (1927) to a senior scientist (over age 60) for career-long service to plant biology, the Kettering Award (1962) for excellence in photosynthesis research, the Charles A. Shull Award (1971) to a scientist under age 40 for novel investigations in plant biology, the Adoph E. Gude award (1983) for outstanding service to plant biology, the Dennis R. Hoagland Award (1985) for research in support of agriculture, and the Excellence in Education Award (1988) for recognition of distinguished plant science teaching.

ASPP/ASPB benefited tremendously from dedicated members who served the Society on a volunteer basis. Many outstanding examples are cited in J.B. Hanson's book and are demonstrated by the recent endowment of several new awards that recognize young plant physiologists, educators, mentors, and those working to diversify fields. Thus, the positive, selfless attitude of ASPP/ASPB members has prevailed to the present time.

PART 2: LEADERSHIP/STAFF/GOVERNANCE

A Work in Progress

Crispin Taylor, Alan Jones, and Kent Chapman

Until the mid-2010s, ASPB's governance structure had remained largely unchanged for most of its history, with a single Executive Committee (ExCom) responsible for steering the organization. Subordinate committees handled specific tasks, such as finance and audit (Board of Trustees), membership, and so on. As time went by, the Society established new committees, including, for example, Women in Plant Biology (originally the Committee on the Status of Women in Plant Physiology), Minority Affairs (now Equity, Diversity, and Inclusion [EDIC]), and Education, committees that brought vital new perspectives to the table. Governance also added new Sections, both special interest (e.g., Primarily Undergraduate Institution, Environmental and Ecological Plant Physiology), and regional (e.g., México).

Indeed, developing appropriate affinity groups, however defined, was an important way to engage a diversity of people around issues of importance to them.

Governance

From both governance and budgetary perspectives, the problem with this historical model was that ASPB's constitution established that for each new committee or section, additional representation was added to the ExCom. As a result, the ExCom grew to comprise more than 25 people, which made decision-making unwieldy and tended toward creating meetings that were consumed by reporting, rather than long-term strategic planning and short-term action-taking. In effect, there was a lack of accountability and strategic direction. As one former participant in many ExCom meetings quipped, "we can spend two hours discussing an expenditure of $2,000 but will pass a $6 million dollar budget in two minutes." In addition, the costs of convening this group in person twice a year – once during the annual Plant Biology meeting and again at the headquarters building in Rockville, MD each spring – were ballooning.

This is not to say expansion of ASPB's sections was a bad idea; it wasn't. Indeed, developing appropriate affinity groups, however defined, was an important way to engage a diversity of people around issues of importance to them. It was also an effective way to create pathways towards organizational leadership. However, both fiscal and governance structures under-pinning growth in ASPB's section portfolio were important to define.

In terms of governance and moving the organization forward, as early as the mid-90s an unofficial workaround was in place: the Society's Executive Director(s) – the key staff position that linked the governance decision-making body to the staff that execute its directives and policies – was meeting regularly with an "operations subcommittee" comprising the three Presidents (past, current, and succeeding) and, sometimes, the Treasurer. By the time Crispin Taylor joined the Society as Executive Director in 2004, this group was the de facto decision-making body, at least for many decisions that needed to be made in between the twice yearly in-person meetings of the ExCom.

After Alan Jones became President-elect in 2012, discussion began in earnest about reimagining ASPB's governance structure. The two primary goals were to retain – indeed, to reinforce the value of –a broadly conceived, inclusive governance committee to inform decision making and help set strategic direction, and to create a more nimble and flexible decision-making body.

Over the next few years, these objectives crystallized into the current administrative structure, in which an expanded and even more inclusive Council replaced the ExCom. Per the reworded bylaws, which were approved in 2015, "The Council is the Society's deliberating body; it provides leadership and guidance in setting the overall direction of the Society via the strategic plan. The council evaluates novel program, product, and service concepts in the context of the strategic plan and through its comprehensive sense of the Society's directions and priorities, makes recommendations to the board of directors accordingly." At the same time, a Board of Directors (BoD) was established. The BoD is a subset of the Council, and it is comprised of the elected officers (the current and succeeding President and the Secretary); the Treasurer; two members elected at large by the membership, including one from outside the US; and an individual – a committee chair or section representative – elected from the Council. The bylaw describing BoD currently reads, "The Board of Directors is the Society's decision-making body; it has authority to act on all matters of the Society as a whole. The Board of Directors is charged with maintaining the activities of the Society in furthering the science of plant biology. Decisions of the BoD require a quorum of five voting members and a minimum of four votes cast in favor of the decision, unless otherwise directed herein." In addition to these structural and functional changes, the rewritten bylaws specify that the president-elect chairs the BoD and the past president chairs the Council.

Following implementation of the new system of governance, there have been other changes in the management of the Society's business. For example, in addition to annual in-person meetings, the BoD now has frequent remote meetings through conference calls, which allows rapid decision making. While the Council continues to meet annually in person, just as the ExCom had done, several years ago the COVID pandemic (2020-2022) accelerated the use of virtual (Zoom) meetings, allowing increased frequency of virtual meetings for both the Council and the BoD.

Another consequence of the change in ASPB governance that accompanied creation of a BoD was a repurposing of sorts for the Board of Trustees (BoT), allowing it to function more explicitly as a finance and audit committee for the BoD. The BoT retains some of its historical functions, including financial oversight of the Society's funds and investments. Its chair, the Society Treasurer, is a member of the BoD, an arrangement that provides the BoD knowledge of investments, audits, the budget process and general financial oversight. Cooperation between the BoD and the BoT has increased in recent years to include a joint meeting that evaluates and approves annual budgets for operations. Going forward, what is needed is a budgeting process informed by strategic plans so that financial activities align with member-focused priorities. Although this governance structure has been in place for several years, arguably it has yet to fulfill its envisioned potential. This is largely because the strategic planning process initiated in 2022 has been bogged down as programmatic priorities competed with fiscal realities.

Strategic Planning

Strategic planning is a work in progress; as of mid-2024 the Society's core ideology – comprising its purpose, values, and mission – is coming together, as are its major thematic objectives. A draft of the core purpose reads "to advance the community of plant biologists", and includes the values of inclusivity, respect, transparency, integrity, and sustainability, among others. The draft of ASPB's mission states that the Society's role is to "nurture, empower, and inform the community of plant biologists and advance plant science research, education, careers, leadership, advocacy, and public engagement." Major strategic goal areas are also coming into focus and currently include the following:

1. **ASPB:** ASPB will fulfill its potential as a leader in the plant science community by securing its organizational health and prosperity and by centering the well-being and growth of a diverse, inclusive, and thriving membership.
2. **Scientific Community:** ASPB will take on a leadership role, with its allies around the world, to promote and adopt technologies, programs, and tools for the global plant science community to advance knowledge in plant biology, to address societal challenges and to further equity and inclusion in science.
3. **Society and Public:** The global society will trust, recognize, and celebrate the impacts of plants and plant science on personal livelihoods, environmental sustainability, and global economies and be inspired to dismantle systemic patterns that propagate inequity.
4. **Educators and Students:** ASPB will support a diverse community of lifelong learners expanding their awareness and knowledge of plant science.

What remains to be done is for the Council to solicit input and consensus from stakeholder groups beyond the Council, including the staff, for clarity and direction; and for the BoD to establish specific objectives, timelines, and budgets for each of these major goals. It is important that each goal is discrete and encompasses the vital objective of ensuring that fiscal well-being is properly addressed.

Finances

Since shortly after the Society's inception in 1924, enabled by a donation from Charles A. Shull and his wife to publish *Plant Physiology*, its single largest revenue stream has come from its scientific publications. Indeed, financial surpluses from journal operations – along with income derived from the Society's investment portfolio – have powered all the other Society functions, from education and outreach to the launch of Plantae and science policy activities. Although this approach has proven successful over the Society's first century, governance, informed by staff analyses and projections, determined that this business model is no longer sustainable. As described in the "Other Publications" section of Part 4, the pending transition to Open Access publication, coupled with smaller annual publication-derived surpluses the Society receives since partnering with Oxford University Press to produce and disseminate *Plant Physiology* and *The Plant Cell*, require that we increase existing revenue streams and develop new ones to ensure the fiscal stability of the Society well into the future. This will involve building on the success of the Centennial Challenge fundraising campaign by nurturing and expanding our already substantial pool of donors, fundraising from foundations, and an expanded portfolio of federal grants.

Regarding ASPB's financial situation, it is important to note that since recruiting CFO Clara Woodall in 2017 our accounting procedures and the depth of our understanding of our fiscal situation – both on an annual basis and over the longer term – have improved enormously. As a result, ASPB has enjoyed a succession of "clean" audits that provide reassurance to governance that staff are providing an accurate picture of our finances and that gifts to ASPB from donors are well managed.

Speaking of donors, in recent years, the BoD chartered the ad hoc Legacy Society Leadership Committee (LSLC) and the Centennial Challenge Committee (CCC). These groups, respectively, developed the concept of the Legacy Society, which was launched with a recruitment drive for "founding members," each of whom donated at least $5,000 to the Society; and the Centennial Challenge – a goal of raising at least $3 million by the time of the Society's Centennial conference in Honolulu.

Both efforts have proven successful. As we reach the Centennial, these fundraising groups are at an inflection point, and a revised charter combining the liaison role of the LSLC and the fundraising efforts of the CCC will be developed in due course. Additional avenues for support, including additional federal grants as well as funding from private philanthropies, will be pursued by staff and volunteers in and outside of governance.

Another current tangible asset, as described in John Hansen's "History of ASPP," is the Gude Plant Science Center, also called "the headquarters building," which was gifted to ASPB (then ASPP) in late 1980. This building, portions of which were constructed in 1916, has served the Society well; however, it has largely stood empty since March 2020 as ASPB's staff has worked remotely since the start of the COVID pandemic in 2020.

Potential sale of the building and the 4.5 acres on which it stands will depend on the urgency of ASPB's fiscal needs, as well as the overall real estate market in the vicinity. In the meantime, the Gude Plant Science Center may offer other, revenue-generating opportunities for the Society that have yet to be explored.

Staff

ASPB continues to accomplish a great deal with a staff that, along with everyone involved in its governance, is dedicated to the Society's success and well-being. These individuals bring expertise and perspectives of value to the organization. As ASPB's staffing structure evolved over time, it has historically been built around four main components: Operations/IT/Finance; Publications; Meetings and Membership; and Executive and Governance. This is largely a consequence of the initial governance structure, which reinforced silos and heavily staffed units that tended to operate with a high degree of autonomy. This lack of integration diminished the efficiency with which new initiatives could be visualized and take hold.

The Gude Plant Science Center in Rockville, MD, pictured in May 2019.

To address this shortcoming, recent staffing changes, including the creation of a VP for Content and Communications with a remit that spans all of the Society's content-generating activities, were implemented to identify and exploit synergies among currently dispersed content streams. The recent addition of ASPB's first VP for Business Development provides an organization-wide heft for efforts to develop and bolster existing and new revenue streams. A sound strategic plan will help in this regard, and further evolution of staffing is possible, in so far as strategic objectives indicate its necessity.

Another governance and leadership innovation that was deployed around the same time as establishment of the Council/BoD structure was elevation of the role of the Executive Director to that of a Chief Executive Officer (CEO). This may seem like a subtle change; however, it is an important difference in the way in which governance and staff interact to identify and pursue the Society's objectives. Put simply, the CEO position reinforced the concept that the professional staff are experts in their designated areas – active participants in governance discussions, as well as the source of novel product and service concepts for governance consideration – and not simply there to carry out the instructions of the BoD. In this construct, BoD, BoT, Council, and staff collaborate to discuss and broadly define directions for the Society, and the staff retain a direct and respected (albeit non-voting) voice at the governance table.

Nominations Committee

One governance innovation of note, which explicitly intended to lower barriers to participation and inclusion in ASPB leadership, was a change in the structure of the nominations committee, whose role it is to identify candidates that stand for election to the BoD. Prior to 2020, the nominations committee simply consisted of the three Presidents, a construct with an arguably narrow perspective that could be perceived as exclusive despite the open call for nominations from ASPB members with which the nominations process begins each year.

After deliberation within the Council and BoD, and engagement with the Society's membership, the Constitution and Bylaws were changed in 2020 such that the Nominations Committee now includes two Presidents (current President and the President-elect as chair), a member elected from the Council, a member appointed by the Early Career Plant Science Section, and a member appointed by the EDIC – this latter participant reflecting an expanded role for the EDIC that was approved at the same time. Input from membership remains in place, although ideas to increase participation by ASPB members in the process of identifying potential future leaders are being explored.

PART 3: WOMEN IN ASPP/ASPB

Judy Callis and Eva Farré

Since publication of the *History of the American Society of Plant Physiologists* about 40 years ago, women have become increasingly involved in plant biology research and have increasingly participated in leadership activities of the American Society of Plant Physiologists, later the American Society of Plant Biologists (ASPP/ASPB). One of the transformative actions by women in the Society was creation of a committee whose primary focus is on issues of special interest and significance to women. This committee's functions are to promote the involvement and recognition of women in the Society, build a mentoring community of women plant biologists who can provide career and life guidance to others, and help mitigate barriers to the equal involvement and success of women as plant biologists.

Elisabeth (Beth) Gantt

Jane Silverthorne — Mary Clutter

Enid MacRobbie

In the early years of ASPP, as was the case in many other areas of science at the time (Kohlstedt, 2006), there were few female members of the Society, and women were not involved in its leadership. But in the late 1970s, with the emergence of a growing number of women doing plant-centric research and increased federal funding for it, there was a significant rise in the number of women members in the Society. Recognition and representation of these women was slow to follow. Prior to 1984, there was limited representation by women in leadership roles, such as membership on the Executive Committee, speakers in symposia at the annual meeting, and recipients of Society-wide awards. From 1985 to 1994, the number of women on the 15-person Executive Committee hovered between one and three. In October 1996, its membership included six women out of a total of 17 members (with the addition of two standing committee chairs). Likewise, only a few women served on the Editorial Board of the Society's founding journal, *Plant Physiology*. Despite their valuable scientific contributions, by 1984 Birgett Vennesland was the only female recipient of a Society-wide award (in 1950) out of more than 100 individuals receiving recognition for the five different prizes and awards given at that time. (Notably, a husband-wife couple, William and Winnifred Klein, were the recipients of the inaugural Alfred Gude Award in 1984.)

Gradually, during the third quarter of the 20th century, women became recognized as leaders in ASPP/ASPB. After Mary Stiller's service to ASPP as the first female Secretary (1973-1975), it was another decade before a second woman, Elisabeth (Beth) Gantt, held that same elected position. Gantt served ASPP in multiple positions. Prior to her service as Secretary and member of the Executive Committee (1985-1987), she served as Chair of the Publications Committee (1983-1985). Subsequently, she was

President-elect (1987-1988) and became the Society's first female President the following year.

Women in ASPP leadership roles became more common in the late 1980s and early 1990s. As a member of the Western Section, Terri Lomax became the first woman Section Representative to the Executive Committee (1990-1995). Mary Helen Goldsmith served a three-year term on the Executive Committee from 1985 to 1988, and she was President-elect in 1990. Deborah (Debby) Delmer joined her on the Executive Committee as an Elected Member-at-large in 1987-1990. As chair of the Publications Committee, Machi Dilworth served on the Executive Committee from 1989 to 1990. Nevertheless, the number of women engaged in Society leadership remained small, and the limited engagement and recognition of women in ASPP concerned many members of the Society.

Johanna Döbereiner

Aruna Kilaru

There was also growing national recognition of constraints on the advancement of women in science, many of whom were a minority in their home department, research group, or government laboratory. Women in ASPP began to realize that the Society should help provide professional guidance and support for women. For all the issues described above, a group of committed women and a small number of progressive male members of ASPP realized there needed to be greater participation of women in ASPP-based activities.

To address these issues, the women of ASPP organized informal gatherings at the annual meetings in the early 1980s. In 1984, at the request of several women including Wendy Boss and Ellen Weaver, a meeting with an assigned room was scheduled in advance of the annual meeting at University of California, Davis. The gathering was announced as "Women in Plant Physiology" in the meeting program, and all interested attendees were invited to participate. People attending the meeting decided to form a coordinating group that was chosen by nomination between the 1984 and 1985 meetings. Subsequently, there were open meetings organized by women members at the annual Society meetings (1985-1988) where decisions were made on how to organize, identify issues of highest priority, and plan for future activities. Among the leaders were Jaleh Daie, Kathryn Edwards, Janet Slovin, Mary Jo Vesper, Mary Helen Goldsmith, and Beth Gantt, in addition to Wendy Boss and Ellen Weaver. One important outcome was development of a proposal for the establishment of a woman-focused standing committee. Such a committee would provide a stable forum with connections to ASPP leadership and create a basis for financial support. At the 1985 annual meeting, the group voted unanimously to support establishment of a standing committee (1985 Meeting of Women in Plant Physiology group meeting minutes), and Ellen Weaver presented the rationale for this proposal a few days later at the Executive Committee meeting (History of ASPB, p. 254). In 1986, after considering the proposal, the Executive Committee voted to establish an ad hoc committee on the "Status of Women in Plant Physiology" (Executive Committee minutes, August, 1986) and, as a member of the Executive Committee, Mary Helen Goldsmith was charged with developing a budget request and recommendations for committee appointments. Mary Jo Vesper was the first chair of the newly formed Committee on the Status of Women in Plant Physiology (CSWIPP). While initially established as an ad hoc committee, CSWIPP subsequently became a standing committee after formal approval by the Executive Committee. In 1995-1996, the chair of this committee became a member of the Executive Committee, with Elizabeth (Beth) Bray serving as the first CSWIPP chair to also hold a seat on the Executive Committee. In 2001, when the Society's name was changed to the American Society of Plant Biologists (ASPB), the name of the committee was changed to Women in Plant Biology (WIPB).

Through the years, CSWIPP began and WIPB has continued to address a number of key issues: 1) How to enhance representation of women in the Society's leadership positions, for example, as committee members, committee chairs, and recipients of the Society's awards; 2) how to mitigate the financial and logistical challenges for women to participate in ASPP activities at the annual meeting, particularly in instances of childcare and the lack of financial support for it; 3) how to improve mentoring of early career ASPP members, particularly women; and 4) how to enhance communication among women members of the Society. The CSWIPP/WIPB engaged in multiple activities that addressed these issues, and in so doing improved the professional and personal lives of ASPB members.

Table 1. Committee Chairs and Members

Year (Oct-Oct)	Chair	Committee Members
1988	Mary Jo Vesper	J. Slovin
1989	J. Slovin	C. Waggoner, S. O'Neill, A. Vasconcelos, J. Radin, J. Verbeke, M.J. Vesper
1990	S. O'Neill	C. Mitchell, T.K. Peterman, C. Goldstein, J. Verbeke
1991	Judith Verbeke	G. Hagan
1992	Judith Verbeke	G. Hagen, R. Fisher, E. Bray
1993	Judith Verbeke	G. Hagan, R. Fisher, A. Harmon, E. Bray, J. Deikman
1994	E. Bray	R. Fisher, A. Harmon, J. Deikman, D. Mohnen, K. Koch
1995	E. Bray	A. Harmon, J. Deikman, D. Mohnen, K. Koch, R. Alscher
1996	R. Alscher	D. Mohnen, K. Koch, S. Heinhorst, E. Hood, C. Henson
1997	R. Alscher	S. Heinhorst, E. Hood, C. Henson, D. Della Penna, R. McClung
1998	Ann Hirsch	E. Hood, C. Henson, D. Della Penna, R. McClung, R. Boston
1999	Ann Hirsch	D. Della Penna, R. McClung, R. Boston, E. Hood, F.C. Guinel-Jefferson
2000	Ann Hirsch	R. Boston, E. Hood, F.C. Guinel-Jefferson, P. Okubara, B. Larkins
2001	Elizabeth Hood	F.C. Guinel-Jefferson, P. Okubara, B. Larkins, K. Wang, M. Tierney
2002	Mary Tierney	P. Okubara, B. Larkins, K. Wang, L.Olsen, L. Feldman
2003	Mary Tierney	K. Wang, L. Olsen, P. Bedinger, D. Schachtman, E. Walker
2004	Laura Olsen	P. Bedinger, D. Schachtman, E. Walker, J. Brusslan, J. Kiss
2005	P. Bedinger	D. Schachtman, E. Walker, J Brusslan, J. Kiss, L. Olsen
2006	Judy Brusslan	J. Kiss, L. Olsen, T. Delaney, M. Laskowski, E. Pilon-Smits
2007	Judy Brusslan	L. Olsen, T. Delaney, M. Laskowski, E. Pilon-Smits, J. Kiss

continued on next page

Table 1. Committee Chairs and Members *(continued)*

Year (Oct-Oct)	Chair	Committee Members
2008	Judy Brusslan	T. Delaney, M. Laskowski, E. Pilon-Smits, J. Kiss, Y-R Lee
2009	Judy Brusslan	J. Kiss, Y-R Lee, D. Bassham, M. Laskowski, M. Neff
2010	Marta Laskowski	Y-R Lee, D. Bassham, M. Neff, W. Peer, C. Wetzel
2011	Marta Laskowski	D. Bassham, M. Neff, W. Peer, C. Wetzel, K. Duncan
2012	Kateri Duncan	W. Peer, C. Wetzel, D. Bassham, C. Bermudez-Kandianis, M. Neff
2013	Kateri Duncan	M. Neff, D. Bassham, M. Otegui, C. Bermudez-Kandianis, E. Wurtzel
2014	Marisa Otegui	M. Neff, D. Bassham, E. Wurtzel, K. Osteryoung, L. Wayne
2015	Marisa Otegui	E. Wurtzel, K. Osteryoung, L. Wayne, L. Tian, E. Farre
2016	Marisa Otegui	K. Osteryoung, L. Wayne, L. Tian, E. Farre, S. Chellamma
2017	Laura Wayne	E. Farre, L. Tian, S. Chellamma, KM. Gillespie, G. Miller
2018	Laura Wayne	S. Chellamma, KM. Gillespie, G. Miller, E. Farre, L. Tian
2019	Laura Wayne	KM. Gillespie, G. Miller, K. Murphy, E. Farre, L. Tian, A. Kilaru
2020	Eva Farre	K. Murphy, S. Chin, L. Tian, A. Kilaru, S. Mafu, J. Mortimer, D. Pater
2021	Eva Farre	S. Chin, M. Rahmati Ishka, A. Kilaru, S. Mafu, J. Mortimer, D. Pater, R. Roston
2022	Eva Farre	M. Rahmati Ishka, B. Alptekin, S. Mafu, J. Mortimer, D. Pater, R. Roston, A. Foudree
2023	Rebecca Roston	B. Alptekin, B. Dhat, R. Melo, L. Murphy, A. Noori, A. Foudree

A subset of ASPP/ASPB members volunteered time and expertise to serve as committee chair and/or as a member, generating ideas and turning them into activities. Below are some of their most impactful successes. Table 1 lists the names of committee chairs and members who dedicated time and effort to design and implement successful activities of the CSWIPP/WIPB.

1. Increased representation of women in Society leadership roles and as award recipients.

Establishment of the CSWIPP, by itself, vastly increased the proportion of women involved in shaping ASPP's activities and vision while providing increased leadership opportunities for women. Inclusion of the CSWIPP/WIPB chair on the Executive Committee created a connection between the committee's activities and Society's leadership. Members of the CSWIPP/WIPB had increased visibility, providing female role models for early career researchers.

CSWIPP members knew that recognition of women as leaders and acknowledgment of their scientific accomplishments were important for their careers outside ASPP. Promotion at academic institutions requires service activities, and Society research awards document national recognition important to appointment and promotion committees. Initially, as the proportion of women in the Society increased, their representation in leadership positions and as recipients of Society-wide awards did not increase in parallel. This disparity was noted during early discussions by members of the CSWIPP, and there was considerable brainstorming about how to affect change.

Historically, the nomination process for elected leadership positions in the Society involved one candidate solicited from the membership at large, and another recommended by the Nominating Committee, which was composed of the Society's President, President-elect, and Past President. An effort was made by the CSWIPP to nominate female candidates through this mechanism, although this approach did not guarantee success. For increased participation on standing committees, the CSWIPP forwarded names of potential candidates to the President-elect, who appoints committee members. The list of Presidents, Secretaries, Elected Members of the Executive Committee (now Council) and members of standing committees showed an increase in female representation after the establishment of CSWIPP; gender representation approached parity by the early 2000s.

In addition to reduced recognition as candidates for the Society's research awards, women were at a disadvantage for some awards because of their specifications. For example, the age requirement of an award was particularly a problem for women. Putting a hold on one's career to take time out for childbearing and rearing or postponing graduate studies (until children were of school age, for example) meant that women with these career trajectories were often not eligible. In 1985, the women's group petitioned that the criteria for these types of awards and any future awards be changed from an absolute age to "years post-degree." The Charles Albert Shull award required the awardee be 45 years or younger. While mention of a specific age has not disappeared completely, the Shull Award now reads, "by a member who is generally under forty-five years of age on January 1st of the year of presentation or is fewer than ten years from the granting of the doctoral degree. Breaks in careers will be considered when addressing the age limit of this award." The award now allows a scientist with a non-traditional career path to be eligible for this award. Currently, all ASPB awards with a time limit include a statement regarding "breaks" so as to be as inclusive as possible while still staying within the spirit of the award. For example, the Early Career Award specifies eligibility as one "who is generally not more than seven years post-PhD on January 1 of the year of the presentation. Breaks in careers will be considered when addressing the time limit of this award." The early women's group and then the CSWIPP played a critical role in bringing this issue to the attention of the Society's leadership.

2. Increased networking and mentoring through annual meeting luncheons and workshops.

Prior to 1985, workshops at the annual meeting focused attention on "hot" topical areas and new techniques, but there was very little consideration of topics such as possible career options or how to navigate a career in plant sciences. Also, general attendance at standing committee meetings was closed. ASPP women organized themselves using a model in which inclusivity was valued and the meetings were open to all. Meetings with speakers, including a lunch, were held even before the formal establishment of CSWIPP. This format continued once CSWIPP was authorized. Currently, the WIPB meets to coordinate activities and sponsors an invited speaker/panel program and workshops at every annual meeting. Notably, these events were and continue to be open to all attendees and typically address topics of broad interest to all Society members.

An important early goal for these events was showcasing women as scientists and role models. Thus, by design the vast majority of speakers at these meetings were women. The luncheon (or dinner) program varied from single speakers to panel discussions with multiple members. Topics ranged from research presentations to discussions of data on gender-based aspects of career advancement. They also included scientists describing their own career experiences and topics of importance to women and young scientists and sometimes even established scientists. For example, the luncheon panel of women in 1990, coordinated by Janet Slovin, featured a panel of speakers addressing grantsmanship and balancing career and personal relationships. Various themes were repeated because first-time attendees were always a significant fraction of the meeting population. The list of topics and speakers at these luncheons/dinners (Table 2) demonstrates the breadth of topics covered over the years.

In addition to mentoring and guiding early career researchers, luncheons/dinners also provided networking opportunities for women, which was strongly promoted both formally and informally by the committee. The program often included organized seating to foster mentorship, with one or two experienced woman scientists seated at each table. Some years the mentor stayed at a table and attendees rotated to different tables; other times the mentors rotated. At some programs, tables were organized by sub-discipline. The committee worked with mentors to ensure a safe and welcoming environment where everyone felt free to ask questions and express their opinion.

Due to the time of the meeting, the venue required meal service, and hence attendance required an additional cost above the annual meeting registration fee. The organizers did not intend that purchase of a meal ticket be a barrier to attendance, so the committee worked hard to obtain funds to subsidize this expense for students and post-doctoral fellows. In 1985, ASPP member Ellen Weaver made a donation to support attendance at the meeting, allowing several women to attend who otherwise would not have.

Table 2. CSWIPP/WIPB Sponsored Speaker Series

Year	Meeting Location	Speaker(s)	Topic/title
1988	Reno, NV	Elisabeth Gantt	Experiences as a female scientist
1989	Toronto, Ontario, Canada	Machi Dilworth, and a USDA representative	Granting and funding agencies
1990	Indianapolis, IN	Panel: Janet Slovin, Wendy Boss, Sharon Long, Marcia Harrison	Experiences as female scientists
1991	Albuquerque, NM	Mary Helen Goldsmith	Service opportunities in ASPB
1992	Pittsburgh, OH	Michelle Shinn	Committee on the Status of Women in Physics
1993	Minneapolis, MN	Vicki Chandler	Research presentation-maize genetics
1994	Portland, OR	Sharmon O'Neill	Research presentation-regulation of flowering
1995	Charlotte, NC	Panel: Patricia Myers, Catherine Didion, Dina Mandoli	Panel discussion on mentoring
1996	San Antonio, TX	Margrete Klein	NSF programs and the advancement of women in science
1997	Vancouver, BC Canada	Sue Rosser	Female friendly science
1998	Madison, WI	Gerhard Sonnert	Women in science and the "two-body-problem" of dual-career couples
1999	Baltimore, MD	Eileen Dowse	Communicating to overcome resistance and increase success
2000	San Diego, CA	Debra Rolison	A Title IX challenge
2001	Providence, RI	Nancy Hopkins	A report on the status of women faculty at MIT leads to new initiatives for faculty diversity
2002	Denver, CO	Nancy Dickey	Will the glass ceiling finally retract in the 21st century?

continued on next page

Table 2. CSWIPP/WIPB Sponsored Speaker Series *(continued)*

Year	Meeting Location	Speaker(s)	Topic/title
2003	Honolulu, HI	Debby Delmer	A woman's life: coping with challenges and change
2004	Orlando, FL	Elsa Murano	Achieving leadership positions in science
2005	Seattle, WA	Judy Verbeke	The World According to Garth
2006	Boston, MA	Wendy Boss	Tapping on the Puzzle Piece
2007	Chicago, IL	Jo Handelsman	The Women Don't Need Fixing: The Role of Institutions in Advancing the Participation of Women in Science
2008	Mérida, Mexico	Patricia Leon	The Good Side of a Macho Culture
2009	Honolulu, HI	Mary Lou Guerinot	Just Do It
2010	Montreal, Quebec CAN	Shirley Malcom	Women's Career Choices: finding the third way
2011	Minneapolis, MN	Karen Plaut	Women in Science: Leading the Way
2012	Austin, TX	Donna Dean (evening)	Mentoring that Works: Building Connections for a Lifetime of Career Options
2013	Providence, RI	Elizabeth Vierling	Do One Thing Every Day that Scares You
2014	Portland, OR	Nancy Ryan Gray	The Key to Success is... on My Keychain Somewhere at the Bottom of My Purse
2015	Minneapolis, MN	Michelle LeBaron	Conflict Resolution in the Workplace
2016	Austin, TX	David Pedulla	Work-Family Policies and Gender Inequality
2017	Honolulu, HI	Machi Dilworth	Challenges and Opportunities for Women in Science
2018	Montreal, Quebec CAN	Michelle Smith	Leadership Development
2019	San Jose, CA	Denneal Jamison-McClung	The UC Davis ADVANCE program
2020	Virtual	Sharlene Weatherwax	Perspectives in Science Leadership and Policy

There were no WIPB-sponsored speakers in 2021, 2022, or 2023.

Subsequently, the Committee sought outside funds to subsidize the cost, and several corporations made donations reducing the ticket price for students and post-doctoral fellows. At times, both ASPP and the Canadian Society of Plant Physiologists (CSPP) helped subsidize the cost. The luncheons/dinners were very successful annual events through 2019; the 2020 meeting was held online because of the COVID pandemic. Attendees reported that they left these programs with new tools and ideas to navigate their scientific careers, and with an expanded network of scientific colleagues.

In addition to speaker/panelist events at committee meetings, CSWIPP/WIPB also sponsored workshops at the annual meeting (Table 3). The committee decided on topics, invited speakers, and coordinated with ASPB staff and conference managers for available times and rooms. The goals of these workshops were to increase opportunities to inform, mentor, and network with members. Topics were chosen based on suggestions and feedback from the women's community. CSWIPP/WIPB chose a speaker or a panel of presenters, with interaction and participation by the attendees. Impactful topics such as work-life balance, negotiating for a position, job opportunities, and other issues of career mentoring were repeated on a regular basis. These workshops have continued to be open to all who wish to attend and are hugely successful.

3. Childcare

Childcare disproportionately impacts women, affecting their ability to travel to conferences. It negatively impacts their ability to network and communicate the results of their research, and it affects development of collaborations and other career opportunities. Childcare at the annual meeting was one of the main topics of conversation among women in the 1980s. CSWIPP organized parent-cooperative daycare

Table 3. CSWIPP/WIPB Sponsored Workshops

Year	Meeting Location	Topic/title
1999	Baltimore, MD	unknown
2000	San Diego, CA	Workshop 1: "Where Are the Jobs?"; Workshop 2: "How to Get the Jobs"
2001	Providence, RI	unknown
2002	Denver, CO	unknown
2003	Honolulu, HI	"Career/Family Balance"
2004	Orlando, FL	Workshop 1: "Career/Family Balance"; Workshop 2: "Getting and Keeping a Job"
2005	Seattle, WA	Workshop 1: "Where Are the Other Jobs?" ; Workshop 2: "Getting and Keeping a Job"
2006	Boston, MA	Workshop 1: "Where Are the Jobs?" ; Workshop 2: "Getting and Keeping a Job"
2007	Chicago, IL	Workshop 1: "Where Are the Jobs?"; Workshop 2: "Getting the Most Out of the Postdoc Experience"
2008	Mérida, Mexico	unknown
2009	Honolulu, HI	Workshop 1: "How to Survive a New Job"; Workshop 2: "Alternate Careers"
2010	Montreal, Quebec	Workshop 1: "Getting the Most Out of Graduate School"; Workshop 2: "Is Industry the Right Choice for You?"
2011	Minneapolis, MN	Workshop 1: "Communication Styles that Promote Success"; Career Workshop 2: "The Art and Science of Successful Negotiation"
2012	Austin, TX	1 Workshop: "Finding a Work/Life Balance"
2013	Providence, RI	1 Workshop: "Networking to Success: Attaining Diverse Careers in Plant Biology" (panel presentation + discussion)
2014	Portland, OR	Workshop 1: "Career Adaptability: How to Set Yourself Up for a Changing Professional Environment"; Workshop 2: "Your Research Career- What's Next?"
2015	Minneapolis, MN	1 Workshop: "Life/Work Balance"
2016	Austin, TX	Workshop 1: "Family Related Policies in the Workplace"
2017	Honolulu, HI	Workshop 1: Breakout tables on various topics
2018	Montreal, Quebec	Joint workshop with MAC (EDIC): "Strategies f Leadership Development"
2019	San Jose, CA	Joint workshop with EDIC: "Implicit Bias"
2020	Remote	Joint workshop with EDIC: "Building Sustaine Commitments to Equity and Inclusion"; Workshop 2: "Careers beyond Academia" (in collaboration with Membership Committe and ECPS)
2021	Remote	Joint workshop with EDIC: "Building Sustaine Commitments to Equity and Inclusion"; Workshop 2: "Careers beyond Academia" (in collaboration with Membership Committe and ECPS)
2022	Portland, OR	Workshop 1: "Science Communication: Enhancing your Delivery, Selling your Science Promoting Yourself"; Workshop 2: "How to be a Diversity Advocat in your Lab"
2023	Savannah, GA	Workshop 1: "How to Create your Mentoring Network"; Workshop 2: "Entering Leadership Roles"

Erin Friedman

at the 1989 joint ASPP/CSPP annual meeting in Toronto using a form completed by interested participants that was included in the meeting registration packet. ASPP/CSPP covered the costs of the daycare site, and parents and volunteers rotated to cover the necessary childcare time slots. John Radin was an early supporter, organizer, and volunteer for this initiative, and at the 1990 meeting he coordinated this facility and recruited additional volunteers. That year the facility took care of nine children ages 1-10. This arrangement lasted about 10 years, but it was difficult to organize and, in the words of one volunteer, "they would rather have been at talks than babysitting and not a lot of parents were comfortable about relying on volunteers being there." However, participation in this activity did develop some lasting relationships among many of the volunteers.

Subsequently, childcare at the annual meeting moved toward a model whereby ASPB reimbursed part of the childcare costs, but parents had to make their own arrangements (May/June 1999 ASPP News). Financial support that allowed this subsidy was provided by a very generous contribution from John Radin.

In 2007, ASPB received a significant donation from the Eli Romanoff estate (Sept/Oct 2007 ASPB News) that was meant to increase diversity in ASPB. Since 2008, it has been allocated to support childcare at the annual meetings. At first, these funds were used to support childcare arrangements made by parents and ASPB helped identify local services. Childcare subsidized by ASPB at the meeting site (e.g., in the convention center) was first arranged in 2010, and nine people signed up. At the Honolulu meeting in 2017, a total of 107 hours of childcare were provided. During the 2020 and 2021 annual meetings, which were held remotely due to the COVID pandemic, these funds were used to support childcare at the attendees' home locations.

4. Travel awards for women to attend the annual meeting.

In 1997, the CSWIPP, together with the Minority Affairs Committee, began organizing travel awards to support students and early career faculty attendance at the annual meeting. As the chair of WIPB, Judy Brusslan led an initiative to create a travel award entitled The Women's Young Investigator Travel Award (WYITA), which is specifically for early-career women; it was first announced in 2008 (ASPB, 2008). The goal was for ASPB to support early career women investigators by providing funds to offset costs, such as travel, housing, and the registration fee for the annual meeting. The WYITA was first awarded in 2009, with funding from ASPB's investment portfolio. These travel grants support women working in both teaching- and research-intensive institutions, as well as those with government or industry positions. Between 2017 and 2020, the WYITA program was supported by a grant from the Sharon Gray Foundation. Since its inception, the WYITA program has supported more than 100 women, with an average of seven awards per year. To continue to recognize the contributions of Sharon Gray, WIPB proposed that from 2021 to 2024 one of the travel awardees be named as a Sharon Gray Women's Young Investigator Travel Awardee. The Executive Committee approved this action in 2020.

5. Additional notable initiatives.

In 2005, Judy Brusslan led a WIPB initiative to sponsor a series of articles published in the ASPB News. The goal of this series was to introduce the broader community to women scientists and to "offer perspectives, ideas, and suggestions for those just starting careers in plant biology, as well as those further along in their careers." The WIPB identified topics and authors. These articles ran from May 2005 to 2008, with an article in every subsequent issue. From 2009 to 2011, three or four articles were published each year, ending with the Nov/Dec 2011 issue. These articles are archived on the ASPB website (https://aspb.org/aspb-news-women-in-plant-biology/).

The WIPB committee was an early user of the Plantae.org website. It created a group page and sponsored or co-sponsored webinars that remain posted; author blog pages are available (https://plantae.org/community/groups/women-in-plant-biology/). It also developed a website recognizing women pioneers in plant biology (https://aspb.org/wipb-pioneer-biographies/) and posted summaries of committee meetings held throughout the year. Recently, the committee initiated a Women in Plant Biology interview series to highlight backgrounds and identities of women at different career stages (https://plantae.org/women-in-plant-biology-interview-series/). WIPB chair Laura Wayne promoted the establishment of the Plantae Mentoring Center (https://jobs.plantae.org/ementor/), which currently contains profiles of 58 mentors.

Laura Wayne

Tessa Burch-Smith

Keiko Torii

Reference

Kohlstedt SG. 2004. Sustaining gains: reflections on women in science and technology in 20th-century United States. NWSA J. 16(1):1–26. DOI: 10.1353/nwsa.2004.0036

PART 4: PUBLICATIONS

Maarten Chrispeels, Brian Larkins, and Ivan Baxter

Plant Physiology: Change was the only constant

Maarten Chrispeels

The past 25 years of *Plant Physiology* have been characterized by constant change, not only of the content but of the method of production and presentation. The journal is now strictly online without a print version. Five Editors-in-Chief have been at the helm:

July 1992–April 2000
Maarten J. Chrispeels
Department of Biology of the University of California, San Diego

May 2000–June 2005
Natasha V. Raikhel
Department of Plant Sciences, University of California, Riverside

September 2005–December 2012
Donald R. Ort
United States Department of Agriculture, University of Illinois at Urbana-Champaign

January 2013–December 2021
Michael Blatt
Regius Professor of Botany in the School of Molecular Biosciences, University of Glasgow, U.K.

January 2022–Present
Yunde Zhao
School of Biological Sciences, University of California, San Diego

In 1987 the ASPP Executive Committee decided to introduce the position of Director of Publications and stated that the editor of *Plant Physiology* would report to the director of publications rather than have a seat on the Executive Committee. At that time, the Executive Committee was looking for an Editor-in-Chief for a new journal, *The Plant Cell*, and decided to offer five-year contracts to all Editors-in-Chief. Martin Gibbs at Brandeis University, who had been at the helm of *Plant Physiology* for 25 years, was given a final five-year contract, and the decision was made to look for a new editor in 1991. Gibbs had done a fantastic job of growing the journal, but it was time to find a new leader. Maarten Chrispeels was selected from among several candidates. As he liaised with Deborah Weiner, the managing editor for *Plant Physiology*, he realized that there was a problem with manuscript editing. When Deb came on board, she was shown a room full of manuscripts approved for publication but otherwise unprocessed. Chrispeels and Weiner convinced the Executive Committee that more staff and funds were needed to publish some "really fat" issues of *Plant Physiology* to get rid of this backlog.

As Chrispeels took over the leadership of the journal, he was asked by his postdoc mentor, Joe Varner, how the readers would know a new wind was blowing at *Plant Physiology*. Chrispeels decided to put an appropriate image associated with one of the papers in the issue on the journal's cover, as was being done for *The Plant Cell*. The first such issue carried a beautiful transmission electron micrograph of a Golgi apparatus by Andrew Staehelin on the cover. A second change was the publication of an editorial (99:797-781, 1992) that indicated how manuscripts would be handled in the future. Chrispeels asked all (over 100) Associate Editors to resign and went to a new system with six Associate Editors and 33 Decision Editors. The makeup of the editorial board made it clear the journal would remain strong in traditional areas, like biochemistry, photosynthesis, and whole plant physiology, but it would also attract papers in molecular genetics, cell biology, and molecular physiology. Two Updates in each monthly issue would update readers on a developing area. The main challenge facing the journal, besides publishing excellent science, was to compete with *Planta*, a strong Springer journal, and *The Plant Journal*, the European counterpart of *Plant Physiology*. Although *Plant Physiology* remained the most-cited journal by far, the issue was the impact factor, a measure of the number of times a paper was cited in the two years after its publication. Rather than evaluate the research of their faculty members, Chairs in many academic departments simply looked at the impact factors of the journals in which their articles appeared.

Going Electronic

In 1997, when Donald R. Ort, a biochemist and crop physiologist who worked on photosynthesis in leaves and plant canopy level, was president of ASPP, he wrote an editorial in the *ASPP News* that both journals would have an electronic version by January 1998. ASPP negotiated with High Wire Press, a nonprofit subsidiary of Stanford University that provided electronic services to 50 scholarly journals. Nancy Winchester was the new director of publications at the time. There were long discussions on how online publication would affect revenue flows to the Society, both from members — who would be able to get access to publications through their library — and from the libraries themselves. Many in the leadership of the Society and many staff members, not quite up on all this electronic "stuff," were anxious. But everything went smoothly.

Chrispeels decided that eight years would be a long enough tenure as Editor-in-Chief, and he was replaced in the summer of 2000 by Natasha Raikhel, who had made a name for herself in plant cell biology. Raikhel added two Associate Editors in the areas of Genome Analysis and Breakthrough Technologies. These were, and still are, new and growing scientific fields. She also named several women as Decision Editors. In January 2001, she edited the 75th anniversary issue of *Plant Physiology*, which contained 42 commentaries that attempted to summarize conceptual breakthroughs in plant biology during the previous 25 years. Three technological advances stood out in 2001: the development of molecular tools, the development and rapid adoption of plant transformation by Agrobacterium tumefaciens, and the widespread adoption of Arabidopsis thaliana as a model organism by hundreds of plant biologists. In her editorial (125:1-3, 2001), she projected that the interconnection between biology and physics, chemistry, and mathematics would lead to new tools to investigate gene function at all levels. She saw this as the greatest challenge facing the journal. She resigned in summer of 2005 at the end of her five-year contract.

In the fall of 2005, Don Ort was appointed Editor-in-Chief. He introduced a new feature called "Founders Reviews." For electronic publishing of *Plant Physiology*, everything had been worked out, but the issue at hand was "open access." Open access meant the publication was freely available to anyone. When he became Editor-in-Chief, Ort wrote an editorial that said it was the "preeminent force driving a change in academic publishing." Should the journal be accessible to all dues-paying members of ASPB, or should it be available for the whole world? Ort was a big proponent of open access. Initially there was much pushback because of the financial uncertainties it created for the Society. As things stood, scientists paid a membership fee that included a subscription to the journal.

There was also a publication fee. Subscription fees for *Plant Physiology* and *The Plant Cell* paid by libraries accounted for over half of ASPB's income. How could ASPB survive without that income? A new financial model was needed, and this issue occupied both the Publications Committee's and the Executive Committee's meetings. Some in the US Congress pushed for a law that would mandate all federally funded research be published in open access journals. ASPB and other scientific societies and most journals published by companies (like Nature) opposed such laws. The issue was resolved by charging authors for open access publication of their manuscript at the time of publication if they wanted it to be open access. The publication fees were included as part of grant requests scientists made to funding agencies. Libraries continued to pay the (often outrageous) subscription fees charged by some journals, although the ASPB journals — packaged into one bundle — remained reasonably priced.

In January 2013, Michael Blatt, a membrane physiologist from Glasgow, took over as Editor-in-Chief. Blatt was keen to rationalize and expand the content of *Plant Physiology* in line with trends in the field. He introduced a streamlined organization of topical headings in the journal and in editorial handling, and he added Ecophysiology and Systems and Synthetic Biology as new focus areas. In 2015, the journal ceased printing paper copies and went strictly to an online format. The demand for hard copies was dwindling and the cost of printing so few copies and mailing them became prohibitive. The reduction in cost would allow ASPB to stop charging authors for color printed figures and make the journal more competitive.

ASPB migrated both journals to e-Journal Press in 2016. This change enabled Blatt to streamline the process of submission and review of manuscripts. Most important, Blatt encouraged correspondence between editors in screening submissions and, later in the review process, in enabling editors and reviewers to consult on decisions. It also allowed for approved articles to be submitted directly to PubMed Central, as required for all NIH-supported research.

A growing concern for the journal was the (often inadvertent) mishandling of image data. In 2018, Blatt initiated action to implement more rigorous assessment of the quality of illustrations and, with the support of ASPB, engaged scientific editors in this process. Three years on, Blatt and the editorial board were pleased to note that, while many examples of mishandling were identified prior to final publication, up through the end of 2021 in every case these instances arose through unfamiliarity with good practice rather than any malicious intent.

An important decision was made in 2020. To secure ongoing financial stability in an age of great uncertainty about the future of scholarly publishing, ASPB entered into an agreement with Oxford University Press (OUP) that it would take over the production and selling/distribution of *Plant Physiology* and *The Plant Cell*. How would this affect submissions? Authors submit articles to *Plant Physiology* primarily because they are part of a community. Would that sense of community decrease and eventually disappear?

All these actions combined to smooth out the process of manuscript submission and handling. In addition, Blatt appointed first one and then two editors to handle reviews (Updates and Topical Reviews) and kept for himself the responsibility of Founders Reviews, which were started by Don Ort.

In January 2022, Yunde Zhao took over as Editor-in-Chief. Zhao is a molecular geneticist who works on the biosynthesis of auxin in Arabidopsis thaliana. He fashioned an editorial board of 13 Associate Editors, 63 Monitoring

Editors, and 29 Assistant Features Editors. The Assistant Features Editors write "News and Views," short research article summaries of research articles appearing in an issue. This greatly expanded the Assistant Feature Editors category that was started by Mike Blatt. To promote the careers of young scientists, Assistant Feature Editors were selected as young rising stars in plant biology.

Submissions and Acceptances

Over the years, there was a gradual increase in the number of research articles (450-575) published in *Plant Physiology*. A drop in submissions in 1995 was caused by a change in the format of the *Plant Gene Register* (report of a nucleotide sequence) from a hard copy in the journal to a strictly electronic form. Nearly every submission to the *Plant Gene Register* was published, and authors clearly liked to see their submissions in print. But such articles are not counted in the total research articles published. Figure 1 shows the number of manuscripts submitted to *Plant Physiology* (B) compared to the number of manuscripts published (C) between 1990 and 2020. A large increase in submissions in 2001–2003 was testimony to the youth and vigor that Natasha Raikhel brought to the position of Editor-in-Chief. She clearly energized the plant biology community about sending their articles to *Plant Physiology*. Authors had to get used to and accept major changes in the presentation of their research in the journal. The gradual rise in research articles published over subsequent years indicates that *Plant Physiology* is keeping up with the changing complexity of plant biology topics in general. In the past 25 years a number of well-established science journals connected to scientific societies (e.g., the *Journal of Biological Chemistry*) have seen dramatic drops in articles published.

Figure 1. Number of Manuscripts Submitted to *Plant Physiology* (B) Compared to Published Manuscripts (C)

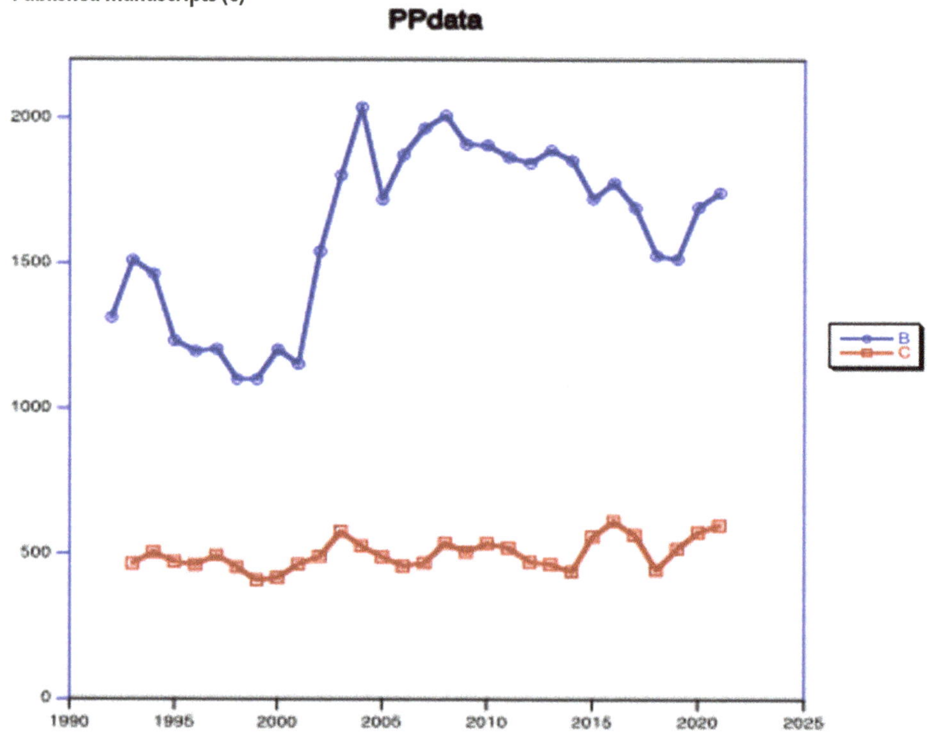

What did a typical issue of *Plant Physiology* look like in 1993? There were 35 research articles on the following topics: Development and Growth Regulation (10%), Cell Biology and Signal Transduction (14.5%), Metabolism and Enzymology (18%), Molecular Biology and Gene Regulation (12%), Environmental Stress (13%), Plant–Microbe Interactions (10%), Bioenergetics (3%), and Molecular Biology and Gene Regulation (12%). In addition, there was a category entitled Plant Gene Register (19%). Each issue had two Updates and a few Short Communications.

What did a typical issue of *Plant Physiology* look like in 2022? The June issue of Volume 189 had the following content: Nine News and Views articles written by Assistant Features Editors, three Letters, one Topical Review, three Breakthrough Technologies, Tools and Resources articles, one Research Report, and 40 Research Articles covering Biochemistry and Metabolism (1), Cell Biology (3), Genes, Development, and Evolution (12), Membranes, Transport, and Energetics (4), Signaling and Response (18), Systems and Synthetic Biology (2), and four Corrections.

As shown by the location of the corresponding author, where are the scientists working who publish articles to *Plant Physiology*? They are found in the United States (14), China (20), Japan (6), India (2), Europe (14), Argentina (1), and Australia (2). Four articles listed two corresponding authors, indicating they were collaborations between two institutions. The enormous increase in articles from China is testimony to the quality of science being carried out in China and the enormous increase in funding for science in China.

What's In a Name: ASPP or ASPB?

In 1999, Chrispeels suggested that we start naming our annual meeting "Plant Biology 2000" and for the future the Society kept this name with a changing date. This was followed by a campaign to change the name of the Society from ASPP (American Society of Plant Physiologists) to ASPB (American Society of Plant Biologists). This change was subject to membership approval. In the January/February 2000 issue, President Debby Delmer assured the membership that there would be extensive discussion and a vote by the membership to decide this question. The exact wording of the name change proposal was formulated by the Constitution and Bylaws Committee, and the proposal appeared on the Fall 2000 membership ballot and passed handily. The new name was then adopted by the Executive Committee.

Some people were dissatisfied with the name of *Plant Physiology* because its content was by now very broad and much broader than it was 75 years earlier at its inception. Various names were proposed, but none of them garnered overwhelming enthusiasm in a vote by the membership. Ultimately, it was felt that name recognition was paramount. So, the journal kept its name.

The Plant Cell

Brian Larkins

The first issue of *The Plant Cell* was published in January 1989. The decision for the American Society of Plant Physiologists (ASPP) to start a second journal that would deal with the emerging field of plant molecular biology was largely a consequence of the vision, enthusiasm and persistence of Larry Vanderhoef, Charlie Arntzen, and Joe Varner. Several people, including those involved in publishing *Plant Physiology*, opposed the idea. The leadership of ASPP had been slow to realize the impact plant molecular biology would have on plant physiology. Although *Plant Physiology*, which began publication in 1926, published a few articles involving plant molecular biology during the 1980s, much of the cutting-edge research was appearing in other journals, including *Plant Molecular Biology*, a new journal published by Kluwer Academic Press and primarily supported by members of the International Society of Plant Molecular Biology. The format, style, and scope of *Plant Physiology* were not the most attractive venue for publishing plant molecular biology research at the time, nor did it have the prestige of *Science, Nature,* or *Cell.* To understand how these circumstances arose and *The Plant Cell* came about, it is instructive to know the history of plant molecular biology research. The following is a brief overview.

What was considered plant molecular biology research in the 1960s and early 1970s focused on gene expression at the level of quantitative changes of DNA, ribosomal RNA, tRNA and, more often, enzyme activity. There was growing evidence of an adenine-rich RNA related to mRNA (Key and Ingle, 1964), but not at the level of mRNA poly(A) tails. There was no research approaching molecular genetics as we understand it today, because there was no easy way to identify a protein-coding gene or its mRNA product. Indeed, there were a variety of studies in genetics (writ-large), plant cell structure and plant development, but these were not at the nucleic acid level of gene expression and regulation.

The structure of DNA was described in 1953, and research shortly thereafter demonstrated what is known as the Central Dogma of biology: DNA→RNA→protein. The molecular mechanisms of the Central Dogma were documented during the 1950s and 1960s by research involving bacteria and their viruses. During this period, research on animal cells began to emerge, particularly as it applied to medicine. It included studies not only on mechanisms related to gene transcription and protein synthesis, but also post-translational regulation of proteins and their targeting to organelles. This research was facilitated by the efficiency of introducing radioactive precursors into animal cells and tissues and the ease with which proteins and nucleic acids could be extracted and characterized. Besides investigations of eukaryotic cells and tissues that produced highly abundant proteins, e.g., hemoglobin, insulin and ovalbumin, model organisms including yeast, Drosophila, Xenopus and sea urchins were widely used for the utility they provided to study development and genetic regulation. Rapid progress and new insights were made possible by financial support for basic research from the National Institutes of Health (NIH) and the National Science Foundation (NSF). Later, in the 1970s, the discovery of DNA restriction endonucleases in bacteria and the demonstration by Cohen and Boyer (1973) of their utility for constructing recombinant DNA molecules, coincident with the development of methods to sequence DNA molecules (Sanger and Colson, 1975; Maxam and Gilbert, 1977), revolutionized genetic research on microbes and animals, and launched the field of biotechnology in the 1980s. There was a report on plant gene cloning in 1979, but this technology was not widely applied to plant research until the early 1980s. Why did it take so long?

Research in plant molecular biology had its origins in the 1960s, but it was slow to develop. Most studies at that time involved characterizing changes in the amount of DNA and RNA as they related to growth and development, the application of hormones and herbicides, or environmental stresses, such as drought. There were studies of enzymes involved in these processes, but not the genes responsible for them. While there were many interesting questions and topics to study in plant biology, there weren't many people applying molecular biology to investigate them. There were a few labs in Europe and the US beginning to study protein and nucleic acid synthesis in plant cells and organelles in the 1960s. The proceedings of an international meeting on this topic in 1971, organized by G.L. Farkas at the Biological Research Institute in Tihany, Hungary, illustrate the limited scope of this research and the technical challenges of plant molecular biology at the time (Farkas, 1972). In contrast to prokaryotic and animal cells, labeling and isolation of intact plant nucleic acids were major challenges. The cell wall made access to plant protoplasts difficult, and the hydrolytic enzymes released from vacuoles created problems for purifying intact nucleic acids and organelles. These and many other circumstances impeded this research:

1. Most plant scientists in Botany and Biology departments were trained primarily in classical botanical sciences: taxonomy, morphology, ecology, mycology, algology, and physiology. Most plant physiologists were primarily trained in biochemistry and physiology, not molecular biology and genetics, and indeed not the molecular genetics of bacteria.
2. Plant scientists in Colleges of Agriculture, certainly those in US Land Grant Institutions, did so-called "applied research" that focused on crop improvement, soil science, weed science and plant breeding. While well-versed in transmission genetics, plant breeders focused primarily on the genotype and phenotype and their relationship to the environment. Agricultural research was primarily supported by the US Department of Agriculture (USDA) and similar agencies around the world, but developing high-yielding, disease resistant germplasm did not generally have a molecular genetic focus.
3. Because research in most plant science departments in Colleges of Agriculture was whole plant oriented, there were limited opportunities for students, particularly graduate students, to be trained in molecular biology and molecular genetics. And, of course, this was a major impediment when it came to the techniques involved in DNA cloning and sequencing.
4. Another major limitation was the need for more infrastructure and funding for plant molecular biology research (Arntzen, 1989). In the 1960s and 1970s, there were few, if any, scientific meetings devoted to this topic. Why would there have been? The primary source of funding for basic plant research in the US was through the NSF, and those funds were rather limited. Also, unlike the situation in animal and medical research, there were no model plant systems that facilitated these investigations. The Maize Genetics Cooperative provided intellectual and material support for basic research on corn, but it occurred primarily at the whole plant phenotypic level. The same was true of research on wheat and soybean genetics. Consequently, when plant DNA was first cloned and sequenced, few people or institutions were prepared to engage in the frontier of plant molecular genetics.

During the 1970s, the field of plant molecular biology began to blossom, with research devoted to a range of topics. Techniques were developed for isolating intact plant mRNA molecules (Larkins et al., 1976), and the development of an easily prepared in vitro protein synthesis system made from wheat seed embryos (Marcus et al., 1974) made it possible to characterize the proteins they encoded. The technique of DNA-DNA reassociation kinetics developed by Britten and Davidson (1974) to characterize animal genomes was beginning to be applied to plants (Goldberg et al., 1973; Walbot and Dure, 1974), revealing their genomes were generally larger and more complex than those of animals. And RNA excess DNA/RNA hybrid-ization studies determined that plants had complex nuclear RNA populations like animal cells and that at least 25,000 diverse genes were required to program a plant organ such as leaves (Goldberg et al., 1978). Although a model plant system had yet to be agreed upon, developing seeds were recognized as model systems to study protein synthesis because of the abundant storage proteins they synthesize (Larkins et al., 1975; Hall et al., 1976). Chloroplast and mitochondrial genomes started to be analyzed (Rochaix, 1972; Bedbrook and Bogorad, 1976; Levings and Pring, 1976), as were nuclear genes encoding some of the proteins they contain (Chua and Schmidt, 1978; Leaver and Forde, 1980). The nature of plant viruses and plant viroid genomes was also being characterized (Hirth, 1980; Sänger, 1980), as were the structure and movement mechanisms of the maize transposable elements (Sanger, 1980) made famous by Nobel Prize winner, Barbara McClintock (1950). It had also become clear that the tumor-inducing property of Agrobacterium tumefaciens was due to a plasmid it contained (Schell and Van Montagu, 1980).

As the field of plant molecular biology grew during the 1970s, several international meetings focused on this research. The first of these were NATO Advanced Study Institutes in 1976 and 1979. A comparison of the research presented in 1976 at the meeting on Nucleic Acids and Protein Synthesis in Plants in Strasbourg, France (Bogorad and Weil, 1977) and the second one in 1979 at Edinburgh, Scotland on Genome Organization and Expression in Plants (Leaver, 1980) is striking and shows how rapidly the scientific understanding of a variety of topics was advancing. It also shows that many young people were entering the field. There was growing awareness that plant molecular biology would impact agriculture, and representatives from several major agricultural research companies, including Monsanto, DuPont, and Ciba-Geigy attended the Edinburgh meeting.

The two NATO-sponsored meetings in Europe provided the impetus to start an annual Plant Molecular Biology Gordon Conference in the US, the first of which was held in 1980. A few years later, starting in 1983, there were also annual winter ARCO/UCLA-sponsored research conferences in Colorado. These meetings brought together scientists from all over the world who were doing cutting-edge plant molecular biology research. These meetings fostered an exchange of scientific information and innovative technical approaches, leading to the creation of the Plant Molecular Biology Association, led by Maureen Hanson, a young Assistant Professor at the University of Virginia. Hanson coordinated the publication of a newsletter, *The Plant Molecular Biology Reporter* (Mary Ann Liebert, Inc.), a monthly magazine containing short articles and technical advice. Eventually, the Plant Molecular Biology Association was superseded by the International Society of Plant Molecular Biologists (ISPMB), led by Leon Dure at the University of Georgia and Rob Schilperoort at Leiden University.In 1981, ISPMB sponsored the publication of a new journal, *Plant Molecular Biology* (PMB),

(Kluwer Academic Press, The Netherlands; currently Springer Netherlands), with Schilperoort as the Editor-in-Chief. PMB quickly became a popular outlet for publishing plant molecular biology research. Then, starting in 1985, ISPMB began holding a Congress every three years: 1986, Savanah, GA; 1988, Jerusalem, Israel; and 1991, Tucson, Arizona.

The importance of plant molecular biology research was also becoming recognized by Federal funding agencies, including NSF, NIH, and the Department of Energy (DOE). At first, this was mainly through the NSF and DOE Basic Energy Sciences Program, but after years of lobbying the USDA to fund basic plant research and support institutions outside of Colleges of Agriculture at Land Grant Universities, the USDA launched the Competitive Grants Research Organization (CRGO) in 1978. The CRGO had a small budget of $15 million, with a third of that targeted to nutrition research, but it had an impact. Plant molecular biology was also capturing the imagination of venture capitalists who envisioned a large amount of money could be made once this research began creating novel traits. Several plant biotechnology companies were started in the early 1980s, including Agrigenetics, Agracetus, Advanced Genetic Systems, Calgene, The International Plant Research Institute, Phytogen, Plant Genetic Systems, Sun Gene, and Zoecon, among others, and new research divisions were also created at traditional agricultural product companies like Monsanto, DuPont, Ciba-Geigy, Pioneer Hi-Bred, and Dow Chemical. Even oil companies, like ARCO, started investing in plant molecular biology research. Until data from research at these companies began to have practical applications, their results were openly published in scientific journals.

The advances in plant molecular biology research of the 1980s led to diversification of the faculty in Agronomy and Horticulture Departments throughout the US and the hiring of molecular plant scientists in Biology and Biochemistry departments worldwide. The impetus to create these positions was partly a consequence of increased research funding by Federal agencies in the US and other countries. The NSF inaugurated a new postdoctoral research program to attract non-plant biologists (specifically those trained in molecular biology) to plant biology research. Later, this program was opened to plant biologists not trained in molecular biology, seeking to move them into areas of plant molecular biology. The NIH began funding some aspects of plant molecular biology research, there were new funding initiatives at NSF, DOE, and eventually, funding of the USDA CRGO increased to $50 million. Some agriculture-related corporations and biotechnology start-up companies also funded labs doing research of special interest to them.

As the plant molecular biology community grew and their investigations spread to a diversity of topics, there was increasing recognition of the importance of focusing on a model plant system that could be used to investigate the complexities of plant physiology. Initially, maize and petunia vied for recognition, but later an appreciation for Arabidopsis carried the day (Somerville and Koornneef, 2003). Its genetics, small size, and short generation time were well known, as was the ease with which EMS-induced mutants affecting diverse aspects of its growth, development, and physiology could be obtained. By the early 1980s, there was a core group of scientists characterizing Arabidopsis mutants that affected many different biochemical and physiological processes. But the utility of Arabidopsis became even more apparent following the demonstration of methods for creating gene-tagged mutants via insertion of the T-DNA of Agrobacterium tumefaciens (Feldman and Marks, 1987; Bechthold et al., 1993).

Arguably, T-DNA turned out to be perhaps more important for understanding plant biology than for producing transgenic crops, as it was the only way to experimentally validate the function of genes and their regulatory sequences.

Realizing the potential for Arabidopsis to uncover the molecular basis of plant growth and development made it an obvious choice for the first plant to have its genome sequenced, a milestone achieved in 2000. After that, as they say, the rest is history. Studies of Arabidopsis went on to explain a myriad of plant biochemical and physiological reactions, from the mechanisms of hormone synthesis and perception to the perception of light and subsequent reactions of photosynthesis and flowering.

Early on, the developments occurring in plant molecular biology research did not escape the notice of senior ASPP members. They were aware that much of the cutting-edge research was not being published in *Plant Physiology*. Furthermore, most of the people doing this research were not members of ASPP. Of the nearly 1,800 people who attended the first ISPB Congress in 1988, less than 20% were members of the Society. Since many of these people were not trained as plant physiologists, nor particularly familiar with ASPP, they were unlikely to become members of the Society. And what would the future of the Society be without them? Clearly, this had serious implications for its future.

In 1986, Larry Vanderhoef approached the President of ASPP, Eric Conn, with the idea of ASPP starting a new journal that focused on plant molecular and cellular biology. This idea was not embraced by the Editor-in-Chief of *Plant Physiology*, nor some members of the Editorial Board, but it was hard to dismiss its potential importance and long-term impact on ASPP. One major concern was how the Society would pay for publishing a second journal; it would constitute a significant new expense, and a source of funding was not obvious. But in 1986, the ASPP Executive Committee voted almost unanimously to start the new journal. The cost would be met by bundling its subscription with that of *Plant Physiology* so that library subscriptions would cover the expense of both journals. It was felt that the reputation of *Plant Physiology* would ensure the subscriptions wouldn't be canceled; besides, the cost of the joint subscription was competitive with the cost of individual subscriptions to privately published journals, such as *Cell, Planta,* and *Plant Molecular Biology.*

A committee composed of Joe Varner (chair), Jack Shannon, Martin Gibbs, Machi Dilworth, Lowell Owens, Bob Goldberg, Rob Horsch, and Mel Josephs (Executive Director of ASPP) was appointed to decide on the title and Editor-in-Chief of the new journal, and devise an editorial structure, editorial board, and scope. There was strong sentiment that Bob Goldberg, an ASPP member and well-known plant molecular biologist at UCLA, was the best person to be the Editor-in-Chief and lead this effort. After he was lobbied by Varner, Arntzen, and his colleague Bernie Phinney, Goldberg agreed to do it. *Cell* was one of the most prestigious journals publishing molecular biology research at the time, and it sometimes even included articles about plants, so it was a good model for what Goldberg envisioned. The title he decided on was *The Plant Cell* (Goldberg et al., 2019). The goal was to attract the highest impact articles in plant molecular biology, as well as cell and developmental biology, and review and publish them in a rapid and efficient manner. This would be accomplished by accelerating the review process with FedEx and Fax machines (previously everything was done by regular mail) and using a centralized management system. The goal was a two to three months turnaround from submission to publication — clearly an attractive feature for those doing the research! Also, the journal would be published with the highest-quality materials, including four-color figures and a full-color cover featuring research described in one of the articles. *The Plant Cell* would also include some features of *Science* and *Nature,* including reviews, commentaries, editorials, and news. Another idea was a yearly issue devoted to a series of reviews on a special topic. This was, to say the least, a significant departure from the way in which *Plant Physiology* and many other plant journals were published. The plan was approved by the Executive Committee in 1987, and the first issue of the journal was published in January 1989 (Goldberg et al., 2019).

To help launch *The Plant Cell,* Goldberg recruited several well-known plant molecular biologists (Brian Larkins, Ralph Quatrano, Rob Horsch, Nam-Hai Chua, Bill Timberlake, and Sam Levings III), and Machi Dilworth and Becky Chasan as Founding News and Views Editors. Judy Grollman, hired as the Managing Editor, oversaw the entire process, from submission to final publication. The first year was not without challenges: the goal was to have at least eight manuscripts each month. However, within three years, the journal routinely published 10-12 articles each month, and submissions came from all over the world! The journal quickly developed an excellent reputation and became the preferred option for publishing plant molecular biology research.

Bob Goldberg completed his term as the inaugural Editor-in-Chief of *The Plant Cell* in 1992 and was followed by Brian Larkins (1993-1998) and Ralph Quatrano (1998-2003). Larkins and Quatrano continued the format and procedures formulated when *The Plant Cell* was started, but they changed the composition of the Editorial Board and the scope of topics published in the journal. Crispin Taylor became the News and Reviews Editor in 1995 when Becky Chasan left, followed by Harry Smith when Crispin replaced Judy Grollman as the Managing Editor when she retired in 1998. Crispin Taylor took a job at *Science* in 2000, but he returned to the Society (renamed as the American Society of Plant Biologists, or ASPB, in 2001) to become ASPB Executive Director in 2004. Beth Staehle was brought in as Managing Editor and Nan Eckardt as News and Reviews Editor in mid-2000. By 2000, the yearly number of manuscripts published in the journal had grown to nearly 200, and its impact factor was more than 11, placing it at the top of plant science-related journals and comparable with several leading general science journals, e.g., *The EMBO Journal, Proceedings of the National Academy of Sciences USA, Current Biology, Genes and Development,* and the *Journal of Cell Biology.*

In July 2003, Richard Jorgensen became the fourth Editor-in-Chief of *The Plant Cell.* He introduced a number of changes in the way manuscripts were reviewed and edited. He significantly expanded the Editorial Board, making it more international and including more women. To maintain the journal's high standards and scientific integrity, he added a statistician to the Editorial Board to perform routine analysis of the data in manuscripts. He also appointed a staff science editor for a final review of all accepted manuscripts. This ensured that uniform publication standards were met and enhanced the written quality of the articles. He decided *The Plant Cell* would publish all topics of plant science that met its standards, and Co-editors were willing to work with authors so they understood what would be required of a resubmission. Also, all decision letters had to be written and signed by the Co-editor making the decision, not the Managing Editor as was previously the case. By the end of his term, it was apparent that a single full-time editor was insufficient to handle the workload, and the next Editor-in-Chief, Cathie Martin (2008-2014), established a team of five freelance science editors under the supervision of Nan Eckardt, who became Senior Features Editor. The team of science editors that started in 2009 is still with the journal today (2024). Several of these editors created a successful scientific editing service (PlantEditors.com) that now handles post-acceptance editing for *The Plant Cell* and *Plant Physiology*.

One of Cathie Martin's important innovations was the creation of "Teaching Tools in Plant Biology" (TTPB). In January 2009, Mary Williams was hired as a features editor to create the TTPB. Currently, there are 50 sets of teaching tools that are freely available and immensely popular: (https://academic.oup.com/plcell/pages/teaching-tools-plant-biology).

In January 2015, Sabeeha Merchant replaced Cathie Martin as Editor-in-Chief, and she made many changes in the structure and membership of the Editorial Board, as well as the way in which the review process worked. She was aware of complaints about the growing length of time for the review process and what was becoming multiple rounds of revision based on reviewers' questions and opinions. Authors complained the reviews were becoming prescriptive about the direction of their research. Merchant felt editors should assess the data presented in a manuscript and decide to accept or reject it based on the merits.

Merchant's Editorial Board consisted of Senior Editors and Reviewing Editors. A manuscript would be assigned to a Senior Editor who handled the pre-review consultation and assigned a Reviewing Editor. She expanded the Editorial Board significantly to avoid overloading the editors, whom she wanted to spend more time carefully assessing manuscripts, both pre- and post-review, and crafting decision letters. At the beginning of her term, there were approximately 35 Co-editors; by the end, there there more than 90 editors, divided among Senior, Reviewing, Guest, and Consulting Editor roles. She increased the Editorial Board's gender diversity and expertise and added more editors from China and Europe.

The pre-review consultation among editors helped decide whether a manuscript merited consideration for publication, while the post-review consultation among editors and reviewers occurred before a final decision was made to accept or reject the manuscript for publication. The pre-review process meant a manuscript could not be declined based on just one person's opinion. Consequently, the portion of manuscripts declined without external review went from less than 10% to approximately 50%; however, the overall acceptance rate remained roughly the same (20-35%, depending on whether one considers resubmission of a substantially revised work as a "new" paper).This system saved time for reviewers and authors. Authors got answers more quickly — and could submit their work elsewhere should they elect to do so — and reviewers didn't waste their time on manuscripts editors determined were unlikely to be accepted for publication. The post-review process was designed to help reviewers and editors agree on revisions that were needed (or explain the reason for declining the manuscript), which helped the editors craft a decision letter that clarified what was needed for acceptance.

Another of Merchant's innovations was the Assistant Features Editor (AFE) program. Starting in 2017, a group of early career researchers (mainly postdocs) were included to gain writing experience and learn about the workings of the Editorial Board. The AFEs, managed by Senior Features Editor Nan Eckardt, wrote "In Brief" articles, in which they highlighted papers in the journal each month, attended editorial board meetings, and participated in pre-review consultations to learn about how peer review was handled.

This program was extremely popular and successful (and was adopted by *Plant Physiology* as well). As of 2023, *The Plant Cell* had 24 AFEs, plus a team of seven Chinese-language AFEs and two Special Content Editors who managed the translation of article summaries into the Chinese language and posted them on *The Plant Cell* WeChat channel.

Blake Meyers is the current Editor-in-Chief of *The Plant Cell*, having taken over for Sabeeha Merchant in January 2020. Meyers continues to refine the management of the journal, from overseeing the review process and communication with authors to monitoring post-acceptance scientific editing. He has maintained many editorial innovations implemented by Sabeeha Merchant, including Assistant Features Editors (postdocs) who write "In Brief" articles for the journal. He increased the diversity of the Editorial Board, appointing more women as Senior Editors, adding people from diverse backgrounds, and extending the global outreach of the Board (https://academic.oup.com/plcell/pages/editorial-board).

When Rich Jorgensen became Editor-in-Chief in 2003, the journal abandoned annual review issues on specialized topics. But Meyers re-embraced the concept and started having focus issues containing sets of in-depth review articles to encourage the submission of research publications in the targeted area. Focus issues to date have been on the Biology of Plant Genomes (April 2021), Cell Biology (January 2022), Plant-Microbe Interactions (May 2022), Plant Genetics: Celebrating the 200th Birth Anniversary of Gregor Mendel (July 2022), Climate Change and Plant Abiotic Stress (January 2023), and RNA Biology (June 2023). Going forward, the plan is to have two or three such issues per year to position the journal as a major driver of progress in plant sciences. These issues will include "vignette-style" reviews addressing outstanding questions pertaining to the focus topic, as well as commentaries addressing best practices in the field.

Since its inception 35 years ago, *The Plant Cell* has endeavored to publish the most interesting and impactful research in plant biology being done around the world. This includes not only molecular, cellular, and developmental biology, but also plant pathology, biochemistry, and molecular genetics. The journal's Editorial Board has grown to include scientists in Europe, Japan, and more recently, China. Twenty years ago, China began making a major investment in plant research, and there has been a marked increase in the quantity and quality of research articles submitted from China. The journal is also experiencing increased submissions of manuscripts from India and South America.

During its first 10 years, *The Plant Cell* averaged about 180-200 articles per year. This number grew to over 300 by 2014, but it is back to around 200 currently. It isn't clear whether the decline in submissions reflects increased scrutiny and expectations of the editors, or perhaps competition for manuscripts from other publications. The popularity of publishing in *The Plant Cell* in 1989-1990 led to a creation of a similar publication, *The Plant Journal* (John Wiley and Sons) in Europe in July 1991. Since then, commercial publishers Cell Press and Nature Press, released like-minded publications: *Molecular Plant* (2008) and *Nature Plants* (2015). Now, plant scientists have multiple high-profile journals in which to publish their research. A journal's impact factor and the cost of publication are important considerations that influence authors' decisions of where to publish their research. *The Plant Cell* has an impact factor of more than 12 (2021-2022 year journal impact factor), which is higher than many plant science journals (*Plant Journal, Plant Physiology, Planta, New Phytologist*), higher than some general biology journals (*Plos Biology* and *Journal of Cell Biology*) and comparable with others (*Proceedings of the National Academy of Sciences USA, EMBO Journal, Genes and Development*). However, it trails that of *Molecular Plant* and *Nature Plants*, perhaps a consequence of these publishers' longstanding reputations.

As described earlier, since its inception *The Plant Cell* has undergone many changes in publication procedures and policies, as well as how the journal is printed. Initially (1989-1991), this was done by Waverly Press, Inc. in Baltimore, MD. For various reasons, in 1992 printing was switched to Capital City Press in Montpelier, VT, and in 2004 it was taken over by Dartmouth Journal Services, a sister company of Capital City Press. In 1998, ASPB partnered with HighWire Press to publish the journal online. ASPP/ASPB is the owner and publisher of its journals, and revenue generated from their library subscriptions provides the bulk of operating funds for the Society. Electronic publication with "Open Access" (OA) profoundly impacted the financial model by which ASPB and other scientific societies operate. Consequently in 2020, ASPB contracted with Oxford University Press for sales, distribution, and production services for *The Plant Cell* and *Plant Physiology*, and to be able to partner in ongoing discussions about how best to deal with OA mandates.

From 2000 to 2020, the journal had four different managing editors: Beth Staehle (2000–2004), John Long (2004–2011), Patti Lockhart 2011–2018, and Jennifer Regala (2018–2020). In 2021, as publication transitioned to Oxford University Press, Chris Lowe joined ASPB as managing editor of both *The Plant Cell* and *Plant Physiology*.

It is fair to say that the decision in 1987 for ASPP to publish a journal focusing on plant molecular biology had a profound effect on the Society. It influenced its vision and its membership. The molecular biology/ molecular genetic approach to plant physiological research has become pervasive. Twenty years after *The Plant Cell* was published, nearly every topic at the annual meeting (Plant Biology) included a molecular approach to answering fundamental questions. It has been said that standards for publication created by *The Plant Cell* raised the bar for other journals, as reflected not only in the expectation for research impact, but also the description and illustration of data. There may be no better evidence for the impact of *The Plant Cell* than the number of similar journals that have been created since its inception.

References

Arntzen CJ. 1989. Funding for plant sciences. The Plant Cell 1, 1-2.

Bechtold N, Ellis J, Pelletier G. 1993. In-planta Agrobacterium-mediated gene transfer by infiltration of adult Arabidopsis thaliana plants. CR Acad Sci Ser. III, 316, 1194-1199.

Bedbrook JR, Bogorad L. 1976. Endonuclease recognition sites mapped on Zea mays chloroplast DNA (chloroplast rRNAs/Sal I/Bam I/EcoRI/hybridization. Proc Natl Acad Sci. USA Vol. 73, 4309-4313.

Blake T. 1989. Will American agricultural research continue to be productive. The Plant Cell. 1, 481.

Britten RJ, Graham DE, Neufeld BR. 1974. Analysis of repeating DNA sequences by reassociation. Methods in Enzymology. 29, 363-418.

Chua N-H, Schmidt. 1978. Post-translational transport into intact chloroplasts of a precursor to the small subunit of ribulose-1,5-bisphosphate carboxylase. Proc Natl Acad Sci. USA 75, 6110-6114.

continued on next page

Cohen, S.N., Chang, AC, Boyer, H.W. and Helling, R.B. (1973). Construction of biologically functional bacterial plasmids in vitro. Proc. Nat. Acad. Sci. USA. 70, 3240–3204.

Farkas, G.L.,ed. (1972) Symposia Biologica Hungarica 13: Nucleic Acids and Proteins in Higher Plants, Akademiai Kiado, Budapest.

Feldmann, K.A. and Marks, M.D. (1987) Agrobacterium-mediated transformation of germinating seeds o Arabidopsis thaliana: a non-tissue culture approach. Mol. Gen. Genet. 208,1-9.

Fink, G.R. (1989) The state of plant molecular biology: Is there a second chance for Ag Schools? Cell 56, 141-142.

Goldberg, R. B., Galau, G. A., Britten, R. J., and Davidson, E. H. (1973). Nonrepetitive DNA sequence representation in sea urchin embryo messenger RNA. Proc. Natl. Acad. Sci. USA 70, 3510-3520.

Goldberg, R. B., Hoschek, G., Kamalay, J. C., and Timberlake, W. E. (1978). Sequence complexity of nuclear and polysomal RNA in leaves of the tobacco plant. Cell 14, 123-131.2

Goldberg, R.B., Larkins, B.A., and Quatrano, R.S. (2019) The Plant Cell: Beginnings. The Plant Cell 1, 3-4.

Hall, T.C., Ma, Y., Buchbinder, B.U. Pyne, J.W., Sun, S.M., and Bliss, F.A. (1978) Messenger RNA for G1 protein of French bean seeds: cell-free translation and product characterization. Proc. Natl. Acad. Sci. USA 75, 3196-3200.

Hirth, L. (1980) Structure of Plant Viral Genomes. In Genome Organization and Expression in Plants, C.J. Leaver, ed., Plenum Press, pp. 497-510.

Kaul, S. et al. (2000) Analysis of the genome sequence of the flowering plant Arabidopsis thaliana. Nature 408, 797-815.

Key, J.L., and Ingle, J. (1964) Requirement for the synthesis of DNA-like RNA for growth of excised plant tissue. Proc. Natl. Acad. Sci. USA 52, 1382-1388.

Larkins, B.A. and Dalby, A. (1975) In vitro synthesis of zein-like protein by maize polyribosomes. Biochemical and Biophysical Research Communications 66, 1042-1047.

Larkins, B.A., Jones, R.A., and Tsai, C.Y. (1976) Isolation and in vitro translation of zein messenger RNA. Biochemistry 15, 5505-5511.

Leaver, C.J., ed., (1979) Genome Organization and Expression in Plants. NATO Advance Study Institute, Plenum Press, NY and London.

Leaver, C.J. and Forde, B.G. (1980) Mitochondrial genome expression in higher plants. In Genome Organization and Expression in Plants, C.J. Leaver, ed., Plenum Press, pp. 407-425.

Levings, C.S. III and Pring, D.R. (1976) Restriction endonuclease analysis of mitochondrial DNA from normal and Texas cytoplasmic male sterile maize. Science 193, 158-160.

Marcus, A., Efron, D. and Weeks, D.P. (1974) The wheat embryo cell-free system. Methods in Enzymology 30, 749-754.

Maxam, A.M. and Gilbert, W. (1977). A new method for sequencing DNA. Proc. Natl. Acad. Sci. U.S.A. 74, 560-564.

McClintock, B. (1950) The origin and behavior of mutable loci in maize. Proc Natl Acad Sci USA 36, 344–355.

Rochaix, J-D., (1972) Cyclization of chloroplast DNA fragments of Chlamydomonas reinhardi. Nature New Bio. 238, 76-78.

Sanger, F. and Coulson, A.R. (1975). A rapid method for determining sequences in DNA by primed synthesis with DNA polymerase. J. Mol. Biol. 94, 441–448.

Sänger, H.L. (1980) Viroids: biology, structure and possible functions. In Genome Organization and Expression in Plants, C.J. Leaver, ed., Plenum Press, pp. 553-601.

Schell, J. and Van Montagu, M. (1980) The Ti-plasmids of Agrobacterium tumefaciens and their role in Crown Gall formation. In Genome Organization and Expression in Plants, C.J. Leaver, ed., Plenum Press, pp. 453-470.

continued on next page

Somerville, C. and Koornneef, M. A fortunate choice: the history of Arabidopsis as a model plant. Nature Reviews Genetics (2003) 3, 883-889.

Starlinger, P. A reexamination of McClintock's "Controlling Elements" in maize in view of recent advances in molecular biology. In Genome Organization and Expression in Plants, C.J. Leaver, ed., Plenum Press, pp. 537-551.

Varner, J.E. (1989) The greening of Gerry Fink. The Plant Cell 1, 479-480.

Walbot, V. and Dure, L.S. (1976) Developmental biochemistry of cotton seed embryogenesis and germination: VII. Characterization of the cotton genome J. Mol. Biol. 101, 503-536.

Warner, J.R., Rich, A. and Hall, C.E. (1962) Electron microscope studies of ribosomal clusters synthesizing hemoglobin. Science 138, 1399-1403.

Zimmerman, J. L., and Goldberg, R. B. (1977). DNA sequence organization in the genome of Nicotiana tabacum. Chromosoma 59, 227-252.

Plant Direct

Ivan Baxter

At the beginning of the 21st century, *Plant Physiology* started focusing on improving its impact factor (IF) by becoming more exclusive about the perceived merit of what it published. However, this meant a large number of manuscripts from Society members no longer had an obvious journal home. To address this, ASPB joined with the Society for Experimental Biology (SEB) and Wiley Press in 2017 to launch *Plant Direct,* a sound science journal for the plant sciences.

The scientific publishing landscape dramatically changed during the two previous decades. Almost all journals became electronically available on the internet, making it easier to disseminate manuscripts around the world. Also, the impact factor of journals influenced authors' choices for where to publish, as well as hiring and promotion decisions, despite clear evidence for the IF's negative effects on the scientific culture. Starting with *PLoS ONE* in 2007, a new type of journal emerged and rapidly expanded. Termed "sound science," these journals applied different criteria to submissions, focusing on a manuscript's technical standards and ethical integrity instead of its scientific novelty and impact. *PLoS ONE* published the most manuscripts in the world in 2011, and today most of the largest journals in the world are sound science journals. These publications have proven to be quite profitable, and their success contributed to the rise of predatory journals that don't require peer review. Scientific societies like ASPB, traditionally conservative and reluctant to risk their publishing revenue streams, were slow to react to the emergence of sound science journals. But, as the "real world" manifestation of plant scientific communities, Societies like ASPB had a unique opportunity to become publishers for most of their community's research, while leveraging profit from the publication process to support their existence.

As *Plant Direct*'s first Editor-in-Chief, Ivan Baxter, working with the management committee and the editorial board, used the opportunity of starting a new journal to take a fresh look at the way journals are run. Many researchers found the traditional publishing process — in which manuscripts pass through multiple stages of review and revision, sometimes spanning multiple journals, each using new reviewers — slow, frustrating, and wasteful of authors' and reviewers' time. Although this process often improves manuscripts, it also greatly delays the availability of science to the scientific community. *Plant Direct* implemented three innovations to improve the speed, efficiency, and openness of the publication process:

1. Promotion of manuscript preprint posting at the time of submission.
2. Consideration of the reviews from *Plant Physiology, The Plant Cell*, and *The Plant Journal* when manuscripts were transferred to *Plant Direct* for consideration.
3. Publication of anonymous reviewer reports upon acceptance for publication.

These features were intended to speed scientific communication in two ways.
The preprint became immediately visible to the scientific community, although readers had to judge the work's technical merit for themselves. This visibility has major benefits for authors, especially junior scientists applying for fellowships, grants, and jobs, by showing their research to the world. It also benefits readers who are not in the author's network by making the work available more rapidly than if it was published in a typical peer-reviewed journal. The *Plant Direct* team thought preprints were such an important benefit to the community that they implemented a discount on publication charges if the manuscript was posted on the preprint server at the time of submission to *Plant Direct*.

Other features of *Plant Direct* were intended to speed the publication process by reducing or eliminating steps that cause delays. Allowing transfer of manuscripts from referring journals, or from the bioRxiv preprint server, saved authors the hassle of entering manuscript metadata multiple times. Utilizing reviews from participating journals allowed the editor to benefit from previous reviews, possibly removing the need to send the manuscript out for another review, thereby reducing reviewer workload. Publication of the reviewers' comments was intended to make the review process more open and informative and help ensure the integrity and transparency of the review process.

With review criteria that are simple to articulate and understand (i.e., no need to define impact and novelty), *Plant Direct* offered an obvious opportunity for early career scientists to become reviewers. To that end, *Plant Direct* launched a reviewer training program in which scientists obtained real-world experience reviewing manuscripts by working with a *Plant Direct* editor.

Plant Direct Progress

Larry York took over as Editor-in-Chief of *Plant Direct* in 2022. In the seven years since the journal's launch, authors from 36 countries published 338 original research papers in *Plant Direct*, along with seven white papers. As demonstrated by the author stories series (see https://plantae.org/plant-direct-qa-with-judy-brusslan/ and https://doi.org/10.1002/pld3.176), the creation of *Plant Direct* provided an opportunity for researchers to rapidly communicate their research to the community. *Plant Direct* papers have been downloaded over 500,000 times and received over 1,600 citations to date. *Plant Direct* authors were able to make an impact with their research, while supporting the activities of two important plant societies, ASPB and SEB.

Through its emphasis on sound science, efficient publication processes, and support for the plant community, *Plant Direct* has made remarkable strides in a relatively short span of time. Looking ahead, the journal's focus on continual improvement, engagement with the community, and collaboration with authors, reviewers, and editors will ensure its continued success in advancing plant science knowledge and fostering scientific communication.

Other Publications

Crispin Taylor

ASPP's/ASPB's scientific journal publication program was part of the Society's original *raison d'etre,* and it has continued to be its single largest revenue source. Nevertheless, the Society has also developed a number of other noteworthy publications over the past 30 years. Their financial impact has been mixed; however, they have increased ASPB's visibility in novel market areas, and they have supported and enhanced aspects of its mission that relate to dissemination of knowledge about plants.

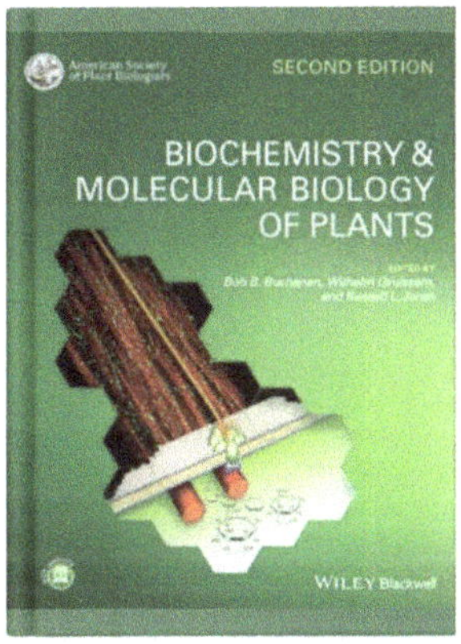

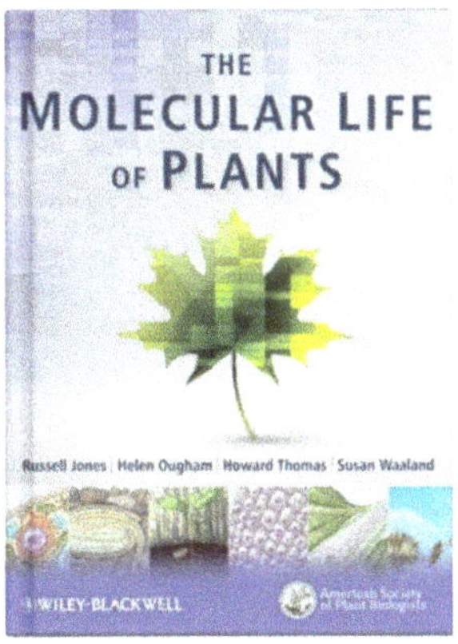

Even as early as 1940, ASPP began publishing monographs; typically, these were a collection of presentations from a symposium or topical meeting. This practice continued well into the 21st century and evolved to incorporate books of higher production value that included figures and tables. In the mid-to-late 1990s, a large, novel project was initiated, i.e., the creation of a comprehensive, graduate-level textbook covering a spectrum of plant biology topics. The book, entitled *Biochemistry and Molecular Biology of Plants,* was edited by Bob Buchanan, Willi Gruissem, and Russell Jones. ASPB invested several years of work before self-publishing the first edition of this text in 2000. Realizing the Society was not able to produce and disseminate a book of this size, ASPB took advantage of an existing business relationship with John Wiley and Sons to distribute the book.

BMBP, as it is now known, was well reviewed and received. Its sales were brisk and necessitated additional print runs that were paid for by ASPB. In 2006, however, ASPB passed control of future print runs and the rights to the book's content to Wiley, which incurred responsibility for subsequent print runs and the cost of developing a second edition. The second edition of BMBP2 was not published until 2015, by which time sale of the first edition and the royalties paid to ASPB by Wiley had trailed off significantly. Nevertheless, the second edition, with its updated text and graphics, had a major impact on the field and sold strongly.

In between publication of the two editions of BMBP, ASPB remained active in the "other publications" arena. A derivative publication from BMPB was an online image library comprising all the original figures in the book, a resource that is now freely available on the ASPB website.

More tangentially related to BMBP was a second textbook project: *The Molecular Life of Plants* (MLOP), edited by Russell Jones, Helen Ougham, Howard Thomas, and Susan Waaland. MLOP was published in 2012 with a target audience of undergraduate classes in plant biology. Like its larger, older sibling (BMBP), it marketed very well.

Embracing the growth in interest and financial support for research using the model plant Arabidopsis thaliana, as well as the possibility of inexpensive online publication, in 2002 ASPB began to publish *The Arabidopsis Book* (TAB), with Chris Somerville and Eliott Meyerowitz as editors. TAB, which is archived at https://www.ncbi.nlm.nih.gov/pmc/journals/1655/,

continued publishing free, online articles through 2019. Initially, interest was high, with this quasi-journal publishing 36 review-type articles in 2002. But only a handful of articles were published each of the following years. Interest and activity increased around 2012, when Keiko Torii was appointed as Editor-in-Chief, but despite Torii's strenuous efforts to generate new content and update previous reviews, TAB struggled because it was not readily accessible outside of the Arabidopsis and ASPB communities. Although it had a lower production cost than other journals, it was not publishing enough articles each year to be considered a serial by the Institute of Scientific Information (ISI) and therefore was ineligible to receive a Journal Impact Factor. Its publication ceased after 2019.

During these years, ASPB supported development and co-publication of a number of books and serial products, including *Current Protocols in Plant Biology*, initially edited by Gary Stacey, in 2016; and *Plant Genes, Genomes, and Genetics* in 2015, another graduate-level textbook that focused on all aspects of plant gene expression, edited by Erich Grotewold, Joe Chappell, and Toby Kellogg. Through ASPB's agreement with Wiley, ASPB members were, and still are, offered a discount on the purchase price of these publications.

The same is true of an agreement created with another non-profit organization, CABI (previously the Commonwealth Agricultural Bureaux International). In 2017, *Maize Kernel Development*, which offers an authoritative guide to understanding and doing research on the development of maize kernels, was edited by Brian Larkins. This book is valuable for those working in the areas of development, genetics, and crop physiology of maize and other cereals.

Finally, we must give a shout out to *My Life as A Plant* (2012) – a kindergarten through fourth grade targeted coloring book that introduces plant science concepts, including photosynthesis, tropisms, biotic and abiotic stress responses, and the geographical distribution of plans to future generations of plant scientists. Designed by (then) ASPB president elect, Alan Jones and former Education Committee chair Jane Ellis, this book has since been translated into more than 20 languages, including a second English dialect. All of these versions are available as pdf downloads on ASPB's website.

PART 5: HISTORY OF FUNDING FOR PLANT BIOLOGY RESEARCH*, **

Machi F. Dilworth

The history of the American Society of Plant Biologists (ASPB) reflects advances in plant biology research. John Hanson wrote in *The History of the American Society of Plant Physiologists (ASPP)*, "Classical botany knew no boundaries – all areas of plant science were included – but with the growth of the science there was the inevitable advent of specialization." Some 60 years after creation of ASPP, the Society was renamed ASPB to reflect the broadening of the field from "physiology" to "biology." Today, a century later, ASPB members' research encompasses "all areas of plant science" and more – Back to the Future!

As in all fields of science and technology, plant biology experienced revolutionary changes after World War II. The arrival of sophisticated techniques and instrumentation, such as electron microscopes, electrophoretic separations, chromatography, and ultracentrifuges, transformed biological research from descriptive to reductionist science in the 1960s. Then, with the arrival of the age of biotechnology followed by the "genomics revolution" and the latest "integrating everything" science, biological sciences made dramatic advances that led to large increases in research funding.

According to Science and Engineering Indicators, which is published biannually by the National Center for Science and Engineering Statistics (NCSES) (Burke, Okrent, & Hale, 2022), the major source of funding for research conducted at academic and not-for profit research organizations in all fields of science is the federal government and the institutions themselves. Private organizations provide additional support, which can often fill in areas for which public funding is hard to justify. In this overview, private funding sources are included only if they offered grants.

Federal Government Funding for Plant Biology Research

Federal government funding for Research and Development (R&D) began with the establishment of Land Grant Universities and enactment of the Morrill Act I of 1862 signed by President Lincoln, the Hatch Act of 1887, and the Morrill Act II of 1890, which were followed by the 1914 Smith Lever Act that created the Cooperative Extension Service of the Land Grant institutions. The Bankhead-Jones Act of 1935 established formula funding for research conducted by the various State Agricultural Experiment Stations (SAES)

* "Plant biology research" is defined as research that addresses questions about plants and plant-associated organisms, and any research that uses a plant and plant-associated organisms as experimental material, regardless of the approach or motivation.

** Descriptions focus on funding that supports research conducted in higher education sectors, as well as public and not-for-profit research institutions where a majority of the ASPB members are located and conduct research.

(NIFA, 2004) at these institutions. Around the time when ASPP was formed (1924), the major source of funding for plant biology research came from the US Department of Agriculture (USDA) and universities' funding by the State. Hanson's book describes the difficulties members experienced in paying the Society dues, annual meeting attendances, and publication fees. It was not until the 1950s, when the federal government started providing research funds through grants and fellowships, that plant biologists were able to pursue research, train students, and publish their discoveries.

National Science Foundation (NSF)

The NSF was established in 1950 specifically to support basic research in all fields of science and train scientists. Plant biology research and researchers were part of NSF's portfolio from the beginning, and NSF has continued to be the major funder of basic plant biology research. Among early recipients of research grants (NSF, 2004) were plant biologists, who investigated a wide range of subdisciplines of plant biology covering the entire spectrum of the plant kingdom.

In 1969, Dr. Elijah (Eli) Romanoff arrived at NSF as Director of the Metabolic Biology Program. Although he had no training in plant biology (he was an endocrinologist), Romanoff recognized the importance of plants and actively supported plant biology research. Around the same time, Anne (Holly) Schauer joined NSF as Associate Director for the Developmental Biology Program, and she increased representation of plant developmental biologists on grant review panels.

In 1974, Dr. Mary E. Clutter joined NSF as Director for the Developmental Biology Program. She worked at NSF for 30 years, steadily rising in the administrative ranks. This period saw revolutionary advances in biological sciences. While Clutter facilitated advances in all non-medical biology fields (Walbot et al., 2020), there is no question that her arrival at NSF changed plant biology research forever. The following lists the major plant biology-specific funding programs/initiatives she established at NSF:

1979: Cold Spring Harbor Plant Molecular Biology Course

This course was originally designed to introduce non-plant biologists to plant biology and traditional plant biologists to molecular biology. The course has continued to this day with a changing focus and target audience, often but not always supported by NSF.

1983-1994: Postdoctoral Research Fellowship in Plant Biology

This program was designed to encourage molecular biologists to become plant biologists and plant biologists to become molecular biologists. A total of 236 fellows were supported. Most of them went on to have successful careers in research, education, and administration that directly contributed to the advances in plant biology.

1990-2000: Arabidopsis Genome Research Initiative

The multinational Arabidopsis Genome Research Project was officially launched in 1990, but Clutter had been working for some time to make sure this initiative would be a national as well as an international program, just like the human genome initiative had become (NSF, 1990). Within the US, the project was supported by NSF, DOE, USDA, and NIH. The major goal, the Arabidopsis genome sequencing project, started in 1996 and was completed in 2000 by six international groups of scientists including three US groups (The Arabidopsis Genome Initiative, 2000). This rapid progress was made possible because the FY1998-2000 Plant Genome Research Program's budget provided more than $30 million toward the cost of sequencing by the US groups. After the completion of the sequencing project, Arabidopsis genome research was fully integrated into the regular NSF programs.

1998: Plant Genome Research Program (PGRP)

The PGRP is part of the National Plant Genome Initiative (NPGI) that was established in 1997 as a coordinated national plant genome research program by the Interagency Working Group (IWG) on Plant Genomes within the Office of Science and Technology Policy (OSTP). The IWG published a long-range plan for the NPGI outlining its needs, goals, objectives, and guiding principles to operate such a program (National Science and Technology Council, 1998). Congress appropriated $40 million in the FY1998 budget to NSF to implement the IWG's long-range plan. Total budget allocation for the Biological Sciences (BIO) Directorate was $324M for FY1997 and $370M for FY1998, indicating that the budget for PGRP represented "new" money. Although NSF was the only agency that received funds specifically designated for the initiative, the IWG made sure that the NSF's PGRP operated as part of the coordinated NPGI (NPGI 2009; 2014). Other funding agencies participated in and contributed to the NPGI in a manner consistent with each agency's mandate. One example of such coordination was the USDA's Agriculture Research Service (ARS), which under the leadership of Dr. Judith St. John provided a permanent home for resources created by PGRP-funded research, including mutant collections and genome data of major crops. PGRP's budget continued to grow and was $94 million in 2005, the year Dr. Clutter retired. The PGRP still exists today, continuously evolving in a manner consistent with other BIO programs.

Its relatively large budget allowed the PGRP to make large awards. A grant of $1 million to $2 million per year for five years was not an exception. This changed the perception of plant biology at research-intensive universities, as their faculty brought "NIH-size" grants to campus, increasing the visibility of plant biology research in general. In 2008, a committee of the National Research Council evaluated the NPGI and documented the impact of the NPGI for the first 10 years (Achievements of the national plant genome initiative and New Horizons in plant biology, 2008).

US Department of Agriculture (USDA)

Since 1950, the USDA increased funding for research while continuing to provide support for the State Agricultural Experiment Stations and the Cooperative Extension Service established in the early 1900s.

1953: Agriculture Research Service (ARS)

The USDA established the ARS with a charge to expand the nation's scientific knowledge and solve agricultural problems. Headquartered in the metropolitan Washington, DC area, the ARS continued to grow to include five major regional research centers and many specialized centers across the US. One of the specialty centers is the Plant Gene Expression Center (PGEC) (PGEC USDA), established in 1986 as a joint venture between ARS's Western Regional Research Center and the Department of Plant Biology at the University of California at Berkeley (UCB). The PGEC is situated in Albany, CA, close to the UCB campus. ARS provides PI salaries, support personnel, and indirect costs associated with the operation of the Center, and all PIs hold a full-time faculty position at UCB.

Some specialty center buildings are located on university campuses, such as the Robert W. Holley Research Center for Agriculture and Health (RHRC), which is located on the campus of Cornell University and houses the plant, soil and nutrition research unit among others. Some ARS scientists are located at universities with a regular faculty appointment and a laboratory in a university building. Both mechanisms enable graduate students and postdocs to contribute to ARS missions. ARS research is organized into 15 National Programs. Plant biology research is primarily supported within the Crop Production and Protection (CPP) program area. CPP's budget for FY2021 was $524 million.

In addition to research, the ARS has a unique role as the nation's custodian of agriculture-related research resources such as germplasms, mutant collections, publications, and databases of many kinds. These resources, managed professionally through the National Agriculture Library (NAL), the National Plant Germplasm System (NPGS), and others are irreplaceable national treasures.

1978: Competitive research grants programs (CRGO)
Perhaps the most significant event for plant biology research at USDA in modern times was the establishment of a competitive research grants program to support basic research that is investigator-initiated and peer-reviewed. While the newly established CRGO struggled a few years in the beginning (Krogmann & Key, 1981), it had become accepted by the traditional agriculture research community by 1985 (Huang, 1988). Over the years, the program was expanded to include additional research mechanisms, such as postdoctoral fellowships. A major expansion took place in 1991, when the program was renamed as the National Research Initiative Competitive Research Program (NRI). In 2008, USDA reorganized its research, education, and extension functions into a new agency, The National Institutes of Food and Agriculture (NIFA). In 2018, all USDA competitive grants programs were consolidated into the Agriculture and Food Research Initiative (AFRI). Currently, AFRI supports six priority research areas, including two that are relevant to the ASPB community: "Plant Health and Production and Plant Products" and "Bioenergy, Natural Resources, and Environment." In its first year of operation, CRGO received an appropriation of $14.4 million and provided $9.6 million for plant biology research. In FY2021, AFRI received an appropriation of $406.7 million and provided $79.5 million for the Plant Health & Production and the Plant Products areas (AFRI, 2021).

2014: Foundation for Food and Agriculture Research (FFAR)
FFAR was authorized by Congress as part of the 2014 Farm Bill and is a non-profit corporation consisting of public-private partnerships. Its mission is to "support audacious science addressing today's food and agriculture challenges." One identified challenge area is directly relevant to plant biology research, "Next Generation Crops." The 2014 Farm Bill provided $200 million for FFAR which had to be matched by non-federal funds. The 2018 Farm Bill further provided $185 million. By 2023, over 300 grants were awarded in all challenge areas, 70% of which were given to academic institutions (FFAR, 2023).

Department of Energy (DOE)

DOE's interest in scientific research stems from a post-World War II idea of developing peaceful uses of atomic energy. Over the years, the nature of energy research went beyond atomic energy. In 1977, President Carter signed the Department of Energy Organization Act, consolidating more than 30 separate energy functions carried out by various government agencies. The reorganization created the Office of Energy Research (renamed the Office of Science in 1998) to manage all basic research programs at DOE. Today, two of the Office's divisions support research in biology, Basic Energy Sciences (BES) and Biological and Environmental Research (BER) (Energy.gov), among the predominantly physics-focused research programs.

1965: Plant Research Laboratory (PRL) at Michigan State University

For the ASPB community, DOE support for plant biology research started in 1959, when the Atomic Energy Commission (AEC – later became DOE) formed an advisory committee of plant biologists to strengthen plant biology research. One of their recommendations led to the establishment of the AEC/MSU Plant Research Laboratory (PRL) ("MSU-DOE Plant Research Laboratory History: The First 50 Years"). Dr. Anton Lang served as the inaugural PRL director and recruited Phil Filner, Hans Kende, Derek Lamport, John Scandalios, Joe Varner, Peter Wolk and Jan Zeevaart, putting the PRL on a solid footing. Today, PRL alumni can be found all over the world, where they contribute to research, development and education. DOE has continuously supported the PRL to this day.

1976: Basic Energy Sciences (BES)

A major boost in plant biology research support happened when Dr. Robert (Bob) Rabson joined DOE as Director for the Energy Biosciences Program, as it was called then (Sussman, 2019). This program's mission was to support energy-related biological research at academic institutions and national laboratories. He alone was responsible to define "energy-related research," and he chose to focus on underlying mechanisms of plant growth and development, including supporting broader group programs with community access such as the Complex Carbohydrate Research Center at the University of Georgia. In addition, the program supported research on microbial energy bioconversion mechanisms and plant-microbe associations (including mechanistic work on plant pathogens). While BES used the standard peer review system, Bob's focus was to support people working on a long-term project. From the late 1980s throughout the 1990s, the program partnered with the Life Sciences Research Foundation (LSRF) ("Home," 2021) to support postdoctoral researchers in plant biology including microbiology associated with bioenergy production. A reorganization in late 2000s within the BES resulted in plant biology research moving to the Chemical Sciences, Geosciences, and Biosciences (CSGB) division, with new programs in nanoscience and newly created large collaborative projects at Energy Frontier Research Centers (EFRCs) (2021).

1986: Biological and Environmental Research (BER)

Since the age of genomics arrived, followed by a big push for developing sources of renewable energy, DOE support for plant biology has expanded beyond BES to include the BER. BER is where the human genome project originated before the NIH took over, and it supports all the fundamental ecosystem research for the DOE. Within BER, plant biology research is well integrated in the Biological Systems Science Division's activities (US DOE Office of Science, 2021).

Examples of BER's support for plant biology research include sequencing of the genomes of many plant and plant-associated organisms through BER's Community Sequencing Program (CSP), and the Facilities Integrating Collaborations for User Science (FICUS) Initiative ("Approved User Proposals," 2013) at the Joint Genome Institute (JGI), as well as support for structural biology for plant protein and mechanistic research at the DOE-supported synchrotrons (Lightsources.org) or molecular imaging tools at the Environmental Molecular Sciences Laboratory.

From 2006 to 2018, DOE-BER partnered with USDA-NIFA to jointly support competitive research through the USDA-DOE Plant Feedstock Genomics for Bioenergy program (2019) leveraging DOE's genomics expertise with USDA's longstanding expertise in plant breeding. BER also initiated support for Bioenergy Research Centers (BRCs) (US Department of Energy Genomic Science Program, 2024), sizeable multidisciplinary, multi-institutional research programs to vertically integrate fundamental knowledge of plants and microbes and accelerate the pathway to improving and scaling up advanced biofuel and bioproduct production processes. In FY2023 the BRCs were awarded $110 million. In FY2022, the BER's Genomics Enabled Plant Biology activity within the Genomic Science program provided $27.4 million for research to characterize gene function in bioenergy crop plants (OSTI, 2022).

National Institutes of Health (NIH)

Establishment of the National Institutes of Health (NIH) extramural grants in the second half of the 20th century marked a significant shift in support for biomedical research in the United States (Schneider, 2015). NIH has supported research on fundamental life processes in plants, including genetics, cell biology, photosynthesis, and metabolism, as these are important to our understanding of life.

1963: National Institute of General Medical Sciences (NIGMS)

NIGMS was established as part of the NIH, and funding programs for basic research, including physical and biological sciences research, was consolidated into the NIGMS. A cursory review of NIGMS's grant award data (going back to 1985) shows 25 ASPB Pioneer Members ("Pioneer Members," 2024) received career-long support from NIGMS, with an additional five occasionally receiving grants.* In addition, many plant biologists were supported by the NIH's training grants during their graduate studies, and many received NIH postdoctoral fellowships. In recent years, the budget for NIH's extramural programs has increased at an unprecedented rate (NIH Office of Budget, 2024). This increase has benefited the plant biology research community. A quick search of the NIH's award database (RePORT > RePORTER, 2024) reveals that in FY2022 NIGMS awarded about 80 plant biology-related grants including eight prestigious Ruth L. Kirschstein National Research Service Award Individual Postdoctoral Fellowships. The average annual size of these research grants was $354,000 (total cost).

Other Federal Government Research Grant Programs

Multiple other agencies of the Federal government support research and their funding programs can be found at Grants.gov ("Grant-making Agencies," 2024). Of those currently listed, the National Aeronautics and Space Administration (NASA), Defense Advanced Research Projects Agency (DARPA), and US Agency for International Development (USAID) have offered research grants in biology, including plant biology.

Most relevant to the ASPB community is NASA's space biology program that has supported research on plant responses to environmental factors, especially gravity, since the 1970s. Plant responses to gravity involve complex processes, from signal sensing and transduction to cell growth and ion transport, and NASA's program contributed to bringing plant physiology researchers from several specialized areas of expertise together. Plant biology is a focus area for the NASA's Space Biology Program (2024).

Private Organizations Funding Plant Biology Research

In addition to the US Federal government, industry and philanthropic organizations fund

* One-pass data check done against the list of ASPB Pioneer members as of September 2023

research and research-related activities, including plant biology, although they rarely support basic research through an open competition. Nevertheless, private entities offer a wide variety of funding opportunities, each with specific goals that fit their mission. Below is a list of organizations that have provided non-string attached support of research and research related activities in plant biology. Only those activities that were open to all qualified plant biology researchers are listed.

1980s

The McKnight Foundation

www.mcknight.org

Established in 1953, the McKnight Foundation is a Minnesota-based family foundation whose support for plant biology research began in 1983. The Plant Biology Research Program granted a total of $18 million for basic plant biology research from 1983 to 1992. The program focused on individual professional development grants for young faculty and institutional training grants for graduate students — two opportunities that were not available to plant biology researchers at the time. The individual faculty fellowship grant could be used at the PI's discretion with no restrictions, which is exactly what a young faculty needs, yet such a grant is very rare, even today. Following this program, in 1993, the McKnight Foundation established the Collaborative Crop Research Program to focus their efforts on supporting crop research in developing countries (CCRP, 2019).

Agrigenetics

Founded in 1975, Agrigenetics was one of the first agricultural biotechnology companies that were established to apply recombinant DNA technologies to plant agriculture (US Congress, Office of Technology Assessment, 1984). In the mid-1980s, Agrigenetics identified 20 plant biologists in and outside of the US whose research was at the forefront of plant biotechnology at the time and provided a total of $50+ million for basic biotechnology research (Kidd, 1995). While the results were to be funneled into the Agrigenetics' own Madison research laboratory, the grant recipients were free to publish their research with a minimal delay. The recipients also enjoyed freedom from the bureaucratic reporting requirements associated with government grants.

This program created an example of how a basic plant biology researcher could partner with profit-making entities without compromising their academic career. More importantly, the program increased awareness among plant biologists in academia about the enormous potential of plant biology research in the rapidly developing biotechnology industry and changed people's perception about getting involved in commercialization of research results. The establishment of Agrigenetics led to the founding of a number of other plant biotechnology companies in the early 1980s, including Agricetus, Advanced Genetic Systems, Calgene, International Plant Research Institute, Phytogen, Plant Genetic Systems, and others. This opened new career opportunities for trained plant biologists. New research divisions were also created at traditional agricultural product companies like DuPont, Ciba-Geigy, and Dow Chemical. Agrigenetics was later purchased by Dow Agriscience and eventually became part of Corteva Agriscience.

1990s

The Rockefeller Foundation

www.rockefellerfoundation.org

Established in 1913, the Foundation's mission has been "to promote the well-being of humanity throughout the world." They have four areas where they provide grants, among which the most relevant to plant biology is Food Security. Their typical grants are given to an organization (often a group of organizations) that conducts or supports research and/or training in agriculture. One of the foci of their food security research has been on rice as rice (Oryza sativa L.) feeds over half of the global population. The foundation staff who are PhD-level scientists identify the area in need of investment. One example is their support of the International Rice Genome Sequencing project. The project was officially launched in 1998 and the rice genome sequence was completed and published in 2005 by the international consortium involving scientists from 11 countries (Sasaki & Burr, 2000). The Rockefeller Foundation was at the table from the beginning and provided support, especially to the participants from the developing countries. The scientific/technical aspect of the project was the most straight-forward part of this project. Coordinating the funders from the various countries was difficult. The role to work through these issues fell on Dr. Gary Toenniessen of the Rockefeller Foundation. His contribution to the success of the International Rice Genome Sequencing project cannot be overstated.

2000s-2010s

Bill and Melinda Gates Foundation (BMGF)
www.gatesfoundation.org

Launched in 2000, BMGF's primary stated goal is "to enhance healthcare and reduce extreme poverty across the world, expand educational opportunities and access to information technology in the US." As part of their support of subsistence farmers in developing countries, BMGF partnered with the NSF's PGRP and created the Basic Research to Enable Agricultural Development (BREAD) program (NSF, 2010). The idea of the BREAD program came in 2006 from Dr. Robert Horsch soon after he took up a new position at BMGF. Actual hard work of establishing the joint NSF-BMGF program was carried out by Dr. Jane Silverthorne at NSF and Dr. Kathy Kahn at BMGF. The BREAD program operated for four cycles in FY2009, FY2011, FY2012, and FY2015, and made a total of 34 awards (NSF, 2015).

The Howard Hughs Medical Institute (HHMI)
www.hhmi.org
The Gordon and Betty Moore Foundation (GBMF)
www.moore.org

HHMI is the second-largest private foundation for biomedical research in the world, and they support basic biomedical research and science education. The HHMI states "Our research programs support people, not projects, and give our scientists the time and freedom to pursue difficult, long-range questions." While their science education programs are aimed at improving undergraduate science education at colleges and universities, a new program was created in 2002 to recognize individual science educators called "HHMI Professors." HHMI Professors are "accomplished research scientists who also are deeply committed to creating effective ways to engage under-graduate students in science." Currently, six prominent plant biologists are among the HHMI Professors.

In 2009, HHMI and the Gordon & Betty Moore Foundation (GBMF) came together and conducted a joint study "From Petioles to Transposons: How Plant Research Is Opening Scientific Frontiers and Solving Global Problems" (Carey, 2011). Based on the findings from the study, Dr. Robert Tijan, President of HHMI, and Dr. Vicki Chandler, Chief Program Officer of the Science Program at GBMF at the time, developed a joint HHMI/GBMF Plant Investigator program, which funded 15 investigators in 2011 for an initial five years (HHMI, 2011). Beginning in 2016, HHMI/GBMF Plant Investigators who met renewal criteria transitioned to the regular HHMI Investigator program, and in subsequent HHMI investigator competitions additional plant scientists have received HHMI Investigator awards.

Lessons Learned From the First 100 Years

In the US, the Federal government is the major funder of basic research at academic and not-for profit institutions. In addition, there are several private funding sources available to support scientific research and education. One great advantage of the US research funding system is the multiplicity of funding sources. Not all funding programs, in fact most funding programs for biology, do not specifically mention "plants" in their titles. Plant biologists can look widely and deeply for funding opportunities. Here, an old saying "Seek and you shall find" does apply! Any program whose purpose is to support biological research and education should be applicable to plant biology. Research programs that address issues such as energy and the environment might not mention the word biology at all, yet they are directly relevant to the field of plant biology. It is important that plant biologists themselves become much more aware of the value of studying plants well beyond advancing the field of biology. Plant biology research can provide solutions to major societal issues. This important advice needs to be conveyed to the next generation of plant biologists.

Federal government funding for basic research increased after WWII, when society (the policy makers) clearly saw the benefits of the investment (Bush, 1945). Support for plant biology research did not grow as fast as that for biomedical research, because the public has a difficult time seeing the direct and immediate benefits of plant biology research. The gap between funding for biomedical research and other areas of biological science will likely continue to widen, as there is no shortage of human ailments and new diseases. In 2009, a National Academy of Sciences committee issued a report, "A New Biology for the 21st Century," in which they identified "four key societal needs: sustainable food production, ecosystem restoration, optimized biofuel production, and improvement in human health." Plant biologists know their research can directly contribute solutions to all of them. This speaks to the necessity of increasing funding for plant biology research. ASPB, as an organization that represents the broadest community of plant biology researchers, is in the best position to lead this effort. In fact, ASPB began this effort in 2011, when it organized a meeting of scientists representing the full spectrum of plant science research – from basic to applied including industry, government, and academia – to develop a consensus plan that would invigorate and guide plant science research over the next decade. In 2015, this was followed by ASPB establishing the Plant Science Research Network (PSRN) (Henkhaus, 2018) with a grant from NSF. The PSRN is a coalition of 14 plant science-related professional associations. Working together, in 2020 the PSRN issued a report, "Plant Science Decadal Vision 2020-2030." Due to the COVID pandemic, follow-up actions in terms of increased funding for plant biology research have not yet occurred. It would be a fitting activity to pursue at the beginning of the ASPB's second Century.

Acknowledgement

I would like to thank Dr. Judy St.John, Dr. Sharlene Weatherwax, Dr. Vicki Chandler, Dr. Brian Larkins, Dr. Debrah Delmer and Dr. Rob Last for their review of factual accuracies and/or help with locating old reference documents and sharing their recollections. In addition, I thank Dr. Beth Gantt and Dr. Maureen Hanson for their helpful comments.

References

Achievements of the national plant genome initiative and New Horizons in plant biology. 2008. Washington, D.C.: National Academies Press.

continued on next page

AFRI. 2021. Agriculture and Food Research Initiative (AFRI) FY 2021-2022 Annual Review.

Approved user proposals. 2013 May 23. DOE Joint Genome Institute. [accessed 2024 Jun 12]. https://jgi.doe.gov/our-projects/csp-plans/.

The Arabidopsis Genome Initiative. 2000. Analysis of the genome sequence of the flowering plant Arabidopsis thaliana. Nature. 408(6814):796–815. doi:10.1038/35048692. http://dx.doi.org/10.1038/35048692.

ARS Home : USDA ARS. Usda.gov. [accessed 2024 Jun 12]. https://www.ars.usda.gov/.

ASPB pioneer members. American Society of Plant Biologists. [accessed 2024 Jun 12]. https://aspb.org/membership/aspb-pioneer-members/.

Basic Energy Sciences Advisory Committee, Department of Energy. 2018. A Remarkable Return on Investment in Fundamental Research. Energy.gov. [accessed 2024 Jun 12]. https://www.energy.gov/sites/default/files/2019/01/f58/bes-at-40.pdf.

Burke A, Okrent A, Hale K. 2022. The state of U.s. science and engineering 2022. Nsf.gov. [accessed 2024 Jun 12]. https://ncses.nsf.gov/pubs/nsb20221/u-s-and-global-research-and-development.

Bush V. 1945 Jul. Science: The Endless Frontier. Nsf.gov. [accessed 2024 Jun 12]. https://www.nsf.gov/about/history/EndlessFrontier_w.pdf.

Walbot V, Silverthorne J, Dilworth M. 2020. Mary E. Clutter. Aspb.org. *ASPB News.* Jan/Feb 2020. https://aspb.org/newsletter/archive/2020/JanFeb2020.pdf#page=15.

Carey J. 2011. From Petioles to Transposons: How Plant Research Is Opening Scientific Frontiers and Solving Global Problems. Hhmi.org. [accessed 2024 Jun 12]. https://www.hhmi.org/sites/default/files/PlantSciences_Backgrounder_060711%5B1%5D.pdf.

Collaborative Crop Research Program. 2019 May 15. Collaborative Crop Research Program History. CCRP. [accessed 2024 Jun 12]. https://www.ccrp.org/resources/collaborative-crop-research-program-history/.

Committee on a New Biology for the 21st Century: Ensuring the United States Leads the Coming Biology Revolution, Board on Life Sciences, Division on Earth and Life Studies, National Research Council. 2009. A new biology for the 21st century. Washington, D.C., DC: National Academies Press.

Energy frontier research centers. 2022 Jan 21. Osti.gov. [accessed 2024 Jun 12]. https://science.osti.gov/bes/efrc.

Environmental Molecular Sciences Laboratory. Pnnl.gov. [accessed 2024 Jun 12]. https://www.emsl.pnnl.gov/.

FFAR. 2023. FFAR Grants Overview. Foundationfar.org. [accessed 2024 Jun 12]. https://foundationfar.org/wp-content/uploads/2023/04/FFAR2023ImpactData4-11-23.pdf.

Grant-making agencies. Grants.gov. [accessed 2024 Jun 12]. https://www.grants.gov/learn-grants/grant-making-agencies/.

Henkhaus N. 2018 Jul 25. About the Plant Science Research Network. Plantae. [accessed 2024 Jun 12]. https://plantae.org/about-the-plant-science-research-network/.

HHMI. 2011 Jun 16. 2011 plant science program HHMI-GBMF investigators. Hhmi.org. [accessed 2024 Jun 12]. https://www.hhmi.org/news/2011-plant-science-program-hhmi-gbmf-investigators.

continued on next page

HHMI. HHMI professors. Hhmi.org. [accessed 2024 Jun 12]. https://www.hhmi.org/programs/hhmi-professors.

History. National Institute of Food and Agriculture. [accessed 2024a Jun 12]. https://www.nifa.usda.gov/about-nifa/who-we-are/history.

History. Energy.gov. [accessed 2024b Jun 12]. https://www.energy.gov/science/history.

Home. 2021 Jan 16. Life Sciences Research Foundation. [accessed 2024 Jun 12]. https://lsrf.org/.

Huang HT. 1988. The USDA's competitive grants program and agricultural research. Agric Hist. 62(2):270–278. http://www.jstor.org/stable/3743299.

Sasaki T, Burr B. 2000. International Rice Genome Sequencing Project: the effort to completely sequence the rice genome. Curr Opin Plant Biol. 3(2):138–142. doi:10.1016/s1369-5266(99)00047-3. http://dx.doi.org/10.1016/s1369-5266(99)00047-3.

Kidd G. 1995. Nature Biotechnology. 13:433.

Krogmann DW, Key J. 1981. The agriculture grants program. Science. 213(4504):178–182. doi:10.1126/science.213.4504.178. http://dx.doi.org/10.1126/science.213.4504.178.

Lightsources.org. Lightsources.org. [accessed 2024 Jun 12]. https://lightsources.org/.

MSU-DOE Plant Research Laboratory History: The First 50 Years (1965-2015). Msu.edu. [accessed 2024 Jun 12]. https://prl.natsci.msu.edu/_assets/files/PRL_History.pdf.

National Science and Technology Council. 1998 Jan. Archives.gov. [accessed 2024 Jun 12]. https://clintonwhitehouse3.archives.gov/WH/EOP/OSTP/NSTC/html/npgireport.html.

National Science Foundation, (NSF). 1990. A Long-range Plan for the Multinational Coordinated Arabidopsis Thaliana Genome Research Project. National Science Foundation.

NIFA. National Institute of Food and Agriculture. Usda.gov. [accessed 2024 Jun 12]. https://www.nifa.usda.gov/.

NIH — office of budget —appropriations history by institute/center (1938 to present). Nih.gov. [accessed 2024 Jun 12]. https://officeofbudget.od.nih.gov/approp_hist.html.

NPGI. 2009. NPGI Five-Year Plan 2009-2014. Nsf.gov. [accessed 2024 Jun 12]. https://www.nsf.gov/bio/pubs/reports/npgi_five_year_plan_2009_2013.pdf.

NPGI. 2014. NPGI Five-Year Plan 2014-2018. Archives.gov. [accessed 2024 Jun 12]. https://obamawhitehouse.archives.gov/sites/default/files/microsites/ostp/NSTC/npgi_five-year_plan_5-2014.pdf.

NSF. 2010 May 12. NSF provides $20 million to support 15 projects through BREAD program. Nsf.gov. [accessed 2024 Jun 12]. https://www.nsf.gov/news/news_summ.jsp?cntn_id=116932&org=NSF&from=news.

NSF. 2015. Basic Research to Enable Agricultural Development (BREAD) Fiscal Year 2015 Awards. Nsf.gov. [accessed 2024 Jun 12]. https://www.nsf.gov/bio/pubs/awards/bread10_fy12.htm.

NSF. Annual reports - history. Nsf.gov. [accessed 2024 Jun 12]. https://www.nsf.gov/about/history/annual-reports.jsp.

OSTI. 2022. Department of energy announces $27.4 million for research to characterize gene function in bioenergy crop plants. Osti.gov. [accessed 2024 Jun 12]. https://science.osti.gov/-/media/ber/pdf/Awards/GenomicsEnabled-Plant-Biology-FOA-2601-Award-List-2022-v2.pdf.

continued on next page

PGEC USDA. Berkeley.edu. [accessed 2024 Jun 12]. https://pgec.berkeley.edu/.

RePORT > RePORTER. Nih.gov. [accessed 2024 Jun 12]. https://reporter.nih.gov/advanced-search.

Schneider WH. 2015. The origin of the medical research grant in the United States: The Rockefeller foundation and the NIH extramural funding program. J Hist Med Allied Sci. 70(2):279–311. doi:10.1093/jhmas/jrt074. http://dx.doi.org/10.1093/jhmas/jrt074.

Space Biology program. 2020 Aug 12. Nasa.gov. [accessed 2024 Jun 12]. https://science.nasa.gov/biological-physical/programs/space-biology.

Sussman MR. 2019 Jun. Bob Rabson. Aspb.org. [accessed 2024 Jun 12]. https://aspb.org/newsletter/archive/2019/MayJun19.pdf#page=28.

U.S. Congress, Office of Technology Assessment, OTA-BA-. 1984. Commercial Biotechnology: An International Analysis. https://govinfo.library.unt.edu/ota/Ota_4/DATA/1984/8407.PDF.

U.S. Department of Energy Genomic Science Program. 2019. Plant Feedstock Genomics for Bioenergy Joint Awards 2006-2018. [accessed 2024 Jun 12]. https://genomicscience.energy.gov/usda-doe-plant-feedstock-genomics-for-bioenergy/.

US Department of Energy Genomic Science Program. Bioenergy Research Centers. genomicscience.energy.gov. [accessed 2024 Jun 12]. https://genomicscience.energy.gov/bioenergy-research-centers/.

US DOE Office of Science. 2021 Apr. Biological Systems Science Division Strategic Plan. Energy.gov. [accessed 2024 Jun 12]. https://genomicscience.energy.gov/wp-content/uploads/2021/09/BSSD_Strategic_Plan_2021.pdf.

Wikipedia contributors. 2024 Mar 26. McKnight Foundation. Wikipedia, The Free Encyclopedia. https://en.wikipedia.org/w/index.php?title=McKnight_Foundation&oldid=1215644723.

PART 6: PHILANTHROPY

An American Society of Plant Physiology/Plant Biology Tradition

Brian Larkins and Crispin Taylor

Since the founding of the Society in 1923-24, a philanthropic spirit has long been the hallmark of ASPP/ASPB members. Donations of more than $20,000 by Charles A. Shull and his wife, Lena Mearle Shull, provided funds to launch the Society and its journal, Plant Physiology. Donations were also used to create the Stephen Hales and Charles Reid Barns Awards.

Since then, multiple donations and bequests have supported and nurtured many aspects of the Society's functions. Included among them was a major gift of land and property from the Gude family that created the current headquarters of the Society in Rockville, MD. Others allowed creation of new awards, including the Robert Rabson Award, the Eric Conn Award, and the ASPB Innovation Prize for Agricultural Technology. In most instances, an initial request to the Society for creating a new award was supported by a major financial donation that was, in turn, leveraged to garner sufficient support to permanently endow it. This was the case with the most recently named award, the ASPB-Carnegie Winslow Briggs Mentorship Award. In this case, an ASPB member offered to match donations garnered by the Carnegie Institution; ultimately, over $120,000 was raised in less than a year.

Donations and bequests have also been used to support other aspects of ASPB's operations. For example, the subsidized child services that ASPB provides at annual meetings was funded initially through a temporarily restricted donation from John Radin and later by a bequest from the estate of Eli Romanoff, a former Director of the NSF Metabolic Biology Program. More recently, bequests to the Society totaling more than $500,000 came from the estates of former NSF Biology Division Director, Mary Clutter, and two long-time members of the Society, Jim Siedow, and Marc Cohn. The donations from Clutter's and Siedow's estates were unrestricted, meaning that ASPB's board and senior staff have authority to direct use of these funds. In the case of the bequest from Siedow's estate, a portion was used to support additional

SURF awards in 2023. Cohn's bequest stipulated that the Society periodically organizes a symposium on seed biology, the donor's lifelong research passion. Other donations to ASPB have come in different forms, including equities.

In 2016, at a gathering of former Presidents during the annual ASPB meeting in Minneapolis, the suggestion was made to launch a formal program to solicit donations to support ASPP; Brian Larkins agreed to lead it. Because several related scientific societies had a "legacy society" for such purposes, Larkins recommended ASPB do the same. Accordingly, in 2017 he contacted a number of long-time ASPB members, inviting them to become a "Founding Member" of the ASPB Legacy Society by making a donation of $5,000. Over the next two years, 135 individuals volunteered to do so and helped launch the ASPB Legacy Society.

The Legacy Society was created with the expectation that the principal of the donations would be held in perpetuity, while revenue generated from their investment would be made available for use by ASPB. No financial ceiling for fund-raising was set; rather, the goal was to build the Legacy Society to a point at which the interest and dividends generated from the permanently restricted donations would become sufficiently large to impact the Society's mission.

Although Brian Larkins and Crispin Taylor (ASPB CEO) initially operated the Legacy Society on an ad hoc basis, in 2018 they petitioned the ASPB Board of Trustees and Board of Directors to formally authorize a "Legacy Society Leadership Committee" (LSLC) to oversee its activities. The committee was charged with a number of responsibilities that included: a) Working with the ASPB Board of Trustees and relevant staff to establish and oversee donations; b)Developing a comprehensive strategy and timeline for future fund raising; c) Participating actively in the process of solicitation of new funds for the Legacy Society and, from time to time, making recommendations regarding whether any outside entities would be required to assist in fundraising; d) Proposing which ASPB staff could be asked to assist the Legacy Society in its efforts and in what ways; e) Preparing a document that clarifies the current fundraising strategy, proposes how the funds raised by the LSLC might be managed, and proposes how money from generated interest/dividends could be used. Final decisions on the appropriation and use of these funds are subject to approval by the Board of Trustees and Board of Directors.

Initially, the LSLC comprised three ASPB members and the ASPB Treasurer, who served as an ex-officio member. Members, appointed by the current President in consultation with the Board of Trustees, serve staggered three-year terms. The most senior member of the committee typically serves as chair. The first LSLC consisted of Ralph Quatrano, Debby Delmer, and Brian Larkins, with Kent Chapman, ASPB Treasurer, ex officio. Subsequently, Rob Last and Jerry Cohen replaced Quatrano and Delmer when they rotated off the committee, and later Rob Last replaced Brian Larkins as Chair. The committee was greatly aided by input from Machi Dilworth, who served on the ASPB Centennial Challenge Committee (described further in the following paragraphs).

In anticipation of ASPB's Centennial, in 2019 the LSLC worked with the ASPB's Board of Directors to launch the "Centennial Challenge," with a goal of raising $3 million in donations. To support this effort, the Centennial Challenge Committee (CCC) was created, which included members of the LSLC, as well as Machi Dilworth, Laura Wayne, and Ali Gardner. Among the activities associated with the Centennial Challenge was recognition of a special group of plant scientists, ASPB Pioneers, many of whom helped make the Society what it had become. These individuals provided the education and research training for many members of the ASPB community, and in some cases played leadership roles in ASPB and/or its journals.

Recognition as a Pioneer came from donations by former graduate students, postdocs, colleagues, family members, and friends who collectively contributed $5,000 or more in support of the individual's nomination. The names of contributors, and in some cases their testimonials, are listed in the link for each Pioneer, which also includes biographical information about them and their contributions to plant biology.

To date, more than 1,800 people, including many ASPB members and the broader plant science community, have donated to recognize ASPB Pioneers, raising more than $1 million.

Another focus of the LSLC and CCC was improving the ASPB's approach to soliciting donations through its website. This involved redesigning the "Support ASPB" webpage, describing ASPB's needs and priorities that merit members' financial support, and honoring those who had done so. A series of "donor circles" was created: Charles Reid Barnes Circle, $5,000-$9,999; Katherine Esau Circle, $10,000-$24,999; George Washington Carver, $25,000-$49,000; Melvin Calvin and Andrew Benson, $50,000-$74,999; Joseph Varner, $75,000-$99,999; Barbara McClintock, more than $100,000. Any individual who donated $5,000 automatically became a member of the Legacy Society, and their donations were tracked over time, allowing them to progress through the donor Circles.

Based on the enthusiastic support and generosity of ASPB members and other plant science supporters, the Legacy Society promises to become a valuable asset to ASPB and help ensure the long-term success of its members. The ASPB Board of Directors invests up to 90% of these donations and uses the dividends from them to provide financial support for ASPB activities that foster professional development and early career-centered programs. The Legacy Society's primary goal is to nurture future generations of plant biologists and, thereby, ensure the health and longevity of the Society. In this way, members contribute back to the Society that has nurtured their professional development and careers.

Part 2

Centennial Reviews from *Plant Physiology*®

https://doi.org/10.1093/plphys/kiae141 PLANT PHYSIOLOGY 2024: 195: 1–3

Plant Physiology®

Celebrating the American Society of Plant Biologists centennial anniversary: A compendium of review articles in plant biology

Nancy A. Eckardt,[1,*] **Blake C. Meyers**,[1,2] **Yunde Zhao**[3,4]

1 The Plant Cell, American Society of Plant Biologists, USA
2 UC Davis Genome Center, University of California Davis, Davis, CA 95616, USA
3 Plant Physiology, American Society of Plant Biologists, USA
4 Department of Cell and Developmental Biology, University California San Diego, La Jolla, CA 92093-0116, USA

*Author for correspondence: neckardt@aspb.org

Editorial

This year we celebrate 100 years of the American Society of Plant Biologists (ASPB; formerly the American Society of Plant Physiologists). A major motivation for the society was the intent to publish a journal devoted to plant physiology (Hanson 1989), and the first issue of *Plant Physiology* was published in January 1926. *The Plant Cell* began publication in January 1989, to fill a gap in the rapidly progressing areas of modern plant molecular cell and developmental biology (Goldberg et al. 2019). To celebrate the 100-year milestone of ASPB and the society's role in the publication of the highest-quality plant science over the ensuing years, we are pleased to present, in this issue of both journals, a compendium of review articles covering a range of topics in plant biology. The list of topics was divided more or less randomly between the 2 journals, and in each case, a more senior author was invited and asked to consider, where feasible, bringing on board one or more early- or mid-career coauthors.

The May 2024 issue of *Plant Physiology* includes 11 articles for this compendium. Two articles delve into aspects of plant metabolism. In *The end game(s) of photosynthetic carbon metabolism*, Sharkey (2024) reviews the last stages of photosynthetic metabolism in leaves that occur after the Calvin–Benson cycle, including major discoveries in sucrose and starch synthesis. Dixon and Dickinson (2024) follow with *A century of studying plant secondary metabolism—From "what?" to "where, how, and why?"* describing the fascinating history of research on small molecules known as plant secondary metabolites and proposing future areas of exploration. Five reviews cover topics in plant development and signaling. Maple et al. (2024) bring us *Flowering time: From physiology, through genetics to mechanism* covering flowering time mutants, natural variation in flowering time, and molecular mechanisms that can inform breeding strategies for new crops and changing environments. In *Light signaling in plants—a selective history*, Huq et al. (2024) review the discovery and characterization of phytochrome and cryptochrome sensory photoreceptors. Shani et al. (2024) cover all things gibberellin: biosynthesis, metabolism, perception, signaling, and transport in *Highlights in gibberellin research: A tale of the dwarf and the slender*. Gasperini and Howe (2024) then guide us through the "blurred distinction between classical phytohormones and other bioactive metabolites" using jasmonate as an example in their review *Phytohormones in a universe of regulatory metabolites: Lessons from jasmonate*. Kim et al. (2024) explore the convergence of signaling pathways in plant responses to dehydration and temperature changes in their review *Regulatory networks in plant responses to drought and cold stress*, with an eye toward engineering climate-resilient crops. Two reviews focus on aspects of plant reproduction. Zhong et al. (2024) review major events in plant reproduction, including gametic differentiation to pollen–pistil interaction, fertilization, and early zygotic events in *From gametes to zygote: Mechanistic advances and emerging possibilities in plant reproduction*. In *Molecular basis and evolutionary drivers of endosperm-based hybridization barriers*, Bente and Köhler

Received March 04, 2024. Accepted March 04, 2024. Advance access publication March 28, 2024.

(2024) focus on the endosperm, which plays a key role in sup-porting embryo growth and germination in flowering plants. Rounding out the set in *Plant Physiology* are 2 reviews cover-ing topics in cell biology and synthetic biology. In *A charged existence: A century of transmembrane ion transport in plants*, Blatt (2024) brings us into the field of electrophysiology, fo-cusing on stomatal guard cells, with an engaging review of the history and current state of research on ion transport across plant membranes. Finally, in *A pilot oral history of plant synthetic biology*, Joshi and Hanson (2024) provide a fas-cinating glimpse into the world of plant synthetic biology, re-vealed through interviews with eight of the pioneering researchers in the field.

The May 2024 issue of *The Plant Cell* includes 17 articles for this compendium. Four articles treat topics in genomics and evo-lution. Paterson and Queitsch (2024) explore the relationship between angiosperm genome organization and botanical di-versity in *Genome organization and botanical diversity*, show-ing how the indeterminate development of plants may have led to some of the major differences in gene regulation be-tween plants and animals. Mabry et al. (2024) seek to extend the usefulness of model species across larger taxonomic groups in their review *Complementing model species with model clades*, focusing on establishing the Brassicales as a model clade. In *Domestication and the evolution of crops: Variable syndromes, complex genetic architectures, and eco-logical entanglements*, Alam and Purugganan (2024) review major themes in crop domestication research as key to devel-oping resource-efficient crops and sustainable cultivation practices. In *Why should we study plant sex chromosomes?*, Charlesworth and Harkess (2024) review the evolution of di-oecious plant species and features of sex chromosomes such as heteromorphism, recombination suppression, dosage compensation, and the repeated evolution and breakdown of dioecy. Two reviews take us into topics in cell biology. Delmer et al. (2024) review major aspects of the 5 major plant cell wall polymers, recognizing "some of the colorful historic-al figures who shaped cell wall research over the past 100 years" in *The plant cell wall—dynamic, strong, and adapt-able—is a natural shapeshifter*. Zhuang et al. (2024) highlight the important technologies and key scientists who contributed to research on plant organelles and the endomembrane system over the last century in *A century journey of organelles research in the plant endomembrane system*. Three reviews cover fundamental topics in plant development. In *Reflections on the ABC model of flower development*, Bowman and Moyroud (2024) review the colorful history of the ABC model of flower development, highlighting "un-solved riddles still hidden in the floral alphabet." Sajeev et al. (2024) review the complex and interlinked processes of seed dormancy and germination in *A commitment for life: Decades of unraveling the molecular mechanisms behind seed dormancy and germination*. In their review *Not so hidden anymore: Advances and challenges in understanding root growth under water deficits*, Voothuluru et al. (2024) explore biophysical and metabolic constraints affecting the interactions of root systems with the soil environment and plants' ability to maintain access to soil water under water deficit conditions. Six more articles explore various topics in phytohormones and signaling. In *An auxin research odys-sey: 1989–2023*, Cohen and Strader (2024) review this "master regulator" of plant processes, the role of auxin in spatial and temporal regulation of plant growth and development, and the role *The Plant Cell* has played in publishing seminal work in this area. In *Cytokinin: From autoclaved DNA to two-component signaling*, Argueso and Kieber (2024) review cyto-kinin function, metabolism, transport, and the nature of cytokinin 2-component signaling pathways. Spoel and Dong (2024) review the major plant defense hormone salicyl-ic acid and its crosstalk with other phytohormones in *Salicylic acid in plant immunity and beyond*. Dodds et al. (2024) review some of the major themes in plant immunity in *Pathogen perception and signaling in plant immunity*, including the gene-for-gene hypothesis, nucleotide-binding/leucine-rich-repeat receptor (NLR) receptors, pattern-triggered and effector-triggered immunity, and efforts in NLR engineering. In *Nitrogen sensing and regulatory networks: It's about time and space*, Shanks et al. (2024) focus on temporal and spatial aspects of nitrogen sensing and signaling and discuss efforts to improve nitrogen-use efficiency in crops. Yang et al. (2024) turn to the essential nutrient phosphorus and review the molecular mechanisms that regulate phosphorous starvation responses and phosphorus transport, sensing, and signaling in *Milestones in understanding transport, sensing, and signal-ing of the plant nutrient phosphorus*. Finally, 2 reviews focus on aspects of transcription and translation. Marathe et al. (2024) explore the understudied area of extra-nuclear tran-scription factors that reside in the cytoplasm in an inactive state, entering the nucleus and becoming transcriptionally active only in response to specific signals, such as from biotic and abiotic stresses, in *Should I stay or should I go? Trafficking of plant extra-nuclear transcription factors*. Wu et al. (2024) provide a comprehensive yet concise and engaging review of how the translation of messenger RNA is regulated in plants, providing an excellent summary and useful illustra-tions of this complex topic in *What, where, and how: Regulation of translation and the translational landscape in plants*.

We acknowledge that this collection is far from a complete representation of the major topics in plant biology. Life often gets in the way of the best intentions, and some invited authors were unable to submit in time to make publication in this issue. Other topics were omitted if recently covered, for example, with extensive reviews in recent or upcoming fo-cus issues in either journal (including e.g. many topics within the areas of cell biology, gene editing, RNA biology, plant im-munity, photosynthesis, and proteolysis). We apologize if your favorite topic is not included. We thank Brian Larkins for the initial ideas and discussions that led to this project. We hope our readers find these articles instructive and in-spirational as avenues into the next 100 years of plant biology research.

Conflict of interest statement. None declared.

References

Alam O, Purugganan MD. Domestication and the evolution of crops: variable syndromes, complex genetic architectures, and ecological entanglements. Plant Cell. 2024:**36**(5):1227–1241. https://doi.org/10.1093/plcell/koae013

Argueso CT, Kieber JJ. Cytokinin: from autoclaved DNA to two-component signaling. Plant Cell. 2024:**36**(5):1429–1450. https://doi.org/10.1093/plcell/koad327

Bente H, Köhler C. Molecular basis and evolutionary drivers of endosperm-based hybridization barriers. Plant Physiol. 2024:**195**(1): 155–169. https://doi.org/10.1093/plphys/kiae050

Blatt MR. A charged existence: a century of transmembrane ion transport in plants. Plant Physiol. 2024:**195**(1):79–110. https://doi.org/10.1093/plphys/kiad630

Bowman JL, Moyroud E. Reflections on the ABC model of flower development. Plant Cell. 2024:**36**(5):1334–1357. https://doi.org/10.1093/plcell/koae044

Charlesworth D, Harkess A. Why should we study plant sex chromosomes? Plant Cell. 2024:**36**(5):1242–1256. https://doi.org/10.1093/plcell/koad278

Cohen JD, Strader LC. An auxin research odyssey: 1989–2023. Plant Cell. 2024:**36**(5):1410–1428. https://doi.org/10.1093/plcell/koae054

Delmer D, Dixon RA, Keegstra K, Mohnen D. The plant cell wall—dynamic, strong, and adaptable—is a natural shapeshifter. Plant Cell. 2024:**36**(5):1257–1311. https://doi.org/10.1093/plcell/koad325

Dixon RA, Dickinson AJ. A century of studying plant secondary metabolism—from "what?" to "where, how, and why?" Plant Physiol. 2024:**195**(1):48–66. https://doi.org/10.1093/plphys/kiad596

Dodds PN, Chen J, Outram MA. Pathogen perception and signaling in plant immunity. Plant Cell. 2024:**36**(5):1465–1481. https://doi.org/10.1093/plcell/koae020

Gasperini D, Howe GA. Phytohormones in a universe of regulatory metabolites: lessons from jasmonate. Plant Physiol. 2024:**195**(1):135–154. https://doi.org/10.1093/plphys/kiae045

Goldberg RB, Larkins BA, Quatrano RS. The plant cell: beginnings. Plant Cell. 2019:**31**(7):1391–1392. https://doi.org/10.1105/tpc.19.00347

Hanson JB. History of the American Society of Plant Physiologists. Am. Soc. Plant Physiol. 1989. https://aspb.org/about/history/

Huq E, Lin C, Quail PH. Light signaling in plants—a selective history. Plant Physiol. 2024:**195**(1):213–231. https://doi.org/10.1093/plphys/kiae110

Joshi J, Hanson AD. A pilot oral history of plant synthetic biology. Plant Physiol. 2024:**195**(1):36–47. https://doi.org/10.1093/plphys/kiad585

Kim J-S, Kidokoro S, Yamaguchi-Shinozaki K, Shinozaki K. Regulatory networks in plant responses to drought and cold stress. Plant Physiol. 2024:**195**(1):170–189. https://doi.org/10.1093/plphys/kiae105

Mabry ME, Abrahams RS, Al-Shehbaz IA, Baker WJ, Barak S, Barker MS, Barrett RL, Beric A, Bhattacharya S, Carey SB, et al. Complementing model species with model clades. Plant Cell. 2024: **36**(5):1205–1226. https://doi.org/10.1093/plcell/koad260

Maple R, Zhu P, Hepworth J, Wang J-W, Dean C. Flowering time: from physiology, through genetics to mechanism. Plant Physiol. 2024: **195**(1):190–212. https://doi.org/10.1093/plphys/kiae109

Marathe S, Grotewold E, Otegui MS. Should I stay or should I go? Trafficking of plant extra-nuclear transcription factors. Plant Cell. 2024:**36**(5):1524–1539. https://doi.org/10.1093/plcell/koad277

Paterson AH, Queitsch C. Genome organization and botanical diversity. Plant Cell. 2024:**36**(5):1186–1204. https://doi.org/10.1093/plcell/koae045

Sajeev N, Koornneef M, Bentsink L. A commitment for *life*: decades of unraveling the molecular mechanisms behind seed dormancy and germination. Plant Cell. 2024:**36**(5):1358–1376. https://doi.org/10.1093/plcell/koad328

Shani E, Hedden P, Sun T-p. Highlights in gibberellin research: a tale of the dwarf and the slender. Plant Physiol. 2024:**195**(1):111–134. https://doi.org/10.1093/plphys/kiae044

Shanks CM, Rothkegel K, Brooks MD, Cheng C-Y, Alvarez JM, Ruffel S, Krouk G, Gutiérrez RA, Coruzzi GM. Nitrogen sensing and regulatory networks: it's about time and space. Plant Cell. 2024:**36**(5): 1482–1503. https://doi.org/10.1093/plcell/koae038

Sharkey TC. The end game(s) of photosynthetic carbon metabolism. Plant Physiol. 2024:**195**(1):67–78. https://doi.org/10.1093/plphys/kiad601

Spoel SH, Dong X. Salicylic acid in plant immunity and beyond. Plant Cell. 2024:**36**(5):1451–1464. https://doi.org/10.1093/plcell/koad329

Voothuluru P, Wu Y, Sharp RE. Not so hidden anymore: advances and challenges in understanding root growth under water deficits. Plant Cell. 2024:**36**(5):1377–1409. https://doi.org/10.1093/plcell/koae055

Wu HL, Jen J, Hsu PY. What, where, and how: regulation of translation and the translational landscape in plants. Plant Cell. 2024:**36**(5): 1540–1564. https://doi.org/10.1093/plcell/koad197

Yang S-Y, Lin W-Y, Hsiao Y-M, Chiou T-J. Milestones in understanding transport, sensing, and signaling of the plant nutrient phosphorus. Plant Cell. 2024:**36**(5):1504–1523. https://doi.org/10.1093/plcell/koad326

Zhong S, Zhao P, Peng X, Li H-J, Duan Q, Cheung AY. From gametes to zygote: mechanistic advances and emerging possibilities in plant reproduction. Plant Physiol. 2024:**195**(1):4–35. https://doi.org/10.1093/plphys/kiae125

Zhuang X, Li R, Jiang L. A century journey of organelles research in the plant endomembrane system. Plant Cell. 2024:**36**(5):1312–1333. https://doi.org/10.1093/plcell/koae004

https://doi.org/10.1093/plphys/kiae125 PLANT PHYSIOLOGY 2024: 195: 4–35

Plant Physiology®

Update

From gametes to zygote: Mechanistic advances and emerging possibilities in plant reproduction

Sheng Zhong,[1,†] **Peng Zhao**,[2,3,†] **Xiongbo Peng**,[2,3,†] **Hong-Ju Li**,[4,†] **Qiaohong Duan**,[5,†] **Alice Y. Cheung**[6,*,†]

1 State Key Laboratory for Protein and Plant Gene Research, Peking-Tsinghua Center for Life Sciences, New Cornerstone Science Laboratory, College of Life Sciences, Peking University, Beijing 100871, China
2 State Key Laboratory of Hybrid Rice, College of Life Sciences, Wuhan University, Wuhan 430072, China
3 Hubei Hongshan Laboratory, Wuhan 430070, China
4 Key Laboratory of Seed Innovation, Institute of Genetics and Developmental Biology, Chinese Academy of Sciences, Center for Molecular Agrobiology, Institute of Genetics and Developmental Biology, Chinese Academy of Sciences, Beijing 100101, China
5 College of Horticulture Science and Engineering, Shandong Agricultural University, Tai'an, Shandong 271018, China
6 Department of Biochemistry and Molecular Biology, Molecular and Cellular Biology Program, Plant Biology Graduate Program, University of Massachusetts, Amherst, MA 01003, USA

*Author for correspondence: acheung@biochem.umass.edu
†All authors contributed equally.

Introduction

Research in angiosperm sexual reproduction is among the most classical botanical studies with a rich knowledge base. The introduction of molecular tools since the late 1970s, especially the development of genomic, transcriptomics, and gene identification tools, and increasingly sensitive biochemical and cell biological methods have culminated in explosive advances in the molecular and mechanistic understanding of reproduction in flowering plants. With outcrossing being crucial for hybrid vigor, understanding the molecular mechanisms of "self-incompatibility" (SI), which prevents inbreeding, led the study of plant reproduction into the molecular era. These efforts had achieved significant milestones (see Broz and Bedinger 2021; Díaz et al. 2021; Goring et al. 2023; Nasrallah 2023) before the molecular nuts and bolts that enable fertilization emerged. Here, we use SI studies from the 1980s and 1990s as the watershed to begin our discussions of mechanistic advances ranging from the developmental standpoints of gamete production to the cell biological perspectives of pollen-pistil interactions that enable fertilization and post-fertilization zygotic development (see Johnson et al. 2019; Hater et al. 2020; Hafidh and Honys 2021; Zhao et al. 2022a) (Fig. 1). We shall also discuss strategies that maximize reproductive success and how they might be deployed to impact agriculture. While many of our examples come from the model plant Arabidopsis, studies in other model or crop species provide a richness that we also tapped into for our discussion. Box 1 provides a glossary for terms; Box 2 looks toward where future efforts would most benefit from current knowledge.

Male and female gametogenesis

Flowering plants produce germ cells via the sequential patterning events of sporogenesis and gametogenesis (Fig. 1, A and B) (Ma 2005; Yang et al. 2010; Hater et al. 2020; Hafidh and Honys 2021; Cai et al. 2022). Here, we focus on cell fate determination of male germline (MG) cells, the generative cell (GC) and its mitotic product sperm cells (SCs) in male gametogenesis, and of the female gametes, the egg and the central cell, in female gametogenesis.

Differentiation of MG cells and vegetative cell

Milestone discoveries on MG cell fate determination can be considered 2-fold. One is the importance of the asymmetric division of a microspore to produce the larger vegetative cell

Received December 12, 2023. Accepted February 13, 2024. Advance access publication March 2, 2024.

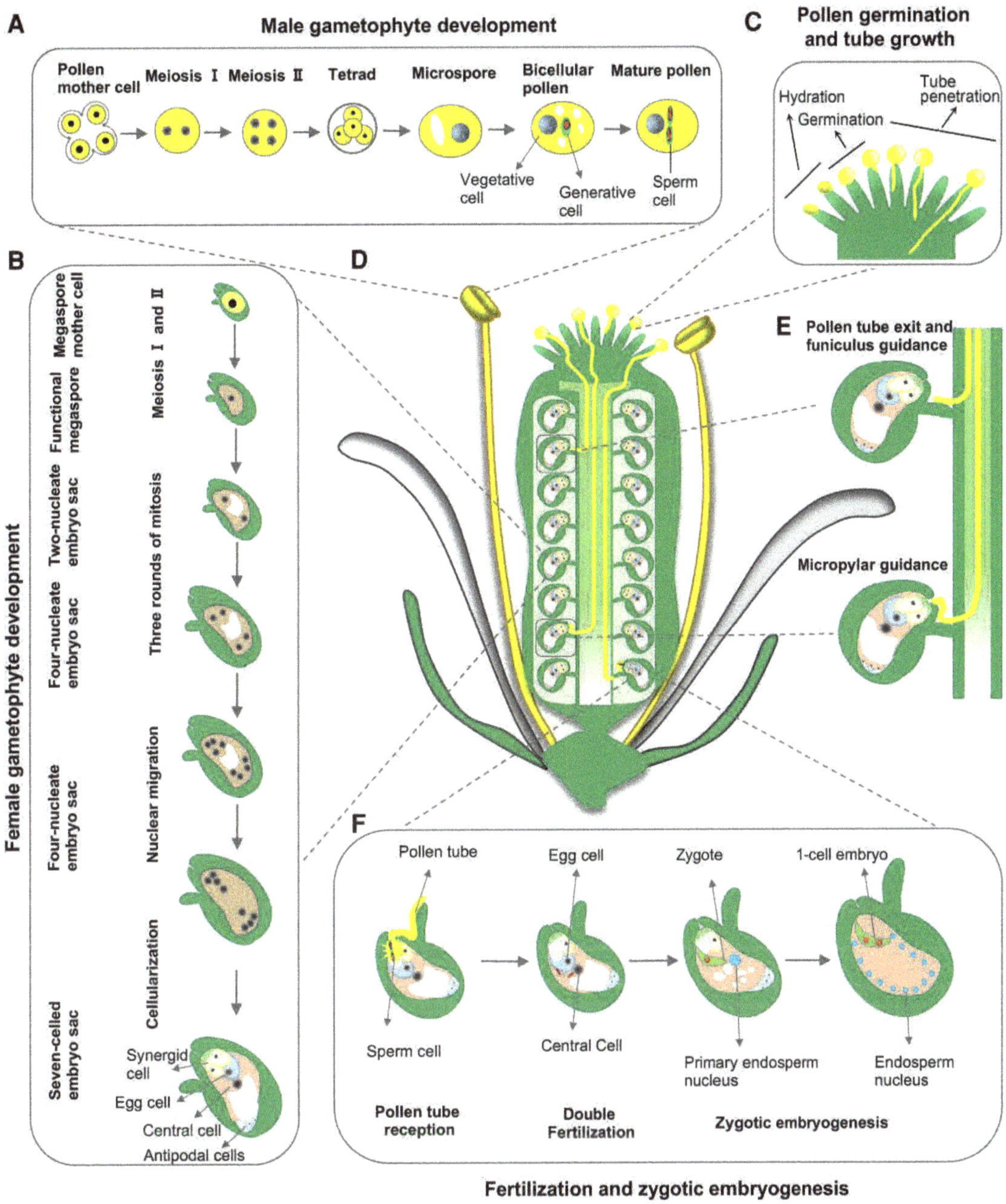

Figure 1. A schematic overview of the reproduction process in flowering plants. The eudicot Arabidopsis is used as a model for illustration. **A)** Male gametogenesis. Two-step meiosis in pollen mother cells produce 4 haploid microspores as a tetrad. Each microspore then undergoes an asymmetrical division to give rise to a large vegetative cell and a small generative cell. The generative cell then undergoes another mitotic division, leading to the production of 2 sperm cells giving rise to tricellular pollen. In some species with bicellular pollen, for example, tobacco, the second mitotic division of the generative cell occurs in the pollen tube. **B)** Female gametogenesis. Meiosis of the megaspore mother cell producing the functional megaspore. The megaspore undergoes 3 rounds of nuclear division, resulting in a syncytial female gametophyte (embryo sac) composed of 8 nuclei. Cytokinesis follows, resulting in a mature female gametophyte comprising the egg cell, central cell, antipodal cells, and synergid cells, each with specialized functions. **C)** Pollen germination and tube growth. After landing on a receptive stigma, pollen grains undergo rehydration and activation to initiate germination, leading to the formation of a pollen tube that develops from the vegetative cell. Hydration is almost instantaneous and temporally well-resolved from the later event of pollen tube penetration into the stigmatic surface. **D)** The pollen tube growth path in the pistil. Sperm cells are immotile and rely on the growth of the pollen tube in the stylar and transmitting tissues to reach the ovules where the target female gametophyte is located. **E)** Pollen tube guidance to ovules. In the transmitting tract, the pollen tubes grow towards the ovules by navigating through a series of female reproductive tissues. Attractants secreted by the ovules regulate this process. **F)** Pollen tube reception and double fertilization. Once the pollen tube reaches its destination, it releases its cargo—2 sperm cells—in a process referred to as pollen tube reception. The egg cell is fertilized by 1 sperm cell to form the zygote, and the central cell is fertilized by another sperm cell to initiate the development of the endosperm. Box 1 provides a glossary for nomenclature and terms. Supplementary Table S1 provides a list of components important for the gametogenesis processes, including some not discussed here.

Box 1. A glossary of nomenclature and terms.

Microsporogenesis and Microgametogenesis
- Diploid pollen mother cell (PMC)
- Male germline (MG) cells
  - a generative cell (GC) and its mitotic twin sperm cells (SCs)
- Male gametophyte (pollen)
  - VC (the pollen tube cell)
  - GC and SCs

Megasporogenesis and Megagametogenesis
- Megaspore mother cell (MMC)
- Functional megaspore (FM)
- Embryo sac (ES), female gametophyte (FG);

The polygonum type is monosporic, 8-nucleate and 7-celled at maturity.
- Synergid cells: twin cells at the entrance to the FG; an arriving pollen tube penetrates one of these
- Egg and central cells: female gametes for double fertilization producing the zygote and endosperm of a future seed
- Antipodal cell: for nutrient provisions to the FG

Pre-zygotic events
- Pistillate structures
  - Stigmatic papilla: pollen receptive surface.
  - Transmitting tract: a specialized tissue with highly secretory cells that produce an extracellular matrix to support pollen tube growth.
  - Septum: a partition or wall-like structure/tissue that divides the locules or compartments.
  - Funicular: a stalk-like structure that connects the ovule to the placenta.
  - Ovules: the target organ for pollen tubes that supports the development of the female gametophyte, the precursors to future seeds after fertilization.
  - Micropyle: a small opening or pore in the integuments of an ovule to allow pollen tube entrance.
- Pollen tube, a single highly polarized cell developed from the pollen grain specialized as a vehicle to transport sperm to the female gametophyte for fertilization.
- Pollen tube guidance, the process that the pollen tube is tightly navigated to the target embryo sac along various female tissues:
  - Funiculus guidance, the process that the pollen tube is signaled to exit the septum, then guided toward an ovule along the funiculus, the stalk that connects the ovule to the septum.
  - Ovular guidance, the process by which the ovule itself plays a role in guiding the pollen tube towards it, typically through chemical attractants or other cues that emanate from the ovule.
- Pollen tube reception, the process where the pollen tube, upon entering the micropyle, establishes contact and interaction with the filiform apparatus, triggering the degeneration of the synergid cell and eventually leading to the pollen tube rupture.
- Receptive synergid, the synergid cell that is penetrated by the arriving pollen tube and becomes degenerative.
- Persistent synergid, the synergid cell that is not penetrated by the arriving pollen tube and plays an important "reserved" role for "fertilization recovery."
- "Polytubey": the phenotype of supernumerary pollen tube penetration of an ovule/the female gametophyte
- Fertilization recovery: the ability of different tissues within the pistil to activate and support growth and entry of additional pollen tubes into the ovule to bring new sperm cells to salvage double fertilization when the egg and/or the central cell fail to be fertilized.

Gametic fusion and early zygotic events
- Gamete recognition: the process by which male and female gametes recognize and fuse with each other during fertilization.
- Gametic fusion: gametic fusion comprises 2 sequential processes: plasmogamy, the fusion of the cytoplasm of male and female gametic cells, and karyogamy, the fusion of the nuclei of male and female gametes during fertilization.
- Zygotic genome activation (ZGA): the onset of de novo transcription from the zygotic genome, which occurs shortly after fertilization in flowering plants.

Elaborating strategies
- Polyspermy control: ensures single pollen tube entry per ovule, maximizing ovule occupancy, seed yield and protecting progeny health.
- Fertilization recovery: salvage mechanisms when fertilization fails to ensure production of some progeny ensuring species survival.
- Reproductive barriers: intraspecific barriers facilitate out-crossing, reducing inbred suppression; interspecific barriers prevent hybridizations that are likely to result in aborted reproduction, or differentiate between self and pollen from closely related species to favor conspecific fertilization and species preservation.
- Apomixis: generating seeds that contain embryos genetically identical with the mother plants.

Box 2. A look to the future.

- Developmentally, how the multiple factors that control cell fate determination during male and female gametogenesis coordinate to regulate the differentiation of VC, GC, and the female gametophyte remains to be elucidated.
- For the prezygotic phase:
  - The emerging mechanistic understanding for stigma interaction with pollen, pollen tube attraction, reception, prevention of polyspermy and fertilization suggest diverse strategies. A cohesive understanding integrating available knowledge is needed to provide solid foundations for future efforts.
  - The in vivo configurations of various FER-related signaling complexes need to be determined. Is there a predominant configuration, or are various possible configurations assembled dynamically in response to different conditions?
  - With FER and related receptor kinases at the core of multiple strategies, the precise contributions from other positive and negative regulatory components need to be defined.
  - Molecular and biochemical understanding of pre-ovular and funiculus guidance of pollen tubes remains limited. This knowledge will provide useful context to formulate working hypotheses to explore how polyspermy blocks and fertilization recovery are orchestrated.
  - The ability to monitor the pollen tube growth and targeting process inside the pistil (e.g. Mizuta et al. 2023) and with intracellular definitions will illuminate how the pollen tube growth machinery responds to pistillate cues in vivo.
  - Envisioning an all-encompassing mechanism on how fertilization recovery is achieved remains challenging; multiple mechanisms likely exist. Investigating the temporal and spatial correlation between the degradation of the receptive synergid cell and the 2 fertilization events should provide more clues for identifying the initiating signals of persistent synergid cell degradation. Understanding how the central cell senses and reacts to the degeneration of both synergid cells activates the transcription and release of its pool of "back-up" pollen tube attractants should provide insights on the hypothesis that accessory cells in the FG had evolved to render more efficient fertilization.
- For the fertilization phase, whether and how the egg cell impacts its own precedence over the central cell for being fertilized requires further investigations. Achieving this will require identifying factors in both the egg cell and central cell that are responsible for gamete recognition, attachment and fusion.
- For post-fertilization, ZGA is an emerging area with the possibility to probe pioneer factors responsible for the first wave of zygotic genome transcription and the initiation of zygotic embryogenesis. Illustration of epigenetic characteristics including DNA methylation, histone methylation, and other modifications in both gametes and early embryos will promote the understanding of mechanisms for ZGA.
- Reproductive barriers maintained at different male-female interaction steps involved distinct strategies, each could also be diverse in mechanisms. Currently, compared with research on postzygotic hybridization barriers (in rice, for example), studies in prezygotic hybridization barriers, especially those with clear molecular barriers, mechanisms, are very limited. Proof-of-principle experiments to overcome interspecific barriers at the stigma appear promising.
- Recent advances in generating synthetic apomixis are promising. Further investigations are necessary to devise a synthetic apomixis strategy for hybrid in different crops, ensuring elevated seed-setting rates and improved cloning seed induction rates. Advancements in haploid induction and MiMe processes offer promising avenues for implementing synthetic apomixis in staple crops beyond rice.

(VC) of the male gametophyte (pollen) and the smaller GC (Fig. 1) (Eady et al. 1995; Twell et al. 2002). The other is the role of the R2R3 transcription factor DUO POLLEN 1 (DUO1) in controlling sperm formation from the GC (Rotman et al. 2005; Brownfield et al. 2009). The asymmetric division of the microspore follows the microtubule-dependent migration of its nucleus to the future germ cell pole. Failing to do so results in symmetric mitosis and both daughter cells adopting the VC fate (Eady et al. 1995; Chen and McCormick 1996; Park et al. 1998; Park et al. 2004; Zeng et al. 2009; Oh et al. 2010; Oh et al. 2016; Lee et al. 2017). Loss of the functionally redundant helix-loop-helix transcription factors *BONOBO1* and *BONOBO2* resulted in the smaller daughter cell from microspore asymmetric division failing to adopt the GC fate, implying a broader genetic network underlies GM cell fate determination (Yamaoka et al. 2018).

Cell cycle progression is critical for GC mitosis to produce 2 SCs. Mutations in several cell cycle regulators blocked GC division, resulting in bicellular pollen harboring only a single MG cell (Iwakawa et al. 2006; Nowack et al. 2006; Kim et al. 2008; Liu et al. 2008b; Gusti et al. 2009). The A-type cyclin-dependent kinase CDKA;1 is crucial for the GC progression through S-phase (Iwakawa et al. 2006; Nowack et al. 2006). Defective chromatin assembly causes cell cycle arrest, producing pollen with a single MG cell (Chen et al. 2008). DUO1 is expressed specifically in the MG and controls the expression of the G2/M regulator CYCLIN B1;1 (Rotman et al. 2005; Brownfield et al. 2009), and single MG cell phenotypes are also detected in a group of *duo* pollen (Durbarry et al. 2005). The single MG cell in pollen with mutations that directly target cell cycle processes differentiates sperm cell character, including the full capacity for fertilization (Chen et al. 2008; Gusti et al. 2009; Aw et al. 2010). Contrarily, the single MG cell in *duo1* lacks fertilization capacity (Rotman et al. 2005), consistent with DUO1 controlling the expression of several MG-specific genes, including the gamete fusogen *GCS1/HAP2* (GENERATIVE CELL SPECIFIC 1/HAPLESS2) on the sperm cell surface required for fertilization (Mori et al. 2006; Borg et al. 2011).

Epigenetic reprogramming, especially the erasure and remodeling of histone methylation, also plays an important role in the differentiation of the VC and MG cells. The chromatin is diffuse in the VC but highly condensed in the GC/SCs. Some histone variants showed VC- or SC-specific expression (Borg and Berger 2015; Jiang et al. 2020). The VC, GC, and SC nuclei also harbor differential levels of various histone modifications, and the histone3 mark H3K27me3 is essential for the VC fate as its erasure resulted in the VC fate shifting toward the MG cell fate (Houben et al. 2011; Borg et al. 2020; Huang and Sun 2022). The replacement of H3.1 by H3.10, silencing of H3K27me3 writers, and activation of H3K27me3 erasers/demethylases together might underlie how the H3K27me3 marks are globally lost from Arabidopsis SCs (Ingouff et al. 2010; Borg et al. 2020).

The mature pollen grain

The complexity of the surface of the mature pollen grain and the diverse patterns of apertures on pollen grains have long intrigued biologists (Furness and Rudall 2004; Quilichini et al. 2014). The development of increasingly sensitive analytical tools has enabled considerable advances in characterizing the protective sporollenin that sculpts the outer wall of pollen grains (Li et al. 2019). Genetics studies, especially of transcriptional cascades controlled by the transcription factor MS188, have uncovered many key regulatory and biosynthetic genes for the pollen cell wall (see Ariizumi and Toriyama 2011; Wang et al. 2018; Dehors et al. 2019; Ma et al. 2021; Yao et al. 2022; Xue et al. 2023).

Pollen from most species germinates extruding the pollen tube through an aperture, although some species can extrude a tube directly through the pollen wall (Edlund et al. 2016). Arabidopsis pollen has 3 equidistantly spaced apertures. Loss of *INAPERTURATE POLLEN 1* (*INP1*), encoding a membrane protein, resulted in mutant pollen with no aperture, though they remained fertile (Dobritsa and Coerper 2012). INP1 marks pollen cell membrane domains early in pollen development, and they are specified to become lightly covered by the pollen wall and differentiate into 3 spatially restricted pollen apertures around the pollen equator (Dobritsa and Coerper 2012; Lee et al. 2018; Zhou and Dobritsa 2019; Lee et al. 2021; Mazuecos-Aguilera et al. 2021). Another family of proteins, MACARONs, function upstream of INP1 and quantitatively determine the number and patterning of pollen apertures (Zhou et al. 2021a).

INP1 is conserved in cereal crops, such as maize and rice, which typically have a single pollen aperture (Christensen et al. 1997). Although the maize INP1 could functionally substitute for the Arabidopsis INP1, the maize inaperturate pollen is sterile (Lee et al. 2018). In rice, interaction between INP1 and a lectin receptor-like kinase is crucial for aperture formation (Zhang et al. 2020b).

Differentiation of synergid cell, egg cell, and central cell

In polygonum-type embryo sac (ES), development is staged between FG (female gametophyte)1 to FG7, starting from a functional megaspore (FM) (Christensen et al. 1997), which undergoes nuclear division, nuclear migration, and cellularization, producing a mature FG with 7 cells (Fig. 1). Each member cell type in the ES has its function to ensure double fertilization (Box 1). The 2 synergid cells attract and receive pollen tubes (Higashiyama 2002). The 2 sperm cells, 1 fuses with the egg and the other with the central cell for a double fertilization to form the zygote and the endosperm of a future seed. Three antipodal cells may function in nutrient transfer, especially in the ES of species with as many as 100 cells before fertilization (Engell 1994). Consistent with their functional differentiation, member cells in the ES have different transcriptomes (Kumlehn et al. 2001; Le et al. 2005; Sprunck et al. 2005; Okuda et al. 2009; Wuest et al. 2010; Ohnishi et al. 2011; Zhao et al. 2011; Anderson et al. 2013; Yu et al. 2016; Song et al. 2020; Susaki et al. 2021). These enabled detailed functional studies and facilitated the development of marker genes for individual cell types, greatly aiding live-imaging of processes pre- and post-pollen tube arrival (Kasahara et al. 2005; Portereiko et al. 2006; Steffen et al. 2007; Okuda et al. 2009; Sprunck et al. 2012; Chen et al. 2017; Tedeschi et al. 2017; Yu et al. 2021). Cell type–specific transcription factors play crucial roles in determining the synergid, egg, or central cell fate. Among these, the transcription factor MYB98 is crucial for the synergid cell fate determination and maintenance, RKD (RWP-RK domain) proteins control the egg cell fate, AGAMOUS-like 80 promotes central cell fate and inhibits MYB98 (Portereiko et al. 2006; Bemer et al. 2008; Steffen et al. 2008; Zhang et al. 2020c), while MYB119 and BYB64 redundantly restrict the central cell fate (Rabiger and Drews 2013).

Interestingly, female gametes also regulate fate specialization of adjacent cells. For example, a mutation in the egg and central cell–expressed LACHESIS (LIS), a homolog of the yeast mRNA splicing factor PRP4, induced the mis-expression of egg-specific genes in the mutant synergid and central cells, and the antipodal cells in the mutant resembled the central cell (Gross-Hardt et al. 2007). A similar phenotype as *lis-1* was also observed in other splicing factor mutants (Moll et al. 2008). *LIS* transcript levels in the egg cell specifically affected the development of all female gametophytic cells (Völz et al. 2012), further supporting an important role for RNA maturation. Cell ablation studies also supported the idea that the egg cell affects synergid cell specification (Lawit et al. 2013; Susaki et al. 2015).

Nuclear positions before cellularization in the ES are important for cell fate determination. In Arabidopsis, nuclei positions at FG5 already mark the post-cellularization destiny of these nuclei (Christensen et al. 1997; Susaki et al. 2021). When the typical nuclear location in the cell destined to be the egg in FG5 was replaced by 2 nuclei, an extra egg cell formed at the expense of a synergid cell (Gross-Hardt et al. 2007; Pagnussat et al. 2007; Kirioukhova et al. 2011). Disrupting the actin cytoskeleton before cellularization impaired nuclear migration and resulted in the formation of 2 functional egg cells. Both could be fertilized, producing double embryos, thus further supporting that the final positions of the nuclei, rather than nuclear lineages, play critical roles in the specification of the egg cell (Sun et al. 2021). Extra rounds of free nuclear divisions in the maize mutant *indeterminate gametophyte 1* ES also resulted in extra synergid, egg, and central cells according to nuclear position in ES (Guo et al. 2004). Auxin and cytokinin are also important for ES cell fate determination (Pagnussat et al. 2009; Yuan et al. 2016; Sun et al. 2021) and might rely on uneven distribution of these hormones along the chalazal to antipodal axis of the ES (Skinner and Sundaresan 2018). Disrupting normal auxin synthesis or distribution pattern impaired egg cell specialization (Pagnussat et al. 2009; Panoli et al. 2015; Sun et al. 2021). Cytokinin signaling, from perception to phosphorylation cascades and downstream response regulators, is also critical for FG development, in particular, central cell specification (Kong et al. 2015; Yuan et al. 2016; Zhu et al. 2022).

Pre-zygotic events

Impressive advances have been made in elucidating key mo-lecular players that enable prezygotic pollen-pistil interac-tions and render fertilization possible (Fig. 1, C–E) (Johnson et al. 2019). An elaborate network for structural and signaling components for the elongating pollen tube has been established for the polarized cell growth process in vitro (Steer and Steer 1989; Cheung and Wu 2008; Hepler and Winship 2015; Michard et al. 2017; Cameron and Geitmann 2018). Here, we discuss several related signal-ing modules important for pollen-pistil cooperation and how their studies have advanced the mechanistic under-standing of pollen tube growth and ovule targeting in the pistil (Figs. 2 and 3).

The FERONIA (FER) receptor kinase and related signaling modules

The establishment of the FER-LORELEI (LRE) signaling module (Fig. 2, B and C) provided an invaluable entry point into the mechanistic dissections of the pre-zygotic processes (Cheung et al. 2022). FER-LRE is a key regulator of pollen tube reception at the pollen tube-FG interface, w here the pollen tube would penetrate the FG, burst and release sperm for fertilization (Fig. 1E; Box 1). FER is a malectin domain–containing receptor kinase (Escobar-Restrepo et al. 2007; Franck et al. 2018a, 2018b; Yang et al. 2021), and LRE is a glycosylphosphatidylinositol-anchored protein (GPI-AP) (Capron et al. 2008; Tsukamoto et al. 2010). As a coreceptor pair (Li et al. 2015a; Xiao et al. 2019), FER and LRE colocalize in the synergid cell membrane that densely packed the filiform apparatus (Box 1), the synergid cell wall gating the entrance to the female gametophyte. FER is 1 of 17 related receptor kinases in Arabidopsis. It is expressed almost ubiquitously except in pollen and plays profound roles for growth and survival (Li et al. 2016; Malivert and Hamant 2023; Cheung 2024; Liu et al. 2024). LRE is expressed almost exclusively in the FG; LRE-like GPI-AP1 (LLG1) is expressed in sporophytic tissues throughout development, including in the stigma, while LLG2 and LLG3 are pollen-specific (Noble et al. 2022).

Studies of FER during early plant development identified multiple basic parameters of FER-LRE/LLG signaling and have been reviewed extensively (Liao et al. 2017; Doblas et al. 2018; Franck et al. 2018a, 2018b; Cheung et al. 2022; Ortiz-Morea et al. 2022). FER-LRE/LLGs are coreceptors (Li et al. 2015a), and RALF (RAPID ALKALINIZATION FACTOR) peptide regulators are their ligands (Haruta et al. 2014; Blackburn et al. 2020; Abarca et al. 2021). FER binds the major cell wall polysaccharide pectin (Feng et al. 2018; Lin et al. 2022). RAC/ROP GTPases and reactive oxygen species (ROS) are mediators of FER-LLG/LRE signaling (Duan et al. 2010; Li et al. 2015a; Song et al. 2021). Here, we focus on the modules that regulate prezygotic pollen-pistil interac-tions (Fig. 2, B and C).

Pollen-stigma interaction

Pollen germination and tube penetration into the stigmatic papilla Compatible pollen germination and tube penetration into the pistil involve an early phase of pollen hydration and germin-ation, that is, extruding a pollen tube, and the later event of pol-len tube penetration into the stigmatic papilla (Fig. 1C). In Arabidopsis, hydration occurs within minutes after pollination (MAP) and is completed around 10 MAP (Rozier et al. 2020; Liu et al. 2021; Huang et al. 2023; Lan et al. 2023a). Pollen tube penetration of the stigmatic papilla is notable around 30 MAP; by 3 hours pollen tubes could clearly be traced along the stigmatic cell surface, and, typically, they have grown into the transmitting tissue (Huang et al. 2023; Lan et al. 2023a) (Figs. 1C and 2D). A RALF-FER-LLG1 module mediating RAC/ ROP signaled-ROS production in the Arabidopsis stigma

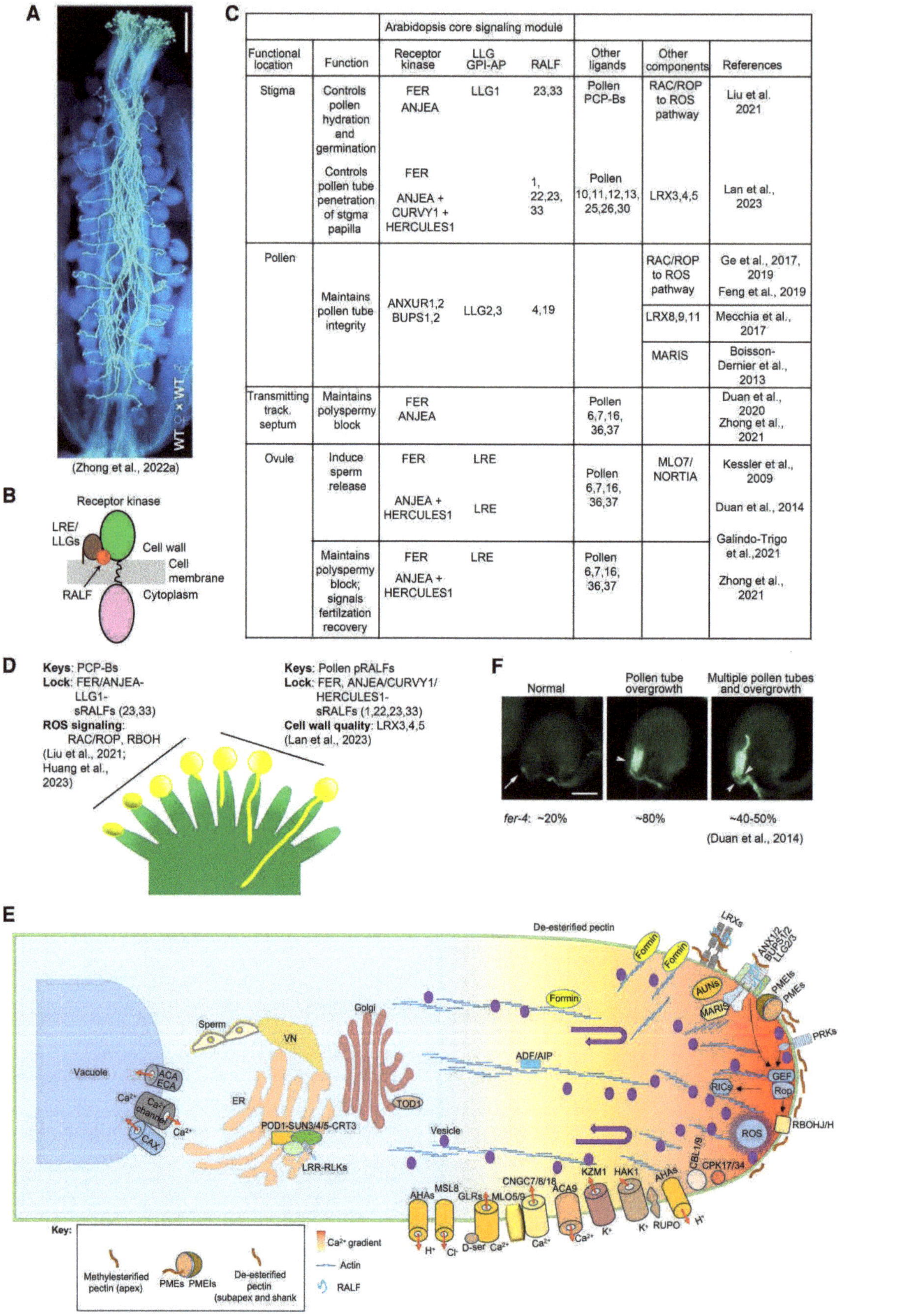

Functional location	Function	Arabidopsis core signaling module: Receptor kinase	LLG GPI-AP	RALF	Other ligands	Other components	References
Stigma	Controls pollen hydration and germination	FER ANJEA	LLG1	23,33	Pollen PCP-Bs	RAC/ROP to ROS pathway	Liu et al. 2021
	Controls pollen tube penetration of stgma papilla	FER ANJEA + CURVY1 + HERCULES1		1, 22,23, 33	Pollen 10,11,12,13, 25,26,30	LRX3,4,5	Lan et al., 2023
Pollen	Maintains pollen tube integrity	ANXUR1,2 BUPS1,2	LLG2,3	4,19		RAC/ROP to ROS pathway	Ge et al., 2017, 2019 Feng et al., 2019
						LRX8,9,11	Mecchia et al., 2017
						MARIS	Boisson-Dernier et al., 2013
Transmitting track, septum	Maintains polyspermy block	FER ANJEA			Pollen 6,7,16, 36,37		Duan et al., 2020 Zhong et al., 2021
Ovule	Induce sperm release	FER	LRE		Pollen 6,7,16, 36,37	MLO7/ NORTIA	Kessler et al., 2009
		ANJEA + HERCULES1	LRE				Duan et al., 2014 Galindo-Trigo et al.,2021
	Maintains polyspermy block; signals fertilzation recovery	FER ANJEA + HERCULES1	LRE		Pollen 6,7,16, 36,37		Zhong et al., 2021

Figure 2. The prezygotic phase of reproduction. **A)** Aniline blue–stained pollinated pistil showing the pollen tube growth path in an Arabidopsis pistil. It highlights pollen tube targeting of ovules and the typical 1 pollen tube: 1 ovule pattern of pollen (reproduced from Zhong et al. 2022a).

(continued)

provides a redox-controlled gate, which is relaxed by compatible pollen, for example, during self-pollination (Liu et al. 2021). Deposition of wild-type pollen suppresses stigmatic ROS, mitigating the gate to allow compatible pollen hydration and germination, followed by tube growth and penetration into the stigmatic papilla. Loss of each of the components of the signaling module, FER, LLG1, RALF33, RAC/ROP or RBOHs (which produce ROS), resulted in the reduction of stigmatic ROS. This accelerated pollen hydration and germination, hastening pollen tube growth into the stigmatic papilla. FER-related ANJEA, which interacted with FER in vitro, and RALF23 also participated in maintaining the stigmatic gate, suggesting mul-tiple configurations in planta are likely for the stigmatic module (Fig. 2, C and D). In search of the pollen trigger that relaxes the FER-anchored gate, Liu et al. (2021) discovered several PCP-B class pollen coat proteins (PCPs) as signals that disengage the RALF23/33-FER/ANJEA-LLG1-signaled, RAC/ROP-mediated, and ROS-controlled stigmatic gate for pollen hydration and germination.

Lan et al. 2023a proposed a more elaborate FER-dependent stigmatic gate. It engages not only FER and ANJEA but also FER-related receptor kinases HERCULES1 and CURVY1, additional RALFs (RALF1,22 along with RALF23,33; referred to as sRALFs), and Leucine-Rich Extensin (LRXs), cell wall proteins with an extensin domain (Bedinger 2018). FER functioned adequately on its own, whereas ANJEA, HERCULES1, and CURVY1 functioned redundantly to maintain the stigmatic gate. The Leucine-rich-repeat (LRR) domains in LRXs bind FER, RALFs, and other membrane proteins and mediate cell wall integrity (Fabrice et al. 2018; Dünser et al. 2019; Herger et al. 2020; Moussu et al. 2020). Using multiplexed CRISPR/Cas mutagenesis, Lan et al. (2023a) identified multiple pollen-expressed RALFs (RALF10,11,12,13,25,26,30; referred to as pRALFs) as triggers to disengage the sRALF-FER/ANJEA/HERCULES1/CURVY1-dependent and LRX-controlled gate.

Molecular and biochemical mechanisms of FER-dependent stigmatic gating

The stigmatic gate for pollen hydration and germination and subsequent pollen tube penetration into the stigmatic papilla are separable events, one unlocked by PCP-Bs (Liu et al. 2021; Goring et al. 2023; Huang et al. 2023; Nasrallah 2023) and the other by pRALFs (Lan et al. 2023a) (Fig. 2, C and D). PCP-Bs are low-molecular-weight, cysteine-rich PCPs that decorate the Brassica pollen surface, acting at pol-len hydration to facilitate germination (Doughty et al. 1998; Wang et al. 2017a; Wang et al. 2023a). Loss of PCP-B peptides, such as PCP-Bγ, induced delays in stigmatic ROS decline, pollen hydration, and germination on wild-type stigmas (Liu et al. 2021). Application of PCP-Bγ was adequate to trigger rapid ROS decline in unpollinated stigmas, suggesting that PCP-Bγ downregulates the stigmatic FER to ROS pathway to facilitate pollen hydration, followed by germination and tube growth into the pistil. Loss of FER compromises cell wall integrity (Feng et al. 2018), which most likely also con-tributed to *fer* mutant stigmas being more readily penetrable by pollen tubes (Liu et al. 2021; Zhang et al. 2021; Huang et al. 2023; Lan et al. 2023a).

For the pRALFs-unlocked gate (Lan et al. 2023a), light micros-copy, scanning, and transmission electron microscopy docu-mented that pollen from the multiplexed loss-of-function *ralf* septuplet mutant (referred to as *ralf sept*) behaved just like wild-type pollen during hydration and germination. The *ralf sept* pollen tubes failed to penetrate the wild-type stigmatic papilla cells (Lan et al. 2023a), but they efficiently grew into mu-tant stigmas lacking FER; or combinations of ANJEA, HERCULES1, and CURVY1; or all 4 of the sRALFs (referred to as *ralf quad*). Mutant stigmas lacking LRX3,4,5 also presented no barrier to *ralf sept* pollen tubes (Lan et al. 2023a).

However, unlike PCP-B–mediated early events of pollen hydration and germination being dependent on the relaxation of FER-controlled, RBOH produced stigmatic ROS (Liu et al. 2021; Huang et al. 2023), *ralf sept* mutant tubes remained unable to penetrate the ROS-deficient *rboh* stigmas (Lan et al. 2023a). However, LRX3,4,5, which interacted with both pRALFs and sRALFs in vitro, are essential for the FER-dependent gate to prevent *ralf sept* mutant tube penetration. Application of RALF33 restored FER to ROS production in the *ralf33* mutant stigma (Liu et al. 2021); it also reimplemented the *ralf quad*-weakened stigmatic gate,

Figure 2. (Continued)
B) A prototypical RALF-FER-LRE/LLG tripartite signaling module. **C)** A list of the tripartite signaling modules playing key roles in the prezygotic phase of the reproductive process. **D)** FER controls pollen germination and pollen tube penetration of the pistil. The schematic shows the components involved in regulating the early events in pollen hydration and the later events of pollen tube penetration of the stigmatic papilla. **E)** An abbreviated schematic representation of an elongating pollen tube. Readers are referred to many overall reviews referred to in the text. Listed here are some well-established components that regulate diverse conditions in the pollen tube. Components that regulate Ca^{2+} dynamics include CPK34/17 and CPK2/20 (Myers et al. 2009; Gutermuth et al. 2013), CPK11/24 (Zhao et al. 2013), CPK32 (Zhou et al. 2014), CBL1/9 (Mähs et al. 2013), CBL2/3 together with CIPK12 (Steinhorst et al. 2015), Ca^{2+} exchanger (Demidchik et al. 2018), and P-type IIA Ca^{2+}-ATPases (García Bossi et al. 2020). Regulators for the dynamics of other ions, such as K^+ (Mouline et al. 2002), are also important. Multiple regulators of cytoskeletons are also required (Chen et al. 2003; Lovy-Wheeler et al. 2006; Cheung and Wu 2008; Ye et al. 2009; Cheung et al. 2010; Zhu et al. 2017; Lan et al. 2018; Liu et al. 2018); microtubules are not displayed. Golgi-resident proteins involved in secretion of membrane proteins and cell wall materials are also essential for pollen tube growth (Chen et al. 2015; Tan et al. 2016; Jia et al. 2018). We refer readers to Supplementary Tables S2 to S4 for an extended list of components contributing to diverse functional needs for the pollen tube growth process in vitro and *in the pistil.* **F)** FER controls sperm release and suppresses polyspermy. Images show *fer-4* ovule phenotypes of pollen tube overgrowth and supernumerary pollen tube entrance. (Reproduced from Duan et al. 2014).

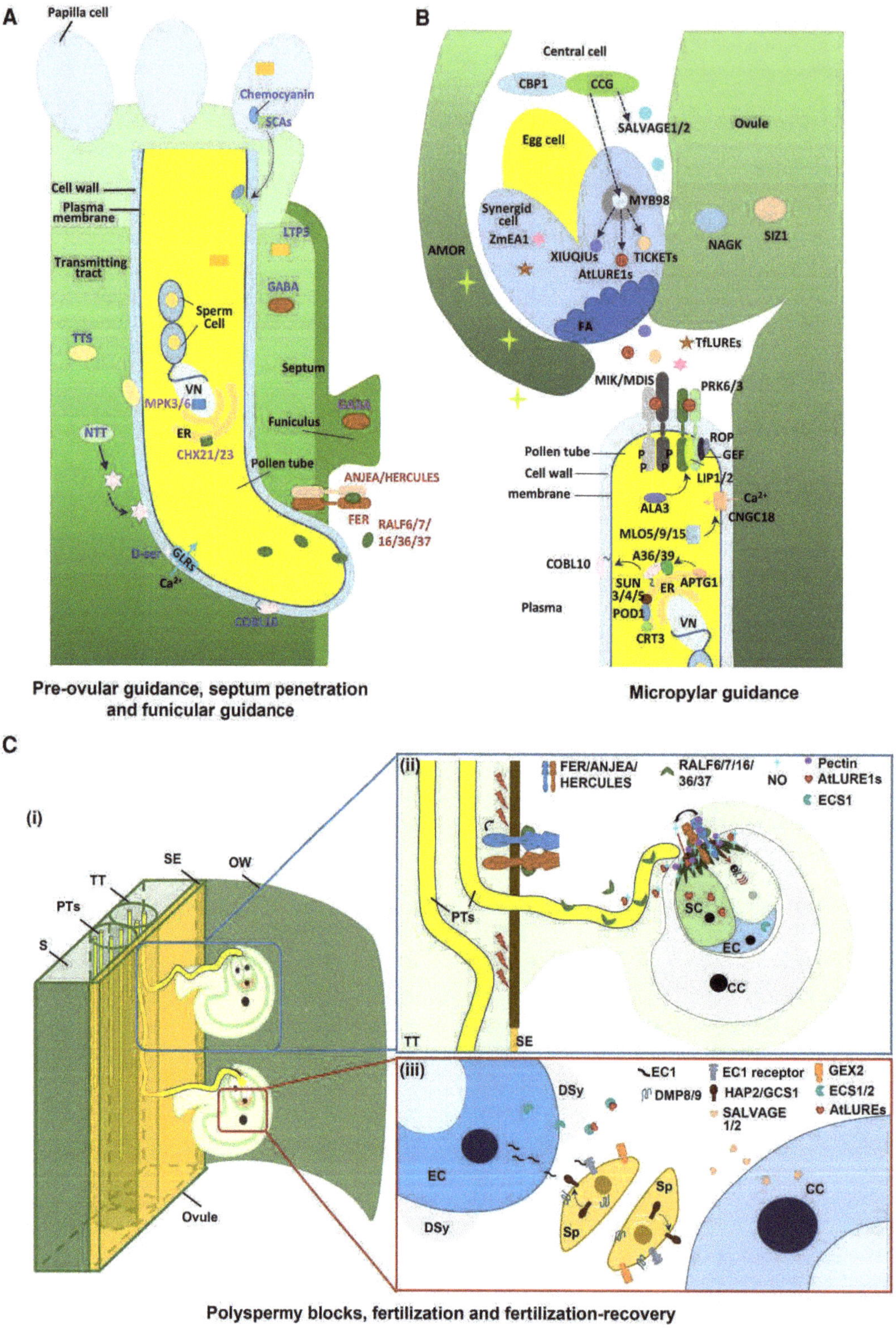

Figure 3. A schematic summary of key components for pollen tube growth in the pistil, pollen tube guidance and post-targeting events. **A)** Key factors involved in pre-ovular guidance (blue), pollen tube penetration at the septum (red), and funicular guidance (purple); the collection is from Arabidopsis, lily (chemocyanin, SCAs, LTPs), and tobacco (TTS). **B)** Key factors involved in micropylar guidance. The collection is mostly from Arabidopsis; AMOR and TfLUREs are from *T. fournieri*, and ZmEA1 from *Z. mays*. **C)** Key factors for polyspermy control, fertilization, and fertilization recovery. (i) A view in the pistil illustrating ovular conditions triggered upon pollen tube arrival (upper ovule and ii) and key molecules that induce the ovular polyspermy block, and an ovule (lower ovule and iii) showing key molecules for gametic fusion (HAP2/GCS1, GEX2 and DMP8/9) and for fertilization recovery (ECS1/2 and SALVAGE1/2). OW, ovule wall; SE, septum; TT, transmitting tract; PTs, pollen tubes; S, style.EC, egg cell; CC, central cell; SC, synergid cell; DSy, degenerated synergid cell; Sp, sperm. Red thunder bolts along the SE, triggers for SE polyspermy block. Not all components here are discussed in the text. Supplementary Tables S3 to S5 provide a more complete list.

restoring it to be nonpenetrable by *ralf sept* tubes (Lan et al. 2023a). However, application of RALF33 had no impact on *lrx3,4,5* stigmas, which remained penetrable by *ralf sept* tubes. The most parsimonious interpretation would be that the FER-dependent stigmatic gate controls 2 separable events. First, it gates the early events of pollen hydration and germination via RBOH-produced ROS, and PCP-B unlocks the gate by triggering the decline of stigmatic ROS (Liu et al. 2021). Then, the FER-dependent, LRX-controlled gate mediates papilla cell wall quality to guard against the later event of pollen tube penetration of the stigma, which is unlocked by pRALFs (Lan et al. 2023a).

Molecularly, key-lock mechanisms were proposed with the stigmatic FER-anchored modules as locks that protect the pistil from intrusion until triggered by keys from compatible pollen (Liu et al. 2021; Zhang et al. 2021; Lan et al. 2023a). PCP-Bγcompeted with RALF33 for binding to the FER extracellular domain in vitro. On the *ralf33* mutant stigma, PCP-Bγ efficiently blocked e xogenous R ALF33 f rom r estoring stigmatic ROS production. Together, these results led to a model w hereby upon pollination, PCP-Bs compete w ith stigma RALF23/33 to disengage the FER to ROS pathway, resulting in pollination-triggered stigmatic ROS decline, relaxation of the stigmatic gate for the early events of pollen hydration, and germination (Fig. 2D) (Liu et al. 2021; Huang et al. 2023).

For the FER-dependent, LRX-controlled stigmatic gate (Lan et al. 2023a), in vitro experiments indicate that pRALF competes with sRALF for binding to the extracellular domain of FER/CURVY1. When both pRALF and sRALF were simultaneously applied to *ralf quad* mutant stigmas, pRALF reversed the inhibition of *ralf sept* pollen tube penetration caused by sRALF treatment. These results suggest that pRALFs compete with sRALFs for FER binding to unlock the FER-dependent gate for pollen tube penetration into the stigmatic cell wall. On the other hand, *ralf sept* pollen tubes could penetrate the stigma of *lrx345*, and pRALF/sRALF both bind the LRX, indicating that LRX3,4,5 are essential for the stigma gate that guards against pollen tube penetration. Interestingly, pRALF and sRALF did not compete for stigma LRX3,4,5. With RALFs being disorder peptides (Xiao et al. 2019; Liu et al. 2024), it would be interesting to resolve what characteristics between the sRALFs and pRALFs rendered them competitive interactors with FER and what rendered them equal partners in their interactions with stigma LRXs. Also interesting to examine is whether there are 2 classes of pollen RALFs, with, for example, RALF4 and 19 interacting with pollen LRXs on the pollen tube surface to mediate tube wall integrity (Mecchia et al. 2017; Sede et al. 2018; Moussu et al. 2020, 2023), and others such as RALF10,11,12,13,25,26,30 interacting with stigma LRXs inducing weakening of wall integrity (Lan et al. 2023a). Considering the roles of LRXs in cell wall integrity (Fabrice et al. 2018; Herger et al. 2019), the establishment and opening of the LRX3,4,5-controlled stigma gate are likely achieved through dynamic changes in the cell wall.

PCP-Bγ, in fact, engages a dual process to secure a hospital stigmatic environment for pollen germination. Huang et al. (2023) discovered that as a prelude to reducing FER-controlled ROS, compatible pollination triggered a rapid and FER-dependent increase in stigmatic NO that rapidly reached a maximum (around 5 minutes after pollination). This rise mirrored the reduction in stigmatic ROS and progress of pollen hydration. Application of PCP-Bγ triggered similar NO increases in unpollinated Arabidopsis stigmas. The mirroring impacts of pollination and PCP-Bγ on stigmatic NO and ROS—that is, when NO is low, ROS is high and vice versa—strongly support a causal relationship between compatible pollination-induced stigmatic NO rise and ROS decline, facilitating pollen hydration and germination to start a pollen tube on its journey in the pistil.

Pollen tube growth in the pistil: an overview

The polarized pollen tube growth process

After penetrating the stigmatic papilla and underlying tissue, the pollen tube grows into the transmitting tract surrounded by ovules (Fig. 1). The polarized cell growth process involves multifaceted coordination of extracellular, cell surface, and cytoplasmic components and is dependent on regulated ion dynamics, redox status, and turgor (Fig. 2E) (Steer and Steer 1989; Cheung et al. 2008; Cheung and Wu 2008; Tavares et al. 2011; Hepler and Winship 2015; Michard et al. 2017; Cameron and Geitmann 2018; Hoffmann et al. 2020; Hayashi and Palmgren 2021). Apically located RAC/ROP GTPases (Guo and Yang 2020; Li et al. 2023c) and a dynamically regulated actin cytoskeleton (Cheung and Wu 2008; Cheung et al. 2010) are crucial for tip-focused growth. Pollen tube growth is self-organized, requiring only Ca^{2+} and Boron in minimal culture media. Here, we focus on the pollen tube grow th process in the pistil, w hich is facilitated and regulated by pistillate tissues (Fig. 3).

The pistil environment

Pistils are architecturally diverse. Studies in lily, tobacco, and tomato, representing open- and solid-style pistils, produced early insights on pistillate contributions to the pollen tube grow th process (Cheung 1995, 1996; Lord and Russell 2002). In tomato, the stigmatic factor LeSTIG supported pollen tube grow th by competing w ith the pollen-expressed LeLAT52 for binding to pollen receptor-like kinases (PRK) LePRK1/2, w hich activate RAC/ROP GTPases (Muschietti et al. 1998; Tang et al. 2002; Tang et al. 2004; Kaothien et al. 2005; Liu et al. 2020). As the highway for pollen tubes, the transmitting tract is spatially accommodating and, being enriched in polysaccharides, glycolipids, glycoproteins, and adhesive molecules, supportive of the metabolically demanding pollen tube growth process as nutrients and provides cell-cell adhesion (Cheung et al. 1995; Park et al. 2000; Nieuwland et al. 2005; Chae and Lord 2011; Pereira et al. 2016a) (Fig. 3A). The pollen transcriptome changes during growth in the pistil, implying adaptive responses to the pistil environment (Qin et al. 2009). The pistillate tissues prime the pollen tube for responding to directional cues from the ovules. A complex glycan, ethyl-glucuronosyl arabinogalactan (AMOR) from

Torenia fournieri ovules, confers pollen tube competency in responding to ovular attractants (Mizukami et al. 2016). Low-molecular-weight, pistil-derived factors are also crucial for pollen germination and tube growth, such as brassinosteroid (Vogler et al. 2014), sulfinated azadecalins (Qin et al. 2011), the amino acid D-serine, which affected pollen tube growth through regulating glutamate receptor-like (GRL) Ca^{2+} channels (Michard et al. 2011; Wudick et al. 2018), and γ-amino butyric acid (GABA) (Palanivelu et al. 2003; Yu et al. 2014). Pistils also express peptides to regulate pollen tube growth, for example, CLE45, protecting them against heat stress through interacting with a pollen-expressed kinase (Endo et al. 2013).

Pistils from Arabidopsis, with their short style and elongated ovary with ovules differentiating from the septum (the epidermis of the transmitting tract), provide excellent specimens to trace pollen tube behavior along its growth pathway (Figs. 1 and 2A). With the vast genetic resource repertoire, the popularized use of semi-in vivo pollen tube growth system (Cheung et al. 1995; Higashiyama et al. 1998; Palanivelu and Preuss 2006) and improvements in histological and live-cell visualization of the pollen tube growth process, studies in Arabidopsis have provided major contributions to our current understanding of the in vivo process discussed below.

Pollen tube growth and exit from the stylar transmitting tissue

The pollen tube extends through cycles of cell wall softening and stiffening at the tip (Hepler et al. 2013). This, combined with turgor exerted from the expanding cell (Hepler and Winship 2015), makes the pollen tube susceptible to rupture. FER-related signaling modules, assembled from pollen counterparts of FER, LRE, and LLG1 and pollen-expressed RALFs (Fig. 2, B and C), are crucial for maintaining pollen tube integrity until it reaches the target FG. The paired ANXUR1 (ANX1)/ANX2 (Boisson-Dernier et al. 2009; Miyazaki et al. 2009) and BUPS1/BUPS2 (BUDDHA'S PAPER SEAL1/2) (Ge et al. 2017) are highly and the most prominently pollen-expressed paralogs of FER. Loss of either pair resulted in male sterility. During pollination, *anx1 anx2*, *bups1 bups2*, and *bups1* pollen tubes displayed tip defects consistent with precocious rupture and were arrested early in the pistils, precluding arrival at the ovules, thus male sterility. In rice, the BUPS/ANX orthologs also control pollen tube integrity (Liu et al. 2016b). A recent study in maize suggests considerable conservation of a FER-like receptor kinase-anchored cell surface module in control pollen tube integrity during growth (Zhou et al. 2024).

ANXs and BUPSs form heteromers, all 4 receptor kinases interacted with LLG2 and LLG3, and the pollen-expressed RALF4 and RALF19 bound to these coreceptors (Ge et al. 2017; Zhu et al. 2018; Feng et al. 2019; Ge et al. 2019). While the precise in vivo configuration remains unclear, these tripartite complexes are located at the pollen tube tip membrane, where they signal the RAC/ROP-controlled ROS production pathway in the pollen tube and the proper deposition of pectin in the tube wall. ANX1/2 promotes exocytosis through the downstream cytoplasmic kinase MARIS and RBOH oxidases RBOHH/J, generating ROS at the tip (Boisson-Dernier et al. 2013; Kaya et al. 2014; Lassig et al. 2014; Boisson-Dernier et al. 2015). Zhou et al. (2021b) suggested that BUPS1 served as a mechano-sensor of the physical barriers from the female tissues to regulate RAC/ROP-controlled ROS production. Furthermore, GPI-AP COBL11 also modulates pollen tube cell wall integrity via interaction with RALF4/19 (Li et al. 2023b), and protein phosphatases ATUNIS1 and ATUNIS2 function in a distinct but converging pathway as ANX1/2 to negatively regulate tip growth (Franck et al. 2018a, 2018b).

The pollen tube cell wall property, particularly its pectin matrix (Cascallares et al. 2020; Du et al. 2022), is important for pollen tube growth. Pectin is secreted in the methylester-ified form at the pollen tube tip and demethylated by suba-pical cell wall–located pectin methylesterases (PMEs) as the tip advances. The de-esterified pectin is crosslinked by Ca^{2+}, mediating a stiffened wall starting from the subapical region. Pectin de-esterification is countered by PME inhibitors (PMEIs). This dynamic pattern facilitates elasticity for growth at the apex and rigidity in distal regions to confine the cell geometry and provide integrity (Fig. 2E). Mutations in *PME* genes, like *VANGUARD* and *PME48*, lead to precocious pollen tube rupture (Jiang et al. 2005; Leroux et al. 2015). Therefore, PMEs, PMEIs, and other pectin-modifying enzymes (Du et al. 2022), FER, ANX/BUPS, and possibly other FER-related receptors are functionally intertwined in impacting physical and biological properties at the cell wall/cell membrane interface (Feng et al. 2018; Herger et al. 2019; Liu et al. 2024).

Extensins are cell wall hydroxyproline-rich glycoproteins. They are crosslinked via tyrosine residues and associate tightly with the cell wall matrix, reinforce the wall, and form a scaffold for pectin assembly (Lamport et al. 2011; Mecchia et al. 2017; Marzol et al. 2018; Moussu et al. 2020; Moussu and Ingram 2023). In Arabidopsis, the extensin-like pollen LRX8/9/10/11 interact with RALF4/19, and loss of these cell wall–linked proteins also causes precocious pollen tube rupture (Mecchia et al. 2017; Fabrice et al. 2018; Sede et al. 2018). Other extensin-like proteins, for example, proline-rich extensin-like receptor kinases, regulate pollen tube growth and impact its cell wall polysaccharide composition (Borassi et al. 2016; Borassi et al. 2021).

LRXs comprise an N-terminal LRR domain that interacts with RALF and a C-terminal pectin-associated extensin domain (Herger et al 2019; Moussu et al. 2020; Moussu et al. 2023). RALF4 inhibition of pollen tube growth depends on the LRXs (Mecchia et al. 2017), suggesting that RALF4-LRX binding stiffens the wall as a brake to control pollen tube growth rate. Interestingly, Moussu et al. (2020) noted that tertiary complexes, that is, receptor kinase-LLG/LRX-RALF, had not been observed. Although pH and redox conditions might impact the interactions between the structurally versatile RALF with neighboring molecules, Moussu et al. suggested that the mutually exclusive binding of RALF peptides to the receptor kinases or LRXs might reflect how RALF peptides regulate 2 parallel, mechanistically distinct but convergent pathways to fine tune pollen tube integrity. Whether a similar scenario could be applicable to the FER-dependent, sRALFs-maintained FER-dependent, LRX-controlled, and pRALFs-unlocked stigmatic gate (Lan et al. 2023a) will be interesting to explore.

RALF-pectin phase-separates in vitro and in vivo to mediate a global regulatory mechanism from the cell surface (Liu et al. 2024). Aligned with the idea of RALF-pectin interaction is the finding that RALF4-LRX8 LRR domain complex interacted with de-esterified pectin through the polycationic C-terminal domain of RALF4, forming a network of puncta with sizes suggestive of heterotetrameric complexes (Moussu et al. 2023). C-terminal–mutated RALF4s were compromised for this inter-action; they were also less able to restore normal growth of *amiralf4 ralf19* mutant pollen tubes. Instead, *amiralf4 ralf19* pollen tubes expressing the C-terminal–mutated RALF4s dis-played a wall quality more akin to control mutant tubes than to wild-type RALF4-complement *amiralf4 ralf19* tubes. Therefore, the ability for RALF4-LRX8-pectin interaction correlated well with pollen tube wall quality and integrity and tube growth property. Given the hypothesis that RALF peptides regulate 2 parallel but mechanistically distinct but convergent pathways in pollen tubes (Moussu et al. 2020) and that the RALF C-terminal interacts with FER (Xiao et al. 2019), whether and how the C-terminal–mutated RALF4s (Moussu et al. 2023) would interact with ANX/BUPS and regulate their signaling will be interesting to explore.

Proper pollen tube growth rate and integrity rely on many parallel and intertwining factors (Figs. 2E and 3; Supplementary Tables S2 and S3). Some of these factors function through balancing the intracellular turgor pressure and cell wall mechanics during pollen germination and oscillatory growth. Disrupting this balance usually results in pollen tube rupture. For instance, the Cl^--gating mechanosensitive-like channel 8 (MSL8) senses membrane tension caused by the osmotic changes during pollen hydration, and loss of its function causes pollen rupture during germination (Hamilton et al. 2015; Hamilton and Haswell 2017). MSL8 intersects exocytosis and possibly the ANX1/2 pathway to achieve osmotic balance and membrane integrity (Zhou et al. 2021b; Wang et al. 2022a; Zhou et al. 2022). At the same time, the apical Ca^{2+} gradient and RAC/ROPs underlie the oscillatory growth through cellular feedback loops (Tunc-Ozdemir et al. 2013; Lichocka et al. 2022), actin-based cytoskeleton assembly/disassembly (Cheung et al. 2010; Qu et al. 2017), and sphingolipid-mediated feedbacks between the turgor pressure and cell wall thickness (Chen et al. 2015).

Pollen tube guidance

Pollen tube emergence from the septum

Architectural and physiochemical properties along the transmitting tract provide an environment for pollen tubes to grow directionally toward their target ovules. Pollen tube guidance (Box 1), further divided into pre-ovular and ovular, is a phenomenon whereby female cues guide a pollen tube to change its growth trajectory, first to reach an ovule, then to target the ovular aperture micropyle and penetrate the FG to deliver sperm (Figs. 1D, 2A and 3) (Higashiyama and Takeuchi 2015; Higashiyama and Yang 2016). The mechanisms for pre-ovular guidance remain unclear, conceivably involving various physicochemical conditions along the pollen tube growth path (Supplementary Table S3) (Crawford et al. 2007; Herrera-Ubaldo et al. 2019). Ovular guidance, studied mainly in dicots in recent years, is further divided into funicular guidance that each pollen tube follows from the septum to reach the ovule, and micropylar guidance, during which an approaching pollen tube responds to attractants produced by the FG and then targets and enters the micropyle.

For funicular guidance (Mizuta and Higashiyama 2018), studies in Arabidopsis showed that the pollen tube engages actively in this event. Loss of 2 pollen-specific and endoplasmic reticulum-located K^+ transporters CATION/PROTON EXCHANGERS 21/23 (CHX21/23) resulted in mutant pollen tubes bypassing all the ovules to reach the bottom of the pistil. Nevertheless, *chx21 chx23* pollen tubes also failed to target the micropyle when exposed directly to ovules (Lu et al. 2011). As K^+/H^+ exchangers, CHX21 CHX23 could function, albeit not been investigated yet, through modifying cation balance and pH, conditions critical for proper pollen tube growth. Loss of a GPI-AP, COBLA-like 10, induced phenotypes similar to that of *chx21 chx23* (Li et al. 2013). Other factors, such as phytosulfokine (Stührwohldt et al. 2015), mitogen-activated protein kinase cascades (Guan et al. 2014), and the shikimate biosynthetic pathway (Wang et al. 2023b) also impact funicular guidance.

Ovular guidance

Synergid cell-produced pollen tube attractant LUREs were first discovered from the architecturally unique *T. fournieri* ovules with exposed ES (Okuda et al. 2009); much mechanistic insight has been advanced since (Higashiyama and Takeuchi 2015; Li et al. 2018; Johnson et al. 2019; Zhou and Dresselhaus 2019). Seven TfLURE homologs from *Arabidopsis thaliana* (*A. thaliana*), AtLURE1s, belong to a subclade among hundreds of cysteine-rich proteins (CRPs) (Takeuchi and Higashiyama 2012). They are secreted by the synergids and diffuse to the micropyle and funicular surfaces. AtLURE1 receptors include 2 receptor-like cytoplasmic kinases (RLCKs), LOST IN POLLEN TUBE GUIDANCE 1 and 2 (LIP1/2), and the pollen tube tip-localized PRK6 (Liu et al. 2013; Takeuchi and Higashiyama 2016; Zhang et al. 2017). PRK6 interacts with LIP1/2, which lack an extracellular domain, and the PRK/RLCK module could perceive AtLURE1s and mediate cytoplasmic signaling. Additional cell surface receptors also partake in ATLURE1 perception. For instance, the LRR-RLKs MDIS1-INTERACTING RECEPTOR LIKE KINASE 1/2 (MIK1/2) and MALE DISCOVERER 1/2 (MDIS1/2) form a heterodimer and participate in AtLURE1-regulated growth (Wang et al. 2016).

How do attractant receptors regulate pollen tube reorientation? PRK6 and several other pollen-expressed homologs from different species, for example, the tomato LePRK1 and LePRK2 (Kaothien et al. 2005), interact with ROPGEFs, which activate RAC/ROPs (Berken et al. 2005; Gu et al. 2006). RAC/ROPs regulate cytoskeleton dynamics, cytosolic calcium gradients, and exocytosis, all crucial for polar cell growth (Gu et al. 2005; Lavy et al. 2007). Modeling and ex-perimental studies showed that pollen tube guidance results from reorienting the ROP1-exocytosis signaling network that regulates tip growth (Luo et al. 2017). During semi-in vivo growth and after exposure to AtLURE1, the symmetric distribution of PRK6 on the pollen tube tip rapidly changed to asym-metric and concentrated more on the side of attractant application. This was followed by the pollen tube redirecting its growth trajectory toward the attractant. Together, evidence strongly supports that pollen tube guidance is a result of AtLURE1-triggered repositioning of the LURE receptor-ROPGEF-RAC/ROP signaling module and the down-stream growth supporting activity at the apical dome (Takeuchi and Higashiyama 2016; Luo et al. 2017).

The central cell also plays an important role in micropylar guidance in Arabidopsis. CENTRAL CELL GUIDANCE (CCG) is a TFIIB domain-containing central cell-specific transcrip-tion factor (Fig. 3). *ccg* mutant ovules display normal ES development but lack pollen tube attraction ability, resulting in severe sterility (Chen et al. 2007). Central cell-located CCG BINDING PROTEIN 1 (CBP1) interacts with CCG to re-cruit the Mediator complex and RNA Pol II (Li et al. 2015b). Several genes, including *MYB98*, crucial for filiform apparatus development, are downregulated in *ccg* ovules, consistent with the CCG-CBP1 complex functioning in pollen tube guidance through the MYB98 signaling pathway.

The pollen tube itself plays a critical role in responding to ovular guidance. For instance, the pollen tube–localized Ca^{2+} channel CNGC18 and GLRs regulate pollen tube guid-ance by modulating Ca^{2+} oscillations at the pollen tube tip (Michard et al. 2011; Gao et al. 2016). In response to ovular signals, CNGC18 was asymmetrically recruited by MLO5/9/ 15 to plasma membrane on the side where the pollen tube senses the ovular signals to induce Ca^{2+} influx and direct the pollen tube on the septum surface to the nearest funicu-lus. Unable to induce this, *mlo5/9* pollen tubes penetrated the septum to emerge on the septum surface but twisted into a pile, apparently not able to decode the ovular signals. The pollen tube endoplasmic reticulum serves a depot in synthesizing, assembling, and chaperoning key molecules to the apical dome membrane to mediate growth in response to micropylar guidance (Li et al. 2011; Dai et al. 2014; Xue et al. 2022).

Due to the architecture of the pistil, monocots rely only on ovular guidance (Lausser and Dresselhaus 2010). For instance, the maize transmitting tract terminates at the chalazal end and the pollen tube grows toward the micropylar region following the ovary anatomy and enters by breaking through the inner epidermis of the ovary wall (Zhou and Dresselhaus 2019). RNAi knock-down of the egg apparatus specific peptide, EGG APPARATUS 1 (ZmEA1), induced a severe micropylar targeting defect and significantly reduced seed sets (Márton et al. 2005).

Pollen tube reception

After entering the micropyle, the pollen tube interacts with one of the synergid cells, the receptive synergid, to undergo pollen tube reception, terminating in pollen tube bursting in the FG to release sperm (Hater et al. 2020; Hafidh and Honys 2021) (Fig. 1; Box 1). The process lasts approximately 30 to 60 min, including a brief halt in pollen tube growth, and is accompanied by the degradation of the receptive synergid (Sandaklie-Nikolova et al. 2007; Hamamura et al. 2011; Ngo et al. 2014). The FER-LRE signaling module located at the filiform apparatus (Box 1) of maturing and receptive ovules (Fig. 1) was identified through their roles in pollen tube reception (Huck et al. 2003; Rotman et al. 2003; Escobar-Restrepo et al. 2007; Capron et al. 2008; Tsukamoto et al. 2010; Duan et al. 2014; Li et al. 2015a; Liu et al. 2016a). Loss of FER or LRE alone prevents pollen tube bursting in the FG, resulting in pollen tube overgrowth (Fig. 2F). HERCULES1 and ANJEA participate redundantly in pollen tube reception. In vitro, they complexed with FER (Galindo-Trigo et al. 2020). Five pollen-expressed RALFs (RALF6/7/16/36/37) function as ligands for the FER/ANJEA/ HERCULES1-LRE complex (Zhong et al. 2022a). The *ralf* quin-tuple mutant pollen tubes failed to burst in wild-type FGs. In vitro, these RALFs interact with each receptor and enhanced interactions between FER and ANJEA/HERK1, suggesting receptor-ligand complexes of varying peptide-receptor kinase-LRE configurations are possible in vivo (Zhong et al. 2022a, 2022b).

A Powdery Mildew Resistance Locus-O (MLO) family protein, NORTIA (NTA)/MLO7, interacts with FER-LRE to participate in pollen tube reception (Kessler et al. 2010). Before pollen tube arrival, NTA/MLO7 is maintained intracellularly in the synergids. Pollen tube arrival induced FER-dependent re-location of NTA/MLO7 to FER-LRE-occupied filiform appar-atus. The approaching pollen tube induced the degeneration of the receptive synergid cell is linked to pollen tube rupture. The approaching pollen tube induced FER-, LRE-, and NTA-dependent changes in synergid cytoplasmic Ca^{2+} dynam-ics (Iwano et al. 2012; Denninger et al. 2014; Ngo et al. 2014; Gao et al. 2022; Ponvert and Johnson 2024). The calcium sta-tus/dynamics in the pollen tube is also important for pollen tube rupture. The plasma membrane–localized Ca^{2+}-ATPase ACA9 plays a pivotal role for cytosolic Ca^{2+} efflux, maintain-ing $[Ca^{2+}]$ within the pollen tube. *aca9* pollen tubes halted upon entering the micropyle, and bursting did not occur (Schiøtt et al. 2004). Moreover, the pollen tube apical $[Ca^{2+}]$ gradient was obliterated just before bursting in vitro and required Ca^{2+} channel activity. Sequestration of pistillate $[Ca^{2+}]$ inhibited FER-mediated pollen tube bursting in the ovules (Duan et al. 2014). Genes regulating synergid cell death also appeared to be involved (Christensen et al. 1997; Leydon et al. 2015; Mendes et al. 2016; Wang et al. 2017b), consistent with pollen tube penetration and sperm release being tightly linked.

In monocots, K^+ plays an important role in pollen tube rupture. The maize pollen tube cytoplasmic membrane localized potassium channel KZM1 functions downstream of the synergid cell-secreted peptide EMBRYO SAC4 (ZmES4). The interaction between ZmES4 and KZM1 induced K^+ influx, the change in osmotic pressure due to water uptake might have led to pollen tube bursts (Amien et al. 2010). In rice, the K^+ transporter OsHAK1/19/20 interacted directly with pollen-expressed FER homolog RUPTURED POLLEN TUBE (RUPO) to regulate between pollen tube integrity and rup-ture (Liu et al. 2016b).

Gamete interaction and fertilization

Fertilization and the subsequent zygote development are pivotal stages in the life cycles of both animals and plants, during which haploid male and female gametes fuse to generate a diploid organism. With immotile sperm and 2 gamete fusion events, angiosperm fertilization is complex. Upon sperm release, the 2 gametic fusions usually occur around a 7-minute time window, probably the time needed for gametic cell recognition, attachment, and activation (Hamamura et al. 2011). Then, 1 sperm cell fuses with the egg cell to form the diploid embryo, and the other fuses with the central cell to generate the triploid endosperm. We discussed recent advances, mainly from studies in Arabidopsis, in the understanding of gametic and early post-zygotic events.

Gamete recognition and attachment

In Arabidopsis, the 2 sperm cells in a pollen grain seem isomorphic, i.e. without preferential fusion of either sperm with the egg cell or central cell (Mori et al. 2006; Liu et al. 2008a, 2008b; Ingouff et al. 2009; Hamamura et al. 2011; Cole et al. 2014; Chen et al. 2022). In contrast, the 2 female gametes are dimorphic, showing distinct fertilization prefer-ence. Typically sperm-egg fusion occurs first, followed by sperm-central cell fusion several minutes later (Denninger et al. 2014; Hamamura et al. 2014). The use of cell-cycle defective mutants that produced a single sperm-like cell but remaining competent for fertilization showed that the egg cell is preferentially fertilized than the central cell (Li et al. 2022a; Wang et al. 2024), and egg cell–specific aspartic endopeptidases as well as small CRPs might partially contribute to the preferential fertilization of egg cell (Jiang et al. 2022a; Wang et al. 2024).

Current knowledge about gamete recognition and attachment (Fig. 3) has mainly derived from GAMETE EXPRESSED 2 (GEX2), a sperm plasma membrane protein, whose expression is controlled by DUO1 (Brownfield et al. 2009; Borg et al. 2011). GEX2 is expressed with an N-terminal signal sequence, a filamin repeat domain followed by a single-pass transmembrane domain at the C terminal. In vivo gamete attachment assay demonstrated that GEX2 is essential for sperm-egg and sperm-central cell attachment (Mori et al. 2014), consistent with both sperm cells having comparable capacity to adhere to either female gamete for fertilization. Three sperm cell endomembrane-associated cation/H^+ exchangers (CHXs) are also required for gamete attachment, which failed in the *chx17/18/19* mutant (Chen et al. 2022). Disturbed osmotic balance in *chx17/18/19* sperm cells and their resultant rupture inside the ES suggest an interplay between turgor and the sperm cell surface.

Gamete activation, fusion, and sperm nucleus migration

After gamete attachment, contributions from the male and female sides are required for the critical events of gamete activation and fusion (Fig. 3). On the female side, a group of egg cell–specific small CRPs, EGG CELL 1 (EC1), are located in vesicles of the unfertilized egg cell. Upon sperm arrival at the sperm-egg fusion sites, the egg cell secretes EC1, and this stimulates the redistribution of the fusogen GCS1/HAP2 from the sperm endomembrane system to its cell surface, a step critical for sperm cell activation and fertilization competence acquisition (Sprunck et al. 2012). *ec1*-RNAi plants had 2 or 4 unfused sperm cells in about one-half of the ESs, consistent with EC1 is essential for gamete fusion. On the male side, GCS1/HAP2, is an evolutionarily conserved protein for gamete fusion in animals and plants and essential for gamete fusion in algae and angiosperms (Mori et al. 2006; Liu et al. 2008a; Cole et al. 2014; Wang et al. 2022d). The Arabidopsis GCS1/ HAP2 functions by interacting via its N terminus with female gamete-expressed proteins, while its positively charged C terminus facilitates fusion between the plasma membranes. The EC1-triggered GCS1/HAP2 translocation is dependent on 2 sperm-specific DOMAIN OF UNKNOWN FUNCTION 679 membrane proteins (DMP8 and DMP9) with functions conserved across seed plants (Wang et al. 2022d). Sperm from *dmp8 dmp9* preferentially fertilized the central cell rather than the egg, consistent with sperm-egg fusion being more dependent on DMP8/9 (Cyprys et al. 2019).

After gamete membrane fusion, gamete nuclei migration is the critical next step for the fusion of parental genomes to generate the zygotic genome. In contrast to the microtubule-assisted female pronuclei movement toward the male nucleus in fertilized animal eggs, an F-actin meshwork supports sperm nuclear migration toward the nucleus of female gametes in Arabidopsis in a RAC/ROP regulated process involving formin-regulated actin nucleation and a plant-specific myosin, XI-G (Kawashima et al. 2014; Ohnishi et al. 2014; Peng et al. 2017; Ali et al. 2020; Ali and Kawashima 2021).

Genome-wide zygotic genome activation (ZGA)

Genome-wide ZGA

The fusion of a large egg cell and a small sperm cell generates the fertilized egg cell, the zygote, and the starting point of a new generation in both animals and plants. In animals, ZGA is usually characterized by a minor and a major activation waves (Tadros and Lipshitz 2009). The animal process may start as early as the 1-cell embryo stage or spread out over longer durations, depending on species (Vastenhouw et al. 2019). Starting about 20 years ago and focusing on the eudicot Arabidopsis and monocots rice and maize, the relatively short history of plant ZGA studies has already yielded major insights. In contrast to animals, a comparative transcriptome analysis of egg cells and zygotes revealed that ZGA in eudicots and monocots occurs shortly after fertilization. For example, in Arabidopsis, ~13% of its genes were already activated before the first zygotic division (Zhao et al. 2017; Dresselhaus and Jürgens 2021; Zhao et al. 2022a; Huang et al. 2023). Notably, among the activated genes are those encoding transcription factors important for zygote and early embryo development, such as WUSCHEL RELATED HOMEOBOX (WOX)2, WOX9, BABY BOOM (BBM), and its homolog PLETHORA 2 (PLT2) (Zhao et al. 2019). In maize, sperm, egg, and zygote transcriptomes at 2 different developmental stages showed that ~9% of its almost 40,000 annotated genes began to be transcribed in zygotes (Chen et al. 2017). On the other hand, only 181 genes are expressed de novo in rice zygotes (Anderson et al. 2017). Regardless, BBM and WOX9 homologs were activated in maize and rice zygotes, suggesting conserved roles in ZGA between eudicots and monocots. Taken together with comparative transcriptomic data from the brown alga Dictyota (Bogaert et al. 2017), evidence strongly supports that plant ZGA occurs at the zygotic stage, and de novo transcription is required for zygote division (Bogaert et al. 2017; Kao and Nodine 2019; Zhao et al. 2019).

Parental contributions to ZGA

Parental genomes merge upon fertilization. In mammals, the genetic information deposited in the egg cell controls early embryogenesis (Tadros and Lipshitz 2009; Vastenhouw et al. 2019). In plants, the paternal and maternal genomes fused quickly to form the zygotic genome after fertilization (Ingouff et al. 2010; Kimata et al. 2016). In Arabidopsis, genome-wide transcriptome analyses using zygotes and early embryos established that maternal and paternal genomes contribute equally to the transcriptomes of Arabidopsis embryos as early as the elongated zygote stage (Zhao et al. 2019; Zhao et al. 2020). In monocots, parental contributions to the transcriptomes of zygotes and early embryos remain largely unknown. Emerging information from hybrid rice zygote at 9 HAP suggests highly asymmetric parental contribution, with maternal alleles contributing to about 98% of transcripts (Anderson et al. 2017).

Elaborations to maximize reproductive success

Angiosperms have evolved multiple male-female interactive strategies to elaborate beyond the required steps to achieve fertilization to maximize progeny production and their vibrancy. While some of these phenomena are long-known, recent efforts illuminated emerging understandings of some fascinating strategies.

Polyspermy block and the pollen tube to one ovule rule

The entry of multiple sperm cells into a single egg (polyspermy) can lead to anomalies in chromosome numbers in the zygote, resulting in severe consequences in animals (van der Ven et al. 1985), and is generally considered disadvantageous for plants (Spielman and Scott 2008; Bianchi and Wright 2016; Tekleyohans et al. 2017). Contrasting the barriers within the animal female reproductive tract to prevent polyspermy (Bianchi and Wright 2016), angiosperms have evolved multiple mechanisms to preclude multiple pollen tube entrance into the same ovule to avert polyspermy.

A strict 1-to-1 relationship between pollen tubes and ovules is established during pollen tube growth in the pistil when they exit the main growth path to target ovules (Johnson et al. 2019; Cheung et al. 2022) (Fig. 2A). Recent studies have unveiled multiple mechanisms behind this "1 pollen tube to 1 ovule" phenomenon. First, pollen tube overgrowth in loss-of-function *fer* mutant pistils is frequently accompanied by supernumerary pollen tube penetration of single ovules (Fig. 2E), as well as bundled exits of multiple pollen tubes from the transmitting tract (Huck et al. 2003; Rotman et al. 2003; Duan et al. 2014; Duan et al. 2020; Zhong et al. 2022a). These dual anomalies implicate the relaxation of an early polyspermy block at the septum (Box 1) and a local block at the entry to the FG in the ovule (Fig. 3C). Genetic and biochemical studies suggest that upon the emergence of the first pollen tube from the septum, 5 pollen tube-secreted RALFs (RALF6/7/16/36/37) interact with septum-located receptors FER/ANJEA/HERCULES1 to establish the early polyspermy block (Zhong et al. 2022a). Various multiplexed *ralf6/7/16/36/37* mutant pollen tubes failed to trigger the block, resulting in increased levels of mul-tiple pollen tubes exiting as bundles and entering single ovules. Thus, the ovule phenotypes in multiplexed *ralf*-pollinated wild-type pistils phenocopied ovules from wild-type pollen tube penetrated *fer* pistils.

The local polyspermy block is triggered upon the entry of the first pollen tube into the ovule and fertilization is assured (Fig. 3C). Thus far, 2 distinct mechanisms involve the inactivation of the pollen tube attractant AtLURE1s (Duan et al. 2020; Yu et al. 2021). The FERONIA-controlled block is mediated by signaling events involving FERONIA, pectin, and pollen tube arrival–triggered nitric oxide (NO) at the fili-form apparatus. NO mediates a 2-prong mechanism to disen-gage the AtLURE1-mediated pollen tube attraction, one to inhibit the activity of the already secreted attractant and the other to prevent its further secretion from the synergids (Duan et al. 2020). Upon fusion with a sperm cell, the egg might also aid in fortifying the polyspermy block by secreting

2 aspartic proteases, ECS1 and ECS2, which would cleave and render any residual attractants inactive (Yu et al. 2021).

Fertilization recovery

Angiosperms have evolved an amazing capacity, referred to as "fertilization recovery" (Box 1), to salvage failed fertilization, for example, because of incompetent sperm or natural calamity has delayed pollination during the peak fertility window. With fertilization recovery, the plant can attract additional pollen tubes to a penetrated but not fertilized ovule or relax the polyspermy control in a dwindling time frame, affording more opportunities to produce offspring (Kasahara et al. 2012; Kasahara et al. 2013; Cheung et al. 2022). Although a naturally occurring process, studies of mutations in *A. thaliana* that impact the ability of sperm cells to fuse with the female gametes enabled systematic characterization and mechanistic elucidation of this fascinating phenomenon (Beale et al. 2012; Kasahara et al. 2012). When the *cdka;1*, *duo1*, *duo3* and *hap2* mutant pollen pollinated wild-type pistils, they induced significantly increased number of ovules penetrated by multiple pollen tubes, a phenotype referred to as "polytubey" (Beale et al. 2012) (Box 1). A variety of mutations that induce supernumerary pollen tubes in an ovule have since been reported (Supplementary Table S5) (Völz et al. 2013; Maruyama et al. 2015; Pereira et al. 2016b; Nagahara et al. 2020; Yu et al. 2021; Zhong et al. 2022a; Meng et al. 2023).

What triggers fertilization recovery? The principle of fertilization recovery assumes that the synergids, egg, or central cell somehow sense fertilization is threatened, thus the need to salvage the situation (Fig. 3C). While multiple me-chanisms likely exist, one obvious requirement is to re-open the early gate created at the septum upon pollen tube exit and the local gate in the ovule generated by the passage of the first pollen tube. Another requirement is to ensure the presence of attractants to guide additional pollen tubes. At the local gate, a pollen tube arrival–triggered FER- and pectin-dependent NO block disengages attractants. Since NO is gaseous, the local gate is easily reversible when NO has dissipated (Duan et al. 2020). Another mechanism ema-nates from the egg cell. When fertilization failed, the egg cell does not secrete the ECS1 and ECS2 proteases, which degrade attractants, thus preserving them and improving the chances to re-implement pollen tube attraction and fertilization recovery (Yu et al. 2021). How the septum polyspermy block is controlled remains to be demonstrated. Studies of RALF6,7,16,36,37 led to a provocative model that links pollen tube burst to fertilization recovery (Zhong et al. 2022a). It po-sits that the pollen RALFs are also required not only for estab-lishing but also maintaining the polyspermy block and that under normal conditions, with the first-arriving pollen tube bursting in the FG, the production and secretion of the pollen RALFs would halt. Without the pollen RALFs for mainten-ance, the septum block re-opens to allow exits of additional pollen tubes to target already penetrated ovules.

Persistent synergid cell represents a crucial component for fertilization recovery and the phenomenon depends on the persistent synergid not being terminally degraded so that it can reactivate attractant production. After the first fertilization attempt fails, the receptive synergid cell degrades (Maruyama and Higashiyama 2016), and the ES relies on the persistent synergid cell to continue secreting attractants to attract additional pollen tubes. It could be argued that the persistent synergid exists for the purpose of fertilization recovery. However, it also poses a risk to the polyspermy block, so it needs to be degraded rapidly upon fertilization. A mechanism that mitigates this threat could be the fusion between the persistent synergid and the endosperm, thus rapidly diluting the persistent cell-produced attractants (Maruyama et al. 2015). Yet fertilization recovery depends on maintaining the attractant factory (Fig. 3C). Therefore, the survival of the persistent synergid cell appears to be an outcome of a delicate balance. Multiple mechanisms for the degradation and survival of the persistent synerdid have emerged. These involve, for example, components of ethylene signaling, though not ethylene per se (Völz et al. 2013; Li et al. 2022a, 2022b, 2022c) and epigenetic regulation (Maruyama et al. 2013). While how these mechanisms integrate into the orchestration of fertilization recovery, they illustrate the connection between the degradation of the persistent synergid cell and the fertilization recovery-induced secondary pollen tube entrance to ovules.

The central cell also plays a crucial role and participates directly in fertilization recovery. When double fertilization fails and both synergids have degenerated after pollen tube entry, the central cell secretes additional peptide pollen tube attractants, SALVAGER1 and SALVAGER2, to attract late-arriving pollen tubes (Dresselhaus and van der Linde 2023; Lan et al. 2023b; Meng et al. 2023). In the absence of functional synergids, as in *myb98* mutant ovules, SALVAGERs and its upstream transcription factor CCG are upregulated, and SALVAGERs retained in the central cell before fertilization are directionally secreted to the micropyle (Meng et al. 2023) (Fig. 3B). These observations suggest that the synergid has an inhibitory effect on the transcription and secretion pathways of SALVAGERs, which would normally be switched on for pollen tube attraction and reception after the synergids have been eliminated or are impaired. Compared with the transient effect of synergid-based fertilization recovery, this central cell-controlled fertilization recovery mechanism acts as a main back-up system for fertilization success (Dresselhaus and van der Linde 2023). These observations set the stage for the elucidation of the multifaceted role of the central cell and its interaction with the synergids (Box 2).

Interspecific reproduction barriers

In natural surroundings, besides responding to compatible or incompatible pollen of their own species (intraspecific) via SI (Díaz et al. 2021; Goring et al. 2023; Nasrallah 2023), pistils also must respond to and set up barriers for pollen of other species (interspecific) (Fig. 4). Interspecific reproductive

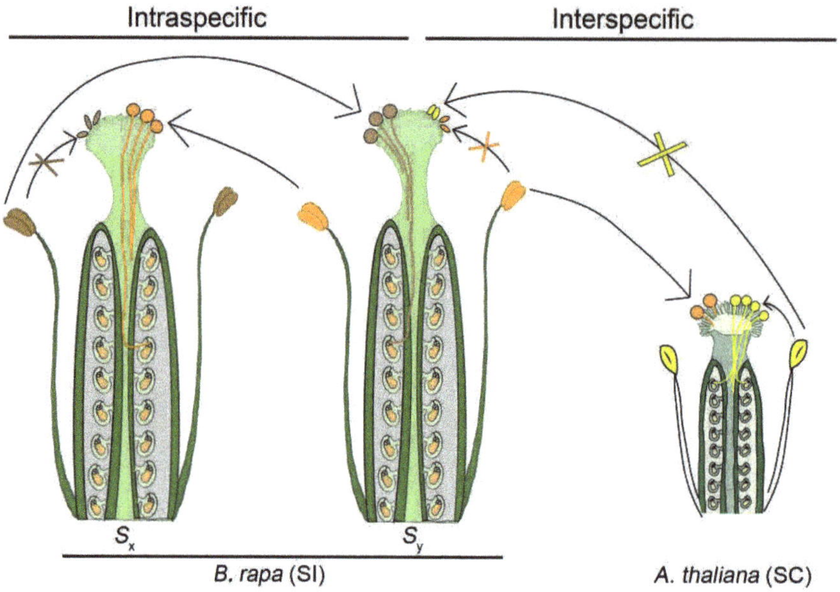

Figure 4. Intra- and interspecific pollination. *B. rapa* and *A. thaliana* are used as representative self-incompatible (SI) and self-compatible (SC) species. For intraspecific pollination, SI species displays incompatibility to self-pollen or pollen with the same S-haplotype (brown X) but are receptive to pollen with distinct haplotype (arrows with no X). Interspecific pollination often follows the SI × SC rule; SI species show unilaterally incompatibility (UI) to pollen of another species (yellow X), while SC species are unilaterally compatible (UC) to interspecific pollen, albeit they often display a notable preference for intraspecific pollen over interspecific pollen (compare yellow tubes and orange tubes in the *A. thaliana* pistil).

barriers prevent crossing between species that are not likely to produce healthy progeny and ensures species integrity. Overcoming hybridization barriers between interspecific species could facilitate species diversification and promote vigor. Interspecific barriers are diverse and genetically complex and could be implemented on multiple levels. Because of their importance in species preservation and diversification, and in agricultural concerns of the introgression of genes encoding high-quality traits, interspecific reproductive barriers are frequent subjects of reviews, for example, in the Brassicaceae (Kitashiba and Nasrallah 2014; Nasrallah 2023) and the Solanaceae (Baek et al. 2015; Bedinger et al. 2017; Broz and Bedinger 2021). Here, we focus our discussions on recent efforts surrounding the self-compatible *A. thaliana* and its related Brassicaceae species.

Stigmatic barriers

Membrane localized-regulators stigmatic gatekeepers. Pollen-pistil interaction studies revealed SI-independent and SI-dependent stigmatic gating mechanisms against interspecific hybridization. In Brassicaceae SI, self-pollen is recognized by the ligand–receptor interaction between the pollen-expressed SCR/SP11 and its receptor, the stigma-expressed, plasma membrane-localized SRK, followed by the activation of downstream signaling pathways in papilla cells that ultimately lead to the rejection of self-pollen (Kitashiba and Nasrallah 2014; Nasrallah 2023). SI-independent interspecific incompatibility was identified by Fujii et al. 2019 in a GWAS screen. Using more than 300 *A. thaliana* lines as recipients of pollen from the distantly related *Malcolmia littorea*, Fujii et al. identified the stigma-specific gene *STIGMATIC PRIVACY 1* (*SPRI1*) in *A. thaliana* as important for interspecific incompatibility. *M. littorea* pollen typically fails to germinate on *A. thaliana* Col-0 stigmas, but it was able to germinate and penetrate various *A. thaliana* ecotypes that harbor mutations in the *SPRI1* gene or Col-0 T-DNA knockouts of *SPRI1*, which encodes a 221-amino-acid stigmatic papilla transmembrane protein. However, *spri1* mutant expressing a functional SRK retained the typical SI capacity, rejecting pollen expressing the cognate SI pollen factor, SP11/SCR, but did not impact the penetration of *M. littorea* pollen tubes into the mutant stigma, indicating that SPRI1 is functionally independent of SI. The underlying mechanism of how SPRI1 regulates interspecific incompatibility remains to be explored.

Redox conditions controlled by the FER to ROS pathway in the Brassicaceae stigma (Liu et al. 2021; Zhang et al. 2021) play a major role in establishing an SI-dependent barrier for incompatible pollen intrusion (Franklin-Tong and Bosch 2021; Zhang et al. 2021; Huang et al. 2023). Like self-pollination in Arabidopsis, the self-compatible (SC) *Brassica rapa* (*B. rapa*) pollen suppresses *B. rapa* stigmatic ROS via PCP-B-triggered stigmatic NO increase and the mirrored decline of FER to ROS signaling, allowing pollen hydration and germination to proceed (Liu et al. 2021). On the other hand, SI pollination induced rapid increases in stigmatic ROS, paralleling the SI response of pollen arrest. Suppressing SRK expression on the stigma relaxed the SI pollen response. It abrogated the SI pollen-induced stigmatic ROS response, establishing stigmatic ROS as an SI-dependent barrier for incompatible pollen. The underlying mechanism involves allele-specific SP11/SCR-enhanced SRK-FER interaction, stimulating the FER to

ROS pathway and fortifying the intraspecific b arrier f or SI pollen.

In nature, plants could be nondiscriminately pollinated, including by sympatric species, some could be closely related. A mechanism similar to that in the Arabidopsis stigma also mediates the *B. rapa* stigmatic barrier for other Brassicaceae species. For stigmas of SI species, an SRK-dependent rise of the FER-controlled stigmatic ROS levels was triggered by interspecific pollen to reject them (Huang et al. 2023; Lan et al. 2023; Song et al. 2023; Wang and Filatov 2023). The FER-controlled ROS stigmatic gate not only blocked pollen from distantly related species but also delayed the hydration and germination of closely related interspecific compatible pollen, resulting in retarded pollen tube penetration of the stigma relative to SC pollen. The tunable stigmatic ROS gate was found to be responsive to PCP-B in a species differential manner, preventing interspecific hybridization and providing a head-start advantage for conspecific pollen to complete self-fertilization before pollen from closely related interspecies have a chance for hybridization, thereby ensuring species integrity.

Cell wall quality as ga tekeeper. The cell walls at the pollen-stigma interface are apparently important for species differentiation. Fujii et al. 2023 reported a zinc-finger transcriptional factor, STIGMATIC PRIVACY 2 (SPRI2), as having a controlling role in interspecies incompatibility in *A. thaliana*. SPRI2 is important in maintaining the expression of many genes related to cell wall organization or biosynthesis, including 2 genes encoding xylan or pectin O-acetyltransferases, TRICHOME BIREFRINGENCE-LIKE (TBL)40 and TBL45, as its direct targets. Loss of SPRI2 or TBLs relaxed the stigmatic barrier, allowing the interspecific pollen from *M. littorea* and *C. rubella* to hydrate and germinate and their tubes to penetrate the mutant pistils (Fujii et al. 2023). Furthermore, Lan et al. (2023a) showed that not only pollen from *ralf sept* mutant Arabidopsis but also pollen from interspecific incompatible species were able to penetrate stigmas from loss of LRX3,4,5 mutants. With LRXs being an integral part of the cell wall matrix (Borassi et al. 2016; Bedinger 2018; Zhao et al. 2018; Herger et al. 2019), the findings from Lan et al. (2023a) directly supports papilla cell wall quality as critical for a pollen tube gating mechanism.

Studies in other incompatibility systems also support that the pollen tube wall property is an important determinant for barrier-crossing capacity. For example, the maize *Teosinte crossing barrier1-s* (*Tcb1-s*) genes confer cross-incompatibility between *Teosinte* and maize. The female *Tcb1-s* prevents growth and fertilization by interspecific pollen and encodes a pistil-expressed pectin methylesterase (PME), a key enzyme in regulating cell wall stiffness and porosity (Lu et al. 2019). The maize *Gametophyte factor 1* (*Ga1*) locus, which controls intraspecific cross-incompatibility between popcorn and sweet corn, involves at least a silk-specific PME and 2 pollen-expressed PMEs (Moran Lauter et al. 2017; Zhang et al. 2018). The results underscore the importance of properly regulated pectin status in the cell wall in maintaining stigmatic incompatibility barriers.

Reproduction barriers at the level of pollen tube attraction

Earlier studies of attractants from Arabidopsis and maize demonstrated interspecific barriers at the level of pollen tube attraction to the FG. For LURE attractants, recombinant LURE1 from *A. thaliana* preferentially attracted *A. thaliana* pollen tubes than those from *A. lyarata* in vitro. In planta, the expression of the Arabidopsis LURE1 in the synergid cells of *T. fournieri* guided the heterologous *A. thaliana* pollen tube to the Torenia FG (Takeuchi and Higashiyama 2012). Expression of the maize attractant EA1 in *A. thaliana* ovules also augmented the frequency of the maize pollen tube targeting the *A. thaliana* ovule micropyle (Márton et al. 2012). Studies of AtLUREs further showed that in semi-in vivo assays *Capsella rubella* (*C. rubella*) pollen tubes expressing the *A. thaliana* PRK6 responded to AtLURE1, unlike wild-type *C. rubella* tubes (Takeuchi and Higashiyama 2016). Implicated in AtLURE1 perception, the Arabidopsis receptor kinase MDIS1, when expressed in *C. rubella* pollen tubes enhanced Capsella pollen tube targeting to *A. thaliana* ovules (Wang et al. 2016). These results indicate that both cross-species ex-pression of attractants or their receptors on the pollen tubes could overcome the barrier in pollen tube guidance.

Pollen tube guidance plays an important role in maintaining interspecific isolation. For example, compared with *A. thaliana* pollen tubes, *A. lyrata* and the more distantly re-lated *Cardamine fluexuosa* pollen tubes were increasingly less able to target *A. thaliana* ovules (Escobar-Restrepo et al. (2007). The contribution from pollen tube attractants is underscored by recent studies of AtLUREs and their related XUIQIUs and TICKETs, all from the CRP810 subclade (Takeuchi and Higashiyama 2012; Meng et al. 2019; Zhong et al. 2019a). The key function of cognate AtLURE1-PRK6 attractant-receptor interaction was determined to be for facilitating the exit of conspecific pollen tubes from the *A. thaliana* septum where interspecific *A. lyrata* pollen tubes emerged considerably later than the self-*A. thaliana* pollen tubes. AtLUREs therefore promote conspecific precedence in fertilization, reducing the chance for interspecific hybridization (Zhong et al. 2019a). Conversely, the self-pollen tubes were significantly delayed in their exits from the *atlure1* null mutant transmitting tract, exits of *A. lyrata* pollen tubes were not affected at all. A member of synergid-secreted non-CRP peptide NPA1 also contribute to species barrier by its species-specific pollen tube attraction activity among sister species in Brassicacea (Wang et al. 2024).

Barrier at the level of pollen tube reception in the ovule

Interspecific barriers also occur in the FG to prevent fertilization by an already penetrated pollen tube from a distant species, for example, in the Rhododendron, interspecific pollen tubes continued growth within the FG, aborting fertilization (Williams et al. 1986). Furthermore, of the tubes attracted to *A. thaliana* ovules, approximately 50% of *A. lyrata* and 70% of

C. fluexusa failed to burst inside the FG, correlated interspecific barrier strength with species isolation (Escobar-Restrepo et al. 2007). Interestingly, ectopic accumulation of NTA/MLO7 at the filiform apparatus suppressed the interspecific barrier resulting in significantly reduced levels (approximately 25%–30%) of *A. lyrata* pollen tube overgrowth in these transformed *A. thaliana* ovules (Ju et al. 2021). *A. lyrata* and interspecific *Olimarabidopsis pumila* pollen tubes, which could not burst in *A. thaliana* FG, nevertheless triggered FER-LRE-dependent changes in *A. thaliana* synergid cell Ca^{2+} oscillation dynamics, suggesting that downstream signaling of the FER-LRE-NORTIA apparatus plays key roles in maintaining the interspecific barrier (Ponvert and Johnson 2024).

Crossing barriers: emerging capability and continuing challenges

Genetic hybridization can rapidly expand species divergence in the ecosphere and is an important agricultural tool to generate plants with improved performance, especially under suboptimal conditions, or improved quality as a food source. Interspecific barriers for pollen invasion have been observed as weakened by genetic lesions impacting gating mechanism (e.g. Jones et al. 2017; Ju et al. 2021; Huang et al. 2023; Lan et al. 2023a) or the ability to cross barriers could be gained by heterologous expression of interspecific determinants (e.g. Márton et al. 2012; Takeuchi and Higashiyama 2012, 2016; Wang et al. 2016). These demonstrations suggest the possibility to manipulate interspecific barriers allowing hybridization of closely or more distantly related species. Two sets of test-of-principal experiments, one in SI *B. rapa* (Huang et al. 2023) and the other in SC *A. thaliana* pistils (Lan et al. 2023a), demonstrated that genetically weakening the stigmatic barrier could indeed be engineered to allow interspecific pollination to proceed. In *B. rapa*, downregulating the functionally linked SRK, FER, or the FER-controlled RBOH during pollination allowed efficient intraspecific SI pollen tube penetration producing developmentally normal embryos. These manipulations also allowed a limited level of cross-fertilization success by closely related *B. oleracea* pollen and more distantly related *B. vulagris*, albeit producing developmentally compromised embryos. In *A. thaliana*, the FER-RALF-dependent, LRX3,4,5-gated stigmatic barrier for interspecific pollen tube penetration could be relaxed by loss of FER alone, CURVY1, ANJEA and HERCULES1 together, sRALFs, or all LRX3,4,5. Moreover, co-applying a pRALF capable of weakening the wild-type stigma during interspecific pollination provided a mentoring effect for interspecific pollen tube penetration, producing hybrid embryos. Together, test of principal experiments demonstrate that fundamental understanding of how pollen and pistil interact to maintain hybridization barrier could contribute to manipulating these barriers to modulate interspecific isolation.

Natural and synthetic apomixis

In angiosperm, over 400 species from different plant phyla can propagate asexually through apomixis (Box 1). Apomixis exists in the gametophytic and sporophytic types. In gametophytic apomixis, apospory is characterized by the direct formation of an ES from somatic ovule cell, while the diplospory refers to the formation of unreduced FG from the MMC by skipping meiosis (Hojsgaard and Hörandl 2019). In both cases, the diploid ES develops into clonal embryo parthenogenetically. Sporophytic apomixis through adventitious embryos is found in some Citrus species and Fortunella genus (Wang et al. 2017c; Wang et al. 2022b). Synthetic apomixis is a newly invented term, defined as the introduction of apomixis into sexual plants through genomic engineering (Xiong et al. 2023). Unlike sexual reproduction, the MMC in synthetic apomicts bypasses meiosis and divides mitotically to form unreduced FG. The unreduced egg cell then undergoes a pseudo-zygotic program to form an embryo without integrating the sperm genome, and the endosperm forms autonomously or differentiates from the fertilized central cell (Xiong et al. 2023; Li et al. 2023a). In the last decade, utilizing the combined genetic system of *Mitosis instead of Meiosis* (*MiMe*) and haploid induction (HI) strategies, synthetic apomixis has been achieved in Arabidopsis and rice (Marimuthu et al. 2011; Khanday et al. 2019; Wang et al. 2019; Xie et al. 2019; Vernet et al. 2022; Liu et al. 2023; Wei et al. 2023; Xiong et al. 2023).

Genes involved in MiMe in Arabidopsis and rice

Meiosis is a specialized form of cell division that reduces the ploidy of germ cells (Mercier et al. 2015). Meiosis differs from mitosis in many aspects, including 2 rounds of chromosome segregation (Meiosis I and II) and the segregation of sister chromatids in Meiosis II (Mercier et al. 2015). The core mechanism of *MiMe* is to inactivate the essential genes required for these unique meiotic processes (dErfurth et al. 2009; dErfurth et al. 2010; Cifuentes et al. 2016; Mieulet et al. 2016; Khanday et al. 2019; Xie et al. 2019; Vernet et al. 2022). In Arabidopsis and rice, homologs of the highly conserved SPO11 with sequence similarity to the A subunit of Archaeal Topoisomerase VI play a key role in the formation of DNA double-strand breaks to initiate meiotic recombination (Stacey et al. 2006; Benyahya et al. 2020; Fayos et al. 2020). Lesions in these *SPO11* homologs abolish meiotic recombination (Li et al. 2023a, 2023b, 2023c; Xiong et al. 2023). Additional factors are also required, including *PUTATIVE RECOMBINATION INITIATION DEFECT 1*, 2, and 3 (*PRD1*, 2 and 3) and *DSB FORMING* (*DFO*) in Arabidopsis (De Muyt et al. 2007; De Muyt et al. 2009; Zhang et al. 2012; Shi et al. 2021; Wang et al. 2022c). In rice, *PRD3*/*HOMOLOGOUS PAIRING ABERRATION IN RICE MEIOSIS1* (*PAIR1*), *CENTRAL REGION COMPONENT 1* (*CRC1*), *P31*^{comet}, and *SOLO DANCERS* (*SDS*) are involved in DSB formation during meiotic recombination (Nonomura et al. 2004; Miao et al. 2013; Wu et al. 2015; Ji et al. 2016). Mutations in these factors hinder recombination events (Mercier et al. 2015). *RECOMBINATION-DEFICIENT 8* (*REC8*) encodes an exclusive meiosis-specific cohesin subunit protein and its impairment causes aberrant meiosis I (Chelysheva et al. 2005; Sun et al. 2013). Other cohesion-related components, such as SISTER-CHROMATID COHESION PROTEIN 3 (SCC3), SHUGOSHIN1 (SGO1), and Bub1-related kinase 1 (BRK1), are also required for cohesion functions in meiosis (Chelysheva et al. 2005; Wang et al. 2012; Zamariola et al. 2013). Genes associated with meiosis progression need to be modified to fully abrogate meiosis II. These include the A-type cyclin *TARDY ASYNCHRONOUS MEIOSIS* (*TAM*), *CDKs*, *THREE DIVISION MUTATION 1* (*TDM1*), and *OMISSION of SECOND DIVISION 1* (*OSD1*) (Dissmeyer et al. 2007; dErfurth et al. 2010; Cifuentes et al. 2016). During meiosis I, TAM interacts with CDKA;1 to suppress the activity of TDM1, a component of the Anaphase Promoting Complex/Cyclosome (APC/C) regulating the exit from meiosis II (Dissmeyer et al. 2007; Cairo et al. 2022). OSD1 is pivotal in initiating meiosis II by governing CDK activity in rice and Arabidopsis (dErfurth et al. 2009; Mieulet et al. 2016). Thus, OSD1, TAM, and TDM1 collectively govern critical transitions within the meiotic cell cycle, and mutations in any of these 3 genes result in the abrogation of meiosis II (dErfurth et al. 2010; Cifuentes et al. 2016).

Gene loci related to HI

HI presents an effective approach to bypass zygote formation. In maize, *quir1* and *quir8* are gene loci related to HI widely used in agricultural breeding (Wang et al. 2017c; Zhong et al. 2019b). *quir1* a mutation in *PLA1/NOT LIKE DAD* (*NLD*)/*MATRILINEAL* (*MTL*). *PLA1/NLD/MTL* encodes a phospholipase A1 localized on the endo-plasma membrane (PM) around the 2 sperm cells, also known as the inner vegetative PM for having originated from the PM of the pollen VC (Dong et al. 2013; Gilles et al. 2017; Kelliher et al. 2017). The *pla1/mtl*/nld haploid inducer line displays a remarkably high rate of sperm DNA fragmentation (Li et al. 2017). Such fragmentation might be caused by perturbed redox status that triggers postfertilization paternal genome elimination, culminating in the generation of maternal haploids (Jiang et al. 2022a). The utilization of *PLA1/MTL/NLD*-based chromosome fragmentation has been extended to generating haploid-inducing lines in other monocot staple crops, such as wheat, rice, and foxtail millet (Cheng et al. 2021; Liu et al. 2023). The depletion of pollen-specific *PHOSPHOLIPASE D3* (*ZmPLD3*) and *PLA1/MTL/NLD* also induces a higher HI rate in maize (Li et al. 2021).

Ectopic expression of *BBM* or its homologs in the egg cell of monocots or dicots induces autonomous transition to embryonic development (Conner et al. 2015; Khanday et al. 2019; Deng et al. 2022). In *Pennisetum squamulatum*, several *BBM-like* (*BBML*) genes in apospory-specific genomic region (ASGR) govern parthenogenesis (Akiyama et al. 2011). Notably, in diverse species, including tobacco, pearl millet, rice, maize, Arabidopsis, *Brassica napus*, and *Solanum lycopersicon*, parthenogenesis is induced when *BBMLs* are ectopically expressed in the egg cells (Conner et al. 2015; Conner et al. 2017; Khanday et al. 2019; Zhang et al. 2020a; Deng et al. 2022). Rice has 4 *OsBBM* homologs; *BBM1,3,4* are exclusively expressed in sperm cells pre-fertilization (Khanday et al. 2019). Ectopic expression of these *OsBBMs* in rice egg cells induces parthenogenesis (Khanday et al. 2019). Depletion of the gamete-fusion-related *DMP8/9* in Arabidopsis or orthologs in multiple plants leads to maternal haploid offspring (Zhong et al. 2019b; Zhong et al. 2020; Li et al. 2022c; Wang et al. 2022b; Zhang et al. 2022a, 2022b; Zhao et al. 2022b; Zhong et al. 2022b; Chen et al. 2023). Except for the fertilization defect, maternal haploid offspring has also been observed in *ec-s1ecs2* double mutants in Arabidopsis and the rice *esc1* mutant (Yu et al. 2021; Jiang et al. 2022b; Jiang and Qu 2023; Zhang et al. 2023).

Combining MiMe and HI in fixation of hybrid rice

Based on the findings discussed above, several strategies have been developed to achieve synthetic apomixis in rice (Khanday et al. 2019; Wang et al. 2019). Using the *MiMe* mutants, *rec8 pair1 osd1*, either combining with *osmtl* or ectopic expression of *OsBBM1* in the rice egg cell resulted in clonal hybrid rice seeds (Khanday et al. 2019; Wang et al. 2019). However, these strategies face challenges, such as low seed yields and clonal seed proportion. Employing a single vector for mutating *PAIR1*, *REC8*, and *OSD1* along with ectopic *OsBBM1* expression has subsequently increased the clonal seed proportion to 90% in a hybrid rice variety (Vernet et al. 2022). Another novel strategy combines *BBM4* ectopic expression in the egg cell with *MiMe*, generating clonal hybrid plants with high seed setting, albeit the clonal seed proportion remained low (Wei et al. 2023). Recent advances are promising for further improvements (Box 2).

Concluding remarks

We hope the discussions here provide a perspective of important advances in plant reproduction culminated in the past 30 years or so through the efforts of a huge research community. We hope some of the more definitive conclusions will guide us to explore imminently approachable problems and tackle challenges that might be considered too grand to contemplate even in the past decade (Box 2). For the vast amount of accumulated knowledge that might require further definitions, we hope they will fuel productive discussions and entice new generations of biologists to join plant reproduction research.

Acknowledgments

We thank H-M Wu, James M-C Liu (U MASS), and Yunyun Cao (SDAU and UMASS) for their comments, Yang Lin (SDAU) for preparing Fig. 4, and Kelly Simpson (UMASS) for assistance in assembling the reference list.

Funding

Work from our laboratories was supported by National Natural Science Foundation of China (#32170343 to HL), CAS Project for Young Scientists in Basic Research

(YSBR-078 to HL); National Natural Science Foundation of China (#32270360) and The Foundation of Hubei Hongshan Laboratory (#2021hszd010; #2022hszd017 to X-BP and PZ); The Science and Technology Department of Hubei Province (#2022CFA071 to PZ); National Natural Science Foundation of China (#32122014; #32070854 to SZ); Shandong Natural Science Foundation (ZR2020KC017 to Q.D.). U.S. National Science Foundation (MCB-1715764, IOS-1645858, IOS-2101467 to AYC and H-M Wu), National Institute of Food and Agriculture (NIFA)/USDA, the Center for Agriculture, Food, and the Environment under project number MAS00525 (to AYC), and National Institute of Health (R01GM147548-01A1).

Conflict of interest statement. None declared

Data availability

There is no new data in this paper.

References

Abarca A, Franck CM, Zipfel C. Family-wide evaluation of RAPID ALKALINIZATION FACTOR peptides. Plant Physiol. 2021:**187**(2): 996–1010. https://doi.org/10.1093/plphys/kiab308

Akiyama Y, Goel S, Conner JA, Hanna WW, Yamada-Akiyama H, Ozias-Akins P. Evolution of the apomixis transmitting chromosome in Pennisetum. BMC Evol Biol. 2011:**11**(1):289. https://doi.org/10.1186/1471-2148-11-289

Ali MF, Fatema U, Peng X, Hacker SW, Maruyama D, Sun M-X, Kawashima T. ARP2/3-independent WAVE/SCAR pathway and class XI myosin control sperm nuclear migration in flowering plants. Proc Natl Acad Sci U S A. 2020:**117**(51):32757–32763. https://doi.org/10.1073/pnas.2015550117

Ali MF, Kawashima T. Formins control dynamics of F-actin in the central cell of Arabidopsis thaliana. Plant Signal Behav. 2021:**16**(8): 1920192. https://doi.org/10.1080/15592324.2021.1920192

Amien S, Kliwer I, Márton ML, Debener T, Geiger D, Becker D, Dresselhaus T. Defensin-like ZmES4 mediates pollen tube burst in maize via opening of the potassium channel KZM1. PLoS Biol. 2010:**8**(6):e1000388. https://doi.org/10.1371/journal.pbio.1000388

Anderson SN, Johnson CS, Chesnut J, Jones DS, Khanday I, Woodhouse M, Li C, Conrad LJ, Russell SD, Sundaresan V. The zygotic transition is initiated in unicellular plant zygotes with asymmetric activation of parental genomes. Dev Cell. 2017:**43**(3):349–358.e4. https://doi.org/10.1016/j.devcel.2017.10.005

Anderson SN, Johnson CS, Jones DS, Conrad LJ, Gou X, Russell SD, Sundaresan V. Transcriptomes of isolated Oryza sativa gametes characterized by deep sequencing: evidence for distinct sex-dependent chromatin and epigenetic states before fertilization. Plant J. 2013:**76**(5):729–741. https://doi.org/10.1111/tpj.12336

Ariizumi T, Toriyama K. Genetic regulation of sporopollenin synthesis and pollen exine development. Annu Rev Plant Biol. 2011:**62**(1):437–460. https://doi.org/10.1146/annurev-arplant-042809-112312

Aw SJ, Hamamura Y, Chen Z, Schnittger A, Berger F. Sperm entry is sufficient to trigger division of the central cell but the paternal genome is required for endosperm development in *Arabidopsis*. Development. 2010:**137**(16):2683–2690. https://doi.org/10.1242/dev.052928

Baek YS, Covey PA, Petersen JJ, Chetelat RT, McClure B, Bedinger PA. Testing the SI × SC rule: Pollen-pistil interactions in interspecific crosses between members of the tomato clade (Solanum section Lycopersicon, Solanaceae). Am J Bot. 2015:**102**:302–311.

Beale KM, Leydon AR, Johnson MA. Gamete fusion is required to block multiple pollen tubes from entering an Arabidopsis ovule. Curr Biol. 2012:**22**(12):1090–1094. https://doi.org/10.1016/j.cub.2012.04.041

Bedinger P. Coordinating cell walls and cell growth: a role for LRX extensin chimeras. Plant Physiol. 2018:**176**(3):1890–1891. https://doi.org/10.1104/pp.18.00084

Bedinger PA, Broz AK, Tovar-Mendez A, McClure B. Pollen-pistil interactions and their role in mate selection. Plant Physiol. 2017:**173**: 79–90.

Bemer M, Wolters-Arts M, Grossniklaus U, Angenent GC. The MADS domain protein DIANA acts together with AGAMOUS-LIKE80 to specify the central cell in Arabidopsis ovules. Plant Cell. 2008:**20**(8): 2088–2101. https://doi.org/10.1105/tpc.108.058958

Benyahya F, Nadaud I, Da Ines O, Rimbert H, White C, Sourdille P. SPO11.2 is essential for programmed double-strand break formation during meiosis in bread wheat (Triticum aestivum L.). Plant J. 2020:**104**(1):30–43. https://doi.org/10.1111/tpj.14903

Berken A, Thomas C, Wittinghofer A. A new family of RhoGEFs activates the Rop molecular switch in plants. Nature. 2005:**436**(7054): 1176–1180. https://doi.org/10.1038/nature03883

Bianchi E, Wright GJ. Sperm meets egg: the genetics of mammalian fertilization. Annu Rev Genet. 2016:**50**(1):93–111. https://doi.org/10.1146/annurev-genet-121415-121834

Blackburn MR, Haruta M, Moura DS. Twenty years of progress in physiological and biochemical investigation of RALF peptides. Plant Physiol. 2020:**182**(4):1657–1666. https://doi.org/10.1104/pp.19.01310

Bogaert KA, Beeckman T, De Clerck O. Two-step cell polarization in algal zygotes. Nat Plants. 2017:**3**(2):16221. https://doi.org/10.1038/nplants.2016.221

Boisson-Dernier A, Franck CM, Lituiev DS, Grossniklaus U. Receptor-like cytoplasmic kinase MARIS functions downstream of CrRLK1L-dependent signaling during tip growth. Proc Natl Acad Sci U S A. 2015:**112**(39):12211–12216. https://doi.org/10.1073/pnas.1512375112

Boisson-Dernier A, Lituiev DS, Nestorova A, Franck CM, Thirugnanarajah S, Grossniklaus U. ANXUR receptor-like kinases coordinate cell wall integrity with growth at the pollen tube tip via NADPH oxidases. PLoS Biol. 2013:**11**(11):e1001719. https://doi.org/10.1371/journal.pbio.1001719

Boisson-Dernier A, Roy S, Kritsas K, Grobei MA, Jaciubek M, Schroeder JI, Grossniklaus U. Disruption of the pollen-expressed FERONIA homologs ANXUR1 and ANXUR2 triggers pollen tube discharge. Development. 2009:**136**(19):3279–3288. https://doi.org/10.1242/dev.040071

Borassi C, Sede AR, Mecchia MA, Mangano S, Marzol E, Denita-Juarez SP, Salgado Salter JD, Velasquez SM, Muschietti JP, Estevez JM. Proline-rich extensin-like receptor kinases PERK5 and PERK12 are involved in pollen tube growth. FEBS Lett. 2021:**595**(20):2593–2607. https://doi.org/10.1002/1873-3468.14185

Borassi C, Sede AR, Mecchia MA, Salgado Salter JD, Marzol E, Muschietti JP, Estevez JM. An update on cell surface proteins containing extensin-motifs. J Exp Bot. 2016:**67**(2):477–487. https://doi.org/10.1093/jxb/erv455

Borg M, Berger F. Chromatin remodelling during male gametophyte development. Plant J. 2015:**83**(1):177–188. https://doi.org/10.1111/tpj.12856

Borg M, Brownfield L, Khatab H, Sidorova A, Lingaya M, Twell D. The R2R3 MYB transcription factor DUO1 activates a male germline-specific regulon essential for sperm cell differentiation in Arabidopsis. Plant Cell. 2011:**23**(2):534–549. https://doi.org/10.1105/tpc.110.081059

Borg M, Jacob Y, Susaki D, LeBlanc C, Buendía D, Axelsson E, Kawashima T, Voigt P, Boavida L, Becker J, et al. Targeted reprogramming of H3K27me3 resets epigenetic memory in plant paternal chromatin. Nat Cell Biol. 2020:**22**(6):621–629. https://doi.org/10.1038/s41556-020-0515-y

Brownfield L, Hafidh S, Borg M, Sidorova A, Mori T, Twell D. A plant germline-specific integrator of sperm specification and cell cycle progression. PLoS Genet. 2009:**5**(3):e1000430. https://doi.org/10.1371/journal.pgen.1000430

Broz AK, Bedinger PA. Pollen-pistil interactions as reproductive barriers. Annu Rev Plant Biol. 2021:**72**(1):615–639. https://doi.org/10.1146/annurev-arplant-080620-102159

Cai H, Ma S, Su H, Liu K, Aslam M, Qin Y. Positional signals establishment in the regulation of female germline specification. Seed Biology. 2022:**1**(1):1–7. https://doi.org/10.48130/SeedBio-2022-0006

Cairo A, Vargova A, Shukla N, Capitao C, Mikulkova P, Valuchova S, Pecinkova J, Bulankova P, Riha K. Meiotic exit in Arabidopsis is driven by P-body-mediated inhibition of translation. Science. 2022:**377**(6606):629–634. https://doi.org/10.1126/science.abo0904

Cameron C, Geitmann A. Cell mechanics of pollen tube growth. Curr Opin Genet Dev. 2018:**51**:11–17. https://doi.org/10.1016/j.gde.2018.03.008

Capron A, Gourgues M, Neiva LS, Faure J-E, Berger F, Pagnussat G, Krishnan A, Alvarez-Mejia C, Vielle-Calzada J-P, Lee Y-R, et al. Maternal control of male-gamete delivery in Arabidopsis involves a putative GPI-anchored protein encoded by the LORELEI gene. Plant Cell. 2008:**20**(11):3038–3049. https://doi.org/10.1105/tpc.108.061713

Cascallares M, Setzes N, Marchetti F, López GA, Distéfano AM, Cainzos M, Zabaleta E, Pagnussat GC. A complex journey: cell wall remodeling, interactions, and integrity during pollen tube growth. Front Plant Sci. 2020:**11**:599247. https://doi.org/10.3389/fpls.2020.599247

Chae K, Lord EM. Pollen tube growth and guidance: roles of small, secreted proteins. Ann Bot. 2011:**108**(4):627–636. https://doi.org/10.1093/aob/mcr015

Chen CY-h, Cheung AY, Wu H-m. Actin-depolymerizing factor mediates Rac/Rop GTPase–regulated pollen tube growth. Plant Cell. 2003:**15**(1):237–249. https://doi.org/10.1105/tpc.007153

Chelysheva L, Diallo S, Vezon D, Gendrot G, Vrielynck N, Belcram K, Rocques N, Márquez-Lema A, Bhatt AM, Horlow C, et al. AtREC8 and AtSCC3 are essential to the monopolar orientation of the kinetochores during meiosis. J Cell Sci. 2005:**118**(Pt 20):4621–4632. https://doi.org/10.1242/jcs.02583

Chen J, Strieder N, Krohn NG, Cyprys P, Sprunck S, Engelmann JC, Dresselhaus T. Zygotic genome activation occurs shortly after fertilization in maize. Plant Cell. 2017:**29**(9):2106–2125. https://doi.org/10.1105/tpc.17.00099

Chen L-Y, Shi D-Q, Zhang W-J, Tang Z-S, Liu J, Yang W-C. The Arabidopsis alkaline ceramidase TOD1 is a key turgor pressure regulator in plant cells. Nat Commun. 2015:**6**(1):6030. https://doi.org/10.1038/ncomms7030

Chen S-Y, Wang L, Jia P-F, Yang W-C, Sze H, Li H-J. Osmoregulation determines sperm cell geometry and integrity for double fertilization in flowering plants. Mol Plant. 2022:**15**(9):1488–1496. https://doi.org/10.1016/j.molp.2022.07.013

Chen X, Li Y, Ai G, Chen J, Guo D, Zhu Z, Zhu X, Tian S, Wang J, Liu M, et al. Creation of a watermelon haploid inducer line via ClDMP3-mediated single fertilization of the central cell. Horti Res. 2023:**10**(6):uhad081. https://doi.org/10.1093/hr/uhad081

Chen Y-C, McCormick S. Sidecar pollen, an Arabidopsis thaliana male gametophytic mutant with aberrant cell divisions during pollen development. Development. 1996:**122**(10):3243–3253. https://doi.org/10.1242/dev.122.10.3243

Chen Y-H, Li H-J, Shi D-Q, Yuan L, Liu J, Sreenivasan R, Baskar R, Grossniklaus U, Yang W-C. The central cell plays a critical role in pollen tube guidance in Arabidopsis. Plant Cell. 2007:**19**(11):3563–3577. https://doi.org/10.1105/tpc.107.053967

Chen Z, Tan JLH, Ingouff M, Sundaresan V, Berger F. Chromatin assembly factor 1 regulates the cell cycle but not cell fate during male gametogenesis in Arabidopsis thaliana. Development. 2008:**135**(1):65–73. https://doi.org/10.1242/dev.010108

Cheng Z, Sun Y, Yang S, Zhi H, Yin T, Ma X, Zhang H, Diao X, Guo Y, Li X, et al. Establishing in planta haploid inducer line by edited SiMTL in foxtail millet (Setaria italica). Plant Biotechnol J. 2021:**19**(6):1089–1091. https://doi.org/10.1111/pbi.13584

Cheung AY. Pollen-pistil interactions in compatible pollination. Proc Natl Acad Sci U S A. 1995:**92**:3077–3080.

Cheung AY. Pollen—pistil interactions during pollen-tube growth. Trends Plant Sci. 1996:**1**:45–51.

Cheung AY. FERONIA: a receptor kinase at the core of a global signaling network. Annu Rev Plant Biol. 2024:**75**. https://doi.org/10.1146/annurev-arplant-102820-103424 Online ahead of print.

Cheung AY, Duan Q, Costa SS, de Graaf BHJ, Di Stilio VS, Feijo J, Wu H-M. The dynamic pollen tube cytoskeleton: live cell studies using actin-binding and microtubule-binding reporter proteins. Mol Plant. 2008:**1**(4):686–702. https://doi.org/10.1093/mp/ssn026

Cheung AY, Duan Q, Li C, James Liu M-C, Wu H-M. Pollen-pistil interactions: it takes two to tangle but a molecular cast of many to deliver. Curr Opin Plant Biol. 2022:**69**:102279. https://doi.org/10.1016/j.pbi.2022.102279

Cheung AY, Niroomand S, Zou Y, Wu H-M. A transmembrane formin nucleates subapical actin assembly and controls tip-focused growth in pollen tubes. Proc Natl Acad Sci U S A. 2010:**107**(37):16390–16395. https://doi.org/10.1073/pnas.1008527107

Cheung AY, Wang H, Wu HM. A floral transmitting tissue-specific glycoprotein attracts pollen tubes and stimulates their growth. Cell. 1995:**82**(3):383–393. https://doi.org/10.1016/0092-8674(95)90427-1

Cheung AY, Wu HM. Structural and signaling networks for the polar cell growth machinery in pollen tubes. Annu Rev Plant Biol. 2008:**59**(1):542–572. https://doi.org/10.1146/annurev.arplant.59.032607.092921

Christensen CA, King EJ, Jordan JR, Drews GN. Megagametogenesis in Arabidopsis wild type and the Gf mutant. Sex Plant Reprod. 1997:**10**(1):49–64. https://doi.org/10.1007/s004970050067

Cifuentes M, Jolivet S, Cromer L, Harashima H, Bulankova P, Renne C, Crismani W, Nomura Y, Nakagami H, Sugimoto K, et al. TDM1 regulation determines the number of meiotic divisions. PLoS Genet. 2016:**12**(2):e1005856. https://doi.org/10.1371/journal.pgen.1005856

Cole ES, Cassidy-Hanley D, Fricke Pinello J, Zeng H, Hsueh M, Kolbin D, Ozzello C, Giddings T Jr, Winey M, Clark TG. Function of the male-gamete-specific fusion protein, HAP2 in a seven-sexed ciliate. Curr Biol. 2014:**24**(18):2168–2173. https://doi.org/10.1016/j.cub.2014.07.064

Conner JA, Mookkan M, Huo H, Chae K, Ozias-Akins P. A parthenogenesis gene of apomict origin elicits embryo formation from unfertilized eggs in a sexual plant. Proc Nat'l Acad Sci U S A. 2015:**112**(36):11205–11210. https://doi.org/10.1073/pnas.1505856112

Conner JA, Podio M, Ozias-Akins P. Haploid embryo production in rice and maize induced by PsASGR-BBML transgenes. Plant Reprod. 2017:**30**(1):41–52. https://doi.org/10.1007/s00497-017-0298-x

Crawford BCW, Ditta G, Yanofsky MF. The NTT gene is required for transmitting-tract development in carpels of Arabidopsis thaliana. Curr Biol. 2007:**17**(13):1101–1108. https://doi.org/10.1016/j.cub.2007.05.079

Cyprys P, Lindemeier M, Sprunck S. Gamete fusion is facilitated by two sperm cell-expressed DUF679 membrane proteins. Nat Plants. 2019:**5**(3):253–257. https://doi.org/10.1038/s41477-019-0382-3

Dai XR, Gao X-Q, Chen GH, Tang LL, Wang H, Zhang XS. ABNORMAL POLLEN TUBE GUIDANCE1, an endoplasmic reticulum-localized mannosyltransferase homolog of GLYCOSYLPHOSPHATIDYLINOSITOL10 in yeast and PHOSPHATIDYLINOSITOL GLYCAN ANCHOR BIOSYNTHESIS B in human, is required for Arabidopsis pollen tube micropylar guidance and embryo development. Plant Physiol. 2014:**165**(4):1544–1556. https://doi.org/10.1104/pp.114.236133.

Dehors J, Mareck A, Kiefer-Meyer M-C, Menu-Bouaouiche L, Lehner A, Mollet J-C. Evolution of cell wall polymers in tip-growing land plant gametophytes: composition, distribution, functional aspects and their remodeling. Front Plant Sci. 2019:**10**:441. https://doi.org/10.3389/fpls.2019.00441

Demidchik V, Shabala S, Isayenkov S, Cuin TA, Pottosin I. Calcium transport across plant membranes: mechanisms and functions. New Phytol. 2018:**220**(1):49–69. https://doi.org/10.1111/nph.15266

De Muyt A, Pereira L, Vezon D, Chelysheva L, Gendrot G, Chambon A, Lainé-Choinard S, Pelletier G, Mercier R, Nogué F, et al. A high throughput genetic screen identifies new early meiotic recombination functions in Arabidopsis thaliana. PLoS Genet. 2009:**5**(9): e1000654. https://doi.org/10.1371/journal.pgen.1000654

De Muyt A, Vezon D, Gendrot G, Gallois J-L, Stevens R, Grelon M. AtPRD1 is required for meiotic double strand break formation in Arabidopsis thaliana. EMBO J. 2007:**26**(18):4126–4137. https://doi.org/10.1038/sj.emboj.7601815

Deng F, Zeng F, Shen Q, Abbas A, Cheng J, Jiang W, Chen G, Shah AN, Holford P, Tanveer M, et al. Molecular evolution and functional modification of plant miRNAs with CRISPR. Trends Plant Sci. 2022:**27**(9):890–907. https://doi.org/10.1016/j.tplants.2022.01.009

Denninger P, Bleckmann A, Lausser A, Vogler F, Ott T, Ehrhardt DW, Frommer WB, Sprunck S, Dresselhaus T, Grossmann G. Male–female communication triggers calcium signatures during fertilization in Arabidopsis. Nat Commun. 2014:**5**(1):4645. https://doi.org/10.1038/ncomms5645

dErfurth I, Cromer L, Jolivet S, Girard C, Horlow C, Sun Y, To JPC, Berchowitz LE, Copenhaver GP, Mercier R. The cyclin-A CYCA1;2/TAM is required for the meiosis I to meiosis II transition and cooperates with OSD1 for the prophase to first meiotic division transition. PLoS Genet. 2010:**6**(6):e1000989. https://doi.org/10.1371/journal.pgen.1000989

dErfurth I, Jolivet S, Froger N, Catrice O, Novatchkova M, Mercier R. Turning meiosis into mitosis. PLoS Biol. 2009:**7**(6):e1000124. https://doi.org/10.1371/journal.pbio.1000124

Díaz JAJ, Zamora YC, García FC. Self-incompatibility in S-RNase-based systems: are there differences in the pollen rejection mechanisms among species? MCDA. 2021:**8**:803–807.

Dissmeyer N, Nowack MK, Pusch S, Stals H, Inzé D, Grini PE, Schnittger A. T-loop phosphorylation of Arabidopsis CDKA;1 is required for its function and can be partially substituted by an aspartate residue. Plant Cell. 2007:**19**(3):972–985. https://doi.org/10.1105/tpc.107.050401

Doblas VG, Gonneau M, Höfte H. Cell wall integrity signaling in plants: malectin-domain kinases and lessons from other kingdoms. Cell Surf. 2018:**3**:1–11. https://doi.org/10.1016/j.tcsw.2018.06.001

Dobritsa AA, Coerper D. The novel plant protein INAPERTURATE POLLEN1 marks distinct cellular domains and controls formation of apertures in the Arabidopsis pollen exine. Plant Cell. 2012:**24**(11):4452–4464. https://doi.org/10.1105/tpc.112.101220

Dong X, Xu X, Miao J, Li L, Zhang D, Mi X, Liu C, Tian X, Melchinger AE, Chen S. Fine mapping of qhir1 influencing in vivo haploid induction in maize. Theor Appl Genet. 2013:**126**(7):1713–1720. https://doi.org/10.1007/s00122-013-2086-9

Doughty J, Dixon S, Hiscock SJ, Willis AC, Parkin IAP, Dickinson HG. PCP-A1, a defensin-like brassica pollen coat protein that binds the S locus glycoprotein, is the product of gametophytic gene expression. Plant Cell. 1998:**10**(8):1333–1347. https://doi.org/10.1105/tpc.10.8.1333

Dresselhaus T, Jürgens G. Comparative embryogenesis in angiosperms: activation and patterning of embryonic cell lineages. Annu Rev Plant Biol. 2021:**72**(1):641–676. https://doi.org/10.1146/annurev-arplant-082520-094112

Dresselhaus T, van der Linde K. Plant reproduction: fertilization SALvaged by the central cell. Curr Biol. 2023:**33**(19):R1013–R1015. https://doi.org/10.1016/j.cub.2023.08.078

Du J, Anderson CT, Xiao C. Dynamics of pectic homogalacturonan in cellular morphogenesis and adhesion, wall integrity sensing and plant development. Nat Plants. 2022:**8**(4):332–340. https://doi.org/10.1038/s41477-022-01120-2

Duan Q, Kita D, Johnson EA, Aggarwal M, Gates L, Wu H-M, Cheung AY. Reactive oxygen species mediate pollen tube rupture to release sperm for fertilization in Arabidopsis. Nat Commun. 2014:**5**(1):3129. https://doi.org/10.1038/ncomms4129

Duan Q, Kita D, Li C, Cheung AY, Wu H-M. FERONIA receptor-like kinase regulates RHO GTPase signaling of root hair development. Proc Natl Acad Sci U S A. 2010:**107**(41):17821–17826. https://doi.org/10.1073/pnas.1005366107

Duan Q, Liu M-CJ, Kita D, Jordan SS, Yeh F-LJ, Yvon R, Carpenter H, Federico AN, Garcia-Valencia LE, Eyles SJ, et al. FERONIA controls pectin- and nitric oxide-mediated male-female interaction. Nature. 2020:**579**(7800):561–566. https://doi.org/10.1038/s41586-020-2106-2

Dünser K, Gupta S, Herger A, Feraru MI, Ringli C, Kleine-Vehn J. Extracellular matrix sensing by FERONIA and Leucine-Rich Repeat Extensins controls vacuolar expansion during cellular elongation in Arabidopsis thaliana. EMBO J. 2019:**38**:e100353.

Durbarry A, Vizir I, Twell D. Male germ line development in Arabidopsis. Duo pollen mutants reveal gametophytic regulators of generative cell cycle progression. Plant Physiol. 2005:**137**(1): 297–307. https://doi.org/10.1104/pp.104.053165

Eady C, Lindsey K, Twell D. The significance of microspore division and division symmetry for vegetative cell-specific transcription and generative cell differentiation. Plant Cell. 1995:**7**(1):65–74. https://doi.org/10.2307/3869838

Edlund AF, Zheng Q, Lowe N, Kuseryk S, Ainsworth KL, Lyles RH, Sibener SJ, Preuss D. Pollen from Arabidopsis thaliana and other Brassicaceae are functionally omniaperturate. Am J Bot. 2016:**103**(6): 1006–1019. https://doi.org/10.3732/ajb.1600031

Endo S, Shinohara H, Matsubayashi Y, Fukuda H. A novel pollen-pistil interaction conferring high-temperature tolerance during reproduction via CLE45 signaling. Curr Biol. 2013:**23**(17):1670–1676. https://doi.org/10.1016/j.cub.2013.06.060

Engell K. Embryology of barley. IV. Ultrastructure of the antipodal cells of Hordeum vulgare L. cv. Bomi before and after fertilization of the egg cell. Sex Plant Reprod. 1994:**7**(6):333–346. https://doi.org/10.1007/BF00230512

Escobar-Restrepo J-M, Huck N, Kessler S, Gagliardini V, Gheyselinck J, Yang W-C, Grossniklaus U. The FERONIA receptor-like kinase mediates male-female interactions during pollen tube reception. Science. 2007:**317**(5838):656–660. https://doi.org/10.1126/science.1143562

Fabrice TN, Vogler H, Draeger C, Munglani G, Gupta S, Herger AG, Knox P, Grossniklaus U, Ringli C. LRX proteins play a crucial role in pollen grain and pollen tube cell wall development. Plant Physiol. 2018:**176**(3):1981–1992. https://doi.org/10.1104/pp.17.01374

Fayos I, Meunier AC, Vernet A, Navarro-Sanz S, Portefaix M, Lartaud M, Bastianelli G, Périn C, Nicolas A, Guiderdoni E. Assessment of the roles of SPO11-2 and SPO11-4 in meiosis in rice using CRISPR/Cas9 mutagenesis. J Exp Bot. 2020:**71**(22):7046–7058. https://doi.org/10.1093/jxb/eraa391

Feng H, Liu C, Fu R, Zhang M, Li H, Shen L, Wei Q, Sun X, Xu L, Ni B, et al. LORELEI-LIKE GPI-ANCHORED PROTEINS 2/3 regulate pollen tube growth as chaperones and coreceptors for ANXUR/BUPS receptor kinases in Arabidopsis. Mol Plant. 2019:**12**(12):1612–1623. https://doi.org/10.1016/j.molp.2019.09.004

Feng W, Kita D, Peaucelle A, Cartwright HN, Doan V, Duan Q, Liu M-C, Maman J, Steinhorst L, Schmitz-Thom I, et al. The FERONIA receptor kinase maintains cell-wall integrity during salt stress through Ca2+ signaling. Curr Biol. 2018:**28**(5):666–675.e5. https://doi.org/10.1016/j.cub.2018.01.023

Franck CM, Westermann J, Boisson-Dernier A. Plant malectin-like receptor kinases: from cell wall integrity to immunity and beyond. Annu Rev Plant Biol. 2018a:**69**(1):301–328. https://doi.org/10.1146/annurev-arplant-042817-040557

Franck CM, Westermann J, Burssner S, Lentz R, Lituiev DS, Boisson-Dernier A. The protein phosphatases ATUNIS1 and ATUNIS2 regulate cell wall integrity in tip-growing cells. Plant Cell. 2018b:**30**(8):1906–1923. https://doi.org/10.1105/tpc.18.00284

Franklin-Tong N, Bosch M. Plant biology: stigmatic ROS decide whether pollen is accepted or rejected. Curr Biol. 2021:**31**(14): R891–R917. https://doi.org/10.1016/j.cub.2021.05.034

Fujii S, Tsuchimatsu T, Kimura Y, Ishida S, Tangpranomkorn S, Shimosato-Asano H, Iwano M, Furukawa S, Itoyama W, Wada Y, et al. A stigmatic gene confers interspecies incompatibility in the Brassicaceae. Nat Plants. 2019:**5**(7):731–741. https://doi.org/10.1038/s41477-019-0444-6

Fujii S, Yamamoto E, Ito S, Tangpranomkorn S, Kimura Y, Miura H, Yamaguchi N, Kato Y, Niidome M, Yoshida A, et al. SHI family transcription factors regulate an interspecific barrier. Nat Plants. 2023:**9**(11):1862–1873. https://doi.org/10.1038/s41477-023-01535-5

Furness CA, Rudall PJ. Pollen aperture evolution—a crucial factor for eudicot success? Trends Plant Sci. 2004:**9**(3):154–158. https://doi.org/10.1016/j.tplants.2004.01.001

Galindo-Trigo S, Blanco-Touriñán N, DeFalco TA, Wells ES, Gray JE, Zipfel C, Smith LM. Cr RLK1L receptor-like kinases HERK1 and ANJEA are female determinants of pollen tube reception. EMBO Rep. 2020:**21**(2):e48466. https://doi.org/10.15252/embr.201948466

Gao Q, Wang C, Xi Y, Shao Q, Li L, Luan S. A receptor-channel trio conducts Ca2+ signalling for pollen tube reception. Nature. 2022:**607**(7919):534–539. https://doi.org/10.1038/s41586-022-04923-7

Gao Q-F, Gu L-L, Wang H-Q, Fei C-F, Fang X, Hussain J, Sun S-J, Dong J-Y, Liu H, Wang Y-F. Cyclic nucleotide-gated channel 18 is an essential Ca2+ channel in pollen tube tips for pollen tube guidance to ovules in Arabidopsis. Proc Natl Acad Sci U S A. 2016:**113**(11): 3096. https://doi.org/10.1073/pnas.1524629113

García Bossi J, Kumar K, Barberini ML, Domínguez GD, Rondón Guerrero YDC, Marino-Buslje C, Obertello M, Muschietti JP, Estevez JM. The role of P-type IIA and P-type IIB Ca2+-ATPases in plant development and growth. J Exp Bot. 2020:**71**(4):1239–1248. https://doi.org/10.1093/jxb/erz521

Ge Z, Bergonci T, Zhao Y, Zou Y, Du S, Liu M-C, Luo X, Ruan H, García-Valencia LE, Zhong S, et al. Arabidopsis pollen tube integrity and sperm release are regulated by RALF-mediated signaling. Science. 2017:**358**(6370):1596–1600. https://doi.org/10.1126/science.aao3642

Ge Z, Zhao Y, Liu M-C, Zhou L-Z, Wang L, Zhong S, Hou S, Jiang J, Liu T, Huang Q, et al. LLG2/3 Are Co-receptors in BUPS/ANX-RALF signaling to regulate Arabidopsis pollen tube integrity. Curr Biol. 2019:**29**(19):3256–3265.e5. https://doi.org/10.1016/j.cub.2019.08.032

Gilles LM, Khaled A, Laffaire J-B, Chaignon S, Gendrot G, Laplaige J, Bergès H, Beydon G, Bayle V, Barret P, et al. Loss of pollen-specific phospholipase NOT LIKE DAD triggers gynogenesis in maize. EMBO J. 2017:**36**(6):707–717. https://doi.org/10.15252/embj.201796603

Goring DR, Bosch M, Franklin-Tong VE. Contrasting self-recognition rejection systems for self-incompatibility in Brassica and Papaver. Curr Biol. 2023:**33**(11):R530–R542. https://doi.org/10.1016/j.cub.2023.03.037

Gross-Hardt R, Kägi C, Baumann N, Moore JM, Baskar R, Gagliano WB, Jürgens G, Grossniklaus U. LACHESIS restricts gametic cell fate in the female gametophyte of Arabidopsis. PLoS Biol. 2007:**5**(3):e47. https://doi.org/10.1371/journal.pbio.0050047

Gu Y, Fu Y, Dowd P, Li S, Vernoud V, Gilroy S, Yang Z. A Rho family GTPase controls actin dynamics and tip growth via two counteracting downstream pathways in pollen tubes. J Cell Biol. 2005:**169**(1): 127–138. https://doi.org/10.1083/jcb.200409140

Gu Y, Li S, Lord EM, Yang Z. Members of a novel class of Arabidopsis Rho guanine nucleotide exchange factors control Rho GTPase-dependent polar growth. Plant Cell. 2006:**18**(2):366–381. https://doi.org/10.1105/tpc.105.036434

Guan Y, Lu J, Xu J, McClure B, Zhang S. Two mitogen-activated protein kinases, mpk3 and mpk6, are required for funicular guidance of pollen tubes in Arabidopsis. Plant Physiol. 2014:**165**:528–533.

Guo F, Huang B-Q, Han Y, Zee S-Y. Fertilization in maize indeterminate gametophyte1 mutant. Protoplasma. 2004:**223**(2–4):111–120. https://doi.org/10.1007/s00709-004-0045-7

Guo J, Yang Z. Exocytosis and endocytosis: coordinating and fine-tuning the polar tip growth domain in pollen tubes. J Exp Bot. 2020:**71**(8):2428–2438. https://doi.org/10.1093/jxb/eraa134

Gusti A, Baumberger N, Nowack M, Pusch S, Eisler H, Potuschak T, De Veylder L, Schnittger A, Genschik P. The Arabidopsis thaliana F-box protein FBL17 is essential for progression through the second mitosis during pollen development. PLoS One. 2009:**4**(3):e4780. https://doi.org/10.1371/journal.pone.0004780

Gutermuth T, Lassig R, Portes M-T, Maierhofer T, Romeis T, Borst J-W, Hedrich R, Feijó JA, Konrad KR. Pollen tube growth regulation by free anions depends on the interaction between the anion channel SLAH3 and calcium-dependent protein kinases CPK2 and CPK20. Plant Cell. 2013:**25**:4525–4543.

Hafidh S, Honys D. Reproduction multitasking: the male gametophyte. Annu Rev Plant Biol. 2021:**72**(1):581–614. https://doi.org/10.1146/annurev-arplant-080620-021907

Hamamura Y, Nishimaki M, Takeuchi H, Geitmann A, Kurihara D, Higashiyama T. Live imaging of calcium spikes during double fertilization in Arabidopsis. Nat Commun. 2014:**5**(1):4722. https://doi.org/10.1038/ncomms5722

Hamamura Y, Saito C, Awai C, Kurihara D, Miyawaki A, Nakagawa T, Kanaoka MM, Sasaki N, Nakano A, Berger F, et al. Live-cell imaging reveals the dynamics of two sperm cells during double fertilization in Arabidopsis thaliana. Curr Biol. 2011:**21**(6):497–502. https://doi.org/10.1016/j.cub.2011.02.013

Hamilton ES, Haswell ES. The tension-sensitive ion transport activity of MSL8 is critical for its function in pollen hydration and germination. Plant Cell Physiol. 2017:**58**(7):1222–1237. https://doi.org/10.1093/pcp/pcw230

Hamilton ES, Jensen GS, Maksaev G, Katims A, Sherp AM, Haswell ES. Mechanosensitive channel MSL8 regulates osmotic forces during pollen hydration and germination. Science. 2015:**350**(6259):438–441. https://doi.org/10.1126/science.aac6014

Haruta M, Sabat G, Stecker K, Minkoff BB, Sussman MR. A peptide hormone and its receptor protein kinase regulate plant cell expansion. Science. 2014:**343**(6169):408–411. https://doi.org/10.1126/science.1244454

Hater F, Nakel T, Groß-Hardt R. Reproductive multitasking: the female gametophyte. Annu Rev Plant Biol. 2020:**71**(1):517–546. https://doi.org/10.1146/annurev-arplant-081519-035943

Hayashi M, Palmgren M. The quest for the central players governing pollen tube growth and guidance. Plant Physiol. 2021:**185**(3): 682–693. https://doi.org/10.1093/plphys/kiaa092

Hepler PK, Rounds CM, Winship LJ. Control of cell wall extensibility during pollen tube growth. Mol Plant. 2013:**6**(4):998–1017. https://doi.org/10.1093/mp/sst103

Hepler PK, Winship LJ. The pollen tube clear zone: clues to the mechanism of polarized growth. J Integr Plant Biol. 2015:**57**(1):79–92. https://doi.org/10.1111/jipb.12315

Herger A, Dünser K, Kleine-Vehn J, Ringli C. Leucine-rich repeat extensin proteins and their role in cell wall sensing. Curr Biol. 2019:**29**(17):R851–R858. https://doi.org/10.1016/j.cub.2019.07.039

Herger A, Gupta S, Kadler G, Franck CM, Boisson-Dernier A, Ringli C. Overlapping functions and protein-protein interactions of LRR-extensins in Arabidopsis. PLoS Genet. 2020:**16**(6):e1008847. https://doi.org/10.1371/journal.pgen.1008847

Herrera-Ubaldo H, Lozano-Sotomayor P, Ezquer I, Di Marzo M, Chávez Montes RA, Gómez-Felipe A, Pablo-Villa J, Diaz-Ramirez D, Ballester P, Ferrándiz C, et al. New roles of NO TRANSMITTING TRACT and SEEDSTICK during medial domain development in Arabidopsis fruits. Development. 2019:**146**(1): dev172395. https://doi.org/10.1242/dev.172395

Higashiyama T. The synergid cell: attractor and acceptor of the pollen tube for double fertilization. J Plant Res. 2002:**115**(2):149–160. https://doi.org/10.1007/s102650200020

Higashiyama T, Kuroiwa H, Kawano S, Kuroiwa T. Guidance in vitro of the pollen tube to the naked embryo sac of torenia fournieri.

Plant Cell. 1998:**10**(12):2019–2031. https://doi.org/10.1105/tpc.10.12.2019

Higashiyama T, Takeuchi H. The mechanism and key molecules involved in pollen tube guidance. Annu Rev Plant Biol. 2015:**66**(1):393–413. https://doi.org/10.1146/annurev-arplant-043014-115635

Higashiyama T, Yang WC. Gametophytic pollen tube guidance: attractant peptides, gametic controls, and receptors. Plant Physiol. 2016:**173**(1):112–121. https://doi.org/10.1104/pp.16.01571

Hoffmann RD, Portes MT, Olsen LI, Damineli DSC, Hayashi M, Nunes CO, Pedersen JT, Lima PT, Campos C, Feijo JA, et al. Plasma membrane H^+-ATPases sustain pollen tube growth and fertilization. Nat Commun. 2020:**11**(1):2395. https://doi.org/10.1038/s41467-020-16253-1

Hojsgaard D, Hörandl E. The rise of apomixis in natural plant populations. Front Plant Sci. 2019:**10**:358. https://doi.org/10.3389/fpls.2019.00358

Houben A, Kumke K, Nagaki K, Hause G. CENH3 distribution and differential chromatin modifications during pollen development in rye (Secale cereale L.). Chromosome Res. 2011:**19**(4):471–480. https://doi.org/10.1007/s10577-011-9207-6

Huang J, Yang L, Yang L, Wu X, Cui X, Zhang L, Hui J, Zhao Y, Yang H, Liu S, et al. Stigma receptors control intraspecies and interspecies barriers in Brassicaceae. Nature. 2023:**614**(7947):303–308. https://doi.org/10.1038/s41586-022-05640-x

Huang XR, Sun MX. H3K27 methylation regulates the fate of two cell lineages in male gametophytes. Plant Cell. 2022:**34**(8):2989–3005. https://doi.org/10.1093/plcell/koac136

Huck N, Moore JM, Federer M, Grossniklaus U. The Arabidopsis mutant feronia disrupts the female gametophytic control of pollen tube reception. Development. 2003:**130**(10):2149–2159. https://doi.org/10.1242/dev.00458

Ingouff M, Rademacher S, Holec S, Šoljić L, Xin N, Readshaw A, Foo SH, Lahouze B, Sprunck S, Berger F. Zygotic resetting of the HISTONE 3 variant repertoire participates in epigenetic reprogramming in Arabidopsis. Curr Biol. 2010:**20**(23):2137–2143. https://doi.org/10.1016/j.cub.2010.11.012

Ingouff M, Sakata T, Li J, Sprunck S, Dresselhaus T, Berger F. The two male gametes share equal ability to fertilize the egg cell in Arabidopsis thaliana. Curr Biol. 2009:**19**(1):R19–R20. https://doi.org/10.1016/j.cub.2008.11.025

Iwakawa H, Shinmyo A, Sekine M. Arabidopsis CDKA;1, a cdc2 homologue, controls proliferation of generative cells in male gametogenesis. Plant J. 2006:**45**(5):819–831. https://doi.org/10.1111/j.1365-313X.2005.02643.x

Iwano M, Ngo QA, Entani T, Shiba H, Nagai T, Miyawaki A, Isogai A, Grossniklaus U, Takayama S. Cytoplasmic Ca2+ changes dynamically during the interaction of the pollen tube with synergid cells. Development. 2012:**139**(22):4202–4209. https://doi.org/10.1242/dev.081208

Ji J, Tang D, Shen Y, Xue Z, Wang H, Shi W, Zhang C, Du G, Li Y, Cheng Z. P31comet, a member of the synaptonemal complex, participates in meiotic DSB formation in rice. Proc Natl Acad Sci U S A. 2016:**113**(38):10577–10582. https://doi.org/10.1073/pnas.1607334113

Jia P-F, Xue Y, Li H-J, Yang W-C. Golgi-localized LOT regulates trans-Golgi network biogenesis and pollen tube growth. Proc Natl Acad Sci U S A. 2018:**115**:12307–12312.

Jiang C, Sun J, Li R, Yan S, Chen W, Guo L, Qin G, Wang P, Luo C, Huang W, et al. A reactive oxygen species burst causes haploid induction in maize. Mol Plant. 2022b:**15**(6):943–955. https://doi.org/10.1016/j.molp.2022.04.001

Jiang D, Borg M, Lorković ZJ, Montgomery SA, Osakabe A, Yelagandula R, Axelsson E, Berger F. The evolution and functional divergence of the histone H2B family in plants. PLoS Genet. 2020:**16**(7):e1008964. https://doi.org/10.1371/journal.pgen.1008964

Jiang J, Qu L-J. Response to Zhang et al., "do egg cell-secreted aspartic proteases promote gamete attachment?" J Integr Plant Biol. 2023:**65**(1):7–9. https://doi.org/10.1111/jipb.13448

Jiang J, Stührwohldt N, Liu T, Huang Q, Li L, Zhang L, Gu H, Fan L, Zhong S, Schaller A, et al. Egg cell-secreted aspartic proteases ECS1/2 promote gamete attachment to prioritize the fertilization of egg cells over central cells in Arabidopsis. J Integr Plant Biol. 2022a:**64**(11):2047–2059. https://doi.org/10.1111/jipb.13371

Jiang L, Yang S-L, Xie L-F, Puah CS, Zhang X-Q, Yang W-C, Sundaresan V, Ye D. VANGUARD1 encodes a pectin methylesterase that enhances pollen tube growth in the Arabidopsis style and transmitting tract. Plant Cell. 2005:**17**(2):584–596. https://doi.org/10.1105/tpc.104.027631

Johnson MA, Harper JF, Palanivelu R. A fruitful journey: pollen tube navigation from germination to fertilization. Annu Rev Plant Biol. 2019:**70**(1):809–837. https://doi.org/10.1146/annurev-arplant-050718-100133

Jones DS, Yuan J, Smith BE, Willoughby AC, Kumimoto EL, Kessler SA. MILDEW RESISTANCE LOCUS O function in pollen tube reception is linked to its oligomerization and subcellular distribution. Plant Physiol. 2017:**175**(1):172–185. https://doi.org/10.1104/pp.17.00523

Ju Y, Yuan J, Jones DS, Zhang W, Staiger CJ, Kessler SA. Polarized NORTIA accumulation in response to pollen tube arrival at synergids promotes fertilization. Dev Cell. 2021:**56**(21):2938–2951.e6. https://doi.org/10.1016/j.devcel.2021.09.026

Kao P, Nodine MD. Transcriptional activation of Arabidopsis zygotes is required for initial cell divisions. Sci Rep. 2019:**9**(1):17159. https://doi.org/10.1038/s41598-019-53704-2

Kaothien P, Ok SH, Shuai B, Wengier D, Cotter R, Kelley D, Kiriakopolos S, Muschietti J, McCormick S. Kinase partner protein interacts with the LePRK1 and LePRK2 receptor kinases and plays a role in polarized pollen tube growth. Plant J. 2005:**42**(4):492–503. https://doi.org/10.1111/j.1365-313X.2005.02388.x

Kasahara RD, Maruyama D, Hamamura Y, Sakakibara T, Twell D, Higashiyama T. Fertilization recovery after defective sperm cell release in Arabidopsis. Curr Biol. 2012:**22**(12):1084–1089. https://doi.org/10.1016/j.cub.2012.03.069

Kasahara RD, Maruyama D, Higashiyama T. Fertilization recovery system is dependent on the number of pollen grains for efficient reproduction in plants. Plant Signal Behav. 2013:**8**(4):e23690. https://doi.org/10.4161/psb.23690

Kasahara RD, Portereiko MF, Sandaklie-Nikolova L, Rabiger DS, Drews GN. MYB98 is required for pollen tube guidance and synergid cell differentiation in Arabidopsis. Plant Cell. 2005:**17**(11):2981–2992. https://doi.org/10.1105/tpc.105.034603

Kawashima T, Maruyama D, Shagirov M, Li J, Hamamura Y, Yelagandula R, Toyama Y, Berger F. Dynamic F-actin movement is essential for fertilization in Arabidopsis thaliana. Elife. 2014:**3**:e04501. https://doi.org/10.7554/eLife.04501

Kaya H, Nakajima R, Iwano M, Kanaoka MM, Kimura S, Takeda S, Kawarazaki T, Senzaki E, Hamamura Y, Higashiyama T, et al. Ca2+-activated reactive oxygen species production by Arabidopsis RbohH and RbohJ is essential for proper pollen tube tip growth. Plant Cell. 2014:**26**(3):1069–1080. https://doi.org/10.1105/tpc.113.120642

Kelliher T, Starr D, Richbourg L, Chintamanani S, Delzer B, Nuccio ML, Green J, Chen Z, McCuiston J, Wang W, et al. MATRILINEAL, a sperm-specific phospholipase, triggers maize haploid induction. Nature. 2017:**542**(7639):105–109. https://doi.org/10.1038/nature20827

Kessler SA, Shimosato-Asano H, Keinath NF, Wuest SE, Ingram G, Panstruga R, Grossniklaus U. Conserved molecular components for pollen tube reception and fungal invasion. Science. 2010:**330**(6006):968–971. https://doi.org/10.1126/science.1195211

Khanday I, Skinner D, Yang B, Mercier R, Sundaresan V. A male-expressed rice embryogenic trigger redirected for asexual propagation through seeds. Nature. 2019:**565**(7737):91–95. https://doi.org/10.1038/s41586-018-0785-8

Kim HJ, Oh SA, Brownfield L, Hong SH, Ryu H, Hwang I, Twell D, Nam HG. Control of plant germline proliferation by SCFFBL17 degradation of cell cycle inhibitors. Nature. 2008:**455**(7216):1134–1137. https://doi.org/10.1038/nature07289

Kimata Y, Higaki T, Kawashima T, Kurihara D, Sato Y, Yamada T, Hasezawa S, Berger F, Higashiyama T, Ueda M. Cytoskeleton dynamics control the first asymmetric cell division in Arabidopsis zygote. Proc Natl Acad Sci U S A. 2016:**113**(49):14157–14162. https://doi.org/10.1073/pnas.1613979113

Kirioukhova O, Johnston AJ, Kleen D, Kägi C, Baskar R, Moore JM, Bäumlein H, Groß-Hardt R, Grossniklaus U. Female gametophytic cell specification and seed development require the function of the putative Arabidopsis INCENP ortholog WYRD. Development. 2011:**138**(16):3409–3420. https://doi.org/10.1242/dev.060384

Kitashiba H, Nasrallah JB. Self-incompatibility in Brassicaceae crops: lessons for interspecific incompatibility. Breed Sci. 2014:**64**(1): 23–37. https://doi.org/10.1270/jsbbs.64.23

Kong J, Lau S, Jürgens G. Twin plants from supernumerary egg cells in Arabidopsis. Curr Biol. 2015:**25**(2):225–230. https://doi.org/10.1016/j.cub.2014.11.021

Kumlehn J, Kirik V, Czihal A, Altschmied L, Matzk F, Lörz H, Bäumlein H. Parthenogenetic egg cells of wheat: cellular and molecular studies. Sex Plant Reprod. 2001:**14**(4):239–243. https://doi.org/10.1007/s00497-001-0115-3

Lamport DTA, Kieliszewski MJ, Chen Y, Cannon MC. Role of the extensin superfamily in primary cell wall architecture. Plant Physiol. 2011:**156**(1):11–19. https://doi.org/10.1104/pp.110.169011

Lan Y, Liu X, Fu Y, Huang S. Arabidopsis class I formins control membrane-originated actin polymerization at pollen tube tips. PLoS Genet. 2018:**14**:e1007789.

Lan Z, Song Z, Wang Z, Li L, Liu Y, Zhi S, Wang R, Wang J, Li Q, Bleckmann A, et al. Antagonistic RALF peptides control an intergeneric hybridization barrier on Brassicaceae stigmas. Cell. 2023a:**186**(22): 4773–4787.e12. https://doi.org/10.1016/j.cell.2023.09.003

Lan Z, Song Z, Zhong S, Qu L-J. The central cell: another opportunity for fertilization recovery in plants. Seed Biology. 2023b:**2**:22. https://doi.org/10.48130/SeedBio-2023-0022.

Lan Z, Zhong S, Qu L-J. Insights into pollen–stigma recognition: self-incompatibility mechanisms serve as interspecies barriers in Brassicaceae? aBIOTECH. 2023:**4**:176–179. https://doi.org/10.1007/s42994-023-00105-9

Lassig R, Gutermuth T, Bey TD, Konrad KR, Romeis T. Pollen tube NAD(P)H oxidases act as a speed control to dampen growth rate oscillations during polarized cell growth. Plant J. 2014:**78**(1):94–106. https://doi.org/10.1111/tpj.12452

Lausser A, Dresselhaus T. Sporophytic control of pollen tube growth and guidance in grasses. Biochem Soc Trans. 2010:**38**(2):631–634. https://doi.org/10.1042/BST0380631

Lavy M, Bloch D, Hazak O, Gutman I, Poraty L, Sorek N, Sternberg H, Yalovsky S. A novel ROP/RAC effector links cell polarity, root-meristem maintenance, and vesicle trafficking. Curr Biol. 2007:**17**(11):947–952. https://doi.org/10.1016/j.cub.2007.04.038

Lawit SJ, Chamberlin MA, Agee A, Caswell ES, Albertsen MC. Transgenic manipulation of plant embryo sacs tracked through cell-type-specific fluorescent markers: cell labeling, cell ablation, and adventitious embryos. Plant Reprod. 2013:**26**(2):125–137. https://doi.org/10.1007/s00497-013-0215-x

Le Q, Gutierrez-Marcos JF, Costa LM, Meyer S, Dickinson HG, Lorz H, Kranz E, Scholten S. Construction and screening of subtracted cDNA libraries from limited populations of plant cells: a comparative analysis of gene expression between maize egg cells and central cells. Plant J. 2005:**44**(1):167–178. https://doi.org/10.1111/j.1365-313X.2005.02518.x

Lee BH, Wang R, Moberg IM, Reeder SH, Amom P, Tan MH, Amstutz K, Chandna P, Helton A, Andrianova EP, et al. A species-specific functional module controls formation of pollen apertures. Nat Plants. 2021:**7**(7):966–978. https://doi.org/10.1038/s41477-021-00951-9

Lee BH, Weber ZT, Zourelidou M, Hofmeister BT, Schmitz RJ, Schwechheimer C, Dobritsa AA. Arabidopsis protein kinase D6PKL3 is involved in the formation of distinct plasma membrane aperture domains on the pollen surface. Plant Cell. 2018:**30**(9): 2038–2056. https://doi.org/10.1105/tpc.18.00442

Lee Y-RJ, Hiwatashi Y, Hotta T, Xie T, Doonan JH, Liu B. The mitotic function of augmin is dependent on its microtubule-associated protein subunit EDE1 in Arabidopsis thaliana. Curr Biol. 2017:**27**(24): 3891–3897.e4. https://doi.org/10.1016/j.cub.2017.11.030

Leroux C, Bouton S, Kiefer-Meyer M-C, Fabrice TN, Mareck A, Guénin S, Fournet F, Ringli C, Pelloux J, Driouich A, et al. PECTIN METHYLESTERASE48 is involved in Arabidopsis pollen grain germination. Plant Physio. 2015:**167**(2):367–380. https://doi.org/10.1104/pp.114.250928

Leydon AR, Tsukamoto T, Dunatunga D, Qin Y, Johnson MA, Palanivelu R. Pollen tube discharge completes the process of synergid degeneration that is initiated by pollen tube-synergid interaction in Arabidopsis. Plant Physiol. 2015:**169**(1):485–496. https://doi.org/10.1104/pp.15.00528

Li C, Wu H-M, Cheung AY. FERONIA and her pals: functions and mechanisms. Plant Physiol. 2016:**171**(4):2379–2392. https://doi.org/10.1104/pp.16.00667

Li C, Yeh FL, Cheung AY, Duan Q, Kita D, Liu MC, Maman J, Luu EJ, Wu BW, Gates L, et al. Glycosylphosphatidylinositol-anchored proteins as chaperones and co-receptors for FERONIA receptor kinase signaling in Arabidopsis. Elife. 2015a:**4**:e06587. https://doi.org/10.7554/eLife.06587

Li E, Zhang Y-L, Qin Z, Xu M, Qiao Q, Li S, Li S-W, Zhang Y. Signaling network controlling ROP-mediated tip growth in Arabidopsis and beyond. Plant Commun. 2023c:**4**(1):100451. https://doi.org/10.1016/j.xplc.2022.100451

Li F-S, Phyo P, Jacobowitz J, Hong M, Weng J-K. The molecular structure of plant sporopollenin. Nat Plants. 2019:**5**(1):41–46. https://doi.org/10.1038/s41477-018-0330-7

Li H, Yang Y, Zhang H, Li C, Du P, Bi M, Chen T, Qian D, Niu Y, Ren H, et al. the Arabidopsis GPI-anchored protein CBL11 is necessary for regulating pollen tube integrity. Cell Rep. 2023b:**42**(11):113353. https://doi.org/10.1016/j.celrep.2023.113353

Li H-J, Meng J-G, Yang W-C. Multilayered signaling pathways for pollen tube growth and guidance. Plant Reprod. 2018:**31**(1):31–41. https://doi.org/10.1007/s00497-018-0324-7

Li H-J, Xue Y, Jia D-J, Wang T, hi D-Q, Liu J, Cui F, Xie Q, Ye D, Yang W-C. POD1 Regulates pollen tube guidance in response to micropylar female signaling and acts in early embryo patterning in Arabidopsis. Plant Cell. 2011:**23**(9):3288–3302. https://doi.org/10.1105/tpc.111.088914

Li H-J, Zhu S-S, Zhang M-X, Wang T, Liang L, Xue Y, Shi D-Q, Liu J, Yang W-C. Arabidopsis CBP1 Is a novel regulator of transcription initiation in central cell-mediated pollen tube guidance. Plant Cell. 2015b:**27**(10):2880–2893. https://doi.org/10.1105/tpc.15.00370

Li L, Hou S, Xiang W, Song Z, Wang Y, Zhang L, Li J, Gu H, Dong J, Dresselhaus T, et al. The egg cell is preferentially fertilized in Arabidopsis double fertilization. J Integr Plant Biol. 2022a:**64**(11): 2039–2046. https://doi.org/10.1111/jipb.13370

Li S, Ge F-R, Xu M, Zhao X-Y, Huang G-Q, Zhou L-Z, Wang J-G, Kombrink A, McCormick S, Zhang XS, et al. Arabidopsis COBRA-LIKE 10, a GPI-anchored protein, mediates directional growth of pollen tubes. Plant J. 2013:**74**(3):486–497. https://doi.org/10.1111/tpj.12139

Li S-Z, Wang J, Jia S-G, Wang K, Li H-J. Synthetic apomixis: from genetic basis to agricultural application. Seed Biol. 2023a:**2**(1):10. https://doi.org/10.48130/SeedBio-2023-0010

Li W, Li Q, Lyu M, Wang Z, Song Z, Zhong S, Gu H, Dong J, Dresselhaus T, Zhong S, et al. Lack of ethylene does not affect reproductive success and synergid cell death in Arabidopsis. Mol Plant. 2022b:**15**(2):354–362. https://doi.org/10.1016/j.molp.2021.11.001

Li X, Meng D, Chen S, Luo H, Zhang Q, Jin W, Yan J. Single nucleus sequencing reveals spermatid chromosome fragmentation as a possible cause of maize haploid induction. Nat Commun. 2017:**8**(1): 991. https://doi.org/10.1038/s41467-017-00969-8

Li Y, Li D, Xiao Q, Wang H, Wen J, Tu J, Shen J, Fu T, Yi B. An in planta haploid induction system in Brassica napus. J Integr Plant Biol. 2022c:**64**(6):1140–1144. https://doi.org/10.1111/jipb.13270

Li Y, Lin Z, Yue Y, Zhao H, Fei X EL, Liu C, Chen S, Lai J, Song W. Loss-of-function alleles of ZmPLD3 cause haploid induction in maize. Nat Plants. 2021:**7**(12):1579–1588. https://doi.org/10.1038/s41477-021-01037-2

Liao H, Tang R, Zhang X, Luan S, Yu F. FERONIA receptor kinase at the crossroads of hormone signaling and stress responses. Plant Cell Physiol. 2017:**58**(7):1143–1150. https://doi.org/10.1093/pcp/pcx048

Lichocka M, Krzymowska M, Górecka M, Hennig J. Arabidopsis annexin 5 is involved in maintenance of pollen membrane integrity and permeability. J Exp Bot. 2022:**73**(1):94–109. https://doi.org/10.1093/jxb/erab419

Lin W, Tang W, Pan X, Huang A, Gao X, Anderson CT, Yang Z. Arabidopsis pavement cell morphogenesis requires FERONIA binding to pectin for activation of ROP GTPase signaling. Curr Biol. 2022:**32**(3):497–507.e4. https://doi.org/10.1016/j.cub.2021.11.030

Liu C, He Z, Zhang Y, Hu F, Li M, Liu Q, Huang Y, Wang J, Zhang W, Wang C, et al. Synthetic apomixis enables stable transgenerational transmission of heterotic phenotypes in hybrid rice. Plant Commun. 2023:**4**(2):100470. https://doi.org/10.1016/j.xplc.2022.100470

Liu C, Shen L, Xiao Y, Vyshedsky D, Peng C, Sun X, Liu Z, Cheng L, Zhang H, Han Z, et al. Pollen PCP-B peptides unlock a stigma peptide-receptor kinase gating mechanism for pollination. Science. 2021:**372**(6538):171–175. https://doi.org/10.1126/science.abc6107

Liu C, Zhang Y, Ren H. Actin polymerization mediated by AtFH5 directs the polarity establishment and vesicle trafficking for pollen germination in Arabidopsis. Mol Plant. 2018:**11**(11):1389–1399. https://doi.org/10.1016/j.molp.2018.09.004

Liu H-K, Li Y-J, Wang S-J, Yuan T-L, Huang W-J, Dong X, Pei J-Q, Zhang D, McCormick S, Tang W-H. Kinase partner protein plays a key role in controlling the speed and shape of pollen tube growth in tomato. Plant Physiol. 2020:**184**(4):1853–1869. https://doi.org/10.1104/pp.20.01081

Liu J, Zhang Y, Qin G, Tsuge T, Sakaguchi N, Luo G, Sun K, Shi D, Aki S, Zheng N, et al. Targeted degradation of the cyclin-dependent kinase inhibitor ICK4/KRP6 by ring-type E3 ligases is essential for mitotic cell cycle progression during Arabidopsis gametogenesis. Plant Cell. 2008b:**20**(6):1538–1554. https://doi.org/10.1105/tpc.108.059741

Liu J, Zhong S, Guo X, Hao L, Wei X, Huang Q, Hou Y, Shi J, Wang C, Gu H, et al. Membrane-bound RLCKs LIP1 and LIP2 Are essential male factors controlling male-female attraction in Arabidopsis. Curr Biol. 2013:**23**(11):993–998. https://doi.org/10.1016/j.cub.2013.04.043

Liu L, Zheng C, Kuang B, Wei L, Yan L, Wang T. Receptor-like kinase RUPO interacts with potassium transporters to regulate pollen tube growth and integrity in rice. PLoS Genet. 2016b:**12**(7):e1006085. https://doi.org/10.1371/journal.pgen.1006085

Liu M-CJ, Yeh F-LJ, Yvon R, Simpson K, Jordan S, Chambers J, Wu H-M, Cheung AY. Extracellular pectin-RALF phase separation mediates FERONIA global signaling function. Cell. 2024:**187**(2):312–330. https://doi.org/10.1016/j.cell.2023.11.038

Liu X, Castro C, Wang Y, Noble J, Ponvert N, Bundy M, Hoel C, Shpak E, Palanivelu R. The role of LORELEI in pollen tube reception at the interface of the synergid cell and pollen tube requires the modified eight-cysteine motif and the receptor-like kinase FERONIA. Plant Cell. 2016a:**28**(5):1035–1052. https://doi.org/10.1105/tpc.15.00703

Liu YJ, Tewari R, Ning J, Blagborough AM, Garbom S, Pei JM, Grishin NV, Steele RE, Sinden RE, Snell WJ, et al. The conserved plant sterility gene HAP2 functions after attachment of fusogenic membranes in Chlamydomonas and Plasmodium gametes. Genes Dev. 2008a:**22**(8):1051–1068. https://doi.org/10.1101/gad.1656508

Lord EM, Russell SD. The mechanisms of pollination and fertilization in plants. Annu Rev Cell Dev Biol. 2002:**18**:81–105.

Lovy-Wheeler A, Kunkel JG, Allwood EG, Hussey PJ, Hepler PK. Oscillatory increases in alkalinity anticipate growth and may regulate actin dynamics in pollen tubes of lily. Plant Cell. 2006:**18**:2182–2193.

Lu Y, Chanroj S, Zulkifli L, Johnson MA, Uozumi N, Cheung A, Sze H. Pollen tubes lacking a pair of K+ transporters fail to target ovules in Arabidopsis. Plant Cell. 2011:**23**(1):81–93. https://doi.org/10.1105/tpc.110.080499

Lu Y, Hokin SA, Kermicle JL, Hartwig T, Evans MMS. A pistil-expressed pectin methylesterase confers cross-incompatibility between strains of Zea mays. Nat Commun. 2019:**10**(1):2304. https://doi.org/10.1038/s41467-019-10259-0

Luo N, Yan A, Liu G, Guo J, Rong D, Kanaoka MM, Xiao Z, Xu G, Higashiyama T, Cui X, et al. Exocytosis-coordinated mechanisms for tip growth underlie pollen tube growth guidance. Nat Commun. 2017:**8**(1):1687. https://doi.org/10.1038/s41467-017-01452-0

Ma H. Molecular genetic analyses of microsporogenesis and microgametogenesis in flowering plants. Annu Rev Plant Biol. 2005:**56**(1): 393–434. https://doi.org/10.1146/annurev.arplant.55.031903.141717

Ma X, Wu Y, Zhang G. Formation pattern and regulatory mechanisms of pollen wall in Arabidopsis. J Plant Physiol. 2021:**260**:153388. https://doi.org/10.1016/j.jplph.2021.153388

Mähs A, Steinhorst L, Han J-P, Shen L-K, Wang Y, Kudla J. The calcineurin B-like Ca2+sensors CBL1 and CBL9 function in pollen germination and pollen tube growth in Arabidopsis. Mol Plant. 2013:**6**(4): 1149–1162. https://doi.org/10.1093/mp/sst095

Malivert A, Hamant O. Why is FERONIA pleiotropic? Nat Plants. 2023:**9**(7):1018–1025. https://doi.org/10.1038/s41477-023-01434-9

Marimuthu MPA, Jolivet S, Ravi M, Pereira L, Davda JN, Cromer L, Wang L, Nogué F, Chan SWL, Siddiqi I, et al. Synthetic clonal reproduction through seeds. Science. 2011:**331**(6019):876. https://doi.org/10.1126/science.1199682

Márton ML, Cordts S, Broadhvest J, Dresselhaus T. Micropylar pollen tube guidance by egg apparatus 1 of maize. Science. 2005:**307**(5709): 573–576. https://doi.org/10.1126/science.1104954

Márton ML, Fastner A, Uebler S, Dresselhaus T. Overcoming hybridization barriers by the secretion of the maize pollen tube attractant ZmEA1 from Arabidopsis ovules. Curr Biol. 2012:**22**(13):1194–1198. https://doi.org/10.1016/j.cub.2012.04.061

Maruyama D, Hamamura Y, Takeuchi H, Susaki D, Nishimaki M, Kurihara D, Kasahara RD, Higashiyama T. Independent control by each female gamete prevents the attraction of multiple pollen tubes. Dev Cell. 2013:**25**(3):317–323. https://doi.org/10.1016/j.devcel.2013.03.013

Maruyama D, Higashiyama T. The end of temptation: the elimination of persistent synergid cell identity. Curr Opin Plant Biol. 2016:**34**: 122–126. https://doi.org/10.1016/j.pbi.2016.10.011

Maruyama D, Völz R, Takeuchi H, Mori T, Igawa T, Kurihara D, Kawashima T, Ueda M, Ito M, Umeda M, et al. Rapid elimination of the persistent synergid through a cell fusion mechanism. Cell. 2015:**161**(4):907–918. https://doi.org/10.1016/j.cell.2015.03.018

Marzol E, Borassi C, Bringas M, Sede A, Rodríguez Garcia DR, Capece L, Estevez JM. Filling the gaps to solve the extensin puzzle. Mol Plant. 2018:**11**(5):645–658. https://doi.org/10.1016/j.molp.2018.03.003

Mazuecos-Aguilera I, Romero-García AT, Klodová B, Honys D, Fernández-Fernández MC, Ben-Menni Schuler S, Dobritsa AA, Suárez-Santiago VN. The role of INAPERTURATE POLLEN1 as a pollen aperture factor is conserved in the basal eudicot Eschscholzia californica (Papaveraceae). Front Plant Sci. 2021:**12**:701286. https://doi.org/10.3389/fpls.2021.701286

Mecchia MA, Santos-Fernandez G, Duss NN, Somoza SC, Boisson-Dernier A, Gagliardini V, Martínez-Bernardini A, Fabrice TN, Ringli C, Muschietti JP, et al. RALF4/19 peptides interact with LRX proteins to control pollen tube growth in Arabidopsis.

Science. 2017:**358**(6370):1600–1603. https://doi.org/10.1126/science.aao5467

Mendes MA, Guerra RF, Castelnovo B, Silva-Velazquez Y, Morandini P, Manrique S, Baumann N, Groß-Hardt R, Dickinson H, Colombo L. Live and let die: a REM complex promotes fertilization through synergid cell death in Arabidopsis. Development. 2016:**143**(15): 2780–2790. https://doi.org/10.1242/dev.134916

Meng J-G, Xu Y-J, Wang W-Q, Yang F, Chen S-Y, Jia P-F, Yang W-C, Li H-J. Central-cell-produced attractants control fertilization recovery. Cell. 2023:**186**(17):3593–3605.e12. https://doi.org/10.1016/j.cell.2023.06.024

Meng J-G, Zhang MX, Yang WC, Li HJ. TICKET attracts pollen tubes and mediates reproductive isolation between relative species in Brassicaceae. Sci China Life Sci. 2019:**62**(11):1413–1419. https://doi.org/10.1007/s11427-019-9833-3

Mercier R, Mézard C, Jenczewski E, Macaisne N, Grelon M. The molecular biology of meiosis in plants. Annu Rev Plant Biol. 2015:**66**(1):297–327. https://doi.org/10.1146/annurev-arplant-050213-035923

Miao C, Tang D, Zhang H, Wang M, Li Y, Tang S, Yu H, Gu M, Cheng Z. Central region component1, a novel synaptonemal complex component, is essential for meiotic recombination initiation in rice. Plant Cell. 2013:**25**(8):2998–3009. https://doi.org/10.1105/tpc.113.113175

Michard E, Lima PT, Borges F, Silva AC, Portes MT, Carvalho JE, Gilliham M, Liu L-H, Obermeyer G, Feijó JA. Glutamate receptor–like genes form Ca2+ channels in pollen tubes and are regulated by pistil D-serine. Science. 2011:**332**(6028):434–437. https://doi.org/10.1126/science.1201101

Michard E, Simon AA, Tavares B, Wudick MM, Feijó JA. Signaling with ions: the keystone for apical cell growth and morphogenesis in pollen tubes. Plant Physiol. 2017:**173**(1):91–111. https://doi.org/10.1104/pp.16.01561

Mieulet D, Jolivet S, Rivard M, Cromer L, Vernet A, Mayonove P, Pereira L, Droc G, Courtois B, Guiderdoni E, et al. Turning rice meiosis into mitosis. Cell Res. 2016:**26**(11):1242–1254. https://doi.org/10.1038/cr.2016.117

Miyazaki S, Murata T, Sakurai-Ozato N, Kubo M, Demura T, Fukuda H, Hasebe M. ANXUR1 and 2, sister genes to FERONIA/SIRENE, are male factors for coordinated fertilization. Curr Biol. 2009:**19**(15): 1327–1331. https://doi.org/10.1016/j.cub.2009.06.064

Mizukami AG, Inatsugi R, Jiao J, Kotake T, Kuwata K, Ootani K, Okuda S, Sankaranarayanan S, Sato Y, Maruyama D, et al. The AMOR arabinogalactan sugar chain induces pollen-tube competency to respond to ovular guidance. Curr Biol. 2016:**26**(8):1091–1097. https://doi.org/10.1016/j.cub.2016.02.040

Mizuta Y, Higashiyama T. Chemical signaling for pollen tube guidance at a glance. J Cell Sci. 2018:**131**(2):jcs208447. https://doi.org/10.1242/jcs.208447

Mizuta Y, Sakakibara D, Nagahara S, Kaneshiro I, Nagae TT, Kurihara D, Higashiyama T. Deep imaging revealed dynamics and signaling in one-to-one pollen tube guidance. 2023:2023.04.19.537439.

Moll C, von Lyncker L, Zimmermann S, Kägi C, Baumann N, Twell D, Grossniklaus U, Gross-Hardt R. CLO/GFA1 and ATO are novel regulators of gametic cell fate in plants. Plant J. 2008:**56**:913–921.

Moran Lauter AN, Muszynski MG, Huffman RD, Scott MP. A pectin methylesterase ZmPme3 is expressed in gametophyte factor1-s (Ga1-s) silks and maps to that locus in maize (Zea mays L.). Front Plant Sci. 2017:**8**:1926. https://doi.org/10.3389/fpls.2017.01926

Mori T, Igawa T, Tamiya G, Miyagishima S, Berger F. Gamete attachment requires GEX2 for successful fertilization in Arabidopsis. Curr Biol. 2014:**24**(2):170–175. https://doi.org/10.1016/j.cub.2013.11.030

Mori T, Kuroiwa H, Higashiyama T, Kuroiwa T. GENERATIVE CELL SPECIFIC 1 is essential for angiosperm fertilization. Nat Cell Biol. 2006:**8**(1):64–71. https://doi.org/10.1038/ncb1345

Mouline K, Véry A-A, Gaymard F, Boucherez J, Pilot G, Devic M, Bouchez D, Thibaud J-B, Sentenac H. Pollen tube development and competitive ability are impaired by disruption of a Shaker K+ channel in Arabidopsis. Genes Dev. 2002:**16**(3):339–350. https://doi.org/10.1101/gad.213902

Moussu S, Broyart C, Santos-Fernandez G, Augustin S, Wehrle S, Grossniklaus U, Santiago J. Structural basis for recognition of RALF peptides by LRX proteins during pollen tube growth. Proc Natl Acad Sci U S A. 2020:**117**(13):7494–7503. https://doi.org/10.1073/pnas.2000100117

Moussu S, Ingram G. The EXTENSIN enigma. Cell Surf. 2023:**9**:100094. https://doi.org/10.1016/j.tcsw.2023.100094

Moussu S, Lee HK, Haas KT, Broyart C, Rathgeb U, De Bellis D, Levasseur T, Schoenaers S, Fernandez GS, Grossniklaus U, et al. Plant cell wall patterning and expansion mediated by protein-peptide-polysaccharide interaction. Science. 2023:**382**(6671): 719–725. https://doi.org/10.1126/science.adi4720

Muschietti J, Eyal Y, McCormick S. Pollen tube localization implies a role in pollen-pistil interactions for the tomato receptor-like protein kinases LePRK1 and LePRK2. Plant Cell. 1998:**10**(3):319–330. https://doi.org/10.1105/tpc.10.3.319

Myers C, Romanowsky SM, Barron YD, Garg S, Azuse CL, Curran A, Davis RM, Hatton J, Harmon AC, Harper JF. Calcium-dependent protein kinases regulate polarized tip growth in pollen tubes. Plant J. 2009:**59**(4):528–439. https://doi.org/10.1111/j.1365-313X.2009.03894.x

Nagahara S, Takeuchi H, Higashiyama T. Polyspermy block in the central cell during double fertilization of Arabidopsis thaliana. Front Plant Sci. 2020:**11**:588700. https://doi.org/10.3389/fpls.2020.588700

Nasrallah JB. Stop and go signals at the stigma–pollen interface of the Brassicaceae. Plant Physiol. 2023:**193**(2):927–948. https://doi.org/10.1093/plphys/kiad301

Ngo QA, Vogler H, Lituiev DS, Nestorova A, Grossniklaus U. A calcium dialog mediated by the FERONIA signal transduction pathway controls plant sperm delivery. Dev Cell. 2014:**29**(4):491–500. https://doi.org/10.1016/j.devcel.2014.04.008

Nieuwland J, Feron R, Huisman BAH, Fasolino A, Hilbers CW, Derksen J, Mariani C. Lipid transfer proteins enhance cell wall extension in tobacco. Plant Cell. 2005:**17**(7):2009–2019. https://doi.org/10.1105/tpc.105.032094

Noble JA, Bielski NV, Liu M-CJ, DeFalco TA, Stegmann M, Nelson ADL, McNamara K, Sullivan B, Dinh KK, Khuu N, et al. Evolutionary analysis of the LORELEI gene family in plants reveals regulatory subfunctionalization. Plant Physiol. 2022:**190**(4): 2539–1556. https://doi.org/10.1093/plphys/kiac444

Nonomura K-I, Nakano M, Fukuda T, Eiguchi M, Miyao A, Hirochika H, Kurata N. The novel gene HOMOLOGOUS PAIRING ABERRATION IN RICE MEIOSIS1 of rice encodes a putative coiled-coil protein required for homologous chromosome pairing in meiosis. Plant Cell. 2004:**16**(4):1008–1020. https://doi.org/10.1105/tpc.020701

Nowack MK, Grini PE, Jakoby MJ, Lafos M, Koncz C, Schnittger A. A positive signal from the fertilization of the egg cell sets off endosperm proliferation in angiosperm embryogenesis. Nat Genet. 2006:**38**(1): 63–67. https://doi.org/10.1038/ng1694

Oh S-A, Jeon J, Park H-J, Grini PE, Twell D, Park SK. Analysis of gemini pollen 3 mutant suggests a broad function of AUGMIN in microtubule organization during sexual reproduction in Arabidopsis. Plant J. 2016:**87**(2):188–201. https://doi.org/10.1111/tpj.13192

Oh SA, Park KS, Twell D, Park SK. The SIDECAR POLLEN gene encodes a microspore-specific LOB/AS2 domain protein required for the correct timing and orientation of asymmetric cell division. Plant J. 2010:**64**(5):839–850. https://doi.org/10.1111/j.1365-313X.2010.04374.x

Ohnishi T, Takanashi H, Mogi M, Takahashi H, Kikuchi S, Yano K, Okamoto T, Fujita M, Kurata N, Tsutsumi N. Distinct gene expression profiles in egg and synergid cells of rice as revealed by cell type-specific microarrays. Plant Physiol. 2011:**155**(2):881–891. https://doi.org/10.1104/pp.110.167502

Ohnishi Y, Hoshino R, Okamoto T. Dynamics of male and female chromatin during karyogamy in rice zygotes. Plant Physiol. 2014:**165**(4):1533–1543. https://doi.org/10.1104/pp.114.236059

Okuda S, Tsutsui H, Shiina K, Sprunck S, Takeuchi H, Yui R, Kasahara RD, Hamamura Y, Mizukami A, Susaki D, et al. Defensin-like polypeptide LUREs are pollen tube attractants secreted from synergid cells. Nature. 2009:**458**(7236):357–361. https://doi.org/10.1038/nature07882

Ortiz-Morea FA, Liu J, Shan L, He P. Malectin-like receptor kinases as protector deities in plant immunity. Nat Plants. 2022:**8**(1):27–37. https://doi.org/10.1038/s41477-021-01028-3

Pagnussat GC, Alandete-Saez M, Bowman JL, Sundaresan V. Auxin-dependent patterning and gamete specification in the Arabidopsis female gametophyte. Science. 2009:**324**(5935): 1684–1689. https://doi.org/10.1126/science.1167324

Pagnussat GC, Yu H-J, Sundaresan V. Cell-fate switch of synergid to egg cell in Arabidopsis eostre mutant embryo sacs arises from misexpression of the BEL1-like homeodomain gene BLH1. Plant Cell. 2007:**19**(11):3578–3592. https://doi.org/10.1105/tpc.107.054890

Palanivelu R, Brass L, Edlund AF, Preuss D. Pollen tube growth and guidance is regulated by POP2, an Arabidopsis gene that controls GABA levels. Cell. 2003:**114**(1):47–59. https://doi.org/10.1016/S0092-8674(03)00479-3

Palanivelu R, Preuss D. Distinct short-range ovule signals attract or repel Arabidopsis thaliana pollen tubes in vitro. BMC Plant Biol. 2006:**6**(1):7. https://doi.org/10.1186/1471-2229-6-7

Panoli A, Martin MV, Alandete-Saez M, Simon M, Neff C, Swarup R, Bellido A, Yuan L, Pagnussat GC, Sundaresan V. Auxin import and local auxin biosynthesis are required for mitotic divisions, cell expansion and cell specification during female gametophyte development in Arabidopsis thaliana. PLoS One. 2015:**10**(5):e0126164. https://doi.org/10.1371/journal.pone.0126164

Park SK, Howden R, Twell D. The Arabidopsis thaliana gametophytic mutation gemini pollen1 disrupts microspore polarity, division asymmetry and pollen cell fate. Development. 1998:**125**(19):3789–3799. https://doi.org/10.1242/dev.125.19.3789

Park SK, Rahman D, Oh SA, Twell D. Gemini pollen 2, a male and female gametophytic cytokinesis defective mutation. Sex Plant Reprod. 2004:**17**(2):63–70. https://doi.org/10.1007/s00497-004-0216-x

Park S-Y, Jauh G-Y, Mollet J-C, Eckard KJ, Nothnagel EA, Walling LL, Lord EM. A lipid transfer–like protein is necessary for lily pollen tube adhesion to an in vitro stylar matrix. Plant Cell. 2000:**12**(1):151–163. https://doi.org/10.1105/tpc.12.1.151

Peng X, Yan T, Sun M. The WASP-Arp2/3 complex signal cascade is involved in actin-dependent sperm nuclei migration during double fertilization in tobacco and maize. Sci Rep. 2017:**7**(1):43161. https://doi.org/10.1038/srep43161

Pereira AM, Lopes AL, Coimbra S. Arabinogalactan proteins as interactors along the crosstalk between the pollen tube and the female tissues. Front Plant Sci. 2016a:**7**:1895. https://doi.org/10.3389/fpls.2016.01895

Pereira AM, Lopes AL, Coimbra S. *JAGGER*, an AGP essential for persistent synergid degeneration and polytubey block in Arabidopsis. Plant Signal Behav. 2016b:**11**(8):e1209616. https://doi.org/10.1080/15592324.2016.1209616

Ponvert N, Johnson MA. Synergid cell calcium oscillations refine understanding of FERONIA/LORELEI signaling during interspecific hybridization. Plant Reprod. 2024:**37**:57–68.

Portereiko MF, Lloyd A, Steffen JG, Punwani JA, Otsuga D, Drews GN. AGL80 is required for central cell and endosperm development in Arabidopsis. Plant Cell. 2006:**18**(8):1862–1872. https://doi.org/10.1105/tpc.106.040824

Qin Y, Leydon AR, Manziello A, Pandey R, Mount D, Denic S, Vasic B, Johnson MA, Palanivelu R. Penetration of the stigma and style elicits a novel transcriptome in pollen tubes, pointing to genes critical for growth in a pistil. PLoS Genet. 2009:**5**(8):e1000621. https://doi.org/10.1371/journal.pgen.1000621

Qin Y, Wysocki RJ, Somogyi A, Feinstein Y, Franco JY, Tsukamoto T, Dunatunga D, Levy C, Smith S, Simpson R, et al. Sulfinylated azadecalins act as functional mimics of a pollen germination stimulant in Arabidopsis pistils. Plant J. 2011:**68**(5):800–815. https://doi.org/10.1111/j.1365-313X.2011.04729.x

Qu X, Zhang R, Zhang M, Diao M, Xue Y, Huang S. Organizational innovation of apical actin filaments drives rapid pollen tube growth and turning. Mol Plant. 2017:**10**(7):930–947. https://doi.org/10.1016/j.molp.2017.05.002

Quilichini TD, Douglas CJ, Samuels AL. New views of tapetum ultrastructure and pollen exine development in Arabidopsis thaliana. Ann Bot. 2014:**114**(6):1189–1201. https://doi.org/10.1093/aob/mcu042

Rabiger DS, Drews GN. MYB64 and MYB119 are required for cellularization and differentiation during female gametogenesis in Arabidopsis thaliana. PLoS Genet. 2013:**9**(9):e1003783. https://doi.org/10.1371/journal.pgen.1003783

Rotman N, Durbarry A, Wardle A, Yang WC, Chaboud A, Faure J-E, Berger F, Twell D. A novel class of MYB factors controls sperm-cell formation in plants. Curr Biol. 2005:**15**(3):244–248. https://doi.org/10.1016/j.cub.2005.01.013

Rotman N, Rozier F, Boavida L, Dumas C, Berger F, Faure J-E. Female control of male gamete delivery during fertilization in Arabidopsis thaliana. Curr Biol. 2003:**13**(5):432–436. https://doi.org/10.1016/S0960-9822(03)00093-9

Rozier F, Riglet L, Kodera C, Bayle V, Durand E, Schnabel J, Gaude T, Fobis-Loisy I. Live-cell imaging of early events following pollen perception in self-incompatible Arabidopsis thaliana. J Exp Bot. 2020:**71**(9):2513–2526. https://doi.org/10.1093/jxb/eraa008

Sandaklie-Nikolova L, Palanivelu R, King EJ, Copenhaver GP, Drews GN. Synergid cell death in Arabidopsis is triggered following direct interaction with the pollen tube. Plant Physiol. 2007:**144**(4): 1753–1762. https://doi.org/10.1104/pp.107.098236

Schiøtt M, Romanowsky SM, Bækgaard L, Jakobsen MK, Palmgren MG, Harper JF. A plant plasma membrane Ca^{2+} pump is required for normal pollen tube growth and fertilization. Proc Natl Acad Sci U S A. 2004:**101**(25):9502–9507. https://doi.org/10.1073/pnas.0401542101

Sede AR, Borassi C, Wengier DL, Mecchia MA, Estevez JM, Muschietti JP. Arabidopsis pollen extensins LRX are required for cell wall integrity during pollen tube growth. FEBS Lett. 2018:**592**(2):233–243. https://doi.org/10.1002/1873-3468.12947

Shi W, Ji J, Xue Z, Zhang F, Miao Y, Yang H, Tang D, Du G, Li Y, Shen Y, et al. PRD1, a homologous recombination initiation factor, is involved in spindle assembly in rice meiosis. New Phytol. 2021:**230**(2):585–600. https://doi.org/10.1111/nph.17178

Skinner DJ, Sundaresan V. Recent advances in understanding female gametophyte development. F1000Res. 2018:**7**:F1000 Faculty Rev-804. https://doi.org/10.12688/f1000research.14508.1

Song Q, Ando A, Jiang N, Ikeda Y, Chen ZJ. Single-cell RNA-seq analysis reveals ploidy-dependent and cell-specific transcriptome changes in Arabidopsis female gametophytes. Genome Biol. 2020:**21**(1):178. https://doi.org/10.1186/s13059-020-02094-0

Song Y, Wilson AJ, Zhang X-C, Thoms D, Sohrabi R, Song S, Geissmann Q, Liu Y, Walgren L, He SY, et al. FERONIA restricts Pseudomonas in the rhizosphere microbiome via regulation of reactive oxygen species. Nat Plants. 2021:**7**(5):644–654. https://doi.org/10.1038/s41477-021-00914-0

Song Z, Zhong S, Qu L-J. FERONIA and reactive oxygen species: regulators in the self-incompatibility response and in interspecific pollination. Mol Hortic. 2023:**3**(1):10. https://doi.org/10.1186/s43897-023-00058-z

Spielman M, Scott Rj. Polyspermy barriers in plants: from preventing to promoting fertilization. Sex Plant Reprod. 2008:**21**(1):53–65. https://doi.org/10.1007/s00497-007-0063-7

Sprunck S, Baumann U, Edwards K, Langridge P, Dresselhaus T. The transcript composition of egg cells changes significantly following fertilization in wheat (Triticum aestivum L.). Plant J. 2005:**41**(5): 660–672. https://doi.org/10.1111/j.1365-313X.2005.02332.x

Sprunck S, Rademacher S, Vogler F, Gheyselinck J, Grossniklaus U, Dresselhaus T. Egg cell-secreted EC1 triggers sperm cell activation during double fertilization. Science. 2012:**338**(6110):1093–1097. https://doi.org/10.1126/science.1223944

Stacey NJ, Kuromori T, Azumi Y, Roberts G, Breuer C, Wada T, Maxwell A, Roberts K, Sugimoto-Shirasu K. Arabidopsis SPO11-2 functions with SPO11-1 in meiotic recombination. Plant J. 2006:**48**(2):206–216. https://doi.org/10.1111/j.1365-313X.2006.02867.x

Steer MW, Steer JM. Pollen tube tip growth. New Phytol. 1989:**111**(3): 323–358. https://doi.org/10.1111/j.1469-8137.1989.tb00697.x

Steffen JG, Kang I-H, Macfarlane J, Drews GN. Identification of genes expressed in the Arabidopsis female gametophyte. Plant J. 2007:**51**(2):281–292. https://doi.org/10.1111/j.1365-313X.2007.03137.x

Steffen JG, Kang I-H, Portereiko MF, Lloyd A, Drews GN. AGL61 interacts with AGL80 and is required for central cell development in Arabidopsis. Plant Physiol. 2008:**148**(1):259–268. https://doi.org/10.1104/pp.108.119404

Steinhorst L, Mähs A, Ischebeck T, Zhang C, Zhang X, Arendt S, Schültke S, Heilmann I, Kudla J. Vacuolar CBL-CIPK12 Ca2 +-sensor-kinase complexes are required for polarized pollen tube growth. Curr Biol. 2015:**25**(11):1475–1482. https://doi.org/10.1016/j.cub.2015.03.053

Stührwohldt N, Dahlke RI, Kutschmar A, Peng X, Sun M-X, Sauter M. Phytosulfokine peptide signaling controls pollen tube growth and funicular pollen tube guidance in Arabidopsis thaliana. Physiol Plant. 2015:**153**(4):643–653. https://doi.org/10.1111/ppl.12270

Sun M, Nishino T, Marko JF. The SMC1-SMC3 cohesin heterodimer structures DNA through supercoiling-dependent loop formation. Nucl Acids Res. 2013:**41**(12):6149–6160. https://doi.org/10.1093/nar/gkt303

Sun Y, Wang X, Pan L, Xie F, Dai B, Sun M, Peng X. Plant egg cell fate determination depends on its exact position in female gametophyte. Proc Natl Acad Sci U S A. 2021:**118**(8):e2017488118. https://doi.org/10.1073/pnas.2017488118

Susaki D, Suzuki T, Maruyama D, Ueda M, Higashiyama T, Kurihara D. Dynamics of the cell fate specifications during female gametophyte development in Arabidopsis. PLoS Biol. 2021:**19**(3):e3001123. https://doi.org/10.1371/journal.pbio.3001123

Susaki D, Takeuchi H, Tsutsui H, Kurihara D, Higashiyama T. Live imaging and laser disruption reveal the dynamics and cell-cell communication during Torenia fournieri female gametophyte development. Plant Cell Physiol. 2015:**56**(5):1031–1041. https://doi.org/10.1093/pcp/pcv031

Tadros W, Lipshitz HD. The maternal-to-zygotic transition: a play in two acts. Development. 2009:**136**(18):3033–3042. https://doi.org/10.1242/dev.033183

Takeuchi H, Higashiyama T. A species-specific cluster of defensin-like genes encodes diffusible pollen tube attractants in Arabidopsis. PLoS Biol. 2012:**10**(12):e1001449. https://doi.org/10.1371/journal.pbio.1001449

Takeuchi H, Higashiyama T. Tip-localized receptors control pollen tube growth and LURE sensing in Arabidopsis. Nature. 2016:**531**(7593):245–248. https://doi.org/10.1038/nature17413

Tan X, Cao K, Liu F, Li Y, Li P, Gao C, Ding Y, Lan Z, Shi Z, Rui Q, et al Arabidopsis COG complex subunits COG3 and COG8 modulate golgi morphology, vesicle trafficking homeostasis and are essential for pollen tube growth. PLoS Genet. 2016:**12**:e1006140.

Tang W, Ezcurra I, Muschietti J, McCormick S. A cysteine-rich extracellular protein, LAT52, interacts with the extracellular domain of the pollen receptor kinase LePRK2. Plant Cell. 2002:**14**(9):2277–2287. https://doi.org/10.1105/tpc.003103

Tang W, Kelley D, Ezcurra I, Cotter R, McCormick S. LeSTIG1, an extracellular binding partner for the pollen receptor kinases LePRK1 and LePRK2, promotes pollen tube growth in vitro. Plant J. 2004:**39**(3): 343–353. https://doi.org/10.1111/j.1365-313X.2004.02139.x

Tavares B, Domingos P, Dias PN, Feijó JA, Bicho A. The essential role of anionic transport in plant cells: the pollen tube as a case study. J Exp Bot. 2011:**62**(7):2273–2298. https://doi.org/10.1093/jxb/err036

Tedeschi F, Rizzo P, Rutten T, Altschmied L, Bäumlein H. RWP-RK domain-containing transcription factors control cell differentiation during female gametophyte development in Arabidopsis. New Phytol. 2017:**213**(4):1909–1924. https://doi.org/10.1111/nph.14293

Tekleyohans DG, Mao Y, Kägi C, Stierhof Y-D, Groß-Hardt R. Polyspermy barriers: a plant perspective. Curr Opin Plant Biol. 2017:**35**:131–137. https://doi.org/10.1016/j.pbi.2016.11.012

Tsukamoto T, Qin Y, Huang Y, Dunatunga D, Palanivelu R. A role for LORELEI, a putative glycosylphosphatidylinositol-anchored protein, in Arabidopsis thaliana double fertilization and early seed development. Plant J. 2010:**62**(4):571–588. https://doi.org/10.1111/j.1365-313X.2010.04177.x

Tunc-Ozdemir M, Rato C, Brown E, Rogers S, Mooneyham A, Frietsch S, Myers CT, Poulsen LR, Malhó R, Harper JF. Cyclic nucleotide gated channels 7 and 8 are essential for male reproductive fertility. PLoS One. 2013:**8**(2):e55277. https://doi.org/10.1371/journal.pone.0055277

Twell D, Park SK, Hawkins TJ, Schubert D, Schmidt R, Smertenko A, Hussey PJ. MOR1/GEM1 has an essential role in the plant-specific cytokinetic phragmoplast. Nat Cell Biol. 2002:**4**(9):711–714. https://doi.org/10.1038/ncb844

van der Ven HH, Al-Hasani S, Diedrich K, Hamerich U, Lehmann F, Krebs D. Polyspermy in in vitro fertilization of human oocytes: frequency and possible causes. Ann N Y Acad Sci. 1985:**442**(1):88–95. https://doi.org/10.1111/j.1749-6632.1985.tb37508.x

Vastenhouw NL, Cao WX, Lipshitz HD. The maternal-to-zygotic transition revisited. Development. 2019:**146**(11):dev161471. https://doi.org/10.1242/dev.161471

Vernet A, Meynard D, Lian Q, Mieulet D, Gibert O, Bissah M, Rivallan R, Autran D, Leblanc O, Meunier AC, et al. High-frequency synthetic apomixis in hybrid rice. Nat Commun. 2022:**13**(1):7963. https://doi.org/10.1038/s41467-022-35679-3

Vogler F, Schmalzl C, Englhart M, Bircheneder M, Sprunck S. Brassinosteroids promote Arabidopsis pollen germination and growth. Plant Reprod. 2014:**27**(3):153–167. https://doi.org/10.1007/s00497-014-0247-x

Völz R, Heydlauff J, Ripper D, von Lyncker L, Groß-Hardt R. Ethylene signaling is required for synergid degeneration and the establishment of a pollen tube block. Dev Cell. 2013:**25**(3):310–316. https://doi.org/10.1016/j.devcel.2013.04.001

Völz R, von Lyncker L, Baumann N, Dresselhaus T, Sprunck S, Gross-Hardt R. LACHESIS-dependent egg-cell signaling regulates the development of female gametophytic cells. Development. 2012:**139**(3):498–502. https://doi.org/10.1242/dev.075234

Wang C, Liu Q, Shen Y, Hua Y, Wang J, Lin J, Wu M, Sun T, Cheng Z, Mercier R, et al. Clonal seeds from hybrid rice by simultaneous genome engineering of meiosis and fertilization genes. Nat Biotechnol. 2019:**37**(3):283–286. https://doi.org/10.1038/s41587-018-0003-0

Wang J-G, Feng C, Liu H-H, Feng Q-N, Li S, Zhang Y. AP1G mediates vacuolar acidification during synergid-controlled pollen tube reception. Proc Natl Acad Sci U S A. 2017b:**114**(24):E4877–E4883. https://doi.org/10.1073/pnas.1617967114

Wang K, Guo ZL, Zhou WT, Zhang C, Zhang ZY, Lou Y, Xiong SX, Yao XZ, Fan JJ, Zhu J, et al. The regulation of sporopollenin biosynthesis genes for rapid pollen wall formation. Plant Physiol. 2018:**178**(1): 283–294. https://doi.org/10.1104/pp.18.00219

Wang L, Clarke LA, Eason RJ, Parker CC, Qi B, Scott RJ, Doughty J. PCP-B class pollen coat proteins are key regulators of the hydration checkpoint in Arabidopsis thaliana pollen-stigma interactions. New Phytol. 2017a:**213**(2):764–777. https://doi.org/10.1111/nph.14162

Wang L, Filatov DA. Mechanisms of prezygotic post-pollination reproductive barriers in plants. Front Plant Sci. 2023:**14**:1230278. https://doi.org/10.3389/FPLS.2023.1230278

Wang L, Lau Y-L, Fan L, Bosch M, Doughty J. Pollen coat proteomes of Arabidopsis thaliana, Arabidopsis lyrata, and Brassica oleracea reveal remarkable diversity of small cysteine-rich proteins at the pollen-stigma interface. Biomolecules. 2023a:**13**(1):157. https://doi.org/10.3390/biom13010157

Wang M, Tang D, Luo Q, Jin Y, Shen Y, Wang K, Cheng Z. BRK1, a Bub1-related kinase, is essential for generating proper tension between homologous kinetochores at metaphase I of rice meiosis. Plant Cell. 2012:**24**(12):4961–4973. https://doi.org/10.1105/tpc.112.105874

Wang N, Song X, Ye J, Zhang S, Cao Z, Zhu C, Hu J, Zhou Y, Huang Y, Cao S, et al. Structural variation and parallel evolution of apomixis in citrus during domestication and diversification. Natl Sci Rev. 2022b:**9**(10):nwac114. https://doi.org/10.1093/nsr/nwac114

Wang N, Xia X, Jiang T, Li L, Zhang P, Niu L, Cheng H, Wang K, Lin H. In planta haploid induction by genome editing of DMP in the model legume Medicago truncatula. Plant Biotech J. 2022c:**20**(1):22–24. https://doi.org/10.1111/pbi.13740

Wang T, Liang L, Xue Y, Jia P-F, Chen W, Zhang M-X, Wang Y-C, Li H-J, Yang W-C. A receptor heteromer mediates the male perception of female attractants in plants. Nature. 2016:**531**(7593):241–244. https://doi.org/10.1038/nature16975

Wang W, Xiong H, Zhao P, Peng X, Sun M-X. DMP8 and 9 regulate HAP2/GCS1 trafficking for the timely acquisition of sperm fusion competence. Proc Natl Acad Sci U S A. 2022d:**119**(45): e2207608119. https://doi.org/10.1073/pnas.2207608119

Wang W-Q, Meng J-G, Yang F, Xu Y-J, Li S-Z, Li H-J. A non-defensin peptide NPA1 attracts pollen tube in Arabidopsis. Seed Biol. 2024:**1**(1):1–11. https://doi.org/10.48130/seedbio-0024-0001

Wang X, Liu X, Yi X, Wang M, Shi W, Li R, Tang W, Zhang L, Sun M, Peng X. The female germ unit is essential for pollen tube funicular guidance in Arabidopsis thaliana. New Phytol. 2023b:**238**(1): 155–168. https://doi.org/10.1111/nph.18686

Wang X, Xu Y, Zhang S, Cao L, Huang Y, Cheng J, Wu G, Tian S, Chen C, Liu Y, et al. Genomic analyses of primitive, wild and cultivated citrus provide insights into asexual reproduction. Nat Genet. 2017c:**49**(5):765–772. https://doi.org/10.1038/ng.3839

Wang Y, Coomey J, Miller K, Jensen GS, Haswell ES. Interactions between a mechanosensitive channel and cell wall integrity signaling influence pollen germination in Arabidopsis thaliana. J Exp Bot. 2022a:**73**(5):1533–1545. https://doi.org/10.1093/jxb/erab525

Wei X, Liu C, Chen X, Lu H, Wang J, Yang S, Wang K. Synthetic apomixis with normal hybrid rice seed production. Mol Plant. 2023:**16**(3):489–492. https://doi.org/10.1016/j.molp.2023.01.005

Williams E, Kaul V, Rouse J, Palser B. Overgrowth of pollen tubes in embryo sacs of rhododendron following interspecific pollinations. Aust J Bot. 1986:**34**(4):413. https://doi.org/10.1071/BT9860413

Wu Z, Ji J, Tang D, Wang H, Shen Y, Shi W, Li Y, Tan X, Cheng Z, Luo Q. OsSDS is essential for DSB formation in rice meiosis. Front Plant Sci. 2015:**6**:21. https://doi.org/10.3389/fpls.2015.00021

Wudick MM, Portes MT, Michard E, Rosas-Santiago P, Lizzio MA, Nunes CO, Campos C, Santa Cruz Damineli D, Carvalho JC, Lima PT, et al. CORNICHON sorting and regulation of GLR channels underlie pollen tube Ca2+ homeostasis. Science. 2018:**360**(6388): 533–536. https://doi.org/10.1126/science.aar6464

Wuest SE, Vijverberg K, Schmidt A, Weiss M, Gheyselinck J, Lohr M, Wellmer F, Rahnenführer J, von Mering C, Grossniklaus U. Arabidopsis female gametophyte gene expression map reveals similarities between plant and animal gametes. Curr Biol. 2010:**20**(6): 506–512. https://doi.org/10.1016/j.cub.2010.01.051

Xiao Y, Stegmann M, Han Z, DeFalco TA, Parys K, Xu L, Belkhadir Y, Zipfel C, Chai J. Mechanisms of RALF peptide perception by a heterotypic receptor complex. Nature. 2019:**572**(7768):270–274. https://doi.org/10.1038/s41586-019-1409-7

Xie E, Li Y, Tang D, Lv Y, Shen Y, Cheng Z. A strategy for generating rice apomixis by gene editing. J Integr Plant Biol. 2019:**61**(8):911–916. https://doi.org/10.1111/jipb.12785

Xiong J, Hu F, Ren J, Huang Y, Liu C, Wang K. Synthetic apomixis: the beginning of a new era. Curr Opin Biotech. 2023:**79**:102877. https://doi.org/10.1016/j.copbio.2022.102877

Xue JS, Qiu S, Jia XL, Shen SY, Shen CW, Wang S, Xu P, Tong Q, Lou YX, Yang NY, et al. Stepwise changes in flavonoids in spores/pollen contributed to terrestrial adaptation of plants. Plant Physiol. 2023:**193**(1):627–642. https://doi.org/10.1093/plphys/kiad313

Xue Y, Meng J-G, Jia P-F, Zhang Z-R, Li H-J, Yang W-C. POD1-SUN-CRT3 chaperone complex guards the ER sorting of LRR receptor kinases in Arabidopsis. Nat Commun. 2022:**13**(1):2703. https://doi.org/10.1038/s41467-022-30179-w

Yamaoka S, Nishihama R, Yoshitake Y, Ishida S, Inoue K, Saito M, Okahashi K, Bao H, Nishida H, Yamaguchi K, et al. Generative cell specification requires transcription factors evolutionarily conserved in land plants. Curr Biol. 2018:**28**(3):479–486.e5. https://doi.org/10.1016/j.cub.2017.12.053

Yang H, Wang D, Guo L, Pan H, Yvon R, Garman S, Wu HM, Cheung AY. Malectin/Malectin-like domain-containing proteins: A repertoire of cell surface molecules with broad functional potential. Cell Surf. 2021:**7**:100056. https://doi.org/10.1016/j.tcsw.2021.100056

Yang W-C, Shi D-Q, Chen Y-H. Female gametophyte development in flowering plants. Annu Rev Plant Biol. 2010:**61**(1):89–108. https://doi.org/10.1146/annurev-arplant-042809-112203

Yao X, Hu W, Yang Z-N. The contributions of sporophytic tapetum to pollen formation. Seed Biology. 2022:**1**(1):1–13. https://doi.org/10.48130/SeedBio-2022-0005

Ye J, Zheng Y, Yan A, Chen N, Wang Z, Huang S, Yang Z. Arabidopsis formin3 directs the formation of actin cables and polarized growth in pollen tubes. Plant Cell. 2009:**21**:3868–3884.

Yu G-H, Zou J, Feng J, Peng X-B, Wu J-Y, Wu Y-L, Palanivelu R, Sun M-X. Exogenous γ-aminobutyric acid (GABA) affects pollen tube growth via modulating putative Ca2+-permeable membrane channels and is coupled to negative regulation on glutamate decarboxylase. J Exp Bot. 2014:**65**(12):3235–3248. https://doi.org/10.1093/jxb/eru171

Yu X, Luo A, Sun M-X. An efficient antipodal cell isolation method for screening of cell type-specific genes in Arabidopsis thaliana. PLoS One. 2016:**11**(11):e0166390. https://doi.org/10.1371/journal.pone.0166390

Yu X, Zhang X, Zhao P, Peng X, Chen H, Bleckmann A, Bazhenova A, Shi C, Dresselhaus T, Sun M. Fertilized egg cells secrete endopeptidases to avoid polytubey. Nature. 2021:**592**(7854):433–437. https://doi.org/10.1038/s41586-021-03387-5

Yuan L, Liu Z, Song X, Johnson C, Yu X, Sundaresan V. The CKI1 histidine kinase specifies the female gametic precursor of the endosperm. Dev Cell. 2016:**37**(1):34–46. https://doi.org/10.1016/j.devcel.2016.03.009

Zamariola L, De Storme N, Tiang CL, Armstrong SJ, Franklin FCH, Geelen D. SGO1 but not SGO2 is required for maintenance of centromere cohesion in Arabidopsis thaliana meiosis. Plant Reprod. 2013:**26**(3):197–208. https://doi.org/10.1007/s00497-013-0231-x

Zeng CJT, Lee Y-RJ, Liu B. The WD40 repeat protein NEDD1 functions in microtubule organization during cell division in Arabidopsis thaliana. Plant Cell. 2009:**21**(4):1129–1140. https://doi.org/10.1105/tpc.109.065953

Zhang C, Song Y, Cheng Z, Wang Y, Zhu J, Ma H, Xu L, Yang Z-N. The Arabidopsis thaliana DSB formation (AtDFO) gene is required for meiotic double-strand break formation. Plant J. 2012:**72**(2): 271–281. https://doi.org/10.1111/j.1365-313X.2012.05075.x

Zhang J, Yin J, Luo J, Tang D, Zhu X, Wang J, Liu Z, Wang P, Zhong Y, Liu C, et al. Construction of homozygous diploid potato through maternal haploid induction. aBIOTECH. 2022a:**3**(3):163–168. https://doi.org/10.1007/s42994-022-00080-7

Zhang L, Huang J, Su S, Wei X, Yang L, Zhao H, Yu J, Wang J, Hui J, Hao S, et al. FERONIA receptor kinase-regulated reactive oxygen species mediate self-incompatibility in Brassica rapa. Curr Biol. 2021:**31**(14):3004–3016.e4. https://doi.org/10.1016/j.cub.2021.04.060

Zhang M-X, Zhu S-S, Xu Y-C, Guo Y-L, Yang W-C, Li H-J. Transcriptional repression specifies the central cell for double fertilization. Proc Natl Acad Sci U S A. 2020c:**117**(11):6231–6236. https://doi.org/10.1073/pnas.1909465117

Zhang X, Liu W, Nagae TT, Takeuchi H, Zhang H, Han Z, Higashiyama T, Chai J. Structural basis for receptor recognition of pollen tube attraction peptides. Nat Commun. 2017:**8**(1):1331. https://doi.org/10.1038/s41467-017-01323-8

Zhang X, Shi C, Li S, Zhang B, Luo P, Peng X, Zhao P, Dresselhaus T, Sun M-X. A female in vivo haploid-induction system via mutagenesis of egg cell-specific peptidases. Mol Plant. 2023:**16**(2):471–480. https://doi.org/10.1016/j.molp.2023.01.001

Zhang X, Zhang L, Zhang J, Jia M, Cao L, Yu J, Zhao D. Haploid induction in allotetraploid tobacco using DMPs mutation. Planta. 2022b:**255**(5):98. https://doi.org/10.1007/s00425-022-03877-4

Zhang X, Zhao G, Tan Q, Yuan H, Betts N, Zhu L, Zhang D, Liang W. Rice pollen aperture formation is regulated by the interplay between OsINP1 and OsDAF1. Nat Plants. 2020b:**6**(4):394–403. https://doi.org/10.1038/s41477-020-0630-6

Zhang Z, Conner J, Guo Y, Ozias-Akins P. Haploidy in tobacco induced by PsASGR-BBML transgenes via parthenogenesis. Genes (Basel). 2020a:**11**(9):1072. https://doi.org/10.3390/genes11091072

Zhang Z, Zhang B, Chen Z, Zhang D, Zhang H, Wang H, Zhang Y, Cai D, Liu J, Xiao S, et al. A PECTIN METHYLESTERASE gene at the maize Ga1 locus confers male function in unilateral cross-incompatibility. Nat Commun. 2018:**9**(1):3678. https://doi.org/10.1038/s41467-018-06139-8

Zhao C, Zayed O, Yu Z, Jiang W, Zhu P, Hsu C-C, Zhang L, Tao WA, Lozano-Durán R, Zhu J-K. Leucine-rich repeat extensin proteins regulate plant salt tolerance in Arabidopsis. Proc Natl Acad Sci U S A. 2018:**115**(51):13123–13128. https://doi.org/10.1073/pnas.1816991115

Zhao J, Xin H, Qu L, Ning J, Peng X, Yan T, Ma L, Li S, Sun M-X. Dynamic changes of transcript profiles after fertilization are associated with de novo transcription and maternal elimination in tobacco zygote, and mark the onset of the maternal-to-zygotic transition. Plant J. 2011:**65**(1):131–145. https://doi.org/10.1111/j.1365-313X.2010.04403.x

Zhao L-N, Shen L-K, Zhang W-Z, Zhang W, Wang Y, Wu W-H. $Ca2^{+}$-dependent protein kinase11 and 24 modulate the activity of the inward rectifying K+ channels in Arabidopsis pollen tubes. Plant Cell. 2013:**25**:649–661.

Zhao P, Begcy K, Dresselhaus T, Sun M-X. Does early embryogenesis in eudicots and monocots involve the same mechanism and molecular players? Plant Physiol. 2017:**173**(1):130–142. https://doi.org/10.1104/pp.16.01406

Zhao P, Shi C, Wang L, Sun M-X. The parental contributions to early plant embryogenesis and the concept of maternal-to-zygotic transition in plants. Curr Opin Plant Biol. 2022a:**65**:102144. https://doi.org/10.1016/j.pbi.2021.102144

Zhao P, Zhou X, Shen K, Liu Z, Cheng T, Liu D, Cheng Y, Peng X, Sun M-X. Two-step maternal-to-zygotic transition with two-phase parental genome contributions. Dev Cell. 2019:**49**(6):882–893.e5. https://doi.org/10.1016/j.devcel.2019.04.016

Zhao P, Zhou X, Zheng Y, Ren Y, Sun M-X. Equal parental contribution to the transcriptome is not equal control of embryogenesis. Nat Plants. 2020:**6**(11):1354–1364. https://doi.org/10.1038/s41477-020-00793-x

Zhao X, Yuan K, Liu Y, Zhang N, Yang L, Zhang Y, Wang Y, Ji J, Fang Z, Han F, et al. In vivo maternal haploid induction based on genome editing of DMP in Brassica oleracea. Plant Biotech J. 2022b:**20**(12):2242–2244. https://doi.org/10.1111/pbi.13934

Zhong S, Li L, Wang Z, Ge Z, Li Q, Bleckmann A, Wang J, Song Z, Shi Y, Liu T, et al. RALF peptide signaling controls the polytubey block in Arabidopsis. Science. 2022a:**375**(6578):290–296. https://doi.org/10.1126/science.abl4683

Zhong S, Liu M, Wang Z, Huang Q, Hou S, Xu YC, Ge Z, Song Z, Huang J, Qiu X, et al. Cysteine-rich peptides promote interspecific genetic isolation in Arabidopsis. Science. 2019a:**364**(6443): eaau9564. https://doi.org/10.1126/science.aau9564

Zhong Y, Chen B, Li M, Wang D, Jiao Y, Qi X, Wang M, Liu Z, Chen C, Wang Y, et al. A DMP-triggered in vivo maternal haploid induction system in the dicotyledonous Arabidopsis. Nat Plants. 2020:**6**(5): 466–472. https://doi.org/10.1038/s41477-020-0658-7

Zhong Y, Liu C, Qi X, Jiao Y, Wang D, Wang Y, Liu Z, Chen C, Chen B, Tian X, et al. Mutation of ZmDMP enhances haploid induction in maize. Nat Plants. 2019b:**5**(6):575–580. https://doi.org/10.1038/s41477-019-0443-7

Zhong Y, Wang Y, Chen B, Liu J, Wang D, Li M, Qi X, Liu C, Boutilier K, Chen S. Establishment of a dmp based maternal haploid induction system for polyploid Brassica napus and Nicotiana tabacum. J Integr Plant Biol. 2022b:**64**(6):1281–1294. https://doi.org/10.1111/jipb.13244

Zhou L, Lan W, Jiang Y, Fang W, Luan S. A calcium-dependent protein kinase interacts with and activates a calcium channel to regulate pollen tube growth. Mol Plant. 2014:**7**(2):369–376. https://doi.org/10.1093/mp/sst125

Zhou L-Z, Dresselhaus T. Friend or foe: signaling mechanisms during double fertilization in flowering seed plants. Curr Top Dev Biol. 2019:**131**:453–496. https://doi.org/10.1016/bs.ctdb.2018.11.013

Zhou L-Z, Wang L, Chen X, Ge Z, Mergner J, Li X, Küster B, Langst G, Qu L-J, Dresselhaus T. The RALF signaling pathway regulates cell wall integrity during pollen tube growth in maize. Plant Cell. 2024:koad324. https://doi.org/10.1093/plcell/koad324 Online ahead of print.

Zhou X, Lu J, Zhang Y, Guo J, Lin W, Van Norman JM, Qin Y, Zhu X, Yang Z. Membrane receptor-mediated mechano-transduction maintains cell integrity during pollen tube growth within the pistil. Dev Cell. 2021b:**56**(7):1030–1042.e6. https://doi.org/10.1016/j.devcel.2021.02.030

Zhou X, Zheng Y, Wang L, Li H, Guo Y, Li M, Sun MX, Zhao P. SYP72 interacts with the mechanosensitive channel MSL8 to protect pollen from hypoosmotic shock during hydration. Nat Commun. 2022:**13**(1):73. https://doi.org/10.1038/s41467-021-27757-9

Zhou Y, Amom P, Reeder SH, Lee BH, Helton A, Dobritsa AA. Members of the ELMOD protein family specify formation of distinct aperture domains on the Arabidopsis pollen surface. Elife. 2021a:**10**: e71061. https://doi.org/10.7554/eLife.71061

Zhou Y, Dobritsa AA. Formation of aperture sites on the pollen surface as a model for development of distinct cellular domains. Plant Sci. 2019:**288**:110222. https://doi.org/10.1016/j.plantsci.2019.110222

Zhu J, Nan Q, Qin T, Qian D, Mao T, Yuan S, Wu X, Niu Y, Bai Q, An L, et al. Higher-ordered actin structures remodeled by Arabidopsis ACTIN-DEPOLYMERIZING FACTOR5 are important for pollen germination and pollen tube growth. Mol Plant. 2017:**10**(8):1065–1081. https://doi.org/10.1016/j.molp.2017.06.001

Zhu L, Chu L-C, Liang Y, Zhang X-Q, Chen L-Q, Ye D. The Arabidopsis CrRLK1L protein kinases BUPS1 and BUPS2 are required for normal growth of pollen tubes in the pistil. Plant J. 2018:**95**(3):474–486. https://doi.org/10.1111/tpj.13963

Zhu M, Tao L, Zhang J, Liu R, Tian H, Hu C, Zhu Y, Li M, Wei Z, Yi J, et al. The type-B response regulators ARR10, ARR12, and ARR18 specify the central cell in Arabidopsis. Plant Cell. 2022:**34**(12):4714–4737. https://doi.org/10.1093/plcell/koac285

https://doi.org/10.1093/plphys/kiad585 PLANT PHYSIOLOGY 2024: 195: 36–47

Plant Physiology®

Topical Review

A pilot oral history of plant synthetic biology

Jaya Joshi [1,*,†] **and Andrew D. Hanson** [2,*,†]

1 Department of Wood Science, University of British Columbia, 2424 Main Mall, Vancouver, BC V6T 1Z4, Canada
2 Horticultural Sciences Department, University of Florida, 2550 Hull Road, Gainesville, FL 32611, USA

*Author for correspondence: jaya.joshi@ubc.ca (J.J.); adha@ufl.edu (A.D.H.)
†Senior authors.
The author responsible for distribution of materials integral to the findings presented in this article in accordance with the policy described in the Instructions for Authors (https://academic.oup.com/plphys/pages/General-Instructions) is: Jaya Joshi (jaya.joshi@ubc.ca), Andrew D. Hanson (adha@ufl.edu).

Abstract

The whole field of synthetic biology (SynBio) is only about 20 years old, and plant SynBio is younger still. Nevertheless, within that short time, SynBio in general has drawn more scientific, philosophical, government, and private-sector interest than anything in biology since the recombinant DNA revolution. Plant SynBio, in particular, is now drawing more and more interest in relation to plants' potential to help solve planetary problems such as carbon capture and storage and replacing fossil fuels and feedstocks. As plant SynBio is so young and so fast-developing, we felt it was too soon to try to analyze its history. Instead, we set out to capture the essence of plant SynBio's origins and early development through interviews with 8 of the field's founders, representing 5 countries and 3 continents. We then distilled these founders' personal recollections and reflections into this review, centering the narrative on timelines for pivotal events, articles, funding programs, and quoting from interviews. We have archived the interview recordings and documented timeline entries. This work provides a resource for future historical scholarship.

Introduction

What synthetic biology is and does

Synthetic biology (SynBio) has been defined in various ways, of which these 2 are typical: "[SynBio] aims to build new properties into living systems so that they do or make something useful" (Davies 2019); and "SynBio is a transformative combination of DNA technology, engineering principles, and computational tools that makes it possible to design new life processes and to repurpose existing natural ones for useful purposes" (Hanson et al. 2019). These and most other definitions stress that SynBio designs, builds, and uses biological components ("parts") that do not exist in nature, and that the end goal is utility, i.e. that at its core, SynBio is primarily an engineering endeavor that aims to solve problems, not a scientific one that seeks to answer questions and to uncover information. To understand SynBio it is therefore essential to grasp the differences between engineering and science. The following pithy quotes—and the articles they come from—sum up key characteristics of engineering:

> Scientists study the world as it is, engineers create the world that never has been (Hammack and Anderson 2022);
>
> The engineering method is the use of heuristics to cause the best change in a poorly understood situation within the available resources (Hammack and Anderson 2022);
>
> Critically different from colleagues in the sciences, the chemical engineer must be willing to attempt a solution in situations in which many important details of the system are undefined or uncertain (Bailey 1995);
>
> In order to achieve reliable [engineering] outcomes it is necessary to avoid jumping to conclusions, to avoid making uninformed guesses where information can be made available. (MacLeod 2012)

The last quote refers particularly to the value of numerate thinking, which emerged in our interviews.

Received July 31, 2023. Accepted September 15, 2023. Advance access publication January 2, 2024

ADVANCES BOX

- Plant SynBio is a distinct, vibrant, diverse, and growing field with huge potential for real-world engineering impact that branched off mainline microbial SynBio but remains integrated with it.
- Plant SynBio was much influenced by the electronic engineering paradigm of the 2000s but has seen its shortcomings and is now hammering out plant-centric tools and ways to think and work.
- Plant SynBio is starting to internalize the essential differences between science (in which most plant SynBio practitioners were trained) and engineering (in which they were not trained).

In the beginning

Although the first use of the term "synthetic biology" in its modern sense can be dated to 1980 (Hirschi et al. 2022), a landmark *Nature* paper in 2000 on construction of a genetic toggle switch in *Escherichia coli* (Gardner et al. 2000) is generally seen as the start of the SynBio field. It is worth noting that this seminal work came out of Professor Jim Collins' group in the Biological Engineering Department at MIT, was done with a microbe, was based on an electronic engineering paradigm, and was funded by the Office of Naval Research. Associations with engineering departments, with electronic concepts instantiated in microbes, with elite universities, and with defense agency support were brand features in SynBio's early development, as discussed further below. The same associations remain strong to this day, although less in plant SynBio than in other areas. Also still strong is the early connection between SynBio and biomimetic chemistry, particularly in the branch of SynBio that seeks to redesign life from artificial building blocks (Benner and Sismour 2005). As artificial life-themed SynBio has little to do with plants we do not cover it in this review. The inception and progress of mainstream (i.e. microbe-centric and engineering-oriented) SynBio is well covered in many reviews, of which the following is an illustrative selection (de Lorenzo and Danchin 2008; de Lorenzo 2017; Erb 2019; Cameron et al. 2014; Clarke and Kitney 2020; Hanczyc 2020; Voigt 2020; Gallup et al. 2021; Volk et al. 2023).

Hallmark principles and vocabulary of SynBio

SynBio has some distinguishing concepts and language. The first of these features—inherent in all forward engineering—is SynBio's prescriptive nature; this sets it apart from traditional descriptive biology (Calladine and ter Meulen 2012; Kurth et al. 2020; Hanson and de Lorenzo 2023). Other features have, over the years, almost become shibboleths (in-group catch phrases) used to set SynBio apart from all other biology; these include the principles of decoupling, abstraction, standardization, modularity, and orthogonality, and their fruits—automation and the industrialization of biology in biofoundry facilities (Endy 2005; Lucks et al. 2008; Chao et al. 2017). See the review by Liu and Stewart (2015) for definitions and explanations of these principles and the review by Holowko et al. (2021) for explanations of what biofoundries are and how they scale up throughput. Also diagnostic of SynBio are the words "platform" and "chassis," which are metaphorical—and philosophically loaded—terms for the organism in which a SynBio project is executed (de Lorenzo 2011; Stewart et al. 2018; Voigt 2020).

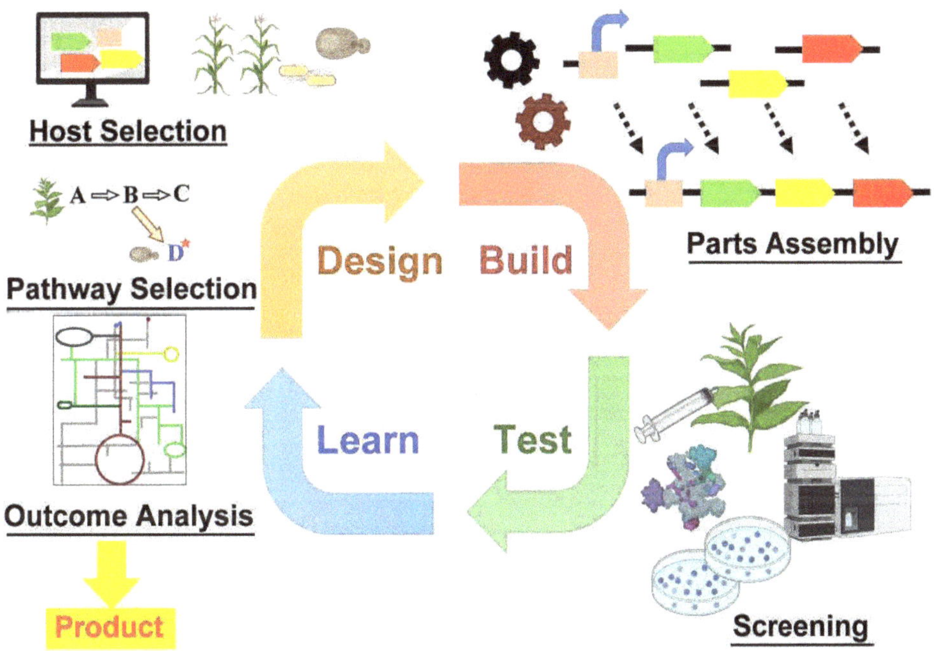

Figure 1. SynBio's iconic DBTL cycle. Note that the cycle outputs products, not just information and understanding.

Another SynBio hallmark is the design-build-test-learn (DBTL) cycle (Fig. 1). This iconic cycle conveys the idea that SynBio projects are iterative, i.e. that a design is first prototyped, then tested, and the test results inform further rounds of improvement until a final product exits the cycle (Poust et al. 2014). The idea of iterative cycles in design is not special to SynBio; it is in the ethos of all engineering (MacLeod 2012) and was explicitly adopted by metabolic engineering when it began (Bailey 1991).

What could also be considered a hallmark of SynBio is the porosity of its borders with other sectors of biology, most of all with metabolic engineering, but also with genetic engineering and, more recently, with genome editing. The continuum between metabolic engineering and metabolic SynBio projects is neatly captured in Fig. 1 of an excellent short review by Erb et al. (2017). This figure's message is that the more the enzyme reactions in a pathway and the pathway itself depart from nature, the more the project becomes SynBio, the culmination being new-to-nature pathways featuring novel reactions based on novel enzyme mechanisms. Plant-related examples of such enzymes and pathways are the late Arren Bar-Even's "daring metabolic designs" to replace Rubisco and the Calvin cycle with alternative reactions (Bar-Even 2018). Lastly, in practical terms, the continuum with metabolic engineering helped plant metabolic biochemists and engineers transition to SynBio (i.e. to "re-invent themselves"), as many of those working in mainline and specialized metabolism have done (Stewart et al. 2018).

Plant SynBio's take-off and (future) flight path

Oral history approach

As said above, SynBio's origin as a distinct field is usually dated to 2000 and began with gene circuits and then metabolic path-ways in microbes, some of which involved parts—genes and en-zymes—taken from plants (e.g. Chen and Weiss 2005; Ro et al. 2006). Plant SynBio sensu stricto (SynBio in which the target for improvement is a plant or photosynthetic bacterium, and what this review mainly covers) took off about a decade later, making it so young that it barely has any history under its belt.

Figure 2. The 8 founding figures in plant SynBio who were interviewed. Box 1 gives microsketches of their careers and contributions to the field (arranged in clockwise order, starting with Claudia Vickers at the top left).

Moreover, plant SynBio's flight path to real-world engineering impact is still largely unclear. Unlike most areas of microbial SynBio, including engineering high-value plant products in yeast or bacteria (Schempp et al. 2017) and engineering plant-associated bacteria (Ke et al. 2021), plant SynBio does not really even have a flight *plan* yet, much less a flight *path*. The nearness to the present and the fluidity of everything about plant SynBio sensu stricto plus its lack of tangible impacts so far led us to decide not to survey the field and attempt an analytical history but instead to give readers a sense of what the early days of Plant SynBio were like through the eyes of representative founders of the field, i.e. to offer a pilot oral history. That we are not historians and did not ourselves witness plant SynBio taxiing along the runway before takeoff also played into this de-cision. For surveys of current activities and frontiers in plant SynBio, we refer readers to special journal issues (e.g. Benning and Sweetlove 2016; Stewart et al. 2018; Hanson et al. 2019; Kumar and Van de Peer 2020; Patron and Burgess 2023; Yang and Reyna-Llorens 2023), and to the reviews in Supplemental Table S1.

We interviewed 8 founding figures who work in 5 of the countries that were among the first to see the potential of SynBio in general and plant SynBio in particular, and to put substantial funding programs in place. Figure 2 introduces these founders, and Box 1 sketches some of their key contributions to plant SynBio. The vacant central space in Fig. 2 symbolically honors the many other founders to whom the field owes much but whom we could not interview in the scope of this project. This applies particularly to plant researchers in Asia and other continents that were not among the very first to move into SynBio but that have contributed greatly to its subsequent flourishing. The interviews were one-on-one, via Zoom. We asked interviewees for their recollections and reflections on a standard set of topics, some of which were open-ended (Box 2 and Outstanding Questions Box). The interview recordings are available in Supplemental Appendix S1. We integrated the interview data into a set of timelines covering pivotal events, articles, and funding programs (Fig. 3 and Supplemental Table S2) and extracted key points from responses on the open-ended topics. Below, we discuss trends and highspots in the timelines and summarize key points from the responses. What else interviewees said on these and other points is in the recordings, which we encourage readers to visit; they are rich in experience, wisdom, and vision—and never boring! The timeline entries in Fig. 3 are partly based on consensus and are not attributed to particular founders; the text section on responses includes attributions. Lastly, a disclaimer: as an oral history based on the lived experiences and perspectives of a small number of people, this review could not be comprehensive or balanced in every aspect. We take full responsibility as authors for the review's shortcomings and ask readers to pardon these.

Timeline trends and highspots

Comparing the events and articles timelines with the funding timeline (Fig. 3) shows that the first 2 tend to have gaps in the

Box 1. Brief biodata for the interviewees

Dr Claudia Vickers has a background in plant molecular biology and engineering and has worked in research and development in universities, government, and industry. As Director of CSIRO's Synthetic Biology Future Science Platform from 2017 to 2022, she took a leading role in fostering development of SynBio in Australia and internationally, at scientific and strategic policy levels, serving as President of Synthetic Biology Australasia and as a founding member of the Global Biofoundries Alliance executive, and cochairing the WEF Synthetic Biology Global Future Council. Her recent research has been in developing biobased industrial products. She now leads her own venture, BioBuilt Solutions.

Dr Nicola Patron is a plant molecular biotechnologist and a national and international leader in plant SynBio. She directs the Earlham Biofoundry, one of the United Kingdom's first foundries and the only one specializing in plant SynBio workflows. She has taken leadership initiative to standardize DNA parts for the plant SynBio community. Her research applies SynBio approaches to understand how phenotypes emerge from network functions and to explore and utilize plant metabolic diversity. Through the creation of cutting-edge plant SynBio tools at the intersections of engineering, computing, biology, and design, she represents the future of SynBio and is an inspiration for diversity in STEM.

Dr Jennifer Nemhauser is a developmental biologist who builds SynBio tools to investigate and manipulate plant signaling dynamics, particularly those centered on auxin and other hormone signaling. Her ultimate goal is to uncover core principles that shape multicellular behaviors across eukaryotes. In addition to her scientific research and advocacy, her voice also reinforces the importance of inclusivity and diversity in the plant SynBio community and beyond. She has been a driving force in encouraging greater representation and equal opportunities for individuals from minoritized communities in STEM, as well as calling for major shifts in the culture of plant science toward equity-centered policies and actions.

Dr Vincent Martin has a microbiology background. He entered the nascent SynBio field as a postdoctoral at UC Berkeley in Jay Keasling's group, where he worked with a group of chemical engineers and cofounded Amyris Inc., which was established to commercialize production of the anti-malarial drug artemisinin in yeast and became a shining SynBio success. He went on to establish the first Synthetic Biology Center in Canada, equipped with a state-of-the-art Genome Foundry to accelerate the DBTL cycle. His research interest is in green biofactories for production of chemicals, pharmaceuticals, and next-generation therapeutics; he is currently working on natural products and organic acid synthesis in yeast.

Dr C. Neal Stewart, Jr is a plant biotechnologist (and songwriter!) with many accomplishments and patents, particularly in crop genetic engineering. He has successfully developed transformation technologies for diverse plants. He became a trailblazer and early adopter of SynBio in plant biotech and agriculture and established University of Tennessee as one of the first US Land-Grant Institutions to take a lead in the field. He currently codirects the University of Tennessee's Center for Agricultural Synthetic Biology. He is an active proponent of plants as SynBio chassis. His current research ranges from agricultural sustainability to bioenergy crops and plants as environmental sensors.

Dr June Medford has a deep background in plant molecular biology and genetic engineering, having been in the Monsanto group that first commercialized transgenics. A very early entrant into plant SynBio, she rose rapidly to become an acclaimed world leader and advocate for SynBio in the plant sciences. She developed some of the first Plant Sentinels, featuring computationally designed sensors and control circuits to detect explosives, hazardous chemicals, and environmental pollutants. Recently, she has worked on a plant synthetic desalination genetic circuit. She has been instrumental in shaping plant SynBio's vision and advancing plant SynBio in government and industrial policies.

Dr Birger Lindberg Møller is a plant biochemist, particularly in the specialized metabolism area. He was among the first to recognize SynBio's potential in plant science and biotechnology, and quickly became a SynBio pioneer and visionary advocate for plant SynBio in Denmark, Europe, and worldwide. He leads the University of Copenhagen's Center for Synthetic Biology, whose mission is to develop knowledge-based solutions for grand challenges such as energy supply, transitioning to green growth, and advanced medical solutions. His research emphasizes harnessing chloroplasts as powerhouses for light-driven synthesis of natural products and bulk compounds, thus reducing reliance on fossil carbon.

Dr Jim Haseloff has a background in engineering and plant biology. He initially pioneered artificial RNA enzymes with targeted substrate specificity. An early SynBio adopter and leading SynBio expert, he quickly moved to the forefront of applying SynBio approaches to plants. His contributions include pioneering new quantitative imaging techniques, genetic circuits for cell-to-cell communication, and advancing the liverwort *Marchantia polymorpha* as a simple model system for bioengineering, for which he has developed DNA tools. He works through the OpenPlant initiative and Biomaker project to provide platforms for accessible technologies, fostering interdisciplinary activities and bringing together biologists, engineers, and social scientists.

Box 2. Interview topics

Interviewees were asked for their recollections and reflections on:

- Key meetings, workshops, or other events in plant SynBio development.
- Key funding agencies and programs and the dates they started.
- Two or three landmark papers or reviews that impacted the field overall, or them personally.
- Researchers whom they saw as founders.
- The suitability of the electronics engineering paradigm.
- Where plant SynBio *is* going—and where it *ought to be* going.
- Meetings, government initiatives, and industry collaborations that will be taking place soon to support the growth of plant SynBio.

middle and that much of the activity in the funding timeline comes in that middle region. Given that the funding timeline refers to when agencies *first started* funding SynBio (Box 2), a pattern emerges: influential meetings and papers in the early 2000s (not on plants) sparked interest inside funding agencies and among a few plant scientists. After lags of various lengths, the agencies started funding programs that were first microbe-centric and then extended to plants and mammals, and some plant scientists became early adopters. Plant-centric SynBio meetings and key papers then followed, about a decade behind the field as a whole. The lag between SynBio's inception and the start of major funding was not-ably short among the US funding agencies, particularly the Office for Naval Research and the Defense Advanced Research Projects Agency (DARPA), i.e. defense agencies, and the Advanced Research Projects Agency-Energy. These agencies remain major players, and all fund sensu stricto plant SynBio research and development. Although intervie-wees did not comment on this, defense agency support for plant SynBio has not been without controversy. One instance is DARPA's 2016 Insect Allies program to use insect

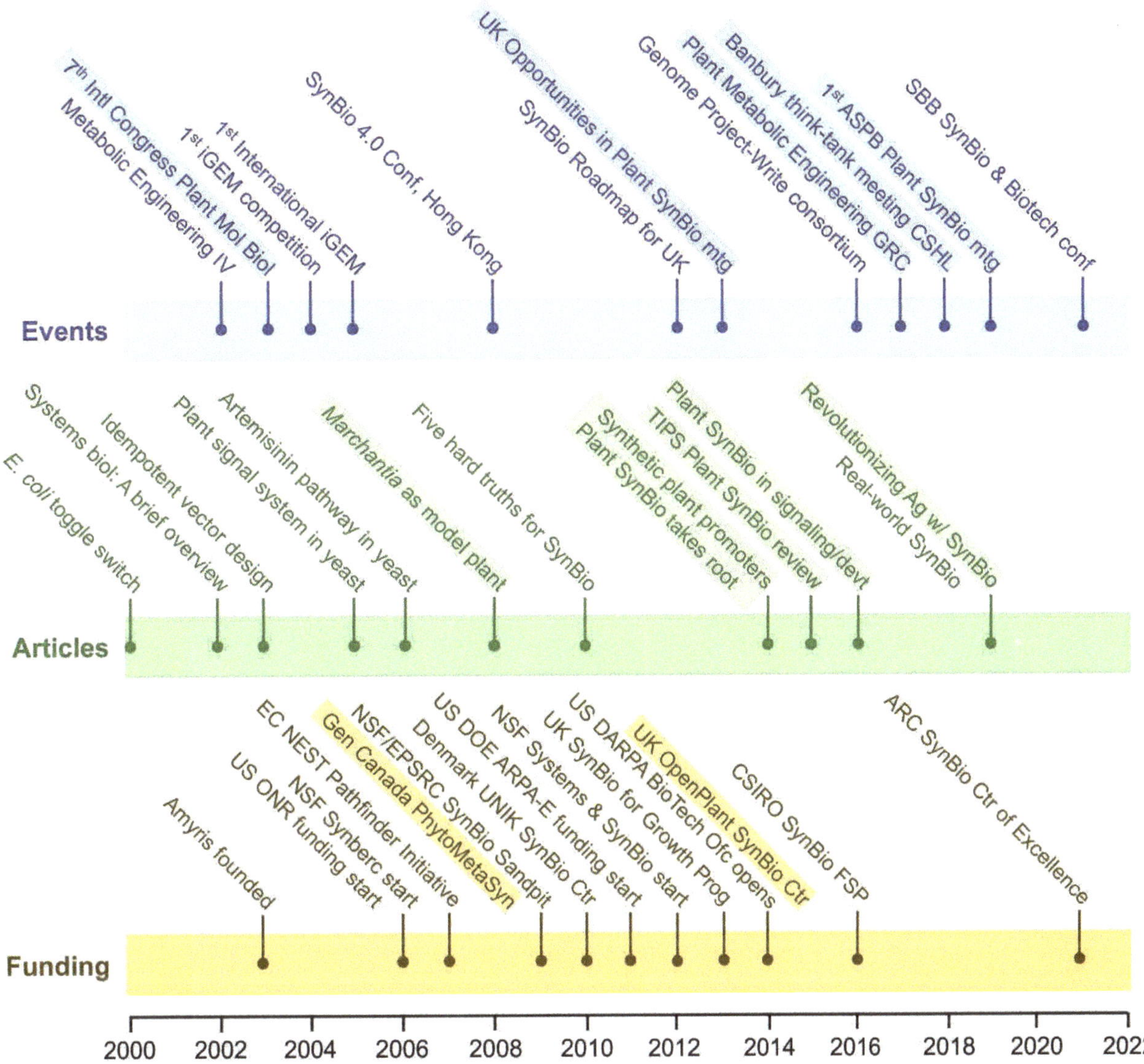

Figure 3. Timelines of selected meetings or other events, published articles, and funding programs that founders identified as pivotal in plant SynBio's evolution. Plant SynBio-specific entries are highlighted. That over half the entries are not from plant SynBio but from mainstream microbial SynBio reflects how much this mainstream inspired plant SynBio intellectually and nurtured it practically. Some dates are approximate. Timeline entries are documented in Supplemental Table S2.

dispersion of genetically modified viruses engineered to edit crop chromosomes in the field (Reeves et al. 2018).

A second trend evident in the timelines is that the earliest key SynBio papers and events, like the archetypal toggle switch paper (Gardner et al. 2000) and the first International Genetically Engineered Machine competition, came out of engineering departments in elite, globally networked institutions in the Northeastern United States and California; those identified in Fig. 3 are MIT, Princeton, and U.C. Berkeley; others include Harvard, Stanford, and Caltech. Although none of the earliest SynBio practitioners at these institutions were plant scientists, some became involved in plant-related work and helped inspire current and future plant scientists; MIT's Chris Voigt (e.g. Geddes et al. 2015) and Berkeley's Adam Arkin (Rubin et al. 2016) are examples. Much inspiration of this sort came out of the US National Science Foundation (NSF)-supported Synthetic Biology Engineering Research Center (SynBERC). This center, based at UC Berkeley and led by Jay Keasling (Synberc 2019), became a focal point for SynBio's spread into plant science.

A third trend concerns which countries outside the United States were nimblest in setting up government funding, research centers, and training programs for SynBio in general and plant SynBio in particular. In Europe, the U.K. and Denmark were particularly quick off the mark. Up to 2009, the United Kingdom was the only European country with an established SynBio funding scheme (Pei et al. 2012), and in 2009 the Danish Ministry of Research launched the UNIK SynBio initiative (Center for Synthetic Biology 2012). Further evidencing early leadership, in 2009 the U.K.

Engineering and Physical Sciences Research Council (EPSRC) held a SynBio "Sandpit" jointly with NSF to address grand challenges (NSF 2008). Plant SynBio was represented at the Sandpit. Max Planck Institutes in Germany were strong centers. Beyond Europe, Australia was quite quick to systematically foster SynBio research and development (CSIRO 2022).

The publication and event highspots that interviewees identified necessarily had individual flavors; not everyone went to the same meetings or read the same things. There was nonetheless a fair amount of consensus on 2 points. First, early SynBio-related articles now held up as icons such as the first and fifth from the left in Fig. 3 truly did change mindsets in the way their citation statistics suggest they did. Second, the sense of a like-minded community born and nurtured in what were often quite small meetings was critical to get plant SynBio out of the construction hangar and onto the runway.

Responses to open-ended questions

Researchers named by interviewees as founding figures

Supplemental Table S3 lists these researchers alphabetically. Europe, Australia, and Canada are strongly represented, particularly for plant SynBio as opposed to microbial SynBio, testifying to plant SynBio's international character. That founders of the whole SynBio field such as Frances Arnold, Chris Voigt, Jay Keasling, Adam Arkin, Tobi Erb, and Victor de Lorenzo appear in this list reflects not only their general influencer status via publications and talks, but also specific contacts and collaborations with several of the interviewees, in some cases via Synberc (see above).

Suitability of the electronics engineering paradigm

Like Victor de Lorenzo (de Lorenzo 2011), Antoine Danchin (Danchin 2012), and Jamie Davies (Davies 2019), all the foun-ders interviewed saw the reigning electronic engineering paradigm—"cells as circuit boards"—as a double-edged sword: useful but of limited applicability, and even conceptually dangerous. The lively individual responses (edited from notes and transcripts) were as follows.

Nicola Patron: The analogy to electrical engineering is really helpful to think about things in a new way. But beware of the analogy because it will fall down really quickly. If you flog this analogy to everything in biology, it's definitely going to break. This is not the idea. Plants are not electronic circuits.

Vincent Martin: I think this analogy has a role to play but we have to be careful of the limitations, e.g. a genetic switch built with a transcription factor and a transcription factor-binding site. The premise would be that that the binding factor only binds to that sequence, and only binds in a very defined way. This is theoretically on/off, but there's a lot of data now coming out showing that transcription factors bind in a lot of different places and they bind to varying degrees to show varying effects. So the idea of being able to have orthologous or isolated circuits in the long run is going to be very, very difficult to do. But that being said, there are still wonderful examples of how these approaches can be used to control metabolism and cell development. So theoretically, it seems that it could be done. Practically, there are examples of successes. Whether or not those successes can scale to more complex systems or more advanced processes, I don't know.

Claudia Vickers: It's a very useful framework as long as you recognize the limitations in the context of biological systems and the additional limitations in the context of plants. As long as you don't stand up and say this is how it works, we have transistors and AND/OR/NOT gates that are always reproduced for X, it's fine. It's what we're aiming for and is an objective to aim for as long as you recognize the challenges that are inherent in it and, in particular, as long as you understand the reasons behind why it's not that black-and-white, why it's not simple, then it's a very valuable paradigm. Because what you seek to achieve using those approaches is important, right? But if you pretend that it actually works that way and if you don't recognize the limitations, then you will fail.

Jim Haseloff: I kind-of agree with the paradigm at one level. If you're trying to put DNAs together, then having modular components that will be easy to string together, and where the relationships between gene, encoded protein, and target sequence are predictable, then there are a lot of useful principles that you can take from electrical engineering. But there are some people who take the electronics analogy to mean that you can deal with biological components as the equivalent of electronic components and draw circuits and make connections that way. You can probably do this at a local scale but, broadly speaking, I think the principles that build biological systems are very different to more-linear electronic systems. There are obviously examples of feedback and other network-type interactions in electronic systems. But *all* the interesting aspects of biological systems are based on networks and more social ways of interacting that couple into self-organization and self-repair. I think one has to recognize that the architecture of biological systems is fundamentally different to most human-engineered systems.

Jennifer Nemhauser: Metaphors are very powerful and it's also important to remember that they are metaphors. People who are trained as engineers really want biology to fit into that paradigm because it makes biology a lot less messy, which is definitely what engineers by training and by personality do. As biologists, our job is to hold complexity as a value to be comfortable with, to acknowledge when we're reducing something to something simple, that we're intentionally making assumptions. We can then remember those assumptions, and add back the complexity in a thoughtful way to challenge the conclusions we've come to. If someone working in plant SynBio finds a particular engineering paradigm useful for helping them think about how to approach their system—great. If not, they can try to find something else that feels more aligned.

Neal Stewart: I think it's a good example of a boilerplate to think about engineering plants, as circuit boards, as devices. But for real biological devices it kind-of breaks down, though. The idea here is that SynBio should be very useful for actually solving real-life problems, and the complexity of gene expression means that "cells as circuit boards" isn't.

June Medford: As a paradigm, it helps to move us forward if we first start out and say, hey, can we make these simple gates? And yes, I would totally agree with a lot of folks who are coming from metabolic engineering and say it's much more complex. Absolutely. But if you don't take it apart and put it in simple steps that your mind can understand, you're never going to get to more complex issues. It is a thinking paradigm to move us forward; eventually we're going to understand much more complex things like the influence of chromatin and epistasis that is so complex. We're going to get there, but we need steps to get there. You don't just go from "poof!" You have to take it in steps.

Birger Møller: This paradigm is helpful and unhelpful at the same time. It is a helpful conceptual frame—but biological units are not Lego bricks. Many pathways are organized in dynamic metabolons based on complex interactions between the modules. This serves to secure essential metabolic channeling. In this context, it is important to work with and learn from Nature. This is the way to build and scale up robust sustainable production systems for the future. Engineered regulatory circuits without context to Nature may work in the lab, but are not likely to be robust enough to work in field conditions.

Where plant SynBio is going—and where it ought to be going

Responses on these topics ranged from the grand strategic level through to operational and tactical levels (MacGregor 1992) and naturally tended to reflect each interviewee's domain expertise and interests. Where plant SynBio ought to be going (or the converse, where it is going but should not be) drew the most comments. The responses were pithy and thought-provoking, and converged on some common themes. Regarding where plant SynBio is currently going, there was a fair consensus that sustainability (particularly in agriculture) and climate change are major current directions. Regarding where it ought (or ought not) to be going, the idea that plant SynBio should "get real" or "come down to earth" came out in various ways, including statements that echo long-standing reservations (e.g. Good and Bell 1980) about the prospects for biofuels. A need for plant SynBio to break with entrenched, traditional ways of thinking and of doing things, and to look outward and engage with the world were other common threads. The responses below are preceded by "Is" or "Ought" in parentheses to make clear which category they refer to.

Nicola Patron: (Is) It is really exciting to see model-informed design of biological processes then getting them into the field to see how they work there happening, and to see trait engineering papers. (Ought) I did expect to see more fundamental biology taking in SynBio approaches.

Vincent Martin: (Is) Genomics is empowering enzyme discovery; the old model of "discover one enzyme → install in microbe → find product" is being overtaken, I don't think this is where it's going. (Ought) Where I think it is going is our capacity to do 2 things: to bring these things to scale if they're going to bring any value to anybody other than academic publications, but also our capacity to generate either natural or non-natural diversity at a much bigger scale than we can do now. We need to diversify the structures that we can produce. At present we are nowhere near matching nature's ability to do this. We need a combinatorial, high-throughput approach.

Claudia Vickers: (Is) Plant SynBio is going into climate change and sustainable agriculture. (Ought) We need to ask what impact SynBio can have that is *scalable* and will have impact at the planetary scale. Two areas that will be most impactful are in food and agriculture and there will be lots of plant SynBio applications that will be important for that. Examples are biobased construction materials and biobased binders. The holy grail of accelerating photosynthesis for carbon sequestration has proven to be very challenging. Plant SynBio should go where we can have planetary impact. There are many, many things, but the first thing you should do is OK, let's do an emissions analysis and a techno-economic analysis and determine is it scalable? Is it fast enough? And will it make a difference?

Jim Haseloff: (Is) "Plant blindness" is a problem. There is undue emphasis on microbial SynBio and new forms of bioproduction, and ignorance of the importance of plant systems and agriculture. I always find this quite curious because it's completely at odds with current practices in human culture, where agriculture is universal and cheap and has a huge capacity for production. Why is there so much emphasis on microbial-based systems, or in vitro-based systems, compared to the obvious importance, for the future, of plant-based SynBio and engineering biology? There is a bias in the plant world from the conservative (for good reason) agendas of large agribusiness companies that contrasts with startup culture for fermentations. Inari Agriculture and Afingen are exceptions. (Ought) For biofuels, even just simple life cycle analysis and economic analysis will tell you this is based on very doubtful premises. A lot of the early decision-making in this area was somewhat corrupted by the politics of bioethanol and energy security; there wasn't really any solid scientific basis to it. With AI-protein folding we can now begin to predict consequences of residue changes in terms of interactions. Up to now, we have confused our (in)ability to understand these systems with the ability to begin engineering them. (See quotes on engineering in the Introduction.) Engineering development via gene regulatory circuits is hard because the circuit must be robust to environmental insults. This could be achieved by integrating environmental inputs to compensate for changes. Neo-organogenesis (building new plant organs) can be combined with metabolic engineering to locate pathway products in a new organ. This can also be the basis for neo-domestication (creating new crops).

Jennifer Nemhauser: (Ought) We need an ethos and genuine commitment to open-source, to DIY, to coproduction of knowledge, to novices and experts having high-quality conversations to find solutions for the food and climate crisis. A plant SynBio Reddit board where researchers can have conversations like that, to build a community up organically? We also need a more inclusive approach to build relationships with indigenous communities, and to train more international communities to find solutions for the food and climate crisis. There's a huge amount of biases left in the grant-reviewing process—there's still more work to do to be done around the people who are reviewing grants, maybe more than the people who are managing the funding. A more cross-disciplinary approach is needed.

Neal Stewart: (Ought) Plant SynBio should focus more on the chloroplast genome, which is more prokaryote-like, more amenable to manipulation. "Kill" (neutralization) switches are needed for containment. We should distinguish between SynBio toys and useful tools. Systems biology is delivering an endless parts list without understanding of how the parts fit together, still less how to manipulate them. We're drowning in uncharacterized parts.

June Medford: (Is) Plant SynBio going towards sustainability of life on earth, we're going to engineer a much more sustainable future! (Ought) ASPB should take more leadership. We should bring in big agribusiness players, e.g. Cargill, Bayer, and Syngenta. We need to help grant program managers to understand our plant SynBio needs and challenges, e.g. that plant research needs more time, so five-year rather than three-year funding.

Birger Møller: (Ought) Biofuels are "not an attractive climate measure" according to the Danish Council on Climate Change, which now "advises against investing in them as a means of achieving the climate goals" (Klima 2023). Two drivers for biofuels enthusiasm are that importing biomass is a way to offshore energy demands while still keeping a green façade and that enzyme companies have a vested interest in biomass valorization. It is important to build a startup culture. Promising research areas to pursue are (i) light-driven production in algae of high-value compounds with the pathways being expressed in the plant powerhouse, the chloroplast; (ii) windmill or solar cell electrically powered CO_2 reduction to acetate and use of the acetate as a supplemental substrate for algal growth under light-limited conditions in environmentally contained algae farms offering dual production of bulk and high-value products (Sørensen et al. 2022); and (iii) exploiting multifunctional P450s as new ways to make high-value molecules hard to obtain by classical chemical synthesis, e.g. triptonide (Hansen et al. 2022) and ginkgolides (Forman et al. 2022).

Meetings, government initiatives, industry collaborations that will support plant SynBio

Although this topic elicited fewer specific responses than the others, there was general satisfaction at how plant SynBio is gaining traction on the world stage (captured in the first quote below). Two interviewees cited regulatory policies as choke points.

Vincent Martin: Right now, there's so much happening. SynBio meetings of various types are now mushrooming all over the world, SynBio is spreading out fast from the Northern California and Boston areas where it used to be concentrated. I love the stuff that's coming out of DTU in Denmark. There's a lot of interesting things out of Australia. SynBio has really taken off in Asia. It's just wonderful, investigators are coming from all over the place as opposed to the field being dominated by a few groups.

Claudia Vickers: The Organization for Economic Co-operation and Development (OECD) Global Forum on Technology in Paris in June 2023 ("Shaping our future at the tech frontier") featured SynBio; I contributed. We need to work on agile regulatory policy that can be updated as technology develops. We need to support startups, tech transfer, scaleup and to break through the issues of conflict-of-interest management, risk aversion, and intellectual property (IP) ownership that complicate getting technologies out of research institutions. We need dramatically more support for role models who demonstrate that such things are possible.

Neal Stewart: Government regulations are getting more precautionary, deeper—it's now a regulatory quagmire. Big Ag has been rather slow to embrace the power of SynBio, other than gene editing. Thus, industry is not in the lead (unlike it was for transgene technology), nor are government laboratories. It's therefore up to universities. More diversity in the funding pool is needed. DARPA still dominates. I wish we had a better name than "synthetic biology," but I don't know what to call it. I think there's a segment of the population who feel or have felt that biotechnology is maybe "playing God." I've never really thought about it in those terms, because I feel that we are simply rearranging some of God's creation, in a very minor way, actually. I think we're utilizing that and recombining it to a degree, to actually do something useful—which is what humans do. They innovate on top of innovation. Designing a "synthetic crop" of some sort might be what biologists in a hundred years end up doing. And here again, I don't think it's "playing God." I think it's just being creative as humans are creative.

Birger Møller: There's a lot of transfer of money to the plant field in Denmark and the Green Transition, whatever this means. Likewise, a few big players decide how the real moneys are spent. Now, SynBio is a buzzword, and everyone suddenly has been doing SynBio their entire life. We should really make sure that funds assigned to SynBio research are spent in the right way, so that we don't have too many mishaps and the technology misses and people start not trusting what is promised.

Take the SynBio tide at the flood

A palpable feeling of being on the edge of a paradigm shift in Thomas Kuhn's deep original sense (Kuhn 1962), and even on the edge of history, pervaded our interviews. The feeling was more intense in some interviews than others, but always pre-sent. The Kuhnian paradigm shift is to a new, *actionable* way to think about and do bioscience. The historical aspect is that putting actionable SynBio ideas and information *to work* in the bioeconomy, agriculture, and the environment can help with many facets of the "global polycrisis"

A

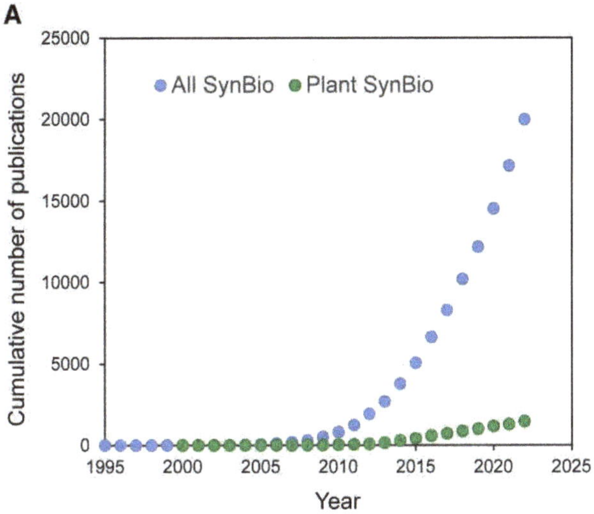

B

2022 Plant SynBio publications		%
Reviews and Perspectives	76	41.5
Bioprospecting and parts characterization	37	20.2
Tools (including modeling)	31	16.9
Plant parts in heterologous host	26	14.2
Engineering in plants	10	5.5
Plant enzyme engineering	3	1.6
Total	**183**	

Figure 4. Trends in publications in SynBio as a whole and in plant SynBio **(A)** and a breakdown of 2022 plant SynBio publications by category **(B)**. Plant SynBio publications in 2022 were 7.3% of the total. Data were extracted from PubMed using the "All Fields" setting and the search terms "synthetic biology" or "synthetic biology" plus "plant." The plant SynBio publications were curated manually.

(Lawrence et al. 2022). The same awareness of being on the brink of something big comes through in the burgeoning plant SynBio literature, which in 2022 reached 7% of total SynBio articles (Fig. 4). Of the 2022 plant SynBio publications, >40% were forward-looking reviews or perspectives; ~8% were original research articles on engineering plants or their enzymes. A 5-to-1 ratio of secondary to primary literature betokens a young field in the grip of a sense of promise.

We detected a second feeling, perhaps an anxious undercurrent, in the interviews. It was signaled by the length and strength of the responses about where plant SynBio *ought* and *ought not* to be going, and made manifest in Claudia Vickers' insistence that "Plant SynBio should go where we can have planetary impact…the first thing you should do is…an emissions analysis and a techno-economic analysis and determine is it scalable? Is it fast enough? And will it make a difference?" The undercurrent of feeling was basically that time is running out if plant SynBio is to fully realize its potential to help fix the planetary problems that will change so much about life in the next 30 years.

There was a somewhat similar sense in microbial SynBio a decade ago, expressed in a hard-hitting *Nature* article ("Five hard truths for synthetic biology"; Kwok 2010). This article

OUTSTANDING QUESTIONS BOX

- How can the science-to-engineering culture shift from descriptive, largely qualitative, curiosity-driven biology to prescriptive, quantitative, problem-solving SynBio be sped up and made easier?
- Which large-scale problems in agricultural, food, energy, and environmental systems are key targets for SynBio, and how can SynBio technology be deployed to deliver workable solutions?
- How do we open SynBio researchers' minds to the urgency of engaging with real-world problems while the chances of finding and being able to implement solutions are still optimal?

quoted SynBio guru George Church as saying: "The field has had its hype phase. Now it needs to deliver." A dozen years on, microbial SynBio still has plenty of hype (Hanson and de Lorenzo 2023), but it has real products too (Voigt 2020). "Hype" is too harsh a term for most current plant SynBio ideation. Maybe "distraction by misinformed popular narratives," or "blue skies thinking unconnected to practical needs," or "visions with a thousand-year timeline when we only have decades," or even "aspiring engineers learning numeracy on the job" would be nearer the mark. In any case, these lines from Shakespeare's play *Julius Caesar* express the tension in our interviews between scientific excitement on one hand and the great but time-limited opportunity to take up real-world engineering challenges on the other.

> *There is a tide in the affairs of [humans] which, taken at the flood, leads on to fortune;*
>
> *Omitted, all the voyage of their life is bound in shallows and in miseries.*
>
> *On such a full sea are we now afloat; And we must take the current when it serves,*
>
> *Or lose our ventures.*

Acknowledgments

We thank the interviewees for the effort that they put into the interviews themselves and into follow-up correspondence. We thank Dr Scott C. Lenaghan for the gift of Fig. 1, created with BioRender.com.

Author contributions

Both authors conducted interviews, analyzed data, and wrote the article.

Supplemental data

The following materials are available in the online version of this article.

Supplemental Table S1. Recent reviews of plant SynBio.
Supplemental Table S2. Documentation of the entries in Fig. 3 plus entries that were not included.

Supplemental Table S3. Researchers named by interviewees as founding figures.

Supplemental Appendix S1. Recordings of the 8 interviews.

Funding

This work was supported primarily by the U.S. Department of Energy, Office of Science, Basic Energy Sciences under Award DE-SC0020153, the USDA National Institute of Food and Agriculture (Hatch project FLA-HOS-005796), and an Endowment from the C.V. Griffin, Sr. Foundation.

Conflict of interest statement. None declared.

Data availability

The data underlying this article are available in the article and in its online supplementary material.

References

Bailey JE. Toward a science of metabolic engineering. Science. 1991:**252**(5013):1668–1675. https://doi.org/10.1126/science.2047876

Bailey JE. Chemical engineering of cellular processes. Chem Eng Sci. 1995:**50**:4091–4108. https://doi.org/10.1016/0009-2509(96)81834-3

Bar-Even A. Daring metabolic designs for enhanced plant carbon fixation. Plant Sci. 2018:**273**:71–83. https://doi.org/10.1016/j.plantsci.2017.12.007

Benner S, Sismour A. Synthetic biology. Nat Rev Genet. 2005:**6**(7):533–543. https://doi.org/10.1038/nrg1637

Benning C, Sweetlove L. Synthetic biology for basic and applied plant research. Plant J. 2016:**87**(1):3–4. https://doi.org/10.1111/tpj.13245

Calladine AM, ter Meulen R. Synthetic biology. In: **Chadwick R**, editor. Encyclopedia of applied ethics. 2nd ed. Salt Lake City (UT): Academic Press; 2012. p. 281–288.

Cameron DE, Bashor CJ, Collins JJ. A brief history of synthetic biology. Nat Rev Microbiol. 2014:**12**(5):381–390. https://doi.org/10.1038/nrmicro3239

Center for Synthetic Biology. 2012 [accessed 2023 Jul 22]. https://synbio.ku.dk/research-presentation/unik-synthetic-biology/.

Chao R, Mishra S, Si T, Zhao H. Engineering biological systems using automated biofoundries. Metab Eng. 2017:**42**:98–108. https://doi.org/10.1016/j.ymben.2017.06.003

Chen MT, Weiss R. Artificial cell-cell communication in yeast *Saccharomyces cerevisiae* using signaling elements from *Arabidopsis thaliana*. Nat Biotechnol. 2005:**23**(12):1551–1555. https://doi.org/10.1038/nbt1162

Clarke L, Kitney R. Developing synthetic biology for industrial biotechnology applications. Biochem Soc Trans. 2020:**48**(1):113–122. https://doi.org/10.1042/BST20190349

CSIRO. 2022 [accessed 2023 Jul 22]. https://research.csiro.au/synthetic-biology-fsp/.

Danchin A. Scaling up synthetic biology: do not forget the chassis. FEBS Lett. 2012:**586**(15):2129–2137. https://doi.org/10.1016/j.febslet.2011.12.024

Davies JA. Real-world synthetic biology: is it founded on an engineering approach, and should it be? Life (Basel). 2019:**9**(1):6. https://doi.org/10.3390/life9010006

de Lorenzo V. Beware of metaphors: chasses and orthogonality in synthetic biology. Bioeng Bugs. 2011:**2**(1):3–7. https://doi.org/10.4161/bbug.2.1.13388

de Lorenzo V. Synthetic microbiology: from analogy to methodology. Microb Biotechnol. 2017:**10**(5):1264–1266. https://doi.org/10.1111/1751-7915.12786

de Lorenzo V, Danchin A. Synthetic biology: discovering new worlds and new words. EMBO Rep. 2008:**9**(9):822–827. https://doi.org/10.1038/embor.2008.159

Endy D. Foundations for engineering biology. Nature. 2005:**438**(7067):449–453. https://doi.org/10.1038/nature04342

Erb TJ. Back to the future: why we need enzymology to build a synthetic metabolism of the future. Beilstein J Org Chem. 2019:**15**:551–557. https://doi.org/10.3762/bjoc.15.49

Erb TJ, Jones PR, Bar-Even A. Synthetic metabolism: metabolic engineering meets enzyme design. Curr Opin Chem Biol. 2017:**37**:56–62. https://doi.org/10.1016/j.cbpa.2016.12.023

Forman V, Luo D, Geu-Flores F, Lemcke R, Nelson DR, Kampranis SC, Staerk D, Møller BL, Pateraki I. A gene cluster in *Ginkgo biloba* encodes unique multifunctional cytochrome P450s that initiate ginkgolide biosynthesis. Nat Commun. 2022:**13**(1):5143. https://doi.org/10.1038/s41467-022-32879-9

Gallup O, Ming H, Ellis T. Ten future challenges for synthetic biology. Eng Biol. 2021:**5**(3):51–59. https://doi.org/10.1049/enb2.12011

Gardner TS, Cantor CR, Collins JJ. Construction of a genetic toggle switch in *Escherichia coli*. Nature. 2000:**403**(6767):339–342. https://doi.org/10.1038/35002131

Geddes BA, Ryu MH, Mus F, Garcia Costas A, Peters JW, Voigt CA, Poole P. Use of plant colonizing bacteria as chassis for transfer of N_2-fixation to cereals. Curr Opin Biotechnol. 2015:**32**:216–222. https://doi.org/10.1016/j.copbio.2015.01.004

Good NE, Bell DH. Photosynthesis, plant productivity, and crop yield. In: **Carlson PS**, editor. The biology of crop productivity. New York: Academic Press; 1980. p. 3–51.

Hammack WS, Anderson JL. Working in the penumbra of understanding. Issues in Science and Technology [2022 February 16]; 2022 [accessed 2023 Jul 20]. https://issues.org/penumbra-engineering-perspective-hammack-anderson/.

Hanczyc MM. Engineering life: a review of synthetic biology. Artif Life. 2020:**26**(2):260–273. https://doi.org/10.1162/artl_a_00318

Hansen NL, Kjaerulff L, Heck QK, Forman V, Staerk D, Møller BL, Andersen-Ranberg J. *Tripterygium wilfordii* cytochrome P450s catalyze the methyl shift and epoxidations in the biosynthesis of triptonide. Nat Commun. 2022:**13**:5011. https://doi.org/10.1038/s41467-022-32667-5

Hanson AD, de Lorenzo V. Synthetic biology—high time to deliver? ACS Synth Biol. 2023:**12**(6):1579–1582. https://doi.org/10.1021/acssynbio.3c00238

Hanson AD, Hibberd JM, Koffas MAG, Kopka J, Wurtzel ET. Focus issue editorial: synthetic biology. Plant Physiol. 2019:**179**(3):772–774. https://doi.org/10.1104/pp.19.00074

Hirschi S, Ward TR, Meier WP, Müller DJ, Fotiadis D. Synthetic biology: bottom-up assembly of molecular systems. Chem Rev. 2022:**122**(21):16294–16328. https://doi.org/10.1021/acs.chemrev.2c00339

Holowko MB, Frow EK, Reid JC, Rourke M, Vickers CE. Building a biofoundry. Synth Biol. 2021:**6**(1):p.ysaa026. https://doi.org/10.1093/synbio/ysaa026

Ke J, Wang B, Yoshikuni Y. Microbiome engineering: synthetic biology of plant-associated microbiomes in sustainable agriculture. Trends Biotechnol. 2021:**39**(3):244–261. https://doi.org/10.1016/j.tibtech.2020.07.008

Klima. 2023 [accessed 2023 Jul 24]. https://www.altinget.dk/klima/artikel/klimaraadet-advarer-mod-yderligere-satsning-paa-biobraendstoffer-ikke-et-attraktivt-klimatiltag (in Danish).

Kuhn TS. The structure of scientific revolutions, 3rd ed. Chicago (IL): University of Chicago Press; 1962.

Kumar D, Van de Peer Y. Plant synthetic biology. Curr Plant Biol. 2020:**24**:100186. https://doi.org/10.1016/j.cpb.2020.100186

Kurth MH, Ali R, Bridges T, Suedel BC, Linkov I. Evaluating resilience co-benefits of engineering with Nature® projects. Front Ecol Evol. 2020:**8**:149. https://doi.org/10.3389/fevo.2020.00149

Kwok R. Five hard truths for synthetic biology. Nature. 2010:**463**(7279): 288–291. https://doi.org/10.1038/463288a

Lawrence M, Janzwood S, Homer-Dixon T. What is a global polycrisis? Version #2. Cascade Institute 2022, No. 2022-4; 2022 [accessed 2023 Jul 24]. https://cascadeinstitute.org/technical-paper/what-is-a-global-polycrisis/.

Liu W, Stewart CN. Plant synthetic biology. Trends Plant Sci. 2015:**20**(5):309–317. https://doi.org/10.1016/j.tplants.2015.02.004

Lucks JB, Qi L, Whitaker WR, Arkin AP. Toward scalable parts families for predictable design of biological circuits. Curr Opin Microbiol. 2008:**11**(6):567–573. https://doi.org/10.1016/j.mib.2008.10.002

MacGregor DA. Future battle: the merging levels of war. Parameters. 1992:**22**:1992–1993. https://apps.dtic.mil/sti/pdfs/ADA528099.pdf

MacLeod IA. The ethos of professional engineering. Journal IESIS, Paper 1665, 154; 2012.

NSF. 2008 [accessed 2023 Jul 22]. https://www.nsf.gov/pubs/2009/nsf09012/nsf09012.jsp?org=NSF.

Patron NJ, Burgess SJ. Editorial overview: engineering plants and plant products for a green bioeconomy. Curr Opin Plant Biol. 2023:**71**: 102346. https://doi.org/10.1016/j.pbi.2023.102346

Pei L, Gaisser S, Schmidt M. Synthetic biology in the view of European public funding organisations. Public Underst Sci. 2012:**21**:149–162. https://doi.org/10.1177/0963662510393624

Poust S, Hagen A, Katz L, Keasling JD. Narrowing the gap between the promise and reality of polyketide synthases as a synthetic biology platform. Curr Opin Biotechnol. 2014:**30**:32–39. https://doi.org/10.1016/j.copbio.2014.04.011

Reeves RG, Voeneky S, Caetano-Anollés D, Beck F, Boëte C. Agricultural research, or a new bioweapon system? Science. 2018:**362**(6410):35–37. https://doi.org/10.1126/science.aat7664

Ro DK, Paradise EM, Ouellet M, Fisher KJ, Newman KL, Ndungu JM, Ho KA, Eachus RA, Ham TS, Kirby J, et al. Production of the antimalarial drug precursor artemisinic acid in engineered yeast. Nature. 2006:**440**(7086):940–943. https://doi.org/10.1038/nature04640

Rubin BE, Wetmore KM, Price MN, Diamond S, Shultzaberger RK, Welkie DG, Simkovsky R, Ota M, Lowe LC, Curtin G, et al. Massive mutant screens to develop a photosynthetic bioproduction platform. In Vitro Cell Dev Biol Plant. 2016:**52**:441.

Schempp FM, Drummond L, Buchhaupt M, Schrader J. Microbial cell factories for the production of terpenoid flavor and fragrance compounds. J Agric Food Chem. 2017:**66**:2247–2258. https://doi.org/10.1021/acs.jafc.7b00473

Sørensen M, Andersen-Ranberg J, Hankamer B, Møller BL. Circular biomanufacturing through harvesting solar energy and CO_2. Trends Plant Sci. 2022:**27**(7):655–673. https://doi.org/10.1016/j.tplants.2022.03.001

Stewart CN, Patron N, Hanson AD, Jez JM. Plant metabolic engineering in the synthetic biology era: plant chassis selection. Plant Cell Rep. 2018:**37**(10):135–1358. https://doi.org/10.1007/s00299-018-2342-1

Synberc. 2019 [accessed 2023 Jul 23]. https://ebrc.org/wp-content/uploads/2019/07/Synberc-10-years-book.pdf.

Voigt CA. Synthetic biology 2020–2030: six commercially-available products that are changing our world. Nat Commun. 2020:**11**(1):6379. https://doi.org/10.1038/s41467-020-20122-2

Volk MJ, Tran VG, Tan SI, Mishra S, Fatma Z, Boob A, Li H, Xue P, Martin TA, Zhao H. Metabolic engineering: methodologies and applications. Chem Rev. 2023:**123**(9):5521–5570. https://doi.org/10.1021/acs.chemrev.2c00403

Yang JS, Reyna-Llorens I. Plant synthetic biology: exploring the frontiers of sustainable agriculture and fundamental plant biology. J Exp Bot. 2023:**74**:3787–3790. https://doi.org/10.1093/jxb/erad220

https://doi.org/10.1093/plphys/kiad596 PLANT PHYSIOLOGY 2024: 195: 48–66

Plant Physiology®

Topical Review

A century of studying plant secondary metabolism—From "what?" to "where, how, and why?"

Richard A. Dixon [1,*] **and Alexandra Jazz Dickinson** [2]

1 BioDiscovery Institute and Department of Biological Sciences, University of North Texas, Denton, TX 76203, USA
2 Department of Cell and Developmental Biology, University of California at San Diego, La Jolla, CA 92093, USA

*Author for correspondence: richard.dixon@unt.edu
R.A.D. and A.J.D. contributed equally to this review.
The author responsible for distribution of materials integral to the findings presented in this article in accordance with the policy described in the Instructions for Authors (https://academic.oup.com/plphys/pages/General-Instructions) is: Richard A. Dixon.

Abstract

Over the past century, early advances in understanding the identity of the chemicals that collectively form a living plant have led scientists to deeper investigations exploring *where* these molecules localize, *how* they are made, and *why* they are synthesized in the first place. Many small molecules are specific to the plant kingdom and have been termed plant secondary metabolites, despite the fact that they can play primary and essential roles in plant structure, development, and response to the environment. The past 100 yr have witnessed elucidation of the structure, function, localization, and biosynthesis of selected plant secondary metabolites. Nevertheless, many mysteries remain about the vast diversity of chemicals produced by plants and their roles in plant biology. From early work characterizing unpurified plant extracts, to modern integration of 'omics technology to discover genes in metabolite biosynthesis and perception, research in plant (bio)chemistry has produced knowledge with substantial benefits for society, including human medicine and agricultural biotechnology. Here, we review the history of this work and offer suggestions for future areas of exploration. We also highlight some of the recently developed technologies that are leading to ongoing research advances.

Introduction

A hundred years ago, scientists were trying to understand the fundamental question, "What makes a plant?" Since then, we have learned a great deal about the chemical constituents of plants and their functions. But, there is still much exploration needed to answer this question; tens of thousands of plant chemicals (known as secondary metabolites, phytochemicals, or natural products) remain to be identified, and many mysteries remain even in some of the most well-studied organisms and pathways.

The term "secondary metabolite" suffers from the implication that the chemical in question is of secondary importance, and the term "specialized metabolite" is sometimes preferred. Even this may have negative connotations, as specialization can be taken to mean "not of general interest," as seen by the relative numbers of papers on plant secondary metabolism compared with, say, plant development, in high-impact general interest journals over the years. This situation is changing, and here, we hope to further explode the myth by pointing out the major advances that the study of plant secondary metabolism has made in the areas of phylogenetics, genome evolution, basic biochemistry, plant defense, ecology, and plant development. Important principles of biology have indeed been learned by the study of plant secondary metabolism (some are highlighted in Fig. 1).

The difference between our current approach to plant secondary metabolism and the work done a century ago is largely a question of techniques and ways of thinking. We therefore also highlight the major technological and

Received July 10, 2023. Accepted August 15, 2023. Advance access publication January 2, 2024

ADVANCES

- High-throughput gene sequencing, MS, and AI technologies are helping to assemble the complement of secondary metabolites across the plant kingdom.
- Studies in plant–insect ecology highlight the evolutionary flexibility of secondary metabolite biosynthesis and the roles of secondary metabolites in multitrophic interactions.
- Coupling genome mining and metabolomics with multigene transient expression systems provide a pipeline for gene discovery in complex pathways.
- The discovery of plant metabolite gene clusters is a major advance in understanding the biosynthesis and evolution of secondary metabolites and plant genome evolution.
- Single-cell "omics" and small molecule biosensors hold future promise in determining the roles of specialized metabolites as signal compounds.

intellectual innovations over the past 100 yr. We discuss the examples of particular interest and offer our predictions for the future of the field. This is a vast area, and our selection of examples cannot encompass all of the important work in this topic. Therefore, we invite interested readers to continue to explore this broad and exciting topic beyond this review.

The era of chemistry and function

A key driver of early phytochemistry was understanding the compounds that gave medicinal plants their healing abilities. As a classic review of phytochemistry states, "The desire to discover the healing virtues of plants or again, to extract deadly poisons from them, is one of the oldest endeavors of mankind" (Burrell 1937). By the early 20th century, organic compounds from plants were generally classified by whether they were commonly found in many plant species (today, many of these compounds could be described as primary metabolites) or whether they were specific to certain plants. Compounds identified as specific to medicinal or poisonous plants were treated with particular interest, and chemical characterization of these types of compounds led to some of the first plant-derived medicinal compounds: aspirin (salicylic acid), morphine, digitoxin, and quinine (Leroux 1830; Fisch 1985; Schmitz 1985; Achan et al. 2011). As characteriza-tion of plant metabolites has improved, subsequent research has shown that many "specific" compounds were in fact much more broadly distributed in plants than previously rea-lized. Salicylic acid is an excellent example of this—originally identified as the medicinal component in willow (*Salix* spp) bark, we know now that salicylic acid is an important defense hormone found across the plant kingdom (Peng et al. 2021).

Early attempts to identify the medicinal components of plants led to the discovery and characterization of many secondary metabolites, including alkaloids, glucosides, anthocyanins, carotenoids, and terpenes (Burrell 1937). Bioactivity assays were frequently used to determine which plants made compounds with "desirable" effects. These assays would apply plant extracts to tissues, animals, or microbes of interest. If favorable biological changes were induced by the extracts, then natural product chemistry techniques would be used to isolate the causative plant metabolite. These assays were frequently integrated with chemical synthesis to identify and refine compounds of interest. For instance, the discovery of aspirin involved treating feverish patients with extracts of willow bark, which had high levels of salicin, a compound convertible to salicylic acid (Montinari et al. 2019; Mahdi et al. 2006). Chemical structure confirmation was achieved by comparing the natural compound with synthesized salicylic acid. Further synthetic modifications resulted in acetylsalicylic acid, the water-soluble compound used in the commercial formulation for aspirin. Over several decades, synthetic analogs of acetylsalicylic acid were generated to create a number of other nonsteroidal anti-inflammatory drugs, including ibuprofen. This integrated natural product and synthetic chemistry approach to medicine has served as a template for the discovery and creation of a wide range of drugs ranging from chemotherapeutics to antibiotics.

Studying plant chemistry in the context of animals and insects became an expanding theme that gave rise to important discoveries in understanding the functions of plant metabolites. Starting in the 1960s, the work of Tony Swain, Jeffrey Harborne, and later Thomas Hartmann and Tom Mabry, among others, led to the rebirth of chemical ecology (Hartmann 2008) and chemotaxonomy (Bate-Smith 1962). Swain, along with Bate-Smith, Haslam, and White, coined the term "polyphenol," and the technique of paper chromatography allowed polyphenols to be used as markers in broad taxonomic studies (Swain 1985). By the mid-20th century, explorations in the field of chemical ecology were leading to the discovery of roles of plant metabolites in regulating insect behavior and biology (Dethier 1941). The 1940s to 1960s also saw the birth of our understanding of inducible (phytoalexin) as opposed to static (phytoanticipin) antimicrobial chemical defenses in plants (Cruickshank and Perrin 1961; Tomiyama et al. 1968; VanEtten et al. 1994), followed by appreciation of the role of compounds such as flavonoids and (later) sinapate esters as protectants against abiotic stress (UV light) (Markham and Mabry 1975; Li et al. 1993; Sheahan 1996).

The desire to understand what controls plant growth and development led to the identification of other groups of bioactive plant secondary metabolites. Although compounds now recognized as plant hormones are believed to be distributed across the plant kingdom, several are structurally related to other secondary metabolites with more specialized distributions and different functions. Thus, auxin is in the same indole family as the Arabidopsis (*Arabidopsis thaliana*) phytoalexin camalexin, gibberellins are diterpenes, abscisic acid is derived from carotenoids, and brassinosteroids are triterpenes. Certain flavonoids

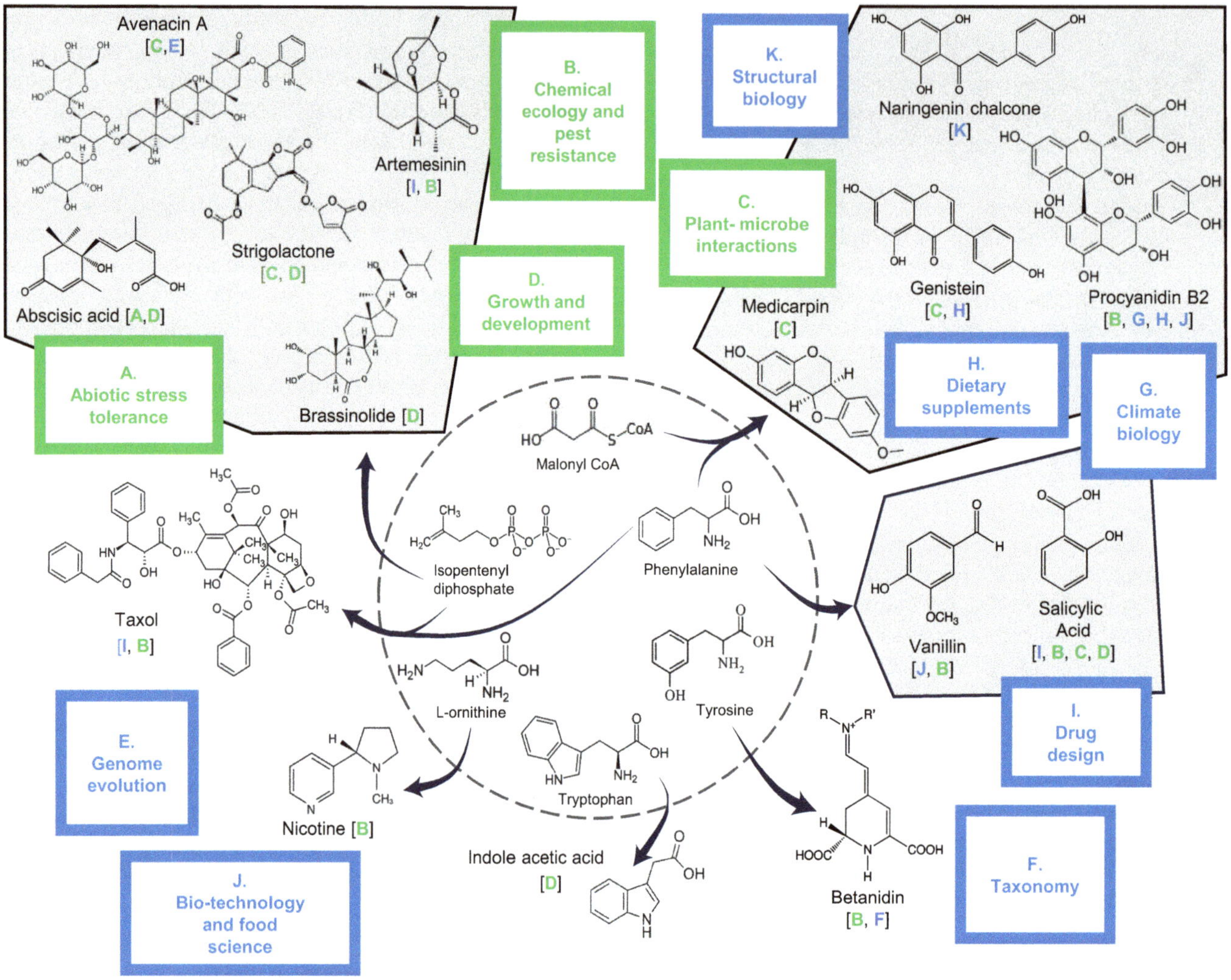

Figure 1. Structures of a selection of plant secondary metabolites, showing their biosynthetic origins, in planta functions, and uses or importance in plant biology. The bracket under each compound indicates the in planta function (green letter referring to a green box) and potential use or area of biology that the study of the compound has impacted (blue letter referring to a blue box). Arrows link precursors from primary metabolism (central circle) to the end products. Salicylic acid can be derived from phenylalanine or isochorismate (shikimate pathway intermediate).

function as inhibitors of auxin transport (Jacobs and Rubery 1988), and a coniferyl alcohol dimer (lignan) was proposed to act as a cytokinin, although this has recently been challenged (Witvrouw et al. 2023). Here, we consider "plant hormones" only from the perspective of where their study provides broader insight into secondary metabolism or has led to the develop-ment of innovative techniques for studying plant metabolism.

Figure 1 summarizes the biological activities of a selection of well-studied plant secondary metabolites and also indicates their usefulness to humans and, where appropriate, to the study of biology in general.

The era of biosynthesis

Model systems for the study of plant secondary metabolism

By their nature and with some exceptions, most plant secondary metabolites are not widely present throughout the plant kingdom. There are also limits to distribution at the level of chemical classes, with, for example, betalains replacing anthocyanins in the Caryophyllales and isoflavonoids being generally, but not exclusively, restricted to the Leguminosae. Multiple model species have therefore been developed to study plant secondary metabolism, especially at the molecular genetic level. Early studies prior to the application of genetic approaches targeted species irrespective of their genetic complexity or tractability, but work soon began to focus on model species with advanced genetic features such as easy transformability and availability of mutant collections. Unfortunately, Arabidopsis, although possessing many of the conserved secondary metabolic pathways such as phenylpropanoid and triterpene, is not always a good model for studying other aspects of metabolism/ecology. For example, although our un-derstanding of the biosynthesis of flavonoid-derived condensed tannins was accelerated through study of the set of Arabidopsis *transparent testa* mutants (Appelhagen et al. 2014), Arabidopsis has since been shown to lack several of the enzymatic steps involved in tannin biosynthesis in other species (Yu et al. 2023). Madagascar periwinkle (*Catharanthus roseus*) has been and continues to be widely studied as a model plant for alkaloid bio-synthesis (Kulagina et al. 2022), and the wild tobacco *Nicotiana alata*, with facile transformation and the existence of distinct field populations, has been an important model for studies in chemical ecology (Kumar et al. 2014). Barrel medic (*Medicago truncatula*), with highly developed genetic resources (Tadege et al. 2008), has served as one model for (iso)flavonoid and tri-terpene biosynthesis (Liu et al. 2018; Ribeiro et al. 2020). With ever-increasing improvements in genomic/transcriptomic re-sources (One Thousand Plant Transcriptomes Initiative 2019) and synthetic biology and rapid transient expression technolo-gies (Owen et al. 2017), plant transformation efficiency is now the major hurdle to investigations of secondary metabolism across the broad diversity of plant species.

Unraveling the biosynthesis of secondary metabolites

Phase 1: Chemistry informs biochemistry. Early work on the biosynthesis of plant secondary metabolites relied heavily on the principles of organic chemistry. The pioneering work of Robinson was among the first to contemplate a role for enzymes in the assembly of complex plant metabolites (Robinson 1917), leading to a Nobel Prize in Chemistry in 1947. Chemical synthesis or semisynthesis was used as a guide to how these molecules might be assembled in the plant under enzymatic control. For example, the chemical assembly of condensed tannins through nucleophilic addition via a carbocation intermediate was proposed as the basis for their biosynthesis from flavonoid precursors (Haslam 1974; Roux and Ferreira 1982). Biosynthetic pathways were also interrogated through labeling experiments, usually utilizing ^{14}C-labeled precursors which had to be custom synthesized. Examples include pioneering studies on the biosynthesis of lignin from L-phenylalanine and coniferyl alcohol (Freudenberg 1965), of vanillin from ferulic acid (Zenk 1965) and of benzylisoquinoline alkaloids from tyrosine (Rueffer and Zenk 1987). At the same time, in vitro assays were being developed to examine whether crude extracts from plants might contain predicted enzyme activities for involvement in the synthesis of a range of different secondary metabolites (e.g. Koukol and Conn 1961; Britsch and Grisebach 1986; Kochs and Grisebach 1986, e.g. in the phenylpropanoid/flavonoid pathway), and pathways were "assembled" based on the ordering of such activities. Figure 1 gives a brief indication of the biosynthetic origins (precursors from primary metabolism) of a selection of secondary metabolites, and Fig. 2 provides a historical timeline for the identification and characterization of selected phenylpropanoids, flavonoids, alkaloids, and their biosynthetic pathways.

Phase 2: Molecular biology enters the race. Large strides in our understanding of the molecular basis of plant secondary metabolism were made in the 1980s following the development of inducible systems such as cell cultures or biotically/abiotically stressed tissues (Hadwiger and Schwochau 1971; Kreuzaler and Hahlbrock 1973; Lawton et al. 1980; Zenk 1991) to which the tools of the emerging discipline of molecular biology could be applied. A comparison of the timelines for discoveries (Fig. 2) and for the adoption of techniques to the study of plant metabolism (Fig. 3) shows the enormous impact of molecular biology on the field. Once a reliable assay had been developed for an enzyme that was induced in parallel with an associated metabolite, it was then possible to purify the enzyme to homogeneity, sequence peptides derived from the enzyme protein, design oligonucleotide probes corresponding to the protein sequence, and screen cDNA or genomic libraries to identify the corresponding gene and provide DNA probes to enable the detection of transcripts on RNA blots (e.g. Kreuzaler et al. 1983; Edwards et al. 1985). In some cases, plant genes were cloned based on their homology to conserved regions in genes from mammalian species (e.g. Learned and Fink 1989).

Experiments, in which a particular metabolite or sets of metabolites were induced as a result of application of a chemical, elicitor molecule, or a light stimulus, were then designed to address the question of whether the induction involved de novo gene expression (Hahlbrock and Grisebach 1979; Hahlbrock and Scheel 1989). Earlier studies on environ-mentally controlled or elicitor-induced plant metabolism had relied on cumbersome techniques such as density label-ing with deuterated water to determine whether enzymes were being synthesized de novo (Filner and Varner 1967; Lawton et al. 1980). Now, antibodies could be used as re-agents for in vitro translation assays to determine whether translatable mRNA activities were being increased as a result of the chemical, environmental, or developmental stimulus (e.g. Kreuzaler et al. 1983), and nuclear run-off experiments could address control at the level of transcription itself (Somssich et al. 1989). The cDNAs could also be used as probes for in situ hybridization experiments to address the cellular localization of the transcripts and, by extension, the pathway (Schmelzer et al. 1988; St Pierre et al. 1999). Together, this work led to the appreciation that plant secondary metabolism was under complex spatial and temporal control, often regulated primarily at the level of gene expression (Hahlbrock and Scheel 1989). For example, transcriptional regulation at the levels of L-phenylalanine ammonia-lyase (PAL) (Edwards et al. 1985), chalcone syn-thase (CHS) (Bell et al. 1984), and 3-hydroxymethylglutaryl coenzyme A reductase (HMGR) (Choi et al. 1992) were proposed as contributing to control of phenylpropanoid, flavonoid, and terpenoid biosynthesis, respectively. It also became apparent, by comparison of levels of transcripts and their corresponding proteins, that posttranslational control could be important, particularly for key entrypoint enzymes into secondary metabolism such as HMGR (Korth et al. 2000).

Phase 3: "Omics" changes the game. Between the mid-1980s and the mid-1990s, many of the key enzymes in the biosynthesis of different classes of plant secondary

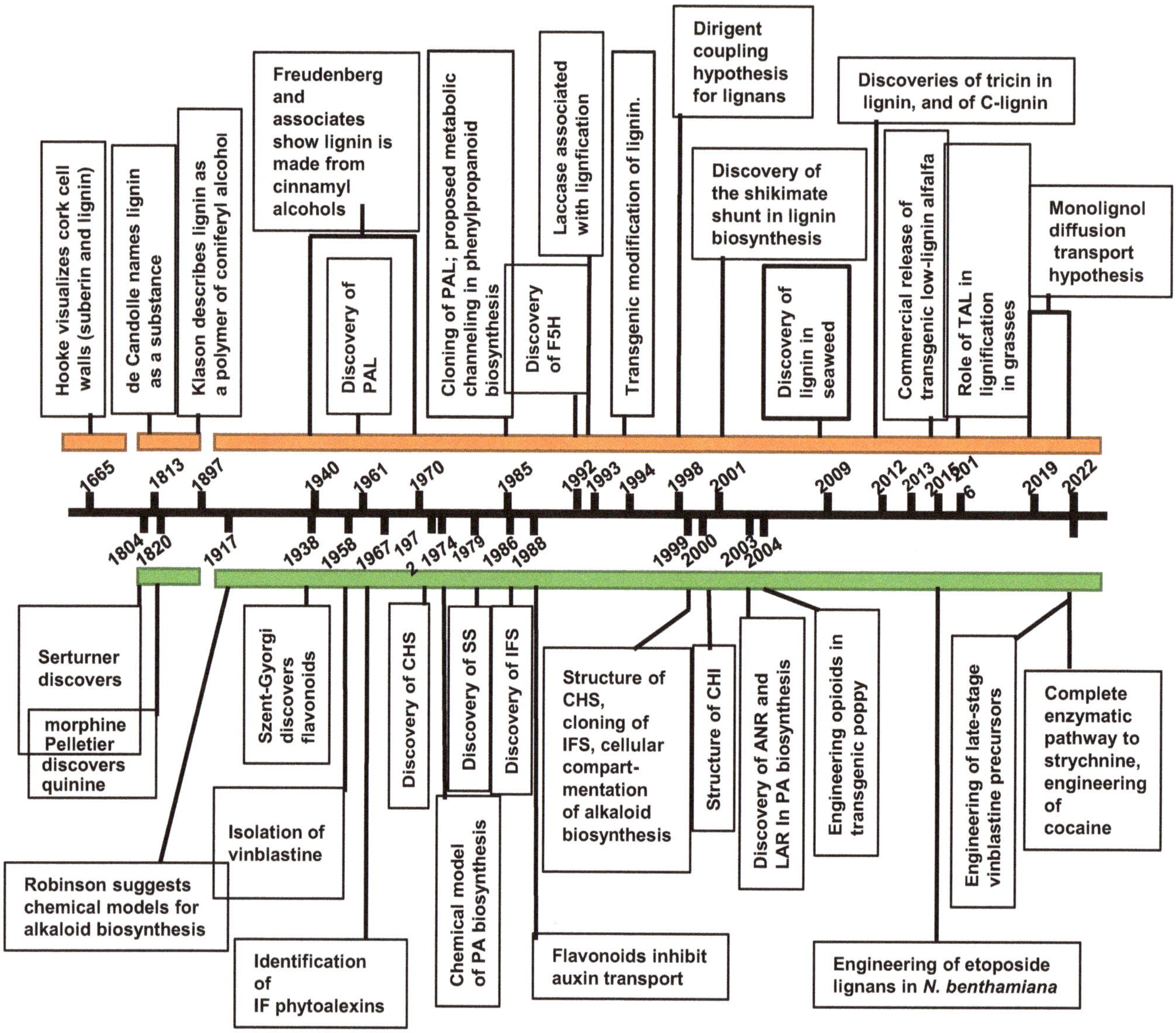

Figure 2. Historical timeline for research on lignin, lignans, flavonoids, and alkaloids, with a focus on phenolic pathways. ANR, anthocyanidin reductase; CHI, chalcone isomerase; CHS, chalcone synthase; F5H, ferulate:coniferaldehyde 5-hydroxylase; IF, isoflavonoid; IFS, isoflavone synthase; LAR, leucoanthocyanidin reductase; PA, proanthocyanidin (condensed tannin); PAL, L-phenylalanine ammonia-lyase; SS, strictosidine synthase; TAL, L-tyrosine ammonia-lyase. The enzymes are referenced further in the text.

metabolite had been identified at the molecular level. However, further progress was often impeded by uncertainty about and unavailability of potential intermediates for later pathway reactions. The advent of RNA and genome sequencing in the early 2000s made available a powerful resource for studies on plant secondary metabolism (Kellner et al. 2015; Wurtzel and Kutchan 2016). It was possible to ob-tain the total complement of all classes of enzymes expressed within an organism, a particular tissue, or in response to a particular biotic or abiotic stimulus, providing a set of candidate genes that fitted predicted expression patterns to test for involvement of their encoded proteins in a "missing reaction" within a biosynthetic pathway. It became clear that specific classes of enzymes had been recruited, through evolution, for the purpose of elaborating plant secondary metabolites (Modolo et al. 2009). These include cytochrome P450 enzymes or 2-oxoglutarate–dependent dioxygenases for hydroxylation reactions, uridine diphosphate glycosyltransferases for addition of sugar moieties, BAHD-family acyltransferases for acyl group transfer, etc. The gene candidates could be selected and expressed heterologously in a microbe (generally *Escherichia coli*, but with yeast [usually *Saccharomyces cerevisiae*] a favored host for membrane-bound enzymes such as the P450s) or transiently expressed in a plant host such as *N. benthamiana* (see below for subsequent developments of this technology) to determine the catalytic activity of the encoded enzyme. At first, recombinant enzymes were tested against a range of predicted substrates, providing tentative in vitro functions, the physiological relevance of which could then be

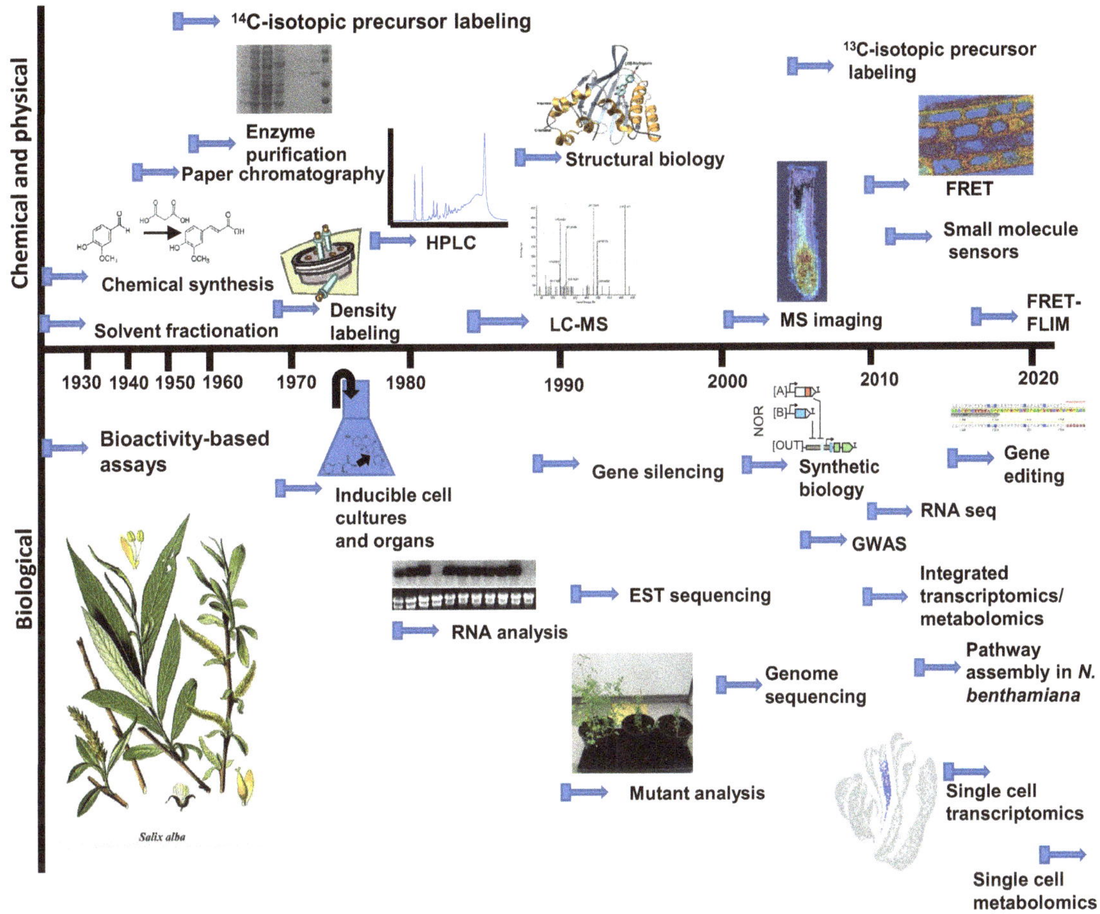

Figure 3. Timeline for the development of techniques that have driven research on plant secondary metabolites. Pictures are from: MS imaging (Zhang et al. 2023); FRET (Herud-Sikimić et al. 2021); mutant analysis (Adiji et al. 2021); synthetic biology (Brophy et al. 2022); and single-cell transcriptomics (Nolan et al. 2023). The picture bottom left is a vintage lithograph showing the white willow, *Salix alba* (original book source: Prof. Dr Otto Wilhelm Thomé, Flora von Deutschland, Österreich und der Schweiz 1885, Gera (Germany)—public domain, https://commons.wikimedia.org/ w/index.php?curid=2358667). The figure from Brophy et al. (2022) is reprinted with permission from AAAS. Other images are our own or licensed under a Creative Commons Attribution License (CC BY, https://creativecommons.org/licenses/by/4.0/) by Zhang et al. (2023), Herud-Sikimić et al. (2021), and Nolan et al. (2023). EST, expressed sequence tag; FLIM, fluorescence lifetime imaging microscopy.

determined by subsequent genetic analysis. This approach was, however, still cumbersome for the elucidation of multistep pathways and still severely limited by the availability of pathway intermediates to test as substrates.

The development of robust metabolomic technologies allowed the interrogation of metabolite/transcript co-occurrence for predicting the involvement of genes in secondary metabolite biosynthesis (Saito and Matsuda 2010). Subsequently, coupling genome mining and metabolomics with multigene transient expression systems in *N. benthamiana* has provided an efficient route to deciphering multistep pathways using a gain-of-function approach. Excellent examples are the discovery of the pathways to the etoposide lignans in mayapple (*Podophyllum peltatum*; Lau and Sattely 2015; Box 1), colchicine in *Gloriosa superba* (Nett et al. 2020), strychnine in *Strychnos nux-vomica* (Hong et al. 2022), momilactone diterpenes in rice (*Oryza sativa*) (De La Peña and Sattely 2021), limonoid triterpenes in *Citrus* and *Meliaceae* species (De La Peña et al. 2023), QS saponins (vaccine adjuvants) in soap bark tree (*Quillaja saponaria*) (Reed et al. 2023), and the late stages of vinblastine synthesis in *C. roseus* (Grzech et al. 2023). In these studies, transcriptome data were generated to provide a catalog of differentially expressed transcripts in tissues producing the end products and candidates were selected based on expression profile and enzyme class predicted from a chemically logical theoretical pathway and single and multigene expression through leaf infiltration of expression constructs was used to build the pathway step-by-step in *N. benthamiana*. LC-MS–based metabolite profiling was used to demonstrate the sequential appearance and disappearance of pathway intermediates as consecutive enzymes were sequentially coexpressed. Because flux from primary metabolism, or earlier branches of secondary metabolism, may not be specifically channeled into the engineered pathway, yields of end products could be quite low. However, it has often been possible

BOX 1

Case study: Discovery of the etoposide pathway

The paper published by Lau and Sattely (2015) is a notable example of coupling transcriptomics, chemical intuition, and metabolite analysis to unravel a complex biosynthetic pathway. The authors were searching for the genes that synthesize etoposide, a chemotherapeutic agent produced naturally by Mayapple. Four genes in this pathway had been previously discovered, over a span of about 17 yr. To further unravel the remaining biosynthetic genes, Lau and Sattely (2015) identified biosynthetic gene candidates by searching for enzymes that were co-expressed in the same tissues and environmental conditions as the known genes in the pathway. Then, they heterologously expressed these genes in *N. benthamiana* and measured the biosynthetic products. This led to the identification of 6 genes that, when combined, produced etoposide aglycone, the immediate precursor to etoposide. More recently, the Sattely lab has demonstrated a semi-semisynthetic approach to producing etoposide precursors at a milligram scale in *N. benthamiana*. Overall, this approach demonstrates a compelling example of how recently developed ideas and technology can rapidly lead to substantial discoveries in complex plant metabolic pathways.

to increase flux into the engineered pathway. For example, activation of the general phenylpropanoid pathway through coexpression of a MYB transcription factor massively increased etoposide formation, to levels higher than in the plant of origin, in *N. benthamiana* coinfiltrated with the structural genes for the pathway (Kim et al. 2022), confirming the utility of *N. benthamiana* as a chassis for commercial pro-duction of high value plant metabolites. In the case of ter-pene production, coexpression of a feedback insensitive form of HMG-CoA reductase increased the yields of triterpenes by 4- to 10-fold (Reed et al. 2017), and momilactone production could be enhanced by rerouting diterpene biosynthesis from the chloroplastic to the cytosolic, high-flux mevalonate pathway (De La Peña and Sattely 2021).

Genetic loss of function provides critical information on the involvement of a gene and its encoded protein in a particular biological function. Candidate genes identified by the approaches above can be functionally characterized by isolation of mutants and impacts on secondary metabolism determined by metabolomics or targeted metabolite profiling (e.g. Adiji et al. 2021 as one of many examples). The advent of widely applicable gene editing technologies removes the requirement for available mutant populations, and virus-induced gene silencing (VIGS) has been and continues to be developed as a rapid approach to gene knock-down. VIGS applies recombinant viruses to transiently downregulate the expression of targeted genes.

This technology has found use in identifying genes involved in the biosynthesis of several types of secondary metabolites, including monoterpene indole alkaloids (MIAs) in opium poppy (*Papaver somniferum*) (Wijekoon and Facchini 2012; Salim et al. 2013; Chen et al. 2020), bisindole alkaloids in *C. roseus* (Liscombe and O'Connor 2011), and acyl sugars in *Solanum* and *Nicotiana* species (Slocombe et al. 2008).

Random assembly of secondary metabolite-derived polymers

Polymers of secondary metabolites in the form of lignin (derived from hydroxycinnamyl alcohols) and condensed tannins (derived from flavonoids) constitute a substantial proportion of the fixed carbon on the planet. Other plant polymers derived wholly or in part from secondary metabolites include suberin (fatty acids, hydroxycinnamic acids associated with lignin), melanin (phenolics), phlobaphenes (flavonoids), and sporopollenin (phenolics, fatty acids, flavonoids). With the exception of the highly inert polymer sporopollenin, the structure of which was only recently discovered (Li et al. 2019), the biosynthesis of the building blocks of these polymers has been intensively studied and reviewed elsewhere (Dixon and Barros 2019; Yu et al. 2023). However, in contrast to other plant polymers such as polysaccharides and proteins, the assembly of the monomeric units into polymers requires less precise enzymatic control. In the case of melanin, the polymer results from the oxidation of vacuolar phenolics and possibly other molecules after rupture of the chloroplast membrane releases polyphenol oxidase (Glagoleva et al. 2020). After many years of debate, it now appears that the monomers that constitute the chains of condensed tannins are linked nonenzymatically via nucleophilic attack between an activated flavan-3-ol or leucoanthocyanidin carbocation extension unit and a flavan-3-ol starter unit (Liu et al. 2016; Jun et al. 2021); genetic analysis has failed to demonstrate any type of polymerase that would yield the observed linkage patterns but rather reveals proteins that appear to control the spatial distribution of starter and extension units (Yu et al. 2023).

In the case of lignin, it has been known for many years that the monolignols (hydroxycinnamyl alcohols) are polymerized via a free radical reaction initiated by the action of laccases and peroxidases. However, the large size of the laccase and peroxidase gene families in plants has hampered genetic analysis of this process, and there has been much debate as to the extent to which the polymerization is purely chemical or requires some type of template (Ralph et al. 2008). As in many things, the truth probably lies somewhere in the middle; the discovery of C-lignin composed only of catechyl alcohol units (Chen et al. 2012) and evidence for the incorporation of structurally diverse nonconventional units into the lignin of grasses (e.g. the flavonoid tricin), and other species (e.g. stilbenes and polyamines) argues against strict proteinaceous control of coupling (Ralph et al. 2019), whereas a picture is now emerging of precise spatial control of the initiation of macromolecular lignin assembly in the Casparian strip of the root involving dirigent proteins, previously implicated in stereoselective lignan coupling (Davin and Lewis 2000; Hosmani et al. 2013; Wang et al. 2022).

The above examples highlight the importance of cell biology in the accumulation of secondary metabolite-derived polymers and the need for unbiased genetic analysis in elucidating the complete biosynthetic machinery.

Biological insights derived from studies on plant secondary metabolism

Secondary metabolite synthesis is under complex spatial control

Secondary metabolites are seldom found throughout all plant tissues but rather are deposited in specific tissues or even specialized organs. Glandular trichomes are common sites of storage of metabolites required on the outer surface of the plant and can be isolated in pure form, facilitating the elucidation of pathways involved in the biosynthesis of their contents through RNA sequencing. A classic example is the interrogation of the peppermint (*Mentha* x *piperita*) glandular trichome transcriptome for genes involved in the biosyn-thesis of menthol and related monoterpenes (Lange et al. 2000). In some cases, secondary metabolites are localized to the leaf epidermis, and transcriptome analysis of the *C. roseus* leaf "epidermone" revealed the presence of transcripts for the complete MIA pathway, including a transporter for leaf sur-face secretion (Murata et al. 2008; Yu and De Luca 2013).

However, not all secondary metabolites are synthesized in the organs in which they ultimately accumulate. A well know example is nicotine in *Nicotiana* species, which accumu-lates in leaves but is synthesized in roots (Zenkner et al. 2019), its transport being facilitated by a purine uptake permease (Kato et al. 2015). Application of single cell metabolomics ana-lysis revealed that, although most MIA precursors are indeed found in the epidermal cells of *C. roseus*, the major MIAs vindoline and serpentine localize to idioblast cells, and their precursor strictosidine accumulates in both cell types (Yamamoto et al. 2019). The biosynthesis of benzyisoquinoline alkaloid biosynthesis in the opium poppy is equally complex: the initial steps of biosynthesis of several of the alkaloids ap-pear to occur in companion cells and sieve elements, with final accumulation of the alkaloids in laticifers (reviewed in Beaudoin and Facchini 2014). Surprisingly, however, the pathway gene transcripts are localized to companion cells, whereas the proteins themselves are found in both sieve elements and laticifers, and shotgun proteomics has confirmed the presence of enzymes specific for the final steps of morphine biosynthesis to be present in laticifers (Onoyovwe et al. 2013). MS imaging techniques (Fig. 3) hold promise for higher resolution understanding of the spatial distributions of the precursors and end products of plant secondary metabolism (Dong and Aharoni 2022). This understanding is assisted by our increasing knowledge of transporters for secondary metabolites (Gani et al. 2021), although in some cases (e.g. monolignol transport) genetic analysis has failed to reveal transport mechanisms, and passive diffusion is becoming increasingly accepted (Perkins et al. 2022). At the same time, the release of floral volatiles to the atmosphere, long believed to occur via passive diffusion, has recently been shown to be mediated by nonspecific lipid transfer proteins (Liao et al. 2023).

Metabolic channeling provides an additional layer of pathway regulation

Metabolic channeling is the passing of intermediates of enzymatic reactions from one enzyme to the next without equilibration into the cellular milieu (cytosol). It provides a means to regulate flux into bifurcating pathways, to enhance catalytic activity, and to protect highly reactive intermediates from undesirable chemical conversion or from causing cellular toxicity (Winkel-Shirley 1999). Although first postulated in mammalian metabolism, many of the best examples of metabolic channeling have been demonstrated, or at least proposed, in plants. Unfortunately, metabolic channeling can be used as a convenient catch-all interpretation when metabolic data do not fit the current paradigm, in some cases perhaps because the model is incorrect. Data supporting metabolic channeling require careful interpretation and usually confirmation via multiple approaches.

Metabolic channeling has been proposed to operate in the synthesis of many types of plant secondary metabolites, including terpenoids, alkaloids, phenylpropanoids, flavonoids, and cyanogenic glycosides (Gutensohn et al. 2022; Jørgensen et al. 2005). Techniques for determining protein–protein interactions have been widely used in these studies. For example, bimolecu-lar fluorescence complementation (BiFC) and coimmunopreci-pitation approaches revealed that soybean (*Glycine max*) isoflavone synthase interacted with CHS, chalcone isomerase (CHI), and cinnamate 4-hydroxylase (C4H), suggesting the for-mation of an isoflavone biosynthesis metabolon (Dastmalchi et al. 2016). But metabolic channeling is not necessarily con-served in the same pathway between different species. For example, snapdragon (*Antirrhinum majus*) flavone synthase II (FNS II) was shown to interact with CHS, CHI, and dihydroflavanol reductase (DFR) by yeast 2-hybrid and BiFC assays (Fujino et al. 2018). In contrast, using similar methods, no interaction between FNS II and CHS could be verified in torenia (*Torenia hybrida*), where FNS II interacted instead with flavonoid 3-hydroxylase (F3H) (Fujino et al. 2018). In a further example from the flavonoid pathway, Förster resonance energy transfer (FRET) detected by fluorescence lifetime imaging microscopy (FLIM) (FRET-FLIM) provided in vivo evidence that DFR competed with flavonol synthase (FLS) in physical binding to CHS in Arabidopsis; this suggests that channeling may operate to de-termine the direction of flux into flavonols and anthocyanins/ proanthocyanidins (Crosby et al. 2011). Metabolic channeling is likely a dynamic process that is potentially dependent upon physiological conditions, an aspect that requires further study.

It was proposed that the membrane-associated cytochrome P450 enzymes C4H, flavanone 3′-hydroxylase (F3′H), and FNS II act to anchor the phenylpropanoid/flavonoid enzyme complex to the cytosolic side of the endoplasmic reticulum (Winkel-Shirley 1999, 2001; Fujino et al. 2018), and C4H, coumaroyl shikimate 3′-hydroxylase (C3′H), and ferulate/coniferaldehyde 5-hydroxylase (F5H) P450s were assumed to operate similarly in the pathway to monolignols. Direct evidence of weak PAL/C4H interactions came from FRET studies in *N. tabacum* (Achnine et al. 2004), and FLIM revealed the consecutive P450s C4H and C3′H to be colocalized on the endoplasmic reticulum (ER) membrane and restricted in their lateral diffusion by association with other ER proteins (Bassard et al. 2012). The operationally soluble enzymes hydroxycinnamoyl CoA:shikimate hydroxycinnamoyl transferase and 4-coumarate:CoA ligase were partially associated with the ER when the 2 P450s were coexpressed. It was subsequently shown that the 3 P450 enzymes of the monolignol pathway, although in spatial proximity to one another on the ER membrane, do not interact directly but rather through the intermediacy of 2 members of the membrane sterol binding protein family, which themselves can form multimers to allow organization of P450 clusters with resultant stabilization of the P450 enzymes (Gou et al. 2018). Although most examples of metabolic channeling in plant secondary metabolism have been proposed to involve complexes anchored to the ER, compartmentation between cytosol and organelles can also occur. During the biosynthesis of MIA alkaloids in *C. roseus*, strictosidine is formed from secologanin and tyrosine by a Pictet–Spengler reaction catalyzed by strictosidine synthase (Treimer and Zenk 1979); many plants contain putative orthologs of strictosidine synthase with functions yet to be determined (Kibble et al. 2010). Strictosidine is then deglycosylated by an enzyme, strictosidine β-glucosidase (SDG), that multimerizes to form a shield protecting the reactive aglycone, which spontaneously reorganizes to yield, among other products, cathenamine and epi-cathenamine, which are downstream intermediates for the formation of MIAs. The enzymes that initiate further conversion of these products may be localized to the nucleus, where they may physically interact with the SDG multimers, forming a nuclear metabolon that is thought to further protect reactive intermediates until more stable compounds are generated for subsequent oxidation by cytochrome P450 enzymes localized on the ER (Stavrinides et al. 2015). Nuclear localization of secondary metabolism warrants further study.

Plant secondary metabolism evolves through a variety of mechanisms

Because many secondary metabolites are involved in the interactions of plants with their environment, their appearance in plant lineages is intimately associated with plant microbe/pest/predator coevolution. In some cases, the same metabolites are made by both the plant and its potential pest/predator. Aside from ecological considerations, this situation is of interest for the light it sheds on the evolutionary flexibility of secondary metabolite biosynthesis. A remarkable example concerns the iridoid sex pheromones, which are synthesized in the hind legs of female pea aphids (*Acyrthosiphon pisum*) by the same sequence of reactions as the iridoids in plants, although 6 enzymes of the insect pathway are unrelated to their plant counterparts (Köllner et al. 2022). The same situation of common intermediates but involvement of different classes of enzymes is seen in the case of the diterpene gibberellins of plants, fungi, and bacteria (Salazar-Cerezo et al. 2018), highlighting the operation of cross-kingdom convergent evolution of secondary biosynthetic pathways. At the same time, phylogenetic studies have revealed that some enzymes of plant secondary metabolism have arisen through horizontal gene transfer, such as the origin of PAL from soil bacteria and fungi during symbioses early in land colonization (Emiliani et al. 2009). A number of enzymes of secondary metabolism have also been coopted during evolution from enzymes of primary metabolism (Sonawane et al. 2020; Lou et al. 2022), an understanding of which can inform rational metabolic engineering.

Some enzymes of secondary metabolism, for example, the terpene synthases, are remarkably promiscuous, based on the contours and dynamics, rather than catalytic residues or mechanism, of their active sites that direct the folding of the tertiary carbocation toward a large range of potential products (Greenhagen et al. 2006; Jia et al. 2019). Subsequent modification of terpene structures largely includes hydroxylation reactions, most of which are catalyzed by cytochrome P450 enzymes, and kingdom-wide phylogenetic analysis has suggested that the CYP716 family has evolved specifically toward triterpenoid biosynthesis in eudicots, with neofunctionalization of the CYP716 family coinciding with the emergence of oxidosqualene cyclases with specific roles in secondary metabolism (Miettinen et al. 2017).

In contrast to the situation in fungi, where secondary metabolic genes are usually clustered together in operons, it was thought, until the late 1990s, that the genes involved in plant secondary metabolism were randomly distributed throughout the genome. This appears true in many cases, including, for example, the genes of general phenylpropanoid metabolism and the lignin and anthocyanin branches. However, the picture changed in 1997, when it was shown that all 5 structural genes (a tryptophan synthase homolog and 4 cytochrome P450s) for the biosynthesis of anti-insect benzoxazinones in grasses were linked in an operon-like structure (Frey et al. 1997). Following the subsequent discovery of the genomic organization of the avenacin pathway in oat (*Avena sativa*), the momilactone pathway in rice, and the thalianol pathway in Arabidopsis (Qi et al. 2004; Wilderman et al. 2004; Field and Osbourn 2008), diterpene and triterpene biosynthesis emerged as a focus area for studies on metabolic diversification through independent evolution of gene clusters in plants. Gene clusters have since been discovered in several other areas of plant secondary metabolism (Bharadwaj et al. 2021), including the biosynthesis of the surface β-diketone waxes in wheat (*Triticum aestivum*) and barley (*Hordeum vulgare*) where the cluster contains a polyketide synthase, a cytochrome, P450 and a hydrolase/carboxylase, functions that were previously considered to be part of a multifunctional protein (Hen-Avivi et al. 2016).

Clustering of genes allows for more highly coordinated coregulation, as well as opportunities for regulation via targeted chromatin remodeling (Osbourn 2010). Analysis of the structure of the thalianol cluster in a wide range of *A. thaliana* accessions and related Arabidopsis species showed that the cluster is largely fixed, with a low frequency of deleterious haplotypes, and that chromosomal inversion can shuttle distant genes into the cluster to enable cluster compaction (Liu et al. 2020). Cross-cluster effects may also play a part in the maintenance of biosynthetic gene clusters. For example, in addition to the momilactone cluster on chromosome 4, rice also contains a gene cluster on chromosome 2 for synthesis of related phytocassane diterpenoids, which contains 2 genes encoding related cytochrome P450 enzymes (CYP76M7/8). CRISPR-Cas9 editing to remove the cluster on chromosome 2 resulted in phytotoxicity due to apparent buildup of momilactone pathway intermediates requiring CYP76M7/8 for further conversion (Li et al. 2022).

Plants also contain examples of highly tandemly duplicated genes, the best known being the leucine-rich repeat disease resistance genes in many plants (van Wersch and Li 2019). The tandem arrangement may provide a "reservoir" for generation of gene diversity through site-directed recombination or ensure cosegregation of genes that are most effective when expressed in concert. In the area of secondary metabolism, the ~100 Arabidopsis small molecule uridine diphosphate glycosyl transferase (*UGT*) genes are scattered throughout the genome but cluster into groups of up to 7 genes. The over 300 *M. truncatula UGT* genes are distributed similarly, but the number of genes in each array is greater, with one locus on chromosome 6 containing 27 tandemly arrayed *UGT* genes and *CYP* genes often cocluster with the *UGT* genes (Young et al. 2011). It has yet to be shown whether this particular organization provides a reservoir for evolutionary adaptation of UGT functions.

It is now becoming apparent that plant secondary metabolism has evolved to include built-in redundancies. This likely provides both flexibility and protection against fluctuations in precursor flux. Examples include dual pathways for *O*-methylation (at the free acid and CoA ester levels) and 3′-hydroxylation (direct or via the "shikimate shunt") in monolignol biosynthesis (Ye et al. 1994; Dixon and Barros 2019; Vanholme et al. 2019) and parallel pathways to epicatechin as a starter or extension unit in condensed tannin biosynthesis (reviewed in Yu et al. 2023). Such phenomena complicate the genetic analysis of pathway structure.

Manipulation of secondary metabolism provides a paradigm for plant ecology

The field of plant chemical ecology was developed from the discoveries in the 1940s to 1980s that characterized secondary metabolites and their biological activities against insect predators and microbial pathogens. The subsequent discoveries of pathway genes have now allowed ecological predictions to be tested in natural environments. A good example is the work of Baldwin and associates with the wild tobacco *N. alata* in habitat in the Great Basin Desert in Utah, which established that release of volatiles from insect larval-attacked leaves can attract predatory insects in nature (Kessler and Baldwin 2001), and that adjacent plants can perceive volatile signals from attacked plants to induce preemptive defense pathways (Baldwin et al. 2006). Planting transgenically modified *N. alata* with strongly reduced nicotine levels in its natural environment confirmed that nicotine does indeed protect the plant from larval feeding (Steppuhn et al. 2004) and demonstrated that insect predators (wolf spiders) prefer larvae that have fed on plants lacking nicotine (Kumar et al. 2014). Furthermore, use of plant-mediated RNAi to downregulate an insect cytochrome P450 implicated in the exhalation of digested nicotine from the spiracles resulted in increased spider attack, showing that larvae themselves utilize nicotine as a defense (Kumar et al. 2014).

The optimal defense theory (McKey 1979) states that the evolution and allocation of defenses is based on both the fitness value and vulnerability of individual plant parts to simultaneously optimize growth and defense. This theory was recently tested for the glucosinolate defense compounds of *A. thaliana*, which naturally exhibit a gradient of distribution from the youngest (highest) to the oldest leaves, where the compounds are synthesized in the lower mature leaves and transported to the younger leaves, which are the preferred tissues for herbivory. Modifying this gradient by knock-out of 2 glucosinolate transporter genes altered the feeding preference of larvae of the generalist insect herbivore African cotton leafworm (*Spodoptera littoralis*) in a way consistent with the tenets of the optimal defense theory (Hunziker et al. 2021). However, glucosinolates appear to have functions beyond protection against herbivory and pathogen attack. Levels of aliphatic glucosinolates in Arabidopsis are regulated by a transcriptional cascade involving auxin-sensitive AUXIN/INDOLE-3-ACETIC ACID-INDUCIBLE (Aux/IAA) repressors, with elevated glucosinolate levels being necessary for correct stomatal regulation under drought (Salehin et al. 2019).

It can be expected that the ecological functions of more and more secondary metabolites will be examined through field studies with genetically modified plants in the coming years.

Cutting-edge technologies are paving the road to discoveries in secondary metabolism research

Emerging technologies are enabling understandings of metabolites and their functional roles in plants. Here, we highlight recently developed or improved technical approaches and how these have influenced the field of plant metabolic biology.

Genome-wide association studies allow unbiased analysis of gene function

Genome-wide association studies (GWAS) are well known in the biomedical field but have now become a powerful tool in many other areas of biology, including plants (Tibbs Cortes et al. 2021). Linking traits to genotypes across populations allows an unbiased approach to discovery of genes impacting important traits, and this approach is expected to have broad impacts for plant secondary metabolism in the future. As one example, in a GWAS analysis of more than 1,000 individual lines of black cottonwood (*Populus trichocarpa*), some single nucleotide polymorphisms (SNPs) associated with lignin content across the *Populus* species range and in plants grown under different environmental conditions resided in a 5-enolpyruvylshikimate-3-phosphate synthase-like (*EPSP*) gene (Xie et al. 2018). EPSP synthase is an enzyme of the shikimate pathway that provides precursors for phenylpropanoid biosynthesis, but the rare alleles were not associated with enzymatic properties but rather structural changes that converted the enzyme to a transcription factor that acted as a repressor of lignin biosynthesis (Xie et al. 2018). An approach in which untargeted, qualitative metabolic traits are subjected to GWAS analysis (QT-GWAS) appears promising in uncovering gene metabolite relations, revealing enzymes of chroman, nucleotide, and neolignan biosynthesis in Arabidopsis (Brouckaert et al. 2023).

To determine the genetic basis for production of defensive terpene metabolites secreted by maize roots, a combination of metabolite-based QTL mapping, GWAS analysis, and use of near-isogenic lines was applied, leading to the identification of a terpene cyclase involved in the synthesis of the non-volatile antibiotic β-costic acid (Ding et al. 2017). Such approaches provide powerful tools for molecular breeding to enhance biotic stress tolerance.

Single-cell transcriptomics is enabling deep characterization of phytohormone signaling leading to the identification of regulators

Another rapidly evolving method that is advancing our understanding of gene function is single-cell sequencing. The recent explosion of use of single-cell transcriptomics has led to the extensive generation of data on thousands of genes underlying plant development and stress response. Several of these studies have already made progress deepening our understanding of small molecule biosynthesis and signaling. For example, single-cell transcriptomic studies during phosphate deficiency revealed cytokinin production in the central vasculature, which has a cascading effect to promote root hair development in the epidermis (Wendrich et al. 2020). In another example, single-cell RNA sequencing of developing Arabidopsis lateral roots revealed camalexin, an indole phytoalexin, as a regulator of lateral root formation (Serrano-Ron et al. 2021). Other studies have found additional roles for auxin, cytokinin, and gibberellin in various developmental processes, including root development (Torii et al. 2020), leaf vein development (Liu et al. 2022), and tissue regeneration (Efroni et al. 2016). In addition, single-cell transcriptomics has been used to analyze genes that respond to phytohormone stimulus. For example, single-cell analysis in brassinosteroid-blind mutants revealed non-cell-autonomous functions of brassinosteroid signaling (Graeff et al. 2021). Another single-cell study of the brassinosteroid response in roots identified responsive transcription factors expressed with high specificity in the elongating cortex (Nolan et al. 2023). In the future, it will be important to delineate the transcriptomic responses to abiotic and biotic stresses at the single cell level and explore methods capable of combining single-cell transcriptomics with metabolomics measurements to address the involvement of unidentified regulatory natural products.

Technological advances to visualize localization patterns of small molecules

Key advances in small molecule biology have been enabled by the development of strategies to visualize phytochemicals across space and time using chemical and genetic-based approaches. The most common technique for visualizing metabolite activity is to design and measure genetically encoded reporters. This technique has been mostly applied to plant hormones in the context of development. Promoter fusions, which typically rely on synthetic promoters that are activated by phytohormone-specific transcription factors, can be used to measure the activities of auxin (Ulmasov et al. 1997), ABA (Wu et al. 2018), brassinosteroids (Chaiwanon and Wang 2015), and cytokinins (Liu and Müller 2017). For example, DR5 is a synthetic promoter element that contains 7 repeats of an auxin-response element (Ulmasov et al. 1997). DR5 has been used extensively to drive various reporters, including GUS, GFP, and luciferase (LUC) to quantify auxin-regulated processes. 6xABRE is another synthetic promoter used to visualize the transcriptional response to ABA (Wu et al. 2018; Fig. 4A). However, it is advised to use multiple strategies to test conclusions using synthetic promoter:reporter lines, as it has been shown that these lines do not always reflect known phytohormone concentrations (Chandler and Werr 2015). A comparison of multiple approaches for the visualization of ABA in plant roots and shoots is given in Fig. 4. Another genetic approach to visualizing phytohormones uses signal-induced degradation to indirectly measure phytohormone levels. This approach is exemplified by R2D2 (ratiometric version of 2 DIIs), which leverages an interaction between auxin, an Aux/IAA DII domain, and a ubiquitin ligase to measure auxin levels (Liao et al. 2015). In this system, the DII domain is fused to the reporter n3xVenus and the mDII domain, which is not degraded by auxin, is fused to nTdTomato. Changes in auxin levels are determined by measuring the ratio of yellow to red fluorescence. FRET sensors are another way to measure small molecule activity. For instance, a FRET reporter for auxin that utilizes an engineered tryptophan repressor protein-binding pocket allows for transient and reversible detection

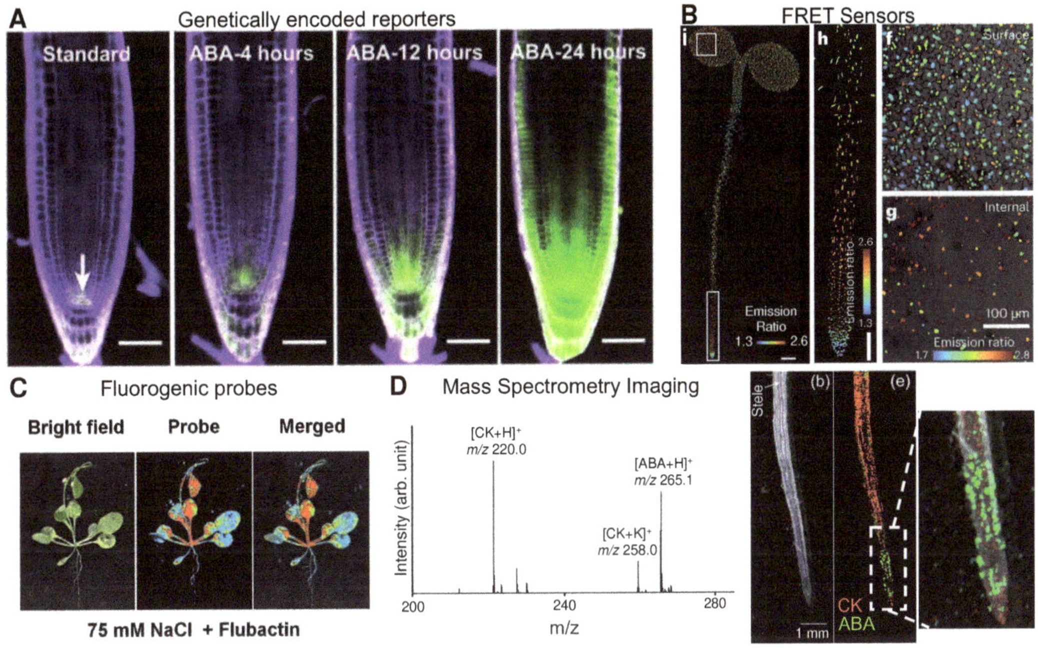

Figure 4. Examples of techniques used to visualize the localization of secondary metabolites. Each technology shown here is designed to image abscisic acid (ABA). **A)** A genetically encoded reporter (Wu et al. 2018); **B)** a FRET sensor (Rowe et al. 2023); **C)** a fluorogenic, small molecule-based probe (Yang et al. 2022), and this image was reprinted with permission from Yang et al. (2022) copyright 2022 American Chemical Society; and **D)** an MALDI MS imaging method for imaging phytohormones cytokinin (CK) and ABA (Shiono et al. 2017), and this image was reprinted with permission from Shiono et al. (2017), copyright 2017 American Chemical Society. Images are licensed under a Creative Commons Attribution License (CC BY, https://creativecommons.org/licenses/by/4.0/) by Wu et al. (2018) and Rowe et al. (2023).

of IAA (Herud-Sikimić et al. 2021). A FRET reporter for ABA relies on the signal-induced interaction between covalently linked ABA receptor domains (such as pyrabactin resistance 1 and PP2C-type phosphatase domains) to measure changes in ABA localization and concentration (Jones et al. 2014; Waadt et al. 2014; Rowe et al. 2023; Fig. 4B).

Currently, there are few examples of biosensors for secondary metabolites beyond the classical hormones, which reflects the general lack of detailed information required to generate these types of reporters. Designing receptor-based biosensors requires atomic scale resolution, which is lacking for most secondary metabolite receptors, if known (Nemhauser and Torii 2016). Likewise, promoter-based biosensors require the identification of highly specific sequences responsive only to the metabolite of interest. Further elucidation of these interactions will expand the use of biosensors and lead to understanding of the localization patterns of secondary metabolites and their potential functions as signal molecules. Global protein–metabolite interactome studies could provide a basis for technology development in this space (Luzarowski et al. 2021).

Direct imaging of metabolites can occasionally be performed if the chemical of interest already has natural fluorescence properties. For example, phenolic compounds such as hydroxycinnamic acids, coumarins, stilbenes, and styrylpyrones are often auto-fluorescent under the proper excitation wavelength (Hutzler et al. 1998). Phenylcoumarans and stilbenes are major chromophore monomers in lignin and have enabled nondestructive imaging of lignin (Maceda and Terrazas 2022). The fluorescence of lignin can be further amplified by dyes or by feeding monolignols with small fluorescent tags (Tobimatsu et al. 2013). Flavonoids are other phenolic metabolites that naturally possess low levels of fluorescence, which can be enhanced with dye conjugates to reveal localization and intensity, as well as uncover their regulatory interactions with signaling pathways and phytohormones (Lewis et al. 2011; Muhlemann et al. 2018). However, the vast majority of metabolites do not have inherent fluorescence. To address this, bioactive compounds with fluorescent tags have been developed to track brassinosteroids (Irani et al. 2012), auxins (Pařízková et al. 2021; Hayashi et al. 2014), and jasmonates (Liu and Sang 2013). While such bioactive fluorescent tags have provided information on the spatiotemporal dynamics of these molecules, they typically cannot be utilized to determine activity (i.e. protein-binding or participation in enzymatic reactions). One strategy for enabling measurements of metabolite activity is the design of synthetic protein-responsive small molecule reporters that change fluorescence properties upon protein interactions. For example, the computationally optimized synthetic compound, flubactin, increases in fluorescence 8-fold upon binding to the ABA pyrabactin resistance 1 receptor (Yang et al. 2022; Fig. 4C). Protein-responsive fluorescent reporters have also been developed for strigolactone (de Saint Germain et al. 2022) and retinaldehyde (Dickinson et al. 2021). These reporters enable access to a key element of secondary metabolite function, interactions with signaling proteins. However, an obvious limitation of pro-fluorescent and fluorescent reporters is that they can only be designed where the identity of a secondary metabolite is known, and the vast majority of secondary metabolites in plants remain unknown.

In order to explore the "black box" of secondary metabolism, chemical imaging technologies have provided avenues for direct measurements of metabolites in situ and in vivo. For example, desorption electrospray ionization and matrix assisted laser desorption/ionization (MALDI) MS imaging has been applied to map the spatial profiles of primary metabolites (Zhang et al. 2023), glucosinolates (Sarsby et al. 2012), lipids (Sarabia et al. 2018), alkaloids (Conceição et al. 2021), and phytohormones (Shiono et al. 2017; Fig. 4D) across plant tissue sections of interest. This work has revealed localization patterns of known metabolites as well as mass to charge signatures of many unknown compounds. MS imaging is capable of measuring relative levels of mass to charge ratios of tens to hundreds of small molecules with spatial resolution ranging from 5 to 250 μm. However, a limitation of this technology is that it cannot perform dynamic measurements. Two chemical imaging techniques that are nondestructive and capable of measuring dynamic information are Raman and NMR microscopy. Raman spectroscopy is a selective technique that measures vibrational modes of molecules. When applied as a microscopic method, it has been used to measure different classes of secondary metabolites (Yang et al. 2017) and cell wall polymers (Gierlinger and Schwanninger 2006) with high spatial resolution (1 to 10 μm). NMR microscopy is related to magnetic resonance imaging (MRI), which has been widely used for whole body imaging in human medical applications. In plants, NMR microscopy is ideally suited to measure water flow, but can also be applied to track C^{13} labeled plant metabolites (Köckenberger et al. 2004).

Advances in plant synthetic biology are deepening our understanding of how to produce and utilize secondary metabolites

Due to the ability of plants to synthesize and store large amounts of metabolites of interest for human health and society, both primary and secondary metabolism are now major targets of plant synthetic biology efforts (Maeda 2019), and these advances have led to engineering of secondary metabolite pathways with applications in medicine, agriculture, and renewable fuels in sustainable ways (Goold et al. 2018; Liu et al. 2023). This research has also been critical for generating deeper understandings of metabolite biosynthesis and regulation. For example, elucidating the biological pathway for colchicine biosynthesis generated understanding of the tissue specificity of alkaloid biosynthesis in medicinal plants in addition to providing engineering strategies for the production of this medicinal compound (Nett et al. 2020). In spite of this and the other examples given earlier, the discovery of natural product biosynthesis pathways still remains a major obstacle for achieving the full potential for engineering secondary metabolites in plant and microbes (Anarat-Cappillino et al. 2014). Advances will rely not only on discovery of pathway genes but also on deployment of tunable synthetic regulatory circuits (Nemhauser and Torii 2016).

Another branch of synthetic biology research is to utilize metabolic pathways to understand functions or to engineer new traits. As previously mentioned, the development of genetically encoded biosensors is a growing area of plant biology research. This includes the development of small molecule reporters to study the localization of signaling components and the creation of plant "sentinels" that produce dramatic phenotypic changes in response to external chemical cues (Jez et al. 2016; Uslu and Grossmann 2016). One of the earliest synthetic biology examples in plants was the engineering of plant metabolism for phytoremediation of toxins (Bizily et al. 2000). An exciting recently developed approach is the synthetic transfer of features of plant secondary metabolite regulation into heterologous systems for precise control of regulatory processes. For example, the auxin-inducible degron system, used to rapidly degrade proteins in organisms that do not endogenously synthesize auxin, has already made important contributions to biological research (Nishimura et al. 2009). Another current theme in synthetic biology is to generate strategies for precise tuning of secondary metabolite localization to improve plant growth or stress response. For example, researchers have rewired phytohormone pathways to control plant development and architecture (Brophy et al. 2022). Finally, there is potential in the future for engineering plant–microbial interactions, which often involve metabolite-based communication, for improving plant growth and stress responses (Ke et al. 2021).

Conclusions

The past century has seen enormous advances in our understanding of the broad repertoire of chemicals elaborated by plants, their localizations, functions, and biosynthetic pathways. These developments have in many ways tracked the technological developments in analytical chemistry and molecular and cellular biology. There is a long history of the use of plant secondary metabolites in human health and society. Continuing the discovery of new molecules and pathways will lead to further societal benefits. At the same time, the study of plant secondary metabolism itself has led to a number of conceptual advances in biology and has raised a number of questions that, even if recognized, could not have been addressed by early workers in the field (see "OUTSTANDING QUESTIONS"). We are entering an exciting period where advances in gene discovery and synthetic biology are opening up possibilities for the sustainable production of a wide range of plant secondary metabolites as replacements for traditional medicines, food ingredients, and fuels. To further our understanding of the functions of secondary metabolites in plant ecosystems, particularly under climate change, it will be necessary to manipulate these compounds in a wide variety of species beyond crops. Recent advances combining VIGS with CRISPR Cas9-based gene editing (e.g. Ellison et al. 2020) promise solutions to the need for species-independent and genotype-independent genetic transformation methods for rapid pathway manipulation, thereby allowing scientists to expand the legacy of the past century's insights into plant secondary metabolism.

OUTSTANDING QUESTIONS

- What are the chemical identities of the myriad of specialized metabolites that have remained uncharacterized?
- What are the genes that synthesize and respond to specialized metabolites, and can we improve the speed and reliability of plant transformation and VIGS to enable their study in any species of interest?
- What are the roles of secondary metabolites in signaling, and can this be addressed in the future through proteome:metabolome-level interaction mapping?
- How do we improve our ability to measure and engineer these metabolites with high spatial precision?
- What translational applications for plant specialized metabolites will be developed in the future?

Acknowledgments

A.J.D. would like to thank Philip Benfey for many years of exemplary mentorship and guidance and for seeing the potential for understanding the chemistry underlying developmental plant biology. R.A.D. thanks Vernon Butt, Derek Bendall, and Chris Lamb for inspiration during the early part of his career.

Funding

Work in the authors' laboratories was funded by the Center for Bioenergy Innovation, a U.S. Department of Energy Research Center supported by the Office of Biological and Environmental Research in the DOE Office of Science (to R.A.D.); the National Science Foundation Division of Integrated Organismal Systems (award number 1456286 to R.A.D.); the National Science Foundation Early-Concept Grants for Exploratory Research program (collaborative EAGER 2028649 to A.J.D.); the National Institutes of Health (NIGMS 1R35GM147216 to A.J.D.); and the Hellman Foundation (to A.J.D.).

Conflict of interest statement. None declared.

References

Achan J, Talisuna AO, Erhart A, Yeka A, Tibenderana JK, Baliraine FN, Rosenthal PJ, D'Alessandro U. Quinine, an old anti-malarial drug in a modern world: role in the treatment of malaria. Malar J. 2011:**10**(1):144. https://doi.org/10.1186/1475-2875-10-144

Achnine L, Blancaflor EB, Rasmussen S, Dixon RA. Colocalization of L-phenylalanine ammonia-lyase and cinnamate 4-hydroxylase for metabolic channeling in phenylpropanoid biosynthesis. Plant Cell. 2004:**16**(11):3098–3109. https://doi.org/10.1105/tpc.104.024406

Adiji O, Docampo-Palacios ML, Alvarez-Hernandez A, Pasinetti GM, Wang X, Dixon RA. UGT84F9 Is the major flavonoid UDP-glucuronosyltransferase in *Medicago truncatula*. Plant Physiol. 2021:**185**(4):1617–1637. https://doi.org/10.1093/plphys/kiab016

Anarat-Cappillino G, Sattely ES. The chemical logic of plant natural product biosynthesis. Curr Opin Plant Biol. 2014:**19**:51–58. https://doi.org/10.1016/j.pbi.2014.03.007

Appelhagen I, Thiedig K, Nordholt N, Schmidt N, Huep G, Sagasser M, Weisshaar B. Update on *transparent testa* mutants from *Arabidopsis thaliana*: characterisation of new alleles from an isogenic collection. Planta. 2014:**240**(5):955–970. https://doi.org/10.1007/s00425-014-2088-0

Baldwin IT, Halitschke R, Paschold A, von Dahl CC, Preston CA. Volatile signaling in plant-plant interactions: "talking trees" in the genomics era. Science. 2006:**311**(5762):812–815. https://doi.org/10.1126/science.1118446

Bassard JE, Richert L, Geerinck J, Renault H, Duval F, Ullmann P, Schmitt M, Meyer E, Mutterer J, Boerjan W, et al. Protein-protein and protein-membrane associations in the lignin pathway. Plant Cell. 2012:**24**(11):4465–4482. https://doi.org/10.1105/tpc.112.102566

Bate-Smith EC. The phenolic constituents of plants and their taxonomic significance. I. Dicotyledons. Bot J Linnean Soc. 1962:**58**(371):95–173. https://doi.org/10.1111/j.1095-8339.1962.tb00890.x

Beaudoin GA, Facchini PJ. Benzylisoquinoline alkaloid biosynthesis in opium poppy. Planta. 2014:**240**(1):19–32. https://doi.org/10.1007/s00425-014-2056-8

Bell JN, Dixon RA, Bailey JA, Rowell PM, Lamb CJ. Differential induction of chalcone synthase mRNA activity at the onset of phytoalexin accumulation in compatible and incompatible plant-pathogen interactions. Proc Natl Acad Sci U S A. 1984:**81**(11):3384–3388. https://doi.org/10.1073/pnas.81.11.3384

Bharadwaj R, Kumar SR, Sharma A, Sathishkumar R. Plant metabolic gene clusters: evolution, organization, and their applications in synthetic biology. Front Plant Sci. 2021:**12**:697318. https://doi.org/10.3389/fpls.2021.697318

Bizily SP, Rugh CL, Meagher RB. Phytodetoxification of hazardous organomercurials by genetically engineered plants. Nat Biotechnol. 2000:**18**(2):213 217. https://doi.org/10.1038/72678

Britsch L, Grisebach H. Purification and characterization of (2S)-flavanone 3-hydroxylase from *Petunia hybrida*. Eur J Biochem. 1986:**156**(3):569–577. https://doi.org/10.1111/j.1432-1033.1986.tb09616.x

Brophy JAN, Magallon KJ, Duan L, Zhong V, Ramachandran P, Kniazev K, Dinneny JR. Synthetic genetic circuits as a means of reprogramming plant roots. Science. 2022:**377**(6607):747–751. https://doi.org/10.1126/science.abo4326

Brouckaert M, Peng M, Höfer R, Houari IE, Darrah C, Storme V, Saeys Y, Vanholme R, Goeminne G, Timokhin VI, et al. QT-GWAS: a novel method for unveiling biosynthetic loci affecting qualitative metabolic traits. Mol Plant. 2023:**16**(7):1212–1227. 10.1016/j.molp.2023.06.004

Burrell RC. Phytochemistry. What it is and how it has developed? J Chem Educ. 1937:**14**(11):520. https://doi.org/10.1021/ed014p520

Chaiwanon J, Wang ZY. Spatiotemporal brassinosteroid signaling and antagonism with auxin pattern stem cell dynamics in Arabidopsis roots. Curr Biol. 2015:**25**(8):1031–1042. https://doi.org/10.1016/j.cub.2015.02.046

Chandler JW, Werr W. Cytokinin–auxin crosstalk in cell type specification. Trends Plant Sci. 2015:**20**(5):291–300. https://doi.org/10.1016/j.tplants.2015.02.003

Chen F, Tobimatsu Y, Havkin-Frenkel D, Dixon RA, Ralph J. A polymer of caffeyl alcohol in plant seeds. Proc Natl Acad Sci U S A. 2012:**109**(5):1772–1177. https://doi.org/10.1073/pnas.1120992109

Chen R, Chen X, Hagel JM, Facchini PJ. Virus-induced gene silencing to investigate alkaloid biosynthesis in Opium Poppy. Meth Mol Biol. 2020:**2172**:75–92. https://doi.org/10.1007/978-1-0716-0751-0_7

Choi D, Ward BL, Bostock RM. Differential induction and suppression of potato 3-hydroxy-3-methylglutaryl coenzyme A reductase genes in response to *Phytophthora infestans* and to its elicitor arachidonic acid. Plant Cell. 1992:**4**(10):1333–1344. https://doi.org/10.1105/tpc.4.10.1333

Conceição RS, Perez CJ, Branco A, Botura MB, Ifa DR. Identification of *Sassafras albidum* alkaloids by high-performance thin-layer chromatography tandem mass spectrometry and mapping by desorption electrospray ionization mass spectrometry imaging. J Mass Spectrom. 2021:**56**(1):e4674. https://doi.org/10.1002/jms.4674

Crosby KC, Pietraszewska-Bogiel A, Gadella TWJ, Winkel BSJ. Förster resonance energy transfer demonstrates a flavonoid metabolon in living plant cells that displays competitive interactions between enzymes. FEBS Lett. 2011:**585**(14):2193–2198. https://doi.org/10.1016/j.febslet.2011.05.066

Cruickshank IAM, Perrin DR. Studies on phytoalexins. III. The isolation, assay, and general properties of a phytoalexin from *Pisum sativum* L. Australian J Biol Sci. 1961:**14**(3):336–348. https://doi.org/10.1071/BI9610336

Dastmalchi M, Bernards MA, Dhaubhadel S. Twin anchors of the soybean isoflavonoid metabolon: evidence for tethering of the complex to the endoplasmic reticulum by IFS and C4H. Plant J. 2016:**85**(6): 689–706. https://doi.org/10.1111/tpj.13137

Davin LB, Lewis NG. Dirigent proteins and dirigent sites explain the mystery of specificity of radical precursor coupling in lignan and lignin biosynthesis. Plant Physiol. 2000:**123**(2):453–462. https://doi.org/10.1104/pp.123.2.453

De La Peña R, Hodgson H, Liu JCT, Stephenson MJ, Martin AC, Owen C, Harkess A, Leebens-Mack J, Jimenez LE, Osbourn A, et al. Complex scaffold remodeling in plant triterpene biosynthesis. Science. 2023:**379**(6630):361–368. https://doi.org/10.1126/science.adf1017

De La Peña R, Sattely ES. Re-routing plant terpene biosynthesis enables momilactone pathway elucidation. Nat Chem Biol. 2021:**17**(2):205–212. https://doi.org/10.1038/s41589-020-00669-3

de Saint Germain A, Clavé G, Schouveiler P, Pillot JP, Singh AV, Chevalier A, Daignan Fornier S, Guillory A, Bonhomme S, Rameau C, et al. Expansion of the strigolactone profluorescent probes repertory: the right probe for the right application. Front Plant Sci. 2022:**13**:887347. https://doi.org/10.3389/fpls.2022.887347

Dethier VG. Chemical factors determining the choice of food plants by *Papilio larvae*. Am Nat. 1941:**75**(756):61–73. https://doi.org/10.1086/280929

Dickinson AJ, Zhang J, Luciano M, Wachsman G, Sandoval E, Schnermann M, Dinneny JR, Benfey PN. A plant lipocalin promotes retinal-mediated oscillatory lateral root initiation. Science. 2021:**373**(6562):1532–1536. https://doi.org/10.1126/science.abf7461

Ding Y, Huffaker A, Köllner TG, Weckwerth P, Robert CAM, Pencer JL, Lipka AE, Schmelz EA. Selinene volatiles are essential precursors for maize defense promoting fungal pathogen resistance. Plant Physiol. 2017:**175**(3):1455–1468. https://doi.org/10.1104/pp.17.00879

Dixon RA, Barros J. Lignin biosynthesis: old roads revisited and new roads explored. Open Biol. 2019:**9**(12):190215. https://doi.org/10.1098/rsob.190215

Dong Y, Aharoni A. Image to insight: exploring natural products through mass spectrometry imaging. Nat Prod Rep. 2022:**39**(7): 1510–1530. https://doi.org/10.1039/D2NP00011C

Edwards K, Cramer CL, Bolwell GP, Dixon RA, Schuch W, Lamb CJ. Rapid transient induction of phenylalanine ammonia-lyase mRNA in elicitor-treated bean cells. Proc Natl Acad Sci U S A. 1985:**82**(20): 6731–6673. https://doi.org/10.1073/pnas.82.20.6731

Efroni I, Mello A, Nawy T, Ip PL, Rahni R, DelRose N, Powers A, Satija R, Birnbaum KD. Root regeneration triggers an embryo-like sequence guided by hormonal interactions. Cell. 2016:**165**(7): 1721–1733. https://doi.org/10.1016/j.cell.2016.04.046

Ellison EE, Nagalakshmi U, Gamo ME, Huang P-j, Dinesh Kumar S, Voytas D. Multiplexed heritable gene editing using RNA viruses and mobile single guide RNAs. Nat Plants. 2020:**6**(6):620–624. https://doi.org/10.1038/s41477-020-0670-y

Emiliani G, Fondi M, Fani R, Gribaldo S. A horizontal gene transfer at the origin of phenylpropanoid metabolism: a key adaptation of plants to land. Biol Direct. 2009:**4**(1):7. https://doi.org/10.1186/1745-6150-4-7

Field B, Osbourn AE. Metabolic diversification-independent assembly of operon-like gene clusters in different plants. Science. 2008:**320**(5875):543–547. https://doi.org/10.1126/science.1154990

Filner P, Varner JE. A test for de novo synthesis of enzymes: density labeling with H_2O^{18} of barley alpha-amylase induced by gibberellic acid. Proc Natl Acad Sci U S A. 1967:**58**(4):1520–1526. https://doi.org/10.1073/pnas.58.4.1520

Fisch C. William withering: an account of foxglove and some of its medical uses. 1785–1985. J Am Coll Cardiol. 1985:**5**(5):1A–2A. https://doi.org/10.1016/S0735-1097(85)80456-3

Freudenberg K. Lignin: its constitution and formation from *p*-hydroxycinnamyl alcohols: lignin is duplicated by dehydrogenation of these alcohols; intermediates explain formation and structure. Science. 1965:**148**(3670):595–600. https://doi.org/10.1126/science.148.3670.595

Frey M, Chomet P, Glawischnig E, Stettner C, Grun S, Winklmair A, Eisenreich W, Bacher A, Meeley RB, Briggs SP, et al. Analysis of a chemical plant defense mechanism in grasses. Science. 1997:**277**(5326): 696–699. https://doi.org/10.1126/science.277.5326.696

Fujino N, Tenma N, Waki T, Ito K, Komatsuzaki Y, Sugiyama K, Yamazaki T, Yoshida S, Hatayama M, Yamashita S, et al. Physical interactions among flavonoid enzymes in snapdragon and torenia reveal the diversity in the flavonoid metabolon organization of different plant species. Plant J. 2018:**94**(2):372–392. https://doi.org/10.1111/tpj.13864

Gani U, Vishwakarma RA, Misra P. Membrane transporters: the key drivers of transport of secondary metabolites in plants. Plant Cell Rep. 2021:**40**(1):1–18. https://doi.org/10.1007/s00299-020-02599-9

Gierlinger N, Schwanninger M. Chemical imaging of poplar wood cell walls by confocal Raman microscopy. Plant Physiol. 2006:**140**(4): 1246–1254. https://doi.org/10.1104/pp.105.066993

Glagoleva AY, Shoeva OY, Khlestkina EK. Melanin pigment in plants: current knowledge and future perspectives. Front Plant Sci. 2020:**11**: 770. https://doi.org/10.3389/fpls.2020.00770

Goold HD, Wright P, Hailstones D. Emerging opportunities for synthetic biology in agriculture. Genes (Basel). 2018:**9**(7):341. https://doi.org/10.3390/genes9070341

Gou M, Ran X, Martin DW, Liu CJ. The scaffold proteins of lignin biosynthetic cytochrome P450 enzymes. Nat Plants. 2018:**4**(5):299–310. https://doi.org/10.1038/s41477-018-0142-9

Graeff M, Rana S, Wendrich JR, Dorier J, Eekhout T, Aliaga Fandino AC, Guex N, Bassel GW, De Rybel B, Hardtke CS. A single-cell morpho-transcriptomic map of brassinosteroid action in the Arabidopsis root. Mol Plant. 2021:**14**(12):1985–1999. https://doi.org/10.1016/j.molp.2021.07.021

Greenhagen BT, O'Maille PE, Noel JP, Chappell J. Identifying and manipulating structural determinates linking catalytic specificities in terpene synthases. Proc Natl Acad Sci U S A. 2006:**103**(26): 9826–9831. https://doi.org/10.1073/pnas.0601605103

Grzech D, Hong B, Caputi L, Sonawane PD, O'Connor SE. Engineering the biosynthesis of late-stage vinblastine precursors precondylocarpine acetate, catharanthine, tabersonine in *Nicotiana benthamiana*. ACS Synth Biol. 2023:**12**(1):27–34. https://doi.org/10.1021/acssynbio.2c00434

Gutensohn M, Hartzell E, Dudareva N. Another level of complexity: the role of metabolic channeling and metabolons in plant terpenoid metabolism. Front Plant Sci. 2022:**13**:954083. https://doi.org/10.3389/fpls.2022.954083

Hadwiger LA, Schwochau ME. Ultraviolet light-induced formation of pisatin and phenylalanine ammonia lyase. Plant Physiol. 1971:**47**(4): 588–590. https://doi.org/10.1104/pp.47.4.588

Hahlbrock K, Grisebach H. Enzymic controls in the biosynthesis of lignin and flavonoids. Annu Rev Plant Biol. 1979:**30**(1):105–130. https://doi.org/10.1146/annurev.pp.30.060179.000541

Hahlbrock K, Scheel D. Physiology and molecular biology of phenylpropanoid metabolism. Annu. Rev Plant Physiol Plant Mol Biol. 1989:**40**(1): 347–369. https://doi.org/10.1146/annurev.pp.40.060189.002023

Hartmann T. The lost origin of chemical ecology in the late 19th century. Proc Natl Acad Sci U S A. 2008:**105**(12):4541–4546. https://doi.org/10.1073/pnas.0709231105

Haslam E. Biogenetically patterned synthesis of procyanidins. J Chem Soc Chem Commun. 1974:**1974**(15):594–595. https://doi.org/10.1039/c39740000594

Hayashi K, Nakamura S, Fukunaga S, Nishimura T, Jenness MK, Murphy AS, Motose H, Nozaki H, Furutani M, Aoyama T. Auxin transport sites are visualized in planta using fluorescent auxin analogs. Proc Natl Acad Sci U S A. 2014:**111**(31):11557–11562. https://doi.org/10.1073/pnas.1408960111

Hen-Avivi S, Savin O, Racovita RC, Lee W-S, Adamski NM, Malitsky S, Almekias-Siegl E, Levy M, Vautrin S, Bergès H, et al. A metabolic gene cluster in the wheat W1 and the barley Cer-cqu loci determines β-diketone biosynthesis and glaucousness. Plant Cell. 2016:**28**(6): 1440–1460. https://doi.org/10.1105/tpc.16.00197

Herud-Sikimić O, Stiel AC, Kolb M, Shanmugaratnam S, Berendzen KW, Feldhaus C, Höcker B, Jürgens G. A biosensor for the direct visualization of auxin. Nature. 2021:**592**(7856):768–772. https://doi.org/10.1038/s41586-021-03425-2

Hong B, Grzech D, Caputi L, Sonawane P, Rodriguez Lopez CE, Kamileen MO, Hernandez Lozada NJ, Grabe V, O'Connor S. Biosynthesis of strychnine. Nature. 2022:**607**(7919):617–622. https://doi.org/10.1038/s41586-022-04950-4

Hosmani PS, Kamiya T, Danku J, Naseer S, Geldner N, Guerinot ML, Salt DE. Dirigent domain-containing protein is part of the machinery required for formation of the lignin-based casparian strip in the root. Proc Natl Acad Sci U S A. 2013:**110**(35):14498–11503. https://doi.org/10.1073/pnas.1308412110

Hunziker P, Lambertz SK, Weber K, Crocoll C, Halkier BA, Schulz A. Herbivore feeding preference corroborates optimal defense theory for specialized metabolites within plants. Proc Natl Acad Sci U S A. 2021:**118**(47):e2111977118. https://doi.org/10.1073/pnas.2111977118

Hutzler P, Fischbach R, Heller W, Jungblut T, Reuber S, Schmitz R, Veit M, Weissenbock G, Schnitzler J. Tissue localization of phenolic compounds in plants by confocal laser scanning microscopy. J Exp Bot. 1998:**49**(323):953–965. https://doi.org/10.1093/jxb/49.323.953

Irani NG, Di Rubbo S, Mylle E, Van den Begin J, Schneider-Pizoń J, Hniliková J, Šíša M, Buyst D, Vilarrasa-Blasi J, Szatmári AM, et al. Fluorescent castasterone reveals BRI1 signaling from the plasma membrane. Nat Chem Biol. 2012:**8**(6):583–589. https://doi.org/10.1038/nchembio.958

Jacobs M, Rubery PH. Naturally occurring auxin transport regulators. Science. 1988:**241**(4863):346–349. https://doi.org/10.1126/science.241.4863.346

Jez JM, Lee SG, Sherp AM. The next green movement: plant biology for the environment and sustainability. Science. 2016:**353**(6305): 1241–1244. https://doi.org/10.1126/science.aag1698

Jia M, Mishra SK, Tufts S, Jernigan RL, Peters RJ. Combinatorial biosynthesis and the basis for substrate promiscuity in class I diterpene synthases. Metab Eng. 2019:**55**:44–58. https://doi.org/10.1016/j.ymben.2019.06.008

Jones AM, Danielson JA, Manojkumar SN, Lanquar V, Grossmann G, Frommer WB. Abscisic acid dynamics in roots detected with genetically encoded FRET sensors. Elife. 2014:**3**:e01741. https://doi.org/10.7554/eLife.01741

Jørgensen K, Rasmussen AV, Morant M, Nielsen AH, Bjarnholt N, Zagrobelny M, Bak S, Møller BL. Metabolon formation and metabolic channeling in the biosynthesis of plant natural products. Curr Opin Plant Biol. 2005:**8**(3):280–291. https://doi.org/10.1016/j.pbi.2005.03.014

Jun JH, Lu N, Docampo-Palacios M, Wang X, Dixon RA. Dual activity of anthocyanidin reductase supports the dominant plant proanthocyanidin extension unit pathway. Sci Adv. 2021:**7**(20):eabg4682. https://doi.org/10.1126/sciadv.abg4682

Kato K, Shitan N, Shoji T, Hashimoto T. Tobacco NUP1 transports both tobacco alkaloids and vitamin B6. Phytochemistry. 2015:**113**: 33–40. https://doi.org/10.1016/j.phytochem.2014.05.011

Ke J, Wang B, Yoshikuni Y. Microbiome engineering: synthetic biology of plant-associated microbiomes in sustainable agriculture. Trends Biotechnol. 2021:**39**(3):244–261. https://doi.org/10.1016/j.tibtech.2020.07.008

Kellner F, Kim J, Clavijo BJ, Hamilton JP, Childs KL, Vaillancourt B, Cepela J, Habermann M, Steuernagel B, Clissold L, et al. Genome-guided investigation of plant natural product biosynthesis. Plant J. 2015:**82**(4):680–692. https://doi.org/10.1111/tpj.12827

Kessler A, Baldwin IT. Defensive function of herbivore-induced plant volatile emissions in nature. Science. 2001:**291**(5511):2141–2144. https://doi.org/10.1126/science.291.5511.2141

Kibble NAJ, Sohani MM, Shirley N, Byrt C, Roessner U, Bacic A, Schmidt O, Schultz CJ. Phylogenetic analysis and functional characterisation of strictosidine synthase-like genes in *Arabidopsis thaliana*. Funct Plant Biol. 2010:**36**(12):1098–1109. https://doi.org/10.1071/FP09104

Kim SS, Wengier DL, Ragland CJ, Sattely ES. Transcriptional reactivation of lignin biosynthesis for the heterologous production of etoposide aglycone in *Nicotiana benthamiana*. ACS Synth Biol. 2022:**11**(10): 3379–3387. https://doi.org/10.1021/acssynbio.2c00289

Kochs G, Grisebach H. Enzymic synthesis of isoflavones. Eur J Biochem. 1986:**155**(2):311–318. https://doi.org/10.1111/j.1432-1033.1986.tb09492.x

Köckenberger W, De Panfilis C, Santoro D, Dahiya P, Rawsthorne S. High resolution NMR microscopy of plants and fungi. J Microsc. 2004:**214**(2):182–189. https://doi.org/10.1111/j.0022-2720.2004.01351.x

Köllner TG, David A, Luck K, Beran F, Kunert G, Zhou J-J, Caputi L, O'Connor SE. Biosynthesis of iridoid sex pheromones in aphids. Proc Natl Acad Sci U S A. 2022:**119**(42):e2211254119. https://doi.org/10.1073/pnas.2211254119

Korth KL, Jaggard DAW, Dixon RA. Developmental and light-regulated post-translational control of 3-hydroxy-3-methylglutaryl-CoA reductase levels in potato. Plant J. 2000:**23**(4):507–516. https://doi.org/10.1046/j.1365-313x.2000.00821.x

Koukol J, Conn EE. The metabolism of aromatic compounds in higher plants. IV. Purification and properties of the phenylalanine deaminase of *Hordeum vulgare*. J Biol Chem. 1961:**236**(10):2692–2698. https://doi.org/10.1016/S0021-9258(19)61721-7

Kreuzaler F, Hahlbrock K. Flavonoid glycosides from illuminated cell suspension cultures of *Petroselinum hortense*. Phytochemistry. 1973:**12**(5):1149–1152. https://doi.org/10.1016/0031-9422(73)85031-9

Kreuzaler F, Ragg H, Fautz E, Kuhn DN, Hahlbrock K. UV induction of chalcone synthase mRNA in cell suspension cultures of *Petroselinum hortense*. Proc Natl Acad Sci U S A. 1983:**80**(9):2591–2593. https://doi.org/10.1073/pnas.80.9.2591

Kulagina N, Méteignier L-V, Papon N, O'Connor SE, Courdavault V. More than a Catharanthus plant: a multicellular and pluri-organelle alkaloid-producing factory. Curr Opin Plant Biol. 2022:**67**:102200. https://doi.org/10.1016/j.pbi.2022.102200

Kumar P, Pandit SS, Steppuhn BI. Natural history-driven, plant-mediated RNAi-based study reveals CYP6B46's role in a nicotine-mediated antipredator herbivore defense. Proc Natl Acad Sci U S A. 2014:**111**(4):1245–1252. https://doi.org/10.1073/pnas.1314848111

Lange BM, Wildung MR, Stauber EJ, Sanchez C, Pouchnik D, Croteau R. Probing essential oil biosynthesis and secretion by functional evaluation of expressed sequence tags from mint glandular trichomes. Proc Natl Acad Sci U S A. 2000:**97**(6):2934–2939. https://doi.org/10.1073/pnas.97.6.2934

Lau W, Sattely ES. Six enzymes from mayapple that complete the biosynthetic pathway to the etoposide aglycone. Science. 2015:**349**(6253):1224–1228. https://doi.org/10.1126/science.aac7202

Lawton MA, Dixon RA, Lamb CJ. Elicitor modulation of L-phenylalanine ammonia-lyase in French bean cell suspension cultures. Biochim Biophys Acta. 1980:**633**(2):162–175. https://doi.org/10.1016/0304-4165(80)90402-X

Learned RM, Fink GR. 3-Hydroxy-3-methylglutaryl-coenzyme A reductase from *Arabidopsis thaliana* is structurally distinct from the yeast and animal enzymes. Proc Natl Acad Sci U S A. 1989:**86**(8): 2779–2783. https://doi.org/10.1073/pnas.86.8.2779

Leroux H. Discovery of salicine. J Chim Med. 1830:**6**:340–432.

Lewis DR, Ramirez MV, Miller ND, Vallabhaneni P, Ray WK, Helm RF, Winkel BS, Muday GK. Auxin and ethylene induce flavonol accumulation through distinct transcriptional networks. Plant Physiol. 2011:**156**(1):144–164. https://doi.org/10.1104/pp.111.172502

Li FS, Phyo P, Jacobowitz J, Hong M, Weng JK. The molecular structure of plant sporopollenin. Nat Plants. 2019:**5**(1):41–46. https://doi.org/10.1038/s41477-018-0330-7

Li J, Ou-Lee TM, Raba R, Amundson RG, Last RL. Arabidopsis flavonoid mutants are hypersensitive to UV-B irradiation. Plant Cell. 1993:**5**(2):171–179. https://doi.org/10.2307/3869583

Li R, Zhang J, Li Z, Peters RJ, Yang B. Dissecting the labdane-related diterpenoid biosynthetic gene clusters in rice reveals directional cross-cluster phytotoxicity. New Phytol. 2022:**233**(2):878–889. https://doi.org/10.1111/nph.17806

Liao P, Maoz I, Shih ML, Lee JH, Huang XQ, Morgan JA, Dudareva N. Emission of floral volatiles is facilitated by cell-wall non-specific lipid transfer proteins. Nat Commun. 2023:**14**(1):330. https://doi.org/10.1038/s41467-023-36027-9

Liao CY, Smet W, Brunoud G, Yoshida S, Vernoux T, Weijers D. Reporters for sensitive and quantitative measurement of auxin response. Nat Methods. 2015:**12**(3):207–210. https://doi.org/10.1038/nmeth.3279

Liscombe DK, O'Connor SE. A virus-induced gene silencing approach to understanding alkaloid metabolism in *Catharanthus roseus*. Phytochemistry. 2011:**72**(16):1969–1977. https://doi.org/10.1016/j.phytochem.2011.07.001

Liu Z, Cheema J, Vigouroux M, Hill L, Reed J, Paajanen P, Yant L, Osbourn A. Formation and diversification of a paradigm biosynthetic gene cluster in plants. Nat Commun. 2020:**11**(1):5354. https://doi.org/10.1038/s41467-020-19153-6

Liu C, Ha CM, Dixon RA. Functional genomics in the study of metabolic pathways in *Medicago truncatula*: an overview. Methods Mol Biol. 2018:**1822**:315–337. https://doi.org/10.1007/978-1-4939-8633-0_20

Liu C, Wang X, Shulaev V, Dixon RA. A role for leucoanthocyanidin reductase in the extension of proanthocyanidins. Nat Plants. 2016:**2**(12):16182. https://doi.org/10.1038/nplants.2016.182

Liu J, Müller B. Imaging TCSn::GFP, a synthetic cytokinin reporter, in *Arabidopsis thaliana*. Methods Mol Biol. 2017:**1497**:81–90. https://doi.org/10.1007/978-1-4939-6469-7_9

Liu S, Sang R. Bioactive fluorescent jasmonate designed by molecular modeling and its migration in tomato visualized by fluorescent molecular imaging. Tetrahedron. 2013:**69**(2):844–848. https://doi.org/10.1016/j.tet.2012.10.105

Liu X, Zhang P, Zhao Q, Huang AC. Making small molecules in plants: a chassis for synthetic biology-based production of plant natural products. J Integr Plant Biol. 2023:**65**(2):417–443. https://doi.org/10.1111/jipb.13330

Liu Z, Wang J, Zhou Y, Zhang Y, Qin A, Yu X, Zhao Z, Wu R, Guo C, Bawa G, et al. Identification of novel regulators required for early development of vein pattern in the cotyledons by single-cell RNA-sequencing. Plant J. 2022:**110**(1):7–22. https://doi.org/10.1111/tpj.15719

Lou Y-R, Pichersky E, Last RL. Deep roots and many branches: origins of plant-specialized metabolic enzymes in general metabolism. Curr Opin Plant Biol. 2022:**66**:102192. https://doi.org/10.1016/j.pbi.2022.102192

Luzarowski M, Vicente R, Kiselev A, Wagner M, Schlossarek D, Erban A, de Souza LP, Childs D, Wojciechowska I, Luzarowska U, et al. Global mapping of protein-metabolite interactions in Saccharomyces cerevisiae reveals that Ser-Leu dipeptide regulates phosphoglycerate kinase activity. Commun Biol. 2021:**4**(1):181. https://doi.org/10.1038/s42003-021-01684-3

Maceda A, Terrazas T. Fluorescence microscopy methods for the analysis and characterization of lignin. Polymers (Basel). 2022:**14**(5):961. https://doi.org/10.3390/polym14050961

Maeda HA. Harnessing evolutionary diversification of primary metabolism for plant synthetic biology. J Biol Chem. 2019:**294**(45): 16549–16566. https://doi.org/10.1074/jbc.REV119.006132

Mahdi JG, Mahdi AJ, Bowen ID. The historical analysis of aspirin discovery, its relation to the willow tree and antiproliferative and anticancer potential. Cell Prolif. 2006:**39**(2):147–155. https://doi.org/10.1111/j.1365-2184.2006.00377.x

Markham KR, Mabry TJ. Ultraviolet-visible and proton magnetic resonance spectroscopy of flavonoids. In: **Harborne HB, Mabry TJ, Mabry H**, editors. The flavonoids. New York (NY): Academic Press; 1975. p. 45–77.

McKey D. The distribution of secondary compounds within plants. In: **Rosenthal GA, Janzen DH**, editors. Herbivores: their interaction with secondary plant metabolites. New York (NY): Academic Press; 1979. p. 55–134.

Miettinen K, Pollier J, Buyst D, Arendt P, Csuk R, Sommerwerk S, Moses T, Mertens J, Sonawane PD, Pauwels L, et al. The ancient CYP716 family is a major contributor to the diversification of eudicot triterpenoid biosynthesis. Nat Commun. 2017:**8**(1):14153. https://doi.org/10.1038/ncomms14153

Modolo LV, Reichert AI, Dixon RA. Introduction to the different classes of biosynthetic enzymes. In: **Osbourn A, Lanzotti V**, editors. Plant-derived natural products, synthesis, function and application. New York (NY): Springer; 2009. p. 143–163.

Montinari MR, Minelli S, De Caterina R. The first 3500 years of aspirin history from its roots—a concise summary. Vascul Pharmacol. 2019:**113**:1–8. https://doi.org/10.1016/j.vph.2018.10.008

Muhlemann JK, Younts TLB, Muday GK. Flavonols control pollen tube growth and integrity by regulating ROS homeostasis during high-temperature stress. Proc Natl Acad Sci U S A. 2018:**115**(47): 11188–11197. https://doi.org/10.1073/pnas.1811492115

Murata J, Roepke J, Gordon H, De Luca V. The leaf epidermome of *Catharanthus roseus* reveals its biochemical specialization. Plant Cell. 2008:**20**(3):524–542. https://doi.org/10.1105/tpc.107.056630

Nemhauser JL, Torii KU. Plant synthetic biology for molecular engineering of signalling and development. Nat Plants. 2016:**2**(3):16010. https://doi.org/10.1038/nplants.2016.10

Nett RS, Lau W, Sattely ES. Discovery and engineering of colchicine alkaloid biosynthesis. Nature. 2020:**584**(7819):148–153. https://doi.org/10.1038/s41586-020-2546-8

Nishimura K, Fukagawa T, Takisawa H, Kakimoto T, Kanemaki M. An auxin-based degron system for the rapid depletion of proteins in nonplant cells. Nat Meth. 2009:**6**(12):917–292. https://doi.org/10.1038/nmeth.1401

Nolan TM, Vukašinović N, Hsu CW, Zhang J, Vanhoutte I, Shahan R, Taylor IW, Greenstreet L, Heitz M, Afanassiev A, et al. Brassinosteroid gene regulatory networks at cellular resolution in the Arabidopsis root. Science. 2023:**379**(6639):eadf4721. https://doi.org/10.1126/science.adf4721

One Thousand Plant Transcriptomes Initiative. One thousand plant transcriptomes and the phylogenomics of green plants. Nature. 2019:**574**(7780):679–685. https://doi.org/10.1038/s41586-019-1693-2

Onoyovwe A, Hagel JM, Chen X, Khan MF, Schriemer D, Facchini PJ. Morphine biosynthesis in opium poppy involves two cell types: sieve elements and laticifers. Plant Cell. 2013:**25**(10):4110–4122. https://doi.org/10.1105/tpc.113.115113

Osbourn A. Gene clusters for secondary metabolic pathways: an emerging theme in plant biology. Plant Physiol. 2010:**154**(2):531–535. https://doi.org/10.1104/pp.110.161315

Owen C, Patron NJ, Huang A, Osbourn A. Harnessing plant metabolic diversity. Curr Opin Chem Biol. 2017:**40**:24–30. https://doi.org/10.1016/j.cbpa.2017.04.015

Pařízková B, Žukauskaitė A, Vain T, Grones P, Raggi S, Kubeš MF, Kieffer M, Doyle SM, Strnad M, Kepinski S, et al. New fluorescent auxin probes visualise tissue-specific and subcellular distributions of auxin in Arabidopsis. New Phytol. 2021:**230**(2):535–549. https://doi.org/10.1111/nph.17183

Peng Y, Yang J, Li X, Zhang Y. Salicylic acid: biosynthesis and signaling. Annu Rev Plant Biol. 2021:**72**(1):761–791. https://doi.org/10.1146/annurev-arplant-081320-092855

Perkins ML, Schuetz M, Unda F, Chen KT, Bally MB, Kulkarni JA, Yan Y, Pico J, Castellarin SD, Mansfield SD, et al. Monolignol export by diffusion down a polymerization-induced concentration gradient. Plant Cell. 2022:**34**(5):2080–2095. https://doi.org/10.1093/plcell/koac051

Qi X, Bakht S, Leggett M, Maxwell C, Melton R, Osbourn A. A gene cluster for secondary metabolism in oat: implications for the evolution of metabolic diversity in plants. Proc Natl Acad Sci U S A. 2004:**101**(21):8233–8238. https://doi.org/10.1073/pnas.0401301101

Ralph J, Brunow G, Harris P, Dixon RA, Boerjan W. Lignification: are lignins biosynthesized via simple combinatorial chemistry or via proteinaceous control and template replication? Rec Adv Polyphenols Res. 2008:**1**:36–66. https://doi.org/10.1002/9781444302400.ch2

Ralph J, Lapierre C, Boerjan W. Lignin structure and its engineering. Curr Opin Biotechnol. 2019:**56**:240–249. https://doi.org/10.1016/j.copbio.2019.02.019

Reed J, Orme A, El-Demerdash A, Owen C, Martin LBB, Misra RC, Kikuchi S, Rejzek M, Martin AC, Harkess A, et al. Elucidation of the pathway for biosynthesis of saponin adjuvants from the soapbark tree. Science. 2023:**379**(6638):1252–1264. https://doi.org/10.1126/science.adf3727

Reed J, Stephenson MJ, Miettinen K, Brouwer B, Leveau A, Brett P, Goss RJM, Goossens A, O'Connell MA, Osbourn A. A translational synthetic biology platform for rapid access to gram-scale quantities of novel drug-like molecules. Metabolic Eng. 2017:**42**:185–193. https://doi.org/10.1016/j.ymben.2017.06.012

Ribeiro B, Lacchini E, Bicalho KU, Mertens J, Arendt P, Vanden Bossche R, Calegario G, Gryffroy L, Ceulemans E, Buitink J, et al. A seed-specific regulator of triterpene saponin biosynthesis in *Medicago truncatula*. Plant Cell. 2020:**32**(6):2020–2042. https://doi.org/10.1105/tpc.19.00609

Robinson R. A theory of the mechanism of the phytochemical synthesis of certain alkaloids. J Chem Soc Trans. 1917:**111**(0):876–899. https://doi.org/10.1039/CT9171100876

Roux D, Ferreira D. Structure and function in the biomimetic synthesis of linear, angular and branched condensed tannins. Pure Appl Chem. 1982:**54**(12):2465–2478. https://doi.org/10.1351/pac198254122465

Rowe J, Grangé-Guermente M, Exposito-Rodriguez M, Wimalasekera R, Lenz MO, Shetty KN, Cutler SR, Jones AM. Next-generation ABACUS biosensors reveal cellular ABA dynamics driving root growth at low aerial humidity. Nat Plants. 2023:**9**(7):1103–1115. https://doi.org/10.1038/s41477-023-01447-4

Rueffer M, Zenk MH. Distant precursors of benzylisoquinoline alkaloids and their enzymatic formation. Z Naturforsch. 1987:**42**(4):319–332. https://doi.org/10.1515/znc-1987-0402

Saito K, Matsuda F. Metabolomics for functional genomics, systems biology, and biotechnology. Annu Rev Plant Biol. 2010:**61**(1):463–489. https://doi.org/10.1146/annurev.arplant.043008.092035

Salazar-Cerezo S, Martínez-Montiel N, García-Sánchez J, Pérez-y-Terrón R, Martínez-Contreras RD. Gibberellin biosynthesis and metabolism: a convergent route for plants, fungi and bacteria. Microbiol Res. 2018:**208**:85–98. https://doi.org/10.1016/j.micres.2018.01.010

Salehin M, Li B, Tang M, Katz E, Song L, Ecker JR, Kliebenstein DJ, Estelle M. Auxin-sensitive Aux/IAA proteins mediate drought tolerance in Arabidopsis by regulating glucosinolate levels. Nat Commun. 2019:**10**(1):4021. https://doi.org/10.1038/s41467-019-12002-1

Salim V, Yu F, Altarejos J, De Luca V. Virus-induced gene silencing identifies *Catharanthus roseus* 7-deoxyloganic acid-7-hydroxylase, a step in iridoid and monoterpene indole alkaloid biosynthesis. Plant J. 2013:**76**(5):754–765. https://doi.org/10.1111/tpj.12330

Sarabia LD, Boughton BA, Rupasinghe T, van de Meene AML, Callahan DL, Hill CB, Roessner U. High-mass-resolution MALDI mass spectrometry imaging reveals detailed spatial distribution of metabolites and lipids in roots of barley seedlings in response to salinity stress. Metabolomics. 2018:**14**(5):63. https://doi.org/10.1007/s11306-018-1359-3

Sarsby J, Towers MW, Stain C, Cramer R, Koroleva OA. Mass spectrometry imaging of glucosinolates in Arabidopsis flowers and siliques. Phytochemistry. 2012:**77**:110–118. https://doi.org/10.1016/j.phytochem.2012.01.026

Schmelzer E, Jahnen W, Hahlbrock K. In situ localization of light-induced chalcone synthase mRNA, chalcone synthase, and flavonoid end products in epidermal cells of parsley leaves. Proc Natl Acad Sci U S A. 1988:**85**(9):2989–2993. https://doi.org/10.1073/pnas.85.9.2989

Schmitz R. Friedrich Wilhelm Serturner and the discovery of morphine. Pharm Hist. 1985:**27**:61–74. PMID: 11611724.

Serrano-Ron L, Perez-Garcia P, Sanchez-Corrionero A, Gude I, Cabrera J, Ip PL, Birnbaum KD, Moreno-Risueno MA. Reconstruction of lateral root formation through single-cell RNA sequencing reveals order of tissue initiation. Mol Plant. 2021:**14**(8):1362–1378. https://doi.org/10.1016/j.molp.2021.05.028

Sheahan JJ. Sinapate esters provide greater UV-B attenuation than flavonoids in *Arabidopsis thaliana* (Brassicaceae). Am J Botany. 1996:**83**(6):679–686. https://doi.org/10.1002/j.1537-2197.1996.tb12757.x

Shiono K, Hashizaki R, Nakanishi T, Sakai T, Yamamoto T, Ogata K, Harada KI, Ohtani H, Katano H, Taira S. Multi-imaging of cytokinin and abscisic acid on the roots of rice (*Oryza sativa*) using matrix-assisted laser desorption/ionization mass spectrometry. J Agric Food Chem. 2017:**65**(35):7624–7628. https://doi.org/10.1021/acs.jafc.7b02255

Slocombe SP, Schauvinhold I, McQuinn RP, Besser K, Welsby NA, Harper A, Aziz N, Li Y, Larson TR, Giovannoni J, et al. Transcriptomic and reverse genetic analyses of branched-chain fatty acid and acyl sugar production in *Solanum pennellii* and *Nicotiana benthamiana*. Plant Physiol. 2008:**148**(4):1830–1846. https://doi.org/10.1104/pp.108.129510

Somssich IE, Bollmann J, Hahlbrock K, Kombrink E, Schulz W. Differential early activation of defense-related genes in elicitor-treated parsley cells. Plant Mol Biol. 1989:**12**(2):227–234. https://doi.org/10.1007/BF00020507

Sonawane PD, Jozwiak A, Panda S, Aharoni A. 'Hijacking' core metabolism: a new panache for the evolution of steroidal glycoalkaloids structural diversity. Curr Opin Plant Biol. 2020:**55**:118–128. https://doi.org/10.1016/j.pbi.2020.03.008

St-Pierre B, Vazquez-Flota F, De Luca V. Multicellular compartmentation of *Catharanthus roseus* alkaloid biosynthesis predicts intercellular translocation of a pathway intermediate. Plant Cell. 1999:**11**(5):887–900. https://doi.org/10.1105/tpc.11.5.887

Stavrinides A, Tatsis EC, Foureau E, Caputi L, Kellner F, Courdavault V, O'Connor SE. Unlocking the diversity of alkaloids in *Catharanthus*

roseus: nuclear localization suggests metabolic channeling in secondary metabolism. Chem Biol. 2015:**22**(3):336–341. https://doi.org/10.1016/j.chembiol.2015.02.006

Steppuhn A, Gase K, Krock B, Halitschke R, Baldwin IT. Nicotine's defensive function in nature. PLoS Biol. 2004:**2**(8):e217. https://doi.org/10.1371/journal.pbio.0020217

Swain T. Plant phenolics: past and future. Annu Proc Phytochem Soc Europe. 1985:**25**:453–468.

Tadege M, Wen J, He J, Tu H, Kwak Y, Eschstruth A, Cayrel A, Endre G, Zhao PX, Chabaud M, et al. Large-scale insertional mutagenesis using the *Tnt1* retrotransposon in the model legume *Medicago truncatula*. Plant J. 2008:**54**(2):335–347. https://doi.org/10.1111/j.1365-313X.2008.03418.x

Tibbs Cortes L, Zhang Z, Yu J. Status and prospects of genome-wide association studies in plants. Plant Genome. 2021:**14**(1):e20077. https://doi.org/10.1002/tpg2.20077

Tobimatsu Y, Wagner A, Donaldson L, Mitra P, Niculaes C, Dima O, Kim JI, Anderson N, Loque D, Boerjan W, et al. Visualization of plant cell wall lignification using fluorescence-tagged monolignols. Plant J. 2013:**76**(3):357–366. https://doi.org/10.1111/tpj.12299

Tomiyama K, Sakuma T, Ishizaka N, Sato N, Katsui N, Tagasugi M, Masamune T. A new antifungal substance isolated from resistant potato tuber tissue. Phytopathol. 1968:**58**:115–116.

Torii K, Kubota A, Araki T, Endo M. Time-series single-cell RNA-seq data reveal auxin fluctuation during endocycle. Plant Cell Physiol. 2020:**61**(2):243–254. https://doi.org/10.1093/pcp/pcz228

Treimer JF, Zenk MH. Purification and properties of strictosidine synthase, the key enzyme in indole alkaloid formation. Eur J Biochem. 1979:**101**(1):225–233. https://doi.org/10.1111/j.1432-1033.1979.tb04235.x

Ulmasov T, Murfett J, Hagen G, Guilfoyle TJ. Aux/IAA proteins repress expression of reporter genes containing natural and highly active synthetic auxin response elements. Plant Cell. 1997:**9**(11): 1963–1971. https://doi.org/10.1105/tpc.9.11.1963

Uslu VV, Grossmann G. The biosensor toolbox for plant developmental biology. Curr Opin Plant Biol. 2016:**29**:138–147. https://doi.org/10.1016/j.pbi.2015.12.001

VanEtten HD, Mansfield JW, Bailey JA, Farmer EE. Two classes of plant antibiotics: phytoalexins versus "phytoanticipins". Plant Cell. 1994:**6**(9):1191–1192. https://doi.org/10.2307/3869817

Vanholme R, De Meester B, Ralph J, Boerjan W. Lignin biosynthesis and its integration into metabolism. Curr Opin Biotechnol. 2019:**56**:230–239. https://doi.org/10.1016/j.copbio.2019.02.018

van Wersch S, Li X. Stronger when together: clustering of plant NLR disease resistance genes. Trends Plant Sci. 2019:**24**(8):688–699. https://doi.org/10.1016/j.tplants.2019.05.005

Waadt R, Hitomi K, Nishimura N, Hitomi C, Adams SR, Getzoff ED, Schroeder JI. FRET-based reporters for the direct visualization of abscisic acid concentration changes and distribution in Arabidopsis. eLife. 2014:**3**:e01739. https://doi.org/10.7554/eLife.01739

Wang Y, Cao Y, Liang X, Zhuang J, Wang X, Qin F, Jiang C. A dirigent family protein confers variation of casparian strip thickness and salt tolerance in maize. Nat Commun. 2022:**13**(1):2222. https://doi.org/10.1038/s41467-022-29809-0

Wendrich JR, Yang B, Vandamme N, Verstaen K, Smet W, Van de Velde C, Minne M, Wybouw B, Mor E, Arents HE, et al. Vascular transcription factors guide plant epidermal responses to limiting phosphate conditions. Science. 2020:**370**(6518):eaay4970. https://doi.org/10.1126/science.aay4970

Wijekoon CP, Facchini PJ. Systematic knockdown of morphine pathway enzymes in opium poppy using virus-induced gene silencing. Plant J. 2012:**69**(6):1052–1063. https://doi.org/10.1111/j.1365-313X.2011.04855.x

Wilderman PR, Xu M, Jin Y, Coates RM, Peters RJ. Identification of syn-imara-7,15-diene synthase reveals functional clustering of terpene synthases involved in rice phytoalexin/allelochemical biosynthesis. Plant Physiol. 2004:**135**(4):2098–2105. https://doi.org/10.1104/pp.104.045971

Winkel-Shirley B. Evidence for enzyme complexes in the phenylpropanoid and flavonoid pathways. Physiol Plant. 1999:**107**(1):142–149. https://doi.org/10.1034/j.1399-3054.1999.100119.x

Winkel-Shirley B. Flavonoid biosynthesis. A colorful model for genetics, biochemistry, cell biology, and biotechnology. Plant Physiol. 2001:**126**(2):485–493. https://doi.org/10.1104/pp.126.2.485

Witvrouw K, Kim H, Vanholme R, Goeminne G, Ralph J, Boerjan W, Vanholme B. Reassessing the claimed cytokinin-substituting activity of dehydrodiconiferyl alcohol glucoside. Proc Natl Acad Sci U S A. 2023:**120**(9):e2123301120. https://doi.org/10.1073/pnas.2123301120

Wu R, Duan L, Pruneda-Paz JL, Oh DH, Pound M, Kay S, Dinneny JR. The 6xABRE synthetic promoter enables the spatiotemporal analysis of ABA-mediated transcriptional regulation. Plant Physiol. 2018:**177**(4):1650–1665. https://doi.org/10.1104/pp.18.00401

Wurtzel E, Kutchan T. Plant metabolism, the diverse chemistry set of the future. Science. 2016:**353**(6305):1232–1236. https://doi.org/10.1126/science.aad2062

Xie M, Muchero W, Bryan AC, Yee K, Guo HB, Zhang J, Tschaplinski TJ, Singan VR, Lindquist E, Payyavula RS, et al. A 5-enolpyruvylshikimate 3-phosphate synthase functions as a transcriptional repressor in Populus. Plant Cell. 2018:**30**(7):1645–1660. https://doi.org/10.1105/tpc.18.00168

Yamamoto K, Takahashi K, Caputi L, Mizuno H, Rodriguez-Lopez CE, Iwasaki T, Ishizaki K, Fukaki H, Ohnishi M, Yamazaki M, et al. The complexity of intercellular localisation of alkaloids revealed by single-cell metabolomics. New Phytol. 2019:**224**(2):848–859. https://doi.org/10.1111/nph.16138

Yang JF, Chen WJ, Zhou LM, Hewage KAH, Fu YX, Chen MX, He B, Pei RJ, Song K, Zhang JH, et al. Real-time fluorescence imaging of the abscisic acid receptor allows nondestructive visualization of plant stress. ACS Appl Mater Interfaces. 2022:**14**(25):28489–28500. https://doi.org/10.1021/acsami.2c02156

Yang Y, Wang X, Zhao C, Tian G, Zhang H, Xiao H, He L, Zheng J. Chemical mapping of essential oils, flavonoids and carotenoids in citrus peels by Raman microscopy. J Food Sci. 2017:**82**(12):2840–2846. https://doi.org/10.1111/1750-3841.13952

Ye Z-H, Kneusel RE, Matern U, Varner JE. An alternative methylation pathway in lignin biosynthesis in Zinnia. Plant Cell. 1994:**6**(10): 1427–1439. https://doi.org/10.1105/tpc.6.10.1427

Young ND, Debellé F, Oldroyd GE, Geurts R, Cannon SB, Udvardi MK, Benedito VA, Mayer KF, Gouzy J, Schoof H, et al. The Medicago genome provides insight into the evolution of rhizobial symbioses. Nature. 2011:**480**(7378):520–524. https://doi.org/10.1038/nature10625

Yu F, De Luca V. ATP-binding cassette transporter controls leaf surface secretion of anticancer drug components in *Catharanthus roseus*. Proc Natl Acad Sci U S A. 2013:**110**(39):15830–15835. https://doi.org/10.1073/pnas.1307504110

Yu K, Song Y, Lin J, Dixon RA. The complexities of proanthocyanidin biosynthesis and its regulation in plants. Plant Commun. 2023:**4**(2): 100498. https://doi.org/10.1016/j.xplc.2022.100498

Zenk MH. Biosynthese von vanillin in *Vanilla planifolia*. Andr Z Pflanzenphysiol. 1965:**53**:404–414.

Zenk MH. Chasing the enzymes of secondary metabolism: plant cell cultures as a pot of gold. Phytochemistry. 1991:**30**(12):3861–3863. https://doi.org/10.1016/0031-9422(91)83424-J

Zenkner FF, Margis-Pinheiro M, Cagliari A. Nicotine biosynthesis in Nicotiana: a metabolic overview. Tobacco Sci. 2019:**56**(1):1–9. https://doi.org/10.3381/18-063

Zhang T, Noll SE, Peng JT, Klair A, Tripka A, Stutzman N, Cheng C, Zare RN, Dickinson AJ. Chemical imaging reveals diverse functions of tricarboxylic acid metabolites in root growth and development. Nat Commun. 2023:**14**(1):2567. https://doi.org/10.1038/s41467-023-38150-z

https://doi.org/10.1093/plphys/kiad601 PLANT PHYSIOLOGY 2024: 195: 67–78

Plant Physiology®

Topical Review

The end game(s) of photosynthetic carbon metabolism

Thomas D. Sharkey *

MSU-DOE Plant Research Laboratory, Plant Resilience Institute, and Department of Biochemistry and Molecular Biology, Michigan State University, East Lansing, MI 48824, USA

*Author for correspondence: tsharkey@msu.edu

The author responsible for distribution of materials integral to the findings presented in this article in accordance with the policy described in the Instructions for Authors (https://academic.oup.com/plphys/pages/General-Instructions) is Thomas D. Sharkey (tsharkey@msu.edu).

Abstract

The year 2024 marks 70 years since the general outline of the carbon pathway in photosynthesis was published. Although several alternative pathways are now known, it is remarkable how many organisms use the reaction sequence described 70 yrs ago, which is now known as the Calvin–Benson cycle or variants such as the Calvin–Benson–Bassham cycle or Benson–Calvin cycle. However, once the carbon has entered the Calvin–Benson cycle and is converted to a 3-carbon sugar, it has many potential fates. This review will examine the last stages of photosynthetic metabolism in leaves. In land plants, this process mostly involves the production of sucrose provided by an endosymbiont (the chloroplast) to its host for use and transport to the rest of the plant. Photosynthetic metabolism also usually involves the synthesis of starch, which helps maintain respiration in the dark and enables the symbiont to supply sugars during both the day and night. Other end products made in the chloroplast are closely tied to photosynthetic CO_2 assimilation. These include serine from photorespiration and various amino acids, fatty acids, isoprenoids, and shikimate pathway products. I also describe 2 pathways that can short circuit parts of the Calvin–Benson cycle. These final processes of photosynthetic metabolism play many important roles in plants.

Introduction

The phrase "path of carbon in photosynthesis" was used in the titles of 22 papers published in the 1940s and 1950s, plus 2 books, and Melvin Calvin's Nobel Lecture (Calvin 1964). The most consequential is the 21st such paper (Bassham et al. 1954), which will be 70 years old in 2024 and is referred to here as "paper 21." This paper has been ci-ted over 320 times (as of August 9, 2023) but laid the groundwork for many thousands of papers. The achievement of Bassham, Benson, and Calvin has been extensively documented (e.g. Nickelsen 2012; Sharkey 2019). Like the Krebs/tricarboxylic acid (TCA)/citric acid cycle, the scheme in paper 21 goes by various names, but here, I will refer to it as the Calvin–Benson cycle. Much current interest in this cycle focuses on sedoheptulose bisphosphatase (SBPase) (Lefebvre et al. 2005; Zhu et al. 2007; Rosenthal et al. 2011; Driever et al. 2017; Simkin et al. 2017). This enzyme is used in preference to transaldolase normally found in the nonoxidative pentose phosphate pathway, which has the effect of directing carbon from trioses to pentoses rather than from pentoses to hexoses and trioses, as is normally depicted for the transaldolase-dependent pentose phosphate pathway (Sharkey 2021).

In 1954, when paper 21 was published, almost nothing was known about how carbon enters the cycle, except that there are 3 sources of pentose and that a pentose (now called RuBP) is carboxylated to generate 3-phosphoglycerate (3-PGA). The carboxylase eventually became known as rubisco (Portis and Parry 2007; Sharkey 2022). A big question about photosynthetic carbon metabolism remained: what are the last steps as carbon leaves the cycle? While paper 21 did not track the phosphorylation status of the intermediates, it became clear that ATP drives the cycle forward. Therefore, the phosphate added to make RuBP must be

Received August 18, 2023. Accepted October 27, 2023. Advance access publication January 2, 2024

ADVANCES BOX

- The last steps of photosynthetic carbon metabolism in the endosymbiont chloroplast provide important precursors for plants.
- The last steps of photosynthetic carbon metabolism are essential in order to release phosphate from the organic phosphate pool for reuse in ATP.
- Photorespiration can make serine as an end product of photosynthetic carbon metabolism although the estimates of 40% serine export seem high.
- A stromal G6P shunt involving the oxidative pentose phosphate pathway consumes carbon from the Calvin–Benson cycle and consumes ATP. It is normally not functioning during the day.
- A cytosolic G6P shunt reinjects free glucose, fructose, and sucrose into the Calvin–Benson cycle helping it refill very quickly after periods of low light.

released so that ATP can be regenerated. All the phosphate added by phosphoglycerate kinase is immediately released by glyceraldehyde 3-phosphate dehydrogenase. One third of the phosphate added by phosphoribulokinase is released by fructose bisphosphatase (FBPase) and one third by SBPase, but the last third must be released during end-product synthesis. This sets a reasonable boundary for photosynthetic carbon metabolism: it begins with CO_2 and ends only when an equivalent amount of carbon in intermediates is dephosphorylated.

Glucose is often regarded as the ultimate product of photosynthetic carbon metabolism, but unphosphorylated glucose only occurs in part of the starch utilization pathway. The bulk of carbon fixed in photosynthesis ends up as either starch or sucrose (Fig. 1). If starch is made, all photosynthetic carbon metabolism is said to occur inside chloroplasts. However, as an endosymbiont, the chloroplast supplies the host cell with reduced carbon, and starch synthesis capacity is typically not sufficient to account for the high rates of CO_2 fixation that could be observed using isolated intact chloroplasts (Walker and Herold 1977).

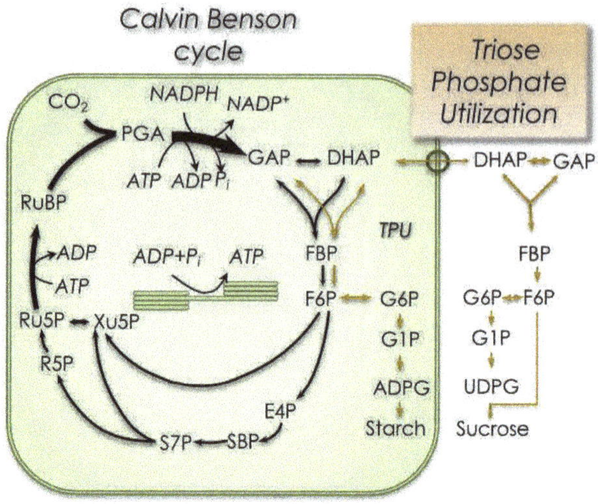

Figure 1. The carbon dioxide that enters the Calvin–Benson cycle is metabolized primarily to starch and sucrose (lines on right side of the figure). Since both sucrose and starch synthesis can be readily traced back to the triose phosphates, this is often called triose phosphate utilization. PGA, 3-phosphoglycerate; TPI, triose phosphate isomerase; GAP, glyceraldehyde 3-phosphate (distinguished from glycerol 3-phosphate denoted G3P); DHAP, dihydroxyacetone phosphate; G6P, glucose 6-phosphate; F6P, fructose 6-phosphate; FBP, fructose 1,6-bisphosphate; E4P, erythrose 4-phosphate; glyco-TDP, glycoaldehyde thiamine diphosphate; SBP, sedoheptulose 1,7-bisphosphate; S7P, sedoheptulose 7-phosphate; R5P, ribose 5-phosphate; Xu5P; xylulose 5-phosphate; Ru5P, ribulose 5-phosphate.

Sucrose metabolism

Carbon export from chloroplasts during CO_2 fixation depends on the obligatory export of triose phosphate and the import of inorganic phosphate (Heldt et al. 1977; Fliege et al. 1978). Phosphate and triose phosphate exchange occurs at a triose phosphate/phosphate antiporter (TPT), an abun-dant protein on the inner envelope of the chloroplast and a member of a family of related phosphate antiporters (Bockwoldt et al. 2019). The evolution of these antiporters is thought to have been an essential step to establishing the endosymbiotic nature of chloroplasts (Weber et al. 2006; Linka and Weber 2010). The export of triose phosphate supports the synthesis of sucrose in the cytosol of plant cells (Bird et al. 1974). As sucrose is a phosphate-free molecule, its production marks the end of photosynthetic carbon metabolism.

Sucrose synthesis involves the formation of sucrose 6-phosphate from fructose 6-phosphate (F6P) and uridine diphosphate (UDP)-glucose by the enzyme sucrose-phosphate synthase (SPS). The sucrose phosphate is readily broken down to sucrose by SPS. Sucrose-phosphate synthase is highly regulated, especially by phosphorylation (Huber and Huber 1996; Hardin et al. 2003). Several sites can be phosphorylated with different effects, and it is dephosphorylated by SPS protein phosphatase, a type 2A protein phosphatase (Huber and Huber 1996). This regulation also involves a 14-3-3 protein; such proteins are often involved in regulating phosphorylation (Bachmann et al. 1996). While SPS activity is lower and more sensitive to metabolite concentrations at night, it is not turned off completely (Jones and Ort 1997), allowing sucrose synthesis to occur during both day and night.

Sucrose is a disaccharide that plays several central roles in plants (Salerno and Curatti 2003). A similar disaccharide, trehalose, is more widespread phylogenetically. The answer to the question "why sucrose?" is unclear (Salerno and Curatti 2003). In plants, trehalose metabolism appears to be specialized for carbohydrate status signaling, especially trehalose

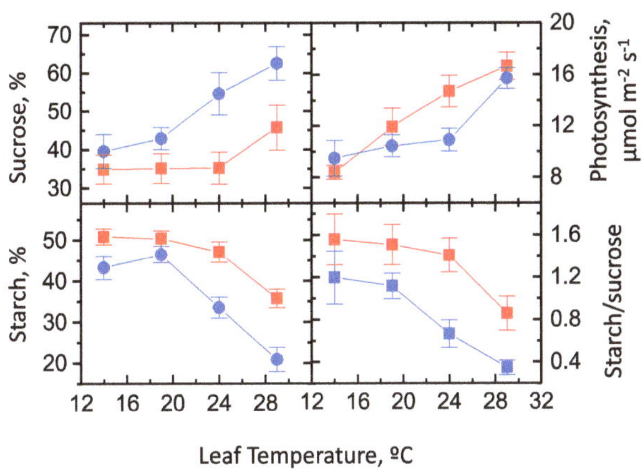

Figure 2. Sucrose and starch synthesis as a function of temperature in tomato with and without a SPS gene from maize. Squares are data from transformed plants, and circles are from untransformed plants. The proportion of sucrose synthesis from CO_2 fixation increases with temperature, while starch synthesis declines. The SPS gene had little effect on the rate of photosynthesis, which increased over a temperature range of 14 °C to 29 °C. Data are from Laporte (1997).

6-phosphate (Lastdrager et al. 2014; Fichtner and Lunn 2021; Peixoto et al. 2021) and its interactions with sucrose nonfermenting-related kinase (SnRK) and target of rapamycin (TOR) regulatory mechanisms (Jamsheer et al. 2019; Ryabova et al. 2019; Sharma et al. 2022).

Starch metabolism

The identification of the TPT explained carbon export from the chloroplast during the day, but what about carbon export at night? Starch accumulates throughout the day in most plants. At low rates of photosynthesis, most carbon is partitioned to sucrose, but as the photosynthetic rate increases, due to either increased light or increased CO_2 availability, more and more carbon is partitioned to starch (Sharkey et al. 1985; Mullen and Koller 1988). The ratio of starch to sucrose is higher in plants growing in short days (Sulpice et al. 2014; Xu et al. 2023) and can be low when leaves are incubated at low temperature (Pollock and Lloyd 1987). The ratio of partitioning to starch versus su-crose can be regulated by activity of SPS (Galtier et al. 1993). Increasing SPS activity by transforming tomato (*Solanum lyco-persicum*) with an SPS gene from maize (*Zea mays*) increased the proportion of carbon going to sucrose and decreased the proportion going to starch (Fig. 2, redrawn from Laporte (1997)). The starch/sucrose ratio declined with increasing tem-perature in both the transformed line and the controls.

Another critical control point for sucrose synthesis is the ratio of FBPase activity in the stroma to that in the cytosol. In the stroma, this regulation involves light activation, among other mechanisms. In the cytosol, FBPase activity is regulated by fructose 2,6-bisphosphate (Stitt 1990). F6P is phosphorylated and dephosphorylated at the 2 position, a process that integrates many different signals. The regulation of FBPase, and its mirror image, i.e., the regulation of phosphofructokinase, prevents a futile cycle in which ATP is used to convert F6P to F1,6BP, and FBPase removes that phosphate, consuming ATP. Regulation involving fructose 2,6-bisphosphate ensures that these reactions are occurring in only 1 direction, toward sucrose during photosynthesis and consuming glucose only when photosynthesis is not occurring. The regulation at FBPase is more complete than the regulation of SPS so that at night, sucrose cannot be produced from triose phosphates but it can be produced from hexose phosphates.

Starch breakdown then supplies the plant with sugars at night. Starch is broken down at a constant rate at night; more than 90% of the carbon present in starch at the end of the day is released from starch at night (Lu et al. 2005; Pokhilko et al. 2014; Flis et al. 2019). As a result, the chloroplast supplies reduced carbon to the host at an almost constant rate day and night (Sulpice et al. 2014).

It was assumed that at night, the sugars in starch broke down to triose phosphates, which were then exported on the TPT. However, deuterium labeling showed that when starch breaks down and is converted to sucrose, the carbon skeletons are never broken down to triose phosphates (Schleucher et al. 1998). Instead, maltose, the product of β-amylase, is the primary export sugar resulting from starch breakdown (Weise et al. 2004). Initial attempts to show a gra-dient for maltose export were unsuccessful, leading to the finding that β-maltose is the primary exported sugar. β-maltose, but not total maltose, shows a significant gradient in favor of export from chloroplasts (Weise et al. 2005).

β-amylase cleaves 2 glucose residues from α-1,4 maltodextrins (glucose chains) to generate β-maltose. However, this enzyme cannot work on maltotriose, only longer maltodextrins. To finish breaking down starch remnants, an enzyme called disproportionating enzyme 1 (DPE1) (1 for the plastidial form) uses maltotriose in this reaction:

$$\text{Glu3} + \text{Glu3} \rightleftharpoons \text{Glu5} + \text{Glu}$$

where Glu3 is maltotriose and Glu5 is maltopentaose. Other reengagements are possible, but the net effect is that maltodextrins can be broken down primarily to produce maltose but some glucose as well. Glucose in the chloroplast can exit through a glucose transporter, which was first reported in 2000 (Weber et al. 2000). A maltose transporter was discovered soon after (Niittylä et al., 2004). Once in the cytosol, the maltose is acted on by DPE2 (Chia et al. 2004; Lu and Sharkey 2004) (Fig. 3). DPE2 is similar to DPE1 but contains a 150 amino acid insert (Steichen et al. 2008). However, unlike DPE1, DPE2 transfers the nonreducing glucose to a soluble heteroglycan (SHG) (Fettke et al. 2006). This large-molecular-weight polysaccharide primarily comprises arabinose, galactose, ribose, and some glucose (Yang and Steup 1990). The glucose transferred to the heteroglycan can be liberated by phosphorolysis. In this way, the energy in the glucose–glucose bond of maltose can be conserved in glucose 1-phosphate (G1P), which can be isomerized to glucose 6-phosphate (G6P) or used to make UDP-glucose.

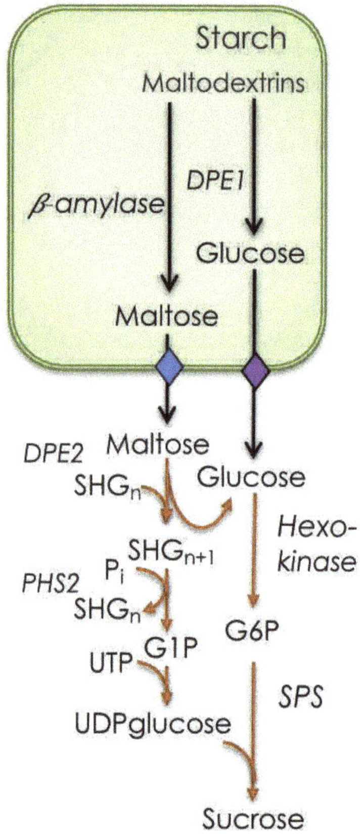

Figure 3. Conversion of starch to sucrose at night. Various starch-degrading enzymes break the starch down to linear maltodextrins, which are hydrolyzed by β-amylase to generate β-maltose. Once the maltodextrin reaches a degree of polymerization of 3, it is acted on by DPE1. Mostly maltose, but some glucose too, leaves the chloroplast through specific transporters. Maltose is acted on by disproportionating enzyme 2 (DPE2) to add 1 glucose to a SHG and glucose. Starch phosphorylase 2 (PHS2) removes a glucose from the SHG to produce G1P. The glucose is acted on by hexokinase to make G6P. SPS generates sucrose phosphate, which is then dephosphorylated to sucrose.

Currently, there is much interest in how starch breakdown is regulated. The remarkably constant rate of starch breakdown at night has attracted much attention (Lu et al. 2005; Graf et al. 2010; Scialdone et al. 2013; Fernandez et al. 2017; Mengin et al. 2017). Transitory starch is phosphorylated by a glucan water dikinase (Ritte et al. 2002) and phosphoglucan water dikinase (Kötting et al. 2005). Without this phosphorylation, starch becomes more difficult to break down. Some plants accumulate nontransitory starch as they age (Chu et al. 2022), possibly due to the accumulation of unphosphorylated starch, which is no longer accessible to the plant, over wks or mos.

Photorespiration

It was recently hypothesized that photorespiration can provide some capacity for end-product synthesis in the form of glycine and serine (Harley and Sharkey 1991; Busch et al. 2018; Fu et al. 2023). It has long been recognized that O_2 inhibits photosynthesis. This is sometimes called the (green) Warburg effect (Kutschera et al. 2020); the red effect is related to cancer metabolism (Liberti and Locasale 2016). Rabinowitch (1945) used the term "photorespiration" to describe metabolism that begins upon illumination and ends upon darkness. Decker (1955) described a postillumination burst of CO_2 release from leaves believed to result from the light-dependent process that releases CO_2. Wilson and Calvin (1955) found that glycolate was produced in their re-actions and that this could be suppressed by using 1% CO_2. Bassham and Kirk (1962) established that the production of glycolate (and glycine) could be affected by O_2. However, the major advance in the field of photorespiration was the dis-covery that the CO_2-fixing enzyme rubisco also fixes O_2 (Bowes et al. 1971; Ogren and Bowes 1971). The 2-phospho-glycolate formed in this reaction is a potent inhibitor of triose phosphate isomerase (Anderson 1971; Flügel et al. 2017; Li et al. 2019). Phosphoglycolate also inhibits phosphofructoki-nase (Kelly and Latzko 1976), but at higher concentrations. Phosphofructokinase inside chloroplasts is not operational during photosynthesis when FBPase is functioning. NADPH provides 1 mechanism for ensuring that phosphofructoki-nase is not active in the light (Cséke et al. 1982). The pathway for metabolizing 2-phosphoglycolate formed by the oxygen-ation of RuBP was proposed by Tolbert (1971).

In the view of photorespiration presented above, this process is not a method for end-product synthesis, since all the carbon that enters the pathway is either released as CO_2 or rephosphorylated to reenter the Calvin–Benson cycle. However, for many years, it appeared as though photorespiration could add capacity for end-product synthesis. An early example was reported by Jolliffe and Tregunna (1973), who showed that under high CO_2 and low temperature conditions, the photosynthetic rate under low O_2 conditions (low photorespiration) was lower than the rate in 21% O_2. Sharkey (1985) proposed that when end-product synthesis limits photosynthesis, CO_2 assimilation becomes insensitive to CO_2 and O_2 (in other words, photorespiration does not decrease the rate of CO_2 assimilation). We now know that photorespiration in fact increases the capacity for photosynthesis when starch and sucrose synthesis rates are maximal. This may be explained by the hypothesis that glycine, and more likely serine, leaves photorespiratory metabolism, providing both reduced nitrogen and carbon for use in leaf metabolism (Busch et al. 2018). If this process occurs at a rate proportional to the rate of photorespiration, and if end-product synthesis sets the upper bound of CO_2 assimilation, then photorespiration increases the upper bound. As a result, as CO_2 levels increase and the rate of photorespiration de-creases, the overall rate of CO_2 assimilation will also decrease.

Under low light conditions, increasing CO_2 or decreasing O_2 levels stimulate photosynthesis, as would be expected if the rate of photorespiration is reduced (Fig. 4). However, as light levels increase, the upper bound of the rate of CO_2 assimilation falls under the control of end-product synthesis, and increased CO_2 or decreased O_2 levels reduce

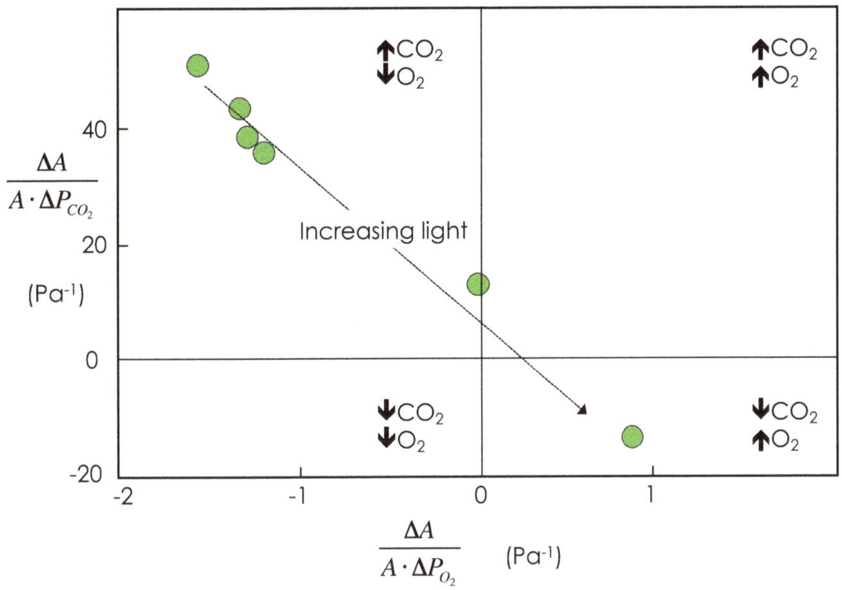

Figure 4. Assimilation-weighted sensitivity to CO_2 (vertical axis) and O_2 (horizontal axis) as affected by increasing light levels. At limiting light levels, the energy cost of photorespiration results in a reduction in CO_2 assimilation capacity. As light levels increase, the upper bound becomes sensitive to end-product synthesis. Photorespiration stimulates CO_2 assimilation by adding end-product synthesis capacity in the form of serine export. Redrawn from Sharkey (1990). Arrows indicate the response of photosynthesis to the indicated gases.

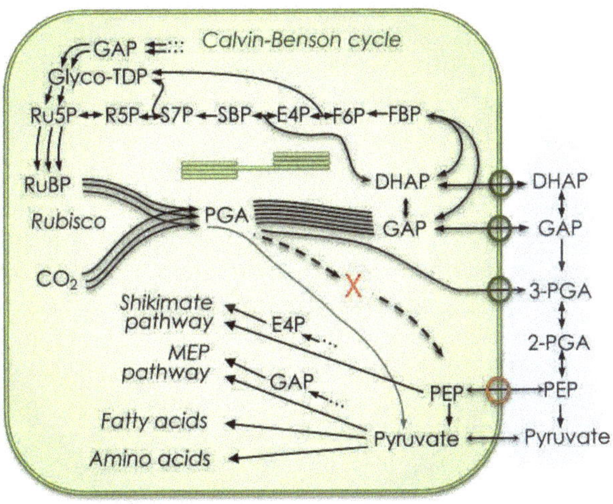

Figure 5. Chloroplast metabolism emphasizing the lower glycolytic pathway. The TPT (circles on the chloroplast membrane) exchanges DHAP, GAP, and PGA. The PEP/phosphate antiporter exchanges PEP for phosphate. The production of PEP inside the chloroplast is restricted by very low phosphoglyceromutase and enolase activities (denoted with X). Pyruvate is generated at a rate of ∼0.5% of total carboxylations, directly providing pyruvate for chloroplast reactions (line connecting rubisco to pyruvate). Additional products of the Calvin–Benson cycle include shikimate pathway products (e.g. aromatic amino acids), isoprenoids, and pentoses. See Raines (2011, 2022) (Fig. 1) for a more complete depiction of export sites from the Calvin–Benson cycle.

photorespiration, thereby reducing the overall rate of CO_2 assimilation. The value of 30% to 40% of carbon diverted from 2-phosphoglycolate to serine found from curve fitting (Busch et al. 2018), and isotopically non–steady-state mass flux analysis (Fu et al. 2023) seems high. It is possible that other metabolisms occur that have the same effect as serine export for nitrogen metabolism.

While serine export can explain many observations of reversed sensitivity, some observations are too extreme to be ac-counted for by serine export from photorespiration. Sharkey and Vassey (1989) determined that the rate of CO_2 assimilation decreased by 20% upon switching from 200 to 20 kPa O_2. This decline was almost entirely the result of a reduced rate of starch synthesis. The amount of stromal G6P fell by 73%, likely due to the inhibition of phosphoglucoisomerase (PGI). This enzyme is known to be out of equilibrium in the stroma (Gerhardt et al. 1987; Schleucher et al. 1999) and can be inhibited by some Calvin–Benson cycle intermediates, especially erythrose 4-phos-phate (Preiser et al. 2020). In this case, the reversed sensitivity to O_2 is not a function of increased end-product synthesis capacity associated with photorespiration. Instead, it is a function of a metabolic "traffic jam" in which increases in the levels of some Calvin–Benson cycle intermediates reduce the capacity for end-product (in this case starch) synthesis.

There is currently substantial interest in devising alternative pathways for metabolizing 2-phosphoglycerate, the metabolite that initiates photorespiration (Peterhänsel and Maurino 2011; Peterhänsel et al. 2013; Dalal et al. 2015; Xin et al. 2015; Engqvist and Maurino 2017; South et al. 2019). On the other hand, some researchers believe that photorespiration plays an important role in nitrogen metabolism (Rachmilevitch et al. 2004; Bloom 2015; Busch et al. 2018) and that plants experience evolutionary pressure to maintain photorespiration.

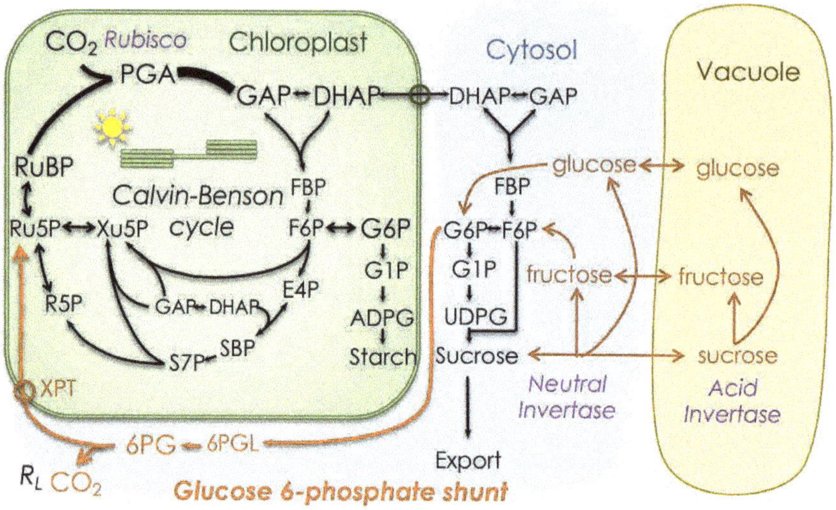

Figure 6. Model of photosynthetic carbon metabolism including sources of unlabeled carbon that is imported into the Calvin–Benson cycle. The import of unlabeled carbon that enters as hexose slows the rate of labeling of Calvin–Benson cycle intermediates. The G6P shunt also releases moderately labeled CO_2 that can be detected as $^{12}CO_2$ emitted into a 99+% $^{13}CO_2$ atmosphere. Abbreviations as in Fig. 1 plus 6PGL, 6-phosphogluconolactone; 6PG, 6-phosphogluconate; XPT, xylulose 5-phosphate/phosphate antiporter (also transports Ru5P). A detailed flux map based on this model is available in Xu et al. (2022).

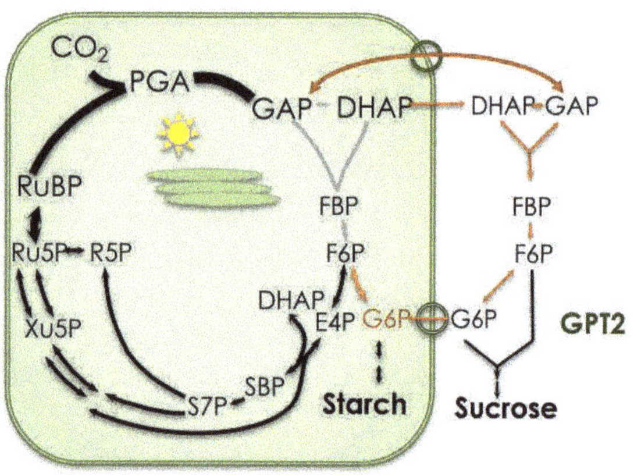

Figure 7. Mechanism for the cytosolic bypass of critical enzymes of the Calvin–Benson cycle. Triose phosphate isomerase, aldolase, and FBPase can be bypassed when the G6P transporter GPT2 is induced. However, the low capacity of PGI causes G6P to build up in the stroma, ultimately activating stromal G6PD, causing a futile cycle that consumes ATP. The ATP is replaced by cyclic electron flow, which has been observed in plants carrying out cytosolic bypass.

The lower glycolysis metabolism of chloroplasts

The reduction step and several steps in the regeneration portion of the Calvin–Benson cycle are gluconeogenic, that is, the reverse of glycolysis, from 3-PGA to F6P and G6P. The lower branch of glycolysis from PGA to phospho*enol*pyruvate (PEP) appears to be absent from photosynthesizing chloroplasts. Phosphoglyceromutase and enolase activities were found to be <1% of the activities in the cytosol (Stitt and ap Rees 1979; Schulze-Siebert et al. 1987; Hoppe et al. 1993) (X in Fig. 5). The TPT (circles on the chloroplast membrane in Fig. 5) exchanges phosphate, dihydroxyacetone phosphate (DHAP), glyceraldehyde 3-phosphate (GAP), and PGA. The metabolism in the stroma and cytosol determines the concentration gradients of these 4 molecules, and they are exchanged based on their concentration gradients. A triose phosphate isomerase can isomerize DHAP and GAP, but whole-leaf triose phosphate concentrations are not in isomerase equilibrium (Li et al. 2019). A non-phosphorylating glyceraldehyde 3-phosphate dehydrogenase (GAPN) that converts GAP to PGA is present and can provide NADPH in the cytosol. This reaction is irreversible. There are also phosphorylating GAPD enzymes, i.e. GAPC1 and GAPC2 (Guo et al. 2014). The amount of PGA that can be converted to GAP depends on the energetic status of the cytosol.

Several biosynthetic pathways require PEP (e.g. the shikimate pathway and its many products). PEP is imported from the cytosol by a PEP/P_i antiporter (bottom ring on the chloroplast membrane in Fig. 5). When the antiporter is absent, as in the Arabidopsis (*Arabidopsis thaliana*) *cue1* (chlorophyll a/b binding protein [CAB] underexpressors) mutants, the plants do not grow well (Voll et al. 2003). Voll et al. (2003) found that the *cue1* mutant could be rescued by either expressing a gene encoding a PEP/P_i antiporter (PPT) or by expressing a pyruvate phosphate dikinase (PPDK) gene in the stroma. The finding that expressing a PPT gene in the mutant background rescued this mutant indicates that the chloroplast does not have sufficient phosphoglycerate mutase to support the requirement for PEP in the chloroplast.

The reaction from PEP to pyruvate catalyzed by pyruvate kinase is not easily reversed, but the PPDK reaction is energetically favorable because it, in essence, uses the energy of 2 ATPs. The rescue of the *cue1* mutant by PPDK means that there is sufficient pyruvate to satisfy the need for PEP inside the chloroplast. Thus, either pyruvate is transported into the chloroplast or the small amount of pyruvate made by ru-bisco (Andrews and Kane 1991) is sufficient for the plant's needs.

Other chloroplast-localized metabolic pathways require pyruvate, e.g. the methylerythritol 4-phosphate (MEP) pathway (the source of carotenoids among other classes of molecules), fatty acid synthesis, and the synthesis of branched chain amino acids. There are 4 possible sources of pyruvate in the chloroplast stroma: PEP that enters through the PPT could be converted to pyruvate by pyruvate kinase inside the chloroplast, pyruvate could be imported from the cytosol by passive permeation, pyruvate could be transported by the BASS2 importer (Furumoto et al. 2011), or pyruvate could be supplied as a side reaction of rubisco (Andrews and Kane 1991). One source of cytosolic pyruvate is malic enzyme acting on malate from the vacuole.

The pyruvate paradox

When $^{13}CO_2$ is fed to leaves, intermediates of the Calvin–Benson cycle are labeled very quickly at first, but the rate of labeling slows considerably when 80% to 90% of the intermediates are labeled. All the intermediates of the Calvin–Benson cycle are labeled with very similar kinetics (Szecowka et al. 2013; Ma et al. 2014; Xu et al. 2021; Xu et al. 2022). This includes 2-PGA (Szecowka et al. 2013), which should only be present in the cytosol, not the chloroplast, and PEP (Hasunuma et al. 2010; Ma et al. 2014; Xu et al. 2022), which should mix with any PEP in the cytosol. However, pyruvate, which should be produced from PEP, is labeled much more slowly (Szecowka et al. 2013; Xu et al. 2022) than all the other molecules shown in Fig. 5. Although whole-leaf pyruvate is relatively unlabeled after 20 min, isoprene, 40% of which is derived from pyruvate, is labeled to the same degree as Calvin–Benson cycle intermediates (Delwiche and Sharkey 1993; Sharkey et al. 2020), indicating that the metabolically active pyruvate pool is labeled to the same degree as Calvin–Benson cycle intermediates. Perhaps a very large amount of pyruvate in the vacuole is labeled very slowly, masking the very rapid labeling of a small, metabolically active pyruvate pool in the chloroplast (and presumably also in the cytosol). It is possible that pyruvate is generated in the chloroplast by rubisco (Andrews and Kane 1991) and that this rubisco-derived pyruvate is heavily labeled and is used preferentially in the MEP pathway and presumably for fatty acid synthesis and other stromal reac-tions requiring pyruvate.

In summary, the intermediates of the Calvin–Benson cycle, starch synthesis precursors (e.g. ADP-glucose) and both PEP and pyruvate, constitute a large metabolically active pool of carbon that is labeled with similar kinetics. Of all these molecules, only pyruvate is also present in a metabolically inactive pool.

Shunts that bypass Calvin–Benson cycle reactions

The radioactive isotope ^{14}C (Ruben and Kamen 1940) played a critical role in elucidating the carbon pathway in photosyn-thesis. However, when 99+% labeled CO_2 is fed to photosyn-thesizing leaves, Calvin–Benson cycle intermediates do not become fully labeled (Mahon et al. 1974; Hasunuma et al. 2010; Szecowka et al. 2013; Ma et al. 2014), indicating that ei-ther nonmetabolic pools of most Calvin–Benson cycle inter-mediates are present or a source of unlabeled carbon enters the cycle.

A model of photosynthetic carbon metabolism has been used in metabolic flux analysis that includes oxidative pentose phosphate pathway reactions forming a shunt that bypasses a substantial portion of the Calvin–Benson cycle (Xu et al. 2022). This model (Fig. 6) can account for the lack of complete labeling and a nonphotorespiratory CO_2 release known as day respiration (R_d) or respiration in the light (R_L).

The G6P shunt occurring in the cytosol appears to operate continuously and at a reasonably constant rate, as assessed by measuring R_L (Tcherkez et al. 2017; Schmiege et al. 2023). A similar shunt can operate in the stroma, but this ap-pears to occur only in response to stress (Sharkey et al. 2020), although a recent report describes a very large stromal shunt that operates under either low or high CO_2 levels (Wieloch et al. 2022). A different shunt can be induced to bypass missing enzymes in the Calvin–Benson cycle. This shunt involves GAP leaving the chloroplast via the TPT, its conversion to G6P, and its reimport through an inducible G6P transporter called GPT2 (Kammerer et al. 1998).

Ordinarily, there is no exchange of G6P across the chloroplast envelope, and large gradients of G6P between the stroma and cytosol have been reported (Gerhardt et al. 1987; Sharkey and Vassey 1989; Schleucher et al. 1998; Schleucher et al. 1999). Keeping the G6P level low in the chloroplast has the advantage of preventing stimulation of the stromal G6P dehydrogenase (G6PD), which would lead to a futile cycle. The stromal G6PD is normally inactive during the day due to the thioredoxin-dependent reduction of a disulfide bridge (Wenderoth et al. 1997; Née et al. 2009; Née et al. 2014; Cardi et al. 2016; Yoshida et al. 2019). However, the deactivation is mostly the result of an increase in K_m rather than a reduction in overall capacity (Scheibe et al. 1989; Née et al. 2014). In addition, G6P can overcome the redox regulation of G6PD (Cossar et al. 1984; Preiser et al. 2019). Therefore, G6PD activity in the stroma can be very sensitive to G6P concentration. This G6P sensitivity likely explains why regulatory mechanisms are in place to limit the G6P concentration in the stroma, including the lack of G6P exchange with the cytosol and the limited capacity at the stromal PGI. However, plants lacking otherwise essential enzymes such as stromal triose phosphate isomerase, aldolase, and FBPase can survive by making use of the cytosolic versions of these enzymes and letting G6P back into the stroma through the inducible GPT2 (Fig. 7). The low capacity of

PGI will cause G6P to build up, stimulating the futile G6P shunt in the stroma. This leads to cyclic electron flow to make up for the ATP lost in the futile cycle. Therefore, although there is evidence for the cytosolic bypass mechanism described here (Gotoh et al. 2010; Livingston et al. 2010; Li et al. 2019), plants that depend on this bypass mechanism are generally compromised.

Conclusion

The core of photosynthetic carbon metabolism is the Calvin–Benson cycle. To keep this cycle going, end products must be generated that result in the release of phosphate for reuse in ATP. Chloroplasts produce many end products. The rates of sucrose, starch, and serine synthesis (from photorespiration) can affect the gas exchange behavior of leaves. These processes, plus some bypass mechanisms, are the end game of photosynthetic CO_2 fixation, but they mark the beginning of the next phase, such as using the resources for maintenance, to build the plant, to flower, and to reproduce. The production of end products of photosynthesis also represents a starting point for designing ways to modify photosynthesis to enhance food, fuel, and fiber production for human use. There has been substantial progress in understanding the end-product metabolism of photosynthesis, but regulation of G6P in the stroma and the pyruvate paradox remain important areas for research (Outstanding questions box).

OUTSTANDING QUESTIONS BOX

- The apparent large flux of serine out of photorespiration requires additional study. What is the fate of all of that serine?
- G6P in the chloroplast appears to be strongly controlled by regulation of G6P isomerase. How this regulation occurs and whether it can be modified to improve how photosynthesis meets human interests is an important area for research.
- Solving the pyruvate paradox also is an outstanding concern that should be investigated. The ability of rubisco to make pyruvate has been reported once but needs to be independently verified, and then, the consequences of this pyruvate production need additional study.

Author contributions

T.D.S. conceived the idea and wrote the paper.

Funding

The ideas presented here draw on my photosynthesis research, which is supported by the Division of Chemical Sciences, Geosciences, and Biosciences, Office of Basic Energy Sciences, of the U.S. Department of Energy (grant DE-FG02-91ER20021), and my isoprene research, which is supported by U.S. National Science Foundation, award number IOS-2022495. I also received partial salary support from Michigan AgBioResearch.

Conflict of interest statement. None declared.

Data availability

No new data were created for this work.

References

Anderson LE. Chloroplast and cytoplasmic enzymes. II. Pea leaf triose phosphate isomerases. Biochim Biophys Acta. 1971:**235**(1):237–244. https://doi.org/10.1016/0005-2744(71)90051-9

Andrews TJ, Kane HJ. Pyruvate is a by-product of catalysis by ribulosebisphosphate carboxylase/oxygenase. J Biol Chem. 1991:**266**(15):9447–9452. https://doi.org/10.1016/S0021-9258(18)92841-3

Bachmann M, Huber JL, Athwal GS, Wu K, Ferl RJ, Huber SC. 14-3-3 proteins associate with the regulatory phosphorylation site of spinach leaf nitrate reductase in an isoform-specific manner and reduce dephosphorylation of Ser-543 by endogenous protein phosphatases. FEBS Lett. 1996:**398**(1):26–30. https://doi.org/10.1016/S0014-5793(96)01188-X

Bassham JA, Benson AA, Kay LD, Harris AZ, Wilson AT, Calvin M. The path of carbon in photosynthesis. XXI. The cyclic regeneration of carbon dioxide acceptor. J Am Chem Soc. 1954:**76**(7):1760–1770. https://doi.org/10.1021/ja01636a012

Bassham JA, Kirk M. The effect of oxygen on the reduction of CO_2 to glycolic acid and other products during photosynthesis by Chlorella. Biochem Biophys Res Commun. 1962:**9**(5):376–380. https://doi.org/10.1016/0006-291X(62)90019-0

Bird IF, Cornelius MJ, Keys AJ, Whittingham CP. Intracellular site of sucrose synthesis in leaves. Phytochemistry 1974:**13**(1):59–64. https://doi.org/10.1016/S0031-9422(00)91267-6

Bloom AJ. Photorespiration and nitrate assimilation: a major intersection between plant carbon and nitrogen. Photosynth Res. 2015:**123**(2):117–128. https://doi.org/10.1007/s11120-014-0056-y

Bockwoldt M, Heiland I, Fischer K. The evolution of the plastid phosphate translocator family. Planta 2019:**250**(1):245–261. https://doi.org/10.1007/s00425-019-03161-y

Bowes G, Ogren WL, Hageman RH. Phosphoglycolate production catalyzed by ribulose diphosphate carboxylase. Biochem Biophys Res Commun. 1971:**45**(3):716–722. https://doi.org/10.1016/0006-291X(71)90475-X

Busch FA, Sage RF, Farquhar GD. Plants increase CO_2 uptake by assimilating nitrogen via the photorespiratory pathway. Nat Plants. 2018:**4**(1):46–54. https://doi.org/10.1038/s41477-017-0065-x

Calvin M. The path of carbon in photosynthesis. In nobel lectures, chemistry 1942–1962. Amsterdam: Elsivier Publishing Company; 1964. http://www.nobelprize.org/nobel_prizes/chemistry/laureates/1961/calvin-lecture.html

Cardi M, Zaffagnini M, De Lillo A, Castiglia D, Chibani K, Gualberto JM, Rouhier N, Jacquot J-P, Esposito S. Plastidic P2 glucose-6P dehydrogenase from poplar is modulated by thioredoxin *m*-type: distinct roles of cysteine residues in redox regulation and NADPH inhibition. Plant Sci. 2016:**252**:257–266. https://doi.org/10.1016/j.plantsci.2016.08.003

Chia T, Thorneycroft D, Chapple A, Messerli G, Chen J, Zeeman SC, Smith SM, Smith AM. A cytosolic glucosyltransferase is required for conversion of starch to sucrose in Arabidopsis leaves at night. Plant J. 2004:**37**(6):853–863. https://doi.org/10.1111/j.1365-313X.2003.02012.x

Chu KL, Koley S, Jenkins LM, Bailey SR, Kambhampati S, Foley K, Arp JJ, Morley SA, Czymmek KJ, Bates PD, et al. Metabolic flux analysis of the non-transitory starch tradeoff for lipid production in mature tobacco leaves. Metab Eng. 2022:**69**:231–248. https://doi.org/10.1016/j.ymben.2021.12.003

Cossar JD, Rowell P, Stewart WDP. Thioredoxin as a modulator of glucose-6-phosphate dehydrogenase in a N_2-fixing cyanobacterium. J Gen Microbiol. 1984:**130**:991–998. https://doi.org/10.1099/00221287-130-4-991

Cséke C, Nishizawa AN, Buchanan BB. Modulation of chloroplast phosphofructokinase by NADPH: a mechanism for linking light to the regulation of glycolysis. Plant Physiol. 1982:**70**(3):658–661. https://doi.org/10.1104/pp.70.3.658

Dalal J, Lopez H, Vasani NB, Hu Z, Swift JE, Yalamanchili R, Dvora M, Lin X, Xie D, Qu R, et al. A photorespiratory bypass increases plant growth and seed yield in biofuel crop *Camelina sativa*. Biotechnol Biofuels. 2015:**8**(1):175. https://doi.org/10.1186/s13068-015-0357-1

Decker JP. A rapid, postillumination deceleration of respiration in green leaves. Plant Physiol. 1955:**30**(1):82–84. https://doi.org/10.1104/pp.30.1.82

Delwiche CF, Sharkey TD. Rapid appearance of ^{13}C in biogenic isoprene when $^{13}CO_2$ is fed to intact leaves. Plant Cell Environ. 1993:**16**(5):587–591. https://doi.org/10.1111/j.1365-3040.1993.tb00907.x

Driever SM, Simkin AJ, Alotaibi S, Fisk SJ, Madgwick PJ, Sparks CA, Jones HD, Lawson T, Parry MAJ, Raines CA. Increased SBPase activity improves photosynthesis and grain yield in wheat grown in greenhouse conditions. Philos Trans R Soc Lond B Biol Sci. 2017:**372**(1730):20160384. https://doi.org/10.1098/rstb.2016.0384

Engqvist MKM, Maurino VG. Metabolic engineering of photorespiration. In: **Fernie AR, Bauwe H, Weber APM,** editors. Photorespiration: methods and protocols. New York (NY): Springer New York; 2017. p. 137–155.

Fernandez O, Ishihara H, George GM, Mengin V, Flis A, Sumner D, Arrivault S, Feil R, Lunn JE, Zeeman SC, et al. Leaf starch turnover occurs in long days and in falling light at the end of the day. Plant Physiol. 2017:**174**(4):2199–2212. https://doi.org/10.1104/pp.17.00601

Fettke J, Chia T, Eckermann N, Smith A, Steup M. A transglucosidase necessary for starch degradation and maltose metabolism in leaves at night acts on cytosolic heteroglycans (SHG). Plant J. 2006:**46**(4):668–684. https://doi.org/10.1111/j.1365-313X.2006.02732.x

Fichtner F, Lunn JE. The role of trehalose 6-phosphate (Tre6P) in plant metabolism and development. Annu Rev Plant Biol. 2021:**72**(1):737–760. https://doi.org/10.1146/annurev-arplant-050718-095929

Fliege R, Flügge U-I, Werdan K, Heldt HW. Specific transport of inorganic phosphate, 3-phosphoglycerate and triosephosphates across the inner membrane of the envelope in spinach chloroplasts. Biochim Biophys Acta. 1978:**502**(2):232–247. https://doi.org/10.1016/0005-2728(78)90045-2

Flis A, Mengin V, Ivakov AA, Mugford ST, Hubberten H-M, Encke B, Krohn N, Höhne M, Feil R, Hoefgen R, et al. Multiple circadian clock outputs regulate diel turnover of carbon and nitrogen reserves. Plant Cell Environ. 2019:**42**(2):549–573. https://doi.org/10.1111/pce.13440

Flügel F, Timm S, Arrivault S, Florian A, Stitt M, Fernie AR, Bauwe H. The photorespiratory metabolite 2-phosphoglycolate regulates photosynthesis and starch accumulation in Arabidopsis. Plant Cell. 2017:**29**(10):2537–2551. https://doi.org/10.1105/tpc.17.00256

Fu X, Gregory LM, Weise SE, Walker BJ. Integrated flux and pool size analysis in plant central metabolism reveals unique roles of glycine and serine during photorespiration. Nat Plants. 2023:**9**(1):169–178. https://doi.org/10.1038/s41477-022-01294-9

Furumoto T, Yamaguchi T, Ohshima-Ichie Y, Nakamura M, Tsuchida-Iwata Y, Shimamura M, Ohnishi J, Hata S, Gowik U, Westhoff P, et al. A plastidial sodium-dependent pyruvate transporter. Nature 2011:**476**(7361):472–475. https://doi.org/10.1038/nature10250

Galtier N, Foyer CH, Huber J, Voelker TA, Huber SC. Effects of elevated sucrose-phosphate synthase activity on photosynthesis, assimilate partitioning, and growth in tomato (Lycopersicon esculentum var UC82B). Plant Physiol. 1993:**101**(2):535–543. https://doi.org/10.1104/pp.101.2.535

Gerhardt R, Stitt M, Heldt HW. Subcellular metabolite levels in spinach leaves: regulation of sucrose synthesis during diurnal alterations in photosynthetic partitioning. Plant Physiol. 1987:**83**(2):399–407. https://doi.org/10.1104/pp.83.2.399

Gotoh E, Matsumoto M, Ogawa K, Kobayashi Y, Tsuyama M. A qualitative analysis of the regulation of cyclic electron flow around photosystem I from the post-illumination chlorophyll fluorescence transient in Arabidopsis: a new platform for the in vivo investigation of the chloroplast redox state. Photosynth Res. 2010:**103**(2):111–123. https://doi.org/10.1007/s11120-009-9525-0

Graf A, Schlereth A, Stitt M, Smith AM. Circadian control of carbohydrate availability for growth in *Arabidopsis* plants at night. Proc Natl Acad Sci U S A. 2010:**107**(20):9458–9463. https://doi.org/10.1073/pnas.0914299107

Guo L, Ma F, Wei F, Fanella B, Allen DK, Wang X. Cytosolic phosphorylating glyceraldehyde-3-phosphate dehydrogenases affect Arabidopsis cellular metabolism and promote seed oil accumulation. Plant Cell. 2014:**26**(7):3023–3035. https://doi.org/10.1105/tpc.114.126946

Hardin SC, Tang GQ, Scholz A, Holtgraewe D, Winter H, Huber SC. Phosphorylation of sucrose synthase at serine 170: occurrence and possible role as a signal for proteolysis. Plant J. 2003:**35**(5):588–603. https://doi.org/10.1046/j.1365-313X.2003.01831.x

Harley PC, Sharkey TD. An improved model of C_3 photosynthesis at high CO_2: reversed O_2 sensitivity explained by lack of glycerate reentry into the chloroplast. Photosynth Res. 1991:**27**(3):169–178. https://doi.org/10.1007/BF00035838

Hasunuma T, Harada K, Miyazawa S-I, Kondo A, Fukusaki E, Miyake C. Metabolic turnover analysis by a combination of in vivo ^{13}C-labelling from $^{13}CO_2$ and metabolic profiling with CE-MS/MS reveals rate-limiting steps of the C_3 photosynthetic pathway in *Nicotiana tabacum* leaves. J Exp Bot. 2010:**61**(4):1041–1051. https://doi.org/10.1093/jxb/erp374

Heldt HW, Chon CJ, Maronde D. Role of orthophosphate and other factors in the regulation of starch formation in leaves and isolated chloroplasts. Plant Physiol. 1977:**59**(6):1146–1155. https://doi.org/10.1104/pp.59.6.1146

Hoppe P, Heintze A, Riedel A, Creuzer C, Schultz G. The plastidic 3-phosphoglycerate → acetyl-CoA pathway in barley leaves and its involvement in the synthesis of amino acids, plastidic isoprenoids and fatty acids during chloroplast development. Planta 1993:**190**(2):253–262. https://doi.org/10.1007/BF00196619

Huber SC, Huber JL. Role and regulation of sucrose-phosphate synthase in higher plants. Annu Rev Plant Physiol Plant Mol Biol. 1996:**47**(1):431–444. https://doi.org/10.1146/annurev.arplant.47.1.431

Jamsheer KM, Jindal S, Laxmi A. Evolution of TOR-SnRK dynamics in green plants and its integration with phytohormone signaling networks. J Exp Bot. 2019:**70**(8):2239–2259. https://doi.org/10.1093/jxb/erz107

Jolliffe PA, Tregunna EB. Environmental regulation of the oxygen effect on apparent photosynthesis in wheat. Can J Bot. 1973:**51**(5):841–853. https://doi.org/10.1139/b73-107

Jones TL, Ort DR. Circadian regulation of sucrose phosphate synthase activity in tomato by protein phosphatase activity. Plant Physiol. 1997:**113**(4):1167–1175. https://doi.org/10.1104/pp.113.4.1167

Kammerer B, Fischer K, Hilpert B, Schubert S, Gutensohn M, Weber A, Flügge UI. Molecular characterization of a carbon transporter in plastids from heterotrophic tissues: the glucose 6-phosphate/phosphate antiporter. Plant Cell. 1998:**10**(1):105–117. https://doi.org/10.1105/tpc.10.1.105

Kelly GJ, Latzko E. Inhibition of spinach-leaf phosphofructokinase by 2-phosphoglycollate. FEBS Lett. 1976:**68**(1):55–58. https://doi.org/10.1016/0014-5793(76)80403-6

Kötting O, Pusch K, Tiessen A, Geigenberger P, Steup M, Ritte G. Identification of a novel enzyme required for starch metabolism in Arabidopsis leaves. The phosphoglucan, water dikinase. Plant Physiol. 2005:**137**(1):242–252. https://doi.org/10.1104/pp.104.055954

Kutschera U, Pieruschka R, Farmer S, Berry JA. The Warburg-effects: basic metabolic processes with reference to cancer development and global photosynthesis. Plant Signal Behav. 2020:**15**(7):1776477. https://doi.org/10.1080/15592324.2020.1776477

Laporte MM. The role of sucrose synthesis in determining patterns of growth, development, and allocation to reproduction in plants. Madison (WI): University of Wisconsin-Madison; 1997.

Lastdrager J, Hanson J, Smeekens S. Sugar signals and the control of plant growth and development. J Exp Bot. 2014:**65**(3):799–807. https://doi.org/10.1093/jxb/ert474

Lefebvre S, Lawson T, Zakhleniuk OV, Lloyd JC, Raines CA, Fryer M. Increased sedoheptulose-1,7-bisphosphatase activity in transgenic tobacco plants stimulates photosynthesis and growth from an early stage in development. Plant Physiol. 2005:**138**(1):451. https://doi.org/10.1104/pp.104.055046

Li J, Weraduwage SM, Preiser AL, Tietz S, Weise SE, Strand DD, Froehlich JE, Kramer DM, Hu J, Sharkey TD. A cytosolic bypass and G6P shunt in plants lacking peroxisomal hydroxypyruvate reductase. Plant Physiol. 2019:**180**(2):783–792. https://doi.org/10.1104/pp.19.00256

Liberti MV, Locasale JW. The Warburg effect: how does it benefit cancer cells? Trends Biochem Sci. 2016:**41**(3):211–218. https://doi.org/10.1016/j.tibs.2015.12.001

Linka N, Weber AP. Intracellular metabolite transporters in plants. Mol Plant. 2010:**3**(1):21–53. https://doi.org/10.1093/mp/ssp108

Livingston AK, Cruz JA, Kohzuma K, Dhingra A, Kramer DM. An *Arabidopsis* mutant with high cyclic electron flow around photosystem I (*hcef*) involving the NADPH dehydrogenase complex. Plant Cell. 2010:**22**(1):221. https://doi.org/10.1105/tpc.109.071084

Lu Y, Gehan JP, Sharkey TD. Daylength and circadian effects on starch degradation and maltose metabolism. Plant Physiol. 2005:**138**(4): 2280–2291. https://doi.org/10.1104/pp.105.061903

Lu Y, Sharkey TD. The role of amylomaltase in maltose metabolism in the cytosol of photosynthetic cells. Planta 2004:**218**(3):466–473. https://doi.org/10.1007/s00425-003-1127-z

Ma F, Jazmin LJ, Young JD, Allen DK. Isotopically nonstationary ^{13}C flux analysis of changes in *Arabidopsis thaliana* leaf metabolism due to high light acclimation. Proc Natl Acad Sci U S A. 2014:**111**(47):16967–16972. https://doi.org/10.1073/pnas.1319485111

Mahon JD, Fock H, Canvin DT. Changes in specific radioactivities of sunflower leaf metabolites during photosynthesis in $^{14}CO_2$ and $^{12}C0_2$ at normal and low oxygen. Planta 1974:**120**(2):125–134. https://doi.org/10.1007/BF00384922

Mengin V, Pyl E-T, Alexandre Moraes T, Sulpice R, Krohn N, Encke B, Stitt M. Photosynthate partitioning to starch in *Arabidopsis thaliana* is insensitive to light intensity but sensitive to photoperiod due to a restriction on growth in the light in short photoperiods. Plant Cell Environ. 2017:**40**(11):2608–2627. https://doi.org/10.1111/pce.13000

Mullen JA, Koller HR. Daytime and nighttime carbon balance and assimilate export in soybean leaves at different photon flux densities. Plant Physiol. 1988:**86**(3):880–884. https://doi.org/10.1104/pp.86.3.880

Née G, Aumont-Nicaise M, Zaffagnini M, Nessler S, Valerio-Lepiniec M, Issakidis-Bourguet E. Redox regulation of chloroplastic G6PDH activity by thioredoxin occurs through structural changes modifying substrate accessibility and cofactor binding. Biochem J. 2014:**457**(1): 117–125. https://doi.org/10.1042/BJ20130337

Née G, Zaffagnini M, Trost P, Issakidis-Bourguet E. Redox regulation of chloroplastic glucose-6-phosphate dehydrogenase: a new role for f-type thioredoxin. FEBS Lett. 2009:**583**(17):2827–2832. https://doi.org/10.1016/j.febslet.2009.07.035

Nickelsen K. The path of carbon in photosynthesis: how to discover a biochemical pathway. Ambix 2012:**59**(3):266–293. https://doi.org/10.1179/174582312X13457672281867

Niittylä T, Messerli G, Trevisan M, Chen J, Smith AM, Zeeman SC. A previously unknown maltose transporter essential for starch degradation in leaves. Science 2004:**303**(5654):87–89. https://doi.org/10.1126/science.1091811

Ogren WL, Bowes G. Ribulose diphosphate carboxylase regulates soybean photorespiration. Nat New Biol. 1971:**230**(13):159–160. https://doi.org/10.1038/newbio230159a0

Peixoto B, Moraes TA, Mengin V, Margalha L, Vicente R, Feil R, Höhne M, Sousa AGG, Lilue J, Stitt M, et al. Impact of the SnRK1 protein kinase on sucrose homeostasis and the transcriptome during the diel cycle. Plant Physiol. 2021:**187**(3):1357–1373. https://doi.org/10.1093/plphys/kiab350

Peterhänsel C, Krause K, Braun HP, Espie GS, Fernie AR, Hanson DT, Keech O, Maurino VG, Mielewczik M, Sage RF. Engineering photorespiration: current state and future possibilities. Plant Biol (Stuttg). 2013:**15**(4):754–758. https://doi.org/10.1111/j.1438-8677.2012.00681.x

Peterhänsel C, Maurino VG. Photorespiration redesigned. Plant Physiol. 2011:**155**(1):49–55. https://doi.org/10.1104/pp.110.165019

Pokhilko A, Flis A, Sulpice R, Stitt M, Ebenhoh O. Adjustment of carbon fluxes to light conditions regulates the daily turnover of starch in plants: a computational model. Mol Biosyst. 2014:**10**(3):613–627. https://doi.org/10.1039/C3MB70459A

Pollock CJ, Lloyd EJ. The effect of low temperature upon starch, sucrose and fructan synthesis in leaves. Ann Bot. 1987:**60**(2):231–235. https://doi.org/10.1093/oxfordjournals.aob.a087441

Portis AR Jr, Parry MA. Discoveries in Rubisco (ribulose 1,5-bisphosphate carboxylase/oxygenase): a historical perspective. Photosyn Res. 2007:**94**(1):121–143. https://doi.org/10.1007/s11120-007-9225-6

Preiser AL, Banerjee A, Weise SE, Renna L, Brandizzi F, Sharkey TD. Phosphoglucoisomerase is an important regulatory enzyme in partitioning carbon out of the Calvin-Benson cycle. Front Plant Sci. 2020:**11**:580726. https://doi.org/10.3389/fpls.2020.580726

Preiser AL, Fisher N, Banerjee A, Sharkey TD. Plastidic glucose-6-phosphate dehydrogenases are regulated to maintain activity in the light. Biochem J. 2019:**476**(10):1539–1551. https://doi.org/10.1042/BCJ20190234

Rabinowitch EI. Photosynthesis and related processes. New York: Interscience Publishers Inc.; 1945.

Rachmilevitch S, Cousins AB, Bloom AJ. Nitrate assimilation in plant shoots depends on photorespiration. Proc Natl Acad Sci U S A. 2004:**101**(31):11506–11510. https://doi.org/10.1073/pnas.0404388101

Raines CA. Increasing photosynthetic carbon assimilation in C_3 plants to improve crop yield: current and future strategies. Plant Physiol. 2011:**155**(1):36–42. https://doi.org/10.1104/pp.110.168559

Raines CA. Improving plant productivity by re-tuning the regeneration of RuBP in the Calvin–Benson–Bassham cycle. New Phytol. 2022:**236**(2):350–356. https://doi.org/10.1111/nph.18394

Ritte G, Lloyd JR, Eckermann N, Rottmann A, Kossmann J, Steup M. The starch-related R1 protein is an alpha-glucan, water dikinase. Proc Natl Acad Sci U S A. 2002:**99**(10):7166–7171. https://doi.org/10.1073/pnas.062053099

Rosenthal DM, Locke AM, Khozaei M, Raines CA, Long SP, Ort DR. Over-expressing the C_3 photosynthesis cycle enzyme sedoheptulose-1-7 bisphosphatase improves photosynthetic carbon gain and yield under fully open air CO_2 fumigation (FACE). BMC Plant Biol. 2011:**11**(1):123. https://doi.org/10.1186/1471-2229-11-123

Ruben S, Kamen MD. Radioactive carbon of long half-life. Phys Rev. 1940:**57**(6):549–549. https://doi.org/10.1103/PhysRev.57.549

Ryabova LA, Robaglia C, Meyer C. Target of rapamycin kinase: central regulatory hub for plant growth and metabolism. J Exp Bot. 2019:**70**(8):2211–2216. https://doi.org/10.1093/jxb/erz108

Salerno GL, Curatti L. Origin of sucrose metabolism in higher plants: when, how and why? Trends Plant Sci. 2003:**8**(2):63–69. https://doi.org/10.1016/S1360-1385(02)00029-8

Scheibe R, Geissler A, Fickenscher K. Chloroplast glucose-6-phosphate dehydrogenase: K_m shift upon light modulation and reduction. Arch Biochem Biophys. 1989:**274**(1):290–297. https://doi.org/10.1016/0003-9861(89)90441-4

Schleucher J, Vanderveer P, Markley JL, Sharkey TD. Intramolecular deuterium distributions reveal disequilibrium of chloroplast phosphoglucose isomerase. Plant Cell Environ. 1999:**22**(5):525–533. https://doi.org/10.1046/j.1365-3040.1999.00440.x

Schleucher J, Vanderveer PJ, Sharkey TD. Export of carbon from chloroplasts at night. Plant Physiol. 1998:**118**(4):1439–1445. https://doi.org/10.1104/pp.118.4.1439

Schmiege SC, Sharkey TD, Walker B, Hammer J, Way DA. Laisk measurements in the nonsteady state: tests in plants exposed to warming and variable CO_2 concentrations. Plant Physiol. 2023:**193**(2): 1045–1057. https://doi.org/10.1093/plphys/kiad305

Schulze-Siebert D, Heintze A, Schultz G. Substrate flow from photosynthetic carbon metabolism to chloroplast isoprenoid synthesis in spinach evidence for a plastidic phosphoglycerate mutase. Z Naturforsch C. 1987:**42**(5):570–580. https://doi.org/10.1515/znc-1987-0513

Scialdone A, Mugford ST, Feike D, Skeffington A, Borrill P, Graf A, Smith AM, Howard M. Arabidopsis plants perform arithmetic division to prevent starvation at night. eLife. 2013:**2**:e00669. https://doi.org/10.7554/eLife.00669

Sharkey TD. O_2-insensitive photosynthesis in C_3 plants: its occurrence and a possible explanation. Plant Physiol. 1985:**78**(1):71–75. https://doi.org/10.1104/pp.78.1.71

Sharkey TD. Feedback limitation of photosynthesis and the physiological role of ribulose bisphosphate carboxylase carbamylation. Botanical Magazine. Tokyo special issue. 1990:**2**:87–105.

Sharkey TD. Discovery of the canonical Calvin–Benson cycle. Photosynth Res. 2019:**140**(2):235–252. https://doi.org/10.1007/s11120-018-0600-2

Sharkey TD. Pentose phosphate pathway reactions in photosynthesizing cells. Cells 2021:**10**(6):1547. https://doi.org/10.3390/cells10061547

Sharkey TD. The discovery of Rubisco. J Exp Bot. 2022:**74**(2):510–519. https://doi.org/10.1093/jxb/erac254

Sharkey TD, Berry JA, Raschke K. Starch and sucrose synthesis in *Phaseolus vulgaris* as affected by light, CO2 and abscisic acid. Plant Physiol. 1985:**77**(3):617–620. https://doi.org/10.1104/pp.77.3.617

Sharkey TD, Preiser AL, Weraduwage SM, Gog L. Source of ^{12}C in Calvin-Benson cycle intermediates and isoprene emitted from plant leaves fed with $^{13}CO_2$. Biochem J. 2020:**477**(17):3237–3252. https://doi.org/10.1042/BCJ20200480

Sharkey TD, Vassey TL. Low oxygen inhibition of photosynthesis is caused by inhibition of starch synthesis. Plant Physiol. 1989:**90**(2): 385–387. https://doi.org/10.1104/pp.90.2.385

Sharma M, Sharma M, Jamsheer K M, Laxmi A. Jasmonic acid coordinates with light, glucose and auxin signalling in regulating branching angle of Arabidopsis lateral roots. Plant Cell Environ. 2022:**45**(5): 1554–1572. https://doi.org/10.1111/pce.14290

Simkin AJ, Lopez-Calcagno PE, Davey PA, Headland LR, Lawson T, Timm S, Bauwe H, Raines CA. Simultaneous stimulation of sedoheptulose 1,7-bisphosphatase, fructose 1,6-bisphophate aldolase and the photorespiratory glycine decarboxylase-H protein increases CO(2) assimilation, vegetative biomass and seed yield in Arabidopsis. Plant Biotechnol J. 2017:**15**(7):805–816. https://doi.org/10.1111/pbi.12676

South PF, Cavanagh AP, Liu HW, Ort DR. Synthetic glycolate metabolism pathways stimulate crop growth and productivity in the field. Science 2019:**363**(6422):eaat9077. https://doi.org/10.1126/science.aat9077

Steichen JM, Petty RV, Sharkey TD. Domain characterization of a 4-α-glucanotransferase essential for maltose metabolism in photosynthetic leaves. J Biol Chem. 2008:**283**(30):20797–20804. https://doi.org/10.1074/jbc.M803051200

Stitt M. Fructose-2,6-bisphosphate as a regulatory molecule in plants. Annu Rev Plant Physiol Plant Mol Biol. 1990:**41**(1):153–185. https://doi.org/10.1146/annurev.pp.41.060190.001101

Stitt M, ap Rees T. Capacities of pea chloroplasts to catalyse the oxidative pentose phosphate pathway and glycolysis. Phytochemistry 1979:**18**(12):1905–1911. https://doi.org/10.1016/S0031-9422(00)82700-4

Sulpice R, Flis A, Ivakov AA, Apelt F, Krohn N, Encke B, Abel C, Feil R, Lunn JE, Stitt M. Arabidopsis coordinates the diurnal regulation of carbon allocation and growth across a wide range of photoperiods. Mol Plant. 2014:**7**(1):137–155. https://doi.org/10.1093/mp/sst127

Szecowka M, Heise R, Tohge T, Nunes-Nesi A, Vosloh D, Huege J, Feil R, Lunn J, Nikoloski Z, Stitt M, et al. Metabolic fluxes in an illuminated *Arabidopsis* rosette. Plant Cell. 2013:**25**(2):694–714. https://doi.org/10.1105/tpc.112.106989

Tcherkez G, Gauthier P, Buckley TN, Busch FA, Barbour MM, Bruhn D, Heskel MA, Gong XY, Crous KY, Griffin K, et al. Leaf day respiration: low CO_2 flux but high significance for metabolism and carbon balance. New Phytol. 2017:**216**(4):986–1001. https://doi.org/10.1111/nph.14816

Tolbert N. Microbodies-peroxisomes and glyoxysomes. Annu Rev Plant Physiol. 1971:**22**(1):45–74. https://doi.org/10.1146/annurev.pp.22.060171.000401

Voll L, Hausler RE, Hecker R, Weber A, Weissenbock G, Fiene G, Waffenschmidt S, Flugge U-I. The phenotype of the Arabidopsis cue1 mutant is not simply caused by a general restriction of the shikimate pathway. Plant J. 2003:**36**(3):301–317. https://doi.org/10.1046/j.1365-313X.2003.01889.x

Walker DA, Herold A. Can the chloroplast support photosynthesis unaided. Plant Cell Physiol. 1977:**SI**:295–310.

Weber AP, Linka M, Bhattacharya D. Single, ancient origin of a plastid metabolite translocator family in Plantae from an endomembrane-derived ancestor. Eukaryot Cell. 2006:**5**(3):609–612. https://doi.org/10.1128/EC.5.3.609-612.2006

Weber A, Servaites JC, Geiger DR, Kofler H, Hille D, Gröner F, Hebbeker U, Flügge UI. Identification, purification, and molecular cloning of a putative plastidic glucose translocator. Plant Cell. 2000:**12**(5):787–801. https://doi.org/10.1105/tpc.12.5.787

Weise SE, Kim KS, Stewart RP, Sharkey TD. β-maltose is the metabolically active anomer of maltose during transitory starch degradation. Plant Physiol. 2005:**137**(2):756–761. https://doi.org/10.1104/pp.104.055996

Weise SE, Weber AP, Sharkey TD. Maltose is the major form of carbon exported from the chloroplast at night. Planta 2004:**218**(3):474–482. https://doi.org/10.1007/s00425-003-1128-y

Wenderoth I, Scheibe R, von Schaewen A. Identification of the cysteine residues involved in redox modification of plant plastidic glucose-6-phosphate dehydrogenase. J Biol Chem. 1997:**272**(43): 26985–26990. https://doi.org/10.1074/jbc.272.43.26985

Wieloch T, Augusti A, Schleucher J. Anaplerotic flux into the Calvin–Benson cycle: hydrogen isotope evidence for in vivo occurrence in C3 metabolism. New Phytol. 2022:**234**(2):405–411. https://doi.org/10.1111/nph.17957

Wilson AT, Calvin M. The photosynthetic cycle. CO_2 dependent transients. J Am Chem Soc. 1955:**77**(22):5948–5957. https://doi.org/10.1021/ja01627a050

Xin C-P, Tholen D, Devloo V, Zhu X-G. The benefits of photorespiratory bypasses: how can they work? Plant Physiol. 2015:**167**(2): 574–585. https://doi.org/10.1104/pp.114.248013

Xu Y, Fu X, Sharkey TD, Shachar-Hill Y, Walker ABJ. The metabolic origins of non-photorespiratory CO_2 release during photosynthesis: a metabolic flux analysis. Plant Physiol. 2021:**186**(1):297–314. https://doi.org/10.1093/plphys/kiab076

Xu Y, Koroma AA, Weise SE, Fu X, Sharkey TD, Shachar-Hill Y. Daylength variation affects growth, photosynthesis, leaf metabolism, partitioning, and metabolic fluxes. Plant Physiol. 2024:**194**(1):475–490. https://doi.org/10.1093/plphys/kiad507

Xu Y, Wieloch T, Kaste JAM, Shachar-Hill Y, Sharkey TD. Reimport of carbon from cytosolic and vacuolar sugar pools into the Calvin-Benson cycle explains photosynthesis labeling anomalies. Proc Natl Acad Sci U S A. 2022:**119**(11):e2121531119. https://doi.org/10.1073/pnas.2121531119

Yang Y, Steup M. Polysaccharide fraction from higher plants which strongly interacts with the cytosolic phosphorylase isozyme. Plant Physiol. 1990:**94**(3):960–969. https://doi.org/10.1104/pp.94.3.960

Yoshida K, Uchikoshi E, Hara S, HisaboriT. Thioredoxin-like2/2-Cys peroxiredoxin redox cascade acts as oxidative activator of glucose-6-phosphate dehydrogenase in chloroplasts. Biochem J 2019: **476**(12):1781–1790. https://doi.org/10.1042/BCJ20190336

Zhu X-G, de Sturler E, Long SP. Optimizing the distribution of resources between enzymes of carbon metabolism can dramatically increase photosynthetic rate: a numerical simulation using an evolutionary algorithm. Plant Physiol. 2007:**145**(2):513–526. https://doi.org/10.1104/pp.107.103713

https://doi.org/10.1093/plphys/kiad630 PLANT PHYSIOLOGY 2024: 195: 79–110

Plant Physiology®

Topical Review

A charged existence: A century of transmembrane ion transport in plants

Michael R. Blatt *

Laboratory of Plant Physiology and Biophysics, University of Glasgow, Bower Building, Glasgow G12 8QQ, UK

*Author for correspondence: michael.blatt@glasgow.ac.uk
The author responsible for distribution of materials integral to the findings presented in this article in accordance with the policy described in the Instructions for Authors (https://academic.oup.com/plphys/pages/General-Instructions) is Michael R. Blatt.

Abstract

If the past century marked the birth of membrane transport as a focus for research in plants, the past 50 years has seen the field mature from arcane interest to a central pillar of plant physiology. Ion transport across plant membranes accounts for roughly 30% of the metabolic energy consumed by a plant cell, and it underpins virtually every aspect of plant biology, from mineral nutrition, cell expansion, and development to auxin polarity, fertilization, plant pathogen defense, and senescence. The means to quantify ion flux through individual transporters, even single channel proteins, became widely available as voltage clamp methods expanded from giant algal cells to the fungus *Neurospora crassa* in the 1970s and the cells of angiosperms in the 1980s. Here, I touch briefly on some key aspects of the development of modern electrophysiology with a focus on the guard cells of stomata, now without dispute the premier plant cell model for ion transport and its regulation. Guard cells have proven to be a crucible for many technical and conceptual developments that have since emerged into the mainstream of plant science. Their study continues to provide fundamental insights and carries much importance for the global challenges that face us today.

Introduction

To posit the past 50 years as marking the transition to modern research on plant membrane transport is understatement. The period encompasses two major revolutions in research, both with a profound impact on our present understanding of plant membranes and their physiology. The first and most obvious is the tectonic shift from traditional biochemical methods to those of molecular genetics. The second, if subtle, is in many ways more profound and is rooted in an awakening within the plant research community to the value of electrophysiology, chiefly of the voltage clamp, in dissecting and quantifying transport. These two drivers have facilitated our understanding of protein structures that support ion flux and its regulation and the molecular mechanics behind subcellular as well as multicellular signal transduction; they underpin the detailed and quantitative understanding we now have of stomatal physiology as well as long-distance signaling in pathogenesis and herbivory; and they are now generating developments in bioengineering that hold real promise for more sustainable ways to live and work within the global environment.

Having researched membrane transport in several model systems over much of this period, I am astounded by the sheer volume of literature now facing anyone starting out, especially those with interests in stomata. When I joined Enid MacRobbie's laboratory in 1983, intent on exploring the electrophysiology of stomatal guard cells, the entire body of literature comprised 4 papers, all empirical observations on guard cell membrane voltage (Pallaghy 1968; Moody and Zeiger 1978; Saftner and Raschke 1981; Ishikawa et al. 1983), and 11 more describing the ion fluxes and contents of guard cells in the closed and open states. Twenty-five years later, in developing the first-generation OnGuard platform to model stomata (Chen et al. 2012; Hills et al. 2012), Adrian Hills, Zhonghua Chen, V.L. (Arieh) Lew, and I sifted through

Received July 27, 2023. Accepted November 1, 2023. Advance access publication January 2, 2024

ADVANCES BOX

- Beginning in the 1980s, adopting voltage clamp and patch clamp methods, plant electrophysiologists established a wealth of quantitative data showing how the dominant pumps, channels, and ion-driven carriers operate in multicellular plants.
- Plant membrane transport is governed by overlapping circuits of H^+ and charge flux that pass through ion pumps, carriers, and channels. These transporters operate in parallel across each membrane and in series between the plasma membrane and each endomembrane.
- Data for stomatal guard cells, and their relevance to foliar gas exchange and photosynthesis, established these cells as the premier model for studies of membrane transport and its regulation that they are today.
- Membrane ion transport in plants is well-described mathematically with flux equations that incorporate voltage as a driver and parameters resolved for all dominant charge transporters in guard cells.
- Model and experiment show an extraordinary degree of entanglement that gives rise to unexpected, often counterintuitive physiology.

more than 8,000 papers, including a core of almost 500 electrophysiological studies. Integrating the voltage clamp, patch clamp and flux data from these publications allowed us to detail the 23 classes of plasma membrane and endomembrane transporters that contribute directly to osmotic solute and water flux for stomatal movements (Blatt et al. 2022). Five years later, in preparing a major review on the topic for this journal (Jezek and Blatt 2017), Mareike Jezek and I confronted as many publications again. Last year alone saw almost 6,000 publications, roughly 500 per month, incorporating "stomata" and "guard cell" in the title and abstract.

While guard cells are now, without dispute, the premier cell model for studies of membrane transport and its regulation in plants, they have also proven to be a crucible for a much greater breadth of research as a transitional ground, a microcosm of methods and concepts emerging into the mainstream of research in plant membrane transport. I need hardly elaborate, too, that these developments are directly relevant to foliar gas exchange, photosynthesis, and global environmental change (Berry et al. 2010; Smith and Dukes 2013; Lawson and Blatt 2014; Blatt et al. 2022). Stomata play a key role in atmospheric modeling and weather prediction (Beljaars et al. 1996; Betts et al. 1996; Berry et al. 2010; Jasechko et al. 2013), and they are at the center of a crisis in fresh water availability that is presently unfolding (UNESCO 2015; Smedley 2023). It would be impossible to cover more than a tiny fraction of the current knowledge pertinent to membrane transport and its place in plant development, nutrition, signal transduction, stress, and environmental physiology. Here, I have chosen to touch on a small selection of topics that are relevant, both historically and as landmarks going forward. I interweave the first sections of this review with some essential conceptual background to electrophysiology before focusing on what has been, and still remains to be, learned from guard cells.

For the interested reader, I recommend several excellent reviews that cover transport integration, voltage-dependent gating of channels and channel structure, and signaling in guard cells (Lawson and Blatt 2014; Maurel et al. 2015; Assmann and Jegla 2016; Jezek and Blatt 2017; Jegla et al. 2018; Lefoulon 2021; Blatt et al. 2022), Ca^{2+} signaling (Hetherington and Brownlee 2004; McAinsh and Pittman 2009; Roelfsema and Hedrich 2010), action potentials, transport, and signaling in Characean algae (Thiel et al. 1997; Shepherd et al. 2002; Beilby and Al Khazaaly 2016, 2017; Beilby et al. 2022), the Venus flytrap (*Dionaea muscipula*) (Sibaoka 1969; Hedrich et al. 2016), pollen and root hairs (Campanoni and Blatt 2007; Zonia and Munnik 2007), and systemic signaling in pathogenesis and herbivory (Choi et al. 2016; Johns et al. 2021; Klejchova et al. 2021). Readers interested in the conceptual approaches and methods of membrane transport research will find much detail in treatments of the all-important kinetics of transport and its inter-pretation [see Sanders (1990) and several useful chapters in *Membrane Transport in Plants* (Blatt 2004)], the practical guides *Microelectrode Techniques* (Standen et al. 1987) and *Microelectrode Methods for Intracellular Recording and Electrophoresis* (Purves 1981), as well as the textbooks *Ion Channels of Excitable Membranes* (Hille 2001) and *Cellular Biophysics* (Weiss 1996). An annotated timeline and short history of some of the key developments relevant to plant membrane transport research will be found in Klejchova et al (2021).

An early focus on membrane voltage

Voltage as a measure of membrane transport is deeply engrained in the collective consciousness of plant physiologists. Prior to the introduction by Cole (1949) of the voltage clamp—a circuit that incorporates the cell membrane in a feedback loop to measure membrane current—voltage was also the principal variable available to the experimenter. Recording membrane voltage required only a high-impedance "follower" amplifier, a set of electrolyte junctions for connection to the cell and bath, and some means to gain electrical access to the inside of a cell. More elaborate circuits to measure membrane conductance by balancing a small stimulus across a Wheatstone bridge required considerable patience and a robust tissue. It is little surprise, then, that membrane voltage was, and in some circles still is, the most commonly cited variable.

Both voltage and conductance measurements appear in the earliest intracellular recordings from plants, predating by two decades similar studies on animal cells (Ling and Gerard 1949; Bretag 2017). The concept of a surrounding cell membrane had been expounded on by Pfeffer in the 1870s (Pfeffer 1877), and a role for the cell membrane in supporting animal "bioelectricity" was recognized early in the 20th century (Bernstein 1902). Yet, it was in plants in the 1920s—around the time of the founding of this journal—that Umrath (1930, 1932), Osterhout (1931), and later Blinks (1942, 1949) were making use of pulled-glass capillaries, microelectrodes filled with salt solutions for electrical contact, in recordings upon insertion into single cells of the giant algae *Chara* and *Nitella* and the marine alga *Valonia* [see also Hope and Walker (1975)]. These cells offered both the size and robustness needed to withstand physical handling and impalements with the relatively coarse microelectrodes of the time, but their use also left doubts at first about whether the findings were applicable to plants more generally. The studies noted a principal dependence on the K^+ concentration, with voltages largely consistent with passive diffusion of the ion across the cell membrane. They also highlighted the dynamic nature of the membrane, its sensitivity to temperature, and its capacity to support large voltage transients, an "irritability" later identified as action potentials, that accompanied substantial increases in membrane conductance. With the origins of the action potentials in nerves resolved to discrete changes in the conductances for Na^+ and K^+ (Hodgkin and Huxley 1952; Hodgkin et al. 1952), what would later be recognized as ion channels, the connection of membrane voltage to passive ionic diffusion was firmly established.

Voltage recordings take on a different, if complementary, perspective as a tool to assess active transport. By convention, biological membranes are referenced to the outside [or, for the tonoplast, to the vacuole volume]. The diffusion of an individual ion is governed by the Nernst Equation:

$$E_x = \frac{RT}{zF}\ln\left\{\frac{[X]_o}{[X]_i}\right\}, \quad (1)$$

which defines the voltage at equilibrium, E_x, for a single ion X. Here, R is the universal gas constant, T is the temperature in °K, z is the charge on ion X, and F is the Faraday constant. In effect, diffusion of ion X progresses until the driving force for diffusion, determined by the ratio of the concentrations inside vs outside, $[X]_i/[X]_o$, is balanced by an equal opposing electrical driving force, E_x, the equilibrium voltage. The high molar value of the Faraday constant (approx. 10^5 Coul/mol) and the small electrical capacitance of biological membranes (approx. 10^{-6} Coul/V.cm^2 = 1 F/cm^2) means that only a very small number of ions need move across a membrane to reach equilibrium and ensures that this voltage builds up exceedingly rapidly, typically over very small fractions of a millisecond.

For multiple ions, this equation expands, for example in the Goldman–Hodgkin–Katz Equation, to include weightings for the differences in relative conductances (permeabilities) for each ionic species, p_x, so that the resting voltage of the membrane

$$V_m = \frac{RT}{F}\ln\left\{\frac{p_K[K^+]_o + p_{Na}[Na^+]_o + p_{Cl}[Cl^-]_i}{p_K[K^+]_i + p_{Na}[Na^+]_i + p_{Cl}[Cl^-]_o}\right\}. \quad (2)$$

Note that V_m is not the same as the ionic equilibrium voltage. Instead, it reflects the balance of ionic fluxes, the voltage at which the sum of all charge movements equals zero. I return to this, most important point shortly.

For a typical plant cell with 150 mM K^+ and 10 mM Cl^- in the cytosol and bathed in a solution of 1 mM KCl outside at 25 °C, the individual equilibrium voltages for K^+ and Cl^- are −126 mV and +58 mV, respectively (Eqn [1]). Provided we can ignore Na^+, as is often the case for plants, and assuming a realistic conductance ratio $p_K/p_{Cl} = 10$, we can expect a resting voltage near −110 mV (Eqn [2]). At −110 mV, neither K^+ nor Cl^- is at equilibrium, and therefore both ions will slowly diffuse out of the cell, driven towards their respective equilibrium voltages. In most plant cells, these fluxes represent an exceedingly small fraction of the total cellular contents—typically less than 0.000001% s^{-1} for the average mesophyll cell—meaning that diffusion alone may support a charge on a membrane for many hours before the voltage collapses to zero. (Such low conductances also mean that a similarly small increase in conductance for another ion, for example opening channels for Ca^{2+} flux for which E_{Ca} is usually far from the resting voltage, will have a substantial impact on the membrane voltage with very little change in the total ion flux.)

Why are these calculations important? A membrane voltage well in excess (negative) of the equilibrium voltages for all of the major ions present is a good indicator of metabolic energy-dependent ion pumping. So, new evidence that surfaced in the 1960s was a direct challenge to the dogma that the voltages on biological membranes are determined solely by passive ionic diffusion. Clifford and Carolyn Slayman (Slayman and Slayman 1962; Slayman 1965) reported membrane voltages from the fungus *Neurospora* of −200 mV, far beyond the most negative equilibrium voltage for K^+ of −25 mV. Most importantly, these studies also showed that the membrane collapsed to voltages near K^+ equilibrium when respiration was inhibited by treatment with azide, dinitrophenol or carbon monoxide. Once it was recognized that small amounts of Ca^{2+} were important to reduce so-called leak conductances, similarly large, inside-negative voltages were reported in plant cells as well (see Spanswick 1981).

Identifying the ion primarily responsible for the membrane voltage is often a process of elimination. The discovery (Kitasato 1968) that the membrane voltage of *Nitella* is strongly dependent on pH and often substantially more negative than all other ionic equilibria was a vital breakthrough; it helped establish the concept of plasma membrane ATPases that pump H^+ out of the cell, with a major proportion of the energy of ATP hydrolysis used to maintain a large transmembrane voltage. The concept of H^+ as the driver ion coupling the "economy" of transport in plants and fungi gained support in the 1970s and 1980s in the wake of Peter Mitchell's ideas of chemiosmosis (Mitchell 1969). Studies of *Neurospora* (Slayman and Slayman 1974; Hansen and Slayman 1978; Blatt et al. 1997) and of microalgae (Schwab and Komor 1978), liverworts (Felle 1981), Characean algae (Beilby and Walker 1981; Sanders and Hansen 1981; Sanders et al. 1985), and angiosperm plants (Ullrich et al. 1984; Basso and Ullrich-Eberius 1987; McCutcheon and Bown 1987; Meharg and Blatt 1995) showed that the membrane voltage depolarized and the extracellular medium alkalinized when uncharged or negatively charged solutes such as glucose, Cl^-, NO^-, and glutamate were taken up, for the anions against the voltage across the membrane. These observations could only be explained if these solutes were transported together with an inward flux of H^+. The findings left no doubt that transport in plant and fungal cells is coupled to the H^+ electrochemical gradient, with the H^+-ATPase generating an electrochemical driving force for H^+ that is then used to drive the coupled transport of other solutes; they also established the H^+-ATPase, and by inference membrane transport generally, as a major consumer of metabolic energy, some 30% to 35% in most estimates [see Klejchova et al. (2021), Jezek and Blatt (2017), and Sanders and Slayman (1982)].

So, is membrane voltage a good indicator of H^+-ATPase activity? The answer is a qualified "yes". When compared with ionic diffusion equilibria, membrane voltage is useful as a "first test" for energy-dependent transport. However, as a readout with which to quantify pump activity, voltage is a problematic measure. Not least, the observations that plant and fungal membranes depolarize in the presence of uncharged and negatively charged solutes clearly demonstrate that voltage reflects the activities of more than just ATP-dependent H^+ pumps (see Current matters, below).

Some pitfalls of the voltage readout

Of course, a major difficulty in quantifying H^+ flux across any membrane is simply that the H^+ is highly buffered through its association and dissociation with H_2O, which is present in aqueous solution at a concentration of roughly 55 mol/L. Therefore, undertaking unidirectional flux measurements is virtually impossible. Yet relying on voltage alone as a readout for pump activity—like net H^+ (pH)—can lead to serious errors of interpretation. Evidence surrounding the concept of the H^+–K^+ exchange ATPase offer a prime example of the danger of relying on voltage as a measure of pump activity. The solution to this problem lies in the use of membrane current, rather than voltage, as a measure of transport. Its resolution was my first major venture into membrane physiology as I started out at the Yale University Medical School in 1980 as a postdoctoral researcher with Clifford Slayman.

When replete with K^+ (and in the presence of small amounts of Ca^{2+}), plant and fungal cells show membrane voltages that are highly sensitive to extracellular H^+ concentrations in the micromolar and submicromolar range (pH 5 to 7) but change very little when extracellular K^+ concentrations are reduced below 1 mM [cf. Spanswick et al. (1967), Blatt (1987b), Beilby and Blatt (1986)]. This H^+ dependence indicates a substantial conductance to H^+ and, as noted above, was a key piece of evidence for the presence of ATP-powered H^+ transport. However, when starved of K^+ so that the cellular K^+ content falls below approximately 60 mM, the membrane voltages of the same cells become hyperpolarized and strongly dependent on extracellular K^+ concentration in the micromolar range. For example, Rodriguez-Navarrro et al. (1986) reported that the membrane voltage of K^+-starved *Neurospora* in the absence of K^+ was typically near −300 mV; adding even 5 μM K^+ outside depolarized the membrane by +70 mV; and K^+ uptake occurred in near one-to-one exchange with H^+. These findings complemented earlier studies of K^+-starved yeast that, likewise, showed acidification of the medium that was stimulated by K^+ additions outside in apparent exchange of H^+ with K^+ (Conway and O'Malley 1944; Conway and Brady 1950).

Why should K^+ starvation lead to large (negative) membrane voltages, and why should adding micromolar K^+ outside evoke such large voltage changes with an exchange of K^+ uptake with H^+ release? One interpretation (Fig. 1A) was that the large voltages were a consequence of enhanced ATPase activity, and the stimulation by K^+ represented a mode of pump operation that, in K^+-starved cells, engages an alternative transport cycle to couple K^+ uptake in exchange with H^+. This concept of a H^+/K^+-ATPase was heavily influenced by contemporary studies of mammalian Na^+/K^+-ATPases and was proposed to incorporate "slippage" that would accommodate electrogenesis as well as H^+ exchange with K^+, subject to the K^+ content within the cell. These ideas, and suggestive biochemical studies with microsomal membrane vesicles (Hodges et al. 1972; Leonard et al. 1973), led to more than a decade of research beginning in the 1970s in an ultimately futile search for this hypothetical plant H^+/K^+ exchange ATPase (Leonard and Hotchkiss 1976; Briskin and Leonard 1982). [As late as the 1980s, the concept that membrane voltages, plant or animal, could be determined by more than ionic diffusion was ridiculed in some circles. The electrogenic nature of the mammalian Na^+/K^+-ATPase, itself, remained a matter of controversy for over 30 years and ended only following studies in the mid-1980s that demonstrated a current carried by the Na^+/K^+-ATPase (DeWeer et al. 1988)].

An alternative explanation for the extreme negative voltages and sensitivity to K^+ outside is that K^+ starvation promotes the expression of transporters that achieve a high affinity for K^+ by coupling its uptake, in symport, to the influx of H^+ (Fig. 1B). Physical laws dictate that in the steady state,

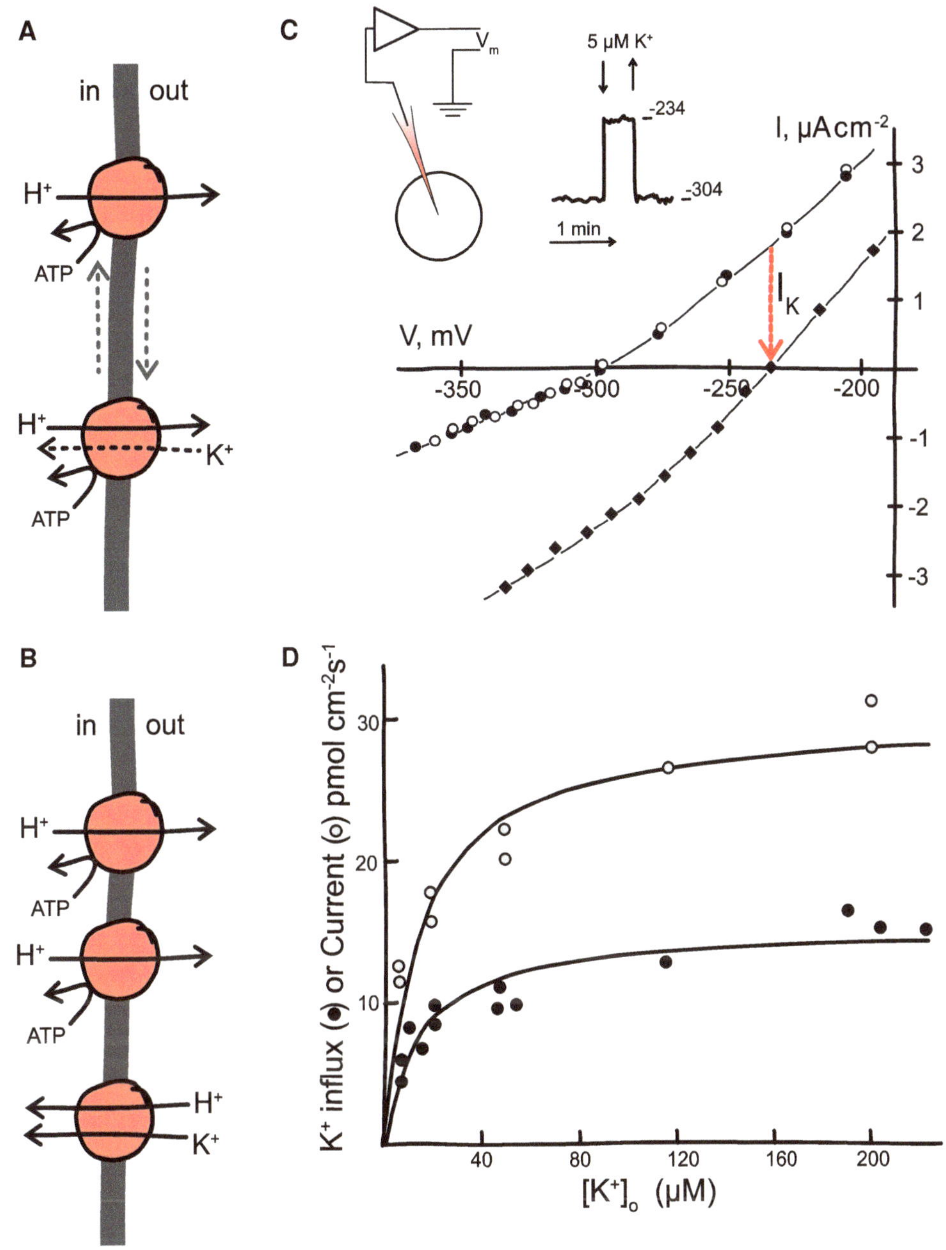

Figure 1. Alternative models for high-affinity K^+ transport and data critical for selecting between them. **A)** The H^+/K^+-ATPase "slippage" model effectively postulated two interchangeable modes of operation for the pump: one that hydrolyses ATP to transport H^+ out of the cell (*above*) and contributes to membrane voltage, and the second that additionally transports H^+ out of the cell in exchange for K^+ uptake (*below*). The second mode was proposed to engage on K^+ starvation and to operate concurrently with the first mode in order to explain membrane hyperpolarization. **B)** The chemiosmotic model proposes that the H^+-ATPase (*above*) and a separate H^+–K^+ symporter (*below*) are coupled across a common membrane. Disabling the symporter by removing K^+ outside would cause the H^+-ATPase to hyperpolarise the membrane toward its equilibrium voltage, offset only by a small background conductance (not shown). In the presence of K^+ outside, the membrane would depolarize, with the current through the H^+-ATPase balanced by the symporter so that two H^+ ions would pass through the H^+-ATPase for every pair of charges, one H^+ and one K^+, returning through the symporter. **C)** Current–voltage (IV) curves recorded from K^+-starved *Neurospora* before (❍), upon adding 5 μM K^+ (◆), and after removing K^+ from the bath (●). The current introduced at the free-running voltage on adding K^+ (I_K) is indicated (red arrow). *Insets*: Simplified circuit for recordings from spherical cells of *Neurospora* and the free-running voltage before, during, and after adding 5 μM K^+ (free-running voltages and time scale indicated by trace). The simplified circuit comprises an amplifier (triangle) that reports the membrane voltage (V_m) from electrode impaled in the spherical cell (circle) referenced to the earthed bath solution. **D)** Net chemical flux ($^{42}K^+$, ●) and current at the free-running voltage (❍) determined from concurrent radiotracer and voltage clamp recordings of *Neurospora* such as shown in **(C)**. Current here converted to units of pmol cm^{-2} s^{-1} for direct comparison with chemical flux. The results show that two charges move across the membrane with each K^+ ion. Data in **(C)** and **(D)** are redrawn from Rodriguez-Navarro et al. (1986).

the net charge movement across a membrane must always be zero at the free-running voltage, a point I noted earlier. So, such a transporter, together with a H^+-ATPase, would explain the observations, provided that H^+ flux through the ATPase was balanced primarily by charge return through H^+-coupled K^+ transport: in other words, for every two H^+ ions pumped out of the cell by the H^+-ATPase, two charges, one H^+ and one K^+, would enter the cell. The overall balance sheet, then, would give a net exchange of H^+ with K^+, acidification of the external medium, a strongly rheogenic uptake of K^+ with two charges moved with each K^+ and, with K^+ outside removed to disable the H^+-K^+ return pathway, an extremely negative membrane voltage driven by the H^+-ATPase.

The idea of H^+-coupled cation symport had been mooted in bacteria (Bakker and Harold 1980) but met with stiff resistance: in the early 1980s, cation uptake by eukaryotes was universally assumed to be driven by voltage only, negative inside the cell, or directly by ATP hydrolysis as is the case for K^+ in animal and insect cells. So, the finding by Rodriguez-Navarro et al (1986) that high-affinity K^+ uptake in K^+-starved *Neurospora* occurred in 1:1 exchange with H^+—and, most telling, that K^+ additions depolarized the membrane—was a direct challenge to this dogma (Fig. 1C). Comparisons of $^{42}K^+$ radiotracer flux with K^+ current showed that two charges enter the cell with each K^+ ion (Fig. 1D), precluding any notion of a channel-mediated K^+ flux (Hedrich and Schroeder 1989). Instead, the instantaneous and reversible depolarizations upon adding K^+ outside indicated that a strongly rheogenic transporter operated in a chemiosmotic cycle with the H^+-ATPase. This interpretation also explained why uptake was inhibited by metabolic blockade to deplete the energy supply.

Ultimately, characterizing any transporter—whether a pump, ion-driven carrier or channel—requires the transport rate to be quantified as a function of the relevant substrate(s). Current is a direct measure of transport rate, whereas voltage is not. Furthermore, for any transporter that moves charge and is therefore subject to membrane voltage, control of the voltage is essential for quantification. The voltage clamp provides the means to this control, and it yields the membrane current associated with each clamp voltage that defines the current-voltage (IV) relation of the membrane. Of course, every cell harbors a number of different transporters that move a net charge across the membrane and will therefore contribute to the total membrane current. So, the challenge for the experimenter is to isolate that fraction of the current carried by the transporter of interest.

One common approach to this challenge is to use an inhibitor to selectively block the transporter and record membrane current at each voltage before, and immediately after, adding the inhibitor. The difference between the two sets of measurements can then be credited to the transporter of interest. Good examples of this approach is the use of Ba^{2+} and the tetraethylammonium cation to block K^+ channels (Hagiwara et al. 1978; Blatt 1988b; Tester 1988; Wegner et al. 1994; Blatt and Gradmann 1997) and of ouabain to inhibit the Na^+/K^+-ATPase (Skou and Norby 1979; Karlish and Stein 1982). In the absence of a selective inhibitor, transport may be blocked by removing a substrate. In the case of high-affinity K^+ transport, for which there is no specific inhibitor, transport can be suppressed by removing K^+ outside. (In practice, K^+ leaching from glass puts a lower limit on $[K^+]_o$ in distilled water of approx. 0.2 µM, but this concentration is almost two orders of magnitude below the apparent $K_{1/2}$ for high-affinity K^+ uptake.) A difficulty with any approach that alters a substrate concentration is that it affects the thermodynamic driving force on the transporter so that current subtractions become distorted near and beyond the equilibrium voltage. However, current subtractions will yield identifiable subsets of relations, so-called difference current-voltage curves (dIVs), that define the limiting transport kinetics and even the (hidden) equilibrium characteristics (Blatt 1986).

For *Neurospora*, the dIVs for high-affinity K^+ uptake showed a strong voltage dependence with the difference current saturating at voltages negative of −250 mV. Using the voltage clamp to control voltage, it was possible to demonstrate that the K^+ current was independent of H^+-ATPase activity and not directly affected by metabolic blockade; in the absence of the voltage clamp, K^+ uptake was inhibited by metabolic blockade, but only because inhibiting the H^+-ATPase led to large depolarizations of the membrane (Blatt et al. 1987; Blatt and Slayman 1987). A decade later, work with purified maize (*Zea mays*) plasma membrane vesicles ruled out a direct stimulation by K^+ of the H^+-ATPase (Gibrat et al. 1990) and drew to a close the hunt for a H^+/K^+-ATPase in plants. Subsequent voltage clamp studies of K^+-starved Arabidopsis (*Arabidopsis thaliana*) roots uncovered evidence for a H^+–K^+ symporter (Maathuis and Sanders 1994) much like that of *Neurospora*.

These high-affinity K^+ transporters are now identified as members of the widely expressed KT/KUP/HAK family that support K^+ nutrition, growth, and stress responses (Quintero and Blatt 1997; Ahn et al. 2004; Osakabe et al. 2013; Véry et al. 2014). Nevertheless, the physiological and kinetic studies with *Neurospora* stand as a landmark in demonstrating a true chemiosmotic cycle that connects high-affinity, H^+-coupled K^+ uptake with the H^+-ATPase at the plasma membrane, and they underline the importance of voltage as a common intermediate shared between the two transporters. Much the same characteristics are now known for the H^+-ATPase of *Chara* (Blatt et al. 1990a) and guard cells (Blatt 1987a; Lohse and Hedrich 1992), and their voltage sensitivities are a key feature defining cellular transport in plants.

Current matters

Why is current important and how does it relate to membrane voltage? Quite simply, transport current is a stoichiometric readout of transport rate and therefore of the kinetic properties of an ion-transporting enzyme. Voltage is both a driving force on transport—the electrical analog of a "substrate" for a transporter that moves charge—and a product of charge movement across the membrane. For high-affinity K^+ uptake by fungi and plants, voltage is a product of the H^+-ATPase and a substrate for K^+ uptake by the H^+–K^+ symporter. Of course, the difference here between voltage as a substrate and as a product is a matter of perspective: voltage will act on the charge flux as a "substrate", and the charge flux in turn will affect the membrane voltage as a "product". For example, the plasma membrane H^+-ATPase uses ATP hydrolysis to pump H^+ out of the cell, thereby contributing to a voltage "product" that is negative inside the cell. At the same time, this voltage also acts as a driver "substrate" that reduces H^+ efflux via the H^+-ATPase by opposing the movement of H^+ out of the cell. Not surprisingly, the plasma membrane H^+-ATPases of fungi and plant cells show a significant dependence on voltage across the physiological voltage range (Gradmann et al. 1978; Sanders et al. 1981; Blatt 1987a; Blatt and Slayman 1987; Blatt et al. 1990a; Chen et al. 2012).

An equally important, if more subtle, aspect of these relations pertains to voltage as an intermediate shared between all charge-carrying transporters in a membrane. As noted above, basic laws of physics dictate that net charge move-ment across a membrane must always sum to zero in the steady state. [The speed at which biological membranes charge, as noted earlier, ensures that a steady state is virtually always maintained. Even during the voltage transient of an action potential, the net charge movement across the mem-brane remains essentially zero, millisecond by millisecond, as charging of the membrane requires only a negligible fraction of the ion flux (Thiel 1995).] So, the charge (ion) flux through each transporter is conjoined to the flux through all other transporters that carry net charge and affect—and are affected by—voltage across the same membrane. An immediate consequence is that changes in voltage, for example membrane hyperpolarization, may arise either from an increase in H^+-ATPase activity or from a decrease in the activity of the balancing current return pathways (Fig. 2). Only by clamping the voltage is this interconnection between transporters bypassed, enabling the experimenter to characterize individual transporters using manipulations such as those I used to isolate the *Neurspora* H^+-K^+ symporter (Blatt et al. 1987; Blatt and Slayman 1987) and the H^+–NO_3^- symporter in Arabidopsis roots (Meharg and Blatt 1995). In other words, without the voltage clamp, changes to any charge-carrying transporter will always influence the activities of other transporters in the same membrane.

The position of membrane voltage as a common intermedi-ate often has consequences that are unexpected, even counterintuitive. One example serves to illustrate this point. The *open stomata2* (*ost2*) mutation in Arabidopsis (Merlot et al. 2007) affects the predominant H $^+$-ATPase AHA1, ren-dering it insensitive to Ca^{2+}. It promotes membrane hyperpo-larization and reduces the dynamic range of stomatal movements. As an intuitive assessment, these observations would suggest that H^+-ATPase activation simply prevents full stomatal closure in the dark or in abscisic acid (ABA), a water-stress signal, by maintaining membrane energization. It was to be expected that stomata of the *ost2* mutant would close more slowly than wild-type stomata. Counterintuitively, however, stomata of the *ost2* mutant also opened more slowly than the wild type. This observation could be explained by the finding that, under voltage clamp, the inward-rectifying K^+ current—the KAT1 K^+ channel activity—in the *ost2* mutant was reduced to 10% or less of the currents recorded in wild-type guard cells (Wang et al. 2017). In other words, the *capacity* for K^+ flux was greatly reduced (see also Modeling guard cells, below).

Why should a mutation that uncouples the control of the H^+-ATPase and promotes its activity slow stomatal opening as well as closing, and why should it greatly reduce the K^+ channel current? The answer lies in the connections to membrane voltage as a shared intermediate between the H^+-ATPase and the R-type anion channels on one hand, and their connections to plasma membrane Ca^{2+} channels and, hence, to cytosolic-free $[Ca^{2+}]$ ($[Ca^{2+}]_i$) and cytosolic pH on the other. Wang et al (2017) found that the *ost2* mutation, by uncoupling the control of the H^+-ATPase, elevated the resting $[Ca^{2+}]_i$ and cytosolic pH. Both the rise in $[Ca^{2+}]_i$ and in pH suppressed the K^+ current to slow opening; and, by hyperpolarizing the membrane and raising pH, the effect was also to suppress activation of the R-type channels (Meyer et al. 2010). The latter are a key to driving $[Ca^{2+}]_i$ and voltage oscillations (Grabov and Blatt 1997, 1998, 1999; Hamilton et al. 2000) that accelerate stomatal closure (Hamilton et al. 2000; Garcia-Mata et al. 2003; Blatt et al. 2007; Minguet-Parramona et al. 2016).

There are several insights to be gained from this study. The first and most obvious is of voltage as a shared intermediate, in this case shared between the H^+-ATPase and Ca^{2+} channels that trigger $[Ca^{2+}]_i$ elevations as well as the K^+ and anion channels. These connections, and changes in the cytosolic H^+ concentration that are expected of H^+-ATPase stimulation, have far-reaching consequences, affecting—and connecting—virtually all of the major transporters localized to both the plasma membrane and tonoplast (Jezek and Blatt 2017; Vialet-Chabrand et al. 2017; Klejchova et al. 2021). Furthermore, as transport processes at these two membranes also share a common pool of solutes within the cytosol, transport across one membrane will also influence, and be influenced by, transport across the other (Horaruang et al. 2020), and transport at both membranes will be affected by metabol-ism (Santelia and Lawson 2016; Jezek and Blatt 2017). As a cor-ollary here, an all-important message is that intuition is often a very poor guide to understanding the physiological impacts of transport and to predicting its behaviors, a point I return to later.

Ion channels and the patch clamp

"A channel is a hole with a very special role" begins the rhyming couplet. Indeed, so special were they viewed that, for a time, ion channels were widely associated with the

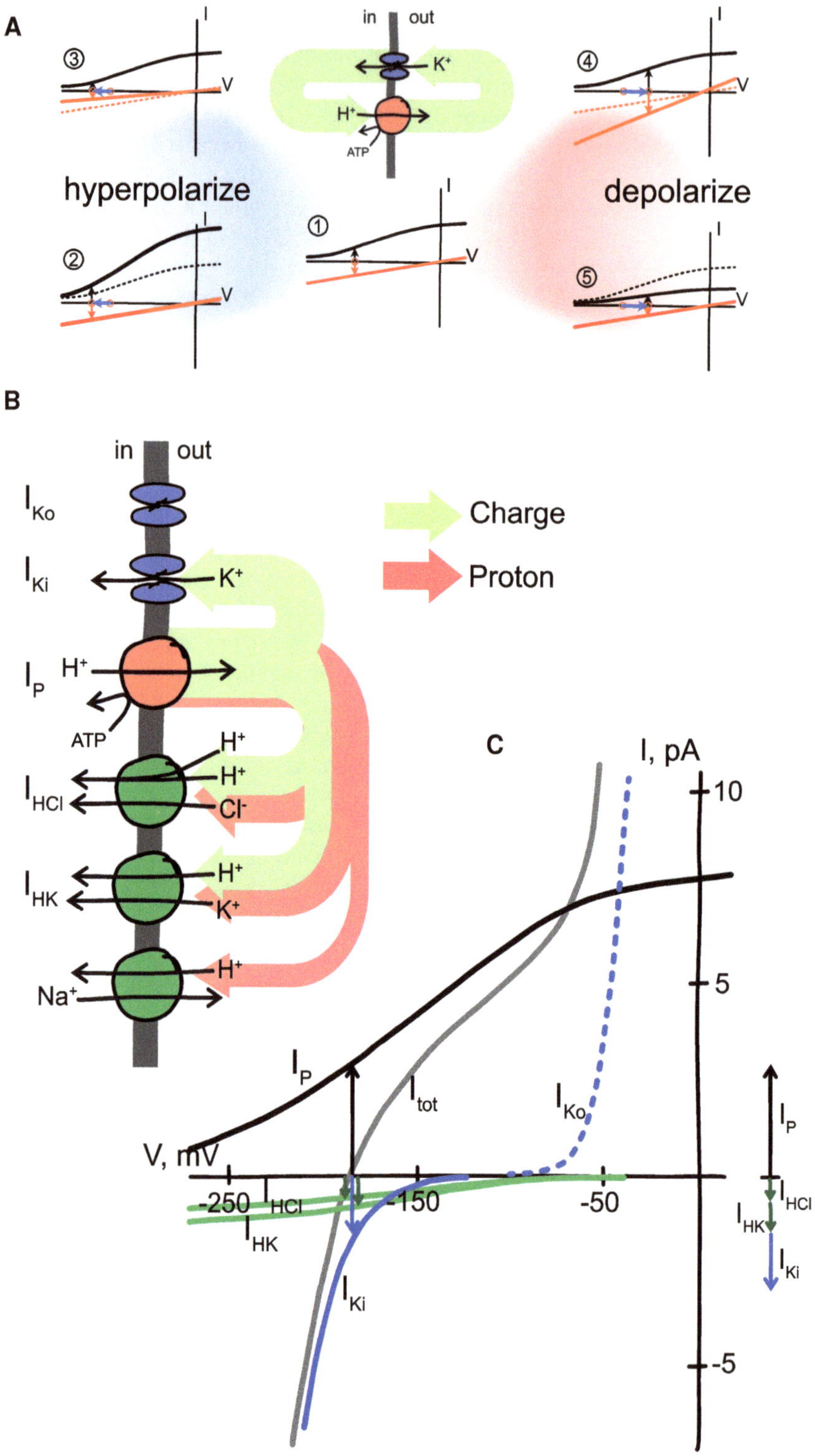

Figure 2. The charge and ionic circuits typical of plant membranes. **A)** A minimal charge circuit for a plant membrane, incorporating a H^+-ATPase and K^+ channel, illustrates the codependence of the two arms of the current cycle, outward via the H^+-ATPase and inward via the channel (*central schematic and green arrows*), and their impacts on membrane voltage. In the steady state, charge movement (current) through the two arms must always be equal in amplitude and opposite in direction across the membrane (vertical *black* and *red arrows*). Thus, from the central current–voltage (IV) plot (1), membrane hyperpolarization is possible (2) by increasing the H^+-ATPase activity (*black curve*), and/or (3) by decreasing the activity of the current return pathway through the K^+ channel (*red curve*). Conversely, membrane depolarization is possible (4) by increasing the activity of the current return pathway through the K^+ channel and/or (5) by decreasing the H^+-ATPase activity. The initial and final free-running voltages in each case is marked by a red circle and the shift in voltage marked by horizontal blue arrows in each plot. Previous H^+-ATPase and K^+ channel activities

(continued)

Characean algae, cells that had come to be synonymous with action potentials in plants. Equally, it was commonly held that channels were absent from most other plant cells, none of which were thought to feature action potentials. Beginning in the mid-1980s, two separate developments closed any debate and would also shift the focus of much of plant transport research for the next decades to address channel activities and their functions.

The first of these developments was the demonstration of single-channel activities in protoplasts of guard cells from the broad bean *Vicia faba* and the mesophyll of wheat, *Triticum aestivum*. Influenced by work with mammalian cells, Schroeder et al (1984) restricted their recordings across a narrow voltage range and to short time intervals, leading to the erroneous conclusions that the K^+ channels of these cells were insensitive to membrane voltage and that their activity did not vary with time. Moran et al (1984) observed a range of discrete, step-like currents in leaf mesophyll protoplasts of wheat, including a conductance of 35 pS with a voltage-dependence to its activity, but did not identify the ions transported. Nonetheless, the findings of quantized activities of single channels in the plasma membranes of wheat leaf cells and *Vicia* guard cells left no doubt about their broader presence within the cells of plants. By the beginning of the 1990s, a number of K^+, Cl^-, and anion-permeable channels were documented in the plasma membranes of marine algae (Bertl et al. 1988), several cell types of land plants (Schroeder et al. 1987; Bush et al. 1988; Hosoi et al. 1988; Moran et al. 1988; Stoeckel and Takeda 1989; Colombo and Cerana 1991; Spalding et al. 1992), and also in mitochondria (Smack and Colombini 1985), chloroplasts (Schoenknecht et al. 1988; Tester and Blatt 1989), and isolated vacuoles (Hedrich et al. 1986; Hedrich and Kurkdjian 1988; Alexandre and Lassalles 1990; Smith et al. 1990).

The second of the developments stems from the gradual recognition that action potentials—and, by inference, the excitabilities unique to ion channels—are a common feature of most, if not all plant cells, not just of the Characean algae, the traps of insectivorous plants, and pulvinar cells of *Mimosa* and its relatives (Sibaoka 1969; Williams and Pickard 1972a, 1972b; Simons 1981). Electrical excitability with bona fide action potentials were described early on in *Neurospora* (Slayman et al. 1976) and subsequently in the leaves of the marine angiosperm *Zostera* (Garrill et al. 1994), in the vasculature of pea, Arabidopsis, barley, and maize (Fromm and Bauer 1994; Herde et al. 1999; Favre et al. 2001; Felle and Zimmermann 2007), and of course in guard cells (Thiel et al. 1992; Gradmann et al. 1993; Blatt and Thiel 1994; Minguet-Parramona et al. 2016).

Gerhard Thiel recorded the original cardiac-like action potentials of guard cells. To the best of my knowledge, these are also the earliest recordings from the cells of any land plant beyond those of the traps of the insectivorous plant *Dionea muscipula*. Gerhard arrived in Cambridge in 1988 as a postdoctoral researcher with Enid MacRobbie. His unorthodox and eclectic approach was the perfect anodyne to the regimented focus my research had followed until then, and we were soon fast friends. Together we demonstrated inositol trisphosphate (IP_3)-mediated control of guard cell K^+ channels via endomembrane Ca^{2+} release (Blatt et al. 1990b). Later, we flagged the puzzle that remains of how a peptide of the auxin-binding protein ABP1 triggers large and near-instantaneous cytosolic alkalinizations that also regulate these K^+ channels (Thiel et al. 1993a). While at Cambridge, Gerhard went on to resolve the single-channel gating characteristics of the anion channels behind the *Chara* action potential (Thiel et al. 1993b, 1997). These studies firmly established the sequence of their activation by $[Ca^{2+}]_i$ in what is now the model for Ca^{2+}-induced Ca^{2+} release that underpins the action potentials and osmotic K^+ and Cl^- release in plant cells.

Many of the early studies of ion channels in land plants took advantage of the new patch clamp methods devised by Neher and Sakmann (1976) and later refined with improvements in pipette manufacture (Hamill et al. 1981). In the so-called whole-cell configuration, the patch clamp yields data equivalent to those of the voltage clamp. Like the comparator at the heart of the voltage clamp circuit, the patch clamp is a current-to-voltage converter. The patch clamp differs from that of the voltage clamp principally in its omission of a follower amplifier input. For the experimenter, therefore,

Figure 2. (Continued)
from the central plot are shown as dotted lines in each case. **B)** Simplified charge and chemical (H^+) circuits of the plasma membrane illustrated with arrows weighted to show the overall balance of chemical (H^+) and charge flux. A similar pair of circuits occurs at the tonoplast. Physical laws require that the net charge flux (*green arrows*) across the membrane in the steady state must sum to zero. [Note: The chemical (H^+, *red arrows*) circuit does not need to balance in the same way.] In other words, at the free-running membrane voltage charge passing out of the cell, here shown as charge movement through the H^+-ATPase (I_P), must be the same as the sum of charges passing back into the cell through the other transporters (active charge-carrying transport, I_{HCl}, I_{Ki}, I_{HK}). As a consequence, a change in charge flux through any one transport pathway will affect the balance of charge movement through all of the other transporters that move charge. Note that one-to-one exchange transport (antiport) of a positively charged solute with H^+, here shown as a Na^+/H^+ antiporter (Quintero et al. 2000; Pardo et al. 2006; Bassil and Blumwald 2014), does not result in net charge movement across the membrane and therefore does not contribute to charge balance. **C)** The simplified charge circuit of the plasma membrane in **(B)** illustrated as current–voltage (IV) curves for each of the transporters to show the *capacity* for current through each transporter as a function of voltage. Total membrane current (I_{tot}) comprises the sum of the transporter currents, here the H^+-ATPase (I_P), the H^+–K^+ (I_{HK}), and H^+–Cl^- (I_{HCl}) symports, and the inward- and outward-rectifying K^+ channels (I_{Ki} and I_{Ko}). At the free-running voltage, the point at which I_{tot} crosses the voltage axis, the vector sum of these currents is zero (*right*). Note that the outward-rectifying K^+ channels contribute to I_{tot} but, because of their gating properties, only at voltages well positive of the free-running voltage.

the patch clamp is always engaged to clamp the membrane to a user-specified voltage (Sigworth 1983). The comparator of the voltage clamp, by contrast, can be bypassed during recordings, allowing the user to record the true free-running voltage across the membrane with the follower amplifier. Nonetheless, patch clamp methods offered access to virtually any cell type in the plant through the expedient of protoplast generation. It circumvented the problems associated with current spread in intact plant tissues that I come to shortly. Exceptions to the patch clamp approach were the use of vesicle fractions fused into planar lipid bilayers to study endomembrane channels, including K^+ channels of chloroplast thylakoids (Tester and Blatt 1989), and my own work on guard cells that drew on the classical, two-electrode voltage clamp (Blatt 1988b, 1990, 1992; Blatt and Clint 1989; Blatt et al. 1990b).

The patch clamp offers other advantages as well, but it also presents certain limitations in studies of ion channels. Among the advantages, the most important is that, in single-channel configuration, the patch clamp enables analysis of the molecular kinetics of protein conformations that determine gating. These details are resolved by analyzing the open and closed lifetime distributions, in other words the time of residence of the channel in each state. Indeed, recordings of single channels in membrane patches are still the only example in which the molecular dynamics of an individual protein molecule can be studied in situ and in real time. Very few researchers have availed themselves of this facility, however, and even today there remain barely a handful of publications that describe the lifetime kinetics of any plant ion channels. Notable are studies of the gating of the KAT1 K^+ channel (Hoshi 1995; Zei and Aldrich 1998; Tang and Hoshi 1999) and its regulation by binding to vesicle-trafficking proteins (Lefoulon et al. 2018), of the stelar SKOR K^+ channel and the K^+ dependence of its gating (Johansson et al. 2006), and of how Ca^{2+} regulates the activity of Ca^+ channels at the guard cell plasma membrane (Hamilton et al. 2000, 2001).

Because patch pipettes are usually prepared with tip diameters of 1 µm or slightly greater, diffusion from the pipette allows rapid dialysis of the cell contents against the much larger volume of the patch pipette. In effect, dialysis with the pipette solution allows for experimental control of the cytosolic contents. However, such dialysis also implies that mechanisms of channel regulation that depend on local and transient changes in vivo, for example in $[Ca^{2+}]_i$, will be masked and can be difficult to assess. A related limitation of patch methods is that dialysis against the pipette often leads to so-called rundown in channel activity as cofactors, essential kinases, and phosphatases, and other regulatory proteins are diluted and lost (Tang and Hoshi 1999; Hamilton et al. 2000; Horvath et al. 2002).

Voltage clamp and the intact cell

Voltage clamp methods with intracellular microelectrodes avoid the need to generate and maintain protoplasts. They also minimize the problems of cellular dialysis. The tip diameters of microelectrodes are commonly near 0.05 µm and lead to diffusion rates typically 300- to 1000-fold less than those of patch pipettes (Blatt and Slayman 1983; Sakmann and Neher 1983; Blatt 1987b). However, work with intact tissues presents a different set of constraints for the electrophysiologist. Of these, the most challenging is to know the extent of the membrane surface that is under the control of the clamp. Within any cell, the spread of current—and, hence, the voltage imposed by the clamp—is determined by the conductivity of the cell contents and its limiting membrane (Jack et al. 1983; Weiss 1996). These characteristics are encapsulated in the so-called space constant that defines how a current, injected at a point in space within a cell, dissipates with distance from that point. As the conductivity of the cytosol is generally three to four orders of magnitude higher than that of the cell membrane, the shape of the cell surface relative to the point of current injection becomes the most important determinant of current flow.

For an accurate assessment of transport, the electrophysiologist needs current to flow equally over the membrane surface. Small cells with a near-spherical shape, for example protoplasts, are ideal for this purpose, as they ensure a uniform distribution of current over the membrane (Fig. 3A). Cells such as root hairs and the giant cells of the alga *Chara* behave like long conductive cables and are more difficult to work with, because current flow across the membrane will fall off with distance from the point of injection. So, the current driving a voltage change on the membrane will decline with distance, resulting in an indeterminant clamp voltage for the membrane as a whole. The space constant for Arabidopsis root hairs, for example, varies between 100 and 300 µm (Meharg et al. 1994). Thus, driving a current to clamp the membrane +100 mV from the free-running (unclamped) voltage at the base of a root hair is likely to achieve only a +30 mV displacement 200 µm away near the root hair tip (Fig. 3B). Of course, the consequence is that transporters at the base of the cell will see a very different membrane voltage from those near the tip, confounding any quantitative interpretation (Smith 1984).

These difficulties are further compounded by plasmodesmata that occur in the majority of multicellular plant tissues. Plasmodesmatal connections present a low-resistance pathway for current flow between c ells t hat e nsure electrical coupling with a spatial dissipation analogous to the current spread along an elongated cell (Spanswick 1972; Goldsmith and Goldsmith 1978). Plasmodesmata are usually distributed over the surfaces of cells without substantial orientational bias. So within the mesophyll of the leaf, for example, these connections create a 3D network (Fig. 3C) over which current spread is virtually impossible to define in practice (Eisenberg and Johnson 1970; Eisenberg et al. 1979; Jack et al. 1983). It is no surprise, then, that voltage clamp studies using intact plant tissues are confined almost entirely to single cells and to cells in tissues that are relatively easy to isolate from their neighbors.

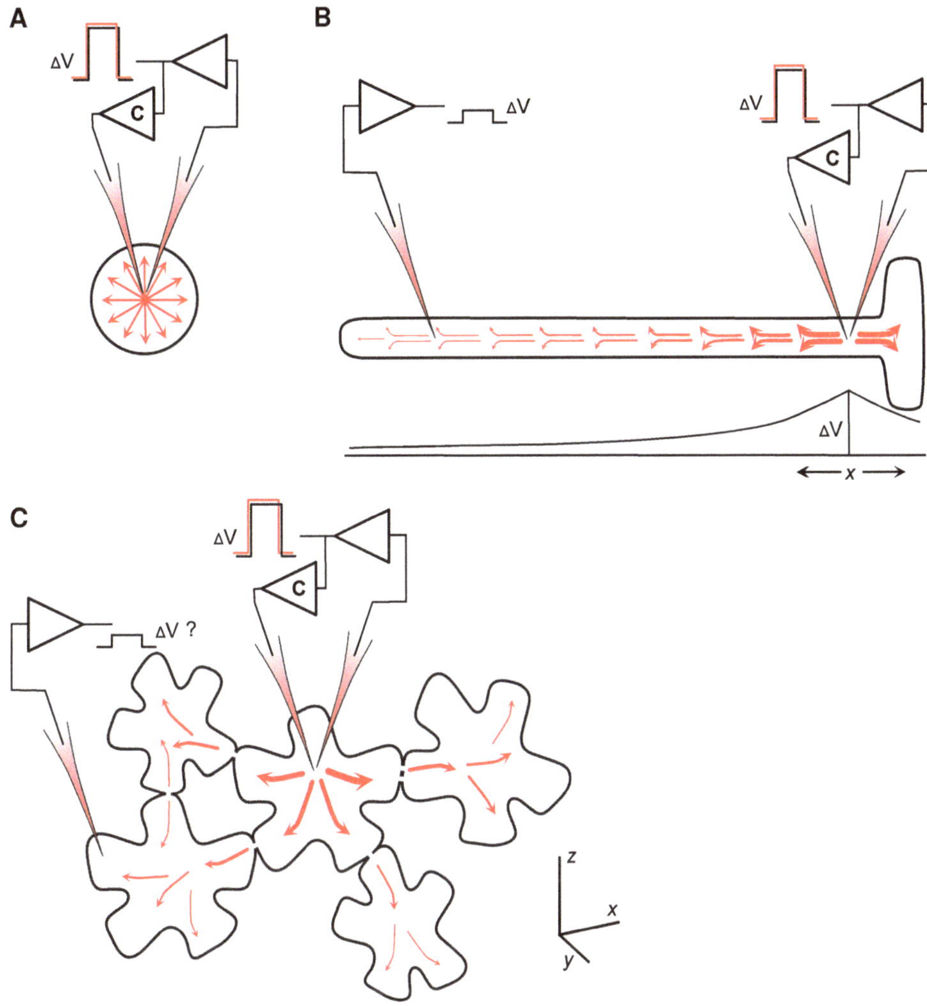

Figure 3. Current spread under voltage clamp in different cell geometries. Cell geometry is a critical factor in voltage clamp studies that make use of microelectrodes for clamp-current injection. Shown in each illustration is the current spread over the cell membrane as red arrows, with the arrow weight indicating the current density. In every case, the membrane is the primary resistance to ground (the external bath). The voltage clamp circuit in each illustration comprises two amplifiers (triangles). A voltage follower amplifier reports the membrane voltage from the right-most electrode; a comparator amplifier, labeled with C, compares this voltage with a command voltage input (not shown) and then returns a current to the cell through the other electrode(s) to correct for any difference between the recorded and command voltages. Also shown is a voltage step (ΔV) reported at the follower amplifier output, with the command step indicated in red and offset for visibility. **A)** Voltage clamp of a spherical or near-spherical cell ensures a uniform current distribution over the cell membrane. **B)** Voltage clamp of a near-cylindrical cell, here shown as a root hair cell, leads to current dissipation along the length of the cell as if it were an infinite cable. Thus, current passing across the membrane is reduced with distance away from the point of injection as current passes across the more proximal surfaces. The decline in the voltage step (ΔV) as the membrane current falls off with distance *x* is shown below. The follower amplifier recording voltage near the root hair tip sees only a fraction of this voltage change. **C)** Voltage clamp of a 3D matrix (*x, y, z* frame, *below*), comprising cells interconnected by plasmodesmata, leads to an ill-defined spread of current over the surfaces of the connected cells. Individual mesophyll cells may present a near-spherical shape, but their interconnections will draw down current locally. Most estimates (Spanswick 1972; Goldsmith and Goldsmith 1978) suggest that 10% to 20% of the current injected into a cell passes to the next through these connections. Thus, the challenge becomes one of defining the cellular interconnections and their spatial distributions.

One way around the problem of current spread in single, elongated cells is to use a linear, rather than a point, source for the current by inserting a fine wire through the length of the cell. This approach works well with the large and robust cells of *Chara* (Beilby and Coster 1979; Beilby 1986; Blatt et al. 1990a). For smaller and delicate cells such as root hairs, the alternative is to measure how the voltage change declines—and, hence, how current spreads—with distance from the point of injection and then correct for this behavior. Usually, the voltage decline can be approximated as a simple exponential function of distance. This approach works for changes imposed by a constant current and is suitable for studies of ion-coupled transport in the steady state (Meharg and Blatt 1995). It proves much more challenging to quantify current flow during transitions that are important for resolving relaxation characteristics of ion channel activation and deactivation. Such analyses require both temporal and spatial knowledge of the membrane voltage (Jack et al. 1983).

The difficulties of current spread are circumvented in intact guard cells, making work with these cells well-suited for the voltage clamp. This was the approach I championed from the start. It built on my experience with *Neurospora* and knowledge of mammalian physiology. Indeed, a long and distinguished history of research on mammalian as well as plant ion channels predates the patch clamp, only a very small part of which I touch on in this review. For those unfamiliar with this history (and for a few pedants who were), at first only the quantized transitions of single-channel recordings were recognized as *bona f ide*evidence of channel activity (see Hedrich and Schroeder 1989). Several of these same laboratories now use the voltage clamp methods that I pioneered for guard cells. Of course, my laboratory also used the patch clamp to resolve the lifetime kinetics and regulation of several ion channels (Hamilton et al. 2000, 2001; Johansson et al. 2006; Lefoulon et al. 2018).

Guard cells are easy to isolate within epidermal peels and to mount and impale with microelectrodes. When mature, these cells lack functional plasmodesmata and are therefore isolated electrically from the neighboring epidermal cells (Wille and Lucas 1984; Erwee et al. 1985; Palevitz and Hepler 1985). Equally important, under the voltage clamp, current from an intracellular microelectrode spreads homogeneously over the membrane surface: analysis of the space constant, even for the large guard cells of *Vicia*, gave values near 300 to 500 µm (Blatt 1987a), more than an order of magnitude greater than the cell dimensions. In other words, voltage clamping these, and the much smaller guard cells of Arabidopsis, with a microelectrode—a point source—is clearly sufficient to ensure a uniform distribution of current over the cell membrane. If the questions were of transport and its regulation in the whole cell rather than of the conformational mechanics of a single channel, then the choice of the voltage clamp with intracellular microelectrodes was, for me, a "no brainer". Why add the difficulties of preparing protoplasts, the uncertainties of their physiology, and the complications of cellular dialysis when measurements were possible from intact cells that were known to retain meaningful responses (Fischer 1968; Humble and Hsiao 1969; Lange et al. 1971; Willmer and Fricker 1996) to environmental and hormonal stimuli?

The guard cell model

From the start, the need to establish the operational boundaries that constrain transport in the guard cell was abundantly clear: What voltages are typical of these cells? Which of the ions prevalent outside dominate the membrane conductance? Which ionic conductances respond to stimuli known to trigger stomatal movements? These are basic questions for which the answers are best derived from intact cells. An initial analysis (Blatt 1987b) showed that the plasma membrane of guard cells of *Vicia*, the model in common use at the time, was dominated by conductances to K^+ and H^+, was largely insensitive to Na^+, and was only weakly dependent on Cl^- and Ca^{2+} outside. The sensitivity to H^+ could be ascribed to H^+-ATPases that, like the *Neurospora* (Sanders et al. 1981; Blatt and Slayman 1987) and *Chara* H^+-ATPases (Kishimoto et al. 1984; Blatt et al. 1990a), transported one H^+ per ATP hydrolyzed and introduced a significant con-ductance over much of the physiological voltage range (Blatt 1987a; Lohse and Hedrich 1992). Likewise, the K^+ conductance was soon recognized to arise from two major K^+ channel currents with opposing voltage sensitivities (Schroeder et al. 1987; Blatt 1988b; Schroeder 1988; Blatt and Clint 1989). These characteristics were supported by analyses with the H^+-ATPase agonist fusicoccin and experi-ments with $^{86}Rb^+$ as a radiotracer for K^+ flux (Blatt 1988a; Blatt and Clint 1989; Clint and Blatt 1989). Of interest, the radiotracer experiments and supporting electrophysiological data (Clint and Blatt 1989) remain the only direct evidence for H^+-coupled K^+ uptake in guard cells and suggest that this pathway can account for as much as half of the cation uptake during stomatal opening.

Detailing the two K^+ currents exposed a major divergence from the biophysical properties of K^+ channels known in animals at the time. The outward-rectifying channels exhibited a dependence on K^+ outside that interacted with the voltage action on the gate, what I perceived as a failsafe to ensure that the channels mediated K^+ efflux for stomatal closure regardless of the K^+ concentration outside (Blatt 1988b; Blatt and Gradmann 1997). This characteristic contrasted with the voltage dependence for gating of the inward-rectifying channels that was insensitive to K^+ outside (Schroeder 1988; Blatt et al. 1990b; Blatt 1992). It was as if the animal membrane had been turned inside-out. Nerve, muscle, and invertebrate eggs incorporated inward-rectifying K^+ channels with a K^+ dependence that displaced channel gating with ex-ternal K^+ concentration (Armstrong and Binstock 1965; Hagiwara and Takahashi 1974) and was later attributed to block by intracellular ions (Lu and MacKinnon 1994; Taglialatela et al. 1994), especially Mg^{2+} (Vandenberg 1987; Agus and Morad 1991); by contrast, the animal outward-rectifying K^+ channels—often identified as delayed rectifiers because of the association with the later, repolariza-tion phase of the nerve and muscle action potential— showed little or no change in gating with external K^+ (Hodgkin and Huxley 1952; Hodgkin et al. 1952; Wu and Haugland 1985), unlike the plant outward-rectifying K^+ channels.

We now know that the inward- and outward-rectifying currents are carried principally by KAT1 and GORK, respectively, in Arabidopsis and their homologs in other species (Mueller-Roeber et al. 1995; Nakamura et al. 1995; Ache et al. 2000; Pilot et al. 2001; Hosy et al. 2003). Similar characteristics show up in the few grass stomata that have been

Table 1. Ion channel and aquaporin structures and gating relevant to guard cells[a]

Channel	Gene	PDB code	Gating factors	Phosphorylated by	Function	Refs
K^+ channels						
KAT1	At5g46240	6V1Y	V, $[Ca^{2+}]_i$, $[H^+]_o$, $[H^+]_i$	CPK3,CPK6,OST1	K^+ uptake	b
AKT1	At2g26650	7T4X,7WSW, 7FCV, 7WM2	V	CBL1/CIPK23	K^+ uptake	c
SKOR	At3g02850	8JET, 8JEU	V, $[H^+]_o$, $[H^+]_i$		K^+ efflux	d
TPC1	At4g03560	5E1J, 7HFK, 5DQQ	V, $[Ca^{2+}]_o$, $[H^+]_o$, $[H^+]_i$		Unknown	e
Anion channels						
SLAC1	At1g12480	7EN0, 7WNQ	V, $[Ca^{2+}]_i$, $[H^+]_i$, $[Ac]_i$	HT1,OST1,CBL1/CIPK6, CPK3, CPK6	Cl^-/NO_3^- efflux	f
ALMT12	At4g17970	7W6K	V, $[Ca^{2+}]_i$, $[H^+]_i$, $[Mal]_o$	OST1,CBL1/CIPK6	Cl^-/Mal efflux	g
Aquaporins						
PIP2;1	At3g53420	1Z98, 3CN5, 4JC6	$[Ca^{2+}]_i$, $[H^+]_o$, $[H^+]_o$	OST1, SnRK2.8	H_2O flux	h
TIP2;1	At3g16240	5I32	$[H^+]_i$		H_2O flux	i

[a]Tabular summary of gene and structural detail available to date for ion and aquaporin channels pertinent to guard cells along with a brief summary of gating factors and protein kinases known to act on channel activity. See Table IX of Jezek and Blatt (2017) for an extensive summary of kinases and phosphatases targeting the channels.

[b]Clark et al. (2020).

[c]Dickinson et al. (2022) and Lu et al. (2022).

[d]Li et al. (2023).

[e]Guo et al. (2016) and Kintzer and Stroud (2016).

[f]Chen et al. (2010a, 2010b), Deng et al. (2021), and Li et al. (2022).

[g]Qin et al. (2022).

[h]Tornroth-Horsefield et al. (2006).

[i]Kirscht et al. (2016).

characterized (Fairley-Grenot and Assmann 1992; Buchsenschutz et al. 2005). Furthermore, we have a good un-derstanding of the mechanics of their gating with voltage and its dependence on the conformation of the so-called voltage sensor domain (Hoshi 1995; Blatt and Gradmann 1997; Marten and Hoshi 1997; Zei and Aldrich 1998; Eisenach et al. 2014; Lefoulon et al. 2018; Lefoulon 2021). This under-standing is supported by the crystal structures (Table 1) for KAT1 (Clark et al. 2020), the related AKT1 channel (Dickinson et al. 2022; Lu et al. 2022) and the SKOR channel (Li et al. 2023) that is near sequence-identical with GORK. These channels are all members of the cyclic-nucleotide binding domain (CNBD) channel subfamily (Jegla et al. 2018) that includes the cardiac pacemaker HCN channels (Moroni et al. 2001; Lee and MacKinnon 2017). How gating in GORK and its homologs is coupled with extracellular K^+ remains a matter of interest, not least because it is likely to underpin the rapid stomatal movements of many plants with C_4 photosynthesis (Alvim 2022) and has implications for the channel protein dynamics (Eisenach et al. 2014). My laboratory recently found that manipulating this coupling yields substantial gains in both water use efficiency and photosynthetic carbon assimilation when light fluctuates, as it commonly does in the environment (Blatt and Alvim 2022; Horaruang et al. 2022).

Early evidence for the function of anion channels in guard cells surfaced as a background conductance that was en-hanced 2-fold within minutes of exposing the cells to abscisic acid, which normally triggers stomatal closure (Blatt 1990; Thiel et al. 1992; Grabov et al. 1997), and later to CO_2 (Brearley et al. 1997). The quasi-linear characteristic of this conductance was consistent with a Ca^{2+}-sensitive leak ob-served in guard cell protoplasts (Schroeder and Hagiwara 1989) but was at odds with the sharp voltage dependence of what was subsequently labeled as GCAC1 (Guard Cell Anion Channel 1), the "rapid activating" or R-type anion channel (Keller et al. 1989; Hedrich et al. 1990; Marten et al. 1992). A definitive study by Schroeder and Keller (1992) clarified the relationship between the two currents, the first a slow, or S-type, anion current that activated and deactivated over 20 to 30 s and the second the rapid, R-type, current that responded to voltage with halftimes of 50 to 100 ms.

The S-type current is now recognized to arise with the SLAC1 channel and its homolog SLAH3 (Negi et al. 2008; Vahisalu et al. 2008; Geiger et al. 2011), members of the struc-turally similar dicarboxylate transporter family (Chen et al. 2010a). The R-type current was associated with ALMT12 (Meyer et al. 2010), a member of the aluminium-sensitive malate transporter family of proteins (Table 1). However, the Arabidopsis *almt12* mutant retains the R-type current (Sasaki et al. 2010), and even double and triple *almt* mutants show a near-wild-type current in vivo (Jaslan et al. 2023). These observations close the door on assertions that the R-type anion current is mediated by *ALMT* gene products and leave its identity unresolved.

Molecular identities aside, the distinctive contributions of the S- and R-type anion currents to stomatal closure was a matter of controversy for some time. Their respective contri-butions were subsequently resolved through a computation-al analysis incorporating the quantitative characteristics of guard cell transport (Chen et al. 2012; Minguet-Parramona et al. 2016). The analysis shows that while the S-type current favors an overall bias to more positive voltages, the R-type current is instrumental in the repetitive membrane depolar-izations, the action potentials, that arise through coupling with K^+ and Ca^{2+} channels and $[Ca^{2+}]_i$ elevations to acceler-ate a rapid loss of K^+, Cl^-, and malate from guard cells (Blatt and Thiel 1994; Grabov and Blatt 1998; Allen et al. 1999; Hamilton et al. 2000; Chen et al. 2010b; Stange et al. 2010).

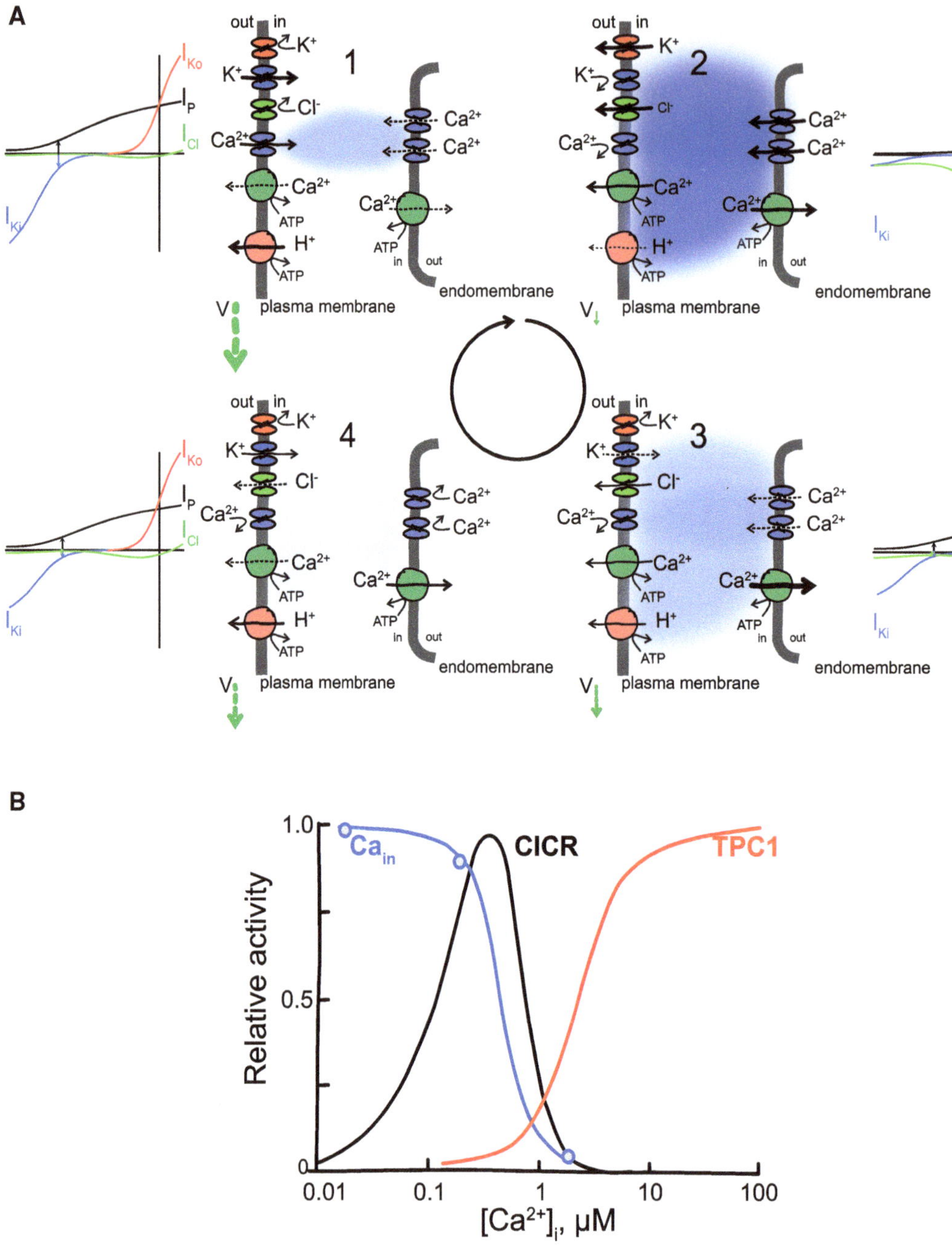

Figure 4. Ca^{2+}-dependent gating of Ca^{2+} channels needed to support Ca^{2+}-induced Ca^{2+} release. **A)** The cycle of Ca^{2+}-induced Ca^{2+} release and its recovery progresses clockwise through four steps in guard cells (see Blatt 2000). Outer insets at each step illustrate the component current-voltage curves for the plasma membrane H^+-ATPase (I_P, *black*), the two K^+ channels (I_{Ki}, *blue*; I_{Ko}, *red*), and the Cl^- channel (I_{Cl}, *green*). The free-running membrane voltage in each is indicated by the position of the opposing arrows along the voltage axis (color-coded to the dominant current in each case) and below each cartoon frame (V, *dark green arrows*). (1) Ca^{2+} influx across the plasma membrane, driven by the H^+-ATPase and membrane hyperpolarisation, triggers endomembrane Ca^{2+}-permeable channels to activate. (2) Endomembrane Ca^{2+} release floods the cytosol to raise $[Ca^{2+}]_i$ to micromolar concentrations, suppressing inward-rectifying K^+ channel and H^+-ATPase activities and promoting Cl^- channel activity, thereby depolarizing the plasma membrane. (3) $[Ca^{2+}]_i$ elevation and membrane depolarization promote Ca^{2+}-ATPases to resequester the cation and remove it to the apoplast. (4) The lowered $[Ca^{2+}]_i$ reduces Cl^- channel activity, promotes inward-rectifying K^+ channel activity and allows the H^+-ATPase to re-activate, thereby repolarising the plasma membrane. The outward-rectifying K^+ channel activity is insensitive to $[Ca^{2+}]_i$ and parallels that of the Cl^- channels, driven principally by the changes in membrane voltage [see also Chen et al (2012) and Blatt (2000)]. **B)** Ca^{2+}-dependent gating char-acteristics of the 13 pS, plasma membrane Ca^{2+} channel (Ca_{in}) is consistent with Ca^{2+}-dependent inhibition expected of a channel that overlaps with, and triggers, Ca^{2+}-induced Ca^{2+} release [data redrawn from Hamilton et al (2000, 2001)]. The Ca^{2+}-dependent gating characteristics of the dominant Ca^{2+} release pathway (CICR) requires both Ca^{2+}-dependent activation at submicromolar $[Ca^{2+}]_i$, overlapping with the activity of the trig-gering Ca^{2+} channels, and Ca^{2+}-dependent inactivation at supramicromolar $[Ca^{2+}]_i$ [curve redrawn from Bezprozvanny et al (1991)]. These char-acteristics are not a feature of TPC1, which shows only Ca^{2+}-dependent activation at $[Ca^{2+}]_i$ beyond the effective range for CICR [curve redrawn from data of Schulz-Lessdorf et al (1995) and Pei et al (1999)].

[It is difficult to reconcile the claim that the *almt12* mutant slows stomatal closure in high CO_2 (Jaslan et al. 2023), given the wild-type characteristics of the current and an analysis of the R-type current in guard cell action potentials and stomatal closure (Chen et al. 2012; Minguet-Parramona et al. 2016); instead, the question arises whether the *almt12* mutant harbors additional T-DNA insertions.]

Guard cell Ca^{2+} channels and $[Ca^{2+}]_i$ oscillations

Remarkably, although their physiological characteristics are well-established, the molecular identities of the Ca^{2+} channels that mediate $[Ca^{2+}]_i$ increases remain one of the major unsolved mysteries of the guard cell. These channels play a central role in triggering the $[Ca^{2+}]_i$ transients that coordinate transport for stomatal closure (Grabov and Blatt 1998; Blatt 2000; Hamilton et al. 2000; Pei et al. 2000; Allen et al. 2001; Koehler and Blatt 2002; Roelfsema and Hedrich 2010; Jezek and Blatt 2017). It was a chance observation that led Alex Grabov and me (Grabov and Blatt 1998) to connect $[Ca^{2+}]_i$ transients in guard cells with membrane voltage. These studies, combining the voltage clamp with the Ca^{2+}-sensitive dye Fura2, showed a steep rise in $[Ca^{2+}]_i$ with membrane hyperpolarization. From these findings, we identified the dominant, voltage-activated Ca^{2+} channel in the guard cell plasma membrane, its activation by ABA (Hamilton et al. 2000, 2001), and its role in triggering endomembrane Ca^{2+} release (Grabov and Blatt 1999; Garcia-Mata et al. 2003). Together with our concurrent work (Grabov and Blatt 1997, 1999) quantifying the kinetics of the inward-rectifying K^+ channels and their dependence on $[Ca^{2+}]_i$, it was soon clear (Fig. 4A) that the voltage and $[Ca^{2+}]_i$ oscillations were coupled within a cycle of K^+ and Cl^- efflux to accelerate solute loss (Blatt 2000). Later studies quantifying the $[Ca^{2+}]_i$-dependence of the guard cell R- and S-type anion channels (Mori et al. 2006; Chen et al. 2010b; Meyer et al. 2010; Stange et al. 2010) only added depth to this conclusion. A central lesson here was that the $[Ca^{2+}]_i$ oscillations "reflect a spectrum of frequencies that emerge from the balance of intrinsic transport activities" in the guard cell and the interactions between Ca^{2+}, K^+, and Cl^- flux (Minguet-Parramona et al. 2016) rather than a unique $[Ca^{2+}]_i$ frequency that "encodes" stomatal closure.

What do we know about the plasma membrane Ca^{2+} channels? Hamilton et al. (2000, 2001) reported a Ca^{2+}-permeable channel with a limiting single-channel conductance of 13 pS, like many Ca^{2+} channels an equal or greater permeability to Ba^{2+}, and a high selectivity (>10:1) for the divalents over K^+ and Cl^-. The channel showed a significant voltage dependence with an apparent gating charge of unity, possibly higher (Koehler and Blatt 2002; Sokolovski et al. 2008), thus distinguishing the Ca^{2+} channel from K^+ channels, including KAT1 and GORK, that show gating charges of 1.8 to 2.2. [In whole-cell analysis, channel gating can be described with two parameters: the gating charge defines the slope of the change in relative activity (relative open probability) with voltage; the midpoint voltage defines the voltage at which this activity is half-maximal. While the channel midpoint voltage is often under ligand or environmental control, the gating charge is a unique "fingerprint" intrinsic to the structure of the channel protein.] The mean open probability of the Ca^{2+} channel was enhanced over 100-fold within 10 to 20 s of adding ABA to excised, inside-out membrane patches, leading to a +40 mV shift in the midpoint voltage for the relative open probability; an equivalent, but slower, response in cell-attached recordings implied that that ABA must traverse the membrane before binding to its intracellular receptor (Hamilton et al. 2000).

Concurrent work by Pei et al. (2000) similarly uncovered a divalent cation-selective current that was enhanced by reactive oxygen species and abscisic acid treatment, but these studies provided little other quantitative detail. Hamilton et al. (2000, 2001), Koehler and Blatt (2002), and Koehler et al (2003) reported channel mean open probabilities around 0.0003 to 0.001 near −120 mV that rose rapidly at more negative voltages and channel activation by reactive oxygen species with a $K_{1/2}$ of 76 μM. Unusually, the channel open probability was strongly enhanced, with a Hill (binding) coefficient near 4, as the external divalent concentration rose above 2 mM; this effect resulted primarily from an increase in the open lifetime of the channel. These studies also showed that phosphorylation of the Ca^{2+} channels was a prerequisite for their subsequent response to ABA. Finally, the channel activity was strongly dependent on internal Ca^{2+}, falling roughly 10-fold as $[Ca^{2+}]_i$ increased from 200 nM to 2 μM (Hamilton et al. 2000). Suppression by $[Ca^{2+}]_i$ was precisely the feedback necessary to support $[Ca^{2+}]_i$ transients and their temporal oscillatory behavior, indicating that the 13 pS channel was responsible for triggering the $[Ca^{2+}]_i$ oscillations observed in vivo.

Could these Ca^{2+} channels be ascribed to one or more assemblies of cyclic nucleotide-gated channel (CNGC) subunits? Possibly, but the evidence to date is not convincing (Table 2). Cyclic nucleotide-gated channels incorporate a structure, broadly similar to plant CNBD K^+ channels, that comprises six transmembrane-spanning domains (Kohler et al. 1999) with sites for calmodulin binding (Kohler and Neuhaus 2000; DeFalco et al. 2016). Certainly, calmodulin binding and its regulation of channel activity might explain the negative feedback evident with $[Ca^{2+}]_i$ elevation (Fischer et al. 2017; Pan et al. 2019; Jarratt-Barnham et al. 2021). However, other characteristics exhibited by CNGCs to date do not match those of the plasma membrane Ca^{2+} channels first described by Hamilton et al (2000, 2001). With the possible exception of CNGC18 (Gao et al. 2014), CNGC channels -including the CNGCs of guard cells (Wang et al. 2013b)—are broadly permeant to monovalent as well as divalent cations (Dietrich et al. 2020; Jarratt-Barnham et al. 2021). Furthermore, CNGCs uniformly show shallow voltage sensitivities with gating charges of 0.2 to 0.4, even in the presence of cytosolic Mg^{2+} (Lemtiri-Chlieh et al.

Table 2. Comparison of guard cell Ca^{2+} channels, candidate channel proteins and their gating[a]

Channel	Gene	Voltage dependence		Action: $[Ca^{2+}]_o$	$[Ca^{2+}]_i$	Other	Refs
		δ	$V_{1/2}$ (mV)				
Plasma membrane							
Ca_{in}	Unknown	1 to 1.4	−140 to −80	+shift $V_{1/2}$	K_{inh}, 350 nM, $h=4$ − shift $V_{1/2}$	ROS act ABA act	c
CNGC5	*CNGC5*	<0.4[b]	<−160[b]	n.d.	CAM-mediated K_{inh}	ROS insens	d
CNGC6	*CNGC6*	<0.4[b]	<−160[b]	n.d.	CAM-mediated K_{inh}	ROS insens	d
CNGC9	*CNGC9*	<0.4[b]	<−160[b]	n.d.	CAM-mediated K_{inh}	ROS insens	d
CNGC12	*CNGC12*	<0.4[b]	<−160[b]	n.d.	CAM-mediated K_{inh}	ROS insens	d
Tonoplast and endomembranes							
CICR	Unknown	3 to 4	−80 to −40	+shift $V_{1/2}$	K_{act}, 400 nM, $h=4$ K_{inh}, 1.2 µM, $h=4$	−shift $V_{1/2}$, HCO_3^-	e
SV	*TPC1*	1 to 1.4	+40 to +90	+shift $V_{1/2}$	K_{act}, 3 µM, $h=4$ no K_{inh}	inhib, pH < 7 $[Mg^{2+}]_i$ act	f
IP_3R	Unknown	~1[b]	−80 to −50[b]	n.d.	no direct K_{act} no direct K_{inh}	$IP_3 > 1$ µM, act	g
IP_6R	Unknown	n.d.	n.d.	n.d.	n.d.	$IP_6 > 1$ µM, act	h
cADPR	Unknown	<1[b]	−120 to −80[b]	n.d.	No direct K_{act} K_{inh}, 500 nM[b]	cADP > 20 nM act	j

[a]act, activation; inh, inhibition; insens, insensitive; K_{act}, activation $K_{1/2}$; K_{inh}, inhibition $K_{1/2}$; *h*, Hill coefficient; $V_{1/2}$, conductance midpoint voltage; ±shift, positive/negative shift in $V_{1/2}$; n.d., not determined.

[b]Estimated from published data.

[c]Grabov and Blatt (1998), Hamilton et al. (2000, 2001), Pei et al. (2000), Koehler and Blatt (2002), Garcia-Mata et al. (2003), Sokolovski et al. (2005, 2008), and Wang et al. (2013a). [d]Wang et al. (2013b), Pan et al. (2019), Dietrich et al. (2020), Lemtiri-Chlieh et al. (2020), and Tan et al. (2023).

[e]Minimal OnGuard parameters established for endomembrane Ca^{2+} release (Chen et al. 2012; Hills et al. 2012; Wang et al. 2017; Jezek et al. 2021; Blatt et al. 2022). [f]Hedrich and Neher (1987), Ward and Schroeder (1994), Schulz-Lessdorf and Hedrich (1995), Pei et al. (1999), Pottosin et al. (2004), Beyhl et al. (2009), and Hedrich and Marten (2011).

[g]Alexandre et al. (1990), Blatt et al. (1990a, 1990b), Gilroy et al. (1990), and Allen et al. (1995).

[h]Lemtiri-Chlieh et al. (2000, 2003).

[j]Allen et al. (1995), Muir et al. (1997), and Leckie et al. (1998).

2020). Tan et al. (2023) reported that a quadruple knockout of the four dominant CNGCs in Arabidopsis guard cells slows the $[Ca^{2+}]_i$ oscillations evoked by ABA. The oscillations are still present, however. Furthermore, reactive oxygen species elicit $[Ca^{2+}]_i$ elevations in the quadruple knockout mutant, much as they do in the wild type. So, yes, CNGCs may contribute to $[Ca^{2+}]_i$ homeostasis in guard cells, but the molecular identity of the 13 pS Ca^{2+} channel most likely lies somewhere else.

Probing the tonoplast

The tonoplast membrane surrounding the guard cell vacuole presents many more challenges. As the vacuole comprises some 80% to 90% of the volume of the guard cell, the major fraction of solute and water flux occurring during stomatal movements must cross both membranes (Willmer and Fricker 1996; Jezek and Blatt 2017). It is no surprise, then, that transport across the tonoplast is coordinated with that of the plasma membrane. Both membranes share a common pool of solutes within the cytosol, and the transport of these solutes, including that of Ca^{2+}, are affected by, and impact on, the activities at both membranes (Horaruang et al. 2020).

There are substantial technical challenges in gaining access to transport across the tonoplast. Analysis in vivo is limited to radiotracer methods (MacRobbie 1995, 2000; MacRobbie and Kurup 2007) that, with few exceptions (Clint and Blatt 1989), yield data only for the sum fluxes of any one ion. Patch clamp methods, by contrast, have provided much quantitative data on the individual channels and pumps activities from isolated vacuoles. These include the biophysical and regulatory characteristics of the slow- (SV) and fast-activating (FV) and the voltage-independent (VK) K^+ permeable channels (Hedrich and Neher 1987; Weiser and Bentrup 1991; Ward and Schroeder 1994; Schulz-Lessdorf and Hedrich 1995; Allen et al. 1998), the SV and VK channels now recognized as the two-pore channel proteins TPC1 and TPK1, respectively (Furuichi et al. 2001; Bihler et al. 2005; Peiter et al. 2005; Gobert et al. 2007). Like work with membrane vesicles, however, patch recordings of the tonoplast require some educated guesswork to define the environment on the cytosolic side of the membrane. As a result, there remain many more mysteries that have yet to be resolved about the components, nature and regulation of tonoplast transport as it operates in vivo.

The growing interest in transport across other endomembranes, especially the endoplasmic reticulum and nuclear en-velope, chloroplasts, and mitochondria (De Angeli et al. 2009; Chanroj et al. 2011; Furumoto et al. 2011; Loro and Costa 2013; Mueller et al. 2014; Maurel et al. 2015; Charpentier et al. 2016; De Col et al. 2017; Leitão et al. 2019), faces very similar challenges. Among others, these compartments con-tribute to the $[Ca^{2+}]_i$ characteristics of plant cells and include the endoplasmic reticulum (Garcia-Mata et al. 2003; Blatt et al. 2007; Bonza et al. 2013), the vacuole (Allen et al. 1995; Beyhl et al. 2009) and, over a higher $[Ca^{2+}]_i$ range, also mitochondria and chloroplasts (McAinsh and Pittman 2009; Loro et al. 2012; Loro and Costa 2013). There is not space here to cover these topics, but they are bound to add substantially to an understanding of cellular dynamics and signaling (Resentini et al. 2021).

Among the defining mysteries, the pathways of endomembrane Ca^{2+} release for $[Ca^{2+}]_i$ transients and oscillations remain to be identified. This is a challenge that, by historical circumstance, is tied hand-in-hand with the puzzle that is the TPC1 (SV) channel. The SV current was originally identified as a Ca^{2+}-activated and cation-permeable channel of 60 pS that rectified strongly into the beet vacuole (Hedrich et al. 1988), by convention hereafter defined as an outward-rectifying current (Bertl et al. 1992). The current was later identified with a tandem-pore channel (TPC) analogous to the mammalian channels carrying the same moniker (Furuichi et al. 2001; Peiter et al. 2005) and has since proven to be a dominant feature of the vacuoles of the cells in virtually all land plants (Hedrich and Marten 2011). TPC1 current is activated by $[Ca^{2+}]_i$ elevation on the cytosolic side of the membrane (Hedrich and Neher 1987) that promotes a permeability also to Ca^{2+} (Ward and Schroeder 1994; Schulz-Lessdorf and Hedrich 1995), but the current is much reduced at the acid pH values typical of the vacuole (Schulz-Lessdorf and Hedrich 1995; Beyhl et al. 2009). Thus, the TPC1 current is recovered at voltages negative of 0 mV only with supramicromolar $[Ca^{2+}]_i$ concentrations (Pottosin and Schonknecht 2007; Hedrich and Marten 2011). As tonoplast voltages normally range from −20 to −70 mV (Spanswick and Williams 1964; Bentrup et al. 1986), the first, and foremost puzzle is simply why the current activates only well outside the physiological voltage range.

It has proven equally difficult to "shoehorn" the $[Ca^{2+}]_i$ dependence of TPC1 into the features essential for evoked Ca^{2+} release (Fig. 4B), the phenomenon of Ca^{2+}-induced Ca^{2+} release (CICR) that underpins guard cell $[Ca^{2+}]_i$ transients (Grabov and Blatt 1999; Blatt 2000; Garcia-Mata et al. 2003; Chen et al. 2012; Minguet-Parramona et al. 2016). Yet, there is unequivocal evidence that the vacuole is a major endomembrane Ca^{2+} store (MacRobbie 1992; Allen and Sanders 1994; Pottosin and Schonknecht 2007; Pittman 2011) and that it contributes to $[Ca^{2+}]_i$ transients in vivo in guard cells in response to changes in light and CO_2 (Jezek et al. 2021). As Mareike Jezek and I noted before (Jezek and Blatt 2017), one of the prerequisites for $[Ca^{2+}]_i$ transients and oscillations is that the pathways for Ca^{2+} entry to the cytosol must be self-limiting; for endomem-brane Ca^{2+} release, this requirement implies a bell-shaped dependence on $[Ca^{2+}]_i$ so that the Ca^{2+} flux shuts down at su-pramicromolar $[Ca^{2+}]_i$ to avoid catastrophic Ca^{2+} release.

The activity of TPC1 only rises with $[Ca^{2+}]_i$ and with an ap-parent $K_{1/2}$ near 3 μM and Hill coefficient of 4 (Hedrich and Neher 1987; Ward and Schroeder 1994; Schulz-Lessdorf and Hedrich 1995), values that also place TPC1 activation outside the range needed for triggered $[Ca^{2+}]_i$ increases (Fig. 4B). [Note that Pei et al. (1999) erroneously quote a $K_{1/2}$ of 227 nM for Ca^{2+} activation of the SV (TPC1) current, even though their data clearly shows a value near 3 μM, an order of magnitude greater (Hills et al. 2012).] Thus, it is difficult to see how the activation of TPC1 might contribute to anything other than cellular collapse. What, then, is the role of TPC1? To date, in vivo studies with the Arabidopsis *tpc1* null mutant and the *fatty acid oxygenation upregulated2* (*fou2*) mutant, which renders TPC1 largely insensitive to vacuolar Ca^{2+}, have failed to uncover a phenotype attributable to $[Ca^{2+}]_i$ signaling (Peiter et al. 2005; Islam et al. 2010; Hedrich and Marten 2011). Thus, a question mark hangs over the function of TPC1, as it has for over 35 years.

All estimates indicate that 95% of the Ca^{2+} flux for $[Ca^{2+}]_i$ transients comes from endomembrane stores (Grabov and Blatt 1999; Chen et al. 2012; Jezek et al. 2021), much as in animals (Bezprozvanny et al. 1991; Hille 2001). So, which channels, then, might facilitate CICR and $[Ca^{2+}]_i$ oscillations in guard cells? Several Ca^{2+}-permeable channels (Table 2) associated with endomembrane stores are activated by submicromolar $[Ca^{2+}]_i$ and by ligands, including IP_3, cyclic ADP-ribose (cADPR), nitric oxide (NO), and inositol hexakisphosphate (IP_6), and have been implicated in Ca^{2+} release (Alexandre and Lassalles 1990; Blatt et al. 1990b; Muir and Sanders 1996; Wu et al. 1997; Leckie et al. 1998; Grabov and Blatt 1999; Garcia-Mata et al. 2003; Lemtiri-Chlieh et al. 2003). Furthermore, there is unequivocal evidence for ryanodine- and IP_3-sensitive Ca^{2+} release behind CICR in guard cells (Blatt et al. 1990b; Gilroy et al. 1990; Grabov and Blatt 1999). However, no Ca^{2+}-permeable channel has yet to be resolved at the tonoplast or at any other endomem-brane with the bell-shaped dependence on $[Ca^{2+}]_i$ required for CICR (Bezprozvanny et al. 1991; Hille 2001). I have no doubt that solving this mystery will open entirely new per-spectives on signaling in plants that extend far beyond the guard cell.

Modeling guard cells

At a major conference in 2004, a delegate asked of me "Why do you work on guard cells when they're so complex?" The delegate, an internationally renowned colleague, had worked for decades on wounding and plant pathogen defence. So, the irony of the situation was not lost on me. [Another colleague once quipped that plant pathogenesis is the study of multiply nested double-negatives—as a linguistic construct, a usage that can be difficult to parse even when not nested (Pinker 2014)]. At the start of this article, I noted that by the turn of this century, the sheer volume of data for transport, especially in guard cells, was overwhelming. I have said very little of signaling and transport regulation, beyond mentioning $[Ca^{2+}]_i$ and reactive oxygen species, and nothing of the regulatory impacts of cytosolic pH, nitrosylation, persulfidation, or phosphorylation, let alone of gene expression and vesicle trafficking for transport protein delivery to, and recycling from, target membranes.

Table 3. A minimal set of solute transporters used by the OnGuard platform[a]

Transporter[b]	Representative gene product[c]	Transported ions	Gating[e]: $[Ca^{2+}]_i$	$[H^+]_i$	Function[f]
Plasma membrane					
H^+-ATPase	AHA1, AHA2	H^+	inhib	substr	Energisation
Ca^{2+}-ATPase	ACA8, ACA10	Ca^{2+}, H^+	substr		Ca^{2+} efflux
H^+–K^+ symport	HAK5	K^+, H^+		substr	K^+ influx, high affinity
H^+-anion symport	NPF7	$Cl^-(NO_3^-)$, H^+		substr	Cl^- influx, high affinity
H^+-Mal symport	ABCB14	Mal, H^+		substr	Mal influx
Ca^{2+} channel (in)[d]	(unknown)	Ca^{2+}	inhib		Ca^{2+} influx, CICR
K^+ channel (in)	KAT1, KAT2	K^+	inhib	activ	K^+ influx
K^+ channel (out)	GORK	K^+		inhib	K^+ efflux
K^+ channel (none)	(unknown)	K^+			K^+ leak
Anion channel (weak)	SLAC1, SLAH3	$Cl^-(NO_3^-)$	activ	activ	Anion efflux
Anion channel (out)	(unknown)	Cl^-, Mal	activ	activ	Anion efflux
Aquaporin	PIP2;1		inhib		Water flux
Tonoplast					
H^+-ATPase	VHA1	H^+		substr	Energisation
H^+-PPase	PVA1	H^+		substr	Energisation
Ca^{2+}-ATPase	ACA4, ACA11	Ca^{2+}, H^+	substr		Ca^{2+} efflux
$Cl^-(NO_3^-)$ channel	ALMT9	$Cl^-(NO_3^-)$			Anion influx
Mal channel	ALMT6	Mal			Mal influx
H^+/Ca^{2+} antiport	CAX2, CAX3	Ca^{2+}, H^+	substr	substr	Ca^{2+} efflux
H^+/Cl^- antiport	CLC1	$Cl^-(NO_3^-)$, H^+		substr	$Cl^-(NO_3^-)$ efflux
Ca^{2+} channel (in)[d]	(unknown)	Ca^{2+}	inhib	inhib	Ca^{2+} influx
K^+ channel (weak)	TPK1	K^+	activ	activ	K^+ influx
K^+ channel (none)	FV	K^+	inhib	inhib	K^+ leak
K^+/Ca^{2+} channel (out)[g]	TPC1	K^+, Ca^{2+}	activ	inhib	(unknown)
Aquaporin	TIP1;1, TIP2;1		inhib		Water flux

[a]Listed transporters are essential for osmotically-active solute, H^+ and Ca^{2+} flux. See Chen et al (2012), Jezek and Blatt (2017), and Jezek et al. (2021) for a complete list of the transporter properties.

[b]Transporters groups by functional characteristics. Channel rectification and charge flux direction is indicated in parentheses.

[c]Representative Arabidopsis gene products, where known.

[d]Channels identified by physiological characteristics, yet to be associated with a gene product.

[e]activ, activation; inhib, inhibition; substr, transported solute (substrate or product).

[f]Ion (chemical) flux direction relative to the cytosol.

[g]Function of TPC1 remains unknown and is included here as a placeholder only.

Of course, signaling processes are intimately connected with our understanding transport and add to this body of data [cf. Kollist et al. (2014), Rodriguez and Shan (2022), Hedrich and Geiger (2017), Lefoulon (2021), Karnik et al. (2017), Chen et al. (2016), Papanatsiou et al. (2015), Pantaleno et al. (2021), and Jezek and Blatt (2017)]. They offer important insights into how guard cells coordinate transport for stomatal movements. Yes, their impacts are multifaceted, but these can be tiered with clear patterns to the physiology. For example, Lefoulon (2021) pointed out that phosphorylation of K^+ channels, almost without exception, affects ensemble current amplitude rather than gating. In other words, phosphorylation alters the pool size of channels available for activation by voltage but does not alter their sensitivity to voltage. The same charac-teristics also to apply to Ca^{2+} and Cl^- channels (Koehler and Blatt 2002; Assmann and Jegla 2016; Eisenach et al. 2017; Jezek and Blatt 2017; Li et al. 2022), and H^+-coupled Cl^- transport is similarly affected by phosphorylation in amplitude only (Wege et al. 2014).

My colleague's comment was valid in one sense, though perhaps not as it was intended: The kinetics of most biologic-al processes are highly non-linear, often with respect to multiple variables, for ion transport including voltage. Interactions between these processes, not least through the entanglements of a common membrane voltage and the pool of cytosolic solutes shared between the plasma membrane and tonoplast (see Current matters, above). These characteristics give rise to physiology that is frequently unexpected or counterintuitive, what is termed "emergent" behavior. Simply put, by the turn of this century, it was obvious that intuition was a very poor guide to understanding how stomata work.

While in Cambridge, I often stopped by the Physiological Laboratories on the Downing site adjacent Plant Sciences to share a coffee with V.L. (Arieh) Lew and discuss the challenges of transport interactions. We continued these discussions long after I left Cambridge for my first substantive posting at Wye College in Kent in 1990. Arieh had developed mechanism-based, mathematical models of the erythrocyte (Lew and Bookchin 1986) that explained much of the phenomenology of Gardos dehydration (Gardos 1958) and of sickle-cell anemia with counterintuitive predictions that were subsequently validated in experiments (Lew and Bookchin 2005; Lew 2022). It was obvious to us the need to approach the problems of stomatal physiology—for example, how guard cells respond to abscisic acid—using similar, mechanism-based strategies. Eventually, we secured funding from the UK Biotechnology and Biological Sciences Research Council and embarked on this project in 2008.

[As an aside, Wye College, part of the University of London, gained an international reputation in large measure from the work of agricultural chemist Ralph Louis Wain CBE, FRS. During the 1950s, 1960s and 1970s while at Wye College, Wain developed, synthesized, and studied the actions of a number of auxin-like herbicides and was instrumental in discovering the role of ABA in regulating foliar transpiration (Cornforth et al. 1966; Wright and Hiron 1969; Hiron and Wright 1973). By chance, my own laboratory while at the College focused on ABA, work that led to our discoveries, noted earlier, of the guard cell Ca^{2+} channel and of voltage and $[Ca^{2+}]_i$ coupling in K^+ and Cl^- flux control (Grabov and Blatt 1997, 1998, 1999; Blatt 2000; Hamilton et al. 2000; Koehler and Blatt 2002; Sokolovski et al. 2008). A second discovery of ABA action through vesicle trafficking (SNARE) proteins (Leyman et al. 1999) led my laboratory at Glasgow to SNARE-K^+ channel binding and their coregulation in cell expansion (Honsbein et al. 2009; Grefen et al. 2010, 2015; Karnik et al. 2017; Lefoulon et al. 2018; Waghmare et al. 2019). Wye College was subsumed within Imperial College London in 2000, shortly before I left to take up the Regius Chair of Botany at the University of Glasgow. The merger was poorly handled and its oversight unsympathetic to Wye and its surrounds. Wye College closed in 2009.]

Stomata are ideally suited to the "bottom-up" approach of mechanistic, computational modeling. Their movements are driven by changes in the osmotic solute contents of the guard cells that are connected to cell volume, turgor, and stomatal aperture. The osmotic solute contents, in turn, are governed by ion and water flux that connect to the relevant ion gradients and permeabilities across each membrane. Finally, the operation of each of the 23 major solutes and water transporters can be described quantitatively within sets of well-established kinetic equations, fully constrained by experimental data and by physical laws of mass and charge conservation (Table 3). Even though our knowledge of transport at the tonoplast was less well-developed, there was sufficient detailed information from experiments (Blatt 2000; Webb et al. 2001; Hetherington and Brownlee 2004; Blatt et al. 2007; McAinsh and Pittman 2009; Assmann and Jegla 2016; Jezek and Blatt 2017) to constrain the range of parameters needed to comply with experimental results (Chen et al. 2012).

Ultimately, for the guard cell, the important outputs are the cell volume and osmolality, water potential and turgor, the voltages across each membrane, the predominant ion concentrations, especially K^+, Cl^-, and malate (Mal), the total and free concentrations of Ca^{2+} and H^+ and their buffering, and the corresponding ion fluxes through each transporter. For such modeling, it is sufficient to know the kinetic relationships that describe a process, even if the structural gene products are not. The same considerations apply to the molecular mechanisms behind their regulation. To understand how a system responds to physiological or experimental perturbation, the only relevant biology is encapsulated in how one model variable is connected to another. For example, we still do not know the relative contributions of the KAT1, KAT2, and KC1 subunits to the inward-rectifying K^+ channels at the plasma membrane of Arabidopsis guard cells (Jeanguenin et al. 2011; Lefoulon 2021). Nevertheless, we know how the ensemble K^+ current depends on voltage and how its amplitude depends on extracellular $[K^+]$, pH, and $[Ca^{2+}]_i$, and we can describe these dependencies in quantitative terms (Blatt 1990, 1992; Lemtiri-Chlieh and MacRobbie 1994; Grabov and Blatt 1997, 1999; Roelfsema and Prins 1997; Gajdanowicz et al. 2009). Such knowledge is sufficient to describe both depend-encies and to model the current using a set of precise, mathematical equations. This approach introduces modules—"black boxes" with adjustable levels of kinetic detail—that may be expanded if, and when, a specific mod-ule becomes the focus of study; it also greatly reduces the computational overhead without a loss in predictive power (Endy and Brent 2001).

In general, the true value of any model rests in its ability not only to reproduce known behaviors but also to predict new ones that can be tested experimentally. Models built on a collection of elements generate system behavior through the interactions that arise between the elements; these interactions often lead to unexpected, or emergent, be-haviors that could not be anticipated from knowledge of the elements in isolation. More often than not, these emergent behaviors become a focus for model testing and validation. Model and experiment are thus complementary tools (Stelling et al. 2004; van Riel 2006), each model representing a hypothesis to be discarded, validated, or refined by compar-ing model predictions with experimental results.

Models built on the original OnGuard platform (Chen et al. 2012; Hills et al. 2012) proved their predictive power (Table 4), among others, by explaining why guard cell $[Ca^{2+}]_i$ rises during the daytime, how Cl^- availability affects vacuolar Mal accumulation (Dodd et al. 2007; Chen et al. 2012), and how net Cl^- and Mal transport is directed to the vacuole during the day and out of the vacuole to the apo-plast at night (Chen et al. 2012; Hills et al. 2012). These models predicted how the constitutively-active H^+-ATPase *ost2* mutation affects Cl^- and Mal accumulation (Blatt et al. 2014) and slows stomatal opening as well as closing (Wang et al. 2017), as I noted earlier (Fig. 5). They also explained why stomata of the *slac1* mutant open as well as close slowly (Wang et al. 2012). These studies uncovered networks con-necting the activity of the K^+ channels with the anion chan-nels and H^+-ATPase. The model outputs include many other testable predictions, including the importance of gating of the outward-rectifying K^+ channels as a target to accelerate stomatal kinetics (Blatt et al. 2014).

With the second generation platform, we introduced foliar transpiration, connecting whole-plant water relations with water and solute transport in the guard cells (Wang et al. 2017). OnGuard2 incorporated vapor equilibration with water in the guard cell wall, harmonizing the concepts of

Table 4. Selected predictions arising from the OnGuard platform[a]

Prediction	Observations	Principal targets	Validated?[a]	Refs
OnGuard1				
slac1	KAT1 inhibition, slow opening $[Ca^{2+}]_i$ and pH_i increases	Ca^{2+} channel, H^+–Cl^- symport, K^+ channels	Yes	b
Elevated $[Ca^{2+}]_i$	Daytime rise in $[Ca^{2+}]_i$	H^+-ATPase, Ca^{2+} channel, Ca^{2+}-ATPases	Historic	c
Cl^- accumulation	preference for Cl^- over Mal accumulation with $[Cl^-]_o$	H^+-Cl^- symport, VCl channel, VMal channel	Historic	d
Tonoplast voltage	Daytime depolarization	All tonoplast transporters	Untested	e
$[Ca^{2+}]_i$ oscillations	$[Ca^{2+}]_i$ oscillations accelerate closure, coupled with solute flux	H^+-ATPase, Ca^{2+} channel, Ca^{2+}-ATPases	Yes	e
OnGuard2				
ost2	Elevated pH_i, slowed opening and closing, elevated $[Ca^{2+}]_i$	H^+-ATPase, $[Ca^{2+}]_i$ increases, Ca^{2+} channel, VCa^{2+} channel, H^+-Cl^- symport	Yes	f
slac1 and water	No effect of *slac1* on VPD-driven stomatal movement		Yes	f
Mal and stress	preference for Mal retention under water/ osmotic stress	VCl channel, SLAC1 channel	Historic	d
OnGuard3				
$[Ca^{2+}]_i$ oscillations	CO_2-evoked $[Ca^{2+}]_i$ rise	VCa^{2+}-ATPase, VCa_{in} channel	Historic	g
Carbon memory	Latency in stomatal closing	VCa^{2+}-ATPase, VCa_{in} channel	Yes	h
CO_2 "locking"	CO_2 precedence over dark in stomatal closure	VCa_{in} channel	Yes	j
K^+ channel engineering	K^+ channel gating as a target to accelerate stomata	GORK K^+ channel	Yes	k

[a]Listed are a small selection of the predictions arising from simulations with the OnGuard platform that are documented in the literature. Predictions are validated through previous publications (historic) or following the modeling (yes); other predictions remain to be examined (untested).

[b]Wang et al. (2012).

[c]Dodd et al. (2005, 2007) and Chen et al. (2012).

[d]Allaway (1973), Van Kirk and Raschke (1977, 1978), Raschke and Schnabl (1978), and Chen et al. (2012).

[e]Chen et al. (2012), Hills et al. (2012), and Wang et al. (2012).

[f]Wang et al. (2017).

[g]Webb et al. (1996), Grabov and Blatt (1997, 1998), Minguet-Parramona et al. (2016), and Jezek et al. (2021).

[h]Jezek et al. (2021).

[j]Blatt et al. (2022).

[k]Blatt and Alvim (2022) and Horaruang et al. (2022).

stomatal vapor equilibration (Peak and Mott 2011) and of liquid water delivery to the mesophyll and guard cells (Buckley et al. 2003, 2017; Rockwell et al. 2014). Apoplastic solute and turgor "exchange" with the surrounding epidermal cells ac-commodated the opposing mechanical pressures of the epi-dermis and the visco-elastic properties of the guard cell wall (Jezek et al. 2019). Finally, with the third-generation platform, OnGuard3, we added CO_2 diffusion from the atmosphere and its fixation through mesophyll photosynthesis (Jezek et al. 2021). Including CO_2 diffusion allowed tests for the min-imum set of targets necessary to explain CO_2 action on sto-matal aperture. It uncovered an emergent "carbon memory" in stomatal responsiveness and correctly predicted its impact on foliar gas exchange and water use efficiency as well as its dependence on endomembrane Ca^{2+} storage and release in the guard cell (Jezek et al. 2021). OnGuard3 has also proven instrumental in guiding the recent bioengineering of the GORK K^+ channel to accelerate stomatal movement for gains in both water use efficiency and photosynthetic yield (Blatt and Alvim 2022; Horaruang et al. 2022).

What do we learn from OnGuard and, more generally, from mechanistic modeling of transport physiology? The foremost lesson, again, is just how poor intuition is as a guide to understanding the consequences of transporters operating in parallel across a common membrane and sharing substrates and products on either side. Simulations such as those possible with OnGuard are therefore keys to resolving the "decisions made" by stomata (Berry et al. 2010) and encoded within the network of transport entanglements. A second, and equally important, lesson is that a very large part of transport physiology is explicable through these entanglements, even without considering additional tiers of regulatory activity, whether of gene expression, vesicle trafficking, or post-translational protein modification. Of course, these additional tiers of control may be important for other reasons, for example in long-term adaptation or in fine-tuning transport. So, a third lesson is simply that modeling based on the mechanics intrinsic to transport is a good place to start and should be an early step in any systematic understanding physiology before refocusing research to address extrinsic regulatory processes.

In short, in recognizing the wealth of emergent behaviors, it becomes evident that much of stomatal physiology simply reflects "a spectrum of behaviors that emerge from the ionic fluxes that play across the guard cell membranes, their entanglement and balance with the metabolic activities of the guard cells" (Blatt et al. 2022). For those seeking fundamental knowledge, the challenge is to resolve these entanglements and the behaviors that they engender. Mechanism-based models are essential tools in this process. They are equally

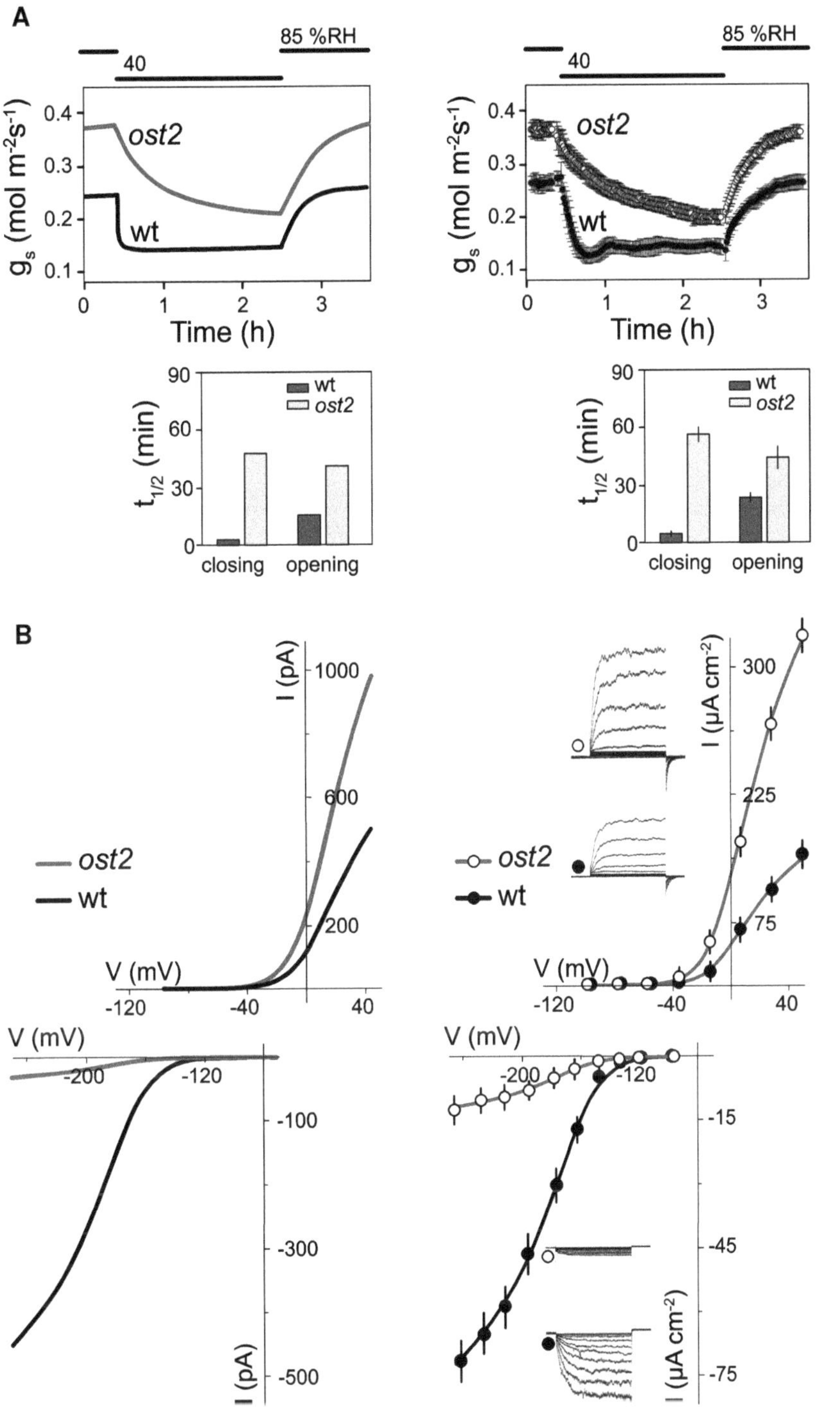

Figure 5. OnGuard correctly predicts slowed stomatal opening and closing in the *ost2* H^+-ATPase mutant in response to steps in relative humidity. OnGuard modeling outputs (*left*) and experimental data (*right*) connect slowed stomatal reopening in the Arabidopsis *ost2* mutant with the greatly reduced activity of the inward-rectifying K^+ channels when compared with wild-type Arabidopsis (wt). Modeling and experimental data consolidated and redrawn from Figs 3, 5, and 6 of Wang et al (2017). **A)** Stomatal conductances (g_s) predicted (*left*) and measured (*right*) as means ± SE of 4 independent experiments. Relative humidity (%RH) steps indicated (*above*). Corresponding halftimes ($t_{1/2}$) for closing and opening (*below*) were calculated by nonlinear least-squares fitting of model outputs and experimental data to a single exponential function. **B)** Current–voltage (IV) curves predicted (*left*) and recorded (*right*) under voltage clamp as means ± SE of 5 independent experiments. Steady-state currents shown are for the outward- (*above*) and inward-rectifying (*below*) K^+ channels. *Insets* (*right*): Representative current relaxations with voltage steps recorded under voltage clamp. Additional data and analysis (Wang et al. 2017) ascribes the enhanced outward-rectifying K^+ current to a 0.3 unit rise in cyto-solic pH and the reduced inward-rectifying K^+ current to this pH rise and an increase in $[Ca^{2+}]_i$ to near 500 nM at rest. Slowed reopening in the *ost2* mutant was predicted as a direct consequence of reducing the inward K^+ current by roughly 90%; the slowed closure was the predicted consequence of hyperpolarizing the membrane and raising pH, thereby suppressing activation of the R-type channels that engage the $[Ca^{2+}]_i$ and voltage oscilla-tions needed to promote the outward-rectifying K^+ channels and accelerate stomatal closure.

important, therefore, as practical guides in the effort towards rational biodesign of membrane transport going forward. If nothing else, the re-engineering of GORK gating (Blatt and Alvim 2022; Horaruang et al. 2022) is a clear example of how such knowledge can be put to use to improve whole-plant water use efficiency and photosynthetic carbon assimilation.

Concluding remarks

The International Workshop on Plant Membrane Biology was held this past year in Taipei, delayed beyond the usual triannual occurrence by COVID. These meetings have a long history going back to 1968 when the first international gathering was held in Schloss Reinhardsbrunn, in what was then East Germany. The workshop title did not appear until the meeting in Liverpool in 1972, and the focus expanded from ion transport to membrane biology with the meeting in Monterey, California, in 1992. The science has similarly progressed from descriptive studies, analyses of the driving forces on ion flux, and questions around their energy dependencies to the quantitative kinetics of individual transporters, their molecular properties, regulation, and trafficking.

Our knowledge of transport protein structure is growing rapidly too (Table 1). Crystal structures for the Arabidopsis plasma membrane H^+-ATPase AHA2 (Pedersen et al. 2007) and spinach aquaporin PIP2;1 (Tornroth-Horsefield et al. 2006) have been followed by detail of the structures, mechanics, and assembly of a bacterial SLAC1 channel homolog (Chen et al. 2010a) and more recently the Arabidopsis SLAC1 (Deng et al. 2021), the tonoplast aquaporin TIP2;1 (Kirscht et al. 2016) and TPC1 (Guo et al. 2016; Kintzer and Stroud 2016), the CNBD K^+ channels KAT1, AKT1, and SKOR (Clark et al. 2020; Dickinson et al. 2022; Lu et al. 2022; Li et al. 2023), as well as the ALMT12 channel (Qin et al. 2022).

With its operation across a common membrane, ion trans-port is intrinsically modular and therefore well-suited to bio-engineering. I have no doubt that the first tentative steps in this direction (Papanatsiou et al. 2019; Zhou et al. 2021; Horaruang et al. 2022) will soon become routine tools in re-search and its application, much as is becoming the case for synthetic photosynthesis (Miller et al. 2020). The means to these advances requires only that we combine kinetic and physiological data with molecular structural information of this kind, and that we use mechanism-based models to guide these efforts.

It is important, too, that we keep in mind from where these advances have come, and not only as a roadmap to past false starts and dead ends. The historical backdrop to research in plant membrane transport also informs on many opportun-ities that remain to be explored, including those set aside of-ten enough because they appeared too difficult to pursue at the time. Recognizing and utilizing this history wisely is all the more urgent in light of the global imperatives of climate change and the need to husband our environment and its re-sources. Rodriguez-Martinez et al. (2022) make the case for solutions to come through molecular biology, but clearly the call applies to biological research generally and its poten-tial for innovation.

OUTSTANDING QUESTIONS BOX

- A role for the slow-vacuolar cation channel TPC1 in plant physiology remains as obscure as it was 35 years on from its discovery.
- The identity of the plasma membrane Ca^{2+} channels that trigger $[Ca^{2+}]_i$ and voltage oscillations in guard cells and close stomata remains a mystery, as the current differs in biophysical characteristics from that of any cyclic nucleotide-gated channel identified to date.
- Similarly, the nature and identities remain obscure for the channels mediating endomembrane Ca^{2+} release and $[Ca^{2+}]_i$ elevations.
- Mechanistic models highlight the entanglements of ion transport, providing explanations for seemingly counterintuitive behaviors without a need for extrinsic regulation. A general challenge, then, is to establish what cannot be understood as a result of these entanglements and the behaviors that emerge from them.

Acknowledgments

Several colleagues kindly read drafts of this article. I thank Rucha Karnik, Carlos Garcia-Mata, Myriam Charpentier, and Gerhard Thiel for their comments and suggestions. I am indebted to the many colleagues with whom I had the pleasure of working over the past five decades, and especially to my mentors Clifford Slayman, Enid MacRobbie and, in absentia, Winslow Briggs and Wolfgang Haupt. Thank you for your patience, guidance and friendship.

Author Contributions

In accordance with journal requirements, M.B. acknowledges his contribution to this review as the sole author. He undertook to assess and interpret the literature, and he conceived and wrote the manuscript in order to share his experience and perceptions with his readers. As a review, there were no tools used, no experiments performed, and no new data analysed.

Funding

I am grateful for continued support from the UK research councils, and funding from the National Science Foundation

(USA), the John Simon Guggenheim Memorial Foundation, the Human Frontiers Science Program, the European Research Council, the Royal Society, and the Leverhulme Trust.

Conflict of interest statement. None declared.

References

Ache P, Becker D, Ivashikina N, Dietrich P, Roelfsema MRG, Hedrich R. GORK, a delayed outward rectifier expressed in guard cells of *Arabidopsis thaliana*, is a K^+-selective, K^+-sensing ion channel. FEBS Lett. 2000:**486**(2):93–98. https://doi.org/10.1016/S0014-5793(00)02248-1

Agus ZS, Morad M. Modulation of cardiac ion channels by magnesium. Annu Rev Physiol. 1991:**53**(1):299–307. https://doi.org/10.1146/annurev.ph.53.030191.001503

Ahn SJ, Shin R, Schachtman DP. Expression of *KT/KUP* genes in Arabidopsis and the role of root hairs in K+ uptake. Plant Physiol. 2004:**134**(3):1135–1145. https://doi.org/10.1104/pp.103.034660

Alexandre J, Lassalles JP. Effect of D-*myo*-inositol-1,4,5-trisphosphate on the electrical properties of the red beet vacuole membrane. Plant Physiol. 1990:**93**(2):837–840. https://doi.org/10.1104/pp.93.2.837

Alexandre J, Lassalles JP, Kado RT. Opening of Ca^{2+} channels in isolated red beet root vacuole membrane by inositol 1,4,5-trisphosphate. Nature. 1990:**343**(6258):567–570. https://doi.org/10.1038/343567a0

Allaway WG. Accumulation of malate in guard cells of *Vicia faba* during stomatal opening. Planta. 1973:**110**(1):63–70. https://doi.org/10.1007/BF00386923

Allen GJ, Amtmann A, Sanders D. Calcium-dependent and calcium-independent K+ mobilization channels in *Vicia faba* guard cell vacuoles. J Exp Bot. 1998:**49**(Special):305–318. https://doi.org/10.1093/jxb/49.Special_Issue.305

Allen GJ, Chu SP, Harrington CL, Schumacher K, Hoffman T, Tang YY, Grill E, Schroeder JI. A defined range of guard cell calcium oscillation parameters encodes stomatal movements. Nature. 2001:**411**(6841):1053–1057. https://doi.org/10.1038/35082575

Allen GJ, Kwak JM, Chu SP, Llopis J, Tsien RY, Harper JF, Schroeder JI. Cameleon calcium indicator reports cytoplasmic calcium dynamics in Arabidopsis guard cells. Plant J. 1999:**19**(6):735–747. https://doi.org/10.1046/j.1365-313x.1999.00574.x

Allen GJ, Muir SR, Sanders D. Release of Ca^{2+} from individual plant vacuoles by both insp(3) and cyclic ADP-ribose. Science. 1995:**268**(5211):735–737. https://doi.org/10.1126/science.7732384

Allen GJ, Sanders D. Two voltage-gated, calcium-release channels coreside in the vacuolar membrane of broad bean guard-cells. Plant Cell. 1994:**6**(5):685–694. https://doi.org/10.2307/3869872

Alvim FSL. Potassium channels of C3 and C4 plants. Glasgow: University of Glasgow; 2022.

Armstrong CM, Binstock L. Anomalous rectification in the squid giant axon injected with tetraethylammonium chloride. J Gen Physiol. 1965:**48**(5):859–872. https://doi.org/10.1085/jgp.48.5.859

Assmann SM, Jegla T. Guard cell sensory systems: recent insights on stomatal responses to light, abscisic acid, and CO_2. Curr Opin Plant Biol. 2016:**33**:157–167. https://doi.org/10.1016/j.pbi.2016.07.003

Bakker EP, Harold FM. Energy coupling to potassium-transport in *Streptococcus-faecalis*—interplay of ATP and the protonmotive force. J Biol Chem. 1980:**255**(2):433–440. https://doi.org/10.1016/S0021-9258(19)86192-6

Bassil E, Blumwald E. The ins and outs of intracellular ion homeostasis: NHX-type cation/H^+ transporters. Curr Opin Plant Biol. 2014:**22**:1–6. https://doi.org/10.1016/j.pbi.2014.08.002

Basso B, Ullrich-Eberius CI. Membrane potential and proton cotransport of alanine and phosphate as affected by permeant weak acids in *Lemna gibba*. Plant Physiol. 1987:**85**(3):674–678. https://doi.org/10.1104/pp.85.3.674

Beilby MJ. Potassium channels and different states of the *Chara* plasmalemma. J Membr Biol. 1986:**89**(3):241–249. https://doi.org/10.1007/BF01870667

Beilby MJ, Al Khazaaly S. Re-modeling *Chara* action potential: I. From thiel model of Ca^{2+} transient to action potential form. AIMS Biophys. 2016:**3**(3):431–449. https://doi.org/10.3934/biophy.2016.3.431

Beilby MJ, Al Khazaaly S. Re-modeling *Chara* action potential: II. The action potential form under salinity stress. AIMS Biophys. 2017:**4**(2):298–315. https://doi.org/10.3934/biophy.2017.2.298

Beilby MJ, Bisson MA, Schneider SC. How Characean algae take up needed and excrete unwanted ions—an overview explaining how insights from electrophysiology are useful to understand the ecology of aquatic macrophytes. Aquatic Bot. 2022:**181**:103542. https://doi.org/10.1016/j.aquabot.2022.103542

Beilby MJ, Blatt MR. Simultaneous measurements of cytoplasmic K^+ concentration and plasma membrane electrical parameters in single membrane samples of *Chara corallina*. Plant Physiol. 1986:**82**(2):417–422. https://doi.org/10.1104/pp.82.2.417

Beilby MJ, Coster HGL. The action potential in *Chara corallina* III. The Hodgkin-Huxley parameters for the plasmalemma. Aust J Plant Physiol. 1979:**6**(3):337–353. https://doi.org/10.1071/PP9790337

Beilby M, Walker NA. Chloride transport in Chara: I. Kinetics and current-voltage curves for a probable proton symport. J Exp Bot. 1981:**32**(1):43–54. https://doi.org/10.1093/jxb/32.1.43

Beljaars ACM, Viterbo P, Miller MJ, Betts AK. The anomalous rainfall over the United States during July 1993: sensitivity to land surface parameterization and soil moisture. Monthly Weather Rev. 1996:**124**(3):362–383. https://doi.org/10.1175/1520-0493(1996)124<0362:TAROTU>2.0.CO;2

Bentrup FW, Gogarten-Boekels M, Hoffmann B, Gogarten J, Baumann C. ATP-dependent acidification and tonoplast hyperpolarization in isolated vacuoles from green suspension cells of *Chenopodium rubrum* L. Proc Natl Acad Sci U S A. 1986:**83**(8):2431–2433. https://doi.org/10.1073/pnas.83.8.2431

Bernstein J. Untersuchungen zur thermodynamik der bioelektrischen strîme. PflÅgers Archiv Eur J Physiol. 1902:**92**(10–12):521–562. https://doi.org/10.1007/BF01790181

Berry JA, Beerling DJ, Franks PJ. Stomata: key players in the earth system, past and present. Curr Opin Plant Biol. 2010:**13**(3):233–240. https://doi.org/10.1016/j.pbi.2010.04.013

Bertl A, Blumwald E, Coronado R, Eisenberg R, Findlay G, Gradmann D, Hille B, Kohler K, Kolb HA, Macrobbie E, et al. Electrical measurements on endomembranes. Science. 1992:**258**(5084):873–874. https://doi.org/10.1126/science.1439795

Bertl A, Klieber HG, Gradmann D. Slow kinetics of a potassium channel in acetabularia. J Membr Biol. 1988:**102**(2):141–152. https://doi.org/10.1007/BF01870452

Betts AK, Ball JH, Beljaars ACM, Miller MJ, Viterbo PA. The land surface-atmosphere interaction: a review based on observational and global modeling perspectives. J Geophys Res. 1996:**101**(D3):7209–7225. https://doi.org/10.1029/95JD02135

Beyhl D, Hortensteiner S, Martinoia E, Farmer EE, Fromm J, Marten I, Hedrich R. The fou2 mutation in the major vacuolar cation channel TPC1 confers tolerance to inhibitory luminal calcium. Plant J. 2009:**58**(5):715–723. https://doi.org/10.1111/j.1365-313X.2009.03820.x

Bezprozvanny I, Watras J, Ehrlich BE. Bell-shaped calcium-response curves of Ins(1,4,5)P3- and calcium-gated channels from endoplasmic reticulum of cerebellum. Nature. 1991:**351**(6329):751–754. https://doi.org/10.1038/351751a0

Bihler H, Eing C, Hebeisen S, Roller A, Czempinski K, Bertl A. TPK1 is a vacuolar ion channel different from the slow-vacuolar cation channel. Plant Physiol. 2005:**139**(1):417–424. https://doi.org/10.1104/pp.105.065599

Blatt MR. Interpretation of steady-state current-voltage curves: consequences and implications of current subtraction in transport studies. J Membr Biol. 1986:**92**(1):91–110. https://doi.org/10.1007/BF01869018

Blatt MR. Electrical characteristics of stomatal guard cells: the contribution of ATP-dependent, "electrogenic" transport revealed by current-voltage and difference-current-voltage analysis. J Membr Biol. 1987a:**98**(3):257–274. https://doi.org/10.1007/BF01871188

Blatt MR. Electrical characteristics of stomatal guard cells: the ionic basis of the membrane potential and the consequence of potassium chloride leakage from microelectrodes. Planta. 1987b:**170**(2): 272–287. https://doi.org/10.1007/BF00397898

Blatt MR. Mechanisms of fusicoccin action: a dominant role for secondary transport in a higher-plant cell. Planta. 1988a:**174**(2): 187–200. https://doi.org/10.1007/BF00394771

Blatt MR. Potassium-dependent bipolar gating of potassium channels in guard cells. J Membr Biol. 1988b:**102**(3):235–246. https://doi.org/10.1007/BF01925717

Blatt MR. Potassium channel currents in intact stomatal guard cells: rapid enhancement by abscisic acid. Planta. 1990:**180**(3):445–455. https://doi.org/10.1007/BF01160403

Blatt MR. K^+ channels of stomatal guard cells: characteristics of the inward rectifier and its control by pH. J Gen Physiol. 1992:**99**(4): 615–644. https://doi.org/10.1085/jgp.99.4.615

Blatt MR. Cellular signaling and volume control in stomatal movements in plants. Annu Rev Cell Dev Biol. 2000:**16**(1):221–241. https://doi.org/10.1146/annurev.cellbio.16.1.221

Blatt MR. Membrane transport in plants. In: **Blatt MR**, editor. Membrane transport in plants. Vol 1. Oxford: Blackwell; 2004. p. 1–256.

Blatt MR, Alvim J. Exploiting a channel voltage 'antenna' for gains in water use efficiency and biomass. Nat Plants. 2022:**8**(11): 1216–1217. https://doi.org/10.1038/s41477-022-01268-x

Blatt MR, Beilby MJ, Tester M. Voltage dependence of the Chara proton pump revealed by current-voltage measurement during rapid metabolic blockade with cyanide. J Membr Biol. 1990a:**114**(3): 205–223. https://doi.org/10.1007/BF01869215

Blatt MR, Clint GM. Mechanisms of fusicoccin action kinetic modification and inactivation of potassium channels in guard cells. Planta. 1989:**178**(4):509–523. https://doi.org/10.1007/BF00963821

Blatt MR, Garcia-Mata C, Sokolovski S. Membrane transport and Ca^{2+} oscillations in guard cells. In: **Mancuso S, Shabala S**, editors. Rhythms in plants. Vol 1. Berlin: Springer; 2007. p. 115–134.

Blatt MR, Gradmann D. K^+-sensitive gating of the K^+ outward rectifier in Vicia guard cells. J Membr Biol. 1997:**158**(3):241–256. https://doi.org/10.1007/s002329900261

Blatt MR, Jezek M, Lew VL, Hills A. What can mechanistic models tell us about guard cells, photosynthesis, and water use efficiency? Trends Plant Sci. 2022:**27**(2):166–179. https://doi.org/10.1016/j.tplants.2021.08.010

Blatt MR, Maurousset L, Meharg AA. High-affinity NO_3^--H^+ cotransport in the fungus Neurospora: induction and control by pH and membrane voltage. J Membr Biol. 1997:**160**(1):59–76. https://doi.org/10.1007/s002329900295

Blatt MR, Rodriguez-Navarro A, Slayman CL. Potassium-proton symport in Neurospora: kinetic control by pH and membrane potential. J Membr Biol. 1987:**98**(2):169–189. https://doi.org/10.1007/BF01872129

Blatt MR, Slayman CL. KCl leakage from microelectrodes and its impact on the membrane parameters of a nonexcitable cell. J Membr Biol. 1983:**72**(3):223–234. https://doi.org/10.1007/BF01870589

Blatt MR, Slayman CL. Role of "active" potassium transport in the regulation of cytoplasmic pH by nonanimal cells. Proc Natl Acad Sci USA. 1987:**84**(9):2737–2741. https://doi.org/10.1073/pnas.84.9.2737

Blatt MR, Thiel G. K^+ channels of stomatal guard cells: bimodal control of the K^+ inward-rectifier evoked by auxin. Plant J. 1994:**5**(1):55–68. https://doi.org/10.1046/j.1365-313X.1994.5010055.x

Blatt MR, Thiel G, Trentham DR. Reversible inactivation of K^+ channels of Vicia stomatal guard cells following the photolysis of caged inositol 1,4,5-trisphosphate. Nature. 1990b:**346**(6286):766–769. https://doi.org/10.1038/346766a0

Blatt MR, Wang Y, Leonhardt N, Hills A. Exploring emergent properties in cellular homeostasis using OnGuard to model K^+ and other ion transport in guard cells. J Plant Physiol. 2014:**171**(9):770–778. https://doi.org/10.1016/j.jplph.2013.09.014

Blinks LR. The relations of temperature to the potassium effect and the bioelectric potential of valonia. J Gen Physiol. 1942:**25**(6):905–916. https://doi.org/10.1085/jgp.25.6.905

Blinks LR. The source of the bioelectric potentials in large plant cells. Proc Natl Acad Sci USA. 1949:**35**(10):566–575. https://doi.org/10.1073/pnas.35.10.566

Bonza MC, Loro G, Behera S, Wong A, Kudla J, Costa A. Analyses of Ca^{2+} accumulation and dynamics in the endoplasmic reticulum of Arabidopsis root cells using a genetically encoded cameleon sensor. Plant Physiol. 2013:**163**(3):1230–1241. https://doi.org/10.1104/pp.113.226050

Brearley J, Venis MA, Blatt MR. The effect of elevated CO_2 concentrations on K^+ and anion channels of *Vicia faba* L. guard cells. Planta. 1997:**203**(2):145–154. https://doi.org/10.1007/s004250050176

Bretag AW. The glass micropipette electrode: a history of its inventors and users to 1950. J General Physiol. 2017:**149**:417–430.

Briskin DP, Leonard RT. Partial characterization of a phosphorylated intermediate associated with the plasma-membrane atpase of corn roots. Proc Natl Acad Sci USA. 1982:**79**(22):6922–6926. https://doi.org/10.1073/pnas.79.22.6922

Buchsenschutz K, Marten I, Becker D, Philippar K, Ache P, Hedrich R. Differential expression of K+ channels between guard cells and subsidiary cells within the maize stomatal complex. Planta. 2005:**222**(6):968–976. https://doi.org/10.1007/s00425-005-0038-6

Buckley TN, John GP, Scoffoni C, Sack L. The sites of evaporation within leaves. Plant Physiol. 2017:**173**(3):1763–1782. https://doi.org/10.1104/pp.16.01605

Buckley TN, Mott KA, Farquhar GD. A hydromechanical and biochemical model of stomatal conductance. Plant Cell Environ. 2003: **26**(10):1767–1785. https://doi.org/10.1046/j.1365-3040.2003.01094.x

Bush DS, Hedrich R, Schroeder JI, Jones RL. Channel-mediated potassium ion flux in barley aleurone protoplasts. Planta. 1988:**176**(3): 368–377. https://doi.org/10.1007/BF00395417

Campanoni P, Blatt MR. Membrane trafficking and polar growth in root hairs and pollen tubes. J Exp Bot. 2007:**58**(1):65–74. https://doi.org/10.1093/jxb/erl059

Chanroj S, Lu YX, Padmanaban S, Nanatani K, Uozumi N, Rao R, Sze H. Plant-specific cation/H^+ exchanger 17 and its homologs are endomembrane K^+ transporters with roles in protein sorting. J Biol Chem. 2011:**286**(39):33931–33941. https://doi.org/10.1074/jbc.M111.252650

Charpentier M, Sun JH, Martins TV, Radhakrishnan GV, Findlay K, Soumpourou E, Thouin J, Very AA, Sanders D, Morris RJ, et al. Nuclear-localized cyclic nucleotide-gated channels mediate symbiotic calcium oscillations. Science. 2016:**352**(6289):1102–1105. https://doi.org/10.1126/science.aae0109

Chen Z-H, Hills A, Baetz U, Amtmann A, Lew VL, Blatt MR. Systems dynamic modeling of the stomatal guard cell predicts emergent behaviors in transport, signaling, and volume control. Plant Physiol. 2012:**159**(3):1235–1251. https://doi.org/10.1104/pp.112.197350

Chen Z-H, Hills A, Lim CK, Blatt MR. Dynamic regulation of guard cell anion channels by cytosolic free Ca^{2+} concentration and protein phosphorylation. Plant J. 2010b:**61**(5):816–825. https://doi.org/10.1111/j.1365-313X.2009.04108.x

Chen Y-H, Hu L, Punta M, Bruni R, Hillerich B, Kloss B, Rost B, Love J, Siegelbaum SA, Hendrickson WA. Homologue structure of the SLAC1 anion channel for closing stomata in leaves. Nature. 2010a:**467**(7319):1074–1080. https://doi.org/10.1038/nature09487

Chen Z-H, Wang Y, Wang J-W, Babla M, Zhao C, Garcia-Mata C, Sani E, Differ C, Mak M, Hills A, et al. Nitrate reductase mutation alters potassium nutrition as well as nitric oxide-mediated control of guard cell ion channels in Arabidopsis. New Phytol. 2016:**209**(4):1456–1469. https://doi.org/10.1111/nph.13714

Choi W-G, Hilleary R, Swanson SJ, Kim S-H, Gilroy S. Rapid, long-distance electrical and calcium signaling in plants. Annu Rev Plant Biol. 2016:**67**(1):287–307. https://doi.org/10.1146/annurev-arplant-043015-112130

Clark MD, Contreras GF, Shen R, Perozo E. Electromechanical coupling in the hyperpolarization-activated K+ channel KAT1. Nature. 2020:**583**(7814):145. https://doi.org/10.1038/s41586-020-2335-4

Clint GM, Blatt MR. Mechanisms of fusicoccin action: evidence for concerted modulations of secondary K^+ transport in a higher-plant cell. Planta. 1989:**178**(4):495–508. https://doi.org/10.1007/BF00963820

Cole KS. Dynamic electrical characteristics of the squid axon membrane. Arch Sci Physiol. 1949:**3**:253–257.

Colombo R, Cerana R. Inward rectifying potassium channels in the plasma membrane of *Arabidopsis thaliana*. Plant Physiol. 1991: **97**(3):1130–1135. https://doi.org/10.1104/pp.97.3.1130

Conway EJ, Brady TG. Biological production of acid and alkali. 1. Quantitative relations of succinic and carbonic acids to the potassium and hydrogen ion exchange in fermenting yeast. Biochem J. 1950:**47**(3): 360–369. https://doi.org/10.1042/bj0470360

Conway EJ, O'Malley E. Nature of the cation exchanges during short-period yeast fermentation. Nature. 1944:**153**(3888):555–556. https://doi.org/10.1038/153555b0

Cornforth JW, Milborrow BV, Ryback G, Rothwell K, Wain RL. Identification of yellow lupin growth inhibitor as (+)-abscisin 2 ((+)-dormin). Nature. 1966:**211**(5050):742–743. https://doi.org/10.1038/211742b0

De Angeli A, Monachello D, Ephritikhine G, Frachisse JM, Thomine S, Gambale F, Barbier-Brygoo H. CLC-mediated anion transport in plant cells. Philos Trans R Soc B-Biol Sci. 2009:**364**(1514):195–201. https://doi.org/10.1098/rstb.2008.0128

De Col V, Fuchs P, Nietzel T, Elsaesser M, Voon CP, Candeo A, Seeliger I, Fricker MD, Grefen C, Moller IM, et al. ATP sensing in living plant cells reveals tissue gradients and stress dynamics of energy physiology. Elife. 2017:**6**:e26770. https://doi.org/10.7554/eLife.26770

DeFalco TA, Marshall CB, Munro K, Kang HG, Moeder W, Ikura M, Snedden WA, Yoshioka K. Multiple calmodulin-binding sites positively and negatively regulate Arabidopsis CYCLIC NUCLEOTIDE-GATED CHANNEL12. Plant Cell. 2016:**28**(7):1738–1751. https://doi.org/10.1105/tpc.15.00870

Deng Y-N, Kashtoh H, Wang Q, Zhen G-X, Li Q-Y, Tang L-H, Gao H-L, Zhang C-R, Qin L, Su M, et al. Structure and activity of SLAC1 channels for stomatal signaling in leaves. Proc Natl Acad Sci USA. 2021:**118**(18):e2015151118. https://doi.org/10.1073/pnas.2015151118

DeWeer P, Gadsby DC, Rakowski RF. Voltage dependence of the Na-K pump. Annu Rev Physiol. 1988:**50**(1):225–241. https://doi.org/10.1146/annurev.ph.50.030188.001301

Dickinson MS, Pourmal S, Gupta M, Bi M, Stroud RM. Symmetry reduction in a hyperpolarization-activated homotetrameric ion channel. Biochemistry. 2022:**61**(20):2177–2181. https://doi.org/10.1021/acs.biochem.1c00654

Dietrich P, Moeder W, Yoshioka K. Plant cyclic nucleotide-gated channels: new insights on their functions and regulation. Plant Physiol. 2020:**184**(1):27–38. https://doi.org/10.1104/pp.20.00425

Dodd AN, Gardner MJ, Hotta CT, Hubbard KE, Dalchau N, Love J, Assie JM, Robertson FC, Jakobsen MK, Goncalves J, et al. The Arabidopsis circadian clock incorporates a cADPR-based feedback loop. Science. 2007:**318**(5857):1789–1792. https://doi.org/10.1126/science.1146757

Dodd AN, Love J, Webb AAR. The plant clock shows its metal: circadian regulation of cytosolic free Ca^{2+}. Trends Plant Sci. 2005:**10**(1): 15–21. https://doi.org/10.1016/j.tplants.2004.12.001

Eisenach C, Baetz U, Huck NV, Zhang J, De Angeli A, Beckers GJM, Martinoia E. ABA-induced stomatal closure involves ALMT4, a phosphorylation-dependent vacuolar anion channel of Arabidopsis. Plant Cell. 2017:**29**(10):2552–2569. https://doi.org/10.1105/tpc.17.00452

Eisenach C, Papanatsiou M, Hillert EK, Blatt MR. Clustering of the K^+ channel GORK of Arabidopsis parallels its gating by extracellular K^+. Plant J. 2014:**78**(2):203–214. https://doi.org/10.1111/tpj.12471

Eisenberg R, Barcilon V, Mathias R. Electrical properties and analysis of spherical syncytia. Biophys J. 1979:**25**(1):151–180. https://doi.org/10.1016/S0006-3495(79)85283-2

Eisenberg RS, Johnson EA. Three-dimensional electrical field problems in physiology. Prog Biophys Mol Biol. 1970:**20**:1–65. https://doi.org/10.1016/0079-6107(70)90013-1

Endy D, Brent R. Modelling cellular behaviour. Nature. 2001:**409**(6818): 391–395. https://doi.org/10.1038/35053181

Erwee MG, Goodwin PB, Van Bel AJE. Cell-cell communication in the leaves of *Commelina cyanea* and other plants. Plant Cell Environ. 1985:**8**(3):173–178. https://doi.org/10.1111/1365-3040.ep11604602

Fairley-Grenot KA, Assmann SM. Whole-cell K^+ current across the plasma membrane of guard cells from a grass: *Zea mays*. Planta. 1992:**186**(2):282–293. https://doi.org/10.1007/BF00196258

Favre P, Greppin H, Agosti RD. Repetitive action potentials induced in *Arabidopsis thaliana* leaves by wounding and potassium chloride application. Plant Physiol Biochem. 2001:**39**(11):961–969. https://doi.org/10.1016/S0981-9428(01)01317-1

Felle H. Steriospecificity and electrogenicity of amino acid transport in *Riccia fluitans*. Planta. 1981:**152**(6):505–512. https://doi.org/10.1007/BF00380821

Felle HH, Zimmermann MR. Systemic signalling in barley through action potentials. Planta. 2007:**226**(1):203–214. https://doi.org/10.1007/s00425-006-0458-y

Fischer RA. Stomatal opening in isolated epidermal strips of *Vicia faba* I. Responses to light and CO_2-free air. Plant Physiol. 1968:**43**(12): 1947–1952. https://doi.org/10.1104/pp.43.12.1947

Fischer C, DeFalco TA, Karia P, Snedden WA, Moeder W, Yoshioka K, Dietrich P. Calmodulin as a Ca^{2+}-sensing subunit of Arabidopsis cyclic nucleotide-gated channel complexes. Plant Cell Physiol. 2017:**58**(7):1208–1221. https://doi.org/10.1093/pcp/pcx052

Fromm J, Bauer T. Action-potentials in maize sieve tubes change phloem translocation. J Exp Bot. 1994:**45**(4):463–469. https://doi.org/10.1093/jxb/45.4.463

Furuichi T, Cunningham KW, Muto S. A putative two pore channel AtTPC1 mediates Ca^{2+} flux in Arabidopsis leaf cells. Plant Cell Physiol. 2001:**42**(9):900–905. https://doi.org/10.1093/pcp/pce145

Furumoto T, Yamaguchi T, Ohshima-Ichie Y, Nakamura M, Tsuchida-Iwata Y, Shimamura M, Ohnishi J, Hata S, Gowik U, Westhoff P, et al. A plastidial sodium-dependent pyruvate transporter. Nature. 2011:**476**(7361):472–475. https://doi.org/10.1038/nature10250

Gajdanowicz P, Garcia-Mata C, Sharma T, Gonzalez W, Morales-Navarro SE, Gonzalez-Nilo FD, Gutowicz J, Mueller-Roeber B, Blatt MR, Dreyer I. Distributed structures determine K^+ and voltage dependent gating of the K_{in} channel KAT1 and the K_{out} channel SKOR. New Phytol. 2009:**182**:380–391. https://doi.org/10.1111/j.1469-8137.2008.02749.x

Gao Q-F, Fei C-F, Dong J-Y, Gu L-L, Wang Y-F. Arabidopsis CNGC18 is a Ca^{2+}-permeable channel. Mol Plant. 2014:**7**(4):739–743. https://doi.org/10.1093/mp/sst174

Garcia-Mata C, Gay R, Sokolovski S, Hills A, Lamattina L, Blatt MR. Nitric oxide regulates K^+ and Cl^- channels in guard cells through a subset of abscisic acid-evoked signaling pathways. Proc Natl Acad Sci USA. 2003:**100**(19):11116–11121. https://doi.org/10.1073/pnas.1434381100

Gardos G. The function of calcium in the potassium permeability of human erythrocytes. Biochim Biophys Acta. 1958:**30**(3):653–654. https://doi.org/10.1016/0006-3002(58)90124-0

Garrill A, Tyerman SD, Findlay GP. Ion channels in the plasma membrane of protoplasts from the halophytic angiosperm *Zostera muelleri*. J Membr Biol. 1994:**142**(3):381–393. https://doi.org/10.1007/BF00233443

Geiger D, Maierhofer T, Al-Rasheid KAS, Scherzer S, Mumm P, Liese A, Ache P, Wellmann C, Marten I, Grill E, et al. Stomatal closure by fast abscisic acid signaling is mediated by the guard cell anion channel SLAH3 and the receptor RCAR1. Sci Signal. 2011:**4**(173):ra32. https://doi.org/10.1126/scisignal.2001346

Gibrat R, Grouzis JP, Rigaud J, Grignon C. Potassium stimulation of corn root plasmalemma ATPase: II. H+-pumping in native and reconstituted vesicles with purified ATPase. Plant Physiol. 1990:**93**(3): 1183–1189. https://doi.org/10.1104/pp.93.3.1183

Gilroy S, Read ND, Trewavas AJ. Elevation of cytoplasmic calcium by caged calcium or caged inositol trisphosphate initiates stomatal closure. Nature. 1990:**346**(6286):769–771. https://doi.org/10.1038/346769a0

Gobert A, Isayenkov S, Voelker C, Czempinski K, Maathuis FJM. The two-pore channel TPK1 gene encodes the vacuolar K+ conductance and plays a role in K+ homeostasis. Proc Natl Acad Sci USA. 2007:**104**(25):10726–10731. https://doi.org/10.1073/pnas.0702595104

Goldsmith T, Goldsmith MHM. The interpretation of intracellular measurements of membrane potential, resistance, and coupling in cells of higher plants. Planta. 1978:**143**(3):267–274. https://doi.org/10.1007/BF00391997

Grabov A, Blatt MR. Parallel control of the inward-rectifier K^+ channel by cytosolic-free Ca^{2+} and pH in Vicia guard cells. Planta. 1997:**201**(1):84–95. https://doi.org/10.1007/BF01258684

Grabov A, Blatt MR. Membrane voltage initiates Ca^{2+} waves and potentiates Ca^{2+} increases with abscisic acid in stomatal guard cells. Proc Natl Acad Sci USA. 1998:**95**(8):4778–4783. https://doi.org/10.1073/pnas.95.8.4778

Grabov A, Blatt MR. A steep dependence of inward-rectifying potassium channels on cytosolic free calcium concentration increase evoked by hyperpolarization in guard cells. Plant Physiol. 1999:**119**(1):277–287. https://doi.org/10.1104/pp.119.1.277

Grabov A, Leung J, Giraudat J, Blatt MR. Alteration of anion channel kinetics in wild-type and abi1-1 transgenic *Nicotiana benthamiana* guard cells by abscisic acid. Plant J. 1997:**12**(1):203–213. https://doi.org/10.1046/j.1365-313X.1997.12010203.x

Gradmann D, Blatt MR, Thiel G. Electrocoupling of ion transporters in plants. J Membr Biol. 1993:**136**(3):327–332. https://doi.org/10.1007/BF00233671

Gradmann D, Hansen UP, Long W, Slayman CL, Warnke J. Current-voltage relationships for the plasma membrane and its principle electrogenic pump in *Neurospora crassa*. I. Steady-state conditions. J Membr Biol. 1978:**29**(4):333–367. https://doi.org/10.1007/BF01869898

Grefen C, Chen ZH, Honsbein A, Donald N, Hills A, Blatt MR. A novel motif essential for SNARE interaction with the K^+ channel KC1 and channel gating in Arabidopsis. Plant Cell. 2010:**22**(9):3076–3092. https://doi.org/10.1105/tpc.110.077768

Grefen C, Karnik R, Larson E, Lefoulon C, Wang Y, Waghmare S, Zhang B, Hills A, Blatt MR. A vesicle-trafficking protein commandeers Kv channel voltage sensors for voltage-dependent secretion. Nat Plants. 2015:**1**(8):15108–15119. https://doi.org/10.1038/nplants.2015.108

Guo JT, Zeng WZ, Chen QF, Lee C, Chen LP, Yang Y, Cang CL, Ren DJ, Jiang YX. Structure of the voltage-gated two-pore channel TPC1 from *Arabidopsis thaliana*. Nature. 2016:**531**(7593):196–201. https://doi.org/10.1038/nature16446

Hagiwara S, Myazaki S, Moody W, Patlak J. Blocking effects of barium and hydrogen ions on the potassium current during anomalous rectification in the starfish egg. J Physiol. 1978:**279**(1):167–185. https://doi.org/10.1113/jphysiol.1978.sp012338

Hagiwara S, Takahashi K. The anomalous rectification and cation selectivity of the membrane of a starfish egg cell. J Membr Biol. 1974:**18**(1):61–80. https://doi.org/10.1007/BF01870103

Hamill OP, Marty A, Neher E, Sakmann B, Sigworth FJ. Improved patch-clamp techniques for high-resolution current recording from cells and cell-free membrane patches. Pfluegers Archiv. 1981: **391**(2):85–100. https://doi.org/10.1007/BF00656997

Hamilton DWA, Hills A, Blatt MR. Extracellular Ba^{2+} and voltage interact to gate Ca^{2+} channels at the plasma membrane of stomatal guard cells. FEBS Lett. 2001:**491**(1-2):99–103. https://doi.org/10.1016/S0014-5793(01)02176-7

Hamilton DWA, Hills A, Kohler B, Blatt MR. Ca^{2+} channels at the plasma membrane of stomatal guard cells are activated by hyperpolarization and abscisic acid. Proc Natl Acad Sci USA. 2000:**97**(9): 4967–4972. https://doi.org/10.1073/pnas.080068897

Hansen UP, Slayman CL. Current-voltage relationships for a clearly electrogenic cotransport system. In: **Hoffman JF**, editor. Membrane transport processes. New York: Raven Press; 1978. p. 141–154.

Hedrich R, Barbier-Brygoo H, Felle H, Fluegge UI, Luettge U, Maathuis FJM, Marx S, Prins HBA, Raschke K, Schnabl H, et al. General mechanisms for solute transport across the tonoplast of plant vacuoles: a patch-clamp survey of ion channels and proton pumps. Bot Acta. 1988:**101**(1):7–13. https://doi.org/10.1111/j.1438-8677.1988.tb00003.x

Hedrich R, Busch H, Raschke K. Ca^{2+} and nucleotide dependent regulation of voltage dependent anion channels in the plasma membrane of guard cells. EMBO J. 1990:**9**(12):3889–3892. https://doi.org/10.1002/j.1460-2075.1990.tb07608.x

Hedrich R, Flugge UI, Fernandez JM. Patch clamp studies of ion transport in isolated plant vacuoles. FEBS Lett. 1986:**204**(2):228–232. https://doi.org/10.1016/0014-5793(86)80817-1

Hedrich R, Geiger D. Biology of SLAC1-type anion channels-from nutrient uptake to stomatal closure. New Phytol. 2017:**216**(1):46–61. https://doi.org/10.1111/nph.14685

Hedrich R, Kurkdjian A. Characterization of an anion-permeable channel from sugar beet vacuoles effect of inhibitors. EMBO J. 1988:**7**(12): 3661–3666. https://doi.org/10.1002/j.1460-2075.1988.tb03247.x

Hedrich R, Marten I. TPC1-SV channels gain shape. Mol Plant. 2011:**4**(3):428–441. https://doi.org/10.1093/mp/ssr017

Hedrich R, Neher E. Cytoplasmic calcium regulates voltage-dependent ion channels in plant vacuoles. Nature. 1987:**329**(6142):833–836. https://doi.org/10.1038/329833a0

Hedrich R, Salvador-Recatala V, Dreyer I. Electrical wiring and long-distance plant communication. Trends Plant Sci. 2016:**21**(5): 376–387. https://doi.org/10.1016/j.tplants.2016.01.016

Hedrich R, Schroeder JI. The physiology of ion channels and electrogenic pumps in higher plants. Annu Rev Plant Physiol Plant Mol Biol. 1989:**40**(1):539–569. https://doi.org/10.1146/annurev.pp.40.060189.002543

Herde O, Cortes HP, Wasternack C, Willmitzer L, Fisahn J. Electric signaling and Pin2 gene expression on different abiotic stimuli depend on a distinct threshold level of endogenous abscisic acid in several abscisic acid-deficient tomato mutants. Plant Physiol. 1999:**119**(1):213–218. https://doi.org/10.1104/pp.119.1.213

Hetherington AM, Brownlee C. The generation of Ca^{2+} signals in plants. Annu Rev Plant Biol. 2004:**55**(1):401–427. https://doi.org/10.1146/annurev.arplant.55.031903.141624

Hille B. Ionic channels of excitable membranes. Vol 3. Sunderland (MA): Sinauer Press; 2001.

Hills A, Chen Z-H, Amtmann A, Blatt MR, Lew VL. Onguard, a computational platform for quantitative kinetic modeling of guard cell physiology. Plant Physiol. 2012:**159**(3):1026–1042. https://doi.org/10.1104/pp.112.197244

Hiron RWP, Wright STC. Role of endogenous abscisic acid in response of plants to stress. J Exp Bot. 1973:**24**(4):769–781. https://doi.org/10.1093/jxb/24.4.769

Hodges TK, Leonard RT, Keenan TW, Bracker CE. Purification of an ion-stimulated adenosine-triphosphatase from plant roots—association with plasma-membranes. Proc Natl Acad Sci USA. 1972:**69**(11):3307–3311. https://doi.org/10.1073/pnas.69.11.3307

Hodgkin AL, Huxley AF. A quantitative description of membrane current and its application to conduction and excitation in nerve. J Physiol. 1952:**117**(4):500–544. https://doi.org/10.1113/jphysiol.1952.sp004764

Hodgkin AL, Huxley AF, Katz B. Measurements of current-voltage relations in the membrane of the giant axon of loligo. J Physiol. 1952:**116**(4):424–448. https://doi.org/10.1113/jphysiol.1952.sp004716

Honsbein A, Sokolovski S, Grefen C, Campanoni P, Pratelli R, Paneque M, Chen ZH, Johansson I, Blatto MR. A tripartite SNARE-K^+ channel complex mediates in channel-dependent K^+ nutrition in Arabidopsis. Plant Cell. 2009:**21**(9):2859–2877. https://doi.org/10.1105/tpc.109.066118

Hope AB, Walker NA. The physiology of giant algal cells. Cambridge: CUP Archive; 1975.

Horaruang W, Hills A, Blatt MR. Communication between the plasma membrane and tonoplast is an emergent property of ion transport. Plant Physiol. 2020:**182**(4):1833–1835. https://doi.org/10.1104/pp.19.01485

Horaruang W, Klejchová M, Carroll W, Silva-Alvim F, Waghmare S, Papanatsiou M, Amtmann A, Hills A, Alvim J, Blatt MR, et al. Engineering a K^+ channel 'sensory antenna' enhances stomatal kinetics, water use efficiency and photosynthesis. Nat Plants. 2022: **8**(11):1262–1274. https://doi.org/10.1038/s41477-022-01255-2

Horvath F, Erdei L, Wodala B, Homann U, Thiel G. K^+ outward rectifying channels as targets of phosphatase inhibitor deltamethrin in Vicia faba guard cells. J Plant Physiol. 2002:**159**(10):1097–1103. https://doi.org/10.1078/0176-1617-00559

Hoshi T. Regulation of voltage-dependence of the KAT1 channel by intracellular factors. J Gen Physiol. 1995:**105**(3):309–328. https://doi.org/10.1085/jgp.105.3.309

Hosoi S, Iino M, Shimazaki K. Outward-rectifying K+ channels in stomatal guard cell protoplasts. Plant Cell Physiol. 1988:**29**:907–911. https://doi.org/10.1093/oxfordjournals.pcp.a077594

Hosy E, Vavasseur A, Mouline K, Dreyer I, Gaymard F, Porée F, Boucherez J, Lebaudy A, Bouchez D, Véry A-A, et al. The Arabidopsis outward K^+ channel GORK is involved in regulation of stomatal movements and plant transpiration. Proc Natl Acad Sci USA. 2003:**100**(9):5549–5554. https://doi.org/10.1073/pnas.0733970100

Humble GD, Hsiao T. Specific requirement for potassium for light-activated opening of stomata in epidermal strips. Plant Physiol. 1969:**44**(2):230–234. https://doi.org/10.1104/pp.44.2.230

Ishikawa H, Aizawa H, Kishira H, Ogawa T, Sakata M. Light-induced changes of membrane potential in guard cells of *Vicia faba*. Plant Cell Physiol. 1983:**24**(4):769–772. https://doi.org/10.1093/oxfordjournals.pcp.a076575

Islam MM, Munemasa S, Hossain MA, Nakamura Y, Mori IC, Murata Y. Roles of AtTPC1, vacuolar two pore channel 1, in Arabidopsis stomatal closure. Plant Cell Physiol. 2010:**51**(2):302–311. https://doi.org/10.1093/pcp/pcq001

Jack JJB, Noble D, Tsien RW. Electric current flow in excitable cells. Vol 1. Oxford: Clarendon Press; 1983.

Jarratt-Barnham E, Wang LM, Ning YZ, Davies JM. The complex story of plant cyclic nucleotide-gated channels. Int J Mol Sci. 2021:**22**(2): 874. https://doi.org/10.3390/ijms22020874

Jasechko S, Sharp ZD, Gibson JJ, Birks SJ, Yi Y, Fawcett PJ. Terrestrial water fluxes dominated by transpiration. Nature. 2013:**496**(7445): 347–351. https://doi.org/10.1038/nature11983

Jaslan J, Marten I, Jakobson L, Arjus T, Deeken R, Sarmiento C, De Angeli A, Brosche M, Kollist H, Hedrich R. ALMT-independent guard cell R-type anion currents. New Phytol. 2023:**239**(6): 2225–2234. https://doi.org/10.1111/nph.19124

Jeanguenin L, Alcon C, Duby G, Boeglin M, Cherel I, Gaillard I, Zimmermann S, Sentenac H, Véry A-A. AtKC1 is a general modulator of Arabidopsis inward Shaker channel activity. Plant J. 2011:**67**(4): 570–582. https://doi.org/10.1111/j.1365-313X.2011.04617.x

Jegla T, Busey G, Assmann SM. Evolution and structural characteristics of plant voltage-gated K+ channels. Plant Cell. 2018:**30**(12): 2898–2909. https://doi.org/10.1105/tpc.18.00523

Jezek M, Blatt MR. The membrane transport system of the guard cell and its integration for stomatal dynamics. Plant Physiol. 2017:**174**(2): 487–519. https://doi.org/10.1104/pp.16.01949

Jezek M, Hills A, Blatt MR, Lew VL. A constraint-relaxation-recovery mechanism for stomatal dynamics. Plant Cell Environ. 2019:**42**(8): 2399–2410. https://doi.org/10.1111/pce.13568

Jezek M, Silva-Alvim F, Hills A, Donald N, Ishka MR, Shadbolt J, He B, Lawson T, Harper JF, Wang Y, et al. Guard cell endomembrane Ca^{2+}-ATPases underpin a 'carbon memory' of photosynthetic assimilation that impacts on water use efficiency. Nat Plants. 2021:**7**(9): 1301–1307. https://doi.org/10.1038/s41477-021-00966-2

Johansson I, Wulfetange K, Poree F, Michard E, Gajdanowicz P, Lacombe B, Sentenac H, Thibaud JB, Mueller-Roeber B, Blatt MR, et al. External K+ modulates the activity of the Arabidopsis potassium channel SKOR via an unusual mechanism. Plant J. 2006:**46**(2): 269–281. https://doi.org/10.1111/j.1365-313X.2006.02690.x

Johns S, Hagihara T, Toyota M, Gilroy S. The fast and the furious: rapid long-range signaling in plants. Plant Physiol. 2021:**185**(3): 694–706. https://doi.org/10.1093/plphys/kiaa098

Karlish SJD, Stein WD. Protein conformational-changes in (Na, K)-ATPase and the role of cation occlusion in active-transport. Ann N Y Acad Sci. 1982:**402**(1):226–238. https://doi.org/10.1111/j.1749-6632.1982.tb25744.x

Karnik R, Waghmare S, Zhang B, Larson E, Lefoulon C, Gonzalez W, Blatt MR. Commandeering channel voltage sensors for secretion, cell turgor, and volume control. Trends Plant Sci. 2017:**11**(1):81–95. https://doi.org/10.1016/j.tplants.2016.10.006

Keller BU, Hedrich R, Raschke K. Voltage-dependent anion channels in the plasma membrane of guard cells. Nature. 1989:**341**(6241): 450–453. https://doi.org/10.1038/341450a0

Kintzer AF, Stroud RM. Structure, inhibition and regulation of two-pore channel TPC1 from *Arabidopsis thaliana*. Nature. 2016:**531**(7593):258–262. https://doi.org/10.1038/nature17194

Kirscht A, Kaptan SS, Bienert GP, Chaumont F, Nissen P, de Groot BL, Kjellbom P, Gourdon P, Johanson U. Crystal structure of an ammonia-permeable aquaporin. PLoS Biol. 2016:**14**(3):e1002411. https://doi.org/10.1371/journal.pbio.1002411

Kishimoto U, Kami-ike N, Takeuchi Y, Ohkawa T. A kinetic analysis of the electrogenic pump of *Chara corallina*: I. Inhibition of the pump by DCCD. J Membr Biol. 1984:**80**(2):175–183. https://doi.org/10.1007/BF01868773

Kitasato H. The influence of H+ on the membrane potential and ion fluxes of *Nitella clavata*. J Gen Physiol. 1968:**52**(1):60–87. https://doi.org/10.1085/jgp.52.1.60

Klejchova M, Silva-Alvim FAL, Blatt MR, Alvim JC. Membrane voltage as a dynamic platform for spatiotemporal signaling, physiological, and developmental regulation. Plant Physiol. 2021:**185**(4):1523–1541. https://doi.org/10.1093/plphys/kiab032

Koehler B, Blatt MR. Protein phosphorylation activates the guard cell Ca^{2+} channel and is a prerequisite for gating by abscisic acid. Plant J. 2002:**32**(2):185–194. https://doi.org/10.1046/j.1365-313X.2002.01414.x

Koehler B, Hills A, Blatt MR. Control of guard cell ion channels by hydrogen peroxide and abscisic acid indicates their action through alternate signaling pathways. Plant Physiol. 2003:**131**(2):385–388. https://doi.org/10.1104/pp.016014

Kohler C, Merkle T, Neuhaus G. Characterisation of a novel gene family of putative cyclic nucleotide- and calmodulin-regulated ion channels in *Arabidopsis thaliana*. Plant J. 1999:**18**(1):97–104. https://doi.org/10.1046/j.1365-313X.1999.00422.x

Kohler C, Neuhaus G. Characterisation of calmodulin binding to cyclic nucleotide-gated ion channels from *Arabidopsis thaliana*. FEBS Lett. 2000:**471**(2-3):133–136. https://doi.org/10.1016/S0014-5793(00)01383-1

Kollist H, Nuhkat M, Roelfsema MRG. Closing gaps: linking elements that control stomatal movement. New Phytol. 2014:**203**(1):44–62. https://doi.org/10.1111/nph.12832

Lange OL, Losch R, Schulze ED, Kappen L. Responses of stomata to changes in humidity. Planta. 1971:**100**(1):76–84. https://doi.org/10.1007/BF00386887

Lawson T, Blatt MR. Stomatal size, speed, and responsiveness impact on photosynthesis and water use efficiency. Plant Physiol. 2014:**164**(4):1556–1570. https://doi.org/10.1104/pp.114.237107

Leckie CP, McAinsh MR, Allen GJ, Sanders D, Hetherington AM. Abscisic acid-induced stomatal closure mediated by cyclic ADP-ribose. Proc Natl Acad Sci USA. 1998:**95**(26):15837–15842. https://doi.org/10.1073/pnas.95.26.15837

Lee C-H, MacKinnon R. Structures of the human HCN1 hyperpolarization-activated channel. Cell. 2017:**168**(1-2):111–120. https://doi.org/10.1016/j.cell.2016.12.023

Lefoulon C. The bare necessities of plant K+ channel regulation. Plant Physiol. 2021:**187**(4):2092–2109. https://doi.org/10.1093/plphys/kiab266

Lefoulon C, Waghmare S, Karnik R, Blatt MR. Gating control and K+ uptake by the KAT1 K+ channel leaveraged through membrane anchoring of the trafficking protein SYP121. Plant Cell Environ. 2018:**41**(11):2668–2677. https://doi.org/10.1111/pce.13392

Leitão N, Dangeville P, Carter R, Charpentier M. Nuclear calcium signatures are associated with root development. Nat Commun. 2019:**10**(1):4865. https://doi.org/10.1038/s41467-019-12845-8

Lemtiri-Chlieh F, Arold ST, Gehring C. Mg^{2+} is a missing link in plant cell Ca^{2+} signalling and homeostasis-a study on *Vicia faba* guard cells. Int J Mol Sci. 2020:**21**(11):3771. https://doi.org/10.3390/ijms21113771

Lemtiri-Chlieh F, MacRobbie EAC. Role of calcium in the modulation of *Vicia* guard cell potassium channels by abscisic acid: a patch-clamp study. J Membr Biol. 1994:**137**(2):99–107. https://doi.org/10.1007/BF00233479

Lemtiri-Chlieh F, MacRobbie EAC, Brearley CA. Inositol hexakisphosphate is a physiological signal regulating the K+-inward rectifying conductance in guard cells. Proc Natl Acad Sci USA. 2000:**97**(15):8687–8692. https://doi.org/10.1073/pnas.140217497

Lemtiri-Chlieh F, MacRobbie EAC, Webb AAR, Manison NF, Brownlee C, Skepper JN, Chen J, Prestwich GD, Brearley CA. Inositol hexakisphosphate mobilizes an endomembrane store of calcium in guard cells. Proc Natl Acad Sci USA. 2003:**100**(17):10091--10095. https://doi.org/10.1073/pnas.1133289100

Leonard RT, Hansen D, Hodges TK. Membrane-bound adenosine-triphosphatase activities of oat roots. Plant Physiol. 1973:**51**(4):749–754. https://doi.org/10.1104/pp.51.4.749

Leonard RT, Hotchkiss CW. Cation-stimulated adenosine-triphosphatase activity and cation-transport in corn roots. Plant Physiol. 1976:**58**(3):331–335. https://doi.org/10.1104/pp.58.3.331

Lew VL. The predictive power of cellular homeostasis models illustrated with a study of the gardos effect. Paracelsus Proc Exp Med. 2022:**1**(2):1–13. https://doi.org/10.33594/000000593

Lew VL, Bookchin RM. Volume, pH, and ion-content regulation in human red-cells—analysis of transient-behavior with an integrated model. J Membr Biol. 1986:**92**(1):57–74. https://doi.org/10.1007/BF01869016

Lew VL, Bookchin RM. Ion transport pathology in the mechanism of sickle cell dehydration. Physiol Rev. 2005:**85**(1):179–200. https://doi.org/10.1152/physrev.00052.2003

Leyman B, Geelen D, Quintero FJ, Blatt MR. A tobacco syntaxin with a role in hormonal control of guard cell ion channels. Science. 1999:**283**(5401):537–540. https://doi.org/10.1126/science.283.5401.537

Li Y, Ding Y, Qu L, Li X, Lai Q, Zhao P, Gao Y, Xiang C, Cang C, Liu X, et al. Structure of the Arabidopsis guard cell anion channel SLAC1 suggests activation mechanism by phosphorylation. Nat Commun. 2022:**13**(1):2511–2511. https://doi.org/10.1038/s41467-022-30253-3

Li S, Wang Y, Wang C, Zhang Y, Sun D, Zhou P, Tian C, Liu S. Cryo-EM structure reveals a symmetry reduction of the plant outward-rectifier potassium channel SKOR. Cell Discov. 2023:**9**(1):67–67. https://doi.org/10.1038/s41421-023-00572-w

Ling G, Gerard RW. The normal membrane potential of frog sartorius fibers. J Cell Comp Physiol. 1949:**34**(3):383–396. https://doi.org/10.1002/jcp.1030340304

Lohse G, Hedrich R. Characterization of the plasma-membrane H^+-ATPase from *Vicia faba* guard cells. Planta. 1992:**188**(2):206–214. https://doi.org/10.1007/BF00216815

Loro G, Costa A. Imaging of mitochondrial and nuclear Ca^{2+} dynamics in Arabidopsis roots. Cold Spring Harb Protoc. 2013:**2013**(8):781–785. https://doi.org/10.1101/pdb.prot073049

Loro G, Drago I, Pozzan T, Lo Schiavo F, Zottini M, Costa A. Targeting of Cameleons to various subcellular compartments reveals a strict cytoplasmic/mitochondrial Ca^{2+} handling relationship in plant cells. Plant J. 2012:**71**(1):1–13. https://doi.org/10.1111/j.1365-313X.2012.04968.x

Lu Z, MacKinnon R. Electrostatic tuning of Mg^{2+} affinity in an inward-rectifier K^+ channel. Nature. 1994:**371**(6494):243–246. https://doi.org/10.1038/371243a0

Lu YM, Yu M, Jia YT, Yang F, Zhang YM, Xu X, Li XM, Yang F, Lei JL, Wang Y, et al. Structural basis for the activity regulation of a potassium channel AKT1 from Arabidopsis. Nat Commun. 2022:**13**(1):5682. https://doi.org/10.1038/s41467-022-33420-8

Maathuis FJM, Sanders D. Mechanism of high-affinity potassium uptake in roots of *Arabidopsis thaliana*. Proc Natl Acad Sci USA. 1994:**91**(20):9272–9276. https://doi.org/10.1073/pnas.91.20.9272

MacRobbie EAC. Calcium and ABA-induced stomatal closure. Philos Trans R Soc Lond B Biol Sci. 1992:**338**(1283):5–18. https://doi.org/10.1098/rstb.1992.0124

MacRobbie EAC. Effects of ABA on 86 Rb^+ fluxes at plasmalemma and tonoplast of stomatal guard cells. Plant J. 1995:**7**(5):835–843. https://doi.org/10.1046/j.1365-313X.1995.07050835.x

MacRobbie EAC. ABA activates multiple Ca^{2+} fluxes in stomatal guard cells, triggering vacuolar K^+ (Rb^+) release. Proc Natl Acad Sci USA. 2000:**97**(22):12361–12368. https://doi.org/10.1073/pnas.220417197

MacRobbie EAC, Kurup S. Signalling mechanisms in the regulation of vacuolar ion release in guard cells. New Phytol. 2007:**175**(4):630–640. https://doi.org/10.1111/j.1469-8137.2007.02131.x

Marten I, Hoshi T. Voltage-dependent gating characteristics of the K^+ channel KAT1 depend on the N and C termini. Proc Natl Acad Sci USA. 1997:**94**(7):3448–3453. https://doi.org/10.1073/pnas.94.7.3448

Marten I, Zeilinger C, Redhead C, Landry DW, Al-Awqati Q, Hedrich R. Identification and modulation of a voltage-dependent anion channel in the plasma membrane of guard cells by high-affinity ligands. EMBO J. 1992:**11**(10):3569–3575. https://doi.org/10.1002/j.1460-2075.1992.tb05440.x

Maurel C, Boursiac Y, Doan-Trung L, Santoni V, Shahzad Z, Verdoucq L. Aquaporins in plants. Physiol Rev. 2015:**95**(4):1321–1358. https://doi.org/10.1152/physrev.00008.2015

McAinsh MR, Pittman JK. Shaping the calcium signature. New Phytol. 2009:**181**(2):275–294. https://doi.org/10.1111/j.1469-8137.2008.02682.x

McCutcheon SL, Bown AW. Evidence for a specific glutamate/H^+ cotransport in isolated mesophyll cells. Plant Physiol. 1987:**83**(3):691–697. https://doi.org/10.1104/pp.83.3.691

Meharg AA, Blatt MR. Nitrate transport in root hairs of *Arabidopsis thaliana*: kinetic control by membrane voltage and pH. J Membr Biol. 1995:**145**(1):49–66. https://doi.org/10.1007/BF00233306

Meharg AA, Maurousset L, Blatt MR. Cable correction of membrane currents recorded from root hairs of *Arabidopsis thaliana* L. J Exp Bot. 1994:**45**(1):1–6. https://doi.org/10.1093/jxb/45.1.1

Merlot S, Leonhardt N, Fenzi F, Valon C, Costa M, Piette L, Vavasseur A, Genty B, Boivin K, Mueller A, et al. Constitutive activation of a plasma membrane H^+-ATPase prevents abscisic acid-mediated stomatal closure. EMBO J. 2007:**26**(13):3216–3226. https://doi.org/10.1038/sj.emboj.7601750

Meyer S, Mumm P, Imes D, Endler A, Weder B, Al-Rasheid KAS, Geiger D, Marten I, Martinoia E, Hedrich R. AtALMT12 represents an R-type anion channel required for stomatal movement in

Arabidopsis guard cells. Plant J. 2010:**63**(6):1054–1062. https://doi.org/10.1111/j.1365-313X.2010.04302.x

Miller TE, Beneyton T, Schwander T, Diehl C, Girault M, McLean R, Chotel T, Claus P, Cortina NS, Baret J-C, et al. Light-powered CO_2 fixation ina chloroplast mimic with natural and synthetic parts. Science. 2020:**368**(6491):649–654. https://doi.org/10.1126/science.aaz6802

Minguet-Parramona C, Wang Y, Hills A, Vialet-Chabrand S, Griffiths H, Rogers S, Lawson T, Lew VL, Blatt MR. An optimal frequency in Ca^{2+} oscillations for stomatal closure is an emergent property of Ion transport in guard cells. Plant Physiol. 2016:**170**(1):32–45. https://doi.org/10.1104/pp.15.01607

Mitchell P. Chemiosmotic coupling and energy transduction. Theor Exp Biophys. 1969:**2**:159–216.

Moody W, Zeiger E. Electrophysiological properties of onion guard cells. Planta. 1978:**139**(2):159–165. https://doi.org/10.1007/BF00387142

Moran N, Ehrenstein G, Iwasa K, Barre C, Mischke C. Ion channels in the plasmalemma of wheat protoplasts. Science. 1984:**226**(4676): 835–838. https://doi.org/10.1126/science.6093255

Moran N, Ehrenstein G, Iwasa K, Mischke C, Bare C, Satter RL. Potassium channels in motor cells of *Sananea saman*: a patch-clamp study. Plant Physiol. 1988:**88**(3):643–648. https://doi.org/10.1104/pp.88.3.643

Mori IC, Murata Y, Yang Y, Munemasa S, Wang Y-F, Andreoli S, Tiriac H, Alonso JM, Harper JF, Ecker JR, et al. CDPKs CPK6 and CPK3 function in ABA regulation of guard cell S-type anion- and Ca^{2+}-permeable channels and stomatal closure. PLoS Biol. 2006:**4**(10):e327. https://doi.org/10.1371/journal.pbio.0040327

Moroni A, Gorza L, Beltrame M, Gravante B, Vaccari T, Bianchi ME, Altomare C, Longhi R, Heurteaux C, Vitadello M, et al. Hyperpolarization-activated cyclic nucleotide-gated channel 1 is a molecular determinant of the cardiac pacemaker current I-f. J Biol Chem. 2001:**276**(31):29233–29241. https://doi.org/10.1074/jbc.M100830200

Mueller-Roeber B, Ellenberg J, Provart N, Willmitzer L, Busch H, Becker D, Dietrich P, Hoth S, Hedrich R. Cloning and electrophysiological analysis of KST1, an inward rectifying K^+ channel expressed in potato guard cells. EMBO J. 1995:**14**(11):2409–2416. https://doi.org/10.1002/j.1460-2075.1995.tb07238.x

Mueller M, Kunz H-H, Schroeder JI, Kemp G, Young HS, Neuhaus HE. Decreased capacity for sodium export out of Arabidopsis chloroplasts impairs salt tolerance, photosynthesis and plant performance. Plant J. 2014:**78**(4):646–658. https://doi.org/10.1111/tpj.12501

Muir SR, Bewell MA, Sanders D, Allen GJ. Ligand-gated Ca^{2+} channels and Ca^{2+} signalling in higher plants. J Exp Bot. 1997:**48**(Special): 589–597. https://doi.org/10.1093/jxb/48.Special_Issue.589

Muir SR, Sanders D. Pharmacology of Ca^{2+} release from red beet microsomes suggests the presence of ryanodine receptor homologs in higher plants. FEBS Lett. 1996:**395**(1):39–42. https://doi.org/10.1016/0014-5793(96)01000-9

Nakamura RL, McKendree WL Jr, Hirsch RE, Sedbrook JC, Gaber RF, Sussman MR. Expression of an Arabidopsis potassium channel gene in guard cells. Plant Physiol. 1995:**109**(2):371–374. https://doi.org/10.1104/pp.109.2.371

Negi J, Matsuda O, Nagasawa T, Oba Y, Takahashi H, Kawai-Yamada M, Uchimiya H, Hashimoto M, Iba K. CO_2 regulator SLAC1 and its homologues are essential for anion homeostasis in plant cells. Nature. 2008:**452**(7186):483–486. https://doi.org/10.1038/nature06720

Neher E, Sakmann B. Single-channel currents recorded from the membrane of denervated frog muscle fibres. Nature. 1976:**260**(5554): 779–802. https://doi.org/10.1038/260799a0

Osakabe Y, Arinaga N, Umezawa T, Katsura S, Nagamachi K, Tanaka H, Ohiraki H, Yamada K, Seo S-U, Abo M, et al. Osmotic stress responses and plant growth controlled by potassium transporters in Arabidopsis. Plant Cell. 2013:**25**(2):609–624. https://doi.org/10.1105/tpc.112.105700

Osterhout WJV. Physiological studies of single plant cells. Biol Rev. 1931:**6**(4):369–411. https://doi.org/10.1111/j.1469-185X.1931.tb01031.x

Palevitz BA, Hepler PK. Changes in dye coupling of stomatal cells of Allium and Commelina demonstrated by microinjection of lucifer yellow. Planta. 1985:**164**(4):473–479. https://doi.org/10.1007/BF00395962

Pallaghy CK. Electrophysiological studies in guard cells of tobacco. Planta. 1968:**80**(2):147. https://doi.org/10.1007/BF00385590

Pan YJ, Chai XY, Gao QF, Zhou LM, Zhang SS, Li LG, Luan S. Dynamic interactions of plant CNGC subunits and calmodulins drive oscillatory Ca^{2+} channel activities. Dev Cell. 2019:**48**(5):710–725.e5. https://doi.org/10.1016/j.devcel.2018.12.025

Pantaleno R, Scuffi D, Garcia-Mata C. Hydrogen sulphide as a guard cell network regulator. New Phytol. 2021:**230**(2):451–456. https://doi.org/10.1111/nph.17113

Papanatsiou M, Petersen J, Henderson L, Wang Y, Christie JM, Blatt MR. Optogenetic manipulation of stomatal kinetics improves carbon assimilation and water use efficiency. Science. 2019:**363**(6434): 1456–1459. https://doi.org/10.1126/science.aaw0046

Papanatsiou M, Scuffi D, Blatt MR, Garcia-Mata C. Hydrogen sulfide regulates inward-rectifying K^+ channels in conjunction with stomatal closure. Plant Physiol. 2015:**168**(1):29–35. https://doi.org/10.1104/pp.114.256057

Pardo JM, Cubero B, Leidi EO, Quintero FJ. Alkali cation exchangers: roles in cellular homeostasis and stress tolerance. J Exp Bot. 2006:**57**(5):1181–1199. https://doi.org/10.1093/jxb/erj114

Peak D, Mott KA. A new, vapour-phase mechanism for stomatal responses to humidity and temperature. Plant Cell Environ. 2011:**34**(1):162–178. https://doi.org/10.1111/j.1365-3040.2010.02234.x

Pedersen BP, Buch-Pedersen MJ, Morth JP, Palmgren MG, Nissen P. Crystal structure of the plasma membrane proton pump. Nature. 2007:**450**(7172):1111–1114. https://doi.org/10.1038/nature06417

Pei Z-M, Murata Y, Benning G, Thomine S, Klusener B, Allen GJ, Grill E, Schroeder JI. Calcium channels activated by hydrogen peroxide mediate abscisic acid signalling in guard cells. Nature. 2000: **406**(6797):731–734. https://doi.org/10.1038/35021067

Pei Z-M, Ward JM, Schroeder JI. Magnesium sensitizes slow vacuolar channels to physiological cytosolic calcium and inhibits fast vacuolar channels in fava bean guard cell vacuoles. Plant Physiol. 1999:**121**(3): 977–986. https://doi.org/10.1104/pp.121.3.977

Peiter E, Maathuis FJM, Mills LN, Knight H, Pelloux M, Hetherington AM, Sanders D. The vacuolar Ca^{2+}-activated channel TPC1 regulates germination and stomatal movement. Nature. 2005:**434**(7031): 404–408. https://doi.org/10.1038/nature03381

Pfeffer W. Osmotische untersuchungen. Vol 1. Leipzig: Wilhelm Engelmann; 1877.

Pilot G, Lacombe B, Gaymard F, Cherel I, Boucherez J, Thibaud JB, Sentenac H. Guard cell inward K+ channel activity in Arabidopsis involves expression of the twin channel subunits KAT1 and KAT2. J Biol Chem. 2001:**276**(5):3215–3221. https://doi.org/10.1074/jbc.M007303200

Pinker S. The sense of style: the thinking person's Guide to writing in the 21st century. New York (NY): Penguin; 2014.

Pittman JK. Vacuolar Ca^{2+} uptake. Cell Calcium. 2011:**50**(2):139–146. https://doi.org/10.1016/j.ceca.2011.01.004

Pottosin I, Martinez-Estevez M, Dobrovinskaya O, Muniz J, Schonknecht G. Mechanism of luminal Ca^{2+} and Mg^{2+} action on the vacuolar slowly activating channels. Planta. 2004:**219**(6): 1057–1070. https://doi.org/10.1007/s00425-004-1293-7

Pottosin II, Schonknecht G. Vacuolar calcium channels. J Exp Bot. 2007:**58**(7):1559–1569. https://doi.org/10.1093/jxb/erm035

Purves JD. Microelectrode methods for intracellular recording and ionophoresis. Vol 1. London: Academic Press; 1981. p. 1–146.

Qin L, Tang L-H, Xu J-S, Zhang X-H, Zhu Y, Zhang C-R, Wang M-H, Liu X-L, Li F, Sun F, et al. Cryo-EM structure and electrophysiological characterization of ALMT from Glycine max reveal a previously uncharacterized class of anion channels. Sci Adv. 2022:**8**(9):eabm3238. https://doi.org/10.1126/sciadv.abm3238

Quintero FJ, Blatt MR. A new family of K^+ transporters from Arabidopsis that are conserved across phyla. FEBS Lett. 1997:**415**(2):206–211. https://doi.org/10.1016/S0014-5793(97)01125-3

Quintero FJ, Blatt MR, Pardo JM. Functional conservation between yeast and plant endosomal Na^+/H^+ antiporters. FEBS Lett. 2000:**471**(2–3):224–228. https://doi.org/10.1016/S0014-5793(00)01412-5

Raschke K, Schnabl H. Availability of chloride affects balance between potassium chloride and potassium malate in guard cells of *Vicia faba* L. Plant Physiol. 1978:**62**(1):84–87. https://doi.org/10.1104/pp.62.1.84

Resentini F, Ruberti C, Grenzi M, Bonza M, Costa A. The signatures of organellar calcium. Plant Physiol. 2021:**187**(4):1985–2004. https://doi.org/10.1093/plphys/kiab189

Rockwell FE, Holbrook NM, Stroock AD. The competition between liquid and vapor transport in transpiring leaves. Plant Physiol. 2014:**164**(4):1741–1758. https://doi.org/10.1104/pp.114.236323

Rodrigues O, Shan LB. Stomata in a state of emergency: H_2O_2 is the target locked. Trends Plant Sci. 2022:**27**(3):274–286. https://doi.org/10.1016/j.tplants.2021.10.002

Rodriguez-Martinez M, Nielsen J, Dupont S, Vamathevan J, Glover BJ, Crosswell LC, Rouse B, Luisi BF, Bowler C, Gasser SM, et al. Molecular biology for green recovery-a call for action. PLoS Biol. 2022:**20**(4):e3001623. https://doi.org/10.1371/journal.pbio.3001623

Rodriguez-Navarro A, Blatt MR, Slayman CL. A potassium-proton symport in *Neurospora crassa*. J Gen Physiol. 1986:**87**(5):649–674. https://doi.org/10.1085/jgp.87.5.649

Roelfsema MR, Hedrich R. Making sense out of Ca^{2+} signals: their role in regulating stomatal movements. Plant Cell Environ. 2010:**33**(3):305–321. https://doi.org/10.1111/j.1365-3040.2009.02075.x

Roelfsema MG, Prins HA. Ion channels in guard cells of *Arabidopsis thaliana* (L) Heynh. Planta. 1997:**202**(1):18–27. https://doi.org/10.1007/s004250050098

Saftner RA, Raschke K. Electrical potentials in stomatal complexes. Plant Physiol. 1981:**67**(6):1124–1132. https://doi.org/10.1104/pp.67.6.1124

Sakmann B, Neher E. Geometric parameters of pipettes and membrane patches. In: **Sakmann B, Neher E**, editors. Single channel recording. Vol 1. New York: Plenum; 1983. p. 37–51.

Sanders D. Kinetic modeling of plant and fungal membrane transport systems. Annu Rev Plant Physiol Plant Mol Biol. 1990:**41**(1):77–108. https://doi.org/10.1146/annurev.pp.41.060190.000453

Sanders D, Hansen UP. Mechanism of Cl^- transport at the plasma membrane of *Chara corallina*: II. Transinhibition and determination of H^+/Cl^- binding order from a reaction kinetic model. J Membr Biol. 1981:**58**(2):139–153. https://doi.org/10.1007/BF01870976

Sanders D, Hansen UP, Slayman CL. Role of the plasma membrane proton pump in pH regulation in non-animal cells. Proc Natl Acad Sci U S A. 1981:**78**(9):5903–5907. https://doi.org/10.1073/pnas.78.9.5903

Sanders D, Slayman CL. Control of intracellular pH: predominant role of oxidative metabolism, not proton transport, in the eukaryotic microorganism Neurospora. J Gen Physiol. 1982:**80**(3):377–402. https://doi.org/10.1085/jgp.80.3.377

Sanders D, Smith FA, Walker NA. Proton-chloride cotransport in Chara: mechanism of enhanced influx after rapid external acidification. Planta. 1985:**163**(3):411–418. https://doi.org/10.1007/BF00395151

Santelia D, Lawson T. Rethinking guard cell metabolism. Plant Physiol. 2016:**172**(3):1371–1392. https://doi.org/10.1104/pp.16.00767

Sasaki T, Mori IC, Furuichi T, Munemasa S, Toyooka K, Matsuoka K, Murata Y, Yamamoto YT. Closing plant stomata requires a homolog of an aluminum-activated malate transporter. Plant Cell Physiol. 2010:**51**(3):354–365. https://doi.org/10.1093/pcp/pcq016

Schoenknecht G, Hedrich R, Junge W, Raschke K. A voltage-dependent chloride channel in the photosynthetic membrane of a higher plant. Nature. 1988:**336**(6199):589–592. https://doi.org/10.1038/336589a0

Schroeder JI. K+ transport properties of K+ channels in the plasma membrane of *Vicia faba* guard cells. J Gen Physiol. 1988:**92**(5):667–683. https://doi.org/10.1085/jgp.92.5.667

Schroeder JI, Hagiwara S. Cytosolic calcium regulates ion channels in the plasma membrane of *Vicia faba* guard cells. Nature. 1989:**338**(6214):427–430. https://doi.org/10.1038/338427a0

Schroeder JI, Hedrich R, Fernandez JM. Potassium-selective single channels in guard cell protoplasts of *Vicia faba* guard cells. Nature. 1984:**312**(5992):361–362. https://doi.org/10.1038/312361a0

Schroeder JI, Keller BU. Two types of anion channel currents in guard cells with distinct voltage regulation. Proc Natl Acad Sci USA. 1992:**89**(11):5025–5029. https://doi.org/10.1073/pnas.89.11.5025

Schroeder JI,Raschke K, Neher E. Voltage dependence of K+ channels in guard-cell protoplasts. Proc Natl Acad Sci USA. 1987:**84**(12):4108–4112. https://doi.org/10.1073/pnas.84.12.4108

Schulz-Lessdorf B, Hedrich R. Protons and calcium modulate SV-type channels in the vacuolar-lysosomal compartment–channel interaction with calmodulin inhibitors. Planta. 1995:**197**(4):655–671. https://doi.org/10.1007/BF00191574

Schwab W, Komor E. A possible mechanistic role of the membrane potential in proton-sugar cotransport of Chlorella. FEBS Lett. 1978:**87**(1):157–160. https://doi.org/10.1016/0014-5793(78)80156-2

Shepherd VA, Beilby MJ, Shimmen T. Mechanosensory ion channels in charophyte cells: the response to touch and salinity stress. Eur Biophys J. 2002:**31**(5):341–355. https://doi.org/10.1007/s00249-002-0222-6

Sibaoka T. Physiology of rapid movements in higher plants. Annu Rev Plant Physiol. 1969:**20**(1):165–184. https://doi.org/10.1146/annurev.pp.20.060169.001121

Sigworth FJ. Electronic design of the patch clamp. In: **Sakmann B, Neher E**, editors. Single channel recording. Vol 1. New York: Plenum; 1983. p. 3–36.

Simons PJ. The role of electricity in plant movements. New Phytol. 1981:**87**(1):11–37. https://doi.org/10.1111/j.1469-8137.1981.tb01687.x

Skou JC, Norby J. Na/K-ATPase: structure and kinetics. New York: Academic Press; 1979.

Slayman CL. Electrical properties of *Neurospora crassa*. Respiration and the intracellular potential. J Gen Physiol. 1965:**49**(1):93–116. https://doi.org/10.1085/jgp.49.1.93

Slayman CL, Long WS, Gradmann D. Action potentials in neurospora-crassa, a mycelial fungus. Biochim Biophys Acta. 1976:**426**(4):732–744. https://doi.org/10.1016/0005-2736(76)90138-3

Slayman CL, Slayman CW. Measurement of membrane potentials in neurospora. Science. 1962:**136**(3519):876–877. https://doi.org/10.1126/science.136.3519.876

Slayman CL, Slayman CW. Depolarization of the plasma membrane of Neurospora during active transport of glucose: evidence for a proton-dependent cotransport system. Proc Natl Acad Sci USA. 1974:**71**(5):1935–1939. https://doi.org/10.1073/pnas.71.5.1935

Smack DP, Colombini M. Voltage-dependent channels found in the membrane fraction of corn mitochondria. Plant Physiol. 1985:**79**(4):1094–1097. https://doi.org/10.1104/pp.79.4.1094

Smedley T. The last drop. London: Pan Macmillan; 2023.

Smith JR. The electrical properties of plant cell membranes. II. Distortion of non-linear current-voltage characteristics induced by the cable properties of *Chara*. Aust J Plant Physiol. 1984:**11**(3):211–224. https://doi.org/10.1071/PP9840211

Smith NG, Dukes JS. Plant respiration and photosynthesis in global-scale models: incorporating acclimation to temperature and CO_2. Glob Change Biol. 2013:**19**(1):45–63. https://doi.org/10.1111/j.1365-2486.2012.02797.x

Smith JAC, Pennington AJ, Martin RJ. A malate-specific ion channel in the vacuolar membrane of cam plants. Plant Physiol. 1990:**93**:32.

Sokolovski S, Hills A, Gay R, Blatt MR. Functional interaction of the SNARE protein NtSyp121 in Ca^{2+} channel gating, Ca^{2+} transients and ABA signalling of stomatal guard cells. Mol Plant. 2008:**1**(2):347–358. https://doi.org/10.1093/mp/ssm029

Sokolovski S, Hills A, Gay R, Garcia-Mata C, Lamattina L, Blatt MR. Protein phosphorylation is a prerequisite for intracellular Ca^{2+} release and ion channel control by nitric oxide and abscisic acid in guard cells. Plant J. 2005:**43**(4):520–529. https://doi.org/10.1111/j.1365-313X.2005.02471.x

Spalding EP, Slayman CL, Goldsmith MHM, Gradmann D, Bertl A. Ion channels in Arabidopsis plasma membrane. Plant Physiol. 1992:**99**(1):96–102. https://doi.org/10.1104/pp.99.1.96

Spanswick RM. Electrical coupling between cells of higher-plants—a direct demonstration of intercellular communication. Planta. 1972:**102**(3):215–223. https://doi.org/10.1007/BF00386892

Spanswick RM. Electrogenic ion pumps. Annu Rev Plant Physiol. 1981:**32**(1):267–281. https://doi.org/10.1146/annurev.pp.32.060181.001411

Spanswick RM, Stolarek J, Williams EJ. Membrane potential of *Nitella translucens*. J Exp Bot. 1967:**18**(1):1. https://doi.org/10.1093/jxb/18.1.1

Spanswick RM, Williams EJ. Electrical potentials and Na, K, and Cl concentrations in the vacuole and cytoplasm of *Nitella translucens*. J Exp Bot. 1964:**15**(2):193–200. https://doi.org/10.1093/jxb/15.2.193

Standen NB, Gray PTA, Whitaker MJ. Microelectrode techniques. Vol 1. Cambridge: Company of Biologists; 1987.

Stange A, Hedrich R, Roelfsema MRG. Ca^{2+}-dependent activation of guard cell anion channels, triggered by hyperpolarization, is promoted by prolonged depolarization. Plant J. 2010:**62**(2):265–276. https://doi.org/10.1111/j.1365-313X.2010.04141.x

Stelling J, Sauer U, Szallasi Z, Doyle FJ, Doyle J. Robustness of cellular functions. Cell. 2004:**118**(6):675–685. https://doi.org/10.1016/j.cell.2004.09.008

Stoeckel H, Takeda K. Calcium-activated voltage-dependent nonselective cation currents in endosperm plasma membrane from higher plants. Proc R Soc Lond B Biol Sci. 1989:**237**(1287):213–231. https://doi.org/10.1098/rspb.1989.0045

Taglialatela M, Wible BA, Caporaso R, Brown AM. Specification of pore properties by the carboxyl-terminus of inwardly rectifying K^+ channels. Science. 1994:**264**(5160):844–847. https://doi.org/10.1126/science.8171340

Tan Y-Q, Yang Y, Shen X, Zhu MJ, Shen JL, Zhang W, Hu HH, Wang Y-F. Multiple cyclic nucleotide-gated channels function as ABA-activated Ca^{2+} channels required for ABA-induced stomatal closure in Arabidopsis. Plant Cell. 2023:**35**(1):239–259. https://doi.org/10.1093/plcell/koac274

Tang XD, Hoshi T. Rundown of the hyperpolarization-activated KAT1 channel involves slowing of the opening transitions regulated by phosphorylation. Biophys J. 1999:**76**(6):3089–3098. https://doi.org/10.1016/S0006-3495(99)77461-8

Tester M. Blockade of potassium channels in the plasmalemma of *Chara corallina* by tetraethylammonium, Ba^{2+}, Na^+ and Cs^+. J Membr Biol. 1988:**105**(1):77–85. https://doi.org/10.1007/BF01871108

Tester M, Blatt MR. Direct measurement of K^+ channels in thylakoid membranes by incorporation of vesicles into planar lipid bilayers. Plant Physiol. 1989:**91**(1):249–252. https://doi.org/10.1104/pp.91.1.249

Thiel G. Dynamics of chloride and potassium currents during the action potential in *Chara* studies with action-potential clamp. Eur Biophys J. 1995:**24**(2):85–92. https://doi.org/10.1007/BF00211403

Thiel G, Blatt MR, Fricker MD, White IR, Millner PA. Modulation of K^+ channels in Vicia stomatal guard cells by peptide homologs to the auxin-binding protein C-terminus. Proc Natl Acad Sci USA. 1993a:**90**(24):11493–11497. https://doi.org/10.1073/pnas.90.24.11493

Thiel G, Homann U, Gradmann D. Microscopic elements of electrical excitation in Chara—transient activity of voltage-dependent Cl^- channels. J Membr Biol. 1993b:**134**(1):53–66. https://doi.org/10.1007/BF00233475

Thiel G, Homann U, Plieth C. Ion channel activity during the action potential in Chara: new insights with new techniques. J Exp Bot. 1997:**48**(Special):609–622. https://doi.org/10.1093/jxb/48.Special_Issue.609

Thiel G, MacRobbie EAC, Blatt MR. Membrane transport in stomatal guard cells: the importance of voltage control. J Membr Biol. 1992:**126**(1):1–18. https://doi.org/10.1007/BF00233456

Tornroth-Horsefield S, Wang Y, Hedfalk K, Johanson U, Karlsson M, Tajkhorshid E, Neutze R, Kjellbom P. Structural mechanism of plant aquaporin gating. Nature. 2006:**439**(7077):688–694. https://doi.org/10.1038/nature04316

Ullrich WR, Larsson M, Larsson CM, Lesch S, Novacky A. Ammonium uptake in *Lemna gibba* G1, related membrane potential changes, and inhibition of anion uptake. Physiol Plant. 1984:**61**(3):369–376. https://doi.org/10.1111/j.1399-3054.1984.tb06342.x

Umrath K. Untersuchungen Åber Plasma und Plasmastrîmung an Characean. IV. Potentialmessungen an *Nitella mucronata* mit besonderer BerÅcksichtigung der Erregungserscheinungen. Protoplasma. 1930:**9**(1):576–597. https://doi.org/10.1007/BF01943373

Umrath K. Die Bildung von Plasmalemma (Plasmahaut) bei *Nitella mucronata*. Protoplasma. 1932:**16**(1):173–188. https://doi.org/10.1007/BF01638804

UNESCO. Water for a sustainable world—UN world water development report. Vol 1. New York: UNESCO; 2015.

Vahisalu T, Kollist H, Wang Y-F, Nishimura N, Chan W-Y, Valerio G, Lamminmaki A, Brosche M, Moldau H, Desikan R, et al. SLAC1 is required for plant guard cell S-type anion channel function in stomatal signalling. Nature. 2008:**452**(7186):487–491. https://doi.org/10.1038/nature06608

Vandenberg CA. Inward rectification of a potassium channel in cardiac ventricular cells depends on internal magnesium ions. Proc Natl Acad Sci USA. 1987:**84**(8):2560–2564. https://doi.org/10.1073/pnas.84.8.2560

Van Kirk CA, Raschke K. Stomatal aperture and malate content of epidermis-effects of chloride and abscisic acid. Plant Physiol. 1977:**59**:96–108.

Van Kirk CA, Raschke K. Presence of chloride reduces malate production in epidermis during stomatal opening. Plant Physiol. 1978:**61**(3):361–364. https://doi.org/10.1104/pp.61.3.361

van Riel NAW. Dynamic modelling and analysis of biochemical networks: mechanism-based models and model-based experiments. Brief Bioinform. 2006:**7**(4):364–374. https://doi.org/10.1093/bib/bbl040

Véry A-A, Nieves-Cordones M, Daly M, Khan I, Fizames C, Sentenac H. Molecular biology of K^+ transport across the plant cell membrane: what do we learn from comparison between plant species? J Plant Physiol. 2014:**171**(9):748–769. https://doi.org/10.1016/j.jplph.2014.01.011

Vialet-Chabrand S, Hills A, Wang Y, Griffiths H, Lew VL, Lawson T, Blatt MR, Rogers S. Global sensitivity analysis of OnGuard models identifies key hubs for transport interaction in stomatal dynamics. Plant Physiol. 2017:**174**(2):680–688. https://doi.org/10.1104/pp.17.00170

Waghmare S, Lefoulon C, Zhang B, Liliekyte E, Donald N, Blatt MR. K + channel-SEC11 binding exchange regulates SNARE assembly for secretory traffic. Plant Physiol. 2019:**181**(3):1096–1113. https://doi.org/10.1104/pp.19.00919

Wang Y, Chen Z-H, Zhang B, Hills A, Blatt MR. PYR/PYL/RCAR abscisic acid receptors regulate K^+ and Cl^- channels through reactive oxygen species-mediated activation of Ca^{2+} channels at the plasma membrane of intact Arabidopsis guard cells. Plant Physiol. 2013a:**163**(2):566–577. https://doi.org/10.1104/pp.113.219758

Wang Y, Hills A, Vialet-Chabrand SR, Papanatsiou M, Griffiths H, Rogers S, Lawson T, Lew V, Blatt MR. Unexpected connections between humidity and ion transport discovered using a model to bridge guard cell- to-leaf scales. Plant Cell. 2017:**29**(11):2921–2139. https://doi.org/10.1105/tpc.17.00694

Wang Y-F, Munemasa S, Nishimura N, Ren H-M, Robert N, Han M, Puzorjova I, Kollist H, Lee S, Mori I, et al. Identification of cyclic GMP-activated nonselective Ca^{2+}-permeable cation channels and associated CNGC5 and CNGC6 genes in Arabidopsis guard cells. Plant Physiol. 2013b:**163**(2):578–590. https://doi.org/10.1104/pp.113.225045

Wang Y, Papanatsiou M, Eisenach C, Karnik R, Williams M, Hills A, Lew VL, Blatt MR. Systems dynamic modelling of a guard cell Cl^- channel mutant uncovers an emergent homeostatic network regulating stomatal transpiration. Plant Physiol. 2012:**160**(4):1956–1972. https://doi.org/10.1104/pp.112.207704

Ward JM, Schroeder JI. Calcium-activated K+ channels and calcium-induced calcium-release by slow vacuolar ion channels in guard-cell vacuoles implicated in the control of stomatal closure. Plant Cell. 1994:**6**(5):669–683. https://doi.org/10.2307/3869871

Webb AAR, Larman MG, Montgomery LT, Taylor JE, Hetherington AM. The role of calcium in ABA-induced gene expression and stomatal movements. Plant J. 2001:**26**(3):351–362. https://doi.org/10.1046/j.1365-313X.2001.01032.x

Webb AAR, McAinsh MR, Mansfield TA, Hetherington AM. Carbon dioxide induces increases in guard cell cytosolic free calcium. Plant J. 1996:**9**(3):297–304. https://doi.org/10.1046/j.1365-313X.1996.09030297.x

Wege S, De Angeli A, Droillard MJ, Kroniewicz L, Merlot S, Cornu D, Gambale F, Martinoia E, Barbier-Brygoo H, Thomine S, et al. Phosphorylation of the vacuolar anion exchanger AtCLCa is required for the stomatal response to abscisic acid. Sci Signal. 2014:**7**(333): ra65. https://doi.org/10.1126/scisignal.2005140

Wegner LH, deBoer AH, Raschke K. Properties of the K+ inward rectifier in the plasma membrane of xylem parenchyma cells from barley roots: effects of TEA, Ba2+ and La3+. J Membr Biol. 1994:**142**(3): 363–379. https://doi.org/10.1007/BF00233442

Weiser T, Bentrup FW. Charybdotoxin blocks cation-channels in the vacuolar membrane of suspension cells of chenopodium-rubrum l. Biochim Biophys Acta. 1991:**1066**(1):109–110. https://doi.org/10.1016/0005-2736(91)90258-A

Weiss TF. Cellular biophysics. Vol 1. Cambridge (MA): MIT Press; 1996.

Wille A, Lucas W. Ultrastructural and histochemical studies on guard cells. Planta. 1984:**160**(2):129–142. https://doi.org/10.1007/BF00392861

Williams SE, Pickard BG. Properties of action potentials in drosera tentacles. Planta. 1972a:**103**(3):222–240. https://doi.org/10.1007/BF00386845

Williams SE, Pickard BG. Receptor potentials and action potentials in drosera tentacles. Planta. 1972b:**103**(3):193–221. https://doi.org/10.1007/BF00386844

Willmer C, Fricker MD. Stomata. Vol 1. London: Chapman and Hall; 1996. p. 1–375.

Wright STC, Hiron RWP. (+)-Abscisic acid, the growth inhibitor induced in wheat leaves by a period of wilting. Nature. 1969:**224**(5220):719–720. https://doi.org/10.1038/224719a0

Wu CF, Haugland FN. Voltage clamp analysis of membrane currents in laval muscle fibers of Drosophila: alteration of potassium currents in Shaker mutants. J Neurosci. 1985:**5**(10):2626–2640. https://doi.org/10.1523/JNEUROSCI.05-10-02626.1985

Wu Y, Kuzma J, Marechal E, Graeff R, Lee HC, Foster R, Chua N-H. Abscisic acid signaling through cyclic ADP-ribose in plants. Science. 1997:**278**(5346):2126–2130. https://doi.org/10.1126/science.278.5346.2126

Zei PC, Aldrich RW. Voltage-dependent gating of single wild-type and S4 mutant KAT1 inward rectifier potassium channels. J Gen Physiol. 1998:**112**(6):679–713. https://doi.org/10.1085/jgp.112.6.679

Zhou Y, Ding MQ, Gao SQ, Yu-Strzelczyk J, Krischke M, Duan XD, Leide J, Riederer M, Mueller MJ, Hedrich R, et al. Optogenetic control of plant growth by a microbial rhodopsin. Nat Plants. 2021:**7**(2): 144–151. https://doi.org/10.1038/s41477-021-00853-w

Zonia L, Munnik T. Life under pressure: hydrostatic pressure in cell growth and function. Trends Plant Sci. 2007:**12**(3):90–97. https://doi.org/10.1016/j.tplants.2007.01.006

https://doi.org/10.1093/plphys/kiae044 PLANT PHYSIOLOGY 2024: 195: 111–134

Plant Physiology®

Topical Review

Highlights in gibberellin research: A tale of the dwarf and the slender

Eilon Shani[1] **Peter Hedden**[2,3] **and Tai-ping Sun**[4,*]

1 School of Plant Sciences and Food Security, Tel Aviv University, Tel Aviv 69978, Israel
2 Laboratory of Growth Regulators, Institute of Experimental Botany and Palacky University, 78371 Olomouc, Czech Republic
3 Sustainable Soils and Crops, Rothamsted Research, Harpenden AL5 2JQ, UK
4 Department of Biology, Duke University, Durham, NC 27708, USA

*Author for correspondence: tps@duke.edu

Abstract

It has been almost a century since biologically active gibberellin (GA) was isolated. Here, we give a historical overview of the early efforts in establishing the GA biosynthesis and catabolism pathway, characterizing the enzymes for GA metabolism, and elucidating their corresponding genes. We then highlight more recent studies that have identified the GA receptors and early GA signaling components (DELLA repressors and F-box activators), determined the molecular mechanism of DELLA-mediated transcription reprograming, and revealed how DELLAs integrate multiple signaling pathways to regulate plant vegetative and reproductive development in response to internal and external cues. Finally, we discuss the GA transporters and their roles in GA-mediated plant development.

Introduction

The gibberellins (GAs) were named for the phytopathogenic fungus *Gibberella fujikuroi*, whose secretions caused abnormal growth and sterility in infected rice (*Oryza sativa*) plants (Kurosawa 1926). The biologically active factor was isolated in impure form from fungal cultures in the 1930s in Japan and named gibberellin A (Yabuta and Sumiki 1938), but this research was not widely known outside of Japan until the late 1940s. The main active component was identified in the 1950s in the UK, USA, and Japan, where it was named gibberellic acid, gibberellin-X, and gibberellin A_3 (GA_3), respectively (Curtis and Cross 1954; Stodola et al. 1955; Takahashi et al. 1955). The structure of gibberellic acid, the name agreed upon by the UK and US groups, or GA_3, was proposed in the late 1950s by chemists working at the ICI Ackers Laboratory, Welwyn, UK (reviewed by Grove 1961). The remarkable effect of this substance on plants stimulated interest in fungal gibberellins (GAs). GAs promote plant growth, particularly to rescue the growth of dwarf mutants of pea (*Pisum sativum*) and maize (*Zea mays*) (Brian et al. 1954; Phinney 1956) and to induce bolting in long-day (LD) rosette species (Lang 1956; Wittwer et al. 1957), prompting the suggestion that they may be en-dogenous plant hormones (Stowe and Yamaki 1957). Plant extracts promoted the growth of mutants in a similar manner to GA_3, reinforcing this hypothesis, which was confirmed by the isolation of 2 mg of GA_1 from 87.3 kg of immature seeds of runner bean (*Phaseolus coccineus*) (Macmillan and Suter 1958). The identity of the isolated bioactive compound was determined by comparing its infrared spectrum with that of authentic GA_1 from *G. fuji-kuroi.* Thus, GA was established as the second endogenous growth regulator (plant hormone) after auxin. The role of GA in plant growth regulation is illustrated in Fig. 1, which compares wild-type and mutant wheat (*Triticum aestivum*) plants with compromised GA-biosynthesis (GA-responsive) or signaling (GA-unresponsive) without or with treatment with GA_3. While GA_3 was the first GA to be discovered, it is a minor form in plants, whereas the major bioactive forms

Received August 01, 2023. Accepted November 06, 2023. Advance access publication January 30, 2024

ADVANCES

- The use of biosensors, gene reporters, single-cell RNA-sequencing, and tissue-specific manipulation of GA metabolism is revealing the cellular distribution of GA biosynthesis and accumulation, and its relevance in plant development.
- The GA signaling repressors DELLAs function as master growth regulators by interacting with regulators in many cellular pathways in response to internal and external cues.
- Besides GA-GID1-induced degradation, DELLA activity is regulated by interacting transcription factors, GA-GID1-independent polyubiquitination and degradation, and other PTMs (glycosylation, SUMOylation, and phosphorylation).
- The movement of GA precursors provides an additional layer of regulation for bioactive GA contents at responding tissues, which is particularly crucial for long-distance communication in coordinating plant growth and development.

are GA_1 or GA_4 (see Fig. 2). GA_3 differs from GA_1 by the presence of a double bond that prevents inactivation by 2β-hydroxylation.

The identification of GA_1 and other GAs in bean seeds was followed by the isolation of other GAs from many plant species. Their structures were determined by conversion to compounds of known structure and/or nuclear magnetic resonance spectroscopy profiles. Later, the development of combined gas chromatography-mass spectrometry for GA analysis allowed them to be identified in plant tissues without the need to obtain pure compounds (Binks et al. 1969). This technique, and later liquid chromatography–mass spectrometry (LC–MS), has enabled GA identification and quantitation to become routine and not just the purview of chemists. The high sensitivity and resolution of ultra high performance LC-MS-MS now allow GAs to be measured in low (mg) amounts of plant tissues.

Following their discovery, there was initially steady but slow progress in elucidating the biosynthetic pathways for GAs in *G. fujikuroi* (the rice pathogen is now reclassified as *Fusarium fujikuroi*) and in plants. In addition, although there was considerable information on the physiological action of GAs on plants, advances in understanding their molecular modes of action were sluggish. By contrast, the application of GAs in agriculture and horticulture developed rapidly with their availability from fungal cultures, with major uses in the production of seedless grapes, to improve skin finish in apples, and many other applications (Rademacher 2015). Furthermore, inhibitors of GA biosynthesis found important applications as plant growth retardants (Rademacher 2000), while semidwarf varieties of major crop species that were key contributors to the Green Revolution were later shown to be defective in GA biosynthesis or action (Phillips 2016). In the last 30 years, with the use of mutants and develop-ments in molecular genetics and genomics, progress in our understanding of both GA metabolism and signaling has accelerated, with details emerging on the biosynthetic reactions, enzymes, genes, and their regulation as well as GA perception and signal transduction. The movement of GAs between cells or over longer distances between organs

Figure 1. GA regulates plant growth and development. Shown are 12-wk-old GA-unresponsive vs -responsive mutants in wheat. All the mutant lines are in cv. Cadenza background. The gain-of-function *Rht-D1b* and *Rht-B1c* (*DELLA*) alleles were introduced into Cadenza from cvs Avalon and Mercia, respectively (Van De Velde et al. 2021). The hypomorphic *gid1* and loss-of-function *ga3ox2* and *rht-1* plants were produced by TILLING after EMS-induced mutagenesis (A.L. Phillips and S.G. Thomas, unpublished data). All plants were untreated (−) or treated (+) twice-weekly with 10 μM GA_3. *Rht* and *gid1* mutants are unresponsive to GA treatment, whereas GA completely rescued the GA-biosynthesis mutant *ga3ox2*. Scale bar = 40 cm.

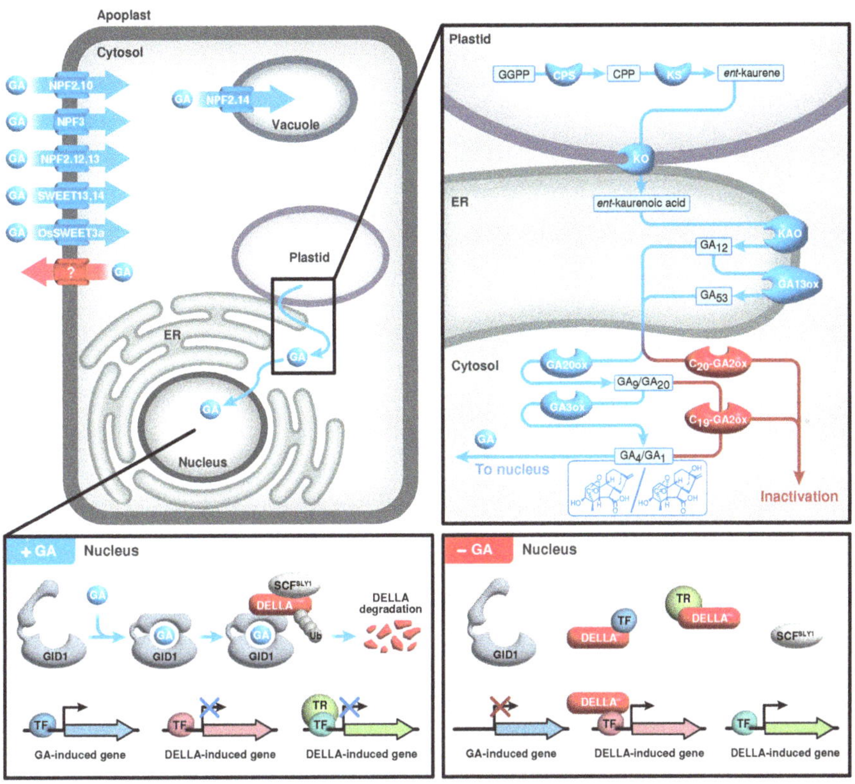

Figure 2. GA metabolism, transport, perception, and signaling in plant cells. GA biosynthesis takes place in three cellular compartments: *ent*-Kaurene is synthesized from GGPP by CPS and KS in the plastid; *ent*-Kaurene is converted to *ent*-kaurenoic acid by KO on the outer plastid membrane, which is connected to the ER; *ent*-Kaurenoic acid is converted to GA_{12} by KAO and GA_{12} to GA_{53} by GA13ox in the ER; GA_{12} and GA_{53} are converted to bioactive GA_4 and GA_1, respectively, by GA20ox and GA3ox in the cytoplasm. GA_4 and GA_1 as well as their immediate precursors GA_9 and GA_{20}, respectively, are oxidized on C-2 by C_{19}-GA2ox, resulting in inactivation, while C_{20}-GA2ox acts on earlier C_{20}-GA precursors. In addition to de novo biosynthesis, GA can be imported into the cell by GA transporters nitrate and peptide transporter families (NPFs) and SWEETs or transported into the vacuole as labeled. GA perception and signaling occur in the nucleus where GA binding to its receptor GID1 (+GA) promotes DELLA degradation via Skp, Cullin, F-box $(SCF)^{SLY1/GID2}$-mediated polyubiquitination and subsequent proteolysis by the 26S proteasome. When GA levels are low (− GA), DELLAs accumulate to high levels. Three distinct modes of DELLA action are shown: (i) DELLA represses transcription by blocking DNA binding and sequestering transcription factors (TF in blue) from target promoters; (ii) DELLA induces transcription by recruiting TFs (in pink); and (iii) DELLA induces transcription by sequestering transcription repressors (TR in green) from target promoters. GGPP, *trans*-geranylgeranyl diphosphate; CPP, *ent*-copalyl diphosphate; CPS, *ent*-copalyl diphosphate synthase; KS, *ent*-kaurene synthase; KO, *ent*-kaurene oxidase; KAO, *ent*-kaurenoic acid oxidase; GA13ox, GA 13-oxidase; GA20ox, GA 20-oxidase; GA3ox, GA 3-oxidase; GA2ox, GA 2-oxidase. GA biosynthesis enzymes are labeled in blue, and the deactivation enzymes are labeled in red. GA, gibberellin; TF, transcription factor; TR, transcription repressors; ER, endoplasmic reticulum.

is an important factor in their function, and recent progress in understanding their transport and transporters is a major development. This review highlights the advances that have contributed to our current understanding of GA metabolism, signaling, and transport and their role in plant development.

GA metabolism

Establishing the GA-biosynthetic pathways

The diterpenoid nature of GAs was demonstrated in the fungus *F. fujikuroi* by the incorporation of ^{14}C-labeled mevalonic acid (MVA) into GA_3 (Birch et al. 1958). Subsequently, the GA-biosynthetic and catabolic pathways were established in plants, primarily using cell-free systems from developing seeds (Graebe et al. 1965; Dennis and West 1967; Graebe et al. 1972, 1974a, b; Kamiya and Graebe 1983) and in the fungus using liquid cultures (Cross et al. 1964; Bearder et al. 1975; Evans and Hanson 1975). The GA metabolism pathway in plants is summarized in Fig. 2, and details can be found in a recent review (Hedden 2020). The diterpene precursor *trans*-geranylgeranyl diphosphate (GGPP), which is formed from MVA in the fungus and mainly via the methylerythritol phosphate (MEP) pathway in plants, is converted in two steps to the tetracyclic diterpene *ent*-kaurene via *ent*-copalyl diphosphate (CPP). *ent*-Kaurene is oxidized to *ent*-kaurenoic acid and then to GA_{12} via several intermediates by two multifunctional cytochrome P450 (CYP450) monooxygenases, *ent*-kaurene oxidase (KO) and *ent*-kaurenoic acid oxidase (KAO), respectively. A third CYP450 converts GA_{12} to GA_{53} by hydroxylation on C-13. These C_{20} intermediates are converted in parallel pathways by soluble dioxygenases to the C_{19}-GAs GA_9 and GA_{20}, respectively, and then to the bioactive phytohormones GA_4 and GA_1. The loss of a C atom (C-20) in the formation of C_{19}-GAs occurs from an aldehyde intermediate (Kamiya and Graebe 1983). In contrast to plants, in which 13-hydroxylation (in GA_{53} formation) occurs early in the pathway and 3β-hydroxylation is the final step, in *F. fujikuroi*, 3β-hydroxylation occurs earlier, while 13-hydroxylation is the last step in the formation of GA_3 (Bearder et al. 1975; Evans and Hanson 1975).

Further metabolism of inactive products is critical to the regulation of GA concentration. The most important inactivation process is 2-oxidation, which can occur on the bioactive GAs and their immediate C_{19} precursors, as well as on earlier C_{20} intermediates. Oxidation of C_{19}-GAs to 2β-hydroxy products is especially strong in late-developing legume seeds, including pea seeds, in which further oxidation on C-2 to GA-catabolites was noted (Sponsel 1983).

GA-biosynthesis mutants

GA-responsive dwarf mutants with lesions in the GA-biosynthetic pathway (Fig. 1) proved extremely useful in understanding GA physiology, identifying the underlying enzymes, and isolating the corresponding genes. Collections of single gene mutants of maize and peas were assembled by Phinney at University of California, Los Angeles (UCLA) and Murfet at Hobart, Tasmania, respectively, and the application of precursors revealed the positions of the lesions in the pathway. Such experiments established that GA_1, but not its biosynthetic precursors, had biological activity in maize (Phinney and Spray 1982) and that *DWARF-1* in maize and *LE* in pea encode 3β-hydroxylases that convert GA_{20} to GA_1 (Ingram et al. 1984; Spray et al. 1984). The *le* mutation corresponded to one of the traits, the difference in stem length, used in Mendel's classical experiments. Once the gene was identified, the mutation was shown to cause an amino acid substitution close to the Fe binding site that reduced enzyme activity (Lester et al. 1997; Martin et al. 1997). The *dwarf-5* mutation alters the activity of *ent*-kaurene synthase (KS) to produce mainly *ent-iso*kaurene, as shown in a cell-free system from maize seedlings (Hedden and Phinney 1979). The *slender* (*sln*) pea mutant illustrates the importance of GA inactivation in regulating GA concentration. The mutation is associated with excessive seedling growth, which decreases later in development (Reid et al. 1992; Ross et al. 1995). The gene, which encodes a GA 2-oxidase, is highly expressed in developing seeds, particularly in the tes-tae (Lester et al. 1999; Martin et al. 1999). Mature pea seeds contain high levels of inactivation products oxidized on C-2 (Sponsel 1983), but *sln* seeds accumulate the precursor GA_{20}, which upon germination is converted to GA_1, causing the overgrowth symptoms (Reid et al. 1992).

In addition, Koornneef and van der Veen (1980) produced 56 independent GA-sensitive Arabidopsis (*Arabidopsis thaliana*) mutants, representing 5 loci, through irradiation- or EMS-induced mutagenesis. Mutations at three loci, named *ga1*, *ga2*, and *ga3*, prevented germination and caused severe dwarfism. By contrast, the other 2 mutations, *ga4* and *ga5*, allowed germination without GA treatment and produced only mild dwarfism. It was later shown that *GA1*, *GA2*, and *GA3*, which encode CPP synthase (CPS), KS, and KO, respectively, are single-copy genes, while the mild phenotypes of *ga4* and *ga5*, with impaired GA 3-oxidation and GA 20-oxidation, respectively, are due to gene redundancy (see below).

Characterizing the enzymes of GA metabolism

Work in Charles West's laboratory at UCLA with cell-free preparations from *Marah macrocarpus* endosperm and *F. fujikuroi* mycelia led to the characterization of the enzymes involved in the conversion of GGPP to *ent*-kaurene. Using the *F. fujikuroi* system, Shechter and West (1969) showed that the conversion occurred in two steps, with CPP as an intermediate. The 2 activities, named activity A for conversion of GGPP to CPP and activity B for conversion of CPP to *ent*-kaurene, were found (after purification) to be present in a single polypeptide (Fall and West 1971). However, they are separate enzymes in *M. macrocarpus* but probably function in association (Duncan and West 1981). Activity A was renamed as the type II terpene cyclase CPS, while activity B, a type I cyclase, was renamed KS (MacMillan 1997). Early indications that these activities were present in plastids (Simcox et al. 1975) were later confirmed (Aach et al. 1995) and further substantiated by the presence of plastid-targeting leader sequences in CPS and KS (Sun and Kamiya 1994; Yamaguchi et al. 1996) and by the plastid localization of enzyme fusions with green fluorescent protein (GFP) (Helliwell et al. 2001b).

The enzyme activities responsible for the middle section of GA biosynthesis, from *ent*-kaurene to GA_{12} and GA_{53} in plants and to GA_{14} (3β-hydroxy GA_{12}) in *F. fujikuroi*, were present in microsomes from cell-free systems from *M. macrocarpus*, pumpkin (*Cucurbita maxima*) endosperm, developing pea cotyledons, and the fungal mycelia and required NADPH (West 1973; Hasson and West 1976; Ropers et al. 1978; Graebe et al. 1980). These enzymes have the properties of CYP450s, which was confirmed when cDNAs encoding the enzymes were isolated (see below). By contrast, the final reactions in the pathway are catalyzed by soluble oxidases requiring Fe^{2+} and are therefore different from the monooxygenases responsible for earlier steps. The pumpkin enzymes were shown to require a small molecule whose identification as 2-oxoglutarate established the enzymes catalyzing GA_{12}-aldehyde 7-oxidation, GA 20-oxidation, 3β-hydroxylation, and 2β-hydroxylation as 2-oxoglutarate-dependent dioxygenases (2-ODDs) (Hedden and Graebe 1982). The soluble 7-oxidase has restricted distribution between plant families, with most plants employing only a monooxygenase for this reaction, while in pumpkin endosperm, this reaction is catalyzed by both monooxygenase and dioxygenase enzymes.

Although work with cell-free homogenates demonstrated the efficient conversion of MVA into GAs, the application of ^{13}C-labeled substrates to Arabidopsis seedlings indicated that *ent*-kaurene and GA_{12} were synthesized mainly from the MEP pathway with a small contribution from the MVA pathway (Kasahara et al. 2002). The extent of cross-over between these pathways, which is dependent on the movement of isoprenoid intermediates into and out of the plastid, may vary among tissues and developmental stages.

Identification of genes encoding GA-metabolic enzymes

The isolation of transcripts and genes encoding the GA-biosynthetic enzymes were major developments that advanced our understanding of the regulation of GA metabolism. Sun et al. (1992) took advantage of the large deletion in the *ga1-3* Arabidopsis mutant to isolate the *GA1* gene by genomic subtraction. Expression of its cDNA in *Escherichia coli* demonstrated that it encodes CPS (Sun and Kamiya 1994). The maize gene *ANTHER EAR1* (*AN1*), which also encodes CPS, was isolated by tagging with the Mutator transposon shortly thereafter (Bensen et al. 1995). After purifying a GA 20-oxidase from pumpkin endosperm and partial amino acid sequencing (Lange 1994), the use of antibodies raised against synthetic peptides led to the isolation of its cDNA from an expression library (Lange et al. 1994). Expression in *E. coli* confirmed its enzymatic activity as oxidizing C-20 mainly to the carboxylic acid. Based on the nucleotide sequence of the pumpkin transcript, three *GA20ox* cDNAs were isolated from Arabidopsis and shown (by expression in *E. coli*) to encode enzymes that convert GA_{12} to the C_{19}-GA, GA_9 (Phillips et al. 1995). The tissue-specific expression patterns of the genes differed, but these genes showed partial redundancy, explaining the mild phenotype of the *ga5* mutant (Rieu et al. 2008b). Their expression was down-regulated by GA, confirming feedback regulation, which had been proposed earlier (Hedden and Croker 1992). A similar approach was used to clone one of these genes, *AtGA20ox1*, which corresponds to *GA5* (Xu et al. 1995). Arabidopsis contains five *GA20ox* genes, but only three of these, *AtGA20ox1,2*, and 3, play major roles in plant development (Plackett et al. 2012).

The cloning of other GA-biosynthetic genes quickly followed. The Arabidopsis *GA4* gene was cloned by T-DNA tagging (Chiang et al. 1995) and confirmed (by heterologous expression) to encode a GA3ox (Williams et al. 1998). Of the four Arabidopsis *GA3ox* genes, two genes, *AtGA3ox1* and *AtGA3ox2*, regulate vegetative growth (Mitchum et al. 2006). Like *AtGA20ox1,2*, and 3, *AtGA3ox1* is down-regulated by GA signaling as part of GA homeostasis (Cowling et al. 1998). *KS* was cloned from pumpkin following the purification of KS protein from cotyledons (Yamaguchi et al. 1996). This led to the isolation of *KS* cDNA from Arabidopsis and the finding (by mutant complementation) that it corresponded to *GA2* (Yamaguchi et al. 1998). Helliwell et al. (1998) demonstrated that *GA3* encodes KO based on the accumulation of *ent*-kaurene in the *ga3-1* mutant and its inability to respond to *ent*-kaurene application. The authors used map-based cloning and random sequencing to isolate *GA3*, which encodes a CYP450, and confirmed its identity by mutant complementation and by demonstrating KO activity after expression in yeast. *KAO* was cloned from barley (*Hordeum vulgare*) in which it corresponds to *GA-RESPONSIVE DWARF5* (*GRD5*), whose mutants accumulate *ent*-kaurenoic acid (Helliwell et al. 2001a). *GRD5* and 2 Arabidopsis homologs encode CYP88A family members, which, after expression in yeast, were shown to catalyze the 3-step conversion of *ent*-kaurenoic acid to GA_{12} via *ent*-7α-hydroxykaurenoic acid and GA_{12}-aldehyde (Helliwell et al. 2001a). The equivalent enzyme in *F. fujikuroi* (CYP68A) also has 3β-hydroxylase activity and produces GA_{14} (Rojas et al. 2001). The two *AtKAO* genes share fully redundant functions, with the double mutant being severely dwarfed (Regnault et al. 2014). Thus, two enzymes, KO and KAO, are required to convert *ent*-kaurene to GA_{12} in plants, with a third CYP enzyme catalyzes the 13-hydroxylation of GA_{12} to GA_{53} (see below). Fusions of AtKO and AtKAO with GFP localized to the outer chloroplast envelope and endoplasmic reticulum (ER), respectively, provide a mechanism for the transit of *ent*-kaurene from plastids to the ER (Helliwell et al. 2001b).

The first cloning of a *GA2ox* cDNA took advantage of the very high GA2ox activity in late-developing *P. coccineus* seeds by functional screening of a cDNA expression library for release of ^3H from [2β, 3β-$^3H_2$]GA_9 (Thomas et al. 1999). The functions of the *P. coccineus* cDNA and three homologous Arabidopsis cDNAs identified in genomic databases were determined by expression in *E. coli*. The enzymes converted C_{19}-GAs to their 2β-hydroxy analogs and, depending on the substrate and paralog, catalyzed further oxidation to GA-catabolites. GA promoted the expression of two of the Arabidopsis *GA2ox* genes, whereas it had the opposite effect on *GA20ox* and *GA3ox* gene expression. Like *GA20ox* and *GA3ox*, the C_{19}-*GA2oxs* form a gene family with five functional members in Arabidopsis (Rieu et al. 2008a). Soon after, similar approaches were used to clone *GA2ox* cDNAs from developing pea seeds, one of which, *PsGA2ox1*, corresponds to *SLN* (Lester et al. 1999; Martin et al. 1999). A second clade of *GA2ox* genes with 2 members was identified from Arabidopsis by activation tagging, encoding enzymes that act on C_{20}-GAs (Schomburg et al. 2003). Two additional members of this clade were identified in Arabidopsis recently (Lange et al. 2020). Apart from the *GA7ox* genes with restricted distribution, angiosperms contain four families of 2-ODDs, GA20ox, GA3ox, C_{19}-GA2ox, and C_{20}-GA2ox, while this last clade is absent from gymnosperms (Yoshida et al. 2020). The C_{19}-*GA2ox* gene family is the largest in most species, with some tissue-specific expression but considerable redundancy.

GA 13-hydroxylase genes were first identified in rice. The encoded monooxygenases CYP714B1 and CYP714B2 converted GA_{12} to GA_{53} following expression in yeast (Magome et al. 2013). Overexpression of these genes in Arabidopsis caused a slight reduction in height, suggesting that 13-hydroxylation is a mild inactivation reaction. Other members of the CYP714 family oxidize 13-deoxy GAs and/or *ent*-kaurenoic acid on C-13 or adjacent C atoms and are also inactivating. Mutation of the rice enzyme CYP714D1, known as ELONGATED UPPERMOST INTERNODE (EUI), which oxidizes the 16,17-double bond to the epoxide, is utilized in hybrid rice production to promote panicle emergence in male-sterile cultivars (Zhu et al. 2006). Arabidopsis, which produces low amounts of 13-hydroxy GAs, except in the seed, contains two *CYP714* genes, *CYP714A1* and *CYP714A2*. Overexpression of CYP714A1, which converts GA_{12} to 16α-carboxy-17-norGA_{12}, causes extreme dwarfism, while CYP714A2 functions mainly as a 12α-hydroxylase, with only low 13-hydroxylase activity, and causes mild dwarfism when overexpressed (Nomura et al. 2013). Notably, AtCYP72A9 was found to 13-hydroxylate GA_{12}, GA_9, and GA_4 (following expression in yeast) and may be the main source of 13-hydroxy GAs in Arabidopsis seeds (He et al. 2019). Its overexpression resulted in dwarfism.

Regulation of GA metabolism

GA biosynthesis and inactivation are tightly regulated by developmental and environmental cues (Yamaguchi and Kamiya 2000; Sun 2008; Yamaguchi 2008; Hedden and Thomas 2012; Hedden 2020; Bouré and Arnaud 2023). Following the identification of genes encoding the metabolic enzymes, there have been numerous reports of their transcriptional regulation, particularly for the 2-ODDs, which limit the production of bioactive GAs. As discussed under GA perception and signaling, their ex-pression is modified by other hormones as well as by numerous environmental factors, including stress (Fig. 3). While there is some evidence for post-transcriptional regulation (Lee and Zeevaart 2007), this process has been little studied for practical reasons. Reports of altered GA levels in GA-response mutants were early indications that GA-metabolism was regulated via GA signaling (reviewed in Hedden and Sponsel 2015). These ob-servations highlighted 20-oxidation as a potential site of regula-tion by GA action, which was confirmed by experiment (Hedden and Croker 1992). Subsequently, as noted above, ex-pression of some *GA20ox* and *GA3ox* genes was found to be re-pressed by GA, while *GA2ox* expression was upregulated (Thomas et al. 1999). Furthermore, down-regulation of *GA INSENSITIVE DWARF1* (*GID1*) receptor genes by GA extended homeostasis to GA signaling (Griffiths et al. 2006). The involvement of the DELLA GA-signaling component in this process is discussed below. A nontranscriptional homeostatic mechanism was revealed from the X-ray crystal structure of OsGA2ox3 (Takehara et al. 2020). In the presence of its substrate GA_4, the enzyme forms a tetramer, thereby increasing its catalytic efficiency.

GA biosynthesis in nonseed plants, fungi, and bacteria

In the evolution of land plants, DELLA-mediated signaling predates its regulation by GA-GID1 (Hernandez-Garcia et al. 2021). The evolution of GA biosynthesis was discussed in detail recently (Yoshida et al. 2020). The bryophyte *Physcomitrium patens* produces *ent*-kaurenoic acid derivatives but not GAs (Miyazaki et al. 2018), whereas the lycophyte *Selaginella moellendorffii* and the fern *Lycodium japonicum* produce GA_4 using 2-ODDs for 20ox and 3ox activities and possess the GA-GID1-DELLA signaling system (Hernandez-Garcia et al. 2019; Yoshida et al. 2020). Notably, while *P. patens* and *L. japonicum* have bifunctional CPS/KS enzymes, *S. moellendorffii* has separate CPS and KS enzymes for *ent*-kaurene synthesis, as in seed plants. GA-inactivation by 2-oxidation is not present in nonseed plants and was acquired just before the establishment of gymnosperms (Yoshida et al. 2020). GA 2β-hydroxylation is also absent from GA-producing fungi and bacteria, whose genes for GA-biosynthesis are clustered in operons and were acquired independently of plants and each other (Hedden et al. 2001; Nett et al. 2017b). The members of both kingdoms employ CYP450s rather than 2-ODDs for 3- and 20-oxidation, while *F. fujikuroi* uses a 2-ODD to convert GA_4 to GA_7 in GA_3 biosynthesis (Bhattacharya et al. 2012). Bacteria have separate CPS and KS enzymes, whereas fungi contain a bifunctional CPS/KS (Morrone et al. 2009). The capacity to produce bioactive GAs appears to be related to pathogenicity. Most symbiotic N-fixing bacteria produce GA_9, allowing the plant to regulate GA_4 production, while the phytopathogenic species *Xanthomonas oryzae* contains an extra gene encoding a 3β-hydroxylase (CYP115) and produces GA_4 (Nagel and Peters 2017; Nett et al. 2017a).

GA perception and signaling

Dwarf and slender mutants with reduced or elevated GA responses

Genetic analyses of mutants displaying altered GA responses and molecular cloning of their corresponding genes have been instrumental in unveiling the long-sought GA receptor (GID1) and its immediate downstream repressors (DELLAs) and activators (F-box proteins GID2/SLEEPY1 [SLY1]). GA-unresponsive recessive mutants exhibit a dark-green, dwarf phenotype that mimics GA biosynthesis mutants, but their growth cannot be restored by GA treatment (Fig. 1). Plants with defects in genes encoding the GA receptor GID1 or the F-box protein GID2/SLY1 belong to this mutant class (McGinnis et al. 2003; Sasaki et al. 2003; Ueguchi-Tanaka et al. 2005). Gain-of-function DELLA mutants are dominant GA-unresponsive dwarves, e.g. *GA insensitive* (*gai*) in Arabidopsis (Koornneef et al. 1985) and *Reduced height* (*Rht*) varieties in wheat (Fig. 1), which were major contributors to the "Green Revolution" by increasing grain yield in the 1960s to 1970s (Börner et al. 1996; Peng et al. 1999). Conversely, "slen-der" mutants with elevated GA responses (e.g. *la cry* in pea and *slender* (*sln*) in barley) are recessive and display a tall and thin stem phenotype, which resembles wild-type plants that have been overdosed with GA (Brian 1957; Foster 1977).

Besides genetic analysis, cereal aleurone was also used extensively to study GA signaling (Lovegrove and Hooley 2000; Sun and Gubler 2004). During seed germination, aleurone cells produce hydrolytic enzymes in response to GA (diffused from the embryo) to degrade the starchy endosperm.

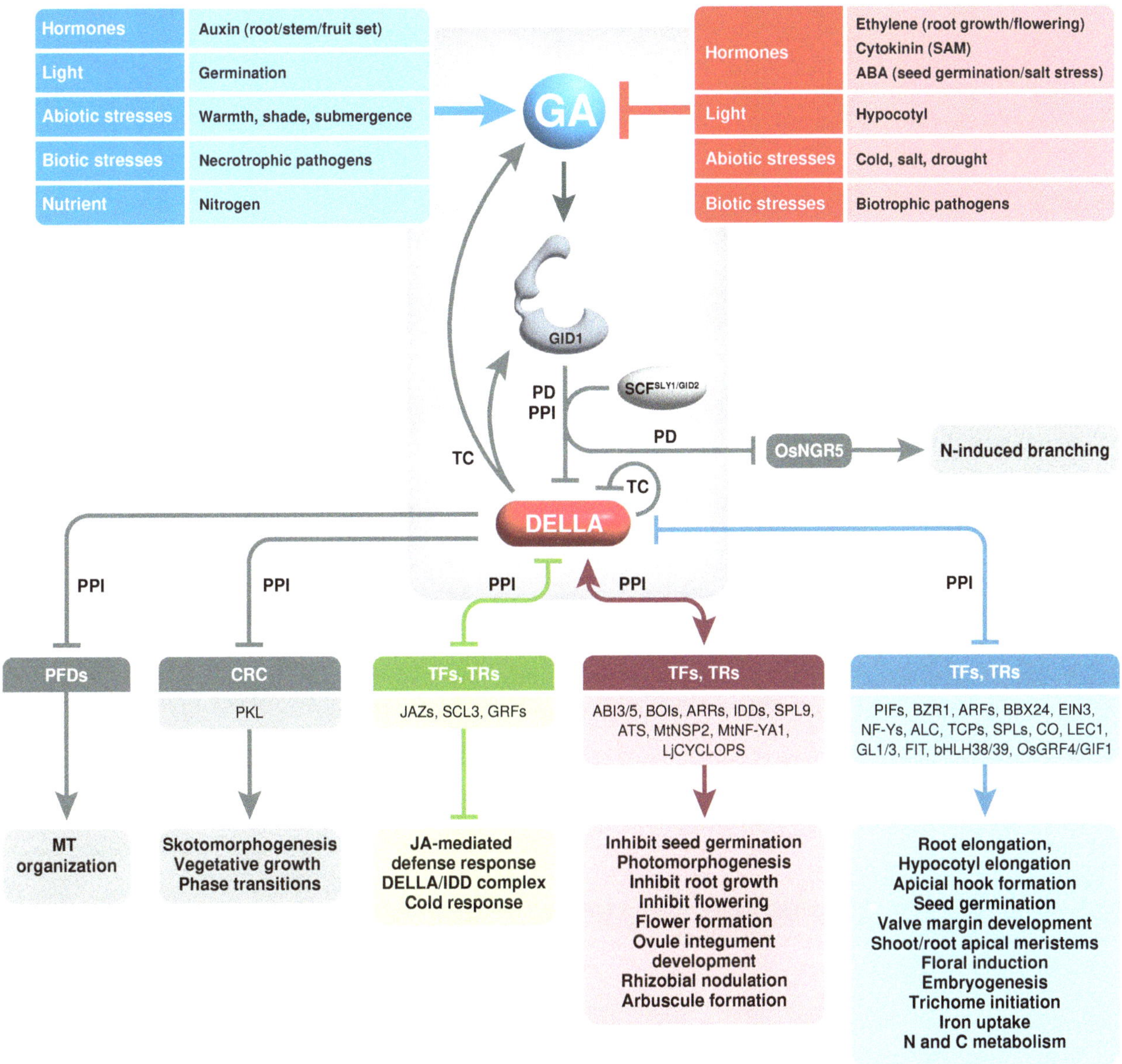

Figure 3. Interaction network between the GA-GID1-DELLA signaling module and various internal and external cues. Signals that increase bioactive GA levels are labeled in blue, while signals that decrease GA levels are shown in red. GA-GID1 triggers DELLA degradation via $SCF^{SLY1/GID2}$-mediated polyubiquitination. GA-GID1 also induces OsNGR5 degradation in a SCF^{GID2}-dependent manner to inhibit nitrogen-induced shoot branching in rice. DELLAs interact antagonistically or additively with a myriad of transcription factors (TFs), transcription regulators (TRs), CRC, and PREFOLDINs (PFDs) to modulate specific developmental processes. Most of the DELLA interactors are Arabidopsis proteins, except those that are labeled (Os, *Oryza sativa*; Mt, *Medicago truncatula*; Lj, *Lotus japonicus*). See Boxes 1 and 2 for details. PD, protein degradation; PPI, protein–protein interaction; TC, transcription; SAM, shoot apical meristem; MT, microtubule; GA, gibberellin; ABA, abscisic acid.

Using this aleurone system, pharmacological assays have identified GA-induced genes and GAMYB, the transcription factor responsible for this transcriptional induction.

GA-GID1 induces $SCF^{SLY1/GID2}$-mediated DELLA degradation

With the development of Arabidopsis as a model system for plant research in the 1980s, the *gai-1* mutant and *repressor of ga1-3* (*rga*) mutants in Arabidopsis were isolated by screening for GA-unresponsive dwarves (Koornneef et al. 1985) or suppressors of the dwarf phenotype of *ga1-3* (Silverstone et al. 1997b), respectively. Intragenic suppressors of the semidominant *gai-1* mutant were generated by Ds transposon insertion mutagenesis (Peng and Harberd 1993), which guided the cloning of *GAI* (Peng et al. 1997). Suppressor mutants of *ga1-3* generated by fast-neutron mutagenesis contained large deletions in the *RGA* locus, which facilitated the cloning of *RGA* by genomic subtraction (Silverstone et al. 1998). The Arabidopsis genome contains five DELLA genes, *RGA*, *GAI*, *RGA-LIKE1* (*RGL1*), *RGL2*, and *RGL3*, which belong to a subfamily of GRAS (for GAI, RGA and SCARECROW) transcription regulators. The N-terminal DELLA domain is unique to the DELLA subfamily, whereas the C-terminal GRAS domain is shared among all GRAS family members (Pysh et al. 1999; Sun and Gubler 2004). Further characterization of orthologs in other species showed that DELLA genes are conserved in plants, including *Rht* in wheat (Peng et al. 1999), *SLENDER RICE1* (*SLR1*) in rice (Ikeda et al. 2001), *SLN1* in barley (Chandler et al. 2002), *D8* and *D9* in maize (Peng et al. 1999; Lawit et al. 2010), *PROCERA* in tomato (*Solanum lycopersicum*) (Jasinski et al. 2008), and *LA* and *CRY* in pea (Weston et al. 2008).

How are DELLA proteins regulated by the GA signal? The *la cry* slender mutant phenotype resembles the effects of GA overdose, leading to the idea that LA and CRY control the production of inhibitors (for GA-induced growth) and suggesting that GAs may function as "inhibitors of inhibitors" to promote growth (Brian 1957). Importantly, an examination of RGA protein levels by immunoblot analysis and confocal microscopy revealed that RGA protein levels in planta rapidly decreased in response to GA treatment (Silverstone et al. 2001). Furthermore, the *gai-1* Arabidopsis mutant encodes a mutant protein with a 17-amino acid deletion within the DELLA motif, suggesting that this mutation turns the GAI protein into a constitutive repressor of GA signaling (Peng et al. 1997). Indeed, deletion of the identical DELLA motif in RGA (rga-Δ17) abolished GA-induced degradation and conferred a GA-unresponsive dwarf phenotype, indicating that the DELLA motif is required for its proteolysis in re-sponse to GA (Dill et al. 2001). Notably, the semidwarf var-ieties of wheat and maize were found to be caused by deletion mutations in the DELLA domain of *Rht* and *D8*, respectively (Peng et al. 1999). Further studies of the recessive GA-unresponsive dwarf mutants Arabidopsis *sly1* and rice *gid2* led to the identification of F-box proteins and associated Skp, Cullin, F-box (SCF) ubiquitin E3 ligase complexes ($SCF^{SLY1/GID2}$) that are responsible for polyubiquitination of DELLA and degradation by the 26S proteasome (Steber et al. 1998; McGinnis et al. 2003; Sasaki et al. 2003).

Although Arabidopsis research led to the breakthrough discoveries of the DELLA repressors and SLY1 (F-box) activators of GA signaling, the GA receptor remained elusive despite the efforts of multiple mutant screens. Eventually, aided by the completion of the rice genome sequence (Goff et al. 2002; Yu et al. 2002), the GA receptor was unveiled by positional cloning of *GA-insensitive dwarf1* (*gid1*) mutants in rice (Ueguchi-Tanaka et al. 2005). Notably, the smaller genome of Arabidopsis contains three *GID1* orthologs (*GID1A*, *GID1B*, and *GID1C*), whose functional redundancy explains why genetic screens failed to identify GA receptors in Arabidopsis (Griffiths et al. 2006; Nakajima et al. 2006). GID1 is localized to both the cytoplasm and nu-cleus and belongs to the hormone-sensitive lipase (HSL) family, although it lacks one of three key catalytic residues for lipase activity. Yeast two-hybrid (Y2H) and in vitro pull-down assays show that GA binding to GID1 promotes GID1-DELLA interactions (Ueguchi-Tanaka et al. 2005) and that the DELLA domain (including DELLA, LEXLE and VHYNP motifs) is essential for its interaction with GID1 (Griffiths et al. 2006; Willige et al. 2007). Molecular details of the GA-AtGID1A-DELLA domain (GAI) complex and GA-OsGID1 determined by X-ray crystallography revealed that bioactive GA is an allosteric inducer of its receptor GID1 (Murase et al. 2008; Shimada et al. 2008). The carboxy-terminal core domain of GID1 forms a GA-binding pocket, and the amino-terminal extension (N-Ex) domain acts as a lid. GA binding induces a conformational switch of its N-Ex to close the GA-binding pocket and creates hydrophobic surfaces for DELLA binding (Fig. 2).

How does GA-GID1-DELLA promote SLY1/GID2 recognition? Mutant and Y2H analyses indicated that SLY1/GID2 interacts with the GRAS domain of the DELLA protein (Dill et al. 2004; Hirano et al. 2010). Yeast 3-hybrid assays further demonstrated that GA-bound GID1 enhances the RGA–SLY1 interaction, suggesting that the GA/GID1-DELLA domain interaction triggers conformational changes in the GRAS domain for SLY1 recognition (Griffiths et al. 2006). Moreover, the GRAS domain of SLR1 was shown to interact with GID1 after the binding of the DELLA domain to further stabilize the GID1-SLR1 complex, which allows efficient recognition by the F-box protein GID2 (Hirano et al. 2010).

Mechanism of DELLA action: transcriptional reprograming via protein–protein interactions with hundreds of transcription factors

DEL L Aɜre nucleus-localized transcription regulators. Transcriptome studies on early GA- and DELLA-responsive genes showed that DELLAs can activate or repress transcription, depending on the target genes (Zentella et al. 2007; Hou et al. 2008). DELLAs do not contain any canonical DNA binding motifs and have not been shown to bind DNA directly. Importantly, chromatin immunoprecipitation (ChIP)-qPCR analysis demonstrated an association of RGA with its target chromatin (Zentella et al. 2007). ChIP-sequencing (seq) analysis identified genome-wide RGA binding sites: 421 associated genes in Arabidopsis seedlings and 2,327 associated genes in the inflorescence meristem (Marin-de la Rosa et al. 2015; Serrano-Mislata et al. 2017). Abundant evidence indicates that DELLAs regulate transcription via antagonistic or additive interactions with a myriad of transcription factors/regulators (Figs. 2 and 3 and Boxes 1 and 2) (Sun 2011; Daviere and Achard 2016; Van De Velde et al. 2017). PHYTOCHROME INTERACTING FACTORS (PIFs), PIF3, and PIF4 were the first reported DELLA-interacting transcription factors (TFs) (de Lucas et al. 2008; Feng et al. 2008). PIF3 and PIF4 are light-responsive bHLH TFs that promote hypocotyl elongation. Genetic analysis and ChIP-qPCR showed that the DEL L A-PIF3/4interaction sequesters PIFs from binding to the promoters of growth-related genes, revealing

BOX 1. DELLA INTERACTORS: TRANSCRIPTION FACTORS/REGULATORS

(1) DELLA-repressed transcription activators include several TFs that promote hypocotyl elongation: PIF3/4 (light signaling regulators), BRASSINAZOL E-RESISTANT1 (BZR1, a brassinosteroid signaling activator) (Bai et al. 2012; Gallego-Bartolome et al. 2012), AUXIN RESPONSE FACTORs (ARFs, auxin signaling activators) (Oh et al. 2014), and BBX24 (B-box zinc finger protein) (Crocco et al. 2015). DEL L Asalso inhibit the activities of ETHYLENE INSENSITIVE3 (EIN3, an ethylene signaling activator) in apical hook formation (An et al. 2012); NUCLEAR FACTOR Ys (NF-Ys) in seed germination and flowering (Hou et al. 2014); ALCATRAZ (ALC, bHLH) in fruit valve margin development (Arnaud et al. 2010); Type I TCPs (TEOSINTE BRANCHED 1 [TB1], CYCLOIDEA [CYC], and PROLIFERATING CELL FACTOR [PCF]) in cell division in shoot and root apical meristems (Daviere et al. 2014; Resentini et al. 2015); GL ABRA1 (GL 1, MYB23) and GL 3 (bHL H) in trichome initiation (Qi et al. 2014); SQUAMOSA PROMOTER BINDING PROTEIN L IKEs (SPL s) and CONSTANS (CO) in floral induction (Yu et al. 2012; Hyun et al. 2016; Xu et al. 2016); L EAFY COTYL EDON1 (LEC1 = NF-YB9) in late embryogenesis (Hu et al. 2018); FIT (bHL H) and bHL H38/39 for iron uptake in the root (Wild et al. 2016); and the GROWTH REGUL ATING FACTOR4 (OsGRF4)/ GRF-Interacting Factor1 (OsGIF1) complex in nitrogen and carbon metabolism and nitrogen uptake (Li et al. 2018).
(2) DELLA-activated transcription factors/regulators include ABSCISIC ACID INSENSITIVE 3 (ABI3) and ABI5 (a bZIP TF), which mediate ABA-inhibited seed germination (Lim et al. 2013); BOTRYTIS SUSCEPTIBLE1 INTERACTORs (BOIs, RING domain protein), which inhibit seed germination, the juvenile-to-adult transition, and floral induction (Park et al. 2013); type-B ARRs, which function in cytokinin-induced de-etiolation and root meristem cell division (Marin-de la Rosa et al. 2015); the INDETERMINATE DOMAIN (IDD) subfamily of C2H2 zinc finger TFs, which regulate root development, inhibit floral induction, and regulate GA homeostasis (Fukazawa et al. 2014; Yoshida et al. 2014); SPL9, which promotes flower formation; ABERRANT TESTA SHAPE (ATS = KANADI4, KAN4), which promotes ovule integument development (Gomez et al. 2016); NODULATION SIGNALING PATHWAY2 (MtNSP2, a GRAS protein) and MtNF-YA1 in *Medicago truncatula*, which promote rhizobial nodulation (Fonouni-Farde et al. 2016; Jin et al. 2016); and LjCYCLOPS in *Lotus japonicus*, which functions in arbuscule formation (Pimprikar et al. 2016).

BOX 2. DELLA INTERACTORS: TRANSCRIPTION REPRESSORS AND OTHERS

(1) DEL L A-repressedtranscription repressors include the jasmonate (JA) signaling repressors JAZs, which promote JA-induced defense responses against herbivory and necrotrophs (Hou et al. 2010). Moreover, the JAZ–DELLA interaction inhibits DELLA-PIF3 to promote plant growth, revealing the role of JAZ/DELLA/PIF in balancing plant defense and growth (Yang et al. 2012). DELLAs also inhibit SCL3 (a GRAS protein) activity by interacting and competing with SCL3 for binding to IDDs (Zhang et al. 2011; Yoshida et al. 2014). SCL3 is an activator of GA signaling whose transcription is induced by DEL LA but repressed by itself. In addition, DEL L As inhibit the activity of GRFs in promoting the expression of cold-induced genes (Lantzouni et al. 2020).
(2) DELLAs interact with CRC, including SWI3C (Sarnowska et al. 2013) and PICKLE (PKL) (Zhang et al. 2014). SWI3C is a core subunit of the Switch (SWI)/Sucrose Nonfermenting (SNF)-type CRC. Transcript analysis suggested that SWI3C promotes the expression of DEL L A-inducedgenes (e.g. *GID1 A* and *SCL3*), although the mechanism is unclear (Sarnowska et al. 2013). An antagonistic interaction between DELLA and PKL regulates GA-induced skotomorphogenesis, vegetative growth, and the phase transition (Zhang et al. 2014; Park et al. 2017).
(3) DELLAs sequester the cochaperones PREFOLDINs (PFDs) to the nucleus, which disrupts microtubule organization in the cytoplasm (Locascio et al. 2013).

molecular crosstalk between light and GA signaling. Extensive studies in the last 15 years have identified 370 potential DELLA-interacting proteins in Arabidopsis by Y2H screens, and over 40 of them have been verified by co-IP and/or genetic analyses (Marin-de la Rosa et al. 2014; Van De Velde et al. 2017; Lantzouni et al. 2020).

DELLAs appear to function as transcription co-activators or corepressors, depending on which transcription factors/ regulators they interact with. Three distinct modes of DELLA action have been reported: (i) DELLA represses transcription by blocking DNA binding and sequestering transcriptional activators (e.g. PIFs) from their target promoters; (ii) DELLA activates transcription by recruiting transcription factors (e.g. ABSCISIC ACID INSENSITIVE 3 [ABI3] and ABI5, ARABIDOPSIS RESPONSE REGULATORs (ARRs), and IDDs); and (iii) DELLA activates transcription by sequestering transcription repressors (e.g. JASMONATE ZIM DOMAINs [JAZs], SCARECROW-LIKE3 [SCL3]) from their target promoters (Figs. 2 and 3) (Daviere and Achard 2016; Thomas et al. 2016; Van De Velde et al. 2017). ChIP-seq analyses showed that RGA binding peaks are enriched near cis-elements for several DELLA-interacting TFs (Marin-de la Rosa et al. 2015; Serrano-Mislata et al. 2017), supporting the notion that DELLAs are recruited to target promoters by their interacting TFs.

Multiple signaling pathways regulate DELLA-mediated plant responses to internal and environmental cues

DELLAs were initially identified as GA signaling repressors, and GA promotes rapid DELLA proteolysis mediated by GID1 and $SCF^{SLY1/GID2}$ (Sun and Gubler 2004). Notably, DELLAs also play a key role in feedback regulation to help maintain GA homeostasis by inducing the transcription of genes encoding GA biosynthetic enzymes and GID1s (Fig. 3) (Zentella et al. 2007). However, extensive studies of DELLA interactors and DELLA-regulated processes in the last 15 years have unveiled a much broader function of DELLAs as master growth regulators that integrate the activities of many signaling pathways in response to developmental and external cues, including biotic and abiotic stress (Thomas et al. 2016; Van De Velde et al. 2017). Boxes 1 and 2, and Fig. 3 highlight the diverse processes regulated by DELLAs and their interactors. Importantly, DELLA activity can be regulated by several mechanisms: (i) altered DELLA stability (GA-GID1 dependent) by modulating GA metabolism; (ii) altered DELLA activity by interacting with TFs/TRs/chromatin-remodeling complexes (CRC); (iii) GA-GID1-independent polyubiquitination and degradation; (iv) other post-translational modifications (PTMs); a nd (v) transcriptional induction of *RGL3* by JA signaling.

(1) Altered DELLA stability (GA-GID1 dependent) by modulating GA metabolism

Factors that decrease bioactive GA levels to increase DELLA accumulation include other phytohormones (ABA, cytokinin, and ethylene), external cues (e.g. light, abiotic stresses [cold, salt, and drought], and biotic stress [biotrophic pathogens]) (Fig. 3). Conversely, factors that increase bio-active GA levels to decrease DELLA accumulation include auxin, and external cues (e.g. nitrogen, light, abiotic stresses [warmth, shade, submergence], and biotic stress [necrotrophic pathogens]). Most of these findings have been discussed in previous reviews (Sun 2011; Colebrook et al. 2014), except for two recent studies showing that drought inhibits GA biosynthesis in the leaf base of wheat seedlings (Ptošková et al. 2022) and that nitrate (a major nitrogen sources) induces root growth by increasing GA biosynthesis, resulting in reduced DELLA protein accumulation in Arabidopsis and wheat (Camut et al. 2021). Recent studies on the semidwarf Green Revolution varieties (GRVs) of wheat (*Rht* alleles encoding dominant DELLAs) and rice (*semi-dwarf1* [*sd1*], defective in GA20ox2) demonstrated that nitrogen use efficiency (NUE) is regulated by the GA signaling pathway (Wang et al. 2021; Liu et al. 2022). The GRVs dramatically increase crop yields, although they exhibit low NUE because SLR1 inhibits nitrogen and carbon metabolism, nitrogen uptake, and assimilation by disrupting the interaction between the key TF OsGRF4 and its coactivator OsGIF1 (Li et al. 2018) (Fig. 3). Notably, elevated DELLA activities in GRVs promote nitrogen-induced tillering (shoot branching). It turns out that GID1 and DELLA can interact with NITROGEN MEDIATED TILLER GROWTH RESPONSE 5 (NGR5), a transcription factor that represses the expression of genes that inhibit tillering (Wu et al. 2020). GA-GID1 promotes NGR5 degradation by the SCF^{GID2}-mediated ubiquitin-proteasome pathway (Fig. 3), while DELLA competes with GID1-NGR5 interaction to stabilize NGR5.

(2) Altered DELLA activity by antagonistic or additive interactions with TFs/TRs

(Boxes 1 and 2, and Fig. 3).

(3) GA/GID1-independent DELLA degradation

Besides $GA/GID1/SCF^{SLY1/GID2}$-dependent proteolysis, DELLA protein stability can be regulated by 3 other pathways (Fig. 4).

CONSTITUTIVE PHOTOMORPHOGENIC1 (COP1) mediates DELLA degradation in response to warm temperatures and shade:

Shade- or warm temperature-induced hypocotyl elongation correlates with elevated bioactive GA levels and a re-duction in DELLA protein levels (Djakovic-Petrovic et al. 2007; Stavang et al. 2009; Arana et al. 2011). Surprisingly, the abundance of the GA-resistant rga-Δ17 protein is also reduced by these environmental cues, suggesting the presence of a GA/GID1-independent mechanism for DELLA degradation (Blanco-Tourinan et al. 2020a). Biochemical and genetic analyses showed that the E3 ubiquitin ligase COP1 and its interacting protein SUPPRESSOR OF phyA-105 proteins (SPAs) play a direct role in the rapid proteolysis of DELLA prior to changes in GA content in the shade or under warm con-ditions. As the liverwort (*Marchantia polymorpha*) gen-ome contains putative orthologs of *COP1* and *DELLA* genes, COP1-mediated DELLA degradation may serve to regulate DELLA activity prior to the GA/GID1-mediated mechanism, which appears later in lycophytes.

FLAVIN-BINDING KELCH REPEAT F-BOX 1 (FKF1) mediates DELLA degradation to promote flowering in LD conditions:

FKF1 promotes flowering under long-day (LD) conditions. The *fkf1* mutant is late flowering and has elevated RGA protein levels even in the GA-deficient background (Yan et al. 2020). co-IP assays in *Nicotiana benthamiana* and in vitro assays showed that FKF1 directly binds to DELLAs and promotes their ubiquitination and degradation. It was proposed that FKF1 regulates the cyclical degradation of DELLA in LDs, but this remains to be verified.

Strigolactone (SL)-D14 mediates DELLA degradation in response to low nitrogen conditions:

As described above, GA reduces NUE by promoting NGR5 degradation. Conversely, SL increases NUE. D53 is a repressor

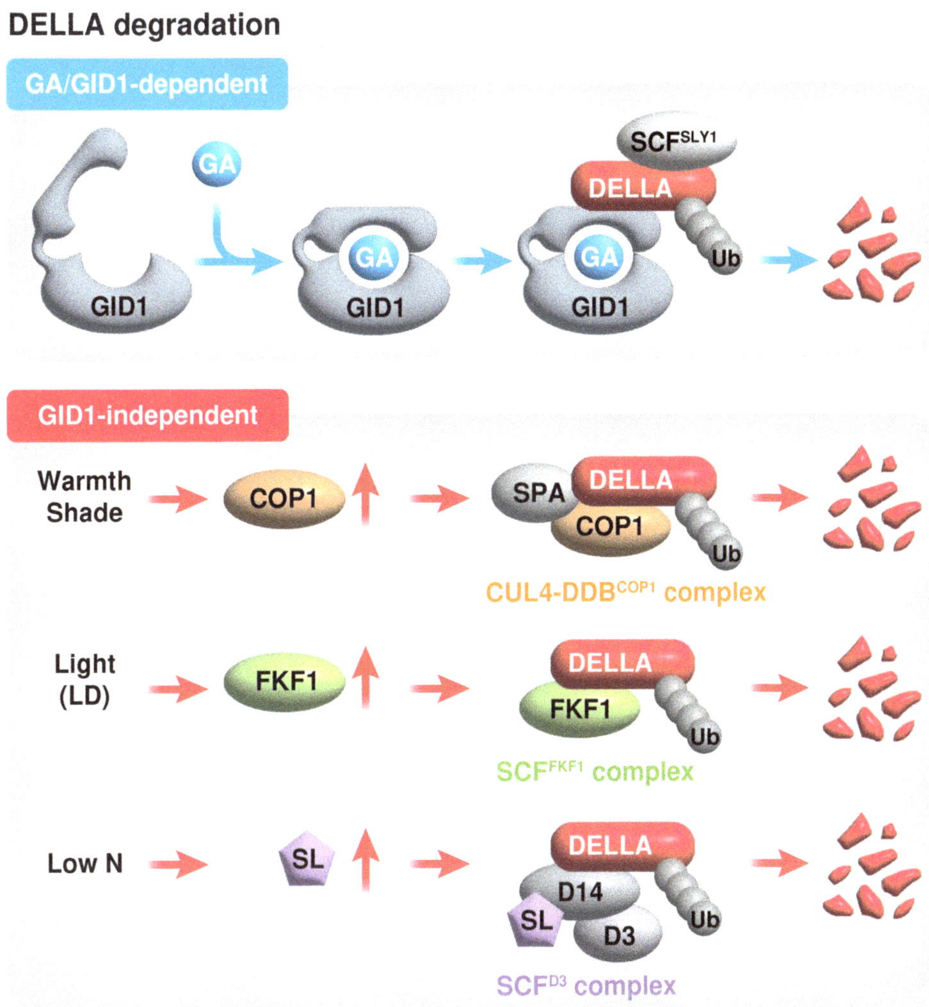

Figure 4. GA/GID1-dependent versus GA/GID1-independent DELLA degradation. GA/GID1-dependent proteolysis of DELLA is mediated by $SCF^{SLY1/GID2}$. DELLA can be destabilized by signals including warmth, shade, LD light, and low nitrogen conditions. COP1 and its associated $CUL4\text{-}DDB^{COP1}$ ubiquitin E3 ligase complex mediate warmth- and shade-induced DELLA degradation in Arabidopsis. The SCF^{FKF1} ubiquitin E3 ligase complex mediates LD-induced DELLA proteolysis in Arabidopsis. Low nitrogen conditions induce the biosynthesis of SL in rice, which binds to its receptor D14 and promotes the D14-DELLA interaction and DELLA degradation mediated by SCF^{D3}. GA, gibberellin; Ub, ubiquitin.

of SL signaling that inhibits the expression of GRF4-induced genes for N metabolism. Low nitrogen conditions induce the biosynthesis of SL, which binds to its receptor D14 and promotes the D14-D53 interaction and subsequent SCF^{D3}-mediated ubiquitination and degradation of D53. Notably, SL also promotes the interaction of D14 with SLR1 (rice DELLA) (Nakamura et al. 2013), which leads to SLR1 degradation mediated by SCF^{D3} (Sun et al. 2023). D53 competes with SLR1 for binding to D14, adding another layer of regulation for DELLA degradation.

(4) Regulation of DELLA Function by PTMs:

In addition to polyubiquitination, which promotes DELLA proteolysis, DELLA activity is also modulated by other PTMs, including Small Ubiquitin-like Modifier (SUMO)-conjugation (SUMOylation), phosphorylation, and *O*-glycosylation (*O*-linked *N*-acetylglucosamine [*O*-GlcNAc] and *O*-fucose modifications) (Blanco-Tourinan et al. 2020b; Sun 2021). Under salt-stress conditions, DELLA SUMOylation is induced due to increased degradation of the SUMO proteases OVERLY TOLERANT TO SALT 1 and 2 (OTS1 OTS2) (Conti et al. 2014). SUMOylated DELLA binds to and sequesters GID1 independently of GA, thereby promoting the accu-mulation of non-SUMO-DELLA and restricting plant growth. Under nonstress conditions, OTS-mediated de-SUMOylation of DELLA promotes the growth of stamen filaments (Campanaro et al. 2016). The role of phosphorylation in DELLA function is not well understood. An early study reported that GA-induced SLR1 degradation in rice occurs independently of phosphorylation (Itoh et al. 2005). However, another study suggested that the phosphorylation of SLR1 by the casein kinase I EARLIER FLOWERING1 (EL1) increases its stability (Dai and Xue 2010). The *el1* mutant displays elevated GA response and early flowering phenotypes, and the

SLR1-YFP (yellow fluorescent protein) protein in *35S:SLR1-YFP el1* transgenic rice was degraded more rapidly after GA treatment than in the wild-type background. These findings suggest that EL1 may inhibit GA signaling by enhancing DELLA stability, although SLR1 phosphorylation by EL1 was only shown in vitro.

The discovery of the role of *O*-glycosylation in regulating DELLA activity came from the characterization of the Arabidopsis *spindly* (*spy*) mutants, which partially rescue the GA-deficient dwarf phenotype caused by a GA biosynthesis inhibitor (paclobutrazol) or a mutation (*ga1*), indicating that SPY is a repressor of GA signaling (Jacobsen and Olszewski 1993; Jacobsen et al. 1996; Silverstone et al. 1997b, 2007). Both SPY and its paralog SECRET AGENT (SEC) in Arabidopsis were predicted to be *O*-GlcNAc transferases (OGTs) based on sequence analysis (Olszewski et al. 2010). Both SPY and SEC contain a tetratricopeptide-repeat (TPR) domain and a putative OGT catalytic domain. Surprisingly, electron transfer dissociation (ETD)-MS/MS and in vitro enzyme assays showed that SPY *O*-fucosylates DELLAs and that SEC *O*-GlcNAcylates DELLAs (Zentella et al. 2016, 2017). Genetic analysis and pulldown assays further showed that *O*-fucosylation of DELLA by SPY enhances DELLA binding to TFs (e.g. BZR1 and PIFs), while *O*-GlcNAcylation of DEL L Aby SEC reduces DEL L Aactivity. As OGT serves as a nutrient sensor in metazoans (Hart 2019), it was proposed that *O*-Fuc and *O*-GlcNAc modifications might modulate DELLA activity and plant growth in response to nutrient availability.

(5) Transcriptional induction of *RGL3* by JA signaling

RGL3 transcription is rapidly induced by MYC2, which is a JA signaling activator (Wild et al. 2012). As RGL3 binds to and sequesters the JA signaling repressors JAZs, elevated expression of *RGL3* in response to the JA signal enhances MYC2 activity to promote JA-mediated resistance to necrotrophic pathogens.

DELLA-independent GA responses

Although DELLAs control almost all GA-regulated processes, a few DELLA-independent GA responses have been reported. SPATULA (SPT), a bHLH TF that is unrelated to DELLAs, also inhibits GA-induced cotyledon expansion and fruit growth (Josse et al. 2011; Fuentes et al. 2012). The SPT-repressed cotyledon expansion is independent of light conditions, which is in contrast to DELLA, whose stability is reduced by red light-induced GA biosynthesis. Notably, DELLAs negatively regulate *SPT* transcript accumulation, which provides a balance between the two classes of repressors (Josse et al. 2011). Another DELLA-independent GA response is GA-induced increases in cytosolic Ca^{2+} levels, although it remains to be determined whether this GA response is mediated by GID1 in the cytoplasm (Okada et al. 2017).

GA transport mechanism

Long-distance GA transport

Nearly half a century ago, initial findings were reported on the movement of GA in plants (Zweig et al. 1961; Chin and Lockhart 1965; Hoad and Bowen 1968). These studies confirmed the presence of GA in the phloem sap and its ability to travel through this medium. Since then, ongoing efforts have been made to understand and measure the movement of GA in plants (Hedden and Sponsel 2015; Binenbaum et al. 2018). GA moves within the plant in both upward (acropetal) and downward (basipetal) directions (Proebsting et al. 1992; Bjorklund et al. 2007; Regnault et al. 2015; Lacombe and Achard 2016). The movement of GA is crucial for various developmental processes in plants (Anfang and Shani 2021; Zhang et al. 2023).

Several studies have attempted to identify the mobile form of GA by perturbing GA biosynthesis at different stages of the pathway. In peas, grafting mutant plants deficient in GA biosynthetic enzymes onto wild-type plants led to an increase in GA content in the shoots compared to non-grafted mutant plants. GA analysis revealed that GA_{20} was the major mobile form in pea plants (Proebsting et al. 1992). Similarly, in Arabidopsis, grafting experiments with mutant plants at different stages of GA biosynthesis identified GA_{12} as the major form transported over long distances through the vasculature (Regnault et al. 2015) (Fig. 5). GA_{12} moves through the xylem from roots to shoots and through the phloem from shoots to roots to regulate plant growth (Regnault et al. 2015). The transport activity is most evident in plants that fail to synthesize GA locally. It was furthermore demonstrated that GA_{12} derived in the roots plays a role in regulating the growth of shoots in response to temperature changes in Arabidopsis (Camut et al. 2019). Most recently, it was re-ported that two GA and ABA transporters (NPF2.12 and NPF2.13) are required for shoot-to-root GA_{12} translocation to regulate endodermal root suberization (summarized be-low) (Binenbaum et al. 2023).

GA movement and localization

The synthesis of active GA is a complex and multistep process involving various intermediate compounds (Fig. 2). This complexity makes it challenging to identify the specific tissues or organs where GAs are produced and localized. Analysis of GA biosynthesis reporter lines indicated that while some tissues show colocalization of GA biosynthesis genes and GA perception genes, there are cases where these two groups of genes do not overlap. For example, GA biosyn-thesis genes are not expressed in the aleurone cells of the rice endosperm, but GA signaling genes are (Kaneko et al. 2003). This spatial separation suggests the need for GA movement within the plant. In addition, the expression levels of genes involved in the GA biosynthesis pathway itself do not always align. For instance, the expression of the late-stage GA biosynthesis genes *AtGA3ox1* and *AtGA3ox2* in germinating embryos differs spatially from that of the early GA biosynthesis

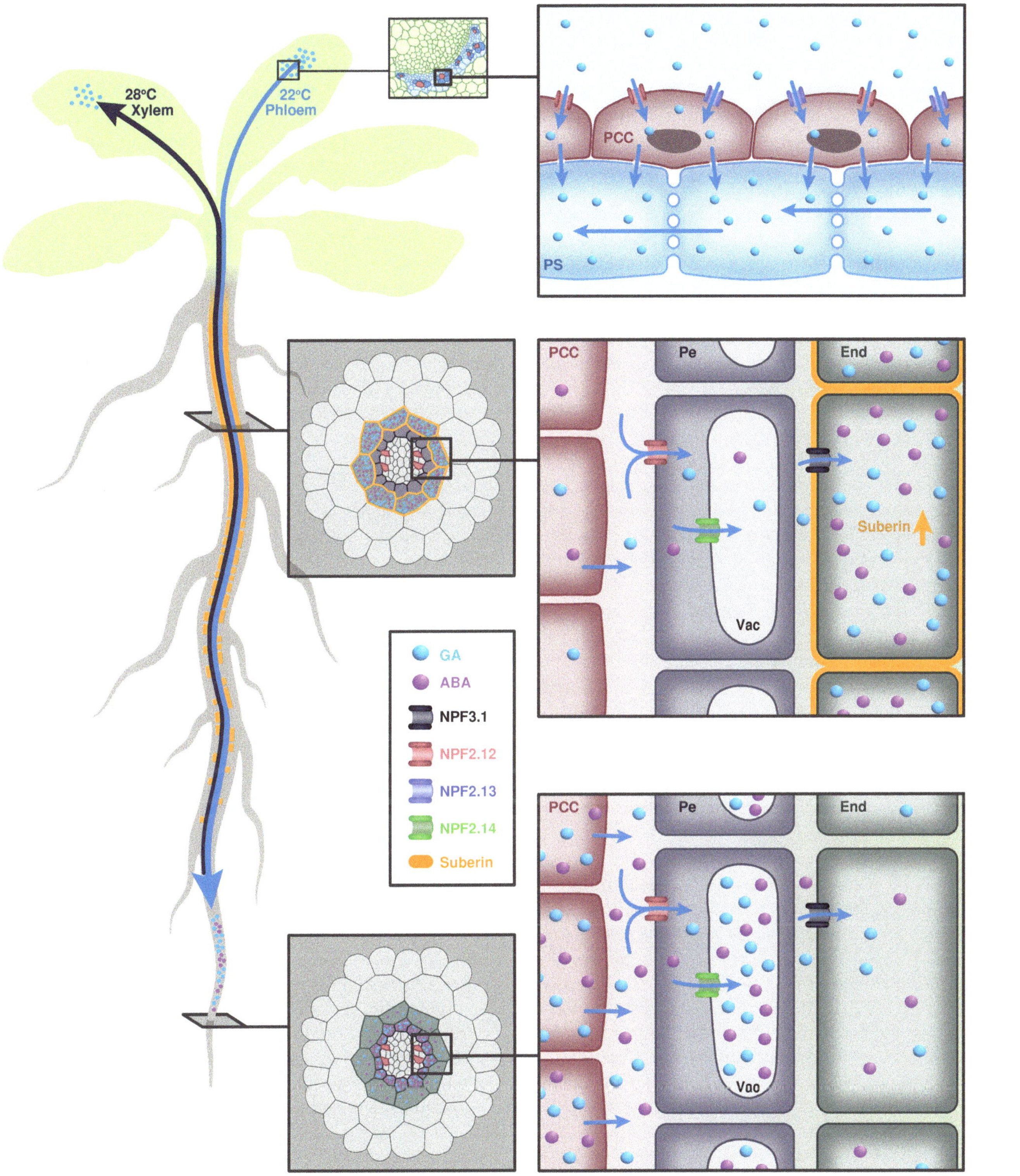

Figure 5. Model of the role of NPF2s activity in regulating shoot-to-root translocation of GA_{12} to promote endodermal suberin deposition. NPF2.12 and NPF2.13 are found in the phloem of the shoot and play vital roles in transporting GA_{12} over long distances from shoot to root. NPF2.12 is active in the membranes of root pericycle cells and facilitates the movement of ABA and GA from the vasculature to the pericycle region. Once inside the pericycle's cytoplasm, NPF2.14 transports these hormones into the vacuole, where they are possibly stored as a reserve for future use. As the root elongates over time and these cells mature, the stored hormones are released from the pericycle vacuole and taken up by the endodermis through the action of NPF3.1. This uptake triggers suberization, a process that forms a protective barrier in the endodermis. In addition, GA_{12} derived in the root, plays a role in regulating the shoot growth in response to ambient temperatures (28°C) in Arabidopsis. PCC, phloem companion cell; PS, phloem sieve; Pe, pericycle; End, endodermis; Vac, vacuole; GA, gibberellin; ABA, abscisic acid.

gene *AtCPS* (Silverstone et al. 1997a; Yamaguchi et al. 2001). Similar patterns are found in different root cell types (Barker et al. 2021). Such differences indicate that the location and movement of GA precursors could play a vital role in regulat-ing GA responses. An interesting example of GA precursor translocation was found in the fern *L. japonicum*, where GA movement is involved in the sex-determining mechan-ism. The study proposed a model in which different stages of prothalli in a colony express different GA-biosynthetic genes, producing specific forms of GA that regulate the for-mation of reproductive structures (Tanaka et al. 2014). In the future, there is a need to measure GA contents (including bioactive GAs and precursors) in different tissues and cell types at the single-cell level to shed light on the GA map with respect to its dynamic movement.

Long and short-distance movement of different GA forms are crucial for plant development (Rizza and Jones 2019; Wexler et al. 2019; Anfang and Shani 2021). For ex-ample, one of the functions of GA is to induce xylem differentiation in the hypocotyl following the floral transition in Arabidopsis. *GA3ox1* mRNA levels increase in the shoot but not the hypocotyl after flowering, sug-gesting GA movement. Mutant plants lacking GA biosyn-thesis, such as the *ga1-3* mutant, exhibit reduced hypocotyl xylem expansion after the flowering stage (Ragni et al. 2011). However, xylem expansion was restored when the mutant plants were grafted onto wild-type root-stocks. This suggests that GA acts as a mobile signal derived from the shoot that triggers xylem expansion in the hypocotyl (Ragni et al. 2011). Similar effects were observed in tobacco (*Nicotiana tabacum*) plants when defoliation occurred, resulting in reduced GA content and growth abnormalities in the stem (Dayan et al. 2012).

Additional reports have described the dependency of certain organs on external sources of GA. For example, GA movement from the embryo scutellum to the aleurone in cereal grains plays a pivotal role in regulating seed germination (Paleg 1960). Upon imbibition, during germination, the scutellum synthesizes GAs, which move to the aleurone layer (Hayashi 1940). In the aleurone, GA triggers the activation of hydrolytic enzymes that break down the stored starch and proteins in the endosperm into simpler forms, providing essential nutrients for the developing embryo (Lovegrove and Hooley 2000; Sun and Gubler 2004). This coordinated GA-mediated movement and response from scutellum to aleurone are fundamental for successful germination and early seedling growth in cereal grains.

In addition, petals rely on anthers as a source of GA for their growth and development (Hu et al. 2008). Studies in Arabidopsis, petunia (*Petunia hybrida*), tobacco, and rice demonstrated that GA produced in the anthers is crucial for pe-tal development (Weiss and Halevy 1989; Itoh et al. 1999, 2001; Hu et al. 2008). These pieces of evidence support the idea that GA acts as a mobile plant hormone and that its movement is necessary for various processes involved in plant growth and development.

In cucumber flowers, specific forms of GA were found to be localized to different floral parts, suggesting their involvement in localized growth regulation. Experiments using deuterated GA provided quantitative support for the production and movement of GA from ovaries to sepals and petals, where it is converted to a bioactive form to regulate organ growth (Lange and Lange 2016). A study characterizing GA biosynthesis sites in roots at the cellular level, coupled with cell-type-specific GA synthesis rescue experiments, indicated GA movement between root cell-files (Barker et al. 2021). Analysis of photocaged bioactive GA_4 released endogenously in Arabidopsis roots allowed the kinetic parameters of its flow to be measured, such as decay length and velocity (Wexler et al. 2019). More comprehensive studies are needed to explore additional developmental stages where GA movement plays a fundamental role by correlating the expression patterns of GA biosynthetic genes with direct measurements of GA levels.

GA transporters

The first evidence of bioactive and regulated GA transport came with the identification of GA transporters from the NPF (Chiba et al. 2015; Saito et al. 2015; David et al. 2016; Tal et al. 2016; Anfang and Shani 2021; Kanstrup and Nour-Eldin 2022). Several NPF transporter proteins, including NPF2.3, NPF2.4, NPF2.5, NPF2.7, NPF2.10, and NPF3.1 have been identified as potential GA transporters using yeast-modified 2-hybrid systems and further confirmed in *Xenopus* oocytes (Chiba et al. 2015; Saito et al. 2015; Wulff et al. 2019). However, the physiological importance of most of these transporters in plants remains to be fully understood. The main challenge in characterizing GA transporters and their physiological function is the lack of apparent GA-mediated phenotypes in the respective mutants. Genetic redundancy plays a substantial role in this shortcoming, as most, if not all, transporter proteins belong to large and robust gene families. Thus, the knockout of one putative GA transporter is compensated for by another family member (Zhang et al. 2023).

Recent research has revealed that a subset of NPF proteins is required for the mechanisms behind long-distance GA transport from shoot to root and its developmental importance (Binenbaum et al. 2023). GA transport plays a critical role in suberin formation in the root. NPF2.12 and NPF2.13 (2 recently identified GA and ABA importers), along with NPF2.14 (a tonoplast importer), coordinate the regulation of suberin formation (Binenbaum et al. 2023). NPF2.12 and NPF2.13 are membrane-localized proteins expressed in leaf phloem companion cells that facilitate the transport of GA_{12} from shoot to root (Fig. 5). Once reaching the root, GA_{12} is converted to GA_4 by the enzymes GA20ox and GA3ox. It is speculated that the bioactive GA and ABA exit the phloem at the phloem unloading zone (Robe and Barberon 2023) located around the root elongation zone (Binenbaum et al. 2023). NPF2.12 is then able to import GA_4 and ABA into the pericycle, and subsequently, the pericycle-specific transporter NPF2.14 transports these phytohormones into the vacuole (Fig. 5) GA and ABA accumulate in the vacuole within the phloem unloading zone located around the root elongation zone, where they are stored dur-ing root maturation and differentiation. Only later in devel-opment are these plant hormones released from the vacuole by an unknown mechanism and are able to be taken into the endodermis by NPF3 to promote suberization. NPF3 is loca-lized to the plasma membrane and imports GAs in a pH-dependent manner (David et al. 2016; Tal et al. 2016) (Fig. 5). These findings indicate that GA and ABA can work in a nonantagonistic manner to regulate plant development. This mechanism highlights the importance of long-distance shoot-to-root movement of GA_{12} and the accumulation of bioactive GA_4 and ABA in the endodermis for regulating endodermal suberization (Binenbaum et al. 2023; Zhang et al. 2023).

In Arabidopsis, the Sugars Will Eventually Be Exported Transporters (SWEET) family members SWEET13 and SWEET14 have been identified as GA transporters (Kanno et al. 2016; Morii et al. 2020). These transporters import GAs, as demonstrated in yeast and oocyte transport assays (Kanno et al. 2016). SWEET13 and SWEET14 function redundantly to regulate anther development, and the application of exogenous GAs to the *sweet13 sweet14* double mutant res-cues the anther's dehiscence defect (Kanno et al. 2016). In rice, OsSWEET3a acts as both a sugar transporter and a GA transporter, playing roles in seed germination and early shoot development (Morii et al. 2020).

GA biosensors and markers

Studies utilizing GA biosensors have provided insights into the transport and localization of GAs, indicating that GAs are highly mobile (Rizza and Jones 2019). Specifically, analysis using the GA perception biosensor GA Perception Sensor 1 (GPS1) revealed that, in the root, the concentration of bioactive GA is highest in the root elongation zone (Rizza et al. 2017). GPS1 represents a pioneering biosensor that utilizes Förster resonance energy transfer (FRET) to detect and track cellular GA levels in planta. This biosensor contains AtGID1C and the N-terminal domain of AtGAI, which are linked to two fluorescent proteins that produce FRET when GA binding to GID1C triggers an intramolecular conformational change. The assessment of a fluorescence emission ratio of nuclear localized-GPS1 (nlsGPS1) enables the precise mapping of endogenous and externally administered GA gradients within various tissue structures at the cellular level. The use of nlsGPS1 live imaging, combined with comprehensive modeling, revealed that a disparity in GA biosynthesis along the roots is accountable for shaping the distribution of GA (Rizza et al. 2021). Another biosensor based on the DELLA protein RGA, named qmRGA (*pRPS5a::RGAmPFYR-VENUS*), provided in planta information on changes in GA responses at the cellular level in the shoot apical meristem, with GA signaling found primarily in cells located between organ primordia (Shi et al. 2021). Furthermore, experiments involving fluorescently labeled bioactive GAs demonstrated the exclusive accumulation of GAs in the root elongation zone (Shani et al. 2013) and in leaf mesophyll cells (Matias-Hernandez et al. 2016), suggesting that GAs move from one tissue to another. The latter process is regulated by two transcription factors, TEMPRANILLO1 (TEM1) and TEM2, which negatively regulate the expression of genes encoding specific GA transporters (GLUCOSINOLATE TRANSPORTER1 (GTR1), NPF3, and NPF2.3) belonging to the NPF family, leading to variable GA accumulation and distribution in mesophyll cells that regulate trichome initiation in the epidermis (Matias-Hernandez et al. 2016).

While the GPS1 FRET biosensor (Rizza et al. 2021) and the qmRGA ratiometric GA signaling biosensor (Shi et al. 2021) report on GA localization based on perception mechanisms, limited progress has been made in generating GA biosensors that are based on transcriptional reporters. Such transcriptional reporters have been widely used in other phytohormone research based on promoters of endogenous phytohormone-induced genes or synthetic transcriptional reporters (Ulmasov et al. 1997; Muller and Sheen 2008; Kim et al. 2011; Okamoto et al. 2013; Liao et al. 2015; Wu et al. 2018). Dayan et al. generated sev-eral β-glucuronidase reporters based on the GA-induced promoters of *EXP1*, *MYB34*, and *GA2OX2* and a synthetic GA-responsive promoter (FK) containing known GA-response *cis*-elements found in promoters of α-amylase genes from cereal crops (Dayan et al. 2012). However, constructing a universal GA reporter that re-flects the broad range of transcriptional regulation (Fig. 3), remains challenging. Such a reporter would need to respond specifically to endogenous GA levels in all tis-sues and cell types, with an appropriate reporter turnover rate. It may be difficult to design GA reporters that re-spond to both DELLA-dependent and -independent pathways.

Despite the progress made, several open questions remain regarding GA transport (see Outstanding questions). One unanswered question is whether there are GA exporters capable of transporting GAs from the cytosol to the apoplast. Currently, no proteins with this function have been identified, but it is believed that such proteins must exist to overcome the GA ion-trapping mechanism. In addition, considering the recent findings regarding GA accumulation in the pericycle vacuole, one could speculate that GA is actively transported out of the vacuole and pericycle cells to reach the endodermis. However, the specific transporters responsible for this process have not yet been identified. Furthermore, the relevance of movement of GA through plasmodesmata and the apoplast remains unclear. A recent discovery demonstrated that the plant hormone ABA travels radially through the plasmodesmata in the root to regulate lateral root branching in response to water stress (air gaps in the soil) (Mehra et al. 2022). Investigating whether a similar mechanism applies to GA in various developmental and

OUTSTANDING QUESTIONS

- How are GA metabolism, transport, and signaling activities regulated in different cell/tissue types to coordinate plant growth and development?
- What is the role of post-transcriptional regulation of GA-metabolic enzymes in determining GA concentration?
- How do DELLAs interact with a myriad of distinct classes of TFs/transcriptional regulators (TRs), and does the tissue/cell-type specific expression of DELLAs and their interacting TFs/TRs determine the selective regulation of a subset of target genes?
- Do GA exporters exist and, if so, what are their developmental roles and specificity?

environmental responses would be intriguing. Further research is required to fully understand the transport and localization of GAs in plants.

Concluding remarks

In the last three decades, substantial progress has been made in elucidating the regulation of GA metabolism and the molecular mechanism of GA perception and early GA signaling. The central role of DELLAs as integrators of multiple signaling pathways has clearly been demonstrated, although the specificity of these key growth regulators in distinct tissue/cell-type requires further investigation (see Outstanding Questions). Recently identified GA transporters and the development of GA biosensors are important advances toward understanding how GA regulates plant growth and development in response to internal and external cues (see Outstanding Questions). The development of methods for analyzing GA content at the single-cell level in combination with existing molecular/genomics/proteomics tools will allow us to achieve this goal in the future.

Acknowledgments

We thank Steve Thomas and Graham Shephard, Rothamsted Research, for providing the plant photo in Fig. 1.

Author contributions

E.S., P.H., and T.-p.S. wrote and edited the manuscript.

Funding

This work was supported by the National Institutes of Health (GM100051 and GM150029 to T.-p.S.), the National Science Foundation (MCB-1818161 to T.-p.S.), the National Institute of Food and Agriculture (2023-67013-39532 to T.-p.S.), the Czech Science Foundation (20-17984S to P.H.), the European Regional Developmental Fund "Centre for Experimental Plant Biology" (CZ.02.1.01/0.0/0.0/16_019/0000738 to P.H.), the Israel Science Foundation (2378/19 and 3419/20 to E.S.), the Human Frontier Science Program (HFSP-LIY000540/2020 to E.S) and the European Research Council (ERC) (757683-RobustHormoneTrans and 101118769-HYDROSENSING to ES).

Conflict of interest statement. None declared.

Data availability

No new data were generated in this review.

References

Aach H, Böse G, Graebe JE. *ent*-Kaurene biosynthesis in a cell-free system from wheat (*Triticum aestivum* L.) seedlings and the localization of *ent*-kaurene synthetase in plastids of 3 species. Planta. 1995:**197**(2): 333–342. https://doi.org/10.1007/BF00202655

An F, Zhang X, Zhu Z, Ji Y, He W, Jiang Z, Li M, Guo H. Coordinated regulation of apical hook development by gibberellins and ethylene in etiolated Arabidopsis seedlings. Cell Res. 2012:**22**(5):915–927. https://doi.org/10.1038/cr.2012.29

Anfang M, Shani E. Transport mechanisms of plant hormones. Curr Opin Plant Biol. 2021:**63**:102055. https://doi.org/10.1016/j.pbi.2021.102055

Arana MV, Marin-de la Rosa N, Maloof JN, Blázquez MA, Alabadí D. Circadian oscillation of gibberellin signaling in Arabidopsis. Proc Natl Acad Sci U S A. 2011:**108**(22):9292–9297. https://doi.org/10.1073/pnas.1101050108

Arnaud N, Girin T, Sorefan K, Fuentes S, Wood TA, Lawrenson T, Sablowski R, Østergaard L. Gibberellins control fruit patterning in *Arabidopsis thaliana*. Genes Dev. 2010:**24**(19):2127–2132. https://doi.org/10.1101/gad.593410

Bai MY, Shang JX, Oh E, Fan M, Bai Y, Zentella R, Sun TP, Wang Z-Y. Brassinosteroid, gibberellin, and phytochrome signalling pathways impinge on a common transcription module in Arabidopsis. Nat Cell Biol. 2012:**14**(8):810–817. https://doi.org/10.1038/ncb2546

Barker R, Fernandez Garcia MN, Powers SJ, Vaughan S, Bennett MJ, Phillips AL, Thomas SG, Hedden P. Mapping sites of gibberellin biosynthesis in the Arabidopsis root tip. New Phytol. 2021:**229**(3): 1521–1534. https://doi.org/10.1111/nph.16967

Bearder JR, MacMillan J, Phinney BO. Fungal products. Part XIV. Metabolic pathways from *ent*-kaurenoic acid to fungal gibberellins in mutant B1-41a of *Gibberella fujikuroi*. J Chem Soc, Perkin Trans. 1975:**1**(8):721–726. https://doi.org/10.1039/p19750000721

Bensen RJ, Johal GS, Crane VC, Tossberg JT, Schnable PS, Meeley RB, Briggs SP. Cloning and characterization of the maize *AN1* gene. Plant Cell. 1995:**7**(1):75–84. https://doi.org/10.1105/tpc.7.1.75

Bhattacharya A, Kourmpetli S, Ward DA, Thomas SG, Gong F, Powers SJ, Carrera E, Taylor B, Gonzalez FND, Tudzynski B, et al. Characterization of the fungal gibberellin desaturase as a 2-oxoglutarate-dependent dioxygenase and its utilization for enhancing plant growth. Plant Physiol. 2012:**160**(2):837–845. https://doi.org/10.1104/pp.112.201756

Binenbaum J, Weinstain R, Shani E. Gibberellin localization and transport in plants. Trends Plant Sci. 2018:**23**(5):410–421. https://doi.org/10.1016/j.tplants.2018.02.005

Binenbaum J, Wulff N, Camut L, Kiradjiev K, Anfang M, Tal I, Vasuki H, Zhang Y, Sakvarelidze-Achard L, Davière J-M, et al. Gibberellin and abscisic acid transporters facilitate endodermal suberin formation in Arabidopsis. Nat Plants. 2023:**9**:785–802. https://doi.org/10.1038/s41477-023-01391-3

Binks R, MacMillan J, Pryce RJ. Plant hormones—VIII: combined gas chromatography-mass spectrometry of methyl esters of gibberellins A_1 to A_{24} and their trimethylsilyl ethers. Phytochemistry. 1969:**8**(1): 271–284. https://doi.org/10.1016/S0031-9422(00)85825-2

Birch AJ, Rickards RW, Smith H. The biosynthesis of gibberellic acid. Proc Chem Soc. 1958:192–193.

Björklund S, Antti H, Uddestrand I, Moritz T, Sundberg B. Cross-talk between gibberellin and auxin in development of Populus wood: gibberellin stimulates polar auxin transport and has a common transcriptome with auxin. Plant J. 2007:**52**(3):499–511. https://doi.org/10.1111/j.1365-313X.2007.03250.x

Blanco-Touriñán N, Legris M, Minguet EG, Costigliolo-Rojas C, Nohales MA, Iniesto E, García-León M, Pacín M, Heucken N, Bloemeir T, et al. COP1 destabilizes DELLA proteins in *Arabidopsis*. Proc Natl Acad Sci U S A. 2020a:**117**(24):13792–13799. https://doi.org/10.1073/pnas.1907969117

Blanco-Touriñán N, Serrano-Mislata A, Alabadí D. Regulation of DELLA proteins by post-translational modifications. Plant Cell Physiol. 2020b:**61**(11):1891–1901. https://doi.org/10.1093/pcp/pcaa113

Börner A, Plaschke J, Korzun V, Worland AJ. The relationships between the dwarfing genes of wheat and rye. Euphytica. 1996:**89**(1): 69–75. https://doi.org/10.1007/BF00015721

Bouré N, Arnaud N. Molecular GA pathways as conserved integrators for adaptive responses. Plant Biol. 2023:**25**(5):649–660. https://doi.org/10.1111/plb.13549

Brian PW. The effects of some microbial metabolic products on plant growth. Symp Soc Exp Biol. 1957:**11**:166–181.

Brian PW, Elson GW, Hemming HG, Radley M. The plant growth promoting properties of gibberellic acid, a metabolic product of the fungus, *Gibberella fujikuroi*. J Sci Food Agric. 1954:**5**(12):602–612. https://doi.org/10.1002/jsfa.2740051210

Campanaro A, Battaglia R, Galbiati M, Sadanandom A, Tonelli C, Conti L. SUMO proteases OTS1 and 2 control filament elongation through a DELLA-dependent mechanism. Plant Reprod. 2016:**29**(4): 287–290. https://doi.org/10.1007/s00497-016-0292-8

Camut L, Gallova B, Jilli L, Sirlin-Josserand M, Carrera E, Sakvarelidze-Achard L, Ruffel S, Krouk G, Thomas SG, Hedden P, et al. Nitrate signaling promotes plant growth by upregulating gibberellin biosynthesis and destabilization of DELLA proteins. Curr Biol. 2021:**31**(22):4971–4982.e4974. https://doi.org/10.1016/j.cub.2021.09.024

Camut L, Regnault T, Sirlin-Josserand M, Sakvarelidze-Achard L, Carrera E, Zumsteg J, Heintz D, Leonhardt N, Lange MJP, Lange T. Root-derived GA_{12} contributes to temperature-induced shoot growth in Arabidopsis. Nat Plants. 2019:**5**(12):1216–1221. https://doi.org/10.1038/s41477-019-0568-8

Chandler PM, Marion-Poll A, Ellis M, Gubler F. Mutants at the *Slender1* locus of barley cv Himalaya: molecular and physiological characterization. Plant Physiol. 2002:**129**(1):181–190. https://doi.org/10.1104/pp.010917

Chiang HH, Hwang I, Goodman HM. Isolation of the *Arabidopsis GA4* locus. Plant Cell. 1995:**7**(2):195–201. https://doi.org/10.1105/tpc.7.2.195

Chiba Y, Shimizu T, Miyakawa S, Kanno Y, Koshiba T, Kamiya Y, Seo M. Identification of *Arabidopsis thaliana* NRT1/PTR family (NPF) proteins capable of transporting plant hormones. J Plant Res. 2015:**128**(4):679–686. https://doi.org/10.1007/s10265-015-0710-2

Chin TY, Lockhart JA. Translocation of applied gibberellin in bean seedlings. Am J Bot. 1965:**52**(8):828–833. https://doi.org/10.1002/j.1537-2197.1965.tb07254.x

Colebrook EH, Thomas SG, Phillips AL, Hedden P. The role of gibberellin signalling in plant responses to abiotic stress. J Exp Biol. 2014:**217**(1):67–75. https://doi.org/10.1242/jeb.089938

Conti L, Nelis S, Zhang C, Woodcock A, Swarup R, Galbiati M, Tonelli C, Napier R, Hedden P, Bennett M, et al. Small ubiquitin-like modifier protein SUMO enables plants to control growth independently of the phytohormone gibberellin. Dev Cell. 2014:**28**(1): 102–110. https://doi.org/10.1016/j.devcel.2013.12.004

Cowling RJ, Kamiya Y, Seto H, Harberd NP. Gibberellin dose-response regulation of *GA4* gene transcript levels in Arabidopsis. Plant Physiol. 1998:**117**(4):1195–1203. https://doi.org/10.1104/pp.117.4.1195

Crocco CD, Locascio A, Escudero CM, Alabadí D, Blázquez MA, Botto JF. The transcriptional regulator BBX24 impairs DELLA activity to promote shade avoidance in Arabidopsis thaliana. Nat Commun. 2015:**6**(1):6202. https://doi.org/10.1038/ncomms7202

Cross BE, Hanson JR, Galt RHB. The biosynthesis of the gibberellins. Part I. (-)-kaurene as a precursor of gibberellic acid. J Chem Soc. 1964:295–300. https://doi.org/10.1039/JR9640000295

Curtis PJ, Cross BE. Gibberellic acid—a new metabolite from the culture filtrates of *Gibberella fujikuroi*. Chem Ind. 1954:1066.

Dai C, Xue HW. Rice early flowering1, a CKI, phosphorylates DELLA protein SLR1 to negatively regulate gibberellin signalling. EMBO J. 2010:**29**(11):1916–1927. https://doi.org/10.1038/emboj.2010.75

David LC, Berquin P, Kanno Y, Seo M, Daniel-Vedele F, Ferrario-Méry S. N availability modulates the role of NPF3.1, a gibberellin transporter, in GA-mediated phenotypes in Arabidopsis. Planta. 2016:**244**(6):1315–1328. https://doi.org/10.1007/s00425-016-2588-1

Davière JM, Achard P. A pivotal role of DELLAs in regulating multiple hormone signals. Mol Plant. 2016:**9**(1):10–20. https://doi.org/10.1016/j.molp.2015.09.011

Davière JM, Wild M, Regnault T, Baumberger N, Eisler H, Genschik P, Achard P. Class I TCP-DELLA interactions in inflorescence shoot apex determine plant height. Curr Biol. 2014:**24**(16):1923–1928. https://doi.org/10.1016/j.cub.2014.07.012

Dayan J, Voronin N, Gong F, Sun TP, Hedden P, Fromm H, Aloni R. Leaf-induced gibberellin signaling is essential for internode elongation, cambial activity, and fiber differentiation in tobacco stems. Plant Cell. 2012:**24**(1):66–79. https://doi.org/10.1105/tpc.111.093096

de Lucas M, Daviere JM, Rodríguez-Falcón M, Pontin M, Iglesias-Pedraz JM, Lorrain S, Fankhauser C, Blázquez MA, Titarenko E, Prat S. A molecular framework for light and gibberellin control of cell elongation. Nature. 2008:**451**(7177):480–484. https://doi.org/10.1038/nature06520

Dennis DT, West CA. Biosynthesis of gibberellins. III. Conversion of (-)-kaurene to (-)-kauren-19-oic acid in endosperm of *Echinocystis macrocarpa* greene. J Biol Chem. 1967:**242**(14):3293–3300. https://doi.org/10.1016/S0021-9258(18)95909-0

Dill A, Jung H-S, Sun TP. The DELLA motif is essential for gibberellin-induced degradation of RGA. Proc Natl Acad Sci U S A. 2001:**98**(24):14162–14167. https://doi.org/10.1073/pnas.251534098

Dill A, Thomas SG, Hu J, Steber CM, Sun TP. The Arabidopsis F-box protein SLEEPY1 targets GA signaling repressors for GA-induced degradation. Plant Cell. 2004:**16**(6):1392–1405. https://doi.org/10.1105/tpc.020958

Djakovic-Petrovic T, de Wit M, Voesenek LA, Pierik R. DELLA protein function in growth responses to canopy signals. Plant J. 2007:**51**(1): 117–126. https://doi.org/10.1111/j.1365-313X.2007.03122.x

Duncan JD, West CA. Properties of kaurene synthetase from *Marah macrocarpus* endosperm—evidence for the participation of separate but interacting enzymes. Plant Physiol. 1981:**68**(5):1128–1134. https://doi.org/10.1104/pp.68.5.1128

Evans R, Hanson JR. Studies in terpenoid biosynthesis. Part XIII. Biosynthetic relationship of gibberellins in *Gibberella fujikuroi*. J Chem Soc Perkin Trans. 1975:**1**(7):663–666. https://doi.org/10.1039/p19750000663

Fall RR, West CA. Purification and properties of kaurene synthetase from *Fusarium moniliforme*. J Biol Chem. 1971:**246**(22):6913–6928. https://doi.org/10.1016/S0021-9258(19)45933-4

Feng S, Martinez C, Gusmaroli G, Wang Y, Zhou J, Wang F, Chen L, Yu L, Iglesias-Pedraz JM, Kircher S, et al. Coordinated regulation of *Arabidopsis thaliana* development by light and gibberellins. Nature. 2008:**451**(7177):475–479. https://doi.org/10.1038/nature06448

Fonouni-Farde C, Tan S, Baudin M, Brault M, Wen J, Mysore KS, Niebel A, Frugier F, Diet A. DELLA-mediated gibberellin signalling regulates nod factor signalling and rhizobial infection. Nat Commun. 2016:**7**(1):12636. https://doi.org/10.1038/ncomms12636

Foster CA. Slender: an accelerated extension growth mutant of barley. Barley Genet Newsl. 1977:**7**:24–27.

Fuentes S, Ljung K, Sorefan K, Alvey E, Harberd NP, Østergaard L. Fruit growth in Arabidopsis occurs via DELLA-dependent and DELLA-independent gibberellin responses. Plant Cell. 2012:**24**(10): 3982–3996. https://doi.org/10.1105/tpc.112.103192

Fukazawa J, Teramura H, Murakoshi S, Nasuno K, Nishida N, Ito T, Yoshida M, Kamiya Y, Yamaguchi S, Takahashi Y. DELLAs function as coactivators of GAI-ASSOCIATED FACTOR1 in regulation of gibberellin homeostasis and signaling in Arabidopsis. Plant Cell. 2014:**26**(7):2920–2938. https://doi.org/10.1105/tpc.114.125690

Gallego-Bartolomé J, Minguet EG, Grau-Enguix F, Abbas M, Locascio A, Thomas SG, Alabadí D, Blázquez MA. Molecular mechanism for the interaction between gibberellin and brassinosteroid signaling pathways in Arabidopsis. Proc Natl Acad Sci U S A. 2012:**109**(33): 13446–13451. https://doi.org/10.1073/pnas.1119992109

Goff SA, Ricke D, Lan TH, Presting G, Wang R, Dunn M, Glazebrook J, Sessions A, Oeller P, Varma H, et al. A draft sequence of the rice genome (*Oryza sativa* L. ssp. *japonica*). Science. 2002:**296**(5565): 92–100. https://doi.org/10.1126/science.1068275

Gomez MD, Ventimilla D, Sacristan R, Perez-Amador MA. Gibberellins regulate ovule integument development by interfering with the transcription factor ATS. Plant Physiol. 2016:**172**(4): 2403–2415. https://doi.org/10.1104/pp.16.01231

Graebe JE, Bowen DH, MacMillan J. Conversion of mevalonic acid into gibberellin A_{12}-aldehyde in a cell-free system from *Cucurbita pepo*. Planta. 1972:**102**(3):261–271. https://doi.org/10.1007/BF00386896

Graebe JE, Dennis DT, Upper CD, West CA. Biosynthesis of gibberellins. I. Biosynthesis of (-)-kaurene, (-)-kauren-19-ol and *trans*-geranylgeraniol in endosperm nucellus of *Echinocystis macrocarpa* greene. J Biol Chem. 1965:**240**(4):1847–1854. https://doi.org/10.1016/S0021-9258(18)97516-2

Graebe JE, Hedden P, Gaskin P, MacMillan J. Biosynthesis of a C_{19}-gibberellin from mevalonic acid in a cell-free system from a higher plant. Planta. 1974a:**120**(3):307–309. https://doi.org/10.1007/BF00390299

Graebe JE, Hedden P, Gaskin P, MacMillan J. Biosynthesis of gibberellins A_{12}, A_{15}, A_{24}, A_{36} and A_{37} by a cell-free system from *Cucurbita maxima*. Phytochem. 1974b:**13**(8):1433–1440. https://doi.org/10.1016/0031-9422(74)80304-3

Graebe JE, Hedden P, Rademacher W. Gibberellin biosynthesis. In: **Lenton JR**, editors. Gibberellins—chemistry, physiology and use, vol 5. Wantage: British Plant Growth Regulator Group; 1980. p. 31–47.

Griffiths J, Murase K, Rieu I, Zentella R, Zhang ZL, Powers SJ, Gong F, Phillips AL, Hedden P, Sun TP, et al. Genetic characterization and functional analysis of the GID1 gibberellin receptors in Arabidopsis. Plant Cell. 2006:**18**(12):3399–3414. https://doi.org/10.1105/tpc.106.047415

Grove JF. The gibberellins. Q Rev. 1961:**15**(1):56–70. https://doi.org/10.1039/qr9611500056

Hart GW. Nutrient regulation of signaling and transcription. J Biol Chem. 2019:**294**(7):2211–2231. https://doi.org/10.1074/jbc.AW119.003226

Hasson EP, West CA. Properties of the system for the mixed-function oxidation of kaurene and kaurene derivatives in microsomes of the immature seed of *Marah macrocarpus*: cofactor requirements. Plant Physiol. 1976:**58**(4):473–478. https://doi.org/10.1104/pp.58.4.473

Hayashi T. Biochemical studies on 'bakanae' fungus of the rice. Part VI. Effect of gibberellin on the activity of amylase in germinated cereal grains. J Agri Chem Soc Japan. 1940:**16**:531–538.

He J, Chen QW, Xin PY, Yuan J, Ma YH, Wang XM, Xu MM, Chu JF, Peters RJ, Wang GD. CYP72A enzymes catalyse 13-hydrolyzation of gibberellins. Nat Plants. 2019:**5**(10):1057–1065. https://doi.org/10.1038/s41477-019-0511-z

Hedden P. The current status of research on gibberellin biosynthesis. Plant Cell Physiol. 2020:**61**(11):1832–1849. https://doi.org/10.1093/pcp/pcaa092

Hedden P, Croker SJ. Regulation of gibberellin biosynthesis in maize seedlings. In: **Karssen CM, Van Loon LC, Vreugdenhil D**, editors. Progress in plant growth regulation: proceedings of the 14th international conference on plant growth substances. Dordrecht: Kluwer; 1992. p. 534–544.

Hedden P, Graebe JE. Cofactor requirements for the soluble oxidases in the metabolism of C_{20}-gibberellins. J Plant Growth Regul. 1982:**1**: 105–116.

Hedden P, Phillips AL, Rojas MC, Carrera E, Tudzynski B. Gibberellin biosynthesis in plants and fungi: a case of convergent evolution? J Plant Growth Regul. 2001:**20**(4):319–331. https://doi.org/10.1007/s003440010037

Hedden P, Phinney BO. Comparison of *ent*-kaurene and *ent*-isokaurene synthesis in cell-free systems from etiolated shoots of normal and *dwarf-5* maize seedlings. Phytochem. 1979:**18**(9): 1475–1479. https://doi.org/10.1016/S0031-9422(00)98478-4

Hedden P, Sponsel V. A century of gibberellin research. J Plant Growth Regul. 2015:**34**(4):740–760. https://doi.org/10.1007/s00344-015-9546-1

Hedden P, Thomas SG. Gibberellin biosynthesis and its regulation. Biochem J. 2012:**444**(1):11–25. https://doi.org/10.1042/BJ20120245

Helliwell CA, Chandler PM, Poole A, Dennis ES, Peacock WJ. The CYP88A cytochrome P450, *ent*-kaurenoic acid oxidase, catalyzes three steps of the gibberellin biosynthesis pathway. Proc Natl Acad Sci U S A. 2001a:**98**(4):2065–2070. https://doi.org/10.1073/pnas.98.4.2065

Helliwell CA, Sheldon CC, Olive MR, Walker AR, Zeevaart JA, Peacock WJ, Dennis ES. Cloning of the *Arabidopsis ent*-kaurene oxidase gene *GA3*. Proc Natl Acad Sci U S A. 1998:**95**(15):9019–9024. https://doi.org/10.1073/pnas.95.15.9019

Helliwell CA, Sullivan JA, Mould RM, Gray JC, Peacock WJ, Dennis ES. A plastid envelope location of *Arabidopsis ent*-kaurene oxidase links the plastid and endoplasmic reticulum steps of the gibberellin biosynthesis pathway. Plant J. 2001b:**28**(2):201–208. https://doi.org/10.1046/j.1365-313X.2001.01150.x

Hernández-García J, Briones-Moreno A, Blázquez MA. Origin and evolution of gibberellin signaling and metabolism in plants. Semin Cell Dev Biol. 2021:**109**:46–54. https://doi.org/10.1016/j.semcdb.2020.04.009

Hernández-García J, Briones-Moreno A, Dumas R, Blázquez MA. Origin of gibberellin-dependent transcriptional regulation by molecular exploitation of a transactivation domain in DELLA proteins. Mol Biol Evol. 2019:**36**(5):908–918. https://doi.org/10.1093/molbev/msz009

Hirano K, Asano K, Tsuji H, Kawamura M, Mori H, Kitano H, Ueguchi-Tanaka M, Matsuoka M. Characterization of the molecular mechanism underlying gibberellin perception complex formation in rice. Plant Cell. 2010:**22**(8):2680–2696. https://doi.org/10.1105/tpc.110.075549

Hoad G, Bowen M. Evidence for gibberellin-like substances in phloem exudate of higher plants. Planta. 1968:**82**(1):22–32. https://doi.org/10.1007/BF00384695

Hou X, Hu WW, Shen L, Lee LY, Tao Z, Han JH, Yu H. Global identification of DELLA target genes during Arabidopsis flower development. Plant Physiol. 2008:**147**(3):1126–1142. https://doi.org/10.1104/pp.108.121301

Hou X, Lee LY, Xia K, Yan Y, Yu H. DELLAs modulate jasmonate signaling via competitive binding to JAZs. Dev Cell. 2010:**19**(6): 884–894. https://doi.org/10.1016/j.devcel.2010.10.024

Hou X, Zhou J, Liu C, Liu L, Shen L, Yu H. Nuclear factor Y-mediated H3K27me3 demethylation of the *SOC1* locus orchestrates flowering responses of Arabidopsis. Nat Commun. 2014:**5**(1):4601. https://doi.org/10.1038/ncomms5601

Hu J, Mitchum MG, Barnaby N, Ayele BT, Ogawa M, Nam E, Lai WC, Hanada A, Alonso JM, Ecker JR, et al. Potential sites of bioactive gibberellin production during reproductive growth in Arabidopsis. Plant Cell. 2008:**20**(2):320336. https://doi.org/10.1105/tpc.107.057752

Hu Y, Zhou L, Huang M, He X, Yang Y, Liu X, Li Y, Hou X. Gibberellins play an essential role in late embryogenesis of Arabidopsis. Nat Plants. 2018:**4**(5):289–298. https://doi.org/10.1038/s41477-018-0143-8

Hyun Y, Richter R, Vincent C, Martinez-Gallegos R, Porri A, Coupland G. Multi-layered regulation of SPL15 and cooperation with SOC1 integrate endogenous flowering pathways at the Arabidopsis shoot meristem. Dev Cell. 2016:**37**(3):254–266. https://doi.org/10.1016/j.devcel.2016.04.001

Ikeda A, Ueguchi-Tanaka M, Sonoda Y, Kitano H, Koshioka M, Futsuhara Y, Matsuoka M, Yamaguchi J. *Slender* rice, a constitutive gibberellin response mutant is caused by a null mutation of the *SLR1* gene, an ortholog of the height-regulating gene *GAI/RGA/RHT/D8*. Plant Cell. 2001:**13**(5):999–1010. https://doi.org/10.1105/tpc.13.5.999

Ingram TJ, Reid JB, Murfet IC, Gaskin P, Willis CL, MacMillan J. Internode length in *Pisum*: the *Le* gene controls the 3β-hydroxylation of gibberellin A_{20} to gibberellin A_1. Planta. 1984:**160**(5):455–463. https://doi.org/10.1007/BF00429763

Itoh H, Sasaki A, Ueguchi-Tanaka M, Ishiyama K, Kobayashi M, Hasegawa Y, Minami E, Ashikari M, Matsuoka M. Dissection of the phosphorylation of rice DELLA protein, SLENDER RICE1. Plant Cell Physiol. 2005:**46**(8):1392–1399. https://doi.org/10.1093/pcp/pci152

Itoh H, Tanaka-Ueguchi M, Kawaide H, Chen X, Kamiya Y, Matsuoka M. The gene encoding tobacco gibberellin 3β-hydroxylase is expressed at the site of GA action during stem elongation and flower organ development. Plant J. 1999:**20**(1): 15–24. https://doi.org/10.1046/j.1365-313X.1999.00568.x

Itoh H, Ueguchi-Tanaka M, Sentoku N, Kitano H, Matsuoka M, Kobayashi M. Cloning and functional analysis of two gibberellin 3β-hydroxylase genes that are differently expressed during the growth of rice. Proc Natl Acad Sci U S A. 2001:**98**(15):8909–8914. https://doi.org/10.1073/pnas.141239398

Jacobsen SE, Binkowski KA, Olszewski NE. SPINDLY, a tetratricopeptide repeat protein involved in gibberellin signal transduction in *Arabidopsis*. Proc Natl Acad Sci U S A. 1996:**93**(17):9292–9296. https://doi.org/10.1073/pnas.93.17.9292

Jacobsen SE, Olszewski NE. Mutations at the *SPINDLY* locus of Arabidopsis alter gibberellin signal transduction. Plant Cell. 1993:**5**(8):887–896. https://doi.org/10.1105/tpc.5.8.887

Jasinski S, Tattersall A, Piazza P, Hay A, Martinez-Garcia JF, Schmitz G, Theres K, McCormick S, Tsiantis M. *PROCERA* encodes a DELLA protein that mediates control of dissected leaf form in tomato. Plant J. 2008:**56**(4):603–612. https://doi.org/10.1111/j.1365-313X.2008.03628.x

Jin Y, Liu H, Luo D, Yu N, Dong W, Wang C, Zhang X, Dai H, Yang J, Wang E. DELLA proteins are common components of symbiotic rhizobial and mycorrhizal signalling pathways. Nat Commun. 2016:**7**(1): 12433. https://doi.org/10.1038/ncomms12433

Josse EM, Gan Y, Bou-Torrent J, Stewart KL, Gilday AD, Jeffree CE, Vaistij FE, Martínez-García JF, Nagy F, Graham IA, et al. A DELLA in disguise: SPATULA restrains the growth of the developing Arabidopsis seedling. Plant Cell. 2011:**23**(4):1337–1351. https://doi.org/10.1105/tpc.110.082594

Kamiya Y, Graebe JE. The biosynthesis of all major pea gibberellins in a cell-free system from *Pisum sativum*. Phytochem. 1983:**22**(3): 681–689. https://doi.org/10.1016/S0031-9422(00)86962-9

Kaneko M, Itoh H, Inukai Y, Sakamoto T, Ueguchi-Tanaka M, Ashikari M, Matsuoka M. Where do gibberellin biosynthesis and gibberellin signaling occur in rice plants? Plant J. 2003:**35**(1): 104–115. https://doi.org/10.1046/j.1365-313X.2003.01780.x

Kanno Y, Oikawa T, Chiba Y, Ishimaru Y, Shimizu T, Sano N, Koshiba T, Kamiya Y, Ueda M, Seo M. AtSWEET13 and AtSWEET14 regulate gibberellin-mediated physiological processes. Nat Commun. 2016:**7**(1):13245. https://doi.org/10.1038/ncomms13245

Kanstrup C, Nour-Eldin HH. The emerging role of the nitrate and peptide transporter family: Npf in plant specialized metabolism. Curr Opin Plant Biol. 2022:**68**:102243. https://doi.org/10.1016/j.pbi.2022.102243

Kasahara H, Hanada A, Kuzuyama T, Takagi M, Kamiya Y, Yamaguchi S. Contribution of the mevalonate and methylerythritol phosphate pathways to the biosynthesis of gibberellins in *Arabidopsis*. J Biol Chem. 2002:**277**(47):45188–45194. https://doi.org/10.1074/jbc.M208659200

Kim TH, Hauser F, Ha T, Xue S, Böhmer M, Nishimura N, Munemasa S, Hubbard K, Peine N, Lee BH, et al. Chemical genetics reveals negative regulation of abscisic acid signaling by a plant immune response pathway. Curr Biol. 2011:**21**(11):990–997. https://doi.org/10.1016/j.cub.2011.04.045

Koornneef M, Elgersma A, Hanhart CJ, van Loenen MEP, van Rijn L, Zeevaart JAD. A gibberellin insensitive mutant of *Arabidopsis thaliana*. Physiol Plant. 1985:**65**(1):33–39. https://doi.org/10.1111/j.1399-3054.1985.tb02355.x

Koornneef M, van der Veen JH. Induction and analysis of gibberellin sensitive mutants in *Arabidopsis thaliana* (L.) heynh. Theor Appl Genet. 1980:**58**(6):257–263. https://doi.org/10.1007/BF00265176

Kurosawa E. Experimental studies on the nature of the substance excreted by the 'bakanae' fungus. Transa Nat Hist Soc Formosa. 1926:**16**:213–227.

Lacombe B, Achard P. Long-distance transport of phytohormones through the plant vascular system. Curr Opin Plant Biol. 2016:**34**: 1–8. https://doi.org/10.1016/j.pbi.2016.06.007

Lang A. Bolting and flowering in biennial *Hyoscyamus niger*, induced by gibberellic acid. Plant Physiol. 1956:**31**(suppl):35.

Lange T. Purification and partial amino-acid-sequence of gibberellin 20-oxidase from *Cucurbita maxima* L. endosperm. Planta. 1994: **195**(1):108–115. https://doi.org/10.1007/BF00206298

Lange T, Hedden P, Graebe JE. Expression cloning of a gibberellin 20-oxidase, a multifunctional enzyme involved in gibberellin biosynthesis. Proc Natl Acad Sci U S A. 1994:**91**(18):8552–8556. https://doi.org/10.1073/pnas.91.18.8552

Lange T, Kramer C, Lange MJP. The class III gibberellin 2-oxidases *AtGA2ox9* and *AtGA2ox10* contribute to cold stress tolerance and fertility. Plant Physiol. 2020:**184**(1):478–486. https://doi.org/10.1104/pp.20.00594

Lange MJP, Lange T. Ovary-derived precursor gibberellin A_9 is essential for female flower development in cucumber. Development. 2016: **143**(23):4425–4429. https://doi.org/10.1242/dev.135947

Lantzouni O, Alkofer A, Falter-Braun P, Schwechheimer C. GROWTH-REGULATING FACTORS interact with DELLAs and regulate growth in cold stress. Plant Cell. 2020:**32**(4):1018–1034. https://doi.org/10.1105/tpc.19.00784

Lawit SJ, Wych HM, Xu D, Kundu S, Tomes DT. Maize DELLA proteins dwarf plant8 and dwarf plant9 as modulators of plant development. Plant Cell Physiol. 2010:**51**(11):1854–1868. https://doi.org/10.1093/pcp/pcq153

Lee DJ, Zeevaart JAD. Regulation of gibberellin 20-oxidase1 expression in spinach by photoperiod. Planta. 2007:**226**(1):35–44. https://doi.org/10.1007/s00425-006-0463-1

Lester DR, Ross JJ, Davies PJ, Reid JB. Mendel's stem length gene (*Le*) encodes a gibberellin 3 beta-hydroxylase. Plant Cell. 1997:**9**(8): 1435–1443. https://doi.org/10.1105/tpc.9.8.1435

Lester DR, Ross JJ, Smith JJ, Elliott RC, Reid JB. Gibberellin 2-oxidation and the *SLN* gene of *Pisum sativum*. Plant J. 1999:**19**(1):65–73. https://doi.org/10.1046/j.1365-313X.1999.00501.x

Li S, Tian Y, Wu K, Ye Y, Yu J, Zhang J, Liu Q, Hu M, Li H, Tong Y, et al. Modulating plant growth-metabolism coordination for sustainable agriculture. Nature. 2018:**560**(7720):595–600. https://doi.org/10.1038/s41586-018-0415-5

Liao CY, Smet W, Brunoud G, Yoshida S, Vernoux T, Weijers D. Reporters for sensitive and quantitative measurement of auxin response. Nat Methods. 2015:**12**(3):207–210. https://doi.org/10.1038/nmeth.3279

Lim S, Park J, Lee N, Jeong J, Toh S, Watanabe A, Kim J, Kang H, Kim DH, Kawakami N, et al. ABA-insensitive3, ABA-insensitive5, and DELLAs interact to activate the expression of *SOMNUS* and other high-temperature-inducible genes in imbibed seeds in Arabidopsis. Plant Cell. 2013:**25**(12):4863–4878. https://doi.org/10.1105/tpc.113.118604

Liu Q, Wu K, Song W, Zhong N, Wu Y, Fu X. Improving crop nitrogen use efficiency toward sustainable green revolution. Annu Rev Plant Biol. 2022:**73**(1):523–551. https://doi.org/10.1146/annurev-arplant-070121-015752

Locascio A, Blázquez MA, Alabadí D. Dynamic regulation of cortical microtubule organization through prefoldin-DELLA interaction. Curr Biol. 2013:**23**(9):804–809. https://doi.org/10.1016/j.cub.2013.03.053

Lovegrove A, Hooley R. Gibberellin and abscisic acid signalling in aleurone. Trends Plant Sci. 2000:**5**(3):102–110. https://doi.org/10.1016/S1360-1385(00)01571-5

MacMillan J. Biosynthesis of the gibberellin plant hormones. Nat Prod Rep. 1997:**14**(3):221–243. https://doi.org/10.1039/np9971400221

Macmillan J, Suter PJ. The occurrence of gibberellin A_1 in higher plants—isolation from the seed of runner bean (*Phaseolus multiflorus*). Naturwissenschaften. 1958:**45**(2):46. https://doi.org/10.1007/BF00635028

Magome H, Nomura T, Hanada A, Takeda-Kamiya N, Ohnishi T, Shinma Y, Katsumata T, Kawaide H, Kamiya Y, Yamaguchi S. CYP714B1 and CYP714B2 encode gibberellin 13-oxidases that reduce gibberellin activity in rice. Proc Natl Acad Sci U S A. 2013:**110**(5):1947–1952. https://doi.org/10.1073/pnas.1215788110

Marin-de la Rosa N, Sotillo B, Miskolczi P, Gibbs DJ, Vicente J, Carbonero P, Onate-Sanchez L, Holdsworth MJ, Bhalerao R, Alabadí D, et al. Large-scale identification of gibberellin-related transcription factors defines group VII ETHYLENE RESPONSE FACTORS as functional DELLA partners. Plant Physiol. 2014:**166**(2):1022–1032. https://doi.org/10.1104/pp.114.244723

Marín-de la Rosa N, Pfeiffer A, Hill K, Locascio A, Bhalerao RP, Miskolczi P, Grønlund AL, Wanchoo-Kohli A, Thomas SG, Bennett MJ, et al. Genome wide binding site analysis reveals transcriptional coactivation of cytokinin-responsive genes by DELLA proteins. PLoS Genet. 2015:**11**(7):e1005337. https://doi.org/10.1371/journal.pgen.1005337

Martin DN, Proebsting WM, Hedden P. Mendel's dwarfing gene: cDNAs from the *Le* alleles and function of the expressed proteins. Proc Natl Acad Sci U S A. 1997:**94**(16):8907–8911. https://doi.org/10.1073/pnas.94.16.8907

Martin DN, Proebsting WM, Hedden P. The *SLENDER* gene of pea encodes a gibberellin 2-oxidase. Plant Physiol. 1999:**121**(3):775–781. https://doi.org/10.1104/pp.121.3.775

Matías-Hernández L, Aguilar-Jaramillo AE, Osnato M, Weinstain R, Shani E, Suárez-López P, Pelaz S. TEMPRANILLO reveals the mesophyll as crucial for epidermal trichome formation. Plant Physiol. 2016:**170**(3):1624–1639. https://doi.org/10.1104/pp.15.01309

McGinnis KM, Thomas SG, Soule JD, Strader LC, Zale JM, Sun TP, Steber CM. The Arabidopsis *SLEEPY1* gene encodes a putative F-box subunit of an SCF E3 ubiquitin ligase. Plant Cell. 2003:**15**(5):1120–1130. https://doi.org/10.1105/tpc.010827

Mehra P, Pandey BK, Melebari D, Banda J, Leftley N, Couvreur V, Rowe J, Anfang M, De Gernier H, Morris E. Hydraulic flux–responsive hormone redistribution determines root branching. Science. 2022:**378**(6621):762–768. https://doi.org/10.1126/science.add3771

Mitchum MG, Yamaguchi S, Hanada A, Kuwahara A, Yoshioka Y, Kato T, Tabata S, Kamiya Y, Sun TP. Distinct and overlapping roles of two gibberellin 3-oxidases in Arabidopsis development. Plant J. 2006:**45**(5):804–818. https://doi.org/10.1111/j.1365-313X.2005.02642.x

Miyazaki S, Hara M, Ito S, Tanaka K, Asami T, Hayashi K, Kawaide H, Nakajima M. An ancestral gibberellin in a moss *Physcomitrella patens*. Mol Plant. 2018:**11**(8):1097–1100. https://doi.org/10.1016/j.molp.2018.03.010

Morii M, Sugihara A, Takehara S, Kanno Y, Kawai K, Hobo T, Hattori M, Yoshimura H, Seo M, Ueguchi-Tanaka M. The dual function of OsSWEET3a as a gibberellin and glucose transporter is important for young shoot development in rice. Plant Cell Physiol. 2020:**61**(11):1935–1945. https://doi.org/10.1093/pcp/pcaa130

Morrone D, Chambers J, Lowry L, Kim G, Anterola A, Bender K, Peters RJ. Gibberellin biosynthesis in bacteria: separate *ent*-copalyl diphosphate and *ent*-kaurene synthases in *Bradyrhizobium japonicum*. FEBS Lett. 2009:**583**(2):475–480. https://doi.org/10.1016/j.febslet.2008.12.052

Müller B, Sheen J. Cytokinin and auxin interaction in root stem-cell specification during early embryogenesis. Nature. 2008:**453**(7198):1094–1097. https://doi.org/10.1038/nature06943

Murase K, Hirano Y, Sun TP, Hakoshima T. Gibberellin-induced DELLA recognition by the gibberellin receptor GID1. Nature. 2008:**456**(7221):459–463. https://doi.org/10.1038/nature07519

Nagel R, Peters RJ. Investigating the phylogenetic range of gibberellin biosynthesis in bacteria. Mol Plant-Microbe Interact. 2017:**30**(4):343–349. https://doi.org/10.1094/MPMI-01-17-0001-R

Nakajima M, Shimada A, Takashi Y, Kim YC, Park SH, Ueguchi-Tanaka M, Suzuki H, Katoh E, Iuchi S, Kobayashi M, et al. Identification and characterization of Arabidopsis gibberellin receptors. Plant J. 2006:**46**(5):880–889. https://doi.org/10.1111/j.1365-313X.2006.02748.x

Nakamura H, Xue YL, Miyakawa T, Hou F, Qin HM, Fukui K, Shi X, Ito E, Ito S, Park SH, et al. Molecular mechanism of strigolactone perception by DWARF14. Nat Commun. 2013:**4**(1):2613. https://doi.org/10.1038/ncomms3613

Nett RS, Contreras T, Peters RJ. Characterization of CYP115 as a gibberellin 3-oxidase indicates that certain rhizobia can produce bioactive gibberellin A_4. ACS Chem Biol. 2017a:**12**(4):912–917. https://doi.org/10.1021/acschembio.6b01038

Nett RS, Montanares M, Marcassa A, Lu X, Nagel R, Charles TC, Hedden P, Rojas MC, Peters RJ. Elucidation of gibberellin biosynthesis in bacteria reveals convergent evolution. Nat Chem Biol. 2017b:**13**(1):69–74. https://doi.org/10.1038/nchembio.2232

Nomura T, Magome H, Hanada A, Takeda-Kamiya N, Mander LN, Kamiya Y, Yamaguchi S. Functional analysis of *Arabidopsis* CYP714a1 and CYP714A2 reveals that they are distinct gibberellin modification enzymes. Plant Cell Physiol. 2013:**54**(11):1837–1851. https://doi.org/10.1093/pcp/pct125

Oh E, Zhu JY, Bai MY, Arenhart RA, Sun Y, Wang ZY. Cell elongation is regulated through a central circuit of interacting transcription factors in the Arabidopsis hypocotyl. Elife. 2014:**3**:e03031. https://doi.org/10.7554/eLife.03031

Okada K, Ito T, Fukazawa J, Takahashi Y. Gibberellin induces an increase in cytosolic ca^{2+} via a DELLA-independent signaling pathway. Plant Physiol. 2017:**175**(4):1536–1542. https://doi.org/10.1104/pp.17.01433

Okamoto M, Peterson FC, Defries A, Park SY, Endo A, Nambara E, Volkman BF, Cutler SR. Activation of dimeric ABA receptors elicits guard cell closure, ABA-regulated gene expression, and drought tolerance. Proc Natl Acad Sci U S A. 2013:**110**(29):12132–12137. https://doi.org/10.1073/pnas.1305919110

Olszewski NE, West CM, Sassi SO, Hartweck LM. *O*-GlcNAc protein modification in plants: evolution and function. Biochim Biophys Acta. 2010:**1800**(2):49–56. https://doi.org/10.1016/j.bbagen.2009.11.016

Paleg LG. Physiological effects of gibberellic acid. II. On starch hydrolyzing enzymes of barley endosperm. Plant Physiol. 1960:**35**(6):902–906. https://doi.org/10.1104/pp.35.6.902

Park J, Nguyen KT, Park E, Jeon JS, Choi G. DELLA proteins and their interacting RING finger proteins repress gibberellin responses by binding to the promoters of a subset of gibberellin-responsive genes in Arabidopsis. Plant Cell. 2013:**25**(3):927–943. https://doi.org/10.1105/tpc.112.108951

Park J, Oh D, Dassanayaka M, Nguyen KT, Ogas J, Choi G, Sun TP. GA signaling requires chromatin remodeler PICKLE to promote vegetative growth and phase transitions. Plant Physiol. 2017:**173**(2):1463–1474. https://doi.org/10.1104/pp.16.01471

Peng J, Carol P, Richards DE, King KE, Cowling RJ, Murphy GP, Harberd NP. The Arabidopsis *GAI* gene defines a signalling pathway that negatively regulates gibberellin responses. Genes Dev. 1997:**11**(23):3194–3205. https://doi.org/10.1101/gad.11.23.3194

Peng J, Harberd NP. Derivative alleles of the Arabidopsis gibberellin-insensitive (*gai*) mutation confer a wild-type phenotype. Plant Cell. 1993:**5**(3):351–360. https://doi.org/10.2307/3869602

Peng J, Richards DE, Hartley NM, Murphy GP, Devos KM, Flintham JE, Beales J, Fish LJ, Worland AJ, Pelica F, et al. 'Green revolution' genes encode mutant gibberellin response modulators. Nature. 1999:**400**(6741):256–261. https://doi.org/10.1038/22307

Phillips AL. Genetic control of gibberellin metabolism and signalling in crop improvement. In: **Hedden P, Thomas SG**, editors. The gibberellins, annual plant reviews, vol 49. Oxford, UK: Wiley Blackwell; 2016. p. 405–430.

Phillips AL, Ward DA, Uknes S, Appleford NEJ, Lange T, Huttly AK, Gaskin P, Graebe JE, Hedden P. Isolation and expression of three gibberellin 20-oxidase cDNA clones from Arabidopsis. Plant Physiol. 1995:**108**(3):1049–1057. https://doi.org/10.1104/pp.108.3.1049

Phinney BO. Growth response of single-gene dwarf mutants in maize to gibberellic acid. Proc Natl Acad Sci U S A. 1956:**42**(4):185–189. https://doi.org/10.1073/pnas.42.4.185

Phinney BO, Spray C. Chemical genetics and the gibberellin pathway in *Zea mays* L. In: **Wareing P**, editors. Plant growth substances 1982. New York: Academic Press, London; 1982. p. 101–110.

Pimprikar P, Carbonnel S, Paries M, Katzer K, Klingl V, Bohmer MJ, Karl L, Floss DS, Harrison MJ, Parniske M, et al. A CCaMK-CYCLOPS-DELLA complex activates transcription of RAM1 to regulate arbuscule branching. Curr Biol. 2016:**26**(8):987–998. https://doi.org/10.1016/j.cub.2016.01.069

Plackett ARG, Powers SJ, Fernandez-Garcia N, Urbanova T, Takebayashi Y, Seo M, Jikumaru Y, Benlloch R, Nilsson O, Ruiz-Rivero O, et al. Analysis of the developmental roles of the Arabidopsis gibberellin 20-oxidases demonstrates that GA20ox1, -2, and -3 are the dominant paralogs. Plant Cell. 2012:**24**(3):941–960. https://doi.org/10.1105/tpc.111.095109

Proebsting WM, Hedden P, Lewis MJ, Croker SJ, Proebsting LN. Gibberellin concentration and transport in genetic lines of pea: effects of grafting. Plant Physiol. 1992:**100**(3):1354–1360. https://doi.org/10.1104/pp.100.3.1354

Ptošková K, Szecówka M, Jaworek P, Tarkowská D, Petřík I, Pavlović I, Novák O, Thomas SG, Phillips AL, Hedden P. Changes in the concentrations and transcripts for gibberellins and other hormones in a growing leaf and roots of wheat seedlings in response to water restriction. BMC Plant Biol. 2022:**22**(1):284. https://doi.org/10.1186/s12870-022-03667-w

Pysh LD, Wysocka-Diller JW, Camilleri C, Bouchez D, Benfey PN. The GRAS gene family in Arabidopsis: sequence characterization and basic expression analysis of the *SCARECROW-LIKE* genes. Plant J. 1999:**18**(1):111–119. https://doi.org/10.1046/j.1365-313X.1999.00431.x

Qi T, Huang H, Wu D, Yan J, Qi Y, Song S, Xie D. Arabidopsis DELLA and JAZ proteins bind the WD-repeat/bHLH/MYB complex to modulate gibberellin and jasmonate signaling synergy. Plant Cell. 2014:**26**(3):1118–1133. https://doi.org/10.1105/tpc.113.121731

Rademacher W. GROWTH RETARDANTS: effects on gibberellin biosynthesis and other metabolic pathways. Annu Rev Plant Physiol Plant Mol Biol. 2000:**51**(1):501–531. https://doi.org/10.1146/annurev.arplant.51.1.501

Rademacher W. Plant growth regulators: backgrounds and uses in plant production. J Plant Growth Regul. 2015:**34**(4):845–872. https://doi.org/10.1007/s00344-015-9541-6

Ragni L, Nieminen K, Pacheco-Villalobos D, Sibout R, Schwechheimer C, Hardtke CS. Mobile gibberellin directly stimulates Arabidopsis hypocotyl xylem expansion. Plant Cell. 2011:**23**(4):1322–1336. https://doi.org/10.1105/tpc.111.084020

Regnault T, Davière JM, Heintz D, Lange T, Achard P. The gibberellin biosynthetic genes *AtKAO1* and *AtKAO2* have overlapping roles throughout Arabidopsis development. Plant J. 2014:**80**(3):462–474. https://doi.org/10.1111/tpj.12648

Regnault T, Davière J-M, Wild M, Sakvarelidze-Achard L, Heintz D, Bergua EC, Diaz IL, Gong F, Hedden P, Achard P. The gibberellin precursor GA_{12} acts as a long-distance growth signal in Arabidopsis. Nat Plants. 2015:**1**(6):15073. https://doi.org/10.1038/nplants.2015.73

Reid JB, Ross JJ, Swain SM. Internode length in *Pisum*—a new, slender mutant with elevated levels of C_{19} gibberellins. Planta. 1992:**188**(4):462–467. https://doi.org/10.1007/BF00197036

Resentini F, Felipo-Benavent A, Colombo L, Blázquez MA, Alabadí D, Masiero S. TCP14 and TCP15 mediate the promotion of seed germination by gibberellins in *Arabidopsis thaliana*. Mol Plant. 2015:**8**(3):482–485. https://doi.org/10.1016/j.molp.2014.11.018

Rieu I, Eriksson S, Powers SJ, Gong F, Griffiths J, Woolley L, Benlloch R, Nilsson O, Thomas SG, Hedden P, et al. Genetic analysis reveals that C_{19}-GA 2-oxidation is a major gibberellin inactivation pathway in Arabidopsis. Plant Cell. 2008a:**20**(9):2420–2436. https://doi.org/10.1105/tpc.108.058818

Rieu I, Ruiz-Rivero O, Fernandez-Garcia N, Griffiths J, Powers SJ, Gong F, Linhartova T, Eriksson S, Nilsson O, Thomas SG, et al. The gibberellin biosynthetic genes *AtGA20ox1* and *AtGA20ox2* act, partially redundantly, to promote growth and development throughout the Arabidopsis life cycle. Plant J. 2008b:**53**(3):488–504. https://doi.org/10.1111/j.1365-313X.2007.03356.x

Rizza A, Jones AM. The makings of a gradient: spatiotemporal distribution of gibberellins in plant development. Curr Opin Plant Biol. 2019:**47**:9–15. https://doi.org/10.1016/j.pbi.2018.08.001

Rizza A, Tang B, Stanley CE, Grossmann G, Owen MR, Band LR, Jones AM. Differential biosynthesis and cellular permeability explain longitudinal gibberellin gradients in growing roots. Proc Natl Acad Sci U S A. 2021:**118**(8):e1921960118. https://doi.org/10.1073/pnas.1921960118

Rizza A, Walia A, Lanquar V, Frommer WB, Jones AM. In vivo gibberellin gradients visualized in rapidly elongating tissues. Nat Plants. 2017:**3**(10):803–813. https://doi.org/10.1038/s41477-017-0021-9

Robe K, Barberon M. NPFs rule suberization. Nat Plants. 2023:**9**(5):689–690. https://doi.org/10.1038/s41477-023-01411-2

Rojas MC, Hedden P, Gaskin P, Tudzynki B. The P450-1 gene of *Gibberella fujikuroi* encodes a multifunctional enzyme in gibberellin biosynthesis. Proc Natl Acad Sci U S A. 2001:**98**(10):5838–5843. https://doi.org/10.1073/pnas.091096298

Ropers HJ, Graebe JE, Gaskin P, Macmillan J. Gibberellin biosynthesis in a cell-free system from immature seeds of *Pisum sativum*. Biochem Biophys Res Commun. 1978:**80**(4):690–697. https://doi.org/10.1016/0006-291X(78)91299-8

Ross JJ, Reid JB, Swain SM, Hasan O, Poole AT, Hedden P, Willis CL. Genetic-regulation of gibberellin deactivation in *Pisum*. Plant J.

1995:**7**(3):513–523. https://doi.org/10.1046/j.1365-313X.1995.7030513.x

Saito H, Oikawa T, Hamamoto S, Ishimaru Y, Kanamori-Sato M, Sasaki-Sekimoto Y, Utsumi T, Chen J, Kanno Y, Masuda S. The jasmonate-responsive GTR1 transporter is required for gibberellin-mediated stamen development in Arabidopsis. Nat Commun. 2015:**6**(1):6095. https://doi.org/10.1038/ncomms7095

Sarnowska EA, Rolicka AT, Bucior E, Cwiek P, Tohge T, Fernie AR, Jikumaru Y, Kamiya Y, Franzen R, Schmelzer E, et al. DELLA-interacting SWI3C core subunit of switch/sucrose nonfermenting chromatin remodeling complex modulates gibberellin responses and hormonal cross talk in Arabidopsis. Plant Physiol. 2013:**163**(1):305–317. https://doi.org/10.1104/pp.113.223933

Sasaki A, Itoh H, Gomi K, Ueguchi-Tanaka M, Ishiyama K, Kobayashi M, Jeong D-H, An G, Kitano J, Ashikari M, et al. Accumulation of phosphorylated repressor for gibberellin signaling in an F-box mutant. Science. 2003:**299**(5614):1896–1898. https://doi.org/10.1126/science.1081077

Schomburg FM, Bizzell CM, Lee DJ, Zeevaart JAD, Amasino RM. Overexpression of a novel class of gibberellin 2-oxidases decreases gibberellin levels and creates dwarf plants. Plant Cell. 2003:**15**(1):151–163. https://doi.org/10.1105/tpc.005975

Serrano-Mislata A, Bencivenga S, Bush M, Schiessl K, Boden S, Sablowski R. DELLA genes restrict inflorescence meristem function independently of plant height. Nat Plants. 2017:**3**(9):749–754. https://doi.org/10.1038/s41477-017-0003-y

Shani E, Weinstain R, Zhang Y, Castillejo C, Kaiserli E, Chory J, Tsien RY, Estelle M. Gibberellins accumulate in the elongating endodermal cells of Arabidopsis root. Proc Natl Acad Sci U S A. 2013:**110**(12):4834–4839. https://doi.org/10.1073/pnas.1300436110

Shechter I, West CA. Biosynthesis of gibberellins. IV. Biosynthesis of cyclic diterpenes from trans-geranylgeranyl pyrophosphate. J Biol Chem. 1969:**244**(12):3200–3209. https://doi.org/10.1016/S0021-9258(18)93114-5

Shi B, Felipo-Benavent A, Cerutti G, Galvan-Ampudia C, Jilli L, Brunoud G, Mutterer J, Vallet E, Sakvarelidze-Achard L, Davière J-M. A quantitative gibberellin signalling biosensor reveals a role for gibberellins in internode specification at the shoot apical meristem. bioRxiv 448154. https://doi.org/10.1101/2021.06.11.448154, 15 June 2021, preprint: not peer reviewed.

Shimada A, Ueguchi-Tanaka M, Nakatsu T, Nakajima M, Naoe Y, Ohmiya H, Kato H, Matsuoka M. Structural basis for gibberellin recognition by its receptor GID1. Nature. 2008:**456**(7221):520–523. https://doi.org/10.1038/nature07546

Silverstone AL, Chang C, Krol E, Sun TP. Developmental regulation of the gibberellin biosynthetic gene *GA1* in *Arabidopsis thaliana*. Plant J. 1997a:**12**(1):9–19. https://doi.org/10.1046/j.1365-313X.1997.12010009.x

Silverstone AL, Ciampaglio CN, Sun TP. The Arabidopsis *RGA* gene encodes a transcriptional regulator repressing the gibberellin signal transduction pathway. Plant Cell. 1998:**10**(2):155–169. https://doi.org/10.1105/tpc.10.2.155

Silverstone AL, Jung H-S, Dill A, Kawaide H, Kamiya Y, Sun TP. Repressing a repressor: gibberellin-induced rapid reduction of the RGA protein in Arabidopsis. Plant Cell. 2001:**13**(7):1555–1566. https://doi.org/10.1105/tpc.010047

Silverstone AL, Mak PYA, Casamitjana Martínez E, Sun TP. The new *RGA* locus encodes a negative regulator of gibberellin response in *Arabidopsis thaliana*. Genetics. 1997b:**146**(3):1087–1099. https://doi.org/10.1093/genetics/146.3.1087

Silverstone AL, Tseng T-S, Swain S, Dill A, Jeong SY, Olszewski NE, Sun TP. Functional analysis of SPINDLY in gibberellin signaling in Arabidopsis. Plant Physiol. 2007:**143**(2):987–1000. https://doi.org/10.1104/pp.106.091025

Simcox PD, Dennis DT, West CA. Kaurene synthetase from plastids of developing plant tissues. Biochem Biophys Res Commun. 1975:**66**(1):166–172. https://doi.org/10.1016/S0006-291X(75)80309-3

Sponsel VM. The localization, metabolism and biological activity of gibberellins in maturing and germinating seeds of *Pisum sativum* cv. Progress No. 9. Planta. 1983:**159**(5):454–468. https://doi.org/10.1007/BF00392082

Spray C, Phinney BO, Gaskin P, Gilmour SJ, MacMillan J. Internode length in *Zea mays* L.—the *dwarf-1* mutation controls the 3β-hydroxylation of gibberellin A_{20} to gibberellin A_1. Planta. 1984:**160**(5):464–468. https://doi.org/10.1007/BF00429764

Stavang JA, Gallego-Bartolomé J, Gómez MD, Yoshida S, Asami T, Olsen JE, García-Martínez JL, Alabadí D, Blázquez MA. Hormonal regulation of temperature-induced growth in Arabidopsis. Plant J. 2009:**60**(4):589–601. https://doi.org/10.1111/j.1365-313X.2009.03983.x

Steber CM, Cooney S, McCourt P. Isolation of the GA-response mutant *sly1* as a suppressor of *ABI1-1* in *Arabidopsis thaliana*. Genetics. 1998:**149**(2):509–521. https://doi.org/10.1093/genetics/149.2.509

Stodola FH, Raper KB, Fennell DI, Conway HF, Sohns VE, Langford CT, Jackson RW. The microbiological production of gibberellins A and X. Arch Biochem Biophys. 1955:**54**(1):240–245. https://doi.org/10.1016/0003-9861(55)90027-8

Stowe BB, Yamaki T. The history and physiological action of the gibberellins. Annu Rev Plant Physiol Plant Mol Biol. 1957:**8**(1):181–216. https://doi.org/10.1146/annurev.pp.08.060157.001145

Sun TP. Gibberellin metabolism, perception and signaling pathways in Arabidopsis. The Arabidopsis Book. 2008:**6**:e0103. https://doi.org/10.1199/tab.0103

Sun TP. The molecular mechanism and evolution of the GA-GID1-DELLA signaling module in plants. Curr Biol. 2011:**21**(9):R338–R345. https://doi.org/10.1016/j.cub.2011.02.036

Sun TP. Novel nucleocytoplasmic protein O-fucosylation by SPINDLY regulates diverse developmental processes in plants. Curr Opin Struct Biol. 2021:**68**:113–121. https://doi.org/10.1016/j.sbi.2020.12.013

Sun TP, Goodman HM, Ausubel FM. Cloning the Arabidopsis *GA1* locus by genomic subtraction. Plant Cell. 1992:**4**(2):119–128. https://doi.org/10.2307/3869565

Sun TP, Gubler F. Molecular mechanism of gibberellin signaling in plants. Annu Rev Plant Biol. 2004:**55**(1):197–223. https://doi.org/10.1146/annurev.arplant.55.031903.141753

Sun H, Guo X, Zhu X, Gu P, Zhang W, Tao W, Wang D, Wu Y, Zhao Q, Xu G, et al. Strigolactone and gibberellin signaling coordinately regulate metabolic adaptations to changes in nitrogen availability in rice. Mol Plant. 2023:**16**(3):588–598. https://doi.org/10.1016/j.molp.2023.01.009

Sun TP, Kamiya Y. The Arabidopsis *GA1* locus encodes the cyclase *ent*-kaurene synthetase A of gibberellin biosynthesis. Plant Cell. 1994:**6**(10):1509–1518. https://doi.org/10.1105/tpc.6.10.1509

Takahashi N, Kitamura H, Kawarada A, Seta Y, Takai M, Tamura S, Sumiki Y. Biochemical studies on "Bakanae" fungus. Part XXXIV. Isolation of gibberellins and their properties. Bull Agr Chem Soc Japan. 1955:**19**:267–277. https://doi.org/10.1080/03758397.1955.10856832

Takehara S, Sakuraba S, Mikama B, Yoshida H, Yoshimura H, Itoh A, Endo M, Watanabe N, Nagae T, Matsuoka M, et al. A common allosteric mechanism regulates homeostatic inactivation of auxin and gibberellin. Nat Commun. 2020:**11**(1):2143. https://doi.org/10.1038/s41467-020-16068-0

Tal I, Zhang Y, Jørgensen ME, Pisanty O, Barbosa IC, Zourelidou M, Regnault T, Crocoll C, Olsen CE, Weinstain R, et al. The Arabidopsis NPF3 protein is a GA transporter. Nat Commun. 2016:**7**(1):11486. https://doi.org/10.1038/ncomms11486

Tanaka J, Yano K, Aya K, Hirano K, Takehara S, Koketsu E, Ordonio RL, Park S-H, Nakajima M, Ueguchi-Tanaka M. Antheridiogen determines sex in ferns via a spatiotemporally split gibberellin synthesis pathway. Science. 2014:**346**(6208):469–473. https://doi.org/10.1126/science.1259923

Thomas SG, Bláquez MA, Alabadí D. DELLA Proteins: master regulators of gibberellin-responsive growth and development. In: **Hedden P, Thomas SG**, editors. Annual plant reviews, vol 49. Chichester: Wiley; 2016. p. 189–228.

Thomas SG, Phillips AL, Hedden P. Molecular cloning and functional expression of gibberellin 2-oxidases, multifunctional enzymes involved in gibberellin deactivation. Proc Natl Acad Sci U S A. 1999:**96**(8):4698–4703. https://doi.org/10.1073/pnas.96.8.4698

Ueguchi-Tanaka M, Ashikari M, Nakajima M, Itoh H, Katoh E, Kobayashi M, Chow TY, Hsing YI, Kitano H, Yamaguchi I, et al. *GIBBERELLIN INSENSITIVE DWARF1* encodes a soluble receptor for gibberellin. Nature. 2005:**437**(7059):693–698. https://doi.org/10.1038/nature04028

Ulmasov T, Murfett J, Hagen G, Guilfoyle TJ. Aux/IAA proteins repress expression of reporter genes containing natural and highly active synthetic auxin response elements. Plant Cell. 1997:**9**(11): 1963–1971. https://doi.org/10.1105/tpc.9.11.1963

Van De Velde K, Ruelens P, Geuten K, Rohde A, Van Der Straeten D. Exploiting DELLA signaling in cereals. Trends Plant Sci. 2017:**22**(10): 880–893. https://doi.org/10.1016/j.tplants.2017.07.010

Van De Velde K, Thomas SG, Heyse F, Kaspar R, Van Der Straeten D, Rohde A. N-terminal truncated RHT-1 proteins generated by translational reinitiation cause semi-dwarfing of wheat green revolution alleles. Mol Plant. 2021:**14**(4):679–687. https://doi.org/10.1016/j.molp.2021.01.002

Wang F, Yoshida H, Matsuoka M. Making the 'green revolution' truly green: improving crop nitrogen use efficiency. Plant Cell Physiol. 2021:**62**(6):942–947. https://doi.org/10.1093/pcp/pcab051

Weiss D, Halevy AH. Stamens and gibberellin in the regulation of corolla pigmentation and growth in *Petunia hybrida*. Planta. 1989:**179**(1):89–96. https://doi.org/10.1007/BF00395775

West CA. Biosynthesis of gibberellins. In: **Milborrow BV**, editors. Biosynthesis and its control in plants. London, New York: Academic Press; 1973. p. 473–482.

Weston DE, Elliott RC, Lester DR, Rameau C, Reid JB, Murfet IC, Ross JJ. The pea DELLA proteins LA and CRY are important regulators of gibberellin synthesis and root growth. Plant Physiol. 2008:**147**(1): 199–205. https://doi.org/10.1104/pp.108.115808

Wexler S, Schayek H, Rajendar K, Tal I, Shani E, Meroz Y, Dobrovetsky R, Weinstain R. Characterizing gibberellin flow *in planta* using photocaged gibberellins. Chem Sci. 2019:**10**(5):1500–1505. https://doi.org/10.1039/C8SC04528C

Wild M, Davière JM, Cheminant S, Regnault T, Baumberger N, Heintz D, Baltz R, Genschik P, Achard P. The Arabidopsis DELLA *RGA-LIKE3* is a direct target of MYC2 and modulates jasmonate signaling responses. Plant Cell. 2012:**24**(8):3307–3319. https://doi.org/10.1105/tpc.112.101428

Wild M, Daviere JM, Regnault T, Sakvarelidze-Achard L, Carrera E, Lopez Diaz I, Cayrel A, Dubeaux G, Vert G, Achard P. Tissue-specific regulation of gibberellin signaling fine-tunes Arabidopsis iron-deficiency responses. Dev Cell. 2016:**37**(2): 190–200. https://doi.org/10.1016/j.devcel.2016.03.022

Williams J, Phillips AL, Gaskin P, Hedden P. Function and substrate specificity of the gibberellin 3β-hydroxylase encoded by the Arabidopsis *GA4* gene. Plant Physiol. 1998:**117**(2):559–563. https://doi.org/10.1104/pp.117.2.559

Willige BC, Ghosh S, Nill C, Zourelidou M, Dohmann EMN, Maier A, Schwechheimer C. The DELLA domain of GA INSENSITIVE mediates the interaction with the GA INSENSITIVE DWARF1A gibberellin receptor of Arabidopsis. Plant Cell. 2007:**19**(4):1209–1220. https://doi.org/10.1105/tpc.107.051441

Wittwer SH, Bukovac MJ, Sell HM, Weller LE. Some effects of gibberellin on flowering and fruit setting. Plant Physiol. 1957:**32**(1):39–41. https://doi.org/10.1104/pp.32.1.39

Wu R, Duan L, Pruneda-Paz JL, Oh DH, Pound M, Kay S, Dinneny JR. The *6xABRE* synthetic promoter enables the spatiotemporal analysis of ABA-mediated transcriptional regulation. Plant Physiol. 2018:**177**(4):1650–1665. https://doi.org/10.1104/pp.18.00401

Wu K, Wang SS, Song WZ, Zhang JQ, Wang Y, Liu Q, Yu JP, Ye YF, Li S, Chen JF, et al. Enhanced sustainable green revolution yield via nitrogen-responsive chromatin modulation in rice. Science. 2020:**367**(6478):eaaz2046. https://doi.org/10.1126/science.aaz2046

Wulff N, Ernst HA, Jørgensen ME, Lambertz S, Maierhofer T, Belew ZM, Crocoll C, Motawia MS, Geiger D, Jørgensen FS. An optimized screen reduces the number of GA transporters and provides insights into nitrate transporter 1/peptide transporter family substrate determinants. Front Plant Sci. 2019:**10**:1106. https://doi.org/10.3389/fpls.2019.01106

Xu YL, Li L, Wu KQ, Peeters AJM, Gage DA, Zeevaart JAD. The *GA5* locus of *Arabidopsis thaliana* encodes a multifunctional gibberellin 20-oxidase—molecular cloning and functional expression. Proc Natl Acad Sci U S A. 1995:**92**(14):6640–6644. https://doi.org/10.1073/pnas.92.14.6640

Xu F, Li T, Xu PB, Li L, Du SS, Lian HL, Yang HQ. DELLA proteins physically interact with CONSTANS to regulate flowering under long days in Arabidopsis. FEBS Lett. 2016:**590**(4):541–549. https://doi.org/10.1002/1873-3468.12076

Yabuta T, Sumiki T. Communication to the editor. J Agr Chem Soc Japan. 1938:**14**:1526.

Yamaguchi S. Gibberellin metabolism and its regulation. Annu Rev Plant Biol. 2008:**59**(1):225–251. https://doi.org/10.1146/annurev.arplant.59.032607.092804

Yamaguchi S, Kamiya Y. Gibberellin biosynthesis: its regulation by endogenous and environmental signals. Plant Cell Physiol. 2000:**41**(3): 251–257. https://doi.org/10.1093/pcp/41.3.251

Yamaguchi S, Kamiya Y, Sun T. Distinct cell-specific expression patterns of early and late gibberellin biosynthetic genes during Arabidopsis seed germination. Plant J. 2001:**28**(4):443–453. https://doi.org/10.1046/j.1365-313X.2001.01168.x

Yamaguchi S, Saito T, Abe H, Yamane H, Murofushi N, Kamiya Y. Molecular cloning and characterization of a cDNA encoding the gibberellin biosynthetic enzyme *ent*-kaurene synthase B from pumpkin (*Cucurbita maxima* L.). Plant J. 1996:**10**(2):203–213. https://doi.org/10.1046/j.1365-313X.1996.10020203.x

Yamaguchi S, Sun T, Kawaide H, Kamiya Y. The *GA2* locus of *Arabidopsis thaliana* encodes *ent*-kaurene synthase of gibberellin biosynthesis. Plant Physiol. 1998:**116**(4):1271–1278. https://doi.org/10.1104/pp.116.4.1271

Yan J, Li X, Zeng B, Zhong M, Yang J, Yang P, Li X, He C, Lin J, Liu X, et al. FKF1 F-box protein promotes flowering in part by negatively regulating DELLA protein stability under long-day photoperiod in Arabidopsis. J Integr Plant Biol. 2020:**62**(11):1717–1740. https://doi.org/10.1111/jipb.12971

Yang DL, Yao J, Mei CS, Tong XH, Zeng LJ, Li Q, Xiao LT, Sun TP, Li J, Deng XW, et al. Plant hormone jasmonate prioritizes defense over growth by interfering with gibberellin signaling cascade. Proc Natl Acad Sci U S A. 2012:**109**(19):E1192–E1200. https://doi.org/10.1073/pnas.1201616109

Yoshida H, Hirano K, Sato T, Mitsuda N, Nomoto M, Maeo K, Koketsu E, Mitani R, Kawamura M, Ishiguro S, et al. DELLA protein functions as a transcriptional activator through the DNA binding of the INDETERMINATE DOMAIN family proteins. Proc Natl Acad Sci U S A. 2014:**111**(21):7861–7866. https://doi.org/10.1073/pnas.1321669111

Yoshida H, Takehara S, Mori M, Ordonio RL, Matsuoka M. Evolution of GA metabolic enzymes in land plants. Plant Cell Physiol. 2020:**61**(11):1919–1934. https://doi.org/10.1093/pcp/pcaa126

Yu S, Galvão VC, Zhang YC, Horrer D, Zhang TQ, Hao YH, Feng YQ, Wang S, Schmid M, Wang JW. Gibberellin regulates the Arabidopsis floral transition through miR156-targeted SQUAMOSA promoter binding-like transcription factors. Plant Cell. 2012:**24**(8):3320–3332. https://doi.org/10.1105/tpc.112.101014

Yu J, Hu S, Wang J, Wong GK, Li S, Liu B, Deng Y, Dai L, Zhou Y, Zhang X, et al. A draft sequence of the rice genome (*Oryza sativa* L. ssp. *indica*). Science. 2002:**296**(5565):79–92. https://doi.org/10.1126/science.1068037

Zentella R, Hu J, Hsieh WP, Matsumoto PA, Dawdy A, Barnhill B, Oldenhof H, Hartweck LM, Maitra S, Thomas SG, et al. *O*-GlcNAcylation of master growth repressor DELLA by SECRET AGENT modulates multiple signaling pathways in Arabidopsis. Genes Dev. 2016:**30**(2):164–176. https://doi.org/10.1101/gad.270587.115

Zentella R, Sui N, Barnhill B, Hsieh WP, Hu J, Shabanowitz J, Boyce M, Olszewski NE, Zhou P, Hunt DF, et al. The Arabidopsis *O*-fucosyltransferase SPINDLY activates nuclear growth repressor DELLA. Nat Chem Biol. 2017:**13**(5):479–485. https://doi.org/10.1038/nchembio.2320

Zentella R, Zhang ZL, Park M, Thomas SG, Endo A, Murase K, Fleet CM, Jikumaru Y, Nambara E, Kamiya Y, et al. Global analysis of DELLA direct targets in early gibberellin signaling in Arabidopsis. Plant Cell. 2007:**19**(10):3037–3057. https://doi.org/10.1105/tpc.107.054999

Zhang Y, Berman A, Shani E. Plant hormone transport and localization: signaling molecules on the move. Annu Rev Plant Biol. 2023:**74**(1): 453–479. https://doi.org/10.1146/annurev-arplant-070722-015329

Zhang D, Jing Y, Jiang Z, Lin R. The chromatin-remodeling factor PICKLE integrates brassinosteroid and gibberellin signaling during skotomorphogenic growth in Arabidopsis. Plant Cell. 2014:**26**(6): 2472–2485. https://doi.org/10.1105/tpc.113.121848

Zhang ZL, Ogawa M, Fleet CM, Zentella R, Hu J, Heo J-O, Lim J, Kamiya Y, Yamaguchi S, Sun TP. SCARECROW-LIKE 3 promotes gibberellin signaling by antagonizing DELLA in Arabidopsis. Proc Natl Acad Sci U S A. 2011:**108**(5):2160–2165. https://doi.org/10.1073/pnas.1012232108

Zhu YY, Nomura T, Xu YH, Zhang YY, Peng Y, Mao BZ, Hanada A, Zhou HC, Wang RX, Li PJ, et al. *ELONGATED UPPERMOST INTERNODE* encodes a cytochrome P450 monooxygenase that epoxidizes gibberellins in a novel deactivation reaction in rice. Plant Cell. 2006:**18**(2):442–456. https://doi.org/10.1105/tpc.105.038455

Zweig G, Yamaguchi S, Mason GW. Translocation of C^{14}-gibberellin in red kidney bean, normal corn, and dwarf corn. Advances Chem. 1961:**28**:122–134. https://doi.org/10.1021/ba-1961-0028.ch013

https://doi.org/10.1093/plphys/kiae045 PLANT PHYSIOLOGY 2024: 195: 135–154

Plant Physiology®

Topical Review

Phytohormones in a universe of regulatory metabolites: lessons from jasmonate

Debora Gasperini [1,*] **and Gregg A. Howe** [2,3,4,*]

1 Department of Molecular Signal Processing, Leibniz Institute of Plant Biochemistry, Halle 06120, Germany
2 Department of Energy-Plant Research Laboratory, Michigan State University, East Lansing, MI 48824, USA
3 Department of Biochemistry and Molecular Biology, Michigan State University, East Lansing, MI 48824, USA
4 Plant Resilience Institute, Michigan State University, East Lansing, MI 42284, USA

*Author for correspondence: debora.gasperini@ipb-halle.de (D.G.), howeg@msu.edu (G.A.H.)
The author responsible for distribution of materials integral to the findings presented in this article in accordance with the policy described in the Instructions for Authors (https://academic.oup.com/plphys/pages/General-Instructions) is: Gregg A. Howe.

Abstract

Small-molecule phytohormones exert control over plant growth, development, and stress responses by coordinating the patterns of gene expression within and between cells. Increasing evidence indicates that currently recognized plant hormones are part of a larger group of regulatory metabolites that have acquired signaling properties during the evolution of land plants. This rich assortment of chemical signals reflects the tremendous diversity of plant secondary metabolism, which offers evolutionary solutions to the daunting challenges of sessility and other unique aspects of plant biology. A major gap in our current understanding of plant regulatory metabolites is the lack of insight into the direct targets of these compounds. Here, we illustrate the blurred distinction between classical phytohormones and other bioactive metabolites by highlighting the major scientific advances that transformed the view of jasmonate from an interesting floral scent to a potent transcriptional regulator. Lessons from jasmonate research generally apply to other phytohormones and thus may help provide a broad understanding of regulatory metabolite–protein interactions. In providing a framework that links small-molecule diversity to transcriptional plasticity, we hope to stimulate future research to explore the evolution, functions, and mechanisms of perception of a broad range of plant regulatory metabolites.

Introduction

Plants use a diverse repertoire of chemical messengers to coordinate their growth and development with informational cues from the environment. Among organic signaling compounds, small-molecule plant hormones (i.e. phytohormones) have been the focus of much research during the past century. Currently recognized phytohormones influence virtually all aspects of growth, development, and environmental acclimation and tend to be broadly conserved among land plants. Evolutionary reconstruction of plant hormone response pathways using phylogenomic approaches has provided important insights into the emergence of small molecules as drivers of phenotypic plasticity and innovations in life history traits (Bowman et al. 2017; Blázquez et al. 2020). Moreover, the molecular mechanisms governing phytohormone homeostasis and action have emerged as key targets for improving the productivity and stress resilience of crops (Hedden 2003; Bailey-Serres et al. 2019).

What is the meaning of "hormone" as it applies to the unique biology of plants? Beyond the limited number of compounds elevated to phytohormone status in textbooks, what other plant metabolites have signaling properties? What are the mechanisms by which these molecules are perceived and transduced to downstream effectors? How do such metabolites modulate transcription, and how are these functions acquired during evolution? Finally, how can this knowledge inform the development of new crop varieties that are more productive,

Received October 27, 2023. Accepted January 05, 2024. Advance access publication January 30, 2024

ADVANCES

- The past century of plant hormone research has helped reveal how low-molecular-weight metabolites modulate transcription to promote plasticity in growth, development, and environmental responses.
- Recent knowledge of the evolutionary origins of phytohormone biosynthesis, perception, and signaling is helping redefine the hormone concept to reflect the unique biology of plants.
- Currently recognized phytohormones are part of a larger group of metabolites that alter gene expression with different biological effects and degrees of conservation within land plants.
- Classical genetic approaches to study jasmonates and other phytohormones provide a roadmap to exploring the activities of other regulatory metabolites.
- Advances in mass spectrometry, molecular modeling, and artificial intelligence-based structural prediction provide new tools to identify regulatory metabolite–protein interactions in plants.

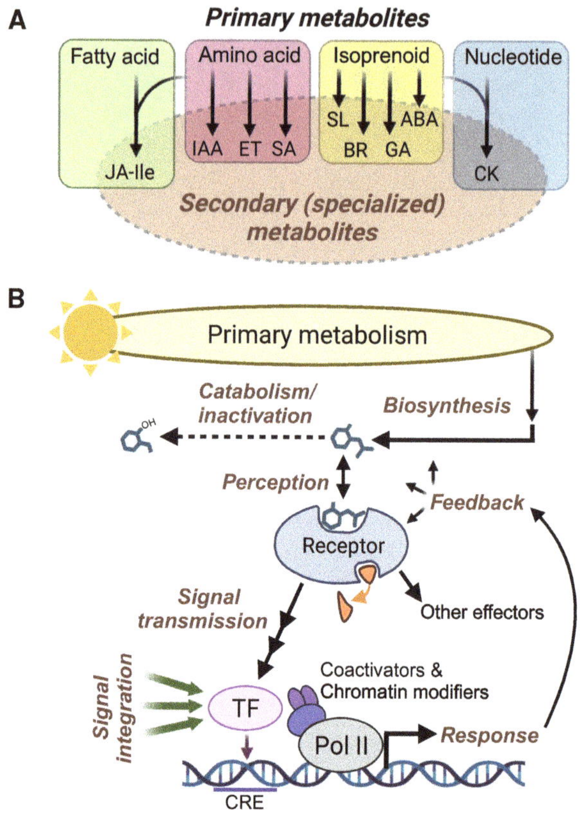

Figure 1. Common themes in plant hormone biosynthesis and signaling. **A)** Phytohormones can be viewed as secondary (i.e. specialized) metabolites that are derived from several major classes of primary metabolites. Some hormones, including JA-Ile and cytokinin (CK), are synthesized from more than one primary metabolite class. Peptide hormones are omitted for simplicity. **B)** Phytohormone-response pathways consist of several highly integrated processes (bold text) that collectively modulate the activity of one or more effectors. Hormone biosynthetic pathways utilize precursors from primary metabolism and give rise to bioactive compounds whose intracellular concentrations are several orders of magnitude below those of primary metabolic precursors. Bioactive hormones (depicted as a generic chemical structure) are subject to oxidation, conjugation, and other catabolic reactions that inactivate the signal (dashed arrow). Hormone perception is mediated by specific proteinaceous receptors. Many receptors are associated with a ligand-induced enzymatic activity (small curved arrow). Ligand binding transmits the signal to a downstream effector such as a TF. Effector TFs bind to specific *cis*-regulatory elements (CRE) in response genes and work together with various coactivators and chromatin-modifying enzymes to control the rate of transcription via RNA polymerase II (Pol II). TFs and other effectors generate physiological responses. These responses are subject to multiple levels of feedback regulation, for example, as a means of preventing the overactivation of a response. Hormone-responsive TFs are subject to many additional layers of regulation, including inputs from multiple developmental and environmental signals (green arrows). As a consequence, the expression of any given target gene is context dependent; effector TFs constitute the apex of the regulatory hierarchy, with hormone-receptor inputs serving a condition-specific role. ABA, abscisic acid; BR, brassinosteroid; ET, ethylene; GA, gibberellin; IAA, indole acetic acid (auxin); SA, salicylic acid; SL, strigolactone.

nutritious, and better adapted to changing climate conditions? These are the main questions that motivated this topical review.

Phytohormones as regulatory metabolites

Here, we develop the hypothesis that currently recognized plant hormones are part of a larger and more diverse group of metabolites that acquired regulatory attributes over the course of embryophyte evolution. This supposition is based on the increasing recognition that many secondary (also referred to as "specialized") metabolites exhibit hormone-like effects on growth, development, and responses to the environment. The mechanisms by which these bioactive compounds act remain largely unknown but are often associated with changes in gene expression. The past century of plant hormone research, combined with new approaches to systematically map protein–metabolite interactions, provides a general roadmap for future progress in this area. We hope that this conceptual framework helps guide future research aimed at understanding the evolution, functions, and mechanisms of action of plant regulatory metabolites beyond currently recognized phytohormones.

Overview of phytohormone actions and origins

The identification of genes required for plant hormone biosynthesis and action resulted largely from the use of genetic approaches applied to the model eudicot species *Arabidopsis* (*Arabidopsis thaliana*). This branch of research greatly advanced our understanding of how "linear" phytohormone-response pathways function at the molecular level, including the discovery of ethylene as the first gaseous signaling molecule (Kende 1998), the elucidation of mechanisms to perceive steroids and auxins at the cell surface (He et al. 2000; Yu et al. 2023), and the paradigm that some hormones act as molecular glue to promote the reorganization of transcription complexes (Dharmasiri et al. 2003; Tan et al. 2007; Sheard et al. 2010). A deep mechanistic understanding of phytohormone-response pathways has also provided a foundation on which to explore more complex questions, such as the evolutionary origins of chemical communication in plants, cross-regulation between phytohormone-response pathways, the ecological relevance of phytohormones, and strategies to engineer economically im-portant plant traits.

Currently recognized plant hormones exert their effects by triggering signal transduction pathways that modulate the activities of specific cellular effectors, most often transcription factors (TFs). It is now evident that the phenotypic outputs of hormone signaling cascades are strongly influenced by synergistic and antagonistic interactions with many other signaling pathways (Bari and Jones 2009; Santner et al. 2009; Altmann et al. 2020). Due to this interconnectivity, phytohormone-response networks regulate gene expression and other physiological processes in a context-specific manner that is shaped by multiple developmental and environmental (e.g. light) cues (Fig. 1). This complex web of interactions likely explains the pleiotropic effects of hormones and why they appear to "do" so many things. A better understanding of how phytohormones and other bioactive metabolites interact to influence gene expression networks has the potential to improve crop performance and thus represents a major goal of modern plant science (Berens et al. 2017).

Plant hormones are derived from nearly all classes of primary metabolites, including isoprenoids, amino acids, fatty acids, and nucleotides (Fig. 1 and Box 1). Given their potent and wide-ranging effects on gene expression, endogenous hormone concentrations are subject to exquisite feedback control at the level of biosynthesis, transport, and catabolism. Bioactive end products of hormone biosynthetic pathways are perceived by specific proteinaceous receptors that, upon ligand binding, transduce the signal to a major TF or other effector (Fig. 1). High-affinity binding of hormones to their cognate receptor accounts for the ability of these metabolites to exert physiological effects at very low concentrations. Elucidation of the 3-dimensional structures of hormone–receptor complexes has yielded unprecedented insight into how ligand binding is linked to rapid changes in effector activity (Takeuchi et al. 2021). Some hormones induce a conformational change (i.e. allosteric regulation) in their receptor, whereas others act as molecular glues to promote protein–protein interactions that transduce the signal to an effector. Ligand–receptor interactions typically activate target effectors through changes in the activity of a receptor-linked enzyme, such as a kinase, phosphatase, or ubiquitin ligase. The integration of hormone perception with enzymatic

Box 1. Plant hormones are found in unexpected places

The tendency of phytohormones to exert broad-spectrum effects (Fig. 2) may reflect their discovery through the use of bioassays designed to identify substances that modulate plant growth and cell division. In this context, it is interesting to note that activity-driven purification of phytohormones employed starting materials not only from plant sources but also complex chemical mixtures from non-plant sources, including coal gas (ethylene), herring sperm DNA (cytokinin), human urine (auxin), and fungal exudates (gibberellins and jasmonates); only later was it determined that these metabolites are endogenous plant compounds. The powerful biological effects of classical hormones may explain why these low-hanging fruits on the larger tree of regulatory metabolites were the first to be identified using bioassay-based approaches biased toward easily measurable phenotypes (e.g. growth) that involve large-scale reprogramming of gene expression. Endogenous regulatory metabolites with more subtle effects on a narrow range of effectors surely exist (Fig. 2), but alternative experimental approaches may be needed to identify them (see Concluding remarks and future perspectives). Many phytohormones are produced by nonplant taxa, including bacteria, fungi, and algae, which lack the canonical hormone response pathways found in land plants. The production of phytohormones in these lineages has fascinating implications for understanding how processes such as horizontal gene transfer and endosymbiosis have shaped the relationships of plants with other groups of organisms (Bari and Jones 2009; Morffy and Strader 2020).

function is further illustrated by the recent discovery of plant hormone receptors having α/β-hydrolase or nucleotide cyclase activity (Takeuchi et al. 2021; Qi et al. 2022).

The advent of functional phylogenomic approaches has spurred remarkable progress in understanding the evolution of phytohormone-response pathways and, more broadly, the understudied interface between metabolism and signaling (Bowman et al. 2017; Blázquez et al. 2020; Rieseberg et al. 2023). Small-molecule plant hormones have features ascribed to both primary and secondary metabolites (Erb and Kliebenstein 2020; Fàbregas and Fernie 2021). Primary metabolites, such as amino acids, fatty acids, and sugars, are relatively abundant compounds (e.g. micromolar range) whose critical roles in the plant lifecycle stem from their use as building blocks that fuel tissue growth and increases in biomass. Although many hormones are also essential for the plant lifecycle, this dependency reflects the signaling

Box 2. Phytohormone signaling pathways are unique

The term "hormone" was coined by animal physiologists to convey the notion of chemical messengers that coordinate the growth and metabolic activity in different parts of the body (Tata 2005). This view was founded on observations that mammalian hormones are produced in specific locations, transported through the bloodstream, and generate physiological effects in target tissues. Plant scientists have generally, but not universally (Trewavas 1981; Bennett and Leyser 2014), embraced the hormone paradigm because it accounts for the potent biological effects of endogenous plant growth and immune regulators (Davies 2010). Beyond a general role in mediating intercellular communication, however, fundamental differences between plant and animal biology suggest that extrapolating hormone signaling mechanisms from animals to plants is naive at best, and at worst hinders a full appreciation of how plants use small molecules as informational cues. Indeed, the independent evolution of multicellularity in plants endowed them with signal communication networks that were likely shaped by unique features of the embryophyte lineage: a thick cell wall that restricts cell mobility; an autotrophic, sessile lifestyle enabled by photosynthetic plastids; increased developmental flexibility (i.e. totipotency) of interdependent above- and belowground organs; and an extraordinary diversity of metabolites to help meet the profound challenges of sessility. While both plants and animals rely on transcriptional programming to specify patterns of cell identity and growth, the signaling components responsible for executing these developmental programs in plants and animals are largely nonhomologous (Meyerowitz 2002). Moreover, many of the storied signaling components that helped to establish the animal hormone paradigm, including G protein-coupled receptors and associated second messengers such as cyclic AMP, either appear to be missing in plants (Urano and Jones 2014) or function in plants in ways that deviate from the animal paradigm (Wong et al. 2023). Plant cells also appear to be more autonomous than their animal counterparts, such that most plant cell types likely have the capacity to both synthesize and perceive a given chemical signal. Accordingly, the classic hormone concept of "action at a distance" does not necessarily apply to plants, even though many plant signaling compounds are transported long distances. If we have learned anything from the past century of phytohormone research, it is that plants use low-molecular-weight metabolites as informational cues in ways that are often unique to the plant kingdom.

properties of these metabolites, which typically accumulate to minuscule levels (e.g. subnanomolar) relative to the primary compounds from which they are derived. Whereas primary metabolites are produced by all organisms and cell types, the distribution of phytohormones in the tree of life is more sporadic and includes recent innovations within land plants (Box 1) (Bowman et al. 2017; Weng et al. 2021; Chini et al. 2023; Rieseberg et al. 2023). These collective observations support the view that small-molecule phytohormones trace their evolutionary origins to specialized metabolites that were co-opted for signaling roles in either the ancestral land plant or within land plants (Bowman et al. 2017; Rieseberg et al. 2023). Accordingly, plants use small-molecule signals in ways that are often unique to the plant kingdom (Box 2).

Phytohormones versus regulatory metabolites: what's the difference?

There is general consensus about the properties that define a phytohormone (Davies 2010). These characteristics include the ability to pleiotropically modulate growth, developmental, and immune-related processes, activity at very low concentrations, rapid biological responses to fluctuating concentrations of the hormone signal, and a high degree of conservation in the plant kingdom. Plant hormones are also perceived by dedicated receptors that transduce the signal to major effectors of physiological change.

Beyond these general features, how should we think about phytohormones conceptually? As is the case for other complex multicellular organisms, the spatial and temporal action of transcriptional regulators (e.g. TFs) plays a dominant role in executing a genetic blueprint that specifies rudimentary plant phenotypes. Phytohormone networks appear to confer plasticity to these hard-wired processes by providing real-time, environmentally dependent inputs that coordinate gene expression patterns between cell types, tissue types, and organs. This conceptual view of phytohormones as conditional modulators of gene expression has implications for understanding the stepwise assembly of hormone-response pathways during evolution (Bowman et al. 2017; Mutte et al. 2018; Blázquez et al. 2020). Increasing evidence indicates that protein components of various phytohormone signaling cascades predated the recruitment of the hormone into the pathway and thus have the capacity to control output responses independently of the hormone (Mutte et al. 2018; Sun et al. 2019; Bunsick et al. 2021). For example, auxin appears to have been recruited as a signal based on its ability to promote protein–protein interactions between preexist-ing components of the pathway, thereby enabling efficient control of TF activity (Mutte et al. 2018; Cao et al. 2022). In this scenario, the major drivers of plant phenotype are ancient transcriptional regulators such as TFs, which over evolutionary time acquired the capacity to receive and integrate information from internal (e.g. metabolic) and external (e.g. light) cues (Fig. 1).

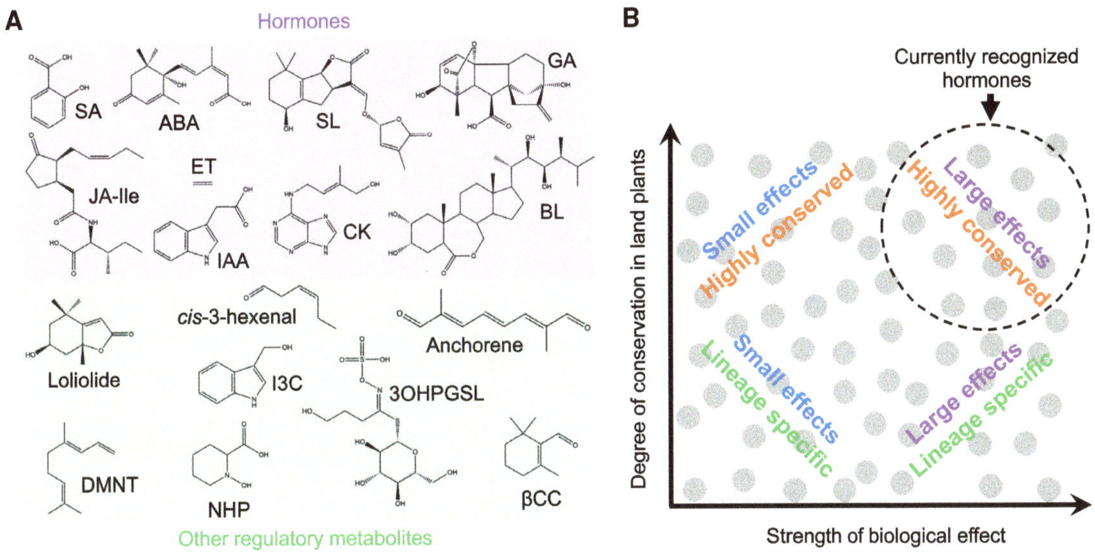

Figure 2. Diversity of plant regulatory metabolites. **A)** Chemical structures of currently recognized phytohormones (top group) and representative regulatory metabolites (bottom group). **B)** Plant regulatory metabolites (gray circles) differ not only in their chemical structures but also in their phylogenetic distribution (*y* axis) and range of biological effects (*x* axis). For example, phytohormone action is typically linked to the differential expression of hundreds (or thousands) of genes, whereas other regulatory metabolites modulate the expression of a smaller set of genes. Owing to their strong biological effects and high degree of conservation in land plants, currently recognized phytohormones tend to occupy the upper right quadrant (dashed circle) of the plot. 3OHPGSL, 3-hydroxypropylglucosinolate; ABA, abscisic acid; BL, brassinolide; CK, cytokinin (*trans*-zeatin); DMNT, (*E*)-3,8-dimethyl-1,4,7-nonatriene; ET, ethylene; GA, gibberellic acid (GA_3); I3C, indole-3-carbinol; IAA, indole-3-acetic acid (auxin); Jasmonic acid (JA)-Ile, jasmonoyl-L-isoleucine; NHP, *N*-hydroxy-pipecolic acid; SA, salicylic acid; SL, strigolactone (strigol); βCC, β-cyclocitral.

A unified phytohormone concept also needs to address the lack of a clear distinction between known hormones and an increasing number of plant metabolites that exhibit regulatory properties (Erb and Kliebenstein 2020). Evolutionary theory predicts that plants possess a diverse array of regulatory metabolites that differ not only in their biosynthetic origin and phylogenetic distribution but also in their range of biological effects (Fig. 2). We propose that hormones should be viewed not as an exclusive group of endogenous mediators but rather as part of a spectrum of compounds that acquired signaling attributes during or after the colonization of land by plants. This framework is consistent with current knowledge of how specialized metabolites in general, and phytohormones in particular, evolved (Bowman et al. 2017; Blázquez et al. 2020; Rieseberg et al. 2023). Phylogenomic analyses suggested that most hormone-response pathways emerged early in the evolution of land plants and, because of their essential roles in the plant lifecycle, were retained in present-day lineages (Bowman et al. 2017). Regulatory metabolites whose distribu-tion is restricted to specific plant taxa were either recruited later in evolution or perhaps discarded from other lineages due to their lack of utility in a particular environment. This evolutionary perspective accounts for the existence of lineage-specific regulatory metabolites (e.g. certain glucosinolate derivatives) that confer subtle—and thus difficult-to-detect—fitness advantages in a given environment (Fig. 2) (Erb and Kliebenstein 2020).

While it is obvious that all small-molecule hormones are regulatory metabolites, it is less clear which of these compounds should be categorized as phytohormones. Dwelling on this ambiguous distinction has limited practical value because the evolutionary tinkering that recruits metabolites into their myriad roles is blind to such classification schemes. One solution to this quandary is to phase out the use of the term "hormone" in favor of a more general term such as regulatory metabolite (Bennett and Leyser 2014). A quick survey of the literature, however, suggests that the plant hormone/phytohormone labels are here to stay. Suffice it to say that currently recognized phytohormone pathways mediate potent and pleiotropic biological effects in a wide range of plant species. Debating whether or not a particular metabolite should be classified as a hormone is less meaningful than determining how that compound functions at a mechanistic level; such insights are needed to obtain a satisfactory understanding of the physiological, ecological, and evolutionary relevance of bioactive metabolites.

The expanding universe of plant regulatory metabolites

Twenty-five years ago, Kende and Zeevaart (1997) predicted that the list of 5 "classical" plant hormones, namely abscisic acid, auxin, cytokinin, gibberellin, and ethylene, would expand as the functions of other bioactive compounds became recog-nized. This expectation has materialized with the general acceptance of brassinosteroid, jasmonate, salicylic acid, and strigolactone as the newest members of the group (Santner et al. 2009; Takeuchi et al. 2021). Beyond these known phytohormones, to what extent do plant metabolites have signaling attributes and how were these properties acquired during evolution? One way to address this question is through the study of presumptive regulatory metabolites whose chemical identity is unknown. The KARRIKIN INSENSITIVE 2 receptor, which per-ceives smoke-derived compounds (i.e. karrikins) that promote seed germination, is thought to perceive an endogenous ligand, but the identity of this compound(s) remains elusive (Waters and Nelson 2023). Another example is the cytochrome P450 KLUH/KLU, which in *Arabidopsis* is involved in the production of a mobile growth-promoting factor that is likely distinct from known phytohormones (Anastasiou et al. 2007).

A second approach to expand our knowledge of plant regulatory metabolites is to identify the molecular targets of endogenous compounds that exhibit signaling properties. The list of such compounds is lengthy and includes peptides, sugars (e.g. trehalose-6-phosphate), nucleotides (e.g. cyclic nucleotides), and amino acids (e.g. Glu). The direct targets of several of these compounds have been identified and studied in detail (Sandoval and Santiago 2020). Below, we highlight a few examples of secondary metabolites whose regulatory properties and modes of action are unknown but are under active investigation. Readers are referred to recent reviews for a more in-depth discussion of bioactive plant compounds originating from secondary metabolism (Erb and Kliebenstein 2020; Durán-Medina et al. 2021; Weng et al. 2021; Cai and Aharoni 2022).

Plant secondary metabolites account for much of the chemical diversity in nature. This large assortment of compounds exerts direct effects on plant-interacting organisms and also promotes adaptation to abiotic stress conditions. Interestingly, an increasing number of specialized metabolites are reported to possess signaling attributes reminiscent of phytohormones (Fig. 2) (Erb and Kliebenstein 2020). Flavonoids and polyamines have long been known for their ability to influence gene expression and other physiological processes (Peer and Murphy 2007; Davies 2010; Pourcel et al. 2013). Glucosinolates comprise a phylogenetically re-stricted group of defense compounds, some of which exert effects on growth and stress resilience (Kim et al. 2015; Malinovsky et al. 2017; Salehin et al. 2019; Katz et al. 2020) (Fig. 2). *N*-hydroxypipecolic acid, β-aminobutyric acid, and indole have received increasing attention for their immuno-modulatory properties (Erb et al. 2015; Cai and Aharoni 2022; Shields et al. 2022). Chemically diverse isoprenoids, including volatile terpenoids, have been intensively studied for their signaling roles in growth, defense, and interplant communi-cation (Zuo et al. 2019; Bai et al. 2021; Ninkovic et al. 2021; Hu 2022; Wang and Erb 2022). Oxidative cleavage of carote-noids is also a rich source of metabolites with regulatory functions, beyond the well-characterized abscisic acid and strigolactone hormones (Wang et al. 2021). Examples include the promotion of leaf stress responses by β-cyclocitral (Dinneny et al. 2019) and β-ionone (Felemban et al. 2023), the mediation of plant rhizosphere interactions by zaxinone (Wang, Haider, et al. 2019), and the control of root architecture by anchorene (Jia et al. 2019). Oxygenated fatty acids (i.e. oxylipins) and their derivatives, including green leaf volatiles (Matsui 2006), azelaic acid (Jung et al. 2009), and 12-oxo-phytodienoic acid (OPDA) (Wang et al. 2020), are also recognized as signaling compounds. These examples high-light the regulatory properties of plant secondary metabolites and raise the central question: what is their mode of action?

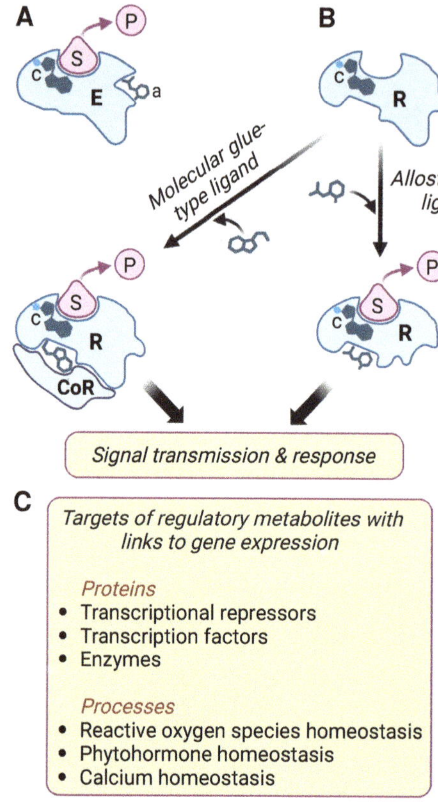

Figure 3. Low-molecular-weight metabolites interact with target proteins to modulate cellular processes that affect gene expression. **A)** Schematic diagram of an enzyme (E) interacting with its substrate (S) to generate a product (P). Enzymes often contain cofactors (c) to assist in catalysis and may also bind allosteric modulators (a) that positively or negatively regulate catalysis through conformational changes in the active site. **B)** Small molecules bind to proteinaceous receptors (R) and transmit information via 2 general mechanisms. First, and analogous to enzyme–metabolite interactions, allosteric-type ligands induce a conformational change in the receptor, which in turn propagates the signal (e.g. ABA, GA, and SL perception). Second, binding of molecular glue-type ligands creates a unique surface that interacts with a coreceptor (CoR) protein (e.g. IAA, JA, and BR perception). Many currently recognized phytohormone receptors possess (or are associated with) an enzymatic activity involved in signal transmission. Some plant receptors, including TIR1 and COI1, are associated with a cofactor (c) that is required for receptor structure and function. **C)** Examples of how gene expression may be modulated by the interactions of specific proteins or cellular processes with low-molecular-weight metabolites other than known phytohormones. See main text for details.

Consistent with their biological activity, most of the above-mentioned bioactive compounds have been shown to modulate nuclear gene expression. Low-molecular-weight metabolites functionally interact with target proteins in well-defined ways, including as enzyme substrates and products, cofactors, allosteric modulators, and molecular glues (Fig. 3). In theory, each of the protein–metabolite interactions has the potential to alter physiological responses through changes in gene expression. Specific examples of how small molecules may regulate transcription include the binding of β-caryophyllene to a TOPLESS-like transcriptional corepressor (Nagashima et al. 2019), the activation of a NIN-like protein TF by nitrate (Liu et al. 2022), the increase in cytosolic Ca^{2+} levels by Z-3-hexenal (Aratani et al. 2023), the perception of β-aminobutyric acid by an aspartyl tRNA synthetase (Luna et al. 2014), and the modulation of reactive oxygen species homeostasis by flavonols (Pourcel et al. 2013; Sugimoto et al. 2022; Daryanavard et al. 2023) (Fig. 3). Regulatory metabolites may also alter gene expression networks by affecting the metabolism or action of known phytohormones. For example, auxin signaling and transport are influenced by certain flavonols (Brown et al. 2001; Peer and Murphy 2007), apocarotenoids (Jia et al. 2019), and glucosinolates (Katz et al. 2015, 2020). Priming of plant defense responses by various volatile signals is also mediated by phytohormones, most notably jasmonate and salicylic acid (Fig. 3) (Engelberth et al. 2004; Dani and Loreto 2022; Hu 2022; Li et al. 2022).

It is also likely that the signaling properties of some secondary metabolites reflect dedicated perception systems that remain to be discovered. In their review of secondary metabolites that exhibit signaling properties, Erb and Kliebenstein (2020) concluded that these compounds affect growth and development by mechanisms that "are barely distinguishable from mechanisms normally assigned to plant hormones." Although this may be true at a conceptual level, studies claiming to identify the direct target of a bioactive plant secondary metabolite are sparse. Additional work is clearly needed to define the physiological relevance and precise modes of action of these and other protein–metabolite interactions.

Lessons from jasmonate

The past century of plant hormone research, together with emerging technologies to elucidate broader networks of plant metabolite–protein interactions, provides a general road map for future efforts to identify and characterize regulatory metabolites. As a case study, we highlight the major experimental approaches and discoveries that led to the current understanding of jasmonate as a lipid-derived signal for promoting transcriptional flexibility. Detailed mechanistic knowledge of the jasmonate response pathway and its evolutionary origins offer valuable insights into how metabolites are recruited into roles that promote transcriptional pliability in dynamic environments. Although we draw mainly on studies performed with *Arabidopsis* as a model system, increasing attention to other model and nonmodel species has provided key advances into our understanding of jasmonate biology. Readers are referred to several recent reviews for a comprehensive discussion of jasmonate biosynthesis, signaling, and evolution (Chini et al. 2016; Wasternack and Song 2017; Howe et al. 2018; Wasternack and Feussner 2018; Wang, Wu, et al. 2019; Monte 2023).

Exogenous jasmonates promote diverse responses

The discovery of jasmonates (collectively referred to as JAs) dates back more than half a century to the identification of methyl-JA (MeJA) as a component of floral scent (Demole et al. 1962). This work laid the foundation for the chemical synthesis of JAs, which are widely used as additives in the food and fragrance industries (Wasternack 2015). Jasmonic acid (JA) was first identified as a plant growth inhibitory substance from cultures of the fungus *Lasiodiplodia theobromae* (Aldridge et al. 1971). This finding foreshadowed the current paradigm that pathogens often produce phytohormones or structurally related compounds to co-opt host plant immunity (Box 1) (Bari and Jones 2009). The availability of synthetic JAs facilitated "spray and pray" approaches to investigate the effects of exogenous compounds in a wide range of plant species. These studies provided evidence that JAs inhibit growth (Aldridge et al. 1971; Staswick et al. 1992), promote senescence (Ueda and Kato 1980), and attract insect pollinators (Baker et al. 1981). Subsequent studies showed that exogenous JAs also induce the accumulation of various stress-related compounds, including proteinase inhibitors (PIs), defensive secondary metabolites, pathogenesis-related proteins, and vegetative storage proteins (Farmer and Ryan 1990; Parthier 1990; Staswick 1990; Gundlach et al. 1992). Newly emerging molecular approaches further demonstrated that low concentrations of exogenous JAs provoke global changes in plant gene expression (Farmer et al. 1992; Sembdner and Parthier 1993; Creelman and Mullet 1997). This early era of jasmonate research established that exogenous JAs elicit growth- and stress-related responses by a mechanism that likely depends on changes in gene expression.

Quantification of endogenous jasmonates

The ability of any exogenous substance to elicit a biological response is insufficient to establish a role for that compound as a regulatory factor. Rather, it is necessary to demonstrate that specific physiological effects coincide with the spatial and temporal accumulation of the metabolite in plant tissues. The discovery that JAs have signaling properties prompted the development of methods to measure these compounds in plant tissues (Creelman and Mullet 1997). Analytical platforms for quantifying JAs range from high-performance liquid chromatography (Yamane et al. 1981) and qualitative radio immunolabeling (Knöfel et al. 1984) to quantitative GC-MS analyses (Müller et al. 2002) and current state-of-the-art UPLC-MS/MS approaches capable of detecting target metabolites in the pmol range (e.g. Balcke et al. 2012). The low abundance of JAs in plant tissues is consistent with the notion that phytohormones act at concentrations several orders of magnitude below that of most primary metabolites. Analysis using increasingly sensitive instruments revealed the chemical diversity of JAs and related oxylipins (i.e. oxygenated fatty acids) in a

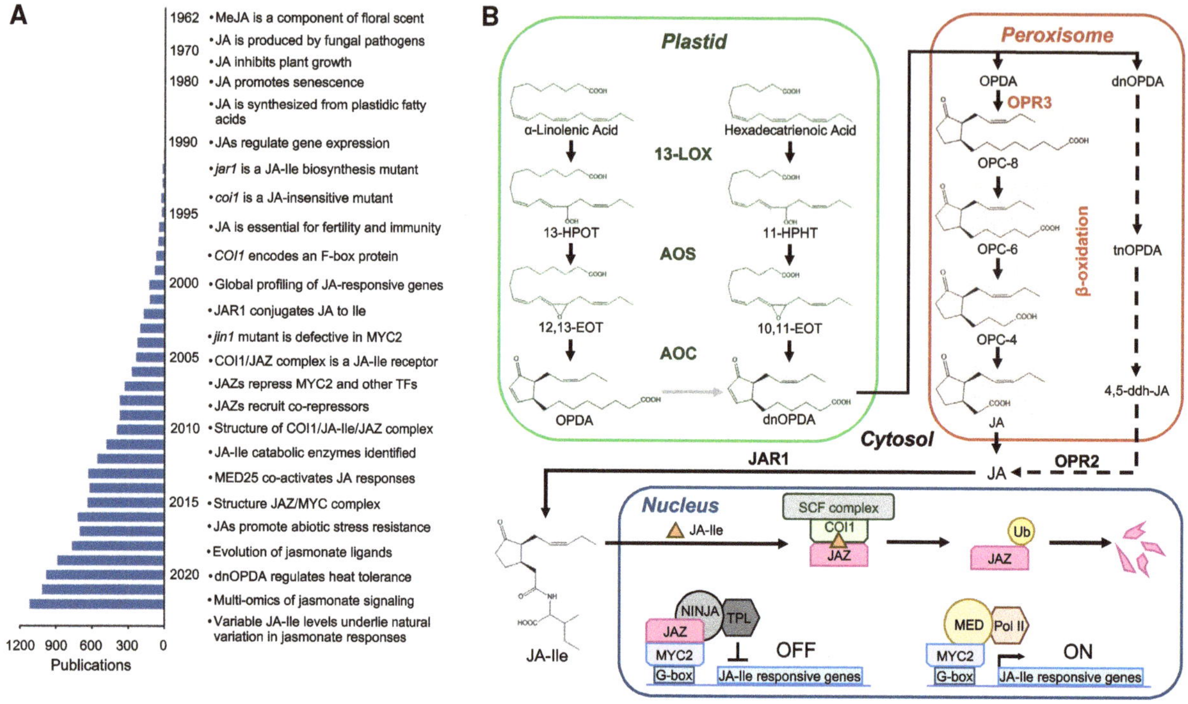

Figure 4. Major advances in elucidating the jasmonate pathway. **A)** Substantial milestones in jasmonate research are listed together with the number of jasmonate-related publications, based on the results of a 2023 PubMed search for "jasmonate." Due to space constraints, not all milestones are included. See Wasternack (2015) for a comprehensive history of jasmonate research. **B)** Scheme of JA-Ile biosynthesis and signaling. See main text for details. 10,11-EOT, (11*S*)-10,11-epoxy-octadecatrienoic acid; 11-HPHT, 11(*S*)-hydroperoxy-hexadecatrienoic acid; 12,13-EOT, (13*S*)-12,13-epoxy-octadecatrienoic acid; 13-HPOT, 13(*S*)-hydroperoxy-octadecatrienoic acid; 13-LOX, 13-LIPOXYGENASE; 4,5-ddh-JA, 4,5-didehydro JA; AOC, ALLENE OXIDE CYCLASE; AOS, ALLENE OXIDE SYNTHASE; COI1, CORONATINE INSENSITIVE 1; dnOPDA, dinor-oxo-phytodienoic acid; JA, (+)-7-*iso*-jasmonic acid; JA-Ile, (+)-7-*iso*-jasmonoyl-isoleucine; JAZ, JAZMONATE ZIM DOMAIN; MED, MEDIATOR complex; NINJA, NOVEL INTERACTOR OF JAZ; OPC-4, 3-oxo-2-(2-(Z)-pentenyl cyclopentane-1-butanoic acid; OPC-6, 3-oxo-2-(2-(Z)-pentenyl)-cyclopentane-1-hexanoic acid); OPC-8, 3-oxo-2-(2-(Z)-pentenyl)-cyclopentane-1-octanoic acid; OPDA, 12-oxo-phytodienoic acid; OPR2, OPDA REDUCTASE 2; OPR3, OXO-PHYTODIENOIC ACID REDUCTASE 3; Pol II, RNA polymerase II; SCF, SKP1-CUL1-F-box; tnOPDA, tetranor-OPDA; TPL, TOPLESS; Ub, ubiquitin.

broad range of plant species (Creelman and Mullet 1997; Wasternack and Feussner 2018).

Elucidation of the jasmonate biosynthetic pathway

A complete understanding of regulatory metabolites requires knowledge of how and where the compound is produced in plant cells. A combination of biochemical and genetic approaches was particularly useful for elucidating the enzymatic reactions that convert polyunsaturated fatty acids to JA and the major receptor-active form of the phytohormone, jasmonoyl-L-isoleucine (JA-Ile) (Browse 2009) (Fig. 4). Classical biochemical fractionation and labeling experiments demonstrated that JA is produced from chloroplastic pools of α-linolenic acid (18:3) via lipoxygenase (LOX) enzymes (Vick and Zimmerman 1984). The elucidation of jasmonate biosyn-thesis has been extensively reviewed (Wasternack 2015), and the pathway is summarized in Fig. 4. The intracellular location of jasmonate biosynthetic enzymes was validated using immunocytochemical and cell biological approaches (Schaller and Weiler 1997; Wasternack and Feussner 2018). Activity- based purification of several jasmonate biosynthetic enzymes from diverse plant species facilitated the molecular cloning of the corresponding genes using information deduced from the amino acid sequences of the purified proteins (e.g. Song and Brash 1991). These studies confirmed the presence of plastidic or peroxisomal targeting sequences in the relevant enzymes. The development of plant genomic resources, including sequencing of the *Arabidopsis* genome (Arabidopsis Genome Initiative 2000), greatly facilitated efforts to identify genes involved in jasmonate biosynthesis and other aspects of jasmonate biology.

Mutants galore

The importance of well-defined mutants (e.g. lines harboring null alleles) for elucidating the jasmonate pathway and under-standing the functions of JAs cannot be overstated. Identification of jasmonate-deficient *Arabidopsis* mutants in forward genetic screens proved instrumental not only for val-idating early models of jasmonate biosynthesis but also for uncovering new steps in the pathway (Schaller and Weiler 1997; Browse 2009; Wasternack and Feussner 2018). This approach was facilitated by the easily recognizable phenotypes of JA deficiency in *Arabidopsis*, including male sterility and increased susceptibility to pathogens and insects. The first jasmonate bio-synthetic mutant to be identified was *jasmonic acid resistant 1* (*jar1*), which displays insensitivity to JA- and MeJA-inhibited root growth (Staswick et al. 1992). Because these JAs strongly elicit responses in wild-type plants, the most straightforward interpretation at the time was that *jar1* is a jasmonate perception mutant. However, map-based cloning of *JAR1* showed that it encodes a JA-conjugating enzyme responsible for the production of JA-Ile, indicating that *jar1* is in fact a biosynthetic mutant that fails to produce a positive regulator (i.e. JA-Ile) of jasmonate responses (Staswick et al. 2002; Staswick and Tiryaki 2004). JAR1 is a member of the GH3 family of acyl acid amido synthetases, which conjugate amino acids to many phytohormones and likely other regulatory metabolites as well (Staswick and Tiryaki 2004; Westfall et al. 2012). Interestingly, a natural sequence polymorphism abrogating the expression of *JAR1* accounts for much of the variation in induced defense responses within wild populations of *Nicotiana attenuata* (Ray et al. 2023).

Another key advance in understanding the biosynthesis and function of JAs was the discovery that severe JA deficiency in *Arabidopsis* results in male sterility due to defects in pollen development and anther dehiscence (Feys et al. 1994; Browse 2009). The reversibility of this phenotype by exogenous JA was first documented in a fatty acid desaturase triple mutant (*fad3 fad7 fad8*) that fails to produce trienoic fatty acid precursors of JA (McConn and Browse 1996). The application of JA to flower buds of the *fad3 fad7 fad8* mutant recovered its fertility, demonstrating that the conversion of 18:3 to JA is essential for normal flower development. Subsequent screens for male-sterile mutants that could be chemically complemented by JA treatment led to the isolation of additional biosynthetic mutants, including those impaired in ALLENE OXIDE SYNTHASE, 12-OXO-PHYTODIENOIC ACID REDUCTASE 3 (OPR3), and DEFECTIVE IN ANTHER DEHISCENCE1 (DAD1) (reviewed by Browse 2009) (Fig. 4). The latter enzyme is a phospholipase A1 that acts in floral tissues to produce fatty acid precursors of JA from plastidial galactolipids (Ishiguro et al. 2001). In parallel with studies on the role of JAs in male fertility, the availability of JA biosynthetic mutants was instrumental in establishing a causal link between JAs and resistance to a broad spectrum of biotic agents, including arthropod herbivores and necrotrophic pathogens (Glazebrook 2005; Howe and Jander 2008; Bari and Jones 2009; Browse 2009).

Functional analysis of genes encoding biosynthetic enzymes and other components of the jasmonate pathway was advanced through the use of reverse genetic approaches that leveraged newly developed molecular tools, most notably the sequenced *Arabidopsis* genome (Arabidopsis Genome Initiative 2000) and sequence-indexed insertional mutants (Alonso et al. 2003). These resources were particularly important for the characterization of jasmonate biosynthetic enzymes encoded by multigene families, as functional redundancy among paralogs often precludes the identification of mutants in forward genetic screens (Schilmiller et al. 2007; Delfin et al. 2022). The *13-LOX* gene family in *Arabidopsis* exemplifies the use of reverse genetics in helping to decipher the subfunctionalization of specific family members. All 4 13-LOX enzymes (LOX2, LOX3, LOX4, and LOX6) in *Arabidopsis* oxygenate 18:3 and contribute to JA production, likely explaining why *Arabidopsis 13-lox* mutants have not been identified in forward genetic screens. Detailed analysis of single and multiorder *lox* mutants showed that each paralog preferentially acts in different tissues and cell types or is involved in other aspects of oxylipin biosynthesis (Bannenberg et al. 2009; Caldelari et al. 2011; Chauvin et al. 2013; Grebner et al. 2013; Mochizuki et al. 2016). Reverse genetic approaches were also useful for uncovering the existence of alternative pathways that operate in the absence of the major JA biosynthetic route. For example, in mutants lacking OPR3, OPDA can be converted to 4,5-di-dehydro-JA in the peroxisome and then reduced to JA in the cytosol by OPR2 (Fig. 4) (Chini et al. 2018; Howe 2018). Although OPR2 contributes weakly to JA production in *Arabidopsis*, recent evidence suggests that this alternative biosynthetic route predated the OPR3-dependent pathway in the evolution of jasmonate biosynthesis (Chini et al. 2023).

A satisfactory understanding of phytohormones and other regulatory metabolites depends on knowledge obtained from diverse plant species. In general, the roles of JAs in promoting senescence, growth inhibition, and broad-spectrum resistance to herbivores and pathogens are broadly conserved across land plants (Monte et al. 2019). However, whereas JAs appear to be essential for normal reproductive development in all plants examined to date, defects in the JA pathway often give rise to reproductive phenotypes that are specific to a particular species or lineage. The rice *extra glume1* (*eg1*) mutant, which is defective in a DAD1 homolog, is impaired in early stages of spikelet and floret development but show normal stamen and pollen maturation (Cai et al. 2014). Disruption of JA biosynthesis in the maize (*Zea mays*) sex determination mutant *tasselseed 1* (*ts1*) results in feminization of the male inflorescence (tassel) but does not affect early spikelet and floret development (Acosta et al. 2009). The lack of jasmonate responsiveness in the tomato (*Solanum lycopersicum*) *jasmonic acid-insensitive 1* (*jai1*) mutant impairs the maternal control of seed development while decreasing but not eliminating pollen viability (Li et al. 2004; Goetz et al. 2012). These examples illustrate how the study of mutants in diverse plants adds important perspectives to our understanding of jasmonate biology.

Regulation of jasmonate biosynthesis and catabolism

A substantial advance toward understanding the regulation of jasmonate biosynthesis was the discovery that various stress-related treatments result in the rapid accumulation of endogenous JAs. Examples of eliciting stimuli include the treatment of cultured plant cells with fungal cell wall fragments (Gundlach et al. 1992) and mechanical wounding of leaves (Creelman et al. 1992; Farmer and Ryan 1992). The latter observation linked groundbreaking work on wound-inducible defensive PIs in tomato to the discovery that exogenous JAs rapidly activate the expression of PI-encoding genes (Green and Ryan 1972; Farmer and Ryan 1990; Farmer et al. 1992). A model in which the perception of wounding and other

Table 1. Distribution of jasmonates in *Arabidopsis*

Tissue	Condition	OPDA[a]	JA[a]	JA-Ile[a]	References
Leaves	Resting	600	<10	<2	Schulze et al. (2019), Reymond et al. (2000), Koo et al. (2009)
Leaves	1 h post leaf wounding	25,000	7,000	400	Schulze et al. (2019), Reymond et al. (2000), Koo et al. (2009)
Roots	Resting	35	<10	<2	Schulze et al. 2019
Roots	1 h Post leaf wounding	100	60	60	Schulze et al. 2019

[a]Metabolite levels are in pmol/g FW tissue.

stress-related signals at the cell surface triggers the production of JA and concomitant defense responses has garnered strong support over the past 3 decades (Farmer and Ryan 1992; Howe 2019). These early insights into the roles of JAs in inducing the expression of genes encoding defensive compounds and their accumulation opened up new avenues of investigation into how this lipid-based signaling pathway controls plant immun-ity (Glazebrook 2005; Karban and Baldwin 2007; Campos et al. 2014; Wang, Wu, et al. 2019).

Owing to its simplicity and effectiveness in inducing jasmonate biosynthesis, mechanical wounding has been used extensively to study the dynamics of jasmonate responses within and between different plant tissues. In general, JA and JA-Ile levels increase within minutes of tissue damage and peak within approximately 1 h (Table 1) (Baldwin et al. 1997; McConn and Browse 1996; Glauser et al. 2009; Koo et al. 2009). Basal levels of OPDA are approximately 500– 600 pmol/g fresh weight (FW) in undamaged *Arabidopsis* leaves and 30–40 pmol/g FW in roots and increase 40-fold in leaves and 6-fold in roots after leaf wounding (Table 1). Although basal JA levels are lower than basal OPDA levels and are often below the limit of quantification in undamaged tissues, JA levels can increase up to 500-fold after mechanical wounding. Peak levels of wound-induced JA-Ile are typically much less (e.g. 10-fold) than peak levels of JA in wounded tis-sues (Table 1). This observation is consistent with the "pro-hormone" concept in which JA availability is a limiting factor for the JAR1-catalyzed formation of JA-Ile (Staswick and Tiryaki 2004; Acosta and Farmer 2010; Koo 2018). The large differences in the relative abundances of JA biosynthetic precursors and intermediates (18:3 > OPDA > JA > JA-Ile) also illustrate how the tightly regulated biosynthesis of these regulatory metabolites depends on an appropriate supply of precursors along the entire biosynthetic route (Table 1).

Precise control over the spatial and temporal accumulation of phytohormones depends not only on biosynthetic en-zymes but also on factors involved in metabolite transport and catabolism (Li et al. 2021). In the case of JAs, several metabolic transformations are crucial for preventing the overactivation of costly jasmonate-mediated stress responses (Koo and Howe 2009; Heitz et al. 2016; Wasternack and Feussner 2018). For example, cytosolic pools of JA are depleted via the combined action of highly specific JA hydroxylases and sulfotransferases (Caarls et al. 2017; Smirnova et al. 2017; Fernández-Milmanda et al. 2020). JA-Ile itself is inactivated both by hydrolysis of the amide bond (Woldemariam et al. 2012; Widemann et al. 2013; Koo 2018) and by cytochrome P450 enzymes that oxidize the pentenyl side chain to generate the corresponding hydroxy and carboxy derivatives (Koo et al. 2011, 2014; Heitz et al. 2012; Aubert et al. 2015). The finding that JA induces the expression of genes encoding these catabolic enzymes is consistent with the notion of negative feedback regulation of JA-Ile levels (Heitz et al. 2016; Koo 2018). Notably, the rapid inducibility of these and other jasmonate-related genes has been exploited through "guilt-by-associated" coexpression approaches to identify components of the jasmonate response pathway (e.g. Koo et al. 2011).

Homeostasis of plant regulatory metabolites also involves processes that control the systemic production of the chemical signal. The jasmonate pathway provides an excellent model to address this question (Gao et al. 2023; Morin et al. 2023). JA and JA-Ile levels increase rapidly not only locally at the site of injury but also in distal unwounded leaves (Glauser et al. 2009; Koo et al. 2009) and roots (Gasperini et al. 2015). Leaf injury provoked by mechanical wounding or insect herbivory stimulates the long-distance translocation of OPDA and JA to mount defense responses in distal organs (Schulze et al. 2019; Li et al. 2020). Concomitant with this burst of JAs is the rapid propagation of slow wave electrical potentials governed by GLUTAMATE RECEPTOR-LIKE (GLR) proteins (Mousavi et al. 2013; Nguyen et al. 2018). Wounding also stimulates the release of extracellular elicitors such as Glu, which binds to specific GLRs to regulate long-distance Ca^{2+} waves (Toyota et al. 2018; Alfieri et al. 2020; Bellandi et al. 2022). Recent studies in *Arabidopsis* further showed that tissue injury results in the generation of aglucone elicitors of the GLR-dependent propagation of slow wave potentials via glucohydrolase enzymes (Gao et al. 2023). Consistent with these findings, mutants defective in specific GLRs (e.g. GLR3.3 and GLR3.6) are impaired in wound-induced slow wave potentials, long-distance increases in cytosolic Ca^{2+} and jasmonate accumulation in distal leaves (Mousavi et al. 2013; Nguyen et al. 2018; Toyota et al. 2018). Specific DAD1-like lipases appear to link wounding-responsive electrical signaling to the activation of JA biosynthesis (Morin et al. 2023).

Elucidation of the core jasmonate signaling pathway

By the mid-1990s, it was becoming increasingly evident that the potent effects of JAs on gene expression involve a specific mechanism that links the perception of these compounds to transcriptional reprogramming (Parthier 1990; Farmer and Ryan 1992; Wasternack and Parthier 1997). However, in the absence of any clues into how JAs function, which experimental approaches were best suited to tackle this issue? Once again, mutants identified in genetic screens provided the way forward. The first major advance in understanding the mode of action of JAs came from the identification of the *Arabidopsis coronatine insensitive 1* (*coi1*) mutant. This mutant fails to respond to coronatine, which is a toxin produced by some patho-genic strains of *Pseudomonas syringae* (Feys et al. 1994). As a structural mimic of JA-Ile, exogenous coronatine elicits physiological responses that are very similar to those induced by ex-ogenous JAs. The finding that *coi1* mutants were insensitive to both coronatine and JA indicated that these compounds signal via a similar mechanism (Feys et al. 1994). Importantly, *coi1* mutants are also male sterile and highly susceptible to attack by necrotrophic pathogens and herbivores (Feys et al. 1994; Thomma et al. 1998; Glazebrook 2005). The phenotypic similarity between JA biosynthesis and *coi1* perception mutants provided compelling genetic evidence that endogenous JAs activate a COI1-dependent signaling pathway leading to the ex-pression of jasmonate-responsive genes.

Map-based cloning studies revealed that *COI1* encodes a member of the large (>500 in *Arabidopsis*) F-box family of proteins (Xie et al. 1998). F-box proteins act as specificity determinants in SCF (Skp1/Cullin/F-box)-type E3 ubiquitin ligase complexes that target specific substrates for ubiquitylation and proteolytic destruction (Shabek and Zheng 2014). Accordingly, it was suggested that COI1 recruits a negative regulator of jasmonate responses for targeted degradation, which in turn promotes fertility and defense responses (Xie et al. 1998). This model is supported by the sequence similarity between COI1 and the closely related F-box protein TRANSPORT INHIBITOR RESPONSE 1 (TIR1) and by emerging evidence that TIR1 is a central component of the auxin perception apparatus (Ruegger et al. 1998). An intensive search for COI1 substrates included genetic screens to identify additional regulators of the pathway (Ellis and Turner 2001; Hilpert et al. 2001; Xu et al. 2001; Jensen et al. 2002; Lorenzo et al. 2004), identification of COI1-interacting proteins (Devoto et al. 2002) and cataloging of JA-responsive genes that are expressed in a COI1-dependent manner (Devoto et al. 2005). Despite these efforts, viable candidate COI1 substrates were not identified, and progress toward understanding the molecular mechanism of jasmonate signaling was at an impasse for nearly a decade (Browse 2009). One notable discovery during this period was the identification of jasmonate-insensitive mutants that are defective in a basic helix–loop–helix-type TF called MYC2 (Berger et al. 1996; Boter et al. 2004; Lorenzo et al. 2004). Although MYC2 and its closely related proteins (e.g. MYC3 and MYC4 in *Arabidopsis*) are regarded as master positive regulators of jasmonate responses (Fernández-Calvo et al. 2011; Kazan and Manners 2013), a satisfactory model to explain how MYC activity is linked to jasmonate and COI1 failed to emerge at this time.

In 2007, the discovery of JASMONATE ZIM DOMAIN (JAZ) proteins as the elusive COI1 substrates was achieved through the analysis of gene coexpression modules (Thines et al. 2007; Yan et al. 2007) and positional cloning of a mutated gene that confers dominant insensitivity to JA (Chini et al. 2007). Thines and coworkers found that 8 related genes encoding uncharacterized proteins containing a ZINC-FINGER PROTEIN EXPRESSED IN INFLORESCENCE MERISTEM (ZIM) motif were rapidly induced by exogenous jasmonate (Thines et al. 2007). Similarly, genome-wide expression profiling showed that several wound-induced transcripts encoding these ZIM domain-containing proteins accumulated in a JA-dependent manner (Yan et al. 2007). Finally, the causative mutation underlying the JA-insensitive *jai3-1* mutant phenotype was shown to reside in a chromosomal region that contains a member (*JAZ3*) of the same gene family (Chini et al. 2007). This plant-specific gene family was named *JAZ*, with 13 recognized members in *Arabidopsis* (Chini et al. 2007; Thines et al. 2007; Yan et al. 2007; Thireault et al. 2015). The size of the rapidly evolving *JAZ* family ranges from one in the common liverwort (*Marchantia polymorpha*) to more than 20 members in many flowering plants (Howe et al. 2018; Monte et al. 2019).

Three key findings associated with the discovery of JAZ helped to fill in the remaining major gaps in understanding the core mechanism of jasmonate signaling. First, mutated JAZ proteins that fail to interact with COI1 strongly repress the expression of jasmonate-responsive genes (Chini et al. 2007; Thines et al. 2007; Yan et al. 2007). Second, JAZ proteins exert repression by binding directly to MYC TFs (Chini et al. 2007) and many other TFs as well (Fernández-Calvo et al. 2011; Chini et al. 2016; Howe et al. 2018). Finally, the physical interaction between COI1 and JAZ is promoted by JA-Ile but not its metabolic precursors or derivatives, including OPDA, JA, and MeJA; it was therefore concluded that a COI1–JAZ complex constitutes the site of JA-Ile perception (Thines et al. 2007). These collective findings provided a basis for a model (Fig. 4) in which MYC-dependent responses are repressed by JAZ in the absence of JA-Ile. In response to stress-related or developmental cues that stimulate JA-Ile accumulation, JA-Ile promotes the SCF^{COI1}-dependent turnover of JAZ and subsequent expression of jasmonate-responsive genes.

Follow-up studies showed that coronatine also promotes the formation of COI1–JAZ complexes and that the C-terminal Jas motif of JAZ harbors a "degron" sequence that mediates the hormone-dependent COI1–JAZ interaction (Katsir et al. 2008; Melotto et al. 2008). Mutation of the degron motif stabilizes JAZ against hormone-dependent degradation, resulting in dominant repression of jasmonate responses (Chini et al. 2007; Thines et al. 2007; Yan et al. 2007; Melotto et al. 2008). Biochemical and structural analyses further showed that the COI1–JAZ interaction is mediated specifically by (+)-7-*iso*-JA-Ile, which is the major JA-Ile stereoisomer in plant tissues (Fonseca et al. 2009; Sheard et al. 2010). In validating and extending the pioneering work on JAR1 (Staswick and Tiryaki 2004), it is now evident that jasmonate responses elicited by exogenous JA and MeJA require metabolic transformation of these compounds to (+)-7-*iso*-JA-Ile in plant tissues.

Structural insights into jasmonate perception and signaling

The discovery that COI1–JAZ complexes are a high-affinity receptor for JA-Ile (Thines et al. 2007) was aided by the molecular glue paradigm, which first emerged from the study of auxin perception (Dharmasiri et al. 2003; Kepinski and Leyser 2005; Tan et al. 2007). The 3-dimensional structure of the ligand-bound COI1–JAZ ternary complex shows that an inositol pyrophosphate (likely $InsP_8$) serves as a COI1 structural cofactor and that the JAZ degron simultaneously interacts with both the bound hormone and COI1 (Fonseca et al. 2009; Sheard et al. 2010; Laha et al. 2015). This model of COI1–JAZ as a high-affinity coreceptor does not exclude the possibility that JA-Ile first binds to COI1 with low affinity, followed by the formation of a high-affinity COI1–JAZ complex. This 2-step model of JA-Ile perception has received experimental support by showing that coronatine weakly interacts (i.e. low micromolar affinity) with COI1 in the absence of JAZ (Yan et al. 2018). Similar results were recently reported for TIR1, which can bind auxin at low micromolar affinity in the absence of an Aux/IAA substrate (Cao et al. 2022). These collective findings support a scenario in which molecular glue-type hormones were recruited as signals based on their ability to promote protein–protein interactions (Fig. 3).

The structural elucidation of MYC3 binding to the Jas domain of JAZ9 further clarified the molecular mechanism by which JAZ proteins repress MYC TFs (Zhang et al. 2015). In the absence of the hormone, the Jas motif assumes an α-helical structure that intercalates within the MYC N-terminal fold, thereby inhibiting the interaction of MYC with the MED25 subunit of the mediator complex (Çevik et al. 2012; Chen et al. 2012; Zhang et al. 2015). JAZ proteins recruit transcriptional corepressors of the TOPLESS family via the adaptor protein NINJA (Pauwels et al. 2010) and, in some cases, directly interact with TOPLESS through an EAR motif (Kagale et al. 2010; Shyu et al. 2012). JA-Ile-induced JAZ degradation releases MYCs from repression while simultaneously facilitating the recruitment of coactivators such as MED25 (Zhang et al. 2015; An et al. 2017) (Fig. 4). Thus, JAZ proteins perform dual functions as transcriptional repressors in the absence of JA-Ile and as coreceptors in the presence of the hormone. Understanding the structural and dynamic behavior of JAZ-associated transcriptional complexes represents an important future challenge (Zhang et al. 2017; Zhou et al. 2023). The prolific ability of JAZs to interact with proteins in other signaling pathways (Robert-Seilaniantz et al. 2011; Song et al. 2014; Berens et al. 2017; Howe et al. 2018; Zhou et al. 2023) provides a window into how regulatory metabolites connect pathways to reconfigure transcriptional programs in a flexible and dynamic manner.

Multiple layers of repression

A recurring theme in hormone biology is the existence of mechanisms to attenuate the amplitude and duration of responses through negative feedback control (Fig. 1). In the case of jasmo-nate, overactivation of the pathway results in stunted growth, decreased fecundity, loss of photosynthetic capacity, and, in its extreme, cell death (Baldwin 1998; Yan et al. 2007; Zhang and Turner 2008; Guo et al. 2018, 2022). Among the major mechan-isms to attenuate these costly responses are the catabolism of bioactive JAs (see above) and de novo synthesis of JAZ repressors. The latter mechanism is based on the observation that *JAZ* genes are rapidly and strongly expressed by MYC2 in response to jasmonate (Chini et al. 2007; Thines et al. 2007; Chung et al. 2008; Moreno et al. 2013; Zander et al. 2020). Effective JAZ-mediated repression relies on the recruitment of various corepressor and adaptor proteins, including NINJA, TOPLESS, and epigenetic modifying enzymes (Pauwels et al. 2010; Howe et al. 2018; Zander 2021). JAZ stabilization resulting from alternative splicing events that eliminate or weaken COI1 interactions, as well as jasmonate-induced expression of factors that compete with MYCs for DNA binding, represent additional layers of negative feedback control (Chung et al. 2010; Nakata et al. 2013; Fonseca et al. 2014; Qi et al. 2015; Zhang et al. 2017; Wu et al. 2020). These many layers of signal attenuation highlight the negative fitness consequences of overactivating jasmonate responses.

Evolution of the jasmonate pathway

Functional phylogenomic approaches based on recently sequenced streptophyte (charophyte algae and land plants) genomes have provided important new insights into processes that shaped the emergence of metabolic and signaling pathways during the colonization of land by plants (Maeda and Fernie 2021; Bowman 2022; Rieseberg et al. 2023). Recent studies on the evolution of JAs have contributed to these efforts by revealing lineage-specific innovations in both JA biosynthetic and signaling pathways. Genome analysis revealed that bryophyte lineages (hornworts, liverworts, and mosses) contain orthologs for all core signaling JA components, including COI1, JAZ, NINJA, TPL, and MYC (Bowman et al. 2017; Monte et al. 2018). Functional analyses of the COI1-JAZ-MYC signaling module showed that the jasmonate pathway likely emerged in the common ancestor of land plants more than 450 million years ago (Bowman et al. 2017; Peñuelas et al. 2019; Chini et al. 2023). Interestingly, however, bryophytes neither synthesize nor respond to JA-Ile, which is the receptor-active form of the hormone in vascular plants. Rather, studies with *M. polymorpha* (Mp) and the lycophyte *Huperzia selago* identified several iso-mers of dnOPDA (dn-cis-OPDA and dn-iso-OPDA) as active ligands in these species (Monte et al. 2018; Kneeshaw et al. 2022). The selectivity of the MpCOI1–JAZ coreceptor for dnOPDA ligands was attributed to a single amino acid change in MpCOI1 that alters the size of the ligand-binding pocket, as well as specific residues in the Jas motif of MpJAZ. A broad survey of the green lineage recently confirmed and extended these insights into the evolutionary trajectory of JAs (Chini et al. 2023).

Functional phylogenomic studies have also provided insight into the extent to which the physiological functions of JAs are conserved in land plants. Transcriptional changes in *M. polymorpha* induced by treatment with dnOPDA are similar to transcriptional changes elicited by JA-Ile in flowering plants, including the induction of genes related to secondary metabolism and wounding responses (Devoto et al. 2005; Chini et al. 2018). However, whereas JA-Ile promotes defense and inhibits growth and fertility in *Arabidopsis*, dnOPDA appears to regulate defense and growth responses but not fertility in *M. polymorpha* (Monte et al. 2018). Given the reduced fertility of OPDA-deficient mutants of the moss *Physcomitrium patens* (Stumpe et al. 2010), additional work is needed to better under-stand the evolutionary basis of the roles of JAs in plant repro-ductive processes.

It is noteworthy that cyclopentenone JAs such as OPDA and dnOPDA are reactive electrophilic compounds that exhibit signaling properties independently of COI1. A recent study provided evidence that these compounds signal through a COI1-independent pathway to promote thermotolerance in diverse species, including *Arabidopsis*, *M. polymorpha*, and the charophyte alga *Klebsormidium nitens* (Monte et al. 2020). These findings suggest an ancestral role for JAs in regulating protection against heat stress, which was likely important for plant colonization of terrestrial habitats. Elucidation of the COI1-independent signaling pathway by which cyclopentanone JAs promote thermotolerance promises to provide new insight into how the larger family of lipid-derived JAs were recruited into distinct signaling pathways during plant evolution.

OUTSTANDING QUESTIONS

- What are the direct targets of the myriad volatile and nonvolatile plant secondary metabolites that exhibit signaling attributes?
- To what extent can untargeted mass spectrometry-based and predictive approaches be used to systematically identify protein–metabolite interactions with a wide range of affinities?
- How do evolutionary forces recruit secondary metabolites into signaling pathways?
- Beyond the small number of plant hormones known to act in the nucleus, to what extent do small molecules interact with and modulate the activities of transcriptional regulators?
- How can knowledge of regulatory metabolites be used to develop crop varieties that are more productive, nutritious, and better adapted to changing climate conditions?

Concluding remarks and future perspectives

The extraordinary diversity of plant metabolism arises from adaptive processes such as the neofunctionalization of duplicated enzyme-encoding genes and increased enzyme promiscuity (Chae et al. 2014; Weng et al. 2021). As a consequence, hundreds of thousands of secondary metabolites constitute an evolving chemical "library" from which to recruit small molecules into roles that promote plant fitness in dynamic and often harsh environments. This process is exemplified by small-molecule plant hormones, which have acquired the ability to modulate cellular processes needed to help the plant adapt its growth and development to changing conditions. Currently recognized phytohormones appear to be part of a larger and more diverse group of chemical mediators whose action is linked in part to transcriptional control. Lessons from the study of the jasmonate pathway and other phytohormones provide a guide to drive future progress in unveiling the physiological roles of regulatory metabolites, their modes of perception and signaling, and their evolutionary history. Current knowledge gaps (see Outstanding Questions) pose exciting challenges for the future of plant science, which may help to inform efforts to modify signaling pathways for crop improvement.

A major bottleneck in understanding how plant metabolites influence gene expression and associated traits is the lack of insight into the direct targets of these compounds. The vast majority of small-molecule receptors in plants were identified through the use of mutants that are unresponsive to a defined eliciting cue, including (pro)hormones and hormone agonists. Following the identification of such mutants, molecular genetic techniques are employed to track down the causal mutation. The strength of this approach lies in linking a specific gene to a robust biological response, as well as its application to any bioactive compound whose chemical identity is known. Moreover, advances in genome-sequencing technology have greatly accelerated the process of identifying the mutated gene underlying a phenotype of interest (e.g. Schneeberger et al. 2009). Limitations of the classical genetic approach include the labor intensiveness of genetic screens and the possibility that perception mutants are difficult to identify due to functional redundancy within receptor-encoding gene families.

These considerations justify future research to systemically investigate regulatory metabolite–protein interactions in plants. Such efforts have historically been hampered by technical challenges related to the transient, low-affinity nature of these bimolecular associations. Nevertheless, recent advances in mass spectrometry have facilitated the development of powerful approaches to identify biologically relevant protein–metabolite interactions in various model organisms (Piazza et al. 2018; Luzarowski and Skirycz 2019). This technology can be applied in either a non-targeted or targeted manner (i.e. identification of proteins that bind to a specific metabolite) and is adaptable for use in plants (Schlossarek et al. 2023). Improvements in molecular modeling and other computational approaches for predicting protein–metabolite interactions, which are widely used for drug discovery (Lavecchia 2015), are also expected to be employed to identify candidate protein targets of plant regulatory metabolites. In the coming years, there is much to look forward to as these new technologies are combined with robust genetic approaches to delve deeper into the expanding universe of plant regulatory metabolites.

Acknowledgments

We thank Erich Grotewold, Danve Castroverde, and anonymous reviewers for helpful comments on the manuscript. We gratefully acknowledge current and past laboratory members for their many contributions to jasmonate research over the years. We also acknowledge the influence of the stimulating research environments in our respective institutions (Leibniz Institute of Plant Biochemistry and MSU-DOE Plant Research Laboratory), where there is a strong legacy of phytohormone research. G.A.H. thanks Hans Kende and Jan Zeevaart for inspiration in the early stage of his career, and the Fulbright U.S. Scholar program for the opportunity to discuss aspects of this work with colleagues in Japan. The graphical abstract and Figs. 1 and 3 were created with BioRender.com.

Author contributions

G.A.H. conceived the structure of the review; D.G. and G.A.H. wrote the paper.

Funding

This work was supported by the Chemical Sciences, Geosciences, and Biosciences Division, Basic Energy Sciences, Office of Science at the U.S. Department of Energy (through grant DE-FG02–91ER20021) and the Michigan AgBioResearch Project MICL02278 to G.A.H. and by the IPB core funding from the Leibniz Association, the German Research Foundation (grants GRK2498/400681449 and GA 2419/3-1) and the European Union (ERC, MECHANOJAS, 101088876) to D.G.

Conflict of interest statement. None declared.

References

Acosta IF, Farmer EE. Jasmonates. Arabidopsis Book. 2010:**8**:e0129. https://doi.org/10.1199/tab.0129

Acosta IF, Laparra H, Romero SP, Schmelz E, Hamberg M, Mottinger JP, Moreno MA, Dellaporta SL. *Tasselseed1* is a lipoxygenase affecting jasmonic acid signaling in sex determination of maize. Science. 2009:**323**(5911):262–265. https://doi.org/10.1126/science.1164645

Aldridge DC, Galt S, Giles D, Turner WB. Metabolites of *Lasiodiplodia theobromae*. J Chem Soc. 1971:**1971**:1623–1627. https://doi.org/10.1039/j39710001623

Alfieri A, Doccula FG, Pederzoli R, Grenzi M, Bonza MC, Luoni L, Candeo A, Romano Armada N, Barbiroli A, Valentini G, et al. The structural bases for agonist diversity in an *Arabidopsis thaliana* glutamate receptor-like channel. Proc Natl Acad Sci U S A. 2020:**117**(1):752–760. https://doi.org/10.1073/pnas.1905142117

Alonso JM, Stepanova AN, Leisse TJ, Kim CJ, Chen H, Shinn P, Stevenson DK, Zimmerman J, Barajas P, Cheuk R, et al. Genome-wide insertional mutagenesis of *Arabidopsis thaliana*. Science. 2003:**301**(5633):653–657. https://doi.org/10.1126/science.1086391

Altmann M, Altmann S, Rodriguez PA, Weller B, Elorduy Vergara L, Palme J, Marín-de la Rosa N, Sauer M, Wenig M, Villaécija-Aguilar JA, et al. Extensive signal integration by the phytohormone protein network. Nature. 2020:**583**(7815):271–276. https://doi.org/10.1038/s41586-020-2460-0

An C, Li L, Zhai Q, You Y, Deng L, Wu F, Chen R, Jiang H, Wang H, Chen Q, et al. Mediator subunit MED25 links the jasmonate receptor to transcriptionally active chromatin. Proc Natl Acad Sci U S A. 2017:**114**(42):E8930-E8939. https://doi.org/10.1073/pnas.1710885114

Anastasiou E, Kenz S, Gerstung M, MacLean D, Timmer J, Fleck C, Lenhard M. Control of plant organ size by KLUH/CYP78A5-dependent intercellular signaling. Dev Cell. 2007:**13**(6):843–856. https://doi.org/10.1016/j.devcel.2007.10.001

Arabidopsis Genome Initiative. Analysis of the genome sequence of the flowering plant *Arabidopsis thaliana*. Nature. 2000:**408**(6814):796–815. https://doi.org/10.1038/35048692

Aratani Y, Uemura T, Hagihara T, Matsui K, Toyota M. Green leaf volatile sensory calcium transduction in *Arabidopsis*. Nat Commun. 2023:**14**(1):6236. https://doi.org/10.1038/s41467-023-41589-9

Aubert Y, Widemann E, Miesch L, Pinot F, Heitz T. CYP94-mediated jasmonoyl-isoleucine hormone oxidation shapes jasmonate profiles and attenuates defence responses to *Botrytis cinerea* infection. J Exp Bot. 2015:**66**(13):3879–3892. https://doi.org/10.1093/jxb/erv190

Bai Y, Fernández-Calvo P, Ritter A, Huang AC, Morales-Herrera S, Bicalho KU, Karady M, Pauwels L, Buyst D, Njo M, et al. Modulation of *Arabidopsis* root growth by specialized triterpenes. New Phytol. 2021:**230**(1):228–243. https://doi.org/10.1111/nph.17144

Bailey-Serres J, Parker JE, Ainsworth EA, Oldroyd GED, Schroeder JI. Genetic strategies for improving crop yields. Nature. 2019:**575**(7781):109–118. https://doi.org/10.1038/s41586-019-1679-0

Baker TC, Nishida R, Roelofs WL. Close-range attraction of female oriental fruit moths to herbal scent of male hairpencils. Science. 1981:**214**(4527):1359–1361. https://doi.org/10.1126/science.214.4527.1359

Balcke GU, Handrick V, Bergau N, Fichtner M, Henning A, Stellmach H, Tissier A, Hause B, Frolov A. An UPLC-MS/MS method for highly sensitive high-throughput analysis of phytohormones in plant tissues. Plant Methods. 2012:**8**(1):47. https://doi.org/10.1186/1746-4811-8-47

Baldwin IT. Jasmonate-induced responses are costly but benefit plants under attack in native populations. Proc Natl Acad Sci U S A. 1998:**95**(14):8113–8118. https://doi.org/10.1073/pnas.95.14.8113

Baldwin IT, Zhang ZP, Diab N, Ohnmeiss TE. Quantification, correlations and manipulations of wound-induced changes in jasmonic acid and nicotine in *Nicotiana sylvestris*. Planta. 1997:**201**(4):397–404. https://doi.org/10.1007/s004250050082

Bannenberg G, Martínez M, Hamberg M, Castresana C. Diversity of the enzymatic activity in the lipoxygenase gene family of *Arabidopsis thaliana*. Lipids. 2009:**44**(2):85–95. https://doi.org/10.1007/s11745-008-3245-7

Bari R, Jones JDG. Role of plant hormones in plant defence responses. Plant Mol Biol. 2009:**69**(4):473–488. https://doi.org/10.1007/s11103-008-9435-0

Bellandi A, Papp D, Breakspear A, Joyce J, Johnston MG, de Keijzer J, Raven EC, Ohtsu M, Vincent TR, Miller AJ, et al. Diffusion and bulk flow of amino acids mediate calcium waves in plants. Sci Adv. 2022:**8**(42):eabo6693. https://doi.org/10.1126/sciadv.abo6693

Bennett T, Leyser O. The auxin question: a philosophical overview. In: **Zažímalová E, Petrášek J, Benková E**, editors. Auxin and its role in plant development. Vienna: Springer; 2014. p. 3–19.

Berens ML, Berry HM, Mine A, Argueso CT, Tsuda K. Evolution of hormone signaling networks in plant defense. Annu Rev Phytopathol. 2017:**55**(1):401–425. https://doi.org/10.1146/annurev-phyto-080516-035544

Berger S, Bell E, Mullet JE. Two methyl jasmonate-insensitive mutants show altered expression of AtVsp in response to methyl jasmonate and wounding. Plant Physiol. 1996:**111**:525–531.

Blázquez MA, Nelson DC, Weijers D. Evolution of plant hormone response pathways. Annu Rev Plant Biol. 2020:**71**(1):327–353. https://doi.org/10.1146/annurev-arplant-050718-100309

Boter M, Ruíz-Rivero O, Abdeen A, Prat S. Conserved MYC transcription factors play a key role in jasmonate signaling both in tomato and *Arabidopsis*. Genes Dev. 2004:**18**(13):1577–1591. https://doi.org/10.1101/gad.297704

Bowman JL. The origin of a land flora. Nat Plants. 2022:**8**(12):1352–1369. https://doi.org/10.1038/s41477-022-01283-y

Bowman JL, Kohchi T, Yamato KT, Jenkins J, Shu S, Ishizaki K, Yamaoka S, Nishihama R, Nakamura Y, Berger F, et al. Insights into land plant evolution garnered from the *Marchantia* polymorpha genome. Cell. 2017:**171**(2):287–304.e15. https://doi.org/10.1016/j.cell.2017.09.030

Brown DE, Rashotte AM, Murphy AS, Normanly J, Tague BW, Peer WA, Taiz L, Muday GK. Flavonoids act as negative regulators of auxin transport in vivo in arabidopsis. Plant Physiol. 2001:**126**(2):524–535. https://doi.org/10.1104/pp.126.2.524

Browse J. The power of mutants for investigating jasmonate biosynthesis and signaling. Phytochemistry. 2009:**70**(13–14):1539–1546. https://doi.org/10.1016/j.phytochem.2009.08.004

Bunsick M, McCullough R, McCourt P, Lumba S. Plant hormone signaling: is upside down right side up? Curr Opin Plant Biol. 2021:**63**:102070. https://doi.org/10.1016/j.pbi.2021.102070

Caarls L, Elberse J, Awwanah M, Ludwig NR, de Vries M, Zeilmaker T, Van Wees SCM, Schuurink RC, Van den Ackerveken G. *Arabidopsis* JASMONATE-INDUCED OXYGENASES down-regulate plant immunity by hydroxylation and inactivation of the hormone jasmonic acid. Proc Natl Acad Sci U S A. 2017:**114**(24):6388–6393. https://doi.org/10.1073/pnas.1701101114

Cai J, Aharoni A. Amino acids and their derivatives mediating defense priming and growth tradeoff. Curr Opin Plant Biol. 2022:**69**:102288. https://doi.org/10.1016/j.pbi.2022.102288

Cai Q, Yuan Z, Chen M, Yin C, Luo Z, Zhao X, Liang W, Hu J, Zhang D. Jasmonic acid regulates spikelet development in rice. Nat Commun. 2014:**5**(1):3476. https://doi.org/10.1038/ncomms4476

Caldelari D, Wang G, Farmer EE, Dong X. *Arabidopsis lox3 lox4* double mutants are male sterile and defective in global proliferative arrest. Plant Mol Biol. 2011:**75**(1–2):25–33. https://doi.org/10.1007/s11103-010-9701-9

Campos ML, Kang J-H, Howe GA. Jasmonate-triggered plant immunity. J Chem Ecol. 2014:**40**(7):657–675. https://doi.org/10.1007/s10886-014-0468-3

Cao S, Kang S, Mao H, Yao J, Gu L, Zheng N. Defining molecular glues with a dual-nanobody cannabidiol sensor. Nat Commun. 2022:**13**(1):815. https://doi.org/10.1038/s41467-022-28507-1

Çevik V, Kidd BN, Zhang P, Hill C, Kiddle S, Denby KJ, Holub EB, Cahill DM, Manners JM, Schenk PM, et al. MEDIATOR25 acts as an integrative hub for the regulation of jasmonate-responsive gene expression in *Arabidopsis*. Plant Physiol. 2012:**160**(1):541–555. https://doi.org/10.1104/pp.112.202697

Chae L, Kim T, Nilo-Poyanco R, Rhee SY. Genomic signatures of specialized metabolism in plants. Science. 2014:**344**(6183):510–513. https://doi.org/10.1126/science.1252076

Chauvin A, Caldelari D, Wolfender J-L, Farmer EE. Four 13-lipoxygenases contribute to rapid jasmonate synthesis in wounded *Arabidopsis thaliana* leaves: a role for lipoxygenase 6 in responses to long-distance wound signals. New Phytol. 2013:**197**(2):566–575. https://doi.org/10.1111/nph.12029

Chen R, Jiang H, Li L, Zhai Q, Qi L, Zhou W, Liu X, Li H, Zheng W, Sun J, et al. The *Arabidopsis* mediator subunit MED25 differentially regulates jasmonate and abscisic acid signaling through interacting with the MYC2 and ABI5 transcription factors. Plant Cell. 2012:**24**(7):2898–2916. https://doi.org/10.1105/tpc.112.098277

Chini A, Fonseca S, Fernández G, Adie B, Chico JM, Lorenzo O, García-Casado G, López-Vidriero I, Lozano FM, Ponce MR, et al. The JAZ family of repressors is the missing link in jasmonate signalling. Nature. 2007:**448**(7154):666–671. https://doi.org/10.1038/nature06006

Chini A, Gimenez-Ibanez S, Goossens A, Solano R. Redundancy and specificity in jasmonate signalling. Curr Opin Plant Biol. 2016:**33**:147–156. https://doi.org/10.1016/j.pbi.2016.07.005

Chini A, Monte I, Zamarreño AM, García-Mina JM, Solano R. Evolution of the jasmonate ligands and their biosynthetic pathways. New Phytol. 2023:**238**(5):2236–2246. https://doi.org/10.1111/nph.18891

Chini A, Monte I, Zamarreño AM, Hamberg M, Lassueur S, Reymond P, Weiss S, Stintzi A, Schaller A, Porzel A, et al. An OPR3-independent pathway uses 4,5-didehydrojasmonate for jasmonate synthesis. Nat Chem Biol. 2018:**14**(2):171–178. https://doi.org/10.1038/nchembio.2540

Chung HS, Cooke TF, Depew CL, Patel LC, Ogawa N, Kobayashi Y, Howe GA. Alternative splicing expands the repertoire of dominant JAZ repressors of jasmonate signaling. Plant J. 2010:**63**(4):613–622. https://doi.org/10.1111/j.1365-313X.2010.04265.x

Chung HS, Koo AJK, Gao X, Jayany S, Thines B, Jones AD, Howe GA. Regulation and function of *Arabidopsis JASMONATE-ZIM*-domain genes in response to wounding and herbivory. Plant Physiol. 2008:**146**(3):952–964. https://doi.org/10.1104/pp.107.115691

Creelman RA, Mullet JE. Biosynthesis and action of jasmonates in plants. Annu Rev Plant Physiol Plant Mol Biol. 1997:**48**(1):355–381. https://doi.org/10.1146/annurev.arplant.48.1.355

Creelman RA, Tierney ML, Mullet JE. Jasmonic acid/methyl jasmonate accumulate in wounded soybean hypocotyls and modulate wound gene expression. Proc Natl Acad Sci U S A. 1992:**89**(11):4938–4941. https://doi.org/10.1073/pnas.89.11.4938

Dani KGS, Loreto F. Plant volatiles as regulators of hormone homeostasis. New Phytol. 2022:**234**(3):804–812. https://doi.org/10.1111/nph.18035

Daryanavard H, Postiglione AE, Mühlemann JK, Muday GK. Flavonols modulate plant development, signaling, and stress responses. Curr Opin Plant Biol. 2023:**72**:102350. https://doi.org/10.1016/j.pbi.2023.102350

Davies PJ. The plant hormones: their nature, occurrence, and functions. In: **Davies PJ**, editor. Plant hormones: biosynthesis, signal transduction, action!. Dordrecht, Netherlands: Springer; 2010. p. 1–15.

Delfin JC, Kanno Y, Seo M, Kitaoka N, Matsuura H, Tohge T, Shimizu T. AtGH3.10 is another jasmonic acid-amido synthetase in *Arabidopsis thaliana*. Plant J. 2022:**110**(4):1082–1096. https://doi.org/10.1111/tpj.15724

Demole E, Lederer E, Mercier D. Isolement et détermination de la structure du jasmonate de méthyle, constituant odorant caractéristique de l'essence de jasmin. Helv Chim Acta. 1962:**45**(2):675–685. https://doi.org/10.1002/hlca.19620450233

Devoto A, Ellis C, Magusin A, Chang H-S, Chilcott C, Zhu T, Turner JG. Expression profiling reveals COI1 to be a key regulator of genes involved in wound- and methyl jasmonate-induced secondary metabolism, defence, and hormone interactions. Plant Mol Biol. 2005:**58**(4):497–513. https://doi.org/10.1007/s11103-005-7306-5

Devoto A, Nieto-Rostro M, Xie D, Ellis C, Harmston R, Patrick E, Davis J, Sherratt L, Coleman M, Turner JG. COI1 links jasmonate signalling and fertility to the SCF ubiquitin-ligase complex in *Arabidopsis*. Plant J. 2002:**32**(4):457–466. https://doi.org/10.1046/j.1365-313X.2002.01432.x

Dharmasiri N, Dharmasiri S, Jones AM, Estelle M. Auxin action in a cell-free system. Curr Biol. 2003:**13**(16):1418–1422. https://doi.org/10.1016/S0960-9822(03)00536-0

Dinneny J, Al-Babili S, Benfey PN. β-Cyclocitral is a conserved root growth regulator. Proc Natl Acad Sci U S A. 2019:**116**(21):10563–10567. https://doi.org/10.1073/pnas.1821445116

Durán-Medina Y, Ruiz-Cortés BE, Guerrero-Largo H, Marsch-Martínez N. Specialized metabolism and development: an unexpected friendship. Curr Opin Plant Biol. 2021:**64**:102142. https://doi.org/10.1016/j.pbi.2021.102142

Ellis C, Turner JG. The *Arabidopsis* mutant *cev1* has constitutively active jasmonate and ethylene signal pathways and enhanced resistance to pathogens. Plant Cell. 2001:**13**(5):1025–1033. https://doi.org/10.1105/tpc.13.5.1025

Engelberth J, Alborn HT, Schmelz EA, Tumlinson JH. Airborne signals prime plants against insect herbivore attack. Proc Natl Acad Sci U S A. 2004:**101**(6):1781–1785. https://doi.org/10.1073/pnas.0308037100

Erb M, Kliebenstein DJ. Plant secondary metabolites as defenses, regulators, and primary metabolites: the blurred functional trichotomy. Plant Physiol. 2020:**184**(1):39–52. https://doi.org/10.1104/pp.20.00433

Erb M, Veyrat N, Robert CAM, Xu H, Frey M, Ton J, Turlings TCJ. Indole is an essential herbivore-induced volatile priming signal in maize. Nat Commun. 2015:**6**(1):6273. https://doi.org/10.1038/ncomms7273

Fàbregas N, Fernie AR. The reliance of phytohormone biosynthesis on primary metabolite precursors. J Plant Physiol. 2021:**268**:153589. https://doi.org/10.1016/j.jplph.2021.153589

Farmer EE, Johnson RR, Ryan CA. Regulation of expression of proteinase inhibitor genes by methyl jasmonate and jasmonic acid. Plant Physiol. 1992:**98**(3):995–1002. https://doi.org/10.1104/pp.98.3.995

Farmer EE, Ryan CA. Interplant communication: airborne methyl jasmonate induces synthesis of proteinase inhibitors in plant leaves. Proc Natl Acad Sci U S A. 1990:**87**(19):7713–7716. https://doi.org/10.1073/pnas.87.19.7713

Farmer EE, Ryan CA. Octadecanoid precursors of jasmonic acid activate the synthesis of wound-inducible proteinase inhibitors. Plant Cell. 1992:**4**(2):129–134. https://doi.org/10.2307/3869566

Felemban A, Moreno JC, Mi J, Ali S, Sham A, AbuQamar SF, Al-Babili S. The apocarotenoid β-ionone regulates the transcriptome of *Arabidopsis thaliana* and increases its resistance against *Botrytis cinerea*. Plant J. 2023:**117**(2):541–560.

Fernández-Calvo P, Chini A, Fernández-Barbero G, Chico J-M, Gimenez-Ibanez S, Geerinck J, Eeckhout D, Schweizer F, Godoy M, Franco-Zorrilla JM, et al. The *Arabidopsis* bHLH transcription factors MYC3 and MYC4 are targets of JAZ repressors and act additively with MYC2 in the activation of jasmonate responses. Plant Cell. 2011:**23**(2):701–715. https://doi.org/10.1105/tpc.110.080788

Fernández-Milmanda GL, Crocco CD, Reichelt M, Mazza CA, Köllner TG, Zhang T, Cargnel MD, Lichy MZ, Fiorucci A-S, Fankhauser C, et al. A light-dependent molecular link between competition cues and defence responses in plants. Nat Plants. 2020:**6**(3):223–230. https://doi.org/10.1038/s41477-020-0604-8

Feys B, Benedetti CE, Penfold CN, Turner JG. *Arabidopsis* mutants selected for resistance to the phytotoxin coronatine are male sterile, insensitive to methyl jasmonate, and resistant to a bacterial pathogen. Plant Cell. 1994:**6**(5):751–759. https://doi.org/10.2307/3869877

Fonseca S, Chini A, Hamberg M, Adie B, Porzel A, Kramell R, Miersch O, Wasternack C, Solano R. (+)-7-iso-Jasmonoyl-L-isoleucine is the endogenous bioactive jasmonate. Nat Chem Biol. 2009:**5**(5):344–350. https://doi.org/10.1038/nchembio.161

Fonseca S, Fernández-Calvo P, Fernández GM, Díez-Díaz M, Gimenez-Ibanez S, López-Vidriero I, Godoy M, Fernández-Barbero G, Van Leene J, De Jaeger G, et al. bHLH003, bHLH013 and bHLH017 are new targets of JAZ repressors negatively regulating JA responses. PLoS One. 2014:**9**(1):e86182. https://doi.org/10.1371/journal.pone.0086182

Gao Y-Q, Jimenez-Sandoval P, Tiwari S, Stolz S, Wang J, Glauser G, Santiago J, Farmer EE. Ricca's factors as mobile proteinaceous effectors of electrical signaling. Cell. 2023:**186**(7):1337–1351.e20. https://doi.org/10.1016/j.cell.2023.02.006

Gasperini D, Chauvin A, Acosta IF, Kurenda A, Stolz S, Chételat A, Wolfender J-L, Farmer EE. Axial and radial oxylipin transport. Plant Physiol. 2015:**169**(3):2244–2254. https://doi.org/10.1104/pp.15.01104

Glauser G, Dubugnon L, Mousavi SAR, Rudaz S, Wolfender J-L, Farmer EE. Velocity estimates for signal propagation leading to systemic jasmonic acid accumulation in wounded *Arabidopsis*. J Biol Chem. 2009:**284**(50):34506–34513. https://doi.org/10.1074/jbc.M109.061432

Glazebrook J. Contrasting mechanisms of defense against biotrophic and necrotrophic pathogens. Annu Rev Phytopathol. 2005:**43**(1):205–227. https://doi.org/10.1146/annurev.phyto.43.040204.135923

Goetz S, Hellwege A, Stenzel I, Kutter C, Hauptmann V, Forner S, McCaig B, Hause G, Miersch O, Wasternack C, et al. Role of *cis*-12-oxo-phytodienoic acid in tomato embryo development. Plant Physiol. 2012:**158**(4):1715–1727. https://doi.org/10.1104/pp.111.192658

Grebner W, Stingl NE, Oenel A, Mueller MJ, Berger S. Lipoxygenase6-dependent oxylipin synthesis in roots is required for abiotic and biotic stress resistance of *Arabidopsis*. Plant Physiol. 2013:**161**(4):2159–2170. https://doi.org/10.1104/pp.113.214544

Green TR, Ryan CA. Wound-induced proteinase inhibitor in plant leaves: a possible defense mechanism against insects. Science. 1972:**175**(4023):776–777. https://doi.org/10.1126/science.175.4023.776

Gundlach H, Müller MJ, Kutchan TM, Zenk MH. Jasmonic acid is a signal transducer in elicitor-induced plant cell cultures. Proc Natl Acad Sci U S A. 1992:**89**(6):2389–2393. https://doi.org/10.1073/pnas.89.6.2389

Guo Q, Major IT, Kapali G, Howe GA. MYC transcription factors coordinate tryptophan-dependent defence responses and compromise seed yield in *Arabidopsis*. New Phytol. 2022:**236**(1):132–145. https://doi.org/10.1111/nph.18293

Guo Q, Yoshida Y, Major IT, Wang K, Sugimoto K, Kapali G, Havko NE, Benning C, Howe GA. JAZ repressors of metabolic defense promote growth and reproductive fitness in *Arabidopsis*. Proc Natl Acad Sci U S A. 2018:**115**(45):E10768-E10777. https://doi.org/10.1073/pnas.1811828115

He Z, Wang ZY, Li J, Zhu Q, Lamb C, Ronald P, Chory J. Perception of brassinosteroids by the extracellular domain of the receptor kinase BRI1. Science. 2000:**288**(5475):2360–2363. https://doi.org/10.1126/science.288.5475.2360

Hedden P. The genes of the green revolution. Trends Genet. 2003:**19**(1):5–9. https://doi.org/10.1016/S0168-9525(02)00009-4

Heitz T, Smirnova E, Widemann E, Aubert Y, Pinot F, Ménard R. The rise and fall of jasmonate biological activities. Subcell Biochem. 2016:**86**:405–426. https://doi.org/10.1007/978-3-319-25979-6_16

Heitz T, Widemann E, Lugan R, Miesch L, Ullmann P, Désaubry L, Holder E, Grausem B, Kandel S, Miesch M, et al. Cytochromes P450 CYP94C1 and CYP94B3 catalyze two successive oxidation steps of plant hormone jasmonoyl-isoleucine for catabolic turnover. J Biol Chem. 2012:**287**(9):6296–6306. https://doi.org/10.1074/jbc.M111.316364

Hilpert B, Bohlmann H, op den Camp RO, Przybyla D, Miersch O, Buchala A, Apel K. Isolation and characterization of signal transduction mutants of *Arabidopsis thaliana* that constitutively activate the octadecanoid pathway and form necrotic microlesions. Plant J. 2001:**26**(4):435–446. https://doi.org/10.1046/j.1365-313X.2001.2641036.x

Howe GA. Plant hormones: metabolic end run to jasmonate. Nat Chem Biol. 2018:**14**(2):109–110. https://doi.org/10.1038/nchembio.2553

Howe GA. Auspicious beginnings for the defense hormone jasmonate. Plant Cell. 2019:**31**(6):1198–1199. https://doi.org/10.1105/tpc.19.00333

Howe GA, Jander G. Plant immunity to insect herbivores. Annu Rev Plant Biol. 2008:**59**(1):41–66. https://doi.org/10.1146/annurev.arplant.59.032607.092825

Howe GA, Major IT, Koo AJ. Modularity in jasmonate signaling for multistress resilience. Annu Rev Plant Biol. 2018:**69**(1):387–415. https://doi.org/10.1146/annurev-arplant-042817-040047

Hu L. Integration of multiple volatile cues into plant defense responses. New Phytol. 2022:**233**(2):618–623. https://doi.org/10.1111/nph.17724

Ishiguro S, Kawai-Oda A, Ueda J, Nishida I, Okada K. The *DEFECTIVE IN ANTHER DEHISCIENCE* gene encodes a novel phospholipase A1 catalyzing the initial step of jasmonic acid biosynthesis, which synchronizes pollen maturation, anther dehiscence, and flower opening in *Arabidopsis*. Plant Cell. 2001:**13**(10):2191–2209. https://doi.org/10.1105/tpc.010192

Jensen AB, Raventos D, Mundy J. Fusion genetic analysis of jasmonate-signalling mutants in *Arabidopsis*. Plant J. 2002:**29**(5):595–606. https://doi.org/10.1046/j.0960-7412.2001.01241.x

Jia K-P, Dickinson AJ, Mi J, Cui G, Xiao TT, Kharbatia NM, Guo X, Sugiono E, Aranda M, Blilou I, et al. Anchorene is a carotenoid-derived regulatory metabolite required for anchor root formation in *Arabidopsis*. Sci Adv. 2019:**5**(11):eaaw6787. https://doi.org/10.1126/sciadv.aaw6787

Jung HW, Tschaplinski TJ, Wang L, Glazebrook J, Greenberg JT. Priming in systemic plant immunity. Science. 2009:**324**(5923):89–91. https://doi.org/10.1126/science.1170025

Kagale S, Links MG, Rozwadowski K. Genome-wide analysis of ethylene-responsive element binding factor-associated amphiphilic repression motif-containing transcriptional regulators in *Arabidopsis*. Plant Physiol. 2010:**152**(3):1109–1134. https://doi.org/10.1104/pp.109.151704

Karban R, Baldwin IT. Induced responses to herbivory. Chicago: University of Chicago Press; 2007.

Katsir L, Schilmiller AL, Staswick PE, He SY, Howe GA. COI1 is a critical component of a receptor for jasmonate and the bacterial virulence factor coronatine. Proc Natl Acad Sci U S A. 2008:**105**(19): 7100–7105. https://doi.org/10.1073/pnas.0802332105

Katz E, Bagchi R, Jeschke V, Rasmussen ARM, Hopper A, Burow M, Estelle M, Kliebenstein DJ. Diverse allyl glucosinolate catabolites independently influence root growth and development. Plant Physiol. 2020:**183**(3):1376–1390. https://doi.org/10.1104/pp.20.00170

Katz E, Nisani S, Yadav BS, Woldemariam MG, Shai B, Obolski U, Ehrlich M, Shani E, Jander G, Chamovitz DA. The glucosinolate breakdown product indole-3-carbinol acts as an auxin antagonist in roots of *Arabidopsis thaliana*. Plant J. 2015:**82**(4):547–555. https://doi.org/10.1111/tpj.12824

Kazan K, Manners JM. MYC2: the master in action. Mol Plant. 2013:**6**(3):686–703. https://doi.org/10.1093/mp/sss128

Kende H. Plant biology and the Nobel Prize. Science. 1998:**282**(5389): 627. https://doi.org/10.1126/science.282.5389.627b

Kende H, Zeevaart J. The five "classical" plant hormones. Plant Cell. 1997:**9**:1197–1210. https://doi.org/10.1105/tpc.9.7.1197

Kepinski S, Leyser O. The *Arabidopsis* F-box protein TIR1 is an auxin receptor. Nature. 2005:**435**(7041):446–451. https://doi.org/10.1038/nature03542

Kim JI, Dolan WL, Anderson NA, Chapple C. Indole glucosinolate biosynthesis limits phenylpropanoid accumulation in *Arabidopsis thaliana*. Plant Cell. 2015:**27**(5):1529–1546. https://doi.org/10.1105/tpc.15.00127

Kneeshaw S, Soriano G, Monte I, Hamberg M, Zamarreño ÁM, García-Mina JM, Franco-Zorrilla JM, Kato N, Ueda M, Rey-Stolle MF, et al. Ligand diversity contributes to the full activation of the jasmonate pathway in *Marchantia polymorpha*. Proc Natl Acad Sci U S A. 2022:**119**(36):e2202930119. https://doi.org/10.1073/pnas.2202930119

Knöfel H-D, Brückner C, Kramell R, Sembdner G, Schreiber K. A radioimmunoassay for jasmonic acid. Biochem Physiol Pflanz. 1984:**179**(4):317–325. https://doi.org/10.1016/S0015-3796(84)80048-7

Koo AJ. Metabolism of the plant hormone jasmonate: a sentinel for tissue damage and master regulator of stress response. Phytochem Rev. 2018:**17**(1):51–80. https://doi.org/10.1007/s11101-017-9510-8

Koo AJK, Cooke TF, Howe GA. Cytochrome P450 CYP94B3 mediates catabolism and inactivation of the plant hormone jasmonoyl-L-isoleucine. Proc Natl Acad Sci U S A. 2011:**108**(22):9298–9303. https://doi.org/10.1073/pnas.1103542108

Koo AJK, Gao X, Jones AD, Howe GA. A rapid wound signal activates the systemic synthesis of bioactive jasmonates in *Arabidopsis*. Plant J. 2009:**59**(6):974–986. https://doi.org/10.1111/j.1365-313X.2009.03924.x

Koo AJK, Howe GA. The wound hormone jasmonate. Phytochemistry. 2009:**70**(13–14):1571–1580. https://doi.org/10.1016/j.phytochem.2009.07.018

Koo AJ, Thireault C, Zemelis S, Poudel AN, Zhang T, Kitaoka N, Brandizzi F, Matsuura H, Howe GA. Endoplasmic reticulum-associated inactivation of the hormone jasmonoyl-L-isoleucine by multiple members of the cytochrome P450 94 family in *Arabidopsis*. J Biol Chem. 2014:**289**(43):29728–29738. https://doi.org/10.1074/jbc.M114.603084

Laha D, Johnen P, Azevedo C, Dynowski M, Weiß M, Capolicchio S, Mao H, Iven T, Steenbergen M, Freyer M, et al. VIH2 regulates the synthesis of inositol pyrophosphate $InsP_8$ and jasmonate-dependent defenses in *Arabidopsis*. Plant Cell. 2015:**27**(4):1082–1097. https://doi.org/10.1105/tpc.114.135160

Lavecchia A. Machine-learning approaches in drug discovery: methods and applications. Drug Discov Today. 2015:**20**(3):318–331. https://doi.org/10.1016/j.drudis.2014.10.012

Li L-L, Li Z, Lou Y, Meiners SJ, Kong C-H. (−)-Loliolide is a general signal of plant stress that activates jasmonate-related responses. New Phytol. 2022:**238**(5):2099–2112. https://doi.org/10.1111/nph.18644

Li M, Wang F, Li S, Yu G, Wang L, Li Q, Zhu X, Li Z, Yuan L, Liu P. Importers drive leaf-to-leaf jasmonic acid transmission in wound-induced systemic immunity. Mol Plant. 2020:**13**(10):1485–1498. https://doi.org/10.1016/j.molp.2020.08.017

Li M, Yu G, Cao C, Liu P. Metabolism, signaling, and transport of jasmonates. Plant Commun. 2021:**2**(5):100231. https://doi.org/10.1016/j.xplc.2021.100231

Li L, Zhao Y, McCaig BC, Wingerd BA, Wang J, Whalon ME, Pichersky E, Howe GA. The tomato homolog of CORONATINE-INSENSITIVE1 is required for the maternal control of seed maturation, jasmonate-signaled defense responses, and glandular trichome development. Plant Cell. 2004:**16**(1):126–143. https://doi.org/10.1105/tpc.017954

Liu K-H, Liu M, Lin Z, Wang Z-F, Chen B, Liu C, Guo A, Konishi M, Yanagisawa S, Wagner G, et al. NIN-like protein 7 transcription factor is a plant nitrate sensor. Science. 2022:**377**(6613):1419–1425. https://doi.org/10.1126/science.add1104

Lorenzo O, Chico JM, Sánchez-Serrano JJ, Solano R. *JASMONATE-INSENSITIVE1* encodes a MYC transcription factor essential to discriminate between different jasmonate-regulated defense responses in *Arabidopsis*. Plant Cell. 2004:**16**(7):1938–1950. https://doi.org/10.1105/tpc.022319

Luna E, van Hulten M, Zhang Y, Berkowitz O, López A, Pétriacq P, Sellwood MA, Chen B, Burrell M, van de Meene A, et al. Plant perception of β-aminobutyric acid is mediated by an aspartyl-tRNA synthetase. Nat Chem Biol. 2014:**10**(6):450–456. https://doi.org/10.1038/nchembio.1520

Luzarowski M, Skirycz A. Emerging strategies for the identification of protein–metabolite interactions. J Exp Bot. 2019:**70**(18):4605–4618. https://doi.org/10.1093/jxb/erz228

Maeda HA, Fernie AR. Evolutionary history of plant metabolism. Annu Rev Plant Biol. 2021:**72**(1):185–216. https://doi.org/10.1146/annurev-arplant-080620-031054

Malinovsky FG, Thomsen M-LF, Nintemann SJ, Jagd LM, Bourgine B, Burow M, Kliebenstein DJ. An evolutionarily young defense metabolite influences the root growth of plants via the ancient TOR signaling pathway. eLife. 2017:**6**:e29353. https://doi.org/10.7554/eLife.29353

Matsui K. Green leaf volatiles: hydroperoxide lyase pathway of oxylipin metabolism. Curr Opin Plant Biol. 2006:**9**(3):274–280. https://doi.org/10.1016/j.pbi.2006.03.002

McConn M, Browse J. The critical requirement for linolenic acid is pollen development, not photosynthesis, in an *Arabidopsis* mutant. Plant Cell. 1996:**8**(3):403–416. https://doi.org/10.2307/3870321

Melotto M, Mecey C, Niu Y, Chung HS, Katsir L, Yao J, Zeng W, Thines B, Staswick P, Browse J, et al. A critical role of two positively charged amino acids in the jas motif of *Arabidopsis* JAZ proteins in mediating coronatine- and jasmonoyl isoleucine-dependent interactions with the COI1 F-box protein. Plant J. 2008:**55**(6):979–988. https://doi.org/10.1111/j.1365-313X.2008.03566.x

Meyerowitz EM. Plants compared to animals: the broadest comparative study of development. Science. 2002:**295**(5559):1482–1485. https://doi.org/10.1126/science.1066609

Mochizuki S, Sugimoto K, Koeduka T, Matsui K. *Arabidopsis* lipoxygenase 2 is essential for formation of green leaf volatiles and five-carbon volatiles. FEBS Lett. 2016:**590**(7):1017–1027. https://doi.org/10.1002/1873-3468.12133

Monte I. Jasmonates and salicylic acid: evolution of defense hormones in land plants. Curr Opin Plant Biol. 2023:**76**:102470. https://doi.org/10.1016/j.pbi.2023.102470

Monte I, Franco-Zorrilla JM, García-Casado G, Zamarreño AM, García-Mina JM, Nishihama R, Kohchi T, Solano R. A single JAZ repressor controls the jasmonate pathway in *Marchantia* polymorpha. Mol Plant. 2019:**12**(2):185–198. https://doi.org/10.1016/j.molp.2018.12.017

Monte I, Ishida S, Zamarreño AM, Hamberg M, Franco-Zorrilla JM, García-Casado G, Gouhier-Darimont C, Reymond P, Takahashi K,

García-Mina JM, et al. Ligand–receptor co-evolution shaped the jasmonate pathway in land plants. Nat Chem Biol. 2018:**14**(5):480–488. https://doi.org/10.1038/s41589-018-0033-4

Monte I, Kneeshaw S, Franco-Zorrilla JM, Chini A, Zamarreño AM, García-Mina JM, Solano R. An ancient COI1-independent function for reactive electrophilic oxylipins in thermotolerance. Curr Biol. 2020:**30**(6):962–971.e3. https://doi.org/10.1016/j.cub.2020.01.023

Moreno JE, Shyu C, Campos ML, Patel LC, Chung HS, Yao J, He SY, Howe GA. Negative feedback control of jasmonate signaling by an alternative splice variant of JAZ10. Plant Physiol. 2013:**162**(2): 1006–1017. https://doi.org/10.1104/pp.113.218164

Morffy N, Strader LC. Old town roads: routes of auxin biosynthesis across kingdoms. Curr Opin Plant Biol. 2020:**55**:21–27. https://doi.org/10.1016/j.pbi.2020.02.002

Morin H, Chételat A, Stolz S, Marcourt L, Glauser G, Wolfender J-L, Farmer EE. Wound-response jasmonate dynamics in the primary vasculature. New Phytol. 2023:**240**(4):1484–1496. https://doi.org/10.1111/nph.19207

Mousavi SAR, Chauvin A, Pascaud F, Kellenberger S, Farmer EE. GLUTAMATE RECEPTOR-LIKE genes mediate leaf-to-leaf wound signalling. Nature. 2013:**500**(7463):422–426. https://doi.org/10.1038/nature12478

Müller A, Düchting P, Weiler EW. A multiplex GC-MS/MS technique for the sensitive and quantitative single-run analysis of acidic phytohormones and related compounds, and its application to *Arabidopsis thaliana*. Planta. 2002:**216**(1):44–56. https://doi.org/10.1007/s00425-002-0866-6

Mutte SK, Kato H, Rothfels C, Melkonian M, Wong GK-S, Weijers D. Origin and evolution of the nuclear auxin response system. eLife. 2018:**7**:e33399. https://doi.org/10.7554/eLife.33399

Nagashima A, Higaki T, Koeduka T, Ishigami K, Hosokawa S, Watanabe H, Matsui K, Hasezawa S, Touhara K. Transcriptional regulators involved in responses to volatile organic compounds in plants. J Biol Chem. 2019:**294**(7):2256–2266. https://doi.org/10.1074/jbc.RA118.005843

Nakata M, Mitsuda N, Herde M, Koo AJK, Moreno JE, Suzuki K, Howe GA, Ohme-Takagi M. A bHLH-type transcription factor, ABA-INDUCIBLE BHLH-TYPE TRANSCRIPTION FACTOR/JA-ASSOCIATED MYC2-LIKE1, acts as a repressor to negatively regulate jasmonate signaling in *Arabidopsis*. Plant Cell. 2013:**25**(5):1641–1656. https://doi.org/10.1105/tpc.113.111112

Nguyen CT, Kurenda A, Stolz S, Chételat A, Farmer EE. Identification of cell populations necessary for leaf-to-leaf electrical signaling in a wounded plant. Proc Natl Acad Sci U S A. 2018:**115**(40):10178–10183. https://doi.org/10.1073/pnas.1807049115

Ninkovic V, Markovic D, Rensing M. Plant volatiles as cues and signals in plant communication. Plant Cell Environ. 2021:**44**(4):1030–1043. https://doi.org/10.1111/pce.13910

Parthier B. Jasmonates: hormonal regulators or stress factors in leaf senescence? J Plant Growth Regul. 1990:**9**(1–4):57. https://doi.org/10.1007/BF02041942

Pauwels L, Barbero GF, Geerinck J, Tilleman S, Grunewald W, Pérez AC, Chico JM, Bossche RV, Sewell J, Gil E, et al. NINJA connects the co-repressor TOPLESS to jasmonate signalling. Nature. 2010:**464**(7289):788–791. https://doi.org/10.1038/nature08854

Peer WA, Murphy AS. Flavonoids and auxin transport: modulators or regulators? Trends Plant Sci. 2007:**12**(12):556–563. https://doi.org/10.1016/j.tplants.2007.10.003

Peñuelas M, Monte I, Schweizer F, Vallat A, Reymond P, García-Casado G, Franco-Zorrilla JM, Solano R. Jasmonate-related MYC transcription factors are functionally conserved in *Marchantia polymorpha*. Plant Cell. 2019:**31**(10):2491–2509. https://doi.org/10.1105/tpc.18.00974

Piazza I, Kochanowski K, Cappelletti V, Fuhrer T, Noor E, Sauer U, Picotti P. A map of protein-metabolite interactions reveals principles of chemical communication. Cell. 2018:**172**(1–2):358–372.e23. https://doi.org/10.1016/j.cell.2017.12.006

Pourcel L, Irani NG, Koo AJK, Bohorquez-Restrepo A, Howe GA, Grotewold E. A chemical complementation approach reveals genes and interactions of flavonoids with other pathways. Plant J. 2013:**74**(3):383–397. https://doi.org/10.1111/tpj.12129

Qi L, Kwiatkowski M, Chen H, Hoermayer L, Sinclair S, Zou M, Del Genio CI, Kubeš MF, Napier R, Jaworski K, et al. Adenylate cyclase activity of TIR1/AFB auxin receptors in plants. Nature. 2022:**611**(7934):133–138. https://doi.org/10.1038/s41586-022-05369-7

Qi T, Wang J, Huang H, Liu B, Gao H, Liu Y, Song S, Xie D. Regulation of jasmonate-induced leaf senescence by antagonism between bHLH subgroup IIIe and IIId factors in *Arabidopsis*. Plant Cell. 2015:**27**(6): 1634–1649. https://doi.org/10.1105/tpc.15.00110

Ray R, Halitschke R, Gase K, Leddy SM, Schuman MC, Rodde N, Baldwin IT. A persistent major mutation in canonical jasmonate signaling is embedded in an herbivory-elicited gene network. Proc Natl Acad Sci U S A. 2023:**120**(35):e2308500120. https://doi.org/10.1073/pnas.2308500120

Reymond P, Weber H, Damond M, Farmer EE. Differential gene expression in response to mechanical wounding and insect feeding in Arabidopsis. Plant Cell. 2000:**12**:707–720.

Rieseberg TP, Dadras A, Fürst-Jansen JMR, Dhabalia Ashok A, Darienko T, de Vries S, Irisarri I, de Vries J. Crossroads in the evolution of plant specialized metabolism. Semin Cell Dev Biol. 2023:**134**: 37–58. https://doi.org/10.1016/j.semcdb.2022.03.004

Robert-Seilaniantz A, Grant M, Jones JDG. Hormone crosstalk in plant disease and defense: more than just jasmonate-salicylate antagonism. Annu Rev Phytopathol. 2011:**49**(1):317–343. https://doi.org/10.1146/annurev-phyto-073009-114447

Ruegger M, Dewey E, Gray WM, Hobbie L, Turner J, Estelle M. The TIR1 protein of *Arabidopsis* functions in auxin response and is related to human SKP2 and yeast grr1p. Genes Dev. 1998:**12**(2):198–207. https://doi.org/10.1101/gad.12.2.198

Salehin M, Li B, Tang M, Katz E, Song L, Ecker JR, Kliebenstein DJ, Estelle M. Auxin-sensitive aux/IAA proteins mediate drought tolerance in Arabidopsis by regulating glucosinolate levels. Nat Commun. 2019:**10**(1):4021. https://doi.org/10.1038/s41467-019-12002-1

Sandoval PJ, Santiago J. In vitro analytical approaches to study plant ligand-receptor interactions. Plant Physiol. 2020:**182**(4):1697–1712. https://doi.org/10.1104/pp.19.01396

Santner A, Calderon-Villalobos LIA, Estelle M. Plant hormones are versatile chemical regulators of plant growth. Nat Chem Biol. 2009:**5**(5):301–307. https://doi.org/10.1038/nchembio.165

Schaller F, Weiler EW. Enzymes of octadecanoid biosynthesis in plants–12-oxo-phytodienoate 10,11-reductase. Eur J Biochem. 1997:**245**(2): 294–299. https://doi.org/10.1111/j.1432-1033.1997.t01-1-00294.x

Schilmiller AL, Koo AJK, Howe GA. Functional diversification of acyl-coenzyme A oxidases in jasmonic acid biosynthesis and action. Plant Physiol. 2007:**143**(2):812–824. https://doi.org/10.1104/pp.106.092916

Schlossarek D, Zhang Y, Sokolowska EM, Fernie AR, Luzarowski M, Skirycz A. Don't let go: co-fractionation mass spectrometry for untargeted mapping of protein-metabolite interactomes. Plant J. 2023:**113**(5):904–914. https://doi.org/10.1111/tpj.16084

Schneeberger K, Ossowski S, Lanz C, Juul T, Petersen AH, Nielsen KL, Jørgensen J-E, Weigel D, Andersen SU. SHOREmap: simultaneous mapping and mutation identification by deep sequencing. Nat Methods. 2009:**6**(8):550–551. https://doi.org/10.1038/nmeth0809-550

Schulze A, Zimmer M, Mielke S, Stellmach H, Melnyk CW, Hause B, Gasperini D. Wound-Induced shoot-to-root relocation of JA-Ile precursors coordinates *Arabidopsis* growth. Mol Plant. 2019:**12**(10): 1383–1394. https://doi.org/10.1016/j.molp.2019.05.013

Sembdner G, Parthier B. The biochemistry and the physiological and molecular actions of jasmonates. Annu Rev Plant Physiol Plant Mol Biol. 1993:**44**(1):569–589. https://doi.org/10.1146/annurev.pp.44.060193.003033

Shabek N, Zheng N. Plant ubiquitin ligases as signaling hubs. Nat Struct Mol Biol. 2014:**21**(4):293–296. https://doi.org/10.1038/nsmb.2804

Sheard LB, Tan X, Mao H, Withers J, Ben-Nissan G. Jasmonate perception by inositol-phosphate-potentiated COI1–JAZ co-receptor. Nature. 2010:**468**(7322):400–405. https://doi.org/10.1038/nature09430

Shields A, Shivnauth V, Castroverde CDM. Salicylic acid and *N*-hydroxypipecolic acid at the fulcrum of the plant immunity-growth equilibrium. Front Plant Sci. 2022:**13**:841688. https://doi.org/10.3389/fpls.2022.841688

Shyu C, Figueroa P, Depew CL, Cooke TF, Sheard LB, Moreno JE, Katsir L, Zheng N, Browse J, Howe GA. JAZ8 lacks a canonical degron and has an EAR motif that mediates transcriptional repression of jasmonate responses in *Arabidopsis*. Plant Cell. 2012:**24**(2):536–550. https://doi.org/10.1105/tpc.111.093005

Smirnova E, Marquis V, Poirier L, Aubert Y, Zumsteg J, Ménard R, Miesch L, Heitz T. Jasmonic acid oxidase 2 hydroxylates jasmonic acid and represses basal defense and resistance responses against *Botrytis cinerea* infection. Mol Plant. 2017:**10**(9):1159–1173. https://doi.org/10.1016/j.molp.2017.07.010

Song WC, Brash AR. Purification of an allene oxide synthase and identification of the enzyme as a cytochrome P-450. Science. 1991:**253**(5021):781–784. https://doi.org/10.1126/science.1876834

Song S, Qi T, Wasternack C, Xie D. Jasmonate signaling and crosstalk with gibberellin and ethylene. Curr Opin Plant Biol. 2014:**21**:112–119. https://doi.org/10.1016/j.pbi.2014.07.005

Staswick PE. Novel regulation of vegetative storage protein genes. Plant Cell. 1990:**2**(1):1–6. https://doi.org/10.2307/3869045

Staswick PE, Su W, Howell SH. Methyl jasmonate inhibition of root growth and induction of a leaf protein are decreased in an *Arabidopsis thaliana* mutant. Proc Natl Acad Sci U S A. 1992:**89**(15):6837–6840. https://doi.org/10.1073/pnas.89.15.6837

Staswick PE, Tiryaki I. The oxylipin signal jasmonic acid is activated by an enzyme that conjugates it to isoleucine in *Arabidopsis*. Plant Cell. 2004:**16**(8):2117–2127. https://doi.org/10.1105/tpc.104.023549

Staswick PE, Tiryaki I, Rowe ML. Jasmonate response locus *JAR1* and several related *Arabidopsis* genes encode enzymes of the firefly luciferase superfamily that show activity on jasmonic, salicylic, and indole-3-acetic acids in an assay for adenylation. Plant Cell. 2002:**14**(6):1405–1415. https://doi.org/10.1105/tpc.000885

Stumpe M, Göbel C, Faltin B, Beike AK, Hause B, Himmelsbach K, Bode J, Kramell R, Wasternack C, Frank W, et al. The moss *Physcomitrella patens* contains cyclopentenones but no jasmonates: mutations in allene oxide cyclase lead to reduced fertility and altered sporophyte morphology. New Phytol. 2010:**188**(3):740–749. https://doi.org/10.1111/j.1469-8137.2010.03406.x

Sugimoto K, Zager JJ, Aubin BS, Lange BM, Howe GA. Flavonoid deficiency disrupts redox homeostasis and terpenoid biosynthesis in glandular trichomes of tomato. Plant Physiol. 2022:**188**(3):1450–1468. https://doi.org/10.1093/plphys/kiab488

Sun Y, Harpazi B, Wijerathna-Yapa A, Merilo E, de Vries J, Michaeli D, Gal M, Cuming AC, Kollist H, Mosquna A. A ligand-independent origin of abscisic acid perception. Proc Natl Acad Sci U S A. 2019:**116**(49):24892–24899. https://doi.org/10.1073/pnas.1914480116

Takeuchi J, Fukui K, Seto Y, Takaoka Y, Okamoto M. Ligand-receptor interactions in plant hormone signaling. Plant J. 2021:**105**(2):290–306. https://doi.org/10.1111/tpj.15115

Tan X, Calderon-Villalobos LIA, Sharon M, Zheng C, Robinson CV, Estelle M, Zheng N. Mechanism of auxin perception by the TIR1 ubiquitin ligase. Nature. 2007:**446**(7136):640–645. https://doi.org/10.1038/nature05731

Tata JR. One hundred years of hormones. EMBO Rep. 2005:**6**(6):490–496. https://doi.org/10.1038/sj.embor.7400444

Thines B, Katsir L, Melotto M, Niu Y, Mandaokar A, Liu G, Nomura K, He SY, Howe GA, Browse J. JAZ repressor proteins are targets of the SCF(COI1) complex during jasmonate signalling. Nature. 2007:**448**(7154):661–665. https://doi.org/10.1038/nature05960

Thireault C, Shyu C, Yoshida Y, St Aubin B, Campos ML, Howe GA. Repression of jasmonate signaling by a non-TIFY JAZ protein in *Arabidopsis*. Plant J. 2015:**82**(4):669–679. https://doi.org/10.1111/tpj.12841

Thomma BP, Eggermont K, Penninckx IA, Mauch-Mani B, Vogelsang R, Cammue BP, Broekaert WF. Separate jasmonate-dependent and salicylate-dependent defense-response pathways in *Arabidopsis* are essential for resistance to distinct microbial pathogens. Proc Natl Acad Sci U S A. 1998:**95**(25):15107–15111. https://doi.org/10.1073/pnas.95.25.15107

Toyota M, Spencer D, Sawai-Toyota S, Jiaqi W, Zhang T, Koo AJ, Howe GA, Gilroy S. Glutamate triggers long-distance, calcium-based plant defense signaling. Science. 2018:**361**(6407):1112–1115. https://doi.org/10.1126/science.aat7744

Trewavas A. How do plant growth substances work? Plant Cell Environ. 1981:**4**(3):203–228. https://doi.org/10.1111/j.1365-3040.1981.tb01048.x

Ueda J, Kato J. Isolation and identification of a senescence-promoting substance from wormwood (*Artemisia absinthium* L). Plant Physiol. 1980:**66**(2):246–249. https://doi.org/10.1104/pp.66.2.246

Urano D, Jones AM. Heterotrimeric G protein-coupled signaling in plants. Annu Rev Plant Biol. 2014:**65**(1):365–384. https://doi.org/10.1146/annurev-arplant-050213-040133

Vick BA, Zimmerman DC. Biosynthesis of jasmonic acid by several plant species. Plant Physiol. 1984:**75**(2):458–461. https://doi.org/10.1104/pp.75.2.458

Wang K-D, Borrego EJ, Kenerley CM, Kolomeits MV. Oxylipins other than jasmonic acid are xylem-resident signals regulating systemic resistance induced by *Trichoderma virens* in maize. Plant Cell. 2020:**32**(1):16–185. https://doi.org/10.1105/tpc.19.00487

Wang L, Erb M. Volatile uptake, transport, perception, and signaling shape a plant's nose. Essays Biochem. 2022:**66**(5):695–702. https://doi.org/10.1042/EBC20210092

Wang JY, Haider I, Jamil M, Fiorilli V, Saito Y, Mi J, Baz L, Kountche BA, Jia K-P, Guo X, et al. The apocarotenoid metabolite zaxinone regulates growth and strigolactone biosynthesis in rice. Nat Commun. 2019:**10**(1):810. https://doi.org/10.1038/s41467-019-08461-1

Wang JY, Lin P-Y, Al-Babili S. On the biosynthesis and evolution of apocarotenoid plant growth regulators. Semin Cell Dev Biol. 2021:**109**:3–11. https://doi.org/10.1016/j.semcdb.2020.07.007

Wang J, Wu D, Wang Y, Xie D. Jasmonate action in plant defense against insects. J Exp Bot. 2019:**70**(13):3391–3400. https://doi.org/10.1093/jxb/erz174

Wasternack C. How jasmonates earned their laurels: past and present. J Plant Growth Regul. 2015:**34**(4):761–794. https://doi.org/10.1007/s00344-015-9526-5

Wasternack C, Feussner I. The oxylipin pathways: biochemistry and function. Annu Rev Plant Biol. 2018:**69**(1):363–386. https://doi.org/10.1146/annurev-arplant-042817-040440

Wasternack C, Parthier B. Jasmonate-signalled plant gene expression. Trends Plant Sci. 1997:**2**(8):302–307. https://doi.org/10.1016/S1360-1385(97)89952-9

Wasternack C, Song S. Jasmonates: biosynthesis, metabolism, and signaling by proteins activating and repressing transciption. J Exp Bot. 2017:**68**(6)1303–1321. https://doi.org/10.1093/jxb/erw443

Waters MT, Nelson DC. Karrikin perception and signalling. New Phytol. 2023:**237**(3):1525–1541. https://doi.org/10.1111/nph.18598

Weng J-K, Lynch JH, Matos JO, Dudareva N. Adaptive mechanisms of plant specialized metabolism connecting chemistry to function. Nat Chem Biol. 2021:**17**(10):1037–1045. https://doi.org/10.1038/s41589-021-00822-6

Westfall CS, Zubieta C, Herrmann J, Kapp U, Nanao MH, Jez JM. Structural basis for prereceptor modulation of plant hormones by GH3 proteins. Science. 2012:**336**(6089):1708–1711. https://doi.org/10.1126/science.1221863

Widemann E, Miesch L, Lugan R, Holder E, Heinrich C, Aubert Y, Miesch M, Pinot F, Heitz T. The amidohydrolases IAR3 and ILL6 contribute to jasmonoyl-isoleucine hormone turnover and generate 12-hydroxyjasmonic acid upon wounding in *Arabidopsis* leaves. J

Biol Chem. 2013:**288**(44):31701–31714. https://doi.org/10.1074/jbc.M113.499228

Woldemariam MG, Onkokesung N, Baldwin IT, Galis I. Jasmonoyl-L-isoleucine hydrolase 1 (JIH1) regulates jasmonoyl-L-isoleucine levels and attenuates plant defenses against herbivores. Plant J. 2012:**72**(5):758–767. https://doi.org/10.1111/j.1365-313X.2012.05117.x

Wong A, Chi W, Yu J, Bi C, Tian X, Yang Y, Gehring C. Plant adenylate cyclases have come full circle. Nat Plants. 2023:**9**(9):1389–1397. https://doi.org/10.1038/s41477-023-01486-x

Wu F, Deng L, Zhai Q, Zhao J, Chen Q, Li C. Mediator subunit MED25 couples alternative splicing of *JAZ* genes with fine-tuning of jasmonate signaling. Plant Cell. 2020:**32**(2):429–448. https://doi.org/10.1105/tpc.19.00583

Xie DX, Feys BF, James S, Nieto-Rostro M, Turner JG. *COI1*: an *Arabidopsis* gene required for jasmonate-regulated defense and fertility. Science. 1998:**280**(5366):1091–1094. https://doi.org/10.1126/science.280.5366.1091

Xu L, Liu F, Wang Z, Peng W, Huang R, Huang D, Xie D. An *Arabidopsis* mutant *cex1* exhibits constant accumulation of jasmonate-regulated *AtVSP*, *Thi2.1* and *PDF1.2*. FEBS Lett. 2001:**494**(3):161–164. https://doi.org/10.1016/S0014-5793(01)02331-6

Yamane H, Takahashi N, Ueda J-I, Kato J. Resolution of (±)-methyl jasmonate by high performance liquid chromatography and the inhibitory effect of (+)-enantiomer on the growth of rice seedlings. Agric Biol Chem. 1981:**45**(7):1709–1711. https://doi.org/10.1080/00021369.1981.10864755

Yan Y, Stolz S, Chételat A, Reymond P, Pagni M, Dubugnon L, Farmer EE. A downstream mediator in the growth repression limb of the jasmonate pathway. Plant Cell. 2007:**19**(8):2470–2483. https://doi.org/10.1105/tpc.107.050708

Yan J, Yao R, Chen L, Li S, Gu M, Nan F, Xie D. Dynamic perception of jasmonates by the F-box protein COI1. Mol Plant. 2018:**11**:1237–1247. https://doi.org/10.1016/j.molp.2018.07.007

Yu Y, Tang W, Lin W, Li W, Zhou X, Li Y, Chen R, Zheng R, Qin G, Cao W, et al. ABLs and TMKs are co-receptors for extracellular auxin. Cell. 2023:**186**(25):5457–5471.e17. https://doi.org/10.1016/j.cell.2023.10.017

Zander M. Many ways to repress! JAZ's agony of choices. Mol Plant. 2021:**14**(5):714–716. https://doi.org/10.1016/j.molp.2021.04.010

Zander M, Lewsey MG, Clark NM, Yin L, Bartlett A, Saldierna Guzmán JP, Hann E, Langford AE, Jow B, Wise A, et al. Integrated multi-omics framework of the plant response to jasmonic acid. Nat Plants. 2020:**6**(3):290–302. https://doi.org/10.1038/s41477-020-0605-7

Zhang F, Ke J, Zhang L, Chen R, Sugimoto K, Howe GA, Xu HE, Zhou M, He SY, Melcher K. Structural insights into alternative splicing-mediated desensitization of jasmonate signaling. Proc Natl Acad Sci U S A. 2017:**114**(7):1720–1725. https://doi.org/10.1073/pnas.1616938114

Zhang Y, Turner JG. Wound-induced endogenous jasmonates stunt plant growth by inhibiting mitosis. PLoS One. 2008:**3**(11):e3699. https://doi.org/10.1371/journal.pone.0003699

Zhang F, Yao J, Ke J, Zhang L, Lam VQ, Xin X-F, Zhou XE, Chen J, Brunzelle J, Griffin PR, et al. Structural basis of JAZ repression of MYC transcription factors in jasmonate signalling. Nature. 2015:**525**(7568):269–273. https://doi.org/10.1038/nature14661

Zhou XE, Zhang Y, Yao J, Zheng J, Zhou Y, He Q, Moreno J, Lam VQ, Cao X, Sugimoto K, et al. Assembly of JAZ-JAZ and JAZ-NINJA complexes in jasmonate signaling. Plant Commun. 2023:**4**(6):100639–. https://doi.org/10.1016/j.xplc.2023.100639

Zuo Z, Weraduwage SM, Lantz AT, Sanchez LM, Weise SE, Wang J, Childs KL, Sharkey TD. Isoprene acts as a signaling molecule in gene networks important for stress responses and plant growth. Plant Physiol. 2019:**180**(1):124–152. https://doi.org/10.1104/pp.18.01391

https://doi.org/10.1093/plphys/kiae050 PLANT PHYSIOLOGY 2024: 195: 155–169

Plant Physiology®

Topical Review

Molecular basis and evolutionary drivers of endosperm-based hybridization barriers

Heinrich Bente[1] and Claudia Köhler[1,2,*]

1 Department of Plant Reproductive Biology and Epigenetics, Max Planck Institute of Molecular Plant Physiology, Potsdam 14476, Germany
2 Department of Plant Biology, Swedish University of Agricultural Sciences and Linnean Center for Plant Biology, Uppsala 75007, Sweden

*Author for correspondence: koehler@mpimp-golm.mpg.de
The author responsible for distribution of materials integral to the findings presented in this article in accordance with the policy described in the Instructions for Authors (https://academic.oup.com/plphys/pages/General-Instructions) is: Claudia Köhler (koehler@mpimp-golm.mpg.de).

Abstract

The endosperm, a transient seed tissue, plays a pivotal role in supporting embryo growth and germination. This unique feature sets flowering plants apart from gymnosperms, marking an evolutionary innovation in the world of seed-bearing plants. Nevertheless, the importance of the endosperm extends beyond its role in providing nutrients to the developing embryo by acting as a versatile protector, preventing hybridization events between distinct species and between individuals with different ploidy. This phenomenon centers on growth and differentiation of the endosperm and the speed at which both processes unfold. Emerging studies underscore the important role played by type I MADS-box transcription factors, including the paternally expressed gene *PHERES1*. These factors, along with downstream signaling pathways involving auxin and abscisic acid, are instrumental in regulating endosperm development and, consequently, the establishment of hybridization barriers. Moreover, mutations in various epigenetic regulators mitigate these barriers, unveiling a complex interplay of pathways involved in their formation. In this review, we discuss the molecular underpinnings of endosperm-based hybridization barriers and their evolutionary drivers.

Introduction

Angiosperms comprise approximately 300,000 extant species and represent the most species-rich plant clade, accounting for approximately 90% of all land plant species (Christenhusz and Byng 2016; Katz 2018). Their rapid rise to ecological dominance in the mid-Cretaceous remained the "abominable mystery" to Charles Darwin, as evident from his 1879 letter to Joseph Hooker (Darwin and Seward 1903). Understanding this enigmatic rise to prominence has remained a central objective in evolutionary biology, necessitating insights into the evolution of reproductive barriers that limit gene flow between populations. Much previous research has focused on compre-hending prefertilization reproductive barriers, which prevent successful zygote formation through mechanisms like pollinator adaptations or pollen–pistil incompatibilities (Widmer et al. 2009; Hernández-Hernández et al. 2021). Recent advances in the field of evolutionary biology have recognized postzygotic mechanisms as substantial drivers of plant speciation (Coughlan 2023). These mechanisms operate at various time points after fertilization, leading to impaired viability of hybrid seeds or seedlings or sterility of the resulting hybrids (He et al. 2023). In this review, we focus on the evolution of postzygotic barriers arising from the endosperm, particularly in the context of interspecies and interploidy hybridizations. We discuss the molecular mechanisms underlying these barriers and explore their impact on plant speciation.

Endosperm development

Seed development in flowering plants starts with double fertilization, whereby 1 of the 2 pollen-delivered sperm cells fertilizes the egg cell, initiating the formation of the embryo, and the other sperm cell fertilizes the central cell, giving rise to the

Received August 31, 2023. Accepted January 06, 2024. Advance access publication January 31, 2024

ADVANCES

- Genetic data substantiate a model wherein AGLs function upstream of a regulatory cascade governing endosperm cellularization, making understanding of *AGL* regulation and their downstream targets of crucial relevance.
- Maternal Pol IV–dependent siRNAs play a pivotal role in suppressing *AGL* expression, thus fostering endosperm cellularization. Conversely, paternal Pol IV–dependent siRNAs exert a negative influence on endosperm cellularization, illustrating the antagonistic functions of maternal and paternal siRNAs in this process.
- The inhibitory effect of auxin on endosperm cellularization underscores the importance of unraveling auxin signaling within the endosperm.
- Endosperm cellularization serves to protect the embryo from desiccation, a process typically occurring during maturation. The precise role of ABA in this process remains in important open question.

endosperm (Dresselhaus et al. 2016). The endosperm is a unique feature of angiosperms, playing a crucial role as a nourishing tissue to support the growth and germination of the embryo (Doll and Ingram 2022). In most angiosperms, the endosperm is a triploid tissue with 2 maternal and 1 paternal genome copies. This triploid constitution arises since the central cell is a homodiploid cell (Dresselhaus et al. 2016). This increased maternal copy number may have evolved to regu-late resource allocation to the embryo, possibly reducing the influence of paternally provided growth-promoting alleles and ensuring equal provisioning of seeds (Haig and Westoby 1989). The ancestral state of the endosperm, ob-served in sister lineages to all angiosperms such as Nymphaeales and Austrobaileyales, is diploid. Resource allo-cation in these species is typically mediated by the perisperm, a nutritive tissue derived from the maternal sporophyte (Floyd and Friedman 2000; Williams and Friedman 2002).

Endosperm development leads to the formation of a fully cellularized tissue surrounding the embryo. However, the process of endosperm cellularization is not uniform across all species. In most angiosperms, the endosperm does not become cellular from the start; instead, nuclear divisions occur initially without cellularization, resulting in the formation of a coenocyte. Subsequently, cellularization takes place after a defined number of mitotic cycles, followed by the differentiation of distinct cell types (Boisnard-Lorig et al. 2001; Sorensen et al. 2002; Berger 2003). This developmental transition is critical for embryo survival; impaired endosperm cellularization in Arabidopsis (*Arabidopsis thaliana*) mutants is generally connected with embryo arrest (Grossniklaus et al. 1998; Kiyosue et al. 1999; Pignocchi et al. 2009; Hehenberger et al. 2012). In contrast to the nuclear type of endosperm development, certain genera like *Solanum*, *Mimulus*, and *Nicotiana* exhibit the cellular endosperm type, where nuclear division and cellularization are coupled events (Lopes and Larkins 1993; Floyd and Friedman 2001). Additionally, an intermediate type of endosperm development exists, known as the helobial endosperm. After the first division of the fertilized central cell, 1 cell follows the nuclear type of development while the other follows the cellular type, resulting in a unique helobial endosperm. Helobial endosperm development is relatively rare but can be found in species of the Cabombaceae, Sabiaceae, and Saxifragaceae families (Geeta 2003).

The endosperm plays a crucial role as a nutrient reservoir for the embryo, and its reserves can be remobilized during embryo development and germination. While in monocots the endosperm is preserved in the mature seed, serving to support embryo germination, in most eudicots, the endosperm is consumed by the developing embryo as it grows (Li and Berger 2012).

The endosperm establishes reproductive isolation

Numerous studies conducted in the last century have reported the failure of producing viable crop hybrids when crossing related plant species due to the widespread occurrence of seed lethality (Brink and Cooper 1947; Williams and White 1976; Gill and Waines 1978; Johnston and Hanneman Jr 1982; Sukno et al. 1999). By histological observations and in vitro embryo rescue experiments, these studies identified abnormal endosperm development as the primary cause of hybrid seed lethality in response to interspe-cies crosses (Brink and Cooper 1947; Valentine and Woodell 1963; Sukno et al. 1999; Dinu et al. 2005; Roy et al. 2011). This phenomenon's prevalence across diverse plant taxa (Fig. 1; Coughlan 2023) suggests that endosperm-based hybrid seed lethality (later on referred to as endosperm-based bar-riers [EBBs]) represents a substantial mode of reproductive isolation in flowering plants.

Similar to interspecies crosses, crosses involving plants with differing numbers of chromosome sets (ploidy) also result in seed lethality, with the outcome dependent on the direction of the cross. To describe this phenomenon, we will employ the term "paternal excess seeds" for seeds resulting from crosses where the paternal parent has a higher ploidy, while "maternal excess seeds" refer to those resulting from the re-ciprocal cross direction. Both types of seeds, whether they ex-hibit maternal or paternal excess, show abnormalities in endosperm development. In species characterized by the nu-clear mode of endosperm development, paternal excess leads to a delay in endosperm cellularization, whereas maternal ex-cess has the opposite effect, causing precocious cellulariza-tion (Scott et al. 1998; Pennington et al. 2008; Sekine et al. 2013). In species with the cellular mode of endosperm devel-opment, paternal excess seeds are characterized by larger and more vacuolated endosperm cells, while maternal excess seeds have smaller and less vacuolated cells (Woodell and

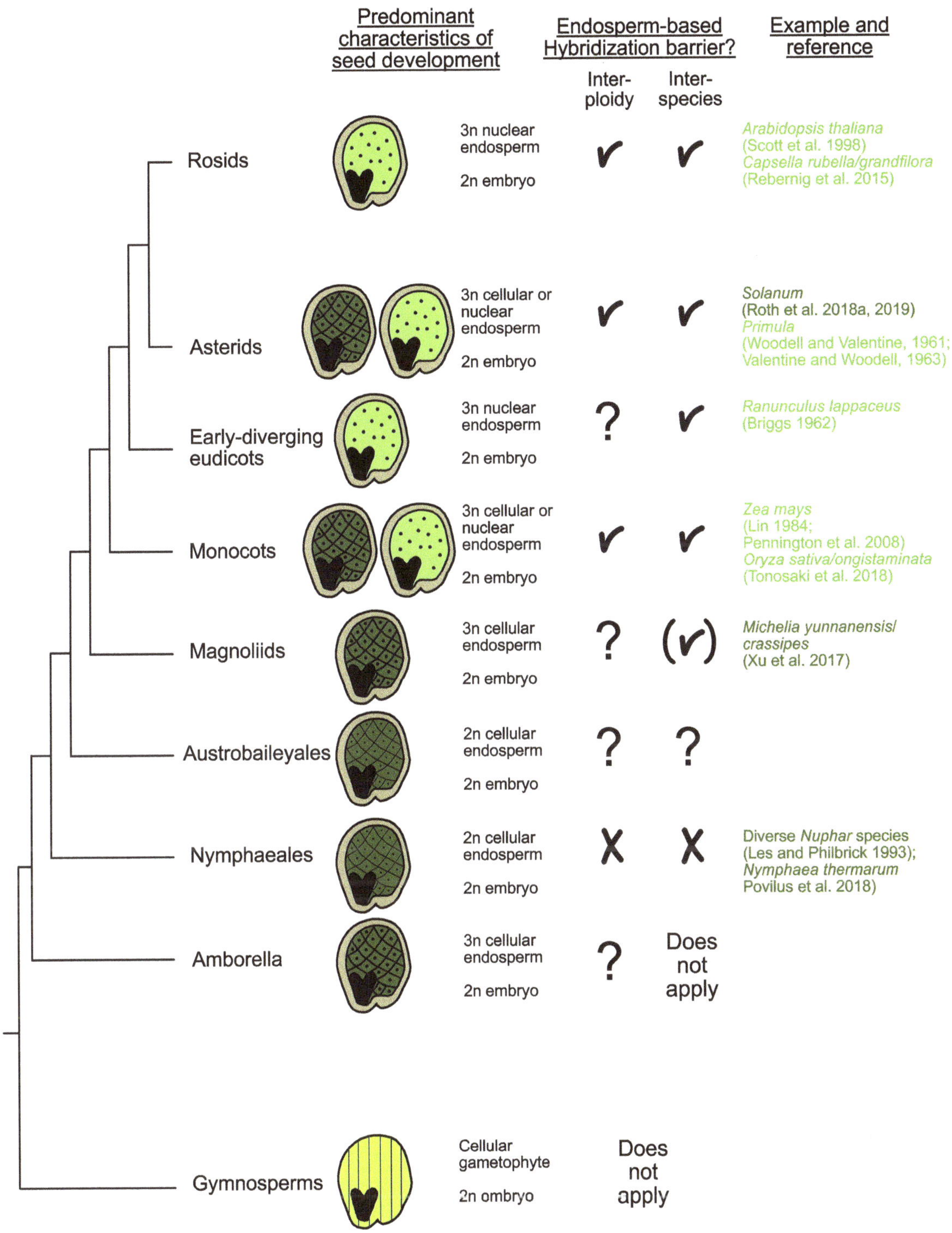

Figure 1. Widespread occurrence of EBBs. The presence of endosperm corresponds closely with the emergence of robust hybridization barriers. In gymnosperms, embryo development is supported by the enlarged female gametophyte (denoted by parallel lines), whereas angiosperms have successfully integrated fertilization with the formation of a nourishing endosperm tissue. The endosperm may initiate as a cellular structure, as seen in ANA (Amborella, Nymphaeales, and Austrobaileyales) species that are sister lineages to all other angiosperms (indicated by the checkerboard pattern), or as a coenocyte where initial nuclear divisions are not synchronized with cellularization (depicted by the dotted pattern). References for each category of barriers are provided; for a more comprehensive exploration of reported barriers, we direct readers to Coughlan (2023). References in light green refer to barriers occurring in nuclear-type endosperm. Dark green references refer to cellular-type barriers. Checkmarks symbolize presence of EBBs. Brackets symbolize that insufficient data are available to determine whether the endosperm is affected in hybrid seeds. Question marks symbolize that no data are available yet.

Valentine 1961; von Wangenheim and Peterson 2004). Maternal and paternal excess-like phenotypes have also been observed in interspecies crosses where the ploidy of the parental plants is alike (Brink and Cooper 1947; Sukno et al. 1999; Dinu et al. 2005; Roy et al. 2011; Baek et al. 2016; Oneal et al. 2016; Roth et al. 2018a; Coughlan et al. 2020). These deviations in endosperm development observed in both interploidy and interspecies crosses suggest a common mechanistic basis for these phenomena. This notion gains substantial support from the observation that interspecies EBBs can be overcome by increasing the ploidy of 1 parent, highlighting the involvement of a dosage-sensitive component in both types of EBBs (Johnston and Hanneman Jr 1982; Lafon-Placette et al. 2017; Tonosaki et al. 2018).

The concept of the endosperm balance number (EBN) provides a theoretical framework for understanding the dosage sensitivity of the endosperm and offers a means to assess hybridization success between different species (Johnston et al. 1980). According to this concept, each species possesses an effective ploidy, which may differ from its actual (karyotypic) ploidy. The effective ploidy is determined through test crosses with a defined species, allowing the assessment of cross-compatibility with other species (Johnston et al. 1980; Städler et al. 2021). Only species with similar EBNs can produce viable seeds by maintaining an effective 2:1 maternal-to-paternal ratio in the endosperm, a characteristic feature in most angiosperms (Lin 1984; Scott et al. 1998; Leblanc et al. 2002). From now on, we will refer to maternal or paternal excess crosses when the maternal or paternal parent has the higher effective ploidy, respectively. The focus however will be on paternal excess crosses, as most molecular research has concentrated on unraveling the molecular mechanisms underlying EBBs originating from these specific types of crosses.

Genomic imprinting: An epigenetic phenomenon underpinning reproductive isolation

The nonreciprocal effects observed in interploidy and interspecies hybridizations defy Mendelian expectations, revealing that parental genomes exert varying influences on offspring. Genomic imprinting, an epigenetic phenomenon that induces alterations in gene expression based on parental origin, explains these nonreciprocal hybridization outcomes. In the context of flowering plants, genomic imprinting primarily occurs within the endosperm and substantially influences its growth (Rodrigues and Zilberman 2015; Satyaki and Gehring 2017; Batista and Köhler 2020).

Imprinted genes are epigenetically modified, a process that typically initiates within the gametes and persists following fertilization. This results in genes that are specifically expressed either maternally (MEGs) or paternally (PEGs). Imprinted genes are enriched for transposable elements (TEs) in their flanking regions (Gehring et al. 2009; Wolff et al. 2011; Rodrigues et al. 2013; Pignatta et al. 2014; Hatorangan et al. 2016), with many of them subject to regulation by the DNA glycosylase DEMETER (DME). DME excises 5′-methylcytosine, and its activity in the central cell of the fe-male gametophyte leads to the activation of maternal alleles for certain MEGs that remain hypomethylated after fertilization (Choi et al. 2002; Pignatta et al. 2014; Park et al. 2016). Conversely, for PEGs, DNA hypomethylation facilitates Polycomb Repressive Complex2 (PRC2)–directed trimethylation at lysine 27 of histone H3 (H3K27me3), resulting in the suppression of maternal alleles of PEGs (Moreno-Romero et al. 2016, 2019).

The relationship between imprinted genes and flanking TEs initially led to the hypothesis that imprinting primarily functions to repress TEs (Gehring et al. 2009; Hsieh et al. 2009). However, recent research challenges this notion, revealing that numerous TEs harbor transcription factor binding sites. This discovery implies that DME-mediated demethylation plays a vital role in kick-starting gene expression by granting access to transcription factors (Schmitz et al. 2013; Batista et al. 2019b; Khouider et al. 2021). Hence, it ap-pears that endosperm imprinting results from a process en-abling the transcriptional activation of MEGs through DME-mediated demethylation, with some of these MEGs being crucial for seed development. Notably, *MEDEA* and *FERTILIZATION INDEPENDENT SEED2* (*FIS2*), encoding subu-nits of the PRC2, are essential MEGs that subsequently con-tribute to the suppression of the maternal alleles of PEGs (Zhang et al. 2014; Moreno-Romero et al. 2019).

Aside from the molecular explanations of how genomic imprinting is established, there are many considerations on the possible evolutionary advantages of this phenomenon. Among those, the "parental conflict" or "kinship theory" (Trivers 1974; Haig and Westoby 1989, 1991) is the most widely accepted evolutionary theory that provides a theoret-ical framework for the selection of imprinted genes. According to this theory, conflict arises because siblings compete for limited maternal resources. Since the maternal parent is equally related to its offspring, balanced provisioning of the progeny maximizes maternal fitness (Haig and Westoby 1989). Multiple paternity increases this conflict, since each offspring is more closely related to itself than to its siblings (Trivers 1974; Haig and Westoby 1989, 1991). This conflict drives the evolution of paternally derived alleles that enhance resource acquisition, while maternally derived alleles are selected to equalize nutrient allocation (Queller 1983; Haig and Westoby 1989, 1991). Consequently, MEGs and PEGs are predicted to be frequently associated with opposing functions in the endosperm, either restricting or promoting endosperm growth (Haig and Westoby 1989, 1991; Costa et al. 2012). Supporting this theory, an increased maternal genome dosage leads to reduced endosperm growth and smaller seeds, whereas an increased paternal genome dosage produces the opposite effect (Brink and Cooper 1947; Scott et al. 1998; von Wangenheim and Peterson 2004; Stoute et al. 2012; Sekine et al. 2013; Rebernig et al. 2015; Roth et al. 2019). Interestingly, the chalazal region of the endosperm is particularly affected by changes in parental genome dosage (Scott et al. 1998; Sandstedt and Sweigart 2022). This region of the endosperm serves as an interface between maternal and endosperm tissues, playing a pivotal role in determining the extent of nutrient flow from the maternal plant to the embryo and endosperm (Povilus and Gehring 2022). PEGs are most highly expressed in the chalazal endo-sperm, pointing at a functional role of PEGs in regulating nu-trient transfer by affecting the growth of the chalazal endosperm (Picard et al. 2021).

The cross-direction–dependent seed phenotypes observed in interspecies and interploidy hybridizations have led to the hypothesis that imprinted genes play a causal role in establishing EBBs (Haig and Westoby 1989, 1991; Moore 2001; Gutierrez-Marcos et al. 2003). Genetic studies provide support for this notion, demonstrating that mutations in several PEGs can suppress triploid seed abortion (referred to as triploid block) in Arabidopsis (Kradolfer et al. 2013b; Wolff et al. 2015; Huang et al. 2017; Batista et al. 2019b). Suppression of the triploid block by mutations in PEGs consistently corresponds with the restoration of endosperm cellularization (Wolff et al. 2015; Batista et al. 2019b), underscoring the critical role of endosperm cellularization in embryo growth. Many PEGs associated with the triploid block are generally highly upregulated in triploid seeds. However, this upregulation is not attributed to a breakdown of imprinting but rather to enhanced expression of the active paternal alleles (Batista et al. 2019b). In contrast, seed arrest in interspecies hybrids is associated with disruption of imprinting in hybrid endosperm (Josefsson et al. 2006; Burkart-Waco et al. 2015; Florez-Rueda et al. 2016; Kinser et al. 2021). Nevertheless, while the cause for the deregulation of imprinted genes in interploidy and interspecies hybrids may differ, the consequences of their deregulation are similar and lead to shared endosperm phenotypes with lethal consequences.

Small RNAs and DNA methylation in the establishment of reproductive isolation

Paternally inherited mutations in the RNA-directed DNA methylation (RdDM) pathway suppress the triploid block (Erdmann et al. 2017; Martinez et al. 2018; Satyaki and Gehring 2019; Wang et al. 2021), similar to mutants in the maintenance methyltransferase MET1 or chemically induced hypomethylation (Schatlowski et al. 2014; Huc et al. 2021). The RdDM pathway initiates with the activity of RNA polymerase IV (Pol IV), which generates short transcripts that are transformed into double-stranded RNAs by RNA-dependent RNA POLYMEREASE 2. These double-stranded RNAs are then cleaved into small interfering RNAs (siRNAs) of 21 to 24 nucleotides (nts) by DICER-like proteins that can be loaded into different ARGONAUTE proteins (Cuerda-Gil and Slotkin 2016). In pollen, Pol IV is required for generating 21- to 22-nt epigenetically activated siRNAs (easiRNAs) as well as 24-nt siRNAs. easiRNAs originate from TEs within the vegetative nucleus and are likely transported to sperm cells afterwards (Slotkin et al. 2009; Martinez et al. 2016). Notably, the formation of easiRNAs relies on microRNA845a (miR845a) and miR845b, 2 abundant pollen miRNAs that target long terminal repeat retrotransposons at their reverse transcription primer binding site. In line with the importance of easiRNAs in the triploid block, a mutation in *MIR845b* can alleviate this block (Borges et al. 2018).

Pol IV–generated siRNAs recruit the de novo DOMAINS REARRANGED METHYLTRANSFERASE2, which establishes DNA methylation in all sequence contexts (Chow et al. 2020). Interestingly, inbreeding of RdDM mutants enhances their ability to suppress the triploid block and is linked with a gradual decrease in DNA methylation over generations. In contrast to other RdDM pathway mutants, *nrpd1* (affected in the large subunit of Pol IV, NUCLEAR RNA POLYMERASE D1) exhibits a suppressive effect in the first generation of inbreeding (Wang et al. 2021), suggesting that the loss of easiRNAs and paternal DNA methylation can independently suppress the triploid block. Recent findings further support this notion, revealing that the depletion of easiRNAs in the vegetative cell of pollen (but not in sperm cells) can suppress the triploid block (Pachamuthu et al. 2023). Whether paternally delivered easiRNAs are causally responsible for DNA methylation changes in the endosperm as previously proposed (Martinez et al. 2018) remains to be investigated.

Despite that paternally inherited *nrpd1* can suppress the triploid block, the transcriptome changes in triploid seeds inheriting *nrpd1* through pollen remain relatively mild. Among the genes that continue to be upregulated in triploid *nrpd1* seeds are several PEGs previously identified to suppress the triploid block when mutated (Martinez et al. 2018; Satyaki and Gehring 2019). There are 2 possible hypotheses to explain this data: either Pol IV operates in a parallel pathway to the known suppressor genes or it acts downstream of these genes. Further research is required to distinguish between these hypotheses.

A link between the dosage of siRNAs and gene deregulation in hybrid endosperm has also been established in *Solanum*, where reciprocal hybrid seeds exhibit reduced siRNA levels corresponding with increased gene expression at specific loci potentially targeted by siRNAs (Florez-Rueda et al. 2021). Similarly, *Capsella* paternal excess hybrid endosperm shows strongly depleted DNA methylation, which associates with chromatin decondensation (Dziasek et al. 2021) and loss of siRNAs that depend on maternal Pol IV function (Dziasek et al. 2023). The loss of chromatin condensation in *Capsella* hybrid endosperm is associated with increased gene expression, particularly near heterochromatic pericentromeric regions, including many targets of type I MADS-box transcription factors (referred to as AGAMOUS-LIKE TFs or AGLs). These AGLs are consistently upregulated in paternal excess interploidy and interspecies hybrids (Josefsson et al. 2006; Erilova et al. 2009; Ishikawa et al. 2010; Burkart-Waco et al. 2013; Sekine et al. 2013; Rebernig et al. 2015; Roth et al. 2019; Bjerkan et al. 2020; Dziasek et al. 2021). AGL upregulation is likely causally relevant in establishing EBBs, since mutants in several *AGLs* show weakened interploidy and interspecies barriers (Josefsson et al. 2006; Walia et al. 2009; Hehenberger et al. 2012; Batista et al. 2019b). Maternal siRNAs inversely correspond with *AGL* expression (Lu et al. 2012; Kirkbride et al. 2019; Dziasek et al. 2023), suggesting a connection between the loss of maternal siRNAs, DNA methylation, and increased expression of *AGLs*. These Pol IV–dependent siRNAs exhibit a proportional increase corresponding to the ploidy of the maternal genome. Consequently, maternal excess in Arabidopsis leads to heightened accumulation of siRNAs, while paternal excess results in diminished siRNA abundance, corresponding with increased *AGL* expression (Lu et al. 2012; Kirkbride et al. 2019). Notably, maternal Pol IV–dependent siRNAs, originating from a few loci in high abundance, have been observed across various flowering plant species. The loci responsible for producing these siRNAs appear to evolve rapidly (Grover et al. 2020), hinting at a potential role in establishing reproductive barriers between species. Where exactly maternal Pol IV–dependent siRNAs are produced remains to be established; while they have been proposed to be predominantly maternally produced in *Brassica* (Grover et al. 2018, 2020), biparental production was found in Arabidopsis (Satyaki and Gehring 2022). Collectively, alterations in the dosage of maternal and paternal Pol IV–dependent siRNAs impact EBBs. However, dependent on their parental origin, these siRNAs are generated from distinct loci and influence different targets, in line with their antagonistic effects on gene expression and endosperm development (Satyaki and Gehring 2022).

Epigenetic regulators as key players in overcoming the triploid block

PHERES1 (PHE1) is a highly upregulated AGL in triploid Arabidopsis seeds that directly binds to the promoter region of PEGs, many of those also being highly upregulated in triploid seeds (Batista et al. 2019b). Among them, *ADMETOS (ADM)*, *SU(VAR)3–9 HOMOLOG 7 (SUVH7)*, *PICKLE RELATED2 (PKR2)*, and *PEG2* have been identified to suppress the triploid block when mutated (Kradolfer et al. 2013a; Wolff et al. 2015; Huang et al. 2017; Fig. 2). ADM physically interacts with SUVH9 and various AT HOOK-LIKE (AHL) proteins. Mutations in *suvh9* and *ahl10*, similar to *adm*, effectively suppress the triploid block, suggesting the 3 genes act in the same pathway (Jiang et al. 2017). SUVH9 has the capacity to bind to methylated DNA, facilitating the recruitment of DNA-dependent RNA polymerase V (Pol V) to chromatin (Johnson et al. 2014; Liu et al. 2014b). Simultaneously, AHLs bind to AT-rich chromosomal regions, inducing increased levels of H3K9me2 via unknown mechanisms (Yun et al. 2012; Xu et al. 2013). Notably, in triploid seeds, ADM-dependent mislocalization of H3K9me2 is targeted at AT-rich small TEs, inversely correlating with CHH methylation. Recent work has shown that the catalytic domain of SUVH9 is essential for its function (Parent et al. 2021), suggesting that AHL10-ADM-SUVH9 may collectively recruit H3K9me2 to TEs in the endosperm, potentially leading to the ectopic recruitment of H3K9me2 in triploid seeds (Jiang et al. 2017). This ectopic recruitment of H3K9me2 inversely correlates with CHH methylation, suggesting that AHL10-ADM-SUVH9–mediated H3K9me2 at specific TEs opposes the RdDM pathway in triploid seeds, in line with previous reports on the inhibitory effect of H3K9me2 on the RdDM pathway (Schoft et al. 2009; Zemach et al. 2013; Gent et al. 2014).

The roles of SUVH7 and PKR2 in relation to the triploid block are yet to be fully elucidated. Given that PKR2 shares homology with PICKLE (PKL), which acts in the RdDM pathway by supporting the accumulation of Pol V–dependent transcripts (Yang et al. 2017), it is plausible that PKR2 serves a related function. Similar to other mutants in the RdDM pathway (Wang et al. 2021), the suppressive effect of *pkr2* appears to require several generations of inbreeding (Huang et al. 2017). The *pkr2* mutant was not initially identified as a suppressor when tested after introgression into the meiotic mutant *omission of second division 1* (*osd1*; Wolff et al. 2015), in contrast to its suppressive effect following inbreeding (Huang et al. 2017).

The suppressive effect of *peg2* on the triploid block has been linked to the ability of the *PEG2* transcript to sequester the TE-derived *siRNA854* (Wang et al. 2018). This suggests that a reduced abundance of *siRNA854* has functional consequences for endosperm development. *SiRNA854* regulates the abundance of the stress granule–associated protein OLIGOURIDYLATE BINDING PROTEIN 1B (UBP1b) by translational repression (McCue et al. 2012), which is consistent with *ubp1b* partially suppressing the triploid block (Wang et al. 2018). UBP1b has been proposed to play a role in the translational repression of *Athila* retrotransposon transcripts (McCue et al. 2012). However, whether this mechanism is mechanistically connected to the triploid block requires further investigation.

Collectively, many of the triploid block suppressors identified in Arabidopsis encode for chromatin regulators with potential roles in TE silencing or the establishment of heterochromatin (Wolff et al. 2015; Jiang et al. 2017; Yang et al. 2017; Satyaki and Gehring 2019; Wang et al. 2021). Similarly, many genes responsible for hybrid incompatibility in *Drosophila* encode for dosage-sensitive heterochromatin-interacting proteins or components of the PIWI pathway that silences TEs (Brideau et al. 2006; Bayes and Malik 2009; Thomae et al. 2013; Parhad et al. 2017). In *Drosophila*, TEs have been implicated in causing intraspecies incompatibility (Castillo and Moyle 2022). Nevertheless, like in plants, the mechanisms through which TEs and deregulated chromatin regulators induce lethality remain to be fully established.

The role of auxin in antagonizing endosperm cellularization

Beyond its interactions with various epigenetic regulators, PHE1 also binds to auxin biosynthesis genes *YUCCA10* (*YUC10*) and *TRYPTOPHAN AMINOTRANSFERASE RELATED*

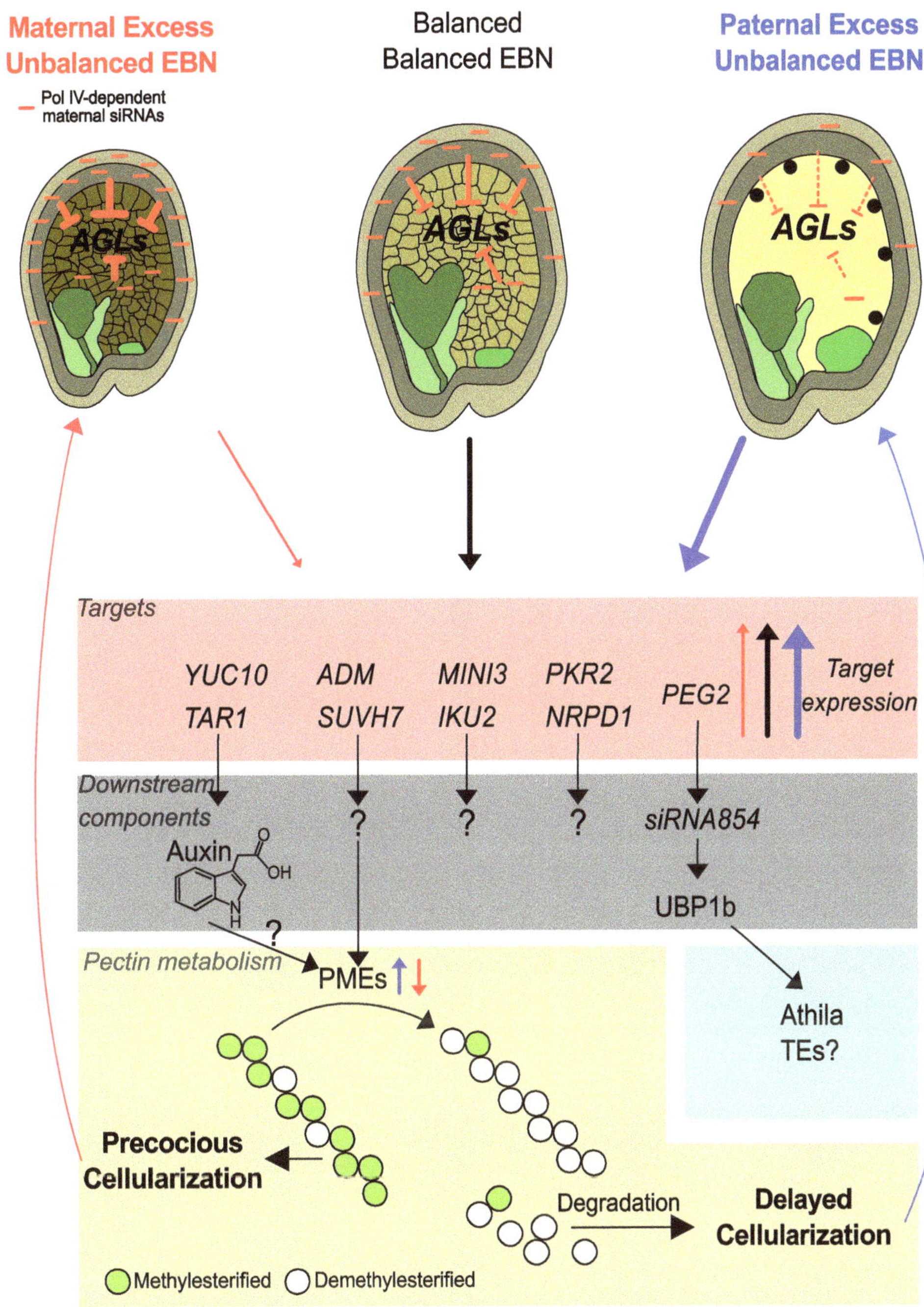

Figure 2. Molecular pathways underlying endosperm-based interploidy and interspecies barriers. Maternal Pol IV–dependent siRNAs act as dosage-dependent repressors of AGL MADS-box genes. Increased dosage of siRNAs in maternal excess crosses corresponds with strong AGL repression (reflected by blunt-ended arrows; thickness of the arrows corresponds to intensity of repression) and early endosperm cellularization, while decreased siRNA dosage in paternal excess crosses corresponds with increased AGL abundance and late endosperm cellularization. High and low AGL dosage causes increased or decreased expression of downstream AGL targets, respectively, including auxin biosynthesis genes. Downstream direct or indirect targets of this signaling cascade are genes involved in pectin metabolism, including pectin methylesterase (PME) genes. Demethylesterification exposes pectin to pectin–degrading enzymes, thereby establishing a connection between the expression level of *AGLs* and the timing of endosperm cellularization. Abundance of AGLs and their downstream targets in response to hybridization are depicted by red (low), gray (balanced), and blue (high) arrows. Light and dark green endosperm regions mark micropylar and chalazal domains, respectively. ADM, ADMETOS; AGLs, AGAMOUS-LIKE genes; EBN, endosperm balance number; IKU2, HAIKU2; MINI3, MINISEED3; NRPD1, NUCLEAR RNA POLYMERASE D1; PEG2, PATERNALLY EXPRESSED GENE2; UBP1, OLIGOURIDYLATE BINDING PROTEIN 1B; PKR2, PICKLE RELATED2; SUVH7, SU(VAR)3–9 HOMOLOG 7; TAR1, TAA1-RELATED1; TEs, transposable elements; YUC10, YUCCA10.

1 (*TAR1*). Activated by PHE1, *YUC10* and *TAR1* are highly overexpressed in triploid seeds, concomitant with increased auxin activity (Batista et al. 2019a, 2019b; Fig. 3).

Elevated auxin levels wield a negative impact on endo-sperm cellularization, establishing a direct link between the upregulation of *PHE1*, increased auxin production, and the failure of endosperm cellularization (Batista et al. 2019a). Interestingly, increasing auxin levels, specifically in the seed coat, also block endosperm cellularization, implicating the seed coat as a target of auxin action that indirectly impacts endosperm cellularization. Notably, in triploid seeds, increased auxin activity is primarily detected in the seed coat rather than the endosperm, hinting at auxin's production in the endo-sperm and its subsequent transportation to the seed coat, drives seed coat expansion (Batista et al. 2019a). AGL62 emerges as a crucial player in regulating auxin transport from the endo-sperm to the seed coat, potentially by modulating the expres-sion of the ABCB-type transporter gene *PGP10* (Figueiredo et al. 2016) and regulating auxin biosynthesis (Guo et al. 2022). Supporting this notion is the observation that the *agl62* mutant exhibits partial suppression of EBBs (Walia et al. 2009; Hehenberger et al. 2012). This intricate regulation of auxin transport between the endosperm and seed coat potentially underpins the mechanism behind the triploid block.

The suppressive effect of *transparent testa* (*tt*) mutants, affecting flavonoid biosynthesis in the innermost layer of the inner integument, offers another potential link between aux-in and the triploid block. Flavonoids impact polar auxin transport by interacting with auxin transporters such as PGPs (Peer and Murphy 2007). Several mutants within the TT pathway, including *transparent testa glabra2* (*ttg2*), *tt4*, *tt7*, *tt8*, and *tt13*, exhibit the ability to suppress the triploid block (Dilkes et al. 2008; Scott et al. 2013; Doughty et al. 2014; Zumajo-Cardona et al. 2023), although the precise mechanisms remain elusive. Intriguingly, the efficiency of this suppression varies among the *tt* mutants, with *tt8* causing complete suppression, while others have a milder effect (Zumajo-Cardona et al. 2023), hinting at the involvement of multiple pathways regulated by *TT* genes.

A plausible model suggests that type I MADS-box transcription factors, such as PHE1 and AGL62, govern auxin homeostasis within the endosperm. This regulation includes binding to the promoter regions of auxin biosynthesis genes, resulting in increased auxin production in direct proportion to the increased dosage of AGLs (Figueiredo et al. 2016; Batista et al. 2019a). Auxin reduces tissue rigidity in the shoot apical meristem by initiating the demethylesterification of homogalacturonan, a component of pectin. This process ren-ders it susceptible to enzymatic depolymerization, facilitating gel formation and degradation (Braybrook and Peaucelle 2013; Wormit and Usadel 2018). Genes linked to demethyles-terification and degradation of pectin are significantly over-expressed in triploid hybrid seeds that fail to undergo endosperm cellularization, suggesting that auxin regulates pectin metabolism with consequences on endosperm cellularization (Rebernig et al. 2015; Wolff et al. 2015).

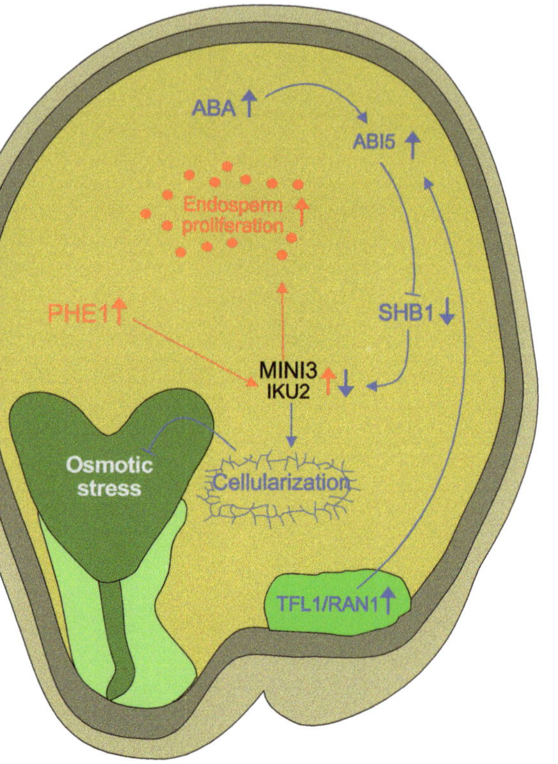

Figure 3. Regulation of endosperm cellularization by antagonizing activities of auxin and ABA. ABA activates the downstream transcription factor ABI5, which, in turn, is stabilized by mobile factors such as TFL1, which is transported from the chalazal region to the peripheral endosperm. The mobilization of TFL1 is facilitated by RAN1. ABI5, under the influence of ABA, exerts its regulatory function by downregulating *SHB1*. SHB1 is a positive regulator of 2 critical genes, *MINI3* and *IKU2*, both of which are direct targets of PHE1. MINI3 and IKU2 act as positive regulators of endosperm proliferation (symbolized by red dots). The repression of *MINI3* and *IKU2* via the ABA pathway likely promotes endosperm cellularization. Failure of endosperm cellularization can trigger an osmotic stress response in the embryo. Red arrows in this network represent the auxin-PHE1 pathway, and blue arrows highlight the ABA signaling pathway. Upward-pointing arrows reflect increased expression or activity, while downward-pointing arrows reflect decreased expression or activity. Arrows with blunt ends reflect repressive activity. Light and dark green endosperm regions mark micropylar and chalazal domains, respectively. ABI5, ABSCISIC ACID-INSENSITIVE5; IKU2, HAIKU2; MINI3, MINISEED3; PHE1, PHERES1; RAN1, Ras-related nuclear GTPase1; SHB1, SHORT HYPOCOTYL UNDER BLUE1; TFL1, TERMINAL FLOWER1.

Abscisic acid facilitates endosperm cellularization and enhances triploid seed viability

The precise timing of endosperm cellularization represents a critical factor for embryo survival, a phenomenon consistently disrupted in interploidy and interspecies hybrids (Muntzing 1933; Brink and Cooper 1947; Woodell and Valentine 1961; Scott et al. 1998; Dilkes et al. 2008; Hehenberger et al. 2012). The reasons underpinning the necessity of endosperm cellularization for embryo survival have long been a matter of conjecture. One proposed hypothesis suggests that the failure of endosperm cellularization results in an inadequate supply of sucrose to the embryo, as the central vacuole remains the primary resource sink in the seed (Lafon-Placette and Kohler 2014). However, recent research suggests an alternative perspective: embryos surrounded by noncellularized endosperm elicit an embryo-specific osmotic stress response akin to the seed maturation process. This implies that endosperm cellularization serves to protect the embryo from desiccation, a process typically occurring during maturation (Kozaki and Aoyanagi 2022; Xu et al. 2023). Supporting this view, mutations in abscisic acid (ABA) biosynthesis and signaling exacerbate the triploid block, while elevated endogenous ABA levels induce endosperm cellularization and suppress the arrest of embryo growth (Xu et al. 2023). Nevertheless, whether the suppressive effect of increased ABA is a consequence of ABA-induced endosperm cellularization or an enhanced tolerance of the embryo to desiccation remains to be established.

Interestingly, exposure to cooler temperatures during early seed maturation leads to ABA retention in the endosperm upon desiccation (Chen et al. 2021). Similarly, cold treatment enhances the viability of Arabidopsis hybrid seeds impaired in endosperm cellularization (Bjerkan et al. 2020), suggesting that the survival of hybrid seeds in response to cold is linked to elevated ABA levels.

The process of ABA-mediated endosperm cellularization hinges on the downregulation of *SHORT HYPOCOTYL UNDER BLUE1* (*SHB1*) by ABSCISIC ACID-INSENSITIVE5 (ABI5). ABI5-deficient plants exhibit delayed endosperm cellularization and develop enlarged seeds (Cheng et al. 2014), which, notably, intensify the triploid block (Xu et al. 2023). SHB1, in turn, upre-gulates *MINISEED3* (*MINI3*) and the LRR receptor kinase gene *HAIKU2* (*IKU2*), both of which are direct targets of PHE1 and exert a positive influence on endosperm proliferation (Garcia et al. 2003; Luo et al. 2005; Batista et al. 2019a).

Recent investigations have unveiled the interaction between ABI5 and TERMINAL FLOWER1 (TFL1), a mobile regulator generated in the chalazal endosperm that then relocates to the peripheral endosperm to stabilize ABI5 (Zhang et al. 2020). The movement of TFL1 is modulated by Ras-related nuclear GTPases (RANs) previously implicated in seed size regulation (Liu et al. 2014a). A plausible model posits that ABA activates ABI5, which is subsequently stabilized by TFL1 and RAN1. ABI5 then represses *SHB1*, which in turn restrains *IKU2* and *MINI3*, recognized regulators of endosperm proliferation (Fig. 3). This intricate network is juxtaposed with PHE1, which binds to *IKU2* and *MINI3*, likely promoting their transcription (Batista et al. 2019b).

These findings underscore the convergence of multiple pathways on MINI3 and IKU2, with 1 pathway positively regulated by auxin and PHE1 and the other negatively regulated by ABA, ABI5, and SHB1. While the loss of MINI3 function does not impact the triploid block (Batista et al. 2019a), a mutation in the MINI3 interaction partner IKU1 enhances the survival of Arabidopsis interspecies hybrids (Burkart-Waco et al. 2013), suggesting that the dosage of components within the IKU1-IKU2-MINI3 pathway plays a pivotal role in erecting hybridization barriers by affecting endosperm cellularization.

Understanding the relevance and drivers of endosperm-mediated reproductive barriers

As elaborated in the preceding sections, the establishment of reproductive barriers mediated by the endosperm is intricately linked to the dysregulation of imprinted genes, particularly PEGs (Wolff et al. 2015; Florez-Rueda et al. 2016; Batista et al. 2019b; Kinser et al. 2021). In line with theoretical expectations, species characterized by outbreeding tendencies are presumed to experience heightened parental conflict, thus predicting an increased abundance or elevated expression of imprinted genes (Brandvain and Haig 2005). Indeed, empirical evidence supports this prediction, as outbreeding species tend to exhibit higher numbers or levels of PEGs compared to their inbreeding relatives (Klosinska et al. 2016; Lafon-Placette et al. 2018).

This augmented prevalence and expression of PEGs in outbreeders could account for the robust EBBs that often manifest between inbreeding and outbreeding species. This phenomenon has been theoretically framed as the "weak inbreeder/strong outbreeder" (WISO) hypothesis (Brandvain and Haig 2005). The validity of the WISO hypothesis is substantiated by extensive evidence, with documented EBBs observed between inbreeders and outbreeders in plant genera such as Arabidopsis, *Arabis*, and *Capsella* (Brandvain and Haig 2005; Lafon-Placette and Kohler 2016; Lafon-Placette et al. 2018; İltaş et al. 2021; Petrén et al. 2023). Similar to differences in breeding mode, also differences in effective popu-lation size can lead to strong reproductive barriers. A decrease in population size is anticipated to diminish con-flicts among siblings as mating partners become highly re-lated, akin to the impact of selfing. Correspondingly, within outbreeding species of *Solanum* and *Mimulus*, a smaller population size (reflected by lower nt diversity) is associated with a lower EBN (Roth et al. 2018a, 2019; Coughlan et al. 2020; Sandstedt et al. 2021; Sandstedt and Sweigart 2022). Interestingly, in *Solanum*, out of 3 tested species, the one with the highest EBN has the highest number of PEGs (Roth et al. 2018b, 2019). This finding underscores a potential connection between number of PEGs, intensity of parental conflict, and strength of EBBs.

Genomic imprinting is a dynamically evolving phenomenon, characterized by low conservation of imprinted loci across species (Waters et al. 2013; Pignatta et al. 2014; Hatorangan et al. 2016; Klosinska et al. 2016; Chen et al. 2018). In alignment with the hypothesized causal role of PEGs in the establishment of reproductive barriers, EBBs themselves exhibit a propensity for rapid evolution (Rebernig et al. 2015; Oneal et al. 2016; Roth et al. 2018a; Coughlan et al. 2020; İltaş et al. 2021). The exploration of the functional role of imprinted genes in the context of interspecies hybridization barriers promises to be an exciting avenue for future research.

OUTSTANDING QUESTIONS

- What is the functional role of paternal Pol IV–dependent siRNAs (easiRNAs) in the establishment of reproductive barriers?
- Are maternal Pol IV–dependent siRNAs transmitted from maternal sporophytic tissues to the endosperm, and what are the underlying mechanisms of this process?
- What are the epistatic relationships among the identified suppressor genes associated with hybridization barriers?
- At what pace do hybridization barriers evolve?

Another compelling research direction lies in the relationship between the strength of EBBs and the mode of endosperm development. Notably, robust barriers are encountered in species characterized by both nuclear and cellular endosperm development, suggesting that the mode of endosperm development itself does not substantially influence barrier strength. Furthermore, the ploidy level of the endosperm does not appear to be the decisive factor, as evidenced by the robust interploidy barriers observed in species like evening primrose (*Oenothera hookeri*), which have a diploid endosperm (von Wangenheim 1962). Instead, the key de-terminant seems to be the role of the endosperm in provisioning the embryo, a role that varies among species. In species where the endosperm plays a pivotal role in embryo provisioning, strong EBBs tend to be prevalent. Conversely, barriers are generally weaker in species where the endosperm has a relatively minor role in this context, such as in Nymphaeales, which rely on the maternal sporophytic nucel-lus for seed storage (Les and Philbrick 1993; Povilus et al. 2018). An intriguing question for future investigation pertains to whether the presence of weak EBBs is a recurring pattern among angiosperms with reduced endosperms.

In conclusion, our comprehension of the genetic underpinnings of EBBs has made substantial progress in recent years, shedding light on intriguing facets while simultaneously unveiling additional inquiries (see Outstanding Questions). Tackling these emerging questions collectively as a scientific community promises to be an exciting endeavor for the future.

Acknowledgments

We thank Gesa Hoffmann for critically reading this manuscript and support in generating the figures. We furthermore thank Yichun Qiu, Katarzyna Dziasek, and Varsha Vasudevan for helpful suggestions to improve the manuscript.

Author contributions

H.B. and C.K. conceived, wrote, and edited the article.

Funding

This work was financially supported by the Max Planck Society.

Conflict of interest statement. None declared.

References

Baek YS, Royer SM, Broz AK, Covey PA, López-Casado G, Nuñez R, Kear PJ, Bonierbale M, Orillo M, van der Knaap E, et al. Interspecific reproductive barriers between sympatric populations of wild tomato species (*Solanum* section *Lycopersicon*). Am J Bot. 2016:**103**(11):1964–1978. https://doi.org/10.3732/ajb.1600356

Batista RA, Figueiredo DD, Santos-Gonzalez J, Köhler C. Auxin regulates endosperm cellularization in *Arabidopsis*. Genes Dev. 2019a:**33**(7–8):466–476. https://doi.org/10.1101/gad.316554.118

Batista RA, Köhler C. Genomic imprinting in plants-revisiting existing models. Genes Dev. 2020:**34**(1–2):24–36. https://doi.org/10.1101/gad.332924.119

Batista RA, Moreno-Romero J, Qiu Y, van Boven J, Santos-Gonzalez J, Figueiredo DD, Köhler C. The MADS-box transcription factor PHERES1 controls imprinting in the endosperm by binding to domesticated transposons. Elife 2019b:**8**:e50541. https://doi.org/10.7554/eLife.50541

Bayes JJ, Malik HS. Altered heterochromatin binding by a hybrid sterility protein in *Drosophila* sibling species. Science 2009:**326**(5959): 1538–1541. https://doi.org/10.1126/science.1181756

Berger F. Endosperm: the crossroad of seed development. Curr Opin Plant Biol. 2003:**6**(1):42–50. https://doi.org/10.1016/S1369526602000043

Bjerkan KN, Hornslien KS, Johannessen IM, Krabberød AK, van Ekelenburg YS, Kalantarian M, Shirzadi R, Comai L, Brysting AK, Bramsiepe J, et al. Genetic variation and temperature affects hybrid barriers during interspecific hybridization. Plant J. 2020:**101**(1): 122–140. https://doi.org/10.1111/tpj.14523

Boisnard-Lorig C, Colon-Carmona A, Bauch M, Hodge S, Doerner P, Bancharel E, Dumas C, Haseloff J, Berger F. Dynamic analyses of the expression of the histone::YFP fusion protein in Arabidopsis show that syncytial endosperm is divided in mitotic domains. Plant Cell. 2001:**13**(3):495–509. https://doi.org/10.1105/tpc.13.3.495

Borges F, Parent JS, van Ex F, Wolff P, Martinez G, Köhler C, Martienssen RA. Transposon-derived small RNAs triggered by miR845 mediate genome dosage response in *Arabidopsis*. Nat Genet. 2018:**50**(2):186–192. https://doi.org/10.1038/s41588-017-0032-5

Brandvain Y, Haig D. Divergent mating systems and parental conflict as a barrier to hybridization in flowering plants. Am Nat. 2005:**166**(3): 330–338. https://doi.org/10.1086/432036

Braybrook SA, Peaucelle A. Mechano-chemical aspects of organ formation in *Arabidopsis thaliana*: the relationship between auxin and pectin. PLoS One 2013:**8**(3):e57813. https://doi.org/10.1371/journal.pone.0057813

Brideau NJ, Flores HA, Wang J, Maheshwari S, Wang X, Barbash DA. Two Dobzhansky-Muller genes interact to cause hybrid lethality in *Drosophila*. Science 2006:**314**(5803):1292–1295. https://doi.org/10.1126/science.1133953

Briggs BG. Interspecific hybridization in the *Ranunculus lappaceus* group. Evolution 1962:**16**(3):372–390. https://doi.org/10.2307/2406286

Brink RA, Cooper DC. The endosperm in seed development. Bot Rev. 1947:**132**(8):423–541. https://doi.org/10.1007/BF02861548

Burkart-Waco D, Ngo K, Dilkes B, Josefsson C, Comai L. Early disruption of maternal-zygotic interaction and activation of defense-like responses in Arabidopsis interspecific crosses. Plant Cell. 2013:**25**(6): 2037–2055. https://doi.org/10.1105/tpc.112.108258

Burkart-Waco D, Ngo K, Lieberman M, Comai L. Perturbation of parentally biased gene expression during interspecific hybridization.

PLoS One 2015:**10**(2):e0117293. https://doi.org/10.1371/journal.pone.0117293

Castillo DM, Moyle LC. Hybrid incompatibility between *Drosophila virilis* and *D. lummei* is stronger in the presence of transposable elements. J Evol Biol. 2022:**35**(10):1319–1334. https://doi.org/10.1111/jeb.14079

Chen C, Li T, Zhu S, Liu Z, Shi Z, Zheng X, Chen R, Huang J, Shen Y, Luo S, et al. Characterization of imprinted genes in rice reveals conservation of regulation and imprinting with other plant species. Plant Physiol. 2018:**177**(4):1754–1771. https://doi.org/10.1104/pp.17.01621

Chen X, Yoong F-Y, O'Neill CM, Penfield S. Temperature during seed maturation controls seed vigour through ABA breakdown in the endosperm and causes a passive effect on DOG1 mRNA levels during entry into quiescence. New Phytol. 2021:**232**(3):1311–1322. https://doi.org/10.1111/nph.17646

Cheng ZJ, Zhao XY, Shao XX, Wang F, Zhou C, Liu YG, Zhang Y, Zhang XS. Abscisic acid regulates early seed development in Arabidopsis by ABI5-mediated transcription of SHORT HYPOCOTYL UNDER BLUE1. Plant Cell. 2014:**26**(3):1053–1068. https://doi.org/10.1105/tpc.113.121566

Choi Y, Gehring M, Johnson L, Hannon M, Harada JJ, Goldberg RB, Jacobsen SE, Fischer RL. DEMETER, a DNA glycosylase domain protein, is required for endosperm gene imprinting and seed viability in *Arabidopsis*. Cell 2002:**110**(1):33–42. https://doi.org/10.1016/S0092-8674(02)00807-3

Chow HT, Chakraborty T, Mosher RA. RNA-directed DNA methylation and sexual reproduction: expanding beyond the seed. Curr Opin Plant Biol. 2020:**54**:11–17. https://doi.org/10.1016/j.pbi.2019.11.006

Christenhusz MJM, Byng JW. The number of known plants species in the world and its annual increase. Phytotaxa 2016:**261**(3):201–217. https://doi.org/10.11646/phytotaxa.261.3.1

Costa LM, Yuan J, Rouster J, Paul W, Dickinson H, Gutierrez-Marcos JF. Maternal control of nutrient allocation in plant seeds by genomic imprinting. Curr Biol. 2012:**22**(2):160–165. https://doi.org/10.1016/j.cub.2011.11.059

Coughlan JM. The role of hybrid seed inviability in angiosperm speciation. Am J Bot. 2023:**110**(3):1–14. https://doi.org/10.1002/ajb2.16135

Coughlan JM, Wilson Brown M, Willis JH. Patterns of hybrid seed inviability in the *Mimulus guttatus* sp. complex reveal a potential role of parental conflict in reproductive isolation. Curr Biol. 2020:**30**(1): 83–93.e85. https://doi.org/10.1016/j.cub.2019.11.023

Cuerda-Gil D, Slotkin RK. Non-canonical RNA-directed DNA methylation. Nat Plants. 2016:**2**(11):16163. https://doi.org/10.1038/nplants.2016.163

Darwin F, Seward AC. More letters of Charles Darwin. A record of his work in a series of hitherto. London: John Murray; 1903. (unpublished letters).

Dilkes BP, Spielman M, Weizbauer R, Watson B, Burkart-Waco D, Scott RJ, Comai L. The maternally expressed WRKY transcription factor TTG2 controls lethality in interploidy crosses of *Arabidopsis*. PLoS Biol. 2008:**6**(12):e308. https://doi.org/10.1371/journal.pbio.0060308

Dinu II, Hayes RJ, Kynast RG, Phillips RL, Thill CA. Novel inter-series hybrids in *Solanum*, section *Petota*. Theor Appl Genet. 2005:**110**(3): 403–415. https://doi.org/10.1007/s00122-004-1782-x

Doll NM, Ingram GC. Embryo-endosperm interactions. Annu Rev Plant Biol. 2022:**73**(1):293–321. https://doi.org/10.1146/annurev-arplant-102820-091838

Doughty J, Aljabri M, Scott RJ. Flavonoids and the regulation of seed size in *Arabidopsis*. Biochem Soc Trans. 2014:**42**(2):364–369. https://doi.org/10.1042/BST20140040

Dresselhaus T, Sprunck S, Wessel GM. Fertilization mechanisms in flowering plants. Curr Biol. 2016:**26**(3):R125–R139. https://doi.org/10.1016/j.cub.2015.12.032

Dziasek K, Santos-González J, Qiu Y, Rigola D, Nijbroek K, Köhler C. Dosage sensitive maternal siRNAs determine hybridization success in *Capsella*. Res Square. 2023. https://doi.org/10.21203/rs.21203.rs-3705839/v3705831

Dziasek K, Simon L, Lafon-Placette C, Laenen B, Wärdig C, Santos-González J, Slotte T, Köhler C. Hybrid seed incompatibility in *Capsella* is connected to chromatin condensation defects in the endosperm. PLoS Genet. 2021:**17**(2):e1009370. https://doi.org/10.1371/journal.pgen.1009370

Erdmann RM, Satyaki PR, Klosinska M, Gehring M. A small RNA pathway mediates allelic dosage in endosperm. Cell Rep. 2017:**21**(12):3364–3372. https://doi.org/10.1016/j.celrep.2017.11.078

Erilova A, Brownfield L, Exner V, Rosa M, Twell D, Mittelsten Scheid O, Hennig L, Köhler C. Imprinting of the Polycomb group gene MEDEA serves as a ploidy sensor in *Arabidopsis*. PLoS Genet. 2009:**5**(9):e1000663. https://doi.org/10.1371/journal.pgen.1000663

Figueiredo DD, Batista RA, Roszak PJ, Hennig L, Köhler C. Auxin production in the endosperm drives seed coat development in *Arabidopsis*. Elife 2016:**5**:e20542. https://doi.org/10.7554/eLife.20542

Florez-Rueda AM, Fiscalini F, Roth M, Grossniklaus U, Städler T. Endosperm and seed transcriptomes reveal possible roles for small RNA pathways in wild tomato hybrid seed failure. Genome Biol Evol. 2021:**13**(8):evab107. https://doi.org/10.1093/gbe/evab107

Florez-Rueda AM, Paris M, Schmidt A, Widmer A, Grossniklaus U, Stadler T. Genomic imprinting in the endosperm is systematically perturbed in abortive hybrid tomato seeds. Mol Biol Evol. 2016:**33**(11):2935–2946. https://doi.org/10.1093/molbev/msw175

Floyd S, Friedman W. Evolution of endosperm developmental patterns among basal flowering plants. Int J Plant Sci. 2000:**161**(S6):S57S81. https://doi.org/10.1086/317579

Floyd SK, Friedman WE. Developmental evolution of endosperm in basal angiosperms: evidence from *Amborella* (Amborellaceae), *Nuphar* (Nymphaeaceae), and *Illicium* (Illiciaceae). Plant Syst Evol. 2001:**228**(3–4):153–169. https://doi.org/10.1007/s006060170026

Garcia D, Saingery V, Chambrier P, Mayer U, Jürgens G, Berger F. Arabidopsis haiku mutants reveal new controls of seed size by endosperm. Plant Physiol. 2003:**131**(4):1661–1670. https://doi.org/10.1104/pp.102.018762

Geeta R. The origin and maintenance of nuclear endosperms: viewing development through a phylogenetic lens. Proc R Soc Lond B Biol Sci. 2003:**270**(1510):29–35. https://doi.org/10.1098/rspb.2002.2206

Gehring M, Bubb KL, Henikoff S. Extensive demethylation of repetitive elements during seed development underlies gene imprinting. Science 2009:**324**(5933):1447–1451. https://doi.org/10.1126/science.1171609

Gent JI, Madzima TF, Bader R, Kent MR, Zhang X, Stam M, McGinnis KM, Dawe RK. Accessible DNA and relative depletion of H3K9me2 at maize loci undergoing RNA-directed DNA methylation. Plant Cell. 2014:**26**(12):4903–4917. https://doi.org/10.1105/tpc.114.130427

Gill BS, Waines JG. Paternal regulation of seed development in wheat hybrids. Theor Appl Genet. 1978:**51**(6):265–270. https://doi.org/10.1007/BF00274813

Grossniklaus U, Vielle-Calzada JP, Hoeppner MA, Gagliano WB. Maternal control of embryogenesis by MEDEA a Polycomb group gene in *Arabidopsis*. Science 1998:**280**(5362):446–450. https://doi.org/10.1126/science.280.5362.446

Grover JW, Burgess D, Kendall T, Baten A, Pokhrel S, King GJ, Meyers BC, Freeling M, Mosher RA. Abundant expression of maternal siRNAs is a conserved feature of seed development. Proc Natl Acad Sci U S A. 2020:**117**(26):15305–15315. https://doi.org/10.1073/pnas.2001332117

Grover JW, Kendall T, Baten A, Burgess D, Freeling M, King GJ, Mosher RA. Maternal components of RNA-directed DNA methylation are required for seed development in *Brassica rapa*. Plant J. 2018:**94**(4):575–582. https://doi.org/10.1111/tpj.13910

Guo L, Luo X, Li M, Joldersma D, Plunkert M, Liu Z. Mechanism of fertilization-induced auxin synthesis in the endosperm for seed and fruit development. Nat Commun. 2022:**13**(1):3985. https://doi.org/10.1038/s41467-022-31656-y

Gutierrez-Marcos JF, Pennington PD, Costa LM, Dickinson HG. Imprinting in the endosperm: a possible role in preventing wide hybridization. Philos Trans R Soc Lond B Biol Sci. 2003:**358**(1434): 1105–1111. https://doi.org/10.1098/rstb.2003.1292

Haig D, Westoby M. Genomic imprinting in endosperm: its effect on seed development in crosses between species, and between different ploidies of the same species, and its implications for the evolution of apomixes. Phil Trans R Soc Lond. 1991:**333**(1266):1–13. https://doi.org/10.1098/rstb.1991.0057

Haig D, Westoby M. Parent-specific gene expression and the triploid endosperm. Am Nat. 1989:**134**(1):147–155. https://doi.org/10.1086/284971

Hatorangan MR, Laenen B, Steige KA, Slotte T, Kohler C. Rapid evolution of genomic imprinting in two species of the Brassicaceae. Plant Cell. 2016:**28**(8):1815–1827. https://doi.org/10.1105/tpc.16.00304

He H, Shiragaki K, Tezuka T. Understanding and overcoming hybrid lethality in seed and seedling stages as barriers to hybridization and gene flow. Front Plant Sci. 2023:**14**:1219417. https://doi.org/10.3389/fpls.2023.1219417

Hehenberger E, Kradolfer D, Köhler C. Endosperm cellularization defines an important developmental transition for embryo development. Development 2012:**139**(11):2031–2039. https://doi.org/10.1242/dev.077057

Hernández-Hernández T, Miller EC, Román-Palacios C, Wiens JJ. Speciation across the tree of life. Biol Rev. 2021:**96**(4):1205–1242. https://doi.org/10.1111/brv.12698

Hsieh TF, Ibarra CA, Silva P, Zemach A, Eshed-Williams L, Fischer RL, Zilberman D. Genome-wide demethylation of Arabidopsis endosperm. Science 2009:**324**(5933):1451–1454. https://doi.org/10.1126/science.1172417

Huang F, Zhu QH, Zhu A, Wu X, Xie L, Wu X, Helliwell C, Chaudhury A, Jean Finnegan E, Luo M. Mutants in the imprinted PICKLE RELATED 2 gene, suppress seed abortion of fertilization independent seed class mutants and paternal excess interploidy crosses in *Arabidopsis*. Plant J. 2017:**90**(2):383–395. https://doi.org/10.1111/tpj.13500

Huc J, Dziasek K, Pachamuthu K, Woh T, Köhler C, Borges F. Bypassing reproductive barriers in hybrid seeds using chemically induced epimutagenesis. Plant Cell. 2021:**34**(3):989–1001. https://doi.org/10.1093/plcell/koab284

İltaş Ö, Svitok M, Cornille A, Schmickl R, Lafon Placette C. Early evolution of reproductive isolation: a case of weak inbreeder/strong outbreeder leads to an intraspecific hybridization barrier in *Arabidopsis lyrata*. Evolution 2021:**75**(6):1466–1476. https://doi.org/10.1111/evo.14240

Ishikawa R, Ohnishi T, Kinoshita Y, Eiguchi M, Kurata N, Kinoshita T. Rice interspecies hybrids show precocious or delayed developmental transitions in the endosperm without change to the rate of syncytial nuclear division. Plant J. 2010:**65**(5):798–806. https://doi.org/10.1111/j.1365-313X.2010.04466.x

Jiang H, Moreno-Romero J, Santos-González J, De Jaeger G, Gevaert K, Van De Slijke E, Köhler C. Ectopic application of the repressive histone modification H3K9me2 establishes post-zygotic reproductive isolation in *Arabidopsis thaliana*. Genes Dev. 2017:**31**(12): 1272–1287. https://doi.org/10.1101/gad.299347.117

Johnson LM, Du J, Hale CJ, Bischof S, Feng S, Chodavarapu RK, Zhong X, Marson G, Pellegrini M, Segal DJ, et al. SRA- and SET-domain-containing proteins link RNA polymerase V occupancy to DNA methylation. Nature 2014:**507**(7490):124–128. https://doi.org/10.1038/nature12931

Johnston SA, Hanneman RE Jr. Manipulations of endosperm balance number overcome crossing barriers between diploid *Solanum* species. Science 1982:**217**(4558):446–448. https://doi.org/10.1126/science.217.4558.446

Johnston SA, Nijs TPM, Peloquin SJ, Hanneman RE Jr. The significance of genic balance to endosperm development in interspecific crosses. Theor Appl Genet. 1980:**57**(1):5–9. https://doi.org/10.1007/BF00276002

Josefsson C, Dilkes B, Comai L. Parent-dependent loss of gene silencing during interspecies hybridization. Curr Biol. 2006:**16**(13): 1322–1328. https://doi.org/10.1016/j.cub.2006.05.045

Katz O. Extending the scope of Darwin's 'abominable mystery': integrative approaches to understanding angiosperm origins and species richness. Ann Bot. 2018:**121**(1):1–8. https://doi.org/10.1093/aob/mcx109

Khouider S, Borges F, LeBlanc C, Ungru A, Schnittger A, Martienssen R, Colot V, Bouyer D. Male fertility in *Arabidopsis* requires active DNA demethylation of genes that control pollen tube function. Nat Commun. 2021:**12**(1):410. https://doi.org/10.1038/s41467-020-20606-1

Kinser TJ, Smith RD, Lawrence AH, Cooley AM, Vallejo-Marín M, Conradi Smith GD, Puzey JR. Endosperm-based incompatibilities in hybrid monkeyflowers. Plant Cell. 2021:**33**(7):2235–2257. https://doi.org/10.1093/plcell/koab117

Kirkbride RC, Lu J, Zhang C, Mosher RA, Baulcombe DC, Chen ZJ. Maternal small RNAs mediate spatial-temporal regulation of gene expression, imprinting, and seed development in *Arabidopsis*. Proc Natl Acad Sci U S A. 2019:**116**(7):2761–2766. https://doi.org/10.1073/pnas.1807621116

Kiyosue T, Ohad N, Yadegari R, Hannon M, Dinneny J, Wells D, Katz A, Margossian L, Harada JJ, Goldberg RB, et al. Control of fertilization-independent endosperm development by the MEDEA Polycomb gene in *Arabidopsis*. Proc Natl Acad Sci U S A. 1999:**96**(7):4186–4191. https://doi.org/10.1073/pnas.96.7.4186

Klosinska M, Picard CL, Gehring M. Conserved imprinting associated with unique epigenetic signatures in the *Arabidopsis* genus. Nat Plants. 2016:**2**(10):16145. https://doi.org/10.1038/nplants.2016.145

Kozaki A, Aoyanagi T. Molecular aspects of seed development controlled by gibberellins and abscisic acids. Int J Mol Sci. 2022:**23**(3): 1876. https://doi.org/10.3390/ijms23031876

Kradolfer D, Hennig L, Köhler C. Increased maternal genome dosage bypasses the requirement of the FIS Polycomb Repressive Complex 2 in *Arabidopsis* seed development. PLoS Genet. 2013a:**9**(1):e1003163. https://doi.org/10.1371/journal.pgen.1003163

Kradolfer D, Wolff P, Jiang H, Siretskiy A, Köhler C. An imprinted gene underlies postzygotic reproductive isolation in *Arabidopsis thaliana*. Dev Cell. 2013b:**26**(5):525–535. https://doi.org/10.1016/j.devcel.2013.08.006

Lafon-Placette C, Hatorangan MR, Steige KA, Cornille A, Lascoux M, Slotte T, Kohler C. Paternally expressed imprinted genes associate with hybridization barriers in *Capsella*. Nat Plants. 2018:**4**(6): 352–357. https://doi.org/10.1038/s41477-018-0161-6

Lafon-Placette C, Johannessen IM, Hornslien KS, Ali MF, Bjerkan KN, Bramsiepe J, Glockle BM, Rebernig CA, Brysting AK, Grini PE, et al. Endosperm-based hybridization barriers explain the pattern of gene flow between *Arabidopsis lyrata* and *Arabidopsis arenosa* in Central Europe. Proc Natl Acad Sci U S A. 2017:**114**(6):E1027–E1035. https://doi.org/10.1073/pnas.1615123114

Lafon-Placette C, Kohler C. Embryo and endosperm, partners in seed development. Curr Opin Plant Biol. 2014:**17**:64–69. https://doi.org/10.1016/j.pbi.2013.11.008

Lafon-Placette C, Kohler C. Endosperm-based postzygotic hybridization barriers: developmental mechanisms and evolutionary drivers. Mol Ecol. 2016:**25**(11):2620–2629. https://doi.org/10.1111/mec.13552

Leblanc O, Pointe C, Hernandez M. Cell cycle progression during endosperm development in zea mays depends on parental dosage effects. Plant J. 2002:**32**(6):1057–1066. https://doi.org/10.1046/j.1365-313X.2002.01491.x

Les DH, Philbrick CT. Studies of hybridization and chromosome number variation in aquatic angiosperms: evolutionary implications. Aquatic Bot. 1993:**44**(2–3):181–228. https://doi.org/10.1016/0304-3770(93)90071-4

Li J, Berger F. Endosperm: food for humankind and fodder for scientific discoveries. New Phytol. 2012:**195**(2):290–305. https://doi.org/10.1111/j.1469-8137.2012.04182.x

Lin BY. Ploidy barrier to endosperm development in maize. Genetics 1984:**107**(1):103–115. https://doi.org/10.1093/genetics/107.1.103

Liu P, Qi M, Wang Y, Chang M, Liu C, Sun M, Yang W, Ren H. Arabidopsis RAN1 mediates seed development through its parental ratio by affecting the onset of endosperm cellularization. Mol Plant. 2014a:**7**(8):1316–1328. https://doi.org/10.1093/mp/ssu041

Liu ZW, Shao CR, Zhang CJ, Zhou JX, Zhang SW, Li L, Chen S, Huang HW, Cai T, He XJ. The SET domain proteins SUVH2 and SUVH9 are required for Pol V occupancy at RNA-directed DNA methylation loci. PLoS Genet. 2014b:**10**(1):e1003948. https://doi.org/10.1371/journal.pgen.1003948

Lopes MA, Larkins BA. Endosperm origin, development and function. Plant Cell. 1993:**5**(10):1383–1399. https://doi.org/10.1105/tpc.5.10.1383

Lu J, Zhang C, Baulcombe DC, Chen ZJ. Maternal siRNAs as regulators of parental genome imbalance and gene expression in endosperm of *Arabidopsis* seeds. Proc Natl Acad Sci U S A. 2012:**109**(14): 5529–5534. https://doi.org/10.1073/pnas.1203094109

Luo M, Dennis ES, Berger F, Peacock WJ, Chaudhury A. MINISEED3 (MINI3), a WRKY family gene, and HAIKU2 (IKU2), a leucine-rich repeat (LRR) KINASE gene, are regulators of seed size in *Arabidopsis*. Proc Natl Acad Sci U S A. 2005:**102**(48):17531–17536. https://doi.org/10.1073/pnas.0508418102

Martinez G, Panda K, Köhler C, Slotkin RK. Silencing in sperm cells is directed by RNA movement from the surrounding nurse cell. Nat Plants. 2016:**2**(4):16030. https://doi.org/10.1038/nplants.2016.30

Martinez G, Wolff P, Wang Z, Moreno-Romero J, Santos-Gonzalez J, Conze LL, DeFraia C, Slotkin RK, Köhler C. Paternal easiRNAs regulate parental genome dosage in *Arabidopsis*. Nat Genet. 2018:**50**(2): 193–198. https://doi.org/10.1038/s41588-017-0033-4

McCue AD, Nuthikattu S, Reeder SH, Slotkin RK. Gene expression and stress response mediated by the epigenetic regulation of a transposable element small RNA. PLoS Genet. 2012:**8**(2):e1002474. https://doi.org/10.1371/journal.pgen.1002474

Moore T. Genetic conflict, genomic imprinting and establishment of the epigenotype in relation to growth. Reproduction 2001:**122**(2): 185–193. https://doi.org/10.1530/rep.0.1220185

Moreno-Romero J, Del Toro-De León G, Yadav VK, Santos-Gonzalez J, Kohler C. Epigenetic signatures associated with imprinted paternally expressed genes in the *Arabidopsis* endosperm. Genome Biol. 2019:**20**(1):41. https://doi.org/10.1186/s13059-019-1652-0

Moreno-Romero J, Jiang H, Santos-Gonzalez J, Kohler C. Parental epigenetic asymmetry of PRC2-mediated histone modifications in the *Arabidopsis* endosperm. EMBO J. 2016:**35**(12):1298–1311. https://doi.org/10.15252/embj.201593534

Muntzing A. Hybrid incompatibility and the origin of polyploidy. Hereditas 1933:**18**(1–2):33–55. https://doi.org/10.1111/j.1601-5223.1933.tb02596.x

Oneal E, Willis JH, Franks RG. Disruption of endosperm development is a major cause of hybrid seed inviability between *Mimulus guttatus* and *Mimulus nudatus*. New Phytol. 2016:**210**(3):1107–1120. https://doi.org/10.1111/nph.13842

Pachamuthu K, Simon M, Borges F. Targeted suppression of siRNA biogenesis in *Arabidopsis* pollen reveals distinct Pol IV activities in the sperm and vegetative cell lineages. Res Square. 2023. https://doi.org/10.21203/rs.21203.rs 2947716/v2947711

Parent J-S, Cahn J, Herridge RP, Grimanelli D, Martienssen RA. Small RNAs guide histone methylation in *Arabidopsis* embryos. Genes Dev. 2021:**35**(11–12):841–846. https://doi.org/10.1101/gad.343871.120

Parhad SS, Tu S, Weng Z, Theurkauf WE. Adaptive evolution leads to cross-species incompatibility in the piRNA transposon silencing machinery. Dev Cell. 2017:**43**(1):60–70.e65. https://doi.org/10.1016/j.devcel.2017.08.012

Park K, Kim MY, Vickers M, Park JS, Hyun Y, Okamoto T, Zilberman D, Fischer RL, Feng X, Choi Y, et al. DNA demethylation is initiated in the central cells of Arabidopsis and rice. Proc Natl Acad Sci U S A. 2016:**113**(52):15138–15143. https://doi.org/10.1073/pnas.1619047114

Peer WA, Murphy AS. Flavonoids and auxin transport: modulators or regulators? Trends Plant Sci. 2007:**12**(12):556–563. https://doi.org/10.1016/j.tplants.2007.10.003

Pennington PD, Costa LM, Gutierrez-Marcos JF, Greenland AJ, Dickinson HG. When genomes collide: aberrant seed development following maize interploidy crosses. Ann Bot. 2008:**101**(6):833–843. https://doi.org/10.1093/aob/mcn017

Petrén H, Thosteman H, Stift M, Toräng P, Ågren J, Friberg M. Differences in mating system and predicted parental conflict affect post-pollination reproductive isolation in a flowering plant. Evolution 2023:**4**(4):1019–1030. https://doi.org/10.1093/evolut/qpad016

Picard CL, Povilus RA, Williams BP, Gehring M. Transcriptional and imprinting complexity in *Arabidopsis* seeds at single-nucleus resolution. Nat Plants. 2021:**7**(6):730–738. https://doi.org/10.1038/s41477-021-00922-0

Pignatta D, Erdmann RM, Scheer E, Picard CL, Bell GW, Gehring M. Natural epigenetic polymorphisms lead to intraspecific variation in *Arabidopsis* gene imprinting. Elife 2014:**3**:e03198. https://doi.org/10.7554/eLife.03198

Pignocchi C, Minns GE, Nesi N, Koumproglou R, Kitsios G, Benning C, Lloyd CW, Doonan JH, Hills MJ. ENDOSPERM DEFECTIVE1 is a novel microtubule-associated protein essential for seed development in Arabidopsis. Plant Cell. 2009:**21**(1):90–105. https://doi.org/10.1105/tpc.108.061812

Povilus RA, Diggle PK, Friedman WE. Evidence for parent-of-origin effects and interparental conflict in seeds of an ancient flowering plant lineage. Proc Biol Sci. 2018:**285**(1872):20172491. https://doi.org/10.1098/rspb.2017.2491

Povilus RA, Gehring M. Maternal-filial transfer structures in endosperm: a nexus of nutritional dynamics and seed development. Curr Opin Plant Biol. 2022:**65**:102121. https://doi.org/10.1016/j.pbi.2021.102121

Queller DC. Kin selection and conflict in seed maturation. J Theor Biol. 1983:**100**(1):153–172. https://doi.org/10.1016/0022-5193(83)90099-1

Rebernig CA, Lafon-Placette C, Hatorangan MR, Slotte T, Kohler C. Non-reciprocal interspecies hybridization barriers in the *Capsella* genus are established in the endosperm. PLoS Genet. 2015:**11**(6): e1005295. https://doi.org/10.1371/journal.pgen.1005295

Rodrigues JA, Ruan R, Nishimura T, Sharma MK, Sharma R, Ronald PC, Fischer RL, Zilberman D. Imprinted expression of genes and small RNA is associated with localized hypomethylation of the maternal genome in rice endosperm. Proc Natl Acad Sci U S A. 2013:**110**(19):7934–7939. https://doi.org/10.1073/pnas.1306164110

Rodrigues JA, Zilberman D. Evolution and function of genomic imprinting in plants. Genes Dev. 2015:**29**(24):2517–2531. https://doi.org/10.1101/gad.269902.115

Roth M, Florez-Rueda AM, Griesser S, Paris M, Städler T. Incidence and developmental timing of endosperm failure in post-zygotic isolation between wild tomato lineages. Ann Bot. 2018a:**121**(1): 107–118. https://doi.org/10.1093/aob/mcx133

Roth M, Florez-Rueda AM, Paris M, Städler T. Wild tomato endosperm transcriptomes reveal common roles of genomic imprinting in both nuclear and cellular endosperm. Plant J. 2018b:**95**(6): 1084–1101. https://doi.org/10.1111/tpj.14012

Roth M, Florez-Rueda AM, Städler T. Differences in effective ploidy drive genome-wide endosperm expression polarization and seed failure in wild tomato hybrids. Genetics 2019:**212**(1):141–152. https://doi.org/10.1534/genetics.119.302056

Roy AK, Malaviya DR, Kaushal P. Generation of interspecific hybrids of *Trifolium* using embryo rescue techniques. Methods Mol Biol. 2011:**710**:141–151. https://doi.org/10.1007/978-1-61737-988-8_12

Sandstedt GD, Sweigart AL. Developmental evidence for parental conflict in driving *Mimulus* species barriers. New Phytol. 2022:**236**(4): 1545–1557. https://doi.org/10.1111/nph.18438

Sandstedt GD, Wu CA, Sweigart AL. Evolution of multiple postzygotic barriers between species of the *Mimulus tilingii* complex. Evolution 2021:**75**(3):600–613. https://doi.org/10.1111/evo.14105

Satyaki PR, Gehring M. DNA methylation and imprinting in plants: machinery and mechanisms. Crit Rev Biochem Mol Biol. 2017:**52**(2):163–175. https://doi.org/10.1080/10409238.2017.1279119

Satyaki PR, Gehring M. Paternally acting canonical RNA-directed DNA methylation pathway genes sensitize Arabidopsis endosperm to paternal genome dosage. Plant Cell. 2019:**31**(7):1563–1578. https://doi.org/10.1105/tpc.19.00047

Satyaki PRV, Gehring M. RNA pol IV induces antagonistic parent-of-origin effects on *Arabidopsis* endosperm. PLoS Biol. 2022:**20**(4):e3001602. https://doi.org/10.1371/journal.pbio.3001602

Schatlowski N, Wolff P, Santos-Gonzalez J, Schoft V, Siretskiy A, Scott R, Tamaru H, Köhler C. Hypomethylated pollen bypasses the interploidy hybridization barrier in Arabidopsis. Plant Cell. 2014:**26**(9):3556–3568. https://doi.org/10.1105/tpc.114.130120

Schmitz RJ, Schultz MD, Urich MA, Nery JR, Pelizzola M, Libiger O, Alix A, McCosh RB, Chen H, Schork NJ, et al. Patterns of population epigenomic diversity. Nature 2013:**495**(7440):193–198. https://doi.org/10.1038/nature11968

Schoft VK, Chumak N, Mosiolek M, Slusarz L, Komnenovic V, Brownfield L, Twell D, Kakutani T, Tamaru H. Induction of RNA-directed DNA methylation upon decondensation of constitutive heterochromatin. EMBO Rep. 2009:**10**(9):1015–1021. https://doi.org/10.1038/embor.2009.152

Scott RJ, Spielman M, Bailey J, Dickinson HG. Parent-of-origin effects on seed development in *Arabidopsis thaliana*. Development 1998:**125**(17):3329–3341. https://doi.org/10.1242/dev.125.17.3329

Scott RJ, Tratt JL, Bolbol A. Seed development in interploidy hybrids. In: **Chen ZJ, Birchler JA**, editors. Polyploid and hybrid genomics. Oxford, UK: Wiley; 2013. p. 271–290.

Sekine D, Ohnishi T, Furuumi H, Ono A, Yamada T, Kurata N, Kinoshita T. Dissection of two major components of the post-zygotic hybridization barrier in rice endosperm. Plant J. 2013:**76**(5):792–799. https://doi.org/10.1111/tpj.12333

Slotkin RK, Vaughn M, Borges F, Tanurdzic M, Becker JD, Feijo JA, Martienssen RA. Epigenetic reprogramming and small RNA silencing of transposable elements in pollen. Cell 2009:**136**(3):461–472. https://doi.org/10.1016/j.cell.2008.12.038

Sorensen MB, Mayer U, Lukowitz W, Robert H, Chambrier P, Jürgens G, Somerville C, Lepiniec L, Berger F. Cellularisation in the endosperm of *Arabidopsis thaliana* is coupled to mitosis and shares multiple components with cytokinesis. Development 2002:**129**(24):5567–5576. https://doi.org/10.1242/dev.00152

Städler T, Florez-Rueda AM, Roth M. A revival of effective ploidy: the asymmetry of parental roles in endosperm-based hybridization barriers. Curr Opin Plant Biol. 2021:**61**:102015. https://doi.org/10.1016/j.pbi.2021.102015

Stoute AI, Varenko V, King GJ, Scott RJ, Kurup S. Parental genome imbalance in *Brassica oleracea* causes asymmetric triploid block. Plant J. 2012:**71**(3):503–516. https://doi.org/10.1111/j.1365-313X.2012.05015.x

Sukno S, Ruso J, Jan CC, Melero-Vara JM, Fernández-martínez JM. Interspecific hybridization between sunflower and wild perennial *Helianthus* species via embryo rescue. Euphytica 1999:**106**(1):69–78. https://doi.org/10.1023/A:1003524822284

Thomae AW, Schade GO, Padeken J, Borath M, Vetter I, Kremmer E, Heun P, Imhof A. A pair of centromeric proteins mediates reproductive isolation in *Drosophila* species. Dev Cell. 2013:**27**(4):412–424. https://doi.org/10.1016/j.devcel.2013.10.001

Tonosaki K, Sekine D, Ohnishi T, Ono A, Furuumi H, Kurata N, Kinoshita T. Overcoming the species hybridization barrier by ploidy manipulation in the genus *Oryza*. Plant J. 2018:**93**(3):534–544. https://doi.org/10.1111/tpj.13803

Trivers RL. Parent-offspring conflict. Am Zool. 1974:**14**(1):249–264. https://doi.org/10.1093/icb/14.1.249

von Wangenheim KH. Zur Ursache Der Abortion Von Samenanlagen In Diploid-Polyploid-Kreuzungen. Zeitschrift für Vererbungslehre. 1962:**93**:319–334. https://doi.org/10.1007/BF00888790

von Wangenheim KH, Peterson HP. Aberrant endosperm development in interploidy crosses reveals a timer of differentiation. Dev Biol. 2004:**270**(2):277–289. https://doi.org/10.1016/j.ydbio.2004.03.014

Valentine DH, Woodell SRJ. Studies in British *Primulas* X. Seed incompatibility in intraspecific and interspecific crosses at diploid and tetraploid levels. New Phytol. 1963:**62**(2):125–143. https://doi.org/10.1111/j.1469-8137.1963.tb06321.x

Walia H, Josefsson C, Dilkes B, Kirkbride R, Harada J, Comai L. Dosage-dependent deregulation of an AGAMOUS-LIKE gene cluster contributes to interspecific incompatibility. Curr Biol. 2009:**19**(13):1128–1132. https://doi.org/10.1016/j.cub.2009.05.068

Wang G, Jiang H, Del Toro-De León G, Martinez G, Kohler C. Sequestration of a transposon-derived siRNA by a target mimic imprinted gene induces postzygotic reproductive isolation in *Arabidopsis*. Dev Cell. 2018:**46**(6):696–705. https://doi.org/10.1016/j.devcel.2018.07.014

Wang Z, Butel N, Santos-González J, Simon L, Wärdig C, Köhler C. Transgenerational effect of mutants in the RNA-directed DNA methylation pathway on the triploid block in *Arabidopsis*. Genome Biol. 2021:**22**(1):141. https://doi.org/10.1186/s13059-021-02359-2

Waters AJ, Bilinski P, Eichten SR, Vaughn MW, Ross-Ibarra J, Gehring M, Springer NM. Comprehensive analysis of imprinted genes in maize reveals allelic variation for imprinting and limited conservation with other species. Proc Natl Acad Sci U S A. 2013:**110**(48):19639–19644. https://doi.org/10.1073/pnas.1309182110

Widmer A, Lexer C, Cozzolino S. Evolution of reproductive isolation in plants. Heredity (Edinb). 2009:**102**(1):31–38. https://doi.org/10.1038/hdy.2008.69

Williams E, White DWR. Early seed development after crossing of *Trifolium ambiguum* and *T. repens*. N Z J Bot. 1976:**14**(4):307–314. https://doi.org/10.1080/0028825X.1976.10428903

Williams JH, Friedman WE. Identification of diploid endosperm in an early angiosperm lineage. Nature 2002:**415**(6871):522–526. https://doi.org/10.1038/415522a

Wolff P, Jiang H, Wang G, Santos-Gonzalez J, Köhler C. Paternally expressed imprinted genes establish postzygotic hybridization barriers in *Arabidopsis thaliana*. Elife 2015:**4**:e10074. https://doi.org/10.7554/eLife.10074

Wolff P, Weinhofer I, Seguin J, Roszak P, Beisel C, Donoghue MT, Spillane C, Nordborg M, Rehmsmeier M, Köhler C. High-resolution analysis of parent-of-origin allelic expression in the *Arabidopsis* endosperm. PLoS Genet. 2011:**7**(6):e1002126. https://doi.org/10.1371/journal.pgen.1002126

Woodell SRJ, Valentine DH. Studies in British *Primulas*. IX. Seed incompatibility in diploid-autotetraploid crosses. New Phytol. 1961:**60**(3):282–294. https://doi.org/10.1111/j.1469-8137.1961.tb06256.x

Wormit A, Usadel B. The multifaceted role of pectin methylesterase inhibitors (PMEIs). Int J Mol Sci. 2018:**19**(10):2878. https://doi.org/10.3390/ijms19102878

Xu H, Li F, Pan Y, Gong X. Interspecific hybridization processes between *Michelia yunnanensis* and *M. crassipes* and embryogenesis of the heterozygote. HortScience 2017:**52**(8):1043–1047. https://doi.org/10.21273/HORTSCI12086-17

Xu W, Sato H, Bente H, Santos-González J, Köhler C. Endosperm cellularization failure induces a dehydration-stress response leading to embryo arrest. Plant Cell. 2023:**35**(2):874–888. https://doi.org/10.1093/plcell/koac337

Xu Y, Wang Y, Stroud H, Gu X, Sun B, Gan ES, Ng KH, Jacobsen SE, He Y, Ito T. A matrix protein silences transposons and repeats through interaction with retinoblastoma-associated proteins. Curr Biol. 2013:**23**(4):345–350. https://doi.org/10.1016/j.cub.2013.01.030

Yang R, Zheng Z, Chen Q, Yang L, Huang H, Miki D, Wu W, Zeng L, Liu J, Zhou JX, et al. The developmental regulator PKL is required to maintain correct DNA methylation patterns at RNA-directed DNA methylation loci. Genome Biol. 2017:**18**(1):103. https://doi.org/10.1186/s13059-017-1226-y

Yun J, Kim YS, Jung JH, Seo PJ, Park CM. The AT-hook motif-containing protein AHL22 regulates flowering initiation by modifying FLOWERING LOCUS T chromatin in *Arabidopsis*. J Biol Chem.

2012:**287**(19):15307–15316. https://doi.org/10.1074/jbc.M111.318477

Zemach A, Kim MY, Hsieh PH, Coleman-Derr D, Eshed-Williams L, Thao K, Harmer SL, Zilberman D. The *Arabidopsis* nucleosome remodeler DDM1 allows DNA methyltransferases to access H1-containing heterochromatin. Cell 2013:**153**(1):193–205. https://doi.org/10.1016/j.cell.2013.02.033

Zhang B, Li C, Li Y, Yu H. Mobile TERMINAL FLOWER1 determines seed size in *Arabidopsis*. Nat Plants. 2020:**6**(9):1146–1157. https://doi.org/10.1038/s41477-020-0749-5

Zhang M, Xie S, Dong X, Zhao X, Zeng B, Chen J, Li H, Yang W, Zhao H, Wang G, et al. Genome-wide high resolution parental-specific DNA and histone methylation maps uncover patterns of imprinting regulation in maize. Genome Res. 2014:**24**(1):167–176. https://doi.org/10.1101/gr.155879.113

Zumajo-Cardona C, Aguirre M, Castillo-Bravo R, Mizzotti C, Di Marzo M, Banfi C, Mendes MA, Spillane C, Colombo L, Ezquer I. Maternal control of triploid seed development by the TRANSPARENT TESTA 8 (TT8) transcription factor in *Arabidopsis thaliana*. Sci Rep. 2023:**13**(1):1316. https://doi.org/10.1038/s41598-023-28252-5

https://doi.org/10.1093/plphys/kiae105 PLANT PHYSIOLOGY 2024: 195: 170–189

Plant Physiology®

Topical Review

Regulatory networks in plant responses to drought and cold stress

June-Sik Kim [1,2,*,†] Satoshi Kidokoro,[3,*,†] Kazuko Yamaguchi-Shinozaki,[4,5] Kazuo Shinozaki [1,6]

1 RIKEN Center for Sustainable Resource Science, 1-7-22 Suehiro-cho, Tsurumi-ku, Yokohama, 230-0045 Japan
2 Institute of Plant Science and Resources, Okayama University, 2-20-1 Chuo, Kurashiki, 710-0046 Japan
3 School of Life Science and Technology, Tokyo Institute of Technology, 4259 Nagatsuta-cho, Midori-ku, Yokohama, 226-8502 Japan
4 Research Institute for Agriculture and Life Sciences, Tokyo University of Agriculture, 1-1-1 Sakuragaoka, Setagaya-ku, Tokyo, 156-8502 Japan
5 Graduate School of Agriculture and Life Science, The University of Tokyo, 1-1-1 Yayoi, Bunkyo-ku, Tokyo, 113-0032 Japan
6 Institute for Advanced Research, Nagoya University, Furo-cho, Chikusa-ku, Nagoya, 464-8601 Japan

*Author for correspondence: june-sik.kim@riken.jp (J.-S.K.), kidokoro.s.aa@m.titech.ac.jp (S.K.)
†These authors contributed equally to this work.

Abstract

Drought and cold represent distinct types of abiotic stress, each initiating unique primary signaling pathways in response to dehydration and temperature changes, respectively. However, a convergence at the gene regulatory level is observed where a common set of stress-responsive genes is activated to mitigate the impacts of both stresses. In this review, we explore these intricate regulatory networks, illustrating how plants coordinate distinct stress signals into a collective transcriptional strategy. We delve into the molecular mechanisms of stress perception, stress signaling, and the activation of gene regulatory pathways, with a focus on insights gained from model species. By elucidating both the shared and distinct aspects of plant responses to drought and cold, we provide insight into the adaptive strategies of plants, paving the way for the engineering of stress-resilient crop varieties that can withstand a changing climate.

Introduction

The substantial rise in atmospheric carbon dioxide levels has contributed to global warming, adversely affecting ecological conditions. This surge is intensifying climate change, leading to frequent droughts, severe temperature swings, and rising soil salinity, all of which detrimentally affect ecosystem health and agricultural output. Indeed, plant growth is greatly affected by water-limited conditions such as drought and salinity, as well as extreme temperature conditions. However, land plants have evolved mechanisms that allow them to withstand these stressors and sustain survival.

Early research on plant responses to abiotic stress focused on the underlying physiology and biochemistry using the techniques available at the time (Iljin 1957; Lyons 1973). Since the 1980s, the focus has shifted to obtaining a sophisticated understanding of the genetic mechanisms behind these responses, driven by continuous technical advances and the adoption of Arabidopsis (*Arabidopsis thaliana*) as a model species. The complete sequencing of the Arabidopsis genome in 2000 (The Arabidopsis Genome Initiative 2000) accelerated the identification of genes activated in response to various stresses, including cold and drought, revealing both significant overlap in gene networks and distinct pathways for each stress type (Maruyama et al. 2012; Zhu 2016; Waadt et al. 2022). Over the past decade, next-generation sequencing has further advanced this field. RNA-seq has become a crucial tool for transcriptome analysis, providing detailed and digitized global expression profiles and allowing the comparison of even subtle changes in expression (Wang et al. 2014). Concurrently, DNA-seq has hastened the creation of reference genomes for a growing number of plant species, facilitating the construction of orthologous groups of functional genes to broaden insights from model species to other plants (Nguyen et al. 2019).

Received August 29, 2023. Accepted February 15, 2024. Advance access publication March 21, 2024.

The accumulated knowledge is being harnessed to develop stress-resilient crops through transgenic DNA technology and genome editing (Bailey-Serres et al. 2019; Hickey et al. 2019; Cooper and Messina 2023).

Plant drought and cold response pathways comprise highly complex sensing, signaling, and response mechanisms specific to each type of stress (Fig. 1). As the 2 stresses do not often co-occur, the pathways for each stress are essentially inde-pendent. Drought responses are largely driven by biosynthesis of and signaling associated with the phytohormone abscisic acid (ABA), with plants utilizing root-to-shoot signal-ing to adapt to dry conditions of varying intensities and dura-tions. Responses to chilling and freezing are more uniform, since temperature shifts are generally simultaneously per-ceived across plant shoots and are connected with diurnal, circadian, and annual rhythms. However, the molecular func-tions of cold- and dehydration-responsive genes significantly overlap. Angiosperms such as Arabidopsis, soybean (*Glycine max*), and rice (*Oryza sativa*) induce genes encoding late embryogenesis abundant (LEA) proteins and molecular chaperones during both drought and cold stresses, while genes related to photosynthesis are typically downregulated (Maruyama et al. 2012). Cold affects plant physiology via temperature-related effects, whereas freezing-induced damage primarily results from dehydration associated with the formation of ice crystals (Thomashow 1999). Thus, plants have developed a unified strategy of gene regulation to coun-ter the effects of high osmolarity during both drought and cold stress (Thomashow 1999; Yamaguchi-Shinozaki and Shinozaki 2006; Zhu 2016).

In this review, we summarize milestone discoveries and recent advances in our understanding of drought and cold stress responses and tolerance, with a primary focus on regulatory systems including gene expression, signaling, and perception. We also discuss notable applications of recent technological advancements in crop breeding. The term "plant" refers to terrestrial angiosperms throughout this review unless otherwise specified.

Plant responses to drought stress

Water scarcity during drought poses significant challenges to plants. Dehydration, resulting from high osmolality or drought stress, leads to a loss of turgidity and pressure against the cell wall, both essential for maintaining plant structure and growth. Energy production falters as photo-synthesis rates drop, causing a metabolic imbalance and the accumulation of harmful compounds. The phytohor-mone ABA plays a crucial role in drought responses, minim-izing water loss through stomatal closure and initiating protective mechanisms against water scarcity (Finkelstein 2013). Drought- and ABA-inducible genes are integral to these protective actions, encoding LEA proteins, chaperones, enzymes for ABA and osmoprotectant biosynthesis, sugar and proline transporters, aquaporins, and detoxification enzymes to neutralize reactive oxygen species (ROS) (Yamaguchi-Shinozaki and Shinozaki 2006; Urano et al. 2009; Zhu 2016; Yoshida and Fernie 2023). *RESPONSIVE TO DEHYDRATION29A* (*RD29A*) and *RD29B*, early identified Arabidopsis genes for LEA proteins, are widely used as stress markers in research due to their sensitive induction by ABA and drought (Yamaguchi-Shinozaki et al. 1992). Other genes inducible by cold or drought stress, such as galactinol synthase (GolS) genes, lead to the production of raffinose and galactinol —soluble carbohydrates that protect against osmotic damage (Taji et al. 2002). Drought-responsive transcription factors (TFs) regulate these stress-protective genes and are modulated by stress signals from ABA and other signaling pathways.

Biosynthesis, catabolism, and transport of ABA in response to drought stress

Endogenous ABA levels increase when plants encounter water-limited conditions, leading to systemic changes that protect them against the stress. The well-established ABA biosynthesis pathway involves catalytic enzymes, including NINE-CIS-EPOXYCAROTENOID DIOXYGENASE (NCED), ABA DEFICIENT2 (ABA2), and ARABIDOPSIS ALDEHYDE OXIDASE3 (AAO3) (Hauser et al. 2011; Finkelstein 2013). NCED, originally identified as Viviporous14 in maize (*Zea mays*), is the rate-limiting enzyme of this pathway (Tan et al. 1997), with notable induction by drought first observed in cowpea (*Vigna unguiculata*) (Iuchi et al. 2000). Of the 5 NCED isoforms in Arabidopsis, *NCED3* transcripts show rapid drought-responsive accumulation (Iuchi et al. 2001; Tan et al. 2003). Loss-of-function *nced3* mutants have reduced endogenous ABA levels, with impaired stomatal closure and plant survival (Iuchi et al. 2001). Other isoforms (such as *NCED5*, *NCED6*, and *NCED9*) are primarily expressed in developing seeds, contributing to seed maturation and dormancy (Martínez-Andújar et al. 2011; Frey et al. 2012).

ABA catabolism maintains ABA homeostasis. Within the cytochrome P450 monooxygenase family 707 (CYP707), a specific subclade of ABA 8′-hydroxylases comprising CYP707A1 to CYP707A4 plays a pivotal role in Arabidopsis (Krochko et al. 1998; Kushiro et al. 2004; Saito et al. 2004). These 4 members collectively regulate seed dormancy and germination, exhibiting inducible gene expression in response to dehydration stress and subsequent rehydration (Kushiro et al. 2004; Saito et al. 2004; Okamoto et al. 2006; Umezawa et al. 2006). The drought-responsive expression of *CYP707A* and *NCED3* genes underscores the precise control of ABA levels through the maintenance of the delicate balance between biosynthesis and catabolism. Moreover, ABA inactivation occurs through conjugation with oxidative catabolites. Notably, ABA glucosyl ester (ABA-GE) has been identified in various tissues of different plant species (Piotrowska and Bajguz 2011). β-Glucosidases hydrolyze ABA-GE to release uncon-jugated ABA from the endoplasmic reticulum (Dietz et al. 2000; Lee et al. 2006). Hence, this pathway likely serves as a rapid ABA release mechanism for an immediate response to drought before the full activation of ABA biosynthesis.

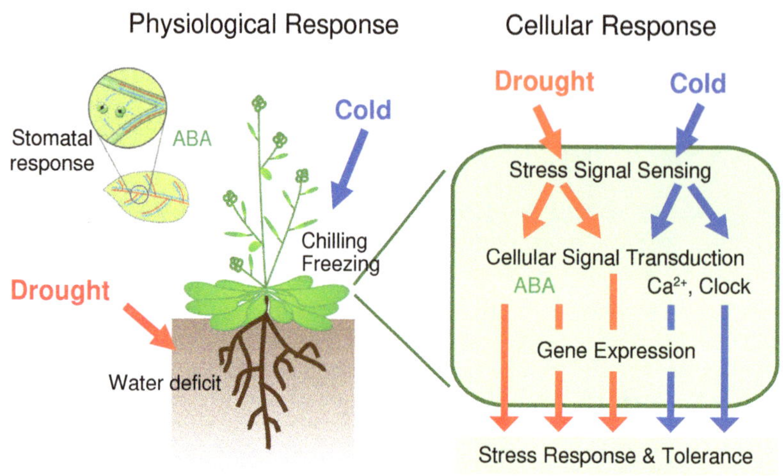

Figure 1. Physiological and cellular responses to drought and cold stress in plants. Under severe drought and cold stress conditions, plants alter their physiology to acquire stress tolerance. During drought stress, ABA plays a crucial role in stomatal closure, which limits water loss. By contrast, cold stress is characterized by chilling or freezing, which severely disrupts plant growth. However, plants have evolved mechanisms to survive these stresses. During cellular stress responses, plant cells directly sense drought and cold stress or receive systemic signals from other parts of the plant. Through multiple interconnected signaling cascades, these stress signals activate the expression of genes specific to each type of stress. These genes encode proteins that enhance plant stress tolerance.

The preferential expression of genes encoding key ABA biosynthesis enzymes including NCED3, ABA2, and AAO3 has been observed in leaf vascular tissues, where water availability is sensed and through which ABA is transported in the plant (Koiwai et al. 2004; Endo et al. 2008; Kuromori et al. 2014). Several transporters involved in ABA trafficking across cellular membranes have been identified (Kuromori et al. 2018). These transporters belong to families such as the ATP-binding cassette (ABC) subfamily, Nitrate transporter1/Peptide transporter family (NPF), and Detoxification efflux carrier (DTX)/Multidrug and toxic compound extrusion family. In particular, Arabidopsis *ABCG25* exhibits ABA- and drought-inducible expression and encodes an ABA exporter restricted to vascular tissues (Kuromori et al. 2010). *ABCG40* shows strong expression in guard cells and facilitates ABA uptake to promote ABA-mediated stomatal closure (Kang et al. 2010). Among the other transporter families, *NPF4.6* is an ABA importer gene predominantly expressed in inflorescence stems and roots (Kanno et al. 2012). In addition, *DTX50* is strongly expressed in rosette leaf vasculature, facilitating ABA-responsive stomatal closure through ABA efflux (Zhang et al. 2014).

ABA perception complex and drought signaling

The established model for ABA perception and signaling in vascular plants involves a series of protein phosphorylation cascades mediated by 3 key components: the ABA receptors PYRABACTIN RESISTANCE1 (PYR1) and PYR1-LIKE (PYR1/PYL, also named REGULATORY COMPONENT OF ABA RECEPTOR [RCAR]), the signaling activator SUCROSE NON-FERMENTING1-RELATED PROTEIN KINASE2 (SnRK2 or SRK2), and the suppressor TYPE 2C PROTEIN PHOSPHATASE (PP2C) (Cutler et al. 2010; Finkelstein 2013) (Fig. 2, A and B).

The discovery of the PYR1/PYL/RCAR family of ABA receptors marked a significant breakthrough in understanding ABA signaling in plants. In Arabidopsis, this family has 12 functionally redundant members, with the prominent PYR1/RCAR11 being the first receptor found to bind ABA (Ma et al. 2009; Park et al. 2009; Santiago et al. 2009). Upon binding to ABA, the receptors PYR1/PYLs/RCAR interact with PP2Cs and inhibit their activity, enabling the phosphorylation and activation of SnRK2s, thereby initiating ABA signaling (Melcher et al. 2009; Miyazono et al. 2009; Nishimura et al. 2009). The search for additional receptor ligands for agricultur-al applications has identified several ABA analogs that exhibit stronger sensitivity and greater receptor selectivity, such as quinabactin (Okamoto et al. 2013) and opabactin (Vaidya et al. 2019).

SnRK2s phosphorylate target proteins, including themselves, and their catalytic activity requires their own phosphorylation (Yoshida et al. 2002; Belin et al. 2006; Soma et al. 2023). The Arabidopsis SnRK2/SRK2 family consists of 10 isoforms of 3 subclasses based on protein homology (Mustilli et al. 2002; Boudsocq et al. 2004). Three SnRK2 isoforms, SnRK2.2/SRK2D, SnRK2.3/SRK2I, and SnRK2.6/SRK2E/OPEN STOMATA1 (OST1), belong to subclass III and redundantly mediate most ABA signal transduction. The triple loss-of-function mutant of subclass III SnRK2 members lacks ABA-responsive stomatal closure, seed maturation, and dormancy and exhibits no ABA-mediated induction of most drought-responsive genes (Fujii and Zhu 2009; Fujita et al. 2009; Nakashima et al. 2009; Umezawa et al. 2009). OST1 is specifically involved in regulating stomatal closure, as its ABA-induced phosphorylation activates the channel SLOW

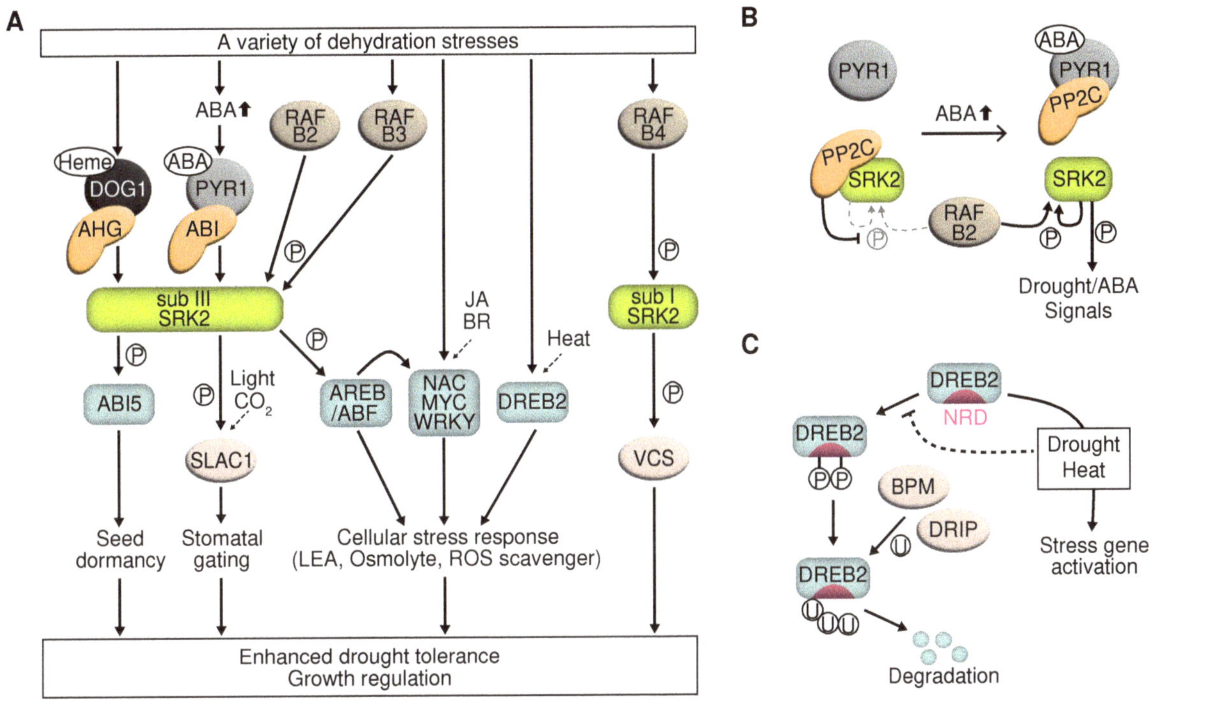

Figure 2. Molecular machinery of drought stress perception, signaling, and responses in vascular plants. **A)** Overview of ABA–dependent and –independent drought signaling pathways. In the ABA-dependent pathway, the ABA-bound receptor complex modulates drought signaling via protein phosphorylation (circled P). ABA signal transduction induces stress responses including seed maturation and dormancy, stomatal gating, and drought-responsive gene expression. There are also ABA-independent branches of drought stress signaling, including the heme-bound DOG1–AHG pathway and DREB2-mediated transcriptional regulation. B-type RAFs are recently discovered components of drought signaling. JA, jasmonic acid; BR, brassinosteroid. **B)** The ternary complex model of ABA perception consists of the receptor (PYR1/PYL/RCAR), the signal activator (subclass III SnRK2/SRK2), and the suppressor (PP2C). B2-type RAFs are recently discovered components that constitutively phosphorylate SnRK2s. **C)** Post-translational modifications of DREB2 in response to stress. Under normal conditions, the nascent DREB2 protein is phosphorylated within its NRD domain and is subsequently targeted for proteasomal degradation via ubiquitination (circled U) mediated by specific E3 ligases (BPMs and DRIPs). Drought and heat stress prevent DREB2 phosphorylation, thereby enhancing its stability and allowing it to activate downstream gene expression.

ANION CHANNEL-ASSOCIATED1, which induces anion efflux from guard cells, resulting in stomatal closure (Mustilli et al. 2002; Geiger et al. 2009). This pathway is also required for other stimuli that induce stomatal closure, such as CO_2 and the phytohormone jasmonate (JA) (Hsu et al. 2018; Munemasa et al. 2019). By contrast, ABA treatment failed to induce or activate any of the 4 subclass I–type SnRK2s (SnRK2.1/SRK2G, SnRK2.4/SRK2A, SnRK2.5/SRK2H, and SnRK2.10/SRK2B), which are rapidly activated by osmotic stress without stress-responsive ABA accumulation (Boudsocq et al. 2004; Yoshida et al. 2006). Subsequent studies unveiled their ABA-independent role under drought stress: subclass I–type SnRK2s redundantly target and activate the mRNA decapping activator VARICOSE in cytosolic processing (P)-bodies, promoting stress-responsive mRNA decay to enhance stress acclimation (Soma et al. 2017; Kawa et al. 2020).

Arabidopsis harbors dozens of PP2C isoforms, with those in subclade A being key players in ABA signaling (Schweighofer et al. 2004), including ABA INSENSITIVE 1 (ABI1) and ABI2 (Koornneef et al. 1984). In the absence of ABA, ABI1 interacts with and dephosphorylates subclass III–type SnRK2s, preventing their activation (Gosti et al. 1999; Yoshida et al. 2006). PP2C subclade A comprises 9 members, including 2 ABIs. At least 6 of these PP2Cs (ABI1, ABI2, HYPERSENSITIVE TO ABA1 [HAB1], HAB2, ABA HYPERSENSITIVE GERMINATION1 [AHG1], and AHG3) target and dephosphorylate the SnRK2s, albeit with slight differences in their physiological roles. Two HABs share similar functions with ABI (Umezawa et al. 2009; Vlad et al. 2009). AHG1 and AHG3 are involved in seed-related responses, promoting and maintaining dormancy through both PYR1/PYL/RCAR-dependent and -independent mechanisms. The eukaryote-conserved heme-binding protein DELAY OF GERMINATION1 (DOG1) plays a critical role in Arabidopsis seed dormancy and the adaptive control of germination (Graeber et al. 2014). Heme-bound DOG1 interacts with AHG1, preventing this PP2C from binding to PYR1/PYLs/RCARs and inactivating sub-class III–type SnRK2s (Nishimura et al. 2018). A similar regulatory mechanism was established for AHG3, which is also modulated by ABA-bound PYR1/PYLs/RCARs (Née et al. 2017). The remaining subclade A PP2C members are 3 homologs: HIGHLY ABA-INDUCED1 (HAI1), HAI2, and HAI3. The genes encoding these subclade members exhibit ABA-inducible expression, and the *hai* triple mutant displays hypersensitivity to ABA (Fujita et al. 2009; Bhaskara et al. 2012). Thus, HAIs likely function in the negative feedback regulation of ABA signaling, although their target proteins and underlying mechanisms remain unclear (Chong et al. 2019).

MAP kinases in ABA and drought signaling

Although subclass III SnRK2s are likely self-activated by autophosphorylation after their release from PP2C-imposed inhibition, an earlier discovery in the bryophyte *Physcomitrium patens* shed light on the involvement of the B-type MAP KINASE KINASE KINASE (M3K) subgroup of Raf-like kinases in SnRK2 activation under drought stress (Saruhashi et al. 2015). B-type Raf-like kinases are divided into 4 subfamilies (B1–B4). Four B3-type kinases (RAF3/M3Kδ1, RAF4/M3Kδ7, RAF5/M3Kδ6, and RAF6/M3Kδ5) phosphorylate and activate subclass III SnRK2s under osmotic stress in Arabidopsis (Katsuta et al. 2020; Lin et al. 2020; Takahashi et al. 2020). Through mutant analysis, 4 B2-RAFs (RAF7, RAF10, RAF11, and RAF12), in addition to B3-RAFs, were identified as activators of subclass III SnRK2s. Phosphorylation by these B2-RAFs and B3-RAFs is essential for the activation of subclass III SnRK2s (Lin et al. 2020). A recent study showed that B2-RAFs are constitutively active and activate subclass III SnRK2s when released from PP2C-imposed inhibition, whereas B3-RAFs are activated by stress (Soma et al. 2023). These findings indicate that autophosphorylation of subclass III SnRKs is not sufficient for ABA responses and that B2-RAFs are essential for subclass III SnRK2 activation in response to ABA, whereas B3-RAFs enhance subclass III SnRK2 activity under drought. By contrast, 3 B4-RAFs (RAF18, RAF20, and RAF24) activate the ABA-insensitive subclass I–type SnRK2s under dehydration conditions (Soma et al. 2020). The upstream signaling mechanisms for the B3-Raf-like and B4-Raf-like kinases remain unclear, leaving open questions about how plants sense drought stress.

AREB/ABF transcription factors function in ABA-dependent gene expression

During plant drought responses, ABA signaling elicits the induction of numerous drought-responsive genes (Fig. 2A). The most highly conserved *cis*-acting element detected in the promoters of these genes is the ABA-RESPONSIVE ELEMENT (ABRE), a variant of the G-box (or E-box) element with an AGCT core (Marcotte et al. 1989; Mundy et al. 1990). The G-box sequence is preferentially recognized by basic leucine zipper (bZIP)-type TFs. Arabidopsis contains 78 bZIPs classified into 13 subgroups, with subgroup A encompassing the ABRE-BINDING PROTEIN/ABRE-BINDING FACTOR (AREB/ABF) members that play central roles in regulating ABA-responsive gene expression (Dröge-Laser et al. 2018). Among the AREBs/ABFs, AREB1/ABF2 and its 3 homologs (AREB2/ABF4, ABF1, and ABF3) regulate ABA- and drought-responsive gene expression in vegetative tissues by activating promoters harboring the ABRE (Choi et al. 2000; Uno et al. 2000). The activity of these TFs is induced by subclass III–type SnRK2-mediated phosphorylation (Fujita et al. 2005; Furihata et al. 2006). These 4 AREB/ABF members are functionally redundant, as evidenced by the more severe phenotypes observed in higher-order mutants compared with single *areb1* mutants (Fujita et al. 2005; Yoshida et al. 2010, 2015). Transcriptome analysis of quadruple *areb/abf* mutants revealed that these 4 members collectively regulate a major portion of ABA-responsive genes under the control of subclass III–type SnRK2s (Yoshida et al. 2015).

ABI5 is another AREB/ABF member that functions exclusively in seed dormancy and early seedling development. Germination in *abi5* mutants is insensitive to ABA treatment, whereas *ABI5*-overexpressing plants exhibit hypersensitivity to ABA (Finkelstein and Lynch 2000; Lopez-Molina et al. 2001). Similar to AREB1/ABF2, ABI5 TF activity requires phosphorylation mediated by subclass III–type SnRK2s (Nakashima et al. 2009; Nishimura et al. 2018). *ABI5* expression is induced by ABA, and ABI5 promotes seed dormancy by suppressing the expression of genes involved in germination, including those related to gibberellin biosynthesis and the production of seed coat mucilage (Finch-Savage and Leubner-Metzger 2006).

DREB transcription factors function in ABA-independent gene expression

Drought-responsive ABA-independent gene regulatory pathways also exist in plants (Fig. 2A). Initially, the DEHYDRATION-RESPONSIVE ELEMENT (DRE) *cis*-acting element was identified from a set of dehydration-responsive genes expressed in leaves (Yamaguchi-Shinozaki and Shinozaki 1994). The same promoter motif was also reported in *COLD-REGULATED* (*COR*) genes and was named C-REPEAT (CRT) (Baker et al. 1994; Jiang et al. 1996). DRE/CRT is thus involved in cold stress–induced gene induction but does not function in ABA-responsive gene expression. Two subclades of APETALA2/ETHYLENE-RESPONSIVE FACTOR (AP2/ERF) TFs induce gene expression by targeting promoters containing DRE/CRT motifs: DRE-BINDING PROTEIN1/CRT-BINDING FACTOR (DREB1/CBF) and DREB2 (Stockinger et al. 1997; Liu et al. 1998; Shinwari et al. 1998). The DREB1/CBF family members are primarily recognized as cold-responsive regulators of gene expression with prominent roles in the chilling response (Liu et al. 1998). We delve further into their regulation and mo-lecular functions in cold responses below.

The DREB2 subclade features a prominent and extensively characterized member, DREB2A. *DREB2A* is expressed differently in response to dehydration and heat stress: heat stress triggers a relatively short and strong induction, whereas drought stress leads to prolonged *DREB2A* transcript accumulation (Sakuma et al. 2006b). These distinct expression patterns involve 2 discrete regions in the *DREB2A* promoter, each independently inducing expression through different upstream TFs (Yoshida et al. 2008; Kim et al. 2011). Moreover, the post-translational regulation of DREB2A is

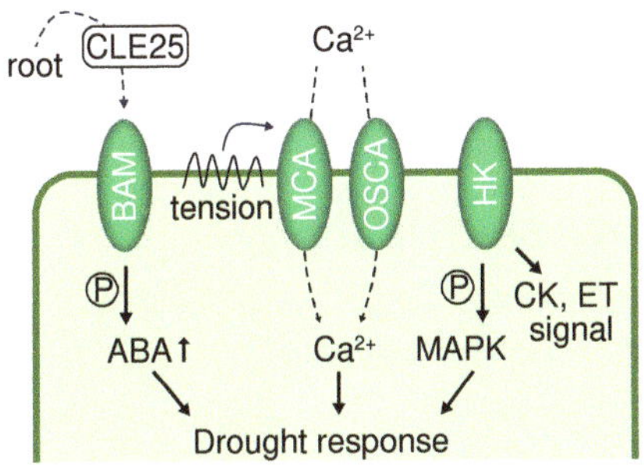

Figure 3. The osmosensor machinery in plants. HK, a homolog of the yeast osmosensor protein Sln1p, activates plant stress responses via MAP kinase (MAPK) signaling. HK also participates in cytokinin (CK) and ethylene (ET) signaling. Two BAM family members (BAM1 and BAM3) form a receptor complex for small CLE25 peptides, conveying the drought stress signal from the root. MCA and OSCA family members are permeable calcium ion (Ca^{2+}) channels whose activation triggers Ca^{2+} uptake and associated stress responses.

crucial for its TF activity (Fig. 2C). DREB2A is highly unstable due to the presence of a serine/threonine-rich negative regulatory domain (NRD) located in the center of the protein (Sakuma et al. 2006a). The NRD undergoes extensive phosphorylation under normal growth conditions (Mizoi et al. 2019), leading to DREB2A degradation, which is facilitated by the ubiquitin E3 ligases BTB/POZ AND MATH DOMAIN (BPM) and DREB2-INTERACTING PROTEIN (DRIP) (Qin et al. 2008; Morimoto et al. 2017). However, heat and dehydration prevent these post-translational modifications, resulting in increased DREB2A stability (Morimoto et al. 2017; Mizoi et al. 2019). Additionally, heat stress–induced SUMOylation targeting the NRD inhibits the DREB2A–BPM interaction (Wang et al. 2020b). Notably, a similar BPM-mediated decay pathway has been identified for other drought-signaling components, including ABI1 (Julian et al. 2019) and HOMEOBOX PROTEIN6 (HB6) (Lechner et al. 2011), indicating the existence of an intricate protein turnover and signal interaction system that facilitates efficient drought responses in plants. Alongside the transcriptional repressor acting on *DREB2A* expression under normal growth conditions (Kim et al. 2012), these multi-level regulatory mechanisms indicate that plants use DREB2A to handle severe drought stress while also minimizing its negative effects on vegetative growth and development (Sakuma et al. 2006a; Qin et al. 2008).

Other transcription factors and phytohormones involved in drought-responsive gene expression

The intricate mechanism of plant response to drought features plant-specific TFs that modulate drought-responsive gene expression (Fig. 2A). Members of the NAC (NAM, ATAF1/2, and CUC) family include STRESS-RESPONSIVE NACs (SNACs) that mediate the expression of abiotic stress–inducible genes (Tran et al. 2004). Among SNACs, ANAC072 (also known as RD26) and its functionally redundant homologs enhance the expression of many drought-responsive genes and regulate ABA-responsive leaf senescence (Fujita et al. 2004; Takasaki et al. 2015; Kamranfar et al. 2018). Similarly, ANAC096 is another SNAC that contributes to plant drought responses by physically interacting with members of the AREB/ABF family (Xu et al. 2013). Additionally, the plant-specific WRKY family is involved in a variety of plant stress responses (Eulgem et al. 2000). Several WRKY family members play roles in drought and ABA signaling. For example, Arabidopsis WRKY63 targets the *AREB1/ABF2* gene and enhances its ABA-inducible expression (Ren et al. 2010). By contrast, a subgroup comprising WRKY18, WRKY40, and WRKY60 represses multiple TFs involved in drought signaling, including AREB1/ABF2 and ABI5 (Shang et al. 2010). Drought- and ABA-inducible *HB6* encodes a TF that negatively regulates ABA-mediated drought responses, creating a feedback loop for efficient and flexible stress adaptation (Himmelbach et al. 2002).

Research on plant drought responses has increasingly focused on crosstalk with phytohormone pathways beyond ABA, although only a few genetic intersections have been established so far (Zhu 2016; Waadt et al. 2022; Yoshida and Fernie 2023). Arabidopsis MYC2, also known as JASMONATE INSENSITIVE1, is a key regulator of JA-responsive gene expression (Lorenzo et al. 2004). MYC2 and its cooperating TF MYB2 were originally identified as being responsible for ABA- and drought-inducible *RD22* expression (Abe et al. 1997, 2003). *MYC2* and *MYB2* also show ABA- and drought-inducible expression, suggesting that certain genetic interactions occur between JA signaling and the drought response (Abe et al. 2003). ANAC072/RD26 modulates brassinosteroid (BR) signaling as well by interacting with the essential BR-responsive TF BRASSINAZOLE RESISTANT1 (BZR1), thus influencing the induction of BR-responsive genes (Ye et al. 2017). Conversely, the BR-signaling kinase BRASSINOSTEROID-INSENSITIVE2, a repressor of BZR1 activity, activates ABA signaling by targeting SnRK2s (Cai et al. 2014). DEEPER ROOTING1 (DRO1) is a plasma membrane (PM)–localized protein of uncertain function, which was initially discovered in rice for its contribution to drought tolerance by altering root developmental patterns (Uga et al. 2013). DRO1 in angiosperms shows auxin-responsive gene expression and modifies root architecture to make the roots extend more deeply into the soil, thereby enhancing drought tolerance through improved water uptake (Guseman et al. 2016).

Osmotic sensors and long-distance signaling in drought stress responses

In nature, plants encounter drought conditions with spatiotemporal heterogeneity, requiring signaling mechanisms to sense and communicate the stress throughout the plant (Fig. 3). Arabidopsis HISTIDINE KINASE1 (AHK1) is a PM-localized protein that is homologous to the yeast (*Saccharomyces cerevisiae*) osmosensor protein Sln1p (Urao et al. 1999). Indeed, ectopic expression of *AHK1* enhanced plant survival under drought (Tran et al. 2007; Wohlbach et al. 2008). Close homologs of AHK1 also function in cytokinin signaling, making it challenging to determine their specific roles as osmosensors (Jeon et al. 2010; Nishiyama et al. 2011). A recent study in the moss *P. patens* revealed that ethylene receptor–related HKs activate SnRK2s via the activation of RAF kinase, providing additional insights into plant HK-mediated osmosensing mechanisms (Toriyama et al. 2022). Calcium ion (Ca^{2+}) channels are also candidate osmosensors. The Ca^{2+}-permeable mechanosensitive channels MID1-COMPLEMENTING ACTIVITY1 (MCA1) and MCA2 sense cell wall tension in Arabidopsis, facilitating adaptive growth under dehydration conditions (Nakagawa et al. 2007; Yoshimura et al. 2021).

Both directionless diffusion and directional transduction occur during systemic signaling arising from stressed tissues. ROS and Ca^{2+} influence ABA signaling and stomatal closure (Pei et al. 2000; Waszczak et al. 2018). RESPIRATORY BURST OXIDASE HOMOLOG D (RBOHD), a key component of the systemic ROS wave, is activated by local ROS waves and subsequent Ca^{2+} accumulation. RBOHD then triggers another ROS burst that is transferred to neighboring cells (Suzuki et al. 2013). In guard cells, the ROS–Ca^{2+} signal cooperates with ABA to stimulate stomatal closure, triggering plant stress acclimation (Suzuki et al. 2013; Zandalinas et al. 2020). The H_2O_2 sensor HYDROGEN-PEROXIDE-INDUCED Ca^{2+} INCREASES1, which is located on the guard cell PM, is responsible for the H_2O_2-responsive Ca^{2+} influx that initiates stomatal closure (Wu et al. 2020). The Ca^{2+} wave is a systemic signaling mech-anism for defense (Lecourieux et al. 2006; Toyota et al. 2018). The Ca^{2+} channel REDUCED HYPEROSMOLALITY-INDUCED Ca^{2+} INCREASE1 functions in guard cells and roots, mediating osmotic stress–inducible Ca^{2+} influx in guard cells (Yuan et al. 2014). Liquid–liquid phase separation is another plausible mechanism for monitoring environmental changes (Eljebbawi et al. 2023). In the context of water availability, the prion-like protein FLOE1 stimulates seed germination in Arabidopsis through a reversible hydration-dependent phase separation process (Dorone et al. 2021).

In addition to directionless diffusion of signals, plants also use short peptides for directional signaling. The CLAVATA3/EMBRYO-SURROUNDING REGION-RELATED (CLE) family consists of mobile peptides that play active roles in development (Fletcher 2020). Thus, the discovery of a role for Arabidopsis CLE25 in drought responses was unexpected. Soil dehydration leads to CLE25 accumulation in roots, which then travels through the leaf vasculature (Fig. 3). At the leaf PM, the receptor-like protein kinases BARELY ANY MERISTEM1 (BAM1) and BAM3 perceive CLE25, inducing *NCED3* expression and ultimately resulting in stomatal clos-ure (Takahashi et al. 2018). Notably, the Arabidopsis TF NGATHA1 mediates the drought-responsive induction of *NCED3* predominantly in the leaf vasculature (Sato et al. 2018). This coincidence suggests that CLE25–BAM ligand formation leads to ABA biosynthesis in leaves. Systematic studies have shown that most *CLE* family members are expressed in roots (Jun et al. 2010), encouraging further exploration of CLE-mediated root–shoot stress communication.

Plant responses to cold stress

Cold stress has severe effects on plant growth and survival. Plant growth and development are negatively affected by chilling temperatures (0–12 °C) due to the inhibition of various vital biochemical reactions, such as decreases in water and nutrient uptake (Lyons 1973). Chilling stress also causes stomatal closure and the subsequent inhibition of photosynthesis and transpiration activity, with negative effects on plant growth similar to those of drought stress (Wilkinson et al. 2001). Furthermore, under freezing temperatures (below 0 °C), ice crystals start to form and damage cell membranes and cell walls. Ice nucleation occurs in intercellular spaces, leading to decreased water potential outside the cells. Consequently, water moves out of the cells to the intercellular space, resulting in severe cellular dehydration.

Plants acquire tolerance to subsequent freezing conditions through temporary exposure to chilling but nonfreezing temperatures. Cold acclimation involves dramatic transcriptional and metabolic changes that protect plant cells from cold-induced damage. These changes during cold acclimation include the accumulation of osmoprotectants and ROS scavenging, which also occur during drought stress responses (Cook et al. 2004; Maruyama et al. 2009; Mittler et al. 2022). Thus, some cold-responsive genes overlap with drought-responsive genes (Maruyama et al. 2012). In addition, cold stress alters the PM composition to mitigate decreased fluidity and freezing injury (Miki et al. 2019). Moreover, nonlethal freezing temperatures (approximately −3 °C) induce cell wall modifications and alter extracellular components, thereby inducing an additional response referred to as subzero acclimation (Takahashi et al. 2019, 2021).

Transcriptional networks involved in cold-responsive gene expression

As described in the previous section, DREB1/CBF proteins, which are found only in angiosperms, function as major transcriptional regulators of cold stress responses by binding to the DRE/CRT elements in the promoters of their target genes (Stockinger et al. 1997; Liu et al. 1998; Shinwari et al. 1998). The Arabidopsis genome contains 3 *DREB1/CBF* genes, which are tandemly arrayed on chromosome 4: *DREB1B/CBF1*, *DREB1A/CBF3*, and *DREB1C/CBF2*. The constitutive expression of *D REB1/CBF* genes in transgenic plants upregulated hundreds of cold- and drought-responsive genes, even under normal growth conditions (Seki et al. 2001; Fowler and Thomashow 2002; Maruyama et al. 2004, 2009), improving tolerance to freezing, drought, and high-salinity stress (Jaglo-Ottosen et al. 1998; Liu et al. 1998; Kasuga et al. 1999). Many of the upregulated genes encode proteins necessary for cold stress tolerance and acclimation, such as LEA proteins, and osmolyte biosynthesis enzymes. Disruption of all 3 *DREB1/CBF* genes by genome editing resulted in dramatically reduced cold-inducible gene expression and tolerance of freezing stress (Jia et al. 2016; Zhao et al. 2016). By contrast, overexpressing *DREB1/CBF* led to the downregulation of growth-related genes and adversely affected plant growth (Kasuga et al. 1999; Sakamoto et al. 2004; Achard et al. 2008; Kudo et al. 2019).

Upstream regulators of cold-responsive *DREB1/CBF* expression

Three *DREB1/CBF* genes in Arabidopsis are rapidly induced by cold stress (Liu et al. 1998; Shinwari et al. 1998). Several TFs have been reported to regulate *D REB1/CBF* expression (Thomashow 2010; Kidokoro et al. 2022). The MYC-like basic helix-loop-helix (bHLH) TF INDUCER OF CBF EXPRESSION1 (ICE1) was identified following the isolation of the Arabidopsis *ice1-1* mutant in a screen for mutations that impair the cold-induced transcription of the firefly luciferase (*LUC*) reporter gene driven by the Arabidopsis *D REB1A/CBF3* promoter (Chinnusamy et al. 2003). In the *ice1-1* mutant, cold-induced expression of endogenous *D REB1A/CBF3* was reduced, whereas *D REB1B/CBF1* and *D REB1C/CBF2* expression was unaffected. It was long believed that ICE1 promotes cold stress tolerance by directly activating *DREB1A/CBF3*. However, *D REB1A/CBF3* repression in *ice1-1* plants is not caused by the mutation originally identified i n ICE1 (R23 6H); rather, it is due to DNA methylation–mediated gene silencing triggered by the inserted T-DNA harboring the *D REB1A/CBF3pro:LUC* reporter construct. Therefore, ICE1 does not regulate *DREB1A/CBF3* expression in response to cold stress (Kidokoro et al. 2020). Nevertheless, ICE1 is identical to SCREAM (SCRM), which regulates stomatal development along with its homolog SCRM2; the *ice1-2 scrm2-1* double mutant exhibits an epidermis devoid of stomata (Kanaoka et al. 2008). The dominant *SCRM* mutant allele *scrm-D* carries the same missense mutation (R236H) in SCRM/ICE1 as *ice1-1*, and both mutants present clear defects in stomatal development (Kanaoka et al. 2008). However, cold stress–induced *D REB1A/CBF3* expression did not decrease in *scrm-D* mutants compared with wild-type plants (Kidokoro et al. 2020). This finding validates the idea that *D REB1A/CBF3* repression in *ice1-1* is not caused by the $ICE1^{R23\ 6H}$ mutation and that ICE1 is not involved in cold stress–responsive *D REB1A/CBF3* expression. There have been many reports on the effect of *ICE1* overexpression on cold stress tolerance in crops, but they need to be carefully reassessed (Kidokoro et al. 2020; Thomashow and Torii 2020).

Other TFs are also crucial regulators of *DREB1/CBF* expression during early cold stress responses, including CALMODULIN-BINDING TRANSCRIPTION ACTIVATOR (CAMTA) TFs and circadian clock–related MYB-like TFs such as REVEILLEs (RVEs), CIRCADIAN CLOCK ASSOCIATED1 (CCA1), and LATE ELONGATED HYPOCOTYL (LHY) (Doherty et al. 2009; Kidokoro et al. 2017, 2021). CAMTAs are CG-1-type TFs that are conserved in plants and other multicellular eukaryotes (Bouché et al. 2002). These proteins contain calmodulin-interacting domains: IQ motifs and a calmodulin-binding domain. The CAMTA TFs recognize the CGCG-box [(A/C/G)CGCG(G/T/C)] and strongly upregulate *DREB1B/CBF1* and *DREB1C/CBF2* (Kidokoro et al. 2017). Unlike CAMTA TFs, circadian clock–related MYB-like TFs bind to a *cis*-acting motif known as the evening element (EE) [AAAATATCT] in the *DREB1A/CBF3* and *DREB1C/CBF2* promoters (Kidokoro et al. 2021). Detailed analysis of these CAMTA and MYB-like TFs revealed that Arabidopsis recognizes cold stress as 2 different signals (rapid or gradual decreases in temperature) and that the expression of *DREB1A/CBF3* genes is differentially induced by these 2 stress signals through 2 different signal transduction pathways (Kidokoro et al. 2017). The CAMTA TFs only respond to a rapid drop in temperature, but the MYB-like TFs recognize both rapid and gradual decreases in temperature. The MYB-like TFs strongly induce *DREB1/CBF* expression in response to cold only in the daytime; by contrast, CAMTA TFs activate *DREB1/CBF* expression regardless of the time of day (Kidokoro et al. 2017).

CAMTA-mediated regulatory pathways

Among Arabidopsis CAMTA family members (CAMTA1– 6), CAMTA3 (also named *Arabidopsis thaliana* SIGNAL-RESPONSIVE GENE1), along with its homologs CAMTA1 and CAMTA2, activate *DREB1B/CBF1* and *DREB1C/CBF2* expression (Doherty et al. 2009; Kim et al. 2013) (Fig. 4). These CAMTA proteins also act as transcriptional repressors of plant immune signaling (Du et al. 2009). This signaling is induced not only by pathogen infection but also by long-term low-temperature conditions and by mechanical stimuli such as raindrops (Kim et al. 2013, 2020; Matsumura et al. 2022). The CAMTA-mediated pathway is likely involved in Ca^{2+} signaling, as these TFs contain calmodulin-binding sites (Bouché et al. 2002). However, the N-terminal 334– amino acid region of CAMTA3, which contains the CG-1 DNA-binding domain and lacks calmodulin-binding sites, is sufficient to induce *DREB1/CBF* gene expression in re-sponse to cold stress and to suppress immune-related genes under normal temperatures (Kim et al. 2017; Chao et al. 2022). Under cold stress conditions, calmodulin-binding sites in CAMTA3 release their negative regulation of the N-terminal region of this protein, leading to immune- responsive gene expression (Kim et al. 2017). The CG-1 DNA-binding domain of CAMTA3 modulates its transcrip-tional activity in response to cold, whereas CAMTA3 stably binds to target promoters under both unstressed and cold stress conditions (Chao et al. 2022). Therefore, the role of the calmodulin-binding sites of CAMTA3 in cold stress–induced transcriptional regulation remains unclear.

In addition to CAMTA3, an analysis of quintuple and sextuple *camta* mutants revealed that CAMTA5 also induces *DREB1/CBF* expression (Kidokoro et al. 2017). CAMTA5 functions in the cold stress response but not in the immune

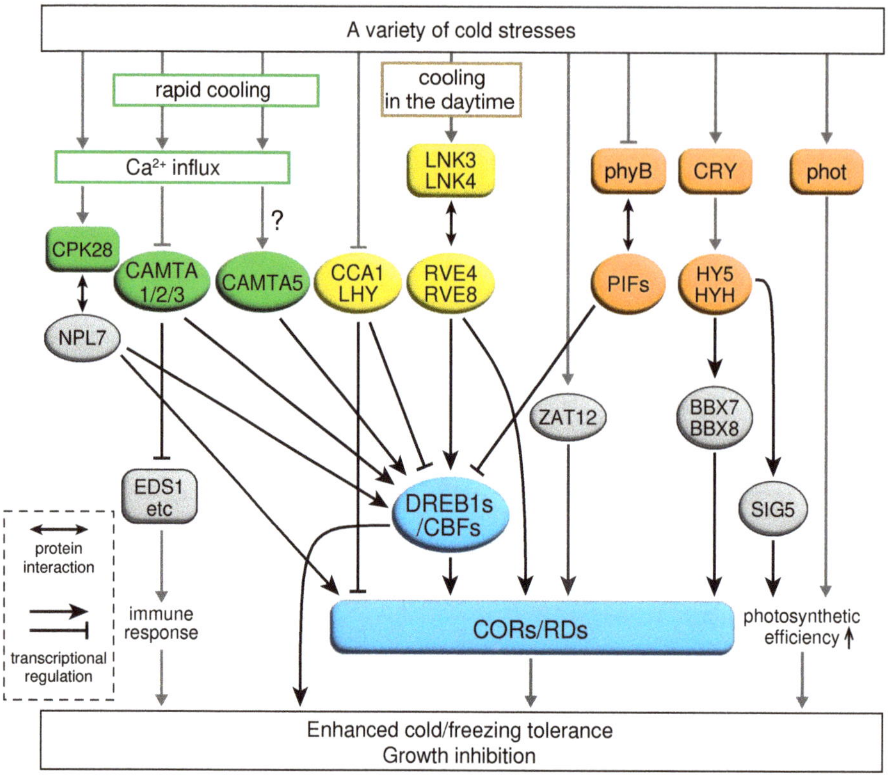

Figure 4. The regulatory network activated under cold stress conditions. In angiosperms, DREB1/CBF transcription factors regulate cold acclimation and stress tolerance acquisition by inducing numerous *COR* and *RD* genes. CAMTA and RVE transcription factors rapidly induce *DREB1/CBF* expression. By contrast, CCA1, LHY, and PIFs repress *DREB1/CBF* expression under normal growth conditions. CAMTA3 is also involved in plant immune responses and inhibits growth under cold stress conditions together with CAMTA1 and CAMTA2. Other transcription factors, such as NPL7, HY5, HYH, and ZAT12, also promote cold stress tolerance. Moreover, several photoreceptors, such as phyB and phot, influence cold stress responses. Notably, phototropins in liverwort perceive cold stress and regulate chloroplast relocation. Circles indicate transcription factors. Yellow, orange, and green represent proteins involved in the circadian clock, light signaling, and calcium (Ca^{2+}) signaling, respectively.

response, whereas CAMTA3 regulates both cold stress and immune responses. Moreover, these CAMTA proteins regulate different target genes during cold stress (Chao et al. 2022), indicating that although both CAMTA3 and CAMTA5 activate *DREB1/CBF* expression, their activation mechanisms may differ. Additional analyses are required to elucidate the Ca^{2+} signaling–regulated functions of CAMTA proteins in cold-responsive gene expression.

Clock-related TF-mediated regulatory pathways

In the Arabidopsis circadian clock, several transcriptional repressors form negative feedback loops and are key oscillators under normal growth conditions. These repressors are CCA1 and LHY, PSEUDO-RESPONSE REGULATORs (PRR9, PRR7, PRR5, and PRR1/TIMING OF CAB EXPRESSION1 [TOC1]), and an evening complex (EC) that consists of LUX ARRHYTHMO (LUX), EARLY FLOWERING3 (ELF3), and ELF4 (Nohales and Kay 2016). CCA1 and LHY levels rise shortly before dawn; these TFs repress evening-phased genes via the evening element (EE) mentioned above and the CCA1-binding site (Nagel et al. 2015; Kamioka et al. 2016). Three RVE proteins, RVE4, RVE6, and RVE8, then activate these genes in the evening (Hsu et al. 2013).

The regulation of cold-inducible gene expression by circadian clock–related MYB-like TFs resembles circadian-controlled gene expression. CCA1 and LHY repress cold-inducible gene expression under normal temperatures, whereas RVEs activate cold-responsive genes under cold stress. RVE4 and RVE8 play dominant roles in cold-inducible gene expression (Kidokoro et al. 2021). RVE6 redundantly and conditionally activates its target cold-inducible genes together with RVE3 and RVE5, when *RVE4* and *RVE8* are not expressed. CCA1, LHY, RVE4, and RVE8 activities are controlled by post-translational modifications. Under normal temperature conditions, CCA1 and LHY localize to the nucleus, whereas RVE4 and RVE8 are located mainly in the cytosol and partly in the nucleus (Kidokoro et al. 2021). During the early cold stress response, CCA1 and LHY are degraded, but RVE4 and RVE8 quickly accumulate in the nucleus to activate cold-inducible transcription.

Truncation of potential PEST sequences in CCA1 prevented its degradation and led to decreased *DREB1A/CBF3* induction under cold stress. This finding suggests that CCA1 and LHY degradation is important for cold-inducible gene expression, in addition to the nuclear accumulation of RVE4 and RVE8. Additionally, RVE4 and RVE8 activate heat-inducible genes such as *ETHYLENE RESPONSE*

FACTOR53 (*ERF53*) and *ERF54* and regulate heat stress tolerance during the day (Li et al. 2019). Taken together, these findings indicate that RVE4 and RVE8 likely activate distinct sets of target genes through different signaling pathways under normal temperatures or cold or heat stress conditions. In response to heat, RVE4 and RVE8 accumulate in the nucleus (Kidokoro et al. 2021). However, *DREB1/CBF* expression is not induced by heat stress, indicating that factors other than the nuclear accumulation of RVE4 and RVE8 are also required to regulate target genes under each temperature condition.

Under normal growth conditions, NIGHT LIGHT-INDUCIBLE AND CLOCK-REGULATED1 (LNK1) and LNK2 interact with RVEs. LNK1 and LNK2 act as RVE coactivators by recruiting the transcriptional machinery to target RVEs to the promoters of circadian-controlled genes (Rugnone et al. 2013; Xie et al. 2014; Ma et al. 2018). Similarly, LNK proteins also function as coactivators in stress-inducible gene expression (Kidokoro et al. 2023). Among the 4 Arabidopsis LNK proteins, LNK3 and LNK4 regulate cold-inducible gene expression. Because RVE4, RVE8, and LNK proteins accumulate in the nucleus during the day, many cold-inducible target genes, including *DREB1/CBF* genes, are strongly induced by cold stress in the daytime (Kidokoro et al. 2021, 2023). Furthermore, LNK3 and LNK4 are specifically phosphorylated under cold conditions, whereas RVEs and LNKs stably interact under all temperature conditions. Therefore, the specific phosphorylation of LNKs may control cold-specific target gene expression. However, LNK1 and LNK2 function under heat stress in addition to normal temperature conditions (Kidokoro et al. 2023). The heat stress response pathway mediated by RVEs–LNK1/LNK2 is independent from another major pathway regulated by the HEAT SHOCK FACTOR1A family (Ohama et al. 2016).

Moreover, PRR proteins recognize the *DREB1/CBF* gene promoters and regulate their expression. The triple loss of function of *PRR9*, *PRR7*, and *PRR5* resulted in the strong upregulation of *DREB1/CBF* genes and the downstream genes of their encoded TFs under normal growth conditions (Nakamichi et al. 2009, 2012). These findings indicate that multiple circadian clock–related TFs regulate *DREB1/CBF* expression.

Other transcription factors regulating the cold stress response

Several TFs that are key regulators of other signaling pathways also control *D REB1/CBF* expression, indicating that the regulation of *D REB1/CBF* genes is complex (Fig. 4) (Kidokoro et al. 2022). PHYTOCHROME-INTERACTING FACTORs (PIFs) contain a putative phytochrome-interacting domain and are involved in photomorphogenesis (Leivar and Quail 2011). Among these bHLH TFs, Arabidopsis PIF3, PIF4, and PIF7 and rice OsPIL16 negatively regulate *DREB1* expression, whereas tomato (*Solanum lycopersicum*) SlPIF4 enhances *DREB1/CBF* expression and cold acclimation (Kidokoro et al. 2009; Lee and Thomashow 2012; Jiang et al. 2017; Wang et al. 2020a). bZIP-type TFs, including ELONGATED HYPOCOTYL5 (HY5), are key components of light signaling. HY5 stability is regulated by E3 ubiquitin ligase complexes, including CONSTITUTIVELY PHOTOMORPHOGENIC1, in a blue light–dependent manner. Under cold stress conditions, HY5 is stabilized and upregulates target genes encoding B-BOX DOMAIN PROTEIN TFs and anthocyanin biosynthesis enzymes (Catalá et al. 2011; Li et al. 2021). HY5 and HY5 HOMOLOG (HYH) also induce *SIGMA FACTOR5* (*SIG5*) expression in a circadian manner (Cano-Ramirez et al. 2023). SIG5 regulates the expression of chloroplast genes such as *psbD* to maintain photo-synthetic efficiency under long-term cold stress. NIN-LIKE PROTEIN7, a key TF involved in nitrate responses, activates the cold-inducible expression of *DREB1/CBF* and their downstream genes (Ding et al. 2022). Moreover, cold stress activates the PM-localized CALCIUM-DEPENDENT PROTEIN KINASE28 (CPK28) in a Ca^{2+}-dependent manner. CPK28 induces a phos-phorylation cascade that triggers the translocation of NPL7 into the nucleus (Ding et al. 2022).

In addition to the DREB1/CBF-mediated pathway, other pathways are also involved in the acquisition of cold tolerance through transcriptional regulation. In particular, the clock-related factors RVE4, RVE8, CCA1, and LHY directly bind to the EE and regulate many *RD* (*RESPONSIVE TO DEHYDRATION*) and *COR* genes (Kidokoro et al. 2021). The EE is enriched in the promoters of cold-inducible genes in several plant species, such as Arabidopsis, soybean, and rice (Mikkelsen and Thomashow 2009; Maruyama et al. 2012). ZINC FINGER OF ARABIDOPSIS THALIANA12 is another TF that upregulates *RD* and *COR* gene expression, whereas it downregulates *DREB1/CBF* expression (Vogel et al. 2005). COR27 and COR28 are negative regulators of freezing tolerance, but their encoding genes are transcriptionally induced by cold (Li et al. 2016). They repress the transcription of even-ing phase genes such as *TOC1* by interacting with and desta-bilizing the RVE8–LNK1/LNK2 complex (Sorkin et al. 2023). The existence of various transcriptional regulatory pathways for cold-inducible gene expression reflects the ability to accli-mate to complex cold stress conditions in nature.

Perception of cold stress

Although many TFs involved in cold stress signaling have been reported, how plants perceive temperature changes and adapt to different climates remain poorly understood (Verslues et al. 2023). Ca^{2+} is a general second messenger involved in various signaling pathways. While plants lack ortho-logs of transient receptor potential cation channels, which are well-known thermosensors in many organisms, including mammals (Kashio and Tominaga 2022), Ca^{2+} signaling plays a crucial role in cold stress responses, including in the DREB1/ CBF-dependent and -independent pathways (Plieth et al. 1999; Yuan et al. 2018).

Cold stress triggers rapid and transient increases in cytosolic Ca^{2+} levels in a cooling rate–dependent and/or tissue-specific manner in plants (Knight and Knight 2000; Hiraki et al. 2019; Lamers et al. 2020). Several PM-localized proteins, such as calcium-permeable mechanosensitive channels (MCA1 and MCA2), cyclic nucleotide-gated ion channels, and Ca^{2+}-permeable transporters (ANNEXINs), mediate increases in cytosolic Ca^{2+} levels under cold stress (Mori et al. 2018; Cui et al. 2020; Liu et al. 2021). In rice, the PM and endoplasmic reticulum–localized protein CHILLING-TOLERANCE DIVERGENCE1 activates Ca^{2+} influx in response to cold stress and induces cold stress tolerance by regulating G-protein signaling (Ma et al. 2015). In cold stress–activated Ca^{2+} signaling pathways, Ca^{2+}/calmodulin-regulated receptor-like kinases and Ca^{2+}-dependent protein kinases (CPKs/CDPKs) positively regulate cold tolerance by activating MAP kinase cascades (Yang et al. 2010a, 2010b; Yuan et al. 2018). In maize, ZmMPK8 phosphorylates type-A Response Regulator1, leading to its degradation and the induction of cold-responsive gene expression to promote chilling tolerance (Zeng et al. 2021).

Several photoreceptors function as thermosensors in plants (Noguchi and Kodama 2022; Jung et al. 2023; Kerbler and Wigge 2023). The red-light photoreceptor PHYTOCHROME B (phyB) regulates photomorphogenesis, light-mediated development, and thermomorphogenesis under normal temperatures in Arabidopsis (Legris et al. 2016; Jung et al. 2023). While red light stimulates conversion between 2 photo-interconvertible forms of phyB, from the inactive Pr state to the active Pfr state, cool temperatures prevent the reversion of phyB from the Pfr state to the Pr state. In Arabidopsis and several other plants, light quality and photoreceptors such as phyB affect cold acclimation and freezing tolerance (Kim et al. 2002; Franklin and Whitelam 2007; Wang et al. 2016; Prerostova et al. 2021). In the liverwort *Marchantia polymorpha*, the blue light receptor phototropin (phot) regulates chloroplast positioning under low-temperature conditions via the temperature-dependent lifetime of its photoactivated chromophore (Fujii et al. 2017). Among the central oscillators of the Arabidopsis circadian clock, the EC functions as a thermosensor. ELF3, an EC component that contains a polyglutamine-rich repeat embedded within a prion-like domain, is a large scaffold protein and a key temperature-sensing component (Jung et al. 2020). ELF4, another EC component, strengthens the DNA-binding activity of the EC at low temperatures (Ezer et al. 2017; Silva et al. 2020). In addition, ELF4 can be transported from shoots to roots to deliver circadian information; low temperatures promote its mobility (Chen et al. 2020). However, how the EC perceives the cold signal and functions at temperatures below 12 °C remain unknown.

Interconnection between drought and cold responses

Exogenous ABA application improves the plant cold tolerance, while cold stress alone only causes a modest increase in ABA levels compared with drought stress (Maruyama et al. 2012). Both stresses lead to stomatal closure, but during cold stress, this mechanism appears to be independent of ABA (Wilkinson et al. 2001). ABA-deficient or -insensitive mutants exhibit reduced freezing stress tolerance, although the cold-inducible expression of *COR* and *RD* genes may not always be affected in these mutants, indicating that a substantial part of the cold response operates independently of ABA (Gilmour and Thomashow 1991). Therefore, ABA contributes to plant cold resilience through its cumulative effects on growth and development, alongside its role in signaling during the drought response (Yoshida and Fernie 2023).

DREB1/CBF and DREB2 TFs are highly homologous isoforms that share the DRE/CRT *cis*-acting element for identifying their target genes. This trait positions them as a pivotal interconnection point in gene regulation of drought and cold stress responses. DREB2 homologs have been widely identified in both mosses and lycophytes as essential components of drought stress signaling. By contrast, DREB1/CBF homologs have been identified exclusively in angiosperms, highlighting a significant evolutionary adaptation (Zhang et al. 2004; Ito et al. 2006; Mizoi et al. 2012; Kidokoro et al. 2015; Li et al. 2020). This specialization suggests that the regulation of cold-responsive genes in plants has evolved to cope with the challenges posed by habitat expansion onto dry land, where plants face reduced water availability and do not receive the temperature buffering provided by aquatic environ-ments. As a result, plants have developed diverse sensing mechanisms that converge on a shared set of protective genes, enabling them to effectively respond to both the combined and specific environmental challenges of cold and drought.

Breeding stress-resilient crops

Improving stress tolerance is a major focus of plant stress biology and crop breeding programs, with many significant advancements already documented (Bailey-Serres et al. 2019; Cooper and Messina 2023). Specifically, in molecular breeding, a range of stress-related genes have been incorporated into transgenic crops to determine their effects on stress resilience during field trials. Targeting the ABA receptor to adjust stress sensitivity via the ectopic expression of PYR1/PYL/RCAR homologs in Arabidopsis and wheat (*Triticum aestivum*) enhanced drought tolerance and water use efficiency with minimal yield losses (Yang et al. 2016; Mega et al. 2019). However, transgenic rice expressing these homologs showed reduced growth under field conditions (Kim et al. 2014). Another strategy for improving tolerance is metabolic engineering, such as the introduction of Arabidopsis GolS2 into rice, which increased osmoprotectant levels and maintained yields under drought (Selvaraj et al. 2017). Stress-signaling TFs, such as DREB1/CBF and its homo-logs, have been successfully used to increase drought toler-ance in crops including wheat, soybean, and rice (Xiao et al. 2009; Saint Pierre et al. 2012; Fuganti-Pagliarini et al. 2017). Although evaluating cold tolerance in field trials is challenging, overexpressing rice *bZIP73* notably improved cold tolerance in rice at the reproductive stage (Liu et al. 2019). Additionally, the overexpression of various fatty acid desaturase genes enhanced cold stress tolerance in crops, with some demonstrating field-level cold resilience (Dar et al. 2017). Commercially available transgenic maize expressing bacterial cold shock protein, which acts as an RNA chaperone, showed improved environmental resilience and yield retention (Nemali et al. 2015).

Recent biotechnological advancements allow for precise and flexible gene modulation and the ability to edit multiple genes simultaneously (Chen et al. 2019; Shi et al. 2023). Therefore, future molecular breeding efforts are expected to involve physiological context-based strategies, moving from the current single-gene interventions (Husaini 2022). For instance, enhancing drought tolerance may involve a variety of changes, such as increasing ABA sensitivity, augmenting leaf wax layers, and reducing stomatal density to conserve water. Similarly, for cold tolerance, adjusting the switch between growth and stress responses through CAMTA- and circadian clock–regulated pathways may offer improvements.

New challenges and perspectives

New technologies have the potential to expand the field of plant stress biology. Recent advancements have ushered in the capability for genome-wide analyses of the transcriptome and epigenome at the single-cell level, granting researchers the ability to dissect the molecular profiles of individual cells implicated in abiotic stress responses, such as guard cells and phloem companion cells (Shaw et al. 2020; Seyfferth et al. 2021). Spatial sequencing is poised to enhance our understanding of single-cell data within the framework of intercellular positioning (Giacomello 2021). Additionally, computational approaches empowered by machine learning are refining the analysis of nuanced developmental changes under complex environmental conditions, complemented by automated phenotyping (Fujita et al. 2018; Lew et al. 2020; Singh et al. 2021). Advanced imaging methods such as hyperspectral imaging and super-resolution microscopy are providing unprecedented spatial and temporal insights into plant stress reactions, enabling a detailed exploration of variability in stress responses and the signaling pathways that mediate stress adaptation across different genotypes and environmental conditions (Galieni et al. 2021; Ovečka et al. 2022). These technological developments are instrumental for researchers seeking to decode the complex networks of plant stress responses and to engineer crops with optimized traits for stress resilience.

Acknowledgments

We apologize to all authors whose work could not be cited here due to space restrictions.

Author contributions

J.-S.K. and S.K. designed the figures and wrote the manuscript draft, based on the data provided by K.S. and K.Y.-S. All authors contributed to the review and editing of the manuscript.

Funding

This work was partly supported by Grants-in-Aid for Scientific Research from the Japan Society for the Promotion of Science (grant numbers JP21K05524 to J.-S.K. and JP21K06209 to S.K.), a Grant-in-Aid for Transformative Research Areas (grant number JP23H04193 to S.K.) from MEXT, and a Fusion Oriented Research for Disruptive Science and Technology grant (grant number JPMJFR2206 to S.K.) from the Japan Science and Technology Agency.

Conflict of interest statement. None declared.

References

Abe H, Urao T, Ito T, Seki M, Shinozaki K, Yamaguchi-Shinozaki K. Arabidopsis AtMYC2 (bHLH) and AtMYB2 (MYB) function as transcriptional activators in abscisic acid signaling. Plant cell. 2003:**15**(1): 63–78. https://doi.org/10.1105/tpc.006130

Abe H, Yamaguchi-Shinozaki K, Urao T, Iwasaki T, Hosokawa H, Shinozaki K. Role of Arabidopsis MYC and MYB homolog in drought- and abscisic acid-regulated gene expression. Plant Cell. 1997:**9**(10):1859–1868. https://doi.org/10.1105/tpc.9.10.1859

Achard P, Gong F, Cheminant S, Alioua M, Hedden P, Genschik P. The cold-inducible CBF1 factor-dependent signaling pathway modulates the accumulation of the growth-repressing DELLA proteins via its effect on gibberellin metabolism. Plant Cell. 2008:**20**(8):2117–2129. https://doi.org/10.1105/tpc.108.058941

Bailey-Serres J, Parker JE, Ainsworth EA, Oldroyd GED, Schroeder JI. Genetic strategies for improving crop yields. Nature. 2019:**575**(7781): 109–118. https://doi.org/10.1038/s41586-019-1679-0

Baker SS, Wilhelm KS, Thomashow MF. The 5′-region of Arabidopsis thaliana cor15a has cis-acting elements that confer cold-, drought- and ABA-regulated gene expression. Plant Mol Biol. 1994:**24**(5): 701–713. https://doi.org/10.1007/BF00029852

Belin C, de Franco P-O, Bourbousse C, Chaignepain S, Schmitter J-M, Vavasseur A, Giraudat J, Barbier-Brygoo H, Thomine S. Identification of features regulating OST1 kinase activity and OST1 function in guard cells. Plant Physiol. 2006:**141**(4):1316–1327. https://doi.org/10.1104/pp.106.079327

Bhaskara GB, Nguyen TT, Verslues PE. Unique drought resistance functions of the highly ABA-induced clade A protein phosphatase 2Cs. Plant Physiol. 2012:**160**(1):379–395. https://doi.org/10.1104/pp.112.202408

Bouché N, Scharlat A, Snedden W, Bouchez D, Fromm H. A novel family of calmodulin-binding transcription activators in multicellular organisms. J Biol Chem. 2002:**277**(24):21851–21861. https://doi.org/10.1074/jbc.M200268200

Boudsocq M, Barbier-Brygoo H, Laurière C. Identification of nine sucrose nonfermenting 1-related protein kinases 2 activated by hyperosmotic and saline stresses in Arabidopsis thaliana. J Biol Chem. 2004:**279**(40):41758 41766. https://doi.org/10.1074/jbc.M405259200

Cai Z, Liu J, Wang H, Yang C, Chen Y, Chen Y, Li Y, Pan S, Dong R, Tang G, et al. GSK3-like kinases positively modulate abscisic acid signaling through phosphorylating subgroup III SnRK2s in Arabidopsis. Proc Natl Acad Sci U S A. 2014:**111**(26):9561–9656. https://doi.org/10.1073/pnas.1316717111

Cano-Ramirez DL, Panter PE, Takemura T, de Fraine TS, de Barros Dantas LL, Dekeya R, Barros-Galvão T, Paajanen P, Bellandi A, Batstone T, et al. Low-temperature and circadian signals are integrated by the sigma factor SIG5. Nat Plants. 2023:**9**(4):661–672. https://doi.org/10.1038/s41477-023-01377-1

Catalá R, Medina J, Salinas J. Integration of low temperature and light signaling during cold acclimation response in Arabidopsis. Proc Natl Acad Sci U S A. 2011:**108**(39):16475–16480. https://doi.org/10.1073/pnas.1107161108

Chao L, Kim Y, Gilmour SJ, Thomashow MF. Temperature modulation of CAMTA3 gene induction activity is mediated through the DNA binding domain. Plant J. 2022:**112**(1):235–248. https://doi.org/10.1111/tpj.15944

Chen K, Wang Y, Zhang R, Zhang H, Gao C. CRISPR/cas genome editing and precision plant breeding in agriculture. Ann Rev Plant Biol. 2019:**70**(1):667–697. https://doi.org/10.1146/annurev-arplant-050718-100049

Chen WW, Takahashi N, Hirata Y, Ronald J, Porco S, Davis SJ, Nusinow DA, Kay SA, Mas P. A mobile ELF4 delivers circadian temperature information from shoots to roots. Nat Plants. 2020:**6**(4): 416–426. https://doi.org/10.1038/s41477-020-0634-2

Chinnusamy V, Ohta M, Kanrar S, Lee B-H, Hong X, Agarwal M, Zhu JK. ICE1: a regulator of cold-induced transcriptome and freezing tolerance in Arabidopsis. Genes Dev. 2003:**17**(8):1043–1054. https://doi.org/10.1101/gad.1077503

Choi H, Hong J, Ha J, Kang J, Kim SY. ABFs, a family of ABA-responsive element binding factors. J Biol Chem. 2000:**275**(3):1723–1730. https://doi.org/10.1074/jbc.275.3.1723

Chong GL, Foo MH, Lin WD, Wong MM, Verslues PE. Highly ABA-induced 1 (HAI1)-interacting protein HIN1 and drought acclimation-enhanced splicing efficiency at intron retention sites. Proc Natl Acad Sci U S A. 2019:**116**(44):22376–22385. https://doi.org/10.1073/pnas.1906244116

Cook D, Fowler S, Fiehn O, Thomashow MF. A prominent role for the CBF cold response pathway in configuring the low-temperature metabolome of Arabidopsis. Proc Natl Acad Sci U S A. 2004:**101**(42): 15243–15248. https://doi.org/10.1073/pnas.0406069101

Cooper M, Messina CD. Breeding crops for drought-affected environments and improved climate resilience. Plant Cell. 2023:**35**(1): 162–186. https://doi.org/10.1093/plcell/koac321

Cui Y, Lu S, Li Z, Cheng J, Hu P, Zhu T, Wang X, Jin M, Wang X, Li L, et al. CYCLIC NUCLEOTIDE-GATED ION CHANNELs 14 and 16 promote tolerance to heat and chilling in rice. Plant Physiol. 2020:**183**(4): 1794–1808. https://doi.org/10.1104/pp.20.00591

Cutler SR, Rodriguez PL, Finkelstein RR, Abrams SR. Abscisic acid: emergence of a core signaling network. Ann Rev Plant Biol. 2010:**61**(1): 651–679. https://doi.org/10.1146/annurev-arplant-042809-112122

Dar AA, Choudhury AR, Kancharla PK, Arumugam N. The FAD2 gene in plants: occurrence, regulation, and role. Front Plant Sci. 2017:**8**:1789. https://doi.org/10.3389/fpls.2017.01789

Dietz KJ, Sauter A, Wichert K, Messdaghi D, Hartung W. Extracellular beta-glucosidase activity in barley involved in the hydrolysis of ABA glucose conjugate in leaves. J Exp Bot. 2000:**51**(346):937–944. https://doi.org/10.1093/jexbot/51.346.937

Ding Y, Yang H, Wu S, Fu D, Li M, Gong Z, Yang S. CPK28-NLP7 module integrates cold-induced ca(2+) signal and transcriptional reprogramming in Arabidopsis. Sci Adv. 2022:**8**(26):eabn7901. https://doi.org/10.1126/sciadv.abn7901

Doherty CJ, Van Buskirk HA, Myers SJ, Thomashow MF. Roles for Arabidopsis CAMTA transcription factors in cold-regulated gene expression and freezing tolerance. Plant Cell. 2009:**21**(3):972–984. https://doi.org/10.1105/tpc.108.063958

Dorone Y, Boeynaems S, Flores E, Jin B, Hateley S, Bossi F, Lazarus E, Pennington JG, Michiels E, De Decker M, et al. A prion-like protein regulator of seed germination undergoes hydration-dependent phase separation. Cell. 2021:**184**(16):4284–4298.e27. https://doi.org/10.1016/j.cell.2021.06.009

Dröge-Laser W, Snoek BL, Snel B, Weiste C. The Arabidopsis bZIP transcription factor family—an update. Curr Opin Plant Biol. 2018:**45**:36–49. https://doi.org/10.1016/j.pbi.2018.05.001

Du L, Ali GS, Simons KA, Hou J, Yang T, Reddy ASN, Poovaiah BW. Ca(2+)/calmodulin regulates salicylic-acid-mediated plant immunity. Nature. 2009:**457**(7233):1154–1158. https://doi.org/10.1038/nature07612

Eljebbawi A, Dolata A, Strotmann VI, Stahl Y. Unlocking nature's (sub)cellular symphony: phase separation in plant meristems. Curr Opin Plant Biol. 2023:**76**:102480. https://doi.org/10.1016/j.pbi.2023.102480

Endo A, Sawada Y, Takahashi H, Okamoto M, Ikegami K, Koiwai H, Seo M, Toyomasu T, Mitsuhashi W, Shinozaki K, et al. Drought induction of Arabidopsis 9-cis-epoxycarotenoid dioxygenase occurs in vascular parenchyma cells. Plant Physiol. 2008:**147**(4):1984–1993. https://doi.org/10.1104/pp.108.116632

Eulgem T, Rushton PJ, Robatzek S, Somssich IE. The WRKY superfamily of plant transcription factors. Trends Plant Sci. 2000:**5**(5):199–206. https://doi.org/10.1016/S1360-1385(00)01600-9

Ezer D, Jung JH, Lan H, Biswas S, Gregoire L, Box MS, Charoensawan V, Cortijo S, Lai X, Stöckle D, et al. The evening complex coordinates environmental and endogenous signals in Arabidopsis. Nat Plants. 2017:**3**(7):17087. https://doi.org/10.1038/nplants.2017.87

Finch-Savage WE, Leubner-Metzger G. Seed dormancy and the control of germination. New Phytol. 2006:**171**(3):501–523. https://doi.org/10.1111/j.1469-8137.2006.01787.x

Finkelstein R. Abscisic acid synthesis and response. Arabidopsis Book. 2013:**11**:e0166. https://doi.org/10.1199/tab.0166

Finkelstein RR, Lynch TJ. The Arabidopsis abscisic acid response gene ABI5 encodes a basic leucine zipper transcription factor. Plant Cell. 2000:**12**(4):599–609. https://doi.org/10.1105/tpc.12.4.599

Fletcher JC. Recent advances in Arabidopsis CLE peptide signaling. Trends Plant Sci. 2020:**25**(10):1005–1016. https://doi.org/10.1016/j.tplants.2020.04.014

Fowler S, Thomashow MF. Arabidopsis transcriptome profiling indicates that multiple regulatory pathways are activated during cold acclimation in addition to the CBF cold response pathway. Plant Cell. 2002:**14**(8):1675–1690. https://doi.org/10.1105/tpc.003483

Franklin KA, Whitelam GC. Light-quality regulation of freezing tolerance in Arabidopsis thaliana. Nat Genet. 2007:**39**(11):1410–1413. https://doi.org/10.1038/ng.2007.3

Frey A, Effroy D, Lefebvre V, Seo M, Perreau F, Berger A, Sechet J, To A, North HM, Marion-Poll A. Epoxycarotenoid cleavage by NCED5 fine-tunes ABA accumulation and affects seed dormancy and drought tolerance with other NCED family members. Plant J. 2012: **70**(3):501–512. https://doi.org/10.1111/j.1365-313X.2011.04887.x

Fuganti-Pagliarini R, Ferreira LC, Rodrigues FA, Molinari HBC, Marin SRR, Molinari MDC, Marcolino-Gomes J, Mertz-Henning LM, Farias JRB, de Oliveira MCN, et al. Characterization of soybean genetically modified for drought tolerance in field conditions. Front Plant Sci. 2017:**8**:448. https://doi.org/10.3389/fpls.2017.00448

Fujii H, Zhu JK. Arabidopsis mutant deficient in 3 abscisic acid-activated protein kinases reveals critical roles in growth, reproduction, and stress. Proc Natl Acad Sci USA. 2009:**106**(20):8380–8385. https://doi.org/10.1073/pnas.0903144106

Fujii Y, Tanaka H, Konno N, Ogasawara Y, Hamashima N, Tamura S, Hasegawa S, Hayasaki Y, Okajima K, Kodama Y. Phototropin perceives temperature based on the lifetime of its photoactivated state. Proc Natl Acad Sci U S A. 2017:**114**(34):9206–9211. https://doi.org/10.1073/pnas.1704462114

Fujita M, Fujita Y, Maruyama K, Seki M, Hiratsu K, Ohme-Takagi M, Tran LSP, Yamaguchi-Shinozaki K, Shinozaki K. A dehydration-induced NAC protein, RD26, is involved in a novel ABA-dependent stress-signaling pathway. Plant J. 2004:**39**(6): 863–876. https://doi.org/10.1111/j.1365-313X.2004.02171.x

Fujita M, Tanabata T, Urano K, Kikuchi S, Shinozaki K. RIPPS: a plant phenotyping system for quantitative evaluation of growth under controlled environmental stress conditions. Plant Cell Physiol. 2018:**59**(10):2030–2038. https://doi.org/10.1093/pcp/pcy122

Fujita Y, Fujita M, Satoh R, Maruyama K, Parvez MM, Seki M, Hiratsu K, Ohme-Takagi M, Shinozaki K, Yamaguchi-Shinozaki K. AREB1 is a transcription activator of novel ABRE-dependent ABA signaling that enhances drought stress tolerance in Arabidopsis. Plant Cell. 2005:**17**(12):3470–3488. https://doi.org/10.1105/tpc.105.035659

Fujita Y, Nakashima K, Yoshida T, Katagiri T, Kidokoro S, Kanamori N, Umezawa T, Fujita M, Maruyama K, Ishiyama K, et al. Three SnRK2 protein kinases are the main positive regulators of abscisic acid signaling in response to water stress in Arabidopsis. Plant Cell Physiol. 2009:**50**(12):2123–2132. https://doi.org/10.1093/pcp/pcp147

Furihata T, Maruyama K, Fujita Y, Umezawa T, Yoshida R, Shinozaki K, Yamaguchi-Shinozaki K. Abscisic acid-dependent multisite phosphorylation regulates the activity of a transcription activator AREB1. Proc Natl Acad Sci USA. 2006:**103**(6):1988–1993. https://doi.org/10.1073/pnas.0505667103

Galieni A, D'Ascenzo N, Stagnari F, Pagnani G, Xie Q, Pisante M. Past and future of plant stress detection: an overview from remote sensing to positron emission tomography. Front Plant Sci. 2021:**11**: 609155. https://doi.org/10.3389/fpls.2020.609155

Geiger D, Scherzer S, Mumm P, Stange A, Marten I, Bauer H, Ache P, Matschi S, Liese A, Al-Rasheid KAS, et al. Activity of guard cell anion channel SLAC1 is controlled by drought-stress signaling kinase-phosphatase pair. Proc Natl Acad Sci USA. 2009:**106**(50): 21425–21430. https://doi.org/10.1073/pnas.0912021106

Giacomello S. A new era for plant science: spatial single-cell transcriptomics. Curr Opin Plant Biol. 2021:**60**:102041. https://doi.org/10.1016/j.pbi.2021.102041

Gilmour SJ, Thomashow MF. Cold acclimation and cold-regulated gene expression in ABA mutants of Arabidopsis thaliana. Plant Mol Biol. 1991:**17**(6):1233–1240. https://doi.org/10.1007/BF00028738

Gosti F, Beaudoin N, Serizet C, Webb AA, Vartanian N, Giraudat J. ABI1 protein phosphatase 2C is a negative regulator of abscisic acid signaling. Plant Cell. 1999:**11**(10):1897–1910. https://doi.org/10.1105/tpc.11.10.1897

Graeber K, Linkies A, Steinbrecher T, Mummenhoff K, Tarkowská D, Turečková V, Ignatz M, Sperber K, Voegele A, de Jong H, et al. DELAY OF GERMINATION 1 mediates a conserved coat-dormancy mechanism for the temperature- and gibberellin-dependent control of seed germination. Proc Natl Acad Sci USA. 2014:**111**(34): E3571–E3580. https://doi.org/10.1073/pnas.1403851111

Guseman JM, Webb K, Srinivasan C, Dardick C. DRO1 influences root system architecture in Arabidopsis and Prunus species. Plant J. 2016:**89**(6):1093–1105. https://doi.org/10.1111/tpj.13470

Hauser F, Waadt R, Schroeder JI. Evolution of abscisic acid synthesis and signaling mechanisms. Curr Biol. 2011:**21**(9):R346–R355. https://doi.org/10.1016/j.cub.2011.03.015

Hickey LT, Hafeez AN, Robinson H, Jackson SA, Leal-Bertioli SCM, Tester M, Gao C, Godwin ID, Hayes BJ, Wulff BBH. Breeding crops to feed 10 billion. Nat Biotechnol. 2019:**37**(7):744–754. https://doi.org/10.1038/s41587-019-0152-9

Himmelbach A, Hoffmann T, Leube M, Höhener B, Grill E. Homeodomain protein ATHB6 is a target of the protein phosphatase ABI1 and regulates hormone responses in Arabidopsis. EMBO J. 2002:**21**(12):3029–3038. https://doi.org/10.1093/emboj/cdf316

Hiraki H, Uemura M, Kawamura Y. Calcium signaling-linked CBF/DREB1 gene expression was induced depending on the temperature fluctuation in the field: views from the natural condition of cold acclimation. Plant Cell Physiol. 2019:**60**(2):303–317. https://doi.org/10.1093/pcp/pcy210

Hsu PY, Devisetty UK, Harmer SL. Accurate timekeeping is controlled by a cycling activator in Arabidopsis. Elife. 2013:**2**:e00473. https://doi.org/10.7554/eLife.00473

Hsu PK, Takahashi Y, Munemasa S, Merilo E, Laanemets K, Waadt R, Pater D, Kollist H, Schroeder JI. Abscisic acid-independent stomatal CO2 signal transduction pathway and convergence of CO2 and ABA signaling downstream of OST1 kinase. Proc Natl Acad Sci USA. 2018:**115**(42):E9971–E9980. https://doi.org/10.1073/pnas.1809204115

Husaini AM. High-value pleiotropic genes for developing multiple stress-tolerant biofortified crops for 21st-century challenges. Heredity (Edinb). 2022:**128**(6):460–472. https://doi.org/10.1038/s41437-022-00500-w

Iljin WS. Drought resistance in plants and physiological process. Annu Rev Plant Physiol. 1957:**8**(1):257–274. https://doi.org/10.1146/annurev.pp.08.060157.001353

Ito Y, Katsura K, Maruyama K, Taji T, Kobayashi M, Seki M, Shinozaki K, Yamaguchi-Shinozaki K. Functional analysis of rice DREB1/CBF-type transcription factors involved in cold-responsive gene expression in transgenic rice. Plant Cell Physiol. 2006:**47**(1): 141–153. https://doi.org/10.1093/pcp/pci230

Iuchi S, Kobayashi M, Taji T, Naramoto M, Seki M, Kato T, Tabata S, Kakubari Y, Yamaguchi-Shinozaki K, Shinozaki K. Regulation of drought tolerance by gene manipulation of 9-cis-epoxycarotenoid dioxygenase, a key enzyme in abscisic acid biosynthesis in Arabidopsis. Plant J. 2001:**27**(4):325–333. https://doi.org/10.1046/j.1365-313x.2001.01096.x

Iuchi S, Kobayashi M, Yamaguchi-Shinozaki K, Shinozaki K. A stress-inducible gene for 9-cis-epoxycarotenoid dioxygenase involved in abscisic acid biosynthesis under water stress in drought-tolerant cowpea. Plant Physiol. 2000:**123**(2):553–562. https://doi.org/10.1104/pp.123.2.553

Jaglo-Ottosen KR, Gilmour SJ, Zarka DG, Schabenberger O, Thomashow MF. Arabidopsis CBF1 overexpression induces COR genes and enhances freezing tolerance. Science. 1998:**280**(5360): 104–106. https://doi.org/10.1126/science.280.5360.104

Jeon J, Kim NY, Kim S, Kang NY, Novák O, Ku SJ, Cho C, Lee DJ, Lee EJ, Strnad M, et al. A subset of cytokinin two-component signaling system plays a role in cold temperature stress response in Arabidopsis. J Biol Chem. 2010:**285**(30):23371–23386. https://doi.org/10.1074/jbc.M109.096644

Jia Y, Ding Y, Shi Y, Zhang X, Gong Z, Yang S. The cbfs triple mutants reveal the essential functions of CBFs in cold acclimation and allow the definition of CBF regulons in Arabidopsis. New Phytol. 2016:**212**(2):345–353. https://doi.org/10.1111/nph.14088

Jiang B, Shi Y, Zhang X, Xin X, Qi L, Guo H, Li J, Yang S. PIF3 is a negative regulator of the CBF pathway and freezing tolerance in Arabidopsis. Proc Natl Acad Sci U S A. 2017:**114**(32):E6695–E6702. https://doi.org/10.1073/pnas.1706226114

Jiang C, Iu B, Singh J. Requirement of a CCGAC cis-acting element for cold induction of the BN115 gene from winter Brassica napus. Plant Mol Biol. 1996:**30**(3):679–684. https://doi.org/10.1007/BF00049344

Julian J, Coego A, Lozano-Juste J, Lechner E, Wu Q, Zhang X, Merilo E, Belda-Palazon B, Park SY, Cutler SR, et al. The MATH-BTB BPM3 and BPM5 subunits of cullin3-RING E3 ubiquitin ligases target PP2CA and other clade A PP2Cs for degradation. Proc Natl Acad Sci U S A. 2019:**116**(31):15725–15734. https://doi.org/10.1073/pnas.1908677116

Jun J, Fiume E, Roeder AHK, Meng L, Sharma VK, Osmont KS, Baker C, Ha CM, Meyerowitz EM, Feldman LJ, et al. Comprehensive analysis of CLE polypeptide signaling gene expression and overexpression activity in Arabidopsis. Plant Physiol. 2010:**154**(4):1721–1736. https://doi.org/10.1104/pp.110.163683

Jung JH, Barbosa AD, Hutin S, Kumita JR, Gao M, Derwort D, Silva CS, Lai X, Pierre E, Geng F, et al. A prion-like domain in ELF3 functions as a thermosensor in Arabidopsis. Nature. 2020:**585**(7824): 256–260. https://doi.org/10.1038/s41586-020-2644-7

Jung JH, Seo PJ, Oh E, Kim J. Temperature perception by plants. Trends Plant Sci. 2023:**28**(8):924–940. https://doi.org/10.1016/j.tplants.2023.03.006

Kamioka M, Takao S, Suzuki T, Taki K, Higashiyama T, Kinoshita T, Nakamichi N. Direct repression of evening genes by CIRCADIAN CLOCK-ASSOCIATED1 in the Arabidopsis circadian clock. Plant Cell. 2016:**28**(3):696–711. https://doi.org/10.1105/tpc.15.00737

Kamranfar I, Xue GP, Tohge T, Sedaghatmehr M, Fernie AR, Balazadeh S, Mueller-Roeber B. Transcription factor RD26 is a key regulator of metabolic reprogramming during dark-induced senescence. New Phytol. 2018:**218**(4):1543–1557. https://doi.org/10.1111/nph.15127

Kanaoka MM, Pillitteri LJ, Fujii H, Yoshida Y, Bogenschutz NL, Takabayashi J, Zhu JK, KU T. SCREAM/ICE1 and SCREAM2 specify three cell-state transitional steps leading to Arabidopsis stomatal differentiation. Plant Cell. 2008:**20**(7):1775–1785. https://doi.org/10.1105/tpc.108.060848

Kang J, Hwang JU, Lee M, Kim YY, Assmann SM, Martinoia E, Lee Y. PDR-type ABC transporter mediates cellular uptake of the phytohormone abscisic acid. Proc Natl Acad Sci U S A. 2010:**107**(5):2355–2360. https://doi.org/10.1073/pnas.0909222107

Kanno Y, Hanada A, Chiba Y, Ichikawa T, Nakazawa M, Matsui M, Koshiba T, Kamiya Y, Seo M. Identification of an abscisic acid transporter by functional screening using the receptor complex as a sensor. Proc Natl Acad Sci U S A. 2012:**109**(24):9653–9658. https://doi.org/10.1073/pnas.1203567109

Kashio M, Tominaga M. TRP channels in thermosensation. Curr Opin Neurobiol. 2022:**75**:102591. https://doi.org/10.1016/j.conb.2022.102591

Kasuga M, Liu Q, Miura S, Yamaguchi-Shinozaki K, Shinozaki K. Improving plant drought, salt, and freezing tolerance by gene transfer of a single stress-inducible transcription factor. Nat Biotechnol. 1999:**17**(3):287–291. https://doi.org/10.1038/7036

Katsuta S, Masuda G, Bak H, Shinozawa A, Kamiyama Y, Umezawa T, Takezawa D, Yotsui I, Taji T, Sakata Y. Arabidopsis Raf-like kinases act as positive regulators of subclass III SnRK2 in osmostress signaling. Plant J. 2020:**103**(2):634–644. https://doi.org/10.1111/tpj.14756

Kawa D, Meyer AJ, Dekker HL, Abd-El-Haliem AM, Gevaert K, Van De Slijke E, Maszkowska J, Bucholc M, Dobrowolska G, De Jaeger G, et al. SnRK2 protein kinases and mRNA decapping machinery control root development and response to salt. Plant Physiol. 2020:**182**(1):361–377. https://doi.org/10.1104/pp.19.00818

Kerbler SM, Wigge PA. Temperature sensing in plants. Annu Rev Plant Biol. 2023:**74**(1):341–366. https://doi.org/10.1146/annurev-arplant-102820-102235

Kidokoro S, Hayashi K, Haraguchi H, Ishikawa T, Soma F, Konoura I, Toda S, Mizoi J, Suzuki T, Shinozaki K, et al. Posttranslational regulation of multiple clock-related transcription factors triggers cold-inducible gene expression in Arabidopsis. Proc Natl Acad Sci U S A. 2021:**118**(10):e2021048118. https://doi.org/10.1073/pnas.2021048118

Kidokoro S, Kim JS, Ishikawa T, Suzuki T, Shinozaki K, Yamaguchi-Shinozaki K. DREB1A/CBF3 is repressed by transgene-induced DNA methylation in the Arabidopsis ice1-1 mutant. Plant Cell. 2020:**32**(4):1035–1048. https://doi.org/10.1105/tpc.19.00532

Kidokoro S, Konoura I, Soma F, Suzuki T, Miyakawa T, Tanokura M, Shinozaki K, Yamaguchi-Shinozaki K. Clock-regulated coactivators selectively control gene expression in response to different temperature stress conditions in Arabidopsis. Proc Natl Acad Sci U S A. 2023:**120**(16):e2216183120. https://doi.org/10.1073/pnas.2216183120

Kidokoro S, Maruyama K, Nakashima K, Imura Y, Narusaka Y, Shinwari ZK, Osakabe Y, Fujita Y, Mizoi J, Shinozaki K, et al. The phytochrome-interacting factor PIF7 negatively regulates DREB1 expression under circadian control in Arabidopsis. Plant Physiol. 2009:**151**(4):2046–2057. https://doi.org/10.1104/pp.109.147033

Kidokoro S, Shinozaki K, Yamaguchi-Shinozaki K. Transcriptional regulatory network of plant cold-stress responses. Trends Plant Sci. 2022:**27**(9):922–935. https://doi.org/10.1016/j.tplants.2022.01.008

Kidokoro S, Watanabe K, Ohori T, Moriwaki T, Maruyama K, Mizoi J, Myint Phyu Sin Htwe N, Fujita Y, Sekita S, Shinozaki K, et al. Soybean DREB1/CBF-type transcription factors function in heat and drought as well as cold stress-responsive gene expression. Plant J. 2015:**81**(3):505–518. https://doi.org/10.1111/tpj.12746

Kidokoro S, Yoneda K, Takasaki H, Takahashi F, Shinozaki K, Yamaguchi-Shinozaki K. Different cold-signaling pathways function in the responses to rapid and gradual decreases in temperature. Plant Cell. 2017:**29**(4):760–774. https://doi.org/10.1105/tpc.16.00669

Kim HJ, Kim YK, Park JY, Kim J. Light signalling mediated by phytochrome plays an important role in cold-induced gene expression through the C-repeat/dehydration responsive element (C/DRE) in Arabidopsis thaliana. Plant J. 2002:**29**(6):693–704. https://doi.org/10.1046/j.1365-313X.2002.01249.x

Kim J, Lee K, Hwang H, Bhatnagar N, Kim DY, Yoon IS, Byun MO, Kim ST, Jung KH, Kim BG, et al. Overexpression of PYL5 in rice enhances drought tolerance, inhibits growth, and modulates gene expression. J Exp Bot. 2014:**65**(2):453–464. https://doi.org/10.1093/jxb/ert397

Kim JS, Mizoi J, Kidokoro S, Maruyama K, Nakajima J, Nakashima K, Mitsuda N, Takiguchi Y, Ohme-Takagi M, Kondou Y, et al. Arabidopsis GROWTH-REGULATING FACTOR7 functions as a transcriptional repressor of abscisic acid- and osmotic stress-responsive genes, including DREB2A. Plant Cell. 2012:**24**(8):3393–3405. https://doi.org/10.1105/tpc.112.100933

Kim JS, Mizoi J, Yoshida T, Fujita Y, Nakajima J, Ohori T, Todaka D, Nakashima K, Hirayama T, Shinozaki K, et al. An ABRE promoter sequence is involved in osmotic stress-responsive expression of the DREB2A gene, which encodes a transcription factor regulating drought-inducible genes in Arabidopsis. Plant Cell Physiol. 2011:**52**(12):2136–2146. https://doi.org/10.1093/pcp/pcr143

Kim Y, Gilmour SJ, Chao L, Park S, Thomashow MF. Arabidopsis CAMTA transcription factors regulate pipecolic acid biosynthesis and priming of immunity genes. Mol Plant. 2020:**13**(1):157–168. https://doi.org/10.1016/j.molp.2019.11.001

Kim Y, Park S, Gilmour SJ, Thomashow MF. Roles of CAMTA transcription factors and salicylic acid in configuring the low-temperature transcriptome and freezing tolerance of Arabidopsis. Plant J. 2013:**75**(3):364–376. https://doi.org/10.1111/tpj.12205

Kim YS, An C, Park S, Gilmour SJ, Wang L, Renna L, Brandizzi F, Grumet R, Thomashow MF. CAMTA-mediated regulation of salicylic acid immunity pathway genes in Arabidopsis exposed to low temperature and pathogen infection. Plant Cell. 2017:**29**(10):2465–2477. https://doi.org/10.1105/tpc.16.00865

Knight H, Knight MR. Imaging spatial and cellular characteristics of low temperature calcium signature after cold acclimation in Arabidopsis. J Exp Bot. 2000:**51**(351):1679–1686. https://doi.org/10.1093/jexbot/51.351.1679

Koiwai H, Nakaminami K, Seo M, Mitsuhashi W, Toyomasu T, Koshiba T. Tissue-specific localization of an abscisic acid biosynthetic enzyme, AAO3, in Arabidopsis. Plant Physiol. 2004:**134**(4):1697–1707. https://doi.org/10.1104/pp.103.036970

Koornneef M, Reuling G, Karssen CM. The isolation and characterization of abscisic acid-insensitive mutants of Arabidopsis thaliana. Physiol Plant. 1984:**61**(3):377–383. https://doi.org/10.1111/j.1399-3054.1984.tb06343.x

Krochko JE, Abrams GD, Loewen MK, Abrams SR, Cutler AJ. (+)-Abscisic acid 8′-hydroxylase is a cytochrome P450 monooxygenase1. Plant Physiol. 1998:**118**(3):849–860. https://doi.org/10.1104/pp.118.3.849

Kudo M, Kidokoro S, Yoshida T, Mizoi J, Kojima M, Takebayashi Y, Sakakibara H, Fernie AR, Shinozaki K, Yamaguchi-Shinozaki K. A gene-stacking approach to overcome the trade-off between drought stress tolerance and growth in Arabidopsis. Plant J. 2019:**97**(2):240–256. https://doi.org/10.1111/tpj.14110

Kuromori T, Miyaji T, Yabuuchi H, Shimizu H, Sugimoto E, Kamiya A, Moriyama Y, Shinozaki K. ABC transporter AtABCG25 is involved in abscisic acid transport and responses. Proc Natl Acad Sci U S A. 2010:**107**(5):2361–2366. https://doi.org/10.1073/pnas.0912516107

Kuromori T, Seo M, Shinozaki K. ABA transport and plant water stress responses. Trends in Plant Science. 2018:**23**(6):513–522. https://doi.org/10.1016/j.tplants.2018.04.001

Kuromori T, Sugimoto E, Shinozaki K. Intertissue signal transfer of abscisic acid from vascular cells to guard cells. Plant Physiol. 2014:**164**(4):1587–1592. https://doi.org/10.1104/pp.114.235556

Kushiro T, Okamoto M, Nakabayashi K, Yamagishi K, Kitamura S, Asami T, Hirai N, Koshiba T, Kamiya Y, Nambara E. The

Arabidopsis cytochrome P450 CYP707A encodes ABA 8′-hydroxylases: key enzymes in ABA catabolism. EMBO J. 2004:**23**(7):1647–1656. https://doi.org/10.1038/sj.emboj.7600121

Lamers J, van der Meer T, Testerink C. How plants sense and respond to stressful environments. Plant Physiol. 2020:**182**(4):1624–1635. https://doi.org/10.1104/pp.19.01464

Lechner E, Leonhardt N, Eisler H, Parmentier Y, Alioua M, Jacquet H, Leung J, Genschik P. MATH/BTB CRL3 receptors target the homeodomain-leucine zipper ATHB6 to modulate abscisic acid signaling. Dev Cell. 2011:**21**(6):1116–1128. https://doi.org/10.1016/j.devcel.2011.10.018

Lecourieux D, Ranjeva R, Pugin A. Calcium in plant defence-signalling pathways. New Phytol. 2006:**171**(2):249–269. https://doi.org/10.1111/j.1469-8137.2006.01777.x

Lee CM, Thomashow MF. Photoperiodic regulation of the C-repeat binding factor (CBF) cold acclimation pathway and freezing tolerance in Arabidopsis thaliana. Proc Natl Acad Sci U S A. 2012:**109**(37): 15054–15059. https://doi.org/10.1073/pnas.1211295109

Lee KH, Piao HL, Kim H-Y, Choi SM, Jiang F, Hartung W, Hwang I, Kwak JM, Lee IJ, Hwang I. Activation of glucosidase via stress-induced polymerization rapidly increases active pools of abscisic acid. Cell. 2006:**126**(6):1109–1120. https://doi.org/10.1016/j.cell.2006.07.034

Legris M, Klose C, Burgie ES, Rojas CCR, Neme M, Hiltbrunner A, Wigge PA, Schäfer E, Vierstra RD, Casal JJ. Phytochrome B integrates light and temperature signals in Arabidopsis. Science. 2016:**354**(6314):897–900. https://doi.org/10.1126/science.aaf5656

Leivar P, Quail PH. PIFs: pivotal components in a cellular signaling hub. Trends Plant Sci. 2011:**16**(1):19–28. https://doi.org/10.1016/j.tplants.2010.08.003

Lew TTS, Sarojam R, Jang IC, Park BS, Naqvi NI, Wong MH, Singh GP, Ram RJ, Shoseyov O, Saito K, et al. Species-independent analytical tools for next-generation agriculture. Nat Plants. 2020:**6**(12): 1408–1417. https://doi.org/10.1038/s41477-020-00808-7

Li B, Gao Z, Liu X, Sun D, Tang W. Transcriptional profiling reveals a time-of-day-specific role of REVEILLE 4/8 in regulating the first wave of heat shock-induced gene expression in Arabidopsis. Plant Cell. 2019:**31**(10):2353–2369. https://doi.org/10.1105/tpc.19.00519

Li W, Chen Y, Ye M, Lu H, Wang D, Chen Q. Evolutionary history of the C-repeat binding factor/dehydration-responsive element-binding 1 (CBF/DREB1) protein family in 43 plant species and characterization of CBF/DREB1 proteins in Solanum tuberosum. BMC Evol Biol. 2020:**20**(1):142. https://doi.org/10.1186/s12862-020-01710-8

Li X, Ma D, Lu SX, Hu X, Huang R, Liang T, Xu T, Tobin EM, Liu H. Blue light- and low temperature-regulated COR27 and COR28 play roles in the Arabidopsis circadian clock. Plant Cell. 2016:**28**(11): 2755–2769. https://doi.org/10.1105/tpc.16.00354

Li Y, Shi Y, Li M, Fu D, Wu S, Li J, Gong Z, Liu H, Yang S. The CRY2-COP1-HY5-BBX7/8 module regulates blue light-dependent cold acclimation in Arabidopsis. Plant Cell. 2021:**33**(11):3555–3573. https://doi.org/10.1093/plcell/koab215

Lin Z, Li Y, Zhang Z, Liu X, Hsu CC, Du Y, Sang T, Zhu C, Wang Y, Satheesh V, et al. A RAF-SnRK2 kinase cascade mediates early osmotic stress signaling in higher plants. Nat Commun. 2020:**11**(1): 613. https://doi.org/10.1038/s41467-020-14477-9

Liu C, Schläppi MR, Mao B, Wang W, Wang A, Chu C. The bZIP73 transcription factor controls rice cold tolerance at the reproductive stage. Plant Biotech J. 2019:**17**(9):1834–1849. https://doi.org/10.1111/pbi.13104

Liu Q, Ding Y, Shi Y, Ma L, Wang Y, Song C, Wilkins KA, Davies JM, Knight H, Knight MR, et al. The calcium transporter ANNEXIN1 mediates cold-induced calcium signaling and freezing tolerance in plants. EMBO J. 2021:**40**(2):e104559. https://doi.org/10.15252/embj.2020104559

Liu Q, Kasuga M, Sakuma Y, Abe H, Miura S, Yamaguchi-Shinozaki K, Shinozaki K. Two transcription factors, DREB1 and DREB2, with an EREBP/AP2 DNA binding domain separate two cellular signal transduction pathways in drought- and low-temperature-responsive gene expression, respectively, in Arabidopsis. Plant Cell. 1998:**10**(8): 1391–1406. https://doi.org/10.1105/tpc.10.8.1391

Lopez-Molina L, Mongrand S, Chua NH. A postgermination developmental arrest checkpoint is mediated by abscisic acid and requires the ABI5 transcription factor in Arabidopsis. Proc Natl Acad Sci U S A. 2001:**98**(8):4782–4787. https://doi.org/10.1073/pnas.081594298

Lorenzo O, Chico JM, Saénchez-Serrano JJ, Solano R. JASMONATE-INSENSITIVE1 encodes a MYC transcription factor essential to discriminate between different jasmonate-regulated defense responses in Arabidopsis. Plant Cell. 2004:**16**(7):1938–1950. https://doi.org/10.1105/tpc.022319

Lyons JM. Chilling injury in plants. Ann Rev Plant Physiol. 1973:**24**(1): 445–466. https://doi.org/10.1146/annurev.pp.24.060173.002305

Ma Y, Dai X, Xu Y, Luo W, Zheng X, Zeng D, Pan Y, Lin X, Liu H, Zhang D, et al. COLD1 confers chilling tolerance in rice. Cell. 2015:**160**(6):1209–1221. https://doi.org/10.1016/j.cell.2015.01.046

Ma Y, Gil S, Grasser KD, Mas P. Targeted recruitment of the basal transcriptional machinery by LNK clock components controls the circadian rhythms of nascent RNAs in Arabidopsis. Plant Cell. 2018:**30**(4): 907–924. https://doi.org/10.1105/tpc.18.00052

Ma Y, Szostkiewicz I, Korte A, Moes D, Yang Y, Christmann A, Grill E. Regulators of PP2C phosphatase activity function as abscisic acid sensors. Science. 2009:**324**(5930):1064–1068. https://doi.org/10.1126/science.1172408

Marcotte WR Jr, Russell SH, Quatrano RS. Abscisic acid-responsive sequences from the em gene of wheat. Plant Cell. 1989:**1**(10):969–976. https://doi.org/10.1105/tpc.1.10.969

Martínez-Andújar C, Ordiz MI, Huang Z, Nonogaki M, Beachy RN, Nonogaki H. Induction of 9-cis-epoxycarotenoid dioxygenase in Arabidopsis thaliana seeds enhances seed dormancy. Proc Natl Acad Sci U S A. 2011:**108**(41):17225–17229. https://doi.org/10.1073/pnas.1112151108

Maruyama K, Sakuma Y, Kasuga M, Ito Y, Seki M, Goda H, Shimada Y, Yoshida S, Shinozaki K, Yamaguchi-Shinozaki K. Identification of cold-inducible downstream genes of the Arabidopsis DREB1A/CBF3 transcriptional factor using two microarray systems. Plant J. 2004:**38**(6):982–993. https://doi.org/10.1111/j.1365-313X.2004.02100.x

Maruyama K, Takeda M, Kidokoro S, Yamada K, Sakuma Y, Urano K, Fujita M, Yoshiwara K, Matsukura S, Morishita Y, et al. Metabolic pathways involved in cold acclimation identified by integrated analysis of metabolites and transcripts regulated by DREB1A and DREB2A. Plant Physiol. 2009:**150**(4):1972–1980. https://doi.org/10.1104/pp.109.135327

Maruyama K, Todaka D, Mizoi J, Yoshida T, Kidokoro S, Matsukura S, Takasaki H, Sakurai T, Yamamoto YY, Yoshiwara K, et al. Identification of cis-acting promoter elements in cold- and dehydration-induced transcriptional pathways in Arabidopsis, rice, and soybean. DNA Res. 2012:**19**(1):37–49. https://doi.org/10.1093/dnares/dsr040

Matsumura M, Nomoto M, Itaya T, Aratani Y, Iwamoto M, Matsuura T, Hayashi Y, Mori T, Skelly MJ, Yamamoto YY, et al. Mechanosensory trichome cells evoke a mechanical stimuli–induced immune response in Arabidopsis thaliana. Nat Commun. 2022:**13**(1): 1216. https://doi.org/10.1038/s41467-022-28813-8

Mega R, Abe F, Kim JS, Tsuboi Y, Tanaka K, Kobayashi H, Sakata Y, Hanada K, Tsujimoto H, Kikuchi J, et al. Tuning water-use efficiency and drought tolerance in wheat using abscisic acid receptors. Nat Plants. 2019:**5**(2):153–159. https://doi.org/10.1038/s41477-019-0361-8

Melcher K, Ng LM, Zhou XE, Soon FF, Xu Y, Suino-Powell KM, Park SY, Weiner JJ, Fujii H, Chinnusamy V, et al. A gate-latch-lock mechanism for hormone signalling by abscisic acid receptors. Nature. 2009:**462**(7273):602–608. https://doi.org/10.1038/nature08613

Miki Y, Takahashi D, Kawamura Y, Uemura M. Temporal proteomics of Arabidopsis plasma membrane during cold- and de-acclimation. J Proteomics. 2019:**197**:71–81. https://doi.org/10.1016/j.jprot.2018.11.008

Mikkelsen MD, Thomashow MF. A role for circadian evening elements in cold-regulated gene expression in Arabidopsis. Plant J. 2009:**60**(2): 328–339. https://doi.org/10.1111/j.1365-313X.2009.03957.x

Mittler R, Zandalinas SI, Fichman Y, Van Breusegem F. Reactive oxygen species signalling in plant stress responses. Nat Rev Mol Cell Biol. 2022:**23**(10):663–679. https://doi.org/10.1038/s41580-022-00499-2

Miyazono K, Miyakawa T, Sawano Y, Kubota K, Kang HJ, Asano A, Miyauchi Y, Takahashi M, Zhi Y, Fujita Y, et al. Structural basis of abscisic acid signalling. Nature. 2009:**462**(7273):609–614. https://doi.org/10.1038/nature08583

Mizoi J, Kanazawa N, Kidokoro S, Takahashi F, Qin F, Morimoto K, Shinozaki K, Yamaguchi-Shinozaki K. Heat-induced inhibition of phosphorylation of the stress-protective transcription factor DREB2A promotes thermotolerance of Arabidopsis thaliana. J Biol Chem. 2019:**294**(3):902–917. https://doi.org/10.1074/jbc.RA118.002662

Mizoi J, Shinozaki K, Yamaguchi-Shinozaki K. AP2/ERF family transcription factors in plant abiotic stress responses. Biochim Biophys Acta. 2012:**1819**(2):86–96. https://doi.org/10.1016/j.bbagrm.2011.08.004

Mori K, Renhu N, Naito M, Nakamura A, Shiba H, Yamamoto T, Suzaki T, Iida H, Miura K. Ca2+-permeable mechanosensitive channels MCA1 and MCA2 mediate cold-induced cytosolic Ca2+ increase and cold tolerance in Arabidopsis. Sci Rep. 2018:**8**(1):550. https://doi.org/10.1038/s41598-017-17483-y

Morimoto K, Ohama N, Kidokoro S, Mizoi J, Takahashi F, Todaka D, Mogami J, Sato H, Qin F, Kim JS, et al. BPM-CUL3 E3 ligase modulates thermotolerance by facilitating negative regulatory domain-mediated degradation of DREB2A in *Arabidopsis*. Proc Natl Acad Sci U S A. 2017:**114**(40):E8528–E8536. https://doi.org/10.1073/pnas.1704189114

Mundy J, Yamaguchi-Shinozaki K, Chua NH. Nuclear proteins bind conserved elements in the abscisic acid-responsive promoter of a rice rab gene. Proc Natl Acad Sci U S A. 1990:**87**(4):1406–1410. https://doi.org/10.1073/pnas.87.4.1406

Munemasa S, Hirao Y, Tanami K, Mimata Y, Nakamura Y, Murata Y. Ethylene inhibits methyl jasmonate-induced stomatal closure by modulating guard cell slow-type anion channel activity via the OPEN STOMATA 1/SnRK2.6 kinase-independent pathway in Arabidopsis. Plant Cell Physiol. 2019:**60**(10):2263–2271. https://doi.org/10.1093/pcp/pcz121

Mustilli AC, Merlot S, Vavasseur A, Fenzi F, Giraudat J. Arabidopsis OST1 protein kinase mediates the regulation of stomatal aperture by abscisic acid and acts upstream of reactive oxygen species production. Plant Cell. 2002:**14**(12):3089–3099. https://doi.org/10.1105/tpc.007906

Nagel DH, Doherty CJ, Pruneda-Paz JL, Schmitz RJ, Ecker JR, SA K. Genome-wide identification of CCA1 targets uncovers an expanded clock network in Arabidopsis. Proc Natl Acad Sci U S A. 2015:**112**(34): E4802–E4810. https://doi.org/10.1073/pnas.1513609112

Nakagawa Y, Katagiri T, Shinozaki K, Qi Z, Tatsumi H, Furuichi T, Kishigami A, Sokabe M, Kojima I, Sato S, et al. Arabidopsis plasma membrane protein crucial for Ca2+ influx and touch sensing in roots. Proc Natl Acad Sci U S A. 2007:**104**(9):3639–3644. https://doi.org/10.1073/pnas.0607703104

Nakamichi N, Kiba T, Kamioka M, Suzuki T, Yamashino T, Higashiyama T, Sakakibara H, Mizuno T. Transcriptional repressor PRR5 directly regulates clock-output pathways. Proc Natl Acad Sci U S A. 2012:**109**(42):17123–17128. https://doi.org/10.1073/pnas.1205156109

Nakamichi N, Kusano M, Fukushima A, Kita M, Ito S, Yamashino T, Saito K, Sakakibara H, Mizuno T. Transcript profiling of an Arabidopsis PSEUDO RESPONSE REGULATOR arrhythmic triple mutant reveals a role for the circadian clock in cold stress response. Plant Cell Physiol. 2009:**50**(3):447–462. https://doi.org/10.1093/pcp/pcp004

Nakashima K, Fujita Y, Kanamori N, Katagiri T, Umezawa T, Kidokoro S, Maruyama K, Yoshida T, Ishiyama K, Kobayashi M, et al. Three Arabidopsis SnRK2 protein kinases, SRK2D/SnRK2.2, SRK2E/SnRK2.6/OST1 and SRK2I/SnRK2.3, involved in ABA signaling are essential for the control of seed development and dormancy. Plant Cell Physiol. 2009:**50**(7):1345–1363. https://doi.org/10.1093/pcp/pcp083

Née G, Kramer K, Nakabayashi K, Yuan B, Xiang Y, Miatton E, Finkemeier I, Soppe WJJ. DELAY OF GERMINATION1 requires PP2C phosphatases of the ABA signalling pathway to control seed dormancy. Nat Commun. 2017:**8**(1):72. https://doi.org/10.1038/s41467-017-00113-6

Nemali KS, Bonin C, Dohleman FG, Reeves WR, Reeves WR, Nelson DE, Castiglioni P, Whitsel JE, Sammons B, Silady RA, et al. Physiological responses related to increased grain yield under drought in the first biotechnology-derived drought-tolerant maize. Plant Cell Environ. 2015:**38**(9):1866–1880. https://doi.org/10.1111/pce.12446

Nguyen KL, Grondin B, Courtois B, Gantet P. Next-generation sequencing accelerates crop gene discovery. Trend Plant Sci. 2019:**24**(3):264–274. https://doi.org/10.1016/j.tplants.2018.11.008

Nishimura N, Hitomi K, Arvai AS, Rambo RP, Hitomi C, Cutler SR, Schroeder JI, Getzoff ED. Structural mechanism of abscisic acid binding and signaling by dimeric PYR1. Science. 2009:**326**(5958): 1373–1379. https://doi.org/10.1126/science.1181829

Nishimura N, Tsuchiya W, Moresco JJ, Hayashi Y, Satoh K, Kaiwa N, Irisa T, Kinoshita T, Schroeder JI, Yates JR, et al. Control of seed dormancy and germination by DOG1-AHG1 PP2C phosphatase complex via binding to heme. Nat Commun. 2018:**9**(1):2132. https://doi.org/10.1038/s41467-018-04437-9

Nishiyama R, Watanabe Y, Fujita Y, Le DT, Kojima M, Werner T, Vankova R, Yamaguchi-Shinozaki K, Shinozaki K, Kakimoto T, et al. Analysis of cytokinin mutants and regulation of cytokinin metabolic genes reveals important regulatory roles of cytokinins in drought, salt and abscisic acid responses, and abscisic acid biosynthesis. Plant Cell. 2011:**23**(6):2169–2183. https://doi.org/10.1105/tpc.111.087395

Noguchi M, Kodama Y. Temperature sensing in plants: on the Dawn of molecular thermosensor research. Plant Cell Physiol. 2022:**63**(6): 737–743. https://doi.org/10.1093/pcp/pcac033

Nohales MA, Kay SA. Molecular mechanisms at the core of the plant circadian oscillator. Nat Struct Mol Biol. 2016:**23**(12):1061–1069. https://doi.org/10.1038/nsmb.3327

Ohama N, Kusakabe K, Mizoi J, Zhao H, Kidokoro S, Koizumi S, Takahashi F, Ishida T, Yanagisawa S, Shinozaki K, et al. The transcriptional cascade in the heat stress response of Arabidopsis is strictly regulated at the level of transcription factor expression. Plant Cell. 2016:**28**(1):181–201. https://doi.org/10.1105/tpc.15.00435

Okamoto M, Kuwahara A, Seo M, Kushiro T, Asami T, Hirai N, Kamiya Y, Koshiba T, Nambara E. CYP707A1 and CYP707A2, which encode abscisic acid 8′-hydroxylases, are indispensable for proper control of seed dormancy and germination in Arabidopsis. Plant Physiol. 2006:**141**(1):97–107. https://doi.org/10.1104/pp.106.079475

Okamoto M, Peterson FC, Defries A, Park S-Y, Endo A, Nambara E, Volkman BF, Cutler SR. Activation of dimeric ABA receptors elicits guard cell closure, ABA-regulated gene expression, and drought tolerance. Proc Natl Acad Sci U S A. 2013:**110**(29):12132–12137. https://doi.org/10.1073/pnas.1305919110

Ovečka M, Sojka J, Tichá M, Komis G, Basheer J, Marchetti C, Šamajová O, Kuběnová L, Šamaj J, et al. Imaging plant cells and organs with light-sheet and super-resolution microscopy. Plant Physiol. 2022:**188**(2):683–702. https://doi.org/10.1093/plphys/kiab349

Park SY, Fung P, Nishimura N, Jensen DR, Fujii H, Zhao Y, Lumba S, Santiago J, Rodrigues A, Chow TF, et al. Abscisic acid inhibits type 2c protein phosphatases via the PYR/PYL family of START proteins. Science. 2009:**324**(5930):1068–1071. https://doi.org/10.1126/science.1173041

Pei ZM, Murata Y, Benning G, Thomine S, Klüsener B, Allen GJ, Grill E, Schroeder JI. Calcium channels activated by hydrogen peroxide mediate abscisic acid signalling in guard cells. Nature. 2000:**406**(6797):731–734. https://doi.org/10.1038/35021067

Piotrowska A, Bajguz A. Conjugates of abscisic acid, brassinosteroids, ethylene, gibberellins, and jasmonates. Phytochem. 2011:**72**(17): 2097–2112. https://doi.org/10.1016/j.phytochem.2011.08.012

Plieth C, Hansen UP, Knight H, Knight MR. Temperature sensing by plants: the primary characteristics of signal perception and calcium response. Plant J. 1999:**18**(5):491–497. https://doi.org/10.1046/j.1365-313X.1999.00471.x

Prerostova S, Dobrev PI, Knirsch V, Jarosova J, Gaudinova A, Zupkova B, Prášil IT, Janda T, Brzobohatý B, Skalák J, et al. Light quality and intensity modulate cold acclimation in Arabidopsis. Int J Mol Sci. 2021:**22**(5):2736. https://doi.org/10.3390/ijms22052736

Qin F, Sakuma Y, Tran LS, Maruyama K, Kidokoro S, Fujita Y, Fujita M, Umezawa T, Sawano Y, Miyazono K, et al. Arabidopsis DREB2A-interacting proteins function as RING E3 ligases and negatively regulate plant drought stress-responsive gene expression. Plant Cell. 2008:**20**(6):1693–1707. https://doi.org/10.1105/tpc.107.057380

Ren X, Chen Z, Liu Y, Zhang H, Zhang M, Liu Q, Hong X, Zhu JK, Gong Z. ABO3, a WRKY transcription factor, mediates plant responses to abscisic acid and drought tolerance in Arabidopsis. Plant J. 2010:**63**(3): 417–429. https://doi.org/10.1111/j.1365-313X.2010.04248.x

Rugnone ML, Faigón Soverna A, Sanchez SE, Schlaen RG, Hernando CE, Seymour DK, Mancini E, Chernomoretz A, Weigel D, Más P, et al. LNK genes integrate light and clock signaling networks at the core of the Arabidopsis oscillator. Proc Natl Acad Sci U S A. 2013:**110**(29):12120–12125. https://doi.org/10.1073/pnas.1302170110

Saint Pierre C, Crossa JL, Bonnett D, Yamaguchi-Shinozaki K, Reynolds MP. Phenotyping transgenic wheat for drought resistance. J Exp Bot. 2012:**63**(5):1799–1808. https://doi.org/10.1093/jxb/err385

Saito S, Hirai N, Matsumoto C, Ohigashi H, Ohta D, Sakata K, Mizutani M. Arabidopsis CYP707As encode (+)-abscisic acid 8′-hydroxylase, a key enzyme in the oxidative catabolism of abscisic acid. Plant Physiol. 2004:**134**(4):1439–1449. https://doi.org/10.1104/pp.103.037614

Sakamoto H, Maruyama K, Sakuma Y, Meshi T, Iwabuchi M, Shinozaki K, Yamaguchi-Shinozaki K. Arabidopsis Cys2/His2-type zinc-finger proteins function as transcription repressors under drought, cold, and high-salinity stress conditions. Plant Physiol. 2004:**136**(1):2734–2746. https://doi.org/10.1104/pp.104.046599

Sakuma Y, Maruyama K, Osakabe Y, Qin F, Seki M, Shinozaki K, Yamaguchi-Shinozaki K. Functional analysis of an Arabidopsis transcription factor, DREB2A, involved in drought-responsive gene expression. Plant Cell. 2006a:**18**(5):1292–1309. https://doi.org/10.1105/tpc.105.035881

Sakuma Y, Maruyama K, Qin F, Osakabe Y, Shinozaki K, Yamaguchi-Shinozaki K. Dual function of an Arabidopsis transcription factor DREB2A in water-stress-responsive and heat-stress-responsive gene expression. Proc Natl Acad Sci U S A. 2006b:**103**(49):18822–18827. https://doi.org/10.1073/pnas.0605639103

Santiago J, Dupeux F, Round A, Antoni R, Park SY, Jamin M, Cutler SR, Rodriguez PL, Márquez JA. The abscisic acid receptor PYR1 in complex with abscisic acid. Nature. 2009:**462**(7273):665–668. https://doi.org/10.1038/nature08591

Saruhashi M, Ghosh TK, Arai K, Ishizaki Y, Hagiwara K, Komatsu K, Shiwa Y, Izumikawa K, Yoshikawa H, Umezawa T, et al. Plant Raf-like kinase integrates abscisic acid and hyperosmotic stress signaling upstream of SNF1-related protein kinase2. Proc Natl Acad Sci U S A. 2015:**112**(46):E6388–E6396. https://doi.org/10.1073/pnas.1511238112

Sato H, Takasaki H, Takahashi F, Suzuki T, Iuchi S, Mitsuda N, Ohme-Takagi M, Ikeda M, Seo M, Yamaguchi-Shinozaki K, et al. Arabidopsis thaliana NGATHA1 transcription factor induces ABA biosynthesis by activating NCED3 gene during dehydration stress. Proc Natl Acad Sci U S A. 2018:**115**(47):E11178–E11187. https://doi.org/10.1073/pnas.1811491115

Schweighofer A, Hirt H, Meskiene I. Plant PP2C phosphatases: emerging functions in stress signaling. Trends Plant Sci. 2004:**9**(5):236–243. https://doi.org/10.1016/j.tplants.2004.03.007

Seki M, Narusaka M, Abe H, Kasuga M, Yamaguchi-Shinozaki K, Carninci P, Hayashizaki Y, Shinozaki K. Monitoring the expression pattern of 1300 Arabidopsis genes under drought and cold stresses by using a full-length cDNA microarray. Plant Cell. 2001:**13**(1): 61–72. https://doi.org/10.1105/tpc.13.1.61

Selvaraj MG, Ishizaki T, Valencia M, Ogawa S, Dedicova B, Ogata T, Yoshiwara K, Maruyama K, Kusano M, Saito K, et al. Overexpression of an Arabidopsis thaliana galactinol synthase gene improves drought tolerance in transgenic rice and increased grain yield in the field. Plant Biotech J. 2017:**15**(11):1465–1477. https://doi.org/10.1111/pbi.12731

Seyfferth C, Renema J, Wendrich JR, Eekhout T, Seurinck R, Vandamme N, Blob B, Saeys Y, Helariutta Y, Birnbaum KD, et al. Advances and opportunities in single-cell transcriptomics for plant research. Annu Rev Plant Biol. 2021:**72**(1):847–866. https://doi.org/10.1146/annurev-arplant-081720-010120

Shang Y, Yan L, Liu ZQ, Cao Z, Mei C, Xin Q, Wu FQ, Wang XF, Du SY, Jiang T, et al. The mg-chelatase H subunit of Arabidopsis antagonizes a group of WRKY transcription repressors to relieve ABA-responsive genes of inhibition. Plant Cell. 2010:**22**(6): 1909–1935. https://doi.org/10.1105/tpc.110.073874

Shaw R, Tian X, Xu J. Single-cell transcriptome analysis in plants: advances and challenges. Mol Plant. 2020:**14**(1):115–126. https://doi.org/10.1016/j.molp.2020.10.012

Shi L, Su J, Cho MJ, Song H, Dong X, Liang Y, Zhang Z. Promoter editing for the genetic improvement of crops. J Exp Bot. 2023:**74**(15): 4349–4366. https://doi.org/10.1093/jxb/erad175

Shinwari ZK, Nakashima K, Miura S, Kasuga M, Seki M, Yamaguchi-Shinozaki K, Shinozaki K. An Arabidopsis gene family encoding DRE/CRT binding proteins involved in low-temperature-responsive gene expression. Biochem Biophys Res Commun. 1998:**250**(1): 161–170. https://doi.org/10.1006/bbrc.1998.9267

Silva CS, Nayak A, Lai X, Hutin S, Hugouvieux V, Jung JH, López-Vidriero I, Franco-Zorrilla JM, Panigrahi KCS, Nanao MH, et al. Molecular mechanisms of evening complex activity in Arabidopsis. Proc Natl Acad Sci U S A. 2020:**117**(12):6901–6909. https://doi.org/10.1073/pnas.1920972117

Singh A, Jones S, Ganapathysubramarian B, Sarkar S, Mueller D, Sandhu K, Nagasubramanian K, et al. Challenges and opportunities in machine-augmented plant stress phenotyping. Trends Plant Sci. 2021:**26**(1):53–69. https://doi.org/10.1016/j.tplants.2020.07.010

Soma F, Mogami J, Yoshida T, Abekura M, Takahashi F, Kidokoro S, Mizoi J, Shinozaki K, Yamaguchi-Shinozaki K. ABA-unresponsive SnRK2 protein kinases regulate mRNA decay under osmotic stress in plants. Nat Plants. 2017:**3**(1):16204. https://doi.org/10.1038/nplants.2016.204

Soma F, Takahashi F, Kidokoro S, Kameoka H, Suzuki T, Uga Y, Shinozaki K, Yamaguchi-Shinozaki K. Constitutively active B2 Raf-like kinases are required for drought-responsive gene expression upstream of ABA-activated SnRK2 kinases. Proc Natl Acad Sci U S A. 2023:**120**(24):e2221863120. https://doi.org/10.1073/pnas.2221863120

Soma F, Takahashi F, Suzuki T, Shinozaki K, Yamaguchi-Shinozaki K. Plant Raf-like kinases regulate the mRNA population upstream of ABA-unresponsive SnRK2 kinases under drought stress. Nat Commun. 2020:**11**(1):1373. https://doi.org/10.1038/s41467-020-15239-3

Sorkin ML, Tzeng SC, King S, Romanowski A, Kahle N, Bindbeutel R, Hiltbrunner A, Yanovsky MJ, Evans BS, Nusinow DA. COLD REGULATED GENE 27 and 28 antagonize the transcriptional activity of the RVE8/LNK1/LNK2 circadian complex. Plant Physiol. 2023:**192**(3):2436–2456. https://doi.org/10.1093/plphys/kiad210

Stockinger EJ, Gilmour SJ, Thomashow MF. Arabidopsis thaliana CBF1 encodes an AP2 domain-containing transcriptional activator that binds to the C-repeat/DRE, a cis-acting DNA regulatory element that stimulates transcription in response to low temperature and water deficit. Proc Natl Acad Sci U S A. 1997:**94**(3):1035–1040. https://doi.org/10.1073/pnas.94.3.1035

Suzuki N, Miller G, Salazar C, Mondal HA, Shulaev E, Cortes DF, Shuman JL, Luo X, Shah J, Schlauch K, et al. Temporal-spatial interaction between reactive oxygen species and abscisic acid regulates rapid systemic acclimation in plants. Plant Cell. 2013:**25**(9):3553–3569. https://doi.org/10.1105/tpc.113.114595

Taji T, Ohsumi C, Iuchi S, Seki M, Kasuga M, Kobayashi M, Yamaguchi-Shinozaki K, Shinozaki K. Important roles of drought- and cold-inducible genes for galactinol synthase in stress tolerance in Arabidopsis thaliana. Plant J. 2002:**29**(4):417–426. https://doi.org/10.1046/j.0960-7412.2001.01227.x

Takahashi D, Gorka M, Erban A, Graf A, Kopka J, Zuther E, Hincha DK. Both cold and sub-zero acclimation induce cell wall modification and changes in the extracellular proteome in Arabidopsis thaliana. Sci Rep. 2019:**9**(1):2289. https://doi.org/10.1038/s41598-019-38688-3

Takahashi D, Willick IR, Kasuga J, Livingston III DP. Responses of the plant cell wall to sub-zero temperatures: a brief update. Plant Cell Physiol. 2021:**62**(12):1858–1866. https://doi.org/10.1093/pcp/pcab103

Takahashi F, Suzuki T, Osakabe Y, Betsuyaku S, Kondo Y, Dohmae N, Fukuda H, Yamaguchi-Shinozaki K, Shinozaki K. A small peptide modulates stomatal control via abscisic acid in long-distance signaling. Nature. 2018:**556**(7700):235–238. https://doi.org/10.1038/s41586-018-0009-2

Takasaki H, Maruyama K, Takahashi F, Fujita M, Yoshida T, Nakashima K, Myouga F, Toyooka K, Yamaguchi-Shinozaki K, Shinozaki K. SNAC-As, stress-responsive NAC transcription factors, mediate ABA-inducible leaf senescence. Plant J. 2015:**84**(6): 1114–1123. https://doi.org/10.1111/tpj.13067

Takahashi Y, Zhang J, Hsu P-K, Ceciliato PHO, Zhang L, Dubeaux G, Munemasa S, Ge C, Zhao Y, Hauser F, et al. MAP3Kinase-dependent SnRK2-kinase activation is required for abscisic acid signal transduction and rapid osmotic stress response. Nat Commun. 2020:**11**(1):12. https://doi.org/10.1038/s41467-019-13875-y

Tan BC, Joseph LM, Deng W-T, Liu L, Li QB, Cline K, McCarty DR. Molecular characterization of the Arabidopsis 9-cis epoxycarotenoid dioxygenase gene family. Plant J. 2003:**35**(1):44–56. https://doi.org/10.1046/j.1365-313X.2003.01786.x

Tan BC, Schwartz SH, Zeevaart JAD, McCarty DR. Genetic control of abscisic acid biosynthesis in maize. Proc Natl Acad Sci U S A. 1997:**94**(22):12235–12240. https://doi.org/10.1073/pnas.94.22.12235

The Arabidopsis Genome Initiative. Analysis of the genome sequence of the flowering plant Arabidopsis thaliana. Nature. 2000:**408**(6814): 796–815. https://doi.org/10.1038/35048692

Thomashow MF. PLANT COLD ACCLIMATION: freezing tolerance genes and regulatory mechanisms. Annu Rev Plant Physiol Plant Mol Biol. 1999:**50**(1):571–599. https://doi.org/10.1146/annurev.arplant.50.1.571

Thomashow MF. Molecular basis of plant cold acclimation: insights gained from studying the CBF cold response pathway. Plant Physiol. 2010:**154**(2):571–577. https://doi.org/10.1104/pp.110.161794

Thomashow MF, Torii KU. SCREAMing twist on the role of ICE1 in freezing tolerance. Plant Cell. 2020:**32**(4):816–819. https://doi.org/10.1105/tpc.20.00124

Toriyama T, Shinozawa A, Yasumura Y, Saruhashi M, Hiraide M, Ito S, Matsuura H, Kuwata K, Yoshida M, Baba T, et al. Sensor histidine kinases mediate ABA and osmostress signaling in the moss Physcomitrium patens. Curr Biol. 2022:**32**(1):164–175.e8. https://doi.org/10.1016/j.cub.2021.10.068

Toyota M, Spencer D, Sawai-Toyota S, Jiaqi W, Zhang T, Koo AJ, Howe GA, Gilroy S. Glutamate triggers long-distance, calcium-based plant defense signaling. Science. 2018:**361**(6407):1112–1115. https://doi.org/10.1126/science.aat7744

Tran L-SP, Nakashima K, Sakuma Y, Simpson SD, Fujita Y, Maruyama K, Fujita M, Seki M, Shinozaki K, Yamaguchi-Shinozaki K. Isolation and functional analysis of Arabidopsis stress-inducible NAC transcription factors that bind to a drought-responsive cis-element in the early responsive to dehydration stress 1 promoter. Plant Cell. 2004:**16**(9): 2481–2498. https://doi.org/10.1105/tpc.104.022699

Tran L-SP, Urao T, Qin F, Maruyama K, Kakimoto T, Shinozaki K, Yamaguchi-Shinozaki K. Functional analysis of AHK1/ATHK1 and cytokinin receptor histidine kinases in response to abscisic acid, drought, and salt stress in Arabidopsis. Proc Natl Acad Sci U S A. 2007:**104**(51):20623–20628. https://doi.org/10.1073/pnas.0706547105

Uga Y, Sugimoto K, Ogawa S, Rane J, Ishitani M, Hara N, Kitomi Y, Inukai Y, Ono K, Kanno N, et al. Control of root system architecture by DEEPER ROOTING 1 increases rice yield under drought conditions. Nat Genet. 2013:**45**(9):1097–1102. https://doi.org/10.1038/ng.2725

Umezawa T, Okamoto M, Kushiro T, Nambara E, Oono Y, Seki M, Kobayashi M, Koshiba T, Kamiya Y, Shinozaki K. CYP707A3, a major ABA 8′-hydroxylase involved in dehydration and rehydration response in Arabidopsis thaliana. Plant J. 2006:**46**(2):171–182. https://doi.org/10.1111/j.1365-313X.2006.02683.x

Umezawa T, Sugiyama N, Mizoguchi M, Hayashi S, Myouga F, Yamaguchi-Shinozaki K, Ishihama Y, Hirayama T, Shinozaki K. Type 2C protein phosphatases directly regulate abscisic acid-activated protein kinases in Arabidopsis. Proc Natl Acad Sci U S A. 2009:**106**(41): 17588–17593. https://doi.org/10.1073/pnas.0907095106

Uno Y, Furihata T, Abe H, Yoshida R, Shinozaki K, Yamaguchi-Shinozaki K. Arabidopsis basic leucine zipper transcription factors involved in an abscisic acid-dependent signal transduction pathway under drought and high-salinity conditions. Proc Natl Acad Sci U S A. 2000:**97**(21):11632–11637. https://doi.org/10.1073/pnas.190309197

Urano K, Maruyama K, Ogata Y, Morishita Y, Takeda M, Sakurai N, Suzuki H, Saito K, Shibata D, Kobayashi M, et al. Characterization of the ABA-regulated global responses to dehydration in Arabidopsis by metabolomics. Plant J. 2009:**57**(6):1065–1078. https://doi.org/10.1111/j.1365-313X.2008.03748.x

Urao T, Yakubov B, Satoh R, Yamaguchi-Shinozaki K, Seki M, Hirayama T, Shinozaki K. A transmembrane hybrid-type histidine kinase in Arabidopsis functions as an osmosensor. Plant Cell. 1999:**11**(9):1743–1754. https://doi.org/10.1105/tpc.11.9.1743

Vaidya AS, Helander JDM, Peterson FC, Elzinga D, Dejonghe W, Kaundal A, Park SY, Xing Z, Mega R, Takeuchi J, et al. Dynamic control of plant water use using designed ABA receptor agonists. Science. 2019:**366**(6464):eaaw8848. https://doi.org/10.1126/science.aaw8848

Verslues PE, Bailey-Serres J, Brodersen C, Buckley TN, Conti L, Christmann A, Dinneny JR, Grill E, Hayes S, Heckman RW, et al. Burning questions for a warming and changing world: 15 unknowns in plant abiotic stress. Plant Cell. 2023:**35**(1):67–108. https://doi.org/10.1093/plcell/koac263

Vlad F, Rubio S, Rodrigues A, Sirichandra C, Belin C, Robert N, Leung J, Rodriguez PL, Laurière C, Merlot S. Protein phosphatases 2C regulate the activation of the snf1-related kinase OST1 by abscisic acid in Arabidopsis. Plant Cell. 2009:**21**(10):3170–3184. https://doi.org/10.1105/tpc.109.069179

Vogel JT, Zarka DG, Van Buskirk HA, Fowler SG, Thomashow MF. Roles of the CBF2 and ZAT12 transcription factors in configuring the low temperature transcriptome of Arabidopsis. Plant J. 2005:**41**(2): 195–211. https://doi.org/10.1111/j.1365-313X.2004.02288.x

Waadt R, Seller CA, Hsu PK, Takahashi Y, Munemasa S, Schroeder JI. Plant hormone regulation of abiotic stress responses. Nat Rev Mol Cell Biol. 2022:**23**(10):680–694. https://doi.org/10.1038/s41580-022-00479-6

Wang C, Gong B, Bushel PR, Thierry-Mieg J, Thierry-Mieg D, Xu J, Fang H, Hong H, Shen J, Su Z, et al. The concordance between RNA-seq and microarray data depends on chemical treatment and transcript abundance. Nat Biotechnol. 2014:**32**(9):926–932. https://doi.org/10.1038/nbt.3001

Wang F, Chen X, Dong S, Jiang X, Wang L, Yu J, Zhou Y. Crosstalk of PIF4 and DELLA modulates CBF transcript and hormone homeostasis in cold response in tomato. Plant Biotechnol J. 2020a:**18**(4): 1041–1055. https://doi.org/10.1111/pbi.13272

Wang F, Guo Z, Li H, Wang M, Onac E, Zhou J, Xia X, Shi K, Yu J, Zhou Y. Phytochrome A and B function antagonistically to regulate cold tolerance via abscisic acid-dependent jasmonate signaling. Plant Physiol. 2016:**170**(1):459–471. https://doi.org/10.1104/pp.15.01171

Wang F, Liu Y, Shi Y, Han D, Wu Y, Ye W, Yang H, Li G, Cui F, Wan S, et al. SUMOylation stabilizes the transcription factor DREB2A to improve plant thermotolerance. Plant Physiol. 2020b:**183**(1):41–50. https://doi.org/10.1104/pp.20.00080

Waszczak C, Carmody M, Kangasjärvi J. Reactive oxygen species in plant signaling. Annu Rev Plant Biol. 2018:**69**(1):209–236. https://doi.org/10.1146/annurev-arplant-042817-040322

Wilkinson S, Clephan AL, Davies WJ. Rapid low temperature-induced stomatal closure occurs in cold-tolerant *Commelina communis* leaves but not in cold-sensitive tobacco leaves, via a mechanism that involves apoplastic calcium but not abscisic acid. Plant Physiol. 2001:**126**(4):1566–1578. https://doi.org/10.1104/pp.126.4.1566

Wohlbach DJ, Quirino BF, Sussman MR. Analysis of the Arabidopsis histidine kinase ATHK1 reveals a connection between vegetative osmotic stress sensing and seed maturation. Plant Cell. 2008:**20**(4): 1101–1117. https://doi.org/10.1105/tpc.107.055871

Wu F, Chi Y, Jiang Z, Xu Y, Xie L, Huang F, Wan D, Ni J, Yuan F, Wu X, et al. Hydrogen peroxide sensor HPCA1 is an LRR receptor kinase in Arabidopsis. Nature. 2020:**578**(7796):577–581. https://doi.org/10.1038/s41586-020-2032-3

Xiao BZ, Chen Z, Xiang CB, Tang N, Zhang QF, Xiong LZ. Evaluation of seven function-known candidate genes for their effects on improving drought resistance of transgenic rice under field conditions. Mol Plant. 2009:**2**(1):73–83. https://doi.org/10.1093/mp/ssn068

Xie Q, Wang P, Liu X, Yuan L, Wang L, Zhang C, Li Y, Xing H, Zhi L, Yue Z, et al. LNK1 and LNK2 are transcriptional coactivators in the Arabidopsis circadian oscillator. Plant Cell. 2014:**26**(7):2843–2857. https://doi.org/10.1105/tpc.114.126573

Xu ZY, Kim SY, Hyeon DY, Kim DH, Dong T, Park Y, Jin JB, Joo SH, Kim SK, Hong JC, et al. The Arabidopsis NAC transcription factor ANAC096 cooperates with bZIP-type transcription factors in dehydration and osmotic stress responses. Plant Cell. 2013:**25**(11): 4708–4724. https://doi.org/10.1105/tpc.113.119099

Yamaguchi-Shinozaki K, Koizumi M, Urao S, Shinozaki K. Molecular cloning and characterization of 9 cDNAs for genes that are responsive to desiccation in Arabidopsis thaliana: sequence analysis of one cDNA clone that encodes a putative transmembrane channel protein. Plant Cell Physiol. 1992:**33**(3):217–224. https://doi.org/10.1093/oxfordjournals.pcp.a078243

Yamaguchi-Shinozaki K, Shinozaki K. A novel cis-acting element in an Arabidopsis gene is involved in responsiveness to drought, low-temperature, or high-salt stress. Plant Cell. 1994:**6**(2):251–264. https://doi.org/10.1105/tpc.6.2.251

Yamaguchi-Shinozaki K, Shinozaki K. Transcriptional regulatory networks in cellular responses and tolerance to dehydration and cold stresses. Annu Rev Plant Biol. 2006:**57**(1):781–803. https://doi.org/10.1146/annurev.arplant.57.032905.105444

Yang T, Chaudhuri S, Yang L, Du L, Poovaiah BW. A calcium/calmodulin-regulated member of the receptor-like kinase family confers cold tolerance in plants. J Biol Chem. 2010a:**285**(10):7119–7126. https://doi.org/10.1074/jbc.M109.035659

Yang T, Shad Ali G, Yang L, Du L, Reddy ASN, Poovaiah BW. Calcium/calmodulin-regulated receptor-like kinase CRLK1 interacts with MEKK1 in plants. Plant Signal Behav. 2010b:**5**(8):991–994. https://doi.org/10.4161/psb.5.8.12225

Yang Z, Liu J, Tischer SV, Christmann A, Windisch W, Schnyder H, Grill E. Leveraging abscisic acid receptors for efficient water use in Arabidopsis. Proc Natl Acad Sci U S A. 2016:**113**(24):6791–6796. https://doi.org/10.1073/pnas.1601954113

Ye H, Liu S, Tang B, Chen J, Xie Z, Nolan TM, Jiang H, Guo H, Lin HY, Li L, et al. RD26 mediates crosstalk between drought and brassinosteroid signalling pathways. Nat Commun. 2017:**8**(1):14573. https://doi.org/10.1038/ncomms14573

Yoshida T, Fernie AR. Hormonal regulation of plant primary metabolism under drought. J Exp Bot. 2023:erad358. Online ahead of print. https://doi.org/10.1093/jxb/erad358

Yoshida T, Fujita Y, Maruyama K, Mogami J, Todaka D, Shinozaki K, Yamaguchi-Shinozaki K. Four Arabidopsis AREB/ABF transcription factors function predominantly in gene expression downstream of SnRK2 kinases in abscisic acid signalling in response to osmotic stress. Plant Cell Environ. 2015:**38**(1):35–49. https://doi.org/10.1111/pce.12351

Yoshida T, Fujita Y, Sayama H, Kidokoro S, Maruyama K, Mizoi J, Shinozaki K, Yamaguchi-Shinozaki K. AREB1, AREB2, and ABF3 are master transcription factors that cooperatively regulate ABRE-dependent ABA signaling involved in drought stress tolerance and require ABA for full activation. Plant J. 2010:**61**(4):672–685. https://doi.org/10.1111/j.1365-313X.2009.04092.x

Yoshida R, Hobo T, Ichimura K, Mizoguchi T, Takahashi F, Aronso J, Ecker JR, Shinozaki K. ABA-activated SnRK2 protein kinase is required for dehydration stress signaling in Arabidopsis. Plant Cell Physiol. 2002:**43**(12):1473–1483. https://doi.org/10.1093/pcp/pcf188

Yoshida T, Sakuma Y, Todaka D, Maruyama K, Qin F, Mizoi J, Kidokoro S, Fujita Y, Shinozaki K, Yamaguchi-Shinozaki K. Functional analysis of an Arabidopsis heat-shock transcription factor HsfA3 in the transcriptional cascade downstream of the DREB2A stress-regulatory system. Biochem Biophys Res Commun. 2008:**368**(3):515–521. https://doi.org/10.1016/j.bbrc.2008.01.134

Yoshida R, Umezawa T, Mizoguchi T, Takahashi S, Takahashi F, Shinozaki K. The regulatory domain of SRK2E/OST1/SnRK2.6 interacts with ABI1 and integrates abscisic acid (ABA) and osmotic stress signals controlling stomatal closure in Arabidopsis. J Biol Chem. 2006:**281**(8):5310–5318. https://doi.org/10.1074/jbc.M509820200

Yoshimura K, Iida K, Iida H. MCAs in Arabidopsis are Ca2+-permeable mechanosensitive channels inherently sensitive to membrane tension. Nat Commun. 2021:**12**(1):6074. https://doi.org/10.1038/s41467-021-26363-z

Yuan F, Yang H, Xue Y, Kong D, Ye R, Li C, Zhang J, Theprungsirikul L, Shrift T, Krichilsky B, et al. OSCA1 mediates osmotic-stress-evoked Ca2+ increases vital for osmosensing in Arabidopsis. Nature. 2014:**514**(7522):367–371. https://doi.org/10.1038/nature13593

Yuan P, Yang T, Poovaiah BW. Calcium signaling-mediated plant response to cold stress. Int J Mol Sci. 2018:**19**(12):3896. https://doi.org/10.3390/ijms19123896

Zandalinas SI, Fichman Y, Devireddy AR, Sengupta S, Azad RK, Mittler R. Systemic signaling during abiotic stress combination in plants. Proc Natl Acad Sci U S A. 2020:**117**(24):13810–13820. https://doi.org/10.1073/pnas.2005077117

Zeng R, Li Z, Shi Y, Fu D, Yin P, Cheng J, Jiang C, Yang S. Natural variation in a type-A response regulator confers maize chilling tolerance. Nat Commun. 2021:**12**(1):4713. https://doi.org/10.1038/s41467-021-25001-y

Zhang H, Zhu H, Pan Y, Yu Y, Luan S, Li L. A DTX/MATE-type transporter facilitates abscisic acid efflux and modulates ABA sensitivity and drought tolerance in Arabidopsis. Mol Plant. 2014:**7**(10): 1522–1532. https://doi.org/10.1093/mp/ssu063

Zhang X, Fowler SG, Cheng H, Lou Y, Rhee SY, Stockinger EJ, Thomashow MF. Freezing-sensitive tomato has a functional CBF cold response pathway, but a CBF regulon that differs from that of freezing-tolerant Arabidopsis. Plant J. 2004:**39**(6):905–919. https://doi.org/10.1111/j.1365-313X.2004.02176.x

Zhao C, Zhang Z, Xie S, Si T, Li Y, Zhu JK. Mutational evidence for the critical role of CBF transcription factors in cold acclimation in Arabidopsis. Plant Physiol. 2016:**171**(4):2744–2759. https://doi.org/10.1104/pp.16.00533

Zhu JK. Abiotic stress signaling and responses in plants. Cell. 2016:**167**(2):313–324. https://doi.org/10.1016/j.cell.2016.08.029

https://doi.org/10.1093/plphys/kiae109 PLANT PHYSIOLOGY 2024: 195: 190–212

Plant Physiology®

Topical Review

Flowering time: From physiology, through genetics to mechanism

Robert Maple [1,†] Pan Zhu [1,†] Jo Hepworth [2] Jia-Wei Wang [3–5,*] Caroline Dean [1,*]

1 Department of Cell and Developmental Biology, John Innes Centre, Norwich Research Park, Norwich, NR4 7UH, UK
2 Department of Biosciences, Durham University, Stockton Road, Durham, DH1 3LE, UK
3 National Key Laboratory of Plant Molecular Genetics (NKLPMG), CAS Center for Excellence in Molecular Plant Sciences (CEMPS), Institute of Plant Physiology and Ecology (SIPPE), Chinese Academy of Sciences (CAS), Shanghai 200032, China
4 School of Life Science and Technology, Shanghai Tech University, Shanghai 201210, China
5 New Cornerstone Science Laboratory, Shanghai 200032, China

*Author for correspondence: jwwang@sippe.ac.cn (J.W.W); caroline.dean@jic.ac.uk (C.D.)
†These authors contributed equally to this work.

Abstract

Plant species have evolved different requirements for environmental/endogenous cues to induce flowering. Originally, these varying requirements were thought to reflect the action of different molecular mechanisms. Thinking changed when genetic and molecular analysis in *Arabidopsis thaliana* revealed that a network of environmental and endogenous signaling input pathways converge to regulate a common set of "floral pathway integrators." Variation in the predominance of the different input pathways within a network can generate the diversity of requirements observed in different species. Many genes identified by flowering time mutants were found to encode general developmental and gene regulators, with their targets having a specific flowering function. Studies of natural variation in flowering were more successful at identifying genes acting as nodes in the network central to adaptation and domestication. Attention has now turned to mechanistic dissection of flowering time gene function and how that has changed during adaptation. This will inform breeding strategies for climate-proof crops and help define which genes act as critical flowering nodes in many other species.

Introduction

Flowering is a major developmental transition in the life cycle of a plant. Correct timing of this process has a huge impact on reproductive success and has thus been of central importance in plant breeding. Different strategies have evolved to ensure correct timing for successful outcrossing, alignment of flowering with pollinators, and sexual reproduction in favorable external conditions (reviewed in Lee et al. 2023). However, different critical factors influence the timing of the transition when environmental conditions change and flowering is a last defense against acute heat/drought stress, as seeds are more likely to survive (reviewed in Takeno 2016).

Early studies focused on flowering physiology—production of transmissible signals from the leaf to the apex, and changes in reproductive competence of the meristem (reviewed in Bernier et al. 1993). There were 3 major theories for flowering control: the "florigen/antiflorigen" concept (Lang 1984), envisioning a specific floral promoter and inhibitor; the "nutrient diversion" hypothesis (Sachs and Hackett 1983), where modification of source/sink relationships in inductive conditions resulted in the shoot apex receiving a better supply of assimilates; and finally, the "multifactorial control" theory (Bernier et al. 1981; Bernier 1988) postulated that multiple signals involving chemicals, assimilates, and phytohormones synergize to induce the floral transition. These early studies still make good reading and suggest many new molecular experiments.

The *Arabidopsis* genetic revolution then transformed thinking: *Arabidopsis thaliana* mutants (Koornneef et al. 1991;

Received October 12, 2023. Accepted February 12, 2024. Advance access publication February 28, 2024.

ADVANCES

- Forward genetics screens in *Arabidopsis thaliana* identified many flowering time mutants and established molecular understanding of the genetic pathways mediating the floral transition.
- Multiple pathways converge on a common set of genes specifying the floral meristem identity, so multiple environmental and endogenous cues can be integrated into the floral transition.
- Natural selection has targeted different nodes of the floral network to adapt flowering time to new environments. Nature has done the mutagenesis for us.
- These have also been selected in the domestication of crop species, although in each species a different node or pathway has been predominantly selected and researched.
- Combining field studies with more accurate simulation in the laboratory is key to understanding flowering time control in natural environments.

Koornneef et al. 1998) with altered flowering time revealed an integrated network of environmental and endogenous input pathways. These converge to quantitatively control the expression of floral pathway integrators, which when expressed above a certain threshold trigger the transition to flowering (Koornneef et al. 1991; Chandler et al. 1996; Koornneef et al. 1998). This then provided the conceptual framework to explain the diversity of physiological requirements in different species. Although much of our understanding of flowering time has come from studies on *A. thaliana*, significant progress has also been made in other species such as rice, wheat, and the model temperate grass *Brachypodium* (Higgins et al. 2010; Tsuji et al. 2013; Cao et al. 2021). We refer the reader to the following reviews (Osnato et al. 2021; Quiroz et al. 2021; Li et al. 2022a) that cover the extensive recent literature on flowering time control. In this short review we focus on how the molecular understanding of environmental signal integration fits into thinking from earlier physiological analyses. We also discuss how current knowledge may facilitate breeding in crops and pose questions for future research.

Arabidopsis: the Rosetta stone?

How did *Arabidopsis* genetics change thinking on flowering time regulation? Forward genetics analyses in *A. thaliana* identified many flowering time mutants (Koornneef et al. 1991). These mutants were categorized into pathways mediating environmental and endogenous cues that promoted the floral transition: maturity, photoperiod, autonomous, vernalization, light quality, and hormonal pathways (Fig. 1). These pathways converge to regulate a common set of genes known as "floral pathway integrators." These include FLOWERING LOCUS T (FT) and SUPPRESSOR OF OVEREXPRESSION OF CONSTANS1 (SOC1). These in turn regulate the expression of the floral meristem identity genes such as *APETALA1* (*AP1*), *AP2*, *FRUITFULL* (*FUL*), *CAULIFLOWER* (*CAL*), and *LEAFY* (*LFY*). Over time, more and more genes (over 300) have been added to this complex interconnected network of *Arabidopsis* flowering time regulators (Bouche et al. 2016). Such a regulatory network can explain how flowering time can vary in different conditions, how different pathways are predominant in different seasons, and, together with analysis of shoot maturation, how diversity of reproductive strategies evolve. It also enabled the previous physiological understanding of the floral transition—production of transmissible signals from the leaf to the apex and changes in meristem competence—to be described in genetic terms.

Transmissible floral-promoting signals

Transmissible signals are key to photoperiodic control of flowering. Systemic signaling mechanisms involving long-range inter-tissue transportation integrate the signals received from different parts of the plant (Lifschitz et al. 2006; Lin et al. 2007; Tamaki et al. 2007). The identification of FT as florigen (Kardailsky et al. 1999; Kobayashi et al. 1999) was a major step forward in our understanding of the macromolecules that move through the phloem together with sugars and hormones to regulate flowering and development (Giaquinta 1983; Corbesier et al. 2007; Molnar et al. 2010; Regnault et al. 2015).

The first photoperiodic regulator, the B-box transcription factor CONSTANS (CO), was cloned (Putterill et al. 1995) and shown to transmit photoperiod information to flowering time control through induction of *FT* in leaves. CO is regulated at the transcriptional and posttranslational level by the circadian clock. *CO* mRNA accumulates during the day, peaking 16 h after dawn during long day (more than 12 h of light [LD]) photoperiods (Suarez-Lopez et al. 2001). In the morning *CO* mRNA levels are repressed by CYCLIN DOF FACTORs (CDFs) and through the day, but FLAVIN BINDING, KELCH REPEAT, F-BOX1 (FKF1), and ZEITLUPE (ZTL) relieve CDF-mediated repression in the evening, permitting mRNA accumulation in the evening (Imaizumi et al. 2005; Song et al. 2014). CO protein is itself targeted at night by CONSTITUTIVE PHOTOMORPHOGENIC1 (COP1) and SUPPRESSOR OF PHYA-105S1 (SPA1) for degradation by the proteosome (Valverde et al. 2004; Laubinger et al. 2006; Jang et al. 2008). This complex regulation culminates in CO accumulating only in LD conditions, when light coincides with the evening, to bind to the promoter of *FT* and activate its transcription. Partly as a result of direct CO regulatory dependence on the circadian clock, the mutations in clock component genes, such as *GIGANTEA* (*GI*), *EARLY FLOWERING3* (*ELF3*), and pseudo response regulators (*PRRs*), also affect flowering time (Hicks et al. 1996; Fowler et al. 1999; Nakamichi et al. 2007). However, many clock components

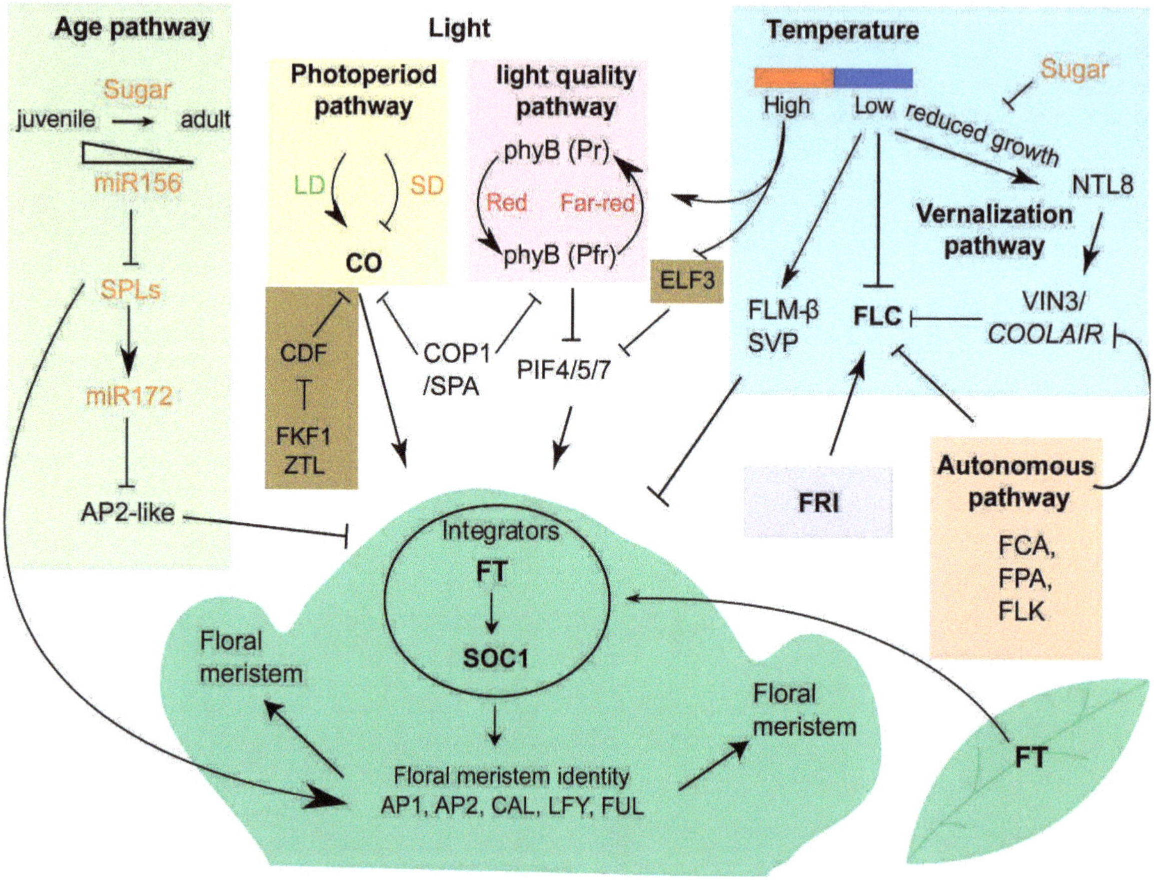

Figure 1. The main genetic pathways controlling flowering time in *Arabidopsis*. Colored boxes highlight different pathways; FRI (purple) and clock components (brown), key integration nodes (FLC, CO, FT, and SOC1), and those with extensive natural variation (FRIGIDA, FLC, and FT) are in bold. Arrows indicate positive and bars represent negative regulatory relationships. Genetic pathways converge on *FT*, encoding a transmissible signaling molecule transported from the leaves to the SAM. The floral pathway integrators (in a circle) and floral meristem identity genes are shown in the green schematic meristem. The influence of sugar on some pathways is indicated. Different pathways are interconnected, for example, photoperiod and light quality and temperature pathways by COP1/SPA, and circadian and high temperature pathways by ELF3.

have direct transcriptional outputs affecting flowering through other pathways and themselves integrate temperature and light signals—mechanisms that are still being elucidated (for more detailed reviews, see Harmer et al. 2022; Maeda and Nakamichi 2022; Nakamichi et al. 2022; Oravec and Greenham 2022).

In *A. thaliana*, *FT* expression is activated in the companion cells (CCs) in the leaf, and FT protein is then loaded into the sieve elements (SEs) and transported to the SAM, where it forms a floral activation complex with the bZIP transcription factors (TFs), FD and FD-related proteins, as well as 14-3-3 proteins to induce the floral transition (Martignago et al. 2023). Another phosphatidylethanolamine-binding (PEPB) protein related to FT, called TERMINAL FLOWER1 (TFL1), antagonizes FT's function by competition for chromatin-bound FD at shared target loci (Goretti et al. 2020; Zhu et al. 2020). Many studies have shown that long-distance FT transportation is not just a result of diffusion but is highly controlled. The transport of FT protein from CCs to SEs is mediated by FT-INTERACTING PROTEIN 1 (FTIP1), QUIRKY (QKY), and SYNTAXIN OF PLANTS121 (SYP121). FTIP1 is an endoplasmic reticulum (ER) membrane protein that mediates the movement of FT protein through a continuous ER network running through the intercellular connections known as plasmodesmata between CCs and SEs (Liu et al. 2012). QKY and SYP121 (MCTP-SNARE Complex) coordinately facilitate FT export from CCs to SEs through the endosomal trafficking pathway (Liu et al. 2019). After entering into the phloem stream, the long-distance trafficking of FT from leaves to the SAM is regulated by a heavy metal–associated domain-containing protein, NaKR1 (Zhu et al. 2016). However, little is known about how FT is unloaded post-phloem and transported to the shoot apex (Yoo et al. 2013).

FT expression is also promoted by PHYTOCHROME-INTERACTING FACTOR 4 (PIF4) and its orthologs (PIF5 and PIF7) (Kumar et al. 2012; Galvao et al. 2019). Under optimal ambient temperatures, red light converts photoreceptor phyB to the active (Pfr) state, which leads to the degradation of PIF4/5/7 and CO. Under elevated temperatures (27 °C), the active state is rapidly converted to the inactive (Pr) state to allow the induction of *PIF4/CO* and sub-sequently *FT* (Kumar et al. 2012; Fernandez et al. 2016). Flowering time can also be regulated by phyB through PHYTOCHROME AND FLOWERING TIME1 (PFT1) to regu-late *FT* transcription (Cerdan and Chory 2003). In addition to *FT* mRNA expression, FT movement from CC to SE is also temperature sensitive; low temperatures facilitate FT sequestration in the cellular membrane of the companion cell due to its phospholipid-binding properties, thus reducing soluble FT levels and delaying flowering (Liu et al. 2020; Susila et al. 2021). FTIP1, QKY, and SYP121 single mutants show dif-ferent temperature responses, indicating ambient tempera-ture may influence different steps of FT trafficking (Liu et al. 2020). So, FT not only integrates environmental signals from different branches of the floral network but is itself directly regulated by these cues.

Reproductive competence

Before the floral transition, the shoot apical meristem (SAM) must first become competent. This competence is thought to be associated with the transition from juvenile to adult vegetative phase; however, this is not true for all species (Poethig 2003; Baurle and Dean 2006; Hyun et al. 2017; Poethig and Fouracre 2024). There are some species that flower without a juvenile to adult vegetative phase change, but in many cases, only the adult plants can respond to diverse environmental cues such as photoperiod or low temperature to flower while some others show varied responses in the juvenile and adult phases (Hyun et al. 2017).

The transition from juvenile to adult phase (reviewed in Poethig and Fouracre 2024) is governed by a decrease in expression of microRNA156 (miR156), which represses the expression of SQUAMOSA PROMOTER BINDING-LIKE (SPL) TFs (Wu and Poethig 2006; Wu et al. 2009; Wang et al. 2009a; Yu et al. 2013; Gao et al. 2022). This decrease defines the length of juvenility; a recent study has revealed that the miR156 decline rate is correlated with developmental age rather than chronological age. Upon seed germination, the onset of cell division in the SAM serves as a trigger for the decline in miR156. Concomitant with cell division, the transcriptional activity of *MIR156C* is gradually attenuated by the deposition of the repressive histone mark trimethylation of lysine 27 of histone 3 (H3K27me3) (Cheng et al. 2021). These results provide a plausible explanation for why the decline in miR156 is unidirectional.

Genetic studies and mis-expression experiments have revealed that miR156 regulates flowering time in both leaf and SAM through distinct mechanisms. In leaves, the miR156/SPL module primarily controls flowering via miR172, which targets 5 AP2-like TFs. *MIR172B* is activated directly by SPL9 (Wu et al. 2009). The 5 AP2-like TFs act as flowering repressors by inhibiting the expression of the florigen gene *FT*, which normally contributes to signaling from leaves to meristem. Overexpression of miR172 leads to early flowering, while the simultaneous mutation of 5 *MIR172* genes results in late flowering, particularly under non-inductive conditions (Aukerman and Sakai 2003; Chen 2004; Lian et al. 2021; Ó'Maoileidigh et al. 2021).

Within the shoot apex itself, miR156-targeted SPLs (mainly SPL15) and FT promote flowering directly by activating a shared set of targets, including *AP1*, *FUL*, *LFY*, and *SOC1* (Wang et al. 2009a; Yamaguchi et al. 2009). Additionally, SPL15 releases the inhibition of flowering by AP2 itself through activating *MIR172A* and *MIR172D* (Lian et al. 2021; Ó'Maoileidigh et al. 2021). These findings demonstrate the highly redundant activities and feed-forward action of the miR156/SPL and FT modules in regulating flowering while also revealing the interplay between meristem competence and photoperiod.

Vernalization removes the floral repressors responsible for reducing the sensitivity of the meristem to inductive signals. In Arabidopsis, the major repressor is *FLOWERING LOCUS C* (*FLC*), whose expression is upregulated by *FRIGIDA* (*FRI*). *FLC*, a MADS-box transcription factor, forms a heterodimer with SHORT VEGETATIVE PHASE (SVP; also a MADS-box TF) to negatively regulate *FT* and *SOC1* and thereby prevent flowering (Michaels and Amasino 1999; Searle et al. 2006). Progressive cold treatment represses *FLC* through a co-transcriptional mechanism involving *FLC* antisense transcripts (known as *COOLAIR*) (see Box 1) and in parallel is epigenetically silenced by Polycomb Repressive Complex 2 (PRC2) with both processes regulated through NTL8 (De Lucia et al. 2008; Csorba et al. 2014; Zhao et al. 2020; Zhao et al. 2021; Nielsen et al. 2024). PRC2 associates with a sense noncoding transcript, *COLDWRAP*, derived from the promoter region of *FLC* (Heo and Sung 2011; Kim et al. 2017), involved in the formation of a repressive intragenic chromatin loop at *FLC* (Kim and Sung 2017). A second sense noncoding, intronic transcript, *COLDAIR*, is also proposed to interact with PRC2 (Heo and Sung 2011), but its identity and function are still not fully resolved. The PRC2-induced epigenetic silencing is mitotically stable when temperatures rise in the spring but is reset every generation (Sheldon et al. 2008; Crevillen et al. 2014). Early forward screens uncovered *VERNALIZATION1* (*VRN1*), *VERNALIZATION2* (*VRN2*), *VERNALIZATION INSENSITIVE3* (*VIN3*), and *VERNALIZATION5* (*VRN5*), all of which compromise the plant's ability to establish or maintain stable *FLC* silencing (Chandler et al. 1996; Greb et al. 2007). During the cold, VIN3 protein accumulates and associates with the nucleation region of *FLC* (Sung and Amasino 2004; Finnegan et al. 2005; Wood et al. 2006), where it functions with the VRN2-PRC2 complex through its interaction with VRN5 (Yang et al. 2017; Franco-Echevarria et al. 2023). PRC2 catalyzes the deposition of H3K27me3 around the nucleation region and when the plant is returned to warm conditions, this modification spreads across the whole locus to silence *FLC*. H3K27me3 is stable through cell division and maintains *FLC* in an epigenetically silenced state (Bastow et al. 2004).

In addition to *FLC*, other floral repressors play important roles in temperature sensitivity of the Arabidopsis floral transition. FLOWERING LOCUS M (FLM/MADS AFFECTING

Box 1. The environmental sensitivity of *FLC* antisense transcription.

A set of cold-induced antisense transcripts, named *COOLAIR*, is expressed at the *FLC* locus in *Arabidopsis thaliana* (Swiezewski et al. 2009). *COOLAIR* initiates immediately downstream of the major sense *FLC* poly (A) site, can transcribe through to the *FLC* promoter, and plays many roles in *FLC* silencing (Csorba et al. 2014; Kim et al. 2017; Nielsen et al. 2024). *COOLAIR* homologs in the semi-perennial relative *Arabis alpina* are induced each winter (Castaings et al. 2014). *COOLAIR* transcripts adopt multiple secondary structures with different conformational dynamics, influenced by temperature (Hawkes et al. 2016; Yang et al. 2022). Monocot *FLC* homologs also show cold-induced antisense tran-scripts (Jiao et al. 2019).

Cold temperature not only promotes *COOLAIR* RNA levels but also affects its processing; promoting the use of a proximal polyadenylation site and enhancing splicing to form a distal *COOLAIR* isoform called Class II.ii (Zhao et al. 2021; Zhu et al. 2021) (Box 1 Fig. A). Several cold-responsive TFs facilitate cold induction of *COOLAIR* including NTL8, CRT/DRE-binding factors (CBFs), and the group-III WRKY transcription factor WRKY63 (Zhao et al. 2020; Hung et al. 2022; Jeon et al. 2023) (Box 1 Fig. A). These factors have distinct cold sensitivities; for example, CBFs are upre-gulated upon short cold (minutes/hours) exposure (Jeon et al., 2023), while NTL8 accumulates over weeks of cold exposure (Zhao et al. 2020). The slow timescale of NTL8 induction is due to an indirect thermosensory mechanism whereby cold slows cell division enabling NTL8 protein accumulation through reduced dilution. Thus, CBFs and NTL8 are likely to be responsible for *COOLAIR* induction at different stages of vernalization. Because components of the growth medium, particularly sugar levels, change plant growth rate and influence protein accumulation, *COOLAIR* expression peaks at different stages when analysed in different laboratories (Box 1 Fig. B).

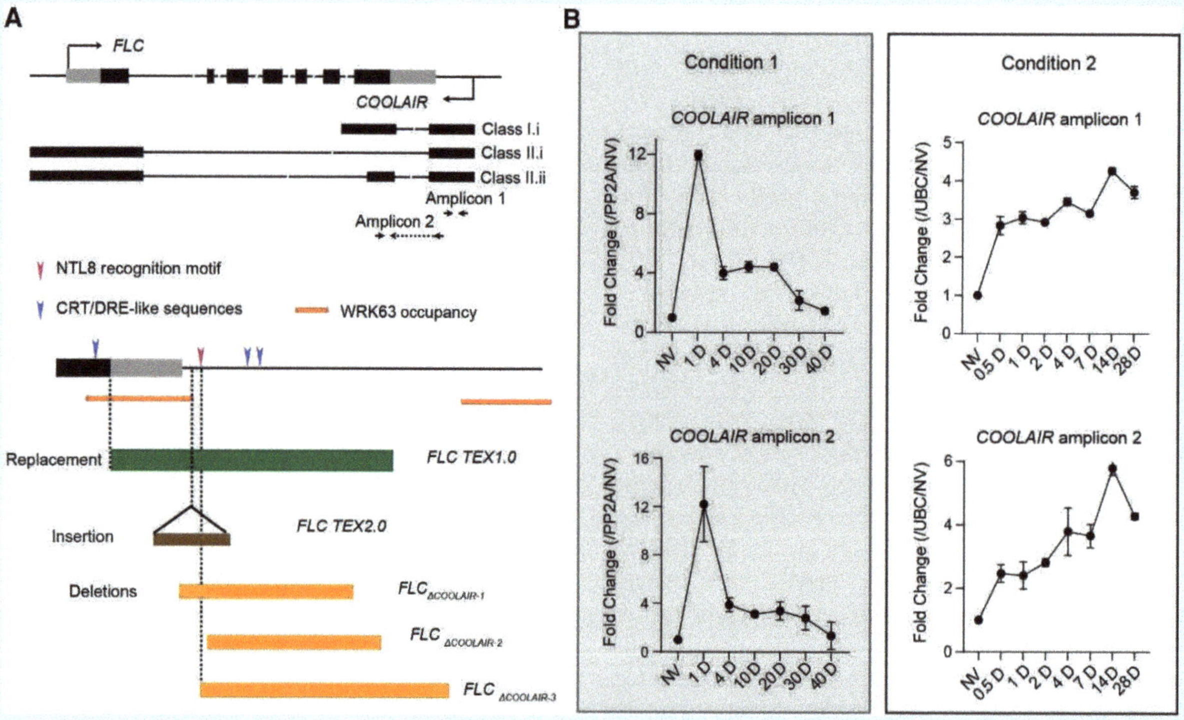

Figure A. Antisense transcription at *FLC* locus. **A)** Schematic illustration of *FLC* gene architecture and *COOLAIR* transcripts. Black lines represent introns, black boxes represent exons, and grey boxes indicate UTR regions. *FLC* and *COOLAIR* transcription start sites are shown by black arrows. The 3′ end of *FLC* is enlarged below to show binding motifs/regions and currently available mutants that disrupt *COOLAIR*. **B)** *COOLAIR* expression profile during cold treatment, measured using Q-RT-PCR and 2 amplicons shown in (A) at different conditions. The data in condition 1 is from Jeon et al., 2023 while the condition 2 is the same as described in Swiezewski et al. 2009.

There are contradictory conclusions on the role of *COOLAIR* in cold-induced *FLC* silencing. These are based on knockdown/out mutants generated using different genetic methods (Zhu et al. 2021; Jeon et al., 2023; Zhang et al. 2023a; Zhu and Dean 2023). None of the antisense mutants entirely remove antisense transcription; when transcription is suppressed from 1 region it initiates in another within the locus (Zhao et al. 2020; Zhu and Dean 2023). In addition, the overlap of the *COOLAIR* transcription start region containing transcription factor binding sites, with the sense transcript *FLC* 3′ UTR end, makes it difficult to completely remove the cold sensitivity of *COOLAIR* (Box 1 Fig. A). The combination of molecular analyses with computational modelling helped explain the contradictory findings. *FLC* is silenced through pathways that function with different dynamics: a *COOLAIR* transcription-mediated pathway capable of fast response; and in parallel a slow Polycomb Repressive Complex 2 (PRC2) switching mechanism that maintains each allele in an epigenetically silenced (Nielsen et al. 2024). The parallel repressive inputs and extensive feedback make the mechanism counter-intuitive but provide great flexibility to the plant to cope with ever-changing seasonal conditions.

FLOWERING1), a MADS-box transcription factor related to FLC, represses *FT* and *SOC1* (Balasubramanian et al. 2006a). The *FLM* locus is transcribed into a number of different isoforms via temperature-dependent alternative splicing mechanisms (Pose et al. 2013; Lutz et al. 2015; Sureshkumar et al. 2016). At low ambient temperatures, transcription of the repressive *FLM-β* isoform, which contains the functional DNA binding domain, is promoted compared with other isoforms, allowing FLM to multimerize with SVP and FLC to repress floral integrators (Lee et al. 2013a, 2013b; Pose et al. 2013). At elevated temperatures, active SVP protein is re-duced via decreased transcription and increased proteasomal degradation, permitting accelerated flowering (Lee et al. 2007; Lee et al. 2013a, 2013b; Jin et al. 2022). However, muta-tions in *FLM* and *SVP* have relatively small effects on *A. thaliana* flowering time.

Natural winter annual *A. thaliana* accessions show a *FRI*/*FLC*-dependent dominant requirement for vernalization, in contrast to the mutation-induced flowering mutants such as *fca* (*flowering locus ca*), *fpa* (*flowering locus pa*), and *flk* (*flowering locus k*) that confer a recessive vernalization requirement—rather like the *Vrn* loci in wheat. These mutants flower late regardless of day length but respond to vernalization or growth in far-red light and were classified into the autonomous floral pathway (Koornneef et al. 1991; Koornneef et al. 1998). The compo-nents typically regulate flowering by limiting *FLC* expression levels. FCA, FPA, and FLK are general RNA binding and 3′ process-ing factors that regulate *FLC* through a transcription-coupled chromatin silencing mechanism (Macknight et al. 1997; Schomburg et al. 2001; Lim et al. 2004; Marquardt et al. 2006). This involves proximal transcription termination linked to deliv-ery of a chromatin environment that affects transcriptional out-put—initiation, processivity, and elongation (Liu et al. 2007; Baurle and Dean 2008). The proximal termination process influ-ences both *COOLAIR* transcripts from *FLC* and the sense *FLC* transcription (Box 2) (Schon et al. 2021; Menon et al. 2023).

Genes identified through flowering time mutants frequently encode general developmental and gene regulators

The molecular analysis of genes in the Arabidopsis integrated flowering network has led to important mechanistic information as to how organisms can perceive different environmental signals and integrate them over time. The surprise has been that despite having relatively specific flowering time mutant phenotypes, many genes were found to encode general developmental and gene regulators. We discuss a few examples below.

FT encodes florigen. FT is a member of the PEBP family that includes *TWIN SISTER OF FT* (*TSF*), *MOTHER OF FT* (*MFT*), and *TFL1* (Yamaguchi et al. 2005; Xi et al. 2010; Hiraoka et al. 2013). In *A. thaliana*, *FT*, *TSF*, and *TFL1* jointly affect shoot architecture through differential activation of axillary meristems, and *FT*, *MFT*, and *TFL1* all affect seed development (Hiraoka et al. 2013; Chen et al. 2014; Zhang et al. 2020a). In other species, the roles for the PEBP family have proliferated: in potato (*Solanum tuberosum*), the *FT-like SELF-PRUNING 6A* (*StSP6A*) gene is activated in leaves under short-day conditions (SDs) due to the inactivation of the FT-like repressor *SELF PRUNING 5G* (*StSP5G*) by unstable StCOL1 (Navarro et al. 2011; Abelenda et al. 2016). Along with 2 other SD-activated transmissible FT-like proteins, StSP3D and FT-like 1 (StFTL1), SP6A protein is then transported via phloem from leaves to stolons, where it forms a floral activation-like complex, termed the tuberigen activation complex, which promotes tuber formation (Teo et al. 2017; Jing et al. 2023). A similar shoot-to-root translocation is found in soybean (*Glycine max*), where the shoot-derived ortholog of *Arabidopsis* FT, GmNN1/FT2a, triggers nodulation upon rhizobial infection (Kong et al. 2010; Sun et al. 2011; Li et al. 2022b). In the root, GmNN1/FT2a interacts with GmNFYA-C to activate symbiotic signaling through the GmNFYA-C-ENOD40 module (Li et al. 2022b). Therefore, the mobile signaling functions of FT-like proteins, first established in flowering time control, appear to be involved in many other environmentally controlled development processes.

Other genes first identified through their flowering time phenotype have now also been found to have additional functions. For instance, miR156-targeted SPL TFs influence leaf mor-phological changes associated with developmental progression (Poethig 2010), male fertility (Xing et al. 2010), nodulation (Wang et al. 2015; Yun et al. 2022), defense against insects (Mao et al. 2017), and anthocyanin biosynthesis (Gou et al. 2011). PFT1 encodes MED25 of the plant Mediator complex and plays an essential role in transcription initiation, regulating jasmonate signaling, biotic and abiotic stress responses, and flowering (Backstrom et al. 2007; Kidd et al. 2009; Inigo et al. 2012).

Many mutants affecting *FLC* expression were found to encode core transcription regulators. These include ELF7 (PAF1) and other Paf1C components, FRI (and FRIc), CAP-BINDING PROTEIN20 (CBP20), and CBP80 (He et al. 2004; Geraldo et al. 2009); splicing regulators: the apoptosis and splicing–associated protein (ASAP) complex and PRE-MRNA

Box 2. Feedback that complicate analysis of *FLC* and its sensitivity to co-transcriptional regulators.

The feedback between transcription and chromatin is central to regulation of expression at *FLOWERING LOCUS C* (*FLC*). These mechanisms are difficult to tease apart without support of computational modelling (Menon et al. 2023). Both transcriptional activation by FRIGIDA and repression by the autonomous pathway involve co- transcriptional pathways that link transcription termination with delivery of a changed chromatin environment. This chromatin environment then feeds back to affect transcriptional output by changing transcription initiation, pro-cessivity and elongation. FRIGIDA promotes *FLC* transcription by acting as an anti-terminator in the developing em-bryo, enhancing usage of distal termination sites for both sense and antisense transcription (Schon et al. 2021). The higher transcription delivers active chromatin modifications to the locus (H3K4me, H3K36me3), which enhance distal site usage (Liu et al. 2010). FCA represses *FLC* transcription by promoting proximal termination of both sense and antisense transcription (Menon et al. 2023). This is linked via FLD-mediated H3K4 demethylation to a changed chro-matin environment that reduces transcriptional output by feeding back to enhance use of the proximal polyadenyla-tion site (Liu et al. 2007; Liu et al. 2010). These feedback result in counter-intuitive outcomes on steady state RNA levels. For example, loss of FCA primarily reduces the relative propensity for proximal termination, but loss of that step affects the chromatin environment at the whole locus, which results in higher transcription of all the *FLC* and *COOLAIR* transcripts, including proximal *COOLAIR* (Liu et al. 2010). Thus, analyzing changes in absolute levels at steady state can give confusing answers. The answer has been to measure the ratio of proximal to distal polyadenylation. However, sufficient sequencing depth is required to reliably measure these low abundance transcripts. 3′RNA sequen-cing approaches provide useful data on polyadenylation of both *FLC* and *COOLAIR* (Box 2 Fig. A-B) (Schon et al. 2021; Menon et al. 2023), and these confirm all the analyses using PCR (Liu et al. 2010).

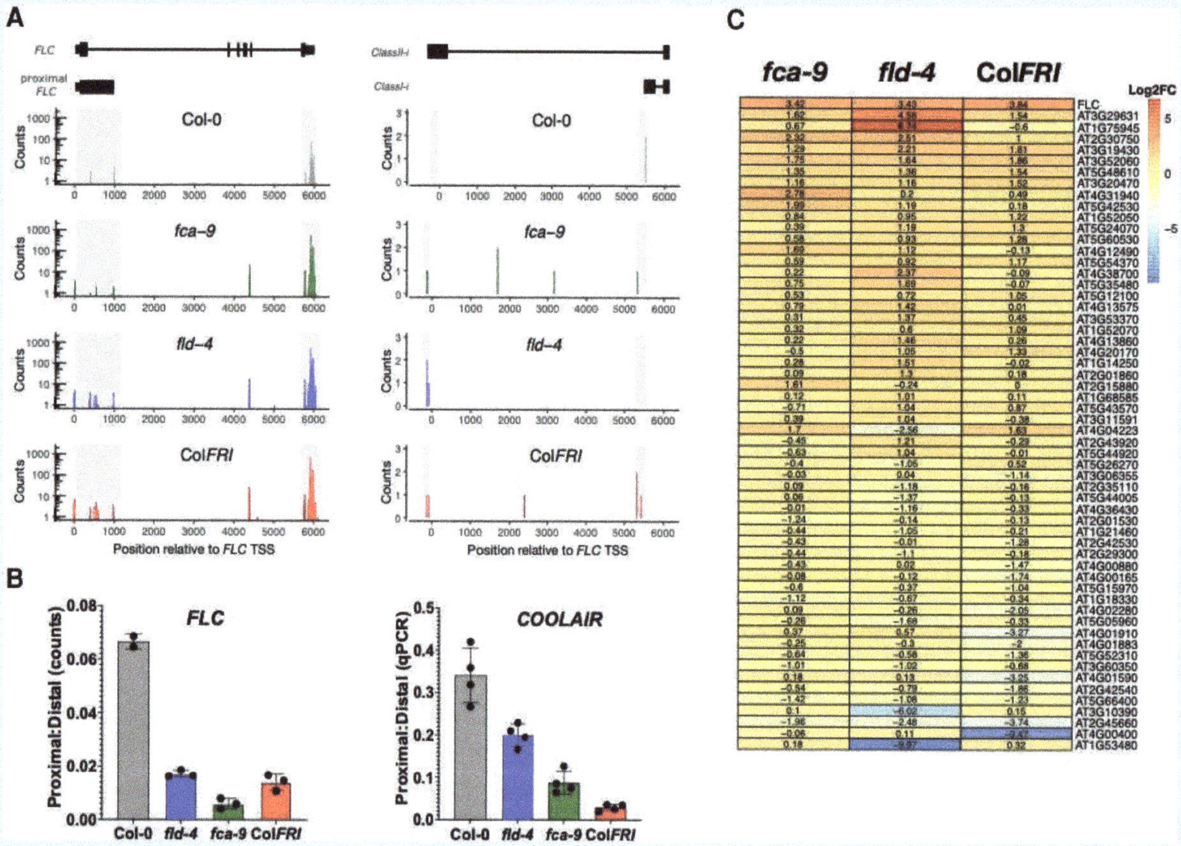

Figure B. Co-transcriptional 3′processing at the *FLC* locus. **A)** RNA 3′ sequencing at the *FLC* locus reveals altered polyadenylation site selection of both the sense (left) and antisense (right) transcripts by *fca-9*, *fld-4* or Col*FRI* compared with Col-0. **B)** The proximal to distal polyadenylation ratio of *FLC* by 3′RNA sequencing matches conventional qPCR analysis of *COOLAIR* polyadenylation ratio. **C)** Differentially expressed genes in *fca-9*, *fld-4* and Col*FRI* compared with Col-0 (padj < 0.05 and Log2FC > 1). Consistently, *FLC* is one of the top upregulated genes.

These 3′ sequencing approaches show the extent of mis-regulation in the Arabidopsis genome when functionality of FRI or autonomous pathway components is disrupted. Several hundred genes are differentially transcribed in these genotypes (with significant *P*-values when compared with Col-0), *fca-9* 104 up, 69 down; *fld-4* 201 up,152 down; ColFRI 511 up, 420 down; padj < 0.05. This agrees with their predicted roles as general co-transcriptional regulators. However, it is striking that *FLC* is in each case the most affected gene, with the highest fold change: 10.7 for *fca-9*, 10.8 for *fld-4* and 14.3 for ColFRI (Box 2 Fig. **C**). What confers this sensitivity to general acting co-transcriptional regulators is a fascinating question, with major implications for why *FLC* has become the node for flowering time variation in the Brassicaceae (Li et al. 2014).

PROCESSING8 (PRP8) (Marquardt et al. 2014; Mikulski et al. 2022); RNA binding proteins: THO/TREX, FCA, and FY (Liu et al. 2007; Xu et al. 2021); 3′ processing factors: CPSF, CstF64, CstF77, ANTHESIS PROMOTING FACTOR1 (APRF1), and TYPE ONE SERINE/THREONINE PROTEIN PHOSPHATASE4 (TOPP4) (Liu et al. 2010; Mateo Bonmati et al. 2023). This raises the question: What makes *FLC* such a sensitive target to generic transcriptional regulators? Analysis of fold changes in genes mis-expressed in *fca* reveal the sen-sitivity of *FLC* regulation compared with other targets (Box 2). Similarly, the vernalization mediators VRN2, VRN5, and VIN3 are PRC2 core and accessory proteins, respectively, and associate widely with loci across the *Arabidopsis* genome (Franco- Echevarria et al. 2023). Redundancy and subfunctionalization between PRC2 forms may explain how some of the mutants have more vernalization-specific phenotypes. However, this relative specificity has provided a nonlethal platform for investigation of the mechanisms of these important general regulators, allowing ongoing insights into conserved regulation of transcriptional processing and epigenetic chromatin memory.

Even the floral repressor *FLC* is not specific to the floral transition. *FLC* is known to bind to many genes and is associated with broader developmental contexts such as seed dormancy, cold tolerance, juvenile-to-adult phase transition, and inflorescence patterning (Chiang et al. 2009; Deng et al. 2011; Huang et al. 2013; Mateos et al. 2017; Auge et al. 2019). Moreover, FRI and FLC increase the stress tolerance of plants during drought and pathogen infection (Wilson et al. 2013; Chen et al. 2018; Chen et al. 2022; Shukla et al. 2022; Xu et al. 2022). The naming of genes based on original forward screens implies a specificity to their function, which in reality is not the case. This convention has caused confusion in the whole field and is not exclusive to the flowering time field.

So, if the genes classified in the regulatory network affecting flowering actually function in many processes important in plant growth and development, which genes are most influential in determining flowering in natural conditions and how have these evolved? The study of natural variation of flowering time has been most helpful in answering these questions.

Natural variation has identified critical nodes driving flowering time diversity

We summarize studies of the genetic basis of natural variation in flowering, focusing on Arabidopsis and its relatives as 1 example, and then contrast those findings with our understanding of the loci selected during domestication of our major crops.

Flowering time variation within Arabidopsis and relatives

Arabidopsis thaliana accessions show considerable variation for many aspects of flowering time. This has enabled adaptation to their wide climate envelope, from the Arctic Circle to near the equator (Hoffmann 2005). Studies of natural variation of flowering time have focused on vernalization requirement (winter vs rapid-cycling habit), vernalization response, or photoperiod sensitivity. Variation in vernalization requirement was first mapped by Klaus Napp-Zinn in the 1950s, who showed that, despite the quantitative nature of flowering time control, *FRI* could be mapped as a single Mendelian locus (Napp-Zinn 1957; Clarke and Dean 1994). Subsequent QTL analyses between a range of winter annual and rapid cycling accessions showed approximately 70% of winter annual/rapid-cycling variation can be accounted for by allelic variation at *FRI* (Le Corre et al. 2002; Gazzani et al. 2003; Shindo et al. 2005; Werner et al. 2005a; Méndez-Vigo et al. 2011; Strange et al. 2011; Ågrena et al., 2013; Kinmonth-Schultz et al. 2021). Molecular analysis has shown loss-of-function *FRI* mutations are a recurrent feature in the evolution of the rapid-cycling habit (Johanson et al. 2000; Shindo et al. 2005). Two amino acid polymorphisms in the central domain of the FRI protein that change stability or subcellular localization also cause early flowering (Zhang et al. 2020b). *FRI* alleles are associated with flowering time plasticity in regions experiencing high annual temperature variation (Fournier-Level et al. 2022), which may be related to its temperature-sensitive properties (Zhu et al. 2021). The evolutionary relevance of FRI variation extends to *Cardamine hirsuta*, where 3 distinct *FRI* loss-of-function alleles associate with early flowering (Baumgarten et al. 2023).

Variation in vernalization response mapped to *FLC* in QTL analysis of variation between different winter annual types (Shindo et al. 2006). Analysis of the *FLC* genomic sequence in the natural accessions revealed ~20 major haplotypes, with 5 widely represented in the worldwide population. These haplotypes are distinguished by noncoding SNPs, which were shown to be causative for the different *FLC* expression levels and response to cold (Li et al. 2014). In many cases 1 noncoding SNP has a substantial effect on the phenotype (Li et al. 2014;

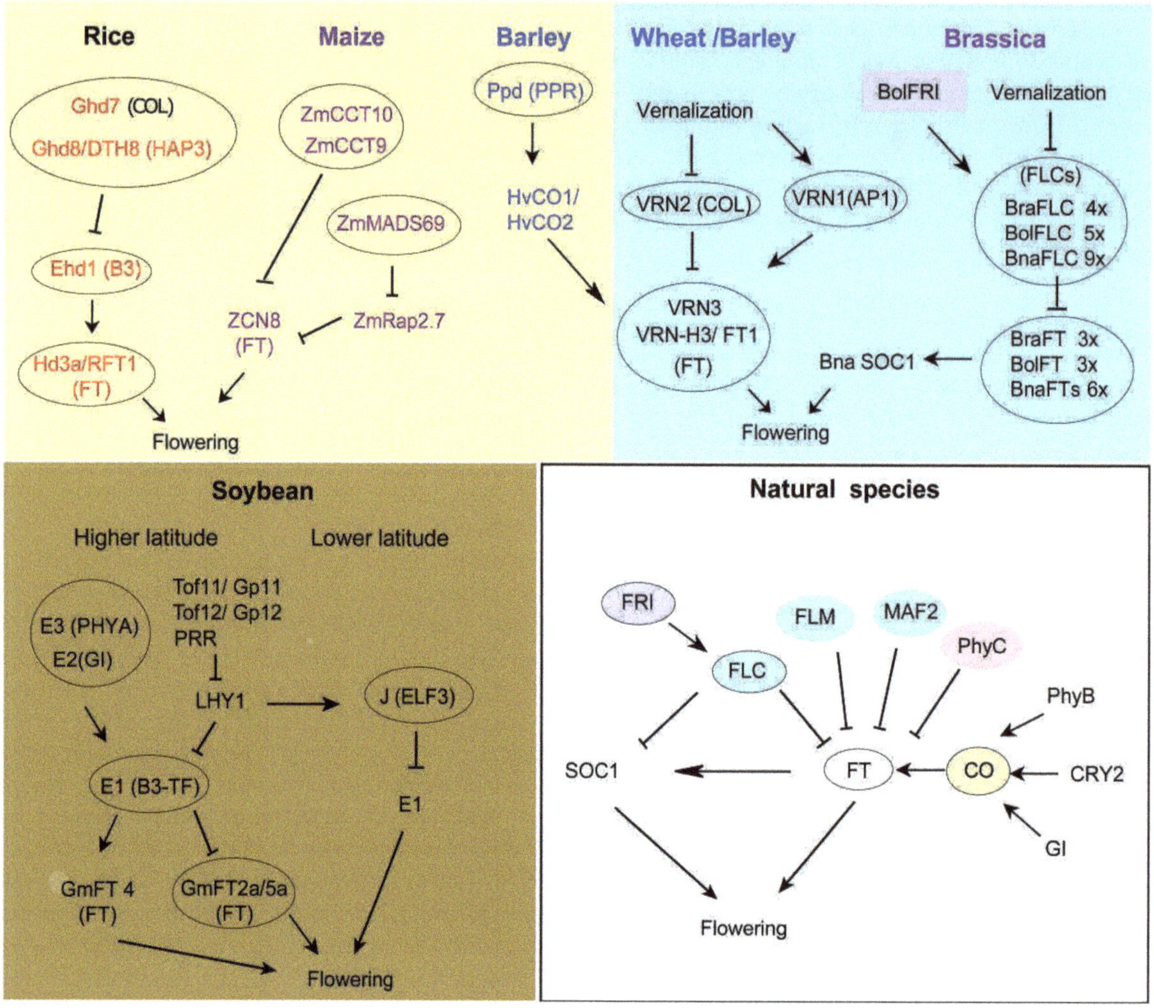

Figure 2. Natural variation in flowering pathways, comparing domesticated crops and natural species. Nodes circled in black in each pathway represent the major nodes with high allelic diversity. Yellow box shows selected photoperiodic regulators in rice (red text), maize (purple text), and barley (blue text). Extensive natural variation also occurs in vernalization regulators in barley (blue box), wheat, and brassica crops. Soybean breeding has predominantly targeted circadian clock components (pale brown box). White box: loci showing natural variation in *A. thaliana*.

Li et al. 2015; Qüesta et al. 2020; Zhu et al. 2023). In natural populations these SNPs accumulate over time, driven by natural selection, but their effect may be reduced or enhanced by epistatic interactions in the genetic background (Neto and Hancock 2023). The different causative SNPs at *FLC* are an excellent tool to dissect mechanism and act through influence on promoter activity, epigenetic silencing, or antisense alternative splicing (Li et al. 2015; Qüesta et al. 2020; Zhu et al. 2023). The major haplotypes of *A. thaliana FLC* have been introgressed into a common background to generate a set of near-isogenic lines. Use of these lines in field experiments across Sweden and the UK revealed that autumnal *FLC* expression determined by the noncoding SNPs was the most important determinant modulating flowering time and fitness in response to different natural fluctuating environments (Hepworth et al. 2020).

The variation in *FLC* epigenetic silencing is also a feature in Arabidopsis relatives. In the annual *A. thaliana*, the floral meristem is determinate and growth ceases, but for *Arabis alpina*, a polycarpic perennial, some meristems remain vegetative for growth in the following year. This requires reactivation of the *FLC* ortholog *PERPETUAL FLOWERING1* (*PEP1*) (Wang et al. 2009b). Unlike *FLC* in *Arabidopsis*, *PEP1* repression by cold is not epigenetically stable (Wang et al. 2009b). In *A. alpina*, miR156 repression of *SPL15* in younger meristems prevents conversion of that meristem to flowering post vernalization, even in long photoperiods. *FLC* reactivation is required to prevent high *FT* expression overriding this, as it can in young *A. thaliana* shoots (Hyun et al. 2019). This strategy for perennial growth is not isolated: in *Arabidopsis halleri*, *Arabidopsis lyrata*, *Cardamine hirsute*, *Capsella rubella*, *A. alpina*, and its close annual relative, *Arabis montbretiana*, *FLC* orthologs also influence flowering in an expression-dependent manner and contribute to nat-ural variation and life histories (Wang et al. 2009b; Aikawa et al. 2010; Guo et al. 2012; Kemi et al. 2013; Kiefer et al. 2017). *FLC cis* variation in 3 orthologs include gene duplications and noncoding changes to promoters and introns

(Albani et al. 2012; Kemi et al. 2013; Kiefer et al. 2017). Conservation of *COOLAIR* sequences and structures across the Brassicaceae suggests the antisense functionality is also conserved (Castaings et al. 2014; Hawkes et al. 2016; Kiefer et al. 2017).

FT is an important contributor for major-effect QTLs underlying flowering time variation in response to environmental conditions, such as light and temperature (Schwartz et al. 2009; Li et al. 2010; Strange et al. 2011). In *A. thaliana*, long photoperiod induction of *FT* expression is controlled by 2 interdependent regulatory regions, with the distance between them essential for the responsiveness (Schwartz et al. 2009; Adrian et al. 2010). One polymorphism in 1 of the *cis* elements, the CCAAT box (C block), in the Ull2-5 accession is causative for impaired *FT* expression pattern in response to LD induction (Strange et al. 2011). Natural promoter length variation at *FT* creates promoter length differences that correlate with longitudinal and latitudinal clines (Liu et al. 2014). The bHLH transcription factor MYC3 competes with CO to repress the expression of *FT* under unfavorable photoperiods (Bao et al. 2019). The MYC3 binding site, the ACGGAT motif, is specifically present in accessions bearing the long *FT* promoter variant situated toward more northern latitudes (Bao et al. 2019). It should be noted that FT can be functionally converted to TFL1 and vice versa by a single amino acid substitution, and such mutations have been selected during crop domestication (Hanzawa et al. 2005). However, it is still unclear how these mutations are linked to the protein function, especially their movement; TFL1 moves within SAM only, outward from the central region (Conti and Bradley 2007), in a manner very different from FT.

Other genes showing natural variation in both *A. thaliana* and relatives include photoperiod and vernalization regulators. *VIN3* underlies GWAS and QTL peaks in *A. thaliana* and *Brassica napus* (Dittmar et al. 2014; Song et al. 2020). The blue light receptor gene *CRYPTOCHROME2* (*CRY2*) is a rare QTL in *A. thaliana*. In a recent study of the evolution of the ruderal weed species *Cardamine occulta*, CRY2 was selected for photoperiod insensitivity, as well as *FLC* (Li et al. 2023a; Li et al. 2023b). In addition, the photoreceptors *PHYTOCHROME C* (*PHYC*), *PHYB*, and *CO* are likely targets to explain phenotypic variation in other studies (Balasubramanian et al. 2006b; Caicedo et al. 2009; Salome et al. 2011; Rosas et al. 2014). Apart from these pathways, variant alleles at *FLM* have also been associated with flowering time in *A. thaliana* (Lutz et al. 2015; Lutz et al. 2017; Kinmonth-Schultz et al. 2023) as well as *MADS AFFECTING FLOWERING2* (*MAF2*, a relative of *FLM*), *SVP*, *GIS5*, and *HUA2* albeit at lower frequency (El-Din El-Assal et al. 2001; Werner et al. 2005b; Wang et al. 2007; Filiault et al. 2008; Schwartz et al. 2009; Méndez-Vigo et al. 2013; Fournier-Level et al. 2022; Kinmonth-Schultz et al. 2023).

In summary, the natural variation in flowering time predominantly influences either light or temperature pathways. Natural variation for alternative (less predictable) environmental conditions such as light quality, annual rainfall, drought, or heat stress have yet to be fully explored (Kobayashi et al. 2013; Yeoh et al. 2017). To date these have been categorized as stress factors and linked indirectly to flowering time (Takeno 2016).

Crop domestication

Flowering time variation has been a major trait during crop domestication. The current growing regions of many crop species are frequently very far from their centers of origin. As such, many aspects of their biology have been bred to accommodate the latitudinal range expansion. Traits such as annual, biennial, and perennial habits; ability to grow in high-density monoculture; and shoot architecture have all been modified as breeders selected for higher yield (Gaudinier and Blackman 2020; Liang et al. 2021). Initial domestication likely selected unconsciously for predictability of flowering time and maximizing the growing period to improve yield. Further crop range expansion required local adaptation of the crop to the new environmental conditions and involved selection for variants alleles (Gaudinier and Blackman 2020). Extreme phenotypes in different cultivars of the same species have been bred: for example, shorter lifecycles to meet strict rotation requirements, or highly delayed bolting to increase storage organ size (for classic Brassica examples, see Cheng et al. 2016; Helal et al. 2016).

Wild rice and maize (*Zea mays*) species are naturally found in low latitudes and exhibit characteristics of short-day plants (Matsuoka et al. 2002; Huang et al. 2012). To adapt to cultivation regions in high latitudes with longer daylight periods, the selection process involved the frequent choice of loss-of-function or weakened alleles of long-day suppressor genes enabling domesticated cultivars to flower early and maximize yield (Izawa 2007; Zhang et al. 2023b). For example, QTL mapping has revealed natural mutations in *Ghd7* and *Ghd8/DTH8* with reduced functions that enable rice to be cultivated in temperate regions. Ghd7 is a CCT domain protein showing homology to *Arabidopsis* CO and CO-LIKE (COL) (Xue et al. 2008), while *Ghd8/DTH8* encodes a HAP3 subunit of a CCAAT-box binding protein the HEME ACTIVATOR PROTEIN (HAP) complex (Fig. 2) (Wei et al. 2010; Yan et al. 2011; Dai et al. 2012). In *A. thaliana*, CO and HAP also form a transcriptional activation complex to modulate *FT* expression (Wenkel et al. 2006; Gnesutta et al. 2017; Lv et al. 2021). Similarly, 2 COL genes, *ZmCCT9* and *ZmCCT10*, have been cloned as flowering-time QTLs in maize (Yang et al. 2013; Huang et al. 2018). Extensive studies on nat-ural variation in rice has identified *Hd16* and *Hd6*, encoding casein kinase I (CKI) and alpha subunit of casein kinase II (CKII alpha), respectively (Takahashi et al. 2001; Hori et al. 2013), and *DTH7*, which encodes a pseudo-response regula-tor protein whose expression is regulated by photoperiod (Fig. 2) (Liu et al. 2013; Gao et al. 2014).

In addition to disrupting LD suppressor genes, mutations that lead to enhanced flowering activators under both LD and SD have also been identified during domestication. For instance, the *FT* ortholog genes *Hd3a* and *RFT1* in rice and *ZCN8* in maize have been preferentially selected at different evolutionary times for local adaptation (Fig. 2) (Kojima et al. 2002; Komiya et al. 2008; Ogiso-Tanaka et al. 2013; Guo et al. 2018). Sequence polymorphisms in the regulatory and coding regions of *RFT1* and *ZCN8* may underlie divergence of flowering time among various cultivars and wild accessions (Fig. 2) (Ogiso-Tanaka et al. 2013; Guo et al. 2018). Moreover, rice *Ehd1*, which encodes a B-type response regulator that does not have a clear ortholog in *Arabidopsis*, promotes flowering under both LD and SD conditions (Doi et al. 2004). Finally, *ZmMADS69*, a MADS-box gene, contributed to the difference in flowering time between maize and its wild ancestor (teosinte) and may have played an important adaptive role during the expansion of maize from the tropics to temperate zones (Fig. 2) (Liang et al. 2019). ZmMADS69 likely downregulates the expression of the flowering time repressor *ZmRap2.7*, thereby alleviating the repression on *ZCN8* and promoting early flowering.

Unlike maize and rice, cultivated soybean has its origin near the temperate Yellow River region (Li et al. 2008). As such, the spread of soybean cultivars involved adaptation to both lower and higher latitudes. Evidence suggests that circadian clock genes were the primary targets for flowering time regulation during domestication (Lu et al. 2022a). Specifically, the adaptation of soybean to higher latitudes was facilitated by naturally occurring loss-of-function mutations in 5 flowering suppressors. Among them, *E1* encodes a legume-specific B3-like transcription factor acting as a suppressor in photoperiod pathway (Xia et al. 2012). E2 and E3 are likely the orthologs of Arabidopsis GI and PHYTOCHROME A (PHYA) (Watanabe et al. 2009; Watanabe et al. 2011). Tof11/Gp11 and Tof12/Gp12, paralogs of Arabidopsis PRR genes, function via homologs of LATE ELONGATED HYPOCOTYL1 (LHY1), a central component of the plant circadian clock. Tof11/Gp11 and Tof12/Gp12 promote *E1* expression and delay flowering under LD (Li et al. 2019; Gong 2020; Lu et al. 2020). The adaptation of soybean to lower latitudes, on the contrary, was driven by an impairment in *J*, the soybean ortholog of *Arabidopsis ELF3* that plays a highly conserved role maintaining circadian rhythms in different species (Fig. 2) (Lu et al. 2017). *J* promotes flowering under SD through repressing *E1*. Loss-of-function mutations in *J* led to an extended vegetative phase and higher yields at lower latitudes (Lu et al. 2017).

Barley (*Hordeum vulgare*) is primarily cultivated in temperate regions. The different varieties of barley can be categorized into 2 main groups based on their response to photoperiod: photoperiod-sensitive and photoperiod-insensitive. The *Ppd-1* gene, an *Arabidopsis PRR* ortholog, controls this photoperiod sensitivity (Turner et al. 2005). The presence of functional alleles of *Ppd-1* results in photoperiod insensitivity, allowing barley plants to flower under both LD and SD conditions (Fig. 2). As such, in higher latitudes with shorter summers, barley varieties with functional *Ppd-1* alleles are particularly advantageous as they enable earlier flowering and maturity. This trait ensures successful reproduction and higher yields in areas where the growing season is limited. By enabling barley to thrive in diverse environments, *Ppd-1* alleles contribute significantly to the cultivation of barley in temperate regions around the world.

Natural variations in vernalization are responsible for the differences in winter or spring growth habit observed in cereals (Kippes et al. 2018; Xu and Chong 2018). In wheat (*Triticum aestivum*), *VRN1* encodes a MADS-box transcription factor that shares homology with *Arabidopsis* AP1 (Yan et al. 2003; Konopatskaia et al. 2016). VRN2 (not the same protein as Arabidopsis VRN2!) and VRN3 show high similarities to *Arabidopsis* COL and FT, respectively (Yan et al. 2004, 2006). VRN2 functions as a floral repressor. Similar to *FLC* in *Arabidopsis*, the expression of *VRN2* is downregulated through vernalization. Loss-of-function alleles of *VRN2* lead to an increased level of *VRN1* and *VRN3*, thereby converting wheat from a winter annual to a spring annual growth habit (Fig. 2) (Yan et al. 2004). Dominant overexpressors of *VRN1* can also cause spring habit by overriding VRN2, and like *FLC*, *VRN1* alleles with differing expression influence both vernalization requirement and environmental sensitivity (Fu et al. 2005; Dixon et al. 2019). The core components of the vernalization pathway—VRN1, VRN2, and VRN3—are conserved between wheat and barley (Yan et al. 2004; Yan et al. 2006; Oliver et al. 2009). Notably, natural allelic variations in *VRN-H3* (*HvFT*) gene also contribute to difference in vernalization requirement in barley (Yan et al. 2006). Moreover, *EPS2/CEN*, which is a homolog of *Arabidopsis TFL1*, is involved in differentiating between winter and spring barley and has been selected and maintained during geographic range extension (Comadran et al. 2012).

Unlike the highly conserved flowering roles of the PEBP proteins, *CO* homologs and clock components, the *FLC* clade of MADS-box genes has not been reported as major flowering time regulators in crops outside of the Brassicaceae (Becker and Theißen 2003; Schilling et al. 2018). Members of other MADS-box clades quantitatively regulate phenology in crops: as well as the cereal *VRN2s* (AP1 family), *ODDSOC2* in barley is also a vernalization regulator but is cereal specific (Greenup et al. 2010). The *DORMANCY ASSOCIATED MADS-BOX* (*DAM*) genes involved in chilling requirement for bud break and bloom time across a range of perennial fruit trees (from apple to pear, peach, and sweet cherry) are most closely related to the SVP clade (Bielenberg et al. 2008; Falavigna et al. 2019; Calle et al. 2020), although an *FLC-like* gene has been found beneath QTL peaks in apple too, along with *AGL24* and *FT* homologs (Allard et al. 2016).

Compared with *Arabidopsis*, the "diploid" Brassica are mesohexaploids, and these diploids have further hybridized to form amphidiploids, which have preferentially retained copies of flowering time genes (Jones et al. 2018). *Brassica rapa* and *Brassica oleracea*, the main vegetable species, have at least 4 and 5 copies of *FLC* respectively, and their hybrid, *B. napus*, thus has at least 9, most of which have been implicated in flowering time variation between different cultivars and which underlie the majority of the main flowering time QTLs, with *FT* homologs accounting for several of the rest (for a comprehensive review, see Schiessl 2020; Song et al. 2020). Critically, these

FLC copies have different expression sensitivity over cold as well as different expression dynamics between alleles, and it is total *FLC* expression, rather than expression of specific *FLC* paralogs, that best explains differences in cold requirement between cultivars (Schiessl 2020; Calderwood et al. 2021a). Given the very wide variation in flowering timing between conspecific Brassica crops, it is likely that the proliferation of *FLC*, each copy with different alleles, has contributed to the evolutionary space that permitted these different domesticated morphs (Calderwood et al. 2021a).

Taken together, while different alleles conferring accelerated or delayed flowering time were selected to aid the adaptation of crops to diverse cultivation areas at various latitudes, the underlying genes involved in these processes appear to be repeatedly selected across different crops. We direct the reader to further literature of major effect genes in other minor crop species (chickpea, Upadhyaya et al. 2015; sunflower, McAssey et al. 2016; apple, Urrestarazu et al. 2017; strawberry, Gaston et al. 2020; litchi, Lu et al. 2022b; flax, Saroha et al. 2022; pepper, Choi et al. 2023). Despite our focus on major crops, the conclusions are likely to extend to other crops. The main pattern observed is that during domestication, there is a notable preference for selection of master regulatory genes within each flowering time pathway—for example, orthologues of *CO* and *COL* in the photoperiod pathway, *FT* and *TFL1* in the florigen-related pathway, *PRR* and *ELF3* in the circadian clock, and *AP1* in the floral-promoting MADS-box gene family. Although the exact reasons for this are yet to be fully understood, variation at these genes may have provided com-patibility with high-density field growth conditions and culti-vation practices aimed at achieving high yield. In support of this notion, it has been shown that *Ghd7* has a broad impact on various traits in rice, including yield, plant height, and heading date (Xue et al. 2008). Similarly, in soybean do-mestication, Tof12/Gp12-dependent acceleration of maturity is associated with reduced dormancy and seed dispersal (Lu et al. 2020). Under this scenario, the aforementioned genes hold great potential as targets for crop breeding and future de novo domestication of wild crop-related species using gen-ome editing approaches (Li et al. 2018; Zsogon et al. 2018; Yu et al. 2021).

In conclusion, a comparison between genes underpinning natural diversity in *Arabidopsis* and relatives compared with those selected during domestication of our major crops reveals crop domestication has utilized fewer loci with reduced allelic diversity. Perhaps nature hedges its bets maintaining high variation in the population, while crop domestication has prized predictability, at the cost of maximizing plant fit-ness in a fluctuating environment.

Importance of *in natura* flowering analysis

A new realization in the field is the importance of undertaking experiments under field conditions; plants have not evolved to grow in the constant conditions we provide in the laboratory (Shimizu et al. 2011; Nishio et al. 2016; Hepworth et al. 2018; Zhao et al. 2021). So-called in natura field experiments are increasingly important for dissection of molecular mechanisms. Natural fluctuations in both photoperiod and temperature averages and ranges influence the plant transcriptome widely and have multifaceted effects on plant fitness (Nagano et al. 2019). Analysis of classic flowering time mutants revealed that under field conditions, many have much fewer phenotypic effects than in the la-boratory (Wilczek et al. 2009; Song et al. 2018; Taylor et al. 2019). Conversely, field conditions may reveal critical me-chanisms considered less important when studied in the laboratory (Brachi et al. 2010). Temperature dynamics in autumn, not winter, are likely to be the critical variable for vernalization (Duncan et al. 2015; Dixon et al. 2019; Hepworth et al. 2020) partly because as temperatures reduce during autumn their fluctuations have different effects on *FLC* silencing depending on their precise range and timing (Antoniou-Kourounioti et al. 2018; Hepworth et al. 2018). Early freezing is one such effect (Zhao et al. 2021). Temperature fluctuations also turn out to be key to the func-tion of the photoperiod pathway in the field, reducing the im-pact of mutations in *FKF1* and *GI* compared with the laboratory but revealing important roles for *ELF3* and *PHYA* in natural conditions (Song et al. 2018; Kinmonth-Schultz et al. 2023). By investigating the expression dynamics of haplotypes in the field and the subsequent fitness of plants carrying these haplotypes, avoidance of precocious flowering in autumn rather than in spring was revealed as a key driver for the vernalization pathway in a sub-artic environment, with high-expression "slow vernalizing" alleles providing protection against precocious flowering (Hepworth et al. 2020). This fits with findings by Fournier-Level et al. (2022) that late-flowering alleles are promoted in environments with high seasonal temperature fluctuation. Most of these studies monitored the behavior of genes in the field itself: however, many then recapitulated these observations in laboratory settings, in order to test and quantify the observed environmental drivers of different molecular responses. This combination of approaches nullifies some of the distorting effects of laboratory investigation, while exploiting its power to verify interactions, and is likely to be of continuing importance to future research on plant environmental sensitivity.

Analyzing flowering in a new species

What lessons have emerged from all these studies that will influence identification of critical nodes for flowering regulation in newly researched species? Research on rice, wheat, barley, and soybean shows that many of the same principles hold, and *A. thaliana* has (and continues to have) lessons for the functioning of fundamental pathways. A favored strategy in crops with the advent of deep sequencing has been to use RNA-seq analysis and GWAS to look for gene expression variation and linked polymorphism. In the original GWAS studies in *A. thaliana*, *FRI* and *FLC* were not detected as significantly associated loci partly due to population structure in the original set of 96 lines (Atwell et al. 2010), although larger studies with more advanced statistical tools have improved on this (Sasaki et al. 2015). Subsequent QTL (Shindo et al. 2005) and molecular analysis (Li et al. 2014) showed that allelic heterogeneity (different haplotypes having the same phenotype) also reduces the statistical significance, so preventing detection by GWAS. Sample size will also influence detection—GWAS and QTL studies require very large sample sizes, and they can only map those differences that are captured between the initial parental strains. The environmental sensitivity of flowering time is also probably one of the largest difficulties in any single GWAS or QTL experiment. Field studies have demonstrated that much selection is conditional—and the power of GWAS studies to investigate GxE inter-actions can be low (Sasaki et al. 2015).

So going forward, this is we learned that will be useful to define flowering time regulators in a new species, enabling us to breed staple crops adapted to climate change, or develop underutilized crops:

1) Make the most of the considerable molecular knowledge of flowering time gene action.
2) Select GWAS on diversity panels (Harper et al. 2012; 1001 Genomes Consortium 2016) to detect a wide variety of natural alleles affecting flowering; but use intercrossed mapping populations to better identify loci of critical nodes such as FLC (Brachi et al. 2010; Song et al. 2020).
3) Once a critical quantitative node has been identified, explore the dynamics of standing variation of that gene across a wide diversity of accessions. SNPs, presence/absence, transposons etc. within haplotypes will assist in generating targeted, quantitative change in flowering responses while reducing interference with pleiotropic functions of these master regulators (Turner et al. 2005; Liu et al. 2014; Bao et al. 2019; Dixon et al. 2019; Song et al. 2020).
4) Dynamic genes require dynamic methods: many of the critical nodes change quantitatively over time, so time-course analyses are essential (Shindo et al. 2006; Duncan et al. 2015; Nagano et al. 2019; Schiessl et al. 2019; Calderwood et al. 2021a), and tools are becoming available for easier comparison of transcriptomics (Calderwood et al. 2021b).
5) Move molecular experimentation into the field early in the research pipeline—but then return to the laboratory to validate and quantify the results.

Knowledge from molecular research has had high barriers to implementation in crops, in part because phenotypic analysis from the laboratory does not always translate into the field (Atwell et al. 2010). One reason for this is lack of knowledge about critical field conditions, such as the temperature profiles required for vernalization (Hepworth et al. 2018; Dixon et al. 2019) or the light patterns that induce *FT in natura* (Song et al. 2018). However, by combining field studies with more accurate simulation in the laboratory, *Arabidopsis* research is developing a range of new methods for understanding, and crucially predicting, how pathways respond and control flowering in the changing field conditions (Antoniou-Kourounioti et al. 2018; Song et al. 2018; Nagano et al. 2019). With climate change challenging the key mechanisms plants rely on for their timing, *Arabidopsis* research remains critical to fundamental knowledge and plant breeding alike.

Mechanistic information from *A. thaliana* has significantly accelerated understanding of flowering time regulation in all plant species. This is readily recognized by breeding companies (Enza Zaden Box 3). Thus, for the timely production of a

Box 3. Impact of flowering time research to advance plant breeding.

Drs. Xana Verweij and Jeroen Rouppe van der Voort Global Biotech director and Research Manager IP/External Projects at Enza Zaden Research and Development.

Flowering time plays a crucial role in breeding. There are numerous examples ranging from breeding for different seasonal product types in e.g. cauliflower and lettuce (Leijten et al. 2018), balancing the switches from vegetative to generative plant growth in peppers, climate zone adaptation in onion types (Lee et al. 2013a, 2013b), to escaping Phytophthora disease pressure in potato (by planting "early varieties") and seed quality and yield (applicable to any seeded crop variety). In addition, the finding that flowering time regulators are involved in many other plant developmental pathways stresses the importance of obtaining a deep understanding of flowering-related processes in food crops. The advent of the application of omics tools to create any type of data, and the increased capabilities for analyzing such data sets finally enables breeders to identify the key loci and allelic variation to breed for. The work done by fundamental research is essential to uncover the spatio-temporal regulation of flowering time laying the foundations of knowledge that can be translated by breeding companies to create predictable and adaptable crop products. We aim to identify floral pathway integrators in different crops which might serve as breeding targets, allowing us to design strategies towards optimal flowering and robust fruit and seed production even under adverse growing conditions. Therefore, it is essential to bring fundamental research and commercial crop breeding objectives closer together. This will guarantee that we work in synergy for the relevant traits that can have a positive impact in our agricultural systems by e.g. developing resilient crops with optimal yield and quality potential and minimal trade-off effects.

OUTSTANDING QUESTIONS

- To what extent have we identified all the pathways regulating flowering time in distinct climates?
- Do the same principles hold broadly across natural plant species and most crops?
- What makes some floral regulators such sensitive targets to general transcriptional/epigenetic regulators?
- How much does noncoding transcription/long noncoding RNA contribute to altering regulatory dynamics in fluctuating natural environments?

range of climate-proof crops, we need to focus on expanding our mechanistic understanding of flowering time gene function in natural environments and those mechanisms that have changed during adaptation. This will be the fastest route to open new opportunities for crop improvement.

Acknowledgments

The authors thank Dr. Antony Dodd, Trine Hvoslef-Eide, and Odd Arne Rognli for their insightful comments on the manuscript.

Author contributions

R.M., P.Z., J.H., J.-W.W., and C.D. wrote the manuscript. R.M. and P.Z. prepared figures.

Funding

This work was supported by the UK Biotechnology and Biological Sciences Research Council (BB/P013511/1), Wellcome Trust (210654/Z/18/Z), Royal Society Professorship to C.D. and the European Research Council (EPISWITCH-833254). The work in the J.-W.W. laboratory is supported by grants from the National Natural Science Foundation of China (32388201; 31721001) and the New Cornerstone Science Foundation through the XPLORER PRIZE.

Conflict of interest statement. None declared.

Data availability

Sequencing data presented in this article is available on the Short Read Archive (SRA) under the project reference PRJNA1088482.

References

1001 Genomes Consortium. 1,135 genomes reveal the global pattern of polymorphism in *Arabidopsis thaliana*. Cell. 2016:**166**(2):481–491. https://doi.org/10.1016/j.cell.2016.05.063

Abelenda JA, Cruz-Oro E, Franco-Zorrilla JM, Prat S. Potato *StCONSTANS-like1* suppresses storage organ formation by directly activating the FT-like *StSP5G* repressor. Curr Biol. 2016:**26**(7): 872–881. https://doi.org/10.1016/j.cub.2016.01.066

Adrian J, Farrona S, Reimer JJ, Albani MC, Coupland G, Turck F. cis-Regulatory elements and chromatin state coordinately control temporal and spatial expression of *FLOWERING LOCUS T* in Arabidopsis. Plant Cell. 2010:**22**(5):1425–1440. https://doi.org/10.1105/tpc.110.074682

Ågrena J, Oakley CG, McKay JK, Lovell JT, Schemske DW. Genetic mapping of adaptation reveals fitness tradeoffs in *Arabidopsis thaliana*. Proc Natl Acad Sci U S A. 2013:**110**(52):21077–21082. https://doi.org/10.1073/pnas.1316773110

Aikawa S, Kobayashi MJ, Satake A, Shimizu KK, Kudoh H. Robust control of the seasonal expression of the *Arabidopsis FLC* gene in a fluctuating environment. Proc Natl Acad Sci U S A. 2010:**107**(25): 11632–11637. https://doi.org/10.1073/pnas.0914293107

Albani MC, Castaings L, Wötzel S, Mateos JL, Wunder J, Wang R, Reymond M, Coupland G. PEP1 of *Arabis alpina* is encoded by two overlapping genes that contribute to natural genetic variation in perennial flowering. PLoS Genet. 2012:**8**(12):e1003130. https://doi.org/10.1371/journal.pgen.1003130

Allard A, Bink MC, Martinez S, Kelner J-J, Legave J-M, Di Guardo M, Di Pierro EA, Laurens F, Van De Weg EW, Costes E. Detecting QTLs and putative candidate genes involved in budbreak and flowering time in an apple multiparental population. J Exp Bot. 2016:**67**(9): 2875–2888. https://doi.org/10.1093/jxb/erw130

Antoniou-Kourounioti RL, Hepworth J, Heckmann A, Duncan S, Questa J, Rosa S, Sall T, Holm S, Dean C, Howard M. Temperature sensing is distributed throughout the regulatory network that controls *FLC* epigenetic silencing in vernalization. Cell Syst. 2018:**7**(6):643–655. https://doi.org/10.1016/j.cels.2018.10.011

Atwell S, Huang YS, Vilhjalmsson BJ, Willems G, Horton M, Li Y, Meng D, Platt A, Tarone AM, Hu TT, et al. Genome-wide association study of 107 phenotypes in *Arabidopsis thaliana* inbred lines. Nature. 2010:**465**(7298):627–631. https://doi.org/10.1038/nature08800

Auge GA, Penfield S, Donohue K. Pleiotropy in developmental regulation by flowering-pathway genes: is it an evolutionary constraint? New Phytol. 2019:**224**(1):55–70. https://doi.org/10.1111/nph.15901

Aukerman MJ, Sakai H. Regulation of flowering time and floral organ identity by a microRNA and its *APETALA2*-like target genes. Plant Cell. 2003:**15**(11):2730–2741. https://doi.org/10.1105/tpc.016238

Backstrom S, Elfving N, Nilsson R, Wingsle G, Bjorklund S. Purification of a plant mediator from *Arabidopsis thaliana* identifies PFT1 as the Med25 subunit. Mol Cell. 2007:**26**(5):717–729. https://doi.org/10.1016/j.molcel.2007.05.007

Balasubramanian S, Sureshkumar S, Agrawal M, Michael TP, Wessinger C, Maloof JN, Clark R, Warthmann N, Chory J, Weigel D. The *PHYTOCHROME C* photoreceptor gene mediates natural variation in flowering and growth responses of *Arabidopsis thaliana*. Nat Genet. 2006b:**38**(6):711–715. https://doi.org/10.1038/ng1818

Balasubramanian S, Sureshkumar S, Lempe J, Weigel D. Potent induction of *Arabidopsis thaliana* flowering by elevated growth temperature. PLoS Genet. 2006a:**2**(7):e106. https://doi.org/10.1371/journal.pgen.0020106

Bao SJ, Hua CM, Huang GQ, Cheng P, Gong XM, Shen LS, Yu H. Molecular basis of natural variation in photoperiodic flowering responses. Dev Cell. 2019:**50**(1):90–101. https://doi.org/10.1016/j.devcel.2019.05.018

Bastow R, Mylne JS, Lister C, Lippman Z, Martienssen RA, Dean C. Vernalization requires epigenetic silencing of *FLC* by histone methylation. Nature. 2004:**427**(6970):164–167. https://doi.org/10.1038/nature02269

Baumgarten L, Pieper B, Song B, Mane S, Lempe J, Lamb J, Cooke EL, Srivastava R, Strutt S, Zanko D, et al. Pan-European study of genotypes and phenotypes in the Arabidopsis relative *Cardamine hirsuta* reveals how adaptation, demography, and development shape diversity patterns. PLoS Biol. 2023:**21**(7):e3002191. https://doi.org/10.1371/journal.pbio.3002191

Baurle I, Dean C. The timing of developmental transitions in plants. Cell. 2006:**125**(4):655–664. https://doi.org/10.1016/j.cell.2006.05.005

Baurle I, Dean C. Differential interactions of the autonomous pathway RRM proteins and chromatin regulators in the silencing of arabidopsis targets. Plos One. 2008:**3**(7):e2733. https://doi.org/10.1371/journal.pone.0002733

Becker A, Theißen G. The major clades of MADS-box genes and their role in the development and evolution of flowering plants. Mol Phylogenet Evol. 2003:**29**(3):464–489. https://doi.org/10.1016/S1055-7903(03)00207-0

Bernier G. The control of floral evocation and morphogenesis. Annu Rev Plant Physiol Plant Mol Biol. 1988:**39**(1):175–219. https://doi.org/10.1146/annurev.pp.39.060188.001135

Bernier G, Havelange A, Houssa C, Petitjean A, Lejeune P. Physiological signals that induce flowering. Plant Cell. 1993:**5**(10):1147–1155. https://doi.org/10.2307/3869768

Bernier G, Jean-Marie K, Sachs RM. The physiology of flowering. Boca Raton (FL): CRC Press; 1981.

Bielenberg DG, Wang Y, Li Z, Zhebentyayeva T, Fan S, Reighard GL, Scorza R, Abbott AG. Sequencing and annotation of the evergrowing locus in peach [*Prunus persica* (L.) batsch] reveals a cluster of six MADS-box transcription factors as candidate genes for regulation of terminal bud formation. Tree Genet Genomes. 2008:**4**(3):495–507. https://doi.org/10.1007/s11295-007-0126-9

Bouche F, Lobet G, Tocquin P, Perilleux C. FLOR-ID: an interactive database of flowering-time gene networks in Arabidopsis thaliana. Nucleic Acids Res. 2016:**44**(D1):D1167–D1171. https://doi.org/10.1093/nar/gkv1054

Brachi B, Faure N, Horton M, Flahauw E, Vazquez A, Nordborg M, Bergelson J, Cuguen J, Roux F. Linkage and association mapping of *Arabidopsis thaliana* flowering time in nature. PLoS Genet. 2010:**6**(5):e1000940. https://doi.org/10.1371/journal.pgen.1000940

Caicedo AL, Richards C, Ehrenreich IM, Purugganan MD. Complex rearrangements lead to novel chimeric gene fusion polymorphisms at the *Arabidopsis thaliana MAF2-5* flowering time gene cluster. Mol Biol Evol. 2009:**26**(3):699–711. https://doi.org/10.1093/molbev/msn300

Calderwood A, Hepworth J, Woodhouse S, Bilham L, Jones DM, Tudor E, Ali M, Dean C, Wells R, Irwin JA, et al. Comparative transcriptomics reveals desynchronisation of gene expression during the floral transition between Arabidopsis and *Brassica rapa* cultivars. Quant Plant Biol. 2021b:**2**:e4. https://doi.org/10.1017/qpb.2021.6

Calderwood A, Lloyd A, Hepworth J, Tudor EH, Jones DM, Woodhouse S, Bilham L, Chinoy C, Williams K, Corke F, et al. Total *FLC* transcript dynamics from divergent paralogue expression explains flowering diversity in *Brassica napus*. New Phytol. 2021a:**229**(6):3534–3548. https://doi.org/10.1111/nph.17131

Calle A, Cai L, Iezzoni A, Wünsch A. Genetic dissection of bloom time in low chilling sweet cherry (*Prunus avium* L.) using a multi-family QTL approach. Front Plant Sci. 2020:**10**:1647. https://doi.org/10.3389/fpls.2019.01647

Cao S, Luo X, Xu D, Tian X, Song J, Xia X, Chu C, He Z. Genetic architecture underlying light and temperature mediated flowering in *Arabidopsis*, rice, and temperate cereals. New Phytol. 2021:**230**(5):1731–1745. https://doi.org/10.1111/nph.17276

Castaings L, Bergonzi S, Albani MC, Kemi U, Savolainen O, Coupland G. Evolutionary conservation of cold-induced antisense RNAs of *FLOWERING LOCUS C* in *Arabidopsis thaliana* perennial relatives. Nat Commun. 2014:**5**(1):4457. https://doi.org/10.1038/ncomms5457

Cerdan PD, Chory J. Regulation of flowering time by light quality. Nature. 2003:**423**(6942):881–885. https://doi.org/10.1038/nature01636

Chandler J, Wilson A, Dean C. *Arabidopsis* mutants showing an altered response to vernalization. Plant J. 1996:**10**(4):637–644. https://doi.org/10.1046/j.1365-313X.1996.10040637.x

Chen L, Hu PC, Lu QQ, Zhang F, Su YH, Ding Y. Vernalization attenuates dehydration tolerance in winter-annual Arabidopsis. Plant Physiol. 2022:**190**(1):732–744. https://doi.org/10.1093/plphys/kiac264

Chen M, MacGregor DR, Dave A, Florance H, Moore K, Paszkiewicz K, Smirnoff N, Graham IA, Penfield S. Maternal temperature history activates *Flowering Locus T* in fruits to control progeny dormancy according to time of year. Proc Natl Acad Sci U S A. 2014:**111**(52):18787–18792. https://doi.org/10.1073/pnas.1412274111

Chen Q, Zheng Y, Luo LD, Yang YP, Hu XY, Kong XX. Functional FRIGIDA allele enhances drought tolerance by regulating the P5CS1 pathway in Arabidopsis thaliana. Biochem Biophys Res Commun. 2018:**495**(1):1102–1107. https://doi.org/10.1016/j.bbrc.2017.11.149

Chen XM. A microRNA as a translational repressor of APETALA2 in Arabidopsis flower development. Science. 2004:**303**(5666):2022–2025. https://doi.org/10.1126/science.1088060

Cheng F, Sun R, Hou X, Zheng H, Zhang F, Zhang Y, Liu B, Liang J, Zhuang M, Liu Y, et al. Subgenome parallel selection is associated with morphotype diversification and convergent crop domestication in *Brassica rapa* and *Brassica oleracea*. Nat Genet. 2016:**48**(10):1218–1224. https://doi.org/10.1038/ng.3634

Cheng YJ, Shang GD, Xu ZG, Yu S, Wu LY, Zhai D, Tian SL, Gao J, Wang L, Wang JW. Cell division in the shoot apical meristem is a trigger for miR156 decline and vegetative phase transition in Arabidopsis. Proc Natl Acad Sci U S A. 2021:**118**(46):e2115667118. https://doi.org/10.1073/pnas.2115667118

Chiang GC, Barua D, Kramer EM, Amasino RM, Donohue K. Major flowering time gene, *FLOWERING LOCUS C*, regulates seed germination in *Arabidopsis thaliana*. Proc Natl Acad Sci U S A. 2009:**106**(28):11661–11666. https://doi.org/10.1073/pnas.0901367106

Choi H, Back S, Kim GW, Lee K, Venkatesh J, Lee HB, Kwon J-K, Kang B-C. Development of a speed breeding protocol with flowering gene investigation in pepper (Capsicum annuum). Front Plant Sci. 2023:**14**:1151765. https://doi.org/10.3389/fpls.2023.1151765

Clarke JH, Dean C. Mapping FRI, a locus controlling flowering time and vernalization response in Arabidopsis thaliana. Mol Gen Genet. 1994:**242**(1):81–89. https://doi.org/10.1007/BF00277351

Comadran J, Kilian B, Russell J, Ramsay L, Stein N, Ganal M, Shaw P, Bayer M, Thomas W, Marshall D, et al. Natural variation in a homolog of *Antirrhinum CENTRORADIALIS* contributed to spring growth habit and environmental adaptation in cultivated barley. Nat Genet. 2012:**44**(12):1388–1392. https://doi.org/10.1038/ng.2447

Conti L, Bradley D. *TERMINAL FLOWER1* is a mobile signal controlling *Arabidopsis* architecture. Plant Cell. 2007:**19**(3):767–778. https://doi.org/10.1105/tpc.106.049767

Corbesier L, Vincent C, Jang S, Fornara F, Fan Q, Searle I, Giakountis A, Farrona S, Gissot L, Turnbull C, et al. FT protein movement contributes to long-distance signaling in floral induction of *Arabidopsis*. Science. 2007:**316**(5827):1030–1033. https://doi.org/10.1126/science.1141752

Crevillen P, Yang H, Cui X, Greeff C, Trick M, Qiu Q, Cao X, Dean C. Epigenetic reprogramming that prevents transgenerational inheritance of the vernalized state. Nature. 2014:**515**(7528):587–590. https://doi.org/10.1038/nature13722

Csorba T, Questa JI, Sun Q, Dean C. Antisense *COOLAIR* mediates the coordinated switching of chromatin states at *FLC* during vernalization. Proc Natl Acad Sci U S A. 2014:**111**(45):16160–16165. https://doi.org/10.1073/pnas.1419030111

Dai XD, Ding YN, Tan LB, Fu YC, Liu FX, Zhu ZF, Sun XY, Sun XW, Gu P, Cai HW, et al. *LHD1*, an allele of *DTH8/Ghd8*, controls late heading date in common wild rice (*Oryza rufipogon*). J Integr Plant Biol. 2012:**54**(10):790–799. https://doi.org/10.1111/j.1744-7909.2012.01166.x

De Lucia F, Crevillen P, Jones AM, Greb T, Dean C. A PHD-polycomb repressive complex 2 triggers the epigenetic silencing of *FLC* during vernalization. Proc Natl Acad Sci U S A. 2008:**105**(44):16831–16836. https://doi.org/10.1073/pnas.0808687105

Deng W, Ying H, Helliwell CA, Taylor JM, Peacock WJ, Dennis ES. FLOWERING LOCUS C (FLC) regulates development pathways throughout the life cycle of Arabidopsis. Proc Natl Acad Sci U S A. 2011:**108**(16):6680–6685. https://doi.org/10.1073/pnas.1103175108

Dittmar EL, Oakley CG, Ågren J, Schemske DW. Flowering time QTL in natural populations of *Arabidopsis thaliana* and implications for their adaptive value. Mol Ecol. 2014:**23**(17):4291–4303. https://doi.org/10.1111/mec.12857

Dixon LE, Karsai I, Kiss T, Adamski NM, Liu Z, Ding Y, Allard V, Boden SA, Griffiths S. *VERNALIZATION1* controls developmental responses of winter wheat under high ambient temperatures. Development. 2019:**146**(3):dev172684. https://doi.org/10.1242/dev.172684

Doi K, Izawa T, Fuse T, Yamanouchi U, Kubo T, Shimatani Z, Yano M, Yoshimura A. *Ehd1*, a B-type response regulator in rice, confers short-day promotion of flowering and controls *FT-like* gene expression independently of *Hd1*. Genes Dev. 2004:**18**(8):926–936. https://doi.org/10.1101/gad.1189604

Duncan S, Holm S, Questa J, Irwin J, Grant A, Dean C. Seasonal shift in timing of vernalization as an adaptation to extreme winter. Elife. 2015:**4**:e06620. https://doi.org/10.7554/eLife.06620

El-Din El-Assal S, Alonso-Blanco C, Peeters AJM, Raz V, Koornneef M. A QTL for flowering time in *Arabidopsis* reveals a novel allele of *CRY2*. Nat Genet. 2001:**29**(4):435–440. https://doi.org/10.1038/ng767

Falavigna VDS, Guitton B, Costes E, Andrés F. I want to (bud) break free: the potential role of *DAM* and *SVP*-like genes in regulating dormancy cycle in temperate fruit trees. Front Plant Sci. 2019:**9**:1990. https://doi.org/10.3389/fpls.2018.01990

Fernandez V, Takahashi Y, Le Gourrierec J, Coupland G. Photoperiodic and thermosensory pathways interact through *CONSTANS* to promote flowering at high temperature under short days. Plant J. 2016:**86**(5):426–440. https://doi.org/10.1111/tpj.13183

Filiault DL, Wessinger CA, Dinneny JR, Lutes J, Borevitz JO, Weigel D, Chory J, Maloof JN. Amino acid polymorphisms in *Arabidopsis* phytochrome B cause differential responses to light. Proc Natl Acad Sci U S A. 2008:**105**(8):3157–3162. https://doi.org/10.1073/pnas.0712174105

Finnegan EJ, Kovac KA, Jaligot E, Sheldon CC, Peacock WJ, Dennis ES. The downregulation of *FLOWERING LOCUS C* (*FLC*) expression in plants with low levels of DNA methylation and by vernalization occurs by distinct mechanisms. Plant J. 2005:**44**(3):420–432. https://doi.org/10.1111/j.1365-313X.2005.02541.x

Fournier-Level A, Taylor MA, Paril JF, Martinez-Berdeja A, Stitzer MC, Cooper MD, Roe JL, Wilczek AM, Schmitt J. Adaptive significance of flowering time variation across natural seasonal environments in *Arabidopsis thaliana*. New Phytol. 2022:**234**(2):719–734. https://doi.org/10.1111/nph.17999

Fowler S, Lee K, Onouchi H, Samach A, Richardson K, Morris B, Coupland G, Putterill J. *GIGANTEA*: a circadian clock-controlled gene that regulates photoperiodic flowering in *Arabidopsis* and encodes a protein with several possible membrane-spanning domains. EMBO J. 1999:**18**(17):4679–4688. https://doi.org/10.1093/emboj/18.17.4679

Franco-Echevarria E, Nielsen M, Schulten A, Cheema J, Morgan TE, Bienz M, Dean C. Distinct accessory roles of *Arabidopsis* VEL proteins in Polycomb silencing. Genes Dev. 2023:**37**(17–18):801. https://doi.org/10.1101/gad.350814.123

Fu D, Szűcs P, Yan L, Helguera M, Skinner JS, Von Zitzewitz J, Hayes PM, Dubcovsky J. Large deletions within the first intron in *VRN-1* are associated with spring growth habit in barley and wheat. Mol Genet Genom. 2005:**273**(1):54–65. https://doi.org/10.1007/s00438-004-1095-4

Galvao VC, Fiorucci AS, Trevisan M, Franco-Zorilla JM, Goyal A, Schmid-Siegert E, Solano R, Fankhauser C. PIF transcription factors link a neighbor threat cue to accelerated reproduction in *Arabidopsis*. Nat Commun. 2019:**10**(1):4005. https://doi.org/10.1038/s41467-019-11882-7

Gao H, Jin MN, Zheng XM, Chen J, Yuan DY, Xin YY, Wang MQ, Huang DY, Zhang Z, Zhou KN, et al. *Days to heading 7*, a major quantitative locus determining photoperiod sensitivity and regional adaptation in rice. Proc Natl Acad Sci U S A. 2014:**111**(46):18337–18342. https://doi.org/10.1073/pnas.1418204111

Gao J, Zhang K, Cheng YJ, Yu S, Shang GD, Wang FX, Wu LY, Xu ZG, Mai YX, Zhao XY, et al. A robust mechanism for resetting juvenility during each generation in *Arabidopsis*. Nat Plants. 2022:**8**(3):257–268. https://doi.org/10.1038/s41477-022-01110-4

Gaston A, Osorio S, Denoyes B, Rothan C. Applying the Solanaceae strategies to strawberry crop improvement. Trends Plant Sci. 2020:**25**(2):130–140. https://doi.org/10.1016/j.tplants.2019.10.003

Gaudinier A, Blackman BK. Evolutionary processes from the perspective of flowering time diversity. New Phytol. 2020:**225**(5):1883–1898. https://doi.org/10.1111/nph.16205

Gazzani S, Gendall AR, Lister C, Dean C. Analysis of the molecular basis of flowering time variation in *Arabidopsis* accessions. Plant Physiol. 2003:**132**(2):1107–1114. https://doi.org/10.1104/pp.103.021212

Geraldo N, Baurle I, Kidou S, Hu X, Dean C. FRIGIDA delays flowering in Arabidopsis via a cotranscriptional mechanism involving direct interaction with the nuclear cap-binding complex. Plant Physiol. 2009:**150**(3):1611–1618. https://doi.org/10.1104/pp.109.137448

Giaquinta RT. Phloem loading of sucrose. Annu Rev Plant Physiol Plant Mol Biol. 1983:**34**(1):347–387. https://doi.org/10.1146/annurev.pp.34.060183.002023

Gnesutta N, Kumimoto RW, Swain S, Chiara M, Siriwardana C, Horner DS, Holt BF III, Mantovani R. CONSTANS imparts dna sequence specificity to the histone fold NF-YB/NF-YC dimer. Plant Cell. 2017:**29**(6):1516–1532. https://doi.org/10.1105/tpc.16.00864

Gong ZZ. Flowering phenology as a core domestication trait in soybean. J Integr Plant Biol. 2020:**62**(5):546–549. https://doi.org/10.1111/jipb.12934

Goretti D, Silvestre M, Collani S, Langenecker T, Mendez C, Madueno F, Schmid M. TERMINAL FLOWER1 functions as a mobile transcriptional cofactor in the shoot apical meristem. Plant Physiol. 2020:**182**(4):2081–2095. https://doi.org/10.1104/pp.19.00867

Gou JY, Felippes FF, Liu CJ, Weigel D, Wang JW. Negative regulation of anthocyanin biosynthesis in *Arabidopsis* by a miR156-targeted SPL transcription factor. Plant Cell. 2011:**23**(4):1512–1522. https://doi.org/10.1105/tpc.111.084525

Greb T, Mylne JS, Crevillen P, Geraldo N, An H, Gendall AR, Dean C. The PHD finger protein VRN5 functions in the epigenetic silencing of *Arabidopsis FLC*. Curr Biol. 2007:**17**(1):73–78. https://doi.org/10.1016/j.cub.2006.11.052

Greenup AG, Sasani S, Oliver SN, Talbot MJ, Dennis ES, Hemming MN, Trevaskis B. *ODDSOC2* is a MADS box floral repressor that is down-regulated by vernalization in temperate cereals. Plant Physiol. 2010:**153**(3):1062–1073. https://doi.org/10.1104/pp.109.152488

Guo Y-L, Todesco M, Hagmann J, Das S, Weigel D. Independent *FLC* mutations as causes of flowering-time variation in *Arabidopsis thaliana* and *Capsella rubella*. Genetics. 2012:**192**(2):729–739. https://doi.org/10.1534/genetics.112.143958

Guo L, Wang XH, Zhao M, Huang C, Li C, Li D, Yang CJ, York AM, Xue W, Xu GH, et al. Stepwise cis-regulatory changes in *ZCN8* contribute to maize flowering-time adaptation. Curr Biol. 2018:**28**(18):3005–3015. https://doi.org/10.1016/j.cub.2018.07.029

Hanzawa Y, Money T, Bradley D. A single amino acid converts a repressor to an activator of flowering. Proc Natl Acad Sci U S A. 2005:**102**(21):7748–7753. https://doi.org/10.1073/pnas.0500932102

Harmer SL, Fankhauser C, Webb AA. Focus on circadian rhythms. Plant Physiol. 2022:**190**(2):921–923. https://doi.org/10.1093/plphys/kiac353

Harper AL, Trick M, Higgins J, Fraser F, Clissold L, Wells R, Hattori C, Werner P, Bancroft I. Associative transcriptomics of traits in the polyploid crop species *Brassica napus*. Nat Biotechnol. 2012:**30**(8):798–802. https://doi.org/10.1038/nbt.2302

Hawkes EJ, Hennelly SP, Novikova IV, Irwin JA, Dean C, Sanbonmatsu KY. *COOLAIR* antisense RNAs form evolutionarily conserved elaborate secondary structures. Cell Rep. 2016:**16**(12):3087–3096. https://doi.org/10.1016/j.celrep.2016.08.045

He Y, Doyle MR, Amasino RM. PAF1-complex-mediated histone methylation of *FLOWERING LOCUS C* chromatin is required for the vernalization-responsive, winter-annual habit in *Arabidopsis*. Genes Dev. 2004:**18**(22):2774–2784. https://doi.org/10.1101/gad.1244504

Helal MU, Islam N, Kadir M, Miah NH. Performance of rapeseed and mustard (*Brassica sp.*) varieties/lines in north-east region (Sylhet) of Bangladesh. Agril Res Technol. 2016:**2**(1):01–06. https://doi.org/10.19080/ARTOAJ.2016.02.555576

Heo JB, Sung S. Vernalization-mediated epigenetic silencing by a long intronic noncoding RNA. Science. 2011:**331**(6013):76–79. https://doi.org/10.1126/science.1197349

Hepworth J, Antoniou-Kourounioti RL, Berggren K, Selga C, Tudor EH, Yates B, Cox D, Collier Harris BR, Irwin JA, Howard M. Natural variation in autumn expression is the major adaptive determinant distinguishing *Arabidopsis FLC* haplotypes. Elife. 2020:**9**:e57671. https://doi.org/10.7554/eLife.57671

Hepworth J, Antoniou-Kourounioti RL, Bloomer RH, Selga C, Berggren K, Cox D, Collier Harris BR, Irwin JA, Holm S, Sall T, et al. Absence of warmth permits epigenetic memory of winter in *Arabidopsis*. Nat Commun. 2018:**9**(1):639. https://doi.org/10.1038/s41467-018-03065-7

Hicks KA, Millar AJ, Carre IA, Somers DE, Straume M, Meeks-Wagner DR, Kay SA. Conditional circadian dysfunction of the *Arabidopsis early-flowering* 3 mutant. Science. 1996:**274**(5288):790–792. https://doi.org/10.1126/science.274.5288.790

Higgins JA, Bailey PC, Laurie DA. Comparative genomics of flowering time pathways using Brachypodium distachyon as a model for the temperate grasses. PLoS One. 2010:**5**(4):e10065. https://doi.org/10.1371/journal.pone.0010065

Hiraoka K, Yamaguchi A, Abe M, Araki T. The florigen genes *FT* and *TSF* modulate lateral shoot outgrowth in *Arabidopsis thaliana*. Plant Cell Physiol. 2013:**54**(3):352–368. https://doi.org/10.1093/pcp/pcs168

Hoffmann MH. Evolution of the realized climatic niche in the genus Arabidopsis (Brassicaceae). Evolution. 2005:**59**:1425–1436. https://doi.org/10.1111/j.0014-3820.2005.tb01793.x

Hori K, Ogiso-Tanaka E, Matsubara K, Yamanouchi U, Ebana K, Yano M. *Hd16*, a gene for casein kinase I, is involved in the control of rice flowering time by modulating the day-length response. Plant J. 2013:**76**(1):36–46. https://doi.org/10.1111/tpj.12268

Huang X, Ding J, Effgen S, Turck F, Koornneef M. Multiple loci and genetic interactions involving flowering time genes regulate stem branching among natural variants of *Arabidopsis*. New Phytol. 2013:**199**(3):843–857. https://doi.org/10.1111/nph.12306

Huang XH, Kurata N, Wei XH, Wang ZX, Wang A, Zhao Q, Zhao Y, Liu KY, Lu HY, Li WJ, et al. A map of rice genome variation reveals the origin of cultivated rice. Nature. 2012:**490**(7421):497–501. https://doi.org/10.1038/nature11532

Huang C, Sun HY, Xu DY, Chen QY, Liang YM, Wang XF, Xu GH, Tian JG, Wang CL, Li D, et al. *ZmCCT9* enhances maize adaptation to higher latitudes. Proc Natl Acad Sci U S A. 2018:**115**(2):E334–E341. https://doi.org/10.1073/pnas.1718058115

Hung F-Y, Shih Y-H, Lin P-Y, Feng Y-R, Li C, Wu K. WRKY63 transcriptional activation of *COOLAIR* and *COLDAIR* regulates vernalization-induced flowering. Plant Physiol. 2022:**190**(1):532–547. https://doi.org/10.1093/plphys/kiac295

Hyun Y, Richter R, Coupland G. Competence to flower: age-controlled sensitivity to environmental cues. Plant Physiol. 2017:**173**(1):36–46. https://doi.org/10.1104/pp.16.01523

Hyun Y, Vincent C, Tilmes V, Bergonzi S, Kiefer C, Richter R, Martinez-Gallegos R, Severing E, Coupland G. A regulatory circuit conferring varied flowering response to cold in annual and perennial plants. Science. 2019:**363**(6425):409–412. https://doi.org/10.1126/science.aau8197

Imaizumi T, Schultz TF, Harmon FG, Ho LA, Kay SA. FKF1 F-box protein mediates cyclic degradation of a repressor of *CONSTANS* in *Arabidopsis*. Science. 2005:**309**(5732):293–297. https://doi.org/10.1126/science.1110586

Inigo S, Alvarez MJ, Strasser B, Califano A, Cerdan PD. PFT1, the MED25 subunit of the plant mediator complex, promotes flowering through *CONSTANS* dependent and independent mechanisms in *Arabidopsis*. Plant J. 2012:**69**(4):601–612. https://doi.org/10.1111/j.1365-313X.2011.04815.x

Izawa T. Adaptation of flowering-time by natural and artificial selection in *Arabidopsis* and rice. J Exp Bot. 2007:**58**(12):3091–3097. https://doi.org/10.1093/jxb/erm159

Jang S, Marchal V, Panigrahi KC, Wenkel S, Soppe W, Deng XW, Valverde F, Coupland G. *Arabidopsis* COP1 shapes the temporal pattern of CO accumulation conferring a photoperiodic flowering response. EMBO J. 2008:**27**(8):1277–1288. https://doi.org/10.1038/emboj.2008.68

Jeon M, Jeong G, Yang Y, Luo X, Jeong D, Kyung J, Hyun Y, He Y, Lee I. Vernalization-triggered expression of the antisense transcript *COOLAIR* is mediated by *CBF* genes. eLife. 2023:**12**. https://doi.org/10.7554/eLife.84594

Jiao F, Pahwa K, Manning M, Dochy N, Geuten K. Cold induced antisense transcription of *FLOWERING LOCUS C* in distant grasses. Front Plant Sci. 2019:**10**. https://doi.org/10.3389/fpls.2019.00072

Jin S, Kim SY, Susila H, Nasim Z, Youn G, Ahn JH. FLOWERING LOCUS M isoforms differentially affect the subcellular localization and stability of SHORT VEGETATIVE PHASE to regulate temperature-responsive flowering in Arabidopsis. Mol Plant. 2022:**15**(11):1696–1709. https://doi.org/10.1016/j.molp.2022.08.007

Jing SL, Jiang P, Sun XM, Yu L, Wang ES, Qin J, Zhang F, Prat S, Song BT. Long-distance control of potato storage organ formation by SELF PRUNING 3D and FLOWERING LOCUS T-like 1. Plant Commun. 2023:**4**(3):100547. https://doi.org/10.1016/j.xplc.2023.100547

Johanson U, West J, Lister C, Michaels S, Amasino R, Dean C. Molecular analysis of FRIGIDA, a major determinant of natural variation in *Arabidopsis* flowering time. Science. 2000:**290**(5490):344–347. https://doi.org/10.1126/science.290.5490.344

Jones DM, Wells R, Pullen N, Trick M, Irwin JA, Morris RJ. Spatio-temporal expression dynamics differ between homologues of flowering time genes in the allopolyploid *Brassica napus*. Plant J. 2018:**96**(1):103–118. https://doi.org/10.1111/tpj.14020

Kardailsky I, Shukla VK, Ahn JH, Dagenais N, Christensen SK, Nguyen JT, Chory J, Harrison MJ, Weigel D. Activation tagging of the floral inducer FT. Science. 1999:**286**(5446):1962–1965. https://doi.org/10.1126/science.286.5446.1962

Kemi U, Niittyvuopio A, Toivainen T, Pasanen A, Quilot-Turion B, Holm K, Lagercrantz U, Savolainen O, Kuittinen H. Role of vernalization and of duplicated *FLOWERING LOCUS C* in the perennial *Arabidopsis lyrata*. New Phytol. 2013:**197**(1):323–335. https://doi.org/10.1111/j.1469-8137.2012.04378.x

Kidd BN, Edgar CI, Kumar KK, Aitken EA, Schenk PM, Manners JM, Kazan K. The mediator complex subunit PFT1 is a key regulator of jasmonate-dependent defense in *Arabidopsis*. Plant Cell. 2009:**21**(8):2237–2252. https://doi.org/10.1105/tpc.109.066910

Kiefer C, Severing E, Karl R, Bergonzi S, Koch M, Tresch A, Coupland G. Divergence of annual and perennial species in the Brassicaceae and the contribution of cis-acting variation at *FLC* orthologues. Mol Ecol. 2017:**26**(13):3437–3457. https://doi.org/10.1111/mec.14084

Kim D-H, Sung S. Vernalization-triggered intragenic chromatin loop formation by long noncoding RNAs. Dev Cell. 2017:**40**(3):302–331. https://doi.org/10.1016/j.devcel.2016.12.021

Kim D-H, Xi Y, Sung S. Modular function of long noncoding RNA, COLDAIR, in the vernalization response. PLoS Genet. 2017:**13**(7):e1006939. https://doi.org/10.1371/journal.pgen.1006939

Kinmonth-Schultz H, Lewandowska-Sabat A, Imaizumi T, Ward JK, Rognli OA, Fjellheim S. Flowering times of wild *Arabidopsis* accessions from across Norway correlate with expression levels of *FT*, *CO*, and *FLC* genes. Front Plant Sci. 2021:**12**:747740. https://doi.org/10.3389/fpls.2021.747740

Kinmonth-Schultz H, Sønstebø JH, Croneberger AJ, Johnsen SS, Leder E, Lewandowska-Sabat A, Imaizumi T, Rognli OA, Vinje H, Ward JK,

et al. Responsiveness to long days for flowering is reduced in *Arabidopsis* by yearly variation in growing season temperatures. Plant Cell Environ. 2023:**46**(11):3337–3352. https://doi.org/10.1111/pce.14632

Kippes N, Guedira M, Lin L, Alvarez MA, Brown-Guedira GL, Dubcovsky J. Single nucleotide polymorphisms in a regulatory site of *VRN-A1* first intron are associated with differences in vernalization requirement in winter wheat. Mol Genet Genom. 2018:**293**(5): 1231–1243. https://doi.org/10.1007/s00438-018-1455-0

Kobayashi Y, Kaya H, Goto K, Iwabuchi M, Araki T. A pair of related genes with antagonistic roles in mediating flowering signals. Science. 1999:**286**(5446):1960–1962. https://doi.org/10.1126/science.286.5446.1960

Kobayashi MJ, Takeuchi Y, Kenta T, Kume T, Diway B, Shimizu KK. Mass flowering of the tropical tree Shorea beccariana was preceded by expression changes in flowering and drought-responsive genes. Mol Ecol. 2013:**22**(18):4767–4782. https://doi.org/10.1111/mec.12344

Kojima S, Takahashi Y, Kobayashi Y, Monna L, Sasaki T, Araki T, Yano M. *Hd3a*, a rice ortholog of the Arabidopsis *FT* gene, promotes transition to flowering downstream of *Hd1* under short-day conditions. Plant Cell Physiol. 2002:**43**(10):1096–1105. https://doi.org/10.1093/pcp/pcf156

Komiya R, Ikegami A, Tamaki S, Yokoi S, Shimamoto K. *Hd3a* and *RFT1* are essential for flowering in rice. Development. 2008:**135**(4): 767–774. https://doi.org/10.1242/dev.008631

Kong FJ, Liu BH, Xia ZJ, Sato S, Kim BM, Watanabe S, Yamada T, Tabata S, Kanazawa A, Harada K, et al. Two coordinately regulated homologs of *FLOWERING LOCUS T* are involved in the control of photoperiodic flowering in soybean. Plant Physiol. 2010:**154**(3): 1220–1231. https://doi.org/10.1104/pp.110.160796

Konopatskaia I, Vavilova V, Kondratenko EY, Blinov A, Goncharov NP. *VRN1* genes variability in tetraploid wheat species with a spring growth habit. BMC Plant Biol. 2016:**16**(S3):93–106. https://doi.org/10.1186/s12870-016-0924-z

Koornneef M, Alonso-Blanco C, Blankestijn-de Vries H, Hanhart CJ, Peeters AJM. Genetic interactions among late-flowering mutants of Arabidopsis. Genetics. 1998:**148**(2):885–892. https://doi.org/10.1093/genetics/148.2.885

Koornneef M, Hanhart CJ, van der Veen JH. A genetic and physiological analysis of late flowering mutants in *Arabidopsis thaliana*. Mol Gen Genet. 1991:**229**(1):57–66. https://doi.org/10.1007/BF00264213

Kumar SV, Lucyshyn D, Jaeger KE, Alos E, Alvey E, Harberd NP, Wigge PA. Transcription factor PIF4 controls the thermosensory activation of flowering. Nature. 2012:**484**(7393):242–245. https://doi.org/10.1038/nature10928

Lang A. The photoperiodic regulation of promotion and inhibition of flower formation. Ber Dtsch Bot Ges. 1984:**97**(1):293–314. https://doi.org/10.1111/j.1438-8677.1984.tb03470.x

Laubinger S, Marchal V, Le Gourrierec J, Wenkel S, Adrian J, Jang S, Kulajta C, Braun H, Coupland G, Hoecker U. Arabidopsis SPA proteins regulate photoperiodic flowering and interact with the floral inducer CONSTANS to regulate its stability. Development. 2006:**133**(16): 3213–3222. https://doi.org/10.1242/dev.02481

Le Corre V, Roux F, Reboud X. DNA polymorphism at the *FRIGIDA* gene in *Arabidopsis thaliana:* extensive nonsynonymous variation is consistent with local selection for flowering time. Mol Biol Evol. 2002:**19**(8):1261–1271. https://doi.org/10.1093/oxfordjournals.molbev.a004187

Lee JH, Ryu HS, Chung KS, Pose D, Kim S, Schmid M, Ahn JH. Regulation of temperature-responsive flowering by MADS-box transcription factor repressors. Science. 2013a:**342**(6158):628–632. https://doi.org/10.1126/science.1241097

Lee JH, Yoo SJ, Park SH, Hwang I, Lee JS, Ahn JH. Role of SVP in the control of flowering time by ambient temperature in *Arabidopsis*. Genes Dev. 2007:**21**(4):397–402. https://doi.org/10.1101/gad.1518407

Lee R, Baldwin S, Kenel F, McCallum J, Macknight R. *FLOWERING LOCUS T* genes control onion bulb formation and flowering. Nat Commun. 2013b:**4**(1):2884. https://doi.org/10.1038/ncomms3884

Lee Z, Kim S, Choi SJ, Joung E, Kwon M, Park HJ, Shim JS. Regulation of flowering time by environmental factors in plants. Plants (Basel). 2023:**12**(21):3680. https://doi.org/10.3390/plants12213680

Leijten W, Koes R, Roobeek I, Frugis G. Translating flowering time from *Arabidopsis thaliana* to Brassicaceae and Asteraceae crop species. Plants (Basel). 2018:**7**(4):111. https://doi.org/10.3390/plants7040111

Li LZ, Chang TG, Kang H, Zhu XG, Wang JW. Simulating evolution trajectory of ruderal weeds by computational modeling. Plant Physiol. 2023a:**193**(2):923–926. https://doi.org/10.1093/plphys/kiad392

Li LZ, Xu ZG, Chang TG, Wang L, Kang H, Zhai D, Zhang LY, Zhang P, Liu HT, Zhu XG, et al. Common evolutionary trajectory of short life-cycle in Brassicaceae ruderal weeds. Nat Commun. 2023b:**14**(1):290. https://doi.org/10.1038/s41467-023-35966-7

Li MW, Liu W, Lam HM, Gendron JM. Characterization of two growth period QTLs reveals modification of *PRR3* genes during soybean domestication. Plant Cell Physiol. 2019:**60**(2):407–420. https://doi.org/10.1093/pcp/pcy215

Li P, Tao Z, Dean C. Phenotypic evolution through variation in splicing of the noncoding RNA *COOLAIR*. Genes Dev. 2015:**29**(7):696–701. https://doi.org/10.1101/gad.258814.115

Li PJ, Filiault D, Box MS, Kerdaffrec E, van Oosterhout C, Wilczek AM, Schmitt J, McMullan M, Bergelson J, Nordborg M, et al. Multiple *FLC* haplotypes defined by independent cis-regulatory variation underpin life history diversity in *Arabidopsis thaliana*. Genes Dev. 2014:**28**(15):1635–1640. https://doi.org/10.1101/gad.245993.114

Li TD, Yang XP, Yu Y, Si XM, Zhai XW, Zhang HW, Dong WX, Gao CX, Xu C. Domestication of wild tomato is accelerated by genome editing. Nat Biotechnol. 2018:**36**(12):1160–1163. https://doi.org/10.1038/nbt.4273

Li X, Liang T, Liu HT. How plants coordinate their development in response to light and temperature signals. Plant Cell. 2022a:**34**(3): 955–966. https://doi.org/10.1093/plcell/koab302

Li XX, Zhou HW, Cheng L, Ma NN, Cui BF, Wang WF, Zhong YJ, Liao H. Shoot-to-root translocated GmNN1/FT2a triggers nodulation and regulates soybean nitrogen nutrition. Plos Biol. 2022b:**20**(8): e3001739. https://doi.org/10.1371/journal.pbio.3001739

Li Y, Huang Y, Bergelson J, Nordborg M, Borevitz JO. Association mapping of local climate-sensitive quantitative trait loci in *Arabidopsis thaliana*. Proc Natl Acad Sci U S A. 2010:**107**(49):21199–-21204. https://doi.org/10.1073/pnas.1007431107

Li YH, Guan RX, Liu ZX, Ma YS, Wang LX, Li LH, Lin FY, Luan WJ, Chen PY, Yan Z, et al. Genetic structure and diversity of cultivated soybean (*Glycine max* (L.) merr.) landraces in China. Theor Appl Genet. 2008:**117**(6):857–871. https://doi.org/10.1007/s00122-008-0825-0

Lian H, Wang L, Ma N, Zhou CM, Han L, Zhang TQ, Wang JW. Redundant and specific roles of individual *MIR172* genes in plant development. Plos Biol. 2021:**19**(2):e3001044. https://doi.org/10.1371/journal.pbio.3001044

Liang YM, Liu Q, Wang XF, Huang C, Xu GH, Hey S, Lin HY, Li C, Xu DY, Wu LS, et al. *ZmMADS69* functions as a flowering activator through the *ZmRap2.7-ZCN8* regulatory module and contributes to maize flowering time adaptation. New Phytol. 2019:**221**(4):2335–2347. https://doi.org/10.1111/nph.15512

Liang YM, Liu HJ, Yan JB, Tian F. Natural variation in crops: realized understanding, continuing promise. Annu Rev Plant Biol. 2021:**72**(1): 357–385. https://doi.org/10.1146/annurev-arplant-080720-090632

Lifschitz E, Eviatar T, Rozman A, Shalit A, Goldshmidt A, Amsellem Z, Alvarez JP, Eshed Y. The tomato *FT* ortholog triggers systemic signals that regulate growth and flowering and substitute for diverse environmental stimuli. Proc Natl Acad Sci U S A. 2006:**103**(16): 6398–6403. https://doi.org/10.1073/pnas.0601620103

Lim MH, Kim J, Kim YS, Chung KS, Seo YH, Lee I, Kim J, Hong CB, Kim HJ, Park CM. A new Arabidopsis gene, *FLK*, encodes an RNA binding protein with K homology motifs and regulates flowering time via *FLOWERING LOCUS C*. Plant Cell. 2004:**16**(3):731–740. https://doi.org/10.1105/tpc.019331

Lin MK, Belanger H, Lee YJ, Varkonyi-Gasic E, Taoka K, Miura E, Xoconostle-Cazares B, Gendler K, Jorgensen RA, Phinney B, et al. *FLOWERING LOCUS T* protein may act as the long-distance florigenic signal in the cucurbits. Plant Cell. 2007:**19**(5):1488–1506. https://doi.org/10.1105/tpc.107.051920

Liu F, Marquardt S, Lister C, Swiezewski S, Dean C. Targeted 3′ processing of antisense transcripts triggers *Arabidopsis FLC* chromatin silencing. Science. 2010:**327**(5961):94–97. https://doi.org/10.1126/science.1180278

Liu FQ, Quesada V, Crevillen P, Baurle I, Swiezewski S, Dean C. The *Arabidopsis* RNA-binding protein FCA requires a lysine-specific demethylase 1 homolog to downregulate *FLC*. Mol Cell. 2007:**28**(3): 398–407. https://doi.org/10.1016/j.molcel.2007.10.018

Liu L, Li CY, Teo ZWN, Zhang B, Yu H. The MCTP-SNARE complex regulates florigen transport in *Arabidopsis*. Plant Cell. 2019:**31**(10): 2475–2490. https://doi.org/10.1105/tpc.18.00960

Liu L, Liu C, Hou XL, Xi WY, Shen LS, Tao Z, Wang Y, Yu H. FTIP1 is an essential regulator required for florigen transport. Plos Biol. 2012:**10**(4):e1001313. https://doi.org/10.1371/journal.pbio.1001313

Liu L, Zhang Y, Yu H. Florigen trafficking integrates photoperiod and temperature signals in *Arabidopsis*. J Integr Plant Biol. 2020:**62**(9): 1385–1398. https://doi.org/10.1111/jipb.13000

Liu LY, Adrian J, Pankin A, Hu JY, Dong X, von Korff M, Turck F. Induced and natural variation of promoter length modulates the photoperiodic response of *FLOWERING LOCUS T*. Nat Commun. 2014:**5**(1):4558. https://doi.org/10.1038/ncomms5558

Liu TM, Liu HY, Zhang H, Xing YZ. Validation and characterization of *Ghd7.1*, a Major quantitative trait locus with pleiotropic effects on spikelets per panicle, plant height, and heading date in rice (*Oryza sativa* L.). J Integr Plant Biol. 2013:**55**(10):917–927. https://doi.org/10.1111/jipb.12070

Lu SJ, Dong LD, Fang C, Liu SL, Kong LP, Cheng Q, Chen LY, Su T, Nan HY, Zhang D, et al. Stepwise selection on homeologous *PRR* genes controlling flowering and maturity during soybean domestication. Nat Genet. 2020:**52**(4):428–436. https://doi.org/10.1038/s41588-020-0604-7

Lu SJ, Fang C, Abe J, Kong FJ, Liu BH. Current overview on the genetic basis of key genes involved in soybean domestication. aBIOTECH. 2022a:**3**(2):126–139. https://doi.org/10.1007/s42994-022-00074-5

Lu X, Lü P, Liu H, Chen H, Pan X, Liu P, Feng L, Zhong S, Zhou B. Identification of chilling accumulation-associated genes for litchi flowering by transcriptome-based genome-wide association studies. Front Plant Sci. 2022b:**13**:819188. https://doi.org/10.3389/fpls.2022.819188

Lu SJ, Zhao XH, Hu YL, Liu SL, Nan HY, Li XM, Fang C, Cao D, Shi XY, Kong LP, et al. Natural variation at the soybean *J* locus improves adaptation to the tropics and enhances yield. Nat Genet. 2017:**49**(5):773–779. https://doi.org/10.1038/ng.3819

Lutz U, Nussbaumer T, Spannagl M, Diener J, Mayer KF, Schwechheimer C. Natural haplotypes of *FLM* non-coding sequences fine-tune flowering time in ambient spring temperatures in *Arabidopsis*. Elife. 2017:**6**:e22114. https://doi.org/10.7554/eLife.22114

Lutz U, Posé D, Pfeifer M, Gundlach H, Hagmann J, Wang C, Weigel D, Mayer KF, Schmid M, Schwechheimer C. Modulation of ambient temperature-dependent flowering in *Arabidopsis thaliana* by natural variation of *FLOWERING LOCUS M*. PLoS Genet. 2015:**11**(10): e1005588. https://doi.org/10.1371/journal.pgen.1005588

Lv XC, Zeng XL, Hu HM, Chen LX, Zhang F, Liu R, Liu Y, Zhou XL, Wang CS, Wu Z, et al. Structural insights into the multivalent binding of the *Arabidopsis FLOWERING LOCUS T* promoter by the CO-NF-Y master transcription factor complex. Plant Cell. 2021:**33**(4):1182–1195. https://doi.org/10.1093/plcell/koab016

Macknight R, Bancroft I, Page T, Lister C, Schmidt R, Love K, Westphal L, Murphy G, Sherson S, Cobbett C, et al. *FCA*, a gene controlling flowering time in *Arabidopsis*, encodes a protein containing RNA-binding domains. Cell. 1997:**89**(5):737–745. https://doi.org/10.1016/S0092-8674(00)80256-1

Maeda AE, Nakamichi N. Plant clock modifications for adapting flowering time to local environments. Plant Physiol. 2022:**190**(2): 952–967. https://doi.org/10.1093/plphys/kiac107

Mao YB, Liu YQ, Chen DY, Chen FY, Fang X, Hong GJ, Wang LJ, Wang JW, Chen XY. Jasmonate response decay and defense metabolite accumulation contributes to age-regulated dynamics of plant insect resistance. Nat Commun. 2017:**8**(1):13925. https://doi.org/10.1038/ncomms13925

Marquardt S, Boss PK, Hadfield J, Dean C. Additional targets of the *Arabidopsis* autonomous pathway members, *FCA* and *FY*. J Exp Bot. 2006:**57**(13):3379–3386. https://doi.org/10.1093/jxb/erl073

Marquardt S, Raitskin O, Wu Z, Liu F, Sun Q, Dean C. Functional consequences of splicing of the antisense transcript *COOLAIR* on *FLC* transcription. Mol Cell. 2014:**54**(1):156–165. https://doi.org/10.1016/j.molcel.2014.03.026

Martignago D, da Silveira Falavigna V, Lombardi A, Gao H, Korwin Kurkowski PK, Galbiati M, Tonelli C, Coupland G, Conti L. The bZIP transcription factor AREB3 mediates FT signalling and floral transition at the Arabidopsis shoot apical meristem. PLoS Genet. 2023:**19**(5): e1010766. https://doi.org/10.1371/journal.pgen.1010766

Mateo Bonmati E, Fang X, Maple R, Fiedler M, Passmore LA, Dean CA. The CPSF phosphatase module links transcription termination to chromatin silencing. bioRxiv 547976. https://doi.org/10.1101/2023.07.06.547976, July 6, 2023. Preprint: not peer reviewed.

Mateos JL, Tilmes V, Madrigal P, Severing E, Richter R, Rijkenberg CW, Krajewski P, Coupland G. Divergence of regulatory networks governed by the orthologous transcription factors FLC and PEP1 in Brassicaceae species. Proc Natl Acad Sci U S A. 2017:**114**(51): E11037–E11046. https://doi.org/10.1073/pnas.1618075114

Matsuoka Y, Vigouroux Y, Goodman MM, Sanchez G J, Buckler E, Doebley J. A single domestication for maize shown by multilocus microsatellite genotyping. Proc Natl Acad Sci U S A. 2002:**99**(9): 6080–6084. https://doi.org/10.1073/pnas.052125199

McAssey EV, Corbi J, Burke JM. Range-wide phenotypic and genetic differentiation in wild sunflower. BMC Plant Biol. 2016:**16**(1):249. https://doi.org/10.1186/s12870-016-0937-7

Méndez-Vigo B, Martínez-Zapater JM, Alonso-Blanco C. The flowering repressor *SVP* underlies a novel *Arabidopsis thaliana* QTL interacting with the genetic background. PLoS Genet. 2013:**9**(1): e1003289. https://doi.org/10.1371/journal.pgen.1003289

Méndez-Vigo B, Picó FX, Ramiro M, Martínez-Zapater JM, Alonso-Blanco C. Altitudinal and climatic adaptation is mediated by flowering traits and *FRI*, *FLC*, and *PHYC* genes in *Arabidopsis*. Plant Physiol. 2011:**157**(4):1942–1955. https://doi.org/10.1104/pp.111.183426

Menon G, Mateo Bonmati E, Reeck S, Maple R, Wu Z, Ietswaart R, Dean C, Howard M. Proximal termination generates a transcriptional state that determines the rate of establishment of Polycomb silencing. bioRxiv. https://doi.org/10.1101/2023.07.06.547969, July 6, 2023. Preprint: not peer reviewed.

Michaels SD, Amasino RM. *FLOWERING LOCUS C* encodes a novel MADS domain protein that acts as a repressor of flowering. Plant Cell. 1999:**11**(5):949–956. https://doi.org/10.1105/tpc.11.5.949

Mikulski P, Wolff P, Lu T, Nielsen M, Echevarria EF, Zhu D, Questa JI, Saalbach G, Martins C, Dean C. VAL1 acts as an assembly platform co-ordinating co-transcriptional repression and chromatin regulation at *Arabidopsis FLC*. Nat Commun. 2022:**13**(1):5542. https://doi.org/10.1038/s41467-022-32897-7

Molnar A, Melnyk CW, Bassett A, Hardcastle TJ, Dunn R, Baulcombe DC. Small silencing RNAs in plants are mobile and direct epigenetic modification in recipient cells. Science. 2010:**328**(5980):872–875. https://doi.org/10.1126/science.1187959

Nagano AJ, Kawagoe T, Sugisaka J, Honjo MN, Iwayama K, Kudoh H. Annual transcriptome dynamics in natural environments reveals plant seasonal adaptation. Nat Plants. 2019:**5**(1):74–83. https://doi.org/10.1038/s41477-018-0338-z

Nakamichi N, Kita M, Niinuma K, Ito S, Yamashino T, Mizoguchi T, Mizuno T. Arabidopsis clock-associated pseudo-response regulators

PRR9, PRR7 and PRR5 coordinately and positively regulate flowering time through the canonical CONSTANS-dependent photoperiodic pathway. Plant Cell Physiol. 2007:**48**(6):822–832. https://doi.org/10.1093/pcp/pcm056

Nakamichi N, Yamaguchi J, Sato A, Fujimoto KJ, Ota E. Chemical biology to dissect molecular mechanisms underlying plant circadian clocks. New Phytol. 2022:**235**(4):1336–1343. https://doi.org/10.1111/nph.18298

Napp-Zinn K. Untersuchungen zur genetik des kältebedürfnisses bei Arabidopsis thaliana. Z Indukt AbstammungsVererbungsl. 1957:**88**: 253–285. https://doi.org/10.1007/BF00308342

Navarro C, Abelenda JA, Cruz-Oro E, Cuellar CA, Tamaki S, Silva J, Shimamoto K, Prat S. Control of flowering and storage organ formation in potato by FLOWERING LOCUS T. Nature. 2011:**478**(7367): 119–U132. https://doi.org/10.1038/nature10431

Neto C, Hancock A. Genetic architecture of flowering time differs between populations with contrasting demographic and selective histories. Mol Biol Evol. 2023:**40**(8):msad185. https://doi.org/10.1093/molbev/msad185

Nielsen M, Menon G, Zhao Y, Mateo-Bonmati E, Wolff P, Zhou S, Howard M, Dean C. *COOLAIR* and PRC2 function in parallel to silence *FLC* during vernalization. Proc Natl Acad Sci U S A. 2024:**121**(4):e2311474121. https://doi.org/10.1073/pnas.2311474121

Nishio H, Buzas DM, Nagano AJ, Suzuki Y, Sugano S, Ito M, Morinaga S-I, Kudoh H. From the laboratory to the field: assaying histone methylation at *FLOWERING LOCUS C* in naturally growing *Arabidopsis halleri*. Genes Genet Syst. 2016:**91**(1):15–26. https://doi.org/10.1266/ggs.15-00071

Ogiso-Tanaka E, Matsubara K, Yamamoto S, Nonoue Y, Wu JZ, Fujisawa H, Ishikubo H, Tanaka T, Ando T, Matsumoto T, et al. Natural variation of the RICE *FLOWERING LOCUS T 1* contributes to flowering time divergence in rice. Plos One. 2013:**8**(10):e75959. https://doi.org/10.1371/journal.pone.0075959

Oliver SN, Finnegan EJ, Dennis ES, Peacock WJ, Trevaskis B. Vernalization-induced flowering in cereals is associated with changes in histone methylation at the *VERNALIZATION1* gene. Proc Natl Acad Sci U S A. 2009:**106**(20):8386–8391. https://doi.org/10.1073/pnas.0903566106

Ó'Maoileidigh DS, van Driel AD, Singh A, Sang Q, Le Bec N, Vincent C, de Olalla EBG, Vayssieres A, Romera Branchat M, Severing E, et al. Systematic analyses of the *MIR172* family members of *Arabidopsis* define their distinct roles in regulation of *APETALA2* during floral transition. Plos Biol. 2021:**19**(2):e3001043. https://doi.org/10.1371/journal.pbio.3001043

Oravec MW, Greenham K. The adaptive nature of the plant circadian clock in natural environments. Plant Physiol. 2022:**190**(2):968–980. https://doi.org/10.1093/plphys/kiac337

Osnato M, Cota I, Nebhnani P, Cereijo U, Pelaz S. Photoperiod control of plant growth: flowering time genes beyond flowering. Front Plant Sci. 2021:**12**:805635. https://doi.org/10.3389/fpls.2021.805635

Poethig RS. Phase change and the regulation of developmental timing in plants. Science. 2003:**301**(5631):334–336. https://doi.org/10.1126/science.1085328

Poethig RS. The past, present, and future of vegetative phase change. Plant Physiol. 2010:**154**(2):541–544. https://doi.org/10.1104/pp.110.161620

Poethig, RS, Fouracre J. Temporal regulation of vegetative phase change in plants. Dev Cell. 2024:**59**(1):4–19. https://doi.org/10.1016/j.devcel.2023.11.010

Pose D, Verhage L, Ott F, Yant L, Mathieu J, Angenent GC, Immink RG, Schmid M. Temperature-dependent regulation of flowering by antagonistic *FLM* variants. Nature. 2013:**503**(7476):414–417. https://doi.org/10.1038/nature12633

Putterill J, Robson F, Lee K, Simon R, Coupland G. The CONSTANS gene of Arabidopsis promotes flowering and encodes a protein showing similarities to zinc finger transcription factors. Cell. 1995:**80**(6): 847–857. https://doi.org/10.1016/0092-8674(95)90288-0

Qüesta JI, Antoniou-Kourounioti RL, Rosa S, Li P, Duncan S, Whittaker C, Howard M, Dean C. Noncoding SNPs influence a distinct phase of Polycomb silencing to destabilize long-term epigenetic memory at *Arabidopsis FLC*. Genes Dev. 2020:**34**(5–6):446–461. https://doi.org/10.1101/gad.333245.119

Quiroz S, Yustis JC, Chavez-Hernandez EC, Martinez T, Sanchez MP, Garay-Arroyo A, Alvarez-Buylla ER, Garcia-Ponce B. Beyond the genetic pathways, flowering regulation complexity in *Arabidopsis thaliana*. Int J Mol Sci. 2021:**22**(11):5716. https://doi.org/10.3390/ijms22115716

Regnault T, Daviere JM, Wild M, Sakvarelidze-Achard L, Heintz D, Carrera Bergua E, Lopez Diaz I, Gong F, Hedden P, Achard P. The gibberellin precursor GA12 acts as a long-distance growth signal in *Arabidopsis*. Nat Plants. 2015:**1**(6):15073. https://doi.org/10.1038/nplants.2015.73

Rosas U, Mei Y, Xie QG, Banta JA, Zhou RW, Seufferheld G, Gerard S, Chou L, Bhambhra N, Parks JD, et al. Variation in *Arabidopsis* flowering time associated with *cis*-regulatory variation in *CONSTANS*. Nat Commun. 2014:**5**(1):3651. https://doi.org/10.1038/ncomms4651

Sachs RM, Hackett WP. Source-sink relationships and flowering. In: **Meudt WJ**, editor. Beltsville symposia in agricultural research. UK: Allanheld, Osmun; 1983. p. 263–272.

Salome PA, Bomblies K, Laitinen RAE, Yant L, Mott R, Weigel D. Genetic architecture of flowering-time variation in *Arabidopsis thaliana*. Genetics. 2011:**188**(2):421–433. https://doi.org/10.1534/genetics.111.126607

Saroha A, Pal D, Gomashe SS, Akash, Kaur V, Ujjainwal S, Rajkumar S, Aravind J, Radhamani J, Kumar R, et al. Identification of QTNs associated with flowering time, maturity, and plant height traits in Linum usitatissimum L. Using genome-wide association study. Front Genet. 2022:**13**:811924. https://doi.org/10.3389/fgene.2022.811924

Sasaki E, Zhang P, Atwell S, Meng D, Nordborg M. "Missing" G x E variation controls flowering time in *Arabidopsis thaliana*. PLoS Genet. 2015:**11**(10):e1005597. https://doi.org/10.1371/journal.pgen.1005597

Schiessl S. Regulation and subfunctionalization of flowering time genes in the allotetraploid oil crop *Brassica napus*. Front Plant Sci. 2020:**11**: 605155. https://doi.org/10.3389/fpls.2020.605155

Schiessl SV, Quezada-Martinez D, Tebartz E, Snowdon RJ, Qian L. The vernalisation regulator *FLOWERING LOCUS C* is differentially expressed in biennial and annual *Brassica napus*. Sci Rep. 2019:**9**(1): 14911. https://doi.org/10.1038/s41598-019-51212-x

Schilling S, Pan S, Kennedy A, Melzer R. MADS-box genes and crop domestication: the jack of all traits. J Exp Bot. 2018:**69**(7): 1447–1469. https://doi.org/10.1093/jxb/erx479

Schomburg FM, Patton DA, Meinke DW, Amasino RM. *FPA*, a gene involved in floral induction in *Arabidopsis*, encodes a protein containing RNA-recognition motifs. Plant Cell. 2001:**13**(6):1427–1436. https://doi.org/10.1105/TPC.010017

Schon M, Baxter C, Xu C, Enugutti B, Nodine MD, Dean C. Antagonistic activities of cotranscriptional regulators within an early developmental window set *FLC* expression level. Proc Natl Acad Sci U S A. 2021:**118**(17):e2102753118. https://doi.org/10.1073/pnas.2102753118

Schwartz C, Balasubramanian S, Warthmann N, Michael TP, Lempe J, Sureshkumar S, Kobayashi Y, Maloof JN, Borevitz JO, Chory J, et al. Cis-regulatory changes at *FLOWERING LOCUS T* mediate natural variation in flowering responses of *Arabidopsis thaliana*. Genetics. 2009:**183**(2):723–732. https://doi.org/10.1534/genetics.109.104984

Searle I, He Y, Turck F, Vincent C, Fornara F, Krober S, Amasino RA, Coupland G. The transcription factor *FLC* confers a flowering response to vernalization by repressing meristem competence and systemic signaling in *Arabidopsis*. Genes Dev. 2006:**20**(7):898–912. https://doi.org/10.1101/gad.373506

Sheldon CC, Hills MJ, Lister C, Dean C, Dennis ES, Peacock WJ. Resetting of *FLOWERING LOCUS C* expression after epigenetic repression by vernalization. Proc Natl Acad Sci U S A. 2008:**105**(6): 2214–2219. https://doi.org/10.1073/pnas.0711453105

Shimizu KK, Kudoh H, Kobayashi MJ. Plant sexual reproduction during climate change: gene function in natura studied by ecological and evolutionary systems biology. Ann Bot. 2011:**108**(4):777–787. https://doi.org/10.1093/aob/mcr180

Shindo C, Aranzana MJ, Lister C, Baxter C, Nicholls C, Nordborg M, Dean C. Role of *FRIGIDA* and *FLOWERING LOCUS C* in determining variation in flowering time of *Arabidopsis*. Plant Physiol. 2005:**138**(2):1163–1173. https://doi.org/10.1104/pp.105.061309

Shindo C, Lister C, Crevillen P, Nordborg M, Dean C. Variation in the epigenetic silencing of *FLC* contributes to natural variation in *Arabidopsis* vernalization response. Genes Dev. 2006:**20**(22):3079–3083. https://doi.org/10.1101/gad.405306

Shukla A, Pagan I, Crevillen P, Alonso-Blanco C, Garcia-Arenal F. A role of flowering genes in the tolerance of *Arabidopsis thaliana* to cucumber mosaic virus. Mol Plant Pathol. 2022:**23**(2):175–187. https://doi.org/10.1111/mpp.13151

Song YH, Estrada DA, Johnson RS, Kim SK, Lee SY, MacCoss MJ, Imaizumi T. Distinct roles of FKF1, Gigantea, and Zeitlupe proteins in the regulation of Constans stability in *Arabidopsis* photoperiodic flowering. Proc Natl Acad Sci U S A. 2014:**111**(49):17672–17677. https://doi.org/10.1073/pnas.1415375111

Song J-M, Guan Z, Hu J, Guo C, Yang Z, Wang S, Liu D, Wang B, Lu S, Zhou R, et al. Eight high-quality genomes reveal pan-genome architecture and ecotype differentiation of *Brassica napus*. Nat Plants. 2020:**6**(1):34–45. https://doi.org/10.1038/s41477-019-0577-7

Song YH, Kubota A, Kwon MS, Covington MF, Lee N, Taagen ER, Laboy Cintron D, Hwang DY, Akiyama R, Hodge SK, et al. Molecular basis of flowering under natural long-day conditions in *Arabidopsis*. Nat Plants. 2018:**4**(10):824–835. https://doi.org/10.1038/s41477-018-0253-3

Strange A, Li P, Lister C, Anderson J, Warthmann N, Shindo C, Irwin J, Nordborg M, Dean C. Major-effect alleles at relatively few loci underlie distinct vernalization and flowering variation in *Arabidopsis accessions*. PLoS One. 2011:**6**(5):e19949. https://doi.org/10.1371/journal.pone.0019949

Suarez-Lopez P, Wheatley K, Robson F, Onouchi H, Valverde F, Coupland G. *CONSTANS* mediates between the circadian clock and the control of flowering in *Arabidopsis*. Nature. 2001:**410**(6832):1116–1120. https://doi.org/10.1038/35074138

Sun HB, Jia Z, Cao D, Jiang BJ, Wu CX, Hou WS, Liu YK, Fei ZH, ZhaoDZ, Han TF. *GmFT2a*, a soybean homolog of *FLOWERING LOCUS T*, is involved in flowering transition and maintenance. Plos One. 2011:**6**(12):e29238. https://doi.org/10.1371/journal.pone.0029238

Sung SB, Amasino RM. Vernalization in *Arabidopsis thaliana* is mediated by the PHD finger protein VIN3. Nature. 2004:**427**(6970):159–164. https://doi.org/10.1038/nature02195

Sureshkumar S, Dent C, Seleznev A, Tasset C, Balasubramanian S. Nonsense-mediated mRNA decay modulates FLM-dependent thermosensory flowering response in *Arabidopsis*. Nat Plants. 2016:**2**(5):16055. https://doi.org/10.1038/nplants.2016.55

Susila H, Juric S, Liu L, Gawarecka K, Chung KS, Jin S, Kim SJ, Nasim Z, Youn G, Suh MC, et al. Florigen sequestration in cellular membranes modulates temperature-responsive flowering. Science. 2021:**373**(6559):1137–1141. https://doi.org/10.1126/science.abh4054

Swiezewski S, Liu F, Magusin A, Dean C. Cold-induced silencing by long antisense transcripts of an Arabidopsis polycomb target. Nature. 2009:**462**(7274):799–802. https://doi.org/10.1038/nature08618

Takahashi Y, Shomura A, Sasaki T, Yano M. *Hd6*, a rice quantitative trait locus involved in photoperiod sensitivity, encodes the alpha subunit of protein kinase CK2. Proc Natl Acad Sci U S A. 2001:**98**(14):7922–7927. https://doi.org/10.1073/pnas.111136798

Takeno K. Stress-induced flowering: the third category of flowering response. J Exp Bot. 2016:**67**(17):4925–4934. https://doi.org/10.1093/jxb/erw272

Tamaki S, Matsuo S, Wong HL, Yokoi S, Shimamoto K. Hd3a protein is a mobile flowering signal in rice. Science. 2007:**316**(5827):1033–1036. https://doi.org/10.1126/science.1141753

Taylor MA, Wilczek AM, Roe JL, Welch SM, Runcie DE, Cooper MD, Schmitt J. Large-effect flowering time mutations reveal conditionally adaptive paths through fitness landscapes in *Arabidopsis thaliana*. Proc Natl Acad Sci U S A. 2019:**116**(36):17890–17899. https://doi.org/10.1073/pnas.1902731116

Teo CJ, Takahashi K, Shimizu K, Shimamoto K, Taoka K. Potato tuber induction is regulated by interactions between components of a tuberigen complex. Plant Cell Physiol. 2017:**58**(2):365–374. https://doi.org/10.1093/pcp/pcw197

Tsuji H, Taoka K-i, Shimamoto K. Florigen in rice: complex gene network for florigen transcription, florigen activation complex, and multiple functions. Curr Opin Plant Biol. 2013:**16**(2):228–235. https://doi.org/10.1016/j.pbi.2013.01.005

Turner A, Beales J, Faure S, Dunford RP, Laurie DA. The pseudo-response regulator *Ppd-H1* provides adaptation to photoperiod in barley. Science. 2005:**310**(5750):1031–1034. https://doi.org/10.1126/science.1117619

Upadhyaya HD, Bajaj D, Das S, Saxena MS, Badoni S, Kumar V, Tripathi S, Gowda C, Sharma S, Tyagi AK, et al. A genome-scale integrated approach aids in genetic dissection of complex flowering time trait in chickpea. Plant Mol Biol. 2015:**89**(4–5):403–420. https://doi.org/10.1007/s11103-015-0377-z

Urrestarazu J, Muranty H, Denancé C, Leforestier D, Ravon E, Guyader A, Guisnel R, Feugey L, Aubourg S, Celton J-M, et al. Genome-wide association mapping of flowering and ripening periods in apple. Front Plant Sci. 2017:**8**:1923. https://doi.org/10.3389/fpls.2017.01923

Valverde F, Mouradov A, Soppe W, Ravenscroft D, Samach A, Coupland G. Photoreceptor regulation of CONSTANS protein in photoperiodic flowering. Science. 2004:**303**(5660):1003–1006. https://doi.org/10.1126/science.1091761

Wang JW, Czech B, Weigel D. miR156-Regulated SPL transcription factors define an endogenous flowering pathway in *Arabidopsis thaliana*. Cell. 2009a:**138**(4):738–749. https://doi.org/10.1016/j.cell.2009.06.014

Wang Q, Sajja U, Rosloski S, Humphrey T, Kim MC, Bomblies K, Weigel D, Grbic V. *HUA2* caused natural variation in shoot morphology of *A. thaliana*. Curr Biol. 2007:**17**(17):1513–1519. https://doi.org/10.1016/j.cub.2007.07.059

Wang R, Farrona S, Vincent C, Joecker A, Schoof H, Turck F, Alonso-Blanco C, Coupland G, Albani MC. *PEP1* regulates perennial flowering in *Arabis alpina*. Nature. 2009b:**459**(7245):423–427. https://doi.org/10.1038/nature07988

Wang Y, Wang Z, Amyot L, Tian L, Xu Z, Gruber MY, Hannoufa A. Ectopic expression of *miR156* represses nodulation and causes morphological and developmental changes in *Lotus japonicus*. Mol Genet Genomics. 2015:**290**(2):471–484. https://doi.org/10.1007/s00438-014-0931-4

Watanabe S, Hideshima R, Xia ZJ, Tsubokura Y, Sato S, Nakamoto Y, Yamanaka N, Takahashi R, Ishimoto M, Anai T, et al. Map-based cloning of the gene associated with the soybean maturity locus *E3*. Genetics. 2009:**182**(4):1251–1262. https://doi.org/10.1534/genetics.108.098772

Watanabe S, Xia ZJ, Hideshima R, Tsubokura Y, Sato S, Yamanaka N, Takahashi R, Anai T, Tabata S, Kitamura K, et al. A map-based cloning strategy employing a residual heterozygous line reveals that the *GIGANTEA* gene is involved in soybean maturity and flowering. Genetics. 2011:**188**(2):395–407. https://doi.org/10.1534/genetics.110.125062

Wei XJ, Xu JF, Guo HN, Jiang L, Chen SH, Yu CY, Zhou ZL, Hu PS, Zhai HQ, Wan JM. *DTH8* suppresses flowering in rice, influencing plant height and yield potential simultaneously. Plant Physiol. 2010:**153**(4):1747–1758. https://doi.org/10.1104/pp.110.156943

Wenkel S, Turck F, Singer K, Gissot L, Le Gourrierec J, Samach A, Coupland G. CONSTANS and the CCAAT box binding complex share a functionally important domain and interact to regulate flowering of Arabidopsis. Plant Cell. 2006:**18**(11):2971–2984. https://doi.org/10.1105/tpc.106.043299

Werner JD, Borevitz JO, Uhlenhaut NH, Ecker JR, Chory J, Weigel D. *FRIGIDA*-independent variation in flowering time of natural *Arabidopsis thaliana* accessions. Genetics. 2005a:**170**(3):1197–1207. https://doi.org/10.1534/genetics.104.036533

Werner JD, Borevitz JO, Warthmann N, Trainer GT, Ecker JR, Chory J, Weigel D. Quantitative trait locus mapping and DNA array hybridization identify an *FLM* deletion as a cause for natural flowering-time variation. Proc Natl Acad Sci U S A. 2005b:**102**(7):2460–2465. https://doi.org/10.1073/pnas.0409474102

Wilczek AM, Roe JL, Knapp MC, Cooper MD, Lopez-Gallego C, Martin LJ, Muir CD, Sim S, Walker A, Anderson J, et al. Effects of genetic perturbation on seasonal life history plasticity. Science. 2009:**323**(5916):930–934. https://doi.org/10.1126/science.1165826

Wilson DC, Carella P, Isaacs M, Cameron RK. The floral transition is not the developmental switch that confers competence for the *Arabidopsis* age-related resistance response to *Pseudomonas syringae* pv. *tomato*. Plant Mol Biol. 2013:**83**(3):235–246. https://doi.org/10.1007/s11103-013-0083-7

Wood CC, Robertson M, Tanner G, Peacock WJ, Dennis ES, Helliwell CA. The *Arabidopsis thaliana* vernalization response requires a polycomb-like protein complex that also includes VERNALIZATION INSENSITIVE 3. Proc Natl Acad Sci U S A. 2006:**103**(39):14631–14636. https://doi.org/10.1073/pnas.0606385103

Wu G, Park MY, Conway SR, Wang JW, Weigel D, Poethig RS. The sequential action of *miR156* and *miR172* regulates developmental timing in *Arabidopsis*. Cell. 2009:**138**(4):750–759. https://doi.org/10.1016/j.cell.2009.06.031

Wu G, Poethig RS. Temporal regulation of shoot development in *Arabidopsis* thaliana by *miR156* and its target *SPL3*. Development. 2006:**133**(18):3539–3547. https://doi.org/10.1242/dev.02521

Xi W, Liu C, Hou X, Yu H. *MOTHER OF FT AND TFL1* regulates seed germination through a negative feedback loop modulating ABA signaling in *Arabidopsis*. Plant Cell. 2010:**22**(6):1733–1748. https://doi.org/10.1105/tpc.109.073072

Xia ZJ, Watanabe S, Yamada T, Tsubokura Y, Nakashima H, Zhai H, Anai T, Sato S, Yamazaki T, Lu SX, et al. Positional cloning and characterization reveal the molecular basis for soybean maturity locus *E1* that regulates photoperiodic flowering. Proc Natl Acad Sci U S A. 2012:**109**(32):E2155–E2164. https://doi.org/10.1073/pnas.1117982109

Xing S, Salinas M, Hohmann S, Berndtgen R, Huijser P. miR156-targeted and nontargeted SBP-box transcription factors act in concert to secure male fertility in *Arabidopsis*. Plant Cell. 2010:**22**(12):3935–3950. https://doi.org/10.1105/tpc.110.079343

Xu C, Fang X, Lu T, Dean C. Antagonistic cotranscriptional regulation through ARGONAUTE1 and the THO/TREX complex orchestrates *FLC* transcriptional output. Proc Natl Acad Sci U S A. 2021:**118**(47): e2113757118. https://doi.org/10.1073/pnas.2113757118

Xu SJ, Chong K. Remembering winter through vernalisation. Nat Plants. 2018:**4**(12):997–1009. https://doi.org/10.1038/s41477-018-0301-z

Xu YY, Zeng RF, Zhou H, Qiu MQ, Gan ZM, Yang YL, Hu SF, Zhou JJ, Hu CG, Zhang JZ. Citrus FRIGIDA cooperates with its interaction partner dehydrin to regulate drought tolerance. Plant J. 2022:**111**(1): 164–182. https://doi.org/10.1111/tpj.15785

Xue WY, Xing YZ, Weng XY, Zhao Y, Tang WJ, Wang L, Zhou HJ, Yu SB, Xu CG, Li XH, et al. Natural variation in *Ghd7* is an important regulator of heading date and yield potential in rice. Nat Genet. 2008:**40**(6):761–767. https://doi.org/10.1038/ng.143

Yamaguchi A, Kobayashi Y, Goto K, Abe M, Araki T. *TWIN SISTER OF FT (TSF)* acts as a floral pathway integrator redundantly with *FT*. Plant Cell Physiol. 2005:**46**(8):1175–1189. https://doi.org/10.1093/pcp/pci151

Yamaguchi A, Wu MF, Yang L, Wu G, Poethig RS, Wagner D. The MicroRNA-regulated SBP-box transcription factor SPL3 is a direct upstream activator of *LEAFY, FRUITFULL*, and *APETALA1*. Dev Cell. 2009:**17**(2):268–278. https://doi.org/10.1016/j.devcel.2009.06.007

Yan L, Fu D, Li C, Blechl A, Tranquilli G, Bonafede M, Sanchez A, Valarik M, Yasuda S, Dubcovsky J. The wheat and barley vernalization gene *VRN3* is an orthologue of *FT*. Proc Natl Acad Sci U S A. 2006:**103**(51): 19581–19586. https://doi.org/10.1073/pnas.0607142103

Yan L, Loukoianov A, Tranquilli G, Helguera M, Fahima T, Dubcovsky J. Positional cloning of the wheat vernalization gene *VRN1*. Proc Natl Acad Sci U S A. 2003:**100**(10):6263–6268. https://doi.org/10.1073/pnas.0937399100

Yan LL, Loukoianov A, Blechl A, Tranquilli G, Ramakrishna W, SanMiguel P, Bennetzen JL, Echenique V, Dubcovsky J. The wheat *VRN2* gene is a flowering repressor down-regulated by vernalization. Science. 2004:**303**(5664):1640–1644. https://doi.org/10.1126/science.1094305

Yan WH, Wang P, Chen HX, Zhou HJ, Li QP, Wang CR, Ding ZH, Zhang YS, Yu SB, Xing YZ, et al. A Major QTL, *ghd8*, plays pleiotropic roles in regulating grain productivity, plant height, and heading date in rice. Mol Plant. 2011:**4**(2):319–330. https://doi.org/10.1093/mp/ssq070

Yang H, Berry S, Olsson TSG, Hartley M, Howard M, Dean C. Distinct phases of polycomb silencing to hold epigenetic memory of cold in *Arabidopsis*. Science. 2017:**357**(6356):1142–1145. https://doi.org/10.1126/science.aan1121

Yang M, Zhu P, Cheema J, Bloomer R, Mikulski P, Liu Q, Zhang Y, Dean C, Ding Y. In vivo single-molecule analysis reveals *COOLAIR* RNA structural diversity. Nature. 2022:**609**(7926):394–399. https://doi.org/10.1038/s41586-022-05135-9

Yang Q, Li Z, Li WQ, Ku LX, Wang C, Ye JR, Li K, Yang N, Li YP, ZhongT, et al. CACTA-like transposable element in *ZmCCT* attenuated photoperiod sensitivity and accelerated the postdomestication spread of maize. Proc Natl Acad Sci U S A. 2013:**110**(42):16969–16974. https://doi.org/10.1073/pnas.1310949110

Yeoh SH, Satake A, Numata S, Ichie T, Lee SL, Basherudin N, Muhammad N, Kondo T, Otani T, Hashim M. Unravelling proximate cues of mass flowering in the tropical forests of south-east Asia from gene expression analyses. Mol Ecol. 2017:**26**(19):5074–5085. https://doi.org/10.1111/mec.14257

Yoo SC, Chen C, Rojas M, Daimon Y, Ham BK, Araki T, Lucas WJ. Phloem long-distance delivery of FLOWERING LOCUS T (FT) to the apex. Plant J. 2013:**75**(3):456–468. https://doi.org/10.1111/tpj.12213

Yu H, Lin T, Meng XB, Du HL, Zhang JK, Liu GF, Chen MJ, Jing YH, Kou LQ, Li XX, et al. A route to de novo domestication of wild allotetraploid rice. Cell. 2021:**184**(5):1156–1170. https://doi.org/10.1016/j.cell.2021.01.013

Yu S, Cao L, Zhou CM, Zhang TQ, Lian H, Sun Y, Wu JQ, Huang JR, Wang GD, Wang JW. Sugar is an endogenous cue for juvenile-to-adult phase transition in plants. Elife. 2013:**2**:e00269. https://doi.org/10.7554/eLife.00269

Yun J, Sun Z, Jiang Q, Wang Y, Wang C, Luo Y, Zhang F, Li X. The miR156b-GmSPL9d module modulates nodulation by targeting multiple core nodulation genes in soybean. New Phytol. 2022:**233**(4): 1881–1899. https://doi.org/10.1111/nph.17899

Zhang B, Li C, Li Y, Yu H. Mobile TERMINAL FLOWER1 determines seed size in *Arabidopsis*. Nat Plants. 2020a:**6**(9):1146–1157. https://doi.org/10.1038/s41477-020-0749-5

Zhang L, Jimenez-Gomez JM. Functional analysis of FRIGIDA using naturally occurring variation in Arabidopsis thaliana. Plant J. 2020b:**103**(1):154–165. https://doi.org/10.1111/tpj.14716

Zhang M, Kong D, Wang H. Genomic landscape of maize domestication and breeding improvement. Seed Biol. 2023b:**2**(1):9. https://doi.org/10.48130/SeedBio-2023-0009

Zhang Z, Luo X, Yang Y, He Y. Cold induction of nuclear FRIGIDA condensation in Arabidopsis. Nature. 2023a:**619**(7969):E27–E32. https://doi.org/10.1038/s41586-023-06189-z

Zhao Y, Antoniou-Kourounioti RL, Calder G, Dean C, Howard M. Temperature-dependent growth contributes to long-term cold sensing. Nature. 2020:**583**(7818):825–829. https://doi.org/10.1038/s41586-020-2485-4

Zhao Y, Zhu P, Hepworth J, Bloomer R, Antoniou-Kourounioti RL, Doughty J, Heckmann A, Xu C, Yang H, Dean C. Natural temperature fluctuations promote *COOLAIR* regulation of *FLC*. Genes Dev. 2021:**35**(11–12):888–898. https://doi.org/10.1101/gad.348362.121

Zhu P, Dean C. Reply to: cold induction of nuclear FRIGIDA condensation in Arabidopsis. Nature. 2023:**619**(7969):E33–E37. https://doi.org/10.1038/s41586-023-06190-6

Zhu P, Lister C, Dean C. Cold-induced Arabidopsis FRIGIDA nuclear condensates for *FLC* repression. Nature. 2021:**599**(7886):657–661. https://doi.org/10.1038/s41586-021-04062-5

Zhu P, Schon M, Questa J, Nodine M, Dean C. Causal role of a promoter polymorphism in natural variation of the Arabidopsis floral repressor gene *FLC*. Curr Biol. 2023:**33**(20):4381–4391. https://doi.org/10.1016/j.cub.2023.08.079

Zhu Y, Liu L, Shen LS, Yu H. NaKR1 regulates long-distance movement of FLOWERING LOCUS T in *Arabidopsis*. Nat Plants. 2016:**2**(6):16075. https://doi.org/10.1038/nplants.2016.75

Zhu Y, Klasfeld S, Jeong CW, Jin R, Goto K, Yamaguchi N, Wagner D. TERMINAL FLOWER 1-FD complex target genes and competition with FLOWERING LOCUS T. Nat Commun. 2020:**11**(1):5118. https://doi.org/10.1038/s41467-020-18782-1

Zsogon A, Cermak T, Naves ER, Notini MM, Edel KH, Weinl S, Freschi L, Voytas DF, Kudla J, Peres LEP. De novo domestication of wild tomato using genome editing. Nat Biotechnol. 2018:**36**(12):1211–1216. https://doi.org/10.1038/nbt.4272

https://doi.org/10.1093/plphys/kiae110 PLANT PHYSIOLOGY 2024: 195: 213–231

Plant Physiology®

Topical Review

Light signaling in plants—a selective history

Enamul Huq [1], Chentao Lin [2], Peter H. Quail [3,4,*]

1 Department of Molecular Biosciences and The Institute for Cellular and Molecular Biology, The University of Texas at Austin, Austin, TX 78712, USA
2 Basic Forestry and Plant Proteomics Research Center, Fujian Agriculture and Forestry University, Fuzhou 350002, China
3 Department of Plant and Microbial Biology, University of California, Berkeley, Berkeley, CA 94720, USA
4 Plant Gene Expression Center, Agricultural Research Service, US Department of Agriculture, Albany, CA 94710, USA

*Author for correspondence: Quail@berkeley.edu

Abstract

In addition to providing the radiant energy that drives photosynthesis, sunlight carries signals that enable plants to grow, develop and adapt optimally to the prevailing environment. Here we trace the path of research that has led to our current understanding of the cellular and molecular mechanisms underlying the plant's capacity to perceive and transduce these signals into appropriate growth and developmental responses. Because a fully comprehensive review was not possible, we have restricted our coverage to the phytochrome and cryptochrome classes of photosensory receptors, while recognizing that the phototropin and UV classes also contribute importantly to the full scope of light-signal monitoring by the plant.

Our story begins almost a century ago. The USDA established a research program in Beltsville, Maryland, in the 1920s, with the goal of identifying environmental factors that impact the growth and productivity of crop plants grown by US farmers. Light was soon defined as a major regulator of multiple facets of plant growth and development, including seed germination, seedling development, and flowering, the latter through a process they called photoperiodism. In 1936, plant physiologist Harry Borthwick was recruited to the USDA facility to form a new group focused on using basic research to define mechanisms underlying such light responses. An extensive history of the research initiated by these efforts, up until about 1990, can be found in Linda Sage's book *Pigment of the Imagination: A History of Phytochrome Research* (Sage 1992).

Phytochrome

Brief early history (1940s to early 1960s)

In 1940, Sterling Hendricks, an eminent photochemist, joined this effort as a collaborator. He suggested using the strategy of performing action spectra, as this would define the light wavelengths (colors) most active in inducing the relevant plant responses. This work clearly identified the red and far-red wavelengths as the most active in eliciting these responses, thereby defining the light-absorption profile of the molecule in the plant responsible for this activity. A key experiment in 1952 provided critical insight into the unique properties of the receptor molecule (Borthwick et al. 1952). It was found that as little as 1 min of red light was sufficient to trigger germination of lettuce seed but that an immediately subsequent minute of far-red light could cancel the red-light effect, thereby blocking germination. This led Hendricks to conclude that the photoreceptor molecule responsible exists in 2 light-interconvertible forms: a red light–absorbing form (which he called Pr) and a far red–absorbing form (christened Pfr). He also concluded that red light converted Pr to Pfr, far-red light converted Pfr back to Pr, and the Pr form was inactive in inducing seed germination, whereas the Pfr form was active (Fig. 1).

Moreover, he proposed that formation of the Pfr form could be detected physically in the plant by the small increase in far-red absorbance intrinsic to that form upon formation; and conversely, the Pr form would exhibit an

Received October 17, 2023. Accepted February 16, 2024. Advance access publication March 2, 2024.

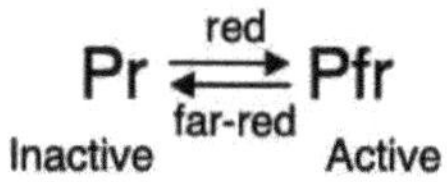

Figure 1. Scheme depicting Hendrick's proposed photoreversible switch-like behavior of the photoreceptor responsible for plant responses to the red/far-red region of the light spectrum. Pr, red-light absorbing form. Pfr, far-red light absorbing form.

increase in red absorbance. Karl Noris, an engineer at the Beltsville facility, was recruited to construct the highly sensitive, specialized spectrophotometer required to detect the extremely small absorbance changes in the plant. Warren Butler, a biophysicist who joined the group in 1956, first succeeded in directly detecting the photoreceptor in dark-grown turnip seedlings in 1959 using Noris's spectrophotometer (Butler et al. 1959). That same day, Harold (Bill) Siegelman, a biochemist who had joined the group in 1957, showed that the activity was also detectable in crude homogenates of the tissue. Butler proposed the name "phytochrome" for the new molecule, and this was officially announced by Borthwick and Hendricks in 1960 (Borthwick and Hendricks 1960).

Siegelman, with the help of Winslow Briggs, went on to provide the first purified preparations of phytochrome, showing that it is a soluble protein with a tetrapyrrole chromophore covalently attached (Siegelman and Firer 1964). This was followed over the next several years by several laboratories that developed additional methods resulting in high-quality preparations of pure phytochrome (Vierstra and Quail 1982) that enabled investigation of the biochem-ical and photochemical properties of the molecule.

The early 1960s to late 1970s

As noted by Furuya, little substantive progress toward defining the primary molecular mechanism of phy action in the cell was made during the late 1960s and 1970s (Furuya 1987). This had to wait until the onset of the molecular biology era in the 1980s. One notable exception is the work of Lee Pratt and colleagues starting in the mid-1970s. They produced antibodies to the phy protein and introduced the use of immunochemical procedures, which permitted in vitro analysis of the phy protein (e.g. immunoprecipitation, western blotting) and in situ behavior of the phy molecule at the tissue and subcellular levels (Mackenzie et al. 1975, 1978).

The early 1980s to the present—photoreceptor molecularly defined

The 1980s saw the dawn of the molecular biology and molecular genetics era in plants. As was the case more broadly, this had enormous impact on the phy field. One early major focus on was to isolate and characterize phytochrome genes from various species, including *Avena sativa*, zucchini, and cucurbita (Hershey et al. 1984). However, *Arabidopsis thaliana* soon became the preferred model system for both genetic and molecular studies. In 1980, Maarten Koornneef pioneered the use of Arabidopsis for obtaining mutants defective in responsiveness to light as a strategy for eventually identifying the photoreceptor molecule responsible (Koornneef et al. 1980). He isolated several long-hypocotyl (*hy*) mutants, some of which would later help achieve this goal. Using a molecular biological approach, Sharrock and Quail described the isolation and characterization of 3 phytochrome genes (*PHYA-PHYC*) from Arabidopsis (Sharrock and Quail 1989), revealing for the first time that there is not 1 but multiple phytochromes in the cell. Two other Arabidopsis phytochrome genes (*PHYD-PHYE*) were later cloned and characterized (Clack et al. 1994). In parallel, mutants in phytochrome genes were isolated for all 5 phytochromes using genetic screens under various conditions. *phyA* mutants were isolated using a 2-step screen, initially selecting long hypocotyl mutants under continuous far-red light and then wild-type–like seedlings under continuous red light from the progeny of the first round [*hy8* for phyA; (Parks and Quail 1993)]. Koornneef's *hy3* mutant yielded *phyB* mutants, and the *hy1* and *hy2* mutants were found to be defective in chromophore biosynthesis (Koornneef et al. 1980). Both *HY3* and *HY8* loci were later cloned, verifying that they encode *PHYB* and *PHYA* genes, respectively (Dehesh et al. 1993; Reed et al. 1993). *phyC* mutants were isolated using a targeted PCR-based screening procedure from various collections of T-DNA insertional mutants and fast neutron deletion libraries (Monte et al. 2003). A *phyD* mutant was isolated by cloning the *PHYD* gene from the WS ecotype, which was later found to have a 14-bp deletion in this ecotype compared with the Col-0 background (Aukerman et al. 1997). Finally, *phyE* mutants were isolated from a 2-step genetic screen for early flowering and elongated rosette internodes in the first round of screen and then selecting mutants displaying attenuated internode elongation and flowering responses under end-of day far-red treatments from the progeny of the first round (Devlin et al. 1998).

The isolation of individual phytochrome mutants helped define the biological roles and photosensory specificity of each phytochrome. Phenotypic characterization showed that although all phytochromes are activated by red light, individual phytochromes have both distinct and overlapping roles in regulating plant development under various light conditions (Li et al. 2011). Strikingly, the phytochrome-deficient quintuple mutant (*phyabcde*) of Arabidopsis showed that phytochromes are dispensable for completion of the Arabidopsis life cycle, although these mutants are severely defective in developmental responses to the prevailing light environment (Strassera et al. 2010; Hua et al. 2013). More recently, in a major paradigm shift, phytochromes have been shown to function not only as a light receptor but also as a temperature sensor in plants (Jung et al. 2016; Legris et al. 2016). These studies highlight the importance of phytochromes as a broad environmental sensor in regulating plant growth and development.

Two major long-term goals in the field have been to solve the structure of the phytochrome molecule and to understand its functional activity at the molecular level. A breakthrough regarding the first goal came with the success of Vierstra and colleagues, initially in solving the crystal structure of the Pr and Pfr forms of the photosensory domains, and later the full-length phytochromes, from bacteria (Wagner et al. 2005; Essen et al. 2008; Ulijasz et al. 2010). Most recently, using cryo-electron microscopy (cryo-EM), these authors have defined the 3D structures of the Pr forms of the full-length Arabidopsis phyA and phyB molecules (Li et al. 2022; Burgie et al. 2023; Zhang et al. 2023). These structures have revealed that both phyA and phyB are asymmetric dimers in their N-terminal photosensory domain, while they have a symmetric dimer configuration in their C-terminal: histidine kinase-related domains (HKRD). One notable difference between the phyA and phyB structures is that, whereas the phyB N-terminal photosensory domain and the HKRD region associate asymmetrically, these associations are absent in phyA. These data suggest that this decoupling between the N- and C-terminal domains in phyA might have functional consequences for Pfr stability and photosensory specificity.

Molecular signaling mechanism defined

In parallel with this structural research, efforts were ongoing in multiple laboratories toward the second goal of understanding how these photoreceptors function inside the plant cell. Early immunocytochemical studies had led to the proposal that phys are constitutively localized in the cytosol (Mackenzie et al. 1975) and that the activated form communicates with the nucleus through second messengers to control gene expression (Millar et al. 1994). In support of this hypothesis, a few second messengers, including Ca2+, calmodulin, and cGMP, had been shown to impact gene expression and chloroplast development (Bowler et al. 1994). However, in 1996, Sakamoto and Nagatani provided initial evidence that phyB localizes to the nucleus in the Pfr form (Sakamoto and Nagatani 1996). Following this study, all phytochromes were shown to be initially in the cytoplasm in the Pr form and to then translocate rapidly into the nucleus in response to light, with varying kinetics and fluence-rate specificity (Kircher et al. 1999, 2002; Yamaguchi et al. 1999).

Nagatani's findings converged in the late 1990s with efforts to identify phy-interacting proteins directed toward the goal of defining the molecular mechanism of light-signal transfer in the cell. The breakthrough came when Ni et al. (Ni et al. 1998, 1999) identified a basic helix-loop-helix (bHLH) transcription factor (Toledo-Ortiz et al. 2003), which they called Phytochrome Interacting Factor 3 (PIF3) and which interacted specifically with the Pfr form of phyB (Martinez-Garcia et al. 2000). These findings established that the light-activated photoreceptor molecule signals directly to the genome via physical interaction with components of the transcriptional machinery, thereby regulating the expression of target genes (Fig. 2). It had been noted by Pratt and colleagues in the late 1970s that light-activated phy formed "speckles" (later called "photobodies") in the nucleus in prolonged irradiation (Mackenzie et al. 1978). In 2004, Schaefer and colleagues showed that light-activated phy rapidly (within minutes) formed photobodies upon nuclear translocation and that constitutively nuclear-localized PIF3 also coalesced together with phy in these photobodies (Bauer et al. 2004). Moreover, they found that this interaction led to rapid degradation of the PIF3 protein.

Using PIF3 as a bait, PIF1 and PIF4, other members of the bHLH family, were also isolated and characterized (Huq and Quail 2002; Huq et al. 2004). When the genome sequence of Arabidopsis became available, a number of additional bHLH PIFs (PIF5-PIF8) were identified and characterized (Khanna et al. 2004, 2007; Leivar et al. 2008a, 2008b; Oh et al. 2020). Among all the PIFs, PIF1 and PIF3 were shown to interact with the Pfr form of phyA and phyB (Huq et al. 2004), whereas all the other PIFs were shown to interact with the Pfr form of phyB. Thus, these transcription factors display differential affinities for the different phys, suggesting the possibility of combinatorial pathway divergence at the genome interface.

In parallel studies, earlier genetic screens to identify mutants in the phy signaling pathways yielded another class of important mutants, named *constitutive photomorphogenic/ de-etiolated/fusca* (*cop/det/fus*) (Chory et al. 1989; Deng et al. 1991). These mutants displayed light-grown phenotypes in darkness, suggesting that these factors act as negative regulators not only of phy signaling but also of other light signaling pathways. Molecular cloning and characterization of the *COP1* gene revealed that this factor is mostly involved in protein deg-radation through the ubiquitin/26S proteasome pathway (Deng et al. 1992). The COP1 protein functions as an E3 ligase for degradation of positively acting components in light signaling pathways (e.g. ELONGATED HYPOCOTYL 5, HY5 and others) (Osterlund et al. 2000). COP1 has been shown to inter-act with SUPPRESSOR OF PHYA-105 1 (SPA1-SPA4) family of proteins (Saijo et al. 2003; Seo et al. 2003). The COP1-SPA complex acts as a substrate adaptor for the CULLIN4-based E3 ligase that degrades positively acting transcription factors in light signaling pathways (Xu et al. 2015; Han et al. 2020). Upon photoactivation, phys reverse the COP1-SPA function in promoting photomorphogenesis in 2 ways. On the one hand, light-activated phys interact with the SPA proteins upon translocation into the nucleus, reorganizing the COP1-SPA complex (Lu et al. 2015; Sheerin et al. 2015) and thereby inhibiting the E3 ligase activity. In addition, in prolonged light, COP1 is excluded from the nucleus (Subramanian et al. 2004), thereby reducing the E3 ligase activity within the nucleus. These dual inhibitory mechanisms result in stabilization and accumulation of the positively acting transcription factors (e.g. HY5, LAF1, HFR1, and others), which then promote photomorphogenesis.

To understand the regulatory functions of the PIFs, *pif* mutants were identified, mainly using reverse-genetic approaches, although *pif4* was isolated by both genetic and reverse-genetic approaches (Huq and Quail 2002). As

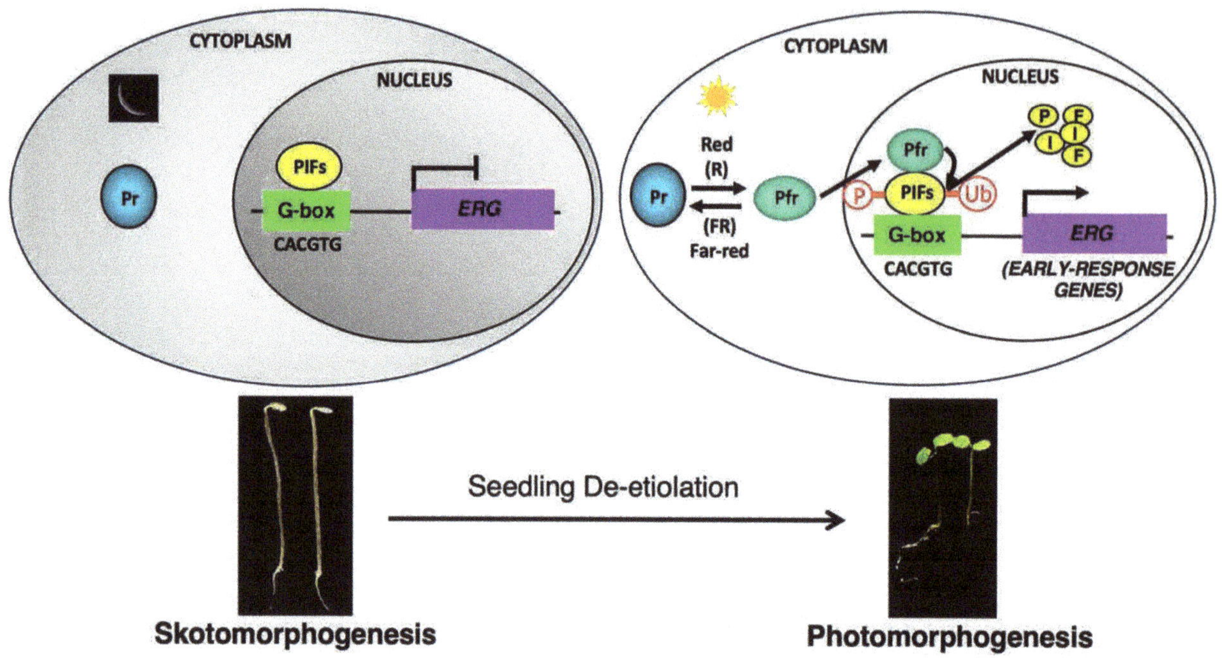

Figure 2. Photoactivated phys directly regulates target-gene expression via molecular interaction with PIF transcription factors. In the dark (left), phytochromes are synthesized in their biologically inactive Pr form, which is localized to the cytosol. PIFs are constitutively nuclear-localized transcription factors that bind to G-box (CACGTG) sequence elements in the promoters of many light-regulated genes. Arabidopsis seedlings undergo etiolated development (skotomorphogenesis). Upon red light illumination (right), phytochromes are converted to the biologically active Pfr form. The Pfr form translocates into the nucleus, binds to PIFs, and induces direct-target gene expression. The phy-PIF interaction within the nucleus results in rapid phosphorylation, ubiquitination, and degradation of PIFs. This light-induced degradation of PIFs results in activation of PIF-repressed genes and suppression of PIF-activated genes. The transcriptional network elicited by light exposure drives photomorphogenic development.

members of the bHLH family, PIFs are highly homologous proteins (Toledo-Ortiz et al. 2003; Castillon et al. 2007; Leivar and Quail 2011). However, phenotypic characteriza-tion of the *pif* mutants showed that individual PIFs regulate specific, as well as overlapping, sets of biological pathways. For example, PIF1 and PIF8 regulate seed germination (Oh et al. 2004, 2020); PIF1 and PIF3 regulate chlorophyll and carotenoid biosynthetic pathways (Huq et al. 2004; Stephenson et al. 2009); PIF4, PIF5, and PIF7 regulate shade avoidance responses (Lorrain et al. 2008; Leivar et al. 2012); and all the PIFs regulate hypocotyl lengths to varying degrees under red light (Huq and Quail 2002; Monte et al. 2004; Leivar et al. 2008a, 2008b). Strikingly, higher-order *pif* mutants (e.g. *pif1 pif3 pif4 pif5*, termed *pifQ*) display short hypocotyl and expanded cotyledons in darkness mimicking light-grown phenotypes (Leivar et al. 2008a, 2008b; Shin et al. 2009), suggesting that PIFs function negatively in regulating these processes.

The array of gene expression changes underlying these morphological phenotypes regulated by the individual phys were first defined in the early 2000s using early microarray technology (Tepperman et al. 2001, 2004, 2006). This was followed later by RNAseq studies showing that a large number of light-controlled genes are regulated by the PIFs by com-paring the *pifQ* mutant in darkness with the gene expression pattern in the light-exposed wild type seedlings (Leivar et al. 2009; Shin et al. 2009). Moreover, PIFs directly regulated a subset of these genes both individually as well as in an overlapping manner (Zhang et al. 2013; Leivar and Monte 2014; Pfeiffer et al. 2014). Thus, the PIFs function as negative regulators whereas the phys function as positive regulators of photomorphogenesis, establishing a "yin-yang" relationship between the 2 signaling partners.

The earlier discovery that rapid, light-induced colocalization of the phy and PIF3 molecules in photobodies induced rapid degradation of the transcription factor (Bauer et al. 2004) supports the proposal that the promotion of photomorphogenesis by light results from reversal of PIF-imposed repression of this process. Subsequently, a number of studies showed that, with the exception of PIF7 (Leivar et al. 2008a, 2008b), all other PIFs are also degraded in light with varying kinetics and fluence-rate dependency (Park et al. 2004; Shen et al. 2005, 2007; Al-Sady et al. 2006; Lorrain et al. 2008). Among all the PIFs, PIF1 is the most light-sensitive, with a half-life less than 1 min, reflecting its strong interaction with both phyA and phyB (Huq et al. 2004; Shen et al. 2008). Thus, the phy-induced degradation of PIFs has formed the foundation of the primary signaling event in phy signaling pathways. In addition, phyB has been shown to inhibit the DNA binding activity of PIF3 by sequestration (Park et al. 2012). In fact, the N-terminal domain of phyB binds to PIF3 and inhibits the DNA binding and transcriptional activation activity of PIF3, while the C-terminal domain of phyB promotes the degradation of PIF3 (Park et al. 2018; Yoo et al. 2021). This is also consistent with the deetiolated phenotypes of the *pifQ* quadruple mutant in darkness (Leivar et al. 2008a, 2008b), suggesting that the removal of PIFs either genetically or by light-induced degradation or inhibition of PIF function by sequestration is sufficient to promote photomorphogenesis.

Early investigation of the mechanism of phy-induced PIF degradation provided evidence that PIFs are first phosphorylated and then ubiquitinated before being degraded through the 26S proteasome pathway (Fig. 2) (Al-Sady et al. 2006). Intense efforts by various laboratories were then focused on identifying the kinases and E3 ubiquitin ligases for PIFs. In 2014, Ni et al. showed that LRB proteins [Light-Response Bric-a-Brack/Tramtrack/Broad (BTB)] act as substrate adaptors for PIF3 degradation under light (Ni et al. 2014). Moreover, LRBs also act as substrate adaptors for phyB, ensuring mutual co-degradation of the phyB-PIF3 complex in the light. In 2015, Zhu et al. described a different E3 ligase for light-induced degradation of PIF1 (Zhu et al. 2015). These authors showed that COP1-SPA proteins, well-known repressors of photomorphogenesis, act as a substrate adaptor complex in a CUL4-based E3 ligase to recruit PIF1 and PIF5 for light-induced degradation (Zhu et al. 2015; Pham et al. 2018). In addition, Dong et al. showed that EIN3-BINDING F BOX PROTEIN 1 (EBF1) and EBF2, which form the $CUL1^{EBF1}$ and $CUL1^{EBF2}$ complexes, respectively, also recruited PIF3 for degradation (Dong et al. 2017). Moreover, a number of other E3 ligases, including COLD TEMPERATURE-GERMINATING 10 (CTG10) (Majee et al. 2018), BLADE-ON-PETIOLE (BOP) (Zhang et al. 2017), and HIGH EXPRESSION OF OSMOTICALLY RESPONSIVE GENES1 (HOS1), were described to recruit various PIFs for light-induced degradation (Kim et al. 2017a, 2017b). Thus, it appears that multiple E3 ligases can recruit PIFs for light-induced degradation to promote photomorphogenesis.

Following the discovery of E3 ligases, progress in identifying kinases for the relevant PIFs was made by various laboratories. The first candidate was phys themselves, as early evidence showed the C-terminal domain of phys displays sequence homology to the histidine kinase families (Quail 1997), and biochemical evidence suggested that phys might function as a serine threonine kinase (Yeh and Lagarias 1998; Shin et al. 2016). However, recent cryo-EM structures did not reveal any kinase-related domain in either phyA and phyB (Li et al. 2022; Burgie et al. 2023; Zhang et al. 2023). Although the phy kinase hypothesis has remained a hot topic for debate in the field, various laboratories were focused on identifying other kinases for PIFs. The first of these kinases is called the Photoregulatory Protein Kinases (PPK1–PPK4), which were previously known as MUT9-Like Kinases (MLKs), that phosphorylate PIF3 in vitro (Ni et al. 2017). Interestingly, like LRBs, PPKs also controlled the degradation of the phyB-PIF3 complex resulting in a hypersensitive phenotype of the higher order *ppk* mutants. Paik et al. showed that SPA1, previously known as an adaptor for COP1, functions as a kinase for PIF1 (Paik et al. 2019). SPA proteins have a serine threonine kinase related domain at the N terminus (Hoecker et al. 1999), but the functional significance of this domain is unknown. A mutant form of SPA1 with a mutation at the kinase domain failed to rescue the *spaQ* phenotypes, whereas the WT SPA1 largely rescued the phenotypes, suggesting that this domain is important for SPA function. Apart from PPKs and SPAs, a few other kinases have been described for PIFs. These include Casein Kinase 2 (CK2) (Bu et al. 2011) (Bernardo-García et al. 2014) and mitogen-activated protein kinase 6 (MPK6) (Xin et al. 2018a). However, these kinases may not be involved in the light-induced phosphor-ylation of the PIFs.

In addition to the phy-PIF module-controlled transcriptional regulation of gene expression described above, phys have been shown to control the post-transcriptional and translational steps in gene regulation (Cheng et al. 2021; Kathare and Huq 2021). In 2014, Shikata and colleagues first showed that phys extensively regulate alternative splicing of pre-mRNAs, where variable splice sites were used in multi-intron pre-mRNAs to generate multiple forms of mRNAs from the same transcript (Shikata et al. 2014). In another study, overaccumulation of phyB, especially in the *lrb* mutant backgrounds, was shown to promote intron retention in the 5′ untranslated region of PIF3 mRNA, resulting in inhibition of PIF3 translation (Dong et al. 2017). Following these studies, a number of splicing factors have been identified by genetic and biochemical approaches in both Arabidopsis (Splicing Factor for Phytochrome Signaling, SFPS; Reduced Red-light Responses in Cry1Cry2 background, RRC1, Suppressor of White Apricot 1, SWAP1, and SMP2) (Xin et al. 2018b; Xin et al. 2019; Kathare et al. 2022; Yan et al. 2022) and *Physcomitrella patens* (heterogeneous nuclear ribonucleo-proteins, PphnRNP-H1; PphnRNP-F1) (Shih et al. 2019; Lin et al. 2020). Strikingly, all of these splicing factors directly interact with phys. However, unlike PIFs, these splicing fac-tors are not degraded under light, suggesting a distinct mech-anism to control their function by phys.

Light signals not only control transcription and post-transcriptional processing of mRNAs but also translation of many mRNAs. The first direct evidence that phys are involved in this process came from identifying a phy-interacting factor named PENTA1 (PNT1) (Paik et al. 2012). PNT1 directly binds to the 5′ untranslated region of the protochlorophyllide reductase A (PORA) mRNA. Light-activated phys interact with PNT1 and repress translation of PORA. In addition, COP1 has been shown to repress translation in darkness (Chen et al. 2018). Light-induced inactivation of COP1 through phys and crys enhances translation of many mRNAs.

Recent findings have unveiled an exciting new direction in defining the functional activities of the phy-PIF duo in the nucleus. The discovery by Chentao Lin and colleagues [(Wang et al. 2021); see below] that light-induced CRY2 photobody formation in the nucleus results from the rapid coalescence of the CRY2 molecules into liquid-liquid droplets highlighted existing long-standing evidence of physical interaction between, and co-occupation of, photobodies by light-activated CRY2 and phys (Yamaguchi et al. 1999; Más et al. 2000), thereby predicting that phy photobody formation likewise results from liquid-liquid phase separation of the photoreceptor molecule (Quail 2021). Direct evidence that light-activated phy is induced to form liquid-liquid phase separated (LLPS) photobodies, that function as signaling hubs of the phys, soon followed (Chen et al. 2022). In parallel, Chory and colleagues have provided elegant evidence of 1 major signaling-hub function of phy-PIF7–harboring photobodies (Willige et al. 2021). These authors show that light-induced, Pfr-activated phosphorylation of PIF7 sequesters the transcription factor in photobodies away from access to its DNA binding sites. Conversely, vegetative shade-induced reduction in Pfr levels results in PIF7 dephosphorylation, releasing it from the photobodies to bind to its target cis-acting DNA sites. The bound PIF7 then ejects the H2A.Z histone variant from chromatin, H3K9 is acetylated, and gene expression of PIF7 target-gene expression is initiated (Willige et al. 2021). Because more than 100 transcription factor genes are directly targeted, an extensive, diverse PIF7-initiated gene network is modulated by this regulatory module.

Kim et al. (2021) have provided additional evidence that phys regulate chromatin remodeling directly and indirectly to control the expression of target genes that promote plant development. They showed that phyB directly interacts with VERNALIZATION INSENSITIVE 3-LIKE1 (VIL1), a component of the Polycomb Repressive Complex 2 (PRC2), in a light-dependent manner (Kim et al. 2021). phyB-VIL1 synergistically represses the expression of growth promoting genes (e.g. ATHB2) through the formation of a chromatin loop in a light-dependent manner that promotes the enrichment of Histone H3 Lys27 trimethylation (H3K27me3), a repressive histone modification. In addition, Gonzalez-Grandio and colleagues have examined whether the rapid activation and repression of PIF direct-target genes are associated with chromatin changes (González-Grandío et al. 2022). These authors found evidence that the light-regulated transcriptional changes and chromatin-remodeling processes might be mechanistically intertwined.

Cryptochromes

A brief early history

Blue and red light are the 2 major spectral regions of sunlight used by plants for photosynthesis, and these 2 regions are also the 2 major wavelengths used for photomorphogenesis. In his book "The Power of Movement in Plants," Charles Darwin described an experiment in his study of the circumnutational movement of cabbage seedlings. In this experiment, "the plants were illuminated by light passing through a solution of bichromate of potassium so as to eliminate heliotropism" (Darwin 1881). Given that heliotropism is primarily a phototropic response and that orange-colored potassium bichromate solution has an absorption spectrum of approximately 310 to 450 nm (from UV-A to blue light) (von Halban et al. 1927), the correlation between removal of UV-A and blue-light photons and abolition of plant heliotropism implied that phototropism is a blue-light response. Indeed, Julius von Sachs, who is considered the father of plant physiology (Kutschera and Niklas 2018), measured the first crude action spectrum of plant phototropic responses and found that it is maximal in the blue-light region (Sachs 1887). Since the discovery of phototropism, plant biologists have made enormous efforts to identify the molecular mechanism underlying this fascinating blue-light response of plants (Butler et al. 1959; Briggs et al. 1983). For example, Winslow Briggs and his students discovered light-induced auxin transport and demonstrated that this transport was lateral within the plant organ in response to the light, causing asymmetric cell elongation, which led in turn to phototropism (Briggs et al. 1957). However, early efforts to biochemical-ly isolate blue-light receptors encountered many technical difficulties (Briggs et al. 1983), such that molecular identification of blue-light receptors had to wait until 1990s (Briggs and Huala 1999). Because no biochemical activity was known for these receptors at the time, the lack of an appropriate readout imposed a technical hurdle difficult to overcome in those days. The biochemical study of plant blue-light receptors was greatly encouraged by a seemingly promising readout called LIAC (light-induced absorption changes), which was originally discovered in the study of photosynthesis (Kok 1957). It was found that extracts of membrane fractions of fungi, algae, and higher plants exhibited LIAC with an action spectrum similar to that of phototropism, as well as photoreduction of flavoproteins, especially the b-type cytochromes (Poff and Butler 1974; Brain et al. 1977; Briggs et al. 1983). Unfortunately, this seemingly promising approach was also fruitless in identifying the molecular nature of plant blue-light sensory receptors, due to the relatively low abundance of these photoreceptors in comparison to the photosynthetic pigments. Although it was correctly predicted from these earlier studies that the blue-light sensory receptors would be flavoproteins that contain a chromophore of either FAD (Flavin Adenine Dinucleotide) or FMN (Flavin Mononucleotide), none of the presently known plant receptors is a cytochrome and none appear to be responsible for the LIAC phenomenon (Briggs 2010). Despite this setback, Jonathan Gressel, inspired by the successes in identifying the phytochromes, proposed the term "cryptochrome" in 1979 for the-yet-to-be molecularly identified pigments having both the absorption and action spectra in the UV-A/blue light region of the spectrum and actually mo-lecularly responsible for the plant photoresponse (Gressel 1979; Senger 1984). The name cryptochrome was coined "because of importance in cryptogamic plants (in the study of blue-light responses) and its cryptic nature" (Gressel 1979).

Molecular identification of cryptochromes

Fourteen years later, Cashmore and colleagues molecularly identified the first bona fide blue-light sensory photoreceptor (Ahmad and Cashmore 1993; Cashmore et al. 1999).

This achievement was made possible by the molecular-genetic methodologies developed in the model plant *Arabidopsis thaliana* (Koornneef et al. 1980; Feldmann 1991; Somerville and Koornneef 2002). In 1980, in addition to the phytochrome mutants described above in the previous section, Maarten Koornneef also isolated the *hy4* mutant that exhibits a long-hypocotyl phenotype in blue light (Koornneef et al. 1980). In 1991, Kenneth Feldmann generated 8,000 Arabidopsis T-DNA insertion-mutant lines, making it much easier to identify genes corresponding to mutants that have visible phenotypes (Feldmann 1991. In 1993, Anthony Cashmore's laboratory screened this mutant population, identified a T-DNA tagged allele of the *hy4* mutant, *hy4-2*, and cloned the *HY4* gene (Ahmad and Cashmore 1993). The *HY4* gene encodes a protein with an N-terminal domain that is approximately 30% identical to DNA photolyases. Interestingly, a *Sinapis alba* photolyase-like gene (SA-Phr1) was reported (Batschauer 1993) a couple of months before the publication of the *HY4* gene. Because DNA photolyase is a blue light-dependent flavoenzyme that repairs cyclobutane pyrimidine dimers of UV-damaged DNA (Malhotra et al. 1992), these results immediately suggested that the HY4 protein is most likely a blue-light receptor responsible for regulating seedling hypocotyl growth (Ahmad and Cashmore 1993). Consistent with this prediction, both the full-length HY4 protein expressed in and purified from insect cells and the N-terminal domain expressed and purified from *E. coli* as the MBP (Maltose Binding Protein) fusion protein were found to contain oxidized FAD that absorbs UV-A and blue light (Lin et al. 1995; Malhotra et al. 1995). It was observed that the HY4 protein is not only required for a blue light-specific response (Koornneef et al. 1980; Ahmad and Cashmore 1993; Ahmad et al. 1995) but also itself absorbs blue light, fulfilling the 2 key criteria of a blue-light sensory receptor or a crypto-chrome; it was named cryptochrome 1, or CRY1 (Lin et al. 1995).

The Arabidopsis genome also encodes a CRY1 homolog that was called CRY2 or PHH1 (Hoffman et al. 1996; Lin et al. 1996). CRY1 and CRY2 share 58% sequence identity in their N-terminal photolyase-like domains and 13% in their C-terminal nonphotolyase domains. Antibodies against CRY2 were prepared and used in a genetic screen to isolate the loss-of-function *cry2* mutants (Guo et al. 1998; Lin et al. 1998). In this experiment, a fast neutron-mutagenized population of M2 seedlings was grown in blue light to select those that grew slightly taller than the wild-type because it was speculated that CRY2 may have overlapping function with CRY1 in mediating blue light inhibition of hypocotyl elongation. The isolates from the first screen were analyzed by immunoblots probed with the anti-CRY2 antibody. This 2-step genetic screen resulted in 2 alleles of the *cry2* mutants that suffer from complete or partial deletion of the *CRY2* gene and absence of the CRY2 protein (Guo et al. 1998; Lin et al. 1998). The *cry2* mutants exhibited 2 abnormal phenotypes. First, the mutant exhibited the long-hypocotyl phenotype compared with wild-type, especially when grown in low intensities of blue light (Lin et al. 1998). Second, this mutant exhibited a later-flowering phenotype compared with the wild type when they were grown in long-day photoperiods or continuous white light (Guo et al. 1998). A screen of the late-flowering mutants known at that time demonstrated that *cry2* is allelic to *fha*, which is one of the many late-flowering mutants previously isolated by Maarten Koornneef (Koornneef et al. 1991). The late-flowering phenotype of the *cry2* mutant is wavelength specific but in a manner more complex than expected. It was known that blue light promotes and red light inhibits flowering in Arabidopsis (Brown and Klein 1971; Eskins 1992). Surprisingly, the *cry2* mutant showed normal flowering time when grown in monochromatic blue light. It turned out that cry2 displayed the late-flowering phenotype only in the presence of both blue light and red light (Guo et al. 1998). Based on this observation, it was proposed that CRY2 mediates blue light inhibition of phyB-dependent suppression of floral initiation (Guo et al. 1998). Indeed, it was demonstrated that phyB mediates red light inhibition of pro-teolysis of the flowering promoter protein CO (CONSTANS), whereas CRY2 mediates blue light suppression of phyB func-tion and thereby of CO degradation. These experiments ex-plained the puzzling wavelength-specific phenotype of the *cry2* mutant and demonstrated the close functional associ-ation of the blue light–receptor cryptochromes and the red light–receptor phytochromes.

Soon after discovery of the Arabidopsis CRY1 and CRY2, the presence of cryptochromes was also reported in animals. It was found that *Drosophila* dCRY acts as the photoreceptor mediating light regulation of the circadian clock in *Drosophila* (Emery et al. 1998; Stanewsky et al. 1998), whereas the mouse mCRYs act as light-independent core compo-nents of the circadian clock in mice (van der Horst et al. 1999). At about the same time, CRYs and phytochromes were shown to mediate light entrainment of the circadian clock in Arabidopsis (Somers et al. 1998). All CRYs discovered by that time showed regulatory activity of gene expression, without DNA photolyase activity. Subsequently, a third type of CRY, referred to as CRY-DASH (Drosophila, Arabidopsis, Synechocystis, Human) (Brudler et al. 2003) or CRY3 (Kleine et al. 2003), was discovered. It was found that CRY-DASH is a DNA-binding protein that affects gene expression (Brudler et al. 2003), and Arabidopsis CRY3 has a signal peptide that targets it into chloroplasts and mitochondria (Kleine et al. 2003). In contrast to the canonical CRYs that have no enzymatic activity repairing UV-damaged DNA, the DASH-type CRYs from bacteria and plants act as light-dependent ssDNA repairing DNA photolyases (Huang et al. 2006; Selby and Sancar 2006). The DASH-type cryptochromes lack an efficient flipping mechanism for repairing cyclobutane pyrimidine dimers within dsDNA, which explains why CRY-DASH cannot repair dsDNA (Pokorny et al. 2008). The function of DASH-type CRYs appears important to the genome stability of bacteria and organelles of eukaryotes because these smaller genomes may have relatively more abundant "melted" ssDNA regions resulting from rapid DNA replication and transcription. In the subsequent sections of this article, CRYs refer only to the canonical cryptochromes, especially Arabidopsis CRY1 and CRY2, that show no enzymatic activity in repairing cyclobutane pyrimidine dimers of DNA but function instead as signaling photoreceptors that regulate gene expression and photomorphogenesis.

Studies of the evolutionary history of photolyase/cryptochrome families indicate that ancient DNA photolyases may have duplicated multiple times during evolution, resulting in expansion of the photolyase/cryptochrome families and divergence of plant and animal CRYs more than 1 billion years ago (Lin and Todo 2005; Kim et al. 2014; Mei and Dvornyk 2015; Deppisch et al. 2022). This is consistent with the hypothesis that CRYs are the first sensory photoreceptors that evolved in plants (Han et al. 2019). It appears that CRYs in different eukaryotes may have arisen independently from different photolyases, which would explain the different modes of action of the different CRYs in different eukaryotes, despite their common role in regulating the circadian clock (Cashmore 2003; Sancar 2016). Up to now, neither canonical CRY nor CRY-DASH DNA photolyase has been found in Archaea (Deppisch et al. 2022). Thus CRYs might not have existed 2 billion years ago when eukaryotes diverged from Archaea (Doolittle 1997; Williams et al. 2013), making the evolutionary history of CRYs about 1 to 2 billion years old. Although the canonical CRYs, such as plant and mammalian CRY1 and CRY2, do not have the catalytic activity repairing cyclobutane pyrimidine dimers of DNA, Arabidopsis CRY1 and CRY2 have been recently reported to promote repairing of DNA double-strand breaks (DSBs) in response to blue light (Guo et al. 2023). Plant CRY1 and CRY2 undergo blue light–dependent interactions with ADA2b, a subunit of the highly conserved GCN5 acetyltransferase that is often recruited to DSBs to facilitate DNA repair. The CRY2-ADA2b interaction enhances the ADA2b interaction with SMC5/6 (Structural Maintenance of Chromosome 5/6) that repairs DSBs by homologous recombination (De Piccoli et al. 2006) and/or DNA loop extrusion mechanisms (Pradhan et al. 2023). Therefore, canonical cryptochrome may have evolved with not only new functions in the regulation of gene expression and the circadian clock by protein-protein interactions (see later) but also reinvented its ancestral function in DNA repair by blue light–dependent protein-protein interactions. Given the similar DSB repairing mechanisms in eukaryotes, it would be interesting to examine whether metazoan CRYs also possess similar functions promoting DNA repairing reactions in response to blue light.

Signal transduction mechanisms of cryptochromes

The CRY signaling mechanisms were recently reviewed (Wang and Lin 2020; Ponnu and Hoecker 2022). Regulation of gene expression is the major cellular function of Arabidopsis CRYs, and the blue light–dependent formation or disintegration of CRY complexes composed of CRYs, CRY-signaling proteins, and CRY regulatory proteins is the major signal transduction mechanism of plant CRYs. For example, plant CRYs are known to mediate blue light promotion of anthocyanin biosynthesis (Sponga et al. 1986), and Arabidopsis CRY1 was shown to mediate blue light–induced expression of the mRNA encoding the key enzyme in anthocyanin biosynthesis, *CHS* (chalcone synthase) (Kubasek et al. 1992; Fuglevand et al. 1996). Soon after the Arabidopsis genome sequence was completed, Arabidopsis CRY1 and CRY2 were demonstrated to mediate blue light–regulated transcriptome changes (Ma et al. 2001; Folta et al. 2003; Ohgishi et al. 2004). Arabidopsis CRY1 (Cashmore et al. 1999) and CRY2 (Guo et al. 1999; Kleiner et al. 1999) are nuclear proteins. CRY2 is a "constitutive" nuclear protein that functions only in the nucleus (Yu et al. 2007a, 2007b), whereas CRY1 exists and functions in both the nucleus and cyto-plasm (Fig. 3) (Wu and Spalding 2007). It is now clear plant CRYs interact with various gene expression regulators, including E3 ubiquitin ligases, transcription factors and co-factors, protein kinases, chromatin remodelers, splicing factors, RNA modifying enzymes, and phytochromes to regulate proteolysis, transcription, chromatin remodeling, mRNA splicing, mRNA modification, and translation, in response to blue light. For example, results of a recent multiple omics analysis indicates that, in the Arabidopsis *cry1cry2* mutant seedlings grown under continuous blue light, at least 7,400 genes showed significant changes of mRNA abundance in comparison to the wild type (Jiang et al. 2023a, 2003b), con-firming that CRYs regulate mRNA expression of approxi-mately one-quarter of the transcriptome (Ma et al. 2001). It is the light-dependent changes in association of the CRY protein complexes that cause the photoresponsive changes in gene expression and eventually photomorphogenic changes of plants (Wang et al. 2018a, 2018b; Wang and Lin 2020; Ponnu and Hoecker 2022).

Regarding proteolysis, similarly to that described for the phys above, the CRY-COP1/SPA-Transcription factor (TF) axis was the first discovered, and probably most extensively studied, mechanism of plant CRY signaling (Yang et al. 2000, 2001; Wang et al. 2001). Current understanding (Han et al. 2020; Wang and Lin 2020; Ponnu and Hoecker 2022) indicates that CRYs interact with the E3 ubiquitin ligase complex $CUL4^{COP1/SPAs}$ to inhibit its activity, resulting in accumulation of the transcription factors, such as HY5, CO, LAF1, HFR1, and BBX21 (Wei and Deng 1996; Lau and Deng 2012; Kim et al. 2017a, 2017b; Han et al. 2020). Because these transcription factors act to promote transcription of genes required for photomorphogenesis, the CRY-dependent inhibition of $CUL4^{COP1/SPAs}$ activity promotes photoresponsive changes of the transcriptome and proteome driving photomorphogenic development of plants. Arabidopsis CRYs interact with both COP1 (Wang et al. 2001; Yang et al. 2001) and the SPA1/COP1 complex in a blue light–dependent manner, imposing light-dependent inhibition of COP1 activity in the dark (Lian et al. 2011; Liu et al. 2011;

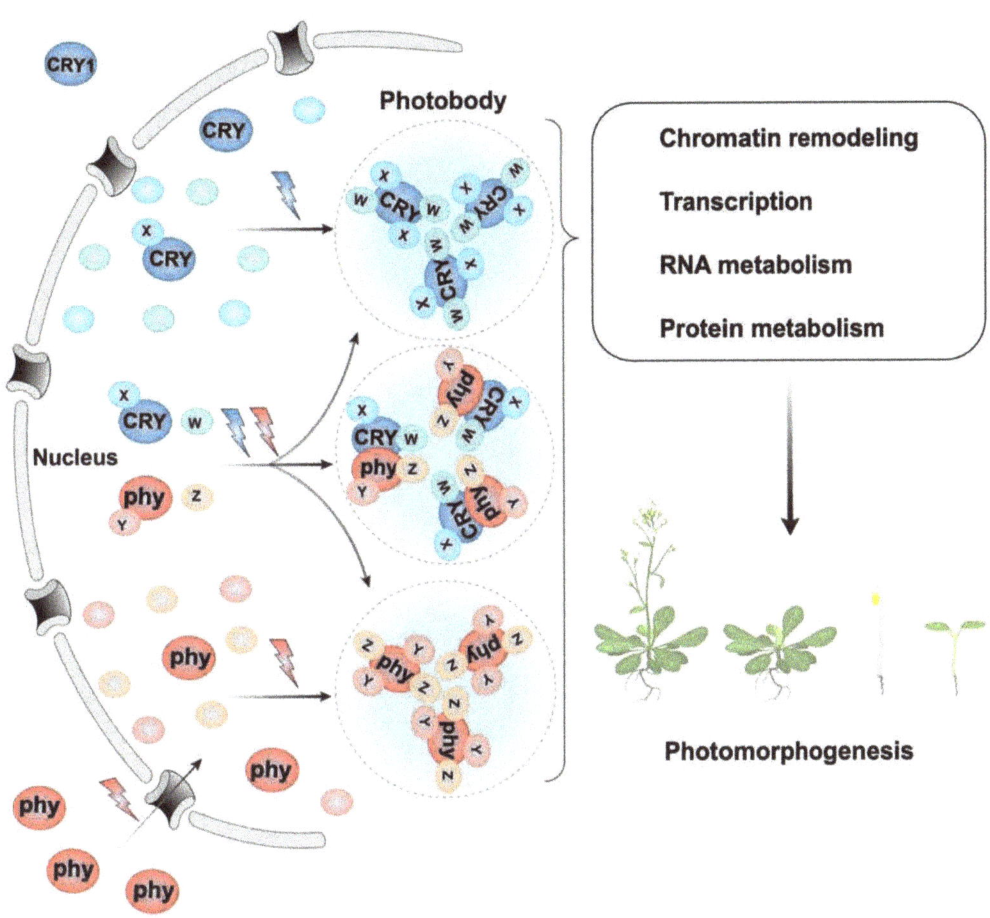

Figure 3. Light-triggered photobody formation by LLPS facilitates convergence of cryptochrome and phytochrome signaling pathways directly at the genome interface. The diagram depicts the nuclear importation of photoactivated phy and the condensation of both the photoactivated phytochromes and CRY by LLPS. This light-induced LLPS of the photoreceptors facilitates protein-protein interactions both between the phy and CRY molecules and with the numerous photoreceptor-bound signaling proteins, denoted W, X, Y, and Z. These multilateral interactions regulate processes that include chromatin remodeling, transcription, RNA metabolism (splicing, modification, degradation, etc), protein metabolism (translation, modification, degradation, etc), and photomorphogenesis.

Holtkotte et al. 2017). SPA1 is one of a 4-member family of COP1-interacting proteins that positively regulate COP1 activity (Hoecker et al. 1998; Lau and Deng 2012; Hoecker 2017; Podolec and Ulm 2018; Han et al. 2020). CRY2 also interacts with SPA1 in a light-dependent manner to inhibit COP1 activity, but photoexcited CRY2 may interact with COP1 in either a SPA-dependent or a SPA-independent manner (Zuo et al. 2011; Holtkotte et al. 2017; Ponnu et al. 2019). COP1 interacts with the VP motif of its substrates to cause polyubiquitination and degradation of the COP1 substrates (Holm et al. 2001). The VP motif in the CCE domain of Arabidopsis CRY1 and CRY2 is required for physiological functions of the CRYs (Lau et al. 2019; Ponnu et al. 2019; Liu et al. 2020) in their action as competitive inhibitors that suppress the $CUL4^{COP1/SPAs}$ activity (Lau et al. 2019; Ponnu et al. 2019).

In addition to interacting with the COP1/SPAs complex to regulate proteolysis, CRYs interact with at least 80 other CRY-interacting proteins to regulate other aspects of gene expression, including transcription, chromatin remodeling, DNA repair, RNA splicing, and RNA modifications. For example, photoexcited CRYs interact with the bHLH transcription factors, called CIBs (CRY-Interacting bHLHs), to regulate flowering time in Arabidopsis or leaf senescence in soybean (Liu et al. 2008, 2013; Meng et al. 2013). CRYs also interact with the phytochrome-signaling bHLH transcription factors PIF4 and PIF5 to regulate shade avoidance and thermoresponses (Ma et al. 2016; Pedmale et al. 2016; He et al. 2022). CRYs interact with the transcriptional repressors PRR9 and TCP22, among others, to regulate the circadian clock (He et al. 2022; Mo et al. 2022). CRYs interact with hormonal responsive transcription regulators, such as IAAs, ARF6, BZR1, BIM1, GID, DELLA, the protein kinase BIN2, and the E3 ubiquitin ligases SAINTs, to modulate auxin, brassinosteroid, gibberellin, and ethylene responses, respect-ively (Wang et al. 2018a, 2018b; Xu et al. 2018; He et al. 2019; Mao et al. 2020; Lee et al. 2021; Xu et al. 2021; Yan et al. 2021; Zhong et al. 2021).

CRYs interact with these CRY-signaling proteins via 2 different domains of CRYs. All CRYs are characterized by a highly conserved N-terminal domain, referred to as PHR for

Photolyase Homologous Region (Lin and Todo 2005), or CNT for CRY N-Terminus (Sang et al. 2005). Plant CRYs also possess a less conserved C-terminal domain unrelated to DNA photolyase, which was originally referred to as CCT (Yang et al. 2000), but was later renamed CCE (Cryptochrome C-terminal Extension) (Yu et al. 2010), to avoid possible confusion caused by the same domain name (IntePro# IPR010402 or Pfam# PF06203) registered for the CCT-family proteins (CONSTANS, CO-LIKE, and TIMING OF CAB1), that also have important functions in light signal transduction (Strayer et al. 2000). PHR is the chromophore-binding domain of CRYs that facilitates their blue light–dependent photoreactivity. Several lines of evidence support a hypothesis that absorption of blue-light photons by FAD changes the conformation of CRYs, including interactions between the PHR and CCE domains, resulting in the "open" conformation of the molecules that alter their protein-protein interactions with the CRY-signaling proteins, thereby altering gene expression (Yang et al. 2000; Partch et al. 2005; Yu et al. 2007a, 2007b; Goett-Zink et al. 2021). Consistent with this model, some CRY-signaling proteins, such as ADA2a/2b, AGB1, ARF6/8, BEE2, BIC1, CIBs, CIS1, CO, HBI1, IAAs, PhyB, TCP2, etc, interact with the PHR domain of CRYs, whereas others, such as ARP6, BES1, BZR1, MTA, RGA, SINATs, TOE1/2, etc, interact with both the PHR domain and the CCE domains (see review by Qu et al. 2023). The light-dependent conformational changes of the 2 different domains of CRYs may explain the complex network of the CRY-signaling proteins. However, the above domain-based classification of the CRY-interacting proteins is based on the reported results that were each analyzed under different conditions, leaving the general mechanisms underlying the structural specificity between CRYs and CRY-interacting proteins unclear at present.

Photoactivation and inactivation of cryptochromes

Like all receptor proteins, the CRY photoreceptors undergo an activation and inactivation cycle upon light absorption. According to our current understanding, CRYs are photoactivated and inactivated/degraded by the following reactions. First, photoexcited CRYs change conformation to oligomerize, forming the CRY homo-tetramer that is physiologically active. This photoresponsive reaction of CRYs is referred to as CRY photoactivation (Sang et al. 2005; Rosenfeldt et al. 2008; Wang et al. 2016; Ma et al. 2020a, 2020b; Shao et al. 2020; Palayam et al. 2021). Photoreduction of the FAD chromophore of plant CRYs (Lin et al. 1995) has been hypothesized to be important for CRY photoexcitation, and its role in the mechanism of CRY activity has been previously reviewed (Liu et al. 2010; Chaves et al. 2011; Ahmad 2016). However, how FAD photoreduction is associated with oligomerization-dependent activation of CRY proteins remains unclear. Second, the oligomerized CRYs transduce the light signal by modulating the CRY complexes with or without changing the affinity between CRYs and CRY-interacting proteins (see below). Third, photoexcited CRYs are inactivated by interacting with BICs (Blue-light Inhibitor of Cryptochromes) that directly inhibit CRY oligomerization (Wang et al. 2016; Ma et al. 2020a, 2020b) and with 4 protein kinases PPKs (PPK1-4) that phosphorylate CRYs at more than 20 serine and threonine residues to not only stimulate CRY activity but also promote CRY ubiquitination and degradation (Shalitin et al. 2002; Liu et al. 2017; Gao et al. 2022). Two distinct E3 ubiquitin ligases, $Cul4^{COP1/SPAs}$ and $Cul3^{LRBs}$, catalyze CRY polyubiquitination to promote CRY degradation by the 26S proteosome (Shalitin et al. 2002; Chen et al. 2021; Liu et al. 2022; Miao et al. 2022). Although CRY1 appears stable upon relatively short exposure to blue light, 2 recent reports showed that Arabidopsis CRY1 is phosphorylated, ubiquitinated, and degraded in response to prolonged exposure to blue light via the same mechanism as CRY2 (Liu et al. 2022; Miao et al. 2022). It is interesting that PPKs, COP1, and SPAs are required for both the CRY function and CRY degradation. Fourth, the oligomerized CRYs undergo spontaneous monomerization in darkness (Liu et al. 2020). Distinct from photoinactivation of CRYs that requires the blue light–dependent CRY-BIC interaction, this light-independent CRY monomerization process presumably results from thermal relaxation, and other reactions, that reverse the CRY conformation, dissolving the CRY oligomers to inactivate the CRY photoreceptors in darkness (Liu et al. 2020).

CRY photobodies and light-induced LLPS

Más and colleagues first showed that blue-light activation of Arabidopsis CRY2 induces its rapid coalescence into nuclear condensates (Más et al. 2000). These condensates were initially called CRY2 nuclear speckles (Más et al. 2000), or CRY2 nuclear bodies (Yu et al. 2009), and, more recently, CRY2 photobodies (Chen and Chory 2011; Zuo et al. 2012; Ozkan-Dagliyan et al. 2013). However, the molecular basis for this coalescence, as for the phy photobodies, remained enigmatic until 2021 when Lin and colleagues discovered that the Arabidopsis CRY2 molecule undergoes LLPS (Wang et al. 2021) to form these biomolecular condensates (Fig. 3).

Recently it has been shown that this CRY2 LLPS formation is dependent on blue light, the FAD chromophore, the Intrinsically Disordered Region (IDR) of the CCE domain, and phosphorylation of the photoreceptor (Wang et al. 2021; Mo et al. 2022; Jiang et al. 2023a, 2023b; Ma et al. 2023). It has also been reported that CRY2 LLPS determines its function in multiple light-dependent responses, including transcriptional regulation, post-transcriptional mRNA methylation, CRY-phy co-action (see below), light regulation of the circadian clock, and light regulation of chlorophyll homeostasis (Wang et al. 2021; Mo et al. 2022; Jiang et al. 2023a, 2023b). Therefore, in addition to the regulation of many blue light-dependent CRY-interacting proteins that change the binding affinity to CRYs in response to blue light, photoexcitation-induced oligomerization is also necessary for the CRY-dependent functions of the CRY-interacting proteins that do not change the binding affinity to CRYs in response to light. Instead, these CRY-interacting proteins depend on the blue light–induced LLPS for photoresponsiveness. For example, the blue light–induced CRY oligomerization and LLPS are necessary for the functions of CRY-interacting proteins, such as COP1 (Wang et al. 2001; Yang et al. 2001), MTA (Wang et al. 2021), TCP22 (Mo et al. 2022), MAC3A (Jiang et al. 2023a, 2023b), FIO1 (Jiang et al. 2023a, 2023b), and probably many more that do not appear to change the CRY-binding affinity in response to light. It has been proposed that the light-induced coalescence of the CRYs increases the local concentration of both the photoreceptor and CRY-interacting proteins within the condensates (CRY photobodies), thereby accelerating the ensuing biochemical reactions (Wang et al. 2021; Jiang et al. 2023a, 2023b).

The available data are consistent with the concept that the IDR of the CRY molecule is responsible for keeping CRYs in the liquid phase during the LLPS process, and that the PHR domain is involved in photoresponsiveness and intermolecular interactions of the CRY complexes (Wang et al. 2021). Indeed, both phy and CRY molecules have not only well-structured but also unstructured IDR-containing regions, suggesting that both photoreceptors utilize this bifurcated structure for their photoresponsive signaling function (Pardi and Nusinow 2021; Wang et al. 2021; Chen et al. 2022).

Although as mentioned above, phototropism was the first blue-light response studied in plants (Darwin 1881; Briggs and Huala 1999; Christie 2007), because of space constraints here, we have omitted detailed discussion of the phototropins, the second major blue-light photosensory receptors molecularly defined, as well as of ZTL/FKF/LKP2 and the UV-light receptors. The phototropins, which sense directional blue light and induce the phototropic plant responses, were discovered by Winslow Briggs and colleagues in the 1990s (Christie et al. 1998). Subsequent progress in this area can be found in a number of periodic reviews (Christie 2007; Briggs 2014; Morrow et al. 2018). The ZTL and related proteins have been reviewed by Christie et al. (2015), and UVR8 by Ulm and colleagues (Rizzini et al 2011; Podolec et al. 2021).

phy-CRY signaling convergence

A variety of genetic and physiological studies over a number of years have indicated that the blue- and red-light photoreceptors function both synergistically and antagonistically in regulating responses to variations in the natural daylight spectrum (Casal 2000; Lin 2000; Folta and Spalding 2001; Usami et al. 2004; Imaizumi and Kay 2006; Nozue et al. 2007; Su et al. 2017). Following identification of the phy and CRY photoreceptor molecules, there were several reports of the physical interaction between various family members of the 2 molecules by co-immunoprecipitation in vitro [phyA and CRY1; (Ahmad et al. 1998)]; phyB and CRY2; (Más et al. 2000) and by co-localization of phyB and CRY2 in nuclear speckles within the cell [by fluorescence resonance energy transfer microscopy (Más et al. 2000)].

In addition, both CRY1 and CRY2 have been shown to physically interact with the phy-interacting PIF transcription factors in a blue light–dependent manner. For example, CRY1 binds to PIF3 and PIF4 (Ma et al. 2016), while CRY2 binds to PIF4 and PIF5 (Pedmale et al. 2016). CRY1 inhibits the transcriptional activation activity of PIF4, thereby reducing the hypocotyl elongation response to both blue light and high ambient temperature (Ma et al. 2016). On the other hand, the interactions of CRY2 with PIF4 and PIF5 result in promotion of hypocotyl elongation under low blue light, a condition mimicking canopy shade where blue light is filtered through the canopy (Pedmale et al. 2016).

As detailed above, phyA, phyB, CRY1, and CRY2 have all been shown to physically interact with the COP1-SPA complex (Yang et al. 2001; Ma et al. 2016; Pedmale et al. 2016). These interactions dissociate this complex (Lian et al. 2011; Liu et al. 2011; Zuo et al. 2011; Lu et al. 2015; Sheerin et al. 2015), disrupting its E3-ligase activity, thereby allowing accumulation of positively acting transcription factors (e.g. HY5, HFR1) that function in both the phy and CRY signaling pathways to promote photomorphogenesis under red/far-red and blue light conditions. Taken together, these studies suggest that the phys and CRYs interact physically both with each other and with common signaling partners to regulate plant development under the prevailing light conditions in nature.

The discovery that both the photoactivated CRY and phy receptors coalesce very rapidly to form photobodies in the nu-cleus by LLPS (Wang et al. 2021; Chen et al. 2022) poises the system to provide a new dimension of understanding of these light-signaling pathways (Fig. 3). Numerous cellular factors de-fined as interactors of one or both photoreceptor molecules have been shown to co-localize with them in photobodies. These factors are known to have a diversity of functional roles in the cell suggesting that the photobodies may function as component-sequestration sites, as well as highly condensed hubs of transcriptional activity, regulated protein modification and degradation, chromatin remodeling, regulated mRNA degradation, circadian clock regulation, and cross-talk with other signaling pathways, such as the plant hormones (Paik et al. 2017; Ronald and Davis 2019; Willige et al. 2021). Overall, the emerging picture is that the phy and CRY sensory photoreceptors nucleate the coalescence of large complexes inside shared photobodies, generating a highly concentrated, dynamic milieux of interacting partners, where numerous multivalent transactions are enhanced.

Acknowledgments

EH: Author's laboratory is supported in part by the National Science Foundation (MCB-2014408 to EH); CL thanks UCLA and NIH for supporting his research from 1996 to 2022, previous members of his laboratory, and Drs. Xu Wang and Guifang Lin for assistance in preparing Fig. 3. Author's current research is supported in part by the Natural Science Foundation of China (32330009 to C.L.) and Fujian Agriculture and Forestry University Research Fund (to C.L.).

P.H.Q. thanks the many wonderful graduate students, postdocs, and lab research assistants who have contributed not only their ingenuity, time, and hard work to the research successes we have had over the years but also their readiness and energy in forming a welcoming and supportive community among the lab members. Special thanks go to Jim Tepperman, my lab manager, colleague, and enduring friend, who contributed so much to the science and community building. The research referred to in this review was supported over various periods by grants from NIH, NSF, DOE, USDA competitive grants, and by the USDA/ARS support of The Plant Gene Expression Center.

Author contributions

EH, CL and PHQ all contributed to researching the literature for, writing and editing this review article.

Conflict of interest statement. None declared.

Data availability

There are no new data associated with this article.

References

Ahmad M. Photocycle and signaling mechanisms of plant cryptochromes. Curr Opin Plant Biol. 2016:**33**:108–115. https://doi.org/10.1016/j.pbi.2016.06.013

Ahmad M, Cashmore AR. HY4 gene of *A. thaliana* encodes a protein with characteristics of a blue-light photoreceptor. Nature. 1993:**366**(6451):162–166. https://doi.org/10.1038/366162a0

Ahmad M, Jarillo JA, Smirnova O, Cashmore AR. The CRY1 blue light photoreceptor of Arabidopsis inter- acts with phytochrome A in vitro. Mol Cell. 1998:**1**(7):939–948. https://doi.org/10.1016/S1097-2765(00)80094-5

Ahmad M, Lin C, Cashmore AR. Mutations throughout an Arabidopsis blue-light photoreceptor impair blue-light-responsive anthocyanin accumulation and inhibition of hypocotyl elongation. Plant J. 1995:**8**(5):653–658. https://doi.org/10.1046/j.1365-313X.1995.08050653.x

Al-Sady B, Ni W, Kircher S, Schafer E, Quail PH. Photoactivated phytochrome induces rapid PIF3 phosphorylation prior to proteasome-mediated degradation. Mol Cell. 2006:**23**(3):439–446. https://doi.org/10.1016/j.molcel.2006.06.011

Aukerman MJ, Hirschfeld M, Wester L, Weaver M, Clack T, Amasino RM, Sharrock RA. A deletion in the PHYD gene of the Arabidopsis Wassilewskija ecotype defines a role for phytochrome D in red/far-red light sensing. Plant Cell. 1997:**9**(8):1317–1326. https://doi.org/10.1105/tpc.9.8.1317

Batschauer A. A plant gene for photolyase: an enzyme catalyzing the repair of UV-light-induced DNA damage. Plant J. 1993:**4**(4):705–709. https://doi.org/10.1046/j.1365-313X.1993.04040705.x

Bauer D, Viczian A, Kircher S, Nobis T, Nitschke R, Kunkel T, Panigrahi KC, Adam E, Fejes E, Schafer E, et al. Constitutive photomorphogenesis 1 and multiple photoreceptors control degradation of phytochrome interacting factor 3, a transcription factor required for light signaling in Arabidopsis. Plant Cell. 2004:**16**(6):1433–1445. https://doi.org/10.1105/tpc.021568

Bernardo-García S, de Lucas M, Martínez C, Espinosa-Ruiz A, Daviere J-M, Prat S. BR-dependent phosphorylation modulates PIF4 transcriptional activity and shapes diurnal hypocotyl growth. Genes Dev. 2014:**28**(15):1681–1694. https://doi.org/10.1101/gad.243675.114

Borthwick HA, Hendricks SB. Photoperiodism in plants: growth is controlled by light and the measurement of night length through reversible reactions of a pigment. Science. 1960:**132**(3435):1223–1228. https://doi.org/10.1126/science.132.3435.1223

Borthwick HA, Hendricks SB, Parker MW, Toole EH, Toole VK. A reversible photoreaction controlling seed germination. Proc Natl Acad Sci U S A. 1952:**38**(8):662–666. https://doi.org/10.1073/pnas.38.8.662

Bowler C, Neuhaus G, Yamagata H, Chua NH. Cyclic GMP and calcium mediate phytochrome phototransduction. Cell. 1994:**77**(1):73–81. https://doi.org/10.1016/0092-8674(94)90236-4

Brain RD, Freeberg JA, Weiss CV, Briggs WR. Blue light-induced absorbance changes in membrane fractions from corn and neurospora. Plant Physiol. 1977:**59**(5):948–952. https://doi.org/10.1104/pp.59.5.948

Briggs WR. A wandering pathway in plant biology: from wildflowers to phototropins to bacterial virulence. Annu Rev Plant Biol. 2010:**61**(1):1–20. https://doi.org/10.1146/annurev-arplant-042809-112326

Briggs WR. Phototropism: some history, some puzzles, and a look ahead. Plant Physiol. 2014:**164**(1):13–23. https://doi.org/10.1104/pp.113.230573

Briggs WR, Huala E. Blue-light photoreceptors in higher plants. Ann Rev Cell Dev Biol. 1999:**15**(1):33–62. https://doi.org/10.1146/annurev.cellbio.15.1.33

Briggs WR, Iino M, Galston AW, Digby J, Firn RD, Wareing PF, Smith H. Blue-light-absorbing photoreceptors in plants. Philos Trans R Soc Lond B Biol Sci. 1983:**303**(1116):347–359. https://doi.org/10.1098/rstb.1983.0098

Briggs WR, Tocher RD, Wilson JF. Phototropic auxin redistribution in corn coleoptiles. Science. 1957:**126**(3266):210–212. https://doi.org/10.1126/science.126.3266.210

Brown JAM, Klein WH. Photomorphogenesis in Arabidopsis thaliana (L.) heynh, threshhold intensity and blue-far-red synergism in floral induction. Plant Physiol. 1971:**47**(3):393–399. https://doi.org/10.1104/pp.47.3.393

Brudler R, Hitomi K, Daiyasu H, Toh H, Kucho K, Ishiura M, Kanehisa M, Roberts VA, Todo T, Tainer JA, et al. Identification of a new cryptochrome class. Structure, function, and evolution. Mol Cell. 2003:**11**(1):59–67. https://doi.org/10.1016/S1097-2765(03)00008-X

Bu Q, Zhu L, Yu L, Dennis M, Lu X, Person M, Tobin E, Browning K, Huq E. Phosphorylation by CK2 enhances the rapid light-induced degradation of PIF1. J Biol Chem. 2011:**286**(14):12066–12074. https://doi.org/10.1074/jbc.M110.186882

Burgie ES, Li H, Gannam ZTK, McLoughlin KE, Vierstra RD, Li H. The structure of Arabidopsis phytochrome A reveals topological and functional diversification among the plant photoreceptor isoforms. Nat Plants. 2023:**9**(7):1116–1129. https://doi.org/10.1038/s41477-023-01435-8

Butler WL, Norris KH, Siegelman HW, Hendricks SB. Detection, assay, and preliminary purification of the pigment controlling photoresponsive development of plants. Proc Natl Acad Sci U S A. 1959:**45**(12):1703–1708. https://doi.org/10.1073/pnas.45.12.1703

Casal J. Phytochromes, cryptochromes, phototropin: photoreceptor interactions in plants. Photochem Photobiol. 2000:**71**(1):1–11. https://doi.org/10.1562/0031-8655(2000)071<0001:PCPPII>2.0.CO;2

Cashmore AR. Cryptochromes: enabling plants and animals to determine circadian time. Cell. 2003:**114**:537–543. https: //doi.org/10.1016/j.cell.2003.08.004.

Cashmore AR, Jarillo JA, Wu YJ, Liu D. Cryptochromes: blue light receptors for plants and animals. Science. 1999:**284**(5415):760–765. https://doi.org/10.1126/science.284.5415.760

Castillon A, Shen H, Huq E. Phytochrome interacting factors: central players in phytochrome-mediated light signaling networks. Trends

Plant Sci. 2007:**12**(11):514–521. https://doi.org/10.1016/j.tplants.2007.10.001

Chaves I, Nijman RM, Biernat MA, Bajek MI, Brand K, da Silva AC, Saito S, Yagita K, Eker AP, van der Horst GT. The potorous CPD photolyase rescues a cryptochrome-deficient mammalian circadian clock. PLoS One. 2011:**6**(8):e23447. https://doi.org/10.1371/journal.pone.0023447

Chen M, Chory J. Phytochrome signaling mechanisms and the control of plant development. Trends Cell Biol. 2011:**21**(11):664–671. https://doi.org/10.1016/j.tcb.2011.07.002

Chen Y, Hu X, Liu S, Su T, Huang H, Ren H, Gao Z, Wang X, Lin D, Wohlschlegel JA, et al. Regulation of Arabidopsis photoreceptor CRY2 by two distinct E3 ubiquitin ligases. Nat Commun. 2021:**12**(1):2155. https://doi.org/10.1038/s41467-021-22410-x

Chen G-H, Liu M-J, Xiong Y, Sheen J, Wu S-H. TOR and RPS6 transmit light signals to enhance protein translation in deetiolating Arabidopsis seedlings. Proc Natl Acad Sci U S A. 2018:**115**(50):12823–12828. https://doi.org/10.1073/pnas.1809526115

Chen D, Lyu M, Kou X, Li J, Yang Z, Gao L, Li Y, Fan L-M, Shi H, Zhong S. Integration of light and temperature sensing by liquid-liquid phase separation of phytochrome B. Mol Cell. 2022:**82**(16):3015–3029.e6. https://doi.org/10.1016/j.molcel.2022.05.026

Cheng M-C, Kathare PK, Paik I, Huq E. Phytochrome signaling networks. Annu Rev Plant Biol. 2021:**72**(1):217–244. https://doi.org/10.1146/annurev-arplant-080620-024221

Chory J, Peto C, Feinbaum R, Pratt L, Ausubel F. Arabidopsis thaliana mutant that develops as a light-grown plant in the absence of light. Cell. 1989:**58**(5):991–999. https://doi.org/10.1016/0092-8674(89)90950-1

Christie JM. Phototropin blue-light receptors. Annu Rev Plant Biol. 2007:**58**(1):21–45. https://doi.org/10.1146/annurev.arplant.58.032806.103951

Christie JM, Blackwood L, Petersen J, Sullivan S. Plant flavoprotein photoreceptors. Plant Cell Physiol. 2015:**56**(3):401–413. https://doi.org/10.1093/pcp/pcu196

Christie JM, Reymond P, Powell GK, Bernasconi P, Raibekas AA, Liscum E, Briggs WR. Arabidopsis NPH1: a flavoprotein with the properties of a photoreceptor for phototropism. Science. 1998:**282**(5394):1698–1701. https://doi.org/10.1126/science.282.5394.1698

Clack T, Mathews S, Sharrock RA. The phytochrome apoprotein family in Arabidopsis is encoded by five genes: the sequences and expression of PHYD and PHYE. Plant Mol Biol. 1994:**25**(3):413–427. https://doi.org/10.1007/BF00043870

Darwin C. The power of movement in plants. New York: D. Appleton and Company; 1881.

Dehesh K, Franci C, Parks BM, Seeley KA, Short TW, Tepperman JM, Quail PH. Arabidopsis HY8 locus encodes phytochrome A. Plant Cell. 1993:**5**(9):1081–1088. https://doi.org/10.1105/tpc.5.9.1081

Deng XW, Caspar T, Quail PH. Cop1: a regulatory locus involved in light-controlled development and gene expression in Arabidopsis. Genes Dev. 1991:**5**(7):1172–1182. https://doi.org/10.1101/gad.5.7.1172

Deng XW, Matsui M, Wei N, Wagner D, Chu AM, Feldman KA, Quail PH. *COP1*, an arabidopsis regulatory gene, encodes a protein with both a zinc-binding motif and a Gβ homologous domain. Cell. 1992:**71**(5):791–801. https://doi.org/10.1016/0092-8674(92)90555-Q

De Piccoli G, Cortes-Ledesma F, Ira G, Torres-Rosell J, Uhle S, Farmer S, Hwang JY, Machin F, Ceschia A, McAleenan A, et al. Smc5-Sm c6 mediate DNA double-strand-break repair by promoting sister-chromatid recombination. Nat Cell Biol. 2006:**8**(9):1032–1034. https://doi.org/10.1038/ncb1466

Deppisch P, Helfrich-Förster C, Senthilan PR. The gain and loss of cryptochrome/photolyase family members during evolution. Genes (Basel). 2022:**13**(9):1613. https://doi.org/10.3390/genes13091613

Devlin PF, Patel SR, Whitelam GC. Phytochrome E influences internode elongation and flowering time in Arabidopsis. Plant Cell. 1998:**10**(9):1479–1487. https://doi.org/10.1105/tpc.10.9.1479

Dong J, Ni W, Yu R, Deng XW, Chen H, Wei N. Light-dependent degradation of PIF3 by SCF-EBF1/2 promotes a photomorphogenic response in Arabidopsis. Curr Biol. 2017:**27**(16):2420–2430. https://doi.org/10.1016/j.cub.2017.06.062

Doolittle WF. Fun with genealogy. Proc Natl Acad Sci U S A. 1997:**94**(24):12751–12753. https://doi.org/10.1073/pnas.94.24.12751

Emery P, So WV, Kaneko M, Hall JC, Rosbash M. CRY, a Drosophila clock and light-regulated cryptochrome, is a major contributor to circadian rhythm resetting and photosensitivity. Cell. 1998:**95**(5):669–679. https://doi.org/10.1016/S0092-8674(00)81637-2

Eskins K. Light-quality effects on Arabidopsis development. Red, blue and far-red regulation of flowering and morphology. Physiol Planta. 1992:**86**(3):439–444. https://doi.org/10.1111/j.1399-3054.1992.tb01341.x

Essen LO, Mailliet J, Hughes J. The structure of a complete phytochrome sensory module in the Pr ground state. Proc Natl Acad Sci U S A. 2008:**105**(38):14709–14714. https://doi.org/10.1073/pnas.0806477105

Feldmann KA. T-DNA insertion mutagenesis in Arabidopsis: mutation spectrum. Plant J. 1991:**1**(1):71–82. https://doi.org/10.1111/j.1365-313X.1991.00071.x

Folta KM, Pontin MA, Karlin-Neumann G, Bottini R, Spalding EP. Genomic and physiological studies of early cryptochrome 1 action demonstrate roles for auxin and gibberellin in the control of hypocotyl growth by blue light. Plant J. 2003:**36**(2):203–214. https://doi.org/10.1046/j.1365-313X.2003.01870.x

Folta KM, Spalding EP. Opposing roles of phytochrome A and phytochrome B in early cryptochrome-mediated growth inhibition. Plant J. 2001:**28**(3):333–340. https://doi.org/10.1046/j.1365-313X.2001.01157.x

Fuglevand G, Jackson JA, Jenkins GI. UV-B, UV-A, and blue light signal transduction pathways interact synergistically to regulate chalcone synthase gene expression in Arabidopsis. Plant Cell. 1996:**8**(12):2347–2357. https://doi.org/10.1105/tpc.8.12.2347

Furuya M. History and insights. In: **Wada KSM, Shimazaki K, Iino M**, editors. Light sensing in plants. Tokyo: Springer; 1987, p. 3–20.

Gao L, Liu Q, Zhong M, Zeng N, Deng W, Li Y, Wang D, Liu S, Wang Q. Blue light-induced phosphorylation of Arabidopsis cryptochrome 1 is essential for its photosensitivity. J Integr Plant Biol. 2022:**64**(9):1724–1738. https://doi.org/10.1111/jipb.13331

Goett-Zink L, Toschke AL, Petersen J, Mittag M, Kottke T. C-terminal extension of a plant cryptochrome dissociates from the β-sheet of the flavin-binding domain. J Phys Chem Lett. 2021:**12**(23):5558–5563. https://doi.org/10.1021/acs.jpclett.1c00844

González-Grandío E, Álamos S, Zhang Y, Dalton-Roesler J, Niyogi KK, García HG, Quail PH. Chromatin changes in phytochrome interacting factor-regulated genes parallel their rapid transcriptional response to light. Front Plant Sci. 2022:**13**:803441. https://doi.org/10.3389/fpls.2022.803441

Gressel J. Blue light photoreception. Photochem Photobiol. 1979:**30**(6):749–754. https://doi.org/10.1111/j.1751-1097.1979.tb07209.x

Guo HW, Duong H, Ma N, Lin CT. The Arabidopsis blue light recep- tor cryptochrome 2 is a nuclear protein regulated by a blue light-dependent post- transcriptional mechanism. Plant J. 1999:**19**:279–287.

Guo T, Liu M, Chen L, Liu Y, Li L, Li Y, Cao X, Mao Z, Wang W, Yang H-Q. Photoexcited cryptochromes interact with ADA2b and SMC5 to promote the repair of DNA double-strand breaks in Arabidopsis. Nat Plants. 2023:**9**(8):1280–1290. https://doi.org/10.1038/s41477-023-01461-6

Guo H, Yang H, Mockler TC, Lin C. Regulation of flowering time by Arabidopsis photoreceptors. Science. 1998:**279**(5355):1360–1363. https://doi.org/10.1126/science.279.5355.1360

Han X, Chang X, Zhang Z, Chen H, He H, Zhong B, Deng XW. Origin and evolution of core components responsible for monitoring light environment changes during plant terrestrialization. Mol Plant. 2019:**12**(6):847–862. https://doi.org/10.1016/j.molp.2019.04.006

Han X, Huang X, Deng XW. The photomorphogenic central repressor COP1: conservation and functional diversification during evolution. Plant Commun. 2020:**1**(3):100044. https://doi.org/10.1016/j.xplc.2020.100044

He G, Liu J, Dong H, Sun J. The blue-light receptor CRY1 interacts with BZR1 and BIN2 to modulate the phosphorylation and nuclear function of BZR1 in repressing BR signaling in Arabidopsis. Mol Plant. 2019:**12**(5):689–703. https://doi.org/10.1016/j.molp.2019.02.001

He Y, Yu Y, Wang X, Qin Y, Su C, Wang L. Aschoff's rule on circadian rhythms orchestrated by blue light sensor CRY2 and clock component PRR9. Nat Commun. 2022:**13**(1):5869. https://doi.org/10.1038/s41467-022-33568-3

Hershey HP, Colbert JT, Lissemore JL, Barker RF, Quail PH. Molecular cloning of cDNA for Avena phytochrome. Proc Natl Acad Sci U S A. 1984:**81**(8):2332–2336. https://doi.org/10.1073/pnas.81.8.2332

Hoecker U. The activities of the E3 ubiquitin ligase COP1/SPA, a key repressor in light signaling. Curr Opin Plant Biol. 2017:**37**:63–69. https://doi.org/10.1016/j.pbi.2017.03.015

Hoecker U, Tepperman JM, Quail PH. SPA1, a WD-repeat protein specific to phytochrome A signal transduction. Science. 1999:**284**(5413):496–499. https://doi.org/10.1126/science.284.5413.496

Hoecker U, Xu Y, Quail PH. SPA1: a new genetic locus involved in phytochrome A-specific signal transduction. Plant Cell. 1998:**10**(1):19–33. https://doi.org/10.1105/tpc.10.1.19

Hoffman PD, Batschauer A, Hays JB. PHH1, a novel gene from Arabidopsis thaliana that encodes a protein similar to plant blue-light photoreceptors and microbial photolyases. Mol Gen Genet. 1996:**253**(1–2):259–265. https://doi.org/10.1007/s004380050321

Holm M, Hardtke CS, Gaudet R, Deng XW. Identification of a structural motif that confers specific interaction with the WD40 repeat domain of Arabidopsis COP1. EMBO J. 2001:**20**(1):118–127. https://doi.org/10.1093/emboj/20.1.118

Holtkotte X, Ponnu J, Ahmad M, Hoecker U. The blue light-induced interaction of cryptochrome 1 with COP1 requires SPA proteins during Arabidopsis light signaling. PLoS Genet. 2017:**13**(10):e1007044. https://doi.org/10.1371/journal.pgen.1007044

Hua W, Franklin KA, Sharrock RA, Jones MA, Harmer SL, Lagarias JC. Unanticipated regulatory roles for Arabidopsis phytochromes revealed by null mutant analysis. Proc Natl Acad Sci U S A. 2013:**110**(4):1542–1547. https://doi.org/10.1073/pnas.1221738110

Huang Y, Baxter R, Smith BS, Partch CL, Colbert CL, Deisenhofer J. Crystal structure of cryptochrome 3 from Arabidopsis thaliana and its implications for photolyase activity. Proc Natl Acad Sci U S A. 2006:**103**(47):17701–17706. https://doi.org/10.1073/pnas.0608554103

Huq E, Al-Sady B, Hudson M, Kim C, Apel K, Quail PH. Phytochrome-interacting factor 1 is a critical bHLH regulator of chlorophyll biosynthesis. Science. 2004:**305**(5692):1937–1941. https://doi.org/10.1126/science.1099728

Huq E, Quail PH. PIF4, a phytochrome-interacting bHLH factor, functions as a negative regulator of phytochrome B signaling in Arabidopsis. EMBO J. 2002:**21**(10):2441–2450. https://doi.org/10.1093/emboj/21.10.2441

Imaizumi T, Kay S. Photoperiodic control of flowering: not only by coincidence. Trends Plant Sci. 2006:**11**(11):550–558. https://doi.org/10.1016/j.tplants.2006.09.004

Jiang B, Zhong Z, Gu L, Zhang X, Wei J, Ye C, Lin G, Qu G, Xiang X, Wen C, et al. Light-induced LLPS of the CRY2/SPA1/FIO1 complex regulating mRNA methylation and chlorophyll homeostasis in Arabidopsis. Nat Plants. 2023a:**9**(12):2042–2058. https://doi.org/10.1038/s41477-023-01580-0

Jiang B, Zhong Z, Su J, Zhu T, Yueh T, Bragasin J, Bu V, Zhou C, LinC, Wang X. Co-condensation with photoexcited cryptochromes facilitates MAC3A to positively control hypocotyl growth in Arabidopsis. Sci Adv. 2023b:**9**(32):eadh4048. https://doi.org/10.1126/sciadv.adh4048

Jung J-H, Domijan M, Klose C, Biswas S, Ezer D, Gao M, Khattak AK, Box MS, Charoensawan V, Cortijo S, et al. Phytochromes function as thermosensors in Arabidopsis. Science. 2016:**354**(6314):886–889. https://doi.org/10.1126/science.aaf6005

Kathare PK, Huq E. Light-regulated pre-mRNA splicing in plants. Curr Opin Plant Biol. 2021:**63**:102037. https://doi.org/10.1016/j.pbi.2021.102037

Kathare PK, Xin R, Ganesan AS, June VM, Reddy ASN, Huq E. SWAP1-SFPS-RRC1 splicing factor complex modulates pre-mRNA splicing to promote photomorphogenesis in Arabidopsis. Proc Natl Acad Sci U S A. 2022:**119**(44):e2214565119. https://doi.org/10.1073/pnas.2214565119

Khanna R, Huq E, Kikis EA, Al-Sady B, Lanzatella C, Quail PH. A novel molecular recognition motif necessary for targeting photoactivated phytochrome signaling to specific basic helix-loop-helix transcription factors. Plant Cell. 2004:**16**(11):3033–3044. https://doi.org/10.1105/tpc.104.025643

Khanna R, Shen Y, Marion CM, Tsuchisaka A, Theologis A, Schäfer E, Quail PH. The basic helix-loop-helix transcription factor PIF5 acts on ethylene biosynthesis and phytochrome signaling by distinct mechanisms. Plant Cell. 2007:**19**(12):3915–3929. https://doi.org/10.1105/tpc.107.051508

Kim J, Bordiya Y, Kathare PK, Zhao B, Zong W, Huq E, Sung S. Phytochrome B triggers light-dependent chromatin remodelling through the PRC2-associated PHD finger protein VIL1. Nat Plants. 2021:**7**(9):1213–1219. https://doi.org/10.1038/s41477-021-00986-y

Kim Y-M, Choi J, Lee H-Y, Lee G-W, Lee Y-H, Choi D. dbCRY: a web-based comparative and evolutionary genomics platform for blue-light receptors. Database (Oxford). 2014:**2014**(0):bau037. https://doi.org/10.1093/database/bau037

Kim JH, Lee H-J, Jung J-H, Lee S, Park C-M. HOS1 facilitates the phytochrome B-mediated inhibition of PIF4 function during hypocotyl growth in Arabidopsis. Mol Plant. 2017a:**10**(2):274–284. https://doi.org/10.1016/j.molp.2016.11.009

Kim JY, Song JT, Seo HS. COP1 regulates plant growth and development in response to light at the post-translational level. J Exp Bot. 2017b:**68**(17):4737–4748. https://doi.org/10.1093/jxb/erx312

Kircher S, Gil P, Kozma-Bognár L, Fejes E, Speth V, Husselstein-Muller T, Bauer D, Ádám É, Schäfer E, Nagy F. Nucleocytoplasmic partitioning of the plant photoreceptors phytochrome A, B, C, D, and E is regulated differentially by light and exhibits a diurnal rhythm. Plant Cell. 2002:**14**(7):1541–1555. https://doi.org/10.1105/tpc.001156

Kircher S, Kozma-Bognar L, Kim L, Adam E, Harter K, Schäfer E, Nagy F. Light quality–dependent nuclear import of the plant photoreceptors phytochrome A and B. Plant Cell. 1999:**11**(8):1445–1456. https://doi.org/10.1105/tpc.11.8.1445

Kleine T, Lockhart P, Batschauer A. An Arabidopsis protein closely related to synechocystis cryptochrome is targeted to organelles. Plant J. 2003:**35**(1):93–103. https://doi.org/10.1046/j.1365-313X.2003.01787.x

Kleiner O, Kircher S, Harter K, Batschauer A. Nuclear localization of the Arabidopsis blue light receptor cryptochrome 2. Plant J. 1999:**19**(3):289–296. https://doi.org/10.1046/j.1365-313X.1999.00535.x

Kok B. Absorption changes induced by the photochemical reaction of photosynthesis. Nature. 1957:**179**(4559):583–584. https://doi.org/10.1038/179583a0

Koornneef M, Hanhart CJ, van der Veen JH. A genetic and physiological analysis of late flowering mutants in *Arabidopsis thaliana*. Mol Gen Genet. 1991:**229**(1):57–66. https://doi.org/10.1007/BF00264213

Koornneef M, Rolff E, Spruit CJP. Genetic control of light-inhibited hypocotyl elongation in Arabidopsis thaliana (L.) Heynh. Zeitschrift für Pflanzenphysiologie. 1980:**100**(2):147–160. https://doi.org/10.1016/S0044-328X(80)80208-X

Kubasek WL, Shirley BW, McKillop A, Goodman HM, Briggs W, Ausubel FM. Regulation of flavonoid biosynthetic genes in germinating Arabidopsis seedlings. Plant Cell. 1992:**4**(10):1229–1236. https://doi.org/10.2307/3869409

Kutschera U, Niklas KJ. Julius sachs (1868): the father of plant physiology. Am J Bot. 2018:**105**(4):656–666. https://doi.org/10.1002/ajb2.1078

Lau OS, Deng XW. The photomorphogenic repressors COP1 and DET1: 20 years later. Trends Plant Sci. 2012:**17**(10):584–593. https://doi.org/10.1016/j.tplants.2012.05.004

Lau K, Podolec R, Chappuis R, Ulm R, Hothorn M. Plant photoreceptors and their signaling components compete for COP1 binding via VP peptide motifs. EMBO J. 2019:**38**(18):e102140. https://doi.org/10.15252/embj.2019102140

Lee HY, Park HL, Park C, Chen Y-C, Yoon GM. Reciprocal antagonistic regulation of E3 ligases controls ACC synthase stability and responses to stress. Proc Natl Acad Sci U S A. 2021:**118**(34):e2011900118. https://doi.org/10.1073/pnas.2011900118

Legris M, Klose C, Burgie ES, Costigliolo C, Neme M, Hiltbrunner A, Wigge PA, Schäfer E, Vierstra RD, Casal JJ. Phytochrome B integrates light and temperature signals in Arabidopsis. Science. 2016:**354**(6314):897–900. https://doi.org/10.1126/science.aaf5656

Leivar P, Monte E. PIFs: systems integrators in plant development. Plant Cell. 2014:**26**(1):56–78. https://doi.org/10.1105/tpc.113.120857

Leivar P, Monte E, Al-Sady B, Carle C, Storer A, Alonso JM, Ecker JR, Quail PH. The Arabidopsis phytochrome-interacting factor PIF7, together with PIF3 and PIF4, regulates responses to prolonged red light by modulating phyB levels. Plant Cell. 2008a:**20**(2):337–352. https://doi.org/10.1105/tpc.107.052142

Leivar P, Monte E, Oka Y, Liu T, Carle C, Castillon A, Huq E, Quail PH. Multiple phytochrome-interacting bHLH transcription factors repress premature seedling photomorphogenesis in darkness. Curr Biol. 2008b:**18**(23):1815–1823. https://doi.org/10.1016/j.cub.2008.10.058

Leivar P, Quail PH. PIFs: pivotal components in a cellular signaling hub. Trends Plant Sci. 2011:**16**(1):19–28. https://doi.org/10.1016/j.tplants.2010.08.003

Leivar P, Tepperman JM, Cohn MM, Monte E, Al-Sady B, Erickson E, Quail PH. Dynamic antagonism between phytochromes and PIF-family bHLH factors generates selective reciprocal responses during deetiolation and shade-avoidance in a rapidly light-responsive transcriptional network. Plant Cell. 2012:**24**(4):1398–1419. https://doi.org/10.1105/tpc.112.095711

Leivar P, Tepperman JM, Monte E, Calderon RH, Liu TL, Quail PH. Definition of early transcriptional circuitry involved in light-induced reversal of PIF-imposed repression of photomorphogenesis in young Arabidopsis seedlings. Plant Cell. 2009:**21**(11):3535–3553. https://doi.org/10.1105/tpc.109.070672

Li H, Burgie ES, Gannam ZTK, Li H, Vierstra RD. Plant phytochrome B is an asymmetric dimer with unique signalling potential. Nature. 2022:**64**(7904):127–133. https://doi.org/10.1038/s41586-022-04529-z

Li J, Li G, Wang H, Deng XW. Phytochrome signaling mechanisms. In: The Arabidopsis book. Rockville, MD: American Society of Plant Biologists; 2011:p. 9.

Lian HL, He SH, Zhang YC, Zhu DM, Zhang JY, Jia KP, Sun SX, Li L, Yang HQ. Blue-light-dependent interaction of cryptochrome 1 with SPA1 defines a dynamic signaling mechanism. Genes Dev. 2011:**25**(10):1023–1028. https://doi.org/10.1101/gad.2025111

Lin C. Photoreceptors and regulation of flowering time. Plant Physiol. 2000:**123**(1):39–50. https://doi.org/10.1104/pp.123.1.39

Lin C, Ahmad M, Chan J, Cashmore AR. CRY2, a second member of the Arabidopsis cryptochrome gene family. Plant Physiol. 1996:**110**(3):1047. https://doi.org/10.1104/pp.110.3.1047

Lin C, Robertson DE, Ahmad M, Raibekas AA, Jorns MS, Dutton PL, Cashmore AR. Association of flavin adenine dinucleotide with the Arabidopsis blue light receptor CRY1. Science. 1995:**269**(5226):968–970. https://doi.org/10.1126/science.7638620

Lin B-Y, Shih C-J, Hsieh H-Y, Chen H-C, Tu S-L. Phytochrome coordinates with a hnRNP to regulate alternative splicing via an exonic splicing silencer. Plant Physiol. 2020:**182**(1):243–254. https://doi.org/10.1104/pp.19.00289

Lin C, Todo T. The cryptochromes. Genome Biol. 2005:**6**(5):220. https://doi.org/10.1186/gb-2005-6-5-220

Lin C, Yang H, Guo H, Mockler T, Chen J, Cashmore AR. Enhancement of blue-light sensitivity of Arabidopsis seedlings by a blue light receptor cryptochrome 2. Proc Natl Acad Sci U S A. 1998:**95**(5):2686–2690. https://doi.org/10.1073/pnas.95.5.2686

Liu Y, Li X, Li K, Liu H, Lin C. Multiple bHLH proteins form heterodimers to mediate CRY2-dependent regulation of flowering-time in Arabidopsis. PLoS Genet. 2013:**9**(10):e1003861. https://doi.org/10.1371/journal.pgen.1003861

Liu B, Liu H, Zhong D, Lin C. Searching for a photocycle of the cryptochrome photoreceptors. Curr Opin Plant Biol. 2010:**13**(5):578–586. https://doi.org/10.1016/j.pbi.2010.09.005

Liu Q, Su T, He W, Ren H, Liu S, Chen Y, Gao L, Hu X, Lu H, Cao S, et al. Photooligomerization determines photosensitivity and photoreactivity of plant cryptochromes. Mol Plant. 2020:**13**(3):398–413. https://doi.org/10.1016/j.molp.2020.01.002

Liu Q, Wang Q, Deng W, Wang X, Piao M, Cai D, Li Y, Barshop WD, Yu X, Zhou T, et al. Molecular basis for blue light-dependent phosphorylation of Arabidopsis cryptochrome 2. Nat Commun. 2017:**8**(1):15234. https://doi.org/10.1038/ncomms15234

Liu H, Yu X, Li K, Klejnot J, Yang H, Lisiero D, Lin C. Photoexcited CRY2 interacts with CIB1 to regulate transcription and floral initiation in Arabidopsis. Science. 2008:**322**(5907):1535–1539. https://doi.org/10.1126/science.1163927

Liu S, Zhang L, Gao L, Chen Z, Bie Y, Zhao Q, Zhang S, Hu X, Liu Q, Wang X, et al. Differential photoregulation of the nuclear and cytoplasmic CRY1 in Arabidopsis. New Phytol. 2022:**234**(4):1332–1346. https://doi.org/10.1111/nph.18007

Liu B, Zuo Z, Liu H, Liu X, Lin C. Arabidopsis cryptochrome 1 interacts with SPA1 to suppress COP1 activity in response to blue light. Genes Dev. 2011:**25**(10):1029–1034. https://doi.org/10.1101/gad.2025011

Lorrain S, Allen T, Duek PD, Whitelam GC, Fankhauser C. Phytochrome-mediated inhibition of shade avoidance involves degradation of growth-promoting bHLH transcription factors. Plant J. 2008:**53**(2):312–323. https://doi.org/10.1111/j.1365-313X.2007.03341.x

Lu X-D, Zhou C-M, Xu P-B, Luo Q, Lian H-L, Yang H-Q. Red-light-dependent interaction of phyB with SPA1 promotes COP1–SPA1 dissociation and photomorphogenic development in Arabidopsis. Mol Plant. 2015:**8**(3):467–478. https://doi.org/10.1016/j.molp.2014.11.025

Ma L, Guan Z, Wang Q, Yan X, Wang J, Wang Z, Cao J, Zhang D, Gong X, Yin P. Structural insights into the photoactivation of Arabidopsis CRY2. Nat Plants. 2020a:**6**(12):1432–1438. https://doi.org/10.1038/s41477-020-00800-1

Ma L, Jia H, Shen A-L, Ding J, Wang X, Wang J, Wan J, Yan J, Zhang D, Dong X, et al. Two determinants influence CRY2 photobody formation and function. Plant Biotechnol J. 2023:**21**(3):460–462. https://doi.org/10.1111/pbi.13978

Ma D, Li X, Guo Y, Chu J, Fang S, Yan C, Noel JP, Liu H. Cryptochrome 1 interacts with PIF4 to regulate high temperature-mediated hypocotyl elongation in response to blue light. Proc Natl Acad Sci U S A. 2016:**113**(1):224–229. https://doi.org/10.1073/pnas.1511437113

Ma L, Li J, Qu L, Hager J, Chen Z, Zhao H, Deng XW. Light control of Arabidopsis development entails coordinated regulation of genome expression and cellular pathways. Plant Cell. 2001:**13**(12):2589–2607. https://doi.org/10.1105/tpc.010229

Ma L, Wang X, Guan Z, Wang L, Wang Y, Zheng L, Gong Z, Shen C, Wang J, Zhang D, et al. Structural insights into BIC-mediated inactivation of Arabidopsis cryptochrome 2. Nat Struct Mol Biol. 2020b:**27**(5):472–479. https://doi.org/10.1038/s41594-020-0410-z

Mackenzie JM Jr, Briggs WR, Pratt LH. Intracellular phytochrome distribution as a function of its molecular form and of its destruction. Am J Bot. 1978:**65**(6):671–676. https://doi.org/10.1002/j.1537-2197.1978.tb06124.x

Mackenzie JM Jr, Coleman R, Briggs W, Pratt L. Reversible redistribution of phytochrome within the cell upon conversion to its physiologically active form. Proc Natl Acad Sci U S A. 1975:**72**(3):799–803. https://doi.org/10.1073/pnas.72.3.799

Majee M, Kumar S, Kathare PK, Wu S, Gingerich D, Nayak NR, Salaita L, Dinkins R, Martin K, Goodin M, et al. KELCH F-BOX protein positively influences Arabidopsis seed germination by targeting PHYTOCHROME-INTERACTING FACTOR1. Proc Natl Acad Sci U S A. 2018:**115**(17):E4120–E4129. https://doi.org/10.1073/pnas.1711919115

Malhotra K, Baer M, Li YF, Sancar GB, Sancar A. Identification of chromophore binding domains of yeast DNA photolyase. J Biol Chem. 1992:**267**(5):2909–2914. https://doi.org/10.1016/S0021-9258(19)50672-X

Malhotra K, Kim ST, Batschauer A, Dawut L, Sancar A. Putative blue-light photoreceptors from *Arabidopsis thaliana* and *Sinapis alba* with a high degree of sequence homology to DNA photolyase contain the two photolyase cofactors but lack DNA repair activity. Biochemistry. 1995:**34**(20):6892–6899. https://doi.org/10.1021/bi00020a037

Mao Z, He S, Xu F, Wei X, Jiang L, Liu Y, Wang W, Li T, Xu P, Du S, et al. Photoexcited CRY1 and phyB interact directly with ARF6 and ARF8 to regulate their DNA-binding activity and auxin-induced hypocotyl elongation in Arabidopsis. New Phytol. 2020:**225**(2):848–865. https://doi.org/10.1111/nph.16194

Martinez-Garcia JF, Huq E, Quail PH. Direct targeting of light signals to a promoter element-bound transcription factor. Science. 2000:**288**(5467):859–863. https://doi.org/10.1126/science.288.5467.859

Más P, Devlin PF, Panda S, Kay SA. Functional interaction of phytochrome B and cryptochrome 2. Nature. 2000:**408**(6809):207–211. https://doi.org/10.1038/35041583

Mei Q, Dvornyk V. Evolutionary history of the photolyase/cryptochrome superfamily in eukaryotes. PLoS One. 2015:**10**(9):e0135940. https://doi.org/10.1371/journal.pone.0135940

Meng Y, Li H, Wang Q, Liu B, Lin C. Blue light-dependent interaction between cryptochrome2 and CIB1 regulates transcription and leaf senescence in soybean. Plant Cell. 2013:**25**(11):4405–4420. https://doi.org/10.1105/tpc.113.116590

Miao L, Zhao J, Yang G, Xu P, Cao X, Du S, Xu F, Jiang L, Zhang S, Wei X, et al. Arabidopsis cryptochrome 1 undergoes COP1 and LRBs-dependent degradation in response to high blue light. New Phytol. 2022:**234**(4):1347–1362. https://doi.org/10.1111/nph.17695

Millar A, McGrath R, Chua N-H. Phytochrome phototransduction pathways. Annu Rev Genet. 1994:**28**(1):325–349. https://doi.org/10.1146/annurev.ge.28.120194.001545

Mo W, Zhang J, Zhang L, Yang Z, Yang L, Yao N, Xiao Y, Li T, Li Y, Zhang G, et al. Arabidopsis cryptochrome 2 forms photobodies with TCP22 under blue light and regulates the circadian clock. Nat Commun. 2022:**13**(1):2631. https://doi.org/10.1038/s41467-022-30231-9

Monte E, Alonso JM, Ecker JR, Zhang Y, Li X, Young J, Austin-Phillips S, Quail PH. Isolation and characterization of phyC mutants in Arabidopsis reveals complex crosstalk between phytochrome signaling pathways. Plant Cell. 2003:**15**(9):1962–1980. https://doi.org/10.1105/tpc.012971

Monte E, Tepperman JM, Al-Sady B, Kaczorowski KA, Alonso JM, Ecker JR, Li X, Zhang Y, Quail PH. The phytochrome-interacting transcription factor, PIF3, acts early, selectively, and positively in light-induced chloroplast development. Proc Natl Acad Sci U S A. 2004:**101**(46):16091–16098. https://doi.org/10.1073/pnas.0407107101

Morrow J, Willenburg KT, Liscum E. Phototropism in land plants: molecules and mechanism from light perception to response. Front Biol. 2018:**13**(5):342–357. https://doi.org/10.1007/s11515-018-1518-y

Ni M, Tepperman JM, Quail PH. PIF3, a phytochrome-interacting factor necessary for normal photoinduced signal transduction, is a novel basic helix-loop-helix protein. Cell. 1998:**95**(5):657–667. https://doi.org/10.1016/S0092-8674(00)81636-0

Ni M, Tepperman JM, Quail PH. Binding of phytochrome B to its nuclear signalling partner PIF3 is reversibly induced by light. Nature. 1999:**400**(6746):781–784. https://doi.org/10.1038/23500

Ni W, Xu S-L, González-Grandío E, Chalkley RJ, Huhmer AFR, Burlingame AL, Wang Z-Y, Quail PH. PPKs mediate direct signal transfer from phytochrome photoreceptors to transcription factor PIF3. Nat Commun. 2017:**8**(1):15236. https://doi.org/10.1038/ncomms15236

Ni W, Xu S-L, Tepperman JM, Stanley DJ, Maltby DA, Gross JD, Burlingame AL, Wang Z-Y, Quail PH. A mutually assured destruction mechanism attenuates light signaling in Arabidopsis. Science. 2014:**344**(6188):1160–1164. https://doi.org/10.1126/science.1250778

Nozue K, Covington MF, Duek PD, Lorrain AA, Fankhauser C, Harmer SL, Maloof JN. Rhythmic growth explained by coincidence between internal and external cues. Nature. 2007:**448**(7151):358–361. https://doi.org/10.1038/nature05946

Oh E, Kim J, Park E, Kim JI, Kang C, Choi G. PIL5, a phytochrome-interacting basic helix-loop-helix protein, is a key negative regulator of seed germination in Arabidopsis thaliana. Plant Cell. 2004:**16**(11):3045–3058. https://doi.org/10.1105/tpc.104.025163

Oh J, Park E, Song K, Bae G, Choi G. PHYTOCHROME INTERACTING FACTOR8 inhibits phytochrome A-mediated far-red light responses in Arabidopsis. Plant Cell. 2020:**32**(1):186–205. https://doi.org/10.1105/tpc.19.00515

Ohgishi M, Saji K, Okada K, Sakai T. Functional analysis of each blue light receptor, cry1, cry2, phot1, and phot2, by using combinatorial multiple mutants in Arabidopsis. Proc Natl Acad Sci U S A. 2004:**101**(8):2223–2228. https://doi.org/10.1073/pnas.0305984101

Osterlund MT, Hardtke CS, Wei N, Deng XW. Targeted destabilization of HY5 during light-regulated development of Arabidopsis. Nature. 2000:**405**(6785):462–466. https://doi.org/10.1038/35013076

Ozkan-Dagliyan I, Chiou Y-Y, Ye R, Hassan BH, Ozturk N, Sancar A. Formation of Arabidopsis cryptochrome 2 photobodies in mammalian nuclei: application as an optogenetic DNA damage checkpoint switch. J Biol Chem. 2013:**288**(32):23244–23251. https://doi.org/10.1074/jbc.M113.493361

Paik I, Chen F, Ngoc Pham V, Zhu L, Kim J-I, Huq E. A phyB-PIF1-SPA1 kinase regulatory complex promotes photomorphogenesis in Arabidopsis. Nat Commun. 2019:**10**(1):4216. https://doi.org/10.1038/s41467-019-12110-y

Paik I, Kathare PK, Kim J-I, Huq E. Expanding roles of PIFs in signal integration from multiple processes. Mol Plant. 2017:**10**(8):1035–1046. https://doi.org/10.1016/j.molp.2017.07.002

Paik I, Yang S, Choi G. Phytochrome regulates translation of mRNA in the cytosol. Proc Natl Acad Sci U S A. 2012:**109**(4):1335–1340. https://doi.org/10.1073/pnas.1109683109

Palayam M, Ganapathy J, Guercio AM, Tal L, Deck SL, Shabek N. Structural insights into photoactivation of plant cryptochrome-2. Commun Biol. 2021:**4**(1):28. https://doi.org/10.1038/s42003-020-01531-x

Pardi SA, Nusinow DA. Out of the dark and into the light: a new view of phytochrome photobodies. Front Plant Sci. 2021:**12**:732947. https://doi.org/10.3389/fpls.2021.732947

Park E, Kim Y, Choi G. Phytochrome B requires PIF degradation and sequestration to induce light responses across a wide range of light conditions. Plant Cell. 2018:**30**(6):1277–1292. https://doi.org/10.1105/tpc.17.00913

Park E, Kim J, Lee Y, Shin J, Oh E, Chung WI, Liu JR, Choi G. Degradation of phytochrome interacting factor 3 in phytochrome-mediated light signaling. Plant Cell Physiol. 2004:**45**(8):968–975. https://doi.org/10.1093/pcp/pch125

Park E, Park J, Kim J, Nagatani A, Lagarias JC, Choi G. Phytochrome B inhibits binding of phytochrome-interacting factors to their target promoters. Plant J. 2012:**72**(4):537–546. https://doi.org/10.1111/j.1365-313X.2012.05114.x

Parks BM, Quail PH. Hy8, a new class of arabidopsis long hypocotyl mutants deficient in functional phytochrome A. Plant Cell. 1993:**5**(1):39–48. https://doi.org/10.1105/tpc.5.1.39

Partch CL, Clarkson MW, Özgür S, Lee AL, Sancar A. Role of structural plasticity in signal transduction by the cryptochrome blue-light photoreceptor. Biochemistry. 2005:**44**(10):3795–3805. https://doi.org/10.1021/bi047545g

Pedmale UV, Huang SC, Zander M, Cole Benjamin J, Hetzel J, Ljung K, Reis Pedro AB, Sridevi P, Nito K, Nery JR, et al. Cryptochromes interact directly with PIFs to control plant growth in limiting blue light. Cell. 2016:**164**(1–2):233–245. https://doi.org/10.1016/j.cell.2015.12.018

Pfeiffer A, Shi H, Tepperman JM, Zhang Y, Quail PH. Combinatorial complexity in a transcriptionally-centered signaling hub in Arabidopsis. Mol Plant. 2014:**7**(11):1598–1618. https://doi.org/10.1093/mp/ssu087

Pham VN, Kathare PK, Huq E. Dynamic regulation of PIF5 by COP1-SPA complex to optimize photomorphogenesis in Arabidopsis. Plant J. 2018:**96**(2):260–273. https://doi.org/10.1111/tpj.14074

Podolec R, Demarsy E, Ulm R. Perception and signaling of ultraviolet-B radiation in plants. Annu Rev Plant Biol. 2021:**72**(1):793–822. https://doi.org/10.1146/annurev-arplant-050718-095946

Podolec R, Ulm R. Photoreceptor-mediated regulation of the COP1/SPA E3 ubiquitin ligase. Curr Opin Plant Biol. 2018:**45**:18–25. https://doi.org/10.1016/j.pbi.2018.04.018

Poff KL, Butler WL. Absorbance changes induced by blue light in Phycomyces blakesleeanus and Dictyostelium discoideum. Nature. 1974:**248**(5451):799–801. https://doi.org/10.1038/248799a0

Pokorny R, Klar T, Hennecke U, Carell T, Batschauer A, Essen L-O. Recognition and repair of UV-lesions in loop structures of duplex DNA by DASH-type cryptochrome. Proc Natl Acad Sci U S A. 2008: **105**(52):21023–21027. https://doi.org/10.1073/pnas.0805830106

Ponnu J, Hoecker U. Signaling mechanisms by Arabidopsis cryptochromes. Front Plant Sci. 2022:**13**:844714. https://doi.org/10.3389/fpls.2022.844714

Ponnu J, Riedel T, Penner E, Schrader A, Hoecker U. Cryptochrome 2 competes with COP1 substrates to repress COP1 ubiquitin ligase activity during Arabidopsis photomorphogenesis. Proc Natl Acad Sci U S A. 2019:**116**(52):27133–27141. https://doi.org/10.1073/pnas.1909181116

Pradhan B, Kanno T, Umeda Igarashi M, Loke MS, Baaske MD, Wong JSK, Jeppsson K, Björkegren C, Kim E. The Smc5/6 complex is a DNA loop-extruding motor. Nature. 2023:**616**(7958):843–848. https://doi.org/10.1038/s41586-023-05963-3

Qu G, Jiang B, Lin C. A dual action mechanism of Arabidopsis cryptochromes. J Integr Plant Biol. 2023. Online ahead of print. https://doi.org/10.1111/jipb.13578.

Quail PH. The phytochromes: a biochemical mechanism of signaling in sight? BioEssays. 1997:**19**(7):571–579. https://doi.org/10.1002/bies.950190708

Quail PH. Photobodies reveal their secret. Nat Plants. 2021:**7**(10): 1326–1327. https://doi.org/10.1038/s41477-021-01010-z

Reed JW, Nagpal P, Poole DS, Furuya M, Chory J. Mutations in the gene for the red/far-red light receptor phytochrome B alter cell elongation and physiological responses throughout Arabidopsis development. Plant Cell. 1993:**5**(2):147–157. https://doi.org/10.1105/tpc.5.2.147

Rizzini L, Favory JJ, Cloix C, Faggionato D, O'Hara A, Kaiserli E, Baumeister R, Schafer E, Nagy F, Jenkins GI, et al. Perception of UV-B by the Arabidopsis UVR8 protein. Science. 2011:**332**(6025): 103–106. https://doi.org/10.1126/science.1200660

Ronald J, Davis S. Focusing on the nuclear and subnuclear dynamics of light and circadian signalling. Plant Cell Environ. 2019:**42**(10): 2871–2884. https://doi.org/10.1111/pce.13634

Rosenfeldt G, Viana RM, Mootz HD, von Arnim AG, Batschauer A. Chemically induced and light-independent cryptochrome photoreceptor activation. Mol Plant. 2008:**1**(1):4–12. https://doi.org/10.1093/mp/ssm002

Sachs JV. Lectures on the physiology of plants. Oxford: Clarendon Press; 1887.

Sage LC. Pigment of the imagination: a history of phytochrome research. Cambrige, MA: Academic Press, Inc.; 1992. p. 562.

Saijo Y, Sullivan JA, Wang H, Yang J, Shen Y, Rubio V, Ma L, Hoecker U, Deng XW. The COP1-SPA1 interaction defines a critical step in phytochrome A-mediated regulation of HY5 activity. Genes Dev. 2003:**17**(21):2642–2647. https://doi.org/10.1101/gad.1122903

Sakamoto K, Nagatani A. Nuclear localization activity of phytochrome B. Plant J. 1996:**10**(5):859–868. https://doi.org/10.1046/j.1365-313X.1996.10050859.x

Sancar A. Mechanisms of DNA repair by photolyase and excision nuclease (Nobel lecture). Angew Chem Int Ed Engl. 2016:**55**(30): 8502–8527. https://doi.org/10.1002/anie.201601524

Sang Y, Li Q-H, Rubio V, Zhang Y-C, Mao J, Deng X-W, Yang H-Q. N-terminal domain–mediated homodimerization is required for photoreceptor activity of Arabidopsis CRYPTOCHROME 1. Plant Cell. 2005:**17**(5):1569–1584. https://doi.org/10.1105/tpc.104.029645

Selby CP, Sancar A. A cryptochrome/photolyase class of enzymes with single-stranded DNA-specific photolyase activity. Proc Natl Acad Sci U S A. 2006:**103**(47):17696–17700. https://doi.org/10.1073/pnas.0607993103

Senger H. Cryptochrome, some terminological thoughts. In: **Senger H**, editor. Blue light effects in biological systems. Berlin, Heidelberg: Springer Berlin Heidelberg; 1984. p. 72–72.

Seo HS, Yang JY, Ishikawa M, Bolle C, Ballesteros ML, Chua NH. LAF1 ubiquitination by COP1 controls photomorphogenesis and is stimulated by SPA1. Nature. 2003:**423**(6943):995–999. https://doi.org/10.1038/nature01696

Shalitin D, Yang H, Mockler TC, Maymon M, Guo H, WhitelamGC, Lin C. Regulation of Arabidopsis cryptochrome 2 by blue-light-dependent phosphorylation. Nature. 2002:**417**(6890): 763–767. https://doi.org/10.1038/nature00815

Shao K, Zhang X, Li X, Hao Y, Huang X, Ma M, Zhang M, Yu F, Liu H, Zhang P. The oligomeric structures of plant cryptochromes. Nat Struct Mol Biol. 2020:**27**(5):480–488. https://doi.org/10.1038/s41594-020-0420-x

Sharrock RA, Quail PH. Novel phytochrome sequences in Arabidopsis thaliana: structure, evolution, and differential expression of a plant regulatory photoreceptor family. Genes Dev. 1989:**3**(11):1745–1757. https://doi.org/10.1101/gad.3.11.1745

Sheerin DJ, Menon C, zur Oven-Krockhaus S, Enderle B, Zhu L, Johnen P, Schleifenbaum F, Stierhof Y-D, Huq E, Hiltbrunner A. Light-activated phytochrome A and B interact with members of the SPA family to promote photomorphogenesis in Arabidopsis by reorganizing the COP1/SPA complex. Plant Cell. 2015:**27**(1): 189–201. https://doi.org/10.1105/tpc.114.134775

Shen Y, Khanna R, Carle CM, Quail PH. Phytochrome induces rapid PIF5 phosphorylation and degradation in response to red-light activation. Plant Physiol. 2007:**145**(3):1043–1051. https://doi.org/10.1104/pp.107.105601

Shen H, Ling Z, Castillon A, Majee M, Downie B, Huq E. Light-induced phosphorylation and degradation of the negative regulator PHYTOCHROME INTERACTING FACTOR 1 depends upon its direct physical interactions with photoactivated phytochromes. Plant Cell. 2008:**20**(6):1586–1602. https://doi.org/10.1105/tpc.108.060020

Shen H, Moon J, Huq E. PIF1 is regulated by light-mediated degradation through the ubiquitin-26S proteasome pathway to optimize seedling photomorphogenesis in Arabidopsis. Plant J. 2005:**44**(6): 1023–1035. https://doi.org/10.1111/j.1365-313X.2005.02606.x

Shih C-J, Chen H-W, Hsieh H-Y, Lai Y-H, Chiu F-Y, Chen Y-R, Tu S-L. Heterogeneous nuclear ribonucleoprotein H1 coordinates with phytochrome and the U1 snRNP complex to regulate alternative splicing in Physcomitrella patens. Plant Cell. 2019:**31**(10):2510–2524. https://doi.org/10.1105/tpc.19.00314

Shikata H, Hanada K, Ushijima T, Nakashima M, Suzuki Y, Matsushita T. Phytochrome controls alternative splicing to mediate light responses in Arabidopsis. Proc Natl Acad Sci U S A. 2014: **111**(52):18781–18786. https://doi.org/10.1073/pnas.1407147112

Shin A-Y, Han Y-J, Baek A, Ahn T, Kim SY, Nguyen TS, Son M, Lee KW, Shen Y, Song P-S, et al. Evidence that phytochrome functions as a protein kinase in plant light signalling. Nat Commun. 2016:**7**(1): 11545. https://doi.org/10.1038/ncomms11545

Shin J, Kim K, Kang H, Zulfugarov IS, Bae G, Lee CH, Lee D, Choi G. Phytochromes promote seedling light responses by inhibiting four negatively-acting phytochrome-interacting factors. Proc Nat Acad Sci U S A. 2009:**106**(18):7660–7665. https://doi.org/10.1073/pnas.0812219106

Siegelman HW, Firer EM. Purification of phytochrome from oat seedlings. Biochemistry. 1964:**3**(3):418–423. https://doi.org/10.1021/bi00891a019

Somers DE, Devlin PF, Kay SA. Phytochromes and cryptochromes in the entrainment of the Arabidopsis circadian clock. Science. 1998: **282**(5393):1488–1490. https://doi.org/10.1126/science.282.5393.1488

Somerville C, Koornneef M. A fortunate choice: the history of Arabidopsis as a model plant. Nat Rev Genet. 2002:**3**(11):883–889. https://doi.org/10.1038/nrg927

Sponga F, Deitzer GF, Mancinelli AL. Cryptochrome, phytochrome, and the photoregulation of anthocyanin production under blue light. Plant Physiol. 1986:**82**(4):952–955. https://doi.org/10.1104/pp.82.4.952

Stanewsky R, Kaneko M, Emery P, Beretta B, Wager-Smith K, Kay SA, Rosbash M, Hall JC. The cryb mutation identifies cryptochrome as a circadian photoreceptor in Drosophila. Cell. 1998:**95**(5):681–692. https://doi.org/10.1016/S0092-8674(00)81638-4

Stephenson PG, Fankhauser C, Terry MJ. PIF3 is a repressor of chloroplast development. Proc Natl Acad Sci U S A. 2009:**106**(18): 7654–7659. https://doi.org/10.1073/pnas.0811684106

Strassera B, Sánchez-Lamasa M, Yanovsky MJ, Casal JJ, Cerdán PD. Arabidopsis thaliana life without phytochromes. Proc Natl Acad Sci U S A. 2010:**107**(10):4776–4781. https://doi.org/10.1073/pnas.0910446107

Strayer C, Oyama T, Schultz TF, Raman R, Somers DE, Mas P, Panda S, Kreps JA, Kay SA. Cloning of the Arabidopsis clock gene TOC1, an autoregulatory response regulator homolog. Science. 2000: **289**(5480):768–771. https://doi.org/10.1126/science.289.5480.768

Su J, Liu B, Liao J, Yang Z, Lin C, Oka Y. Coordination of cryptochrome and phytochrome signals in the regulation of plant light responses. Agronomy. 2017:**7**(1):25. https://doi.org/10.3390/agronomy7010025

Subramanian C, Kim BH, Lyssenko NN, Xu X, Johnson CH, von Arnim AG. The Arabidopsis repressor of light signaling, COP1, is regulated by nuclear exclusion: mutational analysis by bioluminescence resonance energy transfer. Proc Natl Acad Sci U S A. 2004:**101**(17):6798–6802. https://doi.org/10.1073/pnas.0307964101

Tepperman JM, Hudson ME, Khanna R, Zhu T, Chang SH, Wang X, Quail PH. Expression profiling of phyB mutant demonstrates substantial contribution of other phytochromes to red-light-regulated gene expression during seedling de-etiolation. Plant J. 2004:**38**(5): 725–739. https://doi.org/10.1111/j.1365-313X.2004.02084.x

Tepperman JM, Hwang YS, Quail PH. Phya dominates in transduction of red-light signals to rapidly-responding genes at the initiation of *Arabidopsis* seedling deetiolation. Plant J. 2006:**48**(5):728–742. https://doi.org/10.1111/j.1365-1313X.2006.02914.x

Tepperman JM, Zhu T, Chang H-S, Wang X, Quail PH. Multiple transcription-factor genes are early targets of phytochrome A signaling. Proc Natl Acad Sci U S A. 2001:**98**(16):9437–9442. https://doi.org/10.1073/pnas.161300998

Toledo-Ortiz G, Huq E, Quail PH. The Arabidopsis basic/helix-loop-helix transcription factor family. Plant Cell. 2003:**15**(8):1749–1770. https://doi.org/10.1105/tpc.013839

Ulijasz AT, Cornilescu G, Cornilescu CC, Zhang JR, Rivera M, Markley JL, Vierstra RD. Structural basis for the photoconversion of a phytochrome to the activated Pfr form. Nature. 2010: **463**(7278):250–U143. https://doi.org/10.1038/nature08671

Usami T, Mochizuki N, Kondo M, Nishimura M, Nagatani A. Cryptochromes and phytochromes synergistically regulate Arabidopsis root greening under blue light. Plant Cell Physiol. 2004:**45**(12): 1798–1808. https://doi.org/10.1093/pcp/pch205

van der Horst GT, Muijtjens M, Kobayashi K, Takano R, Kanno S, Takao M, de Wit J, Verkerk A, Eker AP, van Leenen D, et al. Mammalian Cry1 and Cry2 are essential for maintenance of circadian rhythms. Nature. 1999:**398**(6728):627–630. https://doi.org/10.1038/19323

Vierstra R, Quail P. Native phytochrome: inhibition of proteolysis yields a homogeneous monomer of 124 kilodaltons from Avena. Proc Natl Acad Sci U S A. 1982:**79**(17):5272–5276. https://doi.org/10.1073/pnas.79.17.5272

von Halban H, Eisenbrand J, Baly ECC. On the measurement of light absorption. Proc R Soc Lond Ser A Contain Pap Math Phys Character. 1927:**116**(773):153–162. https://doi.org/10.1098/rspa.1927.0128

Wagner JR, Brunzelle JS, Forest KT, Vierstra RD. A light-sensing knot revealed by the structure of the chromophore binding domain of phytochrome. Nature. 2005:**438**(7066):325–331. https://doi.org/10.1038/nature04118

Wang X, Jiang B, Gu L, Chen Y, Mora M, Zhu M, Noory E, Wang Q, Lin C. A photoregulatory mechanism of the circadian clock in Arabidopsis. Nat Plants. 2021:**7**(10):1397–1408. https://doi.org/10.1038/s41477-021-01002-z

Wang Q, Lin C. Mechanisms of cryptochrome-mediated photoresponses in plants. Annu Rev Plant Biol. 2020:**71**(1):103–129. https://doi.org/10.1146/annurev-arplant-050718-100300

Wang W, Lu X, Li L, Lian H, Mao Z, Xu P, Guo T, Xu F, Du S, Cao X, et al. Photoexcited CRYPTOCHROME1 interacts with dephosphorylated BES1 to regulate brassinosteroid signaling and photomorphogenesis in Arabidopsis. Plant Cell. 2018b:**30**(9):1989–2005. https://doi.org/10.1105/tpc.17.00994

Wang H, Ma LG, Li JM, Zhao HY, Deng XW. Direct interaction of Arabidopsis cryptochromes with COP1 in light control development. Science. 2001:**294**(5540):154–158. https://doi.org/10.1126/science.1063630

Wang Q, Zuo Z, Wang X, Gu L, Yoshizumi T, Yang Z, Yang L, Liu Q, Liu W, Han Y-J, et al. Photoactivation and inactivation of Arabidopsis cryptochrome 2. Science. 2016:**354**(6310):343–347. https://doi.org/10.1126/science.aaf9030

Wang Q, Zuo Z, Wang X, Liu Q, Gu L, Oka Y, Lin C. Beyond the photocycle-how cryptochromes regulate photoresponses in plants? Curr Opin Plant Biol. 2018a:**45**:120–126. https://doi.org/10.1016/j.pbi.2018.05.014

Wei N, Deng XW. The role of the COP/DET/FUS genes in light control of Arabidopsis seedling development. Plant Physiol. 1996:**112**(3): 871–878. https://doi.org/10.1104/pp.112.3.871

Williams TA, Foster PG, Cox CJ, Embley TM. An archaeal origin of eukaryotes supports only two primary domains of life. Nature. 2013:**504**(7479):231–236. https://doi.org/10.1038/nature12779

Willige BC, Zander M, Yoo CY, Phan A, Garza RM, Trigg SA, He Y, Nery JR, Chen H, Chen M, et al. PHYTOCHROME-INTERACTING FACTORs trigger environmentally responsive chromatin dynamics in plants. Nat Genet. 2021:**53**(7):955–961. https://doi.org/10.1038/s41588-021-00882-3

Wu G, Spalding EP. Separate functions for nuclear and cytoplasmic cryptochrome 1 during photomorphogenesis of Arabidopsis seedlings. Proc Natl Acad Sci U S A. 2007:**104**(47):18813–18818. https://doi.org/10.1073/pnas.0705082104

Xin R, Kathare PK, Huq E. Coordinated regulation of Pre-mRNA splicing by the SFPS-RRC1 complex to promote photomorphogenesis. Plant Cell. 2019:**31**(9):2052–2069. https://doi.org/10.1105/tpc.18.00786

Xin X, Chen W, Wang B, Zhu F, Li Y, Yang H, Li J, Ren D. Arabidopsis MKK10-MPK6 mediates red-light-regulated opening of seedling cotyledons through phosphorylation of PIF3. J Exp Bot. 2018a:**69**(3):423–439. https://doi.org/10.1093/jxb/erx418

Xin R, Zhu L, Salomé PA, Mancini E, Marshall CM, Harmon FG, Yanovsky MJ, Weigel D, Huq E. SPF45-related splicing factor for phytochrome signaling promotes photomorphogenesis by regulating pre-mRNA splicing in Arabidopsis. Proc Natl Acad Sci U S A. 2018b: **114**(33):E7018–E7027. https://doi.org/10.1073/pnas.1706379114

Xu P, Chen H, Li T, Xu F, Mao Z, Cao X, Miao L, Du S, Hua J, Zhao J, et al. Blue light-dependent interactions of CRY1 with GID1 and DELLA proteins regulate gibberellin signaling and photomorphogenesis in Arabidopsis. Plant Cell. 2021:**33**(7):2375–2394. https://doi.org/10.1093/plcell/koab124

Xu F, He S, Zhang J, Mao Z, Wang W, Li T, Hua J, Du S, Xu P, Li L, et al. Photoactivated CRY1 and phyB interact directly with AUX/IAA proteins to inhibit auxin signaling in Arabidopsis. Mol Plant. 2018:**11**(4): 523–541. https://doi.org/10.1016/j.molp.2017.12.003

Xu X, Paik I, Zhu L, Huq E. Illuminating progress in phytochrome-mediated light signaling pathways. Trends Plant Sci. 2015:**20**(10): 641–650. https://doi.org/10.1016/j.tplants.2015.06.010

Yamaguchi R, Nakamura M, Mochizuki N, Kay SA, Nagatani A. Light-dependent translocation of a phytochrome B-GFP fusion protein to the nucleus in transgenic Arabidopsis. J Cell Biol. 1999:**145**(3): 437–445. https://doi.org/10.1083/jcb.145.3.437

Yan T, Heng Y, Wang W, Li J, Deng XW. SWELLMAP 2, a phyB-interacting splicing factor, negatively regulates seedling photomorphogenesis in Arabidopsis. Front Plant Sci. 2022:**13**:836519. https://doi.org/10.3389/fpls.2022.836519

Yan B, Yang Z, He G, Jing Y, Dong H, Ju L, Zhang Y, Zhu Y, Zhou Y, Sun J. The blue light receptor CRY1 interacts with GID1 and DELLA proteins to repress gibberellin signaling and plant growth. Plant Commun. 2021:**2**(6):100245. https://doi.org/10.1016/j.xplc.2021.100245

Yang H-Q, Tang R-H, Cashmore AR. The signaling mechanism of Arabidopsis CRY1 involves direct interaction with COP1. Plant Cell. 2001:**13**(12):2573–2587. https://doi.org/10.1105/tpc.010367

Yang H-Q, Wu Y-J, Tang R-H, Liu D, Liu Y, Cashmore AR. The C termini of Arabidopsis cryptochromes mediate a constitutive light response. Cell. 2000:**103**(5):815–827. https://doi.org/10.1016/S0092-8674(00)00184-7

Yeh KC, Lagarias JC. Eukaryotic phytochromes: light-regulated serine/threonine protein kinases with histidine kinase ancestry. Proc Natl Acad Sci U S A. 1998:**95**(23):13976–13981. https://doi.org/10.1073/pnas.95.23.13976

Yoo CY, He J, Sang Q, Qiu Y, Long L, Kim RJ-A, Chong EG, Hahm J, Morffy N, Zhou P, et al. Direct photoresponsive inhibition of a p53-like transcription activation domain in PIF3 by Arabidopsis phytochrome B. Nat Commun. 2021:**12**(1):5614. https://doi.org/10.1038/s41467-021-25909-5

Yu X, Klejnot J, Zhao X, Shalitin D, Maymon M, Yang H, Lee J, Liu X, Lopez J, Lin C. Arabidopsis cryptochrome 2 completes its posttranslational life cycle in the nucleus. Plant Cell. 2007a:**19**(10): 3146–3156. https://doi.org/10.1105/tpc.107.053017

Yu X, Liu H, Klejnot J, Lin C. The cryptochrome blue-light receptors. In: The Arabidopsis book. Vol 8. Rockville (MD): American Society of Plant Biologists; 2010. p. e0135

Yu X, Sayegh R, Maymon M, Warpeha K, Klejnot J, Yang H, Huang J, Lee J, Kaufman L, Lin C. Formation of nuclear bodies of Arabidopsis CRY2 in response to blue light is associated with its blue light-dependent degradation. Plant Cell. 2009:**21**(1):118–130. https://doi.org/10.1105/tpc.108.061663

Yu X, Shalitin D, Liu X, Maymon M, Klejnot J, Yang H, Lopez J, Zhao X, Bendehakkalu KT, Lin C. Derepression of the NC80 motif is critical for the photoactivation of Arabidopsis CRY2. Proc Natl Acad Sci U S A. 2007b:**104**(17):7289–7294. https://doi.org/10.1073/pnas.0701912104

Zhang B, Holmlund M, Lorrain S, Norberg M, Bakó L, Fankhauser C, Nilsson O. BLADE-ON-PETIOLE proteins act in an E3 ubiquitin ligase complex to regulate PHYTOCHROME INTERACTING FACTOR 4 abundance. Elife. 2017:**6**:e26759. https://doi.org/10.7554/eLife.26759

Zhang Y, Lin X, Ma C, Zhao J, Shang X, Wang Z, Xu B, Gao N, Deng XW, Wang J. Structural insights into plant phytochrome A as a highly sensitized photoreceptor. Cell Res. 2023:**33**(10):806–809. https://doi.org/10.1038/s41422-023-00858-4

Zhang Y, Mayba O, Pfeiffer A, Shi H, Tepperman JM, Speed TP, Quail PH. A quartet of PIF bHLH factors provides a transcriptionally centered signaling hub that regulates seedling morphogenesis through differential expression-patterning of shared target genes in Arabidopsis. PLoS Genet. 2013:**9**(1):e1003244. https://doi.org/10.1371/journal.pgen.1003244

Zhong M, Zeng B, Tang D, Yang J, Qu L, Yan J, Wang X, Li X, Liu X, Zhao X. The blue light receptor CRY1 interacts with GID1 and DELLA proteins to repress GA signaling during photomorphogenesis in Arabidopsis. Mol Plant. 2021:**14**(8):1328–1342. https://doi.org/10.1016/j.molp.2021.05.011

Zhu L, Bu Q, Xu X, Paik I, Huang X, Hoecker U, Deng XW, Huq E. CUL4 forms an E3 ligase with COP1 and SPA to promote light-induced degradation of PIF1. Nat Commun. 2015:**6**(1):7245. https://doi.org/10.1038/ncomms8245

Zuo Z, Liu H, Liu B, Liu X, Lin C. Blue light-dependent interaction of CRY2 with SPA1 regulates COP1 activity and floral initiation in Arabidopsis. Curr Biol. 2011:**21**(10):841–847. https://doi.org/10.1016/j.cub.2011.03.048

Zuo Z-C, Meng Y-Y, Yu X-H, Zhang Z-L, Feng D-S, Sun S-F, Liu B, Lin C-T. A study of the blue-light-dependent phosphorylation, degradation, and photobody formation of Arabidopsis CRY2. Mol Plant. 2012:**5**(3):726–733. https://doi.org/10.1093/mp/sss007

Part 3

Centennial Reviews from

THE

PLANT

CELL

https://doi.org/10.1093/plcell/koae045

THE PLANT CELL

THE PLANT CELL 2024: 36: 1186–1204

Review

Genome organization and botanical diversity

Andrew H. Paterson [1,*] **Christine Queitsch** [2]

1 Plant Genome Mapping Laboratory, University of Georgia, Athens, GA, USA
2 Department of Genome Sciences, University of Washington, Seattle, WA, USA

*Author for correspondence: paterson@uga.edu

Abstract

The rich diversity of angiosperms, both the planet's dominant flora and the cornerstone of agriculture, is integrally intertwined with a distinctive evolutionary history. Here, we explore the interplay between angiosperm genome organization and botanical diversity, empowered by genomic approaches ranging from genetic linkage mapping to analysis of gene regulation. Commonality in the genetic hardware of plants has enabled robust comparative genomics that has provided a broad picture of angiosperm evolution and implicated both general processes and specific elements in contributing to botanical diversity. We argue that the hardware of plant genomes—both in content and in dynamics—has been shaped by selection for rather substantial differences in gene regulation between plants and animals such as maize and human, organisms of comparable genome size and gene number. Their distinctive genome content and dynamics may reflect in part the indeterminate development of plants that puts strikingly different demands on gene regulation than in animals. Repeated polyploidization of plant genomes and multiplication of individual genes together with extensive rearrangement and differential retention provide rich raw material for selection of morphological and/or physiological variations conferring fitness in specific niches, whether natural or artificial. These findings exemplify the burgeoning information available to employ in increasing knowledge of plant biology and in modifying selected plants to better meet human needs.

Introduction

The principle that plants share many commonalities in their underlying genetic "hardware" traces to at least Vavilov, basing his law of homologous series in variation on observations that comparable variant forms tended to appear in different varieties of the same species, different species of the same genus, and different genera of the same family (Vavilov 1922).

How is underlying genetic commonality reconciled with the rich diversity of extant angiosperms? First, commonality is not identity. Most genes have recognizable homologs in most plant species, but with coding and/or regulatory sequence differences ranging from minute (single nucleotide polymorphisms [SNPs]) to massive. Likewise, gene order along the chromosomes ranges from near-identity within a population [although even subtle differences may be of evolutionary importance by reducing gene flow (Rieseberg 2001)] to barely discernible among distant taxa. The progression of such divergence is by no means linear—for example, the genome of sorghum more closely parallels that of rice, separated by ~100 million years, than that of maize, separated by approximately 12 million years (Swigonova *et al.* 2004). Often, striking differences in genome structure of closely related taxa reflect punctuational consequences of a genome duplication, for example in the maize lineage (see below).

Second, similar hardware may use different software. The regulatory cues that determine gene expression are often short and easily mutable. Indeed, regulatory mutations may create interdependence that is a basis for the retention of duplicated genes, for example by reciprocal loss of expression cues that make different members of a duplicated gene pair each essential in different tissues (Force *et al.* 1999, 2005). We elaborate on this important dimension of (plant) diversity below.

Third, polymorphism determines the subset of genes that contribute to standing genetic variation within a taxon. Both random and nonrandom factors determine the persistence of gene polymorphism with functional consequences, the

Received October 16, 2023. Accepted February 07, 2024. Advance access publication February 21, 2024.

former being heavily weighted in gene pools that have experienced evolutionarily recent genetic bottlenecks such as those of most major crops. Only the subset of genes in which functional polymorphism differentiates the progenitors of a study population are accessible by forward genetics, while reverse genetics scans most genes.

Thus, although botanical diversity is a focus of this paper, essential to its study has been commonality in the genetic hardware and software of plants. In 2 hypothetical genomes that differed by only a single nucleotide, the identical surroundings rather than the difference itself would inform its likely phenotypic consequences. Robust comparative genomics, distinctive in plants in accommodating recursive genome duplication (see below), enables information from facile models to be extrapolated across the plant phylogeny as at least an indicator of gene function. This principle has provided for tremendous leveraging of both genome sequence and reverse genetics resources, with deep annotation and hard-won functional data for botanical models accelerating hypothesis formation in diverse taxa. Forward genetics also benefits from comparative genomics, although with comparisons among taxa confounded by factors that determine the persistence of polymorphism with functional consequences.

In the final scientific paper of an illustrious career, G. Ledyard Stebbins articulated *"a plant evolutionist's point of view . . . based on profound differences . . . between the 2 kingdoms."* In particular, he highlighted the lack of a germline, propensity for inbreeding, and greater tolerance of diversity, particularly polyploidy, attributed to the fewer cellular structures of plants than animals (Stebbins 1999). Little more than a year after his paper was published, Stebbins had died when the initial sequence of the first higher plant genome (Arabidopsis Genome Initiative 2000), followed 2 months later by the human genome (Lander *et al.* 2001), kicked off a new era in comparative biology. This era has been enabled by technological improvements that made genome sequencing rapid and routine and empowered by CPUs sufficient to archive and sift voluminous digital data for grains of insight.

More than 2 decades hence, voluminous literature has accumulated on the hardware and software of genomes, how these differ between and among plants and animals, and how they relate to diversity among and polymorphism within plant taxa. Here, we present a plant-centric view, reviewing the broad state of the field based on findings of and connections among a selected subset of studies that focus on plant-specific features. The widespread application of genomic technologies, particularly in agriculture, make truly exhaustive coverage intractable. Overrepresentation herein of our own studies and others in our focal taxa reflects those we are best able to interpret and integrate, with no disrespect intended to the work of many others that would be similarly suitable. We have further stretched our writ to volunteer opinions about areas in which we see promise for future investigations, an act of commission intended to stimulate second-level thinking by our colleagues and successors.

The hardware: using genome structure and composition to probe diversity

Robust DNA sequences provide digital access to the genome of each new organism, filling gaps in knowledge of evolutionary history and the spectrum of botanical diversity. As more taxa are sequenced, diminishing returns are naturally realized in discovery of new genes and macroevolutionary events such as genome duplications. However, closely related taxa with less functional diversity may be favorable systems in which to identify causal genomic differences, just as subtle differences in beak shape related to diet of closely related finches contributed to the theory of natural selection (Darwin 1859).

An end in itself in that it comprises the entire hereditary information of an organism, a genome is also a means to other ends in providing improved tools, as summarized in Table 1. High-quality genome sequences have expedited forward genetics approaches to home in on the region of a causal gene and reverse genetics approaches to identify causative variants. Dramatic improvement in both cost and throughput of sequencing has empowered "evolutionary genetics" to achieve critical mass as a complement to reverse genetics toward determining causative variants and added the potential precision of genome-wide association study (GWAS) to the repertoire of forward genetics. Robust comparative genomics (section IV, below) enables hard-won functional information about specific genes from forward, reverse, evolutionary, and/or association approaches to be extrapolated to newly sequenced genomes as at least an early indicator of gene function.

Forward genetics remains central in the botany toolbox, especially in crop genomes for which target traits tend to be complex (quantitative trait loci). Robust genome sequences obviate the need for laborious "dart-throwing" approaches by which were produced early genetic maps that guided sequence assembly. A variety of reduced representation "genotyping by sequencing" approaches (Davey *et al.* 2011) now permit one to routinely undertake whole-genome scans for trait mapping, and reference genomes expedite digital identification of precisely targeted diagnostic tools such as Kompetitive Allele Specific PCR (Semagn *et al.* 2013) or simple-sequence repeat markers. Genetic maps based on voluminous SNPs permit anchoring of even small sequence contigs as well as orientation of those spanning a recombination event—for example, more than 2 million SNPs in 1,178 positions separated by recombinants anchored 57,270 scaffolds each containing 5 or more mapped SNPs and collectively comprising 14% of the subject genome (Bowers *et al.* 2016). Such rich data also empower GWAS (Ozaki *et al.* 2002) that use physically dense marker sets to take ad-vantage of historical accumulations of recombinations in re-lating a phenotype to a gene(s) in a crop gene pool or natural population.

Reverse genetics has tremendou sly increased plant gene fu nctional knowledge, in particu lar with many thousands

Table 1. Genome-enabled genetics approaches in plants. Most plant genetics studies utilize 1 or more of 4 broad approaches, each suitable in different contexts, with different strengths and limitations, and requiring different resources

Genetics approach	% (subset) of gene set accessed	Rate (source) of false positives?	Rate (source) of false negatives?	Resolution (level) of DNA characterization	Resources required
Forward	Low (polymorphic in single cross)	Low	Medium (small population size)	Low (10 cM marker spacing	Mapping populations (BC, F2, DH, RIL, NIL)
Reverse	High (mutable resulting in phenotype)	Low	Medium (subtle phenotypes)	High (gene space sequence)	Mutant collection
Evolutionary	Medium (polymorphic in species)	High (relatedness of genotypes)	Medium (rare alleles)	High (gene space sequence)	Diversity panel (core or mini-core collection)
GWAS	Medium (polymorphic in species subset)	High (relatedness of genotypes)	Medium (rare alleles)	Medium (deep SNP coverage)	Diversity panel (gene pool sample)

GWAS, genome-wide association study; SNP, single nucleotide polymorphism.

of genes identified using the *Agrobacterium* T-DNA as both a mutagen and a tag (Feldmann *et al.* 1989), taking advantage of high throughput methods (Alonso *et al.* 2003) to produce extensive collections of mutant alleles. Historically, the ability to generate phenotypically interesting mutations has greatly exceeded the ability to identify causative sequence variants. Easily identifiable "tags" such as T-DNA or various transposons (Fladung 2016) and methods for efficient searches for point mutations (McCallum *et al.* 2000; Colbert *et al.* 2001; Till *et al.* 2003; Henikoff *et al.* 2004) have mitigated constraints to identifying causative plant gene mutations. Knockout mutant populations generated by means such as fast neutron mediated mutagenesis that cause mostly deletions of DNA fragments ranging in size from a few base pairs to more than 30 kb (Bruggemann *et al.* 1996; Li *et al.* 2001, 2002) can readily be sequence tagged even by low-coverage methods, although tolerating far fewer mutations per genome (Belfield *et al.* 2012) than EMS populations in which it is more challenging to discern (single-nucleotide) mutations with confidence [e.g. (Addo-Quaye *et al.* 2018)].

Naturally occurring alleles, "biased" in informative ways by the action of selection, have become a powerful complement to both forward and reverse genetics. The ability to generate draft genome sequences enables comprehensive searches for naturally occurring alleles that are statistically associated with a trait, incorporating elements of both forward and re-verse genetics. Demonstrated in 2005 (Ozaki *et al.* 2002), such GWAS have been widely used, taking advantage of the histor-ical accumulation of recombination events since the evolu-tion of an allele. Sequencing of large germplasm collections, such as many collected and phenotyped for numerous crops by CGIAR centers, reveals much of the spectrum of naturally occurring alleles in a gene pool, also evaluating associations of the more abundant ones with traits [e.g. (Morris *et al.* 2013)]. In addition to de novo searches for trait association, phenotypic and passport information for such collections can be of value in testing hypotheses, for example, regarding the geographic distribution of specific candidate alleles (Cuevas et al. 2016).

Germplasm drawn from cultivated gene pools derived recently from small numbers of progenitors, or prepared by crossing strategies that combine broad samples of diversity with appreciable linkage disequilibrium (Yu *et al.* 2008), harnesses naturally occurring alleles and recombinations to improve the precision of trait mapping while mitigating the propensity of GWAS for false-positive associations [e.g. (Buckler *et al.* 2009)]. Such a "nested association mapping" approach is an attractive means to investigate effects of naturally occurring alleles that are too rare to obtain a significant GWAS signal, in that biparental populations placing the allele in hundreds of progeny offer higher statistical power.

While commonality of plant gene content has permitted extensive leveraging of reverse genetics findings across taxa, extrapolation of forward genetics data is constrained by random and nonrandom factors that affect the persistence of functional polymorphism, together with statistical constraints that limit quantitative trait loci mapping by either bi-parental methods [e.g. (Lander and Botstein 1989)] or GWAS (Yu *et al.* 2008). For example, the notion that crosses between wild and elite forms of different crops may segregate for convergent alleles at corresponding loci (Paterson *et al.* 1995) has had both supporting (Lin *et al.* 2012) and conflicting evi-dence (Tang *et al.* 2013). Nonetheless, across numerous traits and taxa, meta-analysis of the collected literature empow-ered by robust comparative genomics is informative in re-vealing "hotspots" for discovery of quantitative trait loci in elite gene pools and in providing diagnostic tools for such hotspots.

Exceptions to the commonality of genetic hardware often offer hypotheses about potential causal agents of specific traits. Genome sequences abound with examples of copy number amplifications that are correlated with distinctive features of particular plant taxa. *Brassica napus*, or canola, has experienced massive expansion of oil biosynthesis genes in an oilseed plant (Chalhoub *et al.* 2014) at almost double the number annotated in soybean (Schmutz *et al.* 2010) and more than double those in oil palm (Singh *et al.* 2013). Striking differences in the cell walls of monocots and dicots

(McCann and Roberts 1991; Carpita and Gibeaut 1993) are correlated with abundance of CesA/Csl cell wall biogenesis gene superfamily members, *Arabidopsis* containing a single Group F GT31 gene, whereas sorghum and rice contain 6 and 10, respectively (Paterson *et al.* 2009). Contraction or complete loss in tomato of several cytochrome P450 subfamilies associated with toxic alkaloid biosynthesis (Sato *et al.* 2012) may be related to the importance of attracting vertebrate frugivores to disperse seeds via fleshy fruits (Howe and Smallwood 1982). Copy number amplifications contribute to the evolution of herbicide resistance in weeds of genetically engineered crops (Patterson *et al.* 2018), potentially with multiple origins of such resistance (Fernández *et al.* 2013), attesting to the speed at which this mechanism permits a plant to respond to new selective pressure. Some such rapid responses appear to be enabled by heritable extranuclear DNA (Molin *et al.* 2020).

While the commonality of genetic hardware among plants is a useful generalization, formation of new genes and loss of existing ones are each continuous, with the content of an extant genome ranging from taxonomically widespread, highly conserved genes to taxonomically restricted "orphan" genes. Percentages of orphan genes in a genome range widely, with 5% to 15% being typical (Arendsee *et al.* 2014) and often declining as additional closely related genomes are sequenced. Age stratification found 4% of *A. thaliana* genes to be species specific, 61.3% to date to the origin of eukaryotes (30.3%, an estimated 1.6 billion years ago) or cellular organisms (31.0%, 2.52 billion years ago), and varying percentages shared with ascending hierarchical taxonomic groups, for example 0.8% originating with the *Arabidopsis* genus and 3.3% with the Brassicaceae family (Arendsee *et al.* 2014).

In that their comparative analysis is constrained by definition, orphan genes are prone to artifactual annotation; however, various types of evidence attest to subsets having important functions. For example, sorghum genes containing 1 functional domain that was absent from rice encoded alpha kafirins that account for most sorghum seed storage protein (Paterson *et al.* 2009) and were found to correspond to maize zeins (Xu and Messing 2008). A recent analysis of 3,553 orthogroups (totaling 5,456 genes) found, with very few exceptions, that those comprised solely of species specific (in the example, *P. alba*) genes in which multiple genes showed molecular signatures of positive selection had features associated with disease resistance (Kong *et al.* 2023).

The software: plant gene regulation in light of genome structure and lifestyle

Gene expression translates the genome's hardware into messenger RNAs and noncoding RNAs and ultimately generates the diversity of plant cell types and tissues and their finely tuned responses to environmental cues and challenges. As in animals, epigenetic marks reinforce patterns of cell-type and condition-specific gene expression and contribute to appropriate transcriptional responses (Sullivan *et al.* 2014; Schmitz *et al.* 2022). Accessible chromatin regions (i.e. regulatory DNA, ACRs) are enriched for trait-associated genetic variation implicated in domestication and evolution (Maurano *et al.* 2012; Sullivan *et al.* 2014; Rodgers-Melnick *et al.* 2016). Although researchers have learned some of the rules that govern the transcription of genomic information in plants, there is much left to be discovered.

Plant gene regulation research lacked the unified, comprehensive, and well-funded ENCODE and ModENCODE projects that annotated transcripts, transcription factor binding sites, histone modifications, ACRs, and long-range regulatory interactions for many human and animal cell lines and tissues (ENCODE 2012; Sanyal *et al.* 2012; Thurman *et al.* 2012). However, during the past decade, the methods developed and employed in these projects and other genome-scale interrogations of animal genomes (Patwardhan *et al.* 2009; Arnold *et al.* 2013; Gasperini *et al.* 2019; Klein *et al.* 2020) have been adapted for use in plants in many studies (Rodríguez-Leal *et al.* 2017; Ricci *et al.* 2019; Jores *et al.* 2020, 2021, 2023; Deng *et al.* 2023; Tan *et al.* 2023). Taken together, these studies have identified ACRs, histone marks associated with various activity states, transcription factor binding sites, and candidate long-range regulatory interactions in diverse plant genomes (Schmitz *et al.* 2022; Jores *et al.* 2023). Too often, however, data interpretation and conclusions appear to be driven by the deep existing knowledge of gene regulation mechanisms in animals without sufficient consideration given to the distinct features of plant genomes and plant life that might be relevant to understand gene regulation in these sessile and largely autotrophic organisms.

Plant and animal genomes differ strikingly in gene size and prevalence of transposable elements, both features of likely importance for gene regulation. The maize and human genomes are of comparable size (2.4 and 3 Gb, respectively), yet the average protein-coding gene size is ~4,000 bp in maize and approximately 32,000 bp in human. This 8-fold difference is largely due to the enormous size of introns in human (human, mean length 6,174 bp, median 1,590 bp in protein coding transcripts; maize, mean length 699 bp, median 145 bp in protein coding transcripts). In human, the genomic "real estate" occupied by introns and intronic ACRs is far greater than in maize (Fig. 1) and impacts the distances that need to be spanned between intronic enhancer elements and their target transcription start sites (TSS; human, distance of 8,530 intronic ACRs to nearest TSS: mean 35,800 bp, median 11,501 bp; maize, distance of 350 intronic ACRs to nearest TSS: mean 8,074 bp, median 4,118 bp).

The first genome-wide plant regulatory landscapes established in *A. thaliana* (Zhang *et al.* 2012; Sullivan *et al.* 2014) show marked differences in the genomic distribution of ACRs compared with those in animals. In *A. thaliana*, less than 5% of ACRs reside in introns, about 37% reside within 400 bps of a TSS, and another 37% were classified as intergenic or distal (with the latter designation not well suited to this compact genome). In the equally compact and

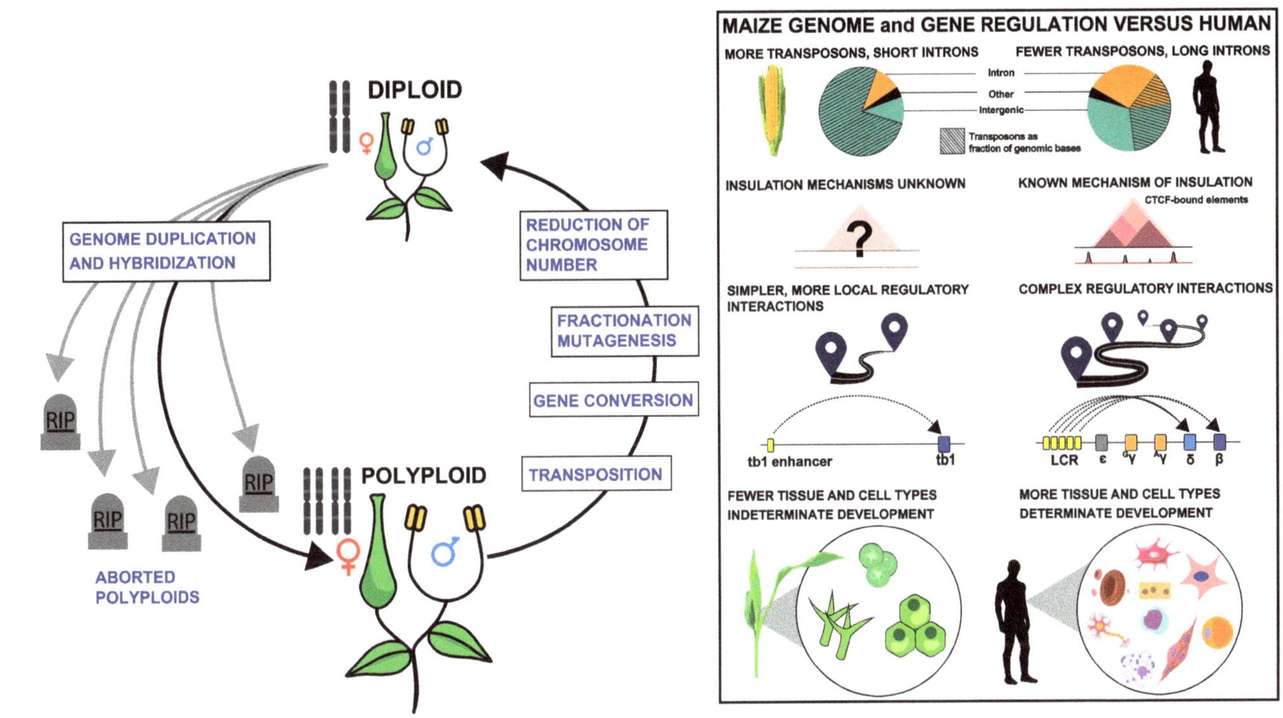

Figure 1. Plant genomes—the hardware—have been shaped by frequent polyploidization events (left figure panel) via hybridization or duplication. Newly formed polyploids face chromosome segregation and gene dosage challenges, leading to many aborted lineages. In a small subset of lineages, selection acts on the variation created via transposition, gene conversion, and fractionation mutagenesis, generally with reduction of chromosome number such that a diploid-like state is restored. Maize provides a particularly good example, now having the same chromosome number as sorghum despite having experienced genome duplication since their divergence approximately 20 million years ago. This cycle has consequences for plant genome structure and gene regulation—the software—as shown here for maize and human (right figure panel). Although both genomes are of comparable size and contain similar numbers of genes, human genes are much longer, largely because their introns are 10 times longer (pie charts, top). Compared with humans, maize genes and the maize regulatory landscape appear much more compact, possibly reducing the need for "expression domains" enabled by the human insulator protein CTCF (middle 2 panels). Although long-distance regulatory interactions exist in plants, including between the maize *tb1* gene and its transposon-derived enhancer, these interactions appear to be generally less complex than those observed in humans, as illustrated here with the beta-globin locus control region. The reduced complexity at this regulatory level is consistent with fewer cell types and the indeterminate, environmentally responsive mode of development in maize vs human (bottom panel). Shorter introns and a generally more compact and less complex regulatory landscape may render plant genomes less susceptible to disruption by the mutational processes following polyploidization.

gene-rich genome of the fly *Drosophila melanogaster*, 31% of ACRs reside in introns, 12% reside near a TSS, and 29% were classified as intergenic (Thomas *et al.* 2011). In human, the 2012 ENCODE study reported that 51% of ACRs reside in introns, 3% near a TSS, and 41% were classified as distal in 125 cell lines and tissue samples (Thurman *et al.* 2012). A 2020 ENCODE study improved on the latter work by vastly increasing the number of analyzed samples and broadening the coverage of primary cells and tissues (Moore *et al.* 2020). Moreover, the study integrated chromatin accessibility with histone modification marks on flanking nucleosomes and annotated CTCF-bound elements. This approach enables ACR annotation as promoters (high H3K4me3 signal, within 200 bp of a TSS), proximal (high H3K27ac, low H3K4me3, within 2 kb of a TSS) or distal enhancers (high H3K27ac, low H3K4me3, outside 2 kb of a TSS), insulator or looping elements (CTCF-bound), and elements of unknown function (high H3K4me3, not within 200 bp of a TSS). Using these annotations, ~4% of ACRs are promoter-like, 15% are proximal enhancer-like elements within 2 kb of a TSS, and 71% are distal enhancer-like elements, many of which are residing in introns. In stark contrast to human, in maize, 29% of ACRs reside within 2 kb of a gene and only 45% are classified as distal ACRs residing outside of this interval (Ricci *et al.* 2019). Moreover, the majority of distal ACRs (51.2%) are depleted of flanking histone modifications, 10.2% show H3K9/K27/K56 acetylation, and 27.5% show histone modifications consistent with transcribed genes, possibly re-presenting unannotated open reading frames.

The absence of the canonical flanking histone modifications marking active enhancers in animals (high H3K 27ac, high H3K4me1, low H3K4me3) is consistent with the absence of functional enhancer RNAs in plants. Enhancer RNAs are short (~200 nt-2 kb) and short-lived transcripts that are pervasively and bidirectionally transcribed from active enhancers (Kim *et al.* 2010; Arner *et al.* 2015; Harrison and Bose 2022). They interact with the histone modifying enzyme complexes Polycomb Repressive Complex 2 and histone acetyltransferases CREB binding protein/p300 to inhibit the deposition of repressive H3K27me3 histone marks and promote the deposition of activating H3K27ac histone marks, respectively, thereby maintaining enhancer chromatin accessibility and promoting further eRNA synthesis (Bose *et al.* 2017). Both enhancer RNAs and H3K27ac marks are used to annotate active animal enhancers (Andersson *et al.* 2014; Core *et al.* 2014; Wu *et al.* 2014; Moore *et al.* 2020). Analyses of nascent bi-directionally transcribed RNAs in *A. thaliana* and maize yielded conflicting results for (Lozano *et al.* 2021) and against the widespread existence of enhancer RNAs (Erhard *et al.* 2015; Hetzel *et al.* 2016; McDonald *et al.* 2023). A recent *A. thaliana* study revisited this controversy by inhibiting exosome-dependent degradation of unstable nascent RNAs and finding only 113 instances of bi-directionally transcription, of which 78 were intronic, and the remaining 35 were intergenic (Thieffry *et al.* 2020). Thus, plants appear to use different mechanisms than animals to maintain enhancer chromatin accessibility and activity or at least additional ones (McDonald *et al.* 2023).

Taken together, there appear to be profound differences in ACR spacing, ACR genomic context, and ACR histone modifications between human and maize. These differences may relate to another notable difference between animal and plant genomes: the absence of the transcriptional repressor CTCF in plant genomes (Fig. 1). In animals, CTCF is involved in regulating the 3D structure of chromatin and forming loops and marking TAD (Topologically Associating Domain) boundaries. CTCF-bound TAD boundaries function as insulators, separating regulatory units within which enhancers and promoters interact (Lupiáñez *et al.* 2015). In animals and human, disruption of TAD boundaries can lead to inappropriate interactions and profound misregulation (Okhovat *et al.* 2023). Although large plant genomes exhibit TAD-like structures, long-range chromatin interactions can span these structures (Dong *et al.* 2017; Liu *et al.* 2017; Mascher *et al.* 2017; Doğan and Liu 2018). As neither the protein(s) nor any sequence motif(s) underpinning boundary activity are known in any plant (Heger and Wiehe 2014), it remains unresolved how plant TAD-like structures arise and how relevant they are for gene regulation (Domb *et al.* 2022; Schmitz *et al.* 2022). Given the differences in spacing and genomic context of maize and human ACRs, the need for strictly enforcing insulation of long-distance enhancer-promoter interactions might be less profound in plants.

It is tempting to speculate that some of the described differences arise in part from the indeterminate development of plants that puts strikingly different demands on gene regulation. Throughout their life, plants continue to form new organs in response to environmental cues, which requires close and constant integration of environmental response and developmental gene expression pathways in ways simply not present in fly, mouse, and human. This constant integration might not be compatible with the strict expression boundaries observed in these animals; instead, it likely requires rapid rewiring across both long- and short-distance regulatory interactions. In human, of the nearly 3 million union ACRs detected in 125 diverse samples, nearly 1 million were specific to 1 sample, nearly 2 million in 2 or more samples, and only 3,692 ACRs were detected in all samples (Thurman *et al.* 2012). ACRs in human are so specific to cell and tissue type that their changing patterns and decreasing numbers along developmental trajectories can be used to reconstruct cell fate and lineage relationships from embryonic stem cells to terminal fates (Stergachis *et al.* 2013). Although direct comparisons are challenging because far fewer plant cell and tissue types have been studied in this way, ACRs in *A. thaliana* and maize appear to be far less dynamic—i.e. cell-type or condition-specific—than those in human. Across 13 diverse *A. thaliana* samples that detected nearly 47,000 union ACRs, only about 2,000 ACRs were specific to 1 sample while nearly 45,000 ACRs were detected in 2 or more samples (Sullivan *et al.* 2019). In maize, only 15% to 21% of the distal ACRs accessible in leaf tissue were inaccessible in inflorescence tissue (Ricci *et al.* 2019). Although single-cell genomics studies in both *A. thaliana* and maize have yielded evidence for larger numbers of dynamic, cell-type-specific ACR (Dorrity *et al.* 2021; Marand *et al.* 2021) cell and tissue identity appears less rigidly engrained in the regulatory landscape of plants than in animals.

However, rigid cell fate determination through altered patterns of ACRs might be less required in plants. Unlike animal development, plant development does not involve movement of cells because cell identity and lineage are established by cell division and position (van den Berg *et al.* 1995). This developmental mode likely minimizes the role of cell-autonomous gene regulation and cell-type-specific epigenomes and allows for scenarios in which transcription factors and other molecules expressed in a particular cell type can act in a concerted fashion with those in neighboring cells. It is tempting to speculate that the less rigid (i.e. less cell and tissue-specific) regulatory landscapes found in plants contribute to their capacity to regenerate fully functional plants from excised tissue or protoplasts (Gaillochet and Lohmann 2015) because there are fewer ACRs that have to be reo-pened to turn terminally differentiated plant cells into stem-cell-like ones. Moreover, coming back to Stebbin's ob-servations about animal and plant kingdom differences, there are likely many fewer distinct cell types in maize than in hu-man or mouse (Fig. 1).

The relative stasis of plant regulatory landscapes may also reflect another phenomenon: the constant integration of environmental cues with developmental gene expression may require that a large number of ACRs are poised for activation, i.e., occupied by trans-acting factors, without active transcription occurring until a signal is perceived. This mode of transcriptional regulation, occurring at enhancers but also immediately downstream of the transcription start site with paused, transcriptionally engaged Pol II molecules awaiting activation, allows for fast response times and is often observed for developmental and environmental response genes (Boehm *et al.* 2003; Creyghton *et al.* 2010; Xi *et al.*

2011). For example, in *A. thaliana*, treatment with the growth hormone auxin results in differential regulation of over 1,200 genes (Lewis *et al.* 2013); however, accessibility increases for only 155 ACRs and decreases for another 167 ACRs (Sullivan *et al.* 2019), in part because auxin-responsive factors are DNA-bound irrespective of auxin treatment. A greater fraction of poised ACRs in plants than in human would result in a comparatively more static regulatory landscape and a weaker correlation between chromatin accessibility and nearby gene expression. Alternatively, the relative stasis of plant regulatory landscapes might be explained by the large expansion of plant transcription factor (TF) families (Shiu *et al.* 2005), with their repressive and activating family members recognizing very similar motifs. While repressive and activating TFs of a family may trade places upon an environmental or developmental stimulus, this dynamic change would be largely invisible at the level of ACRs.

Stebbins highlighted another plant-specific phenomenon —greater tolerance of polyploidy—that is of consequence for the genome structure of angiosperms and gene expression regulation. Polyploidy is well tolerated across maturing tissues of individual plants (endoreduplication) (Lang and Schnittger 2020), and ploidy changes occur frequently along plant lineages (Song and Chen 2015), as discussed in detail below. There is very little accounting in most gene expression studies for the fact that commonly assayed tissues like leaves and roots are highly divergent in ploidy (Bhosale *et al.* 2018; Lang and Schnittger 2020). Single-cell expression studies in *A. thaliana* roots suggest that gene expression overall drops with increasing ploidy and maturity of root hair cells, but the expression of cell-type-specific genes increases (Jean-Baptiste *et al.* 2019). It is unknown to what extent the many additional copies contribute to gene expression in endoreduplicated cells and how regulatory landscapes and interactions among regulatory elements may change in response to endoreduplication.

The frequent changes of ploidy across plant lineages may contribute to shorter genes, shorter introns, less pronounced TAD structures, fewer distal enhancers, and higher transposon content and activity in plants (Fig. 1). Increases in ploidy require rapid dosage compensation and rapid divergence (Conant and Wolfe 2008). Plants have mastered both through RNA-mediated DNA methylation (Song and Chen 2015) and through high transposon activity (Flagel and Wendel 2009). Of course, transposon activity has remodeled both plant and animal genomes and contributed to evolutionary novelty (Flagel and Wendel 2009; Lynch *et al.* 2015) and domestication traits (Studer *et al.* 2011). Nevertheless, the frequent reshuffling of plant genomes during the course of their deep evolutionary history through cycles of polyploidization and returns to a diploid state might contribute to their shorter genes, their more compact distance distribution of regulatory elements, their less cell type–specific landscapes, and their higher genomic transposon content, all which favor local gene regulation environments and disfavor the highly intricate enhancer architecture and large distances observed in the animal *HOX* gene clusters (Montavon and Duboule 2013) or the beta-globin locus control region (Levings and Bungert 2002) (Fig. 1).

In summary, we argue that the hardware of plant genomes —both in content and in dynamics—has contributed to rather substantial differences in gene regulation from that in animals as exemplified by maize and human, organisms of comparable genome size and gene number. The mechanistic underpinnings of several of the phenomena discussed here such as insulation and TAD formation, maintenance of enhancer accessibility, ACR presence, and gene expression in endoreduplicated cells remain unknown and will require technological innovation to be resolved. Specifically, we need to move beyond averaging large numbers of small fragments to infer chromatin states and element activity as we do in the current chromosome conformation and accessibility assays. We favor recently developed single-molecule, long-read assays like Fiber-seq and others (Abdulhay *et al.* 2020; Stergachis *et al.* 2020), which can capture ACRs, TF binding sites, nucleosome position, Pol II footprints, and cytosine methylation along 20-kb fibers. The cited methods use a DNA N^6-adenine methyltransferase to methylate accessible adenines and long-read sequencing to detect this modification at nucleotide resolution, in addition to cytosine methylation. Adenine methylation is extremely sparse in plants (Kong *et al.* 2022), contrary to earlier reports. Assessing gene regulatory landscapes in this way eliminates the need to infer activity states and co-regulation of neighboring regulatory elements (Pliner *et al.* 2018) and eliminates bias from interpreting data across different assay types. Single-molecule regulatory landscapes will give insight into the regulatory activity of formerly inaccessible, highly repetitive regions, including the large regions of plant genomes occupied by transposons, and they will capture the stochasticity of TF occupancy and nucleosome positioning across many fibers corresponding to a large genomic interval. As long-read sequencing becomes more efficient (and thus cheaper) and capable of even longer reads, the promise of visualizing and understanding gene regulation at the level of single molecules across diverse plants and diverse plant cell types cannot be overstated.

A singular history of polyploidization

Stebbins' appreciation of the importance of polyploidy in plant evolution, albeit considerable, was an underestimate. Building on early clues from genetic mapping (Kowalski *et al.* 1994; Paterson *et al.* 1996), its genome sequence revealed that even the relatively small 5-chromosome genome of *A. thaliana*, chosen as the first plant genome to be sequenced in part for its simplicity, retained traces of 2 duplications and 1 triplication (Blanc *et al.* 2000; Arabidopsis Genome Initiative 2000; Lynch and Conery 2000; Paterson *et al.* 2000; Vision *et al.* 2000; Bowers *et al.* 2003). With the further discovery that the first monocot genome sequence, rice, also confirmed clues from cytology (Lawrence 1931) and genetic mapping (Kishimoto *et al.* 1994; Nagamura *et al.* 1995) of genome duplication (Goff *et al.* 2002), the study of duplicated gene relationships advanced from an isolated specialization to an essential element of genetic, evolutionary, and developmental investigations in higher plants.

In contrast to algorithms refined in microbes and/or animals to discern often-faint signal of ancient genome duplications (Bailey *et al.* 2002, 2004; Cheung *et al.* 2003a, 2003b; Tuzun *et al.* 2004), consideration of multiple colinearity was necessary to take full advantage of the relatively strong signals of, often recursive, plant genome duplications. For example, if a chromosomal region is preserved in 3 genomes (A, B, and C), then pair-wise predictions would combinatorially yield 3 inferences about ancestral gene repertoire and order (A-B, B-C, A-C). Polyploidy in 1 or more of the 3 genomes multiply the number of comparisons accordingly. To retrieve maximal information, both consequences of paleopolyploidy within a genome and comprehensive alignments of multiple paleopolyploid genomes to one another, require related pair-wise colinear segment to be combined into one inferred or-der (A-B-C). For example, multi-alignment of *A. thaliana*, *Populus trichocarpa*, and *Carica papaya* genomes revealed evidence of previously unknown "triplicated" structure that was validated empirically by comparison to the *Vitis* genome (Tang *et al.* 2008), a phylogenetic outgroup that had not experienced any more recent polyploidy.

The more than 1,600 green plant genomes now available, representing over 800 species, have yielded a relatively good (though by no means complete) history of plant genome duplications (Fig. 2) and knowledge of their consequences, as follows:

1) Phylogenetic tree topologies for hundreds of putatively orthologous expressed sequence tagged (EST) sequences support the occurrence of 1 genome duplication in the common ancestor of seed plants, and another in the common ancestor of angiosperms (Jiao *et al.* 2011), with the lineage of only 1 extant angiosperm, *Amborella trichopoda*, not known to have preserved additional paleopolyploidy events (Amborela Genome Project 2013);
2) In the monocots, phylogenetic analysis of nested synteny blocks indicates 1 genome duplication in a common ancestor of grasses and commelinids, upon which additional duplications have been superimposed in many lineages (e.g. 1 in oil palm, 2 in grasses, possibly 3 in banana) (Jiao *et al.* 2014) and with a host of still more recent duplications in some, for example, allopolyploidy in a common *Miscanthus-Saccharum* ancestor ~3.8 to 4.6 million years ago (Kim *et al.* 2014);
3) In a common ancestor of eudicots, colinearity information revealed an ancient event first discerned in *Arabidopsis* (Bowers *et al.* 2003) and clearly determined to be a genome triplication in the grape genome (Jaillon *et al.* 2007), upon which a host of additional lineage specific events have been superimposed;
4) While most paleopolyploidizations are duplications (i.e. forming tetraploids), several triplications have been inferred to form hexaploids, for example, in a eudicot ancestor (Jaillon *et al.* 2007) and twice consecutively in the tomato lineage (Sato *et al.* 2012); and a decaploid may have been formed by a complex event in the cotton lineage involving multiple polyploidizations plus hybridization in short succession (Wang *et al.* 2016);
5) Plant genome duplication may not be merely episodic but cyclic in the sense that some fitness benefits gradually deteriorate and favor recursive polyploidization (Chapman *et al.* 2006). However, the interval between successive events varies widely even in closely-related taxa, for example, with nearly 100 million years between a pan-grass duplication and the formation of *Sorghum halepense* (Paterson *et al.* 2020), but only ~1 million years between an allopolyploidy shared by Miscanthus and Saccharum, and a Saccharum-specific autoploidy (Kim *et al.* 2014);
6) Genome duplication is a punctuational event in the evolution of a lineage, triggering changes such that closely related genomes differing by a duplication may be less similar to one another than much more ancient genomes of common ploidy. Study of recently formed natural polyploids (Paterson *et al.* 2012; Chalhoub *et al.* 2014; Liu *et al.* 2014; Chen *et al.* 2016; Zhuang *et al.* 2019) and 1 synthetic polyploid (Zhuang *et al.* 2019) reveals macromolecular processes similar to those well known in synthetic polyploids, including loss and restructuring of low-copy DNA sequences (Song et al. 1995; Feldman et al. 1997; Liu et al. 1998; Ozkan et al. 2001; Shaked et al. 2001; Kashkush et al. 2002; Ozkan et al. 2002; Pires et al. 2004), activation of genes and retrotransposons (O'Neill *et al.* 2002; Kashkush *et al.* 2003; Fontdevila 2005), gene silencing (Chen and Pikaard 1997a, 1997b; Comai *et al.* 2000; Lee and Chen 2001) and out-right loss (Langham *et al.* 2004; Freeling 2009; Schnable *et al.* 2010), and subfunctionalization of gene expression patterns (Adams et al. 2003; Adams et al. 2004; Adams and Wendel 2005).
7) Modern chromosome numbers provide little information about the history of polyploidy in a lineage, as chromosome numbers tend to return to a narrow range following polyploidization, often via joining of homoeologous chromosomes near their termini (Wang *et al.* 2014). For example, the 5 chromosomes of *A. thaliana* are thought to trace to a total of 84 ancestral chromosomes (Wang *et al.* 2014). Reciprocal gene loss in different polyploid individuals or subpopulations, leading to a special case of Bateson–Dobzhansky–Muller incompatibility (Werth and Windham 1991), may favor survival of lineages with low chromosome numbers (Bowers and Paterson 2021).

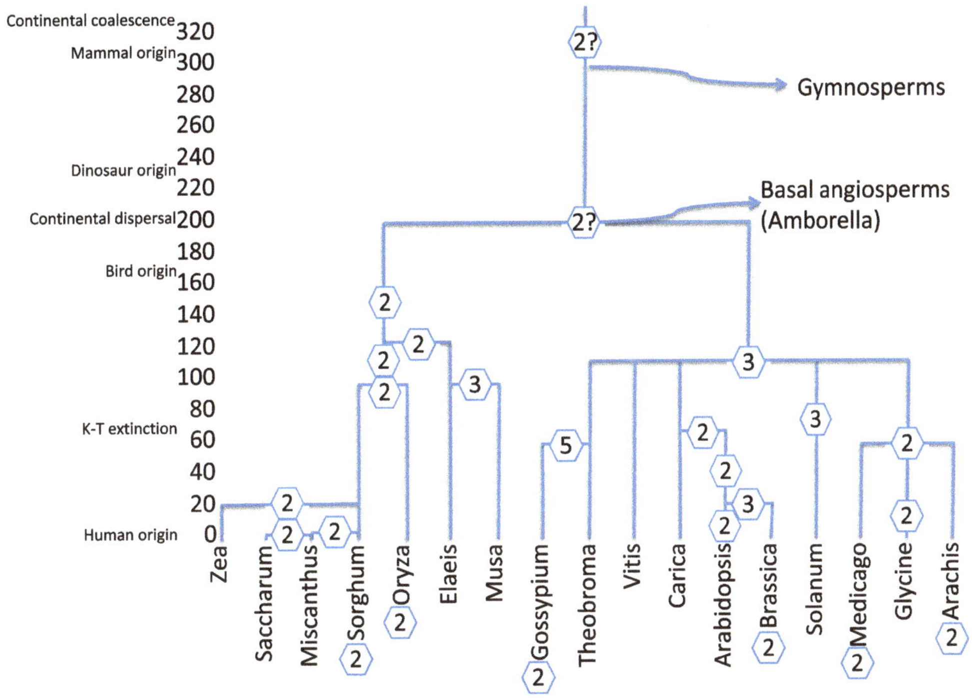

Figure 2. A brief history of plant genome duplication, illustrated in selected taxa. Genome duplication has been central to evolution in plants, arguably to a greater degree than in any other taxon. Duplication of regulatory genes important to seed and flower development appears to be concentrated around 319 and 192 million years ago (MYA) (Jiao *et al.* 2011): the former, roughly coincident with continental coalescence (Rogers and Santosh 2004), suggests a genome duplication in a common ancestor of seed plants; the latter, roughly coincident with continental dispersal (Rogers and Santosh 2004), preceded the diversification of angiosperms. While 1 basal angiosperm, *Amborella trichopoda*, has not preserved further duplications (Amborela Genome Project 2013), others have [e.g. (Chaw *et al.* 2019; Chen *et al.* 2019)]. Early duplication events also shaped the eudicot (Bowers *et al.* 2003; Jaillon *et al.* 2007) and monocot (Jiao *et al.* 2014) lineages. Genomic analyses have revealed many additional prehistoric duplica-tions (Goff *et al.* 2002; Paterson *et al.* 2004; Schmutz *et al.* 2010; D'Hont *et al.* 2012; Sato *et al.* 2012; Kim *et al.* 2014), the most affected lineage being that of Brassica with 36-fold multiplication relative to the angiosperm common ancestor, plus an additional post-Neolithic polyploid formation between extant diploids (Chalhoub *et al.* 2014). Indeed, genus-specific polyploid formations led to some primary cultigens, for example in Arachis (Zhuang *et al.* 2019), Gossypium (Paterson *et al.* 2012), and Medicago (Shen *et al.* 2020); widespread invasives, for example, in sorghum (Paterson *et al.* 2020), and abundant species, for example, in Oryza (Zou *et al.* 2015). Selected outgroups are depicted to delineate timing of salient events. Timing of all events is based on indicated citations, and variation in estimation methods together with resolution of the figure make these timings approximate—readers should refer to original citations for precise estimates.

A natural question is whether the propensity of plants for genome duplication may have contributed to their evolutionary success. Duplication of existing genes is thought to be a primary source of genetic material available for evolution of genes with new functions (Taylor and Raes 2004), and a long-held hypothesis has been that polyploidy is an engine for the evolution of genetic diversity by facilitating "functional divergence" of duplicated genes. Duplication of an entire genome at once retains stoichiometric balance between its constituents (Birchler and Veitia 2007) and has long been recognized as less disruptive to the phenotype and viability than duplication of only a subset such as a single chromosome (Blakeslee *et al.* 1920). Following duplication, 1 member of a gene pair may be free to experience function-al divergence, acquiring unique functionality [neo-functionalization, (Stephens 1951; Ohno 1970)] or the 2 copies subdividing ancestral functions, for example, by reciprocal loss of regulatory cues that render expression of different copies specific to different tissues [subfunctionalization —(Lynch and Force 2000)], with the fitness of the organism insulated by the homeolog. However, in that endoreduplication is well tolerated across maturing tissues of individual plants as noted above, it remains an open question whether expression-based neo- or subfunctionalization associated with polyploidy [e.g. (Adams *et al.* 2003, 2004; Adams and Wendel 2005)] is a striking response to a newly duplicated nucleus or merely a previously undetected adaptation to the frequent occurrence of many additional gene copies.

While diverse examples associate polyploidy with broader zones of adaptation (Kiedrzynski *et al.* 2021) and striking phenotypic changes such as the seedborne epidermal fibers of cotton (Jiang *et al.* 1998; Paterson *et al.* 2012), genomic data have also raised questions about the classical "functional divergence" model. For example, if a primary advantage of polyploidy is the opportunity for the evolution of genes with new functions, then patterns of genetic diversity among strains within paleopolyploid taxa might reveal footprints of selection that are consistent with duplicated genes being relatively free to acquire unique functionality. Contrary to this prediction, SNPs encode less radical amino acid changes in genes for which there exists a duplicated copy at a "paleologous" locus, than in "singleton" genes among both *Arabidopsis* ecotypes and *Oryza* subspecies (Chapman *et al.* 2006). While this does not preclude the possibility that important adaptations have occurred by functional divergence of duplicated genes, it suggests that there may also be a population of genes in which fitness benefits accrue to functional buffering, consistent with remarkably long duration that duplicated yeast genes continue to compensate one another (Gu *et al.* 2003).

Organisms continuously require genetic variation to adapt to constantly fluctuating environments, yet genome duplications are episodic—in lineages that do not preserve whole genome duplications for long time periods, diverse mechanisms have provided the raw genetic material for adaptation. Sorghum and rice, which have not experienced whole genome duplication in an estimated 98 million years (Wang *et al.* 2015), preserve relatively more SNPs in tandem than paralogous duplicated genes (Guo *et al.* 2019). However, maize, which experienced genome duplication shortly after divergence from a common ancestor shared with sorghum as recently as 12 million years ago (Swigonova *et al.* 2004), shows SNP enrichment in its large supply of paralogous duplicates. The proportion of genes showing signatures of recent positive selection is higher in small-scale (tandem and transposed) than genome-scale duplicates in sorghum, but the opposite in maize (Guo *et al.* 2019).

Even complex biochemical pathways, which might be favored by stoichiometric balance resulting from whole-genome duplications, in at least 1 case did not evolve from available whole-genome duplicates but came together subsequently from a series of single-gene duplicates. For example, genome duplication in a common ancestor of grasses putatively provided "spare" copies of genes for the entire photosynthetic pathway; however, the evolution of C4 photosynthesis from C3 ancestors used few if any of these, instead involving single-gene duplications (Wang *et al.* 2009).

Paleopolyploidy results in islands of conserved sequence duplicated in parallel along stretches of homoeologous chromosomes, providing a genomic environment conducive to genetic exchanges between non-allelic loci. Similar DNA sequences may transiently form heteroduplex DNA (Holliday 1964, 1966), in which repair of unmatched bases may allow "gene conversion," rendering similar DNA sequences identical (Galtier 2003). Often implicated in homogenization of small tracts of paralogous DNA sequences, usually between several and several hundred base pairs (Petes and Symington 1991), gene conversion has long been thought to account for the evolution of various multigene families (Sawyer 1989; White and Crowther 2000; Mondragon-Palomino and Gaut 2005) and proximal gene clusters such as rRNA (Brown *et al.* 1972) and histone genes (Ohta 1984). Gene conversion may explain low divergence rates between paralogs produced by ancient large-scale duplication events in yeast (Gao and Innan 2004), appearing to have affected 2% of *Caenorhabditis elegans* duplicated genes (Semple and Wolfe 1999) and 18% of homeologs duplicated before the mouse–rat divergence (Ezawa *et al.* 2006).

Plant whole-genome duplications and associated genome-wide sets of simultaneously duplicated homoeologous genes have facilitated surveys for gene conversion, finding rich evidence of the action of this mechanism with clear temporal and spatial patterns. Complete genome sequences enable searches for footprints of gene conversion using elegant approaches such as phylogenetic incongruity in "gene trees," for example, finding that ancient duplicates within the same genome are more similar to one another than they are to their alleles in a different genome. Gene conversion begins virtually immediately after polyploid formation, being found at appreciable frequency in "synthetic" polyploid *Arachis* a few years after being produced by humans (Zhuang *et al.* 2019) and neopolyploid canola formed in nature a few thousand years ago (Chalhoub *et al.* 2014), with gradually declining frequency over a few million years of evolution of a new lineage but still discernible within the elite gene pools of crops such as cotton (Guo *et al.* 2014).

Striking spatial patterns of gene evolution are evident along plant chromosomes, with some general principles that appear to have persisted for millennia, and intriguing special cases. "Euchromatin" and "heterochromatin," long recognizable cytologically by differential staining, have profoundly different composition and evolution, effectively providing each plant genome with 2 qualitatively different compartments that respectively facilitate different types and rates of evolution. Euchromatin tends to be terminal to the chromosomes, accounts for most reciprocal exchange between orthologs ("conventional" recombination) and is gene-rich with gene orders persisting over long time periods (Bowers *et al.* 2005). Heterochromatin tends to be the physically larger compartment, accounting for much of the difference in DNA content between closely related taxa (Bowers *et al.* 2005), with relative paucity of genes but enrichment for recent single-gene duplications [or perhaps merely less effective removal than in euchromatin (Paterson *et al.* 2009)] and extensive migration of DNA between pericentromeric regions of different chromosomes. On paleo-duplicated chromosomes, regions of euchromatin and heterochromatin closely correspond (Paterson *et al.* 2009), indicating that these respective states have persisted for long time periods, for example, nearly 100 million years in the cereals (Wang *et al.* 2015).

Intriguing and perplexing are the case of rice chromosomes 11 and 12 and their homologs and homeologs in other cereals. Derived from a single ancestral chromosome in the pan-cereal genome duplication (Wang *et al.* 2007), these chromosomes experienced illegitimate recombination that has been temporally restricted in a stepwise manner, independently forming "strata" in different grass lineages with perplexing properties (Wang *et al.* 2011). The pericentromeric region accounts for two-thirds of the gene content differences between this homeologous chromosome pair, yet there is generally low sequence divergence between paleo-duplicated genes. Indeed, a distal region has the greatest DNA similarity between surviving duplicated genes found anywhere in the genome but also the highest concentration of lineage-specific gene pairs found anywhere in these genomes and with a significantly elevated gene evolutionary rate.

In partial summary, the neopolyploidy that has long been evident from classical genetics and cytology [e.g. (Stebbins 1966)] is merely the tip of the iceberg—plants have been experiencing recursive whole-genome duplications throughout their evolutionary history, followed by gene and chromosome losses and rearrangements that obscure cytological evidence but leave signatures of these ancient events in genome sequences. The classical notion of polyploidy as a generator of diversity may be somewhat simplistic in that endoreduplication of some tissues may "preadapt" plants to having multiple functioning gene copies in a cell, that signatures of selection consistent with functional buffering are also evident, and that the availability of duplicated genes for entire pathways was sometimes insufficient to catalyze major events.

Toward harnessing botanical diversity

With a broad range of plant genome sequences now available and the ability to routinely deepen the sample as needed to address fundamental questions and/or applied goals, better knowledge of plant evolutionary history informs and guides utilization of botanical diversity to improve human lives.

Building on early forays showing that plant genomes were not the staid and stable environments we once envisioned but in fact were highly fluid [e.g. (McClintock 1984)], detailed assemblies have highlighted heterogeneity across the genomic landscape. Plant genomes are dynamic and variable environments, broadly comprised of 2 qualitatively different compartments that respectively facilitate different types and rates of evolution, corresponding to classical "euchromatin" and "heterochromatin" (Bowers *et al.* 2005). Particularly striking chromosomal regions show both the most extensive gene loss and the most striking conservation of the remaining genes, putatively reflecting high levels of nonreciprocal exchanges but driven by selective forces that are not yet understood (Wang *et al.* 2011). In sum, a plant genome provides a range of options that may permit the same adaptive need to be met by different means at different times and/or in different lineages.

Much like the genome as a whole, the gene space also includes qualitatively different components that may respectively facilitate different types and rates of evolution. Polyploids have been thought to acquire capabilities that are "more than the sum" of those of their diploid progenitors, with gene duplication (Maere *et al.* 2005) providing material available for divergence to new function (Stephens 1951; Ohno 1970; Force *et al.* 1999; Lynch and Conery 2000; Taylor and Raes 2004), while normal function by the homeolog insulates the fitness of the organism. However, in contrast with the classical "functional divergence" model, some groups of gene families show signatures of selection consistent with "functional buffering" (conservation of ancestral function) (Chapman *et al.* 2006; Paterson *et al.* 2006), with duplicated copies continuing to functionally compensate one another for far longer periods (Gu *et al.* 2003) than are thought to be necessary for "functional divergence" to occur (Lynch and Conery 2000).

In nature, heterogeneity and flexibility, rather than a specific pattern of genome organization, may confer evolutionary success. Evolution occurs at the intersection of genetics and ecology, each the outcome of hosts of multifactorial interactions, and this complexity may be reflected in outcomes incongruous with evolutionary expectations rooted in genomics. For example, the evolution of a complex pathway such as C4 photosynthesis intuitively would have been greatly simplified by stoichiometric balance between its constituent parts (Birchler and Veitia 2007)—the inference that it evolved *not* from available whole-genome duplicates but from single-gene duplications (Wang *et al.* 2009) suggests a series of adaptive steps toward its greater efficiency at ambient CO_2 levels and elevated temperatures (Heckmann *et al.* 2013; Schluter and Weber 2020). Likewise, in lineages that survived long periods during which no whole-genome duplications survived, adaptation utilized raw material provided by other mechanisms (Guo *et al.* 2019).

The spectrum of adaptations observed in nature informs approaches by which plants might be improved to better meet human needs. For example, cotton domestication was associated with conversion of a few dozen genes in the "D" genome (from a progenitor that does not produce spinnable fibers) to the "A" genome sequence, thus doubling copy number of the allele from the progenitor that does produce spinnable fibers (Guo *et al.* 2014). This raises several questions. First, did this doubling confer variation in fiber yield or quality for which domesticates were selected? Second, would doubling of other cotton genes permit further improvements—or does a similar phenomenon contribute to phenotypes of other crops? Genome editing now permits empirical testing of such questions by engineering of conversions not found in nature.

A natural means by which large numbers of single gene-sized chromosomal tracts could be transferred between exotic and elite genotypes for empirical phenotypic evaluation would be of high value. Transfer of favorable traits from wild or exotic relatives by recurrent "backcrossing" to an elite crop cultivar typically includes 10% to 20% of the chromosome carrying the target gene, including hundreds of nearby genes that often impose "linkage drag" from associated undesirable phenotypes. DNA markers expedite this process, especially for target genes with recessive or subtle phenotypes otherwise requiring replicated progeny testing and identifying recombinants that minimized linkage drag. However, better still would be a means to induce such single-allele transfers, for example, in F1 hybrids between an exotic donor and an elite line engineered with appropriate molecular machinery—followed by selfing or backcrossing to permit "screening" large numbers of individual exotic alleles for useful effects in progeny in much the same manner as reverse genetics screens while dramatically reducing the impact of linkage drag. This would especially facilitate evaluation of allelic variation in recombinationally recalcitrant heterochromatin, usually the physically larger compartment of a plant genome. Albeit not involving transfer or gene-sized tracts, methods by which hundreds of novel regulatory alleles can be generated (Rodríguez-Leal *et al.* 2017) foretell similar new opportunities to screen large numbers of candidates for alleles conferring novel functionality.

Polyploidy itself has long been employed to introgress botanical diversity from wild species into cultivated gene pools, but many species combinations fail to produce viable hybrids. Moreover, hybrids from such crosses often form novel alleles per se, by loss and restructuring of low-copy DNA sequences (cited above), which in some cases are associated with phenotypes (Schranz and Osborn 2000; Pires et al. 2004; Schranz and Osborn 2004). Convergent loss of duplicated copies of specific genes following independent duplications that are separated by hundreds of millions of years of evolution may reflect an underlying set of principles of molecular evolution that contribute to the fates of genome duplications (Paterson *et al.* 2006), raising the intriguing hypothesis that persistence of artificial polyploids might be facilitated by silencing 1 copy of these genes. Methods for simultaneously editing dozens of genes at once (Campa *et al.* 2019; Yuan and Gao 2022) or in orthogonal combinations (Cetin *et al.* 2023) may permit empirical tests of this hypothesis, potentially accessing rich additional diversity.

While the creation of "synthetic" tetraploids by humans has been widely attempted to exploit otherwise-inaccessible botanical variation in crop improvement, a recent example reverses the process. The formation of tetraploid *S. halepense* ~1 to 2 million years ago from a natu rally occu rring event merging the genomes of African *S. bicolor* (sorghu m) and Asian *S. propinquum* was the first surviving genome duplication in the sorghum lineage in nearly 100 million years (Wang *et al.* 2015), and its spread across 6 continents has exposed its rich variation to diverse selective forces (Paterson *et al.* 2020). Recent discovery that diploid sorghu ms can be obtained from certain crosses with tetraploid *S. halepense*-derived materials (Cox *et al.* 2017) permits genetic novelty from *S. halepense* to be investigated for contributions to the conventional sorghum gene pool.

In closing, the world has changed since the first author walked the wheat fields as a graduate student, learning empirical whole organism–level methods of plant breeding. It is hard to imagine an alternative that evaluates the voluminous number of interactions between and among hardware, software, and environment that are reflected in the whole organism–level phenotyping essential to making plant breeding decisions. However, the potential inputs into crop improvement programs have snowballed, genomics transcending transgenesis with enriched knowledge of the innate hardware and software of plant genomes and their function, and with new abilities to track genome transmission and alter the hardware and software in designed ways that leave behind no exogenous DNA. Moreover, we have far better understanding of the heterogeneity and flexibility of plant genomes and the forces that have acted to permit them to adapt to habitats from the tropics to near the poles and from sea level to at least 20,100 feet in altitude (Younghusband 1926), forming ramets ranging from floating *Wolffia* spp. plants of 1 mm in length (Raven *et al.* 1992) to *Eucalyptus regnans* trees of 100 m in height and 10 m in trunk diameter. Writing this article was especially interesting—perhaps as was envisioned by the editors who suggested such a partnership, the second ("junior") author challenged the first to reach well beyond the original assignment, with the end result presenting a broad and, we hope, integrative picture of the state of knowledge of botanical diversity at the molecular level. We emphasize once again that this breadth made truly exhaustive coverage intractable, and the overrepresentation of our own studies and others in our focal taxa is with no disrespect intended to the excellent work of many esteemed colleagues that would be similarly suitable.

Acknowledgments

Paterson thanks numerous past and current PGML and other colleagues for influencing his thinking, empowering, and conducting research salient to this paper. Queitsch is grateful to Kerry Bubb for inspiration, discussion, and analyses concerning the "software" of plant genomes. She thanks Sayeh Gorjifard for assistance with creating Fig. 1.

Author contributions

Both authors contributed significantly to the preparation of this review. A.H.P. was the primary author of the "hardware", "polyploidization" and "harnessing" sections. C.Q. was the primary author of the "software" section. A.H.P. and C.Q. were the primary authors of the abstract and introduction sections, also each reviewing and revising the sections written by the other.

Funding

Salient work in the laboratory of A.H.P. was partially funded by many past grants from the US National Science Foundation, US Department of Agriculture, US Agency for International Development, US Department of Energy Joint Genome Institute Community Sequencing Program, Cotton Incorporated, and a University System of Georgia Regents Professorship; and for C.Q., from the US National Science Foundation (RESEARCH-PGR grant no. 1748843, PlantSynBio grant no. 2240888) and the US National Institute of Health (NIGMS MIRA grant no. 1R35GM13953).

Conflict of interest statement. None declared.

Data availability

No new data were generated or analysed in support of this research.

References

Abdulhay NJ, McNally CP, Hsieh LJ, Kasinathan S, Keith A, Estes LS, Karimzadeh M, Underwood JG, Goodarzi H, Narlikar GJ, et al. Massively multiplex single-molecule oligonucleosome footprinting. Elife. 2020:**9**:e59404. https://doi.org/10.7554/eLife.59404

Adams KL, Cronn R, Percifield R, Wendel JF. Genes duplicated by polyploidy show unequal contributions to the transcriptome and organ-specific reciprocal silencing. Proc Natl Acad Sci U S A. 2003:**100**(8):4649–4654. https://doi.org/10.1073/pnas.0630618100

Adams KL, Percifield R, Wendel JF. Organ-specific silencing of duplicated genes in a newly synthesized cotton allotetraploid. Genetics. 2004:**168**(4):2217–2226. https://doi.org/10.1534/genetics.104.033522

Adams KL, Wendel JF. Novel patterns of gene expression in polyploid plants. Trends Genet. 2005:**21**(10):539–543. https://doi.org/10.1016/j.tig.2005.07.009

Addo-Quaye C, Tuinstra M, Carraro N, Weil C, Dilkes BP. Whole-genome sequence accuracy is improved by replication in a population of mutagenized sorghum. 2018:**8**(3):1079–1094. https://doi.org/10.1534/g3.117.300301

Alonso JM, Stepanova AN, Leisse TJ, Kim CJ, Chen H, Shinn P, Stevenson DK, Zimmerman J, Barajas P, Cheuk R, et al. Genome-wide insertional mutagenesis of Arabidopsis thaliana. Science. 2003:**301**(5633):653–657. https://doi.org/10.1126/science.1086391

Amborela Genome Project. The amborella genome and the evolution of flowering plants. Science. 2013:**342**:1241089. https://doi.org/10.1126/science.1241089

Andersson R, Gebhard C, Miguel-Escalada I, Hoof I, Bornholdt J, Boyd M, Chen Y, Zhao X, Schmidl C, Suzuki T, et al. An atlas of active enhancers across human cell types and tissues. Nature. 2014:**507**(7493):455–461. https://doi.org/10.1038/nature12787

Arabidopsis Genome Initiative. Analysis of the genome sequence of the flowering plant *Arabidopsis thaliana*. Nature. 2000:**408**(6814):796–815. https://doi.org/10.1038/35048692

Arendsee ZW, Li L, Wurtele ES. Coming of age: orphan genes in plants. Trends Plant Sci. 2014:**19**(11):698–708. https://doi.org/10.1016/j.tplants.2014.07.003

Arner E, Daub CO, Vitting-Seerup K, Andersson R, Lilje B, Drabløs F, Lennartsson A, Rönnerblad M, Hrydziuszko O, Vitezic M, et al. Transcribed enhancers lead waves of coordinated transcription in transitioning mammalian cells. Science. 2015:**347**(6225):1010–1014. https://doi.org/10.1126/science.1259418

Arnold CD, Gerlach D, Stelzer C, Boryń ŁM, Rath M, Stark A. Genome-wide quantitative enhancer activity maps identified by STARR-seq. Science. 2013:**339**(6123):1074–1077. https://doi.org/10.1126/science.1232542

Bailey JA, Church DM, Ventura M, Rocchi M, Eichler EE. Analysis of segmental duplications and genome assembly in the mouse. Genome Res. 2004:**14**(5):789–801. https://doi.org/10.1101/gr.2238404

Bailey JA, Gu ZP, Clark RA, Reinert K, Samonte RV, Schwartz S, Adams MD, Myers EW, Li PW, Eichler EE. Recent segmental duplications in the human genome. Science. 2002:**297**(5583):1003–1007. https://doi.org/10.1126/science.1072047

Belfield EJ, Gan X, Mithani A, Brown C, Jiang C, Franklin K, Alvey E, Wibowo A, Jung M, Bailey K, et al. Genome-wide analysis of mutations in mutant lineages selected following fast-neutron irradiation mutagenesis of Arabidopsis thaliana. Genome Res. 2012:**22**(7):1306–1315. https://doi.org/10.1101/gr.131474.111

Bhosale R, Boudolf V, Cuevas F, Lu R, Eekhout T, Hu Z, Van Isterdael G, Lambert GM, Xu F, Nowack MK, et al. A spatiotemporal DNA endoploidy map of the arabidopsis root reveals roles for the endocycle in root development and stress adaptation. Plant Cell. 2018:**30**(10):2330–2351. https://doi.org/10.1105/tpc.17.00983

Birchler JA, Veitia RA. The gene balance hypothesis: FROM classical genetics to modern genomics. Plant Cell. 2007:**19**(2):395–402. https://doi.org/10.1105/tpc.106.049338

Blakeslee AF, Belling J, Farnham ME. Chromosomal duplication and Mendelian phenomena in datura mutants. Science. 1920:**52**(1347):388–390. https://doi.org/10.1126/science.52.1347.388

Blanc G, Barakat A, Guyot R, Cooke R, Delseny I. Extensive duplication and reshuffling in the arabidopsis genome. Plant Cell. 2000:**12**(7):1093–1101. https://doi.org/10.1105/tpc.12.7.1093

Boehm AK, Saunders A, Werner J, Lis JT. Transcription factor and polymerase recruitment, modification, and movement on dhsp70 in vivo in the minutes following heat shock. Mol Cell Biol. 2003:**23**(21):7628–7637. https://doi.org/10.1128/MCB.23.21.7628-7637.2003

Bose DA, Donahue G, Reinberg D, Shiekhattar R, Bonasio R, Berger SL. RNA binding to CBP stimulates histone acetylation and transcription. Cell. 2017:**168**(1–2):135–149.e122. https://doi.org/10.1016/j.cell.2016.12.020

Bowers JE, Arias MA, Asher R, Avise JA, Ball RT, Brewer GA, Buss RW, Chen AH, Edwards TM, Estill JC, et al. Comparative physical mapping links conservation of microsynteny to chromosome structure and recombination in grasses. Proc Natl Acad Sci U S A. 2005:**102**(37):13206–13211. https://doi.org/10.1073/pnas.0502365102

Bowers JE, Chapman BA, Rong J, Paterson AH. Unravelling angiosperm genome evolution by phylogenetic analysis of chromosomal duplication events. Nature. 2003:**422**(6930):433–438. https://doi.org/10.1038/nature01521

Bowers JE, Paterson AH. Chromosome number is key to longevity of polyploid lineages. New phytol. 2021:**231**(1):19–28. https://doi.org/10.1111/nph.17361

Bowers JE, Pearl SA, Burke JM. Genetic mapping of millions of SNPs in safflower (Carthamus tinctorius L.) via whole-genome resequencing. G3 (Bethesda). 2016:**3**(6):2203–2211. https://doi.org/10.1534/g3.115.026690

Brown DD, Wensink PC, Jordan E. A comparison of the ribosomal DNA's of Xenopus laevis and Xenopus mulleri: the evolution of tandem genes. J Mol Biol. 1972:**63**(1):57–73. https://doi.org/10.1016/0022-2836(72)90521-9

Bruggemann E, Handwerger K, Essex C, Storz G. Analysis of fast neutron-generated mutants at the Arabidopsis thaliana HY4 locus. Plant J. 1996:**10**(4):755–760. https://doi.org/10.1046/j.1365-313X.1996.10040755.x

Buckler ES, Holland JB, Bradbury PJ, Acharya CB, Brown PJ, Browne C, Ersoz E, Flint-Garcia S, Garcia A, Glaubitz JC, et al. The genetic architecture of maize flowering time. Science. 2009:**325**(5941):714–718. https://doi.org/10.1126/science.1174276

Campa CC, Weisbach NR, Santinha AJ, Incarnato D, Platt RJ. Multiplexed genome engineering by Cas12a and CRISPR arrays encoded on single transcripts. Nat Methods. 2019:**16**(9):887–893. https://doi.org/10.1038/s41592-019-0508-6

Carpita NC, Gibeaut DM. Structural models of primary-cell walls in flowering plants—consistency of molecular-structure with the physical-properties of the walls during growth. Plant J. 1993:**3**(1):1–30. https://doi.org/10.1111/j.1365-313X.1993.tb00007.x

Cetin R, Wegner M, Luwisch L, Saud S, Achmedov T, Susser S, Vera-Gupti A, Muller K, Matthess Y, Quandt E, et al. Optimized metrics for orthogonal combinatorial CRISPR screens. Sci Rep. 2023:**13**(1):7405. https://doi.org/10.1038/s41598-023-34597-8

Chalhoub B, Denoeud F, Liu S, Parkin IAP, Tang H, Wang X, Chiquet J, Belcram H, Tong C, Samans B, et al. Early allopolyploid evolution in the post-neolithic Brassica napus oilseed genome. Science. 2014:**345**(6199):950–953. https://doi.org/10.1126/science.1253435

Chapman BA, Bowers JE, Feltus FA, Paterson AH. Buffering crucial functions by paleologous duplicated genes may impart cyclicality to angiosperm genome duplication. Proc Natl Acad Sci U S A. 2006:**103**(8):2730–2735. https://doi.org/10.1073/pnas.0507782103

Chaw SM, Liu YC, Wu YW, Wang HY, Lin CI, Wu CS, Ke HM, Chang LY, Hsu CY, Yang HT, et al. Stout camphor tree genome fills gaps in understanding of flowering plant genome evolution. Nat Plants. 2019:**5**(1):63–73. https://doi.org/10.1038/s41477-018-0337-0

Chen J, Hao Z, Guang X, Zhao C, Wang P, Xue L, Zhu Q, Yang L, Sheng Y, Zhou Y, et al. Liriodendron genome sheds light on angiosperm phylogeny and species-pair differentiation. Nat Plants. 2019:**5**(1):18–25. https://doi.org/10.1038/s41477-018-0323-6

Chen X, Li H, Pandey MK, Yang Q, Wang X, Garg V, Li H, Chi X, Doddamani D, Hong Y, et al. Draft genome of the peanut A-genome progenitor (*Arachis duranensis*) provides insights into geocarpy, oil biosynthesis and allergens. Proc Natl Acad Sci USA. 2016:**113**(24):6785–6790. https://doi.org/10.1073/pnas.1600899113

Chen ZJ, Pikaard CS. Transcriptional analysis of nucleolar dominance in polyploid plants: biased expression/silencing of progenitor rRNA genes is developmentally regulated in Brassica. Proc Natl Acad Sci U S A. 1997a:**94**(7):3442–3447. https://doi.org/10.1073/pnas.94.7.3442

Chen ZJ, Pikaard CS. Epigenetic silencing of RNA polymerase I transcription: a role for DNA methylation and histone modification in nucleolar dominance. Genes Dev. 1997b:**11**(16):2124–2136. https://doi.org/10.1101/gad.11.16.2124

Cheung J, Estivill X, Khaja R, MacDonald JR, Lau K, Tsui LC, Scherer SW. Genome-wide detection of segmental duplications and potential assembly errors in the human genome sequence. Genome Biol. 2003a:**4**(4):R25. https://doi.org/10.1186/gb-2003-4-4-r25

Cheung J, Wilson MD, Zhang JJ, Khaja R, MacDonald JR, Heng HHQ, Koop BF, Scherer SW. Recent segmental and gene duplications in the mouse genome. Genome Biol. 2003b:**4**(8):R47. https://doi.org/10.1186/gb-2003-4-8-r47

Colbert T, Till BJ, Tompa R, Reynolds S, Steine MN, Yeung AT, McCallum CM, Comai L, Henikoff S. High-throughput screening for induced point mutations. Plant Physiol. 2001:**126**(2):480–484. https://doi.org/10.1104/pp.126.2.480

Comai L, Tyagi AP, Winter K, Holmes-Davis R, Reynolds SH, Stevens Y, Byers B. Phenotypic instability and rapid gene silencing in newly formed arabidopsis allotetraploids. Plant Cell. 2000:**12**(9):1551–1567. https://doi.org/10.1105/tpc.12.9.1551

Conant GC, Wolfe KH. Turning a hobby into a job: how duplicated genes find new functions. Nat Rev Genet. 2008:**9**(12):938–950. https://doi.org/10.1038/nrg2482

Core LJ, Martins AL, Danko CG, Waters CT, Siepel A, Lis JT. Analysis of nascent RNA identifies a unified architecture of initiation regions at mammalian promoters and enhancers. Nat Genet. 2014:**46**(12):1311–1320. https://doi.org/10.1038/ng.3142

Cox S, Nabukalu P, Paterson AH, Kong W, Auckland S, Rainville L, Cox S, Wang S. High proportion of diploid hybrids produced by interspecific diploid × tetraploid Sorghum hybridization. Genet Resour Crop Evol. 2017:**65**:645–657. https://doi.org/10.1007/s10722-017-0562-9

Creyghton MP, Cheng AW, Welstead GG, Kooistra T, Carey BW, Steine EJ, Hanna J, Lodato MA, Frampton GM, Sharp PA, et al. Histone H3K27ac separates active from poised enhancers and predicts developmental state. Proc Natl Acad Sci U S A. 2010:**107**(50):21931–21936. https://doi.org/10.1073/pnas.1016071107

Cuevas HE, Zhou C, Tang H, Khadke PP, Das SK, Lin YR, Ge Z, Clemente T, Upadhyaya HD, Hash CT, et al. The evolution of photoperiod-insensitive flowering in sorghum, a genomic model for panicoid grasses. Mol Biol Evol. 2016:**33**(9):2417–2428. https://doi.org/10.1093/molbev/msw120

Darwin C. On the origin of species. London: Murray; 1859.

Davey JW, Hohenlohe PA, Etter PD, Boone JQ, Catchen JM, Blaxter ML. Genome-wide genetic marker discovery and genotyping using next-generation sequencing. Nat Rev Genet. 2011:**12**(7):499–510. https://doi.org/10.1038/nrg3012

Deng L, Zhou Q, Zhou J, Zhang Q, Jia Z, Zhu G, Cheng S, Cheng L, Yin C, Yang C, et al. 3D organization of regulatory elements for transcriptional regulation in Arabidopsis. Genome Biol. 2023:**24**(1):181. https://doi.org/10.1186/s13059-023-03018-4

D'Hont A, Denoeud F, Aury JM, Baurens FC, Carreel F, Garsmeur O, Noel B, Bocs S, Droc G, Rouard M, et al. The banana (Musa acuminata) genome and the evolution of monocotyledonous plants. Nature. 2012:**488**(7410):213–217. https://doi.org/10.1038/nature11241

Doğan ES, Liu C. Three-dimensional chromatin packing and positioning of plant genomes. Nat Plants. 2018:**4**(8):521–529. https://doi.org/10.1038/s41477-018-0199-5

Domb K, Wang N, Hummel G, Liu C. Spatial features and functional implications of plant 3D genome organization. Annu Rev Plant Biol. 2022:**73**(1):173–200. https://doi.org/10.1146/annurev-arplant-102720-022810

Dong P, Tu X, Chu PY, Lü P, Zhu N, Grierson D, Du B, Li P, Zhong S. 3D chromatin architecture of large plant genomes determined by local A/B compartments. Mol Plant. 2017:**10**(12):1497–1509. https://doi.org/10.1016/j.molp.2017.11.005

Dorrity MW, Alexandre CM, Hamm MO, Vigil AL, Fields S, Queitsch C, Cuperus JT. The regulatory landscape of Arabidopsis thaliana roots at single-cell resolution. Nat Commun. 2021:**12**(1):3334. https://doi.org/10.1038/s41467-021-23675-y

ENCODE Project Consortium. An integrated encyclopedia of DNA elements in the human genome. Nature. 2012:**489**(7414):57–74. https://doi.org/10.1038/nature11247

Erhard KF Jr., Talbot JE, Deans NC, McClish AE, Hollick JB. Nascent transcription affected by RNA polymerase IV in Zea mays. Genetics. 2015:**199**(4):1107–1125. https://doi.org/10.1534/genetics.115.174714

Ezawa K, Oota S, Saitou N. Proceedings of the SMBE Tri-National Young Investigators' Workshop 2005. Genome-wide search of gene conversions in duplicated genes of mouse and rat. Mol Biol Evol. 2006:**23**(5):927–940. https://doi.org/10.1093/molbev/msj093

Feldman M, Liu B, Segal G, Abbo S, Levy AA, Vega JM. Rapid elimination of low-copy DNA sequences in polyploid wheat: a possible mechanism for differentiation of homoeologous chromosomes. Genetics. 1997:**147**(3):1381–1387. https://doi.org/10.1093/genetics/147.3.1381

Feldmann KA, Marks MD, Christianson ML, Quatrano RS. A dwarf mutant of arabidopsis generated by T-Dna insertion mutagenesis. Science. 1989:**243**(4896):1351–1354. https://doi.org/10.1126/science.243.4896.1351

Fernández L, de Haro LA, Distefano AJ, Carolina Martínez M, Lía V, Papa JC, Olea I, Tosto D, Esteban Hopp H. Population genetics structure of glyphosate-resistant Johnsongrass (Sorghum halepense L. Pers) does not support a single origin of the resistance. Ecol Evol. 2013:**3**:3388–3400. https://doi.org/10.1002/ece3.67

Fladung M. Transposon activation tagging in plants for gene function discovery. In: **Botany PI, Luttge U, Canovas F, Matyssek R**, editors. Progress in Botany. vol. 77. Switzerland: Springer Cham; 2016. p. 265–289.

Flagel LE, Wendel JF. Gene duplication and evolutionary novelty in plants. New Phytol. 2009:**183**(3):557–564. https://doi.org/10.1111/j.1469-8137.2009.02923.x

Fontdevila A. Hybrid genome evolution by transposition. Cytogenet Genome Res. 2005:**110**(1–4):49–55. https://doi.org/10.1159/000084937

Force A, Cresko WA, Pickett FB, Proulx SR, Amemiya C, Lynch M. The origin of subfunctions and modular gene regulation. Genetics. 2005:**170**(1):433–446. https://doi.org/10.1534/genetics.104.027607

Force A, Lynch M, Pickett FB, Amores A, Yan YL, Postlethwait J. Preservation of duplicate genes by complementary, degenerative mutations. Genetics. 1999:**151**(4):1531–1545. https://doi.org/10.1093/genetics/151.4.1531

Freeling M. Bias in plant gene content following different sorts of duplication: tandem, whole-genome, segmental, or by transposition. Ann Rev Plant Biol. 2009:**60**(1):433–453. https://doi.org/10.1146/annurev.arplant.043008.092122

Gaillochet C, Lohmann JU. The never-ending story: from pluripotency to plant developmental plasticity. Development. 2015:**142**(13): 2237–2249. https://doi.org/10.1242/dev.117614

Galtier N. Gene conversion drives GC content evolution in mammalian histones. Trends Genet. 2003:**19**(2):65–68. https://doi.org/10.1016/S0168-9525(02)00002-1

Gao LZ, Innan H. Very low gene duplication rate in the yeast genome. Science. 2004:**306**(5700):1367–1370. https://doi.org/10.1126/science.1102033

Gasperini M, Hill AJ, McFaline-Figueroa JL, Martin B, Kim S, Zhang MD, Jackson D, Leith A, Schreiber J, Noble WS, et al. A genome-wide framework for mapping gene regulation via cellular genetic screens. Cell. 2019:**176**(1–2):377–390.e319. https://doi.org/10.1016/j.cell.2018.11.029

Goff SA, Ricke D, Lan TH, Presting G, Wang RL, Dunn M, Glazebrook J, Sessions A, Oeller P, Varma H, et al. A draft sequence of the rice genome (Oryza sativa L. ssp japonica). Science. 2002:**296**(5565): 92–100. https://doi.org/10.1126/science.1068275

Gu ZL, Steinmetz LM, Gu X, Scharfe C, Davis RW, Li WH. Role of duplicate genes in genetic robustness against null mutations. Nature. 2003:**421**(6918):63–66. https://doi.org/10.1038/nature01198

Guo H, Jiao Y, Tan X, Wang X, Huang X, Jin H, Paterson AH. Gene duplication and genetic innovation in cereal genomes. Genome Res. 2019:**29**(2):261–269. https://doi.org/10.1101/gr.237511.118

Guo H, Wang X, Gundlach H, Mayer KFX, Peterson DG, Scheffler BE, Chee PW, Paterson AH. Extensive and biased intergenomic nonreciprocal DNA exchanges shaped a nascent polyploid genome, gossypium (cotton). Genetics. 2014:**197**(4):1153–1163. https://doi.org/10.1534/genetics.114.166124

Harrison LJ, Bose D. Enhancer RNAs step forward: new insights into enhancer function. Development. 2022:**149**(16):dev200398. https://doi.org/10.1242/dev.200398

Heckmann D, Schulze S, Denton A, Westhoff P, Weber APM, Lercher MJ. Predicting C4 photosynthesis evolution: modular, individually adaptive steps on a mount fuji fitness landscape. Cell. 2013:**153**(7):1579–1588. https://doi.org/10.1016/j.cell.2013.04.058

Heger P, Wiehe T. New tools in the box: an evolutionary synopsis of chromatin insulators. Trends Genet. 2014:**30**(5):161–171. https://doi.org/10.1016/j.tig.2014.03.004

Henikoff S, Till BJ, Comai L. TILLING. Traditional mutagenesis meets functional genomics. Plant Physiol. 2004:**135**(2):630–636. https://doi.org/10.1104/pp.104.041061

Hetzel J, Duttke SH, Benner C, Chory J. Nascent RNA sequencing reveals distinct features in plant transcription. Proc Natl Acad Sci U S A. 2016:**113**(43):12316–12321. https://doi.org/10.1073/pnas.1603217113

Holliday R. Mechanism for gene conversion in fungi. Genet Res. 1964:**5**(2):282–304. https://doi.org/10.1017/S0016672300001233

Holliday R. Studies on mitotic gene conversion in ustilago. Genet Res. 1966:**8**(3):323–337. https://doi.org/10.1017/S0016672300010181

Howe HF, Smallwood J. Ecology of seed dispersal. Annu Rev Ecol Syst. 1982:**13**(1):201–228. https://doi.org/10.1146/annurev.es.13.110182.001221

Jaillon O, Aury JM, Noel B, Policriti A, Clepet C, Casagrande A, Choisne N, Aubourg S, Vitulo N, Jubin C, et al. The grapevine genome sequence suggests ancestral hexaploidization in major angiosperm phyla. Nature. 2007:**449**(7161):463–467. https://doi.org/10.1038/nature06148

Jean-Baptiste K, McFaline-Figueroa JL, Alexandre CM, Dorrity MW, Saunders L, Bubb KL, Trapnell C, Fields S, Queitsch C, Cuperus JT. Dynamics of gene expression in single root cells of Arabidopsis thaliana. Plant Cell. 2019:**31**(5):993–1011. https://doi.org/10.1105/tpc.18.00785

Jiang CX, Wright RJ, El-Zik KM, Paterson AH. Polyploid formation created unique avenues for response to selection in Gossypium (cotton). Proc Natl Acad Sci U S A. 1998:**95**(8):4419–4424. https://doi.org/10.1073/pnas.95.8.4419

Jiao Y, Li J, Tang H, Paterson AH. Integrated syntenic and phylogenomic analyses reveal a new ancient genome duplication in monocots. Plant Cell. 2014:**26**(7):2792–2802. https://doi.org/10.1105/tpc.114.127597

Jiao Y, Wickett NJ, Ayyampalayam S, Chanderbali AS, Landherr L, Ralph PE, Tomsho LP, Hu Y, Liang H, Soltis PS, et al. Ancestral polyploidy in seed plants and angiosperms. Nature. 2011:**473**(7345): 97–100. https://doi.org/10.1038/nature09916

Jores T, Hamm M, Cuperus JT, Queitsch C. Frontiers and techniques in plant gene regulation. Curr Opin Plant Biol. 2023:**75**:102403. https://doi.org/10.1016/j.pbi.2023.102403

Jores T, Tonnies J, Dorrity MW, Cuperus JT, Fields S, Queitsch C. Identification of plant enhancers and their constituent elements by STARR-seq in tobacco leaves. Plant Cell. 2020:**32**(7):2120–2131. https://doi.org/10.1105/tpc.20.00155

Jores T, Tonnies J, Wrightsman T, Buckler ES, Cuperus JT, Fields S, Queitsch C. Synthetic promoter designs enabled by a comprehensive analysis of plant core promoters. Nat Plants. 2021:**7**(6):842–855. https://doi.org/10.1038/s41477-021-00932-y

Kashkush K, Feldman M, Levy AA. Gene loss, silencing and activation in a newly synthesized wheat allotetraploid. Genetics. 2002:**160**(4): 1651–1659. https://doi.org/10.1093/genetics/160.4.1651

Kashkush K, Feldman M, Levy AA. Transcriptional activation of retrotransposons alters the expression of adjacent genes in wheat. Nat Genet. 2003:**33**(1):102–106. https://doi.org/10.1038/ng1063

Kiedrzynski M, Zielinska KM, Jedrzecczyk I, Kiedrzynska E, Toomczyk PP, Rewicz A, Rewers M, Indreica A, Bednarska I, Stupar V, et al. Tetraploids expanded beyond the mountain niche of their diploid ancestors in the mixed-ploidy grass Festuca amethystina L. Sci Rep. 2021:**11**(1):18735. https://doi.org/10.1038/s41598-021-97767-6

Kim C, Wang X, Lee T-H, Jakob K, Lee G-J, Paterson AH. Comparative analysis of two saccharinae genera, miscanthus and saccharum reveals a shared whole-genome duplication but different evolutionary fate. Plant Cell. 2014:**26**(6):2420–2429. https://doi.org/10.1105/tpc.114.125583

Kim TK, Hemberg M, Gray JM, Costa AM, Bear DM, Wu J, Harmin DA, Laptewicz M, Barbara-Haley K, Kuersten S, et al. Widespread transcription at neuronal activity-regulated enhancers. Nature. 2010:**465**(7295):182–187. https://doi.org/10.1038/nature09033

Kishimoto N, Higo H, Abe K, Arai S, Saito A, Higo K. Identification of the duplicated segments in rice chromosomes 1 and 5 by linkage analysis of cDNA markers of known functions. Theor Appl Genet. 1994:**88**(6–7):722–726. https://doi.org/10.1007/BF01253976

Klein JC, Agarwal V, Inoue F, Keith A, Martin B, Kircher M, Ahituv N, Shendure J. A systematic evaluation of the design and context dependencies of massively parallel reporter assays. Nat Methods. 2020:**17**(11):1083–1091. https://doi.org/10.1038/s41592-020-0965-y

Kong W, Liu M, Felker P, Ewens M, Bessega C, Pometti C, Wang J, Peng X, Teng J, Wang J, et al. Genome and evolution of Prosopis alba Griseb., a drought and salinity tolerant tree legume crop for arid climates. Plants People Planet. 2023:**5**(6):933–947. https://doi.org/10.1002/ppp3.10404

Kong Y, Cao L, Deikus G, Fan Y, Mead EA, Lai W, Zhang Y, Yong R, Sebra R, Wang H, et al. Critical assessment of DNA adenine methylation in eukaryotes using quantitative deconvolution. Science. 2022:**375**(6580):515–522. https://doi.org/10.1126/science.abe7489

Kowalski S, Lan T-H, Feldmann K, Paterson A. Comparative mapping of *Arabidopsis thaliana* and *Brassica oleracea* chromosomes reveals islands of conserved gene order. Genetics. 1994:**138**(2):499–510. https://doi.org/10.1093/genetics/138.2.499

Lander E, Botstein D. Mapping Mendelian factors underlying quantitative traits using RFLP linkage maps. Genetics. 1989:**121**(1): 185–199. https://doi.org/10.1093/genetics/121.1.185

Lander ES, Linton LM, Birren B, Nusbaum C, Zody MC, Baldwin J, Devon K, Dewar K, Doyle M, FitzHugh W, et al. Initial sequencing and analysis of the human genome. Nature. 2001:**409**(6822):860–921. https://doi.org/10.1038/35057062

Lang L, Schnittger A. Endoreplication—a means to an end in cell growth and stress response. Curr Opin Plant Biol. 2020:**54**:85–92. https://doi.org/10.1016/j.pbi.2020.02.006

Langham RJ, Walsh J, Dunn M, Ko C, Goff SA, Freeling M. Genomic duplication, fractionation and the origin of regulatory novelty. Genetics. 2004:**166**(2):935–945. https://doi.org/10.1093/genetics/166.2.935

Lawrence WJC. The secondary association of chromosomes. Cytologia (Tokyo). 1931:**2**(4):352–384. https://doi.org/10.1508/cytologia.2.352

Lee HS, Chen ZJ. Protein-coding genes are epigenetically regulated in Arabidopsis polyploids. Proc Natl Acad Sci U S A. 2001:**98**(12):6753–6758. https://doi.org/10.1073/pnas.121064698

Levings PP, Bungert J. The human beta-globin locus control region. Eur J Biochem. 2002:**269**(6):1589–1599. https://doi.org/10.1046/j.1432-1327.2002.02797.x

Lewis DR, Olex AL, Lundy SR, Turkett WH, Fetrow JS, Muday GK. A kinetic analysis of the auxin transcriptome reveals cell wall remodeling proteins that modulate lateral root development in Arabidopsis. Plant Cell. 2013:**25**(9):3329–3346. https://doi.org/10.1105/tpc.113.114868

Li X, Lassner M, Zhang Y. Deleteagene: a fast neutron deletion mutagenesis-based gene knockout system for plants. Comp Funct Genomics. 2002:**3**(2):158–160. https://doi.org/10.1002/cfg.148

Li X, Song Y, Century K, Straight S, Ronald PC, Dong X, Lassner M, Zhang Y. A fast neutron deletion mutagenesis-based reverse genetics system for plants. Plant J. 2001:**27**(3):235–242. https://doi.org/10.1046/j.1365-313x.2001.01084.x

Lin Z, Li X, Shannon LM, Yeh C-T, Wang ML, Bai G, Peng Z, Li J, Trick HN, Clemente TE, et al. Parallel domestication of the Shattering1 genes in cereals. Nat Genet. 2012:**44**(6):720–724. https://doi.org/10.1038/ng.2281

Liu B, Vega JM, Feldman M. Rapid genomic changes in newly synthesized amphiploids of triticum and aegilops. II. changes in low-copy coding DNA sequences. Genome. 1998:**41**(4):535–542. https://doi.org/10.1139/g98-052

Liu C, Cheng YJ, Wang JW, Weigel D. Prominent topologically associated domains differentiate global chromatin packing in rice from Arabidopsis. Nat Plants. 2017:**3**(9):742–748. https://doi.org/10.1038/s41477-017-0005-9

Liu S, Liu Y, Yang Y, Tong C, Edwards D, Parkin I, Zhao M, Ma J, Yu J, Huang S, et al. The Brassica oleracea genome reveals the asymmetrical evolution of polyploid genomes. Nat Commun. 2014:**5**(1):3930. https://doi.org/10.1038/ncomms4930

Lozano R, Booth GT, Omar BY, Li B, Buckler ES, Lis JT, Del Carpio DP, Jannink JL. RNA polymerase mapping in plants identifies intergenic regulatory elements enriched in causal variants. G3 (Bethesda). 2021:**11**(11):jkab273. https://doi.org/10.1093/g3journal/jkab273

Lupiáñez DG, Kraft K, Heinrich V, Krawitz P, Brancati F, Klopocki E, Horn D, Kayserili H, Opitz JM, Laxova R, et al. Disruptions of topological chromatin domains cause pathogenic rewiring of gene-enhancer interactions. Cell. 2015:**161**(5):1012–1025. https://doi.org/10.1016/j.cell.2015.04.004

Lynch M, Conery JS. The evolutionary fate and consequences of duplicate genes. Science. 2000:**290**(5494):1151–1155. https://doi.org/10.1126/science.290.5494.1151

Lynch M, Force A. The probability of duplicate gene preservation by subfunctionalization. Genetics. 2000:**154**(1):459–473. https://doi.org/10.1093/genetics/154.1.459

Lynch VJ, Nnamani MC, Kapusta A, Brayer K, Plaza SL, Mazur EC, Emera D, Sheikh SZ, Grützner F, Bauersachs S, et al. Ancient transposable elements transformed the uterine regulatory landscape and transcriptome during the evolution of mammalian pregnancy. Cell Rep. 2015:**10**(4):551–561. https://doi.org/10.1016/j.celrep.2014.12.052

Maere S, De Bodt S, Raes J, Casneuf T, Van Montagu M, Kuiper M, Van de Peer Y. Modeling gene and genome duplications in eukaryotes. Proc Natl Acad Sci U S A. 2005:**102**(15):5454–5459. https://doi.org/10.1073/pnas.0501102102

Marand AP, Chen Z, Gallavotti A, Schmitz RJ. A cis-regulatory atlas in maize at single-cell resolution. Cell. 2021:**184**(11):3041–3055.e3021. https://doi.org/10.1016/j.cell.2021.04.014

Mascher M, Gundlach H, Himmelbach A, Beier S, Twardziok SO, Wicker T, Radchuk V, Dockter C, Hedley PE, Russell J, et al. A chromosome conformation capture ordered sequence of the barley genome. Nature. 2017:**544**(7651):427–433. https://doi.org/10.1038/nature22043

Maurano MT, Humbert R, Rynes E, Thurman RE, Haugen E, Wang H, Reynolds AP, Sandstrom R, Qu H, Brody J, et al. Systematic localization of common disease-associated variation in regulatory DNA. Science. 2012:**337**(6099):1190–1195. https://doi.org/10.1126/science.1222794

McCallum CM, Comai L, Greene EA, Henikoff S. Targeting induced local lesions in genomes (TILLING) for plant functional genomics. Plant Physiol. 2000:**123**(2):439–442. https://doi.org/10.1104/pp.123.2.439

McCann MC, Roberts K. Architecture of the primary cell wall. In: **Lloyd CW**, editor. The cytoskeletal basis of plant growth and form. New York: Academic Press; 1991. p. 109–129.

McClintock B. The significance of responses of the genome to challenge. Science. 1984:**226**(4676):792–801. https://doi.org/10.1126/science.15739260

McDonald BR, Picard C, Brabb IM, Savenkova MI, Schmitz RJ, Jacobsen SE, and Duttke SH. Enhancers associated with unstable RNAs are rare in plants. bioRxiv 2023.09.25.559415. https://doi.org/10.1101/2023.09.25.559415, 2023, preprint: not peer reviewed.

Molin WT, Yaguchi A, Blenner M, Saski CA. The EccDNA replicon: a heritable, extranuclear vehicle that enables gene amplification and glyphosate resistance in amaranthus palmeri. Plant Cell. 2020:**32**(7):2132–2140. https://doi.org/10.1105/tpc.20.00099

Mondragon-Palomino M, Gaut BS. Gene conversion and the evolution of three leucine-rich repeat gene families in Arabidopsis thaliana. Mol Biol Evol. 2005:**22**(12):2444–2456. https://doi.org/10.1093/molbev/msi241

Montavon T, Duboule D. Chromatin organization and global regulation of Hox gene clusters. Philos Trans R Soc Lond B Biol Sci. 2013:**368**(1620):20120367. https://doi.org/10.1098/rstb.2012.0367

Moore JE, Purcaro MJ, Pratt HE, Epstein CB, Shoresh N, Adrian J, Kawli T, Davis CA, Dobin A, Kaul R, et al. Expanded encyclopaedias of DNA elements in the human and mouse genomes. Nature. 2020:**583**(7818):699–710. https://doi.org/10.1038/s41586-020-2493-4

Morris GP, Ramu P, Deshpande SP, Hash CT, Shah T, Upadhyaya HD, Riera-Lizarazu O, Brown PJ, Acharya CB, Mitchell SE, et al. Population genomic and genome-wide association studies of agroclimatic traits in sorghum. Proc Natl Acad Sci U S A. 2013:**110**(2):453–458. https://doi.org/10.1073/pnas.1215985110

Nagamura Y, Inoue T, Antonio B, Shimano T, Kajiya H, Shomura A, Lin S, Kuboki Y, Harushima Y, Kurata N, et al. Conservation of duplicated segments between rice chromosomes 11 and 12. Breed Sci. 1995:**45**:373–376. https://doi.org/10.1270/JSBBS1951.45.373

Ohno S. Evolution by gene duplication. Berlin: Springer; 1970.

Ohta T. Some models of gene conversion for treating the evolution of multigene families. Genetics. 1984:**106**(3):517–528. https://doi.org/10.1093/genetics/106.3.517

Okhovat M, VanCampen J, Nevonen KA, Harshman L, Li W, Layman CE, Ward S, Herrera J, Wells J, Sheng RR, et al. TAD evolutionary and functional characterization reveals diversity in mammalian TAD boundary properties and function. Nat Commun. 2023:**14**(1):8111. https://doi.org/10.1038/s41467-023-43841-8

O'Neill RJW, O'Neill MJ, Graves JAM. Undermethylation associated with retroelement activation and chromosome remodelling in an interspecific mammalian hybrid. Nature. 2002:**420**:106–106. https://doi.org/10.1038/nature01162

Ozaki K, Ohnishi Y, Iida A, Yamada R, Tsunoda T, Sato H, Sato H, Hori M, Nakamura Y, Tanaka T. Functional SNPs in the lymphotoxin-α gene that are associated with susceptibility to myocardial infarction. Nat Genet. 2002:**32**(4):650–654. https://doi.org/10.1038/ng1047

Ozkan H, Levy AA, Feldman M. Allopolyploidy-induced rapid genome evolution in the wheat (Aegilops-Triticum) group. Plant Cell. 2001:**13**(8):1735–1747. https://doi.org/10.1105/tpc.010082

Ozkan H, Levy AA, Feldman M. Rapid differentiation of homeologous chromosomes in newly-formed allopolyploid wheat. Isr J Plant Sci. 2002:**50**(1):S65–S76. https://doi.org/10.1560/E282-PV55-G4XT-DRWJ

Paterson A, Bowers J, Burow M, Draye X, Elsik C, Jiang C, Katsar C, Lan T, Lin Y, Ming R, et al. Comparative genomics of plant chromosomes. Plant Cell. 2000:**12**(9):1523–1539. https://doi.org/10.1105/tpc.12.9.1523

Paterson AH, Bowers JE, Bruggmann R, Dubchak I, Grimwood J, Gundlach H, Haberer G, Hellsten U, Mitros T, Poliakov A, et al. The Sorghum bicolor genome and the diversification of grasses. Nature. 2009:**457**(7229):551–556. https://doi.org/10.1038/nature07723

Paterson AH, Bowers JE, Chapman BA. Ancient polyploidization predating divergence of the cereals, and its consequences for comparative genomics. Proc Natl Acad Sci U S A. 2004:**101**(26):9903–9908. https://doi.org/10.1073/pnas.0307901101

Paterson AH, Chapman BA, Kissinger JC, Bowers JE, Feltus FA, Estill JC. Many gene and domain families have convergent fates following independent whole-genome duplication events in Arabidopsis, Oryza, Saccharomyces and Tetraodon. Trends Genet. 2006:**22**(11):597–602. https://doi.org/10.1016/j.tig.2006.09.003

Paterson AH, Kong W, Johnston RM, Nabukalu P, Wu G, Poehlman W, Goff VH, Isaacs K, Lee T-H, Guo H, et al. The evolution of an invasive plant, Sorghum halepense L. ('johnsongrass'). Front Genet. 2020:**11**:317. https://doi.org/10.3389/fgene.2020.00317

Paterson AH, Lan TH, Reischmann KP, Chang C, Lin YR, Liu SC, Burow MD, Kowalski SP, Katsar CS, DelMonte TA, et al. Toward a unified genetic map of higher plants, transcending the monocot-dicot divergence. Nat Genet. 1996:**14**(4):380–382. https://doi.org/10.1038/ng1296-380

Paterson AH, Lin YR, Li ZK, Schertz KF, Doebley JF, Pinson SRM, Liu SC, Stansel JW, Irvine JE. Convergent domestication of cereal crops by independent mutations at corresponding genetic loci. Science. 1995:**269**(5231):1714–1718. https://doi.org/10.1126/science.269.5231.1714

Paterson AH, Wendel J, Gundlach F, Guo H, Jenkins H, Jin J, Llewellyn D, Showmaker D, Shu KC, Udall S, et al. Repeated polyploidization of Gossypium genomes and the evolution of spinnable cotton fibres. Nature. 2012:**492**(7429):423–427. https://doi.org/10.1038/nature11798

Patterson EL, Pettinga DJ, Ravet K, Neve P, Gaines TA. Glyphosate resistance and EPSPS gene duplication: convergent evolution in multiple plant species. J Hered. 2018:**109**(2):117–125. https://doi.org/10.1093/jhered/esx087

Patwardhan RP, Lee C, Litvin O, Young DL, Pe'er D, Shendure J. High-resolution analysis of DNA regulatory elements by synthetic saturation mutagenesis. Nat Biotechnol. 2009:**27**(12):1173–1175. https://doi.org/10.1038/nbt.1589

Petes TD, Symington LS. The molecular and cellular biology of the yeast Saccharomyces: genome dynamics, protein synthesis and energetics. In: **Broach JJ, Pringle J**, editors. Recombination in yeast. Cold Spring Harbor (NY): Cold Spring Harbor Press; 1991.

Pires JC, Zhao JW, Schranz ME, Leon EJ, Quijada PA, Lukens LN, Osborn TC. Flowering time divergence and genomic rearrangements in resynthesized Brassica polyploids (Brassicaceae). Biol J Linn Soc. 2004:**82**(4):675–688. https://doi.org/10.1111/j.1095-8312.2004.00350.x

Pliner HA, Packer JS, McFaline-Figueroa JL, Cusanovich DA, Daza RM, Aghamirzaie D, Srivatsan S, Qiu X, Jackson D, Minkina A, et al. Cicero predicts cis-regulatory DNA interactions from single-cell chromatin accessibility data. Mol Cell. 2018:**71**(5):858–871.e858. https://doi.org/10.1016/j.molcel.2018.06.044

Raven P, Evert R, Eichhorn S. Biology of plants. New York: Worth Publishers, Inc.; 1992.

Ricci WA, Lu Z, Ji L, Marand AP, Ethridge CL, Murphy NG, Noshay JM, Galli M, Mejía-Guerra MK, Colomé-Tatché M, et al. Widespread long-range cis-regulatory elements in the maize genome. Nat Plants. 2019:**5**(12):1237–1249. https://doi.org/10.1038/s41477-019-0547-0

Rieseberg LH. Chromosomal rearrangements and speciation. Trends Ecol Evol. 2001:**16**(7):351–358. https://doi.org/10.1016/S0169-5347(01)02187-5

Rodgers-Melnick E, Vera DL, Bass HW, Buckler ES. Open chromatin reveals the functional maize genome. Proc Natl Acad Sci U S A. 2016:**113**(22):E3177-3184. https://doi.org/10.1073/pnas.1525244113

Rodríguez-Leal D, Lemmon ZH, Man J, Bartlett ME, Lippman ZB. Engineering quantitative trait variation for crop improvement by genome editing. Cell. 2017:**171**(2):470–480.e478. https://doi.org/10.1016/j.cell.2017.08.030

Rogers JJW, Santosh M. Continents and supercontinents. Oxford: Oxford University Press; 2004.

Sanyal A, Lajoie BR, Jain G, Dekker J. The long-range interaction landscape of gene promoters. Nature. 2012:**489**(7414):109–113. https://doi.org/10.1038/nature11279

Sato S, Tabata S, Hirakawa H, Asamizu E, Shirasawa K, Isobe S, Kaneko T, Nakamura Y, Shibata D, Aoki K, et al. The tomato genome sequence provides insights into fleshy fruit evolution. Nature. 2012:**485**(7400):635–641. https://doi.org/10.1038/nature11119

Sawyer S. Statistical tests for detecting gene conversion. Mol Biol Evol. 1989:**6**(5):526–538. https://doi.org/10.1093/oxfordjournals.molbev.a040567

Schluter U, Weber APM. Regulation and evolution of C4 photosynthesis. Annu Rev Plant Biol. 2020:**71**(1):183–215. https://doi.org/10.1146/annurev-arplant-042916-040915

Schmitz RJ, Grotewold E, Stam M. Cis-regulatory sequences in plants: their importance, discovery, and future challenges. Plant Cell. 2022:**34**(2):718–741. https://doi.org/10.1093/plcell/koab281

Schmutz J, Cannon SB, Schlueter J, Ma JX, Mitros T, Nelson W, Hyten DL, Song QJ, Thelen JJ, Cheng JL, et al. Genome sequence of the palaeopolyploid soybean. Nature. 2010:**463**(7278):178–183. https://doi.org/10.1038/nature08670

Schnable JC, Springer N, Freeling M. Biased gene loss following tetraploidy in maize reflects genome dominance and ongoing selection. Proc Natl Acad Sci U S A. 2010:**108**(10):4069–4074. https://doi.org/10.1073/pnas.1101368108

Schranz ME, Osborn TC. Novel flowering time variation in the resynthesized polyploid Brassica napus. J Hered. 2000:**91**(3):242–246. https://doi.org/10.1093/jhered/91.3.242

Schranz ME, Osborn TC. De novo variation in life-history traits and responses to growth conditions of resynthesized polyploid Brassica napus (Brassicaceae). Am J Bot. 2004:**91**(2):174–183. https://doi.org/10.3732/ajb.91.2.174

Semagn K, Babu R, Hearne S, Olsen M. Single nucleotide polymorphism genotyping using kompetitive allele specific PCR (KASP): overview of the technology and its application in crop improvement. Mol Breed. 2013:**33**(1):1–14. https://doi.org/10.1007/s11032-013-9917-x

Semple C, Wolfe KH. Gene duplication and gene conversion in the caenorhabditis elegans genome. J Mol Evol. 1999:**48**(5):555–564. https://doi.org/10.1007/PL00006498

Shaked H, Kashkush K, Ozkan H, Feldman M, Levy AA. Sequence elimination and cytosine methylation are rapid and reproducible responses of the genome to wide hybridization and allopolyploidy in wheat. Plant Cell. 2001:**13**(8):1749–1759. https://doi.org/10.1105/TPC.010083

Shen C, Du H, Chen Z, Lu H, Zhu F, Chen H, Meng X, Liu Q, Liu P, Zheng L, et al. The chromosome-level genome sequence of the autotetraploid alfalfa and resequencing of core germplasms provide genomic resources for alfalfa research. Mol Plant. 2020:**13**(9): 1250–1261. https://doi.org/10.1016/j.molp.2020.07.003

Shiu SH, Shih MC, Li WH. Transcription factor families have much higher expansion rates in plants than in animals. Plant Physiol. 2005:**139**(1):18–26. https://doi.org/10.1104/pp.105.065110

Singh R, Ong-Abdullah M, Low ET, Manaf MA, Rosli R, Nookiah R, Ooi LC, Ooi SE, Chan KL, Halim MA, et al. Oil palm genome sequence reveals divergence of interfertile species in old and new worlds. Nature. 2013:**500**(7462):335–339. https://doi.org/10.1038/nature12309

Song KM, Lu P, Tang KL, Osborn TC. Rapid genome change in synthetic polyploids of brassica and its implications for polyploid evolution. Proc Natl Acad Sci U S A. 1995:**92**(17):7719–7723. https://doi.org/10.1073/pnas.92.17.7719

Song Q, Chen ZJ. Epigenetic and developmental regulation in plant polyploids. Curr Opin Plant Biol. 2015:**24**:101–109. https://doi.org/10.1016/j.pbi.2015.02.007

Stebbins G. Chromosomal variation and evolution; polyploidy and chromosome size and number shed light on evolutionary processes in higher plants. Science. 1966:**152**(3728):1463–1469. https://doi.org/10.1126/science.152.3728.1463

Stebbins GL. A brief summary of my ideas on evolution. Am J Bot. 1999:**86**(8):1207–1208. https://doi.org/10.2307/2656985

Stephens S. Possible significance of duplications in evolution. Adv Genet. 1951:**4**:247–265. https://doi.org/10.1016/S0065-2660(08)60237-0

Stergachis AB, Debo BM, Haugen E, Churchman LS, Stamatoyannopoulos JA. Single-molecule regulatory architectures captured by chromatin fiber sequencing. Science. 2020:**368**(6498): 1449–1454. https://doi.org/10.1126/science.aaz1646

Stergachis AB, Neph S, Reynolds A, Humbert R, Miller B, Paige SL, Vernot B, Cheng JB, Thurman RE, Sandstrom R, et al. Developmental fate and cellular maturity encoded in human regulatory DNA landscapes. Cell. 2013:**154**(4):888–903. https://doi.org/10.1016/j.cell.2013.07.020

Studer A, Zhao Q, Ross-Ibarra J, Doebley J. Identification of a functional transposon insertion in the maize domestication gene tb1. Nat Genet. 2011:**43**(11):1160–1163. https://doi.org/10.1038/ng.942

Sullivan AM, Arsovski AA, Lempe J, Bubb KL, Weirauch MT, Sabo PJ, Sandstrom R, Thurman RE, Neph S, Reynolds AP, et al. Mapping and dynamics of regulatory DNA and transcription factor networks in A. thaliana. Cell Rep. 2014:**8**(6):2015–2030. https://doi.org/10.1016/j.celrep.2014.08.019

Sullivan AM, Arsovski AA, Thompson A, Sandstrom R, Thurman RE, Neph S, Johnson AK, Sullivan ST, Sabo PJ, Neri FV, et al. Mapping and dynamics of regulatory DNA in maturing arabidopsis thaliana siliques. Front Plant Sci. 2019:**10**:1434. https://doi.org/10.3389/fpls.2019.01434

Swigonova Z, Lai JS, Ma JX, Ramakrishna W, Llaca V, Bennetzen JL, Messing J. Close split of sorghum and maize genome progenitors. Genome Res. 2004:**14**(10a):1916–1923. https://doi.org/10.1101/gr.2332504

Tan Y, Yan X, Sun J, Wan J, Li X, Huang Y, Li L, Niu L, Hou C. Genome-wide enhancer identification by massively parallel reporter assay in Arabidopsis. Plant J. 2023:**116**(1):234–250. https://doi.org/10.1111/tpj.16373

Tang H, Bowers JE, Wang X, Ming R, Alam M, Paterson AH. Synteny and colinearity in plant genomes. Science. 2008:**320**(5875):486–488. https://doi.org/10.1126/science.1153917

Tang H, Cuevas HE, Das S, Sezen UU, Zhou C, Guo H, Goff VH, Ge Z, Clemente TE, Paterson AH. Seed shattering in a wild sorghum is conferred by a locus unrelated to domestication. Proc Natl Acad Sci U S A. 2013:**110**(39):15824–15829. https://doi.org/10.1073/pnas.1305213110

Taylor JS, Raes J. Duplication and divergence: the evolution of new genes and old ideas. Ann Rev Genet. 2004:**38**(1):615–643. https://doi.org/10.1146/annurev.genet.38.072902.092831

Thieffry A, Vigh ML, Bornholdt J, Ivanov M, Brodersen P, Sandelin A. Characterization of Arabidopsis thaliana promoter bidirectionality and antisense RNAs by inactivation of nuclear RNA decay pathways. Plant Cell. 2020:**32**(6):1845–1867. https://doi.org/10.1105/tpc.19.00815

Thomas S, Li XY, Sabo PJ, Sandstrom R, Thurman RE, Canfield TK, Giste E, Fisher W, Hammonds A, Celniker SE, et al. Dynamic reprogramming of chromatin accessibility during Drosophila embryo development. Genome Biol. 2011:**12**(5):R43. https://doi.org/10.1186/gb-2011-12-5-r43

Thurman RE, Rynes E, Humbert R, Vierstra J, Maurano MT, Haugen E, Sheffield NC, Stergachis AB, Wang H, Vernot B, et al. The accessible chromatin landscape of the human genome. Nature. 2012:**489**(7414):75–82. https://doi.org/10.1038/nature11232

Till BJ, Reynolds SH, Greene EA, Codomo CA, Enns LC, Johnson JE, Burtner C, Odden AR, Young K, Taylor NE, et al. Large-scale discovery of induced point mutations with high-throughput TILLING. Genome Res. 2003:**13**(3):524–530. https://doi.org/10.1101/gr.977903

Tuzun E, Bailey JA, Eichler EE. Recent segmental duplications in the working draft assembly of the brown Norway rat. Genome Res. 2004:**14**(4):493–506. https://doi.org/10.1101/gr.1907504

van den Berg C, Willemsen V, Hage W, Weisbeek P, Scheres B. Cell fate in the Arabidopsis root meristem determined by directional signalling. Nature. 1995:**378**(6552):62–65. https://doi.org/10.1038/378062a0

Vavilov N. The law of homologous series in variation. J Genet. 1922:**12**(1):1. https://doi.org/10.1007/BF02983073

Vision TJ, Brown DG, Tanksley SD. The origins of genomic duplications in Arabidopsis. Science. 2000:**290**(5499):2114–2117. https://doi.org/10.1126/science.290.5499.2114

Wang X, Gowik U, Tang H, Bowers JE, Westhoff P, Paterson AH. Comparative genomic analysis of C4 photosynthetic pathway evolution in grasses. Genome Biol. 2009:**10**(6):R68. https://doi.org/10.1186/gb-2009-10-6-r68

Wang X, Guo H, Wang J, Liu T, Lei T, Lee T-H, Li J, Tang H, Paterson AH. Comparative genomic de-convolution of the cotton genome revealed a decaploid ancestor and widespread chromosomal fractionation. New Phytol. 2016:**209**(3):1252–1263. https://doi.org/10.1111/nph.13689

Wang X, Tang H, Bowers JE, Feltus FA, Paterson AH. Extensive concerted evolution of rice paralogs and the road to regaining independence. Genetics. 2007:**177**(3):1753–1763. https://doi.org/10.1534/genetics.107.073197

Wang X, Tang H, Paterson AH. Seventy million years of concerted evolution of a homoeologous chromosome pair, in parallel in major Poaceae lineages. Plant Cell. 2011:**23**(1):27–37. https://doi.org/10.1105/tpc.110.080622

Wang X, Wang J, Guo H, Jin D, Lee T-H, Liu T, Paterson AH. Genome alignment spanning major Poaceae lineages reveals heterogeneous evolutionary rates and alters inferred dates for key evolutionary events. Mol Plant. 2015:**8**(6):885–898. https://doi.org/10.1016/j.molp.2015.04.004

Wang X, Wang Z, Guo H, Zhang L, Wang L, Li J, Jin D, Paterson AH. Telomere-centric genome repatterning determines recurring chromosome number reductions during the evolution of eukaryotes. New Phytol. 2014:**205**(1):378–389. https://doi.org/10.1111/nph.12985

Werth CR, Windham MD. A model for divergent, allopatric speciation of polyploid pteridophytes resulting from silencing of duplicate-gene expression. Am Nat. 1991:**137**(4):515–526. https://doi.org/10.1086/285180

White ME, Crowther BI. Gene conversions may obscure actin gene family relationships. J Mol Evol. 2000:**50**(2):170–174. https://doi.org/10.1007/s002399910018

Wu H, Nord AS, Akiyama JA, Shoukry M, Afzal V, Rubin EM, Pennacchio LA, Visel A. Tissue-specific RNA expression marks distant-acting developmental enhancers. PLoS Genet. 2014:**10**(9): e1004610. https://doi.org/10.1371/journal.pgen.1004610

Xi Q, Wang Z, Zaromytidou AI, Zhang XH, Chow-Tsang LF, Liu JX, Kim H, Barlas A, Manova-Todorova K, Kaartinen V, et al. A poised chromatin platform for TGF-β access to master regulators. Cell. 2011:**147**(7):1511–1524. https://doi.org/10.1016/j.cell.2011.11.032

Xu J-H, Messing J. Organization of the prolamin gene family provides insight into the evolution of the maize genome and gene duplications in grass species. Proc Natl Acad Sci USA. 2008:**105**(38):14330–14335. https://doi.org/10.1073/pnas.0807026105

Younghusband F. The epic of Mount Everest. London: E.P. Publishing; 1926.

Yu JM, Holland JB, McMullen MD, Buckler ES. Genetic design and statistical power of nested association mapping in maize. Genetics. 2008:**178**(1):539–551. https://doi.org/10.1534/genetics.107.074245

Yuan Q, Gao X. Multiplex base- and prime-editing with drive-and-process CRISPR arrays. Nat Commun. 2022:**13**(1):2771. https://doi.org/10.1038/s41467-022-30514-1

Zhang W, Zhang T, Wu Y, Jiang J. Genome-wide identification of regulatory DNA elements and protein-binding footprints using signatures of open chromatin in Arabidopsis. Plant Cell. 2012:**24**(7):2719–2731. https://doi.org/10.1105/tpc.112.098061

Zhuang W, Chen H, Yang M, Wang J, Pandey MK, Zhang C, Chang WC, Zhang L, Zhang X, Tang R, et al. The genome of cultivated peanut provides insight into legume karyotypes, polyploid evolution and crop domestication. Nat Genet. 2019:**51**(5):865–876. https://doi.org/10.1038/s41588-019-0402-2

Zou XH, Du YS, Tang L, Xu XW, Doyle JJ, Sang T, Ge S. Multiple origins of BBCC allopolyploid species in the rice genus (Oryza). Sci Rep. 2015:**5**(1):14876. https://doi.org/10.1038/srep14876

https://doi.org/10.1093/plcell/koad260

THE PLANT CELL

THE PLANT CELL 2024: 36: 1205–1226

Review

Complementing model species with model clades

Makenzie E. Mabry [1], R. Shawn Abrahams [2,3], Ihsan A. Al-Shehbaz [4], William J. Baker [5], Simon Barak [6], Michael S. Barker [7], Russell L. Barrett [8], Aleksandra Beric [9,10], Samik Bhattacharya [11], Sarah B. Carey [12], Gavin C. Conant [13], John G. Conran [14], Maheshi Dassanayake [15], Patrick P. Edger [16], Jocelyn C. Hall [17], Yue Hao [18], Kasper P. Hendriks [11,19], Julian M. Hibberd [20], Graham J. King [21], Daniel J. Kliebenstein [22], Marcus A. Koch [23], Ilia J. Leitch [5], Frederic Lens [19,24], Martin A. Lysak [25], Alex C. McAlvay [26], Michael T.W. McKibben [7], Francesco Mercati [27], Richard C. Moore [28], Klaus Mummenhoff [11], Daniel J. Murphy [29], Lachezar A. Nikolov [30], Michael Pisias [31], Eric H. Roalson [32], M. Eric Schranz [33], Shawn K. Thomas [34,35], Qingyi Yu [36], Alan Yocca [12], J. Chris Pires [37] and Alex E. Harkess [12,*]

1 Florida Museum of Natural History, University of Florida, Gainesville, FL 32611, USA
2 Department of Ecology and Evolutionary Biology, Yale University, New Haven, CT 06520, USA
3 Department of Biochemistry, Purdue University, West Lafayette, IN 47906, USA
4 Missouri Botanical Garden, Shaw Boulevard, St. Louis, MO 63110, USA
5 Royal Botanic Gardens, Kew, Richmond, Surrey TW9 3AE, UK
6 Ben-Gurion University of the Negev, French Associates Institute for Agriculture and Biotechnology of Drylands, Jacob Blaustein Institutes for Desert Research, Midreshet Ben-Gurion, 8499000, Israel
7 Department of Ecology and Evolutionary Biology, University of Arizona, Tucson, AZ 85721, USA
8 National Herbarium of New South Wales, Australian Botanic Garden, Locked Bag 6002, Mount Annan, NSW 2567, Australia
9 Department of Psychiatry, Washington University in Saint Louis School of Medicine, St. Louis, MO 63110, USA
10 NeuroGenomics and Informatics Center, Washington University in Saint Louis School of Medicine, St. Louis, MO 63108, USA
11 Department of Biology, Botany, University of Osnabrück, D-49076 Osnabrück, Germany
12 HudsonAlpha Institute for Biotechnology, Huntsville, AL 35806, USA
13 Department of Biological Sciences, Bioinformatics Research Center, Program in Genetics, North Carolina State University, Raleigh, NC 27695, USA
14 ACEBB and SGC, School of Biological Sciences, The University of Adelaide, Adelaide, SA 5005, Australia
15 Department of Biological Sciences, Louisiana State University, Baton Rouge, LA 70803, USA
16 Department of Horticulture, Michigan State University, East Lansing, MI 48864, USA
17 Department of Biological Sciences, University of Alberta, Edmonton, Alberta T6G 2E9, Canada
18 Cancer and Cell Biology Division, Translational Genomics Research Institute, Phoenix, AZ 85004, USA
19 Functional Traits, Naturalis Biodiversity Center, PO Box 9517, Leiden 2300 RA, the Netherlands
20 Department of Plant Sciences, University of Cambridge, Cambridge, CB2 1TN, UK
21 Southern Cross Plant Science, Southern Cross University, Lismore, NSW 2480, Australia
22 Department of Plant Sciences, UC Davis, Davis, CA 95616, USA
23 Centre for Organismal Studies (COS), Heidelberg University, 69120 Heidelberg, Germany
24 Institute of Biology Leiden, Plant Sciences, Leiden University, 2333 BE Leiden, the Netherlands
25 CEITEC, and NCBR, Faculty of Science, Masaryk University, 625 00 Brno, Czech Republic
26 Institute of Economic Botany, New York Botanical Garden, The Bronx, NY 10458, USA
27 National Research Council (CNR), Institute of Biosciences and Bioresource (IBBR), Palermo 90129, Italy
28 Biology Department, Miami University, Oxford, OH 45056, USA
29 Royal Botanic Gardens Victoria, Melbourne, VIC 3004, Australia

Received May 09, 2023. Accepted September 22, 2023. Advance access publication October 12, 2023

30 Department of Biology, Indiana University, Bloomington, IN 47405, USA
31 Division of Plant Sciences, University of Missouri, Columbia, MO 65211, USA
32 School of Biological Sciences, Washington State University, Pullman, WA 99164-4236, USA
33 Biosystematics Group, Wageningen University, 6708 PB Wageningen, the Netherlands
34 Division of Biological Sciences, University of Missouri, Columbia, MO 65211, USA
35 Bioinformatics and Analytics Core, University of Missouri, Columbia, MO 65211, USA
36 Daniel K. Inouye U.S. Pacific Basin Agricultural Research Center, Agricultural Research Service, United States Department of Agriculture, Hilo, HI 96720, USA
37 Department of Soil and Crop Sciences, Colorado State University, Fort Collins, CO 80523-1170, USA

*Author for correspondence: aharkess@hudsonalpha.org

Abstract

Model species continue to underpin groundbreaking plant science research. At the same time, the phylogenetic resolution of the land plant tree of life continues to improve. The intersection of these 2 research paths creates a unique opportunity to further extend the usefulness of model species across larger taxonomic groups. Here we promote the utility of the *Arabidopsis thaliana* model species, especially the ability to connect its genetic and functional resources, to species across the entire Brassicales order. We focus on the utility of using genomics and phylogenomics to bridge the evolution and diversification of several traits across the Brassicales to the resources in *Arabidopsis*, thereby extending scope from a model species by establishing a "model clade." These Brassicales-wide traits are discussed in the context of both the model species *Arabidopsis* and the family Brassicaceae. We promote the utility of such a "model clade" and make suggestions for building global networks to support future studies in the model order Brassicales.

Introduction

Model species are essential in plant biology. For example, Barbara McClintock carefully tracked corn (*Zea mays*) kernel variation to characterize transposable elements and chromo-some segregation (McClintock 1941, 1950). The genome of *Arabidopsis thaliana* (The Arabidopsis Genome Initiative 2000) and foundational mutant and germplasm collections like the *Arabidopsis* T-DNA Insertional Mutant Collection (O'Malley et al. 2015) have driven our understanding of gene function in all plants. However, the plant kingdom is vast, comprising nearly 500,000 species and spanning a billion years of evolutionary time (Nic Lughadha et al. 2016). Collaborations among systematists, herbaria, and botanical gardens continue to drive advances in building the plant tree of life (Johnson et al. 2019; One Thousand Plant Transcriptomes Initiative 2019), in part due to rapid advances and the decreasing cost of genome and transcriptome sequencing technologies. These technological advancements enable us to build resources that complement and elevate the usefulness of a model species, encompassing larger taxonomic groups to now produce "model clades" of plants. As species relationships become clearer, the plant biology community is in an ideal position to leverage comparative genomics across these clades to unravel the evolution and function of genes, regulatory networks, repetitive elements, and noncoding DNA. Comparative biologists and phylogeneticists have long been interdisciplinarily working on these problems, for example, to examine floral organ evolution across the order Ranunculales (Damerval and Becker 2017) and the grasses (Schrager-Lavelle et al. 2017). The development of accessible comparative genome and gene family evolution platforms like Phytozome (Goodstein et al. 2012), CoGE (Lyons 2008), Ensembl Plants (Yates et al. 2022), and GENESPACE (Lovell et al. 2018, 2022) allow the placement of stand-alone genome assemblies in a larger evolutionary and phylogenetic context, creating an opportune time for plant biologists to identify clades with diverse phenotypes that can be anchored to a closely related model species.

The order Brassicales exhibits extensive diversity of species, traits, and environmental adaptations across a broad geographical distribution and includes several crops like canola, caper, broccoli, kale, papaya, rapeseed, and saltwort. Within the Brassicales, one can find recurrent evolution and diversification of complex traits involving genomics (e.g. hybridization, polyploidy, repeat sequence turnover, speciation, and reproductive systems), physiology (e.g. extremophytism, photosynthesis modes), morphology (e.g. woodiness, tuberization), metabolomics (e.g. oil production, secondary metabolite diversification), and environment (e.g. wet and hyperarid habitats). To understand the genetic underpinnings of these traits, the predicted functions of orthologous genes in any understudied Brassicales species can be anchored to the massive mutant collections in the Brassicaceae "model species" *A. thaliana*. In this review, we highlight the utility of all Brassicales for benefiting from comparative genomics to understand trait variation and evolution in the context of producing a "model clade" of plants that builds out from a "model species." We frame each major section with 3 key points. First, we describe what is known about a given trait from studying the family Brassicaceae, often derived from research using *A. thaliana*. Second, we describe what is known about the diversity or variation of a given trait across the order Brassicales. Third, we discuss how leveraging comparative genomics across the order will lead to a better understanding of gene, genome, and trait evolution.

Systematics and diversification of the Brassicales

Systematics across the Brassicaceae

The Brassicales comprise ca. 4,700 species in 18 families (Table 1; Magallon et al. 1999; Swanepoel et al. 2020), with the Brassicaceae, by far the largest family, comprising ca. 4,140 species in 349 genera (Koch et al. 2018; Walden et al. 2020a; German et al. 2023; Hendriks et al. 2023; Table 1). Brassicaceae systematics has long been complicated by high levels of parallel and convergent evolution of morphological characters, likely reflecting recent and rapid radiations (Huang et al. 2016; Guo et al. 2017; Walden et al. 2020a; Zuo et al. 2022). Nearly all species are assigned to 1 of 58 mono-phyletic tribes (German et al. 2023; Hendriks et al. 2023), which in turn are assigned to 2 subfamilies: the small mono-generic Aethionemoideae and the large Brassicoideae (98.6% of species; German et al. 2023). Additionally, the 5 main lineages within Brassicoideae recently received supertribe status (German et al. 2023). While assignment of tribes to supertribes is more or less consistent between reconstruction methods, relationships among supertribes vary. For instance, in plastid phylogenies, the Camelinodae (formerly lineage I) consistently diverges first (Fig. 1; Huang et al. 2016; Guo et al. 2017; Mabry et al. 2020; Walden et al. 2020a; Hendriks et al. 2023), whereas nuclear phylogenies show the Hesperodae (formerly lineage III) diverging first (Fig. 1; Huang et al. 2016; Nikolov et al. 2019; Mabry et al. 2020; Hendriks et al. 2023). With the availability of whole genome sequences from all main lineages and advances in synteny and coalescent-based methods for phylogenetic inference, additional evolutionary hypotheses have been proposed recently (Walden et al. 2020b; Zou et al. 2020). For example, the similarity of the *Arabis alpina* genome structure with that of *Aethionema* (Aethionemeae) may indicate an earlier diverging position of the lineage leading to Arabideae (Walden et al. 2020b) and potentially Alysseae (both super-tribe Arabodae, formerly lineage IV; Hendriks et al. 2023). Such contrasting gene and species tree topologies suggest complex evolutionary histories and lead to natural complictions in achieving a stable classification based on the principle of monophyly at many taxonomic levels (Forsythe et al. 2020).

Systematics across the Brassicales

The morphological, biochemical, and diverse genomic composition across the Brassicales permits comparative investigation of evolutionary novelties and macroevolutionary patterns (Fig. 2). However, these approaches rely on first constructing a well-supported species tree. Most of the families across the

Table 1. The distribution of families within the Brassicales, along with the noted genera and species numbers[a]

Family	Genera/Species	Geographical Distribution
Akaniaceae	2/2	*Akania* – Australia *Bretschneidera* – Asia
Bataceae	1/2	Tropical America, Austral-Asia
Brassicaceae	349/4,140	Cosmopolitan (mostly temperate)
Capparaceae	16/480	Cosmopolitan (mostly tropical)
Caricaceae	6/34	Tropical Africa and America (mostly New World)
Cleomaceae	26+/270	Cosmopolitan
Emblingiaceae	1/1	Southwest Australia
Gyrostemonaceae	5/18+	Australia
Koeberliniaceae	1/2	North, Central, and South America
Limnanthaceae	2/8	North America
Moringaceae	1/12	Africa, Madagascar, and Asia
Pentadiplandraceae	1/1	Western Africa
Resedaceae	8/96	North Africa, Eurasia, North America
Salvadoraceae	3/11	Africa to Southeast Asia
Setchellanthaceae	1/1	Mexico, North America
Tiganophytaceae	1/1	Namibia, Africa
Tovariaceae	1/2	Tropical America
Tropaeolaceae	1/105	North and South America

[a]Genera and species numbers from the Angiosperm Phylogeny Website (Stevens 2001). The recently proposed family Borthwickiaceae (found in China and Myanmar) is included within Resedaceae.

Brassicales, including Borthwickiaceae (Zhao et al. 2015) and the recently described Tiganophytaceae (Swanepoel et al. 2020), are only minimally sampled at the species level. Relationships across the order have been established with phylotranscriptomics (Edger et al. 2015, 2018a; Mabry et al. 2020), target sequence capture (Baker et al. 2022), and more traditional molecular systematic approaches based on fewer genes and larger taxonomic sampling (Rodman et al. 1993, 1996; Cardinal-McTeague et al. 2016). Many relationships among the Brassicales are well supported (Hall et al. 2002, 2004; Ronse De Craene and Haston 2006; Hall 2008; Edger et al. 2015, 2018a; Cardinal-McTeague et al. 2016; Mabry et al. 2020); however, some key nodes remain unresolved, which may be the consequence of rapid diversification and/or introgression (Edger et al. 2018a). Due to a large number of extant Brassicales species, building a high-resolution species tree requires both a unified research community effort as well as creative approaches to integrating multiple types of data.

The future of Brassicales systematics research

A complete tree of all Brassicales species, with multiple representative sampling for each species, is desirable and tractable, esp ecially given the growing p op ularityof reduced rep resentation methods such as targeted sequence capture (Dodsworth et al. 2019; Baker et al. 2021). Targeted sequence cap ture can be ap p liedvery effectively even to degraded DNA obtained from herbarium sp ecimens (Brewer et al. 2019), which vastly increases op p ortunitiesfor extensive

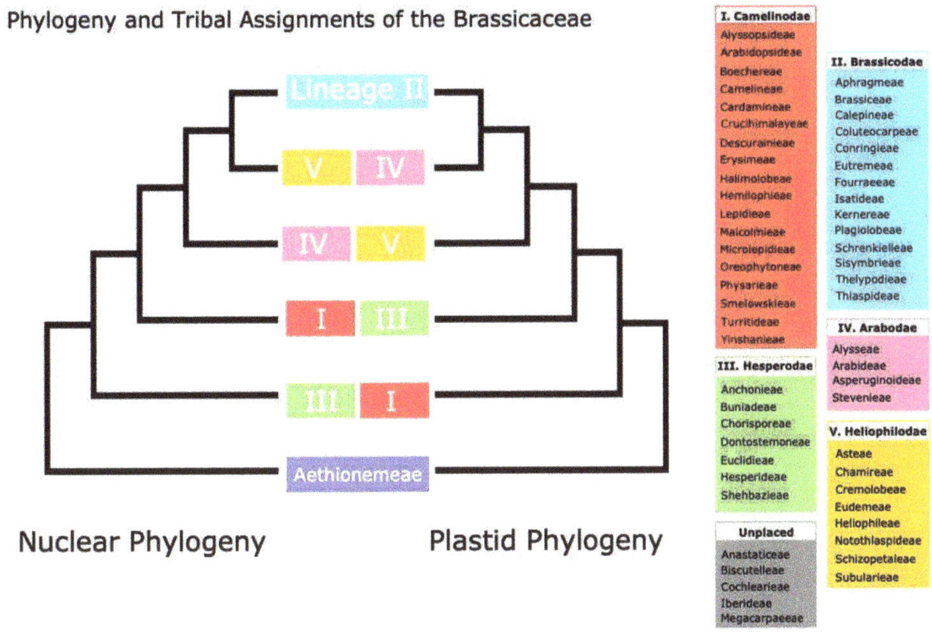

Figure 1. Phylogeny indicating the lineage relationships of the Brassicaceae using nuclear and plastid data. Placed tribes are noted with corresponding color and numbering. Summary, classification, and relationships based on Hendriks et al. (2023).

species sampling. These methods have proven useful for understanding species relationships at the order level across angiosperms (Antonelli et al. 2021; Lee et al. 2021; Maurin et al. 2021; Thomas et al. 2021; Zuntini et al. 2021), and efforts are already underway to build comprehensive phylogenomic datasets across the Brassicales by employing a universal sequence capture toolkit (Johnson et al. 2019; Baker et al. 2022). The combination of universal and lineage-specific toolkits (Nikolov et al. 2019; Eserman et al. 2021; Hendriks et al. 2021) offers great promise for a complete species phylogeny of the order. Further, a Department of Energy Joint Genome Institute Community Sequencing Project (https://jgi.doe.gov/csp-2021-brassicales-genome-initiative/) is currently funded to assemble and annotate chromosome-scale assemblies for at least 26 species, spanning every family in the order. The future for the phylogenomic community studying the Brassicales likely resides in a dual approach, integrating reduced representation genomic data with high-quality, chromosome-scale genomes and annotations.

Polyploidy and Diploidization Dynamics

Polyploidy across the Brassicaceae

Whole-genome duplications (WGDs) are prevalent throughout plant evolution and are often associated with the origin of novel traits and shifts in diversification (Schranz et al. 2012; Van de Peer et al. 2021). Early studies suggested that *Brassica* species (Brassiceae) were hexaploids relative to *A. thaliana* (Arabidopsideae) and that their genomes experienced chromosomal rearrangements during their return to the diploid state (Lagercrantz 1998; Babula et al. 2003). These results were confirmed by later studies and the sequencing of *Brassica* genomes (Lysak et al. 2005, 2007; Parkin et al. 2005; Schranz et al. 2006; *Brassica rapa Genome Sequencing Project Consortium 2011*; Fig. 3). This hexaploidy or whole-genome triplication (WGT) occurred through 2 successive hybridizations creating first an allotetraploid and eventually an allohexaploid (Cheng et al. 2012; Tang et al. 2012). However, the timing and placement of the hexaploidy on a resolved phylogeny remains elusive (Lysak et al. 2007; *Brassica rapa* Genome Sequencing Project Consortium 2011; Arias and Pires 2012; Cardinal-McTeague et al. 2016; Huang et al. 2016). More than 13 other independent, genus- or tribe-specific, polyploidy events have been identified across the Brassicaceae (Mandáková et al. 2017; Huang et al. 2020; Walden et al. 2020b; Hendriks et al. 2023). A study of nearly one-half of all Brassicaceae species found that 50% of these are recent polyploids (i.e. still recognized as polyploids based on cytological data) with at least 1 hidden state (characters that are not observed for the purpose of the model but influence the diversification rate (Beaulieu and O'Meara 2016), which impacts the net diversification of diploid and polyploid species (Román-Palacios et al. 2020). Another study by Walden et al. (2020a) suggests these unassessed traits from Román-Palacios et al. (2020) may be associated with morphological diversity.

Polyploidy across the Brassicales

Multiple studies have found ancient WGDs spread more broadly across the Brassicales (e.g. At-β event; Fig. 2; Barker et al. 2009; Edger et al. 2015, 2018a; One Thousand Plant Transcriptomes Initiative 2019; Mabry et al. 2020). Many families have also experienced family-specific WGDs, including the Brassicaceae (At-α; Barker et al. 2009; Haudry et al. 2013;

Figure 2. Phylogeny of 17 families in the Brassicales and character trait matrix. The presence of a given trait is marked by a filled dark-green square. Traits that are unknown or absent are in light green. Possible locations of At-β are marked in blue stars, with family-specific whole-genome duplications in orange. Traits displayed are: A) C4, B) extremophytism, C) glucosinolates, D) ancestral woodiness, and E) dioecy. Additional proposed families, which are not included here, still need to be analyzed using multiple single-copy nuclear genes (e.g. Tiganophytaceae and Borthwickiaceae). Areas of the phylogeny with no or low support are indicated by thin branch lines. Topology and support based on Edger et al. (2018a).

Edger et al. 2015, 2018a; One Thousand Plant Transcriptomes Initiative 2019; Mabry et al. 2020), Cleomaceae (Cs-α or Th-α; Schranz and Mitchell-Olds 2006; Barker et al. 2009; Bayat et al. 2018; Mabry et al. 2020; Hoang et al. 2023), and Resedaceae (Rs-α; Lysak 2018; One Thousand Plant Transcriptomes Initiative 2019; Mabry et al. 2020). Family-specific WGDs have also been identified in Gyrostemonaceae, Koeberliniaceae, and Tropaeolaceae (One Thousand Plant Transcriptomes Initiative 2019).

The future of Brassicales polyploidy research

Understanding polyploidy and diploidization in the Brassicales is highly relevant and impactful for several reasons. Firstly, the Brassicales are ideal for studying these processes due to their relatively small genomes ($N = 342$ taxa, mean = 761.0 Mb/1C, median = 537.0 Mb/1C, mode = 392.0 Mb/1C; based on data from the Plant DNA C-values database release 7.1 together with additional data not yet incorporated; Pellicer and Leitch (2020) and Leitch et al., release 7.1, April 2019). We can also produce high-quality genomes that allow for fine-scale comparisons (Kreiner et al. 2017), making them well-suited for in-depth genomic analysis. This relevance stems from the potential insights that can be gained into broader biological phenomena.

The impact of this research extends to various areas. Firstly, resolving the current struggles in understanding gene loss and retention after polyploidy in Brassicales can provide critical insights into the evolution of complex genomes. While

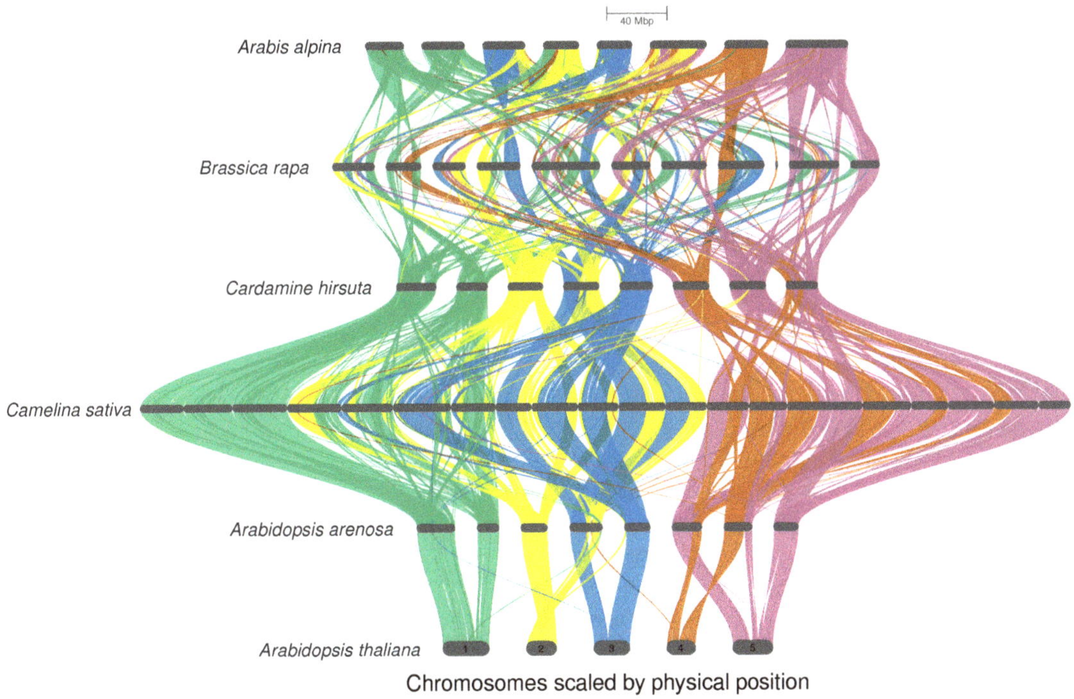

Figure 3. Whole-genome synteny of chromosome-scale Brassicaceae genomes. The synteny plot was generated using GENESPACE v0.9.1 (Lovell et al. 2022) with "diamondMode" and "orthofinderMode" set to fast, using *A. thaliana* (Arabidopsideae; Lamesch et al. 2012), *A. arenosa* (Arabidopsideae; Barragan et al. 2021), *B. rapa* (Brassiceae; v1.3; downloaded from Phytozome), *Camelina sativa* (Camelineae; Kagale et al. 2014), *Cardamine hirsuta* (Cardamineae; Gan et al. 2016), and *A. alpina* (Arabideae; Willing et al. 2015) genome annotations. The 4 diploid genomes (*A. arenosa, A. thaliana, A. alpina,* and *C. hirsuta*) show 1:3 syntenic relationships with the younger hexaploid *C. sativa* genome ($n = 20$) and evolutionarily older hexaploid genome of *B. rapa* ($n = 10$).

gene loss after polyploidy is common, it is not random, and the fate of duplicated genes is determined by the interplay of many forces, including dosage-induced constraints (Birchler and Veitia 2007, 2012; Conant 2014; Conant et al. 2014; Hao et al. 2022), differentially methylated transposable element (TE) density patterns among subgenomes (Hollister and Gaut 2009; Freeling et al. 2012), changes in regulatory networks (Blischak et al. 2018; Hu and Wendel 2019), and epigenetic changes after hybridization (Cheng et al. 2016; Zhao et al. 2017). These processes yield diploidized genomes with genes retained in different copy numbers and patterns of variation (Li et al. 2021). This understanding is not only significant for advancing our knowledge of plant genetics but also has implications for broader evolutionary biology.

Secondly, the knowledge gained from studying polyploidy and diploidization in Brassicales can have practical applications in agriculture. Genes retained after polyploidy harbor more genetic diversity than other genes and were important during the domestication of *Brassica rapa* (Qi et al. 2021). Therefore, a better understanding of the processes involved can lead to more efficient breeding strategies and the development of improved crop varieties.

Thirdly, resolving these struggles can open up new avenues for future research. It can enable the development of advanced tools and methods, such as POInT (the Polyploid Orthology Inference Tool), a likelihood method for modeling subgenome partitioning and biased fractionation (Conant and Wolfe 2008), which can then be applied not only to Brassicales but also to other polyploid systems. Inferring the biased pattern of gene loss has been challenging for several reasons, including genomic rearrangement and homoeologous exchanges (Xiong et al. 2011; Mandáková et al. 2017; Edger et al. 2018b; Bird et al. 2020, 2023). Orthology inference using gene synteny can reli-ably differentiate subgenomes in polyploids (Lyons et al. 2008; Haug-Baltzell et al. 2017). POInT has been used to confirm that biased fractionation cannot be attributed to artifacts in synteny block inference (Conant 2020). Integrating tools like POInT and GENESPACE (Lovell et al. 2022) can incorporate synteny into phylogenetically robust models of evolution. These approaches are promising for studying the diploidization process, particularly as the wider Brassicales community generates more chromosome-scale genome assemblies from across the order.

Epigenetics, small RNAS, and repetitive elements

Epigenetics, small RNAs, and transposable elements across the Brassicaceae

Much of our knowledge of small RNA (sRNA) biology has been derived from mutant screens in *A. thaliana*, including studies of TE mobilization and repression pathways (Miura et al. 2001; Gendrel et al. 2002; Lippman et al. 2004; Martinez et al. 2017). Plant genomes are typically dominated by TEs that can move throughout the genome using a variety of mechanisms (Wicker et al. 2007). Although cycles of TE turnover are crucial for shaping plant genomes, most TEs are kept in a transcriptionally repressed state via DECREASED DNA METHYLATION 1 (DDM1) and RNA-directed DNA methylation pathways (Zemach et al. 2013; Matzke and Mosher 2014; Li et al. 2015; Gallego-Bartolomé et al. 2019). The loss of DDM1 leads to loss of DNA methylation, accompanied by TE mobilization and developmental defects (Kakutani et al. 1999; Miura et al. 2001). The *A. thaliana* genome, which was the first plant genome to be sequenced (The Arabidopsis Genome Initiative 2000), has been paramount to the dissection of these pathways; however, there is immense variation in DNA methylation levels within the Brassicaceae (Seymour et al. 2014; Bewick et al. 2016), such as the loss of gene body methylation in *Eutrema salsugineum* (Eutremeae) due to a loss of the *CHROMOMETHYLASE* 3 gene (Bewick et al. 2016). A recent study by Lu et al. (2019) investigated epigenetic variation across 13 angiosperm species, 2 of which are in the Brassicaceae (*A. thaliana* and *E. salsugineum*). They found that species-specific accessible chromatin regions were strongly enriched for TE sequences. Therefore, species-specific TEs may regulate gene expression across the order, emphasizing the importance of sequencing across the clade.

Epigenetics, sRNAs, and TEs across the Brassicales

In the Brassicales, sRNAs have been p oorly samp led in non-Brassicaceae species, though there are several emerging studies related to classes of sRNAs. One example is research on phased, secondary, small interfering RNAs (phasiRNAs), which display an intriguing pattern of loss in the Brassicales. In *A. thaliana*, phasiRNAs are expressed in vegetative tissues (Fei et al. 2013); additionally, 21-nt and 24-nt reproductive phasiRNAs, triggered by miR2118 and miR2275, respectively, have been found broadly across monocots (Kakrana et al. 2018; Patel et al. 2018) and eudicots (Xia et al. 2019) and shown to play a role in anther fertility in corn and rice (Teng et al. 2020). Intriguingly, miR2275 is not present in any of the sampled Brassicales (Polydore et al. 2018; Xia et al. 2019). The loss of miR2275 has been reported in several other lineages, including legumes and Solanaceous species, allowing us to place the Brassicales in a larger context of sRNA loss, thus enabling comparisons across the tree of life to understand the anther fertility phenotype compared with other monocots and eudicots. Other floral-related microRNAs in the Brassicales are implicated in sex determination and sexual dimorphism in papaya (Aryal et al. 2014). Further examples of lineage-specific sRNA evolution in the Brassicales are tandem microRNA loci, miR826, and miR5090, involved in nitrogen starvation adaptation (He et al. 2014).

With multiple documented WGDs, the Brassicales are a model clade for testing the interplay between polyploidy, genome size, and repetitive element proliferation. A recent study characterized TE abundances from low-coverage genome shotgun data in 71 phylogenetically diverse taxa across the order, placing transposon content in the context of WGD and phylogeny (Beric et al. 2021); surprisingly, little connec-tion was found between phylogeny and TE expansion or con-traction, nor was there any correlation between patterns of WGD and TE abundance. Although WGDs might be ex-pected to lead to global TE reactivation, TE purging mechan-isms seem to work efficiently in the Brassicales (Hawkins et al. 2009). Interestingly, the elimination of TEs is less effective in some Brassicales clades, resulting in large diploid genomes (1,500–4,200 Mb) that exceed the modal C-value of the Brassicales by 4- to 10-fold (Hloušková et al. 2019; Zuo et al. 2022).

The future of Brassicales epigenetics, sRNA, and TE research

The future of epigenetics, sRNA, and repetitive element studies within the Brassicales hinges on our ability to generate diverse sRNA and methylome sequencing datasets and anchor analyses onto chromosome-scale assemblies in a comparative genomic context. Genomes from across the order will provide opportunities for investigation of how sRNA biogenesis and methylation pathways of *A. thaliana* function in phylogenetically distant relatives, as well as provide insight into polyploidy, transposon biology, and lineage-specific s RNA e mergence a nd loss.

Reproductive biology

Reproductive biology across the Brassicaceae

Thirty years ago, homeotic genes discovered in *A. thaliana* formed the foundation of the "ABC model" (Bowman et al. 1991a, 1991b; Drews et al. 1991). *A. thaliana* is also a model for understanding the evolutionary genetic underpinnings of the loss of self-incompatibility (SI) which promotes out-crossing. SI is a widespread trait found in 40% of flowering plant species and is based on recognition of self-pollen (e.g. Brassicaceae, Papaveraceae), or non–self-pollen (e.g. Solanaceae; Igic et al. 2008; Fujii et al. 2016). The mechanisms underlying sporophytic SI in the Brassicaceae were discovered in *Brassica* and involve a pollen-specific ligand (*S-LOCUS PROTEIN 11*; or *S-LOCUS CYS-RICH PROTEIN*, *SCR*) and pistil-specific receptor (*S LOCUS RECEPTOR KINASE*) tightly linked at the S-locus (Stein et al. 1991; Schopfer et al. 1999; Takasaki et al. 2000; Takayama et al. 2000). The loss of SI arose in *A. thaliana* at least 3 times (Shimizu et al. 2008; Tsuchimatsu et al. 2017) through loss-of-function mutations in *SCR*, followed by rearrangements of the S-locus (Shimizu et al. 2008; Tsuchimatsu et al. 2010, 2017). The loss of SI in the Brassicaceae has occurred independently in at least 5 other species (Shimizu and Tsuchimatsu 2015; Nasrallah 2017). Further, there is an epigenetic component to SI plasticity. In *Brassica*, both DNA methylation (Shiba et al. 2006) and 24-nt sRNAs (Tarutani et al. 2010) target recessive alleles in allelic pollen S-determinants. Both genomic and ecological factors can influence the repeated evolution and breakdown of SI, so understanding the diversity of pathways leading to SI loss can be gained by studying systems across the Brassicales, although to date most involve the primary loss of the male-specific determinant of SI (Shimizu and Tsuchimatsu 2015).

Reproductive biology across the Brassicales

Most species in the Brassicales are hermaphroditic or monoecious (Renner 2014). However, roughly 1% are dioecious, where unisexual flowers develop on separate plants (Renner 2014). Dioecy has evolved in 7 different families, representing at least 5 origins (Fig. 2; Renner 2014; Soza et al. 2014). Although structurally hermaphroditic, Cleomaceae commonly vary the ratio of functionally male to functionally female flowers according to environmental conditions (Zohoungbogbo et al. 2018).

In some Brassicales, sex chromosomes have been identified, with the most detailed studies in papaya (Caricaceae). The papaya Y is defined by the presence of an approximately 10-Mb nonrecombining sex-determining region that is only a fraction (approximately 10%) of the entire sex chromosome (Liu et al. 2004; Na et al. 2012; Wang et al. 2012; Yue et al. 2022); this is similar to other plant sex chromosomes like asparagus and kiwifruit (Akagi et al. 2018; Harkess et al. 2020). The closely related genus *Vasconcellea* also has evidence of XY chromosomes; however, evidence suggests this is an independent origin (Wang et al. 2012; Carvalho and Renner 2015). Although papaya is almost exclusively dioecious in the wild, it is often described as trioecious because most cultivars are essentially gynodioecious (Brown et al. 2012; Fuentes and Santamaría 2014).

The Brassicales are also an excellent model for the evolution of diverse fruit structures. The Brassicaceae fruits exhibit the greatest diversity in size and shape (Ronse De Craene and Haston 2006); however, fruit characters are highly homoplastic (Franzke et al. 2011), with clear evidence of parallel evolution of indehiscent fruits (Mühlhausen et al. 2013) and independent origins of dehiscence in Capparaceae and Cleomaceae (Hall et al. 2002). Moreover, the diverse fruit characters in the Brassicaceae play major ecological roles in fruit/seed dispersal (Sperber et al. 2017; Arshad et al. 2019; Bhattacharya et al. 2019a; Nichols et al. 2020). The recent evidence of plasticity in *Aethionema* in response to stress, with corresponding glucosinolate allocation to fruit tissues (Bhattacharya et al. 2019b), demands a holistic evo-devo analysis among other members of the Brassicales to understand the genetic and epigenetic basis of diversity in fruit characters.

The future of Brassicales reproductive biology research

The repeated evolution of dioecy, SI, and diverse fruit types across the Brassicales makes it a powerful model order for comparative developmental and genomic analyses. For instance, are the same sterility genes or pathways involved in dioecious flower development across the Brassicales? Do the same genes control fruit shape and dehiscence? These comparisons will be valuable because they may be potential targets for controlling sex and fruit characteristics in breeding programs for the many economically important traits found across the order. Moreover, the many origins of sex chromosomes will provide further insight into their early evolution from ancestral autosomes (Carey et al. 2021).

Specialized metabolite diversity

Specialized metabolite diversity across the Brassicaceae

Glucosinolates, or mustard oils, are a clade-defining characteristic of plants in the Brassicales. The presence of major ecological model systems within this group, including *Arabidopsis*, *Boechera* (Boechereae), *Brassica*, and *Streptanthus* (Thelypodieae), has allowed mechanistic ecology studies to identify the forces shaping specialized diversity in the Brassicaceae. Competition experiments between *A. thaliana* accessions from central and northern Europe in the presence of multiple herbivores showed structural variation in the glucosinolate *METHYLTHIOALKYLMALATE* (MAM) locus linked to the presence or absence of 2 different aphids across the continent (Kroymann et al. 2003; Züst et al. 2012). Recreating natural variation in isogenic lines with field trials in both *Arabidopsis* and *Boechera* showed specialized metabolism is under strong selective pressure that fluctuates locally, creating local bet-hedging (Schranz et al. 2009; Manzaneda et al. 2010; Kerwin et al. 2015, 2017). Manipulating glucosinolates within *Brassica* showed how intraspecific glucosinolate variation shaped inter- and intraspecific competition and enabled stochastic variation in species occupancy across a landscape (Lankau and Strauss 2007). Combining ecological parameters, biochemistry, and phylogenetics within *Streptanthus* has been used to understand how specialized metabolism is shaped within the family in relation to serpentine soils (Cacho et al. 2015, 2021). Interestingly, the above ecological model systems and other less studied Brassicales can co-occur within the same or similar environments. This raises the potential to develop studies investigating how specialized metabolism within one species may influence the specialized metabolism within another.

Specialized metabolite diversity across the Brassicales

Extending the ecology work via the combination of genomics and chemistry across the entire order Brassicales has begun to illustrate how specialized metabolite pathways evolve. Glucosinolates are estimated to have first occurred at the base of the Brassicales approximately 108 MYA and span the phylogeny from the leafy greens of the Brassicaceae to the seeds of the Caricaceae, where they co-occur with the hypothesized evolutionary progenitor cyanogenic glycosides (Olafsdottir et al. 2002). Glucosinolates then diversified further approximately 43 MYA with the evolution of the aliphatic glucosinolates (Beekwilder et al. 2008; Zhang et al.

2015; Czerniawski et al. 2021). Some glucosinolates, like 2-hydroxy-2methylbutyl glucosinolate (glucocleomin), are family-specific innovations (e.g. Cleomaceae), and others like 3-benzoyloxybutyl glucosinolate (glucomalcomiin) appear to be relatively species specific (e.g. *A. thaliana*). In contrast, glucosinolates like methyl glucosinolate (glucocapparin) are found in the Cleomaceae and the Capparaceae but not in the Brassicaceae, suggesting a loss or replacement in the Brassicaceae that has yet to be characterized genetically.

Although some of these innovations in glucosinolate amino acid substrate types have been mapped on the phylogeny (Edger et al. 2018a; Kumar et al. 2019; Petersen et al. 2019), the underlying genomic mechanisms influencing these transitions have yet to be phylogenetically characterized. What is clear is that the evolution of the glucosinolate pathway has involved whole-genome, local duplication, and gene transposition events, each at different levels. In the glucosinolate pathway, whole-genome duplications gave rise to variation in the CYP79 family that determines the specific amino acid used for the glucosinolate backbone (Edger et al. 2015, 2018a) and the origin of the MAM locus derived from primary metabolism (Abrahams et al. 2020). In contrast, local structural diversity, duplications, and inversions create enzymatic variation leading to novel enzymes that create the structural modifications to the glucosinolate backbone, determining the ultimate biological activity (Kroymann et al. 2003; Hansen et al. 2008; Chan et al. 2010).

The future of Brassicales specialized metabolite diversity research

Moving beyond glucosinolates to other specialized pathways, the identified diversity and innovations at different time scales, both within and between metabolic pathways, allow for unique comparisons using the Brassicales. For example, is there a difference in how older and younger genes evolve within a pathway and is this difference consistent across pathways? In addition to investigating how pathways are gained, the Brassicales provide the ability to study how pathways can be lost and independently recreated (Haribal et al. 2001; Rajniak et al. 2015). Additional questions related to specialized metabolism can also be addressed when sampling across the Brassicales. For example, does the evolution of cardiac glycosides within the *Erysimum* lineage (Erysimeae) influence the selective pressure on specialized metabolism in other species when they coexist or within the same species when the 2 pathways coexist (Züst et al. 2020)? Developing extensive functional, biochemical, genetic, and genomic resources within the Brassicales will allow testing these questions about how specialized metabolism is driven to novelty.

The evolution of woodiness

Woodiness across the Brassicaceae

The initiation and activation of a vascular cambium can lead to massive secondary growth in woody plants, in which the wood cylinder is the main component in stems and roots of shrubs, trees, and lianas. The herbaceous *A. thaliana* is an excellent model for elucidating the genetic controls of wood formation. Woody mutants show that the genetic mechanism turning on the wood pathway in stems can be simple, for example by knocking out 2 flowering control genes (Melzer et al. 2008; Lens et al. 2012; Davin et al. 2016) or by overexpressing a single gene that is downstream of these 2 flowering control genes (Rahimi et al. 2022). Interestingly, about 10% of Brassicaceae species have developed into woody shrubs, treelets, and even lianas under natural conditions, and these approximately 400 woody species are the result of nearly 100 independent transitions from herbaceousness toward phylogenetically derived woodiness (e.g. Mohammadin et al. 2017; Zizka et al. 2022; Lens et al., unpublished dataset). This makes the Brassicaceae a perfect model to better understand why flowering plants became woody during evolutionary history.

Woodiness across the Brassicales

Woodiness represents the ancestral state in the Brassicales (Figs. 2 and 4). Across the order, evolutionary transitions from ancestral woodiness toward herbaceousness and reversals back to phylogenetically derived woodiness have repeatedly occurred. Transitions toward herbaceousness evolved at least 7 times, leading to, among others, the predominantly herbaceous sister families Brassicaceae (Al-Shehbaz 1984; Franzke et al. 2011) and Cleomaceae (Byng 2014; Patchell et al. 2014) that include by far the majority of perennial or annual herbaceous species within the order. Other families including herbs are Emblingiaceae (Hall et al. 2004; Byng 2014), Gyrostemonaceae (Byng 2014), Limnanthaceae (Meyers et al. 2010; Edger et al. 2018a), Resedaceae (Martín-Bravo et al. 2007), Tovariaceae (Hall et al. 2004), and Tropaeolaceae (Andersson and Andersson 2000). Within the predominantly herbaceous Brassicales clades, we also identified approximately 100 evolutionary reversals from herbaceousness toward derived woodiness (e.g. Lens et al., unpublished dataset), accounting for about 15% of the total number of reversals within flowering plants. Nearly all of these independent reversals are in the Brassicaceae, but they also occurred in Cleomaceae (at least 4 transitions; Patchell et al. 2014) and Resedaceae (at least 3; Martín-Bravo et al. 2007).

Although the tallest Capparaceae trees reach 40 m in height, other trees in the order are typically much smaller (e.g. Akaniaceae, other Capparaceae; Bayer and Appel 2003; Mercado Gómez and Escalante 2018). Across the entire order, multiple invasions into seasonally dry (Brassicaceae, Gyrostemonaceae, Resedaceae), (hyper)arid (Brassicaceae, Cleomaceae, Gyrostemonaceae, Koeberliniaceae, Moringaceae, Salvadoraceae, Setchellanthaceae, Tiganophytaceae), and temperate habitats (Brassicaceae, Cleomaceae, Limnanthaceae) gave rise to a general reduction in plant size, leading to small shrubs and herbs in many lineages (Al-Shehbaz 1984; Martín-Bravo et al. 2007; Byng 2014;

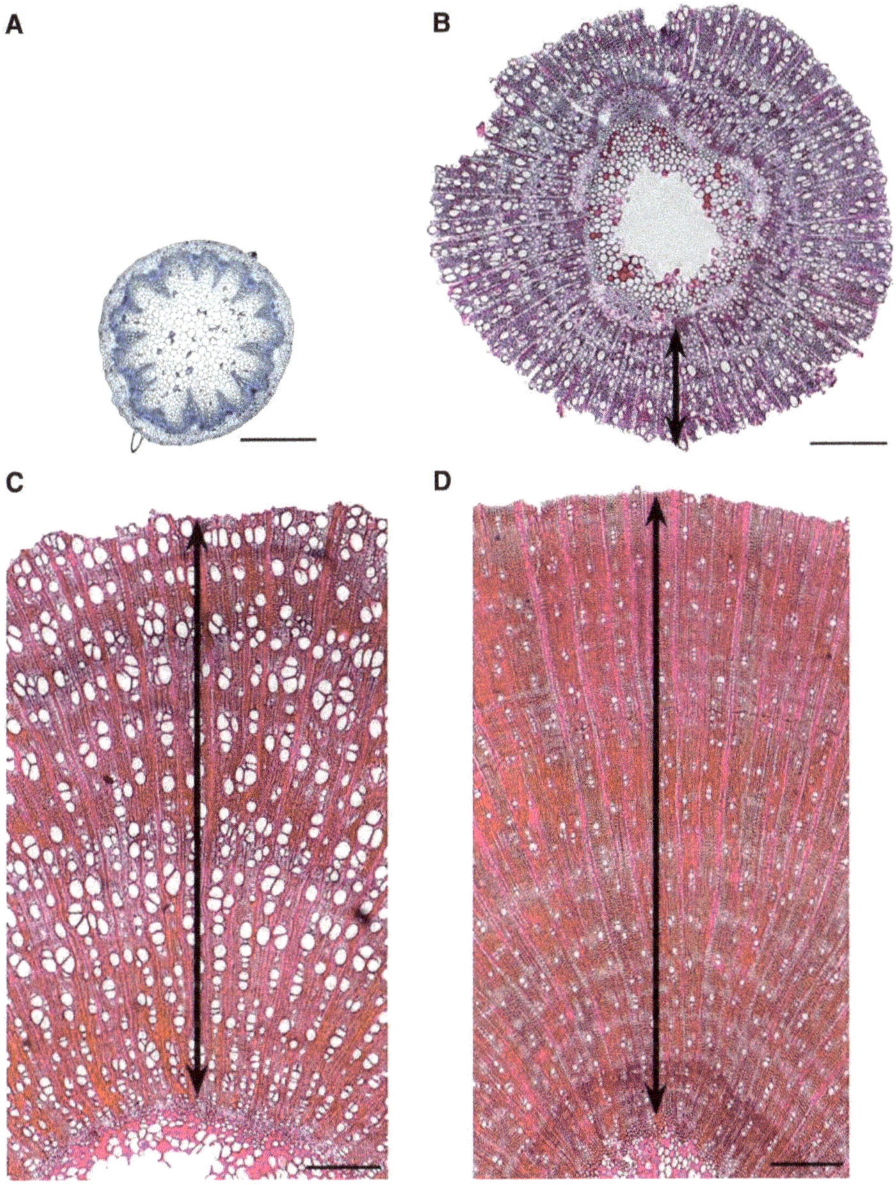

Figure 4. Light microscope cross sections through Brassicales stems showing the difference between: herbaceous **A)** and woody stems **B–D)**. Double-pointed arrows indicate the wood cylinder. All images are at the same magnification (scale bar = 500 µm). **A)** Basal inflorescence stem part of *A. thaliana* (Brassicaceae). **B)** *Reseda* sp. (Resedaceae). **C)** *C. spinosa* (Capparaceae). **D)** *Sinapidendron angustifolium* (Brassicaceae).

Salariato et al. 2016; Swanepoel et al. 2020). It is noteworthy that woody Gyrostemonaceae, which may reach 10 m, remain relatively short-lived (usually <10 years) in arid areas (Baker et al. 2005). The successful attempts to establish and diversify into (hyper)arid habitats is unusual across the largely tropical rosids, making the Brassicales in general and the species-rich Brassicaceae, in particular, stand out as ecologic-ally distinct lineages (Cornwell et al. 2014).

The future of Brassicales woodiness research

Ongoing efforts to start building a species-dense Brassicales phylogeny (Nikolov et al. 2019; Hendriks et al. 2023), required to identify all the species pairs with contrasting life forms, makes the Brassicales a new, representative model clade to investigate the environmental and genetic drivers of the multiple woodiness shifts. The emerging phylogenetic framework will inform the first step toward reconstructing the evolution of niche dimensions and allow for solid statistical testing of hypotheses explaining why plants became woody (Lens et al. 2013; Dória et al. 2018; Zizka et al. 2022). Moreover, identification of the approximately 100 contrasting life form pairs in the Brassicales will serve as a great tool to help discover key regulatory genes controlling wood formation and subsequently validate these results in *A. thaliana* (Melzer et al. 2008; Lens et al. 2012; Davin et al. 2016; Rahimi et al. 2022). Using newly developed approaches such as phylogenetic inter-species genome-wide association studies (Kiefer et al. 2019), we will be able to assess whether the same genes are involved in the independent transitions, thereby contributing to our general understanding of parallel and convergent evolution as a principal evolutionary phenomenon (Elmer and Meyer 2011).

Photosynthesis

Photosynthesis across the Brassicaceae

Modifications to C_3 photosynthesis have repeatedly occurred in angiosperms and typically are characterized by spatial (C_4) or temporal (crassulacean acid metabolism) sequestration of CO_2 fixation (Edwards 2019). Within the Brassicaceae, so far only *Moricandia* (Brassiceae) has been reported to include 5 C_3-C_4 intermediate species as well as 3 C_3 species, whereas no species has been shown to have C_4 photosynthesis (Apel et al. 1997; Schlüter et al. 2017). Schlüter et al. (2017) suggest various constraints on the path to C_4, including anatomical limitations to efficient metabolite exchange, limitation in nutrients other than carbon, and limited pressure to reduce photorespiration as these species inhabit cooler climates than those occupied by C_4 species. Other Brassicaceae lineages deviate from typical C_3 physiology (e.g. carbon isotope ratios not typical of C_3), including *Diplotaxis tenuifolia* (Brassiceae; Apel et al. 1997; Ueno et al. 2003, 2006), *Brassica gravinae* (Brassiceae; Ueno 2011), and *Hirschfeldia incana* (Brassiceae; Garassino et al. 2022). Such species and close relatives need detailed analysis to understand how C_3 photosynthesis has been modified.

Photosynthesis across the Brassicales

In the Brassicales, at least 4 instances of partial- or full-C_4 photosynthesis have evolved. In Cleomaceae, C_4 NAD-dependent malic enzyme (NAD-ME) photosynthesis has been extensively characterized in *Gynandropsis gynandra* (Marshall et al. 2007; Voznesenskaya et al. 2007; Bräutigam et al. 2011; Brown et al. 2011; Koteyeva et al. 2011, 2014; Kajala et al. 2012; Williams et al. 2016; Huang et al. 2021). This research indicates that C_4 photosynthesis has evolved from components found in C_3 species. Identification of a bipartite transcription factor module driving gene expression in the *A. thaliana* bundle sheath provides insight into *trans*-factors important for C_4 photosynthesis (Dickinson et al. 2020). Significant natural variation of C_4 traits exist in *Gynandropsis* (Reeves et al. 2018), and the *G. gynandra* genome has gene families retained in duplicate following polyploidy that facilitated C_4 evolution (Hoang et al. 2023). The *Coalisina* clade (Cleomaceae; Roalson and Hall 2017) includes C_3, intermediate, and C_4 species (Voznesenskaya et al. 2007). *Coalisina angustifolia* has NAD-ME C_4, whereas *Coalisina paradoxa* has been characterized as intermediate/C_2 (Koteyeva et al. 2011). Only 1 taxon within the *C. angustifolia* complex has been fully characterized (*C. angustifolia* subsp. *angustifolia*; Voznesenskaya et al. 2007). Denser sampling of this clade would inform us about the C_3 to C_4 transition as it appears to include C_3, C_3-C_4 intermediate, and fully C_4 species and would provide an interesting comparative lineage to *Moricandia*, where there are C_3 and C_3-C_4 intermediates but no known C_4 species. *Areocleome* is the least studied of Cleomaceae C_4 lineages and includes at least 1 species restricted to arid Australia (Barrett et al. 2017). *Areocleome oxalidea* has NAD-ME-type C_4 physiology (Voznesenskaya et al. 2007; Koteyeva et al. 2011), but the underlying C_4 genetics remain unstudied. The sister genus *Arivela* may have C_3-C_4 intermediate characteristics as it is sympatric with *Areocleome* and more widespread in the monsoonal tropics. Other Cleomaceae lineages, including *Cleomella sparsifolia*, *Tarenaya siliculifera*, *Sieruela allamanii*, and *S. gallaensis*, appear to deviate from typical C_3 physiology (Voznesenskaya et al. 2007; Parma et al. 2021). These species significantly differ in their distributions from dry southwestern North America to tropical Brazil, and tropical to subtropical Africa, providing opportunities to consider C_3 pathway modifications across environments.

The future of Brassicales photosynthesis research

Understanding the modifications in photosynthetic pathways and carbon-concentrating mechanisms holds significant relevance and impact in several key aspects. Gaining insights into these fundamental physiological processes is crucial for advancing our comprehension of plant biology and evolution. This knowledge helps us unravel the intricacies of how plants have adapted to various environmental conditions, contributing to a deeper understanding of the natural world.

Photosynthetic pathway modifications occur in numerous lineages and in some cases co-occur and shift between the 2 major modification types (Holtum et al. 2017; Edwards 2019; Winter et al. 2019). Further, C_4 photosynthesis has multiple biochemical forms (Sage 2004). As our understanding of these basic physiological processes has expanded, 4 foundational hypotheses have solidified: (1) there have been multiple independent origins of C_4/CAM (Sage et al. 2011); (2) independent origins of carbon-concentrating mechanisms can occur through the co-option of similar molecular components (Christin et al. 2007; Brown et al. 2011), and these components exist in the ancestral C_3 state (Brown et al. 2011; Williams et al. 2016; Reyna-Llorens et al. 2018); (3) precursor anatomical modifications that appear unrelated to photosynthesis act as facilitators of biochemical adaptations (Christin et al. 2013; Williams et al. 2013); and (4) a number of carbon-concentrating mechanisms do not fully fit our concepts of C_4/CAM (e.g. C_2; Lundgren 2020). Our understand-ing of carbon-concentrating adaptations has become better attuned to these nuances as knowledge on independent origins has accumulated.

Resolving the current struggles in this field would open up exciting future directions for research. Access to genomes and a comprehensive understanding of photosynthetic pathway modifications would enable researchers to explore the genetic and molecular underpinnings of these processes in even greater detail. This, in turn, could lead to the discovery of novel molecular components and mechanisms that drive photosynthetic adaptations. The relevance of this research extends to agriculture and environmental conservation. By understanding the multiple independent origins of C_4/CAM photosynthesis and the co-option of molecular components, we can potentially engineer crops to be more efficient in photosynthesis and better adapted to changing environmental conditions. This has implications for food security and sustainable agriculture, particularly in the face of climate change. Additionally, the recognition of carbon concentrating mechanisms that do not fit traditional C_4/CAM concepts, such as C_2, presents an intriguing avenue for further investigation. Unraveling the molecular basis of these mechanisms could challenge and expand our current understanding of photosynthesis and carbon fixation in plants.

Extremophytism and stress tolerance

Extremophytism across the Brassicaceae

Many Brassicaceae (and Brassicales) species thrive in extreme environments including desert, arctic, and saline ecosystems. To survive in such habitats, extremophytes possess genetic adaptations for tolerating severe abiotic stresses. Such adaptations are exemplified in 2 halophytic Brassicaceae, *Schrenkiella parvula* (Schrenkielleae) and *Eutrema salsugineum* (Eutremeae), quintessential models (Zhu 2015) for studying tolerance to ionic and nutrient stresses (Kazachkova et al. 2018; Pantha et al. 2021; Tran et al. 2021; Wang et al. 2021). Curated genomes, multiple tissues, and environmental response “omics” datasets have been generated for both species as well as transformation protocols that facilitate independent genetic analyses (Oh et al. 2010, 2014; Dassanayake et al. 2011; Wu et al. 2012; Wang et al. 2019; Pantha et al. 2021; Tran et al. 2022; Wijesinghege et al. 2022a, 2022b; Sun et al. 2022).

Because multiple evolutionary paths can provide unique solutions to different environmental challenges, the Brassicaceae from various extreme environments are being studied including: (1) the desert species *Anastatica hierochuntica* (Anastaticeae), which is tolerant to heat, low nutrient, and salt stress (Eshel et al. 2017, 2022); (2) the metal hyperaccumulator *Arabidopsis halleri* (Arabidopsideae; Briskine et al. 2017; Honjo and Kudoh 2019); and (3) 3 Arctic species, *Cardamine bellidifolia* (Cardamineae), *Cochlearia groenlandica* (Cochlearieae), and *Draba nivalis* (Arabideae; Birkeland et al. 2020). Together, these models should provide a window into evolutionary adaptations that confer stress tolerance and fa-cilitate an extremophyte lifestyle.

Extremophytism across the Brassicales

Beyond the Brassicaceae, only a few studies have examined extremophyte Brassicales. For instance, *Batis maritima* (Bataceae), a perennial, succulent halophytic shrub that inhabits mangrove swamps, salt marshes, and salt flats (Marcone 2003; Debez et al. 2010), tolerates up to 1 M NaCl and has evolved different salt tolerance mechanisms compared with the 2 annual Brassicaceae halophyte models, *E. salsugineum* and *S. parvula*. For example, *B. maritima* accumulates high levels of shoot Na^+ while *E. salsugineum* minimizes entry of Na^+ (Kant et al. 2006; Debez et al. 2010).

Drought responses were investigated in the stress-tolerant, evergreen, African tree species, *Dobera glabra* (Salvadoraceae). When compared with co-occurring non-Brassicales species, *D. glabra* displayed the lowest water potential, illustrating its ability to thrive in drought-prone areas (Gebrekirstos et al. 2014).

Molecular studies of non-Brassicaceae Brassicales are extremely sparse. Mercati et al. (2019) reported a de novo transcriptome assembly from the drought-tolerant, xerophytic crop, *Capparis spinosa* (Capparaceae), and they generated the first set of SSR markers distinguishing subspecies of *C. spinosa*. A chromosome-level *C. spinosa* var. *herbacea* genome assembly showed expansion of gene families involved in photosynthesis and response to abscisic acid (Wang et al. 2021). Phylogenetic analysis of a Brassicales-specific gene encoding *PROTEIN PHOSPHATASE7-LIKE* (*PP7L*) involved in chloroplast development and abiotic stress tolerance suggested that *PP7L* evolved about 43–92 MYA, after the divergence of Caricaceae from the other Brassicales families but before the separation of Cleomaceae from Brassicaceae (Xu et al. 2019).

The future of Brassicales extremophyte research

Core stress response pathways exist in all land plants. Yet, selective pressures imposed by extreme environments have driven novel evolutionary innovations to modify these core pathways or add alternatives. These evolutionary innovations are being revealed by comparative analyses of the premier model species *A. thaliana* (“stress-sensitive”) with the Brassicaceae extremophyte models. Yet these Brassicaceae species are all annuals, and alternative stress tolerance strategies could be found in perennial extremophytes that survive harsh stresses over multiple years. Thus, the gap in the adaptive trait space left by the Brassicaceae models can be narrowed by studies of perennial Brassicales extremophytes, which possess the added advantage of shared ancestry with *A. thaliana*, many of whose core stress response pathways have been identified.

The lack of genomic resources for the Brassicales extremophytes has limited research into their adaptations to extreme environments. Emerging technologies in large-scale and single-molecule/cell sequencing offer unprecedented molecular tools that require little prior genetic information to produce primary genomic resources. Single-molecule/hyperspectral imaging and phenomics can then facilitate linkage of novel genomic variation to unique extremophyte traits. Thus, the broader phylogenetic footprint capturing the life history strategies of the Brassicales extremophytes could facilitate new discoveries for sustainable crop production, land reclamation, and biodiversity conservation in this era of climate change and global threats to the environment and food insecurity.

Supporting model orders with global communities

Issues identified not only in the determination of relationships among or within the Brassicales but also in analysis of all the phenotypes discussed here create many opportunities to leverage the trait diversity across the order. To fully take advantage of these opportunities for both those currently investigating this diverse plant order and those who wish to, there must be a cohesive development of a global network for plant material, data generated, and researchers. Turner-Hissong et al. (2020) highlight ways to curate community efforts in organizing not only germplasm and data but also outreach and broader impacts. Although these suggestions are primarily directed at crop systems, many ideas are useful here too. For example, there is a need to develop reliable stock sources or live genetic material, which researchers can both order and submit samples, to increase reproducibility across studies. Some examples of this exist, such as the *Arabidopsis* Biological Resource Center (ABRC; https://abrc.osu.edu/), the Universidad Politécnica de Madrid Plant Germplasm Bank (UPM-PGB; which includes the César Gómez Campo Collection), the USDA Germplasm Resources Information Network (GRIN), and the Royal Botanic Gardens (Kew) Millennium Seed Bank and its DNA and tissue bank collections. However, one must first know about these resources to utilize them. However, even with these resources available, cultivating them demands substantial time, effort, funding, and space. Numerous species lack established protocols, particularly those accessible to the public, on optimizing plant growth for generating sufficient seeds for trait-focused studies. It is essential to meticulously document the origins of genetic resources and acknowledge the knowledge's source by returning it to the country of origin.

Following germplasm curation, data curation of all types is crucial for continual progress. As we generate these large, multi-omic datasets, there is an increasing need for scalable, systematic community efforts to organize these resources, all of which will require standards for data collection and curation and ideally utilize established open-access repositories. Fortunately, there are already databases in existence, such as BrassiBase (Koch et al. 2018), which consolidate information encompassing taxonomy, systematics, evolution, available germplasm resources, and specimen collections and could be expanded to include all of the Brassicales in the future. Furthermore, networks like the Multinational *Brassica* Genome Project (MBGP; brassica.info) could play a pivotal role if expanded to Brassicales. Currently, MBGP organizes annual meetings at international conferences and is dedicated to establishing a *Brassica* Information System, which includes the formulation of standards for describing data entities and experimental resources. Additionally, platforms like Phytozome, which serves as a central hub for accessing, visualizing, and analyzing JGI-sequenced plant genomes, along with the recently published and updated JGI Gene Atlas (Sreedasyam et al. 2023), can greatly contribute to the development of a cohesive community focused on the Brassicales model clade. These types of databases provide a place where not only data can be accessed, but researchers can also keep others updated on current projects, results, and future plans. Centralizing access to these plant materials and data generated would significantly decrease the barriers for scientists, especially early career scientists and those with less access to modern resources, and ultimately encourage equity and recruit new and diverse researchers to study the model order Brassicales. Nonetheless, persistent challenges exist, including a decline in taxonomic expertise, restricted access to materials from several small endemic families, and a noticeable bias toward collaboration with researchers primarily from the global North.

Greater accessibility to plant genomics at lower costs has expanded global collaboration opportunities. However, it is important to acknowledge that the discovery of genes and pathways in plant species can have significant economic implications, so these initiatives must also address the growing complexity surrounding genetic resources and intellectual property rights. There are growing calls for decolonization of science in the global North that extracts materials and knowledge from the global South and Indigenous groups without reciprocity (Radcliffe 2017; Armstrong and McAlvay 2019; Baker et al. 2019; Carroll et al. 2020; The First Nations Information Governance Centre 2022). Within the order Brassicales, *Pentadiplandra brazzeana* was at the center of a biopiracy controversy when scientists from the University of Wisconsin patented a protein, brazzein, extracted from the berries of Gabonese plants as a sweetening agent with no benefit-sharing agreement with Gabon or Gabonese people (Blakeney 2019; Dwivedy et al. 2019). A growing number of countries have implemented the Nagoya Protocol on Access and Benefit-sharing (Buck and Hamilton 2011) and other measures to govern the exchange and use of biological materials. Moving forward, we urge all researchers working on Brassicales collection and sequencing efforts to consider genuine collaboration with Indigenous groups and scientists in the global South to work to align project goals with local interests (Baker et al. 2019). We understand the need to build a foundation of respect, reciprocity, and accountability within our growing collaborative efforts in keeping with the principles of Indigenous data sovereignty and equitable benefit sharing. The authors here commit to these core principles throughout our contacts and research.

Acknowledgments

The authors would like to acknowledge the effort from the Joint Genome Institute (JGI) Community Science Program which inspired ideas for this review paper.

Author contributions

All authors were involved in the conception, writing, and editing of this manuscript.

Funding

MD and SB would like to acknowledge the support provided by the NSF-BSF-IOS-EDGE (National Science Foundation, USA, and The United States-Israel Binational Science Foundation) 1923589/2019610, The Goldinger Trust Jewish Fund for the Future awarded to SB, and Department of Energy BER-DE-SC0020358 awarded to MD. AH would like to acknowledge the support provided by NSF-IOS-PGRP CAREER (National Science Foundation, USA) Award 2239530.

Conflict of interest statement. None declared.

References

Abrahams RS, Pires JC, Schranz ME. Genomic origin and diversification of the glucosinolate MAM locus. Front Plant Sci. 2020:**11**:711. https://doi.org/10.3389/fpls.2020.00711

Akagi T, Henry IM, Ohtani H, Morimoto T, Beppu K, Kataoka I, Tao R. A Y-encoded suppressor of feminization arose via lineage-specific duplication of a cytokinin response regulator in kiwifruit. Plant Cell. 2018:**30**(4):780–795. https://doi.org/10.1105/tpc.17.00787

Al-Shehbaz IA. The tribes of Cruciferae (Brassicaceae) in the southeastern United States. J Arnold Arbor. 1984:**65**(3):343–373. https://doi.org/10.5962/p.36696

Andersson L, Andersson S. A molecular phylogeny of *Tropaeolaceae* and its systematic implications. Taxon. 2000:**49**(4):721–736. https://doi.org/10.2307/1223973

Antonelli A, Clarkson JJ, Kainulainen K, Maurin O, Brewer GE, Davis AP, Epitawalage N, Goyder DJ, Livshultz T, Persson C, et al. Settling a family feud: a high-level phylogenomic framework for the Gentianales based on 353 nuclear genes and partial plastomes. Am J Bot. 2021:**108**(7):1143–1165. https://doi.org/10.1002/ajb2.1697

Apel P, Horstmann C, Pfeffer M. The *Moricandia* syndrome in species of the Brassicaceae - evolutionary aspects. Photosynthetica. 1997:**33**(2):205–215. https://doi.org/10.1023/A:1022108229082

Arias T, Pires JC. A fully resolved chloroplast phylogeny of the brassica crops and wild relatives (Brassicaceae: Brassiceae): novel clades and potential taxonomic implications. Taxon. 2012:**61**(5):980–988. https://doi.org/10.1002/tax.615005

Armstrong CG, McAlvay AC. Introduction to special section on action ethnobiology. J Ethnobiol. 2019:**39**(1):3–13. https://doi.org/10.2993/0278-0771-39.1.3

Arshad W, Sperber K, Steinbrecher T, Nichols B, Jansen VAA, Leubner-Metzger G, Mummenhoff K. Dispersal biophysics and adaptive significance of dimorphic diaspores in the annual *Aethionema arabicum* (Brassicaceae). New Phytol. 2019:**221**(3):1434–1446. https://doi.org/10.1111/nph.15490

Aryal R, Jagadeeswaran G, Zheng Y, Yu Q, Sunkar R, Ming R. Sex specific expression and distribution of small RNAs in papaya. BMC Genomics. 2014:**15**(1):20. https://doi.org/10.1186/1471-2164-15-20

Babula D, Kaczmarek M, Barakat A, Delseny M, Quiros CF, Sadowski J. Chromosomal mapping of Brassica oleracea based on ESTs from Arabidopsis thaliana: complexity of the comparative map. Mol Gen Genomics. 2003:**268**(5):656–665. https://doi.org/10.1007/s00438-002-0782-2

Baker WJ, Bailey P, Barber V, Barker A, Bellot S, Bishop D, Botigué LR, Brewer G, Carruthers T, Clarkson JJ, et al. A comprehensive phylogenomic platform for exploring the angiosperm tree of life. Syst Biol. 2022:**71**(2):301–319. https://doi.org/10.1093/sysbio/syab035

Baker WJ, Dodsworth S, Forest F, Graham SW, Johnson MG, McDonnell A, Pokorny L, Tate JA, Wicke S, Wickett NJ. Exploring Angiosperms353: an open, community toolkit for collaborative phylogenomic research on flowering plants. Am J Bot. 2021:**108**(7):1059–1065. https://doi.org/10.1002/ajb2.1703

Baker K, Eichhorn MP, Griffiths M. Decolonizing field ecology. Biotropica. 2019:**51**(3):288–292. https://doi.org/10.1111/btp.12663

Baker KS, Steadman KJ, Plummer JA, Dixon KW. Seed dormancy and germination responses of nine Australian fire ephemerals. Plant Soil. 2005:**277**(1–2):345–358. https://doi.org/10.1007/s11104-005-7971-9

Barker MS, Vogel H, Schranz ME. Paleopolyploidy in the Brassicales: analyses of the *Cleome* transcriptome elucidate the history of genome duplications in *Arabidopsis* and other Brassicales. Genome Biol Evol. 2009:**1**:391–399. https://doi.org/10.1093/gbe/evp040

Barragan AC, Collenberg M, Schwab R, Kerstens M, Bezrukov I, Bemm F, Požárová D, Kolář F, Weigel D. Homozygosity at its limit: inbreeding depression in wild Arabidopsis arenosa populations. bioRxiv. 2021. https://doi.org/10.1101/2021.01.24.427284, 25 January 2021, preprint: not peer reviewed.

Barrett RL, Roalson EH, Ottewell K, Byrne M, Govindwar SP, Yadav SR, Tamboli AS, Gholave AR. Resolving generic boundaries in Indian-Australasian Cleomaceae: circumscription of *Areocleome*, *Arivela*, and *Corynandra* as distinct genera. Syst Bot. 2017:**42**(4):694–708. https://doi.org/10.1600/036364417X696401

Bayat S, Schranz ME, Roalson EH, Hall JC. Lessons from Cleomaceae, the sister of crucifers. Trends Plant Sci. 2018:**23**(9):808–821. https://doi.org/10.1016/j.tplants.2018.06.010

Bayer C, Appel O. Bataceae. In: **Kubitzki K, Bayer C**, editors. Flowering plants · dicotyledons: Malvales, Capparales and non-Betalain Caryophyllales. Berlin, Heidelberg: Springer Berlin Heidelberg; 2003. p. 30–32.

Beaulieu JM, O'Meara BC. Detecting hidden diversification shifts in models of trait-dependent speciation and extinction. Syst Biol. 2016:**65**(4):583–601. https://doi.org/10.1093/sysbio/syw022

Beekwilder J, van Leeuwen W, van Dam NM, Bertossi M, Grandi V, Mizzi L, Soloviev M, Szabados L, Molthoff JW, Schipper B, et al. The impact of the absence of aliphatic glucosinolates on insect herbivory in *Arabidopsis*. PLoS One. 2008:**3**(4):e2068. https://doi.org/10.1371/journal.pone.0002068

Beric A, Mabry ME, Harkess AE, Brose J, Schranz ME, Conant GC, Edger PP, Meyers BC, Pires JC. Comparative phylogenetics of repetitive elements in a diverse order of flowering plants (Brassicales). G3 (Bethesda). 2021:**11**:jkab140. https://doi.org/10.1093/g3journal/jkab140

Bewick AJ, Ji L, Niederhuth CE, Willing E-M, Hofmeister BT, Shi X, Wang L, Lu Z, Rohr NA, Hartwig B, et al. On the origin and evolutionary consequences of gene body DNA methylation. Proc Natl Acad Sci USA. 2016:**113**(32):9111–9116. https://doi.org/10.1073/pnas.1604666113

Bhattacharya S, Mayland-Quellhorst S, Müller C, Mummenhoff K. Two-tier morpho-chemical defence tactic in *Aethionema* via fruit morph plasticity and glucosinolates allocation in diaspores. Plant Cell Environ. 2019b:**42**(4):1381–1392. https://doi.org/10.1111/pce.13462

Bhattacharya S, Sperber K, Özüdoğru B, Leubner-Metzger G, Mummenhoff K. Naturally-primed life strategy plasticity of dimorphic *Aethionema arabicum* facilitates optimal habitat colonization. Sci Rep. 2019a:**9**(1):16108. https://doi.org/10.1038/s41598-019-52520-y

Birchler JA, Veitia RA. The gene balance hypothesis: from classical genetics to modern genomics. Plant Cell. 2007:**19**(2):395–402. https://doi.org/10.1105/tpc.106.049338

Birchler JA, Veitia RA. Gene balance hypothesis: connecting issues of dosage sensitivity across biological disciplines. Proc Natl Acad Sci USA. 2012:**109**(37):14746–14753. https://doi.org/10.1073/pnas.1207726109

Bird KA, Niederhuth C, Ou S, Gehan M, Pires JC, Xiong Z, VanBuren R, Edger PP. Replaying the evolutionary tape to investigate subgenome dominance in allopolyploid *Brassica napus*. New Phytol. 2020:**230**(1):354–371. https://doi.org/10.1111/nph.17137

Bird KA, Pires JC, VanBuren R, Xiong Z, Edger PP. Dosage-sensitivity shapes how genes transcriptionally respond to allopolyploidy and homoeologous exchange in resynthesized *Brassica napus*. Genetics. 2023:**225**(1):iyad114. https://doi.org/10.1093/genetics/iyad114

Birkeland S, Gustafsson ALS, Brysting AK, Brochmann C, Nowak MD. Multiple genetic trajectories to extreme abiotic stress adaptation in arctic Brassicaceae. Mol Biol Evol. 2020:**37**(7):2052–2068. https://doi.org/10.1093/molbev/msaa068

Blakeney M. Remedying the misappropriation of genetic resources. In: **Singh HB Keswani C, Singh SP**, editors. Intellectual property issues in microbiology. Singapore: Springer Singapore; 2019. p. 147–161.

Blischak PD, Mabry ME, Conant GC, Pires JC. Integrating networks, phylogenomics, and population genomics for the study of polyploidy. Annu Rev Ecol Evol Syst. 2018:**49**(1):253–278. https://doi.org/10.1146/annurev-ecolsys-121415-032302

Bowman JL, Drews GN, Meyerowitz EM. Expression of the *Arabidopsis* floral homeotic gene AGAMOUS is restricted to specific cell types late in flower development. Plant Cell. 1991a:**3**(8):749–758. https://doi.org/10.1105/tpc.3.8.749

Bowman JL, Smyth DR, Meyerowitz EM. Genetic interactions among floral homeotic genes of *Arabidopsis*. Development. 1991b:**112**(1): 1–20. https://doi.org/10.1242/dev.112.1.1

Brassica rapa Genome Sequencing Project Consortium. The genome of the mesopolyploid crop species Brassica rapa. Nat Genetics. 2011:**43**(10):1035–1039. https://doi.org/10.1038/ng.919

Bräutigam A, Kajala K, Wullenweber J, Sommer M, Gagneul D, Weber KL, Carr KM, Gowik U, Maß J, Lercher MJ, et al. An mRNA blueprint for C_4 photosynthesis derived from comparative transcriptomics of closely related C_3 and C_4 species. Plant Physiol. 2011:**155**(1):142–156. https://doi.org/10.1104/pp.110.159442

Brewer GE, Clarkson JJ, Maurin O, Zuntini AR, Barber V, Bellot S, Biggs N, Cowan RS, Davies NMJ, Dodsworth S, et al. Factors affecting targeted sequencing of 353 nuclear genes from herbarium specimens spanning the diversity of angiosperms. Front Plant Sci. 2019:**10**: e01102. https://doi.org/10.3389/fpls.2019.01102

Briskine RV, Paape T, Shimizu-Inatsugi R, Nishiyama T, Akama S, Sese J, Shimizu KK. Genome assembly and annotation of *Arabidopsis halleri*, a model for heavy metal hyperaccumulation and evolutionary ecology. Mol Ecol Resour. 2017:**17**(5):1025–1036. https://doi.org/10.1111/1755-0998.12604

Brown JE, Bauman JM, Lawrie JF, Rocha OJ, Moore RC. The structure of morphological and genetic diversity in natural populations of *Carica papaya* (Caricaceae) in Costa Rica. Biotropica. 2012:**44**(2): 179–188. https://doi.org/10.1111/j.1744-7429.2011.00779.x

Brown NJ, Newell CA, Stanley S, Chen JE, Perrin AJ, Kajala K, Hibberd JM. Independent and parallel recruitment of preexisting mechanisms underlying C_4 photosynthesis. Science. 2011:**331**(6023):1436–1439. https://doi.org/10.1126/science.1201248

Buck M, Hamilton C. The Nagoya protocol on access to genetic resources and the fair and equitable sharing of benefits arising from their utilization to the convention on biological diversity. Rev Eur Community Int Environ Law. 2011:**20**(1):47–61. https://doi.org/10.1111/j.1467-9388.2011.00703.x

Byng JW. The flowering plants handbook: a practical guide to families and genera of the world. United Kingdom: Plant Gateway Ltd; 2014.

Cacho NI, Kliebenstein DJ, Strauss SY. Macroevolutionary patterns of glucosinolate defense and tests of defense-escalation and resource availability hypotheses. New Phytol. 2015:**208**(3):915–927. https://doi.org/10.1111/nph.13561

Cacho NI, McIntyre PJ, Kliebenstein DJ, Strauss SY. Genome size evolution is associated with climate seasonality and glucosinolates, but not life history, soil nutrients or range size, across a clade of mustards. Ann Bot. 2021:**127**(7):887–902. https://doi.org/10.1093/aob/mcab028

Cardinal-McTeague WM, Sytsma KJ, Hall JC. Biogeography and diversification of Brassicales: a 103 million year tale. Mol Phylogenet Evol. 2016:**99**:204–224. https://doi.org/10.1016/j.ympev.2016.02.021

Carey S, Yu Q, Harkess A. The diversity of plant sex chromosomes highlighted through advances in genome sequencing. Genes. 2021:**12**(3):381. https://doi.org/10.3390/genes12030381

Carroll SR, Garba I, Figueroa-Rodríguez OL, Holbrook J, Lovett R, Materechera S, Parsons M, Raseroka K, Rodriguez-Lonebear D, Rowe R, et al. The CARE principles for indigenous data governance. Data Sci J. 2020:**19**:43. https://doi.org/10.5334/dsj-2020-043

Carvalho FA, Renner SS. IV. A dated phylogeny of the papaya family (Caricaceae) reveals the crop's closest relatives and the family's biogeographic history. In: **Carvalho FA** ed., Molecular phylogeny, biogeography and an e-monograph of the papaya family (Caricaceae) as an example of taxonomy in the electronic age. Wiesbaden, Germany: Springer; 2015. p. 49–81.

Chan EKF, Rowe HC, Kliebenstein DJ. Understanding the evolution of defense metabolites in *Arabidopsis thaliana* using genome-wide association mapping. Genetics. 2010:**185**(3):991–1007. https://doi.org/10.1534/genetics.109.108522

Cheng F, Sun C, Wu J, Schnable J, Woodhouse MR, Liang J, Cai C, Freeling M, Wang X. Epigenetic regulation of subgenome dominance following whole genome triplication in *Brassica rapa*. New Phytol. 2016:**211**(1):288–299. https://doi.org/10.1111/nph.13884

Cheng F, Wu J, Fang L, Sun S, Liu B, Lin K, Bonnema G, Wang X. Biased gene fractionation and dominant gene expression among the subgenomes of *Brassica rapa*. PLoS One. 2012:**7**(5):e36442. https://doi.org/10.1371/journal.pone.0036442

Christin P-A, Osborne CP, Chatelet DS, Columbus JT, Besnard G, Hodkinson TR, Garrison LM, Vorontsova MS, Edwards EJ. Anatomical enablers and the evolution of C_4 photosynthesis in grasses. Proc Natl Acad Sci USA. 2013:**110**(4):1381–1386. https://doi.org/10.1073/pnas.1216777110

Christin P-A, Salamin N, Savolainen V, Duvall MR, Besnard G. C_4 photosynthesis evolved in grasses via parallel adaptive genetic changes. Curr Biol. 2007:**17**(14):1241–1247. https://doi.org/10.1016/j.cub.2007.06.036

Conant GC. Comparative genomics as a time machine: how relative gene dosage and metabolic requirements shaped the time-dependent resolution of yeast polyploidy. Mol Biol Evol. 2014:**31**(12):3184–3193. https://doi.org/10.1093/molbev/msu250

Conant GC. The lasting after-effects of an ancient polyploidy on the genomes of teleosts. PLoS One. 2020:**15**(4):e0231356. https://doi.org/10.1371/journal.pone.0231356

Conant GC, Birchler JA, Pires JC. Dosage, duplication, and diploidization: clarifying the interplay of multiple models for duplicate gene evolution over time. Curr Opin Plant Biol. 2014:**19**:91–98. https://doi.org/10.1016/j.pbi.2014.05.008

Conant GC, Wolfe KH. Probabilistic cross-species inference of orthologous genomic regions created by whole-genome duplication in yeast. Genetics. 2008:**179**(3):1681–1692. https://doi.org/10.1534/genetics.107.074450

Cornwell WK, Westoby M, Falster DS, FitzJohn RG, O'Meara BC, Pennell MW, McGlinn DJ, Eastman JM, Moles AT, Reich PB, et al. Functional distinctiveness of major plant lineages. J Ecol. 2014:**102**(2):345–356. https://doi.org/10.1111/1365-2745.12208

Czerniawski P, Piasecka A, Bednarek P. Evolutionary changes in the glucosinolate biosynthetic capacity in species representing *Capsella, Camelina* and *Neslia* genera. Phytochemistry. 2021:**181**: 112571. https://doi.org/10.1016/j.phytochem.2020.112571

Damerval C, Becker A. Genetics of flower development in Ranunculales - a new, basal eudicot model order for studying flower evolution. New Phytol. 2017:**216**(2):361–366. https://doi.org/10.1111/nph.14401

Dassanayake M, Oh D-H, Haas JS, Hernandez A, Hong H, Ali S, Yun D-J, Bressan RA, Zhu J-K, Bohnert HJ, et al. The genome of the extremophile crucifer *Thellungiella parvula*. Nat Genet. 2011:**43**(9): 913–918. https://doi.org/10.1038/ng.889

Davin N, Edger PP, Hefer CA, Mizrachi E, Schuetz M, Smets E, Myburg AA, Douglas CJ, Schranz ME, Lens F. Functional network analysis of genes differentially expressed during xylogenesis in *soc1ful* woody *Arabidopsis* plants. Plant J. 2016:**86**(5):376–390. https://doi.org/10.1111/tpj.13157

Debez A, Saadaoui D, Slama I, Huchzermeyer B, Abdelly C. Responses of *Batis maritima* plants challenged with up to two-fold seawater NaCl salinity. J Plant Nutr Soil Sci. 2010:**173**(2):291–299. https://doi.org/10.1002/jpln.200900222

Dickinson PJ, Kneřová J, Szecówka M, Stevenson SR, Burgess SJ, Mulvey H, Bågman A-M, Gaudinier A, Brady SM, Hibberd JM. A bipartite transcription factor module controlling expression in the bundle sheath of *Arabidopsis thaliana*. Nat Plants. 2020:**6**(12): 1468–1479. https://doi.org/10.1038/s41477-020-00805-w

Dodsworth S, Pokorny L, Johnson MG, Kim JT, Maurin O, Wickett NJ, Forest F, Baker WJ. Hyb-Seq for flowering plant systematics. Trends Plant Sci. 2019:**24**(10):887–891. https://doi.org/10.1016/j.tplants.2019.07.011

Dória LC, Podadera DS, del Arco M, Chauvin T, Smets E, Delzon S, Lens F. Insular woody daisies (*Argyranthemum*, Asteraceae) are more resistant to drought-induced hydraulic failure than their herbaceous relatives. Funct Ecol. 2018:**32**(6):1467–1478. https://doi.org/10.1111/1365-2435.13085

Drews GN, Bowman JL, Meyerowitz EM. Negative regulation of the *Arabidopsis* homeotic gene AGAMOUS by the APETALA2 product. Cell. 1991:**65**(6):991–1002. https://doi.org/10.1016/0092-8674(91)90551-9

Dwivedy AK, Singh VK, Das S, Chaudhari AK, Upadhyay N, Singh A, Singh A, Dubey NK. Biodiversity bioprospection with respect to medicinal plants. In: Kumar Patra J, Das G, Kumar S, Thatoi H, editors. Ethnopharmacology and biodiversity of medicinal plants. Boca Raton: Apple Academic Press; 2019. p. 3–30.

Edger PP, Hall JC, Harkess A, Tang M, Coombs J, Mohammadin S, Schranz ME, Xiong Z, Leebens-Mack J, Meyers BC, et al. Brassicales Phylogeny inferred from 72 plastid genes: a reanalysis of the phylogenetic localization of two paleopolyploid events and origin of novel chemical defenses. Am J Bot. 2018a:**105**(3):463–469. https://doi.org/10.1002/ajb2.1040

Edger PP, Heidel-Fischer HM, Bekaert M, Rota J, Glöckner G, Platts AE, Heckel DG, Der JP, Wafula EK, Tang M, et al. The butterfly plant arms-race escalated by gene and genome duplications. Proc Natl Acad Sci USA. 2015:**112**(27):8362–8366. https://doi.org/10.1073/pnas.1503926112

Edger PP, McKain MR, Bird KA, VanBuren R. Subgenome assignment in allopolyploids: challenges and future directions. Curr Opin Plant Biol. 2018b:**42**:76–80. https://doi.org/10.1016/j.pbi.2018.03.006

Edwards EJ. Evolutionary trajectories, accessibility and other metaphors: the case of C_4 and CAM photosynthesis. New Phytol. 2019:**223**(4):1742–1755. https://doi.org/10.1111/nph.15851

Elmer KR, Meyer A. Adaptation in the age of ecological genomics: insights from parallelism and convergence. Trends Ecol Evol. 2011:**26**(6):298–306. https://doi.org/10.1016/j.tree.2011.02.008

Eserman LA, Thomas SK, Coffey EED, Leebens-Mack JH. Target sequence capture in orchids: developing a kit to sequence hundreds of single-copy loci. Appl Plant Sci. 2021:**9**(7):e11416. https://doi.org/10.1002/aps3.11416

Eshel G, Duppen N, Wang G, Oh D-H, Kazachkova Y, Herzyk P, Amtmann A, Gordon M, Chalifa-Caspi V, Oscar MA, et al. Positive selection and heat-response transcriptomes reveal adaptive features of the Brassicaceae desert model, *Anastatica hierochuntica*. New Phytol. 2022:**236**(3):1006–1026. https://doi.org/10.1111/nph.18411

Eshel G, Shaked R, Kazachkova Y, Khan A, Eppel A, Cisneros A, Acuna T, Gutterman Y, Tel-Zur N, Rachmilevitch S, et al. *Anastatica hierochuntica*, an *Arabidopsis* desert relative, is tolerant to multiple abiotic stresses and exhibits Species-specific and common stress tolerance strategies with its halophytic relative, *Eutrema (Thellungiella) salsugineum*. Front Plant Sci. 2017:**7**:1992. https://doi.org/10.3389/fpls.2016.01992

Fei Q, Xia R, Meyers BC. Phased, secondary, small interfering RNAs in posttranscriptional regulatory networks. Plant Cell. 2013:**25**(7): 2400–2415. https://doi.org/10.1105/tpc.113.114652

Forsythe ES, Nelson ADL, Beilstein MA. Biased gene retention in the face of introgression obscures species relationships. Genome Biol Evol. 2020:**12**(9):1646–1663. https://doi.org/10.1093/gbe/evaa149

Franzke A, Lysak MA, Al-Shehbaz IA, Koch MA, Mummenhoff K. Cabbage family affairs: the evolutionary history of Brassicaceae. Trends Plant Sci. 2011:**16**(2):108–116. https://doi.org/10.1016/j.tplants.2010.11.005

Freeling M, Woodhouse MR, Subramaniam S, Turco G, Lisch D, Schnable JC. Fractionation mutagenesis and similar consequences of mechanisms removing dispensable or less-expressed DNA in plants. Curr Opin Plant Biol. 2012:**15**(2):131–139. https://doi.org/10.1016/j.pbi.2012.01.015

Fuentes G, Santamaría JM. Papaya (*Carica papaya* L.): origin, domestication, and production. In: **Ming R, Moore PH**, editors. Genetics and genomics of papaya. New York (NY): Springer New York; 2014. p. 3–15.

Fujii S, Kubo K-I, Takayama S. Non-self- and self-recognition models in plant self-incompatibility. Nat Plants. 2016:**2**(9):16130. https://doi.org/10.1038/nplants.2016.130

Gallego-Bartolomé J, Liu W, Kuo PH, Feng S, Ghoshal B, Gardiner J, Zhao JM-C, Park SY, Chory J, Jacobsen SE. Co-targeting RNA polymerases IV and V promotes efficient de novo DNA methylation in *Arabidopsis*. Cell. 2019:**176**(5):1068–1082.e19. https://doi.org/10.1016/j.cell.2019.01.029

Gan X, Hay A, Kwantes M, Haberer G, Hallab A, Ioio RD, Hofhuis H, Pieper B, Cartolano M, Neumann U, et al. Erratum: the *Cardamine hirsuta* genome offers insight into the evolution of morphological diversity. Nat Plants. 2016:**2**(12):16189. https://doi.org/10.1038/nplants.2016.189

Garassino F, Wijfjes RY, Boesten R, Reyes Marquez F, Becker FFM, Clapero V, van den Hatert I, Holmer R, Schranz ME, Harbinson J, et al. The genome sequence of *Hirschfeldia incana*, a new Brassicaceae model to improve photosynthetic light-use efficiency. Plant J. 2022:**112**(5):1298–1315. https://doi.org/10.1111/tpj.16005

Gebrekirstos A, Teketay D, Mitlöhner R. Responses of *Dobera glabra* and eight co-occurring species to drought and salinity stress at a savanna-scrub ecotone: implications in the face of climate change. Open J For. 2014:**4**(4):327–337. https://doi.org/10.4236/ojf.2014.44039

Gendrel A-V, Lippman Z, Yordan C, Colot V, Martienssen RA. Dependence of heterochromatic histone H3 methylation patterns on the *Arabidopsis gene DDM1*. Science. 2002:**297**(5588): 1871–1873. https://doi.org/10.1126/science.1074950

German DA, Hendriks KP, Koch MA, Lens F, Lysak MA, Bailey CD, Mummenhoff K, Al-Shehbaz IA. An updated classification of the Brassicaceae (Cruciferae). PhytoKeys. 2023:**220**:127–144. https://doi.org/10.3897/phytokeys.220.97724

Goodstein DM, Shu S, Howson R, Neupane R, Hayes RD, Fazo J, Mitros T, Dirks W, Hellsten U, Putnam N, et al. Phytozome: a comparative platform for green plant genomics. Nucleic Acids Res. 2012:**40**(D1):D1178–D1186. https://doi.org/10.1093/nar/gkr944

Guo X, Liu J, Hao G, Zhang L, Mao K, Wang X, Zhang D, Ma T, Hu Q, Al-Shehbaz IA, et al. Plastome phylogeny and early diversification of Brassicaceae. BMC Genom. 2017:**18**(1):176. https://doi.org/10.1186/s12864-017-3555-3

Hall JC. Systematics of Capparaceae and Cleomaceae: an evaluation of the generic delimitations of *Capparis* and *Cleome* using plastid DNA sequence data. Botany. 2008:**86**(7):682–696. https://doi.org/10.1139/B08-026

Hall JC, Iltis HH, Sytsma KJ. Molecular phylogenetics of core Brassicales, placement of orphan genera *Emblingia*, *Forchhammeria*, *Tirania*, and character evolution. Syst Bot. 2004:**29**(3):654–669. https://doi.org/10.1600/0363644041744491

Hall JC, Sytsma KJ, Iltis HH. Phylogeny of Capparaceae and Brassicaceae based on chloroplast sequence data. Am J Bot. 2002:**89**(11):1826–1842. https://doi.org/10.3732/ajb.89.11.1826

Hansen BG, Kerwin RE, Ober JA, Lambrix VM, Mitchell-Olds T, Gershenzon J, Halkier BA, Kliebenstein DJ. A novel 2-oxoacid-dependent dioxygenase involved in the formation of the goiterogenic 2-hydroxybut-3-enyl glucosinolate and generalist insect resistance in *Arabidopsis*. Plant Physiol. 2008:**148**(4):2096–2108. https://doi.org/10.1104/pp.108.129981

Hao Y, Fleming J, Petterson J, Lyons E, Edger PP, Pires JC, Thorne JL, Conant GC. Convergent evolution of polyploid genomes from across the eukaryotic tree of life. G3 (Bethesda). 2022:**12**(6):jkac094. https://doi.org/10.1093/g3journal/jkac094

Haribal M, Yang Z, Attygalle AB, Renwick JA, Meinwald J. A cyanoallyl glucoside from *Alliaria petiolata*, as a feeding deterrent for larvae of *Pieris napi oleracea*. J Nat Prod. 2001:**64**(4):440–443. https://doi.org/10.1021/np000534d

Harkess A, Huang K, van der Hulst R, Tissen B, Caplan JL, Koppula A, Batish M, Meyers BC, Leebens-Mack J. Sex determination by two Y-linked genes in garden asparagus. Plant Cell. 2020:**32**(6): 1790–1796. https://doi.org/10.1105/tpc.19.00859

Haudry A, Platts AE, Vello E, Hoen DR, Leclercq M, Williamson RJ, Forczek E, Joly-Lopez Z, Steffen JG, Hazzouri KM, et al. An atlas of over 90,000 conserved noncoding sequences provides insight into crucifer regulatory regions. Nat Genet. 2013:**45**(8):891–898. https://doi.org/10.1038/ng.2684

Haug-Baltzell A, Stephens SA, Davey S, Scheidegger CE, Lyons E. Synmap2 and SynMap3D: web-based whole-genome synteny browsers. Bioinformatics. 2017:**33**(14):2197–2198. https://doi.org/10.1093/bioinformatics/btx144

Hawkins JS, Proulx SR, Rapp RA, Wendel JF. Rapid DNA loss as a counterbalance to genome expansion through retrotransposon proliferation in plants. Proc Natl Acad Sci USA. 2009:**106**(42): 17811–17816. https://doi.org/10.1073/pnas.0904339106

He H, Liang G, Li Y, Wang F, Yu D. Two young microRNAs originating from target duplication mediate nitrogen starvation adaptation via regulation of glucosinolate synthesis in *Arabidopsis thaliana*. Plant Physiol. 2014:**164**(2):853–865. https://doi.org/10.1104/pp.113.228635

Hendriks KP, Kiefer C, Al-Shehbaz IA, Bailey CD, Hooft van Huysduynen A, Nikolov LA, Nauheimer L, Zuntini AR, German DA, Franzke A, et al. Global Brassicaceae phylogeny based on filtering of 1,000-gene dataset. Curr Biol. 2023:**33**(19):4052–4068.e6. https://doi.org/10.1016/j.cub.2023.08.026

Hendriks KP, Mandáková T, Hay NM, Ly E, Hooft van Huysduynen A, Tamrakar R, Thomas SK, Toro-Núñez O, Pires JC, Nikolov LA, et al. The best of both worlds: combining lineage-specific and universal bait sets in target-enrichment hybridization reactions. Appl Plant Sci. 2021:**9**(7):e11438. https://doi.org/10.1002/aps3.11438

Hloušková P, Mandáková T, Pouch M, Trávníček P, Lysak MA. The large genome size variation in the *Hesperis* clade was shaped by the prevalent proliferation of DNA repeats and rarer genome downsizing. Ann Bot. 2019:**124**(1):103–120. https://doi.org/10.1093/aob/mcz036

Hoang NV, Deedi Sogbohossou EO, Xiong W, Simpson CJC, Singh P, Walden N, van den Bergh E, Becker FFM, Li F, Zhu X-G, et al. The *Gynandropsis gynandra* genome provides insights into whole-genome duplications and the evolution of C_4 photosynthesis in Cleomaceae. Plant Cell. 2023:**35**(5):1334–1359. https://doi.org/10.1093/plcell/koad018

Hollister JD, Gaut BS. Epigenetic silencing of transposable elements: a trade-off between reduced transposition and deleterious effects on neighboring gene expression. Genome Res. 2009:**19**(8):1419–1428. https://doi.org/10.1101/gr.091678.109

Holtum JAM, Hancock LP, Edwards EJ, Winter K. Optional use of CAM photosynthesis in two C_4 species, *Portulaca cyclophylla* and *Portulaca digyna*. J Plant Physiol. 2017:**214**:91–96. https://doi.org/10.1016/j.jplph.2017.01.010

Honjo MN, Kudoh H. *Arabidopsis halleri*: a perennial model system for studying population differentiation and local adaptation. AoB Plants. 2019:**11**(6):plz076. https://doi.org/10.1093/aobpla/plz076

Hu G, Wendel JF. *Cis*-trans controls and regulatory novelty accompanying allopolyploidization. New Phytol. 2019:**221**(4):1691–1700. https://doi.org/10.1111/nph.15515

Huang X-C, German DA, Koch MA. Temporal patterns of diversification in Brassicaceae demonstrate decoupling of rate shifts and mesopolyploidization events. Ann Bot. 2020:**125**(1):29–47. https://doi.org/10.1093/aob/mcz123

Huang C-F, Liu W-Y, Lu M-YJ, Chen Y-H, Ku MSB, Li W-H. Whole-Genome duplication facilitated the evolution of C_4 photosynthesis in *Gynandropsis gynandra*. Mol Biol Evol. 2021:**38**(11): 4715–4731. https://doi.org/10.1093/molbev/msab200

Huang C-H, Sun R, Hu Yi, Zeng L, Zhang N, Cai L, Zhang Q, Koch MA, Al-Shehbaz I, Edger PP, et al. Resolution of Brassicaceae phylogeny using nuclear genes uncovers nested radiations and supports convergent morphological evolution. Mol Biol Evol. 2016:**33**(2):394–412. https://doi.org/10.1093/molbev/msv226

Igic B, Lande R, Kohn JR. Loss of self-incompatibility and its evolutionary consequences. Int J Plant Sci. 2008:**169**(1):93–104. https://doi.org/10.1086/523362

Johnson MG, Pokorny L, Dodsworth S, Botigué LR, Cowan RS, Devault A, Eiserhardt WL, Epitawalage N, Forest F, Kim JT, et al. A universal probe set for targeted sequencing of 353 nuclear genes from any flowering plant designed using k-medoids clustering. Syst Biol. 2019:**68**(4):594–606. https://doi.org/10.1093/sysbio/syy086

Kagale S, Koh C, Nixon J, Bollina V, Clarke WE, Tuteja R, Spillane C, Robinson SJ, Links MG, Clarke C, et al. The emerging biofuel crop *Camelina sativa* retains a highly undifferentiated hexaploid genome structure. Nat Commun. 2014:**5**(1):3706. https://doi.org/10.1038/ncomms4706

Kajala K, Brown NJ, Williams BP, Borrill P, Taylor LE, Hibberd JM. Multiple *Arabidopsis* genes primed for recruitment into C_4 photosynthesis. Plant J. 2012:**69**(1):47–56. https://doi.org/10.1111/j.1365-313X.2011.04769.x

Kakrana A, Mathioni SM, Huang K, Hammond R, Vandivier L, Patel P, Arikit S, Shevchenko O, Harkess AE, Kingham B, et al. Plant 24-nt reproductive phasiRNAs from intramolecular duplex mRNAs in diverse monocots. Genome Res. 2018:**28**(9):1333–1344. https://doi.org/10.1101/gr.228163.117

Kakutani T, Munakata K, Richards EJ, Hirochika H. Meiotically and mitotically stable inheritance of DNA hypomethylation induced by ddm1 mutation of *Arabidopsis thaliana*. Genetics. 1999:**151**(2): 831–838. https://doi.org/10.1093/genetics/151.2.831

Kant S, Kant P, Raveh E, Barak S. Evidence that differential gene expression between the halophyte, *Thellungiella halophila*, and *Arabidopsis thaliana* is responsible for higher levels of the compatible osmolyte proline and tight control of Na+ uptake in *T. halophila*. Plant Cell Env. 2006:**29**(7):1220–1234. https://doi.org/10.1111/j.1365-3040.2006.01502.x

Kazachkova Y, Eshel G, Pantha P, Cheeseman JM, Dassanayake M, Barak S. Halophytism: what have we learnt from *Arabidopsis thaliana* relative model systems? Plant Physiol. 2018:**178**(3):972–988. https://doi.org/10.1104/pp.18.00863

Kerwin R, Feusier J, Corwin J, Rubin M, Lin C, Muok A, Larson B, Li B, Joseph B, Francisco M, et al. Natural genetic variation in *Arabidopsis thaliana* defense metabolism genes modulates field fitness. Elife. 2015:**4**: e05604 https://doi.org/10.7554/eLife.05604

Kerwin RE, Feusier J, Muok A, Lin C, Larson B, Copeland D, Corwin JA, Rubin MJ, Francisco M, Li B, et al. Epistasis × environment interactions among *Arabidopsis thaliana* glucosinolate genes impact complex traits and fitness in the field. New Phytol. 2017:**215**(3): 1249–1263. https://doi.org/10.1111/nph.14646

Kiefer C, Willing E-M, Jiao W-B, Sun H, Piednoël M, Hümann U, Hartwig B, Koch MA, Schneeberger K. Interspecies association mapping links reduced CG to TG substitution rates to the loss of gene-body methylation. Nat Plants. 2019:**5**(8):846–855. https://doi.org/10.1038/s41477-019-0486-9

Koch MA, German DA, Kiefer M, Franzke A. Database taxonomics as key to modern plant biology. Trends Plant Sci. 2018:**23**(1):4–6. https://doi.org/10.1016/j.tplants.2017.10.005

Koteyeva NK, Voznesenskaya EV, Cousins AB, Edwards GE. Differentiation of C_4 photosynthesis along a leaf developmental gradient in two *Cleome* species having different forms of Kranz anatomy. J Exp Bot. 2014:**65**(13):3525–3541. https://doi.org/10.1093/jxb/eru042

Koteyeva NK, Voznesenskaya EV, Roalson EH, Edwards GE. Diversity in forms of C_4 in the genus *Cleome* (Cleomaceae). Ann Bot. 2011:**107**(2):269–283. https://doi.org/10.1093/aob/mcq239

Kreiner JM, Kron P, Husband BC. Frequency and maintenance of unreduced gametes in natural plant populations: associations with reproductive mode, life history and genome size. New Phytol. 2017:**214**(2):879–889. https://doi.org/10.1111/nph.14423

Kroymann J, Donnerhacke S, Schnabelrauch D, Mitchell-Olds T. Evolutionary dynamics of an *Arabidopsis* insect resistance quantitative trait locus. Proc Natl Acad Sci USA. 2003:**100**(Suppl 2): 14587–14592. https://doi.org/10.1073/pnas.1734046100

Kumar R, Lee SG, Augustine R, Reichelt M, Vassão DG, Palavalli MH, Allen A, Gershenzon J, Jez JM, Bisht NC. Molecular basis of the evolution of methylthioalkylmalate synthase and the diversity of methionine-derived glucosinolates. Plant Cell. 2019:**31**(7): 1633–1647. https://doi.org/10.1105/tpc.19.00046

Lagercrantz U. Comparative mapping between *Arabidopsis thaliana* and *Brassica nigra* indicates that Brassica genomes have evolved through extensive genome replication accompanied by chromosome fusions and frequent rearrangements. Genetics. 1998:**150**(3): 1217–1228. https://doi.org/10.1093/genetics/150.3.1217

Lamesch P, Berardini TZ, Li D, Swarbreck D, Wilks C, Sasidharan R, Muller R, Dreher K, Alexander DL, Garcia-Hernandez M, et al. The *Arabidopsis* information resource (TAIR): improved gene annotation and new tools. Nucleic Acids Res. 2012:**40**(D1):D1202–D1210. https://doi.org/10.1093/nar/gkr1090

Lankau RA, Strauss SY. Mutual feedbacks maintain both genetic and species diversity in a plant community. Science. 2007:**317**(5844): 1561–1563. https://doi.org/10.1126/science.1147455

Lee AK, Gilman IS, Srivastav M, Lerner AD, Donoghue MJ, Clement WL. Reconstructing Dipsacales phylogeny using Angiosperms353: issues and insights. Am J Bot. 2021:**108**(7):1122–1142. https://doi.org/10.1002/ajb2.1695

Leitch IJ, Johnston E, Pellicer J, Hidalgo O, Bennett . Plant DNA C-values Database. Plant DNA C-values Database. Release 7.1, April 2019.

Lens F, Smets E, Melzer S. Stem anatomy supports *Arabidopsis thaliana* as a model for insular woodiness. New Phytol. 2012:**193**(1): 12–17. https://doi.org/10.1111/j.1469-8137.2011.03888.x

Lens F, Tixier A, Cochard H, Sperry JS, Jansen S, Herbette S. Embolism resistance as a key mechanism to understand adaptive plant strategies. Curr Opin Plant Biol. 2013:**16**(3):287–292. https://doi.org/10.1016/j.pbi.2013.02.005

Li ZQ, Li JX, Li HJ, Shi ZH, Zhang GF. Overexpression of TsApx1 from *Thellungiella salsuginea* improves abiotic stress tolerance in transgenic *Arabidopsis thaliana*. Biol Plant. 2015:**59**(3):497–506. https://doi.org/10.1007/s10535-015-0533-y

Li Z, McKibben MTW, Finch GS, Blischak PD, Sutherland BL, Barker MS. Patterns and processes of diploidization in land plants. Annu Rev Plant Biol. 2021:**72**(1):387–410. https://doi.org/10.1146/annurev-arplant-050718-100344

Lippman Z, Gendrel A-V, Black M, Vaughn MW, Dedhia N, Richard McCombie W, Lavine K, Mittal V, May B, Kasschau KD, et al. Role of transposable elements in heterochromatin and epigenetic control. Nature. 2004:**430**(6998):471–476. https://doi.org/10.1038/nature02651

Liu Z, Moore PH, Ma H, Ackerman CM, Ragiba M, Yu Q, Pearl HM, Kim MS, Charlton JW, Stiles JI, et al. A primitive Y chromosome in papaya marks incipient sex chromosome evolution. Nature. 2004:**427**(6972):348–352. https://doi.org/10.1038/nature02228

Lovell JT, Jenkins J, Lowry DB, Mamidi S, Sreedasyam A, Weng X, Barry K, Bonnette J, Campitelli B, Daum C, et al. The genomic landscape of molecular responses to natural drought stress in *Panicum hallii*. Nat Commun. 2018:**9**(1):5213. https://doi.org/10.1038/s41467-018-07669-x

Lovell JT, Sreedasyam A, Schranz ME, Wilson M, Carlson JW, Harkess A, Emms D, Goodstein DM, Schmutz J. GENESPACE Tracks regions of interest and gene copy number variation across multiple genomes. Elife. 2022:**11**: e78526. https://doi.org/10.7554/eLife.78526

Lu Z, Marand AP, Ricci WA, Ethridge CL, Zhang X, Schmitz RJ. The prevalence, evolution and chromatin signatures of plant regulatory elements. Nat Plants. 2019:**5**(12):1250–1259. https://doi.org/10.1038/s41477-019-0548-z

Lundgren MR. C_2 photosynthesis: a promising route towards crop improvement? New Phytol. 2020:**228**(6):1734–1740. https://doi.org/10.1111/nph.16494

Lyons EH. Coge, a new kind of comparative genomics platform: insights into the evolution of plant genomes. Berkeley, CA: University of California; 2008.

Lyons E, Pedersen B, Kane J, Freeling M. The value of nonmodel genomes and an example using SynMap within CoGe to dissect the hexaploidy that predates the rosids. Tropical Plant Biology. 2008:**1**(3–4):181–190. https://doi.org/10.1007/s12042-008-9017-y

Lysak MA. Brassicales: an update on chromosomal evolution and ancient polyploidy. Plant Syst Evol. 2018:**304**(6):757–762. https://doi.org/10.1007/s00606-018-1507-2

Lysak MA, Cheung K, Kitschke M, Bures P. Ancestral chromosomal blocks are triplicated in Brassiceae species with varying chromosome number and genome size. Plant Physiol. 2007:**145**(2):402–410. https://doi.org/10.1104/pp.107.104380

Lysak MA, Koch MA, Pecinka A, Schubert I. Chromosome triplication found across the tribe Brassiceae. Genome Res. 2005:**15**(4):516–525. https://doi.org/10.1101/gr.3531105

Mabry ME, Brose JM, Blischak PD, Sutherland B, Dismukes WT, Bottoms CA, Edger PP, Washburn JD, An H, Hall JC, et al. Phylogeny and multiple independent whole-genome duplication events in the Brassicales. Am J Bot. 2020:**107**(8):1148–1164. https://doi.org/10.1002/ajb2.1514

Magallon S, Crane PR, Herendeen PS. Phylogenetic pattern, diversity, and diversification of eudicots. Ann Missouri Bot Gard. 1999:**86**(2): 297. https://doi.org/10.2307/2666180

Mandáková T, Li Z, Barker MS, Lysak MA. Diverse genome organization following 13 independent mesopolyploid events in Brassicaceae contrasts with convergent patterns of gene retention. Plant J. 2017:**91**(1):3–21. https://doi.org/10.1111/tpj.13553

Manzaneda AJ, Prasad KVSK, Mitchell-Olds T. Variation and fitness costs for tolerance to different types of herbivore damage in *Boechera stricta* genotypes with contrasting glucosinolate structures. New Phytol. 2010:**188**(2):464–477. https://doi.org/10.1111/j.1469-8137.2010.03385.x

Marcone MF. *Batis maritima* (Saltwort/Beachwort): a nutritious, halophytic, seed bearings, perennial shrub for cultivation and recovery of otherwise unproductive agricultural land affected by salinity. Food Res Int. 2003:**36**(2):123–130. https://doi.org/10.1016/S0963-9969(02)00117-5

Marshall DM, Muhaidat R, Brown NJ, Liu Z, Stanley S, Griffiths H, Sage RF, Hibberd JM. *Cleome*, a genus closely related to *Arabidopsis*, contains species spanning a developmental progression from C(3) to C(4) photosynthesis. Plant J. 2007:**51**(5):886–896. https://doi.org/10.1111/j.1365-313X.2007.03188.x

Martinez G, Choudury SG, Keith Slotkin R. tRNA-derived small RNAs target transposable element transcripts. Nucleic Acids Res. 2017:**45**(9):5142–5152. https://doi.org/10.1093/nar/gkx103

Martín-Bravo S, Meimberg H, Luceño M, Märkl W, Valcárcel V, Bräuchler C, Vargas P, Heubl G. Molecular systematics and biogeography of Resedaceae based on ITS and trnL-F sequences. Mol Phylogenet Evol. 2007:**44**(3):1105–1120. https://doi.org/10.1016/j.ympev.2006.12.016

Matzke MA, Mosher RA. RNA-directed DNA methylation: an epigenetic pathway of increasing complexity. Nat Rev Genet. 2014:**15**(6): 394–408. https://doi.org/10.1038/nrg3683

Maurin O, Anest A, Bellot S, Biffin E., Brewer G., Charles-Dominique T, Cowan RS, Dodsworth S, Epitawalage N, Gallego B, et al. A nuclear phylogenomic study of the angiosperm order Myrtales, exploring the potential and limitations of the universal Angiosperms353 probe set. Am J Bot. 2021:**108**(7):1087–1111. https://doi.org/10.1002/ajb2.1699

McClintock B. The stability of broken ends of chromosomes in *Zea mays*. Genetics. 1941:**26**(2):234–282. https://doi.org/10.1093/genetics/26.2.234

McClintock B. The origin and behavior of mutable loci in maize. Proc Natl Acad Sci USA 1950:**36**(6):344–355. https://doi.org/10.1073/pnas.36.6.344

Melzer S, Lens F, Gennen J, Vanneste S, Rohde A, Beeckman T. Flowering-time genes modulate meristem determinacy and growth form in *Arabidopsis thaliana*. Nat Genet. 2008:**40**(12):1489–1492. https://doi.org/10.1038/ng.253

Mercado Gómez JD, Escalante T. Areas of endemism of the neotropical species of Capparaceae. Biol J Linn Soc Lond. 2018:**126**(3): 507–520. https://doi.org/10.1093/biolinnean/bly186

Mercati F, Fontana I, Gristina AS, Martorana A, El Nagar M, De Michele R, Fici S, Carimi F. Transcriptome analysis and codominant markers development in caper, a drought tolerant orphan crop with medicinal value. Sci Rep. 2019:**9**(1):10411. https://doi.org/10.1038/s41598-019-46613-x

Meyers SC, Liston A, Meinke R. A molecular phylogeny of *Limnanthes* (Limnanthaceae) and investigation of an anomalous *Limnanthes* population from California, U.S.A. Syst Bot. 2010:**35**(3):552–558. https://doi.org/10.1600/036364410792495854

Miura A, Yonebayashi S, Watanabe K, Toyama T, Shimada H, Kakutani T. Mobilization of transposons by a mutation abolishing full DNA methylation in *Arabidopsis*. Nature. 2001:**411**(6834) 212–214. https://doi.org/10.1038/35075612

Mohammadin S, Peterse K, van de Kerke SJ, Chatrou LW, Dönmez AA, Mummenhoff K, Pires JC, Edger PP, Al-Shehbaz IA, Schranz ME. Anatolian origins and diversification of *Aethionema*, the sister lineage of the core Brassicaceae. Am J Bot. 2017:**104**(7):1042–1054. https://doi.org/10.3732/ajb.1700091

Mühlhausen A, Lenser T, Mummenhoff K, Theißen G. Evidence that an evolutionary transition from dehiscent to indehiscent fruits in *Lepidium* (Brassicaceae) was caused by a change in the control of valve margin identity genes. Plant J. 2013:**73**(5):824–835. https://doi.org/10.1111/tpj.12079

Na J-K, Wang J, Murray JE, Gschwend AR, Zhang W, Yu Q, Pérez RN, Feltus FA, Chen C, Kubat Z, et al. Construction of physical maps for the sex-specific regions of papaya sex chromosomes. BMC Genomics. 2012:**13**(1):176. https://doi.org/10.1186/1471-2164-13-176

Nasrallah JB. Plant mating systems: self-incompatibility and evolutionary transitions to self-fertility in the mustard family. Curr Opin Genet Dev. 2017:**47**:54–60. https://doi.org/10.1016/j.gde.2017.08.005

Nichols BS, Leubner-Metzger G, Jansen VAA. Between a rock and a hard place: adaptive sensing and site-specific dispersal. Ecol Lett. 2020:**23**(9):1370–1379. https://doi.org/10.1111/ele.13564

Nic Lughadha EN, Govaerts R, Belyaeva I, Black N, Lindon H, Allkin R, Magill RE, Nicolson N. Counting counts: revised estimates of numbers of accepted species of flowering plants, seed plants, vascular plants and land plants with a review of other recent estimates. Phytotaxa. 2016:**272**(1):82–88. https://doi.org/10.11646/phytotaxa.272.1.5

Nikolov LA, Shushkov P, Nevado B, Gan X, Al-Shehbaz IA, Filatov D, Bailey CD, Tsiantis M. Resolving the backbone of the Brassicaceae phylogeny for investigating trait diversity. New Phytol. 2019:**222**(3): 1638–1651. https://doi.org/10.1111/nph.15732

Oh D-H, Dassanayake M, Haas JS, Kropornika A, Wright C, d'Urzo MP, Hong H, Ali S, Hernandez A, Lambert GM, et al. Genome structures and halophyte-specific gene expression of the extremophile *Thellungiella parvula* in comparison with *Thellungiella salsuginea* (*Thellungiella halophila*) and *Arabidopsis*. Plant Physiol. 2010:**154**(3): 1040–1052. https://doi.org/10.1104/pp.110.163923

Oh D-H, Hong H, Lee SY, Yun D-J, Bohnert HJ, Dassanayake M. Genome structures and transcriptomes signify niche adaptation for the multi-ion tolerant extremophyte *Schrenkiella parvula*. Plant Physiol. 2014:**164**(4): 2123–2138. https://doi.org/10.1104/pp.113.233551

Olafsdottir ES, Bolt Jørgensen L, Jaroszewski JW. Cyanogenesis in glucosinolate-producing plants: *Carica papaya* and *Carica quercifolia*. Phytochemistry. 2002:**60**(3):269–273. https://doi.org/10.1016/S0031-9422(02)00106-1

O'Malley RC, Barragan CC, Ecker JR. A user's guide to the *Arabidopsis* T-DNA insertion mutant collections. Methods Mol Biol. 2015:**1284**: 323–342. https://doi.org/10.1007/978-1-4939-2444-8_16

One Thousand Plant Transcriptomes Initiative. One thousand plant transcriptomes and the phylogenomics of green plants. Nature. 2019:**574**:679–685. https://doi.org/10.1038/s41586-019-1693-2

Pantha P, Oh D-H, Longstreth D, Dassanayake M. Living with high potassium: a balance between nutrient acquisition and stress signaling during K-induced salt stress. bioRxiv. 2021. https://doi.org/10.1101/2021.07.01.450778, 07 December 2021, preprint: not peer reviewed.

Parkin IAP, Gulden SM, Sharpe AG, Lukens L, Trick M, Osborn TC, Lydiate DJ. Segmental structure of the *Brassica napus* genome based on comparative analysis with *Arabidopsis thaliana*. Genetics. 2005:**171**(2):765–781. https://doi.org/10.1534/genetics.105.042093

Parma DF, Vaz MGMV, Falquetto P, Silva JC, Clarindo WR, Westhoff P, van Velzen R, Schlüter U, Araújo WL, Schranz ME, et al. New insights into the evolution of C_4 photosynthesis offered by the *Tarenaya* cluster of Cleomaceae. Front Plant Sci. 2021:**12**:756505. https://doi.org/10.3389/fpls.2021.756505

Patchell MJ, Roalson EH, Hall JC. Resolved phylogeny of Cleomaceae based on all three genomes. Taxon. 2014:**63**:315–328. https://doi.org/10.12705/632.17

Patel P, Mathioni S, Kakrana A, Shatkay H, Meyers BC. Reproductive phasiRNAs in grasses are compositionally distinct from other classes of small RNAs. New Phytol. 2018:**220**(3):851–864. https://doi.org/10.1111/nph.15349

Pellicer J, Leitch IJ. The plant DNA C-values database (release 7.1): an updated online repository of plant genome size data for comparative studies. New Phytologist. 2020:**226**(2):301–305. https://doi.org/10.1111/nph.16261

Petersen A, Hansen LG, Mirza N, Crocoll C, Mirza O, Halkier BA. Changing substrate specificity and iteration of amino acid chain elongation in glucosinolate biosynthesis through targeted mutagenesis of Arabidopsis methylthioalkylmalate synthase 1. Biosci Rep. 2019:**39**(7):BSR20190446. https://doi.org/10.1042/BSR20190446

Polydore S, Lunardon A, Axtell MJ. Several phased siRNA annotation methods can frequently misidentify 24 nucleotide siRNA-dominated PHAS loci. Plant Direct. 2018:**2**(12):e00101. https://doi.org/10.1002/pld3.101

Qi X, An H, Hall TE, Di C, Blischak PD, McKibben MTW, Hao Y, Conant GC, Pires JC, Barker MS. Genes derived from ancient polyploidy have higher genetic diversity and are associated with domestication in *Brassica rapa*. New Phytol. 2021:**230**(1):372–386. https://doi.org/10.1111/nph.17194

Radcliffe SA. Decolonising geographical knowledges. Trans Inst Br Geogr. 2017:**42**(3):329–333. https://doi.org/10.1111/tran.12195

Rahimi A, Karami O, Lestari AD, de Werk T, Amakorová P, Shi D, Novák O, Greb T, Offringa R. Control of cambium initiation and activity in *Arabidopsis* by the transcriptional regulator AHL15. Curr Biol. 2022:**32**(8):1764–1775.e3. https://doi.org/10.1016/j.cub.2022.02.060

Rajniak J, Barco B, Clay NK, Sattely ES. A new cyanogenic metabolite in *Arabidopsis* required for inducible pathogen defence. Nature. 2015:**525**(7569):376–379. https://doi.org/10.1038/nature14907

Reeves G, Singh P, Rossberg TA, Deedi Sogbohossou EO, Schranz ME, Hibberd JM. Natural variation within a species for traits underpinning C_4 photosynthesis. Plant Physiol. 2018:**177**(2):504–512. https://doi.org/10.1104/pp.18.00168

Renner SS. The relative and absolute frequencies of angiosperm sexual systems: dioecy, monoecy, gynodioecy, and an updated online database. Am J Bot. 2014:**101**(10):1588–1596. https://doi.org/10.3732/ajb.1400196

Reyna-Llorens I, Burgess SJ, Reeves G, Singh P, Stevenson SR, Williams BP, Stanley S, Hibberd JM. Ancient duons may underpin spatial patterning of gene expression in C_4 leaves. Proc Natl Acad Sci. 2018:**115**(8):1931–1936. https://doi.org/10.1073/pnas.1720576115

Roalson EH, Hall JC. New generic concepts for African Cleomaceae. Syst Bot. 2017:**42**(4):925–942. https://doi.org/10.1600/036364417X696393

Rodman JE, Karol KG, Price RA, Sytsma KJ. Molecules, morphology, and Dahlgren's expanded order Capparales. Syst Bot. 1996:**21**(3): 289. https://doi.org/10.2307/2419660

Rodman J, Price RA, Karol K, Conti E, Systma KJ, Palmer JD. Nucleotide sequences of the rbcL gene indicate monophyly of mustard oil plants. Ann Missouri Bot Gard. 1993:**80**(3):686. https://doi.org/10.2307/2399854

Román-Palacios C, Molina-Henao YF, Barker MS. Polyploids increase overall diversity despite higher turnover than diploids in the Brassicaceae. Proc Biol Sci 2020:**287**(1934):20200962. https://doi.org/10.1098/rspb.2020.0962

Ronse De Craene LP, Haston E. The systematic relationships of glucosinolate-producing plants and related families: a cladistic investigation based on morphological and molecular characters. Bot J Linn Soc. 2006:**151**(4):453–494. https://doi.org/10.1111/j.1095-8339.2006.00580.x

Sage RF. The evolution of C_4 photosynthesis. New Phytol. 2004:**161**(2): 341–370. https://doi.org/10.1111/j.1469-8137.2004.00974.x

Sage RF, Christin P-A, Edwards EJ. The C 4 plant lineages of planet earth. J Exp Bot. 2011:**62**(9):3155–3169. https://doi.org/10.1093/jxb/err048

Salariato DL, Zuloaga FO, Franzke A, Mummenhoff K, Al-Shehbaz IA. Diversification patterns in the CES clade (Brassicaceae tribes Cremolobeae, Eudemeae, Schizopetaleae) in Andean South America. Bot J Linn Soc. 2016:**181**(4):543–566. https://doi.org/10.1111/boj.12430

Schlüter U, Bräutigam A, Gowik U, Melzer M, Christin P-A, Kurz S, Mettler-Altmann T, Weber AP. Photosynthesis in C_3-C_4 intermediate *Moricandia* species. J Exp Bot. 2017:**68**(2):191–206. https://doi.org/10.1093/jxb/erw391

Schopfer CR, Nasrallah ME, Nasrallah JB. The male determinant of self-incompatibility in *Brassica*. Science. 1999:**286**(5445):1697–1700. https://doi.org/10.1126/science.286.5445.1697

Schrager-Lavelle A, Klein H, Fisher A, Bartlett M. Grass flowers: an untapped resource for floral evo-devo. J Syst Evol. 2017:**55**(6): 525–541. https://doi.org/10.1111/jse.12251

Schranz ME, Lysak MA, Mitchell-Olds T. The ABC's of comparative genomics in the Brassicaceae: building blocks of crucifer genomes. Trends Plant Sci. 2006:**11**(11):535–542. https://doi.org/10.1016/j.tplants.2006.09.002

Schranz ME, Manzaneda AJ, Windsor AJ, Clauss MJ, Mitchell-Olds T. Ecological genomics of *Boechera stricta*: identification of a QTL controlling the allocation of methionine- vs branched-chain amino acid-derived glucosinolates and levels of insect herbivory. Heredity. 2009:**102**(5):465–474. https://doi.org/10.1038/hdy.2009.12

Schranz ME, Mitchell-Olds T. Independent ancient polyploidy events in the sister families Brassicaceae and Cleomaceae. Plant Cell. 2006:**18**(5):1152–1165. https://doi.org/10.1105/tpc.106.041111

Schranz ME, Mohammadin S, Edger PP. Ancient whole genome duplications, novelty and diversification: the WGD radiation lag-time model. Curr Opin Plant Biol. 2012:**15**(2):147–153. https://doi.org/10.1016/j.pbi.2012.03.011

Seymour DK, Koenig D, Hagmann J, Becker C, Weigel D. Evolution of DNA methylation patterns in the Brassicaceae is driven by differences in genome organization. PLoS Genet. 2014:**10**(11):e1004785. https://doi.org/10.1371/journal.pgen.1004785

Shiba H, Kakizaki T, Iwano M, Tarutani Y, Watanabe M, Isogai A, Takayama S. Dominance relationships between self-incompatibility alleles controlled by DNA methylation. Nat Genet. 2006:**38**(3): 297–299. https://doi.org/10.1038/ng1734

Shimizu KK, Shimizu-Inatsugi R, Tsuchimatsu T, Purugganan MD. Independent origins of self-compatibility in *Arabidopsis thaliana*. Mol Ecol. 2008:**17**(2):704–714. https://doi.org/10.1111/j.1365-294X.2007.03605.x

Shimizu KK, Tsuchimatsu T. Evolution of selfing: recurrent patterns in molecular adaptation. Ann Rev Ecol. 2015:**46**(1):593–622. https://doi.org/10.1146/annurev-ecolsys-112414-054249

Soza VL, Le Huynh V, Di Stilio VS. Pattern and process in the evolution of the sole dioecious member of Brassicaceae. Evodevo. 2014:**5**(1):42. https://doi.org/10.1186/2041-9139-5-42

Sperber K, Steinbrecher T, Graeber K, Scherer G, Clausing S, Wiegand N, Hourston JE, Kurre R, Leubner-Metzger G, Mummenhoff K. Fruit fracture biomechanics and the release of *Lepidium didymum* pericarp-imposed mechanical dormancy by fungi. Nat Commun. 2017:**8**(1):1868. https://doi.org/10.1038/s41467-017-02051-9

Sreedasyam A, Plott C, Hossain MS, Lovell JT, Grimwood J, Jenkins JW, Daum C, Barry K, Carlson J, Shu S, et al. JGI Plant gene atlas: an updateable transcriptome resource to improve functional gene descriptions across the plant kingdom. Nucleic Acid Res. 2023:**51**(16): 8383–8401. https://doi.org/10.1093/nar/gkad616

Stein JC, Howlett B, Boyes DC, Nasrallah ME, Nasrallah JB. Molecular cloning of a putative receptor protein kinase gene encoded at the self-incompatibility locus of *Brassica oleracea*. Proc Natl Acad Sci USA. 1991:**88**(19):8816–8820. https://doi.org/10.1073/pnas.88.19.8816

Stevens PF. Angiosperm phylogeny website. Version 14, July 2017. 2001; [accessed 2021 Jan 10] http://www.mobot.org/MOBOT/research/APweb/.

Sun Y, Oh D-H, Duan L, Ramachandran P, Ramirez A, Bartlett A, Tran K-N, Wang G, Dassanayake M, Dinneny JR. Divergence in the ABA gene regulatory network underlies differential growth control. Nat Plants. 2022:**8**(5):549–560. https://doi.org/10.1038/s41477-022-01139-5

Swanepoel W, Chase MW, Christenhusz MJM, Maurin O, Forest F, Van Wyk AE. From the frying pan: an unusual dwarf shrub from Namibia turns out to be a new Brassicalean family. Phytotaxa. 2020:**439**(3):171–185. https://doi.org/10.11646/phytotaxa.439.3.1

Takasaki T, Hatakeyama K, Suzuki G, Watanabe M, Isogai A, Hinata K. The S receptor kinase determines self-incompatibility in *Brassica* stigma. Nature. 2000:**403**(6772):913–916. https://doi.org/10.1038/35002628

Takayama S, Shiba H, Iwano M, Shimosato H, Che FS, Kai N, Watanabe M, Suzuki G, Hinata K, Isogai A. The pollen determinant of self-incompatibility in *Brassica campestris*. Proc Natl Acad Sci USA. 2000:**97**(4):1920–1925. https://doi.org/10.1073/pnas.040556397

Tang H, Woodhouse MR, Cheng F, Schnable JC, Pedersen BS, Conant G, Wang X, Freeling M, Pires JC. Altered patterns of fractionation and exon deletions in *Brassica rapa* support a two-step model of paleohexaploidy. Genetics. 2012:**190**(4):1563–1574. https://doi.org/10.1534/genetics.111.137349

Tarutani Y, Shiba H, Iwano M, Kakizaki T, Suzuki G, Watanabe M, Isogai A, Takayama S. Trans-acting small RNA determines dominance relationships in *Brassica* self-incompatibility. Nature. 2010:**466**(7309):983–986. https://doi.org/10.1038/nature09308

Teng C, Zhang H, Hammond R, Huang K, Meyers BC, Walbot V. Dicer-like 5 deficiency confers temperature-sensitive male sterility in maize. Nat Commun. 2020:**11**(1):1–9. https://doi.org/10.1038/s41467-020-16634-6

The Arabidopsis Genome Initiative. Analysis of the genome sequence of the flowering plant *Arabidopsis thaliana*. Nature. 2000:**408**(6814): 796–815. https://doi.org/10.1038/35048692

The First Nations Information Governance Centre. The First Nations Information Governance Centre; 2022 [accessed 2022 Feb 1]. https://fnigc.ca/.

Thomas SK, Liu X, Du Z-Y, Dong Y, Cummings A, Pokorny L, Xiang Q-YJ, Leebens-Mack JH. Comprehending Cornales: phylogenetic reconstruction of the order using the Angiosperms353 probe set. Am J Bot. 2021:**108**(7):1112–1121. https://doi.org/10.1002/ajb2.1696

Tran K-N, Pantha P, Wang G, Kumar N, Wijesinghege C, Oh DH, Duppen N, He J, Li H, Hong H, et al. Balancing growth amidst salinity stress – lifestyle perspectives from the extremophyte model. bioRxiv. 2021. https://doi.org/10.1101/2021.08.27.457575, 2021, preprint: not peer reviewed.

Tran K-N, Wang G, Oh D-H, Larkin JC, Smith AP, Dassanayake M. Multiple paths lead to salt tolerance - pre-adaptation vs dynamic responses from two closely related extremophytes. bioRxiv. 2022. https://doi.org/10.1101/2021.10.23.465591, 08 February, 2022, preprint: not peer reviewed.

Tsuchimatsu T, Goubet PM, Gallina S, Holl A-C, Fobis-Loisy I, Bergès H, Marande W, Prat E, Meng D, Long Q, et al. Patterns of polymorphism at the self-incompatibility locus in 1,083 *Arabidopsis thaliana* genomes. Mol Biol Evol. 2017:**34**(8):1878–1889. https://doi.org/10.1093/molbev/msx122

Tsuchimatsu T, Suwabe K, Shimizu-Inatsugi R, Isokawa S, Pavlidis P, Städler T, Suzuki G, Takayama S, Watanabe M, Shimizu KK. Evolution of self-compatibility in *Arabidopsis* by a mutation in the male specificity gene. Nature. 2010:**464**(7293):1342–1346. https://doi.org/10.1038/nature08927

Turner-Hissong SD, Mabry ME, Beissinger TM, Ross-Ibarra J, Pires JC. Evolutionary insights into plant breeding. Curr Opin Plant Biol. 2020:**54**:93–100. https://doi.org/10.1016/j.pbi.2020.03.003

Ueno O. Structural and biochemical characterization of the C_3-C_4 intermediate *Brassica gravinae* and relatives, with particular reference to cellular distribution of rubisco. J Exp Bot. 2011:**62**(15): 5347–5355. https://doi.org/10.1093/jxb/err187

Ueno O, Bang SW, Wada Y, Kondo A, Ishihara K, Kaneko Y, Matsuzawa Y. Structural and biochemical dissection of photorespiration in hybrids differing in genome constitution between *Diplotaxis tenuifolia* (C_3-C_4) and radish (C_3). Plant Physiol. 2003:**132**(3): 1550–1559. https://doi.org/10.1104/pp.103.021329

Ueno O, Wada Y, Wakai M, Bang SW. Evidence from photosynthetic characteristics for the hybrid origin of *Diplotaxis muralis* from a C_3-C_4 intermediate and a C_3 species. Plant Biol. 2006:**8**(2):253–259. https://doi.org/10.1055/s-2005-873050

Van de Peer Y, Ashman T-L, Soltis PS, Soltis DE. Polyploidy: an evolutionary and ecological force in stressful times. Plant Cell. 2021:**33**(1):11–26. https://doi.org/10.1093/plcell/koaa015

Voznesenskaya EV, Koteyeva NK, Chuong SDX, Ivanova AN, Barroca J, Craven LA, Edwards GE. Physiological, anatomical and biochemical characterisation of photosynthetic types in genus *Cleome* (Cleomaceae). Funct Plant Biol. 2007:**34**(4):247–267. https://doi.org/10.1071/FP06287

Walden N, German DA, Wolf EM, Kiefer M, Rigault P, Huang X-C, Kiefer C, Schmickl R, Franzke A, Neuffer B, et al. Nested whole-genome duplications coincide with diversification and high morphological disparity in Brassicaceae. Nat Commun. 2020a:**11**(1):1–12. https://doi.org/10.1038/s41467-020-17605-7

Walden N, Nguyen T-P, Mandáková T, Lysak MA, Schranz ME. Genomic blocks in *Aethionema arabicum* support Arabideae as next diverging clade in Brassicaceae. Front Plant Sci. 2020b:**11**:719. https://doi.org/10.3389/fpls.2020.00719

Wang G, DiTusa SF, Oh D-H, Herrmann AD, Mendoza-Cozatl DG, O'Neill MA, Smith AP, Dassanayake M. Cross species multi-omics reveals cell wall sequestration and elevated global transcript abundance as mechanisms of boron tolerance in plants. New Phytol. 2021:**230**(5):1985–2000. https://doi.org/10.1111/nph.17295

Wang J, Na J-K, Yu Q, Gschwend AR, Han J, Zeng F, Aryal R, VanBuren R, Murray JE, Zhang W, et al. Sequencing papaya X and yh chromosomes reveals molecular basis of incipient sex chromosome evolution. Proc Natl Acad Sci USA. 2012:**109**(34): 13710–13715. https://doi.org/10.1073/pnas.1207833109

Wang G, Pantha P, Tran K-NK-N, Oh D-HD-H, Dassanayake M. Plant growth and Agrobacterium-mediated floral-dip transformation of the extremophyte *Schrenkiella parvula*. J Vis Exp. 2019:**143**:e58544.

Wicker T, Sabot F, Hua-Van A, Bennetzen JL, Capy P, Chalhoub B, Flavell A, Leroy P, Morgante M, Panaud O, et al. A unified classification system for eukaryotic transposable elements. Nat Rev Genet. 2007:**8**:973–982. https://doi.org/10.1038/nrg2165

Wijesinghege C, Tran K-N, Dassanayake M. Alternative splicing preferentially increases transcript diversity associated with stress responses in the extremophyte Schrenkiella parvula. bioRxiv. 2022a. https://doi.org/10.1101/2022.10.13.512046, 17 October 2022, preprint: not peer reviewed.

Wijesinghege C, Wang G, Pantha P, Tran K-N, Dassanayake M. Spatiotemporal gene expression atlas of the extremophyte Schrenkiella parvula. bioRxiv. 2022b. https://doi.org/10.1101/2022.10.24.513627, 26 October 2022, preprint: not peer reviewed.

Williams BP, Burgess SJ, Reyna-Llorens I, Knerova J, Aubry S, Stanley S, Hibberd JM. An untranslated *cis*-element regulates the accumulation of multiple C_4 enzymes in *Gynandropsis gynandra* mesophyll cells. Plant Cell. 2016:**28**(2):454–465. https://doi.org/10.1105/tpc.15.00570

Williams BP, Johnston IG, Covshoff S, Hibberd JM. Phenotypic landscape inference reveals multiple evolutionary paths to C_4 photosynthesis. Elife. 2013:**2**:e00961. https://doi.org/10.7554/eLife.00961

Willing E-M, Rawat V, Mandáková T, Maumus F, James GV, Nordström KJV, Becker C, Warthmann N, Chica C, Szarzynska B, et al. Genome expansion of *Arabis alpina* linked with retrotransposition and reduced symmetric DNA methylation. Nat Plants. 2015:**1**(2):1–7. https://doi.org/10.1038/nplants.2014.23

Winter K, Sage RF, Edwards EJ, Virgo A, Holtum JAM. Facultative crassulacean acid metabolism in a C_3–C_4 intermediate. J Exp Bot. 2019:**70**(22):6571–6579. https://doi.org/10.1093/jxb/erz085

Wu H-J, Zhang Z, Wang J-Y, Oh D-H Dassanayake M, Liu B, Huang Q, Sun H-X, Xia R, Wu Y, et al. Insights into salt tolerance from the genome of *Thellungiella salsuginea*. Proc Natl Acad Sci USA. 2012:**109**(30): 12219–12224. https://doi.org/10.1073/pnas.1209954109

Xia R, Chen C, Pokhrel S, Ma W, Huang K, Patel P, Wang F, Xu J, Liu Z, Li J, et al. 24-nt Reproductive phasiRNAs are broadly present in angiosperms. Nat Commun. 2019:**10**:627. https://doi.org/10.1038/s41467-019-08543-0

Xiong Z, Gaeta RT, Pires JC. Homoeologous shuffling and chromosome compensation maintain genome balance in resynthesized allopolyploid *Brassica napus*. Proc Natl Acad Sci USA. 2011:**108**(19): 7908–7913. https://doi.org/10.1073/pnas.1014138108

Xu D, Marino G, Klingl A, Enderle B, Monte E, Kurth J, Hiltbrunner A, Leister D, Kleine T. Extrachloroplastic PP7L functions in chloroplast development and abiotic stress tolerance. Plant Physiol. 2019:**180**(1):323–341. https://doi.org/10.1104/pp.19.00070

Yates AD, Allen J, Amode RM, Azov AG, Barba M, Becerra A, Bhai J, Campbell LI, Carbajo Martinez M, Chakiachvili M, et al. Ensembl genomes 2022: an expanding genome resource for non-vertebrates. Nucleic Acids Res. 2022:**50**(D1):D996–D1003. https://doi.org/10.1093/nar/gkab1007

Yue J, VanBuren R, Liu J, Fang J, Zhang X, Liao Z, Wai CM, Xu X, Chen S, Zhang S, et al. Sunup and sunset genomes revealed impact of particle bombardment mediated transformation and domestication history in papaya. Nat Genet. 2022:**54**(5):715–724. https://doi.org/10.1038/s41588-022-01068-1

Zemach A, Yvonne Kim M, Hsieh P-H, Coleman-Derr D, Eshed-Williams L, Thao K, Harmer SL, Zilberman D. The nucleosome remodeler DDM1 allows DNA methyltransferases to access H1-containing heterochromatin. Cell. 2013:**153**(1):193–205. https://doi.org/10.1016/j.cell.2013.02.033

Zhang J, Wang X, Cheng F, Wu J, Liang J, Yang W, Wang X. Lineage-specific evolution of methylthioalkylmalate synthases (MAMs) involved in glucosinolates biosynthesis. Front Plant Sci. 2015:**6**:18. https://doi.org/10.3389/fpls.2015.00018

Zhao H, Lu L, Su J, Chen Z. Phylogeny of Stixeae and Borthwickiaceae based on morphological and molecular data. Chin Bull Bot. 2015:**50**: 473–481. https://doi.org/10.11983/CBB14152

Zhao M, Zhang B, Lisch D, Ma J. Patterns and consequences of subgenome differentiation provide insights into the nature of paleopolyploidy in plants. Plant Cell. 2017:**29**(12):2974–2994. https://doi.org/10.1105/tpc.17.00595

Zhu JK. The next top models. Cell. 2015:**163**(1):18–20. https://doi.org/10.1016/j.cell.2015.09.005

Zizka A, Onstein RE, Rozzi R, Weigelt P, Kreft H, Steinbauer MJ, Bruelheide H, Lens F. The evolution of insular woodiness. Proc Natl Acad Sci. 2022:**119**(37):e2208629119. https://doi.org/10.1073/pnas.2208629119

Zohoungbogbo HPF, Houdegbe CA, Sogbohossou DEO, Tossou MG, Maundu P, Schranz EM, Van Deynze A, Zoundjihekpon J, Achigan-Dako EG. Andromonoecy in *Gynandropsis gynandra* (L.) briq. (Cleomaceae) and effects on fruit and seed production. Genet Resour Crop Evol. 2018:**65**(8):2231–2239. https://doi.org/10.1007/s10722-018-0687-5

Zou Z, Li M, Jia R, Zhao H, He P, Zhang Y, Guo A. Genes encoding light-harvesting chlorophyll a/b-binding proteins in papaya (*Carica papaya* L.) and insight into lineage-specific evolution in Brassicaceae. Gene. 2020:**748**:144685. https://doi.org/10.1016/j.gene.2020.144685

Zuntini AR, Frankel LP, Pokorny L, Forest F, Baker WJ. A comprehensive phylogenomic study of the monocot order Commelinales, with a new classification of Commelinaceae. Am J Bot. 2021:**108**(7): 1066–1086. https://doi.org/10.1002/ajb2.1698

Zuo S, Mandáková T, Kubová M, Lysak MA. Genomes, repeatomes and interphase chromosome organization in the meadowfoam family (Limnanthaceae, Brassicales). Plant J. 2022:**110**(5):1462–1475. https://doi.org/10.1111/tpj.15750

Züst T, Heichinger C, Grossniklaus U, Harrington R, Kliebenstein DJ, Turnbull LA. Natural enemies drive geographic variation in plant defenses. Science. 2012:**338**(6103):116–119. https://doi.org/10.1126/science.1226397

Züst T, Strickler SR, Powell AF, Mabry ME, An H, Mirzaei M, York T, Holland CK, Kumar P, Erb M, et al. Independent evolution of ancestral and novel defenses in a genus of toxic plants (*Erysimum*, Brassicaceae). Elife. 2020:**9**:e51712. https://doi.org/10.7554/eLife.51712

https://doi.org/10.1093/plcell/koae013

THE PLANT CELL

THE PLANT CELL 2024: 36: 1227–1241

Review

Domestication and the evolution of crops: variable syndromes, complex genetic architectures, and ecological entanglements

Ornob Alam[1] and Michael D. Purugganan [1,2,3,*]

1 Center for Genomics and Systems Biology, New York University, New York, NY 10003, USA
2 Center for Genomics and Systems Biology, New York University Abu Dhabi, Abu Dhabi, United Arab Emirates
3 Institute for the Study of the Ancient World, New York University, New York, NY, 10028, USA

*Author for correspondence: mp132@nyu.edu
The author responsible for distribution of materials integral to the findings presented in this article in accordance with the policy described in the Instructions for Authors (https://academic.oup.com/plcell/pages/General-Instructions) is: Michael D. Purugganan (mp132@nyu.edu)

Abstract

Domestication can be considered a specialized mutualism in which a domesticator exerts control over the reproduction or propagation (fitness) of a domesticated species to gain resources or services. The evolution of crops by human-associated selection provides a powerful set of models to study recent evolutionary adaptations and their genetic bases. Moreover, the domestication and dispersal of crops such as rice, maize, and wheat during the Holocene transformed human social and political organization by serving as the key mechanism by which human societies fed themselves. Here we review major themes and identify emerging questions in three fundamental areas of crop domestication research: domestication phenotypes and syndromes, genetic architecture underlying crop evolution, and the ecology of domestication. Current insights on the domestication syndrome in crops largely come from research on cereal crops such as rice and maize, and recent work indicates distinct domestication phenotypes can arise from different domestication histories. While early studies on the genetics of domestication often identified single large-effect loci underlying major domestication traits, emerging evidence supports polygenic bases for many canonical traits such as shattering and plant architecture. Adaptation in human-constructed environments also influenced ecological traits in domesticates such as resource acquisition rates and interactions with other organisms such as root mycorrhizal fungi and pollinators. Understanding the ecological context of domestication will be key to developing resource-efficient crops and implementing more sustainable land management and cultivation practices.

Introduction

Domestication can be considered a specialized mutualism in which a domesticator exerts control over the reproduction or propagation (i.e. fitness) of a domesticated species to gain resources or services, resulting in a unique coevolutionary dynamic (Zeder 2014; Zeder 2015; Purugganan 2022). The process of domestication provides powerful models to study recent evolutionary adaptations and their genetic bases (Ross-Ibarra Morrell and Gaut 2007). Charles Darwin used domesticated crops to understand variation and selection (Darwin 1868), and over the last 150 years, crops have served as models in both genetic and evolutionary studies (Andersson and Purugganan 2022). The domestication and dispersal of crop plant species such as rice, maize, and wheat beginning about 13,000 years ago also helped bring about dramatic changes in human social and political organization (Diamond 2002; Purugganan and Fuller 2009); today, crops are the foundation of modern human societies by providing the majority of food resources—as well as other products—necessary for the survival of the global population.

The nature of domestication has been the subject of intense debate, and over the last two decades our

Received September 07, 2023. Accepted December 14, 2023. Advance access publication January 18, 2024 understanding of this process has advanced considerably (Abbo et al. 2022; Allaby et al. 2022). Although earlier studies had suggested that domestication could have been a rapid process (Abbo et al. 2011), archaeological evidence increasingly indicates that early domestication was largely a protracted, often landscape-level process characterized by weak, unconscious selection occurring over long time per-iods (Purugganan and Fuller 2011; Meyer et al. 2012; Arranz-Otaegui et al. 2016; Allaby et al. 2022). Consistent with a protracted transition from wild to domesticated forms, recent genomic studies in multiple domesticated crops have shown a gradual to negligible decline in effective population size during early domestication (Gross et al. 2014; Meyer et al. 2016; Allaby Ware, and Kistler 2019; Trucchi et al. 2021; Zhou et al. 2017) rather than a rapid or instantaneous bottleneck that might be expected if domestication involved sampling from a wild population once or a small number of times followed by strong selection (Purugganan 2019). For example, a recent study on ancient domesticated common beans from South America showed that the beans—with ages ranging from 600 to 2,500 years old—maintained equivalent levels of genetic diversity to modern wild popula-tions, suggesting that the reduced diversity in modern culti-vars is a consequence of more recent human cultivation practices rather than processes during the early domestica-tion period (Trucchi et al. 2021).

The discovery of different, independent Neolithic centers of domestication (Vavilov 1994) also provides the opportun-ity to explore the process of domestication as parallel natural experiments (Gepts 2003). Overall, domestication of crops appears to have occurred in at least 24 different geographic areas, the earliest being in the Fertile Crescent, China, and the Americas (Purugganan and Fuller 2011). Multiple cereal crop species from different domestication centers such as rice, maize, wheat, and sorghum share to varying degrees a set of traits, including non-shattering seeds, increased seed size, reduced seed dormancy, loss of outcrossing, and apical dominance, which are collectively referred to as the domestication syndrome (Hammer 1984; Allaby 2014; Preece et al. 2017). Adaptations such as non-shattering seeds, shoot architecture, and seed dormancy have been shown to at least partially arise from changes in the same underlying genes—*Sh1* (Lin et al. 2012), *Tb1* (Ramsay et al. 2011; Remigereau et al. 2011), and *G* (M. Wang et al. 2018), respectively—in multiple species. Instances of convergent evolution of domestication phenotypes, and sometimes genotypes, are in line with do-mestication being a distinct evolutionary process among species sometimes leading to shared features and outcomes (Lenser and Theißen 2013). Finally, the process of domestication is also often characterized by gene flow or hybridization between different species or populations, as observed in wheat (Wang et al. 2022), citrus (Wu et al. 2014), maize (Kistler et al. 2020), rice (Choi et al. 2017; Wu et al. 2023), and others. Hexaploid bread wheat, for instance, arose from hybridization between tetraploid emmer wheats—domesticated near the onset of agriculture in the Fertile Crescent more than 10,000 years ago—and wild goatgrass (Dvorak et al. 1998; International Wheat Genome Sequencing Consortium 2014).

Over the last few decades, there have been important strides in our understanding of domestication and crop evolution, although our understanding remains incomplete and multiple issues are unresolved. For this review, we identify and focus on three fundamental areas of crop domestication research that are transforming long-held ideas on the process of domestication. Our first area of focus is on domestication phenotypes and the nature of domestication syndromes. Cereal crop traits heavily inform traditional conceptions of the crop domestication syndrome, but many such traits are not universal across crops (Meyer et al. 2012). In fact, it has been challenging to define a consistent set of domestication phenotypes for other types of crop domesticates such as vegetatively propagated field crops like sugarcane, potato, and cassava (Denham et al. 2020). Understanding variation in domestication traits and syndromes across crops will be key to identifying shared and heterogeneous aspects of the domestication process. Recent comparative analyses have begun to define distinct domestication syndrome traits (Fuller 2018) or reframe the domestication syndrome as more general tendencies (Denham et al. 2020) for specific crop types.

The second area we explore is the genetic architecture underlying domestication and crop evolution. Early studies on genetic loci associated with domestication identified several large-effect mutations associated with phenotypes such as non-shattering seeds and apical dominance (Li et al. 2006; Whipple et al. 2011), which shaped a relatively simple view of the domestication process. It is, however, becoming increasingly clear that many domestication phenotypes have complex and polygenic genetic architectures that often require high-resolution genome-wide association studies (Xue et al. 2016) or observation in a wild genomic background (Ishikawa et al. 2020) to be revealed. In addition, the concept of a domestication locus remains nebulous with regard to when mutations need to have arisen or become selected. Ancient DNA, when available, (Jaenicke-Després et al. 2003; Vallebueno-Estrada et al. 2016) and new methodological advances in dating allele ages and selective sweeps (Speidel et al. 2019; Stern et al. 2019) are likely to be useful in dissecting the evolutionary genetics of the domestication process while distinguishing it from later selection during crop diversification and recent breeding.

Finally, we examine the ecological context of domestication. The milieu and impact of domestication extends beyond the domesticator and domesticated species to their environmental and ecological context. Domestication studies have focused on phenotypic traits that appear to be directly favorable to humans, but there have been relatively fewer studies on selection on ecological traits within human-constructed environments during the protracted domestication period (Chen et al. 2015; Milla 2023). Ecological traits such as growth and resource acquisition rates and interactions with microbes, pollinators, and other organisms have

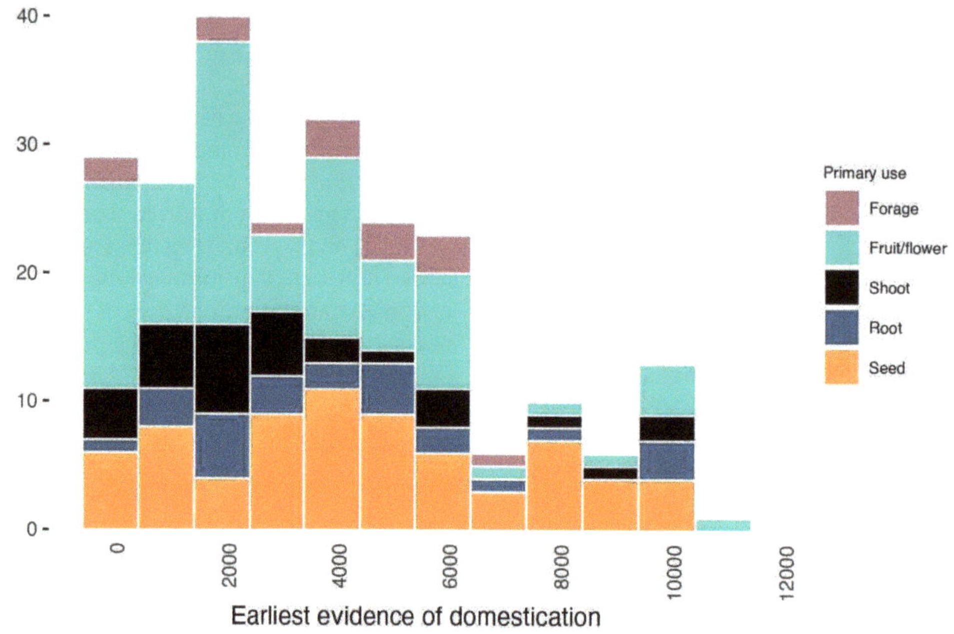

Figure 1. Domestication events through time. Number of domestication events in 1,000-year windows (the label "2000" corresponds to the window between 2,000 and 3,000 years ago). Crops are further grouped by the primary part of the plant used. Based on 235 species with dates for earliest evidence of domestication in the Crop Origins and Phylo Food database reported in Milla 2020.

implications for land management and sustainability of cultivation practices, and a better understanding of the ecology of domestication can help to both raise yield and more sustainably manage local ecosystems.

The domestication syndrome and beyond

Crop plant species evolve under domestication as a direct result of human-associated selection or as adaptations to the human-constructed environments developed for their growth and exploitation. The traditionally defined domestication syndrome in crops comprises a set of common phenotypic traits that were selected across different species during the transition from wild to domesticate, facilitating the mutualistic interaction between crops and humans and allowing for growth in human-constructed agricultural niches (Allaby 2014). For instance, multiple cereal crops including rice, wheat, and sorghum exhibit loss of shattering to facilitate harvesting (Z. Lin et al. 2012), as well as increased grain (Gegas et al. 2010; Zuo and Li 2014; Han et al. 2015) and plant size relative to their respective wild progenitors (Milla et al. 2014; Kluyver et al. 2017; Preece et al. 2018; Allaby et al. 2022).

Analysis of 203 domesticated crop species—including cereals, fruit trees, tubers, and leafy vegetables, among others—revealed that domestication traits vary among crop species depending on biology, the parts of the plant used by humans, and local climates (Meyer et al. 2012). Changes in secondary metabolites linked to changes in flavor, color, and toxicity constituted the most universal domestication-related trait among all the considered species (Meyer et al. 2012). Crops from arid climates such as the Near East tended to undergo changes in seed morphology and shattering during domestication, whereas crops from humid climates such as Near Oceania did not (Meyer et al. 2012). Trees and other perennials tended to be domesticated later than annuals (evident in Fig. 1), while the rate of domestication—measured in terms of the time between first exploitation of wild ancestor and first evidence of domestication for a crop—accelerated with more recently domesticated species, possibly as agriculture became increasingly established (Meyer et al. 2012). In general, the number of domestications attempted increased dramatically starting from about 6,000 to 7,000 years ago (Fig. 1) (Milla 2020).

The degree of completeness of domestication and the number of domestication traits that arise depend on the patterns of human interaction with the domesticate species and their wild ancestors, and these interactions may have prolonged pre-domestication or pre-cultivation histories. Pre-agricultural societies often engaged in some form of ecosystem management, influencing and maintaining the growth of multiple species and foraging what they needed (Rowley-Conwy and Layton 2011). Ecosystems managed for human use, which have been described as domesticated landscapes, can provide the context for greater investment in the care of particularly useful species and include intentional planting and the removal of competitors; these may impose unconscious and conscious selection pressures for domestication even in the absence of cultivation (Levis et al. 2018; Clement et al. 2021). Archaeobotanical studies in the Neotropics have found evidence for incipient or incomplete domestication, occurring as far back as more than 10,000 years ago (Clement et al. 2021). In Amazonia, there is evidence of increasing use and abundance of various palm species starting about 12,000 years ago (Roosevelt et al. 1996; Morcote-Ríos et al. 2021), with species such as the Brazil nut subsequently going on to be domesticated and cultivated (Pärssinen et al. 2021). Major Mesoamerican agricultural crops such as maize and squash had also begun to be domesticated earlier than 9,000 years ago, even though recognizable large-scale food production systems start appearing in the archaeological record only about 4,000 years ago (Clement et al. 2021).

A more nuanced view of domestication traits as continuous (rather than discrete) and possibly unique to the utilization history of individual species is thus emerging. Early landscape management and small-scale cultivation practices pre-dating the origin of agricultural societies likely helped create a continuum of domestication traits rather than a complete domes-tication syndrome (Clement et al. 2021). Domestication may remain incomplete or incipient for long periods, as has been noted in the amaranth species *Amaranthus caudatus* in South America (Stetter et al. 2017) and fonio millet in West Africa (Abrouk et al. 2020). In addition to weak selection pressures likely experienced in managed ecosystems, genetic constraints such as indirect antagonistic pleiotropy, lack of standing genetic variation, and continuous gene flow from wild populations can impact the rate at which domestication traits are fixed (Stetter 2020).

Understanding variation in domestication traits and syndromes across crops will be key to identifying shared and heterogeneous aspects of the domestication process across groups as divergent as cereals, vegetables, fruit trees, and tubers. Indeed, insights on the domestication syndrome in crops largely come from research on major cereal crops such as rice, wheat, and maize (Meyer et al. 2012), and much less is known about what phenotypes or traits constitute domestication syndrome(s) in other types of crop domesticates. Recently, it has been proposed that specific domesticated crops are associated with seven nonexclusive domestication pathways: (i) palms and nut trees through ecosystem engineering or management, (ii) ruderal plants that initially thrived in human-disturbed environments, (iii) tubers, (iv) grain crops, (v) segetal or former weedy species like oats, (vi) fiber crops, and (vii) fruit trees (Fuller et al. 2023). Such a framework opens up avenues to explore domestication traits and evolutionary dynamics within each pathway and shared across pathways. For instance, grain crops, nut and fruit trees, and tuber crops all tend to show increases in yield of the edible portion, such as through the production of swollen regions along the root system in the root tubers cassava and sweet potato (Denham et al. 2020). On the other hand, non-shattering of seeds or pods to retain seeds for harvesting is more unique to grain crops (cereals and legumes), while fused or bunched vegetative storage organs tend to have evolved in fruit- and nut-bearing trees such as bananas, figs, and breadfruit (Denham et al. 2020). In practice, the domestication pathways defined by Fuller et al. may leave the classification of certain species (for example, agave, tea, and leafy greens such as spinach) ambiguous, and require knowledge of the early domestication or pre-cultivation histories of crops. A cursory look at domestication syndrome traits in crops grouped by the primary part of the plant used—which is correlated with the pathways—reveals different distributions of the traits (Fig. 2).

Similar to the domestication trait continuum (Clement et al. 2021), the idea of domestication tendencies rather than traits

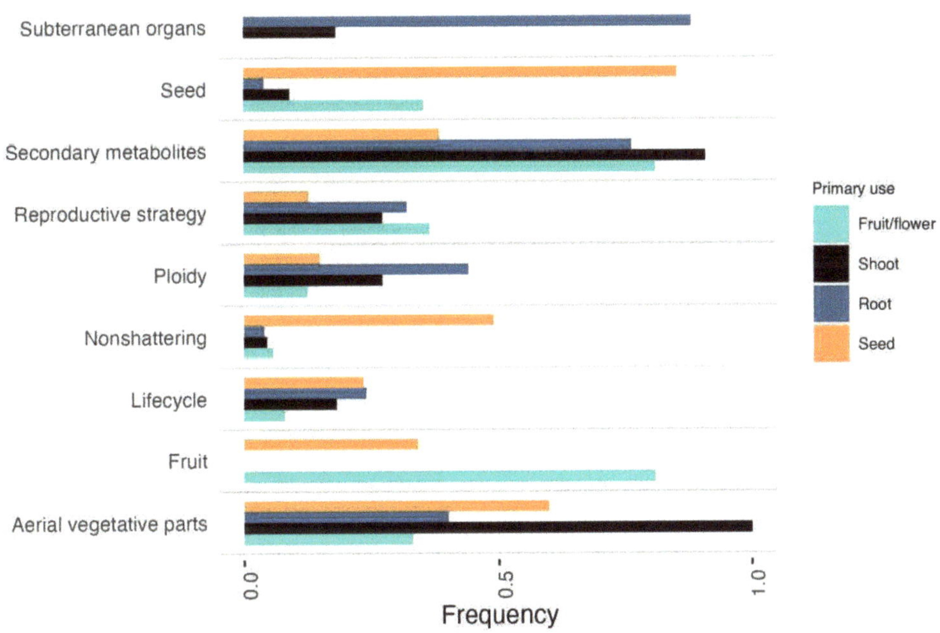

Figure 2. Domestication syndrome traits in crops grouped by the primary part of the plant used. Data on domestication syndrome traits from dataset reported in Meyer et al. 2012 and crop use data from Milla 2020. Based on 182 species identified as either domesticated or semi-domesticated in Meyer et al. 2012 that are present in Milla 2020 and not used primarily as animal feed.

has been invoked for vegetatively propagated field crops such as potato, manioc, and many fruits, as phenotypic signatures of domestication in these species appear more variable than in cereals (Denham et al. 2020; Fuller et al. 2023). More exhaustive study of domestication traits in these crops can help elucidate the process of their domestication. For instance, fewer phenotypic shifts during domestication in perennial fruit trees have been linked to less severe bottleneck effects in the history of perennials compared to annuals; bottleneck effects have a smaller number of generations to accrue in trees, which tend to have longer generation times than cereals (Gaut et al. 2015). Moving forward, evaluating a wider range of possible domestication phenotypes as quantitative traits that differ in distributions or mean values between domesticates and wild progenitors (Casas et al. 2007; Xue et al. 2016) and mapping them along categories like the recently defined domestication pathways (Fuller et al. 2023) may help organize the evident complexity and heterogeneity of crop domestication traits.

Finally, understanding domestication phenotypes in each domesticate is a prerequisite to subsequent exploration of their underlying genetics. The genetic architecture of domes-tication traits—and perhaps tendencies—as well as the order in which genetic variants associated with the traits arose are key to understanding the dynamics and selective landscape of the domestication process.

In search of domestication loci

A major challenge in plant biology has been the identification of genes associated with crop domestication, especially those that underlie key species-specific traits that distinguish the domesticate from their wild ancestor. Across the major crop species, top-down and bottom-up approaches to un-derstanding the genetic architecture of domestication have yielded differing views of the number of genes involved in crop domestication (Kantar et al. 2017).

Top-down methods seek to identify genetic loci underpinning observed domestication traits by genetic mapping of known domestication phenotypes. The study of the genes and genetic architecture underlying domestication traits have advanced significantly with the advent of genetic mapping, primarily QTL mapping approaches (Paterson 2002; Jantasuriyarat et al. 2004; Li et al. 2006; Liu et al. 2007; Wills and Burke 2007) coupled eventually with genome sequencing (Doebley et al. 2006; Meyer and Purugganan 2013; Andersson and Purugganan 2022). These have led to the identification of canonical domestication loci such as *Sh4* for non-shattering in rice (Li et al. 2006), *Tb1* for plant and inflorescence architectures in maize (Doebley et al. 1995), and *Q* for threshability and other traits in wheat (Kato et al. 2003; Jantasuriyarat et al. 2004). Taking a trait-first view of domestication often revealed a small number of strong-effect alleles driving phenotypes. For instance, a single amino acid substitution in *Sh4* dramatically reduces shattering in domesticated rice (Li et al. 2006).

The few domestication loci that have been identified provide some insight into the type of causative mutations and genes associated with key traits (Doebley et al. 2006; Meyer and Purugganan 2013; Andersson and Purugganan 2022). The causative mutations span all kinds of genetic changes, including single-nucleotide polymorphisms (SNPs), insertion/deletions (including transposable element insertions), and larger structural variants. Interestingly, loss-of-function mutations including premature stop codons and frameshifts constitute the most common type of mutation in domestication genes (Meyer and Purugganan 2013). While coding region mutations are widely observed, several cases of *cis*-regulatory mutations (such as in *Tb1*) (R.-L. Wang et al. 1999) have also been identified.

An alternative, bottom-up approach attempts to detect signatures of selective sweeps or highly differentiated loci in genome-wide assays in domesticated populations com-pared to the wild ancestor, independent of associated traits (Kantar et al. 2017). Early whole-genome studies used selec-tion scans on SNP data to identify putative domestication loci (Huang et al. 2012; Kantar et al. 2017). These studies of-ten found a larger number of putative domestication loci than top-down methods, ranging from 55 in rice (Huang et al. 2012) to 484 in maize (Hufford et al. 2012). More recently, top-down approaches have been implemented using crop pangenomes, which allow identification of presence-absence variants (PAVs) and large structural variants and thus provide a more accurate reflection of genome-wide diversity (Gong et al. 2023). In a pangenome analysis of rice (Zhao et al. 2018), a selection scan based on comparing nucleotide diversity between wild and domesticated accessions identi-fied similar putative domestication loci to an earlier study that used whole genomes (Huang et al. 2012) but also detected six new domestication loci. A pangenome study on the foxtail millet identified 4,582 domestication-selected PAVs and mapped them to more than 1,400 genes (He et al. 2023). SNP-based selection scans detected only 22.4% of these genes, suggesting that PAV frequencies may be a complementary approach to finding genes under positive selection (He et al. 2023). A major limitation of the bottom-up approach—with pangenomes or otherwise—is that dissec-tion of the functional consequences of genomic regions un-der selection remains a nontrivial experimental endeavor.

Despite the discovery of several large-effect variants affecting domestication-related traits (Doebley et al. 2006), there is evidence suggesting that domestication may be driven largely by polygenic selection acting on a larger number of loci (Fig. 3, Supplementary Table S1) (Jantasuriyarat et al. 2004; Xue et al. 2016; Kantar et al. 2017; Studer et al. 2017). Even for a trait as well studied as shattering in rice, recent work indicates that there may be more complex, polygenic architecture at play than previously thought (Zhou et al. 2012; Ishikawa et al. 2022). An early study in wheat revealed that multiple genes or loci affected traits such as threshability, spike compactness, and spike length, even though some loci such as *Tg* and *Q* explained large percentages of variation in certain traits (Jantasuriyarat et al. 2004). More recently, a large-scale genome-wide association

Figure 3. Domestication traits and loci. Examples of domestication traits and associated genes or loci across diverse crop types. A combination of large- and small-effect quantitative trait loci contribute to most domestication-related phenotypes. More information on the loci can be found in Supplementary Table S1.

study (GWAS) analysis of domestication phenotypes such as shank length, cob length, and kernel row number in maize revealed small-effect polygenic variants underlying these traits located in genomic regions that had not been previously linked to domestication (Xue et al. 2016). In rice, 27 years of cultivation of landrace varieties under on-farm conservation conditions led to selection at around 180 loci each in *japonica* and *indica* subspecies, indicating continuous polygenic selection even in a shorter timeframe (Cui et al. 2020).

The number of domestication loci has implications for the strength of selection during domestication and the pace at which domestication occurs. A large number of loci implies weaker selection, as strong selection on many variants would likely lead to excessive loss of competing lineages and population collapse (Allaby et al. 2015). It has been estimated that there is an upper limit of 50 to 100 loci that can be simultaneously under selection without leading to the demise of the population (Allaby et al. 2015). This is lower than the number of putative domestication loci reported based on selective sweep detection in some studies (Hufford et al. 2012; Kantar et al. 2017), but many of these detected sweeps may be false positives or driven by more recent selective pressures. Regardless of the upper limit of the number of domestication loci, weak polygenic selection during domestication is supported by the slow rates of change in domestication phenotypes such as grain size and shattering that are observable in the archaeological record (Purugganan and Fuller 2011).

One key question has been the timing of the origin and selection of mutations associated with crop phenotypic evolution. Domestication studies often draw a distinction between domestication loci that were selected early on as wild crops were brought into cultivation and diversification or improvement loci that arose later as the crops spread to different regions and underwent more conscious selection (Doebley et al. 2006; Meyer and Purugganan 2013; Kantar et al. 2017). To have been involved in domestication, a mutation must not only be experimentally linked to a phenotypic trait or effect, but it must generally also have arisen early during the transition to clearly differentiated domesticated populations. Genetic studies of putative domestication loci can lack temporal resolution to differentiate between domestication versus improvement or diversification loci that may have been selected at different time periods during the evolution of the crop species, although population-specific loci under selection can reveal some diversification loci. Here, the emerging area of ancient DNA studies provides the most direct avenue to investigate the timeline of origin of known domestication alleles. In an early ancient DNA study on maize from Mexico and New Mexico, domestication alleles in *Tb1* and *Pbf*—which is involved in regulating seed storage proteins—were found to be present in maize by 4,400 years ago, while *Su1*—a starch metabolism gene containing variants associated with sweet corn varieties (Tracy et al. 2006)—had not undergone selection at one of its modern-day high frequency alleles by about 2,000 years ago

(Jaenicke-Després et al. 2003). A more recent study reported that the oldest maize cob from San Marcos cave in Tehuacán, Mexico, dated to more than 5,000 years ago, showed genomic signs of incomplete domestication even though it was morphologically indistinguishable from modern landraces (Vallebueno-Estrada et al. 2016). While *Tb1* had undergone strong selection and resembled modern maize across its coding and regulatory regions, *Tga1*, a canonical domestication locus conferring naked kernels in maize (H. Wang et al. 2015), still carried many wild alleles that no longer appear in modern maize, suggesting ongoing selection at the time.

Despite its advantages, ancient DNA sampling is sparse, and it is not yet available for many species including rice. In such cases, population genetic approaches that leverage haplotype reconstructions and gene genealogies can help to temporally constrain allele origins or selection to relative evolutionary timeframes or phases. A recent study, for example, used a panel of interploidy wheat accessions to infer ancestral haplotype blocks, allowing the tracing of histories of genomic loci associated with specific ancestries (Z. Wang et al. 2022). This analysis revealed a stepwise recruitment of domestica-tion alleles, with *TaBtr1-3A/-3B* alleles for brittle rachis be-coming fixed in the early domesticated emmer and alleles such as *TaTg-2B* for tenacious glume and *TaQ-5A* (or *Q*) rising to high frequencies later in free-threshing tetraploid wheat and tetraploid durum wheat. Hexaploid wheat subsequently received these domestication loci from its tetraploid contributor. A study in rice reconstructed the genealogies of orthologous genes found in assemblies of wild and domesticated varieties and identified genes under positive selection in early and late phases of domestication (Lu et al. 2022); the late phase followed the emergence of *indica* and *japonica* lineages, and thus represents crop diversification after the initial domestication. Even though this approach used little genome-wide (101 loci) and population-wide information, it was still able to identify canonical loci such as *Sh4*, *PROG1*, and *Hd3a* as under selection during early domestication.

Novel methods that infer and interpret ancestral recombination graphs (ARGs) or their simplified representations (Speidel et al. 2019; Stern Wilton and Nielsen 2019) to date the origin or rise in frequency of specific variants provide powerful possible ways—as yet untapped—to temporally validate domestication loci. An ARG captures the complete coalescent history of every non-recombining block in the genome. In cases where domestication loci have been function-ally characterized, ARGs hold promise in resolving the dynam-ics of the early domestication process and investigating hypotheses such as the sequential recruitment of certain traits and alleles (Ishikawa et al. 2022b). Although ARG-based dating approaches are yet to be applied in domesticates, studies in *Arabidopsis* provide us with a roadmap (Fulgione et al. 2022; Tergemina et al. 2022); in a population of *Arabidopsis* that colonized the base of an active volcano, a mutation in a transporter gene *IRT1* that increased manganese but reduced iron in leaves was shown to have undergone a hard selective sweep about 3,000 years ago, followed by the collective near-fixation of multiple tandem duplicates at NRAMP1 that helped recover iron levels (Tergemina et al. 2022).

Given our evolving understanding of the domestication process, identification of putative domestication loci must incorporate methods to detect multilocus adaptations and estimate timing of allelic origins. This remains a challenge in domesticates that do not yet have the same kind of genomic resources as rice, maize, and wheat.

Domestication and ecological traits

The process of domestication has frequently been viewed through an evolutionary or a genetic lens; less attention has been paid to the ecology of the domestication process. Ecology is often broadly defined as the study of interrelationships among organisms and their environment within an ecosystem (Dice 1955; Friederichs 1958), which in the context of domesticated crops would include canonical domestication traits such as non-shattering seeds that shape interactions between crops and humans. We define ecological traits in domesticated crops in terms of their nonhuman context, including traits such as growth and resource acquisition rates that influence interactions among themselves and their immediate environment, as well as the nature of their interactions with pests, pathogens, and symbionts. For most crops, it remains relatively poorly understood how such ecological traits change during domestication (Chen et al. 2015; Milla 2023); notable exceptions include studies on adaptations to climate (Hung et al. 2012; C. Liu et al. 2018) and photoperiod (Cong Li et al. 2020; C. Liu et al. 2018), and these tend to be linked to diversification and geographical expansion of crops after domestication. Changes in plant density, annual tillage, phenology, fertiliza-tion, and irrigation associated with cultivation likely altered the selection landscape for many ecological traits (Chen et al. 2015; Milla 2023), and recent ecological work, as well as genomic studies of interspecies interactions, have begun to investigate the ecology of domestication.

Understanding the nature of selection on ecological traits during cultivation may provide insight into the early stages of domestication in human-constructed niches. A study of several Fertile Crescent crop domesticates including barley, emmer wheat, and chickpea showed that they had, on average, ~50% higher yield compared to their wild progenitors, driven by increased plant size and seed mass and reduced chaff, and no differences in growth rate and duration, total seed number, and reproductive biomass (Preece et al. 2017); this increased plant size may have arisen from selection for light competition (Milla et al. 2014). A similar pattern was observed for seed traits in a meta-analysis of 49 grain crops that showed an increase in seed mass but no difference in seed number, which may reflect conscious selection for improved nutrition (Garibaldi et al. 2021) or unconscious selection driven, for instance, by light competition as larger seeds are associated with increased plant size (Gómez-Fernández and Milla 2022). Unconscious selection for larger seeds is further supported by the fact that vegetable crops that are not used by humans for their seed also appear to have undergone selection for larger seeds (Kluyver et al. 2017). The magnitude of increase in seed size or mass varies substantially among domesticated crops, suggesting heterogeneity in the nature of selection on the trait during domestication (Preece et al. 2017; Garibaldi et al. 2021).

Domestication also influenced root traits in crops. Maize (Burton et al. 2013), barley (Grando and Ceccarelli 1995), and emmer wheat (Golan et al. 2018) form a higher number of seminal roots—which constitute the initial root system supporting young seedlings—compared to their wild progenitors. This is possibly linked to selection for increased soil nutrient uptake (Perkins and Lynch 2021) or relaxation of selection related to water stress under cultivated conditions (Golan et al. 2018). In wild emmer, for example, two additional suppressed root primordia that do not initially develop into seminal roots can be reactivated after water stress to promote seedling recov-ery (Golan et al. 2018), a drought adaptation lost in domesti-cated wheat that could be of use when breeding crops for future climates with increased frequency of drought (Hari et al. 2020). Certain root traits can also reveal fast versus slow resource-acquisitive strategies; for instance, a fast strategy under increased soil fertility is expected to be associated with lower structural investment in roots and reduced root tissue density (Ryser and Lambers 1995). A 2019 study found that while domesticated crops such as wheat, barley, maize, rice, to-mato, lentil, and chickpea showed an expected increase in total plant dry mass, root traits such as root thickness and root tissue density evolved modestly and in different directions in different crops during domestication (Martín-Robles et al. 2019). Wild progenitors of domesticated crops had thicker and less dense roots—indicating faster resource acquisition—than other wild herbaceous plants, but similar to what has been observed across a wider range of plant species grown in highly fertile soils (Martín-Robles et al. 2019). This suggests that among wild herbaceous plants, crop progenitors were pre-adapted to grow-ing in agricultural conditions (Martín-Robles et al. 2019).

Comparative growth experiments have been used to explore the question of why only certain wild crop progenitors were selected for subsequent domestication over the last 12,000 years. One study comparing three wild crop progenitors to six wild gathered species found that the crop progenitors have higher seed mass and germinate faster (Cunniff et al. 2014). A later comparison of 13 wild grass species, including progenitors of Fertile Crescent domesticates, showed that while stands of the cereal progenitors do not produce a higher seed yield per ground area than closely related wild species, the progeni-tors have a higher seed yield per tiller (Preece et al. 2018). This was driven by larger individual seed size with no reduction in seed number per tiller compared to the related wild species, although it should be noted that a subsequent meta-analysis incorporating more progenitor and non-progenitor wild crops did not find a difference in per-seed mass between cereal progenitors and other wild species (Garibaldi et al. 2021). The evidence for progenitor filtering or pre-adaptation thus appears to be inconsistent and may be dependent on the exact wild species being compared. Nevertheless, the ecological traits of wild progenitors of crops domesticates compared to other wild plants remains an underexplored topic. Understanding whether certain types of wild plants are pre-adapted to being domesticated (Preece et al. 2018) can help inform strategies for de novo domestication of crops to tackle food insecurity and climate change; sampling from a wider set of wild crops could lead to more plant services for humans (Molina-Venegas et al. 2021).

Crop-species interactions and the ecology of crop evolution

Domestication also leads to changes in crop traits that alter interactions of crops with other species or organisms (Chen et al. 2015). In squash, for example, domestication led to increased floral attractiveness to pollinators compared to the wild progenitor (Glasser et al. 2023). Selection for increased yield has been hypothesized to have a negative effect on plant defenses against insect herbivory due to reallocation of metabolic resources from defense to growth (Rosenthal and Dirzo 1997). Reduced defenses have been observed in multiple domesticated species (Rosenthal and Dirzo 1997; Rodriguez-Saona et al. 2011; Turcotte et al. 2014; Chen et al. 2015; Soltis et al. 2018; Fernandez et al. 2021). There is some evidence for a negative relationship between growth and defenses reported in a study on wild teosinte and domesticated maize (Rosenthal and Dirzo 1997), although other studies failed to show a consistent effect (Turcotte et al. 2014; Whitehead et al. 2017).

Direct selection against defense-associated metabolites during domestication may also underlie observations of reduced defenses against herbivory (Moreira et al. 2018). Domesticated cabbage has reduced glucosinolates associated with both induced and constitutive defense with no corresponding tradeoff with leaf area, which is a measure of growth (Moreira et al. 2018). Glucosinolates and other secondary metabolites are known to influence flavor and taste (Bell et al. 2018), and selection for palatability can lead to changes in secondary metabolites during domestication (Johns and Alonso 1990; Q. Lin et al. 2023), sometimes with unintended side-effects (Fernández-Marín et al. 2014; Alseekh et al. 2021). Direct defense traits—such as toxic or distasteful secondary metabolites and physical barriers such as trichomes—tend to be reduced in harvested organs as opposed to vegetative tissues, supporting a role for direct selection against the associated metabolites (Whitehead et al. 2017). However, a meta-analysis that additionally incorporated indirect defense traits such as volatile organic compounds that recruit predators of herbivores and induced defenses that are produced in response to herbivore damage found similar reduction of defense traits in both harvested and vegetative tissues (Fernandez et al. 2021). This may suggest greater support for indirect selection via reallocation of resources away from defense, or different selection pressures on direct versus indirect defense traits. Finally, unintended evolutionary phenotypes can also arise through antagonistic pleiotropy, hitchhiking of deleterious alleles, and negative epistasis (Singh and van der Knaap 2022).

Understanding the impact of crop domestication on other interacting species including the domesticator species will help fully contextualize the ecology and coevolutionary dynamics of the process of domestication. There is evidence of genomic adaptations in human populations after domestication, including to high-starch diets composed of millets and/or rice in East Asia (Raj et al. 2019; Landini et al. 2021) and to potato in the Andes (Jorgensen et al. 2023). Crop domesticates that played an important role in the transition from hunter-gatherer to more sedentary, agriculture-based societies are likely to have shaped human evolution by influencing not just diet but also population density and thereby pathogen exposure and dynamics (Mathieson et al. 2015).

Field experiments in regions of origin of crop domesticates where wild diversity is still present can be informative about changes in their ecological context (Chen et al. 2015). One such study showed that *O. sativa* fields contained about half the number of arthropod taxa than ecosystems dominated by *Oryza rufipogon* (Chen et al. 2013). Domestication can also affect patterns of gene flow in insect species; populations of the bean beetle species *Acanthoscelides obvelatus* associated with wild beans show genetic differentiation by geographical distance, but populations associated with cultivated beans do not, possibly as a result of long-distance movement associated with human exchanges (Alvarez et al. 2007).

Domestication affects the relationship between plants and their microbial symbionts (Porter and Sachs 2020). The diversity of seed microbiota has been observed to be higher in cultivated cereals such as bread wheat, einkorn wheat, durum wheat, and barley compared to their wild progenitors, while co-occurrence network analysis suggests a greater number of microbe–microbe interactions in wild progenitors (Abdullaeva et al. 2021). Seed endophytes and host cereals show phylogenetic congruence, suggesting host–microbe coevolution during domestication (Abdullaeva et al. 2021). Studies on the root microbiome of crops have reported contrasting effects of domestication on diversity (Gutierrez and Grillo 2022). The root arbuscular mycorrhizal (AM) symbiosis is of particular interest, as it plays important roles in plant nutrition and growth and could contribute to the development of more sustainable agroecosystems with reduced fertilizer use (Basu et al. 2018). One major analysis of AM symbiosis in domesticates reported that across 14 crops, including spinach, sunflower, and oat, the benefit to plant growth from AM fungal colonization declines compared to wild progenitors in the presence of phosphorus fertilization, but it remains similar under low phosphorus conditions (Martín-Robles et al. 2018). The AM symbiosis involves the fungi providing phosphorus—among other nutrients—to the plant in exchange for carbohydrates, and it is possible that during cultivation in fertilized soil, the fungi no longer provide an essential service (Martín-Robles et al. 2018). This may have selected for reduced responsiveness to mycorrhiza in cultivated crops under high fertilization. Interestingly, wheat landraces exhibit greater responsiveness to AM symbiosis compared to elite cultivars, suggesting an impact of recent cultivation practices (Manske 1990). Among tomato and wild relatives, increase in plant growth in response to the beneficial soil microbe *Trichoderma* decreased through the domestication gradient along wild relatives, landraces, and cultivars (Jaiswal et al. 2020). Cultivars treated with *Trichoderma* exhibited lower suppression of the necrotic fungus *Botrytis cinerea* than wild relatives and landraces, again suggesting an impact of recent breeding practices on host–microbe interactions (Jaiswal et al. 2020).

The distinct environmental context of domesticates in cultivated fields compared to wild progenitors also led to changes in plant species density and diversity (Chen et al. 2015), and thus likely altered plant–plant interactions. Species mixtures often exhibit higher productivity compared to each species in monoculture (Cardinale et al. 2007). However, evolutionary selection for enhanced net facilitative plant interactions is likely more pronounced in mixtures with a mixture coexistence history rather than a history of growth in monoculture (Schöb et al. 2018), suggesting that the benefits of multispecies mixtures may be reduced under sustained monoculture cultivation conditions. In line with this expectation, a study on eight major crops including sunflower, tomato, and sorghum alongside their wild progenitors demonstrated that mixtures of domesticates have reduced enhancement in productivity compared to mixtures of wild progenitors (Chacón-Labella et al. 2019).

Genetics of crop ecological traits

As ecological traits such as resource acquisition and interactions with other species become clearly established as domestication phenotypes, the genetic bases of these traits will be of great interest not only to understand cultivation-mediated selection processes and outcomes during domestication, but also to optimize ecological performance of crops (Thrall et al. 2011; Bargués-Ribera and Gokhale 2020). Studies in non-domesticated plants have established the heritability and genetic bases of ecological traits such as aboveground community productivity (Lojewski et al. 2009; Wuest and Niklaus 2018), and they point the way toward similar investigations in domesticated crops where genetic loci underlying increased productivity from crop rotations and intercropping would be of great interest (Litrico and Violle 2015). While modern crop breeding and cultivation frequently employ monocultures, there is increasing recognition that mixtures of crops may constitute more robust and resilient agroecosystems (Litrico and Violle 2015; McAlvay et al. 2022), for instance, against pathogens (McDonald and Stukenbrock 2016) and future climate challenges such as increased water stress (Natarajan and Willey 1986) under drought conditions. Comparisons with wild progenitors can facilitate reintroduction of traits that were lost during domestication such as water stress–induced seedling recovery in wild emmer (Golan et al. 2018) and increased productivity from species mixtures of wild progenitors (Chacón-Labella et al. 2019).

One example of genetic analysis on a key crop ecological trait is of nitrogen use efficiency, which has been studied in several crop species including rice (Gao et al. 2019), wheat (Shi et al. 2022), and maize (Ertiro et al. 2020). In rice, a variant in the *OsNR2* gene drives increased nitrate reductase activity and nitrogen use efficiency in the *indica* subspecies, with an accompanying increase in grain yield (Gao et al. 2019). Intriguingly, *japonica* rice—the earliest rice domesticate that contributed many domestication alleles to *indica* rice during its later origin —possibly underwent directional selection in the same gene toward reduced nitrogen use efficiency compared to its wild progenitor *Oryza rufipogon* (Gao et al. 2019). Improved resistance to lodging—the flattening of plants by wind or rain due to nitrogen-promoted stem elongation—in rice varieties selected during the Green Revolution is associated with reduced plant height and nitrogen use efficiency, allowing farm-ers to use large amounts of fertilizer without risk of lodging (Q. Liu et al. 2022). Given that the gene responsible for dwarfism in rice during the Green Revolution, *Sd1*, also underwent selection toward shorter height during *japonica* domestication (Asano et al. 2011), it appears plausible that similar or equivalent selection pressures to those encountered during the Green Revolution—such as highly fertile soils—led to reduced plant height and nitrogen use efficiency during *japonica* domestication. Disentangling the genetic bases of these traits will be key to developing sustainable crops with high nitrogen efficiency without the associated risk of lodging, allowing reduced fertilizer use.

GWAS-based and population genomic approaches have been used to explore evolutionary responses to crop domestication—or to coevolution with domesticates—in interacting species beyond humans. A study in rice found 47 virulence-associated genes in the pathogen *Xanthomonas oryzae* coevolving with 318 rice genes, which included many known resistance genes (Zhang et al. 2021). Several genes in both species showed signatures of positive selection consistent with rapid host–pathogen coevolution. In the agricultural weed *Echinochloa crus-galli* or barnyardgrass, 87 putative plant architecture-related genes have been found to be under selec-tion during the evolution of Vavilovian mimicry to domesti-cated rice (Ye et al. 2019). The domestication and subsequent dispersal of squash or *Cucurbita* species led to the geographical expansion of the wild pollinator species *Eucera (Peponapis) prui-nosa* through the Americas, whose genomes show signatures of selection likely associated with adaptation to novel agricultural niches (López-Uribe et al. 2016; Pope et al. 2023). GWAS studies have identified genetic loci controlling variation in the leaf mi-crobiome in rice (Roman-Reyna et al. 2020) and maize (Wallace et al. 2018) and in the root microbiome in sorghum (Deng et al. 2021), revealing genes involved in plant defense and cell wall function or degradation. These findings invite future investiga-tions of how these loci have evolved in domesticates compared to wild progenitors, and how they may be used to select for microbiome-related outcomes during breeding.

Concluding remarks

Although the study of domestication has a long history that traces back to Darwin, we are beginning to finally get a better understanding of this unique plant evolutionary process, due in part to collaborative efforts by plant biologists, evolutionary geneticists, molecular biologists, archaeologists, and anthropologists. The last few decades have been particularly productive, spurred in part by advances in genomic sciences as well as archaeobotany. Nevertheless, there are still gaps in our understanding of crop domestication. Much of the work in this area has focused on cereal crop species, and this has skewed our knowledge of crop domestication; more efforts are needed to study other crop species, including legumes, vegetable species, and perennial fruit trees. Greater emphasis should be placed on identifying and characterizing more domestication loci, both at the functional and evolutionary genetic levels. A combination of traditional trait-based and selection approaches as well as novel ARG-based dating methods can help reveal more about genomic loci underlying domestication. Some effort should also be made with examining post-domestication evolution, particularly those associated with crop dispersal and adaptation to novel environments.

Domestication may be a process of great evolutionary interest, but crops also underlie our food security, and their study is at the forefront of challenges imposed by climate change (Raza et al. 2019). There has been recent interest in the idea of domesticating new crop species, including through rapid de novo domestication of climate-resilient crops by gene editing, to help human agriculture adapt to future agricultural challenges (Yu et al. 2021; Gutaker et al. 2022). This can also be coupled with more attention to neglected crop species, many of which are confined to local or regional areas, but which may hold promise for future widespread use. In many of these efforts, knowledge of genomic variation associated with desirable domestication-related traits will be key to the success of such approaches, Finally, the ecological context of domestication can not only expand our definition of domestication syndromes in different crops, but also help develop more sustainable agricultural management practices (Milla 2023) to help protect or improve yields in changing environments (Battisti et al. 2018).

Supplementary data

The following materials are available in the online version of this article.

Supplementary Table S1. Details on domestication loci and traits shown in Fig. 3.

Funding

This work was partly supported by grants from the US National Science Foundation Plant Genome Research Program, Zegar Family Foundation, and NYU Abu Dhabi Research Institute.

Conflict of interest statement. None declared.

Data availability

The data underlying the figures in this article are cited within the article and in its online supplementary material.

References

Abbo S, Gopher A, Bar-Gal GK. Plant domestication and the origins of agriculture in the Ancient Near East. Cambridge: Cambridge University Press; 2022.

Abbo S, Rachamim E, Zehavi Y, Zezak I, Lev-Yadun S, Gopher A. Experimental growing of wild pea in Israel and its bearing on near eastern plant domestication. Ann Bot. 2011:**107**(8):1399–1404. https://doi.org/10.1093/aob/mcr081

Abdullaeva Y, Ambika MB, Honermeier B, Schnell S, Cardinale M. Domestication affects the composition, diversity, and co-occurrence of the cereal seed microbiota. J Adv Res. 2021:**31**:75–86. https://doi.org/10.1016/j.jare.2020.12.008

Abrouk M, Ahmed HI, Cubry P, Šimoníková D, Cauet S, Pailles Y, Bettgenhaeuser J, Gapa L, Scarcelli N, Couderc M, et al. Fonio millet genome unlocks African orphan crop diversity for agriculture in a changing climate. Nat Commun. 2020:**11**(1):4488. https://doi.org/10.1038/s41467-020-18329-4

Allaby RG. Domestication syndrome in plants. In: **Smith C**, editors. Encyclopedia of global archaeology. New York (NY): Springer New York; 2014. p. 2182–2184.

Allaby RG, Kitchen JL, Fuller DQ. Surprisingly low limits of selection in plant domestication. Evol Bioinf. 2015:**11**(Suppl 2):41–51. https://doi.org/10.4137/EBO.S33495

Allaby RG, Stevens CJ, Kistler L, Fuller DQ. Emerging evidence of plant domestication as a landscape-level process. Trends Ecol Evol. 2022:**37**(3):268–279. https://doi.org/10.1016/j.tree.2021.11.002

Allaby RG, Ware RL, Kistler L. A re-evaluation of the domestication bottleneck from archaeogenomic evidence. Evol Appl. 2019:**12**(1): 29–37. https://doi.org/10.1111/eva.12680

Alseekh S, Scossa F, Wen W, et al. Domestication of crop metabolomes: desired and unintended consequences. Trends Plant Sci. 2021:**26**(6):650–661. https://doi.org/10.1016/j.tplants.2021.02.005

Alvarez N, Hossaert-McKey M, Restoux G, Delgado-Salinas A, Benrey B. Anthropogenic effects on population genetics of phytophagous insects associated with domesticated plants. Evolution. 2007:**61**(12): 2986–2096. https://doi.org/10.1111/j.1558-5646.2007.00235.x.

Andersson L, Purugganan MD. Molecular genetic variation of animals and plants under domestication. Proc Natl Acad Sci USA. 2022:**119**(30):e2122150119. https://doi.org/10.1073/pnas.2122150119

Arranz-Otaegui A, Colledge S, Zapata L, Teira-Mayolini LC, Ibáñez JJ. Regional diversity on the timing for the initial appearance of cereal cultivation and domestication in southwest Asia. Proc Natl Acad Sci. 2016:**113**(49):14001–14006. https://doi.org/10.1073/pnas.1612797113

Asano K, Yamasaki M, Takuno S, Miura K, Katagiri S, Ito T, Doi K, Wu J, Ebana K, Matsumoto T, et al. Artificial selection for a green revolution gene during japonica rice domestication. Proc Natl Acad Sci USA. 2011:**108**(27):11034–11039. https://doi.org/10.1073/pnas.1019490108

Bargués-Ribera M, Gokhale C. Eco-evolutionary agriculture: host-pathogen dynamics in crop rotations. PLoS Comp Biol. 2020:**16**(1):e1007546. https://doi.org/10.1371/journal.pcbi.1007546

Basu S, Rabara RC, Negi S. AMF: the future prospect for sustainable agriculture. Physiol Mol Plant Path. 2018:**102**:36–45. https://doi.org/10.1016/j.pmpp.2017.11.007

Battisti R, Sentelhas PC, Parker PS, Nendel C, Câmara GMDS, Farias JRB, Basso CJ.et al. Assessment of crop-management strategies to improve soybean resilience to climate change in southern Brazil. Crop Pasture Sci. 2018:**69**(2):154–162. https://doi.org/10.1071/CP17293

Bell L, Oloyede OO, Lignou S, Wagstaff C, Methven L. Taste and flavor perceptions of glucosinolates, isothiocyanates, and related compounds. Mol Nutr Food Res. 2018:**62**(18):e1700990. https://doi.org/10.1002/mnfr.201700990

Burton A, Brown K, Lynch J. Phenotypic diversity of root anatomical and architectural traits in Zea species. Crop Sci. 2013:**53**(3): 1042–1055. https://doi.org/10.2135/cropsci2012.07.0440

Cardinale BJ, Wright JP, Cadotte MW, Carroll IT, Hector A, Srivastava DS, Loreau M, Weis JJ. Impacts of plant diversity on biomass production increase through time because of species complementarity. Proc Natl Acad Sci USA. 2007:**104**(46):18123–18128. https://doi.org/10.1073/pnas.0709069104

Casas A, Otero-Arnaiz A, Pérez-Negrón E, Valiente-Banuet A. In situ management and domestication of plants in Mesoamerica. Ann Bot. 2007:**100**(5):1101–1115. https://doi.org/10.1093/aob/mcm126

Chacón-Labella J, García Palacios Pablo, Matesanz Silvia, Schöb Christian, Milla R. Plant domestication disrupts biodiversity effects across major crop types. Ecol Lett. 2019:**22**(9):1472–1482. https://doi.org/10.1111/ele.13336

Chen Y, Gols R, Benrey B. Crop domestication and its impact on naturally selected trophic interactions. Annu Rev Entomol. 2015:**60**(1): 35–58. https://doi.org/10.1146/annurev-ento-010814-020601

Chen YH, Langellotto G, Barrion AT, Cuong NL. Cultivation of domesticated rice alters arthropod biodiversity and community composition. Ann Entomol Soc Am. 2013:**106**(1):100–110. https://doi.org/10.1603/AN12082

Choi JY, latts AE, Fuller DQ, Hsing Y-I, Wing RA, Purugganan MD. The rice paradox: multiple origins but single domestication in Asian rice. Mol Biol Evol. 2017:**34**(4):969–979. https://doi.org/10.1093/molbev/msx049

Clement CR, Casas A, Parra-Rondinel FA, Levis C, Peroni N, Hanazaki N, Cortés-Zárraga L, Rangel-Landa S, Alves RP, Ferreira MJ, et al. Disentangling domestication from food production systems in the neotropics. Quaternary. 2021:**4**(1):4. https://doi.org/10.3390/quat4010004

Cui D, Lu H, Tang C, Li J, A X, Yu T, Ma X, Zhang E, Wang Y, Cao G, et al. Genomic analyses reveal selection footprints in rice landraces grown under on-farm conservation conditions during a short-term period of domestication. Evol Appl. 2020:**13**(2):290–302. https://doi.org/10.1111/eva.12866

Cunniff J, Wilkinson S, Charles M, Jones G, Rees M, Osborne CP. Functional traits differ between cereal crop progenitors and other wild grasses gathered in the neolithic Fertile crescent. PLoS One. 2014:**9**(1):e87586. https://doi.org/10.1371/journal.pone.0087586

Deng S, Caddell DF, Xu G, Dahlen L, Washington L. Genome wide association study reveals plant loci controlling heritability of the rhizosphere microbiome. ISME J. 2021:**15**(11):3181–3194. https://doi.org/10.1038/s41396-021-00993-z

Darwin CR. The variation of animals and plants under domestication. 1st ed. Vol. 1 (first issue). London: John Murray; 1868.

Denham T, Barton H, Castillo C, Crowther A, Dotte-Sarout E, Florin SA, Pritchard J, Barron A, Zhang Y, Fuller DQ, et al. The domestication syndrome in vegetatively propagated field crops. Ann Bot. 2020:**125**(4):581–597. https://doi.org/10.1093/aob/mcz212

Diamond J. Evolution, consequences and future of plant and animal domestication. Nature. 2002:**418**(6898):700–707. https://doi.org/10.1038/nature01019

Dice LR. What is ecology? Sci Monthly. 1955:**80**:346–351. http://www.jstor.org/stable/21546

Doebley JF, Gaut BS, Smith BD. The molecular genetics of crop domestication. Cell. 2006:**127**(7):1309–1321. https://doi.org/10.1016/j.cell.2006.12.006

Doebley J, Stec A, Gustus C. Teosinte branched1 and the origin of maize: evidence for epistasis and the evolution of dominance. Genetics. 1995:**141**(1):333–346. https://doi.org/10.1093/genetics/141.1.333

Dvorak J, Luo MC, Yang ZL, Zhang HB. The structure of the Aegilops tauschii genepool and the evolution of hexaploid wheat. Theor Appl Genet. 1998:**97**(4):657–670. https://doi.org/10.1007/s001220050942

Ertiro BT, Labuschagne M, Olsen M, Das B, Prasanna BM, Gowda M. Genetic dissection of nitrogen use efficiency in tropical maize through genome-wide association and genomic prediction. Front Plant Sci. 2020:**11**:474. DOI 10.3389/fpls.2020.00474

Fernández-Marín B, Milla R, Martín-Robles N, Arc E, Kranner I, Becerril JM, García-Plazaola JI. Side-effects of domestication: cultivated legume seeds contain similar tocopherols and fatty acids but less carotenoids than their wild counterparts. BMC Plant Biol. 2014:**14**(1):1599. https://doi.org/10.1186/s12870-014-0385-1

Fernandez AR, Sáez A, Quintero C, Gleiser G, Aizen MA. Intentional and unintentional selection during plant domestication: herbivore damage, plant defensive traits and nutritional quality of fruit and seed crops. New Phytol. 2021:**231**(4):1586–1598. https://doi.org/10.1111/nph.17452

Friederichs K. A definition of ecology and some thoughts about basic concepts. Ecology. 1958:**39**(1):154–159. https://doi.org/10.2307/1929981

Fulgione A, Neto C, Elfarargi AF, Tergemina E, Ansari S, Göktay M, Dinis H, Döring N, Flood PJ, Rodriguez-Pacheco S, et al. Parallel reduction in flowering time from de novo mutations enable evolutionary rescue in colonizing lineages. Nat Commun. 2022:**13**(1):1461. https://doi.org/10.1038/s41467-022-28800-z

Fuller DQ. Long and attenuated: comparative trends in the domestication of tree fruits. Veg Hist Archaeobot. 2018:**27**(1):165–176. https://doi.org/10.1007/s00334-017-0659-2

Fuller DQ, Denham T, Allaby R. Plant domestication and agricultural ecologies. Curr Biol. 2023:**33**(11):R636–R649. https://doi.org/10.1016/j.cub.2023.04.038

Gao Z, Wang Y, Chen G, Zhang A, Yang S, Shang L, Wang D, Ruan B, Liu C, Jiang H, et al. The indica nitrate reductase gene OsNR2 allele enhances rice yield potential and nitrogen use efficiency. Nat Commun. 2019:**10**(1):5207. https://doi.org/10.1038/s41467-019-13110-8

Garibaldi LA, Aizen MA, Sáez A, Gleiser G, Strelin MM, Harder LD. The influences of progenitor filtering, domestication selection and the boundaries of nature on the domestication of grain crops. Func Ecol. 2021:**35**(9):1998–2011. https://doi.org/10.1111/1365-2435.13819

Gaut BS, Díez CM, Morrell PL. Genomics and the contrasting dynamics of annual and perennial domestication. Trends Genet. 2015:**31**(12):709–719. https://doi.org/10.1016/j.tig.2015.10.002

Gegas VC, Nazari A, Griffiths S, Simmonds J, Fish L, Orford S, Sayers L, Doonan JH, Snape JW. A genetic framework for grain size and shape variation in wheat. Plant Cell. 2010:**22**(4):1046–1056. https://doi.org/10.1105/tpc.110.074153

Gepts P. Crop domestication as a long-term selection experiment. In: **Janick J**, editor. Plant breeding reviews, vol 24, part 2. New York: Wiley; 2003. p. 1–44. https://doi.org/10.1002/9780470650288.ch1

Glasser SK, de Santiago-Hernández MH, Delgado-Carrillo O, Espino LAV, Pérez AC, González-Rodríguez A, Lira-Saade R, Quesada M. Influence of plant domestication on plant-pollinator interactions: floral attributes and floral visitor communities in wild and cultivated squash plants. Am J Bot. 2023:**110**(5):e16170. https://doi.org/10.1002/ajb2.16170

Golan G, Hendel E, Méndez Espitia GE, Schwartz N, Peleg Z. Activation of seminal root primordia during wheat domestication reveals underlying mechanisms of plant resilience. Plant Cell Environ. 2018:**41**(4):755–766. https://doi.org/10.1111/pce.13138

Gómez-Fernández A, Milla R. How seeds and growth dynamics influence plant size and yield: integrating trait relationships into ontogeny. J Ecol. 2022:**110**(11):2684–2700. https://doi.org/10.1111/1365-2745.13979

Gong Y, Li Y, Liu X, Ma Y, Jiang L. A review of the pangenome: how it affects our understanding of genomic variation, selection and breeding in domestic animals? J Animal Sci Biotech. 2023:**14**(1):73. https://doi.org/10.1186/s40104-023-00860-1

Grando S, Ceccarelli S. Seminal root morphology and coleoptile length in wild (Hordeum vulgare ssp. spontaneum) and cultivated (Hordeum vulgare ssp. vulgare) barley. Euphytica. 1995:**86**(1):73–80. https://doi.org/10.1007/BF00035941

Gross BL, Henk AD, Richards CM, Fazio G, Volk GM. Genetic diversity in Malus ×domestica (Rosaceae) through time in response to domestication. Am J Bot. 2014:**101**(10):1770–1779. https://doi.org/10.3732/ajb.1400297

Gutaker RM, Chater CCC, Brinton J, Castillo-Lorenzo E, Breman E, Pironon S. Scaling up neodomestication for climate-ready crops. Curr Opin Plant Biol. 2022:**66**:102169. https://doi.org/10.1016/j.pbi.2021.102169

Gutierrez A, Grillo MA. Effects of domestication on plant–microbiome interactions. Plant Cell Physiol. 2022:**63**(11):1654–1666. https://doi.org/10.1093/pcp/pcac108

Hammer K. Das domestikationssyndrom. Kulturpflanze. 1984:**32**(1): 11–34. https://doi.org/10.1007/BF02098682

Han L, Chen J, Mace ES, Liu Y, Zhu M, Yuyama N, Jordan DR, Cai H. Fine mapping of qGW1, a major QTL for grain weight in sorghum. Theor Appl Genet. 2015:**128**(9):1813–1825. https://doi.org/10.1007/s00122-015-2549-2

Hari V, Rakovec O, Markonis Y, Hanel M, Kumar R. Increased future occurrences of the exceptional 2018–2019 central European drought under global warming. Sci Rep. 2020:**10**:12207. https://doi.org/10.1038/s41598-020-68872-9

He Q, Tang S, Zhi H, Chen J, Zhang J, Liang H, Alam O, Li H, Zhang H, Xing L, et al. A graph-based genome and pan-genome variation of the model plant Setaria. Nat Genet. 2023:**55**(7):1232–1242. https://doi.org/10.1038/s41588-023-01423-w

Huang X, Kurata N, Wei X, Wang Z-X, Wang A, Zhao Q, Zhao Y, Liu K, Lu H, Li W, et al. A map of rice genome variation reveals the origin of cultivated rice. Nature. 2012:**490**(7421):497–501. https://doi.org/10.1038/nature11532

Hufford MB, Xu X, van Heerwaarden J, Pyhäjärvi T, Chia J-M, Cartwright RA, Elshire RJ, Glaubitz JC, Guill KE, Kaeppler SM, et al. Comparative population genomics of maize domestication and improvement. Nat Genet. 2012:**44**(7):808–811. https://doi.org/10.1038/ng.2309

Hung HY, Shannon LM., Tian F, Bradbury PJ, Chen C, Flint-Garcia SA, McMullen MD, Ware D, Buckler ES, Doebley JF, et al. ZmCCT and the genetic basis of day-length adaptation underlying the postdomestication spread of maize. Proc Natl Acad Sci USA. 2012:**109**(28): E1913–E1921. https://doi.org/10.1073/pnas.1203189109

International Wheat Genome Sequencing Consortium (IWGSC). A chromosome-based draft sequence of the hexaploid bread wheat (Triticum aestivum) genome. Science. 2014:**345**(6194):1251788. https://doi.org/10.1126/science.1251788

Ishikawa R, Castillo CC, Fuller DQ. Genetic evaluation of domestication-related traits in rice: implications for the archaeobotany of rice origins. Archael Anthro Sci. 2020:**12**(8):197. https://doi.org/10.1007/s12520-020-01112-3

Ishikawa R, Castillo CC, Htun TM, Numaguchi K, Inoue K, Oka Y, Ogasawara M, Sugiyama S, Takama N, Orn C, et al. A stepwise route to domesticate rice by controlling seed shattering and panicle shape. Proc Natl Acad Sci USA. 2022:**119**(26):e2121692119. https://doi.org/10.1073/pnas.2121692119

Jaenicke-Després V, Buckler ES, Smith BD, Gilbert MTP, Cooper A, Doebley J, Pääbo S. Early allelic selection in maize as revealed by ancient DNA. Science. 2003:**302**(5648):1206–1208. https://doi.org/10.1126/science.1089056

Jaiswal AK, Mengiste TD, Myers JR, Egel DS, Hoagland LA. Tomato domestication attenuated responsiveness to a beneficial soil microbe for plant growth promotion and induction of systemic resistance to foliar pathogens. Front Microbiol. 2020:**11**:604566. doi 10.3389/fmicb.2020.604566

Jantasuriyarat C, Vales MI, Watson CJW, Riera-Lizarazu O. Identification and mapping of genetic loci affecting the free-threshing habit and spike compactness in wheat (Triticum aestivum L.). Theor Appl Genet. 2004:**108**(2):261–273. https://doi.org/10.1007/s00122-003-1432-8

Johns T, Alonso JG. Glycoalkaloid change during the domestication of the potato, Solanum section petota. Euphytica. 1990:**50**(3):203–210. https://doi.org/10.1007/BF00023646

Jorgensen K, Garcia OA, Kiyamu M, Brutsaert TD, Bigham AW. Genetic adaptations to potato starch digestion in the Peruvian Andes. Am J Biol Anthropol. 2023:**180**:162–172. https://doi.org/10.1002/ajpa.24656

Kantar MB, Nashoba AR, Anderson JE, Blackman BK, Rieseberg LH. The genetics and genomics of plant domestication. BioScience. 2017:**67**(11):971–982. https://doi.org/10.1093/biosci/bix114

Kato K, Sonokawa R, Miura R, Sawada S. Dwarfing effect associated with the threshability gene Q on wheat chromosome 5A. Plant Breed. 2003:**122**(6):489–492. https://doi.org/10.1111/j.1439-0523.2003.00886.x

Kistler L, Thakar HB, VanDerwarker AM, Domic A, Bergström A, George RJ, Harper TK, Allaby RG, Hirth K, Kennett DJ. Archaeological central American maize genomes suggest ancient gene flow from South America. Proc Natl Acad Sci USA. 2020:**117**(52): 33124–33129. https://doi.org/10.1073/pnas.2015560117

Kluyver TA, Jones G, Pujol B, Bennett C, Mockford EJ, Charles M, Rees M, Osborne CP. Unconscious selection drove seed enlargement in vegetable crops. Evol Lett. 2017:**1**(2):64–72. https://doi.org/10.1002/evl3.6

Landini A, Yu S, Gnecchi-Ruscone GA, Abondio P, Ojeda-Granados C, Sarno S, De Fanti S, Gentilini D, Di Blasio AM, Jin H, et al. Genomic adaptations to cereal-based diets contribute to mitigate metabolic risk in some human populations of east Asian ancestry. Evol Appl. 2021:**14**(2):297–313. https://doi.org/10.1111/eva.13090

Lenser T, Theißen G. Molecular mechanisms involved in convergent crop domestication. Trends Plant Sci. 2013:**18**(12):704–714. https://doi.org/10.1016/j.tplants.2013.08.007

Levis C, Flores B, Moreira P, Luize BG, Alves R, Franco-Moraes J, Lins J, Konings E, Pena Claros M, Bongers F, et al. How people domesticated Amazonian forests. Front Ecol Evol. 2018:**5**:1-21. https://doi.org/10.3389/fevo.2017.00171

Li C, Li Y-H, Li Y, Lu H, Hong H, Tian Y, Li H, Zhao T, Zhou X, Liu J, et al. A domestication-associated gene GmPRR3b regulates the circadian clock and flowering time in soybean. Mol Plant. 2020:**13**(5): 745–759. https://doi.org/10.1016/j.molp.2020.01.014

Li C, Zhou A, Sang T. Rice domestication by reducing shattering. Science. 2006:**311**(5769):1936–1939. https://doi.org/10.1126/science.1123604

Lin Q, Chen J, Liu X, Wang B, Zhao Y, Liao L, Allan AC, Sun C, Duan Y, Li X, et al. A metabolic perspective of selection for fruit quality related to apple domestication and improvement. Genome Biol. 2023:**24**(1): 95. https://doi.org/10.1186/s13059-023-02945-6

Lin Z, Li X, Shannon LM, Yeh C-T, Wang ML, Bai G, Peng Z, Li J, Trick HN, Clemente TE, et al. Parallel domestication of the shattering1 genes in cereals. Nat Genet. 2012:**44**(6):720–724. https://doi.org/10.1038/ng.2281

Litrico I, Violle C. Diversity in plant breeding: a new conceptual framework. Trends Plant Sci. 2015:**20**(10):604–613. https://doi.org/10.1016/j.tplants.2015.07.007

Liu B, Fujita T, Yan Z-H, Sakamoto S, Xu D, Abe J. QTL mapping of domestication-related traits in soybean (Glycine max). Ann Bot. 2007:**100**(5):1027–1038. https://doi.org/10.1093/aob/mcm149

Liu C, Ou S, Mao B, Tang J, Wang W, Wang H, Cao S, Schläppi MR, Zhao B, Xiao G, et al. Early selection of bZIP73 facilitated adaptation of japonica rice to cold climates. Nat Commun. 2018:**9**(1):3302. https://doi.org/10.1038/s41467-018-05753-w

Liu Q, Wu K, Song W, Zhong N, Wu Y, Fu X. Improving crop nitrogen use efficiency toward sustainable green revolution. Annu Rev Plant Biol. 2022:**73**(1):523–551. https://doi.org/10.1146/annurev-arplant-070121-015752

Lojewski NR, Fischer DG, Bailey JK., Schweitzer JA, Whitham TG, Hart SC. Genetic basis of aboveground productivity in two native populus species and their hybrids. Tree Physiol. 2009:**29**(9):1133–1142. https://doi.org/10.1093/treephys/tpp046

López-Uribe MM, Cane JH, Minckley RL, Danforth BN. Crop domestication facilitated rapid geographical expansion of a specialist pollinator, the squash bee peponapis pruinosa. Proc Biol Sci. 2016:**283**(1833):20160443. https://doi.org/10.1098/rspb.2016.0443

Lu Y, Xu Y, Li N. Early domestication history of Asian rice revealed by mutations and genome-wide analysis of gene genealogies. Rice. 2022:**15**(1):11. https://doi.org/10.1186/s12284-022-00556-6

Manske GGB. Genetical analysis of the efficiency of VA mycorrhiza with spring wheat. Agric Ecosyst Environ. 1990:**29**(1–4):273–280. https://doi.org/10.1016/0167-8809(90)90287-N

Martín-Robles N, Lehmann A, Seco E, Aroca R, Rillig MC, Milla R. Impacts of domestication on the arbuscular mycorrhizal symbiosis of 27 crop species. New Phytol. 2018:**218**(1):322–334. https://doi.org/10.1111/nph.14962

Martín-Robles N, Morente-López J, Freschet GT, Poorter H, Roumet C, Milla R. Root traits of herbaceous crops: pre-adaptation to cultivation or evolution under domestication? Funct Ecol. 2019:**33**(2): 273–285. https://doi.org/10.1111/1365-2435.13231

Mathieson I, Lazaridis I, Rohland N, Mallick S, Patterson N, Roodenberg SA, Harney E, Stewardson K, Fernandes D, Novak M, et al. Genome-wide patterns of selection in 230 ancient Eurasians. Nature. 2015:**528**(7583):499–503. https://doi.org/10.1038/nature16152

McAlvay AC, DiPaola A, D'Andrea AC, Ruelle ML, Mosulishvili M, Halstead P, Power AG. Cereal species mixtures: an ancient practice with potential for climate resilience: a review. Agron Sustain Dev. 2022:**42**(5):100. https://doi.org/10.1007/s13593-022-00832-1

McDonald BA, Stukenbrock EH. Rapid emergence of pathogens in agro-ecosystems: global threats to agricultural sustainability and food security. Philos Trans R Soc Lond B Biol Sci. 2016:**371**(1709): 20160026. https://doi.org/10.1098/rstb.2016.0026

Meyer RS, Choi JY, Sanches M, Plessis A, Flowers JM, Amas J, Dorph K, Barretto A, Gross B, Fuller DQ, et al. Domestication history and geographical adaptation inferred from a SNP map of African rice. Nat Genet. 2016:**48**(9):1083–1088. https://doi.org/10.1038/ng.3633

Meyer RS, DuVal AE, Jensen HR. Patterns and processes in crop domestication: an historical review and quantitative analysis of 203 global food crops. New Phytol. 2012:**196**(1):29–48. https://doi.org/10.1111/j.1469-8137.2012.04253.x

Meyer RS, Purugganan MD. Evolution of crop species: genetics of domestication and diversification. Nat Rev Genet. 2013:**14**(12):840–852. https://doi.org/10.1038/nrg3605

Milla R. Crop origins and phylo food: a database and a phylogenetic tree to stimulate comparative analyses on the origins of food crops. Global Ecol Biogeogr. 2020:**29**(4):606–614. https://doi.org/10.1111/geb.13057

Milla R. Phenotypic evolution of agricultural crops. Funct Ecol. 2023:**37**(4):976–988. https://doi.org/10.1111/1365-2435.14278

Milla R, Morente-López J, Alonso-Rodrigo JM, Martín-Robles N, Stuart Chapin F. Shifts and disruptions in resource-use trait syndromes during the evolution of herbaceous crops. Proc Biol Sci. 2014:**281**(1793):20141429. https://doi.org/10.1098/rspb.2014.1429

Molina-Venegas R, Rodríguez MÁ., Pardo-de-Santayana M, Ronquillo C, Mabberley DJ. Maximum levels of global phylogenetic diversity efficiently capture plant services for humankind. Nat Ecol Evol. 2021:**5**(5):583–588. https://doi.org/10.1038/s41559-021-01414-2

Morcote-Ríos G, Aceituno FJ, Iriarte J, Robinson M, Chaparro-Cárdenas JL. Colonisation and early peopling of the Colombian Amazon during the late Pleistocene and the early Holocene: new evidence from la Serranía la Lindosa. Quater Int. 2021:**578**:5–19. https://doi.org/10.1016/j.quaint.2020.04.026

Moreira X, Abdala-Roberts L, Gols R, Francisco M. Plant domestication decreases both constitutive and induced chemical defences by direct selection against defensive traits. Sci Rep. 2018:**8**(1):12678. https://doi.org/10.1038/s41598-018-31041-0

Natarajan M, Willey RW. The effects of water stress on yield advantages of intercropping systems. Field Crops Res. 1986:**13**:117–131. https://doi.org/10.1016/0378-4290(86)90015-8

Pärssinen M, Ferreira E, Virtanen PK, Ranzi A. Domestication in motion: macrofossils of pre-colonial Brazilian nuts, palms and other Amazonian planted tree species found in the upper Purus. Environ Archaeol. 2021:**26**(3):309–322. https://doi.org/10.1080/14614103.2020.1765295

Paterson AH. What has QTL mapping taught US about plant domestication? New Phytol. 2002:**154**(3):591–608. https://doi.org/10.1046/j.1469-8137.2002.00420.x

Perkins AC, Lynch JP. Increased seminal root number associated with domestication improves nitrogen and phosphorus acquisition in maize seedlings. Ann Bot. 2021:**128**(4):453–468. https://doi.org/10.1093/aob/mcab074

Pope NS, Singh A, Childers AK, Kapheim KM, Evans JD, López-Uribe MM. The expansion of agriculture has shaped the recent evolutionary history of a specialized squash pollinator. Proc Natl Acad Sci USA. 2023:**120**(15):e2208116120. https://doi.org/10.1073/pnas.2208116120

Porter SS, Sachs JL. Agriculture and the disruption of plant–microbial symbiosis. Trends Ecol Evol. 2020:**35**(5):426–439. https://doi.org/10.1016/j.tree.2020.01.006

Preece C, Clamp NF, Warham G, Charles M, Rees M, Jones G, Osborne CP. Cereal progenitors differ in stand harvest characteristics from related wild grasses. J Ecol. 2018:**106**(3):1286–1297. https://doi.org/10.1111/1365-2745.12905

Preece C, Livarda A, Christin P-A, Wallace M, Martin G, Charles M, Jones G, Rees M, Osborne CP. How did the domestication of Fertile Crescent grain crops increase their yields? Funct Ecol. 2017:**31**(2):387–397. https://doi.org/10.1111/1365-2435.12760

Purugganan MD. Evolutionary insights into the nature of plant domestication. Curr Biol. 2019:**29**(14):R705–R714. https://doi.org/10.1016/j.cub.2019.05.053

Purugganan MD. What is domestication? Trends Ecol Evol. 2022:**37**(8):663–671. https://doi.org/10.1016/j.tree.2022.04.006

Purugganan MD, Fuller DQ. The nature of selection during plant domestication. Nature. 2009:**457**(7231):843–848. https://doi.org/10.1038/nature07895

Purugganan MD, Fuller DQ. Archaeological data reveal slow rates of evolution during plant domestication. Evolution. 2011:**65**(1):171–183. https://doi.org/10.1111/j.1558-5646.2010.01093.x

Raj SM, Pei A, Foll M, Schlamp F, Excoffier L, Fuller DQ, Kivisild T, Clark AG. 2019. Reconstruction of nine thousand years of agriculture-based diet and impact on human genetic diversity in Asia. bioRxiv 747709. https://doi.org/10.1101/747709. preprint: not peer reviewed.

Ramsay L, Comadran J, Druka A, Marshall David F, Thomas William T B, Macaulay M, MacKenzie K, Simpson C, Fuller J, Bonar N, et al. INTERMEDIUM-C, a modifier of lateral spikelet fertility in barley, is an ortholog of the maize domestication gene TEOSINTE BRANCHED 1. Nat Genet. 2011:**43**(2):169–172. https://doi.org/10.1038/ng.745

Raza A, Razzaq Ali, Mehmood Sundas, Zou X, Zhang X, Lv Y, Xu J. Impact of climate change on crops adaptation and strategies to tackle its outcome: a review. Plants. 2019:**8**(2):34. https://doi.org/10.3390/plants8020034

Remigereau M-S, Lakis G, Rekima S, Leveugle M, Fontaine MC, Langin T, Sarr A, Robert T. Cereal domestication and evolution of branching: evidence for soft selection in the tb1 orthologue of pearl millet (Pennisetum glaucum). PLoS One. 2011:**6**(7):e22404. https://doi.org/10.1371/journal.pone.0022404

Rodriguez-Saona C, Vorsa N, Singh AP, Johnson-Cicalese J, Szendrei Z, Mescher MC, Frost CJ. Tracing the history of plant traits under domestication in cranberries: potential consequences on anti-herbivore defences. J Exp Bot. 2011:**62**(8):2633–2644. https://doi.org/10.1093/jxb/erq466

Roman-Reyna V, Pinili Dale, Borja Frances N, Quibod IL, Groen SC, Alexandrov N, Mauleon R, Oliva R. Characterization of the leaf microbiome from whole-genome sequencing data of the 3000 rice genomes project. Rice. 2020:**13**(1):72. https://doi.org/10.1186/s12284-020-00432-1

Roosevelt A, Lima da Costa M, Machado CL, Michab M, Mercier N, Valladas H, Feathers J, Barnett W, Imazio da Silveira M, Henderson A, et al. Paleoindian cave dwellers in the Amazon: the peopling of the Americas. Science. 1996:**272**(5260):373–384. https://doi.org/10.1126/science.272.5260.373

Rosenthal J, Dirzo R. Effects of life history, domestication and agronomic selection on plant defence against insects: evidence from maizes and wild relatives. Evol Ecol. 1997:**11**(3):337–355. https://doi.org/10.1023/A:1018420504439

Ross-Ibarra J, Morrell PL, Gaut BS. Plant domestication, a unique opportunity to identify the genetic basis of adaptation. Proc Natl Acad Sci USA. 2007:**104**(suppl_1):8641–8648. https://doi.org/10.1073/pnas.0700643104

Rowley-Conwy P, Layton R. Foraging and farming as niche construction: stable and unstable adaptations. Philos Trans R Soc B Biol Sci. 2011:**366**(1566):849–862. https://doi.org/10.1098/rstb.2010.0307

Ryser P, Lambers H. Root and leaf attributes accounting for the performance of fast- and slow-growing grasses at different nutrient supply. Plant Soil. 1995:**170**(2):251–265. https://doi.org/10.1007/BF00010478

Schöb C, Brooker RW, Zuppinger-Dingley D. Evolution of facilitation requires diverse communities. Nat Ecol Evol. 2018:**2**(9):1381–1385. https://doi.org/10.1038/s41559-018-0623-2

Shi H, Chen M, Gao L, Wang Y, Bai Y, Yan H, Xu C, Zhou Y, Xu Z, Chen J, et al. Genome-wide association study of agronomic traits related to nitrogen use efficiency in wheat. Theor Appl Genet. 2022:**135**(12):4289–4302. https://doi.org/10.1007/s00122-022-04218-5

Singh J, van der Knaap E. Unintended consequences of plant domestication. Plant Cell Physiol. 2022:**63**(11):1573–1583. https://doi.org/10.1093/pcp/pcac083

Soltis NE, Atwell S, Shi G, Fordyce R, Gwinner R, Gao D, Shafi A, Kliebenstein DJ. 2018. Crop domestication and pathogen virulence: interactions of tomato and Botrytis genetic diversity. bioRxiv 255992. https://doi.org/10.1101/255992. preprint: not peer reviewed.

Speidel L, Forest M, Shi S, Myers SR. A method for genome-wide genealogy estimation for thousands of samples. Nat Genet. 2019:**51**(9):1321–1329. https://doi.org/10.1038/s41588-019-0484-x

Stern AJ, Wilton PR, Nielsen R. An approximate full-likelihood method for inferring selection and allele frequency trajectories from DNA sequence data. PLoS Genet. 2019:**15**(9):e1008384. https://doi.org/10.1371/journal.pgen.1008384

Stetter M. Limits and constraints to crop domestication. Am J Bot. 2020:**107**(12):1617–1621. https://doi.org/10.1002/ajb2.1585

Stetter M, Müller T, Schmid KJ. Genomic and phenotypic evidence for an incomplete domestication of South American grain amaranth (Amaranthus caudatus). Mol Ecol. 2017:**26**(3):871–886. https://doi.org/10.1111/mec.13974

Studer AJ, Wang H, Doebley JF. Selection during maize domestication targeted a gene network controlling plant and inflorescence architecture. Genetics. 2017:**207**(2):755–765. https://doi.org/10.1534/genetics.117.300071

Tergemina E, Elfarargi AF, Flis P, Fulgione A, Göktay M, Neto C, Scholle M, Flood PJ, Xerri S-A, Zicola J, et al. A two-step adaptive walk rewires nutrient transport in a challenging edaphic environment. Sci Adv. 2022:**8**(20):eabm9385. https://doi.org/10.1126/sciadv.abm9385

Thrall PH, Oakeshott JG, Fitt G, Southerton S, Burdon JJ, Sheppard A, Russell RJ, Zalucki M, Heino M, Ford Denison R. Evolution in agriculture: the application of evolutionary approaches to the management of biotic interactions in agro-ecosystems. Evol Appl. 2011:**4**(2):200–215. https://doi.org/10.1111/j.1752-4571.2010.00179.x

Tracy WF, Whitt SR, Buckler ES. Recurrent mutation and genome evolution: example of sugary1 and the origin of sweet maize. Crop Sci. 2006:**46**(S1):S-49. https://doi.org/10.2135/cropsci2006-03-0149tpg

Trucchi E, Benazzo A, Lari M, Iob A, Vai S, Nanni L, Bellucci E, Bitocchi E, Raffini F, Xu C, et al. Ancient genomes reveal early Andean farmers selected common beans while preserving diversity. Nat Plants. 2021:**7**(2):123–128. https://doi.org/10.1038/s41477-021-00848-7

Turcotte MM, Turley NE, Johnson M. The impact of domestication on resistance to two generalist herbivores across 29 independent domestication events. New Phytol. 2014:**204**(3):671–681. https://doi.org/10.1111/nph.12935

Vallebueno-Estrada M, Rodríguez-Arévalo I, Rougon-Cardoso A, Martínez González J, García Cook A, Montiel R, Vielle-Calzada J-P. The earliest maize from San Marcos Tehuacán is a partial domesticate with genomic evidence of inbreeding. Proc Natl Acad Sci USA. 2016:**113**(49):14151–14156. https://doi.org/10.1073/pnas.1609701113

Vavilov N. Origin and geography of cultivated plants (translated by Doris Löve). Arch Nat Hist. 1994:**21**:142–142. https://doi.org/10.3366/anh.1994.21.1.142a

Wallace JG, Kremling KA, Kovar LL, Buckler ES. Quantitative genetics of the maize leaf microbiome. Phytobiomes J. 2018:**2**(4):208–224. https://doi.org/10.1094/PBIOMES-02-18-0008-R

Wang M, Li W, Fang C, Xu F, Liu Y, Wang Z, Yang R, Zhang M, Liu S, Lu S, et al. Parallel selection on a dormancy gene during domestication of crops from multiple families. Nature Genet. 2018:**50**(10): 1435–1441. https://doi.org/10.1038/s41588-018-0229-2

Wang RL, Stec A, Hey J, Lukens L, Doebley J. The limits of selection during maize domestication. Nature. 1999:**398**(6724):236–239. https://doi.org/10.1038/18435

Wang H, Studer AJ, Zhao Q, Meeley R, Doebley JF. Evidence that the origin of naked kernels during maize domestication was caused by a single amino acid substitution in tga1. Genetics. 2015:**200**(3): 965–974. https://doi.org/10.1534/genetics.115.175752

Wang Z, Wang W, Xie X, Wang Y, Yang Z, Peng H, Xin M, Yao Y, Hu Z, Liu J, et al. Dispersed emergence and protracted domestication of polyploid wheat uncovered by mosaic ancestral haploblock inference. Nat Commun. 2022:**13**(1):3891. https://doi.org/10.1038/s41467-022-31581-0

Whipple CJ, Kebrom TH, Weber AL, Yang F, Hall D, Meeley R, Schmidt R, Doebley J, Brutnell TP, Jackson DP. Grassy tillers1 promotes apical dominance in maize and responds to shade signals in the grasses. Proc Natl Acad Sci USA. 2011:**108**(33):E506–E512. https://doi.org/10.1073/pnas.1102819108

Whitehead SR, Turcotte M, Poveda K. Domestication impacts on plant-herbivore interactions: a meta-analysis. Philos Trans R Soc Lond B Biol Sci. 2017:**372**(1712):20160034. doi 10.1098/rstb.2016.0034

Wills DM, Burke JM. Quantitative trait locus analysis of the early domestication of sunflower. Genetics. 2007:**176**(4):2589–2599. https://doi.org/10.1534/genetics.107.075333

Wu GA, Prochnik S, Jenkins J, Salse J, Hellsten U, Murat F, Perrier X, Ruiz M, Scalabrin S, Terol J, et al. Sequencing of diverse mandarin, pummelo and orange genomes reveals complex history of admixture during citrus domestication. Nat Biotech. 2014:**32**(7):656–662. https://doi.org/10.1038/nbt.2906

Wu D, Xie L, Sun Y, Huang Y, Jia L, Dong C, Shen E, Ye C-Y, Qian Q, Fan L, et al. A syntelog-based pan-genome provides insights into rice domestication and de-domestication. Genome Biol. 2023:**24**(1):179. https://doi.org/10.1186/s13059-023-03017-5

Wuest SE, Niklaus PA. A plant biodiversity effect resolved to a single chromosomal region. Nat Ecol Evol. 2018:**2**(12):1933–1939. https://doi.org/10.1038/s41559-018-0708-y

Xue S, Bradbury PJ, Casstevens T, Holland JB. Genetic architecture of domestication-related traits in maize. Genetics. 2016:**204**(1):99–113. https://doi.org/10.1534/genetics.116.191106

Ye C-Y, Tang W, Wu D, Jia L, Qiu J, Chen M, Mao L, Lin F, Xu H, Yu X, et al. Genomic evidence of human selection on vavilovian mimicry. Nat Ecol Evol. 2019:**3**(10):1474–1482. https://doi.org/10.1038/s41559-019-0976-1

Yu H, Lin T, Meng X, Du H, Zhang J, Liu G, Chen M, Jing Y, Kou L, Li X, et al. A route to de novo domestication of wild allotetraploid rice. Cell. 2021:**184**(5):1156–1170.e14. https://doi.org/10.1016/j.cell.2021.01.013

Zeder M. Domestication: definition and overview. In: **Smith C**, editors. Encyclopedia of global Archaeology. New York (NY): Springer New York; 2014. p. 2184–2194.

Zeder M. Core questions in domestication research. Proc Natl Acad Sci USA. 2015:**112**(11):3191–3198. https://doi.org/10.1073/pnas.1501711112

Zhang F, Hu Z, Wu Z, Lu J, Shi Y, Xu J, Wang X, Wang J, Zhang F, Wang M, et al. Reciprocal adaptation of rice and Xanthomonas oryzae pv. oryzae: cross-species 2d GWAS reveals the underlying genetics. Plant Cell. 2021:**33**(8):2538–2561. https://doi.org/10.1093/plcell/koab146

Zhao Q, Feng Q, Lu H, Li Y, Wang A, Tian Q, Zhan Q, Lu Y, Zhang L, Huang T, et al. Pan-genome analysis highlights the extent of genomic variation in cultivated and wild rice. Nat Genet. 2018:**50**(2): 278–284. https://doi.org/10.1038/s41588-018-0041-z

Zhou Y, Lu D, Li C, Luo J, Zhu B-F, Zhu J, Shangguan Y, Wang Z, Sang T, Zhou B, et al. Genetic control of seed shattering in rice by the APETALA2 transcription factor shattering abortion1. Plant Cell. 2012:**24**(3):1034–1048. https://doi.org/10.1105/tpc.111.094383

Zhou Y, Massonnet M, Sanjak JS, Cantu D, Gaut BS. Evolutionary genomics of grape (Vitis vinifera ssp. vinifera) domestication. Proc Natl Acad Sci USA. 2017:**114**(44):11715–11720. https://doi.org/10.1073/pnas.1709257114

Zuo J, Li J. Molecular genetic dissection of quantitative trait loci regulating rice grain size. Ann Rev Genet. 2014:**48**(1):99–118. doi:10.1146/annurev-genet-120213-092138

https://doi.org/10.1093/plcell/koad278

THE PLANT CELL

THE PLANT CELL 2024: 36: 1242–1256

Why should we study plant sex chromosomes?

Deborah Charlesworth [1,*] and Alex Harkess [2]

1 Institute of Ecology and Evolution, University of Edinburgh, Edinburgh EH9 3FL, UK
2 HudsonAlpha Institute for Biotechnology, Huntsville, AL 35806, USA

*Author for correspondence: Deborah.Charlesworth@ed.ac.uk

Abstract

Understanding plant sex chromosomes involves studying interactions between developmental and physiological genetics, genome evolution, and evolutionary ecology. We focus on areas of overlap between these. Ideas about how species with separate sexes (dioecious species, in plant terminology) can evolve are even more relevant to plants than to most animal taxa because dioecy has evolved many times from ancestral functionally hermaphroditic populations, often recently. One aim of studying plant sex chromosomes is to discover how separate males and females evolved from ancestors with no such genetic sex-determining polymorphism, and the diversity in the genetic control of maleness vs femaleness. Different systems share some interesting features, and their differences help to understand why completely sex-linked regions may evolve. In some dioecious plants, the sex-determining genome regions are physically small. In others, regions without crossing over have evolved sometimes extensive regions with properties very similar to those of the familiar animal sex chromosomes. The differences also affect the evolutionary changes possible when the environment (or pollination environment, for angiosperms) changes, as dioecy is an ecologically risky strategy for sessile organisms. Dioecious plants have repeatedly reverted to cosexuality, and hermaphroditic strains of fruit crops such as papaya and grapes are desired by plant breeders. Sex-linked regions are predicted to become enriched in genes with sex differences in expression, especially when higher expression benefits one sex function but harms the other. Such trade-offs may be important for understanding other plant developmental and physiological processes and have direct applications in plant breeding.

Introduction

Now that genomes can be sequenced and assembly is becoming increasingly feasible, even for large genomes (albeit with challenges discussed later), genetic maps can be related to physical maps of the arrangement and nature of sequences, allowing new progress on many classical genetic problems. Among these are questions about sex chromosomes, which may evolve when cosexual species (hermaphroditic angiosperms, with "perfect" flowers, or monoecious ones, like maize) evolve separate sexed individuals. We discuss 2 questions embodied in our title: what questions make sex chromosomes interesting, and why plants are good for studying them. We focus primarily on angiosperms because their genomes tend to be smaller than those of gymnosperms whose genomes are massive and, despite the latest sequencing technologies, still highly challenging to assemble, though data are starting to be published (e.g. Gong and Filatov 2022; Liu et al. 2022). We briefly mention the advantages of studies in species with extended haploid stages, whose genomes are again often small enough for reliable sequencing and where the published studies highlight some interesting questions about sex chromosome evolution.

Plant and animal genetic sex-determining systems share interesting genomic features that demand understanding, but the angiosperms include systems that evolved more recently than the best-studied animal systems, making their evolution more accessible to study. Moreover, widely diverse flowering plant groups appear to have independently evolved separate sexes (dioecy) from cosexual ancestors, allowing consistent changes to be discerned, despite some interesting differences.

Received August 02, 2023. Accepted October 10, 2023. Advance access publication January 2, 2024

Table 1. Information about dioecious angiosperms. The table shows the range of clades in which dioecy is found, and illustrates that sex chromosome heteromorphism (species with names in bold font) is also found in distantly related taxa

Clade	Species	Heterogametic sex	Probable system	Genetic degeneration
Monocotyledon	*Phoenix dactylifera* (date palm)	Male	2 genes	Unknown
	Asparagus	Male	2 genes	Unknown
	***Dioscorea* (yam)**	Male	Unknown	Unknown
Caryophyllales	***Rumex* species (sorrel)**	Male	Unknown	Yes
	***Silene latifolia* and close relatives**	Male[a]	2 genes	Yes
	Simmondsia chinensis (jojoba)	Male	Unknown	Unknown
	Spinach	Male	One gene (insertion)	Unknown
Asterid	Diospyros (persimmon)	Male	One gene (duplication)	Unknown
	Actinidia chinensis (kiwifruit)	Male	2 genes	Unknown
	Solanum appendiculatum	Male	Unknown	Unknown
Rosid	***Cannabis sativa* (hemp), *Humulus lupulus* (hops)**	Male	Unknown	Yes
	Pistacia vera	Female	Unknown	Unknown
	Coccinia grandis	Male	2 genes?[b]	Unknown
	Viscum	Male	Unknown	Unknown
	Hippophae rhamnoides	Male	Unknown	Unknown
	Carica papaya	Male	2 genes	Unknown
	Datisca cannabina	Female	Unknown	Unknown
	Vitis	Male	2 genes	Unknown
	Mercurialis	Male	Unknown	Unknown
	Populus, Salix		One gene (duplication)	Unknown
	Fragaria	Female	Unknown	Unknown
	Morus atropurpurea	Male	Unknown	Unknown
	Morella (or *Myrica*) *rubra*	Female	Unknown	Unknown

Heteromorphic sex chromosome systems are probably overrepresented because, until recently, cytogenetics provided the chief evidence about plant sex chromosomes. A total of 23 species or genera are listed, of which 17 currently have genetic maps and genome sequences (further details, including references, are in Supplemental Table S1). Only 4 of these cases were not discussed in Westergaard's (1958) review; some of the species in this table are not the same as in his list, but they are in the same genus or family. Sex-determining systems with 1 or with 2 genes are found in different clades, but, for most species, the system is unknown.
[a]Female heterogamety is also known in the *Otites* clade.
[b]There is evidence for a 2-gene system in the genus *Ecballium*.

An estimated 5% to 10% of flowering plants have genetic sex determination (Yampolsky and Yampolsky 1922; Garcia et al. 2023), implying that ~30,000 species, in taxa scattered across the angiosperms, have sex-linked genome regions. Early cytological and genetic studies (reviewed by Westergaard 1958) suggested that the sizes of completely sex-linked regions, and degrees of differentiation between the 2 members of the sex chromosome pair, must differ greatly in different angiosperm species, contrasting with the situation in the mammals, a group of similar age (Sauquet et al. 2022; Carlisle et al. 2023). More recent studies are starting to provide details of plant sex determination, and the genomic characteristics of the genome regions involved, and when they evolved, as will be described below.

As shown in Table 1, some angiosperms have giant-sized sex chromosomes that resemble the ancient sex chromosomes of Eutherian mammals, birds, and Dipteran insects in being "heteromorphic," with 1 member of the chromosome pair being repeat-rich and gene poor and sometimes heterochromatic (Fig. 1). The results described in reviews of angiosperms (Yampolsky and Yampolsky 1922; Ming et al. 2011) are probably biased toward species with heteromorphism visible to microscopists. The only heteromorphic system discovered since 1958 appears to be in *Pistacia* (Sola-Campoy et al. 2015; Palmer et al. 2022), though more will probably be discovered as genomes of further dioecious plants are sequenced or through other evidence (see below).

Heteromorphism and differentiation evolve when the 2 members of the sex chromosome pair do not undergo genetic recombination (at least in parts carrying the sex-determining gene or genes; regions in which the sex chromosome pair recombine, are termed "pseudo-autosomal regions" or PARs) and are not differentiated. Many species with heteromorphic sex chromosomes probably evolved separate sexes long ago. This is especially likely if high proportions of genes have been deleted, creating hemizygosity (a haploid copy number) in one sex while the other sex has diploid coverage. In the human XX female/XY male system, the Y is highly degenerated and lacks almost all genes carried on the X (Sayres and Makova 2013). In the modern bird ZZ male/ZW female system, females are similarly hemizygous for most Z-linked genes (as reviewed by Bull 1983). Among angiosperm species studied since Westergaard's (Westergaard 1958) review (Table 1), several with heteromorphic sex chromosomes show evidence for degeneration, including a *Pistacia* species with female heterogamety, whose W is partly or largely nonrecombining (Palmer et al. 2022). Genome sequencing should help to determine species' heterogametic sex and identify the degenerated sex-linked regions, if any. One major reason for studying sex chromosomes is to understand the genomic consequences of absence of recombination (which are shared with other genome regions that rarely recombine) and the processes leading to Y-X or W-Z differentiation. The degeneration just mentioned as well as the accumulation of repetitive sequences

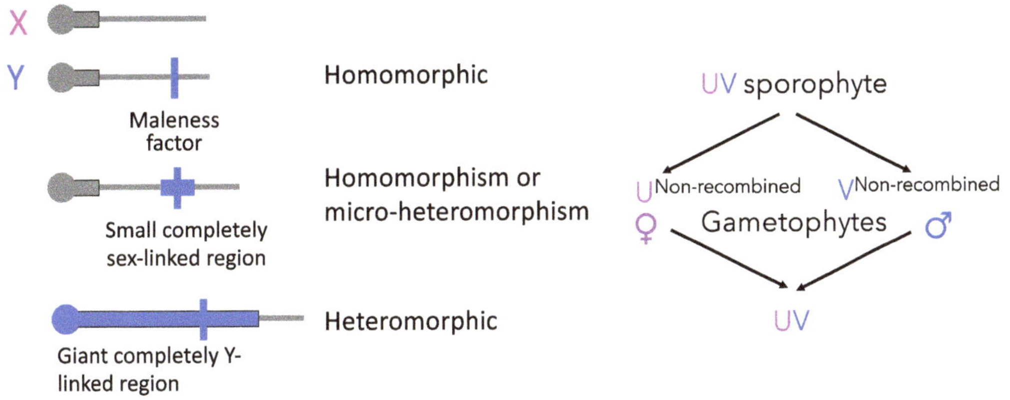

Figure 1. Examples of 3 different kinds of Y-linked regions observed in angiosperms with male heterogamety to illustrate concepts described in the text. The top diagram shows an X chromosome, which is similar to the ancestral state before genetic sex determination evolved; its centromere is indicated at the left (circle), with an adjoining pericentromeric region that rarely recombines. The 3 diagrams below show different types of "Y chromosomes," with thin horizontal lines indicating recombining regions (PARs) and thick ones nonrecombining regions. The top Y is identical with the X (homomorphism), except for presence of a single male-determining factor, which is always heterozygous in males. In the second, the male-determining locus is within a small completely nonrecombining region that may include several other genes (which, again will always be heterozygous with the X in males); this situation is also likely to be classified as homomorphic because the differentiated region is a small part of the chromosome. The bottom diagram shows a strongly heteromorphic situation, with the male-determining locus in a large completely nonrecombining region across most of the chromosome arm, including the centromere, and a larger size of the arm, compared with the homologous X arm.

and rearrangements (see below) are consequences of a lack of recombination of the Y (and W) linked regions, whereas X (and Z) regions recombine and change from their ancestral state much less.

Plants whose haploid phase is dominant are also important for testing ideas about such genome evolutionary processes, as similar processes should operate in very different organisms. Many bryophytes have separate male and female gametophytes, and sex chromosomes were described over 100 years ago (Allen 1919; Garcia et al. 2023). They have nonrecombining regions, and, in some species, genetic PARs have been detected (McDaniel et al. 2007). As recombination occurs in the sporophyte, which is heterozygous for the sex chromosomes, both members of the sex chromosome pair are predicted to degenerate (Bull 1978), and this prediction can now be tested (as will be outlined below). To emphasize this difference from diploid XY and ZW pairs, bryophyte female- and male-determining chromosomes are termed U and V, respectively.

Many flowering plant species do not, however, have heteromorphic sex chromosomes, with detectable differentiation (the same is true in the UV systems of plants such as bryophytes with haploid-dominant life cycles, as discussed later). Their completely sex-linked regions must therefore be small (Fig. 1). Table 1 includes examples, and Westergaard's (Westergaard 1958) review included at least 13 other species or genera. Homomorphism may be more common than heteromorphism in the angiosperms (Garcia et al. 2023). Such plants may help understand why sex-determining genes are often (but not invariably) within extensive nonrecombining regions. Below, we discuss estimation of the ages of plant sex chromosomes (including for young systems that may not yet have evolved heteromorphism).

Genome sequencing now allows different types of sex-determining regions to be distinguished (reviewed by Carey et al. 2021a) and suggests when recombination stopped. One answer to our question about why plant sex chromosomes are interesting is that some homomorphic plants may be in early stages of sex chromosome evolution. This makes them excellent for identifying candidate sex-determining genes within their physically small sex-linked regions. If these genes are within nonrecombining regions, these young systems will also be ideal for discovering the mechanism(s) that created such regions (e.g. chromosomal inversions, as proposed by Lahn and Page 1999) and for describing the earliest consequences of an absence of recombination.

Among the angiosperm species in Table 1, 18 are cultivated plants. Another answer to the same question is therefore that research on sex chromosomes is important for crop breeding (Masuda and Akagi 2023). Flower development can also often be affected by environmental influences so that individuals may not always be completely male or fe-male. So-called inconstant or leaky males express some fe-male function under suitable conditions (Darwin 1877; Käfer and Mousset 2022), probably including genetic background influences. Females are less often inconstant, but such variability has been found in *Mercurialis annua*; in experiments in which females were grown in the absence of males, females reliably expressing male functions rapidly appeared, suggesting presence of major genetic factors (Cossard and Pannell 2021).

Nevertheless many plants exhibit clear discontinuities in gender, and major sex-determining factors can be mapped to defined genomic locations (Veltsos et al. 2019). It is often important to determine individuals' sexes in long-lived species, including many fruit crops, where only a few males are needed for pollination, or for species like asparagus, where males are desired (Harkess et al. 2015), or Ginkgo trees, where the fleshy-coated seeds smell bad, so that males are desired unless the fruits are to be used. This can be done using genotypes of variable sites in or closely linked to a sex- determining locus. Another application is to produce func-tionally hermaphroditic strains of fruit trees and shrubs, such as papaya and grapes, avoiding the need for separate males as pollinators. In some plants, this might involve ma-nipulating the sex-determining system once the genes in-volved are known.

We first outline some recent advances in understanding the developmental properties of sex-determining genes. These are important for understanding how a situation can arise in which the 2 sexes differ at a locus with a major effect on whether individuals develop as female or male, starting from an ancestral cosexual population without such a genetic polymorphism. Then we describe how knowledge of the genes involved illuminates the observed recombination patterns in sex-linked regions.

Sex-determining genes

Flowering plants are excellent for studying how new sex-determining genes arise. As will be seen, the mutations involved in evolving separate sexes are not just loss-of-function mutations producing females (male sterile) and males (female sterile). Both the evolutionary and developmental processes are more complex than this and often involve trade-offs between different phenotype options; male sterility may thus sometimes result from increased female functioning or vice versa.

Discovering the genes involved in plant sex-determining systems is challenging because dioecy evolved independently in different flowering plant lineages, so homology in different species is rarely a useful guide. If a species' sex chromosomes are homomorphic, there may be only a small fully sex-linked genome region, or it may have appeared too recently to have evolved sequence differentiation from its homolog, making it hard to detect or difficult to separate the sequences from the 2 sex-linked regions, as in *Amborella trichopoda* (Käfer et al. 2021). Even identifying a species' heterogametic sex is often difficult. On the other hand, heteromorphic, highly repetitive sequence densities create problems assembling highly differentiated sex chromosomes. Validating sex-determining candidates in nonmodel plants is also challenging. If these are within nonrecombining regions, genetics cannot help identify the genes, and the number of candidates for testing by transgenic experiments can be large. We discuss the technical challenges in more detail later in this review, as it is helpful to first understand important features of sex-linked regions and some of the questions relating to them.

For separate sexes to evolve from a cosexual ancestor requires at least 2 mutations, and at least 1 of them must show trade-offs. Before describing the genes that have been discovered in specific plant sex-determining systems, which seem to confirm these predictions, we first explain how the predictions emerge. Many evolutionary changes, such as changes in morphology or physiology, involve advantageous mutations spreading in populations. Adaptive characteristics such as higher activity of an enzyme, or earlier flowering, usually involve many successive changes that improve a plant function, not single mutations. Each advantageous mutation spreads throughout a population (in evolutionary terminology, becomes "fixed" in all individuals), improving the adaptation. The evolution of sex determination is an especially interesting evolutionary process, because (unlike the scenario just outlined) both sexes cannot evolve by a single mutation: to change from cosexuality to genetically controlled dioecy, at least 2 mutations are required (Charlesworth and Charlesworth 1978; Pannell and Gerchen 2018). Moreover, although mutations suppressing female or male function will rarely be advantageous, such mutations must often be involved in the evolution of dioecy, as appearance of female or male individuals involves losing functions expressed by the cosexual ancestor. This apparent puzzle is resolved if loss of female functions can allow better male functioning, or vice versa, through a developmental or other trade-off (Charlesworth and Charlesworth 1978) such that "male" plants can produce more offspring than cosexuals; Darwin called this "compensation" (Darwin 1877). Trade-offs may be common in many plant (and animal) developmental decisions, for example, in ceasing to grow more leaves and starting to flower, and are of interest for understanding plant development and functioning.

The requirement for 2 mutations to create separate females and males implies that the genes involved must suitably interact during flower development. Mutations may be advantageous only in a suitable genetic background, and this may influence the order in which mutations are able to spread in a species. In the case of sex-determining mutations, males are much more likely to spread in a population after females have first become frequent (Charlesworth and Charlesworth 1978). A final difference from many other kinds of advantageous mutations is that, if a population with genetic sex determination has evolved, at least 1 of the sex-determining mutations must have different frequencies in male and female individuals. If male heterogamety evolves, males have a male-determining factor that is absent in females (unless an X autosome balance system has evolved). Such "polymorphism" makes the genetic differences "visible" to genetic analyses. As will be explained below, either 1 or both of the 2 mutations may establish polymorphisms.

The Westergaard (1958) review of available genetic information first suggested that 2 mutations were indeed involved in sex determination in several plant species. He used "deletion mapping" of cytologically visible Y chromosome features in *Silene latifolia* to infer the location of deletions causing hermaphroditism, demonstrating the presence of a Y-linked actively femaleness suppressing factor (termed *SuF* here, though "gynoecium-suppressing factor," or GSFY is

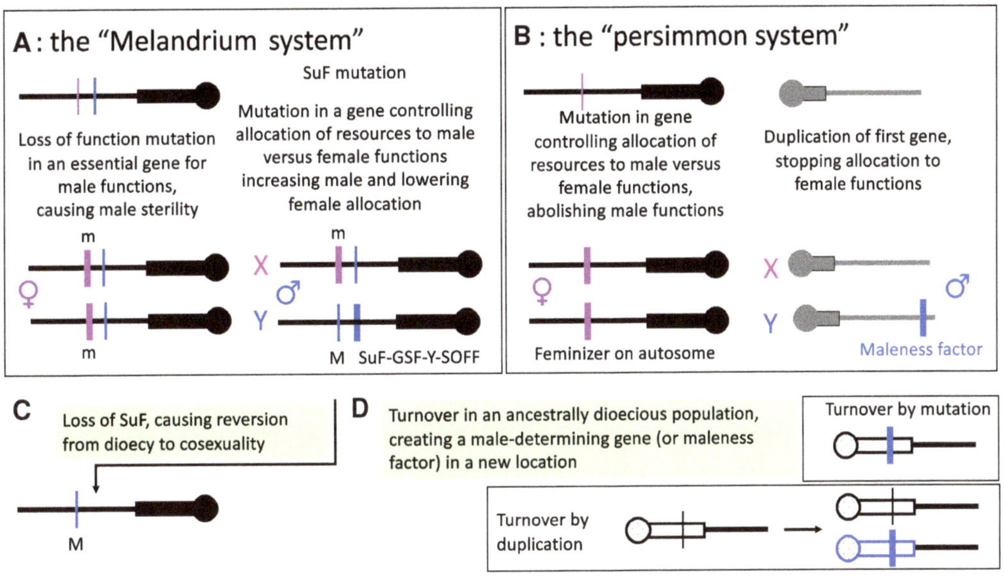

Figure 2. De novo evolution of separate sexes in an initially hermaphrodite or monoecious species, reversions, and turnovers. **A)** The pathway involving an initial M —> m male-sterility mutation in the hermaphroditic ancestor (in a stamen-promoting factor, or SPF), and a second mutation (or succession of mutations), denoted by SuF, GSF, or SOFF, suppresses female functions and creates males. Close linkage is predicted between the 2 mutations that define the Y-linked region, as observed. **B)** The pathway leading to the "persimmon system" in which femaleness is due to a mutation that causes active suppression of maleness (not to a loss of function mutation causing male sterility) and in which the second mutation is an unlinked duplication that creates males by suppressing the action of the allele that created females. **C)** Reversion to co-sexuality is possible under the Silene system by simple loss of the femaleness suppressor function, but in the persimmon system this is less likely (see main text), though turnover events can occur. In the pathway shown in part **D**, a new M factor can appear on any chromosome, including one that was an autosome in the ancestrally dioecious population, or in a new location on a sex chromosome. It can appear by transposition, inserting a duplicate of a progenitor gene from an ancestral location into a new one, or by in situ change of a gene, by a mutation that produces a male-determining function.

also used). Deletions producing flowers with neither male nor female organs revealed 2 essential male function genes in other Y regions (Westergaard 1958; Kazama et al. 2016). Westergaard called this class of system the "Melandrium trigger" (as the genus *Silene* was then named Melandrium); we use "Melandrium pathway" for this pathway from co-sexuality to dioecy (Fig. 2A).

Based on maize mutations, Jones (1932) proposed a single-factor sex-determining system. This "maize trigger" did not fit observations from the dioecious plant species with segregation data available to Westergaard. However, a single-gene system was discovered in the persimmon (Akagi et al. 2014), so we term it the "persimmon system" (see Fig. 2B, and explanation below). Table 1 indicates species with 2-gene or single-gene systems, when evidence is available.

Both systems involve an actively male-promoting factor. In both, this factor defines a Y-linked genome region and triggers downstream developmental differences between male and female inflorescences and flowers and sometimes other "secondary" sex differences, such as fiber or chemical content in some dioecious species. Also, as mentioned above, females are generally expected to evolve before males in both pathways in Fig. 2. There are important differences, however. A single sex-determining factor is, by definition, a "master sex-determining factor": its presence or absence controls whether an individual develops as a male or a female. The sex that develops in its absence is termed the "default sex," and loss of the factor changes individuals to this sex; for example, deletion or inactivation of the maleness factor in the persimmon system can produce females. In contrast, the Melandrium system has no master sex-determining factor whose loss changes the sex and no default sex. Instead, loss of the maleness factor creates co-sexual individuals (Fig. 2C). Thus, in asparagus, mutations changing from male to female involve deletion of both Y-linked factors (Harkess et al. 2020).

Plants with 2-gene systems and identification of the factors

Candidates for 2 primary sex-determining genes have been identified in *Actinidia* (A kagi et al. 2019), *Asparagus* (Harkess et al. 2020), and *V itis* (Massonnet et al. 2020). Consistent with the prediction that the mutations should be in closely linked genes, these species have physically small Y-linked regions (unlike *S. latifolia*). In papaya, co-sexual domesticated strains are controlled by an hermaphrodite-determining Y, Yh, suggesting that papaya also has a Melandrium type of sex determination. A candidate Y-linked *SuF* gene (a homolog of a gene encoding the MA DS-box protein *SHORT V EGETATIV EPHASE*) has been identified, w hose Y h c opy d oes n ot e ncode a complete protein (Ueno et al. 2015; Chae et al. 2021). It appears to have no X-linked homolog. No candidate has yet been identified for the mutation that created females.

In *S. latifolia*, the candidate *SuF* (or *GSFY*) is a CLAVATA *CLV3*-like gene (Kazama et al. 2022) with a highly diverged and dysfunctional X-linked homolog (termed *GSFX*), and the candidate *M* gene (whose mutation created females) is *SlWUS1*, an X-linked *WUSCHEL*-like sequence whose Y copy has been lost (Kazama et al. 2012). Disbalance of the WUSCHEL-CLAVATA feedback loop was proposed to have caused gynoecium suppression in males rather than females reflecting a loss-of-function mutation in an essential male function gene.

The Diospyros system

The persimmon has a single-factor system (Fig. 2B; Akagi et al. 2014). Individuals carrying a factor named OGI are males, and females are the default sex. OGI is a duplication of an autosomal gene MeGI. Assuming that dioecy evolved from a co-sexual ancestor (rather than a turnover, see below), it cannot have involved this duplication alone, as individuals with only MeGI are female, and a wholly female population is impossible. More likely, the MeGI gene controlled the balance between male and female functions in the ancestral co-sexual. A mutant allele could have changed this balance to complete femaleness (rather than femaleness reflecting a loss-of function male sterility mutation). If loss of male function was favored (e.g. by allowing higher female fertility and/or if self-pollination was disfavored), females could increase in frequency; the presence of females would then favor a male-promoting mutation (Charlesworth and Charlesworth 1978). As this second mutation became more common, the advantage of the MeGI allele would also increase, and the hypothetical ancestral allele would disappear from the population, leaving OGI as the master sex-determining factor.

In principle, mutations producing females and males from an ancestral co-sexual population could occur in a single gene that controlled the balance between male and female functions. One mutation could, like MeGI, produce complete femaleness and the other complete maleness. No such systems has yet been discovered.

Changes in sex-determining loci

Movements between genomic locations

After separate sexes have been established, the genetic sex-determining locus may transpose to a new genome location. Related species therefore do not always have homologous sex-linked regions. In *Fragaria*, in which females are the heterogametic sex, genome regions as large as 31 kb, including multiple genes, have moved (as reviewed by Cauret et al. 2022). In the genus *Actinidia*, movements of 3 genes have occurred between different chromosomes, including the 2 sex-determining genes (Akagi et al. 2023). This strongly supports the conclusion that both the sex-determining genes that have been identified are required for the male/female difference (the Melandrium system).

Such changes can occur only if the sex-determining genes are within a physically small genome region. It may therefore be difficult to distinguish between such movements vs evolution of new single-gene systems (turnover events, described below). In *Salix* and *Populus* species, small Y-linked regions are found on different chromosomes in different species (Tuskan et al. 2012). In 2 *Populus* species, the candidate maleness factor is a partial duplication of an autosomal gene similar to the *ARR17* gene, which is important in development of female flower structures in *Arabidopsis thaliana* (Müller et al. 2020; Xue et al. 2020). The ARR17-like partial duplicates probably act as maleness factors by suppressing processes required for femaleness; indeed, experimental knockout changed the sex from male to female, the default state (Müller et al. 2020). This suggests a mechanism like the persimmon system, again with the autosomal progenitor of the duplicates being targeted by the maleness factor. Females may have arisen first, as proposed in Fig. 2B. If so, related outgroup species that have not evolved dioecy should have the ancestral state of this gene. Comparisons of the persimmon MeGI gene (or the Populus functional *ARR17* progenitor gene) with orthologs in other species may indicate how females evolved. However, most *Salicaceae* and most *Diospyros* species are di-oecious, so co-sexual outgroups for such comparisons may be unavailable.

Breakdown of dioecy, turnover events

Plant sex-determining systems are often ephemeral. Loss of separate sexes is frequently advantageous in sessile organisms, especially in colonizing situations where self-fertilization assures reproductive success when conspecific individuals or suitable pollinators are scarce (though inbreeding depression may lower progeny survival or fertility). "Secondary" co-sexuality is common in both angiosperms and bryophytes, but empirical studies of breakdown of dioecy unfortunately remain scarce. Even in the largely dioecious Salicaceae, a few species are co-sexual (Rohwer and Kubitzki 1984; Charlesworth 1985; Mirski 2014).

Under the Melandrium system, co-sexuality can readily evolve by loss of the femaleness suppressing function, *SuF*, or its deletion (see Fig. 2C above). As already mentioned, this has been achieved in papaya breeding (Storey 1953; Weingartner and Moore 2012; VanBuren et al. 2015). Co-sexuals may replace both unisexual forms. If, however, the Y chromosome is degenerated and homozygotes are inviable, as in papaya, co-sexuals may merely replace males, cre-ating a "gynodioecious" population of X/Yh co-sexuals and X/X females (Crossman and Charlesworth 2014). Two-gene systems can also break down to androdioecy, with XY males, and XX (co-sexual) females that can produce some pollen (Charlesworth 1984), as in the genus *Mercurialis* (Gerchen et al. 2022). Turnovers in the genus *Silene* may have involved reversions to co-sexuality, and re-evolution of dioecy, including changing between XY and ZW systems (Balounova et al. 2019; Martin et al. 2019).

In contrast, a persimmon system male-determiner may be replaced with a new single-gene system in a turnover event but seems less likely to revert to the co-sexual state, because the gene that initially changed to produce females must actively suppress male functions. To revert to co-sexuality, this action must therefore be prevented. For example, a different gene involved in controlling the balance between male and female functions might produce a mutation that restores female functions. In the polyploid monoecious persimmon, *Diospyros kaki*, restoration of female functions indeed involves a different gene, a small-Myb RADIALIS-like gene, *DkRAD* (Masuda et al. 2022), and *MeGI* has evolved a epigenetic *cis*-regulatory switch that controls its expression during development, allowing appropriate production of both male and female flowers (Akagi et al. 2018; Masuda et al. 2022); *OGI* is largely silenced by a SINE-like sequence insertion into the promoter region (Akagi et al. 2018), perhaps as a secondary effect after this new system evolved.

Populations that have lost dioecy can therefore help suggest whether a plant group has a 2- or single-gene system, even if the genes have not been identified. In Table 1, 2-gene systems are suggested by such information, including for some Cucurbitaceae. Similarly, observations a gynodioecious marginal population of an otherwise dioecious *Bursera* species suggest that this species has a 2-gene system and a degenerated Y chromosome (Crossman and Charlesworth 2014).

For species with single-gene sex determination, it may be difficult to determine whether this evolved de novo from a co-sexual ancestor or by a turnover event. In principle, the persimmon or Salicaceae XY systems could have evolved when a new male-determining factor arose, replacing a preexisting system. For some taxa, it may be possible to exclude this possibility using information about related outgroup species. However, frequent changes (as in the Salicaceae) may make this problematic; the male-determining factor mentioned above may have moved its location, or dioecy could have been lost and later reevolved through independent duplications creating new, but similar, maleness factors (turnover events). This is plausible because only genes with suitable functions in development are likely to be able to evolve into maleness determiners. In fish, master sex-determining genes include various genes involved in gonad development and its hormonal control, and some genes have male-determining functions in distantly related lineages (Pan et al. 2021). Overall, with so many ways in which they can arise, the existence of single-gene systems should be unsurprising, and it will be interesting in the future to discover which of the different possible mechanisms is involved in specific taxa.

Heterogamety can also change, for example, in both poplars (Tuskan et al. 2012) and willows (Yang et al. 2021). Intriguingly, duplicates of response regulator genes are found within both Y- and W-linked regions in different Salicaceae species (Yang et al. 2021), but it is not known whether they function as sex-determining genes in both XY and ZW systems and, if so, how this is possible. Now that the direction of changes can be inferred from phylogenies and sex ratios can be assessed using sex-linked molecular markers (allowing tests for involvement of segregation distorters, which may often be involved), it will be interesting to compare these systems with theoretical models for such changes (Bull and Charnov 1977). For example, if a femaleness determining mu-tant can override the maleness factors in the ancestral XY + AA genotype, XY females will mate with males, creating YY individuals; this may prevent the change if the Y is degener-ated and YY genotypes are inviable.

Genetic degeneration and chromosome heteromorphism

Once formed, sex-linked regions evolve differently from autosomal regions. Young flowering plant sex-linked regions, including regions like the parts of the *S. latifolia* Y chromosome that became Y-linked only recently (Campos et al. 2016), are ideal for studying the steps involved in sex chromosome evolution. These include accumulation of repetitive transposable element (TE) and satellite sequences, which are predicted to be the first consequences of loss of recombination, because repetitive sequences in such regions do not recombine with similar sequences in other locations and cause problems in meiosis (Charlesworth et al. 1994). Enlarged Y-linked regions can therefore evolve. If high TE densities evolve, suppression of TE activity may also lead to heterochromatin formation, further suppressing recombination (Underwood et al. 2018). Although these processes are most pronounced for Y-linked regions, X-linked regions also recombine less often than autosomes because they recombine only in females. They may therefore evolve similarly, to a smaller extent, especially as TE accumulation may be sex specific, due to differences during male vs female meiosis (reviewed in Jesionek et al. 2021).

In completely sex-linked regions, the ability of natural selection to eliminate deleterious mutations is also weakened, especially for Y-linked mutations that remain heterozygous because they cannot recombine onto the X. Selection against such mutations will generally be weaker than in homozygotes, so they can rise to intermediate frequencies in the population of Y chromosomes, by genetic drift, especially as this population includes only 1 Y chromosome for every 3 X chromosomes (if the sex ratio is 1:1). Genes in Y-linked regions may therefore become nonfunctional pseudogenes or be deleted (reviewed by Bachtrog 2008). Although accumulation of repetitive sequences is not the same as degeneration, it may contribute to altered gene expression. TE mobilization (involving chromosome breaks) can also cause Y chromosome rearrangements, including deletions.

In diploid plants, regions with major gene losses are detectable by coverage analysis of male and female genome sequences: some or all X-linked genes are present in males at one-half the dosage in females (hemizygosity). Table 1 indicates species with evidence for degeneration, all of which are heteromorphic. Degeneration appears to be nearly complete in *Rumex rothschildianus* (Crowson et al. 2017) and also seems likely in *Rumex* species that have evolved X-autosome balance sex-determination systems (Smith 1969). Evidently, therefore, selection in the haploid life cycle stage cannot prevent degeneration. However, the available results do not yet reliably quantify gene losses from Y-linked regions. Most angiosperm species, even those with known heteromorphic sex chromosomes, await study using complete genome sequencing to better quantify deleterious mutations in Y-linked genes and compare with results for X-linked genes.

The nonrecombining parts of the U and V chromosomes of plants with extended haploid stages (or "haploid plants") are expected to evolve differently from autosomal regions. Repetitive sequences are expected to accumulate but should do so similarly for U and V chromosomes because unlike Y or W chromosomes, they both recombine only in the sporophyte stage (Bull 1978). This prediction is supported by the *Ceratodon purpureus* genome sequence; the U and V are both greatly, and similarly, enriched in repetitive sequences (Carey et al. 2021b) even though U-V sequence divergence is small. Similarly, Bull also predicted that U and V chromosomes should both degenerate to similar extents.

In haploid plants, selection opposing genetic degeneration is predicted to be stronger than in diploids because the sex chromosome pair seems likely to carry many genes essential for gametophytes of either sex (Bull 1978). Detecting degeneration in UV systems is, however, difficult. In XY systems, genes present on the X and absent from the Y can be inferred to have been lost from the Y, but similar U and V degeneration prevents any such inference. In *Marchantia polymorpha*, the U and V chromosomes both carry few genes, suggesting that they are both roughly equally degenerated, as predicted. Intriguingly, however, the observations do not support the predicted slight degeneration. Between 1,900 and 3,000 genes are identified on each of the 8 *M. polymorpha* autosomes compared with only 47 apparently functional U-linked genes, 93 on the V chromosome, and 22 shared by both members of the pair, including the female-determining factor, or feminizer, which appears to be retained because it has another essential function in males during the haploid stage (Montgomery et al. 2020; Iwasaki et al. 2021). Because chromosome sizes and genes contents greatly differ, these small numbers of genes are difficult to interpret, but they suggest gene losses. This need not, however, reflect genetic degeneration but could instead indicate selection favoring non–sex-chromosomal locations for genes with essential functions in the gametophyte stages, leading to genes moving to autosomes over the vast evolutionary time since these sex-linked regions evolved. Movements from sex-linked regions to autosomes have recently been detected in bryophytes (Singh et al. 2023).

As mentioned above, the U and V sizes greatly differ in some species (Allen 1919). Such differences are unexpected and are not yet understood, but genome sequencing should help understand their causes. Current bryophyte sequences are only from species whose U and V have similar sizes. In the polyploid *Sphagnum angustifolium*, the UV pair may have evolved independently (perhaps from an ancestral monoicous state like that of some related species) (Healey et al. 2023). It is unknown how haploid plants can evolve to be monoicous and express both sex functions and whether this involves a stage in which gametophytes carry both U and V chromosomes, as suggested for the monoicous species *Ricciocarpos natans* (Singh et al. 2023). Interestingly, the *S. angustifolium* U and V are both small and derived from a larger ancestral chromosome (Healey et al. 2023). This suggests a sex chromosome miniaturization process convergent with that in *M. polymorpha*. In contrast, 2 mosses, *C. purpureus* (Carey et al. 2021b) and *Syntrichia caninervis* (Silva et al. 2021), have large, gene-rich U and V regions. These reflect fusions with at least 1 autosome (ancestral element D), perhaps due to their history of polyploidy followed by reduction in the total number of chromosomes (Lang et al. 2018). Other bryophytes have undergone fusions of an autosome with both the U and V, or with just one of them (Renner et al. 2017; Sousa et al. 2021), which could explain U-V size differences.

Fusions between autosomes and 1 or both of the sex chromosome pair also happen in angiosperms (reviewed in Westergaard 1958), including cannabis and hops, and in the genus *Viscum*, with as many as 5 Y chromosomes (Barlow and Wiens 1976). For example, X-autosome fusions have created visibly XY1Y2 karyotypes in *Rumex* species (Smith 1964; Hough et al. 2014; Grabowska-Joachimiak et al. 2015). If there is a completely Y-linked region, fusion of an autosome with an Y or X region other than the PAR can create sex chromosome heteromorphism in a completely different way from the processes described above. However, even if a completely Y-linked region exists, most loci on an autosome or autosome arm that fuses with it will not stop recombining unless recombination is absent in males. Studies of such "instant heteromorphic systems" can test for new fully Y-linked regions and determine whether proximity to fully Y-linked regions directly prevents recombination (e.g. by hindering pairing with the homologous chromosome near the breakpoint). Such newly nonrecombining regions (including young strata, see below) are also interesting for studying the time course of changes after genome regions becomes completely sex linked.

How old are nonrecombining sex-determining regions?

Sex chromosome turnovers can make the ages of plant sex chromosomes unclear. Direct age estimates, based on estimating sequence divergence between genes within completely sex-linked regions and present on both the X and the Y (termed gametologs) are therefore important. However, the Y-linked region may be small and carry only a few genes, or degeneration of Y-linked genes may mean that few gene pairs are available, resulting in uncertain estimates. Nevertheless, enough gametolog pairs have been found to demonstrate that synonymous site divergence can be as high as 25% (or even more for some genes) in plants with heteromorphic XY pairs, including *S. latifolia* (Yue et al. 2023), and species of Cannabis and hops (Prentout et al. 2021).

Divergence is much higher between UV gametologs of the bryophyte *M. polymorpha* (Yamato et al. 2007), indicating a very ancient origination time without subsequent loss of the separate sexed state (Iwasaki et al. 2021; Singh et al. 2023). However, the haploid gametophytes of many bryophyte species produce gametes of both sexes (monoicy), indicating that sex-determining systems can change. New systems could therefore evolve in turnovers like those described above in angiosperms.

We have explained that completely sex-linked regions differentiate and may become heteromorphic but not why sex-determining genes are often within physically extended nonrecombining regions. The Melandrium pathway for the evolution of separate sexes can readily explain this because crossovers between the 2 genes involved, *SuF* and *M* factors (see above), will produce neuter individuals, such as the *SuF m/m* genotype. This creates selection for closer linkage between the 2 genes (Fig. 3A); because *SuF* factors increase male and suppress female functions, this is termed the sexually antagonistic (SA) polymorphism hypothesis.

However, a lack of recombination does not necessarily imply that suppressed recombination has evolved. Many plant genomes include large pericentromeric regions that rarely recombine (Brazier and Glémin 2022), and sex-determining genes could evolve within them (Fig. 3B). Genetic and physical mapping has revealed that the *Rumex hastatulus* sex-linked region coincides with such a region (Rifkin et al. 2021). The same is likely for part of the *S. latifolia* Y-linked and *Pistacia* W-linked regions (Bergero et al. 2013; Palmer et al. 2022). When a large, repeat-rich Y-linked region arises in this way, the X may also be repeat rich, reflecting the ancestral state.

Duplications can create nonrecombining regions by yet another mechanism (Fig. 3C). The example of the OGI insertion in the persimmon has already been mentioned. Several *Spinacia* species have Y-linked regions in the same genome location and sharing at least 1 gene, though their sizes differ (She et al. 2022). No corresponding sequences are detected in females (Ma et al. 2022; She et al. 2023), suggesting that a duplication resulted in a new male-determining factor and Y-specific region (as diagrammed in Fig. 3D). Because the region is absent from the X, recombination cannot occur with the X. This insertion is within a large pericentromeric region like those just mentioned, instantly creating a large region that recombines rarely, at most, with the male-determining factor.

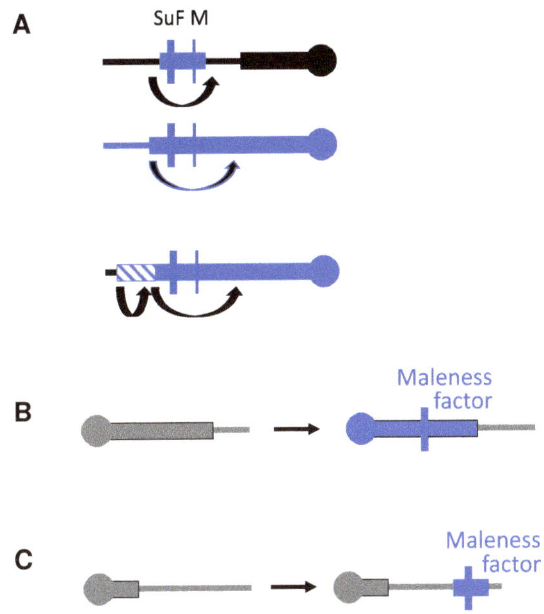

Figure 3. Mechanisms causing lack of recombination. As in Figs. 1 and 2, thin horizontal lines indicate chromosome arms, and thick bars that include the male-determining factors indicate complete Y linkage. **A)** Recombination suppression events creating an old evolutionary stratum. The event may involve inversion of part of the chromosome (the X or the Y, as shown), or control by a genetic recombination modifier. **B)** Evolution of a sex-determining locus in a pericentromeric region that already recombined rarely in 1 or both sexes. The processes in B and C can include turnovers causing appearance of a male-determining locus in a new genomic location. **C)** Insertion of a maleness factor creating a small Y-linked region (in a chromosome that remains morphologically similar to the X). The small Y-specific region may locally hinder recombination with the X.

Evolutionary strata and ongoing recombination suppression

Importantly, however, suppressed recombination must have evolved in some organisms because separate events have been discovered, creating "evolutionary strata" of Y-X divergence. Plants with such evidence for creation of new nonrecombining regions include 3 of the heteromorphic systems in Table 1, *S. latifolia* and its close relatives (Campos et al. 2016; Yue et al. 2023), whose homologous chromosome recombines in *Silene vulgaris* (Bergero et al. 2013), papaya (Wang et al. 2012), and probably the Cannabaceae (Prentout et al. 2021). It remains unclear how many dioecious angiosperms have evolved strata. The mechanism of recombination suppression is also not known. When evolutionary strata were first discovered in the human XY pair, it was suggested that chromosomal inversions could be responsible (Lahn and Page 1999). However, evidence for inversions is still scanty, and the sole example in plants is in papaya (Wang et al. 2012). A large *S. latifolia* Y-linked region is also inverted compared with the homologous chromosome of *S. vulgaris* (Bergero et al. 2013), but it is not clear when this rearrangement occurred; inversions after recombination stopped or, in the *S. vulgaris* lineage, are not excluded.

Until recently, models to explain sex-linked regions by successive changes suppressing recombination proposed that the situation was similar to the SA polymorphism hypothesis outlined above. Later mutations controlling sexually dimorphic traits might also have SA effects, so suppressed recombination might evolve if a single gene system persists long enough to allow maleness-enhancing SA polymorphisms to accumulate in genes closely or completely linked to the "primary" sex-determining locus, largely restricting the male beneficial alleles to males (Jordan and Charlesworth 2012). Polymorphisms controlling such "secondary" sexual characters create selection for closer linkage with an existing fully sex-linked region. Deletion mapping in *S. latifolia* indeed found 2 Y-linked genes whose deletion impairs (Westergaard 1948; Kazama et al. 2016), 1 of which may enhance male functions. A similar situation might explain the apparent spread of a male sterility factor after female sterility in Asparagus, the opposite of the order proposed by the Melandrium path-way (Pannell and Gerchen 2018). Studying the developmen-tal time course of expression of Y-linked genes, and the flower bud tissues in which they are expressed, can now test whether they are likely to be involved in primary sex determination or in later reproductive stages. It will also be interesting to test whether, for genes that are expressed in both sexes, expression of Y-linked alleles differs from that of their X-linked counterparts. Such differences could reflect advan-tageous alleles in males with SA effects in females (or else genetic degeneration).

However, recombination might also be lost without any such SA selection or indeed any form of selection that is specifically related to sex determination. A completely sex-linked region that arises will carry deleterious mutations. Particularly if such a region is small, it may carry fewer mutations than average, favoring its initial spread throughout the Y chromosome population (Olito and Abbott 2023). Although autosomal inversions without other selective advantages are expected almost invariably to be eliminated, because of subsequent deleterious mutations within the region, some Y-linked inversions (or new Y-linked regions created by another mechanism) may escape elimination; this is because Y-linkage ensures that subsequent deleterious mutations do not become homozygous (Jay et al. 2022). If enough small inversions arise, a few lucky ones may survive and enlarge such a Y-linked region with many small strata in an apparently continuous loss of recombination process. Lenormand and Roze (2022) proposed another model, also involving successive small, lucky inversions, with each inversion favored by a special form of SA selection acting in XY heterozygotes: expression of Y-linked genes with deleterious mutations decreases, and expression of the mutation-free X-linked alleles increases in a type of dosage compensation process. This model requires asymmetry in degeneration between the sex chromosomes (see above) and therefore cannot operate in haploid plants. In a final alternative model for recombination suppression, sequence divergence in a completely sex-linked region might affect the recombination process near the boundary with the adjacent PAR, causing recombination to be continuously lost and the sex-linked region enlarged, without involving any type of selection (Jeffries et al. 2021).

These ideas may be empirically testable in dioecious flowering plants. Taxa with homomorphic sex chromosomes retained over long evolutionary times would suggest that SA polymorphisms are rarely established in plant PARs. They would also argue against the sequence divergence mechanism just outlined. Species with very recently evolved nonrecombining regions may allow tests of the other ideas. First, genome sequence assemblies can show whether new stratum formation has involved an inversion or other discontinuous change. Second, such species can be used to test whether genes in a new stratum show sex differences in expression reflecting the proposed evolution of dosage compensation alongside recombination suppression rather than after degeneration has evolved. Initial data from *S. latifolia* (Campos et al. 2016; Filatov 2022) detect no such expression pattern. On the other hand, because control of expression by sex-specific trans-acting factors is required and species in which dioecy is not very long established may not yet have such controls, most dioecious plants may be unsuitable for testing this idea; however, this argument suggests that the process is unlikely in plants.

Dosage compensation

Degenerated Y-linked regions in diploid species may evolve dosage compensation. Degeneration causes expression in males to deviate from the optimal level (which is likely to have evolved in co-sexual ancestors of dioecious plants and which, for many genes, may still be similar in females). Also, for a few genes, expression must be balanced with that from autosomal genes. Control of expression should therefore evolve in males to restore levels nearer the optimum (reviewed in Vicoso and Charlesworth 2006). Lenormand and Roze (2022) proposed that expression of individual genes might be controlled as their Y-linked alleles acquire deleterious mutations. Plants are of special interest for studying dosage compensation because they do not have a long history of ancestors with sexually dimorphic control of gene expression (including dosage compensation). Currently, dosage compensation has been tested in few plant species (Muyle et al. 2012; Fruchard et al. 2020). Moreover, the gene sets studied include a mixture of hemizygous genes (ideal for such tests) and genes with different expression from the Y and X alleles, making the results hard to interpret. It remains unclear whether compensation has evolved in re-sponse to degeneration of sex-linked alleles because it can occur automatically for genes with mutations that decrease expression (Guo and Birchler 1994; Krasovec et al. 2019).

Sex chromosomes and plant genomics

Genome sequences of dioecious plants across angiosperm phylogeny can help identify new sex determining genes, adding to understanding of the diverse genetic factors affecting male and female fertility across angiosperms, complementing mutant library screening within single species. The involvement of sterility mutations in some, if not all, sex-determining systems suggests that, as in fish (Pan et al. 2021), there may be many mechanisms involving different mutations that disrupt male or female functions. Identifying these possibly hundreds of independent sex-determining mutations may reveal involvement of genes not previously suspected of functioning in flower development. For instance, the sole functional annotation of the dominant SUPPRESSOR OF FEMALE FUNCTION (*SOFF*) in garden asparagus is Domain of Unknown Function 247 (Harkess et al. 2020), and orthologs of the *OGI-MEGI* suppression mechanism in persimmons had not been implicated in fertility before their sex determining actions were discovered (Akagi et al. 2014).

Contiguous assemblies of dioecious species' sex-linked regions are most useful if both members of a sex chromosome pair are included, so it is tempting to choose its heterogametic sex (XY male or ZW female) for sequencing. However, as noted above, it is often unknown whether a given dioecious species has male or female heterogamety and thus whether to sequence a male or a female individual. This is especially true for species with homomorphic sex chromosomes with small fully sex-linked regions, where cytological examination of chromosomes is not helpful unless a linkage map is available.

In species with highly degenerated Y- or W-linked regions, the heterogametic sex can be determined from a genome sequence of 2 individuals (or even just 1 of the heterogametic sex, using comparisons with autosomes). Importantly, short-read sequencing of multiple individuals of a species is no longer prohibitively expensive and can be used in analyses to identify sex-specific k-mers (Carvalho and Clark 2013) and determine heterogamety (even if no degeneration has occurred, or degeneration is slight, and coverage is similar in both sexes) by directly identifying Y- or W-linked variants. Provided that the sex-linked region is large enough and there has been enough time for variants to accumulate, this approach can yield reliable results, for example, in spinach (Ma et al. 2022; She et al. 2023). Single-nucleotide polymorphisms, the presence or absence of repetitive elements or genes, including copy number variation, can also serve to detect such associations, and analysis methods are being developed, including pangenome graphs (e.g. He et al. 2023) to represent structural differences between 2 differentiated sex chromosomes. No such analysis has yet been published for a dioecious plant, but this approach will help analyses of gene movements from autosomes into sex-linked regions and movements in the opposite direction, both of which occur in animals (reviewed in Vicoso and Charlesworth 2006) and probably also in plants (see above). Given an assembly, k-mer analysis may give a clear signal across an extensive genome region, distinguishing sex-linked from autosomal and pseudo-autosomal regions. Even assembly-free k-mer analysis can identify the heterogametic sex. There may still be an ascertainment bias toward species with at least moderate Y-X (or W-Z) differentiation across large genome regions, or "micro-heteromorphism," but it will be much less than under complete reliance on cytogenetic observations.

It may thus be possible to assess the relative frequencies of male vs female heterogamety using living material or worldwide herbaria. Sometimes it is possible to assemble short contigs for the sex-linked regions and identify putatively sex-linked genes without expensive long-read genome assembly. Such genomic exploration of fully sex-linked regions in nonmodel species may suggest classes of genes that are enriched in sex-linked regions, test whether certain stages or tissue types or cell layers important for anther and ovule development produce sterility mutations more than others, and reveal new male sterility mutations for hybrid crop breeding.

Heteromorphic sex chromosomes in both plants and animals are often still challenging to assemble and analyze and are often excluded from nuclear genome assemblies (Pinto et al. 2023). If sex linkage is ignored in genome-wide association studies, this can cause mis-inferences, and evolutionary analyses can be compromised. For example, Y-X sequence divergence creates unusually high within-species diversity in completely sex-linked regions and closely linked sequences. Recent advances in genome assembly and scaffolding, such as >99% accurate PacBio HiFi reads, >100-kb ultra-long Oxford Nanopore reads and, more recently, >99% accurate Oxford Nanopore duplex reads (where both strands of a single DNA molecule are read) are now producing gapless genome assemblies, including the first for the human X and Y sex chromosomes (Nurk et al. 2022; Rhie et al. 2022), whereas the first human genome sequence particularly struggled to assemble either sex chromosome (International Human Genome Sequencing Consortium 2001). Such sequence assemblies will also be possible in plants.

A particular problem for sex-linked regions of nonmodel species is that (as just outlined for sex-specific k-mer analysis) the X and Y (or Z and W) versions (or haplotypes) of the region need to be distinguished. In genetical terms, the 2 haplotypes must be assembled separately, or "phased," to study the differences between them. Fully phased assembly of sex chromosomes is difficult without knowing whether the sex-linked region in a species of interest is homomorphic (in which case, it may be small, with sequence divergence too slight for reliable phasing) or heteromorphic (in which case 1 haplotype in the region may be repetitive and/or rearranged). Manual curation, although laborious, can succeed in resolving the 2 haplotypes. Sequencing both sexes of a species can assist phasing, especially using multiple individuals of each sex. For individual gene sequences, this can provide Y-X divergence time estimates, even when sequence divergence is slight (Campos et al. 2016). However, haplotype switching and chimerism in high-throughput sequencing may hinder assembly of highly contiguous sex chromosomes. With sequences from multiple individuals within a species, it will become possible to detect differences in Y chromosomes of different individuals, including copy number variation, variation in the sizes of arrays such as satellite arrays, and large deletions in highly degenerated regions, causing variable Y or W chromosome sizes (Puterova et al. 2018).

We already mentioned the value of identifying sex-determining genes and sex-linked markers to allow early selection of individuals of the desired sex in long-lived dioecious crop plants. In transformable dioecious species, there may be opportunities for improvement of desired functions. For example, in a fruit crop species, rather than using genetic markers to remove the sex that is needed only for pollination, sex ratio distorters could be identified and used to modify sex ratios. Moreover, since 1 member of a sex-linked region is sex specific (Y-linked in species with male heterogamety or W-linked under female heterogamety), transgenes could be placed in a fully sex-linked region to ensure transmission exclusively through females, minimizing escape though the transgenic plant's pollen.

Conclusions

Rapid advances are currently being made in discovering sex-linked regions in dioecious plant genomes. Detailed studies of more species may reveal such regions in diverse species. This may distinguish species with extensive sex-linked regions and without them, and those whose sex-linked regions coincide with ancestrally rarely recombining pericentromeric regions, from others in which recombination has become suppressed as dioecy evolved. It may soon become possible to classify the sex-determining system of flowering plants into those with 2-gene or 1-gene systems. The species so far studied may not, of course, include all possible types of sex-determining system. For example, no system has yet been discovered in which 2 mutations within the same gene create the 2 sexes.

With defined questions in mind, future studies can choose the most suitable species for study. For instance, the Cucurbitacae are a large group of important crops, and many species have genetic sex-determination (while others are monoecious). They may have sex chromosomes of differing ages, but it is currently not even known whether dioecy evolved several times independently in this plant family. Although Cucurbitacae often have large genomes, approaches can now sequence and assemble them, as just discussed. Flowering plants are ideal for studying the time course of genome evolution and evolution of nonrecombining regions. If a succession of inversions is responsible, as recently proposed (Jay et al. 2022; Lenormand and Roze 2022), they should be readily detected in young evolutionary strata where later rearrangements have not yet obscured earlier ones. In addition, plants offer multiple independent instances of loss of recombination, at different times in the past, allowing studies of accumulation of TE and other repetitive sequences and genetic degeneration. Current data, although interesting, have served more to define the interesting questions, and new advances in sequencing hold promise that future studies will answer some of them.

Supplemental data

The following materials are available in the online version of this article.

Supplemental Table S1. Information about dioecious angiosperms, showing the range of clades in which dioecy is found, and illustrating that sex chromosome heteromorphism (species with names in bold font) is also found in distantly related taxa.

Conflict of interest statement. None declared.

References

Akagi T, Henry IM, Ohtani H, Morimoto T, Morimoto K, Beppu K, Kataoka I, Tao R. A Y-encoded suppressor of feminization arose via lineage-specific duplication of a cytokinin response regulator in kiwifruit. Plant Cell. 2018:**30**:780–795. https://doi.org/10.1105/tpc.17.00787

Akagi T, Henry IM, Tao R, Comai L. A Y-chromosome–encoded small RNA acts as a sex determinant in persimmons. Science. 2014: **346**(6209):646–650. https://doi.org/10.1126/science.1257225

Akagi T, Pilkington SM, Varkonyi-Gasic E, Henry IM, Sugano SS, Sonoda M, Firl A, McNeilage MA, Douglas MJ, Wang T, et al. Two Y-chromosome-encoded genes determine sex in kiwifruit. Nat Plants. 2019:**5**(8):801–809. https://doi.org/10.1038/s41477-019-0489-6

Akagi T, Varkonyi-Gasic E, Shirasawa K, Catanach A, Henry IM, Mertten D, Datson P, Masuda K, Fujita N, Kuwada E, et al. Recurrent neo-sex chromosome evolution in kiwifruit. Nat Plants. 2023:**9**(3):393–402. https://doi.org/10.1038/s41477-023-01361-9

Allen CE. The basis of sex inheritance in *Sphaerocarpos*. Proc Amer Philos Soc. 1919:**58**:289–316.

Bachtrog D. The temporal dynamics of processes underlying Y chromosome degeneration. Genetics. 2008:**179**(3):1513–1525. https://doi.org/10.1534/genetics.107.084012

Balounova V, Gogela R, Cegan R, Cangren P, Zluvova J, Safar J, Kovacova V, Bergero R, Hobza R, Vyskot B, et al. Evolution of sex determination and heterogamety changes in section Otites of the genus *Silene*. Sci Rep. 2019:**9**(1):1045. https://doi.org/10.1038/s41598-018-37412-x

Barlow BA, Wiens D. Translocation heterozygosity and sex ratio in *Viscum fischeri*. Heredity. 1976:**37**(1):27–40. https://doi.org/10.1038/hdy.1976.63

Bergero R, Qiu S, Forrest A, Borthwick H, Charlesworth D. Expansion of the pseudoautosomal region and ongoing recombination suppression in the *Silene latifolia* sex chromosomes. Genetics. 2013:**194**(3): 673–686. https://doi.org/10.1534/genetics.113.150755

Brazier T, Glémin S. Diversity and determinants of recombination landscapes in flowering plants. PLoS Genet. 2022:**18**(8):e1010141. https://doi.org/10.1371/journal.pgen.1010141

Bull JJ, Charnov EL. Changes in the heterogametic mechanism of sex determination. Heredity. 1977:**39**(1):1–14. https://doi.org/10.1038/hdy.1977.38

Bull JJ. Evolution of sex determining mechanisms. Menlo Park (CA): Benjamin/Cummings; 1983.

Bull JJ. Sex-chromosomes in haploid dioecy—unique contrast to Mullers theory for diploid dioecy. Am Nat. 1978:**112**(983):245–250. https://doi.org/10.1086/283267

Campos J, Qiu S, Guirao-Rico S, Bergero R, Charlesworth D. Recombination changes at the boundaries of fully and partially sex-linked regions between closely related Silene species pairs. Heredity. 2016:**118**(4):395–403. https://doi.org/10.1038/hdy.2016.113

Carey S, Yu QY, Harkess A. The diversity of plant sex chromosomes highlighted through advances in genome sequencing. Genes (Basel). 2021a:**12**(3):381. https://doi.org/10.3390/genes12030381

Carey SB, Jenkins J, Lovell JT, Maumus F, Sreedasyam A, Payton AC, Shu S, Tiley GP, Fernandez-Pozo N, Healey A, et al. Gene-rich UV sex chromosomes harbor conserved regulators of sexual development. Sci Adv. 2021b:**7**(27):eabh2488. https://doi.org/10.1126/sciadv.abh2488

Carlisle E, Janis CM, Pisan D, Donoghue PCJ, Silvestro D. A timescale for placental mammal diversification based on Bayesian modeling of the fossil record. Curr Biol. 2023:**33**(15):3073–3082.e3. https://doi.org/10.1016/j.cub.2023.06.016

Carvalho A, Clark AG. Efficient identification of Y chromosome sequences in the human and Drosophila genomes. Genome Res. 2013:**23**(11):1894–1907. https://doi.org/10.1101/gr.156034.113

Cauret CMS, Mortimer SME, Roberti MC, Ashman T-L, Liston A. Chromosome-scale assembly with a phased sex-determining region resolves features of early Z and W chromosome differentiation in a wild octoploid strawberry. G3 (Bethesda). 2022:**12**(8):jkac139. https://doi.org/10.1093/g3journal/jkac139

Chae T, Harkess AA, Moore RC. Sex-linked gene expression and the emergence of hermaphrodites in *Carica papaya*. Am J Bot. 2021:**108**(6):1029–1041. https://doi.org/10.1002/ajb2.1689

Charlesworth D. Androdioecy and the evolution of dioecy. Biol J Linn Soc. 1984:**23**(4):333–348. https://doi.org/10.1111/j.1095-8312.1984.tb01683.x

Charlesworth, D. Distribution of dioecy and self-incompatibility in angiosperms. In: **Greenwood PJ, Slatkin M**, editors. Evolution—essays in honour of John Maynard Smith. Cambridge: Cambridge University Press; 1985. p. 237–268.

Charlesworth B, Charlesworth D. A model for the evolution of dioecy and gynodioecy. Am Nat. 1978:**112**(988):975–997. https://doi.org/10.1086/283342

Charlesworth B, Sniegowski P, Stephan W. The evolutionary dynamics of repetitive DNA in eukaryotes. Nature. 1994:**371**:215–220. https://doi.org/10.1038/371215a0

Cossard GG, Pannell JR. Enhanced leaky sex expression in response to pollen limitation in the dioecious plant *Mercurialis annua*. J Evol Biol. 2021:**34**(2):416–422. https://doi.org/10.1111/jeb.13720

Crossman A, Charlesworth D. Breakdown of dioecy: models where males acquire cosexual functions. Evolution. 2014:**68**(2):426–440. https://doi.org/10.1111/evo.12283

Crowson D, Barrett SCH, Wright SI. Purifying and positive selection influence patterns of gene loss and gene expression in the evolution of a plant sex chromosome system. Mol Bio Evol. 2017:**34**(5): 1140–1154. https://doi.org/10.1093/molbev/msx064

Darwin CR. The different forms of flowers on plants of the same species. London: John Murray; 1877.

Filatov DA. Recent expansion of the non-recombining sex-linked region on Silene latifolia sex chromosomes. J Evol Biol. 2022:**35**(12): 1696–1708. https://doi.org/10.1111/jeb.14063

Fruchard C, Badouin H, Latrasse D, Devani RS, Muyle A, Rhoné B, Renner SS, Banerjee AK, Bendahmane A, Marais GAB. Evidence for dosage compensation in *Coccinia grandis*, a plant with a highly heteromorphic XY system. Genes (Basel). 2020:**11**(7):787. https://doi.org/10.3390/genes11070787

Garcia S, Janousek B, Pascual-Díaz JP, Renner SS. Sex-chrom v. 2.0: a database of green plant species with sex chromosomes. Chromosoma. 2023:**132**(1):55–58. https://doi.org/10.1007/s00412-023-00786-7

Gerchen J, Veltsos P, Pannell J. Recurrent allopolyploidization, Y-chromosome introgression and the evolution of sexual systems in the plant genus Mercurialis. Phil Trans R Soc B. 2022:**377**(1850): 20210224. https://doi.org/10.1098/rstb.2021.0224

Gong W, Filatov DA. Evolution of the sex-determining region in *Ginkgo biloba*. Philos Trans R Soc B Biol Sci. 2022:**377**(1850):20210229. https://doi.org/10.1098/rstb.2021.0229

Grabowska-Joachimiak A, Kula A, Książczyk T, Chojnicka J, Sliwinska E, Joachimiak A. Chromosome landmarks and autosome-sex chromosome translocations in *Rumex hastatulus*, a plant with XX/XY1Y2 sex chromosome system. Chromosome Res. 2015(2):**23**: 187–197. https://doi.org/10.1007/s10577-014-9446-4

Guo M, Birchler JA. Trans-acting dosage effects on the expression of model gene systems in maize aneuploids. Science. 1994:**266**(5193): 1999–2002. https://doi.org/10.1126/science.266.5193.1999

Harkess A, Huang K, van der Hulst R, Tissen B, Caplan JL, Koppula A, Batish M, Meyers BC, and Leebens-Mack JH. Sex determination by two Y-linked genes in garden asparagus. Plant Cell. 2020:**32**(6): 1790–1796. https://doi.org/10.1105/tpc.19.00859

Harkess A, Mercati F, Shan H, Sunseri F, Falavigna A, Leebens-Mack J. Sex-biased gene expression in dioecious garden asparagus (Asparagus officinalis). New Phytologist. 2015:**207**:883–892. https://doi.org/10.1111/nph.13389

He Q, Tang S, Zhi H, Chen J, Zhang J, Liang H, Alam O, Li H, Zhang H, Xing L, et al. A graph-based genome and pan-genome variation of the model plant Setaria. Nat Genet. 2023:**55**(7):1232–1242. https://doi.org/10.1038/s41588-023-01423-w

Healey AL, Piatkowski B, Lovell JT, Sreedasyam A, Carey SB, Mamidi S, Shu S, Plott C, Jenkins J, Lawrence T, et al. Newly identified sex chromosomes in the Sphagnum (peat moss) genome alter carbon sequestration and ecosystem dynamics. Nat Plants. 2023:**9**(2):238–254. https://doi.org/10.1038/s41477-022-01333-5

Hough J, Hollister JD, Wang W, Barrett SCH, Otto SP. Genetic degeneration of old and young Y chromosomes in the flowering plant *Rumex hastatulus*. Proc Natl Acad Sci U S A. 2014:**111**(21): 7713–7718. https://doi.org/10.1073/pnas.1319227111

Iwasaki M, Kajiwara T, Yasui Y, Yoshitake Y, Miyazaki M, Kawamura S, Suetsugu N, Nishihama R, Yamaoka S, Wanke D, et al. Identification of the sex-determining factor in the liverwort *Marchantia polymorpha* reveals unique evolution of sex chromosomes in a haploid system. Curr Biol. 2021:**31**(24):5522–5532. https://doi.org/10.1016/j.cub.2021.10.023

Jay P, Tezenas E, Giraud T. Sheltering of deleterious mutations explains the stepwise extension of recombination suppression on sex chromosomes and other supergenes. PLoS Biol. 2022:**20**(7): e3001698. https://doi.org/10.1371/journal.pbio.3001698

Jeffries D, Gerchen J, Scharmann M, Pannell J. A neutral model for the loss of recombination on sex chromosomes. Phil Trans R Soc B. 2021:**376**(1832):20200096. https://doi.org/10.1098/rstb.2020.0096

Jesionek W, Bodláková M, Kubát Z, Čegan R, Vyskot B, Vrána J, Šafář J, Puterova J, Hobza R. Fundamentally different repetitive element composition of sex chromosomes in *Rumex acetosa*. Ann Bot. 2021:**127**(1):33–47. https://doi.org/10.1093/aob/mcaa160

Jones DF. The interaction of specific genes in doecious maize. In: Proceedings of the 6th International Congress of Genetics; Ithaca, New York. 1932:104–107.

Jordan C, Charlesworth D. The potential for sexually antagonistic polymorphism in different genome regions. Evolution. 2012:**66**(2): 505–516. https://doi.org/10.1111/j.1558-5646.2011.01448.x

Käfer J, Bewick A, Andres-Robin A, Lapetoule G, Harkess A, Caïus J, Fogliani B, Gâteblé G, Ralph P, dePamphilis CW, et al. A derived ZW chromosome system in Amborella trichopoda, representing the sister lineage to all other extant flowering plants. New Phytol. 2021:**233**(4):1636–1642. https://doi.org/10.1111/nph.17662

Käfer J, Mousset S. Labile sex expression in angiosperm species with sex chromosomes. Philos Trans R Soc Lond B. 2022:**377**(1850):20210216. https://doi.org/10.1098/rstb.2021.0216

Kazama Y, Ishii K, Aonuma W, Ikeda T, Kawamoto H, Koizumi A, Filatov D, Chibalina M, Charlesworth D, Bergero R, et al. A new physical mapping approach refines the sex-determining gene positions on the Silene latifolia Y-chromosome. Sci Rep. 2016:**6**(1): 18917. https://doi.org/10.1038/srep18917

Kazama Y, Kitoh M, Kobayashi T, Ishii K, Krasovec M, Yasui Y, Abe T, Kawano S, Filatov DA. A CLAVATA3-like gene acts as a gynoecium suppression function in white campion. Mol Bio Evol. 2022:**39**(10):msac195. https://doi.org/10.1093/molbev/msac195

Kazama Y, Nishihara K, Bergero R, Fujiwara MT, Abe T, Charlesworth D, Kawano S. SlWUS1: an X-linked gene having no homologous Y-linked copy in *Silene latifolia*. G3 (Bethesda). 2012:**2**(10):1269–1278. https://doi.org/10.1534/g3.112.003749

Krasovec M, Kazama Y, Ishii K, Abe T, Filatov DA. Immediate dosage compensation is triggered by the deletion of Y-linked genes in *Silene latifolia*. Curr Biol. 2019:**29**(13):2214–2221.e4. https://doi.org/10.1016/j.cub.2019.05.060

International Human Genome Sequencing Consortium. Initial sequencing and analysis of the human genome. Nature. 2001:**409**: 860–921. https://doi.org/10.1038/35057062

Lahn BT, Page DC. Four evolutionary strata on the human X chromosome. Science. 1999:**286**(5441):964–967. https://doi.org/10.1126/science.286.5441.964

Lang D, Ullrich KK, Murat F, Fuchs J, Jenkins J, Haas FB, Piednoel M, Gundlach H, Van Bel M, Meyberg R, et al. The Physcomitrella patens chromosome-scale assembly reveals moss genome structure and evolution. Plant J. 2018:**93**(3):515–533. https://doi.org/10.1111/tpj.13801

Lenormand T, Roze D. Y recombination arrest and degeneration in the absence of sexual dimorphism. Science. 2022:**375**(6581):663–666. https://doi.org/10.1126/science.abj1813

Liu Y, Wang S, Li L, Yang T, Dong S, Wei T, Wu S, Liu Y, Gong Y, Feng X, et al. The Cycas genome and the early evolution of seed plants. Nat Plants. 2022:**8**(4):389–401. https://doi.org/10.1038/s41477-022-01129-7

Ma X, Yu LA, Fatima M, Wadlington WH, Hulse-Kemp AM, Zhang X, Zhang S, Xu X, Wang J, Huang H, et al. The spinach YY genome reveals sex chromosome evolution, domestication, and introgression history of the species. Genome Biol. 2022:**23**(1):75. https://doi.org/10.1186/s13059-022-02633-x

Martin H, Carpentier F, Gallina S, Godé C, Schmitt E, Muyle A, Marais G. Evolution of young sex chromosomes in two dioecious sister plant species with distinct sex determination systems. Genome Biol Evol. 2019:**11**(2):350–361. https://doi.org/10.1093/gbe/evz001

Massonnet M, Cochetel N, Minio A, Vondras AM, Muyle A, Lin J, Garcia JF, Zhou Y, Delledonne M, Riaz S, et al. The genetic basis of sex determination in grapevines (Vitis spp.). Nat Commun. 2020:**11**(1):2902. https://doi.org/10.1038/s41467-020-16700-z

Masuda K, Akagi T. Evolution of sex in crops: recurrent scrap and rebuild. Breeding Sci. 2023:**73**(2):95–107. https://doi.org/10.1270/jsbbs.22082

Masuda K, Ikeda Y, Matsuura T, Kawakatsu T, Tao R, Kubo Y, Ushijima K, Henry IM, Akagi T. Reinvention of hermaphroditism via activation of a RADIALIS-like gene in hexaploid persimmon. Nat Plants. 2022:**8**(3):217–224. https://doi.org/10.1038/s41477-022-01107-z

McDaniel SF, Willis JH, Shaw AJ. A linkage map reveals a complex basis for segregation distortion in an interpopulation cross in the moss *Ceratodon purpureus*. Genetics. 2007:**176**(4):2489–2500. https://doi.org/10.1534/genetics.107.075424

Ming R, Bendahmane A, Renner S. Sex chromosomes in land plants. Ann Rev Plant Bio. 2011:**62**(1):485–514. https://doi.org/10.1146/annurev-arplant-042110-103914

Mirski P. Exceptions from dioecy and sex lability in genus Salix. Dendrobiology. 2014:**71**:167–173. https://doi.org/10.12657/denbio.071.017

Montgomery SA, Tanizawa Y, Galik B, Yamato KT, Liu C, Berger F. Chromatin organization in early land plants reveals an ancestral association between H3K27me3, transposons, and constitutive heterochromatin. Curr Biol. 2020:**30**(4):573–588.e7. https://doi.org/10.1016/j.cub.2019.12.015

Müller N, Kersten B, Montalvão A, Mähler N, Bernhardsson C, Bräutigam K, Lorenzo Z, Hoenicka H, Kumar V, Mader M, et al. A single gene underlies the dynamic evolution of poplar sex determination. Nat Plants. 2020:**6**(6):630–637. https://doi.org/10.1038/s41477-020-0672-9

Muyle A, Zemp N, Deschamps C, Mousset S, Widmer A, Marais G. Rapid de novo evolution of X chromosome dosage compensation in *Silene latifolia*, a plant with young sex chromosomes. PLoS Biol. 2012:**10**(4):e1001308. https://doi.org/10.1371/journal.pbio.1001308

Nurk S, Koren S, Rhie A, Rautiainen M, Bzikadze AV, Mikheenko A, Vollger MR, Altemose N, Uralsky L, Gershman A, et al. The complete sequence of a human genome. Science. 2022:**376**(6588):44–53. https://doi.org/10.1126/science.abj6987

Olito C, Abbott JK. The evolution of suppressed recombination between sex chromosomes and the lengths of evolutionary strata. Evolution. 2023:**77**(4):1077–1090. https://doi.org/10.1093/evolut/qpad023

Palmer W, Jacygrad E, Sagayaradj S, Cavanaugh K, Han R, Bertier L, Beede B, Kafkas S, Golino D, Preece J, et al. Genome assembly and association tests identify interacting loci associated with vigor, precocity, and sex in interspecific pistachio rootstocks. G3 (Bethesda). 2022:**13**(2):jkac317. https://doi.org/10.1093/g3journal/jkac317

Pan Q, Kay T, Depincé A, Adolfi M, Schartl M, Guiguen Y, Herpin A. Evolution of master sex determiners: TGF-β signalling pathways at regulatory crossroads. Phil Trans R Soc B. 2021:**376**(1832): 20200091. https://doi.org/10.1098/rstb.2020.0091

Pannell J, Gerchen J. Sex determination: sterility genes out of sequence. Curr Biol. 2018:**28**(2):R66–R88. https://doi.org/10.1016/j.cub.2017.11.052

Pinto B, O'Connor B, Schatz M, Zarate S, Wilson M. Concerning the eXclusion in human genomics: the choice of sex chromosome representation in the human genome drastically affects number of identified variants. G3 (Bethesda). 2023:**13**(10):jkad169. https://doi.org/10.1093/g3journal/jkad169

Prentout D, Stajner N, Cerenak A, Tricou T, Brochier-Armanet C, Jakse J, Käfer J, Marais GAB. Plant genera Cannabis and Humulus share the same pair of well-differentiated sex chromosomes. New Phytol. 2021:**231**(4):1599–1611. https://doi.org/10.1111/nph.17456

Puterova J, Kubat Z, Kejnovsky E, Jesionek W, Cizkova J, Vyskot B, Hobza R. The slowdown of Y chromosome expansion in dioecious *Silene latifolia* due to DNA loss and male-specific silencing of retrotransposons. BMC Genomics. 2018:**19**(1):153. https://doi.org/10.1186/s12864-018-4547-7

Renner SS, Heinrichs J, Sousa A. The sex chromosomes of bryophytes: recent insights, open questions, and reinvestigations of *Frullania dilatata* and *Plagiochila asplenioides*. J Syst Evol. 2017:**55**(4):333–339. https://doi.org/10.1111/jse.12266

Rhie A, Nurk S, Cechova M, Hoyt SJ, Taylor DJ, Altemose N, Hook PW, Koren S, Rautiainen M, Alexandrov IA, et al. The complete sequence of a human Y chromosome. Nature. 2022:**621**(7978): 344–354. https://doi.org/10.1038/s41586-023-06457-y

Rifkin J, Beaudry F, Humphries Z, Choudhury B, Wright S, Barrett S. Widespread recombination suppression facilitates plant sex chromosome evolution. Mol Bio Evol. 2021:**38**(3):1018–1030. https://doi.org/10.1093/molbev/msaa271

Rohwer J, Kubitzki K. Salix martiana, a regularly hermaphrodite willow. Plant Syst Evol. 1984:**144**(2):99–101. https://doi.org/10.1007/BF00986668

Sauquet H, Ramírez-Barahona S, Magallón S. What is the age of flowering plants? J Exp Bot. 2022:**73**(12):3840–3853. https://doi.org/10.1093/jxb/erac130

Sayres M, Makova K. Gene survival and death on the human Y chromosome. Mol Bio Evol. 2013:**30**(4):781–787. https://doi.org/10.1093/molbev/mss267

She H, Liu ZY, Li SF, Xu ZS, Zhang HL, Cheng F, Wu J, Wang XW, Deng CL, Charlesworth D, et al. Evolution of the spinach sex-linked region within a rarely recombining pericentromeric region. Plant Physiol. 2023:**193**(2):1263–1280. https://doi.org/10.1093/plphys/kiad389

She H, Xu Z, Zhang H, Wu J, Wang X, Liu Z, Qian W. Remarkable divergence of the sex-linked region between two wild spinach progenitors. *Spinacia turkestanica* and *Spinacia tetrandra*. Biology (Basel). 2022:**11**(8):1138. https://doi.org/10.3390/biology11081138

Silva A, Gao B, Fisher K, Mishler B, Ekwealor J, Stark L, Li X, Zhang D, Bowker M, Brinda J, et al. To dry perchance to live: insights from the genome of the desiccation-tolerant biocrust moss *Syntrichia caninervis*. Plant J. 2021:**105**(5):1339–1356. https://doi.org/10.1111/tpj.15116

Singh S, Davies KM, Chagné D, Bowman JL. The fate of sex chromosomes during the evolution of monoicy from dioicy in liverworts. Curr Biol. 2023:**33**(17):3597–3609.e3. https://doi.org/10.1016/j.cub.2023.07.023

Smith BW. Evolution of sex-determining mechanisms in *Rumex*. Chromosomes Today. 1969:**2**:172–182.

Smith BW. The evolving karyotype of *Rumex hastatulus*. Evolution. 1964:**18**(1):93–104. https://doi.org/10.2307/2406423

Sola-Campoy PJ, Robles F, Schwarzacher T, Ruiz Rejón C, de la Herrán R, Navajas-Pérez R. The molecular cytogenetic characterization of pistachio (*Pistacia vera* l.) suggests the arrest of recombination in the largest heteropycnotic pai. PLoS One. 2015:**10**(12):e0143861. https://doi.org/10.1371/journal.pone.0143861

Sousa A, Schubert V, Renner SS. Centromere organization and UU/V sex chromosome behavior in a liverwort. Plant J. 2021:**106**(1):133–141. https://doi.org/10.1111/tpj.15150

Storey WB. Genetics of the papaya. J Hered. 1953:**44**(2):70–78. https://doi.org/10.1093/oxfordjournals.jhered.a106358

Tuskan G, Yin T, DiFazio S, Faivre-Rampant P, Gaudet M, Harfouche A, Jorge V, Labbe J, Ranjan P, Sabatti M, et al. The obscure events contributing to the evolution of an incipient sex chromosome in *Populus*: a retrospective working hypothesis. Tree Genet Genomes. 2012:**8**(3):559–571. https://doi.org/10.1007/s11295-012-0495-6

Ueno H, Urasaki N, Natsume S, Yoshida K, Tarora K, Shudo A, Terauchi R, Matsumura H. Genome sequence comparison reveals a candidate gene involved in male–hermaphrodite differentiation in papaya (*Carica papaya*) trees. Mol Genet Genomics. 2015:**290**(2):661–670. https://doi.org/10.1007/s00438-014-0955-9

Underwood C, Choi K, Lambing C, Zhao X, Serra H, Borges F, Simorowski J, Ernst E, Jacob Y, Henderson I, et al. Epigenetic activation of meiotic recombination near *Arabidopsis thaliana* centromeres via loss of H3K9me2 and non-CG DNA methylation. Genome Res. 2018:**28**(4):519–531. https://doi.org/10.1101/gr.227116.117

VanBuren R, Zeng F, Chen C, Zhang J, Wai C, Han J, Aryal R, Gschwend AR, Wang J, Na J, et al. Origin and domestication of papaya Yh chromosome. Genome Res. 2015:**25**(4):524–533. https://doi.org/10.1101/gr.183905.114

Veltsos P, Ridout K, Toups M, González-Martínez S, Muyle A, Emery O, Rastas P, Hudzieczek V, Vyskot B, Marais G, et al. Early sex-chromosome evolution in the diploid dioecious plant *Mercurialis annua*. Genetics. 2019:**212**(3):815–835. https://doi.org/10.1534/genetics.119.302045

Vicoso B, Charlesworth B. Evolution on the X chromosome: unusual patterns and processes. Nature Rev Genet. 2006:**7**(8):645–653. https://doi.org/10.1038/nrg1914

Wang J, Na J, Yu Q, Gschwend AR, Han J, Zeng F, Aryal R, VanBuren R, Murray JE, Zhang W, et al. Sequencing papaya X and Yh chromosomes reveals molecular basis of incipient sex chromosome evolution. Proc Natl Acad Sci U S A. 2012:**109**(34):13710–13715. https://doi.org/10.1073/pnas.1207833109

Weingartner LA, Moore RC. Contrasting patterns of X/Y polymorphism distinguish *Carica papaya* from other sex-chromosome systems. Mol Biol Evol. 2012:**29**(12):3909–3920. https://doi.org/10.1093/molbev/mss196

Westergaard M. The mechanism of sex determination in dioecious plants. Adv Genet. 1958:**9**:217–281. https://doi.org/10.1016/S0065-2660(08)60163-7

Westergaard M. The relation between chromosome constitution and sex in the offspring of triploid Melandrium. Hereditas. 1948:**34**(3):257–279.

Xue L, Wu H, Chen Y, Li X, Hou J, Lu J, Wei S, Dai X, Olson MS, Liu J, et al. Evidences for a role of two Y-specific genes in sex determination in *Populus deltoides*. Nat Commun. 2020:**11**(1):5893. https://doi.org/10.1038/s41467-020-19559-2

Yamato KT, Ishizaki K, Fujisawa M, Okada S, Nakayama S, Fujishita M, Bando H, Yodoya K, Hayashi K, Bando T, et al. Gene organization of the liverwort Y chromosome reveals distinct sex chromosome evolution in a haploid system. Proc Natl Acad Sci U S A. 2007:**104**(150):6472–6477. https://doi.org/10.1073/pnas.0609054104

Yampolsky C, Yampolsky H. Distribution of sex forms in the phanerogamic flora. Biblio Genet. 1922:**3**:1–62.

Yang W, Wang D, Li Y, Zhang Z, Tong S, Li M, Zhang X, Zhang L, Ren L, Ma X, et al. A general model to explain repeated turnovers of sex determination in the Salicaceae. Mol Bio Evol. 2021:**38**(3):968–980. https://doi.org/10.1093/molbev/msaa261

Yue J, Krasovec M, Kazama Y, Zhang X, Xie W, Zhang S, Xu X, Kan B, Ming R, Filatov DA. The origin and evolution of sex chromosomes, revealed by sequencing of the *Silene latifolia* female genome. Curr Biol. 2023:**33**(12):2504–2514.e3. https://doi.org/10.1016/j.cub.2023.05.046

https://doi.org/10.1093/plcell/koad325 THE PLANT CELL 2024: 36: 1257–1311

THE PLANT CELL

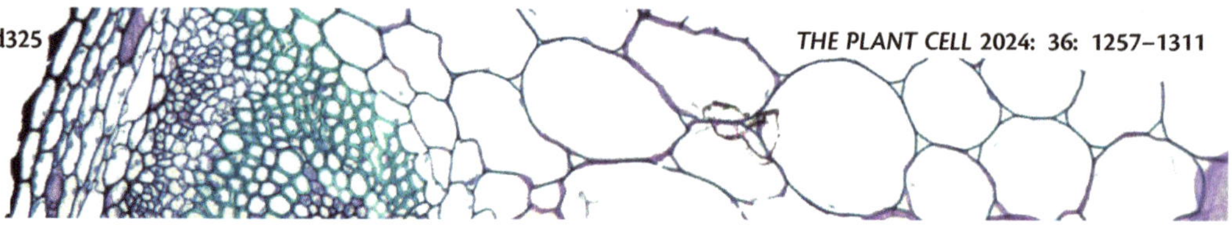

Review

The plant cell wall—dynamic, strong, and adaptable—is a natural shapeshifter

Deborah Delmer [1,*] **Richard A. Dixon** [2,*] **Kenneth Keegstra** [3,*] **and Debra Mohnen** [4,*]

1 Section of Plant Biology, University of California Davis, Davis, CA 95616, USA
2 BioDiscovery Institute and Department of Biological Sciences, University of North Texas, Denton, TX 76203, USA
3 MSU-DOE Plant Research Laboratory, Michigan State University, East Lansing, MI 48823, USA
4 Complex Carbohydrate Research Center and Department of Biochemistry and Molecular Biology, University of Georgia, Athens, GA 30602, USA

*Author for correspondence: dpdelmer@ucdavis.edu (D.D.), richard.dixon@unt.edu (R.A.D.), keegstra@msu.edu (K.K.), dmohnen@ccrc.uga.edu (D.M.)

Abstract

Mythology is replete with good and evil shapeshifters, who, by definition, display great adaptability and assume many different forms—with several even turning themselves into trees. Cell walls certainly fit this definition as they can undergo subtle or dramatic changes in structure, assume many shapes, and perform many functions. In this review, we cover the evolution of knowledge of the structures, biosynthesis, and functions of the 5 major cell wall polymer types that range from deceptively simple to fiendishly complex. Along the way, we recognize some of the colorful historical figures who shaped cell wall research over the past 100 years. The shapeshifter analogy emerges more clearly as we examine the evolving proposals for how cell walls are constructed to allow growth while remaining strong, the complex signaling involved in maintaining cell wall integrity and defense against disease, and the ways cell walls adapt as they progress from birth, through growth to maturation, and in the end, often function long after cell death. We predict the next century of progress will include deciphering cell type–specific wall polymers; regulation at all levels of polymer production, crosslinks, and architecture; and how walls respond to developmental and environmental signals to drive plant success in diverse environments.

Introduction

In 1665, Robert Hooke peered through his primitive microscope at a slice of cork and described little boxes he called "cellula"—rooms that monks inhabited. These "cellula" were dead cells, and all that remained visible to him were their cell walls. For centuries after, much of what we knew about cell walls was derived from what was visible with our own eyes, either viewed directly or through the ever-improving technique of light microscopy. Form helps predict function, and we now know that cell walls play a key role in determining cell, tissue, and organ shapes. Botanists, amazed by the diverse types of cells found in plants, watched the ways in which cells with thin "primary" cell walls (PCWs) first begin to form a new wall at the time of cell division and how cell and overall plant growth is tied to the expansion of such thin walls. Cell diversity is also intimately linked to the way cells mature, building "secondary" cell walls (SCWs) that thicken to form structures that provide strength to stems to with-stand loss of turgor, protect against water loss and diseases, and function in transport of water and minerals from roots to leaves. The overall shape of a plant and its organs, in particular flowers, was found to be one of best predicters of ancestry, and taxonomy became one of the most popular (and controversial) fields for botanists extending well into the early 20th century. Smith (1962) provides an absolutely delightful and colorful history of botany including the hilarious struggles to develop the field of botanical taxonomy.

In this centennial celebration year recognizing the founding of ASPB, we have aimed to provide a historical context to

Received July 17, 2023. Accepted December 19, 2023. Advance access publication February 1, 2024

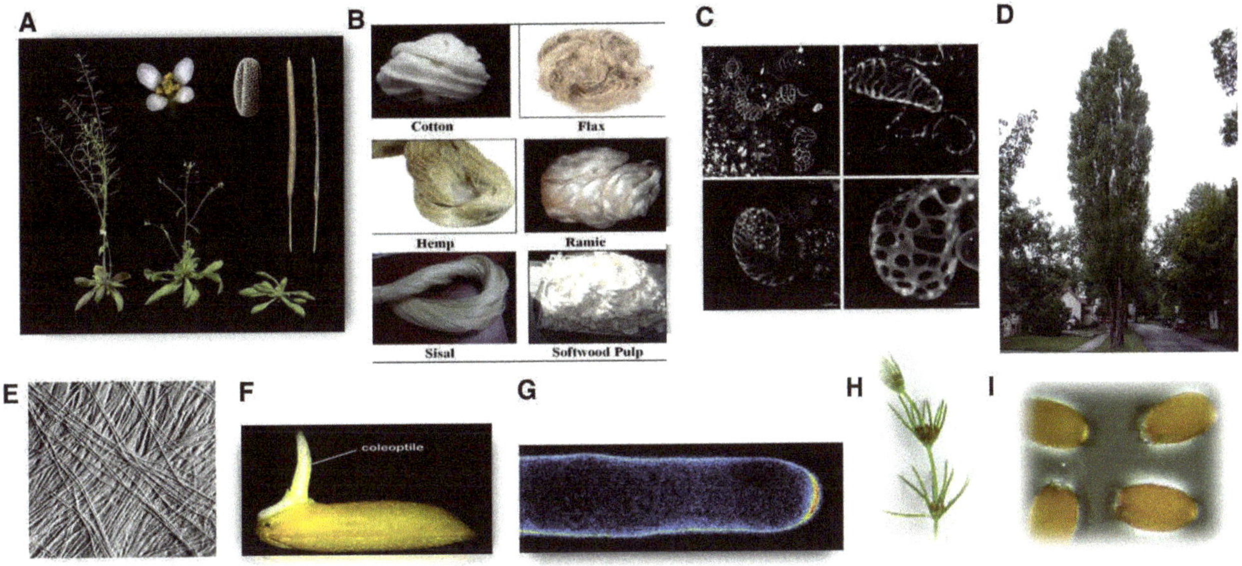

Figure 1. Some model systems that are especially useful for the study of plant cell walls. **A)** *Arabidopsis thaliana*; its small size, short generation time, small genome, and ease of mutant selection make this a favorite for genetic studies (Liepman et al. 2010). **B)** Fiber cells that have SCWs that are cellulose-rich and nearly free—or with much reduced levels—of hemicelluloses and lignin include cotton fibers (actually epidermal hairs), bast fibers (especially from flax and ramie), and fibers of tension wood and are favorites for studying cellulose. Ramie (*Boehemeria nivea*) cellulose is favored by x-ray crystallographers for its clear diffraction patterns; cotton fibers are single cells that show synchronous development within a single boll, and their mature SCWs have >90% cellulose (Haigler et al. 2012). **C)** Suspension-cultured *Zinnia elegans* cells can be induced to differentiate into tracheary elements, making them a good model for study of xylem biogenesis (Demura et al. 2013). **D)** Hybrids of poplar (*Populus*) have become models of choice for trees due to ease of propagation, molecular breeding, transformation, and large genome resources (Tuskan et al. 2004). **E)** Onion epidermal cells (shown here are surface cellulose microfibrils [MFs]) can be easily isolated as a living, hormone-responsive monolayer, especially useful for studying PCW expansion (Suslov et al. 2009; Zhang et al. 2021c). **F)** A popular model for grass walls of very different Type II PCW structure is the maize coleoptile (Carpita 1984). **G)** Pollen tubes of Lily have been used to elucidate the oscillating nature of tip growth (McKenna et al. 2009) and discovery of key signaling cascades (Li et al. 2016a). **H)** Charophycean algae have played a key role in studying the role of pectins in diffusive cell wall elongation (Proseus and Boyer 2012b). **I)** Some mucilage-secreting cells (MSCs) such as those from Arabidopsis, are rich in pectins, whereas others are rich in only one kind of hemicellulose such as xyloglucan, xylan, or glucomannan, making MSCs valuable for studies on biosynthesis and wall polymer interactions (Arsovski et al. 2010). All figures are licensed through the Creative Commons and used with permision.

the process of discovery and to emphasize milestones that made a difference in moving the field forward. The first part of the review presents the progression of learning about the structure, biosynthesis, and functions of the 5 major types of wall polymers: cellulose, hemicelluloses, pectins, wall-associated proteins, and lignin. The second part aims to integrate this knowledge of individual polymers and delve deeper into some of the fascinating challenges faced by plant cell walls. We examine models that have attempted to explain the still-challenging problem of how the polymers found in PCWs can be assembled into a structure strong enough to resist high turgor pressure while allowing expansion to permit growth. Recognizing that walls are not passive structures but constantly communicating and adapting to serve the needs of the cells they protect, we then discuss the rapidly expanding field of cell wall signaling. We end with an examination of the life of the cell wall from birth to survival even beyond the death of their parent cells.

Because each of us has focused most of their career on one of the major wall polymers, we consider this a rare opportunity to integrate our collective knowledge gathered over many years. Aware that new ideas do not always pan out, we also take this opportunity to offer a few new ideas of our own and end with our assessment of major challenges for the future of cell wall research.

Not all cells that build walls do so the same way, so we must emphasize the importance of using a variety of model systems (Fig. 1), *Arabidopsis thaliana*, certainly, but our discussion just might include some of the botanists' favorite very odd-shaped cells.

The 5 major polymers of the plant cell wall

Cellulose

Common claims that cellulose "is the most abundant organic compound on earth" and that taking density differences into consideration, it is "stronger than steel" are actually true (see McNamara et al. 2015). Yet cellulose, the cell wall polymer with the simplest basic structure—one type of sugar joined by 1 type of linkage—is anything but simple given the variety of structures that can be assembled from individual β-1,4-glucan chains. We will see that this feature allows it to interact, not only with itself but with other noncellulosic polysaccharides, and its presence is so important for wall structure that a decrease in cellulose level in the wall sends out distress signals that alert the cell to quickly do whatever possible to correct this situation. Beyond its role in the plant, all these properties make cellulose a key component of timber, textiles, paper, and a major feedstock for chemicals; and the fact that it is difficult to digest influences agriculture and global nutrition.

Structure of cellulose

The early studies. Early botanists often equated the wall with the term cellulose because of its abundance in many cell types. Anselme Payen (1838), "the father of cellulose chemistry," isolated the major sugar of cell walls and was surprised to find that its elemental composition was the same as that of starch. Payen recognized that cellulose could be broken down to glucose but had no idea of its structure nor how the monomers were aggregated to create a polymer so very different from starch. Decades later, the sequential efforts of some towering figures in the emerging fields of sugar and polymer chemistry led to the conclusion that cellulose is a polymer of thousands of glucose residues linked in β-1–4 configuration (Fischer 1902; Haworth et al. 1927, 1931; Staudinger 1932). The concept of a polymer was novel to chemistry at the time, as the length of 1 type of polymer could vary considerably even as the chemical formula stayed the same; but by 1936, the idea became generally accepted (Kanaya and Kaji 2016). Estimates of chain lengths ranged from 100s to many thousands of residues depending on the source and how the cellulose was isolated (Delmer 1983).

The Swiss scientist Albert Frey-Wyssling defined a new field he termed "submicroscopic morphology" that led to several insightful predictions concerning the highly oriented and crystalline properties of cellulose (Frey-Wyssling and Ambronn 1927; Frey-Wyssling 1939). Using the newly intro-duced electron microscope, Frey-Wyssling's predictions were confirmed, and he and Fritz Muhlethaler provided our best early images of cellulose fibrils in native cell walls. One major conclusion was that fibrils of varying widths consisted of smaller fibrils of about 35 Å width, corresponding roughly to a 6×6 array of chains (Mühlethaler 1960; Frey-Wyssling and Mühlethaler 1963). This so-called "elementary fibril" (or microfibril [MF]) is now considered a fundamental unit of cellulose, although the predicted number of chains has been reduced over time. The Federal Institute of Technology Zurich under Frey-Wyssling also pioneered the technique of freeze-fracture (Moor et al. 1961) that led much later to the first images of cellulose synthase com-plexes (CSCs).

It took more than 60 years to answer the key question of whether the chains of native cellulose (cellulose I) were parallel (reducing ends all in the same direction) or anti-parallel (Kroon-Batenburg and Kroon 1997; French 2000; Zugenmaier 2021). Better computer modeling improved analyses of diffraction patterns, and, working independently with the same sample of the alga Valonia native cellulose I, Gardner and Blackwell (1974) and Sarko and Muggli (1974)—and later Woodcock and Sarko (1980) with fibers of ramie (*Boehmeria nivea*)—all concluded that the chains were indeed parallel. This was confirmed by other methods (Hieta et al. 1984; Chanzy and Henrissat 1983), and the parallel chain arrangement for cellulose I is now the accepted model. Cellulose II, produced by alkali treatment of native cellulose, is a more stable form, and the early findings of Meyer and Misch (1937) that the chains are in anti-parallel arrangement were confirmed later by Kolpak and Blackwell (1976) and Buleon and Chanzy (1978).

New technologies provide new insights. It took crystallographers so long to agree on the crystal structure of native cellulose largely because native cellulose can exist in many crystalline forms that show only subtle differences in x-ray diffraction. There has since been a dramatic rise in the number of new technologies that can be applied to study cellulose structure (Jarvis 2013, 2018). According to Jarvis (2018), "Broadly, NMR spectroscopy is most informative about chain conformation; crystallography, about chain packing; Fourier transform infrared spectroscopy, about hydrogen bonding; atomic force microscopy (AFM), about microfibril dimensions; small-angle scattering, about the aggregation of MFs." Using cross-polarization magic angle NMR, Atalla and Vanderhart (1984) demonstrated that native cellulose can exist in at least 2 crystalline forms: Iα, found in native bacterial and algal cellulose, and Iβ, found in tuni-cates and higher plants. Different conformations at C-6 in the crystalline core (Fig. 2A) versus surface MFs (Fig. 2B) can create altered patterns for hydrogen bonding that alter the potential for surface chains to interact with other MFs or other polymers.

The widely cited work of Nishiyama et al. (2002) showed 2 parallel chains in pure Iβ having slightly different conformations, and neutron scattering supported 2 different types of intrachain hydrogen bonding between 1 glucose residue with each of its neighbors that stabilizes a co-planar orientation of glucose. Strikingly, no hint of inter-sheet O—H...O bonds was found, indicating that sheets of cellulose must be held together primarily by van der Waals forces along with some weak C—H...O bonds (see also Notley et al. 2004; McNamara et al. 2015). Thus, the glucan chains have a hydrophilic and a hydrophobic face that allow for very strong interchain interaction, although each individual interaction is weak.

Surprisingly, the cellulose of maize (*Zea mays*) PCWs consists of up to 7 different allomorphs, only 1 of which is almost identical to either Iα or Iβ (Wang et al. 2016b). Some allo-morphs show differences on the surface of the MF and were suggested to be possible targets for interaction with other polymers such as pectins. Others show interior chains

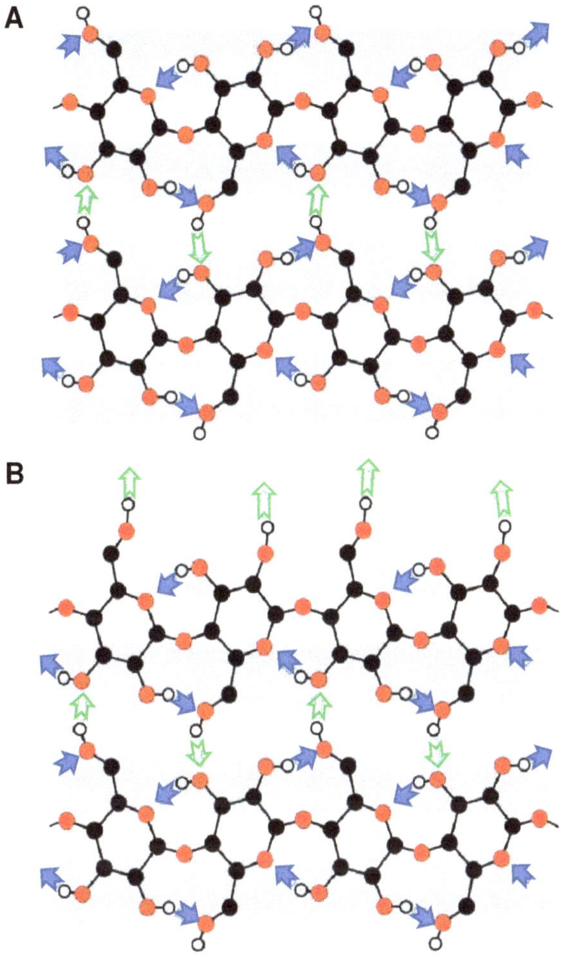

Figure 2. Hydrogen bonding schemes for cellulose. **A)** A 2-chain segment of "crystalline" cellulose with all C-6 in the *tg* conformation, permitting a line of intramolecular hydrogen bonds (shaded arrows) that run along each side of each cellulose chain (lightly shaded horizontal bands). Intermolecular hydrogen bonds are shown as open arrows. **B)** Two-chain segment of "surface" cellulose with the edge of the upper chain having C-6 in the *gt* conformation, so that the line of intramolecular hydrogen bonding is interrupted and there is an increased number of transversely oriented, intermolecular hydrogen bond sites (top row of open arrows) that allows for interaction with water, other MFs or other polymers. Modified from Jarvis (2018) by permission of the author.

associated with surface chains, whereas another is embedded in the core of the MF. Yet another is predicted to be on the surface but poorly hydrated and was suggested as a possible target for expansins (see below). How the surfaces of MFs interact with water, with themselves to create larger bundles or with other wall polymers, is critical to overall wall structure and mode of expansion (see below) and is still an ongoing field of investigation (Cosgrove 2022; Jarvis 2023).

Biosynthesis of cellulose

It is hard to document all the progress that is being made in this field, so it is fortunate that many details have been well-reviewed by others (Wallace and Somerville 2014; McNamara et al. 2015; Turner and Kumar 2018; Wilson et al. 2021). In particular, we leave detailed discussion of CESA (cellulose synthase) trafficking, some aspects of regulation, and the evolution of the genes/proteins involved to others (Haigler and Roberts 2019; Lampugnani et al. 2019; Polko and Kieber 2019). Hopefully our historical perspective provides examples that might inform future work.

The early years. Luis Leloir, an Argentine biochemist, physician, and pioneer of carbohydrate biosynthesis, made several truly milestone discoveries. Key was the discovery in the 1950s of nucleoside diphosphate sugars and recognition that the free energy of hydrolysis releasing the sugar was sufficient for them to serve as donors of sugars for polysaccharide synthesis. Discovery of UDP-Glc in 1950 was only part of many that led to the Nobel Prize (see Leloir 1983). UDP-Glc is a key substrate in polysaccharide biosynthesis, either directly for cellulose, callose (β-1,3-glucan), and hemicellulosic glucans or indirectly as precursor to other NDP-sugars that are themselves substrates for synthesis of noncellulosic polysaccharides (Bar-Peled and O'Neill 2011).

The obligate aerobic bacterium *Acetobacter xylinum*, lacking flagellae, secretes cellulose in order to float and gain oxygen in liquid environments (Hestrin 1962). Glaser (1958) showed that it uses UDP-Glc as substrate for cellulose synthesis in vitro, albeit at quite low rates. Moshe Benziman's group later achieved high rates in vitro through the important discovery of cyclic-di-GMP as a potent activator (Aloni et al. 1982; Ross et al. 1987). One gene (*BcsA*) in a 4-gene operon encodes the catalytic subunit (Saxena et al. 1990); the other genes (Saxena et al. 1994; Römling and Galperin 2015) encode proteins that assist in binding of c-di-GMP to BcsA (*BcsB*), create a pore for MF extrusion through the outer membrane (*BcsC*), or have endoglucanase activity (*BcsD*) that, along with other proteins (Abidi et al. 2022), are important for determining crystallinity of the product.

Another prominent biochemist, Zev Hassid, believed in 1957 that it would only be a short time before his laboratory showed how cellulose was synthesized in plants. He set 2 of his students to work using mung beans expecting to find that UDP-Glc was substrate. The students returned with news that the product was not cellulose but callose (β-1-3-glucan). Hassid could not believe it, and the poor students had to repeat their results several more times before it was published (Feingold et al. 1958; the students, Liz Neufeld and David Feingold, became very well-known polysaccharide biochemists in spite of abandoning cellulose synthesis as a target!). This frustration was repeated by many other laboratories, finding that the bulk of the product in vitro was all or mostly callose (Robinson and Quader 1981; Delmer 1983). Callose synthesis in vitro generally requires 2 activators; micromolar levels of Ca^{2+} and any 1 of a variety of β-glucosides (Hayashi et al. 1987), and assays for cellulose synthase often inadvertently contained both because cellobiose was often supplied and micromolar Ca^{2+} is present in plant extracts if not specifically chelated. Vincent Bulone stands out as one who often reported successful in vitro synthesis of cellulose and deserves much credit for providing protocols for assays of glycosyltransferases (GTs) in plants (Brown et al. 2012). Callose synthases are very stable enzymes in contrast to the quite labile plant CSCs. More recently, Oehme et al. (2019) and Purushotham et al. (2022) showed that only minor changes in amino acid sequences and conformation of glycosyltransferase-2 (GT2) enzymes as classified by the Carbohydrate-Active Enzyme Database (CAZy) (Cantarel et al. 2009; Drula et al. 2022) can lead to β-1,3, as opposed to β-1,4-linkages. One wonders why plants, having evolved CSLF and CSLH enzymes (also GT2 proteins, see Hemicellulose section) that can make glucan backbones with both linkages, have never evolved a callose synthase analogous to the GT2 bacterial β-1,3-glucan (curdlan) synthase (Oehme et al. 2019). Instead, plant callose synthases evolved from yeast/fungi and are quite different in structure (Schneider et al. 2016).

Early contributions from microscopy. Preston (1964)—another force in the world of cell wall research—speculated that a plant CSC is a complex of catalytic subunits in which the number of subunits dictates the number of chains produced in a single MF. Malcolm Brown—a genius at capturing images of the process of cellulose synthesis—showed pictures of *A. xylinum* extruding cellulose from a linear array of pores (Brown et al. 1976). Other freeze-fracture images supported Preston's vision, showing huge multi-subunit "terminal complexes" (TCs) at the ends of very large growing MFs in algae like Oocystis and Valonia (Montezinos and Brown 1976; Itoh and Brown 1984; Brown 1996). Surprisingly, some of these TCs were aligned next to each other and were clearly moving in opposite directions, creating an arrangement that could lead to more stable antiparallel interactions between the MFs; this bi-directional mobility was also later seen for plant CSCs (Paredez et al. 2006). Arabidopsis mutants defective in phosphorylation of the complex lose this property, which affects MF bundling and stem phenotype (Chen et al. 2016). This raises the question of whether antiparallel regions of MF bundling might be the "hot-spots" for expansin function proposed by Cosgrove (2022) (see la-ter). Jarvis (2023) comments that it is not known to what extent MFs bundle in either the parallel or more stable antiparallel orientations, but the issue would certainly seem to deserve further study (Makarem et al. 2020). For plants, a more modest complex was observed in the plasma membrane of maize (Mueller and Brown 1980) and tracheary elements of cress (Herth 1985) showing 6-fold symmetry and was termed a rosette (probably by Andrew Staehelin according to Nick Carpita). These generally appear as solitary struc-tures in plants, and Giddings et al. (1980) demonstrated remarkable arrays of up to 175 rosettes in the plasma membrane of the alga Micrasterias.

CesA genes in plants. The first plant *CESA* gene was not identified until 1996 (Pear et al. 1996). In cotton (*Gossypium hirsutum*) fibers, the rate of cellulose synthesis increases more than 100-fold during the transition from PCW to SCW (Meinert and Delmer 1977), and random cDNA sequencing of a transition stage cDNA library found 2 very similar cDNAs containing scattered sequences characteristic of processive GT2 enzymes and other regions showing strong sequence similarity to the transmembrane helices (TMHs) of the BcsA protein (Pear et al. 1996). But the cotton genes also possessed an N-terminal domain (NTD) encoding a ring-finger domain, a highly plant-conserved region (PCR) and another region of variable sequence (HVR, later called CSR for class-specific region). The latter 2 insertions split the conserved sequences required for catalytic activity, thus explaining why the genes had not been discovered earlier. The cotton genes were strongly induced at the onset of SCW synthesis (Pear et al. 1996), but no genetic evidence was presented to confirm that they encoded cellulose synthases.

Chris Somerville and Herman Hofte had the vision to identify Arabidopsis (Fig. 1) as a model plant in which specific mutations in cellulose synthesis could be easily selected. Williamson's laboratory in Australia isolated a temperature-sensitive mutant of Arabidopsis termed *rsw1* impaired in the synthesis of PCW cellulose (Arioli et al. 1998). It quickly became evident that the *RSW1* locus had a very close relationship to that of the cotton fiber genes and encoded what we now refer to as *AtCESA1* that functions in synthesis of PCW cellulose. And so...off to the races! Richmond and Somerville (2000) showed that Arabidopsis has 10 *CESA* genes. Also identified were a number of related cellulose-synthase-like (*CSL*) families, some of which function in hemicellulose synthesis (see below). The *CSLD* family members are closest to the *CESAs* (Doblin et al. 2002), and reports increasingly suggest that at least some CSLD proteins serve an alternate function to the CESAs for synthesis of cellulose (see below). CSLD3 was recently verified to possess β-1,4-glucan synthase activity and is most similar in structure to CESA6 (Yang et al. 2020b).

A complete CSC in vivo involves more than just CESA proteins (Lampugnani et al. 2019). An endoglucanase, KORRIGAN, analogous to BcsD, may play a role in cleaving disordered chains during synthesis (Nicol et al. 1998; Vain et al. 2014), although its role remains enigmatic because a mutation in the predicted catalytic site still complemented the *kor* mutant phenotype (Lei et al. 2014). Also unclear is the exact function of COBRA, a glycosylphosphatidylinositol (GPI)-anchored protein that is apparently associated with the CSC and affects crystallinity (Schindelman et al. 2001; Roudier et al. 2005). A better understanding of the functions of KOR and COBRA is needed. Furthermore, many scientists seem to have forgotten that freeze-fracture studies of rosettes showed protein globules on the opposite face of the PM that matched in position the central core of the rosettes on the opposite membrane face (Giddings et al. 1980; Mueller and Brown 1980), suggesting the central globule as 1 possible location for accessory proteins. CESAs may also be regulated through phosphorylation and S-acylation that may affect activity, rosette structure, and/or recycling (Kumar et al. 2016; Polko and Kieber 2019).

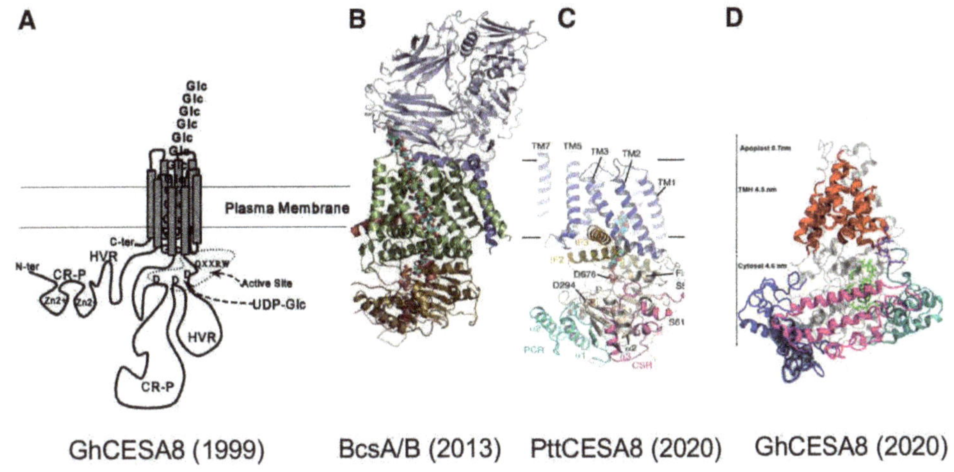

Figure 3. Comparison of models of cellulose synthase catalytic subunits. **A)** Cartoon model taken from Delmer (1999) shows how the active site in plants is recreated to resemble that of BcsA by looping out of CSR (HVR) and PCR (CR-P) domains. **B)** BcsA/B structure determined by X-ray diffraction from its crystal structure (Morgan et al. 2013). **C** and **D)** PttCESA8 modeled from cryo-electron microscopy of reconstituted recombinant protein (Purushotham et al. 2020) and cotton fiber GhCESA8 (original name GhCESA-1) computationally modeled from the sequence (Singh et al. 2020). All models possess an elongated shape with a central catalytic domain (brown in BcsA; wheat in PttCESA8; grey and green in GhCESA) and a channel comprised of TMHs (green, blue, red, moving L to R) through which the growing glucan chain is extruded through the inner/plasma membrane. The PCR (aqua in PttCESA8; magenta in GhCESA8), CSR domains (magenta in poplar; aqua in cotton), and NTD (shown in blue only in GhCESA8) are plant specific and not found in BcsA. The BcsB domain associated with BcsA (blue gray) is not found in plants.

Comparisons of bacterial and plant CESA structures. Study of *CESA* mutants of Arabidopsis has led to 2 surprising but important conclusions: the rosette responsible for synthesis of the fundamental MF is comprised of 2 classes defined largely, but not exclusively, by the homology of their CSRs; and each class contains 3 necessary but non-identical subunits, usually (but not always) in 1:1:1 ratio. Class I proteins (CESA1, CESA3, and any one of CESA 2, 5, 6, or 9) are used primarily for PCW synthesis, whereas Class 2 (CESA 4, 7, and 8) are necessary for SCW synthesis (Turner and Kumar 2018). For clarity, the names of all plant homologs of the Arabidopsis CESAs in other plants have been changed in this review to match those used for Arabidopsis.

A significant milestone came when Jochen Zimmer's group crystallized a catalytically active bacterial BcsA-BcsB protein complex and determined its structure to 3.25-A resolution (Morgan et al. 2013; McNamara et al. 2015); a cello-oligosaccharide remained in the structure during crystallization. In this proposed structure, BcsA contains 4 TMHs in the amino-terminal region and 4 in the carboxy-terminal region, separated by a cytoplasmic intracellular loop that forms the catalytic domain that is protected by a cyclic-di-GMP–regulated gating loop that controls access to the active site (Morgan et al. 2014). The channel formed by TMHs 3–8 accommodates the translocating glucan. The overall structure resembles other membrane-integrated, processive GT2s such as hyaluronic acid synthase and chitin synthase (Bi et al. 2015).

Significant advances have been made in producing catalytically active recombinant CESAs; PpCESA5 from moss (*Physcomitrella patens*) and PttCESA8 from hybrid poplar (*Populus tremula* × *tremuloides*) synthesized very thin β-1,4-glucan fibrils in vitro in the absence of added primer (Purushotham et al. 2016; Cho et al. 2017). When isolated from insect cells, PttCESA8 reconstituted into structures that largely ranged from monomer to trimer, and a homotri-mer fraction was size-selected and analyzed by cryo-electron microscopy (Purushotham et al. 2020). In Fig. 3, a model of the proposed monomer of PttCESA8 is compared with that of other proposed catalytic subunit structures, including BcsA, and, for historical purposes, the original crude model (Delmer 1999) that was based on simple domain structure analyses.

Although differing in many details, all models show that the TMHs form transmembrane channels, and the catalytic domain of plant CESAs is reconstructed by looping out into the cytoplasm of the PCR and CSR domains that are pro-posed to function as sites for CESA-CESA interactions and/or necessary accessory proteins. Possible modes of regulation of the plant gating loop are under study (Olek et al. 2023; Verma et al. 2023). Binding of c-di-GMP regulates catalytic activity through the gating loop in bacteria but not plants (Morgan et al. 2014), and this loop is predicted to collapse into the region of the active site when the CSC is inactive in order to maintain PM membrane potential. Older work by Bacic and Delmer (1981) and Delmer et al. (1982) indicated that glucan synthesis in membrane vesicles of cotton or intact *A. xylinum* is inhibited by collapsing the delta psi (electric component), but not delta, pH component of the membrane potential and perhaps offers some clues in this regard.

Initiation of synthesis without added primer (see above) may put to rest the suggestion that sitosterol glucoside can serve as the primer for cellulose synthesis in cotton fibers

(Peng et al. 2002). Cotton fibers clearly make sterol-di-tri- and tetra-cello-oligosaccharides, but attempts to demonstrate their initiator activity in Arabidopsis have failed. However, it does appear that CSCs function in sterol-rich portions of the plasma membrane (Schrick et al. 2012; Turner and Kumar 2018), another promising area for further study.

MF structure predicts rosette structure (or vice versa)? Current thinking considers 2 different models for the way in which the 3 different CESAs required for synthesis of PCW or SCW assemble into rosettes. One model proposes that each of 2 of the 6 lobes are comprised of CESA homotrimers that alternate positions within the rosette; alternatively, each of the 6 lobes is assembled into identical heterotrimers, wherein each CESA in the lobe has a unique role. As of this writing, there is no clear answer, but homotrimers may be more likely because recombinant SCW CESA8 (and more recently CESA7; Zhang et al. 2021b) can so easily form active homotrimers, although no complete rosette has yet been assembled. Confusingly, recombinant rice (*Oryza sativa*) CESA8 CatD domain can show clear redox-dependent dimerization (Olek et al. 2014), whereas Vandavasi et al. (2016) obtained trimers from a similar recombinant AtCESA1. Analysis of the PCR crystal structure (Rushton et al. 2017; Zhang et al. 2021b) fitted well with the trimeric structure of PttCESA8s proposed by Purushotham et al. (2020). However, studies of the CSR domain by Scavuzzo-Duggan et al. (2018) and Olek et al. (2023) suggest dimerization through key cysteine residues. Kurek et al. (2002) showed redox-dependent formation of either homo- or heterodimers of the ring fingers of cotton fiber CESAs, and the NTD domain is clearly essential for activity, although it has not proven possible to model trimers with dimerized ring fingers.

The answers may lie in the way trimers are assembled into rosettes. Wilson et al. (2021) further examined the idea that redox-dependent CSR dimerization is one stabilizing link between trimers. Examining their models for a dimer of trimers, one can see several possibilities where ring-finger dimers could also form and act as stabilizing trimer-trimer associations, a possibility that deserves further consideration. Dimers clearly predominate in decomposing CSCs from some cells such as cotton fibers (Kurek et al. 2002; Wen et al. 2022), and it was reported that when GhCESA7 was knocked out, the CSC degraded to GhCESA4–8 dimers, while knocking out GhCESA8 led to GhCESA4–7 dimers (Wen et al. 2022). CESAs 4, 7, and 8 can dimerize with each other in a redox-dependent manner, and this interaction was proposed to occur before assembly into a complete rosette (Altanassov et al. 2009). Whether one gets dimers or trimers in decomposing rosettes may depend on the relative stability of the redox-dependent linkages found in trimers vs possible dimerizations that link trimers into rosettes.

The current conclusion is that the basic PCW MF is usually comp rised of 18 chains (Jarvis 2013; Newman et al. 2013; Nixon et al. 2016; Ye et al. 2018; Haigler and Roberts 2019; Song et al. 2020). Models suggest assembly into MFs might be based on 5 layers of chains in a 34,443 arrangement (Kubicki et al. 2018) or 6 layers of 234,432 chains (Song et al. 2020). However, a 24-chain model has been convincingly proposed for spruce (Fernandes et al. 2011), and recent work strongly supports this model for other wood SCW MFs as well (Tai et al. 2023). Haigler and Roberts (2019) have written an excellent analysis of the evolution of rosettes and TCs and the relationship between structure of the MFs and the rosettes.

The role of cortical microtubules in orientation of MFs. Perhaps the most exciting contribution from microscopy that opened a whole new approach to studying cellulose synthesis is the early work of Paredez et al. (2006). A DNA construct encoding AtCESA6 was fused to a fluorescent protein and was active when transformed into a mutant *cesA6* background, and mobile CSC complexes were visualized in hypocotyl epidermal cells. They moved at a constant speed corresponding to an addition of 300 to 1,000 Glc residues per minute and displayed a short residence time in the PM of 7 to 10 min—roughly the amount of time to synthesize 1 PCW glucan chain. Movement was often bidirectional and paralleled the orientation of aligned cortical microtubules beneath the PM. One can only imagine how the early pioneers of cytoskeletal biology would have reacted to actually watching the process of MT-aligned MF synthesis in real time. Although the residence time of PCW CesAs is short, other work indicates that the CesA proteins are quite stable, suggesting they may be recycled (Hill et al. 2018). In contrast, SCW CesAs of cotton fibers showed half-lives of less than 30 minutes (Jacob-Wilk et al. 2006).

The residence time of a CSC was calculated to exceed that of the underlying dynamic MTs, and it was observed often that the MT would disappear, but the CSC would continue its forward motion in the same direction, confirming that CSC movement is not assisted by some MT-motor protein mechanism but rather by the force of polymerization of the MFs. Later, Chan and Coen (2020) studied CSC movement after MT disappearance and confirmed that CSCs continued in the same direction for the residence time of the CSC, after which new CSCs appeared that followed the same trails as previous ones; this pattern continued apparently guided by the orientation of the nascent MFs in the wall. This kind of behavior has been seen in other situations in vivo; for example, the pattern of cellulose deposition during xylem vessel development is initially determined by MT orientation but persists long after MT disruption (Roberts et al. 2004; Schneider et al. 2017; Watanabe et al. 2018). Such observations have occasionally been used as an argument that MTs may not be so important for directing CSC motion. However, the MT-directed path takes preference because, when a CSC encounters an MT, the CSC preferentially begins following the new MT orientation. These studies clearly support the notion that the primary mechanism for

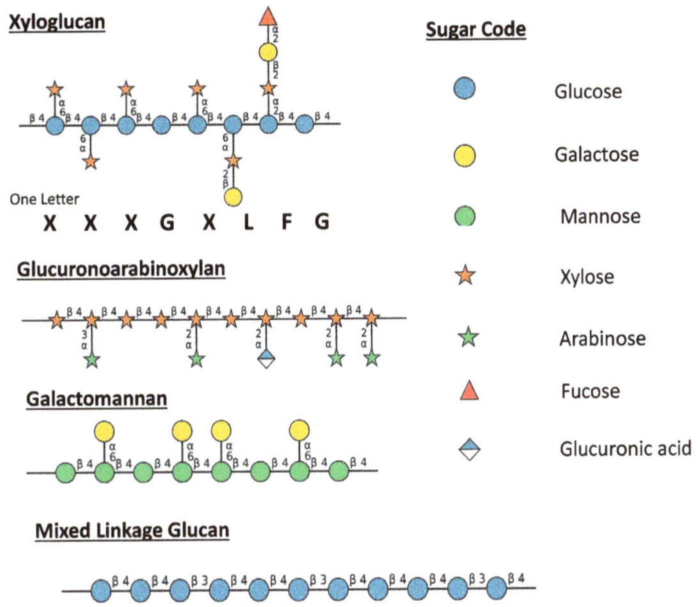

Figure 4. Representative structures of hemicelluloses. The xyloglucan fragment is typical of dicot PCWs. The galactomannan is typical of those found in legume seeds. The glucuronoxyloarabinan is typical of those found in grass PCWs. The structures were drawn using software described by Cheng et al. (2017).

orientation of MFs, a major event in the anisotropic elongation of plant cells, is controlled by the orientation of cortical MTs. But what controls reorientation of the MTs? One intriguing finding is that the localized de-methylesterification of pectins in the wall predicts and precedes the pattern adopted by the MTs (Peaucelle et al. 2015). Other evidence indicates that actin microfilaments interact with MTs and may play a special role in targeting of CSC and MF orientation (especially in SCW synthesis) (Lei et al. 2012). MT orientation may occur by sensing the direction of tensile stress (Hamant et al. 2019), and recent work indicates that the tethering of CSCs to MTs disturbs this stress-induced MT reorganization (Schneider et al. 2022).

The approach conceived in 2006 to follow rosette movement in vivo has also revealed important information on CSC trafficking (Lampugnani et al. 2019). CSCs assemble in the Golgi and are transported to the PM, where they become tethered to cortical MTs through binding of the CESA NTD to POM-POM2/Cellulose Synthase Interacting proteins (Gu et al. 2010; Bringmann et al. 2012). Once activated, the CSCs stall after 7 to 10 min and then are endocytosed. CESAs then transiently accumulate in small CESA compartments or are degraded in the lytic vacuole. While active, the forward movement of the CSC might be expected to displace the MTs from their PM location, but this has been shown to be prevented by cortical MT uncoupling proteins that regulate this interaction (Liu et al. 2016). Other older data may still be relevant: the cellulose synthesis inhibitor 2,6-dichlorobenzonitrile (DCB) somehow may act through disruption of MT organization, and an 18- to 20-kD MT-associated protein (MAP20) binds an active photoaffinity analog of DCB (Delmer et al 1987; Rajangam et al. 2008), consistent with a complex interaction between synthesis of cellulose and orientation of MTs (Schneider et al. 2017). However, genetic evidence that MAP20 is the target of DCB action in vivo is still lacking. Also fascinating is the finding that although kinesins do not control CSC movement, they are involved in movement on MTs of noncellulosic polysaccharide cargos from the Golgi to the PM, and 1 specific kinesin, referred to as Fragile Fiber1, has recently been shown to interact with cortical MT uncoupling proteins. This interaction is proposed to facilitate the localized deposition of matrix polymers in proximity to the sites where cellulose is deposited, representing a potentially important way to coordinate the deposition of cellulose with that of noncellulosic polymers (Ganguly et al. 2020).

Hemicelluloses

Structures of hemicelluloses

Hemicellulosic polysaccharides consist of a diverse collection of different polymers including xyloglucans (XyG), xylans, mannans, and mixed-linkage glucans (MLGs) (Fig. 4). Providing a precise definition of a hemicellulose is difficult. The common feature is that they bind to cellulose. Most have a backbone with β-1,4-linked sugars, but as the name implies, MLG also has β-1,3-linkages in the backbone. The backbones of most hemicelluloses have additional sugars and/or acetyl groups attached, but again MLG is the exception (Fig. 4).

When considering matrix polysaccharides, both hemicelluloses and pectins, it is important to remember that the structures depicted in figures are a representation of many different molecules. Polysaccharides are different from nucleic acids and proteins, where the sequence of monomers is determined by the genetic code. In contrast, the structure of polysaccharides is determined by the specificity of the enzymes that synthesize them, and not all individual molecules in a collection of polysaccharides have the same structure. Another difference is that sugars have multiple hydroxyl residues where a neighboring sugar, or an acetyl residue, can be attached. Thus, determining the structure of a polysaccharide requires complex chemical and analytical methods.

Xyloglucans. Among hemicelluloses, XyGs have the most complex structures (Fig. 4). Like cellulose, they have a backbone of β-1,4–linked glucosyl residues but with α-linked xylosyl residues attached in a regular pattern to the 6-position of most glucosyl residues (Fig. 4). Many xylose residues are further substituted, most frequently with galactosyl or galactosyl plus fucose residues. A shorthand notation for the substitution pattern of each location along the chain uses a single letter designation (Fig. 4) (Fry et al. 1993). Other substitution patterns are found in XyGs from various plant species, with the patterns sometimes varying among different tissues within a species (Hayashi 1989; Scheller and Ulvskov 2010; Schultink et al. 2014; Pauly and Keegstra 2016). XyG is the most abundant hemicellulose in the PCWs of gymnosperms, dicots, and noncommelinid monocots. Plants with XyG as the most abundant hemicellulose are said to have Type I primary walls (Carpita and Gibeaut 1993).

Xylans. Xylans have a pentose-based backbone of β-1,4–linked xylosyl residues linked in a way that allows binding to cellulose. The backbone is substituted with α-linked arabinose residues, α-linked glucuronic acid resides, and variable numbers of acetyl esters (Fig. 4). Xylans, as opposed to XyGs, are the most abundant hemicellulose in the PCWs of grasses; these walls are categorized as Type II (Carpita and Gibeaut 1993). The distribution of the side-chain residues varies depending on the plant species, the plant tissue, and the developmental stage of the tissue (Scheller and Ulvskov 2010). In addition, xylans from gymnosperms and selected angios-perms have a unique tetrasaccharide located at the reducing end of the polysaccharide, although the function of this sequence of 4 sugars is unknown (Smith et al. 2017). The distribution and patterns of substitution along the xylan backbone has functional significance. For example, the backbone substitution patterns of SCW xylans impact their interactions with cellulose (Busse-Wicher et al. 2016; Grantham et al. 2017). More recently, Tryfona et al. (2023) demonstrated that grass cell walls have xylans with at least 3 different substitution patterns and suggest that these patterns influence the way in which xylans interact with other wall components. For example, an arabinoxylan with evenly distributed arabinofuranosyl residues and lacking glucuronic acid facilitates binding to cellulose while other substitution patterns may facilitate interactions with other wall components such as lignin (Tryfona et al. 2023). Xylans are the most abundant hemicellulose in the SCWs of all angiosperms. Because of their abundance in SCWs of woody plants, there are massive quantities of xylans in the biosphere; it has been estimated that 10 billion tons of carbon are incorporated into xylans annually (Smith et al. 2017).

Mannans. Mannans are taxonomically the most widely distributed and evolutionarily the most ancient of the hemicelluloses (Voiniciuc 2022). They contain either a β-1,4–linked mannan backbone or a backbone consisting of both β-1,4–linked mannose and glucose. Some have been found to have alternating mannose and glucose in the backbone (Yu et al. 2022). The backbone sugars are substituted with α-linked galactose attached to the 6-position of backbone mannose (Fig. 4). Some mannans have very few or no side chains, whereas other galactomannans are heavily substituted with galactose, thereby producing polysaccharides with very different physical properties. For example, the un-substituted mannans in ivory nut (from *Phytelephas* species) are insoluble and can be used as carving material for artwork. In some algae, unsubstituted mannans can substitute for cellulose (Scheller and Ulvskov 2010). On the other hand, highly substituted galactomannans are water soluble and can produce viscous solutions with food product and industrial applications (Sharma et al. 2022). Mannans are generally very minor components of PCWs and SCWs of most angiosperms but are frequently the most abundant hemicellulose in the SCWs of gymnosperms.

Mixed-linked glucans (MLGs). MLG has a simple structure with no side chain residues (Fig. 4). It is less widely distributed in the plant kingdom than other hemicelluloses, being found mainly in the primary cell walls of the grasses plus in the cell walls of a few lower plants and some fungi (Burton and Fincher 2009). The β-1,3 linkages create kinks in an otherwise straight polymer so that the frequency and spatial distribu-tion of β-1,3 linkages determines the physical properties of the polymers (Burton and Fincher 2009).

Hemicelluloses as storage compounds in plant seeds

Grant Reid and his colleagues significantly contributed to our understanding of plants that use hemicelluloses as reserve polysaccharides in their seeds. In his 1985 review, he pointed out that studies began late in the 19th century, when "...it was clearly recognized by botanists that the massively thickened cell walls present in many seeds contained reserve substances" (Reid 1985). Different plant species utilize different polysaccharides as storage reserves; for example, tamarind (*Tamarindus indica*), nasturtium (*Tropaeolum majus*), and many other species store XyG as a reserve polysaccharide. On the other hand, fenugreek (*Trigonella foenum-graecum*), coffee (*Coffea arabica*), and many other species store polymers in the mannan family, especially galactomannans that have received significant attention because of their industrial and medical applications (Sharma et al. 2022). Their unique rheological properties cause them to have wide use in the food, pharmaceutical, and cosmetic industries, as well as in hydraulic fracturing fluid (Sharma et al. 2022). *Dendrobium catenatum*, containing large quantities of glucomannan, has been extensively studied for centuries and utilized as both food and a Chinese herbal medicine (Qi et al. 2022).

MLG is also found as a reserve polysaccharide in the seeds of cereal grains (Morrall and Briggs 1978). Use of MLG as a reserve polysaccharide has been studied in detail in the model grass *Brachypodium distachyon*, where it is present in thick endosperm cell walls and is mobilized during seedling germination (Francin-Allami et al. 2023). Although xylans are also found in seeds of some plants, the arabinoxylan from the seed husks of *Plantago ovata* has received the most attention (Belorio and Gomez 2022). This psyllium mucilage consists of a highly branched β-1,4–linked xylan backbone (Fischer et al. 2004) and assists *P. ovata* seeds with hydration and gemination. It has been widely used as fiber in the human diet.

As methods for investigating polysaccharide structure were developed in the middle of the 20th century, it was recognized that storage polysaccharides, primarily from seeds (Kooiman 1957) but also from other organs, were similar in structure to hemicelluloses from plant cell walls. For example, XyG from seeds (Kooiman 1967), extracellular polysaccharides of suspension-cultured sycamore (*Acer pseudoplatanus*) cells, and PCWs (Becker et al. 1964; Aspinall et al. 1969; Bauer et al. 1973) all had similar structures.

Because seed reserve and mucilage polysaccharides can be isolated more easily and using milder conditions, they have provided valuable insight into the structure of the various hemicelluloses. As described below, these seed systems have been exploited to identify the genes and proteins involved in the biosynthesis of cell wall polysaccharides. Because the presence of wall polysaccharides as reserve polymers has arisen independently several times during evolution, analysis of the seed systems provides an opportunity to understand how polysaccharide synthesis is regulated. For example, the entire pathway from sucrose to galactomannan in fenugreek is upregulated during seed formation (Wang et al. 2012). Such systems should provide opportunities to identify transcription factors controlling this process.

Biosynthesis of hemicelluloses

The biosynthesis of matrix polysaccharides, both hemicelluloses and pectins, takes place in the Golgi. It involves a series of complex processes that generally require a different biosynthetic enzyme for every different linkage found in each polymer (Lerouxel et al. 2006). Results over the last 2 decades have revealed that plant cells employ 3 different strategies for the synthesis of matrix polysaccharide backbones. The first is exemplified by MLG that is synthesized by CSLFs, proteins that are closely related to CESA and CSLD proteins (Yin et al. 2014; Little et al. 2018). CSLF is a glycan synthase (Purushotham et al. 2022) that operates similarly to the CESA proteins described earlier. The second strategy is illu-strated by mannans and XyG, which are synthesized by CSLA and CSLC, respectively (Dhugga et al. 2004; Cocuron et al. 2007). These proteins are also glycan synthases but are distant relatives of CESA (Yin et al. 2014; Little et al. 2018). They also differ from the CESA, CSLD, and CSLF proteins in that they operate in concert with GTs that add side chains to the growing backbone (Pauly and Keegstra 2016; Zabotina et al. 2021; Voiniciuc 2022). The third strategy is fundamentally different in that the backbone of xylans and pectic polysaccharides are synthesized by complexes of glycosyltransferase that operate in the lumen of the Golgi (Ye and Zhong 2022; Anders et al. 2023). In these cases, complexes of GTs assemble the backbone (Engle et al. 2022; Ye and Zhong 2022; Anders et al. 2023) while different GTs add the side chain residues when present (Smith et al. 2017; Ye and Zhong 2022).

Over the past 2 decades, investigators have exploited the systems that produce large quantities of individual polysaccharides described above to identify the genes and proteins responsible for their synthesis in a manner analogous to the studies with cotton that led to the identification of the CESA proteins (Pear et al. 1996). Dhugga and colleagues (2004) used developing guar (*Cyamopsis tetragonoloba*) seeds that store large quantities of galactomannan to identify the *CSLA* genes responsible for the synthesis of the mannan backbone. Cocuron et al (2007) used developing nasturtium (*Tropaeolum* spp.) seeds that produce large quantities of XyG to identify the *CSLC* genes responsible for XyG backbone synthesis. For the synthesis of the backbone of rhamnogalactur-onan I (RG-I) (see section on pectins below), Takenaka et al (2018) investigated mutant Arabidopsis plants that had defects in mucilage production to identify a rhamnosyltransfer-ase required for RG-I synthesis. This strategy has also been used to identify the GTs required for the addition of side chain sugar residues (Edwards et al. 1999; Jensen et al. 2012, 2013).

Because GTs are mostly type II membrane proteins where the active site is in the Golgi lumen, they utilize nucleotide sugars present inside the lumen. Thus, Golgi-localized sugar nucleotide transporters are required during polysaccharide synthesis. As noted earlier, the various NDP sugars are made from UDP-Glc via a complex set of pathways present in the cytoplasm or the Golgi membranes (Bar-Peled and O'Neill 2011). On the other hand, glycan synthases, including most CSL proteins, have multiple membrane spans; these enzymes accept sugars from nucleotide sugars in the cytosol and, during synthesis, transport the nascent polymer to the lumen of the Golgi, where GTs add side chain sugars as needed. The newly synthesized polysaccharides move from the Golgi lumen to the cell surface via membrane vesicles, which fuse with the plasma membrane, thereby releasing the polymers into the wall matrix. However, very little is known about how the Golgi-synthesized polysaccharides are assembled into a functional cell wall. A summary of the biosynthesis of each type of hemicellulose is presented below.

Mannans. Soon after the discovery of mannan synthase (Dhugga et al. 2004), Liepman et al. (2005) demonstrated that several members of the Arabidopsis *CSLA* gene family had mannan or glucomannan synthase activity when expressed in cultured insect cells. Subsequently, *CSLA* gene family members from other species were shown to be responsible for mannan biosynthesis, and it has been postulated that all *CSLA* genes encode mannan synthase (Liepman and Cavalier 2012; Voiniciuc 2022). Of note is that GDP-Man and GDP-Glc, as opposed to UDP-Glc, are used for backbone synthesis.

Xyloglucan. Expression of cDNA clones from both nasturtium and Arabidopsis *CSLC* genes in *Pichia pastoris* resulted in the cells producing β-1,4–linked glucan (Cocuron et al. 2007), leading to the conclusion that both likely encode the glucan synthase that makes the XyG backbone. Because *Pichia* does not produce UDP-xyl, it was not possible to confirm these glucan synthases were responsible for XyG biosynthesis, although several lines of evidence supported this conclusion. Later, Kim et al. (2020) demonstrated that a quintuple mutant with disruptions in all 5 Arabidopsis *CSLC* genes has undetectable XyG levels, thereby providing compelling evidence that *CSLC* genes are responsible for XyG biosynthesis. The phenotypes of these and other mutants lacking XyG are described below.

The biosynthesis of mannans and XyGs is similar in many ways (Liepman and Cavalier 2012). With respect to backbone synthesis, the *CSLA* and *CSLC* gene families have sequence similarities and a common evolutionary origin while being evolutionarily distinct from other subgroups in the CESA superfamily (Yin et al. 2014; Little et al. 2018). Little et al. (2018) suggest this lack of connection reflects "...the likely dual endosymbiotic origin of the superfamily" with *CSLA*/*CSLC* coming from 1 event, whereas *CESA*, *CSLD*, and *CSLF* came from a different event. Side chains attached to the 6-position of each polymer are added by enzymes present in CAZy family GT34 that have both sequence and biochemical similarities (Edwards et al. 1999; Faik et al. 2002; Cavalier and Keegstra 2006). Yu et al. (2022) have elaborated on the similarities in the structure and biosynthesis of the 2 polymers, with evidence that they may have similar functions in that mannans may be able to substitute for XyG in mutants lacking XyG.

As noted above, XyG has a complex array of side chain substitutions. Work in the last 2 decades has produced considerable knowledge about the genes and proteins responsible for the addition of these side chains. Because this work is well documented in several recent reviews (Scheller and Ulvskov 2010; Schultink et al. 2014; Pauly and Keegstra 2016; Julian and Zabotina 2022), here we briefly highlight the important role played by genetics, both forward and reverse, in identifying not only the genes and proteins responsible for side chain addition but also the functions of the polysaccharides and their side chains.

The story begins in Chris Somerville's laboratory, where several Arabidopsis mutants with alterations in cell wall com-position were identified (Reiter et al. 1997). One of them, *mur2*, a mutant with reduced levels of fucose, led to identi-fication of the gene responsible for the addition of the fucose side chains on XyG (Perrin et al. 1999; Vanzin et al. 2002). A second, the *mur3* mutant, was later shown to encode a galac-tosyltransferase that adds galactose to XyG at the location where fucose is attached (Fig. 4) (Madson et al. 2003; Kong et al. 2015). Reverse genetics was also used to isolate mutant plants with disruption of the genes encoding XyG xylosyl-transfereases (XXT) (Cavalier et al. 2008; Zabotina et al. 2012). As described in more detail in a later section, these mutant lines also aided in evaluating the function of XyG.

Mixed-linked glucans (MLGs). Early studies on the biosynthesis of MLG focused on confirming that it was made by a unique enzyme. As noted above, early studies on cellulose biosynthesis identified enzyme preparations that made both β-1,3 and β-1,4 linkages. The question was whether MLG was made by the combined action of 2 different enzymes or a unique enzyme that made both linkages. Bruce Stone was a memorable contributor to plant cell wall biochemistry. Both funny and feisty, he possessed encyclopedic knowledge of β-glucans in plants and managed to exhaust 2 sets of authors who worked with him writing 2 volumes on the subject. Bruce and his colleagues resolved the question of 1 or 2 enzymes by identifying and partially characterizing an enzyme activity that produced MLG (Smith and Stone 1973; Henry and Stone 1982a, 1982b). Many years later, Stone was coauthor of a manuscript reporting the identifica-tion of the *CSLF* genes and proteins responsible for MLG bio-synthesis (Burton et al. 2006). Doblin et al. (2009) provided evidence that the grass-specific *CSLH* genes also encode pro-teins capable of MLG synthesis. Both the *CSLF* and *CSLH* genes are more closely related to *CESA* genes than to *CLSA* and *CSLC* genes (Yin et al. 2014; Little et al. 2018). Although many questions remain about the details of MLG biosynthesis, recent in vitro studies using the barley (*Hordeum vulgare*) CSLF protein have demonstrated that 1 enzyme can synthesize both linkages found in MLG. Molecular modeling and mutagenesis identified a region of the protein near the cytoplasmic side of the glucan channel that controls the frequency of β-1,3 links in the polymer product (Purushotham et al. 2022).

Xylan. As noted above, xylan biosynthesis is very different from the biosynthesis of the other hemicelluloses. Most of the genes encoding xylan synthase proteins were identified through forward genetic screens as mutations that altered xylem function, called *irregular xylem* or *irx* mutants. The xylan backbone is synthesized by a complex of GTs and related proteins that either have a single transmembrane domain or that lack a transmembrane domain. The proteins encoded by these genes belong to families GT47 and GT43 in the CAZy classification scheme. Xylan backbone synthesis requires the cooperation of 3 different nonredundant proteins that make up the xylan synthase complex. They include IRX10, or related proteins such as IRX10L, belonging to CAZy family GT47 plus IRX9 and IRX14, or related proteins belonging to CAZy family GT43 (Smith et al. 2017; Ye and Zhong 2022; Anders et al. 2023). The exact role of each of these proteins in the xylan synthase complex is still not completely resolved. The emerging consensus is that IRX10 and related proteins have xylosyltransferase activity (Urbanowicz et al. 2014; Jensen et al. 2018) and provide the active site that links together the xylosyl resides in the backbone. Most, or maybe all, IRX10 and related proteins lack a transmembrane do-main. They are thought to associate with the IRX9 and/or IRX14 proteins, which are type II integral Golgi membrane proteins with a single transmembrane domain. It is not yet clear whether the family GT43 proteins have xylosyltransferase activity or whether they have other functions in the xylan synthase complex. Various members of these protein families are utilized depending on the plant species, the plant tissue, and whether primary or secondary wall xylan is being synthesized (Smith et al. 2017; Ye and Zhong 2022; Anders et al. 2023). Anders et al. (2023) postulate that different members of the gene families are involved in xylan synthesis in PCWs than those in SCWs, similar to the situation with the *CESA* family members.

In addition to the various proteins involved in backbone synthesis, Golgi membranes contain several different GTs involved in adding the side chains found on xylans. These include glucuronosyltransferases in CAZy family GT8 (Mortimer et al. 2010; Oikawa et al. 2010) and arabinosyltransferases in GT61 (Ye and Zhong 2022). Because xylans, es-pecially those in SCWs, are heavily acetylated, Golgi membranes also contain several different enzymes, including acetyltransferases, involved in xylan acetyl esterification as well as acetyl esterification of other matrix polysaccharides (Smith et al. 2017; Pauly and Ramirez 2018; Julian and Zabotina 2022), including pectins as described in the next section.

Pectins

Structures of pectins

History of pectin and its complexity. Pectin (*pectique*), from the Greek πηκτικοί (to coagulate), was first used by Henri Braconnot to describe the acidic viscous substances with gelling properties isolated from multiple tissues of more than 15 plant species (Braconnot 1825a, 1825b; Wisniak 2007). Braconnot noted that the ability of a very small amount of pectin to gelatinize a large mass of sugar water provided it industrial uses such as in the confectionary business (Braconnot 1825b; Wisniak 2007). These properties led to the establishment of the pectin industry in the early 1900s (Kertesz 1951), and numerous studies on the chemical basis of the gelling properties of pectin from commercial sources such as citrus and apple fruit followed. Braconnot also recognized that pectin's ability to bind and remove toxic metals imparted medicinal properties (Braconnot 1825a), and increasing numbers of industrial, food, and medical applications for pectin have since been explored, including its use in edible coatings on foods, antimicrobial bio-based films, nanoparticles, and as healing agents and cancer treatments (Willats et al. 2006; Freitas et al. 2021). The diverse functional characteristics of pectins led Braconnot to propose that a substance so "universally distributed in plants" "must play a role of great importance in the plant and should be studied by plant physiologists" (Braconnot 1825b, p178). Since then, plant biologists and chemists have confirmed multiple functions for this most structurally complex of the plant cell wall glycans, including roles in plant growth and cell expansion (Proseus and Boyer 2012a, 2012b; Daher et al. 2018; Haas et al. 2020), cell shape (Amsbury et al. 2016; Haas et al. 2020; Lin et al. 2022b), organ size and texture (Zhang et al. 2021a), cell-to-cell adhesion (Marry et al. 2006; Yang et al. 2020a), cell wall porosity (Baron-Epel et al. 1988; Fleischer et al. 1999), cell wall structure (Pérez García 2011), signaling (Feng et al. 2018; Lin et al. 2022b; Du et al. 2022), fruit ripening (De Vries et al. 1981; Paniagua et al. 2014), organ abscission (Daher and Braybrook 2015), pollen development (Mollet et al. 2013), and plant defense (Bacete et al. 2018; Molina et al. 2021; Liu et al. 2023). How does a single class of cell wall glycan provide such diverse functions? This question continues to beset plant biologists.

Understanding pectins' multiple functions requires knowledge of (1) their primary, secondary, tertiary, and quaternary structure (Diener et al. 2019); (2) covalent and noncovalent interactions between individual pectic polymers and between pectins and other wall polymers; (3) pectins' roles in cell type-specific wall architecture; and (4) the specific pectic polymeric structure associated with each functional response. In no case is this information currently known. While we know a great deal about the different general types of pectic glycans present in cell wall extracts from different tissues (reviewed by Mohnen et al. 2024), we lack informa-tion about the number and structure of pectic polymers in different specific cell types and how they are integrated into cell wall architecture. Furthermore, pectins that are synthesized and inserted into the wall are subsequently modified during development and in response to biotic and abiotic stress by large families of pectin-modifying en-zymes, including pectin hydrolases (Yang et al. 2018), lyases (Molina-Hidalgo et al. 2013; Leng et al. 2017), methylesterases (Willats et al. 2006; Wolf et al. 2009), acetylesterases (Philippe et al. 2017), and proteins that modify or inhibit these en-zymes (Federici et al. 2006; Hocq et al. 2017b). Importantly, the different classes of enzymes include proteins that uniquely target the different pectic backbones homogalacturonan (HG) and rhamnogalacturonan (see below).

The complexity of pectin structure. All pectic polymers have 1 or 2 backbones: HG and/or rhamnogalacturonan. HG is a

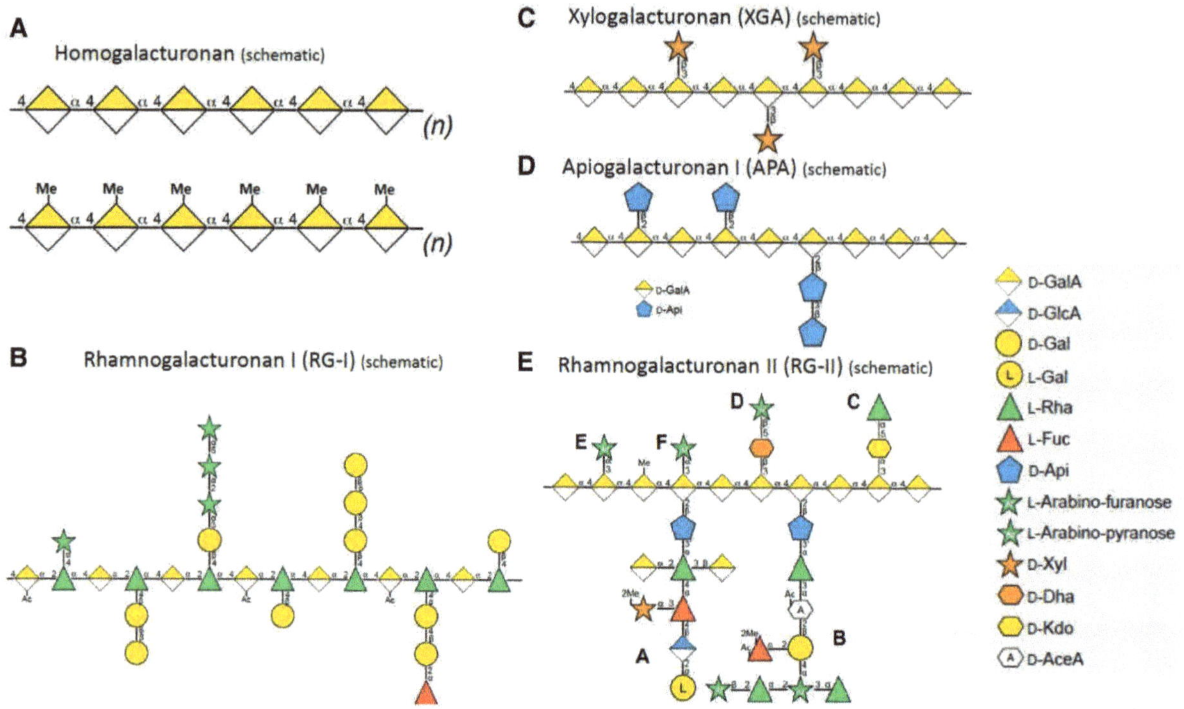

Figure 5. Glycan symbol structure representations of the main pectic glycan domains. **A)** HG. **B)** RG-I. **C)** XGA. **D)** APA. **E)** RG-II. The structures were drawn using Symbol Nomenclature for Glycans from (SNFG) https://www.ncbi.nlm.nih.gov/glycans/snfg.html#nomn.

homopolymer of α-D-1,4–linked galacturonic acid that may have low, partial, or almost full methyl-esterification of the C-6 carboxyl (Bosch and Hepler 2005) (Fig. 5A). It has been proposed that in vivo HG is synthesized in a highly methyl-esterified form; however, the extent and mechanism of esterification of HG synthesized by the different HG biosynthetic enzymes in vivo remain to be determined. HG biosynthetic GALACTURONOSYLTRANSFERASES (GAUTs) can synthesize polymeric HG with no methyl-esterification in vitro (Amos et al. 2018; Engle et al. 2022), indicating that HG methyl-esterification is not a requirement for enzymatic synthesis of HG in vitro (reviewed by Mohnen et al. 2024). Yet, many biological effects are associated with the methyl-esterification status of pectins, and immunolabeling studies indicate that many cell walls have high or changing levels of HG methylester content during development (Wolf et al. 2009).

Each GalA residue in HG has a carboxylic acid group at the C6 position that will be highly ionized and negatively charged at pH above 3.5. Thus, polymeric HG is a polyanion and can interact with positive ions such as calcium or positive patches on proteins. Methylesterification of the carboxyl group by pectin methyltransferases (PMTs) reduces the charge and increases the hydrophobicity of each GalA residue and of methylesterified regions of HG. De-esterification of HG has been associated with changes in wall structure as regions of non-esterified HG can interact to yield HG-Ca^{++}-HG crosslinks and such cross-linking can affect the structure and conformation of the HG polymer and of the pectin network in the wall. An understanding of HG structure during development will require knowledge of the extent and pattern of methyl (and acetyl) esterification in different HG-containing polymers as they are inserted into the wall and subsequent modification of HG once it is deposited in the wall. The degree and pattern of HG esterification and HG size in the wall is dependent on cell type, developmental stage, and status of biotic or abiotic stress because these affect the expression of HG degradative and modifying enzymes. These HG-modifying enzymes include pectin methylesterases (PMEs) (Pelloux et al. 2007), PME inhibitors (PMEIs) (Coculo and Lionetti 2022), acetyl esterases (Sénéchal et al. 2014), and polygalacturonases (Kim et al. 2006), which hydrolyze the methyl/acetyl esters and/or the HG backbone, respectively (Amsbury et al. 2016; Liu et al. 2023). Changes in HG methyl esterification and/or HG size are associated with numerous HG functions, including cell: cell adhesion (Kohorn et al. 2021a), cell elongation/shape/ growth (Daher et al. 2018; Haas et al. 2020), and fruit ripening (Paniagua et al. 2014).

There is no definitive evidence for pure HG as a separate polymer in plant cell walls, although pure HG is synthesized in vitro (Fig. 6A) (Sterling et al. 2006; Atmodjo et al. 2011; Amos et al. 2018; Engle et al. 2022). Rather, HG is generally isolated covalently attached to RG-I (Fig. 6B) or to the substi-tuted HGs rhamnogalacturonan II (RG-II) (Fig. 6E), xylogalacturonan (XGA) (Fig. 6C), and apiogalacturonan (Fig. 5D). All substituted HGs have an HG backbone and the requirement for endopolygalacturonase (EPGase) digestion of cell walls to

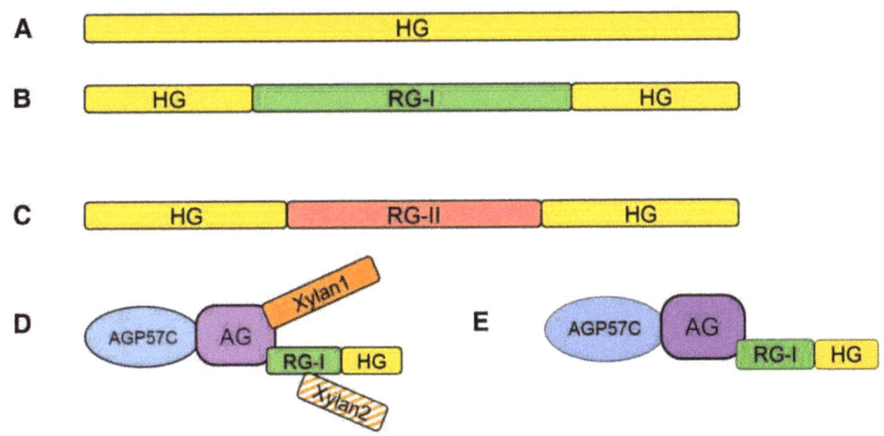

Figure 6. Examples of structurally confirmed homoglycans, heteroglycans, and glycoconjugates that contain the different pectic glycan regions. **A)** HG. **B)** The heteroglycan HG-RG-I-HG. **C)** The heteroglycan HG-RG-II-HG. **D)** The proteoglycan pectic AGP APAP1. **E)** The proteoglycan pectic AGP AGP-RG-I.

release RG-II and RG-I from the walls has often been interpreted as evidence that RG-II and RG-I exist within an HG polymer and possibly as 1 covalently interconnected heteropolymer in the wall. However, this has not been structurally confirmed, and the architecture of pectins and the different pectic heteroglycans and glycoconjugates in the wall remains a critical question.

RG-II (Fig. 5E) is the most structurally complex of the pectic polymers, with 13 different sugars (Darvill et al. 1978; Kobayashi et al. 1996; Bar-Peled et al. 2012; Ndeh et al. 2017) organized into 6 highly conserved side branches on an HG backbone. Why do plants spend so much energy to make this complex glycan? At least part of the answer is likely the structural role it played during the transition of plants from an aqueous environment onto land (Matsunaga et al. 2004), including wall strength-ening via RG-II dimers formed by borate diester complex formation between the side chain A apiosyl residues (Kobayashi et al. 1996; O'Neill et al. 1996; Begum and Fry 2022). Indeed, the dwarf (Reuhs et al. 2004) and cell wall structural (Shi et al. 2017) phenotypes of RG-II mutants support a necessary role for RG-II in cell wall architecture and cell/plant growth (O'Neill et al. 2001).

RG-I is the only pectic polymer not built on an HG backbone (Fig. 6B). RG-I has a GalA-Rha disaccharide repeat backbone $[-\alpha-1,4\text{-D-GalA-}\alpha-1,2\text{-L-Rha-}]_{(n)}$ with about one-half of the rhamnose backbone residues substituted by single or branched galactose, arabinose, or other sugars (Lau et al. 1985) (Fig. 5B). The degree and type of branching is tissue and developmental stage specific (Kaczmarska et al. 2022, 2023). For example, Arabidopsis seed mucilage RG-I is unbranched (Fabrissin et al. 2019; McGee et al. 2021), whereas RG-I in pea cotyledons is initially arabinosylated and acquires significant β-1,4-galactan side branching only in late development (McCartney et al. 2000). The precise order and location of the side branches is not known for any of the pectic polymers, a point that should be kept in mind when viewing schematic representations of pectic heteroglycan structures.

Conformation of pectins. Structure dictates function. Beyond pectins' primary structures (which are already very complex), their secondary (e.g. degree and type of helix formation), tertiary (single molecule 3D structure), and quaternary (polymer-polymer association) structures lead to their vast diversity of biological and industrial functions. Yet, our understanding of these higher levels of pectin structure is unclear (Zdunek et al. 2021). In the mid-1930s, Bonner concluded—based on X-ray diffraction and viscometry studies—that pectins existed as long chains (Bonner 1936). It was known that pectins isolated from plants ranged from high to low methyl esterification content and that the amount of free carboxyl groups affected pectins' colloidal properties (Bonner 1936). Subsequent x-ray diffraction analysis of isolated methyl esterified and de-esterified pectate in the absence and presence of calcium revealed a fibrillar 3_1 (3 residues per turn) helical quaternary conformation for various forms of HG, including sodium pectate, pectic acid, methylesterified HG (pectinic acid), and calcium salt– bridged HG (Walkinshaw and Arnott 1981a, 1981b). Shortly thereafter, Rees and colleagues presented evidence for a 2_1 conformation (2 residues per turn) of calcium pectate dimers, naming this conformation the egg box structure (Morris et al. 1982). More recent modeling suggests further conformational complexity (Braccini and Pérez 2001; Pérez et al. 2003; Braccini et al 2005), and the conformation of HG in vivo remains a matter of debate. Regarding the substituted HG RG-II, Pérez and colleagues provided evidence for a compact, flat, disc-like RG-II monomer structure with compaction of each monomer during dimerization yielding RG-II dimer conformations akin to 2 flattened, parallel disks stacked on top of each other (Pérez et al. 2003). Recent studies of RG-I backbone oligosaccharides indicated a 3_1 right-

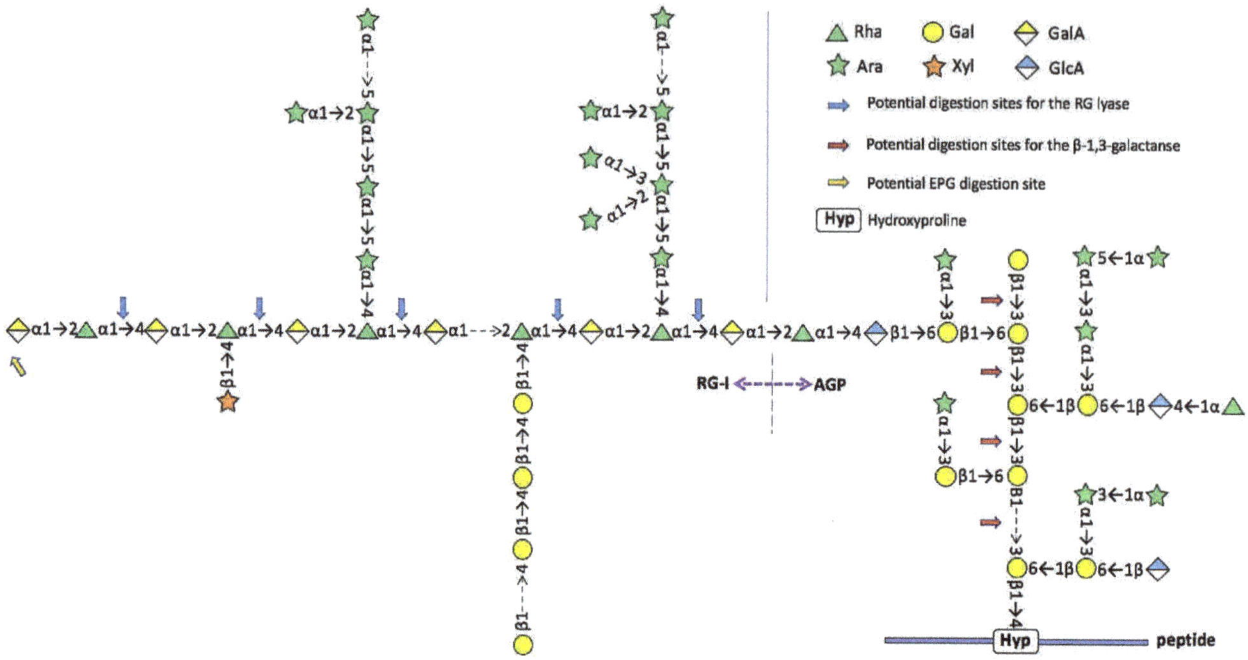

Figure 7. Glycan symbol structure representation of pectic AGP isolated from cell walls of Arabidopsis suspension culture cells. Pectic AGP RG-I-AGP from suspension culture cell walls (from Tan et al. 2023b). The structures were drawn using Symbol Nomenclature for Glycans from (SNFG) as described in Fig. 5.

handed helical structure with 2 backbone disaccharide repeats per turn (Scanlan et al. 2022). Clearly, future studies to clarify the in vivo and in vitro conformation(s) of pectins are critical to delineate pectin structure/function relationships.

Numerous reports over many years have indicated that pectins may be attached to proteins or hemicelluloses (reviewed in Mohnen et al. 2024). Selvendran (1985) presented evidence for pectin-hydroxyproline and serine-rich protein complexes in cell walls of parenchymatous tissues from multiple species. Later, Mort and colleagues reported pectin-protein cross-links in cotton that appeared to be between RG-I and extensin (Qi et al. 1995). The identification of a covalent connection between the arabinogalactan domain of AGP57C and the backbone of RG-I in the pectic AGP APAP1 (arabinoxylan pectin arabinogalactan protein 1) isolated from the medium of Arabidopsis suspension cultures (Tan et al. 2013) provided a biochemical confirmation of a covalent linkage between pectins and protein. APAP1 had short xylan chains linked to the arabinogalactan domain and to Rha in the RG-I backbone. The RG-I backbone was elongated on the nonreducing end with HG (Fig. 6, D and E). Importantly, in a recent study of RG-I isolated from the cell walls of Arabidopsis suspension cells, a pectic AGP named RG-I-AGP was identified as the most abundant form of RG-I present in those walls (Tan et al. 2023b). Because the pectic AGP was isolated by the classical EPGase digestion of the cell walls followed by size exclusion chromatography, the results indicated that the bulk of classically defined RG-I in the walls was covalently linked to protein (Tan et al. 2023b) (Fig. 7). The RG-I-AGP was similar to APAP1; however, it did not have covalently linked xylan chains but rather single xylose residues attached to the RG-I backbone. Efforts to delineate whether all RG-I is synthesized as pectic AGP and, if so, how much is connected to the plasma membrane via GPI-anchoring of the AGP protein core versus how much is synthesized as a free heteroglycan not attached to protein [e.g. existing between HG domains in an HG-RG-HG glycan (Nakamura et al. 2002)] is critical to understanding pectin structure and cell wall architecture. It is noteworthy that the negatively charged family of pectic heteroglycans and glycoconjugates have structural and functional analogies to animal glycosaminoglycans and proteoglycans (Varki et al. 2022).

Biosynthesis of pectins

HG is synthesized by the CAZy GT8 GAUT family in Arabidopsis with GAUTs 1, 4, 10, 11, 13, 14, and possibly more functioning as catalytic HG biosynthetic GAUTs and 3 (GAUTs 5, 6, 7) functioning as Golgi-anchoring subunits of GAUT1 in GAUT1 protein complexes (Lund et al. 2020; Engle et al. 2022). All catalytic GAUTs synthesize polymeric HG in vitro by acceptor-dependent mechanisms, but some such as GAUTs 13, 14, and 1 can also de novo synthesize HG using only UDP-GalA as both the donor and acceptor substrate (Sterling et al. 2006; Atmodjo et al. 2011; Amos et al. 2018; Engle et al. 2022). The role of GAUT4 in wood formation in trees and secondary wall formation in grasses was probed due to its high expression levels in switchgrass (*Panicun virgatum*) stems and during transition from primary wall to wood formation in poplar (*Populus trichocarpa*). Downregulation of GAUT4 in switchgrass, poplar, and rice resulted in greatly reduced amounts of HG and RG-II (Biswal et al. 2018), indicating that in herbaceous dicots, grasses, and trees, GAUT4 may synthesize the bulk of HG and the HG backbone for RG-II. These results provide support for the existence of heteroglycan HG-RG-II-HG glycans (Fig. 6C). Interestingly, there was no reduction in RG-I in these plants, indicating that the synthesis of RG-I is not dependent on the synthesis of the bulk of HG, nor on the synthesis of RG-II. Interrogation of GAUT4 activity in a species with a single GAUT4 homolog, such as Arabidopsis, is challenging due to apparent lethality upon loss of gene function (Caffall et al. 2009). However, expansion of the GAUT family in poplar and switchgrass to 2 and 6 GAUT4 homologs, respectively, allowed knockdown expression of the most highly expressed GAUT4 homolog and analysis of phenotypes. The increase in cell size in the GAUT4 knockdown plants (Biswal et al. 2018) supports proposed roles for the pectin cross-linked matrix in plant cell growth and expansion (Proseus and Boyer 2012a, 2012b; Haas et al. 2020, 2021) (see Models of Cell Wall Organization section below).

Progress has been made in identifying some enzymes involved in synthesis of the substituted HGs. Xylogalacturonan Deficient 1 was identified as a GT47 xylosyltransferase that adds β-Xyl onto the 3-position of GalA in the HG backbone (Jensen et al. 2008), providing tools to explore the biological significance of XGA substitution within the HG backbone. Although originally believed to be most prevalent in fruits, XGA has been identified in many cell types (Zandleven et al. 2007). Immunological evidence suggests it may be associated with plant cell separation and detachment (Willats et al. 2004); however, a complete characterization of the epitope recognized by the antibody will be required to substantiate the functional interpretation of the immunolabeling results, a caution associated with many immunolabeling results. Little is known about the synthesis of the most universal substituted HG, RG-II. Three xylosyltransferases that add α-xylose to the 3-position of L-Fuc (RGXT) (GT77) in RG-II chain A have been identified (Egelund et al. 2006, 2008), but genes encoding the other RG-II biosynthetic enzymes remain to be identified.

The methylesterification state of HG dramatically affects its in vivo and industrial properties, with the pattern and degree of esterification affecting pectins' gelling properties, ability to form pectate calcium ionic crosslinks, and sensitivity to enzymatic cleavage by polygalacturonases (Willats et al. 2006; Wolf et al. 2009). Correspondingly, considerable effort has gone into searches for the methyltransferases that esterify the HG backbone. However, although many mutants affected in methyl ester content have been identified, only with the recent biochemical confirmation of the HG methyltransferase activity of *Quasimodo2* has the door been opened to understanding this crucial process (Du et al. 2020a). Critical experiments for the future include determining whether methyltransferases function in complexes with HG biosynthetic enzymes and the mechanisms for HG methyl esterification levels and patterning during synthesis.

The identification of the biosynthetic RG-I backbone rhamnosyltransferase (RRT) (GT106) (Takenaka et al. 2018) and galacturonosyltransferase (RGGAT) (GT116) (Amos et al. 2022) families that function together to synthesize the RG-I rhamnogalacturonan backbone (Fig. 6B) opens the door to in vitro synthesis of complex pectic heteroglycans and pectic AGP glycoconjugates such as APAP1 (Tan et al. 2013) and RG-I-AGP (Tan et al. 2023b) (Fig. 7D), as well as to efforts to reconstitute the galactose- and arabinose-rich side chains.

Complex RG-I synthesis will require identification of the enzymes that initiate side branch formation on the backbone, enzymes that thus far have remained elusive. The identification of the β-1,4 galactan synthases (GALS) (GT92) that extend the galactose side branches (Liwanag et al. 2012; Ebert et al. 2018; Laursen et al. 2018) was an important step forward and provided catalysts for study of type II galactan synthesis and structure. Interestingly, GALS1 was demonstrated to be a bifunctional enzyme that also can add arabinose to β-1,4-galactan chains, an activity hypothesized to function in chain termination (Laursen et al. 2018; Prabhakar et al. 2023). The 3D crystal structure of GalS1 from *P. trichocarpa* revealed a modular protein structure with an N-terminal CBM95 RG-I backbone binding domain, C-terminal GT92 GT-A fold catalytic domain, and stem region involved in homodimer formation (Prabhakar et al. 2023). Mutagenesis informed by comparison of GT-A fold sequences to the GT92 family identified a novel histidine as the catalytic base and other amino acids involved in β-1,4-galactan syn-thesis and acceptor binding. The approaches used in this study, including docking with molecular dynamics simula-tions, exemplify structure-function and acceptor specificity tools that may accelerate future progress on identifying and characterizing the enzymes involved in plant cell wall synthesis.

The combination of genetic loss-of-function phenotype with enzyme activity determined by in vitro heterologous ex-pression of encoded protein sequence is the most definitive evidence for biochemical function of cell wall GTs (Amos and Mohnen 2019). Two putative arabinosyltransferases (ARAD1 and ARAD2) (GT47) that may add α−1,5-linked Ara onto RG-I side branches have been identified via studies of the *ara-binan deficient 1* mutant and molecular complementation (Harholt et al. 2006, 2012). Although biochemical confirm-ation of enzymatic activity has not yet been provided, these enzymes provide tools to attempt in vitro reconstitution of complex RG-I via coexpression with RG-I backbone synthesizing enzymes.

The identification of pectin biosynthetic genes has enabled the generation of cell wall glycan–modified mutant/variants and their probing using nanoscale imaging to test the hypothesis that pectins interact with cellulose and that

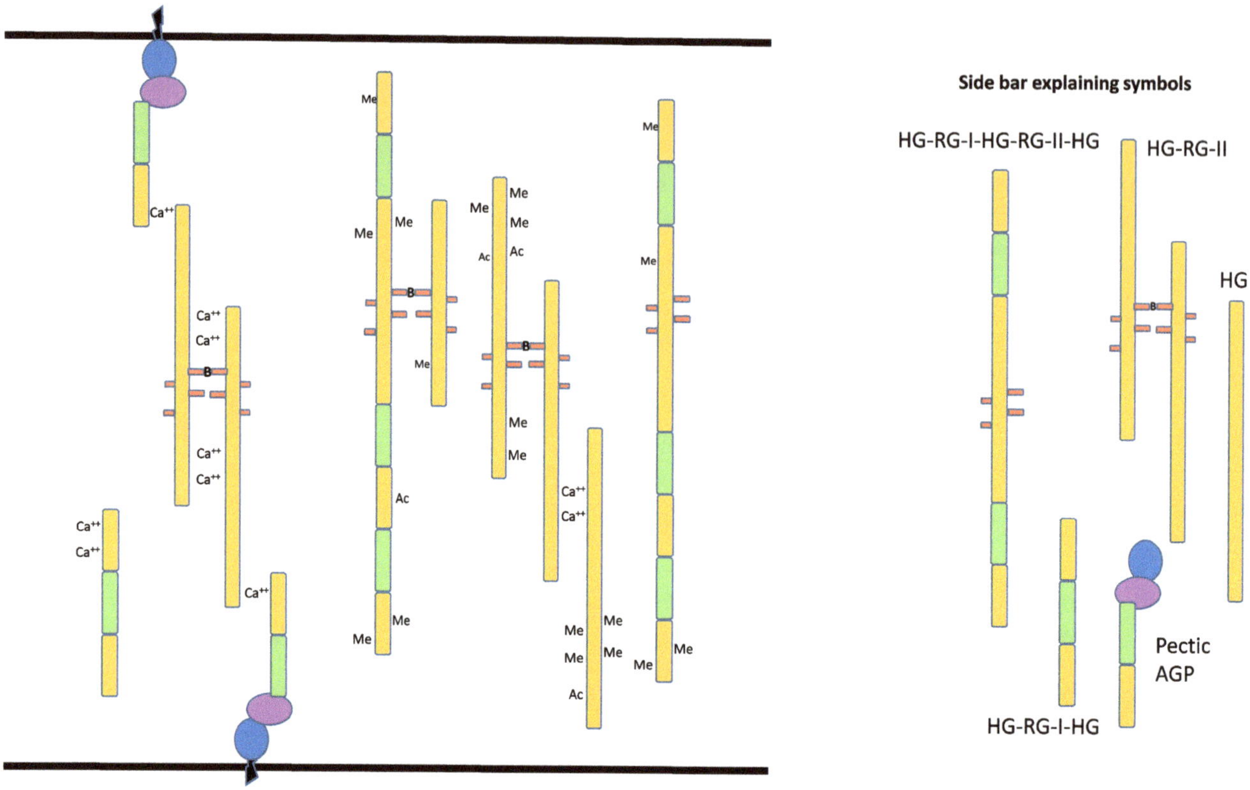

Figure 8. Pectic heteroglycan and glycoconjugate model. In this testable hypothetical model, pectin in the primary wall is proposed to consist of 5 different pectic polymeric structures: a small amount of HG, the pectic heteroglycan HG-RG-II-HG; the pectic heteroglycan HG-RG-I-HG, the pectic glycoconjugate pectic-AGP, and the possible complex pectic heteroglycan HG-RG-I-HG-RG-II-HG representing the covalent connection of the separate glycan and heteroglycan polymers. These 5 pectic polymers may interact via ionic interactions and covalent boron diester cross-linking. Representative methyl and acetyl esterification of the HG backbone is shown but no specific pattern is implied. No side glycan or single sugar side branching is shown but is implied as in Fig. 5. RG-II dimers could occur between any 2 adjacent RG-II molecules. The model is not meant to be quantitatively accurate but rather to denote possible heteroglycan and glycoconjugate structures.

hemicelluloses and pectins share load-bearing functions in the wall (Pérez García 2011). Coexpression in Arabidopsis stems of cytosolic UDP-Glc epimerase, Golgi localized UDP-Rha/UDP-Gal transporter, and RG-I GALS1 increased RG-I galactan content by 50% (Aznar et al. 2018). Multidimensional solid-state NMR analysis of C^{13}-labeled high galactan-expressing Arabidopsis stems showed an increased proximity between galactan and cellulose in the secondary walls, leading the authors to hypothesize that increased galactan chain lengths led to increased mobile contacts between galactan and cellulose and longer spatial contacts between xylan and pectins in Arabidopsis SCWs (Gao et al. 2023). More research is required to understand the degree and importance of molecular interactions between pectins and cellulose in the cell wall, critical information for generation of meaningful cell wall models (see below).

Functions of pectins in vivo

Hypothetical model for pectin structure in the wall. A biologically meaningful interpretation of pectin mutant phenotypes requires, at minimum, knowing how many different types of pectic polymers exist in the wall and how they are covalently and noncovalently connected to each other and to other polymers. At present, pectic polymer models do not provide this information. Thus, a conundrum confronts us. On the one hand, making a model that is structurally inaccurate can lead to erroneous molecular genetic conclusions. On the other hand, without a model, functional hypotheses cannot be made.

Testing hypotheses of pectin function requires knowing which pectic glycan (Fig. 6A), heteroglycan (Fig. 6, B and C), or glycoconjugate (Fig. 6, D and E) is affected in the mutant/variant of interest. For example, the homogalacturon structure depicted in Fig. 6A has been shown to be synthesized in vitro as a nonmethylesterified homopolymer by GAUTs 10, 11, 13, 14, and the GAUT1:7 complex (Amos et al. 2018; Engle et al. 2022; Mohnen et al. 2024), and yet, there is scant evidence that pure homogalacturon exists as an isolated polysaccharide in the wall.

To develop a more pectin polymer-based vocabulary for discussing pectin structures and pectin mutant phenotypes, a working hypothetical model for pectin polymeric structures in plant cell walls is proposed. The pectic heteroglycan

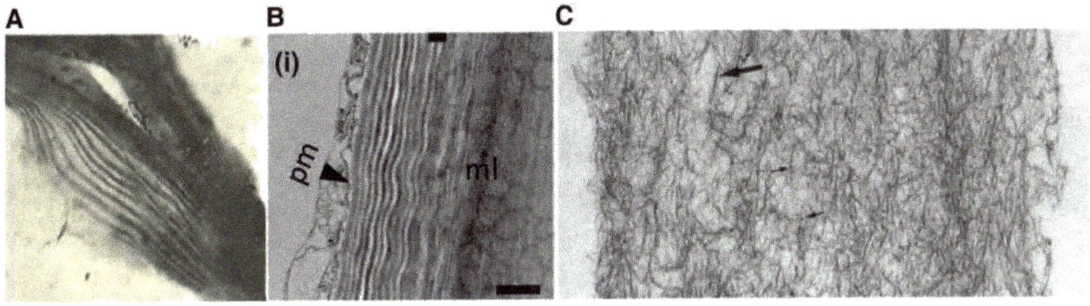

Figure 9. Microscopic evidence for alternating pectin-cellulose layers in collenchyma cells and fibrillary nature of pectin. **A)** A single tomato collenchyma cell wall after treatment with potassium iodide and sulfuric acid to remove pectin and expose cellulose lamellae, ×300 magnification (from Fig. 10 from Anderson 1927). **B)** TEM images of longitudinal sections of celery petiole collenchyma cells showing lamellae of cellulose MFs stained with uranyl acetate and lead citrate. Pm: plasma membrane (from Fig. 1i from Chen et al. 2019). **C)** Fast-freeze, deep-etch rotary shadowed (FDR) replica image of tomato suspension cells adapted to growth on 12 μM DCB showing pectin fibers in wall (from Fig. 3E from Wells et al. 1994).

and glycoconjugate model (Fig. 8) depicts pectins as consisting of 5 different possible glycans, heteroglycans, and glycoconjguates. These include the covalently linked pectic heteroglycan HG:RG-I-HG-RG-II-HG, the separate heteroglycans HG-RG-I-HG and HG-RG-II-HG, the glycan HG by itself, and the pectic glycoconjugate pectic-AGP. The pectic polymers interact via ionic HG-calcium salt bridges and covalent RG-II borate diester cross-links. The pectic AGP is bound to the plasma membrane via a GPI anchor on the AGP protein core (although cleavage by phosphodiesterases may solubilize this into the wall). Representative methyl and acetyl esterification of the HG backbone is shown, but no specific pattern is implied. It is likely that the degree and pattern of methyl and acetyl esterification is development-, tissue-, cell-type-, and stress-level dependent. This model is supported by the observation that RG-I and RG-II are released from cell walls by treatment with EPGase and the supposition that although it is possible that they may all be covalently connected in a single network, it is also possible that they exist as separate heteropolymers in the wall (reviewed in Mohnen et al. 2024). Some of the RG-I and HG are also de-picted as pectic AGPs due to the observation that RG-I in Arabidopsis exists in this form (Tan et al. 2013, Tan et al. 2023b). The backbones of the polymers are depicted, but it is understood that RG-I will have diverse side chains depend-ing on the tissue and developmental stage. Future studies of pectin synthesis and mutant phenotypes are required to test the hypothesized pectic polymeric structures.

Pectin location. Plant biologists soon recognized that pectins were the main components in the middle lamella between cells and were also present in the PCW, along with cellulose, hemicelluloses, and to a lesser extent lignin (Mangin 1893a, 1893b; Kerr and Bailey 1934). Some, including Mangin (1893b) and Bonner (1936), suggested that pectins in the middle lamella existed as calcium pectate, whereas others concluded that they existed as "protopectin" and/or a pectin-protein complex (Sloep 1928; Kertesz 1951). More recent results with antibodies recognizing RG-I (Moore and Staehelin 1988) and transgenic plants expressing reduced levels of pectin metabolic enzymes clearly show that the pectin in the middle lamella contains not only HG but also RG-I (Molina-Hidalgo et al. 2013).

Although less appreciated, considerable data indicate a function for pectins in walls that span the primary-secondary wall definition. There are multiple examples of alternating layers of pectins and cellulose in so-called collenchymatous secondary walls (Anderson 1927; Bonner 1936; Chen et al. 2019), a location consistent with GAUT1 expression in both primary and secondary (xylem) walls in Arabidopsis (Atmodjo et al. 2011). A microscopic study of freshly cut tomato (*Solanum lycopersicum*) (Anderson 1927) provided early evidence for a lamellar wall structure and cellulose-pectin interactions, including alternating layers of protopectin and cellulose in collenchymatic secondary walls (Fig. 9A), a finding confirmed in celery collenchyma walls, walls defined by the authors as thickened PCWs because the cells were still elongating (Chen et al. 2019). Solid-state NMR revealed reduced mobility for pectins and cellulose as the walls matured and expansion ceased (Chen et al. 2019) (Fig. 9B).

It has been proposed that the specialized secondary walls in fibers in G layers of tension wood, extraxylary flax (*Linum usitatissimum*), and hemp (*Cannabis sativa*) fibers, and fibers of climbing plants are different from other secondary walls in that they have a very high cellulose content (85% to 90%), a very low amount of lignin and xylan, and the presence of a unique RG-I highly substituted with galactan (Gorshkova et al. 2022). In such fibers, RG-I is first synthesized with long galactan side chains and inserted between cellulose MFs, preventing lateral association of neighboring MFs. Over time, β-1,4-galactosidase cleaves the galactan chains and RG-I lyase cleaves the backbone, enabling lateral association of the cellulose fibrils and leading to tension that can result in contractile properties of the fibers. These results indicate that pectins have roles in unique secondary walls and in promoting stems to bend and climb.

Pectins in cell:cell adhesion and fruit ripening. Fremy (1847, 1848) proposed the existence of pectose/protopectin in vegetables, fruits, and roots as an insoluble pectin that could be converted into the soluble pectins (reviewed in Bonner 1936) and concluded that what most people studied as pectin was a portion of the original protopectin in plant tissues. Fremy (1848) and Carre (1922) also attributed the thinning of fruit cell walls during ripening to the "decomposition of the insoluble protopectin and the formation of soluble pectin" (Bonner 1936; Kertesz 1951), and enzymes were identified that catalyzed such degradation. Thus, already in the 1800s and early 1900s, functions of pectins were recognized in cell:cell adhesion and fruit texture, and it was understood that easily extractable pectins represented only a portion of pectins in the wall. Both HG (Shedletzsky et al. 1992; Liners et al. 1994; Sobry et al. 2005) and RG-I (Yang et al. 2020a) have since been confirmed to function in cell:cell adhesion, leading to the question of whether they function together in a given cell type or have unique roles in different cell types and tissues. The differential abundance of HG and RG-I in growing vs nongrowing cells, respectively, would support the latter (Mohnen et al. 2024).

Pectin fibers. For more than 50 years, extensive evidence has indicated the existence of pectin fibrils or filaments, yet most modern models of pectins do not represent such structures. In 1951, Roelofsen and Kreger showed, using polarization microscopy, diffraction, and electron microscopy, that pectins in collenchyma cell walls of butterbur (*Petasites hybridus*) existed as oriented fibrils and not, as had been previously held, as amorphous and optically isotropic (Roelofsen and Kreger 1951). Staehelin and Pickett-Heaps (1975) reported pectinase-sensitive fibrils with a 3- to 13-nm diameter in the Chlorophyceaen green alga Scenedesmus. Shortly there-after, Leppard and Colvin (1972) identified electron opaque ruthenium red-staining pectin fibrils in root epidermal cells. More recently, pectic nanofibers have been demonstrated in Arabidopsis epidermal pavement cells, and roles for the fibers in cell expansion have been proposed (Haas et al. 2020). The evidence for pectic fibers highlights the question of the tertiary and quaternary structure of pectins, and more specifically HG. As there is increasing evidence for pectins associating with and between cellulose MFs and fibers (Zykwinska et al. 2005; Lin et al. 2016), the importance of understanding pec-tin structure and architecture in the wall and in association with cellulose is increasingly important. Recently, resonant X-ray scattering of onion epidermal cell walls revealed a 20-nm spacing of cellulose MFs with a pectin matrix between the fibrils (Ye et al. 2018).

Pectins in cell growth and expansion. Tomato cells adapted to growth on DCB are able to divide and expand with a "virtual absence of a cellulose-xyloglucan network" (Shedletzky et al. 1990). Subsequent studies of the DCB-adapted cells revealed enhanced wall-membrane connections and an accompanying reduced ability to plasmolyze the cells, an ability that could be recovered by treatment of the cells with 50 mM CDTA and a pure endopolygalacturonase. These results implicated calcium-homogalacturonan salt bridges as load-bearing components that could support growth of the DCB-treated cells (Shedletzky et al. 1992). Imaging of the DCB-adapted cells provided a visualization of a fibrillar pectin network in the absence of the cellulose-xyloglucan network (Wells et al. 1994) (Fig. 9C). The pectin load-bearing hypothesis was supported by lysis of tobacco and tomato DCB-adapted cells by overnight treatment with CDTA, a lysis that did not occur in non-adapted cells. Although the tensile strength of the DCB-adapted cells was less than one-half that of normal cells, as expected due to the drastic reduction in cellulose, the ability of the calcium cross-linked pectate (i.e. HG) to provide sufficient load-bearing to keep the cell intact and enable growth provided strong evidence for a role of a pectin cross-linked matrix in cell expansion and growth in dicot cells (Shedletzky et al 1992). Further evidence of a role for pectins in cell expansion is presented below (see Models of Cell Wall Organization section).

Wall-associated proteins

Types of cell wall proteins

Plant cell walls contain a diverse collection of proteins, glycoproteins, and proteoglycans that perform an array of different functions (Albersheim et al. 2011). These can be organized into 2 general classes, insoluble and soluble wall proteins, with each class further subdivided into multiple categories (Albersheim et al. 2011). The insoluble proteins are tightly bound to the cell wall and are generally considered to play structural roles; these include large families of hydroxyproline-rich proteins (HRGPs), extensins (EXTs), proline-rich proteins (PRPs), and glycine-rich proteins (Dougall and Shimbayashi 1960; Lamport and Northcote 1960) and have been extensively reviewed (Varner and Lin 1989; Showalter 1993; Kieliszewski and Lamport 1994; Ringli et al. 2001; Showalter et al. 2010; Lamport et al. 2011; Moussu and Ingram 2023). The soluble proteins include en-zymes, lectins, transport- and defense-related proteins, as well as hydroxyproline-rich arabinogalactan proteins (AGPs) (Silva et al. 2020). The AGPs may be insoluble or sol-uble, depending on whether they are covalently attached to wall polysaccharides (reviewed in Tan et al. 2023a, 2023b). Cell wall enzymes include glycan exo- and endo-hydrolases and lyases, transglycosidases, methyl and acetyl esterases, peroxidases and laccases involved in lignin polymerization, enzyme inhibitors, lipid transport proteins functioning in cutin and wax synthesis, and proteases.

Another way to classify the large superfamily of plant cell wall HRGP proteins is via their level of glycosylation, with AGPs being hyperglycosylated, extensins moderately glycosylated, and PRPs lightly glycosylated. The AGP proteins contain noncontiguous Hyp residues and are about 90% carbohydrate by mass, while the extensins have regions of contiguous Hyp residues such as Ser-Pro-Pro-Pro-Pro motifs that become hydroxylated and arabinosylated with 1 to 4 Ara*f* residues (Kieliszewski et al 1992a, 1992b; Petersen et al. 2021). The "Hyp-contiguity hypothesis" developed by Kieliszewski and Lamport (1994) proposes that contiguous Hyp residues [Ser-Hyp4] (as found in extensins) are sites for short arabinose oligosaccharide addition, whereas repetitive noncontiguous Hyp residues (as found in AGPs) are sites for Type II arabinogalactan attachment.

Because it is difficult to identify the large classes of insoluble proteins by BLAST analyses of amino acid sequence due to overlapping protein sequence motifs and arrangements, Showalter et al. (2010) developed a multifaceted bioinformatics approach and identified 166 Arabidopsis HRGPs that grouped into 85 AGPs, 59 EXTs, 18 PRPs, and 4 AGP/EXT hybrids. In poplar, with its expanded genome and large amount of woody secondary walls, a similar approach identified 271 HRGPs, including 162 AGPs, 60 EXTs, and 49 PRPs. The bioinformatics analyses identified and distinguished EXTs with variations of the X-Hyp_4 sequences including SP3, SP4, and SP5. There are also EXT hybrid proteins with additional domains such as the PERKS (proline-rich extension-like receptor kinases), formins with an EXT motif and actin binding domain, hybrid-EXTs such as the LRR-EXTs (leucine rich repeat EXTs), and hybrid AGP-EXT (Borassi et al. 2016). Many of the prolines in the repetitive domains of HRGPs are hydroxylated, and these may, depending on the repetitive amino acid motif, be heavily or partially glycosylated. Pro-Hyp-Val-Tyr-Lys regions in HRGPs appear to be involved in intermolecular crosslinking, adhesion, and cohesion, and Tyr-X-Tyr-X in isodityrosine and pulcherosine covalent cross-linking and self-assembly catalyzed by class III peroxidases (Kieliszewski and Lamport 1994; Showalter et al. 2010; Mishler-Elmore et al. 2021). Analysis of coexpressed GTs, prolyl 4-hydroxylases, and peroxidases have provided candidate enzymes involved in glycosylation, proline hydroxylation, and crosslinking, respectively, and progress is being made in delineating the biosynthetic pathways for these post-translational modifications that are critical for HRGP function (Silva et al. 2020).

The AGPs are a diverse, highly glycosylated, multifunctional subgroup of HRGPs. Tony Bacic and his colleagues in Australia were major contributors to our understanding of this interesting group of complex glycoproteins (Ellis et al. 2010). They are found across all plant species, including algae, bryophytes, and ferns (Seifert and Roberts 2006; Silva et al. 2020; Ma and Johnson 2023; Tan et al. 2023a). AGPS are classified as classical AGPs, lysine-rich AGPs, arabinogalactan peptides, fasciclin-like AGPs, plastocyanin AGPs, and chimeric AGPs (Showalter et al. 2010; Shafee et al. 2020). Of the 85 AGPs in Arabidopsis, 56 were predicted to have a C-terminal GPI anchor sequence (Showalter et al. 2010) directing them to the outer leaflet of the plasma membrane, where it is hypothesized that they remain tethered unless cleaved from the wall by phospholipases. The term GPI anchor is a misnomer because the plant glycolipid contains an inositol ceramide rather than phosphatidylinositol as found in animals (Silva et al. 2020).

The AGPs are glycosylated on selected Hyp residues by large β-1,3-galactan and β-1,6 branched side chains (Type II arabinogalactans) that can be further decorated with sugars such as Ara*f*, Rha, Gal, Fuc, and GlcA. Some AGPs, such as Arabidopsis AGP57C, also have covalently attached pectic glycan RG-I, which is further elongated with HG producing so-called pectic AGPs, and some of these also contain covalently linked xylan (Tan et al. 2013, Tan et al. 2023b) (see Fig. 6). Classic AGPs can be precipitated with β-Yariv reagent, which has been used to detect AGPs in the wall, but pectic AGPs are not reactive with Yariv reagent due to their complex glycosylation. AGP glycosylation is species and tissue dependent with potential multiple AG chains per protein core and with 65 to 142 sugar residues per AG chain reported (Tan et al. 2023a). Although progress continues in defining specific AGP glycosyl structures, we know very little about the variation of specific AG side chain structures within a given cell type or in different tissues or differences in glycosylation of different AGP protein cores. AGP glycan structural diversity is due to the specificity of different GTs and to any glycan processing by hydrolases that may occur within the Golgi complex, as recently suggested for Golgi localized CAZy GH43 β-1,3-galactosyl hydrolase (Nibbering et al. 2020).

Functions of cell wall proteins in vivo

The AGPs are the largest family of cell wall proteins and perhaps the most multi-functional, with roles including plant growth and development, cell expansion, division, signaling, embryogenesis, vascular and gametophyte development, and biotic and abiotic stress response (Seifert and Roberts 2006; Silva et al. 2020; Lin et al. 2022a; Leszczuk et al. 2023; Ma and Johnson 2023; Tan et al. 2023a). Mechanisms for how they may carry out these functions include mediation of information between the cell wall, plasma membrane, and cytoplasm in part due to their anchoring to the plasma membrane and covalent and/or noncovalent interactions with wall carbohydrates and/or proteins. The amount of glucuronidation of the AGP has been associated with calcium binding, signaling, and salt tolerance (Lopez-Hernandez et al. 2020); covalent connection of AGPs to pectin has been implicated in the establishment and maintenance of wall structure and cell expansion (Tan et al. 2013; Biswal et al. 2018; Tan et al. 2023b). AGP signaling mechanisms include indirect signaling through effects on wall properties, action as co-receptors, and activity of the arabinogalactan carbohydrates as signaling molecules (reviewed in Seifert and Roberts 2006; Ma and Johnson 2023). For example, an arabinogalactan 4-methyl-GlcA-β-1,6-Gal terminal AGP epitope from *Torenia fournieri* mature ovules, known as AMOR, has been shown to induce competency of pollen tubes to respond to attractant peptide signals from the synergid cell, thereby identifying specific AGP glycans as signals required for guidance of pollen tubes and pollination (Mizukami et al. 2016), although the specific AGP(s) on

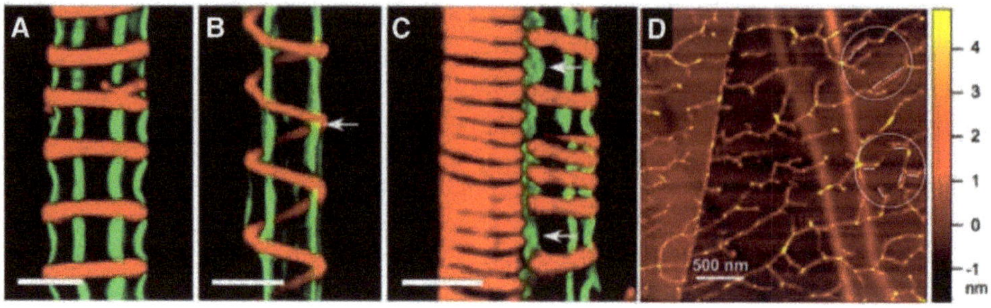

Figure 10. Evidence for roles of structural cell wall proteins in wall structure, assembly, and architecture. **A to C:** Confocal microscopy 3D reconstruction of protoxylem (PX) in apical region of soybean hypocotyl immunolocalized with anti-GRP1.8 polyclonal rabbit antibody (green fluorescence) and lignin (red fluorescence). **A)** PX with ring-shaped lignified SCW and GRP structure. **B)** PX with helical lignified SCW and GRP structure present during passive PX elongation. **C)** Two PXs with younger one (left) with closely spaced SCW and no GRP structure and older one (right) with more widely spaced SCW and GRP structure (taken from Ryser et al. 2003, Fig. 1, A–C). **D)** In vitro self-assembled Arabidopsis EXT3 network imaged by AFM (taken from Cannon et al. 2008, Fig. 5).

which the AG signaling glycan is present remain to be determined. A mechanistic understanding of the roles of AGPs in cell wall structural integrity and cell wall signaling, however, remains fragmentary for most AGPs. A key area of research for the future is to understand how different AGPs and other HRGPs interact in the wall with each other and with carbohydrates, lignin, and proteins, and in those cases where the AGPs are direct signaling molecules to identify their ligand and receptor/co-receptor partners and signaling pathways. Physiological roles for some of the AGPs are just now being identified through molecular genetic analyses of biosynthetic mutants (reviewed in Moreira et al. 2023).

As a group, the structural wall proteins are thought to be part of a cell wall network that functions within or is associated with the cellulose-hemicellulose-pectin carbohydrate network, thereby providing or contributing to mechanical strength and possibly to assembly or architecture of the wall (Fig. 10). It has been proposed that EXTs with their positive charge may interact with pectins to form a protein-carbohydrate polymer gel-like matrix and possible scaffold with the potential to respond to pH and ionic conditions in the wall (Tierney and Varner 1987; Lamport et al. 2011). The repetitive motifs and structures of the EXTs (and other HRGPs) certainly support structural roles (Mishler-Elmore et al. 2021), but the difficulty of extracting them from walls due to cross-linking with the protein, carbohydrate, and lignin components has made this difficult to prove (Qi et al. 1995; Nuñez et al. 2009). In no case has a specific structural role of EXTs in the wall been unambiguously confirmed (Moussu and Ingram 2023). However, recently LEUCINE-RICH REPEAT EXTENSIN (LRX8) in pollen tubes was shown to form a complex with the peptide RAPID ALKANIZATION FACTOR 4 (RALF4), thereby exposing a positive patch on RALF4 that bound HG and stabilized a reticulate cell wall architecture (Moussu et al. 2023) that strengthened the pollen tube wall during growth. A similar case for root hairs involving LRX1 and RALF22 is discussed in more detail in the signaling section (Schoenaers et al. 2023).

Transcript expression analyses suggest that most EXTs, about one-half the AGPs, and a few of the PRPs have organ-specific expression, suggesting that some HRGPs may have roles within most cells whereas others have cell-, tissue-, or organ-specific function. Immunolabeling of wall proteins has revealed cell type and sub-cell wall–specific labeling for some HRGPs. For example, the Arabidopsis proline-rich cell wall protein AtPRP3 is preferentially expressed in root hair–bearing epidermal cells at the root/shoot junction and in the root differentiation zone of light-grown seedlings (Bernhardt and Tierney 2000). Molecular genetic analyses support a role for AtPRP3 in cell wall structure in differentiating root hairs (Bernhardt and Tierney 2000; Albersheim et al. 2011).

Other HRGPs are localized to SCW-rich tissues. In etiolated soybean hypocotyls, bean GRP 1.8 localizes to protoxylem, the water-conducting tissue of young and elongating tissues (Keller et al. 1989; Ryser 2003). Deposition of GRP 1.8 begins in cell corners between protoxylem elements and develops further to interconnect secondary walls between adjacent protoxylem and form a GRP structural wall between 2 dead protoxylem cells. It also connects protoxylem elements and xylem parenchyma cells in a 3-dimensional GRP network that stabilizes the whole protoxylem by interconnecting ring and spiral-shaped secondary wall thickenings of the protoxylem and the middle lamella of living xylem parenchyma cells, preventing premature collapse of the protoxylem. Confocal laser scanning microscopy of pectinase-softened hypocotyls using anti-GRP1.8 antibodies identified the GRP structure as a distinct structural element arranged in the longitudinal axis of the protoxylem elements (Fig. 10, A–C) (Ryser et al. 2003) that is proposed to function in stabilizing cell corners and anchoring and stabilizing protoxylem elements (Ryser 2003). The GRP structural element is closely associated with an RG-I and HG pectin network (Ryser 2003; Ryser et al. 2003), providing further evidence for a function of pectins in secondary walls and a role for passive elongation of protoxylem cells (Fig. 10B).

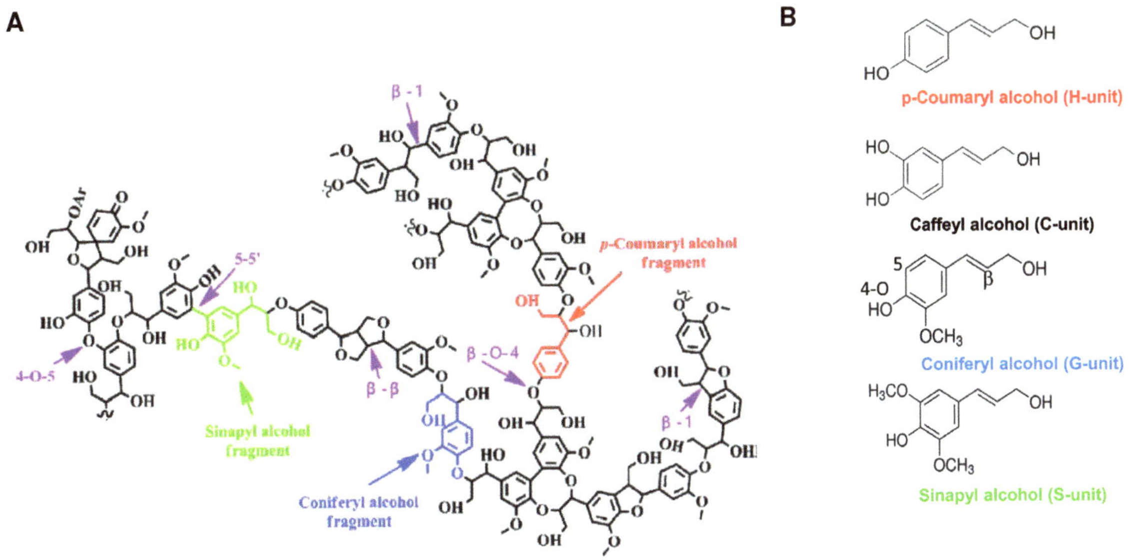

Figure 11. Structure, monomer composition, and linkages of lignin. **A)** A generic lignin molecule showing the major monomer types and the varieties of linkage type. β-*O*-4 linkages are the most common and the easiest to break. Carbon-carbon linkages such as 5-5′ are hard to break and contribute to the recalcitrance of lignin toward degradation. Reproduced from Zakzeski et al. (2010). **B)** The 4 naturally occurring monolignols.

In addition to providing mechanical strength to walls, EXTs are proposed to self-assemble into a network structure that may serve as a positively charged scaffold for pectin in the wall (Fig. 10D) (Cannon et al. 2008) (see "Life of a Cell Wall" below), be involved in wall repair and response to microbial interactions, and be induced when plants are wounded and/or attacked by pathogens (Petersen et al. 2021). Although the name extensin was coined by Lamport (1963) with the proposition that such proteins may be involved in cell expansion, there are many reports in which secretion of extensin into the walls is correlated with the cessation of cell growth (Sadava et al. 1973; Mishler-Elmore et al. 2021). EXTs have also been associated with stresses, including ethylene, wounding, and cold and hot temperatures (Swords and Staehelin 1989).

Lignin

Structure and function of lignin

Lignin is a major structural component of plant SCWs. As mentioned above, the cell walls Robert Hooke first described in the 1660s were from cork, and their shape was immortalized by the recalcitrant layers of what was later shown to be suberin, a complex polymer of polyunsaturated fatty acids esterified to phenolic units that is likely anchored to lignin. Suberin makes up 40% of the cork cell wall and lignin comprises another 20%. The term lignin was coined by the Swiss taxonomist Augustin Pyramus de Candolle in 1813. He had isolated a fibrous plant material that was insoluble in water but soluble in alkaline solutions and named it lignin based on the Latin name for wood (lignum). The underlying chemical structure of lignin as a biopolymer of coniferyl alcohol units joined through ether linkages was proposed by Peter Klason in 1897, and the residue after total acid hydrolysis of the carbohydrate components of wood is still termed Klason lignin. In a series of seminal papers between 1940 and 1970, Freudenberg and colleagues established, through chemical analysis and precursor labeling experiments, that lignin was formed primarily from hydroxyphenyl, coniferyl, and sinapyl alcohols (Fig. 11) derived from the amino acid L-phenylalanine linked through free radical coupling (dehydrogenation) (Freudenberg 1959; Freudenberg and Neish 1968). Studies in spruce and magnolia showed that gymnosperm lignins are composed almost exclusively of guaiacyl (G) units, whereas angiosperm lignins are primarily derived from G and syringyl (S) units, with a much lower proportion of hydroxyphenyl (H) units. As a result of their assembly by non-enzymatic polymerization, there is no exact structure for lignin chains. Earlier models suggested very large 3-dimensional structures with a high degree of cross-linking (Freudenberg and Neish 1968), whereas later models favor shorter, more linear, less branched chains (Ralph et al. 2019). The extent of cross-linking will depend on the S/G ratio based on chemical coupling preferences that can be illustrated with in vitro monolignol polymerization systems (van Parijs et al. 2010).

The linkages between the H, G, and S monolignols can be determined by analysis of partial degradation products or, more conveniently, by NMR (Marita et al. 1999). The most common linkage is the β-*O*-4 linkage between the free 4-OH group of the H, G, or S units and the side chain C2 position of another unit (Fig. 11A), and cleavage at this linkage is the basis for the determination of lignin content and composition by the commonly used technique of thioacidolysis (Lapierre et al. 1985). The several other linkages in lignin are not cleaved by this method. This highlights the problems

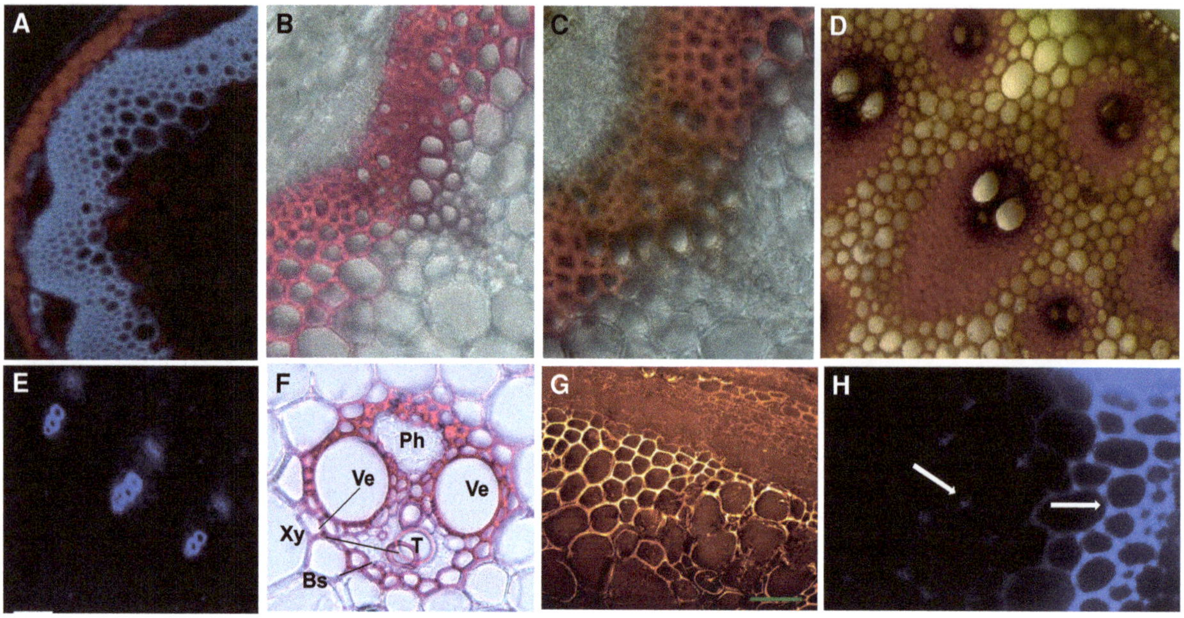

Figure 12. Lignin detection in situ. **A)** Cross section of an Arabidopsis stem viewed under UV light. Lignin autofluorescence is blue, chlorophyll autofluorescence red. **B)** Higher magnification of a cross-section of an Arabidopsis stem stained with phloroglucinol-HCl. **C)** As in **B)** but stained with Maule-reagent. The S-lignin–rich interfascicular fibers stain red, whereas the G-lignin–rich vessel elements stain yellow-brown. A to C are from Wang et al. (2010). **D)** Phloroglucinol staining of a cross-section of a transgenic maize stem with ectopic lignification, showing the different organization of fibers compared with the dicot Arabidopsis. **E)** Cross section of a switchgrass stem viewed under UV light. **F)** Higher magnification of the vascular tissue of a switchgrass stem stained with phloroglucinol-HCl. Bs, bundle sheath cell; Ph, phloem; Ve, vascular element; xy, xylem. D to F are from Gallego-Giraldo et al. (2015). **G)** Forward coherent anti-Stokes Raman scattering (F-CARS) microscopy of a cross section through the epidermal tissue, cortex, and inter-fascicular fiber region of a wild-type alfalfa (*Medicago sativa*) stem. The highest CARS signal is in vascular tissue and can be quantified. From Zeng et al. (2010). **H)** Cross-section through the stem of an alfalfa line undergoing ectopic lignification as a result of downregulation of a WRKY transcription factor. Arrows show lignification in the cell corners of lignified and lignifying cells (L. Gallego-Giraldo and R. A. Dixon, unpublished results).

in the quantitative measurement of lignin; the methods with the most chemical precision are less able to sample the whole molecule, whereas those that rely on simple gravimetric or UV/VIS spectroscopic measurement are prone to interference from contaminants. Conveniently, lignin shows blue autofluorescence under UV light, although similar fluorescence is shown by wall-bound hydroxycinnamic acids. Lignin localization in plant tissues can also be determined by Weisner (phloroglucinol) or Maule (permanganate) staining, the latter showing some specificity for S-lignin, although neither is fully specific or quantitative (Lewis and Yamamoto 1990). Raman spectroscopy provides a means to both visualize and quantify lignin under the microscope (Agarwal and Ralph 1997; Zeng et al. 2010). These various approaches to lignin visualization, along with examples of lignification patterns in stems of both monocots and dicots, are shown in Fig. 12.

It is perhaps surprising that more is not known of the interactions between lignin and the other cell wall components. The current picture is informed by both direct measurements and inferences from indirect approaches such as genetic modification of lignin or other cell wall polysaccharides. It has been generally assumed that lignin assembles on hemicelluloses, and sequential subcritical water extraction of hemicelluloses from birchwood revealed the presence of specific xylan domains that appeared to interact with lignin (Martínez-Abad et al. 2018). This has been confirmed by solid-state NMR measurements of maize stems, where the lignin was shown to be present in hydrophobic nanodomains with electrostatic interactions with the polar motifs of xylans that are not strongly associated with cellulose (Kang et al. 2019).

Sequential glycome profiling of cell wall residues of stems of transgenic plants with altered lignin content and/or composition has shown that both rhamnogalacturonan and xylan epitopes become more water-soluble upon reducing lignin content (Gallego-Giraldo et al. 2018, 2020). Furthermore, biomass from stems of Arabidopsis with reduced lignin content becomes an effective substrate for growth of strains of the thermophilic bacterium *Caldicellulosiruptor bescii* lacking the pectinase gene cluster that is normally required for growth on plant biomass (Gallego-Giraldo et al. 2020). These studies suggest that either lignin is directly associated with pectins and xylan, thereby protecting their accessibility to mild solvent, or that lignin modification brings about active cell wall remodeling leading to greater solubility of other wall polymers, as further discussed below.

It is generally accepted that lignin is found in all higher plants but not in the mosses. Lignin has been identified in at least 1 red alga, where it has been ascribed a role in strengthening cell walls in the genicular tissue linking articulated fronds (Martone et al. 2009). Through studies on delignification of fibers, it has been shown that lignin contributes to tensile strength but has limited impact on bending strength (Wang et al. 2017). Other than during the early stages of lignin deposition in the middle lamella, only plant SCWs are lignified, and such walls are found in xylem, interfascicular fibers, Casparian strip in roots, and some seed coats. Lignin is also found in various other cell types in smaller amounts, including in the walls of anthers and in cotton fibers (Gao et al. 2019). The cell biology of lignification has been well reviewed elsewhere (Barros et al. 2015). Based on the location of lignified walls and more recent studies demonstrating the impairments resulting from genetically engineered reductions in lignin content, the primary functions of lignin are presumed to be the imparting of physical strength, allowing plants to maintain an upright habit or to permit dehiscence of fruits and seeds; preserving a hydrophobic environment to allow passage of water in vessels; and sealing of damaged tissues to prevent pathogen ingress.

The chemical recalcitrance of lignin associated with these functions has been an impediment to the utilization of lignocellulosic materials for conversion of the cellulosic and hemicellulosic cell wall components to liquid biofuels (Li et al. 2016a, 2016b, 2016c; De Meester et al. 2022) and for the digestibility of forage crops (Barros-Rios et al. 2018). At the same time, the realization that lignin could replace petroleum products as a feedstock for the synthesis of biomaterials, high value chemicals, and fuels has led to the concept of lignin valorization (Ragauskas et al. 2014; Martins et al. 2022; Ullah et al. 2022). Together, these concepts have inspired great effort to better understand lignin biosynthesis and its regulation over the past 20 years, which has been the subject of a number of review articles (Dixon and Barros 2019; Ohtani and Demura 2019; Ralph et al. 2019; Vanholme et al. 2019; De Meester et al. 2022).

Biosynthesis of lignin

Early models of lignin biosynthesis envisaged a metabolic grid encompassing multiple routes from the first committed precursor, *trans*-cinnamic acid, to the H, G, and S monolignols. Subsequent models, based primarily on the substrate specificities of recombinant pathway enzymes, suggested a more defined set of routes through the grid (Humphreys et al. 1999), but the complexity and flexibility of the pathway has become more apparent recently through new genetic evidence and the application of kinetic and metabolic flux analysis (Vanholme et al. 2013; Wang et al. 2018; Barros et al. 2022). It is now clear that, in grasses, a significant proportion of the flux into lignin originates from tyrosine rather than phenylalanine (Barros et al. 2016). Figure 13 illustrates the currently accepted routes to the monolignols. The genes encoding all the enzymatic steps have been isolated and most of their functions validated through genetic approaches (Dixon and Barros 2019; Vanholme et al. 2019).

Unlike the other cell wall polymers, lignin shows a degree of flexibility as to its monomeric constituents. This first became apparent from the discovery of noncanonical units such the flavanol tricin, stilbenes, or feruloyl putresceine in the lignin from some species (Ralph et al. 2019), stimulating attempts to introduce additional units with more labile bonds to allow for easier cell wall processing. This was achieved via the engineering of pathways for the formation of artificial monolignol ferulate and the diferuloyl methane curcumin, which were transported to the apoplast and incorporated into lignin, resulting in enhanced saccharification of biomass (Wilkerson et al. 2014; Oyarce et al. 2019). Incorporation of a pathway for formation of *p*-hydroxybenzoate (*p*HBA) resulted in a large increase in esterified *p*HBA in poplar lignin, with potential as a "clip-off" molecule for processing to downstream high value chemicals (Mottiar et al. 2023).

Lignin is polymerized in the apoplast after the pectic, hemicellulosic, and cellulosic components have been laid down following their transport through the Golgi system (pectin and hemicellulose) or synthesis by a trans-membrane molecular "machine" (cellulose). This raises a number of questions concerning the transport of lignin monomers and the initiation of their polymerization. Based on the presence of high levels of coniferin (coniferyl alcohol β-D-glucoside) in xylem sap and its temporal association with lignification in the developing xylem of coniferous trees, along with results of radiolabeling experiments (Terashima and Fukushima 1988), it was proposed that monolignols were transported to the apoplast as their glycosides utilizing membrane transporters and then underwent conversion back to the aglycones through the activity of an apoplastic glycosidase (Tsuyama and Takabe 2015). Biochemical studies suggested the presence of ATP-dependent transporters for coniferin and other monolignol glycosides in membrane preparations from a number of woody species as well as Arabidopsis (Miao and Liu 2010; Tsuyama et al. 2013), but, to date, the only transporter characterized at the molecular level is a low- affinity H-monomer transporter (Alejandro et al. 2012). There is still no genetic evidence for the existence of G- or S-unit transporters, and an alternative hypothesis of passive diffusion down a concentration gradient has been proposed, supported by theoretical thermodynamic models (Vermaas et al. 2019) and by the observation that polymerization to lignin appears to be required for movement of monolignols out of the cytosol to the apoplast (Perkins et al. 2019, 2022; Zhuo et al. 2022). Glycosylation of small monolignol oligomers appears to be required for their intracellular transport from cytosol to vacuole (Dima et al. 2015). A passive diffusion mechanism, or highly promiscuous transporters, are consistent with the ability of "unnatural" monolignols to pass to the apoplast and be incorporated into lignin (Wilkerson et al. 2014; Ralph et al. 2019). The "good neighbor" hypothesis

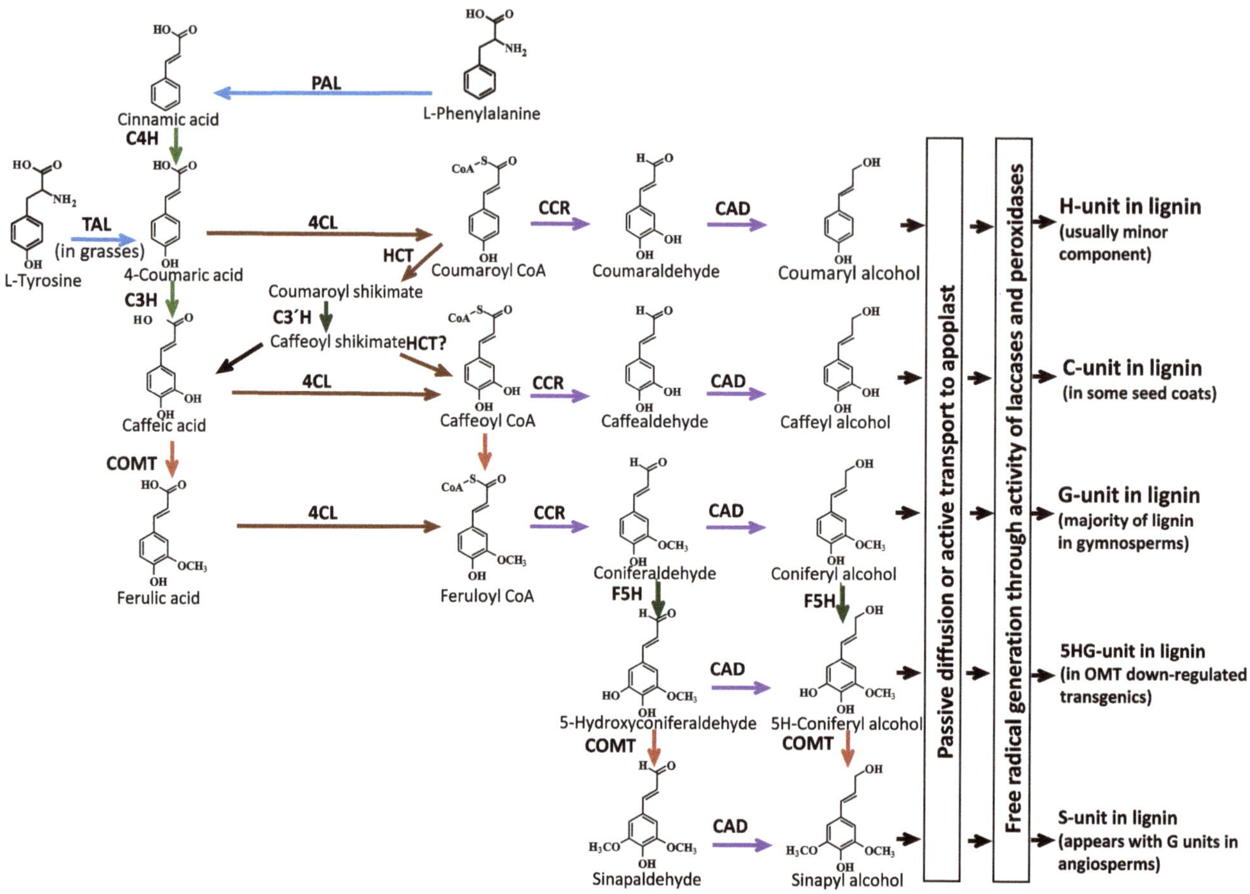

Figure 13. Biosynthetic origins of lignin building blocks, showing enzymatic reactions involved in lignin biosynthesis from primary metabolism (phenylalanine and tyrosine). Arrows are color-coded to show different classes of reactions: blue, deamination; green, hydroxylation; red, *O*-methylation; brown, CoA activation; purple, reduction. Enzymes are PAL, L-phenylalanine ammonia-lyase; TAL, L-tyrosine ammonia-lyase; C4H, cinnamic acid 4-hydroxylase; 4CL, (hydroxy)cinnamate CoA ligase; HCT, hydroxycinnamoyl CoA: shikimate hydroxycinnamoyl transferase (HCT); C3′H, coumaroyl shikimate 3′-hydroxylase; CSE, caffeoyl shikimate esterase, C3H, coumarate 3-hydroxylase; COMT, caffeic acid/5-hydroxyconiferaldehyde 3/5-*O*-methyltransferase; CCoAOMT, caffeoyl CoA 3-*O*-methyltransferase; CAD, cinnamoyl CoA reductase (CCR); CAD, cinnamyl alcohol dehydrogenase; F5H, ferulic acid/coniferaldehyde 5-hydroxylase.

suggests that monolignols may be generated in parenchyma cells adjacent to lignifying xylem elements, then diffusing into and through the apoplast, where lignification is initiated at the cell corners (Fig. 12H) and/or in the middle lamella. In contrast, lignification in interfascicular fibers appears to be cell autonomous (Smith et al. 2013).

The polymerization of monolignols was at first thought to be initiated by the activity of cell wall peroxidases (Freudenberg and Neish 1968), consistent with the cellular localization of these enzymes and the well-studied formation of lignin "dehydrogenation polymers" generated in vitro from monolignols, peroxidase and hydrogen peroxide (e.g. van Parijs et al. 2010). Because the peroxidases are encoded by large gene families (73 members in Arabidopsis) with potential redundancy, it has been difficult to ascribe functions by genetic experiments. It was later proposed that another group of copper oxidases, the laccases, were also likely involved in monolignol polymerization (Sterjiades et al. 1992; Bao et al. 1993). This suggestion has received considerable support from genetic loss of function experiments. For example, lignification is completely absent in stems of the Arabidopsis *lac4 lac11 lac17* triple mutant (Zhao et al. 2013). Genetic analysis has, however, now confirmed roles in lignification for some peroxidases (e.g. Zhang et al. 2022a), and the current model proposes involvement of both enzymes in most lignin biosynthesis, with an exception in the Casparian strip of the root, where peroxidase alone is believed to be necessary (Lee et al. 2013).

Assuming passive transport of monolignols to the apoplast, the localization of the enzymes that initiate free radical formation for polymerization could be critical for determining the site of initiation of lignification. Lignification often starts in the cell corners (Turlapati et al. 2011), and Arabidopsis PEROXIDASE 64 localizes to this region, whereas LACCASE 4 is immobile and localized to the thick secondary wall (Chou et al. 2018). In view of the lack of lignin in the *lac* triple mutant, this suggests that simple initiation of polymerization in the cell corners is insufficient to prime growing lignin chains throughout the wall and that other initiation sites are likely present. Understanding how lignification is initiated in different cell types is a critical question for the future. The pioneering work of Niko Geldner's laboratory has defined the Casparian strip as model for understanding the cell biology and biochemistry of lignin initiation (Barbosa et al. 2019).

Unlike most other plant polymers, lignin is assembled nonenzymatically. The potential lack of any enzymatic control over such an important process led some experts in the field to suggest that the free-radical-based chain elongation might require some kind of template to allow for correct assembly (Davin and Lewis 2000). This view was inspired by the seminal finding of Norman Lewis' group that oxidative coup-ling of 2 coniferyl alcohol units to form dimeric lignans was under strict stereochemical control in most plants and that this control was engendered by the co-action of a so-called dirigent protein with the laccase that generated the monolignol radicals (Davin et al. 1997). Dirigent proteins are trimeric proteins with no identified catalytic reaction, but which direct the stereochemistry of radical-based coupling (Davin et al. 1997). They have been found in nearly all plant species, localize to the cell wall, and are often associated with disease resistance (Paniagua et al. 2017). For several years, a heated debate took place as to whether monolignol assembly for lignin (as opposed to lignans) was under chemical control only, as promoted by John Ralph, or required some kind of template that was composed of an array of dirigent sites (supported by Norman Lewis, Simo Sarkannen, and others). The arguments on both sides can be found in Ralph et al. (2008). Although the pervading view now is that lignin chain assembly is essentially purely chemical (and lignin clearly does not possess any recognizable stereochemistry), the idea of a function for dirigent proteins in some aspects of lignification (initiation or localization) is gaining support. For example, maize mutants lacking a specific dirigent protein displayed impaired deposition of lignin in the Casparian strip, leading to reduced salt tolerance (Wang et al. 2022a, 2022b).

The template hypothesis also posited that the H, G, and S monolignols were the only components that could be assembled into "true" lignin, consistent with steric limitations for a template-driven process. The recent discovery that a wide variety of chemically enabled components (tricin, stilbenes, coniferyl, ferulate, curcumin, etc.) can be incorporated into lignin appears to undermine this tenet of the template hypothesis but still leaves open the possibility of specific proteinaceous initiation sites for lignin chain growth. The flavonol tricin has been proposed to serve as an initiation site for lignification in the grasses, although the effects of manipulating its levels through genetic means differ depending on species (Lam et al. 2021). Nevertheless, the flexibility of lignin structure pro-vides an unprecedented opportunity to fine-tune the composition of a major cell wall structural polymer to either assist biomass degradation or promote its valorization.

Multiple structures involved in many functions

Models of cell wall organization

Studies of cell wall architecture by Roberts and McCann in the early 1990s produced a series of fast-freeze, deep etch, rotary-shadowing images of onion parenchymatous cell walls (McCann et al 1990) that greatly influenced thinking about how the cell wall is arranged. Such images revealed multilamellate 100-nm-thick primary walls with 10-nm pores and cellulose MFs (possibly with associated hemicellulose) of 5- to 12-nm diameter. Importantly, the measured lengths of the isolated polysaccharides were in the 700 nm range, enough to span the wall and enable cell to cell wall cross-links. Already it was recognized that the "spaghetti type" models of the primary cell wall were problematic. "It is clear that the classical description of cellulose microfibrils embedded in an amorphous matrix (Varner and Lin 1989) is conceptually problematic; microfibrils are cross-linked together; all wall components are long rod-like fibres; only removal of cellulose with cellulase gives an image that could be described as an amorphous matrix. It is the nature of the cross-links that remains to be determined" (McCann et al. 1990, p330).

Any comprehensive model of wall structure must answer the crucial unsolved problem of how the various components found in plant cell walls are organized into a matrix that carries out the diverse functions that walls perform. Despite many efforts, with most models focusing on how growth is accomplished, the current "models" are inadequate. Most are descriptive diagrams that aim to depict important structure-function relationships. To be useful, such models should be sufficiently specific that they allow interpretation of experimental data and allow the formulation of predictions that can be experimentally tested. Only recently have models become sufficiently sophisticated that they allow quantitative explanations and predictions (Zhang et al. 2021c).

Over the years, one of the major uses for cell wall models was to explain the reorganization of wall components during elongation growth, especially auxin stimulated growth (see Du et al. 2020b for a recent review). More than half a century ago, it was demonstrated that elongation could be stimulated in many systems by acidic conditions (Rayle and Cleland 1970). It is now understood that auxin stimulates a plasma membrane–localized proton pump that acidifies the cell wall (Du et al. 2020b). What is still not completely resolved is how low pH reorganizes wall components to allow growth.

One of the first efforts to describe such structure/function relationships was the "multi-net" hypothesis put forward in the early 1950s by Roelofsen and Houwink (1953). Its main emphasis was to explain the changes in orientation of cellulose MFs as plant cells undergo anisotropic elongation. They observed that elementary fibrils near the plasma membrane of cotton hairs, and other cells, were oriented transversely, but as cells elongated, the MFs gradually changed their orientation and were mainly axial in outer layers of the primary walls.

A significant step forward came in the early 1970s with the models emerging from the laboratory of Peter Albersheim, one of the great pioneers of cell wall biochemistry; they incorporated growing knowledge of the structures of wall matrix polysaccharides. One model emphasized the cellulose/XyG/RG-I network as a major source of wall strength that resists the forces of turgor pressure but can be reorganized to allow expansion as cells grow (Albersheim 1975). An early version of this model also included experimentally supported connections of pectic molecules to cell wall proteins (Keegstra et al. 1973). However, the proposed XyG to RG-I connection could not be identified, so the emphasis in subsequent models switched to a cellulose/hemicellulose network (Hayashi 1989). Further refinements were made by others, including describing the differences between Type I and Type II walls (Carpita and Gibeaut 1993), by making the dimensions of the primary wall thickness (50–100 nm) proportional to that of the plasma membrane (7–8 nm) (McCann and Roberts 1991), or by showing the various components in the proper proportions to cellulose content (Somerville et al. 2004). All these models postulate that the strength of the wall is determined by a cellulose/hemicellulose network. Many authors have called this model the tethered network hypothesis (Fig. 14A) (reviewed in Cosgrove 2022). But, as described below, several lines of evidence now argue against its validity (Cosgrove 2014, 2022), so new ideas are needed.

The tethered network hypothesis popularly featured XyG, the most abundant hemicellulose in the primary walls of many plant species (Keegstra et al. 1973; Hayashi 1989; Albersheim et al. 2011; Hayashi and Kaida 2011). In the context of XyG tethers connecting MFs (Fig. 14A), several ideas were considered as to how the network was reorganized during growth. One popular idea was the action of XyG transglucosylase hydrolase (XTH) proteins that cleave or reorganize the XyG tethers (Rose et al. 2002). More recent evidence argues against a role for XTHs in cell expansion, but they may have roles in plant responses to biotic and abiotic stresses (Ishida and Yokoyama 2022). Another option for reorganizing the network came with the breakthrough discovery of expansins, proteins that stimulate cell expansion in a pH-dependent manner without breaking covalent bonds (McQueen-Mason et al. 1992). Expansin proteins are encoded by a large family of genes in plants (Sampedro and Cosgrove 2005). Despite a great deal of molecular information about expansin proteins, the molecular details of how they achieve elongation growth remains unclear (Cosgrove 2015, 2016).

An Arabidopsis double mutant (*xxt1/xxt2*; Cavalier et al. 2008) and a triple mutant (*xxt1/xxt2/xxt5*; Zabotina et al. 2012) with disruptions of genes responsible for adding xylose to the XyG backbone have XyG levels below the limit of detection. But these mutant plants grow and develop with only minor defects, casting doubt on the validity of the hypothesis

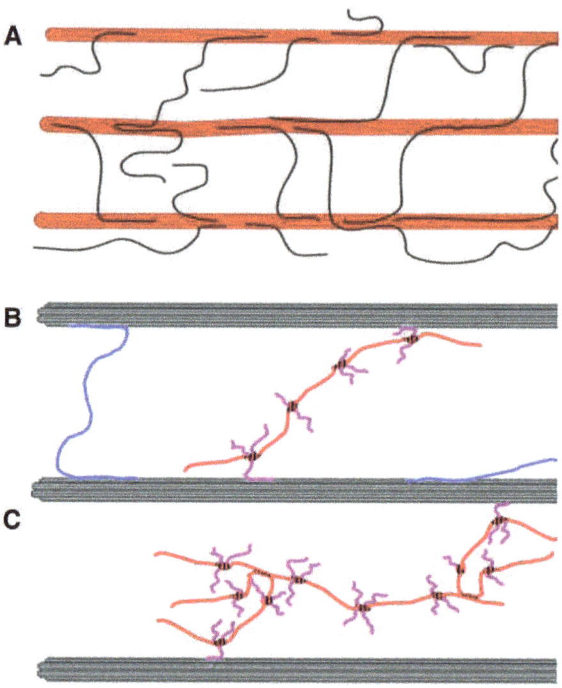

Figure 14. Variations of the tethered network hypothesis using different matrix polysaccharides. **A)** The tethered network hypothesis shows CMFs (thick horizontal rods) to be well spaced and connected by XyG (thin lines). Taken from Cosgrove (2022). **B)** CMFs (thick horizontal rods) linked by both XyG (thin undecorated lines) and pectin (thin line decorated with side chains). **C)** CMFs (thick horizontal rods) linked via a network of pectic polysaccharides (thin lines decorated with side chains). Panels B and C are taken from Zykwinska et al. (2007).

that XyG plays an important role in defining the mechanical properties of primary cell walls. More recently, an Arabidopsis quintuple mutant with disruptions of all 5 *CSLC* genes, encoding proteins that synthesize the glucan backbone of XyG, has undetectable levels of XyG but can grow and develop with minor altered phenotypes, including altered root hairs and pollen tube growth (Kim et al. 2020). It is noteworthy that the presence of aberrant forms of XyG is more detrimental to plant growth and development than the lack of XyG. The Arabidopsis *mur3-3* mutant that lacks fucose and has very low levels of galactose on XyG ex-hibits severe dwarfism and other growth phenotypes (Kong et al. 2015; Xu et al. 2017). But crossing the *mur3-3* mutant to the *xxt1/xxt2* double mutant lacking XyG restores the nor-mal size of the triple mutant plants (Kong et al. 2015). These findings emphasize the importance of the physical properties of wall polysaccharides in determining polymer interactions. Several groups have utilized the Arabidopsis mutants lack-ing XyG to investigate cell wall functions related to expansion growth. The walls in *xxt1/xxt2* mutant plants were examined in detail and found to be more extensible than wild-type walls using stress/strain assays but less extensible in creep mediated by α-expansin (Park and Cosgrove 2012b). The same group used multiple different endoglucanases to treat tissues from wild type and *xxt1/xxt2* mutant plants followed by examination of wall biomechanical properties. Based on their observations, they concluded: "Our results are incompatible with the common depiction of xyloglucan as a load-bearing tether spanning the 20- to 40-nm spacing between cellulose microfibrils...." To explain their observations, they postulated that "...XyG-mediated control of wall extension may be restricted to relatively inaccessible contact surfaces between microfibrils of the primary cell wall" (Park and Cosgrove 2012a, p1933. Later publications referred to these contact surfaces as "biomechanical hotspots" (Cosgrove 2022). One version of the hypothesis assumes that residual XyG is present in the walls of the mutant plants, which has not been demonstrated. Another possibility for the "biomechanical hotspots" is that they consist of direct interactions between cellulose MFs as described by Cosgrove (2022). In this case, the role of matrix polysaccharides, whether XyG, other hemicelluloses, or pectins, is to prevent MF from coalescing into bundles that form a rigid inflexible network. A different strategy was used by Kuki et al. (2020), who prepared protoplasts from the leaves of Arabidopsis *xxt1/xxt2* mutant plants. As protoplasts regenerated cell walls, they observed "...only a slight difference in the structure of cellulose microfibril network between *xxt1/xxt2* and wild-type (WT) protoplasts." From these and other observations, they concluded that "...xyloglucan is not essential for the initial assembly of the cellulose network, and the cellulose network formed in the absence of xyloglucan provides sufficient tensile strength to the primary cell wall regenerated from protoplasts." (Kuki et al. 2020, p. 1)

One possible explanation of the ability of plants to survive the lack of XyG is that other wall molecules can substitute for it, that is, functional redundancy. Sowinski et al. (2022) explored this possibility by performing a detailed analysis of the composition of the *xxt1/xxt2* double and *xxt1/xxt2/xxt5* triple mutants. They observed a small increase in the levels of glucomannan, but significant increases in the levels of pectic polysaccharides. Interestingly, these increases were not caused by any major increases in transcripts responsible for biosynthesis of these polymers. Yu et al (2022) have suggested that a specific galactoglucomannan may function like XyG. Mutants lacking the ability to make this mannan are nearly normal, but when crossed with the *xxt1/xxt2* mutant, the triple mutant has more severe phenotypes than mutants lacking either polymer (Yu et al. 2022) but is still able to grow and complete its life cycle.

Despite the evidence against the tethered network involving cellulose and XyG, there are several reasons why it has proven difficult to abandon this model. First, the structure of XyG is highly conserved in all land plants, arguing that it has an important conserved function. Second, XyG is abun-dant in plants with Type I PCWs, although its abundance var-ies greatly from one species to another and sometimes within the tissues of a single species, and its abundance is low in plants with Type II PCWs. Yet despite the abundance of XyG and its conserved structure, no one (including the authors) have yet proposed a plausible hypothesis for its role in PCWs.

Another version of the tethered network hypothesis proposes that pectic polysaccharides either share duties with XyG (Fig. 14B) or replace it (Fig. 14C) in forming tethers between cellulose MFs (Zykwinska et al. 2007). Although this hypothesis has not received the same level of attention, we argue that this hypothesis, or some variation of it, deserves attention considering the phenotypes of the XyG-deficient mutants.

During his thesis work in the laboratory of James Bonner at Caltech, Peter Albersheim explored the possibility that auxin enhances cell elongation by changes in the metabolism of pectins (Albersheim and Bonner 1959; Jansen et al. 1960). They focused on changes in the extent that galacturonic acid is methyl esterified during auxin treatment of oat coleoptiles. Although primitive by today's standards, this early work examined some of the same ideas for how pectins might regulate cell expansion that are being pursued today.

Boyer and colleagues studied the growing internodes of *Chara corallina*, a green alga with a higher plant-like cell wall (Domozych et al. 2010). In a series of experiments spanning 13 years, they provided compelling evidence that in *Chara* calcium pectate wall polymers provide the critical interactions that regulate cell expansion (Proseus et al. 1999, 2000; Proseus and Boyer 2005, 2007, 2012a, 2012b). They proposed a "calcium pectate cycle" model describing calcium-pectate (HG) as the load-bearing component in the *Chara* wall.

In their review of cell wall mechanics and plant cell growth, Peaucelle et al. (2012) made the compelling argument that the mechanism of cell wall extensibility in Charophycean algae, that is, the coupling of cell wall synthesis to cell wall extensibility through calcium-exchange between existing and newly synthesized calcium-pectate, also exists in terrestrial plants and called that mechanism the "ancient process." They further proposed that land plants have at least 2 load-bearing constituents, a calcium-pectate component, and a cellulose-xyloglucan network and that the latter is associated with an expansin-dependent process. They proposed that the more recent expansin-based process "amplifies" the calcium-pectate mechanism and "allows faster growth."

These ideas were explored in a series of studies examining the elongation of dark-grown Arabidopsis hypocotyls (Pelletier et al. 2010; Peaucelle et al. 2015; Daher et al. 2018). In one study, AFM was used along with antibodies that distinguish ME-HG from de-esterified HG (DE-HG) in plants where expression of PME and PMEIs had been manipulated. Although space constraints preclude a detailed explanation of the study, the authors concluded that "...growth symmetry breaking is controlled at a cellular scale by bipolar pectin de-methylesterification..." (Peaucelle et al. 2015, p. 1746. In a follow-up study using the same experimental system, Daher et al. (2018) also found asymmetry in wall elasticity in cells that was coincident with growth anisotropy and correlated with changes in pectin HG de-methylesterification. They concluded that pectin asymmetry in the hypocotyl epidermis contributes to anisotropic growth.

Jönsson et al. (2021) studied the bending of the apical hook after emergence of the hypocotyl. This process is known to be regulated by auxin and involves differential growth on the 2 sides of the hypocotyl. Again, space constraints preclude a complete description of the experiments, but the authors found a "...spatial correlation between asymmetric auxin distribution, methyl esterified HG pectin, and mechanical properties of the epidermal layer of the hypocotyl in Arabidopsis." (Jönsson et al. 2021, p. 1154)

In a ground-breaking study of expansion in Arabidopsis epidermal pavement cells, Haas et al. (2020) used super-resolution 3-dimensional direct stochastic optical reconstruction microscopy, cryo-scanning electron microscopy, and HG-specific antibodies that distinguish ME-HG versus DE-HG to show that in cotyledon anticlinal walls HG is organized as HG nanofilaments proposed to be similar to the pectate multichain helical crystalline fibers described by Walkinshaw and Arnott (1981a, 1981b). They further showed that the dimensions of DE-HG nanofilaments are 1.4 times greater than the ME-HG nanofilaments and proposed that de-methylesterification of HG nanofilament quaternary structure leads to pavement cell expansion. They call this the "expanding beam" model of pavement cell expansion. They further tested this model through development of a 3D nonlinear finite element method model, which predicted the pavement cell growth observed. Further tests of the model via in silico variation in level of methyl esters on HG by overexpression of PME and PMEI showed that, as expected, overexpression of PMEI led to growth inhibition.

More recently, Haas et al. (2021) hypothesized that enzymatic de-methylesterification of HG to induce calcium-crosslinked HG in the cell wall, in addition to turgor pressure, drives phase separation in cell wall architecture and associated plant cell expansion. Depending on the methylesterification state, HG is either highly partially or relatively poorly charged. As HG appears to be highly methylesterified in vivo when it is synthesized and inserted into the wall, its de-esterification by wall-located PMEs can lead to increases in charge and, when calcium is present, salt bridges between calcium and adjacent HG chains. This HG demethylesterification leads to associated changes in HG quaternary structure and volume transition (Haas et al. 2020) akin to phase separation.

Based on the evidence summarized above and explained in more detail in various reviews (Peaucelle et al. 2012; Daher and Braybrook 2015; Braybrook and Jönsson 2016; Chebli and Geitmann 2017; Hocq et al. 2017a; Haas et al. 2021; Höfte 2023), pectins play an important role in defining cell shape and regulating growth. In addition, increasing amounts of NMR data provide evidence for abundant interactions between cellulose and pectic polysaccharides (Pérez García 2011; Wang et al. 2015; Wang and Hong 2016; Phyo et al. 2019; Temple et al. 2022). Furthermore, genetic evidence supports an important role for pectins. In contrast to the various mutants lacking XyG that have only minor phenotypes, several pectin-deficient mutants have severe phenotypes (Bouton et al. 2002; Mouille et al. 2007; Du et al. 2020a) and others may be lethal (Caffall et al. 2009). The important conclusion is that pectic polysaccharides must be included in models that attempt to explain how wall structure is reorganized during growth. More detailed knowledge about pectin structure as it exists in PCWs is critical to understanding cell growth.

Current efforts to generate a model of the PCW have similarities to the parable of a group of blind people trying to describe an elephant with each person describing their part: a leg, a trunk, a tusk, etc. The problem is made worse by the recognition that PCWs are dynamic structures with many different functions and stages in their life, as described in the following sections.

Cell wall signaling

Perhaps the biggest surprise over the past few decades has been the realization that the cell wall is not the passive structure we assumed but is actively engaged in signaling processes. Plants are constantly under changing stresses, be they abiotic (e.g. salinity, drought, tensile) or biotic. Because degradation of plant cell walls is often a necessity for establishing infection, it seems intuitive that the cell wall would possess sensing mechanisms for cell wall integrity (CWI) that could recognize damage associated with either pathogen ingress or wounding. Even the act of cell elongation involves processes of wall loosening that may create tensile stress (Wolf 2022). Historically, the field of cell wall signaling developed from 3 distinct approaches: the identification of mobile signals within the cell wall such as cell wall–derived oligosaccharide elicitors of plant defense responses (e.g. Ayers et al. 1976); the manipulation of the level of specific wall polymer types that trigger compensatory changes in wall composition (e.g. Hematy et al. 2007); and genetic screens for genes underlying developmental phenotypes, which revealed several key plasma membrane signaling receptors (e.g. Boisson-Dernier et al. 2009; Guo et al. 2009, Duan et al. 2010).

Communication between the extracellular matrix and the cytosol is largely brought about through the binding of ligands that activate PM-localized receptor-like kinases (RLKs), with their ligand-binding domains located within the wall matrix (Wolf 2022). Plants contain large numbers of such transmembrane receptors, many of which, such as the LRR-receptor kinase resistance (R) genes or wall-associated kinases (WAKs), are associated with the rec-ognition of pathogens (McHale et al. 2006; Cho et al. 2022). Other mechano-sensitive receptors sense physical features of the PM-wall continuum such as turgor or tensile stress (Hamant et al. 2019; Bacete and Hamann 2020), whereas others, such as the class referred to as CrRLK1 receptors, physically interact with cell wall components such as pectin or peptides (Bacete et al. 2018; Zhu et al 2021). Remodeling of the cell wall is the process that closes out the cycle of cell wall signaling. The involvement of jasmonic acid (JA) and salicylic acid (SA), and their cross-talk with other hormones such as ABA and ethylene in control of cell wall remodeling, is well-reviewed elsewhere (Vlot et al. 2009; Tenhaken 2014;

Shigenaga et al. 2017; Ali and Baek 2020; Anderson and Kieber 2020; Liu and Timko 2021).

With no intention of covering this complex field in depth, we focus primarily on initial events that occur within the cell wall and the cell wall-PM interface. We discuss causes of loss of CWI, examples of the roles of tensile stresses in growth/ CWI signaling, elicitors and some key receptors and co-receptors, and end with examples of signaling in SCWs. A number of recent reviews cover the broader field in more detail (Bacete et al. 2018; Rui and Dinneny 2019; Vaahtera et al. 2019; Anderson and Kieber 2020; Bacete and Hamann 2020; Wolf 2022; Colin et al. 2023). Unless indicated otherwise, studies were carried out in Arabidopsis and focused on PCWs. Because the field is moving so fast, it is almost guaranteed that, by the time this review appears, there will have been exciting new breakthroughs.

Causes of loss of CWI

One of the most dramatic early examples of the amazing plasticity of cell walls in response to stress is that of Shedletzky et al. (1990, 1992) who showed that cultured cells of tomato, tobacco, or the grass Lolium can slowly adapt and compensate for the loss of cellulose caused by growth on the cellulose synthesis inhibitor DCB. The dicot cells compensated by creating walls comprised almost exclusively of pectin and proteins, while the grass, lacking much pectin, largely enhanced phenolic cross-linking of arabinoxylan. Similar studies with cultured Arabidopsis cells using ISX [that inhibits CSC activity but not that of the CSLD proteins that also synthesize cellulose (Yang et al. 2020b; Wu et al. 2023)], showed less dramatic decreases in cellulose, similar increases in pectin, but also upregulation of CSLD5 and a collagen-like, gly-rich protein (Manfield et al. 2004).

Because cellulose is a major load-bearing polymer in PCWs, it is not surprising that short-term inhibition of its synthesis also has notable consequences (Wang et al. 2016a; Kesten et al. 2017; Wolf 2022). These include swelling of expanding cells, inhibition of elongation, changes in levels and composition of pectins, and sometimes but not always, deposition of callose or ectopic lignin. Natural causes of cellulose synthesis inhibition might include tensile stresses that affect mechanosensitive calcium channels (MCAs), disruption of CSC-MT associations, salt/osmotic stress, or many poorly understood conditions relating to the extreme lability of rosette CSCs. Plants can also detect stress-induced alterations in pectin content or structures due to the presence of heavy metals (Muschitz et al. 2015) or high salinity, which can trigger demethylesterification of loosely bound pectins, MT degradation, and loss of cellulose (Gigli-Bisceglia et al. 2022; Colin et al. 2023). The middle lamella is rich in DE-HG, and mutations in *qua1*, a GAUT, and *qua2*, a PMT, can affect cellulose synthesis and other phenotypes due to increased cell separation (Du et al. 2020a). Cell wall–degrading enzymes from pathogens play a key role in releasing small signaling molecules and altering CWI, especially through attack on cellulose and pectins. There is less evidence for sensing in response to altered levels of PCW XyG or structural proteins, although LRXs and Gly-rich proteins have been implicated in signaling (Park et al. 2001; Manfield et al., 2004; Herger et al. 2019). As described later, engineered changes to lignin in SCWs can surprisingly also lead to large effects.

Oligosaccharin elicitors as mobile signal molecules

Studies on defense signaling began with the pioneering work of Peter Albersheim and his students and post-docs on elicitor-active oligosaccharides from plants and pathogenic fungi (reviewed by Côté and Hahn 1994). Not particularly welcomed coming from a young intruder from the cell wall world into the phytopathology community, the Albersheim hypothesis has nevertheless proven largely valid. This work developed from the elicitor hypothesis for induction of plant defense responses by fungal cell wall components but was extended to include plant developmental responses (Augur et al. 1992) and evidence that pectin-derived fragments sig-nal loss of CWI beyond that caused by pathogens (Wolf 2022). Elicitor-active cell wall–derived oligosaccharides in-clude hepta-glucosides from the cell walls of *Phytophthora sojae*, a range of oligogalacturonides (OGAs) from various sources, xyloglucan oligosaccharides (Ayers et al. 1976; Davis et al. 1986; Fry 1986, 1994), and even a grass-specific MLG-derived elicitor (Yang et al. 2021). Depending on how they are generated, such elicitor molecules are now generally referred to as Damage Associated Molecular Patterns or Pathogen Associated Molecular Patterns (Boutrot and Zipfel 2017; Bacete et al. 2018). The early studies on oligosac-charins were essentially pharmacological, and, although pro-viding reagents for studying defense gene induction, did not provide understanding of their biological significance or mechanism of action. Furthermore, methods for accurately determining the complete repertoire of wall-released mole-cules were in their infancy. Recent improvements in liquid chromatography-mass spectrometry are now making it easier to find in vivo players and link them to their in vivo bioactivities (Voxeur et al. 2019; Liu et al. 2023).

Recent work on oligosaccharins showed that a family of enzymes called lytic monosaccharide oxygenases (LPMOs) found in pathogens can cleave recalcitrant and crystalline polysaccharides such as cellulose through oxidative attack. One group of LPMOs (AA9) was shown to produce oxidized cellobiose and larger oxidized cellodextrins, and a recent study demonstrated that plants perceive the unoxidized and oxidized forms through different signaling pathways and that a combination of both cellobiose and oxidized cellobiose provided better resistance to necrotrophic infection than either alone (Zarattini et al. 2021).

Peptide elicitors

One major class of peptide elicitors that has been recently recognized as central to cell wall signaling is a family of peptides called RALFs. The first of these was discovered accidentally by

Clarence "Bud" Ryan during isolation of the small systemin peptide that induces wound responses in Solanaceous plants; during isolation, a different, larger 5-KDa peptide was also discovered that caused even more rapid alkalinization of the apoplast than systemin (Pearce et al. 2001). RALF peptides are now known to have diverse signaling roles related to alteration of apoplastic pH and growth and/or defense (Haruta et al. 2014; Xiao et al. 2019; Blackburn et al. 2020). Some RALF peptides are generated and activated by proteolysis of propeptides, whereas others have no propeptide sequence or cleavage site, characteristics that can affect the nature of their signaling activity (Stegmann et al. 2017). Another small family of peptides in Arabidopsis referred to as Peps also plays a key role in immune and other types of signaling (Bartels and Boller 2015; Dressano et al. 2020) leading to classic responses such as elevation of intracellular Ca^{2+}, ROS, and production of defense-related hormones such as JA and SA. In a positive feedback loop, they are generated from propeptides by Ca^{2+}-dependent metacaspases (Shen et al. 2019).

Plasma membrane receptors

To understand the receptors for CWI signaling that have been discovered in recent years, we first need to meet a cast of classical Greek and Etruscan characters, starting with FERONIA (FER), named after the Etruscan goddess of fertility, and others to be discussed later such as THESEUS1 (THE1), ERULUS (ERU), and HERCULES (HERK). These are members of a clade of 16 PM-localized proteins referred to as CrRLK1Ls that are within the superfamily of RLKs (Cheung and Wu 2011; Nissen et al. 2016; Zhu et al. 2021). CrRLK1L PM-localized receptors are characterized by their extracellular motifs resembling sugar-binding domains called malectins coupled to an intracellular kinase domain (Nissen et al. 2016). Depending on what ligands are bound, the receptors often interact through phosphorylation of guanine nucleotide exchange factors (GEFs) that enhance the downstream activity of plant-specific small GTPases called ROPs (Berken 2006). Further ROP signaling can lead to many other possible downstream events such as activation or inhibition (depending on the CrRL1K/GEF/ROP combination) of NADPH oxidase to produce ROS and of Ca^{2+} channels that when activated elevate $[Ca^{2+}{}_{cyt}]$ that can promote exocytosis of wall precursors and activate other downstream signals that lead to production of JA or SA (Chen et al. 2020a, 2020b; Tang et al. 2022). In turn, these hormones can lead to upregulated expression of proteins involved in defense responses or cell wall remodeling (Vlot et al. 2009; Ali and Baek 2020; Liu and Timko 2021).

FERONIA—a scaffold on which to create diverse signaling pathways. FER was first discovered in studies of pollen/stigma interactions (Huck et al. 2003). Because it has since been implicated in signaling for a multitude of pathways, FER has been described as a "scaffold" that can interact with different ligands (e.g. RALF peptides, OGAs) and/or co-receptor combinations to control different signaling pathways. The remarkable number of pleiotropic responses signaled by 1 FER protein and its partners are reviewed in depth elsewhere (Cheung and Wu 2011; Li et al. 2016a; Vaahtera et al. 2019; Chen et al. 2020a, 2020b; Xie et al. 2022). FER can play a role in inhibiting immune signaling in several ways. Figure 15A shows 1 example of how certain ligand and protein partners inhibit the outcome of PAMP-mediated ROS production that is signaled through a FLS2/BR11/BAK1 complex. Inhibition occurs when the pro- peptide of RALF23 is cleaved, and the resulting RALF23 inter-acts with the GPI-anchored co-receptor protein LLG1 (Li et al. 2015) that in turn interacts with FER, and this complex in turn inhibits PAMP-induced ROS production and subsequent immune signaling (Stegmann et al. 2017; Xiao et al. 2019) (Fig. 15A).

Two other examples compare FER's partner combinations for control of the auxin-regulated growth of elongating root cells with the complex and differing configurations used to regulate tip growth of root hairs. A study by Mike Sussman's group showed that when FER-LLG1 associates with RALF1, it leads to suppression of cell elongation through alkalinization of the cell wall through FER phosphorylation-induced inhibition of the PM H^+-ATPase AHA_2 (Haruta et al. 2014; Li et al. 2015) (Fig. 15B). Root hair tip growth is a somewhat different process that shows oscillations in which the growth phase is promoted by lowered apoplastic pH and elevated ROS gradients as well as the creation of a tip-focused $[Ca^{2+}{}_{cyt}]$ gradient stimulating actin reorganization and polar exocytosis of wall polymers to the tip (Zhang et al. 2022a, 2022b). One of FER's roles involves its association with LLG1 in root hair tips, activation of a specific associated ROP that further activates NADPH oxidase leading to apoplastic ROS production (Duan et al. 2010; Li et al. 2015; Zhang et al. 2022a, 2022b) (Fig. 15C; right). Another CrRKL1 root hair receptor called ERULUS is auxin induced, positively controls the phosphorylation of FER, and affects pectin dynamics through activation of PME activities (Schoenaers et al. 2018). Ground-breaking work that followed (Schoenaers et al. 2023) showed that RALF22 is expressed specifically in root hairs and primarily located in rings in the cell wall attached to a LRX1 (Fig. 15C, right). These studies also indicate that RALF 22 is not only a signaling peptide but also functions as part of a structural component that regulates the state of HG condensation. From their results, the authors envision an oscillation in which ME-HG, RALF22, and LRX1 are secreted at the tip during the low-pH wall-loosening phase, then RALF22 interacts with FER-LLG1 to inhibit the H + ATPase, and wall pH rises leading to the growth inhibition phase with increased PME activity that leads to a rise in DE-HG. DE-HG then binds to a separate complex of RALF22 with LRX1, leading to wall stiffening and formation of the ring structure that fixes the direction of expansion. The authors hypothesize that the wall stiffening may be a mechanosensory signal to the FER-LLG1-RALF22 complex as part of the growth cycle. More recently, a process strikingly analogous to the novel work described by Schoenaers et al. (2023) for Arabidopsis root hairs was reported for pollen

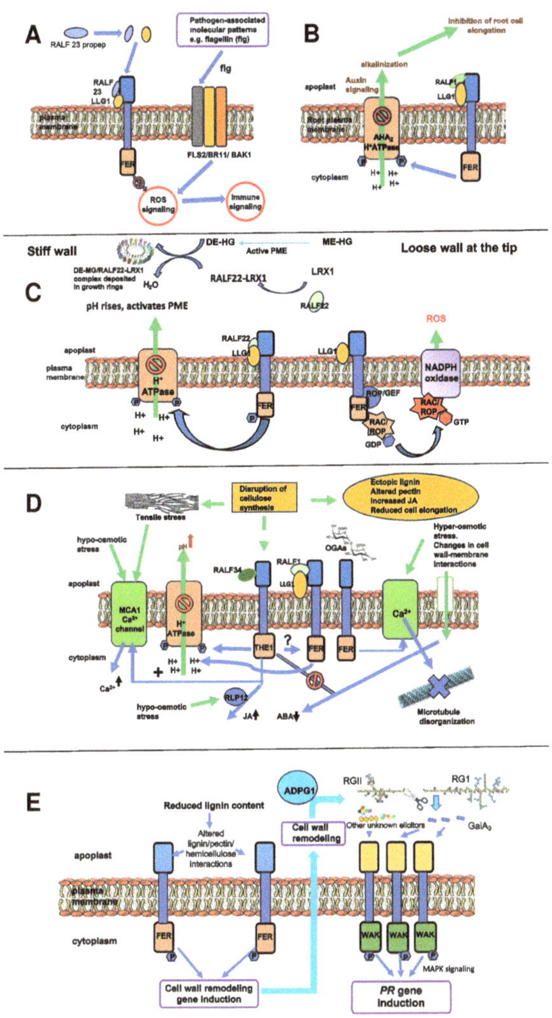

Figure 15. Examples of how variation in ligands and receptor/co-receptor combinations control cell wall signaling outcomes. **A)** A FER-LLG1-RALF23 combination inhibits a PAMP-mediated ROS production and subsequent immunity signaling (Stegmann et al. 2017; Xiao et al. 2019). **B)** The work of Haruta et al. (2014) demonstrated that RALF1 binding to FER complex (that was later shown to include co-receptor LLG1) leads to phosphorylation and inhibition of AHA_2 PM-ATPase. This leads to alkalinization of the cell wall and inhibition of root cell elongation. Later work indicates other RALF1-binding proteins may also be involved (not shown, see Dressano et al. 2017). **C)** The role of FER and an LRX1-DE-HG complex in regulating root hair elong-ation. Loss of function of ERULUS also regulates FER and AHA10 phos-phorylation (not shown). Relevant references are Duan et al. 2010, Li et al. 2015, Schoenaers et al. 2018 and 2023. **D)** Examples of different roles for ligands and/or mechano-sensing by THE1, FER, and Ca^{2+} chan-nels for responses to CWI and for the complex regulation involving RLP12 of hypo- (through JA) and hyper-osmotic stresses (through ABA). Relevant references are Hematy et al. (2007), Engelsdorf et al. (2018), Feng et al. (2018), Hamant et al. (2019), and Bacete et al. (2022). **E)** Induction of PR proteins as a result of engineered reduction of lignin content. FER senses altered wall integrity to induce the poly-galacturonase ADG1, which releases $GalA_3$ and other elicitors from pec-tin; these are recognized by WAKs resulting in induction of PR1 and PR10 (Gallego-Giraldo et al. 2020; Liu et al. 2023).

tubes by Moussu et al. (2023; see also Mohnen 2023). In pollen tubes, 2 molecules of RALF4 were shown to interact with 2 LRX8 molecules to form a positively charged heterotetramer at the pollen tube tip that can then interact with DE-HG to form organized structural complexes in the shank wall that play a very similar role to that in root hairs determining wall strength and shape.

Only in retrospect, through collectively re-examining the examples chosen for Fig. 15, did it occur to us that the pleiotropic character of FER must relate to its key role in balancing the strength of the outcomes of signaling pathways for growth (Fig. 15, B and C), hypo-osmotic (tensile) versus hyper-osmotic stress (Fig. 15D), and defense (Fig. 15, A and E). In a recent thought-provoking paper, Malivert and Hamant (2023) discuss the many different phenotypes gen-erated by *fer* mutants to try to identify a unifying feature that could underlie FER functions. They discuss many interesting options but seem to favor the idea that FER is fundamentally a mechanosensor modulated by different co-receptors and ligands such as pectins/OGAs and RALF peptides leading to changes in pH, Ca^{2+} and/or ROS that function to modulate the mechanical integrity of the cell.

Other receptors involved in CWI signaling. In a study that has defined an important experimental system for interrogating cell wall signaling, Herman Hofte and colleagues recognized that a number of different mutants of Arabidopsis impaired in cellulose synthesis either through mutations in CesAs or as-sociated proteins such as KORRIGAN or by use of the cellulose synthase inhibitor ISX all showed a set of characteristic re-sponses that include production of ectopic lignin, alterations in levels and structure of pectins, reduction in cell elongation, as well as increases in JA production. In a search for suppressor mutations of *Procuste* (*cesA6*), they identified a new, widely ex-pressed CrRLKL1 that they termed THESEUS1 (THE1), "a char-acter from Greek mythology who slew the brigand Procustes" (Hematy et al. 2007) (Fig. 15D). Importantly, they found that loss of THE1 sensing could attenuate the growth effects and production of ectopic lignin caused by the *cesA6* mutation and some other, but not all, *cesA* mutations (in 2007, it was not known that ISX does not inhibit the CSLD CSCs, and this needs to be kept in mind when reading papers that as-sume ISX is completely inhibiting all the cellulose synthesis). Gonneau et al. (2018) have shown that binding of RALF34 to THE1 promotes cytosolic $[Ca^{2+}{}_{cyt}]$ transients and cell wall alkalinization similar to that observed for RALF1 and FER (Fig. 15D). The *fer 1-4* or the *the1/fer 1-4 double* mutant lacked such responses to both RALF1 and RALF34 as well as the same effects of both peptides on upregulation of specific genes, sug-gesting THE1 may act upstream of FER or in parallel in a FER-dependent way.

Mechano-sensing is increasingly recognized as playing a key and perhaps essential role in CWI sensing (Hamant et al. 2019; Bacete and Hamann 2020). Loss of cellulose synthesis can create tensile stress that can affect mechano-sensitive Ca^{2+} channels, and THE1 acts upstream of the MCA1 ion channel that senses hypo-osmotic stress (Engelsdorf et al. 2018), while FER, through binding of OGAs to its malectin-like domains, promotes activity of a different Ca^{2+} channel that opens in response to hyper-osmotic conditions (Feng et al. 2018) (Fig. 15D). Such stress can also lead to microtubule disorganization that is controlled by FER-dependent and -independent pathways (Hamant et al. 2019; Malivert et al. 2021) (Fig. 15D). An exciting new study has tied together the functioning of THE1 with the interactions between sensing of CWI affected by swelling and PM stretching akin to hypo-osmotic stress, as opposed to drought or salinity that cause hyper-osmotic stress that causes membrane contrac-tion and plasmolysis. Bacete et al. (2022) showed that in ISX-treated cells, THE1 modulates both cell wall stiffness as measured by Brillouin microscopy and synthesis of JA, associated with CWI sensing. In contrast, treatment with sorbitol (osmotic stress) stimulated not JA but ABA production, while, with both ISX and sorbitol, JA synthesis occurred but ABA synthesis was inhibited. Stimulation of JA synthesis was under control of a protein called RLP12 that was only expressed in the presence of both sorbitol and ISX (Fig. 15D). Also important was that sorbitol-induced ABA production was not observed in protoplasts, indicating that it involves PM wall connections—most likely the Hecht's threads observed in plasmolyzed cells (Yoneda et al. 2020). Overall, it appears that THE1 is a positive regulator of JA synthesis and negative regulator of ABA synthesis (Fig. 15D). THE1 can also work in concert with another receptor kinase, HERCULES, and FER to mitigate the stress of heavy metals on cell elongation (Richter et al. 2017) and with the LRR kinase FEI2 to induce JA synthesis associated with CWI. Because THE1 is not required for pattern-triggered immunity, it is likely specifically involved in CWI signaling (Engelsdorf et al. 2018). However, Pogorelko et al. (2013) and Bacete et al. (2018) have made clear in their reviews that there are many examples of significant cross-talk between CWI and defense signaling that are not reviewed here.

A direct role of oligosaccharins in defense signaling first became apparent from the important early discovery that they are sensed through their interaction with wall-associated receptor kinases called WAKs (Kohorn et al. 2006, 2021b), leading to induced defenses through activation of downstream MAPK1 signaling pathways (Fig. 15E). In addition to WAKs, many other receptors (including FER; see Fig. 15A) are involved in immune signaling (Narváez-Barragán et al. 2022). The pathogen-related expression pattern of a cluster of 5 other *CrRLK1L* genes suggests a specific role for other CrRLK1Ls in the interaction with pathogens (Lindner et al. 2012). Another interesting example involves the Arabidopsis immune peptides called Peps that are widely distributed in the plant kingdom (Lori et al. 2015); the PROPEP genes are induced by signals such as OGAs, Peps, and defense hormones. Peps bind to a receptor complex of PEPR and other co-receptors, setting off a series of classic immune responses and/or repressing CWI signaling (Shen et al. 2019; Dressano et al. 2020).

Signaling involving SCWs

Although mostly studied in the context of PCWs, signaling may also be triggered following modification of SCWs. For example, reduced growth associated with ectopic activation of defense responses, particularly synthesis of pathogenesis-related (PR) proteins, has sometimes confounded attempts to reduce recalcitrance of SCWs to digestion through reduction of lignin content or modification of lignin composition (Gallego-Giraldo et al. 2011, 2018). Defense trade-offs have been suggested as 1 possible explanation. This is not always the case, because some suppressor mutants of the growth reduction in plants with reduced lignin levels still exhibit defense gene expression (reviewed in Ha et al. 2021). The recent demonstration that some PR proteins are themselves cell wall localized, and function via generating small peptide defense signals (Chen et al. 2020a, 2020b), suggests that PR protein induction could be viewed as part of a broader program of cell wall remodeling that occurs following genetic modification of cell wall polymers.

Although pectin is considered a quantitatively minor component in SCWs and xylan the main wall component interacting with lignin (Kang et al. 2019), emerging evidence points to the possibility that pectin is a critical part of an underlying SCW structural scaffold that encompasses multiple wall components, including xylan and lignin (Hao and Mohnen 2014; Li et al. 2019). Modification of pectin through downregulation of GAUT4 increases both growth of and saccharification efficiency in switchgrass, the latter being a characteristic result of lignin reduction in SCWs (Biswal et al. 2018). Conversely, reduced lignification (e.g. in the *cinnamoyl CoA reductase 1* mutant of Arabidopsis) causes solubilization of pectic elicitors that activate production of a suite of *PR* genes (Gallego-Giraldo et al. 2020). This process involves the FERONIA-dependent transcriptional activation of ADPG1 (Liu et al. 2023), a polygalacturonase normally associated with anther dehiscence, a process that is also dependent on correct lignin deposition (Dai et al. 2019). The ectopic expression of ADPG1 is not itself responsible for the initial release of pectic material from the wall but rather with its processing to elicitor-active moieties, which are then recognized by WAK receptors leading to induced defenses (Gallego-Giraldo et al. 2020; Liu et al. 2023) (Fig. 15E).

Coordinating signal pathways for growth, development, and stress

Clearly activities that occur within the cell wall sit at 1 nexus that controls the regulation of growth, development, and responses to stress. All the processes of cell wall biosynthesis and secretion and polymer modifications and interactions, when functioning normally, are essential to carry out the complex programs involved in plant growth and development. This section has addressed the ways in which cell walls sense when normal growth is disrupted and set in motion processes designed to mitigate such stresses and that these processes rely on similar or sometimes identical ligands, receptors, co-receptors, and downstream processes used for growth. The wall as a shapeshifter is made even more clear in the final section that will examine the changing nature of the cell wall as it plays a key role in coordinated patterns of cell division, expansion, maturation, and programmed cell death (PCD). Again, we shall see that hormones play key roles as input and output signals, and the levels of both intracellular and extracellular Ca^{2+}, ROS, and pH are recurring key actors. Understanding how the signals involved in all these events are coordinated is certainly one of the future challenges facing cell wall signaling.

The life of a cell wall

Although not often discussed, cell walls have 4 distinct life stages: birth at the cell plate, tip, or diffusive growth, SCW maturation, and sometimes prolonged death long after that of their parent cells. Cell plate, tip growth, and maturation show some interesting and rather surprising similarities in terms of wall deposition. The outlier—the process of expansion growth—is one of most-studied processes in all of plant biology and has always seemed different and perhaps the most complex. This section examines these stages, placing emphasis on similarities and differences in wall structure and function between the life stages.

Birth at the cell plate

Cell walls are born during the process of cell division which involves building a new wall where none existed before. Obvious places of birth are stem and root apical meristems, but there are other important meristems such as those in leaf primordia and the cambium. Work by Peter Hepler on the dynamic behavior of the cytoskeleton recognized the importance of temporal and spatial changes in intracellular Ca^{2+} to cell plate formation and tip growth (Hepler 2005). Andrew Staehelin, who coupled a strong intellect to interpretation of his outstanding EM images, contributed greatly to our understanding of membrane dynamics and secretory processes. Hepler and Staehelin's work and several seminal reviews laid the foundation for our current understanding of cell plate formation (Samuels et al. 1995; Staehelin and Hepler 1996).

The initial establishment of a Ca^{2+} gradient focused on the developing cell plate is essential to the process, and condi-tions that dissipate this gradient disrupt cell plate formation (Staehelin and Hepler 1996). Ca^{2+} is important for cytoskeletal dynamics, vesicle fusion, callose synthesis, and pectin structure, all relevant to cell birth. During anaphase the spindle MT array gives way to a new MT array that defines the central position of the cell plate; actin microfilaments also form in this region and both are oriented with their plus ends to the center. This growing structure, called the phragmoplast, becomes a complex array of MTs, actin microfilaments, ER- and Golgi-derived vesicles (Samuels et al. 1995), and/or PM-derived endocytotic vesicles (Baluska et al. 2003) that delivers materials necessary for wall formation and is dismantled upon completion of the new wall. Vesicles fuse to form tubular structures that become the new PMs separated by the newly deposited wall.

Extensin3 (EXT3) was proposed to be deposited and cross-linked to form a scaffold on which to build the cell plate (Cannon et al. 2008). This has recently been questioned (Doll et al. 2022), although it is clear that, in the presence of extensin peroxidase, EXT3 undergoes self-recognition and polymerization into rope-like and cross-linked dendritic assemblies via end-on and lateral association, as revealed by AFM (Fig. 10D). ME-HG is also deposited early and is gradually demethylated, but whether DE-HG interacts with the positively charged EXT3 or some other similarly charged protein to form a "scaffold" requires further examination. Moore and Staehelin (1988) detected xyloglucan showing abundant labeling early in vesicles surrounding the developing plate and also within the plate. Callose is then detected in the well-developed tubular structures in a rare case where it serves as a wall component instead of defense polymer. Callose is by far the most abundant polymer of cell plates, and studies with an inhibitor of synthesis indicate that it is absolutely essential (Park et al. 2014; Drakakaki 2015). Figure 16 shows an impressive image of a developing plate from Arabidopsis that we suggest appears to show 2 different layers that might represent the early scaffold followed by a callose-rich layer.

Cellulose deposition begins sometime after callose synthesis, and the process differs from that in elongating cells because it involves CSLD proteins. In Arabidopsis cell plates, Gu et al. (2016) reported early deposition of cellulose by AtCSLD5 while AtCESA3 was more active at a much later stage, and once the new wall merges with the parent wall, CesA proteins can be seen migrating to the new wall; other reports indicate some overlap in CSLD and CESA expression (Miart et al. 2014; Chen et al. 2018). Wu et al. (2023) have shown that, in moss protonema that display both tip growth and cell divisions, PpCSLD6 (and not CesA) is active in both tips and cell plates; it was also sensitive to DCB but not ISX providing an easy way to distinguish CSLD from CESA activity.

At present, it is not clear why CSLDs are the preferred cellulose synthases for both cell plate and tip growth. One possibility is that CSLDs, but not CESAs, are tolerant of the high Ca^{2+} gradient at the center of cell plates and also at the tip of root hairs and pollen tubes. It may also be important that the products of CSLD synthesis are less oriented and/or less crystalline (Yang et al. 2020b). There can be interactions between β-1,3-glucans and cellulose that result in changed properties in hydrogels or aqueous solutions (Abou-Saleh et al. 2018). Such a hybrid network may be more easily produced when the cellulose MFs are more disordered and less crystalline and such a new composite may be more suitable for the new wall—an area rich for further investigation. As the cell plate matures and joins the parent wall, it is reported that callose is digested (Drakakaki 2015), although it is not

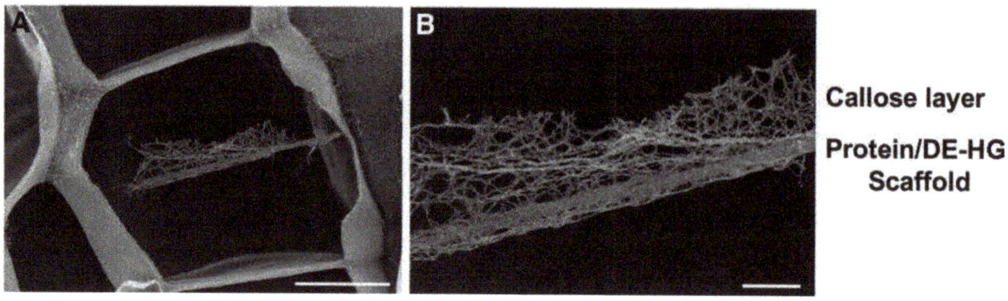

Figure 16. The developing cell plate in Arabidopsis. **A)** The direction of the Ca^{2+} gradient. **B)** An expanded version of **A)**. Modified from Fujita and Wasteneys (2014) to indicate the calcium gradient and the positioning of a proposed location of a callose layer and a scaffold.

entirely clear whether some or all may be masked by the beginning of "normal" wall synthesis.

Growth

Tip-growing cells. The major targets for study, the pollen tube of Lily and both the pollen tube and root hair of Arabidopsis, have been excellently reviewed (Cheung and Wu 2011; Rounds and Bezanilla 2013; Johnson et al. 2019; Çetinbaş-Genç et al. 2022). This discussion focuses mostly on root hairs and some of the interesting similarities with cell plate formation. The deposition of wall polymers at the tip also occurs where no wall exists and relies upon the establishment of a [$Ca^{2+}{}_{cyt}$] tip-focused gradient but with rapid (~30 s) extracellular oscillations that parallel those of protons and ROS but lag slightly behind the auxin-induced oscillations in growth (Monshausen et al. 2007; Zhang et al. 2022b). Wall construction begins with deposition of ME-HG at the tip that gradually becomes DE-HG through the action of PMEs. Beat Keller's group (Baumberger et al. 2001) discovered that the leucine-rich extensin LRX1 becomes insolubilized in root hair cell walls, providing a very early suggestion that it may function as a scaffold along with DE-HG. Xyloglucan is also deposited early, and it is interesting that an unusual acidic xyloglucan (that potentially could also interact with LRX1) is deposited in root hairs but not pollen tubes (Pena et al. 2012). As for cell plates, callose deposition begins later, but with some differences: root hairs are relatively richer in cellulose, although they do deposit some callose, while pollen tubes and cell plate walls are very rich in callose [although some is found in callose plugs in pollen tubes (Kapoor and Geitmann 2023)]. Pollen tubes were one of the first places where CSLDs were shown to function (Doblin et al. 2001) but, in the same year, Favery et al. (2001) showed that the KOJAK gene that encodes AtCSLD3 is necessary for root hair growth. Subsequently it was shown that AtCSLD2 and AtCSLD3 but not CESAs function in cellulose synthesis in root hairs of Arabidopsis (Bernal et al. 2008: Park et al. 2011).

Signaling networks for tip growth are complex and involve many of the players described for CWI signaling as well as those used for diffusive growth (Kroeger et al. 2008; Zhang et al. 2022b). As previously discussed (see Fig. 15C), Schoenaers et al. (2023) provide perhaps the best description yet of oscillations in signals and wall structure that promote either wall-loosening or stiffening and correlate with oscillations in growth in root hairs.

Expansion (diffusive) growth. Studies of expansion growth began with the discovery of auxin by Fritz Went and have continued on for decades by other pioneers such as James Lockhart, Bob Cleland, and Dan Cosgrove (see Du et al. 2020b for recent review). To reiterate briefly, the driving force for growth is turgor, and, if cellulose MFs are aligned perpendicular to the axis of growth, the cell elongates. Auxin promotes wall acidification through activation of PM H^+-ATPase (AHA2) that leads to "acid growth" through wall-loosening that creates new "space" that is filled in by deposition of new wall polymers. Wall remodeling (stiffening) that follows is assumed to involve mechanisms that work in opposition to wall-loosening.

To take a new look at expansion growth, we ask whether it is fundamentally not so very different from tip growth, as has also been implied by others (Wolf and Hofte 2014; Wolf et al. 2014; Höfte 2015; Jobert et al. 2023). It seems not too difficult to imagine such growth as an oscillation between a loose wall and a stiff wall. In a surely oversimplified scenario for diffusive growth, the loose wall has a lower pH, is relatively enriched in newly synthesized polymers (and therefore should initially have a high ratio of ME-HG to DE-HG), has a low level of cross-linking of pectic and/or protein polymers, and newly deposited cellulose is kept from interacting with itself and other polymers by expansin. The stiff wall follows in rapid succession as pH rises, ME-DG is converted to DE-HG promoted by PME, much more stiffening occurs through Ca^{2+}-bridges and/or HG condensation through interaction with positively-charged proteins, peroxidase-mediated activity, and borate cross-links for RGII. The concept of this "loose wall" and the "stiff wall" could easily be consistent with oscillations in Ca^2, ROS (for crosslinking and redox control), new wall polymer deposition, and wall pH (activity of PME vs PMEI, expansin, and THE1 and FER are also all pH regulated). This concept is consistent with FER and THE1 being positive regulators of cell wall stiffness (Höfte 2015; Bacete et al. 2022). But we do not see these 2 different types of walls when we isolate growing tissues because we isolate a mixture of loose and stiff walls.

If this oversimplified scenario has any truth to it, then the larger questions are what feedback loops involving auxin and/or other signals regulate the oscillations, what is the period, where are they spatially located, and why do we not see the process as a growth oscillation? Perhaps because it is not as localized as is tip growth? Perhaps it might be regulated by oscillations of only seconds to minutes and localized to a multitude of growth sites? There is 1 apparent oscillation occurring in multiple places that we can actually see at a microscopic level in elongating hypocotyls: CESAs have a residence time and period of activity in the PM of 7 to 10 minutes after which they become inactive and are internalized where they are targeted for degradation or recycling. Later new and recycled CESAs are inserted along the same MT-aligned path. Assuming deposition of other wall polymers is also locally co-ordinated with cellulose deposition (Ganguly et al. 2020), this is a highly localized oscillation in wall deposition that would occur in a multitude of sites that would be difficult to detect. Hepler and colleagues argued that a measured oscillation in exocytosis of wall materials in tip-growing pollen was a driv-ing force for oscillations in wall thickness and growth (McKenna et al. 2009). A cycle driven by an oscillation in aux-in action leading to oscillations in wall pH, internal Ca^{2+} le-vels and ROS, might fit with an assumption that CESAs are inhibited by Ca^{2+} (something that needs direct investigation) and would require the substitution of CSLDs for CESAs when the Ca^{2+} gradient is tip focused or high at the cell plate and also promote callose synthesis. For such diffuse growth there may be no need for an LRX1-DE-HG scaffold complex or callose. LRX1 is specific for tip growth (Carol and Dolan 2002), and perhaps there is some mechanism to keep callose synthesis inhibited, 1 possibility being its association with the abundant Ca^{2+} -binding annexin (Andrawis et al. 1993).

Exciting recent work indicates feedback loops ranging from seconds to minutes between auxin, wall pH, Ca^{2+}, and pectin remodeling, reviewed by Jobert et al. (2023). They suggest a possibility of special regulation within microdomains wherein there are feedback loops between the state of auxin transporters and cell wall structures. It takes a few minutes for auxin's effects on transcription to show up [e.g. as upregulation of key regulators such as ERU, H^+ ATPases, and small auxin up-regulated RNAs (Du et al. 2020b; Jobert et al. 2023)]. But auxin can also cause very rapid (seconds) fluxes in wall pH in roots that are in opposition to the canonical regulation of AHA2 by auxin carried out through transcriptional regulation (Li et al. 2021).

If we could stop the process at one point in the cycle, would we then see a wall structure that is largely loose or stiff? Perhaps DCB-adapted cells with their pectin-based walls (Shedletzky et al. 1990, 1992) or the alternating pectin/cellu-lose wall layers in celery collenchyma cells (see Fig. 9) offer clues. Or perhaps walls with disturbances on a larger spatial scale such as during hypocotyl hook formation or gravitrop-ism (Jobert et al. 2023). Stay tuned; acid growth has never seemed so exciting!

Maturation

Cell wall maturation is defined here as the production of a SCW that is traditionally said to consist of S-1, S-2, and S-3 layers. The process is assumed to begin when elongation of a cell ceases, but there is sometimes an interesting overlap between these 2 processes during deposition of the S-1 layer. This transition cell wall (TCW) is the most interesting, and the one to which we give most attention. In view of space limitations, we leave discussion of many other aspects of SCW development, including regulation of lignification, to others (Turner et al. 2007; Hobson et al. 2010; Li et al. 2016c; Meents et al. 2018; Rao and Dixon 2018; Liu et al. 2021). In some ways, the making of a TCW is akin to deposition of a new wall because it is different from the PCW and is laid down without integration of wall material into the PCW. Thus, maturation might use the TCW as a scaffold upon which to deposit new material.

The cotton fiber is a unique cell type and has many advantages for study of the development of cell walls (Haigler et al. 2012). It originates from epidermal cells of the ovule. However, evidence indicates that it displays both tip growth and diffusive growth along the length of the fiber (Qin and Zhu 2011). Fiber cells within a boll develop synchronously from the day of anthesis; fibers elongate (up to about 2 cm) and deposit a typical dicot PCW comprised mostly of cellulose, xyloglucan and pectins (Meinert and Delmer 1977). At about 16 days post-anthesis, TCW synthesis begins. The fiber continues to elongate while beginning the deposition of this new TCW called the "winding layer" that is analogous to the S-1 layer of wood cells. From about 24 days onward, the fiber deposits almost exclusively cellulose in S-2 and S-3 layers resulting in a wall that is >90% cellulose by weight at maturity.

The idea that the TCW serves as a scaffold occurred to us because we realized that there are some interesting similarities to wall deposition in cell plates and tip-growing cells. The beginning of the TCW is characterized by a remodeling of pectin resulting in a net loss of uronic acids from walls that occurs at the same time the middle lamella between fibers degrades (Meinert and Delmer 1977; Singh et al. 2009). The expression of 2 Rac/ROP GTPases occurs precisely during this TCW phase of fiber growth (Delmer et al. 1995). Potikha et al. (1999) demonstrated H_20_2 production associated with the TCW stage, and early addition of H_20_2 caused premature initiation of TCW formation while an inhibitor of NADPH oxidase inhibited initiation. Strikingly, these observations are very similar to those observed for initiation of tracheary element differentiation in Zinnia cell cultures (Karlsson et al. 2005). Roberts and Haigler (1989) also presented evidence for a rise in Ca^{2+} levels in tracheary elements during the transition, and significant callose deposition is initiated early in the TCW stage in cotton fibers concomitant with first signs of SCW cellulose, a key component of the winding layer and showing no evidence of turnover (Maltby et al. 1979). As far as we can tell, there is no report of callose in S1-layers of wood, although it is found in tracheids of compression wood where it is called Larcinan (Boyd 1978) and in the G-layer of flax phloem fibers (Ibragimova et al. 2017). Primary wall CesAs apparently continue to be active with a gradual transition to secondary wall CesAs during the TCW transition (Tuttle et al. 2015; Watanabe et al. 2018). CslDs are expressed at all stages of fiber development and their role, if any, needs clarification (Li et al. 2017). In sum, it appears that the cotton fiber winding layer (as a scaffold?) is initiated by elevation of levels of intracellular Ca^{2+}/ROS, production of DE-HG followed by a callose-rich layer mixed with cellulose perhaps made by both PCW and SCW CesAs. There is other very interesting evidence that there may be a pectin-rich scaffold in cells of wood since transgenic modification of poplar to make less HG leads to acceleration of growth and improvement in di-gestibility of the walls (Biswal et al. 2018). This may also relate to the finding that overexpression of an enzyme that degrades RGI enhances degradation of the middle lamella and also enhances growth in poplar (Yang et al. 2020a).

As for the later synthesis of the S-2 and S-3 layers, synthesis of xylans, glucomannans and lignin have already been discussed. CSCs increase significantly in numbers and velocity of move-ment and become more aggregated (Watanabe et al. 2015). Cotton fibers, bast fibers, and tracheary elements all share the same primary transcriptional regulation and the same set of genes is involved in SCW cellulose synthesis. An early transcriptome analysis to identify a common gene network for SCW synthesis in cells such as cotton fibers, xylem, and cellulose-rich tension wood identified 52 genes that met a very wide range of criteria (Ko et al. 2006). Among the genes on the list are NAC and myb TFs, genes encoding enzymes for all the major SCW polymers and, interestingly, also pectin synthesis. Not surprisingly, the cellulose-rich cotton and bast fi-bers all downregulate the transcription factors that induce de-position of lignin and hemicelluloses (Tuttle et al. 2015).

SCW synthesis represents a major carbon sink in maturing cells (Verbančič et al. 2018; Pottier et al. 2023). As discussed in those reviews, the source of the UDP-Glc that is the substrate for both cellulose and callose synthesis still remains controversial for some. For cotton fibers that make only callose and cellulose and little or no starch during SCW synthesis, the carbon comes from sucrose that enters the fibers through plasmodesmata. Amor et al. (1995) showed that 80% of the highly expressed sucrose synthase (SuSy) becomes associated with the plasma membrane fraction during SCW callose and cellulose synthesis, and Salnikov et al. (2001) also showed similar PM localization in developing zinnia tracheary elements. Amor et al. (1995) and Haigler et al. (2001) proposed this SuSy may channel carbon directly to cellulose or callose, a pathway that saves 1 ATP worth of energy for each Glc residue polymerized com-pared with the alternate pathway involving invertase and UDP-Glc pyrophosphorylase, In addition, recycling of the UDP (that inhibits glucan synthases) back to PM-SuSy would prevent its localized accumulation. This work also showed that supplied UDP-Glc can support callose but not cellulose synthesis in permeabilized fibers while sucrose can supply carbon for synthesis of both, another argument for possible in vivo channeling of UDP-Glc directly to the active site of CSCs. This PM-SuSy is most likely the same isoform found to be up-regulated by Arpat et al. (2004), and Brill et al. (2011) also showed strong up-regulation of a special SuSy in PM and wall fractions of cotton fibers starting during the TCW stage. However, as discussed by Verbančič et al. (2018) and Pottier et al. (2023), this role for SuSuy has been questioned, primarily due to mutant analyses showing that SuSy clearly is not absolutely necessary in Arabidopsis (Wang et al. 2022a, 2022b), and mutant analysis of cytoplasmic invertases, that may be important for more reasons than supplying carbon for cellulose synthesis, are essential for normal seedling growth (Barratt et al. 2009). However, both SuSy and UDP-Glc pyrophosphorylase are found in high levels during both PCW and SCW phases and both catalyze reversible reactions, suggesting important roles for each and that one should also be available to substitute when the other is absent. Verbančič et al. (2018) and Pottier et al. (2023) do conclude that there is certainly strong evidence to support an important role for PM-localized SuSy in SCWs. Winter and Huber (2010) have shown that SuSy becomes associated with actin upon Ca-dependent phosphorylation and, its association with actin directs CSCs to appropriate sites on the PM during SCW (Lei et al. 2012)]. Whether such associations might relate to possible channeling of carbon directly through the gating loop of CSCs in vivo deserves further study.

One key step has been missing from the life of the secondary wall—what are the signals that turn on the transcription factors? Clearly hormones are involved (Haigler et al. 2012; Demura 2013; Meents et al. 2018). Just as we were finishing writing this review, a landmark paper appeared that opens up new avenues for this area of research. Cai et al. (2023) have shown that different variants of WAK10, in combination with a brassinosteroid receptor and modulated by binding of OGAs of different sizes and affinities, control the expression of CesA 4,7, and 8 genes, but not lignin, in rice. This process leads to either taller or shorter rice with less or more secondary wall cellulose. Furthermore, breeding of rice has gradually selected a whole set of WAK10 variants leading to shorter or longer rice. This may tie in with the appearance of OGAs at the TCW transition that can bind to and activate WAKs. All of this invokes shades of the Green Revolution and suggests new approaches for regulation of synthesis of cell walls in other crops and particularly those used for biofuels.

Death

Just as the cell signaling community seems to revere Greek gods, the PCD community has fun asking "Was it murder or suicide? What were the murder weapons? Where did they put the corpse?" The emerging answers seem to be: "It was suicide, and, as we have seen for life, the chief weapons are Ca^{2+} and ROS (and probably FER as well) and, as for the corpse, it gets eaten." Intracellular Ca^{2+} and ROS levels certainly rise as an early event of cell death (Roberts and Haigler 1989), and it may be that the regulation of SCW synthesis and PCD are coordinated. A 40-kD serine protease appears at initiation of SCW formation and builds up over time until it appears to serve as the same signal to induce PCD (Groover and Jones 1999). As for the corpse, in some cases the entire cell including the cell wall is digested, but the walls of cells like cotton fibers, bast fibers, vessels, and fiber cells of xylem all survive—sometimes for millennia—in the cotton robes and wooden caskets found in the tombs of the pharaohs.

There are 2 kinds of PCD: one that is developmentally programmed such as that seen for tracheary elements, and the other that is involved in defense against pathogens (pPCD). pPCD can be of the type that is involved in the hypersensitive response (HR-pPCD) to biotrophic and hemi-biotrophic pathogens, whereas another form might be called nec-pPCD in which necrotrophic pathogens digest everything, including walls. Being able to understand the differences between these is very important for devising proper defense strategies, but that gets beyond the scope of this review. The details of plant PCD are all still emerging; the reviews of Huysmans et al. (2017) and Ustun et al. (2017) summarize the overall state of our knowledge.

In sum, it is quite remarkable how the same signaling pathways can be modified in one way or another during the life of the plant and how important the cell wall is for that life. But the life story of most cell walls ends well. Most walls survive for a very long time, carrying out their many functions serving not only their parent plants but the entire global ecosystem.

Challenges for the next 100 years

- What structures do the different wall polymer types have at the time they are synthesized and deposited in walls of specific cell types?
- What are the interactions between and among polymer, and how do these interactions change, both temporally and spatially, for walls with different structures and functions?
- How are all these processes regulated and coordinated?

Acknowledgments

The authors thank the cell wall community that has inspired the work described in this review and, of course, our patient spouses.

Author contributions

All authors contributed significanlty to the preparation of this review. D.D. was the primary author of the "cellulose" and "life of a cell wall" sections, R.A.D. was the primary author of the "lignin" section, K.K. was the primary author of the "hemicellulose" and "model" sections, and D.M. was the primary author of the "pectin" and "wall-associated protein" sections. D.D. and R.A.D. were the primary authors of the "signaling" section.

Funding

Work in the laboratories of R.D. and D.M. was partially funded from The Center for Bioenergy Innovation, a US Department of Energy Research Center supported by the Office of Biological and Environmental Research in the DOE Office of Science and, for D.M., partially from the US Department of Energy, Office of Science, Basic Energy Sciences, Chemical Sciences, Geosciences and Biosciences Division under award no. DE-SC0015662.

Conflict of interest statement. None declared.

Data availability

This historical review contains no new data.

References

Abidi W, Decossas M, Torres-Sánchez L, Puygrenier L, Létoffé S, Ghigo J-M, Krasteva PV. Bacterial crystalline cellulose secretion via a supramolecular BcsHD scaffold. Sci Adv. 2022:**8**(50):eadd1170. https://doi.org/10.1126/sciadv.add1170

Abou-Saleh RH, Hernandez-Gomez MC, Amsbury S, Paniagua C, Bourdon M, Miyashima S, Helariutta Y, Fuller M, Budtova T, Connell SD, et al. Interactions between callose and cellulose revealed through the analysis of biopolymer mixtures. Nat Commun. 2018:**9**(1):4538. https://doi.org/10.1038/s41467-018-06820-y

Agarwal UP, Ralph SA. FT-Raman spectroscopy of wood: identifying contributions of lignin and carbohydrate polymers in the spectrum of black spruce (*Picea mariana*). Appl Spectroscopy. 1997:**51**(11): 1648–1655. https://doi.org/10.1366/0003702971939316

Albersheim P. The walls of growing plant cells. Sci Am. 1975:**232**(4): 80–95. https://doi.org/10.1038/scientificamerican0475-80

Albersheim P, Bonner J. Metabolism and hormonal control of pectic substances. J Biol Chem. 1959:**234**(12):3105–3108. https://doi.org/10.1016/S0021-9258(18)69630-9

Albersheim P, Darvill A, Roberts K, Sederoff R, Staehelin A. Plant cell walls. London: Garland Science, Taylor & Francis; 2011, 430 pp.

Alejandro S, Lee Y, Tohge T, Sudre D, Osorio S, Park J, Bovet L, Lee Y, Geldner N, Fernie AR, et al. AtABCG29 is a monolignol transporter involved in lignin biosynthesis. Curr Biol. 2012:**22**(13):1207–1212. https://doi.org/10.1016/j.cub.2012.04.064

Ali MS, Baek K-H. Jasmonic acid signaling pathway in response to abiotic stresses in plants. Int J Mol Sci. 2020:**21**(2):621. https://doi.org/10.3390/ijms21020621

Aloni Y, Delmer DP, Benziman M. Achievement of high rates of in vitro synthesis of 1, 4-beta-D-glucan: activation by cooperative interaction of the *Acetobacter xylinum* enzyme system with GTP, polyethylene glycol, and a protein factor. Proc Natl Acad Sci U S A. 1982:**79**(21):6448–6452. https://doi.org/10.1073/pnas.79.21.6448

Altanassov II, Pittman JK, Turner SR. Elucidating the mechanisms of assembly and subunit interaction of the cellulose synthase complex of Arabidopsis secondary cell walls. J Biol Chem. 2009:**284**(6): 3833–3841. https://doi.org/10.1074/jbc.M807456200

Amor Y, Haigler CH, Johnson S, Wainscott M, Delmer DP. A membrane-associated form of sucrose synthase and its potential role in synthesis of cellulose and callose in plants. Proc Natl Acad Sci U S A. 1995:**92**(20):9353–9357. https://doi.org/10.1073/pnas.92.20.9353

Amos RA, Atmodjo MA, Huang C, Gao Z, Venkat A, Taujale R, Kannan N, Moremen KW, Mohnen D. Polymerization of the backbone of the pectic polysaccharide rhamnogalacturonan I. Nat Plants. 2022:**8**(11):1289–1303. https://doi.org/10.1038/s41477-022-01270-3

Amos RA, Mohnen D. Critical review of plant cell wall matrix polysaccharide glycosyltransferase activities verified by heterologous protein expression. Front Plant Sci. 2019:**10**:915. https://doi.org/10.3389/fpls.2019.00915

Amos RA, Pattathil S, Yang JY, Atmodjo MA, Urbanowicz BR, Moremen KW, Mohnen D. A two-phase model for the non-processive biosynthesis of homogalacturonan polysaccharides by the GAUT1:GAUT7 complex. J Biol Chem. 2018:**293**(49):19047–19063. https://doi.org/10.1074/jbc.RA118.004463

Amsbury S, Hunt L, Elhaddad N, Baillie A, Lundgren M, Verhertbruggen Y, Scheller HV, Knox JP, Fleming AJ, Gray JE. Stomatal function requires pectin de-methyl-esterification of the guard cell wall. Curr Biol. 2016:**26**(21):2899–2906. https://doi.org/10.1016/j.cub.2016.08.021

Anders N, Frederick L, Wilson L, Sorieul M, Nikolovski N, Dupree P. β-1,4-xylan backbone synthesis in higher plants: how complex can it be? Front Plant Sci. 2023:**13**:1076298. https://doi.org/10.3389/fpls.2022.1076298

Anderson DB. Uber die struktur der kollenchymzellwand auf grund mikrochemischer untersuchungen. Sitzungsberichte der Akademie der Wissenschaften Mathematisch-Naturwissenschaftliche Klasse. 1927:**136**:429–329. Available at: https://www.zobodat.at/pdf/SBAWW_136_0429-0439.pdf

Anderson CT, Kieber JJ. Dynamic construction, perception, and remodeling of plant cell walls. Annu Rev Plant Biol. 2020:**71**(1):39–69. https://doi.org/10.1146/annurev-arplant-081519-035846

Andrawis A, Solomon M, Delmer DP. Cotton fiber annexins: a potential role in the regulation of callose synthase. Plant J. 1993:**3**(6):763–772. https://doi.org/10.1111/j.1365-313X.1993.00763.x

Arioli A, Peng L, Betzner A, Burn J, Wittke W, Herth W, Camilleri C, Hofte H, Plazinski J, Birch R, et al. Molecular analysis of cellulose biosynthesis in *Arabidopsis*. Science. 1998:**279**(5351):717–720. https://doi.org/10.1126/science.279.5351.717

Arpat AB, Waugh M, Sullivan JP, Gonzales M, Frisch D, Main D, Wood T, Leslie A, Wing RA, Wilkins TA. Functional genomics of cell elongation in developing cotton fibers. Plant Mol Biol. 2004:**54**(6):911–929. https://doi.org/10.1007/s11103-004-0392-y

Arsovski AA, Haughn GW, Western TL. Seed coat mucilage cells of *Arabidopsis thaliana* as a model for plant cell wall research. Plant Signal Behav. 2010:**5**(7):796–801. https://doi.org/10.4161/psb.5.7.11773

Aspinall GO, Molloy JA, Craig JWT. Extracellular polysaccharides from suspension-cultured sycamore cells. Can J Biochem. 1969:**47**(11):1063–1070. https://doi.org/10.1139/o69-170

Atalla RH, Vanderhart DL. Native cellulose: a composite of two distinct crystalline forms. Science. 1984:**223**(4633):283–285. https://doi.org/10.1126/science.223.4633.283

Atmodjo MA, Sakuragi Y, Zhu X, Burrell JA, Mohanty SS, Atwood JA III, Orlando R, Scheller HV, Mohnen D. GAUT1:GAUT7 are the core of a plant cell wall pectin biosynthetic homogalacturonan:galacturonosyltransferase complex. Proc Natl Acad Sci U S A. 2011:**108**(50):20225–20230. https://doi.org/10.1073/pnas.1112816108

Augur C, Yu L, Sakai K, Ogawa T, Sinaÿ P, Darvill AG, Albersheim P. Further studies of the ability of xyloglucan oligosaccharides to inhibit auxin-stimulated growth. Plant Physiol. 1992:**99**(1):180–185. https://doi.org/10.1104/pp.99.1.180

Ayers AR, Valent B, Ebel J, Albersheim P. Host-pathogen interactions. XI. Composition and structure of wall-released elicitor fractions. Plant Physiol. 1976:**57**(5):766–774. https://doi.org/10.1104/pp.57.5.766

Aznar A, Chalvin C, Shih PM, Maimann M, Ebert B, Birdseye DS, Loqué D, Scheller HV. Gene stacking of multiple traits for high yield of fermentable sugars in plant biomass. Biotechnol Biofuels. 2018:**11**:2. https://doi.org/10.1186/s13068-017-1007-6

Bacete L, Hamann T. The role of mechanoperception in plant cell wall integrity maintenance. Plants. 2020:**9**(5):574–592. https://doi.org/10.3390/plants9050574

Bacete L, Mélida H, Miedes E, Molina A. Plant cell wall-mediated immunity: cell wall changes trigger disease resistance responses. Plant J. 2018:**93**(4):614–636. https://doi.org/10.1111/tpj.13807

Bacete L, Schulz J, Engelsdorf T, Bartosova Z, Vaahtera L, Yan G, Gerhold JM, Tichá T, Øvstebø C, Gigli-Bisceglia N, et al. THESEUS1 modulates cell wall stiffness and abscisic acid production in *Arabidopsis thaliana*. Proc Natl Acad Sci U S A. 2022:**119**(1):e2119258119. https://doi.org/10.1073/pnas.2119258119

Bacic A, Delmer DP. Stimulation of membrane-associated polysaccharide synthetases by a membrane potential in developing cotton fibers. Planta. 1981:**152**(4):346–351. https://doi.org/10.1007/BF00388260

Baluska F, Samaj J, Wojtaszek P, Volkmann D, Menzel D. Cytoskeleton-plasma membrane-cell wall continuum in plants. Emerging links revisited. Plant Physiol. 2003:**133**(2):482–491. https://doi.org/10.1104/pp.103.027250

Bao W, O'Malley DM, Whetten R, Sederoff RR. A laccase associated with lignification in loblolly pine xylem. Science. 1993:**260**(5108):672–674. https://doi.org/10.1126/science.260.5108.672

Bar-Peled M, O'Neill MA. Plant nucleotide sugar formation, interconversion, and salvage by sugar recycling. Annu Rev Plant Biol. 2011:**62**(1):127–155. https://doi.org/10.1146/annurev-arplant-042110-103918

Bar-Peled M, Urbanowicz BR, O'Neill MA. The synthesis and origin of the pectic polysaccharide rhamnogalacturonan II—insights from nucleotide sugar formation and diversity. Front Plant Sci. 2012:**3**:92. https://doi.org/10.3389/fpls.2012.00092

Barbosa ICR, Rojas-Murcia N, Geldner N. The Casparian strip—one ring to bring cell biology to lignification? Curr Opin Biotechnol. 2019:**56**:121–129. https://doi.org/10.1016/j.copbio.2018.10.004

Baron-Epel O, Gharyal PK, Schindler M. Pectins as mediators of wall porosity in soybean cells. Planta. 1988:**175**(3):389–395. https://doi.org/10.1007/BF00396345

Barratt DHP, Derbyshire P, Findlay K, Pike M, Wellner N, Lunn J, Feil R, Simpson C, Maule AJ, Smith AM. Normal growth of *Arabidopsis* requires cytosolic invertase but not sucrose synthase. Proc Natl Acad Sci U S A. 2009:**106**(31):13124–13129. https://doi.org/10.1073/pnas.0900689106

Barros-Rios J, Temple S, Dixon RA. Development and commercialization of reduced lignin alfalfa. Curr Opin Biotechnol. 2018:**56**:48–54. https://doi.org/10.1016/j.copbio.2018.09.003

Barros J, Serk H, Granlund I, Pesquet E. The cell biology of lignification in higher plants. Ann Bot. 2015:**115**(7):1053–1074. https://doi.org/10.1093/aob/mcv046

Barros J, Serrani-Yarce JC, Chen F, Baxter D, Venables BJ, Dixon RA. Role of bifunctional ammonia-lyase in grass cell wall biosynthesis. Nat Plants. 2016:**2**(6):16050. https://doi.org/10.1038/nplants.2016.50

Barros J, Shrestha HK, Serrani-Yarce JC, Engle N, Abraham PE, Tschaplinski TJ, Hettich RL, Dixon RA. Proteomic and metabolic disturbances in lignin modified *Brachypodium distachyon*. Plant Cell. 2022:**34**(9):3339–3363. https://doi.org/10.1093/plcell/koac171

Bartels S, Boller T. Quo vadis, pep? Plant elicitor peptides at the crossroads of immunity, stress, and development. J Exp Bot. 2015:**66**(17):5183–5193. https://doi.org/10.1093/jxb/erv180

Bauer WD, Talmadge KW, Keegstra K, Albersheim P. The structure of plant cell walls II. The hemicellulose of the walls of suspension-cultured sycamore cells. Plant Physiol. 1973:**51**(1):174–187. https://doi.org/10.1104/pp.51.1.174

Baumberger N, Ringli C, Keller B. The chimeric leucine-rich repeat/extensin cell wall protein LRX1 is required for root hair morphogenesis in *Arabidopsis thaliana*. Genes Dev. 2001:**5**(9):1128–1139. https://doi.org/10.1101/gad.200201

Becker GE, Hui PA, Albersheim P. Synthesis of extracellular polysaccharides by suspensions of *Acer pseudoplatanus* cells. Plant Physiol. 1964:**39**(6):913–920. https://doi.org/10.1104/pp.39.6.913

Begum RA, Fry SC. Boron bridging of rhamnogalacturonan-II in Rosa and Arabidopsis cell cultures occurs mainly in the endo-membrane system and continues at a reduced rate after secretion. Ann Bot. 2022:**130**(5):703–715. https://doi.org/10.1093/aob/mcac119

Belorio M, Gomez M. Psyllium: a useful functional ingredient in food systems. Crit Rev Food Sci Nutr. 2022:**62**(2):527–538. https://doi.org/10.1080/10408398.2020.1822276

Berken A. ROPs in the spotlight of plant signal transduction. Cell Mol Life Sci. 2006:**63**(21):2446–2459. https://doi.org/10.1007/s00018-006-6197-1

Bernal AJ, Yoo C-M, Mutwil M, Jensen JK, Hou G, Blaukopf C, Sorensen I, Blancaflor EB, Scheller HV, Willats WGT. Functional analysis of the cellulose synthase-like genes CSLD1, CSLD2, and CSLD4 in tip-growing Arabidopsis cells. Plant Physiol. 2008:**148**(3): 1238–1253. https://doi.org/10.1104/pp.108.121939

Bernhardt C, Tierney ML. Expression of AtPRP3, a proline-rich structural cell wall protein from Arabidopsis, is regulated by cell-type-specific developmental pathways involved in root hair formation. Plant Physiol. 2000:**122**(3):705–714. https://doi.org/10.1104/pp.122.3.705

Bi Y, Hubbard C, Purushotham P, Zimmer J. Insights into the structure and function of membrane-integrated processive glycosyltransferases. Curr Opin Struct Biol. 2015:**34**:78–86. https://doi.org/10.1016/j.sbi.2015.07.008

Biswal AK, Atmodjo MA, Li M, Baxter HL, Yoo CG, Pu Y, Lee Y-C, Mazarei M, Black IM, Zhang J-Y, et al. Sugar release and growth of biofuel crops are improved by downregulation of pectin biosynthesis. Nat Biotechnol. 2018:**36**(3):249–257. https://doi.org/10.1038/nbt.4067

Blackburn MR, Haruta M, Moura DS. Twenty years of progress in physiological and biochemical investigation of RALF peptides. Plant Physiol. 2020:**182**(4):1657–1666. https://doi.org/10.1104/pp.19.01310

Boisson-Dernier A, Roy S, Kritsas K, Grobei MA, Jaciubek M, Schroeder JI, Grossniklaus U. Disruption of the pollen-expressed FERONIA homologs ANXUR1 and ANXUR2 triggers pollen tube discharge. Development. 2009:**136**(19):3279–3288. https://doi.org/10.1242/dev.040071

Bonner J. The chemistry and physiology of the pectins. Bot Rev. 1936:**2**(10):475–497. https://doi.org/10.1007/BF02869919

Borassi C, Sede AR, Mecchia MA, Salter JDS, Marzol E, Muschietti JP, Estevez JM. An update on cell surface proteins containing extensin-motifs. J Exp Bot. 2016:**67**(2):477–487. https://doi.org/10.1093/jxb/erv455

Bosch M, Hepler PK. Pectin methylesterases and pectin dynamics in pollen tubes. Plant Cell. 2005:**17**(12):3219–3226. https://doi.org/10.1105/tpc.105.037473

Bouton S, Leboeuf E, Mouille G, Leydecker M-T, Talbotec J, Granier F, Lahaye M, Höfte H, Truong H-N. QUASIMODO1 encodes a putative membrane-bound glycosyltransferase required for normal pectin synthesis and cell adhesion in Arabidopsis. Plant Cell. 2002:**14**(10): 2577–2590. https://doi.org/10.1105/tpc.004259

Boutrot F, Zipfel C. Function, discovery, and exploitation of plant pattern recognition receptors for broad-spectrum disease resistance. Annu Rev Phytopathol. 2017:**55**(1):257–286. https://doi.org/10.1146/annurev-phyto-080614-120106

Boyd JD. Significance of larcinan in compression wood tracheids. Wood Sci Technol. 1978:**12**(1):23–35. https://doi.org/10.1007/BF00390008

Braccini I, Pérez S. Molecular basis of Ca^{2+}-induced gelation in alginates and pectins: the egg-box model revisited. Biomacromolecules. 2001:**2**(4): 1089–1096. https://doi.org/10.1021/bm010008g

Braccini I, Rodriguez-Carvajal MA, Pérez S. Chain-chain interactions for methyl polygalacturonate: models for high methyl-esterified pectin junction zones. Biomacromolecules. 2005:**6**(3):1322–1328. https://doi.org/10.1021/bm049457h

Braconnot H. Nouvelles observations sur l'acide pectique. Ann Chim. 1825a:**30**:96–102.

Braconnot H. Recherches sur un nouvel acide universellement répandu dans tous les végétaux. Ann Chim. 1825b:**28**(2):173–178.

Braybrook SA, Jnsson H. Shifting foundations: the mechanical cell wall and development. Curr Opin Plant Biol. 2016:**29**:115–120. https://doi.org/10.1016/j.pbi.2015.12.009

Brill E, van Thournout M, White RG, Llewellyn D, Campbell PM, Engelen S, Ruan YL, Arioli T, Furbank RT. A novel isoform of sucrose synthase is targeted to the cell wall during secondary cell wall synthesis in cotton fiber. Plant Physiol. 2011:**157**(1):40–54. https://doi.org/10.1104/pp.111.178574

Bringmann M, Li E, Sampathkumar A, Kocabek T, Hauser M-T, Persson S. POM-POM2/cellulose synthase interacting1 is essential for the functional association of cellulose synthase and microtubules in Arabidopsis. Plant Cell. 2012:**24**(1):163–177. https://doi.org/10.1105/tpc.111.093575

Brown RM. The biosynthesis of cellulose. J Macromol Sci Part A. 1996:**33**(10):1345–1373. https://doi.org/10.1080/10601329608014912

Brown C, Leijon F, Bulone V. Radiometric and spectrophotometric in vitro assays of glycosyltransferases involved in plant cell wall carbohydrate biosynthesis. Nat Protocols. 2012:**7**(9):1634–1650. https://doi.org/10.1038/nprot.2012.089

Brown RM Jr, Wilison JHM, Richardson CL. Cellulose biosynthesis in *Acetobacter xylinum*: visualization of the site of synthesis and direct measurement of the in vivo process. Proc Natl Acad Sci U S A. 1976:**73**(12):4565–4569. https://doi.org/10.1073/pnas.73.12.4565

Buleon A, Chanzy H. Single crystals of cellulose II. J Polym Sci: Polym Phys Ed. 1978:**16**:833–839. https://doi.org/10.1002/pol.1978.180160508

Burton RA, Fincher GB. (1,3;1,4)-β-D-glucan in cell walls of the *Poaceae*, lower plants, and fungi: a tale of two linkages. Mol Plant. 2009:**2**(5):873–882. https://doi.org/10.1093/mp/ssp063

Burton RA, Wilson SM, Hrmova M, Harvey AJ, Shirley NJ, Medhurst A, Stone BA, Newbigin N, Bacic A, Fincher GB. Cellulose synthase-like *CslF* genes mediate the synthesis of cell wall (1,3;1,4)-β-D-glucans. Science. 2006:**311**(5769):1940–1942. https://doi.org/10.1126/science.1122975

Busse-Wicher M, Li A, Silveira RL, Pereira CS, Tryfona T, Gomes TCF, Skaf MS, Dupree P. Evolution of xylan substitution patterns in gymnosperms and angiosperms: implications for xylan interaction with cellulose. Plant Physiol. 2016:**171**(4):2418–2431. https://doi.org/10.1104/pp.16.00539

Caffall KH, Pattahil S, Phillips SE, Hahn MG, Mohnen D. *Arabidopsis thaliana* T-DNA mutants implicate *GAUT* genes in the biosynthesis of pectin and xylan in cell walls and seed testa. Mol Plant. 2009:**2**(5):1000–1014. https://doi.org/10.1093/mp/ssp062

Cai W, Hong J, Liu Z, Wang W, Zhang J, An G, Liang W, Persson S, Zhang D. A receptor-like kinase controls the amplitude of secondary cell wall synthesis in rice. Curr Biol. 2023:**33**(3):498–506.e6. https://doi.org/10.1016/j.cub.2022.12.035

Cannon MC, Terneus K, Hall Q, Tan L, Wang Y, Wegenhart BL, Chen L, Lamport DT, Chen Y, Kieliszewski MJ. Self-assembly of the plant cell wall requires an extensin scaffold. Proc Natl Acad Sci U S A. 2008:**105**(6):2226–2231. https://doi.org/10.1073/pnas.0711980105

Cantarel BL, Coutinho PM, Rancurel C, Bernard T, Lombard V, Henrissat B. The carbohydrate-active EnZymes database (CAZy): an expert resource for glycogenomics. Nucleic Acids Res. 2009:**37**(Database):D233–D238. https://doi.org/10.1093/nar/gkn663

Carol RJ, Dolan L. Building a hair: tip growth in *Arabidopsis thaliana* root hairs. Philos Trans R Soc Lond B Biol Sci. 2002:**357**(1422): 815–821. https://doi.org/10.1098/rstb.2002.1092

Carpita NC. Cell wall development in maize coleoptiles. Plant Physiol. 1984:**76**(1):205–212. https://doi.org/10.1104/pp.76.1.205

Carpita NC, Gibeaut DM. Structural models of primary cell walls in the flowering plants: consistency of molecular structure with the physical properties of the walls during growth. Plant J. 1993:**3**(1):1–30. https://doi.org/10.1111/j.1365-313X.1993.tb00007.x

Carre M. An investigation of the pectic constituents of stored fruits. Biochem J. 1922:**16**(5):704. https://doi.org/10.1042/bj0160704

Cavalier DM, Keegstra K. Two xyloglucan xylosyltransferases catalyze the addition of multiple xylosyl residues to cellohexaose. J Biol Chem. 2006:**281**(45):34197–34207. https://doi.org/10.1074/jbc.M606379200

Cavalier DM, Lerouxel O, Neumetzler L, Yamauchi K, Reinecke A, Freshour G, Zabotina OA, Hahn MG, Burgert I, Pauly M, et al. Disrupting two *Arabidopsis thaliana* xylosyltransferase genes results in plants deficient in xyloglucan, a major primary cell wall component. Plant Cell. 2008:**20**(6):1519–1537. https://doi.org/10.1105/tpc.108.059873

Çetinbaş-Genç A, Conti V, Cai G. Let's shape again: the concerted molecular action that builds the pollen tube. Plant Reprod. 2022:**35**(2):77–103. https://doi.org/10.1007/s00497-022-00437-4

Chan J, Coen E. Interaction between autonomous and microtubule guidance systems controls cellulose synthase trajectories. Curr Biol. 2020:**30**(5):941–947. https://doi.org/10.1016/j.cub.2019.12.066

Chanzy H, Henrissat B. Electron microscopy study of the enzymic hydrolysis of *Valonia* cellulose. Carbohydr Polym. 1983:**3**(3):161–173. https://doi.org/10.1016/0144-8617(83)90016-4

Chebli Y, Geitmann A. Cellular growth in plants requires regulation of cell wall biochemistry. Curr Opin Cell Biol. 2017:**44**:28–35. https://doi.org/10.1016/j.ceb.2017.01.002

Chen Y-L, Fan K-T, Hung S-C, Chen Y-R. The role of peptides cleaved from protein precursors in eliciting plant stress reactions. New Phytol. 2020b:**225**(6):2267–2282. https://doi.org/10.1111/nph.16241

Chen S, Jia H, Zhao H, Liu D, Liu Y, Liu B, Bauer S, Somerville CR. Anisotropic cell expansion is affected through the bidirectional mobility of cellulose synthase complexes and phosphorylation at two critical residues on CESA3. Plant Physiol. 2016:**171**(1):242–250. https://doi.org/10.1104/pp.15.01874

Chen D, Melton LD, McGillivray DJ, Ryan TM, Harris PJ. Changes in the orientations of cellulose microfibrils during the development of collenchyma cell walls of celery (*Apium graveolens* L). Planta. 2019:**250**(6):1819–1832. https://doi.org/10.1007/s00425-019-03262-8

Chen HW, Persson S, Grebe M, McFarlane HE. Cellulose synthesis during cell plate assembly. Physiol Plant. 2018:**164**(1):17–26. https://doi.org/10.1111/ppl.12703

Chen J, Zhu S, Ming Z, Liu X, Yu F. FERONIA cytoplasmic domain: node of varied signal outputs. aBIOTECH. 2020a:**1**(2):135–146. https://doi.org/10.1007/s42994-020-00017-y

Cheng K, Zhou Y, Neelamegham S. DrawGlycan-SNFG: a robust tool to render glycans and glycopeptides with fragmentation information. Glycobiology. 2017:**27**(3):200–205. https://doi.org/10.1093/glycob/cww115

Cheung AY, Wu HM. THESEUS 1, FERONIA and relatives: a family of cell wall-sensing receptor kinases? Curr Opin Plant Biol. 2011:**14**(6):632–641. https://doi.org/10.1016/j.pbi.2011.09.001

Cho H, Lee J, Oh E. Leucine-rich repeat receptor-like proteins in plants: structure, function, and signaling. J Plant Biol. 2022:**66**(2):99–107. https://doi.org/10.1007/s12374-022-09374-1

Cho SH, Purushotham P, Fang C, Maranas C, Diaz-Moreno SM, Bulone V, Zimmer J, Kumar M, Nixon BT. Synthesis and self-assembly of cellulose microfibrils from reconstituted cellulose synthase. Plant Physiol. 2017:**175**(1):146–156. https://doi.org/10.1104/pp.17.00619

Chou EY, Schuetz M, Hoffmann N, Watanabe Y, Sibout R, Samuels AL. Distribution, mobility, and anchoring of lignin-related oxidative enzymes in Arabidopsis secondary cell walls. J Exp Bot. 2018:**69**(8):1849–1859. https://doi.org/10.1093/jxb/ery067

Coculo D, Lionetti V. The plant invertase/pectin methylesterase inhibitor superfamily. Front Plant Sci. 2022:**13**:863892. https://doi.org/10.3389/fpls.2022.863892

Cocuron J-C, Lerouxel O, Drakakaki G, Alonso AP, Liepman AH, Keegstra K, Raikhel N, Wilkerson CG. A gene from the cellulose synthase-like C family encodes a β-1,4 glucan synthase. Proc Natl Acad Sci U S A. 2007:**104**(20):8550–8555. https://doi.org/10.1073/pnas.0703133104

Colin L, Ruhnow F, Zhu J-K, Zhao C, Zhao Y, Persson S. The cell biology of primary cell walls during salt stress. Plant Cell. 2023:**35**(1):201–217. https://doi.org/10.1093/plcell/koac292

Cosgrove DJ. Re-constructing our models of cellulose and primary cell wall assembly. Curr Opin Plant Biol. 2014:**22**:122–131. https://doi.org/10.1016/j.pbi.2014.11.001

Cosgrove DJ. Plant expansins: diversity and interactions with plant cell walls. Curr Opin Plant Biol. 2015:**25**:162–172. https://doi.org/10.1016/j.pbi.2015.05.014

Cosgrove DJ. Plant cell wall extensibility: connecting plant cell growth with cell wall structure, mechanics, and the action of wall modifying enzymes. J Exp Bot. 2016:**67**(2):463–476. https://doi.org/10.1093/jxb/erv511

Cosgrove DJ. Building an extensible cell wall. Plant Physiol. 2022:**189**(3):1246–1277. https://doi.org/10.1093/plphys/kiac184

Côté F, Hahn MG. Oligosaccharins: structures and signal transduction. Plant Mol Biol. 1994:**26**(5):1379–1411. https://doi.org/10.1007/BF00016481

Daher FB, Braybrook SA. How to let go: pectin and plant cell adhesion. Front Plant Sci. 2015:**6**:523. https://doi.org/10.3389/fpls.2015.00523

Daher FB, Chen Y, Bozorg B, Clough J, Jönsson H, Braybrook SA. Anisotropic growth is achieved through the additive mechanical effect of material anisotropy and elastic asymmetry. eLife. 2018:**7**:e38161. https://doi.org/10.7554/eLife.38161

Dai S-Y, Hsu W-H, Yang C-H. The gene ANTHER DEHISCENCE REPRESSOR (ADR) controls male fertility by suppressing the ROS accumulation and anther cell wall thickening in Arabidopsis. Sci Rep. 2019:**9**(1):5112. https://doi.org/10.1038/s41598-019-41382-z

Darvill AG, McNeil M, Albersheim P. Structure of plant cell walls: VIII. A new pectic polysaccharide. Plant Physiol. 1978:**62**(3):418–422. https://doi.org/10.1104/pp.62.3.418

Davin LB, Lewis NG. Dirigent proteins and dirigent sites explain the mystery of specificity of radical precursor coupling in lignan and lignin biosynthesis. Plant Physiol. 2000:**123**(2):453–462. https://doi.org/10.1104/pp.123.2.453

Davin LB, Wang HB, Crowell AL, Bedgar DL, Martin DM, Sarkanen S, Lewis NG. Stereoselective bimolecular phenoxy radical coupling by an auxiliary (dirigent) protein without an active center. Science. 1997:**275**(5298):362–366. https://doi.org/10.1126/science.275.5298.362

Davis KR, Darvill AG, Albersheim P, Dell A. Host-pathogen interactions. XXIX. Oligogalacturonides released from sodium polypectate by endopolygalacturonic acid lyase are elicitors of phytoalexins in soybean. Plant Physiol. 1986:**80**(2):568–577. https://doi.org/10.1104/pp.80.2.568

Delmer DP. Biosynthesis of cellulose. Adv Carbohydrate Chem Biochem. 1983:**41**:105–153. https://doi.org/10.1016/S0065-2318(08)60057-8

Delmer DP. Cellulose biosynthesis: exciting times for a difficult field of study. Annu Rev Plant Biol. 1999:**50**(1):245–276. https://doi.org/10.1146/annurev.arplant.50.1.245

Delmer DP, Benziman M, Padan E. Requirement for a membrane potential for cellulose synthesis in intact cells of *Acetobacter xylinum*. Proc Natl Acad Sci U S A. 1982:**79**(17):5282–5286. https://doi.org/10.1073/pnas.79.17.5282

Delmer DP, Pear JR, Andrawis A, Stalker DM. Genes encoding small GTP-binding proteins analogous to mammalian rac are preferentially expressed in developing cotton fibers. Mol Gen Genet. 1995:**248**(1):43–51. https://doi.org/10.1007/BF02456612

Delmer DP, Read SM, Cooper G. Identification of a receptor protein in cotton fibers for the herbicide 2, 6-dichlorobenzonitrile. Plant Physio. 1987:**84**:415–420. https://doi.org/10.1104/pp.84.2.415

De Meester B, Vanholme R, Mota T, Boerjan W. Lignin engineering in forest trees: from gene discovery to field trials. Plant Commun. 2022:**3**(6):100465. https://doi.org/10.1016/j.xplc.2022.100465

Demura T. Tracheary element differentiation. Plant Biotechnol Rep. 2013:**8**(1):17–21. https://doi.org/10.1007/s11816-013-0293-0

De Vries JA, Voragen AGJ, Rombouts FM, Pilnik W. Extraction and purification of pectins from Alcohol Insoluble Solids from ripe and unripe apples. Carbohydr Polym. 1981:**1**(2):117–127. https://doi.org/10.1016/0144-8617(81)90004-7

Dhugga KS, Barreirro R, Whistten B, Stecca K, Hazebroek J, Randhawa GS, Dolan M, Kinney AJ, Tomes D, Nichols S, et al. Guar seed (-mannan synthase is a member of the cellulose synthase super gene family. Science. 2004:**303**(5656):363–366. https://doi.org/10.1126/science.1090908

Diener M, Adamcik J, Sánchez-Ferrer A, Jaedig F, Schefer L, Mezzenga R. Primary, secondary, tertiary and quaternary structure levels in linear polysaccharides: from random coil, to single helix to supramolecular assembly. Biomacromolecules. 2019:**20**(4):1731–1739. https://doi.org/10.1021/acs.biomac.9b00087

Dima O, Morreel K, Vanholme B, Kim H, Ralph J, Boerjan W. Small glycosylated lignin oligomers are stored in Arabidopsis leaf vacuoles. Plant Cell. 2015:**27**(3):695–710. https://doi.org/10.1105/tpc.114.134643

Dixon RA, Barros J. Lignin biosynthesis- old roads revisited and new roads explored. Open Biol. 2019:**9**(12):190215. https://doi.org/10.1098/rsob.190215

Doblin MS, De Melis L, Newbigin E, Bacic A, Read SM. Pollen tubes of *Nicotiana alata* express two genes from different β-glucan synthase families. Plant Physiol. 2001:**125**(4):2040–2052. https://doi.org/10.1104/pp.125.4.2040

Doblin MS, Kurek I, Jacob-Wilk D, Delmer DP. Cellulose biosynthesis in plants: from genes to rosettes. Plant Cell Physiol. 2002:**43**(12):1407–1420. https://doi.org/10.1093/pcp/pcf164

Doblin MS, Pettolino FA, Wilson SM, Campbell R, Burton RA, Fincher GB, Newbigin E, Bacic A. A barley *cellulose synthase-like CSLH* gene mediates (1,3:1,4)-β-D-glucan synthesis in transgenic *Arabidopsis*. Proc Natl Acad Sci U S A. 2009:**106**(14):5996–6001. https://doi.org/10.1073/pnas.0902019106

Doll NM, Berenguer E, Truskina J, Ingram G. AtEXT3 is not essential for early embryogenesis or plant viability in Arabidopsis. New Phytol. 2022:**236**(5):1629–1633. https://doi.org/10.1111/nph.18452

Domozych DS, Sørensen I, Pettolino FA, Bacic A, Willats WGT. The cell wall polymers of the Charophycean green alga *Chara corallina*: immunobinding and biochemical screening. Int. J Plant Sci. 2010:**171**(4):345–361. https://doi.org/10.1086/651227

Dougall DK, Shimbayashi K. Factors affecting growth of tobacco callus tissue and its incorporation of tyrosine. Plant Physiol. 1960:**35**(3):396–404. https://doi.org/10.1104/pp.35.3.396

Drakakaki G. Polysaccharide deposition during cytokinesis: challenges and future perspectives. Plant Sci. 2015:**236**:177–184. https://doi.org/10.1016/j.plantsci.2015.03.018

Dressano K, Ceciliato PHO, Silva AL, Guerrero-Abad JC, Bergonci T, Ortiz-Morea FA, Burger M, Silva-Filho MC, Moura DS. BAK1 is involved in AtRALF1-induced inhibition of root cell expansion. PLoS Genet. 2017:**13**(10):e1007053. https://doi.org/10.1371/journal.pgen.1007053

Dressano K, Weckwerth PR, Poretsky E, Takahashi Y, Villarreal C, Shen Z, Schroeder JI, Briggs SP, Huffaker A. Dynamic regulation of Pep-induced immunity through post-translational control of defence transcript splicing. Nat Plants. 2020:**6**(8):1008–1019. https://doi.org/10.1038/s41477-020-0724-1

Drula E, Garron M-L, Dogan S, Lombard V, Henrissat B, Terrapon N. The carbohydrate-active enzyme database: functions and literature. Nucleic Acids Res. 2022:**50**(D1):D571–D577. https://doi.org/10.1093/nar/gkab1045

Du J, Anderson CT, Xiao C. Dynamics of pectic homogalacturonan in cellular morphogenesis and adhesion, wall integrity sensing and plant development. Nat Plants. 2022:**8**(4):332–340. https://doi.org/10.1038/s41477-022-01120-2

Du J, Kirui A, Huang S, Wang L, Barnes WJ, Kiemle SN, Zheng Y, Rui Y, Ruan M, Qi S, et al. Mutations in the pectin methyltransferase QUASIMODO2 influence cellulose biosynthesis and wall integrity in Arabidopsis. Plant Cell. 2020a:**32**(11):3576–3597. https://doi.org/10.1105/tpc.20.00252

Du M, Spalding EP, Gray WM. Rapid auxin-mediated cell expansion. Ann Rev Plant Biol. 2020b:**71**(1):379–402. https://doi.org/10.1146/annurev-arplant-073019-025907

Duan Q, Kita D, Li C, Cheung AY, Wu H-M. FERONIA receptor-like kinase regulates RHO GTPase signaling of root hair development. Proc Natl Acad Sci U S A. 2010:**107**(41):17821–17826. https://doi.org/10.1073/pnas.1005366107

Ebert B, Birdseye D, Liwanag AJM, Laursen T, Rennie EA, Guo X, Catena M, Rautengarten C, Stonebloom SH, Gluza P, et al. The three members of the Arabidopsis glycosyltransferase family 92 are functional β-1,4-galactan synthases. Plant Cell Physiol. 2018:**59**(12):2624–2636. https://doi.org/10.1093/pcp/pcy180

Edwards ME, Dickson CA, Chengappa S, Sidebottom C, Gidley MJ, Reid JSG. Molecular characterization of a membrane-bound galactosyltransferase of plant cell wall matrix polysaccharide biosynthesis. Plant J. 1999:**19**(6):691–697. https://doi.org/10.1046/j.1365-313x.1999.00566.x

Egelund J, Damager I, Faber K, Olsen CE, Ulvskov P, Petersen BL. Functional characterisation of a putative rhamnogalacturonan II specific xylosyltransferase. FEBS Lett. 2008:**582**(21-22):3217–3222. https://doi.org/10.1016/j.febslet.2008.08.015

Egelund J, Petersen BL, Motawia MS, Damager I, Faik A, Olsen CE, Ishii T, Clausen H, Ulvskov P, Geshi N. *Arabidopsis thaliana* RGXT1 and RGXT2 encode Golgi-localized (1,3)-α-D-xylosyltransferases involved in the synthesis of pectic rhamnogalacturonan-II. Plant Cell. 2006:**18**(10):2593–2607. https://doi.org/10.1105/tpc.105.036566

Ellis M, Egelund J, Schultz CJ, Bacic A. Arabinogalactan-proteins: key regulators at the cell surface? Plant Physiol. 2010:**153**(2):403–419. https://doi.org/10.1104/pp.110.156000

Engelsdorf T, Gigli-Bisceglia N, Veerabagu M, McKenna JF, Vaahtera L, Augstein F, Van der Does D, Zipfel C, Hamann T. The plant cell wall integrity maintenance and immune signaling systems cooperate to control stress responses in *Arabidopsis thaliana*. Sci Signal. 2018:**11**(536):eaao3070. https://doi.org/10.1126/scisignal.aao3070

Engle KA, Amos RA, Yang JY, Glushka J, Atmodjo MA, Tan L, Huang C, Moremen KW, Mohnen D. Multiple Arabidopsis galacturonosyltransferases (GAUTs) synthesize polymeric homogalacturonan by oligosaccharide acceptor-dependent or *de novo* synthesis. Plant J. 2022:**109**(6):1441–1456. https://doi.org/10.1111/tpj.15640

Fabrissin I, Cueff G, Berger A, Granier F, Sallé C, Poulain D, Ralet MC, North HM. Natural variation reveals a key role for rhamnogalacturonan I in seed outer mucilage and underlying genes. Plant Physiol. 2019:**181**(4):1498–1518. https://doi.org/10.1104/pp.19.00763

Faik A, Price NJ, Raikhel NV, Keegstra K. An *Arabidopsis* gene encoding an (-xylosyltransferase involved in xyloglucan biosynthesis. Proc Natl Acad Sci U S A. 2002:**99**(11):7797–7802. https://doi.org/10.1073/pnas.102644799

Favery B, Ryan E, Foreman J, Linstead P, Boudonck K, Steer M, Shaw P, Dolan L. KOJAK encodes a cellulose synthase-like protein required for root hair cell morphogenesis in Arabidopsis. Genes Dev. 2001:**15**(1):79–89. https://doi.org/10.1101/gad.188801

Federici L, Matteo AD, Fernandez-Recio J, Tsernoglou D, Cervone F. Polygalacturonase inhibiting proteins: players in plant innate immunity? Trends Plant Sci. 2006:**11**(2):65–70. https://doi.org/10.1016/j.tplants.2005.12.005

Feingold DS, Neufeld EF, Hassid WZ. Synthesis of a β-1, 3-linked glucan by extracts of *Phaseolus aureus* seedlings. J Biol Chem. 1958:**233**(4): 783–788. https://doi.org/10.1016/S0021-9258(18)64655-1

Feng W, Kita D, Peaucelle A, Cartwright HN, Doan V, Duan Q, Liu M-C, Maman J, Steinhorst L, Schmitz-Thom I, et al. The FERONIA receptor kinase maintains cell-wall integrity during salt stress through Ca^{2+} signaling. Curr Biol. 2018:**28**(5):666–675.e5. https://doi.org/10.1016/j.cub.2018.01.023

Fernandes AN, Thomas LH, Altaner CM, Callow P, Forsyth VT, Apperley DC, Kennedy CJ, Jarvis MC. Nanostructure of cellulose microfibrils in spruce wood. Proc Natl Acad Sci U S A. 2011:**108**(47): E1195–E1203. https://doi.org/10.1073/pnas.1108942108

Fischer E. Syntheses in the purine and sugar group. Nobel Prize Lect. 1902:**12**:22–35.

Fischer MH, Yu N, Gray GR, Ralph J, Anderson L, Marlett JA. The gel-forming polysaccharide of psyllium husk (*Plantago ovata* Forsk). Carbohydr Res. 2004:**339**(11):2009–2017. https://doi.org/10.1016/j.carres.2004.05.023

Fleischer A, O'Neill MA, Ehwald R. The pore size of non-graminaceous plant cell walls is rapidly decreased by borate ester cross-linking of the pectic polysaccharide rhamnogalacturonan II. Plant Physiol. 1999:**121**(3):829–838. https://doi.org/10.1104/pp.121.3.829

Francin-Allami M, Bouder A, Geairon A, Alvardo C, Le-Bot L, Daniel S, Shao M, Laudencia-Chingcuanco D, Vogel JP, Guillon F, et al. Mixed-linkage glucan is the main carbohydrate source and starch is an alternative source during *Brachypodium* grain germination. Int J Mol Sci. 2023:**24**(7):6821. https://doi.org/10.3390/ijms24076821

Freitas CMP, Coimbra JSR, Souza VGL, Sousa RCS. Structure and applications of pectin in food, biomedical, and pharmaceutical industry: a review. Coatings. 2021:**11**(8):922. https://doi.org/10.3390/coatings11080922

Fremy E. Recherches sur les matieres gelatineuses des fruits. Compt Rend. 1847:**24**:1076.

Fremy ME. On the ripening of fruits and the gelatious bodies of vegetables. London Edinburgh Dublin Philos Mag J Sci. 1848:**33**(223): 394–396.

French AD. The structure and mechanism of biosynthesis of cellulose. Part I Structure. In: **Kung S-D Yang S-F**, editors. Discoveries in plant biology. Vol. 3. Singapore: World Scientific; 2000. p. 1975–2001. https://doi.org/10.1142/9789812813503_0008.

Freudenberg K. Biosynthesis and constitution of lignin. Nature. 1959:**183**(4669):1152–1155. https://doi.org/10.1038/1831152a0

Freudenberg K, Neish AC. Constitution and biosynthesis of lignin (molecular biology, biochemistry, and biophysics). New York: Springer-Verlag; 1968. 132 pp.

Frey-Wyssling A. The submicroscopic structure of cell walls. Sci Prog. 1939:**34**:249–262.

Frey-Wyssling A, Ambronn H. Das polarisationmikroskop. Ber dt. Bot Ges. 1927:**45**:60–71. https://doi.org/10.1111/j.1438-8677.1927.tb03752.x

Frey-Wyssling A, Mühlethaler K. Die elementarfibrillen der cellulose. Die Makromolekulare Chemie. 1963:**62**(1):25–30. https://doi.org/10.1002/macp.1963.020620103

Fry SC. In-vivo formation of xyloglucan nonasaccharide: a possible biologically active cell-wall fragment. Planta. 1986:**169**(3):443–453. https://doi.org/10.1007/BF00392143

Fry SC. Oligosaccharins as plant growth regulators. Biochem Soc Symp. 1994:**60**:5–14.

Fry SC, York WS, Albersheim P, Darvill A, Hayashi T, Joseleau J-P, Kato Y, Lorences EP, Maclachlan GA, McNeil M, et al. An unambiguous nomenclature for xyloglucan-derived oligosaccharides. Physiol Plant. 1993:**89**(1):1–3. https://doi.org/10.1111/j.1399-3054.1993.tb01778.x

Fujita M, Wasteneys GO. A survey of cellulose microfibril patterns in dividing, expanding, and differentiating cells of *Arabidopsis thaliana*. Protoplasma. 2014:**251**(3):687–698. https://doi.org/10.1007/s00709-013-0571-2

Gallego-Giraldo L, Jikumaru Y, Kamiya Y, Tang Y, Dixon RA. Selective lignin down-regulation leads to constitutive defense response expression in alfalfa (*Medicago sativa* L.). New Phytol. 2011:**190**(3):627–639. https://doi.org/10.1111/j.1469-8137.2010.03621.x

Gallego-Giraldo L, Liu C, Pose-Albacete S, Pattathil S, Peralta AG, Young J, Westphaling J, Hahn M, Rao X, de Meester B, et al. ARABIDOPSIS DEHISCENCE ZONE POLYGALACTURONASE 1 (ADPG1) releases latent defense signals in stems with reduced lignin content. Proc Natl Acad Sci U S A. 2020:**117**(6):3281–3290. https://doi.org/10.1073/pnas.1914422117

Gallego-Giraldo L, Pose-Albacete S, Pattathil S, Peralta AG, Hahn M, Ayre BG, Sunuwar H, Hernandez J, Patel M, Shah J, et al. Elicitors and defense gene induction in plants with altered lignin compositions. New Phytol. 2018:**219**(4):1235–1251. https://doi.org/10.1111/nph.15258

Gallego-Giraldo L, Shadle G, Shen H, Barros-Rios J, Corrales SF, Wang H, Dixon RA. Combining enhanced biomass density with reduced lignin level for improved forage quality. Plant Biotech J. 2015:**14**(3):895–904. https://doi.org/10.1111/pbi.12439

Ganguly A, Zhu C, Chen W, Dixit R. FRA1 kinesin modulates the lateral stability of cortical microtubules through cellulose synthase-microtubule uncoupling proteins. Plant Cell. 2020:**32**(8):2508–2524. https://doi.org/10.1105/tpc.19.00700

Gao Y, Lipton AS, Munson CR, Ma Y, Johnson KL, Murray DT, Scheller HV, Mortimer JC. Elongated galactan side-chains mediate cellulose-pectin interactions in engineered Arabidopsis secondary cell walls. Plant J. 2023:**115**(2):529–545. https://doi.org/10.1111/tpj.16242

Gao Z, Sun W, Wang J, Zhao C, Zuo K. GhbHLH18 negatively regulates fiber strength and length by enhancing lignin biosynthesis in cotton fibers. Plant Sci. 2019:**286**:7–16. https://doi.org/10.1016/j.plantsci.2019.05.020

Gardner K, Blackwell J. The structure of native cellulose. Biopolymers. 1974:**13**(10):1975–2001. https://doi.org/10.1002/bip.1974.360131005

Giddings TH, Brower TL, Staehelin LA. Visualization of the particle complexes in the plasma membrane of *Micrasterias denticulata* associated with the formation of cellulose fibrils in primary and secondary cell walls. J Cell Biol. 1980:**84**(2):327–339. https://doi.org/10.1083/jcb.84.2.327

Gigli-Bisceglia N, van Zelm E, Huo W, Lamers J, Testerink C. Arabidopsis root responses to salinity depend on pectin modification and cell wall sensing. Development. 2022:**149**(12):dev200363. https://doi.org/10.1242/dev.200363

Glaser L. The synthesis of cellulose in cell-free extracts of *Acetobacter xylinum*. J Biol Chem. 1958:**232**(2):627–636. https://doi.org/10.1016/S0021-9258(19)77383-9

Gonneau M, Desprez T, Martin M, Doblas VG, Bacete L, Miart F, Sormani R, Hematy K, Renou J, Landrein B, et al. Receptor kinase THESEUS1 is a rapid alkalinization factor 34 receptor in Arabidopsis. Curr Biol. 2018:**28**(15):2452–2458. https://doi.org/10.1016/j.cub.2018.05.075

Gorshkova T, Petrova A, Mikshina P. Review: tertiary cell wall of plant fibers as a source of inspiration in material design. Carbohydr Polym. 2022:**295**:119849. https://doi.org/10.1016/j.carbpol.2022.119849

Grantham NJ, Wurman-Rodrich J, Terrett OM, Lyczakowski JJ, Stott K, Iuga D, Simmons TJ, Durand-Tardif M, Brown SP, Dupree R, et al. An even pattern of xylan substitution is critical for interaction with cellulose in plant cell walls. Nat Plants. 2017:**3**(11):859–865. https://doi.org/10.1038/s41477-017-0030-8

Groover A, Jones AM. Tracheary element differentiation uses a novel mechanism coordinating programmed cell death and secondary cell wall synthesis. Plant Physiol. 1999:**119**(2):375–384. https://doi.org/10.1104/pp.119.2.375

Gu F, Bringmann M, Combs JR, Yang J, Bergmann DC, Nielsen E. Arabidopsis CSLD5 functions in cell plate formation in a cell cycle-dependent manner. Plant Cell. 2016:**28**(7):1722–1737. https://doi.org/10.1105/tpc.16.00203

Gu Y, Kaplinsky N, Bringmann M, Cobb A, Carroll A, Sampathkumar A, Baskin T I, Persson S, Somerville CR. Identification of a cellulose synthase-associated protein required for cellulose biosynthesis. Proc Natl Acad Sci U S A. 2010:**107**(29): 12866–12871. https://doi.org/10.1073/pnas.1007092107

Guo H, Li L, Ye H, Yu X, Algreen A, Yin Y. Three related receptor-like kinases are required for optimal cell elongation in Arabidopsis thaliana. Proc Natl Acad Sci U S A. 2009:**106**(18):7648–7653. https://doi.org/10.1073/pnas.0812346106

Ha CM, Rao X, Saxena G, Dixon RA. Growth-defense trade-offs as a result of lignin pathway engineering. New Phytol. 2021:**231**(1): 60–74. https://doi.org/10.1111/nph.17383

Haas KT, Wightman R, Meyerowitz EM, Peaucelle A. Pectin homogalacturonan nanofilament expansion drives morphogenesis in plant epidermal cells. Science. 2020:**367**(6481):1003–1007. https://doi.org/10.1126/science.aaz5103

Haas KT, Wightman R, Peaucelle A, Höfte H. The role of pectin phase separation in plant cell wall assembly and growth. Cell Surface. 2021:**7**:100054. https://doi.org/10.106/j.tcsw.2021.100054

Haigler CH, Betancur L, Stiff MR, Tuttle JR. Cotton fiber: a powerful single-cell model for cell wall and cellulose research. Front Plant Sci. 2012:**3**:104. https://doi.org/10.3389/fpls.2012.00104

Haigler CH, Ivanova-Datcheva M, Hogan PS, Salnikov, Vv, Hwang S, Martin K, Delmer DP. Carbon partitioning to cellulose synthesis. Plant Mol Biol. 2001:47:29–51. https://doi.org/10.1007/978-94-010-0668-2_3.

Haigler CH, Roberts AW. Structure/function relationships in the rosette cellulose synthesis complex illuminated by an evolutionary perspective. Cellulose. 2019:**26**(1):227–247. https://doi.org/10.1007/s10570-018-2157-9

Hamant O, Inoue D, Bouchez D, Dumais J, Mjolsness E. Are microtubules tension sensors? Nat Commun. 2019:**10**(1):2360–2373. https://doi.org/10.1038/s41467-019-10207-y

Hao Z, Mohnen D. A review of xylan and lignin biosynthesis: foundation for studying Arabidopsis irregular xylem mutants with pleiotropic phenotypes. Crit Rev Biochem Mol Biol. 2014:**49**(3):212–241. https://doi.org/10.3109/10409238.2014.889651

Harholt J, Jensen JK, Sorensen SO, Orfila C, Pauly M, Scheller HV. ARABINAN DEFICIENT 1 is a putative arabinosyltransferase involved in biosynthesis of pectic arabinan in Arabidopsis. Plant Physiol. 2006:**140**(1):49–58. https://doi.org/10.1104/pp.105.072744

Harholt J, Jensen JK, Verhertbruggen Y, Søgaard C, Bernard S, Nafisi M, Poulsen CP, Geshi N, Sakuragi Y, Driouich A, et al. ARAD proteins associated with pectic arabinan biosynthesis form complexes when transiently overexpressed in planta. Planta. 2012:**236**(1): 115–128. https://doi.org/10.1007/s00425-012-1592-3

Haruta M, Sabat G, Stecker K, Minkoff BB, Sussman M. A peptide hormone and its receptor protein kinase regulate plant cell expansion. Science. 2014:**343**(6169):408–411. https://doi.org/10.1126/science.1244454

Haworth WN, Hirst EL, Thomas HA. CXVI.—Polysaccharides. Part VII. Isolation of octamethyl cellobiose, hendecamethyl cellotriose, and a methylated cellodextrin (cellotetrose?) as crystalline products of the acetolysis of cellulose derivatives. J Chem Soc (Resumed). 1931:(0):824–829. https://doi.org/10.1039/JR9310000824

Haworth WN, Long CW, Plant JHG. CCCLXXVI.—The constitution of the disaccharide. Part XVI. Cellobiose. J Chem Soc (Resumed). 1927:(0):2809–2814. https://doi.org/10.1039/JR9270002809

Hayashi T. Xyloglucans in the primary cell wall. Ann Rev Plant Physiol Plant Mol Biol. 1989:**40**(1):139–168. https://doi.org/10.1146/annurev.pp.40.060189.001035

Hayashi T, Kaida R. Functions of xyloglucan in plant cells. Mol Plant. 2011:**4**(1):17–24. https://doi.org/10.1093/mp/ssq063

Hayashi T, Read SM, Bussell J, Thelen M, Lin F-C, Brown RM Jr, Delmer DP. UDP-Glucose: (1-3)-β-glucan synthases from mung bean and cotton. Diifferential effects of Ca^{2+} and Mg^{2+} on enzyme properties and structure of the reaction product. Plant Physiol. 1987:**83**(4):1054–1062. https://doi.org/10.1104/pp.83.4.1054

Hematy K, Sado PE, Van Tuinen A, Rochange S, Desnos T, Balzergue S, Pelletier S, Renou JP, Hofte H. A receptor-like kinase mediates the response of Arabidopsis cells to the inhibition of cellulose synthesis. Curr Biol. 2007:**17**(11):922–931. https://doi.org/10.1016/j.cub.2007.05.018

Henry RJ, Stone BA. Factors influencing β-glucan synthesis by particulate enzymes from suspension-cultured *Lolium multiflorum* endosperm cells. Plant Physiol. 1982a:**69**(3):632–636. https://doi.org/10.1104/pp.69.3.632

Henry RJ, Stone BA. Solubilization of β-glucan synthases from the membrane of cultured ryegrass endosperm cells. Biochem J. 1982b:**203**(3):629–636. https://doi.org/10.1042/bj2030629

Hepler P. Calcium: a central regulator of plant growth and development. Plant Cell. 2005:**17**(8):2142–2155. https://doi.org/10.1105/tpc.105.032508

Herger A, Dunser K, Kleine-Vehn J, Ringli C. Leucine-rich repeat extensin proteins and their role in cell wall sensing. Curr Biol. 2019:**29**(17):R851–R858. https://doi.org/10.1016/j.cub.2019.07.039

Herth W. Plasma-membrane rosettes involved in localized wall thickening during xylem vessel formation of *Lepidium sativum* L. Planta. 1985:**163**(1):12–21. https://doi.org/10.1007/BF00391020

Hestrin S. Synthesis of polymer homopolysaccharides. In: **Gunsalus RY, Stanier IC**, editors. The bacteria: A treatise on structure and function (Volume IIII Biosynthesis). iii. New York: Academic Press; 1962. p. 373–388.

Hieta K, Kuga S, Usuda M. Electron staining of reducing ends evidences a parallel-chain structure in *Valonia* cellulose. Biopolymers. 1984:**23**(10): 1807–1810. https://doi.org/10.1002/bip.360231002

Hill JL Jr, Josephs C, Barnes WJ, Anderson CT, Tien M. Longevity in vivo of primary cell wall cellulose synthases. Plant Mol Biol. 2018:**96**(3):279–289. https://doi.org/10.1007/s11103-017-0695-4

Hobson N, Roach MJ, Deyholos MK. Gene expression in tension wood and bast fibres. Russian J Plant Physiol. 2010:**57**(3):321–327. https://doi.org/10.1134/S1021443710030039

Hocq L, Pelloux J, Lefebvre V. Connecting homogalacturonan-type pectin remodeling to acid growth. Trends Plant Sci. 2017a:**22**(1): 20–29. https://doi.org/10.1016/j.tplants.2016.10.009

Hocq L, Sénéchal F, Lefebvre F, Lehner A, Domon JM, Mollet JC, Dehors J, Pageau K, Marcelo P, Guérineau F, et al. Combined experimental and computational approaches reveal distinct pH dependence of pectin methylesterase inhibitors. Plant Physiol. 2017b:**173**(2):1075–1093. https://doi.org/10.1104/pp.16.01790

Höfte H. The yin and yang of cell wall integrity control: brassinosteroid and FERONIA signaling. Plant Cell Physiol. 2015:**56**(2):224–231. https://doi.org/10.1093/pcp/pcu182

Höfte H. A dive into the cell wall with Arabidopsis. Comp Rend Biol. 2023:**345**(4):1–20. https://doi.org/10.5802/crbiol.101

Huck N, Moore JM, Federer M, Grossniklaus U. The Arabidopsis mutant feronia disrupts the female gametophytic control of pollen tube reception. Development. 2003:**130**(10):2149–2159. https://doi.org/10.1242/dev.00458

Humphreys JM, Hemm MR, Chapple C. New routes for lignin biosynthesis defined by biochemical characterization of recombinant ferulate 5-hydroxylase, a multifunctional cytochrome P450-dependent monooxygenase. Proc Natl Acad Sci U S A. 1999:**96**(18):10045–10050. https://doi.org/10.1073/pnas.96.18.10045

Huysmans M, Lema AS, Coll NS, Nowack MK. Dying two deaths—programmed cell death regulation in development and disease. Curr Opin Plant Biol. 2017:**35**:37–44. https://doi.org/10.1016/j.pbi.2016.11.005

Ibragimova NN, Ageeva MV, Gorshkova TA. Development of gravitropic response: unusual behavior of flax phloem G-fibers. Protoplasma. 2017:**254**(2):749–762. https://doi.org/10.1007/s00709-016-0985-8

Ishida K, Yokoyama R. Reconsidering the function of the xyloglucan endotransglucosyl/ hydrolase family. J Plant Res. 2022:**135**(2): 145–156. https://doi.org/10.1007/s10265-021-01361-w

Itoh T, Brown RM Jr. The assembly of cellulose microfibrils in *Valonia macrophysa* Kutz. Planta. 1984:**160**(4):372–381. https://doi.org/10.1007/BF00393419

Jacob-Wilk D, Kurek I, Hogan P, Delmer DP. The cotton fiber zinc-binding domain of cellulose synthase A1 from Gossypium hirsutum displays rapid turnover in vitro and in vivo. Proc Natl Acad Sci U S A. 2006:**103**(32):12191–12196. https://doi.org/10.1073/pnas.0605098103

Jansen EF, Jang R, Albersheim P, Bonner J. Pectic metabolism of growing cell walls. Plant Physiol. 1960:**35**(1):87–97. https://doi.org/10.1104/pp.35.1.87

Jarvis MC. Cellulose biosynthesis: counting the chains. Plant Physiol. 2013:**163**(4):1485–1486. https://doi.org/10.1104/pp.113.231092

Jarvis MC. Structure of native cellulose microfibrils, the starting point for nanocellulose manufacture. Philos Trans R Soc A. 2018:**376**(2112): 2017045. https://doi.org/10.1098/rsta.2017.0045

Jarvis MC. Hydrogen bonding and other non-covalent interactions at the surfaces of cellulose microfibrils. Cellulose. 2023:**30**(2):667–687. https://doi.org/10.1007/s10570-022-04954-3

Jensen JK, Busse-Wicher M, Poulsen CP, Fangel JU, Smith PJ, Yang J-Y, Peña M-J, Dinesen MH, Martens HJ, Melkonian M, et al. Identification of an algal xylan synthase indicates that there is functional orthology between algal and plant cell wall biosynthesis. New Phytol. 2018:**218**(3):1049–1060. https://doi.org/10.1111/nph.15050

Jensen JK, Johnson N, Wilkerson CG. Discovery of diversity in xylan biosynthesis genes by transcriptional profiling of a heteroxylan containing mucilaginous tissue. Front Plant Sci. 2013:**4**:183. https://doi.org/10.3389/fpls.2013.001

Jensen JK, Schultink A, Keegstra K, Wilkerson CG, Pauly M. RNA-Seq analysis of developing nasturtium seeds (*Tropaeolum majus*): identification and characterization of an additional galactosyltransferase involved in xyloglucan biosynthesis. Mol Plant. 2012:**5**(5):984–992. https://doi.org/10.1093/mp/sss032

Jensen JK, Sorensen SO, Harholt J, Geshi N, Sakuragi Y, Moller I, Zandleven J, Bernal AJ, Jensen NB, Sørensen C, et al. Identification of a xylogalacturonan xylosyltransferase involved in pectin biosynthesis in Arabidopsis. Plant Cell. 2008:**20**(5): 1289–1302. https://doi.org/10.1105/tpc.107.050906

Jobert F, Yadav S, Robert S. Auxin as an architect of the pectin matrix. J Exp Bot. 2023:**74**(22):6933–:6949. https://doi.org/10.1093/jxb/erad174

Johnson MA, Harper JF, Palanivelu R. A fruitful journey: pollen tube navigation from germination to fertilization. Annu Rev Plant Biol. 2019:**70**(1):809–837. https://doi.org/10.1146/annurev-arplant-050718-100133

Jönsson K, Lathe RS, Kierzkowski D, Routier-Kierzkowska A-L, Hamant O, Bhalearo RP. Mechanochemical feedback mediates tissue bending required for seedling emergence. Curr Biol. 2021:**31**(6): 1154–1164. https://doi.org/10.1016/j.cub.2020.12.016

Julian JD, Zabotina OA. Xyloglucan biosynthesis: from genes to proteins and their functions. Front Pl Sci. 2022:**13**:920494. https://doi.org/10.3389/fpls.2022.920494

Kaczmarska A, Pieczywek PM, Cybulska J, Zdunek A. Structure and functionality of rhamnogalacturonan I in the cell wall and in solution: a review. Carbohydr Polym. 2022:**278**:118909. https://doi.org/10.1016/j.carbpol.2021.118909

Kaczmarska A, Pieczywek PM, Cybulska J, Zdunek A. A mini-review on the plant sources and methods for extraction of rhamnogalacturonan I. Food Chem. 2023:**403**:134378. https://doi.org/10.1016/j.foodchem.2022.134378

Kanaya T, Kaji K. History of fiber structure. In: High-performance and specialty fibers. **The Society of Fiber Science and Technology Japan**, editors. Tokyo: Springer; 2016. p. 3–21. https://doi.org/10.1007/978-4-431-55203-1_1

Kang X, Kirui A, Dickwella Widanage MC, Mentick-Vigier F, Cosgrove DJ, Wang T. Lignin-polysaccharide interactions in plant secondary cell walls revealed by solid-state NMR. Nat Commun. 2019:**10**(1):347. https://doi.org/10.1038/s41467-018-08252-0

Kapoor K, Geitmann A. Pollen tube invasive growth is promoted by callose. Plant Reprod. 2023:**36**(2):157–171. https://doi.org/10.1007/s00497-023-00458-7

Karlsson M, Melzer M, Prokhorenko I, Johansson T, Wingsle G. Hydrogen peroxide and expression of hipl-superoxide dismutase are associated with the development of secondary cell walls in *Zinnia elegans*. J Exp Bot. 2005:**56**(418):2085–2093. https://doi.org/10.1093/jxb/eri207

Keegstra K, Talmadge KW, Bauer WD, Albersheim P. The structure of plant cell walls III. A model of the walls of suspension-cultured sycamore cells based on the interconnections of the macromolecular components. Plant Physiol. 1973:**51**(1):188–196. https://doi.org/10.1104/pp.51.1.188

Keller B, Templeton MD, Lamb CJ. Specific localization of a plant cell wall glycine-rich protein in protoxylem cells of the vascular system. Proc Natl Acad Sci U S A. 1989:**86**(5):1529–1533. https://doi.org/10.1073/pnas.86.5.1529

Kerr T, Bailey IW. The cambium and its derivative tissues: No. X. Structure, optical properties and chemical composition of the so-called middle lamella. J Arnold Arboretum (Harvard University). 1934:**15**(4): 327–349. https://doi.org/10.5962/p.185316

Kertesz ZI. The pectic substances. New York: Interscience Publishers; 1951, p. 1–628.

Kesten C, Menna A, Sánchez-Rodríguez C. Regulation of cellulose synthesis in response to stress. Curr Opin Plant Biol. 2017:**40**:106–113. https://doi.org/10.1016/j.pbi.2017.08.010

Kieliszewski MJ, de Zacks R, Leykam JF, Lamport DTA. A repetitive proline-rich protein from the gymnosperm Douglas fir is a hydroxyproline-rich glycoprotein. Plant Physiol. 1992a:**98**(3): 919–926. https://doi.org/10.1104/pp.98.3.919

Kieliszewski MJ, Kamyab A, Leykam JF, Lamport DTA. A histidine-rich extensin from *Zea mays* is an arabinogalactan protein. Plant Physiol. 1992b:**99**(2):538–547. https://doi.org/10.1104/pp.99.2.538

Kieliszewski MJ, Lamport DTA. Extensive: repetitive motifs, functional sites, post-translational codes, and phylogeny. Plant J. 1994:**5**(2): 157–172. https://doi.org/10.1046/j.1365-313X.1994.05020157.x

Kim SJ, Chandrasekar B, Rea AC, Danhof L, Zemelis-Durfee S, Thrower N, Shepard ZS, Pauly M, Brandizzi F, Keegstra K. The synthesis of xyloglucan, an abundant plant cell wall polysaccharide, requires CSLC function. Proc Natl Acad Sci U S A. 2020:**117**(33): 20316–20324. https://doi.org/10.1073/pnas.2007245117

Kim J, Shiu SH, Thoma S, Li WH, Patterson SE. Patterns of expansion and expression divergence in the plant polygalacturonase gene family. Genome Biol. 2006:**7**(9):R87. https://doi.org/10.1186/gb-2006-7-9-r87

Ko JH, Beers EP, Han KH. Global comparative transcriptome analysis identifies gene network regulating secondary xylem development in *Arabidopsis thaliana*. Mol Genet Genomics. 2006:**276**(6): 517–531. https://doi.org/10.1007/s00438-006-0157-1

Kobayashi M, Matoh T, Azuma J. Two chains of rhamnogalacturonan II are cross-linked by borate-diol ester bonds in higher plant cell walls. Plant Physiol. 1996:**110**(3):1017–1020. https://doi.org/10.1104/pp.110.3.1017

Kohorn BD, Dexter-Meldrum J, Zorensky FDH, Chabout S, Mouille G, Kohorn S. Pectin dependent cell adhesion restored by a mutant microtubule organizing membrane protein. Plants. 2021a:**10**(4):690. https://doi.org/10.3390/plants10040690

Kohorn BD, Greed BE, Mouille G, Verger S, Kohorn SL. Effects of Arabidopsis wall associated kinase mutations on ESMERALDA1 and elicitor induced ROS. PLoS One. 2021b:**16**(5):e0251922. https://doi.org/10.1371/journal.pone.0251922

Kohorn BD, Kobayashi M, Johansen S, Riese J, Huang LF, Koch K, Fu S, Dotson A, Byers N. An Arabidopsis cell wall-associated kinase required for invertase activity and cell growth. Plant J. 2006:**46**(2): 307–316. https://doi.org/10.1111/j.1365-313X.2006.02695.x

Kolpak F, Blackwell J. Determination of the structure of cellulose II. Macromolecules. 1976:**9**(2):273–278. https://doi.org/10.1021/ma60050a019

Kong Y, Pena MJ, Renna L, Avci U, Pattathil S, Tuomivaara ST, Li X, Reiter W-D, Brandizzi F, Hahn M, et al. Galactose-depleted xyloglucan is dysfunctional and leads to dwarfism in *Arabidopsis*. Plant Physiol. 2015:**167**(4):1296–1306. https://doi.org/10.1104/pp.114.255943

Kooiman P. Amyloids of seed plants. Nature. 1957:**179**(4550):107–109. https://doi.org/10.1038/179107a0

Kooiman P. The constitution of the amyloid from seeds of *Annona Muricata L.* Phytochemistry. 1967:**4**(12):1665–1673. https://doi.org/10.1016/S0031-9422(00)82900-3

Kroeger JH, Geitmann A, Grant M. Model for calcium dependent oscillatory growth in pollen tubes. J Theor Biol. 2008:**253**(2):363–374. https://doi.org/10.1016/j.jtbi.2008.02.042

Kroon-Batenburg LMJ, Kroon J. The crystal and molecular structures of cellulose I and II. Glycoconjugate J. 1997:**14**(5):677–690. https://doi.org/10.1023/A:1018509231331

Kubicki JD, Yang H, Sawada D, O'Neill H, Oehme D, Cosgrove D. The shape of native plant cellulose microfibrils. Sci Rep. 2018:**8**(1):13983. https://doi.org/10.1038/s41598-018-32211-w

Kuki H, Yokoyama R, Kuroha T, Nishitani K. Xyloglucan is not essential for the formation and integrity of the cellulose network in the primary cell wall regenerated from Arabidopsis protoplasts. Plants. 2020:**9**(5):629. https://doi.org/10.3390/plants9050629

Kumar M, Wightman R, Atanassov I, Gupta A, Hurst CH, Hemsley PA, Turner S. S-Acylation of the celluose synthase complex is essential for its plasma membrane localization. Science. 2016:**353**(6295): 166–169. https://doi.org/10.1126/science.aaf4009

Kurek I, Kawagoe Y, Jacob-Wilk D, Doblin M, Delmer D. Dimerization of cotton fiber cellulose synthase catalytic subunits occurs via oxidation of the zinc-binding domains. Proc Natl Acad Sci U S A. 2022:**99**(17): 11109–11114. https://doi.org/10.1073/pnas.162077099

Lam PY, Lui ACW, Wang L, Liu H, Umezawa T, Tobimatsu Y, Lo C. Tricin biosynthesis and bioengineering. Front Plant Sci. 2021:**12**: 733198. https://doi.org/10.3389/fpls.2021.733198

Lamport DTA. Oxygen fixation into hydroxyproline of plant cell wall protein. J Biol Chem. 1963:**238**(4):1438–1440. https://doi.org/10.1016/S0021-9258(18)81202-9

Lamport DTA, Kieliszewski MJ, Chen Y, Cannon MC. Role of the extension superfamily in primary cell wall architecture. Plant Physiol. 2011:**156**(1):11–19. https://doi.org/10.1104/pp.110.169011

Lamport DTA, Northcote DH. Hydroxyproline in primary cell walls of higher plants. Nature. 1960:**188**(4751):665–666. https://doi.org/10.1038/188665b0

Lampugnani ER, Flores-Sandoval E, Tan QW, Mutwil M, Bowman JL, Persson S. Cellulose synthesis—central components and their evolutionary relationships. Trends Plant Sci. 2019:**24**(5):402–412. https://doi.org/10.1016/j.tplants.2019.02.011

Lapierre C, Monties B, Rolando C, de Chirale L. Thioacidolysis of lignin: comparison with acidolysis. J Wood Chem Technol. 1985:**5**(2): 277–292. https://doi.org/10.1080/02773818508085193

Lau JM, McNeil M, Darvill AG, Albersheim P. Structure of the backbone of rhamnogalacturonan I, a pectic polysaccharide in the primary cell walls of plants. Carbohydr Res. 1985:**137**:111–125. https://doi.org/10.1016/0008-6215(85)85153-3

Laursen T, Stonebloom SH, Pidatala VR, Birdseye DS, Clausen MH, Mortimer JC, Scheller HV. Bifunctional glycosyltransferases catalyze both extension and termination of pectic galactan oligosaccharides. Plant J. 2018:**94**(2):340–351. https://doi.org/10.1111/tpj.13860

Lee Y, Rubio MC, Alassimone J, Geldner N. A mechanism for localized lignin deposition in the endodermis. Cell. 2013:**153**(2):402–412. https://doi.org/10.1016/j.cell.2013.02.045

Lei L, Li S, Gu Y. Cellulose synthase complexes: composition and regulation. Front Plant Sci. 2012:**3**:75. https://doi.org/10.3389/fpls.2012.00075

Lei L, Zhang T, Strasser R, Lee CM, Gonneau M, Mach L, Vernhettes S, Kim SH, Cosgrove DJ, Li S, et al. The jiaoyao1 mutant is an allele of korrigan1 that abolishes endoglucanase activity and affects the organization of both cellulose microfibrils and microtubules in Arabidopsis. Plant Cell. 2014:**26**(6):2601–2616. https://doi.org/10.1105/tpc.114.126193

Leloir LF. Far away and long ago. Ann Rev Biochem. 1983:**52**(1):1–15. https://doi.org/10.1146/annurev.bi.52.070183.000245

Leng Y, Yang Y, Ren D, Huang L, Dai L, Wang Y, Chen L, Tu Z, Gao Y, Li X, et al. A rice PECTATE LYASE-LIKE gene is required for plant growth and leaf senescence. Plant Physiol. 2017:**172**(2):1151–1166. https://doi.org/10.1104/pp.16.01625

Leppard GG, Colvin JR. Electron-opaque fibrils and granules in and between the cell walls of higher plants. J Cell Biol. 1972:**53**(3):695–703. https://doi.org/10.1083/jcb.53.3.695

Lerouxel O, Cavalier DM, Liepman AH, Keegstra K. Biosynthesis of plant cell wall polysaccharides-a complex process. Curr Opin Plant Biol. 2006:**9**(6):621–630. https://doi.org/10.1016/j.pbi.2006.09.009

Leszczuk A, Kalaitzis P, Kulik J, Zdunek A. Review: structure and modifications of arabinogalactan proteins (AGPs). BMC Plant Biol. 2023:**23**(1):45. https://doi.org/10.1186/s12870-023-04066-5

Lewis NG, Yamamoto E. Lignin: occurrence, biogenesis and biodegradation. Annu Rev Plant Physiol Plant Mol Biol. 1990:**41**(1):455–496. https://doi.org/10.1146/annurev.pp.41.060190.002323

Li Z, Fernie A R, Persson S. Transition of primary to secondary cell wall synthesis. Sci Bull. 2016c:**61**(11):838–846. https://doi.org/10.1007/s11434-016-1061-7

Li M, Pu Y, Ragauskas AJ. Current understanding of the correlation of lignin structure with biomass recalcitrance. Front Chem. 2016b:**4**:45. https://doi.org/10.3389/fchem.2016.00045

Li L, Verstraeten I, Roosjen M, Takahashi K, Rodriguez L, Merrin J, Chen J, Shabala L, Smet W, Ren H, et al. Cell surface and intracellular auxin signalling for H(+) fluxes in root growth. Nature. 2021: **599**(7884):273–277. https://doi.org/10.1038/s41586-021-04037-6

Li C, Wu HM, Cheung AY. FERONIA and her pals: functions and mechanisms. Plant Physiol. 2016a:**171**(4):2379–2392. https://doi.org/10.1104/pp.16.00667

Li Y, Yang T, Dai D, Hu Y, Guo X, Guo H. Evolution, gene expression profiling and 3D modeling of CSLD proteins in cotton. BMC Plant Biol. 2017:**17**(1):119. https://doi.org/10.1186/s12870-017-1063-x

Li C, Yeh F-L, Cheung AY, Duan Q, Kita D, Liu M-C, Maman J, Luu EJ, Wu BW, Gates L, et al. Glycosylphosphatidylinositol-anchored proteins as chaperones and co-receptors for FERONIA receptor kinase signaling in Arabidopsis. Elife. 2015:**4**:e06587. https://doi.org/10.7554/eLife.06587

Li M, Yoo CG, Pu Y, Biswal AK, Tolbert AK, Mohnen D, Ragauskas AJ. Downregulation of pectin biosynthesis gene *GAUT4* leads to reduced ferulate and lignin-carbohydrate cross-linking in switchgrass. Commun Biol. 2019:**2**(1):22. https://doi.org/10.1038/s42003-018-0265-6

Liepman AH, Cavalier DM. The CELLULOSE SYNTHASE-LIKE A and CELLULOSE SYNTHASE-LIKE C families: recent advances and future perspectives. Front Plant Sci. 2012:**3**:109. https://doi.org/10.3389/fpls.2012.00109

Liepman AH, Wightman R, Geshi N, Turner SR, Scheller HV. Arabidopsis—a powerful model system for plant cell wall research. Plant J. 2010:**61**(6):1107–1121. https://doi.org/10.1111/j.1365-313X.2010.04161.x

Liepman AH, Wilkerson CG, Keegstra K. Expression of cellulose synthase-like (Csl) genes in insect cells reveals that CslA family members encode mannan synthases. Proc Natl Acad Sci U S A. 2005:**102**(6):2221–2226. https://doi.org/10.1073/pnas.0409179102

Lin D, Lopez-Sanchez P, Gidley MJ. Interactions of pectins with cellulose during its synthesis in the absence of calcium. Food Hydrocoll. 2016:**52**:57–68. https://doi.org/10.1016/j.foodhyd.2015.06.004

Lin S, Miao Y, Huang H, Zhang Y, Huang L, Cao J. Arabinogalactan proteins: focus on the role in cellulose synthesis and deposition during plant cell wall biogenesis. Int J Mol Sci. 2022a:**23**(12):6578. https://doi.org/10.3390/ijms23126578

Lin W, Tang W, Pan X, Huang A, Gao X, Anderson CT, Yang Z. Arabidopsis pavement cell morphogenesis requires FERONIA binding to pectin for activation of ROP GTPase signaling. Curr Biol. 2022b:**32**(3):497–507. https://doi.org/10.1016/j.cub.2021.11.030

Lindner H, Müller LM, Boisson-Dernier A, Grossniklaus U. CrRLK1L receptor-like kinases: not just another brick in the wall. Curr Opin Plant Biol. 2012:**15**(6):659–669. https://doi.org/10.1016/j.pbi.2012.07.003

Liners F, Gaspar T, Van Cutsem P. Acetyl- and methyl-esterification of pectins of friable and compact sugar-beet calli: consequences for intercellular adhesion. Planta. 1994:**192**(4):545–556. https://doi.org/10.1007/BF00203593

Little A, Schwerdt JG, Shirley NJ, Khor SF, Neumann K, O'Donovan LA, Lahnstein J, Collins HM, Henderson M, Fincher GB, et al. Revised phylogeny of the *Cellulose Synthase* gene superfamily: insights into cell wall evolution. Plant Physiol. 2018:**177**(3):1124–1141. https://doi.org/10.1104/pp.17.01718

Liu X, Cui H, Zhang B, Song M, Chen S, Xiao C, Tang Y, Liesche J. Reduced pectin content of cell walls prevents stress-induced root cell elongation in Arabidopsis. J Exp Bot. 2021:**72**(4):1073–1084. https://doi.org/10.1093/jxb/eraa533

Liu Z, Schneider R, Kesten C, Zhang Y, Somssich M, Zhang Y, Fernie AR, Persson S. Cellulose-microtubule uncoupling proteins prevent lateral displacement of microtubules during cellulose synthesis in Arabidopsis. Dev Cell. 2016:**38**(3):305–315. https://doi.org/10.1016/j.devcel.2016.06.032

Liu H, Timko MP. Jasmonic acid signaling and molecular crosstalk with other phytohormones. Int J Mol Sci. 2021:**22**(6):2914. https://doi.org/10.3390/ijms22062914

Liu C, Yu H, Voxeur A, Rao X, Dixon RA. FERONIA and wall-associated kinases coordinate the defense response induced by cell wall engineering in Arabidopsis. Sci Adv. 2023:**9**(10):eadf7714. https://doi.org/10.1126/sciadv.adf7714

Liwanag AJM, Ebert B, Verhertbruggen Y, Rennie EA, Rautengarten C, Oikawa A, Andersen MCF, Clausen MH, Scheller HV. Pectin biosynthesis: GALS1 in *Arabidopsis thaliana* is a β-1,4-galactan β-1,4-galactosyltransferase. Plant Cell. 2012:**24**(12):5024–5036. https://doi.org/10.1105/tpc.112.106625

Lopez-Hernandez F, Tryfona T, Rizza A, Yu XL, Harris MOB, Webb AR, Kotake T, Dupree P. Calcium binding by arabinogalactan polysaccharides is important for normal plant development. Plant Cell. 2020:**32**(10):3346–3369. https://doi.org/10.1105/tpc.20.00027

Lori M, van Verk MC, Hander T, Schatowitz H, Klauser D, Flury P, Gehring CA, Boller T, Bartels S. Evolutionary divergence of the plant elicitor peptides (peps) and their receptors: interfamily incompatibility of perception but compatibility of downstream signalling. J Exp Bot. 2015:**66**(17):5315–5325. https://doi.org/10.1093/jxb/erv236

Lund CH, Stenbæk A, Atmodjo MA, Rasmussen RE, Moller IE, Erstad SM, Biswal AK, Mohnen D, Mravec J, Sakuragi Y. Pectin synthesis and pollen tube growth in Arabidopsis involves three GAUT1 Golgi-anchoring proteins: gAUT5, GAUT6 and GAUT7. Front Plant Sci. 2020:**11**:585774. https://doi.org/10.3389/fpls.2020.585774

Ma Y, Johnson K. Arabinogalactan proteins—multifunctional glycoproteins of the plant cell wall. Cell Surf. 2023:**9**:100102. https://doi.org/10.1016/j.tcsw.2023.100102

Madson M, Dunand C, Li X, Verma R, Vanzin GF, Caplan J, Shoue DA, Carpita NC, Reiter W-D. The MUR3 gene of Arabidopsis encodes a xyloglucan galactosyltransferase that is evolutionarily related to animal exostosins. Plant Cell. 2003:**15**(7):1662–1670. https://doi.org/10.1105/tpc.009837

Makarem M, Nishiyama Y, Xin X, Durachko DM, Gu Y, Cosgrove DJ, Kim SH. Distinguishing mesoscale polar order (unidirectional vs bidirectional) of cellulose microfibrils in plant cell walls using sum frequency generation spectroscopy. J Phys Chem B. 2020:**124**(37):8071–8081. https://doi.org/10.1021/acs.jpcb.0c07076

Malivert A, Erguvan O, Chevallie A, Dehem A, Friaud R, Liu M, Martin M, Peyraud T, Hamant O, Verger S. FERONIA and microtubules independently contribute to mechanical integrity in the Arabidopsis shoot. PloS Biol. 2021:**19**:e3001454. https://doi.org/10.1371/journal.pbio.3001454

Malivert A, Hamant O. Why is FERONIA pleiotropic? Nat Plants. 2023:**9**(7):1018–1025. https://doi.org/10.1038/s41477-023-01434-9

Maltby D, Carpita NC, Montezinos D, Kulow C, Delmer DP. β-1, 3-glucan in developing cotton fibers: structure, localization, and relationship of synthesis to that of secondary wall cellulose. Plant Physiol. 1979:**63**(6):1158–1164. https://doi.org/10.1104/pp.63.6.1158

Manfield IW, Orfila C, Mccartney L, Harholt J, Bernal AJ, Scheller HV, Gilmartin PM, Mikkelsen JD, Paul Knox J, Willats WG. Novel cell wall architecture of isoxaben-habituated Arabidopsis suspension-cultured cells: global transcript profiling and cellular analysis. Plant J. 2004:**40**:260–275. https://doi.org/10.1111/j.1365-313X.2004.02208.x

Mangin LA. Sur l'emploi du rouge de ruthénium en anatomie végétale. C R Acad Sci (Paris). 1893a:**116**:653–656.

Mangin ML. Nouvelles observations sur la membrane. Bull Soc Bot de France. 1893b:**40**(7):273–280. https://doi.org/10.1080/00378941.1893.10828818

Marita JM, Ralph J, Hatfield RD, Chapple C. NMR characterization of lignins in Arabidopsis altered in the activity of ferulate 5-hydroxylase. Proc Natl Acad Sci U S A. 1999:**96**(22):12328–12332. https://doi.org/10.1073/pnas.96.22.12328

Marry M, Roberts K, Jopson SJ, Huxham IM, Jarvis MC, Corsar J, Robertson E, McCann MC. Cell–cell adhesion in fresh sugar-beet root parenchyma requires both pectin esters and calcium cross-links. Physiol Plant. 2006:**126**(2):243–256. https://doi.org/10.1111/j.1399-3054.2006.00591.x

Martins MM, Carvalheiro F, Gírio F. An overview of lignin pathways of valorization: from isolation to refining and conversion into value-added products. (Biomass Conv Bioref. 2022). Germany: Springer-Verlag GmbH; 2022. https://doi.org/10.1007/s13399-022-02701-z

Martínez-Abad A, Giummarella N, Lawoko M, Vilaplana F. Differences in extractability under subcritical water reveal interconnected hemicellulose and lignin recalcitrance in birch hardwoods. Green Chem. 2018:**20**(11):2534–2546. https://doi.org/10.1039/C8GC00385H

Martone PT, Estevez JM, Lu F, Ruel K, Denny MW, Somerville C, Ralph J. Discovery of lignin in seaweed reveals convergent evolution of cell-wall architecture. Curr Biol. 2009:**19**(2):169–175. https://doi.org/10.1016/j.cub.2008.12.031

Matsunaga T, Ishii T, Matsumoto S, Higuchi M, Darvill A, Albersheim P, O'Neill MA. Occurrence of the primary cell wall polysaccharide rhamnogalacturonan II in pteridophytes, lycophytes, and bryophytes. Implications for the evolution of vascular plants. Plant Physiol. 2004:**134**(1):339–351. https://doi.org/10.1104/pp.103.030072

McCann M, Roberts K. Architecture of the primary cell wall. In: **Lloyd CW**, editor. Cytoskeletal basis of plant growth and form. London: Academic Press; 1991. p. 109–129.

McCann M, Wells B, Roberts K. Direct visualization of cross-links in the primary cell wall. J Cell Sci. 1990:**96**(2):323–334. https://doi.org/10.1242/jcs.96.2.323

McCartney L, Ormerod AP, Gidley MJ, Knox JP. Temporal and spatial regulation of pectic (1-->4)-(-D-GALACTAN in cell walls of developing pea cotyledons: implications for mechanical properties. Plant J. 2000:**22**(2):105–113. https://doi.org/10.1046/j.1365-313x.2000.00719.x

McGee R, Dean GH, Wu D, Zhang Y, Mansfield SD, Haughn GW. Pectin modification in seed coat mucilage by in vivo expression of rhamnogalacturonan-I and homogalacturonan-degrading enzymes. Plant Cell Physiol. 2021:**62**(12):1912–1926. https://doi.org/10.1093/pcp/pcab077

McHale L, Tan X, Koehl P, Michelmore RW. Plant NBS-LRR proteins: adaptable guards. Genome Biol. 2006:**7**(4):212. https://doi.org/10.1186/gb-2006-7-4-212

McKenna ST, Kunkel JG, Bosch M, Rounds CM, Vidali L, Winship LJ, Hepler PK. Exocytosis precedes and predicts the increase in growth in oscillating pollen tubes. Plant Cell. 2009:**21**(10):3026–3040. https://doi.org/10.1105/tpc.109.069260

McNamara JT, Morgan JLW, Zimmer J. A molecular description of cellulose biosynthesis. Annu Rev Biochem. 2015:**84**(1):895–921. https://doi.org/10.1146/annurev-biochem-060614-033930

McQueen-Mason S, Durachko DM, Cosgrove DJ. Two endogenous proteins that induce cell wall expansion in plants. Plant Cell. 1992:**4**(11):1425–1433. https://doi.org/10.1105/tpc.4.11.1425

Meents MJ, Watanabe Y, Samuels AL. The cell biology of secondary cell wall biosynthesis. Ann Bot. 2018:**121**(6):1107–1125. https://doi.org/10.1093/aob/mcy005

Meinert MC, Delmer DP. Changes in biochemical composition of the cell wall of the cotton fiber during development. Plant Physiol. 1977:**59**(6): 1088–1097. https://doi.org/10.1104/pp.59.6.1088

Meyer KH, Misch L. Positions des atomes dans le nouveau modele spatial de la cellulose. Helv Chim Acta. 1937:**20**(1):232–244. https://doi.org/10.1002/hlca.19370200134

Miao YC, Liu CJ. ATP-binding cassette-like transporters are involved in the transport of lignin precursors across plasma and vacuolar membranes. Proc Natl Acad Sci U S A. 2010:**107**(52):22728–22733. https://doi.org/10.1073/pnas.1007747108

Miart F, Desprez T, Biot E, Morin H, Belcram K, Hofte H, Gonneau M, Vernhettes S. Spatio-temporal analysis of cellulose synthesis during cell plate formation in Arabidopsis. Plant J. 2014:**77**(1):71–84. https://doi.org/10.1111/tpj.12362

Mishler-Elmore JW, Zhou Y, Sukul A, Oblak M, Tan L, Faik A, Held MA. Extensins: self-assembly crosslinking, and the role of peroxidases. Front Plant Sci. 2021:**12**:664738. https://doi.org/10.3389/fpls.2021.664738

Mizukami AG, Inatsugi R, Jiao J, Kotake T, Kuwata K, Ootani K, Okuda S, Sankaranarayanan S, Sato Y, Maruyama D, et al. The AMOR arabinogalactan sugar chain induces pollen-tube competency to respond to ovular guidance. Curr Biol. 2016:**26**(8):1091–1097. https://doi.org/10.1016/j.cub.2016.02.040

Mohnen D. A signaling peptide locks pollen tube walls. Science. 2023:**382**(6671):648–649. https://doi.org/10.1126/science.adl1198

Mohnen D, Atmodjo MA, Jayawardhane P. Reconsidering pectin structure: A historical perspective informed by identity and function of the biosynthetic enzymes. In: **Geitmann A**, editor. The plant cell wall - research milestones and conceptual insights. Boca Raton (FL): CRC Press, Taylor & Francis Group; 2024. p. 94–126. https://doi.org/10.1201/9781003178309.

Molina-Hidalgo FJ, Franco AR, Villatoro C, Medina-Puche L, Mercado JA, Hidalgo MA, Monfort A, Caballero JL, Muñoz-Blanco J, Blanco-Portales R. The strawberry (*Fragaria×ananassa*) fruit-specific *rhamnogalacturonate lyase 1* (*FaRGLyase1*) gene encodes an enzyme involved in the degradation of cell-wall middle lamellae. J Exp Bot. 2013:**64**(6):1471–1483. https://doi.org/10.1093/jxb/ers386

Molina A, Miedes E, Bacete L, Rodríguez T, Mélida H, Denancé N, Sánchez-Vallet A, Rivière MP, López G, Freydier A, et al. Arabidopsis cell wall composition determines disease resistance specificity and fitness. Proc Natl Acad Sci U S A. 2021:**118**(5): e2010243118. https://doi.org/10.1073/pnas.2010243118

Mollet JC, Leroux C, Dardelle F, Lehner A. Cell wall composition, biosynthesis and remodeling during pollen tube growth. Plants (Basel). 2013:**2**(1):107–147. https://doi.org/10.3390/plants2010107

Monshausen GB, Bibikova TN, Messerli MA, Shi C, Gilroy S. Oscillations in extracellular pH and reactive oxygen species modulate tip growth of Arabidopsis root hairs. Proc Natl Acad Sci U S A. 2007:**104**(52):20996–21001. https://doi.org/10.1073/pnas.0708586104

Montezinos D, Brown M Jr. Surface architecture of the plant cell: biogenesis of the cell wall, with special emphasis on the role of the plasma membrane in cellulose biosynthesis. J Supramol Struct. 1976:**5**(3): 277–290. https://doi.org/10.1002/jss.400050303

Moor H, Muhlethaler K, Waldner H, Frey-Wyssling A. A new freezing-ultramicrotome. J Biophys Biochem Cytol. 1961:**10**(1): 1–13. https://doi.org/10.1083/jcb.10.1.1

Moore PJ, Staehelin LA. Immunogold localization of the cell-wall-matrix polysaccharides rhamnogalacturonan I and xyloglucan during cell expansion and cytokinesis in *Trifolium pratense* L.; implication for secretory pathways. Planta. 1988:**174**(4):433–445. https://doi.org/10.1007/BF00634471

Moreira D, Kaur D, Pereira AM, Held MA, Showalter AM, Coimbra S. Type II arabinogalactans initiated by hydroxyproline-*O*-galactosyltransferases play important roles in pollen-pistil interactions. Plant J. 2023:**114**(2):371–389. https://doi.org/10.1111/tpj.16141

Morgan JL, McNamara JT, Zimmer J. Mechanism of activation of bacterial cellulose synthase by cyclic di-GMP. Nat Struct Mol Biol. 2014:**21**(5):489–496. https://doi.org/10.1038/nsmb.2803

Morgan JL, Strumillo J, Zimmer J. Crystallographic snapshot of cellulose synthesis and membrane translocation. Nature. 2013:**493**(7431): 181–186. https://doi.org/10.1038/nature11744

Morrall P, Briggs DE. Changes in cell wall polysaccharides of germinating barley grains. Phytochemistry. 1978:**17**(9):1495–1502. https://doi.org/10.1016/S0031-9422(00)94628-4

Morris ER, Powell DA, Gidley MJ, Rees DA. Conformation and interactions of pectins. I. Polymorphism between gel and solid stated of calcium polygalacturonate. J Mol Biol. 1982:**155**(4):507–516. https://doi.org/10.1016/0022-2836(82)90484-3

Mortimer JC, Miles GP, Brown DM, Zhang Z, Segura MP, Weimar T, Yu X, Seffen KA, Stephens E, Turner SR, et al. Absence of branches from xylan in Arabidopsis *gux* mutants reveals potential for simplification of lignocellulosic biomass. Proc Natl Acad Sci U S A. 2010:**107**(40):17409–17414. https://doi.org/10.1073/pnas.1005456107

Mottiar Y, Karlen SD, Goacher RE, Ralph J, Mansfield SD. Metabolic engineering of *p*-hydroxybenzoate in poplar lignin. Plant Biotechnol J. 2023:**21**(1):176–188. https://doi.org/10.1111/pbi.13935

Mouille G, Ralet M-C, Cavelier C, Eland C, Effroy D, Hematy K, McCartney L, Truong H-N, Gaudon V, Tibault J-F, et al. Homogalacturonan synthesis in *Arabidopsis thaliana* requires a Golgi-localized protein with a putative methyltransferase domain. Plant J. 2007:**50**(4):605–614. https://doi.org/10.1111/j.1365-313X.2007.03086.x

Moussu S, Ingram G. The EXTENSIN enigma. Cell Surf. 2023:**9**:100094. https://doi.org/10.1016/j.tcsw.2023.100094

Moussu S, Lee H-K, Haas KT, Broyart C, Rathgeb U, De Bellis D, Levasseur T, Schoenaers S, Fernandez GS, Grossniklasu U, et al. Plant cell wall patterning and expansion mediated by protein-peptide-polysaccharide interaction. Science. 2023:**382**(6671):719–725. https://doi.org/10.1126/science.adi4720

Mueller SC, Brown RM Jr. Evidence for an intra-membrane component associated with a cellulose-microfibril-synthesizing complex in higher plants. J Cell Biol. 1980:**84**(2):315–326. https://doi.org/10.1083/jcb.84.2.315

Mühlethaler K. Die feinstruktur der zellulosemikrofibrillen. Beih Z Schweiz Forstver. 1960:**30**:55–64.

Muschitz A, Riou C, Mollet J-C, Gloaguen V, Faugeron C. Modifications of cell wall pectin in tomato cell suspension in response to cadmium and zinc. Acta Physiol Plant. 2015:**37**:1–11. https://doi.org/10.1007/s11738-015-2000-y

Nakamura A, Furuta H, Maeda H, Takao T, Nagamatsu Y. Structural studies by stepwise enzymatic degradation of the main backbone of soybean soluble polysaccharides consisting of galacturonan and rhamnogalacturonan. Biosci Biotechnol Biochem. 2002:**66**(6): 1301–1313. https://doi.org/10.1271/bbb.66.1301

Narváez-Barragán DA, Tovar-Herrera OE, Guevara-García A, Serrano M, Martinez-Anaya C. Mechanisms of plant cell wall surveillance in response to pathogens, cell wall-derived ligands and the effect of expansins to infection resistance or susceptibility. Front Plant Sci. 2022:**13**: 969343. https://doi.org/10.3389/fpls.2022.969343

Ndeh D, Rogowski A, Cartmell A, Luis AS, Baslé A, Gray J, Venditto I, Briggs J, Zhang X, Labourel A, et al. Complex pectin metabolism by gut bacteria reveals novel catalytic functions. Nature. 2017:**544**(7648):65–70. https://doi.org/10.1038/nature21725

Newman RH, Hill SJ, Harris PJ. Wide-angle x-ray scattering and solid-state nuclear magnetic resonance data combined to test models for cellulose microfibrils in mung bean cell walls. Plant Physiol. 2013:**163**(4):1558–1567. https://doi.org/10.1104/pp.113.228262

Nibbering P, Petersen BL, Motawia MS, Jørgensen B, Ulvskov P, Niittylä T. Golgi-localized exo-(1,3-galactosidases involved in cell expansion and root growth in Arabidopsis. J Biol Chem. 2020:**295**(31): 10581–10592. https://doi.org/10.1074/jbc.RA120.013878

Nicol F, His I, Jauneau A, Vernhettes S, Canut H, Höfte H. A plasma membrane-bound putative endo-1,4-β-D-glucanase is required for normal wall assembly and cell elongation in Arabidopsis. EMBO J. 1998:**17**(19):5563–5576. https://doi.org/10.1093/emboj/17.19.5563

Nishiyama Y, Langan P, Chanzy H. Crystal structure and hydrogen-bonding system in cellulose iβ from synchrotron X-ray and neutron fiber diffraction. J Am Chem Soc. 2002:**124**(31):9074–9082. https://doi.org/10.1021/ja0257319

Nissen KS, Willats WGT, Malinovsky FG. Understanding CrRLK1L function: cell walls and growth control. Trends Plant Sci. 2016:**21**(6):516–527. https://doi.org/10.1016/j.tplants.2015.12.004

Nixon BT, Mansouri K, Singh A, Du J, Davis JK, Lee JG, Slabaugh E, Vandavasi VG, O'Neill H, Roberts EM, et al. Comparative structural and computational analysis supports eighteen cellulose synthases in the plant cellulose synthesis complex. Sci Rep. 2016:**6**(1):28696. https://doi.org/10.1038/srep28696

Notley SM, Pettersson B, Wågberg L. Direct measurement of attractive van der Waals' forces between regenerated cellulose surfaces in an aqueous environment. J Am Chem Soc. 2004:**126**(43):13930–13931. https://doi.org/10.1021/ja045992d

Nuñez A, Fishman ML, Fortis LL, Cooke PH, Hotchkiss AT Jr. Identification of extensin protein associated with sugar beet pectin. J Agric Food Chem. 2009:**57**(22):10951–10958. https://doi.org/10.1021/jf902162t

Oehme DP, Shafee T, Downton MT, Bacic A, Doblin MS. Differences in protein structural regions that impact functional specificity in GT2 family beta-glucan synthases. PLoS One. 2019:**14**(10):e0224442. https://doi.org/10.1371/journal.pone.0224442

Ohtani M, Demura T. The quest for transcriptional hubs of lignin biosynthesis: beyond the NAC-MYB-gene regulatory network model. Curr Opin Biotechnol. 2019:**56**:82–87. https://doi.org/10.1016/j.copbio.2018.10.002

Oikawa A, Joshi HJ, Rennie EA, Ebert B, Manisseri C, Heazlewood JL, Scheller HV. An integrative approach to the identification of Arabidopsis and rice genes involved in xylan and secondary wall development. PLoS One. 2010:**5**(11):e15481. https://doi.org/10.1371/journal.pone.0015481

Olek AT, Rayon C, Makowski L, Kim HR, Ciesielski P, Badger J, Paul LN, Ghosh S, Kihara D, Crowley M, et al. The structure of the catalytic domain of a plant cellulose synthase and its assembly into dimers. Plant Cell. 2014:**26**(7):2996–3009. https://doi.org/10.1105/tpc.114.126862

Olek AT, Rushton PS, Kihara D, Ciesielski P, Aryal UK, Zhang Z, Stauffacher CV, McCann MC, Carpita NC. Essential amino acids in the plant-conserved and class-specific regions of cellulose synthases. Plant Physiol. 2023:**191**(1):142–160. https://doi.org/10.1093/plphys/kiac479

O'Neill MA, Eberhard S, Albersheim P, Darvill AG. Requirement of borate cross-linking of cell wall rhamnogalactaruonan II for Arabidopsis growth. Science. 2001:**294**(5543):846–849. https://doi.org/10.1126/science.1062319

O'Neill MA, Warrenfeltz D, Kates K, Pellerin P, Doco T, Darvill AG, Albersheim P. Rhamnogalacturonan-II, a pectic polysaccharide in the walls of growing plant cells, forms a dimer that is covalently cross-linked by a borate ester. In vitro conditions for the formation and hydrolysis of the dimer. J Biol Chem. 1996:**271**(37):22923–22930. https://doi.org/10.1074/jbc.271.37.22923

Oyarce P, De Meester B, Fonseca F, de Vries L, Goeminne G, Pallidis A, De Rycke R, Tsuji Y, Li Y, Van den Bosch S, et al. Introducing curcumin biosynthesis in Arabidopsis enhances lignocellulosic biomass processing. Nat Plants. 2019:**5**(2):225–237. https://doi.org/10.1038/s41477-018-0350-3

Paniagua C, Bilkova A, Jackson P, Dabravolski S, Riber W, Didi V, Houser J, Gigli-Bisceglia N, Wimmerova M, Budínská E, et al. Dirigent proteins in plants: modulating cell wall metabolism during abiotic and biotic stress exposure. J Exp Bot. 2017:**68**(13): 3287–3301. https://doi.org/10.1093/jxb/erx141

Paniagua C, Posé S, Morris VJ, Kirby AR, Quesada MA, Mercado JA. Fruit softening and pectin disassembly: an overview of nanostructural pectin modifications assessed by atomic force microscopy. Ann Bot. 2014:**114**(6):1375–1383. https://doi.org/10.1093/aob/mcu149

Paredez AR, Somerville CR, Ehrhardt DW. Visualization of cellulose synthase demonstrates functional association with microtubules. Science. 2006:**312**(5779):1491–1495. https://doi.org/10.1126/science.1126551

Park AR, Cho SK, Yun UJ, Jin MY, Lee SH, Sachetto-Martins G, Park OK. Interaction of the Arabidopsis receptor protein kinase Wak1 with a glycine-rich protein, AtGRP-3. J Biol Chem. 2001:**276**(28): 26688–26693. https://doi.org/10.1074/jbc.M101283200

Park YB, Cosgrove DJ. A revised architecture of primary cell walls based on biochemical changes induced by substrate-specific endoglucanases. Plant Physiol. 2012a:**158**(4):1933–1943. https://doi.org/10.1104/pp.111.192880

Park YB, Cosgrove DJ. Changes in cell wall biomechanical properties in the xyloglucan-deficient *xxt1/xxt2* mutant of *Arabidopsis*. Plant Physiol. 2012b:**158**(1):465–475. https://doi.org/10.1104/pp.111.189779

Park E, Diaz-Moreno SM, Davis DJ, Wilkop TE, Bulone V, Drakakaki G. Endosidin 7 specifically arrests late cytokinesis and inhibits callose biosynthesis, revealing distinct trafficking events during cell plate maturation. Plant Physiol. 2014:**165**(3):1019–1034. https://doi.org/10.1104/pp.114.241497

Park S, Szumlanski AL, Gu F, Guo F, Nielsen E. A role for CSLD3 during cell-wall synthesis in apical plasma membranes of tip-growing root-hair cells. Nat Cell Biol. 2011:**13**(8):973–980. https://doi.org/10.1038/ncb2294

Pauly M, Keegstra K. Biosynthesis of the plant cell wall matrix polysaccharide xyloglucan. Annu Rev Plant Biol. 2016:**67**(1):235–259. https://doi.org/10.1146/annurev-arplant-043015-112222

Pauly M, Ramirez V. New insights into wall polysaccharide *O*-acetylation. Front Plant Sci. 2018:**9**:1210. https://doi.org/10.3389/fpls.2018.01210

Payen A. Memoire sur le composition du tissu propre des plantes et du ligneux. Compt Rend. 1838:**7**:1052–1056.

Pear JR, Kawagoe Y, Schreckengost WE, Delmer DP, Stalker DM. Higher plants contain homologs of the bacterial *celA* genes encoding the catalytic subunit of cellulose synthase. Proc Natl Acad Sci U S A. 1996:**93**(22):12637–12642. https://doi.org/10.1073/pnas.93.22.12637

Pearce G, Moura DS, Stratmann J, Ryan CA Jr. RALF, a 5-kDa ubiquitous polypeptide in plants, arrests root growth and development. Proc Natl Acad Sci U S A. 2001:**98**(22):12843–12847. https://doi.org/10.1073/pnas.201416998

Peaucelle A, Braybrook S, Höfte H. Cell wall mechanics and growth control in plants: the role of pectins revisited. Front Plant Sci. 2012:**3**:121. https://doi.org/10.3389/fpls.2012.00121

Peaucelle A, Wightman R, Hofte H. The control of growth symmetry breaking in the Arabidopsis hypocotyl. Curr Biol. 2015:**25**(13): 1746–1752. https://doi.org/10.1016/j.cub.2015.05.022

Pelletier S, Van Orden J, Wolf S, Vissenberg K, Delacourt J, Ndong YA, Pelloux J, Bischoff V, Urbain A, Mouille G, et al. A role for pectin de-methylesterification in a developmentally regulated growth acceleration in dark-grown Arabidopsis hypocotyls. New Phytol. 2010:**188**(3): 726–739. https://doi.org/10.1111/j.1469-8137.2010.03409.x

Pelloux J, Rustérucci C, Mellerowicz EJ. New insights into pectin methylesterase structure and function. Trends Plant Sci. 2007:**12**(6):267–277. https://doi.org/10.1016/j.tplants.2007.04.001

Pena MJ, Kong Y, York WS, O'Neill MA. A galacturonic acid-containing xyloglucan is involved in Arabidopsis root hair tip growth. Plant Cell. 2012:**24**(11):4511–4524. https://doi.org/10.1105/tpc.112.103390

Peng L, Kawagoe Y, Hogan P, Delmer D. Sitosterol-β-glucoside as primer for cellulose synthesis in plants. Science. 2002:**295**(5552): 147–150. https://doi.org/10.1126/science.1064281

Pérez S, Rodríguez-Caravajal MA, Doco T. A complex plant cell wall polysaccharide: rhamnogalacturonan II. A structure in quest of a function. Biochimie. 2003:**85**(1-2):109–121. https://doi.org/10.1016/S0300-9084(03)00053-1

Pérez García M, Zhang Y, Hayes J, Salazar A, Zabotina OA, Hong M. Structure and interactions of plant cell-wall polysaccharides by two and three-dimensional magic-angle-spinning solid-state NMR. Biochemistry. 2011:**50**(6):989–1000. https://doi.org/10.1021/bi101795q

Perkins ML, Schuetz M, Unda F, Chen KT, Bally MB, Kulkarni JA, Yan Y, Pico J, Castellarin SD, Mansfield SD, et al. Monolignol export by diffusion down a polymerization-induced concentration gradient. Plant Cell. 2022:**34**(5):2080–2095. https://doi.org/10.1093/plcell/koac051

Perkins M, Smith RA, Samuels L. The transport of monomers during lignification in plants: anything goes but how? Curr Opin Biotechnol. 2019:**56**:69–74. https://doi.org/10.1016/j.copbio.2018.09.011

Perrin RM, DeRocher AE, Bar-Peled M, Zeng W, Norambuena L, Orellana A, Raikhel NV, Keegstra K. Xyloglucan fucosyltransferase, an enzyme involved in plant cell wall biosynthesis. Science. 1999:**284**(5422): 1976–1979. https://doi.org/10.1126/science.284.5422.1976

Petersen BL, Macalister CA, Ulvskov P. Plant Protein O-Arabinosylation. Front Plant Sci. 2021:**12**:645219. https://doi.org/10.3389/fpls.2021.645219

Philippe F, Pelloux J, Rayon C. Plant pectin acetylesterase structure and function: new insights from bioinformatics analysis. BMC Genomics. 2017:**18**(1):456. https://doi.org/10.1186/s12864-017-3833-0

Phyo P, Gu Y, Hong M. Impact of acidic pH on plant cell wall polysaccharide structure and dynamics: insights into the mechanism of acid growth in plants from solid-state NMR. Cellulose. 2019:**26**(1): 291–304. https://doi.org/10.1007/s10570-018-2094-7

Pogorelko G, Lionetti V, Bellincampi D, Zabotina O. Cell wall integrity: targeted post-synthetic modifications to reveal its role in plant growth and defense against pathogens. Plant Signal Behav. 2013:**8**(9):e25435. https://doi.org/10.4161/psb.25435

Polko JK, Kieber JJ. The regulation of cellulose biosynthesis in plants. Plant Cell. 2019:**31**(2):282–296. https://doi.org/10.1105/tpc.18.00760

Potikha TS, Collins CC, Johnson DI, Delmer DP, Levine A. The involvement of hydrogen peroxide in the differentiation of secondary walls in cotton fibers. Plant Physiol. 1999:**119**(3):849–858. https://doi.org/10.1104/pp.119.3.849

Pottier D, Roitsch T, Persson S. Cell wall regulation by carbon allocation and sugar signaling. Cell Surf. 2023:**9**:100096. https://doi.org/10.1016/j.tcsw.2023.100096

Prabhakar PK, Pereira JH, Taujale R, Shao W, Bharadwaj VS, Chapla D, Yang JY, Bomble YJ, Moremen KW, Kannan N, et al. Structural and biochemical insight into a modular β-1,4-galactan synthase in plants. Nat Plants. 2023:**9**(3):486–500. https://doi.org/10.1038/s41477-023-01358-4

Preston R. The formation of wood in forest trees. NY, USA: Academic Press; 1964, p. 169–187.

Proseus TE, Boyer JS. Turgor pressure moves polysaccharides into growing cell walls of *Chara corallina*. Ann Bot. 2005:**95**(6):967–979. https://doi.org/10.1093/aob/mci113

Proseus TE, Boyer JS. Tension required for pectate chemistry to control growth in *Chara corallina*. J Exp Bot. 2007:**58**(15-16): 4283–4292. https://doi.org/10.1093/jxb/erm318

Proseus TE, Boyer JS. Calcium deprivation disrupts enlargement of *Chara corallina* cells: further evidence for the calcium pectate cycle. J Exp Bot. 2012a:**63**(10):3953–3958. https://doi.org/10.1093/jxb/ers089

Proseus TE, Boyer JS. Pectate chemistry links cell expansion to wall deposition in *Chara corallina*. Plant Signal Behav. 2012b:**7**(11): 1490–1492. https://doi.org/10.4161/psb.21777

Proseus TE, Ortega JKE, Boyer JS. Separating growth from elastic deformation during cell enlargement. Plant Physiol. 1999:**119**(2): 775–784. https://doi.org/10.1104/pp.119.2.775

Proseus TE, Zhu G-L, Boyer JS. Turgor, temperature and the growth of plant cells: using *Chara corallina* as a model system. J Exp Bot. 2000:**51**(350):1481–1494. https://doi.org/10.1093/jexbot/51.350.1481

Purushotham P, Cho SH, Diaz-Moreno SM, Kumar M, Nixon BT, Bulone V, Zimmer J. A single heterologously expressed plant cellulose synthase isoform is sufficient for cellulose microfibril formation in vitro. Proc Natl Acad Sci U S A. 2016:**113**(40):11360–11365. https://doi.org/10.1073/pnas.1606210113

Purushotham P, Ho R, Yu L, Fincher GB, Bulone V, Zimmer J. Mechanism of mixed-linkage glucan biosynthesis by barley cellulose synthase-like CslF6 (1,3;1,4)-β-glucan synthase. Sci Adv. 2022:**8**(45): eadd1596. https://doi.org/10.1126/sciadv.add1596

Purushotham P, Ho R, Zimmer J. Architecture of a catalytically active homotrimeric plant cellulose synthase complex. Science. 2020:**369**(6507): 1089–1094. https://doi.org/10.1126/science.abb2978

Qi X, Behrens BX, West PR, Mort AJ. Solubilization and partial characterization of extensin fragments from cell walls of cotton suspension cultures. Evidence for a covalent cross-link between extensin and pectin. Plant Physiol. 1995:**108**(4):1691–1701. https://doi.org/10.1104/pp.108.4.1691

Qi L, Shi Y, Li C, Liu J, Chong S-L, Lim K-J, Si J, Han Z, Chen D. Glucomannan in *Dendrobium catenatum:* bioactivities, biosynthesis and perspective. Genes (Basel). 2022:**13**(11):1957. https://doi.org/10.3390/genes13111957

Qin Y-M, Zhu Y-X. How cotton fibers elongate: a tale of linear cell-growth mode. Curr Opin Plant Biol. 2011:**14**(1):106–111. https://doi.org/10.1016/j.pbi.2010.09.010

Ragauskas AJ, Beckham GT, Biddy MJ, Chandra R, Chen F, Davis MF, Davison BH, Dixon RA, Gilna P, Keller M, et al. Lignin valorization: improving lignin processing in the biorefinery. Science. 2014:**344**(6185):1246843. https://doi.org/10.1126/science.1246843

Rajangam AS, Kumar M, Aspeborg H, Guerriero G, Arvestad L, Pansri P, Brown CJL, Hober S, Blomqvist K, Divne C, et al. MAP20, a microtubule-associated protein in the secondary cell walls of hybrid aspen, is a target of the cellulose synthesis inhibitor 2,6-dichlorobenzonitrile. Plant Physiol. 2008:**148**(3):1283–1294. https://doi.org/10.1104/pp.108.121913

Ralph J, Brunow G, Harris P, Dixon RA, Boerjan W. Lignification: are lignins biosynthesized via simple combinatorial chemistry or via proteinaceous control and template replication? Rec Adv Polyphenols Res. 2008:**1**:36–66. https://doi.org/10.1002/9781444302400.ch2

Ralph J, Lapierre C, Boerjan W. Lignin structure and its engineering. Curr Opin Biotechnol. 2019:**56**:240–249. https://doi.org/10.1016/j.copbio.2019.02.019

Rao X, Dixon RA. Current models for transcriptional regulation of secondary cell wall biosynthesis in grasses. Front Plant Sci. 2018:**9**:399. https://doi.org/10.3389/fpls.2018.00399

Rayle DL, Cleland R. Enhancement of wall loosening and elongation by acid solutions. Plant Physiol. 1970:**46**(2):250–253. https://doi.org/10.1104/pp.46.2.250

Reid JSG. Cell wall storage carbohydrates in seeds-biochemistry of the seed "gums" and "hemicelluloses". Adv Bot Res. 1985:**11**:125–155. https://doi.org/10.1016/S0065-2296(08)60170-6

Reiter WD, Chapple C, Somerville CR. Mutants of *Arabidopsis thaliana* with altered cell wall polysaccharide composition. Plant J. 1997: **12**(2):335–345. https://doi.org/10.1046/j.1365-313X.1997.12020335.x

Reuhs BL, Glenn J, Stephens SB, Kim JS, Christie DB, Glushka JG, Zablackis E, Albersheim P, Darvill AG. O'Neill MA. l-galactose replaces l-fucose in the pectic polysaccharide rhamnogalacturonan II synthesized by the l-fucose-deficient *mur1* Arabidopsis mutant. Planta. 2004:**219**(1): 147–157. https://doi.org/10.1007/s00425-004-1205-x

Richmond TA, Somerville CR. The cellulose synthase superfamily. Plant Physiol. 2000:**124**(2):495–498. https://doi.org/10.1104/pp.124.2.495

Richter J, Ploderer M, Mongelard G, Gutierrez L, Hauser MT. Role of CrRLK1L cell wall sensors HERCULES1 and 2, THESEUS1, and FERONIA in growth adaptation triggered by heavy metals and trace elements. Front Plant Sci. 2017:**8**:1554. https://doi.org/10.3389/fpls.2017.01554

Ringli C, Keller B, Ryser U. Glycine-rich proteins as structural components of plant cell walls. Cell Mol Life Sci. 2001:**58**(10):1430–1441. https://doi.org/10.1007/PL00000786

Roberts AW, Frost AO, Roberts EM, Haigler CH. Roles of microtubules and cellulose microfibril assembly in the localization of secondary-cell-wall deposition in developing tracheary elements. Protoplasma. 2004:**224**(3-4): 217–229. https://doi.org/10.1007/s00709-004-0064-4

Roberts AW, Haigler CH. Rise in chlorotetracycline fluorescence accompanies tracheary element differentiation in suspension cultures of Zinnia. Protoplasma. 1989:**152**(1):37–45. https://doi.org/10.1007/BF01354238

Robinson DG, Quader H. Towards cellulose synthesis in vitro from plant extracts. J Theor Biol. 1981:**92**(4):483–495. https://doi.org/10.1016/0022-5193(81)90261-7

Roelofsen PA, Houwink AL. Architecture and growth of the primary cell wall in some plant hairs and in the Phycomyces sprangiophore. Acta Bot Neerl. 1953:**2**(2):218–225. https://doi.org/10.1111/j.1438-8677.1953.tb00272.x

Roelofsen PA, Kreger DR. The submicroscopic structure of pectin in collenchyma cell-walls. J Exp Bot. 1951:**2**(3):332–343. https://doi.org/10.1093/jxb/2.3.332

Römling U, Galperin MY. Bacterial cellulose biosynthesis: diversity of operons, subunits, products, and functions. Trends Microbiol. 2015:**23**(9):545–557. https://doi.org/10.1016/j.tim.2015.05.005

Rose JKC, Braam J, Fry SC, Nishitani K. The XTH family of enzymes involved in xyloglucan endotransglucosylation and endohydrolysis: current perspectives and a new unifying nomenclature. Plant Cell Physiol. 2002:**43**(12):1421–1435. https://doi.org/10.1093/pcp/pcf171

Ross P, Weinhouse H, Aloni Y, Michaeli D, Weinberger-Ohana P, Mayer R, Braun S, de Vroomt E, van der Mareit J, van Boomt H, et al. Regulation of cellulose synthesis in *Acetobacter xylinum* by cyclic diguanylic acid. Nature. 1987:**327**(6101):279–281. https://doi.org/10.1038/325279a0

Roudier F, Fernandez AG, Fujita M, Himmelspach R, Borner GHH, Schindelman G, Song S, Baskin TI, Dupree P, Wasteneys GO, et al. COBRA, an Arabidopsis extracellular glycosyl-phosphatidyl inositol-anchored protein, specifically controls highly anisotropic expansion through its involvement in cellulose microfibril orientation. Plant Cell. 2005:**17**(6):1749–1763. https://doi.org/10.1105/tpc.105.031732

Rounds CM, Bezanilla M. Growth mechanisms in tip-growing plant cells. Annu Rev Plant Biol. 2013:**64**(1):243–265. https://doi.org/10.1146/annurev-arplant-050312-120150

Rui Y, Dinneny JR. A wall with integrity: surveillance and maintenance of the plant cell wall under stress. New Phytol. 2019:**225**(4): 1428–1439. https://doi.org/10.1111/nph.16166

Rushton PS, Olek AT, Makowski L, Badger J, Steussy CN, Carpita NC, Stauffacher CV. Rice cellulose synthaseA8 plant-conserved region is a coiled-coil at the catalytic core entrance. Plant Physiol. 2017:**173**(1): 482–494. https://doi.org/10.1104/pp.16.00739

Ryser U. Protoxylem: the deposition of a network containing glycine-rich cell wall proteins starts in the cell corners in close association with the pectins of the middle lamella. Planta. 2003:**216**(5): 854–864. https://doi.org/10.1007/s00425-002-0938-7

Ryser U, Schorderet M, Guyot R, Keller B. A new structural element containing glycine-rich proteins and rhamnogalacturonan I in the protoxylem of seed plants. J Cell Sci. 2003:**117**(7):1179–1190. https://doi.org/10.1242/jcs.00966

Sadava D, Walker F, Chrispeels MJ. Hydroxyproline-rich cell wall protein (extension): biosynthesis and accumulation in growing pea epicotyls. Dev Biol. 1973:**30**(1):42–48. https://doi.org/10.1016/0012-1606(73)90046-8

Salnikov VV, Grimson MJ, Delmer DP, Haigler CH. Sucrose synthase localizes to cellulose synthesis sites in tracheary elements. Phytochemistry. 2001:**57**(6):823–833. https://doi.org/10.1016/S0031-9422(01)00045-0

Sampedro J, Cosgrove DJ. The expansin superfamily. Genome Biol. 2005:**6**(12):242. https://doi.org/10.1186/gb-2005-6-12-242

Samuels AL, Giddings TH Jr, Staehelin LA. Cytokinesis in tobacco BY-2 and root tip cells: a new model of cell plate formation in higher plants. J Cell Biol. 1995:**130**(6):1345–1357. https://doi.org/10.1083/jcb.130.6.1345

Sarko A, Muggli R. Packing analysis of carbohydrates and polysaccharides. III. Valonia cellulose and cellulose II. Macromolecules. 1974:**7**(4): 486–494. https://doi.org/10.1021/ma60040a016

Saxena IM, Kudlicka K, Okuda K, Brown RM Jr. Characterization of genes in the cellulose-synthesizing operon (acs operon) of *Acetobacter xylinum*: implications for cellulose crystallization. J Bacteriol. 1994:**176**(18):5735–5752. https://doi.org/10.1128/jb.176.18.5735-5752.1994

Saxena IM, Lin FC, Brown RM. Cloning and sequencing of the cellulose synthase catalytic subunit gene of *Acetobacter xylinum*. Plant Mol Biol. 1990:**15**(5):673–683. https://doi.org/10.1007/BF00016118

Scanlan EM, Mackeen MM, Wormald MR, Davis BG. Synthesis and solution-phase conformation of the RG-I fragment of the plant polysaccharide pectin reveals a modification-modulated assembly mechanism. J Am Chem Soc. 2022:**132**(21):7238–7239. https://doi.org/10.1021/ja9090963

Scavuzzo-Duggan TR, Chaves AM, Singh A, Sethaphong L, Slabaugh E, Yingling YG, Haigler CH, Roberts AW. Cellulose synthase 'class specific regions' are intrinsically disordered and functionally undifferentiated. J Integr Plant Biol. 2018:**60**(6):481–497. https://doi.org/10.1111/jipb.12637

Scheller HV, Ulvskov P. Hemicelluloses. Annu Rev Plant Biol. 2010:**61**(1):263–289. https://doi.org/10.1146/annurev-arplant-042809-112315

Schindelman G, Morikami A, Jung J, Baskin TI, Carpita NC, Derbyshire P, McCann MC, Benfey PN. COBRA encodes a putative GPI-anchored protein, which is polarly localized and necessary for oriented cell expansion in Arabidopsis. Genes Dev. 2001:**15**(9): 1115–1127. https://doi.org/10.1101/gad.879101

Schneider R, Ehrhardt DW, Meyerowitz EM, Sampathkumar A. Tethering of cellulose synthase to microtubules dampens mechano-induced cytoskeletal organization in Arabidopsis pavement cells. Nat Plants. 2022:**8**(9):1064–1073. https://doi.org/10.1038/s41477-022-01218-7

Schneider R, Hanak T, Persson S, Voigt CA. Cellulose and callose synthesis and organization in focus, what's new? Curr Opin Plant Biol. 2016:**34**:9–16. https://doi.org/10.1016/j.pbi.2016.07.007

Schneider R, Tang L, Lampugnani ER, Barkwill S, Lathe R, Zhang Y, McFarlane HE, Pesquet E, Niittyla T, Mansfield SD, et al. Two complementary mechanisms underpin cell wall patterning during xylem vessel development. Plant Cell. 2017:**29**(10):2433–2449. https://doi.org/10.1105/tpc.17.00309

Schoenaers S, Balcerowicz D, Breen G, Hill K, Zdanio M, Mouille G, Holman TJ, Oh J, Wilson MH, Nikonorova N, et al. The auxin-regulated CrRLK1L kinase ERULUS controls cell wall composition during root hair tip growth. Curr Biol. 2018:**28**(5):722–732. https://doi.org/10.1016/j.cub.2018.01.050

Schoenaers S, Lee HK, Gonneau M, Faucher E, Levasseur T, Akary E, Claeijs N, Moussu S, Broyart C, Balcerowicz D, et al. A pectin-binding peptide with a structural and signaling role in the assembly of the plant cell wall. BioRxiv. 2023. https://doi.org/10.1101/2023.08.23.554416, 23 August 2023, preprint: not peer reviewed.

Schrick K, Debolt S, Bulone V. Deciphering the molecular functions of sterols in cellulose biosynthesis. Front Plant Sci. 2012:**3**:84. https://doi.org/10.3389/fpls.2012.00084

Schultink A, Liu L, Zhu L, Pauly M. Structural diversity and function of xyloglucan sidechain substituents. Plants. 2014:**3**(4):526–542. https://doi.org/10.3390/plants3040526

Seifert GJ, Roberts K. The biology of arabinogalactan proteins. Annu Rev Plant Biol. 2006:**58**(1):137–161. https://doi.org/10.1146/annurev.arplant.58.032806.103801

Selvendran RR. Developments in the chemistry and biochemistry of pectic and hemicellulosic polymers. J Cell Sci Suppl. 1985:**2**(Suppl 2):51–88. https://doi.org/10.1242/jcs.1985.Supplement_2.4

Sénéchal F, Wattier C, Rustérucci C, Pelloux J. Homogalacturonan-modifying enzymes: structure, expression, and roles in plants. J Exp Bot. 2014:**65**(18):5125–5160. https://doi.org/10.1093/jxb/eru272

Shafee T, Bacic A, Johnson K, Wilke C. Evolution of sequence-diverse disordered regions in a protein family: order within the chaos. Mol Biol Evol. 2020:**37**(8):2155–2172. https://doi.org/10.1093/molbev/msaa096

Sharma P, Sharma S, Ramakrishna G, Srivastava H, Gaikwad K. A comprehensive review on leguminous galactomannans: structural analysis, functional properties, biosynthesis process and industrial applications. Critical Rev Food Sci Nutr. 2022:**62**(2):443–465. https://doi.org/10.1080/10408398.2020.1819196

Shedletzky E, Shmuel M, Delmer DP, Lamport DTA. Adaptation and growth of tomato cells on the herbicide 2,6-dichlorobenzonitrile leads to production of unique cell walls virtually lacking a cellulose-xyloglucan network. Plant Physiol. 1990:**94**(3):980–987. https://doi.org/10.1104/pp.94.3.980

Shedletzky E, Shmuel M, Trainin T, Kalman S, Delmer D. Cell wall structure in cells adapted to growth on the cellulose-synthesis inhibitor 2, 6-dichlorobenzonitrile: a comparison between two dicotyledonous plants and a graminaceous monocot. Plant Physiol. 1992:**100**(1):120–130. https://doi.org/10.1104/pp.100.1.120

Shen W, Liu J, Li JF. Type-II metacaspases mediate the processing of plant elicitor peptides in Arabidopsis. Mol Plant. 2019:**12**(11):1524–1533. https://doi.org/10.1016/j.molp.2019.08.003

Shi D-C, Wang J, Hu R-B, Zhou G-K, O'Neill MA, Kong Y-Z. Boron-bridged RG-II and calcium are required to maintain the pectin network of the Arabidopsis seed mucilage ultrastructure. Plant Mol Biol. 2017:**94**(3):267–280. https://doi.org/10.1007/s11103-017-0606-8

Shigenaga AM, Berens ML, Tsuda K, Argueso CT. Towards engineering of hormonal crosstalk in plant immunity. Curr Opin Plant Biol. 2017:**38**:164–172. https://doi.org/10.1016/j.pbi.2017.04.021

Showalter AM. Structure and function of plant cell wall proteins. Plant Cell. 1993:**5**(1):9–23. https://doi.org/10.1105/tpc.5.1.9

Showalter AM, Keppler B, Lichtenberg J, Gu D, Welch LR. A bioinformatics approach to the identification, classification, and analysis of hydroxyproline-rich glycoproteins. Plant Physiol. 2010:**153**(2):485–513. https://doi.org/10.1104/pp.110.156554

Silva J, Ferraz R, Dupree P, Showalter AM, Coimbra S. Three decades of advances in arabinogalactan-protein biosynthesis. Front Plant Sci. 2020:**11**:610377. https://doi.org/10.3389/fpls.2020.610377

Singh B, Avci U, Eichler Inwood SE, Grimson MJ, Landgraf J, Mohnen D, Sorensen I, Wilkerson CG, Willats WG, Haigler CH. A specialized outer layer of the primary cell wall joins elongating cotton fibers into tissue-like bundles. Plant Physiol. 2009:**150**(2):684–699. https://doi.org/10.1104/pp.109.135459

Singh A, Kwansa AL, Kim HS, Williams JT, Yang H, Li NK, Kubicki JD, Roberts AW, Haigler CH, Yingling YG. In silico structure prediction of full-length cotton cellulose synthase protein (GhCESA1) and its hierarchical complexes. Cellulose. 2020:**27**(10):5597–5616. https://doi.org/10.1007/s10570-020-03194-7

Sloep AC. Onderzoekingen over pektinestoffen en hare enzymatische ontleding [PhD Diss.] University of Delft; 1928.

Smith HC. The plasma membrane with notes on the history of botany. Circulation. 1962:**26**:987–1012. https://doi.org/10.1161/01.CIR.26.5.987

Smith RA, Schuetz M, Roach M, Mansfield SD, Ellis B, Samuels L. Neighboring parenchyma cells contribute to Arabidopsis xylem lignification, while lignification of interfascicular fibers is cell autonomous. Plant Cell. 2013:**25**(10):3988–3999. https://doi.org/10.1105/tpc.113.117176

Smith MM, Stone BA. β-Glucan synthesis by cell-free extracts from *Lolium multiforme* endosperm. Biochim Biophys Acta. 1973:**313**(1):72–94. https://doi.org/10.1016/0304-4165(73)90189-X

Smith PJ, Wang H-T, York WS, Peña MJ, Urbanowicz BR. Designer biomass for next-generation biorefineries: leveraging recent insights into xylan structure and biosynthesis. Biotechnol Biofuels. 2017:**10**:286. https://doi.org/10.1186/s13068-017-0973-z

Sobry S, Havelange A, Van Cutsem P. Immunocytochemistry of pectins in shoot apical meristems: consequences for intercellular adhesion. Protoplasma. 2005:**225**(1-2):15–22. https://doi.org/10.1007/s00709-005-0089-3

Somerville C, Bauer S, Brinistool G, Facette M, Hamann T, Milne J, Osborne E, Paradez A, Persson S, Raab T, et al. Toward a systems approach to understanding plant cell walls. Science. 2004:**306**(5705):2206–2211. https://doi.org/10.1126/science.1102765

Song B, Zhao S, Shen W, Collings C, Ding S-Y. Direct measurement of plant cellulose microfibril and bundles in native cell walls. Front Plant Sci. 2020:**11**:479. https://doi.org/10.3389/fpls.2020.00479

Sowinski EE, Westman BM, Redmond CR, Kong Y, Olek AT, Olek J, McCann MC, Carpita NC. Lack of xyloglucan in the cell walls of the Arabidopsis *xxt1/xxt2* mutant results in specific increases in homogalacturonan and glucomannan. Plant J. 2022:**110**(1):212–227. https://doi.org/10.1111/tpj.15666

Staehelin LA, Hepler PK. Cytokinesis in higher plants. Cell. 1996:**94**(6):821–824. https://doi.org/10.1016/S0092-8674(00)81060-0

Staehelin LA, Pickett-Heaps JD. The ultrastructure of *Scenedesmus* (*Chlorophyceae*). I. Species with the "reticulate" or "warty" type ornamental layer. J Phycol. 1975:**11**:163–185. https://doi.org/10.1111/j.1529-8817.1975.tb02765.x

Staudinger H. Der Aufbau der Hochmolecularen Organischen Verbindungen. die Naturwissenschaften. In Naturwissenschaften 1932: Berlin: Springer, pp. 84–89.

Stegmann M, Monaghan J, Smakowska-Luzan E, Rovenich H, Lehner A, Holton N, Belkhadir Y, Zipfel C. The receptor kinase FER is a RALF-regulated scaffold controlling plant immune signaling. Science. 2017:**355**(6322):287–289. https://doi.org/10.1126/science.aal2541

Sterjiades R, Dean JFD, Eriksson K-EL. Laccase from sycamore maple (*Acer pseudoplatanus*) polymerizes monolignols. Plant Physiol. 1992:**99**(3):1162–1168. https://doi.org/10.1104/pp.99.3.1162

Sterling JD, Atmodjo MA, Inwood SE, Kolli VSK, Quigley HF, Hahn MG, Mohnen D. Functional identification of an Arabidopsis pectin biosynthetic homogalacturonan galacturonosyltransferase. Proc Natl Acad Sci U S A. 2006:**103**(13):5236–5241. https://doi.org/10.1073/pnas.0600120103

Suslov D, Verbelen JP, Vissenberg K. Onion epidermis as a new model to study the control of growth anisotropy in higher plants. J Exp Bot. 2009:**60**(14):4175–4187. https://doi.org/10.1093/jxb/erp251

Swords KMM, Staehelin LA. Analysis of extensin structure in plant cell walls. In: **Linskens HF, Jackson JF**, editors. Plant fibers, modern methods of plant analysis. Vol. 10. Germany: Springer-Verlag; 1989. p. 219–233.

Tai H-C, Chang C-H, Cai W, Lin J-H, Huang S-J, Lin Q-Y, Yuan EC-Y, Li S-L, Lin Y-CJ, Chan JCC, et al. Wood cellulose microfibrils have a 24-chain core-shell nanostructure in seed plants. Nat Plants. 2023:**9**(7):1154–1168. https://doi.org/10.1038/s41477-023-01430-z

Takenaka Y, Kato K, Ogawa-Ohnishi M, Tsuruhama K, Kajiura H, Yagyu K, Takeda A, Takeda Y, Kunieda T, Hara-Nishimura I, et al. Pectin RG-I rhamnosyltransferases represent a novel plant-specific glycosyltransferase family. Nat Plants. 2018:**4**(9):669–676. https://doi.org/10.1038/s41477-018-0217-7

Tan L, Eberhard S, Pattathil S, Warder C, Glushka J, Yuan C, Hao Z, Zhu X, Avci U, Miller JS, et al. An Arabidopsis cell wall proteoglycan consists of pectin and arabinoxylan covalently linked to an arabinogalactan protein. Plant Cell. 2013:**25**(1):270–287. https://doi.org/10.1105/tpc.112.107334

Tan L, Xu J, Held M, Lamport DTA, Kieliszewski M. Structures of repetitive serinehydroxyproline glycomodule expressed by Arabidopsis cell suspension cultures. Plants (Basel). 2023a:**12**:1036.

Tan L, Zhang L, Black I, Glushka J, Urbanowicz B, Heiss C, Azadi P. Most of the rhamnogalacturonan-I from cultured Arabidopsis cell walls is covalently linked to arabinogalactan-protein. Carbohydr Polym. 2023b:**301**(Part B):120340. https://doi.org/10.1016/j.carbpol.2022.120340

Tang W, Lin W, Zhou X, Guo J, Dang X, Li B, Li D, Yang Z. Mechano-transduction via the pectin-FERONIA complex activates ROP6 GTPase signaling in Arabidopsis pavement cell morphogenesis. Curr Biol. 2022:**32**(3):508–517.e3. https://doi.org/10.1016/j.cub.2021.11.031

Temple H, Phyo P, Yang W, Lyczakowski JJ, Echevarria-Poza A, Yakunin I, Parra-Rojas JP, Terrett OM, Saez-Aguayo S, Dupree R, et al. Golgi-localized putative S-adenosyl methionine transporters required for cell wall polysaccharide methylation. Nat Plants. 2022:**8**(6):656–669. https://doi.org/10.1038/s41477-022-01156-4

Tenhaken R. Cell wall remodeling under abiotic stress. Front Plant Sci. 2014:**5**:771. https://doi.org/10.3389/fpls.2014.00771

Terashima N, Fukushima K. Heterogeneity in formation of lignin. XI. An autoradiographic study of the heterogeneous formation and structure of pine lignin. Wood Sci Technol. 1988:**22**(3):259–270. https://doi.org/10.1007/BF00386021

Tierney ML, Varner JE. The extensins. Plant Physiol. 1987:**84**(1):1–2. https://doi.org/10.1104/pp.84.1.1

Tryfona T, Bourdon M, Marques RD, Busse-Wicher M, Vilaplana F, Stott K, Dupree P. Grass xylan structural variation functional specialization and distinctive interaction with cellulose and lignin. Plant J. 2023:**113**(5):1004–1020. https://doi.org/10.1111tpj.16096

Tsuyama T, Kawai R, Shitan N, Matoh T, Sugiyama J, Yoshinaga A, Takabe K, Fujita M, Yazaki K. Proton-dependent coniferin transport, a common major transport event in differentiating xylem tissue of woody plants. Plant Physiol. 2013:**162**(2):918–926. https://doi.org/10.1104/pp.113.214957

Tsuyama T, Takabe K. Coniferin β-glucosidase is ionically bound to cell wall in differentiating xylem of poplar. J Wood Sci. 2015:**61**(4): 438–444. https://doi.org/10.1007/s10086-015-1486-7

Turlapati PV, Kim K-W, Davin LB, Lewis NG. The laccase multigene family in *Arabidopsis thaliana*: towards addressing the mystery of their gene function(s). Planta. 2011:**233**(3):439–470. https://doi.org/10.1007/s00425-010-1298-3

Turner S, Gallois P, Brown D. Tracheary element differentiation. Annu Rev Plant Biol. 2007:**58**(1):407–433. https://doi.org/10.1146/annurev.arplant.57.032905.105236

Turner S, Kumar M. Cellulose synthase complex organization and cellulose microfibril structure. Philos Trans A Math Phys Eng Sci. 2018:**376**(2112):20170048. https://doi.org/10.1098/rsta.2017.0048

Tuskan GA, DiFazio SP, Teichmann T. Poplar genomics is getting popular: the impact of the poplar genome project on tree research. Plant Biol (Stuttg). 2004:**6**(1):2–4. https://doi.org/10.1055/s-2003-44715

Tuttle JR, Nah G, Duke MV, Alexander DC, Guan X, Song Q, Chen ZJ, Scheffler BE, Haigler CH. Metabolomic and transcriptomic insights into how cotton fiber transitions to secondary wall synthesis, represses lignification, and prolongs elongation. BMC Genomics. 2015:**16**(1):477. https://doi.org/10.1186/s12864-015-1708-9

Ullah M, Liu P, Xie S, Sun S. Recent advancements and challenges in lignin valorization: green routes towards sustainable bioproducts. Molecules. 2022:**27**(18):6055. https://doi.org/10.3390/molecules27186055

Urbanowicz BR, Peña MJ, Moniz HA, Moremen KW, York WS. Two Arabidopsis proteins synthesize acetylated xylan in vitro. Plant J. 2014:**80**(2):197–206. https://doi.org/10.1111/tpj.12643

Ustun S, Hafren A, Hofius D. Autophagy as a mediator of life and death in plants. Curr Opin Plant Biol. 2017:**40**:122–130. https://doi.org/10.1016/j.pbi.2017.08.011

Vaahtera L, Schulz J, Hamann T. Cell wall integrity maintenance during plant development and interaction with the environment. Nat Plants. 2019:**5**(9):924–932. https://doi.org/10.1038/s41477-019-0502-0

Vain T, Crowell EF, Timpano H, Biot E, Desprez T, Mansoori N, Trindade LM, Pagant S, Robert S, Hofte H, et al. The cellulase KORRIGAN is part of the cellulose synthase complex. Plant Physiol. 2014:**165**(4):1521–1532. https://doi.org/10.1104/pp.114.241216

Vandavasi VG, Putnam DK, Zhang Q, Petridis L, Heller WT, Nixon BT, Haigler CH, Kalluri U, Coates L, Langan P, et al. A structural study of CESA1 catalytic domain of Arabidopsis cellulose synthesis complex: evidence for CESA trimers. Plant Physiol. 2016:**170**(1): 123–135. https://doi.org/10.1104/pp.15.01356

Vanholme R, Cesarino I, Rataj K, Xiao Y, Sundin L, Goeminne G, Kim H, Cross J, Morreel K, Araujo P, et al. Caffeoyl shikimate esterase (CSE) is an enzyme in the lignin biosynthetic pathway in Arabidopsis. Science. 2013:**341**(6150):1103–1106. https://doi.org/10.1126/science.1241602

Vanholme R, De Meester B, Ralph J, Boerjan W. Lignin biosynthesis and its integration into metabolism. Curr Opin Biotechnol. 2019:**56**:230–239. https://doi.org/10.1016/j.copbio.2019.02.018

van Parijs FRD, Morreel K, Ralph J, Boerjan W, Merks RMH. Modeling lignin polymerization. I. Simulation model of dehydrogenation polymers. Plant Physiol. 2010:**53**(3):1332–1344. https://doi.org/10.1104/pp.110.154468

Vanzin GF, Madson M, Carpita NC, Raikhel NV, Keegstra K, Reiter W-D. The *mur2* mutant of *Arabidopsis thaliana* lacks fucosylated xyloglucan because of a lesion in the fucosyltransferase AtFUT1. Proc Natl Acad Sci U S A. 2002:**99**(5):3340–3345. https://doi.org/10.1073/pnas.052450699

Varki A, Cummings RD, Esko JD, Stanley P, Hart GW, Aebi M, Mohnen D, Kinoshita T, Packer NH, Prestegard JH, et al. Essentials of glycobiology. Cold Spring Harbor (NY): Cold Spring Harbor Laboratory Press; 2022. p. 1–859.

Varner JE, Lin L-S. Plant cell wall architecture. Cell. 1989:**56**(2): 231–239. https://doi.org/10.1016/0092-8674(89)90896-9

Verbančič J, Lunn JE, Stitt M, Persson S. Carbon supply and the regulation of cell wall synthesis. Mol Plant. 2018:**11**(1):75–94. https://doi.org/10.1016/j.molp.2017.10.004

Verma P, Kwansa AL, Ho R, Yingling YG, Zimmer J. Insights into substrate coordination and glycosyl transfer of poplar cellulose synthase-8. bioRxiv. 2023. https://doi.org/10.1101/2023.02.07.527505, 7 February 2023, preprint: not peer reviewed.

Vermaas JV, Dixon RA, Chen F, Mansfield SD, Boerjan W, Ralph J, Crowley MF, Beckham GT. Passive membrane transport of lignin-related compounds. Proc Natl Acad Sci U S A. 2019:**116**(46):23117–23123. https://doi.org/10.1073/pnas.1904643116

Vlot AC, Dempsey DMA, Klessig DF. Salicylic acid, a multifaceted hormone to combat disease. Annu Rev Phytopathol. 2009:**47**(1): 177–206. https://doi.org/10.1146/annurev.phyto.050908.135202

Voiniciuc C. Modern mannan: a hemicellulose's journey. New Phytol. 2022:**234**(4):1175–1184. https://doi.org/10.1111/nph.18091

Voxeur A, Habrylo O, Guénin S, Miart F, Soulié MC, Rihouey C, Pau-Roblot C, Domon JM, Gutierrez L, Pelloux J, et al. Oligogalacturonide production upon *Arabidopsis thaliana–Botrytis cinerea* interaction. Proc Natl Acad Sci U S A. 2019:**116**(39):19743–19752. https://doi.org/10.1073/pnas.1900317116

Walkinshaw MD, Arnott S. Conformations and interactions of pectins. I. X-ray diffraction analyses of sodium pectate in neutral and acidified forms. J Mol Biol. 1981a:**53**(4):1055–1073. https://doi.org/10.1016/0022-2836(81)90467-8

Walkinshaw MD, Arnott S. Conformations and interactions of pectins. II. Models for junction zones in pectinic acid and calcium pectate gels. J Mol Biol. 1981b:**153**(4):1075–1085. https://doi.org/10.1016/0022-2836(81)90468-X

Wallace IS, Somerville CR. Plant cell wall patterning and cell shape. Wiley Online; 2014. p. 65–95. https://doi.org/10.1002/9781118647363.ch3.

Wang Y, Alonso AP, Wilkerson CG, Keegstra K. Deep EST profiling of developing fenugreek endosperm to investigate galactomannan biosynthesis and its regulation. Plant Mol Biol. 2012:**79**(3):243–258. https://doi.org/10.1007/s11103-012-9909-y

Wang H, Avci U, Nakashima J, Hahn MG, Chen F, Dixon RA. Mutation of WRKY transcription factors initiates pith secondary wall formation and increases stem biomass in dicotyledonous plants. Proc Natl Acad Sci U S A. 2010:**107**(51):22338–22343. https://doi.org/10.1073/pnas.1016436107

Wang Y, Cao Y, Liang X, Zhuang J, Wang X, Qin F, Jiang C. A dirigent family protein confers variation of Casparian strip thickness and salt tolerance in maize. Nat Commun. 2022b:**13**(1):2222. https://doi.org/10.1038/s41467-022-29809-0

Wang T, Hong M. Solid-state NMR investigations of cellulose structure and interactions with matrix polysaccharides in plant primary cell walls. J Exp Bot. 2016:**67**(2):503–514. https://doi.org/10.1093/jxb/erv416

Wang JP, Matthews ML, Williams CM, Shi R, Yang C, Tunlaya-Anukit S, Chen H-C, Li Q, Liu J, Lin C-Y, et al. Improving wood properties for wood utilization through multi-omics integration in lignin biosynthesis. Nat Commun. 2018:**9**(1):1579. https://doi.org/10.1038/s41467-018-03863-z

Wang T, McFarlane H E, Persson S. The impact of abiotic factors on cellulose synthesis. J Exp Bot. 2016a:**67**(2):543–552. https://doi.org/10.1093/jxb/erv488

Wang T, Park YB, Cosgrove DJ, Hong M. Cellulose-pectin spatial contacts are inherent to never-dried Arabidopsis primary cell walls: evidence from solid state nuclear magnetic resonance. Plant Physiol. 2015:**168**(3):871–884. https://doi.org/10.1104/pp.15.00665

Wang W, Viljamaa S, Hodek O, Moritz T, Niittyla T. Sucrose synthase activity is not required for cellulose biosynthesis in Arabidopsis. Plant J. 2022a:**110**(5):1493–1497. https://doi.org/10.1111/tpj.15752

Wang Q, Xiao S, Shi S, Cai L. Mechanical strength, thermal stability, and hydrophobicity of fiber materials after removal of residual lignin. BioRes. 2017:**13**(1):71–85. https://doi.org/10.15376/biores.13.1.71-85

Wang T, Yang H, Kubicki JD, Hong M. Cellulose structural polymorphism in plant primary cell walls investigated by high-field 2D solid-state NMR spectroscopy and density functional theory calculations. BioMacromolecules. 2016b:**17**(6):2010–2022. https://doi.org/10.1021/acs.biomac.6b00441

Watanabe Y, Meents MJ, McDonnell M, Barkwill S, Sampathkumar A, Cartwright HN, Demura T, Ehrhardt DW, Samuels L, Mansfield SD. Visualization of cellulose synthases in Arabidopsis secondary cell walls. Science. 2015:**350**(6257):198–203. https://doi.org/10.1126/science.aac7446

Watanabe Y, Schneider R, Barkwill S, Gonzales-Vigil E, Hill JL Jr, Samuels AL, Persson S, Mansfield SD. Cellulose synthase complexes display distinct dynamic behaviors during xylem transdifferentiation. Proc Natl Acad Sci U S A. 2018:**115**(27):E6366–E6374. https://doi.org/10.1073/pnas.1802113115

Wells B, McCann MC, Shedletzky E, Delmer D, Roberts K. Structural features of cell walls from tomato cells adapted to grow on the herbicide 2,6-dichlorobenzonitrile. J Microscopy. 1994:**173**(2):155–164. https://doi.org/10.1111/j.1365-2818.1994.tb03438.x

Wen X, Zhai Y, Zhang L, Chen Y, Zhu Z, Chen G, Wang K, Zhu Y. Molecular studies of cellulose synthase supercomplex from cotton fiber reveal its unique biochemical properties. Sci China Life Sci. 2022:**65**(9):1776–1793. https://doi.org/10.1007/s11427-022-2083-9

Wilkerson CG, Mansfield SD, Lu F, Withers S, Park J-Y, Karlen SD, Gonzales-Vigil E, Padmakshan D, Unda F, Rencoret J, et al. Monolignol ferulate transferase introduces chemically labile linkages into the lignin backbone. Science. 2014:**344**(6179):90–93. https://doi.org/10.1126/science.1250161

Willats WGT, Knox JP, Mikkelsen JD. Pectin: new insights into an old polymer are starting to gel. Trends Food Sci Technol. 2006:**17**(3):97–104. https://doi.org/10.1016/j.tifs.2005.10.008

Willats WGT, McCartney L, Steele-King CG, Marcus SE, Mort A, Huisman M, van Alebeek GJ, Schols HA, Voragen AGJ, Le Goff A, et al. A xylogalacturonan epitope is specifically associated with plant cell detachment. Planta. 2004:**218**(4):673–681. https://doi.org/10.1007/s00425-003-1147-8

Wilson TH, Kumar M, Turner SR. The molecular basis of plant cellulose synthase complex organisation and assembly. Biochem Soc Trans. 2021:**49**(1):379–391. https://doi.org/10.1042/BST20200697

Winter H and Huber SC. Regulation of sucrose metabolism in higher plants: localization and regulation of activity of key enzymes. Crit Rev Plant Sci. 2010:**19**:31–67.

Wisniak J. Henri Braconnot. Revista CENIC Ciencias Químicas. 2007:**38**(2):345–355.

Wolf S. Cell wall signaling in plant development and defense. Annu Rev Plant Biol. 2022:**73**(1):323–353. https://doi.org/10.1146/annurev-arplant-102820-095312

Wolf S, Hofte H. Growth control: a saga of cell walls, ROS, and peptide receptors. Plant Cell. 2014:**26**(5):1848–1856. https://doi.org/10.1105/tpc.114.125518

Wolf S, Mouille G, Pelloux J. Homogalacturonan methyl-esterification and plant development. Mol Plant. 2009:**2**(5):851–860. https://doi.org/10.1093/mp/ssp066

Wolf S, van der Does D, Ladwig F, Sticht C, Kolbeck A, Schurholz AK, Augustin S, Keinath N, Rausch T, Greiner S, et al. A receptor-like protein mediates the response to pectin modification by activating brassinosteroid signaling. Proc Natl Acad Sci U S A. 2014:**111**(42):15261–15266. https://doi.org/10.1073/pnas.1322979111

Woodcock C, Sarko A. Packing analysis of carbohydrates and polysaccharides. 11. Molecular and crystal structure of native ramie cellulose. Macromolecules. 1980:**13**(5):1183–1187. https://doi.org/10.1021/ma60077a030

Wu S-Z, Chaves AM, Li R, Roberts AW, Bezanilla M. Cellulose synthase-like D movement in the plasma membrane requires enzymatic activity. J Cell Biol. 2023:**222**(6):e202212117. https://doi.org/10.1083/jcb.202212117

Xiao Y, Stegmann M, Han Z, DeFalco TA, Parys K, Xu L, Belkhadir Y, Zipfel C, Chai J. Mechanisms of RALF peptide perception by a heterotypic receptor complex. Nature. 2019:**572**(7768):270–274. https://doi.org/10.1038/s41586-019-1409-7

Xie Y, Sun P, Li Z, Zhang F, You C, Zhang Z. FERONIA receptor kinase integrates with hormone signaling to regulate plant growth, development, and responses to environmental stimuli. Int J Mol Sci. 2022:**23**(7):3730. https://doi.org/10.3390/ijms23073730

Xu Z, Wang M, Shi D, Zhou G, Niu T, Hahn MG, O'Neill MA, Kong Y. DGE-seq analysis of MUR3-related *Arabidopsis* mutants provide insight into how dysfunctional xyloglucan affects cell elongation. Plant Sci. 2017:**258**:156–159. https://doi.org/10.1016/j.plantsci.2017.01.005

Yang J, Bak G, Burgin T, Barnes WJ, Mayes HB, Pen MJ, Urbanowicz BR, Nielsen E. Biochemical and genetic analysis identify CSLD3 as a β-1,4-glucan synthase that functions during plant cell wall synthesis.

Plant Cell. 2020b:**32**(5):1749–1767. https://doi.org/10.1105/tpc.19.00637

Yang H, Benatti MR, Karve RA, Fox A, Meilan R, Carpita NC, McCann MC. Rhamnogalacturonan-I is a determinant of cell-cell adhesion in poplar wood. Plant Biotechnol J. 2020a:**18**(4):1027–1040. https://doi.org/10.1111/pbi.13271

Yang C, Liu R, Pang J, Ren B, Zhou H, Wang G, Wang E, Liu J. Poaceae-specific cell wall-derived oligosaccharides activate plant immunity via OsCERK1 during *Magnaporthe oryzae* infection in rice. Nat Commun. 2021:**12**(1):2178. https://doi.org/10.1038/s41467-021-22456-x

Yang Y, Yu Y, Liang Y, Anderson CT, Cao J. A profusion of molecular scissors for pectins: classification, expression, and functions of plant polygalacturonases. Front Plant Sci. 2018:**9**:1208. https://doi.org/10.3389/fpls.2018.01208

Ye D, Kiemle SN, Rongpipi S, Wang X, Wang C, Cosgrove DJ, Gomez EW, Gomez ED. Resonant soft X-ray scattering reveals cellulose microfibril spacing in plant primary cell walls. Sci Rep. 2018:**8**(1):12449. https://doi.org/10.1038/s41598-018-31024-1

Ye Z-H, Zhong R. Outstanding questions on xylan bioxynthesis. Plant Sci. 2022:**325**:111476. https://doi.org/10.1016/j.plantsci.2022.111476

Yin Y, Johns MA, Cao H, Rupani M. A survey of plant and algal genomes and transcriptomes reveals new insights into the evolution and function of the cellulose synthase superfamily. BMC Genomics. 2014:**15**(1):260. https://doi.org/10.1186/1471-2164-15-260

Yoneda A, Ohtani M, Katagiri D, Hosokawa Y, Demura T. Hechtian strands transmit cell wall integrity signals in plant cells. Plants. 2020:**9**(5):604. https://doi.org/10.3390/plants9050604

Yu L, Yoshimi Y, Cresswell R, Wightman R, Lyczakowski JJ, Wilson LFL, Ishida K, Stott K, Yu X, Charalambous S, et al. Eudicot primary cell wall glucomannan is related in synthesis, structure, and function to xyloglucan. Plant Cell. 2022:**34**(11):4600–4622. https://doi.org/10.1093/plcell/koac238

Zabotina OA, Avci U, Cavalier D, Pattathil S, Chou Y-H, Eberhard S, Danhof L, Keegstra K, Hahn MG. Mutations in multiple *XXT* genes of Arabidopsis reveal the complexity of xyloglucan biosynthesis. Plant Physiol. 2012:**159**(4):1367–1384. https://doi.org/10.1104/pp.112.198119

Zabotina OA, Zhang N, Weerts R. Polysaccharide biosynthesis: glycosyltransferases and their complexes. Front Plant Sci. 2021:**12**:625307. https://doi.org/10.3389/fpls.2021.625307

Zakzeski J, Bruijnincx PC, Jongerius AL, Weckhuysen BM. The catalytic valorization of lignin for the production of renewable chemicals. Chem Rev. 2010:**110**(6):3552–3599. https://doi.org/10.1021/cr900354u

Zandleven J, Sørensen SO, Harholt J, Beldman G, Schols HA, Scheller HV, Voragen AGJ. XGA exists in cell walls from various tissues of *Arabidopsis thaliana*. Phytochemistry. 2007:**68**(8):1219–1226. https://doi.org/10.1016/j.phytochem.2007.01.016

Zarattini M, Corso M, Kadowaki MA, Monclaro A, Magri S, Milanese I, Jolivet S, de Godoy MO, Hermans C, Fagard M, et al. LPMO-oxidized cellulose oligosaccharides evoke immunity in Arabidopsis conferring resistance towards necrotrophic fungus *B. cinerea*. Commun Biol. 2021:**4**(1):727. https://doi.org/10.1038/s42003-021-02226-7

Zdunek A, Pieczywek PM, Cybulska J. The primary, secondary, and structures of higher levels of pectin polysaccharides. Compr Rev Food Sci Food Safety. 2021:**20**(1):1101–1117. https://doi.org/10.1111/1541-4337.12689

Zeng Y, Saar BG, Friedrich MG, Chen F, Liu Y-S, Dixon RA, Himmel ME, Xie XS, Ding S-Y. Imaging lignin-down-regulated alfalfa using Coherent Anti-Stoke Raman Scattering Microscopy. BioEnerg Res. 2010:**3**(3):272–277. https://doi.org/10.1007/s12155-010-9079-1

Zhang X, Bian A, Li T, Ren L, Li L, Su Y, Zhang Q. ROS and calcium oscillations are required for polarized root hair growth. Plant Signal Behav. 2022b:**17**(1):2106410. https://doi.org/10.1080/15592324.2022.2106410

Zhang H, Guo Z, Zhuang Y, Suo Y, Du J, Gao Z, Pan J, Li L, Wang T, Xiao L, et al. MicroRNA775 regulates intrinsic leaf size and reduces cell wall pectin levels by targeting a galactosyltransferase gene in Arabidopsis. Plant Cell. 2021a:**33**(3):581–602. https://doi.org/10.1093/plcell/koaa049

Zhang J, Liu Y, Li C, Yin B, Liu X, Guo X, Zhang C, Liu D, Hwang I, Li H, et al. PtomtAPX is an autonomous lignification peroxidase during the earliest stage of secondary wall formation in *Populus tomentosa* carr. Nat Plants. 2022a:**8**(7):828–839. https://doi.org/10.1038/s41477-022-01181-3

Zhang X, Xue Y, Guan Z, Zhou C, Nie Y, Men S, Wang Q, Shen C, Zhang D, Jin S, et al. Structural insights into homotrimeric assembly of cellulose synthase CesA7 from *Gossypium hirsutum*. Plant Biotechnol J. 2021b:**19**(8):1579–1587. https://doi.org/10.1111/pbi.13571

Zhang Y, Yu J, Wang X, Durachko DM, Zhang S, Cosgrove DJ. Molecular insights into the complex mechanics of plant epidermal cell walls. Science. 2021c:**372**(6543):706–711. https://doi.org/10.1126/science.abf2824

Zhao Q, Nakashima J, Chen F, Yin Y, Fu C, Yun J, Shao H, Wang X, Wang Z-Y, Dixon RA. LACCASE is necessary and non-redundant with PEROXIDASE for lignin polymerization during vascular development in *Arabidopsis thaliana*. Plant Cell. 2013:**25**(10):3976–3987. https://doi.org/10.1105/tpc.113.117770

Zhu S, Fu Q, Xu F, Zheng H, Yu, F. New paradigms in cell adaptation: decades of discoveries on the CrRLK1L receptor kinase signalling network. New Phytol. 2021:**232**:1168–1183. doi: 10.1111/nph.17683

Zhuo C, Wang X, Docampo-Palacios M, Xiao X, Sanders BC, Engle NL, Tschaplinski TJ, Hendry J, Maranas C, Chen F, et al. Developmental changes in lignin composition are driven by both monolignol supply and laccase specificity. Sci Adv. 2022:**8**(10):eabm8145. https://doi.org/10.1126/sciadv.abm8145

Zugenmaier P. Order in cellulosics: historical review of crystal structure research on cellulose. Carbohydr Polym. 2021:**254**:117417. https://doi.org/10.1016/j.carbpol.2020.117417

Zykwinska AW, Ralet MC, Garnier CD, Thibault JF. Evidence for in vitro binding of pectin side chains to cellulose. Plant Physiol. 2005:**139**(1):397–407. https://doi.org/10.1104/pp.105.065912

Zykwinska A, Thibault J-F, Ralet M-C. Organization of pectic arabinan and galactan side chains in association with cellulose microfibrils in primary cell walls and related models envisaged. J Exp Bot. 2007:**58**(7):1795–1802. https://doi.org/10.1093/jxb/erm037

https://doi.org/10.1093/plcell/koae004 THE PLANT CELL 2024: 36: 1312–1333

A century journey of organelles research in the plant endomembrane system

Xiaohong Zhuang [1,*] Ruixi Li [2,*] and Liwen Jiang [1,3,4,*]

1 School of Life Sciences, Centre for Cell & Developmental Biology and State Key Laboratory of Agrobiotechnology, The Chinese University of Hong Kong, Shatin, New Territories, Hong Kong, China
2 Key Laboratory of Molecular Design for Plant Cell Factory of Guangdong Higher Education Institutes, Institute of Plant and Food Science, School of Life Sciences, Southern University of Science and Technology, Shenzhen 518055, China
3 Institute of Plant Molecular Biology and Agricultural Biotechnology, The Chinese University of Hong Kong, Shatin, Hong Kong, China
4 CUHK Shenzhen Research Institute, Shenzhen 518057, China

*Author for correspondence: xhzhuang@cuhk.edu.hk (X.Z.), lirx@sustech.edu.cn (R.L.), and ljiang@cuhk.edu.hk (L.J.)
The authors responsible for distribution of materials integral to the findings presented in this article in accordance with the policy described in the Instructions for Authors (www.plantcell.org) are: Xiaohong Zhuang (xhzhuang@cuhk.edu.hk), Ruixi Li (lirx@sustech.edu.cn), and Liwen Jiang (ljiang@cuhk.edu.hk).

Abstract

We are entering an exciting century in the study of the plant organelles in the endomembrane system. Over the past century, especially within the past 50 years, tremendous advancements have been made in the complex plant cell to generate a much clearer and informative picture of plant organelles, including the molecular/morphological features, dynamic/spatial behavior, and physiological functions. Importantly, all these discoveries and achievements in the identification and characterization of organelles in the endomembrane system would not have been possible without: (1) the innovations and timely applications of various state-of-art cell biology tools and technologies for organelle biology research; (2) the continuous efforts in developing and characterizing new organelle markers by the plant biology community; and (3) the landmark studies on the identification and characterization of the elusive organelles. While molecular aspects and results for individual organelles have been extensively reviewed, the development of the techniques for organelle research in plant cell biology is less appreciated. As one of the ASPB Centennial Reviews on "organelle biology," here we aim to take a journey across a century of organelle biology research in plants by highlighting the important tools (or landmark technologies) and key scientists that contributed to visualize organelles. We then highlight the landmark studies leading to the identification and characterization of individual organelles in the plant endomembrane systems.

Introduction

Organelles are intracellular membrane-bound compartments that perform specific functions in the cell, and they need to be defined both functionally and morphologically. Plant cells contain an endomembrane system for the secretory pathway with several functionally distinct organelles including the endoplasmic reticulum (ER), the Golgi apparatus, trans-Golgi network (TGN), prevacuolar compartment (PVC), and vacuole. Similarly, the plant endocytic pathway from the plasma membrane (PM) to vacuole consists of endosomal compartments including early endosome (EE) and late endosome (LE) (Fig. 1).

Over the past decades, cumulative results of organelle research on protein trafficking and organelle biogenesis have now generated a better picture about the transport pathways, organelle identity, and function of the endomembrane system as well as crosstalk with other pathways and/or other organelles that are important for the regulation of plant growth and development (Aniento et al. 2022). The current model for plant cells identifies the TGN as an EE merging the secretory and endocytic pathways. The PVC, also known as

Received August 21, 2023. Accepted January 09, 2024. Advance access publication January 16, 2024

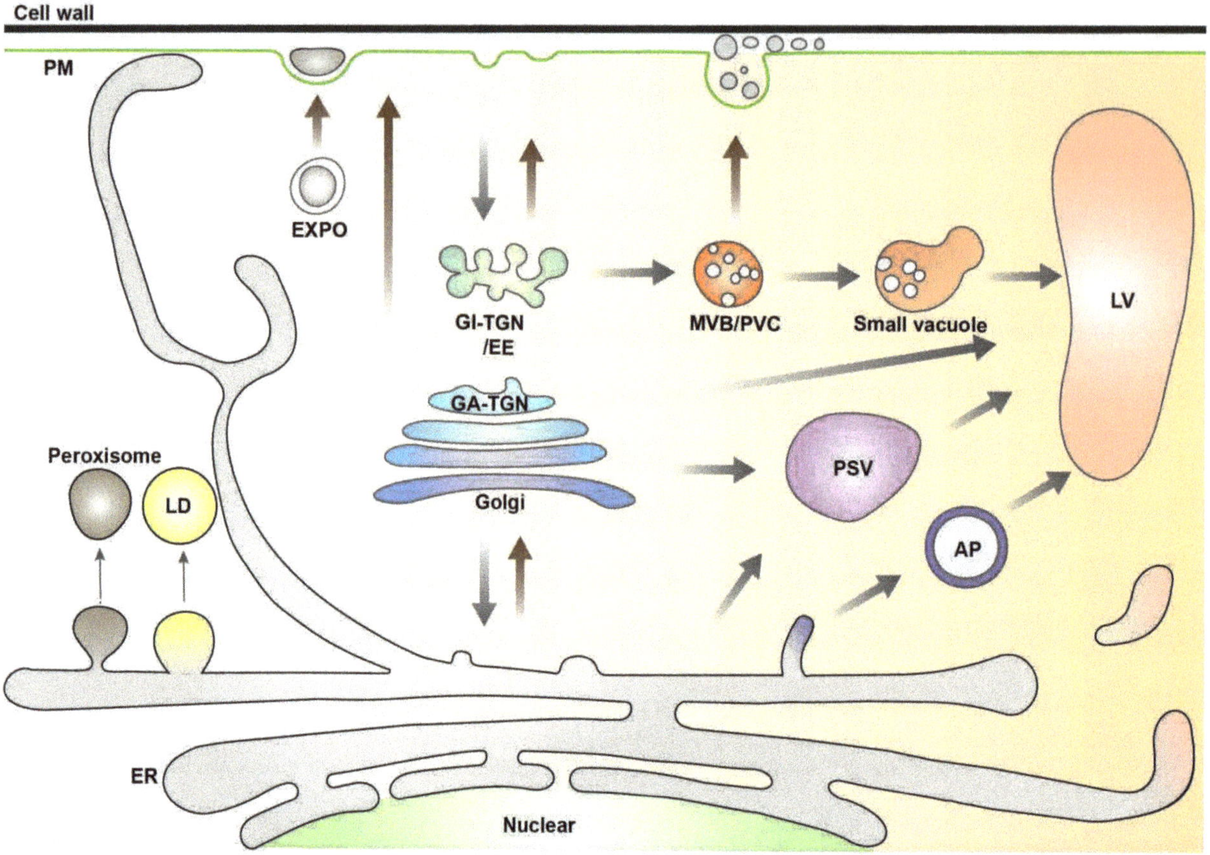

Figure 1. Overview of organelles in the plant endomembrane system. The plant endomembrane system is composed of membrane-enclosed organelles with distinct functions and morphologies, including the ER, Golgi, GA/GI-TGN, MVB/PVC, PSV, and LV. These structures are formed and maintained through different vesicle-mediated transport pathways or membrane interactions, which allow the exchange of materials (e.g. lipids, proteins). In addition, autophagosome, peroxisome and LD also originate from the endoplasmic reticulum, with dynamic interactions with other organelles. AP, autophagosome; ER, endoplasmic reticulum; EVs, extracellular vesicles; DV, dense vesicle; EXPO, exocyst positive organelle; GI-TGN/EE, Golgi-independent trans-Golgi network/early endosome; GA-TGN/EE, Golgi-associated trans-Golgi network/early endosome; MVB/PVC, multivesicular body/prevacuolar compartment; LD, lipid droplet; LV, lytic vacuole; PM, plasma membrane; PSV, protein storage vacuole; SV, small vacuole.

the multivesicular body (MVB), serves as a LE leading to vacuoles (Fig. 1). Emerging evidence suggests that the ER has dynamic interactions and functions with other organelles including the PM for ER-PM contacts and autophagosomes for their formation and progression, as well as for the biogenesis of peroxisome and lipid droplet (LD) (Fig. 1).

All these discoveries and achievements in the identification and characterization of organelles in the endomembrane system would not be possible without (1) the development, evolution, and timely applications of various cell biology tools and technologies for organelle biology research; (2) continuous ef-forts in developing and characterizing new organelle markers; and (3) landmark studies on the identification and character-ization of these elusive organelles.

Thus, as one of the ASPB Centennial Reviews on organelle biology, we first take a journey across a century of organelle biology research in plants by highlighting the important tools and landmark technologies developed over the past decades for visualizing organelles in the plant endomembrane system (Fig. 2). We then highlight the landmark studies leading to the identification and characterization of individual organelles in the plant endomembrane systems. To this end, we point to future studies in answer outstanding questions in plant organelle biology.

Landmark technology in plant organelle biology research

Since the first observation of the cell in a thin slide of cork by Robert Hooke in 1665, our understanding of the organelles would not have been possible if not for technological developments. In 1902, Gottlieb Haberlandt proposed the concept of an in vitro system for plant tissue/cell culture from single somatic cells (Haberlandt 1902). Almost 50 years later, suspension culture cell lines have been successfully developed, which provide a fast-growing and highly synchronous system for live cell imaging, ultrastructural and biochemical analysis in individual cells, including the tobacco Bright Yellow 2 (BY-2) cell line (Nagata et al. 1992). Over the past decades,

Visualizing the organelles in plant cells

	Before 1980	1981-1990	1991-2000	2001-2010	2011-2020	after 2020
Light microscopy	• **Neutral red for vacuole staining** (Guilliermond, 1929.)	• **DiOC6 for ER staining** (Kenebel et al., 1990)	• **BFA for Golgi analysis** (Satiat-Jeunemaitre and Hawes, 1992) • **GFP reporter for visualizing ER and Golgi** (Boevink et al., 1996, 1998) • **BCECF for vacuole staining** (Matsuoka et al., 1997; Mitsuhashi et al., 2000) • **BY-2 system for studying subcellular locations** (Nagata et al., 1992)	• **FM4-64 method** (Ueda et al., 2001; Emans et al., 2002) • **Wortmannin** (Tse et al. , 2004) • **Photoactivatable fluorescent proteins** (Runions et al., 2006; Brown et al., 2010; Mathur et al., 2010) • **Concanamycin A and Bafilomycin A** (Robinson et al., 2004) • **MDC (Contento** et al., 2005) • **Optical tweezers** (Sparkes et al, 2009) • **VAEM and TIRFM** (Wang et al., 2006; Konopka et al., 2008)	• **pH sensors** (Shen et al., 2013; Martiniere et al., 2013) • **Phosphoinositide and electrostatic sensors** (Simon et al., 2014, 2016) • **Nanobody** (Künzl et al., 2016) • **Spinning disk confocal microscopy and SCLIM** (Ito et al., 2012; Zhuang et al., 2013; Uemura et al., 2014; Gadeyne et al., 2014) • **sptPALM** (Hosy et al.,2015; Platre et al. 2019) • **SIM** (Yu et al., 2017) • **SPT and FCS/FCCS** (Li et al., 2011, 2012; Wang et al., 2013) • **Light sheet microscopy** (Berson et al., 2014)	• **4D SCLIM** (Shimizu et al., 2021) • **STED and expansion microscopy** (Fougère et al., 2023)
Electron microscopy	• **Chemical fixation EM** (Chrispeels and Vatter, 1963; Halperin and Jensen,1967; Mesquita,1969; Wheeler and Hanchey, 1971; Fowke et al., 1974; Harris and Chrispeels, 1975; Larkins and Hurkman, 1978; Robinson,1980) • **Freeze-etch EM** (Matile and Moor, 1968; Staehelin and Kiermayer, 1970)	• **High-pressure freezing** (Craig et al, 1987; Staehelin, et al., 1990) • Immunogold EM (Holwerda et al., 1990)	• **Immunogold EM** (Bednarek et al., 1991; Satiat-Jeunemaitre and Hawes, 1992; Hara-Nishimura et al. 1998; Pimple et al, 2000)	• **3D-Tomography** (Otegui et al., 2001; 2006; Donohoe et al., 2007; Kang et al., 2011)	• **Cryo-FIB and Cryo-EM/ET** (Engel et al., 2015; Bykov et al. 2017) • **CLEM** (Bell et al., 2013; Marion et al., 2017; Wang et al., 2019a) • **Whole-cell 3D tomography** (Cui et al., 2019) • **SBF-SEM** (Scheuring et al. 2016)	• **STEM tomography** (Johnson et al., 2021, 2022) • **CLEM** (Chambaud et al., 2022; Dragwidge et al., 2022; Gomez et al., 2022; Jiang et al., 2022; Luo et al., 2023) • **Cryo-EM/ET** (Li et al., 2021; De Bellis et al., 2022) • **FIB-SEM** (Guo et al., 2023)
Organelle isolation & Reconstitution	• **Membrane fractionation for ER and Golgi** (Ray et al., 1969; Huang and Beevers, 1971; Moore and Beevers, 1974) • **Isolation of Golgi** (Morre and Mollenhauer, 1964) • **Isolation of PBs** (Youle and Huang, 1976; Tully and Beevers, 1976)	• **Isolation of PSV** (Hara-Nishimura et al., 1982) • **Isolation of SVs** (Taiz et al, 1983) • **Isolation of lytic vacuole** (Wilkins et al., 1990)	• **Isolation of lytic vacuole** (Bednarek et al., 1991; Ahmed et al., 2000) • **Isolation of PSV** (Kirsch et al., 1994; Paris et al., 1996) • **Isolation of CCVs** (Lin et al., 1992; Kirsch et al., 1994) • **Isolation of PAC vesicles** (Hara-Nishimura et al. 1998)	• **COPI *in vitro* reconstitution** (Pimple et al., 2000) • **Proteomics analysis of vacuole** (Carter et al., 2004) • **PSV/MVB purification** (Tse et al. , 2004; Otegui et al., 2006)	• **Proteomics analysis of TGN** (Drakakaki et al., 2012) • **Proteomics analysis of Golgi** (Parsons et al., 2012; Nikilovski et al., 2014) • **Proteomics analysis of EVs** (Rutter and Innes, 2017)	• **COPII *In vitro* reconstitution** (Li et al., 2021) • **Proteomics analysis of CCVs** (Dahhan et al., 2022)

Figure 2. Timeline of landmark technologies developed over the past decades for visualizing organelles in the plant endomembrane system.

plant cell biology research has profited from great advances in three categories of techniques: light microscopy (LM), electron microscopy (EM), and organelle isolation and reconstitution methods. Here we highlight their landmark applications in plant cells historically (Fig. 2).

For LM-based organelle studies, fluorescent dyes were the first probes to be applied to visualize organelles in plant cells at the subcellular level. These include neutral red and 2′,7′-bis-(2-carboxyethyl)-5-(and-6)-carboxyfluorescein (BCECF) for vacuoles (Guilliermond 1929; Wilkins et al. 1990; Matsuoka et al. 1997; Mitsuhashi et al. 2000), DiOC6(3) (3,3′-dihexyloxacarbocyanine Iodide) for the ER (Knebel et al. 1990), and monodansylcadaverine (MDC) for autophagosomes (Contento et al. 2005), as well as FM4-64 for endosomes and the endocytic PM to vacuole pathway (Ueda et al. 2001; Emans et al. 2002). Subsequently, GFP-tagged organelle markers, first demonstrated for ER and Golgi apparatus (Boevink et al. 1996, 1998), were developed and used in plant cells (Geldner et al. 2009). During the 2010s, photoconvertible fluorescent proteins were developed and used to track the dynamics of organelles (Runions et al. 2006; Brown et al. 2010; Mathur et al. 2010), whereas different biosensors were also generated, including pH sensors for measuring organelle pH (Martiniere et al. 2013; Shen et al. 2013) as well as phosphoinositide and electrostatic sensors for monitoring biophysical properties of membranes (Simon et al. 2014, 2016). Nanobody-conjugated reporters have enabled the detection of transient interactions between different sets of organelles (Kunzl et al. 2016). As a classic system in plant cell biology, tobacco BY-2 suspension cells have also been widely used for studying subcellular locations and sorting routes by transforming fluorescent proteins (Brandizzi et al. 2003). In addition, various inhibitors are in common use, including brefeldin A (BFA) to trap COPI coatomer (Satiatjeunemaitre and Hawes 1992), wortmannin to inhibit PI(3)K (phosphotidylinositol-3-kinase) (Matsuoka et al. 1995; Tse et al. 2004), and concanamycin A or bafilomycin A to block H^+ pumps (Robinson et al. 2004).

Advanced microscopical techniques have also helped researchers investigate the subcellular localization, stoichiom-etry, and degree of interaction between organelles. For example, optical tweezer is a fantastic tool to visualize ER–Golgi interactions (Sparkes et al. 2009). Total internal reflection fluorescence microscopy (TIRFM) and a TIRF-like technique called variable angle epifluorescence microscopy (VAEM) are now used to study protein dynamics on the plasma membrane (PM), transport vesicles of endocytosis and exocytosis as well as ER exist sites close to the PM in plant cells (Wang et al. 2006; Konopka et al. 2008; Li et al. 2011; Takagi et al. 2020). Using single-particle tracking (SPT) systems coupled with fluorescence correlation spectroscopy (FCS), the dynamics of aquaporin and lipid rafts on the PM under physiological conditions have been nicely revealed (Li et al. 2011, 2012). The fluorescence cross-correlation spectroscopy (FCCS) facilities quantitative analysis of mobility and interaction of different proteins (Wang et al. 2013, 2018). Spinning disk microscopy and a more recently developed 4D super-resolution confocal live imaging microscopy (SCLIM) based on the spinning disk microscopy have become a useful tool for dissecting the dynamic behavior of endocytic vesicles, distinct subdomains for ER-Golgi or Golgi-TGN transport with detailed spatial and tem-poral information (Viotti et al. 2010; Ito et al. 2012; Gadeyne et al. 2014; Uemura et al. 2014). Development of super-resolution microscopy (SRM) has opened a door to visualize organelles and dynamics at 10 to 100 nm resolution. Light sheet microscopy has enabled high-speed and long-term live imaging of plant organelles (Berson et al. 2014). Technically, SRM is always combined with a SPT system for quantitative analysis. Structured illumination microscopy (SIM) has also been applied in plant cells to track the dynamics of *Arabidopsis* FLOTILLIN1 (AtFLOT1) on the PM during defense response (Yu et al. 2017). Using another superresolution single-molecule imaging method, photo-activated localization microscopy (PALM), phosphatidylserine molecules, together with the nano-organization of Rho GTPase, have been revealed and quantified on the PM nanodomains under developmental con-trol (Platre et al. 2019). Very recently, a further development of stimulated emission depletion (STED) and expansion microscopy in *Arabidopsis* plants enables live-cell super-resolution imaging of a cis-Golgi tubular intermediate in connecting ER and Golgi transport (Fougère et al. 2023).

Although we are seeing more and more insights into cell structure and organelles with the development of LM- and SRM- based live cell imaging techniques, if we look back through the history of plant cell biology, our view of plant cell structures and organelles has been greatly revolutionized and expanded by the development of EM-based technology (Zhang et al. 2021; Weiner et al. 2022). SRM with 10 to 100 nm resolution still has limitations in revealing information on the spatial connectivity and arrangements of cellular structures (Zhang et al. 2021). In the 1960s, the successful application of chemical fixation using glutaraldehyde (GDH) for EM analysis opened a new door to look at the ultrastructure of plant cells (Chrispeels and Vatter 1963; Halperin and Jensen 1967). In particular, provacuoles were suggested to arise from the rough ER as they display prolongations continuous with the rough ER and the nuclear envelope, as well as a very close proximity to the PM (Mesquita 1969). Other alternative fixatives, including osmium tetroxide and uranyl acetate, were later used to observe the ultrastructure of pinocytosis and cell regeneration (Wheeler and Hanchey 1971; Fowke et al. 1974; Larkins and Hurkman 1978; Harris and Chrispeels 1980). Chris Hawes first applied immunogold labeling with JIM84 antibodies to document morphological changes in the Golgi upon BFA treatment (Satiatjeunemaitre and Hawes 1992). In addition, as the samples were embedded in resin, one can use the serial section transmission EM (ssTEM) method for 3D reconstruction by collecting images for a series of sections. This method was first applied by David G. Robinson to reveal a Golgi dictyosome in which occasional membrane continuities between the Golgi and the ER were observed (Robinson 1980).

To overcome possible artifacts arising during the chemical fixation, the freeze-fracture technique was introduced in plant cell research to visualize the morphology of organelles such as the vacuole (Matile and Moor 1968) and Golgi (Staehelin and Kiermayer 1970). Due to the limits of resolution with this technique, it is more suitable for the analysis of surface structures. Later, Andrew Staehelin greatly improved this technique by using high-pressure freezing/freeze substitution (HPF/FS) for intact plant tissues, allowing a clear differentiation of ER membrane and Golgi stacks in plant root cells (Craig et al. 1987; Staehelin et al. 1990). In addition, immunocytochemistry electron microscopy using antibodies and gold particles has also been used to identify the organelles more precisely at the ultrastructure level (Holwerda et al. 1990; Bednarek and Raikhel 1991; Satiat-Jeunemaitre et al. 1996; Hara-Nishimura et al. 1998). Since then, the combination of HPF and immunogold labeling has become a standard for the characterization of different subcellular structures and reporters, namely COPI, PVC/MVB, and TGN (Pimpl et al. 2000; Otegui et al. 2001, 2006; Tse et al. 2004; Dettmer et al. 2006; Lam et al. 2007). Meanwhile, the samples can also be subjected to electron tomography (ET) and were first used to study cytokinesis in *Arabidopsis* endosperm (Otegui et al. 2001). The application of ET has provided fascin-ating insights into the versatile features and organization of plant organelles. In particular, ET analysis on the Golgi, TGN, and MVB in different plant species has unveiled unex-pected structural beauty and complexity of intermediate compartments that are hardly noticed using conventional TEM techniques (Otegui et al. 2006; Donohoe et al. 2007; Kang et al. 2011; Buono et al. 2017). Recently, ET has been ap-plied to whole root tip cells to reveal a novel intermediate compartment during MVB maturation into the vacuole (Cui et al. 2019b). It should be noted that a typical TEM tomo-gram is produced using semithin sections (~200 to 400 nm), which are tedious to collect to provide enough tomograms for 3D reconstruction of large organelles or cells. Very recent-ly, a concept for volume electron microscopy (vEM) has emerged for large-volume organelle ($>10^6$ μm^3) 3D imaging at nanometer resolution (Kievits et al. 2022). In comparison, scanning-transmission electron microscopy (STEM) tomography can handle thicker sections (e.g. 1,000 nm) with tilted images at high resolution and has been recently used to reveal the arrangement of clathrin-coated vesicles (CCVs) (Johnson et al. 2021, 2022). On the other hand, serial block-face scanning EM (SBF-SEM) by using a ultramicrotome can be applied for 3D reconstruction of plant organelles (Scheuring et al. 2016; Pain et al. 2019; Paterlini et al. 2020), which greatly shortens the sample handling time and expands the limit of sample thickness. In contrast, focused ion beam scanning electron microscopy (FIB-SEM), which uses a gallium ion beam instead for section milling, has recently been successfully used to reveal the interaction between virus particles and the ER network in virus-infected plant cells (Guo et al. 2023). This demonstrates the feasibility of using FIB-SEM on plant samples for large volume 3D imaging.

Correlative light-electron microscopy (CLEM) is another new technique that integrates LM with the TEM or ET (Wang et al. 2019a). This technique was recently applied to detect autophagosome, TGN, plasmodesmata, and biomolecular condensates in plant cells (Bell et al. 2013; Marion et al. 2017; Wang et al. 2019a; Chambaud et al. 2022; Dragwidge et al. 2022; Gomez et al. 2022; Jiang et al. 2022; Luo et al. 2023). Cryo-fixation, which significantly reduces artifacts by plunge-freezing the samples without chemical fixatives, truly preserves the architecture of structures in situ. This method was recently used to investigate purified COPII vesicles from an in vitro reconstitution experiment using *Arabidopsis* cells (Li et al. 2021a). However, when combined with ET, Cryo-FIB milling with SEM imaging will be required to obtain lamellae with suitable thickness (usually <500 nm) for further subtomogram averaging (Villa et al. 2013). So far, Cryo-ET has only been applied to certain plant cell types, such as pollen, the unicellular alga Chlamydomonas, cell walls, and protoplasts (Engel et al. 2015; Bykov et al. 2017; Liu et al. 2021; Nicolas et al. 2022; Sanchez Carrillo et al. 2023). Future development in sample preparation, data collection, and algorithms for image analysis and reconstruction will be essential for revealing the complexity of organelle networks at nanoscale in intact plant cells and tissues.

While the LM and EM help to define the identity of organelles inside the cells, biochemical assays are needed to purify organelles and provide a parallel strategy to characterize their functions. Using centrifugation to isolate subcellular components in plants began early in the 20th century (Andrews 1903), whereas by the middle of 20th century, sugar-based density gradient and rate-zonal sedimentation further improved separation efficiency. Purification of ER and Golgi (Morre and Mollenhauer 1964; Ray et al. 1969; Huang and Beevers 1971) and protein bodies (Tully and Beevers 1976; Youle and Huang 1976) from plants has been achieved in a number of pioneer studies 50 years ago. Groundbreaking works, led by Ikuko Hara-Nishimura, Natasha Raikhel, and John Rogers, respectively, have success-fully isolated lytic vacuole (LV) and protein storage vacuole (PSV) from different plant species and uncovered distinct molecular determines for sorting and targeting to the two different types of vacuoles (Hara-Nishimura et al. 1982, 1998; Wilkins et al. 1990; Bednarek and Raikhel 1991; Kirsch et al. 1994; Paris et al. 1996). Around the same period, secre-tory vesicles (SVs) and MVBs have also been successfully purified and characterized (Taiz et al. 1983; Tse et al. 2004; Otegui et al. 2006). The development of proteomics technologies in concert with the isolation methods to identify the organelle-associated proteins has further expanded our knowledge on the composition of organelles under physiological conditions. Landmark studies include the proteomics analysis of purified vacuoles (Carter et al. 2004), TGN (Drakakaki et al. 2012), Golgi (Parsons et al. 2012; Nikolovski et al. 2014), extracellular vesicles (EVs) (Rutter and Innes 2017), and CCVs (Dahhan et al. 2022). Albeit highly challenging in plant cell biology research, cell-free organelle reconstitution offers the possibility to perform quantitative and qualitative experiments in a controlled environment. The successful reconstitution of COPI (Pimpl et al. 2000) and COPII vesicles (Li et al. 2021a) provided new insights and tools into the mechanisms of vesicular transport in plants.

ER and Golgi: a close but distant relationship

The ER is a remarkable dynamic network as a primary source of lipids and proteins in plant cells (Staehelin 1997; Brandizzi 2021). In terms of morphological features, there are two typical types of ER, the tubules and sheet-like cisternae, which are interconnected and linked to the nuclear envelope (Staehelin 1997; Pain et al. 2019). In plants, ER is considered a single network crossing the whole plant body, which is connected by a special subdomain of ER spanning the cell wall named plasmodesmata (Haywood et al. 2002). Interconversion of the ER cisternae into tubules is highly dynamic, forming by typical three-way junctions through tubule–tubule fusion or extension from the preexisting junctions (Sparkes et al. 2009). By far, several conserved and plant-specific ER-shaping regulators have been discovered (Brandizzi 2021).

Majority proteins are transported from ER to the Golgi apparatus. Unlike single Golgi cisternae distribution in yeasts or Golgi stacks accumulation at the perinuclear as ribbon in animals, the individual Golgi apparatus exhibits a unique polar architecture as an intact mobile multi-disc stack adjacent to the ER in plant cells (Boevink et al. 1998). In the current model, the Golgi apparatus can be divided into three subdomains: ER adjacent cis-Golgi, middle-Golgi, and trans-Golgi (Fig. 3, A and B). Pioneer works, particularly from groups led by Chris Hawes, David G. Robinson, and Andrew Staehelin, have contributed significantly to our current understanding of Golgi structure and function by the application and development of multiple tools (Staehelin et al. 1990; Hawes 2005; Robinson 2020). For example, Chris Hawes's group developed the first Golgi markers, including JIM 84 antibody for immunolocalization (Fitchette et al. 1999) and STtmd–GFP (Boevink et al. 1998) for live cell imaging (Fig. 3C), and first demonstrated the reversible versatility of the Golgi architecture at the EM level using the drug BFA (Satiat-Jeunemaitre et al. 1996). Golgi morphology might vary in different cell types and different plant species, which is suggested to be related to cell wall composition. For example, dilated slime polysac-charide containing SVs are characteristic for many root cap cells. In developing cotyledons, a special type of vesicle termed dense vesicles (DVs; roughly 120 to 150 nm in diameter) are formed for the transport of globulins to the PSV (Hillmer et al. 2001). In the green alga Chlamydomonas, the Golgi stacks have been recently visualized using cryo-EM, with greatly increased cisternal numbers that are adjacent to the ER subdomain (Engel et al. 2015). How the Golgi retains its stacked morphology in plant cells remains unknown. Golgins, a protein family with long coiled-coil domain anchored in the Golgi membrane, have been identified as a tether in maintaining ER–Golgi connection, but it has also been suggested that other tethers might be involved (Osterrieder et al. 2017; Vieira et al. 2020).

Models for Golgi function in plants have been a matter of debate. Basically, there are two models: vesicle transport and cisternae maturation (Hawes 2005; Robinson et al. 2015). In the vesicle transport model, Golgi stacks maintain a relative stable structure by receiving vesicles from the ER. In the Golgi maturation model, COPI vesicles, dominantly residing on the cis-Golgi, are believed to retrieve the enzymes or other receptors back to the cis-Golgi (Pimpl et al. 2000; Hillmer et al. 2001). 3D ET analysis further revealed that two different types of COPI vesicles, COPI1a (with a thicker coat ~12 nm) and COPI1b (with a thinner coat ~6 nm), are participating in inter-Golgi transport (Donohoe et al. 2007). Plant genomes comprise expanded COPI and COPII subunits, suggesting that plants might form various forms of COPI and COPII vesicles to accommodate different cargoes or in response to different environmental stimuli. Indeed, recent in vitro reconstitution and cryofixation studies have elegantly revealed the structures of COPI and COPII vesicles in situ and giant COPII vesicles under abiotic stress condition (Donohoe et al. 2007; Bykov et al. 2017; Li et al. 2021a). However, due to the extreme close dynamic association be-tween ER and Golgi, whether tubule structures formed for ER-Golgi transport in plant cells to function as an intermediate counterpart the ER-Golgi intermediate compartments (ERGIC) in mammalian cells remains an enigma (Robinson et al. 2015). Increasing evidence suggests that a tubular inter-mediate might exist, including irregular cis-Golgi cisternae with variable size and density revealed by 3D ET (Staehelin and Kang 2008), physical contacts by optical tweezer analysis (Sparkes et al. 2009), and Golgi regeneration from the ER exit site upon BFA treatment captured by 3D time-lapse imaging (Ito et al. 2012). A strong clue for the nature and dynamic architecture of such an intermediate compartment was obtained by an elegant combination of STED and expansion microscopy in a recent study (Fougère et al. 2023). With 4 times the physical expansion of the Golgi, two distinguishable cis-Golgi subpopulations were evidently revealed, showing a Golgi-independent tubulo-vesiculated network intertwines with the Golgi-dependent cisternae (Fougère et al. 2023). It will be interesting for future studies to explore whether the COPI and COPII vesicles are incorporated into this tubulo-vesiculated network.

TGN/EE: a central hub for sorting and secretion

After modification within the Golgi cisternae, cargoes are exported out of the Golgi to other destinations, such as the PM and vacuole (Fig. 1). Meanwhile, PM cargoes or receptors are internalized into the cytoplasm through endocytosis to modulate various signaling pathways. The identification of plant intermediate compartments belonging to the secretory

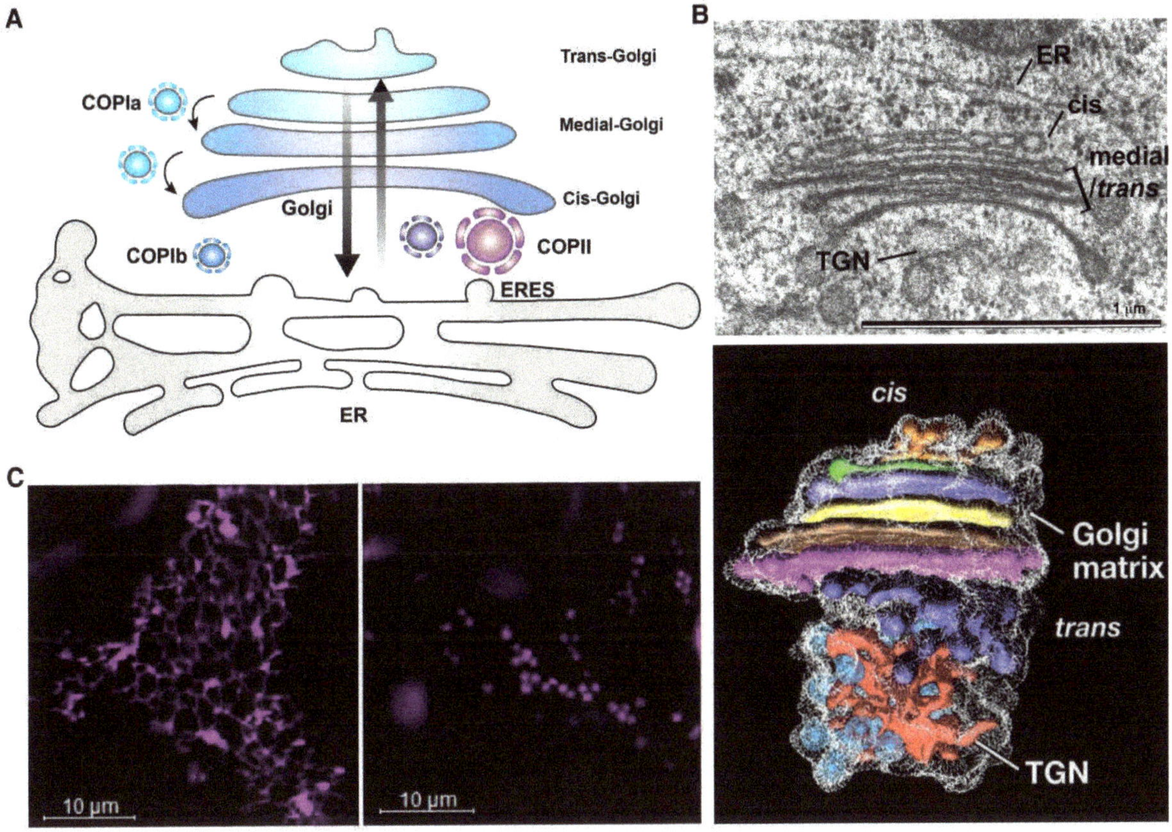

Figure 3. ER and Golgi in plant cells. **A)** Overview of bidirectional transport between the ER and Golgi in plant cells. Membrane lipids and proteins are synthesized in the and transported out of the ER through vesicles pinched off from a specific ER subdomain, also known as ERES. Trafficking occurs in both directions: the forward direction mediated by COPII vesicles to the Golgi, and the reverse direction mediated by COPI vesicles back to the ER. Different sizes of COPII vesicles are formed to transport different cargos. Cargoes are directed to their appropriate destinations, passing from the closest Golgi membranes to the ER, called the cis-Golgi, the medial-Golgi, and reaching the trans-Golgi. COPI, coat protein complex I; COPII, coat protein complex II; ERES, ER exit site. **B)** Electron micrograph and 3D ET revealing the architecture of Golgi complex in a high-pressure-frozen and freeze-substituted BY-2 cell (reprinted from Staehelin and Kang 2008, Figs. 1B and 5. **C)** Confocal images showing the expression patterns of fluorescent tagged ER (HDEL) and Golgi (STtmd) markers in the *Arabidopsis* leaf.

and endocytic pathways within the plant endomembrane system has been challenging because these two pathways had been largely studied separately. Thus, a direct comparison of protein markers between these pathways for specific post-Golgi compartments is necessary for illustrating the nature of endosomal and secretory compartments in plant cells. Early EM work has suggested a "displaced" TGN, named partially coated reticulum (PCR), as an endosome in plant cells (Pesacreta and Lucas 1985; Hillmer et al. 1988). Several studies have laid down the foundation for the subsequent identification and characterization of TGN as an EE merging the secretory and endocytic pathways in plant cells (Dettmer et al. 2006; Lam et al. 2007) (Fig. 4A).

Morphologically, the TGN is often seen under the TEM as a cluster of vesicular and tubular structures that are usually found close to the *trans* side of the Golgi stacks (Dettmer et al. 2006; Lam et al. 2007). Distinct from mammalian cells and yeast, the Golgi apparatus and TGN in plant cells are two distinct compartments as visualized by confocal immunofluorescence (Dettmer et al. 2006; Lam et al. 2007). The Arabidopsis V-ATPase subunit VHA-a1 locates to the TGN by immunogold-TEM analysis in root cells (Fig. 4B) (Dettmer et al. 2006). Interestingly, VHA-a1-GFP was also shown to colocalize with the TGN marker SYP41-mRFP but separate from the late endosomal marker AtRABF2b (Ara7) (Ueda et al. 2004) in double transgenic *Arabidopsis* plants co-expressing these two markers. Further FM4-64 uptake studies in root cells of transgenic plants showed that the internalized FM4-64 first reached the VHA-a1-GFP positive TGN before reaching the AtRABF2b positive endosomes, indicating the early endosomal nature of VHA-a1-labeled TGN in *Arabidopsis* cells (Dettmer et al. 2006).

Another study on the dynamics and subcellular localization of the rice SCAMP1 (secretory carrier protein 1) in trans-genic tobacco BY-2 cells also demonstrated that the TGN is an EE (Lam et al. 2007). Transgenic tobacco BY-2 cells expres-sing SCAMP1-YFP fusion showed both PM and cytosolic punctate patterns, whereby the SCAMP1-positive puncta

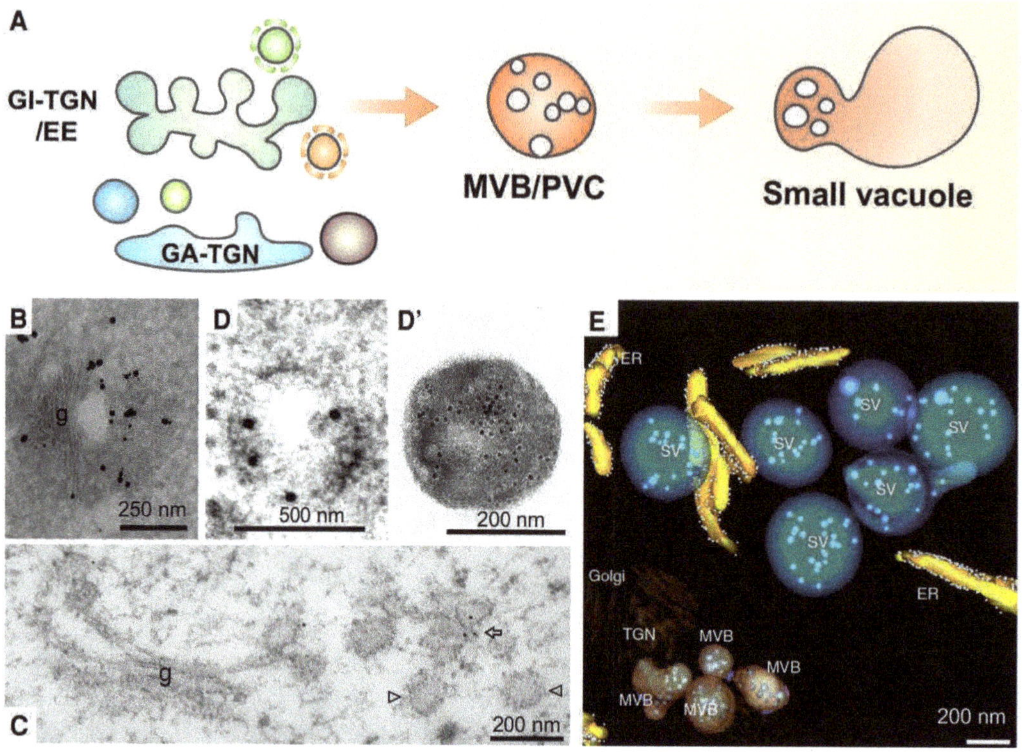

Figure 4. Biogenesis of *trans*-Golgi network (TGN) and multivesicular body/prevacuolar compartment (MVB/PVC). **A)** Golgi-associated (GA)-TGN matures and becomes a free Golgi-independent- (GI) TGN adjacent to the Golgi. GA-TGN or GI-TGN derived vesicles (listed in the box at the bottom right) directly or indirectly transport to MVB/PSV, which later transforms into a small vacuole (SV). **B)** Immunogold localization of *Arabidopsis* VHA-a1 reveals TGN as early endosome in plant cells (reprinted from Dettmer et al. 2006, Fig. 2H). **C)** Immunogold localization of rice SCAMP1 defines TGN as early endosome in plant cells. g, Golgi (reprinted from Lam et al. 2007, Figure 12). **D)** Immunogold labeling of MBV/PVC by vacuolar sorting receptor (VSR) antibodies. (reprinted from Tse et al. 2004, Figs. 6E and 13E). **E)** 3D ET reveals the architecture of MVB transition into SV (reprinted from Cui et al. 2019a, Fig. 1B).

are separate from both Golgi and PVC markers in transgenic tobacco BY-2 cells. An FM4-64 or AM4 to 64 uptake study was performed on transgenic tobacco BY-2 cells expressing SCAMP1-YFP or the PVC marker GFP-BP-80. It was demonstrated that the SCAMP1 and GFP-BP-80/PVC puncta represent EE and LE, respectively, as the internalized FM4-64 or AM4-64 colocalized with the SCAMP1 puncta first before reaching PVCs. Furthermore, immunogold-TEM analysis with HPF/FS samples and specific antibodies identified the SCAMP1-positive tubular-vesicular structures at the *trans*-Golgi, which are distinct from the Golgi stacks but positively labeled with clathrin antibodies on the vesicle coats (Fig. 4C) (Lam et al. 2007). Interestingly, in transgenic BY-2 cells expressing the PVC marker GFP-BP-80, wortmannin treatment caused the redistribution of SCAMP1 from EE to the PVC as visualized by confocal imaging, and fusion between TGN and the enlarged PVC/MVB was confirmed by immunogold-TEM observation (Lam et al. 2007).

Nowadays, advanced imaging analysis has recognized that the Golgi-associated (GA) and Golgi-independent (GI) TGN have diversified functions in protein sorting. Spinning disk confocal and SRM analysis have demonstrated that GI-TGNs are highly dynamic (Viotti et al. 2010; Uemura et al. 2014). By combining 3D ET analysis of the GA- and GI-TGNs, it has been proposed that the GI-TGNs become separated from GA-TGNs via a maturation process (Kang et al. 2011). Different subregions on GI-TGN have been revealed by using 4D SCLIM, implying a complexity of the plant TGN subdomains for different cargo sorting (Shimizu et al. 2021). Various types of vesicles arise from the GA- and GI-TGN. Among them, adaptor protein-mediated vesicles represent the best characterized, including AP1, AP3, and AP4 (see Reynolds et al. 2018 for a nice review). Application of proteomics to characterize these vesicles have provided insightful information for the functions and biogenesis of different types of vesicles from the TGN (Drakakaki et al. 2012; Dahhan et al. 2022). Nevertheless, the determinants for the different subdomains on the dynamic TGNs and the specific cargoes remain elusive. Although cargoes can be delivered by vesicle transport to the late endosomes, recent evidence showed that intraluminal vesicles (ILVs) can be readily observed within the TGN/EE adjacent to the Golgi (Scheuring et al. 2011; Cui et al. 2019b), suggesting a transformation process from TGN/EE into late endosomes (MVB/PVC).

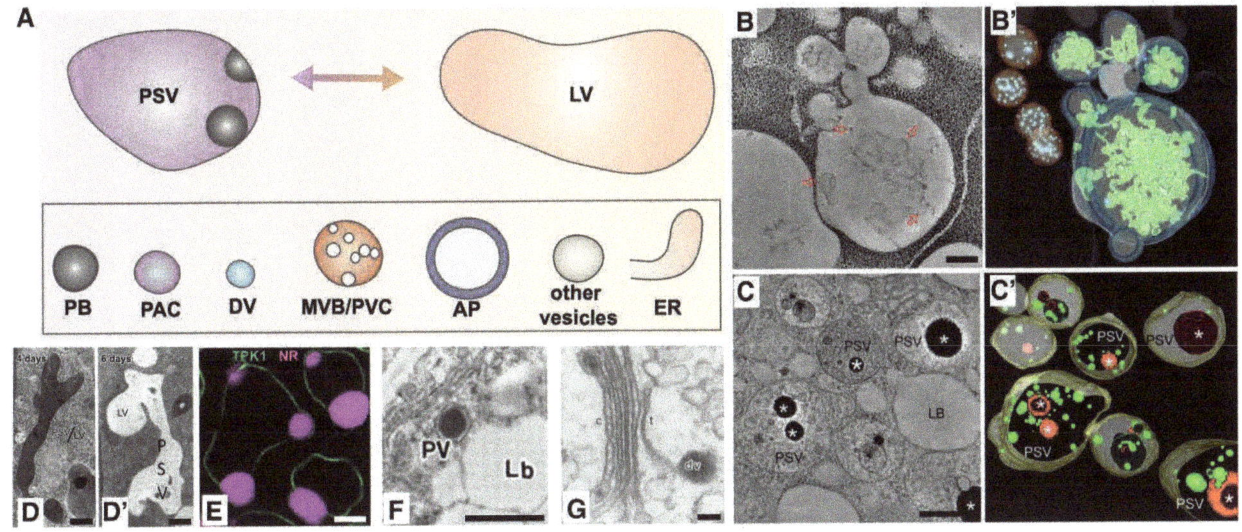

Figure 5. Lytic vacuole (LV) and protein storage vacuole (PSV) in plant cells. **A)** Different membrane sources of LVs are proposed, including the ER, multivesicular bodies/prevacuolar compartment (MVB/PVC), autophagosome (AP), and other vesicles (e.g. AP-3-derived vesicles). The biogenesis of PSVs is species-specific and tissue-dependent. The precursor compartments of PSVs include precursor–accumulation (PAC) vesicle, dense vesicle (DV), protein body (PB), and MVB/PVC. PSVs and LVs can be convertible with each other. **B)** LVs in *Arabidopsis* root cortex cells are generated mainly by MVB maturation and subsequent fusion between small vacuoles (SVs) as revealed by 3D ET. Bar, 200 nm (reprinted from Cui et al. 2019b, Fig. 3b). **C)** Storage proteins in maize aleurone layer are sorted to PBs and translocated to PSVs in a microphagy-like process. Images are representative 3D ET analysis of PBs within the PSVs. Bar, 500 nm (reprinted from Reyes et al. 2011, Fig. 6B); **D)** EM data reveal that PSVs are converted to LVs during seed germination in tobacco root cortex cells. Bar, 500 nm (reprinted from Zheng and Staehelin 2011, Fig. 3G and 3H); **E)** PSVs in *Arabidopsis* embryo are converted from embryonic vacuoles through phase separation and wetting-based membrane remodeling. Bar, 2.5 μm (reprinted from Kusumaatmaja et al. 2021, Fig. 1D); **F)** Seed storage proteins in pumpkin seed cotyledons are sorted to PACs before transport to PSVs. Bar, 500 nm (reprinted from Hara-Nishimura et al. 1998, Fig. 5A); **G)** DVs are generated at the periphery of Golgi cisternae in pea cotyledon. Bar, 100 nm (reprinted from Robinson et al. 1997, Fig. 1B). MVB, multivesicular bodies; PV, precursor-accumulation vesicles; PSV, protein storage vacuoles; LV, lytic vacuole; PB, protein body; SV, small vacuole; DV, dense vesicle; AP, autophagosome; LB, lipid body.

MVB as PVC: a stepwise coordination into maturity

Functionally and morphologically, the identification and characterization of plant intermediate membrane-bound organelles have been challenging tasks due to the complexity and multiple pathways of the plant endomembrane system (Bethke and Jones 2000). Pioneer works conducted independently by Ikuko Hara-Nishimura, Natasha Raikhel, and John Rogers' groups have contributed significantly to the characterization and identification of plant PVCs as an intermediate compartment for the lytic vacuoles in plant cells. Conserved peptide domains (termed vacuolar sorting determinants, VSDs) are both necessary and sufficient for targeting soluble vacuolar proteins to vacuoles in plant cells (Chrispeels and Raikhel 1992; Okita and Rogers 1996). Using an affinity column conjugated with a synthetic peptide containing the VSD Asn-Pro-Ile-Arg (NPIR) from a cysteine protease aleurain, Kirsch et al. (1994) isolated the first plant vacuolar sorting receptor (VSR) termed BP-80 (binding protein 80 kD) from lysates of pea CCV membranes. This study led to the subsequent cloning and characterization of VSRs in plants, including the pea BP-80 (Paris et al. 1997), the *Arabidopsis* epidermal growth factor receptor-like protein (AtELP1) or AtVSR1 (Ahmed et al. 1997), the pumpkin PV-72 (Shimada et al. 1997), and the Arabidopsis receptor homology region transmembrane domain ring H2 motif protein (AtRMR1) (Park et al. 2005). Further immunogold-TEM analysis with specific monoclonal antibodies showed that BP-80 is localized to the Golgi cisterna and a "prevacuole" with features of about 250 nm vacuolar structure separating from but fusing with a vacuole in pea root tips (Paris et al. 1997). Similarly, immunogold-TEM studies on cryosection of *Arabidopsis* roots showed that AtELP localizes to the trans-Golgi and post-Golgi membranes as well as colocalizing with AtEPP12p, a t-SNARE protein and a PVC marker in yeast, on electron-dense structures of 100 nm in diameter representing putative PVCs (Sanderfoot et al. 1998). These results are consistent with the role of VSRs in mediating protein traffic between Golgi/TGN and vacuole in plant cells.

Two subsequent studies led to the morphological identification of PVCs in plant cells. The first study used confocal immunofluorescence with several specific VSR antibodies to compare the subcellular localization of VSRs to that of the well-established Golgi markers in the same plant cells, showing that the VSR-marked organelles are largely (more than 90%) separated from the Golgi organelles (Li et al. 2002). Since VSRs are known to mediate protein trafficking via a PVC between Golgi/TGN and vacuole, these results demonstrated that VSRs are predominantly concentrated on the PVCs at steady state albeit being recycled back from PVC to TGN/Golgi (Li et al. 2002). Thus, the subcellular localization of

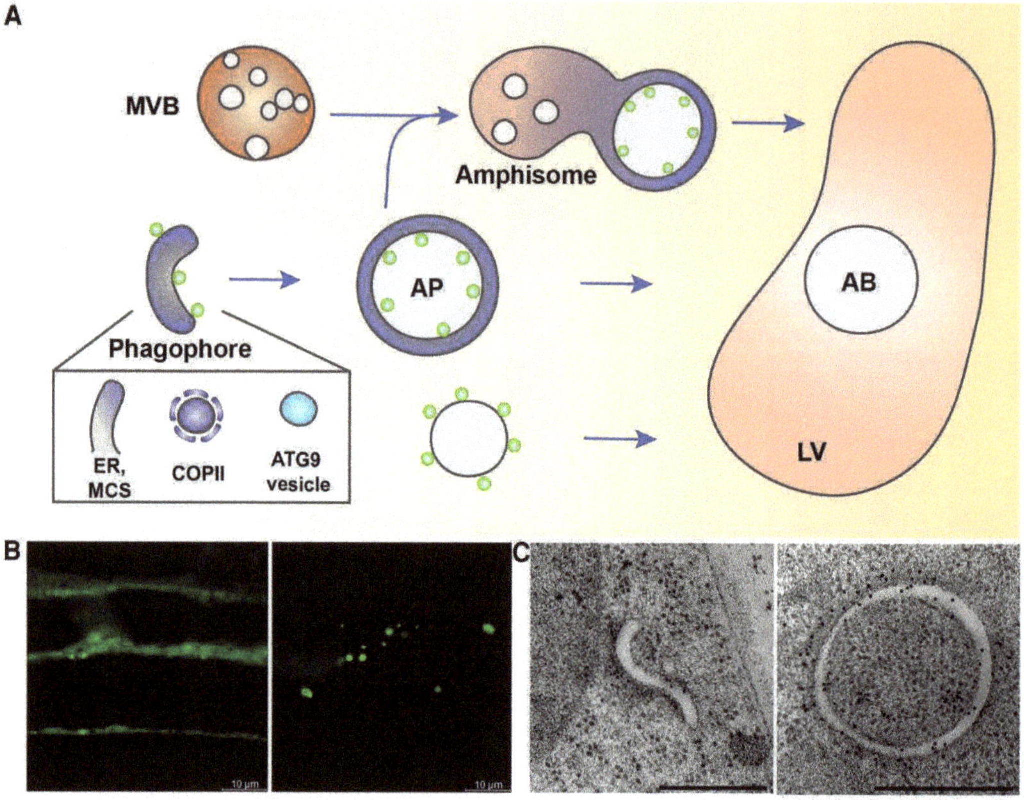

Figure 6. Working model of autophagosome formation in plant cells. **A)** Double-membrane autophagosome is initiated from an isolation membrane sac named as phagophore, which might fuse with MVB/PVC and develop into an intermediate compartment named amphisome, before its delivery into the lytic vacuole (LV). After fusion, the inner part is released into the vacuole, becoming the so-called autophagic body (AB). The phagophore may originate from a preexisting membrane source from other organelles or vesicles, such as ER, MCS, COPII vesicles and ATG9 vesicles. The ATG8 family proteins are associated with the completed autophagosome until its delivery into the vacuole. In addition, ATG8-associated single-membrane vesicle has been reported in mediating cargo degradation. **B)** Expression pattern of autophagosome markers (YFP-ATG8e) before (left)/after (right) autophagic induction in *Arabidopsis* root cells. **C)** Representative autophagosome structures revealed by immunogold labeling in *Arabidopsis* root cells. Bar, 500 nm (reprinted from Zhuang et al. 2013, Figs. 4A and 4F).

VSRs on PVCs at steady state would be a good tool to further identify the PVCs. The second study used two approaches: transgenic tobacco BY-2 cells expressing the Golgi reporter (GONST1-YFP) or the PVC reporter (YFP-BP-80), which showed full colocalization with the endogenous VSRs (Jiang and Rogers 1998), for both live-cell imaging and immunofluorescent analysis, and immunogold-TEM with VSR antibodies for ultra-structure analysis (Tse et al. 2004). Indeed, confocal immunofluorescent analysis with specific Golgi and VSR or PVC marker antibodies showed that GONST1-YFP is largely separated from PVC markers, while YFP-BP-80 is largely colocalized with the PVC markers but separated from the Golgi mar-kers. Interestingly, wortmannin treatment at 16.5 or 33 μM induced YFP-BP-80 PVC puncta, but not the GONST1-YFP Golgi puncta, to enlarge, appearing as small vacuoles in live-cell imaging of transgenic BY-2 cells. Furthermore, immunogold- TEM analysis using HPF/FS-prepared BY-2 cells showed that VSR antibodies specifically labeled MVBs with a diameter of 200 to 500 nm with the gold particles being restricted to the outer membranes of MVBs. Structural TEM analysis further confirmed the presence of small vacuoles of diameter 500 to 800 nm with a small number of 70 nm microvesicles, known as ILVs of MVBs (Fig. 4D) (Tse et al. 2004). A subsequent immunogold-TEM analysis on developing mung bean cotyledon demonstrated that wortmannin induces the homotypic fusion of PVC/MVB, which appears to be the major membrane source for PVC enlargement (Wang et al. 2009). Recently, whole-cell ET analysis on *Arabidopsis* root cortex cells suggests a model of MVB-MVB fusion into small vacuoles (Fig. 4E) (Cui et al. 2019b).

Putative PVCs for the PSV have also been identified and characterized mainly by TEM and/or immunogold-TEM analysis, including the MVBs (size 0.5 to several mm) containing storage proteins as PVCs for PSVs in developing pea seeds (Robinson et al. 1998) and in developing *Arabidopsis* embryos (Otegui et al. 2006), the precursor-accumulating (PAC) vesicles in developing pumpkin seeds (Hara-Nishimura et al. 1998) as well as the dark intrinsic protein (DIP) organelles containing RMR proteins in developing tobacco seeds (Jiang and Rogers 1998) (Fig. 5).

The vacuole: a transport and storage destination

Plant cells contain two types of vacuoles, namely LVs in vegetative tissues and the PSVs in seeds. The term "vacuole" was first given by the French biologist Félix Dujardin to describe the blank space of contractile vesicles in protozoa (Dujardin 1841). The definition was later adopted by the plant biologist De Vries, who proposed that vacuoles were transformed from plastid-like precursors (De Vries 1885). The early systematic study of LVs was carried out by French botanist Guilliermond in the early 20th century, who examined the pattern of anthocyanin formation in living rose petals by using Neutral Red staining (Guilliermond 1929). Studies on vacuoles were further advanced with the arrival of the EM (Buvat 1957, 1958; Matile and Moor 1968). Using BCECF-based confocal 3D reconstruction analysis in root erpidermal cells, a model of interconnecting vacuoles was proposed (Viotti et al. 2013; Kolb et al. 2015). In contrast, a recent whole-cell 3D ET analysis with nanometer resolution in root cortex cells has supported a different model with separated vacuoles and MVB-derived small vacuoles (Cui et al. 2019b).

A longstanding question in vacuole study is the origin of LVs. Four potential upstream compartments have been proposed to be the origins of LVs, including ER, MVB-derived small vacuoles, adaptor protein complex-3 (AP-3)-derived vesicles, and autophagosomes (Fig. 5A). For the ER-derived vacuole membrane model, early evidence came from elec-tron micrographs showing the gradual transition of rough ER into sphere-like structures reminiscent of LVs (Mesquita 1969). This was supported by the similar membrane texture of ER and tonoplast observed by freeze-etching electron microscopy (Matile and Moor 1968). 2D TEM analysis suggested that at least part of vacuole biogenesis in root cells of *Arabidopsis* depended on a yet-to-be-clarified Golgi-bypass route directly from the ER to provacuoles, which were proposed to be the precursor of LVs (Viotti et al. 2013). The notion that MVBs are precursors of LVs was supported by whole-cell 3D ET analysis, which identified small vacuoles as nascent vacuoles in early developmental cortical cells (Cui et al. 2019b). These authors also showed that the LVs were generated by MVB-to-SV transition and subsequent fu-sions of SVs (Cui et al. 2019b) (Fig. 5B). The AP-3 also contributes to LV biogenesis via an MVB-bypass pathway (Feraru et al. 2010; Zwiewka et al. 2011), as *pat2*, a loss-of-function mutant of the AP-3β adaptin, displays obvious defects in vacuole morphology and impairs the conversion from PSV to LV (Feraru et al. 2010). Autophagy is also considered to play a role in LV biogenesis because autophagosomes fuse with the tonoplast during autophagy. However, no obvious defects in vacuole morphology are observed in typical *atg* mutants of the autophagy pathway, questioning the signifi-cance of autophagy in LVs biogenesis. The concept of "autophagic vacuole" was raised based on the EM observation that provacuoles cooperate to drive a programmed cellular autophagy leading to young vacuoles, a process similar to microautophagy (Marty 1978).

The PSV is a special form of vacuole receiving a large amount of seed storage proteins (SSPs) in seeds of flowering plants. The biogenesis of PSV is species-specific and tissue-specific, and even depends on different developmental stages (Zheng and Staehelin 2011; Zheng et al. 2022). While PSVs in *Arabidopsis* embryos develop by remodeling from the embryonic vacuoles during seed maturation (Feeney et al. 2018; Zheng et al. 2022), PSVs in cereals are stored mainly in endosperm tissues. Four major SSP-containing precursor compartments have been reported in different plant species, including PAC vesicles, PB, DV, and PVC (Fig. 5A). ER-derived PAC vesicles were first purified and characterized by Hara-Nishimura's group in developing seeds of pumpkin (*Cucurbita maxima*) and castor bean (*Ricinus communis*) (Hara-Nishimura et al. 1998) (Fig. 5F), and later have also been found in soybean and rice (Mori et al. 2004; Takahashi et al. 2005). PAC vesicles are densely packed with storage proteins with a size around 200 to 400 nm, which are thought to be derived from ER and directly transport from ER to the PSVs bypass Golgi apparatus (Hara-Nishimura et al. 1998). In developing wheat endosperm and the aleurone layer of maize endosperm, PBs are the precursor compartments of SSPs before being transported to PSVs (Levanony et al. 1992; Reyes et al. 2011). PBs are also derived from ER and delivered to the PSV through a microautophagy-like process (Fig. 5C) (Reyes et al. 2011; Ding et al. 2022). In maize starchy endosperm cells, PBs are, however, stuck in ER subdomains (Li et al. 1993; Choi et al. 2000). The common feature of all these com-partments is their dense core with SSP aggregates and relatively large size. However, the Golgi-dependent sorting routes also regulate PSV formation. Immunogold-TEM analysis demonstrated that in *Arabidopsis* embryos, SSPs were sorted into MVBs together with VSRs before their fusion with the PSVs as the enzymatic processing of SSP precursors already occurred in the MVBs (Otegui et al. 2006). Another typical post-Golgi compartment unique to PSV biogenesis are DVs with electron-dense contents without a discernible coat (Fig. 5G). DVs were first discovered by TEM observations in common beans (Chrispeels 1983) and later identified in many other plant species (Krishnan et al. 1986; Otegui et al. 2006). DVs are formed at the peripheral region of cis-Golgi with SSP aggregates, which mature along with the Golgi cisternae and bud off from the TGN (Hohl et al. 1996; Robinson et al. 1997). Thus, both Golgi-independent and Golgi-dependent sorting routes contribute to PSV biogenesis.

The factor determining which precursor compartments SSPs are delivered is still not clear. However, current knowledge suggests that the biophysical property of SSPs drives the membrane remodeling, which ultimately promotes the biogenesis of precursor compartments or PSVs. For example, while prolamins in wheat follow the Golgi-dependent traffic to the PSVs at early stages of seed development, their aggregation inside the ER facilitates PACs formation at later stages (Tosi et al. 2009). Even for Golgi-derived DVs, their formation may also depend on the physical status of cargo proteins. It was postulated that the storage proteins use an aggregation mechanism for sorting and travel in DVs (Hohl et al. 1996). In *Arabidopsis* embryos, the conversion from embryonic vacuoles to PSVs depends on the liquid droplets containing storage proteins formed by phase separation (Fig. 5E). These phase-separated droplets wet the contacting tonoplast and trigger the LV-to-PSV transition (Kusumaatmaja et al. 2021). Therefore, to some extent, the interplay of cargo oligomerization and/or phase separation with membrane remolding may partially explain the phenomenon of multiple precursor compartments during PSV biogenesis.

Although PSVs are enriched in seeds and LVs are distributed mainly in post-embryonic tissues, they can be converted into each other. TEM observations at different stages of seed development clearly support that PSVs in cotyledons are converted from embryonic vacuoles (Zheng and Staehelin 2011) (Fig. 5D). Under certain conditions, LVs can also be transformed back into PSVs (Feeney et al. 2013). The difference between the two types of vacuole is not restricted to their morphology but also depends on their associated membrane proteins. Different aquaporins (tonoplast intrinsic proteins, TIPs) have been identified as markers for either LVs or PSVs (Hofte et al. 1992; Paris et al. 1996). While α-TIP labels PSVs, γ-TIP is found mainly at tonoplast of LVs (Paris et al. 1996). Therefore, the conversion between LVs and PSVs and the maintenance of their organelle identity is a programmed process.

Biogenesis of autophagosome: a shortcut into the vacuole?

In textbooks, most endomembrane compartments in plants are relatively stable. Unlike other organelles, autophagosomes can only occasionally be observed under normal conditions, but they emerge dramatically upon stress induction (Fig. 6). Formation of an autophagosome starts from the assembly of an isolation membrane referred to as the phagophore, which then expands and eventually seals to form a double-membrane vesicular structure (Liu and Bassham 2012) (Fig. 6A). Pioneer works on characterizing the ATG ma-chinery in *Arabidopsis* have significantly advanced our un-derstanding of autophagy in plants (Doelling et al. 2002; Hanaoka et al. 2002). During autophagosome formation, a series of ATG proteins is recruited for phagophore assembly, expansion, and maturation. Among them, a central protein, ATG8, which is anchored to phosphatidylethanolamine (PE) on the membrane through a series of ubiquitin-like con-jugation steps, is thus widely used as a reporter of autophagosomal-related structure (Fig. 6B).

In the currently favored model, it was proposed that de novo-formed autophagosomes arise from different subdomains of the ER network. Several studies have observed phagophore-like structures in close association with the ER membrane. Using live cell imaging and immunogold-TEM analysis, it was shown that phagophore membranes are in close proximity to the ER (Zhuang et al. 2013) (Fig. 6C). By using an ATG5 reporter together with three-dimensional reconstructions, it was proposed that the phagophore was assembled on the cortical ER, but whether there was a direct connection with the ER is unclear (Le Bars et al. 2014). Later, with the application of 3D ET, it was demonstrated that a membrane continuum does indeed exist between the ATG8-labeled structures and the ER membrane in the *atg9* mutant under ER stress, providing clear evidence that ER serves as one of the membrane origins for autophagosome formation (Zhuang et al. 2017). Of note, recent studies have also highlighted that ER membrane contact sites might func-tion as hotspots for phagophore assembly, including ER-PM and ER-mitochondria MCSs, showing that disruption in these MCSs suppress autophagic activity (Wang et al. 2014, 2019b; Li et al. 2022).

After autophagosome initiation, it was suggested that the expansion of the phagophore membrane requires membrane sources from the endomembrane system. Strong evidence was obtained by several recent studies, showing several molecular machineries in connecting the endomembrane system for autophagosome formation. ATG9 vesicles, derived from the Golgi, have long been considered as one membrane source for phagophore expansion (Zhuang et al. 2017). Using spinning disk confocal, it was observed that ATG9 vesicles transiently interact with autophagosome membrane, displaying a "kiss-and run" dynamic (Zhuang et al. 2017). A recent study using 3D ET reveals that vesicle-like structures were trapped on open phagophore membrane in an Arabidopsis *atg2* mutant, implying that vesicle transport event occurs during phagophore development (Luo et al. 2023). Although it was observed that ATG2 deficiency leads to accumulation of ATG9 signals along the autophagosome structures (Luo et al. 2023), whether other vesicles also contribute to phagophore expansion remains unclear. Recently, several studies suggested that COPII vesicles might play a role in this process. It was first noted that Sar1D, a SAR1 GTPase family that mediates COPII vesicle budding from the ER membrane, interacts with ATG8, likely to redirect spe-cific COPII populations to the phagophore for autophagosome formation (Zeng et al. 2021). In line with this finding, it was shown that another SAR1 GTPase family, Sar1B, interacts with FYVE-domain containing protein FYVE2/CFS1 to function in autophagosome maturation (Kim et al. 2022). Noticeably, FYVE2/CFS1 recruits Sar1B to autophagosome by direct binding to ATG8 (Kim et al. 2022). More recently, it was suggested that FYVE2/CFS1-ATG8 interaction might participate in autophagosome and endosome fusion for the formation of amphisome (Zhao et al. 2022). Intriguingly, a previous study reported that FYVE domain protein required for endosomal sorting 1 (FREE1)/FYVE1, a subunit of ESCRT complex in mediating PVC/MVB biogenesis, also functions in autophagosome maturation via interacting with SH3P2

(Gao et al. 2015). Depletion of FREE1 leads to strong association between autophagosome and endosome (Gao et al. 2015). Additionally, FREE1 also regulate autophagosome closure by directly binding with ATG8 (Zeng et al. 2023). By far, the sequential order of autophagosome closure and autophagosome-endosome fusion in plant cells remains unclear, raising a question of whether FREE1 functions independently or corporately in these two processes.

A classical view of canonical autophagy is that cargoes are sequestered by the double-membrane autophagosome, which is morphologically distinguishable from other organelles. Interestingly, it was recently observed that ATG8 proteins are recruited to a single-membrane vesicle derived from the Golgi and interact with clathrin in plant cells upon heat shock stress in *Arabidopsis* (Zhou et al. 2023). Using immunogold EM and 3D ET, it was nicely revealed that clathrin proteins are recruited to buds or tubules of the swollen Golgi cisterns (Zhou et al. 2023). It was suggested that the interaction between ATG8 with clathrin facilitates the recruitment of clathrin proteins for Golgi fission during heat shock recovery, as suppression of ATG8 conjugation to the membrane impairs the recruitment of clathrin, while ATG8-positive puncta were accumulated in a clathrin-deficient mutant. Intriguingly, signals of ATG8 fusion proteins were observed on the vacuole membrane upon PI3P inhibitor treatment (Zhou et al. 2023). It will be interesting to further investigate how these ATG8-decrorated single-membrane vesicles fuse with the vacuole.

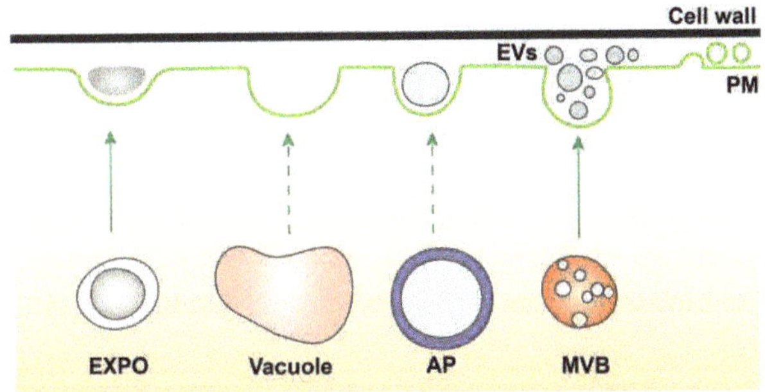

Figure 7. Biogenesis of extracellular vesicles (EVs) in plant cells. Current studies support that EVs in plant cells are derived mainly from the multivesicular body (MVB) and the double-membrane organelle exocyst-positive organelle (EXPO). Lytic vacuole may also contribute to EV release upon pathogen infection. Whether secretory autophagosome and direct budding from plasma membrane regulate EVs production remains to be determined. EV, extracellular vesicle; EXPO, exocyst-positive organelle; AP, autophagosome; MVB, multivesicular body.

Extracellular vesicle (EV): a potential superstar still at its infancy

EVs are membrane-enclosed vesicles released by cells into the extracellular space. While the roles of EVs in mammalian cells are widely reported (van Niel et al. 2018), functional studies on EVs in plants is still very limited. In plants, EVs of variable sizes were first observed in suspension carrot cells by EM in the 1960s (Halperin and Jensen 1967). Later, exosome-like nanoparticles were isolated from grape juice by differential centrifugation and were further confirmed by Cryo-EM (Ju et al. 2013). EVs were also detected in other plant species, including the extracellular fluids of sunflower seeds (Regente et al. 2017), coconut milk (Zhao et al. 2018), and the germinating medium of olive pollen grains (Prado et al. 2014).

Based on our current knowledge, multiple pathways for EV biogenesis have been proposed in plants (Fig. 7). One is from MVB-derived exosomes. Exosome-like structures were first reported in *Arabidopsis* leaf tissue after infection by barley powdery mildew fungus (An et al. 2006). These structures were embedded within the so-called paramural bodies (PMBs), which resembled MVBs with multiple intraluminal vesicles (An et al. 2006). However, the relationship between PMB and MVB is not clear. A recent study revealed the potential role of exosomes in cross-kingdom communications (Cai et al. 2018). Host cells in *Arabidopsis* secreted exosome-like EVs that were taken up by the fungal pathogen *Botrytis cinerea*. These vesicles colocalized with the MVB marker ARA6 and were labeled by TET8 (Tetraspanin-8), a structurally related protein of CD63. Because CD63 is a well-established marker of exosomes in mammalian cells, the TET8-labled EVs were also considered to be exosomes (Cai et al. 2018). Another work detected the fusion of MVBs with PM and the release of their ILVs to the extracellular space under TEM after virus infection in *Nicotiana benthamiana* leaves (Movahed et al. 2019). Using FIB 3D ET, EVs were clearly observed in the cell wall (Movahed et al. 2019). Therefore, MVB-derived exosomes are actively involved in the plant response to pathogens.

The other pathway for EV biogenesis is through the plant-specific compartment exocyst-positive organelle (EXPO) (Wang et al. 2010). EXPOs are labeled by Exo70E2 but do not colocalize with the conventional organelle markers. Unlike MVB-derived exosomes, EXPOs exist without pathogen infection in special types of plant tissues, including the root tips, root hair cells, and pollen grains. Immunogold-TEM reveal that EXPOs are double-membrane structures, which fuse with the PM to release their internal vesicles to the extracellular space (Wang et al. 2010). However, the physiological roles of EXPO-derived EVs remained to be further explored.

The lytic vacuole may also contribute to EVs formation in plants. Viral infection can trigger vacuole collapse (Hatsugai et al. 2004), and bacterial infection can induce vacuole fusion with the PM (Hatsugai et al. 2009). Under both conditions, the internal vesicles within the vacuole can be released to generate the EVs. Whether secretory autophagosomes and direct budding from the PM also regulate EV production remains to be explored.

The most studied function of EVs in plants is their roles in biotic responses. During arbuscular mycorrhizal symbioses, EVs and tubular structures are abundant in the interstitial apoplastic matrix, possibly facilitating the movement of molecules between the symbiotic partners (Ivanov et al. 2019; Roth et al. 2019). EVs are also prevalent in plant extracellular fluids without biotic interactions, but their roles in developmental regulation remain largely unknown. A recent study showed that EVs accumulate specifically in suberizing cells during root suberization, most likely playing a role in delivering monomeric suberin precursors to the extracellular space in order to polymerize into suberin lamellae (De Bellis et al. 2022). In animal cells, EVs are important carriers of signaling molecules for intercellular communications (van Niel et al. 2018; Sanchez et al. 2021; Lee and Kim 2022). EVs have been reported in different plant species, and the identification of small RNAs in these EVs (Cai et al. 2018; Baldrich et al. 2019) make them attractive candidates as signal carriers for many physiological processes.

Organelle membrane contact sites in the endomembrane system: so many unknowns

Organelle interactions have long been recognized in plant cells under the microscope (Staehelin 1997). The ER tubules were found in close association with actin bundles, as well as the PM and vacuole membrane (Staehelin 1997). The term of membrane contact sites (MCS) was first defined in animal cells as regions between two organelles in close proximity (~10 to 80 nm) for non-vesicular communication (e.g. exchange of signals and metabolites), but without the occurrence of membrane fusion (Scorrano et al. 2019; Prinz et al. 2020). Although mitochondria and chloroplasts are not considered parts of the endomembrane system, the formation of MCSs provides another platform for linking them with the endomembrane system. It should be noted that homotypic interactions between organelles followed by membrane fusion are not considered a MCS (e.g. mitochondria–mitochondria interaction) (Scorrano et al. 2019; Prinz et al. 2020) and will not be discussed here. With the development of genetic and advanced imaging tools, our understanding of the functions and types of MCSs in plant cells is just beginning (Fig. 8). Nowadays, MCSs have been implicated to function as critical hubs in the regulation of cellular metabolism for plant development, immunity, and stress response (Lee et al. 2019; Lee et al. 2020; Ruiz-Lopez et al. 2021; Li et al. 2022; Ye et al. 2022; Zhang et al. 2023). Pioneer work using optical tweezers provide the direct evidence for ER-chloroplast MCSs (Andersson et al. 2007). The development of GFP markers combined with optical tweezers reveals tight physical contacts between the ER and the Golgi (Sparkes et al. 2009) (Fig. 8B). Later, the application of photo-convertible fluorescence has enabled the observation of dy-namic communications between different organelles, such as mitochondria and chloroplast (Schattat et al. 2012; Mathur 2021) (Fig. 8C). By dissecting the dynamic behavior, the molecular identity/composition, unexpected novel functions of MCSs have been discovered for inter-organellar communication and organelle biogenesis, including peroxisomes and chloroplasts as well as lipid droplet (LD)-PM tethering (Oikawa et al. 2015; Krawczyk et al. 2022).

MCS between the ER and PM represents the best documented MCS within the plant endomembrane system by far. EPCs have been reported to function in signaling transduction, endocytosis, autophagy, and ER network maintenance, while several regulators in mediating the EPCs have recently been uncovered (Saravanan et al. 2009; Wang et al. 2014; Siao et al. 2016; Lee et al. 2019, 2020; Ye et al. 2022). OSBP, an oxysterol-binding protein, and OSBP-related proteins (ORPs) have been established as key regulators in mediating sterol trafficking from the ER to the PM (Subra et al. 2023). Twelve homologs of ORPs have been found in *Arabidopsis* plant genomes, but their specific roles are unclear (Skirpan et al. 2006). In yeast and mammalian cells, syntaxin-like vesicle-associated membrane protein (VAMP)-associated proteins (VAPs) have been implicated in interactions with OSBP and certain ORPs as tethers be-tween ER and other organelles. Plant VAP homologs (PVAs) are also found, which are grouped in the VAP33 sub-family of SNAREs (Sutter et al. 2006). By analysis of plant VAP and ORP homologs, PVA12, later referred to as VAP27-3, forms a PVA12-ORP3a tether which was first reported to par-ticipate in ER and Golgi sterol trafficking (Saravanan et al. 2009). Moreover, plant-specific actin-binding proteins (the NETWORKED family) NET3C and the clathrin-related protein Pan1/AtEH1 both interact with VAP27 and contribute to autophagosome formation (Wang et al. 2014; Wang et al. 2019b). These results suggest that VAP27 acts as a mo-lecular hub on the ER subdomains to provide a platform to drive autophagosome formation. In *Arabidopsis*, synaptotag-min 1 (SYT1)-enriched EPCs (S-EPCs) represent another type of ER-PM contacts, and evidence is accumulating to support its role in modulating the plasmodesmata during virus infec-tion (Siao et al. 2016). Recently, using an artificial tether named MAPPER, SYT1-dependent EPCs have been demonstrated to mediate ER-PM and PM-LD connectivity in response to abiotic stress in *Arabidopsis* (Lee et al. 2019, 2020; Ruiz-Lopez et al. 2021) (Fig. 8D). Fluorescent protein complementation methods, which are commonly used for protein interaction analysis, have also been applied to visualize ER-related MCSs in plant cells (Cui et al. 2019a; Li et al. 2020).

A new field of MCS-mediated inter-organelle communication has emerged, and many outstanding questions await to be explored. ER-mitochondria MCSs represent another emerging key platform in controlling organelle dynamics and biogenesis in plant cells. However, the molecular compositions for ER-mitochondria MCSs remain elusive. Recently, TraB family proteins, which localize to the outer mitochondrial membrane, have been identified as a tether component of ER-mitochondria MCSs (Li et al. 2022). TraB interacts with the ER-localized VAP27 and ATG8, whereas mitophagy activity is suppressed in the absent of TraB (Li et al. 2022). Accordingly, it was proposed that TRB1-VAP27-dependent

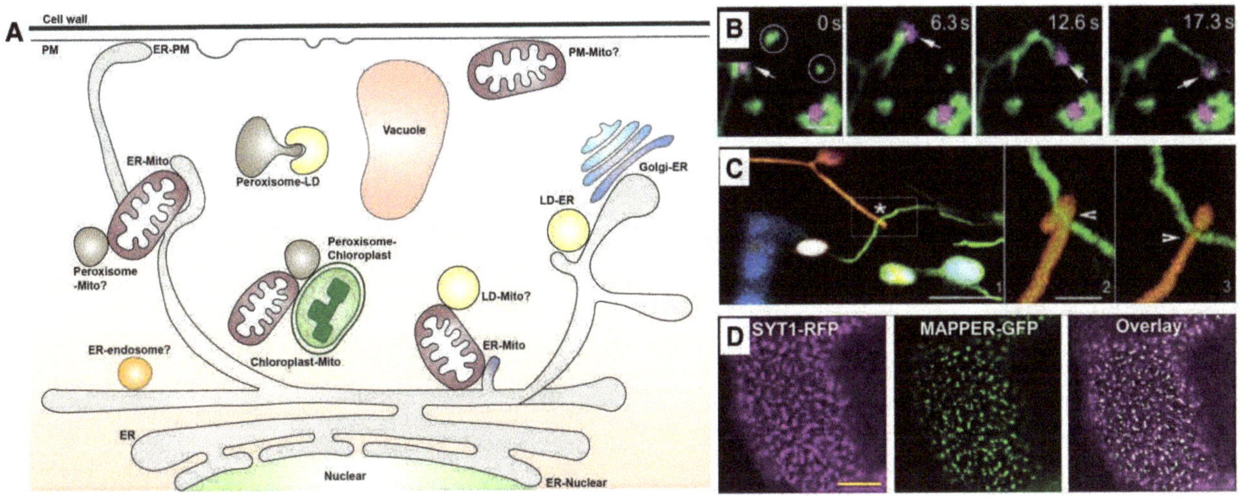

Figure 8. Organelle membrane contacts in plant cells. **A)** Various membrane contact sites have been reported in plants, including contacts for ER–PM, ER–mitochondrial, ER–Golgi, ER–chloroplast, ER-LD, peroxisome-chloroplast, chloroplast and mitochondria, peroxisome, and LD. **B)** Optical trapping of a Golgi body (arrow) showing production of a new 3-way junction by pulling the Golgi body to the anchor points (white circles). Bar, 2 μm (reprinted from Sparkes et al. 2009, Fig. 3F to I). **C)** Photoconvertible probe using mEosFP revealing interaction between a photoconverted stromule and a nonphotoconverted stromule without membrane fusion. Bars, 10 μm in (C1), 2.5 μm in (C2) and (C3) (reprinted from Schattat et al. 2012, Fig. 2F). **D)** A plant artificial marker MAPPER labels the ER–PM membrane contacts. Bar, 20 μm (reprinted from Lee et al. 2019, Fig. 1F).

ER-mitochondria MCSs might contribute to mitophagy by aiding autophagosome formation. Although many types of MCSs for communication with the plant endomembrane system have been observed, their molecular compositions remain to be identified and characterized. Future efforts in developing standard/novel experimental approaches to define and measure the compositions of MCSs are certainly needed.

Conclusions and perspectives

Understanding of the biogenesis and function of plant organelles in the endomembrane system is fundamental to plant cell biology research. The morphologies, functions, and compositions of these organelles are versatile in different cell types, development stages, and environmental stimuli. They are dynamic and exchangeable to facilitate an acute response for plant adaptation and survival. With the development of advancing imaging and omics tools, the spatial and temporal landscape of plant organelles can now be revealed at high resolution, especially for the transitional elements (e.g. transport vesicles, membrane contacts). For instance, the recently developed SRM, cryo-EM, whole-cell ET, 4D SCLIM, and STED have been successfully applied to plant cells to uncover novel intermediates in multiple pathways, including ER-Golgi, Golgi-TGN, and MBV-vacuole transport (Bykov et al. 2017; Cui et al. 2019b; Shimizu et al. 2021; Li et al. 2021a; Fougère et al. 2023). Lipidomic and proteomic analysis of purified organelles/vesicles have also shed new light into the biogenesis and function of plant transport vesicles, such as COPII and EVs (Rutter and Innes 2017; Li et al. 2021a). Nevertheless, new tools and reporters for plant cell research still need to be developed for future systematic investigations into plant organelle architecture and dynamic behavior. Quantitative analysis of the architecture and dynamics of plant organelles by computational modeling is also necessary (Pain et al. 2019; Zhang et al. 2021; Li et al. 2021b). Except for the well-established membrane-shaping domains, biomolecular condensates on the membrane surface provides an efficient yet structure-independent mechanism for inducing membrane curvature, which is just beginning to draw attention in the plant field (Dragwidge and Van Damme 2023; Li and Pang 2023). Nowadays, single-cell technologies have also been greatly revolutionizing research in the plant endomembrane system. To gain further insights into different organelles in various cell types, future efforts are required to integrate the spatial and temporal information of the organelles at the single-cell level for the establishment of the "plant cell atlas" (Rhee et al. 2019).

Looking back, what have we learned from past research? Plant cell biologists are often critical and skeptical about new findings not being in line with conventional views. Our research experience shows that plant cells also have unique regulators, pathways, and organelles that are distinct from the animal model, making the plant cell research both interesting and rewarding.

Looking ahead, through the gradual development and application of advanced technologies in plant cell biology, we may be able to address some of the outstanding questions in organelle biology in plants.

1) Interconversion of tubules and cisternae creates the dynamic ER network. Do different ER membrane shapers coordinate in a sequential order or function independently for tubules or cisternae transformation?

2) Cisternal growth and fission has been observed for Golgi regeneration during mitosis. How are the size and fission of the Golgi cisternae controlled?
3) Are there any TGN subdomains for endocytic recycling as plant cells do not have recycling endosomes?
4) MVBs with typical ILVs are morphologically indistinguishable but with different destinations, like the vacuole and PM. What are the determinants of their fates for fusion with the vacuole or PM?
5) How does membrane remodeling coordinate with vesicle sorting and organelle biogenesis? What are the possible effects of biomolecule condensates on membrane remodeling and membrane geometry?
6) What triggers the contents of the EV to be released? How do EVs cross the cell wall for fusion with other cells? Do EVs in plant cells undergo long-distance transport in vascular tissues?
7) Exactly how cargo sequestration is coupled to the occurrence of phagophores during autophagy remains unclear in plant cells. How do autophagosomes integrate into the endocytic-vacuolar system for cargo delivery and degradation in the vacuole?
8) Diverse modes of MCSs have been observed. How do we characterize and distinguish the physiological functions of novel MCSs in different plant cell types?

These and many other questions will keep organelle research in the plant endomembrane system challenging in the next decades. Innovations of super-resolution and Cryo-ET are becoming popular tools in cell biology, but they still have limited application when using plant tissues as experimental material. Undoubtedly, a comprehensive understanding of the landscape for the plant organelles will provide tremendous opportunities for future biotechnological applications.

Acknowledgments

We apologize to researchers whose work has not been included in this review owing to space limitations. This work was supported by grants from the Research Grants Council of Hong Kong (AoE/M-05/12, CUHK14101219, 14106823, C4033-19E, C4002-20W, C4002-21EF, C2009-19G, C2003-22WF, R4005-18, and Senior Research Fellow Scheme SRFS2122-4S01), the Chinese University of Hong Kong (CUHK) Research Committee, and CAS-Croucher Funding Scheme for Joint Laboratories to L.J; the National Natural Science Foundation of China (32222087), the Research Grants Council of Hong Kong (N_CUHK405/20, 24108820, 14106622, 14110923), and the Chinese University of Hong Kong (CUHK) Research Committee to X.Z; and the National Natural Science Foundation of China (32070193, 32270295), Guangdong Innovation Research Team Fund (2016ZT06S172), and the Shenzhen Sci-Tech Fund (JCYJ20210324105004011 and 20220815103813001) to R.L.

Conflict of interest statement. The authors declare no competing interests.

Data availability

No new data were generated or analysed in support of this research.

References

Ahmed SU, Bar-Peled M, Raikhel NV. Cloning and subcellular location of an Arabidopsis receptor-like protein that shares common features with protein-sorting receptors of eukaryotic cells. Plant Physiol. 1997:**114**(1):325–336. https://doi.org/10.1104/pp.114.1.325

An Q, Huckelhoven R, Kogel KH, van Bel AJ. Multivesicular bodies participate in a cell wall-associated defence response in barley leaves attacked by the pathogenic powdery mildew fungus. Cell Microbiol. 2006:**8**(6):1009–1019. https://doi.org/10.1111/j.1462-5822.2006.00683.x

Andersson MX, Goksor M, Sandelius AS. Optical manipulation reveals strong attracting forces at membrane contact sites between endoplasmic reticulum and chloroplasts. J Biol Chem. 2007:**282**(2): 1170–1174. https://doi.org/10.1074/jbc.M608124200

Andrews FM. Die Wirkung der Zentrifugalkraft auf Pflanzen. Jahrb Wiss Bot. 1903:**38**:1–40.

Aniento F, Sanchez de Medina Hernandez V, Dagdas Y, Rojas-Pierce M, Russinova E. Molecular mechanisms of endomembrane trafficking in plants. Plant Cell. 2022:**34**(1):146–173. https://doi.org/10.1093/plcell/koab235

Baldrich P, Rutter BD, Karimi HZ, Podicheti R, Meyers BC, Innes RW. Plant extracellular vesicles contain diverse small RNA species and are enriched in 10-to 17-nucleotide "tiny" RNAs. Plant Cell. 2019:**31**(2): 315–324. https://doi.org/10.1105/tpc.18.00872

Bednarek SY, Raikhel NV. The barley lectin carboxyl-terminal propeptide is a vacuolar protein sorting determinant in plants. Plant Cell. 1991:**3**(11):1195–1206. https://doi.org/10.1105/tpc.3.11.1195

Bell K, Mitchell S, Paultre D, Posch M, Oparka K. Correlative imaging of fluorescent proteins in resin-embedded plant material. Plant Physiol. 2013:**161**(4):1595–1603. https://doi.org/10.1104/pp.112.212365

Berson T, von Wangenheim D, Takac T, Samajova O, Rosero A, Ovecka M, Komis G, Stelzer EHK, Samaj J. Trans-Golgi network localized small GTPase RabA1d is involved in cell plate formation and oscillatory root hair growth. BMC Plant Biol. 2014:**14**(1):252. https://doi.org/10.1186/s12870-014-0252-0

Bethke PC, Jones RL. Vacuoles and prevacuolar compartments. Curr Opin Plant Biol. 2000:**3**:469–475. https://doi.org/10.1016/S1369-5266(00)00115-1

Boevink P, Oparka K, Santa Cruz S, Martin B, Betteridge A, Hawes C. Stacks on tracks: the plant Golgi apparatus traffics on an actin/ER network. Plant J. 1998:**15**(3):441–447. https://doi.org/10.1046/j.1365-313X.1998.00208.x

Boevink P, SantaCruz S, Hawes C, Harris N, Oparka KJ. Virus-mediated delivery of the green fluorescent protein to the endoplasmic reticulum of plant cells. Plant J. 1996:**10**(5):935–941. https://doi.org/10.1046/j.1365-313X.1996.10050935.x

Brandizzi F. Maintaining the structural and functional homeostasis of the plant endoplasmic reticulum. Dev Cell. 2021:**56**(7):919–932. https://doi.org/10.1016/j.devcel.2021.02.008

Brandizzi F, Irons S, Kearns A, Hawes C. BY-2 cells: culture and transformation for live cell imaging. Curr Protoc Cell Biol. 2003: **Chapter 1**: Unit 1.7. https://doi.org/10.1002/0471143030.cb0107s19

Brown SC, Bolte S, Gaudin M, Pereira C, Marion J, Soler MN, Satiat-Jeunemaitre B. Exploring plant endomembrane dynamics using the photoconvertible protein Kaede. Plant J. 2010:**63**(4): 696–711. https://doi.org/10.1111/j.1365-313X.2010.04272.x

Buono RA, Leier A, Paez-Valencia J, Pennington J, Goodman K, Miller N, Ahlquist P, Marquez-Lago TT, Otegui MS. ESCRT-mediated vesicle concatenation in plant endosomes. J Cell Biol. 2017:**216**(7):2167–2177. https://doi.org/10.1083/jcb.201612040

Buvat R. Relations entre l'ergastoplasme et l'appareil vacuolaire. C R Acad Sci. 1957:**245**:350–352.

Buvat R. Researches sur les infra-structures du cytoplasme dans les cellules du meristeme apical, des ebauches, foliares et des feuilles develpopees d'elodea canadensis. Ann Sci Nat Bot Biol. 1958:**19**:121–161.

Bykov YS, Schaffer M, Dodonova SO, Albert S, Plitzko JM, Baumeister W, Engel BD, Briggs JA. The structure of the COPI coat determined within the cell. Elife. 2017:**6**:e32493. https://doi.org/10.7554/eLife.32493

Cai Q, Qiao L, Wang M, He B, Lin FM, Palmquist J, Huang SD, Jin H. Plants send small RNAs in extracellular vesicles to fungal pathogen to silence virulence genes. Science. 2018:**360**(6393):1126–1129. https://doi.org/10.1126/science.aar4142

Carter C, Pan S, Zouhar J, Avila EL, Girke T, Raikhel NV. The vegetative vacuole proteome of *Arabidopsis thaliana* reveals predicted and unexpected proteins. Plant Cell. 2004:**16**(12):3285–3303. https://doi.org/10.1105/tpc.104.027078

Chambaud C, Cookson SJ, Ollat N, Bayer E, Brocard L. A correlative light electron microscopy approach reveals plasmodesmata ultrastructure at the graft interface. Plant Physiol. 2022:**188**(1):44–55. https://doi.org/10.1093/plphys/kiab485

Choi SB, Wang C, Muench DG, Ozawa K, Franceschi VR, Wu Y, Okita TW. Messenger RNA targeting of rice seed storage proteins to specific ER subdomains. Nature. 2000:**407**(6805):765–767. https://doi.org/10.1038/35037633

Chrispeels MJ. The Golgi-apparatus mediates the transport of phytohemagglutinin to the protein bodies in bean cotyledons. Planta. 1983:**158**(2):140–151. https://doi.org/10.1007/BF00397707

Chrispeels MJ, Raikhel NV. Short peptide domains target proteins to plant vacuoles. Cell. 1992:**68**(4):613–616. https://doi.org/10.1016/0092-8674(92)90134-X

Chrispeels MJ, Vatter AE. Preservation of the ultrastructure of plant cells using glutaric acid dialdehyde as a fixative. Nature. 1963:**200**(4907):711–712. https://doi.org/10.1038/200711a0

Contento AL, Xiong Y, Bassham DC. Visualization of autophagy in Arabidopsis using the fluorescent dye monodansylcadaverine and a GFP-AtATG8e fusion protein. Plant J. 2005:**42**(4):598–608. https://doi.org/10.1111/j.1365-313X.2005.02396.x

Craig S, Gilkey JC, Staehelin LA. Improved specimen support cups and auxiliary devices for the Balzers high pressure freezing apparatus. J Microsc. 1987:**148**(1):103–106. https://doi.org/10.1111/j.1365-2818.1987.tb02857.x

Cui Y, Cao WH, He YL, Zhao Q, Wakazaki M, Zhuang XH, Gao JY, Zeng YL, Gao CJ, Ding Y, et al. A whole-cell electron tomography model of vacuole biogenesis in Arabidopsis root cells. Nat Plants. 2019b:**5**(1):95–105. https://doi.org/10.1038/s41477-018-0328-1

Cui Y, Zhang X, Yu M, Zhu Y, Xing J, Lin J. Techniques for detecting protein-protein interactions in living cells: principles, limitations, and recent progress. Sci China Life Sci. 2019a:**62**(5):619–632. https://doi.org/10.1007/s11427-018-9500-7

Dahhan DA, Reynolds GD, Cardenas JJ, Eeckhout D, Johnson A, Yperman K, Kaufmann WA, Vang N, Yan X, Hwang I, et al. Proteomic characterization of isolated Arabidopsis clathrin-coated vesicles reveals evolutionarily conserved and plant-specific components. Plant Cell. 2022:**34**(6):2150–2173. https://doi.org/10.1093/plcell/koac071

De Bellis D, Kalmbach L, Marhavy P, Daraspe J, Geldner N, Barberon M. Extracellular vesiculo-tubular structures associated with suberin deposition in plant cell walls. Nat Commun. 2022:**13**(1):1489. https://doi.org/10.1038/s41467-022-29110-0

Dettmer J, Hong-Hermesdorf A, Stierhof YD, Schumacher K. Vacuolar H+-ATPase activity is required for endocytic and secretory trafficking in Arabidopsis. Plant Cell. 2006:**18**(3):715–730. https://doi.org/10.1105/tpc.105.037978

De Vries H. Plasmolytische Studien über die Wand der Vakuolen. Jarb Wiss Bot. 1885:**16**:465–598.

Ding XX, Zhang XG, Paez-Valencia J, McLoughlin F, Reyes FC, Morohashi K, Grotewold E, Vierstra RD, Otegui MS. Microautophagy mediates vacuolar delivery of storage proteins in maize aleurone cells. Front Plant Sci. 2022:**13**:833612. https://doi.org/10.3389/fpls.2022.833612

Doelling JH, Walker JM, Friedman EM, Thompson AR, Vierstra RD. The APG8/12-activating enzyme APG7 is required for proper nutrient recycling and senescence in *Arabidopsis thaliana*. J Biol Chem. 2002:**277**(36):33105–33114. https://doi.org/10.1074/jbc.M204630200

Donohoe BS, Kang BH, Staehelin LA. Identification and characterization of COPIa- and COPIb-type vesicle classes associated with plant and algal Golgi. Proc Natl Acad Sci U S A. 2007:**104**(1):163–168. https://doi.org/10.1073/pnas.0609818104

Dragwidge JM, Van Damme D. Protein phase separation in plant membrane biology: more than just a compartmentalization strategy. Plant Cell. 2023:**35**(9):3162–3172. https://doi.org/10.1093/plcell/koad177

Dragwidge JM, Wang Y, Brocard L, Meyer AD, Hudeček R, Eeckhout D, Chambaud C, Pejchar P, Potocký M, Vandorpe M, et al. AtEH/Pan1 proteins drive phase separation of the TPLATE complex and clathrin polymerisation during plant endocytosis. bioRxiv 484738. https://doi.org/10.1101/2022.03.17.484738, 17 March 2022, preprint: not peer reviewed.

Drakakaki G, van de Ven W, Pan S, Miao Y, Wang J, Keinath NF, Weatherly B, Jiang L, Schumacher K, Hicks G, et al. Isolation and proteomic analysis of the SYP61 compartment reveal its role in exocytic trafficking in Arabidopsis. Cell Res. 2012:**22**(2):413–424. https://doi.org/10.1038/cr.2011.129

Dujardin F. Histoire naturelle des zoophytes: infusoires. Paris: Librairie Encyclopédique de Roret; 1841.

Emans N, Zimmermann S, Fischer R. Uptake of a fluorescent marker in plant cells is sensitive to brefeldin A and wortmannin. Plant Cell. 2002:**14**(1):71–86. https://doi.org/10.1105/tpc.010339

Engel BD, Schaffer M, Albert S, Asano S, Plitzko JM, Baumeister W. In situ structural analysis of Golgi intracisternal protein arrays. Proc Natl Acad Sci U S A. 2015:**112**(36):11264–11269. https://doi.org/10.1073/pnas.1515337112

Feeney M, Frigerio L, Cui YH, Menassa R. Following vegetative to embryonic cellular changes in leaves of Arabidopsis overexpressing LEAFY COTYLEDON2. Plant Physiol. 2013:**162**(4):1881–1896. https://doi.org/10.1104/pp.113.220996

Feeney M, Kittelmann M, Menassa R, Hawes C, Frigerio L. Protein storage vacuoles originate from remodeled preexisting vacuoles in *Arabidopsis thaliana*. Plant Physiol. 2018:**177**(1):241–254. https://doi.org/10.1104/pp.18.00010

Feraru E, Paciorek T, Feraru MI, Zwiewka M, De Groodt R, De Rycke R, Kleine-Vehn J, Friml J. The AP-3 beta adaptin mediates the biogenesis and function of lytic vacuoles in Arabidopsis. Plant Cell. 2010:**22**(8):2812–2824. https://doi.org/10.1105/tpc.110.075424

Fitchette AC, Cabanes-Macheteau M, Marvin L, Martin B, Satiat-Jeunemaitre B, Gomord V, Crooks K, Lerouge P, Faye L, Hawes C. Biosynthesis and immunolocalization of Lewis a-containing N-glycans in the plant cell. Plant Physiol. 1999:**121**(2):333–344. https://doi.org/10.1104/pp.121.2.333

Fougère L, Grison M, Laquel P, Montrazi M, Cordelières F, Fernández-Monreal M, Poujol C, Uemura T, Nakano A, Ito Y, et al. ER-to-Golgi trafficking via a dynamic intermediate cis-Golgi tubular network in Arabidopsis. bioRxiv 563925. https://doi.org/10.1101/2023.10.27.563925, 27 October 2023, preprint: not peer reviewed.

Fowke LC, Bech-Hansen CW, Gamborg OL. Electron microscopic observations of cell regeneration from cultured protoplasts of Ammi visnaga. Protoplasma. 1974:**79**(1–2):235–248. https://doi.org/10.1007/BF02055792

Gadeyne A, Sanchez-Rodriguez C, Vanneste S, Di Rubbo S, Zauber H, Vanneste K, Van Leene J, De Winne N, Eeckhout D, Persiau G, et al. The TPLATE adaptor complex drives clathrin-mediated endocytosis in plants. Cell. 2014:**156**(4):691–704. https://doi.org/10.1016/j.cell.2014.01.039

Gao C, Zhuang X, Cui Y, Fu X, He Y, Zhao Q, Zeng Y, Shen J, Luo M, Jiang L. Dual roles of an Arabidopsis ESCRT component FREE1 in regulating vacuolar protein transport and autophagic degradation. Proc Natl Acad Sci U S A. 2015:**112**(6):1886–1891. https://doi.org/10.1073/pnas.1421271112

Geldner N, Denervaud-Tendon V, Hyman DL, Mayer U, Stierhof YD, Chory J. Rapid, combinatorial analysis of membrane compartments in intact plants with a multicolor marker set. Plant J. 2009:**59**(1):169–178. https://doi.org/10.1111/j.1365-313X.2009.03851.x

Gomez RE, Chambaud C, Lupette J, Castets J, Pascal S, Brocard L, Noack L, Jaillais Y, Joubes J, Bernard A. Phosphatidylinositol-4-phosphate controls autophagosome formation in *Arabidopsis thaliana*. Nat Commun. 2022:**13**(1):4385. https://doi.org/10.1038/s41467-022-32109-2

Guilliermond A. The recent development of our idea of the vacuome of plant cells. Am J Bot. 1929:**16**(1):1–22. https://doi.org/10.1002/j.1537-2197.1929.tb09459.x

Guo J, Wang G, Xie L, Wang X, Feng L, Guo W, Tao X, Humbel BM, Zhang Z, Hong J. Three-dimensional analysis of membrane structures associated with tomato spotted wilt virus infection. Plant Cell Environ. 2023:**46**(2):650–664. https://doi.org/10.1111/pce.14511

Haberlandt G. Culturversuche mit isolierten Pflanzenzellen. Sitz-Ber Mat Nat Kl Kais Akad Wiss Wien. 1902:**111**:69–92.

Halperin W, Jensen WA. Ultrastructural changes during growth and embryogenesis in carrot cell cultures. J Ultrastruct Res. 1967:**18**(3–4):428–443. https://doi.org/10.1016/S0022-5320(67)80128-X

Hanaoka H, Noda T, Shirano Y, Kato T, Hayashi H, Shibata D, Tabata S, Ohsumi Y. Leaf senescence and starvation-induced chlorosis are accelerated by the disruption of an Arabidopsis autophagy gene. Plant Physiol. 2002:**129**(3):1181–1193. https://doi.org/10.1104/pp.011024

Hara-Nishimura I, Nishimura M, Matsubara H, Akazawa T. Suborganellar localization of proteinase catalyzing the limited hydrolysis of pumpkin globulin. Plant Physiol. 1982:**70**(3):699–703. https://doi.org/10.1104/pp.70.3.699

Hara-Nishimura I, Shimada T, Hatano K, Takeuchi Y, Nishimura M. Transport of storage proteins to protein storage vacuoles is mediated by large precursor-accumulating vesicles. Plant Cell. 1998:**10**(5):825–836. https://doi.org/10.1105/tpc.10.5.825

Harris N, Chrispeels MJ. The endoplasmic reticulum of mung-bean cotyledons: quantitative morphology of cisternal and tubular ER during seedling growth. Planta. 1980:**148**(3):293–303. https://doi.org/10.1007/BF00380041

Hatsugai N, Iwasaki S, Tamura K, Kondo M, Fuji K, Ogasawara K, Nishimura M, Hara-Nishimura I. A novel membrane fusion-mediated plant immunity against bacterial pathogens. Gene Dev. 2009:**23**(21):2496–2506. https://doi.org/10.1101/gad.1825209

Hatsugai N, Kuroyanagi M, Yamada K, Meshi T, Tsuda S, Kondo M, Nishimura M, Hara-Nishimura I. A plant vacuolar protease, VPE, mediates virus-induced hypersensitive cell death. Science. 2004:**305**(5685):855–858. https://doi.org/10.1126/science.1099859

Hawes C. Cell biology of the plant Golgi apparatus. New Phytol. 2005:**165**(1):29–44. https://doi.org/10.1111/j.1469-8137.2004.01218.x

Haywood V, Kragler F, Lucas WJ. Plasmodesmata: pathways for protein and ribonucleoprotein signaling. Plant Cell. 2002:**14**(suppl 1):S303–S325. https://doi.org/10.1105/tpc.000778

Hillmer S, Freundt H, Robinson DG. The partially coated reticulum and its relationship to the Golgi-apparatus in higher-plant cells. Eur J Cell Biol. 1988:**47**:206–212.

Hillmer S, Movafeghi A, Robinson DG, Hinz G. Vacuolar storage proteins are sorted in the cis-cisternae of the pea cotyledon Golgi apparatus. J Cell Biol. 2001:**152**(1):41–50. https://doi.org/10.1083/jcb.152.1.41

Hofte H, Hubbard L, Reizer J, Ludevid D, Herman EM, Chrispeels MJ. Vegetative and seed-specific forms of tonoplast intrinsic protein in the vacuolar membrane of Arabidopsis-thaliana. Plant Physiol. 1992:**99**(2):561–570. https://doi.org/10.1104/pp.99.2.561

Hohl I, Robinson DG, Chrispeels MJ, Hinz G. Transport of storage proteins to the vacuole is mediated by vesicles without a clathrin coat. J Cell Sci. 1996:**109**(10):2539–2550. https://doi.org/10.1242/jcs.109.10.2539

Holwerda BC, Galvin NJ, Baranski TJ, Rogers JC. In vitro processing of aleurain, a barley vacuolar thiol protease. Plant Cell. 1990:**2**(11):1091–1106. https://doi.org/10.2307/3869262

Huang AH, Beevers H. Isolation of microbodies from plant tissues. Plant Physiol. 1971:**48**(5):637–641. https://doi.org/10.1104/pp.48.5.637

Ito Y, Uemura T, Shoda K, Fujimoto M, Ueda T, Nakano A. cis-Golgi proteins accumulate near the ER exit sites and act as the scaffold for Golgi regeneration after brefeldin A treatment in tobacco BY-2 cells. Mol Biol Cell. 2012:**23**(16):3203–3214. https://doi.org/10.1091/mbc.e12-01-0034

Ivanov S, Austin J, Berg RH, Harrison MJ. Extensive membrane systems at the host-arbuscular mycorrhizal fungus interface. Nat Plants. 2019:**5**(2):194. -+. https://doi.org/10.1038/s41477-019-0364-5

Jiang D, He Y, Zhou X, Cao Z, Pang L, Zhong S, Jiang L, Li R. Arabidopsis HOPS subunit VPS41 carries out plant-specific roles in vacuolar transport and vegetative growth. Plant Physiol. 2022:**189**(3):1416–1434. https://doi.org/10.1093/plphys/kiac167

Jiang L, Rogers JC. Integral membrane protein sorting to vacuoles in plant cells: evidence for two pathways. J Cell Biol. 1998:**143**(5):1183–1199. https://doi.org/10.1083/jcb.143.5.1183

Johnson A, Dahhan DA, Gnyliukh N, Kaufmann WA, Zheden V, Costanzo T, Mahou P, Hrtyan M, Wang J, Aguilera-Servin J, et al. The TPLATE complex mediates membrane bending during plant clathrin-mediated endocytosis. Proc Natl Acad Sci U S A. 2021:**118**(51):e2113046118. https://doi.org/10.1073/pnas.2113046118

Johnson A, Kaufmann WA, Sommer C, Costanzo T, Dahhan DA, Bednarek SY, Friml J. Three-dimensional visualization of planta clathrin-coated vesicles at ultrastructural resolution. Mol Plant. 2022:**15**(10):1533–1542. https://doi.org/10.1016/j.molp.2022.09.003

Ju S, Mu J, Dokland T, Zhuang X, Wang Q, Jiang H, Xiang X, Deng ZB, Wang B, Zhang L, et al. Grape exosome-like nanoparticles induce intestinal stem cells and protect mice from DSS-induced colitis. Mol Ther. 2013:**21**(7):1345–1357. https://doi.org/10.1038/mt.2013.64

Kang BH, Nielsen E, Preuss ML, Mastronarde D, Staehelin LA. Electron tomography of RabA4b- and PI-4Kbeta1-labeled trans Golgi network compartments in Arabidopsis. Traffic. 2011:**12**(3):313–329. https://doi.org/10.1111/j.1600-0854.2010.01146.x

Kievits AJ, Lane R, Carroll EC, Hoogenboom JP. How innovations in methodology offer new prospects for volume electron microscopy. J Microsc. 2022:**287**(3):114–137. https://doi.org/10.1111/jmi.13134

Kim JH, Lee HN, Huang X, Jung H, Otegui MS, Li F, Chung T. FYVE2, a phosphatidylinositol 3-phosphate effector, interacts with the COPII machinery to control autophagosome formation in Arabidopsis. Plant Cell. 2022:**34**(1):351–373. https://doi.org/10.1093/plcell/koab263

Kirsch T, Paris N, Butler JM, Beevers L, Rogers JC. Purification and initial characterization of a potential plant vacuolar targeting receptor. Proc Natl Acad Sci U S A. 1994:**91**(8):3403–3407. https://doi.org/10.1073/pnas.91.8.3403

Knebel W, Quader H, Schnepf E. Mobile and immobile endoplasmic reticulum in onion bulb epidermis cells: short- and long-term observations with a confocal laser scanning microscope. Eur J Cell Biol. 1990:**52**:328–340.

Kolb C, Nagel MK, Kalinowska K, Hagmann J, Ichikawa M, Anzenberger F, Alkofer A, Sato MH, Braun P, Isono E. FYVE1 is essential for vacuole biogenesis and intracellular trafficking in Arabidopsis. Plant Physiol. 2015:**167**(4):1361–1373. https://doi.org/10.1104/pp.114.253377

Konopka CA, Backues SK, Bednarek SY. Dynamics of Arabidopsis dynamin-related protein 1C and a clathrin light chain at the plasma membrane. Plant Cell. 2008:**20**(5):1363–1380. https://doi.org/10.1105/tpc.108.059428

Krawczyk HE, Sun S, Doner NM, Yan Q, Lim MSS, Scholz P, Niemeyer PW, Schmitt K, Valerius O, Pleskot R, et al. Seed lipid droplet protein1, seed lipid droplet protein2, and lipid droplet plasma membrane adaptor mediate lipid droplet-plasma membrane tethering. Plant Cell. 2022:**34**(6):2424–2448. https://doi.org/10.1093/plcell/koac095

Krishnan HB, Franceschi VR, Okita TW. Immunochemical studies on the role of the Golgi-complex in protein-body formation in rice seeds. Planta. 1986:**169**(4):471–480. https://doi.org/10.1007/BF00392095

Kunzl F, Fruholz S, Fassler F, Li B, Pimpl P. Receptor-mediated sorting of soluble vacuolar proteins ends at the trans-Golgi network/early endosome. Nat Plants. 2016:**2**(4):16017. https://doi.org/10.1038/nplants.2016.17

Kusumaatmaja H, May AI, Feeney M, McKenna JF, Mizushima N, Frigerio L, Knorr RL. Wetting of phase-separated droplets on plant vacuole membranes leads to a competition between tonoplast budding and nanotube formation. Proc Natl Acad Sci U S A. 2021:**118**(36):e2024109118. https://doi.org/10.1073/pnas.2024109118

Lam SK, Siu CL, Hillmer S, Jang S, An G, Robinson DG, Jiang L. Rice SCAMP1 defines clathrin-coated, trans-Golgi-located tubular-vesicular structures as an early endosome in tobacco BY-2 cells. Plant Cell. 2007:**19**(1):296–319. https://doi.org/10.1105/tpc.106.045708

Larkins BA, Hurkman WJ. Synthesis and deposition of zein in protein bodies of maize endosperm. Plant Physiol. 1978:**62**(2):256–263. https://doi.org/10.1104/pp.62.2.256

Le Bars R, Marion J, Le Borgne R, Satiat-Jeunemaitre B, Bianchi MW. ATG5 defines a phagophore domain connected to the endoplasmic reticulum during autophagosome formation in plants. Nat Commun. 2014:**5**(1):4121. https://doi.org/10.1038/ncomms5121

Lee Y, Kim JH. The emerging roles of extracellular vesicles as intercellular messengers in liver physiology and pathology. Clin Mol Hepatol. 2022:**28**(4):706–724. https://doi.org/10.3350/cmh.2021.0390

Lee E, Santana BVN, Samuels E, Benitez-Fuente F, Corsi E, Botella MA, Perez-Sancho J, Vanneste S, Friml J, Macho A, et al. Rare earth elements induce cytoskeleton-dependent and PI4P-associated rearrangement of SYT1/SYT5 endoplasmic reticulum-plasma membrane contact site complexes in Arabidopsis. J Exp Bot. 2020:**71**(14):3986–3998. https://doi.org/10.1093/jxb/eraa138

Lee E, Vanneste S, Perez-Sancho J, Benitez-Fuente F, Strelau M, Macho AP, Botella MA, Friml J, Rosado A. Ionic stress enhances ER-PM connectivity via phosphoinositide-associated SYT1 contact site expansion in Arabidopsis. Proc Natl Acad Sci U S A. 2019:**116**(4):1420–1429. https://doi.org/10.1073/pnas.1818099116

Levanony H, Rubin R, Altschuler Y, Galili G. Evidence for a novel route of wheat storage proteins to vacuoles. J Cell Biol. 1992:**119**(5):1117–1128. https://doi.org/10.1083/jcb.119.5.1117

Li C, Duckney P, Zhang T, Fu Y, Li X, Kroon J, De Jaeger G, Cheng Y, Hussey PJ, Wang P. Trab family proteins are components of ER-mitochondrial contact sites and regulate ER-mitochondrial interactions and mitophagy. Nat Commun. 2022:**13**(1):5658. https://doi.org/10.1038/s41467-022-33402-w

Li RL, Liu P, Wan YL, Chen T, Wang QL, Mettbach U, Baluska F, Samaj J, Fang XH, Lucas WJ, et al. A membrane microdomain-associated protein, Flot1, is involved in a clathrin-independent endocytic pathway and is required for seedling development. Plant Cell. 2012:**24**(5):2105–2122. https://doi.org/10.1105/tpc.112.095695

Li R, Pang L. Comparing the effects of proteins with IDRs on membrane system in yeast, mammalian cells, and the model plant Arabidopsis. Curr Opin Plant Biol. 2023:**74**:102375. https://doi.org/10.1016/j.pbi.2023.102375

Li J, Peng J, Jiang X, Rea AC, Peng J, Hu J. DeepLearnMOR: a deep-learning framework for fluorescence image-based classification of organelle morphology. Plant Physiol. 2021b:**186**(4):1786–1799. https://doi.org/10.1093/plphys/kiab223

Li YB, Rogers SW, Tse YC, Lo SW, Sun SS, Jauh GY, Jiang L. BP-80 and homologs are concentrated on post-Golgi, probable lytic prevacuolar compartments. Plant Cell Physiol. 2002:**43**(7):726–742. https://doi.org/10.1093/pcp/pcf085

Li XJ, Wang XH, Yang Y, Li RL, He QH, Fang XH, Luu DT, Maurel C, Lin JX. Single-molecule analysis of PIP2;1 dynamics and partitioning reveals multiple modes of Arabidopsis plasma membrane aquaporin regulation. Plant Cell. 2011:**23**(10):3780–3797. https://doi.org/10.1105/tpc.111.091454

Li X, Wu Y, Zhang DZ, Gillikin JW, Boston RS, Franceschi VR, Okita TW. Rice prolamine protein body biogenesis: a BiP-mediated process. Science. 1993:**262**(5136):1054–1056. https://doi.org/10.1126/science.8235623

Li T, Xiao Z, Li H, Liu C, Shen W, Gao C. A combinatorial reporter set to visualize the membrane contact sites between endoplasmic reticulum and other organelles in plant cell. Front Plant Sci. 2020:**11**:1280. https://doi.org/10.3389/fpls.2020.01280

Li B, Zeng Y, Cao W, Zhang W, Cheng L, Yin H, Wu Q, Wang X, Huang Y, Lau WCY, et al. A distinct giant coat protein complex II vesicle population in *Arabidopsis thaliana*. Nat Plants. 2021a:**7**(10):1335–1346. https://doi.org/10.1038/s41477-021-00997-9

Liu Y, Bassham DC. Autophagy: pathways for self-eating in plant cells. Annu Rev Plant Biol. 2012:**63**(1):215–237. https://doi.org/10.1146/annurev-arplant-042811-105441

Liu Z, Gao J, Cui Y, Klumpe S, Xiang Y, Erdmann PS, Jiang L. Membrane imaging in the plant endomembrane system. Plant Physiol. 2021:**185**(3):562–576. https://doi.org/10.1093/plphys/kiaa040

Luo M, Law KC, He Y, Chung KK, Po MK, Feng L, Chung KP, Gao C, Zhuang X, Jiang L. Arabidopsis AUTOPHAGY-RELATED2 is essential for ATG18a and ATG9 trafficking during autophagosome closure. Plant Physiol. 2023:**193**(1):304–321. https://doi.org/10.1093/plphys/kiad287

Marion J, Le Bars R, Satiat-Jeunemaitre B, Boulogne C. Optimizing CLEM protocols for plants cells: GMA embedding and cryosections as alternatives for preservation of GFP fluorescence in Arabidopsis roots. J Struct Biol. 2017:**198**(3):196–202. https://doi.org/10.1016/j.jsb.2017.03.008

Martiniere A, Bassil E, Jublanc E, Alcon C, Reguera M, Sentenac H, Blumwald E, Paris N. In vivo intracellular pH measurements in tobacco and Arabidopsis reveal an unexpected pH gradient in the endomembrane system. Plant Cell. 2013:**25**(10):4028–4043. https://doi.org/10.1105/tpc.113.116897

Marty F. Cytochemical studies on GERL, provacuoles, and vacuoles in root meristematic cells of Euphorbia. Proc Natl Acad Sci U S A. 1978:**75**(2):852–856. https://doi.org/10.1073/pnas.75.2.852

Mathur J. Organelle extensions in plant cells. Plant Physiol. 2021:**185**(3):593–607. https://doi.org/10.1093/plphys/kiaa055

Mathur J, Radhamony R, Sinclair AM, Donoso A, Dunn N, Roach E, Radford D, Mohaghegh PS, Logan DC, Kokolic K, et al. mEosFP-based green-to-red photoconvertible subcellular probes for plants. Plant Physiol. 2010:**154**(4):1573–1587. https://doi.org/10.1104/pp.110.165431

Matile P, Moor H. Vacuolation—origin and development of lysosomal apparatus in root-tip cells. Planta. 1968:**80**(2):159. https://doi.org/10.1007/BF00385592

Matsuoka K, Bassham DC, Raikhel NV, Nakamura K. Different sensitivity to wortmannin of two vacuolar sorting signals indicates the presence of distinct sorting machineries in tobacco cells. J Cell Biol. 1995:**130**(6):1307–1318. https://doi.org/10.1083/jcb.130.6.1307

Matsuoka K, Higuchi T, Maeshima M, Nakamura K. A vacuolar-type H+-ATPase in a nonvacuolar organelle is required for the sorting of soluble vacuolar protein precursors in tobacco cells. Plant Cell. 1997:**9**(4):533–546. https://doi.org/10.2307/3870505

Mesquita JF. Electron microscope study of the origin and development of the vacuoles in root-tip cells of Lupinus albus L. J Ultrastruct Res. 1969:**26**(3–4):242–250. https://doi.org/10.1016/S0022-5320(69)80004-3

Mitsuhashi N, Shimada T, Mano S, Nishimura M, Hara-Nishimura I. Characterization of organelles in the vacuolar-sorting pathway by visualization with GFP in tobacco BY-2 cells. Plant Cell Physiol. 2000:**41**(9):993–1001. https://doi.org/10.1093/pcp/pcd040

Mori T, Maruyama N, Nishizawa K, Higasa T, Yagasaki K, Ishimoto M, Utsumi S. The composition of newly synthesized proteins in the endoplasmic reticulum determines the transport pathways of soybean seed storage proteins. Plant J. 2004:**40**(2):238–249. https://doi.org/10.1111/j.1365-313X.2004.02204.x

Morre DJ, Mollenhauer HH. Isolation of the Golgi apparatus from plant cells. J Cell Biol. 1964:**23**(2):295–305. https://doi.org/10.1083/jcb.23.2.295

Movahed N, Cabanillas DG, Wan J, Vali H, Laliberte JF, Zheng HQ. Turnip mosaic virus components are released into the extracellular space by vesicles in infected leaves. Plant Physiol. 2019:**180**(3):1375–1388. https://doi.org/10.1104/pp.19.00381

Nagata T, Nemoto A, Hasezawa S. Tobacco BY-2 cell line as the "HeLa" cell in the cell biology of higher plants. Int Rev Cytol. 1992:**132**:1–30. https://doi.org/10.1016/S0074-7696(08)62452-3

Nicolas WJ, Fassler F, Dutka P, Schur FKM, Jensen G, Meyerowitz E. Cryo-electron tomography of the onion cell wall shows bimodally oriented cellulose fibers and reticulated homogalacturonan networks. Curr Biol. 2022:**32**(11):2375–2389.e6. https://doi.org/10.1016/j.cub.2022.04.024

Nikolovski N, Shliaha PV, Gatto L, Dupree P, Lilley KS. Label-free protein quantification for plant Golgi protein localization and abundance. Plant Physiol. 2014:**166**(2):1033–1043. https://doi.org/10.1104/pp.114.245589

Oikawa K, Matsunaga S, Mano S, Kondo M, Yamada K, Hayashi M, Kagawa T, Kadota A, Sakamoto W, Higashi S, et al. Physical interaction between peroxisomes and chloroplasts elucidated by in situ laser analysis. Nat Plants. 2015:**1**(4):15035. https://doi.org/10.1038/nplants.2015.35

Okita TW, Rogers JC. Compartmentation of proteins in the endomembrane system of plant cells. Annu Rev Plant Physiol Plant Mol Biol. 1996:**47**(1):327–350. https://doi.org/10.1146/annurev.arplant.47.1.327

Osterrieder A, Sparkes IA, Botchway SW, Ward A, Ketelaar T, de Ruijter N, Hawes C. Stacks off tracks: a role for the golgin AtCASP in plant endoplasmic reticulum-Golgi apparatus tethering. J Exp Bot. 2017:**68**(13):3339–3350. https://doi.org/10.1093/jxb/erx167

Otegui MS, Herder R, Schulze J, Jung R, Staehelin LA. The proteolytic processing of seed storage proteins in Arabidopsis embryo cells starts in the multivesicular bodies. Plant Cell. 2006:**18**(10):2567–2581. https://doi.org/10.1105/tpc.106.040931

Otegui MS, Mastronarde DN, Kang BH, Bednarek SY, Staehelin LA. Three-dimensional analysis of syncytial-type cell plates during endosperm cellularization visualized by high resolution electron tomography. Plant Cell. 2001:**13**(9):2033–2051. https://doi.org/10.1105/tpc.010150

Pain C, Kriechbaumer V, Kittelmann M, Hawes C, Fricker M. Quantitative analysis of plant ER architecture and dynamics. Nat Commun. 2019:**10**(1):984. https://doi.org/10.1038/s41467-019-08893-9

Paris N, Rogers SW, Jiang L, Kirsch T, Beevers L, Phillips TE, Rogers JC. Molecular cloning and further characterization of a probable plant vacuolar sorting receptor. Plant Physiol. 1997:**115**(1):29–39. https://doi.org/10.1104/pp.115.1.29

Paris N, Stanley CM, Jones RL, Rogers JC. Plant cells contain two functionally distinct vacuolar compartments. Cell. 1996:**85**(4):563–572. https://doi.org/10.1016/S0092-8674(00)81256-8

Park M, Lee D, Lee GJ, Hwang I. AtRMR1 functions as a cargo receptor for protein trafficking to the protein storage vacuole. J Cell Biol. 2005:**170**(5):757–767. https://doi.org/10.1083/jcb.200504112

Parsons HT, Christiansen K, Knierim B, Carroll A, Ito J, Batth TS, Smith-Moritz AM, Morrison S, McInerney P, Hadi MZ, et al. Isolation and proteomic characterization of the Arabidopsis Golgi defines functional and novel components involved in plant cell wall biosynthesis. Plant Physiol. 2012:**159**(1):12–26. https://doi.org/10.1104/pp.111.193151

Paterlini A, Belevich I, Jokitalo E, Helariutta Y. Computational tools for serial block electron microscopy reveal plasmodesmata distributions and wall environments. Plant Physiol. 2020:**184**(1):53–64. https://doi.org/10.1104/pp.20.00396

Pesacreta TC, Lucas WJ. Presence of a partially-coated reticulum in angiosperms. Protoplasma. 1985:**125**(3):173–184. https://doi.org/10.1007/BF01281235

Pimpl P, Movafeghi A, Coughlan S, Denecke J, Hillmer S, Robinson DG. In situ localization and in vitro induction of plant COPI-coated vesicles. Plant Cell. 2000:**12**(11):2219–2235. https://doi.org/10.1105/tpc.12.11.2219

Platre MP, Bayle V, Armengot L, Bareille J, Marques-Bueno MD, Creff A, Maneta-Peyret L, Fiche JB, Nollmann M, Miege C, et al. Developmental control of plant Rho GTPase nano-organization by the lipid phosphatidylserine. Science. 2019:**364**(6435):57. https://doi.org/10.1126/science.aav9959

Prado N, Alche Jde D, Casado-Vela J, Mas S, Villalba M, Rodriguez R, Batanero E. Nanovesicles are secreted during pollen germination and pollen tube growth: a possible role in fertilization. Mol Plant. 2014:**7**(3):573–577. https://doi.org/10.1093/mp/sst153

Prinz WA, Toulmay A, Balla T. The functional universe of membrane contact sites. Nat Rev Mol Cell Biol. 2020:**21**(1):7–24. https://doi.org/10.1038/s41580-019-0180-9

Ray PM, Shininger TL, Ray MM. Isolation of β-glucan synthetase particles from plant cells and identification with Golgi membranes. Proc Natl Acad Sci U S A. 1969:**64**(2):605–612. https://doi.org/10.1073/pnas.64.2.605

Regente M, Pinedo M, San Clemente H, Balliau T, Jamet E, de la Canal L. Plant extracellular vesicles are incorporated by a fungal pathogen and inhibit its growth. J Exp Bot. 2017:**68**(20):5485–5495. https://doi.org/10.1093/jxb/erx355

Reyes FC, Chung T, Holding D, Jung R, Vierstra R, Otegui MS. Delivery of prolamins to the protein storage vacuole in maize aleurone cells. Plant Cell. 2011:**23**(2):769–784. https://doi.org/10.1105/tpc.110.082156

Reynolds GD, Wang C, Pan J, Bednarek SY. Inroads into internalization: five years of endocytic exploration. Plant Physiol. 2018:**176**:208–218. https://doi.org/10.1104/pp.17.01117

Rhee SY, Birnbaum KD, Ehrhardt DW. Towards building a plant cell atlas. Trends Plant Sci. 2019:**24**(4):303–310. https://doi.org/10.1016/j.tplants.2019.01.006

Robinson DG. Dictyosome-endoplasmic reticulum associations in higher plant cells? A serial-section analysis. Eur J Cell Biol. 1980:**23**:22–36.

Robinson DG. Plant Golgi ultrastructure. J Microsc. 2020:**280**(2):111–121. https://doi.org/10.1111/jmi.12899

Robinson DG, Albrecht S, Moriysu Y. The V-ATPase inhibitors concanamycin A and bafilomycin A lead to Golgi swelling in tobacco BY-2 cells. Protoplasma. 2004:**224**(3–4):255–260. https://doi.org/10.1007/s00709-004-0070-6

Robinson DG, Baumer M, Hinz G, Hohl I. Ultrastructure of the pea cotyledon Golgi apparatus: origin of dense vesicles and the action of brefeldin A. Protoplasma. 1997:**200**(3–4):198–209. https://doi.org/10.1007/BF01283295

Robinson DG, Brandizzi F, Hawes C, Nakano A. Vesicles versus tubes: is endoplasmic reticulum-Golgi transport in plants fundamentally different from other eukaryotes? Plant Physiol. 2015:**168**(2):393–406. https://doi.org/10.1104/pp.15.00124

Robinson DG, Hinz G, Holstein SE. The molecular characterization of transport vesicles. Plant Mol Biol. 1998:**38**(1–2):49–76. https://doi.org/10.1023/A:1006025802445

Roth R, Hillmer S, Funaya C, Chiapello M, Schumacher K, Lo Presti L, Kahmann R, Paszkowski U. Arbuscular cell invasion coincides with extracellular vesicles and membrane tubules. Nat Plants. 2019:**5**(2):204–211. https://doi.org/10.1038/s41477-019-0365-4

Ruiz-Lopez N, Perez-Sancho J, Del Valle AE, Haslam RP, Vanneste S, Catala R, Perea-Resa C, Damme DV, Garcia-Hernandez S, Albert A, et al. Synaptotagmins at the endoplasmic reticulum-plasma membrane contact sites maintain diacylglycerol homeostasis during abiotic stress. Plant Cell. 2021:**33**(7):2431–2453. https://doi.org/10.1093/plcell/koab122

Runions J, Brach T, Kuhner S, Hawes C. Photoactivation of GFP reveals protein dynamics within the endoplasmic reticulum membrane. J Exp Bot. 2006:**57**(1):43–50. https://doi.org/10.1093/jxb/eri289

Rutter BD, Innes RW. Extracellular vesicles isolated from the leaf apoplast carry stress-response proteins. Plant Physiol. 2017:**173**(1): 728–741. https://doi.org/10.1104/pp.16.01253

Sanchez GB, Bunn KE, Pua HH, Rafat M. Extracellular vesicles: mediators of intercellular communication in tissue injury and disease. Cell Commun Signal. 2021:**19**:104. https://doi.org/10.1186/s12964-021-00787-y

Sanchez Carrillo IB, Hoffmann PC, Barff T, Beck M, Germain H. Preparing *Arabidopsis thaliana* root protoplasts for cryo electron tomography. Front Plant Sci. 2023:**14**:1261180. https://doi.org/10.3389/fpls.2023.1261180

Sanderfoot AA, Ahmed SU, Marty-Mazars D, Rapoport I, Kirchhausen T, Marty F, Raikhel NV. A putative vacuolar cargo receptor partially colocalizes with AtPEP12p on a prevacuolar compartment in Arabidopsis roots. Proc Natl Acad Sci U S A. 1998:**95**(17):9920–9925. https://doi.org/10.1073/pnas.95.17.9920

Saravanan RS, Slabaugh E, Singh VR, Lapidus LJ, Haas T, Brandizzi F. The targeting of the oxysterol-binding protein ORP3a to the endoplasmic reticulum relies on the plant VAP33 homolog PVA12. Plant J. 2009:**58**(5):817–830. https://doi.org/10.1111/j.1365-313X.2009.03815.x

Satiat-Jeunemaitre B, Cole L, Bourett T, Howard R, Hawes C. Brefeldin A effects in plant and fungal cells: something new about vesicle trafficking? J Microsc. 1996:**181**(2):162–177. https://doi.org/10.1046/j.1365-2818.1996.112393.x

Satiatjeunemaitre B, Hawes C. Redistribution of a Golgi glycoprotein in plant-cells treated with Brefeldin-A. J Cell Sci. 1992:**103**(4): 1153–1166. https://doi.org/10.1242/jcs.103.4.1153

Schattat MH, Griffiths S, Mathur N, Barton K, Wozny MR, Dunn N, Greenwood JS, Mathur J. Differential coloring reveals that plastids do not form networks for exchanging macromolecules. Plant Cell. 2012:**24**(4):1465–1477. https://doi.org/10.1105/tpc.111.095398

Scheuring D, Lofke C, Kruger F, Kittelmann M, Eisa A, Hughes L, Smith RS, Hawes C, Schumacher K, Kleine-Vehn J. Actin-dependent vacuolar occupancy of the cell determines auxin-induced growth repression. Proc Natl Acad Sci U S A. 2016:**113**(2):452–457. https://doi.org/10.1073/pnas.1517445113

Scheuring D, Viotti C, Kruger F, Kunzl F, Sturm S, Bubeck J, Hillmer S, Frigerio L, Robinson DG, Pimpl P, et al. Multivesicular bodies mature from the trans-Golgi network/early endosome in Arabidopsis. Plant Cell. 2011:**23**(9):3463–3481. https://doi.org/10.1105/tpc.111.086918

Scorrano L, De Matteis MA, Emr S, Giordano F, Hajnoczky G, Kornmann B, Lackner LL, Levine TP, Pellegrini L, Reinisch K, et al. Coming together to define membrane contact sites. Nat Commun. 2019:**10**(1):1287. https://doi.org/10.1038/s41467-019-09253-3

Shen J, Zeng Y, Zhuang X, Sun L, Yao X, Pimpl P, Jiang L. Organelle pH in the Arabidopsis endomembrane system. Mol Plant. 2013:**6**(5): 1419–1437. https://doi.org/10.1093/mp/sst079

Shimada T, Kuroyanagi M, Nishimura M, Hara-Nishimura I. A pumpkin 72-kDa membrane protein of precursor-accumulating vesicles has characteristics of a vacuolar sorting receptor. Plant Cell Physiol. 1997:**38**(12):1414–1420. https://doi.org/10.1093/oxfordjournals.pcp.a029138

Shimizu Y, Takagi J, Ito E, Ito Y, Ebine K, Komatsu Y, Goto Y, Sato M, Toyooka K, Ueda T, et al. Cargo sorting zones in the trans-Golgi network visualized by super-resolution confocal live imaging microscopy in plants. Nat Commun. 2021:**12**(1):1901. https://doi.org/10.1038/s41467-021-22267-0

Siao W, Wang P, Voigt B, Hussey PJ, Baluska F. Arabidopsis SYT1 maintains stability of cortical endoplasmic reticulum networks and VAP27-1-enriched endoplasmic reticulum-plasma membrane contact sites. J Exp Bot. 2016:**67**(21):6161–6171. https://doi.org/10.1093/jxb/erw381

Simon ML, Platre MP, Assil S, van Wijk R, Chen WY, Chory J, Dreux M, Munnik T, Jaillais Y. A multi-colour/multi-affinity marker set to visualize phosphoinositide dynamics in Arabidopsis. Plant J. 2014:**77**(2):322–337. https://doi.org/10.1111/tpj.12358

Simon MLA, Platre MP, Marques-Bueno MM, Armengot L, Stanislas T, Bayle V, Caillaud MC, Jaillais Y. A PtdIns(4)P-driven electrostatic field controls cell membrane identity and signalling in plants. Nat Plants. 2016:**2**(7):16089. https://doi.org/10.1038/nplants.2016.89

Skirpan AL, Dowd PE, Sijacic P, Jaworski CJ, Gilroy S, Kao TH. Identification and characterization of PiORP1, a Petunia oxysterol-binding-protein related protein involved in receptor-kinase mediated signaling in pollen, and analysis of the ORP gene family in Arabidopsis. Plant Mol Biol. 2006:**61**(4–5):553–565. https://doi.org/10.1007/s11103-006-0030-y

Sparkes IA, Ketelaar T, de Ruijter NC, Hawes C. Grab a Golgi: laser trapping of Golgi bodies reveals in vivo interactions with the endoplasmic reticulum. Traffic. 2009:**10**(5):567–571. https://doi.org/10.1111/j.1600-0854.2009.00891.x

Staehelin LA. The plant ER: a dynamic organelle composed of a large number of discrete functional domains. Plant J. 1997:**11**(6): 1151–1165. https://doi.org/10.1046/j.1365-313X.1997.11061151.x

Staehelin LA, Giddings TH Jr, Kiss JZ, Sack FD. Macromolecular differentiation of Golgi stacks in root tips of Arabidopsis and Nicotiana seedlings as visualized in high pressure frozen and freeze-substituted samples. Protoplasma. 1990:**157**(1–3):75–91. https://doi.org/10.1007/BF01322640

Staehelin LA, Kang BH. Nanoscale architecture of endoplasmic reticulum export sites and of Golgi membranes as determined by electron tomography. Plant Physiol. 2008:**147**(4):1454–1468. https://doi.org/10.1104/pp.108.120618

Staehelin LA, Kiermayer O. Membrane differentiation in the Golgi complex of Micrasterias denticulata Breb. visualized by freeze-etching. J Cell Sci. 1970:**7**(3):787–792. https://doi.org/10.1242/jcs.7.3.787

Subra M, Antonny B, Mesmin B. New insights into the OSBP–VAP cycle. Curr Opin Cell Biol. 2023:**82**:102172. https://doi.org/10.1016/j.ceb.2023.102172

Sutter JU, Campanoni P, Blatt MR, Paneque M. Setting SNAREs in a different wood. Traffic. 2006:**7**(6):627–638. https://doi.org/10.1111/j.1600-0854.2006.00414.x

Taiz L, Murry M, Robinson DG. Identification of secretory vesicles in homogenates of pea stem segments. Planta. 1983:**158**(6):534–539. https://doi.org/10.1007/BF00397244

Takagi J, Kimori Y, Shimada T, Hara-Nishimura I. Dynamic capture and release of endoplasmic reticulum exit sites by Golgi stacks in Arabidopsis. iScience. 2020:**23**(7):101265. https://doi.org/10.1016/j.isci.2020.101265

Takahashi H, Saito Y, Kitagawa T, Morita S, Masumura T, Tanaka K. A novel vesicle derived directly from endoplasmic reticulum is involved in the transport of vacuolar storage proteins in rice endosperm. Plant Cell Physiol. 2005:**46**(1):245–249. https://doi.org/10.1093/pcp/pci019

Tosi P, Parker M, Gritsch CS, Carzaniga R, Martin B, Shewry PR. Trafficking of storage proteins in developing grain of wheat. J Exp Bot. 2009:**60**(3):979–991. https://doi.org/10.1093/jxb/ern346

Tse YC, Mo B, Hillmer S, Zhao M, Lo SW, Robinson DG, Jiang L. Identification of multivesicular bodies as prevacuolar compartments in Nicotiana tabacum BY-2 cells. Plant Cell. 2004:**16**(3):672–693. https://doi.org/10.1105/tpc.019703

Tully RE, Beevers H. Protein bodies of castor bean endosperm: isolation, fractionation, and the characterization of protein components. Plant Physiol. 1976:**58**(6):710–716. https://doi.org/10.1104/pp.58.6.710

Ueda T, Uemura T, Sato MH, Nakano A. Functional differentiation of endosomes in Arabidopsis cells. Plant J. 2004:**40**(5):783–789. https://doi.org/10.1111/j.1365-313X.2004.02249.x

Ueda T, Yamaguchi M, Uchimiya H, Nakano A. Ara6, a plant-unique novel type Rab GTPase, functions in the endocytic pathway of

Arabidopsis thaliana. EMBO J. 2001:**20**(17):4730–4741. https://doi.org/10.1093/emboj/20.17.4730

Uemura T, Suda Y, Ueda T, Nakano A. Dynamic behavior of the trans-Golgi network in root tissues of Arabidopsis revealed by super-resolution live imaging. Plant Cell Physiol. 2014:**55**(4):694–703. https://doi.org/10.1093/pcp/pcu010

van Niel G, D'Angelo G, Raposo G. Shedding light on the cell biology of extracellular vesicles. Nat Rev Mol Cell Biol. 2018:**19**(4):213–228. https://doi.org/10.1038/nrm.2017.125

Vieira V, Pain C, Wojcik S, Spatola Rossi T, Denecke J, Osterrieder A, Hawes C, Kriechbaumer V. Living on the edge: the role of Atgolgin-84A at the plant ER-Golgi interface. J Microsc. 2020:**280**(2):158–173. https://doi.org/10.1111/jmi.12946

Villa E, Schaffer M, Plitzko JM, Baumeister W. Opening windows into the cell: focused-ion-beam milling for cryo-electron tomography. Curr Opin Struct Biol. 2013:**23**(5):771–777. https://doi.org/10.1016/j.sbi.2013.08.006

Viotti C, Bubeck J, Stierhof YD, Krebs M, Langhans M, van den Berg W, van Dongen W, Richter S, Geldner N, Takano J, et al. Endocytic and secretory traffic in Arabidopsis merge in the trans-Golgi network/early endosome, an independent and highly dynamic organelle. Plant Cell. 2010:**22**(4):1344–1357. https://doi.org/10.1105/tpc.109.072637

Viotti C, Kruger F, Krebs M, Neubert C, Fink F, Lupanga U, Scheuring D, Boutte Y, Frescatada-Rosa M, Wolfenstetter S, et al. The endoplasmic reticulum is the main membrane source for biogenesis of the lytic vacuole in Arabidopsis. Plant Cell. 2013:**25**(9):3434–3449. https://doi.org/10.1105/tpc.113.114827

Wang J, Cai Y, Miao Y, Lam SK, Jiang L. Wortmannin induces homotypic fusion of plant prevacuolar compartments. J Exp Bot. 2009:**60**(11):3075–3083. https://doi.org/10.1093/jxb/erp136

Wang J, Ding Y, Wang J, Hillmer S, Miao Y, Lo SW, Wang X, Robinson DG, Jiang L. EXPO, an exocyst-positive organelle distinct from multivesicular endosomes and autophagosomes, mediates cytosol to cell wall exocytosis in Arabidopsis and tobacco cells. Plant Cell. 2010:**22**(12):4009–4030. https://doi.org/10.1105/tpc.110.080697

Wang P, Hawkins TJ, Richardson C, Cummins I, Deeks MJ, Sparkes I, Hawes C, Hussey PJ. The plant cytoskeleton, NET3C, and VAP27 mediate the link between the plasma membrane and endoplasmic reticulum. Curr Biol. 2014:**24**(12):1397–1405. https://doi.org/10.1016/j.cub.2014.05.003

Wang P, Liang Z, Kang BH. Electron tomography of plant organelles and the outlook for correlative microscopic approaches. New Phytol. 2019a:**223**(4):1756–1761. https://doi.org/10.1111/nph.15882

Wang P, Pleskot R, Zang J, Winkler J, Wang J, Yperman K, Zhang T, Wang K, Gong J, Guan Y, et al. Plant AtEH/Pan1 proteins drive autophagosome formation at ER-PM contact sites with actin and endocytic machinery. Nat Commun. 2019b:**10**(1):5132. https://doi.org/10.1038/s41467-019-12782-6

Wang XH, Teng Y, Wang QL, Li XJ, Sheng XY, Zheng MZ, Samaj J, Baluska F, Lin JX. Imaging of dynamic secretory vesicles in living pollen tubes of Picea meyeri using evanescent wave microscopy. Plant Physiol. 2006:**141**(4):1591–1603. https://doi.org/10.1104/pp.106.080168

Wang L, Xue Y, Xing J, Song K, Lin J. Exploring the spatiotemporal organization of membrane proteins in living plant cells. Annu Rev Plant Biol. 2018:**69**(1):525–551. https://doi.org/10.1146/annurev-arplant-042817-040233

Wang Q, Zhao Y, Luo W, Li R, He Q, Fang X, Michele RD, Ast C, von Wiren N, Lin J. Single-particle analysis reveals shutoff control of the Arabidopsis ammonium transporter AMT1;3 by clustering and internalization. Proc Natl Acad Sci U S A. 2013:**110**(32):13204–13209. https://doi.org/10.1073/pnas.1301160110

Weiner E, Pinskey JM, Nicastro D, Otegui MS. Electron microscopy for imaging organelles in plants and algae. Plant Physiol. 2022:**188**(2): 713–725. https://doi.org/10.1093/plphys/kiab449

Wheeler H, Hanchey P. Pinocytosis and membrane dilation in uranyl-treated plant roots. Science. 1971:**171**(3966):68–71. https://doi.org/10.1126/science.171.3966.68

Wilkins TA, Bednarek SY, Raikhel NV. Role of propeptide glycan in post-translational processing and transport of barley lectin to vacuoles in transgenic tobacco. Plant Cell. 1990:**2**(4):301–313. https://doi.org/10.1105/tpc.2.4.301

Ye H, Gao J, Liang Z, Lin Y, Yu Q, Huang S, Jiang L. Arabidopsis ORP2A mediates ER-autophagosomal membrane contact sites and regulates PI3P in plant autophagy. Proc Natl Acad Sci U S A. 2022:**119**(43): e2205314119. https://doi.org/10.1073/pnas.2205314119

Youle RJ, Huang AH. Protein bodies from the endosperm of castor bean: subfractionation, protein components, lectins, and changes during germination. Plant Physiol. 1976:**58**(6):703–709. https://doi.org/10.1104/pp.58.6.703

Yu M, Liu HJ, Dong ZY, Xiao JW, Su BD, Fan LS, Komis G, Samaj J, Lin JX, Li RL. The dynamics and endocytosis of Flot1 protein in response to flg22 in Arabidopsis. J Plant Physiol. 2017:**215**:73–84. https://doi.org/10.1016/j.jplph.2017.05.010

Zeng Y, Li B, Huang S, Li H, Cao W, Chen Y, Liu G, Li Z, Yang C, Feng L, et al. The plant unique ESCRT component FREE1 regulates autophagosome closure. Nat Commun. 2023:**14**(1):1768. https://doi.org/10.1038/s41467-023-37185-6

Zeng Y, Li B, Ji C, Feng L, Niu F, Deng C, Chen S, Lin Y, Cheung KCP, Shen J, et al. A unique AtSar1D-AtRabD2a nexus modulates autophagosome biogenesis in *Arabidopsis thaliana*. Proc Natl Acad Sci U S A. 2021:**118**(17):e2021293118. https://doi.org/10.1073/pnas.2021293118

Zhang T, Li Y, Li C, Zang J, Gao E, Kroon JT, Qu X, Hussey PJ, Wang P. Exo84c interacts with VAP27 to regulate exocytotic compartment degradation and stigma senescence. Nat Commun. 2023:**14**(1): 4888. https://doi.org/10.1038/s41467-023-40729-5

Zhang X, Man Y, Zhuang XH, Shen JB, Zhang Y, Cui YN, Yu M, Xing JJ, Wang GC, Lian N, et al. Plant multiscale networks: charting plant connectivity by multi-level analysis and imaging techniques. Sci China Life Sci. 2021:**64**(9):1392–1422. https://doi.org/10.1007/s11427-020-1910-1

Zhao J, Bui MT, Ma J, Kunzl F, Picchianti L, De La Concepcion JC, Chen Y, Petsangouraki S, Mohseni A, Garcia-Leon M, et al. Plant autophagosomes mature into amphisomes prior to their delivery to the central vacuole. J Cell Biol. 2022:**221**(12):e202203139. https://doi.org/10.1083/jcb.202203139

Zhao Z, Yu S, Li M, Gui X, Li P. Isolation of exosome-like nanoparticles and analysis of microRNAs derived from coconut water based on small RNA high-throughput sequencing. J Agric Food Chem. 2018:**66**(11):2749–2757. https://doi.org/10.1021/acs.jafc.7b05614

Zheng H, Staehelin LA. Protein storage vacuoles are transformed into lytic vacuoles in root meristematic cells of germinating seedlings by multiple, cell type-specific mechanisms. Plant Physiol. 2011:**155**(4): 2023–2035. https://doi.org/10.1104/pp.110.170159

Zheng P, Zheng C, Otegui MS, Li F. Endomembrane mediated-trafficking of seed storage proteins: from Arabidopsis to cereal crops. J Exp Bot. 2022:**73**(5):1312–1326. https://doi.org/10.1093/jxb/erab519

Zhou J, Ma J, Yang C, Zhu X, Li J, Zheng X, Li X, Chen S, Feng L, Wang P, et al. A non-canonical role of ATG8 in Golgi recovery from heat stress in plants. Nat Plants. 2023:**9**(5):749–765. https://doi.org/10.1038/s41477-023-01398-w

Zhuang XH, Chung KP, Cui Y, Lin WL, Gao CJ, Kang BH, Jiang LW. ATG9 regulates autophagosome progression from the endoplasmic reticulum in. Proc Natl Acad Sci U S A. 2017:**114**(3):E426–E435. https://doi.org/10.1073/pnas.1616299114

Zhuang XH, Wang H, Lam SK, Gao CJ, Wang XF, Cai Y, Jiang LW. A BAR-domain protein SH3P2, which binds to phosphatidylinositol 3-phosphate and ATG8, regulates autophagosome formation in Arabidopsis. Plant Cell. 2013:**25**(11):4596–4615. https://doi.org/10.1105/tpc.113.118307

Zwiewka M, Feraru E, Moller B, Hwang I, Feraru MI, Kleine-Vehn J, Weijers D, Friml J. The AP-3 adaptor complex is required for vacuolar function in Arabidopsis. Cell Res. 2011:**21**(12):1711–1722. https://doi.org/10.1038/cr.2011.99

https://doi.org/10.1093/plcell/koae044

THE PLANT CELL

THE PLANT CELL 2024: 36: 1334–1357

Review

Reflections on the ABC model of flower development

John L. Bowman [1,2,*] and Edwige Moyroud [3,4]

1 School of Biological Sciences, Monash University, Melbourne, VIC 3800, Australia
2 ARC Centre of Excellence for Plant Success in Nature and Agriculture, Monash University, Melbourne, VIC 3800, Australia
3 The Sainsbury Laboratory, Cambridge University, Cambridge CB2 1LR, UK
4 Department of Genetics, University of Cambridge, Cambridge CB2 3EJ, UK

*Author for correspondence: john.bowman@monash.edu
The author responsible for distribution of materials integral to the findings presented in this article in accordance with the policy described in the Instructions for Authors (https://academic.oup.com/plcell/pages/General-Instructions) is: John L. Bowman john.bowman@monash.edu.

Abstract

The formulation of the ABC model by a handful of pioneer plant developmental geneticists was a seminal event in the quest to answer a seemingly simple question: how are flowers formed? Fast forward 30 years and this elegant model has generated a vibrant and diverse community, capturing the imagination of developmental and evolutionary biologists, structuralists, biochemists and molecular biologists alike. Together they have managed to solve many floral mysteries, uncovering the regulatory processes that generate the characteristic spatio-temporal expression patterns of floral homeotic genes, elucidating some of the mechanisms allowing ABC genes to specify distinct organ identities, revealing how evolution tinkers with the ABC to generate morphological diversity, and even shining a light on the origins of the floral gene regulatory network itself. Here we retrace the history of the ABC model, from its genesis to its current form, highlighting specific milestones along the way before drawing attention to some of the unsolved riddles still hidden in the floral alphabet.

History

Humans have long been fascinated by flowers, perhaps initially due to their presence as a sign or harbinger of food. Thus, it is not surprising that variations in flower structure are described in the oldest surviving botanical literature. For example, Theophrastus wrote (circa 300 BC), "Among roses there are many differences, in the number of petals, … in beauty of colour, and in sweetness of scent. Most have five petals, but some have twelve or twenty, and some a great many more than these; for there are some, they say, which are even called "hundred-petalled." [vol II, book VI 3 to 5, p39 (Theophrastus 1916)]. A few hundred years later, Pliny also described (possibly the same) rose variants: "The essential points of difference in the rose are the number of the petals, the comparative number of thorns on the stem, the colour, and the smell. The number of the petals, which is never less than five, goes on increasing in amount, till we find one variety with as many as a hundred, and thence known as the "centifolia" [Book XXI Chap 10 (Plinius the Secondus 1634)]. As China is the center of diversification of the genus *Rosa*, some variants likely predate the Greek and Roman descriptions (Wang 2007). Following the Renaissance, when herbals describing medicinal plants proliferated during the age of exploration, descriptions and drawings of "double flowers" also multiplied, not only for roses, for example, the double-flowered rose depicted on page 93 in Dodoens' *Florum, et Coronariarum Odoratarumque Nonnullarum Herbarum Historia* (Dodoens 1568), but for other species as well. Gerard (Gerard et al. 1597) depicted no fewer than 15 species with double-flowered variants, including in the Brassicaceae (Fig. 1A), the latter being included in some earlier publications (L'Obel 1581; Dodoens 1583). Centuries later, double flowers of *Aquilegia vulgaris* were examined in greater detail with the finding that some stamens are converted into petals with nectaries, with petals from one whorl neatly fitting into those of another (Moquin-Tandon 1841).

Received August 15, 2023. Accepted February 07, 2024. Advance access publication February 12, 2024.

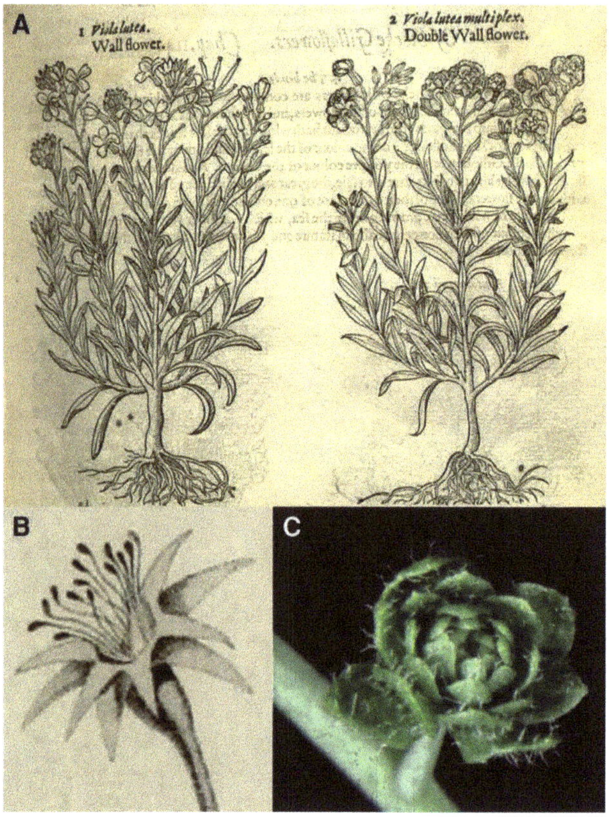

Figure 1. Early descriptions of flower variants. **A)** Single- (left) and double-flowered (right) wallflowers, *Erysimum cheiri* (Gerard et al. 1597). **B)** An apple mutant in which the petals are transformed to another whorl of sepals and the stamens are transformed into carpels. This tree was used for cross-fertilization, where "the ladies and young girls of Saint-Valery compete to 'make their apple' [that's the expression they use]. A hermaphrodite flower, picked in dry weather from any apple tree, is applied to each flower and left there until fertilization is completed and it falls off naturally. Then a ribbon of colour is attached to the fertilized bouquet so that, when autumn comes, each person can recognize the fruit created by their own hand" (Tillette de Clermont-Tonnere 1825). **C)** An *apetala2-1 pistillata-1 agamous-1* flower consisting of an indeterminate number of whorls of leaf-like organs. (A,B, Biodiversity Heritage Library; C, photo, J.L.B.).

While double-flowered blossoms are conspicuous, other less striking flower variants were also described centuries ago (Fig. 1B). For example, an apple tree (*Malus*) in M. Aux's orchard in Saint-Valery-sur-Somme that was "similar in every way to the common apple tree in terms of leaves and flower arrangement, differs in a very remarkable way in the structure of its flowers and fruits. A tomentose peduncle supports a flower composed of a calyx with ten folioles fused at the base, arranged in two alternating rows, the inner ones slightly shorter. The corolla and stamens are absent; the styles, fourteen in number, slightly hairy at the base, are surmounted by a very vigorous oblique stigma" (Tillette de Clermont-Tonnere 1825). Usually in apple flowers "the calyx is simple, with five leaflets; here, it is clearly double, with the inner part similar to the outer part, persisting like it and cannot be assimilated to a corolla. In the former, there are five styles and stigmas; in ours, there are fourteen. Ordinary apples have only five compartments; these have fourteen in two parallel rows, representing two apples fused end to end" (Tillette de Clermont-Tonnere 1825). This phenotype was maintained after grafting onto a normal apple tree, but unfortunately, the phenotype of progeny derived from crosses between mutant flowers and wild-type stamens was not reported (Tillette de Clermont-Tonnere 1825). A similar, albeit less severe, phenotype was noted in a *Rhododendron* variant, known as Misome-sho in Japanese and described in 1692, that exhibits a sepaloid corolla and altered stamen development (Cheon et al. 2018).

It is clear from the descriptions of the *Malus* and *Aquilegia* flowers that it was recognized that there had been a homeotic transformation of one organ type into another (Tillette de Clermont-Tonnere 1825; Moquin-Tandon 1841), although the term "homeotic" was not coined until later. For example, Moquin-Tandon's (1841, p. 386) interpretation of the mutant *Malus* flowers, "It is possible that, due to the absence of stamens, the androecium has transformed into a row of carpels that are fused to the ordinary carpels. In that case, the phenomenon in question would fall into the category of organic metamorphoses". Later, Masters, in his *Vegetable Teratology*, uses the *Malus* example to illustrate his concept of metamorphy (Masters 1869). Bateson subsequently replaced Masters' metamorphy with the term we know today, homeosis (Bateson 1894).

While anatomical and morphological descriptions of "normal" flower development and variations thereof abounded (see Moquin-Tandon 1841; Payer 1857; Masters 1869; Penzig 1890-4; Wordsell 1915-6), the connection to heritability was not often made. Perhaps some of the reluctance stemmed from observations such as made by Darwin: "Such flowers as the carnation, common tulip, and hyacinth, which are believed to be descended, each from a single wild form, present innumerable varieties, differing almost exclusively in the size, form, and colour of the flowers. These and some other anciently cultivated plants which have been long propagated by offsets, pipings, bulbs, & c., become so excessively variable, that almost each new plant raised from seed forms a new variety, 'all of which to describe particularly,' as old Gerarde wrote in 1597, 'were to roll Sisyphus's stone, or to number the sands'" (Darwin 1876). However, heritability is alluded to occasionally. Darwin himself described a number of heritable floral variants, including a poppy in which the stamens are converted into pistils [de Candolle quoted in (Darwin 1876)], and several authors noted the inheritance of a number of peloric floral varieties, including that of *Antirrhinum majus* (Naudin 1865; Godron 1874; Darwin 1876). And much earlier, Parkinson noted that while double-flowered plants never bore seed, if double-flowered plants are desired, one should collect plants from single-flowered plants in the vicinity of the double-flowered plant, and sometimes a proportion of these would be double-flowered even though the original single-flowered parents were indistinguishable (Parkinson 1629). Following the rediscovery of Mendel in the early 20th century, it was rapidly determined that a *Matthiola* double-flower phenotype was recessive with respect to the single-flower phenotype (Bateson et al. 1905), explaining why double flowers, which are entirely male and female sterile, only arise from single-flowered seed stocks, apparently at random, according to breeders in the days before genetics. Furthermore, as noted by some early breeders, single-flowered plants bearing flowers with extra petals may indicate plants that we now know as heterozygotes (Bateson et al. 1905).

For a majority of the 20th century, genetics and developmental biology of plants continued in distinct trajectories despite having merged in the study of animals in the mid-20th century (Meyerowitz et al. 1989; Prunet and Meyerowitz 2016). Thus, early models of flower development typically employed hormonal or biophysical signals or forces, rather than genes, that would influence floral organ identity. Perhaps the earliest model was proposed by Goethe where he viewed floral organs as modified leaves, with their modification from leaves into floral organs directed by ascending sap that became more refined as it made its way to the apex (Goethe 1790). Abnormal organs could be viewed as a failure of refinement of the sap. Other models relied on similar relay mechanisms such that one type of floral organ would induce the next inner set of organs. In some models the mechanism was proposed to be biomechanical (Green 1988), while in others, genetic programs were invoked (Heslop-Harrison 1963). However, these models failed to explain existing floral varieties where organ types were absent or in inappropriate positions. In contrast, models based on concentric rings of hormonal activity, avoiding the inter-whorl communication, were better equipped to explain known floral mutant phenotypes but invoked shifts in the positions of organ primordia rather than changes in the concentric rings of hormonal activity (Brieger 1935; Holder 1979).

Genetics

But I (J.L.B.) did not know most of the above as I peered through the dissecting microscope in a basement annex of the Kerckhoff building at Caltech in early 1990 when I was a PhD student in Elliot Meyerowitz's laboratory. I was searching among Arabidopsis plants for a triple mutant in a family segregating for *apetala2-1*, *agamous-1*, and *pistillata-1*. While its phenotype could be predicted based on already constructed double mutant combinations (Bowman et al. 1989), there was a thrill of observation and contemplating that you have knowledge no one else in the world possessed (Fig. 1C). It was late it the evening, and all members of Elliot's lab had left, and so I sauntered down the hallway to share the discovery with my mate, Joe Minor, who was working on sea urchins in the neighboring laboratory.

These observations were the culmination of experiments begun in the early 1980s, initially by Maarten Koornneef, who was constructing an Arabidopsis genetic map and who isolated several mutants affecting floral architecture (Koornneef et al. 1983). Elliot obtained these stocks from Maarten and, subsequently, Bob Pruitt, a previous PhD student in Elliot's laboratory, and Elliot thought the mutants exhibited homeotic phenotypes. When I joined the laboratory in late 1986, I continued the construction of multiple mu-tants, a process initiated by Bob and Elliot. At first, we had solitary mutant alleles of each of 5 loci [*APETALA1* (*AP1*), *APETALA2* (*AP2*), *APEPALA3* (*AP3*), *PISTILLATA* (*PI*), *AGAMOUS* (*AG*)]. These were chosen based on their homeotic phenotypes, although *AP1* was initially left aside due to its broader role that we now know as determining floral meristem (FM) identity. The chosen loci reflected an early decision to focus on the roles in organ identity, ignoring the effects of the mutations on organ number (e.g. organ initiation or fusion in the case of carpels) and determinacy, even though it was clear the chosen genes might have functions additional to those directly involved in organ identity. In January 1988, David Smyth joined the laboratory as a sabbatical visitor, and we set about looking at wild-type flower development using scanning electron microscopy (Smyth et al. 1990). Given the size of Arabidopsis flowers, this technique was critical to interpreting mutant phenotypes, although it was not without its challenges (see Bowman 2013). David also undertook a mutant screen to isolate additional mutant alleles of the above genes as well as identify other loci (Smyth 2023). The additional alleles were key to estimating whether alleles were likely to be null or hypomorphic, influencing the construction of the model. One early incarnation was drawn on the laboratory fridge door, where we debated whether the A/C boundary was sharp or fuzzy—and based on strong *ap2* alleles (Komaki et al. 1988; Kunst et al. 1989; Bowman et al. 1991), we later drew the boundary sharp. In comparison to older models, the idea of concentric rings of activity was a common theme, only now each concentric ring was defined by specific gene activities (Bowman et al. 1989).

Simply stated, the ABC model postulates 3 distinct gene activities (A, B, and C), each of which are present in 2 adjacent whorls of the Arabidopsis flower, act alone and in combination to specify the 4 types of floral organ (Fig. 2, A to D) (Bowman et al. 1991). A second postulate was that the A and C activities are mutually antagonistic, a prerequisite to be consistent with the single mutant phenotypes (Fig. 2, E to I). The model immediately suggests phenotypes for most double and triple mutants (Fig. 2, J to M). While most double mutant combinations were additive, the key double mutant was *ap2 ag*, where both class A and C are missing, leaving only B present in whorls 2 and 3, and no homeotic function in whorls 1 and 4, both situations not found in any whorl in wild-type flowers (Fig. 2L). The *ap2 ag* phenotype could be described as carpelloid sepals in the first and fourth whorls and stamenoid petals in the second and third whorls, both being new organ types not observed in wild type (Bowman et al. 1989, 1991). Two conclusions can be drawn from the observed phenotype. First, *agamous* mutations have phenotypic effects in the outer whorls in an *ap2* background and vice versa, indicating that the 2 gene activities are in fact mutually

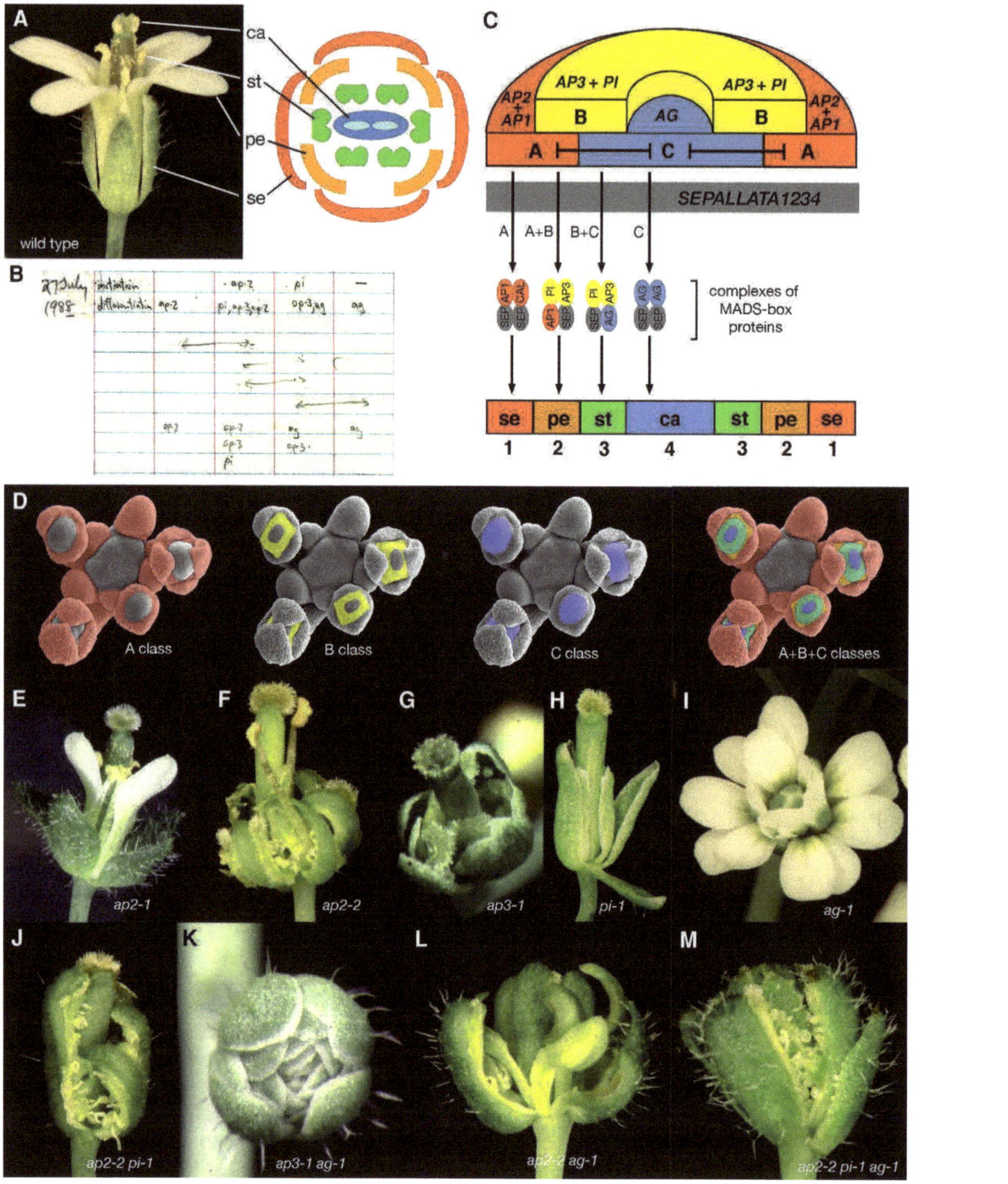

Figure 2. The ABC model and mutant phenotypes on which it is based. **A)** Wild-type Arabidopsis flower (left) and floral diagram (right). se, sepal; pe, petal; st, stamen; ca, carpel; floral structure: se, pe, st, ca. **B)** Early version of the ABC model from mid-1988. **C)** ABC model in its current incarnation; the ABC activities are whorl specific while those of SEPALLATA are throughout the flower. "Dimers of dimers," or quartets, of MIKC^{C}-type MADS box proteins determine organ identity. **D)** Early expression of the ABC genes mapped onto developing flowers. **E)** *apetala2-1*: floral structure: le (leaf), pe, st, ca. **F)** *apetala2-2*: floral structure: ca, missing, st, ca. **G)** *apetala3-1*: floral structure: se, se, ca, ca. **H)** *pistillata-1*. floral structure: se, se, ca (most fused with the central carpels), ca. **I)** *agamous-1*: floral structure: (se, pe, pe)$_n$—the flower meristem is indeterminate with the organs repeating. **J)** *apetala2-2 pistillata-1*: floral structure: ca, missing, missing, ca—in some cases all organs are fused into a single structure. **K)** *apetala3-1 agamous-1*: floral structure: (se, se, se)$_n$. **L)** *apetala2-2 agamous −1*: floral structure: (le/ca (carpelloid leaf), st/pe (stamenoid petal), st/pe)$_n$. **M)** *apetala2-2 pistillata-1 agamous-1*: floral structure: (le/ca (carpelloid leaf), le/ca, le/ca)$_n$. All photos by J.L.B.; A (Lee et al. 2005b); D, F, H, J, L, M (Bowman 2001; copyright 2001, John Wiley & Sons, Ltd., reproduced with permission); G (Bowman et al. 1989); K (Bowman et al. 1991; copyright 1991, Company of Biologists, reproduced with permission).

antagonistic. Second, that the organs in the middle 2 whorls are neither wild-type petals nor wild-type stamens indicates that B class alone in Arabidopsis does promote one or the other. Intriguingly, the cells of these organs are also intermediate between wild-type petals and wild-type stamen cells and thus formed a new cell type. This is in contrast to cases where we observed mosaic organs with distinct boundaries between known cell types in single or multiple mutant combinations (Bowman et al. 1989, 1991), suggesting that floral organ identities were determined largely cell autonomously. One mutant allele of *APETALA3*, *ap3-1*, and all alleles of *APETALA2* were found to be temperature sensitive, allowing the timing of gene activity to be elucidated. Consistent with a role in organ identity specification, gene function was required during the time of organ initiation and early development, with the sensitive period of stamens being quite abbreviated while that of petals was more prolonged (Bowman et al. 1989).

Concomitant with our work in Arabidopsis, 2 groups working with *Antirrhinum* (Enrico Coen's group at the John Innes Centre in Norwich and Suzanna Schwarz-Sommer's group at the Max Planck Institute in Cologne) identified mutants with similar phenotypes and independently came up with a BC model of flower development (Carpenter and Coen 1990; Schwarz-Sommer et al. 1990), the major difference being the lack of *Antirrhinum* A function genes. While it might be difficult to fathom among the current generation of young geneticists, the ABC/BC models were constructed en-tirely with genetics (no molecular biology or genomics) and could have been formulated decades or even a century before [see (Bowman et al. 2012)]—we even named one room in Elliot's laboratory where histology was performed "the 19th-century room." Thus, one question was, why did it take so long? And a second question, why did multiple groups converge on the same methodology and largely similar conclusions almost simultaneously? Such convergence is a common theme in science and must indicate that the time was ripe for discovery, either via a technical or conceptual advance, the latter being key in the present case. I was person-ally influenced by Ed Lewis, whose laboratory was upstairs and whose seminal work in *Drosophila* on the identification and characterization of the bithorax complex led to a model whereby multiple genes act in combination to specify the identity of the thoracic and abdominal segments of the fly body (Lewis 1978)—I recall with both fondness and horror that one of my first experiences as a graduate student was to present this paper in a journal club with Ed in the audience, who "gently" corrected me when I went astray. Thus, it did not seem such a large jump to think about a similar model for the concentric whorls of the flower.

Molecules

The mid-1980s were a time of revolutionary insight into the molecular bases of animal development, with the discovery that genes of the *bithorax* complex and *Antennapedia* encoded paralogous proteins now known as homeobox proteins (Garber et al. 1983; McGinnis et al. 1984b; Scott and Weiner 1984; Karch et al. 1985) and that homologous genes were widespread in animals (McGinnis et al. 1984a). It was immediately speculated that homologs might exist in plants and these might also encode homeotic functions, despite the independent evolution of multicellular plants and animals from a single-celled ancestor. While homeobox genes were subsequently identified in plants (Vollbrecht et al. 1991), the first ABC gene cloned was *DEFICIENS*, a B class gene in *Antirrhinum*, and rather than being similar to homeobox proteins, it had sequence similarity with the known transcription factors MCM1 of yeast and SRF of mammals (Sommer et al. 1990). Some months later, the cloning of the Arabidopsis C class gene, *AGAMOUS*, was reported, and this too was homologous with MCM1, DEF, and SRF (Yanofsky et al. 1990), giving rise to the moniker of "MADS"-box for the putative DNA binding sequence of the gene family (Schwarz-Sommer et al. 1990). Given that 2 of the ABC genes encoded MADS-box genes, it was immediate-ly speculated that other ABC genes may also encode homolo-gous proteins. While the first genes cloned, *DEFICIENS* and *AGAMOUS*, were identified via transposon/T-DNA tagging approaches, cloning of subsequent genes took advantage of homology-based approaches. Although one could not scan genome sequences at the time, low stringency Southern blots and library screens provided a fruitful approach to clone paralogs (Fig. 3A). Thus, the remaining B and C class genes in both Arabidopsis and *Antirrhinum* were quickly identified, either by homology to MADS-box genes in general or via assumption of orthology between the 2 species (Jack et al. 1992; Tröbner et al. 1992; Bradley et al. 1993; Goto and Meyerowitz 1994). However, in retrospect the *Antirrhinum* C-class gene *PLENA* turned out to be a ortholog of Arabidopsis *SHATTERPROOF1/2* genes, which are sister paralogs of *AGAMOUS* (Davies et al. 1999). Later, the Arabidopsis A class gene *APETALA1* was also cloned based on MADS-box homology (Mandel et al. 1992b). The *Antirrhinum AP1* ortholog, *SQUAMOSA*, was not recognized as an A class gene but rather was proposed to play a role in flower meristem identity (Huijser et al. 1992), a theme we will return to shortly. It was also soon recognized that plant genomes harbor a large contingent of MADS-box genes that extended well beyond the ABC genes [e.g. (Ma et al. 1991; Pnueli et al. 1991)], and these would subsequently be targets of future reverse genetic approaches (Fig. 3B).

The cloning of the genes enabled the examination of the ABC model from a molecular perspective, with expression patterns of B and C class genes largely congruent with mutant phenotypes (Yanofsky et al. 1990; Jack et al. 1992; Schwarz-Sommer et al. 1992; Tröbner et al. 1992; Bradley et al. 1993; Goto and Meyerowitz 1994). Critically, in Arabidopsis, the expression pattern of *AGAMOUS* expanded into all floral whorls in the A-class mutant *apetala2*, supporting the hypothesized antagonism between these 2 classes (Drews et al. 1991). Having the cloned genes in hand also allowed further tests by creating gain-of-function alleles in

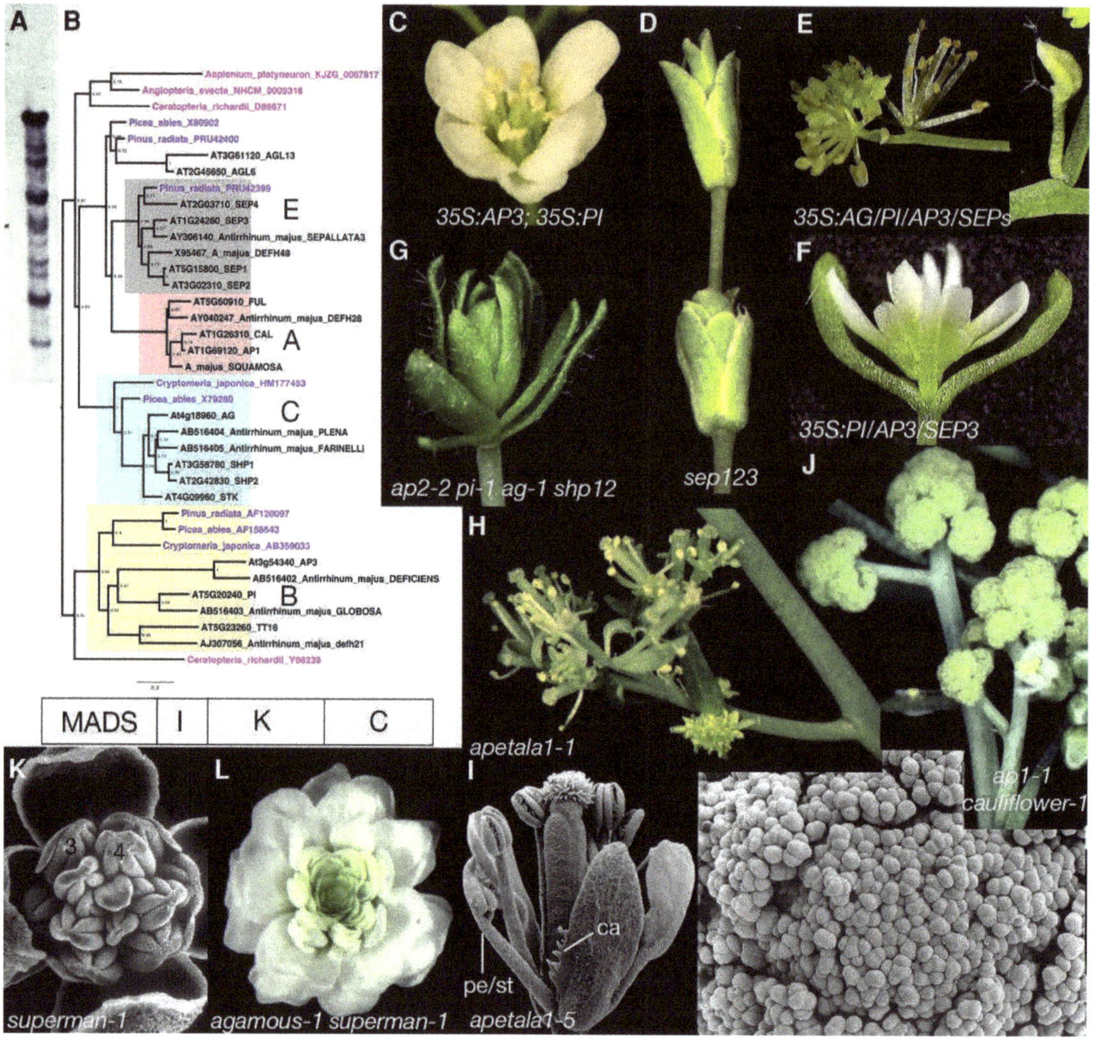

Figure 3. Molecular and genetic extensions of the ABC model. **A)** Low stringency Southern blot using *AGAMOUS* as a probe on Arabidopsis DNA cut with *Eco*RI. **B)** Simplified MADS-box gene phylogeny: A-class, red; B-class, yellow; C-class, blue; E-class, grey; gymnosperm sequences, purple; fern sequences, pink. In both B and C clades, additional angiosperm paralogs are present and often have functions related to the B and C-class genes. Gymnosperm orthologs exist (or existed) for all classes, and the split between the B-class and A/C/E classes occurred in the common ancestor of euphyllophytes. The MIKC^{C} protein structure is indicated at the bottom; MADS, DNA binding domain; I, intervening domain; K, keratin-related domain; C, carboxyl terminal domain. **C)** *35S:APETALA3; 35S:PISTILLATA*: floral structure: pe, pe, st, st. **D)** *sepallata123*: floral structure: (se, se, se)$_n$. **E)** *35S:AG/PI/AP3/SEP3*: floral structure: (st, st, st)$_n$; leaves are transformed into stamens. **F)** *35S:PI/AP3/SEP3*: leaves, but not cotyledons, are transformed into petals. **G)** *apetala2-2 pistillata-1 agamous-1 shatterproof12*: floral structure: (le, le, le)$_n$. **H)** *apetala1-1*: floral structure: br (bract-like), pe (or missing), st, ca; first whorl bracts may also have flower meristem in their axils, and ectopic flower meristems can arise. The photo depicts a "single" basal flower, with later produced flowers having decreasing indeterminacy. **I)** *apetala1-5*: floral structure: br (or br/ca), pe (or pe/st), st, ca. this flower has both a stamenoid petal (pe/st) and carpel (ca) tissue at the margin of a first whorl bract. **J)** *apetala1-1 cauliflower-1*: when grown in nonfloral inductive conditions, what should have been a flower is only a collection of inflorescence meristems that reiterate a spiral initiation pattern (lower panel). **K)** Additional whorls of stamens (4) develop interior to the third whorl (3) in *superman* flowers; floral structure: se, pe, st, st, (st). **L)** Flowers of *superman agamous* double mutants have an outer whorl of sepals and an indeterminate number of whorls of petals; floral structure: se, pe, pe, pe, etc. E-F courtesy of Koji Goto; all other photos by JLB except G by Ji-Young Lee; C (Bowman 2001; copyright 2001, John Wiley & Sons, Ltd., reproduced with permission); D, G (Lee et al. 2005b); I (Bowman et al. 1993; copyright 1993, Company of Biologists, reproduced with permission).

Arabidopsis (disappointingly, *Antirrhinum* is still difficult to transform). Thus, it was noted that ectopic expression of the *Brassica napus AGAMOUS* gene throughout the flower was sufficient to transform the identity of the outer whorls of tobacco flowers into carpels and stamens, respectively, upholding one tenet of the model and also providing additional support for universal ABC gene activity across eudicots (Mandel et al. 1992a). Likewise, ectopic expression of both *APETALA3* and *PISTILLATA* was sufficient to produce homeotic transformations in the first and fourth whorls (Krizek and Meyerowitz 1996) (Fig. 3C). Given that *APETALA3* and *PISTILLATA* were required for B-class activity [they act as an obligate heterodimer (Schwarz-Sommer et al. 1992; Goto and Meyerowitz 1994; Davies et al. 1996; Riechmann et al. 1996)], one surprise was that ectopic expression of *PISTILLATA* alone partially transformed sepals into petals and ectopic expression of *APETALA3* alone partially transformed carpels into stamens. These observations were reconciled by careful examination of gene expression patterns, with *APETALA3* expressed during early sepal development and *PISTILLATA* expression detected in the fourth whorl. Thus, where the 2 gene expression patterns overlap determines B-class activity, and it is possible that the "sloppy" expression patterns could facilitate evolution of perianth phenotypes. Notably, where B- or C-class genes were expressed throughout the entire plant, the leaves were curled and small, but no transformation of leaves into floral organs was observed (Mandel et al. 1992a; Krizek and Meyerowitz 1996).

Converting leaves into floral organs

A major advance at the end of beginning of this century was the discovery of E-class MADS box function. Pivotal was the continued investigation of MADS-box paralogs in Arabidopsis and the discovery that a triple mutant of 3 paralogs resulted in a flower with all sepals (Fig. 3D), implying that this class of gene was required for B- and C-class function (Pelaz et al. 2000). Inclusion of mutations in a fourth paralog resulted in flowers with all leaf-like organs, indicating their requirement for all 4 floral organ types (Ditta et al. 2004). The genes, named *SEPALLATA*, were expressed through the flower and did not appear to provide any organ identity specification activity on their own because their ectopic expression alone is not sufficient to convert leaves to floral organ identity (Honma and Goto 2001); for this reason, I argued that this should be called F-class, for Flower. Remarkably, ectopic expression of ABC genes along with *SEPALLATA* genes was sufficient to convert leaves into floral organs (Honma and Goto 2001; Pelaz et al. 2001). Thus, the MADS-box genes are necessary for the specification of floral organ identity and are sufficient (along with E-class) to convert leaves into floral organs, supporting ideas of the evolution of floral organs from leaf-like ancestors (Fig. 3, E and F). It was noted earlier that DEFICIENS, GLOBOSA, and SQUAMOSA proteins could form a ternary complex (Egea-Cortines et al. 1999), and ternary and quaternary complexes of Arabidopsis proteins including ABC and SEPALLATA MADS-box proteins (e.g. B + C + SEP and A + B + SEP) were observed, prompting the hypothesis that these complexes form the molecular basis of the ABC model (Honma and Goto 2001). These observations were summarized in the "quartet" model wherein MADS-box proteins act as "dimers of dimers' in their binding to regulatory sequences of target genes (Theissen and Saedler 2001) (see "*Good things come in fours*" section below).

That the ABC triple mutant produced carpelloid leaves while the organs of the *sepallata1234* quadruple were more leaf-like suggested that additional MADS-box genes active in the ABC triple mutant background contributed to carpel differentiation. Two *AGAMOUS*-related paralogs, *SHATTERPROOF1/2*, were found to be ectopically expressed in the *apetala2 agamous* background and shown to be responsible for AGAMOUS-independent carpel development (Savidge et al. 1995; Pinyopich et al. 2003). Thus, flowers of the pentuple mutant, *apetala2 pistillata agamous shatterproof1/2* (Fig. 3G), consist entirely of leaf-like organs (Lee et al. 2005a).

A molecular explanation for the indeterminacy of *agamous* flowers was provided by the observation that WUSCHEL, which acts to maintain stem cells in the flower meristem, is repressed directly by AGAMOUS (Sun et al. 2009; Liu et al. 2011). The repression of *WUSCHEL* may also explain other determinacy effects observed when carpels develop in inappropriate whorls, such as the reduced numbers of organs in genotypes containing *apetala2* null alleles. Finally, that carpels promote fusion and determinacy is supported by reduced determinacy in genotypes, where carpel identity is reduced by introduction *SPATULA* or *CRABS CLAW* mutant alleles (Alvarez and Smyth 1999).

The curious case of A-class and subfunctionalization of B- and C-classes

When venturing away from Arabidopsis, genetic manipulations did not always yield the phenotypes expected if function was strictly conserved with Arabidopsis ABC orthologs. Diving into the complex histories of plant genomes revealed that gene/genome duplications, followed by neo- and subfunctionalization or loss, are the source of many variations observed.

The conservation, or even the existence, of A-class has been debated, with the most common hypothesis being that A-class is a continuation of FM identity (e.g. Causier et al. 2010), as is also observed in Arabidopsis. In Arabidopsis, A-class was defined as having 2 functions in the perianth—the repression of C-class and an organ identity role (Bowman et al. 1991). That there exists a distinct A-class activity is supported by the observation that organs of the middle whorls of *apetala2 agamous* flowers are petal/stamen hybrids, and thus, B-class plus SEPALLATA does not specify petals. Furthermore, sepals are not leaves (or bracts), and thus their identity must be specified by some activity, even if it overlaps with the activity that promotes FM identity.

A-class activity was proposed to be composed of the MADS-box gene *APETALA1* and *APETALA2* (Irish and Sussex 1990; Bowman et al. 1991, 1993), encoding an AP2-type transcription factor (Jofuku et al. 1994). In null *apetala1* alleles sepals are converted into bract-like organs (Fig. 3H), and petals are usually missing (Irish and Sussex 1990; Bowman et al. 1993). However, in hypomorphic *apetala1* alleles (Fig. 3I) carpelloid tissue can be formed on the first whorl organs and the petals may be partially transformed into stamens (Bowman et al. 1993), a phenotype often neglected in later reviews of the ABC model. A role for *APETALA1* in organ identity could also be masked by the action of *CAULIFLOWER*, a closely related paralog (Kempin et al. 1995): the double mutant *ap1 cal* produces inflorescence meristems instead of flower ones (Fig. 3J), indicating both genes have a role in FM identity and precluding observation of potential organ identity defects (Bowman et al. 1993). Notably, the *sepallata1234* quadruple mutant does not appear to have defects in FM specification, although *SEPALLATA4* may have redundant roles with *APETALA1* in this regard (Ditta et al. 2004).

Consistent with mutual antagonism and A and C classes in Arabidopsis, *AGAMOUS* mRNA is detected in all 4 whorls of *apetala2* mutants (Drews et al. 1991), and *APETALA1* mRNA expands to the inner whorls in *agamous* flowers (Gustafson-Brown et al. 1994). However, *AGAMOUS* mRNA also accumulates in the inner whorls of *agamous* flowers (Gustafson-Brown et al. 1994), and *APETALA2* mRNA can be found at some levels in all 4 whorls of wild-type flowers (Jofuku et al. 1994), indicating that their mutual antagonism is complex. The negative regulation of *APETALA2* by miR172 provides one mechanism to restrict AP2 activity post-transcriptionally (Aukerman and Sakai 2003; Chen 2004). Expression of miR172 in the central whorls during early flower development appears to provide quantitative regulation of AP2 activity, with the balance of *AGAMOUS* and *APETALA2* activity, especially in the third whorl, proposed to influence organ identity (Wollmann et al. 2010).

Outside of Arabidopsis, findings in asterids demonstrated that floral homeotic phenotypes are incompatible with a straightforward transfer of Arabidopsis A ortholog function (Cartolano et al. 2007; Morel et al. 2017). The AP2 family comprises *APETALA2*-type and TOE-type clades (Wang et al. 2016). Petunia *AP2A/ROB1* and *Antirrhinum LIP1/2* genes are *APETALA2* orthologs, but unlike *APETALA2*, these are unable to repress the C-function. Instead, it is another miRNA, called *BLIND* in *Petunia* and *FIS* in *Antirrhinum*, that tackles this task by targeting NF-YA transcription factors. *BLIND* is helped by *BLIND ENHANCER* (*BEN*), and only the double mutant *blind ben* mimics the expected A class phenotype (Morel et al. 2017). *BEN* is a TOE-type AP2 whose members control flowering time, not organ identity, in Arabidopsis. *BEN* was also shown to work with the 3 Petunia AP2-type genes called *ROBs* (*Repressor Of B function*) so that *BEN* and *ROBs* together repress B function (but not C-function) in the sepal whorl. To summarize, in *Petunia* A function relies on a combination of a related AP2 gene and an unrelated microRNA (*BEN* + *BLIND*); reciprocally, the *APETALA2* orthologs (*ROB* genes) do not re-press C function but rather team up with *BEN* to repress B function.

Compared to A class, the molecular agents fulfilling B and C functions appear broadly conserved across angiosperms. This difference in extent of conservation may be related to the antiquity of the functions, with B and C evolving at the dawn of seed plants, while A-function appears to be an angiosperm novelty (Wellmer et al. 2014). Nonetheless, variations involving related paralogs exist. For example, the B gene *APETALA3* is only 1 of 2 old twins: a duplication created *APETALA3* and *TM6* somewhere near the base of eudicots (Kramer et al. 1998; Kim et al. 2004), but Arabidopsis and *Antirrhinum* have both lost their *TM6* ortholog. So the original floral development genetic models only tell part of the story, and B function can be more complicated in other spe-cies. In *Petunia* the *APETALA3* ortholog (Ph*DEF*) and Ph*TM6* redundantly control stamen identity, but petals are specified by Ph*DEF* only. Ph*tm6* mutants have a wild-type phenotype, whereas in a Ph*def* mutant petals are replaced by sepals but the stamens are normal (Vandenbussche et al. 2004; Rijpkema et al. 2006). A similar logic applies to the PISTILLATA lineage, which represents the other leg of the B function in Arabidopsis: the *Petunia* genome hosts 2 PI orthologs acting redundantly to spe-cify petal and stamen identity (Vandenbussche et al. 2004), with one paralog having acquired an additional role controlling stamen to petal tube fusion. Hence, the expected B phenotype is only obtained when both paralogs are defective. Such duplications also play important roles for petal elaboration (see the "Beyond eudicots" section below), impacting not only organ identity specification but morphogenesis as well (Kramer et al. 2007; Sharma and Kramer 2013; Hsu et al. 2015, 2021).

Evolution within the MADS-box gene family also affected how C-function is executed. Duplications yielded 3 genes in the "C-lineage": *AGAMOUS*, *SHATTERPROOF1/2* (*PLENA*), and *SEEDSTICK*, with the first duplication producing *SEEDSTICK* occurring in the ancestral seed plant and the other occurring within eudicots [e.g. (Carvalho et al. 2018)]. In *Petunia* both *AGAMOUS* and *SHATTERPROOF1/2* orthologs must be inactivated to produce typical C-class mutant phenotype (Kapoor et al. 2002; Heijmans et al. 2012). In Arabidopsis, *AGAMOUS* fulfils the C function with *SHATTERPROOF1/2* controlling aspects of carpel and fruit development (Liljegren et al. 2000). In contrast, *PLENA* executes the C function in snapdragon, while the *AGAMOUS* ortholog, *FARINELLI*, only plays a secondary role during stamen development (Bradley et al. 1993; Davies et al. 1999). The third paralog of the clade, *SEEDSTICK*, directs ovule development (Pinyopich et al. 2003) and is sometimes referred to as D class that is unrelated to floral organ identity. However, even *SEEDSTICK* alone can promote aspects of carpel development in the absence *AGAMOUS* (Favaro et al. 2003), and perhaps its role in ovule development reflects an ancestral function of this clade since ovules predate flowers.

Regarding E-class, in rice (Ohmori et al. 2009; Li et al. 2010), *Petunia* (Rijpkema et al. 2009), orchid (Hsu et al. 2015, 2021), and maize (Thompson et al. 2009) *SEPALLATA* orthologs and also their close paralogs, the *AGL6* genes, can all deliver E function, adding another layer of redundancy and complexity to the ABC story (for review, see Dreni and Zhang 2016).

To accommodate these variations, it was proposed that the model be referred to as the (A)B(C) model (Causier et al. 2010; Theissen et al. 2016) with (A) accounting for A and E activities and (C) combining C and D roles. However, now that we have a broader evolutionary perspective of gene evolution within the ABCE clades (Fig. 3B) combined with functional analyses (above and later), most described variation within core eudicots is derived from vagaries in subfunctionalization of paralogous genes rather than any fundamental differences. It remains to be seen whether exceptions will occur in species with "noncanonical" perianths, such ANA-grade and magnoliid angiosperms (Wellmer et al. 2014).

Paddling upstream and downstream

As all big discoveries, the formulation of the ABC model immediately raised difficult questions: what regulatory processes generate the spatio-temporal expression patterns central to the homeotic function of ABC genes? How can the combinations uncovered by genetic analyses transpose into molecular mechanisms allowing different protein complexes to activate distinct developmental programs? How do the regulatory activities of TF-encoding ABC genes translate into morphogenic effects that shape the different types of floral organs, specifying a range of often organ-specific cell fates? Although we still lack a high-resolution picture, the last 20 years of floral development research have added many pieces to this incomplete puzzle. Our understanding of the complex networks behind the characteristic ABC expression domains has been synthesized regularly over the last 2 decades (Zahn et al. 2006; Causier et al. 2009; Pelayo et al. 2021). In the interest of brevity, we only summarize here the main messages and highlight recent breakthroughs toward solving long-lasting riddles.

Swimming upstream

Early work rapidly established that regulation of ABC genes expression is a multi-layered process integrating inputs from hormonal signaling pathways with the activity of transcription factors, histone modifying enzymes, and chromatin remodeling complexes and roles for post-transcriptional mechanisms often involving mobiles noncoding RNAs and RNA processing factors (e.g. Cheng et al. 2003; Chen 2004; Zhao et al. 2007; Yamaguchi et al. 2009; Wu et al. 2012, 2018; Jung et al. 2014; Sacharowski et al. 2015; Poyatos-Pertíñez 2016; Morel et al. 2017; Lee et al. 2022; Lin et al. 2023). These investigations exposed a 3-step waltz regulating ABC genes: upstream factors initiate transcription across broad territories (step 1), followed by refinement (step 2), and then maintenance (step 3) of precise expression domains. Step 2 and 3 rely extensively on cross regulatory interactions and feedback loops between floral homeotic genes in a man-ner reminiscent of the processes establishing discreet stripes of pair-rule gene expression in flies: combined inputs of maternal determinants and GAP genes first trigger expression of pair-rule genes in broad territories, which then get refined through cross regulations between the pair-rule genes themselves (Edgar et al. 1989; Pankratz and Jäckle 1990) (Fig. 4). Flower patterning follows a similar logic: the broad expression domains of ABC genes in nascent floral meristems largely overlap at first, but cross inhibition between A and C later renders their expression mutually exclusive while positive feedback loops between *APETALA3* and *PISTILLATA* enhance and eventually restrict any significant B genes transcription to the subset of meristematic cells that initially expressed both (Fig. 4). Additionally, B-class genes may act to directly repress genes required for carpel development, providing another layer of cross-regulation (Wuest et al. 2012).

Cadastral genes help define and maintain sharp boundaries between the expression domains of floral homeotic genes: for example, *SUPERMAN* and *RABBIT EARS* encode C2H2 zinc finger repressors that turn off *APETALA3* in whorl 4 (Sakai et al. 1995; Prunet et al. 2017) and *AGAMOUS* in whorl 2 (Krizek et al. 2006; Huang et al. 2012), respectively. In *super-man* mutants, it appears as if the flower meristem is "stuck" in the production of third whorls (Fig. 3, K and L) (Bowman et al. 1992; Baum et al. 2001). These boundaries are no-mans-land for ABC gene expression that keep organ iden-tity apart but also physical borders characterized by a lack of growth, allowing morphogenesis to proceed independently in each whorl to produce discreet organ types.

Many regulators of ABC genes are not exclusive agents of floral organs but instead participate in the control of other developmental processes. A key finding was the discovery that molecular players inducing floral transition directly participate in the activation of floral homeotic selectors. One of those elements, LEAFY (LFY), stands out because it can activate the transcription of A, B, C, and E genes by binding directly to their regulatory regions (Parcy et al. 1998; Busch et al. 1999; Wagner et al. 1999; Honma and Goto 2001; Lohmann et al. 2001; Lamb et al. 2002; Benlloch et al. 2011; Moyroud et al. 2011; Winter et al. 2011; Rieu et al. 2023). Because LFY is a central regulator of the reproductive transition, these results elegantly connect induction of flowering to specifica-tion of floral organ identity but also yield an initial paradox: how can a regulator expressed throughout the floral buds activate different floral homeotic genes in spatially distinct territories? With its uniform expression across floral meristems, *LFY* seemed at first ill-suited to such task.

Watch this space

A possible solution lies in LFY's ability to activate transcription of selected genes only when interacting with specific partners confined to distinct subdomains of the floral meristem. The model proposes WUSCHEL as a cofactor of LFY to promote C gene expression in the inner whorls, UNUSUAL FLORAL ORGAN (UFO) to turn on B genes in whorl 2 and 3 (Levin and Meyerowitz 1995; Wilkinson and Haughn 1995; Lee et al. 1997; Parcy et al. 1998; Honma and Goto 2000; Lenhard et al. 2001), and an unknown cofactor to acti-vate A expression (Parcy et al. 1998; Lohmann et al. 2001) (Fig. 4). However, mechanistic evidence supporting such syn-ergies were initially lacking. The UFO-LEAFY partnership was especially puzzling: fusing a strong activation VP16 domain to LEAFY is sufficient to increase *APETALA1* and *AGAMOUS* expression but fails to affect *APETALA3* transcription, suggesting that the cofactor is also needed for LFY binding to a B gene promoter (Parcy et al. 1998). However, *UFO* encodes

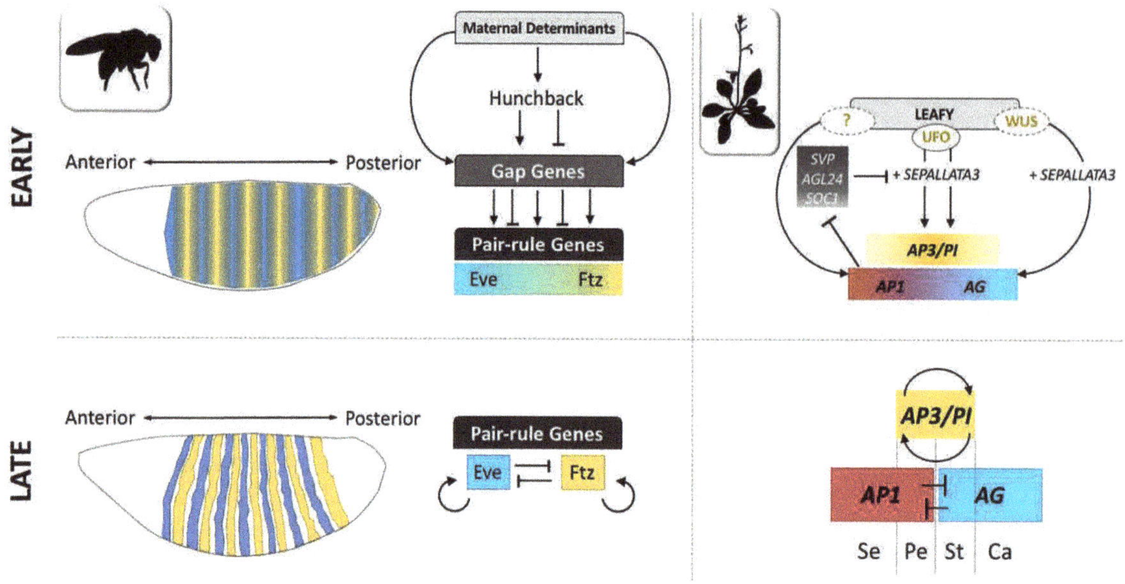

Figure 4. Parallels of developmental gene network architectures. Patterning of pair-rule genes expression [such as *Even-skipped* (*Eve*) and *Fushi tarazu* (*Ftz*)] along the antero-posterior axis of the fly body involves the induction of broad overlapping expression domains by various combination of upstream regulators (GAP genes) at early stages, followed by establishment of discreet stripes through cross-regulation between the different pair-rule genes. Similarly, the overlapping expression domains of ABC genes are refined through the action of positive feedback (i.e. between *APETALA3* (*AP3*) and *PISTILLATA* (*PI*) B class genes) and cross inhibition (i.e. between A and C class genes). Feedforward loops are also common regulatory motifs to control the timing of events patterning the fly body or the floral meristem of Arabidopsis: in Drosophila, maternal determinants such as Bicoid activate the expression of Hunchback and together they pattern GAP gene expression. In Arabidopsis LFY promotes *APETALA1* (*AP1*) transcription during floral induction, which inhibits expression of *SEPALLATA3* repressors (AGL24/SVP/SOC1), preventing precocious expression of B and C class genes. LEAFY and UNUSUAL FLORAL ORGAN (UFO) physically interact (Chae et al. 2008); but whether the partnerships LEAFY—WUSCHEL (WUS) and LEAFY—unknown factor (?) that regulate the expression of A and C class genes respectively also require direct protein-protein interaction remains unclear (Lohmann et al. 2001; Zhao et al. 2018). Se, sepal; Pe, petal, St, stamen; Ca, carpel.

an F-box protein involved in ubiquitination of proteins heading for degradation and, as such, not obviously connected to protein-DNA interactions.

Results obtained a decade later show that LFY DNA binding domain and UFO interact physically in vivo—not via the F-box that enrolls UFO in the SCF complex but via a C-term Kelch-type β-propeller domain and that this interaction is necessary to recruit UFO to *APETALA3* promoter (Chae et al. 2008), but the mystery surrounding LFY-UFO teamwork remained incomplete for the next 15 years. By coupling functional investigation with biochemical approaches, Rieu and colleagues confirmed that the contribution of UFO to the activation of B gene does not require ubiquitination. Instead, the 3D visualization of the LFY-UFO complex bound to DNA inferred from cryoEM data and AlphaFold2 predictions showed that UFO contact both DNA and LFY and thus constitute a bona fide cofactor. Remarkably, UFO tweaks the DNA binding properties of LFY so that it can now stick to DNA sites (named LUBS for LFY-UFO Binding sites) devoid of the canonical LFY cis-elements. In other words, UFO acts as a fairy godmother that turns DNA motifs unable to recruit LFY effectively into strong binding sites. Neither UFO nor LFY can attach to such LUBS by themselves because their affinity is too weak. This elucidates the molecular causes of the LFY-UFO synergy proposed nearly 25 years ago and also explains why previous studies failed to find strong LFY binding sites in the *PISTILLATA* promoter (Rieu et al. 2023). Beyond solving the mystery of B gene regulation by LFY, this uncovered a novel mechanism where an F-box protein directly regulates transcription by modulating the DNA binding specificity of a central regulator of development. Hence, 30 years on research into the ABC model keeps yielding breakthroughs reaching far beyond a mere understanding of flower formation.

A question of tempo

The timing of ABC expression also needs to be carefully balanced: a risk inherent to being controlled by FM identity factor like LFY is that homeotic genes could be expressed too early, leading to incomplete floral organ differentiation and precocious meristem termination. The structure of the regulatory network orchestrating ABC expression provides a safeguard by embedding a regulatory delay in the form of a feedforward loop between *LFY*, *AP1*, and *SEP3*, once again, not unlike feedforward loops at work during the early stages of drosophila body patterning (Fig. 4). *SEP3* and *LFY* together activate B and C gene transcription (Liu et al. 2009; Winter et al. 2011). *LFY* and *APETALA1* activate each other during FM specification (Goslin et al. 2017), but SEP3 is absent at that stage as its expression is repressed by SVP, AGL24 and

SOC1 (Liljegren et al. 1999; Wagner et al. 1999). APETALA1 accumulation later allows repression of SVP/AGL24/SOC1 triggering SEP3 production and eventually B and C gene expression (Liu et al. 2009).

ABC genes must also keep quiet during the vegetative phase at the risk of interfering with correct leaf development; this is not simply due to the absence of LFY but also necessitates active silencing of floral homeotic genes (Goodrich et al. 1997). Recent studies have started to identify specific factors involved in recruiting polycomb complexes to those loci: for instance, 2 Zinc Fingers were shown to recruit PRC2 to induce heterochromatin formation at *APETALA1*, *APETALA3*, and *AGAMOUS* in Arabidopsis (Hu et al. 2023). Those factors are repressed by LFY and APETALA1, allowing release of vegetative inhibition once floral induction is set in motion.

High maintenance

ABC gene expression is initiated and spatially polished in young FM but must also be maintained at later stages. Failure to do so impairs floral morphogenesis, prompting floral organ identity switch and incomplete differentiation of floral organs. The mechanisms responsible for expression maintenance differ from those accounting for ABC initiation, and piecing together findings from the last 20 years builds a clearer picture of those late processes: for instance, *AG* is eventually switched on even when LFY is absent (Weigel and Meyerowitz 1993), and this late AG expression relies in part on post-transcriptional processes involving HUA and HEN proteins (Western et al. 2002). When those are non-functional, stamen become mosaic organs exhibiting charac-teristics of both petal and sepal tissues, whereas carpels become sepal-like.

Transcription maintenance also engages feedback loops: for example, SEP3 enhances its expression both directly by binding to its own promoter but also indirectly by repressing, together with APETALA1, the transcription of its repressors *AGL24*, *SOC1*, and *SVP* (Kaufmann et al. 2009). Maintenance also involves epigenetic processes—for instance, *ARABIDOPSIS TRITHORAX 1*, a plant homolog of the fly trithorax, is necessary to sustain high enough expression levels of A, C, and to a lesser extent B genes (Alvarez-Venegas et al. 2003).

A similar logic applies to B genes: if their expression is turned off prematurely, patches of sepaloid tissues appear in petals. Examining this behavior in a group of aquatic early diverging angiosperms uncovered a possible role for environmental control: in water lilies, exposure to light, even late in development, allows sepal-like features to develop within otherwise petal-like organs (Warner et al. 2008, 2009). Such observations led to the formulation of the Mosaic theory, according to which genetic programs that elaborate characteristics of sepal and petal tissues evolved early in angiosperms history but were mostly controlled by environmental inputs such as light or mechanical cues. Later on, those programs were placed under the sole control of the floral organ identity network and lost their sensitivity to environmental inputs. This represents a candidate case to backup the evo-devo scenario arguing that plasticity could lead to the evolution of novelty (here the production of a differentiated perianth with clear sepal and petal identities) when followed by accommodation and assimilation (Sommer 2020).

Floating downstream

Another key question is how combinations of ABC proteins lead to floral organ differentiation. One approach to identifying downstream targets was to isolate mutants affecting only a single organ type, e.g. carpels (Alvarez and Smyth 1999). For example, *CRABS CLAW* mutants alter carpel development and whereas some effects are independent of *AGAMOUS* activity, others seem to rely on *AGAMOUS* (Alvarez and Smyth 1999). Indeed, when molecular approaches were available, *CRABS CLAW* was found to be a direct downstream target of AGAMOUS (Gómez-Mena et al. 2005; Ó'Maoiléidigh et al. 2013), with requisite dual CArG boxes (MADS-box binding site) for binding of MADS-quartets required for its expression (Lee et al. 2005a). *CRABS CLAW*'s role in carpel development is conserved across angiosperms, with the rice ortholog, *DROOPING LEAF*, indispensable for carpel specification (Yamaguchi et al. 2004).

CRABS CLAW was also found to be required for nectary development (Bowman and Smyth 1999). Arabidopsis nectaries are normally found at the base of stamens and appear to be a third whorl organ independent of the identity of the floral organ occupying that whorl (Baum et al. 2001; Lee et al. 2005a). Such a role for *CRABS CLAW* was found to be widespread in eudicots regardless of nectary position within the flower, or even elsewhere in the plant, leading to speculation that nectaries came under *CRABS CLAW* control when they became associated with reproductive organs within the flower (Lee et al. 2005b). However, my (J.L.B.) foray into flower development was coming to an end as we discovered *CRABS CLAW* was a member of a small gene family, the YABBY gene family (Bowman and Smyth 1999; Siegfried et al. 1999), and enhancer mutagenesis revealed a role for this family and another, the KANADI gene family, in establishing organ polarity not only in the floral organs but also leaves (Eshed et al. 1999, 2001).

Disappointingly, 20 years later our understanding of what lays downstream of ABC remains surprisingly fragmentary: genome-wide binding approaches (SELEX-seq, ChIP-seq, and Dap-seq) have successfully identified in vivo binding sites of ABC regulators associated with genes controlling cell differentiation, division, and expansion, participating in hormonal signaling, or coding for other TFs and even chromatin remodeling and assembly factors (Kaufmann et al. 2009, 2010b; Yant et al. 2010; Wuest et al. 2012; Ó'Maoiléidigh et al. 2013, Pajoro et al. 2014; Chen et al. 2018; Lai et al. 2020). However, picking out bona fide regulatory *cis*-elements from inactive binding sites remains challenging and although many ABC direct targets have been predicted, only a relatively small number have been validated. Despite this, remarkable progress has been made uncovering the mode of action of the floral quartets [recently reviewed in (Goslin et al. 2023; Käppel et al. 2023)] and in particular toward understanding how closely related MADS proteins with very similar DNA binding properties identify targets accurately and selectively among thousands possible targets. Teamwork is key and ABC proteins achieve specificity by following the metaphor of the Swiss cheese model: many leaky layers, which individually would confer very imperfect selectivity, can forge a highly specific system when compounded with each other.

Good things come in fours

Nearly all ABC genes encode MIKC^{C}-type MADS-box TFs, which probably originated from a duplication of MEF2-type genes in the ancestral land plant (Qiu et al. 2023). These TFs display 4 characteristic domains: a MADS-box domain (M), followed by an Intervening (I), a Keratin-like (K), and a C-terminal (C) domains (Ma et al. 1991) (Fig. 3B). Over the last 10 years, SEPALLATA3 has established itself as a model MIKC^{C}-MADS protein for structural investigations (Puranik et al. 2014; Lai et al. 2021). The MADS domain is made of alpha helices exposed by beta-structures and constitutes a well-conserved DNA binding domain (DBD) allowing floral regulators to bind CArG box motifs (CC[A/T]$_6$GG) as dimers. Although the I domain does not directly contact DNA, the MADS domain is unable to bind effectively to its target without it (Lai et al. 2021). The I motif stabilizes the weak interaction between 2 MADS domains, affecting the conformation of the dimer and hence tuning the strength and specificity of the binding to a given locus (Lai et al. 2021). As relatively ill-defined motifs, the CArG boxes are found at high frequency across any plant genome: about 15,000 unique sites in the genome of Arabidopsis match (CC[A/T]$_6$GG) and over 300,000 are detected when 1 mismatch is allowed (Folter and Angenent 2006). ABC proteins have come up with an original strategy to deal with this conundrum common to many TFs.

First, ChIP-seq data independently obtained for A- (Kaufmann et al. 2010b; Pajoro et al. 2014), C- (Ó'Maoiléidigh et al. 2013), B- (Wuest et al. 2012), and E-genes (Kaufmann et al. 2009; Pajoro et al. 2014) revealed that the DNA motif bound by MIKC^{C}-MADS proteins is likely longer (16 bp) than the 10 nucleotides of the original CArG-box. Although this partly explains why not all CArG-box are genuine binding sites, it does not account for differences in specificity between the possible quartets. Mapping genome-wide binding sites of ABC proteins also showed that specific CArG-boxes are not unique targets of a given MADS TF dimer (most CArG-boxes can be bound by most dimers and vice versa) but that preference orders for distinct CArG boxes vary between dimers. For example, DAP-seq data show that SEP3-SEP3 and SEP3-AG dimers have different binding specificities (Lai et al. 2020). Quantitative differences exist not only due to CArG-box base content but also to shape readout (e.g. the width of the minor groove) primarily performed by the N-term extension of the MADS- domain (Käppel et al. 2018) that contributes to specificity of target genes (Muiño et al. 2014; Smaczniak et al. 2017; Aerts et al. 2018).

The second part of the solution to the specificity riddle lies in the physical interactions required to generate quartets: the ABC model quickly led to the assumption that complexes involving 4 floral MADS proteins were needed to trigger specific gene expression, a hypothesis known as the quartet model (Honma and Goto 2001; Theissen and Saedler 2001). Several milestones anchoring this model in reality were subsequentially reached. Those include the reconstitution of quartet formation in vitro (Melzer and Theissen 2009), the elucidation of the structure and role of each subdomain in DNA recognition and tetramer formation (Puranik et al. 2014; Lai et al. 2021), and the demonstration that quartets can induce DNA looping with different quartets exhibiting unique preferences for specific spacing and orientation of the 2 CArG boxes (Melzer et al. 2009; Mendes et al. 2013; Jetha et al. 2014). In vitro (Veron et al. 2007; Zhang et al. 2018) but also in vivo data (Abraham-Juárez et al. 2020; Mao et al. 2021) indicate that quartet composition (e.g. BCE) is mostly conserved across flowering plants. The ability to form quartets rests on the coiled-coil domains of the K region (Puranik et al. 2014), which is made of 2 amphiphatic alpha helixes: the first helix along with the N-term portion of the second one strengthens the dimer. The C-term of the second helix serves as interface for interaction between 2 DNA-bound dimers, mediating tetramerization.

Tetramerization enhances specificity but could also be central to quartet transcriptional activity (Mendes et al. 2013; Hugouvieux et al. 2018): tetramer formation facilitates cooperative binding known to trigger sharp on/off transcriptional responses, well-suited to control critical developmen-tal events (Theissen and Melzer 2007; Kaufmann et al. 2010a). Multimer assembly also creates new interfaces, expanding possibilities to recruit partners such as cofactors or epigenetic regulators (reviewed by Goslin et al. 2023)—how floral quartets recruit chromatin modifying factors remains to be fully understood, but examples are emerging: *STERILE TASSEL SILKY EAR1* (*STS1*) and *SILKY1* (*SI1*) are *PISTILLATA* and *APETALA3* maize orthologs, respectively. IP-MS assays in maize revealed STS1 homodimers and STS1-SI1 heterodimers both interact in vivo with CHR126b, a chromatin remodeling complex ATPase. STS1 homodimers can also contact other chromatin remodeling and scaffolding factors such as CHR12, BRAHMA1, and FRIGIDA-LIKE PROTEIN4a, indicating that floral complex composition directly influences recruitment of different partners (Abraham-Juárez et al. 2020).

Tetramerization appears to be necessary for floral meristem determinacy, and correct organogenesis of floral organs in Arabidopsis as development of petal, stamen, and carpels is severely impaired or even abolished when SEPALLATA3 completely lacks the interface required for tetramer formation (Hugouvieux et al. 2023). However, cooperative 2-site binding to DNA by quartets of Arabidopsis SEPALLATA3-AGAMOUS does not impact the regulation of all target genes equally: it is necessary to trigger floral meristem termination but is not essential to floral organogenesis since expression of SEP$^{\Delta tet}$, a SEPALLATA3 splice variant that exhibits impaired but not abolished tetramerization, suffices to restore petal and stamen development in a *sepallata1/2/3* triple mutant (Hugouvieux et al. 2018). Similarly, AGAMOUS can interact with histone modifiers and chromatin remodelers such as CHR11/CHR17 to activate *YUCCA4* and promote carpel development, but this does not require interaction between a quartet and chromatin modifying agents and might instead result from direct interaction between AGAMOUS and CHR11 [reviewed in (Pelayo et al. 2021)]. Thus, ABC proteins can regulate transcription and contribute to the production of floral organs outside of the floral quartet framework.

ABC through the evolutionary lens

When one examines flower architecture across the quarter-million species of angiosperms, a common theme is organ order. Although flowers differ enormously in numbers, shapes, positions and even numbers of whorls of organs, the organ order (sepals, petals, stamens, carpels) is nearly constant (although missing organ types is a common theme). That similar phenotypes were observed in Arabidopsis, a rosid, and *Antirrhinum*, an asterid, immediately suggested that their common ancestor (that of core eudicots) and perhaps all angiosperms utilized a common genetic program, consistent with a singular origin of flowers (Coen and Meyerowitz 1991). Yet could a universal and relatively simple model underpin the diversity of floral forms seen across the angiosperms? And what is the evolutionary origin of the floral network? To account for various floral morphologies, modifications of ABC model by shifting/sliding or fading boundaries have been proposed (van Tunen et al. 1993; Bowman 1997; Kramer et al. 2003; Buzgo et al. 2004), and we consider only a few variations here.

Variations on a theme

One early line of investigation concerned how the ABC model might apply to grass flowers, which have a highly derived perianth architecture. Maize *silky1* flowers exhibited transformations of stamens into carpels and lodicules into palea/lemma-like organs—a canonical B-class mutant phenotype—and subsequent cloning of the gene revealed *silky1* to encode an *APETALA3* ortholog (Ambrose et al. 2000). Thus, lodicules were interpreted to be petal homologs and the palea/lemma to be sepal homologs (Fig. 5A). C-function was also found to be conserved, although it was split across 2 paralogs (Mena et al. 1996). Similar results were found in rice (Kyozuka et al. 2000), indicating the ABC model applies to grass flowers, despite their divergent perianths. A minor variation of the ABC model was proposed for tulip flowers based on a B-class mutant in which both outer whorls of tepals (perianth organs not differentiated into sepals and petals) were transformed into sepal-like organs not normally found in tulips (Fig. 5B), and stamens were transformed into carpels (van Tunen et al. 1993), with gene expression patterns of ABC orthologs supporting this hypothesis (Kanno et al. 2003).

Transient assays, such as virus-induced gene silencing, facilitated functional investigations in a wider range of systems. Combined with modern -omics approaches, these not only painted a more holistic view of how floral organ identity is specified by the ABC but also started to unveil the molecular bases accounting for morphological variation. Two cases of parallel evolution exemplify sub- and neofunctionalization of B-class genes in the evolution of flower novelty. [I (J.L.B.) once remarked that researchers in the field followed a similar path —they multiplied and diversified via sub- and neofunctionalization...] In the rununculid *Aquilegia*, diversification of *APETALA3* orthologs results in 3 distinct combinations that promote petal, stamen, and staminode identity (Kramer et al. 2007; Sharma and Kramer 2013). In this case a novel or-gan, the staminode, is produced, whose ecological function is unknown, although they likely play a role in pollination (Fig. 5C). Another ranunculid, *Thalictrum*, exhibits expansion of B-class activity into the outer whorl that results in petaloid sepals in a parallel evolution of this phenotype with tulips (Martínez-Gómez et al. 2020).

Sub- and neofunctionalization of B-class is also observed in orchids, where there has been a diversification of *APETALA3* orthologs into 4 distinct clades (Fig. 5D). In this case, the perianth organs are specified by a "perianth code," with each distinct organ type having a different qualitative, and sometimes quantitative, suite of B and E activities that compete with each other, making individual tepals morphologically distinct from each other: sepal/petal vs. labellum (Mondragon-Palomino and Theissen 2011; Pan et al. 2011; Hsu et al. 2015). Such complexes also appear to have gained function beyond specifying organ identity: quartet compo-nents seem to vary spatially with exquisite precision so that subregions within an organ could harbor different multi-mers to sculpt and pattern a given organ section (Fig. 6) —for instance, interfering with the formation of specific quartets in moth orchids prevented curved edge formation and red anthocyanin spots production at the base of lateral sepal (Hsu et al. 2021).

Downregulation of the B-class gene GGLO1 in the Asteraceae family plant *Gerbera hybrida* resulted in transformation of second whorl organs into ones bearing pappus bristles, indicating that the pappus bristles of wild-type Asteraceae flowers represent sepals modified to facilitate seed dispersal (Yu et al. 1999) (Fig. 5E). Another proposed modification of the ABC model is one where the boundaries between the different classes are indistinct, that is, fading/fuzzy borders (Buzgo et al. 2004; Soltis et al. 2007), and which may apply to species where the distinction between sepals and petals or petals and stamens can be vague (Fig. 5F). Lastly, the only known angiosperm in which the organ order

Figure 5. Evolutionary conservation of the ABC model and its variants. **A)** The outermost whorl of grass flowers is occupied by the palea (p) and lemma (l) and the second whorl by lodicules, which swell to facilitate flower opening. **B)** The outer 2 whorls of tulips are tepals, which are petal-like, and B-class gene expression is in all whorls except the innermost carpels. **C)** A proliferation of *APETALA3* paralogs have sub- and neofunctionalized to produce 3 distinct compositions of B-class, with 1 set promoting the development of staminodes (sd), modified stamen-like organs that may act to deter herbivory (Hodges and Tucker 2005). **D)** Orchid also possesses a proliferation of *APETALA3* paralogs that act in combination to promote differentiation of the outer tepals (ot), inner lateral tepals (ilt), and labellum (la) that comprise the perianth. **E)** The capitulum of Asteraceae consists of central disc florets (df) and peripheral ray florets (rf), with the fertile disc florets developing a conspicuous pappus ring (p), homologous with sepals, in their outer whorls. **F)** In some species, organ identity transitions may not be abrupt, such as with petal and stamen identity in *Opuntia*. **G)** *Lacandonia schismatica* is the only angiosperm where the order of organs is transposed, with stamens (st) developing interior to carpels (ca). *L. schismatica* plants (arrow) grow on the Lacandon jungle floor, are small (10 cm), achlorophyllous, and saprophytic. **H)** Pines have separate male and female cones, with C-class being expressed in both and B-class limited to male cones. SEM of *L. schismatica* courtesy of Barbara Ambrose; all other photos by J.L.B.

of the reproductive organs is transposed is *Lacandonia schismatica* (Fig. 5G), a saprophytic plant growing only in the Lacandon jungle of southern Mexico. In this species the B- and C-class gene expression patterns mirror organ identity (Alvarez-Buylla et al. 2010). Closely related to *L. schismatica* is the dioecious *Triuris brevistylus* (Espinosa-Matias et al. 2012), whose female flowers resemble those of *L. schismatica* except for the central stamens, providing a possible evolutionary pathway to the unique floral architecture of *L. schismatica*.

Another early avenue of excitement was whether the ABC genes would be involved in the evolution of monoecy or dioecy in angiosperms. Now that a number of dioecious

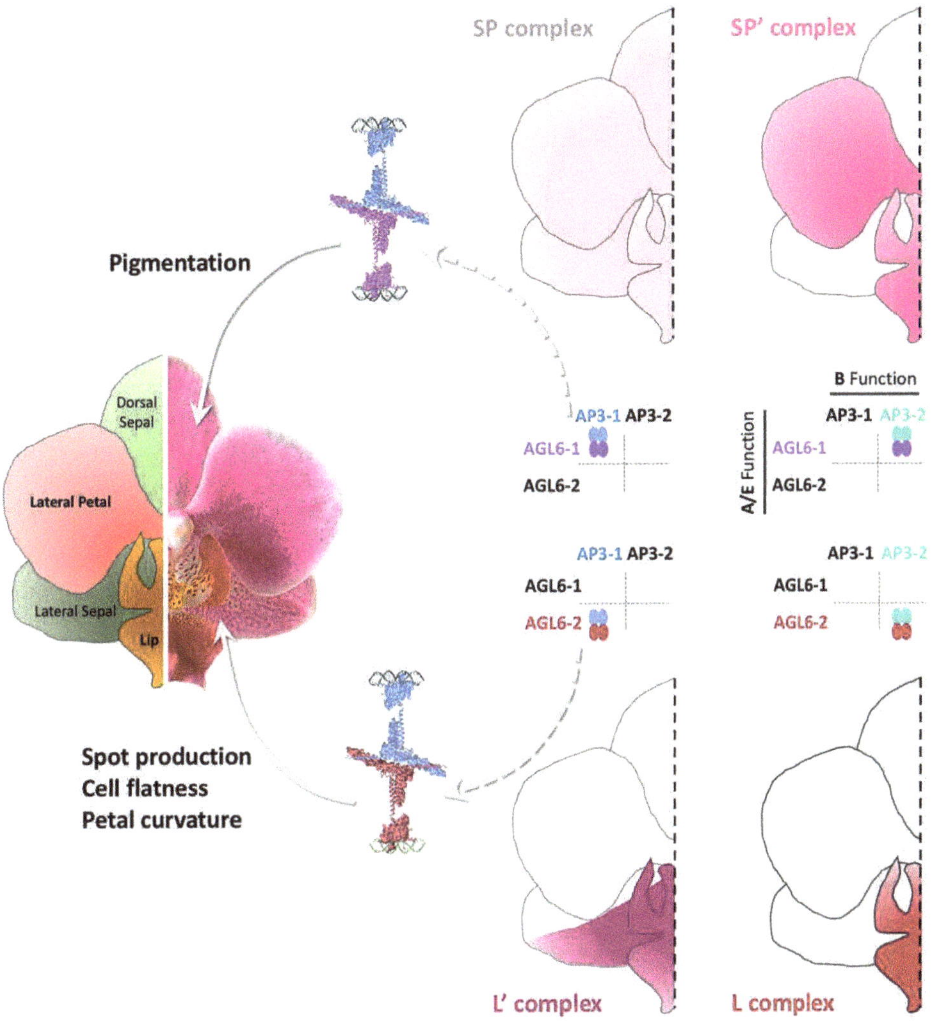

Figure 6. Paralog sub- and neofunctionalization generates diversity. Paralogs of A/E (*AGL6-1, AGL6-2*) and B (*AP3-1, AP3-2*) class genes in orchids have acquired distinct expression patterns, allowing the formation of different quartets across the perianth. This "P code" (Hsu et al. 2015) can explain the activation of different targets genes in different organs within a whorl (i.e. lateral petals vs. lip) or even between different part of the same organ (lower and upper region of the lateral petals): the SP and L" complexes comprise a dimer of AP3-1 (light blue) combines with a dimer of AGL6-1 (purple) or AGL6-2 (red) respectively. Those can bind to promoters harboring distinct CArG box shape, spacing, and orientation, allowing the 2 quartets to regulate different morphological features. The SP complex can trigger uniform pigment production across the petal/sepal epidermis while L" induces spot formation and promotes the production of flat epidermis cells associated with a curvature of the lower lateral petal region (Photo, modified from Zygomorf1 by Christer Johansson via Wikimedia Commons https://creativecommons.org/licenses/by-sa/3.0/).

angiosperms have been molecularly characterized, it seems clear that the ABC genes are not the sex determining genes [as speculated in (Bowman 1997)] but rather that the sex determination mechanisms may converge on changes in B-class gene activity [reviewed in (Montalvão et al. 2021; Zhang et al. 2022)]. One potential exception may be in a gymnosperm, *Cycas*, where *GGM13* (which resides in the phylogenetic sister group to B-class genes) is located in the male determining region of the Y chromosome (Liu et al. 2022). Finally, it must be stressed that not all floral diversity originates from modification of floral MADS-box genes or their downstream targets. For example, change in floral ground plan, such as number and position of floral organ primordia appears largely independent of ABC gene activity (Smyth 2018).

ANA grade and basal eudicots

ABC homologs tend to exhibit broader expression patterns in early diverging angiosperms (Kim et al. 2005; Yoo et al. 2010; Li et al. 2015; Moschin et al. 2021). In vitro studies indicate those are capable of forming floral quartets, but protein-protein interactions are more promiscuous in ANA grade species and early diverging eudicots compared to core eudicots (Liu et al. 2010; Melzer et al. 2014; Li et al. 2015; Rümpler et al. 2018). Hence, the elaboration of the ABC complexes during flowering plant diversification may be best understood by drawing parallels with the formation of a neural network: during animal brain development a large number of neurons is produced (i.e. extensive duplication of

MIKC^{C}-MADS TFs), then as many synapses as possible form between those neurones producing innumerable connections (i.e. almost all combinations of protein-protein interactions between promiscuous MADS-box genes are possible as expression patterns largely overlap). Finally, pruning ensues to selectively eliminate many connections [i.e. paralogous MADS are lost, protein-protein interaction become exclusive between those that remain (Melzer et al. 2014; Li et al. 2015), and expression patterns are sharpened]. Taken together these processes certainly contributed to sculpt the ABC model of eudicots as we know it but also likely introduced the robustness and canalization needed for reliable floral development.

Gymnosperms

While flowers are an angiosperm innovation, ABC-type MADS box genes predate the evolution of flowers (Tandre et al. 1995). Gymnosperm C-class gene expression was detected in both developing female and male cones (Tandre et al. 1998), whereas B-class gene expression is limited to male cones (Mouradov et al. 1999; Sundstrom et al. 1999), suggesting that the interplay between B and C components of the ABC model, key to sex determination, evolved in the ancestral seed plant (Fig. 5H). However, functional investigations in gymnosperms remain difficult, and progress has been limited to heterologous expression, in vitro biochemical assays, and characterization of a handful of natural variants (Carlsbecker et al. 2013). That gymnosperm genomes possess genes similar to A- or E-class angiosperm genes, some of which are expressed in cones (Carlsbecker et al. 2013; Gramzow et al. 2014), suggests that quartets of MADS-box proteins may act in gymnosperms as well (Theißen et al. 2016). Indeed, gymnosperm MIKC^{C}-type proteins can form quartet complexes, with C-homotetramers and BC-heterotetramers able to loop DNA and E proteins (AGL6) not essential to complex formation, at least in vitro (Wang et al. 2010). Whether BC and BCE complexes are both functionally relevant in vivo and/or are capable of activating different sets of genes cannot yet be tested. However, reconstruction and characterization of an ancestral version of BCE complexes suggested that E was necessary for quartet formation specifying male identity in the ancestral seed plant (Ruelens et al. 2017). Some regulatory players accounting for the spatio-temporal expression patterns typical of angiosperms ABC also predate flower evolution: B-gene promoters in *Welwitschia* can recruit LFY to high-affinity binding sites also present in regulatory regions of other conifer B-genes (Moyroud et al. 2017). LFY may already rely on its cofactor UFO to regulate gymnosperm B gene expression because the LFY-UFO connection is likely ancient, but functional UFO genes remain to be identified in nonflowering plants (Rieu et al. 2023).

Thus, key components required to build a flower were already present in the ancestral seed plant, so that the floral ABC network likely derived from a minimal gene regulatory circuit that controlled reproductive organ identity in the last common ancestor of seed plants. In this view the origin of the flower primarily required the construction of an elaborate perianth surrounding the reproductive organs and the evolution of the carpel enclosing the ovules, with the subsequent development of a fruit enclosing the seeds. In this respect, the continued generation of MADS-box gene paralogs was instrumental in the evolution and diversification of angiosperm fruit morphologies (e.g. Gu et al. 1998; Ferrándiz and Fourquin 2014; Karlova et al. 2014; Schilling et al. 2018; Liu et al. 2020).

Origins

The origin of MIKC^{C}-MADS-box genes lies in a duplication in the ancestral land plant, and while most bryophyte species have effectively only a single MIKC^{C}-MADS-box gene (Bowman et al. 2017; Koshimizu et al. 2018), the seeds of ABCE origins can be seen in the ancestral euphyllophyte (Fig. 3B) (Henschel et al. 2002; Kofuji et al. 2003; Wellmer et al. 2014; Qiu et al. 2023). Land plant MIKC^{C}-MADS pro-teins can form quartets in vitro, with a duplication of se-quence within the K domain critical for forming the interface required for tetramer formation (Rümpler et al. 2023). Hence, quartet emergence predates the MADS- boxes learning their ABCs. The ability to form tetramers com-bined with expansion and diversification of MADS-box genes in the ancestral seed plant, and more dramatically within the ancestral angiosperm, enhanced the combinatorial power of MIKC^{C} complexes (Rümpler et al. 2023) and may have contributed to the evolution of tissue and organ types during development. In the ancestral land plant, sex was determined during the haploid generation (Bowman 2022), but in seed plants the sporophyte largely controls gametophyte development, including sex expression. In seedless vascular plants, MIKC^{C}-MADS genes are broadly expressed, often in both generations [e.g. (Hasebe et al. 1998; Ambrose et al. 2021)]. One intriguing speculation is that MIKC^{C}-MADS genes were co-opted into a role in determining sporophyte sex during tracheophyte evolution as the sporophyte was assuming control over gametophyte differentiation. And peering deeper in time, given the role of MCM1 in the yeast life cycle, an idea worth exploring is whether land plant MADS-box genes have retained vestiges of a fundamental role in the eukaryotic life cycle (Herskowitz 1989), as have land plant homeodomain genes (Dierschke et al. 2021; Hisanaga et al. 2021).

Conclusion: a short and biased selection of floral riddles

Thirty years ago, we could not have predicted what a cottage industry the characterization of ABC MADS-box genes in various angiosperms would become, contributing to a deeper and broader understanding of evo-devo processes that have shaped floral architecture. Reciprocally, they also accentuate remaining gaps in our knowledge. For instance, a given quartet is expected to activate different sets of genes at different timepoints as organ primordia initiate, grow, and differentiate, but whether this is the case and how it is achieved remain obscure. Equally, to generate the diversity of floral forms genes regulated by the ABC genes is likely to differ somewhat between species. However, a systematic examination of ABC targets conservation between species is still lacking. Even if a full picture of those targets becomes available, this still does not explain how gene activity translates into morphogenetic effects: here, high-resolution live-imaging coupled with computational simulations and theoretical modeling [see e.g. (Refahi et al. 2021)] could be instrumental in understanding how homeotic regulators interact with gene regulatory networks that specify organ polarity, floral symmetry, and tissue layer identity to sculpt undifferentiated primordia into floral organs with characteristic shape, size, patterned cell types, and metabolic properties. Research into the origins of the ABC genes established that the gene regulatory networks controlling flower development stem from recycling of an ancient circuit at work in the reproductive structures of last common ancestor of seed plants. However, what rewiring occurred along the lineage leading to flowering plants so that discreet B and B + C expression domains could be established enclosed by a sterile perianth remains enigmatic.

These puzzles represent only a handful of the remaining questions and by no means constitute an exhaustive list, but we hope they will spark the curiosity of future flower enthusiasts to continue the successful adventure of the ABC.

Acknowledgments

J.L.B. extends special thanks Elliot Meyerowitz and David Smyth as his "flower mentors" from 1986-95, as they provided both the direction and freedom to pursue curiosity driven discovery research, a theme continuing throughout my entire career. E.M. thanks the members of her laboratory for useful discussions.

Author contributions

Both authors contributed to the outline and preparation of this review. JLB was the primary author for the History, Genetics, Molecules and Converting leaves into floral organs sections; EM was the primary author of the Swimming upstream and Floating downstream sections; both authors contributed to the Curious case and A and subfunctionalization of B- and C-classes, and the ABC through the evolutionary lens sections.

Funding

Work in J.L.B.'s laboratory is funded by the Australian Research Council (DP200100225, DP210101423, CE200100015). E.M. thanks the Gatsby Charitable Foundation and the Isaac Newton/Wellcome Trust ISSF for funding.

Conflict of interest statement. None declared.

References

Abraham-Juárez MJ, Schrager-Lavelle A, Man J, Whipple C, Handakumbura P, Babbitt C, Bartlett M. Evolutionary variation in MADS box dimerization affects floral development and protein abundance in maize. Plant Cell. 2020:**32**(11):3408–3424. https://doi.org/10.1105/tpc.20.00300

Aerts N, de Bruijn S, van Mourik H, Angenent GC, van Dijk ADJ. Comparative analysis of binding patterns of MADS-domain proteins in Arabidopsis thaliana. BMC Plant Biol. 2018:**18**(1):131. https://doi.org/10.1186/s12870-018-1348-8

Alvarez-Buylla ER, Ambrose BA, Flores-Sandoval E, Vergara-Silva F, Englund M, Garay-Arroyo A, Garcia-Ponce B, de la Torre-Barcena E, Espinosa-Matias S, Martinez E, et al. B-function expression in the flower center underlies the homeotic phenotype of Lacandonia schismatica (Triuridaceae). Plant Cell. 2010:**22**(11):3543–3559. https://doi.org/10.1105/tpc.109.069153

Alvarez-Venegas R, Pien S, Sadder M, Witmer X, Grossniklaus U, Avramova Z. ATX-1, an Arabidopsis homolog of trithorax, activates flower homeotic genes. Curr Biol. 2003:**13**(8):627–637. https://doi.org/10.1016/S0960-9822(03)00243-4

Alvarez J, Smyth DR. CRABS CLAW and SPATULA, two Arabidopsis genes that control carpel development in parallel with AGAMOUS. Development. 1999:**126**(11):2377–2386. https://doi.org/10.1242/dev.126.11.2377

Ambrose BA, Lerner DR, Ciceri P, Padilla CM, Yanofsky MF, Schmidt RJ. Molecular and genetic analyses of the silky1 gene reveal conservation in floral organ specification between eudicots and monocots. Mol Cell. 2000:**5**(3):569–579. https://doi.org/10.1016/S1097-2765(00)80450-5

Ambrose BA, Smalls TL, Zumajo-Cardona C. All type II classic MADS-box genes in the lycophyte Selaginella moellendorffii are broadly yet discretely expressed in vegetative and reproductive tissues. Evol Dev. 2021:**23**(3):215–230. https://doi.org/10.1111/ede.12375

Aukerman MJ, Sakai H. Regulation of flowering time and floral organ identity by a microRNA and its APETALA2-like target genes. Plant Cell. 2003:**15**(11):2730–2741. https://doi.org/10.1105/tpc.016238

Bateson W. Materials for the study of variation treated with especial regard to discontinuity in the origin of species. London: Macmillan; 1894.

Bateson W, Saunders ER, Punnett RC. Reports to the evolution committee, report II. Experimental studies in the physiology of heredity-matthiola. London: Royal Society; 1905.

Baum SF, Eshed Y, Bowman JL. The Arabidopsis nectary is an ABC-independent floral structure. Development. 2001:**128**(22): 4657–4667. https://doi.org/10.1242/dev.128.22.4657

Benlloch R, Kim MC, Sayou C, Thévenon E, Parcy F, Nilsson O. Integrating long-day flowering signals: a LEAFY binding site is essential for proper photoperiodic activation of APETALA1. Plant J. 2011:**67**(6): 1094–1102. https://doi.org/10.1111/j.1365-313X.2011.04660.x

Bowman JL. Evolutionary conservation of angiosperm flower development at the molecular and genetic levels. J Biosci. 1997:**22**(4): 515–527. https://doi.org/10.1007/BF02703197

Bowman JL. eLS encyclopedia of life sciences. John Wiley & Sons, Ltd; 2001. p. A734.

Bowman JL. My favourite flowering image. J Exp Bot. 2013:**64**(18): 5779–5782. https://doi.org/10.1093/jxb/erq044

Bowman JL, Alvarez J, Weigel D, Meyerowitz EM, Smyth DR. Control of flower development in Arabidopsis thaliana by APETALA1 and interacting genes. Development. 1993:**119**(3):721–743. https://doi.org/10.1242/dev.119.3.721

Bowman JL, Kohchi T, Yamato KT, Jenkins J, Shu S, Ishizaki K, Yamaoka S, Nishihama R, Nakamura Y, Berger F, et al. Insights into land plant evolution garnered from the Marchantia polymorpha genome. Cell. 2017:**171**(2):287–304.e15. https://doi.org/10.1016/j.cell.2017.09.030

Bowman JL, Sakai H, Jack T, Weigel D, Mayer U, Meyerowitz EM. SUPERMAN, a regulator of floral homeotic genes in Arabidopsis.

Development. 1992:**114**(3):599–615. https://doi.org/10.1242/dev.114.3.599

Bowman JL, Smyth DR. CRABS CLAW, a gene that regulates carpel and nectary development in Arabidopsis, encodes a novel protein with zinc finger and helix-loop-helix domains. Development. 1999:**126**(11):2387–2396. https://doi.org/10.1242/dev.126.11.2387

Bowman JL, Smyth DR, Meyerowitz EM. Genes directing flower development in Arabidopsis. Plant Cell. 1989:**1**(1):37–52. https://doi.org/10.1105/tpc.1.1.37

Bowman JL, Smyth DR, Meyerowitz EM. Genetic interactions among floral homeotic genes. Development. 1991:**112**(1):1–20. https://doi.org/10.1242/dev.112.1.1

Bowman JL, Smyth DR, Meyerowitz EM. The ABC model of flower development: then and now. Development. 2012:**139**(22):4095–4098. https://doi.org/10.1242/dev.083972

Bradley D, Carpenter R, Sommer H, Hartley N, Coen E. Complementary floral homeotic phenotypes result from opposite orientations of a transposon at the plena locus of Antirrhinum. Cell. 1993:**72**(1):85–95. https://doi.org/10.1016/0092-8674(93)90052-R

Brieger FG. The developmental mechanics of normal and abnormal flowers in Primula. Proc Linn Soc Lond. 1935:**147**:126–130. https://doi.org/10.1111/j.1095-8312.1934.tb01264.x

Busch MA, Bomblies K, Weigel D. Activation of a floral homeotic gene in Arabidopsis. Science. 1999:**285**(5427):585–587. https://doi.org/10.1126/science.285.5427.585

Buzgo M, Soltis DE, Soltis PS. Floral developmental morphology of Amborella trichopoda (Amborellaceae). Int J Plant Sci. 2004:**165**(6):925–947. https://doi.org/10.1086/424024

Carlsbecker A, Sundström JF, Englund M, Uddenberg D, Izquierdo L, Kvarnheden A, Vergara-Silva F, Engström P. Molecular control of normal and acrocona mutant seed cone development in Norway spruce (Picea abies) and the evolution of conifer ovule-bearing organs. New Phytologist. 2013:**200**(1):261–275. https://doi.org/10.1111/nph.12360

Carpenter R, Coen ES. Floral homeotic mutations produced by transposon-mutagenesis in antirrhinum-majus. Gene Dev. 1990:**4**(9):1483–1493. https://doi.org/10.1101/gad.4.9.1483

Cartolano M, Castillo R, Efremova N, Kuckenberg M, Zethof J, Gerats T, Schwarz-Sommer Z, Vandenbussche M. A conserved microRNA module exerts homeotic control over Petunia hybrida and Antirrhinum majus floral organ identity. Nat Genet. 2007:**39**(7):901–905. https://doi.org/10.1038/ng2056

Carvalho DS, Schnable JC, Almeida AMR. Integrating phylogenetic and network approaches to study gene family evolution: the case of the AGAMOUS family of floral genes. Evol Bioinform Online. 2018:**14**:117693431876468. https://doi.org/10.1177/1176934318764683

Causier B, Bradley D, Cook H, Davies B. Conserved intragenic elements were critical for the evolution of the floral C-function. Plant J. 2009:**58**(1):41–52. https://doi.org/10.1111/j.1365-313X.2008.03759.x

Causier B, Schwarz-Sommer Z, Davies B. Floral organ identity: 20 years of ABCs. Semin Cell Dev Biol. 2010:**21**(1):73–79. https://doi.org/10.1016/j.semcdb.2009.10.005

Chae E, Tan QK-G, Hill TA, Irish VF. An Arabidopsis F-box protein acts as a transcriptional co-factor to regulate floral development. Development. 2008:**135**(7):1235–1245. https://doi.org/10.1242/dev.015842

Chen D, Yan W, Fu LY, Kaufmann K. Architecture of gene regulatory networks controlling flower development in Arabidopsis thaliana. Nat Commun. 2018:**9**(1):4534.

Chen X. A microRNA as a translational repressor of APETALA2 in Arabidopsis flower development. Science. 2004:**303**(5666):2022–2025. https://doi.org/10.1126/science.1088060

Cheng Y, Kato N, Wang W, Li J, Chen X. Two RNA binding proteins, HEN4 and HUA1, act in the processing of AGAMOUS Pre-mRNA in Arabidopsis thaliana. Dev Cell. 2003:**4**(1):53–66. https://doi.org/10.1016/S1534-5807(02)00399-4

Cheon K-S, Nakatsuka A, Tasaki K, Kobayashi N. Long-lasting corolla cultivars in Japanese azaleas: a mutant AP3/DEF homolog identified in traditional azalea cultivars from more than 300 years ago. Front Plant Sci. 2018:**8**:2239. https://doi.org/10.3389/fpls.2017.02239

Coen E, Meyerowitz EM. The war of the whorls: genetic interactions controlling flower development. Nature. 1991:**353**(6339):31–37. https://doi.org/10.1038/353031a0

Darwin C. The variation of animals and plants under domestication. 2nd ed. New York: D. Appleton; 1876.

Davies B, Egea-Cortines M, Silva EDA, Saedler H, Sommer H. Multiple interactions amongst floral homeotic MADS box proteins. EMBO J. 1996:**15**(16):4330–4343. https://doi.org/10.1002/j.1460-2075.1996.tb00807.x

Davies B, Motte P, Keck E, Saedler H, Sommer H, Schwarz-Sommer Z. PLENA and FARINELLI: redundancy and regulatory interactions between two antirrhinum MADS-box factors controlling flower development. EMBO J. 1999:**18**(14):4023–4034. https://doi.org/10.1093/emboj/18.14.4023

Dierschke T, Flores-Sandoval E, Rast-Somssich MI, Altoff F, Zachgo S, Bowman JL. Gamete expression of TALE class HD genes activates the diploid sporophyte program in Marchantia polymorpha. eLife. 2021:**10**:e57088. https://doi.org/10.7554/eLife.57088

Ditta G, Pinyopich A, Robles P, Pelaz S, Yanofsky MF. The SEP4 gene of Arabidopsis thaliana functions in floral organ and meristem identity. Curr Biol. 2004:**14**(21):1935–1940. https://doi.org/10.1016/j.cub.2004.10.028

Dodoens R. Florum, et coronariarum odoratarumque nonnullarum herbarum historia. Antwerp: Christophori Plantini; 1568.

Dodoens R. Stirpium historiae pemptades sex sive libri XXX. Antwerp: Christophori Plantini; 1583.

Dreni L, Zhang D. Flower development: the evolutionary history and functions of the AGL6 subfamily MADS-box genes. J Exp Bot. 2016:**67**(6):1625–1638. https://doi.org/10.1093/jxb/erw046

Drews GN, Bowman JL, Meyerowitz EM. Negative regulation of the Arabidopsis homeotic gene AGAMOUS by the APETALA2 product. Cell. 1991:**65**(6):991–1002. https://doi.org/10.1016/0092-8674(91)90551-9

Edgar BA, Odell GM, Schubiger G. A genetic switch, based on negative regulation, sharpens stripes in Drosophila embryos. Dev Genet. 1989:**10**(3):124–114. https://doi.org/10.1002/dvg.1020100303

Egea-Cortines M, Saedler H, Sommer H. Ternary complex formation between the MADS-box proteins SQUAMOSA, DEFICIENS and GLOBOSA is involved in the control of floral architecture in Antirrhinum majus. EMBO J. 1999:**18**(19):5370–5379. https://doi.org/10.1093/emboj/18.19.5370

Eshed Y, Baum SF, Bowman JL. Distinct mechanisms promote polarity establishment in carpels of Arabidopsis. Cell. 1999:**99**(2):199–209. https://doi.org/10.1016/S0092-8674(00)81651-7

Eshed Y, Baum SF, Perea JV, Bowman JL. Establishment of polarity in lateral organs of plants. Curr Biol. 2001:**11**(16):1251–1260. https://doi.org/10.1016/S0960-9822(01)00392-X

Espinosa-Matias S, Vergara-Silva F, Vazquez-Santana S, Martinez-Zurita E, Marquez-Guzman J. Complex patterns of morphogenesis, embryology, and reproduction in Triuris brevistylis, a species of Triuridaceae (Pandanales) closely related to Lacandonia schismatica. Botany. 2012:**90**(11):1133–1151. https://doi.org/10.1139/b2012-060

Favaro R, Pinyopich A, Battaglia R, Kooiker M, Borghi L, Ditta G, Yanofsky MF, Kater MM, Colombo L. MADS-box protein complexes control carpel and ovule development in Arabidopsis. Plant Cell. 2003:**15**(11):2603–2611. https://doi.org/10.1105/tpc.015123

Ferrándiz C, Fourquin C. Role of the FUL–SHP network in the evolution of fruit morphology and function. J Exp Bot. 2014:**65**(16):4505–4513. https://doi.org/10.1093/jxb/ert479

Folter SD, Angenent GC. Trans meets cis in MADS science. Trends Plant Sci. 2006:**11**(5):224–231. https://doi.org/10.1016/j.tplants.2006.03.008

Garber RL, Kuroiwa A, Gehring WJ. Genomic and cDNA clones of the homeotic locus antennapedia in Drosophila. EMBO J. 1983:**2**(11):2027–2036. https://doi.org/10.1002/j.1460-2075.1983.tb01696.x

Gerard J, Dodoens R, Norton J, Priest R. The herball, or, generall historie of plantes/gathered by John Gerarde of London, master in chirurgerie. London: John Norton; 1597.

Godron D-A. Des races végétales qui doivent leur ongine a une monstruosité. Mémoires de l'Académie de Stanislas Académie de Nancy. 1874:**6**:77–95.

Goethe JWV. Versuch die metamorphose der Pflanzen zu erklaren. Gotha: C. W. Ettinger; 1790.

Gómez-Mena C, Folter SD, Costa MMR, Angenent GC, Sablowski R. Transcriptional program controlled by the floral homeotic gene AGAMOUS during early organogenesis. Development. 2005:**132**(3): 429–438. https://doi.org/10.1242/dev.01600

Goodrich J, Puangsomlee P, Martin M, Long D, Meyerowitz EM, Coupland G. A polycomb-group gene regulates homeotic gene expression in Arabidopsis. Nature. 1997:**386**(6620):44–51. https://doi.org/10.1038/386044a0

Goslin K, Finocchio A, Wellmer F. Floral homeotic factors: a question of specificity. Plants. 2023:**12**(5):1128. https://doi.org/10.3390/plants12051128

Goslin K, Zheng B, Serrano-Mislata A, Rae L, Ryan PT, Kwaśniewska K, Thomson B, Ó'Maoiléidigh DS, Madueño F, Wellmer F, et al. Transcription factor interplay between LEAFY and APETALA1/CAULIFLOWER during floral initiation. Plant Physiol. 2017:**174**(2): 1097–1109. https://doi.org/10.1104/pp.17.00098

Goto K, Meyerowitz EM. Function and regulation of the Arabidopsis floral homeotic gene PISTILLATA. Genes Dev. 1994:**8**(13): 1548–1560. https://doi.org/10.1101/gad.8.13.1548

Gramzow L, Weilandt L, Theißen G. MADS goes genomic in conifers: towards determining the ancestral set of MADS-box genes in seed plants. Ann Bot. 2014:**114**(7):1407–1429. https://doi.org/10.1093/aob/mcu066

Green PB. A theory for inflorescence development and flower formation based on morphological and biophysical analysis in Echeveria. Planta. 1988:**175**(2):153–169. https://doi.org/10.1007/BF00392424

Gu Q, Ferrándiz C, Yanofsky MF, Martienssen R. The FRUITFULL MADS-box gene mediates cell differentiation during Arabidopsis fruit development. Development. 1998:**125**(8):1509–1517. https://doi.org/10.1242/dev.125.8.1509

Gustafson-Brown C, Savidge B, Yanofsky MF. Regulation of the Arabidopsis floral homeotic gene APETALA1. Cell. 1994:**76**(1): 131–143. https://doi.org/10.1016/0092-8674(94)90178-3

Hasebe M, Wen CK, Kato M, Banks JA. Characterization of MADS homeotic genes in the fern Ceratopteris richardii. Proc Natl Acad Sci U S A. 1998:**95**(11):6222–6227. https://doi.org/10.1073/pnas.95.11.6222

Heijmans K, Ament K, Rijpkema AS, Zethof J, Wolters-Arts M, Gerats T, Vandenbussche M. Redefining C and D In the petunia ABC. Plant Cell. 2012:**24**(6):2305–2317. https://doi.org/10.1105/tpc.112.097030

Henschel K, Kofuji R, Hasebe M, Saedler H, Münster T, Theißen G. Two ancient classes of MIKC-type MADS-box genes are present in the moss Physcomitrella patens. Mol Biol Evol. 2002:**19**(6):801–814. https://doi.org/10.1093/oxfordjournals.molbev.a004137

Herskowitz I. A regulatory hierarchy for cell specialization in yeast. Nature. 1989:**342**:749–757. https://doi.org/10.1038/342749a0

Heslop-Harrison J. Sex expression in flowering plants. Brookhaven Symp Biol. 1963:**16**:109–125.

Hisanaga T, Fujimoto S, Cui Y, Sato K, Sano R, Yamaoka S, Kohchi T, Berger F, Nakajima K. Deep evolutionary origin of gamete-directed zygote activation by KNOX/BELL transcription factors in green plants. Elife. 2021:**10**:e57090. https://doi.org/10.7554/eLife.57090

Holder N. Positional information and pattern formation in plant morphogenesis and a mechanism for the involvement of plant hormones. J Theor Biol. 1979:**77**(2):195–212. https://doi.org/10.1016/0022-5193(79)90307-2

Honma T, Goto K. The Arabidopsis floral homeotic gene PISTILLATA is regulated by discrete cis-elements responsive to induction and maintenance signals. Development. 2000:**127**(10):2021–2030. https://doi.org/10.1242/dev.127.10.2021

Honma T, Goto K. Complexes of MADS-box proteins are sufficient to convert leaves into floral organs. Nature. 2001:**409**(6819):525–529. https://doi.org/10.1038/35054083

Hsu H-F, Chen W-H, Shen Y-H, Hsu W-H, Mao W-T, Yang C-H. Multifunctional evolution of B and AGL6 MADS box genes in orchids. Nat Commun. 2021:**12**(1):902. https://doi.org/10.1038/s41467-021-21229-w

Hsu H-F, Hsu W-H, Lee Y-I, Mao W-T, Yang J-Y, Li J-Y, Yang C-H. Model for perianth formation in orchids. Nat Plants. 2015:**1**(5):1–8. https://doi.org/10.1038/nplants.2015.46

Hu T, Li X, Du L, Manuela D, Xu M. LEAFY and APETALA1 down-regulate ZINC FINGER PROTEIN 1 and 8 to release their repression on class B and C floral homeotic genes. Proc Natl Acad Sci U S A. 2023:**120**(22):e2221181120. https://doi.org/10.1073/pnas.2221181120

Huang T, López-Giráldez F, Townsend JP, Irish VF. RBE controls microRNA164 expression to effect floral organogenesis. Development. 2012:**139**(12):2161–2169. https://doi.org/10.1242/dev.075069

Hugouvieux V, Blanc-Mathieu R, Paul M, Janeau A, Xu X, Lucas J, Lai X, Galien A, Yan W, Nanao M, et al. SEPALLATA-driven MADS transcription factor tetramerization is required for inner whorl floral organ development. bioRxiv 541941. https://doi.org/10.1101/2023.05.23.541941, 23 May 2023, preprint: not peer reviewed.

Hugouvieux V, Silva CS, Jourdain A, Stigliani A, Charras Q, Conn V, Conn SJ, Carles CC, Parcy F, Zubieta C. Tetramerization of MADS family transcription factors SEPALLATA3 and AGAMOUS is required for floral meristem determinacy in Arabidopsis. Nucleic Acids Res. 2018:**46**(10):4966–4977. https://doi.org/10.1093/nar/gky205

Huijser P, Klein J, Lonnig W-E, Meijer H, Saedler H, Sommer H. Bracteomania, an inflorescence anomaly, is caused by the loss of function of the MADS-box gene squamosa in Antirrhinum majus. EMBO J. 1992:**11**(4):1239–1249. https://doi.org/10.1002/j.1460-2075.1992.tb05168.x

Irish VF, Sussex IM. Function of the apetala1-1 gene during Arabidopsis floral development. Plant Cell. 1990:**2**(8):741–751. https://doi.org/10.1105/tpc.2.8.741

Jack T, Brockman LL, Meyerowitz EM. The homeotic gene APETALA3 of Arabidopsis thaliana encodes a MADS box and is expressed in petals and stamens. Cell. 1992:**68**(4):683–697. https://doi.org/10.1016/0092-8674(92)90144-2

Jetha K, Theißen G, Melzer R. Arabidopsis SEPALLATA proteins differ in cooperative DNA-binding during the formation of floral quartet-like complexes. Nucleic Acids Res. 2014:**42**(17):10927–10942. https://doi.org/10.1093/nar/gku755

Jofuku KD, den Boer BGW, Van Montagu M, Okamuro JK. Control of Arabidopsis flower and seed development by the homeotic gene APETALA2. Plant Cell. 1994:**6**(9):1211–1225. https://doi.org/10.1105/tpc.6.9.1211

John JL. Origin of a land flora. Nat Plants. 2022:**8**:1352–1369. https://doi.org/10.1038/s41477-022-01283-y

Jung J-H, Lee S, Yun J, Lee M, Park C-M. The miR172 target TOE3 represses AGAMOUS expression during Arabidopsis floral patterning. Plant Sci. 2014:**215–216**:29–38. https://doi.org/10.1016/j.plantsci.2014.09.010

Kanno A, Saeki H, Kameya T, Saedler H, Theißen G. Heterotopic expression of class B floral homeotic gene supports a modified ABC model for tulip (Tulipa gesneriana). Plant Mol Biol. 2003:**52**(4): 831–841. https://doi.org/10.1023/A:1025070827979

Kapoor M, Tsuda S, Tanaka Y, Mayama T, Okuyama Y, Tsuchimoto S, Takatsuji H. Role of petunia pMADS3 in determination of floral organ and meristem identity, as revealed by its loss of function. Plant J. 2002:**32**(1):115–127. https://doi.org/10.1046/j.1365-313X.2002.01402.x

Käppel S, Melzer R, Rümpler F, Gafert C, Theißen G. The floral homeotic protein SEPALLATA3 recognizes target DNA sequences by shape readout involving a conserved arginine residue in the MADS-domain. Plant J. 2018:**95**(2):341–357. https://doi.org/10.1111/tpj.13954

Käppel S, Rümpler F, Theißen G. Cracking the floral quartet code: how do multimers of MIKCC-type MADS-domain transcription factors recognize their target genes? Int J Mol Sci. 2023:**24**(9):8253. https://doi.org/10.3390/ijms24098253

Karch F, Weiffenbach B, Peifer M, Bender W, Duncan I, Celniker S, Crosby M, Lewis EB, Karch F. The abdominal region of the bithorax complex. Cell. 1985:**43**(1):81–96. https://doi.org/10.1016/0092-8674(85)90014-5

Karlova R, Chapman N, David K, Angenent GC, Seymour GB, Maagd RAD. Transcriptional control of fleshy fruit development and ripening. J Exp Bot. 2014:**65**(16):4527–4541. https://doi.org/10.1093/jxb/eru316

Kaufmann K, Muiño JM, Jauregui R, Airoldi CA, Smaczniak C, Krajewski P, Angenent GC. Target genes of the MADS transcription factor SEPALLATA3: integration of developmental and hormonal pathways in the Arabidopsis flower. PLoS Biol. 2009:**7**(4):e1000090. https://doi.org/10.1371/journal.pbio.1000090

Kaufmann K, Pajoro A, Angenent GC. Regulation of transcription in plants: mechanisms controlling developmental switches. Nat Rev Genet. 2010a:**11**(12):830–842. https://doi.org/10.1038/nrg2885

Kaufmann K, Wellmer F, Muiño JM, Ferrier T, Wuest SE, Kumar V, Serrano-Mislata A, Madueño F, Krajewski P, Meyerowitz EM, et al. Orchestration of floral initiation by APETALA1. Science. 2010b:**328**(5974):85–89. https://doi.org/10.1126/science.1185244

Kempin SA, Savidge B, Yanofsky MF. Molecular basis of the cauliflower phenotype in Arabidopsis. Science. 1995:**267**(5197):522–525. https://doi.org/10.1126/science.7824951

Kim S, Koh J, Yoo M-J, Kong H, Hu Y, Ma H, Soltis PS, Soltis DE. Expression of floral MADS-box genes in basal angiosperms: implications for the evolution of floral regulators. Plant J. 2005:**43**(5):724–744. https://doi.org/10.1111/j.1365-313X.2005.02487.x

Kim S, Yoo M-J, Albert VA, Farris JS, Soltis PS, Soltis DE. Phylogeny and diversification of B-function MADS-box genes in angiosperms: evolutionary and functional implications of a 260-million-year-old duplication. Am J Bot. 2004:**91**(12):2102–2118. https://doi.org/10.3732/ajb.91.12.2102

Kofuji R, Sumikawa N, Yamasaki M, Kondo K, Ueda K, Ito M, Hasebe M. Evolution and divergence of the MADS-box gene family based on genome-wide expression analyses. Mol Biol Evol. 2003:**20**(12):1963–1977. https://doi.org/10.1093/molbev/msg216

Komaki MK, Okada K, Nishino E, Shimura Y. Isolation and characterization of novel mutants of Arabidopsis thaliana defective in flower development. Development. 1988:**104**(2):195–203. https://doi.org/10.1242/dev.104.2.195

Koornneef M, Eden JV, Hanhart CJ, Stam P, Braaksma FJ, Feenstra WJ. Linkage map of Arabidopsis thaliana. J Hered. 1983:**74**(4):265–272. https://doi.org/10.1093/oxfordjournals.jhered.a109781

Koshimizu S, Kofuji R, Sasaki-Sekimoto Y, Kikkawa M, Shimojima M, Ohta H, Shigenobu S, Kabeya Y, Hiwatashi Y, Tamada Y, et al. Physcomitrella MADS-box genes regulate water supply and sperm movement for fertilization. Nat Plants. 2018:**4**(1):36–45. https://doi.org/10.1038/s41477-017-0082-9

Kramer EM, Di Stilio VS, Schluter PM. Complex patterns of gene duplication in the APETALA3 and PISTILLATA lineages of the Ranunculaceae. Int J Plant Sci. 2003:**164**(1):1–11. https://doi.org/10.1086/344694

Kramer EM, Dorit RL, Irish VF. Molecular evolution of genes controlling petal and stamen development: duplication and divergence within the APETALA3 and PISTILLATA MADS-box gene lineages. Genetics. 1998:**149**(2):765–783. https://doi.org/10.1093/genetics/149.2.765

Kramer EM, Holappa L, Gould B, Jaramillo MA, Setnikov D, Santiago PM. Elaboration of B gene function to include the identity of novel floral organs in the lower eudicot aquilegia. Plant Cell. 2007:**19**(3):750–766. https://doi.org/10.1105/tpc.107.050385

Krizek BA, Lewis MW, Fletcher JC. RABBIT EARS is a second-whorl repressor of AGAMOUS that maintains spatial boundaries in Arabidopsis flowers. Plant J. 2006:**45**(3):369–383. https://doi.org/10.1111/j.1365-313X.2005.02633.x

Krizek BA, Meyerowitz EM. The Arabidopsis homeotic genes APETALA3 and PISTILLATA are sufficient to provide the B class organ identity function. Development. 1996:**122**(1):11–22. https://doi.org/10.1242/dev.122.1.11

Kunst L, Klenz JE, Martinez-Zapater J, Haughn GW. AP2 gene determines the identity of perianth organs in flowers of Arabidopsis thaliana. Plant Cell. 1989:**1**(12):1195–1208. https://doi.org/10.2307/3868917

Kyozuka J, Kobayashi T, Morita M, Shimamoto K. Spatially and temporally regulated expression of rice MADS box genes with similarity to Arabidopsis class A, B and C genes. Plant Cell Physiol. 2000:**41**(6):710–718. https://doi.org/10.1093/pcp/41.6.710

Lai X, Stigliani A, Lucas J, Hugouvieux V, Parcy F, Zubieta C. Genome-wide binding of SEPALLATA3 and AGAMOUS complexes determined by sequential DNA-affinity purification sequencing. Nucleic Acids Res. 2020:**48**(17):9637–9648. https://doi.org/10.1093/nar/gkaa729

Lai X, Vega-Léon R, Hugouvieux V, Blanc-Mathieu R, Wal FVD, Lucas J, Silva CS, Jourdain A, Muino JM, Nanao MH, et al. The intervening domain is required for DNA-binding and functional identity of plant MADS transcription factors. Nat Commun. 2021:**12**(1):4760. https://doi.org/10.1038/s41467-021-24978-w

Lamb RS, Hill TA, Tan QK-G, Irish VF. Regulation of APETALA3 floral homeotic gene expression by meristem identity genes. Development. 2002:**129**(9):2079–2086. https://doi.org/10.1242/dev.129.9.2079

Lee J-Y, Baum SF, Alvarez J, Patel A, Chitwood DH, Bowman JL. Activation of CRABS CLAW in the nectaries and carpels of Arabidopsis. Plant Cell. 2005a:**17**(1):25–36. https://doi.org/10.1105/tpc.104.026666

Lee J-Y, Baum SF, Oh S-H, Jiang CZ, Chen JC, Bowman JL. Recruitment of CRABS CLAW to promote nectary development within the eudicot clade. Development. 2005b:**132**(22):5021–5032. https://doi.org/10.1242/dev.02067

Lee U-S, Bieluszewski T, Xiao J, Yamaguchi A, Wagner D. H2a.Z contributes to trithorax activity at the AGAMOUS locus. Mol Plant. 2022:**15**(2):207–210. https://doi.org/10.1016/j.molp.2022.01.005

Lee I, Wolfe DS, Nilsson O, Weigel D. A LEAFY co-regulator encoded by UNUSUAL FLORAL ORGANS. Curr Biol. 1997:**7**(2):95–104. https://doi.org/10.1016/S0960-9822(06)00053-4

Lenhard M, Bohnert A, Jürgens G, Laux T. Termination of stem cell maintenance in Arabidopsis floral meristems by interactions between WUSCHEL and AGAMOUS. Cell. 2001:**105**(6):805–814. https://doi.org/10.1016/S0092-8674(01)00390-7

Levin JZ, Meyerowitz EM. UFO: an Arabidopsis gene involved in both floral meristem and floral organ development. Plant Cell. 1995:**7**(5):529–548. https://doi.org/10.1105/tpc.7.5.529

Lewis EB. A gene complex controlling segmentation in Drosophila. Nature. 1978:**276**(5688):565–570. https://doi.org/10.1038/276565a0

Li H, Liang W, Jia R, Yin C, Zong J, Kong H, Zhang D. The AGL6-like gene OsMADS6 regulates floral organ and meristem identities in rice. Cell Res. 2010:**20**(3):299–313. https://doi.org/10.1038/cr.2009.143

Li L, Yu X-X, Guo C-C, Duan X-S, Shan H-Y, Zhang R, Xu G-X, Kong H-Z. Interactions among proteins of floral MADS-box genes in nuphar pumila (Nymphaeaceae) and the most recent common ancestor of extant angiosperms help understand the underlying mechanisms of the origin of the flower. J Syst Evol. 2015:**53**(4):285–296. https://doi.org/10.1111/jse.12148

Liljegren SJ, Ditta GS, Eshed Y, Savidge B, Bowman JL, Yanofsky MF. Control of fruit dehiscence in Arabidopsis by the SHATTERPROOF MADS-box genes. Nature. 2000:**404**(6779):766–770. https://doi.org/10.1038/35008089

Liljegren SJ, Gustafson-Brown C, Pinyopich A, Ditta GS, Yanofsky MF. Interactions among APETALA1, LEAFY, and TERMINAL FLOWER1 specify meristem fate. Plant Cell. 1999:**11**(6):1007–1018. https://doi.org/10.1105/tpc.11.6.1007

Lin X, Yuan T, Guo H, Guo Y, Yamaguchi N, Wang S, Zhang D, Qi D, Li J, Chen Q, et al. The regulation of chromatin configuration at

AGAMOUS locus by LFR-SYD-containing complex is critical for reproductive organ development in Arabidopsis. Plant J. 2023:**116**(2): 478–496. https://doi.org/10.1111/tpj.16385

Liu X, Kim YJ, Müller R, Yumul RE, Liu C, Pan Y, Cao X, Goodrich J, Chen X. AGAMOUS terminates floral stem cell maintenance in Arabidopsis by directly repressing WUSCHEL through recruitment of polycomb group proteins. Plant Cell. 2011:**23**(10):3654–3670. https://doi.org/10.1105/tpc.111.091538

Liu Z, Ma H, Jung S, Main D, Guo L. Developmental mechanisms of fleshy fruit diversity in Rosaceae. Annu Rev Plant Biol. 2020:**71**(1): 547–573. https://doi.org/10.1146/annurev-arplant-111119-021700

Liu Y, Wang S, Li L, Yang T, Dong S, Wei T, Wu S, Liu Y, Gong Y, Feng X, et al. The Cycas genome and the early evolution of seed plants. Nat Plants. 2022:**8**(4):389–401. https://doi.org/10.1038/s41477-022-01129-7

Liu C, Xi W, Shen L, Tan C, Yu H. Regulation of floral patterning by flowering time genes. Dev Cell. 2009:**16**(5):711–722. https://doi.org/10.1016/j.devcel.2009.03.011

Liu C, Zhang J, Zhang N, Shan H, Su K, Zhang J, Meng Z, Kong H, Chen Z. Interactions among proteins of floral MADS-box genes in basal eudicots: implications for evolution of the regulatory network for flower development. Mol Biol Evol. 2010:**27**(7):1598–1611. https://doi.org/10.1093/molbev/msq044

L'Obel MD. Kruydtboeck oft beschrÿuinghe van allerleye ghewassen, kruyderen, hesteren, ende gheboomten. Antwerpen: Christoffel Plantyn; 1581.

Lohmann JU, Hong RL, Hobe M, Busch MA, Parcy F, Simon R, Weigel D. A molecular link between stem cell regulation and floral patterning in Arabidopsis. Cell. 2001:**105**(6):793–803. https://doi.org/10.1016/S0092-8674(01)00384-1

Ma H, Yanofsky MF, Meyerowitz EM. AGL1-AGL6, an Arabidopsis gene family with similarity to floral homeotic and transcription factor genes. Gene Dev. 1991:**5**(3):484–495. https://doi.org/10.1101/gad.5.3.484

Mandel MA, Bowman JL, Kempin SA, Ma H, Meyerowitz EM, Yanofsky MF. Manipulationn of flower structure in transgenic tobacco. Cell. 1992a:**71**(1):133–143. https://doi.org/10.1016/0092-8674(92)90272-E

Mandel MA, Gustafson-Brown C, Savidge B, Yanofsky MF. Molecular characterization of the Arabidopsis floral homeotic gene APETALA1. Nature. 1992b:**360**(6401):273–277. https://doi.org/10.1038/360273a0

Mao W-T, Hsu W-H, Li J-Y, Yang C-H. Distance-based measurement determines the coexistence of B protein hetero- and homodimers in lily tepal and stamen tetrameric complexes. Plant J. 2021:**105**(5): 1357–1373. https://doi.org/10.1111/tpj.15117

Martínez-Gómez J, Galimba KD, Coté EY, Sullivan AM, Stilio VSD. Spontaneous homeotic mutants and genetic control of floral organ identity in a ranunculid. Evol Dev. 2020:**23**(3):197–214. https://doi.org/10.1111/ede.12357

Masters MT. Vegetable teratology: an account of the principle deviations from the usual construction of plants. London: Ray Society; 1869.

McGinnis W, Garber RL, Wirz J, Kuroiwa A, Gehring WJ. A homologous protein-coding sequence in Drosophila homeotic genes and its conservation in other metazoans. Cell. 1984a:**37**(2):403–408. https://doi.org/10.1016/0092-8674(84)90370-2

McGinnis W, Levine MS, Hafen E, Kuroiwa A, Gehring WJ. A conserved DNA sequence in homoeotic genes of the Drosophila antennapedia and bithorax complexes. Nature. 1984b:**308**(5958):428–433. https://doi.org/10.1038/308428a0

Melzer R, Härter A, Rümpler F, Kim S, Soltis PS, Soltis DE, Theißen G. DEF- and GLO-like proteins may have lost most of their interaction partners during angiosperm evolution. Ann Bot. 2014:**114**(7):1431–1443. https://doi.org/10.1093/aob/mcu094

Melzer R, Theissen G. Reconstitution of 'floral quartets' in vitro involving class B and class E floral homeotic proteins. Nucleic Acids Res. 2009:**37**(8):2723–2736. https://doi.org/10.1093/nar/gkp129

Melzer R, Verelst W, Theißen G. The class E floral homeotic protein SEPALLATA3 is sufficient to loop DNA in 'floral quartet'-like complexes in vitro. Nucleic Acids Res. 2009:**37**(1):144–157. https://doi.org/10.1093/nar/gkn900

Mena M, Ambrose BA, Meeley RB, Briggs SP, Yanofsky MF, Schmidt RJ. Diversification of C-function activity in maize flower development. Science. 1996:**274**(5292):1537–1540. https://doi.org/10.1126/science.274.5292.1537

Mendes MA, Guerra RF, Berns MC, Manzo C, Masiero S, Finzi L, Kater MM, Colombo L. MADS domain transcription factors mediate short-range DNA looping that is essential for target gene expression in Arabidopsis. Plant Cell. 2013:**25**(7):2560–2572. https://doi.org/10.1105/tpc.112.108688

Meyerowitz EM, Smyth DR, Bowman JL. Abnormal flowers and pattern formation in floral development. Development. 1989:**106**(2): 209–217. https://doi.org/10.1242/dev.106.2.209

Mondragon-Palomino M, Theissen G. Conserved differential expression of paralogous DEFICIENS- and GLOBOSA-like MADS-box genes in the flowers of Orchidaceae: refining the 'orchid code'. Plant J. 2011:**66**(6):1008–1019. https://doi.org/10.1111/j.1365-313X.2011.04560.x

Montalvão APL, Kersten B, Fladung M, Müller NA. The diversity and dynamics of sex determination in dioecious plants. Front Plant Sci. 2021:**11**:580488. https://doi.org/10.3389/fpls.2020.580488

Moquin-Tandon A. Elements de tiratologie vigitale, ou histoire abrigie des anomalies de l'Organisation dans les vigitaux. Paris: J.-P. Loss; 1841.

Morel P, Heijmans K, Rozier F, Zethof J, Chamot S, Bento SR, Vialette-Guiraud A, Chambrier P, Trehin C, Vandenbusschea M. Divergence of the floral A-function between an asterid and a rosid species. Plant Cell. 2017:**29**(7):1605–1621. https://doi.org/10.1105/tpc.17.00098

Moschin S, Nigris S, Ezquer I, Masiero S, Cagnin S, Cortese E, Colombo L, Casadoro G, Baldan B. Expression and functional analyses of nymphaea caerulea MADS-box genes contribute to clarify the complex flower patterning of water lilies. Front Plant Sci. 2021:**12**:730270. https://doi.org/10.3389/fpls.2021.730270

Mouradov A, Hamdorf B, Teasdale RD, Kim JT, Winter KU, Theissen G. A DEF/GLO-like MADS-box gene from a gymnosperm: pinus radiata contains an ortholog angiosperm B class floral homeotic genes. Dev Genet. 1999:**25**(3):245–252. https://doi.org/10.1002/(SICI)1520-6408(1999)25:3<245::AID-DVG7>3.0.CO;2-N

Moyroud E, Minguet EG, Ott F, Yant L, Posé D, Monniaux M, Blanchet S, Bastien O, Thévenon E, Weigel D, et al. Prediction of regulatory interactions from genome sequences using a biophysical model for the Arabidopsis LEAFY transcription factor. Plant Cell. 2011:**23**(4):1293–1306. https://doi.org/10.1105/tpc.111.083329

Moyroud E, Monniaux M, Thévenon E, Dumas R, Scutt CP, Frohlich MW, Parcy F. A link between LEAFY and B-gene homologues in Welwitschia mirabilis sheds light on ancestral mechanisms prefiguring floral development. New Phytol. 2017:**216**(2):469–481. https://doi.org/10.1111/nph.14483

Muiño JM, Smaczniak C, Angenent GC, Kaufmann K, van Dijk ADJ. Structural determinants of DNA recognition by plant MADS-domain transcription factors. Nucleic Acids Res. 2014:**42**(4):2138–2146. https://doi.org/10.1093/nar/gkt1172

Naudin C. Nouvelles récherches sur l'hybridité dans les végétaux. Nouvelles archives du Muséum d'histoire naturelle. 1865:**1**:25–176.

Ohmori S, Kimizu M, Sugita M, Miyao A, Hirochika H, Uchida E, Nagato Y, Yoshida H. MOSAIC FLORAL ORGANS1, an AGL6-like MADS box gene, regulates floral organ identity and meristem fate in rice. Plant Cell. 2009:**21**(10):3008–3025. https://doi.org/10.1105/tpc.109.068742

Ó'Maoiléidigh DS, Wuest SE, Rae L, Raganelli A, Ryan PT, Kwaśniewska K, Das P, Lohan AJ, Loftus B, Graciet E, et al. Control of reproductive floral organ identity specification in Arabidopsis by the C function regulator AGAMOUS. Plant Cell. 2013:**25**(7):2482–2503. https://doi.org/10.1105/tpc.113.113209

Pajoro A, Madrigal P, Muiño JM, Matus JT, Jin J, Mecchia MA, Debernardi JM, Palatnik JF, Balazadeh S, Arif M, et al. Dynamics of chromatin accessibility and gene regulation by MADS-domain transcription factors in flower development. Genome Biol. 2014:**15**(3):R41. https://doi.org/10.1186/gb-2014-15-3-r41

Pan Z-J, Cheng C-C, Tsai W-C, Chung M-C, Chen W-H, Hu J-M, ChenH-H. The duplicated B-class MADS-box genes display dualistic characters in orchid floral organ identity and growth. Plant Cell Physiol. 2011:**52**(9):1515–1531. https://doi.org/10.1093/pcp/pcr092

Pankratz MJ, Jäckle H. Making stripes in the Drosophila embryo. Trends Genet. 1990:**6**:287–292. https://doi.org/10.1016/0168-9525(90)90234-W

Parcy F, Nilsson O, Busch MA, Lee I, Weigel D. A genetic framework for floral patterning. Nature. 1998:**395**(6702):561–566. https://doi.org/10.1038/26903

Parkinson J. Paradisi in sole paradisus terrestris. London: Humfrey Lownes and Robert Young; 1629.

Payer J-B. Traité d'organogénie comparée de la fleur. Paris: Libraire de Victor Masson; 1857.

Pelayo MA, Yamaguchi N, Ito T. One factor, many systems: the floral homeotic protein AGAMOUS and its epigenetic regulatory mechanisms. Curr Opin Plant Biol. 2021:**61**:102009. https://doi.org/10.1016/j.pbi.2021.102009

Pelaz S, Ditta GS, Baumann E, Wisman E, Yanofsky MF. B and C floral organ identity functions require SEPALLATA MADS-box genes. Nature. 2000:**405**(6783):200–203. https://doi.org/10.1038/35012103

Pelaz S, Tapia-López R, Alvarez-Buylla ER, Yanofsky MF. Conversion of leaves into petals in Arabidopsis. Curr Biol. 2001:**11**(3):182–184. https://doi.org/10.1016/S0960-9822(01)00024-0

Penzig O. Pflanzen-teratologie systematisch geordnet. Genua: A. Ciminago; 1890–4.

Pinyopich A, Ditta GS, Savidge B, Liljegren SJ, Baumann E, Wisman E, Yanofsky MF. Assessing the redundancy of MADS-box genes during carpel and ovule development. Nature. 2003:**424**(6944):85–88. https://doi.org/10.1038/nature01741

Plinius the Secondus TE. The history of the world, commonly called the naturall historie of C. Plinius Secundus. London: A. Islip; 1634.

Pnueli L, Abu-Abeid M, Zamir D, Nacken W, Schwarz-Sommer Z, Lifschitz E. The MADS box gene family in tomato: temporal expression during floral development, conserved secondary structures and homology with homeotic genes from Antirrhinum and Arabidopsis. Plant J. 1991:**1**(2):255–266. https://doi.org/10.1111/j.1365-313X.1991.00255.x

Poyatos-Pertíñez S, Quinet M, Ortíz-Atienza A, Yuste-Lisbona FJ, Pons C, Giménez E, Angosto T, Granell A, Capel J, Lozano R. A factor linking floral organ identity and growth revealed by characterization of the tomato mutant unfinished flower development (ufd). Front Plant Sci. 2016:**7**:1648. https://doi.org/10.3389/fpls.2016.01648

Prunet N, Meyerowitz EM. Genetics and plant development. Comptes Rendus Biol. 2016:**339**(7–8):240–246. https://doi.org/10.1016/j.crvi.2016.05.003

Prunet N, Yang W, Das P, Meyerowitz EM, Jack TP. SUPERMAN prevents class B gene expression and promotes stem cell termination in the fourth whorl of Arabidopsis thaliana flowers. Proc Natl Acad Sci U S A. 2017:**114**(27):7166–7171. https://doi.org/10.1073/pnas.1705977114

Puranik S, Acajjaoui S, Conn S, Costa L, Conn V, Vial A, Marcellin R, Melzer R, Brown E, Hart D, et al. Structural basis for the oligomerization of the MADS domain transcription factor SEPALLATA3 in Arabidopsis. Plant Cell. 2014:**26**(9):3603–3615. https://doi.org/10.1105/tpc.114.127910

Qiu Y, Li Z, Walther D, Köhler C. Updated phylogeny and protein structure predictions revise the hypothesis on the origin of MADS-box transcription factors in land plants. Mol Biol Evol. 2023:**40**(9):msad194. https://doi.org/10.1101/2023.01.10.523452

Refahi Y, Zardilis A, Michelin G, Wightman R, Leggio B, Legrand J, Faure E, Vachez L, Armezzani A, Risson A-E, et al. A multiscale analysis of early flower development in Arabidopsis provides an integrated view of molecular regulation and growth control. Dev Cell. 2021:**56**(4):540–556. https://doi.org/10.1016/j.devcel.2021.01.019

Riechmann JL, Krizek BA, Meyerowitz EM. Dimerization specificity of Arabidopsis MADS domain homeotic proteins APETALA1, APETALA3, PISTILLATA, and AGAMOUS. Proc Natl Acad Sci U S A. 1996:**93**(10):4793–4798. https://doi.org/10.1073/pnas.93.10.4793

Rieu P, Turchi L, Thévenon E, Zarkadas E, Nanao M, Chahtane H, Tichtinsky G, Lucas J, Blanc-Mathieu R, Zubieta C, et al. The F-box protein UFO controls flower development by redirecting the master transcription factor LEAFY to new cis-elements. Nat Plants. 2023:**9**(2):315–329. https://doi.org/10.1038/s41477-022-01336-2

Rijpkema AS, Royaert S, Zethof J, van der Weerden G, Gerats T, Vandenbussche M. Analysis of the petunia TM6 MADS box gene reveals functional divergence within the DEF/AP3 lineage. Plant Cell. 2006:**18**(8):1819–1832. https://doi.org/10.1105/tpc.106.042937

Rijpkema AS, Zethof J, Gerats T, Vandenbussche M. The petunia AGL6 gene has a SEPALLATA-like function in floral patterning. Plant J. 2009:**60**(1):1–9. https://doi.org/10.1111/j.1365-313X.2009.03917.x

Ruelens P, Zhang Z, van Mourik H, Maere S, Kaufmann K, Geuten K. The origin of floral organ identity quartets. Plant Cell. 2017:**29**(2):229–242. https://doi.org/10.1105/tpc.16.00366

Rümpler F, Tessari C, Gramzow L, Gafert C, Blohs M, Theißen G. The origin of floral quartet formation—ancient exon duplications shaped the evolution of MIKC-type MADS-domain transcription factor interactions. Mol Biol Evol. 2023:**40**(5):msad088. https://doi.org/10.1093/molbev/msad088

Rümpler F, Theißen G, Melzer R. A conserved leucine zipper-like motif accounts for strong tetramerization capabilities of SEPALLATA-like MADS-domain transcription factors. J Exp Bot. 2018:**69**(8):1943–1954. https://doi.org/10.1093/jxb/ery063

Sacharowski SP, Gratkowska DM, Sarnowska EA, Kondrak P, Jancewicz I, Porri A, Bucior E, Rolicka AT, Franzen R, Kowalczyk J, et al. SWP73 subunits of Arabidopsis SWI/SNF chromatin remodeling complexes play distinct roles in leaf and flower development. Plant Cell. 2015:**27**(7):1889–1906. https://doi.org/10.1105/tpc.15.00233

Sakai H, Medrano LJ, Meyerowitz EM. Role of SUPERMAN in maintaining Arabidopsis floral whorl boundaries. Nature. 1995:**378**(6553):199–203. https://doi.org/10.1038/378199a0

Savidge B, Rounsley SD, Yanofsky MF. Temporal relationship between the transcription of two Arabidopsis MADS box genes and the floral organ identity genes. Plant Cell. 1995:**7**(6):721–733. https://doi.org/10.1105/tpc.7.6.721

Schilling S, Pan S, Kennedy A, Melzer R. MADS-box genes and crop domestication: the jack of all traits. J Exp Bot. 2018:**69**(7):1447–1469. https://doi.org/10.1093/jxb/erx479

Schwarz-Sommer Z, Hue I, Huijser P, Flor PJ, Hansen R, Tetens F, Lönnig WE, Saedler H, Sommer H. Characterization of the antirrhinum floral homeotic MADS-box gene deficiens: evidence for DNA binding and autoregulation of its persistent expression throughout flower development. EMBO J. 1992:**11**(1):251–263. https://doi.org/10.1002/j.1460-2075.1992.tb05048.x

Schwarz-Sommer Z, Huijser P, Nacken W, Saedler H, Sommer H. Genetic-control of flower development by homeotic genes in Antirrhinum-majus. Science. 1990:**250**(4983):931–936. https://doi.org/10.1126/science.250.4983.931

Scott MP, Weiner AJ. Structural relationships among genes that control development: sequence homology between the antennapedia, ultrabithorax, and fushi tarazu loci of Drosophila. Proc Natl Acad Sci U S A. 1984:**81**(13):4115–4119. https://doi.org/10.1073/pnas.81.13.4115

Sharma B, Kramer E. Sub- and neo-functionalization of APETALA3 paralogs have contributed to the evolution of novel floral organ identity in Aquilegia (columbine, Ranunculaceae). New Phytol. 2013:**197**(3):949–957. https://doi.org/10.1111/nph.12078

Siegfried KR, Eshed Y, Baum SF, Otsuga D, Drews GN, Bowman JL. Members of the YABBY gene family specify abaxial cell fate in

Arabidopsis. Development. 1999:**126**(18):4117–4128. https://doi.org/10.1242/dev.126.18.4117

Smaczniak C, Muiño JM, Chen D, Angenent GC, Kaufmann K. Differences in DNA binding specificity of floral homeotic protein complexes predict organ-specific target genes. Plant Cell. 2017:**29**(8):1822–1835. https://doi.org/10.1105/tpc.17.00145

Smyth DR. Evolution and genetic control of the floral ground plan. New Phytol. 2018:**220**(1):70–86. https://doi.org/10.1111/nph.15282

Smyth DR. How flower development genes were identified using forward genetic screens in Arabidopsis thaliana. Genetics. 2023:**224**(4):iyad102. https://doi.org/10.1093/genetics/iyad102

Smyth DR, Bowman JL, Meyerowitz EM. Early flower development in Arabidopsis. Plant Cell. 1990:**2**(8):755–768. https://doi.org/10.1105/tpc.2.8.755

Soltis DE, Chanderbali AS, Kim S, Buzco T, Soltis PS. The ABC model and its applicability to basal angiosperms. Ann Bot. 2007:**100**(2):155–163. https://doi.org/10.1093/aob/mcm117

Sommer RJ. Phenotypic plasticity: from theory and genetics to current and future challenges. Genetics. 2020:**215**(1):1–13. https://doi.org/10.1534/genetics.120.303163

Sommer H, Beltrán J-P, Huijser P, Pape H, Lonnig W-E, Saedler H, Schwarz-Sommer Z. Deficiens, a homeotic gene involved in the control of flower morphogenesis in Antirrhinum majus: the protein shows homology to transcription factors. EMBO J. 1990:**9**(3):605–613. https://doi.org/10.1002/j.1460-2075.1990.tb08152.x

Sun B, Xu Y, Ng K-H, Ito T. A timing mechanism for stem cell maintenance and differentiation in the Arabidopsis floral meristem. Gene Dev. 2009:**23**(15):1791–1804. https://doi.org/10.1101/gad.1800409

Sundstrom J, Carlsbecker A, Svensson ME, Svenson M, Johanson U, Theissen G, Engstrom P. MADS-box genes active in developing pollen cones of Norway spruce (Picea abies) are homologous to the B-class floral homeotic genes in angiosperms. Dev Genet. 1999:**25**(3):253–266. https://doi.org/10.1002/(SICI)1520-6408(1999)25:3<253::AID-DVG8>3.0.CO;2-P

Tandre K, Albert VA, Sundgås A, Engström P. Conifer homologues to genes that control floral development in angiosperms. Plant Mol Biol. 1995:**27**(1):69–78. https://doi.org/10.1007/BF00019179

Tandre K, Svenson M, Svensson ME, Engström P. Conservation of gene structure and activity in the regulation of reproductive organ development of conifers and angiosperms. Plant J. 1998:**15**(5):615–623. https://doi.org/10.1046/j.1365-313x.1998.00236.x

Theissen G, Melzer R. Molecular mechanisms underlying origin and diversification of the angiosperm flower. Ann Bot. 2007:**100**(3):603–619. https://doi.org/10.1093/aob/mcm143

Theissen G, Melzer R, Rümpler F. MADS-domain transcription factors and the floral quartet model of flower development: linking plant development and evolution. Development. 2016:**143**(18):3259–3271. https://doi.org/10.1242/dev.134080

Theissen G, Saedler H. Plant biology. Floral quartets. Nature. 2001:**409**(6819):469–471. https://doi.org/10.1038/35054172

Theophrastus. Enquiry into plants and minor works on odours and weather signs. London: William Heinemann; 1916.

Thompson BE, Bartling L, Whipple C, Hall DH, Sakai H, Schmidt R, Hake S. bearded-ear encodes a MADS box transcription factor critical for maize floral development. Plant Cell. 2009:**21**(9):2578–2590. https://doi.org/10.1105/tpc.109.067751

Tillette de Clermont-Tonnere M. Sur une variété femelle du pommier commun. Mémoires de la Société linnéenne de Paris. 1825:**3**:164–167.

Tröbner L, Ramirez P, Motte I, Hue P, Huijser W, Lönnig H, Saedler H, Sommer H, Schwarz-Sommer Z. Globosa—a homeotic gene which interacts with deficiens in the control of antirrhinum floral organogenesis. EMBO J. 1992:**11**(13):4693–4704. https://doi.org/10.1002/j.1460-2075.1992.tb05574.x

Tucker SC, Hodges SA. Floral ontogeny of *Aquilegia, Semiaquilegia,* and *Enemion* (Ranunculaceae). Int J Plant Sci.. 2005:**166**(4):557–574. https://doi.org/10.1086/429848

van Tunen AJ, Eikelboom W, Angenent G. Floral organogenesis in tulipa. Flower Newslett. 1993:**16**:33–38.

Vandenbussche M, Zethof J, Royaert S, Weterings K, Gerats T. The duplicated B-class heterodimer model: whorl-specific effects and complex genetic interactions in petunia hybrida flower development. Plant Cell. 2004:**16**(3):741–754. https://doi.org/10.1105/tpc.019166

Veron AS, Kaufmann K, Bornberg-Bauer E. Evidence of interaction network evolution by whole-genome duplications: a case study in MADS-box proteins. Mol Biol Evol. 2007:**24**(3):670–678. https://doi.org/10.1093/molbev/msl197

Vollbrecht E, Veit B, Sinha N, Hake S. The developmental gene Knotted-1 is a member of a maize homeobox gene family. Nature. 1991:**350**(6315):241–243. https://doi.org/10.1038/350241a0

Wagner D, Sablowski RWM, Meyerowitz EM. Transcriptional activation of APETALA1 by LEAFY. Science. 1999:**285**(5427):582–584. https://doi.org/10.1126/science.285.5427.582

Wang GL. A study on the history of Chinese roses from ancient works and images. Acta Hortic. 2007:**751**(751):347–356. https://doi.org/10.17660/ActaHortic.2007.751.44

Wang P, Cheng T, Lu M, Liu G, Li M, Shi J, Lu Y, Laux T, Chen J. Expansion and functional divergence of AP2 group genes in spermatophytes determined by molecular evolution and Arabidopsis mutant analysis. Front Plant Sci. 2016:**7**. https://doi.org/10.3389/fpls.2016.01383

Wang Y-Q, Melzer R, Theißen G. Molecular interactions of orthologues of floral homeotic proteins from the gymnosperm Gnetum gnemon provide a clue to the evolutionary origin of 'floral quartets'. Plant J. 2010:**64**(2):177–190. https://doi.org/10.1111/j.1365-313X.2010.04325.x

Warner KA, Rudall PJ, Frohlich MW. Differentiation of perianth organs in Nymphaeales. TAXON. 2008:**57**(4):1096–1109. https://doi.org/10.1002/tax.574006

Warner KA, Rudall PJ, Frohlich MW. Environmental control of sepalness and petalness in perianth organs of waterlilies: a new Mosaic theory for the evolutionary origin of a differentiated perianth. J Exp Bot. 2009:**60**(12):3559–3574. https://doi.org/10.1093/jxb/erp202

Weigel D, Meyerowitz EM. Activation of floral homeotic genes in Arabidopsis. Science. 1993:**261**(5129):1723–1726. https://doi.org/10.1126/science.261.5129.1723

Wellmer F, Bowman JL, Davies B, Ferrándiz C, Fletcher JC, Franks RG, Graciet E, Gregis V, Ito T, Jack TP, et al. Flower development: open questions and future directions. In: **Riechmann JL, Wellmer F,** editors. Flower development, methods and protocols. New York: Humana Press; 2014. p. 103–124.

Western TL, Cheng Y, Liu J, Chen X. HUA ENHANCER2, a putative DExH-box RNA helicase, maintains homeotic B and C gene expression in Arabidopsis. Development. 2002:**129**(7):1569–1581. https://doi.org/10.1242/dev.129.7.1569

Wilkinson MD, Haughn GW. UNUSUAL FLORAL ORGANS controls meristem identity and organ primordia fate in Arabidopsis. Plant Cell. 1995:**7**(9):1485–1499. https://doi.org/10.2307/3870137

Winter CM, Austin RS, Blanvillain-Baufumé S, Reback MA, Monniaux M, Wu M-F, Sang Y, Yamaguchi A, Yamaguchi N, Parker JE, et al. LEAFY target genes reveal floral regulatory logic, cis motifs, and a link to biotic stimulus response. Dev Cell. 2011:**20**(4):430–443. https://doi.org/10.1016/j.devcel.2011.03.019

Wollmann H, Mica E, Todesco M, Long JA, Weigel D. On reconciling the interactions between APETALA2, miR172 and AGAMOUS with the ABC model of flower development. Development. 2010:**137**(21):3633–3642. https://doi.org/10.1242/dev.036673

Wordsell WC. The principles of plant teratology. London: Ray Society; 1915–6.

Wu H-W, Deng S, Xu H, Mao H-Z, Liu J, Niu Q-W, Wang H, Chua N-H. A noncoding RNA transcribed from the AGAMOUS (AG) second intron binds to CURLY LEAF and represses AG expression in leaves. New Phytol. 2018:**219**(4):1480–1491. https://doi.org/10.1111/nph.15231

Wu M-F, Sang Y, Bezhani S, Yamaguchi N, Han S-K, Li Z, Su Y, Slewinski TL, Wagner D. SWI2/SNF2 chromatin remodeling ATPases overcome polycomb repression and control floral organ identity with the LEAFY and SEPALLATA3 transcription factors. Proc Natl Acad Sci U S A. 2012:**109**(9):3576–3581. https://doi.org/10.1073/pnas.1113409109

Wuest SE, O'Maoileidigh DS, Rae L, Kwasniewska K, Raganelli A, Hanczaryk K, Lohan AJ, Loftus B, Graciet E, Wellmer F. Molecular basis for the specification of floral organs by APETALA3 and PISTILLATA. Proc Natl Acad Sci U S A. 2012:**109**(33):13452–-13457. https://doi.org/10.1073/pnas.1207075109

Yamaguchi T, Nagasawa N, Kawasaki S, Matsuoka M, Nagato Y, Hirano HY. The YABBY gene DROOPING LEAF regulates carpel specification and midrib development in Oryza sativa. Plant Cell. 2004:**16**(2):500–509. https://doi.org/10.1105/tpc.018044

Yamaguchi A, Wu M-F, Yang L, Wu G, Poethig RS, Wagner D. The microRNA-regulated SBP-box transcription factor SPL3 is a direct upstream activator of LEAFY, FRUITFULL, and APETALA1. Dev Cell. 2009:**17**(2):268–278. https://doi.org/10.1016/j.devcel.2009.06.007

Yanofsky MF, Ma H, Bowman JL, Drews GN, Feldmann KA, Meyerowitz EM. The protein encoded by the Arabidopsis homeotic gene agamous resembles transcription factors. Nature. 1990:**346**(6279): 35–40. https://doi.org/10.1038/346035a0

Yant L, Mathieu J, Dinh TT, Ott F, Lanz C, Wollmann H, Chen X, Schmid M. Orchestration of the floral transition and floral development in Arabidopsis by the bifunctional transcription factor APETALA2. Plant Cell. 2010:**22**(7):2156–2170. https://doi.org/10.1105/tpc.110.075606

Yoo M, Soltis PS, Soltis DE. Expression of floral MADS-box genes in two divergent water lilies: nymphaeales and nelumbo. Int J Plant Sci. 2010:**171**(2):121–146. https://doi.org/10.1086/648986

Yu D, Kotilainen M, Pöllänen E, Mehto M, Elomaa P, Helariutta Y, Albert VA, Teeri TH. Organ identity genes and modified patterns of flower development in Gerbera hybrida (Asteraceae). Plant J. 1999:**17**(1):51–62. https://doi.org/10.1046/j.1365-313X.1999.00351.x

Zahn LM, Feng B, Ma H. Beyond the ABC-model: regulation of floral homeotic genes. In: Soltis DE, Leebens-Mack JH and Soltis PS eds., Advances in botanical research, developmental genetics of the flower. London: Academic Press; 2006. p. 163–207.

Zhang Z, Coenen H, Ruelens P, Hazarika RR, Al Hindi T, Oguis GK, Vandeperre A, van Noort V, Geuten K. Resurrected protein interaction networks reveal the innovation potential of ancient whole-genome duplication. Plant Cell. 2018:**30**(11):2741–2760. https://doi.org/10.1105/tpc.18.00409

Zhang X, Pan L, Guo W, Li Y, Wang W. A convergent mechanism of sex determination in dioecious plants: distinct sex-determining genes display converged regulation on floral B-class genes. Front Plant Sci. 2022:**13**:953445. https://doi.org/10.3389/fpls.2022.953445

Zhao W, Chen Z, Liu X, Che G, Gu R, Zhao J, Wang Z, Hou Y, Zhang X. CsLFY is required for shoot meristem maintenance via interaction with WUSCHEL in cucumber (Cucumis sativus). New Phytol. 2018:**218**(1):344–356. https://doi.org/10.1111/nph.14954

Zhao L, Kim Y, Dinh TT, Chen X. Mir172 regulates stem cell fate and defines the inner boundary of APETALA3 and PISTILLATA expression domain in Arabidopsis floral meristems. Plant J. 2007:**51**(5):840–849. https://doi.org/10.1111/j.1365-313X.2007.03181.x

https://doi.org/10.1093/plcell/koad328

THE PLANT CELL

THE PLANT CELL 2024: 36: 1358–1376

Review

A commitment for *life*: Decades of unraveling the molecular mechanisms behind seed dormancy and germination

Nikita Sajeev,[1] **Maarten Koornneef**[2,3] **and Leónie Bentsink**[1,*]

1 Wageningen Seed Science Centre, Laboratory of Plant Physiology, Wageningen University, 6708PB Wageningen, the Netherlands
2 Laboratory of Genetics, Wageningen University, 6708PB Wageningen, the Netherlands
3 Max Planck Institute for Plant Breeding Research, Former Department of Plant Breeding and Genetics, Koeln 50829, Germany

*Author for correspondence: leonie.bentsink@wur.nl

Abstract

Seeds are unique time capsules that can switch between 2 complex and highly interlinked stages: seed dormancy and germination. Dormancy contributes to the survival of plants because it allows to delay germination to optimal conditions. The switch between dormancy and germination occurs in response to developmental and environmental cues. In this review we provide a comprehensive overview of studies that have helped to unravel the molecular mechanisms underlying dormancy and germination over the last decades. Genetic and physiological studies provided a strong foundation for this field of research and revealed the critical role of the plant hormones abscisic acid and gibberellins in the regulation of dormancy and germination, and later natural variation studies together with quantitative genetics identified previously unknown genetic components that control these processes. Omics technologies like transcriptome, proteome, and translatomics analysis allowed us to mechanistically dissect these processes and identify new components in the regulation of seed dormancy and germination.

Introduction

Seeds are complex entities that encompass all the genetic information needed to give rise to a new plant. Structurally, seeds are composed of 3 basic components: the embryo, which will form the future seedling; the energy-rich and nutritious tissue surrounding the embryo called the endosperm; and a protective layer on the outside called the testa or seed coat (Bewley et al. 2012) (Fig. 1). The seed coat also functions in producing protective metabolites, regulation of nutrient transport to the embryo, and defining the final seed size and shape (Francoz et al. 2018).

The process by which plants establish their next generation is called seed germination. This establishment is a primal goal for plants. Furthermore, seeds can travel for thousands of miles or wait for hundreds of years to germinate. This ability to remain in a quiescent state is called seed dormancy (Bewley 1997).

In the past decades numerous studies have been performed to unravel the mechanisms of seed dormancy and germination; however, we still know little about how emergence is inhibited in dormant seeds and how precisely the embryo emerges from the nondormant seed. Because reproduction is a crucial process in the plant life cycle, the genes regulating seed dormancy and germination are under enormous evolutionary selection pressure. It is challenging to measure dormancy and germination independently because this very much depends on the environmental context in which these measurements are made. An additional complication is that germination is often measured at a population level; when a batch of seeds germinates 100% in optimal conditions, the seeds are said to be nondormant.

In this review, we will attempt to untangle and shed light on the highly inter-linked seed dormancy and germination processes by reporting the most important scientific breakthroughs through time in this field. Here, we emphasize the use of genetic mutants, natural variation, and various omics studies that helped us understand the mechanisms

Received August 24, 2023. Accepted December 19, 2023. Advance access publication January 12, 2024

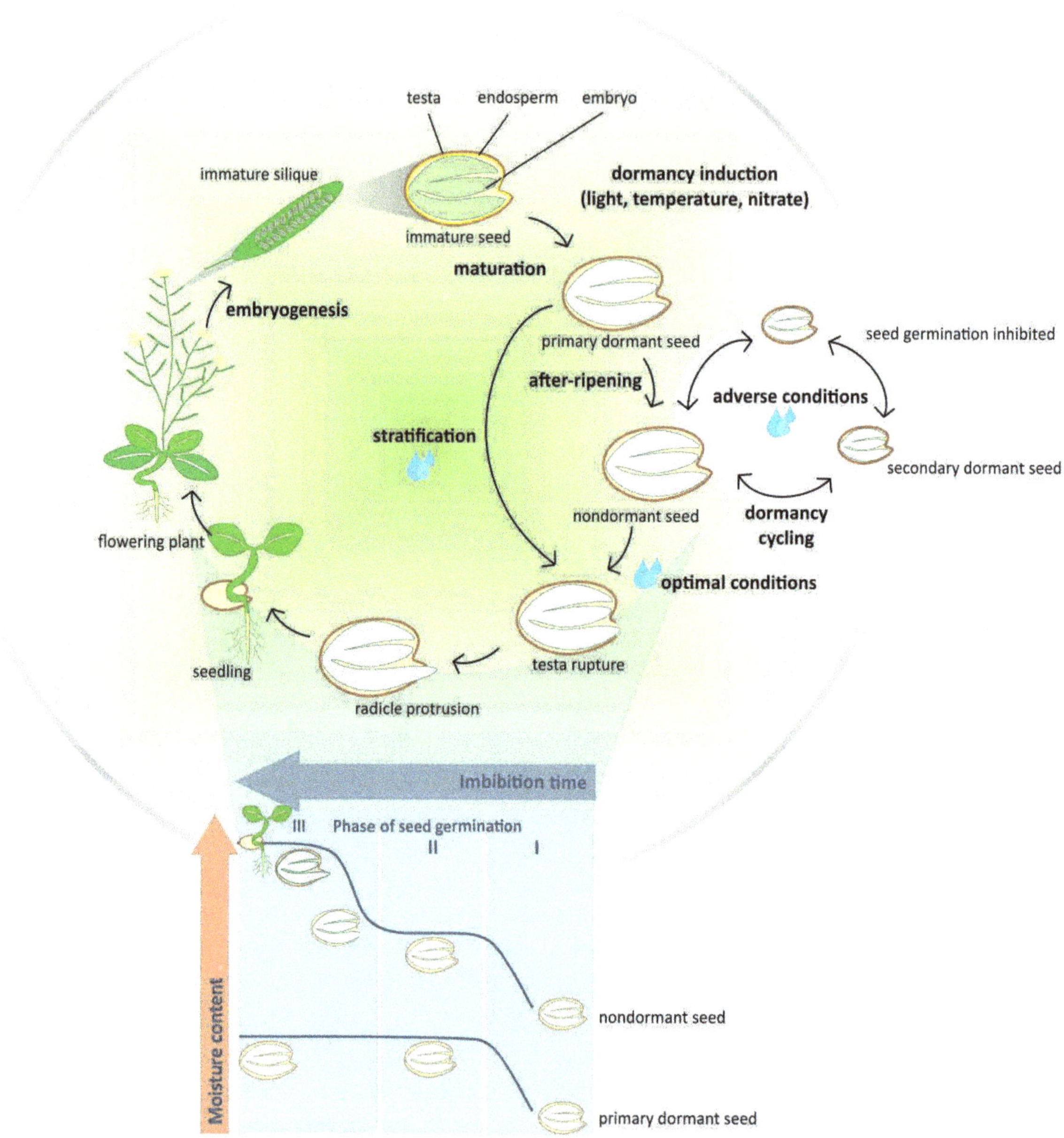

Figure 1. The life cycle of a seed. The seed life cycle begins with the embryogenesis phase occurring in the ovary of the flower of the mother plant. The ovary develops into a silique. Here the immature seed consisting of the testa, the endosperm and the embryo, which will form the future seedling, is developing. This stage is followed by seed maturation during which dormancy, desiccation tolerance, and seed longevity are induced. The induction of dormancy is dependent on genetic factors as well as environmental factors like light, temperature, and nitrate availability that the mother plant experiences. Seed maturation ends with a dry mature seed in which primary dormancy has been established. After-ripening or stratification can release dormancy and make the seeds nondormant. Germination of a nondormant seed can be inhibited, or the seed can reinter a dormant state if they are imbibed in adverse environmental conditions. If the nondormant seed is imbibed in optimal conditions, germination progresses through the different stages of germination, including TR and RP, which leads to the establishment of the seedling. Germination can be divided into 3 phases of water uptake. Stage 1 involves rapid uptake of water, which plateaus at stage II. Dormant seeds will remain in stage II and not progress to stage III of water uptake, whereas nondormant seeds will continue water uptake in stage III and complete germination. Imbibed stages indicated by the blue drops and the blue shades in the lower panel of the figure.

and regulators of seed dormancy and germination. We will mostly focus on the model plant *Arabidopsis thaliana* because most comprehensive studies have been and are still performed with this species.

The seed life cycle

The seed life cycle (Fig. 1) begins with the embryogenesis phase, in which a double fertilization event results in a single zygote. This zygote expands and grows in cell size and number to fill the embryo sac with the mature embryo (Bewley et al. 2012). The embryo growth is then arrested and followed by the maturation phase, in which the main goal is to accumulate resources and properties needed for successful germination and seedling establishment (Nonogaki et al. 2010). These resources can include mRNAs, sugars, and storage proteins required to kickstart germination. The maturation phase is also the critical phase wherein important traits like desiccation tolerance are established by intense moisture loss and seed dormancy and seed longevity are acquired. At the end of this phase, the seed is in a quiescent dry stage (Gutierrez et al. 2007; Holdsworth et al. 2008). Environmental conditions like light and temperature during the maturation phase, also termed as maternal or parental environment, are critical in determining the dormancy levels and the timing of germination in nature (Biere 1991; Fernandez-Pascual et al. 2013).

Seed dormancy: a blessing in disguise

Seed dormancy is defined as the inability of seeds to germinate in favorable conditions. Dormancy enables the mother plants to avoid risking germination of their entire progeny in an unpredictable and dynamic environment. Nikolaeva (1969) was the first seed biologist to develop a comprehensive classification scheme for the various types of seed dormancy. Dormancy can be broadly classified into 5 types of dormancies: physical, morphological, morphophysiological, physiological, and combinational dormancy (Fig. 2; Baskin and Baskin 2004). The most commonly studied type of dormancy is physiological dormancy. This type of dormancy is prevalent in gymnosperms and major classes of angiosperms, including Arabidopsis. The level of dormancy in seeds is high-est at the end of the maturation phase and is called primary dormancy. Primary dormancy in Arabidopsis seeds can be re-leased or alleviated by factors like after-ripening (AR; seed storage in the dry state) or stratification (cold or warm treat-ment in the imbibed state). The dormancy status also changes in response to environmental signals such as tem-perature, nitrate, and light quality exposure during seed de-velopment. Moreover, the sensitivity of the mature dry seeds to these signals depends on the AR time (Finch-Savage and Leubner-Metzger 2006; Nee et al. 2017a). If adverse environ-mental conditions prevail for extended periods of time, then after-ripened (AR) seeds can re-enter dormancy. This cycling of dormancy is known as secondary dormancy and can only occur when the seed is imbibed (Fig. 1) (Donohue 2005; Buijs 2020).

Seed germination: the birth of a seedling

When nondormant seeds are imbibed in optimal conditions, seeds are "woken up," so to say, and germination-promoting processes are initiated. By optimal conditions during imbibition of Arabidopsis seeds, we mean the presence of sufficient moisture, oxygen, optimal temperature, light quality, and absence of environmental stressors such as salinity, drought, cold, or heat stress. Regarding light quality, red light and its perception by photoreceptors called phytochromes is an essential factor for promoting seed germination (Longo et al. 2020). Germination itself is divided into 3 phases (Fig. 1). In Phase I there is tremendous uptake of water by the seed lead-ing to swelling, followed by a phase wherein water uptake plateaus (Phase II). Dormant seeds will not surpass this stage, whereas nondormant seeds will move on to Phase III, in which water uptake is resumed and the seed coat or the testa will rupture (TR). This is a critical physiologically visible stage during seed germination. After TR, cell elongation and expansion take place, which will eventually lead to the embryonic root or radicle to protrude through the endosperm (radicle protrusion [RP]). Once RP has occurred, germination is said to be complete (germination "sensu stricto"). The phase between TR and RP is a very sensitive one because the commitment to germinate into a seedling cannot be reversed once RP has occurred. This could be because seeds drastically lose their desiccation tolerance ability upon RP (Maia et al. 2011). Hence, nondormant seeds can stop or slow germination before RP if environmental conditions do support successful seedling establishment. Therefore, the final commitment to germination is made by integrating signals from the environment, and this process is tightly regulated at the genetic level.

Hormonal control of seed germination and dormancy

The absolute requirement of abscisic acid (ABA) and gibberellins (GAs) for the regulation of seed dormancy and germination was revealed in Arabidopsis and tomato mutants defective in the biosynthesis of these hormones in the early 1980s (Fig. 2) (Koornneef and van der Veen 1980; Groot and Karssen 1987). Most of these mutants were initially isolated on the basis of their altered seed germination phenotype. The requirement of GA for germination depends on the ABA concentration within seeds, as could be concluded from the observation that GA-deficient mutants can germin-ate in the absence of ABA. The latter led to the isolation of ABA-deficient mutants as revertants of GA-deficient mu-tants, which can germinate without GA or on high concen-trations of the GA biosynthesis inhibitor paclobutrazol (Koornneef et al. 1982; Léon-Kloosterziel et al. 1996). The bio-synthesis mutants (especially those in Arabidopsis, maize, and rice) were crucial in elucidating the biochemical steps and genes of the biosynthetic pathways for both plant hor-mones. For this the chemical analysis of the mutants was im-portant. The cloning of the underlying genes was based on the mutants using map-based cloning and transposon tag-ging, also described as forward genetics. In the meantime, the signal transduction steps were also identified. Due to the fact that many of these steps are controlled by multiple genes, this required additional procedures such as chemical genetics and biochemical interaction studies [reviewed by Cutler et al. (2010)]. At this stage, increasing use was made

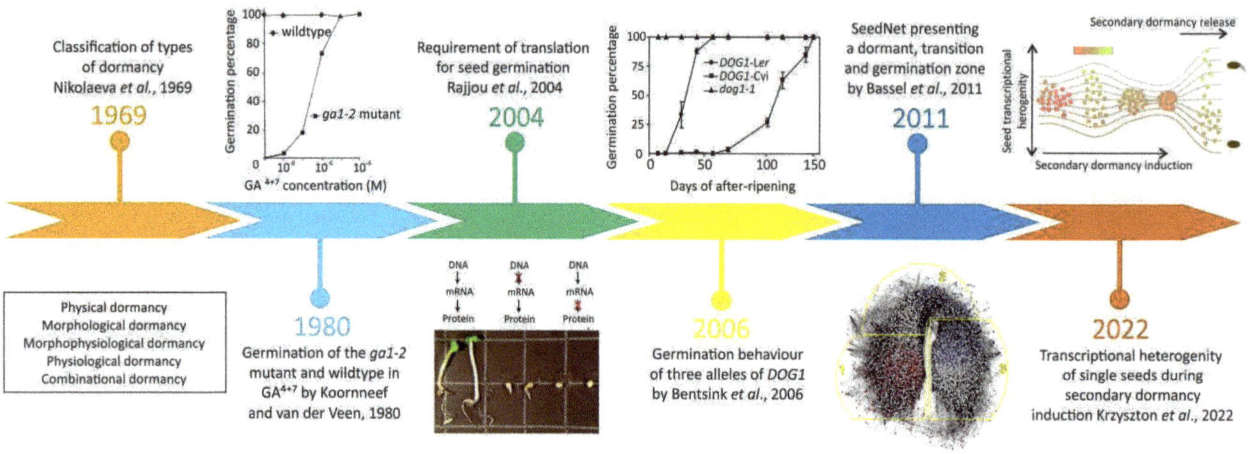

Figure 2. Milestones in the seed dormancy and germination research field through time.This figure graphically represents the biggest insights in the seed dormancy and germination field according to our interpretation. The first milestone was achieved in 1969, when Nikolaeva (1969) classified the 5 types of dormancies. This was followed by the breakthrough study of Koornneef and van der Veen (1980), where they concluded that the ABA concentration within the seeds determines the level of seed dormancy. The graph shows the germination percentage of GA-deficient mutants *ga1-2* and the wild-type seeds in increasing concentrations of exogenous GA. In 2004 it was discovered that the translation, but not transcription, is absolutely required for germination to occur (Rajjou et al. 2004). The next milestone arrived when Bentsink et al. (2006) cloned *DOG1*, the major quantitative trait locus governing seed dormancy in Arabidopsis and fine mapped it to the gene *DOG1* (At5G45830). The graph here shows the germination percentages of 3 *DOG1* alleles Cvi, *Ler*, and the mutant (*dog1-1*) in Arabidopsis during 0 to 150 days of AR. The graph shows the contribution of DOG1 to seed dormancy, as the *dog1-1* germinates 100% without a need for AR. In 2011, Bassel et al. generated a network model of global transcriptional interactions called SeedNet. This network represents 2 state-dependent sets of interactions that associate with dormancy and germination but also an intermediate transition region between these processes. The most recent milestone presented is from 2022, when Krzyszton et al. (2022) revealed transcriptional heterogeneity by performing RNA-seq on individual seeds during secondary dormancy induction. The figure shows that the heterogeneity is the lowest in very dormant seeds and increases when dormancy is broken.

of reverse genetics of candidate genes and the characterization of these mutants and double mutants with seed dormancy tests. Overall, the regulation of seed dormancy and germination and especially the role of ABA and GA in this has been extensively reviewed (Finch-Savage and Leubner-Metzger 2006; Finkelstein et al. 2008; Holdsworth et al. 2008; Bewley et al. 2012; Graeber et al. 2012; Shu et al. 2016). Below we aim at providing an overview of the essential roles of these 2 hormones. Gene interactions during the induction of seed dormancy in immature seeds and during dormancy release in the imbibed state are schematically presented in Fig. 3.

ABA, a repressor of germination and inducer of seed dormancy

In addition to not needing GAs for germination, ABA-deficient *aba1* mutants showed an absence of AR requirement in the shallow dormant Arabidopsis accessions Landsberg *erecta* (*Ler*) and Columbia (Col; Koornneef et al. 1982). Severe ABA biosynthesis mutants (*aba1*, *aba2*, *aba3*, and the *nced6/9* double mutant) show an absolute lack of seed dormancy (Lefebvre et al. 2006). The ABA concentration in seeds is, in addition to the control of ABA biosynthesis, also determined by conversion of ABA into phasic acid leading to inactivation of ABA. Responsible for this conversion are the ABA-8′ hydroxylase(s), which are 2 P450 cytochromes encoded by CYP707A1 and A2. These *cyp707a* mutants have an increased dormancy phenotype (Kushiro et al. 2004).

ABA levels in mature seeds are the result of synthesis during seed maturation. Depending on the species, the ABA levels peak at 2 moments (Kozaki and Aoyanagi 2022). Detailed analysis of ABA concentrations during imbibition showed that although ABA levels initially decrease, they increase at later imbibition stages in dormant seeds (Ali-Rachedi et al. 2004). Moreover, it is ABA produced and released by the endosperm that represses embryo germination, as was revealed by a "seed coat bedding" assay (Lee et al. 2010). The expression of the DELLA protein RGL2 is required for this response. The importance of ABA in maintaining the dormant state in imbibed seeds is also confirmed by the germination-promoting effect of the ABA biosynthesis inhibitor fluoridone when applied during the imbibition of dormant seeds. ABA in seeds is mainly synthesized in the endosperm and transported to the embryo by 4 ATP-binding cassette (ABC) transporter genes (Kang et al. 2015). AtABCG25 and AtABCG31 export ABA from the endosperm, whereas AtABCG30 and AtABCG40 import ABA into the embryo (Kang et al. 2015). The role of ABA transported from the mother plant during dormancy induction is minor, as was shown by ABA-deficient mother plants with ABA-producing embryos and endosperms obtained by reciprocal crossing of *aba1* mutants with its wild type (Karssen et al. 1983).

Seed-specific ABA signaling

The complex regulation of the ABA signaling pathway in seeds, with many gene copies present for its major steps,

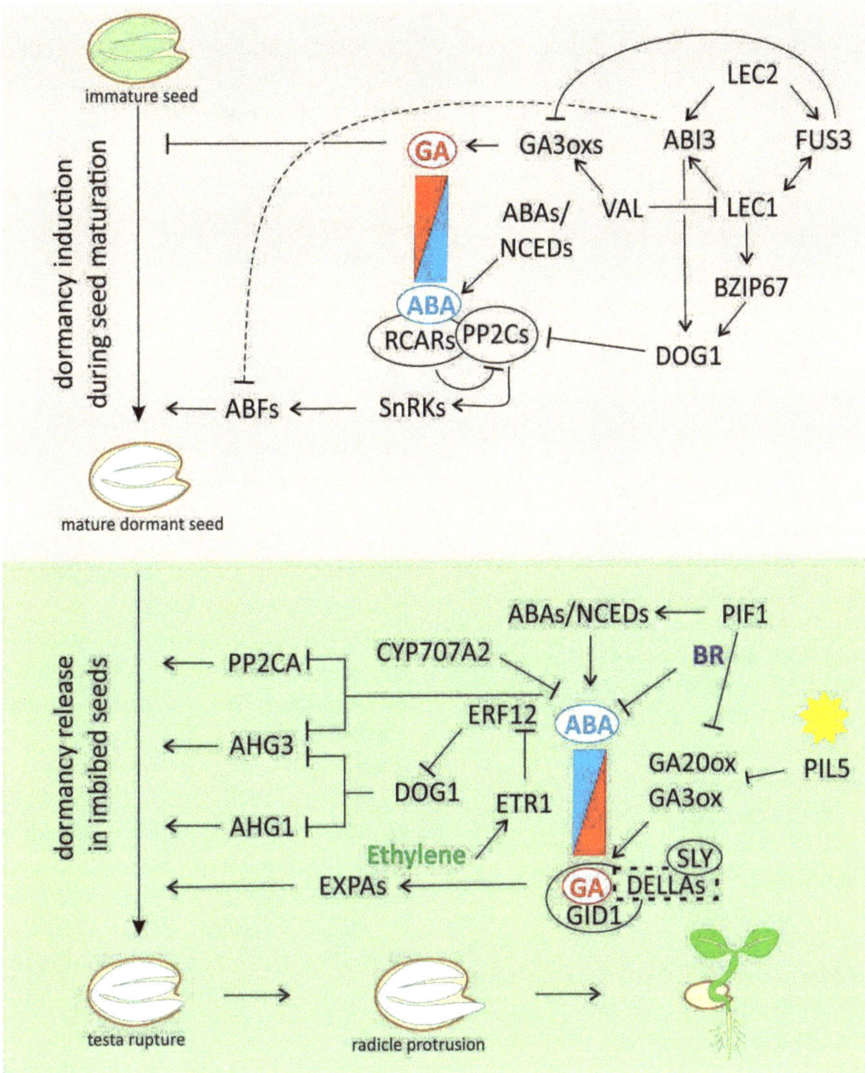

Figure 3. The regulation of seed dormancy and germination during seed maturation and dormancy release in imbibed seeds. The figure represents a schematic overview of gene interactions during induction of seed dormancy in the immature seeds and during dormancy release in the imbibed state. Key for both stages is the balance between concentrations of GA and ABA, indicated by the red/blue rectangle. Both hormones are sensed during seeds maturation and dormancy release; however, in the scheme, factors involved in the perception of ABA (RCARs and PP2Cs) are indicated in more detail the upperpart of the figure, which represents factors involved in dormancy induction during seed maturation. Those involved in GA perception (GID1, DELLAs, and SLY) are depicted in the lower panel, which shows the interactions during the dormancy release in the imbibed seeds. Arrows indicate a promotive effect, bars a repressive effect. The line between ABI3 and ABFs is dashed because the role of ABI3 in ABA perception has remained unclear so far. The indicated interactions are also described in the main text.

has recently been described by Née and Krüger (2023). Mutant screens for seeds germinating in the presence of wild type inhibiting concentrations of ABA identified several genes, among which were ABA insensitive 3 (*ABI3*), *ABI4*, and *ABI5*. These genes are important seed-specific downstream ABA targets and are considered repressors of germination (Koornneef et al. 1984; Finkelstein 1994). Based on their spe-cific expression during seed development and the early stages of germination, they are considered to have seed-specific roles.

Studies using strong *ABI3* mutants revealed, in addition to a lack of dormancy, many important defects during seed development, including the absence of chlorophyll degradation and desiccation intolerance (Ooms et al. 1993; Nambara et al. 1994). The *ABI3* gene together with the seed maturation genes *LEAFY COTYLEDON 1* and 2 (*LEC1/2*) and *FUSCA3* (*FUS3*) form the LAFL (*LEC1*, *ABI3*, *FUS3*, *LEC2*) regulatory network (Carbonero et al. 2017). These genes are essential for seed maturation, and single gene mutants show partial overlap in their phenotypes (Raz et al. 2001; Jia et al. 2013). All these genes (except *LEC1*) are members of the AFL (ABI3/FUS3/LEC2) subfamily of B3 transcription factors (TFs) and bind G-box like sequences in the promoters of seed maturation genes, which are bound by basic leucine zipper (bZIP) TFs (Boulard et al. 2017; Carbonero et al. 2017). ABI3 has many direct targets, including ABI5, DELLAs, and the bHLH type TF PIL5 (PHYTOCHROME INTERACTING FACTOR-LIKE 5) (Tian et al. 2020). ABI3, ABI5, and DELLAs together activate the expression of SOMNUS, which encodes a C3H type zinc finger protein that regulates the expression of genes involved in the synthesis of ABA and in the catabolism of GA in imbibed seeds, thus inhibiting seed germination (Park et al. 2011; Lim et al. 2013). The molecular interaction of ABI3 with the primary ABA signaling pathway remains unclear and is based only on the extreme ABA insensitivity of its mutant. The latter is not observed for the *fus3* mutants despite that they displayed a similar absence of dormancy as the *abi3-3* mutant (Keith et al. 1994).

ABI4 and *ABI5* were also identified in a mutant screen for ABA-insensitive mutants and thereby seemed also to be candidates in the ABA signal transduction (Finkelstein 1994). Their mutant phenotypes were much less severe than that of *abi3* mutants; however, they showed a strong reduction of ABA/ABI3-induced genes, such as the downregulation of the LATE EMBRYO ABUNDANT proteins *AtEM1* and *AtEM6* gene expression (Finkelstein 1994; Carles et al. 2002). Conversely, the *abi4* and 5 mutants did not show clear seed dormancy phenotypes. The *ABI4* gene, reviewed by Chandrasekaran et al. (2020), encodes an APETALA2/ethylene response factor type TF that affects many processes by regulating gene expression, including that of the ABA degradation genes ABA 8′ hydroxylase(s) (*Cyp707A1* and *Cyp707A2*), as well as genes involved in sugar signaling. Additionally, it downregulates several GA biosynthesis genes. The *abi4* mutant has a small effect on the rate of germination, and the mutant is not affected in its seed development phenotype (Shu et al. 2013).

ABI5, reviewed by Skubacz et al. (2016), encodes a bZIP TF, which can both bind to the ABA Responsive Element in promotors and interact with ABI3 via its B1 domain. The protein also interacts with AFPs (ABA 5 binding proteins), which al-lows interaction with other proteins (Lynch et al. 2017). Overexpression of AFP1 and AFP2 resulted in extreme ABA tolerance and desiccation-intolerant seeds resembling strong *abi3* mutant alleles and suggests that these genes are major actors in seed-specific ABA signaling. Deng et al. (2023) de-monstrated that AFP2 negatively regulates primary seed dor-mancy by suppressing the expression of the major regulator of seed dormancy, *DELAY OF GERMINATION 1* (*DOG1*). The TF WRKY36 can interact with AFP2 to transcriptionally co-repress TOPLESS RELATED PROTEIN 2 and epigenetically silence *DOG1* by reducing histone acetylation. The *abi5* mu-tants themselves show tolerance to germination inhibition by NaCl and osmotic stress (Carles et al. 2002). ABI5 can be phosphorylated by SnRK2 proteins as part of the ABA sig-nal transduction chain. The phosphorylation of ABI5 enables interactions with other proteins, including EM1 promotors, suggesting a role in the acquisition of desiccation tolerance in seeds. Furthermore, ABI5 can interact with other hormones, including GA, jasmonic acid, and auxin pathways (Skubacz et al. 2016). Overall, it seems that ABI5 is crucial for part of the ABA transduction chain in seeds, early germination, and its response to stresses.

GA promoters of seed germination

GAs are essential for germination because GA biosynthesis mutants such as *ga1/ga2/ga3* require application of exogenous GAs for germination (Fig. 2, Koornneef and van der Veen 1980). The GA requirement for germination is further indicated by the observation that GA biosynthesis inhibitors such as paclobutrazol and unicazol inhibit germination. GA biosynthesis and GA signaling in seeds has been reviewed by Urbanova and Leubner-Metzger (2018). Active GA binds to the receptor GID1 (encoded by *AtGID1a*, *AtGID1b*, and *AtGID1c*), and together they repress the DELLA proteins. The latter are repressors of GA signaling, and especially RGA2 is the seed-specific representative of this group of 5 proteins (Lee et al. 2002).

The role of GA seems mainly to overcome restraints that are partially related to (ABA induced) dormancy but also to overcome the resistance of the embryo surrounding tissues, including that conferred by the integuments (Debeaujon and Koornneef 2000) and the endosperm (Groot et al. 1988). The latter aspect relates to the induction of genes encoding cell wall–degrading enzymes by GA [reviewed by Nonogaki (2019)]. Focusing on expansin mediate cell-wall loosening, Xu et al. (2020) elucidated the pathway by which DELLA proteins and nitric oxide (NO) control TFs (including TCP14/15, RAP2.2/2.3/2.12, and ZML1), which in turn control expansin gene expression.

Hormonal crosstalk during dormancy and germination

ABA and GA do not regulate seed dormancy and germination on their own. Where interactions with jasmonic acid and auxin were already mentioned, brassinolides (BRs) and ethylene are also known to affect these traits. BRs stimulate germination by promoting TR and endosperm rupture in Arabidopsis in a similar way as GAs do. BRs can rescue the germination phenotypes of the GA biosynthetic mutant *ga1-3* and the GA-insensitive mutant *sleepy1* (Steber and McCourt 2001). Germination does not entirely depend on GA response but rather on GA and BR working in parallel. Two key basic helix–loop–helix transcription factors in the BR signaling pathway, BRASSIN OSTEROID ENHANCED EXPRESSION 2 (BEE2) and BEE 2 IN TERACTIN Gwith INCREASED LEAF INCLINATION 1 BINDING bHLH1 (HBI1), are involved in the regulation of endosperm rupture. Expression of BEE2 and HBI1 was induced in response to BR and GA treatment (Zhong et al. 2021). Furthermore, BRI1-EMS-SUPPRESSOR 1 (BES1) affecting BR signaling was shown to physically interact with ABI5 to hinder its DNA binding capacity, attenuating the ABA-mediated suppression of seed germination by lowering the expression of ABI5 targets. BIN2, a repressor of BR signaling, promotes the ABA responses (Zhao et al. 2019). BES1 and DELLAs were reported to mediate the crosstalk between BRs and GAs during cell elongation in Arabidopsis [reviewed by Ross and Quittenden (2016)].

Ethylene is another promoter of germination. The loss-of-function mutant of one of the ethylene receptors ETHYLENE RESPONSE FACTOR 1 (ETR1) was identified by its reduced dormancy phenotype and therefore named (*reduced dormancy* 3; Li et al. 2019). It was shown that ETR1 represses the expression of *ERF12*, which recruits TOPLESS to form a repressor complex and binds to the DOG1 promoter, thereby inhibiting the expression of *DOG1* and leading to reduced seed dormancy by ethylene.

Another mechanism controlled by EIN2 (ETHYLENE INSENSITIVE 2), a component of ethylene signal transduction, was recently described by Guo et al. (2023). The authors concluded that EIN2 can function independently from the canonical ethylene pathway acting via EIN3/EIL1, by inhibition of a histone H3 acetylase encoded by *HLS1* (*HOOKLESS1*). When active, HLS1 acetylates H3 histones in the region of *ABI5* and *ABI3*, which represses their transcription. In the absence of EIN2, the expression of these ABA signaling molecules is upregulated and germination is inhibited.

Auxin has been shown to function both positively and negatively in seed germination depending on its dose. It has also been reported that high doses of auxins enhance seed dormancy by activating ABI3 mediated by the AUXIN RESPONSE FACTOR 10 TF (Liu et al. 2013) and IAA8 (Hussain et al. 2020). This leads to enhanced expression of *ABI3*, although not by binding to the *ABI3* promotor. *IAA8* expression is upregulated by cold and by ROS signaling and was found to associate with the ABI3 promotor (Hussain et al. 2020). In both cases the auxin effects were relatively minor in affecting the rate of germination. Wang et al. (2016) emphasized the specific role of auxin in the RP and thereby germination rate. They noticed that this is, to a large extent, controlled by the auxin transporter AUX1. *AUX1* expression is negatively controlled by the histone acetylation genes *SNL1/SNL2*, which decrease in expression upon after dormancy breakage.

The power of natural variation, among others, a DOG1 story

Seed germination is a quantitative trait for which ample genetic variation is present in nature. Especially seed dormancy, which together with the timing of flower initiation is a main factor determining where and when a seed should germinate, has shown to be a key trait defining local adaptation. In European accessions a North-South cline was identified, where (primary) dormancy is lower in the North compared with the South. A similar pattern was found for accessions collected in the Iberian Peninsula (Debieu et al. 2013; Vidigal et al. 2016). Studies on the Iberian population also show that dormancy is generally lower at higher altitudes. Kronholm et al. (2012) described genetic signatures of local adaptation as well as a negative correlation between dormancy and summer precipitation. These clines are likely explained by the shorter growing season (in the Northern regions and higher altitudes) and the riskier environment with a lower chance of survival in the Southern regions and lower altitudes. The importance of the right dormancy levels is illustrated in the work of Huang et al. (2010). They showed with a garden experiment that genetic variation for fitness and germination phenology are associated with the genetic regions that were earlier detected for primary seed dormancy by testing AR requirement in the laboratory.

The use of natural variation for the identification of seed dormancy loci has led to the identification of the key regulator of seed dormancy, *DOG1*. Natural variation for dormancy was first explored in the *Ler* Col recombinant population (RIL) (Van der Schaar et al. 1997). However, the use of both these accessions with low dormancy levels only revealed minor quantitative trait loci (QTL). A breakthrough was made when more dormant accessions were used for these genetic approaches. The *Ler*, Cape Verde Island (Cvi) RIL population yielded 7 QTL that together explained more than 60% of the total variance (Alonso-Blanco et al. 2003). The main QTL identified in this study was the aforementioned *DOG1* QTL. Experiments with different RIL populations mainly resulted in the identification of the already known DOG loci (Meng et al. 2008; Bentsink et al. 2010; Amiguet-Vercher et al. 2015; Footitt et al. 2020). Moreover, genome-wide association studies using Arabidopsis accessions from all over the world also identified the already known QTL (He 2014; Martínez-Berdeja et al. 2020), sometimes with a few additional loci (Chen et al. 2023).

So far we have been discussing mostly natural variation for seed dormancy, measured as AR requirement. However, there is also genetic variation for germinability in cold and dark (Meng et al. 2008) and germination variability (Abley et al. 2021), which is referred to as the heterogeneity in germination that occurs in genetically identical plants that have been grown in the same condition. The level of germination variability differs between Arabidopsis accessions, and QTL analyses using MAGIC lines revealed QTL in the *DOG1* and *DOG6* regions (Abley et al. 2021). The major Cold-tolerant Dark Germination QTL occurred at positions for which no QTL have been cloned so far (Meng et al. 2008).

Identifying the genes underlying the DOGs: an ongoing search

The identified *DOG* QTL underly genes of a different nature; so far *DOG1*, *DOG6*, and *DOG18* are cloned, and these encode, respectively, an unknown protein, a NAC transcription factor, and a pseudo phosphatase (RDO5) (Bentsink et al. 2006; Xiang et al. 2016; Song et al. 2021). Although we can only speculate about the nature of the remaining *DOG* loci, genetic analyses and transcriptional profiling have indicated that they control dormancy by different additive genetic and molecular pathways (Bentsink et al. 2010). Recently, several reviews have appeared that describe the regulation of dormancy via DOG1 (Nee et al. 2017a; Carrillo-Barral et al. 2020; Li et al. 2020; Soppe and Bentsink 2020; Iwasaki et al. 2022). *DOG1* was identified using a combined genetic approach involving the fine-mapping of the identified QTL region and a mutagenesis analyses in the background on the near isogenic line (NIL) that contained the *DOG1* Cvi allele (Bentsink et al. 2006). The nondormant mutant identified in the NIL background could not be genetically separated from the *DOG1* dormant allele and turned out to be a mutation in the DOG1 gene. Knowing that knocking-out DOG1 would lead to loss of dormancy allowed for screening knockout lines in the fine-mapped region of the QTL. As such, At5G45830, a gene of unknown function, was identified as the gene encoding *DOG1*. The availability of 3 alleles (Cvi, L*er*, and the mutant *dog1-1*), which showed very strong dormancy phenotypes, was of great importance for the later studies (Fig. 2; Bentsink et al. 2006). It is the amount of DOG1 protein that determines the level of seed dormancy, and loss of function of the protein that occurs during AR leads to a reduction of seed dormancy (Nakabayashi et al. 2012). DOG1 interacts with clade A PP2C phosphatases, the ABA-HYPERSENSITIVE GERMINATION 1 and 3 (AHG1 and 3). DOG1 likely suppresses the action of AHG1 and 3 and as such inhibits the release of seed dormancy (Nee et al. 2017b).

There is a clear link between ABA and DOG1, which involves the above described interaction with the AHG proteins but is also based on the fact that DOG1 cannot confer dormancy in a ABA-deficient/-insensitive background (Bentsink et al. 2006). The fact that DOG1 is an α-helical heme-binding protein is another link to ABA (Nishimura et al. 2018). Heme binding proteins, which are reported to function as sensors for oxygen and NO, are required to counteract ABA during dormancy release (Arc et al. 2013). DOG1 is post-translationally modified as the result of AR (Nakabayashi et al. 2012). It is hypothesized that DOG1-bound heme could cause non-enzymatic oxidative cellular damage by promoting ROS formation during seed dry storage; this would be more severe in the presence of larger amounts of DOG1 protein (Chiabrando et al. 2014).

DOG1 expression is induced during seed maturation (Bentsink et al. 2006). In-depth analyses revealed that during seed development, DOG1 genetically interacts with ABI3 and as such affects other seed maturation traits as well (Dekkers et al. 2016b). Seeds of the *dog1* mutant have reduced expression of genes encoding LATE EMBRYOGENESIS ABUNDANT and HEAT SHOCK PROTEINs and contain reduced amounts of certain primary metabolites and oligosaccharides. Interestingly, the *abi3-1 dog1-1* double mutant has green seeds, which phenocopies the more severe *abi3* mutants. Based on sequence alignment, DOG1 belongs to a family of 6 members, the DOG1-Like (DOGL) genes. None of these proteins affect seed dormancy; however, DOGL4 strongly induced the expression of seed maturation–specific genes, including the major seed reserve proteins ALBUMIN, CRUCIFERIN, and OLEOSIN (Sall et al. 2019).

The regulation of *DOG1* has intensively been studied. Lower temperatures during seed maturation led to increased levels of DOG1 transcript and protein (Nakabayashi et al. 2012). This effect is mediated by an increased expression of the bZIP transcription factor bZIP67, which transactivates *DOG1* expression. Genetic variation in the bZIP67 binding sites of the DOG1 promoter provides a mechanism to explain natural genetic variation in *DOG1* expression (Bryant et al. 2019). Moreover, low temperatures, when perceived during seed maturation, are known to increase DOG1 expression, as well as dormancy (Nakabayashi et al. 2012; He et al. 2014). Gene-swapping experiments between Arabidopsis and the nondormant *Lepidium sativa* showed that the DOG1 function is conserved among these Brassicaceae (Graeber et al. 2014). Whereas biomechanical analyses that allow to investigate the material properties of the endosperm so far have never left the pilot phase in Arabidopsis, these are common ground in *Lepidium* (Muller et al. 2006; Voegele et al. 2012). Such analyses reveal that DOG1 inhibits the expression of cell-wall remodeling genes via temperature-dependent adjustments of key enzymes of the GA biosynthetic pathway (Graeber et al. 2014). How a role for *DOG1* in the endosperm aligns with *DOG1* expression in the embryo remains unknown.

In addition to external factors regulating the expression of DOG1, several mechanisms that involve the DOG1 gene structure have been identified to regulate DOG1 expression. These include the following: 1) alternative splicing and selective polyadenylation resulting in short DOG1 and long DOG1 isoforms (Cyrek et al. 2016; Fedak et al. 2016); 2) long non-coding RNAs in the 5′ region of the gene, which are referred to as PUPPIES and promote DOG1 expression in high-salt conditions (Montez et al. 2023), and an antisense RNA located in the 3′ region of the gene referred to as *1GOD* that negatively affects *DOG1* expression (Cyrek et al. 2016; Kowalczyk et al. 2017); and 3) epigenetic regulation as summarized by Ding et al. (2022).

REDUCED DORMANCY 5 (*RDO5*), originally identified in the same genetic screen as the *dog1-1* mutant and underlying the natural variation at the *DOG18* locus (Xiang et al. 2016), has also been identified to bind to DOG1. RDO5 is annotated as a so-called pseudo phosphatase; although it belongs to the PP2C phosphatase family, it lacks phosphatase activity (Xiang et al. 2014). RDO5, similarly to DOG1, is localized to the nucleus. Three members of the PUM family of RNA binding proteins were upregulated in the *rdo5* mutant, suggesting that these genes promote germination. The PUM proteins are, in contrast to DOG1 or RDO5, localized in the cytoplasm; therefore the regulation of *PUM*s by RDO5 is indirect. In agreement with a role for PUMs in the translational regulation of seed germination, these might be downstream factors that regulate the onset of germination and not dormancy specifically. Downregulation of PUM9 and 11 by RNAi in the *rdo5* mutant background resulted in increased dormancy levels. PUMs are RNA binding proteins and are therefore discussed in more detail in the section "Mechanisms underlying translational control: The role of RNA binding proteins."

DOG6 encodes the NAC transcription factor *ANAC60* (Song et al. 2021, 2022). Different natural alleles of *DOG6* determine the localization of the protein due to the presence or absence of a membrane binding domain. Absence of the membrane binding domain allows the protein to enter the nucleus and as such attenuate seed dormancy.

Exploring transcriptome profiles

Transcriptome analyses have been a preferred method to address seed dormancy– and seed germination–related questions. In the early days, transcript profiling was based on hybridizing mRNAs to cDNA libraries that were constructed from specific tissues. One of the first reports revealing transcriptional changes during seed germination was that of Harada et al. (1988), who made use of nitrocellulose arrays containing a seedling cDNA library. These authors reported on the temporal and spatial specificity of these transcripts in *Brassica napus* cotyledons and seedling axes based on hybridization analyses of the identified clones. The nitrocellulose carriers used for hybridization of mRNA to cDNAs were replaced for coated slides in 1995 and around 2015 for RNA sequencing (RNA-seq). Before the switch to microarrays was made, Bove et al. (2005) reported a cDNA-AFLP study during dormancy breaking in *Nicotiana plumbaginifolia* seeds. Of the 15,000 cDNA fragments identified, 1,020 responded differentially to dormancy breaking or maintenance. The first seed germination–related study involving gene expression analysis using microarrays also dates from 2005 (Nakabayashi et al. 2005). This work performed on Arabidopsis seeds led to the identification of 12,470 seed-stored mRNAs. More than 10,000 of these transcripts were also detected at the different imbibition time-points investigated (6, 12, and 24 hours after imbibition). Differential expressed genes include those related to metabolism, transcription, cell cycle, and DNA processing. Moreover, the analyses of seeds at different stages during imbibition revealed multiple regulatory mechanisms, including epigenetic chromatin structures, co-regulated gene clusters, and cis-acting elements. The first RNA-seq study on seeds revealed 50% more differentially expressed genes then initially reported by Nakabayashi et al. (2005) based on microarray analyses (Narsai et al. 2017). In this section we will discuss transcriptome analyses in dormant and germinating seeds and focus mostly on genome-wide analyses.

Differences between dormant and nondormant seed transcriptomes

Seeds are full of stored mRNAs for which translation is considered important for germination to occur. The importance of stored transcripts for seed germination raised questions about the changes in the transcriptome, particularly when seeds transition from the dormant to the nondormant state during seed AR (dry storage). Although it has been debated whether transcriptional changes, especially upregulation, can occur in dry seeds, studies supporting such findings have been reported for *N. plumbaginifolia*, barley, and Arabidopsis seeds (Bove et al. 2005; Finch-Savage et al. 2007; Leymarie et al. 2007). More specifically, it has been suggested that specific mRNAs are being oxidized during AR. In sunflower seeds 24 oxidized transcripts have been identified that belong to the gene ontology classes metabolic processes, response to stress, and transport. Specific genes identified include protein phosphatase 2C PPH1, mitogen-activated protein kinase phosphatase 1, and phenyl ammonia lyase 1 (Bazin et al. 2011).

More obvious dormancy-specific transcriptomes can be revealed when seeds are investigated in the imbibed state. The comparison of dormant and AR Cvi seeds revealed an upregulation of genes related to translation, including genes that encode ribosomal proteins, translation initiation, and elongation factors in AR seeds (Cadman et al. 2006). This suggests that translation is turned down in dormant seeds, which has been confirmed by the analyses of translation by means of polysome profiling. In imbibed dormant seeds, translation starts like it does in AR seeds; however, it does not reach the same level as it does in nondormant seeds (Bai et al. 2018). Another attempt to identify differences between dormant and AR seeds revealed a set of 30 genes with similar expression profiles in the different physiological stages investigated; among these genes was the key dormancy regulator *DOG1* (Finch-Savage et al. 2007). Although all these 30 genes in theory have the potential to be regulators of seed dormancy or seed germination, they are not reported as such. Only ERF1, which was associated with temperature-sensitive germination in lettuce and improved germination in saline conditions in rice, was described to be germination related (Yoong et al. 2016).

Investigating the temporal changes in dormant and nondormant seeds revealed that transcriptional patterns already differ at 12 hours of imbibition, when nondormant seeds prepare for germination (Dekkers et al. 2016a). After the initial imbibition, which leads to large transcriptional changes (up and down) in both dormant and nondormant seeds (approximately 1,000 and 2,000, respectively), most changes occur in nondormant seeds. Up to 2,000 transcripts change when comparing each time point with the previous, whereas in the dormant seeds this is limited to less than 100. The transcriptional profiles that discriminate dormant and nondormant seeds are found both in the embryo and endosperm and show great overlap with transcriptional changes in seeds in the soil seed bank that cycle in and out of dormancy, suggesting that laboratory-based experiments are a good alternative to study natural processes (Buijs et al. 2020). Transcriptional differences between physiological stages can be used to reveal genes that regulate processes of interest. An example of this is a reverse genetics approach that was chosen to identify the role of genes that are differentially expressed in 24-hour imbibed dormant and AR seeds of 5 dif-ferent genotypes (*Ler* and 4 *DOG* NILs) (Yazdanpanah et al. 2017). Analyses in imbibed dormant and AR seeds in the same 5 genotypes revealed that dormancy maintenance and germination are likely to be very conserved processes (Bentsink et al. 2010). Of the total 3,689 differentially expressed genes, 1,720 were dormancy and 1,969 germination upregulated; of these 45 and 25 were common between all genotypes. Among these genes were the earlier identified germination related GA3OX2, PIF6, and ASG5. Moreover, for 50% of the genes for which T-DNA mutants were available, a germination phenotype could be observed.

Insights from network analyses

The initial transcriptome analyses did not yield new key regulators of seed dormancy and germination, and this became a reason for Bassel et al. (2011) to perform a coexpression analysis on publicly available expression data from imbibed mature seeds. The condition-dependent network SeedNet (http://netvis.ico2s.org/dev/seednet/#/) consisted of 1,583 transcripts associated with germination and 1,844 transcripts associated with nongermination (Fig. 2). The germination and nongermination subdomains were connected by a do-main referred to as the intermediate transition region, which was enriched for genes involved in cellular phase transitions, including *SERRATE* and *EARLY FLOWERING SHORT DAYS*. Knockouts of the latter show both a flowering time and ger-mination phenotype (Soppe et al. 1999). Overall, this work shows that network studies not only have the potential to identify co-regulated genes but also possible phase transitions.

Transcriptional phase transitions during seed germination were also identified by (Dekkers et al. 2013), who investigated spatial and temporal transcriptome changes during seed ger-mination. Arabidopsis seeds were dissected into 4 sections: cotyledons, radicle and hypocotyl, micropylar and chalazal endosperm, and peripheral endosperm. These tissues were investigated from the start of seed imbibition until endo-sperm rupture at 10 time-points, of which at TR (25 HAI) and endosperm rupture (38 HAI), the transcriptomes of rup-tured and nonruptured seeds had been separated. Whereas at 25 hours after imbibition (HAI) the transcriptome of the nonruptured seeds do not differ from those at 20 HAI, there are a large number of differential expressed genes, mostly up-regulated, in the TR seeds compared with the nonruptured in both the embryo and endosperm tissues. These transcrip-tional changes separate germination into 2 transcriptional phases: before and after TR. The transcriptional changes over time have been captured by a coexpression network (Dekkers et al. 2013). The latter showed that coexpression networks form highly confident templates for hypothesis generation, which as such can be further scrutinized to unravel molecular networks. An example of this is the interaction mapping study performed for the cell wall–loosening EXPANSIN (EXPA) gene family (Xu et al. 2020). The EXPA gene family is associated with cell wall modification, and the expression of several family members is induced during Arabidopsis seed germination. One of these expansins, *EXPA9* is regulated by the transcription factors *TCP14* and *TCP15*. Xu et al. (2020) show that the reduced germination phenotype of the *tcp14 tcp15* double mutant can be partially overcome by EXPA9 ectopic expression.

New developments in transcriptome analyses Recently, we have entered a new era of transcriptome ana-lyses: single-cell sequencing. For seeds this is not yet straight-forward; however, a first attempt that includes imbibed and germinated seeds has recently been presented (Lee et al. 2023). In single-cell sequence analyses, cell types are annotated based on a priori known cell type–specific markers. For seeds, such a bona fide cell type–specific data set is not yet available; therefore the researchers made use of spatial transcriptome technologies, Slide-seqV2. The Slide-seqV2 method is based on a combination of detecting RNAs at a spatial resolution of 10 μm (Slide-seq) with single-cell trajectory analysis tools (Rodriques et al. 2019; Stickels et al. 2021). Using this methodology, it was revealed that the identified single-cell clusters spatially mapped to the cotyledons, root tip region, epidermis, seed coat, and provasculature. As such this dataset has the potential to function as foundation for future spatial transcriptome analyses in seeds. Important to consider when performing detailed transcriptome ana-lyses is the material used for these investigations. Seeds are highly dynamic in their response to the environment, the changes they show over time (e.g. the differential expression when comparing TR vs nonruptured seeds from the same imbibition time-point) and on top of that show a large germination variability. With this we refer for example to the variation that occurs in a population of seeds harvested from a single plant (Mitchell et al. 2017). This variability likely contributes to the seed's role as a survival structure for seed plants and is nicely demonstrated by single-seed RNA-seq analyses performed on seeds during secondary dormancy induction (Krzyszton et al. 2022). Transcriptional variability is the lowest in seeds that have been exposed to the secondary dormancy, inducing stress, and is the largest in seeds that have recovered from this stress and have started germination (Fig. 2). Single-seed RNA-seq in combination with single-cell analyses might help to reveal the molecular basis of variability in seed pools at a cellular resolution.

Epigenetic control of seed dormancy and germination

An important aspect of gene regulation is that controlled by epigenetic mechanisms. It has been reported that this mechanism, which includes DNA methylation of cytosines, histone, and in general chromatin modifications, controls important genes during seed development that affect dormancy and subsequent seed germination (Sato et al. 2021;

Ding et al. 2022; Iwasaki et al. 2022; Sato and Köhler 2022). The function of many regulators of these processes had been described before using methods such as methylC-sequencing and ChIP-sequencing applied on mutants of these regulators. Often the respective mutants were used to test their effect on the above-mentioned seed development, seed dormancy, and germination processes leading to conclusions on the role of epigenetic mechanisms.

Epigenetic regulation of seed maturation

Epigenetic control in seeds starts with the differential gene expression in the 2 gametes that together form the zygote and quicky develops into an embryo. This differential parental silencing distinguishes between maternal expressed genes (MEGs) and paternally expressed genes (PEGs). Two histone demethylases, *RELATIVE OF EARLY FLOWERING 6* (*REF6*) *and EARLY FLOWERING 6* (*ELF6*), control endosperm dormancy by activating the MEGs (Chen et al. 2020; Sato and Köhler 2022). A recent study analyzing natural variation for seed dor-mancy in Arabidopsis identified that VERNALIZATION5/ VIN3-LIKE 3 (VEL3) maintains seed dormancy by establishing an epigenetic state in the central cell that primes the depth of primary seed dormancy, which is established later during seed maturation (Chen et al. 2023). VEL3 is required for deacetylation and the tri-methylation of lysine 27 on histone H3 (H3K27me3) established in the central cell. The authors pro-pose that VEL3 maintains a repressive epigenetic state at ORE1 (OBSOLETE1) and other genes with a role in suppressing programmed cell death and inducing seed dormancy (Chen et al. 2023).

The role of epigenetic control during seed maturation included that of the major regulators of seed maturation LEC1, LEC2, FUS3, and ABI3 (Ding et al. 2022). Genetic studies subsequently have uncovered several loss-of-function mutations of chromatin-related genes that lead to ectopic expression of embryo-associated genes. These include mutants of the *HISTON 2B MONO-UBIQUINATION* genes *HUB1* and *HUB2*, resulting in reduced seed dormancy and, among others, downregulation of the dormancy gene *DOG1* (Liu et al. 2007). The proteins encoded by HUB1 and HUB2 most likely bind the polymerase II-associated factor 1 com-plex (PAF1C). This complex modulates the local structure of chromatin during transcription elongation and affects methylation of histone H3 at lysine 4 and 36 (H3K4me and H3K36me), respectively, which are activating epigenetic marks for transcription. Interestingly, the isolation and genetic analysis of additional factors associated with PAF1C, like REDUCED DORMANCY 2 in Arabidopsis, confirmed the important role of transcription elongation factors in seed maturation (Liu et al. 2007). Mutations in the *KRYPTONITE* (*KYP*/ *SU(VAR)3-9 HOMOLOG 4* (*SUVH4*) gene, encoding the histone methyltransferase for H3K9me2, cause increased seed dormancy. In accordance, KYP/SUVH4-overexpressing Arabidopsis plants show decreased dormancy (Zheng et al. 2012). This indicates that this repressive chromatin mark influences seed maturation. Potential direct targets of KYP/ SUVH4 are the seed maturation genes *ABI3* and *DOG1*, which are upregulated in the *kyp-2* mutant.

DELAY OF GERMINATION-LIKE 4 (DOGL4) is a central factor mediating reserve accumulation in seeds without strongly affecting seed dormancy and other processes controlled by DOG1 (Sall et al. 2019). Zhu et al. (2018) described that the DNA demethylase ROS1 regulates the imprinting of *DOGL4*. *DOGL4* is expressed from the MEG and displays preferential methylation and suppression of the paternal allele. ROS1 negatively regulates imprinting by demethylating the paternal allele, preventing its hypermethylation and complete silencing. The dormancy effect of the *dogl4* mutant described by these authors is relatively weak compared with what is described for the *dog1* mutants (Sall et al. 2019). Epigenetic control of the dormancy master regulator *DOG1* was recently summarized by Ding et al. (2022).

The process of epigenetic silencing during germination involves the POLYCOMB REPRESSIVE COMPLEX 2 (PRC2), which plays a role in the repression of flowering in young seedlings by mediating H3K27me3. Mutants in FERTILIZATION INDEPENDENT ENDOSPERM, an essential component of the PRC2 complex, displayed genome-wide abolishment of H3K27me3 and exhibited increased seed dor-mancy and germination defects (Bouyer et al. 2011). Therefore, it was concluded that PRC2 is required for termin-ation of the embryonic transcriptional program to promote the phase transition from embryo to autotrophic seedling. PRC2 sustains the balance between ABA and GA responsiveness via H3K27me3-mediated inhibition of positive ABA and negative GA regulators in maturing seeds. Moreover, the seed dormancy regulator DOG1 is repressed through PRC2-catalyzed H3K27-trimethylation at its locus (Bouyer et al. 2011).

Histone acetylation is another modification that plays a role in silencing of embryonic traits and activating genes associated with seedling establishment and growth. In Arabidopsis, HDA6 and HDA19 act redundantly to repress embryonic properties after germination (Tanaka et al. 2008). The function of *HDA9* is opposite to that of its homologous genes *HDA6* and *HDA19* (van Zanten et al. 2014). HDA9 negatively influences germination and is involved in the suppression of seedling traits in dry seeds. The HDA9 transcript is abundant in dry seeds and becomes reduced during imbibition in wild-type seeds.

Epigenetic effects during seed germination

Dry seeds contain nuclei with reduced size and highly compacted chromatin; this phenomenon occurs during maturation and is therefore related to the acquisition of desiccation tolerance and seems independent from the dormancy state (van Zanten et al. 2012). Transcriptional reprogramming during germination requires repression of embryonic properties and activation of genes involved in the progression into photoautotrophic growth. Independent papers demonstrated roles for chromatin-remodeling factors and epigenetic signaling mechanisms in these processes. Seedlings of mutants lacking the CHD3 class SWI/SNF chromatin-remodeling factor PICKLE (PKL) display embryonic properties, that is, "pickle" roots, with a swollen and greenish distal root tip and accumulation of neutral lipid bodies (Ogas et al. 1999). PKL represses the expression of *ABI3* and *ABI5* in an ABA-dependent manner, as *pkl* mutants showed high *ABI3* and *ABI5* transcript levels after ABA treatment and hypersensitivity to ABA-mediated repression of germination (Perruc et al. 2007). The abundance of these epigenetic silencing marks was reduced in *pkl* mutants compared with the corresponding wild types, which was further enhanced by the application of ABA (Perruc et al. 2007). A direct role for PKL in determining levels of H3K27me3 at these repressed loci during germination was demonstrated (Zhang et al. 2012). Taken together, PKL controls ABA-mediated repression of germination by affecting the abundance of silencing epigenetic histone marks associated with the *ABI3* and *ABI5* loci. A detailed time course of germination using RNA seq showed that changes in gene expression and large-scale demethylation was observed toward the end of germination during the transitions from an embryo-like to a vegetative seedling state (Narsai et al. 2017).

Translational regulation in seeds

As mentioned earlier, during seed maturation there is selective accumulation of mRNA also known as "stored mRNAs" or "long-lived mRNAs" (Ajtkhozhin et al. 1976; Ishibashi et al. 1990; Kuligowski et al. 1991). These stored mRNAs serve as a reservoir for rapid translation into proteins during germination (Ishibashi et al. 1990; Kuligowski et al. 1991; Rajjou et al. 2004; Sano et al. 2012). When nondormant Arabidopsis seeds were imbibed in a solution containing the transcriptional inhibitor α-amanitin, germination could still take place (Fig. 2). This indicated that the long-lived or stored mRNAs within the dry nondormant seed are sufficient to produce proteins required for successful germination. However, when Arabidopsis and rice seeds were imbibed in a solution containing the translational inhibitor cycloheximide, germination was blocked (Fig. 2; Rajjou et al. 2004; Sano et al. 2012, 2015). Therefore, the process of translation is crucial for germination to occur.

With the advent of technologies like ribosome profiling, studying the translational status of all mRNAs in seeds became possible. This technique employs sucrose-based fractions to separate mRNAs based on their association to no ribosomes (free mRNA), 1 ribosome (monosome), or multiple ribosomes (polysomes). mRNAs present in the polysome fraction most likely indicate active translation of those mRNAs into proteins. Using this technique, it was observed that dormant and nondormant seeds show differential association of mRNAs to polysomes (Layat et al. 2014). In the dormant embryos, negative regulators of germination seemed to be associated with polysomes, for example, transcription factors promoting ABA responses like ABI2 and LEA proteins, which should disappear during germination. Moreover, Bai et al. (2018) showed that translation was more abundant in AR Arabidopsis seeds than in dormant seeds at 24HAI. The addition of the transcriptional inhibitor cordycepin resulted in similar levels of translation in dormant and AR seeds, suggesting that transcriptional changes are required for the increased translational activity, which is seen in the nondormant seeds in control conditions. Based on this it was concluded that dormancy maintenance or alleviation is regulated at the transcriptional level (Bai et al. 2018).

Translational regulation, the control of the protein levels produced from its mRNA, is a highly efficient and reversible way of fine-tuning gene expression in response to developmental and environmental (Sablok et al. 2017; Chantarachot and Bailey-Serres 2018). To identify the translational efficiency of transcript one can determine the ratio of the mRNA abundance in the polysome fraction to the total RNA abundance of a given transcript between 2 timepoints or treatments (King and Gerber 2016). Using this approach, selective translation during seed maturation was reported (Bai et al. 2023). Among the translationally regulated genes are ABI2 and ABI5 (Finkelstein 1994; Finkelstein and Lynch 2000). It has also been reported that within the mature dry seed, most stored mRNAs are associated to monosomes and many are translationally upregulated upon imbibition. This indicates that these mRNAs are important for germination. mRNAs associated with polysomes in the dry seed are degraded during germination, suggesting that these are remnants from the maturation program and thus no longer needed for germination (Bai et al. 2020).

The first major study that implicated a role for translational regulation during germination came from Galland et al. (2014). In their study, they observed that differential accumulation of proteins occurred during early germination. The authors compared the transcript abundance, the rate of protein synthesis, and the protein abundance of 158 newly synthesized radiolabeled proteins during Arabidopsis seed germination. The protein abundance of these proteins, which included proteins involved in mRNA translation (10% of the identified proteins), Late Embryogenesis Abundant, and Heat Shock Proteins, correlated very well with the rate of protein synthesis and poorly with the amount of its transcript, thereby indicating that translational regulation occurs during germination for newly synthesized proteins (Galland et al. 2014). Another study performed in rice seeds showed that from the 20 proteins that were synthesized upon from the stored/long-lived mRNAs present dry seeds, only 12 of these were upregulated 2 days after imbibition, and the rest were upregulated 4 days after imbibition. Similarly, in Arabidopsis seeds, using ribosome profiling, Bai et al. (2017) showed that there is extensive sequence-specific translational regulation occurring during 2 shifts of germination, namely the hydration translational shift (HTS) and the germination translational shift (GTS). The HTS shift spans the transition from a dry seed to an imbibed seed, whereas the GTS marks the phase between 2 physiologically visible stages of germination, namely, TR and RP. The pools of mRNA that were translationally regulated were different in these 2 shifts, indicating independent mechanisms (Bai et al. 2017). The purpose of this regulation only in these 2 shifts of germination has not yet been established. However, during the HTS, the seed goes from a completely dry to a fully hydrated imbibed seed. Several metabolic processes must be kick-started, and translation is initiated selectively from stored mRNA. This could be the reason why translational regulation occurs during the HTS, whereas during the GTS, the seed must make the all or nothing commitment to germinate and establish a seedling. This is a complex process based on several factors including a dynamically changing environment, which is why extensive translational regulation takes place during the GTS. Several mRNA features like upstream ORFs, transcript length, GC content, and secondary structures were correlated with an upregulation of translation.

Mechanisms underlying translational control: the role of RNA binding proteins

The precise mechanism involved in the differential translation of mRNAs during germination is not known; however, RNA binding proteins (RBPs) are emerging as key players in the translational control of gene expression in plants (Hentze et al. 2018; Cho et al. 2019; Lou et al. 2020; Sajeev et al. 2022). RBPs are proteins that can bind target mRNAs through their RNA binding domain and regulate the fate and functions of their target mRNA (Lorković 2009; Li et al. 2014; Köster et al. 2017; Lou et al. 2020). Examples of these domains include RNA recognition motif, Pumilio/FBF do-main, DEAD box domain, the K homology domain, and the Pentatricopeptide repeat motif.

RBPs have been implicated in seed development and performance (extensively reviewed by Lou et al. 2020). Some examples include RBPs like GLYCINE RICH PROTEIN 2 that can promote seed germination in stress conditions like low temperature via an ABA-independent pathway, whereas the overexpression of RNA DEAD box (RHs) helicases like RH9 and RH29 inhibit germination in salt conditions (Kim et al. 2005, 2007). APUM9, a conventional RBP from the PUMILIO family, plays a role in promoting germination in optimal conditions possibly by targeting ABA signaling genes for degradation (Xiang et al. 2014; Nyikó et al. 2019; Huang et al. 2021), and APUM5 plays a role in inhibiting seed germination in salt and osmotic stress (Huh and Paek 2014). The RBPs VARICOSE (VCS) and EXORIBONUCLEASE4 also play a role in seed dormancy and germination. The mutant seeds of *VCS* were less dormant, and those of *xrn-4* were more dormant (Basbouss-Serhal et al. 2017). Transcriptomic analyses of the *vcs* mutant seeds showed higher abundance of transcripts pro-moting germination like *WRKY60* and *WRKY63*, which are TFs that negatively regulate ABA signaling, while *xrn-4* seeds dis-played higher transcript levels of genes involved in dormancy maintenance like ABA response elements *ABF3* and *ABP4* (Basbouss-Serhal et al. 2017). Interestingly, VCS and XRN are also components of higher-order RBP-mRNA complexes called processing bodies (P-bodies). These are membrane-less cyto-plasmic granules that can transiently store or decay translationally repressed mRNAs (Maldonado-Bonilla 2014; Jang et al. 2020). Jang et al. (2019) reported that P-bodies control translation of certain mRNAs and are required for optimal seed to seedling transition.

Recently, the first repertoire of RBPs in germinating seeds that could provide potential candidates for future research was reported. This study employed an mRNA interactome capture approach that led to the identification of more than 200 high confidence RBPs present during the GTS (the phase between TR and RP) in seeds (GTS-RBPs) and 30 GTS-RBPs to be present specifically in seeds (Sajeev et al. 2022). Some of the identified RBPs were known to have a seed germination and dormancy phenotype. For in-stance, TUDOR SN1/2 has been shown to promote seed ger-mination under salt stress by regulating the mRNA levels of a key GA biosynthesis enzyme, GA20ox3 (Maldonado-Bonilla 2014), and EIN2 has been reported to play a strong role in al-leviating dormancy (Beaudoin et al. 2000; Ghassemian et al. 2000). This study also identified an RBP from the hyaluro-nan/mRNA, namely, AT5G47210, and showed that it plays a role in inhibiting germination under optimal conditions via an unknown mechanism (Sajeev et al. 2022).

Concluding remarks and perspectives

In this review we provided an overview of the research that led to the current knowledge on the regulation of seed dormancy and germination in the past century. Still major questions remain to be solved. At the level of inducing and overcoming seed dormancy, ABA and DOG1 are major components. Although a lot is known about the ABA perception, the role of ABI3 in this has remained unclear. The same holds for the mechanism of the balance between ABA and GA. On the whole, genetic and natural variation studies in Arabidopsis have contributed significantly to our current knowledge on the regulation of seed dormancy and germination. Moreover, although not yet fully proven, there are strong indications that oxidative processes during AR change the dormancy state, which upon seed imbibition allow the seeds to germinate. The onset of germination itself seems more complicated. Nondormant seeds do not germinate if the environment is not optimal; however, the on/off switch for the decision to germinate has not yet been identified. Factors that could play a role in this are downstream genes of GA signaling, like TCPs and EXPAs; however, seeds contain many copies of these genes, and knocking out a single component generally only results in a small change in germination rate (Xu et al. 2020). Is this redundancy required to allow spreading the risk associated with germination in time and space, even in a single population, and does that mean that such an on/off switch does not exist? Or do we so far not have the insights or tools to identify this component? The use of single-cell analyses and mathematical modeling at the (sub)cellular scale might allow us to address this question. The component(s) that we are looking for may very well lie at the level of translation, as can be concluded from the strong translational control of seed germination at 2 critical physiological phases during seed germination (Bai et al. 2017). Employing techniques like interactome capture and individual-nucleotide resolution UV crosslinking and immunoprecipitation in seeds will advance our understanding and contribute to revealing the role of translational regulation in the control of seed germination. So far, technical difficulties have hampered advances in applying these high-tech protocols in seeds.

Acknowledgments

N.S. and L.B. were supported the project Seeds4Ever: Protection of stored mRNAs to ensure seed survival (with project number 17047) of the research programme Vici which is financed by the Dutch Research Council (NWO).

Author contributions

N.S., M.K., and L.B. conceptualized the review outline, and wrote the original draft, L.B. prepared the figures, M.K. and L.B. reviewed and edited the manuscript.

Conflict of interest statement. None declared.

References

Abley K, Formosa-Jordan P, Tavares H, Chan EYT, Afsharinafar M, Leyser O, Locke JCW. An ABA-GA bistable switch can account for natural variation in the variability of Arabidopsis seed germination time. eLife. 2021:**10**:e59485. https://doi.org/10.7554/eLife.59485

Ajtkhozhin MA, Doschanov KJ, Akhanov AU. Informosomes as a stored form of mRNA in wheat embryos. FEBS Lett. 1976:**66**(1): 124–126. https://doi.org/10.1016/0014-5793(76)80600-X

Ali-Rachedi S, Bouinot D, Wagner MH, Bonnet M, Sotta B, Grappin P, Jullien M. Changes in endogenous abscisic acid levels during dormancy release and maintenance of mature seeds: studies with the Cape Verde Islands ecotype, the dormant model of *Arabidopsis thaliana*. Planta. 2004:**219**(3):479–488. https://doi.org/10.1007/s00425-004-1251-4

Alonso-Blanco C, Bentsink L, Hanhart CJ, Vries HBE, Koornneef M. Analysis of natural allelic variation at seed dormancy loci of *Arabidopsis thaliana*. Genetics. 2003:**164**(2):711–729. https://doi.org/10.1093/genetics/164.2.711

Amiguet-Vercher A, Santuari L, Gonzalez-Guzman M, Depuydt S, Rodriguez PL, Hardtke CS. The IBO germination quantitative trait locus encodes a phosphatase 2C-related variant with a nonsynonymous amino acid change that interferes with abscisic acid signaling. New Phytol. 2015:**205**(3):1076–1082. https://doi.org/10.1111/nph.13225

Arc E, Galland M, Godin B, Cueff G, Rajjou L. Nitric oxide implication in the control of seed dormancy and germination. Front Plant Sci. 2013:**4**:346. https://doi.org/10.3389/fpls.2013.00346

Bai B, Novak O, Ljung K, Hanson J, Bentsink L. Combined transcriptome and translatome analyses reveal a role for tryptophan-dependent auxin biosynthesis in the control of DOG1-dependent seed dormancy. New Phytol. 2018:**217**(3):1077–1085. https://doi.org/10.1111/nph.14885

Bai B, Peviani A, van der Horst S, Gamm M, Snel B, Bentsink L, Hanson J. Extensive translational regulation during seed germination revealed by polysomal profiling. New Phytol. 2017:**214**(1):233–244. https://doi.org/10.1111/nph.14355

Bai B, Schiffthaler B, van der Horst S, Willems L, Vergara A, Karlström J, Mähler N, Delhomme N, Bentsink L, Hanson J. SeedTransNet: a directional translational network revealing regulatory patterns during seed maturation and germination. J Exp Bot. 2023:**74**(7):2416–2432. https://doi.org/10.1093/jxb/erac394

Bai B, van der Horst S, Cordewener JHG, America T, Hanson J, Bentsink L. Seed-Stored mRNAs that are specifically associated to monosomes are translationally regulated during germination. Plant Physiol. 2020:**182**(1):378–392. https://doi.org/10.1104/pp.19.00644

Basbouss-Serhal I, Pateyron S, Cochet F, Leymarie J, Bailly C. 5′ to 3′ mRNA decay contributes to the regulation of arabidopsis seed germination by dormancy. Plant Physiol. 2017:**173**(3):1709–1723. https://doi.org/10.1104/pp.16.01933

Baskin JM, Baskin CC. A classification system for seed dormancy. Seed Sci Res. 2004:**14**(1):1–16. https://doi.org/10.1079/SSR2003150

Bassel GW, Lan H, Glaab E, Gibbs DJ, Gerjets T, Krasnogor N, Bonner AJ, Holdsworth MJ, Provart NJ. Genome-wide network model capturing seed germination reveals coordinated regulation of plant cellular phase transitions. Proc Natl Acad Sci U S A. 2011:**108**(23): 9709–9714. https://doi.org/10.1073/pnas.1100958108

Bazin J, Langlade N, Vincourt P, Arribat S, Balzergue S, El-Maarouf-Bouteau H, Bailly C. Targeted mRNA oxidation regulates sunflower seed dormancy alleviation during dry after-ripening. Plant Cell. 2011:**23**(6):2196–2208. https://doi.org/10.1105/tpc.111.086694

Beaudoin N, Serizet C, Gosti F, Giraudat J. Interactions between abscisic acid and ethylene signaling cascades. Plant Cell. 2000:**12**(7): 1103–1115. https://doi.org/10.1105/tpc.12.7.1103

Bentsink L, Hanson J, Hanhart CJ, Blankestijn-de Vries H, Coltrane C, Keizer P, El-Lithy M, Alonso-Blanco C, Teresa de Andres M, Reymond M, et al. Natural variation for seed dormancy in Arabidopsis is regulated by additive genetic and molecular pathways. Proc Natl Acad Sci U S A. 2010:**107**(9):4264–4269. https://doi.org/10.1073/pnas.1000410107

Bentsink L, Jowett J, Hanhart CJ, Koornneef M. Cloning of *DOG1*, a quantitative trait locus controlling seed dormancy in Arabidopsis. Proc Natl Acad Sci U S A. 2006:**103**(45):17042–17047. https://doi.org/10.1073/pnas.0607877103

Bewley JD. Seed germination and dormancy. Plant Cell. 1997:**9**: 1055–1066. https://doi.org/10.1105/tpc.9.7.1055

Bewley JD, Bradford KJ, Hilhorst HW, Nonogaki H. Seeds. New York: Springer; 2012.

Biere A. Parental effects in Lychnis flos-cuculi. I: Seed size, germination and seedling performance in a controlled environment. J Evol Biol. 1991:**4**(3):447–465. https://doi.org/10.1046/j.1420-9101.1991.4030447.x

Boulard C, Fatihi A, Lepiniec L, Dubreucq B. Regulation and evolution of the interaction of the seed B3 transcription factors with NF-Y subunits. Biochim Biophys Acta Gene Regul Mech. 2017:**1860**(10): 1069–1078. https://doi.org/10.1016/j.bbagrm.2017.08.008

Bouyer D, Roudier F, Heese M, Andersen ED, Gey D, Nowack MK, Goodrich J, Renou J-P, Grini PE, Colot V, et al. Polycomb Repressive Complex 2 controls the embryo-to-seedling phase transition. PLoS Genet. 2011:**7**(3):e1002014. https://doi.org/10.1371/journal.pgen.1002014

Bove J, Lucas P, Godin B, Oge L, Jullien M, Grappin P. Gene expression analysis by cDNA-AFLP highlights a set of new signaling networks and translational control during seed dormancy breaking in *Nicotiana plumbaginifolia*. Plant Mol Biol. 2005:**57**(4):593–612. https://doi.org/10.1007/s11103-005-0953-8

Bryant FM, Hughes D, Hassani-Pak K, Eastmond PJ. Basic LEUCINE ZIPPER TRANSCRIPTION FACTOR67 transactivates DELAY OF GERMINATION1 to establish primary seed dormancy in Arabidopsis. Plant Cell. 2019:**31**(6):1276–1288. https://doi.org/10.1105/tpc.18.00892

Buijs G. A perspective on secondary seed dormancy in *Arabidopsis thaliana*. Plants (Basel). 2020:**9**(6):749. https://doi.org/10.3390/plants9060749

Buijs G, Vogelzang A, Nijveen H, Bentsink L. Dormancy cycling: translation-related transcripts are the main difference between dormant and non-dormant seeds in the field. Plant J. 2020:**102**(2): 327–339. https://doi.org/10.1111/tpj.14626

Cadman CS, Toorop PE, Hilhorst HW, Finch-Savage WE. Gene expression profiles of Arabidopsis Cvi seeds during dormancy cycling indicate a common underlying dormancy control mechanism. Plant J. 2006:**46**(5):805–822. https://doi.org/10.1111/j.1365-313X.2006.02738.x

Carbonero P, Iglesias-Fernández R, Vicente-Carbajosa J. The AFL subfamily of B3 transcription factors: evolution and function in angiosperm seeds. J Exp Bot. 2017:**68**(4):871–880. https://doi.org/10.1093/jxb/erw458

Carles C, Bies-Etheve N, Aspart L, Léon-Kloosterziel KM, Koornneef M, Echeverria M, Delseny M. Regulation of *Arabidopsis thaliana* Em genes: role of ABI5. Plant J. 2002:**30**(3):373–383. https://doi.org/10.1046/j.1365-313X.2002.01295.x

Carrillo-Barral N, Rodríguez-Gacio MDC, Matilla AJ. *Delay of Germination-1 (DOG1)*: a key to understanding seed dormancy. Plants (Basel). 2020:**9**(4):480. https://doi.org/10.3390/plants9040480

Chandrasekaran U, Luo X, Zhou W, Shu K. Multifaceted signaling networks mediated by *Abscisic Acid Insensitive 4*. Plant Commun. 2020:**1**(3):100040. https://doi.org/10.1016/j.xplc.2020.100040

Chantarachot T, Bailey-Serres J. Polysomes, Stress Granules, and Processing Bodies: a dynamic triumvirate controlling cytoplasmic mRNA fate and function. Plant Physiol. 2018:**176**(1):254–269. https://doi.org/10.1104/pp.17.01468

Chen H, Tong J, Fu W, Liang Z, Ruan J, Yu Y, Song X, Yuan L, Xiao L, Liu J, et al. The H3K27me3 demethylase RELATIVE OF EARLY FLOWERING6 suppresses seed dormancy by inducing abscisic acid catabolism. Plant Physiol. 2020:**184**(4):1969–1978. https://doi.org/10.1104/pp.20.01255

Chen X, MacGregor DR, Stefanato FL, Zhang N, Barros-Galvão T, Penfield S. A VEL3 histone deacetylase complex establishes a maternal epigenetic state controlling progeny seed dormancy. Nat Commun. 2023:**14**(1):2220. https://doi.org/10.1038/s41467-023-37805-1

Chiabrando D, Vinchi F, Fiorito V, Mercurio S, Tolosano E. Heme in pathophysiology: a matter of scavenging, metabolism and trafficking across cell membranes. Front Pharmacol. 2014:**5**:61. https://doi.org/10.3389/fphar.2014.00061

Cho H, Cho HS, Hwang I. Emerging roles of RNA-binding proteins in plant development. Curr Plant Biol. 2019:**51**:51–57. https://doi.org/10.1016/j.pbi.2019.03.016

Cutler SR, Rodriguez PL, Finkelstein RR, Abrams SR. Abscisic acid: emergence of a core signaling network. Annu Rev Plant Biol. 2010: **61**(1):651–679. https://doi.org/10.1146/annurev-arplant-042809-112122

Cyrek M, Fedak H, Ciesielski A, Guo Y, Sliwa A, Brzezniak L, Krzyczmonik K, Pietras Z, Kaczanowski S, Liu F, et al. Seed dormancy in Arabidopsis is controlled by alternative polyadenylation of DOG1. Plant Physiol. 2016:**170**(2):947–955. https://doi.org/10.1104/pp.15.01483

Debeaujon I, Koornneef M. Gibberelin requirement for *Arabidopsis thaliana* seed germination is determined both by testa characteristics and embryonic ABA. Plant Physiol. 2000:**122**(2):415–424. https://doi.org/10.1104/pp.122.2.415

Debieu M, Tang C, Stich B, Sikosek T, Effgen S, Josephs E, Schmitt J, Nordborg M, Koornneef M, de Meaux J. Co-variation between seed dormancy, growth rate and flowering time changes with latitude in *Arabidopsis thaliana*. PLoS One. 2013:**8**(5):e61075. https://doi.org/10.1371/journal.pone.0061075

Dekkers BJ, He H, Hanson J, Willems LA, Jamar DC, Cueff G, Rajjou L, Hilhorst HW, Bentsink L. The Arabidopsis *DELAY OF GERMINATION 1* gene affects *ABSCISIC ACID INSENSITIVE 5* (*ABI5*) expression and genetically interacts with *ABI3* during Arabidopsis seed development. Plant J. 2016b:**85**(4):451–465. https://doi.org/10.1111/tpj.13118

Dekkers BJ, Pearce SP, van Bolderen-Veldkamp RP, Holdsworth MJ, Bentsink L. Dormant and after-ripened *Arabidopsis thaliana* seeds are distinguished by early transcriptional differences in the imbibed state. Front Plant Sci. 2016a:**7**:1323. https://doi.org/10.3389/fpls.2016.01323

Dekkers BJ, Pearce S, van Bolderen-Veldkamp RP, Marshall A, Widera P, Gilbert J, Drost HG, Bassel GW, Muller K, King JR, et al. Transcriptional dynamics of two seed compartments with opposing roles in Arabidopsis seed germination. Plant Physiol. 2013:**163**(1):205–215. https://doi.org/10.1104/pp.113.223511

Deng G, Sun H, Hu Y, Yang Y, Li P, Chen Y, Zhu Y, Zhou Y, Huang J, Neill SJ, et al. A transcription factor WRKY36 interacts with AFP2 to break primary seed dormancy by progressively silencing DOG1 in Arabidopsis. New Phytol. 2023:**238**(2):688–704. https://doi.org/10.1111/nph.18750

Ding X, Jia X, Xiang Y, Jiang W. Histone modification and chromatin remodeling during the seed life cycle. Front Plant Sci. 2022:**13**:865361. https://doi.org/10.3389/fpls.2022.865361

Donohue K. Seeds and seasons: interpreting germination timing in the field. Seed Sci Res. 2005:**15**(3):175–187. https://doi.org/10.1079/SSR2005208

Fedak H, Palusinska M, Krzyczmonik K, Brzezniak L, Yatusevich R, Pietras Z, Kaczanowski S, Swiezewski S. Control of seed dormancy in Arabidopsis by a cis-acting noncoding antisense transcript. Proc Natl Acad Sci U S A. 2016:**113**(48):E7846–E7855. https://doi.org/10.1073/pnas.1608827113

Fernandez-Pascual E, Jimenez-Alfaro B, Caujape-Castells J, Jaen-Molina R, Diaz TE. A local dormancy cline is related to the seed maturation environment, population genetic composition and climate. Ann Bot. 2013:**112**(5):937–945. https://doi.org/10.1093/aob/mct154

Finch-Savage WE, Cadman CS, Toorop PE, Lynn JR, Hilhorst HW. Seed dormancy release in Arabidopsis cvi by dry after-ripening, low temperature, nitrate and light shows common quantitative patterns of gene expression directed by environmentally specific sensing. Plant J. 2007:**51**(1):60–78. https://doi.org/10.1111/j.1365-313X.2007.03118.x

Finch-Savage WE, Leubner-Metzger G. Seed dormancy and the control of germination. New Phytol. 2006:**171**(3):501–523. https://doi.org/10.1111/j.1469-8137.2006.01787.x

Finkelstein RR. Mutations at two new Arabidopsis ABA response loci are similar to the abi3 mutations. Plant J. 1994:**5**(6):765–771. https://doi.org/10.1046/j.1365-313X.1994.5060765.x

Finkelstein RR, Lynch TJ. The Arabidopsis abscisic acid response gene ABI5 encodes a basic leucine zipper transcription factor. Plant Cell. 2000:**12**(4):599–609. https://doi.org/10.1105/tpc.12.4.599

Finkelstein R, Reeves W, Ariizumi T, Steber C. Molecular aspects of seed dormancy. Annu Rev Plant Biol. 2008:**59**(1):387–415. https://doi.org/10.1146/annurev.arplant.59.032607.092740

Footitt S, Walley PG, Lynn JR, Hambidge AJ, Penfield S, Finch-Savage WE. Trait analysis reveals DOG1 determines initial depth of seed dormancy, but not changes during dormancy cycling that result in seedling emergence timing. New Phytol. 2020:**225**(5):2035–2047. https://doi.org/10.1111/nph.16081

Francoz E, Lepiniec L, North HM. Seed coats as an alternative molecular factory: thinking outside the box. Plant Reprod. 2018:**31**(3): 327–342. https://doi.org/10.1007/s00497-018-0345-2

Galland M, Huguet R, Arc E, Cueff G, Job D, Rajjou L. Dynamic proteomics emphasizes the importance of selective mRNA translation and protein turnover during Arabidopsis seed germination. Mol Cell Proteomics. 2014:**13**(1):252–268. https://doi.org/10.1074/mcp.M113.032227

Ghassemian M, Nambara E, Cutler S, Kawaide H, Kamiya Y, McCourt P. Regulation of abscisic acid signaling by the ethylene response pathway in Arabidopsis. Plant Cell. 2000:**12**(7):1117–1126. https://doi.org/10.1105/tpc.12.7.1117

Graeber K, Linkies A, Steinbrecher T, Mummenhoff K, Tarkowská D, Turečková V, Ignatz M, Sperber K, Voegele A, de Jong H, et al.

DELAY OF GERMINATION 1 mediates a conserved coat-dormancy mechanism for the temperature- and gibberellin-dependent control of seed germination. Proc Natl Acad Sci U S A. 2014:**111**(34): E3571–E3580. https://doi.org/10.1073/pnas.1403851111

Graeber K, Nakabayashi K, Miatton E, Leubner-Metzger G, Soppe WJ. Molecular mechanisms of seed dormancy. Plant Cell Environ. 2012:**35**(10):1769–1786. https://doi.org/10.1111/j.1365-3040.2012.02542.x

Groot SP, Karssen CM. Gibberellins regulate seed germination in tomato by endosperm weakening: a study with gibberellin-deficient mutants. Planta. 1987:**171**(4):525–531. https://doi.org/10.1007/BF00392302

Groot SP, Kieliszewska-Rokicka B, Vermeer E, Karssen CM. Gibberellin-induced hydrolysis of endosperm cell walls in gibberellin-deficient tomato seeds prior to radicle protrusion. Planta. 1988: **174**(4):500–504. https://doi.org/10.1007/BF00634479

Guo R, Wen X, Zhang W, Huang L, Peng Y, Jin L, Han H, Zhang L, Li W, Guo H. Arabidopsis EIN2 represses ABA responses during germination and early seedling growth by inactivating HLS1 protein independently of the canonical ethylene pathway. Plant J. 2023:**115**(6): 1514–1527. https://doi.org/10.1111/tpj.16335

Gutierrez L, Van Wuytswinkel O, Castelain M, Bellini C. Combined networks regulating seed maturation. Trends Plant Sci. 2007:**12**(7): 294–300. https://doi.org/10.1016/j.tplants.2007.06.003

Harada JJ, Baden CS, Comai L. Spatially regulated genes expressed during seed germination and postgerminative development are activated during embryogeny. Mol Gen Genet. 1988:**212**(3):466–473. https://doi.org/10.1007/BF00330851

He H. Environmental regulation of seed performance. Wageningen, The Netherlands: Wageningen University; 2014.

He H, de Souza Vidigal D, Snoek LB, Schnabel S, Nijveen H, Hilhorst H, Bentsink L. Interaction between parental environment and genotype affects plant and seed performance in Arabidopsis. J Exp Bot. 2014:**65**(22):6603–6615. https://doi.org/10.1093/jxb/eru378

Hentze MW, Castello A, Schwarzl T, Preiss T. A brave new world of RNA-binding proteins. Nat Rev Mol Cell Biol. 2018:**19**(5):327–341. https://doi.org/10.1038/nrm.2017.130

Holdsworth MJ, Bentsink L, Soppe WJJ. Molecular networks regulating Arabidopsis seed maturation, after-ripening, dormancy and germination. New Phytol. 2008:**179**(1):33–54. https://doi.org/10.1111/j.1469-8137.2008.02437.x

Huang R, Liu M, Gong G, Wu P, Patra B, Yuan L, Qin H, Wang X, Wang G, Liao H, et al. The Pumilio RNA-binding protein APUM24 regulates seed maturation by fine-tuning the BPM-WRI1 module in Arabidopsis. J Integr Plant Biol. 2021:**63**(7):1240–1259. https://doi.org/10.1111/jipb.13092

Huang X, Schmitt J, Dorn L, Griffith C, Effgen S, Takao S, Koornneef M, Donohue K. The earliest stages of adaptation in an experimental plant population: strong selection on QTLs for seed dormancy. Mol Ecol. 2010:**19**(7):1335–1351. https://doi.org/10.1111/j.1365-294X.2010.04557.x

Huh SU, Paek KH. APUM5, encoding a Pumilio RNA binding protein, negatively regulates abiotic stress responsive gene expression. BMC Plant Biol. 2014:**14**(1):75. https://doi.org/10.1186/1471-2229-14-75

Hussain S, Kim SH, Bahk S, Ali A, Nguyen XC, Yun D-J, Chung WS. The auxin signaling repressor IAA8 promotes seed germination through down-regulation of ABI3 transcription in Arabidopsis. Front Plant Sci. 2020:**11**:111. https://doi.org/10.3389/fpls.2020.00111

Ishibashi N, Yamauchi D, Minamikawa T. Stored mRNA in cotyledons of *Vigna unguiculata* seeds: nucleotide sequence of cloned cDNA for a stored mRNA and induction of its synthesis by precocious germination. Plant Mol Biol. 1990:**15**(1):59–64. https://doi.org/10.1007/BF00017724

Iwasaki M, Penfield S, Lopez-Molina L. Parental and environmental control of seed dormancy in *Arabidopsis thaliana*. Annu Rev Plant Biol. 2022:**73**(1):355–378. https://doi.org/10.1146/annurev-arplant-102820-090750

Jang GJ, Jang JC, Wu SH. Dynamics and functions of stress granules and processing bodies in plants. Plants (Basel). 2020:**9**(9):1122. https://doi.org/10.3390/plants9091122

Jang GJ, Yang JY, Hsieh HL, Wu SH. Processing bodies control the selective translation for optimal development of Arabidopsis young seedlings. Proc Natl Acad Sci U S A. 2019:**116**(13):6451–6456. https://doi.org/10.1073/pnas.1900084116

Jia H, McCarty DR, Suzuki M. Distinct roles of LAFL network genes in promoting the embryonic seedling fate in the absence of VAL repression. Plant Physiol. 2013:**163**(3):1293–1305. https://doi.org/10.1104/pp.113.220988

Kang J, Yim S, Choi H, Kim A, Lee KP, Lopez-Molina L, Martinoia E, Lee Y. Abscisic acid transporters cooperate to control seed germination. Nat Commun. 2015:**6**(1):8113. https://doi.org/10.1038/ncomms9113

Karssen CM, Brinkhorst-van der Swan DLC, Breekland AE, Koornneef M. Induction of dormancy during seed development by endogenous abscisic acid: studies on abscisic acid deficient genotypes of *Arabidopsis thaliana* (L.) Heynh. Planta. 1983:**157**(2):158–165. https://doi.org/10.1007/BF00393650

Keith K, Kraml M, Dengler NG, McCourt P. Fusca3: a heterochronic mutation affectting late embryo development in Arabidopsis. Plant Cell. 1994:**6**(5):589–600. https://doi.org/10.2307/3869865

Kim Y-O, Kim JS, Kang H. Cold-inducible zinc finger-containing glycine-rich RNA-binding protein contributes to the enhancement of freezing tolerance in *Arabidopsis thaliana*. Plant J. 2005:**42**(6): 890–900. https://doi.org/10.1111/j.1365-313X.2005.02420.x

Kim YO, Pan S, Jung CH, Kang H. A zinc finger-containing glycine-rich RNA-binding protein, atRZ-1a, has a negative impact on seed germination and seedling growth of *Arabidopsis thaliana* under salt or drought stress conditions. Plant Cell Physiol. 2007:**48**(8):1170–1181. https://doi.org/10.1093/pcp/pcm087

King HA, Gerber AP. Translatome profiling: methods for genome-scale analysis of mRNA translation. Brief Funct Genomics. 2016:**15**(1): 22–31. https://doi.org/10.1093/bfgp/elu045

Koornneef M, Jorna ML, Brinkhorst-van der Swan DLC, Karssen CM. The isolation of abscisic acid (ABA) deficient mutants by selection of induced revertants in non-germinating gibberillin sensitive lines of *Arabidopsis thaliana* (L.) Heynh. Theor Appl Genet. 1982:**61**(4): 385–393. https://doi.org/10.1007/BF00272861

Koornneef M, Reuling G, Karssen CM. The isolation and characterization of abscisic acid-insensitive mutants of Arabidopsis thaliana. Physiol Plant. 1984:**61**(3):377–383. https://doi.org/10.1111/j.1399-3054.1984.tb06343.x

Koornneef M, van der Veen JH. Induction and analysis of gibberellin sensitive mutants in *Arabidopsis thaliana* (L.) heynh. Theor Appl Genet. 1980:**58**(6):257–263. https://doi.org/10.1007/BF00265176

Köster T, Marondedze C, Meyer K, Staiger D. RNA-Binding proteins revisited—the emerging Arabidopsis mRNA interactome. Trends Plant Sci. 2017:**22**(6):512–526. https://doi.org/10.1016/j.tplants.2017.03.009

Kowalczyk J, Palusinska M, Wroblewska-Swiniarska A, Pietras Z, Szewc L, Dolata J, Jarmolowski A, Swiezewski S. Alternative polyadenylation of the sense transcript controls antisense transcription of *DELAY OF GERMINATION 1* in Arabidopsis. Mol Plant. 2017:**10**(10):1349–1352. https://doi.org/10.1016/j.molp.2017.07.011

Kozaki A, Aoyanagi T. Molecular aspects of seed development controlled by gibberellins and abscisic acids. Int J Mol Sci. 2022:**23**(3): 1876. https://doi.org/10.3390/ijms23031876

Kronholm I, Picó FX, Alonso-Blanco C, Goudet J, Meaux JD. Genetic basis of adaptation in *Arabidopsis thaliana*: local adaptation at the seed dormancy QTL DOG1. Evolution. 2012:**66**:2287–2302.

Krzyszton M, Yatusevich R, Wrona M, Sacharowski S, Adamska D, Swiezewski S. Single seeds exhibit transcriptional heterogeneity during secondary dormancy induction. Plant Physiol. 2022:**190**(1): 211–225. https://doi.org/10.1093/plphys/kiac265

Kuligowski J, Ferrand M, Chenou E. Stored mRNA in early embryos of a fern *Marsilea vestita*: a paternal and maternal origin. Mol Reprod Dev. 1991:**30**(1):27–33. https://doi.org/10.1002/mrd.1080300104

Kushiro T, Okamoto M, Nakabayashi K, Yamagishi K, Kitamura S, Asami T, Hirai N, Koshiba T, Kamiya Y, Nambara E. The Arabidopsis cytochrome P450 CYP707A encodes ABA 8′-hydroxylases: key enzymes in ABA catabolism. EMBO J. 2004:**23**(7):1647–1656. https://doi.org/10.1038/sj.emboj.7600121

Layat E, Leymarie J, El-Maarouf-Bouteau H, Caius J, Langlade N, Bailly C. Translatome profiling in dormant and nondormant sunflower (*Helianthus annuus*) seeds highlights post-transcriptional regulation of germination. New Phytol. 2014:**204**(4):864–872. https://doi.org/10.1111/nph.13002

Lee KP, Piskurewicz U, Tureckova V, Strnad M, Lopez-Molina L. A seed coat bedding assay shows that RGL2-dependent release of abscisic acid by the endosperm controls embryo growth in Arabidopsis dormant seeds. Proc Natl Acad Sci U S A. 2010:**107**(44):19108–19113. https://doi.org/10.1073/pnas.1012896107

Lee S, Cheng H, King KE, Wang W, He Y, Hussain A, Lo J, Harberd NP, Peng J. Gibberellin regulates Arabidopsis seed germination via RGL2, a GAI/RGA-like gene whose expression is up-regulated following imbibition. Genes Dev. 2002:**16**(5):646–658. https://doi.org/10.1101/gad.969002

Lee TA, Nobori T, Illouz-Eliaz N, Xu J, Jow B, Nery JR, Ecker JR. A Single-nucleus atlas of seed-to-seed development in Arabidopsis. bioRxiv. 2023:2023.2003.2023.533992. https://doi.org/10.1101/2023.03.23.533992

Lefebvre V, North H, Frey A, Sotta B, Seo M, Okamoto M, Nambara E, Marion-Poll A. Functional analysis of Arabidopsis *NCED6* and *NCED9* genes indicates that ABA synthesized in the endosperm is involved in the induction of seed dormancy. Plant J. 2006:**45**(3):309–319. https://doi.org/10.1111/j.1365-313X.2005.02622.x

Léon-Kloosterziel KM, Alvarez Gil M, Ruijs GJ, Jacobsen SE, Olszewski NE, Schwartz SH, Zeevaart JAD, Koornneef M. Isolation and characterization of abscisic acid-deficient Arabidopsis mutants at two new loci. Plant J. 1996:**10**(4):655–661. https://doi.org/10.1046/j.1365-313X.1996.10040655.x

Leymarie J, Bruneaux E, Gibot-Leclerc S, Corbineau F. Identification of transcripts potentially involved in barley seed germination and dormancy using cDNA-AFLP. J Exp Bot. 2007:**58**(3):425–437. https://doi.org/10.1093/jxb/erl211

Li X, Chen T, Li Y, Wang Z, Cao H, Chen F, Li Y, Soppe WJJ, Li W, Liu Y. ETR1/RDO3 regulates seed dormancy by relieving the inhibitory effect of the ERF12-TPL complex on *DELAY OF GERMINATION1* expression. Plant Cell. 2019:**31**(4):832–847. https://doi.org/10.1105/tpc.18.00449

Li X, Kazan H, Lipshitz HD, Morris QD. Finding the target sites of RNA-binding proteins. WIREs RNA. 2014:**5**(1):111–130. https://doi.org/10.1002/wrna.1201

Li P, Ni H, Songbei Y, Wei J, Hu X. Teaching an old dog a new trick: multifaceted strategies to control primary seed germination by *DELAY OF GERMINATION 1* (*DOG1*). Phyton (B Aires). 2020:**89**(1): 1–12. https://doi.org/10.32604/phyton.2020.09817

Lim S, Park J, Lee N, Jeong J, Toh S, Watanabe A, Kim J, Kang H, Kim DH, Kawakami N, et al. *ABA-insensitive3*, *ABA-insensitive5*, and *DELLAs* interact to activate the expression of *SOMNUS* and other high-temperature-inducible genes in imbibed seeds in Arabidopsis. Plant Cell. 2013:**25**(12):4863–4878. https://doi.org/10.1105/tpc.113.118604

Liu X, Zhang H, Zhao Y, Feng Z, Li Q, Yang H-Q, Luan S, Li J, He Z-H. Auxin controls seed dormancy through stimulation of abscisic acid signaling by inducing ARF-mediated *ABI3* activation in Arabidopsis. Proc Natl Acad Sci U S A. 2013:**110**(38):15485–15490. https://doi.org/10.1073/pnas.1304651110

Liu Y, Koornneef M, Soppe WJ. The absence of histone H2B monoubiquitination in the Arabidopsis *hub1* (*rdo4*) mutant reveals a role for chromatin remodeling in seed dormancy. Plant Cell. 2007:**19**(2): 433–444. https://doi.org/10.1105/tpc.106.049221

Longo C, Holness S, De Angelis V, Lepri A, Occhigrossi S, Ruta V, Vittorioso P. From the outside to the inside: new insights on the main factors that guide seed dormancy and germination. Genes (Basel). 2020:**12**(1):52. https://doi.org/10.3390/genes12010052

Lorković ZJ. Role of plant RNA-binding proteins in development, stress response and genome organization. Trends Plant Sci. 2009:**14**(4): 229–236. https://doi.org/10.1016/j.tplants.2009.01.007

Lou L, Ding L, Wang T, Xiang Y. Emerging roles of RNA-binding proteins in seed development and performance. Int J Mol Sci. 2020:**21**(18):6822. https://doi.org/10.3390/ijms21186822

Lynch TJ, Erickson BJ, Miller DR, Finkelstein RR. ABI5-binding proteins (AFPs) alter transcription of ABA-induced genes via a variety of interactions with chromatin modifiers. Plant Mol Biol. 2017:**93**(4–5): 403–418. https://doi.org/10.1007/s11103-016-0569-1

Maia J, Dekkers BJ, Provart NJ, Ligterink W, Hilhorst HW. The re-establishment of desiccation tolerance in germinated *Arabidopsis thaliana* seeds and its associated transcriptome. PLoS One. 2011:**6**(12):e29123. https://doi.org/10.1371/journal.pone.0029123

Maldonado-Bonilla LD. Composition and function of P bodies in *Arabidopsis thaliana*. Front Plant Sci. 2014:**5**:201. https://doi.org/10.3389/fpls.2014.00201

Martínez-Berdeja A, Stitzer MC, Taylor MA, Okada M, Ezcurra E, Runcie DE, Schmitt J. Functional variants of DOG1 control seed chilling responses and variation in seasonal life-history strategies in *Arabidopsis thaliana*. Proc Natl Acad Sci U S A. 2020:**117**(5): 2526–2534. https://doi.org/10.1073/pnas.1912451117

Meng PH, Macquet A, Loudet O, Marion-Poll A, North H. Analysis of natural allelic variation controlling *Arabidopsis thaliana* seed germinability in response to cold and dark: indentification of three major quantitative trait loci. Mol Plant. 2008:**1**(1):145–154. https://doi.org/10.1093/mp/ssm014

Mitchell J, Johnston IG, Bassel GW. Variability in seeds: biological, ecological, and agricultural implications. J Exp Bot. 2017:**68**(4):809–817. https://doi.org/10.1093/jxb/erw397

Montez M, Majchrowska M, Krzyszton M, Bokota G, Sacharowski S, Wrona M, Yatusevich R, Massana F, Plewczynski D, Swiezewski S. Promoter-pervasive transcription causes RNA polymerase II pausing to boost *DOG1* expression in response to salt. EMBO J. 2023:**42**(5): e112443. https://doi.org/10.15252/embj.2022112443

Muller K, Tintelnot S, Leubner-Metzger G. Endosperm-limited Brassicaceae seed germination: abscisic acid inhibits embryo-induced endosperm weakening of *Lepidium sativum* (cress) and endosperm rupture of cress and *Arabidopsis thaliana*. Plant Cell Physiol. 2006:**47**(7):864–877. https://doi.org/10.1093/pcp/pcj059

Nakabayashi K, Bartsch M, Xiang Y, Miatton E, Pellengahr S, Yano R, Seo M, Soppe WJJ. The time required for dormancy release in Arabidopsis is determined by DELAY OF GERMINATION1 protein levels in freshly harvested seeds. Plant Cell. 2012:**24**(7):2826–2838. https://doi.org/10.1105/tpc.112.100214

Nakabayashi K, Okamoto M, Koshiba T, Kamiya Y, Nambara E. Genome-wide profiling of stored mRNA in *Arabidopsis thaliana* seed germination: epigenetic and genetic regulation of transcription in seed. Plant J. 2005:**41**(5):697–709. https://doi.org/10.1111/j.1365-313X.2005.02337.x

Nambara E, Keith K, McCourt P, Naito S. Isolation of an internal deletion mutant of the *Arabidopsis thaliana ABI3* gene. Plant Cell Physiol. 1994:**35**:509–513. https://doi.org/10.1093/oxfordjournals.pcp.a078623

Narsai R, Gouil Q, Secco D, Srivastava A, Karpievitch YV, Liew LC, Lister R, Lewsey MG, Whelan J. Extensive transcriptomic and epigenomic remodelling occurs during *Arabidopsis thaliana* germination. Genome Biol. 2017:**18**(1):172. https://doi.org/10.1186/s13059-017-1302-3

Nee G, Xiang Y, Soppe WJ. The release of dormancy, a wake-up call for seeds to germinate. Curr Opin Plant Biol. 2017a:**35**:8–14. https://doi.org/10.1016/j.pbi.2016.09.002

Nee G, Kramer K, Nakabayashi K, Yuan B, Xiang Y, Miatton E, Finkemeier I, Soppe WJJ. DELAY OF GERMINATION1 requires PP2C phosphatases of the ABA signalling pathway to control seed dormancy. Nat Commun. 2017b:**8**(1):72. https://doi.org/10.1038/s41467-017-00113-6

Née G, Krüger T. Dry side of the core: a meta-analysis addressing the original nature of the ABA signalosome at the onset of seed imbibition. Front Plant Sci. 2023:**14**:1192652. https://doi.org/10.3389/fpls.2023.1192652

Nikolaeva MG. Physiology of deep dormancy in seeds. Leningrad, USSR: Nauka; 1969.

Nishimura N, Tsuchiya W, Moresco JJ, Hayashi Y, Satoh K, Kaiwa N, Irisa T, Kinoshita T, Schroeder JI, Yates JR, 3rd et al. Control of seed dormancy and germination by DOG1-AHG1 PP2C phosphatase complex via binding to heme. Nat Commun. 2018:**9**(1):2132. https://doi.org/10.1038/s41467-018-04437-9

Nonogaki H. Seed germination and dormancy: the classic story, new puzzles, and evolution. J Integr Plant Biol. 2019:**61**(5):541–563. https://doi.org/10.1111/jipb.12762

Nonogaki H, Bassel GW, Bewley JD. Germination-still a mystery. Plant Sci. 2010:**179**(6):574–581. https://doi.org/10.1016/j.plantsci.2010.02.010

Nyikó T, Auber A, Bucher E. Functional and molecular characterization of the conserved Arabidopsis PUMILIO protein, APUM9. Plant Mol Biol. 2019:**100**(1–2):199–214. https://doi.org/10.1007/s11103-019-00853-7

Ogas J, Kaufmann S, Henderson J, Somerville C. PICKLE is a CHD3 chromatin-remodeling factor that regulates the transition from embryonic to vegetative development in Arabidopsis. Proc Natl Acad Sci U S A. 1999:**96**(24):13839–13844. https://doi.org/10.1073/pnas.96.24.13839

Ooms JJJ, Léon-Kloosterziel KM, Bartels D, Koornneef M, Karssen CM. Acquisition of desiccation tolerance and longevity in seeds of *Arabidopsis thaliana*—a comparative study using abscisic acid- insensitive *abi3* mutants. Plant Physiol. 1993:**102**(4):1185–1191. https://doi.org/10.1104/pp.102.4.1185

Park J, Lee N, Kim W, Lim S, Choi G. *ABI3* and *PIL5* collaboratively activate the expression of *SOMNUS* by directly binding to its promoter in imbibed Arabidopsis seeds. Plant Cell. 2011:**23**(4):1404–1415. https://doi.org/10.1105/tpc.110.080721

Perruc E, Kinoshita N, Lopez-Molina L. The role of chromatin-remodeling factor PKL in balancing osmotic stress responses during Arabidopsis seed germination. Plant J. 2007:**52**(5):927–936. https://doi.org/10.1111/j.1365-313X.2007.03288.x

Rajjou L, Gallardo K, Debeaujon I, Vandekerckhove J, Job C, Job D. The effect of alpha-amanitin on the Arabidopsis seed proteome highlights the distinct roles of stored and neosynthesized mRNAs during germination. Plant Physiol. 2004:**134**(4):1598–1613. https://doi.org/10.1104/pp.103.036293

Raz V, Bergervoet JH, Koornneef M. Sequential steps for developmental arrest in Arabidopsis seeds. Development. 2001:**128**(2):243–252. https://doi.org/10.1242/dev.128.2.243

Rodriques SG, Stickels RR, Goeva A, Martin CA, Murray E, Vanderburg CR, Welch J, Chen LM, Chen F, Macosko EZ. Slide-seq: a scalable technology for measuring genome-wide expression at high spatial resolution. Science. 2019:**363**(6434):1463–1467. https://doi.org/10.1126/science.aaw1219

Ross JJ, Quittenden LJ. Interactions between brassinosteroids and gibberellins: synthesis or signaling? Plant Cell. 2016:**28**(4):829–832. https://doi.org/10.1105/tpc.15.00917

Sablok G, Powell JJ, Kazan K. Emerging roles and landscape of translating mRNAs in plants. Front Plant Sci. 2017:**8**:1443. https://doi.org/10.3389/fpls.2017.01443

Sajeev N, Baral A, America AHP, Willems LAJ, Merret R, Bentsink L. The mRNA-binding proteome of a critical phase transition during Arabidopsis seed germination. New Phytol. 2022:**233**(1):251–264. https://doi.org/10.1111/nph.17800

Sall K, Dekkers BJW, Nonogaki M, Katsuragawa Y, Koyari R, Hendrix D, Willems LAJ, Bentsink L, Nonogaki H. DELAY OF GERMINATION1-LIKE4 acts as an inducer of seed reserve accumulation. Plant J. 2019:**100**(1):7–19. https://doi.org/10.1111/tpj.14485

Sano N, Ono H, Murata K, Yamada T, Hirasawa T, Kanekatsu M. Accumulation of long-lived mRNAs associated with germination in embryos during seed development of rice. J Exp Bot. 2015:**66**(13): 4035–4046. https://doi.org/10.1093/jxb/erv209

Sano N, Permana H, Kumada R, Shinozaki Y, Tanabata T, Yamada T, Hirasawa T, Kanekatsu M. Proteomic analysis of embryonic proteins synthesized from long-lived mRNAs during germination of rice seeds. Plant Cell Physiol. 2012:**53**(4):687–698. https://doi.org/10.1093/pcp/pcs024

Sato H, Köhler C. Genomic imprinting regulates establishment and release of seed dormancy. Curr Opin Plant Biol. 2022:**69**:102264. https://doi.org/10.1016/j.pbi.2022.102264

Sato H, Santos-González J, Köhler C. Combinations of maternal-specific repressive epigenetic marks in the endosperm control seed dormancy. eLife. 2021:**10**:e64593. https://doi.org/10.7554/eLife.64593

Shu K, Liu XD, Xie Q, He ZH. Two faces of one seed: hormonal regulation of dormancy and germination. Mol Plant. 2016:**9**(1):34–45. https://doi.org/10.1016/j.molp.2015.08.010

Shu K, Zhang H, Wang S, Chen M, Wu Y, Tang S, Liu C, Feng Y, Cao X, Xie Q. ABI4 regulates primary seed dormancy by regulating the biogenesis of abscisic acid and gibberellins in arabidopsis. PLoS Genet. 2013:**9**(6):e1003577. https://doi.org/10.1371/journal.pgen.1003577

Skubacz A, Daszkowska-Golec A, Szarejko I. The role and regulation of ABI5 (ABA-insensitive 5) in plant development, abiotic stress responses and phytohormone crosstalk. Front Plant Sci. 2016:**7**:1884. https://doi.org/10.3389/fpls.2016.01884

Song S, He H, Gühl K, van Bolderen-Veldkamp M, Buijs G, Willems L, Bentsink L. *DELAY OF GERMINATION 6* encoding the *ANAC060* transcription factor, inhibits seed dormancy. bioRxiv. 2021:2021.05.03.442418. https://doi.org/10.1101/2021.05.03.442418

Song S, Willems LAJ, Jiao A, Zhao T, Eric Schranz M, Bentsink L. The membrane associated NAC transcription factors ANAC060 and ANAC040 are functionally redundant in the inhibition of seed dormancy in *Arabidopsis thaliana*. J Exp Bot. 2022:**73**(16):5514–5528. https://doi.org/10.1093/jxb/erac232

Soppe W, Bentsink L, Koornneef M. The early-flowering mutant efs is involved in the autonomous promotion pathway of *Arabidopsis thaliana*. Development. 1999:**126**(21):4763–4770. https://doi.org/10.1242/dev.126.21.4763

Soppe WJJ, Bentsink L. Seed dormancy back on track; its definition and regulation by DOG1. New Phytol. 2020:**228**(3):816–819. https://doi.org/10.1111/nph.16592

Steber CM, McCourt P. A role for brassinosteriods in germination in Arabidopsis. Plant Physiol. 2001:**125**(2):763–769. https://doi.org/10.1104/pp.125.2.763

Stickels RR, Murray E, Kumar P, Li J, Marshall JL, Di Bella DJ, Arlotta P, Macosko EZ, Chen F. Highly sensitive spatial transcriptomics at near-cellular resolution with Slide-seqV2. Nature Biotech. 2021:**39**(3):313–319. https://doi.org/10.1038/s41587-020-0739-1

Tanaka M, Kikuchi A, Kamada H. The Arabidopsis histone deacetylases HDA6 and HDA19 contribute to the repression of embryonic properties after germination. Plant Physiol. 2008:**146**(1):149–161. https://doi.org/10.1104/pp.107.111674

Tian R, Wang F, Zheng Q, Niza V, Downie AB, Perry SE. Direct and indirect targets of the Arabidopsis seed transcription factor ABSCISIC ACID INSENSITIVE3. Plant J. 2020:**103**(5):1679–1694. https://doi.org/10.1111/tpj.14854

Urbanova T, Leubner-Metzger G. Gibberellins and seed germination. Annual plant reviews online; 2018. p. 253–284. https://onlinelibrary.wiley.com/doi/abs/10.1002/9781119312994.apr0538

Van der Schaar W, Alonso-Blanco C, Léon-Kloosterziel KM, Jansen RC, van Ooijen JW, Koornneef M. QTL analysis of seed dormancy in Arabidopsis using recombinant inbred lines and MQM mapping. Heredity (Edinburgh). 1997:**79**(2):190–200. https://doi.org/10.1038/hdy.1997.142

van Zanten M, Carles A, Li Y, Soppe W. Control and consequences of chromatin compaction during seed maturation in *Arabidopsis thaliana*. Plant Signal Behav. 2012:**7**(3):338–341. https://doi.org/10.4161/psb.19281

van Zanten M, Zoll C, Wang Z, Philipp C, Carles A, Li Y, Kornet NG, Liu Y, Soppe WJ. *HISTONE DEACETYLASE* 9 represses seedling traits in Arabidopsis thaliana dry seeds. Plant J. 2014:**80**(3):475–488. https://doi.org/10.1111/tpj.12646

Vidigal DS, Marques AC, Willems LA, Buijs G, Mendez-Vigo B, Hilhorst HW, Bentsink L, Pico FX, Alonso-Blanco C. Altitudinal and climatic associations of seed dormancy and flowering traits evidence adaptation of annual life cycle timing in *Arabidopsis thaliana*. Plant Cell Environ. 2016:**39**(8):1737–1748. https://doi.org/10.1111/pce.12734

Voegele A, Graeber K, Oracz K, Tarkowská D, Jacquemoud D, Turečková V, Urbanová T, Strnad M, Leubner-Metzger G. Embryo growth, testa permeability, and endosperm weakening are major targets for the environmentally regulated inhibition of *Lepidium sativum* seed germination by myrigalone A. J Exp Bot. 2012:**63**(14):5337–5350. https://doi.org/10.1093/jxb/ers197

Wang Z, Chen F, Li X, Cao H, Ding M, Zhang C, Zuo J, Xu C, Xu J, Deng X, et al. Arabidopsis seed germination speed is controlled by SNL histone deacetylase-binding factor-mediated regulation of AUX1. Nat Commun. 2016:**7**(1):13412. https://doi.org/10.1038/ncomms13412

Xiang Y, Nakabayashi K, Ding J, He F, Bentsink L, Soppe WJ. *Reduced Dormancy5* encodes a protein phosphatase 2C that is required for seed dormancy in Arabidopsis. Plant Cell. 2014:**26**(11):4362–4375. https://doi.org/10.1105/tpc.114.132811

Xiang Y, Song BX, Nee G, Kramer K, Finkemeier I, Soppe WJJ. Sequence polymorphisms at the *REDUCED DORMANCY5* pseudophosphatase underlie natural variation in Arabidopsis dormancy. Plant Physiol. 2016:**171**(4):2659–2670. https://doi.org/10.1104/pp.16.00525

Xu H, Lantzouni O, Bruggink T, Benjamins R, Lanfermeijer F, Denby K, Schwechheimer C, Bassel GW. A molecular signal integration network underpinning Arabidopsis seed germination. Curr Biol. 2020:**30**(19):3703–3712.e3704. https://doi.org/10.1016/j.cub.2020.07.012

Yazdanpanah F, Hanson J, Hilhorst HWM, Bentsink L. Differentially expressed genes during the imbibition of dormant and after-ripened seeds—a reverse genetics approach. BMC Plant Biol. 2017:**17**(1):151. https://doi.org/10.1186/s12870-017-1098-z

Yoong FY, O'Brien LK, Truco MJ, Huo H, Sideman R, Hayes R, Michelmore RW, Bradford KJ. Genetic variation for thermotolerance in lettuce seed germination is associated with temperature-sensitive regulation of ETHYLENE RESPONSE FACTOR1 (ERF1). Plant Physiol. 2016:**170**(1):472–488. https://doi.org/10.1104/pp.15.01251

Zhang H, Bishop B, Ringenberg W, Muir WM, Ogas J. The CHD3 remodeler PICKLE associates with genes enriched for trimethylation of histone H3 lysine 27. Plant Physiol. 2012:**159**(1):418–432. https://doi.org/10.1104/pp.112.194878

Zhao X, Dou L, Gong Z, Wang X, Mao T. BES1 hinders ABSCISIC ACID INSENSITIVE5 and promotes seed germination in Arabidopsis. New Phytol. 2019:**221**(2):908–918. https://doi.org/10.1111/nph.15437

Zheng J, Chen F, Wang Z, Cao H, Li X, Deng X, Soppe WJ, Li Y, Liu Y. A novel role for histone methyltransferase KYP/SUVH4 in the control of Arabidopsis primary seed dormancy. New Phytol. 2012:**193**(3):605–616. https://doi.org/10.1111/j.1469-8137.2011.03969.x

Zhong C, Patra B, Tang Y, Li X, Yuan L, Wang X. A transcriptional hub integrating gibberellin–brassinosteroid signals to promote seed germination in Arabidopsis. J Exp Bot. 2021:**72**(13):4708–4720. https://doi.org/10.1093/jxb/erab192

Zhu H, Xie W, Xu D, Miki D, Tang K, Huang CF, Zhu JK. DNA demethylase ROS1 negatively regulates the imprinting of DOGL4 and seed dormancy in *Arabidopsis thaliana*. Proc Natl Acad Sci U S A. 2018:**115**(42):E9962–E9970. https://doi.org/10.1073/pnas.1812847115

https://doi.org/10.1093/plcell/koae055

THE PLANT CELL

THE PLANT CELL 2024: 36: 1377–1409

Review

Not so hidden anymore: Advances and challenges in understanding root growth under water deficits

Priya Voothuluru [1,2], **Yajun Wu** [3], **Robert E. Sharp** [1,2,*]

1 Division of Plant Science and Technology, University of Missouri, Columbia, MO 65211, USA
2 Interdisciplinary Plant Group, University of Missouri, Columbia, MO 65211, USA
3 Department of Biology and Microbiology, South Dakota State University, Brookings, SD 57007, USA

*Author for correspondence: SharpR@missouri.edu
The author responsible for distribution of materials integral to the findings presented in this article in accordance with the policy described in the Instructions for Authors (https://academic.oup.com/plcell/pages/General-Instructions): Robert E. Sharp (sharpr@missouri.edu).

Abstract

Limited water availability is a major environmental factor constraining plant development and crop yields. One of the prominent adaptations of plants to water deficits is the maintenance of root growth that enables sustained access to soil water. Despite early recognition of the adaptive significance of root growth maintenance under water deficits, progress in understanding has been hampered by the inherent complexity of root systems and their interactions with the soil environment. We highlight selected milestones in the understanding of root growth responses to water deficits, with emphasis on founding studies that have shaped current knowledge and set the stage for further investigation. We revisit the concept of integrated biophysical and metabolic regulation of plant growth and use this framework to review central growth-regulatory processes occurring within root growth zones under water stress at subcellular to organ scales. Key topics include the primary processes of modifications of cell wall–yielding properties and osmotic adjustment, as well as regulatory roles of abscisic acid and its interactions with other hormones. We include consideration of long-recognized responses for which detailed mechanistic understanding has been elusive until recently, for example hydrotropism, and identify gaps in knowledge, ongoing challenges, and opportunities for future research.

Introduction

Limited availability of arable land worldwide creates a pressing need for substantial enhancements of agricultural productivity to satisfy the projected demands for food, feed, fiber, and energy in the near future (Fedoroff et al. 2010). In addition to land limitations, unpredictable changes in climate are creating conditions detrimental to plant growth and crop productivity. Among the stressors, droughts are a major environmental constraint on plant development that adversely affect crop yields and are likely to worsen in many areas of the world (Boyer 1982; Bailey-Serres et al. 2019). To achieve enhanced plant productivity under drought conditions while reducing the environmental footprint of production agriculture (Campbell et al. 2017; Springmann et al. 2018; Pareek et al. 2020), it is critical to elucidate the physiological and molecular mechanisms that regulate plant growth and development under water limitation (Boyer et al. 2013; Tardieu et al. 2018; Bailey-Serres et al. 2019).

It is well known that the growth of different plant organs responds differentially to water deficits. Typically, growth of aerial tissues is reduced or arrested, whereas growth of the root system is relatively maintained or even enhanced under water-limited conditions (Fig. 1; Sharp and Davies 1979; Westgate and Boyer 1985; Sharp et al. 1988). These observations underlie the increased ratio of root to shoot development regarded to be a key adaptive response of plants growing under water-limited conditions (Hsiao 1973; Meyer and Boyer 1981; Sharp and Davies 1989; Hsiao and Xu 2000). The high sensitivities of both vegetative and repro-ductive shoot growth responses to water deficits (Fig. 1A) are considered to be adaptive rather than injurious effects that

Received December 14, 2023. Accepted February 15, 2024. Advance access publication February 21, 2024.

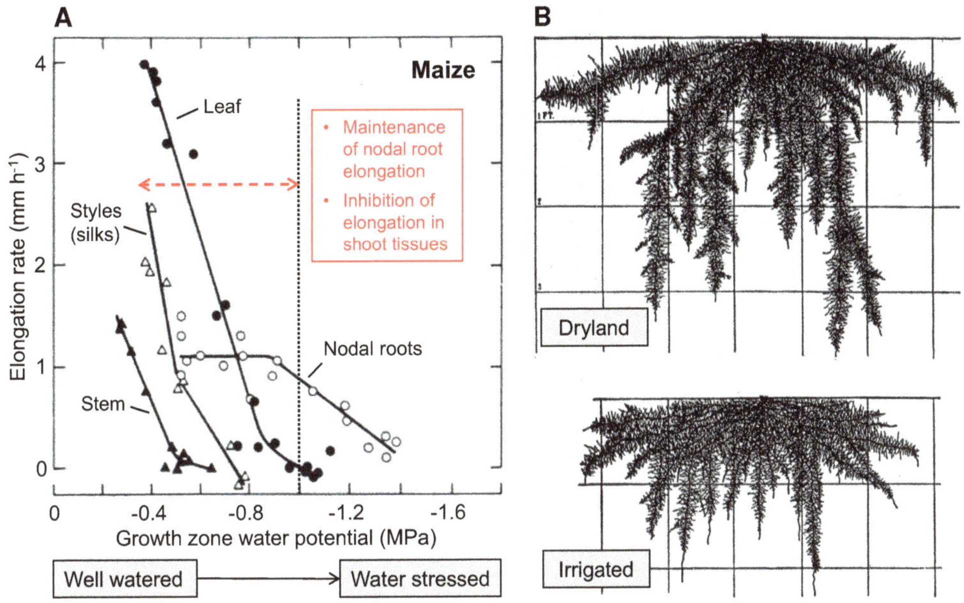

Figure 1. Maintenance of root growth under water deficit conditions. **A)** Comparative responses of elongation rate in different organs of maize to the development of water stress during soil drying. Nodal root elongation continued at growth zone water potentials that caused complete inhibition of elongation in vegetative and reproductive shoot tissues. Because growth responses were determined as a function of the water potentials of the growing tissues, the differential sensitivities reflect inherent differences in how cellular physiology responds to water stress in the different organs. **B)** More extensive root system development in maize plants when grown under soil drying (dryland) compared with irrigated conditions. A modified from Westgate and Boyer (1985), Figure 1, by permission of Springer Nature. B reproduced from Weaver (1926), Figure 87, p 189, by permission of John Wiley and Sons.

are beneficial for plant fitness and survival in an ecological context but tend to reduce yield in an agricultural context (Skirycz and Inzé 2010; Tardieu et al. 2018; Turc and Tardieu 2018). These tradeoffs between survival and growth and the regulatory mechanisms that determine shoot growth responses to water limitation have been studied extensively over the past 50 years, and interested readers are referred to comprehensive reviews on the topic (Hsiao 1973; Skirycz et al. 2011; Tardieu 2012; Claeys and Inzé 2013; Nelissen et al. 2018; Tardieu et al. 2018; Turc and Tardieu 2018).

Increased root system growth under water limitation in several crop and wild species was documented by Weaver (1926) a century ago in a seminal body of work on root development under field conditions. For example, compared with irrigated conditions, maize plants were observed to develop a root system that grew deeper and was more heavily branched under soil-drying conditions (Fig. 1B). These observations in maize and other crops led Weaver (1926, pp 1, 90) to comment on the importance of studying roots, stating that "Frequently, half—and often much more—of every crop plant is invisible. This portion consists entirely or largely of roots which extend far into the soil.... Since roots absorb water and nutrients, a knowledge of their development, extent, and activities and how these are modified by the changes in the environment are necessary for a scientific understanding of plant production." Development of more extensive rooting under water-limited conditions not only reflects the continued growth of root apices into regions of moist soil; in some circumstances, roots must grow through soil that is already dry to reach soil with available water. The ability of roots to grow into and through dry soil has attracted the attention of plant physiologists for many decades (Hendrickson and Veihmeyer 1931; Hunter and Kelley 1946; Portas and Taylor 1976), and it has been shown that certain types of roots—including the primary root of seedlings (see Figs. 2 and 3) and the shoot-borne nodal roots of grasses (Fig. 1A)—have the ability to continue growing at low tissue water potentials that completely inhibit shoot growth (Sharp and Davies 1979; Westgate and Boyer 1985; Sharp et al. 1988; Yamaguchi and Sharp 2010).

Despite early recognition of the adaptive significance of root developmental responses to water-limited conditions, progress in understanding the underlying physiological and genetic control mechanisms has been hampered by the inherent complexity of root systems and their interaction with the soil environment. Different root types, including the primary, seminal, and nodal root axes and their subtending lateral roots, exhibit varying responses to water deficits and are physiologically and genetically distinct (Hochholdinger et al. 2004, 2018; Ahmed et al. 2016, 2018; Waidmann et al. 2020; Freschet et al. 2021). Together with other factors, including root hair production, root exudation, and microbial interactions, this diversity collectively enables optimal root system development and function but complicates experimental investigation. The growth of the root system is further impacted by the spatial and temporal dynamics of soil water and nutrient availability as well as variability in other physical, chemical, and biological properties within the soil matrix. With heightened appreciation of the critical importance yet understudied nature of root development and function (Russell 1977; Eshel and Beeckman 2013; Gregory 2021), and with advances in experimental approaches and measurement techniques, the past several decades have seen increasingly intensive research on root growth responses to water def-icits (Hsiao 1973; Pritchard 1994; Sharp et al. 2004; Ober and Sharp 2007, 2013; Yamaguchi and Sharp 2010; Gowda et al. 2011; Lynch 2013, 2018; Dinneny 2019; Karlova et al. 2021). The development of thermodynamically based methods to measure soil and plant tissue water status in the 1960s was of particular importance in studies of plant responses to water deficits. These techniques allowed precise quantification of experimental conditions and repeatability of plant responses (Slatyer and Taylor 1960; Boyer and Knipling 1965; Scholander et al. 1965; Boyer 1995; Kramer and Boyer 1995; Juenger and Verslues 2023), enabling characterization of the diverse growth responses of different root types, as well as between roots and shoots, to water deficit conditions (Westgate and Boyer 1985; Sharp et al. 1988; Hsiao and Xu 2000; Dowd et al. 2019).

Selected milestones in the understanding of root growth re-sponses to water-limited conditions are the focus of this ASPB Centennial Review, with emphasis on founding studies that have shaped current knowledge and set the stage for further investigation. We first highlight how characteristics of root system architecture (RSA) benefit plant performance under water limitation. We then address the variety of root growth responses that determine the RSA for exploration of the soil profile. These growth responses occur within a relatively small volume of tissue that constitutes the root growth zone at the individual root apices, and it is the mechanisms that control cell production in the meristem and the rate, duration, and direction of cell expansion within the root growth zones that ultimately establish the growth and dimensions of the entire root system. We revisit the concept of integrated bio-physical and metabolic control of plant growth and use this framework to review key growth-regulatory processes occurring within root growth zones under water stress at subcellular, cellular, and tissue scales. Whereas knowledge of the molecular regulation of many aspects of root development is advanced (Motte et al. 2019), we focus on long-recognized physiological responses to water deficits for which, in many cases, detailed mechanistic understanding remains limited. Lastly, we propose future avenues for research to increase understanding of root growth under water limitation and, consequently, for enhanced opportunities to improve crop productivity under drought conditions.

Features of RSA for improved drought tolerance

The root system functions to provide both anchorage and the absorption of water and nutrients necessary for plant growth. The response of deeper rooting in water-limited environments (Fig. 1B) can enable access of water from the subsoil (Klepper et al. 1973; Sharp and Davies 1985; Sponchiado et al. 1989; Lopes and Reynolds 2010), which, by maintaining water avail-ability through to reproductive development, can have a major impact on yield sustainability. For example, in Australian wheat production, an additional 10 cm of rooting depth can result in a 10% to 20% increase in grain yield (Kirkegaard et al. 2007; Lilley and Kirkegaard 2007). Moreover, in elegant reciprocal grafting experiments with common bean lines varying in drought tolerance, White and Castillo (1989, 1992) demonstrated that under soil-drying conditions, the deep-rooting phenotype was genetically determined by the rootstock and, rather than the shoot phenotypes, was of greater importance for maintaining yield. Interestingly, in contrast, grafting experiments with potato cultivars led to opposite conclusions on the relative importance of the scion and rootstock in determining root system development under soil-drying conditions (Jefferies 1993). Accordingly, further studies are warranted to investigate root- vs shoot-sourced regulation of root growth responses to water deficits.

Lynch (2013) proposed that an ideotype of steep-angled roots that explore deeper layers of the soil profile and are anatomically cheaper to build and maintain will enhance drought tolerance and referred to such root system characteristics as a "steep, deep and cheap" ideotype. Several studies have demonstrated that under terminal drought conditions, this ideotype of RSA contributes to drought tolerance in rice (Uga et al. 2011, 2013), maize (Zhu et al. 2010; Chimungu et al. 2014a, 2014b), wheat (Gabay et al. 2021, 2023), chick pea (Kashiwagi et al. 2015), and common bean (Strock et al. 2019). The studies in rice are particularly noteworthy because they show mechanistic understanding of the deep rooting phenotype and demonstrate its impact on increasing grain yield in the field under water deficit conditions (Uga et al. 2011, 2013). The quantitative trait locus *DEEPER ROOTING 1* (*DRO1*) was identified in a rice recombinant in-bred line population, which accounted for 67% of variation in the deep-rooting phenotype (Uga et al. 2011). Further analyses of the *DRO1* locus found that it is negatively regulated by auxin and is involved in strong gravitropic response of the roots, leading to steeper growth angle and deeper root phenotypes of *DRO1*-containing lines compared with the shallow-rooting control lines (Uga et al. 2013). Phylogenetic analyses revealed that *DRO1* homologs are present in a wide range of plant species, and they have been demonstrated to modify RSA in *Arabidopsis*, *Medicago*, and *Prunus* species in addition to rice (Ge and Chen 2016; Guseman et al. 2017; Kitomi et al. 2020; Uga 2021). A similar multidisciplinary approach was used to understand variation in RSA and drought tolerance in wheat, and these studies found that RSA was regulated by dosage of the genes involved in jasmonic acid biosynthesis (Gabay et al. 2021, 2023). These studies indicate that the mechanistic understanding of genes and gene pro-ducts regulating RSA can be effectively used to obtain root phenotypes that improve drought tolerance in plants.

Although the steep, deep, and cheap ideotype was favorable in the terminal drought conditions imposed in these studies, it might not be suitable in dry environments where intermittent rainfall provides water in the upper layers of the soil profile. In a study of root phenotypes of diverse maize genotypes in which stress was imposed by providing 50% of the water required for optimal growth, Klein et al. (2020) found that the steep, deep, and cheap ideotype was not favored. The irrigation provided by the center pivot system used in the study is analogous to intermittent rainfall during the growing season that provides water to the upper layers of the soil profile, and the results demonstrated that maize (and likely other crops) has genetic variability in root growth characteristics that are suited for drought tolerance in these conditions. Favored phenotypes included thicker nodal roots, increased lateral root branching, larger proportion of stele, numerous metaxylem elements, larger cortical cells, and aerenchyma formation; the authors suggested that having multiple phenotypes integrated within an ideotype was necessary for adapting to environments with intermittent rainfall. Therefore, it is important to consider the attributes of RSA and individual root phenotypes in the context of particular drought scenarios and the potential tradeoffs of those features in optimal environmental conditions (Tardieu 2012; Lynch 2018; Tardieu et al. 2018; Verslues et al. 2023). In fact, it has been suggested that an average root system growth phenotype combined with developmental plasticity to environmental changes (rather than constitutive expression of traits) is more productive in a variety of environments and stress conditions (Sandhu et al. 2016; Strock et al. 2019; Schneider and Lynch 2020). Developmental plasticity has many potential ecological and physiological benefits for reducing inputs in production agriculture and is particularly important in low-input systems where water and nutrients are more variable (Weaver 1926; O'Toole and Bland 1987; Sponchiado et al. 1989; Fukai and Cooper 1995; Dardanelli et al. 1997; Sandhu et al. 2016; Strock et al. 2019; Schneider and Lynch 2020; Woods et al. 2022). Therefore, it is vital to discover the genetic and physiological mechanisms that regulate the integrated RSA phenotypes that optimize plant performance under water-limited conditions (Schneider and Lynch 2020; Uga 2021; Lynch 2022; Verslues et al. 2023).

Integrated biophysical and metabolic regulation of plant cell expansion

In his seminal review on plant responses to water stress, Hsiao (1973, p 536) emphasized that "with the shift of attention to metabolic and molecular aspects of stress physiology in the mid-1960s, the importance of water uptake and the resulting turgor as a physical force needed for cell growth has at times been almost overlooked or ignored." Arguably, the same statement could still be made today, and the role of turgor as well as mechanisms of turgor regulation and turgor sensing remain important areas for further investigation (Ali et al. 2023). The Lockhart model of plant cell expansion (Lockhart 1965) was originally formulated to describe the growth of single cells and, with caution, is also useful to gain an understanding of multicellular organ growth (Boyer 1985; Spollen and Sharp 1991; Pritchard 1994; Dumais 2021). In simplified form, the model describes the interdependence of expansive growth (G) on turgor and cell wall–yielding properties:

$$G = m\,(P\text{-}Y),$$

where m is the cell wall extensibility, P the cellular turgor, and Y the yield threshold turgor (i.e. the minimum turgor required for irreversible cell wall extension). An elaborated equation additionally considers the hydraulic resistance and associated water potential gradient required to drive water flow through tissues and into the growing cells ("growth-induced" or "growth-sustaining" water potential gradients; Molz and Boyer 1978; Silk and Wagner 1980), which, by lowering the cellular water potential, reduces the magnitude of turgor that develops for a given osmotic potential (Lockhart 1965; Boyer 1985; Passioura and Boyer 2003). The Lockhart equation conceptualizes that cell expansion occurs only when the internal pressure exerted on the wall is large enough to exceed the yield threshold, resulting in wall yielding at a rate dependent on the extensibility. More precisely, cell wall metabolism first results in wall relaxation (Cosgrove 2016), which relaxes stress and thus lowers turgor and consequently water potential inside the cell (because turgor is a component of water potential). This generates a water potential gradient that, in water-sufficient situations, drives water flow into the cell, resulting in turgor restoration, cell wall yielding, and cell expansion. The cells also take up or generate metabolites to maintain their osmotic concentration and reinforce the primary cell wall. These processes continue until cells reach their final size as secondary cell wall deposition leads to wall stiffening and growth cessation.

The Lockhart equation indicates that a decrease in turgor under water-limited conditions will result in a decrease in growth rate. However, the equation also illustrates that, theoretically, cell expansion can be regulated under water stress by 2 key mechanisms: first, by modifying cell wall–yielding properties, and second, by manipulating the processes of turgor maintenance. In a pioneering study of the dynamic relationship of plant cell expansion to turgor, Green et al. (1971) provided evidence for metabolic as well as physical control of plant cell expansion. Using the large cells of the alga *Nitella* as an experimentally amenable system, it was shown that a small stepwise decrease in turgor caused essentially immedi-ate cessation of cell elongation. (Turgor was measured in situ using an ingenious inserted capillary method, and decreases in turgor were imposed by lowering the external water potential; Green 1968.) However, the original elongation rate resumed within approximately 30 min while the cell remained at the decreased turgor. Conversely, imposed increases in turgor caused very high but short-lived increases in elongation rate, followed by deceleration to the original rate. These

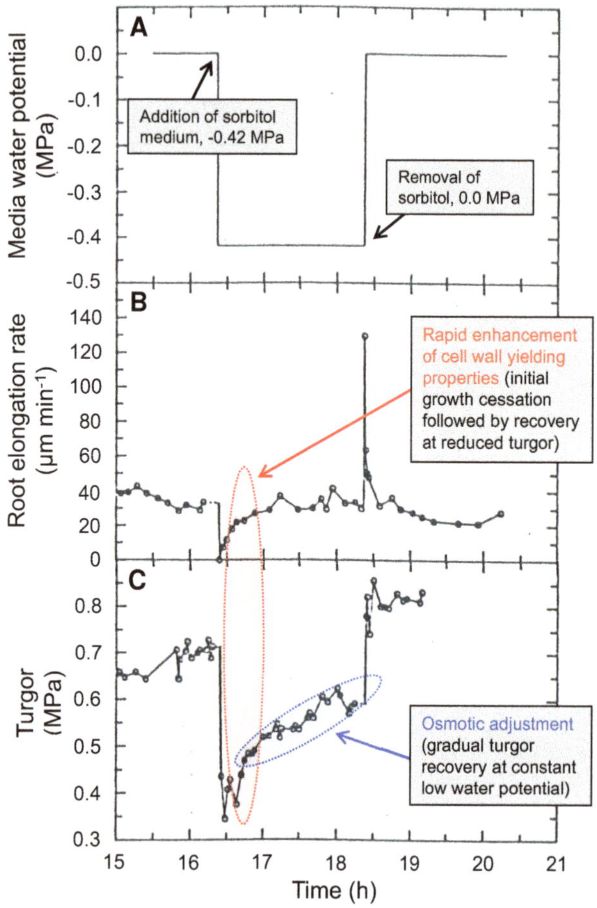

Figure 2. Cell wall–yielding properties are enhanced in maize primary roots after water stress imposition. Maize seedlings were grown in solution at a water potential of approximately 0 MPa (0.1 mM $CaCl_2$). Root elongation rate was monitored with a position transducer, and turgor of surface cells in the central region of the growth zone was measured every few minutes using a pressure microprobe. **A)** A stepwise decrease in media water potential was imposed by addition of −0.42 MPa sorbitol, and after 2 hours the sorbitol was removed. **B)** The abrupt decrease in water potential caused essentially immediate cessation of root elongation, as well as decrease in root turgor **(C)**. Elongation recovered to the well-watered rate within an hour after the onset of stress, whereas turgor recovery as a result of osmotic adjustment was more gradual and turgor did not reach the well-watered level for the duration of the stress treatment. Full recovery of elongation with only partial turgor recovery indicates that cell wall–yielding properties were rapidly enhanced in response to water stress; in terms of the Lockhart model (see text), either the yield threshold decreased or the extensibility increased, or both. Conversely, removal of water stress caused a short-lived spike in root elongation followed by deceleration to the original rate, again pointing to compensatory adjustments of cell wall–yielding properties. Modified from Hsiao and Jing (1987), Figure 7, by permission of ASPB.

observations supported the concept of a yield threshold turgor for cell wall extension that is subject to compensatory and presumably metabolic adjustments (as well as, possibly, changes in wall extensibility) following changes in turgor.

The subsequent development of the pressure microprobe (Husken et al. 1978) allowed similar experiments to be conducted in the much smaller cells within the growth zones of roots and leaves of higher plants. The results indicated that roots exhibit a high capacity for enhanced cell wall yielding and rapid growth resumption in response to turgor decreases (Hsiao and Jing 1987; Frensch and Hsiao 1995; Hsiao and Xu 2000; also see Kuzmanoff and Evans 1981); the original results from Hsiao and Jing (1987) for the maize primary root are shown in Fig. 2. Interestingly, this experiment also showed that turgor began to recover (Fig. 2C) while the low water potential treatment remained constant (Fig. 2A); this observation is indicative of solute accumulation by the process of osmotic adjustment, a process first shown to occur in water-stressed roots by Greacen and Oh (1972). Notably, root elongation remained constant during this phase (Fig. 2B), again pointing to compensatory adjustments of cell wall–yielding properties as turgor increased. Moreover, as a result of the osmotic adjustment, turgor increased to higher levels than in well-watered roots (Fig. 2C) following removal of water stress (Fig. 2A). This resulted in a large spike in root elongation followed by rapid deceleration to the original rate (Fig. 2B), indicating that wall-yielding properties were rapidly moderated. In contrast to the findings with water-stressed roots, studies of leaves indicated that cell wall–yielding properties either did not increase substantially after low water potential imposition or, with longer-term stress exposure, actually decreased, leading to growth inhibition despite turgor maintenance by osmotic adjustment (Matthews et al. 1984; Hsiao and Jing 1987; Serpe and Matthews 1992; Hsiao and Xu 2000).

The Green et al. (1971) study provided the foundation for investigation of the metabolic regulation of plant cell wall expansion, particularly in response to water deficit conditions. The influence of this work was recognized in 2010 by the American Society of Plant Biologists, who included the paper among a "Classics Collection" of 25 papers in *Plant Physiology* that played a key role in shaping modern plant biology research. Modifications of cell wall–yielding properties and osmotic adjustment are now established as primary processes contributing to the ability of roots to maintain growth under water-stressed conditions and are addressed in detail in later sections.

Kinematic approaches to study root growth responses

Among plant organs, roots have a relatively simple growth zone in terms of organization, and root growth zones therefore have been used for many decades as models to study various aspects of plant growth (Sinnott 1939; Erickson and Sax 1956; Goodwin and Avers 1956; Erickson and Silk 1980; Beemster and Baskin 1998; Brady et al. 2005). Cells are first formed in the apical meristem by division of stem cells and then continue to divide for several cycles while cell elongation simultaneously pushes the apex through the soil and

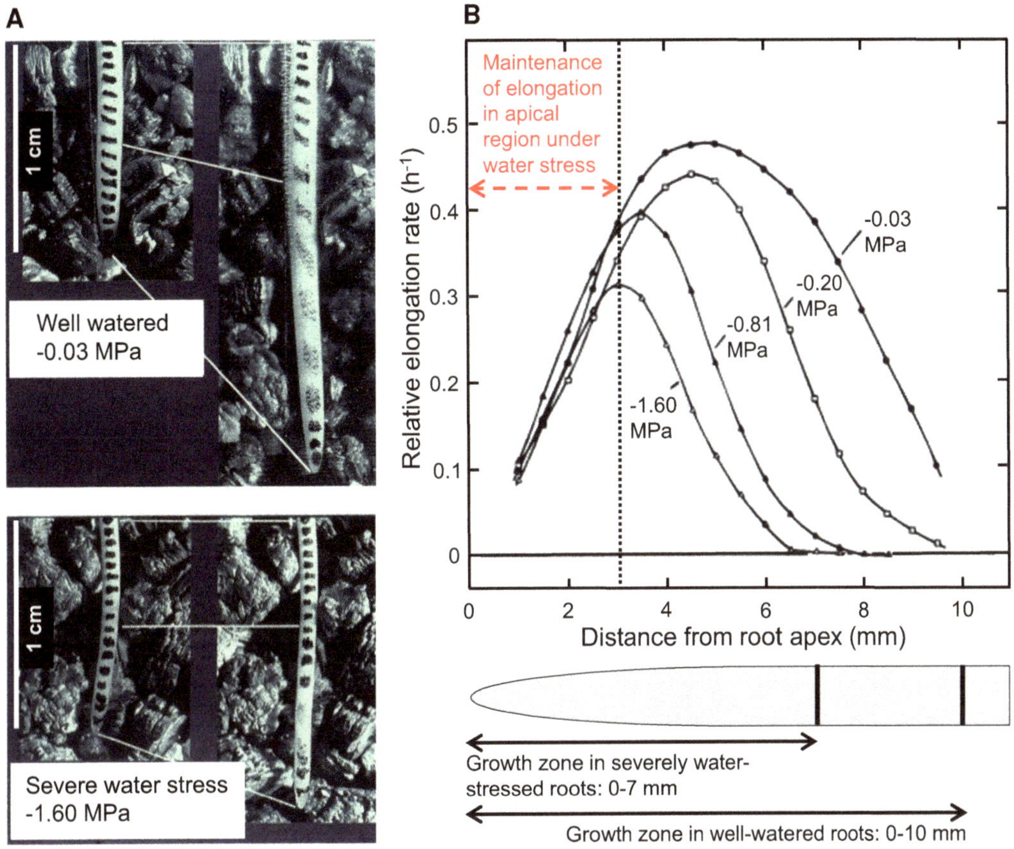

Figure 3. Kinematic analysis reveals spatially differential responses of tissue expansion to water stress within the growth zone of the maize primary root. Maize seedlings were grown under well-watered conditions (water potential of −0.03 MPa) or at mild (−0.20 MPa), moderate (−0.81 MPa), or severe (−1.60 MPa) water stress (obtained by adjusting the vermiculite media water content). When the primary roots were approximately 5 cm long, the apical 10-mm region was marked at approximately 0.6-mm intervals for temporal analysis of mark displacement away from the apex. **A)** Displacement of marks during 3.5 h after marking for representative roots growing under well-watered or severe water stress conditions. White lines indicate vertical displacement of the root apices and of marks originally located at 5 and 10 mm from the apex. In well-watered roots, mark separation, and hence tissue expansion, occurred throughout the apical 10 mm, whereas in severely water-stressed roots, mark separation was confined to the apical 5 mm. Water-stressed roots were also substantially thinner than well-watered controls, indicating inhibition of radial expansion. **B)** Time-lapse analysis of mark displacement during 1 h after marking was used to calculate the distribution of relative elongation rate as a function of distance from the root apex. In all water stress treatments, local elongation rates in the apical 3 mm were maintained at the well-watered rate, whereas elongation was increasingly inhibited with increasing water stress as cells were displaced further from the apex, resulting in progressive shortening of the growth zone. Modified from Sharp et al. (1988), Figures 3 and 5, by permission of ASPB.

displaces older cells away from the apex. The cells continue to expand as they exit the meristem and traverse the growth zone. Hence, as root growth occurs, cells are progressively located at increasing distances from the root apex (Fig. 3A) and experience increasing displacement velocities (Erickson and Silk 1980; Sharp et al. 1988; Baskin et al. 2020). As the cells expand, they typically develop an anisotropic growth pattern due to longitudinal expansion being favored over radial expansion (Liang et al. 1997), resulting in the cylindrical geometry of most roots. The end of the growth zone is marked by cells that have stopped expanding and undergo processes of maturation, including the development of secondary cell wall thickening.

The overall rate of root elongation is determined by the rate of cell production from the meristem and the rate and duration of cell elongation. However, the apparent simplicity of this relationship belies the realization that within the growth zone there is massive spatial and temporal heterogeneity of cellular growth rates and, thus, of underlying growth-regulatory processes. This heterogeneity occurs in the course of normal development (Erickson and Sax 1956; Beemster and Baskin 1998) and in response to various environmental conditions, including, for example, water limitation (Sharp et al. 1988) and soil mechanical resist-ance (Croser et al. 1999). To obtain detailed analyses of spatiotemporal growth patterns within plant growth zones, powerful kinematic approaches that apply concepts of fluid dynamics to tissue expansion were pioneered by Ralph Erickson and Wendy Silk (Erickson and Sax 1956; Erickson 1976; Silk and Erickson 1979; Silk 1984;

Walter et al. 2009). These analyses revealed that in a typical root growth zone, the local elongation rate dramatically accelerates as cells move out of the meristem, reaches a peak in the central region, and then decelerates before growth cessation (Fig. 3B; Erickson and Sax 1956; Goodwin and Avers 1956). Moreover, this large range of tissue expansion rates occurs within a brief frame of space and time. For ex-ample, in the maize primary root growing under well- watered conditions, around 8 hours is required for a cell exiting the meristem to be displaced to the end of the growth zone, located at approximately 10 mm from the apex, during which time the relative elongation rate (longitudinal strain rate) accelerates to a remarkably high peak value of almost 50% h^{-1} (i.e. a tissue element at this location would double in length in an hour; Fig. 3B) and then abruptly decreases (Sharp et al. 1988).

Knowledge of cell expansion patterns within plant growth zones allows comparison of local effects to putative local causes and, thereby, facilitates investigation of underlying regulatory processes (Hsiao et al. 1985; Walter et al. 2009). In the maize primary root, kinematic analyses revealed that cell elongation is differentially responsive to water stress in different regions of the growth zone (Fig. 3). Remarkably, in the apical region that encompasses the meristem, longitudin-al expansion is maintained even under severe water stress (at tissue water potentials as low as −1.6 MPa). In contrast, elongation is progressively inhibited compared with well-watered roots as cells are displaced further from the apex, resulting in decreased final cell lengths and a shortened growth zone (Sharp et al. 1988; Saab et al. 1992; Fan and Neumann 2004; Voothuluru et al. 2020). Similar findings were reported in primary roots of several other species, including wheat (Pritchard et al. 1991), pine (Triboulot et al. 1995), soybean (Yamaguchi et al. 2010), and cotton (Kang et al. 2022), and have also been observed in water-stressed leaves (Durand et al. 1995; Skirycz et al. 2011; Avramova et al. 2015). On the other hand, cell production decreased substantially in roots growing under moderate to severe water stress conditions (Fraser et al. 1990; Voothuluru et al. 2020; Kang et al. 2022; Verslues and Longkumer 2022). The possible adaptive advantage of inhibited cell production in water-stressed roots is discussed in a later section.

Interestingly, the degree of growth anisotropy was also shown to be altered in water-stressed maize primary roots (Sharp et al. 1988; Liang et al. 1997). In contrast to the maintenance of longitudinal expansion in the apical region of the growth zone (Fig. 3B), radial expansion was inhibited, resulting in substantially thinner roots compared with well- watered controls (Fig. 3A). These results indicate that effects of water stress on root expansion in length and width are regulated independently, although the control mechanisms underlying this differential response are not understood (Baskin et al. 1999). Thinner roots in water-stressed com-pared with well-watered plants have been reported in several species (Taylor and Ratliff 1969; van der Weele et al. 2000). It should be noted, however, that whether roots become thinner under water stress depends on soil properties. Many soils increase in mechanical resistance as they dry, and a common response to physical impedance is root swelling (Moss et al. 1988; Pandey et al. 2021; Huang et al. 2022).

Preferential maintenance of cell elongation in the apical region of the growth zone under water deficit conditions, together with root thinning, likely represents a coordinated adaptive response that enables the root to concentrate its use of limited resources to sustain adequate water and solute transport to the vital apical region that includes the meristem and thereby to continue exploration of the soil for water at minimum cost (Sharp et al. 1990; Voetberg and Sharp 1991; Verslues and Sharp 1999; Wiegers et al. 2009; Voothuluru et al. 2020). As described in the following sections, these findings provided a powerful underpinning to investigate the complex network of physiological and molecular processes involved in the regulation of root elongation under water stress conditions (Sharp et al. 2004; Yamaguchi and Sharp 2010, Ober and Sharp 2013).

Historically, kinematic analyses have been an important tool to link cellular growth heterogeneity in root growth zones with spatial variation in, for example, hormones (aux-in [IAA]: Hejnovicz 1961; Goodwin 1972; abscisic acid [ABA]: Saab et al. 1992; Ober and Sharp 2003; gibberellin: Band et al. 2012), cell wall proteins (Wu et al. 1994, 1996; Zhu et al. 2007), apoplastic pH (Peters and Felle 1999; Winch and Pritchard 1999; Fan and Neumann 2004) and reactive oxygen species (ROS) (Voothuluru et al. 2020), and other growth-regulatory factors. Further, the application of growth kinematics provides a powerful approach to ascertain rates of associated developmental processes (Silk et al. 1984, 1986; Sharp et al. 1990; Voetberg and Sharp 1991; Silk and Bogeat-Triboulot 2014), as detailed below with regard to osmotic adjustment. However, despite early recognition of the importance of characterizing how growth patterns are altered by environmental variation (Goodwin and Avers 1956), relatively few studies of root stress biology have taken advantage of these approaches. Whereas original tissue marking and time-lapse photographic techniques were laborious (Fig. 3A), modern tools such as computational video image analysis combined with microscopy techniques have enabled kinematic growth analyses to be obtained with relative ease (Silk et al. 1989; van der Weele et al. 2003; Basu et al. 2007).

Cell wall changes impacting root growth under water deficits

The original indications of enhanced cell wall yielding in water-stressed roots, as described above, were based on the temporal responses of turgor and root growth to low water potential imposition (Fig. 2), and similar inferences were made from comparisons of relative elongation rate and turgor profiles within the growth zone of roots growing under steady water stress conditions (Spollen and Sharp 1991;

Triboulot et al. 1995). However, those studies did not provide direct assessments of cell wall–yielding properties or the biochemical basis for stress-induced changes. Taking advantage of the kinematic growth analysis in water-stressed maize primary roots (Fig. 3), Wu et al. (1996) demonstrated that the maintenance of elongation in the apical region and the premature deceleration and cessation of elongation in the basal region of the growth zone were associated with differential responses of wall-yielding properties. Although substantial osmotic adjustment occurred (see following section), this was insufficient to maintain turgor, which was decreased by over 50% throughout the growth zone (Fig. 4A, inset). Accordingly, the maintenance of elongation in the apical region indicated that longitudinal cell wall yielding was enhanced, which was confirmed by demonstration of substantially increased acid-induced extension in water-stressed compared with well-watered roots (Fig. 4A). Acid-induced growth of plant cell walls has long been recognized (Rayle and Cleland 1992; Peters and Felle 1999) and is mediated at least partly by wall-loosening expansin proteins (McQueen-Mason et al. 1992; Cosgrove 2000, 2022). Consistently, expansin activity (Fig. 4B; Wu et al. 1996) and transcript levels of several expansin genes (Wu et al. 2001; Kang et al. 2023) were markedly increased in the apical region of water-stressed roots, as was activity of the "wall remodeling" enzyme (Cosgrove 2022) xyloglucan endotransglycosylase/hydrolase (XTH) (Fig. 4D; Wu et al. 1994). Interestingly, susceptibility to exogenous expansins also increased in the apical region (Fig. 4C), suggesting that stress-induced modifications of cell wall structure or composition facilitated expansin accessibility (Wu et al. 1996). In contrast, acid-induced extension was greatly decreased in the basal region of growth inhibition (Fig. 4A). Although extractable expansin activity also increased in this region (probably reflecting maintained activity as cells were displaced from the apical region), the minimal extensibility was likely attributable to compositional changes resulting in wall stiffening (Wu et al. 1996; Fan et al. 2006; Yamaguchi and Sharp 2010). Water stress–induced cell wall compositional changes are discussed further below.

Correlations between profiles of apoplastic pH and longitudinal expansion have been demonstrated in root growth zones, with more acidic regions coinciding with peak expansion rates (Peters and Felle 1999). Much evidence indicates that the pH profile is metabolically regulated and causally related to the growth rate distribution (e.g. Staal et al. 2011; Xu et al. 2013). In a study of cell wall pH regulation in water-stressed maize primary roots, Fan and Neumann (2004) found that spatial profiles of root surface acidification (proton efflux) and epidermal cell wall pH correlated with the region-specific growth responses. While the apical region of growth maintenance exhibited profiles similar to the well-watered control, the basal region of growth inhibition showed decreased acidification and a higher pH. Importantly, addition of acidic buffer partially restored growth in the basal region, indicating that the stress-induced increase in wall pH was functionally related to the inhibition of growth in this region. Because apoplastic acidification is important for activation of cell wall loosening proteins (Cosgrove 2000; Hager 2003), the pH profile is likely involved in differentially regulating wall loosening and, thereby, contributing to the spatial growth pattern in water-stressed roots.

To gain a more comprehensive understanding of how cell wall protein composition changes in response to water stress in different regions of the maize primary root growth zone, a proteomics analysis of water-soluble and lightly ionically bound cell wall proteins was conducted by Zhu et al. (2007). The results showed predominantly region-specific changes in several functional categories, suggesting the involvement of multiple processes in the growth responses. Notably, an increase in apoplastic ROS was predicted particularly in the apical region of growth maintenance, which was confirmed by imaging techniques (Zhu et al. 2007; Voothuluru and Sharp 2013). As discussed by Voothuluru et al. (2020), apoplastic ROS may have wall loosening or tightening effects and could also be involved in signaling processes that effect cell production. To investigate these possibilities, root growth characteristics of transgenic maize lines (constitutively expressing a wheat *oxalate oxidase*) with altered apoplastic ROS levels were evaluated. The results revealed a complex picture with apoplastic ROS modulating elongation differentially in well-watered (promoted) or water-stressed (inhibited) roots, in both cases via effects on both cell production and spatial profiles of cell elongation, as discussed in a later section.

Root growth depends on the coordinated and integrated expansion of all cells within the organ, but the biomechanical and biochemical properties of individual tissues may be predominant in regulating and/or limiting the overall rate of elongation. As described above, extensibility assays (Fig. 4A) indicated there are differential changes in wall composition in the apical and basal regions of the growth zone of water-stressed maize primary roots that contribute to the growth pattern, and evidence suggests that these changes oc-cur in a tissue-specific manner. Fan et al. (2006) tested the hypothesis that water stress-induced alterations in wall- linked phenolic compounds are linked with the inhibition of elongation in the basal region of the growth zone. Results from Fourier transform infrared spectroscopy in-dicated region-specific changes in phenolic composition (Fig. 4F), and progressive accumulation of lignin with increas-ing distance from the apex was observed primarily in stelar tissues in correlation with inhibition of mechanical extensi-bility of root segments. A similar spatial pattern of water stress–induced lignification was observed in the soybean pri-mary root (Yamaguchi et al. 2010). When the growth zone of well-watered or water-stressed maize roots was bisected, the roots curved inward as they grew, suggesting that the inner tissues were limiting root elongation (as reported over a century ago in *Vicia faba* L. roots by Darwin and Acton 1909). However, Pritchard and Tomos (1993) found in well-watered roots that extensibility of the separated stele was higher than that of the cortical sleeve and therefore suggested that

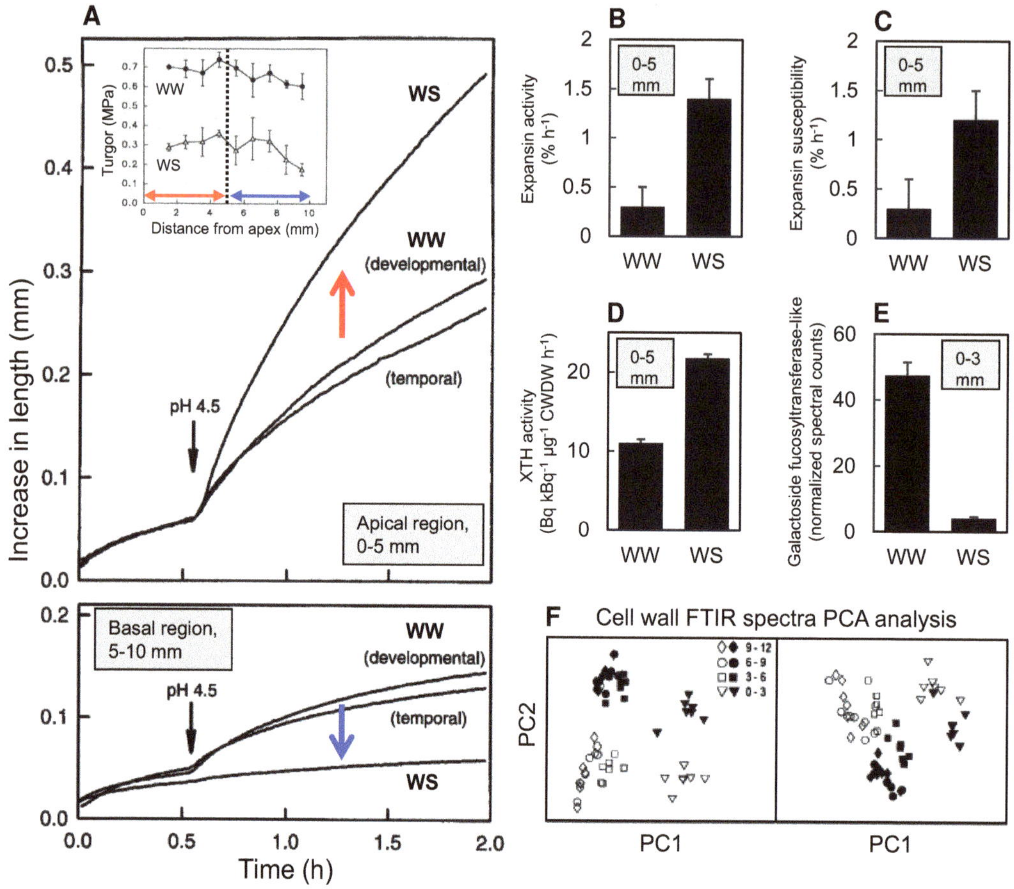

Figure 4. Changes in cell wall–yielding properties and composition in the growth zone of water-stressed maize primary roots. **A)** Acid-induced extension was enhanced in the apical region (0 to 5 mm from the apex) and almost completely inhibited in the basal region (5–10 mm) of the growth zone in water-stressed roots (WS; vermiculite water potential of −1.6 MPa) compared with well-watered (WW) developmental (roots of the same length) and temporal (roots of the same age) controls. The increased acid-induced extension in the apical region is thought to play an important role in maintaining elongation in this region (Fig. 3) despite substantially decreased turgor (inset) due to incomplete osmotic adjustment (see Fig. 6). Conversely, inhibition of acid-induced extension in the basal region, together with decreased turgor, likely contributes to premature slowing and cessation of elongation as cells are displaced through this region (Fig. 3). The apical region of water-stressed compared with well-watered roots also showed large increases in **(B)** expansin activity (change in slope of acid-induced extension of heat-killed cucumber hypocotyl wall preparations following addition of maize root tip expansin extract), **(C)** expansin susceptibility (change in slope of acid-induced extension of heat-killed maize root tip wall preparations following addition of cucumber expansin extract), and **(D)** XTH activity per unit of cell wall dry weight (CWDW), as well as **(E)** decreased abundance of galactoside 2-α-1-fucosyltransferase-like protein. **F)** Principal component analysis (PCA) of cell wall Fourier transform infrared (FTIR) spectra showed that different regions of the growth zone (delineated as mm from the apex) of water-stressed roots (black symbols, water potential of −0.5 MPa imposed with PEG 6000 in solution culture) are compositionally different compared with respective regions of well- watered controls (open symbols). Left and right panels show different wavenumber ranges. **A**, **B**, and **C** modified from Wu et al. (1996), Figures 1, 3B, 6A; **A** inset modified from Spollen and Sharp (1991), Figure 2B; **D** modified from Wu et al. (1994), Figure 3C; **E** modified from Voothuluru et al. (2016), Figure 6; **F** reproduced from Fan et al. (2006), Figure 3; **A–D** and **F** by permission of ASPB, **E** by permission of John Wiley and Sons.

properties of the endodermis and/or the inner layers of the cortex, rather than stelar tissues, are rate limiting for root elongation. Consistently, there is evidence that the endodermis may play a key role in hormone-mediated control of root growth (Dinneny 2014). For example, the endodermis was shown to be the primary GA/DELLA-responsive tissue regulating root growth in *Arabidopsis* (Úbeda-Tomás et al. 2008). On the other hand, evidence suggests that the properties of the epidermis and/or cortex are also important in determining *Arabidopsis* root elongation (Dyson et al. 2014; Vaseva et al. 2018; Verslues and Longkumer 2022), similarly to the more well-characterized regulation of shoot growth (Kutschera and Briggs 1988; Wakabayashi et al. 1989; Peters and Tomos 2000; Passioura and Boyer 2003; Kutschera and Niklas 2007; Savaldi-Goldstein et al. 2007).

Along with the increases in lignin, there is evidence suggesting that water-stressed maize primary roots have differential accumulation of cell wall–bound ferulates in the apical and basal regions of the growth zone (Fan et al. 2006; Spollen et al. 2008; Yamaguchi and Sharp 2010). Ferulates and other hydroxycinnamates are abundant in the cell walls of monocotyledonous plants and have a role in cross-linking wall polysaccharides, including hemicellulose, xylans, pectins, and lignin (Vogel 2008; Vermerris et al. 2010; Hatfield et al. 2018). In mature root and shoot tissues, decreased wall extensibility strongly correlates with increases in cell wall–bound ferulates (Tan et al. 1992; MacAdam and Grabber 2002; Azuma et al. 2005). Therefore, increased accumulation of ferulates in the basal region of the growth zone in water-stressed roots could be involved in enhanced cell wall cross-linking, thereby decreasing cell wall extensibility and elongation. Substantial modifications of cell wall composition in the growth zone of water-stressed maize primary roots were also suggested by cell wall and plasma membrane–enriched proteome analyses (Zhu et al. 2007; Yamaguchi and Sharp 2010; Voothuluru et al. 2016) together with transcriptome studies (Spollen et al. 2008; Opitz et al. 2016; Kang et al. 2023). In particular, modification of xyloglucan composition in the apical region was implicated by the spatial patterns of abundance of enzymes involved in xyloglucan biosynthesis (Fig. 4E). Xyloglucan forms load-bearing associations with cellulose microfibrils, and the potential structural modifications in xyloglucan composition likely impact the susceptibility of cell walls to wall-loosening and remodeling proteins. Along with previous reports of high levels of XTH activity in the root and shoot growth zones of several species (Fry et al. 1992; Wu et al. 1994), the evidence collectively indicates that cell wall remodeling from xyloglucan composition may have an important role in growth regulation of water-stressed roots.

The specific examples of lignin, ferulate, and xyloglucan compositional changes as well as the findings of substantial modifications of cell wall extensibility in water-stressed roots indicate that combining kinematic growth analyses with comprehensive cell wall compositional analyses will enable deciphering of the functional role of various cell wall components. Furthermore, it will be important to ascertain whether stress-induced changes occur in a cell type-specific manner and how these changes impact tissue-level and organ-level growth characteristics. The technical challenge of obtaining the requisite amounts of tissues for cell wall compositional analysis has prevented detailed characterization of the relationship with root growth responses to water stress (Wu and Cosgrove 2000; Cosgrove 2016). Recent developments in glycome profiling and immunohistochemistry (Pattathil et al. 2010) as well as nanoimaging and nanomechanical tech-niques (Kozlova et al. 2019; Coste et al. 2020) provide unpre-cedented opportunities for studying plant cell walls from subcellular to organ scales (Bou Daher et al. 2018; Sampathkumar et al. 2019; Petrova et al. 2021). This multi- scale approach will unveil how components interact within the cell wall matrix and how they impact cell expansion and root growth under normal and water deficit conditions. In the long term, knowledge from these studies will pave the way to selectively alter cell wall components in a tissue-specific manner to promote stress-responsive growth in plants and en-hance agricultural productivity under water deficits.

Osmotic adjustment in roots growing under water deficits

As discussed above, roots exhibit a high capacity to enhance cell wall–yielding properties under water deficit conditions, allowing elongation to continue despite substantial decreases in turgor (Figs. 2 and 4A). However, if tissue water potentials continue to decline with further soil drying, turgor may decrease below the lower limit to which the yield threshold can be adjusted, in which case cell expansion can no longer occur. Accordingly, a second key mechanism for growth maintenance in water-stressed tissues is turgor maintenance by osmotic adjustment (Morgan 1984; Blum 2017; Turner 2018). With osmotic adjustment, increases in cellular solute concentrations (by processes other than dehydration, which does not result in turgor maintenance) lower the osmotic potential and thereby maintain the osmotic driving force for water uptake. A number of early studies reported increased concentrations of sugars and other solutes in root and shoot tissues under water stress conditions and recognized that these changes may positively correlate with drought resistance (Martin et al. 1931; Eaton and Ergle 1948; Iljin 1957). Subsequent studies established that roots, especially the growth zone, have a high capacity for turgor maintenance by osmotic adjustment (Fig. 5) and that this response is associated with continued root elongation under water deficit conditions (Greacen and Oh 1972; Sharp and Davies 1979; Westgate and Boyer 1985). Further, it has been shown that roots can exhibit more pronounced osmot-ic adjustment than leaves under equivalent water stress conditions (Fig. 5; Sharp and Davies 1979; Hsiao and Xu 2000).

To understand the regulation of osmotic adjustment in growing regions, it is important to recognize that increases in solute concentration can occur by 2 distinct overall processes. First, there may be increases in the net rate of solute deposition (encompassing uptake, import, local generation, utilization), which could contribute to growth maintenance. Second, if tissue volume expansion is inhibited, this will reduce rates of water uptake (water represents about 90% of volume increases) and, therefore, of solute dilution. Indeed, this distinction combined with the sensitivity of shoot growth to water stress (Fig. 1A) contributed to early controversy about the potential benefits of osmotic adjustment (Turner and Jones 1980; Steponkus et al. 1982; Munns 1988; Serraj and Sinclair 2002). For example, Wilson and Ludlow (1983, p 536) suggested: "It seems somewhat contradictory to consider osmotic adjustment as a benefit to maintaining growth if the contributing solutes only increase in concentration because growth (and hence solute demand) has slowed down." This view led in turn to the important question of whether selection pressure for enhanced osmotic adjustment might result in reduced growth potential (Quisenberry et al. 1984).

In the case of roots, the observed association of osmotic adjustment with maintenance of elongation under water stress appeared to satisfy this concern (Greacen and Oh

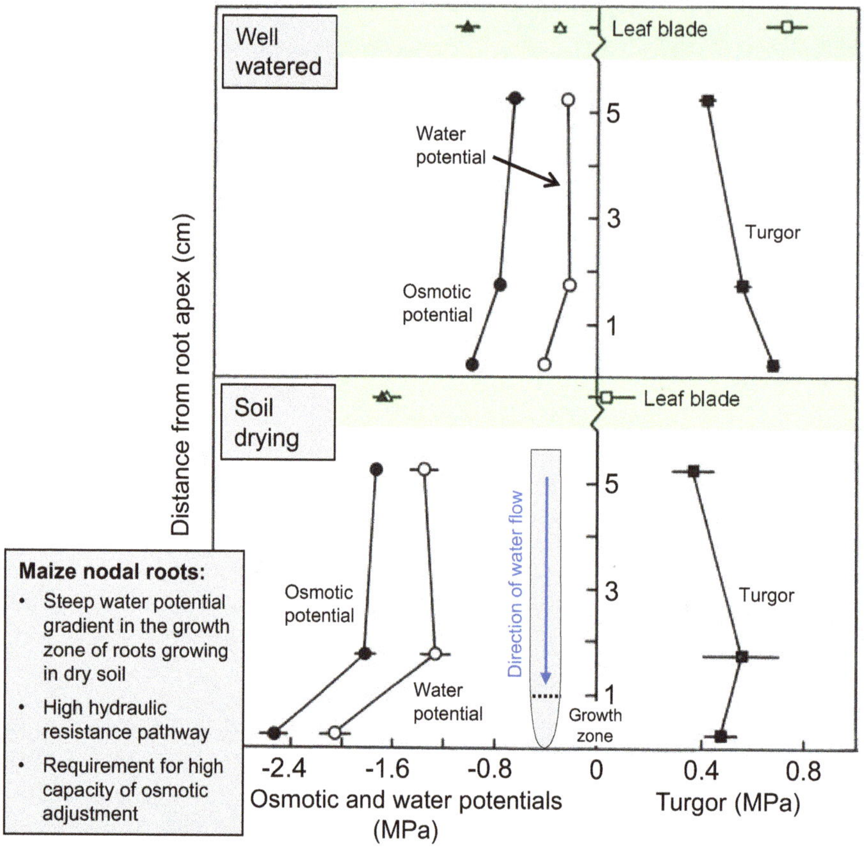

Figure 5. Osmotic adjustment in maize nodal roots under soil-drying conditions. Water potential, osmotic potential, and turgor in the growth zone and mature regions of nodal roots and in mature leaf blade tissues (green shading) of maize plants growing under well-watered or soil-drying conditions. In roots growing through dry soil, a steep "growth-induced" water potential gradient developed between the growth zone and adjacent mature region due to axial delivery of water and the hydraulic resistance of nonvascularized root tip tissues. This necessitates a high capacity for osmotic adjustment to maintain turgor in the root growth zone. The leaf blade was completely wilted in the same plants. Modified from Sharp and Davies (1979), Figure 6, by permission of Springer Nature.

1972; Sharp and Davies 1979; Westgate and Boyer 1985). However, quantitative assessment of this question in the growth zone of the maize primary root revealed a more complex picture, showing that osmotic adjustment involves a multifaceted interplay of morphogenic and metabolic responses (Sharp et al. 1988, 1990; Voetberg and Sharp 1991). While substantial decreases in osmotic potential occurred throughout the growth zone with increasing levels of water stress (Fig. 6A), different solutes were major contributors in the apical region where elongation was maintained vs the basal region of growth inhibition. The apical region showed a dramatic increase in proline concentration that contributed up to 50% of the decrease in osmotic potential (Fig. 6B; Voetberg and Sharp 1991). The particular use of proline for osmotic adjustment in the only slightly vacuolated cells of the apical region is consistent with its function as a cytoplas-mic solute that is compatible with metabolism at high concentrations (Yancey 2005; Verslues and Sharma 2010). In contrast, hexoses increased minimally in the apical region but greatly in the basal region, suggesting a primarily vacuolar compartmentation, where they accounted for up to 60% of the adjustment (Fig. 6C; Sharp et al. 1990). The kinematic analysis of tissue expansion rate profiles (Fig. 3) was used to calculate spatial distributions of solute and water deposition rates using the continuity equation from fluid dynamics (Silk 1984), which revealed that the increased concentrations of these 2 solutes occurred by contrasting mechanisms (Sharp et al. 1990; Voetberg and Sharp 1991). In the apical region, the increase in proline resulted primarily from a large stimulation of the rate of proline deposition (by as much as 10-fold; Fig. 6D) in combination with an approximately 50% decrease in water deposition that was due specifically to the root thinning response as described above (Fig. 3A). In the basal region, in contrast, the large increase in hexose concentration occurred despite the fact that the rate of hexose deposition greatly decreased (Fig. 6E); this result is explained because the rate of water deposition decreased to an even greater extent due to the inhibition of both longitudinal and radial expansion in this region (Fig. 3).

The proline results shown in Fig. 6D provided the first demonstration that increased solute deposition rates can make a major contribution to the osmotic adjustment of

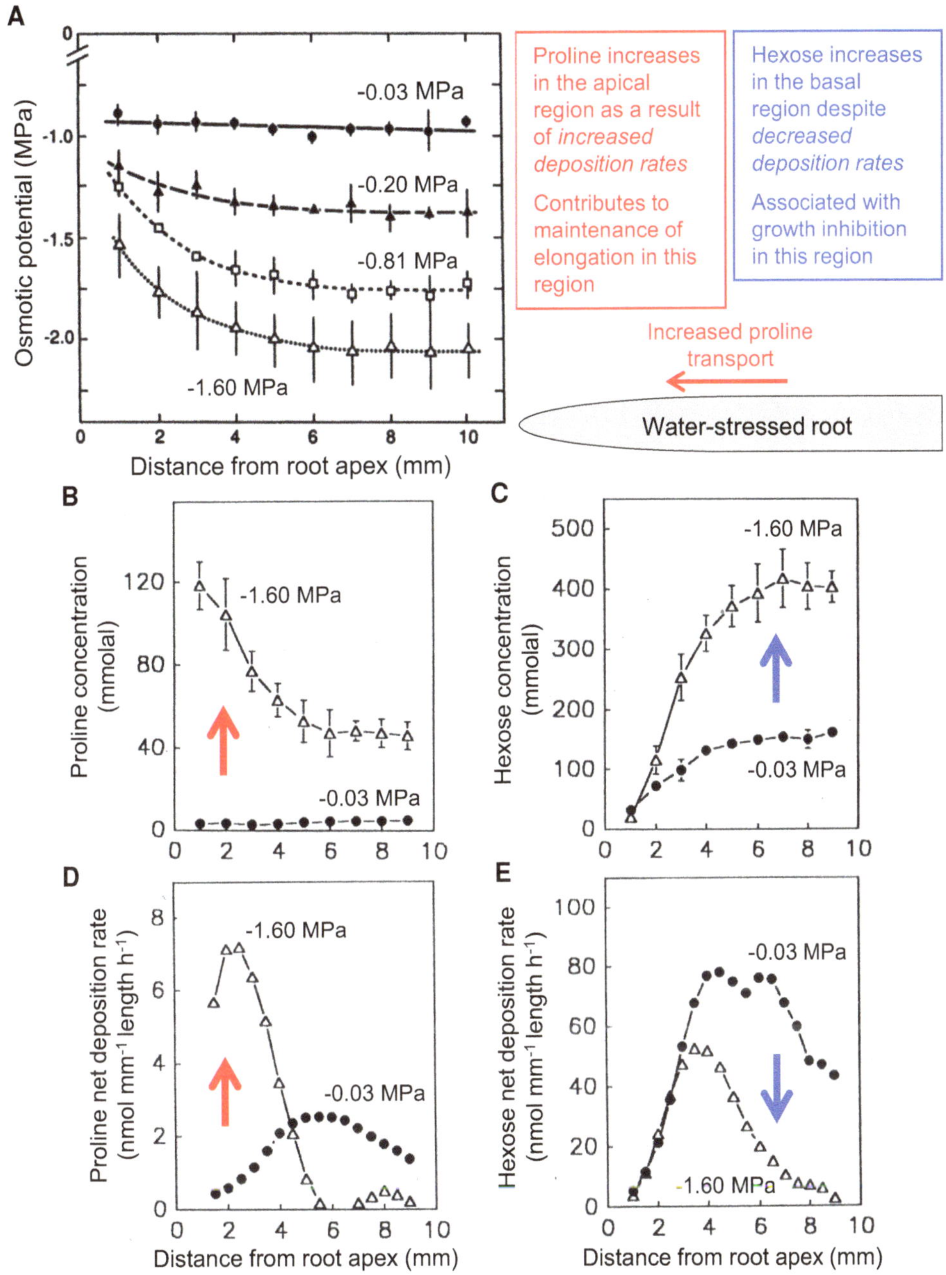

Figure 6. Different solutes contribute to osmotic adjustment in the apical and basal regions of the maize primary root growth zone. **A)** Spatial distribution of osmotic potential in the apical 10 mm of roots growing under well-watered conditions (water potential of −0.03 MPa) or at mild (−0.20 MPa), moderate (−0.81 MPa), or severe (−1.60 MPa) water stress. In the apical region where elongation is maintained in water-stressed roots (Fig. 3), increased proline concentrations **(B)** resulted primarily from increased net rates of proline deposition **(D)**. In the basal region of growth inhibition (Fig. 3), conversely, increased hexose concentrations in water-stressed roots **(C)** occurred despite decreased net rates of hexose deposition **(E)** because tissue expansion, and hence water deposition, decreased to a greater extent. **A**, **C**, and **E** modified from Sharp et al. (1990), Figures 2, 4A, 6C; **B** and **D** modified from Voetberg and Sharp (1991), Figures 1A, 2A; by permission of ASPB.

growing regions and, accordingly, this response is likely to be critical for the maintenance of elongation in the apical region of water-stressed roots (Fig. 3). Subsequent studies indicated that the response is attributable to increased rates of proline import from more basal regions of the root and/or the seed (Verslues and Sharp 1999; Raymond and Smirnoff 2002). In related observations, increased levels of proline in phloem sap (Girousse et al. 1996; Lee et al. 2009) and induction of proline transporter expression (Rentsch et al. 1996; Lehmann et al. 2010) were reported in water-stressed plants. Further analysis of proline metabolism showed that under water limitation, plants can coordinate the metabolism and transport of proline in shoot and root tissues to optimize growth and redox regulation (Verslues et al. 2023). In photosynthesizing shoot tissues, proline is synthesized to regenerate oxidized NADP pools (Sharma et al. 2011). The synthesized proline can then be used for osmotic adjustment in mature tissues or transported to growing tissues for osmotic adjustment and catabolism (Sharma et al. 2011; Bhaskara et al. 2015; Verslues et al. 2023).

In contrast, because the dramatic increase in hexose concentration in the basal part of the root growth zone (Fig. 6C) was associated with inhibition of elongation in this region, this response is not so obviously adaptive—although it is emphasized that in the absence of the increased solute levels the tissues would have become significantly dehydrated. As discussed by Sharp et al. (1990), increases in solute concentrations are not an inevitable result when root growth is restricted by adverse conditions. Accordingly, it was concluded that the osmotic adjustment throughout the root growth zone likely represents an important and highly regulated process involving selective increases of specific solutes in combination with modulation of the growth pattern. More recently, multiomics analyses have provided additional insights into region-specific osmotic regulation within the growth zone of water-stressed maize and cotton primary roots (Spollen et al. 2008; Voothuluru et al. 2016; Kang et al. 2022, 2023).

An important aspect of osmotic adjustment and continuation of cell expansion in water-stressed plants is the maintenance of sink strength in growing tissues. When the root growth zone cannot obtain water from the surrounding soil, for example, during elongation into dry regions or across air gaps (see later section on lateral roots), water is delivered to the growth zone axially via the xylem and/or phloem from regions with greater water availability (upper layers after rainfall or irrigation, or via hydraulic lift from roots in deeper and wetter layers) (Boyer et al. 2010). However, functional xylem does not develop until some distance beyond the growth zone (McCully 1995), whereas phloem occurs closer to the apex and is understood to supply much of the water, along with sugars and other solutes, to support continued cell expansion under these circumstances (Bret-Harte and Silk 1994; Wiegers et al. 2009; Boyer et al. 2010; Rostamza et al. 2013). Water must then flow radially and apically via symplastic and/or apoplastic routes to the expanding cells. Due to the hydraulic resistance to water flow across the nonvascularized root tip tissues, the required growth-induced water potential gradients can be large in this situation (Fig. 5). In dry soil conditions especially, this necessitates a substantially higher capacity for osmotic adjustment in the growth zone than in adjacent mature tissues (Fig. 5; Sharp and Davies 1979; Westgate and Boyer 1985), which, in turn, requires high concentration gradients to drive growth-sustaining solute fluxes. Accordingly, mechanisms that lessen the magni-tude of the water potential gradient are predicted to facilitate continued root elongation under water stress. The shortening of the growth zone and thinning of water-stressed roots (Fig. 3) could play adaptive roles in this regard because of the associated decrease in volume of expanding tissue. Additionally, a modeling study of root tip hydraulics indicated an advantage of more apical phloem differentiation (Wiegers et al. 2009). Modulation of aquaporin-regulated water transport across cell membranes within the growth zone (Chaumont et al. 1998; Hachez et al. 2006; Gambetta et al. 2013) could also play an important role. Notably, increased abundance of a TIP aquaporin was observed in the growth zone of water-stressed maize primary roots specifically in the apical region where elongation was maintained (Voothuluru et al. 2016), and upregulated expression of several aquaporin genes in the growth zone of water-stressed maize roots has also been reported (Poroyko et al. 2007; Opitz et al. 2016).

At the same time, species or genotypes that are able to maintain or enhance source-sink allocation to roots are likely to be better adapted to growing in water-limited conditions (Hsiao and Xu 2000). For example, in a comparison of 2 maize lines with differing abilities for primary root growth maintenance under water stress, the more tolerant line exhibited greater osmotic adjustment and accumulated more sugars and proline in the root growth zone (Velázquez-Márquez et al. 2015). Notably, a plasma membrane–enriched proteomics analysis revealed that 2 sugar transporters increased in abundance in the maize primary root growth zone under water stress conditions (Voothuluru et al. 2016). In related observations in *Arabidopsis*, sugar transporter loss-of-function mutants exhibited impaired primary and lateral root development under control and water stress conditions (Valifard et al. 2021), and conversely, enhanced activity of sucrose transporters was found to increase sucrose levels in the phloem and roots together with increased root growth and enhanced root/shoot ratio under water stress (Chen et al. 2022). These studies indicate that manipulation of source strength can enhance long-distance sugar transport and promote root growth under water stress. However, the authors did not study the accumulation of sugars and osmotic adjustment in developing root tissues, and further studies are needed to assess whether an integrated phenotype of enhanced sugar transport from source tissues and maintenance of osmotic adjustment as well as enhanced sink strength within the root growth zone underlies genetic variability of root development under water-limited conditions.

Studies of several plant species have provided unequivocal evidence that osmotic adjustment is linked to increased yield under water limitation (Blum 2017; Turner 2018). Osmotic adjustment in the roots, by enabling continued root growth and exploration of the soil for water, is recognized as a likely contributing factor (Serraj and Sinclair 2002). Indeed, several early reports showed a greater depth of soil water extraction, indicative of greater rooting depths, in lines selected for high osmotic adjustment in the leaves of wheat (Morgan and Condon 1986; Morgan 1995), sorghum (Wright and Smith 1983; Tangpremsri et al. 1991), and maize (Chimenti et al. 2006). However, it is possible that the enhanced root growth resulted from greater assimilate availability and diversion to the root system due to leaf osmotic adjustment sustaining photosynthesis (Wright et al. 1983; Mervyn and Ludlow 1987) rather than from osmotic adjustment in the roots themselves. Definitive selection or metabolic engineering studies of the importance of osmotic adjustment in roots are lacking, reflecting the fact that over 50 years after first being reported (Greacen and Oh 1972) osmotic adjustment in roots remains an understudied area of investigation.

Reduced cell production in root meristems under water deficits

Organ growth can be regulated by impacting cell production and/or cell elongation processes. The previous sections focused on cell wall changes and osmotic adjustment as processes that contribute to the ability of roots to continue cell elongation under water-limited conditions. Interestingly, several studies showed that although local elongation rates were maintained in the apical region of the growth zone that encompasses the meristem, cell production was substantially reduced (Fraser et al. 1990; Saab et al. 1992; Sacks et al. 1997; Voothuluru et al. 2020; Kang et al. 2022). Reductions in cell production occurred even under moderate water stress (Longkumer et al. 2022), indicating that the effects were likely not due to insufficient availability of resources or cellular damage but, rather, represent an active restriction of meristem activity under water-limited conditions (Verslues and Longkumer 2022). This response could have an adaptive advantage for roots growing under water stress. Decreased cell production combined with maintenance of local elongation results in longer cells in the apical region of the growth zone (Sacks et al. 1997; Voothuluru et al. 2020). This response likely facilitates symplastic translocation of solutes from the phloem to the expanding cells because of the smaller number of plasmodesmata that have to be traversed, thereby helping to promote osmotic adjustment and the maintenance of cell elongation (Bret-Harte and Silk 1994; Sacks et al. 1997; Wiegers et al. 2009; Voothuluru et al. 2020). In addition, given the tendency for shortening of the growth zone in water-stressed roots (Fig. 3) and because fewer cells require less space for expansion, decreased cell production may be part of a coordinated response to reduce the energy costs of continued root elongation.

Nevertheless, it is possible that in an agricultural setting, the downregulation of cell production may be too sensitive to water stress and that overcoming this restriction could improve root growth under water-limited conditions (Verslues and Longkumer 2022). Indeed, recent studies showed that by increasing cell production using chemical or genetic approaches, it was possible to increase root elongation under water-stressed conditions. As mentioned earlier, water-stressed maize primary roots exhibit an increase in apoplastic ROS specifically in the apical region of the growth zone (Zhu et al. 2007; Voothuluru and Sharp 2013). Voothuluru et al. (2020) showed that decreasing ROS levels using scavenger treatments resulted in increased root elongation compared with control roots via promotion of both cell production and the spatial profile of cell elongation. In *Arabidopsis*, Longkumer et al. (2022) reported that mutant and transgenic lines with modified EGR-MASP1 protein stoichiometry (Clade E Growth-Regulating 2 protein phosphatase and Microtubule-Associated Stress Protein 1) exhibited enhanced meristem size and root elongation under water stress conditions. In both cases, the results showed that cell production was normally downregulated under water stress, with the result that root elongation was more inhibited than would otherwise have been the case. It remains unclear how signals from the apoplast/cytosol impact the cell cycle process in the nucleus (Voothuluru et al. 2020; Longkumer et al. 2022), and further studies are needed to identify the direct regulators of cell production and meristem size under water-limited conditions.

In contrast to the evidence for reduced cell production in moderately to severely water-stressed roots, it was reported that cell production was stimulated in the primary root of *Arabidopsis* under mild stress conditions (van der Weele et al. 2000). Also, a recent report showed that enhancement of maize lateral root length under mild water deficit was associated with sustained rates of cell production compared with well-watered controls (Dowd et al. 2020), as discussed in a later section. A future challenge will be to investigate whether specific manipulations of meristem size, cell production, and elongation in different root types can be achieved to optimize root system development under water-limited conditions that are relevant to production agriculture scenarios.

Hydrotropism and root growth responses to heterogenous soil water availability

In addition to the influence of water deficits on the rate and duration of elongation in different root types, effects of soil drying on root architecture are also determined by modifications of the direction of root elongation. For example, nodal root axes of maize (Nakamoto 1993), sorghum, and millet (Rostamza et al. 2013) exhibit phenotypic plasticity to grow more vertically in dry soil conditions, relating to the steep-angled root system ideotype discussed earlier in this review (Lynch 2013; Uga et al. 2013). Besides the possibility of enhanced gravity sensing and response, another potential contributing process to such responses is the phenomenon of hydrotropism, whereby to varying degrees the tips of plant roots can sense the moisture gradient of their surroundings and grow toward wetter areas. Hydrotropism was first observed over 150 years ago (reviewed in Takahashi 1997; Cassab et al. 2013; Dietrich 2018). Sachs (1872) and Molisch (1883) germinated pea, maize, and other seeds in a wet matrix that was suspended in the air. As soon as young roots grew out of the matrix and into the air (due to gravitropism), the roots curved and grew nearly horizontally along the bottom of the wet matrix. These results showed that the roots defied gravitropism to grow toward a water source. However, despite the potential importance of hydrotropism for water acquisition and drought tolerance, the physiological and molecular mechanisms of control have not been extensively studied.

Owing mainly to the efforts of a small number of research groups, we now have a basic understanding of the hydrotropic response (Cassab et al. 2013; Moriwaki et al. 2013; Dietrich 2018; Wang et al. 2020a). Like other responses to environmental sig-nals, hydrotropic responses can be roughly divided into 3 stages: signal sensing, transduction, and the final response. Where in the root the moisture gradient is sensed remains a topic of active investigation, as detailed below. The signal then triggers complex biochemical and physiological changes in the elongating cells of the growth zone, including differential modifications of cell wall–yielding properties in the drier and wetter sides of the root. Kinematic analysis of the maize primary root revealed that these changes result in slower cell elongation on both the drier and wetter sides of the root compared with control roots but more so on the wetter side, and thus the root curves toward the water source (Wang et al. 2020a). Interestingly, this analysis also showed that relative to control roots, cell production rate was enhanced on the drier side of the root and inhibited on the wetter side during hydrotropic bending.

The location along the root that is able to perceive a moisture gradient is a controversial topic (Dietrich 2018; Wang et al. 2020a). Conflicting results have been obtained depending on the species or different approaches used within the same species. Early studies of several species suggested that the very tip of the root is responsible, particularly the root cap (e.g. Takahashi and Scott 1993; Takano et al. 1995). However, a study of the *Arabidopsis* primary root by Dietrich et al. (2017) showed that removal of the root cap and meristem using a microdissection or laser ablation technique did not prevent the hydrotropic response, and their results indicated that the elongation zone is able to perceive the moisture gradient via a cortex-specific mechanism. In contrast, a recent study of the maize primary root by Wang et al. (2020a) used a nondestructive approach to establish a moisture gradient at specific locations along the root. The results demonstrated that the very tip of the root (apical 1.5 mm, including the cap) was the most sensitive to the moisture gradient, whereas establishing the gradient only in the zone of elongation resulted in a weaker response. This nondestructive approach presents an opportunity to conduct a wider species survey to investigate the extent to which moisture sensing in hydrotropism varies among different species.

The mechanism by which roots sense a moisture gradient remains unknown. It is probable that the response is associated with changes in cellular water status and water transport in the root tissues, and mechanosensitive ion channels (Hamilton et al. 2015) that respond to cell turgor and volume changes have been suggested as potential sensing mechanisms (Dietrich 2018). In a study of pea primary roots exhibiting hydrotropic bending (Hirasawa et al. 1997), independent turgor measurements of the 2 halves of the growth zone that were facing or facing away from the hydrostimulant (achieved by splitting the root along its axis) did not reveal differences. However, turgor was calculated indirectly from water and osmotic potential measurements of the bulk tissues and lacked spatial resolution along or across the root. Direct turgor measurements of individual cells in the outer tissue layers using a pressure microprobe, and with spatial resolution along the root, are needed to definitively address this question. Alternatively, root cells may possess specific receptors capable of directly sensing the presence of water molecules in the surrounding environment.

Many studies have depicted a complex web of signal transduction pathways involved in hydrotropism. Several hormones, including IAA, ABA, cytokinins, and brassinosteroids, have been reported to be involved (Quiroz-Figueroa et al. 2010; Dietrich et al. 2017; Miao et al. 2018, 2021; Chang et al. 2019; Wang et al. 2020a). Additionally, other regulatory factors, including calcium signaling (Takano et al. 1997; Shkolnik et al. 2016), H^+-ATPases (Miao et al. 2021), and ROS (Krieger et al. 2016), have also been found to influence hydrotropic responses. For further information on the molecular and signaling processes involved in hydrotropism, interested readers are referred to a comprehensive review by Dietrich (2018). It is not yet known how these signal transduction pathways ultimately regulate the cell production and cell elongation responses that result in hydrotropic bending.

The growth direction of roots is influenced by other processes that may interact with hydrotropism, for example, the interaction of hydrotropism with gravitropism (Takahashi et al. 2009; Dietrich 2018). Among the different hormones that are involved in hydrotropic bending, it has long been proposed that IAA plays an important role. Whether an asymmetric distribution of IAA is necessary for hydrotropic bending, however, remains an active area of investigation. Shkolnik et al. (2016) reported that asymmetric distribution of IAA, based on the fluorescence intensity of an IAA reporter protein, is not required for hydrotropism in *Arabidopsis* because an asymmetric IAA distribution was not observed before bending. Other studies, using indirect approaches such as IAA transporter inhibitors or expression of IAA responsive genes, showed that the asymmetric distribution of IAA is required but is species dependent (Nakajima et al. 2017; Fujii et al. 2018). In maize primary roots, Wang et al. (2020a) quantified hormones directly in the drier and wetter halves of the root tip during hydrotropic bending and found that IAA content, alongside ABA, was higher on the dry side compared with the wet side and that the asymmetric IAA distribution occurred before bending. The higher IAA concentration on the dry side was surprising because a higher concentration of IAA on the lower side of the root during gravitropism results in the inhibition of root cell elongation (Evans 1991; Swarup and Bennett 2009; Konstantinova et al. 2021). This example suggests that the growth control mechanisms involved in gravitropic and hydrotropic curvature may employ different mechanisms, even though both involve IAA.

Because signal sensing, transduction, and root bending encompass the entire growth zone and involve many cell types on both sides of the root, further studies will need to be conducted with spatial resolution and at the tissue- or cell-specific level. Among the tools currently available, single-cell transcriptome analysis has become a mature technology (Rodriques et al. 2019; Ryu et al. 2019) that will allow for the construction of transcript response maps along the root in different cell types and at different time points. With the aid of such comprehensive approaches combined with gene function analysis, our understanding of the complex molecular mechanisms underlying root hydrotropism is likely to advance significantly. There is also a need to develop techniques that are more sensitive than those currently available so that the responses of different root types, including lateral roots, can be characterized. These advances will allow assessment of genetically controlled traits and the biological significance of hydrotropism, particularly its potential role in enhanced water acquisition and drought tolerance.

Role of ABA in root growth responses under water deficits

The involvement of plant hormones in the regulation of plant growth responses to water deficits has been investigated for many decades (Vaadia 1976; Davies and Zhang 1991; Wilkinson and Davies 2002; Waadt et al. 2022). Even for the focus of this review on root growth responses, comprehensive coverage of the roles of different hormones, and their many interactions, is beyond the scope of the article. Instead, we provide a brief history of key discoveries of the role of ABA in root growth regulation under water deficit conditions. Among the hormones, ABA has received the most attention in this regard. However, despite the long-standing interest in the involvement of ABA in root (and shoot) growth regulation, its roles have been challenging to decipher (Sharp 2002; Humplik et al. 2017; Li et al. 2017).

Early interest in the involvement of ABA in regulating plant growth responses to water stress was stimulated because first, it often accumulates in water-stressed tissues in correlation with growth inhibition, and second, it commonly inhi-bits growth when applied to nonstressed plants. Accordingly, ABA was frequently cited as a potential cause of reductions in root and shoot growth under water-stressed conditions (Quarrie and Jones 1977; Creelman et al. 1990; reviewed in Trewavas and Jones 1991). An example is provided by the ob-servation that in well-watered maize seedlings, primary root elongation is progressively inhibited when increasing concentrations of exogenous ABA are applied (Fig. 7A; Sharp et al. 1994). However, interpretation of such findings assumes that effects of applied ABA on growth of nonstressed plants are similar to those of endogenous ABA accumulation in water-stressed plants, which may not be the case. To bypass this concern, mutants, transgenics, or chemical inhibitors can be used to examine the effects of decreasing ABA synthesis or sensitivity on the growth of water-stressed plants. However, despite the availability of ABA-deficient mutants as early as the 1970s (Imber and Tal 1970), this approach was not taken until a study of ABA-deficient maize seedlings by Saab et al. (1990, 1992). Both the *vp5* mutant, which is deficient in carotenoid (and ABA) biosynthesis, and fluridone, an inhibitor of carotenoid (and ABA) biosynthesis, were used to reduce ABA levels under water stress. To ensure that inhibition of carotenoid synthesis, rather than ABA itself, was not a cause of observed growth responses, confirmatory experiments were subsequently conducted (Sharp 2002) using the *vp14* mutant (mutated in a *9-cis-epoxycarotenoid dioxygenase* [*NCED*] gene), which is impaired in the synthesis of xanthoxin that represents the first committed step in ABA biosynthesis and is considered rate-limiting for water stress-induced ABA production (Tan et al. 1997; Qin and Zeevaart 1999). Also, the experiments were performed under conditions of min-imal transpiration (darkness and near-saturation humidity) to avoid the typical "wiltiness" of ABA-deficient plants (due to impaired stomatal function) that can confound interpretation of growth responses (Sharp et al. 2000; Sharp 2002). All 3 approaches showed that reduced ABA levels substantially inhibited primary root elongation under water stress, with a common relationship of growth inhibition to ABA deficiency in the root growth zone (Fig. 7A). Notably, the inhibition of root growth involved impairment of the normal ability to maintain cell elongation in the apical region of the growth zone (Fig. 3), resulting in further shortening of the growth zone toward the apex (Saab et al. 1992; Sharp et al. 1994). Root elongation was restored when growth zone ABA content was returned to normal levels by applying exogenous ABA (Fig. 7B; Sharp et al. 1994). These experiments revealed that rather than acting as a growth inhibitor, ABA accumulation is required for the maintenance of primary root elongation in water-stressed maize seedlings. Notably, this conclusion could not be inferred by applying ABA to nonstressed seedlings, which resulted in growth inhibition over the same range of tissue ABA levels (Fig. 7A). Accordingly, the results reveal that the root growth response to ABA accumulation was altered by the water-limited environment.

Subsequent studies have similarly reported that under water stress conditions, primary root elongation was inhibited in an ABA-deficient tomato mutant (*notabilis*, also a *NCED* mutation; Zhang et al. 2022) and in fluridone-treated *Arabidopsis* and rice seedlings (Xu et al. 2013). In addition, several studies have shown that ABA, generally at low concentrations, can promote root growth in well-watered plants, whereas higher concentrations are generally inhibitory (Li et al. 2017; Miao et al. 2021). Based on these observations, Li et al. (2017) speculated that the biphasic response to ap-plied ABA may be causally related to the promotion of root elongation that has occasionally been reported under mild water stress conditions (Triboulot et al. 1995; Maia et al. 2013), whereas root growth eventually becomes inhibited under more severe water stress (Figs. 1A and 3; Westgate and Boyer 1985; Sebastian et al. 2016). However, it is important to note that the necessity for ABA accumulation in water-stressed maize primary roots (Fig. 7, A and B) was demonstrated under severe water stress conditions (water potential of −1.6 MPa), and the highest levels of ABA occurred in the apical region of the growth zone, where local elongation rates were maintained (Fig. 3B; Saab et al. 1992). It

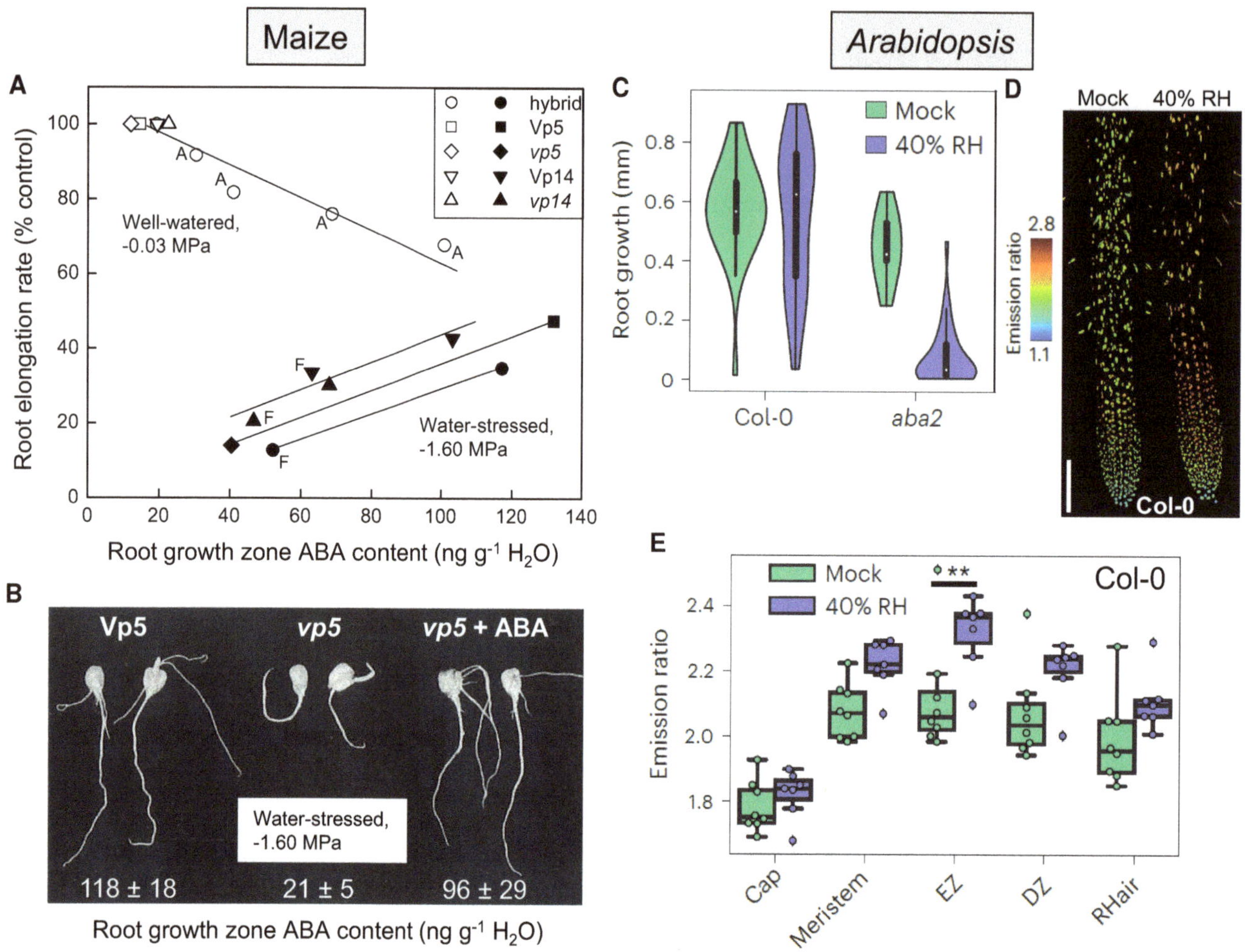

Figure 7. Increased endogenous ABA levels are necessary for maintaining root growth under water deficit conditions. **A)** Maize primary root elongation rate as a function of ABA content in the growth zone (apical 10 mm) for various genotypes growing in vermiculite under well-watered (water potential of −0.03 MPa, open symbols) or water-stressed (water potential of −1.60 MPa, closed symbols) conditions. In well-watered roots, the growth zone ABA content of hybrid (cv. FR27 × FRMo17) seedlings was raised above the normal level by adding various concentrations of ABA (A) to the vermiculite, which caused progressive inhibition of root elongation. Conversely, in water-stressed roots, the growth zone ABA content was decreased below the normal level by treatment with fluridone (F) or by using the *vp5* or *vp14* mutants (ABA deficient), which resulted in inhibition of root elongation with a common relationship of growth inhibition to ABA deficiency. Data are plotted as a percentage of the rate for the same genotype at high water potential. **B)** Recovery of elongation in water-stressed roots of the *vp5* maize mutant when growth zone ABA content was restored by applying exogenous ABA. **C)** *Arabidopsis* primary root growth was greatly reduced in the *aba2* mutant (ABA deficient) but not in the Col-0 wild-type when grown under conditions of low aerial relative humidity (40% RH) compared with a high humidity control (Mock) treatment. The roots were growing in well-watered soil. **D)** ABA accumulation in the growth zone of wild-type roots in the low humidity treatment was visualized using the ABACUS2s ABA biosensor. **E)** Relative quantification of the emission ratio signal in **(D)** in various regions of the root showed that the elongation zone (EZ) accumulates more ABA when grown in the low humidity treatment. DZ, differentiation/maturation zone; RHair, root hair zone. **A** reproduced from Sharp (2002), Figure 2, by permission of John Wiley and Sons; **B** modified from Sharp et al. (1994), Plate 2 and Table 1, by permission of Oxford University Press; **C** to **E**, modified from Rowe et al. (2023), Figure 4, CC BY 4.0.

should also be noted that in the ABA-deficient roots in which elongation under water stress was impaired, ABA levels remained much higher than in well-watered plants (Fig. 7A).

Interestingly, a recent study using the ABA-deficient *aba2* mutant of *Arabidopsis* (impaired in the conversion of xanthoxin to ABA-aldehyde) showed that increased levels of ABA in the primary root growth zone are also required for growth maintenance under conditions of low (40%) aerial relative humidity but where the roots were growing in well-watered soil (Rowe et al. 2023). ABA accumulation in the root growth zone of the wild-type (Col-0) in the low-humidity treatment was visualized using the recently developed ABACUS2s ABA biosensor (Fig. 7, D and E). The water status of the root growth zone was not measured, but effects were likely to have been minimal due to the relative hydraulic isolation of the apical region of roots growing in wet soil from lower water potentials in more shootward locations (Zwieniecki et al. 2003; Wiegers et al. 2009). In this situation, the accumulated ABA in the root growth zone likely represents delivery of shoot-sourced ABA via the phloem (McAdam et al. 2016). As shown in Fig. 7C, the low–aerial humidity treatment caused severe inhibition of root elongation in the *aba2* mutant, whereas there was no effect in the wild-type control. Thus, the results showed that increased levels of ABA in the root growth zone are required to maintain root growth even under very mild water stress conditions.

These findings raise the question of why increased levels of ABA are required for root growth maintenance under plant water deficit conditions regardless of whether the water status of the root tissues themselves changes minimally or substantially. Continuing the study of ABA deficiency (*vp5*, *vp14*, fluridone treatment) in severely water-stressed maize pri-mary roots described above, it was shown that an important role of ABA accumulation is to prevent excess ethylene production that would otherwise cause root growth inhibition (Spollen et al. 2000; Sharp 2002). In ABA-deficient seedlings under water stress, rates of ethylene evolution increased in correlation with both the degree of ABA deficiency and the inhibition of root elongation, and moreover, root elongation was restored by each of 3 inhibitors of ethylene synthesis or action (Spollen et al. 2000). These findings were consistent with an early observation that ABA-deficient tomato mutants exhibit increased ethylene production (Tal et al. 1979) and supported the idea initially suggested by Wright (1980) that ABA accumulation in water-stressed plants may function to restrict stress-induced ethylene production. This hypothesis was based on observations that pretreatment with exogenous ABA prevented wilting-induced increases in ethylene in wheat leaves. More recently, ABA-ethylene interactions have been shown to be involved in various growth responses of plants to water deficits and other abiotic stress conditions (Yang et al. 2004; Rowe et al. 2016; Valluru et al. 2016; Huang et al. 2022). Interestingly, it was recently reported that ethylene is involved in modulating shoot responses to high aerial humidity in *Arabidopsis* (Jiang et al. 2024). Accordingly, it is tempting to speculate that ABA-ethylene interactions may also be involved in the above-described ABA dependency of root growth under low–aerial humidity conditions (Fig. 7, C–E).

Research over several decades has demonstrated that ABA also plays regulatory roles in several other processes contributing to root growth responses to water deficits that were described in earlier sections of this review (Fig. 8). These processes include proline accumulation for osmotic adjustment (Ober and Sharp 1994; Sharma et al. 2011), shoot-to-root sugar transport and accumulation (Chen et al. 2022; Gong and Yang 2022), cell wall XTH activity (Wu et al. 1994), plasma membrane H^+-ATPase activity and cell wall pH regulation (Ober and Sharp 2003; Xu et al. 2013; Miao et al. 2021), ROS regulation (Zhang et al. (2014), hydraulic conductivity including aquaporin activity (Hose et al. 2000; Shahzad et al. 2024), and hydrotropism (Miao et al. 2021). In addition, ABA is involved in the regulation of suberin deposition in the root exodermis and/or endodermis (Wang et al. 2020b; Shiono et al. 2022) and can thereby impact root/soil hydraulics (Baxter et al. 2009; Kreszies et al. 2019; Cantó-Pastor et al. 2024). In many cases, ABA's function involves interactions with IAA and other hormones as well as ethylene (Xu et al. 2013; Rowe et al. 2016; Zhang et al. 2022). Further work is needed to fully decipher the interplay between ABA and other plant hormones in coordinately regulating the cellular processes involved in root growth and RSA development under water deficits.

Growth of lateral roots and root hairs under water deficits

Although this review has focused on the growth responses of root axes to water deficit conditions, the responses of lateral roots are also essential to understand. Lateral roots comprise the bulk of the root system's length, and their spatial and temporal distribution within the soil matrix has major impacts on the ability of the plant to forage for water and nutrients (Russell 1977; Ahmed et al. 2016). Hence, considerable research has focused on understanding how the production and elongation of lateral roots is affected in plants growing under various adverse environmental conditions (Waidmann et al. 2020). A brief synopsis of key responses of lateral root development to water deficits follows.

Lateral roots develop in the maturation zone of primary, sem-inal, and nodal root axes and subsequently undergo higher- order branching. In terminal drought conditions, evidence in maize indicates that the development of fewer and longer lat-eral roots on deeper nodal roots can be more efficient than shorter and numerous lateral roots that are more widely dis-tributed throughout the root system (Zhan et al. 2015). Conversely, analysis of root growth phenotypes in diverse maize lines showed that lateral root branching is a plastic response and that more prolific lateral rooting can be beneficial under intermittent irrigation conditions (Klein et al. 2020). Indeed, in-creased lateral root proliferation under soil-drying conditions (Fig. 1B) was reported a century ago by Weaver (1926). Several later studies reported that promotion of lateral root elongation and/or number occurs specifically in response to mild water deficits, whereas inhibition of lateral root develop-ment generally follows as stress becomes more severe (Read and Bartlett 1972; Ito et al. 2006; Kano et al. 2011; Dowd et al. 2019). A biphasic response of lateral root development to increasing water deficits is rational from the perspective of water uptake; growth promotion under mild water deficits facilitates access to moisture in regions where water is still available, whereas in drier soil continued lateral root development is less effective in obtaining water and, therefore, maintenance of axial growth is prioritized to access deeper and wetter soil layers. It should be noted that although many other studies have concluded that water deficits inhibit lateral root development, imposed stress levels may have been too severe or the rate of dry-down too rapid to characterize the phase of growth promotion, for which high-resolution studies at mild stress le-vels are necessary (Dowd et al. 2019).

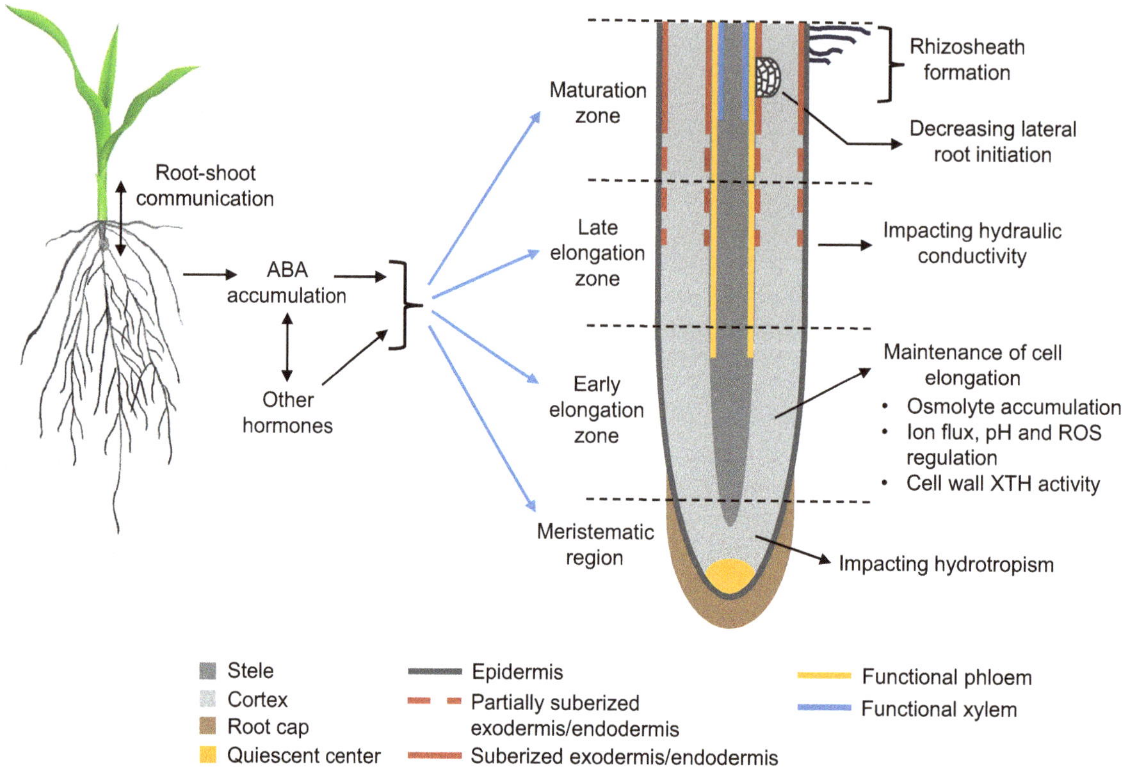

Figure 8. Schematic summarizing the effects of ABA accumulation on diverse cellular processes in different regions of roots growing under water stress conditions.

Although much is known about the regulation of lateral root formation and elongation (Motte et al. 2019; Waidmann et al. 2020), studies of mechanisms underlying increased lateral root growth under water deficits are limited. A significant aspect of lateral root development is their determinate growth pattern, whereby the meristem is genetically programmed to stop cell production at a particular stage of development, resulting in finite growth durations and root lengths (Varney and McCully 1991; Passot et al. 2018, Dowd et al. 2020). As a root approaches determinacy, exhaustion of the apical meristem results in progressive shortening of the growth zone and cell maturation closer to the apex. Dowd et al. (2020) used kinematic growth analyses to demonstrate that enhanced elongation of maize (cv. FR697) first-order lateral roots from the primary root of mildly water-stressed plants (Fig. 9A) was attributable to a delay in the determinate growth program. This was evident from sustained rates of cell flux (approximating the rate of cell production from the meristem; Fig. 9B) and repression of decreases in cell elongation and growth zone length (Fig. 9C) that occurred over time in roots of well-watered plants. Further, large genotypic variation in these responses was evident, because a contrasting genotype (B73) that did not exhibit lateral root growth promotion under water deficits also did not exhibit any changes in the determinate growth program. Interestingly, in FR697 (but not B73), mild water deficits also suppressed lateral root thinning that accompanied the progression of determinacy in well-watered roots (Dowd et al. 2020). This contrasts with water stress–induced thinning of the maize primary root (Fig. 3A). As discussed above, thinning of water-stressed root axes is thought to be adaptive, enabling the root to efficiently maintain elongation and exploration of deeper soil. The contrasting suppression of thinning in lateral roots, along with the maintenance of elongation and thus of volumetric expansion, may help to maintain root-soil contact and thereby facilitate continued water uptake from the surrounding soil.

The cellular and genetic mechanisms underlying the interaction of water deficits with lateral root determinacy are not known. A number of transcription factors, auxin transport and signaling processes, folate metabolism, and other processes are involved in regulating indeterminate-to-determinate root development (Shishkova et al. 2008; Lucas et al 2011; Reyes-Hernández et al. 2014; Rodriguez-Alonso et al. 2018). Future studies of the regulation of delayed determinacy in lateral roots may provide new opportunities to enhance root system developmental plasticity under water deficits. It is also likely that, as in primary roots, mechanisms including changes in cell wall–yielding properties, osmotic adjustment, hormonal regulation, and other processes reviewed above are also important in lateral root growth promotion under water deficits.

Another example of the plasticity of lateral root development with varying water availability is provided by the responses of lateral root formation to transient or local

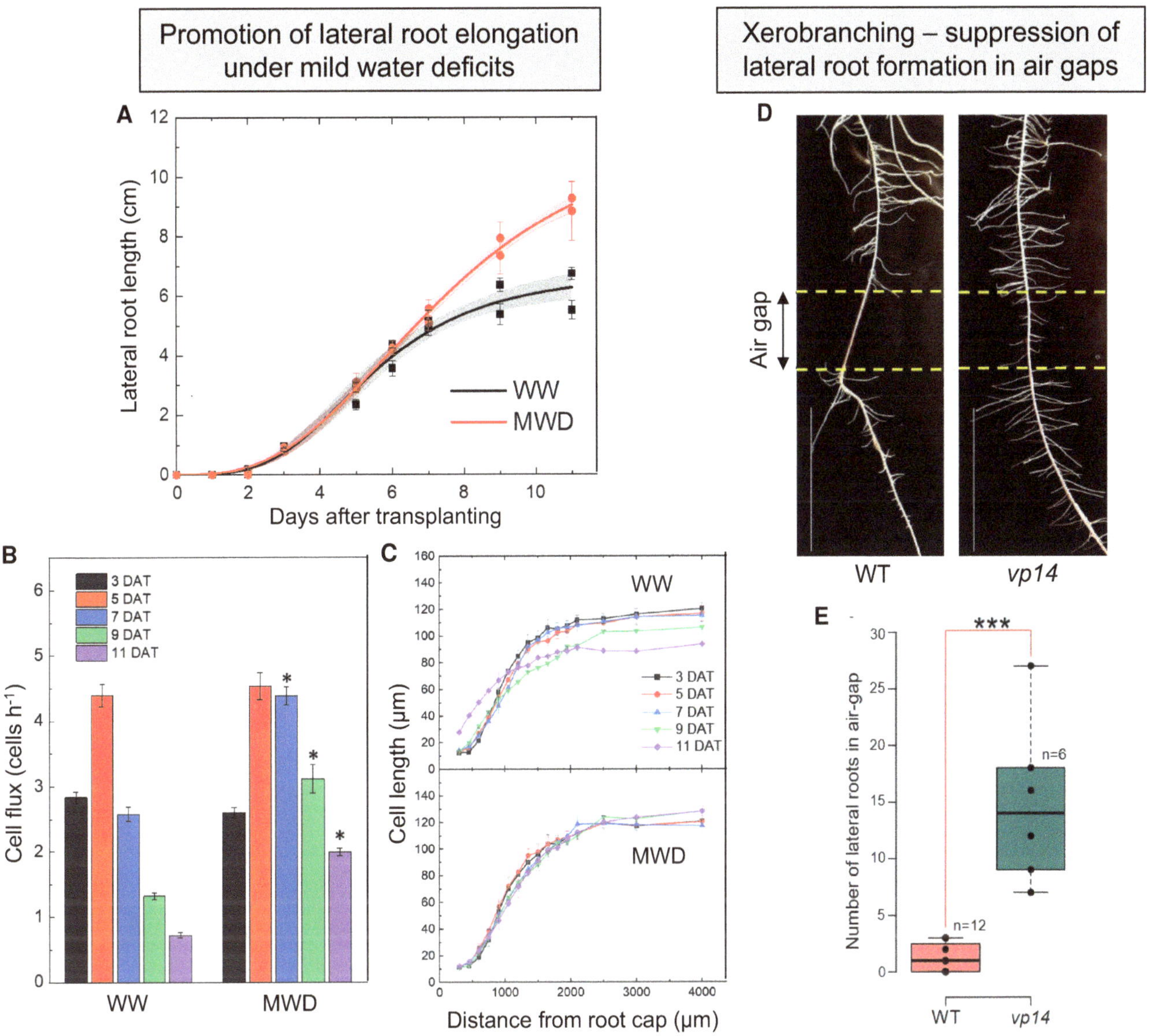

Figure 9. Impact of water availability on lateral root development in the maize primary root system. **A)** Average length, **(B)** cell flux, and **(C)** cortical cell length profiles along the growth zone of the 10 longest first-order lateral roots from the upper 15 cm of the primary root system of maize (cv. FR697) seedlings during 11 days after transplanting (DAT) to well-watered (WW) or mild water deficit (MWD, water potential of −0.28 MPa) conditions. Promotion of lateral root length in the MWD treatment was associated with delayed determinacy compared with WW roots, as evident from sustained rates of cell flux (the rate within a file that cells leave the growth zone, which under steady growth conditions equals the rate of cell production from the meristem) and repression of changes in cortical cell length profile, final cell length, and length of the growth zone that occurred in the WW roots over the course of the experiment. **D)** A xerobranching response is triggered in wild-type (WT) maize when growing root tips lose contact with water, for example, when growing across an air gap, causing repression of lateral root formation until the roots reenter moist conditions. **E)** The ABA-deficient mutant *vp14* produced a significantly higher number of lateral roots in the air gap compared with the wild-type. **A** to **C** modified from Dowd et al. (2020), Figures 3A, 5, 4, by permission of John Wiley and Sons. **D** and **E** reproduced from Mehra et al. (2022), Figure S3, by permission of AAAS.

heterogeneity of soil water. Interestingly, in contrast to the above-described promotion of lateral root development in response to mild soil water deficits, lateral root formation can be completely inhibited under otherwise moist conditions when the root axis temporarily loses contact with moisture, for example, during growth across air gaps (Fig. 9, D and E), in a process known as xerobranching (Orman-Ligeza et al. 2018). It was recently shown that the xerobranching response is regulated by hydraulic flux-responsive redistribution of ABA and IAA within the apical region of the axial root (Mehra et al. 2022). When roots enter an air gap, the phloem rather than the surrounding soil becomes the main source of water to the root growth zone, as described above. This reversal of the direction of water flow also alters the flow of phloem-derived ABA between the inner and outer tissues, triggering closure of plasmodesmata, which, in turn, decreases the inward symplastic movement of IAA and thereby inhibits lateral root formation. ABA-deficient mutants are disrupted in the xerobranching response, for example, in the *vp14* mutant of maize (Fig. 9, D and E). When the root axis regains contact with moist soil, the changes in ABA and IAA flows are attenuated and normal lateral root branching resumes.

The xerobranching response is phenotypically similar to another response of lateral root formation termed hydropatterning (Bao et al. 2014). Under conditions of heterogeneity in water availability around the circumference of the axial root, lateral root formation occurs preferentially on the root surfaces in contact with moisture and is inhibited on air-exposed surfaces. Interestingly, however, although hydropatterning also involves auxin signaling (Orosa-Puente et al. 2018), the response is independent of ABA signaling, distinguishing it mechanistically from xerobranching as well as from other root growth responses to water stress described above (Fig. 7). Nevertheless, evidence suggests that modification of internal growth-induced water potential gradients that arise from the heterogeneity in water availability around the axial root growth zone are involved in the hydropatterning signaling mechanism (Robbins and Dinneny 2018).

In addition to lateral roots, root hairs, originating from epidermal cells, greatly increase the absorbing surface area and enhance root-soil contact, and there is evidence that longer and denser root hair phenotypes are more beneficial in reducing the water potential gradient across the soil-root interface than shorter and sparer phenotypes (Carminati et al. 2017; Burak et al. 2021; Marin et al. 2021; Cai et al. 2022). Accordingly, root hairs have long been assumed to enhance root water uptake, particularly in drier soils, although experimental evidence has been contradictory (Cai and Ahmed 2022). In addition to differences between species and in soil structural parameters, recent evidence suggests that in dry soil conditions, variable loss of root hair turgidity and shrinkage may explain some of the conflicting results (Duddek et al. 2022, 2023). Although shrinkage diminishes their effectiveness, the results nevertheless indicated that root hairs can facilitate water uptake under a range of low soil water potential conditions. In this regard, it is important to evaluate the capacity for osmotic adjustment and turgor maintenance in root hairs.

Root hairs are also an important determinant of rhizosheath formation, which can also positively influence root water uptake (North and Nobel 1997). Soil water deficits have been reported to increase the length of root hairs and enhance rhizosheath formation, and evidence indicates that these responses involve ABA and IAA signaling (Zhang et al. 2020a, 2021). Root hair density has also been shown to be influenced by heterogenous water availability although, intriguingly, with an opposite response to the above-described xerobranching and hydropatterning responses of lateral root formation. In a field study of wheat, White and Kirkegaard (2010) showed that root hair density was greatest where root-soil contact within pores in the soil matrix was minimal. Similarly, in their hydropatterning studies, Bao et al. (2014) observed that root hair development was much greater on the air-exposed root surface compared with surfaces in contact with water. Perhaps the promotion of root hair density concurrent with inhibition of lateral root formation in both situations facilitates maintenance of root-soil contact with minimal metabolic cost.

Future studies are needed to assess whether there is genetic variability within a species that impacts root hair density and length and whether this variability can improve drought tolerance (Cai et al. 2022). Given that there is considerable knowledge about the genes involved in root hair production and expansion (Li et al. 2016; Salazar-Henao et al. 2016; Zhang et al. 2020b), genetic targets with enhanced root hair pheno-types might improve root water uptake under drought conditions.

Conclusions, challenges, and avenues for future research

This review has highlighted several key advances in the understanding of root growth responses to water deficits that have been made over the past century. As summarized in Fig. 10, in many cases these responses are spatially variable within different regions of the root growth zone. Certainly, as emphasized in the title of the article, roots are not so hidden anymore due to rapid acceleration of interest in root development and function among the international plant biology community (Ephrath et al. 2020). Despite these advances, there are still challenges and gaps in understanding as well as important avenues for future research.

Changes in many processes during water stress occur either sequentially or simultaneously. In many cases, the interrelationships between different responses remain poorly understood, and many of the changes may occur indirectly or secondarily. The importance of gaining greater insight into this question was highlighted by Hsiao et al. (1976, p 497) in their conclusion that "... the causes of growth responses under water stress probably will not be understood until the sequence of physiological events developing as water stress sets in is better known." This knowledge is critical to decipher causal vs consequential components of root growth responses. Further, responses to water stress occur across subcellular, cellular, tissue, and organ scales and can differ depending on the stress severity and stage of development of the plant. It is essential to demonstrate that effects observed at different levels of organization are important for the regulation of root growth responses at the whole-plant level under various water deficit conditions.

In addition to changes in the responses of individual root growth and overall RSA under water deficits per se, it is important to understand how plants respond to other stresses that co-occur with soil drying under field conditions. For example, roots generally experience increased soil strength,

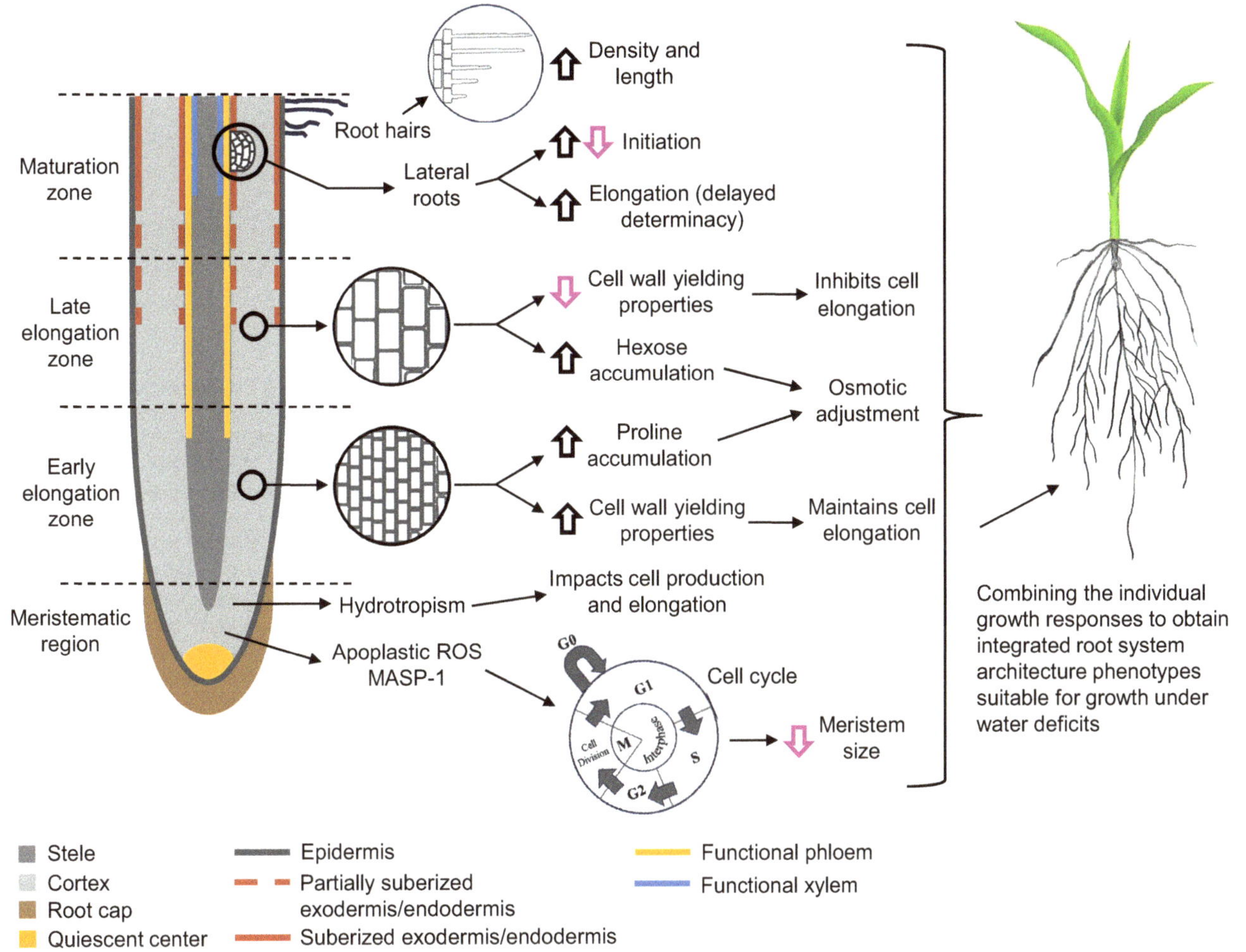

Figure 10. Schematic summarizing the effects of water stress on diverse cellular processes in different regions of roots growing under water stress conditions that eventually determine root system architecture.

which negatively impacts root elongation, simultaneously with decreased water availability (Bengough et al. 2011; Correa et al. 2019). To date, effects of water stress and soil strength on root growth, and the mechanisms underlying the responses, have generally been studied separately, although a few studies have varied soil water content and soil strength to assess the simultaneous impacts of both stresses on root elongation (Greacen and Oh 1972; Mirreh and Ketcheson 1973; Veen and Boone 1990). Importantly, the results suggest that mechanical impedance can be an important limitation to root elongation even in moderately dry soils (Bengough et al. 2011). Accordingly, it is important to examine potential interactions between mechanisms that determine the growth responses to the 2 stresses. Several mechanisms underlying root growth responses to water deficits are also involved in root growth regulation in response to soil strength. Importantly, although some mechanisms are common, for example osmotic adjustment (Greacen and Oh 1972), other processes play contrasting roles. For example, although enhanced cell wall loosening is associated with the maintenance of elongation in the apical region of water-stressed maize primary roots (Fig. 4), Schneider et al. (2021) reported that maize genotypes with thicker and more heavily lignified cortical cell walls (multiseriate cortical sclerenchyma) were better able to penetrate high-strength soils. In another example, it was recently shown that ethylene induces the synthesis of both IAA and ABA to regulate inhibition of elongation and promotion of radial expansion in rice primary roots growing in compacted soil (Huang et al. 2022). This functional pattern contrasts with the role of ABA in preventing excess ethylene production and thereby maintaining elongation in maize primary roots under water stress (Fig. 7, A and B), along with the root thinning response shown in Fig. 3A. Studies are needed to investigate how these and other regulatory processes are impacted when the 2 stresses co-occur.

A second abiotic stress factor that co-occurs with water deficits is high temperature, and future climate change scenarios predict increasing frequencies of combined drought and heat waves that can severely impact plant productivity (Zandalinas et al. 2021; Bheemanahalli et al. 2022). Recent studies with soybean plants subjected to water deficit, heat, and their combination found that leaves, flowers, and pods respond differentially to the stress combination (Sinha et al. 2023a, 2023b). Although root growth responses were not evaluated, it is likely that root tissues acclimate and respond differentially to the individual and combined stressors.

Another key area for intensified future research is the role of root exudation, root-rhizosphere and root-microbiome interactions in root growth responses to water deficits (McCully 1999; Schnepf et al. 2022). Root exudation and mucilage secretion are considered important processes for root growth, particularly in dry and hardening soil conditions (Watt et al. 1994; Bengough and McKenzie 1997; Iijima et al. 2004), and are thought to impact rhizosphere hydraulic properties (McCully and Boyer 1997; Bais et al. 2006; Kroener et al. 2014; Naylor and Coleman-Derr 2018). There is considerable evidence suggesting that root-rhizosphere interactions, particularly with arbuscular mycorrhizal fungi, enable water and nutrient absorption in normal and water deficit conditions (Augé 2001; Bárzana et al. 2014; Augé et al. 2015). Differential spatial and temporal root exudation is hypothesized to impact the diversity and strength of microbial associations (Marschner et al. 2001; Farrar et al. 2003; Watt et al. 2003; Bais et al. 2006; Voothuluru et al. 2018). However, mechanisms involved in root exudation remain largely unknown (Volkov and Schwenke 2020; Williams and de Vries 2020). A deeper mechanistic understanding of root exudation processes and their modulation by environmental stimuli is important to improve our knowledge of beneficial plant-microbial interactions and to manipulate na-tive microbial communities to enhance root and whole plant growth under water deficit conditions.

The coming decades offer tremendously exciting opportunities to build further on the strong foundation of understanding of the diversity of root growth responses to water deficits that has been summarized in this review. Ultimately, deciphering how plants integrate the effects of combined abiotic stresses and biotic interactions to coordinately regulate the diversity of spatially and temporally variable root growth responses will enable strategies for developing integrated phenotypes with RSA suitable for specific drought scenarios.

Acknowledgments

We apologize to colleagues whose work we were unable to cite due to space limitations. We thank Vicki Bryan for assistance with preparation of the manuscript. R.E.S. would like to thank William J. Davies (Lancaster University, UK), John S. Boyer (University of Missouri), Theodore C. Hsiao (University of California, Davis; 1931-2023), and Wendy K. Silk (University of California, Davis) for their mentorship and inspiration over many years.

Author contributions

All authors contributed to the conception, writing, and editing of this review.

Funding

This work was supported by a National Science Foundation EAGER grant (IOS-2318661) to P.V.

Conflict of interest statement. None declared.

Data availability

This historical review contains no new data.

References

Ahmed MA, Zarebanadkouki M, Kaestner A, Carminati A. Measurements of water uptake of maize roots: the key function of lateral roots. Plant Soil. 2016:**398**(1-2):59–77. https://doi.org/10.1007/s11104-015-2639-6

Ahmed MA, Zarebanadkouki M, Muenier F, Javaux M, Kaestner A, Carminati A. Root type matters: measurement of water uptake by seminal, crown and lateral roots in maize. J Exp Bot. 2018:**69**(5): 1199–1206. https://doi.org/10.1093/jxb/erx439

Ali O, Cheddadi I, Landrein B, Long Y. Revisiting the relationship between turgor pressure and plant cell growth. New Phytol. 2023:**238**(1):62–69. https://doi.org/10.1111/nph.18683

Augé RM. Water relations, drought and vesicular-arbuscular mycorrhizal symbiosis. Mycorrhiza. 2001:**11**(1):3–42. https://doi.org/10.1007/s005720100097

Augé RM, Toler HD, Saxton AM. Arbuscular mycorrhizal symbiosis alters stomatal conductance of host plants more under drought than under amply watered conditions: a meta-analysis. Mycorrhiza. 2015:**25**(1):13–24. https://doi.org/10.1007/s00572-014-0585-4

Avramova V, AbdElgawad H, Zhang Z, Fotschki B, Casadevall R, Vergauwen L, Knapen D, Taleisnik E, Guisez Y, Asard H, et al. Drought induces distinct growth response, protection, and recovery mechanisms in the maize leaf growth zone. Plant Physiol. 2015:**169**(2):1382–1396. https://doi.org/10.1104/pp.15.00276

Azuma T, Okita N, Nanmori T, Yasuda T. Changes in cell wall-bound phenolic acids in the internodes of submerged floating rice. Plant Prod Sci. 2005:**8**(4):441–446. https://doi.org/10.1626/pps.8.441

Bailey-Serres J, Parker JE, Ainsworth EA, Oldroyd GED, Schroeder JI. Genetic strategies for improving crop yields. Nature. 2019:**575**(7781): 109–118. https://doi.org/10.1038/s41586-019-1679-0

Bais HP, Weir TL, Perry LG, Gilroy S, Vivanco JM. The role of root exudates in rhizosphere interactions with plants and other organisms. Annu Rev Plant Biol. 2006:**57**(1):233–266. https://doi.org/10.1146/annurev.arplant.57.032905.105159

Band LR, Úbeda-Tomás S, Dyson RJ, Middleton AM, Hodgman TC, Owen MR, Jensen OE, Bennett MJ, King JR. Growth-induced hormone dilution can explain the dynamics of plant root cell elongation. Proc Natl Acad Sci U S A. 2012:**109**(19):7577–7582. https://doi.org/10.1073/pnas.1113632109

Bao Y, Aggarwal P, Robbins NE 2nd, Sturrock CJ, Thompson MC, Tan HQ, Tham C, Duan L, Rodriguez PL, Vernoux T, et al. Plant roots use a patterning mechanism to position lateral root branches toward available water. Proc Natl Acad Sci U S A. 2014:**111**(25): 9319–9324. https://doi.org/10.1073/pnas.1400966111

Bárzana G, Aroca R, Bienert GP, Chaumont F, Ruiz-Lozano JM. New insights into the regulation of aquaporins by the arbuscular mycorrhizal symbiosis in maize plants under drought stress and possible implications for plant performance. Mol Plant Microbe Interact. 2014:**27**(4):349–363. https://doi.org/10.1094/MPMI-09-13-0268-R

Baskin TI, Meekes HTHM, Liang BM, Sharp RE. Regulation of growth anisotropy in well-watered and water-stressed maize roots. II. Role of cortical microtubules and cellulose microfibrils. Plant Physiol. 1999:**119**(2):681–692. https://doi.org/10.1104/pp.119.2.681

Baskin TI, Preston S, Zelinsky E, Yang X, Elmali M, Bellos D, Wells DM, Bennett MJ. Positioning the root elongation zone is saltatory and receives input from the shoot. iScience. 2020:**23**(7):101309. https://doi.org/10.1016/j.isci.2020.101309

Basu P, Pal A, Lynch JP, Brown KM. A novel image-analysis technique for kinematic study of growth and curvature. Plant Physiol. 2007:**145**(2):305–316. https://doi.org/10.1104/pp.107.103226

Baxter I, Hosmani PS, Rus A, Lahner B, Borewitz JO, Muthukumar B, Mickelbart MV, Schreiber L, Franke RB, Salt DE. Root suberin forms an extracellular barrier that affects water relations and mineral nutrition in Arabidopsis. PLoS Genet. 2009:**5**(5):e1000492. https://doi.org/10.1371/journal.pgen.1000492

Beemster GTS, Baskin TI. Analysis of cell division and elongation underlying the developmental acceleration of root growth in *Arabidopsis thaliana*. Plant Physiol. 1998:**116**(4):1515–1526. https://doi.org/10.1104/pp.116.4.1515

Bengough AG, McKenzie BM. Sloughing of root cap cells decreases the frictional resistance to maize (*Zea mays* L.) root growth. J Exp Bot. 1997:**48**(4):885–893. https://doi.org/10.1093/jxb/48.4.885

Bengough AG, McKenzie BM, Hallett PD, Valentine TA. Root elongation, water stress, and mechanical impedance: a review of limiting stresses and beneficial root tip traits. J Exp Bot. 2011:**62**(1):59–68. https://doi.org/10.1093/jxb/erq350

Bhaskara GB, Yang T-H, Verslues PE. Dynamic proline metabolism: importance and regulation in water limited environments. Front Plant Sci. 2015:**6**:484. https://doi.org/10.3389/fpls.2015.00484

Bheemanahalli R, Ramamoorthy P, Poudel S, Samiappan S, Wijewar-dane N, Reddy KR. Effects of drought and heat stresses during reproductive stage on pollen germination, yield, and leaf reflectance properties in maize (*Zea mays* L.). Plant Direct. 2022:**6**(8): e434. https://doi.org/10.1002/pld3.434

Blum A. Osmotic adjustment is a prime drought stress adaptive engine in support of plant production. Plant Cell Environ. 2017:**40**(1):4–10. https://doi.org/10.1111/pce.12800

Bou Daher F, Chen Y, Bozorg B, Clough J, Jönsson H, Braybrook SA. Anisotropic growth is achieved through the additive mechanical effect of material anisotropy and elastic asymmetry. elife. 2018:**7**: e38161. https://doi.org/10.7554/eLife.38161

Boyer JS. Plant productivity and environment. Science. 1982:**218**(4571): 443–448. https://doi.org/10.1126/science.218.4571.443

Boyer JS. Water transport. Annu Rev Plant Physiol. 1985:**36**(1):473–516. https://doi.org/10.1146/annurev.pp.36.060185.002353

Boyer JS. Measuring the water status of plants and soils. San Diego (CA): Academic Press; 1995.

Boyer JS, Byrne P, Cassman KG, Cooper M, Delmer D, Greene T, Gruis F, Habben J, Hausmann N, Kenny N, et al. The US drought of 2012 in perspective: A call to action. Glob Food Security. 2013:**2**(3):139–143. https://doi.org/10.1016/j.gfs.2013.08.002

Boyer JS, Knipling EB. Isopiestic technique for measuring leaf water potentials with a thermocouple psychrometer. Proc Natl Acad Sci U S A. 1965:**54**(4):1044–1051. PMCID: PMC219791 https://www.ncbi.nlm.nih.gov/pmc/articles/PMC219791/

Boyer JS, Silk WK, Watt M. Path of water for root growth. Funct Plant Biol. 2010:**37**(12):1105–1116. https://doi.org/10.1071/FP10108

Brady SM, David A, Orlando DA, Lee J, Wang J-Y, Koch J, Dinneny JR, Mace D, Ohler U, Benfey PN. A high-resolution root spatiotemporal map reveals dominant expression patterns. Science. 2005:**318**(5851): 801–806. https://doi.org/10.1126/science.1146265

Bret-Harte MS, Silk WK. Nonvascular, symplastic diffusion of sucrose cannot satisfy the carbon demands of growth in the primary root tip of *Zea mays* L. Plant Physiol. 1994:**105**(1):19–33. https://doi.org/10.1104/pp.105.1.19

Burak E, Quinton JN, Dodd IC. Root hairs are the most important root trait for rhizosheath formation of barley (*Hordeum vulgare*), maize (*Zea mays*) and *Lotus japonicus* (Gifu). Ann Bot. 2021:**128**(1): 45–57. https://doi.org/10.1093/aob/mcab029

Cai G, Ahmed MA. The role of root hairs in water uptake: recent advances and future perspectives. J Exp Bot. 2022:**73**(11):3330–3338. https://doi.org/10.1093/jxb/erac114

Cai G, Ahmed MA, Abdalla M, Carminati A. Root hydraulic phenotypes impacting water uptake in drying soils. Plant Cell Environ. 2022:**45**(3):650–663. https://doi.org/10.1111/pce.14259

Campbell BM, Beare DJ, Bennett EM, Hall-Spencer JM, Ingram JSI, Jaramillo F, Ortiz R, Ramankutty N, Sayer JA, Shindell D. Agriculture production as a major driver of the Earth system exceeding planetary boundaries. Ecol Soc. 2017:**22**(4):8. https://doi.org/10.5751/ES-09595-220408

Cantó-Pastor A, Kajala K, Shaar-Moshe L, Manzano C, Timilsena P, De Bellis D, Gray S, Holbein J, Yang H, Mohammad S, et al. A suberized exodermis is required for tomato drought tolerance. Nat Plants. 2024:**10**(1):118–130. https://doi.org/10.1038/s41477-023-01567-x

Carminati A, Passioura JB, Zarebanadkouki M, Ahmed MA, Ryan PR, Watt M, Delhaize E. Root hairs enable high transpiration rates in drying soils. New Phytol. 2017:**216**(3):771–781. https://doi.org/10.1111/nph.14715

Cassab GI, Eapen D, Campos ME. Root hydrotropism: an update. Am J Bot. 2013:**100**(1):14–24. https://doi.org/10.3732/ajb.1200306

Chang J, Li X, Fu W, Wang J, Yong Y, Shi H, Ding Z, Kui H, Gou X, He K, et al. Asymmetric distribution of cytokinins determines root hydrotropism in *Arabidopsis thaliana*. Cell Res. 2019:**29**(12): 984–993. https://doi.org/10.1038/s41422-019-0239-3

Chaumont F, Barrieu F, Herman EM, Chrispeels MJ. Characterization of a maize tonoplast aquaporin expressed in zones of cell division and elongation. Plant Physiol. 1998:**117**(4):1143–1152. https://doi.org/10.1104/pp.117.4.1143

Chen Q, Hu T, Li X, Song C-P, Zhu J-K, Chen L, Zhao Y. Phosphorylation of SWEET sucrose transporters regulates plant root:shoot ratio under drought. Nat Plants. 2022:**8**(1):68–77. https://doi.org/10.1038/s41477-021-01040-7

Chimenti CA, Marcantonio M, Hall AJ. Divergent selection for osmotic adjustment results in improved drought tolerance in maize (*Zea mays* L.) in both early growth and flowering stages. Field Crops Res. 2006:**95**(2-3):305–315. https://doi.org/10.1016/j.fcr.2005.04.003

Chimungu JG, Brown KM, Lynch JP. Reduced root cortical cell file number improves drought tolerance in maize. Plant Physiol. 2014a:**166**(4):1943–1955. https://doi.org/10.1104/pp.114.249037

Chimungu JG, Brown KM, Lynch JP. Large root cortical cell size improves drought tolerance in maize (*Zea mays* L.). Plant Physiol. 2014b:**166**(4):2166–2178. https://doi.org/10.1104/pp.114.250449

Claeys H, Inzé D. The agony of choice: how plants balance growth and survival under water-limiting conditions. Plant Physiol. 2013:**162**(4): 1768–1779. https://doi.org/10.1104/pp.113.220921

Correa J, Postma JA, Watt M, Wojciechowski T. Soil compaction and the architectural plasticity of root systems. J Exp Bot. 2019:**70**(21): 6019–6134. https://doi.org/10.1093/jxb/erz383

Cosgrove DJ. Loosening of plant cell walls by expansins. Nature. 2000:**407**(6802):321–326. https://doi.org/10.1038/35030000

Cosgrove DJ. Plant cell wall extensibility: connecting plant cell growth with cell wall structure, mechanics, and the action of wall-modifying enzymes. J Exp Bot. 2016:**67**(2):463–476. https://doi.org/10.1093/jxb/erv511

Cosgrove DJ. Building an extensible cell wall. Plant Physiol. 2022:**189**(3): 1246–1277. https://doi.org/10.1093/plphys/kiac184

Coste R, Pernes M, Tetard L, Molinari M, Chabbert B. Effect of the interplay of composition and environmental humidity on the nanomechanical properties of hemp fibers. ACS Sust Chem Eng. 2020:**8**(16):6381–6390. https://doi.org/10.1021/acssuschemeng.0c00566

Creelman RA, Mason HS, Bensen RJ, Boyer JS, Mullet JE. Water deficit and abscisic acid cause differential inhibition of shoot versus root growth in soybean seedlings: analysis of growth, sugar accumulation, and gene expression. Plant Physiol. 1990:**92**(1):205–214. https://doi.org/10.1104/pp.92.1.205

Croser C, Bengough AG, Pritchard J. The effect of mechanical impedance on root growth in pea (*Pisum sativum*). I. Rates of cell flux, mitosis, and strain during recovery. Physiol Plant. 1999:**107**(3):277–286. https://doi.org/10.1034/j.1399-3054.1999.100304.x

Dardanelli JL, Bachmeier OA, Sereno R, Gil R. Rooting depth and soil water extraction patterns of different crops in a silty loam Haplustoll. Field Crops Res. 1997:**54**(1):29–38. https://doi.org/10.1016/S0378-4290(97)00017-8

Darwin F, Acton EH. Practical physiology of plants. Cambridge: Cambridge University Press; 1909.

Davies WJ, Zhang J. Root signals and the regulation of growth and development of plants in drying soil. Annu Rev Plant Physiol Plant Mol Biol. 1991:**42**(1):55–76. https://doi.org/10.1146/annurev.pp.42.060191.000415

Dietrich D. Hydrotropism: how roots search for water. J Exp Bot. 2018:**69**(11):2759–2771. https://doi.org/10.1093/jxb/ery034

Dietrich D, Pang L, Kobayashi A, Fozard JA, Boudolf V, Bhosale R, Antoni R, Nguyen T, Hiratsuka S, Fujii N, et al. Root hydrotropism is controlled via a cortex-specific growth mechanism. Nat Plants. 2017:**3**:17057. https://doi.org/10.1038/nplants.2017.57

Dinneny JR. A gateway with a guard: how the endodermis regulates growth through hormone signaling. Plant Sci. 2014:**214**:14–19. https://doi.org/10.1016/j.plantsci.2013.09.009

Dinneny JR. Developmental responses to water and salinity in root systems. Annu Rev Cell Dev Biol. 2019:**35**(1):239–257. https://doi.org/10.1146/annurev-cellbio-100617-062949

Dowd TG, Braun DM, Sharp RE. Maize lateral root developmental plasticity induced by mild water stress. I: Genotypic variation across a high-resolution series of water potentials. Plant Cell Environ. 2019:**42**(7):2259–2273. https://doi.org/10.1111/pce.13399

Dowd TG, Braun DM, Sharp RE. Maize lateral root developmental plasticity induced by mild water stress. II: Genotype-specific spatiotemporal effects on determinate development. Plant Cell Environ. 2020:**43**(10):2409–2427. https://doi.org/10.1111/pce.13840

Duddek P, Ahmed MA, Javaux M, Vanderborght J, Lovric G, KingA, Carminati A. The effect of root hairs on root water uptake is determined by root-soil contact and root hair shrinkage. New Phytol. 2023:**240**(6):2484–2497. https://doi.org/10.1111/nph.19144

Duddek P, Carminati A, Koebernick N, Ohmann L, Lovric G, Delzon S, Rodriguez-Dominguez CM, King A, Ahmed MA. The impact of drought-induced root and root hair shrinkage on root-soil contact. Plant Physiol. 2022:**189**(3):1232–1236. https://doi.org/10.1093/plphys/kiac144

Dumais J. Mechanics and hydraulics of pollen tube growth. New Phytol. 2021:**232**(4):1549–1565. https://doi.org/10.1111/nph.17722

Durand J-L, Onillon B, Schnyder H, Rademacher I. Drought effects on cellular and spatial parameters of leaf growth in tall fescue. J Exp Bot. 1995:**46**(9):1147–1155. https://doi.org/10.1093/jxb/46.9.1147

Dyson RJ, Vizcay-Barrena G, Band LR, Fernandes AN, French AP, Fozaed JA, Hodgman TC, Kenobi K, Pridmore TP, Stout M, et al. Mechanical modelling quantifies the functional importance of outer tissue layers during root elongation and bending. New Phytol. 2014:**202**(4):1212–1222. https://doi.org/10.1111/nph.12764

Eaton FM, Ergle DR. Carbohydrate accumulation in the cotton plant at low moisture levels. Plant Physiol. 1948:**23**(2):169–187. https://doi.org/10.1104/pp.23.2.169

Ephrath JE, Klien T, Sharp RE, Lazarovitch N. Exposing the hidden half: root research at the forefront of science. Plant Soil. 2020:**447**(1-2):1–5. https://doi.org/10.1007/s11104-019-04417-y

Erickson RO. Modelling of plant growth. Annu Rev Plant Physiol. 1976:**27**(1):407–434. https://doi.org/10.1146/annurev.pp.27.060176.002203

Erickson RO, Sax KB. Elemental growth rate of the primary root of *Zea mays*. Proc Am Philos Soc (Philadelphia). 1956:**100**:487–498. https://www.jstor.org/stable/3143682

Erickson RO, Silk WK. The kinematics of plant growth. Sci Am. 1980:**242**(5):134–151. https://doi.org/10.1038/scientificamerican0580-134

Eshel A, Beeckman T, editors. Plant roots: the hidden half. 4th ed. Boca Raton, FL:CRC Press; 2013, p 848.

Evans ML. Gravitropism: interaction of sensitivity modulation and effector redistribution. Plant Physiol. 1991:**95**(1):1–5. https://doi.org/10.1104/pp.95.1.1

Fan L, Linker R, Gepstein S, Tanimoto E, Yamamoto R, Neumann PM. Progressive inhibition by water deficit of cell wall extensibility and growth along the elongation zone of maize roots is related to increased lignin metabolism and progressive stelar accumulation of wall phenolics. Plant Physiol. 2006:**140**(2):603–612. https://doi.org/10.1104/pp.105.073130

Fan L, Neumann PM. The spatially variable inhibition by water deficit of maize root growth correlates with altered profiles of proton flux and cell wall pH. Plant Physiol. 2004:**135**(4):2291–2300. https://doi.org/10.1104/pp.104.041426

Farrar J, Hawes M, Jones D, Lindow S. How roots control the flux of carbon to the rhizosphere. Ecology. 2003:**84**(4):827–837. https://doi.org/10.1890/0012-9658(2003)084[0827:HRCTFO]2.0.CO;2

Fedoroff NV. The past, present and future of crop genetic modification. New Biotechnol. 2010:**27**(5):461–465. https://doi.org/10.1016/j.nbt.2009.12.004

Fraser TE, Silk WK, Rost TL. Effects of low water potential on cortical cell length in growing regions of maize roots. Plant Physiol. 1990:**93**(2):648–651. https://doi.org/10.1104/pp.93.2.648

Frensch J, Hsiao TC. Rapid response of the yield threshold and turgor regulation during adjustment of root growth to water stress in *Zea mays*. Plant Physiol. 1995:**108**(1):303–312. https://doi.org/10.1104/pp.108.1.303

Freschet GT, Pagès L, Iversen CM, Comas LH, Rewald B, Roumet C, Klimešová J, Zadworny M, Poorter H, Postma JA, et al. A starting guide to root ecology: strengthening ecological concepts and standardising root classification, sampling, processing and trait measurements. New Phytol. 2021:**232**(3):973–1122. https://doi.org/10.1111/nph.17572

Fry SC, Smith RC, Renwick KF, Martin DJ, Hodge SK, Matthews KJ. Xyloglucan endotransglycosylase, a new wall-loosening enzyme activity from plants. Biochem J. 1992:**282**(3):821–828. https://doi.org/10.1042/bj2820821

Fujii N, Miyabayashi S, Sugita T, Kobayashi A, Yamazaki C, Miyazawa Y, Kamada M, Kasahara H, Osada I, Shimazu T, et al. Root-tip-mediated inhibition of hydrotropism is accompanied with the suppression of asymmetric expression of auxin-inducible genes in response to moisture gradients in cucumber roots. PLoS One. 2018:**13**(1):e0189827. https://doi.org/10.1371/journal.pone.0189827

Fukai S, Cooper M. Development of drought-resistant cultivars using physio-morphological traits in rice. Field Crop Res. 1995:**40**(2):67–86. https://doi.org/10.1016/0378-4290(94)00096-U

Gabay G, Wang H, Zhang J, Moriconi JI, Burguener GF, Gualano LD, Howell T, Lukaszewski A, Staskawicz B, Cho MJ, et al. Dosage differences in *12-OXOPHYTODIENOATE REDUCTASE* genes modulate wheat root growth. Nat Commun. 2023:**14**(1):539. https://doi.org/10.1038/s41467-023-36248-y

Gabay G, Zhang J, Burguener GF, Howell T, Wang H, Fahima T, Lukaszewski A, Moriconi JI, Santa Maria GE, Dubcovsky J. Structural rearrangements in wheat (1BS)-rye (1RS) recombinant chromosomes affect gene dosage and root length. Plant Genome. 2021:**14**(1):e20079. https://doi.org/10.1002/tpg2.20079

Gambetta GA, Fei J, Rost TL, Knipfer T, Matthews MA, Shackel KA, Walker MA, McElrone AJ. Water uptake along the length of grapevine fine roots: developmental anatomy, tissue-specific aquaporin expression, and pathways of water transport. Plant Physiol. 2013:**163**(3):1254–1265. https://doi.org/10.1104/pp.113.221283

Ge L, Chen R. Negative gravitropism in plant roots. Nat Plants. 2016:**2**(11):16155. https://doi.org/10.1038/nplants.2016.155

Girousse C, Bournoville R, Bonnemain J-L. Water deficit-induced changes in concentrations in proline and some other amino acids in the phloem sap of alfalfa. Plant Physiol. 1996:**111**(1):109–113. https://doi.org/10.1104/pp.111.1.109

Gong Z, Yang S. Drought meets SWEET. Nat Plants. 2022:**8**(1):25–26. https://doi.org/10.1038/s41477-021-01032-7

Goodwin RH. Studies on roots. V. Effects of indoleacetic acid on the standard root growth pattern of *Phleum pratense*. Bot Gaz. 1972: **133**(3):224–229. https://doi.org/10.1086/336637

Goodwin RH, Avers CJ. Studies on roots. III. An analysis of root growth in Phleum pratense using photomicrographic records. Am J Bot. 1956:**43**(7):479–487. https://doi.org/10.1002/j.1537-2197.1956.tb10521.x

Gowda VR, Henry A, Yamauchi A, Shashidhar HE, Serraj R. Root biology and genetic improvement for drought avoidance in rice. Field Crops Res. 2011:**122**(1):1–13. https://doi.org/10.1016/j.fcr.2011.03.001

Greacen EL, Oh JS. Physics of root growth. Nature. 1972:**235**(53):24–25. https://doi.org/10.1038/newbio235024a0

Green PB. Growth physics in *Nitella*: a method for continuous *in vivo* analysis of extensibility based on a micro-manometer technique for turgor pressure. Plant Physiol. 1968:**43**(8):1169–1184. https://doi.org/10.1104/pp.43.8.1169

Green PB, Erickson RO, Buggy J. Metabolic and physical control of cell elongation rate—*in vivo* studies in *Nitella*. Plant Physiol. 1971:**47**(3): 423–430. https://doi.org/10.1104/pp.47.3.423

Gregory PJ, editor. Understanding and improving crop root function. Cambridge: Burleigh Dodds Science Publishing; 2021. 440 p.

Guseman JM, Webb K, Srinivasan C, Dardick C. *DRO1* influences root system architecture in Arabidopsis and Prunus species. Plant J. 2017:**89**(6):1093–1105. https://doi.org/10.1111/tpj.13470

Hachez C, Moshelion M, Zelazny E, Cavez D, Chaumont F. Localization and quantification of plasma membrane aquaporin expression in maize primary root: a clue to understanding their role as cellular plumbers. Plant Mol Biol. 2006:**62**(1-2):305–323. https://doi.org/10.1007/s11103-006-9022-1

Hager A. Role of the plasma membrane H+-ATPase in auxin-induced elongation growth: historical and new aspects. J Plant Res. 2003:**116**(6):483–505. https://doi.org/10.1007/s10265-003-0110-x

Hamilton ES, Schlegel AM, Haswell ES. United in diversity: mechanosensitive ion channels in plants. Ann Rev Plant Biol. 2015:**66**(1): 113–137. https://doi.org/10.1146/annurev-arplant-043014-114700

Hatfield RD, Jung H, Marita JM, Kim H. Cell wall characteristics of a maize mutant selected for decreased ferulates. Am J Plant Sci. 2018:**9**(3):446–466. https://doi.org/10.4236/ajps.2018.93034

Hejnovicz Z. The response of the different parts of the cell elongation zone in root to external β-indolylacetic acid. Acta Soc Bot Pol. 1961:**30**(1):25–42. https://doi.org/10.5586/asbp.1961.003

Hendrickson AH, Veihmeyer FJ. Influence of dry soil on root extension. Plant Physiol. 1931:**6**(3):567–576. https://doi.org/10.1104/pp.6.3.567

Hirasawa T, Takahashi H, Suge H, Ishihara K. Water potential, turgor and cell wall properties in elongating tissues of the hydrotropically bending roots of pea (*Pisum sativum* L.). Plant Cell Environ. 1997:**20**(3): 381–386. https://doi.org/10.1046/j.1365-3040.1997.d01-70.x

Hochholdinger F, Woll K, Sauer MA, Dembinsky D. Genetic dissection of root formation in maize (*Zea mays*) reveals root-type specific developmental programmes. Ann Bot. 2004:**93**(4):359–368. https://doi.org/10.1093/aob/mch056

Hochholdinger F, Yu P, Marcon C. Genetic control of root system development in maize. Trends Plant Sci. 2018:**23**(1):79–88. https://doi.org/10.1016/j.tplants.2017.10.004

Hose E, Steudle E, Hartung W. Abscisic acid and hydraulic conductivity of maize roots: a study using cell- and root-pressure probes. Planta. 2000:**211**(6):874–882. https://doi.org/10.1007/s004250000412

Hsiao TC. Plant responses to water stress. Annu Rev Plant Physiol. 1973:**24**(1):519–570. https://doi.org/10.1146/annurev.pp.24.060173.002511

Hsiao TC, Acevedo E, Fereres E, Henderson DW. Water stress, growth, and osmotic adjustment. Philos Trans R Soc Lond B Biol Sci. 1976:**273**(927):479–500. https://doi.org/10.1098/rstb.1976.0026

Hsiao TC, Jing J. Leaf and root expansive growth in response to water deficits. In: **Cosgrove DJ, Knievel DP**, editors. Physiology of cell expansion during plant growth: proceedings of the second annual Penn State symposium in plant physiology (May 21–23, 1987). Rockville (MD): American Society of Plant Physiologists; 1987. p. 180–192.

Hsiao TC, Silk WK, Jing J. Leaf growth and water deficits: biophysical effects. In: **Baker NR, Davies WJ, Ong CK**, editors. Control of leaf growth, SEB seminar 27. Cambridge: Cambridge University Press; 1985. p. 239–266.

Hsiao TC, Xu L-K. Sensitivity of growth of roots versus leaves to water stress: biophysical analysis and relation to water transport. J Exp Bot. 2000:**51**(350):1595–1616. https://doi.org/10.1093/jexbot/51.350.1595

Huang G, Kilic A, Karady M, Zhang J, Mehra P, Song X, Sturrock CJ, Zhu W, Qin H, Hartman S, et al. Ethylene inhibits rice root elongation in compacted soil via ABA- and auxin-mediated mechanisms. Proc Natl Acad Sci U S A. 2022:**119**(30):e2201072119. https://doi.org/10.1073/pnas.2201072119

Humplík JF, Bergougnoux V, Van Volkenburgh E. To stimulate or inhibit? That is the question for the function of abscisic acid. Trends Plant Sci. 2017:**22**(10):830–841. https://doi.org/10.1016/j.tplants.2017.07.009

Hunter AS, Kelley OJ. The extension of plant roots into dry soil. Plant Physiol. 1946:**21**(4):445–451. https://doi.org/10.1104/pp.21.4.445

Husken D, Steudle E, Zimmermann U. Pressure probe technique for measuring water relations of cells in higher plants. Plant Physiol. 1978:**61**(2):158–163. https://doi.org/10.1104/pp.61.2.158

Iijima M, Higuchi T, Barlow PW. Contribution of root cap mucilage and presence of an intact root cap in maize (*Zea mays*) to the reduction of soil mechanical impedance. Ann Bot. 2004:**94**(3):473–477. https://doi.org/10.1093/aob/mch166

Iljin WS. Drought resistance in plants and physiological processes. Annu Rev Plant Physiol. 1957:**8**(1):257–274. https://doi.org/10.1146/annurev.pp.08.060157.001353

Imber D, Tal M. Phenotypic reversion of *flacca*, a wilty mutant of tomato, by abscisic acid. Science. 1970:**169**(3945):592–593. https://doi.org/10.1126/science.169.3945.592

Ito K, Tanakamaru K, Morita S, Abe J, Inanaga S. Lateral root development, including responses to soil drying, of maize (*Zea mays*) and wheat (*Triticum aestivum*) seminal roots. Physiol Plant. 2006:**127**(2):260–267. https://doi.org/10.1111/j.1399-3054.2006.00657.x

Jefferies RA. Cultivar responses to water stress in potato: effects of shoot and roots. New Phytol. 1993:**123**(3):491–498. https://doi.org/10.1111/j.1469-8137.1993.tb03761.x

Jiang Z, Yao L, Zhu X, Hao G, Ding Y, Zhao H, Wang S, Wen C-K, Xu X, Xin X-F. Ethylene signaling modulates air humidity responses in plants. Plant J. 2024:**117**(3):653–668. https://doi.org/10.1111/tpj.16556

Juenger TE, Verslues PE. Time for a drought experiment: do you know your plants' water status? Plant Cell. 2023:**35**(1):10–23. https://doi.org/10.1093/plcell/koac324

Kang J, Sen S, Oliver MJ, Sharp RE. Comparative transcriptomics reveal metabolic rather than genetic control of divergent antioxidant metabolism in the primary root elongation zone of water-stressed cotton and maize. Antioxidants (Basel). 2023:**12**(2):287. https://doi.org/10.3390/antiox12020287

Kang J, Voothuluru P, Hoyos-Miernyk E, Alexander D, Oliver MJ, Sharp RE. Antioxidant metabolism underlies different metabolic strategies for primary root growth maintenance under water stress in cotton and maize. Antioxidants (Basel). 2022:**11**(5):820. https://doi.org/10.3390/antiox11050820

Kano M, Inukai Y, Kitano H, Yamauchi A. Root plasticity as the key root trait for adaptation to various intensities of drought stress in rice. Plant Soil. 2011:**342**(1-2):117–128. https://doi.org/10.1007/s11104-010-0675-9

Karlova R, Boer D, Hayes S, Testerink C. Root plasticity under abiotic stress. Plant Physiol. 2021:**187**(3):1057–1070. https://doi.org/10.1093/plphys/kiab392

Kashiwagi J, Krishnamurthy L, Purushothaman R, Upadhyaya HD, Gaur PM, Gowda CLL, Ito O, Varshney RK. Scope for improvement of yield under drought through the root traits in chickpea (*Cicer arietinum* L). Field Crop Res. 2015:**170**:47–54. https://doi.org/10.1016/j.fcr.2014.10.003

Kirkegaard JA, Lilley JM, Howe GN, Graham JM. Impact of subsoil water use on wheat yield. Aust J Agric Res. 2007:**58**(4):303–315. https://doi.org/10.1071/AR06285

Kitomi Y, Hanzawa E, Kuya N, Inoue H, Hara N, Kawai S, Kanno N, Endo M, Sugimoto K, Yamazaki T, et al. Root angle modifications by the *DRO1* homolog improve rice yields in saline paddy fields. Proc Natl Acad Sci U S A. 2020:**117**(35):21242–21250. https://doi.org/10.1073/pnas.2005911117

Klein SP, Schneider HM, Perkins AC, Brown KM, Lynch JP. Multiple integrated root phenotypes are associated with improved drought tolerance. Plant Physiol. 2020:**83**(3):1011–1025. https://doi.org/10.1104/pp.20.00211

Klepper B, Taylor HM, Huck MG, Fiscus EL. Water relations and growth of cotton in drying soil. Agron J. 1973:**65**(2):307–310. https://doi.org/10.2134/agronj1973.00021962006500020036x

Konstantinova N, Korbei B, Luschnig C. Auxin and root gravitropism: addressing basic cellular processes by exploiting a defined growth response. Int J Mol Sci. 2021:**22**(5):2749. https://doi.org/10.3390/ijms22052749

Kozlova L, Petrova A, Ananchenko B, Gorshkova T. Assessment of primary cell wall nanomechanical properties in internal cells of non-fixed maize roots. Plants (Basel). 2019:**8**(6):172. https://doi.org/10.3390/plants8060172

Kramer PJ, Boyer JS. Water relations of plants and soils. San Diego, (CA): Academic Press; 1995.

Kreszies T, Shellakkutti N, Osthoff A, Yu P, Baldauf JA, Zeisler-Diehl VV, Ranathunge K, Hochholdinger F, Schreiber L. Osmotic stress enhances suberization of apoplastic barriers in barley seminal roots: analysis of chemical, transcriptomic and physiological responses. New Phytol. 2019:**221**(1):180–194. https://doi.org/10.1111/nph.15351

Krieger G, Shkolnik D, Miller G, Fromm H. Reactive oxygen species tune root tropic responses. Plant Physiol. 2016:**172**(2):1209–1220. https://doi.org/10.1104/pp.16.00660

Kroener E, Zarebanadkouki M, Kaestner A, Carminati A. Non-equilibrium water dynamics in the rhizosphere: how mucilage affects water flow in soils. Water Resour Res. 2014:**50**(8):6479–6495. https://doi.org/10.1002/2013WR014756

Kutschera U, Briggs WR. Growth, *in vivo* extensibility, and tissue tension in developing pea internodes. Plant Physiol. 1988:**86**(1):306–311. https://doi.org/10.1104/pp.86.1.306

Kutschera U, Niklas KJ. The epidermal-growth-control theory of stem elongation: an old and a new perspective. J Plant Physiol. 2007:**164**(11):1395–1409. https://doi.org/10.1016/j.jplph.2007.08.002

Kuzmanoff KM, Evans ML. Kinetics of adaptation to osmotic stress in lentil (*Lens culinaris* Med.) roots. Plant Physiol. 1981:**68**(1):244–247. https://doi.org/10.1104/pp.68.1.244

Lee B-R, Jin YL, Avice J-C, Cliquet J-B, Ourry A, Kim T-H. Increased proline loading to phloem and its effects on nitrogen uptake and assimilation in water-stressed white clover (*Trifolium repens*). New Phytol. 2009:**182**(3):654–663. https://doi.org/10.1111/j.1469-8137.2009.02795.x

Lehmann S, Funck D, Szabados L, Rentsch D. Proline metabolism and transport in plant development. Amino Acids. 2010:**39**(4):949–962. https://doi.org/10.1007/s00726-010-0525-3

Li L, Hey S, Liu S, Liu Q, McNinch C, Hu H-C, Wen T-J, Marcon C, Paschold A, Bruce W, et al. Characterization of maize *roothairless6* which encodes a D-type cellulose synthase and controls the switch from bulge formation to tip growth. Sci Rep. 2016:**6**(1):34395. https://doi.org/10.1038/srep34395

Li X, Chen L, Forde BG, Davies WJ. The biphasic root growth response to abscisic acid in Arabidopsis involves interaction with ethylene and auxin signalling pathways. Front Plant Sci. 2017:**8**:1493. https://doi.org/10.3389/fpls.2017.01493

Liang BM, Sharp RE, Baskin TI. Regulation of growth anisotropy in well-watered and water-stressed maize roots. I. Spatial distribution of longitudinal, radial and tangential expansion rates. Plant Physiol. 1997:**115**(1):101–111. https://doi.org/10.1104/pp.115.1.101

Lilley JM, Kirkegaard JA. Seasonal variations in the value of subsoil water to wheat: simulation studies in southern New South Wales. Aust J Agric Res. 2007:**58**(12):1115–1128. https://doi.org/10.1071/AR07046

Lockhart JA. An analysis of irreversible plant cell elongation. J Theor Biol. 1965:**8**(2):264–275. https://doi.org/10.1016/0022-5193(65)90077-9

Longkumer T, Chen C-Y, Biancucci M, Bhaskara GB, Verslues PE. Spatial differences in stoichiometry of EGR phosphatase and microtubule-associated stress protein 1 control root meristem activity during drought stress. Plant Cell. 2022:**34**(2):742–758. https://doi.org/10.1093/plcell/koab290

Lopes MS, Reynolds MP. Partitioning of assimilates to deeper roots is associated with cooler canopies and increased yield under drought in wheat. Funct Plant Biol. 2010:**37**(2):147–156. https://doi.org/10.1071/FP09121

Lucas M, Swarup R, Paponov IA, Swarup K, Casimiro I, Lake D, Peret B, Zappala S, Mairhofer S, Whitworth M, et al. SHORT-ROOT regulates primary, lateral and adventitious root development in Arabidopsis. Plant Physiol. 2011:**155**(1):384–398. https://doi.org/10.1104/pp.110.165126

Lynch JP. Steep, cheap and deep: an ideotype to optimize water and N acquisition by maize root systems. Ann Bot. 2013:**112**(2):347–357. https://doi.org/10.1093/aob/mcs293

Lynch JP. Rightsizing root phenotypes for drought resistance. J Exp Bot. 2018:**69**(13):3279–3292. https://doi.org/10.1093/jxb/ery048

Lynch JP. Harnessing root architecture to address global challenges. Plant J. 2022:**109**(2):415–431. https://doi.org/10.1111/tpj.15560

MacAdam JW, Grabber JH. Relationship of growth cessation with the formation of diferulate cross-links and *p*-coumaroylated lignins in tall fescue leaf blades. Planta. 2002:**215**(5):785–793. https://doi.org/10.1007/s00425-002-0812-7

Maia JM, Voigt EL, Ferreira-Silva SL, Fontenele AV, Macêdo CEC, Silveira JAG. Differences in cowpea root growth triggered by salinity and dehydration are associated with oxidative modulation involving types I and III peroxidases and apoplastic ascorbate. J Plant Growth Regul. 2013:**32**(2):376–387. https://doi.org/10.1007/s00344-012-9308-2

Marin M, Feeney DS, Brown LK, Naveed M, Ruiz S, Koebernick N, Bengough AG, Hallett PD, Roose T, Puértolas J, et al. Significance of root hairs for plant performance under contrasting field conditions and water deficit. Ann Bot. 2021:**128**(1):1–16. https://doi.org/10.1093/aob/mcaa181

Marschner P, Yang C-H, Lieberei R, Crowley DE. Soil and plant specific effects on bacterial community composition in the rhizosphere. Soil Biol Biochem. 2001:**33**(11):1437–1445. https://doi.org/10.1016/S0038-0717(01)00052-9

Martin JH, Harris JA, Jones ID. Freezing-point depression and specific conductivity of sorghum tissue fluids. J Agric Res. 1931:**42**:57–69.

Matthews MA, Van Volkenburgh E, Boyer JS. Acclimation of leaf growth to low water potentials in sunflower. Plant Cell Environ. 1984:**7**(3):199–206. https://doi.org/10.1111/1365-3040.ep11614641

McAdam SAM, Brodribb TJ, Ross JJ. Shoot-derived abscisic acid promotes root growth. Plant Cell Environ. 2016:**39**(3):652–659. https://doi.org/10.1111/pce.12669

McCully M. How do real roots work? Some new views of root structure. Plant Physiol. 1995:**109**(1):1–6. https://doi.org/10.1104/pp.109.1.1

McCully ME. ROOTS IN SOIL: unearthing the complexities of roots and their rhizospheres. Annu Rev Plant Physiol Plant Mol Biol. 1999:**50**(1): 695–718. https://doi.org/10.1146/annurev.arplant.50.1.695

McCully ME, Boyer JS. The expansion of maize root-cap mucilage during hydration. 3. Changes in water potential and water content. Physiol Plant. 1997:**99**(1):169–177. https://doi.org/10.1111/j.1399-3054.1997.tb03445.x

McQueen-Mason S, Durachko DM, Cosgrove DJ. Two endogenous proteins that induce cell wall extension in plants. Plant Cell. 1992:**4**(11):1425–1433. https://doi.org/10.1105/tpc.4.11.1425

Mehra P, Pandey BK, Melebari D, Banda J, Leftley N, Couvreur V, Rowe J, Anfang M, De Gernier H, Morris E, et al. Hydraulic flux-responsive hormone redistribution determines root branching. Science. 2022:**378**(6621):762–768. https://doi.org/10.1126/science.add3771

Mervyn M, Ludlow MM. Contribution of osmotic adjustment to the maintenance of photosynthesis during water stress. In: **Biggins J**, editor. Progress in photosynthesis research. New York: Springer; 1987. p. 161–168.

Meyer RF, Boyer JS. Osmoregulation, solute distribution, and growth in soybean seedlings having low water potentials. Planta. 1981:**151**(5): 482–489. https://doi.org/10.1007/BF00386543

Miao R, Wang M, Yuan W, Ren Y, Li Y, Zhang N, Zhang J, Kronzucker HJ, Xu W. Comparative analysis of Arabidopsis ecotypes reveals a role for brassinosteroids in root hydrotropism. Plant Physiol. 2018:**176**(4): 27202736. https://doi.org/10.1104/pp.17.01563

Miao R, Yuan W, Wang Y, Garcia-Maquilon I, Dang X, Li Y, Zhang J, Zhu Y, Rodriguez PL, Xu W. Low ABA concentration promotes root growth and hydrotropism through relief of ABA INSENSITIVE 1-mediated inhibition of plasma membrane H^+-ATPase 2. Sci Adv. 2021:**7**(12):eabd4113. https://doi.org/10.1126/sciadv.abd4113

Mirreh HF, Ketcheson JW. Influence of soil water matric potential and resistance to penetration on corn root elongation. Can J Soil Sci. 1973:**53**(4):383–388. https://doi.org/10.4141/cjss73-055

Molisch H. Untersuchungen über den hydrotropismus. Sitzungsberichte Akad Wiss Wien. 1883:**88**:897–943. https://www.zobodat.at/pdf/SBAWW_88_0897-0943.pdf

Molz FJ, Boyer JS. Growth-induced water potentials in plant cells and tissues. Plant Physiol. 1978:**62**(3):423–429. https://doi.org/10.1104/pp.62.3.423

Morgan JM. Osmoregulation and water stress in higher plants. Annu Rev Plant Physiol. 1984:**35**(1):299–319. https://doi.org/10.1146/annurev.pp.35.060184.001503

Morgan JM. Growth and yield of wheat lines with differing osmoregulative capacity at high soil water deficit in seasons of varying evaporative demand. Field Crops Res. 1995:**40**(3):143–152. https://doi.org/10.1016/0378-4290(94)00100-Q

Morgan JM, Condon AG. Water use, grain yield, and osmoregulation in wheat. Aust J Plant Physiol. 1986:**13**(4):523–532. https://doi.org/10.1071/PP9860523

Moriwaki T, Miyazawa Y, Kobayashi A, Takahashi H. Molecular mechanisms of hydrotropism in seedling roots of *Arabidopsis thaliana* (Brassicaceae). Am J Bot. 2013:**100**(1):25–34. https://doi.org/10.3732/ajb.1200419

Moss GI, Hall KC, Jackson MB. Ethylene and the responses of roots of maize (*Zea mays* L.) to physical impedance. New Phytol. 1988:**109**(3): 303–311. https://doi.org/10.1111/j.1469-8137.1988.tb04199.x

Motte H, Vanneste S, Beeckman T. Molecular and environmental regulation of root development. Annu Rev Plant Biol. 2019:**70**(1): 465–488. https://doi.org/10.1146/annurev-arplant-050718-100423

Munns R. Why measure osmotic adjustment? Aust J Plant Physiol. 1988:**15**:717–726. https://doi.org/10.1071/PP9880717

Nakajima Y, Nara Y, Kobayashi A, Sugita T, Miyazawa Y, Fujii N, Takahashi H. Auxin transport and response requirements for root hydrotropism differ between plant species. J Exp Bot. 2017:**68**(13): 3441–3456. https://doi.org/10.1093/jxb/erx193

Nakamoto T. Effect of soil water content on the gravitropic behavior of nodal roots in maize. Plant Soil. 1993:**152**(2):261–267. https://doi.org/10.1007/BF00029096

Naylor D, Coleman-Derr D. Drought stress and root-associated bacterial communities. Front Plant Sci. 2018:**8**:2223. https://doi.org/10.3389/fpls.2017.02223

Nelissen H, Sun X-H, Rymen B, Jikumaru Y, Kojima M, Takebayashi Y, Abbeloos R, Demuynck K, Storme V, Vuylsteke M, et al. The reduction in maize leaf growth under mild drought affects the transition between cell division and cell expansion and cannot be restored by elevated gibberellic acid levels. Plant Biotechnol J. 2018:**16**(2): 615–627. https://doi.org/10.1111/pbi.12801

North GB, Nobel PS. Drought-induced changes in soil contact and hydraulic conductivity for roots of *Opuntia ficus-indica* with and without rhizosheaths. Plant Soil. 1997:**191**(2):249–258. https://doi.org/10.1023/A:1004213728734

Ober ES, Sharp RE. Proline accumulation in maize (*Zea mays* L.) primary roots at low water potentials. I. Requirement for increased levels of abscisic acid. Plant Physiol. 1994:**105**(3):981–987. https://doi.org/10.1104/pp.105.3.981

Ober ES, Sharp RE. Electrophysiological responses of maize roots to low water potentials: relationship to growth and ABA accumulation. J Exp Bot. 2003:**54**(383):813–824. https://doi.org/10.1093/jxb/erg060

Ober ES, Sharp RE. Regulation of root growth responses to water deficit. In: **Jenks MA, Hasegawa PM, Jain SM**, editors. Advances in molecular breeding toward drought and salt tolerant crops. Dortrecht, (The Netherlands): Springer; 2007. p. 33–53.

Ober ES, Sharp RE. Maintaining root growth in drying soil: a review of progress and gaps in understanding. In: **Eshel A, Beekman T**, editors. Plant roots: the hidden half. 4th ed. New York: CRC Press; 2013. Chapter 35. p. 1–11.

Opitz N, Marcon C, Paschold A, Malik WA, Lithio A, Brandt R, Piepho H-P, Nettleton D, Hochholdinger F. Extensive tissue-specific transcriptomic plasticity in maize primary roots upon water deficit. J Exp Bot. 2016:**67**(4):1095–1107. https://doi.org/10.1093/jxb/erv453

Orman-Ligeza B, Morris EC, Parizot B, Lavigne T, Babé A, Ligeza A, Klein S, Sturrock C, Xuan W, Novák O, et al. The xerobranching response represses lateral root formation when roots are not in contact with water. Curr Biol. 2018:**28**(19):3165–3173. https://doi.org/10.1016/j.cub.2018.07.074

Orosa-Puente B, Leftley N, von Wangenheim D, Banda J, Srivastava AK, Hill K, Truskina J, Bhosale R, Morris E, Srivastava M, et al. Root branching toward water involves posttranslational modification of transcription factor ARF7. Science. 2018:**362**(6421): 1407–1410. https://doi.org/10.1126/science.aau3956

O'Toole JC, Bland WL. Genotypic variation in crop plant root systems. Adv Agron. 1987:**41**:91–145. https://doi.org/10.1016/S0065-2113(08)60803-2

Pandey BK, Huang G, Bhosale R, Hartman S, Sturrock CJ, Jose L, Martin OC, Karady M, Voesenek LACJ, Ljung K, et al. Plant roots sense soil compaction through restricted ethylene diffusion. Science. 2021:**371**(6526):276–280. https://doi.org/10.1126/science.abf3013

Pareek A, Dhankher OP, Foyer CH. Mitigating the impact of climate change on plant productivity and ecosystem sustainability. J Exp Bot. 2020:**71**(2):451–456. https://doi.org/10.1093/jxb/erz518

Passioura JB, Boyer JS. Tissues stresses and resistance to water flow conspire to uncouple the water potential of the epidermis from that of the xylem in elongating plant stems. Funct Plant Biol. 2003:**30**(3):325–334. https://doi.org/10.1071/FP02202

Passot S, Moreno-Ortega B, Moukouanga D, Balsera C, Guyomarc'h S, Lucas M, Lobet G, Laplaze L, Muller B, Guédon Y. A new phenotyping pipeline reveals three types of lateral roots and a random branching pattern in two cereals. Plant Physiol. 2018:**177**(3): 896–910. https://doi.org/10.1104/pp.17.01648

Pattathil S, Avci U, Baldwin D, Swennes AG, McGill JA, Popper Z, Bootten T, Albert A, Davis RH, Chennareddy C, et al. A comprehensive toolkit of plant cell wall glycan-directed monoclonal antibodies. Plant Physiol. 2010:**153**(2):514–525. https://doi.org/10.1104/pp.109.151985

Peters WS, Felle HH. The correlation of profiles of surface pH and elongation growth in maize roots. Plant Physiol. 1999:**121**(3):905–912. https://doi.org/10.1104/pp.121.3.905

Peters WS, Tomos AD. The mechanistic state of "inner tissue" in the growing zone of sunflower hypocotyls and the regulation of its growth rate following excision. Plant Physiol. 2000:**123**(2):605–612. https://doi.org/10.1104/pp.123.2.605

Petrova A, Gorshkova T, Kozlova L. Gradients of cell wall nano-mechanical properties along and across elongating primary roots of maize. J Exp Bot. 2021:**72**(5):1764–1781. https://doi.org/10.1093/jxb/eraa561

Poroyko V, Spollen WG, Hejlek LG, Hernandez AG, LeNoble ME, Davis G, Nguyen HT, Springer GK, Sharp RE, Bohnert HJ. Comparing regional transcript profiles from maize primary roots under well-watered and low water potential conditions. J Exp Bot. 2007:**58**(2):279–289. https://doi.org/10.1093/jxb/erl119

Portas CAM, Taylor HM. Growth and survival of young plant roots in dry soil. Soil Sci. 1976:**121**(3):170–175. https://doi.org/10.1097/00010694-197603000-00005

Pritchard J. Tansley review No. 68. The control of cell expansion in roots. New Phytol. 1994:**127**(1):3–26. https://doi.org/10.1111/j.1469-8137.1994.tb04255.x

Pritchard J, Tomos AD. Biophysics of root expansion growth under water stressed and non-stressed conditions. In: **Smith JAC, Griffiths H**, editors. Plant responses to water deficits-from cell to community. SEB Environmental Plant Biology Series. Abingdon (UK): BIOS Scientific Publishers; 1993. p. 53–72.

Pritchard J, Wyn Jones RG, Tomos AD. Turgor, growth and rheological gradients of wheat roots following osmotic stress. J Exp Bot. 1991:**42**(8):1043–1049. https://doi.org/10.1093/jxb/42.8.1043

Qin X, Zeevaart JAD. The 9-*cis*-epoxycarotenoid cleavage reaction in the key regulatory step of abscisic acid biosynthesis in water-stressed bean. Proc Natl Acad Sci U S A. 1999:**96**(26):15354–15361. https://doi.org/10.1073/pnas.96.26.15354

Quarrie SA, Jones HG. Effects of abscisic acid and water stress on development and morphology of wheat. J Exp Bot. 1977:**28**(1):192–203. https://doi.org/10.1093/jxb/28.1.192

Quiroz-Figueroa F, Rodríguez-Acosta A, Salazar-Blas A, Hernández-Domínguez E, Campos ME, Kitahata N, Asami T, Galaz-Avalos RM, Cassab GI. Accumulation of high levels of ABA regulates the pleiotropic response of the *nhr1 Arabidopsis* mutant. J Plant Biol. 2010:**53**(1):32–44. https://doi.org/10.1007/s12374-009-9083-1

Quisenberry JE, Cartwright GB, McMichael BL. Genetic relationship between turgor maintenance and growth in cotton germplasm. Crop Sci. 1984:**24**(3):479–482. https://doi.org/10.2135/cropsci1984.0011183X002400030011x

Rayle DL, Cleland RE. The Acid Growth Theory of auxin-induced cell elongation is alive and well. Plant Physiol. 1992:**99**(4):1271–1274. https://doi.org/10.1104/pp.99.4.1271

Raymond MJ, Smirnoff N. Proline metabolism and transport in maize seedlings at low water potential. Ann Bot. 2002:**89**(7):813–823. https://doi.org/10.1093/aob/mcf082

Read DJ, Bartlett EM. The physiology of drought resistance in the soybean plant (*Glycine max*). I. The relationship between drought resistance and growth. J Appl Ecol. 1972:**9**(2):487–499. https://doi.org/10.2307/2402447

Rentsch D, Hirner B, Schmelzer E, Frommer WB. Salt stress-induced proline transporters and salt stress-repressed broad specificity amino acid permeases identified by suppression of a yeast amino acid permease-targeting mutant. Plant Cell. 1996:**8**(8):1437–1446. https://doi.org/10.1105/tpc.8.8.1437

Reyes-Hernández BJ, Srivastava AC, Ugartechea-Chirino Y, Shishkova S, Ramos-Parra PA, Lira-Ruan V, Díaz de la Garza RI, Dong G, Moon J-C, Blancafor EB, et al. The root indeterminacy-to-determinacy developmental switch is operated through a folate-dependent pathway in *Arabidopsis thaliana*. New Phytol. 2014:**202**(4):1223–1236. https://doi.org/10.1111/nph.12757

Robbins NE 2nd, Dinneny JR. Growth is required for perception of water availability to pattern root branches in plants. Proc Natl Acad Sci U S A. 2018:**115**(4):E822–E831. https://doi.org/10.1073/pnas.1710709115

Rodriguez-Alonso G, Matvienko M, López-Valle ML, Lázaro-Mixteco PE, Napsucialy-Mendivil S, Dubrovsky JG, Shishkova S. Transcriptomics insights into the genetic regulation of root apical meristem exhaustion and determinate primary root growth in *Pachycereus pringlei* (Cactaceae). Sci Rep. 2018:**8**(1):8529. https://doi.org/10.1038/s41598-018-26897-1

Rodriques SG, Stickels RR, Goeva A, Martin CA, Murray E, Vanderburg CR, Welch J, Chen LM, Chen F, Macosko EZ. Slide-seq: a scalable technology for measuring genome-wide expression at high spatial resolution. Science. 2019:**363**(6434):1463–1467. https://doi.org/10.1126/science.aaw1219

Rostamza M, Richards RA, Watt M. Response of millet and sorghum to a varying water supply around the primary and nodal roots. Ann Bot. 2013:**112**(2):439–446. https://doi.org/10.1093/aob/mct099

Rowe J, Grangé-Guermente M, Exposito-Rodiguez M, Wimalasekera R, Lenz MO, Shetty KN, Cutler SR, Jones AM. Next-generation ABACUS biosensors reveal cellular ABA dynamics driving root growth at low aerial humidity. Nat Plants. 2023:**9**(7):1103–1115. https://doi.org/10.1038/s41477-023-01447-4

Rowe JH, Topping JF, Liu J, Lindsey K. Abscisic acid regulates root growth under osmotic stress conditions via an interacting hormonal network with cytokinin, ethylene and auxin. New Phytol. 2016:**211**(1):225–239. https://doi.org/10.1111/nph.13882

Russell RS. Plant root systems: their function and interaction with the soil. London: McGraw-Hill Book Company (UK) Ltd.; 1977. 298 p.

Ryu KH, Huang L, Kang HM, Schiefelbein J. Single-cell RNA sequencing resolves molecular relationships among individual plant cells. Plant Physiol. 2019:**179**(4):1444–1456. https://doi.org/10.1104/pp.18.01482

Saab IN, Sharp RE, Pritchard J. Effect of inhibition of abscisic acid accumulation on the spatial distribution of elongation in the primary root and mesocotyl of maize at low water potentials. Plant Physiol. 1992:**99**(1):26–33. https://doi.org/10.1104/pp.99.1.26

Saab IN, Sharp RE, Pritchard J, Voetberg GS. Increased endogenous abscisic acid maintains primary root growth and inhibits shoot growth of maize seedlings at low water potentials. Plant Physiol. 1990:**93**(4):1329–1336. https://doi.org/10.1104/pp.93.4.1329

Sachs J. Ablenkung der Wurzel von ihrer normalen Wachstumsrichtung durch feutchte Körper. Arb D Bot Inst Würzburg. 1872:**1**:209–222.

Sacks MM, Silk WK, Burman P. Effects of water stress on cortical cell division within the apical meristem of primary roots of maize. Plant Physiol. 1997:**114**(2):519–527. https://doi.org/10.1104/pp.114.2.519

Salazar-Henao JE, Vélez-Bermúdez IC, Schmidt W. The regulation and plasticity of root hair patterning and morphogenesis. Development. 2016:**143**(11):1848–1858. https://doi.org/10.1242/dev.132845

Sampathkumar A, Peaucelle A, Fujita M, Schuster C, Persson S, Wasteneys GO, Meyerowitz EM. Primary wall cellulose synthase regulates shoot apical meristem mechanics and growth. Development. 2019:**146**(10):dev179036. https://doi.org/10.1242/dev.179036

Sandhu N, Raman KA, Torres RO, Audebert A, Dardou A, Kumar A, Henry A. Rice root architectural plasticity traits and genetic regions for adaptability to variable cultivation and stress conditions. Plant Physiol. 2016:**171**(4):2562–2576. https://doi.org/10.1104/pp.16.00705

Savaldi-Goldstein S, Peto C, Chory J. The epidermis both drives and restricts plant shoot growth. Nature. 2007:**446**(7132):199–202. https://doi.org/10.1038/nature05618

Schneider HM, Lynch JP. Should root plasticity be a crop breeding target? Front Plant Sci. 2020:**11**:546. https://doi.org/10.3389/fpls.2020.00546

Schneider HM, Strock CF, Hanlon MT, Vanhees DJ, Perkins AC, Ajmera IB, Sidhu JS, Mooney SJ, Brown KM, Lynch JP. Multiseriate cortical sclerenchyma enhance root penetration in compacted soils. Proc Natl Acad Sci U S A. 2021:**118**(6):e2012087118. https://doi.org/10.1073/pnas.2012087118

Schnepf A, Carminati A, Ahmed MA, Ani M, Benard P, Bentz J, Bonkowski M, Knott M, Diehl D, Duddek P, et al. Linking rhizosphere processes across scales: opinion. Plant Soil. 2022:**478**(1–2): 5–42. https://doi.org/10.1007/s11104-022-05306-7

Scholander PF, Bradstreet ED, Hemmingsen EA, Hammel HT. Sap pressure in vascular plants: negative hydrostatic pressure can be measured in plants. Science. 1965:**148**(3668):339–346. https://doi.org/10.1126/science.148.3668.339

Sebastian J, Yee M-C, Goudinho Viana W, Rellán-Álvarez R, Feldman M, Priest HD, Trontin C, Lee T, Jiang H, Baxter I, et al. Grasses suppress shoot-borne roots to conserve water during drought. Proc Natl Acad Sci U S A. 2016:**113**(31):8861–8866. https://doi.org/10.1073/pnas.1604021113

Serpe MD, Matthews MA. Rapid changes in cell wall yielding of elongating *Begonia argenteo-guttata* L. leaves in response to changes in plant water status. Plant Physiol. 1992:**100**(4):1852–1857. https://doi.org/10.1104/pp.100.4.1852

Serraj R, Sinclair TR. Osmolyte accumulation: can it really help increase crop yield under drought conditions? Plant Cell Environ. 2002:**25**(2):333–341. https://doi.org/10.1046/j.1365-3040.2002.00754.x

Shahzad Z, Tournaire-Roux C, Canut M, Adamo M, Roeder J, Verdoucq L, Martinière A, Amtmann A, Santoni V, Grill E, et al. Protein kinase SnRK2.4 is a key regulator of aquaporins and root hydraulics in Arabidopsis. Plant J. 2024:**117**(1):264–279. https://doi.org/10.1111/tpj.16494

Sharma S, Villamor JG, Verslues PE. Essential role of tissue-specific proline synthesis and catabolism in growth and redox balance at low water potential. Plant Physiol. 2011:**157**(1):292–304. https://doi.org/10.1104/pp.111.183210

Sharp RE. Interaction with ethylene: changing views on the role of abscisic acid in root and shoot growth responses to water stress. Plant Cell Environ. 2002:**25**(2):211–222. https://doi.org/10.1046/j.1365-3040.2002.00798.x

Sharp RE, Davies WJ. Solute regulation and growth by roots and shoots of water-stressed maize plants. Planta. 1979:**146**(3):319–326. https://doi.org/10.1007/BF00384589

Sharp RE, Davies WJ. Root growth and water uptake by maize plants in drying soil. J Exp Bot. 1985:**36**(9):1441–1456. https://doi.org/10.1093/jxb/36.9.1441

Sharp RE, Davies WJ. Regulation of growth and development of plants growing with a restricted supply of water. In: **Jones HG, Flowers TL, Jones MB**, editors. Plants under stress. Cambridge: Cambridge University Press; 1989. p. 71–93.

Sharp RE, Hsiao TC, Silk WK. Growth of the maize primary root at low water potentials. II. Role of growth and deposition of hexose and potassium in osmotic adjustment. Plant Physiol. 1990:**93**(4):1337–1346. https://doi.org/10.1104/pp.93.4.1337

Sharp RE, LeNoble ME, Else MA, Thorne ET, Gherardi F. Endogenous ABA maintains shoot growth in tomato independently of effects on plant water balance: evidence for an interaction with ethylene. J Exp Bot. 2000:**51**(350):1575–1584. https://doi.org/10.1093/jexbot/51.350.1575

Sharp RE, Poroyko V, Hejlek LG, Spollen WG, Springer GK, Bohnert HJ, Nguyen HT. Root growth maintenance during water deficits: physiology to functional genomics. J Exp Bot. 2004:**55**(407): 2343–2351. https://doi.org/10.1093/jxb/erh276

Sharp RE, Silk WK, Hsiao TC. Growth of the maize primary root at low water potentials. I. Spatial distribution of expansive growth. Plant Physiol. 1988:**87**(1):50–57. https://doi.org/10.1104/pp.87.1.50

Sharp RE, Wu Y, Voetberg GS, Saab IN, LeNoble ME. Confirmation that abscisic acid accumulation is required for maize primary root elongation at low water potentials. J Exp Bot. 1994:**45**(Special_Issue): 1743–1751. https://doi.org/10.1093/jxb/45.Special_Issue.1743

Shiono K, Yoshikawa M, Kreszies T, Yamada S, Hojo Y, Matsuura T, Mori IC, Schreiber L, Yoshioka T. Abscisic acid is required for exodermal suberization to form a barrier to radial oxygen loss in the adventitious roots of rice (*Oryza sativa*). New Phytol. 2022:**233**(2): 655–669. https://doi.org/10.1111/nph.17751

Shishkova S, Rost TL, Dubrovsky JG. Determinate root growth and meristem maintenance in angiosperms. Ann Bot. 2008:**101**(3): 319–340. https://doi.org/10.1093/aob/mcm251

Shkolnik D, Krieger G, Nuriel R, Fromm H. Hydrotropism: root bending does not require auxin redistribution. Mol Plant. 2016:**9**(5): 757–759. https://doi.org/10.1016/j.molp.2016.02.001

Shkolnik D, Nuriel R, Bonza MC, Costa A, Fromm H. MIZ1 regulates ECA1 to generate a slow, long-distance phloem-transmitted Ca^{2+} signal essential for root water tracking in *Arabidopsis*. Proc Natl Acad Sci U S A. 2018:**115**(31):8031–8036. https://doi.org/10.1073/pnas.1804130115

Silk WK. Quantitative descriptions of development. Annu Rev Plant Physiol. 1984:**35**(1):479–518. https://doi.org/10.1146/annurev.pp.35.060184.002403

Silk WK, Bogeat-Triboulot M-B. Deposition rates in growing tissue: Implications for physiology, molecular biology, and response to environmental variation. Plant Soil. 2014:**374**(1–2):1–17. https://doi.org/10.1007/s11104-013-1726-9

Silk WK, Erickson RO. Kinematics of plant growth. J Theor Biol. 1979:**76**(4):481–501. https://doi.org/10.1016/0022-5193(79)90014-6

Silk WK, Hsiao TC, Diedenhoffen U, Matson C. Spatial distributions of potassium, solutes, and their deposition rates in the growth zone of the primary corn root. Plant Physiol. 1986:**82**(3):853–858. https://doi.org/10.1104/pp.82.3.853

Silk WK, Lord EM, Eckard KJ. Growth patterns inferred from anatomical records. Empirical tests using longisections of roots of *Zea mays* L. Plant Physiol. 1989:**90**(2):708–713. https://doi.org/10.1104/pp.90.2.708

Silk WK, Wagner KK. Growth-sustaining water potential distributions in the primary corn root: A NONCOMPARTMENTED CONTINUUM MODEL. Plant Physiol. 1980:**66**(5):859–863. https://doi.org/10.1104/pp.66.5.859

Silk WK, Walker RC, Labavitch J. Uronide deposition rates in the primary root of *Zea mays*. Plant Physiol. 1984:**74**(3):721–726. https://doi.org/10.1104/pp.74.3.721

Sinha R, Induri SP, Peláez-Vico MÁ, Tukuli A, Shostak B, Zandalinas SI, Joshi T, Fritschi FB, Mittler R. The transcriptome of soybean reproductive tissues subjected to water deficit, heat stress, and a combination of water deficit and heat stress. Plant J. 2023a:**116**(4): 1064–1080. https://doi.org/10.1111/tpj.16222

Sinha R, Shostak B, Induri SP, Sen S, Zandalinas SI, Joshi T, Fritschi FB, Mittler R. Differential transpiration between pods and leaves during stress combination in soybean. Plant Physiol. 2023b:**192**(2): 753–766. https://doi.org/10.1093/plphys/kiad114

Sinnott EW. Growth and differentiation in living plant meristems. Proc Natl Acad Sci U S A. 1939:**25**(2):55–58. https://doi.org/10.1073/pnas.25.2.55

Skirycz A, Claeys H, De Bodt S, Oikawa A, Shinoda S, Andriankaja M, Maleux K, Eloy NB, Coppens F, Yoo S-D, et al. Pause-and-stop: the effects of osmotic stress on cell proliferation during early leaf development in *Arabidopsis* and a role for ethylene signaling in cell cycle arrest. Plant Cell. 2011:**23**(5):1876–1888. https://doi.org/10.1105/tpc.111.084160

Skirycz A, Inzé D. More from less: plant growth under limited water. Curr Opin Biotechnol. 2010:**21**(2):197–203. https://doi.org/10.1016/j.copbio.2010.03.002

Slatyer RO, Taylor SA. Terminology in plant- and soil-water relations. Nature. 1960:**187**(4741):922–924. https://doi.org/10.1038/187922a0

Spollen WG, LeNoble ME, Samuels TD, Bernstein N, Sharp RE. ABA accumulation maintains primary root elongation at low water potentials by restricting ethylene production. Plant Physiol. 2000:**122**(3): 967–976. https://doi.org/10.1104/pp.122.3.967

Spollen WG, Sharp RE. Spatial distribution of turgor and root growth at low water potentials. Plant Physiol. 1991:**96**(2):438–443. https://doi.org/10.1104/pp.96.2.438

Spollen WG, Tao W, Valliyodan B, Chen K, Hejlek LG, Kim JJ, Lenoble ME, Zhu J, Bohnert HJ, Henderson D, et al. Spatial distribution of transcript changes in the maize primary root elongation zone at low water potential. BMC Plant Biol. 2008:**8**(1):32. https://doi.org/10.1186/1471-2229-8-32

Sponchiado BN, White JW, Castillo JA, Jones PG. Root growth of four common bean cultivars in relation to drought tolerance in environments with contrasting soil types. Exp Agric. 1989:**25**(2):249–257. https://doi.org/10.1017/S0014479700016756

Springmann M, Clark M, Mason-D'Croz D, Wiebe K, Bodirsky BL, Lassaletta L, de Vries W, Vermeulen SJ, Herrero M, Carlson KM, et al. Options for keeping the food system within environmental limits. Nature. 2018:**562**(7728):519–525. https://doi.org/10.1038/s41586-018-0594-0

Staal M, De Cnodder T, Simon D, Vandenbussche F, Van Der Straeten D, Verbelen J-P, Elzenga T, Vissenberg K. Apoplastic alkalinization is instrumental for the inhibition of cell elongation in the Arabidopsis root by the ethylene precursor 1-aminocyclopropane-1-carboxylic acid. Plant Physiol. 2011:**155**(4):2049–2055. https://doi.org/10.1104/pp.110.168476

Steponkus PL, Shahan KW, Cutler JM. Osmotic adjustment in rice. In: Drought resistance in crops with emphasis on rice. Los Banos, Leguna, Philippines: IRRI; 1982. p. 181–194.

Strock CF, Burridge J, Massas ASF, Beaver J, Beebe S, Camilo SA, Fourie D, Jochua C, Miguel M, Miklas PN, et al. Seedling root architecture and its relationship with seed yield across diverse environments in *Phaseolus vulgaris*. Field Crops Res. 2019:**237**:53–64. https://doi.org/10.1016/j.fcr.2019.04.012

Swarup R, Bennett MJ. Root gravitropism. In: **Beeckman T**, editor. Root development. Annual Plant Reviews. Vol. 37. Oxford, UK: Blackwell; 2009. p. 157–174.

Takahashi H. Hydrotropism: The current state of our knowledge. J Plant Res. 1997:**110**(2):163–169. https://doi.org/10.1007/BF02509304

Takahashi H, Miyazawa Y, Fujii N. Hormonal interactions during root tropic growth: hydrotropism versus gravitropism. Plant Mol Biol. 2009:**69**(4):489–502. https://doi.org/10.1007/s11103-008-9438-x

Takahashi H, Scott TK. Intensity of hydrostimulation for the induction of root hydrotropism and its sensing by the root cap. Plant Cell Environ. 1993:**16**(1):99–103. https://doi.org/10.1111/j.1365-3040.1993.tb00850.x

Takano M, Takahashi H, Hirasawa T, Suge H. Hydrotropism in roots —sensing of a gradient in water potential by the root cap. Planta. 1995:**197**(2):410–413. https://doi.org/10.1007/BF00202664

Takano M, Takahashi H, Suge H. Calcium requirement for the induction of hydrotropism and enhancement of calcium-induced curvature by water stress in primary roots of pea, *Pisum sativum* L. Plant Cell Physiol. 1997:**38**(4):385–391. https://doi.org/10.1093/oxfordjournals.pcp.a029180

Tal M, Imber D, Erez A, Epstein E. Abnormal stomatal behavior and hormonal imbalance in *flacca*, a wilty mutant of tomato: V. Effect of abscisic acid on indoleacetic acid metabolism and ethylene evolution. Plant Physiol. 1979:**63**(6):1044–1048. https://doi.org/10.1104/pp.63.6.1044

Tan BC, Schwartz SH, Zeevaart JAD, McCarty DR. Genetic control of abscisic acid biosynthesis in maize. Proc Natl Acad Sci U S A. 1997:**94**(22):12235–12240. https://doi.org/10.1073/pnas.94.22.12235

Tan K-S, Hoson T, Masuda Y, Kamisaka S. Involvement of cell wall-bound diferulic acid in light-induced decrease in growth-rate and cell-wall extensibility of *Oryza* coleoptiles. Plant Cell Physiol. 1992:**33**(2):103–108. https://doi.org/10.1093/oxfordjournals.pcp.a078227

Tangpremsri T, Fukai S, Fischer KS, Henzell RG. Genotypic variation in osmotic adjustment in grain sorghum. II. Relation with some growth attributes. Aust J Agric Res. 1991:**42**(5):759–767. https://doi.org/10.1071/AR9910759

Tardieu F. Any trait or trait-related allele can confer drought tolerance: just design the right drought scenario. J Exp Bot. 2012:**63**(1):25–31. https://doi.org/10.1093/jxb/err269

Tardieu F, Simonneau T, Muller B. The physiological basis of drought tolerance in crop plants: a scenario-dependent probabilistic approach. Annu Rev Plant Biol. 2018:**69**(1):733–759. https://doi.org/10.1146/annurev-arplant-042817-040218

Taylor HM, Ratliff LF. Root elongation rates of cotton and peanuts as a function of soil strength and soil water content. Soil Sci. 1969:**108**(2):113–119. https://doi.org/10.1097/00010694-196908000-00006

Trewavas AJ, Jones HG. An assessment of the role of ABA in plant development. In: **Davies WJ, Jones HG**, editors. Abscisic acid: physiology and biochemistry. Oxford: Bios Scientific Publishers; 1991. p. 169–188.

Triboulot M-B, Pritchard J, Tomos D. Stimulation and inhibition of pine root growth by osmotic stress. New Phytol. 1995:**130**(2):169–175. https://doi.org/10.1111/j.1469-8137.1995.tb03038.x

Turc O, Tardieu F. Drought affects abortion of reproductive organs by exacerbating developmentally driven processes via expansive growth and hydraulics. J Exp Bot. 2018:**69**(13):3245–3254. https://doi.org/10.1093/jxb/ery078

Turner NC. Turgor maintenance by osmotic adjustment: 40 years of progress. J Exp Bot. 2018:**69**(13):3223–3233. https://doi.org/10.1093/jxb/ery181

Turner NC, Jones MM. Turgor maintenance by osmotic adjustment: a review and evaluation. In: **Turner NC, Kramer PJ**, editors. Adaptation of plants to water and high temperature stress. New York: Wiley-Interscience; 1980. p. 87–103.

Ubeda-Tomás S, Swarup R, Coates J, Swarup K, Laplaze L, Beemster GTS, Hedden P, Bhalerao R, Bennett MJ. Root growth in *Arabidopsis* requires gibberellin/DELLA signalling in the endodermis. Nat Cell Biol. 2008:**10**(5):625–628. https://doi.org/10.1038/ncb1726

Uga Y. Challenges to design-oriented breeding of root system architecture adapted to climate change. Breed Sci. 2021:**71**(1):3–12. https://doi.org/10.1270/jsbbs.20118

Uga Y, Okuno K, Yano M. *Dro1*, a major QTL involved in deep rooting of rice under upland field conditions. J Exp Bot. 2011:**62**(8):2485–2494. https://doi.org/10.1093/jxb/erq429

Uga Y, Sugimoto K, Ogawa S, Rane J, Ishitani M, Hara N, Kitomi Y, Inukai Y, Ono K, Kanno N, et al. Control of root system architecture by *DEEPER ROOTING 1* increases rice yield under drought conditions. Nat Genet. 2013:**45**(9):1097–1102. https://doi.org/10.1038/ng.2725

Vaadia Y. Plant hormones and water stress. Philos Trans R Soc Lond B. 1976:**273**(927):513–522. https://doi.org/10.1098/rstb.1976.0028

Valifard M, Le Hir R, Müller J, Scheuring D, Neuhaus HE, Pommerrenig B. Vacuolar fructose transporter SWEET17 is critical for root development and drought tolerance. Plant Physiol. 2021:**187**(4):2716–2730. https://doi.org/10.1093/plphys/kiab436

Valluru R, Davies WJ, Reynolds MP, Dodd IC. Foliar abscisic acid-to-ethylene accumulation and response regulate shoot growth sensitivity to mild drought in wheat. Front Plant Sci. 2016:**7**:461. https://doi.org/10.3389/fpls.2016.00461

van der Weele CM, Jiang HS, Palaniappan KK, Ivanov VB, Palaniappan K, Baskin TI. A new algorithm for computational image analysis of deformable motion at high spatial and temporal resolution applied to root growth. Roughly uniform elongation in the meristem and also, after an abrupt acceleration, in the elongation zone. Plant Physiol. 2003:**132**(3):1138–1148. https://doi.org/10.1104/pp.103.021345

van der Weele CM, Spollen WG, Sharp RE, Baskin TI. Growth of *Arabidopsis thaliana* seedlings under water deficit studied by control of water potential in nutrient-agar media. J Exp Bot. 2000:**51**(350):1555–1562. https://doi.org/10.1093/jexbot/51.350.1555

Varney GT, McCully ME. The branch roots of *Zea*. II. Developmental loss of the apical meristem in field-grown roots. New Phytol. 1991:**118**(4):535–546. https://doi.org/10.1111/j.1469-8137.1991.tb00993.x

Vaseva II, Qudeimat E, Potuschak T, Du Y, Genschik P, Vandenbussche F, Van Der Straeten D. The plant hormone ethylene restricts *Arabidopsis* growth via the epidermis. Proc Natl Acad Sci U S A. 2018:**115**(17):E4130–E4139. https://doi.org/10.1073/pnas.1717649115

Veen BW, Boone FR. The influence of mechanical resistance and soil water on the growth of seminal roots of maize. Soil Tillage Res. 1990:**16**(1-2):219–226. https://doi.org/10.1016/0167-1987(90)90031-8

Velázquez-Márquez S, Conde-Martínez V, Trejo C, Delgado-Alvarado A, Carballo A, Suárez R, Mascorro JO, Trujillo AR. Effects of water deficit on radicle apex elongation and solute accumulation in *Zea mays* L. Plant Physiol Biochem. 2015:**96**:29–37. https://doi.org/10.1016/j.plaphy.2015.07.006

Vermerris W, Sherman DM, McIntyre LM. Phenotypic plasticity in cell walls of maize *brown midrib* mutants is limited by lignin composition. J Exp Bot. 2010:**61**(9):2479–2490. https://doi.org/10.1093/jxb/erq093

Verslues PE, Bailey-Serres J, Brodersen C, Buckley TN, Conti L, Christmann A, Dinneny JR, Grill E, Hayes S, Heckman RW, et al. Burning questions for a warming and changing world: 15 unknowns in plant abiotic stress. Plant Cell. 2023:**35**(1):67–108. https://doi.org/10.1093/plcell/koac263

Verslues PE, Longkumer T. Size and activity of the root meristem: a key for drought resistance and a key model of drought-related signaling. Physiol Plant. 2022:**174**(1):e13622. https://doi.org/10.1111/ppl.13622

Verslues PE, Sharma S. Proline metabolism and its implications for plant-environment interaction. In: The Arabidopsis book 8. Rockwille (MD): American Society of Plant Biologists; 2010. p. e0140.

Verslues PE, Sharp RE. Proline accumulation in maize (*Zea mays* L.) primary roots at low water potentials. II. Metabolic source of increased proline deposition in the elongation zone. Plant Physiol. 1999:**119**(4):1349–1360. https://doi.org/10.1104/pp.119.4.1349

Voetberg GS, Sharp RE. Growth of the maize primary root at low water potentials. III. Role of increased proline deposition in osmotic adjustment. Plant Physiol. 1991:**96**(4):1125–1130. https://doi.org/10.1104/pp.96.4.1125

Vogel J. Unique aspects of the grass cell wall. Curr Opin Plant Biol. 2008:**11**(3):301–307. https://doi.org/10.1016/j.pbi.2008.03.002

Volkov V, Schwenke H. A quest for mechanisms of plant root exudation brings new results and models, 300 years after Hales. Plants. 2020:**10**(1):38. https://doi.org/10.3390/plants10010038

Voothuluru P, Anderson JC, Sharp RE, Peck SC. Plasma membrane proteomics in the maize primary root growth zone: novel insights into root growth adaptation to water stress. Plant Cell Environ. 2016:**39**(9):2043–2054. https://doi.org/10.1111/pce.12778

Voothuluru P, Braun DM, Boyer JS. An in vivo imaging assay detects spatial variability in glucose release from plant roots. Plant Physiol. 2018:**178**(3):1002–1010. https://doi.org/10.1104/pp.18.00614

Voothuluru P, Mäkelä P, Zhu J, Yamaguchi M, Cho I-J, Oliver MJ, Simmonds J, Sharp RE. Apoplastic hydrogen peroxide in the growth zone of the maize primary root. Increased levels differentially modulate root elongation under well-watered and water-stressed conditions. Front Plant Sci. 2020:**11**:392. https://doi.org/10.3389/fpls.2020.00392

Voothuluru P, Sharp RE. Apoplastic hydrogen peroxide in the growth zone of the maize primary root under water stress. I. Increased levels are specific to the apical region of growth maintenance. J Exp Bot. 2013:**64**(5):1223–1233. https://doi.org/10.1093/jxb/ers277

Waadt R, Seller CA, Hsu PK, Takahashi Y, Munemasa S, Schroeder JI. Plant hormone regulation of abiotic stress responses. Nat Rev Mol Cell Biol. 2022:**23**(10):680–694. https://doi.org/10.1038/s41580-022-00479-6

Waidmann S, Sarkel E, Kleine-Vehn J. Same same, but different: growth responses of primary and lateral roots. J Exp Bot. 2020:**71**(8):2397–2411. https://doi.org/10.1093/jxb/eraa027

Walter A, Silk W, Schurr U. Environmental effects on spatial and temporal patterns of leaf and root growth. Annu Rev Plant Biol. 2009:**60**(1):279–304. https://doi.org/10.1146/annurev.arplant.59.032607.092819

Wakabayashi K, Sakurai N, Kuraishi S. Role of the outer tissues in abscisic acid-mediated growth suppression of etiolated squash hypocotyl segments. Physiol Plant. 1989:**75**(2):151–156. https://doi.org/10.1111/j.1399-3054.1989.tb06162.x

Wang C, Wang H, Li P, Li H, Xu C, Cohen H, Aharoni A, Wu S. Developmental programs interact with abscisic acid to coordinate root suberization in *Arabidopsis*. Plant J. 2020b:**104**(1):241–251. https://doi.org/10.1111/tpj.14920

Wang Y, Afeworki Y, Geng S, Kanchupati P, Gu M, Martins C, Rude B, Tefera H, Kim Y, Ge X, et al. Hydrotropism in the primary roots of maize. New Phytol. 2020a:**226**(6):1796–1808. https://doi.org/10.1111/nph.16472

Watt M, McCully ME, Canny MJ. Formation and stabilization of rhizosheaths of *Zea mays* L. Effect of soil water content. Plant Physiol. 1994:**106**(1):179–186. https://doi.org/10.1104/pp.106.1.179

Watt M, McCully ME, Kirkegaard JA. Soil strength and rate of root elongation alter the accumulation of *Pseudomonas* spp. and other bacteria in the rhizosphere of wheat. Funct Plant Biol. 2003:**30**(5):483–491. https://doi.org/10.1071/FP03045

Weaver JE. Root development of field crops. New York: McGraw-Hill; 1926. 291 p.

Westgate ME, Boyer JS. Osmotic adjustment and the inhibition of leaf, root, stem and silk growth at low water potentials in maize. Planta. 1985:**164**(4):540–549. https://doi.org/10.1007/BF00395973

White JW, Castillo JA. Relative effect of root and shoot genotypes on yield of common bean under drought stress. Crop Sci. 1989:**29**(2):360–362. https://doi.org/10.2135/cropsci1989.0011183X002900020026x

White JW, Castillo JA. Evaluation of diverse shoot genotypes on selected root genotypes of common bean under soil water deficits. Crop Sci. 1992:**32**(3):762–765. https://doi.org/10.2135/cropsci1992.0011183X003200030037x

White RG, Kirkegaard JA. The distribution and abundance of wheat roots in a dense, structured subsoil—implications for water uptake. Plant Cell Environ. 2010:**33**(2):133–148. https://doi.org/10.1111/j.1365-3040.2009.02059.x

Wiegers BS, Cheer AY, Silk WK. Modeling the hydraulics of root growth in three dimensions with phloem water sources. Plant Physiol. 2009:**150**(4):2092–2103. https://doi.org/10.1104/pp.109.138198

Wilkinson S, Davies WJ. ABA-based chemical signalling: the co-ordination of responses to stress in plants. Plant Cell Environ. 2002:**25**(2):195–210. https://doi.org/10.1046/j.0016-8025.2001.00824.x

Williams A, de Vries FT. Plant root exudation under drought: implications for ecosystem functioning. New Phytol. 2020:**225**(5):1899–1905. https://doi.org/10.1111/nph.16223

Wilson JR, Ludlow MM. Time trends of solute accumulation and the influence of potassium fertilizer on osmotic adjustment of water-stressed leaves of three tropical grasses. Aust J Plant Physiol. 1983:**10**(6):523–537. https://doi.org/10.1071/PP9830523

Winch S, Pritchard J. Acid-induced wall loosening is confined to the accelerating region of the root growing zone. J Exp Bot. 1999:**50**(338):1481–1487. https://doi.org/10.1093/jxb/50.338.1481

Woods P, Lehner KR, Hein K, Mullen JL, McKay JK. Root pulling force across drought in maize reveals genotype by environment interactions and candidate genes. Front Plant Sci. 2022:**13**:883209. https://doi.org/10.3389/fpls.2022.883209

Wright GC, Smith RCG. Differences between two grain sorghum genotypes in adaptation to drought stress. II. Root water uptake and water use. Aust J Agric Res. 1983:**34**(6):627–636. https://doi.org/10.1071/AR9830627

Wright GC, Smith RCG, Morgan JM. Differences between two grain sorghum genotypes in adaptation to drought stress. III. Physiological responses. Aust J Agric Res. 1983:**34**(6):637–651. https://doi.org/10.1071/AR9830637

Wright STC. The effect of plant growth regulator treatments on the levels of ethylene emanating from excised turgid and wilted wheat leaves. Planta. 1980:**148**(4):381–388. https://doi.org/10.1007/BF00388127

Wu Y, Cosgrove DJ. Adaptation of roots to low water potentials by changes in cell wall extensibility and cell wall proteins. J Exp Bot. 2000:**51**(350):1543–1553. https://doi.org/10.1093/jexbot/51.350.1543

Wu Y, Sharp RE, Durachko DM, Cosgrove DJ. Growth maintenance of the maize primary root at low water potentials involves increases in cell-wall extension properties, expansin activity, and wall susceptibility to expansins. Plant Physiol. 1996:**111**(3):765–772. https://doi.org/10.1104/pp.111.3.765

Wu Y, Spollen WG, Sharp RE, Hetherington PR, Fry SC. Root growth maintenance at low water potentials. Increased activity of xyloglucan endotransglycosylase and its possible regulation by abscisic acid. Plant Physiol. 1994:**106**(2):607–615. https://doi.org/10.1104/pp.106.2.607

Wu Y, Thorne ET, Sharp RE, Cosgrove DJ. Modification of expansin transcript levels in the maize primary root at low water potentials. Plant Physiol. 2001:**126**(4):1471–1479. https://doi.org/10.1104/pp.126.4.1471

Xu W, Jia L, Shi W, Liang J, Zhou F, Li Q, Zhang J. Abscisic acid accumulation modulates auxin transport in the root tip to enhance proton secretion for maintaining root growth under moderate water stress. New Phytol. 2013:**197**(1):139–150. https://doi.org/10.1111/nph.12004

Yamaguchi M, Sharp RE. Complexity and coordination of root growth at low water potentials: recent advances from transcriptomic and proteomic analyses. Plant Cell Environ. 2010:**33**(4):590–603. https://doi.org/10.1111/j.1365-3040.2009.02064.x

Yamaguchi M, Valliyodan B, Zhang J, LeNoble ME, Yu O, Rogers EE, Nguyen HT, Sharp RE. Regulation of growth response to water stress in the soybean primary root. I. Proteomic analysis reveals region-specific regulation of phenylpropanoid metabolism and control of free iron in the elongation zone. Plant Cell Environ. 2010:**33**(2):223–243. https://doi.org/10.1111/j.1365-3040.2009.02073.x

Yancey PH. Organic osmolytes as compatible, metabolic and counteracting cytoprotectants in high osmolarity and other stresses. J Exp Bot. 2005:**208**(15):2819–2830. https://doi.org/10.1242/jeb.01730

Yang JC, Zhang JH, Ye YX, Wang ZQ, Zhu QS, Liu LJ. Involvement of abscisic acid and ethylene in the responses of rice grains to water stress during filling. Plant Cell Environ. 2004:**27**(8):1055–1064. https://doi.org/10.1111/j.1365-3040.2004.01210.x

Zandalinas SI, Fritschi FB, Mittler R. Global warming, climate change, and environmental pollution: recipe for a multifactorial stress combination disaster. Trends Plant Sci. 2021:**26**(6):588–599. https://doi.org/10.1016/j.tplants.2021.02.011

Zhan A, Schneider H, Lynch JP. Reduced lateral root branching density improves drought tolerance in maize. Plant Physiol. 2015:**168**(4):1603–1615. https://doi.org/10.1104/pp.15.00187

Zhang C, Bousquet A, Harris JM. Abscisic acid and LATERAL ROOT ORGAN DEFECTIVE/NUMEROUS INFECTIONS AND POLY PHENOLICS modulate root elongation via reactive oxygen species in *Medicago truncatula*. Plant Physiol. 2014:**166**(2):644–658. https://doi.org/10.1104/pp.114.248542

Zhang X, Mi Y, Mao H, Liu S, Chen L, Qin F. Genetic variation in *ZmTIP1* contributes to root hair elongation and drought tolerance in maize. Plant Biotechnol J. 2020b:**18**(5):1271–1283. https://doi.org/10.1111/pbi.13290

Zhang Y, Du H, Gui Y, Xu F, Liu J, Zhang J, Xu W. Moderate water stress in rice induces rhizosheath formation associated with abscisic acid and auxin responses. J Exp Bot. 2020a:**71**(9):2740–2751. https://doi.org/10.1093/jxb/eraa021

Zhang Y, Xu F, Ding Y, Du H, Zhang Q, Dang X, Cao Y, Dodd IC, Xu W. Abscisic acid mediates barley rhizosheath formation under mild soil drying by promoting root hair growth and auxin response. Plant Cell Environ. 2021:**44**(6):1935–1945. https://doi.org/10.1111/pce.14036

Zhang Q, Yuan W, Wang Q, Cao Y, Xu F, Dodd IC, Xu W. ABA regulation of root growth during soil drying and recovery can involve auxin response. Plant Cell Environ. 2022:**45**(3):871–883. https://doi.org/10.1111/pce.14137

Zhu J, Alvarez S, Marsh EL, Lenoble ME, Cho I-J, Sivaguru M, Chen S, Nguyen HT, Wu Y, Schachtman DP, et al. Cell wall proteome in the maize primary root elongation zone. II. Region-specific changes in water soluble and lightly ionically bound proteins under water deficit. Plant Physiol. 2007:**145**(4):1533–1548. https://doi.org/10.1104/pp.107.107250

Zhu J, Brown KM, Lynch JP. Root cortical aerenchyma improves the drought tolerance of maize (*Zea mays* L.). Plant Cell Environ. 2010:**33**(5):740–749. https://doi.org/10.1111/j.1365-3040.2009.02099.x

Zwieniecki MA, Thompson MV, Holbrook NM. Understanding the hydraulics of porous pipes: tradeoffs between water uptake and root length utilization. J Plant Growth Reg. 2003:**21**(4):315–323. https://doi.org/10.1007/s00344-003-0008-9

https://doi.org/10.1093/plcell/koae054

THE PLANT CELL

THE PLANT CELL 2024: 36: 1410–1428

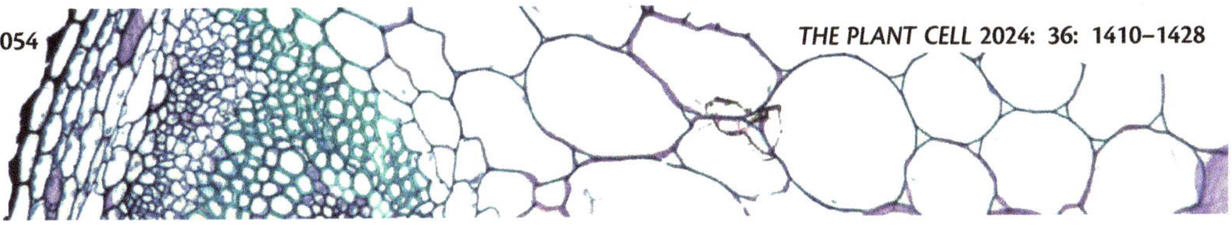

Review

An auxin research odyssey: 1989–2023

Jerry D. Cohen [1,*] **Lucia C. Strader** [2,*]

1 Department of Horticultural Science and the Microbial and Plant Genomics Institute, University of Minnesota, Saint Paul, MN 55108, USA
2 Department of Biology, Duke University, Durham, NC 27008, USA

*Author for correspondence: cohen047@umn.edu (J.D.C.), lucia.strader@duke.edu (L.C.S.)
The author responsible for distribution of materials integral to the findings presented in this article in accordance with the policy described in the Instructions for Authors (https://academic.oup.com/plcell/pages/General-Instructions) is: Jerry D. Cohen (cohen047@umn.edu).

Abstract

The phytohormone auxin is at times called the master regulator of plant processes and has been shown to be a central player in embryo development, the establishment of the polar axis, early aspects of seedling growth, as well as growth and organ formation during later stages of plant development. *The Plant Cell* has been key, since the inception of the journal, to developing an understanding of auxin biology. Auxin-regulated plant growth control is accomplished by both changes in the levels of active hormones and the sensitivity of plant tissues to these concentration changes. In this historical review, we chart auxin research as it has progressed in key areas and highlight the role *The Plant Cell* played in these scientific developments. We focus on understanding auxin-responsive genes, transcription factors, reporter constructs, perception, and signal transduction processes. Auxin metabolism is discussed from the development of tryptophan auxotrophic mutants, the molecular biology of conjugate formation and hydrolysis, indole-3-butyric acid metabolism and transport, and key steps in indole-3-acetic acid biosynthesis, catabolism, and transport. This progress leads to an expectation of a more comprehensive understanding of the systems biology of auxin and the spatial and temporal regulation of cellular growth and development.

Introduction

The first issue of *The Plant Cell* in 1989 began a new focus for the American Society of Plant Biologists into understanding plant function at the molecular level. Auxin, as a central regu-lator of plant growth and development (Carrillo-Carrasco et al. 2023), had already evolved into the molecular age, but the emergence of the new journal with a molecular focus became a critical place to exchange advances in this dynamic area of research (Enders and Strader 2015).

Prelude to plant auxin molecular biology

The chemistry of auxin and its bioactivity filled the auxin literature in the first part of the 20th century. Early reports (see Silberger and Skoog 1953) showed that indole-3-acetic acid (IAA) affected the RNA and DNA contents of plants at concentrations optimal for cell enlargement. During the next decades, the 1960s and 1970s, many advances focused on mechanisms of action of plant hormones. Much of this research dealt with the effect of auxin on nucleic acid metabolism, protein synthesis, and the synthesis of specific enzymes (Carpenter and Cherry 1966; Neumann and Palmer 1971). Although a biophysical mechanism for the rapid effect was put forth in the 1970s, it was still felt that the mechanism of hormone action would involve the control of the production of mRNA and the de novo synthesis of enzymes (Vanderhoef and Dute 1981; Theologis and Ray 1982; Zurfluh and Guilfoyle 1982).

Auxin production in the crown gall system joined the molecular age before similar auxin studies in plants, reflecting the advances in bacterial genetics. The description of the large gall-associated plasmid in *Agrobacterium tumefaciens* (Zaenen et al. 1974) was followed by the demonstration that the gall-forming part of the plasmid integrates into the plant cell chromosome (Chilton et al. 1977; Thomashow et al. 1980). Further analysis of the tumor-inducing part of the plasmid uncovered the auxin biosynthesis pathway genes transferred to the plant (Klee et al. 1984; Thomashow et al. 1984, 1986) and provided a complete

Received December 12, 2023. Accepted February 07, 2024. Advance access publication February 21, 2024.

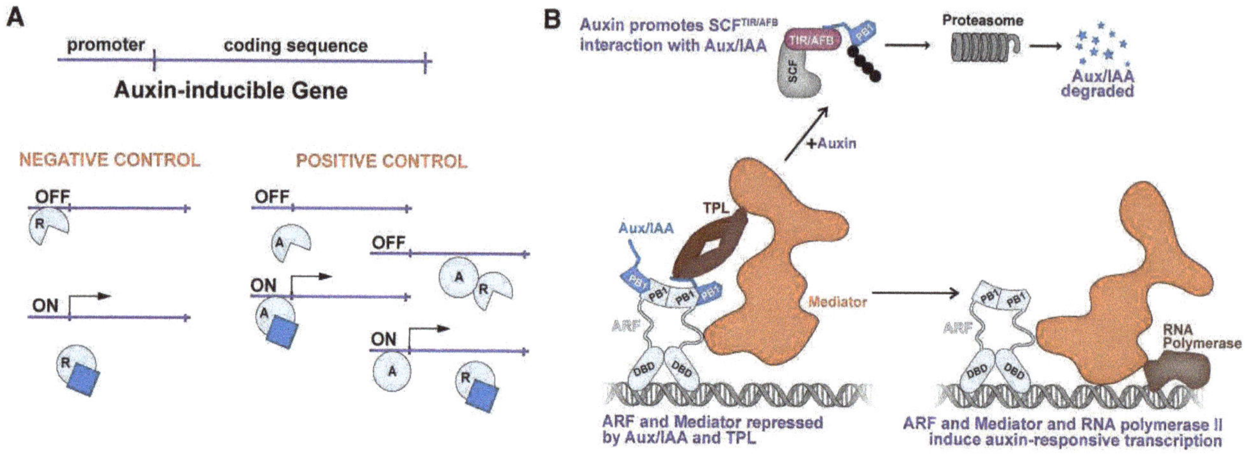

Figure 1. Auxin signaling. **A)** Models proposed in the late 1980s for how the regulation of the auxin-inducible genes might be envisioned (Redrawn from Theologis 1986). "R" indicates a repressor and "A" an activator protein. **B)** A current diagram of auxin signaling, illustrating the remarkable progress in the last 35 years of research. A historically interesting progression of models for the regulation of auxin-inducible genes by derepression, as envisioned in 1985, 1995, and 2005, is diagrammed in Abel and Theologis (2010).

pathway for IAA biosynthesis from tryptophan. It would take more than 25 years before the genes encoding the first described IAA pathway in plants became fully known (Phillips et al. 2011).

Auxin response genes and their promoters

The genes that received early attention were those that were found to be specifically induced by active auxins. Induction was often within minutes of exposure to the hormone and was induced by auxins in the absence of new protein synthesis. These genes were called early, or primary, auxin response genes, and they fell into 3 major groups: Aux/IAAs, SAURs and GH3s. Small auxin up RNA gene clusters were described by McClure et al. (1989), with some responding within 10 minutes of auxin exposure. A common aspect of "all things auxin" is the redundancy of signaling machinery in vascular plants. Thus, over more than 2 decades, it was subsequently shown that in Arabidopsis there are 23 ARF genes, 6 TIR1/AFBs, and 29 Aux/IAA repressors (Du et al. 2020); and other plant species have similar numbers (Ulmasov et al. 1997; Guilfoyle and Hagen 2001; Liscum and Reed 2002).

Aux/IAA genes

The description of multiple copies of *Auxin/INDOLE 3-ACETIC ACID* (*Aux/IAA*) genes per species began before the wide adaption of Arabidopsis as a reference species; *Aux/IAA* were described in several laboratories by differential hybridization in auxin treated vs untreated (or auxin-depleted) hypocotyls or epicotyls in soybean and pea (Hagen and Guilfoyle 2002). The observation that these mRNAs were induced by both active auxins and protein synthesis inhibitors, typically cycloheximide (Guilfoyle 1999), was noteworthy and foreshadowed our current understanding of the nuclear auxin response system. Articles in *The Plant Cell* played a key role in the development of our understanding of the function of the Aux/IAA proteins. Tiwari et al. (2001) showed that *Aux/IAA* genes encode short-lived nuclear proteins that are potent transcriptional repressors, and they identified 4 conserved domains that regulated repressor activity. They concluded that "early auxin response genes are regulated by auxin-modulated stabilities of Aux/IAA proteins" (Tiwari et al. 2001, p 2809), a key aspect of our modern understanding of auxin response regulation. The use of the Arabidopsis genetic resources allowed rapid advances in our understanding of Aux/IAA protein family, and studies of protein turnover and stability showed heterogeneity in both physical properties and the conserved domains. Dreher et al. (2006) linked changes in domain II with variation in both Aux/IAA basal stability and auxin-regulated stability and subsequent auxin responses. Aux/IAA protein stability is now known as the central regulator for auxin signaling processes involving gene expression (Fig. 1) and is the current topic of intense research (reviewed in Figueiredo and Strader 2022).

SAUR genes

The *SMALL, AUXIN-UPREGULATED RNAs*, most often referred to as *SAURs*, were first identified from legume stem sections following 2,4-D treatment (McClure and Guilfoyle 1987) and are among the most rapidly turned over mRNAs found in plants, with half-lives of 10 to 50 min. These RNAs were shown to be induced in only a few minutes after auxin treatment, and induction is not dependent on de novo protein biosynthesis (Abel et al. 1994). The physiological significance of *SAUR* expression in general was clearly shown as an auxin effect by the demonstration of differential expression in gravitropic hypocotyls (McClure and Guilfoyle 1989; Li et al. 1991). *SAURs* were also demonstrated to be present in elongating cells, strongly expressed in epidermal and cortical cells and rapidly increased by auxin treatments (Gee et al. 1991). *SAUR* genes were an early focus in auxin response studies as one of the best characterized gene families induced by auxin. *SAURs* are frequently a large family in flowering plants, with 81 *SAURs* (with 2 pseudogenes) in Arabidopsis (Hagen and Guilfoyle 2002) and similarly large numbers in other plant species (Ren and Gray 2015). Inhibition of protein synthesis, often by treatment with cycloheximide, showed superinduction of *SAURs*, both without exogenous auxin but with a stronger effect after auxin treatment. This increase in transcript levels was through an increase in stabilities of these mRNAs (Li et al. 1994). Although *SAURs* have been studied both in terms of regulation of turnover (Newman et al. 1993; Gil and Green 1996) and control of plant function (Stortenbeker and Bemer 2019), SAUR functions had remained elusive presumably due to the extensive genetic redundancy. Recently, however, decades-long studies of SAUR function at several omics levels have resulted in mechanistic models for SAUR function (Spartz et al. 2014). Now, several intersecting studies have shown that SAURs inhibit the type PP2C-D subfamily of type 2C protein phosphatases to activate plasma membrane H^+-ATPases, resulting in cell expansion (Ren and Gray 2015). In particular, the PP2C.D2, PP2C.D5, and PP2C.D6 protein phosphatases are primarily involved in such cell expansion regulation (Ren et al. 2018), and mutations in a conserved motif of PP2C.D phosphatase result in constitutive activity irrespective of SAUR (Wong et al. 2019).

Gretchen Hagen 3 (GH3) genes

Amino acid conjugates of IAA were first described by Andreae and Good (1955), who found IAA-aspartate as the major product formed after application of IAA to pea stems and roots. These early investigations established several key aspects of auxin amino acid conjugation. First, only active auxins formed separate conjugates; second, conjugation became more rapid over treatment time; third, pretreatment with active auxins induced the conjugation of all applied aux-ins (reviewed in Cohen and Bandurski 1982). Primarily through the work of the laboratories of Brenner, Mumma, and Hamilton, it was shown that a number of different amino acids other than aspartate could form auxin amino acid con-jugates (reviewed in Cohen and Bandurski 1982), and many of these were later found naturally without treatment with auxins (reviewed in Slovin et al. 1999).

The *GH3* gene was first identified in soybean as an early response gene, which is transcriptionally induced in less than 30 min by treatment of auxin (Hagen et al. 1984). Several of the early studies focused on the observation that in contrast to *SAUR* gene expression, *GH3* gene expression was induced by auxin in both hypocotyl and leaf tissues (McClure et al. 1989). In soybean, *GH3* expression occurred in the inner cortex and protoxylem ridges of roots, transient expression was shown during flower and pod development (Gee et al. 1991), and tissue-specific expression has been noted in other species (Peat et al. 2012). Importantly, the observation that different tissues responded rapidly to exogenous auxin by ex-pressing hormone-responsive *GH3* genes (Gee et al. 1991) was informative in driving future studies in context-specific auxin responses (Singh et al. 2015). Careful analysis of the promoter region of *GH3* demonstrated that it is composed of several auxin response elements (AuxREs). These AuxRE are additive in their ability to form a strongly auxin inducible promoter, and auxin inducibility is due to the TGTCTC con-served DNA element (Ulmasov et al. 1995).

These studies of *GH3* promoter function and AuxRE sequence lead directly to the assembly of a highly active synthetic auxin promoter referred to as *DR5* (Ulmasov et al. 1997). In its original description, a DR5-GUS reporter was constructed containing 7 repeats of a 11-nucleotide sequence, including the TGTCTC element cloned upstream of a −46 CaMV 35 s promoter with a TMV 5′ leader (Ulmasov et al. 1997). The DR5-GUS reporter has been an im-portant addition to the tools available for visualization of auxin signaling in planta. Since its first report there have been variations, for example, using luciferase and fluorescent proteins to report. Following the crystallization of the DNA binding domain of AUXIN RESPONSE FACTORs (ARFs), the preferred binding site of both ARF1 and ARF5 appeared to be TGTCGG rather than the "canonical" AuxRE TGTCTC first reported (Boer et al. 2014). These promoters monitor the transcriptional output in the presence of auxin and have been used at the single cell level. Over the last several years significant improvements have been made in auxin re-porters, and these were recently reviewed in *The Plant Cell* (Jedličková et al. 2022). When properly employed, these advances can provide a wealth of information on auxin responses, responsiveness, and cellular distribution un-approachable by other procedures currently available.

The importance of GH3 in advancing our understanding of auxin signaling has been immense, but for many years 2 "problems" in our understanding of auxin remained unsolved: 1) What does the protein encoded by *GH3* do? 2) How are the IAA amino acid conjugates made? These questions were answered when Staswick showed that the *GH3*-like genes encode the GH3 family of acyl acid-amido synthetases, including the substrate specific IAA-amido synthetases (Staswick et al. 2002, 2005; Chen et al. 2010). Several GH3-like proteins form amino acid amide conjugates with IAA, jasmonic acid, or salicylic acid, indicating that they play a generalized role in various hormone metabolic pathways. This discovery linking a critical function to the GH3 gene products had an important role in our current understanding of both auxin signaling and metabolic regulation.

ARFs

The ARFs are transcription factors with diverse but often overlapping functions. The identification of the AuxRE TGTCTC sequence enabled the initial screen for transcription factors that recognized the element. Using a yeast 1-hybrid screen baited with a very responsive, synthetic AuxRE containing a palindromic repeat of the TGTCTC element,

ARF1 was isolated (Ulmasov et al. 1997; Guilfoyle and Hagen 2001). These initial studies provided evidence for 3 additional ARFs, a finding important for defining auxin gene induction at this stage in the progression in our understanding of auxin signaling (Ulmasov et al. 1997). A total of 9 cDNA clones for ARF-like proteins were isolated by cDNA library screens (Guilfoyle et al. 1998). The completion of the first Arabidopsis genome extended the number of ARFs in Arabidopsis to 23, including 1 pseudogene, ARF23, that lacks a N-terminal B3-type DNA binding domain (DBD) (Li et al. 2016; Guilfoyle and Hagen 2001). These early studies provided significant information of ARF function, including by using truncated versions of ARF1 that probed the minimal DBD required to bind to AuxRE and the formation of dimers of ARF as well as Aux/IAA dimerized protein.

Since these early landmark discoveries of ARF transcription factors and the description of their interactions with the Aux/IAA, the description of these genes and their encoded multidomain proteins has revealed much about how auxin is perceived as well as how other signaling systems use related transcription factors. Our understanding of auxin signaling continues to benefit from studies focused on the ARFs. ARF proteins have an N-terminal DBD that has a composite dimerization domain, which has 2 separate subdomains that fold into a single unit: a B3-type DNA-interaction domain and a domain of unknown function (Boer et al. 2014; Mutte et al. 2018). The C-terminal PB1 domain in both ARF and Aux/IAA proteins mediates homo- and hetero- oligomerization among these family members (Korasick et al. 2014; Nanao et al. 2014). "Activator" ARF proteins can elicit a transcription response by recruiting chromatin remodelers (Wu et al. 2015) or components of the Mediator complex (Ito et al. 2016; Leydon et al. 2021)

Heterogeneity among the ARFs has made uncovering functions difficult but also contributes to an understanding of the multiplicity of responses auxin elicits. For example, recently it was shown that a specific E3 ubiquitin ligase directly interacts with ARF19 and ARF7 to regulate their accumulation, con-densation, and partitioning between the nucleus and cyto-plasm (Powers et al. 2019; Jing et al. 2022), unexpected findings that both explain aspects of protein-protein interac-tions and advance new hypotheses about regulatory pro-cesses at play. Indeed, regulation of ARF stability is likely to be important, as "activating" and "repressing" ARFs compete for the same binding sites to regulate gene expression (Lavy et al. 2016; Kato et al. 2020).

TIR1

The importance of the ubiquitin–proteasome pathway in auxin response was recognized well before the function of TRANSPORT INHIBITOR RESPONSE1 (TIR1) and the related AUXIN SIGNALING F BOXs (AFBs) in auxin molecular recognition was elucidated. The isolation of *auxin resistant1* (*axr1*) mutants of Arabidopsis (Lincoln et al. 1990) derived from a large-scale screen for auxin resistance provided an important step. In parallel, isolation of the *tir1* mutant was from a screen for mutants resistant to auxin transport inhibitors (Ruegger et al. 1998). *AXR1* was shown to encode a protein that was related to a similar yeast protein that was known to function to activate the ubiquitin-related protein, whereas *TIR1* was shown to encode a protein related to human SKP2 (Ruegger et al. 1998). Mutations in *TIR1* had a synergistic interaction with mutations in *AXR1*, suggesting that the 2 genes function in overlapping pathways and thus providing strong evidence that both were in the ubiquitin–proteasome pathway (Gray and Estelle 2000). TIR1 was then shown to dir-ectly interact with the Aux/IAA proteins and promote their degradation (Gray et al. 2001; Dharmasiri et al. 2003).

Fifteen years from the appearance of that first paper in *The Plant Cell* (Lincoln et al. 1990), 3 leading-edge papers were published (Dharmasiri et al. 2005a, 2005b; Kepinski and Leyser 2005), showing that the Arabidopsis F-box protein TIR1 is an auxin receptor. Following these reports, the authors suggested that there are additional auxin receptors, including TIR1-related proteins such as AFB1, 2, 3 (Dharmasiri et al. 2005b); others also suggested the possibility of a plasma membrane/extracellular receptor (Badescu and Napier 2006), forward-looking perspectives that seems to have predicted many of our current models. The Arabidopsis genome encodes 6 TIR1/AFB proteins, and genetic analysis shows that the loss of all 6 results in lethal early embryo defects; however, there is pronounced overlap in the function of even the distantly related TIR1/AFB genes. Within the land plants, TIR1/AFB proteins fall into 4 phylogenetic clades: (1) TIR1/AFB1, (2) AFB2/3, (3) AFB4/5, and (4) AFB6 (Prigge et al. 2020). The 6 Arabidopsis members fall into only the first 3 clades (Shimizu-Mitao and Kakimoto 2014). AFB1, however, appears to have a specific function related to both root gravitropism and the inhibition of root growth by auxin (Prigge et al. 2020) while also displaying roles in the cytoplasm to regulate rapid auxin responses (Chen et al. 2023; Dubey et al. 2023). AFB5 shows a different auxin-binding chemical specificity relative to TIR1 (Quareshy et al. 2018), and AFB4 and AFB5 are important in the response to the auxinic herbicide picloram (Prigge et al. 2016). These studies suggest that as more information is obtained about the TIR1/AFB proteins, additional specificities and developmental differences may emerge (Du et al. 2022).

This foundational work coalesced on a model in which nuclear auxin signaling is dominated by a repression-derepression paradigm, with the earliest targets of auxin transcription acting to attenuate auxin signaling (Aux/IAA repressors) or conjugate IAA to reduce the pool of free auxin (GH3s). Thus, it seems likely that a sustained auxin response will be necessary to overcome these immediate feedback responses to result in long-term effects of auxin on organogenesis and development. Contemporary research has focused on understanding the molecular basis of auxin transcriptional response.

Several studies have contributed to our understanding of the molecular basis of nuclear auxin responses. Structures of the ARF DNA binding domain (Boer et al. 2014) and

PB1 interaction domain (Korasick et al. 2014; Nanao et al. 2014) have enabled studies on ARF binding specificity to distinct cis-regulatory elements and their spacing and an understanding of ARF regulation (reviewed in Morffy and Strader 2022). Recapitulation of the auxin signaling module in yeast (Havens et al. 2012) further enabled interrogation of ARF DNA binding specificity (Galli et al. 2018; Lanctot et al. 2020).

The structure of TIR1 revealed that this E3 forms the co-receptor for auxin, along with the degron from an Aux/ IAA protein (Tan et al. 2007), providing—to our knowledge —the first molecular model of a ligand acting as a "molecular glue" between 2 proteins. Later studies have focused on roles for the extensive intrinsically disordered regions of the Aux/ IAAs and their contributions to Aux/IAA interaction with TIR1 (Niemeyer et al. 2020; Ramans-Harborough et al. 2023). These studies are driving forward our understanding of the complicated interactions among the co-receptors and are revealing insights into how auxin responses are tuned.

Auxin metabolism

Tryptophan

In several ways, insightful research into auxin biosynthesis lagged behind that of auxin signaling. Most significantly in 1989, we were at an early stage in even being able to measure auxin biosynthesis and no clear pathways, enzymes or genes, were developed candidates (Cohen et al. 1990). This lack of understanding was to quickly change, however, with ad-vances in studies of tryptophan biosynthesis. Last et al. (1991) described a mutant in tryptophan synthase β (TSβ), 1 of 2 closely related *TSβ* genes (*TSβ1* and *TSβ2*) in Arabidopsis. Both *TSβs* contain 4 introns and a chloroplast leader sequence, and both genes are transcribed; however, *TSβ1* was shown to produce >90% of *TSβ* mRNA in leaf tis-sue. The *trp2-1* mutant, defective in TSβ1, was a conditional mutant able to grow under low light but requiring exogenous tryptophan at higher light. Niyogi and Fink (1992) described 2 anthranilate synthase genes, important because the shunt to anthranilate is the committed step in tryptophan biosyn-thesis. Studies using such mutants allowed investigation of not only IAA but also of other tryptophan-derived com-pounds such as the indolic phytoalexin camalexin (Zhao and Last 1996). An important lesson originating in the cama-lexin studies was to point to the clear need to study not just end-products but entire pathways and metabolic networks. Soon after *trp2-1* was described in Arabidopsis, the maize auxotrophic mutant *orange pericarp* was described, where both copies of the *TSβs* were knocked out, resulting in plants that without tryptophan supplementation were lethal after 1 to 2 weeks post germination (Wright et al. 1992). Both *TSβ* mutants, in maize and Arabidopsis, proved critical for reveal-ing labeling studies of IAA biosynthesis from early precursors (Wright et al. 1991; Normanly et al. 1993). These advances in our understanding of the biochemistry and molecular genetics of tryptophan biosynthesis would, in the coming decades, prove to be invaluable in the furtherance of advances in aux-in biosynthesis as well (Radwanski and Last 1995).

Auxin conjugates

Auxin conjugate biology, however, relative to the study of auxin biosynthesis itself was a richly advancing area of research in the early days of *The Plant Cell*. The first IAA con-jugating enzyme to be cloned, iaglu, UDP-glycosyltransferase, was from maize (Szerszen et al. 1994) and was a testament to decades of foundational biochemical and analytical studies (Slovin et al. 1999). Although it is clear that biochemistry and molecular biology remained codependent sciences, the sensitivity of the community on such as relationship showed at times in the pages of *The Plant Cell* (Varner 1989; Venis 1997). Nevertheless, as discussed earlier, the finding that *GH3* was encoding an IAA-amido synthetase (Staswick et al. 2005) later showed the value of such pairing of subdisciplines.

The IAA amino acid story had been complicated because of the number of different amino acids and their differential utilization by plants for IAA release (Bialek et al. 1983). Most IAA conjugates, when applied to plants, can mimic the ef-fects of free IAA on plant growth due predominately to their slow hydrolysis (Hangarter and Good 1981; Magnus et al. 1992). This observation enabled the identification of Arabidopsis mutants that respond by root growth in the presence of inhibitory levels of applied IAA conjugates (Bartel and Fink 1995)—typically plants are selected that show normal root growth when exposed to concentration of conjugate that stunts roots in the wild-type plants (Davies et al. 1999). IAA-conjugate hydrolases release free IAA from their conjugated forms and thus appear to be in-volved in regulating free IAA levels. The first gene isolated by this strategy, *ILR1* (indoleacetyl-leucine [IALeu]-resistant, but also resistant to IAPhe; Bartel and Fink 1995), was fol-lowed by reports of a number of related mutants resistant to IAPhe (*icr1*, *icr2*; Campanella et al. 1996), IAAla (*iar3*, Davies et al. 1999) that are likely also defective in IAA-AA hy-drolases. Seven IAA-conjugate hydrolases have been reported from Arabidopsis—AtILR1, AtIAR3, AtILL1, AtILL2, AtILL3, AtILL5, and AtILL6—and many have been tested for activity against a robust number of IAA amino acids (LeClere et al. 2002). *ILL5* has been postulated to be a pseudogene (Davies et al. 1999; Widemann et al. 2013). Early reports that IAAsp and IAGlu were only inactive catabolic metabolites have been shown to be incorrect, and several IAA-conjugate hydrolases from different species can hydrolyze these conjugates (LeClere et al. 2002; Tang et al. 2019). However, they can also be directly and irreversibly oxidized to their oxindole derivatives as an alternative metabolic fate (Tuominen et al. 1994; Hayashi et al. 2021; Müller et al. 2021; Guo et al. 2022). Indoleacetyl-L-tryptophan (IAtrp) is a naturally occurring conjugate in both Arabidopsis and soy-bean (Staswick 2009; Yu et al. 2014) and inhibits several as-pects of auxin, including inhibition of root growth and gravitropism. A dominant mutant, called *tryptophan conjugate response1-D*, was identified that conferred resistance to the jasmonic acid-tryptophan conjugate, but not to the IAtrp, and was shown to be due to a single amino acid substitution in the IAA cellular permease AUX1 (Staswick et al. 2017). Apparently, of the 4 Arabidopsis ILR1-like proteins tested, only ILL2 had slight activity toward IAtrp (LeClere et al. 2002), but few other reports have tested for IAtrp hydrolase activity. Because IAtrp is an inhibitor of auxin response, it is possible that some auxin conjugate hydrolases will have the dual role of releasing active IAA and removing a response regulator. Although it has been commonly considered that auxin conjugates are inactive and that observed bioassay activity be attributed to release of an active auxin by hydrolysis, this may be an oversimplification. For example, it is unlikely that IAtrp antiauxin activity is the result of hydrolysis to IAA, and several reports linking auxin conjugates to their own biological activity related to abiotic and biotic stress have been reported (Oetiker and Aeschbacher 1997; González-Lamothe et al. 2012). Other studies have linked GH3-mediated auxin conjugation with stress adaptation (Park et al. 2007), defense response and disease resistance (Ding et al. 2008; Domingo et al. 2009), and grape ripening (Böttcher et al. 2010). Several mechanisms of action have been proposed for auxin conjugates. One possible mechanism of action is through direct perception, as is the case for the active form of jasmonic acid, jasmonoyl-L-isoleucine—but there is no support for this proposed mechanism. A second possible mechanism of action is through "delivery" of the auxin precursor to a specific cell or cellular compartment for release of the active auxin molecule, as the auxin precursor indole-3-butyric acid (IBA) may function. Auxin conjugates may function where the amino acid serves as a kind of biological ZIP code as was proposed over 40 years ago (Cohen and Bandurski 1982), or the conjugates may function at the intersection between release of the free active form and catabolic processes through ring oxidation (Hayashi et al. 2021).

Glucose, inositol, and other sugar conjugates are fundamentally different from the amino acid conjugates. While amino acid conjugates are strong amide bonds, similar to the peptide bonds in proteins, the chemistry of IAA-1-O-glucose (IAglc) is very different. IAglc is formed by the action of UDP-glycosyltransferase that catalyzes the transfer of the monosaccharide from, for example, UDP-glucose to IAA. The reaction is considered to yield the "product" of IAglc, although the formation of UDP-glucose from IAglc is favored with a $K_{eq} = 10^{-1}$; but the reaction proceeds because of the massive amount of UDP-glucose relative to IAA in the cell (Slovin et al. 1999). This, however, means that IAAglc can be considered an "IAA" donor to form esters with suitable sugars, cyclitols, or alcohols (Michalczuk and Bandurski 1982; Starzyńska and Kowalczyk 2012) or to glycoproteins (Jakubowska and Kowalczyk 2004; Ostrowski et al. 2015). UDP-glycosyltransferases have many diverse functions in plants (Ostrowski and Jakubowska 2014), and many of the early studies showed them essential for plant growth but without clear knowledge of the target compound (Woo et al. 1999). Other studies aimed at understanding the phenotype of stress responses discovered novel auxin related substrates such as IBA (Tognetti et al. 2010; Jin et al. 2013), indole-3-pyruvate (Chen et al. 2020), and 2-oxindole-3-acetic acid (Tanaka et al. 2014; Mateo-Bonmatí et al. 2021). Much of the early studies on IAglc formation was accomplished using maize, where IAglc serves as a metabolic precursor to IAA-*myo*-inositol, its glycosides, and higher molecular weight IAA-glucans (Slovin et al. 1999). IAglc is also found in Arabidopsis and in other plants (Tam et al. 2000; Porco et al. 2016; Pěnčík et al. 2018; Brunoni et al. 2020).

When Szerszen et al. (1994) cloned the iaglu gene, a UDP-glucosyltransferase, the presence of IAglc and IAA es-ters in the seed endosperm often led to the speculation that sugar conjugates might be specific to monocots. However, UDPG:indole-3-acetyl-β-D-glucosyltransferase (IAA-UGT) was reported in Arabidopsis by Jackson et al. (2001). Because IAglc and UDP-glucose are in equilibrium in situ, the role of IAA-UGT and IAglc seems more complex than for the amino acid conjugates. Several reports show that overexpression of IAA-UGT can result in an auxin deficient phenotype (Liu et al. 2019; Yu et al. 2019; Aoi et al. 2020); however, earlier reports suggested that IAA-UGT has a regulatory role where IAglc may serve as a metabolic regulator (Iyer et al. 2005; Ludwig-Müller et al. 2005). The broadening roles of these components of auxin metabolism are still evolving, and their role in auxin regulation remains an active area of inquiry.

Indole-3-butyric acid

Research in the 1930s on collections of auxin-like compounds resulted in the adoption of IBA as the standard for rooting of cuttings for propagation. This work further established the now long-standing link between auxin and auxin-like compounds and root development. IBA was initial-ly assumed to be a synthetic auxin, not an endogenous auxin, and in many bioassays it was also shown to be a weaker auxin than IAA (Woodward and Bartel 2005; Simon et al. 2013). A decade or more after its first uses for rooting, there appeared a number of early reports of IBA in plants, and it was later clearly identified by gas chromatography-mass spectrometry in kernels and leaves of maize and subsequently in at least 7 other plant species, including Arabidopsis (Epstein and Ludwig-Müller 1993). The processes by which IBA is synthe-sized in plants (Ludwig-Müller 2007) needs further clarifica-tion (Damodaran and Strader 2019).

The biochemical conversion of IBA to IAA has been demonstrated in a variety of plants using mass spectrometry tracer methods (see e.g. Kreiser et al. 2016). The isolation and characterization of IBA-resistant mutants or mutants resistant to other extended side chain auxins (2,4-dichlorophenoxybutyric acid (2,4-DB); Hayashi et al. 1998) were critical for progress on understanding IBA. Selections for plants that respond normally to IAA or synthetic auxins has allowed the isolation of mutants defective in IBA responses, peroxisomal β-oxidation, and peroxisome development (Zolman et al. 2000; Woodward and Bartel 2005; Baker et al. 2006; Damodaran and Strader 2019). The peroxisomal conversion of IBA to IAA involves 6 steps, all essentially analogous to β-oxidation of fatty acids: activation (thioesterification), oxidation, hydration, dehydration, thiolysis, and hydrolysis (Adham et al. 2005; Spiess and Zolman 2013; Rinaldi et al. 2016; Jawahir and Zolman 2021). IBA and 2,4-DB are both converted to their 2-carbon side chain analog, IAA or 2,4-D, indicating a less restrictive chemical process. Following β-oxidation, the IAA or auxin-like product that is generated is exported from the peroxisomes. The genetic screens for Arabidopsis mutants resistant to applied IBA (initially named *ibr, IBA resistant* or *ped, peroxisome defective*) defined an important aspect of our understanding of auxin metabolism and the special role that IBA plays. For example, the analysis of *ech2* and other *ibr* mutants demonstrated that IBA-derived IAA apparently has a specific role in root cell expansion (Strader et al. 2010b) as well as cell expansion of root hairs and cells in cotyledons (Strader et al. 2010a, 2011). IBA is, in biological tests, active only upon its conversion to IAA (Strader et al. 2010b, 2011), indicating it is an auxin precursor rather than a weak auxin as was originally hypothesized. This finding was confirmed when the 4-carbon side chain of IBA was shown to inhibit formation of the TIR1/AFB-Aux/IAA coreceptor complex required for auxin activity (Lee et al. 2014; Uzunova et al. 2016).

Understanding IBA transport was initially compared with IAA polar auxin transport (PAT), where significant metabolism of the applied compound typically does not occur. In contrast, when [^{3}H]IBA was studied in plant segments, it was found to be metabolized to [^{3}H]IAA as it moved through the tissue (Ruzicka et al. 2010). More detailed metabolic ana-lysis was possible using the stable isotope [^{13}C]-labeling and mass spectrometry analysis (Liu et al. 2012a) in place of the radioactive [^{3}H]-tracers. With a modified agar block PAT experiment, a significant portion of the label detected in the agar receiver block was [$^{13}C_1$]IAA derived from [$^{13}C_1$]IBA. The IAA-PAT inhibitor N-1-naphthylphthalamic acid (NPA) inhibited the movement of [$^{13}C_1$]IAA into the receiver. The transport [^{13}C]-labeled free IBA was much lower than the transport of [^{13}C]-labeled free IAA, and IBA movement was not inhibited by NPA. IBA applied was metabolized to IAA and to IBA conjugates during the transport while applied IAA remained as free IAA throughout transport. IBA transport in the *nakr1* mutant, defective in normal phloem function (Tian et al. 2010), was unaffected. IBA transport was also unaffected in the *ibr1 ibr3 ibr10* triple mutant (Strader et al. 2010b), but conversion to IAA was inhibited and the PAT of IAA thus suppressed (Liu et al. 2012a). IBA movement itself accounted for less than 1% of the bulk transport from exogenous IBA; thus it is not likely that longer distant transport contributes substantially to IBA pools at sites distant from where the IBA is applied. These results show that by using specific stable labeled compounds and mass spectrometry analysis, the movement and metabolic changes of the labeled compounds can be accurately profiled in mutant and wild-type plants and suggests the value of monitoring of metabolic networks in contrast to single compound studies (Liu et al. 2012b).

Transport proteins that are active with IBA remains an incomplete story, but several things are clear. First, IBA and IAA have distinctly different efflux carriers and although there is some indication that like-AUX1 protein LAX3 could be involved in IBA and IAA uptake, this function is not fully resolved (Damodaran and Strader 2019). Conversion of IBA to IAA by β-oxidation requires import of IBA into the peroxisome by PXA1 (Zolman et al. 2001), which also has a role in fatty acid metabolism in dark-treated plants after depletion of starch reserves (Kunz et al. 2009). Conversion also requires ATP, supplied by plant peroxisomal adenylate nucleotide carrier AtPNC gene products (Linka et al. 2008). The IBA efflux process is better studied, and the role Pleiotropic Drug Resistance (PDR) proteins seems clear (Damodaran and Strader 2019). Studies in the early 2000s showed that ABCG36 and ABCG37 ATP-binding cassette transporters were involved in IBA efflux from root tissues (Stein et al. 2006; Strader et al. 2008; Strader and Bartel 2009; Ruzicka et al. 2010). ABCG36 and ABCG37 are found on the outward-facing plasma membrane of root cells suggestive of export to the external media (Strader and Bartel 2009; Ruzicka et al. 2010). Vacuolar movement of IBA involves a protein TOB1 (Michniewicz et al. 2019) that is part of the nitrate or di/tri-peptide transporter family and is a part of the major facilita-tor superfamily of transporters (Quistgaard et al. 2016). Likewise, NPF7.3 mediates IBA uptake to regulate gravitrop-ism (Watanabe et al. 2020) As detailed in their more compre-hensive review, Damodaran and Strader (2019) note description of these transporters has illustrated the involve-ment of IBA transport in control of aspects of plant develop-ment but "much remains to be uncovered."

IAA biosynthesis

While many reactions regulate auxin and plant responses to auxin, the question of the initial perception and how auxin is made have been the focus of research for over a century (Enders and Strader 2015). Early *Plant Cell* papers focused on using *Agrobacterium* genes to design reporter or expression systems for understanding development and light regulation (Lincoln et al. 1990; Karlin-Neurnann et al. 1991; Romano et al. 1993). The birth of *The Plant Cell* coincided with the discovery of the first tryptophan auxotrophic or near auxotrophic mutants (Walbot 1991). These studies were critical because they established the concept that IAA biosynthesis was complex and interwoven with early indole biochemistry in plant metabolism (Tivendale et al. 2014). Before these studies, IAA biosynthesis was envisioned as either a single linear route originating with tryptophan or one with minor differences in the intermediates involved—similar to what was the case in bacterial IAA biosynthesis (Glick et al. 1999). The developmental consequences of multiple pathways, one involving a tryptophan direct route and one independent of tryptophan, was apparent during stages of plant embryogenesis (Laux and Jürgens 1997) where both pathways are essential (Cooke et al. 1993; Wang et al. 2015) and where large changes in IAA levels are observed following fertilization (Ribnicky et al. 2002). The tryptophan independent pathway to IAA has been linked, in part, to indole-generating enzymes, such as indole synthase in Arabidopsis (Zhang et al. 2008) and indole-3-glycerol phosphate lyase in maize (Kriechbaumer et al. 2008) but remains an import-ant topic of current research (Tillmann et al. 2022).

A flavin monooxygenase was first identified as being an enzyme involved in auxin biosynthesis by an activation tag screen in Arabidopsis. This screen found a novel phenotype that suggested an auxin overproduction phenotype, with long hypocotyls and epinastic cotyledons when grown in the light, an Arabidopsis plant reminiscent of the *Yucca schidigera* of the Mojave Desert (Zhao et al. 2001). When first characterized the function of this enzyme, termed YUCCA, was incorrectly assigned based on tryptophan ana-log feeding experiments that suggested it hydroxylated the amino group of tryptamine. This assigned YUCCA, at least temporarily, to the tryptamine branch of the IAA biosyn-thesis pathway.

During this same time the role of indole-3-pyruvate (IPyA) in IAA biosynthesis was advancing (Tivendale et al. 2014). First, IPyA was isolated from tomato and Arabidopsis (Cooney and Nonhebel 1989; Tam and Normanly 1998). Second, the gene for the tryptophan aminotransferase TAA1 was isolated and shown to be developmentally import-ant (Stepanova et al. 2008; Tao et al. 2008). The first sugges-tion that YUCCA may not be in the tryptamine pathway but rather linked together with TAA1 is a single pathway was Strader and Bartel (2008), who observed "Alternatively, it is tempting to take the genetic evidence at face value—the same terminal phenotypes suggest that the same process is disrupted—and speculate that the TAA1/TAR and YUCCA enzymes might actually act in the same pathway" (p 339). This observation was confirmed in a *Plant Cell* paper describ-ing studies of the maize TAA1 co-ortholog mutant *vanishing tassel2* (*vt2*) that shared many phenotypic similarities with *sparse inflorescence1* (*spi1*) monooxygenase mutants in maize. The more complex phenotype of such mutants in maize helped in this characterization, but analysis of auxin levels in the mutants proved decisive. Both *spi1* and *vt2* single mutants exhibited a reduction in free auxin levels, but the *spi1 vt2* double mutants did not have a further reduction compared with *vt2* single mutants (Phillips et al. 2011). This work was followed by 3 papers several months later providing additional confirmation in studies in Arabidopsis (Mashiguchi et al. 2011; Stepanova et al. 2011; Won et al. 2011) and later by the description of an additional tryptophan amino transferase, VAS1, linking IAA and ethylene biosynthesis (Zheng et al. 2013; Pieck et al. 2015). Together, these studies established a fully known pathway of IAA biosynthesis, a research study that began over a half century before it was finally accomplished (Tivendale et al. 2014; Morffy and Strader 2020).

The elucidation of the IPyA pathway has tempted some analysis to downplay research into the role of other intermediates and genes. However, assumptions of the "importance" of other possible routes cannot be deduced by phenotypic analysis alone and especially at only 1 life stage (Di et al. 2016; Solanki and Shukla 2023). There are 3 pathways in plants that apparently lead to IAA production from tryptophan: (a) the indole-3-pyruvic acid (IPyA) pathway, (b) the indole-3-acetaldoxime (IAOx) pathway, and (c) the indole-3-acetamide (IAM) pathway (Fig. 2). There are also variations on these 3 pathways: for example the possible involvement of indole-3-acetonitrile. Evidence for these pathways is not easily dismissed, as, for example, *superroot 1* and 2 are auxin overproducers that also alter glucosinolate metabolism. These mutations lead to the accumulation of endogenous free IAA, which induces development of adventitious roots on the hypocotyl but it seems unlikely the IPyA pathway is directly involved (Boerjan et al. 1995; King et al. 1995; Barlier et al. 2000; Sorin et al. 2005). Studies on sites of IAA biosynthesis in roots suggest that more than 1 tissue produces IAA and also provide evidence for multiple pathways (Ljung et al. 2005). In Brassicaceae, entry into the IAOx pathway involves specific cytochrome P450 enzymes, CYP79B2 and CYP79B3 (Hull et al. 2000; Zhao et al. 2002; Sugawara et al. 2009), and overexpression of these enzymes leads to indole-3-acetonitrile accumulation (Nonhebel et al. 2011). A role for indole-3-acetonitrile and nitrilases was also previously suggested (Bartel and Fink 1994; Normanly et al. 1997), but how they are involved with IAA metabolism seems complex (Lehmann et al. 2017). New tools have come into use for studies to resolve these complex and redundant biochemical events. In addition to genomic methods, including coexpression analysis (Abu-Zaitoon et al. 2022, 2023), new analytical network level analysis has been enabled by stable isotope methods for the measurement of compounds and the metabolic flux through a pathway, and chemical biology has been enabled by a suite of inhibitors of specific processes (Tillmann et al. 2021, Cohen et al. 2022). These promise to continue the progress to solve the unanswered metabolic questions.

IAA oxidation

For a plant hormone to be regulatory, it must turn over to maintain signaling control (Fig. 3). Thus, the catabolism of IAA is central to its function and thus has had a long history of research (Thimann 1934). Because IAA can be degraded by plant peroxidases, the focus before 1980 was largely on IAA peroxidation resulting in decarboxylation (Reinecke 1990; Normanly et al. 1995), although oxidized forms with a carboxylic acid retained were described (Kinashi et al. 1976; Tsurumi and Wada 1980; Nonhebel et al. 1985a). An early example of metabolic flux analysis, however, changed this (Epstein et al. 1980). Quantitative flux analysis led to experiments resulting in the isolation of the ring-oxidized 2-oxo-3-indolineacetic acid (oxIAA) as the major degradation product and identified this first irreversible step in

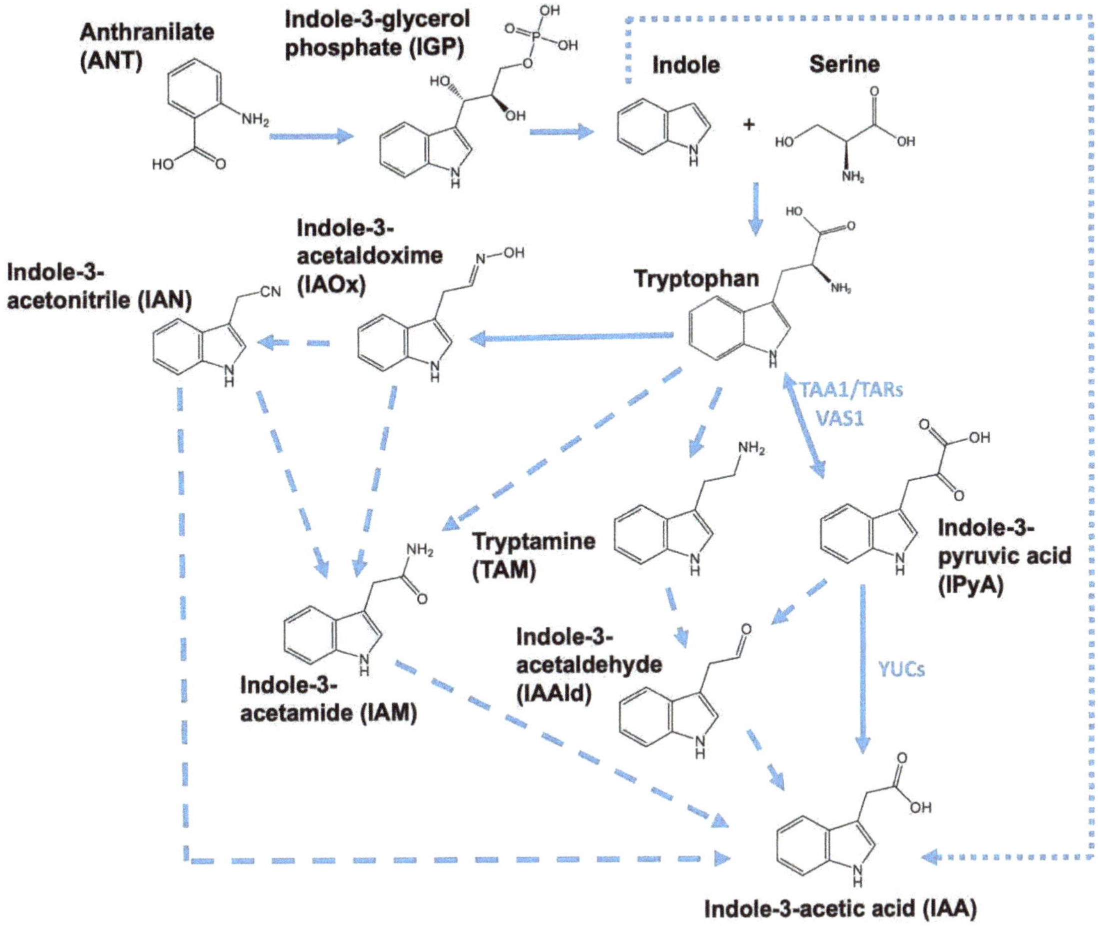

Figure 2. The major proposed routes of IAA biosynthesis showing the IPyA 2 step pathway consisting of an amino transferase step (TAA1/TARs or VAS1) and a monooxygenase reaction (YUC), the indole (tryptophan-independent) option, and possible alternative steps from tryptophan. Figure modified from an original drawn by Dr. Molly Tillmann.

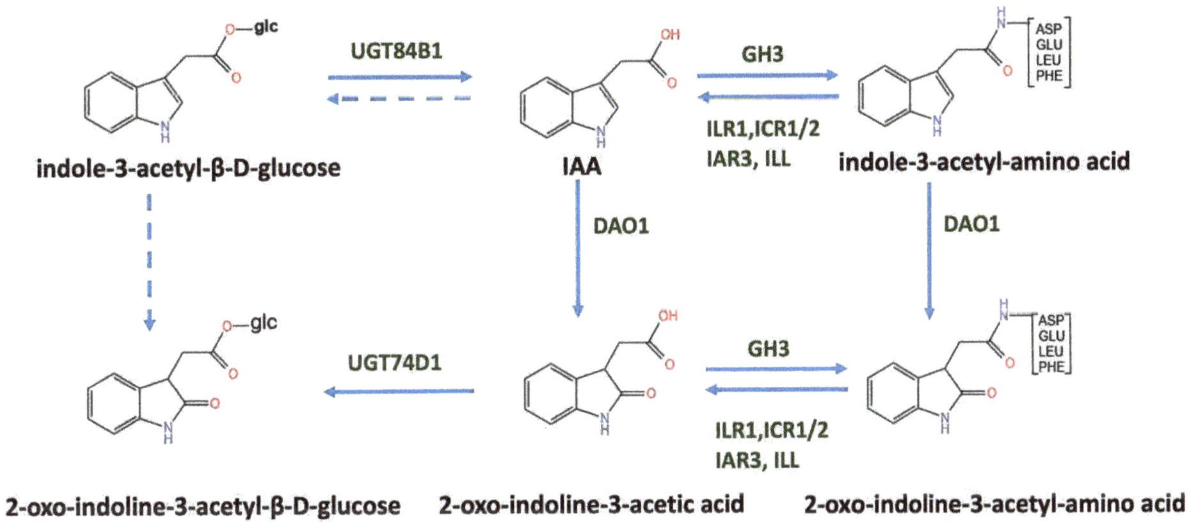

Figure 3. A simplified diagram of the first steps in IAA inactivation pathways as currently envisioned. Dashed arrows represent metabolic steps for which a potential gene function has not been reported. IAA-AA and oxIAA-AA represent amide-linked conjugates with amino acids. The enzymes responsible for IAA metabolism shown in black boxes are as follows: DAO1, Dioxygenase for Auxin Oxidation 1; GH3, Gretchen Hagen 3; UGT84B1/UGT74D1, UDP-glucosyltransferase 84B1/74D1; ILR1, IAA-Leu-Resistance 1; ICR1/2, IAA-conjugate resistance 1 and 2; IAR3, IAA-Ala-resistant; ILL, ILR1-like. Updated and redrawn from Hladík et al. (2023).

IAA catabolism (Reinecke and Bandurski 1981). Quantitative studies showed oxIAA was endogenous in both maize endosperm and shoot tissues and was present in amounts similar to that of IAA (Reinecke and Bandurski 1983). An in vitro enzyme assay for the conversion of IAA to oxIAA was described (Reinecke and Bandurski 1988), and in vivo conversion of exogenous IAA was confirmed in pine seedlings (Ernstsen et al. 1987). Once formed, oxIAA is not the terminal step in catabolism (Nonhebel 1986). In maize, oxIAA was shown to be further metabolized by another oxidation step oxidation to 7-hydroxy-2-oxo-indoline-3-acetic acid and then glycosylated to 7-hydroxy-2-oxo-indoline-3-acetic acid-7^1-O-β-D-glucopyranoside (7-OH-oxIAA-glc) (Nonhebel and Bandurski 1984; Nonhebel et al. 1985b). When 5-^{3}H-7-OH-oxIAA-glc was fed to corn seedlings, it released ^3H_2O, indicating that the 5-position on the ring is labile and may be also oxidized (Lewer and Bandurski 1987). In other plants, IAA-Asp was oxidized into either 3-hydroxy-2-oxo-indoline-3-acetylaspartic acid or 3-(O-β-glucosyl)-2-oxo-indoline-3-acetylaspartic acid in *Vicia faba* (Tsurumi and Wada 1985, 1986a, 1986b) or formed oxIAA-aspartate in poplar (Plüss et al. 1989).

Early studies of oxIAA catabolites showed they were inactive at promoting growth in auxin bioassays (Weis 1966, Sembdner et al. 1981, Reinecke and Bandurski 1983), although there was 1 contradictory report of auxin activity for oxIAA in pea (Galston and Chen 1965). More recent studies, however, confirm the findings that oxIAA lacks auxin activity (Pěnčík et al. 2013; Peer et al. 2013). Also, oxIAA does not induce the expression of the IAA reporters DR5rev:GFP or 2XD0:GUS (Peer et al. 2013), does not interact with the TIR1/AFB auxin receptors (Pěnčík et al. 2013), and is not transported like IAA (Kubeš et al. 2012; Peer et al. 2013).

The genes underpinning IAA degradation were first reported for the gene *DAO* from a male sterile mutant of rice, where the gene product was shown to be capable of converting IAA into oxIAA (Zhao et al. 2013; Stepanova and Alonso 2016) and its subsequent glucosylation by the UDP-glucosyltransferase 74D1 (Tanaka et al. 2014). Soon after the activity was confirmed in rice, functional orthologs of this gene, *AtDAO1* and *AtDAO2* were described in the Arabidopsis genome (Porco et al. 2016; Zhang et al. 2016). Mellor et al. (2016) used a metabolic model to study the *dao1-1* mutant and found the best fit to be a link among IAA oxidation with GH3-catalyzed conjugation and biosynthesis to explain the levels of IAA found. Laskowski and ten Tusscher (2017) built on that model in their Plant Cell Perspective to postulate a mechanism for cell-autonomous oscillations in the induction priming for lateral root formation. The central importance of both oxidation and conjugation has recently been reexamined in new light of reports linking auxin conjugation to aspartate and glutamate with direct DAO dioxygenase catalyzed ring oxidation of the conjugate (Tuominen et al. 1994; Hayashi et al. 2021; Müller et al. 2021; Guo et al. 2022). This established a postulated model building on the one proposed by Mellor et al. (2016) with revised DAO substrate specificity data where IAA levels can be regulated by biosynthesis, direct IAA oxidation, IAA conjugation, IAasp and IAglu hydrolysis, and IAasp/IAglu oxidation—2 ways in and 3 ways out (Hayashi et al. 2021; Ross and Gélinas-Marion 2021). The model continues to evolve, however, as oxidation of other IAA amino acid conjugates has been observed, with both oxIAleu and oxIAphe reported in plants other than Arabidopsis (Hladík et al. 2023) and the relative importance of the pathway may be species dependent (Brunoni et al. 2023). Finally, there have been reported N-linked derivatives and other positions on the indole ring for oxidation that are likely to be catabolic end products (Kai et al. 2007a, 2007b).

Auxin transport

The ability for IAA to move in a polarized direction ("polar auxin transport" [PAT]) lead to the discovery of a "mobile signal," as first described by the defining studies of Theophil Ciesielski (1872) for gravitropism and expanded to phototropic behavior by Charles and Frances Darwin (1881). This arguably was the origin of the concept of hormo-nal signaling in eukaryotic biology. The fundamental role of PAT begins with the embryonic establishment of the polar axis (Laux and Jürgens 1997), establishing ultimately a shoot apical meristem and a root apical meristem and the root cap (Cooke et al. 1993). Early studies of plasma membrane proteins brought focus on possible IAA transporters as well as auxin binding proteins that could be involved in auxin movement (Sussman and Harper 1989; Cox and Muday 1994). The process of understanding these membrane proteins was advanced by studies of the phenotypic similarities of *pin1-1* mutants and plants treated with PAT inhibitors (Okada et al. 1991; Liu et al. 1993). Similarly, PAT inhibitors with other mutants, such as those involved in meristem function, were employed to study other aspects of development (Casimiro et al. 2001). An important link was also established during this timeperiod between early embryo development and PAT using both mutant analysis as well as PAT inhibitors (Cooke et al. 1993; Liu et al. 1993; West and Harada 1993; Zimmerman 1993; Yadegari et al. 1994; Fischer et al. 1997). More recent data (Radoeva et al. 2019) has begun to sort out the complexity of the transcriptional regulation of the auxin relationships during embryo development.

Over the last 25 years many of the molecular details of auxin uptake (AUX1/LAX family) and auxin efflux (PIN family) have been revealed (Bennett et al. 1996; Gälweiler et al. 1998; Luschnig et al. 1998; Marchant et al. 2002; Muday and Murphy 2002). Also, several multiple drug-like resistant PGP family proteins from the ABCB transporter superfamily, which are also involved in transport of auxin across the plasma membrane, have been described (Sidler et al. 1998; Noh et al. 2001; Geisler et al. 2005). An important characteristic of some of the proteins involved with PAT, in particular the PINs, show asymmetric localization at the cell membrane within auxin transport-competent cells (Marhava 2022). How multiple efflux carriers interact has remained an important question since multiple classes of their proteins were described. Recently a systems biology model showed that ABCBs act independently of PINs but PIN-mediated auxin efflux is by a co-dependent process when PINs are co-localized with ABCBs, supporting a model that encompasses interactive ABCB–PIN proteins (Mellor et al. 2022).

Studies of PAT were greatly facilitated by compounds noted generically as phytotropins (Rubery 1990), the most widely used being NPA, first described as an inhibitor for gravitropism (Ching et al. 1956). Studies using NPA have been critical not only for studies of auxin transport genes, but the search for TIR genes by the Estelle group yielded 7 genes involved in diverse auxin related processes (Ruegger et al. 1997): perception, transport, and biosynthesis.

Summary

Since the establishment of *The Plant Cell* in 1989, our understanding of all aspects of auxin biology has advanced in notable ways. Many questions remain, and our ability to use auxin to control plant development is still actively studied. However, our knowledge of auxin-induced gene expression moved our conceptualization of auxin response forward and set the stage for understanding the role for the TIR1-AUX/IAA in auxin perception. Fundamental studies of SAUR were prerequisite for understanding protein phosphatases that activate plasma membrane H^+-ATPases, resulting in cell expansion by "acid growth" as long ago postulated (Rayle and Cleland 1992). We know how auxin is made, at least by 1 pathway, and we have methods to measure the complete metabolic network. We also know how auxin is turned over by ring oxidation of either IAA itself or of a conjugated form. We also know the genes and proteins involved with PAT. However, many open questions remain. For example, whereas we understand how Class-A "activator" ARFs are regulated and are linked to auxin response through the repression-depression model, our understanding of the regulation of Class-B and C "repressor" ARFs is limited. The mechanism and role of rapid auxin responses in the absence of gene expression are an emerging area of active study. Further, the TIR/AFB family was recently shown to have adenylate cyclase activity (Qi et al. 2022), raising the possibility that there is more to uncover in auxin's influence through this family.

Knowledge of auxin is becoming more robust as we understand protein function at molecular levels and this allows development of models to study both transcriptional and metabolic networks, but these need to become much more robust to allow an understanding of auxin regulation of the most fundamental aspects of plant development.

Funding

The work in JDC's laboratory is supported by National Science Foundation MCB 2225057, USDA National Institute of Food and Agriculture, No. 2019-51181-30025 and 2023-67013-39533, Minnesota Department of Agriculture grant FB#6 and by funds from the Minnesota Agricultural Experiment Station and the Gordon and Margaret Bailey Endowment for Environmental Horticulture. The work in LCS's laboratory is supported by National Institutes of Health R35 GM136338 and National Science Foundation PGRP BIO-2112056.

Conflict of interest statement. None declared.

Data availability

Data availability is not applicable to this article as no new data were created or analysed in this work.

References

Abel S, Oeller PW, Theologis A. Early auxin-induced genes encode short-lived nuclear proteins. Proc Natl Acad Sci U S A. 1994:**91**(1): 326–330. https://doi.org/10.1073/pnas.91.1.326

Abel S, Theologis A. Odyssey of auxin. Cold Spring Harb Perspect Biol. 2010:**2**(10):a004572. https://doi.org/10.1101/cshperspect.a004572

Abu-Zaitoon YM, Abu-Zaiton A, Tawaha ARA, Fandi KG, Alnaimat SM, Pati S, Almomani FA. Evidence from co-expression analysis for the involvement of amidase and INS in the tryptophan-independent pathway of IAA synthesis in Arabidopsis. Appl Biochem Biotechnol. 2022:**194**(10):4673–4682. https://doi.org/10.1007/s12010-022-04047-8

Abu-Zaitoon YM, Al-Ramamneh EADM, Al Tawaha AR, Alnaimat SM, Almomani FA. Comparative coexpression analysis of indole synthase and tryptophan synthase A reveals the independent production of auxin via the cytosolic free indole. Plants. 2023:**12**(8):1687. https://doi.org/10.3390/plants12081687

Adham AR, Zolman BK, Millius A, Bartel B. Mutations in Arabidopsis acyl-CoA oxidase genes reveal distinct and overlapping roles in β-oxidation. Plant J. 2005:**41**(6):859–874. https://doi.org/10.1111/j.1365-313X.2005.02343.x

Andreae WA, Good NE. The formation of indoleacetylaspartic acid in pea seedlings. Plant Physiol. 1955:**30**(4):380–382. https://doi.org/10.1104/pp.30.4.380

Aoi Y, Hira H, Hayakawa Y, Liu H, Fukui K, Dai X, Tanaka K, Hayashi KI, Zhao Y, Kasahara H. UDP-glucosyltransferase UGT84B1 regulates the levels of indole-3-acetic acid and phenylacetic acid in Arabidopsis. Biochem Biophys Res Commun. 2020:**532**(2):244–250. https://doi.org/10.1016/j.bbrc.2020.08.026

Badescu GO, Napier RM. Receptors for auxin: will it all end in TIRs? Trends Plant Sci. 2006:**11**(5):217–223. https://doi.org/10.1016/j.tplants.2006.03.001

Baker A, Graham IA, Holdsworth M, Smith SM, Theodoulou FL. Chewing the fat: β-oxidation in signalling and development. Trends Plant Sci. 2006:**11**(3):124–132. https://doi.org/10.1016/j.tplants.2006.01.005

Barlier I, Kowalczyk M, Marchant A, Ljung K, Bhalerao R, Bennett M, Sandberg G, Bellini C. The SUR2 gene of Arabidopsis thaliana encodes the cytochrome P450 CYP83B1, a modulator of auxin homeostasis. Proc Natl Acad Sci U S A. 2000:**97**(26):14819–14824. https://doi.org/10.1073/pnas.260502697

Bartel B, Fink GR. Differential regulation of an auxin-producing nitrilase gene family in *Arabidopsis thaliana*. Proc Natl Acad Sci U S A. 1994:**91**(14):6649–6653. https://doi.org/10.1073/pnas.91.14.6649

Bartel B, Fink GR. ILR1, an amidohydrolase that releases active indole-3-acetic acid from conjugates. Science. 1995:**268**(5218): 1745–1748. https://doi.org/10.1126/science.7792599

Bennett MJ, Marchant A, Green HG, May ST, Ward SP, Millner PA, Walker AR, Schulz B, Feldmann KA. Arabidopsis AUX1 gene: a permease-like regulator of root gravitropism. Science. 1996:**273**(5277): 948–950. https://doi.org/10.1126/science.273.5277.948

Bialek K, Meudt WJ, Cohen JD. Indole-3-acetic acid (IAA) and IAA conjugates applied to bean stem sections: IAA content and the growth response. Plant Physiol. 1983:**73**(1):130–134. https://doi.org/10.1104/pp.73.1.130

Boer DR, Freire-Rios A, van den Berg WAM, Saaki T, Manfield IW, Kepinski S, López-Vidrieo I, Franco-Zorrilla JM, de Vries Sacco C, Solano R, et al. Structural basis for DNA binding specificity by the auxin-dependent ARF transcription factors. Cell. 2014:**156**(3): 577–589. https://doi.org/10.1016/j.cell.2013.12.027

Boerjan W, Cervera MT, Delarue M, Beeckman T, Dewitte W, Bellini C, Caboche M, Van Onckelen H, Van Montagu M, Inzé D. Superroot, a recessive mutation in Arabidopsis, confers auxin overproduction. Plant Cell. 1995:**7**(9):1405–1419. https://doi.org/10.1105/tpc.7.9.1405

Böttcher C, Keyzers RA, Boss PK, Davies C. Sequestration of auxin by the indole-3-acetic acid-amido synthetase GH3-1 in grape berry (*Vitis vinifera* L.) and the proposed role of auxin conjugation during ripening. J Exp Bot. 2010:**61**(13):3615–3625. https://doi.org/10.1093/jxb/erq174

Brunoni F, Collani S, Casanova-Saez R, Simura J, Karady M, Schmid M, Ljung K, Bellini C. Conifers exhibit a characteristic inactivation of auxin to maintain tissue homeostasis. New Phytol. 2020:**226**(6): 1753–1765. https://doi.org/10.1111/nph.16463

Brunoni F, Pěnčík A, Žukauskaitė A, Ament A, Kopečná M, Collani S, Kopečný D, Novák O. Amino acid conjugation of oxIAA is a secondary metabolic regulation involved in auxin homeostasis. New Phytol. 2023:**238**(6):2264–2270. https://doi.org/10.1111/nph.18887

Campanella JJ, Ludwig-Mueller J, Town CD. Isolation and characterization of mutants of *Arabidopsis thaliana* with increased resistance to growth inhibition by indoleacetic acid-amino acid conjugates. Plant Physiol. 1996:**112**(2):735–745. https://doi.org/10.1104/pp.112.2.735

Carpenter WJ, Cherry JH. Effects of protein inhibitors and auxin on nucleic acid metabolism in peanut cotyledons. Plant Physiol. 1966:**41**(6):919–922. https://doi.org/10.1104/pp.41.6.919

Carrillo-Carrasco VP, Hernandez-Garcia J, Mutte SK, Weijers D. The birth of a giant: evolutionary insights into the origin of auxin responses in plants. EMBO J. 2023:**42**(6):e113018. https://doi.org/10.15252/embj.2022113018

Casimiro I, Marchant A, Bhalerao RP, Beeckman T, Dhooge S, Swarup R, Graham N, Inzé D, Sandberg G, Casero PJ, et al. Auxin transport promotes Arabidopsis lateral root initiation. Plant Cell. 2001:**13**(4):843–852. https://doi.org/10.1105/tpc.13.4.843

Chen H, Li L, Zou M, Qi L, Friml J. Distinct functions of TIR1 and AFB1 receptors in auxin signaling. Mol Plant. 2023:**16**(7):1117–1119. https://doi.org/10.1016/j.molp.2023.06.007

Chen L, Huang XX, Zhao SM, Xiao DW, Xiao LT, Tong JH, Wang WS, Li YJ, Ding Z, Hou BK. IPya glucosylation mediates light and temperature signaling to regulate auxin-dependent hypocotyl elongation in Arabidopsis. Proc Natl Acad Sci U S A. 2020:**117**(12):6910–6917. https://doi.org/10.1073/pnas.2000172117

Chen Q, Westfall CS, Hicks LM, Wang S, Jez JM. Kinetic basis for the conjugation of auxin by a GH3 family indole-acetic acid-amido synthetase. J Biol Chem. 2010:**285**(39):29780–29786. https://doi.org/10.1074/jbc.M110.146431

Chilton M-D, Drummond MH, Merlo DJ, Sciaky D, Montoya AL, Gordon MP, Nester EW. Stable incorporation of plasmid DNA into higher plant cells: the molecular basis of crown gall tumorigenesis. Cell. 1977:**11**(2):263–271. https://doi.org/10.1016/0092-8674(77)90043-5

Ching T-MT, Hamilton RH, Bandurski RS. Selective inhibition of the geotropic response by n-1-naphthylphthalamic acid. Physiol Plant. 1956:**9**(4):546–558. https://doi.org/10.1111/j.1399-3054.1956.tb07818.x

Ciesielski T. Untersuchungen über die Abwärtskrümmung der Wurzel. Beitr Biol Pflanzen. 1872:**1**:1–30. https://www.zobodat.at/pdf/Beitr-Biologie-Pflanzen_1_2_0002-0030.pdf

Cohen JD, Bandurski RS. Chemistry and physiology of the bound auxins. Annu Rev Plant Physiol. 1982:**33**(1):403–430. https://doi.org/10.1146/annurev.pp.33.060182.002155

Cohen JD, Bialek K, Slovin JP, Baldi BG, Chen K-H. Development of genetic and analytical systems for studies of auxin metabolism. In: **Pharis RP, Rood SB**, editors. Plant growth substances 1988. Berlin: Springer Verlag; 1990. p. 45–56.

Cohen JD, Tang Q, Hegeman AD. Chapter nine—using targeted metabolomics to elucidate the indole auxin network in plants. Methods Enzymol. 2022:**676**:239–278. https://doi.org/10.1016/bs.mie.2022.07.038

Cooke TJ, Racusen RH, Cohen JD. The role of auxin in plant embryogenesis. Plant Cell. 1993:**5**(11):1494–1495. https://doi.org/10.1105/tpc.5.11.1494

Cooney TP, Nonhebel HM. The measurement and mass spectral identification of indole-3-pyruvate from tomato shoots. Biochem Biophys Res Commun. 1989:**162**(2):761–766. https://doi.org/10.1016/0006-291x(89)92375-9

Cox DN, Muday GK. NPA binding activity is peripheral to the plasma membrane and is associated with the cytoskeleton. Plant Cell. 1994:**6**(12):1941–1953. https://doi.org/10.1105/tpc.6.12.1941

Damodaran S, Strader LC. Indole-3-butyric acid metabolism and transport in *Arabidopsis thaliana*. Front Plant Sci. 2019:**10**:851. https://doi.org/10.3389/fpls.2019.00851

Darwin C, Darwin F. The power of movement in plants. NY: Appleton and Company; 1881, p 592.

Davies RT, Goetz DH, Lasswell J, Anderson MN, Bartel B. IAR3 encodes an auxin conjugate hydrolase from Arabidopsis. Plant Cell. 1999:**11**(3):365–376. https://doi.org/10.1105/tpc.11.3.365

Dharmasiri N, Dharmasiri S, Estelle M. The F-box protein TIR1 is an auxin receptor. Nature. 2005a:**435**(7041):441–445. https://doi.org/10.1038/nature03543

Dharmasiri N, Dharmasiri S, Jones AM, Estelle M. Auxin action in a cell-free system. Curr Biol. 2003:**13**(16):1418–1422. https://doi.org/10.1016/S0960-9822(03)00536-0

Dharmasiri N, Dharmasiri S, Weijers D, Lechner E, Yamada M, Hobbie L, Ehrismann JS, Jurgens G, Estelle M. Plant development is regulated by a family of auxin receptor F box proteins. Dev Cell. 2005b:**9**(1):109–119. https://doi.org/10.1016/j.devcel.2005.05.014

Di D-W, Zhang C, Luo P, An C-W, Guo G-Q. The biosynthesis of auxin: how many paths truly lead to IAA? Plant Growth Regul. 2016:**78**(3): 275–285. https://doi.org/10.1007/s10725-015-0103-5

Ding X, Cao Y, Huang L, Zhao J, Xu C, Li X, Wang S. Activation of the indole-3-acetic acid–amido synthetase GH3-8 suppresses expansin expression and promotes salicylate-and jasmonate-independent basal immunity in rice. Plant Cell. 2008:**20**(1):228–240. https://doi.org/10.1105/tpc.107.055657

Domingo C, Andrés F, Tharreau D, Iglesias DJ, Talón M. Constitutive expression of OsGH3.1 reduces auxin content and enhances defense response and resistance to a fungal pathogen in rice. Mol Plant Microbe Interact. 2009:**22**(2):201–210. https://doi.org/10.1094/MPMI-22-2-0201

Dreher KA, Brown J, Saw RE, Callis J. The Arabidopsis aux/IAA protein family has diversified in degradation and auxin responsiveness. Plant Cell. 2006:**18**(3):699–714. https://doi.org/10.1105/tpc.105.039172

Du W, Lu Y, Li Q, Luo S, Shen S, Li N, Chen X. TIR1/AFB proteins: active players in abiotic and biotic stress signaling. Front Plant Sci. 2022:**13**:1083409. https://doi.org/10.3389/fpls.2022.1083409

Du M, Spalding EP, Gray WM. Rapid auxin-mediated cell expansion. Annu Rev Plant Biol. 2020:**71**(1):379–402. https://doi.org/10.1146/annurev-arplant-073019-025907

Dubey SM, Han S, Stutzman N, Prigge MJ, Medvecká E, Platre MP, Busch W, Fendrych M, Estelle M. The AFB1 auxin receptor controls the cytoplasmic auxin response pathway in *Arabidopsis thaliana*. Mol Plant. 2023:**16**(7):1120–1130. https://doi.org/10.1101/2023.01.04.522696

Enders TA, Strader LC. Auxin activity: past, present, and future. Am J Bot. 2015:**102**(2):180–196. https://doi.org/10.3732/ajb.1400285

Epstein E, Cohen JD, Bandurski RS. Concentration and metabolic turnover of indoles in germinating kernels of *Zea mays* L. Plant Physiol. 1980:**65**(3):415–421. https://doi.org/10.1104/pp.65.3.415

Epstein E, Ludwig-Müller J. Indole-3-butyric acid in plants: occurrence, synthesis, metabolism and transport. Physiol Plant. 1993:**88**(2): 382–389. https://doi.org/10.1111/j.1399-3054.1993.tb05513.x

Ernstsen A, Sandberg G, Lundström K. Identification of oxindole-3-acetic acid, and metabolic conversion of indole-3-acetic acid to oxindole-3-acetic acid in *Pinus sylvestris* seeds. Planta. 1987:**172**(1):47–52. https://doi.org/10.1007/BF00403027

Figueiredo MRA, Strader LC. Intrinsic and extrinsic regulators of aux/IAA protein degradation dynamics. Trends Biochem Sci. 2022:**47**(10): 865–874. https://doi.org/10.1016/j.tibs.2022.06.004

Fischer C, Speth V, Fleig-Eberenz S, Neuhaus G. Induction of zygotic polyembryos in wheat: influence of auxin polar transport. Plant Cell. 1997:**9**(10):1767–1780. https://doi.org/10.1105/tpc.9.10.1767

Galli M, Khakhar A, Lu Z, Chen Z, Sen S, Joshi T, Nemhauser JL, Schmitz RJ, Gallavotti A. The DNA binding landscape of the maize AUXIN RESPONSE FACTOR family. Nat Commun. 2018:**9**(1):4526. https://doi.org/10.1038/s41467-018-06977-6

Galston AW, Chen HR. Auxin activity of isatin and oxindole-3-acetic acid. Plant Physiol. 1965:**40**(4):699–705. https://doi.org/10.1104/pp.40.4.699

Gälweiler L, Guan C, Müller A, Wisman E, Mendgen K, Yephremov A, Palme K. Regulation of polar auxin transport by AtPIN1 in Arabidopsis vascular tissue. Science. 1998:**282**(5397):2226–2230. https://doi.org/10.1126/science.282.5397.2226

Gee MA, Hagen G, Guilfoyle TJ. Tissue-specific and organ-specific expression of soybean auxin-responsive transcripts GH3 and SAURs. Plant Cell. 1991:**3**(4):419–430. https://doi.org/10.1105/tpc.3.4.419

Geisler M, Blakeslee JJ, Bouchard R, Lee OR, Vincenzetti V, Bandyopadhyay A, Titapiwatanakun B, Peer WA, Bailly A, Richards EL, et al. Cellular efflux of auxin catalyzed by the Arabidopsis MDR/PGP transporter AtPGP1. Plant J. 2005:**445**(2): 179–194. https://doi.org/10.1111/j.1365-313X.2005.02519.x

Gil P, Green PJ. Multiple regions of the Arabidopsis SAUR-AC1 gene control transcript abundance: the 3′ untranslated region functions as an mRNA instability determinant. EMBO J. 1996:**15**(7):1678–1686. https://doi.org/10.1002/j.1460-2075.1996.tb00513.x

Glick BR, Patten CL, Holguin G, Penrose DM eds. Chapter 4: Auxin production. Biochemical and genetic mechanisms used by plant growth promoting Bacteria. London: Imperial College Press; 1999. p. 86–133.

González-Lamothe R, El Oirdi M, Brisson N, Bouarab K. The conjugated auxin indole-3-acetic acid-aspartic acid promotes plant disease development. Plant Cell. 2012:**24**(2):762–777. https://doi.org/10.1105/tpc.111.095190

Gray WM, Estelle M. Function of the ubiquitin-proteasome pathway in auxin response. Trends Biochem Sci. 2000:**25**(3):133–138. https://doi.org/10.1016/s0968-0004(00)01544-9

Gray WM, Kepinski S, Rouse D, Leyser O, Estelle M. Auxin regulates SCF^{TIR1}-dependent degradation of AUX/IAA proteins. Nature. 2001:**414**(6861):271–276. https://doi.org/10.1038/35104500

Guilfoyle TJ. Auxin-regulated genes and promoters. New Compr Biochem. 1999:**33**:423–459. https://doi.org/10.1016/S0167-7306(08)60499-8

Guilfoyle TJ, Hagen G. Auxin response factors. J Plant Growth Regul. 2001:**20**(3):281–291. https://doi.org/10.1007/s003440010026

Guilfoyle TJ, Ulmasov T, Hagen G. The ARF family of transcription factors and their role in plant hormone-responsive transcription. Cell Mol Life Sci. 1998:**54**(7):619–627. https://doi.org/10.1007/s000180050190

Guo R, Hu Y, Aoi Y, Hira H, Ge C, Dai X, Kasahara H, Zhao Y. Local conjugation of auxin by the GH3 amido synthetases is required for normal development of roots and flowers in Arabidopsis. Bioch Biophys Res Comm. 2022:**589**:16–22. https://doi.org/10.1016/j.bbrc.2021.11.109

Hagen G, Guilfoyle T. Auxin-responsive gene expression: genes, promoters and regulatory factors. Plant Mol Biol. 2002:**49**(3/4): 373–385. https://doi.org/10.1023/A:1015207114117

Hagen G, Kleinschmidt A, Guilfoyle T. Auxin-regulated gene expression in intact soybean hypocotyl and excised hypocotyl sections. Planta. 1984:**162**(2):147–153. https://doi.org/10.1007/BF00410211

Hangarter RP, Good NE. Evidence that IAA conjugates are slow-release sources of free IAA in plant tissues. Plant Physiol. 1981:**68**(6): 1424–1427. https://doi.org/10.1104/pp.68.6.1424

Havens KA, Guseman JM, Jang SS, Pierre-Jerome E, Bolten N, Klavins E, Nemhauser JL. A synthetic approach reveals extensive tunability of auxin signaling. Plant Physiol. 2012:**160**(1):135–142. https://doi.org/10.1104/pp.112.202184

Hayashi K, Arai K, Aoi Y, Tanaka Y, Hira H, Guo R, Hu Y, Ge C, Zhao Y, Kasahara H, et al. The main oxidative inactivation pathway of the plant hormone auxin. Nat Commun. 2021:**12**(1):6752. https://doi.org/10.1038/s41467-021-27020-1

Hayashi M, Toriyama K, Kondo M, Nishimura M. 2,4-Dichlorophenoxybutyric acid-resistant mutants of Arabidopsis have defects in glyoxysomal fatty acid beta-oxidation. Plant Cell. 1998:**10**(2):183–195. https://doi.org/10.1105/tpc.10.2.183

Hladík P, Petřík I, Žukauskaitė A, Novák O, Pěnčík A. Metabolic profiles of 2-oxindole-3-acetyl-amino acid conjugates differ in various plant species. Front Plant Sci. 2023:**14**:1217421. https://doi.org/10.3389/fpls.2023.1217421

Hull AK, Vij R, Celenza JL. Arabidopsis cytochrome P450s that catalyze the first step of tryptophan-dependent indole-3-acetic acid biosynthesis. Proc Natl Acad Sci U S A. 2000:**97**(5):2379–2384. https://doi.org/10.1073/pnas.040569997

Ito J, Fukaki H, Onoda M, Li L, Li C, Tasaka M, Furutani M. Auxin-dependent compositional change in mediator in ARF7- and ARF19-mediated transcription. Proc Natl Acad Sci U S A. 2016:**113**(23):6562–6567. https://doi.org/10.1073/pnas.1600739113

Iyer M, Slovin JP, Epstein E, Cohen JD. Transgenic tomato plants with a modified ability to synthesize indole-3-acetyl-β-1-O-D-glucose. J Plant Growth Regul. 2005:**24**(2):142–152. https://doi.org/10.1007/s00344-004-0007-5

Jackson RG, Lim EK, Li Y, Kowalczyk M, Sandberg G, Hoggett J, Ashford DA, Bowles DJ. Identification and biochemical characterization of an Arabidopsis indole-3-acetic acid glucosyltransferase. J Biol Chem. 2001:**276**(6):4350–4356. https://doi.org/10.1074/jbc.M006185200

Jakubowska A, Kowalczyk S. The auxin conjugate 1-O-indole-3-acetyl-beta-D-glucose is synthesized in immature legume seeds by IAGlc synthase and may be used for modification of some high molecular weight compounds. J Exp Bot. 2004:**55**(398): 791–801. https://doi.org/10.1093/jxb/erh086

Jawahir V, Zolman BK. Long chain acyl CoA synthetase 4 catalyzes the first step in peroxisomal indole-3-butyric acid to IAA conversion. Plant Physiol. 2021:**185**(1):120–136. https://doi.org/10.1093/plphys/kiaa002

Jedličková V, Ebrahimi Naghani S, Robert HS. On the trail of auxin: reporters and sensors. Plant Cell. 2022:**34**(9):3200–3213. https://doi.org/10.1093/plcell/koac179

Jin SH, Ma XM, Han P, Wang B, Sun YG, Zhang GZ, Li YJ, Hou BK. UGT74D1 is a novel auxin glycosyltransferase from *Arabidopsis thaliana*. PLoS One. 2013:**8**(4):e61705. https://doi.org/10.1371/journal.pone.0061705

Jing H, Korasick DA, Emenecker RJ, Morffy N, Wilkinson EG, Powers SK, Strader LC. Regulation of AUXIN RESPONSE FACTOR condensation and nucleo-cytoplasmic partitioning. Nat Commun. 2022:**13**(1): 4015. https://doi.org/10.1038/s41467-022-31628-2

Kai K, Horita J, Wakasa K, Miyagawa H. Three oxidative metabolites of indole-3-acetic acid from *Arabidopsis thaliana*. Phytochemistry. 2007a:**68**(12):1651–1663. https://doi.org/10.1016/j.phytochem.2007.04.030

Kai K, Wakasa K, Miyagawa H. Metabolism of indole-3-acetic acid in rice: identification and characterization of N-beta-D-glucopyranosyl indole-3-acetic acid and its conjugates. Phytochemistry. 2007b:**68**(20): 2512–2522. https://doi.org/10.1016/j.phytochem.2007.05.040

Karlin-Neurnann GA, Brusslan JA, Tobin EM. Phytochrome control of the tms2 gene in transgenic Arabidopsis: a strategy for selecting mutants in the signal transduction pathway. Plant Cell. 1991:**3**(6): 573–582. https://doi.org/10.1105/tpc.3.6.573

Kato H, Mutte SK, Suzuki H, Crespo I, Das S, Radoeva T, Fontana M, Yoshitake Y, Hainiwa E, van den Berg W. Design principles of a minimal auxin response system. Nat Plants. 2020:**6**(5):473–482. https://doi.org/10.1038/s41477-020-0662-y

Kepinski S, Leyser O. The Arabidopsis F-box protein TIR1 is an auxin receptor. Nature. 2005:**435**(7041):446–451. https://doi.org/10.1038/nature03542

Kinashi H, Suzuki Y, Takeuchi S, Kawarada A. Possible metabolic intermediates from IAA to β-acid in rice bran. Agric Biol Chem. 1976:**40**(12):2465–2470. https://doi.org/10.1080/00021369.1976.10862405

King JJ, Stimart DP, Fisher RH, Bleecker AB. A mutation altering auxin homeostasis and plant morphology in Arabidopsis. Plant Cell. 1995:**7**(12):2023–2037. https://doi.org/10.1105/tpc.7.12.2023

Klee H, Montoya A, Horodyski F, Lichtenstein C, Garfinkel D, Fuller S, Flores C, Peschon J, Nester E, Gordon M. Nucleotide sequences of the *tms* genes of the pTiA6NC octopine Ti-plasmid: two gene products involved in plant tumorigenesis. Proc Natl Acad Sci U S A. 1984:**81**(6): 1728–1732. https://doi.org/10.1073/pnas.81.6.1728

Korasick DA, Westfall CS, Lee SG, Nanao MH, Dumas R, Hagen G, Guilfoyle TJ, Jez JM, Strader LC. Molecular basis for auxin response factor protein interaction and the control of auxin response repression. Proc Natl Acad Sci U S A. 2014:**111**(14):5427–5432. https://doi.org/10.1073/pnas.1400074111

Kreiser M, Giblin C, Murphy R, Fiesel P, Braun L, Johnson G, Wyse D, Cohen JD. Conversion of indole-3-butyric acid to indole-3-acetic acid in shoot tissue of hazelnut (*Corylus*) and elm (*Ulmus*). J Plant Growth Regul. 2016:**35**(3):710–721. https://doi.org/10.1007/s00344-016-9574-5

Kriechbaumer V, Weigang L, Fießelmann A, Letzel T, Frey M, Gierl A, Glawischnig E. Characterisation of the tryptophan synthase alpha subunit in maize. BMC Plant Biol. 2008:**8**(1):44. https://doi.org/10.1186/1471-2229-8-44

Kubeš M, Yang H, Richter GL, Cheng Y, Młodzińska E, Wang X, Blakeslee JJ, Carraro N, Petrášek J, Zažímalová E, et al. The Arabidopsis concentration-dependent influx/efflux transporter ABCB4 regulates cellular auxin levels in the root epidermis. Plant J. 2012:**69**(4):640–654. https://doi.org/10.1111/j.1365-313X.2011.04818.x

Kunz H-H, Scharnewski M, Feussner K, Feussner I, Flügge U-I, FuldaM, Gierth M. The ABC transporter PXA1 and peroxisomal β-oxidation are vital for metabolism in mature leaves of Arabidopsis during extended darkness. Plant Cell. 2009:**21**(9): 2733–2749. https://doi.org/10.1105/tpc.108.064857

Lanctot A, Taylor-Teeples M, Oki EA, Nemhauser JL. Specificity in auxin responses is not explained by the promoter preferences of activator ARFs. Plant Physiol. 2020:**182**(4):1533–1536. https://doi.org/10.1104/pp.19.01474

Laskowski M, ten Tusscher KH. Periodic lateral root priming: what makes it tick? Plant Cell. 2017:**29**(3):432–444. https://doi.org/10.1105/tpc.16.00638

Last RL, Bissinger PH, Mahoney DJ, Radwanski ER, Fink GR. Tryptophan mutants in Arabidopsis: the consequences of duplicated tryptophan synthase beta genes. Plant Cell. 1991:**3**(4):345–358. https://doi.org/10.1105/tpc.3.4.345

Laux T, Jürgens G. Embryogenesis: a new start in life. Plant Cell. 1997:**9**(7):989–1000. https://doi.org/10.1105/tpc.9.7.989

Lavy M, Prigge MJ, Tao S, Shain S, Kuo A, Kirchsteiger K, Estelle M. Constitutive auxin response in *Physcomitrella* reveals complex interactions between aux/IAA and ARF proteins. eLife. 2016:**5**:e13325. https://doi.org/10.7554/eLife.13325

LeClere S, Tellez R, Rampey RA, Matsuda SP, Bartel B. Characterization of a family of IAA-amino acid conjugate hydrolases from Arabidopsis. J Biol Chem. 2002:**277**(23):20446–20452. https://doi.org/10.1074/jbc.M111955200

Lee S, Sundaram S, Armitage L, Evans JP, Hawkes T, Kepinski S, Ferro N, Napier RM. Defining binding efficiency and specificity of auxins for SCF(TIR1/AFB)-aux/IAA co-receptor complex formation. ACS Chem Biol. 2014:**9**(3):673–682. https://doi.org/10.1021/cb400618m

Lehmann T, Janowitz T, Sánchez-Parra B, Pérez Alonso M-M, Trompetter I, Piotrowski M, Pollmann S. Arabidopsis NITRILASE 1 contributes to the regulation of root growth and development through modulation of auxin biosynthesis in seedlings. Front Plant Sci. 2017:**8**:36. https://doi.org/10.3389/fpls.2017.00036

Lewer P, Bandurski RS. Occurrence and metabolism of 7-hydroxy-2-indolinone-3-acetic acid in *Zea mays*. Phytochemistry. 1987:**26**(5): 1247–1250. https://doi.org/10.1016/s0031-9422(00)81790-2

Leydon AR, Wang W, Gala HP, Gilmour S, Juarez-Solis S, Zahler ML, Zemke JE, Zheng N, Nemhauser JL. Repression by the Arabidopsis TOPLESS corepressor requires association with the core mediator complex. eLife. 2021:**10**:e66739. https://doi.org/10.7554/eLife.66739

Li Y, Hagen G, Guilfoyle TJ. An auxin-responsive promoter is differentially induced by auxin gradients during tropisms. Plant Cell. 1991:**3**(11):1167–1175. https://doi.org/10.1105/tpc.3.11.1167

Li Y, Strabala TJ, Hagen G, Guilfoyle TJ. The soybean SAUR open reading frame contains a cis element responsible for cycloheximide-induced mRNA accumulation. Plant Mol Biol. 1994:**24**(5):715–723. https://doi.org/10.1007/BF00029853

Li S-B, Xie Z-Z, Hu C-G, Zhang J-Z. A review of auxin response factors (ARFs) in plants. Front Plant Sci. 2016:**7**:47. https://doi.org/10.3389/fpls.2016.00047

Lincoln C, Britton JH, Estelle M. Growth and development of the axr1 mutants of *Arabidopsis*. Plant Cell. 1990:**2**(11):1071–1080. https://doi.org/10.1105/tpc.2.11.1071

Linka N, Theodoulou FL, Haslam RP, Linka M, Napier JA, Neuhaus HE, Weber A. Peroxisomal ATP import is essential for seedling development in *Arabidopsis thaliana*. Plant Cell. 2008:**20**(12):3241–3257. https://doi.org/10.1105/tpc.108.062042

Liscum E, Reed JW. Genetics of aux/IAA and ARF action in plant growth and development. Plant Mol Biol. 2002:**49**(3/4):387–400. https://doi.org/10.1023/A:1015255030047

Liu C, Xu Z, Chua NH. Auxin polar transport is essential for the establishment of bilateral symmetry during early plant embryogenesis. Plant Cell. 1993:**5**(6):621–630. https://doi.org/10.1105/tpc.5.6.621

Liu Q, Chen TT, Xiao DW, Zhao SM, Lin JS, Wang T, Li YJ, Hou BK. OsIAGT1 is a glucosyltransferase gene involved in the glucose conjugation of auxins in rice. Rice (N Y). 2019:**12**(1):92. https://doi.org/10.1186/s12284-019-0357-z

Liu X, Barkawi L, Gardner G, Cohen JD. Transport of indole-3-butyric acid and indole-3-acetic acid in Arabidopsis hypocotyls using stable isotope labeling. Plant Physiol. 2012a:**158**(4):1988–2000. https://doi.org/10.1104/pp.111.191288

Liu X, Hegeman AD, Gardner G, Cohen JD. Protocol: high-throughput and quantitative assays of auxin and auxin precursors from minute tissue samples. Plant Methods. 2012b:**8**(1):31. https://doi.org/10.1186/1746-4811-8-31

Ljung K, Hull AK, Celenza J, Yamada M, Estelle M, Normanly J, Sandberg G. Sites and regulation of auxin biosynthesis in Arabidopsis roots. Plant Cell. 2005:**17**(4):1090–1104. https://doi.org/10.1105/tpc.104.029272

Ludwig-Müller J. Indole-3-butyric acid synthesis in ecotypes and mutants of *Arabidopsis thaliana* under different growth conditions. J Plant Physiol. 2007:**164**(1):47–59. https://doi.org/10.1016/j.jplph.2005.10.008

Ludwig-Müller J, Walz A, Slovin JP, Epstein E, Cohen JD, Dong W, Town CD. Overexpression of maize IAGLU in *Arabidopsis thaliana* alters plant growth and sensitivity to IAA but not IBA and 2,4-D. J Plant Growth Regul. 2005:**24**(2):127–141. https://doi.org/10.1007/s00344-004-0006-6

Luschnig C, Gaxiola RA, Grisafi P, Fink GR. EIR1, a root-specific protein involved in auxin transport, is required for gravitropism in *Arabidopsis thaliana*. Genes Dev. 1998:**125**(14):2175–2187. https://doi.org/10.1101/gad.12.14.2175

Magnus V, Nigović B, Hangarter RP, Good NE. N-(indol-3-ylacetyl) amino acids as sources of auxin in plant tissue culture. J Plant Growth Regul. 1992:**11**(1):19–28. https://doi.org/10.1007/BF00193839

Marchant A, Bhalerao R, Casimiro I, Eklöf J, Casero PJ, Bennett M, Sandberg G. AUX1 promotes lateral root formation by facilitating indole-3-acetic acid distribution between sink and source tissues in the Arabidopsis seedling. Plant Cell. 2002:**14**(3):589–597. https://doi.org/10.1105/tpc.010354

Marhava P. Recent developments in the understanding of PIN polarity. New Phytol. 2022:**233**(2):624–630. https://doi.org/10.1111/nph.17867

Mashiguchi K, Tanaka K, Sakai T, Sugawara S, Kawaide H, Natsume M, Hanada A, Yaeno T, Shirasu K, Yao H, et al. The main auxin biosynthesis pathway in Arabidopsis. Proc Natl Acad Sci U S A. 2011:**108**(45):18512–18517. https://doi.org/10.1073/pnas.1108434108

Mateo-Bonmatí E, Casanova-Sáez R, Šimura J, Ljung K. Broadening the roles of UDP-glycosyltransferases in auxin homeostasis and plant development. New Phytol. 2021:**232**(2):642–654. https://doi.org/10.1111/nph.17633

McClure BA, Guilfoyle T. Characterization of a class of small auxin-inducible soybean polyadenylated RNAs. Plant Mol Biol. 1987:**9**(6):611–623. https://doi.org/10.1007/BF00020537

McClure BA, Guilfoyle T. Rapid redistribution of auxin-regulated RNAs during gravitropism. Science. 1989:**243**(4887):91–93. https://doi.org/10.1126/science.11540631

McClure BA, Hagen G, Brown CS, Gee MA, Guilfoyle TJ. Transcription, organization, and sequence of an auxin-regulated gene cluster in soybean. Plant Cell. 1989:**1**(2):229–239. https://doi.org/10.1105/tpc.1.2.229

Mellor N, Band LR, Pěnčík A, Novák O, Rashed A, Holman T, Wilson MH, Voß U, Bishopp A, King JR, et al. Dynamic regulation of auxin oxidase and conjugating enzymes AtDAO1 and GH3 modulates auxin homeostasis. Proc Natl Acad Sci U S A. 2016:**113**(39):11022–11027. https://doi.org/10.1073/pnas.1604458113

Mellor NL, Voß U, Ware A, Janes G, Barrack D, Bishopp A, Bennett MJ, Geisler M, Wells DM, Band LR. Systems approaches reveal that ABCB and PIN proteins mediate co-dependent auxin efflux. Plant Cell. 2022:**34**(6):2309–2327. https://doi.org/10.1093/plcell/koac086

Michalczuk L, Bandurski RS. Enzymic synthesis of 1-O-indol-3-ylacetyl-beta-D-glucose and indol-3-ylacetyl-*myo*-inositol. Biochem J. 1982:**207**(2):273–281. https://doi.org/10.1042/bj2070273

Michniewicz M, Ho CH, Enders TA, Floro E, Damodaran S, Gunther LK, Powers SK, Frick EM, Topp CN, Frommer WB, et al. TRANSPORTER OF IBA1 links auxin and cytokinin to influence root architecture. Dev Cell. 2019:**50**(5):599–609.e4. https://doi.org/10.1016/j.devcel.2019.06.010

Morffy N, Strader LC. Old town roads: routes of auxin biosynthesis across kingdoms. Curr Opin Plant Biol. 2020:**55**:21–27. https://doi.org/10.1016/j.pbi.2020.02.002

Morffy N, Strader LC. Structural aspects of auxin signaling. Cold Spring Harbor Perspect Biol. 2022:**14**(1):a039883. https://doi.org/10.1101/cshperspect.a039883

Muday GK, Murphy AS. An emerging model of auxin transport regulation. Plant Cell. 2002:**14**(2):293–299. https://doi.org/10.1105/tpc.140230

Müller K, Dobrev PI, Pěnčik A, Hošek P, Vondráková Z, Filepová R, Malínská K, Brunoni F, Helusova L, Moravec T, et al. DIOXYGENASE FOR AUXIN OXIDATION 1 catalyzes the oxidation of IAA amino acid conjugates. Plant Physiol. 2021:**187**(1):103–115. https://doi.org/10.1093/plphys/kiab242

Mutte SK, Kato H, Rothfels C, Melkonian M, Wong K-SG, Weijers D. Origin and evolution of the nuclear auxin response system. eLife. 2018:**7**:e33399. https://doi.org/10.7554/eLife.33399

Nanao MH, Vinos-Poyo T, Brunoud G, Thévenon E, Mazzoleni M, Mast D, Lainé S, Wang S, Hagen G, Li H, et al. Structural basis for oligomerization of auxin transcriptional regulators. Nat Commun. 2014:**5**(1):3617. https://doi.org/10.1038/ncomms4617

Neumann PM, Palmer JM. Early effects of auxin on growth and nucleic-acid metabolism in sterile pea stems. J Exp Bot. 1971:**22**(4):915–921. https://doi.org/10.1093/jxb/22.4.915

Newman TC, Ohme-Takagi M, Taylor CB, Green PJ. DST sequences, highly conserved among plant SAUR genes, target reporter transcripts for rapid decay in tobacco. Plant Cell. 1993:**5**(6):701–714. https://doi.org/10.1105/tpc.5.6.701

Niemeyer M, Moreno Castillo E, Ihling CH, Iacobucci C, Wilde V, Hellmuth A, Hoehenwarter W, Samodelov SL, Zurbriggen MD, Kastritis PL, et al. Flexibility of intrinsically disordered degrons in AUX/IAA proteins reinforces auxin co-receptor assemblies. Nat Commun. 2020:**11**(1):2277. https://doi.org/10.1038/s41467-020-16147-2

Niyogi KK, Fink GR. Two anthranilate synthase genes in Arabidopsis: defense-related regulation of the tryptophan pathway. Plant Cell. 1992:**4**(6):721–733. https://doi.org/10.1105/tpc.4.6.721

Noh B, Murphy AS, Spalding EP. Multidrug resistance-like genes of Arabidopsis required for auxin transport and auxin-mediated development. Plant Cell. 2001:**13**(11):2441–2454. https://doi.org/10.1105/tpc.010350

Nonhebel HM. Measurement of the rates of oxindole-3-acetic acid turnover, and indole-3-acetic acid oxidation in *Zea mays* seedlings. J Exp Bot. 1986:**37**(11):1691–1697. https://doi.org/10.1093/jxb/37.11.1691

Nonhebel HM, Bandurski RS. Oxidation of indole-3-acetic acid and oxindole-3-acetic acid to 2,3-dihydro-7-hydroxy-2-oxo-1H indole-3-acetic acid-7′-O-beta-D-glucopyranoside in *Zea mays* seedlings. Plant Physiol. 1984:**76**(4):979–983. https://doi.org/10.1104/pp.76.4.979

Nonhebel HM, Hillman JR, Crozier A, Wilkins MB. Metabolism of [^{14}C]indole-3-acetic acid by coleoptiles of *Zea mays* L. J Exp Bot. 1985a:**36**(1):99–109. https://doi.org/10.1093/jxb/36.1.99

Nonhebel HM, Kruse LI, Bandurski RS. Indole-3-acetic acid catabolism in *Zea mays* seedlings. Metabolic conversion of oxindole-3-acetic acid to 7-hydroxy-2-oxindole-3-acetic acid 7′-O-beta-D-glucopyranoside. J Biol Chem. 1985b:**260**(23):12685–12689. https://doi.org/10.1016/S0021-9258(17)38927-5

Nonhebel H, Yuan Y, Al-Amier H, Pieck M, Akor E, Ahamed A, Cohen JD, Celenza JL, Normanly J. Redirection of tryptophan metabolism in tobacco by ectopic expression of an Arabidopsis indolic glucosinolate biosynthetic gene. Phytochemistry. 2011:**72**(1):37–48. https://doi.org/10.1016/j.phytochem.2010.10.018

Normanly J, Cohen JD, Fink GR. *Arabidopsis thaliana* auxotrophs reveal a tryptophan independent biosynthetic pathway for indole-3-acetic acid. Proc Natl Acad Sci U S A. 1993:**90**(21):10355–10359. https://doi.org/10.1073/pnas.90.21.10355

Normanly J, Grisafi P, Fink GR, Bartel B. Arabidopsis mutants resistant to the auxin effects of indole-3-acetonitrile are defective in the nitrilase encoded by the NIT1 gene. Plant Cell. 1997:**9**(10):1781–1790. https://doi.org/10.1105/tpc.9.10.1781

Normanly J, Slovin JP, Cohen JD. Rethinking auxin biosynthesis and metabolism. Plant Physiol. 1995:**107**(2):323–329. https://doi.org/10.1104/pp.107.2.323

Oetiker JH, Aeschbacher G. Temperature-sensitive plant cells with shunted indole-3-acetic acid conjugation. Plant Physiol. 1997:**114**(4):1385–1395. https://doi.org/10.1104/pp.114.4.1385

Okada K, Ueda J, Komaki MK, Bell CJ, Shimura Y. Requirement of the auxin polar transport system in early stages of Arabidopsis floral bud formation. Plant Cell. 1991:**3**(7):677–684. https://doi.org/10.1105/tpc.3.7.677

Ostrowski M, Hetmann A, Jakubowska A. Indole-3-acetic acid UDP-glucosyltransferase from immature seeds of pea is involved in modification of glycoproteins. Phytochemistry. 2015:**117**:25–33. https://doi.org/10.1016/j.phytochem.2015.05.023

Ostrowski M, Jakubowska A. UDP-glycosyltransferases of plant hormones. Med J Cell Biol. 2014:**4**:43–60. https://doi.org/10.2478/acb-2014-0003

Park JE, Park JY, Kim YS, Staswick PE, Jeon J, Yun J, Kim SY, Kim J, Lee YH, Park CM. GH3-mediated auxin homeostasis links growth regulation with stress adaptation response in Arabidopsis.

J Biol Chem. 2007:**282**(13):10036–10046. https://doi.org/10.1074/jbc.M610524200

Peat TS, Böttcher C, Newman J, Lucent D, Cowieson N, Davies C. Crystal structure of an indole-3-acetic acid amido synthetase from grapevine involved in auxin homeostasis. Plant Cell. 2012:**24**(11): 4525–4538. https://doi.org/10.1105/tpc.112.102921

Peer WA, Cheng Y, Murphy AS. Evidence of oxidative attenuation of auxin signalling. J Exp Bot. 2013:**64**(9):2629–2639. https://doi.org/10.1093/jxb/ert152

Pěnčík A, Casanova-Sáez R, Pilarova V, Zukauskaite A, Pinto R, Micol JL, Ljung K, Novák O. Ultra-rapid auxin metabolite profiling for high-throughput mutant screening in Arabidopsis. J Exp Bot. 2018:**69**(10):2569–2579. https://doi.org/1093/jxb/ery084

Pěnčík A, Simonovik B, Petersson SV, Henyková E, Simon S, Greenham K, Zhang Y, Kowalczyk M, Estelle M, Zazímalová E, et al. Regulation of auxin homeostasis and gradients in Arabidopsis roots through the formation of the indole-3-acetic acid catabolite 2-oxindole-3-acetic acid. Plant Cell. 2013:**25**(10): 3858–3870. https://doi.org/10.1105/tpc.113.114421

Phillips KA, Skirpan AL, Liu X, Christensen A, Slewinski TL, Hudson C, Barazesh S, Cohen JD, Malcomber S, Mcsteen P. *Vanishing tassel2* encodes a grass-specific tryptophan aminotransferase required for vegetative and reproductive development in maize. Plant Cell. 2011:**23**(2):550–566. https://doi.org/10.1105/tpc.110.075267

Pieck M, Yuan Y, Godfrey J, Fisher C, Zolj S, Vaughan D, Thomas N, Wu C, Ramos J, Lee N, et al. Auxin and tryptophan homeostasis are facilitated by the ISS1/VAS1 aromatic aminotransferase in Arabidopsis. Genetics. 2015:**201**(1):185–199. https://doi.org/10.1534/genetics.115.180356

Plüss R, Jenny T, Meier H. IAA-induced adventitious root formation in greenwood cuttings of *Populus tremula* and formation of 2-indolone-3-acetylaspartic acid, a new metabolite of exogenously applied indole-3-acetic acid. Physiol Plant. 1989:**75**(1):89–96. https://doi.org/10.1111/j.1399-3054.1989.tb02068.x

Porco S, Pěnčík A, Rashed A, Voß U, Casanova-Sáez R, Bishopp A, Golebiowska A, Bhosale R, Swarup R, Swarup K, et al. Dioxygenase-encoding AtDAO1 gene controls IAA oxidation and homeostasis in Arabidopsis. Proc Natl Acad Sci U S A. 2016:**113**(39): 11016–11021. https://doi.org/10.1073/pnas.1604375113

Powers SK, Holehouse AS, Korasick DA, Schreiber KH, Clark N, Jing H, Emenecker R, Han S, Tycksen E, Hwang I, et al. Nucleo-cytoplasmic partitioning of ARF proteins controls auxin responses in *Arabidopsis thaliana*. Mol Cell. 2019:**76**(1):1–14. https://doi.org/10.1016/j.molcel.2019.06.044

Prigge MJ, Greenham K, Zhang Y, Santner A, Castillejo C, Mutka AM, O'Malley RC, Ecker JR, Kunkel BN, Estelle M. The Arabidopsis auxin receptor F-box proteins AFB4 and AFB5 are required for response to the synthetic auxin picloram. G3 (Bethesda). 2016:**6**(5):1383–1390. https://doi.org/10.1534/g3.115.025585

Prigge MJ, Platre M, Kadakia N, Zhang Y, Greenham K, Szutu W, Pandey BK, Bhosale RA, Bennett MJ, Busch W, et al. Genetic analysis of the Arabidopsis TIR1/AFB auxin receptors reveals both overlapping and specialized functions. eLife. 2020:**9**:e54740. https://doi.org/10.7554/eLife.54740

Qi L, Kwiatkowski M, Chen H, Hoermayer L, Sinclair S, Zou M, Del Genio CI, Kubeš MF, Napier R, Jaworski K, et al. Adenylate cyclase activity of TIR1/AFB auxin receptors in plants. Nature. 2022: **611**(7934):133–138. https://doi.org/10.1038/s41586-022-05369-7

Quareshy M, Prusinska J, Kieffer M, Fukui K, Pardal AJ, Lehmann S, Schafer P, Del Genio C, Kepinski S, Hayashi K, et al. The tetrazole analogue of the auxin indole-3-acetic acid binds preferentially to TIR1 and not AFB5. ACS Chem Biol. 2018:**13**(9):2585–2594. https://doi.org/10.1021/acschembio.8b00527

Quistgaard E, Löw C, Guettou F, Nordlund P. Understanding transport by the major facilitator superfamily (MFS): structures pave the way. Nature Rev Mol Cell Biol. 2016:**17**(2):123–132. https://doi.org/10.1038/nrm.2015.25

Radoeva T, Lokerse AS, Llavata-Peris CI, Wendrich JR, Xiang D, Liao CY, Vlaar L, Boekschoten M, Hooiveld G, Datla R, et al. A robust auxin response network controls embryo and suspensor development through a basic helix loop helix transcriptional module. Plant Cell. 2019:**31**(1):52–67. https://doi.org/10.1105/tpc.18.00518

Radwanski ER, Last RL. Tryptophan biosynthesis and metabolism: biochemical and molecular genetics. Plant Cell. 1995:**7**(7):921–934. https://doi.org/10.1105/tpc.7.7.921

Ramans-Harborough S, Kalverda AP, Manfield IW, Thompson GS, Kieffer M, Uzunova V, Quareshy M, Prusinska JM, Roychoudhry S, Hayashi KI, et al. Intrinsic disorder and conformational coexistence in auxin coreceptors. Proc Natl Acad Sci U S A. 2023:**120**(40): e2221286120. https://doi.org/10.1073/pnas.2221286120

Rayle DL, Cleland RE. The acid growth theory of auxin-induced cell elongation is alive and well. Plant Physiol. 1992:**99**(4):1271–1274. https://doi.org/10.1104/pp.99.4.1271

Reinecke D. The oxindole-3-acetic acid pathway in Zea mays. In: **Pharis RP, Rood SB**, editors. Plant growth substances 1988. Berlin, Heidelberg: Springer; 1990. p. 367–373.

Reinecke DM, Bandurski RS. Metabolic conversion of ^{14}C-indole-3-acetic acid to ^{14}C-oxindole-3-acetic acid. Biochem Biophys Res Commun. 1981:**103**(2):429–433. https://doi.org/10.1016/0006-291x(81)90470-8

Reinecke DM, Bandurski RS. Oxindole-3-acetic acid, an indole-3-acetic acid catabolite in *Zea mays*. Plant Physiol. 1983:**71**(1):211–213. https://doi.org/10.1104/pp.71.1.211

Reinecke DM, Bandurski RS. Oxidation of indole-3-acetic acid to oxindole-3-acetic acid by an enzyme preparation from *Zea mays*. Plant Physiol. 1988:**86**(3):868–872. https://doi.org/10.1104/pp.86.3.868

Ren H, Gray WM. SAUR proteins as effectors of hormonal and environmental signals in plant growth. Mol Plant. 2015:**8**(8):1153–1164. https://doi.org/10.1016/j.molp.2015.05.003

Ren H, Park MY, Spartz AK, Wong JH, Gray WM. A subset of plasma membrane-localized PP2C. D phosphatases negatively regulate SAUR-mediated cell expansion in *Arabidopsis*. PLoS Genet. 2018:**14**(6):e1007455. https://doi.org/10.1371/journal.pgen.1007455

Ribnicky DM, Cohen JD, Hu W-S, Cooke TJ. An auxin surge following fertilization in carrots: a mechanism for regulating plant totipotency. Planta. 2002:**214**(4):505–509. https://doi.org/10.1007/s004250100639

Rinaldi MA, Patel AB, Park J, Lee K, Strader LC, Bartel B. The roles of β-oxidation and cofactor homeostasis in peroxisome distribution and function in *Arabidopsis thaliana*. Genetics. 2016:**204**(3): 1089–1115. https://doi.org/10.1534/genetics.116.193169

Romano CP, Cooper MI, Klee HJ. Uncoupling auxin and ethylene effects in transgenic tobacco and Arabidopsis plants. Plant Cell. 1993:**5**(2):181–189. https://doi.org/10.1105/tpc.5.2.181

Ross JJ, Gélinas-Marion A. Two pathways become one. Nat Plants. 2021:**7**(12):1546–1547. https://doi.org/10.1038/s41477-021-01036-3

Rubery PH. Phytotropins: receptors and endogenous ligands. Symp Soc Exp Biol. 1990:**44**:119–146. PMID:2130510

Ruegger M, Dewey E, Gray WM, Hobbie L, Turner J, Estelle M. The TIR1 protein of Arabidopsis functions in auxin response and is related to human SKP2 and yeast grr1p. Genes Dev. 1998:**12**(2):198–207. https://doi.org/10.1101/gad.12.2.198

Ruegger M, Dewey E, Hobbie L, Brown D, Bernasconi P, Turner J, Muday G, Estelle M. Reduced naphthylphthalamic acid binding in the *tir3* mutant of Arabidopsis is associated with a reduction in polar auxin transport and diverse morphological defects. Plant Cell. 1997:**9**(5):745–757. https://doi.org/10.1105/tpc.9.5.745

Ruzicka K, Strader LC, Bailly A, Yang H, Blakeslee J, Langowski L, Nejedlá E, Fujita H, Itoh H, Syono K, et al. Arabidopsis PIS1 encodes the ABCG37 transporter of auxinic compounds including the auxin precursor indole-3-butyric acid. Proc Natl Acad Sci U S A. 2010:**107**(23): 10749–10753. https://doi.org/10.1073/pnas.1005878107

Sembdner G, Gross D, Liebisch H-W, Schneider G. Biosynthesis and metabolism of plant hormones. In: **MacMillan J**, editor. Hormonal regulation of development. I. Molecular aspects of plant hormones (encyclopedia of plant physiology vol. 9). Berlin: Springer; 1981. p. 281–444.

Shimizu-Mitao Y, Kakimoto T. Auxin sensitivities of all Arabidopsis aux/IAAs for degradation in the presence of every TIR1/AFB. Plant Cell Physiol. 2014:**55**(8):1450–1459. https://doi.org/10.1093/pcp/pcu077

Sidler M, Hassa P, Hasan S, Ringli C, Dudler R. Involvement of an ABC transporter in a developmental pathway regulating hypocotyl cell elongation in the light. Plant Cell. 1998:**10**(10):1623–1636. https://doi.org/10.1105/tpc.10.10.1623

Silberger J, Skoog F. Changes induced by indole-acetic acid in nucleic acid content and growth of tobacco pith tissue. Science. 1953:**118**(3068):443–444. https://doi.org/10.1126/science.118.3068.443

Simon S, Kubeš M, Baster P, Robert S, Dobrev PI, Friml J, Petrášek J, Zažímalová E. Defining the selectivity of processes along the auxin response chain: a study using auxin analogues. New Phytol. 2013:**200**(4):1034–1048. https://doi.org/10.1111/nph.12437

Singh VK, Jain M, Garg R. Genome-wide analysis and expression profiling suggest diverse roles of GH3 genes during development and abiotic stress responses in legumes. Front Plant Sci. 2015:**5**:789. https://doi.org/10.3389/fpls.2014.00789

Slovin JP, Bandurski RS, Cohen JD. Auxin. In: **Hooykaas PJJ, Hall MA, Libbenga KR**, editors. Biochemistry and molecular biology of plant hormones. Amsterdam: Elsevier; 1999. p. 115–140.

Solanki M, Shukla LI. Recent advances in auxin biosynthesis and homeostasis. 3 Biotech. 2023:**13**(9):290. https://doi.org/10.1007/s13205-023-03709-6

Sorin C, Bussell JD, Camus I, Ljung K, Kowalczyk M, Geiss G, McKhann H, Garcion C, Vaucheret H, Sandberg G, et al. Auxin and light control of adventitious rooting in Arabidopsis require ARGONAUTE1. Plant Cell. 2005:**17**(5):1343–1359. https://doi.org/10.1105/tpc.105.031625

Spartz AK, Ren H, Park MY, Grandt KN, Lee SH, Murphy AS, Sussman MR, Overvoorde PJ, Gray WM. SAUR inhibition of PP2C-D phosphatases activates plasma membrane H^+-ATPases to promote cell expansion in *Arabidopsis*. Plant Cell. 2014:**26**(5):2129–2142. https://doi.org/10.1105/tpc.114.126037

Spiess GM, Zolman BK. Peroxisomes as a source of auxin signaling molecules. In: **del Río L**, editor. Peroxisomes and their key role in cellular signaling and metabolism. Subcellular biochemistry, vol 69. Dordrecht: Springer; 2013. p. 257–281.

Starzyńska E, Kowalczyk S. Novel 1-O-indole-3-acetyl-β-D-glucose-dependent acyltransferase transferring indoleacetyl moiety to some mono- di- and oligosaccharides. Acta Physiol Plant. 2012:**34**(1):53–63. https://doi.org/10.1007/s11738-011-0804-y

Staswick PE. The tryptophan conjugates of jasmonic and indole-3-acetic acids are endogenous auxin inhibitors. Plant Physiol. 2009:**150**(3):1310–1321. https://doi.org/10.1104/pp.109.138529

Staswick P, Rowe M, Spalding EP, Splitt BL. Jasmonoyl-L-tryptophan disrupts IAA activity through the AUX1 auxin permease. Front Plant Sci. 2017:**8**:736. https://doi.org/10.3389/fpls.2017.00736

Staswick PE, Serban B, Rowe M, Tiryaki I, Maldonado MT, Maldonado MC, Suza W. Characterization of an Arabidopsis enzyme family that conjugates amino acids to indole-3-acetic acid. Plant Cell. 2005:**17**(2):616–627. https://doi.org/10.1105/tpc.104.026690

Staswick PE, Tiryaki I, Rowe M. The jasmonate response locus JAR1 and several related Arabidopsis genes encode enzymes of the firefly luciferase superfamily that show activity on jasmonic, salicylic, and indole-3-acetic acids in an assay for adenylation. Plant Cell. 2002:**14**(6):1405–1415. https://doi.org/10.1105/tpc.000885

Stein M, Dittgen J, Sánchez-Rodríguez C, Hou B-H, Molina A, Schulze-Lefert P, Lipka V, Somerville S. Arabidopsis PEN3/PDR8, an ATP binding cassette transporter, contributes to nonhost resistance to inappropriate pathogens that enter by direct penetration. Plant Cell. 2006:**18**(3):731–746. https://doi.org/10.1105/tpc.105.038372

Stepanova AN, Alonso JM. Auxin catabolism unplugged: role of IAA oxidation in auxin homeostasis. Proc Natl Acad Sci U S A. 2016:**113**(39):10742–10744. https://doi.org/10.1073/pnas.1613506113

Stepanova AN, Robertson-Hoyt J, Yun J, Benavente LM, Xie DY, Dolezal K, Schlereth A, Jürgens G, Alonso JM. TAA1-mediated auxin biosynthesis is essential for hormone crosstalk and plant development. Cell. 2008:**133**(1):177–191. https://doi.org/10.1016/j.cell.2008.01.047

Stepanova AN, Yun J, Robles LM, Novak O, He W, Guo H, Ljung K, Alonso JM. The Arabidopsis YUCCA1 flavin monooxygenase functions in the indole-3-pyruvic acid branch of auxin biosynthesis. Plant Cell. 2011:**23**(11):3961–3973. https://doi.org/10.1105/tpc.111.088047

Stortenbeker N, Bemer M. The SAUR gene family: the plant's toolbox for adaptation of growth and development. J Exp Bot. 2019:**70**(1):17–27. https://doi.org/10.1093/jxb/ery332

Strader L, Bartel B. A new path to auxin. Nat Chem Biol. 2008:**4**(6):337–339. https://doi.org/10.1038/nchembio0608-337

Strader LC, Bartel B. The Arabidopsis pleiotropic drug resistance8/ABCG36 ATP binding cassette transporter modulates sensitivity to the auxin precursor indole-3-butyric acid. Plant Cell. 2009:**21**(7):1992–2007. https://doi.org/10.1105/tpc.109.065821

Strader LC, Chen GL, Bartel B. Ethylene directs auxin to control root cell expansion. Plant J. 2010a:**64**(5):874–884. https://doi.org/10.1111/j.1365-313X.2010.04373.x

Strader LC, Culler AH, Cohen JD, Bartel B. Conversion of endogenous indole-3-butyric acid to indole-3-acetic acid drives cell expansion in Arabidopsis seedlings. Plant Physiol. 2010b:**153**(4):1577–1586. https://doi.org/10.1104/pp.110.157461

Strader LC, Monroe-Augustus M, Rogers KC, Lin GL, Bartel B. Arabidopsis IBA response5 suppressors separate responses to various hormones. Genetics. 2008:**180**(4):2019–2031. https://doi.org/10.1534/genetics.108.091512

Strader LC, Wheeler DL, Christensen SE, Berens JC, Cohen JD, Rampey RA, Bartel B. Multiple facets of Arabidopsis seedling development require indole-3-butyric acid-derived auxin. Plant Cell. 2011:**23**(3):984–999. https://doi.org/10.1105/tpc.111.083071

Sugawara S, Hishiyama S, Jikumaru Y, Hanada A, Nishimura T, Koshiba T, Zhao Y, Kamiya Y, Kasahara H. Biochemical analyses of indole-3-acetaldoxime-dependent auxin biosynthesis in Arabidopsis. Proc Natl Acad Sci U S A. 2009:**106**(13):5430–5435. https://doi.org/10.1073/pnas.0811226106

Sussman MR, Harper JF. Molecular biology of the plasma membrane of higher plants. Plant Cell. 1989:**1**(10):953–960. https://doi.org/10.1105/tpc.1.10.953

Szerszen JB, Szczyglowski K, Bandurski RS. *Iaglu*, A gene from *Zea mays* involved in conjugation of growth hormone indole-3-acetic acid. Science. 1994:**265**(5179):1699–1701. https://doi.org/10.1126/science.8085154

Tam YY, Epstein E, Normanly J. Characterization of auxin conjugates in Arabidopsis. Low steady-state levels of indole-3-acetyl-aspartate, indole-3-acetyl-glutamate, and indole-3-acetyl-glucose. Plant Physiol. 2000:**123**(2):589–596. https://doi.org/10.1104/pp.123.2.589

Tam YY, Normanly J. Determination of indole-3-pyruvic acid levels in *Arabidopsis thaliana* by gas chromatography-selected ion monitoring-mass spectrometry. J Chromatogr A. 1998:**800**(1):101–108. https://doi.org/10.1016/s0021-9673(97)01051-0

Tan X, Calderon-Villalobos LI, Sharon M, Zheng C, Robinson CV, Estelle M, Zheng N. Mechanism of auxin perception by the TIR1 ubiquitin ligase. Nature. 2007:**446**(7136):640–645. https://doi.org/10.1038/nature05731

Tanaka K, Hayashi K, Natsume M, Kamiya Y, Sakakibara H, Kawaide H, Kasahara H. UGT74D1 catalyzes the glucosylation of 2-oxindole-3-acetic acid in the auxin metabolic pathway in Arabidopsis. Plant Cell Physiol. 2014:**55**(1):218–228. https://doi.org/10.1093/PCP/PCT173

Tang Q, Yu P, Tillmann M, Cohen JD, Slovin JP. Indole-3-acetylaspartate and indole-3-acetylglutamate, the IAA-amide conjugates in the diploid strawberry achene, are hydrolyzed in growing seedlings. Planta. 2019:**249**(4):1073–1085. https://doi.org/10.1007/s00425-018-3061-0

Tao Y, Ferrer JL, Ljung K, Pojer F, Hong F, Long JA, Li L, Moreno JE, Bowman ME, Ivans LJ, et al. Rapid synthesis of auxin via a new tryptophan-dependent pathway is required for shade avoidance in plants. Cell. 2008:**133**(1):164–176. https://doi.org/10.1016/j.cell.2008.01.049

Theologis A. Rapid gene regulation by auxin. Annu Rev Plant Physiol. 1986:**37**(1):407–438. https://doi.org/10.1146/annurev.pp.37.060186.002203

Theologis A, Ray PM. Early auxin-regulated polyadenylated mRNA sequences in pea stem tissue. Proc Natl Acad Sci U S A. 1982:**79**(2):418–421. https://doi.org/10.1073/pnas.79.2.418

Thimann KV. Studies on the growth hormone of plants. VI. The distribution of the growth substance in plant tissues. J Gen Phys. 1934:**18**(1):23–34. https://doi.org/10.1085/jgp.18.1.23

Thomashow MF, Hughly S, Buchlolz WG, Thomashow LS. Molecular basis for the auxin-independent phenotype of crown gall tumor tissue. Science. 1986:**231**(4738):616–618. https://doi.org/10.1126/science.3511528

Thomashow MF, Nutter R, Montoya AL, Gordon MP, Nester EW. Integration and organization of Ti plasmid sequences in crown gall tumors. Cell. 1980:**19**(3):729–739. https://doi.org/10.1016/S0092-8674(80)80049-3

Thomashow LS, Reeves S, Thomashow MF. Crown gall oncogenesis: evidence that a T-DNA gene from the Agrobacterium Ti plasmid pTiA6 encodes an enzyme that catalyzes synthesis of indole-3-acetic acid. Proc Natl Acad Sci US A. 1984:**81**(16):5071–5075. https://doi.org/10.1073/pnas.81.16.5071

Tian H, Baxter IR, Lahner B, Reinders A, Salt DE, Ward JM. Arabidopsis NPCC6/NaKR1 is a phloem mobile metal binding protein necessary for phloem function and root meristem maintenance. Plant Cell. 2010:**22**(12):3963–3979. https://doi.org/10.1105/tpc.110.080010

Tillmann MA, Tang Q, Cohen JD. Protocol: analytical methods for visualizing the indolic precursor network leading to auxin biosynthesis. Plant Methods. 2021:**17**(1):63. https://doi.org/10.1186/s13007-021-00763-0

Tillmann MA, Tang Q, Gardner G, Cohen JD. Complexity of the auxin biosynthetic network in Arabidopsis hypocotyls is revealed by multiple stable-labeled precursors. Phytochemistry. 2022:**200**:113219. https://doi.org/10.1016/j.phytochem.2022.113219

Tivendale ND, Ross JJ, Cohen JD. The shifting paradigms of auxin biosynthesis. Trends Plant Sci. 2014:**19**(1):44–51. https://doi.org/10.1016/j.tplants.2013.09.012

Tiwari SB, Wang XJ, Hagen G, Guilfoyle TJ. AUX/IAA proteins are active repressors, and their stability and activity are modulated by auxin. Plant Cell. 2001:**13**(12):2809–2822. https://doi.org/10.1105/tpc.010289

Tognetti VB, Van Aken O, Morreel K, Vandenbroucke K, van de Cotte B, De Clercq I, Chiwocha S, Fenske R, Prinsen E, Boerjan W, et al. Perturbation of indole-3-butyric acid homeostasis by the UDP-glucosyltransferase UGT74E2 modulates Arabidopsis architecture and water stress tolerance. Plant Cell. 2010:**22**(8):2660–2679. https://doi.org/10.1105/tpc.109.071316

Tsurumi S, Wada S. Metabolism of indole-3-acetic acid and natural occurrence of dioxindole-3-acetic acid derivatives in *Vicia* roots. Plant Cell Physiol. 1980:**21**(8):1515–1525. https://doi.org/10.1093/pcp/21.8.1515

Tsurumi S, Wada S. Identification of 3-(O-β-glucosyl)-2-indolone-3-acetylaspartic acid as a new indole-3-acetic acid metabolite in *Vicia* seedlings. Plant Physiol. 1985:**79**(3):667–671. https://doi.org/10.1104/pp.79.3.667

Tsurumi S, Wada S. Dioxindole-3-acetic acid conjugates formation from indole-3-acetylaspartic acid in *Vicia* seedlings. Plant Cell Physiol. 1986a:**27**(8):1513–1522. https://doi.org/10.1093/oxfordjournals.pcp.a077252

Tsurumi S, Wada S. Identification of 3-hydroxy-2-indolone-3-acetylaspartic acid as a new indole-3-acetic acid metabolite in *Vicia* roots. Plant Cell Physiol. 1986b:**27**(3):559–562. https://doi.org/10.1093/oxfordjournals.pcp.a077133

Tuominen H, Östin A, Sandberg G, Sundberg B. A novel metabolic pathway for indole-3-acetic acid in apical shoots of *Populus tremula* (L.)×*Populus tremuloides* (michx.). Plant Physiol. 1994:**106**(4):1511–1520. https://doi.org/10.1104/pp.106.4.1511

Ulmasov T, Liu ZB, Hagen G, Guilfoyle TJ. Composite structure of auxin response elements. Plant Cell. 1995:**7**(10):1611–1623. https://doi.org/10.1105/tpc.7.10.1611

Ulmasov T, Murfett J, Hagen G, Guilfoyle TJ. Aux/IAA proteins repress expression of reporter genes containing natural and highly active synthetic auxin response elements. Plant Cell. 1997:**9**(11):1963–1971. https://doi.org/10.1105/tpc.9.11.1963

Uzunova VV, Quareshy M, Del Genio CI, Napier RM. Tomographic docking suggests the mechanism of auxin receptor TIR1 selectivity. Open Biol. 2016:**6**(10):160139. https://doi.org/10.1098/rsob.160139

Vanderhoef LN, Dute RR. Auxin-regulated wall loosening and sustained growth in elongation. Plant Physiol. 1981:**67**(1):146–149. https://doi.org/10.1104/pp.67.1.146

Varner JE. The greening of Gerry fink. Plant Cell. 1989:**1**(5):479–480. https://doi.org/10.1105/tpc.1.5.479

Venis MA. When is a hormone really a hormone? Plant Cell. 1997:**9**(6):839–840. https://doi.org/10.1105/tpc.9.6.839

Walbot V. Maize mutants for the 21st century. Plant Cell. 1991:**3**(9):851–856. https://doi.org/10.1105/tpc.3.9.851

Wang B, Chu J, Yu T, Xu Q, Sun X, Yuan J, Xiong G, Wang G, Wang Y, Li J. Tryptophan-independent auxin biosynthesis contributes to early embryogenesis in Arabidopsis. Proc Natl Acad Sci U S A. 2015:**112**(15):4821–4826. https://doi.org/10.1073/pnas.1503998112

Watanabe S, Takahashi N, Kanno Y, Suzuki H, Aoi Y, Takeda-Kamiya N, Toyooka K, Kasahara H, Hayashi KI, Umeda M, et al. The Arabidopsis NRT1/PTR FAMILY protein NPF7.3/NRT1.5 is an indole-3-butyric acid transporter involved in root gravitropism. Proc Natl Acad Sci U S A. 2020:**117**(49):31500–31509. https://doi.org/10.1073/pnas.2013305117

Weis J. Effect of oxindoles on the growth of tobacco tissue cultures. Nature. 1966:**211**(5054):1216–1217. https://doi.org/10.1038/2111216a0

West M, Harada JJ. Embryogenesis in higher plants: an overview. Plant Cell. 1993:**5**(10):1361–1369. https://doi.org/10.1105/tpc.5.10.1361

Widemann E, Miesch L, Lugan R, Holder E, Heinrich C, Aubert Y, Miesch M, Pinot F, Heitz T. The amidohydrolases IAR3 and ILL6 contribute to jasmonoyl-isoleucine hormone turnover and generate 12-hydroxyjasmonic acid upon wounding in Arabidopsis leaves. J Biol Chem. 2013:**288**(44):31701–31714. https://doi.org/10.1074/jbc.M113.499228

Won C, Shen X, Mashiguchi K, Zheng Z, Dai X, Cheng Y, Kasahara H, Kamiya Y, Chory J, Zhao Y. Conversion of tryptophan to indole-3-acetic acid by TRYPTOPHAN AMINOTRANSFERASES OF ARABIDOPSIS and YUCCAs in Arabidopsis. Proc Natl Acad Sci U S A. 2011:**108**(45):18518–18523. https://doi.org/10.1073/pnas.1108436108

Wong JH, Spartz AK, Park MY, Du M, Gray WM. Mutation of a conserved motif of PP2C. D phosphatases confers SAUR immunity and constitutive activity. Plant Physiol. 2019:**181**(1):353–366. https://doi.org/10.1104/pp.19.00496

Woo H-H, Orbach MJ, Hirsch AM, Hawes MC. Meristem-localized inducible expression of a UDP-glycosyltransferase gene is essential for growth and development in pea and alfalfa. Plant Cell. 1999:**11**(12):2303–2315. https://doi.org/10.1105/tpc.11.12.2303

Woodward AW, Bartel B. Auxin: regulation, action, and interaction. Ann Bot. 2005:**95**(5):707–735. https://doi.org/10.1093/aob/mci083

Wright AD, Moehlenkamp CA, Perrot GH, Neuffer MG, Cone KC. The maize auxotrophic mutant orange pericarp is defective in duplicate genes for tryptophan synthase beta. Plant Cell. 1992:**4**(6): 711–719. https://doi.org/10.1105/tpc.4.6.711

Wright AD, Sampson MB, Neuffer MG, Michalczuk L, Slovin JP, Cohen JD. Indole-3-acetic acid biosynthesis in the mutant maize orange pericarp, a tryptophan auxotroph. Science. 1991:**254**(5034): 998–1000. https://doi.org/10.1126/science.254.5034.998

Wu MF, Yamaguchi N, Xiao J, Bargmann B, Estelle M, Sang Y, Wagner D. Auxin-regulated chromatin switch directs acquisition of flower primordium founder fate. eLife. 2015:**4**:e09269. https://doi.org/10.7554/eLife.09269

Yadegari R, Paiva G, Laux T, Koltunow AM, Apuya N, Zimmerman JL, Fischer RL, Harada JJ, Goldberg RB. Cell differentiation and morphogenesis are uncoupled in Arabidopsis raspberry embryos. Plant Cell. 1994:**6**(12):1713–1729. https://doi.org/10.1105/tpc.6.12.1713

Yu P, Hegeman AD, Cohen JD. A facile means for the identification of indolic compounds from plant tissues. Plant J. 2014:**79**(6):1065–1075. https://doi.org/10.1111/tpj.12607

Yu XL, Wang HY, Leung DWM, He ZD, Zhang JJ, Peng XX, Liu EE. Overexpression of OsIAAGLU reveals a role for IAA-glucose conjugation in modulating rice plant architecture. Plant Cell Rep. 2019:**38**(6): 731–739. https://doi.org/10.1007/s00299-019-02402-4

Zaenen I, Van Larabeke N, Teuchy H, Van Montagu M, Schell J. Supercoiled circular DNA in crown gall inducing *Agrobacterium* strains. J Mol Biol. 1974:**86**(1):109–127. https://doi.org/10.1016/S0022-2836(74)80011-2

Zhang J, Lin JE, Harris C, Campos Mastrotti Pereira F, Wu F, Blakeslee JJ, Peer WA. DAO1 catalyzes temporal and tissue-specific oxidative inactivation of auxin in *Arabidopsis thaliana*. Proc Natl Acad Sci U S A. 2016:**113**(39):11010–11015. https://doi.org/10.1073/pnas.1604769113

Zhang R, Wang B, Ouyang J, Li J, Wang Y. Arabidopsis indole synthase, a homolog of tryptophan synthase alpha, is an enzyme involved in the trp-independent indole-containing metabolite biosynthesis. J Integrative Plant Biol. 2008:**50**(9):1070–1077. https://doi.org/10.1111/j.1744-7909.2008.00729.x

Zhao J, Last RL. Coordinate regulation of the tryptophan biosynthetic pathway and indolic phytoalexin accumulation in Arabidopsis. Plant Cell. 1996:**8**(12):2235–2244. https://doi.org/10.1105/tpc.8.12.2235

Zhao Y, Christensen SK, Fankhauser C, Cashman JR, Cohen JD, Weigel D, Chory J. A role for flavin monooxygenase-like enzymes in auxin biosynthesis. Science. 2001:**291**(5502):306–309. https://doi.org/10.1126/science.291.5502.306

Zhao Y, Hull AK, Gupta NR, Goss KA, Alonso J, Ecker JR, Normanly J, Chory J, Celenza JL. Trp-dependent auxin biosynthesis in Arabidopsis: involvement of cytochrome P450s CYP79B2 and CYP79B3. Genes Dev. 2002:**16**(23):3100–3112. https://doi.org/10.1101/gad.1035402

Zhao Z, Zhang Y, Liu X, Zhang X, Liu S, Yu X, Ren Y, Zheng X, Zhou K, Jiang L, et al. A role for a dioxygenase in auxin metabolism and reproductive development in rice. Dev Cell. 2013:**27**(1):113–122. https://doi.org/10.1016/j.devcel.2013.09.005

Zheng Z, Guo Y, Novák O, Dai X, Zhao Y, Ljung K, Noel JP, Chory J. Coordination of auxin and ethylene biosynthesis by the aminotransferase VAS1. Nat Chem Biol. 2013:**9**(4):244–246. https://doi.org/10.1038/nchembio.1178

Zimmerman JL. Somatic embryogenesis: a model for early development in higher plants. Plant Cell. 1993:**5**(10):1411–1423. https://doi.org/10.1105/tpc.5.10.1411

Zolman BK, Silva ID, Bartel B. The Arabidopsis pxa1 mutant is defective in an ATP-binding cassette transporter-like protein required for peroxisomal fatty acid β-oxidation. Plant Physiol. 2001:**127**(3): 1266–1278. https://doi.org/10.1104/pp.010550

Zolman BK, Yoder A, Bartel B. Genetic analysis of indole-3-butyric acid responses in Arabidopsis thaliana reveals four mutant classes. Genetics. 2000:**156**(3):1323–1337. https://doi.org/10.1093/genetics/156.3.1323

Zurfluh LL, Guilfoyle TJ. Auxin-induced changes in the population of translatable messanger RNA in elongating sections of soybean hypocotyls. Plant Physiol. 1982:**69**(2):332–337. https://doi.org/10.1104/pp.69.2.332

https://doi.org/10.1093/plcell/koad327

THE PLANT CELL

THE PLANT CELL 2024: 36: 1429–1450

Cytokinin: From autoclaved DNA to two-component signaling

Cristiana T. Argueso [1] **and Joseph J. Kieber** [2,*]

1 Department of Agricultural Biology, Colorado State University, Fort Collins, CO 80523, USA
2 Department of Biology, University of North Carolina, Chapel Hill, NC 27599, USA

*Author for correspondence: jkieber@unc.edu
The author responsible for distribution of materials integral to the findings presented in this article in accordance with the policy described in the Instructions for Authors (www.plantcell.org) is: Joseph J. Kieber (jkieber@unc.edu).

Abstract

Since its first identification in the 1950s as a regulator of cell division, cytokinin has been linked to many physiological processes in plants, spanning growth and development and various responses to the environment. Studies from the last two and one-half decades have revealed the pathways underlying the biosynthesis and metabolism of cytokinin and have elucidated the mechanisms of its perception and signaling, which reflects an ancient signaling system evolved from two-component elements in bacteria. Mutants in the genes encoding elements involved in these processes have helped refine our understanding of cytokinin functions in plants. Further, recent advances have provided insight into the mechanisms of intracellular and long-distance cytokinin transport and the identification of several proteins that operate downstream of cytokinin signaling. Here, we review these processes through a historical lens, providing an overview of cytokinin metabolism, transport, signaling, and functions in higher plants.

Introduction

In the early part of the 20th century, the Austrian botanist Gottlieb Haberlandt, the "father of plant tissue culture," postulated the existence of "growth enzymes," substances that are released from one set of cells to stimulate the growth and development of other cells (Haberlandt 1902, 1913). In the 1950s, cytokinin was identified as one such factor because it promoted cell division and shoot initiation in cultured cells in concert with another phytohormone, auxin (Miller et al. 1955, 1956). In recent decades, remarkable progress has been made in our understanding of the biology of cytokinin. Here, we highlight recent findings as well as the major break-throughs that led to our understanding of how plants synthe-size, degrade, and transport cytokinins, the signaling pathway underlying the perception of cytokinin, and the myriad roles that this fascinating phytohormone plays in plant growth and development and responses to biotic and abiotic stress. We also discuss how modulation of cytokinin content has been used to improve various aspects of different crops, including enhanced drought tolerance, and yield.

Cytokinin biosynthesis and metabolism

Cytokinins were first identified by Carlos Miller and others in Folke Skoog's laboratory in a search for factors that could promote the growth of plant cells in culture (for an excellent review of the discovery of cytokinin, see Amasino 2005). High activity was found in an old sample of herring sperm DNA, and subsequent work using autoclaved DNA identified the active component as an adenine derivative (6-furfural-aminopurine) that is a degradation product of DNA, which they named kinetin (Miller et al. 1955; Miller et al. 1956). Similar to other cytokinins, kinetin is an N^6-substituted adenine derivative (Fig. 1). In addition to its ability to promote cell division in concert with auxin, different ratios of kinetin to auxin in media were found to promote shoot or root organogenesis, or undifferentiated growth (Skoog and Miller 1957). The first endogenous cytokinin from plants was identified from immature corn kernels and named zeatin (Letham et al. 1964; Jameson 2023). Transzeatin has an isoprene derivative as the N^6 side chain and is the most common naturally occurring cytokinin in

Received August 18, 2023. Accepted November 03, 2023. Advance access publication January 2, 2024

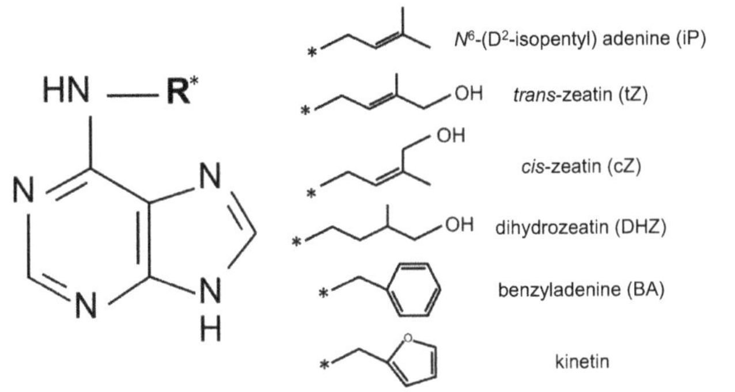

Figure 1. Structures of various cytokinins. Cytokinins are adenine derivatives. The general structure of the adenosine backbone of cytokinin is indicated to the left. Various sides chains, shown on the right, can be attached at the N^6 position of the adenine ring (indicated by an R*), leading to cytokinin species with differing biological activity. Note that BA and kinetin are not naturally occurring cytokinins.

higher plants, which also contain N^6-(Δ2-isopentenyl) adenine (iP) and dihydrozeatin (DHZ), as well as various derivatives of these (Fig. 1). Cytokinins with aromatic sides chains, including kinetin, have also been identified as minor cytokinins from some plant species (Strnad 1997). The various forms of cytokinin differ in their in planta stability, their transport characteristics throughout the plant, and their binding affinity for different cytokinin receptors (Kamada-Nobusada and Sakakibara 2009; Hluska et al. 2021).

tRNA from most organisms, except Archea, contain cis-zeatin as a modified base (Lindner et al. 2014). It was thus postulated that breakdown of this prenylated tRNA was the primary source of cytokinin in plants. However, recent work has shown that plants lack the ability to convert cis-zeatin to trans-zeatin (Hošek et al. 2020), and in many plants, cis-zeatin is not an active cytokinin (Gajdošová et al. 2011). Thus, tRNA is most likely not a significant source of cytokinin in such plants. However, in some plants, including many crop species (e.g. rice, maize, pea, and potato), the cis form of zeatin is found at high levels (Gajdošová et al. 2011; Kudo et al. 2012; Hluska et al. 2021) and is capable of binding to and activating a subset of cytokinin receptors in some species (e.g. Yonekura-Sakakibara et al. 2004; Lomin et al. 2011). The source of this cis-zeatin in these species may be breakdown of tRNA, though an enzyme involved in the breakdown of prenylated tRNA has not yet been identified. Alternatively, there may be a distinct biosynthetic pathway for cis-zeatin.

For most cytokinins, the first step in the biosynthetic pathway involves the linking of an isoprene moiety from dimethylallyl pyrophosphate (DMAPP) to the N^6 position of adenosine to form N^6-(Δ^2-isopentenyl) adenosine (iP) (Fig. 2). The first enzyme (isopentyl transferase [IPT]) catalyzing such a reaction was found in *Dictyostelium discoideum* (Taya et al. 1978), and, subsequently, a gene present on the transfer-DNA from *Agrobacterium tumefaciens* was shown to encode an enzyme with similar activity (Akiyoshi et al. 1984). In 2001, two groups identified nine *IPT* genes in the Arabidopsis genome and found that in addition to the tRNA-IPTs, plants possess a unique clade of *IPT* genes that the research groups subsequently demonstrated were involved in cytokinin biosynthesis (Kakimoto 2001; Takei et al. 2001) (see Table 1 for a list of genes relevant to cytokinin function). An important difference between the plant and bacterial enzymes is that plant IPTs use ATP/ADP rather than AMP as the adenosine source in the reaction. A recent phylogenetic analysis suggests that the tRNA-IPTs can be traced back to the last common ancestor of eukaryotes and that the IPTs involved in cytokinin biosynthesis in plants are derived from tRNA-IPTs (Wang 2020c).

Isotope-labeling experiments in Arabidopsis demonstrated that the primary source of the DMAPP used as the prenyl donor for the IPT reaction comes from the methylerythritol phosphate pathway (Kasahara et al. 2004), which is localized to plastids. Consistent with this, four of the Arabidopsis IPTs involved in cytokinin biosynthesis are localized to plastids, though two others are localized to the cytosol (IPT4) and mitochondria (IPT7) (Miyawaki et al. 2004; Takei et al. 2004a). A minor portion of DMAPP for cytokinin biosynthesis likely also comes from mevalonic acid because lovastatin, an inhibitor of mevalonic acid biosynthesis, can reduce cytokinin content (Laureys et al. 1999). The crystal structures for both a plant and an *A. tumefaciens* IPT have been solved, and this analysis revealed that the prenylation reaction proceeds via an SN2-reaction mechanism (Sugawara et al. 2008; Chu et al. 2010). Interestingly, the plant enzyme has two basic residues (Lys275 and Lys220) that interact with ATP to neutralize the charge of the β- and γ-phosphate, but these residues are replaced by acidic amino acids (Asp221 and Asp171) in bacterial IPTs, which likely explains the difference in the adenosine utilized in bacterial vs plant IPTs.

The products of the IPT reaction are iP ribosides, which are converted to trans-zeatin ribosides by hydroxylation of the prenyl side chain, a reaction catalyzed by the CYP735A cytochrome P450 monooxygenases (Takei et al. 2004b). Disruption of the genes encoding these CYP735As in Arabidopsis and rice eliminates trans-zeatin biosynthesis, allowing one to define specific roles of trans-zeatin type cytokinins in regulating growth and development. In the case of Arabidopsis, the *cyp735a1/2* double mutant shows reduced shoot growth but little effect on root growth (Kiba et al. 2013). This may reflect the observation that ARABIDOPSIS HISTIDINE KINASE3 (AHK3) is the cytokinin receptor most highly expressed in shoots (Higuchi et al. 2004; Nishimura et al. 2004), and it has a higher affinity for trans-zeatin than iP (Romanov et al. 2006; Stolz et al. 2011; Lomin et al. 2012). Disruption of the CYP735As in rice resulted in decreased growth of both roots and shoots, with substantial reductions in leaf size and rate of leaf formation, fewer tillers, delayed flowering, and highly reduced inflorescences (Kiba et al. 2023). This suggests that in rice, trans-zeatin is a positive regulator of shoot and inflorescence meristem function, consistent with analyses of cytokinin receptor mutants in rice (Burr et al. 2020). The enzyme responsible for the conversion

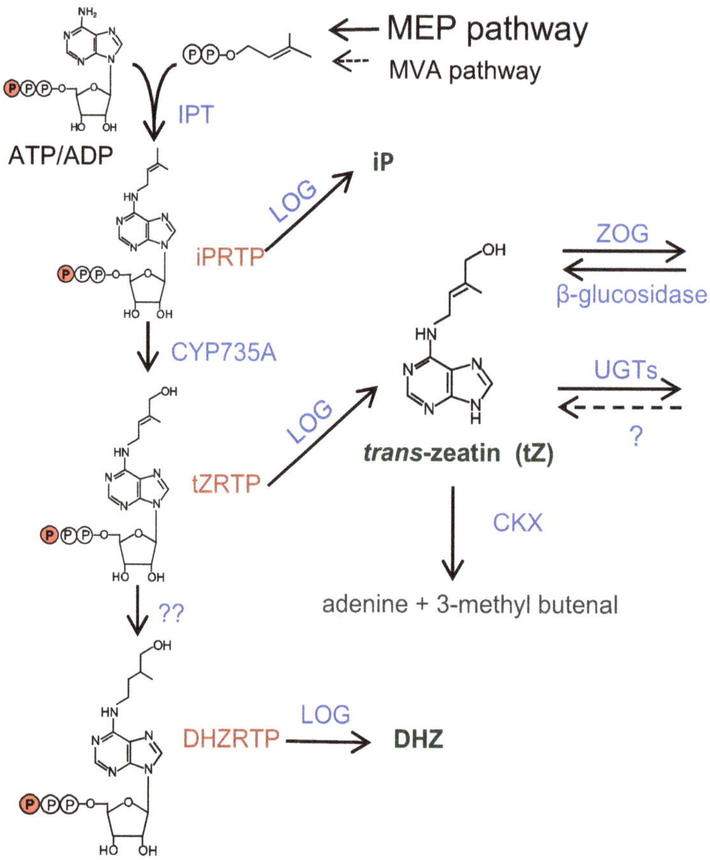

Figure 2. Generalized pathway for cytokinin biosynthesis and catabolism. Enzymes catalyzing each step are shown in blue. Inactive cytokinin ribosides are shown in red, and the active free base forms indicated in green. The source of the dimethylallyl pyrophosphate (DMPP) side chain of cytokinin is primarily from the methylerythritol phosphate (MEP) pathway, with minor contributions from mevalonic acid (MVA) pathway. The structure of the free base form of only zeatin is indicated; for the structures of iP and DHZ, see Fig. 1. Further, the conjugation and degradation of only trans-zeatin is indicated; both iP and DHZ type cytokinin are also subject to these processes to differing extents. The formation of cis-zeatin is not shown. See text for additional details.

of trans-zeatin to dihydrozeatin has not yet been identified in plants.

Biologically active cytokinins are produced by the removal of the riboside groups from iP riboside 5′-monophosphate, trans-zeatin riboside 5′-monophosphate, or dihydrozeatin riboside 5′-monophosphate to form the active, free-base cytokinins, a reaction catalyzed by the LONELYGUY (LOG) family of enzymes (Chen et al. 2022). LOG was first identified in a screen for rice mutants that failed to maintain shoot meri-stem function (Kurakawa et al. 2007). The LOG name comes from the phenotype of the spikelets in the mutants, which often contain only a single stamen and no pistil (Kurakawa et al. 2007). LOG is present as a small gene family in most angiosperms, as well as in moss (*Physcomitrium patens*) and liverworts (*Selagenella moellendorffii*), and are also found as single copy genes in green algae (Chen et al. 2022). Disruption of multiple *LOG* genes in Arabidopsis leads to se-vere retardation of shoot and root growth and defects in the maintenance of the apical meristems, with *LOG4* playing a major role in the shoot apical meristem (SAM) and *LOG3* and *LOG4* in the root (Tokunaga et al. 2012). *LOG4* expression is restricted to the L1 layer of the SAM and may supply apically derived cytokinin to underlying cell layers of the SAM to help pattern the expression of WUSCHEL (WUS), a key regulator of stem cell function within the stem cell niche (Chickarmane et al. 2012). Overexpression of *OsLOGL5* in rice reduced primary root growth, tiller number, and yield (Wang et al. 2020a). In contrast, mutations that alter the C-terminal domain of *OsLOGL5* did not significantly affect the growth or morphology of the plants but led to significantly increased yield in the field (Wang et al. 2020a), though the effects of these C-terminal mutations on *OsLOGL5* function have not been examined.

Cytokinin levels can be decreased through either conjugation to sugars, especially glucose, or irreversible cleavage by cytokinin oxidases. In terms of conjugation, glucose can be linked to the nitrogen at primarily the N^7 or N^9 position of the purine ring or to the oxygen on the side chains of iP, DHZ, and zeatin (both cis and trans) (Chen et al. 2021). Glucosyl conjugates are generally inactive in bioassays and fail to bind to cytokinin receptors (Spichal et al. 2004). Conjugations at the *N* positions were thought to be irreversible, but recent findings are changing that notion as metabolic studies on Arabidopsis seedlings have shown that trans-zeatin

Table 1. List of important genes in cytokinin metabolism, signaling, and transport

Gene/Protein	Name	Type	Function
HK	HISTIDINE KINASE	Signaling	Cytokinin receptors
HP/HPT	HISTIDINE PHOSPHOTRANSFER PROTEINS	Signaling	Phosphorelay proteins from receptors to RRs
PHP	PSEUDO- HISTIDINE PHOSPHOTRANSFER PROTEINS	Signaling	Negative regulators; lack conserved His that is target of phosphorylation
Type-B RR	TYPE-B RESPONSE REGULATOR	Signaling	Primary transcription factors; positive elements in the cytokinin response
Type-A RR	TYPE-A RESPONSE REGULATOR	Signaling	Rapidly induced by cytokinin; act as negative regulators of the cytokinin response
IPT	ISOPENTENYL TRANSFERASE	Biosynthesis	Biosynthetic enzyme; links isoprene moiety to N^6 position of adenosine
CYP735A	CYTOCHROME P450 MONOOXYGENASES	Biosynthesis	Conversion of iP ribosides to trans-zeatin
LOG	LONELY GUY	Biosynthesis	Cytokinin riboside 5′-monophosphate phosphoribohydrolase; activity generates active free-base forms of cytokinins
CKX	CYTOKININ OXIDASE	Degradation	FAD-dependent amine oxidases, cleavage of N^6-side chains; degrades cytokinins
UGT76C1 and 2	UDP-GLUCOSYL TRANSFERASE 76C1 and 2	Inactivation	*N*-glycosyltransferases
OsZOG1	ZEATIN O-GLUCOSYLTRANSFERASE 1	Inactivation	*O*-glycosyltransferases
ENT	EQUILIBRATIVE NUCLEOSIDE TRANSPORTERS	Transport	Cytokinin transporter
ABCGs	ATP-BINDING CASSETE TRANSPORTER G	Transport	Cytokinin transporter
PUPs	PURINE PERMEASES	Transport	Cytokinin transporter
HvSWEET11	SUGARS WILL EVENTUALLY BE EXPORTED TRANSPORTERS	Transport	Cytokinin and sugar transporter
AZG	AZAGUANINE PURINE TRANSPORTER	Transport	Purine and cytokinin transporter

N-glucosides (but not those of iP) are rapidly converted to trans-zeatin (Hošek et al. 2020). The *O*-glycosylated forms can also be converted into active cytokinins by β-glucosidases (Brzobohaty et al. 1993). Genes encoding enzymes catalyzing *N* or *O* (or both) glycosylation of cytokinin have been identified (Chen et al. 2021). In Arabidopsis, the glycosyltransferases UGT76C1 and UGT76C2 can glucosylate multiple different cytokinins at the N^7 or N^9 positions, and UGT85A1 glycosylates cytokinin at the oxygen on the side chain. As expected, disruption of *UGT76C2* led to a significant decrease in cytokinin *N*-glucoside levels, and overexpression had the opposite effect (Wang et al. 2011). The lines with altered *UGT76C2* function did not result in any obvious morphological alterations in plants, though there were some modest changes in the expression of cytokinin-regulated genes (Wang et al. 2011). The lack of effect on growth and development may be due to compensatory changes in cytokinin metabolism in these lines that keep the active, free-base forms within a physiologically suitable range. In contrast, knockdown of a rice cytokinin O-glucosyltransferase (OscZOG1) had substantial effects on growth and development, including increased lateral roots, tillering, panicle branching, grain number per panicle, and seed size; overexpression of OscZOG1 had the opposite effects, concomitant with an increased level of cis-zeatin O-glucoside (Shang et al. 2016). These results are consistent with the idea that conjugation of cytokinin plays an important role in regulating active cytokinin levels in vivo.

A major regulator of cytokinin content in plant tissues is its irreversible degradation by cytokinin oxidases (CKX), which are FAD-dependent amine oxidases that cleave the N^6-side chains from a subset of cytokinins (Werner et al. 2006). Cytokinin oxidase was first cloned from *Zea mays* kernels (Houba-Hérin et al. 1999; Morris et al. 1999) and has since been identified in all land plants. Phylogenetic analysis suggests that the *CKX* genes from land plants are derived from a single chlamydial ancestral gene (Wang et al. 2020b). The substrate preferences vary among CKX isoforms, but in general the free bases and their ribosides are the preferred substrates; dihydrozeatin and aromatic cytokinins such as 6-benzylaminopurine (BA) are generally resistant to cleavage (Galuszka et al. 2007). The expression of multiple *CKX* genes is induced in response to elevated cytokinin, acting as a nega-tive feedback mechanism to reduce cytokinin levels (Bhargava et al. 2013; Polko et al. 2021).

Alteration of CKX function has substantial potential to improve various agronomic traits (reviewed in Jameson and Song 2016; Jameson and Song 2020) and, indeed, various *CKX* genes have been disrupted in multiple crop species using overexpression, RNAi, TILLING, or, more recently, CRISPR/Cas9 (Mandal et al. 2022). Species targeted include rice, wheat, barley, rapeseed, and potato (Jameson and Song 2016). In general, CKX loss-of-function lines tend to show increased yields, while lines overexpressing CKX often have larger root systems and are more drought tolerant. For example, in a recent study in rapeseed (an allotetraploid), disruption of six of the eight copies of two *CKX* genes by TILLING resulted in increased seed yield from both greenhouse and field-grown plants, likely as a result of a more active inflorescence meristem leading to more flowers (Schwarz et al. 2020). Disruption of *Gn1a*, which encodes a CKX enzyme, in the indica subspecies of rice results in a substantial increase in grain yield (Ashikari et al. 2005; Li et al. 2016). Interestingly, overexpression of CKX in multiple crops leads to a higher content of various micro- and macro-elements in the shoots, likely due at least in part to increased size of the root systems. For example, the concentration of zinc, which is deficient in the diet of nearly 25% of the world's population, was significantly increased in the seeds of transgenic barley plants overexpressing CKX (Ramireddy et al. 2018). Likewise, in rapeseed, overexpression of a *CKX* gene resulted in plants that displayed an increase in the content of multiple macro- and microelements in shoots (Nehnevajova et al. 2019). In maize, overexpression of a *CKX* gene specifically in roots resulted in a larger root system without affecting the size of the shoot, and an increase in the content of various minerals, including K, P, and Zn in the shoot and in some lines, resulted in an increase in Cu, Zn, and Mn in seeds (Ramireddy et al. 2021). Similar results were also obtained with overexpression of a *CKX* gene in barley (Ramireddy et al. 2018).

Transport of cytokinins

Cytokinins are synthesized in both roots and shoots and then can be transported short distances to neighboring cells or long distances to other tissues (Sakakibara 2006; Zhang et al. 2023). Trans-zeatin and trans-zeatin riboside cytokinins are generally synthesized in the roots and transported to shoots through the xylem, whereas iP and *cis*-zeatin type cytokinins are mostly synthesized in the shoots and transported to the roots through the phloem (Sakakibara 2006). Although the systemic, mobile nature of cytokinins was known for many years, the identity of the transporters involved in their translocation have only recently come to light.

In the early 2000s, a series of genetic and biochemical studies identified members of the EQUILIBRATIVE NUCLEOSIDE TRANSPORTERS (ENT) family as nucleoside transporters with broad substrate specificity, with some members being able to transport cytokinins (Möhlmann et al. 2001; Hirose et al. 2005). ENT proteins typically localize to the plasma membrane and can transport nucleosides down their concentration gradients (Hyde et al. 2001). A mutation in *ENT8* was identified as a suppressor of an Arabidopsis *IPT* overexpression line, and disruption of this gene resulted in hyposensitivity to exogenous isopentenyladenine and trans-zeatin ribosides, as well as a reduced uptake of isopentenyladenine riboside (Sun et al. 2005). In rice, OsENT2 was shown to transport cytokinin ribosides when expressed in yeast, suggesting it may also participate in cytokinin riboside uptake, though no in planta transport assays have been done for this ENT (Hirose et al. 2008). These studies suggest that a sub-set of ENTs is likely involved in the long-distance transloca-tion from roots to shoots of the riboside forms of isopentenyladenine and trans-zeatin.

Another class of transporters acting in the long-distance movement of cytokinin belongs to the ATP-BINDING CASSETTE TRANSPORTER G (ABCG) family. Disruption of *ABCG14* resulted in phenotypes consistent with reduced cytokinin function, suggesting that it might act as a cytokinin transporter (Zhang et al. 2014). ABCG proteins transport a variety of substrates, including auxin. Through genetic and biochemical analyses, ABCG14 was shown to be a PM-localized protein that functions in the uptake of trans-zeatin and DHZ-type cytokinins (Ko et al. 2014; Zhang et al. 2014). ABCG14 is highly expressed in roots, and quantification of cytokinins in both shoot and roots of the *abcg14* mutant showed a hyperaccumulation of cytokinins in roots compared with wild-type plants, suggesting a function in the root to shoot transport of cytokinin (Ko et al. 2014; Zhang et al. 2014). Grafting experiments confirmed that ABCG14 functions as a cytokinin transporter between roots and shoots (Ko et al. 2014). ABCG14 plays a role in both the loading of cytokinin into the xylem in the roots (Ko et al. 2014; Zhang et al. 2014) and the unloading of cytokinin from the phloem in shoots via an apoplastic pathway (Zhao et al. 2021). In rice, OsABCG18 plays a similar role in root to shoot transport of cytokinin (Kim et al. 2020). Another ABCG transporter, ABCG11, has been proposed as a cytokinin transporter due to cytokinin-associated phenotypes displayed by an *abcg11* loss-of-function mutant (Yang et al. 2022); however, biochemical evidence of transporter activity is lacking. Likewise, disruption of the genes encoding 3 non-intrinsic ABC transporters (ABCI19, ABCI20, and ABCI21) resulted in cytokinin hypersensitivity in Arabidopsis (Kim et al. 2020). These ABCI proteins were localized to the ER and found to be induced by light, though, like ABCG11, these have not yet been shown to transport cytokinin.

Another class of cytokinin transporters belongs to the PURINE PERMEASES (PUPs) family, which likely reflects the fact that cytokinins are purine derivatives. There are 21 *PUP* genes in Arabidopsis (Jelesko 2012), of which *AtPUP1* was the first shown to encode a transporter of cytokinin (Gillissen et al. 2000). A survey of the other *PUP* genes led to the identification *PUP14* as the member with highest expression in a variety of plant tissues and strongest genetic effect on cytokinin responses (Zürcher et al. 2016). PUP14 was localized to the PM, and its activity was inversely correlated with cytokinin content in the apoplast, thus leading to the hypothesis that its physiological function was to attenuate cytokinin signaling by transporting cytokinins from the apoplast into cells, reducing the signal available for PM localized cytokinin receptors (Zürcher et al. 2016). Given the observations that the affinity of this transporter is significantly lower than that of the receptors for cytokinin, that the bulk of the cytokinin receptors are found in the ER, and that a rice PUP cytokinin transporter (OsPUP1) has been localized to the ER (Xiao et al. 2019), a counter-hypothesis is that the PUPs function by transporting cytokinins intracellularly between ER, the cytosol, and the apoplast, thus regulating cytokinin signaling by depleting cytokinin from ER-localized receptors (Romanov et al. 2018).

Recently, a member of the SWEET (SUGARS WILL EVENTUALLY BE EXPORTED TRANSPORTERS) family of sugar transporters in barley (*Hv*SWEET11b) was shown to function as a dual transporter of both sugars and cytokinins in developing barley grains (Radchuk et al. 2023). Given that cytokinin regulates sink-source relationships, which involves sugar transport, this directly links the transport of cytokinins to one of the physiological processes they control (Radchuk et al. 2023).

Finally, two transporters in the azaguanine purine transporter family (AZG) family have been linked to cytokinin transport. AZG2 transported purines and cytokinin with high affinity (Tessi et al. 2021) and localized to both the ER and the plasma membrane. AZG2 plays a role in lateral root development. Interestingly, AZG1, which was also shown to transport cytokinin, was identified as a protein that co-purified with the PIN1 auxin transporter in native gels, indicating that it may play a role in the interaction between auxin and cytokinin (Tessi et al. 2023).

Deciphering the signaling pathway

Starting in the mid-1970s, multiple groups reported the biochemical purification of cytokinin binding proteins (Takegami and Yoshida 1975; Polya and Davis 1978; Sussman and Kende 1978; Moore 1979; Keim and Fox 1980), including binding to isolated plant ribosomes (Fox and Erion 1975), but these approaches failed to identify the molecular nature of the cytokinin receptor. Recent decades have seen remarkable progress in our understanding of the molecular mechanisms underlying cytokinin perception and signaling. In the mid-1990s, 2 independent lines of inves-tigation converged on the same signaling pathway. The CKI1 histidine kinase was identified in a screen for genes that when overexpressed in cultured cells conferred the ability to form shoots in the absence of exogenous cytokinin (Kakimoto 1996). In parallel, two genes encoding response regulators (ARR5 and ARR7) were identified in a screen for cytokinin primary response genes (Brandstatter and Kieber 1998). Histidine Kinases (HKs) and Response Regulators (RRs) comprise the elements of two-component signaling pathways in prokaryotic organisms (Stock Stock and Mottonen 1990; Stock Robinson and Goudreau 2000). In their simplest form, two-component systems are comprised of an often membrane-localized sensor HK that perceives some environmental stimuli (e.g. phosphate levels). Upon perception of the signal, the HK autophosphorylates on a His residue and then in turn transfers the phosphate to an Asp residue within the receiver domain of an RR protein. Phosphorylation of the RR regulates its activity, which most often involves direct regulation of transcription of target genes. Connecting both a HK (CKI1) and two RRs (ARR5 and ARR7) to the response to cytokinin strongly suggested that cytokinin perception involves a two-component–like pathway, which subsequent studies confirmed. However, in the case of cytokinin (and other eukaryotic two-component pathways), the response pathway is actually similar to phosphorelays, which involve an extra pair of phosphotransfers between His and

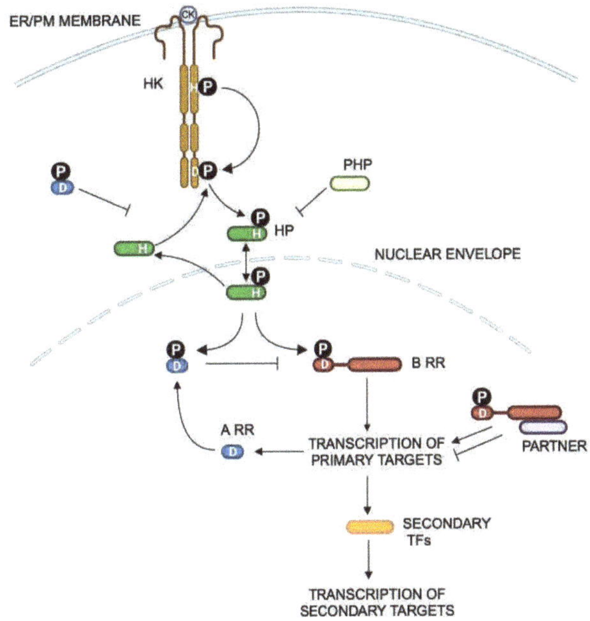

Figure 3. Canonical cytokinin signaling pathway. Cytokinin binds to the CHASE domain in the HK receptors and initiates a phosphorelay as indicated by the P. See text for additional details.

Asp residues (Swanson et al. 1994; Appleby et al. 1996; Schaller et al. 2011). In these systems, there is a receiver domain fused to the HK receptor protein, and when the HK is activated, an Asp residue within this receiver domain accepts the transferred phosphate in an intramolecular reaction. A small histidine-containing phosphotransfer protein then shuttles the phosphate from the HK receiver domain to an Asp residue on the downstream RR proteins (Fig. 3).

While CKI1 was the first HK to be linked to cytokinin signaling, it lacks a cytokinin binding domain and does not act as a cytokinin receptor, though it may feed into the downstream signaling elements (Dobisova et al. 2017). Authentic cytokinin HK receptors were identified in genetic screens for loss-of-function mutants impaired in cytokinin responses in cultured cells [e.g. shoot formation in tissue culture (Inoue et al. 2001)], by reverse genetic analysis of HK homologs in Arabidopsis (Ueguchi et al. 2001), by transient expression in Arabidopsis protoplasts (Hwang and Sheen 2001), by using bioassays in *Escherichia coli* or yeast, and/or by direct binding assays (Inoue et al. 2001; Suzuki et al. 2001; Ueguchi et al. 2001; Yamada et al. 2001). Plant cytokinin HK receptors are generally encoded by small gene families (e.g. three in Arabidopsis and four in rice) and share a similar structure, with two or three transmembrane domains flanking the cytokinin binding domain, a cytosolic HK domain fused to both a functional receiver domain as well as a degenerate receiver domain. Genetic analysis has demonstrated that the His kinase activity of AHK4 in Arabidopsis is required for cytokinin signaling (Mähönen et al. 2006a), as is the Asp residue that receives the phosphate within the functional receiver domain (Hodgens et al. 2020). Interestingly, multiple missense mutations that alter residues within or near the first transmembrane domain of AHK4 also eliminate its function, which may reflect disruption of its membrane insertion (Hodgens et al. 2020).

Cytokinin binds to a CHASE (cyclases/histidine kinases associated sensor extracellular) domain in these HK receptors, an evolutionarily ancient domain that binds a wide variety of ligands, including other adenine derivatives and peptides (Anantharaman and Aravind 2001). Cytokinin-binding, CHASE domain–containing HK cytokinin receptors have been identified in a bryophyte (*P. patens*), a lycophyte (*S. moellendorffii*), a gymnosperm (*Picea abies*), and all angiosperms examined (Lomin et al. 2021), which suggests that the ability to perceive cytokinin arose early in the evolution of land plants. In the case of *P. patens*, the CHASE domain HKs are essential for the response to exogenous cytokinin (von Schwartzenberg et al. 2016), even though *P. patens* contains only a tRNA-IPT and thus these receptors may reflect a response to endogenous cytokinin derived solely from breakdown of prenylated tRNA. The various cytokinin HK receptors in higher plants have different spatial patterns of expression and bind to different cytokinin species with distinct affinities (Spichal et al. 2004; Romanov et al. 2006; Stolz et al. 2011; Lomin et al. 2012, 2021). Thus, although the cytokinin receptors display varying degrees of genetic redundancy in the species that have been examined, they also likely have distinct functionalities.

An unresolved question in the field is the subcellular localization of these receptors. Early studies using analysis of AHK fusions to fluorescent proteins suggested a plasma membrane localization (Kim et al. 2006). However, subsequent biochemical analysis indicated that all three AHK cytokinin receptors localize to the endoplasmic reticulum (ER) (Wulfetange et al. 2011). This was based on analysis of high-affinity cytokinin binding to membrane fractions purified by two-phase partitioning in Arabidopsis lines harboring a single functional AHK receptor, as well as localization of fluorescent fusion proteins in transiently transformed tobacco leaves. Furthermore, aqueous two-phase partitioning of microsomes revealed that epitope-tagged AHK2 and AHK3 colocalized with ER membranes (Wulfetange et al. 2011). Likewise, a second study confirmed that fluorescently labeled AHK3 and AHK4 localized to the ER in both tobacco leaves and stably transformed Arabidopsis, and the authors showed that these fusion proteins complemented an *ahk2 ahk3* loss-of-function mutant (Caesar et al. 2011). Likewise, the maize ZmHK1 protein was also localized to the ER (Lomin et al. 2011). Finally, the pH optima of cytokinin binding to the Arabidopsis receptors is more consistent with binding in the ER lumen than in the apoplast. This issue was extensively discussed in an excellent review (Romanov et al. 2018). However, recent results indicate that a subset of the cytokinin receptors may actually reside in the plasma membrane (Antoniadi et al. 2020). These authors found that immobilized cytokinin (i.e. covalently linked to beads), which cannot enter cells, was capable of activating all three AHK receptors in Arabidopsis protoplasts as measured by both activation of a cytokinin-responsive reporter (Liu and Müller 2017) and induction of Cytokinin Response Factors (*CRF*) transcript levels (Rashotte et al. 2006). Further, super-resolution microscopy of AHK4-GFP and AHK3-GFP fusion proteins indicated that a fraction of the signal is present at the cell surface (Antoniadi et al. 2020). Finally, more recent analysis of cytokinin receptors across the plant kingdom has revealed a wide range of pH optima for cytokinin binding, consistent with different subcellular localizations (Lomin et al. 2021). Thus, it seems that cytokinin receptors in higher plants are primarily localized to the ER with cytokinin binding occurring in the ER lumen, but a fraction of the functional receptors may be at the plasma membrane, binding cytokinin in the apoplast.

Histidine-containing phosphotransfer proteins

The Authentic Histidine-containing Phosphotransfer proteins (AHPs) act downstream of the cytokinin receptors. AHPs are found in unicellular algae, moss, lycophytes, gymnosperms, and angiosperms, with the AHPs from land plants forming a distinct phylogenic clade from those in algae (Pils and Heyl 2009; Rashotte 2021). The AHPs are often found as small gene families in higher plants (e.g. five in Arabidopsis and two in rice), and at least in Arabidopsis, these AHPs have overlapping functions as positive elements in cytokinin signaling (Hutchison et al. 2006). These proteins shuttle between the cytosol and the nucleus to mediate the transfer of a phosphoryl group from the receiver domain of the activated HK receptors to the receiver domain of an RR protein (Punwani et al. 2010). Their phosphotransferase activity is negatively regulated by nitric oxide–induced S-nitrosylation, suggesting a link between redox signaling and cytokinin function (Feng et al. 2013). In addition to their role in cytokinin signaling, the AHPs also act downstream of other HKs, including CKI1 and likely AHK1 in Arabidopsis. For example, disruption of multiple AHPs results in loss of central cell and antipodal cell fates coupled with a gain of egg cell or synergid cell attributes during female gametophyte development in Arabidopsis (Liu et al. 2017), similar to the phenotypes of *cki1* loss-of-function mutations (Yuan et al. 2016).

The AHPs contain an essential His that is the target of phosphotransfer from the upstream AHKs. In addition to the AHPs, plant genomes also encode proteins similar to AHPs that lack the His residue required for phosphorelay. These so-called Pseudo Histidine Phosphotransfer proteins (PHPs) are negative regulators of cytokinin signaling, likely acting in a dominant negative manner to block AHP phos-photransfer activity (Mähönen et al. 2006b). AHP6 (an Arabidopsis PHP) mediates cross talk between auxin and cytokinin in multiple developmental events, including root vascular patterning, the formation of passage cells, lateral root organogenesis, proliferation of the inflorescence meri-stem, shoot phyllotaxy, and gynoecium development

(Mähönen et al. 2006b; Moreira et al. 2013; Besnard et al. 2014; Müller et al. 2017; Andersen et al. 2018). For example, in vascular development, bisymmetric domains of auxin and cytokinin ultimately lead to the formation of phloem and xylem (Mähönen et al. 2006b) (see below). Similar to this, in the shoot apical meristem, leaves initiate at positions of auxin maxima, and the robustness of this phyllotaxy is enhanced by a spatio-temporal pattern of AHP6 that regulates fields of cytokinin signaling (Besnard et al. 2014).

Phylogenic analysis indicates that the PHPs are present in gymnosperms and early diverging angiosperms but not in bryophytes (Vaughan-Hirsch et al. 2021). The monocot and dicot PHPs fall into two distinct clades, with the monocots having a Gln at the position of the conserved His residue and dicots having an Asn at this position. This suggests that PHPs evolved twice independently from AHPs: one event in monocots event giving rise to a to a PHP(Gln) class and a second in eudicots giving rise to the PHP(Asn) class. Consistent with their independent evolutionary origin, rice PHPs appear to have functions distinct from their Arabidopsis counterparts (Vaughan-Hirsch et al. 2021).

Response regulators

The final step of the cytokinin-regulated phosphorelay is the transfer of the phosphate from the AHPs to an Asp residue on the receiver domain of response regulators (RRs). There are two main types of RRs in plants. The type-B RRs are characterized by the presence of a receiver domain and a large C-terminal extension harboring a Myb-like DNA binding domain. Type-B RRs are generally not transcriptionally regulated by cytokinin and act as positive regulators of cytokinin signaling. In contrast, type-A RRs, which form a distinct phylogenetic clade from the type-B RRs, lack a DNA binding domain, are generally transcriptionally induced by cytokinin and act as negative regulators of cytokinin signaling. Most phylogenetic analyses suggest that type-A and type-B RRs originated in the Charophyte algae and are found throughout the land plant lineage, from the liverwort *Marchantia polymorpha* and the moss *Physcomitrella patens*, through Gymnosperms and Angiosperms (Pils and Heyl 2009; Rashotte 2021).

Type-B RRs are DNA-binding transcription factors that mediate the initial transcriptional response to cytokinin (Sakai et al. 2000; Hwang and Sheen 2001; Sakai et al. 2001; Imamura et al. 2003; Mason et al. 2005; Argyros et al. 2008; Hill et al. 2013). As with other elements in the cytokinin signaling pathway, type-Bs are generally encoded by a small gene family in higher plants (e.g. 11 in Arabidopsis and 13 in rice, though not all have been shown to be involved in cytokinin signaling). There is functional overlap in the gene family in Arabidopsis; single mutants have at most modest effects, but disruption of multiple type-B RRs leads to cytokinin insensitivity, effects on growth and development similar to disruption of the cytokinin HK receptors, and, in the highest combination mutants, the near absence of a transcriptional response to cytokinin (Mason et al. 2005; Argyros et al. 2008). The receiver domain of the type-B RRs negatively regulates their transcriptional activity (Sakai et al. 2001; Liang et al. 2012), and phosphorylation of this domain at the conserved Asp residue results in increased binding of type-B RRs to their genomic targets (Zubo et al. 2017; Xie et al. 2018). Type-B RRs were found to bind upstream of genes that were both activated and repressed by cytokinin, and disruption of type-B RRs block the cytokinin regulation of both classes of genes (Mason et al. 2005; Argyros et al. 2008), suggesting that type-B RRs are necessary for both gene activation and repression in response to cytokinin. Type-B RR protein stability is regulated via the KISS ME DEADLY F-box proteins (Kim et al. 2013).

A large number of transcription factors are regulated in re-sponse to cytokinin (Argueso et al. 2010; Brenner et al. 2012; Bhargava et al. 2013), suggesting that type-B RRs operate at the top of a transcriptional cascade, with the succeeding waves of transcription being regulated by these inducible transcription factors (Fig. 3). Various other transcription factors act in concert with type-B RRs and contribute to differences in transcriptomic responses to cytokinin (Leuendorf and Schmülling 2021). These partner transcription factors likely modulate the target genes to which activated type-B RRs bind, either through direct or indirect interaction. Many such type-B RR interacting partners have been identi-fied, including CRFs, DELLA, BPCs, NPR1, TCPs, HY5, TGA3, HD-ZIP IIIs, and EIN3 (Rashotte et al. 2006; Choi et al. 2010; Efroni et al. 2013; Marín-de la Rosa et al. 2015; Raines et al. 2016b; Shanks et al. 2016; Yan et al. 2017; Zhang et al. 2017; Hodgens et al. 2020).

The type-A RRs contain a receiver domain but, unlike the type-B RRs, lack a DNA binding domain and most are transcriptionally induced in response to cytokinin (D'Agostino et al. 2000; Tsai et al. 2012). Genetic analyses indicate that ten type-A RRs in Arabidopsis function as negative regulators of cytokinin signaling, thus acting as a negative feedback loop to dampen cytokinin signaling (Kiba et al. 2003; To et al. 2004; Leibfried et al. 2005; Lee et al. 2007; To et al. 2007). The mechanism by which type-A RRs negatively regulate cytokinin signaling is not fully understood but likely involves both competition with the type-B RRs for phosphotransfer from the upstream AHPs and phospho-specific interactions with regulatory proteins (To et al. 2007; Shanks et al. 2018).

The stability of type-A RR proteins also plays an important role in their regulation. In response to elevated cytokinin, type-A RRs are phosphorylated, and this inhibits their degradation by the 26S proteosome (To et al. 2007). This stabilization acts synergistically with the transcriptional upregulation of type-A RRs to highly increase RR protein levels in response to elevated cytokinin. Intriguingly, in contrast to unphosphorylated type-A RRs, the phosphorylated forms are targeted for degradation by autophagy in an EXO70D-dependent manner (Acheampong et al. 2020). The relative contributions of 26S proteome vs autophagic protein turnover is likely influenced by multiple regulatory inputs and differs across the type-A RR family members. These processes likely complement each other to optimize type-A RR protein levels, and hence cytokinin responsiveness, in response to various developmental and environmental cues.

Another negative feedback loop in cytokinin signaling in addition to the type-A RRs is provided by the cytokinin induction of the EAR motif-containing transcriptional repressor TIE1/TIE2 (He et al. 2022). ARR1 induces transcription of *TIE1* and *TIE2* by directly binding to their regulatory regions in the genome. TIE1/2 also interact directly with type-B RRs to represses transcription of their target genes in Arabidopsis roots, thus dampening the response to elevated cytokinin.

A number of cytokinin-responsive reporter transgenes have been constructed based on the transcriptional induc-tion of the type-A RRs. The first were fusions of the promoter of the Arabidopsis *ARR5* gene to either a GUS or GFP report-er (D'Agostino et al. 2000; Lohar et al. 2004). Although these reporters are useful, they suffer from lack of specificity and universality because ARR5 is likely regulated by other regulatory inputs and is not induced in all cell types. A synthetic reporter was developed comprised of concatemerized type-B binding sites fused to a GFP reporter called *TCS::GFP* (Müller and Sheen 2008). Although useful for reporting cyto-kinin signaling, *TCS::GFP* showed little or no expression in developmental context known to involve cytokinin (e.g. the shoot apical meristem [SAM]) and was found to rapidly silence in transgenic plants. An improved version, called *TCSn::GFP*, was developed with more robust expression that is more consistent with known cytokinin functions in Arabidopsis and which is stable over several generations (Zürcher et al. 2013). *TCSn::GFP* is a valuable tool to querying the endogenous state of cytokinin signaling, and it or derivatives have been used in multiple plant species, including Arabidopsis (Zürcher et al. 2013), rice (Tao et al. 2017), tomato (Steiner et al. 2020), and *Medicago truncatula* (Fonouni-Farde et al. 2017).

Role of cytokinin in plant development and responses to stress

Cytokinin exerts a profound influence on nearly all aspects of plant development, orchestrating intricate processes that de-termine growth, differentiation, and overall morphogenesis (Fig. 4). In addition, cytokinin also plays a pivotal role in regu-lating several physiological processes that help maintain optimal plant development in response to environmental cues. These processes were extensively reviewed previously (Argueso et al. 2009; Hwang et al. 2012; Kieber and Schaller 2014; Schaller et al. 2015; Kieber and Schaller 2018; Wybouw and De Rybel 2019; Liu et al. 2020), and so here we simply highlight a subset of the various responses regu-lated by cytokinin, focusing on seminal work that elucidated these responses, the more recent findings, and the molecular mechanisms at play.

Cell division

Cytokinin plays a crucial role in cell division and cell cycle control, orchestrating the growth and development of plants (Schaller et al. 2014). After this hormone was discovered by its ability to promote cell division, studies showed that oscil-lations in cytokinin levels accompanied the progression of the cell cycle, peaking at phase transitions (Hartig and Beck 2005), suggesting a link between cytokinin and cell cycle con-trol. Further work demonstrated that cytokinin regulates the expression of CYCD3 D-type cyclins (Riou-Khamlichi et al. 1999), which in turn regulate the transition from the G_1 to the S phase of the cell cycle. Cytokinin-induced shoot regeneration in vitro was also found to be dependent on cytokinin-regulated CYCD3s (Dewitte et al. 2007). The bHLH transcription factor CYTOKININ-RESPONSIVE GROWTH REGULATOR (CKG) was recently identified as a mediator of the G_1/S transition in Arabidopsis (Park et al. 2021). *CKG* is transcriptionally induced by cytokinin, and *ckg* mutants show altered expression of cell cycle regulators and display a delay in the G_1/S transition. Although the stud-ies described above indicated a role for cytokinin in the G_1/S transition via the action of CYCD3s and CKG, evidence for a role for cytokinin in the G_2/M transition also exists. Treatment of synchronized BY-2 tobacco cell cultures with the cytokinin biosynthesis inhibitor lovastatin blocks the G_2/M transition (Hartig and Beck 2005), and *ahk2*, *ahk3*, *ahk4* triple mutants show a delayed G_2/M transition in roots (Higuchi et al. 2004).

Cell cycle control by cytokinin regulates the activity and maintenance of meristems, ensuring sustained growth and proper development. SAM size is directly correlated to cytokinin content (Kurakawa et al. 2007; Bartrina et al. 2011), and the spatial distribution of cytokinin is highly regulated within the SAM (Gordon et al. 2009; Chickarmane et al. 2012). Cytokinin also directly regulates the activity of the homeodomain transcription factor WUS, which controls stem cell fate and maintenance in the SAM (Lopes et al. 2021). *WUS* expression is induced by cytokinin, and *WUS* is a direct transcription target of type-B ARR1, ARR10, and ARR12 (Meng et al. 2017; Zhang et al. 2017; Zubo et al. 2017; Xie et al. 2018). In turn, WUS represses type-A *RR* transcription (Leibfried et al. 2005) and cytokinin biosynthesis by downregulating *LOG4* (Chickarmane et al. 2012). This feedback loop between WUS and cytokinin signaling has been hypothesized to control cyto-kinin content and signaling in the SAM and therefore SAM ac-tivity. The mechanisms by which cytokinin regulates cell division in the SAM involve CYCD3, which is a direct transcrip-tional target of the cytokinin-regulated SAM transcription fac-tor SHOOT MERISTEMLESS (Scofield et al. 2013), and MYB3R4 transcription factor, which promotes the G_2/M transition in dividing cells of the SAM (Yang et al. 2021a).

The role of cytokinin in the root apical meristem (RAM) is opposite of that in the SAM. Increased cytokinin content leads to a smaller RAM, and mutants with reduced cytokinin

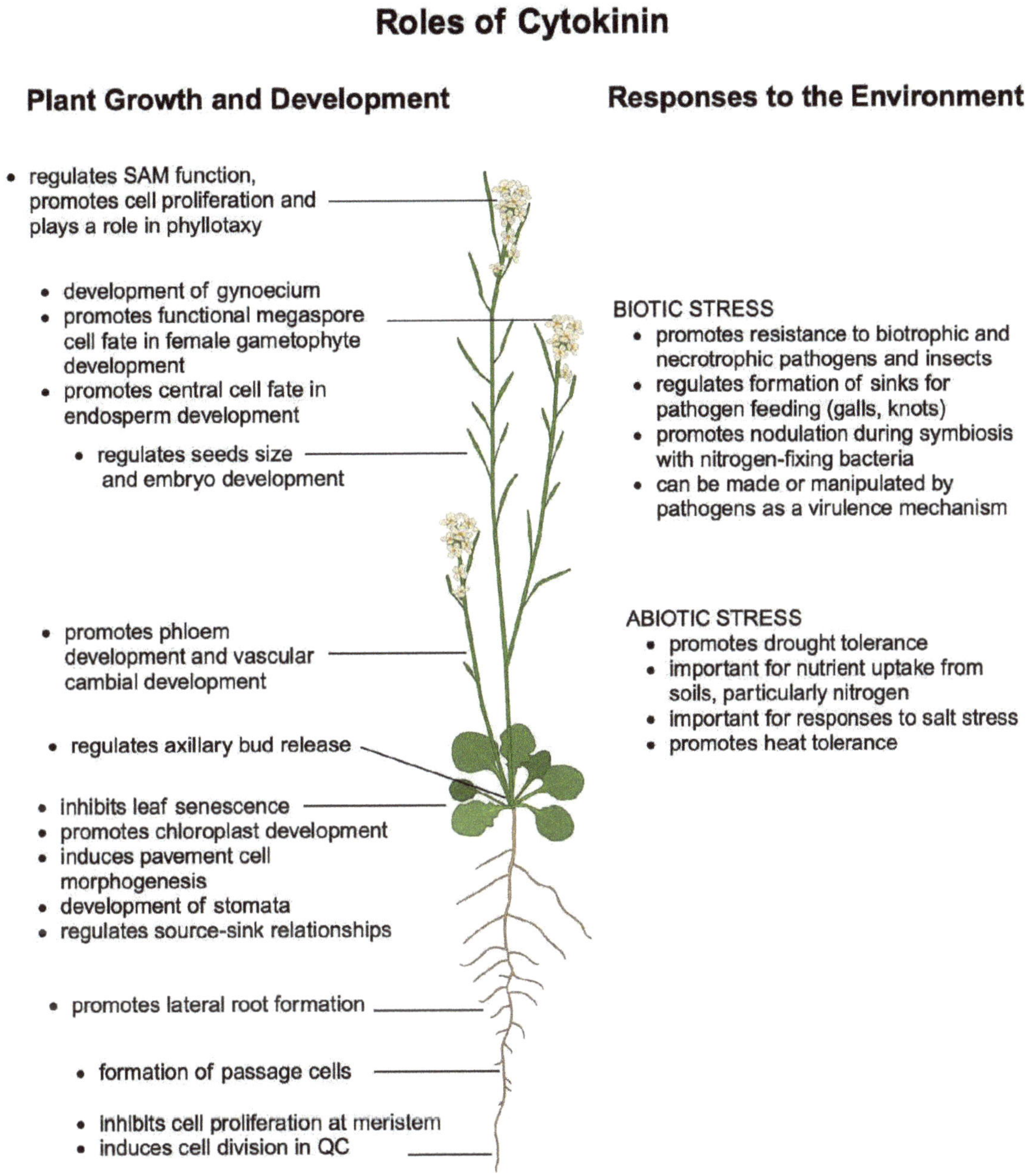

Figure 4. Roles of cytokinins in plant growth and responses to the environment. See text for additional details.

content (e.g. *ipt3*, *ipt5*, *ipt7* triple mutant) have an enlarged RAM (Dello Ioio et al. 2007). Cytokinin reduces RAM size by promoting the exit of cells from the meristem, acting via an incoherent regulatory loop between IPT7 and the HD-ZIPIII transcription factor PHABULOSA (Dello Ioio et al. 2012). This negative effect of cytokinin on the RAM is in part mediated by its antagonistic interaction with auxin, via transcriptional upregulation of auxin signaling repressors (Dello Ioio et al. 2008; Moubayidin et al. 2010; Perilli et al. 2013) and altered regulation of auxin efflux and influx carriers (Marhavý et al. 2011; Zhang et al. 2013), preventing auxin-induced cell division. Cytokinin can directly regulate cell division in the RAM independently of auxin. This is mediated by the CCS52A1 E3 ligase, whose gene expression is directly regulated by ARR2 and whose activity induces endoreplication by degradation of mitotic cyclins (Takahashi et al. 2013).

After cells exit the RAM in the Arabidopsis root, they stop dividing and rapidly elongate in a region called the elongation zone and then finally stop growing and mature into their final cell forms in the differentiation zone (Petricka et al. 2012). In addition to its role in inducing the exit of cells from the root meristem, cytokinin also promotes cessation of cell growth in the distal elongation zone at least in part by increasing the rigidity of the cell wall (Liu et al. 2022).

In rice, disruption of cytokinin function by disruption of cytokinin biosynthesis (LOG genes) or perception (HK cytokinin receptors) results in a drastic effect on both the inflorescence and the spikelet meristems (Kurakawa et al. 2007; Burr et al. 2020). In Arabidopsis, disruption of all three *HK* cytokinin receptors or of all seven *LOG* genes results in a substantial reduction in the size of the SAM (Higuchi et al. 2004; Nishimura et al. 2004; Riefler et al. 2006; Tokunaga et al. 2012), though this may be at least partially a secondary consequence of defects in the vasculature in these mutants. Further, unlike in rice, there are few, if any, effects on the floral meristem in cytokinin signaling mutants in Arabidopsis. In terms of type-A RRs, disruption of a single type-A RR gene (*ABPHYL1*) results in an enlarged meristem in maize (Jackson and Hake 1999), but no such effect is observed even in multiple type-A RR mutants in Arabidopsis (To et al. 2004). Overall, the results suggest that cytokinin may play a more pronounced role in regulating SAM function in monocots compared with dicots.

Vascular and cambium development

The plant vascular system is essential for the survival of plants on land. The main vascular tissues, namely xylem and phloem, originate from procambial stem cells that differenti-ate into mature vascular tissue. A role for cytokinin in vascu-lar morphogenesis was first revealed by the identification of the *wooden leg* allele of *AHK4* (Scheres et al. 1995), which dis-plays a pronounced reduction in the number of cell files in the vascular bundles, a phenotype that is even more pro-nounced in *ahk2,3,4* triple mutants (Riefler et al. 2006). Vascular morphogenesis and differentiation require precise control of cytokinin function in concert with auxin. A feed-back loop that helps define the cellular zones of auxin/ cytokinin function in the developing vasculature involves the auxin transcription factor MONOPTEROS/AUXIN RESPONSE FACTOR 5, which increases the expression of *AHP6*, a negative regulator of cytokinin signaling. This leads to the inhibition of cytokinin signaling in specific domains of the vascular system, regulating meristematic activity of the procambium and the formation of xylem and phloem (Mähönen et al. 2006a, 2006b). Cytokinins also induce a bi-symmetric distribution of the PIN-FORMED (PIN) auxin efflux proteins, resulting in accumulation of auxin in the central domain of protoxylem (Bishopp et al. 2011). In addition to its effects on AHP6, MP can also directly activate the expression of the transcription factor TARGET OF MONOPTEROS 5 (TOM5), which together with the transcription factor LONESOME HIGHWAY (LHW), regulates the expression of *LOG3* and *LOG4*, locally increasing cytokinin levels in the protoxylem, and regulating vascular morphogen-esis and growth (De Rybel et al. 2013). TOM5 and LHW also regulate the cell mobile transcription factor SHORTROOT, which can induce *CKX3* expression and further regulate vas-cular development through localized cytokinin signaling do-mains (Yang et al. 2021b).

As plants grow, their stems and roots thicken, a process known as secondary growth, which results from cambial meristem activity and cell proliferation. Studies in Arabidopsis and poplar demonstrated that *ipt* mutants have decreased root and stem thickening, as well as reduced cambial activity, establishing a role for cytokinin in promoting cambium development (Matsumoto-Kitano et al. 2008; Nieminen et al. 2008). The genes encoding the transcription factor AINTEGUMENTA and the cyclin CYCD3;1 are both transcriptionally upregulated by cytokinin in the cambium (Randall et al. 2015). Both genes are required for proper root and stem thickening, implicating them in a conserved regulatory module controlling vascular development (Randall et al. 2015). Recently, a study in Arabidopsis identified members of the transcription factor family LATERAL ORGAN BOUNDARIES DOMAIN (LBD) as regulators of secondary growth initiation (Ye et al. 2021). *LBD3* and *LBD4* are transcriptionally induced in response to cytokinin and initiate activation of the cambium meristematic cells. Other members, *LBD1* and *LBD11*, are transcriptionally induced by cytokinin later and together contribute to further radial growth and cambial stem cell maintenance. These LBD transcription factors were also determined genetically to inhibit cytokinin signaling, forming a feedback loop that regulates secondary growth (Ye et al. 2021). THE AT-HOOK MOTIF CONTAINING NUCLEAR LOCALIZED 15 transcription factor can also induce cytokinin biosynthesis locally and stimulate cambial activity (Rahimi et al. 2022).

Female gametophyte

During female gametogenesis in plants, the haploid female gametophyte (FG) develops within the ovule, surrounded by the sporophyte, to eventually generate megaspore cells. The development of the FG follows a spatial orientation along the chalazal/micropylar axis of the sporophyte. Cytokinin signaling has been shown to be essential in both the sporophyte (Kinoshita-Tsujimura and Kakimoto 2011) and in the gametophyte (Cheng et al. 2013) for female megaspore formation. In strong alleles of *ahk* receptor triple mutants, FG development is impaired, and the functional megaspore fails to form (Cheng et al. 2013). The effect of cytokinin on FG development was demonstrated to be associated with altered cell cycle regulation (Zhang et al. 2022). Interestingly, even though the requirement for cytokinin signaling through HK receptors for FG development has been demonstrated, cytokinin-independent signaling through two-component elements mediated by CKI1 and multiple AHPs is also required for FG development as *cki1* mutant plants display specific defects in FG development (Hejátko et al. 2009) that can be phenocopied by an *ahp2, ahp3, ahp5* triple mutant (Liu et al. 2017). CKI is also involved in central cell specification that generates the endosperm (Yuan et al. 2016) through a mechanism that appears to involve activation of the cytokinin signaling pathway via the AHPs (Liu et al. 2017), ultimately activating multiple type-B RRs (Zhu et al. 2022).

Leaf senescence

One of the most well-known effects of cytokinin is its anti-senescence activity. During developmental leaf senescence, which is age dependent and genetically programmed, as well as stress-induced senescence, nutrients are remobilized from leaves to other parts of the plant, where they can be reutilized for growth or storage. Levels of senescence are antagonistically correlated with levels of cytokinin content and signaling (Zwack and Rashotte 2013).

Application of cytokinin to plants delays leaf senescence (Gan and Amasino 1996). In addition, increasing endogenous cytokinin levels through transgenic approaches can also reduce senescence when properly regulated (Kant et al. 2015; Glanz-Idan et al. 2022). This was elegantly demonstrated using the promoter of a senescence-associated gene, *SAG12*, to drive the expression of an *IPT* gene in tobacco plants (Gan and Amasino 1995). Using this transgenic approach, cytokinin levels were increased specifically in cells undergoing onset of senescence, and transgenic plants expressing *SAG12::IPT7* showed decreased developmental as well as stress-induced senescence (Gan and Amasino 1995).

At the level of signaling, a genetic screen for mutants with delayed senescence identified a gain-of-function allele of *AHK3* as an important regulator of senescence in Arabidopsis (Kim et al. 2006). AHK3 functions in senescence by the activation of the type-B ARR2, whose transcriptional function is required for the anti-senescence activity (Kim et al. 2006). Downstream of type-B RR function, the CRF transcription factors were also found to have a role in senescence regulation. Higher-order mutant *crf1,3,5,6* plants display fas-ter developmental senescence (Raines et al. 2016a), while overexpression of *CRF6* delays stress-related senescence (Zwack and Rashotte 2013).

The mechanisms underlying the reduction of leaf senes-cence by cytokinin are relatively unknown. The strongest association has been with cell wall-invertases (CWINV). CWINVs function by cleaving sucrose molecules into glucose and fructose, which then become available for metabolic processes associated with plant growth. A delay in develop-mental or stress-induced senescence is associated with increased CWINV enzymatic activity and gene expression (Gan and Amasino 1995). Moreover, expression of a *CWINV* driven by a cytokinin inducible promoter also correlates with decreased senescence, and that can be reverted by the use of a chemical inhibitor of CWINV activity (Balibrea Lara et al. 2004).

Biotic stress

A role for cytokinin in plant-pathogen interaction was long proposed, especially in the relationship between plants and biotrophic pathogens in which exchange of nutrients is required (Walters and McRoberts 2006). This is the case in the formation of the so-called "green islands," which are formed around sites of biotrophic pathogen colonization, an area of reduced senescence that supports pathogen growth. Further studies in Arabidopsis showed that applica-tion of high concentrations of cytokinin, as well as genetic ap-proaches increasing cytokinin content and signaling (Choi et al. 2010; Argueso et al. 2012), do not activate defense directly but rather prime the plants for stronger defense in a variety of plant-pathogen systems (Albrecht and Argueso 2017). This priming effect by cytokinin is also observed in biocontrol agents such as rhizobacteria, which produce cytokinin to activate plant defense and reduce disease (Großkinsky et al. 2016). Molecular mechanisms of immunity activation against biotrophs by cytokinin include formation of a transcriptional complex between the type-B ARR2, the salicylic acid co-receptor NPR1, and the TGA3 transcription factor (Choi et al. 2010). Type-A RRs also participate by negatively regulating immunity in a phosphorylation-dependent manner (Argueso et al. 2012). Stomatal immunity is also par-tially controlled by cytokinin via regulation of peroxidases (Arnaud et al. 2017).

Apart from a role in immunity, cytokinin is also important in interactions of susceptibility between plants and pathogens. In such cases, increased cytokinin production or signaling initiated by pathogens leads to localized regions of cell proliferation, which become the galls, tumors, and knots that are often symptoms of plant diseases (McIntyre et al. 2021). The crown gall disease caused by *A. tumefaciens* is a classic example of this type of interaction (Akiyoshi et al. 1984), but many other pathogens such as fungi (Malinowski et al. 2016), nematodes (Siddique et al. 2015), and even parasitic plants (Spallek et al. 2017) can also manipulate cytokinin biosynthesis or signaling to create regions of cell division at sites of infection. Such regions of cell proliferation function as sinks, to which metabolites are rerouted to, away from source leaves, aiding in pathogen growth and multiplication in a process involving sugars and amino acid transporters that are also regulated by cytokinin (McIntyre et al. 2021).

Recently, a function for cytokinin in immunity against necrotrophic pathogens was also revealed (Gupta et al. 2020; Li et al. 2021; Liu et al. 2023). In tomato, application of micromolar amounts of cytokinin had a protective effect against the necrotrophic fungal pathogen *Botrytis cinerea*, through mechanisms that include the expression of defense marker genes as well as production of ethylene (Gupta et al. 2020). Interestingly, cytokinin can also be produced and sensed by *Botrytis*, having a role in fungal development and energy utilization, which underscores the complex relationship of this hormone in plant-pathogen interactions (Gupta et al. 2021).

Finally, cytokinin also plays a role in symbiotic interactions, particularly in the interactions of legume plants and rhizobia bacteria during nodule formation for nitrogen fixation. Evidence supporting a major role of cytokinin in nodulation comes from the fact that *Medicago truncatula* and *Lotus japonicus* mutants impaired in cytokinin signaling display reduced nodule formation in response to rhizobia infection (Plet et al. 2011; Held et al. 2014), while gain-of-function mutations in these genes leads to nodulation in the absence of rhizobia. Nodule formation is initiated by perception of bacteria-secreted nodulation (Nod) factors by plant cell surface LysM-type receptor kinases. After Nod factor perception, cytokinin biosynthesis and signaling are increased locally in the root, mostly in the root cortex (Lohar et al. 2004). This localized cytokinin accumulation induces the expression of key nodulation transcription factors, including NODULATION SIGNALING PATHWAY (NSP) 1 and 2, by binding of type-B RRs to their promoters to induce their transcription (Ariel et al. 2012). NSP1 and NSP2 then regulate genes that lead to cortical cell dedifferentiation and proliferation for nodule formation, such as *NODULE INCEPTION 1*. Interestingly, legumes also inhibit nodulation in order to balance the carbon costs of nodule formation with their needs for nitrogen fixation, and this suppression of nodulation is also mediated by type-B RRs (Chen et al. 2022). In soybeans the type-B RR GmRRB1d can physically interact with NSP2, interfering with its transcriptional activity, resulting in reduced nodulation (Chen et al. 2022).

Drought and salt tolerance

Cytokinin regulates a myriad of responses to abiotic stress (Argueso et al. 2009; Zwack and Rashotte 2015; Mandal et al. 2022), including the responses to drought and elevated salinity. Pioneering studies from the Blumwald laboratory demonstrated that increased levels of endogenous cytokinin via a transgene harboring an *IPT* gene under the control of a senescence/stress-inducible promoter could confer drought tolerance in tobacco (Rivero et al. 2007) and other plant species (Peleg et al. 2011; Reguera et al. 2013, Décima Oneto et al. 2016; Beznec et al. 2021). The drought tolerance in these lines may be mediated by the increased expression and activity of antioxidant enzymes, which may act to suppress the accumulation of reactive oxygen species during drought stress (Xu et al. 2016). In contrast to these findings, Arabidopsis *ipt* mutants show increased drought tolerance (Tran et al. 2007), and similar results are observed in plants overexpres-sing *CKX* genes (Lubovská et al. 2014). These seemingly opposing roles for cytokinin in drought tolerance maybe be due to different mechanisms of action and specific cellular contexts. In Arabidopsis, a negative role for cytokinin in drought tolerance occurs through a crosstalk with abscisic acid signaling (Huang et al. 2018). SnRK2 kinases, which participate in abscisic acid signaling, phosphorylate type-A RRs, stabilizing their protein levels and thus inhibiting cytokinin signaling. Conversely, type-B RRs physically interact with SnRK2 kinases to inhibit their function (Huang et al. 2018). Through this mutual interaction, these hormones act to fine tune growth and drought responses.

Cytokinin also appears to affect tolerance to salinity, though this also appears to be a complex interaction. In Arabidopsis, *ahk2* and *ahk3* mutants were found to be more salt tolerant than the wild type (Tran et al. 2007). Consistent with this, Arabidopsis lines with reduced cytokinin content (*CKX* overexpression or *ipt* loss-of-function) displayed enhanced salt tolerance (Nishiyama et al. 2011), and elevation of endogenous cytokinin via inducible expression of an *IPT* gene resulted in hypersensitivity to salt (Wang et al. 2015). In tomato, elevation of cytokinin biosynthesis in roots resulted in enhanced salt tolerance (Ghanem et al. 2011). These results suggest that cytokinin negatively regulates salt tolerance. However, a different study reported that *ahk2* mutants, but not mutations in the other 2 AHK receptors, displayed hypersensitivity to salt stress (Kumar and Verslues 2015). Likewise, in rice, reduced expression of *AHPs* via RNAi resulted in hypersensitivity to elevated salinity (Sun et al. 2014), and disruption of 2 type-A RR genes in rice displayed increased salinity tolerance (Wang et al. 2019), suggesting that cytokinin positively regulated salinity tolerance in rice. Consistent with this, reduced expression of *OsCKX2* (via RNAi) resulted in increased endogenous cytokinin and strong salt tolerance in vegetative shoots, which lead to a sig-nificant increase in yield in the presence of high salinity (Joshi et al. 2018). In contrast, overexpression of a cytokinin glucosyltransferase in rice, which would decrease active cytokinin levels, also enhanced salt tolerance in adult plants, though it led to increased salt sensitivity in young seedlings (Li et al. 2020). Modulation of cytokinin redistribution in rice such that there were elevated levels of cytokinin specifically in roots resulted in increased salt tolerance (Yin et al. 2020). Overall, these and other studies suggest a complex relationship between cytokinin function and salinity tolerance, with perhaps differing effects depending on the tissues in which cytokinin function is altered and on the particular plant species examined.

A recent study revealed that salt stress induced degradation of multiple type-B RRs in Arabidopsis via phosphorylation by 2 MAP kinases (MPK3/6) (Yan et al. 2021), suggesting that these MAPKs modulate salt sensitivity by reducing cytokinin sensitivity by targeted degradation of type-B RRs.

Nutrient uptake

Cytokinin influences the uptake and distribution of essential nutrients, such as nitrogen, phosphorus, and potassium (Pavlů et al. 2018). This can enhance nutrient acquisition efficiency, critical for sustaining growth during environmental-ly stressful conditions. Of these essential nutrients, nitrogen, in the form of nitrate, is one of the most growth limiting nutrients and is tightly linked to cytokinin levels and signaling. In response to nitrate supply, the expression of *IPT* genes is increased (Miyawaki et al. 2004), leading to the accumulation of cytokinin in the roots via the action of *IPT3* (Takei et al. 2004a). The increased cytokinin levels in the roots results in changes in root architecture (Ruffel et al. 2011) and also changes in expression of nitrate transporters (Lezhneva et al. 2014), fine-tunning root growth and nitrate uptake to the environmental conditions present. Cytokinins produced in the roots are then translocated to shoots via the xylem, mainly via the ABCG14 cytokinin transporter (Poitout et al. 2018). In the shoot, cytokinin influences multiple aspects of plant growth, including glutamine/glutamate metabolism (Poitout et al. 2018) and SAM activity (Landrein et al.

2018). Thus, levels of cytokinin within the plant function as a long-distance systemic signal of the nitrogen status, modulating plant growth. A possible mechanism mediating this long-distance signal has been recently elucidated: NPL7, a master transcriptional regulator of nitrate signaling, binds to the promoters of the cytokinin transporters *ABCG14*, *PUP14* and *PUP18* (Alvarez et al. 2020), inducing cytokinin transport to shoots. NLP7 is also required for the cytokinin-dependent expression of *CRFs* in the shoot, which in turn regulate expression of *PIN* auxin efflux transporters that mediate shoot growth in response to nitrate (Abualia et al. 2022).

Conclusions

Cytokinin plays pleiotropic roles throughout plant development, with often very distinct roles and effects in different tissues, at different times in development, and in response to various biotic and abiotic factors. An unresolved question remains: What is the molecular basis for the specificity of cytokinin? That is, how does this particular phytohormone have such distinct effects on different cells and tissues throughout the plant and across development? Differences in the cellular context in which the cytokinin signal is perceived likely account for the specificity. The cellular context is defined by the suite of cytokinin signaling elements expressed in a target cell (multiple paralogs of each signaling element in most plants), the activity of various partner transcription factors (multiple such factors identified as described above), the epigenetic state of the genome, and other interacting signaling pathways impinging on the cell. Differences in chromatin accessibility have been linked to differences in the effect of cytokinin on particular target genes in roots and shoots, and cytokinin also alters the accessibility of chromatin throughout the Arabidopsis genome in a type-B RR-dependent manner (Potter et al. 2018). Single cell approaches may shed light on the specifics of various interactions and specificity involved in cytokinin signaling.

Much of this review has focused on results in the model plant Arabidopsis, with some brief highlights from other species. Cytokinin function is being analyzed in many other plant species, including monocots such as maize and rice, as well as the lower plant *Marchantia polymorpha* (Aki et al. 2019). The future will likely reveal new aspects of cytokinin function be-yond the Arabidopsis paradigm.

There are undoubtably additional inputs into cytokinin signaling yet to be identified, as well as missing components involved in cytokinin metabolism (e.g. what is the enzyme involved in the biosynthesis of dihydrozeatin? What is the source of cis-zeatin?). Further, we are only just beginning to understand cytokinin transport throughout the plant and within the cell. Finally, the modulation of cytokinin function to improve various aspects of agriculture is just starting. Although the past decades have seen remarkable progress in our understanding of cytokinin biology, the future should continue to reveal the secrets of this remarkable signaling molecule.

Acknowledgments

We apologize to our colleagues whose work we could not discuss due to length considerations. We thank members of the Kieber laboratory for helpful comments on the text.

Author contributions

Both JJK and CTA contributed to the writing of this manuscript.

Funding

Work in the authors laboratory is generously supported by grants for the National Science Foundation (IOS IOS-2126144 to JJK and MCB-1818211 to CA), the USDA (2023-67013-39413 to JJK), and the USDA National Institute of Food and Agriculture HATCH Project COL00781 (COL0781A) to CA.

Conflict of interest statement. None declared.

References

Abualia R, Ötvös K, Novák O, Bouguyon E, Domanegg K, Krapp A, Nacry P, Gojon A, Lacombe B, Benková E. Molecular framework integrating nitrate sensing in root and auxin-guided shoot adaptive responses. Proc Natl Acad Sci USA. 2022:**119**(31):e2122460119. https://doi.org/10.1073/pnas.2122460119

Acheampong AK, Shanks C, Cheng CY, Schaller GE, Dagdas Y, Kieber JJ. EXO70D isoforms mediate selective autophagic degradation of type-A ARR proteins to regulate cytokinin sensitivity. Proc Natl Acad Sci USA. 2020:**117**(43):27034–27043. https://doi.org/10.1073/pnas.2013161117

Aki SS, Mikami T, Naramoto S, Nishihama R, Ishizaki K, Kojima M, Takebayashi Y, Sakakibara H, Kyozuka J, Kohchi T, et al. Cytokinin signaling is essential for organ formation in *Marchantia polymorpha*. Plant Cell Physiol. 2019:**60**(8):1842–1854. https://doi.org/10.1093/pcp/pcz100

Akiyoshi D, Klee H, Amasino RM, Nester EW, Gordon MP. T-DNA of *Agrobacterium tumefaciens* encodes an enzyme of cytokinin biosynthesis. Proc Natl Acad Sci USA. 1984:**81**(19):5994–5998. https://doi.org/10.1073/pnas.81.19.5994

Albrecht T, Argueso CT. Should I fight or should I grow now? The role of cytokinins in plant growth and immunity and in the growth-defence trade-off. Ann Bot. 2017:**119**(5):725–735. https://doi.org/10.1093/aob/mcw211

Alvarez JM, Schinke AL, Brooks MD, Pasquino A, Leonelli L, Varala K, Safi A, Krouk G, Krapp A, Coruzzi GM. Transient genome-wide interactions of the master transcription factor NLP7 initiate a rapid nitrogen-response cascade. Nat Commun. 2020:**11**(1):1157. https://doi.org/10.1038/s41467-020-14979-6

Amasino R. 1955: kinetin arrives: the 50th anniversary of a new plant hormone. Plant Physiol. 2005:**138**(3):1177–1184. https://doi.org/10.1104/pp.104.900160

Anantharaman V, Aravind L. The CHASE domain: a predicted ligand-binding module in plant cytokinin receptors and other eukaryotic and bacterial receptors. Trends Biochem Sci. 2001:**26**(10):579–582. https://doi.org/10.1016/S0968-0004(01)01968-5

Andersen TG, Naseer S, Ursache R, Wybouw B, Smet W, De Rybel B, Vermeer JEM, Geldner N. Diffusible repression of cytokinin signalling produces endodermal symmetry and passage cells. Nature. 2018:**555**(7697):529–533. https://doi.org/10.1038/nature25976

Antoniadi I, Novák O, Gelová Z, Johnson A, Plíhal O, Simerský R, Mik V, Vain T, Mateo-Bonmatí E, Karady M, et al. Cell-surface receptors enable perception of extracellular cytokinins. Nat Commun. 2020:**11**(1):4284. https://doi.org/10.1038/s41467-020-17700-9

Appleby JL, Parkinson JS, Bourret RB. Signal transduction via the multi-step phosphorelay: not necessarily a road less traveled. Cell. 1996:**86**(6):845–848. https://doi.org/10.1016/S0092-8674(00)80158-0

Argueso CT, Ferreira FJ, Epple P, To JPC, Hutchison CE, Mathews DE, Schaller GE, Dangl JL, Kieber JJ. Two-component elements mediate interactions between cytokinin and salicylic acid in plant immunity. PLoS Genet. 2012:**8**(1):e1002448. https://doi.org/10.1371/journal.pgen.1002448

Argueso CT, Ferreira FJ, Kieber JJ. Environmental perception avenues: the interaction of cytokinin and environmental response pathways. Plant Cell Environ. 2009:**32**(9):1147–1160. https://doi.org/10.1111/j.1365-3040.2009.01940.x

Argueso CT, Raines T, Kieber JJ. Cytokinin signaling and transcriptional networks. Curr Opin Plant Biol. 2010:**13**(5):533–559. https://doi.org/10.1016/j.pbi.2010.08.006

Argyros RD, Mathews DE, Chiang YH, Palmer CM, Thibault DM, Etheridge N, Argyros DA, Mason MG, Kieber JJ, Schaller GE. Type B response regulators of Arabidopsis play key roles in cytokinin signaling and plant development. Plant Cell. 2008:**20**(8):2102–2116. https://doi.org/10.1105/tpc.108.059584

Ariel F, Brault-Hernandez M, Laffont C, Huault E, Brault M, Plet J, Moison M, Blanchet S, Ichanté JL, Chabaud M, et al. Two direct targets of cytokinin signaling regulate symbiotic nodulation in *Medicago truncatula*. Plant Cell. 2012:**24**(9):3838–3852. https://doi.org/10.1105/tpc.112.103267

Arnaud D, Lee S, Takebayashi Y, Choi D, Choi J, Sakakibara H, Hwang I. Cytokinin-mediated regulation of reactive oxygen species homeostasis modulates stomatal immunity in Arabidopsis. Plant Cell. 2017:**29**(3):543–559. https://doi.org/10.1105/tpc.16.00583

Ashikari M, Sakakibara H, Lin S, Yamamoto T, Takashi T, Nishimura A, Angeles ER, Qian Q, Kitano H, Matsuoka M. Cytokinin oxidase regulates rice grain production. Science. 2005:**309**(5735):741–745. https://doi.org/10.1126/science.1113373

Balibrea Lara ME, Gonzalez Garcia MC, Fatima T, Ehneß R, Lee TK, Proels R, Tanner W, Roitsch T. Extracellular invertase is an essential component of cytokinin-mediated delay of senescence. Plant Cell. 2004:**16**(5):1276–1287. https://doi.org/10.1105/tpc.018929

Bartrina I, Otto E, Strnad M, Werner T, Schmülling T. Cytokinin regulates the activity of reproductive meristems, flower organ size, ovule formation, and thus seed yield in *Arabidopsis thaliana*. Plant Cell. 2011:**23**(1):69–80. https://doi.org/10.1105/tpc.110.079079

Besnard F, Refahi Y, Morin V, Marteaux B, Brunoud G, Chambrier P, Rozier F, Mirabet V, Legrand J, Lainé S, et al. Cytokinin signalling inhibitory fields provide robustness to phyllotaxis. Nature. 2014:**505**(7483):417–421. https://doi.org/10.1038/nature12791

Beznec A, Faccio P, Miralles DJ, Abeledo LG, Oneto CD, Garibotto MB, Gonzalez G, Moreyra F, Elizondo M, Ruíz M, et al. Stress-induced expression of IPT gene in transgenic wheat reduces grain yield penalty under drought. J Genet Eng Biotechnol. 2021:**19**(1):67. https://doi.org/10.1186/s43141-021-00171-w

Bhargava A, Clabaugh I, To JP, Maxwell BB, Chiang YH, Schaller GE, Loraine A, Kieber JJ. Identification of cytokinin-responsive genes using microarray meta-analysis and RNA-Seq in Arabidopsis. Plant Physiol. 2013:**162**(1):272–294. https://doi.org/10.1104/pp.113.217026

Bishopp A, Help H, El-Showk S, Weijers D, Scheres B, Friml J, Benková E, Mähönen A, Helariutta Y. A mutually inhibitory interaction between auxin and cytokinin specifies vascular pattern in roots. Curr Biol. 2011:**21**(11):927–932. https://doi.org/10.1016/j.cub.2011.04.049

Brandstatter I, Kieber JJ. Two genes with similarity to bacterial response regulators are rapidly and specifically induced by cytokinin in Arabidopsis. Plant Cell. 1998:**10**(6):1009–1020. https://doi.org/10.1105/tpc.10.6.1009

Brenner WG, Ramireddy E, Heyl A, Schmülling T. Gene regulation by cytokinin in Arabidopsis. Front Plant Sci. 2012:**3**:8. https://doi.org/10.3389/fpls.2012.00008

Brzobohaty B, Moore I, Kristoffersen P, Bako L, Campos N, Schell J, Palme K. Release of active cytokinin by a b-glucosidase localized to the maize root meristem. Science. 1993:**262**(5136):1051–1054. https://doi.org/10.1126/science.8235622

Burr CA, Sun J, Yamburenko MV, Willoughby A, Hodgens C, Boeshore SL, Elmore A, Atkinson J, Nimchuk ZL, Bishopp A, et al. The HK5 and HK6 cytokinin receptors mediate diverse developmental pathways in rice. Development. 2020:**147**(20):dev191734. https://doi.org/10.1242/dev.191734

Caesar K, Thamm AMK, Witthöft J, Elgass K, Huppenberger P, Grefen C, Horak J, Harter K. Evidence for the localization of the Arabidopsis cytokinin receptors AHK3 and AHK4 in the endoplasmic reticulum. J Exp Bot. 2011:**62**(15):5571–5580. https://doi.org/10.1093/jxb/err238

Chen L, Jameson GB, Guo Y, Song J, Jameson PE. The LONELY GUY gene family: from mosses to wheat, the key to the formation of active cytokinins in plants. Plant Biotechnol J. 2022:**20**(4):625–645. https://doi.org/10.1111/pbi.13783

Chen L, Zhao J, Song J, Jameson PE. Cytokinin glucosyl transferases, key regulators of cytokinin homeostasis, have potential value for wheat improvement. Plant Biotechnol J. 2021:**19**(5):878–896. https://doi.org/10.1111/pbi.13595

Cheng C-Y, Mathews DE, Eric Schaller G, Kieber JJ. Cytokinin-dependent specification of the functional megaspore in the Arabidopsis female gametophyte. Plant J. 2013:**73**(6):929–940. https://doi.org/10.1111/tpj.12084

Chickarmane VS, Gordon SP, Tarr PT, Heisler MG, Meyerowitz EM. Cytokinin signaling as a positional cue for patterning the apical–basal axis of the growing Arabidopsis shoot meristem. Proc Natl Acad Sci U S A. 2012:**109**(10):4002–4007. https://doi.org/10.1073/pnas.1200636109

Choi J, Huh SU, Kojima M, Sakakibara H, Paek KH, Hwang I. The cytokinin-activated transcription factor ARR2 promotes plant immunity via TGA3/NPR1-dependent salicylic acid signaling in Arabidopsis. Dev Cell. 2010:**19**(2):284–295. https://doi.org/10.1016/j.devcel.2010.07.011

Chu HM, Ko TP, Wang AH. Crystal structure and substrate specificity of plant adenylate isopentenyltransferase from *Humulus lupulus*: distinctive binding affinity for purine and pyrimidine nucleotides. Nucl Acids Res. 2010:**38**(5):1738–1748. https://doi.org/10.1093/nar/gkp1093

D'Agostino I, Deruère J, Kieber JJ. Characterization of the response of the Arabidopsis *ARR* gene family to cytokinin. Plant Physiol. 2000:**124**(4):1706–1717. https://doi.org/10.1104/pp.124.4.1706

Décima Oneto C, Otegui ME, Baroli I, Beznec A, Faccio P, Bossio E, Blumwald E, Lewi D. Water deficit stress tolerance in maize conferred by expression of an isopentenyltransferase (IPT) gene driven by a stress- and maturation-induced promoter. J Biotechnol. 2016:**220**:66–77. https://doi.org/10.1016/j.jbiotec.2016.01.014

Dello Ioio R, Galinha C, Fletcher AG, Grigg SP, Molnar A, Willemsen V, Scheres B, Sabatini S, Baulcombe D, Maini PK, et al. A PHABULOSA/cytokinin feedback loop controls root growth in Arabidopsis. Curr Biol. 2012:**22**(18):1699–1704. https://doi.org/10.1016/j.cub.2012.07.005

Dello Ioio R, Linhares FS, Scacchi E, Casamitjana-Martinez E, Heidstra R, Costantino P, Sabatini S. Cytokinins determine Arabidopsis root-meristem size by controlling cell differentiation. Curr Biol. 2007:**17**(8):678–682. https://doi.org/10.1016/j.cub.2007.02.047

Dello Ioio R, Nakamura K, Moubayidin L, Perilli S, Taniguchi M, Morita MT, Aoyama T, Costantino P, Sabatini S. A genetic framework for the control of cell division and differentiation in the root meristem. Science. 2008:**322**(5906):1380–1384. https://doi.org/10.1126/science.1164147

De Rybel B, Möller B, Yoshida S, Grabowicz I, Barbier de Reuille P, Boeren S, Smith RS, Borst JW, Weijers D. A bHLH complex controls embryonic vascular tissue establishment and indeterminate growth in Arabidopsis. Dev Cell. 2013:**24**(4):426–437. https://doi.org/10.1016/j.devcel.2012.12.013

Dewitte W, Scofield S, Alcasabas AA, Maughan SC, Menges M, Braun N, Collins C, Nieuwland J, Prinsen E, Sundaresan V, et al. Arabidopsis CYCD3 D-type cyclins link cell proliferation and endocycles and are rate-limiting for cytokinin responses. Proc Natl Acad Sci U S A. 2007:**104**(36):14537–14542. https://doi.org/10.1073/pnas.0704166104

Dobisova T, Hrdinova V, Cuesta C, Michlickova S, Urbankova I, Hejatkova R, Zadnikova P, Pernisova M, Benkova E, Hejatko J. Light controls cytokinin signaling via transcriptional regulation of constitutively active sensor histidine kinase CKI1. Plant Physiol. 2017:**174**(1):387–404. https://doi.org/10.1104/pp.16.01964

Efroni I, Han S-K, Kim HJ, Wu M-F, Steiner E, Birnbaum KD, Hong JC, Eshed Y, Wagner D. Regulation of leaf maturation by chromatin-mediated modulation of cytokinin responses. Dev Cell. 2013:**24**(4):438–445. https://doi.org/10.1016/j.devcel.2013.01.019

Feng J, Wang C, Chen Q, Chen H, Ren B, Li X, Zuo J. S-nitrosylation of phosphotransfer proteins represses cytokinin signaling. Nat Commun. 2013:**4**(1):1529. https://doi.org/10.1038/ncomms2541

Fonouni-Farde C, Kisiala A, Brault M, Emery RJN, Diet A, Frugier F. DELLA1-Mediated gibberellin signaling regulates cytokinin-dependent symbiotic nodulation. Plant Physiol. 2017:**175**(4):1795–1806. https://doi.org/10.1104/pp.17.00919

Fox JE, Erion JL. A cytokinin binding protein from higher plant ribosomes. Biochem Biophys Res Commun. 1975:**64**(2):694–700. https://doi.org/10.1016/0006-291X(75)90376-9

Gajdošová S, Spíchal L, Kamínek M, Hoyerová K, Novák O, Dobrev PI, Galuszka P, Klíma P, Gaudinová A, Žižková E, et al. Distribution, biological activities, metabolism, and the conceivable function of *cis*-zeatin-type cytokinins in plants. J Exp Bot. 2011:**62**(8):2827–2840. https://doi.org/10.1093/jxb/erq457

Galuszka P, Popelková H, Werner T, Frébortová J, Pospíšilová H, Mik V, Köllmer I, Schmülling T, Frébort I. Biochemical characterization of cytokinin oxidases/dehydrogenases from *Arabidopsis thaliana* expressed in *Nicotiana tabacum* L. J Plant Growth Regul. 2007:**26**(3):255–267. https://doi.org/10.1007/s00344-007-9008-5

Gan S, Amasino RM. Inhibition of leaf senescence by autoregulated production of cytokinin. Science. 1995:**270**(5244):1986–1988. https://doi.org/10.1126/science.270.5244.1986

Gan S, Amasino RM. Cytokinins in plant senescence: from spray and pray to clone and play. Bioessays. 1996:**18**(7):557–565. https://doi.org/10.1002/bies.950180707

Ghanem ME, Albacete A, Smigocki AC, Frébort I, Pospísilová H, Martínez-Andújar C, Acosta M, Sánchez-Bravo J, Lutts S, Dodd IC, et al. Root-synthesized cytokinins improve shoot growth and fruit yield in salinized tomato (Solanum lycopersicum L.) plants. J Exp Bot. 2011:**62**(1):125–140. https://doi.org/10.1093/jxb/erq266

Gillissen B, Bürkle L, André B, Kühn C, Rentsch D, Brandl B, Frommer WB. A new family of high-affinity transporters for adenine, cytosine, and purine derivatives in *Arabidopsis*. Plant Cell. 2000:**12**(2):291–300. https://doi.org/10.1105/tpc.12.2.291

Glanz-Idan N, Lach M, Tarkowski P, Vrobel O, Wolf S. Delayed leaf senescence by upregulation of cytokinin biosynthesis specifically in tomato roots. Front Plant Sci. 2022:**13**:922106. https://doi.org/10.3389/fpls.2022.922106

Gordon SP, Chickarmane VS, Ohno C, Meyerowitz EM. Multiple feedback loops through cytokinin signaling control stem cell number within the Arabidopsis shoot meristem. Proc Natl Acad Sci U S A. 2009:**106**(38):16529–16534. https://doi.org/10.1073/pnas.0908122106

Großkinsky DK, Tafner R, Moreno MV, Stenglein SA, García de Salamone IE, Nelson LM, Novák O, Strnad M, van der Graaff E, Roitsch T. Cytokinin production by Pseudomonas fluorescens G20-18 determines biocontrol activity against Pseudomonas syringae in Arabidopsis. Sci Rep. 2016:**6**(1):23310. https://doi.org/10.1038/srep23310

Gupta R, Leibman-Markus M, Pizarro L, Bar M. Cytokinin induces bacterial pathogen resistance in tomato. Plant Path. 2021:**70**(2):318–325. https://doi.org/10.1111/ppa.13279

Gupta R, Pizarro L, Leibman-Markus M, Marash I, Bar M. Cytokinin response induces immunity and fungal pathogen resistance, and modulates trafficking of the PRR LeEIX2 in tomato. Mol Plant Pathol. 2020:**21**(10):1287–1306. https://doi.org/10.1111/mpp.12978

Haberlandt G. Culturversuche mit isolierten pflanzenzellen. Sitz-Ber Mat Nat Kl Kais Akad Wiss Wien. 1902:**111**:69–91. https://biostor.org/reference/221963

Haberlandt G. Zur physiologie der zellteilung. Sitzungsber. Voll XX. Wien: Sitzungsberichte der Preussischen Akademie der Wissenschaften. 1913; p318–345.

Hartig K, Beck E. Endogenous cytokinin oscillations control cell cycle progression of tobacco BY-2 cells. Plant Biol (Stuttg). 2005:**7**(1):33–40. https://doi.org/10.1055/s-2004-830474

He Q, Yuan R, Zhang T, An F, Wang N, Lan J, Wang X, Zhang Z, Pan Y, Wang X, et al. Arabidopsis TIE1 and TIE2 transcriptional repressors dampen cytokinin response during root development. Sci Adv. 2022:**8**(36):eabn5057. https://doi.org/10.1126/sciadv.abn5057

Hejátko J, Ryu H, Kim GT, Dobešová R, Choi S, Choi SM, Souček P, Horák J, Pekárová B, Palme K, et al. The histidine kinases CYTOKININ-INDEPENDENT1 and ARABIDOPSIS HISTIDINE KINASE2 and 3 regulate vascular tissue development in Arabidopsis shoots. Plant Cell. 2009:**21**(7):2008–2021. https://doi.org/10.1105/tpc.109.066696

Held M, Hou H, Miri M, Huynh C, Ross L, Hossain MS, Sato S, Tabata S, Perry J, Wang TL, et al. *Lotus japonicus* cytokinin receptors work partially redundantly to mediate nodule formation. Plant Cell. 2014:**26**(2):678–694. https://doi.org/10.1105/tpc.113.119362

Higuchi M, Pischke MS, Mähönen AP, Miyawaki K, Hashimoto Y, Seki M, Kobayashi M, Shinozaki K, Kato T, Tabata S, et al. In planta functions of the Arabidopsis cytokinin receptor family. Proc Natl Acad Sci U S A. 2004:**101**(23):8821–8826. https://doi.org/10.1073/pnas.0402887101

Hill K, Mathews DE, Kim HJ, Street IH, Wildes SL, Chiang YH, Mason MG, Alonso JM, Ecker JR, Kieber JJ, et al. Functional characterization of type-B response regulators in the Arabidopsis cytokinin response. Plant Physiol. 2013:**162**(1):212–224. https://doi.org/10.1104/pp.112.208736

Hirose N, Makita N, Yamaya T, Sakakibara H. Functional characterization and expression analysis of a gene, OsENT2, encoding an equilibrative nucleoside transporter in rice suggest a function in cytokinin transport. Plant Physiol. 2005:**138**(1):196–206. https://doi.org/10.1104/pp.105.060137

Hirose N, Takei K, Kuroha T, Kamada-Nobusada T, Hayashi H, Sakakibara H. Regulation of cytokinin biosynthesis, compartmentalization and translocation. J Exp Bot. 2008:**59**(1):75–83. https://doi.org/10.1093/jxb/erm157

Hluska T, Hlusková L, Emery RJN. The hulks and the deadpools of the cytokinin universe: a dual strategy for cytokinin production, translocation, and signal transduction. Biomolecules. 2021:**11**(2):209. https://doi.org/10.3390/biom11020209

Hodgens C, Chang N, Schaller GE, Kieber JJ. Mutagenomics: a rapid, high-throughput method to identify causative mutations from a genetic screen. Plant Physiol. 2020:**184**(4):1658–1673. https://doi.org/10.1104/pp.20.00609

Hošek P, Hoyerová K, Kiran NS, Dobrev PI, Zahajská L, Filepová R, Motyka V, Müller K, Kamínek M. Distinct metabolism of N-glucosides of isopentenyladenine and trans-zeatin determines cytokinin metabolic spectrum in Arabidopsis. New Phytol. 2020:**225**(6):2423–2438. https://doi.org/10.1111/nph.16310

Houba-Hérin N, Pethe C, d'Alayer J, Laloue M. Cytokinin oxidase from *Zea mays*: purification, cDNA cloning and expression in moss protoplasts. Plant J. 1999:**17**(6):615–626. https://doi.org/10.1046/j.1365-313X.1999.00408.x

Huang X, Hou L, Meng J, You H, Li Z, Gong Z, Yang S, Shi Y. The antagonistic action of abscisic acid and cytokinin signaling mediates drought stress response in Arabidopsis. Mol Plant. 2018:**11**(7):970–982. https://doi.org/10.1016/j.molp.2018.05.001

Hutchison CE, Li J, Argueso C, Gonzalez M, Lee E, Lewis MW, Maxwell BB, Perdue TD, Schaller GE, Alonso JM, et al. The Arabidopsis histidine phosphotransfer proteins are redundant positive regulators of cytokinin signaling. Plant Cell. 2006:**18**(11):3073–3087. https://doi.org/10.1105/tpc.106.045674

Hwang I, Sheen J. Two-component circuitry in *Arabidopsis* signal transduction. Nature. 2001:**413**(6854):383–389. https://doi.org/10.1038/35096500

Hwang I, Sheen J, Müller B. Cytokinin signaling networks. Annu Rev Plant Biol. 2012:**63**(1):353–380. https://doi.org/10.1146/annurev-arplant-042811-105503

Hyde RJ, Cass CE, Young JD, Baldwin SA. The ENT family of eukaryote nucleoside and nucleobase transporters: recent advances in the investigation of structure/function relationships and the identification of novel isoforms. Mol Membr Biol. 2001:**18**(1):53–63. https://doi.org/10.1080/09687680118799

Imamura A, Kiba T, Tajima Y, Yamashino T, Mizuno T. In vivo and in vitro characterization of the ARR11 response regulator implicated in the his-to-asp phosphorelay signal transduction in *Arabidopsis thaliana*. Plant Cell Physiol. 2003:**44**(2):122–131. https://doi.org/10.1093/pcp/pcg014

Inoue T, Higuchi M, Hashimoto Y, Seki M, Kobayashi M, Kato T, Tabata S, Shinozaki K, Kakimoto T. Identification of CRE1 as a cytokinin receptor from *Arabidopsis*. Nature. 2001:**409**(6823):1060–1063. https://doi.org/10.1038/35059117

Jackson D, Hake S. Control of phyllotaxy in maize by the abphyl1 gene. Development. 1999:**126**(2):315–323. https://doi.org/10.1242/dev.126.2.315

Jameson PE. Zeatin: the 60th anniversary of its identification. Plant Physiol. 2023:**192**(1):34–55. https://doi.org/10.1093/plphys/kiad094

Jameson PE, Song J. Cytokinin: a key driver of seed yield. J Exp Bot. 2016:**67**(3):593–606. https://doi.org/10.1093/jxb/erv461

Jameson PE, Song J. Will cytokinins underpin the second 'green revolution'? J Exp Bot. 2020:**71**(22):6872–6875. https://doi.org/10.1093/jxb/eraa447

Jelesko JG. An expanding role for purine uptake permease-like transporters in plant secondary metabolism. Front Plant Sci. 2012:**3**:78. https://doi.org/10.3389/fpls.2012.00078

Joshi R, Sahoo KK, Tripathi AK, Kumar R, Gupta BK, Pareek A, Singla-Pareek SL. Knockdown of an inflorescence meristem-specific cytokinin oxidase—osCKX2 in rice reduces yield penalty under salinity stress condition. Plant Cell Environ. 2018:**41**(5):936–946. https://doi.org/10.1111/pce.12947

Kakimoto T. CKI1, a histidine kinase homolog implicated in cytokinin signal transduction. Science. 1996:**274**(5289):982–985. https://doi.org/10.1126/science.274.5289.982

Kakimoto T. Identification of plant cytokinin biosynthetic enzymes as dimethylallyl diphosphate: aTP/ADP isopentenyltransferases. Plant Cell Physiol. 2001:**42**(7):677–685. https://doi.org/10.1093/pcp/pce112

Kamada-Nobusada T, Sakakibara H. Molecular basis for cytokinin biosynthesis. Phytochemistry. 2009:**70**(4):444–449. https://doi.org/10.1016/j.phytochem.2009.02.007

Kant S, Burch D, Badenhorst P, Palanisamy R, Mason J, Spangenberg G. Regulated expression of a cytokinin biosynthesis gene IPT delays leaf senescence and improves yield under rainfed and irrigated conditions in canola (*Brassica napus* L.). PLoS One. 2015:**10**(1):e0116349. https://doi.org/10.1371/journal.pone.0116349

Kasahara H, Takei K, Ueda N, Hishiyama S, Yamaya T, Kamiya Y, Yamaguchi S, Sakakibara H. Distinct isoprenoid origins of *cis*- and *trans*-zeatin biosyntheses in *Arabidopsis*. J Biol Chem. 2004:**279**(14):14049–14054. https://doi.org/10.1074/jbc.M314195200

Keim P, Fox JE. Interaction of a radiolabeled cytokinin photoaffinity probe with a receptor protein. Biochem Biophys Res Commun. 1980:**96**(3):1325–1334. https://doi.org/10.1016/0006-291X(80)90096-0

Kiba T, Mizutani K, Nakahara A, Takebayashi Y, Kojima M, Hobo T, Osakabe Y, Osakabe K, Sakakibara H. The trans-zeatin-type side-chain modification of cytokinins controls rice growth. Plant Physiol. 2023:**192**(3):2457–2474. https://doi.org/10.1093/plphys/kiad197

Kiba T, Takei K, Kojima M, Sakakibara H. Side-chain modification of cytokinins controls shoot growth in *Arabidopsis*. Dev Cell. 2013:**27**(4):452–461. https://doi.org/10.1016/j.devcel.2013.10.004

Kiba T, Yamada H, Sato S, Kato T, Tabata S, Yamashino T, Mizuno T. The type-A response regulator, *ARR15*, acts as a negative regulator in the cytokinin-mediated signal transduction in *Arabidopsis thaliana*. Plant Cell Physiol. 2003:**44**(8):868–874. https://doi.org/10.1093/pcp/pcg108

Kieber JJ, Schaller GE. Cytokinins. Arabidopsis Book. 2014:**12**:e0168. https://doi.org/10.1199/tab.0168

Kieber JJ, Schaller GE. Cytokinin signaling in plant development. Development. 2018:**145**(4):dev149344. https://doi.org/10.1242/dev.149344

Kim A, Chen J, Khare D, Jin JY, Yamaoka Y, Maeshima M, Zhao Y, Martinoia E, Hwang JU, Lee Y. Non-intrinsic ATP-binding cassette proteins ABCI19, ABCI20 and ABCI21 modulate cytokinin response at the endoplasmic reticulum in Arabidopsis thaliana. Plant Cell Rep. 2020:**39**(4):473–487. https://doi.org/10.1007/s00299-019-02503-0

Kim HJ, Chiang Y-H, Kieber JJ, Schaller GE. SCF(KMD) controls cytokinin signaling by regulating the degradation of type-B response regulators. Proc Natl Acad Sci U S A. 2013:**110**(24):10028–10033. https://doi.org/10.1073/pnas.1300403110

Kim HJ, Ryu H, Hong SH, Woo HR, Lim PO, Lee IC, Sheen J, Nam HG, Hwang I. Cytokinin-mediated control of leaf longevity by AHK3 through phosphorylation of ARR2 in Arabidopsis. Proc Natl Acad Sci U S A. 2006:**103**(3):814–819. https://doi.org/10.1073/pnas.0505150103

Kinoshita-Tsujimura K, Kakimoto T. Cytokinin receptors in sporophytes are essential for male and female functions in *Arabidopsis thaliana*. Plant Signal Behav. 2011:**6**(1):66–71. https://doi.org/10.4161/psb.6.1.13999

Ko D, Kang J, Kiba T, Park J, Kojima M, Do J, Kim KY, Kwon M, Endler A, Song W-Y, et al. *Arabidopsis* ABCG14 is essential for the root-to-shoot translocation of cytokinin. Proc Natl Acad Sci USA. 2014:**111**(19):7150–7155. https://doi.org/10.1073/pnas.1321519111

Kudo T, Makita N, Kojima M, Tokunaga H, Sakakibara H. Cytokinin activity of cis-zeatin and phenotypic alterations induced by overexpression of putative cis-zeatin-O-glucosyltransferase in rice. Plant Physiol. 2012:**160**(1):319–331. https://doi.org/10.1104/pp.112.196733

Kumar MN, Verslues PE. Stress physiology functions of the Arabidopsis histidine kinase cytokinin receptors. Physiol Plant. 2015:**154**(3):369–380. https://doi.org/10.1111/ppl.12290

Kurakawa T, Ueda N, Maekawa M, Kobayashi K, Kojima M, Nagato Y, Sakakibara H, Kyozuka J. Direct control of shoot meristem activity by a cytokinin-activating enzyme. Nature. 2007:**445**(7128):652–655. https://doi.org/10.1038/nature05504

Landrein B, Formosa-Jordan P, Malivert A, Schuster C, Melnyk CW, Yang W, Turnbull C, Meyerowitz EM, Locke JCW, Jönsson H. Nitrate modulates stem cell dynamics in Arabidopsis shoot meristems through cytokinins. Proc Natl Acad Sci U S A. 2018:**115**(6):1382–1387. https://doi.org/10.1073/pnas.1718670115

Laureys F, Smetsa R, Lenjoub M, Van Bockstaeleb D, Inzéc D, Van Onckelena H. A low content in zeatin type cytokinins is not restrictive for the occurrence of G1/S transition in tobacco BY-2 cells. FEBS

Lett. 1999:**460**(1):123–128. https://doi.org/10.1016/S0014-5793(99)01264-8

Lee DJ, Park JY, Ku SJ, Ha YM, Kim S, Kim MD, Oh MH, Kim J. Genome-wide expression profiling of *ARABIDOPSIS RESPONSE REGULATOR 7(ARR7)* overexpression in cytokinin response. Mol Genet Genomics. 2007:**277**(2):115–137. https://doi.org/10.1007/s00438-006-0177-x

Leibfried A, To JPC, Stehling SK, Busch W, Demar M, Kieber JJ, Lohmann JU. WUSCHEL controls meristem size by direct transcriptional regulation of cytokinin inducible response regulators. Nature. 2005:**438**(7071):1172–1175. https://doi.org/10.1038/nature04270

Letham DS, Shannon JS, McDonald IR. The structure of zeatin, a factor inducing cell division. Proc Chem Soc. 1964:**8**:230–231.

Leuendorf JE, Schmülling T. Meeting at the DNA: specifying cytokinin responses through transcription factor Complex formation. Plants (Basel). 2021:**10**(7):1458. https://doi.org/10.3390/plants10071458

Lezhneva L, Kiba T, Feria-Bourrellier AB, Lafouge F, Boutet-Mercey S, Zoufan P, Sakakibara H, Daniel-Vedele F, Krapp A. The Arabidopsis nitrate transporter NRT2.5 plays a role in nitrate acquisition and remobilization in nitrogen-starved plants. Plant J. 2014:**80**(2):230–241. https://doi.org/10.1111/tpj.12626

Li M, Li X, Zhou Z, Wu P, Fang M, Pan X, Lin Q, Luo W, Wu G, Li H. Reassessment of the four yield-related genes Gn1a, DEP1, GS3, and IPA1 in rice using a CRISPR/cas9 system. Front Plant Sci. 2016:**7**:377. https://doi.org/10.3389/fpls.2016.00377

Li Y, Liu F, Li P, Wang T, Zheng C, Hou B. An Arabidopsis cytokinin-modifying glycosyltransferase UGT76C2 improves drought and salt tolerance in rice. Front Plant Sci. 2020:**11**:560696. https://doi.org/10.3389/fpls.2020.560696

Li B, Wang R, Wang S, Zhang J, Chang L. Diversified regulation of cytokinin levels and signaling during *Botrytis cinerea* infection in Arabidopsis. Front Plant Sci. 2021:**12**:584042. https://doi.org/10.3389/fpls.2021.584042

Liang Y, Wang X, Hong S, Li Y, Zuo J. Deletion of the initial 45 residues of ARR18 induces cytokinin response in Arabidopsis. J Genet Genomics. 2012:**39**(1):37–46. https://doi.org/10.1016/j.jgg.2011.12.004

Lindner AC, Lang D, Seifert M, Podlešáková K, Novák O, Strnad M, Reski R, von Schwartzenberg K. Isopentenyltransferase-1 (IPT1) knockout in Physcomitrella together with phylogenetic analyses of IPTs provide insights into evolution of plant cytokinin biosynthesis. J Exp Bot. 2014:**65**(9):2533–2543. https://doi.org/10.1093/jxb/eru142

Liu J, Müller B. Imaging TCSn::GFP, a synthetic cytokinin reporter, in Arabidopsis thaliana. Methods Mol Biol. 2017:**1497**:81–90. https://doi.org/10.1007/978-1-4939-6469-7_9

Liu S, Strauss S, Adibi M, Mosca G, Yoshida S, Dello Ioio R, Runions A, Andersen TG, Grossmann G, Huijser P, et al. Cytokinin promotes growth cessation in the Arabidopsis root. Curr Biol. 2022:**32**(9):1974–1985.e1973. https://doi.org/10.1016/j.cub.2022.03.019

Liu Z, Yuan L, Song X, Yu X, Sundaresan V. AHP2, AHP3, and AHP5 act downstream of CKI1 in Arabidopsis female gametophyte development. J Exp Bot. 2017:**68**(13):3365–3373. https://doi.org/10.1093/jxb/erx181

Liu Y, Zhang M, Meng Z, Wang B, Chen M. Research progress on the roles of cytokinin in plant response to stress. Int J Mol Sci. 2020:**21**:6574. https://doi.org/10.3390/ijms22010021

Liu X, Zhou X, Li D, Hong B, Gao J, Zhang Z. Rose WRKY13 promotes disease protection to Botrytis by enhancing cytokinin content and reducing abscisic acid signaling. Plant Physiol. 2023:**191**(1):679–693. https://doi.org/10.1093/plphys/kiac495

Lohar DP, Schaff JE, Laskey JG, Kieber JJ, Bilyeu KD, Bird DM. Cytokinins play opposite roles in lateral root formation, and nematode and rhizobial symbioses. Plant J. 2004:**38**:203–214. https://doi.org/10.1111/j.1365-313X.2004.02038.x

Lomin SN, Krivosheev DM, Steklov MY, Osolodkin DI, Romanov GA. Receptor properties and features of cytokinin signaling. Acta Naturae. 2012:**4**(3):31–45. https://doi.org/10.32607/20758251-2012-4-3-31-45

Lomin SN, Savelieva EM, Arkhipov DV, Pashkovskiy PP, Myakushina YA, Heyl A, Romanov GA. Cytokinin perception in ancient plants beyond angiospermae. Int J Mol Sci. 2021:**22**(23):13077. https://doi.org/10.3390/ijms222313077

Lomin SN, Yonekura-Sakakibara K, Romanov GA, Sakakibara H. Ligand-binding properties and subcellular localization of maize cytokinin receptors. J Exp Bot. 2011:**62**(14):5149–5159. https://doi.org/10.1093/jxb/err220

Lopes FL, Galvan-Ampudia C, Landrein B. WUSCHEL in the shoot apical meristem: old player, new tricks. J Exp Bot. 2021:**72**(5):1527–1535. https://doi.org/10.1093/jxb/eraa572

Lubovská Z, Dobrá J, Štorchová H, Wilhelmová N, Vanková R. Cytokinin oxidase/dehydrogenase overexpression modifies antioxidant defense against heat, drought and their combination in *Nicotiana tabacum* plants. J Plant Physiol. 2014:**171**(17):1625–1633. https://doi.org/10.1016/j.jplph.2014.06.021

Mähönen AP, Bishopp A, Higuchi M, Nieminen KM, Kinoshita K, Törmäkangas K, Ikeda Y, Oka A, Kakimoto T, Helariutta Y. Cytokinin signaling and its inhibitor AHP6 regulate cell fate during vascular development. Science. 2006b:**311**(5757):94–98. https://doi.org/10.1126/science.1118875

Mähönen AP, Higuchi M, Törmäkangas K, Miyawaki K, Pischke MS, Sussman MR, Helariutta Y, Kakimoto T. Cytokinins regulate a bidirectional phosphorelay network in Arabidopsis. Curr Biol. 2006a:**16**(11):1116–1122. https://doi.org/10.1016/j.cub.2006.04.030

Malinowski R, Novák O, Borhan MH, Spichal L, Strnad M, Rolfe SA. The role of cytokinins in clubroot disease. Eur J Plant Pathol. 2016:**145**(3):543–557. https://doi.org/10.1007/s10658-015-0845-y

Mandal S, Ghorai M, Anand U, Samanta D, Kant N, Mishra T, Rahman MH, Jha NK, Jha SK, Lal MK, et al. Cytokinin and abiotic stress tolerance -what has been accomplished and the way forward? Front Genet. 2022:**13**:943025. https://doi.org/10.3389/fgene.2022.943025

Marhavý P, Bielach A, Abas L, Abuzeineh A, Duclercq J, Tanaka H, Pařezová M, Petrášek J, Friml J, Kleine-Vehn J, et al. Cytokinin modulates endocytic trafficking of PIN1 auxin efflux carrier to control plant organogenesis. Dev Cell. 2011:**21**(4):796–804. https://doi.org/10.1016/j.devcel.2011.08.014

Marín-de la Rosa N, Pfeiffer A, Hill K, Locascio A, Bhalerao RP, Miskolczi P, Grønlund AL, Wanchoo-Kohli A, Thomas SG, Bennett MJ, et al. Genome wide binding site analysis reveals transcriptional coactivation of cytokinin-responsive genes by DELLA proteins. PLoS Genet. 2015:**11**(7):e1005337. https://doi.org/10.1371/journal.pgen.1005337

Mason MG, Mathews DE, Argyros DA, Maxwell BB, Kieber JJ, Alonso JM, Ecker JRS, Schaller GE. Multiple type-B response regulators mediate cytokinin signal transduction in Arabidopsis. Plant Cell. 2005:**17**(11):3007–3018. https://doi.org/10.1105/tpc.105.035451

Matsumoto-Kitano M, Kusumoto T, Tarkowski P, Kinoshita-Tsujimura K, Václavíková K, Miyawaki K, Kakimoto T. Cytokinins are central regulators of cambial activity. Proc Natl Acad Sci USA. 2008:**105**(50):20027–20031. https://doi.org/10.1073/pnas.0805619105

McIntyre KE, Bush DR, Argueso CT. Cytokinin regulation of source-sink relationships in plant-pathogen interactions. Front Plant Sci. 2021:**12**:677585. https://doi.org/10.3389/fpls.2021.677585

Meng WJ, Cheng ZJ, Sang YL, Zhang MM, Rong XF, Wang ZW, Tang YY, Zhang XS. Type-B ARABIDOPSIS RESPONSE REGULATORs specify the shoot stem cell niche by dual regulation of *WUSCHEL*. Plant Cell. 2017:**29**(6):1357–1372. https://doi.org/10.1105/tpc.16.00640

Miller CO, Skoog F, Okomura FS, von Saltza MH, Strong FM. Isolation, structure and synthesis of kinetin, a substance promoting cell division. J Am Chem Soc. 1956:**78**(7):1345–1350. https://doi.org/10.1021/ja01588a032

Miller CO, Skoog F, Von Saltza MH, Strong F. Kinetin, a cell division factor from deoxyribonucleic acid. J Am Chem Soc. 1955:**77**(5):1392. https://doi.org/10.1021/ja01610a105

Miyawaki K, Matsumoto-Kitano M, Kakimoto T. Expression of cytokinin biosynthetic isopentyltransferase genes in Arabidopsis: tissue specificity and regulation by auxin, cytokinin, and nitrate. Plant J. 2004:**37**(1):128–138. https://doi.org/10.1046/j.1365-313X.2003.01945.x

Möhlmann T, Mezher Z, Schwerdtfeger G, Neuhaus HE. Characterisation of a concentrative type of adenosine transporter from Arabidopsis thaliana (ENT1. At). FEBS Lett. 2001:**509**(3): 370–374. https://doi.org/10.1016/S0014-5793(01)03195-7

Moore FH. A cytokinin-binding protein from wheat germ: isolation by affinity chromatography and properties. Plant Physiol. 1979:**64**(4): 594–599. https://doi.org/10.1104/pp.64.4.594

Moreira S, Bishopp A, Carvalho H, Campilho A. AHP6 inhibits cytokinin signaling to regulate the orientation of pericycle cell division during lateral root initiation. PLoS One. 2013:**8**(2):e56370. https://doi.org/10.1371/journal.pone.0056370

Morris RO, Bilyeu KD, Laskey JG, Cheikh NN. Isolation of a gene encoding a glycosylated cytokinin oxidase from maize. Biochem Biophys Res Comm. 1999:**255**(2):328–333. https://doi.org/10.1006/bbrc.1999.0199

Moubayidin L, Perilli S, Dello Ioio R, Di Mambro R, Costantino P, Sabatini S. The rate of cell differentiation controls the Arabidopsis root meristem growth phase. Curr Biol. 2010:**20**(12):1138–1143. https://doi.org/10.1016/j.cub.2010.05.035

Müller CJ, Larsson E, Spíchal L, Sundberg E. Cytokinin-auxin crosstalk in the gynoecial primordium ensures correct domain patterning. Plant Physiol. 2017:**175**:1144–1157. https://doi.org/10.1104/pp.17.00805

Müller B, Sheen J. Cytokinin and auxin interaction in root stem-cell specification during early embryogenesis. Nature. 2008:**453**(7198): 1094–1097. https://doi.org/10.1038/nature06943

Nehnevajova E, Ramireddy E, Stolz A, Gerdemann-Knörck M, Novák O, Strnad M, Schmülling T. Root enhancement in cytokinin-deficient oilseed rape causes leaf mineral enrichment, increases the chlorophyll concentration under nutrient limitation and enhances the phytoremediation capacity. BMC Plant Biol. 2019:**19**(1):83. https://doi.org/10.1186/s12870-019-1657-6

Nieminen K, Immanen J, Laxell M, Kauppinen L, Tarkowski P, Dolezal K, Tähtiharju S, Elo A, Decourteix M, Ljung K, et al. Cytokinin signaling regulates cambial development in poplar. Proc Natl Acad Sci U S A. 2008:**105**(50):20032–20037. https://doi.org/10.1073/pnas.0805617106

Nishimura C, Ohashi Y, Sato S, Kato T, Tabata S, Ueguchi C. Histidine kinase homologs that act as cytokinin receptors possess overlapping functions in the regulation of shoot and root growth in Arabidopsis. Plant Cell. 2004:**16**(6):1365–1377. https://doi.org/10.1105/tpc.021477

Nishiyama R, Watanabe Y, Fujita Y, Le DT, Kojima M, Werner T, Vankova R, Yamaguchi-Shinozaki K, Shinozaki K, Kakimoto T, et al. Analysis of cytokinin mutants and regulation of cytokinin metabolic genes reveals important regulatory roles of cytokinins in drought, salt and abscisic acid responses, and abscisic acid biosynthesis. Plant Cell. 2011:**23**(6):2169–2183. https://doi.org/10.1105/tpc.111.087395

Park J, Lee S, Park G, Cho H, Choi D, Umeda M, Choi Y, Hwang D, Hwang I. CYTOKININ-RESPONSIVE GROWTH REGULATOR regulates cell expansion and cytokinin-mediated cell cycle progression. Plant Physiol. 2021:**186**(3):1734–1746. https://doi.org/10.1093/plphys/kiab180

Pavlů J, Novák J, Koukalová V, Luklová M, Brzobohatý B, Černý M. Cytokinin at the crossroads of abiotic stress signalling pathways. Int J Mol Sci. 2018:**19**(8):2450. https://doi.org/10.3390/ijms19082450

Peleg Z, Reguera M, Tumimbang E, Walia H, Blumwald E. Cytokinin-mediated source/sink modifications improve drought tolerance and increase grain yield in rice under water-stress. Plant Biotechnol J. 2011:**9**(7):747–758. https://doi.org/10.1111/j.1467-7652.2010.00584.x

Perilli S, Perez-Perez JM, Di Mambro R, Peris CL, Díaz-Triviño S, Del Bianco M, Pierdonati E, Moubayidin L, Cruz-Ramírez A, Costantino P, et al. RETINOBLASTOMA-RELATED protein stimulates cell differentiation in the Arabidopsis root meristem by interacting with cytokinin signaling. Plant Cell. 2013:**25**(11):4469–4478. https://doi.org/10.1105/tpc.113.116632

Petricka JJ, Winter CM, Benfey PN. Control of Arabidopsis root development. Annu Rev Plant Biol. 2012:**63**(1):563–590. https://doi.org/10.1146/annurev-arplant-042811-105501

Pils B, Heyl A. Unraveling the evolution of cytokinin signaling. Plant Physiol. 2009:**151**(2):782–791. https://doi.org/10.1104/pp.109.139188

Plet J, Wasson A, Ariel F, Le Signor C, Baker D, Mathesius U, Crespi M, Frugier F. MtCRE1-dependent cytokinin signaling integrates bacterial and plant cues to coordinate symbiotic nodule organogenesis in *Medicago truncatula*. Plant J. 2011:**65**(4):622–633. https://doi.org/10.1111/j.1365-313X.2010.04447.x

Poitout A, Crabos A, Petřík I, Novák O, Krouk G, Lacombe B, Ruffel S. Responses to systemic nitrogen signaling in Arabidopsis roots involve trans-zeatin in shoots. Plant Cell. 2018:**30**(6):1243–1257. https://doi.org/10.1105/tpc.18.00011

Polko JK, Potter KC, Burr CA, Schaller GE, Kieber JJ. Meta-analysis of transcriptomic studies of cytokinin-treated rice roots defines a core set of cytokinin response genes. Plant J. 2021:**107**(5):1387–1402. https://doi.org/10.1111/tpj.15386

Polya GM, Davis AW. Properties of a high-affinity cytokinin-binding protein from wheat germ. Planta. 1978:**139**(2):139–147. https://doi.org/10.1007/BF00387139

Potter KC, Wang J, Schaller GE, Kieber JJ. Cytokinin modulates context-dependent chromatin accessibility through the type-B response regulators. Nat Plants. 2018:**4**(12):1102–1111. https://doi.org/10.1038/s41477-018-0290-y

Punwani JA, Hutchison CE, Schaller GE, Kieber JJ. The subcellular distribution of the Arabidopsis histidine phosphotransfer proteins is independent of cytokinin signaling. Plant J. 2010:**62**(3):473–482. https://doi.org/10.1111/j.1365-313X.2010.04165.x

Radchuk V, Belew ZM, Gündel A, Mayer S, Hilo A, Hensel G, Sharma R, Neumann K, Ortleb S, Wagner S, et al. SWEET11b transports both sugar and cytokinin in developing barley grains. Plant Cell. 2023:**35**(6):2186–2207. https://doi.org/10.1093/plcell/koad055

Rahimi A, Karami O, Lestari AD, de Werk T, Amakorová P, Shi D, Novák O, Greb T, Offringa R. Control of cambium initiation and activity in Arabidopsis by the transcriptional regulator AHL15. Curr Biol. 2022:**32**(8):1764–1775.e3. https://doi.org/10.1016/j.cub.2022.02.060

Raines T, Blakley IC, Tsai Y-C, Worthen JM, Franco-Zorrilla JM, Solano R, Schaller GE, Loraine AE, Kieber JJ. Characterization of the cytokinin-responsive transcriptome in rice. BMC Plant Biol. 2016a:**16**(1):260. https://doi.org/10.1186/s12870-016-0932-z

Raines T, Shanks C, Cheng C-Y, McPherson D, Argueso CT, Kim HJ, Franco-Zorrilla JM, López-Vidriero I, Solano R, Vaňková R, et al. The cytokinin response factors modulate root and shoot growth and promote leaf senescence in Arabidopsis. Plant J. 2016b:**85**(1): 134–147. https://doi.org/10.1111/tpj.13097

Ramireddy E, Hosseini SA, Eggert K, Gillandt S, Gnad H, von Wirén N, Schmülling T. Root engineering in barley: increasing cytokinin degradation produces a larger root system, mineral enrichment in the shoot and improved drought tolerance. Plant Physiol. 2018:**177**(3):1078–1095. https://doi.org/10.1104/pp.18.00199

Ramireddy E, Nelissen H, Leuendorf JE, Van Lijsebettens M, Inzé D, Schmülling T. Root engineering in maize by increasing cytokinin degradation causes enhanced root growth and leaf mineral enrichment. Plant Mol Biol. 2021:**106**(6):555–567. https://doi.org/10.1007/s11103-021-01173-5

Randall RS, Miyashima S, Blomster T, Zhang J, Elo A, Karlberg A, Immanen J, Nieminen K, Lee JY, Kakimoto T, et al.

AINTEGUMENTA and the D-type cyclin CYCD3; 1 regulate root secondary growth and respond to cytokinins. Biol Open. 2015:**4**(10): 1229–1236. https://doi.org/10.1242/bio.013128

Rashotte AM. The evolution of cytokinin signaling and its role in development before angiosperms. Semin Cell Dev Biol. 2021:**109**:31–38. https://doi.org/10.1016/j.semcdb.2020.06.010

Rashotte AM, Mason MG, Hutchison CE, Ferreira FJ, Schaller GE, Kieber JJ. A subset of Arabidopsis AP2 transcription factors mediates cytokinin responses in concert with a two-component pathway. Proc Natl Acad Sci. U S A. 2006:**103**(29):11081–11085. https://doi.org/10.1073/pnas.0602038103

Reguera M, Peleg Z, Abdel-Tawab YM, Tumimbang EB, Delatorre CA, Blumwald E. Stress-induced cytokinin synthesis increases drought tolerance through the coordinated regulation of carbon and nitrogen assimilation in rice. Plant Physiol. 2013:**163**(4): 1609–1622. https://doi.org/10.1104/pp.113.227702

Riefler M, Novak O, Strnad M, Schmülling T. Arabidopsis cytokinin receptor mutants reveal functions in shoot growth, leaf senescence, seed size, germination, root development, and cytokinin metabolism. Plant Cell. 2006:**18**(1):40–54. https://doi.org/10.1105/tpc.105.037796

Riou-Khamlichi C, Huntley R, Jacqmard A, Murray JA. Cytokinin activation of Arabidopsis cell division through a D-type cyclin. Science. 1999:**283**(5407):1541–1544. https://doi.org/10.1126/science.283.5407.1541

Rivero RM, Kojima M, Gepstein A, Sakakibara H, Mittler R, Gepstein S, Blumwald E. Delayed leaf senescence induces extreme drought tolerance in a flowering plant. Proc Natl Acad Sci U S A. 2007: **104**(49):19631–19636. https://doi.org/10.1073/pnas.0709453104

Romanov GA, Lomin SN, Schmülling T. Biochemical characteristics and ligand-binding properties of Arabidopsis cytokinin receptor AHK3 compared to CRE1/AHK4 as revealed by a direct binding assay. J Exp Bot. 2006:**57**(15):4051–4058. https://doi.org/10.1093/jxb/erl179

Romanov GA, Lomin SN, Schmülling T. Cytokinin signaling: from the ER or from the PM? That is the question! New Phytol. 2018:**218**(1): 41–53. https://doi.org/10.1111/nph.14991

Ruffel S, Krouk G, Ristova D, Shasha D, Birnbaum KD, Coruzzi GM. Nitrogen economics of root foraging: transitive closure of the nitrate–cytokinin relay and distinct systemic signaling for N supply vs. Demand. Proc Natl Acad Sci U S A. 2011:**108**(45):18524–18529. https://doi.org/10.1073/pnas.1108684108

Sakai H, Aoyama T, Oka A. Arabidopsis ARR1 and ARR2 response regulators operate as transcriptional activators. Plant J. 2000:**24**(6): 703–711. https://doi.org/10.1111/j.1365-313X.2000.00909.x

Sakai H, Honma T, Aoyama T, Sato S, Kato T, Tabata S, Oka A. ARR1, a transcription factor for genes immediately responsive to cytokinins. Science. 2001:**294**(5546):1519–1521. https://doi.org/10.1126/science.1065201

Sakakibara H. Cytokinins: activity, biosynthesis, and translocation. Annu Rev Plant Biol. 2006:**57**(1):431–449. https://doi.org/10.1146/annurev.arplant.57.032905.105231

Schaller GE, Bishopp A, Kieber JJ. The yin-yang of hormones: cytokinin and auxin interactions in plant development. Plant Cell. 2015:**27**(1): 44–63. https://doi.org/10.1105/tpc.114.133595

Schaller GE, Shiu S-H, Armitage JP. Two-component systems and their co-option for eukaryotic signal transduction. Curr Biol. 2011:**21**(9):R320–R330. https://doi.org/10.1016/j.cub.2011.02.045

Schaller GE, Street IH, Kieber JJ. Cytokinin and the cell cycle. Curr Opin Plant Biol. 2014:**21C**:7–15. https://doi.org/10.1016/j.pbi.2014.05.015

Scheres B, Di Laurenzo L, Willemsen V, Hauser M-T, Janmaat K, Weisbeek P, Benfey PN. Mutations affecting the radial organization of the Arabidopsis root display specific defects throughout the embryonic axis. Development. 1995:**121**(1):53–62. https://doi.org/10.1242/dev.121.1.53

Schwarz I, Scheirlinck MT, Otto E, Bartrina I, Schmidt RC, Schmülling T. Cytokinin regulates the activity of the inflorescence meristem and components of seed yield in oilseed rape. J Exp Bot. 2020:**71**(22):7146–7159. https://doi.org/10.1093/jxb/eraa419

Scofield S, Dewitte W, Nieuwland J, Murray JAH. The Arabidopsis homeobox gene SHOOT MERISTEMLESS has cellular and meristem-organisational roles with differential requirements for cytokinin and CYCD3 activity. Plant J. 2013:**75**(1):53–66. https://doi.org/10.1111/tpj.12198

Shang X-L, Xie R-R, Tian H, Wang Q-L, Guo F-Q. Putative zeatin O-glucosyltransferase OscZOG1 regulates root and shoot development and formation of agronomic traits in rice. J Integr Plant Biol. 2016:**58**(7):627–641. https://doi.org/10.1111/jipb.12444

Shanks CM, Hecker A, Cheng C-Y, Brand L, Collani S, Schmid M, Schaller GE, Wanke D, Harter K, Kieber JJ. Role of BASIC PENTACYSTEINE transcription factors in a subset of cytokinin signaling responses. Plant J. 2018:**95**(3):458–473. https://doi.org/10.1111/tpj.13962

Shanks CM, Rice H, Hubo Y, Schaller E, Hewezi T, Kieber JJ. The role of cytokinin during infection of *Arabidopsis thaliana* by *Heterodera schachtii*. Mol Plant Microbe Interact. 2016:**29**(1):57–68. https://doi.org/10.1094/MPMI-07-15-0156-R

Siddique S, Radakovic ZS, De La Torre CM, Chronis D, Novák O, Ramireddy E, Holbein J, Matera C, Hutten M, Gutbrod P, et al. A parasitic nematode releases cytokinin that controls cell division and orchestrates feeding site formation in host plants. Proc Natl Acad Sci USA. 2015:**112**(41):12669–12674. https://doi.org/10.1073/pnas.1503657112

Skoog F, Miller C. Chemical regulation of growth and organ formation in plant tissue cultured in vitro. Symp Soc Exp Biol. 1957:**11**:118–130.

Spallek T, Melnyk CW, Wakatake T, Zhang J, Sakamoto Y, Kiba T, Yoshida S, Matsunaga S, Sakakibara H, Shirasu K. Interspecies hormonal control of host root morphology by parasitic plants. Proc Natl Acad Sci U S A. 2017:**114**(20):5283–5288. https://doi.org/10.1073/pnas.1619078114

Spichal L, Rakova NY, Riefler M, Mizuno T, Romanov GA, Strnad M, Schmülling T. Two cytokinin receptors of *Arabidopsis thaliana*, CRE1/AHK4 and AHK3, differ in their ligand specificity in a bacterial assay. Plant Cell Physiol. 2004:**45**(9):1299–1305. https://doi.org/10.1093/pcp/pch132

Steiner E, Israeli A, Gupta R, Shwartz I, Nir I, Leibman-Markus M, Tal L, Farber M, Amsalem Z, Ori N, et al. Characterization of the cytokinin sensor *TCSv2* in Arabidopsis and tomato. Plant Methods. 2020:**16**(1):152. https://doi.org/10.1186/s13007-020-00694-2

Stock AM, Robinson VL, Goudreau PN. Two-component signal transduction. Annu Rev Biochem. 2000:**69**(1):183–215. https://doi.org/10.1146/annurev.biochem.69.1.183

Stock JF, Stock AN, Mottonen JM. Signal transduction in bacteria. Nature. 1990:**344**(6265):395–400. https://doi.org/10.1038/344395a0

Stolz A, Riefler M, Lomin SN, Achazi K, Romanov GA, Schmülling T. The specificity of cytokinin signalling in Arabidopsis thaliana is mediated by differing ligand affinities and expression profiles of the receptors. Plant J. 2011:**67**(1):157–168. https://doi.org/10.1111/j.1365-313X.2011.04584.x

Strnad M. The aromatic cytokinins. Physiol Plant. 1997:**101**(4): 674–688. https://doi.org/10.1111/j.1399-3054.1997.tb01052.x

Sugawara H, Ueda N, Kojima M, Makita N, Yamaya T, Sakakibara H. Structural insight into the reaction mechanism and evolution of cytokinin biosynthesis. Proc Natl Acad Sci U S A. 2008:**105**(7): 2734–2739. https://doi.org/10.1073/pnas.0707374105

Sun J, Hirose N, Wang X, Wen P, Xue L, Sakakibara H, Zuo J. Arabidopsis SOI33/AtENT8 gene encodes a putative equilibrative nucleoside transporter that is involved in cytokinin transport in planta. J Integr Plant Biol. 2005:**47**(5):588–603. https://doi.org/10.1111/j.1744-7909.2005.00104.x

Sun L, Zhang Q, Wu J, Zhang L, Jiao X, Zhang S, Zhang Z, Sun D, LuT, Sun Y. Two rice authentic histidine phosphotransfer proteins, OsAHP1 and OsAHP2, mediate cytokinin signaling and stress responses in rice. Plant Physiol. 2014:**165**(1):335–345. https://doi.org/10.1104/pp.113.232629

Sussman MR, Kende H. In vitro cytokinin binding to a particulate fraction of tobacco cells. Planta. 1978:**140**(3):251–259. https://doi.org/10.1007/BF00390256

Suzuki T, Miwa K, Ishikawa K, Yamada H, Aiba H, Mizuno T. The Arabidopsis sensor his-kinase, AHK4, can respond to cytokinins. Plant Cell Physiol. 2001:**42**(2):107–113. https://doi.org/10.1093/pcp/pce037

Swanson RV, Alex LA, Simon MI. Histidine and aspartate phosphorylation: two-component systems and the limits of homology. Trends Biochem Sci. 1994:**19**(11):485–490. https://doi.org/10.1016/0968-0004(94)90135-X

Takahashi N, Kajihara T, Okamura C, Kim Y, Katagiri Y, Okushima Y, Matsunaga S, Hwang I, Umeda M. Cytokinins control endocycle onset by promoting the expression of an APC/C activator in Arabidopsis roots. Curr Biol. 2013:**23**(18):1812–1817. https://doi.org/10.1016/j.cub.2013.07.051

Takegami T, Yoshida K. Isolation and purification of cytokinin binding protein from tobacco leaves by affinity column chromatography. Biochem Biophys Res Commun. 1975:**67**(2):782–789. https://doi.org/10.1016/0006-291X(75)90881-5

Takei K, Sakakibara H, Sugiyama T. Identification of genes encoding adenylate isopentenyltransferase, a cytokinin biosynthesis enzyme, in *Arabidopsis thaliana*. J Biol Chem. 2001:**276**(28):26405–26410. https://doi.org/10.1074/jbc.M102130200

Takei K, Ueda N, Aoki K, Kuromori T, Hirayama T, Shinozaki K, Yamaya T, Sakakibara H. *AtIPT3* is a key determinant of nitrate-dependent cytokinin biosynthesis in *Arabidopsis*. Plant Cell Physiol. 2004a:**45**(8):1053–1062. https://doi.org/10.1093/pcp/pch119

Takei K, Yamaya T, Sakakibara H. Arabidopsis *CYP735A1* and *CYP735A2* encode cytokinin hydroxylases that catalyze the biosynthesis of *trans*-zeatin. J Biol Chem. 2004b:**279**(40):41866–41872. https://doi.org/10.1074/jbc.M406337200

Tao J, Sun H, Gu P, Liang Z, Chen X, Lou J, Xu G, Zhang Y. A sensitive synthetic reporter for visualizing cytokinin signaling output in rice. Plant Methods. 2017:**13**(1):89. https://doi.org/10.1186/s13007-017-0232-0

Taya Y, Tanaka Y, Nishimura S. 5′-AMP is a direct precursor of cytokinin in *Dictyostelium discoidum*. Nature. 1978:**271**(5645):545–547. https://doi.org/10.1038/271545a0

Tessi TM, Brumm S, Winklbauer E, Schumacher B, Pettinari G, Lescano I, González CA, Wanke D, Maurino VG, Harter K, et al. Arabidopsis AZG2 transports cytokinins in vivo and regulates lateral root emergence. New Phytol. 2021:**229**(2):979–993. https://doi.org/10.1111/nph.16943

Tessi TM, Maurino VG, Shahriari M, Meissner E, Novak O, Pasternak T, Schumacher BS, Ditengou F, Li Z, Duerr J, et al. AZG1 is a cytokinin transporter that interacts with auxin transporter PIN1 and regulates the root stress response. New Phytol. 2023:**238**(5):1924–1941. https://doi.org/10.1111/nph.18879

To JP, Deruère J, Maxwell BB, Morris VF, Hutchison CE, Schaller GE, Kieber JJ. Cytokinin regulates type-A Arabidopsis response regulator activity and protein stability via two-component phosphorelay. Plant Cell. 2007:**19**(12):3901–3914. https://doi.org/10.1105/tpc.107.052662

To JPC, Haberer G, Ferreira FJ, Deruère J, Mason MG, Schaller GE, Alonso JM, Ecker JR, Kieber JJ. Type-A ARRs are partially redundant negative regulators of cytokinin signaling in Arabidopsis. Plant Cell. 2004:**16**(3):658–671. https://doi.org/10.1105/tpc.018978

Tokunaga H, Kojima M, Kuroha T, Ishida T, Sugimoto K, Kiba T, Sakakibara H. Arabidopsis lonely guy (LOG) multiple mutants reveal a central role of the LOG-dependent pathway in cytokinin activation. Plant J. 2012:**69**(2):355–365. https://doi.org/10.1111/j.1365-313X.2011.04795.x

Tran LSP, Urao T, Qin F, Maruyama K, Kakimoto T, Shinozaki K, Yamaguchi-Shinozaki K. Functional analysis of AHK1/ATHK1 and cytokinin receptor histidine kinases in response to abscisic acid, drought, and salt stress in Arabidopsis. Proc Natl Acad Sci U S A. 2007:**104**(51):20623–20628. https://doi.org/10.1073/pnas.0706547105

Tsai YC, Weir NR, Hill K, Zhang W, Kim HJ, Shiu SH, Schaller E, Kieber JJ. Characterization of genes involved in cytokinin signaling and metabolism from rice. Plant Physiol. 2012:**158**(4):1666–1684. https://doi.org/10.1104/pp.111.192765

Ueguchi C, Sato S, Kato T, Tabata S. The *AHK4* gene involved in the cytokinin-signaling pathway as a direct receptor molecule in *Arabidopsis thaliana*. Plant Cell Physiol. 2001:**42**(7):751–755. https://doi.org/10.1093/pcp/pce094

Vaughan-Hirsch J, Tallerday EJ, Burr CA, Hodgens C, Boeshore SL, Beaver K, Melling A, Sari K, Kerr ID, Šimura J, et al. Function of the pseudo phosphotransfer proteins has diverged between rice and Arabidopsis. Plant J. 2021:**106**(1):159–173. https://doi.org/10.1111/tpj.15156

von Schwartzenberg K, Lindner AC, Gruhn N, Šimura J, Novák O, Strnad M, Gonneau M, Nogué F, Heyl A. CHASE domain-containing receptors play an essential role in the cytokinin response of the moss *Physcomitrella patens*. J Exp Bot. 2016:**67**(3):667–679. https://doi.org/10.1093/jxb/erv479

Walters DR, McRoberts N. Plants and biotrophs: a pivotal role for cytokinins? Trends Plant Sci. 2006:**11**(12):581–586. https://doi.org/10.1016/j.tplants.2006.10.003

Wang X, Ding J, Lin S, Liu D, Gu T, Wu H, Trigiano RN, McAvoy R, Huang J, Li Y. Evolution and roles of cytokinin genes in angiosperms 2: do ancient CKXs play housekeeping roles while non-ancient CKXs play regulatory roles? Hortic Res. 2020b:**7**(1):29. https://doi.org/10.1038/s41438-020-0246-z

Wang W-C, Lin T-C, Kieber J, Tsai Y-C. Response regulators 9 and 10 negatively regulate salinity tolerance in rice. Plant Cell Physiol. 2019:**60**(11):2549–2563. https://doi.org/10.1093/pcp/pcz149

Wang X, Lin S, Liu D, Gan L, McAvoy R, Ding J, Li Y. Evolution and roles of cytokinin genes in angiosperms 1: do ancient IPTs play housekeeping while non-ancient IPTs play regulatory roles? Hort Res. 2020c:**7**(1):28. https://doi.org/10.1038/s41438-019-0211-x

Wang J, Ma XM, Kojima M, Sakakibara H, Hou BK. *N*-glucosyltransferase UGT76C2 is involved in cytokinin homeostasis and cytokinin response in *Arabidopsis thaliana*. Plant Cell Physiol. 2011:**52**(12):2200–2201. https://doi.org/10.1093/pcp/pcr152

Wang Y, Shen W, Chan Z, Wu Y. Endogenous cytokinin overproduction modulates ROS homeostasis and decreases salt stress resistance in *Arabidopsis thaliana*. Front Plant Sci. 2015:**6**:1004. https://doi.org/10.3389/fpls.2015.01004

Wang C, Wang G, Gao Y, Lu G, Habben JE, Mao G, Chen G, Wang J, Yang F, Zhao X, et al. A cytokinin-activation enzyme-like gene improves grain yield under various field conditions in rice. Plant Mol Biol. 2020a:**102**(4–5):373–388. https://doi.org/10.1007/s11103-019-00952-5

Werner T, Köllmer I, Bartrina I, Holst K, Schmülling T. New insights into the biology of cytokinin degradation. Plant Biol (Stuttg). 2006:**8**(3):371–381. https://doi.org/10.1055/s-2006-923928

Wulfetange K, Lomin SN, Romanov GA, Stolz A, Heyl A, Schmülling T. The cytokinin receptors of Arabidopsis are located mainly to the endoplasmic reticulum. Plant Physiol. 2011:**156**(4):1808–1818. https://doi.org/10.1104/pp.111.180539

Wybouw B, De Rybel B. Cytokinin—a developing story. Trends Plant Sci. 2019:**24**(2):177–185. https://doi.org/10.1016/j.tplants.2018.10.012

Xiao Y, Liu D, Zhang G, Gao S, Liu L, Xu F, Che R, Wang Y, Tong H, Chu C. Big grain3, encoding a purine permease, regulates grain size via modulating cytokinin transport in rice. J Integr Plant Biol. 2019:**61**(5):581–597. https://doi.org/10.1111/jipb.12727

Xie M, Chen H, Huang L, O'Neil RC, Shokhirev MN, Ecker JR. A B-ARR-mediated cytokinin transcriptional network directs hormone cross-regulation and shoot development. Nat Commun. 2018:**9**(1):1604. https://doi.org/10.1038/s41467-018-03921-6

Xu Y, Burgess P, Huang B. Enhancing cytokinin synthesis by overexpressing ipt alleviated drought inhibition of root growth through activating ROS-scavenging systems in *Agrostis stolonifera*. J Exp Bot. 2016:**67**(6):1979–1992 https://doi.org/10.1093/jxb/erw019

Yamada H, Suzuki T, Terada K, Takei K, Ishikawa K, Miwa K, Yamashino T, Mizuno T. The Arabidopsis AHK4 histidine kinase is a cytokinin-binding receptor that transduces cytokinin signals across the membrane. Plant Cell Physiol. 2001:**41**(9):1017–1023. https://doi.org/10.1093/pcp/pce127

Yan Z, Liu X, Ljung K, Li S, Zhao W, Yang F, Wang M, Tao Y. Type B response regulators act as central integrators in transcriptional control of the auxin biosynthesis enzyme TAA1. Plant Physiol. 2017:**175**(3):1438–1454. https://doi.org/10.1104/pp.17.00878

Yan Z, Wang J, Wang F, Xie C, Lv B, Yu Z, Dai S, Liu X, Xia G, Tian H, et al. MPK3/6-induced degradation of ARR1/10/12 promotes salt tolerance in Arabidopsis. EMBO Rep. 2021:**22**(10):e52457. https://doi.org/10.15252/embr.202152457

Yang W, Cortijo S, Korsbo N, Roszak P, Schiessl K, Gurzadyan A, Wightman R, Jönsson H, Meyerowitz E. Molecular mechanism of cytokinin-activated cell division in Arabidopsis. Science. 2021a: **371**(6536):1350–1355. https://doi.org/10.1126/science.abe2305

Yang B, Minne M, Brunoni F, Plačková L, Petřík I, Sun Y, Nolf J, Smet W, Verstaen K, Wendrich JR, et al. Non-cell autonomous and spatiotemporal signalling from a tissue organizer orchestrates root vascular development. Nat Plants. 2021b:**7**(11):1485–1494. https://doi.org/10.1038/s41477-021-01017-6

Yang Q, Zhang J, Kojima M, Takebayashi Y, Uragami T, Kiba T, Sakakibara H, Lee Y. ABCG11 modulates cytokinin responses in Arabidopsis thaliana. Front Plant Sci. 2022:**13**:976267. https://doi.org/10.3389/fpls.2022.976267

Ye L, Wang X, Lyu M, Siligato R, Eswaran G, Vainio L, Blomster T, Zhang J, Mähönen AP. Cytokinins initiate secondary growth in the Arabidopsis root through a set of LBD genes. Curr Biol. 2021:**31**(15):3365–3373. https://doi.org/10.1016/j.cub.2021.05.036

Yin W, Xiao Y, Niu M, Meng W, Li L, Zhang X, Liu D, Zhang G, Qian Y, Sun Z, et al. ARGONAUTE2 enhances grain length and salt tolerance by activating BIG GRAIN3 to modulate cytokinin distribution in rice. Plant Cell. 2020:**32**(7):2292–2306. https://doi.org/10.1105/tpc.19.00542

Yonekura-Sakakibara K, Kojima M, Yamaya T, Sakakibara H. Molecular characterization of cytokinin-responsive histidine kinases in maize. Differential ligand preferences and response to cis-zeatin. Plant Physiol. 2004:**134**(4):1654–1661. https://doi.org/10.1104/pp.103.037176

Yuan L, Liu Z, Song X, Johnson C, Yu X, Sundaresan V. The CKI1 histidine kinase specifies the female gametic precursor of the endosperm. Dev Cell. 2016:**37**(1):34–46. https://doi.org/10.1016/j.devcel.2016.03.009

Zhang Y, Berman A, Shani E. Plant hormone transport and localization: signaling molecules on the move. Annu Rev Plant Biol. 2023:**74**(1):453–479. https://doi.org/10.1146/annurev-arplant-070722-015329

Zhang T-Q, Lian H, Zhou C-M, Xu L, Jiao Y, Wang J-W. A two-step model for de novo activation of *WUSCHEL* during plant shoot regeneration. Plant Cell. 2017:**29**(5):1073–1087. https://doi.org/10.1105/tpc.16.00863

Zhang K, Novak O, Wei Z, Gou M, Zhang X, Yu Y, Yang H, Cai Y, Strnad M, Liu CJ. Arabidopsis ABCG14 protein controls the acropetal translocation of root-synthesized cytokinins. Nat Commun. 2014:**5**(1):3274. https://doi.org/10.1038/ncomms4274

Zhang J, Pai Q, Yue L, Wu X, Liu H, Wang W. Cytokinin regulates female gametophyte development by cell cycle modulation in Arabidopsis thaliana. Plant Sci. 2022:**324**:111419. https://doi.org/10.1016/j.plantsci.2022.111419

Zhang W, Ranjan Swarup R, Bennett M, Schaller GE, Kieber JJ. Cytokinin induces cell division in the quiescent center of the Arabidopsis root apical meristem. Curr Biol. 2013:**23**(20): 1979–1989. https://doi.org/10.1016/j.cub.2013.08.008

Zhao J, Ding B, Zhu E, Deng X, Zhang M, Zhang P, Wang L, Dai Y, Xiao S, Zhang C, et al. Phloem unloading via the apoplastic pathway is essential for shoot distribution of root-synthesized cytokinins. Plant Physiol. 2021:**186**(4):2111–2123. https://doi.org/10.1093/plphys/kiab188

Zhu M, Tao L, Zhang J, Liu R, Tian H, Hu C, Zhu Y, Li M, Wei Z, Yi J, et al. The type-B response regulators ARR10, ARR12, and ARR18 specify the central cell in Arabidopsis. Plant Cell. 2022:**34**(12):4714–4737. https://doi.org/10.1093/plcell/koac285

Zubo YO, Blakley IC, Yamburenko MV, Worthen JM, Street IH, Franco-Zorrilla JM, Zhang W, Hill K, Raines T, Solano R, et al. Cytokinin induces genome-wide binding of the type-B response regulator ARR10 to regulate growth and development in Arabidopsis. Proc Natl Acad Sci U S A. 2017:**114**(29):E5995–E6004. https://doi.org/10.1073/pnas.1620749114

Zürcher E, Liu J, di Donato M, Geisler M, Müller B. Plant development regulated by cytokinin sinks. Science. 2016:**353**(6303):1027–1030. https://doi.org/10.1126/science.aaf7254

Zürcher E, Tavor-Deslex D, Lituiev D, Enkerli K, Tarr PT, Müller B. A robust and sensitive synthetic sensor to monitor the transcriptional output of the cytokinin signaling network in planta. Plant Physiol. 2013:**161**(3):1066–1075. https://doi.org/10.1104/pp.112.211763

Zwack PJ, Rashotte AM. Cytokinin inhibition of leaf senescence. Plant Signal Behav. 2013:**8**(7):e24737. https://doi.org/10.4161/psb.24737

Zwack PJ, Rashotte AM. Interactions between cytokinin signalling and abiotic stress responses. J Exp Bot. 2015:**66**(16):4863–4871. https://doi.org/10.1093/jxb/erv172

https://doi.org/10.1093/plcell/koad329

THE PLANT CELL

THE PLANT CELL 2024: 36: 1451–1464

Review

Salicylic acid in plant immunity and beyond

Steven H. Spoel [1,*] **and Xinnian Dong** [2,*]

1 Institute of Molecular Plant Sciences, School of Biological Sciences, University of Edinburgh, The King's Buildings, Edinburgh EH9 3BF, UK
2 Department of Biology, Howard Hughes Medical Institute, Duke University, Durham, NC 27708, USA

*Author for correspondence: steven.spoel@ed.ac.uk (S.H.S.), xdong@duke.edu (X.D.)
The authors responsible for distribution of materials integral to the findings presented in this article in accordance with the policy described in the Instructions for Authors (https://academic.oup.com/plcell/pages/General-Instructions) are: Steven Spoel (steven.spoel@ed.ac.uk) and Xinnian Dong (xdong@duke.edu).

Abstract

As the most widely used herbal medicine in human history and a major defence hormone in plants against a broad spectrum of pathogens and abiotic stresses, salicylic acid (SA) has attracted major research interest. With applications of modern technologies over the past 30 years, studies of the effects of SA on plant growth, development, and defence have revealed many new research frontiers and continue to deliver surprises. In this review, we provide an update on recent advances in our understanding of SA metabolism, perception, and signal transduction mechanisms in plant immunity. An overarching theme emerges that SA executes its many functions through intricate regulation at multiple steps: SA biosynthesis is regulated both locally and systemically, while its perception occurs through multiple cellular targets, including metabolic enzymes, redox regulators, transcription cofactors, and, most recently, an RNA-binding protein. Moreover, SA orchestrates a complex series of post-translational modifications of downstream signaling components and promotes the formation of biomolecular condensates that function as cellular signalling hubs. SA also impacts wider cellular functions through crosstalk with other plant hormones. Looking into the future, we propose new areas for exploration of SA functions, which will undoubtedly uncover more surprises for many years to come.

Salicylic acid (SA): a multifaceted plant hormone with a deep history in human civilizations

For over 4,000 years, humanity has been familiar with the plant hormone SA and its derivatives. As documented in ancient inscriptions and texts, willow (Latin name *Salix*) and myrtle, which contain substantial amounts of SA derivatives, were used as herbal medicines by ancient Assyrians, Egyptians, and Chinese civilizations, while Hippocrates was said to have made use of willow bark extracts for reducing fever and pain. Similarly, SA-rich meadowsweet was a sacred medicinal herb utilized by Druids of the ancient Celts. However, it was not until the early 19th century that SA and its derivative, salicin, were isolated from willow bark by Johann Andreas Buchner, Henri Leroux, and Raffaele Piria, and from meadowsweet by Löwig and Weidmann, respectively (Norn et al. 2009). This eventually led to the discovery of SA-derived aspirin as a major innovation in modern medi-cine for the alleviation of pain and prevention of cardiovas-cular disease and even cancer (Norn et al. 2009).

A large body of evidence from the last 50 years demonstrates that plants also utilize SA as a multifaceted endogenous agent for healing plant ailments. It plays important roles in mitigating abiotic stresses, including heat, cold, drought, UV radiation, heavy metals, and osmotic shock (Rivas-San Vicente and Plasencia 2011). Moreover, SA is both a direct and indirect regulator of plant development, influencing processes such as seed germination, growth, photosynthesis, thermogenesis, flowering, and senescence (Rivas-San Vicente and Plasencia 2011). But SA is best known for its role in orchestrating plant immune responses.

Received September 19, 2023. Accepted December 19, 2023. Advance access publication January 2, 2024

Plant immune systems are composed of several layers of sophisticated mechanisms. Conserved microbe-associated molecular patterns are detected by cell surface-localized pattern recognition receptors (PRRs) that activate pattern-triggered immunity (PTI). PTI induces diverse cell signaling events, including the production of reactive oxygen species (ROS) and biosynthesis of SA, as well as other defense hormones (Couto and Zipfel 2016). Collectively, these signaling events lead to induction of immune genes to confer resistance. However, adapted pathogens promote their virulence by suppressing PTI responses through secretion of effectors directly into the host. To negate this, plants have evolved intracellular nucleotide-binding domain leucine-rich repeat (NLR) immune receptors that detect the presence of pathogen effectors and launch effector-triggered immunity (ETI) (Jones et al. 2016). Immune pathways activated by pattern recognition receptors and NLRs display mutual potentiation. Consequently, simultaneous detection of the pathogen through both receptors leads to stronger immune responses that are often associated with programmed cell death (PCD) of the infected tissues, thereby isolating and killing the invading pathogen (Ngou et al. 2021, 2022; Yuan et al. 2021). Such responses can induce SA synthesis in both local and systemic tissues, where SA may have contrasting roles as a cell death agonist and as a cell survival signal to confer long-lasting protection throughout the plant against a wide variety of pathogens (Rate and Greenberg 2001; Torres et al. 2005; Lu et al. 2009; Fu et al. 2012; Zavaliev et al. 2020). This inducible immune mechanism, characterized by A. Frank Ross as systemic acquired resistance (SAR) in 1961 (Ross 1961), has since been extensively studied due to its potential as an immunizing strategy to protect crops against a broad spectrum of pathogens (White 1979; Ward et al. 1991; Friedrich et al. 1996; Gorlach et al. 1996; Lawton et al. 1996; Shimono et al. 2012; Fu and Dong 2013; Vlot et al. 2021). Surprisingly, although SA is synthesized locally upon infection and detected throughout the plant (Malamy et al. 1990; Metraux et al. 1990; Gaffney et al. 1993), grafting experiments demonstrated that de novo synthesis of SA in systemic tissues is required for the establishment of SAR (Gaffney et al. 1993; Vernooij et al. 1994).

In the past 30 years, significant progress has been made in the understanding of SA biosynthesis in both local and systemic tissues, SA perception by metabolic enzymes and receptor proteins, SA-induced biomolecular condensates in different plant tissues and subcellular compartments, and SA crosstalk with other hormones and metabolites. As a complementary update to previously published reviews (Ding and Ding 2020; Lefevere et al. 2020) and in celebration of ASPB's 100th birthday, we focus on more recent breakthroughs in our understanding of the functions of SA in plants.

Precise regulation of local and systemic SA biosynthesis

Although basal SA levels differ between organs and plant species, pathogen-induced biosynthesis of SA is an almost universal step toward the establishment of immunity. Plants have evolved 2 major pathways for SA biosynthesis catalyzed by PHENYLALANINE AMMONIA-LYASE (PAL) and ISOCHORISMATE SYNTHASE (ICS), respectively. Different species utilize these pathways to varying degrees. For example, whereas the PAL pathway is more prevalent in rice, the ICS pathway is predominantly used in *Arabidopsis* (Lefevere et al. 2020; Wu et al. 2023). Both pathways utilize chorismate produced in chloroplasts, which in case of the PAL pathway is exported to the cytoplasm, where it is converted to phenylalanine in a multistep process. PAL then converts phenylalanine to trans-cinnamic acid from which SA is generated through the sequential action of ABNORMAL INFLORESCENCE MERISTEM 1 (AIM1) and a yet-to-be-identified benzoic acid hydrolase. In rice, mutating *PAL4* and *PAL6* reduce SA accumulation and enhance susceptibility to various pathogens, indicating these 2 genes play a predominant role in pathogen-induced SA biosynthesis (Duan et al. 2014; Tonnessen et al. 2015). By contrast, *Arabidopsis* encodes for 2 ICS enzymes, with ICS1 being responsible for the accumulation of SA in leaf tissues in response to biotic or abiotic stress (Wildermuth et al. 2001; Garcion et al. 2008). Recent findings revealed that the ICS1-catalyzed conversion of chorismate to isochorismate is the only step of the SA biosynthesis pathway that is localized in the chloroplasts (Rekhter et al. 2019). Isochorismate is then transported from the chloroplast to the cytosol by the MATE transporter protein ENHANCED DISEASE SUSCEPTIBILITY 5 (EDS5), mutants of which show severely decreased SA accumulation upon pathogen infection (Nawrath and Métraux 1999; Nawrath et al. 2002). In the cytosol, isochorismate is conjugated to glutamate by the amidotransferase avrPphB SUSCEPTIBLE 3 (PBS3), where the resulting product, isochorismate-9-glutamate, is converted to SA either by the acyltransferase ENHANCED PSEUDOMONAS SUSCEPTIBILITY 1 or by spontaneous decomposition (Rekhter et al. 2019; Torrens-Spence et al. 2019).

Although these advances revealed most steps in the SA biosynthesis pathways of plants, relatively little is known about how the enzymes in these pathways are regulated. Reports show that PAL isozymes are ubiquitinated by a SKP1-CULLIN1-F-BOX (SCF)-type E3 ligase and targeted for proteasome-mediated degradation, suggesting post-translational regulation of enzyme activities could be a key regulatory step in SA biosynthesis (Zhang et al. 2013, 2015).

In contrast to the regulation of enzymatic activities, transcriptional regulation of SA synthesis genes has been extensively studied and demonstrated as a major control point in the biosynthesis of SA. The 2 most-studied regulators of SA biosynthesis genes are the related transcriptional activators SAR DEFICIENT 1 (SARD1) and CALMODULIN BINDING PROTEIN 60 g (CBP60g). Mutants of these transcription factors (TFs) fail to induce SA synthesis upon infection and are deficient in both local and systemic immune responses (Wang et al. 2009; Zhang et al. 2010). These TFs were found to bind to the *ICS1* promoter and also associate with the promoters of *EDS5* and *PBS3*, suggesting that they have a regulatory role in the entire SA biosynthesis pathway (Wang et al. 2011; Truman and Glazebrook 2012; Sun et al. 2015). SARD1 and CBP60g also have functions beyond SA biosynthesis because they were found to bind promoters of key signaling proteins downstream of SA, as well as promoters of positive and negative regulators of PTI, ETI, and SAR (Sun et al. 2015). Curiously, another related member of the CBP60 family, CBP60b, was shown to control the expression of *SARD1*, suggesting that CBP60 family members also fine-tune each other's expression levels (Huang et al. 2021b). The biological relevance of the CBP60 family is perhaps best illustrated by the fact that the vascular pathogen *Verticillium dahlia* secretes an effector that directly inhibits CBP60g transcriptional activity to promote its virulence in both *Arabidopsis* and cotton plants (Qin et al. 2018). In the plant pathogen arms race, however, it is thought that members of the CBP60 family have influenced each other's evolution to generate a robust immune regulatory module that is more resilient to perturbation by pathogen effectors (Zheng et al. 2022). Because transcription of both *SARD1* and *CBP60g* genes are induced upon pathogen infection, it raises the question: how are these activators induced upon pathogen challenge? One possible mechanism is activation of CBP60g by PTI-induced calcium transients, because CBP60g's calmodulin-binding domain is required for accumulation of SA and pathogen resistance (Wang et al. 2009). Moreover, calcium transients may also regulate the activities of CALMODULIN-BINDING TRANSCRIPTION ACTIVATOR (CAMTA) TFs that suppress SA biosynthesis via direct repres-sion of *CBP60g* and *SARD1* gene expression (O'Malley et al. 2016; Kim et al. 2020; Sun et al. 2020).

The answer to the question above may also lie with other TFs, including WRKYs, TEOSINTE BRANCHED1/CYCLOIDEA/PCF (TCP) family TFs, NO APICAL MERISTEM ARABIDOPSIS TRANSCRIPTION ACTIVATION FACTOR AND CUP-SHAPED COTYLEDON (NAC) family TFs, DP-E2F-LIKE1 (DEL1), ethylene-responsive ETHYLENE INSENSITIVE3 (EIN3), and ETHYLENE INSENSITIVE3-LIKE1 (EIL1) TFs, all of which have been shown to activate or repress the expression of SA biosynthesis genes (Chen et al. 2009; van Verk et al. 2011; Zheng et al. 2012, 2015; Wang et al. 2015). Of particular interest is NTM1-LIKE 9, a potentially membrane-bound TF that, upon activation, may be cleaved and released to induce *ICS1* gene expression specifically in guard cells, where SA biosynthesis is required for stomatal closure in response to pathogen attack (Zheng et al. 2015). Moreover, TCP21, also known as CHE (CCA1 HIKING EXPEDITION), is a circadian clock TF that regulates daily rhythms of basal SA biosynthesis by binding to the *ICS1* promoter (Zheng et al. 2015). Curiously, CHE is required for pathogen-induced SA biosynthesis only in systemic tissues during the establishment of SAR. CHE contains a conserved, redox-sensitive cysteine residue in its noncanonical basic helix-loop-helix DNA-binding domain (Viola et al. 2013). Upon local induction, H_2O_2 derived from NADPH oxidases, such as RESPIRATORY BURST OXIDASE HOMOLOG PROTEIN D (RBOHD), functions as a mobile signal to induce *S*-sulfenylation (-SOH group) of this cysteine residue in systemic tissues, which enhances CHE's binding affinity for the *ICS1* promoter and stimulates SA synthesis (Cao et al. 2023). CHE is also modified in local tissues, but here, higher H_2O_2 levels lead to further oxidation of the cysteine residue to the *S*-sulfinylated state (-SO_2H group) and possibly even to the *S*-sulfonated state (-SO_3H group), neither of which appears to induce CHE's binding affinity to the promoter of *ICS1*. Thus, a gradient of mobile H_2O_2 combined with concentration-dependent sensing by CHE regulates spatial SA biosynthesis in the establishment of SAR. It is tempting to hypothesize that plants modulate whether and how strongly to turn on SAR based on the severity of the local infection as reflected by the H_2O_2 levels. Although the discovery of the H_2O_2-CHE signaling pathway addressed the 30-year-old question of how a local infection leads to de novo SA synthesis in systemic tissues during SAR, the mechanism by which pathogens and other environmental stresses initiate local SA synthesis remains to be fully elucidated.

SA is perceived directly by multiple target proteins to influence distinct cellular processes

How is SA, a simple phenolic molecule, perceived in the cell to establish immunity? The profound impacts of SA on plant and human physiology suggest that it is likely to have many target proteins. Indeed, screens utilizing SA analogs in combination with protein arrays or crosslinking revealed nearly 100 SA-binding proteins (SABPs) (Tian et al. 2012; Moreau et al. 2013; Manohar et al. 2015). Several of these have already been confirmed to play a role in SA signaling and plant de-fense, supporting the notion that there is unlikely a single re-ceptor for SA or a single paradigm for SA perception, but rather many ways by which SA directly alters cellular responses through binding to metabolic enzymes, redox regulators, and transcription cofactors.

Metabolic enzymes that bind SA include the β-carbonic anhydrase SABP3, the acyl acid amido synthetase GH3, the amidotransferase PBS3 described above, glyceraldehyde 3-phosphate dehydrogenases, alpha-ketoglutarate dehydrogenase, and thimet oligopeptidases (TOP1 and TOP2) (Pokotylo et al. 2019). SA binding to these enzymes often influences their activities, suggesting SA can reprogram cellular metabolism. However, the biological consequences of such interactions have yet to be demonstrated. Therefore, these SABPs may be important gateways toward understanding the role of SA in metabolic reprogramming that favors im-mune responses over other cellular activities.

SA also targets redox regulators. Treatment with SA or SA analogs has been shown to trigger rapid cellular oxidation followed by reduction as measured by changes in reduced vs oxidized glutathione levels (Mou et al. 2003; Spoel and Loake 2011). Consistent with this observation, the earliest identified cellular target of SA was the ROS-scavenging enzyme CATALASE 2 (CAT2). Binding of SA reduces the H_2O_2-detoxifying activity of CAT2, allowing H_2O_2 to accumulate upon pathogen infection and function as a secondary messenger in SAR (Chen et al. 1993; Conrath et al. 1995; Wang et al. 2014; Cao et al. 2023). Moreover, it was shown that SA-mediated suppression of CAT2 activity leads to inhibitory *S*-sulfenylation of TRYPTOPHAN SYNTHETASE B SUBUNIT 1 (TSB1) involved in auxin biosynthesis, and additionally, CAT2 inhibition may also limit jasmonic acid (JA) biosynthesis (Yuan et al. 2017). Thus, redox regulation through direct SA binding is likely to coordinate the plant hormone signaling network. Besides CAT2, SA also inhibits the enzymatic activities of GST enzymes in vitro (Tian et al. 2012). Although the biological relevance of this remains unknown, GST enzymes are involved in the production of antimicrobial compounds, cellular detoxification, and hormone transport (Gullner et al. 2018), suggesting the possible influence of SA on these processes. Further, SA binds to the chloroplast-localized THIOREDOXIN (TRX) *m*1 (Manohar et al. 2015). Although the effect of SA on this enzyme remains unknown, the TRX family of oxidoreductases play key roles in immunity by fine-tuning cellular redox homeostasis, enabling oxidative signaling, and protecting proteins from the dam-aging effects of hyper-oxidation (Mata-Pérez and Spoel 2019; Bleau and Spoel 2021).

Besides enzymes, SA binds to members of the NONEXPRESSOR OF *PR* GENE (NPR) family of transcriptional cofactors (Fu et al. 2012; Manohar et al. 2015; Ding et al. 2018; Wang et al. 2020; Kumar et al. 2022). NPR1 is arguably the most important cellular target/receptor of SA, as illustrated by its repeated identification in genetic screens as an essential component for SA-mediated gene expression and resistance (Cao et al. 1994, 1997; Delaney et al. 1995; Glazebrook et al. 1996; Ryals et al. 1997; Shah et al. 1997). NPR1 acts as a major coactivator that establishes a transcriptional activation complex consisting of NPR1, TGA TFs, and histone acetyltransferases (Fan and Dong 2002; Jin et al. 2018). NPR1 associates with numerous gene promoters to reprogram the expression of thousands of genes (Wang et al. 2005, 2006; Skelly et al. 2019; Nomoto et al. 2021). Although nuclear translocation of NPR1 is induced through SA-triggered redox changes (Mou et al. 2003; Tada et al. 2008), whether NPR1's transcriptional coactivator activity requires direct association with SA has been hotly debated (Fu et al. 2012; Manohar et al. 2015; Ding et al. 2018; Wang et al. 2020; Kumar et al. 2022), because NPR1 exhibited much lower SA-binding activity in side-by-side comparisons with its paralogs, NPR3 and NPR4 (Fu et al. 2012; Wang et al. 2020). The crystal structure of the NPR4 SA-binding core (SBC) revealed that SA is located in an enclosed hydrophobic pocket made of 4 α-helices, indicating that SA may induce a conformational change to accommodate its binding to NPR4 (Wang et al. 2020). Although NPR1 has nearly identical hormone-binding residues as NPR4, nonconserved residues in the SBC appear to be responsible for its low SA-binding activity (Wang et al. 2020). Recent cryo-EM and crystal structure analyses demonstrate that the active form of full-length NPR1 is a bird-shaped homodimer with interacting N-terminal BROAD-COMPLEX, TRAMTRACK AND BRIC-À-BRAC (BTB) domains forming the "body" of the bird and each extending "wing" consisting of a Kelch helix bundle, 4 ankyrin repeats (ANK) and a disordered C-terminal domain (Kumar et al. 2022). The presence of SA induces folding of the C-terminal SA-binding domain (SBD) in vitro and promotes its docking onto the ANK repeat domain. Crosslinking of this docked conformation by artificially engineering a disulfide bond at the interface resulted in enhanced activation of the NPR1 target gene, *PR1*. This finding provides the first structural evidence for a direct role of SA in inducing conformational changes in NPR1 to promote its transcriptional activity. However, it remains to be elucidated why docking of the SA-bound SBD onto ANKs helps enhance TGA activity because there is no direct contact between the ANK-docked SBD and TGA (Kumar et al. 2022). A possible explanation is that a post-translational modification or an unknown chaperone might be required for enhancing in vivo SA binding to NPR1 and bridging the docked SBD with TGA, thus bringing the transcriptional machinery to the TFs.

In contrast to NPR1, NPR3 and NPR4 proteins are negative regulators of SAR and exhibit significantly higher SA binding activities in vitro (Zhang et al. 2006; Fu et al. 2012). SA binding to NPR3 and NPR4 controls their interactions with NPR1 in opposite ways: whereas SA disrupts the NPR4-NPR1 complex, it promotes interaction between NPR3 and NPR1. The SA dependency of these interactions suggests that the C-terminal half of the protein, containing the ANK-repeat domain and the SBD domain, is likely involved (Fu et al. 2012; Wu et al. 2012; Wang et al. 2020). What are the functional consequences of these unusual SA-dependent interactions between NPR3/4 and NPR1 proteins? The domain structures and organization of NPR proteins are typical for substrate adaptors utilized by CULLIN 3-RING LIGASE (CRL3). These substrate adaptors typically contain a BTB domain that interacts with CRL3, whereas an additional protein-protein interaction domain (e.g. the ANK-repeat domain in NPR proteins) recruits the substrate for (poly)ubiquitination, which in many cases leads to substrate degradation by the proteasome (Petroski and Deshaies 2005). Indeed, in the nucleus, NPR3 and NPR4 were shown to function as adaptors for CRL3 to target NPR1 and JASMONATE ZIM DOMAIN (JAZ) corepressors of the JA signaling pathway for degradation in an SA-regulated manner (Fu et al. 2012; Liu et al. 2016). Because SA levels taper off with distance from the site of infection, NPR3/NPR4 provide a way to spatially regulate the level of NPR1: at the site of infection, NPR1 is ubiquitinated and degraded to remove its inhibitory effect on cell death, whereas in adjacent and more distal tissues, NPR1 accumulates to promote cell survival and SAR (Fig. 1) (Fu et al. 2012; Liu et al. 2016). This is a unique mode of action in which one CRL3 substrate adaptor targets another for ubiquitination and degradation in a ligand concentration-dependent fashion to ensure appropriate regulation of cell fate upon pathogen challenge.

Alternative to their CRL3 substrate adaptor function, NPR3 and NPR4 have also been proposed to function as transcriptional corepressors. The C termini of NPR3 and NPR4 contain an ETHYLENE-RESPONSIVE ELEMENT BINDING FACTOR-ASSOCIATED AMPHIPATHIC REPRESSION (EAR) motif widely found in transcriptional (co)repressors. Sequence alignment showed that the EAR motif is uniquely absent in NPR1 proteins of *Brassicaceae* but not NPR1s of other plant lineages (Wang et al. 2020). Moreover, SA binding to NPR4 does not appear to affect the EAR motif conformation or the interaction of NPR3 and NPR4 with TGA TFs (Ding et al. 2018; Wang et al. 2020). Therefore, how SA alleviates the proposed NPR4 transcription corepressor activity remains to be investigated.

SA-mediated post-translational regulation of NPR1 activities

The multifaceted functions of SA are reflected not only in its various cellular targets, but also in the complex post-translational modifications (PTMs) of its downstream signaling component NPR1 (Fig. 2). Although NPR1 is localized to both the cytoplasm and the nucleus, its conformation is different in these cellular compartments. In the cytoplasm, the majority of NPR1 are linked through intermolecular disulfide bonds to form a homo-oligomer (Mou et al. 2003). Mutational and subsequent structural analyses revealed that at least 2 cysteines are involved in generating this oligomer: Cys82 faces its counterpart at the dimer interface, whereas 2 surface-exposed Cys156 residues of the dimer are juxtaposed at the interface of the tetramer (Kumar et al. 2022). Importantly, the solvent exposed Cys156 can be S-nitrosylated (i.e. covalent attachment of nitric oxide), which facilitates further oxidation to a disulfide bond (Tada et al. 2008). SA-induced cellular reduction along with the action of the oxidoreductases TRX*h*3 and TRX*h*5 reduce the NPR1 oligomer, promoting its nuclear translocation (Fig. 2).

Nuclear translocation of NPR1 also requires its phosphorylation at Ser589 and possibly Thr373 by the kinase SnRK2.8 (SNF1-Related Protein Kinase 2.8) and dephosphorylation at Ser55/59 by an unknown phosphatase (Lee et al. 2015; Saleh et al. 2015). In the nucleus, NPR1 undergoes a series of additional PTMs that alter its coactivator behavior (Fig. 2). First, dephosphorylation of Ser55/59 promotes its interaction with the ubiquitin-like modifier, SUMO3, via a SUMO-interaction motif. The resulting SUMOylated NPR1 preferentially associates with TGA3 TF to be recruited to the chromatin (Saleh et al. 2015). Moreover, SUMOylation of NPR1 is a prerequisite for subsequent phosphorylation of a Ser11/15-containing phosphodegron (Saleh et al. 2015), which recruits a CRL3 ligase to ubiquitinate NPR1 (Spoel et al. 2009). Intriguingly, ubiquitination and subsequent degradation of NPR1 were found to enhance its transcription cofactor activity instead of

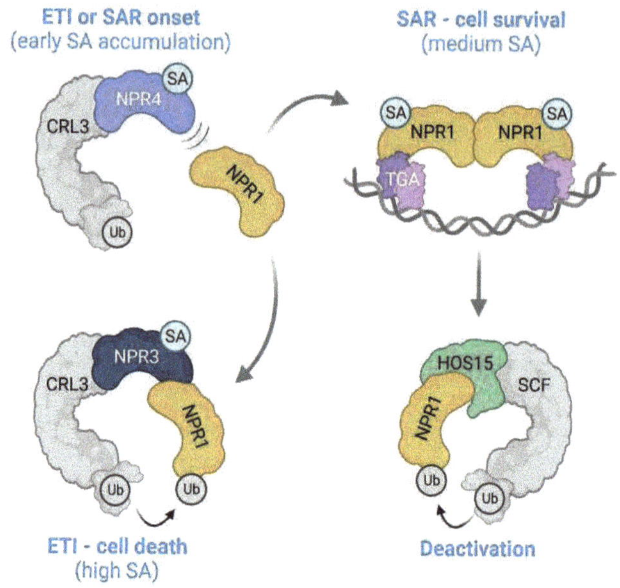

Figure 1. NPR1 is regulated by multiple E3 ubiquitin ligases. Under steady-state conditions, NPR1 is targeted for degradation by a $CRL3^{NPR4}$ ubiquitin ligase to avoid untimely activation of immunity. At the early onset of immunity, SA begins to accumulate and binds to NPR4, which prevents this protein from interacting with NPR1 (top left). This allows NPR1 to activate gene expression and promote cell survival during SAR (top right). Activation of ETI leads to much higher levels of SA, which promote recruitment of NPR1 to a $CRL3^{NPR3}$ ubiquitin ligase that targets NPR1 for degradation, thereby permitting cell death to occur (bottom left). Alternatively, the transcriptionally competent state of NPR1 in SAR can be deactivated by the SCF^{HOS15} ubiquitin ligase (bottom right). Created with BioRender.com.

inhibiting it (Spoel et al. 2009). This conundrum is resolved by the discovery of an intricate ubiquitination relay on NPR1 (Fig. 2). First, CRL3 decorates NPR1 with multi-monoubiquitin or short ubiquitin chains that promote its association with target promoters and enhance its coactivator activity without triggering degradation (Skelly et al. 2019). Therefore, ubiquitination and/or SUMOylation of NPR1 are both possible intramolecular chaperones that may induce conformational changes in NPR1 to facilitate SA binding to the SBD pocket. To test this hypothesis, the site(s) of these PTMs in NPR1 need to be identified. Moreover, after serving its transcriptional coactivator function, NPR1 is further ubiquitinated by the E4 ligase UBE4/MUSE3, which deactivates NPR1 and renders it a substrate for the proteasome (Skelly et al. 2019). Upon arrival at the proteasome, the proteasome-associated HECT-type ligases UPL3 and UPL4 further ubiquitinate NPR1, which promotes proteasome processivity by preventing stalling during its degradation (Wang et al. 2022; Wang and Spoel 2022).

Importantly, cells have several "go/no-go" decision points along the ubiquitin ligase relay that decide the fate of NPR1. First, an SCF ligase containing the F-box protein HIGH EXPRESSION OF OSMOTICALLY RESPONSIVE GENES

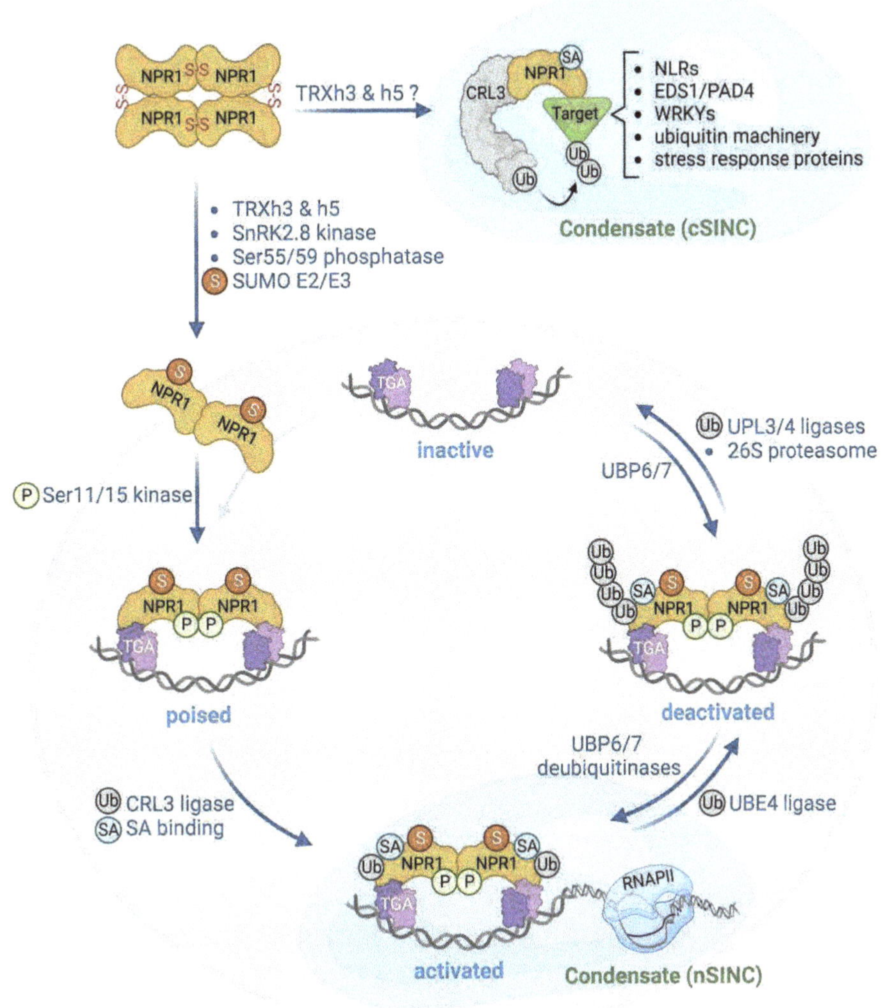

Figure 2. Diverse PTMs dynamically regulate NPR1 localization and activities. In resting cells, NPR1 resides in the cytoplasm as a disulfide-linked (S-S) oligomer. Activation of immunity leads to the TRX*h*3- and TRX*h*5-mediated reduction of NPR1 oligomers and nuclear translocation of NPR1 promoted by SnRK2.8-mediated phosphorylation. Moreover, dephosphorylation of NPR1 at Ser55/59 and its SUMOylation both stimulate localization of NPR1 to the nuclear condensate (nSINC). SUMOylation also promotes NPR1's association with TGA transcription factors and is a prerequisite for phosphorylation (P) at Ser11/15, which stimulates NPR1's transcriptional activity by recruiting a CRL3 ligase that (mono)ubiquitinates NPR1. Sumo (S) and/or ubiquitin (Ub) may act as a molecular chaperone for SA binding, leading to a transcriptionally active NPR1-TGA complex. Eventually, ubiquitin chain elongation by an UBE4 ligase inactivates NPR1 and targets it to the proteasome. At the proteasome UPL3/4 ligases further decorate NPR1 with ubiquitin, which prevents its stalling during degradation and promotes proteasome processivity. NPR1 can be rescued from degradation and returned to its transcriptionally active state by the activities of proteasome-associated UBP6/7 deubiquitinases. In addition to its nuclear function, high SA levels, as found in tissues surrounding ETI-induced cell death, lead NPR1 to form cytoplasmic condensates (cSINCs), where it serves as a $CRL3^{NPR1}$ ligase that targets various cell death-inducing immune regulators for degradation to promote cell survival. Created with BioRender.com.

15 (HOS15) may counteract CRL3-mediated activation of NPR1 by preferentially associating with the Ser11/15 phosphorylated isoform and targeting it directly to the proteasome without activating gene expression (Fig. 2) (Shen et al. 2020). Second, the proteasome-associated deubiquitinases UBP6 and UBP7 reverse ubiquitination by trimming NPR1 ubiquitin chains en bloc, thereby returning NPR1 back to its transcriptionally active state (Fig. 2) (Skelly et al. 2019). Taken together, these findings illustrate how SA utilizes a variety of PTMs to dynamically regulate NPR1 coactivator activity (Fig. 2). Nonetheless, further work is required to understand how SA controls activities of the enzymes that "write" these PTMs.

SA signaling via biomolecular condensates

The large number of cellular targets and profound physiological impacts of SA on plant cells call for centralized organization and dynamic regulatory mechanisms. Such mechanisms have presented themselves to us under the microscope as

SA-induced NPR1-GFP nuclear condensates (nSINCs) (Saleh et al. 2015). However, interest in the potential functional importance of these biomolecular condensates was only raised after the serendipitous discovery that higher concentrations of SA also trigger the formation of SA-induced NPR1 condensates in the cytoplasm (cSINCs) (Zavaliev et al. 2020). These cSINCs, which are unique to NPR1 among the NPR paralogs, contain components of the ubiquitin machinery and numerous stress-responsive proteins, including 10 NLR intracellular immune receptors and their downstream signaling components, ENHANCED DISEASE SUSCEPTIBILITY 1 (EDS1) and PHYTOALEXIN-DEFICIENT 4 (PAD4), required for ETI-associated PCD, as well as redox enzymes and DNA damage response proteins. Transition of NPR1 into cSINCs is required for the formation of a $CRL3^{NPR1}$ ligase complex that targets at least some of these cSINC-containing proteins for ubiquitin-mediated degradation (Fig. 2). cSINCs form only in tissues adjacent to ETI-induced cell death zones where SA concentrations are high. Thus, it is hypothesized that the cSINC-localized $CRL3^{NPR1}$ ligase plays an essential role in promoting cell survival in tissues neighboring ETI-induced cell death by sequestering/degrading the PCD-promoting immune receptors and signaling components.

SA was also found to induce guanylate-binding protein-like GTPases (GBPLs) that assemble condensates in the nucleus. Upon immune activation, catalytically active GBPL3 was found to translocate into the nucleus, where it localizes to the nuclear pore complex and initiates the formation of GBPL defense-activated condensates (GDACs) (Huang et al. 2021a; Tang et al. 2022). GDACs sequester major immune gene promoters, including those of *ICS1*, *EDS5*, *CBP60g*, *SARD1*, and *NPR1*, to a local environment enriched in transcriptional coactivators of the Mediator complex and the RNA polymerase II machinery (Huang et al. 2021a; Kim et al. 2022). Because GBPL3 does not contain any transcriptional activation domains, it appears to act as an SA-induced nucleator to recruit the transcription machinery for reprogramming the transcriptome in response to patho-gen threat. It is plausible that the SA-induced NPR1 conden-sates observed in the nucleus serve a similar hub function as GBPL3 condensates, but instead of regulating SA biosynthesis genes, nSINCs may target downstream SA-responsive genes. Moreover, a recent report demonstrated that upon infection by virulent and avirulent pathogens, MOS4-associated complex (MAC) components form nuclear condensates proposed to activate immunity by sequestering negative regulators of plant defense (Jia et al. 2023). Whether SA plays a role in this process requires further investigation.

In addition to the regulation of defense transcription and protein homeostasis, SA and other phenolic acids have been shown to induce the formation of stress granules in root cells through direct interaction with RNA-BINDING PROTEIN 47B (RBP47B) to inhibit global translation, a mechanism used by plants to suppress growth of their neighboring plants (Xie et al. 2023). Collectively, these findings indicate that SA can orchestrate distinct immune responses by organizing corresponding cellular machineries in close quarters to increase reaction efficiencies and kinetics, as reported for other biological processes in eukaryotic organisms (Banani et al. 2017). The ability of SA-induced NPR1 to form both nu-clear and cytoplasmic condensates with distinct components and biological functions provides a promising new research direction to explore the conditions and regulatory mechanisms of biomolecular condensates. So far, mutating the *redox-sensitive intrinsically disordered region* 3 (*rdr3*) in NPR1 abolished both nSINC and cSINC formation as well as SA-mediated gene expression and resistance (Zavaliev et al. 2020).

Synergistic and antagonistic crosstalk between SA and other cellular signals

It is well-known that the establishment of SAR requires de novo SA synthesis in systemic tissue (Gaffney et al. 1993; Vernooij et al. 1994). Recent discovery of H_2O_2 as a mobile signal for activation of CHE in systemic tissue to induce SA synthesis (Cao et al. 2023) raises questions about the individual functions of this signal and previously reported systemic signals, including N-hydroxy-pipecolic acid (NHP) (Hartmann et al. 2018), azelaic acid (Jung et al. 2009), glycerol-3-phosphate (Chanda et al. 2011), nitric oxide (Wang et al. 2014), dehydroabietinal (Chaturvedi et al. 2012), monoterpenes (Riedlmeier et al. 2017; Wenig et al. 2019), trans-acting small interfering RNAs (tasi-RNAs) (Shine et al. 2022), and extracellular nicotinamide adenine dinucleotide (phosphate) [eNAD(P)] (Wang et al. 2019; Li et al. 2023) and their possible relationships with SA in inducing SAR. So far, there is strong evidence for synergistic interactions between SA and NHP. Pathogen-induced SA accumulation in local tissues first triggers NHP synthesis, which then feedback amplifies SA synthesis in systemic tissues. Moreover, both signals are inactivated by the same glycosyltransferase (Bauer et al. 2021; Cai et al. 2021; Holmes et al. 2021; Mohnike et al. 2021). Recent work demonstrates that NPR1 is required for NHP-induced SAR and associated transcriptional reprogramming (Yildiz et al. 2021). Similarly, genetic analyses suggest that NPR1-interacting TGA TFs are also required for NHP-induced SAR and SAR-related gene expression (Yildiz et al. 2023). Thus, NHP appears to utilize the SA-responsive NPR1/TGA regulatory module to establish SAR. The striking structural similarity between SA and NHP suggests that NPR1 might be a receptor not only for SA but also for NHP. However, isothermal titration calorimetry assays indicated that NHP was unable to bind to recombinant NPR1 derived from insect cells (Nair et al. 2021). Whether the SA and NHP signals converge on NPR1 or further downstream on NPR1-dependent target genes involving NHP-responsive transcriptional regulators remains to be discovered. Moreover, it has yet to be investigated if, like SA, NHP utilizes similar post-translational strategies to regulate NPR1 coactivator activity.

Reciprocal antagonism between the SA and JA signals has been widely reported and plays an important role in shaping

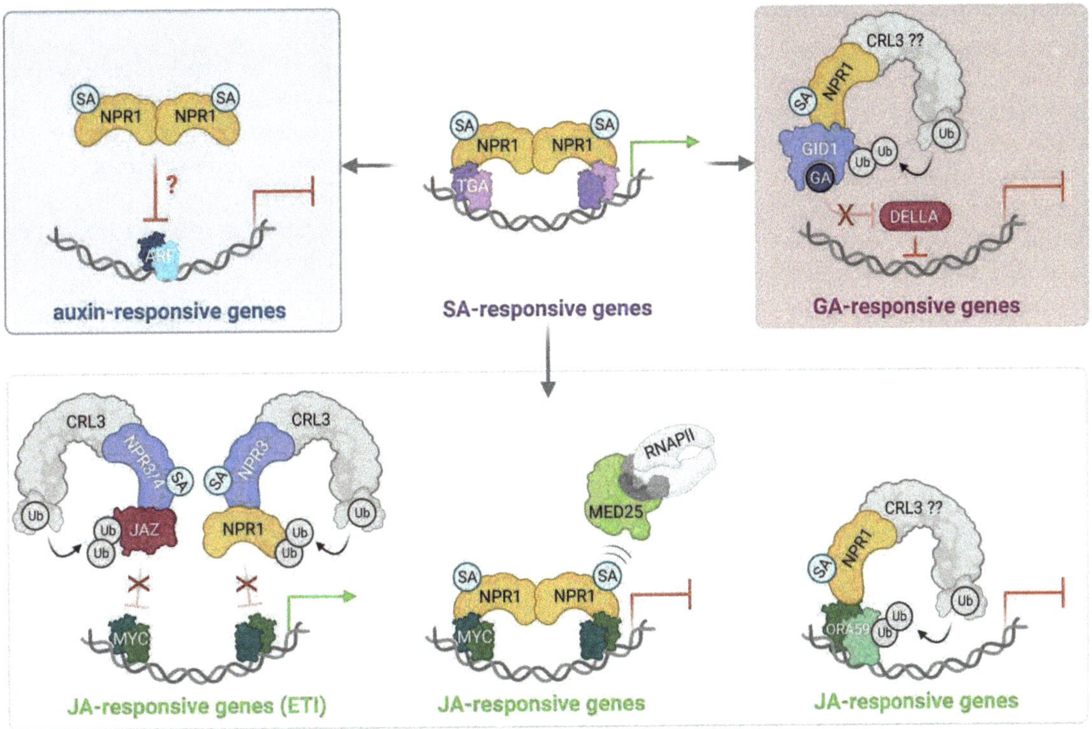

Figure 3. NPR proteins mediate crosstalk between SA and other hormones. While NPR1 is essential for activation of SA-responsive genes during SAR (top center), it can also function as a potent inhibitor of JA-, GA-, and possibly auxin-responsive gene expression. NPR1 inhibits JA-responsive gene expression either by degrading ORA59 activators as part of a CRL3^{NPR1} ligase (bottom right) or by blocking MYC activators' access to Mediator components (MED25) and associated RNA Polymerase II (RNAPII) complex (bottom center). In contrast to NPR1, both NPR3 and NPR4 activate JA-responsive genes during ETI by serving as a CRL3$^{NPR3/4}$ ligase to degrade the JA repressors JAZ and NPR1 (bottom left). Auxin-responsive genes are also inhibited by SA, but whether this process is dependent on NPR1 remains unknown (top left). Lastly, a probable CRL3^{NPR1} ligase inhibits GA signaling by targeting the GA receptor GID1 for degradation, which blocks the removal of DELLA suppressors from GA-responsive genes (top right). Created with BioRender.com.

the outcome of plant-pathogen interactions (Pieterse et al. 2009). For example, strains of the biotrophic leaf pathogen *Pseudomonas syringae* promote their virulence by utilizing the JA-mimicking toxin coronatine that functions as a suppressor of SA-mediated immune responses against this pathogen (Brooks et al. 2005; Zheng et al. 2012; Geng et al. 2014). Some plant hosts have turned the tables by developing strategies to block the virulence-promoting activity of coronatine. For example, *Arabidopsis* employs SA signaling itself to antagonize coronatine- or JA-induced responses, with SA-induced redox changes and glutathione biosynthesis playing a key role in the suppression of JA signaling (Koornneef et al. 2008; Pieterse et al. 2009). While SA suppresses the expression of several JA biosynthesis genes, SA also antagonizes signaling downstream of JA biosynthesis, which is dependent on NPR1 (Spoel et al. 2003; Leon-Reyes et al. 2010). SA-induced NPR1 likely utilizes different me-chanisms to suppress JA-responsive genes. First, SA and NPR1 induce the expression of GRX480, a member of the glu-taredoxin family that interacts with TGA TFs. This GRX480/ TGA complex may associate with a subset of JA-responsive genes and suppress their expression (Ndamukong et al. 2007). Second, SA suppresses genes that contain the JA-responsive GCC-box motif, which is bound by members of the APETALA2/ETHYLENE RESPONSE FACTOR (AP2/ERF) TF family. SA was reported to inhibit JA-responsive accumulation of the AP2/ERF transcriptional activator ORA59, indicating that SA antagonizes JA signaling downstream of the nuclear JA receptor complex, SCF^{COI1}-JAZ (Van der Does et al. 2013). It is therefore plausible that an SA-induced CRL3$^{NPR1/3/4}$ ubiquitin ligase targets ORA59 for proteasome-mediated degradation (Fig. 3). Lastly, SA-induced NPR1 was recently shown to be recruited to G-box motifs highly overrepresented in JA-responsive promoters, where it associates with MYC transcriptional activators (Nomoto et al. 2021). Like JAZ corepressors, NPR1 interacts with the same N-terminal region of MYC activators and competitively prevents recruitment of the Mediator complex, thereby blocking JA-mediated activation of gene expression. Thus, NPR1 is a versatile SA-responsive transcriptional cofactor that can be deployed either as a coactivator or as a corepressor depending on transcriptional context (Fig. 3).

The outcome of crosstalk between the SA and JA signals is modulated by a number of different factors, including their concentrations, spatial distributions, temporal effects, and even the presence of other hormones. For example, the presence of ethylene renders antagonisms between SA and JA signaling independent of NPR1 (Leon-Reyes et al. 2009). Because ethylene is produced during specific plant-pathogen interactions and abiotic stress responses, plant cells may use ethylene to direct NPR1 activity to specific immune pathways. Additionally, spatial regulation of crosstalk ensures that SA and JA are only antagonistic at the site of infection and not in distal systemic tissues (Spoel et al. 2007). This ensures plants can defend themselves against simultaneous attacks by multiple pathogens with different lifestyles. Spatial regulation may be conferred by gradients of hormone concentrations, which determine if SA and JA interactions are synergistic, antagonistic, or absent (Mur et al. 2006). On the other hand, temporal regulation of basal SA and JA levels by the circadian clock to peak at dawn and dusk, respectively, also indicates that plants have evolved mechanisms to avoid the antagonistic effects of these 2 defense hormones (Goodspeed et al. 2013). In fact, during ETI, high concentrations of SA and JA accumulate without apparent antagonism, because in this scenario JA synthesis/responses are not initiated through the canonical SCF^{COI1} ubiquitin ligase, but rather through the SA receptors NPR3 and NPR4 (Liu et al. 2016). These SA receptors were found to target JAZ corepressors for proteasome-mediated degradation, presumably through their ability to form $CRL3^{NPR3/4}$ ubiquitin ligases, which was required for immune receptor-mediated ETI (Fig. 3) (Fu et al. 2012). These findings suggest that SA not only induces ETI and associated local defense responses but also boosts JA responses to prevent vulnerability to necro-trophic pathogens and/or insects.

Hormones other than JA are also antagonized by SA. Many pathogens synthesize auxin-like molecules or alter the host's auxin homeostasis to enhance their virulence (Spoel and Dong 2008). It has been shown that SA strongly inhibits auxin signaling by preventing the degradation of auxin-related transcriptional corepressors, thereby limiting the activation of auxin-responsive gene expression (Wang et al. 2007). The exact molecular mechanisms by which SA inhibits auxin signaling are unknown, but it is plausible that NPR1 mediates this antagonism. Regardless, this crosstalk plays an important role in suppressing pathogen virulence. In addition, NPR1 was also found to suppress gibberellin (GA) signaling by interacting with the GA receptor GA INSENSITIVE DWARF 1 (GID1), possibly as part of a $CRL3^{NPR1}$ ubiquitin ligase that targets GID1 for proteasome-mediated degradation (Yu et al. 2022). NPR1-mediated degradation of GID1 enhances the stability of downstream DELLA transcriptional corepres-sors, thereby suppressing GA-responsive gene expression and associated plant growth responses (Fig. 3).

Collectively, these reports on crosstalk between SA and other hormones begin to paint a picture in which the NPR family of SA receptors play a central role. Their dual function as transcriptional cofactors and as substrate adaptors for CRL3 ubiquitin ligases alter the activities or stabilities of other transcriptional activators and corepressors, thereby extending their reach far beyond the regulation of only SA responses into the realms of other plant hormones (Fig. 3).

Future outlook

As described in this review, the profound impacts of SA on plant and animal physiology, as revealed through years of studies, match its functional complexity and sophisticated regulatory mechanisms. However, there is still a lot more to learn about this small molecule that is full of wonders. Some major questions that need to be addressed include the following: (1) How is SA synthesis initiated in local tissue upon different pathogen and abiotic challenges? It is equally plausible that a common mechanism is used in response to all stimuli or distinct mechanisms are triggered by different stimuli; (2) How does SA exert its distinct functions spatially and temporally? Methods with higher resolutions, such as single-cell sequencing (Nobori et al. 2023; Zhu et al. 2023) and high-affinity biosensors (Chen et al. 2019; Yang et al. 2020), are now becoming available and may soon provide detailed answers; (3) What is the PTM and the associated enzyme or chaperone that helps SA bind to NPR1 to activate the defense transcriptome? (4) How does SA regulate the activities of enzymes that decorate NPR1 and possibly other NPR receptors with PTMs? (5) How does SA control the formation of different biomolecular condensates with distinct biological functions in various plant tissues and subcellular compartments? (6) How is SA perceived in plant species such as rice where the role of NPR proteins is less pronounced? Instead, rice utilizes the transcriptional activator WRKY45 to activate immune gene expression (Shimono et al. 2007; Nakayama et al. 2013). Similar to *Arabidopsis* NPR1, the transcriptional activity of rice WRKY45 is also regulated by phosphorylation and the nuclear ubiquitin-mediated proteasome (Matsushita et al. 2013; Ueno et al. 2015, 2017; Adams and Spoel 2018). Future genetic and biochemical screens may reveal how rice and other species perceive pathogen-induced SA accumulation and transduce signals. By addressing all these questions, we will be able to better understand the multifaceted functions of SA to improve plant and human lives.

Acknowledgments

We thank Lijun Cao, Sargis Karapetyan, Beatriz Orosa-Puente, Soo Park, Jordan Powers, Mindy Sponsel, Zhishuo Wang, Xing Zhang, and Pei Zhou for discussions and suggestions for the review.

Author contributions

S.H.S. and X.D. conceptualized the manuscript outline. S.H.S. wrote the original draft and prepared the figures, and S.H.S. and X.D. reviewed and edited the manuscript.

Funding

This work was supported by the European Research Council (ERC) under the European Union's Horizon 2020 research and innovation program, grant agreement no. 101001137 (to S.H.S.), Biotechnology and Biological Sciences Research Council (BBSRC) grant BB/S016767/1 (to S.H.S.), National Institutes of Health grant NIH 1R35GM118036 (to X.D.), National Science Foundation grants NSF IOS-1645589 and IOS-2041378 (to X.D.); and funding from the Howard Hughes Medical Institute (to X.D.).

Conflict of interest statement: X.D. is a founder of Upstream Biotechnology Inc. and a member of its scientific advisory board, as well as a scientific advisory board member of Inari Agriculture Inc. and Aferna Bio.

Data availability

No new data were generated or analysed in support of this work.

References

Adams EHG, Spoel SH. The ubiquitin-proteasome system as a regulator of plant immunity. J Exp Bot. 2018:**69**(19):4529–4537. https://doi.org/10.1093/jxb/ery216

Banani SF, Lee HO, Hyman AA, Rosen MK. Biomolecular condensates: organizers of cellular biochemistry. Nat Rev Mol Cell Biol. 2017:**18**(5): 285–298. https://doi.org/10.1038/nrm.2017.7

Bauer S, Mekonnen DW, Hartmann M, Yildiz I, Janowski R, Lange B, Geist B, Zeier J, Schaffner AR. UGT76B1, a promiscuous hub of small molecule-based immune signaling, glucosylates N-hydroxypipecolic acid, and balances plant immunity. Plant Cell. 2021:**33**(3):714–734. https://doi.org/10.1093/plcell/koaa044

Bleau JR, Spoel SH. Selective redox signaling shapes plant-pathogen interactions. Plant Physiol. 2021:**186**(1):53–65. https://doi.org/10.1093/plphys/kiaa088

Brooks DM, Bender CL, Kunkel BN. The Pseudomonas syringae phytotoxin coronatine promotes virulence by overcoming salicylic acid-dependent defences in Arabidopsis thaliana. Mol Plant Pathol. 2005:**6**(6):629–639. https://doi.org/10.1111/j.1364-3703.2005.00311.x

Cai J, Jozwiak A, Holoidovsky L, Meijler MM, Meir S, Rogachev I, Aharoni A. Glycosylation of N-hydroxy-pipecolic acid equilibrates between systemic acquired resistance response and plant growth. Mol Plant. 2021:**14**(3):440–455. https://doi.org/10.1016/j.molp.2020.12.018

Cao H, Bowling SA, Gordon AS, Dong X. Characterization of an Arabidopsis mutant that is nonresponsive to inducers of systemic acquired-resistance. Plant Cell. 1994:**6**(11):1583–1592. https://doi.org/10.2307/3869945

Cao H, Glazebrook J, Clarke JD, Volko S, Dong X. The Arabidopsis NPR1 gene that controls systemic acquired resistance encodes a novel protein containing ankyrin repeats. Cell. 1997:**88**(1):57–63. https://doi.org/10.1016/S0092-8674(00)81858-9

Cao L, Yoo H, Chen T, Mwimba M, Zhang X, Dong X. H_2O_2 sulfenylates CHE linking local infection to establishment of systemic acquired resistance. bioRxiv. 2023:07.27.550865. Preprint. https://doi.org/10.1101/2023.1107.1127.550865

Chanda B, Xia Y, Mandal MK, Yu K, Sekine KT, Gao QM, Selote D, Hu Y, Stromberg A, Navarre D, et al. Glycerol-3-phosphate is a critical mobile inducer of systemic immunity in plants. Nat Genet. 2011:**43**(5):421–427. https://doi.org/10.1038/ng.798

Chaturvedi R, Venables B, Petros RA, Nalam V, Li M, Wang X, Takemoto LJ, Shah J. An abietane diterpenoid is a potent activator of systemic acquired resistance. Plant J. 2012:**71**(1):161–172. https://doi.org/10.1111/j.1365-313X.2012.04981.x

Chen C, Feng S, Zhou M, Ji C, Que L, Wang W. Development of a structure-switching aptamer-based nanosensor for salicylic acid detection. Biosens Bioelectron. 2019:**140**:111342. https://doi.org/10.1016/j.bios.2019.111342

Chen H, Xue L, Chintamanani S, Germain H, Lin H, Cui H, Cai R, Zuo J, Tang X, Li X, et al. ETHYLENE INSENSITIVE3 and ETHYLENE INSENSITIVE3-LIKE1 repress SALICYLIC ACID INDUCTION DEFICIENT2 expression to negatively regulate plant innate immunity in Arabidopsis. Plant Cell. 2009:**21**(8):2527–2540. https://doi.org/10.1105/tpc.108.065193

Chen Z, Silva H, Klessig DF. Active oxygen species in the induction of plant systemic acquired resistance by salicylic acid. Science. 1993:**262**(5141):1883–1886. https://doi.org/10.1126/science.8266079

Conrath U, Chen Z, Ricigliano JR, Klessig DF. Two inducers of plant defense responses, 2,6-dichloroisonicotinec acid and salicylic acid, inhibit catalase activity in tobacco. Proc Natl Acad Sci U S A. 1995:**92**(16):7143–7147. https://doi.org/10.1073/pnas.92.16.7143

Couto D, Zipfel C. Regulation of pattern recognition receptor signalling in plants. Nat Rev. Immunol. 2016:**16**(9):537–552. https://doi.org/10.1038/nri.2016.77

Delaney TP, Friedrich L, Ryals JA. Arabidopsis signal transduction mutant defective in chemically and biologically induced disease resistance. Proc Natl Acad Sci U S A. 1995:**92**(14):6602–6606. https://doi.org/10.1073/pnas.92.14.6602

Ding P, Ding Y. Stories of salicylic acid: a plant defense hormone. Trends Plant Sci. 2020:**25**(6):549–565. https://doi.org/10.1016/j.tplants.2020.01.004

Ding Y, Sun T, Ao K, Peng Y, Zhang Y, Li X, Zhang Y. Opposite roles of salicylic acid receptors NPR1 and NPR3/NPR4 in transcriptional regulation of plant immunity. Cell. 2018:**173**(6):1454–1467.e15. https://doi.org/10.1016/j.cell.2018.03.044

Duan L, Liu H, Li X, Xiao J, Wang S. Multiple phytohormones and phytoalexins are involved in disease resistance to Magnaporthe oryzae invaded from roots in rice. Physiol Plant. 2014:**152**(3):486–500. https://doi.org/10.1111/ppl.12192

Fan W, Dong X. In vivo interaction between NPR1 and transcription factor TGA2 leads to salicylic acid-mediated gene activation in Arabidopsis. Plant Cell. 2002:**14**(6):1377–1389. https://doi.org/10.1105/tpc.001628

Friedrich L, Lawton K, Ruess W, Masner P, Specker N, Gut Rella M, Meier B, Dincher S, Staub T, Métraux J-P, et al. A benzothiadiazole derivate induces systemic acquired resistance in tobacco. Plant J. 1996:**10**(1):61–70. https://doi.org/10.1046/j.1365-313X.1996.10010061.x

Fu ZQ, Dong X. Systemic acquired resistance: turning local infection into global defense. Annu Rev Plant Biol. 2013:**64**(1):839–863. https://doi.org/10.1146/annurev-arplant-042811-105606

Fu ZQ, Yan S, Saleh A, Wang W, Ruble J, Oka N, Mohan R, Spoel SH, Tada Y, Zheng N, et al. NPR3 and NPR4 are receptors for the immune signal salicylic acid in plants. Nature. 2012:**486**(7402): 228–232. https://doi.org/10.1038/nature11162

Gaffney T, Friedrich L, Vernooij B, Negrotto D, Nye G, Uknes S, Ward E, Kessman H, Ryals J. Requirement of salicylic acid for the induction of systemic acquired resistance. Science. 1993:**261**(5122): 754–756. https://doi.org/10.1126/science.261.5122.754

Garcion C, Lohmann A, Lamodiere E, Catinot J, Buchala A, Doermann P, Metraux JP. Characterization and biological function of the ISOCHORISMATE SYNTHASE2 gene of Arabidopsis. Plant Physiol. 2008:**147**(3):1279–1287. https://doi.org/10.1104/pp.108.119420

Geng X, Jin L, Shimada M, Kim MG, Mackey D. The phytotoxin coronatine is a multifunctional component of the virulence armament of *Pseudomonas syringae*. Planta. 2014:**240**(6):1149–1165. https://doi.org/10.1007/s00425-014-2151-x

Glazebrook J, Rogers EE, Ausubel FM. Isolation of Arabidopsis mutants with enhanced disease susceptibility by direct screening. Genetics. 1996:**143**(2):973–982. https://doi.org/10.1093/genetics/143.2.973

Goodspeed D, Chehab EW, Covington MF, Braam J. Circadian control of jasmonates and salicylates: the clock role in plant defense. Plant Signal Behav. 2013:**8**(2):e23123. https://doi.org/10.4161/psb.23123

Gorlach J, Volrath S, Knauf-Beiter G, Hengy G, Beckhove U, Kogel KH, Oostendorp M, Staub T, Ward E, Kessmann H, et al. Benzothiadiazole, a novel class of inducers of systemic acquired resistance, activates gene expression and disease resistance in wheat. Plant Cell. 1996:**8**(4):629–643. https://doi.org/10.1105/tpc.8.4.629

Gullner G, Komives T, Kiraly L, Schroder P. Glutathione S-transferase enzymes in plant-pathogen interactions. Front Plant Sci. 2018:**9**:1836. https://doi.org/10.3389/fpls.2018.01836

Hartmann M, Zeier T, Bernsdorff F, Reichel-Deland V, Kim D, Hohmann M, Scholten N, Schuck S, Bräutigam A, Hölzel T, et al. Flavin monooxygenase-generated N-hydroxypipecolic acid is a critical element of plant systemic immunity. Cell. 2018:**173**(2): 456–469.e16. https://doi.org/10.1016/j.cell.2018.02.049

Holmes EC, Chen YC, Mudgett MB, Sattely ES. Arabidopsis UGT76B1 glycosylates N-hydroxy-pipecolic acid and inactivates systemic acquired resistance in tomato. Plant Cell. 2021:**33**(3):750–765. https://doi.org/10.1093/plcell/koaa052

Huang S, Zhu S, Kumar P, MacMicking JD. A phase-separated nuclear GBPL circuit controls immunity in plants. Nature. 2021a:**594**(7863): 424–429. https://doi.org/10.1038/s41586-021-03572-6

Huang W, Wu Z, Tian H, Li X, Zhang Y. Arabidopsis CALMODULIN-BINDING PROTEIN 60b plays dual roles in plant immunity. Plant Commun. 2021b:**2**(6):100213. https://doi.org/10.1016/j.xplc.2021.100213

Jia M, Chen X, Shi X, Fang Y, Gu Y. Nuclear transport receptor KA120 regulates molecular condensation of MAC3 to coordinate plant immune activation. Cell Host Microbe. 2023:**31**(10):1685–1699.e7. https://doi.org/10.1016/j.chom.2023.08.015

Jin H, Choi SM, Kang MJ, Yun SH, Kwon DJ, Noh YS, Noh B. Salicylic acid-induced transcriptional reprogramming by the HAC-NPR1-TGA histone acetyltransferase complex in Arabidopsis. Nucl. Acids Res. 2018:**46**:11712–11725. https://doi.org/10.1093/nar/gky847.

Jones JD, Vance RE, Dangl JL. Intracellular innate immune surveillance devices in plants and animals. Science. 2016:**354**(6316):aaf6395. https://doi.org/10.1126/science.aaf6395

Jung HW, Tschaplinski TJ, Wang L, Glazebrook J, Greenberg JT. Priming in systemic plant immunity. Science. 2009:**324**(5923): 89–91. https://doi.org/10.1126/science.1170025

Kim JH, Castroverde CDM, Huang S, Li C, Hilleary R, Seroka A, Sohrabi R, Medina-Yerena D, Huot B, Wang J, et al. Increasing the resilience of plant immunity to a warming climate. Nature. 2022:**607**(7918):339–344. https://doi.org/10.1038/s41586-022-04902-y

Kim Y, Gilmour SJ, Chao L, Park S, Thomashow MF. Arabidopsis CAMTA transcription factors regulate pipecolic acid biosynthesis and priming of immunity genes. Mol Plant. 2020:**13**(1):157–168. https://doi.org/10.1016/j.molp.2019.11.001

Koornneef A, Leon-Reyes A, Ritsema T, Verhage A, Den Otter FC, Van Loon LC, Pieterse CMJ. Kinetics of salicylate-mediated suppression of jasmonate signaling reveal a role for redox modulation. Plant Physiol. 2008:**147**(3):1358–1368. https://doi.org/10.1104/pp.108.121392

Kumar S, Zavaliev R, Wu Q, Zhou Y, Cheng J, Dillard L, Powers J, Withers J, Zhao J, Guan Z, et al. Structural basis of NPR1 in activating plant immunity. Nature. 2022:**605**(7910):561–566. https://doi.org/10.1038/s41586-022-04699-w

Lawton KA, Friedrich L, Hunt M, Weymann K, Delaney T, Kessmann H, Staub T, Ryals J. Benzothiadiazole induces disease resistance in Arabidopsis by activation of the systemic acquired resistance signal transduction pathway. Plant J. 1996:**10**(1):71–82. https://doi.org/10.1046/j.1365-313X.1996.10010071.x

Lee HJ, Park YJ, Seo PJ, Kim JH, Sim HJ, Kim SG, Park CM. Systemic immunity requires SnRK2.8-mediated nuclear import of NPR1 in Arabidopsis. Plant Cell. 2015:**27**(12):3425–3438. https://doi.org/10.1105/tpc.15.00371

Lefevere H, Bauters L, Gheysen G. Salicylic acid biosynthesis in plants. Front Plant Sci. 2020:**11**:338. https://doi.org/10.3389/fpls.2020.00338

Leon-Reyes A, Spoel SH, De Lange ES, Abe H, Kobayashi M, Tsuda S, Millenaar FF, Welschen RAM, Ritsema T, Pieterse CMJ. Ethylene modulates the role of NONEXPRESSOR OF PATHOGENESIS-RELATED GENES1 in cross talk between salicylate and jasmonate signaling. Plant Physiol. 2009:**149**(4):1797–1809. https://doi.org/10.1104/pp.108.133926

Leon-Reyes A, Van der Does D, De Lange ES, Delker C, Wasternack C, Van Wees SC, Ritsema T, Pieterse CM. Salicylate-mediated suppression of jasmonate-responsive gene expression in Arabidopsis is targeted downstream of the jasmonate biosynthesis pathway. Planta. 2010:**232**(6):1423–1432. https://doi.org/10.1007/s00425-010-1265-z

Li Q, Zhou M, Chhajed S, Yu F, Chen S, Zhang Y, Mou Z. N-hydroxypipecolic acid triggers systemic acquired resistance through extracellular NAD(P). Nat Commun. 2023:**14**(1):6848. https://doi.org/10.1038/s41467-023-42629-0

Liu L, Sonbol FM, Huot B, Gu Y, Withers J, Mwimba M, Yao J, He SY, Dong X. Salicylic acid receptors activate jasmonic acid signalling through a non-canonical pathway to promote effector-triggered immunity. Nat Commun. 2016:**7**(1):13099. https://doi.org/10.1038/ncomms13099

Lu H, Salimian S, Gamelin E, Wang G, Fedorowski J, LaCourse W, Greenberg JT. Genetic analysis of *acd6-1* reveals complex defense networks and leads to identification of novel defense genes in Arabidopsis. Plant J. 2009:**58**(3):401–412. https://doi.org/10.1111/j.1365-313X.2009.03791.x

Malamy J, Carr JP, Klessig DF, Raskin I. Salicylic acid: a likely endogenous signal in the resistance response of tobacco to viral infection. Science. 1990:**250**(4983):1002–1004. https://doi.org/10.1126/science.250.4983.1002

Manohar M, Tian M, Moreau M, Park SW, Choi HW, Fei Z, Friso G, Asif M, Manosalva P, von Dahl CC, et al. Identification of multiple salicylic acid-binding proteins using two high throughput screens. Front Plant Sci. 2015:**5**:777. https://doi.org/10.3389/fpls.2014.00777

Mata-Pérez C, Spoel SH. Thioredoxin-mediated redox signalling in plant immunity. Plant Sci. 2019:**279**:27–33. https://doi.org/10.1016/j.plantsci.2018.05.001

Matsushita A, Inoue H, Goto S, Nakayama A, Sugano S, Hayashi N, Takatsuji H. Nuclear ubiquitin proteasome degradation affects WRKY45 function in the rice defense program. Plant J. 2013:**73**(2): 302–313. https://doi.org/10.1111/tpj.12035

Metraux JP, Signer H, Ryals J, Ward E, Wyssbenz M, Gaudin J, Raschdorf K, Schmid E, Blum W, Inverardi B. Increase in salicylic-acid at the onset of systemic acquired-resistance in cucumber. Science. 1990:**250**(4983):1004–1006. https://doi.org/10.1126/science.250.4983.1004

Mohnike L, Rekhter D, Huang W, Feussner K, Tian H, Herrfurth C, Zhang Y, Feussner I. The glycosyltransferase UGT76B1 modulates N-hydroxy-pipecolic acid homeostasis and plant immunity. Plant Cell. 2021:**33**(3):735–749. https://doi.org/10.1093/plcell/koaa045

Moreau M, Westlake T, Zampogna G, Popescu G, Tian M, Noutsos C, Popescu S. The Arabidopsis oligopeptidases TOP1 and TOP2 are salicylic acid targets that modulate SA-mediated signaling and the immune response. Plant J. 2013:**76**(4):603–614. https://doi.org/10.1111/tpj.12320

Mou Z, Fan WH, Dong XN. Inducers of plant systemic acquired resistance regulate NPR1 function through redox changes. Cell. 2003:**113**(7): 935–944. https://doi.org/10.1016/S0092-8674(03)00429-X

Mur LA, Kenton P, Atzorn R, Miersch O, Wasternack C. The outcomes of concentration-specific interactions between salicylate and jasmonate signaling include synergy, antagonism, and oxidative stress leading to cell death. Plant Physiol. 2006:**140**(1):249–262. https://doi.org/10.1104/pp.105.072348

Nair A, Goyal I, Voß E, Mrozek P, Prajapati S, Thurow C, Tietze L, Tittmann K, Gatz C. N-hydroxypipecolic acid-induced transcription requires the salicylic acid signaling pathway at basal SA levels. Plant Physiol. 2021:**187**(4):2803–2819. https://doi.org/10.1093/plphys/kiab433

Nakayama A, Fukushima S, Goto S, Matsushita A, Shimono M, Sugano S, Jiang CJ, Akagi A, Yamazaki M, Inoue H, et al.

Genome-wide identification of WRKY45-regulated genes that mediate benzothiadiazole-induced defense responses in rice. BMC Plant Biol. 2013:**13**(1):150. https://doi.org/10.1186/1471-2229-13-150

Nawrath C, Métraux J-P. Salicylic acid induction–deficient mutants of Arabidopsis express *PR-2* and *PR-5* and accumulate high levels of camalexin after pathogen inoculation. Plant Cell. 1999:**11**(8): 1393–1404. https://doi.org/10.1105/tpc.11.8.1393

Nawrath C, Heck S, Parinthawong N, Métraux JP. EDS5, an essential component of salicylic acid-dependent signaling for disease resistance in Arabidopsis, is a member of the MATE transporter family. Plant Cell. 2002:**14**(1):275–286. https://doi.org/10.1105/tpc.010376

Ndamukong I, Abdallat AA, Thurow C, Fode B, Zander M, Weigel R, Gatz C. SA-inducible Arabidopsis glutaredoxin interacts with TGA factors and suppresses JA-responsive *PDF1.2* transcription. Plant J. 2007:**50**(1):128–139. https://doi.org/10.1111/j.1365-313X.2007.03039.x

Ngou BPM, Ding P, Jones JDG. Thirty years of resistance: zig-zag through the plant immune system. Plant Cell. 2022:**34**(5): 1447–1478. https://doi.org/10.1093/plcell/koac041

Ngou BPM, Ahn HK, Ding P, Jones JDG. Mutual potentiation of plant immunity by cell-surface and intracellular receptors. Nature. 2021:**592**(7852):110–115. https://doi.org/10.1038/s41586-021-03315-7

Nobori T, Oliva M, Lister R, Ecker JR. Multiplexed single-cell 3D spatial gene expression analysis in plant tissue using PHYTOMap. Nat Plants. 2023:**9**(7):1026–1033. https://doi.org/10.1038/s41477-023-01439-4

Nomoto M, Skelly MJ, Itaya T, Mori T, Suzuki T, Matsushita T, Tokizawa M, Kuwata K, Mori H, Yamamoto YY, et al. Suppression of MYC transcription activators by the immune cofactor NPR1 fine-tunes plant immune responses. Cell Rep. 2021:**37**(11): 110125. https://doi.org/10.1016/j.celrep.2021.110125

Norn S, Permin H, Kruse PR, Kruse E. Fra pilebark til acetylsalicylsyre [from willow bark to acetylsalicylic acid]. Dan Medicinhist Arbog. 2009:**37**:79–98.

O'Malley RC, Huang SC, Song L, Lewsey MG, Bartlett A, Nery JR, Galli M, Gallavotti A, Ecker JR. Cistrome and epicistrome features shape the regulatory DNA landscape. Cell. 2016:**165**(5):1280–1292. https://doi.org/10.1016/j.cell.2016.04.038

Petroski MD, Deshaies RJ. Function and regulation of cullin-RING ubiquitin ligases. Nat Rev Mol Cell Biol. 2005:**6**(1):9–20. https://doi.org/10.1038/nrm1547

Pieterse CMJ, Leon-Reyes A, Van der Ent S, Van Wees SCM. Networking by small-molecule hormones in plant immunity. Nat Chem Biol. 2009:**5**:308–316. https://doi.org/10.1038/nchembio.164.

Pokotylo I, Kravets V, Ruelland E. Salicylic acid binding proteins (SABPs): the hidden forefront of salicylic acid signalling. Int J Mol Sci. 2019:**20**(18):4377. https://doi.org/10.3390/ijms20184377

Qin J, Wang K, Sun L, Xing H, Wang S, Li L, Chen S, Guo HS, Zhang J. The plant-specific transcription factors CBP60g and SARD1 are targeted by a Verticillium secretory protein VdSCP41 to modulate immunity. eLife. 2018:**7**:e34902. https://doi.org/10.7554/eLife.34902

Rate DN, Greenberg JT. The *Arabidopsis aberrant growth and death2* mutant shows resistance to *Pseudomonas syringae* and reveals a role for NPR1 in suppressing hypersensitive cell death. Plant J. 2001:**27**(3): 203–211. https://doi.org/10.1046/j.0960-7412.2001.1075umedoc.x

Rekhter D, Lüdke D, Ding Y, Feussner K, Zienkiewicz K, Lipka V, Wiermer M, Zhang Y, Feussner I. Isochorismate-derived biosynthesis of the plant stress hormone salicylic acid. Science. 2019:**365**(6452):498–502. https://doi.org/10.1126/science.aaw1720

Riedlmeier M, Ghirardo A, Wenig M, Knappe C, Koch K, Georgii E, Dey S, Parker JE, Schnitzler JP, Vlot AC. Monoterpenes support systemic acquired resistance within and between plants. Plant Cell. 2017:**29**(6):1440–1459. https://doi.org/10.1105/tpc.16.00898

Rivas-San Vicente M, Plasencia J. Salicylic acid beyond defence: its role in plant growth and development. J Exp Bot. 2011:**62**(10):3321–3338. https://doi.org/10.1093/jxb/err031

Ross AF. Systemic acquired resistance induced by localized virus infections in plants. Virology. 1961:**14**(3):340–358. https://doi.org/10.1016/0042-6822(61)90319-1

Ryals J, Weymann K, Lawton K, Friedrich L, Ellis D, Steiner HY, Johnson J, Delaney TP, Jesse T, Vos P, et al. The Arabidopsis NIM1 protein shows homology to the mammalian transcription factor inhibitor IkB. Plant Cell. 1997:**9**(3):425–439. https://doi.org/10.1105/tpc.9.3.425

Saleh A, Withers J, Mohan R, Marqués J, Gu Y, Yan S, Zavaliev R, Nomoto M, Tada Y, Dong X. Posttranslational modifications of the master transcriptional regulator NPR1 enable dynamic but tight control of plant immune responses. Cell Host Microbe. 2015:**18**(2): 169–182. https://doi.org/10.1016/j.chom.2015.07.005

Shah J, Tsui F, Klessig DF. Characterization of a salicylic acid-insensitive mutant (*sai1*) of *Arabidopsis thaliana*, identified in a selective screen utilizing the SA-inducible expression of the *tms2* gene. Mol Plant-Microbe Interact. 1997:**10**(1):69–78. https://doi.org/10.1094/MPMI.1997.10.1.69

Shen M, Lim CJ, Park J, Kim JE, Baek D, Nam J, Lee SY, Pardo JM, Kim WY, Mackey D, et al. HOS15 is a transcriptional corepressor of NPR1-mediated gene activation of plant immunity. Proc Natl Acad Sci U S A. 2020:**117**(48):30805–30815. https://doi.org/10.1073/pnas.2016049117

Shimono M, Sugano S, Nakayama A, Jiang CJ, Ono K, Toki S, Takatsuji H. Rice WRKY45 plays a crucial role in benzothiadiazole-inducible blast resistance. Plant Cell. 2007:**19**(6): 2064–2076. https://doi.org/10.1105/tpc.106.046250

Shimono M, Koga H, Akagi A, Hayashi N, Goto S, Sawada M, Kurihara T, Matsushita A, Sugano S, Jiang CJ, et al. Rice WRKY45 plays important roles in fungal and bacterial disease resistance. Mol Plant Pathol. 2012:**13**(1):83–94. https://doi.org/10.1111/j.1364-3703.2011.00732.x

Shine MB, Zhang K, Liu H, Lim GH, Xia F, Yu K, Hunt AG, Kachroo A, Kachroo P. Phased small RNA-mediated systemic signaling in plants. Sci Adv. 2022:**8**(25):eabm8791. https://doi.org/10.1126/sciadv.abm8791

Skelly MJ, Furniss JJ, Grey H, Wong KW, Spoel SH. Dynamic ubiquitination determines transcriptional activity of the plant immune coactivator NPR1. eLife. 2019:**8**:e47005. https://doi.org/10.7554/eLife.47005

Spoel SH, Dong X. Making sense of hormone crosstalk during plant immune responses. Cell Host Microbe. 2008:**3**(6):348–351. https://doi.org/10.1016/j.chom.2008.05.009

Spoel SH, Loake GJ. Redox-based protein modifications: the missing link in plant immune signalling. Curr Opin Plant Biol. 2011:**14**(4): 358–364. https://doi.org/10.1016/j.pbi.2011.03.007

Spoel SH, Johnson JS, Dong X. Regulation of tradeoffs between plant defenses against pathogens with different lifestyles. Proc Natl Acad Sci U S A. 2007:**104**(47):18842–18847. https://doi.org/10.1073/pnas.0708139104

Spoel SH, Mou Z, Tada Y, Spivey NW, Genschik P, Dong X. Proteasome-mediated turnover of the transcription coactivator NPR1 plays dual roles in regulating plant immunity. Cell. 2009:**137**(5):860–872. https://doi.org/10.1016/j.cell.2009.03.038

Spoel SH, Koornneef A, Claessens SMC, Korzelius JP, Van Pelt JA, Mueller MJ, Buchala AJ, Métraux J-P, Brown R, Kazan K, et al. NPR1 modulates cross-talk between salicylate- and jasmonate-dependent defense pathways through a novel function in the cytosol. Plant Cell. 2003:**15**(3):760–770. https://doi.org/10.1105/tpc.009159

Sun T, Huang J, Xu Y, Verma V, Jing B, Sun Y, Ruiz Orduna A, Tian H, Huang X, Xia S, et al. Redundant CAMTA transcription factors negatively regulate the biosynthesis of salicylic acid and N-hydroxypipecolic acid by modulating the expression of SARD1 and CBP60g. Mol Plant. 2020:**13**(1):144–156. https://doi.org/10.1016/j.molp.2019.10.016

Sun T, Zhang Y, Li Y, Zhang Q, Ding Y, Zhang Y. ChIP-seq reveals broad roles of SARD1 and CBP60g in regulating plant immunity. Nat Commun. 2015:**6**(1):10159. https://doi.org/10.1038/ncomms10159

Tada Y, Spoel SH, Pajerowska-Mukhtar K, Mou Z, Song J, Wang C, Zuo J, Dong X. Plant immunity requires conformational changes of NPR1 via *S*-nitrosylation and thioredoxins. Science. 2008:**321**(5891): 952–956. https://doi.org/10.1126/science.1156970

Tang Y, Ho MI, Kang BH, Gu Y. GBPL3 localizes to the nuclear pore complex and functionally connects the nuclear basket with the nucleoskeleton in plants. PLoS Biol. 2022:**20**(10):e3001831. https://doi.org/10.1371/journal.pbio.3001831

Tian M, von Dahl CC, Liu PP, Friso G, van Wijk KJ, Klessig DF. The combined use of photoaffinity labeling and surface plasmon resonance-based technology identifies multiple salicylic acid-binding proteins. Plant J. 2012:**72**(6):1027–1038. https://doi.org/10.1111/tpj.12016

Tonnessen BW, Manosalva P, Lang JM, Baraoidan M, Bordeos A, Mauleon R, Oard J, Hulbert S, Leung H, Leach JE. Rice phenylalanine ammonia-lyase gene OsPAL4 is associated with broad spectrum disease resistance. Plant Mol Biol. 2015:**87**(3):273–286. https://doi.org/10.1007/s11103-014-0275-9

Torrens-Spence MP, Bobokalonova A, Carballo V, Glinkerman CM, Pluskal T, Shen A, Weng JK. PBS3 and EPS1 complete salicylic acid biosynthesis from isochorismate in Arabidopsis. Mol Plant. 2019:**12**(12):1577–1586. https://doi.org/10.1016/j.molp.2019.11.005

Torres MA, Jones JD, Dangl JL. Pathogen-induced, NADPH oxidase-derived reactive oxygen intermediates suppress spread of cell death in Arabidopsis thaliana. Nature Genet. 2005:**37**(10):1130–1134. https://doi.org/10.1038/ng1639

Truman W, Glazebrook J. Co-expression analysis identifies putative targets for CBP60g and SARD1 regulation. BMC Plant Biol. 2012:**12**(1):216. https://doi.org/10.1186/1471-2229-12-216

Ueno Y, Matsushita A, Inoue H, Yoshida R, Jiang CJ, Takatsuji H. WRKY45 phosphorylation at threonine 266 acts negatively on WRKY45-dependent blast resistance in rice. Plant Signal Behav. 2017:**12**(8):e1356968. https://doi.org/10.1080/15592324.2017.1356968

Ueno Y, Yoshida R, Kishi-Kaboshi M, Matsushita A, Jiang CJ, Goto S, Takahashi A, Hirochika H, Takatsuji H. Abiotic stresses antagonize the rice defence pathway through the tyrosine-dephosphorylation of OsMPK6. PLoS Pathog. 2015:**11**(10):e1005231. https://doi.org/10.1371/journal.ppat.1005231

Van der Does D, Leon-Reyes A, Koornneef A, Van Verk MC, Rodenburg N, Pauwels L, Goossens A, Korbes AP, Memelink J, Ritsema T, et al. Salicylic acid suppresses jasmonic acid signaling downstream of SCF^{COI1}-JAZ by targeting GCC promoter motifs via transcription factor ORA59. Plant Cell. 2013:**25**(2):744–761. https://doi.org/10.1105/tpc.112.108548

van Verk MC, Bol JF, Linthorst HJ. WRKY transcription factors involved in activation of SA biosynthesis genes. BMC Plant Biol. 2011:**11**(1):89. https://doi.org/10.1186/1471-2229-11-89

Vernooij B, Friedrich L, Morse A, Reist R, Kolditz-Jawhar R, Ward E, Uknes S, Kessmann H, Ryals J. Salicylic acid is not the translocated signal responsible for inducing systemic acquired resistance but is required in signal transduction. Plant Cell. 1994:**6**(7):959–965. https://doi.org/10.2307/3870006

Viola IL, Guttlein LN, Gonzalez DH. Redox modulation of plant developmental regulators from the class I TCP transcription factor family. Plant Physiol. 2013:**162**(3):1434–1447. https://doi.org/10.1104/pp.113.216416

Vlot AC, Sales JH, Lenk M, Bauer K, Brambilla A, Sommer A, Chen Y, Wenig M, Nayem S. Systemic propagation of immunity in plants. New Phytol. 2021:**229**(3):1234–1250. https://doi.org/10.1111/nph.16953

Wang C, El-Shetehy M, Shine MB, Yu K, Navarre D, Wendehenne D, Kachroo A, Kachroo P. Free radicals mediate systemic acquired resistance. Cell Rep. 2014:**7**(2):348–355. https://doi.org/10.1016/j.celrep.2014.03.032

Wang C, Huang X, Li Q, Zhang Y, Li JL, Mou Z. Extracellular pyridine nucleotides trigger plant systemic immunity through a lectin receptor kinase/BAK1 complex. Nat Commun. 2019:**10**(1):4810. https://doi.org/10.1038/s41467-019-12781-7

Wang D, Amornsiripanitch N, Dong X. A genomic approach to identify regulatory nodes in the transcriptional network of systemic acquired resistance in plants. PLoS Pathog. 2006:**2**(11):e123. https://doi.org/10.1371/journal.ppat.0020123

Wang D, Pajerowska-Mukhtar K, Culler AH, Dong X. Salicylic acid inhibits pathogen growth in plants through repression of the auxin signaling pathway. Curr Biol. 2007:**17**(20):1784–1790. https://doi.org/10.1016/j.cub.2007.09.025

Wang D, Weaver ND, Kesarwani M, Dong X. Induction of protein secretory pathway is required for systemic acquired resistance. Science. 2005:**308**(5724):1036–1040. https://doi.org/10.1126/science.1108791

Wang L, Tsuda K, Sato M, Cohen JD, Katagiri F, Glazebrook J. Arabidopsis CaM binding protein CBP60g contributes to MAMP-induced SA accumulation and is involved in disease resistance against *Pseudomonas syringae*. PLoS Pathog. 2009:**5**(2):e1000301. https://doi.org/10.1371/journal.ppat.1000301

Wang L, Tsuda K, Truman W, Sato M, Nguyen le V, Katagiri F, Glazebrook J. CBP60g and SARD1 play partially redundant critical roles in salicylic acid signaling. Plant J. 2011:**67**(6):1029–1041. https://doi.org/10.1111/j.1365-313X.2011.04655.x

Wang W, Withers J, Li H, Zwack PJ, Rusnac DV, Shi H, Liu L, Yan S, Hinds TR, Guttman M, et al. Structural basis of salicylic acid perception by Arabidopsis NPR proteins. Nature. 2020:**586**(7828):311–316. https://doi.org/10.1038/s41586-020-2596-y

Wang X, Gao J, Zhu Z, Dong X, Wang X, Ren G, Zhou X, Kuai B. TCP transcription factors are critical for the coordinated regulation of isochorismate synthase 1 expression in Arabidopsis thaliana. Plant J. 2015:**82**(1):151–162. https://doi.org/10.1111/tpj.12803

Wang Z, Orosa-Puente B, Nomoto M, Grey H, Potuschak T, Matsuura T, Mori IC, Tada Y, Genschik P, Spoel SH. Proteasome-associated ubiquitin ligase relays target plant hormone-specific transcriptional activators. Sci Adv. 2022:**8**(42):eabn4466. https://doi.org/10.1126/sciadv.abn4466

Wang Z, Spoel SH. HECT ubiquitin ligases as accessory proteins of the plant proteasome. Essays Biochem. 2022:**66**(2):135–145. https://doi.org/10.1042/EBC20210064

Ward ER, Uknes SJ, Williams SC, Dincher SS, Wiederhold DL, Alexander DC, Ahlgoy P, Metraux JP, Ryals JA. Coordinated gene activity in response to agents that induce systemic acquired resistance. Plant Cell. 1991:**3**(10):1085–1094. https://doi.org/10.2307/3869297

Wenig M, Ghirardo A, Sales JH, Pabst ES, Breitenbach HH, Antritter F, Weber B, Lange B, Lenk M, Cameron RK, et al. Systemic acquired resistance networks amplify airborne defense cues. Nat Commun. 2019:**10**(1):3813. https://doi.org/10.1038/s41467-019-11798-2

White RF. Acetylsalicylic acid (aspirin) induces resistance to tobacco mosaic virus in tobacco. Virology. 1979:**99**(2):410–412. https://doi.org/10.1016/0042-6822(79)90019-9

Wildermuth MC, Dewdney J, Wu G, Ausubel FM. Isochorismate synthase is required to synthesize salicylic acid for plant defence. Nature. 2001:**414**(6863):562–565. https://doi.org/10.1038/35107108

Wu J, Zhu W, Zhao Q. Salicylic acid biosynthesis is not from phenylalanine in Arabidopsis. J Integr Plant Biol. 2023:**65**(4):881–887. https://doi.org/10.1111/jipb.13410

Wu Y, Zhang D, Chu JY, Boyle P, Wang Y, Brindle ID, De Luca V, Despres C. The *Arabidopsis* NPR1 protein is a receptor for the plant defense hormone salicylic acid. Cell Rep. 2012:**1**(6):639–647. https://doi.org/10.1016/j.celrep.2012.05.008

Xie Z, Zhao S, Li Y, Deng Y, Shi Y, Chen X, Li Y, Li H, Chen C, Wang X, et al. Phenolic acid-induced phase separation and translation inhibition mediate plant interspecific competition. Nat Plants. 2023:**9**(9):1481–1499. https://doi.org/10.1038/s41477-023-01499-6

Yang L, Chen D, Wang X, Luo B, Wang C, Gao G, Li H, Li A, Chen L. Ratiometric electrochemical sensor for accurate detection of salicylic acid in leaves of living plants. RSC Adv. 2020:**10**(64):38841–38846. https://doi.org/10.1039/D0RA05813K

Yildiz I, Gross M, Moser D, Petzsch P, Köhrer K, Zeier J. N-hydroxypipecolic acid induces systemic acquired resistance and transcriptional reprogramming via TGA transcription factors. Plant Cell Environ. 2023:**46**(6):1900–1920. https://doi.org/10.1111/pce.14572

Yildiz I, Mantz M, Hartmann M, Zeier T, Kessel J, Thurow C, Gatz C, Petzsch P, Köhrer K, Zeier J. The mobile SAR signal

N-hydroxypipecolic acid induces NPR1-dependent transcriptional reprogramming and immune priming. Plant Physiol. 2021:**186**(3): 1679–1705. https://doi.org/10.1093/plphys/kiab166

Yu X, Cui X, Wu C, Shi S, Yan S. Salicylic acid inhibits gibberellin signaling through receptor interactions. Mol Plant. 2022:**15**(11): 1759–1771. https://doi.org/10.1016/j.molp.2022.10.001

Yuan HM, Liu WC, Lu YT. CATALASE2 coordinates SA-mediated repression of both auxin accumulation and JA biosynthesis in plant defenses. Cell Host Microbe. 2017:**21**(2):143–155. https://doi.org/10.1016/j.chom.2017.01.007

Yuan M, Jiang Z, Bi G, Nomura K, Liu M, Wang Y, Cai B, Zhou JM, He SY, Xin XF. Pattern-recognition receptors are required for NLR-mediated plant immunity. Nature. 2021:**592**(7852):105–109. https://doi.org/10.1038/s41586-021-03316-6

Zavaliev R, Mohan R, Chen T, Dong X. Formation of NPR1 condensates promotes cell survival during the plant immune response. Cell. 2020:**182**(5):1093–1108.e18. https://doi.org/10.1016/j.cell.2020.07.016

Zhang X, Gou M, Guo C, Yang H, Liu CJ. Down-regulation of Kelch domain-containing F-box protein in Arabidopsis enhances the production of (poly)phenols and tolerance to ultraviolet radiation. Plant Physiol. 2015:**167**(2):337–350. https://doi.org/10.1104/pp.114.249136

Zhang X, Gou M, Liu CJ. Arabidopsis Kelch repeat F-box proteins regulate phenylpropanoid biosynthesis via controlling the turnover of phenylalanine ammonia-lyase. Plant Cell. 2013:**25**(12):4994–5010. https://doi.org/10.1105/tpc.113.119644

Zhang Y, Cheng YT, Qu N, Zhao Q, Bi D, Li X. Negative regulation of defense responses in Arabidopsis by two *NPR1* paralogs. Plant J. 2006:**48**(5):647–656. https://doi.org/10.1111/j.1365-313X.2006.02903.x

Zhang Y, Xu S, Ding P, Wang D, Cheng YT, He J, Gao M, Xu F, Li Y, Zhu Z, et al. Control of salicylic acid synthesis and systemic acquired resistance by two members of a plant-specific family of transcription factors. Proc Natl Acad Sci U S A. 2010:**107**(42):18220–18225. https://doi.org/10.1073/pnas.1005225107

Zheng Q, Majsec K, Katagiri F. Pathogen-driven coevolution across the CBP60 plant immune regulator subfamilies confers resilience on the regulator module. New Phytol. 2022:**233**(1):479–495. https://doi.org/10.1111/nph.17769

Zheng XY, Spivey NW, Zeng W, Liu PP, Fu ZQ, Klessig DF, He SY, Dong X. Coronatine promotes Pseudomonas syringae virulence in plants by activating a signaling cascade that inhibits salicylic acid accumulation. Cell Host Microbe. 2012:**11**(6):587–596. https://doi.org/10.1016/j.chom.2012.04.014

Zheng XY, Zhou M, Yoo H, Pruneda-Paz JL, Spivey NW, Kay SA, Dong X. Spatial and temporal regulation of biosynthesis of the plant immune signal salicylic acid. Proc Natl Acad Sci U S A. 2015:**112**(30): 9166–9173. https://doi.org/10.1073/pnas.1511182112

Zhu J, Lolle S, Tang A, Guel B, Kvitko B, Cole B, Coaker G. Single-cell profiling of Arabidopsis leaves to *Pseudomonas syringae* infection. Cell Rep. 2023:**42**(7):112676. https://doi.org/10.1016/j.celrep.2023.112676

https://doi.org/10.1093/plcell/koae020

THE PLANT CELL

THE PLANT CELL 2024: 36: 1465–1481

Pathogen perception and signaling in plant immunity

Peter N. Dodds,* **Jian Chen** and **Megan A. Outram**

Commonwealth Scientific and Industrial Research Organization, Agriculture and Food, Canberra, ACT 2601, Australia

*Author for correspondence: peter.dodds@csiro.au

Review

Abstract

Plant diseases are a constant and serious threat to agriculture and ecological biodiversity. Plants possess a sophisticated innate immunity system capable of detecting and responding to pathogen infection to prevent disease. Our understanding of this system has grown enormously over the past century. Early genetic descriptions of plant disease resistance and pathogen virulence were embodied in the gene-for-gene hypothesis, while physiological studies identified pathogen-derived elicitors that could trigger defense responses in plant cells and tissues. Molecular studies of these phenomena have now coalesced into an integrated model of plant immunity involving cell surface and intracellular detection of specific pathogen-derived molecules and proteins culminating in the induction of various cellular responses. Extracellular and intracellular receptors engage distinct signaling processes but converge on many similar outputs with substantial evidence now for integration of these pathways into interdependent networks controlling disease outcomes. Many of the molecular details of pathogen recognition and signaling processes are now known, providing opportunities for bioengineering to enhance plant protection from disease. Here we provide an overview of the current understanding of the main principles of plant immunity, with an emphasis on the key scientific milestones leading to these insights.

Introduction

Like all complex organisms, plants are exposed to the threat of diseases caused by microbial pathogen infection. Plant diseases have profound effects on both crop production and natural biodiversity, the impacts of which have been documented throughout human history, such as the rusts plaguing cereal crops and requiring sacrifices to the gods in Roman days (Zadoks 1981) and the Irish potato famine driving mass migra-tion in the 1800s. The severity of plant disease impacts has mo-tivated research over the last hundred years or so, enormously expanding our understanding of how plants detect and defend against pathogen infection. Here, we discuss the current mod-els of how the plant immune system functions and how this un-derstanding has emerged from key scientific advances in plant–pathogen interaction research.

Genetic control of resistance to plant disease was first described in the early 1900s (Biffen 1905), followed by observations that isolates of a pathogen species could be divided into different "physiologic races" that differed in their ability to infect certain resistant host plant lines (Stakman and Levine 1922). These two concepts were synthesized in the gene-for-gene hypothesis (Flor 1942), which described the outcome of plant–pathogen interactions as equally dependent on host and pathogen genotypes. In this model, disease resistance results from recognition events between the pro-ducts of plant resistance (*R*) genes and pathogen avirulence (*Avr*) genes, while lack of recognition allows disease progression. This basic genetic concept has been found to apply to many different plant diseases caused by bacteria, fungi, oomy-cetes, and viruses and even to interactions with invertebrate pests and parasitic plants. Gene-for-gene resistance has underpinned plant breeding for most of the past century, as *R* genes have been used by breeders to control crop diseases, and it has also been a cornerstone of plant immunity research (Dodds 2023). However, other types of recognition events outside of gene-for-gene interactions were also detected. These stemmed from the identification of pathogen derived "elicitor" molecules, such as bacterial lipopolysaccharides

Received November 02, 2023. Accepted January 16, 2024. Advance access publication January 23, 2024

(LPS) or fungal cell wall components, that caused the induction of defense responses when applied to plant cells or tissues (Cruickshank and Perrin 1968; Graham et al. 1977). However, the contribution of these cellular recognition events to disease outcomes was not understood, because natural variation for the capacity to respond to these elicitors was not detected in plant species. Furthermore, the elicitors themselves were often common to a variety of pathogens, including those able to successfully cause disease. Our understanding of plant immune responses to pathogens remained at this broad phenomenological level until the advent of molecular genetics toward the end of the 20th century, which facilitated the identification of many of the key genes and proteins underlying gene-for-gene and elicitor recognition interactions. This provided the first insights into the molecular mechanisms of plant–pathogen recognition and allowed the development of a unified model of how these interactions trigger defense responses in plant immunity.

Overview of plant immune recognition systems

The key milestones (Table 1) in developing a molecular model of plant immunity were, first, the identification of *Avr* genes from bacterial and fungal pathogens in the 1980s (Staskawicz et al. 1984; de Wit et al. 1985), followed by the first plant *R* genes in the 1990s (Martin et al. 1993; Bent et al. 1994; Jones et al. 1994; Mindrinos et al. 1994; Whitham et al. 1994; Grant et al. 1995; Lawrence et al. 1995; Song et al. 1995). Subsequently, the first plant receptor for a pathogen elicitor was identified in 2000 (Gómez-Gómez and Boller 2000). These findings led to a key conceptual model that has defined plant immunity re-search in the current century, which is the division of pathogen responses into two components: Pattern-triggered and Effector-triggered immunity (PTI and ETI) (Jones and Dangl 2006). These two components broadly correspond to the elicitor-induced responses (PTI) and gene-for-gene resistance (ETI) phenomena that had been previously studied, and this model provided a synthesis of these two sets of observations as representing two parallel components of an overall immune system (Dodds and Rathjen 2010; Couto and Zipfel 2016; Ngou et al. 2022). Under this model (Fig. 1), cell surface located receptors detect conserved molecular patterns associated with microbial infection and trigger defense responses that can provide a broad-spectrum resistance to a wide range of pathogens. However, many adapted pathogens deliver effector proteins into plant cells that act to facilitate infection, including by direct suppression of PTI. Intracellular immune receptors, usually of the nucleotide-binding/leucine-rich-repeat receptor (NB-LRR or NLR) class, can recognize these effector proteins to induce another layer of defense, resulting in resistance that is often associated with the hypersensitive response involving localized cell death.

The PTI and ETI model essentially divides plant immunity into two components; recognition of conserved patterns derived from pathogen structural components or recognition of variable effector proteins. This is a very useful conceptual framework that explains many broad observations of plant–pathogen interactions that had been long known. These include the apocryphal notion that most plants are resistant to most pathogens, which can be explained by the inability of non-adapted pathogens to effectively suppress PTI and evade recognition. It also explains the contrast between the conservation of many elicitor responses across diverse plant species compared to the high genetic variation within species that is inherent to gene-for-gene resistance, with the latter resulting from antagonistic co-evolution between genes encoding interacting effector and receptor proteins in the pathogen and host, respectively. However, from a functional perspective, a more mechanistically relevant divide is between immunity triggered by extracellular or intracellular receptors (Fig. 1). Many infection-associated ligands available in the extracellular space fit the description of conserved patterns, including pathogen/microbe-associated molecular patterns (PAMPs/MAMPs) but also plant-derived signals such as damage-associated molecular patterns (DAMPs) and phytocytokines or other signaling molecules. Nevertheless, there are also apoplastic effector proteins that are recognized by cell surface receptors, with both components showing the high genetic variability characteristic of canonical gene-for-gene interactions. Also, some classically defined PAMPs, such as flagellin, have diversified in different bacteria to escape recognition. In contrast, most intracellularly available ligands are effectors delivered by bacteria, fungi, oomycetes, or animal invaders, but some NLRs recognize essential components of virus particles. Cell surface and intracellular immunity act together to induce defense responses, leading to inhibition of pathogen infection. Although the two fields emerged in parallel and have largely been treated as separate systems, there is increasing evidence of shared pathways and crosstalk indicative of an overall integrated plant immune system (Bernoux et al. 2022; Ngou et al. 2022).

Extracellular perception: monitoring the perimeter

Although cell surface receptors are often referred to as pattern-recognition receptors, or PRRs, for their role in PAMP perception, the first such receptors identified were encoded by the *Cf* resistance genes from tomato and actually detect apoplastic effectors of the fungus *Fulvia fulva* (formerly *Cladosporium fulvum*) in a gene-for-gene resistance system (Jones et al. 1994; Dixon et al. 1998). While many subsequently identified cell surface receptors do detect conserved molecular patterns, such as pathogen-derived elicitors, further examples of membrane receptors that detect apoplastic effector proteins in gene-for-gene interactions have also been found (Ngou et al. 2022; Zhang et al. 2023). Thus, we have avoided using PRR as a generic term for this class of receptor. Although elicitors had been identified much earlier as inducers of plant defense responses, the contribution of these responses to immunity was not properly recognized until the corresponding receptors were identified and their genetic knockout mutations shown

Table 1. Key milestones and timeline in understanding plant immunity

	Advance	Key papers	Decade: 1900s	1910s	1920s	1930s	1940s	1950s	1960s	1970s	1980s	1990s	2000s	2010s	2020s
Genetic and physiological observations	Genetic control of resistance	Biffen (1905)	X												
	Pathogen race differentiation	Stakman and Levine (1922)			X										
	Gene-for-gene model	Flor (1942)					X								
	Pathogen elicitors of defense	Cruikshank and Perrin (1968) and Graham et al. (1977)							X	X					
Identifying key immunity genes	Bacterial/Fungal Avr genes	Staskawicz et al. (1984) and De Wit et al. (1985)									X				
	R genes	Martin et al. (1993), Bent et al. (1994), Jones et al. (1994), Mindrinos et al. (1994) and Whitham et al. (1994)										X			
	RLKs as PAMP detectors	Gomez-Gomez and Boller (2000)											X		
	Role of RLKs in resistance	Zipfel et al. (2004)											X		
	Fungal/oomycete cytosolic Avrs	Orbach et al. (2000), Allen et al. (2004), Dodds et al. (2004), Shan et al. (2004), Armstrong et al. (2005) and Rehmany et al. (2005)											X		
	EDS1 signaling genes	Parker et al. (1996), Falk et al. (1999), Jirage et al. (1999), Grant et al. (2003), Feys et al. (2005), Peart et al. (2005)										X	X		
Defining immune receptor functions	Guard recognition	van der Biezen and Jones (1998)										X			
	Direct recognition	Dodds et al. (2006) and Ueda et al. (2006)											X		
	Role of SERK1 co-receptors (BAK1)	Chinchilla et al. (2007) and Heese et al. (2007)											X		
	Roles of RLCK kinases (e.g. BIK1)	Rowland et al. (2004) and Veronese et al. (2006)											X		
	SOBIR—connects RLPs to BAK1	Liebrand et al. (2013) and Albert et al. (2015)												X	
	RLCK substrates	Kadota et al. (2014), Bi et al. (2018), Tian et al. (2019), Wang et al. (2019c) and Thor et al. (2020)												X	
	LRR controls NLR recognition	Ellis et al. (1999), Dodds et al. (2001) and Shen et al. (2003)											X		
	NB domain ATP/ADP switch	Tameling et al. (2006), Williams et al. (2011) and Bernoux et al. (2016)											X		
	TIR/CC signaling functions	Frost et al. (2004), Swiderski et al. (2009), Collier et al. (2011) and Maekawa et al. (2011)											X	X	
	Integrated decoys	Cesari et al. (2013, 2014b), Le Roux et al. (2015) and Sarris et al. (2015)												X	
	CC-NLR Resistosomes/Ca^{2+} channel	Wang et al. (2019a, 2019b), Bi et al. (2021), Jacob et al. (2021) and Förderer et al. (2022)													X
	TIR domain NADase activity	Horsefield et al. (2019) and Wan et al. (2019)													X
	TIR-NLR Resistosomes	Martin et al. (2020) and Ma et al. (2020)													X
	TIR-derived ligands of the EDS1 family	Huang et al. (2022) and Jia et al. (2022)													X

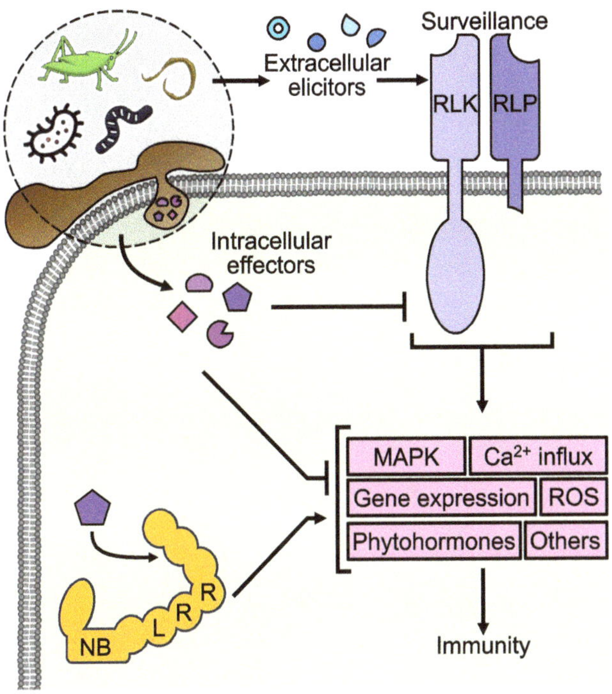

Figure 1. Schematic overview of the plant immune system. Plants use plasma membrane-localized extracellular immunity receptors (RLK/RLP) that can detect extracellular elicitors such as conserved molecular patterns associated with infection or apoplastic effectors and trigger downstream defense responses. Invading pathogens can deliver proteinaceous effector molecules into the plant to suppress defense signaling. Intracellular immunity receptors, predominantly nucleotide-binding leucine-rich-repeat (NLR) proteins, can recognize specific effectors and activate defense. RLK/RLP and NLR immune responses converge on similar downstream events, including reactive oxygen species (ROS) production, defense gene expression, Ca^{2+} influx, mitogen-activated protein kinase (MAPK) activation, and defense hormone production. NLR responses are generally stronger and are more often associated with cell death than RLK/RLP responses.

to have a direct effect on infection outcomes. The first identified PAMP receptor was FLS2 (Flagellin Sensing 2) from *Arabidopsis thaliana* (Gómez-Gómez and Boller 2000), which detects flagellin 22 (flg22), a 22 amino acid peptide derived from bacterial flagellin proteins, by direct binding (Chinchilla et al. 2006). Using genetic knockouts of FLS2, Zipfel et al. (2004) showed that flagellin perception mediated a quantifiable decrease in infection levels of a virulent strain of *Pseudomonas syringae*, which was the key observation linking elicitor responses to plant immunity function. Subsequently, the EF-Tu receptor (EFR) of *Arabidopsis* was shown to detect a fragment of the bacterial elongation factor protein EF-Tu and contribute to restriction of *Agrobacterium* growth and T-DNA delivery (Zipfel et al. 2006). There has since been a substantial expansion in the number of known cell surface immune receptors, with more than 70 characterized functionally from various plants (Ngou et al. 2022).

Cell surface receptors can be broadly classified into two categories: transmembrane receptor-like kinases (RLKs) and transmembrane receptor-like proteins (RLPs). RLKs and RLPs contain a variable extracellular domain (ECD) at the N-terminus, which is responsible for ligand recognition, and either a single-pass transmembrane domain that anchors the protein in the plasma membrane, or in some RLPs a conjugated lipid anchor. The defining difference between RLKs and RLPs is the presence of an intracellular kinase domain in the C-terminal region of RLKs. The ECDs of RLKs/RLPs comprise several different classes (Fig. 2), which are associated with recognition of different types of ligands (Bentham et al. 2020; Ge et al. 2022). The most abundant class of RLKs contain a leucine-rich repeat (LRR) domain and are involved mainly in recognition of proteins and peptides, including plant-derived peptide hormones (phytocytokines). These include the classic FLS2 and EFR proteins, as well as numerous other receptors recognizing conserved peptides from other pathogen proteins, such as the rice OsXa21, which detects a tyrosine-sulfated peptide from *Xanthomonas* spp. (Pruitt et al. 2015), and potato PERU, which senses a conserved peptide of a *Phytophthora* cell wall glycoprotein (Torres Ascurra et al. 2023). Other LRR-RLKs perceive plant-generated phytocytokine signals, such as the *Arabidopsis* PEPR1/2 receptors that detect peptides released by metacaspase activity as part of defense responses (Sun et al. 2013; Saijo et al. 2018). Similarly, most characterized RLPs contain LRR-type extracellular domains, and this class includes most of the receptors recognizing apoplastic effectors, such as the tomato Cf proteins detecting *F. fulvum* Avrs (Jones et al. 1994; Dixon et al. 1998) and canola LepR3 detecting AvrLm1 from *Leptosphaeria maculans* (Larkan et al. 2013). Members of this class also detect more conserved peptide elicitors such as the oomycete elicitin protein INF1 recognized by *Sm*ELR (Du et al. 2015).

Other types of ECDs include the lysin motif (LysM) domain, which is involved mainly in recognition of polysaccharides. These include chitin oligomers derived from fungal cell walls, which are detected by the AtLYM2/4/5 proteins in *Arabidopsis* (Wan et al. 2008) and various LysM-RLKs in other plants (Ngou et al. 2022). The AtLYM1/3 receptors are LysM-RLPs and detect peptidoglycans from bacteria (Willmann et al. 2011). Lectin domains are often involved in de-tecting lipid-derived molecules, such as 3-hydroxy decanoic acid, a component of LPS perceived by *At*LORE (Kutschera et al. 2019). However, the tomato lectin-RLK I-3 detects the apoplastic protein effector Avr3 from *F. oxysporum* (Catanzariti et al. 2015), and the *Arabidopsis* AtDORN1 and AtLecRK-1.8 detect extracellular ATP or NAD molecules as signals of cellular damage (Chen et al. 2017a; Wang et al. 2017). RLKs with an epidermal growth factor (EGF) domain include AtWAK1/2, which perceive fungal-derived oligo galacturonides (Brutus et al. 2010), and wheat Snn1, which is activated by the necrotrophic effector protein *Sn*Tox1 from *Parastagonospora nodorum* (Shi et al. 2016).

The activation and downstream signaling pathways of RLKs/RLPs have been well characterized (DeFalco and Zipfel 2021). A range of physiological responses to elicitors have been described, including an influx of calcium (Ca^{2+}), a reactive oxygen

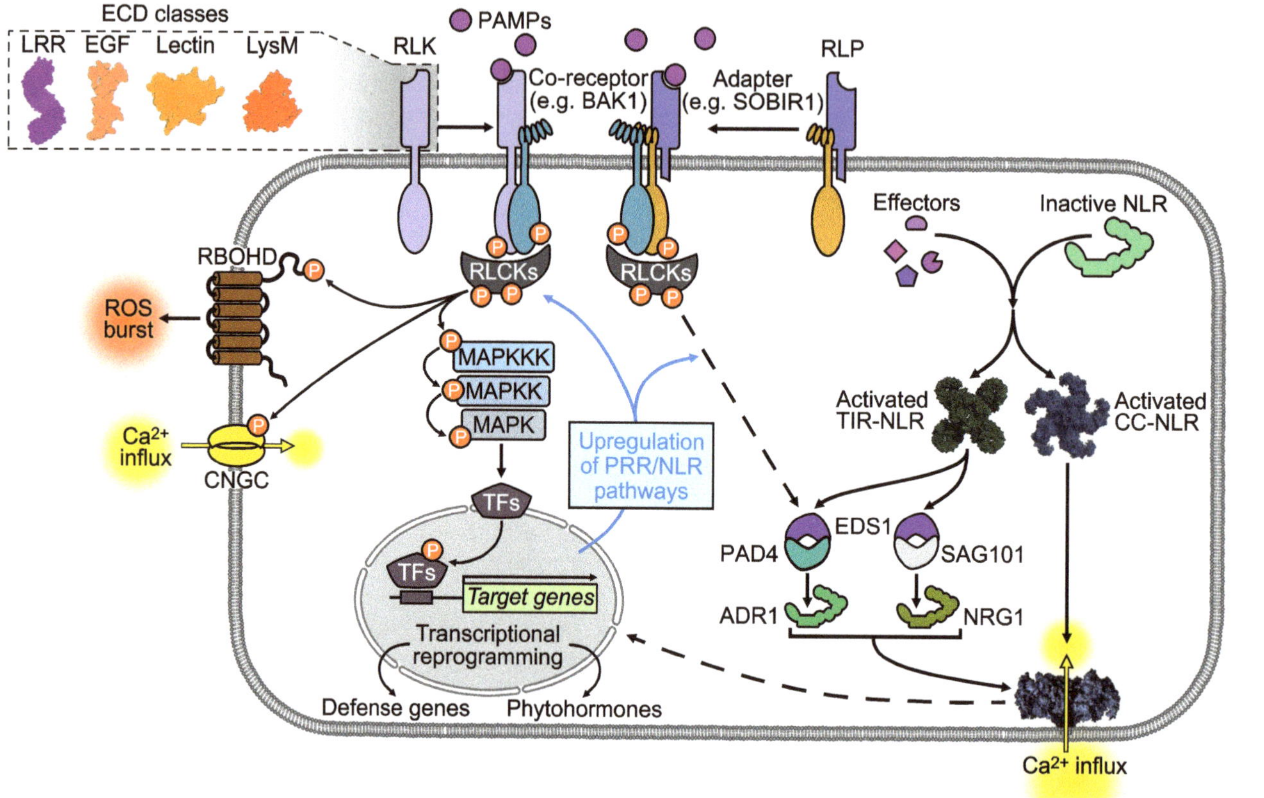

Figure 2. RLK/RLP and NLR signaling pathways. Receptor-like kinases (RLK) and receptor-like proteins (RLP) detect apoplastic ligands, such as pathogen-associated molecular patterns (PAMPs), through their extracellular domains (ECD), which may be of several classes (left panel, LRR—leucine-rich (PDB: 4MNA), EGF—epidermal growth factor (AlphaFold2- A0A7S9C0X0$^{285-374}$), LysM—lysin motif (PDB: 4EBY), Lectin (AlphaFold2—A0A0D4BT35$^{119-201}$). This recognition leads to the association of the RLK or RLP with co-receptors (e.g. BAK1) and recruitment of downstream RLCKs (receptor-like cytoplasmic kinases) through phosphorylation events mediated by the intracellular kinase domains. Activated RLCKs subsequently phosphorylate multiple downstream components, including RBOHD, MAPKs, and CNGCs, which leads to the triggering of reactive oxygen species (ROS) production, Ca2+ influx, MAPK cascade activation, and, ultimately, transcriptional reprogramming. Intracellular nucleotide-binding leucine-rich repeat receptors (NLRs) detect pathogen effectors delivered into the plant cell. Recognition leads to resistosome formation. CC-NLRs form pentameric resistosomes (7XC2) in which the CC domains form a Ca^{2+} channel. TIR-NLRs form tetrameric resistosomes (6J5T) leading to activation of the TIR domain NADase catalytic activity. Small molecules produced by the TIR domains bind to and activate EDS1 heterodimers, which in turn activate CC_R-NLRs (NGR1 and ADR1) forming resistosome-like complexes leading to an influx of Ca^{2+}. RLK/RLP signaling and NLR signaling appear to converge at multiple points and share similar defense activation outputs. Solid arrows indicate known signal transduction mechanisms, and dashed lines indicate that the underlying mechanisms are unknown.

(ROS) burst, activation of mitogen-activated protein kinase (MAPK) cascades, callose deposition to reinforce cell walls, de-fense gene upregulation, and plant hormone changes (Fig. 1). Detailed signaling pathways linking most of these events back to ligand recognition have now been defined (Fig. 2). Key milestones in defining these pathways included the identification of BAK1, a small LRR-RLK of the somatic embryogenesis receptor-like kinase (SERK) family (Chinchilla et al. 2007; Heese et al. 2007), and BIK1, an intracellular kinase of the receptor-like cytoplasmic kinase (RLCK) VII family (Veronese et al. 2006), as required for PAMP-triggered signaling. The Cf-4/Cf-9 RLPs were also shown to require the RLCK-VII pro-tein ACIK1 for defense activation (Rowland et al. 2005). After ligand recognition, LRR-RLKs form heterocomplexes with co-receptor proteins of the SERK family, principally BAK1. Structure determination of the FLS2-flg22-BAK1 complex showed that the flagellin ligand binds principally to the FLS2 LRR domain, but also contacts BAK1 to cement the co-receptor interaction (Sun et al. 2013). This interaction stimulates self- and cross-phosphorylation of their respective intracellular domains, leading to activation of the BAK1 kinase domain. Extensive analyses of these phosphorylation events have revealed a complex code in which multiple phosphosites in the kinase domains control complex activation, kinase cata-lytic activity, protein turnover, and endosomal trafficking (DeFalco and Zipfel 2021).

LRR-type RLPs lack an intracellular kinase domain but form constitutive complexes with the adapter protein SOBIR1 (an LRR-RLK), which bridges their interaction with BAK1 after ligand detection to also activate BAK1 by phosphorylation (Liebrand et al. 2013; Albert et al. 2015) (Fig. 2). The activated receptor/BAK1 (or other co-receptor) kinase complexes also associate with and phosphorylate members of the intracellular RLCK-VII family, such as BIK1, which act as the next-level executors of the immune signal (Couto and Zipfel 2016). Similarly, LysM-type RLKs and RLPs interact with the LysM-RLK co-receptor CERK1 (chitin elicitor receptor kinase 1) to activate signaling via RLCKs (Miya et al. 2007). The formation of receptor/co-receptor/RLCK complexes is regulated both positively and negatively by a variety of processes including accessory RLKs, phosphatases, and ubiquitination to fine-tune responses (DeFalco and Zipfel 2021; Manhaes et al. 2021). Although the precise co-receptor and RLCK components and regulatory interactions vary for different RLKs/RLPs, they appear to function similarly in activation and signaling of immunity.

The next steps in cell surface–initiated signaling involve phosphorylation of a variety of downstream targets by RLCKs to activate several parallel response pathways (Liang and Zhou 2018), although BAK1 and other co-receptors can also phosphorylate some substrates. The rapid oxidative burst initiated upon elicitor perception was shown to be principally mediated by RBOHD, a membrane-localized NADPH oxidase enzyme (Torres et al. 2002) (Fig. 2). RBOHD is a direct substrate of BIK1 (Kadota et al. 2014; Li et al 2014) and other RCLKs (Kadota et al. 2015), and its phosphorylation activates its enzymatic function to generate ROS. RBOHD is also phosphorylated by AtDORN1 upon eATP perception (Chen et al. 2017a). In addition, RLCKs directly phosphorylate cyclic nucleotide gated calcium channel proteins in *Arabidopsis* (Tian et al. 2019) and rice (Wang et al. 2019c), leading to the calcium ion influx associated with PTI. RCLKS also directly phosphor-ylate specific guard cell ion channel proteins to trigger stoma-tal closure (Liu et al. 2019; Thor et al. 2020) and are responsible for initiating MAP kinase cascades through phosphorylation of MAPKKK5 and MEKK1 (Bi et al. 2018). MAPK activity re-sults in phosphorylation of transcription factors, such as WRKY proteins, which are responsible for transcriptional re-programming ultimately leading to production of defense hormones, jasmonic acid, ethylene, and salicylic acid (Lal et al. 2018).

Intracellular perception: policing the interior

The classical gene-for-gene interactions between host *R* genes and pathogen *Avr* genes had often been interpreted in terms of a receptor-ligand model in which R proteins act as receptors to directly detect pathogen Avr proteins or their enzymatic products (Keen 1990). This was confirmed as a general principle via the identification of the underlying host and pathogen genes and their functional analysis, although several unexpected variations in modes of recognition have been elucidated. As noted previously, most of the first plant *R* genes identified in the early 1990s encode intracellular NLR proteins (Bent et al. 1994; Mindrinos et al. 1994; Whitham et al. 1994; Grant et al. 1995; Lawrence et al. 1995). This family of proteins is highly diversified between and within species as they have adapted to recognize rapidly evolving pathogen factors. Hundreds of different NLRs have now been identified that confer disease resistance in a wide variety of plant species (Kourelis et al. 2021; Ngou et al. 2022).

NLR receptors detect intracellular signals, mostly corresponding to effector proteins delivered by pathogens into plant cells during infection. The concept of pathogen effector delivery itself emerged from the study of *Avr* genes. These were first identified from bacterial pathogens (Staskawicz et al. 1984), with subsequent work revealing that the encoded Avr proteins are part of a suite of proteins delivered into host cells via the bacterial type III secretion system, known as type III effectors (Kvitko and Collmer 2023). Many of these effectors specifically target components of the RLK/RLP immune system including co-receptors, RLCKs, and MAPKs to suppress these defense responses (Fig. 1). Although the first Avr proteins identified from the apoplastic fungus *F. fulvum* (de Wit et al. 1985) were secreted proteins detected extracellularly by RLPs, this turned out to be the exception rather than the rule for biotrophic and hemibiotrophic fungal and oomycete pathogens. Subsequently, *Avr-Pita* was identified from the hemibiotroph *Magnaporthe oryzae*, which propagates inside host cells, and encoded a small secreted protein (Orbach et al. 2000) that is delivered into the host cytoplasm during infection (Khang et al. 2010). Identification of *Avr* genes from several biotrophic fungi and oomycetes followed, showing the same trend of encoding small proteins secreted from specialized infection structures being recognized inside plant cells by NLR proteins (Allen et al. 2004; Dodds et al. 2004; Shan et al. 2004; Armstrong et al. 2005; Rehmany et al. 2005). Thus the paradigm of bacterial pathogens delivering a suite of effector proteins into host cells during infection was extended to filamentous (oomycete and fungal) pathogens during biotrophic growth (Dong and Ma 2021; Figueroa et al. 2021; Sperschneider and Dodds 2022). To date, more than 140 NLR-Avr protein pairs have been identified across a wide variety of plant and pathogen species (Ngou et al. 2022).

NLR-mediated effector recognition

Although gene-for-gene recognition had implied a receptor-ligand interaction, the first R-Avr pairs isolated from bacterial disease systems did not appear to conform to this model, since no direct interactions could be detected between the proteins. Instead, a "Guard" model was proposed, initially to explain recognition of the AvrPto effector of *P. syringae*, which required two host R proteins, Pto and Prf (Van der Biezen and Jones 1998). In this case, AvrPto interacted directly with Pto, a protein kinase, but not with Prf, an NLR, so it appeared that the role of the Prf NLR was to monitor the state of the Pto protein kinase to detect perturbations introduced by bacterial effectors targeting this kinase. This indirect recognition model was subsequently verified in detail for the RPS2 NLR, which responds to the degradation of the RIN4 protein by the *P. syringae* effector AvrRpt2 (Mackey et al. 2002, 2003; Axtell and Staskawicz 2003). Likewise, the *P. syringae* effectors AvrB or AvrRpm1 induce phosphorylation of RIN4, which is detected by another NLR protein,

RPM1 (Chung et al. 2011). A modification of this model is the "decoy" concept, where some host proteins have evolved specific functions as mimics of effector virulence targets to allow activation of associated NLRs (van der Hoorn and Kamoun 2008). Such indirect recognition is very common in resistance to bacterial pathogens, whose effectors often direct enzymatic modifications of their host targets (Ngou et al. 2022). Indirect recognition has also been observed for cell surface receptors, such as the tomato Cf-2, which recog-nizes the *F. fulva* effector Avr2 through its action on the apo-plastic protease Rcr3 (Rooney et al. 2005).

In contrast, direct NLR-effector recognition predominates in resistance to biotrophic fungi and oomycetes (Saur et al. 2021; Chen et al. 2022; Ngou et al. 2022). This was established as a significant mode of recognition by observations of direct binding determining the recognition specificity of the flax NLR proteins, L5, L6, and L7, for variants of the corresponding flax rust fungus (*Melampsora lini*) effector AvrL567 (Dodds et al. 2006), and between the tobacco N protein and the p50 helicase protein of tobacco mosaic virus (TMV) (Ueda et al. 2006). Many further examples of direct NLR-effector recognition have been characterized, mostly for biotrophic fungi and oomycetes, including between the *Arabidopsis* NLR RPP1 and the downy mildew effector ATR1 (Krasileva et al. 2010), the flax M protein and flax rust AvrM (Catanzariti et al. 2010), barley MLA NLRs and powdery mildew AvrMla proteins (Saur et al. 2019), and wheat Sr27, Sr35, and Sr50 NLRs with stem rust AvrSr27, AvrSr35, and AvrSr50 proteins, respectively (Chen et al. 2017b, 2023a; Salcedo et al. 2017; Forderer et al. 2022; Zhao et al. 2022).

Some NLRs combine elements of direct interaction and decoy-based recognition through the fusion of additional do-mains, termed integrated decoys (IDs), into the NLR receptor protein (Cesari et al. 2014a). These IDs present a binding site for the pathogen effector that mimics its natural target. Most NLR-IDs function as part of a pair of genetically linked NLRs, with the NLR-ID being the "sensor" responsible for effector detection, while the second NLR is a "helper" that executes the signaling response. The first such described NLR pairs were the rice Pik-1/ Pik-2 (Ashikawa et al. 2008) and RGA4/5 (Okuyama et al. 2011; Cesari et al. 2013) genes and the *Arabidopsis* RRS1/ RPS4 genes (Narusaka et al. 2009), although the latter two had previously been identified independently as single genes. The sensor proteins Pik-1 and RGA5 can recognize different effec-tors that bind to their HMA (heavy-metal-associated) IDs and trigger immunity through the executor NLRs Pik-2 and RGA4, respectively (Cesari et al. 2014b; Maqbool et al. 2015). The RRS1 protein recognizes two unrelated effectors, *P. syringae* AvrRps4 and *R. solanacearum* PopP2, through an ID that mimics their WRKY transcription factor targets and activates an RPS4 mediated immune response (Le Roux et al. 2015; Sarris et al. 2015). In the Solanaceae family, an ancient NLR gene pair has expanded into a large family of sensor and helper NLRs that function as a network (Wu et al. 2017). This family includes most of the CNLs previously defined as disease resistance genes in the Solanaceae, such as Rx, R3a, and Prf, which, in fact, act as sensors and signal through one or more helper NLRs of the NRC family (NLR-required for cell death) to confer disease resistance (Wu et al. 2016, 2017; Adachi et al. 2019).

NLR protein function

Plant NLRs share common NB and LRR domains but can be grouped further into two classes depending on their N-terminal domain: TIR-NLRs containing a toll/interleukin-1 receptor and resistance protein domain; or CC-NLRs containing a coiled-coil domain. Each NLR domain contributes different functions. Early experiments involving domain swaps between allelic variants of NLRs showed that the LRR domain was the main determinant of recognition specificity (Ellis et al. 1999; Dodds et al. 2001; Shen et al. 2003; Rairdan and Moffett 2006), and subsequent studies confirmed this as the inter-action site for certain directly recognized Avr proteins (Dodds et al. 2006; Krasileva et al. 2010). Tameling et al.(2006) first showed that the central NB domain of the tomato Rx NLR has an ATP hydrolysis activity, and an analysis of various inactivating and autoactivating mutations suggested that the activation state of NLRs is determined by a switch mediated by the NB binding to either ADP (the "OFF" state) or ATP (the "ON" state). Likewise, Williams et al. (2011) found that an autoactive variant of the flax M protein preferentially bound to ATP, while Bernoux et al. (2016) found that activation of the flax L6 protein relied on perturbations to the equilibrium between ADP and ATP bound states. The TIR and CC domains contribute the signaling function in NLRs, as overexpression of these truncated domains can often trigger cell death responses without their corresponding effectors (known as autoactivity). This was first recognized for the flax L10 TIR domain (Frost et al. 2004), and then shown for numerous other TIRs (Swiderski et al. 2009). Moreover, the TIR domains of RPP1, L6, and ROQ1 were found to be dispensable for binding of these NLRs to their corresponding effectors (Krasileva et al. 2010; Bernoux et al. 2011; Qi et al. 2018), indicating a function downstream of recognition. Cell death signaling by isolated CC domains was first observed for the barley powdery mildew resistance protein Mla10 (Maekawa et al. 2011) and *Arabidopsis* NRG1 and ADR1 (Collier et al. 2011) and later confirmed in many other CC-NLRs (Wang et al. 2015; Cesari et al. 2016; Wróblewski et al. 2018).

Several lines of research suggested that activated NLRs assemble into oligomeric complexes to signal. First, the tobacco N protein was found to self-associate upon recognition of the TMV helicase protein (Mestre and Baulcombe 2006). Subsequently, it was shown that the isolated TIR and CC domains of L6 and Mla10, respectively, each required self-association for their cell death signaling function (Bernoux et al. 2011; Maekawa et al. 2011). This has also been reported for numerous other TIR and CC domains (Williams et al. 2014; Casey et al. 2016; Cesari et al. 2016; Zhang et al. 2017). Furthermore, induced proximity of TIR domains by fusion to constitutively or inducibly oligo-meric proteins was sufficient to activate their signaling function (Horsefield et al. 2019; Duxbury et al. 2020; Bernoux et al. 2023). TIR domains were also shown to possess an NADase activity that can cleave NAD^+ or $NADP^+$ in vitro and is dependent on self-association, suggesting activation of the enzymatic function by oligomerization (Horsefield et al. 2019; Wan et al. 2019).

In the last few years, cryo-EM structural analysis of NLR complexes have provided a detailed synthesis of NLR immune receptor activation confirming and explaining many of the above functional observations (Burdett et al. 2019; Outram et al. 2022; Chai et al. 2023; Förderer and Kourelis 2023) (Fig. 2). The first breakthrough came with the elucidation of the *Arabidopsis* ZAR1 NLR structure in its inactive state, an intermediate pre-activation state, and a fully activated state, providing a clear picture of how this protein transitions as a result of effector recognition (Wang et al. 2019a, 2019b). ZAR1 interacts with several RLCK-VII proteins to indirectly detect multiple bacterial effector proteins that target these immune signaling components. The inactive ZAR1 structure was solved in complex with the RLCK protein resistance-related kinase 1 (RKS1) and contained ADP bound in the NB domain (Wang et al. 2019b). The bacterial effector AvrAC uridylylates another RLCK-like protein, PBL2, and the $PBL2^{UMP}$ then interacts with RKS1 to form a ZAR1-RKS1-$PBL2^{UMP}$ intermediate complex, in which a rotation has occurred within the NB domain of ZAR1 along with the substitution of the bound ADP by ATP (Wang et al. 2019b). This conformational change is driven by a steric clash between the positions of the ZAR1 NB domain in the inactivate state and the activation segment of RKS1 after binding to $PBL2^{UMP}$. The ZAR1-RKS1-$PBL2^{UMP}$ complex then forms a pentameric complex termed a resistosome (Wang et al. 2019a). The CC domain undergoes a substantial conformational change in the active pentamer, forming a funnel-shaped structure as the five CC domains come together. The activated ZAR1 resistosome associates with the plasma membrane and forms a Ca^{2+} permeable channel that is necessary for immunity signaling (Wang et al. 2019a; Bi et al. 2021). A similar pentameric ATP-bound resistosome structure was observed for the activated Sr35 NLR receptor from wheat, in this case directly bound to its corresponding effector ligand, AvrSr35 (Forderer et al. 2022; Zhao et al. 2022). The *N. benthamiana* NRC2 and NRC4 proteins also self-associate and form resistosome-like complexes after activation of sensor NLRs, suggesting a fundamentally similar operation of helper CC-NLRs in this network (Ahn et al. 2023; Contreras et al. 2023).

Cryo-EM structures of two activated TIR-NLRs also showed formation of oligomeric resistosome complexes, but in these cases involving a tetramer architecture (Ma et al. 2020; Martin et al. 2020). The ROQ1 and RPP1 TIR-NLRs both direct-ly interact with their corresponding effectors, XopQ1 and ATR1, respectively, through their LRR domains as well as a short domain located C-terminal to the LRR. In these tetrameric complexes, the TIR domains self-associate in such a way that the NADase active sites of two TIR subunits are exposed by the asymmetric interactions between these domains (Ma et al. 2020; Martin et al. 2020). This explains how effector-driven TIR-NLR oligomerization leads to signaling activation via this catalytic function. Similar to the CC-NLR resistosomes, the active ROQ1 complex has ATP bound to the NB (Martin et al. 2020), although the RPP1 resistosome is bound to ADP (Ma et al. 2020).

The EDS1 pathway connects TIR-NLRs to CC-NLR outputs

Early genetic attempts to define signaling pathways downstream of NLRs by mutagenesis identified only a few bona fide signaling pathway genes. The first of these was defined by the *eds1* (enhanced disease susceptibility) mutation in *Arabidopsis*, which abolished disease resistance mediated by various TIR-NLRs (Parker et al. 1996), with the wild-type *EDS1* gene encoding a lipase-like protein (Falk et al. 1999). Another key mutant identified the related protein PAD4 (phytoalexin deficient 4) (Jirage et al. 1999), which physically interacts with EDS1 (Feys et al. 2001), while a third protein (SAG101) in this family was identified in a EDS1 protein interaction screen (Feys et al. 2005). EDS1 forms mutually exclusive heterodimers with either PAD4 or SAG101, which are necessary for TIR-NLR-mediated plant immunity (Wagner et al. 2013).

A second layer of signaling proteins downstream of the EDS1-family proteins was also identified in genetic screens: NRG1 (N required gene) as required for TMV resistance mediated by the N protein in *N. benthamiana* (Peart et al. 2005); and ADR1 (Activated disease resistance) through activation tagging leading to enhanced resistance to virulent pathogens (Grant et al. 2003). These are part of two small CC-NLR families that are conserved in many different plant lineages (Shao et al. 2016) and whose CC domains are related to the *Arabidopsis* RPW8 protein and often designated as CC_R domains (Collier et al. 2011). NRG1 and ADR1 work with either the EDS1-SAG101 or EDS1-PAD4 heterocomplexes, respectively, to activate immunity (Qi et al. 2018; Castel et al. 2019; Lapin et al. 2019; Wu et al. 2019). Structural analysis of EDS1/PAD4 and EDS1-SAG101 heterodimers after TIR-NLR signaling activation showed that they each detect different small molecules produced by TIR enzymatic activity: 2′-(5″-phosphoribosyl)-5′-adenosine mono- or di-phosphate (pRib-AMP/ADP) in the case of EDS1/PAD4 and ADP-ribosylated ATP or ADP-ribosylated ADP ribose (ADPr-ATP/di-ADPR) for EDS1/SAG101 (Huang et al. 2022; Jia et al. 2022). Direct interaction of TIR domains with EDS1 complexes may facilitate efficient transfer of these short half- life metabolites (Chen et al. 2024). Binding of these TIR-derived molecules allosterically promotes interaction of EDS1/PAD4 and EDS1/SAG101 heterodimers with ADR1 or NRG1, respectively (Sun et al. 2021; Huang et al. 2022; Jia et al. 2022). Like other CC-NLRs, activated *At*NRG1.1 oligo-merizes and forms puncta in the plasma membrane, and both *At*NRG1.1 and *At*ADR1 can mediate Ca^{2+} influx after ac-tivation (Jacob et al. 2021; Feehan et al. 2023). Thus, TIR-NLR signaling culminates in activation of a CC-NLR resistosome (Fig. 2). How the CC domain-mediated Ca2+ channel activity leads to activation of NLR-induced responses is not under-stood, but these include many of the same outputs as in RLK/RLP-triggered immunity, and emerging evidence suggests extensive interactions between these pathways contribute to the final outcomes of infection.

An integrated plant immune signaling system

The RLK/RLP and NLR-mediated immunity fields emerged in parallel, with most research independently focused on one or the other system. Although it has been clear that both could be activated and contribute to outcomes during infection, they have generally been viewed as independent, with a common view that cell surface receptors offer a first line of defense, which, if breached, could be compensated by the stronger response triggered by intracellular receptors. Nevertheless, cell surface receptor and NLR responses converge on many similar defense responses, such as callose deposition, ROS production, Ca^{2+} influx of calcium, MAPK activation, defense gene upregulation, and plant hormone production, although the NLR responses are generally considered to be greater in magnitude (Peng et al. 2018). This convergence in signaling output has not been well understood, but there is increasing evidence of integrated signaling and crosstalk between the pathways and that an effective immune response to pathogen infection requires both systems acting synergistically (Bernoux et al. 2022; Rhodes et al. 2022). One point of connection between these systems is the EDS1/PAD4-ADR1 pathway, which in addition to acting downstream of TIR-NLRs is also required for LRR-RLP signaling in *Arabidopsis* (Pruitt et al. 2021). Consistent with this, these proteins directly associate with the SOBIR1 co-receptor, although it is still unclear how a signal is propagated between them (Fig. 2). EDS1, PAD4, and ADR1 are also partially required for RLK-induced signaling in *Arabidopsis*. However, Zonnchen et al. (2022) found that RLK/RLP responses in *N. benthamiana* leaves were independent of the EDS1 pathway genes, while Wang et al. (2023) found a specific role of EDS1/PAD4/ADR1 in flg22-induced stomatal closure in both *N. benthamiana* and *Arabidopsis*. While EDS1, PAD4, and ADR homologs are present in all plant species, SAG101 and NRG1 homologs are absent in monocots and some dicot plants, which also lack TIR-NLRs, consistent with a broader role of the EDS1/PAD4/ADR node in non-TIR im-munity signaling (Collier et al. 2011; Shao et al. 2016).

There is also evidence that RLK/RLP signaling is required for NLR outputs, with knockout mutations of co-receptors and RLCKs substantially inhibiting bacterial resistance mediated by the *Arabidopsis* NLRs RPS2 or RPS4/RRS1 (Ngou et al. 2021; Yuan et al. 2021). Similarly, expression of bacterial effectors recognized by the RPS4/RRS1 or RPP4 TIR-NLRs using inducible promoters in the absence of the pathogen did not trigger cell death, which required co-delivery of PAMPs (Ngou et al. 2021). The CNGC2/4 Ca^{2+} channels activated by RLK/RLP signaling also play a role in NLR responses, since their mutations in the *dnd1/2* mutants of *Arabidopsis* lead to loss of NLR-mediated cell death but not bacterial resistance (Clough et al. 2000; Jurkowski et al. 2004). Similarly, MAPK3 and 6 also contribute to salicylic acid production and photosynthetic inhibition triggered by NLRs (Su et al. 2018; Lang et al. 2022).

Part of the explanation for the mutually reinforcing relationship between RLK/RLP and NLR signaling is that both responses lead to transcriptional upregulation and post-transcriptional modification of key components in both signaling pathways. NLR signaling results in enhanced expression of RLKs, RLPs, their co-receptors, and other signaling partners; similarly, NLRs and EDS1 pathway genes are also upregulated by RLK/RLP signaling (Navarro et al. 2004; Ngou et al. 2021; Yuan et al. 2021). Feehan et al. (2023) found that NRG1 was recruited to form complexes with EDS1-SAG101 after TIR-NLR activation in *Arabidopsis* using chemically induced transient expression. However, these complexes did not self-associate into an oligo-meric resistosome unless the nonpathogenic bacterium *Pseudomonas fluorescens* was also infiltrated, suggesting a requirement for an RLK/RLP activation. This suggests a more direct functional connection between these signaling pathways, although the mechanism behind this phenomenon remains to be determined. On the other hand, NLRs provide virus resist-ance in the absence of known PAMP triggers. Taken together, it is clear that cell surface and intracellular immunity are not entirely separate pathways, but there is much work to be done to dissect how they interact to generate effective disease resist-ance responses.

NLR engineering

Although new resistance genes continue to be identified, the recognition repertoire available for breeding in the gene pools of crop species is limited. Thus, a long-term goal of plant immunity research has been to develop the capacity to engineer bespoke disease resistance genes to respond to rapidly evolving pathogens. Several approaches have already emerged from advances in understanding NLR biology and are illustrated in Fig. 3 (Marchal et al. 2022). Since many NLRs recognize effec-tors or decoy proteins through their LRR domain (Chen et al. 2022), recognition specificity can be transferred between re-lated NLRs through swapping whole LRRs (Ellis et al. 1999; Dodds et al. 2001; Bauer et al. 2021) or selected sites within the LRR (Tamborski et al. 2023). Random mutagenesis of the LRR has also been used to expand recognition specificity (Farnham and Baulcombe 2006; Segretin et al. 2014). The recent structural biology advances in understanding the molecular basis of NLR resistosome provide opportunities for targeted engineering approaches. For example, following structure resolution of the AvSr35-Sr35 resistosome, Forderer and colleagues (2022) showed that exchange of the interacting residues from the ascending lateral side of the LRR domain allowed transfer of AvrSr35 recognition to a related NLR (Forderer et al. 2022). Unexpectedly, transfer of the whole Sr35 LRR domain to the distantly related barley HvMLA10 and HvMLA13 NLRs could also confer AvrSr35 specificity, indicating that such exchanges are not necessarily limited to highly sequence-similar NLRs. The recent releases of artificial intelligence tools, such as AlphaFold2, that can predict protein structures individually and in complex should greatly expand the capacity to conduct such targeted

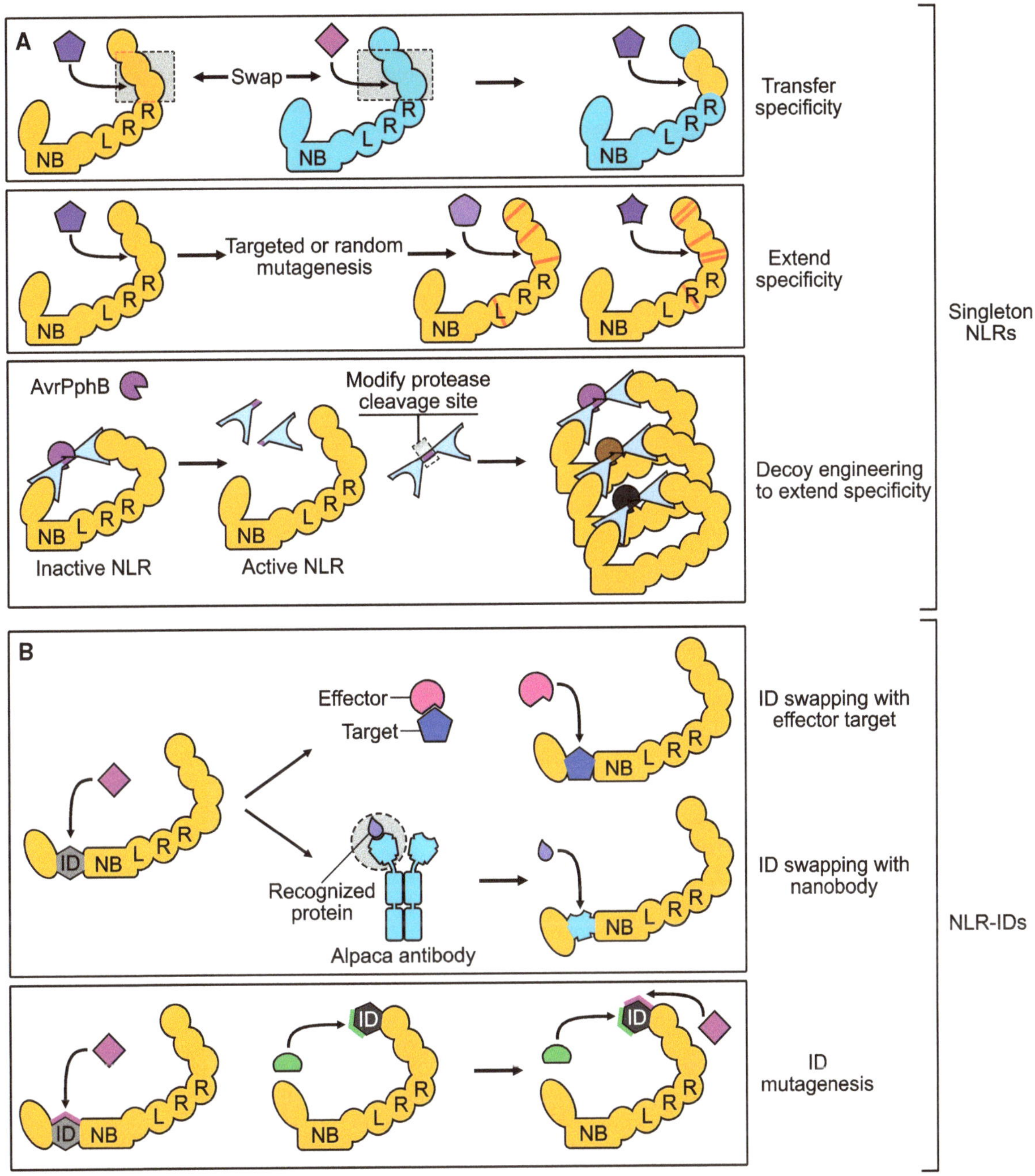

Figure 3. Overview of NLR engineering approaches. **A)** Three broad approaches have been undertaken with respect to engineering singleton NLRs: transferring specificity between two closely related NLRs by swapping the residues conferring effector recognition, targeted mutagenesis, or random mutagenesis to expand the effector recognition profile and decoy engineering. The most prominent example of the latter has been altering the specificity of RPS5 via modification of the decoy protein PBS1 by introducing different protease cleavage sites. **B)** Structure-guided mutagenesis of integrated domain (ID)-containing NLRs has allowed the transfer of recognition and expanded specificity through restructuring the ID domain by swapping ID domains with effector targets can extend specificities to virulent effectors; replacing ID domains with nanobodies specific to secreted effector proteins has the potential to produce made-to-order resistance genes.

engineering in the absence of experimentally determined structures. While the majority of resistance engineering approaches have focused to date on NLRs, modification of effector targets or decoy proteins presents another option. For example, alteration of the AvrPphB protease cleavage site in its target PBS1 allowed its guard NLR (RPS5) to recognize other pathogen proteases (Kim et al. 2016; Helm et al. 2019). Since PBS1 is conserved across flowering plants, this approach could have wide applications (Pottinger et al. 2020). Substituting host effector targets with versions not recognized by the effector can also lead to enhanced resistance (McLellan et al. 2022).

Although the activation mechanism of NLR-IDs and their helper NLRs still remains unclear, this protein architecture offers inherent advantages and flexibility for precise engineering through changes to the ID domain, with the rice Pik-1/Pik-2 and RGA4/RGA5 pairs being extensively tested. For example, using structure guided mutagenesis, Cesari et al. (2022) introduced residues from the AVR-PikD binding interface of the Pikp-1 HMA domain to the RGA5 HMA domain, resulting in the latter gaining recognition for AVR-PikD and retaining recognition of its cognate effectors AVR1-CO39 and AVR-Pia in *N. benthamiana* expression assays. However, transgenic rice plants showed resistance only to *M. oryzae* strains expressing AVR1-CO39 and AVR-Pia, not AVR-PikD, indicating potential discrepancies between heterologous expression and native disease systems. In another approach, the RGA5 HMA domain was engineered to recognize AVR-Pib, which is structurally similar to the recognized effectors AVR1-CO39 and AVR-Pia, and showed both specific recognition leading to cell death in *N. benthamiana* and disease resistance in transgenic rice (Liu et al. 2021). Similarly, structure comparison between the Pikp and Pikm HMA domains provided information to engin-eer PikP and expand its recognition profiles to unrecognized AVR-Pik variants (De la Concepcion et al. 2019). Maidment et al. (2023) replaced the HMA domain of Pikp-1 with the HMA domain from OsHIPP19, a target protein of multiple AVR-Piks, allowing recognition of two variants not recognized by native Pik-1. In an even more ambitious approach, Kourelis et al. (2023) showed that the HMA domain of Pikm-1 can be swapped with single-domain heavy-chain antibodies (nanobodies) allowing the engineered proteins, dubbed "Pikobodies," to recognize protein ligands detected by the inserted nanobodies. Stable transgenic *N. benthamiana* plants expressing Pikobodies targeted to fluorescent proteins showed resistance to *Potato virus X* expressing these proteins (Kourelis et al. 2023). Given the capacity to generate nanobodies detect-ing any targeted protein, this approach holds great promise for designing new-to-nature resistance specificities.

Conclusions

Plant immunity research has advanced enormously building on twin bases of genetic studies of variation in disease outcomes mediated by plant and pathogen genotypes and cell biology studies of plant responses to pathogen-derived molecules. These research streams have coalesced into a broad under-standing of the main principles of plant immunity, involving in-tegration of various immune receptor detection functions into defense signaling pathways. There remain considerable areas of uncertainty, especially in resolving the connections between surface and intracellular perception signaling pathways, which are the focus of current research. One new approach being brought to bear is the use of single-cell analysis (Delannoy et al. 2023; Zhu et al. 2023) to resolve spatial differences in cel-lular responses within plant tissues and to discriminate be-tween cells responding directly to pathogen interactions and those responding to secondary signals, such as DAMPs and phytocytokines induced in interacting cells. The advances in understanding plant immunity mechanisms have provided the theoretical framework for new efforts to engineer immune receptors with new-to-nature recognition capacities to enhance the resistance options available to agriculture, which will also be an exciting avenue to explore in the coming years.

Funding

This work was supported by the Commonwealth Scientific and Industrial Research Organization Research Office.

Conflict of interest statement. None declared.

References

Adachi H, Contreras MP, Harant A, Wu CH, Derevnina L, Sakai T, Duggan C, Moratto E, Bozkurt TO, Maqbool A, et al. An N-terminal motif in NLR immune receptors is functionally conserved across distantly related plant species. Elife. 2019:**8**:e49956. https://doi.org/10.7554/eLife.49956

Ahn HK, Lin X, Olave-Achury AC, Derevnina L, Contreras MP, Kourelis J, Wu CH, Kamoun S, Jones JDG. Effector-dependent activation and oligomerization of plant NRC class helper NLRs by sensor NLR immune receptors Rpi-amr3 and Rpi-amr1. EMBO J. 2023:**42**(5): e111484. https://doi.org/10.15252/embj.2022111484

Albert I, Böhm H, Albert M, Feiler CE, Imkampe J, Wallmeroth N, Brancato C, Raaymakers TM, Oome S, Zhang HQ, et al. An RLP23-SOBIR1-BAK1 complex mediates NLP-triggered immunity. Nat Plants. 2015:**1**(10):15140. https://doi.org/10.1038/nplants.2015.140

Allen RL, Bittner-Eddy PD, Grenvitte-Briggs LJ, Meitz JC, Rehmany AP, Rose LE, Beynon JL. Host-parasite coevolutionary conflict between Arabidopsis and downy mildew. Science. 2004:**306**(5703): 1957–1960. https://doi.org/10.1126/science.1104022

Armstrong MR, Whisson SC, Pritchard L, Bos JIB, Venter E, Avrova AO, Rehmany AP, Bohme U, Brooks K, Cherevach I, et al. An ancestral oomycete locus contains late blight avirulence gene *Avr3a*, encoding a protein that is recognized in the host cytoplasm. Proc Natl Acad Sci U S A. 2005:**102**(21):7766–7771. https://doi.org/10.1073/pnas.0500113102

Ashikawa I, Hayashi N, Yamane H, Kanamori H, Wu JZ, Matsumoto T, Ono K, Yano M. Two adjacent nucleotide-binding site-leucine-rich repeat class genes are required to confer Pikm-specific rice blast resistance. Genetics. 2008:**180**(4):2267–2276. https://doi.org/10.1534/genetics.108.095034

Axtell MJ, Staskawicz BJ. Initiation of RPS2-specified disease resistance in Arabidopsis is coupled to the AvrRpt2-directed elimination of RIN4. Cell. 2003:**112**(3):369–377. https://doi.org/10.1016/S0092-8674(03)00036-9

Bauer S, Yu D, Lawson AW, Saur IML, Frantzeskakis L, Kracher B, Logemann E, Chai J, Maekawa T, Schulze-Lefert P. The leucine-rich repeats in allelic barley MLA immune receptors define specificity towards sequence-unrelated powdery mildew avirulence effectors with a predicted common RNase-like fold. PLoS Pathog. 2021:**17**: e1009223. https://doi.org/10.1371/journal.ppat.1009223

Bent AF, Kunkel BN, Dahlbeck D, Brown KL, Schmidt R, Giraudat J, Leung J, Staskawicz BJ. RPS2 of *Arabidopsis thaliana*: a leucine-rich repeat class of plant disease resistance genes. Science. 1994:**265**(5180): 1856–1860. https://doi.org/10.1126/science.8091210

Bentham AR, De la Concepcion JC, Mukhi N, Zdrzalek R, Draeger M, Gorenkin D, Hughes RK, Banfield MJ. A molecular roadmap to the plant immune system. J Biol Chem. 2020:**295**(44):14916–14935. https://doi.org/10.1074/jbc.REV120.010852

Bernoux M, Burdett H, Williams SJ, Zhang X, Chen C, Newell K, Lawrence GJ, Kobe B, Ellis JG, Anderson PA, et al. Comparative analysis of the flax immune receptors L6 and L7 suggests an equilibrium-based switch activation model. Plant Cell. 2016:**28**(1): 146–159. https://doi.org/10.1105/tpc.15.00303

Bernoux M, Chen J, Zhang XX, Newell K, Hu J, Deslandes L, Dodds P. Subcellular localization requirements and specificities for plant immune receptor toll-interleukin-1 receptor signaling. Plant J. 2023:**114**(6):1319–1337. https://doi.org/10.1111/tpj.16195

Bernoux M, Ve T, Williams S, Warren C, Hatters D, Valkov E, Zhang X, Ellis JG, Kobe B, Dodds PN. Structural and functional analysis of a plant resistance protein TIR domain reveals interfaces for self-association, signaling, and autoregulation. Cell Host Microbe. 2011:**9**(3):200–211. https://doi.org/10.1016/j.chom.2011.02.009

Bernoux M, Zetzsche H, Stuttmann J. Connecting the dots between cell surface- and intracellular-triggered immune pathways in plants. Curr Opin Plant Biol. 2022:**69**:102276. https://doi.org/10.1016/j.pbi.2022.102276

Bi G, Su M, Li N, Liang Y, Dang S, Xu J, Hu M, Wang J, Zou M, Deng Y, et al. The ZAR1 resistosome is a calcium-permeable channel triggering plant immune signaling. Cell. 2021:**184**(13):3528–3541.e12. https://doi.org/10.1016/j.cell.2021.05.003

Bi GZ, Zhou ZY, Wang WB, Li L, Rao SF, Wu Y, Zhang XJ, Menke FLH, Chen S, Zhou J-M. Receptor-like cytoplasmic kinases directly link diverse pattern recognition receptors to the activation of mitogen-activated protein kinase cascades in Arabidopsis. Plant Cell. 2018:**30**(7):1543–1561. https://doi.org/10.1105/tpc.17.00981

Biffen RH. Mendefs laws of inheritance and wheat breeding. J Agric Sci. 1905:**1**(1):4–48. https://doi.org/10.1017/S0021859600000137

Brutus A, Sicilia F, Macone A, Cervone F, De Lorenzo G. A domain swap approach reveals a role of the plant wall-associated kinase 1 (WAK1) as a receptor of oligogalacturonides. Proc Nat Acad Sci U S A. 2010:**107**(20):9452–9457. https://doi.org/10.1073/pnas.1000675107

Burdett H, Bentham AR, Williams SJ, Dodds PN, Anderson PA, Banfield MJ, Kobe B. The plant "resistosome": structural insights into immune signaling. Cell Host Microbe. 2019:**26**(2):193–201. https://doi.org/10.1016/j.chom.2019.07.020

Casey LW, Lavrencic P, Bentham AR, Cesari S, Ericsson DJ, Croll T, Turk D, Anderson PA, Mark AE, Dodds PN, et al. The CC domain structure from the wheat stem rust resistance protein Sr33 challenges paradigms for dimerization in plant NLR proteins. Proc Natl Acad Sci U S A. 2016:**113**(45):12856–12861. https://doi.org/10.1073/pnas.1609922113

Castel B, Ngou P-M, Cevik V, Redkar A, Kim D-S, Yang Y, Ding P, Jones JDG. Diverse NLR immune receptors activate defence via the RPW8-NLR NRG1. New Phytol. 2019:**222**(2):966–980. https://doi.org/10.1111/nph.15659

Catanzariti A, Dodds P, Ve T, Kobe B, Ellis J, Staskawicz B. The AvrM effector from flax rust has a structured C-terminal domain and interacts directly with the M resistance protein. Mol Plant Microbe Interact. 2010:**23**(1):49–57. https://doi.org/10.1094/MPMI-23-1-0049

Catanzariti AM, Lim GTT, Jones DA. The tomato I-3 gene: a novel gene for resistance to Fusarium wilt disease. New Phytol. 2015:**207**(1):106–118. https://doi.org/10.1111/nph.13348

Cesari S, Bernoux M, Moncuquet P, Kroj T, Dodds PN. A novel conserved mechanism for plant NLR protein pairs: the "integrated decoy" hypothesis. Front Plant Sci. 2014a:**5**:606. https://doi.org/10.3389/fpls.2014.00606

Cesari S, Kanzaki H, Fujiwara T, Bernoux M, Chalvon V, Kawano Y, Shimamoto K, Dodds P, Terauchi R, Kroj T. The NB-LRR proteins RGA4 and RGA5 interact functionally and physically to confer disease resistance. EMBO J. 2014b:**33**(17):1941–1959. https://doi.org/10.15252/embj.201487923

Cesari S, Moore J, Chen C, Webb D, Periyannan S, Mago R, Bernoux M, Lagudah ES, Dodds PN. Cytosolic activation of cell death and stem rust resistance by cereal MLA-family CC-NLR proteins. Proc Natl Acad Sci U S A. 2016:**113**(36):10204–10209. https://doi.org/10.1073/pnas.1605483113

Cesari S, Thilliez G, Ribot C, Chalvon V, Michel C, Jauneau A, Rivas S, Alaux L, Kanzaki H, Okuyama Y, et al. The rice resistance protein pair RGA4/RGA5 recognizes the *Magnaporthe oryzae* effectors AVR-Pia and AVR1-CO39 by direct binding. Plant Cell. 2013:**25**(4): 1463–1481. https://doi.org/10.1105/tpc.112.107201

Cesari S, Xi Y, Declerck N, Chalvon V, Mammri L, Pugnière M, Henriquet C, de Guillen K, Chochois V, Padilla A, et al. New recognition specificity in a plant immune receptor by molecular engineering of its integrated domain. Nat Commun. 2022:**13**(1):1524. https://doi.org/10.1038/s41467-022-29196-6

Chai JJ, Song W, Parker JE. New biochemical principles for NLR immunity in plants. Mol Plant Microbe Interact. 2023:**36**(8):468–475. https://doi.org/10.1094/MPMI-05-23-0073-HH

Chen DQ, Cao YR, Li H, Kim D, Ahsan N, Thelen J, Stacey G. Extracellular ATP elicits DORN1-mediated RBOHD phosphorylation to regulate stomatal aperture. Nat Commun. 2017a:**8**(1):2265. https://doi.org/10.1038/s41467-017-02340-3

Chen J, Luo M, Hands P, Rolland V, Zhang J, Li Z, Outram M, Dodds P, Ayliffe M. A split GAL4 RUBY assay for visual in planta detection of protein-protein interactions. Plant J. 2023:**114**(6):1209–1226. https://doi.org/10.1111/tpj.16234

Chen J, Upadhyaya NM, Ortiz D, Sperschneider J, Li F, Bouton C, Breen S, Dong C, Xu B, Zhang X, et al. Loss of AvrSr50 by somatic exchange in stem rust leads to virulence for Sr50 resistance in wheat. Science. 2017b:**358**(6370):1607–1610. https://doi.org/10.1126/science.aao4810

Chen J, Zhang X, Bernoux M, Rathjen JP, Dodds PN. Plant Toll/interleukin-1 receptor/resistance protein domains physically associate with enhanced disease susceptibility1 family proteins in immune signaling. iScience. 2024:108817. https://doi.org/10.1016/j.isci.2024.108817

Chen J, Zhang X, Rathjen JP, Dodds PN. Direct recognition of pathogen effectors by plant NLR immune receptors and downstream signalling. Essays Biochem. 2022:**66**(5):471–483. https://doi.org/10.1042/EBC20210072

Chinchilla D, Bauer Z, Regenass M, Boller T, Felix G. The Arabidopsis receptor kinase FLS2 binds flg22 and determines the specificity of flagellin perception. Plant Cell. 2006:**18**(2):465–476. https://doi.org/10.1105/tpc.105.036574

Chinchilla D, Zipfel C, Robatzek S, Kemmerling B, Nürnberger T, Jones JDG, Felix G, Boller T. A flagellin-induced complex of the receptor FLS2 and BAK1 initiates plant defence. Nature. 2007:**448**(7152): 497–U412. https://doi.org/10.1038/nature05999

Chung EH, da Cunha L, Wu AJ, Gao ZY, Cherkis K, Afzal AJ, Mackey D, Dangl JL. Specific threonine phosphorylation of a host target by two unrelated type III effectors activates a host innate immune receptor in plants. Cell Host Microbe. 2011:**9**(2):125–136. https://doi.org/10.1016/j.chom.2011.01.009

Clough SJ, Fengler KA, Yu IC, Lippok B, Smith RK, Bent AF. The Arabidopsis dnd1 "defense, no death" gene encodes a mutated cyclic nucleotide-gated ion channel. Proc Natl Acad Sci U S A. 2000:**97**(16): 9323–9328. https://doi.org/10.1073/pnas.150005697

Collier SM, Hamel LP, Moffett P. Cell death mediated by the N-terminal domains of a unique and highly conserved class of NB-LRR protein. Mol Plant Microbe Interact. 2011:**24**(8):918–931. https://doi.org/10.1094/MPMI-03-11-0050

Contreras MP, Pai H, Tumtas Y, Duggan C, Yuen ELH, Cruces AV, Kourelis J, Ahn HK, Lee KT, Wu CH, et al. Sensor NLR immune proteins activate oligomerization of their NRC helpers in response to plant pathogens. EMBO J. 2023:**42**(5):e111519. https://doi.org/10.15252/embj.2022111519

Couto D, Zipfel C. Regulation of pattern recognition receptor signalling in plants. NatRev Immunol. 2016:**16**(9):537–552. https://doi.org/10.1038/nri.2016.77

Cruickshank IAM, Perrin DR. The isolation and partial characterization of monilicolin A, a polypeptide with phaseollin-inducing activity from Monilinia Fructicola. Life Sci. 1968:**7**(10):449–458. https://doi.org/10.1016/0024-3205(68)90061-1

DeFalco TA, Zipfel C. Molecular mechanisms of early plant pattern-triggered immune signaling. Mol Cell. 2021:**81**(17):3449–3467. https://doi.org/10.1016/j.molcel.2021.07.029

De la Concepcion JC, Franceschetti M, MacLean D, Terauchi R, Kamoun S, Banfield MJ. Protein engineering expands the effector recognition profile of a rice NLR immune receptor. Elife. 2019:**8**: e47713. https://doi.org/10.7554/eLife.47713

Delannoy E, Batardiere B, Pateyron S, Soubigou-Taconnat L, Chiquet J, Colcombet J, Lang J. Cell specialization and coordination in Arabidopsis leaves upon pathogenic attack revealed by scRNA-seq. Plant Commun. 2023:**4**(5):100676. https://doi.org/10.1016/j.xplc.2023.100676

de Wit PJGM, Hofman AE, Velthuis GCM, Kuc JA. Isolation and characterization of an elicitor of necrosis isolated from intercellular fluids of compatible interactions of Cladosporium-Fulvum (Syn Fulvia-Fulva) and tomato. Plant Physiol. 1985:**77**(3):642–647. https://doi.org/10.1104/pp.77.3.642

Dixon MS, Hatzixanthis K, Jones DA, Harrison K, Jones JDG. The tomato disease resistance gene and six homologs show pronounced allelic variation in leucine-rich repeat copy number. Plant Cell. 1998:**10**(11):1915–1925. https://doi.org/10.1105/tpc.10.11.1915

Dodds P, Lawrence G, Ellis J. Six amino acid changes confined to the leucine-rich repeat beta-strand/beta-turn motif determine the difference between the P and P2 rust resistance specificities in flax. Plant Cell. 2001:**13**(1):163–178. https://doi.org/10.1105/tpc.13.1.163

Dodds P, Rathjen J. Plant immunity: towards an integrated view of plant-pathogen interactions. Nat Rev Genet. 2010:**11**(8):539–548. https://doi.org/10.1038/nrg2812

Dodds PN. From gene-for-gene to resistosomes: Flor's enduring legacy. Mol Plant Microbe Interact. 2023:**36**(8):461–467. https://doi.org/10.1094/MPMI-06-23-0081-HH

Dodds PN, Lawrence GJ, Catanzariti AM, Ayliffe MA, Ellis JG. The *Melampsora lini* AvrL567 avirulence genes are expressed in haustoria and their products are recognized inside plant cells. Plant Cell. 2004:**16**(3):755–768. https://doi.org/10.1105/tpc.020040

Dodds PN, Lawrence GJ, Catanzariti AM, Teh T, Wang CI, Ayliffe MA, Kobe B, Ellis JG. Direct protein interaction underlies gene-for-gene specificity and coevolution of the flax resistance genes and flax rust avirulence genes. Proc Natl Acad Sci U S A. 2006:**103**(23):8888–8893. https://doi.org/10.1073/pnas.0602577103

Dong S, Ma W. How to win a tug-of-war: the adaptive evolution of *Phytophthora* effectors. Curr Opin Plant Biol. 2021:**62**:102027. https://doi.org/10.1016/j.pbi.2021.102027

Du J, Verzaux E, Chaparro-Garcia A, Bijsterbosch G, Keizer LCP, Zhou J, Liebrand TWH, Xie C, Govers F, Robatzek S, et al. Elicitin recognition confers enhanced resistance to *Phytophthora infestans* in potato. Nat Plants. 2015:**1**(4):15034. https://doi.org/10.1038/nplants.2015.34

Duxbury Z, Wang S, MacKenzie CI, Tenthorey JL, Zhang X, Huh SU, Hu L, Hill L, Ngou PM, Ding P, et al. Induced proximity of a TIR signaling domain on a plant-mammalian NLR chimera activates defense in plants. Proc Natl Acad Sci U S A. 2020:**117**(31):18832–18839. https://doi.org/10.1073/pnas.2001185117

Ellis J, Lawrence G, Luck J, Dodds P. Identification of regions in alleles of the flax rust resistance gene L that determine differences in gene-for-gene specificity. Plant Cell. 1999:**11**(3):495–506. https://doi.org/10.1105/tpc.11.3.495

Falk A, Feys BJ, Frost LN, Jones JD, Daniels MJ, Parker JE. EDS1, an essential component of R gene-mediated disease resistance in Arabidopsis has homology to eukaryotic lipases. Proc Natl Acad Sci U S A. 1999:**96**(6):3292–3297. https://doi.org/10.1073/pnas.96.6.3292

Farnham G, Baulcombe DC. Artificial evolution extends the spectrum of viruses that are targeted by a disease-resistance gene from potato. Proc Natl Acad Sci U S A. 2006:**103**:18828–18833. https://doi.org/10.3389/fmicb.2016.00600

Feehan JM, Wang J, Sun X, Choi J, Ahn HK, Ngou BPM, Parker JE, Jones JDG. Oligomerization of a plant helper NLR requires cell-surface and intracellular immune receptor activation. Proc Natl Acad Sci U S A. 2023:**120**(11):e2210406120. https://doi.org/10.1073/pnas.2210406120

Feys BJ, Moisan LJ, Newman MA, Parker JE. Direct interaction between the Arabidopsis disease resistance signaling proteins, EDS1 and PAD4. EMBO J. 2001:**20**(19):5400–5411. https://doi.org/10.1093/emboj/20.19.5400

Feys BJ, Wiermer M, Bhat RA, Moisan LJ, Medina-Escobar N, Neu C, Cabral A, Parker JE. Arabidopsis SENESCENCE-ASSOCIATED GENE101 stabilizes and signals within an ENHANCED DISEASE SUSCEPTIBILITY1 complex in plant innate immunity. Plant Cell. 2005:**17**(9):2601–2613. https://doi.org/10.1105/tpc.105.033910

Figueroa M, Ortiz D, Henningsen EC. Tactics of host manipulation by intracellular effectors from plant pathogenic fungi. Curr Opin Plant Biol. 2021:**62**:102054. https://doi.org/10.1016/j.pbi.2021.102054

Flor HH. Inheritance of pathogenicity in *Melampsora lini*. Phytopathology. 1942:**32**:653–669.

Förderer A, Kourelis J. NLR immune receptors: structure and function in plant disease resistance. Biochem Soc Trans. 2023:**51**(4):1473–1483. https://doi.org/10.1042/BST20221087

Forderer A, Li E, Lawson AW, Deng Y-N, Sun Y, Logemann E, Zhang X, Wen J, Han Z, Chang J, et al. A wheat resistosome defines common principles of immune receptor channels. Nature. 2022:**610**(7932): 532–539. https://doi.org/10.1038/s41586-022-05231-w

Frost D, Way H, Howles P, Luck J, Manners J, Hardham A, Finnegan J, Ellis J. Tobacco transgenic for the flax rust resistance gene L expresses allele-specific activation of defense responses. Mol Plant Microbe Interact. 2004:**17**(2):224–232. https://doi.org/10.1094/MPMI.2004.17.2.224

Ge D, Yeo I-C, Shan L. Knowing me, knowing you: self and non-self recognition in plant immunity. Essays Biochem. 2022:**66**(5):447–458. https://doi.org/10.1042/EBC20210095

Gómez-Gómez L, Boller T. FLS2: an LRR receptor-like kinase involved in the perception of the bacterial elicitor flagellin in Arabidopsis. Mol Cell. 2000:**5**(6):1003–1011. https://doi.org/10.1016/S1097-2765(00)80265-8

Graham TL, Sequeira L, Huang TSR. Bacterial lipopolysaccharides as inducers of disease resistance in tobacco. Appl Environ Microb. 1977:**34**(4):424–432. https://doi.org/10.1128/aem.34.4.424-432.1977

Grant JJ, Chini A, Basu D, Loake GJ. Targeted activation tagging of the Arabidopsis NBS-LRR gene, ADR1, conveys resistance to virulent pathogens. Mol Plant Microbe Interact. 2003:**16**(8):669–680. https://doi.org/10.1094/MPMI.2003.16.8.669

Grant MR, Godiard L, Straube E, Ashfield T, Lewald J, Sattler A, Innes RW, Dangl JL. Structure of the Arabidopsis RPM1 gene enabling dual specificity disease resistance. Science. 1995:**269**(5225): 843–846. https://doi.org/10.1126/science.7638602

Heese A, Hann DR, Gimenez-Ibanez S, Jones AME, He K, Li J, Schroeder JI, Peck SC, Rathjen JP. The receptor-like kinase SERK3/BAK1 is a central regulator of innate immunity in plants. Proc Natl Acad Sci U S A. 2007:**104**(29):12217–12222. https://doi.org/10.1073/pnas.0705306104

Helm M, Qi M, Sarkar S, Yu H, Whitham SA, Innes RW. Engineering a decoy substrate in soybean to enable recognition of the soybean mosaic virus NIa protease. Mol Plant Microbe Interact. 2019:**32**(6): 760–769. https://doi.org/10.1094/MPMI-12-18-0324-R

Horsefield S, Burdett H, Zhang X, Manik MK, Shi Y, Chen J, Qi T, Gilley J, Lai JS, Rank MX, et al. NAD(+) cleavage activity by animal and plant TIR domains in cell death pathways. Science. 2019:**365**(6455):793–799. https://doi.org/10.1126/science.aax1911

Huang S, Jia A, Song W, Hessler G, Meng Y, Sun Y, Xu L, Laessle H, Jirschitzka J, Ma S, et al. Identification and receptor mechanism of TIR-catalyzed small molecules in plant immunity. Science. 2022:**377**(6605):eabq3297. https://doi.org/10.1126/science.abq3297

Jacob P, Kim NH, Wu F, El-Kasmi F, Chi Y, Walton WG, Furzer OJ, Lietzan AD, Sunil S, Kempthorn K, et al. Plant "helper" immune receptors are Ca(2+)-permeable nonselective cation channels. Science. 2021:**373**(6553):420–425. https://doi.org/10.1126/science.abg7917

Jia A, Huang S, Song W, Wang J, Meng Y, Sun Y, Xu L, Laessle H, Jirschitzka J, Hou J, et al. TIR-catalyzed ADP-ribosylation reactions produce signaling molecules for plant immunity. Science. 2022:**377**(6605):eabq8180. https://doi.org/10.1126/science.abq8180

Jirage D, Tootle TL, Reuber TL, Frost LN, Feys BJ, Parker JE, Ausubel FM, Glazebrook J. Arabidopsis thaliana PAD4 encodes a lipase-like gene that is important for salicylic acid signaling. Proc Natl Acad Sci U S A. 1999:**96**(23):13583–13588. https://doi.org/10.1073/pnas.96.23.13583

Jones DA, Thomas CM, Hammondkosack KE, Balintkurti PJ, Jones JDG. Isolation of the tomato Cf-9 gene for resistance to Cladosporium-Fulvum by transposon tagging. Science. 1994:**266**(5186):789–793. https://doi.org/10.1126/science.7973631

Jones JD, Dangl JL. The plant immune system. Nature. 2006:**444**(7117): 323–329. https://doi.org/10.1038/nature05286

Jurkowski GI, Smith RK, Yu IC, Ham JH, Sharma SB, Klessig DF, Fengler KA, Bent AF. Arabidopsis DND2, a second cyclic nucleotide-gated ion channel gene for which mutation causes the "defense, no death" phenotype. Mol Plant Microbe Interact. 2004:**17**(5):511–520. https://doi.org/10.1094/MPMI.2004.17.5.511

Kadota Y, Shirasu K, Zipfel C. Regulation of the NADPH oxidase RBOHD during plant immunity. Plant Cell Physiol. 2015:**56**(8): 1472–1480. https://doi.org/10.1093/pcp/pcv063

Kadota Y, Sklenar J, Derbyshire P, Stransfeld L, Asai S, Ntoukakis V, Jones JDG, Shirasu K, Menke F, Jones A, et al. Direct regulation of the NADPH oxidase RBOHD by the PRR-associated kinase BIK1 during plant immunity. Mol Cell. 2014:**54**(1):43–55. https://doi.org/10.1016/j.molcel.2014.02.021

Keen NT. Gene-for-gene complementarity in plant-pathogen interactions. Annu Rev Genet. 1990:**24**(1):447–463. https://doi.org/10.1146/annurev.ge.24.120190.002311

Khang CH, Berruyer R, Giraldo MC, Kankanala P, Park SY, Czymmek K, Kang S, Valent B. Translocation of *Magnaporther oryzae* effectors into rice cells and their subsequent cell-to-cell movement. Plant Cell. 2010:**22**(4):1388–1403. https://doi.org/10.1105/tpc.109.069666

Kim SH, Qi D, Ashfield T, Helm M, Innes RW. Using decoys to expand the recognition specificity of a plant disease resistance protein. Science. 2016:**351**(6274):684–687. https://doi.org/10.1126/science.aad3436

Kourelis J, Marchal C, Posbeyikian A, Harant A, Kamoun S. NLR immune receptor-nanobody fusions confer plant disease resistance. Science. 2023:**379**(6635):934–939. https://doi.org/10.1126/science.abn4116

Kourelis J, Sakai T, Adachi H, Kamoun S. RefPlantNLR is a comprehensive collection of experimentally validated plant disease resistance proteins from the NLR family. PLoS Biol. 2021:**19**(10):e3001124. https://doi.org/10.1371/journal.pbio.3001124

Krasileva KV, Dahlbeck D, Staskawicz BJ. Activation of an Arabidopsis resistance protein is specified by the in planta association of its leucine-rich repeat domain with the cognate oomycete effector. Plant Cell. 2010:**22**(7):2444–2458. https://doi.org/10.1105/tpc.110.075358

Kutschera A, Dawid C, Gisch N, Schmid C, Raasch L, Gerster T, Schäffer M, Smakowska-Luzan E, Belkhadir Y, Vlot AC, et al. Bacterial medium-chain 3-hydroxy fatty acid metabolites trigger immunity in plants. Science. 2019:**364**(6436):178–181. https://doi.org/10.1126/science.aau1279

Kvitko BH, Collmer A. Discovery of the Hrp type III secretion system in phytopathogenic bacteria: how investigation of hypersensitive cell death in plants led to a novel protein injector system and a world of inter-organismal molecular interactions within plant cells. Phytopathology. 2023:**113**(4):626–636. https://doi.org/10.1094/PHYTO-08-22-0292-KD

Lal NK, Nagalakshmi U, Hurlburt NK, Flores R, Bak A, Sone P, Ma XY, Song GY, Walley J, Shan LB, et al. The receptor-like cytoplasmic kinase BIK1 localizes to the nucleus and regulates defense hormone expression during plant innate immunity. Cell Host Microbe. 2018:**23**(4):485–497. https://doi.org/10.1016/j.chom.2018.03.010

Lang J, Genot B, Bigeard J, Colcombet J. MPK3 and MPK6 control salicylic acid signaling by up-regulating NLR receptors during pattern- and effector-triggered immunity. J Exp Bot. 2022:**73**(7):2190–2205. https://doi.org/10.1093/jxb/erab544

Lapin D, Kovacova V, Sun X, Dongus JA, Bhandari D, von Born P, Bautor J, Guarneri N, Rzemieniewski J, Stuttmann J, et al. A coevolved EDS1-SAG101-NRG1 module mediates cell death signaling by TIR-domain immune receptors. Plant Cell. 2019:**31**(10):2430–2455. https://doi.org/10.1105/tpc.19.00118

Larkan NJ, Lydiate DJ, Parkin IAP, Nelson MN, Epp DJ, Cowling WA, Rimmer SR, Borhan MH. The blackleg resistance gene encodes a receptor-like protein triggered by the effector AVRLM1. New Phytol. 2013:**197**(2):595–605. https://doi.org/10.1111/nph.12043

Lawrence GJ, Finnegan EJ, Ayliffe MA, Ellis JG. The L6 gene for flax rust resistance is related to the Arabidopsis bacterial resistance gene RPS2 and the tobacco viral resistance gene N. Plant Cell. 1995:**7**(8):1195–1206. https://doi.org/10.1105/tpc.7.8.1195

Le Roux C, Huet G, Jauneau A, Camborde L, Trémousaygue D, Kraut A, Zhou BB, Levaillant M, Adachi H, Yoshioka H, et al. A receptor pair with an integrated decoy converts pathogen disabling of transcription factors to immunity. Cell. 2015:**161**(5):1074–1088. https://doi.org/10.1016/j.cell.2015.04.025

Li L, Li M, Yu L, Zhou Z, Liang X, Liu Z, Cai G, Gao L, Zhang X, Wang Y, et al. The FLS2-associated kinase BIK1 directly phosphorylates the NADPH oxidase RbohD to control plant immunity. Cell Host Microbe. 2014:**15**:329–338. https://doi.org/10.1016/j.chom.2014.02.009

Liang X, Zhou J-M. Receptor-like cytoplasmic kinases: central players in plant receptor kinase-mediated signaling. Annu Rev Plant Biol. 2018:**69**(1):267–299. https://doi.org/10.1146/annurev-arplant-042817-040540

Liebrand TWH, van den Berg GCM, Zhang Z, Smit P, Cordewener JHG, America AHP, Sklenar J, Jones AME, Tameling WIL, Robatzek S, et al. Receptor-like kinase SOBIR1/EVR interacts with receptor-like proteins in plant immunity against fungal infection. Proc Natl Acad Sci U S A. 2013:**110**(24):10010–10015. https://doi.org/10.1073/pnas.1220015110

Liu Y, Maierhofer T, Rybak K, Sklenar J, Breakspear A, Johnston MG, Fliegmann J, Huang SG, Roelfsema MRG, Felix G, et al. Anion channel SLAH3 is a regulatory target of chitin receptor-associated kinase PBL27 in microbial stomatal closure. Elife. 2019:**8**:e44474. https://doi.org/10.7554/eLife.44474

Ma S, Lapin D, Liu L, Sun Y, Song W, Zhang X, Logemann E, Yu D, Wang J, Jirschitzka J, et al. Direct pathogen-induced assembly of an NLR immune receptor complex to form a holoenzyme. Science. 2020:**370**(6521):abe3069. https://doi.org/10.1126/science.abe3069

Mackey D, Belkhadir Y, Alonso JM, Ecker JR, Dangl JL. Arabidopsis RIN4 is a target of the type III virulence effector AvrRpt2 and modulates RPS2-mediated resistance. Cell. 2003:**112**(3):379–389. https://doi.org/10.1016/S0092-8674(03)00040-0

Mackey D, Holt BF, Wiig A, Dangl JL. RIN4 interacts with Pseudomonas syringae type III effector molecules and is required for RPM1-mediated resistance in Arabidopsis. Cell. 2002:**108**(6): 743–754. https://doi.org/10.1016/S0092-8674(02)00661-X

Maekawa T, Cheng W, Spiridon LN, Toller A, Lukasik E, Saijo Y, Liu PY, Shen Q-H, Micluta MA, Somssich IE, et al. Coiled-coil domain-dependent homodimerization of intracellular barley immune receptors defines a minimal functional module for triggering cell death. Cell Host Microbe. 2011:**9**(3):187–199. https://doi.org/10.1016/j.chom.2011.02.008

Maidment JHR, Shimizu M, Bentham AR, Vera S, Franceschetti M, Longya A, Stevenson CEM, De la Concepcion JC, Białas A, Kamoun S, et al. Effector target-guided engineering of an integrated domain expands the disease resistance profile of a rice NLR immune receptor. eLife. 2023:**12**:e81123. https://doi.org/10.7554/eLife.81123

Manhaes AMED, Ortiz-Morea FA, He P, Shan LB. Plant plasma membrane-resident receptors: surveillance for infections and coordination for growth and development. J Integr Plant Biol. 2021:**63**(1): 79–101. https://doi.org/10.1111/jipb.13051

Maqbool A, Saitoh H, Franceschetti M, Stevenson CEM, Uemura A, Kanzaki H, Kamoun S, Terauchi R, Banfield MJ. Structural basis of pathogen recognition by an integrated HMA domain in a plant NLR immune receptor. Elife. 2015:**4**:e08709. https://doi.org/10.7554/eLife.08709

Marchal C, Pai H, Kamoun S, Kourelis J. Emerging principles in the design of bioengineered made-to-order plant immune receptors. Curr Opin Plant Biol. 2022:**70**:102311. https://doi.org/10.1016/j.pbi.2022.102311

Martin GB, Brommonschenkel SH, Chunwongse J, Frary A, Ganal MW, Spivey R, Wu T, Earle ED, Tanksley SD. Map-based cloning of a protein kinase gene conferring disease resistance in tomato. Science. 1993:**262**(5138):1432–1436. https://doi.org/10.1126/science.7902614

Martin R, Qi T, Zhang H, Liu F, King M, Toth C, Nogales E, Staskawicz BJ. Structure of the activated ROQ1 resistosome directly recognizing the pathogen effector XopQ. Science. 2020:**370**(6521): abd9993. https://doi.org/10.1126/science.abd9993

McLellan H, Harvey SE, Steinbrenner J, Armstrong MR, He Q, Clewes R, Pritchard L, Wang W, Wang S, Nussbaumer T, et al. Exploiting breakdown in nonhost effector-target interactions to boost host disease resistance. Proc Natl Acad Sci U S A. 2022:**119**(35):e2114064119. https://doi.org/10.1073/pnas.2114064119

Mestre P, Baulcombe DC. Elicitor-mediated oligomerization of the tobacco N disease resistance protein. Plant Cell. 2006:**18**(2):491–501. https://doi.org/10.1105/tpc.105.037234

Mindrinos M, Katagiri F, Yu GL, Ausubel FM. The *A. thaliana* disease resistance gene RPS2 encodes a protein containing a nucleotide-binding site and leucine-rich repeats. Cell. 1994:**78**(6):1089–1099. https://doi.org/10.1016/0092-8674(94)90282-8

Miya A, Albert P, Shinya T, Desaki Y, Ichimura K, Shirasu K, Narusaka Y, Kawakami N, Kaku H, Shibuya N. CERK1, a LysM receptor kinase, is essential for chitin elicitor signaling in Arabidopsis. Proc Natl Acad Sci U S A. 2007:**104**(49):19613–19618. https://doi.org/10.1073/pnas.0705147104

Narusaka M, Shirasu K, Noutoshi Y, Kubo Y, Shiraishi T, Iwabuchi M, Narusaka Y. RRS1 and RPS4 provide a dual resistance-gene system against fungal and bacterial pathogens. Plant J. 2009:**60**(2): 218–226. https://doi.org/10.1111/j.1365-313X.2009.03949.x

Navarro L, Zipfel C, Rowland O, Keller I, Robatzek S, Boller T, Jones JDG. The transcriptional innate immune response to flg22. Interplay and overlap with Avr gene-dependent defense responses and bacterial pathogenesis. Plant Physiol. 2004:**135**(2):1113–1128. https://doi.org/10.1104/pp.103.036749

Ngou BPM, Ahn H-K, Ding P, Jones JDG. Mutual potentiation of plant immunity by cell-surface and intracellular receptors. Nature. 2021:**592**(7852):110115. https://doi.org/10.1038/s41586-021-03315-7

Ngou BPM, Ding PT, Jones JDG. Thirty years of resistance: zig-zag through the plant immune system. Plant Cell. 2022:**34**(5): 1447–1478. https://doi.org/10.1093/plcell/koac041

Okuyama Y, Kanzaki H, Abe A, Yoshida K, Tamiru M, Saitoh H, Fujibe T, Matsumura H, Shenton M, Galam DC, et al. A multifaceted genomics approach allows the isolation of the rice-blast resistance gene consisting of two adjacent NBS-LRR protein genes. Plant J. 2011:**66**(3):467–479. https://doi.org/10.1111/j.1365-313X.2011.04502.x

Orbach MJ, Farrall L, Sweigard JA, Chumley FG, Valent B. A telomeric avirulence gene determines efficacy for the rice blast resistance gene Pi-ta. Plant Cell. 2000:**12**(11):2019–2032. https://doi.org/10.1105/tpc.12.11.2019

Outram MA, Figueroa M, Sperschneider J, Williams SJ, Dodds PN. Seeing is believing: exploiting advances in structural biology to understand and engineer plant immunity. Curr Opin Plant Biol. 2022:**67**:102210. https://doi.org/10.1016/j.pbi.2022.102210

Parker JE, Holub EB, Frost LN, Falk A, Gunn ND, Daniels MJ. Characterization of eds1, a mutation in Arabidopsis suppressing resistance to Peronospora parasitica specified by several different RPP genes. Plant Cell. 1996:**8**(11):2033–2046. https://doi.org/10.1105/tpc.8.11.2033

Peart JR, Mestre P, Lu R, Malcuit I, Baulcombe DC. NRG1, a CC-NB-LRR protein, together with N, a TIR-NB-LRR protein, mediates resistance against tobacco mosaic virus. Curr Biol. 2005:**15**(10): 968–973. https://doi.org/10.1016/j.cub.2005.04.053

Peng YJ, van Wersch R, Zhang YL. Convergent and divergent signaling in PAMP-triggered immunity and effector-triggered immunity. Mol Plant Microbe Interact. 2018:**31**(4):403–409. https://doi.org/10.1094/MPMI-06-17-0145-CR

Pottinger SE, Bak A, Margets A, Helm M, Tang L, Casteel C, Innes RW. Optimizing the PBS1 decoy system to confer resistance to potyvirus infection in *Arabidopsis* and soybean. Mol Plant Microbe Interact. 2020:**33**(7):932–944. https://doi.org/10.1094/MPMI-07-19-0190-R

Pruitt RN, Locci F, Wanke F, Zhang L, Saile SC, Joe A, Karelina D, Hua C, Frohlich K, Wan WL, et al. The EDS1-PAD4-ADR1 node mediates Arabidopsis pattern-triggered immunity. Nature. 2021:**598**(7881): 495–499. https://doi.org/10.1038/s41586-021-03829-0

Pruitt RN, Schwessinger B, Joe A, Thomas N, Liu F, Albert M, Robinson MR, Chan LJG, Luu DD, Chen H, et al. The rice immune receptor XA21 recognizes a tyrosine-sulfated protein from a Gram-negative bacterium. Sci Adv. 2015:**1**(6):1500245. https://doi.org/10.1126/sciadv.1500245

Qi T, Seong K, Thomazella DPT, Kim JR, Pham J, Seo E, Cho MJ, Schultink A, Staskawicz BJ. NRG1 functions downstream of EDS1 to regulate TIR-NLR-mediated plant immunity in *Nicotiana benthamiana*. Proc Natl Acad Sci U S A. 2018:**115**(46):E10979–E10987. https://doi.org/10.1073/pnas.1814856115

Rairdan GJ, Moffett P. Distinct domains in the ARC region of the potato resistance protein Rx mediate LRR binding and inhibition of activation. Plant Cell. 2006:**18**(8):2082–2093. https://doi.org/10.1105/tpc.106.042747

Rehmany AP, Gordon A, Rose LE, Allen RL, Armstrong MR, Whisson SC, Kamoun S, Tyler BM, Birch PRJ, Beynon JL. Differential recognition of highly divergent downy mildew avirulence gene alleles by RPP1 resistance genes from two Arabidopsis lines. Plant Cell. 2005:**17**(6):1839–1850. https://doi.org/10.1105/tpc.105.031807

Rhodes J, Zipfel C, Jones JDG, Ngou BPM. Concerted actions of PRR- and NLR-mediated immunity. Essays Biochem. 2022:**66**(5):501–511. https://doi.org/10.1042/EBC20220067

Rooney HC, Van't Klooster JW, van der Hoorn RA, Joosten MH, Jones JD, de Wit PJ. Cladosporium Avr2 inhibits tomato Rcr3 protease required for Cf-2-dependent disease resistance. Science. 2005:**308**(5729):1783–1786. https://doi.org/10.1126/science.1111404

Rowland O, Ludwig AA, Merrick CJ, Baillieul F, Tracy FE, Durrant WE, Fritz-Laylin L, Nekrasov V, Sjölander K, Yoshioka H, et al. Functional analysis of Avr9/Cf-9 rapidly elicited genes identifies a protein kinase, ACIK1, that is essential for full Cf-9-dependent disease resistance in tomato. Plant Cell. 2005:**17**(1):295–310. https://doi.org/10.1105/tpc.104.026013

Saijo Y, Loo EP, Yasuda S. Pattern recognition receptors and signaling in plant-microbe interactions. Plant J. 2018:**93**(4):592–613. https://doi.org/10.1111/tpj.13808

Salcedo A, Rutter W, Wang S, Akhunova A, Bolus S, Chao S, Anderson N, De Soto MF, Rouse M, Szabo L, et al. Variation in the AvrSr35 gene determines *Sr35* resistance against wheat stem rust race Ug99. Science. 2017:**358**(6370):1604–1606. https://doi.org/10.1126/science.aao7294

Sarris PF, Duxbury Z, Huh SU, Ma Y, Segonzac C, Sklenar J, Derbyshire P, Cevik V, Rallapalli G, Saucet SB, et al. A plant immune receptor detects pathogen effectors that target WRKY transcription factors. Cell. 2015:**161**(5):1089–1100. https://doi.org/10.1016/j.cell.2015.04.024

Saur IM, Bauer S, Kracher B, Lu X, Franzeskakis L, Muller MC, Sabelleck B, Kummel F, Panstruga R, Maekawa T, et al. Multiple pairs of allelic MLA immune receptor-powdery mildew AVRA effectors argue for a direct recognition mechanism. Elife. 2019:**8**:44471. https://doi.org/10.7554/eLife.44471

Saur IML, Panstruga R, Schulze-Lefert P. NOD-like receptor-mediated plant immunity: from structure to cell death. Nat Rev Immunol. 2021:**21**(5):305–318. https://doi.org/10.1038/s41577-020-00473-z

Segretin ME, Pais M, Franceschetti M, Chaparro-Garcia A, Bos JIB, Banfield MJ, Kamoun S. Single amino acid mutations in the potato immune receptor R3a expand response to effectors. Mol Plant

Microbe Interact. 2014:**27**:624–637. https://doi.org/10.1094/Mpmi-02-14-0040-R

Shan W, Cao M, Leung D, Tyler BM. The *Avr1b* locus of *Phytophthora sojae* encodes an elicitor and a regulator required for avirulence on soybean plants carrying resistance gene *Rps*1b. Mol Plant Microbe Interact. 2004:**17**(4):394–403. https://doi.org/10.1094/MPMI.2004.17.4.394

Shao Z-Q, Xue J-Y, Wu P, Zhang Y-M, Wu Y, Hang Y-Y, Wang B, Chen J-Q. Large-scale analyses of angiosperm nucleotide-binding site-leucine-rich repeat genes reveal three anciently diverged classes with distinct evolutionary patterns. Plant Physiol. 2016:**170**(4): 2095–2109. https://doi.org/10.1104/pp.15.01487

Shen Q-H, Zhou F, Bieri S, Haizel T, Shirasu K, Schulze-Lefert P. Recognition specificity and RAR1/SGT1 dependence in barley Mla disease resistance genes to the powdery mildew fungus. Plant Cell. 2003:**15**(3):732–744. https://doi.org/10.1105/tpc.009258

Shi GJ, Zhang ZC, Friesen TL, Raats D, Fahima T, Brueggeman RS, Lu SW, Trick HN, Liu ZH, Chao W, et al. The hijacking of a receptor kinase-driven pathway by a wheat fungal pathogen leads to disease. Sci Adv. 2016:**2**(10):1600822. https://doi.org/10.1126/sciadv.1600822

Song W-Y, Wang G-L, Chen L-L, Kim H-S, Pi L-Y, Holsten T, Gardner J, Wang B, Zhai W-X, Zhu L-H, et al. A receptor kinase-like protein encoded by the rice disease resistance gene, Xa21. Science. 1995:**270**(5243): 1804–1806. https://doi.org/10.1126/science.270.5243.1804

Sperschneider J, Dodds PN. Effectorp 3.0: prediction of apoplastic and cytoplasmic effectors in fungi and oomycetes. Mol Plant Microbe Interact. 2022:**35**(2):146–156. https://doi.org/10.1094/MPMI-08-21-0201-R

Stakman EC, Levine MN. The determination of biologic forms of *Puccinia graminis* on *Triticum* spp. Minnesota Agric Station Tech Bull. 1922:**8**:1–10. https://hdl.handle.net/11299/203941

Staskawicz B, Dahlbeck D, Keen NT. Cloned avirulence gene of Pseudomonas syringae pv. glycinea determines race-specific incompatibility on Glycine max (L.) Merr. Proc Natl Acad Sci U S A. 1984:**81**(19):6024–6028. https://doi.org/10.1073/pnas.81.19.6024

Su JB, Yang LY, Zhu QK, Wu HJ, He Y, Liu YD, Xu J, Jiang DA, Zhang SQ. Active photosynthetic inhibition mediated by MPK3/MPK6 is critical to effector-triggered immunity. PLoS Biol. 2018:**16**(5): e2004122. https://doi.org/10.1371/journal.pbio.2004122

Sun X, Lapin D, Feehan JM, Stolze SC, Kramer K, Dongus JA, Rzemieniewski J, Blanvillain-Baufumé S, Harzen A, Bautor J, et al. Pathogen effector recognition-dependent association of NRG1 with EDS1 and SAG101 in TNL receptor immunity. Nat Commun. 2021:**12**(1):3335. https://doi.org/10.1038/s41467-021-23614-x

Sun Y, Li L, Macho AP, Han Z, Hu Z, Zipfel C, Zhou JM, Chai J. Structural basis for flg22-induced activation of the Arabidopsis FLS2-BAK1 immune complex. Science. 2013:**342**(6158):624–628. https://doi.org/10.1126/science.1243825

Swiderski MR, Birker D, Jones JD. The TIR domain of TIR-NB-LRR resistance proteins is a signaling domain involved in cell death induction. Mol Plant Microbe Interact. 2009:**22**(2):157–165. https://doi.org/10.1094/MPMI-22-2-0157

Tamborski J, Seong K, Liu F, Staskawicz B, Krasileva K. Altering specificity and auto-activity of plant immune receptors Sr33 and Sr50 via a rational engineering approach. Mol Plant Microbe Interact. 2023:**36**(7):434–446. https://doi.org/10.1094/MPMI-07-22-0154-R

Tameling WIL, Vossen JH, Albrecht M, Lengauer T, Berden JA, Haring MA, Cornelissen BJC, Takken FLW. Mutations in the NB-ARC domain of I-2 that impair ATP hydrolysis cause autoactivation. Plant Physiol. 2006:**140**(4):1233–1245. https://doi.org/10.1104/pp.105.073510

Thor K, Jiang SS, Michard E, George J, Scherzer S, Huang SG, Dindas J, Derbyshire P, Leitao N, DeFalco TA, et al. The calcium-permeable channel OSCA1.3 regulates plant stomatal immunity (vol 585, pg 569, 2020). Nature. 2020:**588**(7836):E4. https://doi.org/10.1038/s41586-020-2954-9

Tian W, Hou CC, Ren ZJ, Wang C, Zhao FG, Dahlbeck D, Hu SP, Zhang LY, Niu Q, Li LG, et al. A calmodulin-gated calcium channel links pathogen patterns to plant immunity. Nature. 2019:**572**(7767): 131–135. https://doi.org/10.1038/s41586-019-1413-y

Torres MA, Dangl JL, Jones JDG. Arabidopsis gp91phox homologues AtrbohD and AtrbohF are required for accumulation of reactive oxygen intermediates in the plant defense response. Proc Natl Acad Sci U S Am. 2002:**99**(1):517–522. https://doi.org/10.1073/pnas.012452499

Torres Ascurra YC, Zhang L, Toghani A, Hua C, Rangegowda NJ, Posbeyikian A, Pai H, Lin X, Wolters PJ, Wouters D, et al. Functional diversification of a wild potato immune receptor at its center of origin. Science. 2023:**381**(6660):891–897. https://doi.org/10.1126/science.adg5261

Ueda H, Yamaguchi Y, Sano H. Direct interaction between the tobacco mosaic virus helicase domain and the ATP-bound resistance protein, N factor during the hypersensitive response in tobacco plants. Plant Mol Biol. 2006:**61**(1-2):31–45. https://doi.org/10.1007/s11103-005-5817-8

Van der Biezen EA, Jones JD. Plant disease-resistance proteins and the gene-for-gene concept. Trends Biochem Sci. 1998:**23**(12):454–456. https://doi.org/10.1016/S0968-0004(98)01311-5

van der Hoorn RA, Kamoun S. From guard to decoy: a new model for perception of plant pathogen effectors. Plant Cell. 2008:**20**(8): 2009–2017. https://doi.org/10.1105/tpc.108.060194

Veronese P, Nakagami H, Bluhm B, AbuQamar S, Chen X, Salmeron J, Dietrich RA, Hirt H, Mengiste T. The membrane-anchored BOTRYTIS-INDUCED KINASE 1 plays distinct roles in resistance to necrotrophic and biotrophic pathogens. Plant Cell. 2006:**18**(1): 257–273. https://doi.org/10.1105/tpc.105.035576

Wagner S, Stuttmann J, Rietz S, Guerois R, Brunstein E, Bautor J, Niefind K, Parker JE. Structural basis for signaling by exclusive EDS1 heteromeric complexes with SAG101 or PAD4 in plant innate immunity. Cell Host Microbe. 2013:**14**(6):619–630. https://doi.org/10.1016/j.chom.2013.11.006

Wan L, Essuman K, Anderson RG, Sasaki Y, Monteiro F, Chung EH, Osborne Nishimura E, DiAntonio A, Milbrandt J, Dangl JL, et al. TIR domains of plant immune receptors are NAD(+)-cleaving enzymes that promote cell death. Science. 2019:**365**(6455):799–803. https://doi.org/10.1126/science.aax1771

Wan JR, Zhang XC, Neece D, Ramonell KM, Clough S, Kim SY, Stacey MG, Stacey G. A LysM receptor-like kinase plays a critical role in chitin signaling and fungal resistance in. Plant Cell. 2008:**20**(2):471–481. https://doi.org/10.1105/tpc.107.056754

Wang CG, Zhou MQ, Zhang XD, Yao J, Zhang YP, Mou ZL. A lectin receptor kinase as a potential sensor for extracellular nicotinamide adenine dinucleotide in. Elife. 2017:**6**:25474. https://doi.org/10.7554/eLife.25474

Wang J, Hu M, Wang J, Qi J, Han Z, Wang G, Qi Y, Wang HW, Zhou JM, Chai J. Reconstitution and structure of a plant NLR resistosome conferring immunity. Science. 2019a:**364**(6435):aav5870. https://doi.org/10.1126/science.aav5870

Wang GF, Ji J, El-Kasmi F, Dangl JL, Johal G, Balint-Kurti PJ. Molecular and functional analyses of a maize autoactive NB-LRR protein identify precise structural requirements for activity. PLoS Pathog. 2015:**11**(2):e1004674. https://doi.org/10.1371/journal.ppat.1004674

Wang JC, Liu X, Zhang A, Ren YL, Wu FQ, Wang G, Xu Y, Lei CL, Zhu SS, Pan T, et al. A cyclic nucleotide-gated channel mediates cytoplasmic calcium elevation and disease resistance in rice. Cell Res. 2019c:**29**(10):820–831. https://doi.org/10.1038/s41422-019-0219-7

Wang H, Song S, Gao S, Yu Q, Zhang H, Cui X, Fan J, Xin X, Liu Y, Staskawicz B. The NLR immune receptor ADR1 and lipase-like proteins EDS1 and PAD4 mediate stomatal immunity in *Nicotiana benthamiana* and Arabidopsis. Plant Cell. 2023:**36**:427–446. https://doi.org/10.1093/plcell/koad270

Wang J, Wang J, Hu M, Wu S, Qi J, Wang G, Han Z, Qi Y, Gao N, Wang H-W, et al. Ligand-triggered allosteric ADP release primes a plant NLR complex. Science. 2019b:**364**(6435):aav5868. https://doi.org/10.1126/science.aav5868

Whitham S, Dinesh-Kumar SP, Choi D, Hehl R, Corr C, Baker B. The product of the tobacco mosaic virus resistance gene N: similarity to toll and the interleukin-1 receptor. Cell. 1994:**78**(6):1101–1115. https://doi.org/10.1016/0092-8674(94)90283-6

Williams SJ, Sohn KH, Wan L, Bernoux M, Sarris PF, Segonzac C, Ve T, Ma Y, Saucet SB, Ericsson DJ, et al. Structural basis for assembly and function of a heterodimeric plant immune receptor. Science. 2014:**344**(6181):299–303. https://doi.org/10.1126/science.1247357

Williams SJ, Sornaraj P, deCourcy-Ireland E, Menz RI, Kobe B, Ellis JG, Dodds PN, Anderson PA. An autoactive mutant of the M flax rust resistance protein has a preference for binding ATP, whereas wild-type M protein binds ADP. Mol Plant Microbe Interact. 2011:**24**(8):897–906. https://doi.org/10.1094/MPMI-03-11-0052

Willmann R, Lajunen HM, Erbs G, Newman MA, Kolb D, Tsuda K, Katagiri F, Fliegmann J, Bono JJ, Cullimore JV, et al. Arabidopsis lysin-motif proteins LYM1 LYM3 CERK1 mediate bacterial peptidoglycan sensing and immunity to bacterial infection. Proc Natl Acad Sci U S A. 2011:**108**(49):19824–19829. https://doi.org/10.1073/pnas.1112862108

Wróblewski T, Spiridon L, Martin EC, Petrescu AJ, Cavanaugh K, Truco MJP, Xu HQ, Gozdowski D, Pawlowski K, Michelmore RW, et al. Genome-wide functional analyses of plant coiled-coil NLR-type pathogen receptors reveal essential roles of their N-terminal domain in oligomerization, networking, and immunity. PLoS Biol. 2018:**16**(12): e2005821. https://doi.org/10.1371/journal.pbio.2005821

Wu CH, Abd-El-Haliem A, Bozkurt TO, Belhaj K, Terauchi R, Vossen JH, Kamoun S. NLR network mediates immunity to diverse plant pathogens. Proc Natl Acad Sci U S A. 2017:**114**(30):8113–8118. https://doi.org/10.1073/pnas.1702041114

Wu CH, Belhaj K, Bozkurt TO, Birk MS, Kamoun S. Helper NLR proteins NRC2a/b and NRC3 but not NRC1 are required for Pto-mediated cell death and resistance in *Nicotiana benthamiana*. New Phytol. 2016:**209**(4):1344–1352. https://doi.org/10.1111/nph.13764

Wu Z, Li M, Dong OX, Xia S, Liang W, Bao Y, Wasteneys G, Li X. Differential regulation of TNL-mediated immune signaling by redundant helper CNLs. New Phytol. 2019:**222**(2):938–953. https://doi.org/10.1111/nph.15665

Yuan M, Jiang Z, Bi G, Nomura K, Liu M, Wang Y, Cai B, Zhou JM, He SY, Xin XF. Pattern-recognition receptors are required for NLR-mediated plant immunity. Nature. 2021:**592**(7852):105–109. https://doi.org/10.1038/s41586-021-03316-6

Zadoks JC. Cereal rusts, dogs and stars in antiquity. Garcia de Orta. 1981:**9**:13–20.

Zhang LS, Hua CL, Janocha D, Fliegmann J, Nürnberger T. Plant cell surface immune receptors-novel insights into function and evolution. Curr Opin Plant Biol. 2023:**74**:102384. https://doi.org/10.1016/j.pbi.2023.102384

Zhang X, Bernoux M, Bentham AR, Newman TE, Ve T, Casey LW, Raaymakers TM, Hu J, Croll TI, Schreiber KJ, et al. Multiple functional self-association interfaces in plant TIR domains. Proc Natl Acad Sci U S A. 2017:**114**(10):E2046–E2052. https://doi.org/10.1073/pnas.1621248114

Zhao Y-B, Liu M-X, Chen T-T, Ma XM, Li Z-K, Zheng ZC, Zheng S-R, Chen LF, Li Y-Z, Tang L-R, et al. Pathogen effector AvrSr35 triggers Sr35 resistosome assembly via a direct recognition mechanism. Sci Adv. 2022:**8**(36):eabq5108. https://doi.org/10.1126/sciadv.abq5108

Zhu J, Lolle S, Tang AD, Guel B, Kvitko B, Cole B, Coaker G. Single-cell profiling of Arabidopsis leaves to *Pseudomonas syringae* infection. Cell Rep. 2023:**42**(7):112676. https://doi.org/10.1016/j.celrep.2023.112676

Zipfel C, Kunze G, Chinchilla D, Caniard A, Jones JD, Boller T, Felix G. Perception of the bacterial PAMP EF-Tu by the receptor EFR restricts Agrobacterium-mediated transformation. Cell. 2006:**125**(4): 749–760. https://doi.org/10.1016/j.cell.2006.03.037

Zipfel C, Robatzek S, Navarro L, Oakeley EJ, Jones JDG, Felix G, Boller T. Bacterial disease resistance in Arabidopsis through flagellin perception. Nature. 2004:**428**(6984):764–767. https://doi.org/10.1038/nature02485

Zönnchen J, Gantner J, Lapin D, Barthel K, Eschen-Lippold L, Erickson JL, Villanueva SL, Zantop S, Kretschmer C, Joosten MHAJ, et al. EDS1 complexes are not required for PRR responses and execute TNL-ETI from the nucleus in *Nicotiana benthamiana*. New Phytol. 2022:**236**(6):2249–2264. https://doi.org/10.1111/nph.18511

https://doi.org/10.1093/plcell/koae038

THE PLANT CELL

THE PLANT CELL 2024: 36: 1482–1503

Review

Nitrogen sensing and regulatory networks: it's about time and space

Carly M. Shanks [1,†] Karin Rothkegel [2,3,†] Matthew D. Brooks [4] Chia-Yi Cheng [5] José M. Alvarez [2,6] Sandrine Ruffel [7] Gabriel Krouk [7] Rodrigo A. Gutiérrez [2,3,*,‡] and Gloria M. Coruzzi [1,*,‡]

1 Department of Biology, Center for Genomics and Systems Biology, New York University, New York, NY 10003, USA
2 Agencia Nacional de Investigación y Desarrollo-Millennium Science Initiative Program, Millennium Institute for Integrative Biology (iBio), 7500565 Santiago, Chile
3 Center for Genome Regulation (CRG), Institute of Ecology and Biodiversity (IEB), Facultad de Ciencias Biológicas, Pontificia Universidad Católica de Chile, 8331010 Santiago, Chile
4 Global Change and Photosynthesis Research Unit, USDA-ARS, Urbana, IL 61801, USA
5 Department of Life Science, National Taiwan University, Taipei 10663, Taiwan
6 Centro de Biotecnología Vegetal, Facultad de Ciencias, Universidad Andrés Bello, 8370035 Santiago, Chile
7 Institute for Plant Sciences of Montpellier (IPSiM), Centre National de la Recherche Scientifique (CNRS), Institut National de Recherche pour l'Agriculture, l'Alimentation, et l'Environnement (INRAE), Université de Montpellier, Montpellier 34090, France

*Author for correspondence: gloria.coruzzi@nyu.edu (G.M.C.), rgutierrez@bio.puc.cl (R.A.G.)
†Co-first authors.
‡Co-corresponding authors.

Abstract

A plant's response to external and internal nitrogen signals/status relies on sensing and signaling mechanisms that operate across spatial and temporal dimensions. From a comprehensive systems biology perspective, this involves integrating nitrogen responses in different cell types and over long distances to ensure organ coordination in real time and yield practical applications. In this prospective review, we focus on novel aspects of nitrogen (N) sensing/signaling uncovered using temporal and spatial systems biology approaches, largely in the model Arabidopsis. The temporal aspects span: transcriptional responses to N-dose mediated by Michaelis-Menten kinetics, the role of the master NLP7 transcription factor as a nitrate sensor, its nitrate-dependent TF nuclear retention, its "hit-and-run" mode of target gene regulation, and temporal transcriptional cascade identified by "network walking." Spatial aspects of N-sensing/signaling have been uncovered in cell type-specific studies in roots and in root-to-shoot communication. We explore new approaches using single-cell sequencing data, trajectory inference, and pseudotime analysis as well as machine learning and artificial intelligence approaches. Finally, unveiling the mechanisms underlying the spatial dynamics of nitrogen sensing/signaling networks across species from model to crop could pave the way for translational studies to improve nitrogen-use efficiency in crops. Such outcomes could potentially reduce the detrimental effects of excessive fertilizer usage on groundwater pollution and greenhouse gas emissions.

Introduction

This special *Plant Cell* issue commemorates 100 years of the American Society of Plant Biologists. Our review focuses on the spatial dynamics of nitrogen sensing and signaling networks: it's about time. Aptly, more than 100 years ago, Michalis-Menten's classic paper published in 1913 sought "to achieve the final aim of kinetic research; namely, to

Received August 02, 2023. Accepted January 08, 2024. Advance access publication February 14, 2024 obtain knowledge of the nature of a reaction from a study of its progress" (Michaelis and Menten 1913; Michaelis et al. 2011). Inspired by this, our review focuses on systems biology studies conducted in real time and space. Such studies have uncovered the mechanisms by which plants sense and respond to nitrogen (N) signals within minutes to evoke changes in N-signaling networks in specific cell types that influence plant growth and development. Such discoveries of N-sensing/signaling reside in Pasteur's quadrant, a field of inquiry that aims to gain a fundamental understanding of a scientific problem while also providing immediate societal benefits (Stokes 1997)—in this case, improvements in nitrogen use efficiency (NUE).

The advent of synthetic fertilizers has brought significant advantages to agricultural practices by boosting crop yield, but at both economic and environmental costs (Menegat et al. 2022). Approximately half of applied fertilizers are effectively used by plants, while the remaining portion is prone to runoff, resulting in groundwater contamination and eutrophication (Bijay-Singh and Craswell 2021). Moreover, excess fertilizer application can lead to the production of nitrous oxide, a potent greenhouse gas (Mahmud et al. 2020; Menegat et al. 2022). Considering these challenges, a key objective of nitrogen research is to develop plants with enhanced NUE. Achieving this objective would not only reduce the need for excessive fertilizer usage but also support optimal plant growth in nitrogen-limited soils worldwide.

Nitrogen—the rate limiting element for plant growth—is often found in the soil as nitrate (NO_3^-) and/or ammonium (NH_4^+). Organic forms such as amino acids and urea can also play important roles in specific contexts (Yang et al. 2021b). Nitrate—the main form of nitrogen found in aerobic soils—also acts as a N-signal sensed by a nitrate transceptor in roots (Crawford and Forde 2002). As such, nitrate sensing/signaling has been widely studied by using biochemical, molecular genomics, genetics, and systems biology approaches (Krouk et al. 2010; Gaudinier et al. 2018; Wang et al. 2018; Vidal et al. 2020; Lamig et al. 2022). Herein, we explore studies that use systems biology approaches to examine the temporal and spatial mechanisms behind nitrogen sensing and signaling, largely in the model Arabidopsis. We especially highlight progress in this area published after the "Nitrate in 2020" *Plant Cell* Review, which includes an extensive timeline of milestone publications on nitrate signaling up to 2020 (Vidal et al. 2020). In addition to studies that explore the primary N-response (nitrate sensing/signaling), we include temporal studies that examine the plant response to ammonium nitrate, the source of nitrogen in the widely used Murashige and Skoog cell culture medium (Murashige and Skoog 1962; Varala et al. 2018; Brooks et al. 2019; Swift et al. 2020; Alvarez et al. 2021). Plants respond differently to sole sources of nitrate versus ammonium; therefore, we recommend referring to the following excellent reviews for details on specific ammonium responses not covered herein (Hachiya and Sakakibara 2017; Liu and Von Wirén 2017).

For further insights into advances in nitrogen sensing/signaling, we recommend recent reviews that encompass other aspects such as nitrogen transport (Tegeder and Masclaux-Daubresse 2018; Wang et al. 2018), local and systemic nitrogen signaling (Zhang et al. 2020), post-translational modifications and nitrogen signaling components (Muratore et al. 2021; Wang et al. 2021a), nitrogen-dependent developmental responses (Weber and Burow 2018; Fredes et al. 2019), nitrogen regulation of root system architecture (Jia and von Wirén 2020; Hu et al. 2021), nitrogen interactions with other nutrients (Oldroyd and Leyser 2020; Li et al. 2021), nitrogen and hormone interactions (Sakakibara 2021; Xing et al. 2023), and nitrogen responses under abiotic stress (Araus et al. 2020; Plett et al. 2020).

This review briefly touches on nitrogen response networks in crops (Ueda et al. 2020) and translational studies of nitrogen signaling networks—from model to crop (Obertello et al. 2015; Cheng et al. 2021). We also recommend more extensive recent reviews on crops for a comprehensive understanding of nitrogen signaling in agricultural contexts (Jia and von Wirén 2020; Hou et al. 2021; Sandhu et al. 2021; Gao et al. 2022; Hu et al. 2023).

This review focuses on novel insights gained from systems biology approaches to uncover the temporal and spatial mechanisms of nitrogen sensing and signaling. These include the discovery that Michaelis-Menten kinetics mediates N-dose dependent transcriptome responses (Swift et al. 2020). This finding echoes earlier time-based studies that showed that Michaelis-Menten kinetics mediate N-dose dependent nitrogen uptake (Ho et al. 2009; McNickle and Brown 2014) and plant growth responses (Lana et al. 2005). Furthermore, we examine recent studies that have uncovered the time-dependent mechanisms involving the master transcription factor (TF) NLP7, as a nitrate sensor (Liu et al. 2022), the nitrate-dependent nuclear localization of NLP7/6 (Marchive et al. 2013; Guan et al. 2017; Liu et al. 2017; Cheng et al. 2023), the "hit-and-run" model of transient inter-actions of NLP7-target genes (Alvarez et al. 2020), and the regulation that NLP7 exerts over a temporal cascade of downstream TF2s—uncovered using a method called "network walking" (Alvarez et al. 2020; Brooks et al. 2019, 2020).

We also explore new spatial approaches that can identify how nitrogen sensing, transport, and signaling are governed by cell type specificity in different organs. This includes the cell type-specific signaling responses to nitrate (Chen et al. 2022; Contreras-López et al. 2022), as well as studies that examine nitrate root-to-shoot communication and how plants integrate the shoot and root nitrogen status to systematically regulate nutrient uptake in the roots (Tabata et al. 2014; Ohkubo et al. 2017; Ota et al. 2020; Abualia et al. 2022).

To facilitate and inspire future advances in nitrogen research in space and time, we review advancements in single-cell sequencing technology that can be applied to plant N-sensing/signaling (Rich-Griffin et al. 2020; Cole et al. 2021). For example, the use of single cell (sc) RNA-seq assays

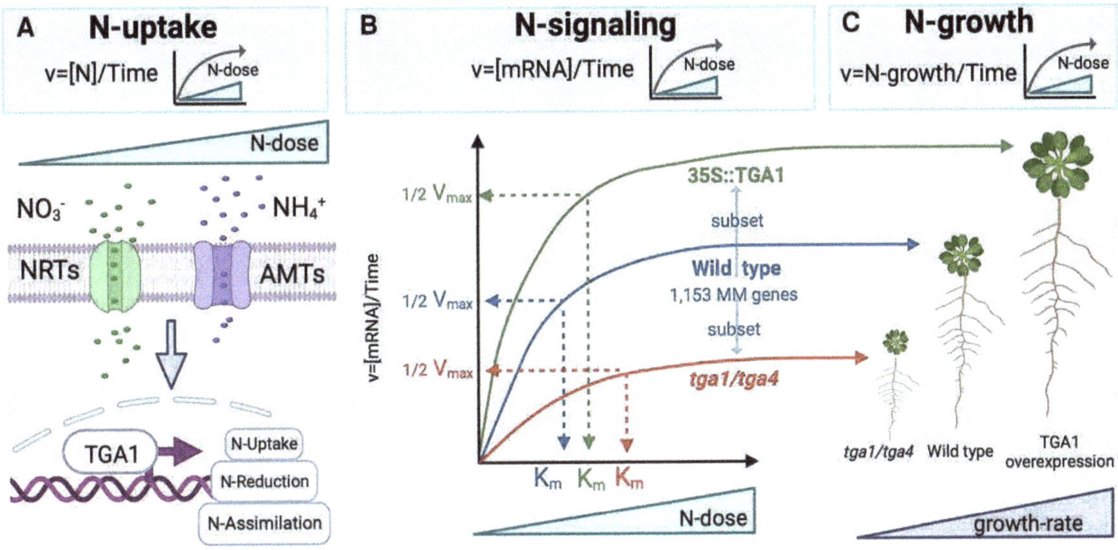

Figure 1. The N-dose-dependent regulation of N-uptake, N-signaling, and N-growth follows Michaelis-Menten (MM) kinetics. **A)** The rate of N-uptake by NRTs and AMTs is regulated by MM kinetics (Ho et al. 2009; McNickle and Brown 2014). **B)** Swift et al. (2020) demonstrated that the transcriptional response to N-dose also follows MM kinetics in Arabidopsis wild-type plants (Swift et al. 2020). Moreover, TGA1 overexpression and *tga1/4* mutant analysis revealed that a portion of this MM-mediated N-dose transcriptional response is mediated by the master transcription factor TGA1, which affects plant growth rate (Swift et al. 2020). **C)** N-dose-regulated growth responses measured by biomass is also regulated by MM kinetics (Lana et al. 2005). Thus, transcriptome kinetics responding to changes in N-dose has the potential to enhance plant growth. Figure adapted from Swift et al. (2020). Figure created with BioRender.com.

could enable (i) tracking of the nitrogen signal from root-to-shoot and (ii) determination of cell fate trajectories using pseudotime analysis (Denyer et al. 2019; Shahan et al. 2022; Nolan et al. 2023). Additionally, we explore how to use computational methods such as machine learning of the gene-to-NUE trait across a model and crop (Cheng et al. 2021) and how artificial intelligence (Gao et al. 2021) may augment experimental nitrogen research endeavors.

Overall, this ASPB Centennial review provides an overview the spatiotemporal dynamics of nitrogen sensing and signaling as an integrated system in plants. These temporal based systems biology approaches can also be applied to study any sensing and signaling network in plant and crop biology.

Nitrogen dose sensing as a function of time

How an organism senses and responds to changes in nitrogen nutrient dose is a basic unanswered question in biology with special relevance to agriculture. Exploiting time to uncover mechanisms underlying N-sensing/signaling in plants derives inspiration from the now classic Michaelis-Menten (MM) paper, which aimed "to obtain knowledge of the nature of a reaction from a study of its progress" (Michaelis and Menten 1913; Michaelis et al. 2011). Importantly, MM kinetics have also previously been shown to mediate nitrogen uptake (Fig. 1A; Ho et al. 2009; McNickle and Brown 2014) and plant growth (Fig. 1C; Lana et al. 2005). Inspired by this, Swift et al (2020) applied the MM kinetic concept to study the molecular basis for N-dose sensing in Arabidopsis, exposing seedlings to a matrix of four increasing N-doses of ammonium nitrate over five time points (Ahmed 2020; Akmakjian and Bailey-Serres 2020; Swift et al. 2020). Modeling of the result-ing RNA-seq data revealed that 3,818 genes increased or decreased their expression in proportion to N-dose over time. Moreover, they found that for a subset of these genes, the N-dose-dependent gene responses mirror simple enzyme kinetics described by Michaelis-Menten (MM) in 1913, where changing levels of enzyme abundance will affect the maximum rate of reaction (V_{max}; Michaelis and Menten 1913; Michaelis et al. 2011; Swift et al. 2020). Specifically, N-dose response genes whose expression pattern significantly fit the MM model allowed for estimating the maximum rate of transcript change (V_{max}), as well as the N-dose at which half of V_{max} was achieved (K_m; Fig. 1B; Swift et al. 2020). Indeed, the classic MM kinetic model was able to explain the expression of 30% of N-dose responsive genes in Arabidopsis (1,153 MM modeled N-dose responsive genes; Fig. 1B), whereas the remaining 70% genes could be explained by more complex kinetics and/or other regulatory mechanisms (Swift et al. 2020). This finding suggests that transcription factors (TFs) that regulate MM response genes can be analogized as catalytic enzymes in the MM model since they establish the rates at which transcription takes place in response to N-dose (Swift et al. 2020). To support this, in vivo studies showed that the overexpression of TGA1, an early N-responsive TF, led to an increase in V_{max} of N-dose responsive mRNAs (Fig. 1B), which was translated as an accelerated plant growth in response to N (Fig. 1C; Swift et al.

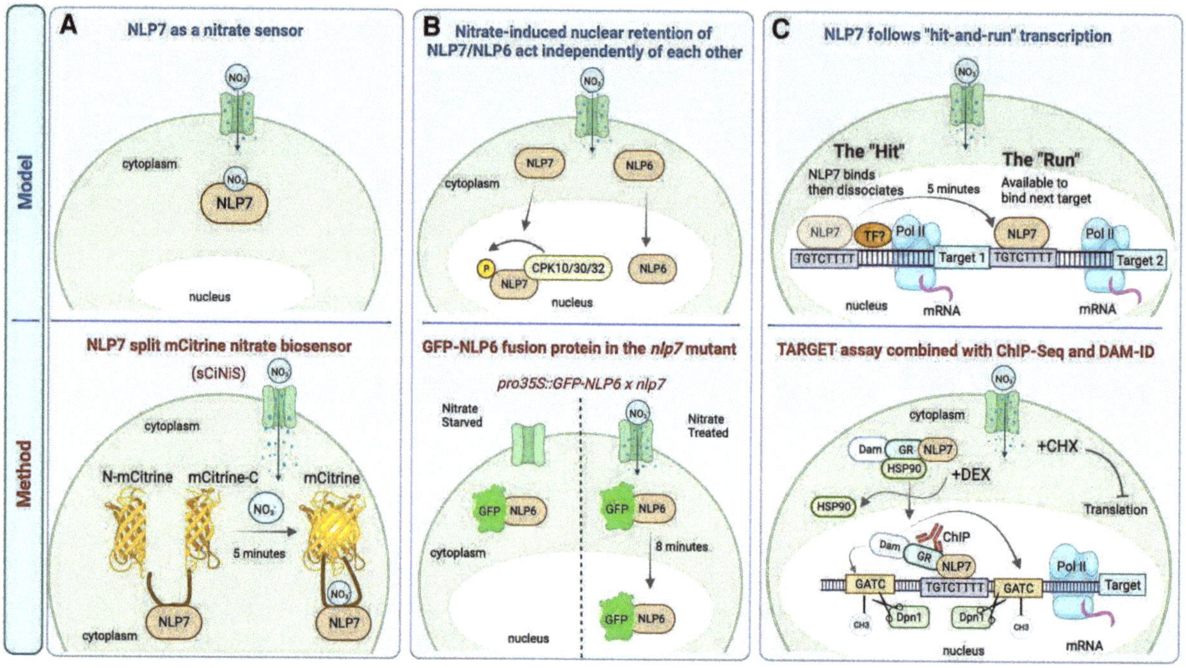

Figure 2. Time- and space-dependent modes of action for NLP7 as a master regulator of nitrate signaling. **A)** NLP7 binds to nitrate and acts as a nitrate sensor as determined using the genetically encoded split mCitrine-NLP7 nitrate biosensor (sCiNiS) assay (Liu et al. 2022). Fluorescent signal was detected 5 min after nitrate treatment in both mesophyll and primary root tip cells (Liu et al. 2022). **B)** Both NLP7 and NLP6 accumulate in the nucleus in response to nitrate as determined with TF-fusion proteins expressed in their respective mutant backgrounds, showing that accumulation of either TF in the nucleus is independent of each other, but dependent on nitrate (Marchive et al. 2013; Guan et al. 2017; Liu et al. 2017; Cheng et al. 2023). **C)** The "hit-and-run" model of transcription posits that a pioneer TF transiently binds to the promoter of a target gene to open the chromatin and allow for other partner TFs to bind the promoter, thereby making NLP7 available to bind the next target gene (Para et al. 2014; Alvarez et al. 2020). The TARGET assay combined with ChIP-seq and DamID was used to identify these highly transient NLP7 target genes (Alvarez et al. 2020). Figure created with BioRender.com.

2020). Uncovering the molecular mechanisms that underlie the transcriptome kinetics responding to changes in N-dose now connects N-uptake (transport) to output (biomass), and thus has the potential to enhance plant growth and improve N-use efficiency in crops (Fig. 1).

Time- and space-dependent modes of action for NLP7 as a master regulator of nitrate signaling

NLP7 is a master regulator of the early nitrate response, acting as a transcriptional regulator of genes involved with nitrate transport, nitrate assimilation, and signal transduction (Marchive et al. 2013; Alvarez et al. 2020). New time-based studies have shown that NLP7 is not only a master transcription factor for mediating nitrate responses, but it can also bind nitrate and act as an intracellular nitrate sensor, as identified using the split mCitrine-NLP7 nitrate biosensor (sCiNiS; Liu et al. 2022; Fig. 2A). The fluorescence signal for NLP7 binding of nitrate was detected after 5 min of nitrate treatment in mesophyll cells of cotyledons and also in primary root tips, showing that NLP7 acts as an intracellular nitrate sensor to initiate nitrate responses (Liu et al. 2022). The nitrate-binding domain on NLP7 is an evolutionarily ancient domain that is conserved among plant NLPs and bacterial nitrate sensors such as NreA (Niemann et al. 2014). Nitrate directly interacts with NLP7 through its amino terminus, inducing its conformational change to activate transcription (Liu et al. 2022; Fig. 2A).

The role of NLP7 as a nitrate sensor is an additional level of NLP7 regulation to the known post-translational modifications that regulate NLP7 in the nucleus in response to nitrate (Fig. 2B; Marchive et al. 2013; Guan et al. 2017; Liu et al. 2017). Once nitrate is transported inside the cell by NRT1.1, a rapid wave of Ca + 2 causes the activation of group III calcium-sensor protein kinases (CPKs) in seconds, which in turn phosphorylates NLP7 to retain it in the nucleus, activating early nitrate-response genes within minutes (Marchive et al. 2013; Liu et al. 2017; Fig. 2B). A recent study by Cheng et al. (2023) found that NLP7 and another NLP family mem-ber, NLP6, are both retained in the nucleus in response to ni-trate (Cheng et al. 2023; Fig. 2B). Moreover, they showed that nitrate-dependent nuclear accumulation of NLP7 and NLP6 act independently of each other. To do this, they constructed translational fusion proteins for both GFP-NLP6 and GFP-NLP7, expressed in *nlp7* or *nlp6* mutant background, respectively, accumulated in the nucleus in response to nitrate

Table 1 . Systems biology techniques and tools applied to studying TFs and their targets involved in nitrogen sensing and signaling

Technique/tool	Description	References
TARGET (TF → direct target regulation in plant cells)	Transient Assay Reporting Genome-wide Effects of Transcription factors (TARGET) is a plant cell-based assay used to identify direct TF target gene interactions with timed nuclear entry of the TF (Fig. 2).	Bargmann et al. (2013), Brooks et al. (2019), Alvarez et al. (2020); Brooks et al. (2023)
DamID-Seq (TF → Target interaction in vivo)	DNA adenine methyltransferase identification (DamID) uses DNA methylation of promoters to detect highly transient TF-DNA binding interactions.	Steensel and Henikoff (2000), Alvarez et al. (2020)
DAP-seq (TF → target interaction in vitro)	DNA affinity purification sequencing (DAP-seq) is a high-throughput TF-DNA binding assay that uses genomic DNA and TFs expressed in vitro.	O'Malley et al. (2016)
Precision/Recall (AUPR) (Validation of inferred TF → target interactions)	Precision/Recall (PR) analysis with area under precision recall (AUPR) curve uses validated TF-target gene data (TF-binding and/or regulation data) to determine the PR of predicted TF-target genes in gene regulatory networks (GRNs) (Fig. 3A).	Brooks et al. (2019), Shanks et al. (2022), Brooks et al. (2021)
Network Walking (TF_1 → direct TF_{2s} → indirect TF_1 targets)	Network walking is a GRN method that charts a path from the direct target genes of a TF_1 to their indirect target genes via a TF_2 (Fig. 3B).	Brooks et al. (2019, 2021)

and in the absence of either endogenous NLP7 or NLP6 proteins (NLP6 experimental setup is shown as an example; Fig. 2B; Cheng et al. 2023). While previous reports show that NLP7 and NLP6 heterodimerize in the cytosol in re-sponse to nitrate (Guan et al. 2017), Cheng et al. (2023) show that the nuclear retention of NLP7 and NLP6 in re-sponse to nitrate is independent of each other (Fig. 2B; Guan et al. 2017; Cheng et al. 2023).

Recent studies also implicate NLP7 in initiating a cascade of early N-responsive downstream transcription factors (Alvarez et al. 2020). Specifically, gene expression changes in response to nitrogen occur rapidly (minutes to hours) and are divided into primary and secondary responses (Medici and Krouk 2014; Alvarez et al. 2021). Primary N-response genes are (i) rapidly induced by nitrate (minutes),(ii) do not require de novo protein synthesis, and (iii) are typ-ically involved in nitrate transport, assimilation, and signaling (Medici and Krouk 2014). Secondary N-response genes are in-duced later (hours) and depend on the transcriptional pro-ducts of the primary response genes. How the primary and secondary nitrogen response is regulated was recently re-vealed to involve rapid, transient protein–DNA interactions by TFs that follow the "hit-and-run" model of regulation (Fig. 2C), which includes the TFs bZIP1 (Para et al. 2014; Doidy et al. 2016) and NIN LIKE PROTEIN 7 (NLP7; Alvarez et al. 2020). As a pioneer or triggering TF, NLP7 is at the top of the nitrate signaling hierarchy following the "hit-and-run" model of transcriptional control (Fig. 2C). It was shown that the transient TF2 targets of NLP7 initiate a temporal cascade of genome-wide changes in the nitrate re-sponse in planta (Marchive et al. 2013; Alvarez et al. 2021).

The "hit-and-run" model suggests that a TF trigger/pioneer can form a stable transcriptional complex (the "hit"), allow-ing transcription to continue even after the initiating TF is no longer bound (the "run"; Fig. 2C; Schaffner 1988; Para et al. 2014; Doidy et al. 2016; Alvarez et al. 2020). Genome-wide evidence for the "hit-and-run" model of transcription for transient TF–target gene interactions was validated for two master TFs involved in the nitrate response, first identified with bZIP1 and more recent evidence for NLP7 (Para et al. 2014; Doidy et al. 2016; Alvarez et al. 2020). Time-series ChIP-seq experiments showed that bZIP1 and NLP7 were transiently bound to early nitrate-response genes, and 4-thiol-uracil labeling of nacent mRNA confirmed the ac-tive transcription of these hit-and-run targets (Para et al. 2014; Doidy et al. 2016; Alvarez et al. 2020).

Importantly, the plant cell-based TARGET assay (Transient Assay Reporting Genome-wide Effects of Transcription factors) used in these studies can capture early and transient TF-target gene regulation events often undetected *in planta* (Para et al. 2014; Doidy et al. 2016; Brooks et al. 2019; Alvarez et al. 2020). The TARGET TF-assay involves transient expression of a TF fused to a glucocorticoid receptor (GR) in plant cell protoplasts. The TF-GR protein is held in the cytoplasm by HSP90 binding to the GR domain (Bargmann et al. 2013; Fig. 2C, Table 1). The addition of the GR-ligand dexamethasone (DEX) displaces HSP90 binding, allowing nuclear entry of the TF-GR fusion protein. When DEX treatment is performed in the presence of cycloheximide, to inhibit the synthesis of proteins encoded by direct target genes (e.g. TF2), direct targets of a TF can be identified with RNA-seq, compared to empty vector (Fig. 2C, Table 1; Bargmann et al. 2013; Brooks et al. 2019, 2023).

The plant cell-based TARGET TF perturbation system allowed the identification of transient NLP7 targets that were undetected by time-series chromatin immunoprecipitation (ChIP; Alvarez et al. 2020). By coupling the DNA adenine methyltransferase identification (DamID) method (Gutierrez-Triana et al. 2016) to the TARGET TF perturbation system, it was possible to capture NLP7 binding to highly transient targets that were missed by time-course ChIP (Fig. 2C, Table 1; Alvarez et al. 2020). DamID uses a fusion protein of DNA adenine methyltransferase (Dam) to detect TF-DNA binding events by leaving a stable methylation mark at the adenine base in the $GA^{me}TC$ sequences near (within 1 kb) to protein-DNA binding sites as soon as the

TF touches down on the promoter (e.g. even transiently). This adenine methylation at GA^{me}TC allows for the binding and DNA cleavage using the DpnI restriction enzyme. DpnI fragments are mapped to the promoter regions to identify genes "touched" by the TF. Thus, this DNA methylation approach overcomes the limitations of biochemical methods such as ChIP-seq and other antibody-based techniques that are biased for stably bound TF-DNA interactions (Fig. 2C, Table 1; Steensel and Henikoff 2000; Alvarez et al. 2020). Using the TARGET and DamID-Seq methods, the study by Alvarez et al. (2020) confirmed that transient interactions of NLP7 initiate active transcription of its targets, consistent with a "hit-and-run" transcription model (Fig. 2; Alvarez et al. 2020). Overall, the multiple levels of NLP7 regulation highlight the important role of NLP7 in primary nitrate response to ensure a fast and broad adaptation by the plant to fluctuating nitrate levels (Fig. 2).

Temporal nitrogen response networks: generation and validation

In addition to the master TFs discussed above, TGA1, bZIP1, and NLP7, which are critical for signaling N-dose over time, gene network analysis studies and mutant screens have identified 40-plus TFs that are involved in propagating the nitrate signal (for review, see Vidal et al. 2020). Thus, we must understand the temporal regulatory connections between these TFs and the nitrate-responsive genes they control to obtain a complete temporal picture of nitrate signaling events. Combining computational and experimental approaches that consider time in gene expression analysis has proven to be a powerful approach to uncovering the temporal me-chanisms of transcriptional responses in plants.

The goal of gene regulatory network (GRN) inference models is to connect a regulator (i.e. TF) to each of the genes it regulates in the genome. As causality moves forward in time, time-series experiments are a valuable resource to infer GRN models that can predict TF-target gene relationships at future untested time points, a main goal of systems biology. To account for the different times of captured gene regulation in time-series data, specialized network inference algorithms have been developed to account for the added factor of time in the data and can be based on correlation: time-lagged, random forest (DynGenie3 and Outpredict), and other regression models (Huynh-Thu and Geurts 2018; Nguyen and Braun 2018; Cirrone et al. 2020).

Over the last 10 years, several studies have exploited time-dependent responses to investigate nitrogen signaling networks in Arabidopsis using fine-scale time series (Krouk et al. 2010; Patterson et al. 2016; Walker et al. 2017; Varala et al. 2018; Brooks et al. 2019; Alvarez et al. 2021). In multiple of these time-based N-response network studies, a state-space model, which is a model that uses first-order differential or difference equations to describe a system, called Dynamic Factor Graph, was applied to fine-scale nitrogen response time series (i.e. many time points close together) datasets to predict regulatory interactions between N-responsive TFs and N-responsive genes in shoots (Varala et al. 2018) or roots (Krouk et al. 2010; Brooks et al. 2019). In general, state-space models are algorithms that model dynamic data (e.g. gene expression) by assuming that data are generated from underlying "hidden states" (Krouk et al. 2010; Brooks et al. 2019). In the case of time series experiments, the data of gene expression consider several time points (e.g. every 5 min, from 5 to 20 min) as consecutive hidden states that form a Markov chain (a mathematical system that statistically modulates random processes). Consequently, each transition in the Markov chain corresponds to a stationary (time) dynamic model. The resulting GRNs revealed the temporal networks operating in each tissue and implicated a hierarchy to the TFs involved (Vidal et al. 2020).

In the first fine-scale time-series study of nitrate signaling, Krouk et al. examined very early (3 to 20 min) gene expression responses to nitrate supply in roots (Krouk et al. 2010), whereas 20 min is the earliest time point that had previously been examined at the genomic level (Wang et al. 2000). The Krouk et al. (2010) study demonstrated that nitrate-triggered gene expression responses occur within as early as 3 min, and that transient changes are missed if plants are only sampled at later time points. In a subsequent study, Varala et al. performed a time-series experiment that included ammonium nitrate treatments across early-to-late time points (starting from 5 min for up to 120 min) and identified 2,737 genes responding to nitrogen as a function of time (NxTime) response genes in shoots (Varala et al. 2018) and 1,458 NxTime response genes in roots (Brooks et al. 2019). Moreover, the concept of "just-in-time" (JIT) analysis developed and deployed in these two studies identified the first time point that a gene was induced >1.4 fold by N-treatment. The JIT analysis bins NxTime genes that are differentially regulated by N for the first time point in the time-series experiment. Analysis of these JIT genes uncovered not only a temporal cascade of enriched cis-elements at each consecutive time point but also GO terms resulting from N-signaling that evolves over time (Varala et al. 2018; Brooks et al. 2019).

A strength of the N-response time-series GRN generated in shoots (Varala et al. 2018) and roots (Brooks et al. 2019) was in assessing the precision and recall accuracy of the TF→target gene GRN predictions using AUPR (area under the precision-recall curve) analysis (Fig. 3A; Table 1). The benefit of AUPR analysis is that it uses validated TF–target interactions to empirically determine precision cutoffs for the TF-target gene predictions in the GRN. By contrast, other methods arbitrarily select the top 1% to 10% of interactions —to prune the GRNs for higher-confidence TF-target edge predictions (Fig. 3A). To conduct the AUPR analysis, the TF–target gene interactions predicted by the GRN for each TF and target gene are ranked based on an edge score computed by each network inference method (Fig. 3A). Next, to determine the accuracy of these predictions, the inferred

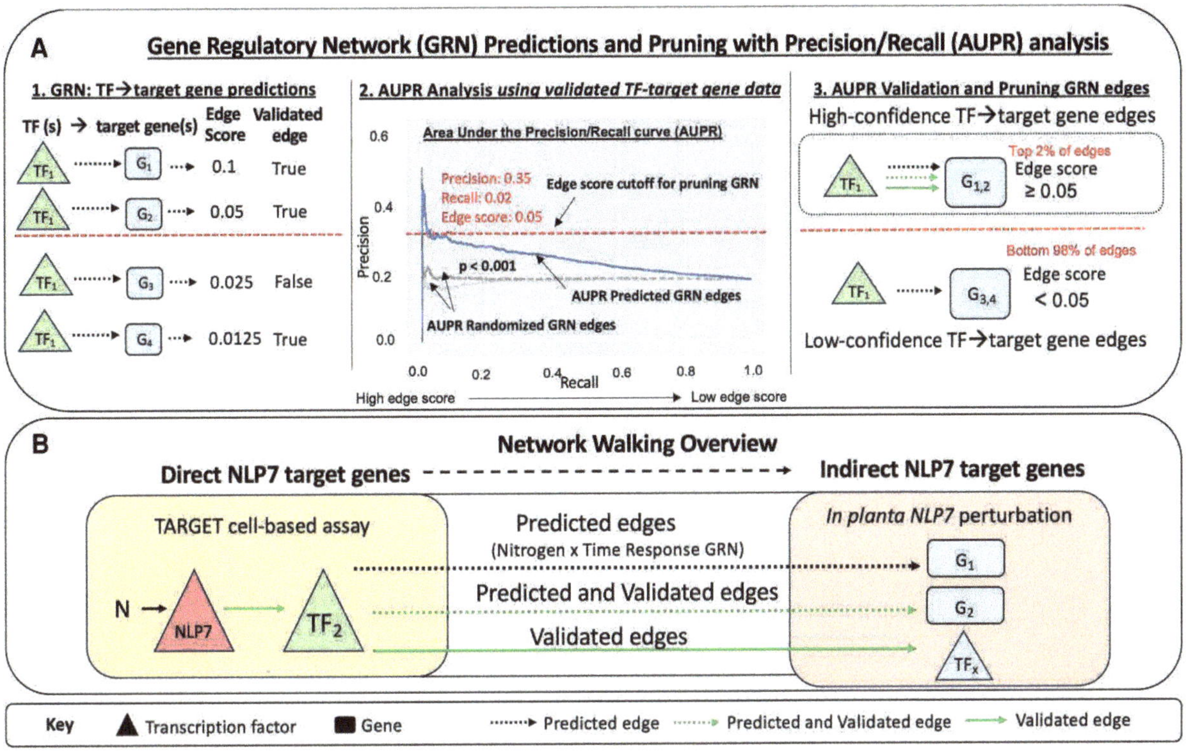

Figure 3. Determining high-confidence GRNs by AUPR and use in network walking. **A)** (A1) The predicted TF–target gene interactions are first ranked according to edge score, and then compared to validated TF–target gene interaction data to calculate precision and recall. (A2) The values are then plotted on the AUPR curve to select a cutoff TF-target edge score. The edges in the predicted network (blue line) were significantly more likely to be true (i.e. validated) edges than when the edge order ranking was randomized (gray lines). The graph is a screenshot from the automated AUPR analysis feature in connectf.org (Brooks et al. 2021). (A3) The edge score cutoff is used to "prune" the network for high-confidence interactions. **B)** Network walking charts a path between direct to indirect target genes of a TF_1 via TF_2s (Brooks et al. 2019, 2021). In this example, the TF NLP7 directly regulates TF_2s as identified with the TARGET cell-based assay (Alvarez et al. 2020). The target genes for each TF_2 can be determined using predicted GRN edges from the NxTime network and/or using validation data from methods such as the TF-TARGET assay and/or TF-target binding by DAP-seq (Table 1). Bottom panel adapted from Brooks et al. (2019).

TF–target gene interactions are compared with validated TF–target gene interactions, for a subset of TFs in the network, as determined by methods such as DAP-seq for TF-target binding interactions in vitro (O'Malley et al. 2016) or direct TF-target gene regulation based on the TARGET TF assay in protoplasts (Table 1, Fig. 3A; Varala et al. 2018; Brooks et al. 2019). This analysis then determines which predicted TF-target edges are supported by experimentally validated data. The validation data are then used to calculate the precision and recall for predicted TF–target gene interactions in the GRN (Schrynemackers et al. 2013). These values are used to produce the AUPR curve, which is then used to select a cutoff edge score for the GRN predictions (Fig. 3A). The selected cutoff edge score from the AUPR curve is used as a threshold to "prune" for high-confidence edge predictions in the GRN. Using the above outlined time-series N response GRNs as examples for AUPR analysis, ConnecTF (connectf.org) is a web-based platform that offers automated AUPR functions where researchers can upload their own networks, select a precision cutoff, and download the high-confidence TF-target gene predictions (Table 1, Fig. 3A; Brooks et al. 2021).

The use of validated TF-target data to prune GRNs for high-confidence TF-target gene predictions is important for identifying key regulatory control points in the absence of comprehensive validated TF-target data. Although there is now experimentally validated TF-target gene binding and regulation data for over 500 Arabidopsis TFs, primarily from DAP-seq (O'Malley et al. 2016), this is still only approximately a quarter of all predicted TFs in Arabidopsis (see data housed in ConnecTF, Brooks et al 2021). This means that for any given signaling pathway studied, it is likely that most of the TFs involved do not have validated target genes. For example, the N-response time-course experiments described above revealed that 326 TFs respond to N-treatment in roots and/or shoots (Varala et al. 2018), but only 95 of those TFs have experimentally validated target genes. Furthermore, for the TFs that lack experimental validation data, it is unclear if the TFs are activators or repressors of their target genes in the network. To address this question, the authors of Hummel et al. (2023) used synthetic biology approaches coupled with this systems biology analysis to determine which TFs in the Varala et al. NxTime GRN are activators or repressors of N-responses by using the reporter genes, nitrate reductase 1 (NR1) and nitrite reductase 1 (NIR1; Varala et al. 2018; Hummel et al. 2023).

A further complication to interpreting TF signaling pathways is that *in planta* gene expression responses in TF mutants and constitutive TF overexpressors reflect both direct and indirect effects of the TF being perturbed. To determine how the N-responsive TFs work to propagate the N signal in a temporal network, Brooks et al. developed a "network walking" approach to chart a temporal network path for a TF of interest (Fig. 3B, Table 1; Brooks et al. 2019, 2021). Network walking connects direct target genes of a focus TF of interest (i.e. genes identified in plant cells with the TARGET TF assay) with their indirect target genes (i.e. genes identified only *in planta*) via their directly regulated TF_2s (Fig. 3B). In the network walking approach, the TF_2s directly regulated by the focus TF being perturbed are then used to explain the response of the indirect target genes *in planta*. Because many of these TF_2s lack experimental data, the high-confidence TF-target gene predictions from the time-series inferred network are crucial to identify the most important TF_2s that mediate the N-response signaling pathway and guide further studies (Fig. 3B; Brooks et al. 2019, 2021). For example, the network walking approach was used to chart a path between direct and indirect target genes for the N-response TFs, TGA1 (Brooks et al. 2019), CRF4 (Brooks et al. 2019), and NLP7 (Alvarez et al. 2020; Brooks et al. 2021).

Learning nitrogen-dependent gene regulatory networks at a temporal level has helped to unravel how shoots integrate multiple root-derived signals. The fine-scale (i.e. many time-points close together) time-series N-response data from Varala et al. have been particularly useful for shoot and root network comparisons as gene expression was measured from both organs for the same sets of Arabidopsis plants (Varala et al. 2018). This study found a significant overlap between shoot and root N-responsive genes and TFs, and yet a large set of genes were also specific to each organ (Varala et al. 2018). The timing of expression between the overlapping shoot and root genes often differed, suggesting that the N-responsive signaling networks had some degree of organ specificity. Additionally, a subset of N-responsive TFs displayed organ specificity in their N-responsive target genes using the TARGET assay (Brooks et al. 2019). For example, CRF4 regulated early N-responses specifically in the shoot, whereas LBD37 regulated N-responses specifically in the root (Brooks et al. 2019). Furthermore, these fine-scale time-series N-response data (Varala et al. 2018) were used to identify the causal relationship of N-responsive genes between organs using Granger-causal analysis (Heerah et al. 2021). Using this analysis, Heerah et al. predicted 1,007 root- and shoot-expressed genes that influenced gene expression in the other organ (Heerah et al. 2021). Interestingly, the list of predicted genes included a significant number (384 genes) of causal genes that are known or predicted mobile transcripts (Heerah et al. 2021). These GRN findings show a coordination between root and shoot N-responses that can be used to determine how these responses coordinat physiological outcomes.

The dynamics of nitrogen responses in specific cell types

The plant's ability to sense and respond to the fluctuating N status of the soil is governed by cell type-specific responses (Jia and von Wirén 2020; Liu et al. 2020; Hu et al. 2021). Three studies have examined cell type-specific nitrogen responses in roots by treating GFP-marked cell lines with nitrogen followed by FACS and transcriptomic analysis (Gifford et al. 2008; Walker et al. 2017; Contreras-López et al. 2022). Consistently, these studies found that nitrate responses in the root are largely cell type-specific and highlight the need for routine cell type studies, as whole root studies will miss a significant portion of the plant response to an environmental stimulus such as nitrate.

The most recent study to examine cell type-specific nitrate responses identified 5,231 differentially expressed genes and a rapid transcriptome reprogramming, with 1,572 genes responding early 12 min after nitrate treatment (Contreras-López et al. 2022). Moreover, 42.5% of regulated genes were localized in the endodermis cell type, suggesting that the endodermis might have a role as a regulatory hub for nitrate signaling since it is embedded with the Casparian strip, being a nutritional checkpoint for the vascular system (Palmgren 2018; Contreras-López et al. 2022). Analysis of gene ontology (GO) terms found that nitrate responses ini-tiate in the epidermis and cortex as outermost cell types, fol-lowed by innermost cell types in later time points (Fig. 4A), which is in line with nitrate uptake and transport (O'Brien et al. 2016; Contreras-López et al. 2022). The first biological processes to be enriched include "response to carbohydrate stimulus," "glycolysis, "response to reactive oxygen species," "response to lipid," "response to abscisic acid," and "response to nitrate," which are initiated from the epidermis and then move toward inner cells (Fig. 4A). For instance, the nitrate transceptor *NRT1.1/NPF6.3* and *NRT2.1* are rapidly induced by nitrate in the epidermis and then within all cell types at later time points (Fig. 4B; Contreras-López et al. 2022). This is in line with previous studies that identified the cell type specificity of nitrate transporters in the epidermis (Tegeder and Masclaux-Daubresse 2018; Wang et al. 2018; Lhamo and Luan 2021; Muratore et al. 2021). At later times, expres-sion of genes involved in "nitrate assimilation" are enriched from inner cell types toward the epidermis (Fig. 4B), while expression of genes involved in "root system develop-ment" are localized in epidermis and endodermis, consistent with their role in lateral root growth and root hair development, showing a localized and transient response to nitrate (Ramakrishna et al. 2019; Liu et al. 2020; Fig. 4B). Furthermore, by integrating the spatiotemporal transcriptomic data with TF-target gene interactions, Contreras-

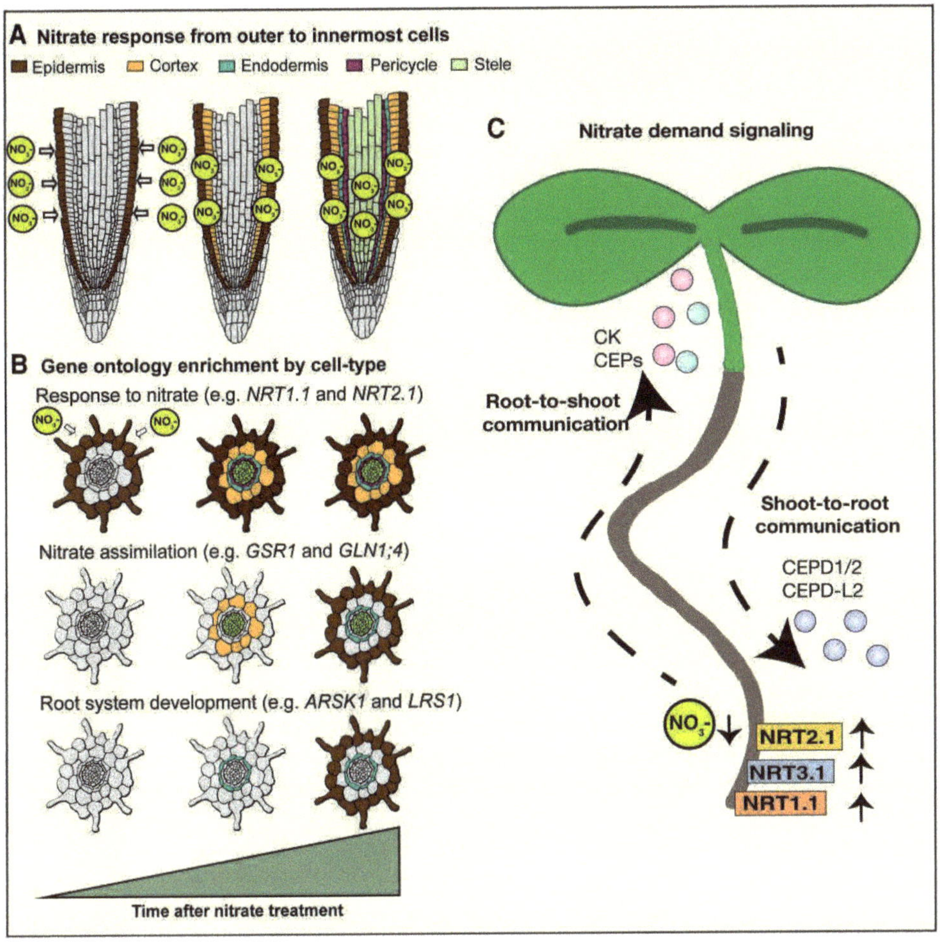

Figure 4. Spatiotemporal responses after nitrate treatments in Arabidopsis root cells are highly dynamic and localized. **A)** During nitrate treatments, the first cell type to respond is epidermis, followed by cortex. Consistent with their outermost location and first layers of nitrate acquisition. At later times of treatment, nitrate responses are present in all major root cell types (Contreras-López et al. 2022). **B)** Transverse view of root cells shows gene ontology (GO) enrichment after nitrate treatments. The first enriched GO term is "response to nitrate," moving from epidermis toward innermost cell types. At later times, "nitrate assimilation" and "root system development" go from inner to outermost cell types. Transcriptomic analysis and GO terms were obtained from sorted root cells by Contreras-López et al. (2022). **C)** Nitrate-demand signaling model. When roots are grown on limited nitrate levels, C-terminally encoded peptides (CEPs) and tZ-type cytokinins (CK) are translocated to the shoot, increasing the expression levels of *CEPD1/2* and *CEPD-L2*. In turn, shoot-derived CEPD1/2 and CEPD-L2 descend back to the root and increase the expression of nitrate trans-porters *NRT3.1* and *NRT1.1/NPF6.3* and *NRT2.1* to compensate for the lack of nitrate in the soil. This highly coordinated system results in plant growth adaptation according to the changing nutrient levels (Tabata et al. 2014; Ohkubo et al. 2017; Ota et al. 2020).

López et al. found that 62% of TF–target interactions were predicted to occur in the endodermis, being an important cell type for transcriptional regulation. The transcription factors ABF2 and ABF3, previously investigated for their role in ABA-mediated signal transduction, were revealed to be master regulators of nitrogen responses in the endodermis, displaying lateral root growth inhibition in *abf2*, *abf3*, and *abf2/3* plants in response to nitrate (Contreras-López et al. 2022). This phenotype is, in part, due to an altered development of lateral root primordium. Overall, these results highlight the importance of spatiotemporal analysis to uncover how the nitrate signal is dynamically propagated in the root and reveal new molecular mechanisms controlling nitrogen responses in specific cell types, which otherwise would be missed.

Complementary with the identification of cell-type transcriptional responses to nitrate over time, the authors of Chen et al. developed a nitrate biosensor to visualize the spatial and temporal distribution of nitrate in the Arabidopsis root (Chen et al. 2022). To accomplish this, Förster resonance energy transfer (FRET) sensors were developed as a (i) fusion fluorescent protein possessing a sensor domain (FRET acceptor protein) and a (ii) fused FRET donor fluorescent protein. Once the donor protein is excited, energy is transferred to the FRET acceptor protein. When the sensor domain from the acceptor protein interacts with its target molecule, a conformational change occurs. This conformational change, in return, alters the efficiency of energy transferred from the FRET fusion donor protein to the FRET fusion acceptor protein. Hence, by measuring the ratio change, as the change between the fluorescence intensity of the donor and the acceptor protein, it is possible to report the concentration of the target (Chen et al. 2022). The FRET sensor developed by Chen et al. used the bacterial protein NasR (NitraMeter3.0), which is a soluble receptor protein containing a nitrate and nitrite sensing domain as a FRET acceptor protein fused to a modified Aphrodite (edAFP) protein. On the other hand, a modified cyan fluorescent protein was used as a FRET donor protein (edeCFP; Chen et al. 2022). When Arabidopsis plants expressing the nitrate biosensor were exposed to exogenous nitrate treatments for 5 min, the fluorescence emission ratios increased in the epidermis, cortex, pericycle, and stele cells, with the highest emissions ratio increase in the cortex cells, suggesting a higher nitrate uptake or transport function in this cell type. These results are in line with the reports of nitrate import and signaling in multiple root cell types (Gifford et al. 2008; Walker et al. 2017; Contreras-López et al. 2022). Additionally, the mutant for the nitrate transceptor, *nrt.1.1/nfp6.3*, displayed lower emission ratios in all root zones, supporting its role as a major nitrate transporter (Chen et al. 2022). The emissions ratio of the endodermal cell layer remained high when roots were grown under low nitrate conditions and increased slowly compared to other cell types, which coincides with the previous result of the endodermis as a nitrate regulatory hub for plants to respond and adapt to their environment (Chen et al. 2022; Contreras-López et al. 2022).

In addition to cell type-specific nitrate responses in the root regulating plant growth and development, there are also nitrate responses localized in the shoot. For example, NRT1.1/NPF6.3 and NLP7 drive stomatal opening by controlling the entry of nitrate into guard cells, resulting in nitrate-induced depolarization and increased nitrate levels during stomatal opening (Guo et al. 2003; Castaings et al. 2009). Indeed, *nrt1.1/npf6.3* and *nlp7* plants are impaired in nitrate content, reducing stomatal opening and water loss, resulting in improved drought tolerance (Guo et al. 2003; Castaings et al. 2009; Araus et al. 2020). However, the role of nitrate signaling mediated by NRT1.1/NPF6.3 and NLP7 in the control of stomatal opening remains to be elucidated.

These studies have shown us how cell type-specific nitrate responses can modulate root and shoot growth, raising the need to implement single-cell level approaches to understand organ-level plasticity. As an in silico approach at single-cell resolution, Lhamo and Luan (2021) profiled putative nitrate transporters in root cell types to understand nitrate uptake and translocation from the soil (Denyer et al. 2019; Ryu et al. 2019; Lhamo and Luan 2021). The dual-affinity transporter *NRT1.1/NPF6.3* and high-affinity transporters *NRT2.1*, *NRT2.2*, *NRT2.4*, and *NRT2.5* were highly expressed in epidermis and root cap cells, concomitant with the role of sensing NO_3^- changes in soil (Ho et al. 2009) and uptake function, respectively (Lhamo and Luan 2021). *NPF1.1* and *NPF1.2* were expressed in procambium cells, indicating that they could be participating in loading NO_3^- to phloem and xylem cells in developing roots (Jouannet et al. 2015; Lhamo and Luan 2021). These results indicate that in the future we will be able to generate maps of local and systemic nitrate sensing/signaling from root to shoot and vice versa by using single-cell approaches.

Nitrogen responses across organs: root-to-shoot communication

Because nitrogen availability in the soil changes constantly, plants have developed communication systems that regulate nitrate uptake from the root according to the nutritional state of the soil and the plant. In response to changes in nitrate availability, the nitrate-response targets the fast reallocation of resources to rebalance biomass between below- and above-ground organs, as well as the regulation of physiological activities such as root nitrate transport. For example, heterogeneous nitrate supply leads to greater development, growth, and nitrate transport stimulation in the roots that are locally exposed to nitrate (Ruffel et al. 2011). Such integrated/adapted responses result from a combination of (i) continuous and long-distance exchange of signals through the vascular system and (ii) organ-specific GRNs. Currently, the challenge is to understand how these multiple signals interact and converge toward regulating central physiological and developmental processes in respective organs.

Nitrate-related long-distance signals are just starting to be understood and as of now belong to the following classes of molecules: hormones (e.g. root-to-shoot trans-zeatin cytokinin signal; Poitout et al. 2018), small peptides (e.g. root-to-shoot C-terminally encoded peptides; CEP; Tabata et al. 2014), and microRNAs (so far only functionally characterized in legumes to regulate the nodulation; Gautrat et al. 2021). The coordination of root/shoot communication and growth responses may rely on other types of long-distance signals that remain to be characterized, such as ions or metabolites. However, connections between the known systemic signals and the discovered nitrate-related local and systemic signaling pathways are starting to be proposed, thereby improving our mechanistic understanding of the nitrate signaling network. In addition to the nitrogen-regulated systemic response of root activity, it has been known for a long time that root nitrate supply is a main input for shoot growth. Interestingly, recent findings indicate that once again, multiple signals likely co-exist to properly coordinate the activity of this aerial organ. Several studies have reported that nitrate in the root induces the synthesis of cytokinins (CK) like tZ-type, which are then translocated to the shoot (Osugi et al. 2017; Poitout et al. 2018). Interestingly, this "CK neo-synthesis" in the root has recently been shown to be under the control of the master nitrate sig-naling regulator NLP7, contributing with CK translocation to the shoot and upregulation of cytokinin response factors (CRF; Abualia et al. 2022). In this manner, CRFs directly induce the expression of auxin transporters (*PINs*), resulting in auxin transport and shoot growth (Abualia et al. 2022).

A very sophisticated example of interaction between systemic signals and nitrate signaling is the case of CEP peptides. CEPs act as root-derived peptides that ascend the nitrate starving signal to the shoot, where the production of a nitrate descending signal induces the expression of root nitrate transporters *NRT2.1*, *NRT3.1* and the transceptor *NRT1.1/NPF6.3* to compensate for the lack of nitrate in the soil (Fig. 4C; Tabata et al. 2014). Years later, Ohkubo et al. identified the nitrate descending signal as a polypeptide named CEP Downstream 1 (CEPD1) and CEPD2 (Fig. 4C; Ohkubo et al. 2017). In 2020, it was also established that CEPD-like 2 (CEPD-L2), together with CEPD1 and CEPD2, contributes to the nitrate demand systemic signaling (Ota et al. 2020). Interaction between tZ-type CK and these peptides also occurs as a response to nitrate starvation. Indeed, the presence of root-synthesized tZ-type CK is necessary in shoots to induce maximal expression levels of shoot-to-root CEPD1/2 and CEPD-L2 peptides, which are mainly induced by the nitrate starvation signal (Fig. 4C; Ota et al. 2020). Moreover, CEPD-L2 positively regulates the expression of high-affinity nitrate transporters and NRT1.5, which loads nitrate into the xylem layer, demonstrating that these peptides have an important role in nitrate uptake and translocation to the shoot under starving conditions (Ota et al. 2020). Altogether, these new findings illustrate that shoots can also perceive and integrate nitrate-related signals, first by receiving a nitrate status signal (e.g. starvation) from the root and responding by sending another signal back to the root (e.g. peptides) to optimize their activity. For future studies, moving on from organ-specific nitrate-responses to cell-specific nitrate-response networks using single-cell data will aid in our understanding of root-shoot-root communication in response to N.

New aspects of N-response in time and space: single-cell analysis

Single-cell RNA-sequencing (scRNA-seq) has emerged as an important tool to better understand dynamic cellular processes such as spatiotemporal gene expression and developmental trajectories from heterogeneous cell populations in a single snapshot (Fig. 5). As outlined above, previous studies that have examined the response of nitrate in specific cell types over time have relied on the use of GFP-marked cell lines followed by bulk RNA-seq (Walker et al. 2017; Contreras-López et al. 2022). The use of single-cell sequencing over GFP-marked lines offers the following advantages: (i) to profile nitrate responses in all the cell types composing an organ, (ii) to track nitrate responses in cell types according to their developmental time in a single snapshot, (iii) to examine nitrate responses in all cell types over time in a single experiment without the need for individual experiments for each GFP-marked cell line, and (iv) to examine the effect of nitrate specific cell types of mutants without the need to develop multiple mutant lines crossed with specific GFP-marked cell lines. Despite these promising advantages, single-cell (sc) RNA-sequencing data are still highly sparse due to cell dropouts. To capture the dynamic cell-specific response to nitrate that regulates multiple plant developmental processes, future studies should examine the cell type-specific responses over multiple time points using single-cell sequencing.

The root has been widely used as a model for scRNA-seq due to its wide characterization, availability of reporter lines and cell type marker genes (Denyer et al. 2019; Jean-Baptiste et al. 2019; Ryu et al. 2019; Zhang et al. 2019; Shahan et al. 2022). Indeed, the root is distributed in different developmental zones, including the less differentiated meristem zone, elongation zone, and most differentiated maturation zone (Fig. 5A). Therefore, we can analyze a gradient of cell differentiation from the root in a single experiment (Fig. 5B). Single-cell transcriptomes offer the unique opportunity to generate computational "developmental trajectories" (Fig. 5C), in which we can order cell type progression from the beginning of cell fate until the final development of ma-ture cell types. Once the developmental trajectory of a spe-cific cell type is established, gene expression throughout development as "pseudotime" can be graphed (Fig. 5D). For example, Denyer et al. observed that during trichoblast development, genes expressed at the beginning of cell fate (e.g. meristematic cells) were enriched for biological processes such as DNA replication, cell proliferation, and ribosomal functions, whereas more differentiated trichoblast cells were enriched in expression of genes controlling unidimensional growth, root hair elongation, and maturation (Fig. 5C; Denyer et al. 2019). Furthermore, pseudotime trajec-tories coupled with GRNs also contributed to identifying the developmental time-regulated TFs that modulate the expression of target genes in a spatiotemporal manner (Denyer et al. 2019). Using the same trichoblast developmental trajec-tory, we find that NRT1.1/NPF6.3 expression exhibits a grad-ual increase as cells become differentiated into mature trichoblasts, which supports the function of nitrate signaling and uptake in this cell type (Fig. 5, C and D). These current single-cell transcriptomic profiles represent plants grown in standard MS media; therefore, expression profiles under changing nitrate conditions need to be examined in future studies.

Multiple studies in Arabidopsis have created root and/or shoot cell atlases from scRNA-seq data with web-based platforms to examine the cell-specific expression profile for genes of interest (Table 2). To develop a comprehensive Arabidopsis root single-cell expression atlas, Shahan et al. analyzed single-cell data for 110,427 Arabidopsis root cells, including root data from previously published single-cell studies, to determine cell-specific gene expression over developmental time in each cell type (Table 2; Denyer et al. 2019; Ryu et al. 2019; Shahan et al. 2022). In addition to single-cell studies, single-nuclei (sn) sequencing has also been used as an alternative method to avoid the limitations of developing protoplasts. To provide a holistic view of the Arabidopsis

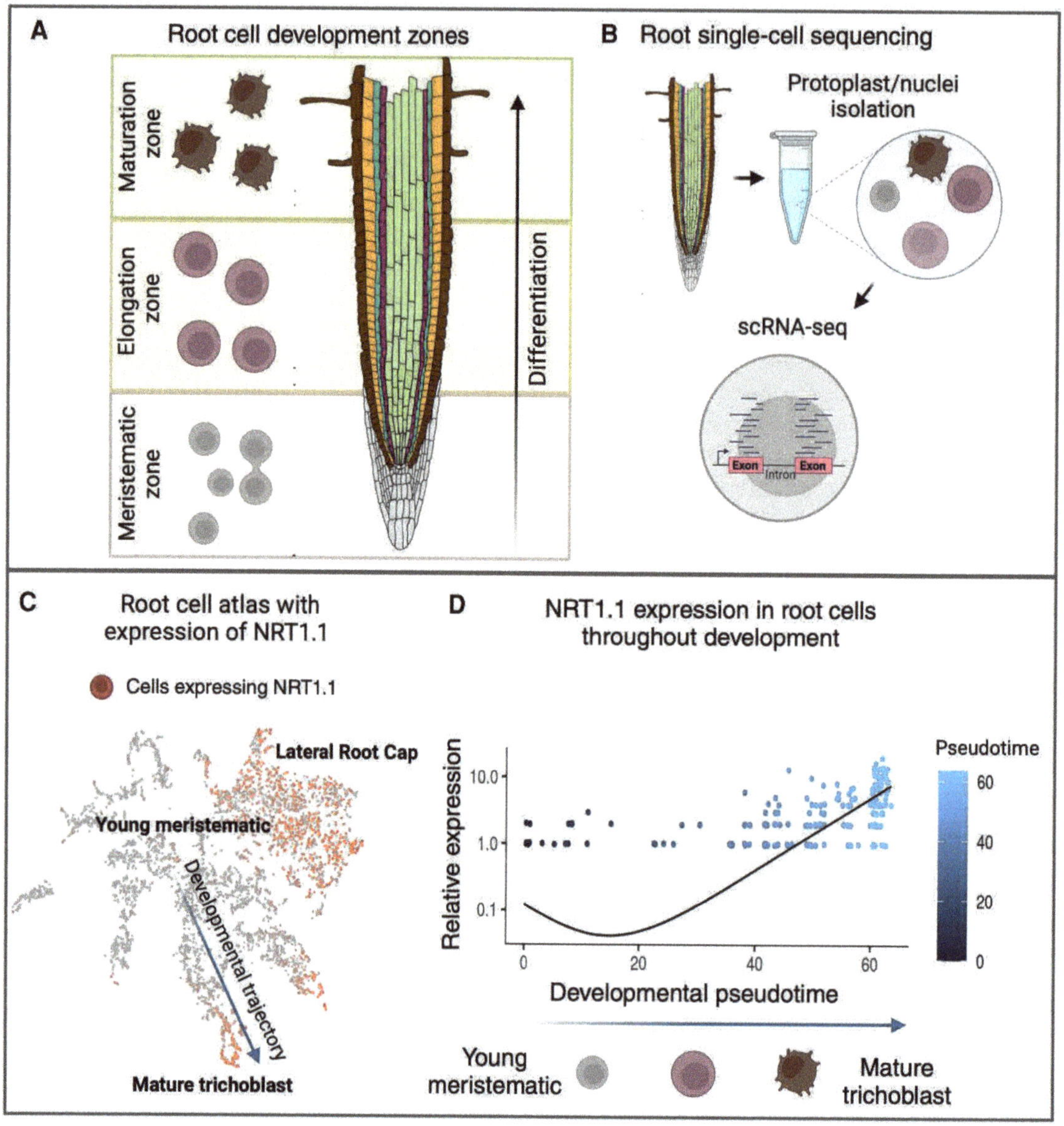

Figure 5. Investigating spatiotemporal gene expression using single-cell RNA sequencing in Arabidopsis thaliana. **A)** Longitudinal view of the root shows different developmental zones from young (meristematic) to mature (maturation zone), which is used as a model for single-cell analysis to construct developmental trajectories in a single experiment (Denyer et al. 2019; Rich-Griffin et al. 2020). **B)** Thousands of protoplasts or nuclei at different developmental stages are used for single-cell library construction (Swift et al. 2022). **C)** Computational analysis of scRNA-data allows the construction of "developmental trajectories" of root cells expressing a gene of interest (red dots), NRT1.1/NPF6.3 using the Plant scRNA-seq Browser with representative screenshots from this tool (Denyer et al. 2019; Ma et al. 2020) (Table 2). **D)** "Pseudotime" expression of *NRT1.1/NPF6.3* from young meristematic cells to mature cells show that NRT1.1 expression is highly expressed in differentiated trichoblast (Denyer et al. 2019; Ma et al. 2020).

transcriptional response over plant development, Lee et al. analyzed single-nuclei data for 801,276 nuclei that represented seed-to-seed development across all major organs during the Arabidopsis life cycle (Lee et al. 2023; Table 2). Furthermore, other researchers have developed web-based interfaces to explore gene expression in smaller-scale single-cell experiments, which include analysis of hormone-treated tissues and/or developmental trajectory analysis (Table 2; Ma et al. 2020; Denyer et al. 2019; Ryu et al. 2019; Wendrich et al. 2020b; Kim et al. 2021; Graeff et al. 2021).

The accessibility to single-cell datasets such as these provides useful tools to form hypotheses on nitrate signaling and explore the spatiotemporal profiles of nitrate-responsive genes in roots/shoots under non-stress cell conditions (Fig. 5). For instance, to understand nitrate uptake and translocation from the soil at single-cell resolution, Lhamo and Luan profiled putative nitrate transporters in root cell types using published single-cell transcriptomic data from Arabidopsis (Denyer et al. 2019; Ryu et al. 2019; Lhamo and Luan 2021). Additionally, we can use the tools outlined in

Table 2. Web browsers available to visualize cell type-specific gene expression from single cell/nuclei datasets in Arabidopsis

Name	Description	URL	References
Arabi Atlas data	Single-cell gene expression in specific root cell types over developmental time for each cell type	https://phytozome-next.jgi.doe.gov/tools/scrna/	Denyer et al. (2019), Ryu et al. (2019), Goodstein et al. (2012) Shahan et al. (2022)
Arabidopsis Developmental Atlas Viewer	Single nuclei transcriptome data for cell type-specific expression of genes throughout plant development stages	http://arabidopsisdevatlas.salk.edu/	Lee et al. (2023)
Plant sc Atlas	Resource and visualization tools for multiple single cell datasets in roots and shoots	https://bioit3.irc.ugent.be/plant-sc-atlas/	VIB Ghent University, Wendrich et al. (2020b), Graeff et al. (2021), Yang et al. (2021a), Nguyen et al. (2023)
The Plant scRNA-seq Browser	Single cell gene expression in roots and shoot cell types including trichoblast, atrichoblast, and cortex pseudotime	https://www.zmbp-resources.uni-tuebingen.de/timmermans/plant-single-cell-browser/	Ma et al. (2020), Denyer et al. (2019), Kim et al. (2021)

Table 2 to examine the expression profiles of genes of interest for nitrate response, such as the transceptor *NRT1.1/NPF6.3*, in roots (Fig. 5, C and D). We find that *NRT1.1/NPF6.3* is expressed in trichoblast, atrichoblast, and lateral root cap, which is also supported with previous studies (Guo et al. 2003; Yang et al. 2008; Denyer et al. 2019). This is also in line with the function of NRT1.1/NPF6.3 in nitrate uptake and signaling; the location of NRT1.1/NPF6.3 expression in the outer cell layers allows for easy access to nitrate in the soil (Contreras-López et al. 2022).

Building on the "pseudotime," cell type-specific gene expression can be monitored in "real time" by adding time-series response data taken throughout development or in response to environmental stimuli (Swift et al. 2022). For example, Nolan et al. generated brassinosteroid (BR) treated time-series expression data of 210,856 single-cell transcriptomes and used GRNs analysis to identify new TFs that activate cell wall-regulated genes in cortex cells to promote elongation (Nolan et al. 2023). Using this approach, it is possible to determine the dynamic nature of how nitrate regulates developmental processes such as lateral root and root hair development in time and space. We propose future studies that incorporate both single-cell profiling and time-series nitrate response assays. Using time-series nitrogen response single-cell data can create cell type-specific developmental trajectories and GRNs to identify the TFs that regu-late cell development during early and late nitrate responses, together with their predicted target genes—thus, enhancing our understanding of how the plant assimilates and responds to nitrate supply to engineer plants with enhanced NUE at the single cell type level.

Following pseudotime and GRN analysis, scRNA-seq of mutant plants will provide new layers of information on nitrate transport and signaling. To date, there are a few studies including single-cell sequencing of mutants to follow changes in developmental trajectories from Arabidopsis roots (Shahan et al. 2022; Nolan et al. 2023). For instance, *SHR* and *SCR* are important TFs for cell identity and differentiation, whereas *BRI1* is an important receptor for BR signaling (Shahan et al. 2022; Nolan et al. 2023). The analysis of *shr* and *scr* single-cell transcriptomes revealed that there is a putative loss of pericycle identity in *shr* mutant and a putative trans-differentiation from cortex to endodermis cells in the scr mutant (Nolan et al. 2023), while Nolan et al. (2023) generated cell-specific *bri1* CRISPR mutants by guiding Cas9 expression into cortex or epidermis cells using as background a *bri1* plant complemented with pBRI1:BRI1:mCitrine. The cortex mutant lines displayed shorter cortex cells in the mature zone but not in the meristem zone, whereas the epidermis mutant lines presented shorter cortex cells in both zones, suggesting that BR signaling is necessary in both the epidermis and cortex to promote cell expansion by modulating cell- wall genes in the elongation zone (Nolan et al. 2023). These results demonstrate that scRNA-seq can address cell identity and signaling pathways in the context of space and time, being also an interesting approach for future N-response studies.

Together with scRNA-seq, single-cell sequencing of accessible chromatin sites will contribute to more resolutive GRNs and will shed light on TFs involved in spatial and temporal regulation during nitrogen responses. Cell type-specific regulation of gene expression is in part modulated by a dynamic chromatin state that responds to development and environment. Changes in the chromatin landscape are reversible and affect the binding of regulatory proteins, such as TFs, controlling gene expression. With single-cell assay for transposase-accessible chromatin (scATAC-seq) as a strategy to uncover putative TF-binding sites during nitrate responses, we can obtain highly resolutive GRNs considering cellular heterogeneity, resulting in cell type-specific accessibility variance and TF-target regulation (Buenrostro et al. 2015). Moreover, together with scRNA-seq approaches, it is possible to correlate chromatin regulation over gene expression across cell types as determined recently by the human ENCODE project and Arabidopsis roots (Buenrostro et al. 2015; Gulko and Siepel 2019; Dorrity et al. 2021; Farmer et al. 2021). Single-cell ATAC-seq in roots has also revealed more accessible and dynamic sites than bulk ATAC-seq, suggesting that a more resolutive network needs to be accompanied by single-cell expression of TFs (Dorrity et al. 2021). Indeed, by integrating snATAC-seq and snRNA-seq from Arabidopsis roots, Farmer et al. led to the identification of

11,858 genes overlapping with chromatin-accessible sites. A high correlation ($P < 10E-05$) was obtained when comparing sn/scRNA-seq and snATAC-seq of root marker genes (Farmer et al. 2021). Overall, in the future, both open chromatin sites and gene expression profiles could be used as biological markers for cell type identity and differentiation level under different nitrogen conditions or time series.

Model-to-crop: nitrogen sensing and signaling and its impact on agricultural outcomes

How can we use results from model species such as Arabidopsis to have an impact on NUE in crops? In reference to this question, a follow-up document to the United States White House's Executive Order 14081, a set of goals for "Harnessing Research and Development to Further Societal Goals" was established and includes multiple goals to improve NUE in agricultural practices (The White House Office of Science and Technology Policy 2023). Specifically, the goals in the White House report are highly relevant to current and future NUE studies, which include reducing nitrogen emissions in agriculture by engineering plants with increased nitrogen use efficiency, improving fertilizer practices, and manipulating plant microbiomes to produce plants capable of growing in nutrient-poor land. To accomplish these goals, we review current research that focuses on (i) how we can apply our knowledge of nitrogen in the model system Arabidopsis to crop species and (ii) understanding how microbial communities affect nitrogen uptake and availability in agriculture.

Advances in our base knowledge of nitrogen uptake, transport, and signaling have greatly benefited from studies in the model Arabidopsis; however, we must develop ways to apply this knowledge base to crops. With the substantial progress in omics technology and the availability of bulk- and single-cell RNA-seq datasets accumulated in the past two decades, cross-species comparisons of genetic information have been gaining momentum (Table 3; Katari et al. 2010; Alvarez et al. 2021; Chen et al. 2021; Fu et al. 2022; Xu et al. 2022). The interspecific comparisons, which can be made at the level of genomic sequences, gene expression, co-expression networks, expression atlas, expression quantitative trait loci (eQTL), and gene regulatory networks, open the possibility of knowledge transferring from model organisms to economically important species. For example, in Obertello et al., a cross-species N-regulatory network between rice and Arabidopsis was constructed by (i) identifying the N-responsive differentially expressed genes in one species as the starting point; (ii) constructing the edges using metabolic interactions, protein-protein interactions, and correlated expression; and (iii) only retaining the nodes whose orthologs were also N-responsive in another species (Obertello et al. 2015). The identified rice candidate TFs targeting the conserved N-regulatory network, including *HYPERSENSITIVITY TO LOW PHOSPHATE-ELICITED PRIMARY ROOT SHORTENING 1*/NITRATE-INDUCIBLE GARP-TYPE TRANSCRIPTIONAL REPRESSOR (HRS1/NIGT1; *HRS1 HOMOLOGs* (HHO), OsHHO3 and OsHHO4) and TGA (OsbZIP11) transcription factors were later functionally validated as key regulators of N-deficiency responses *in planta* (Ueda et al. 2020).

In more recent rice studies, nitrogen-dose sensing in the field was examined as the interaction between nitrogen (N) and water (W; Swift et al. 2019). It was discovered that nitrogen dose is sensed as either moles (N-moles), molarity (N/W), or the synergistic interaction with nitrogen and water (N × W; Swift et al. 2019). Notably, it is the interaction between N and W (N/W or N × W) that positively correlates with phenotypic outcomes such as grain yield and water-use efficiency in the field (Swift et al. 2019). These conclusions were determined using linear models that analyzed RNA-seq and phenotypic data from rice exposed to a factorial matrix of N-by-W conditions of different rice varieties in both laboratory and field conditions (Swift et al. 2019). Using this N-by-W expression and phenotype dataset, Shanks et al. 2022 identified the TFs, OsbZIP23, and Oshox22 as regulators of NUE grain yield (NUEg) by developing gene regulatory networks that linked TFs to target genes to field NUEg phenotypes (Swift et al. 2019; Shanks et al. 2022).

Another widely used approach to uncover gene-to-trait relations from the network perspective is by associating co-expression network modules with either functional information or phenotypes. Cross-species weighted gene co-expression network analysis (WGCNA; Langfelder and Horvath 2008) has identified an N-regulatory network conserved between maize and sorghum, despite the difference in genome size and phylogenetic distance between these two species (Du et al. 2020). Notably, *ZmHPP*, a top-ranked TF in regulating the conserved N-regulatory modules, has an Arabidopsis homolog ATNITR2;2 (AT3G47980) that is also a hub gene (nodes with the most edges) in another co-expression network module for nitrogen signaling (Canales et al. 2014). Overall, these results consistently support the importance of conserved N-regulatory networks in regulating the phenotypic response.

Although tools exist to aid in orthology analysis (Table 3; Huerta-Cepas et al. 2019), the cross-species analyses are limited by the ortholog conversion methods, most of which assume the sequence-based orthologs to have similar functions, the so-called ortholog conjecture, which is arguably applicable to all species as orthology inference tends to be more complicated than a straightforward one-to-one relationship due to genome duplication (Gabaldón and Koonin 2013). This phenomenon is especially prominent in plant species that follow a pattern of genome evolution involving polyploidization followed by the loss or partial retention of duplicated genes (Wendel et al. 2016). As a result, plant genomes have highly complex gene families, making the identification of a single ortholog conceptually impossible. Considering that gene expression can serve as a proxy

Table 3 . Web-based platforms for analysis of genomic data across plant species

Name	Description	URL	References
VirtualPlant	Resource that integrates plant genomic data with visualization and analysis tools	http://virtualplant.bio.nyu.edu/cgi-bin/vpweb/	Katari et al. (2010)
EggNOG	Database for orthology relationships, gene evolutionary relationships and functional annotations for multiple species	http://eggnog5.embl.de/#/app/home	Huerta-Cepas et al. (2019)
ConnecTF	Platform that integrates genome-wide studies to develop and validate networks with AUPR analysis	https://connectf.org	Brooks et al. (2021)
ChIP-Hub	Application to explore the plant regulome and epigenome	https://biobigdata.nju.edu.cn/ChIPHub/	Fu et al. (2022)
Plant Single Cell Hub	Single-cell data repository for plant species	http://jinlab.hzau.edu.cn/PsctH/	Xu et al. (2022)
PlantscRNAdb	Database and browser single-cell expression from multiple species including marker gene selection and analysis for specific cell types	http://ibi.zju.edu.cn/plantscrnadb/index.php	Chen et al. (2021)

for gene function, a novel type of approach has been proposed to classify genes based on not only sequence similarity (genes in the same orthogroup) but also expression patterns (Das et al. 2016). Kasianov et al. constructed a pipeline that trained an XGBoost-based machine learning classifier using developmental transcriptome profiles and sequence-based ortholog information and was able to provide functional correspondence between genes from phylogenetically distinct species (Arabidopsis-maize and Arabidopsis-buckwheat; Kasianov et al. 2023). This method has the potential to enable the selection of functionally similar orthologs, which is approximated by the expression patterns, even for species with distinct morphologies. Furthermore, conservative gene function is not always found between species, which is why mutant analysis and validation must be performed.

In Cheng et al. (2021), the authors exploited the evolutionary conservation of the N-responses across species to identify genes of importance to NUE in a model (Arabidopsis) and crop (maize; Cheng et al. 2021). Specifically, this study integrated the sequence-based and expression-based similarity in orthologs with machine learning modeling to infer gene-to-NUE phenotype relations in Arabidopsis and maize. To do this, the N-response differentially expressed genes (N-DEGs) conserved in Arabidopsis and maize, as determined by sequenced-based orthology, were used as gene features (predictors) in the machine learning, gradient boosting-based method XGBoost (Chen and Guestrin 2016). The output of XGBoost gives each gene feature an importance score for how that gene contributes to a trait (i.e. NUE; Cheng et al. 2021). The top-ranked TFs in the XGBoost models that were important in predicting gene-to-NUE relations were functionally validated using the T-DNA loss-of-function mutants in Arabidopsis. Remarkably, a model-to-crop validation was performed using maize *nfya3* mutant whose Arabidopsis ortholog NF-YA6 (AT3G14020) is the top TF in Arabidopsis XGBoost model and displayed enhanced grain NUE compared to wild-type counterparts in the field experiment (Cheng et al. 2021). In addition, the gene regulatory networks were constructed using a random forest-based algorithm GENIE3 to identify the TFs regulating the conserved N-DEGs predictive of NUE. Functional validation using Arabidopsis mutants defective in the TF hubs (nodes with the most edges) displayed higher NUE (Cheng et al. 2021). These results demonstrated the utility of cross-species transcriptome analyses in optimizing machine learning models and constructing GRNs.

Given the importance of developing high-confidence cross-species networks, the ConnecTF platform (connectf.org) was developed for researchers to perform interactive and automated precision/recall analysis (AUPR) on their uploaded networks, as well as to build and visualize networks and compare validated datasets for one or more TFs in Arabidopsis, rice, and maize (Table 3) (Brooks et al. 2021). To facilitate these analyses, the open-source ConnecTF web platform includes the validated TF-target gene data generated using the plant cell-based TARGET system, along with published *in planta* TF perturbation data, ChIP-seq, and DAP-seq data. Examples of how ConnecTF can be used to develop high-confidence networks using AUPR analysis have been published for Arabidopsis (Brooks et al. 2021) and rice (Shanks et al. 2022).

In addition to cross-species networks and machine learning, single-cell RNA-seq data can be used to uncover conserved regulatory programs and key regulators in specific cell types between species (Table 3; Chen et al. 2021; Xu et al. 2022). Cell type-specific N-responses in crops have been examined on plants grown under low N stress conditions using single-cell RNA-sequencing (scRNA-seq). Like Arabidopsis, studies in crops such as maize and rice also show cell type-specific responses to N in the root using single-cell sequencing technology (Wang et al. 2021b; Li et al. 2022). For example, in maize studies, the nitrate assimilation genes such as ZmGS2 (glutamine synthetase 2) and ZmNAR2.1 (high-affinity nitrate transporter) were induced specifically in the epidermis (Li et al. 2022). One limitation of single-cell studies in crop species compared to Arabidopsis is the lack of well-defined cell-specific marker genes derived from cell type-specific studies and transcriptomes. To address this issue, databases that include single-cell studies in multiple plant species offer tools to identify marker genes for specific cell types in multiple crop species (Table 3; Chen et al. 2021; Xu et al. 2022). Additionally, the new PHYTOMap (plant hybridization-based targeted observation of gene expression map) technique was developed as a fluorescence in situ hybridization method for whole-mount tissue that can be applied to single-cell data for spatial analysis of gene expression and to define marker genes for specific cell types (Nobori et al. 2023). This technique developed in Arabidopsis has the potential to be applied to crops to assist in cluster gene annotation from single-cell studies and marker gene validation without the need to de-velop time-consuming transgenic reporter lines (Nobori et al. 2023). Furthermore, this method is developing rapidly and can be applied to identify the spatial cell type-specific nitrogen response. How plant host–microbe interactions are af-fected by nitrogen levels in the soil is another area of nitrogen research that can benefit from single-cell technology as the colonization of microbes is a cell type-specific response (Cole et al. 2021). For example, multiple studies have examined how nodulation in legumes is initiated in the cortex cells (Walker et al. 2017; Mahmud et al. 2020). During nodulation, legumes will recruit microbes that can perform biological nitrogen fixation (BNF), in which microbes use the enzyme nitrogenase to catalyze the conversion of abundant N_2 gas in the atmosphere into ammonia that can be assimilated by the host plant via glutamine synthase to form glutamine (Gautrat et al. 2021). To reduce the reliance on excess synthetic fertilizers, future research is examining how to harness nitrogen fixation by microbes and apply this process to benefit nonlegume crops such as maize, rice, and wheat (Mahmud et al. 2020; Wen et al. 2021). One way this can be accomplished is by engineering non-legumes to form no-dules with N-fixing bacteria (Mahmud et al. 2020; Wen et al. 2021). Alternatively, crops can be engineered to secrete spe-cific root exudates, which is a mixture of sugars, amino acids, fatty acids, and vitamins, to either recruit beneficial microbes that help improve nitrogen acquisition or to inhibit the colonization of microbes that might compete with nutrient acquisition (Hartman and Tringe 2019). In addition to engineering the crop itself, there is also an effort to use specific microbial inoculants as biofertilizers (Klimasmith and Kent 2022). This can be done either by adding microbes in the soil that will form endosymbiotic relationships with the plant to increase nitrogen fixation or by adding free-living microbes that offer greater nitrogen availability to the plant in the soil (Cole et al. 2021; Klimasmith and Kent 2022).

Nitrogen research in the age of artificial intelligence (AI)

Since Alan Turing first proposed the concept of machines capable of self-learning and self-instruction in 1950, the field of artificial intelligence (AI) has experienced explosive growth. AI, in general, refers to the ability of machines to simulate the intelligence observed in complex organisms.

Increasingly, AI is intersecting with the field of biology. Machine learning and in particular AI, mainly in the form of Deep Neural Net (DNN; Lecun et al. 2015), are transforming the way science is performed. Although traditional model-driven methods still play a valuable role in analyzing biological data, they often lack the capacity to effectively harness vast amounts of available data, including big data, to extract information, forecast data behavior, and comprehend complex data relationships. Particularly over the last decade, we have seen a dramatic increase in the number of large, highly complex datasets being generated from biological experiments, quantifying molecular variables such as gene, protein, and metabolite abundance, microbiome composition, and population-wide genetic variation, to name just a few. Community efforts across research disciplines are regularly generating petabytes of data. The nitrogen signaling field has been very active to use and develop new algorithms to decipher the obvious complexity of the nitrogen signaling pathway even before the rise of the DNN. The nitrogen community has been particularly keen on using several mathematical models to decipher and predict the actual interactions between transcription factors and their target genes or to model solute transport and developmental processes. These approaches ranged from linear models (Gutiérrez et al. 2007; Krouk et al. 2009; Ristova et al. 2016), state-space modeling (Krouk et al. 2010; Brooks et al. 2019; Alvarez et al. 2020), to "ordinary differential equations" (ODEs) embedded in organ and tissue models (Boer et al. 2020; Ötvös et al. 2021). DNNs are very good at image classification and segmentation as they were originally developed for computer vision (Lecun et al. 2015). This is why an important trend is now rising as it relates to the measure of N content on plants from image analysis. Most of the time these approaches use multispectral images to classify the N content of different crops, including maize (Nguyen et al. 2023; Wijewardane et al. 2023), cotton (Xiao et al. 2022), or sorghum (Wijewardane et al. 2023).

Impressive advancements in AI in biology have been made, for example, in precise identification of the three- dimensional structure of biological molecules, such as AlphaFold, a critical task with significant implications for bio-logical research (Jumper et al. 2021; Lin et al. 2023) and now widely used in biology. But the potential for AI to replicate the capacities of living systems, particularly human intelli-gence, represents a significant achievement and a turning point in how science is performed. AI is now capable of ob-ject recognition and decision-making, using cognitive and perceptual abilities akin to those observed in biological systems. A relatively more recent branch of neural networks called "natural language processing" (NLP), originally devel-oped to understand human languages (including translating one language into another), has been applied to biological questions and to published articles or written information and sequence-related data, including DNA or protein se-quences. In the context of NLP, self-supervised large language models, such as "generative pre-trained transformer 3" (GPT-3; Brown et al. 2020) or "pathways language model" (PaLM; Chowdhery et al. 2022), have demonstrated impres-sive abilities to extract meaningful pieces of world knowledge from being exposed to an extremely large quantity of text (billions of words). A relevant challenge is related to how to access the knowledge encoded by the internal representation of a large AI model. Surprisingly, recent research (Gao et al. 2021) has found that it is possible to steer these large models to output relevant knowledge from a novel target task using just a prompt. Specifically, by using a prompt that provides the model with a human language description or several examples of what one wants them to do, the model can output meaningful knowledge related to a target task. This learning strategy, referred to as contextual prompting, offers a new degree of control to selectively access the knowledge encoded in the internal representations of a large language model. Nevertheless, it remains to be assessed how useful and in which way tools such as NLP models or ChatGPT, trained specifically with scientific literature, will be for scientific research.

AI-based algorithms and programs continue to emerge with diverse applications from basic research to precision farming. Precision farming has the potential to revolutionize various agricultural practices, ranging from soil management and water analysis to accurate modeling of fertilizer requirements, as well as the optimization of pesticides, insecticides, herbicides, yield projections, and overall crop management. These advancements in AI intervention can play a pivotal role in meeting the increasing demands for food from a growing global population. Early prediction and identification of agricultural problems, as well as optimization of production practices, are key areas that can greatly benefit from AI applications. Such approaches not only have the potential to save significant costs but also mitigate environmental impacts, leading to more sustainable agricultural practices.

Conclusions

In this review, we examine how studies of nitrogen sensing and signaling over time and space have begun to uncover the underlying dynamic regulatory networks that mediate changes in plant metabolism and development. We also explore how emerging experimental and computational techniques can be applied to advance nitrogen research. This includes leveraging new experimental approaches such as single-cell sequencing to unravel the relay of nitrogen signaling in specific cell types over time and its effect on crucial processes, such as cell differentiation, to modulate organ development. On the other hand, computational methods such as machine learning and artificial intelligence will augment experimental nitrogen research endeavors to uncover the mechanisms by which plants sense and respond to nitrogen sources in their environment. We also highlight how model-to-crop translational studies can be used for practical gain in enhancing NUE in agriculture.

Author contributions

CS, KR, RG, and GC outlined the review topics, current research areas, and prepared figures and tables. CS, KR, MB, CYC, JMA, SR, GK, RG, and GC contributed to writing review topic sections, editing the paper, and approving the submitted version.

Funding

This work was supported by the NSF-PGRP IOS-1840761 grant to GC and JMA, the Zegar Family Foundation (A16-0051) grant to GC, the NIH RO1-GM121753 grant to GC, the NIH NIGMS Fellowship F32GM139299 to CS, the ANID-FONDECYT Postdoctorado 3220033 to KR, and the MOST research grant (NSTC 112-2313-B-002-018-MY3) to CYC. RG's research is supported by ANID-Millennium Science Initiative Program-Millennium Institute for Integrative Biology (iBio) (ICN17_022), the Center for Genome Regulation (ICN2021_044), and the FONDECYT 1180759. JMA's research is supported by the ANID-Millennium Science Initiative Program-Millennium Institute for Integrative Biology (iBio), the ANID FONDECYT 1210389. SR research is supported by the Metaprogram DIGIT-BIO from INRAE and the ANR Plan de Relance TravelPep. GK's research is supported by MUSE University of Montpellier research grant (AI3P: Artificial Intelligence to Predict Phenotypes in Plants), the ANR (the French National Research Agency), under the "Investissements d'avenir" program with the reference ANR-16-IDEX-0006.

Conflict of interest statement. None declared.

References

Abualia R, Ötvös K, Novák O, Bouguyon E, Domanegg K, Krapp A, Nacry P, Gojon A, Lacombe B, Benková E. Molecular framework integrating nitrate sensing in root and auxin-guided shoot adaptive responses. Proc Natl Acad Sci USA. 2022:**119**(32):1–11. https://doi.org/10.1073/pnas.2122460119

Ahmed F. Qnas with Gloria M. Coruzzi. Proc Natl Acad Sci USA. 2020:**117**(23):12508–12509. https://doi.org/10.1073/pnas.2007147117

Akmakjian GZ, Bailey-Serres J. Nitrogen-responsive transcription factor kinetics meter plant growth. Proc Natl Acad Sci USA. 2020:**117**(24):13196–13198. https://doi.org/10.1073/pnas.2007441117

Alvarez JM, Brooks MD, Swift J, Coruzzi GM. Time-based systems biology approaches to capture and model dynamic gene regulatory networks. Annu Rev Plant Biol. 2021:**72**(1):105–131. https://doi.org/10.1146/annurev-arplant-081320-090914

Alvarez JM, Schinke AL, Brooks MD, Pasquino A, Leonelli L, Varala K, Safi A, Krouk G, Krapp A, Coruzzi GM. Transient genome-wide interactions of the master transcription factor NLP7 initiate a rapid nitrogen-response cascade. Nat Commun. 2020:**11**(1):1–13. https://doi.org/10.1038/s41467-020-14979-6

Araus V, Swift J, Alvarez JM, Henry A, Coruzzi GM. A balancing act: how plants integrate nitrogen and water signals. J Exp Bot. 2020:**71**(15):4442–4451. https://doi.org/10.1093/jxb/eraa054

Bargmann BOR, Marshall-Colon A, Efroni I, Ruffel S, Birnbaum KD, Coruzzi GM, Krouk G. TARGET: a transient transformation system for genome-wide transcription factor target discovery. Mol Plant. 2013:**6**(3):978–980. https://doi.org/10.1093/mp/sst010

Bijay-Singh, Craswell E. Fertilizers and nitrate pollution of surface and ground water: an increasingly pervasive global problem. SN Appl Sci. 2021:**3**(4):1–24. https://doi.org/10.1007/s42452-021-04521-8

Boer MD, Santos Teixeira J, Ten Tusscher KH. Modeling of root nitrate responses suggests preferential foraging arises from the integration of demand, supply and local presence signals. Front Plant Sci. 2020:**11**:1–17. https://doi.org/10.3389/fpls.2020.00708

Brooks MD, Cirrone J, Pasquino AV, Alvarez JM, Swift J, Mittal S, Juang CL, Varala K, Gutiérrez RA, Krouk G, et al. Network walking charts transcriptional dynamics of nitrogen signaling by integrating validated and predicted genome-wide interactions. Nat Commun. 2019:**10**(1):1–13. https://doi.org/10.1038/s41467-019-09522-1

Brooks MD, Juang C-L, Katari MS, Alvarez JM, Pasquino AV, Shih H-J, Huang J, Shanks C, Cirrone J, Coruzzi GM. ConnecTF: a platform to build gene networks by integrating transcription factor-target gene interactions. bioRxiv: 2020.07.07.191627. https://doi.org/10.1101/2020.07.07.191627, 2020.

Brooks MD, Juang CL, Katari MS, Alvarez JM, Pasquino A, Shih HJ, Huang J, Shanks C, Cirrone J, Coruzzi GM. ConnecTF: a platform to integrate transcription factor-gene interactions and validate regulatory networks. Plant Physiol. 2021:**185**(1):49–66. https://doi.org/10.1093/plphys/kiaa012

Brooks MD, Reed KM, Krouk G, Coruzzi GM, Bargmann BOR. The TARGET system: rapid identification of direct targets of transcription factors by gene regulation in plant cells. Methods Mol Biol. 2023:**2594**:1–12. https://doi.org/10.1007/978-1-0716-2815-7_1

Brown TB, Mann B, Ryder N, Subbiah M, Kaplan J, Dhariwal P, Neelakantan A, Shyam P, Sastry G, Askell A, et al. Language models are few-shot learners—special version. Adv Neural Inf Process Syst. 2020.

Buenrostro JD, Wu B, Litzenburger UM, Ruff D, Gonzales ML, Snyder MP, Chang HY, Greenleaf WJ. Single-cell chromatin accessibility reveals principles of regulatory variation. Nature. 2015: **523**(7561):486–490. https://doi.org/10.1038/nature14590

Canales J, Moyano T, Villarroel E, Gutiérrez R. Systems analysis of transcriptome data provides new hypotheses about Arabidopsis root response to nitrate treatments. Front Plant Sci. 2014:**5**:22. https://doi.org/10.3389/fpls.2014.00022

Castaings L, Camargo A, Pocholle D, Gaudon V, Texier Y, Boutet-Mercey S, Taconnat L, Renou JP, Daniel-Vedele F, Fernandez E, et al. The nodule inception-like protein 7 modulates nitrate sensing and metabolism in Arabidopsis. Plant J. 2009:**57**(3): 426–435. https://doi.org/10.1111/j.1365-313X.2008.03695.x

Chen YN, Cartwright HN, Ho CH. In vivo visualization of nitrate dynamics using a genetically encoded fluorescent biosensor. Sci Adv. 2022:**8**(42):1–11. https://doi.org/10.1126/sciadv.abq4915

Chen T, Guestrin C. XGBoost: a scalable tree boosting system. In: Proceedings of the 29th ACM SIGKDD Conference on Knowledge Discovery and Data Mining; 2016. p. 785–794.

Chen H, Yin X, Guo L, Yao J, Ding Y, Xu X, Liu L, Zhu QH, Chu Q, Fan L. PlantscRNAdb: a database for plant single-cell RNA analysis. Mol Plant. 2021:**14**(6):855–857. https://doi.org/10.1016/j.molp.2021.05.002

Cheng YH, Durand M, Brehaut V, Hsu FC, Kelemen Z, Texier Y, Krapp A, Tsay YF. Interplay between NIN-LIKE PROTEINs 6 and 7 in nitrate signaling. Plant Physiol. 2023:**192**(4):3049–3068. https://doi.org/10.1093/plphys/kiad242

Cheng CY, Li Y, Varala K, Bubert J, Huang J, Kim GJ, Halim J, Arp J, Shih H-JS, Levinson G, et al. Evolutionarily informed machine learning enhances the power of predictive gene-to-phenotype relationships. Nat Commun. 2021:**12**(1):1–15. https://doi.org/10.1038/s41467-021-25893-w

Chowdhery A, Narang S, Devlin J, Bosma M, Mishra G, Roberts A, Barham P, Chung HW, Sutton C, Gehrmann S, et al. PaLM: scaling language modeling with pathways; 2022. pp. 1–87.

Cirrone J, Brooks MD, Bonneau R, Coruzzi GM, Shasha DE. OutPredict: multiple datasets can improve prediction of expression and inference of causality. Sci Rep. 2020:**10**(1):6804. https://doi.org/10.1038/s41598-020-63347-3

Cole B, Bergmann D, Blaby-Haas CE, Blaby IK, Bouchard KE, Brady SM, Ciobanu D, Coleman-Derr D, Leiboff S, Mortimer JC, et al. Plant single-cell solutions for energy and the environment. Commun Biol. 2021:**4**:962. https://doi.org/10.1038/s42003-021-02477-4

Contreras-López O, Vidal EA, Riveras E, Alvarez JM, Moyano TC, Sparks EE, Medina J, Pasquino A, Benfey PN, Coruzzi GM, et al. Spatiotemporal analysis identifies ABF2 and ABF3 as key hubs of endodermal response to nitrate. Proc Natl Acad Sci USA. 2022:**119**: 1–17. https://doi.org/10.1073/pnas.2107879119

Crawford NM, Forde BG. Molecular and developmental biology of inorganic nitrogen nutrition. Arab B. 2002:**1**:e0011.

Das M, Haberer G, Panda A, Laha SD, Ghosh TC, Schäffner AR. Expression pattern similarities support the prediction of orthologs retaining common functions after gene duplication events. Plant Physiol. 2016:**171**(4):2343–2357. https://doi.org/10.1104/pp.15.01207

Denyer T, Ma X, Klesen S, Scacchi E, Nieselt K, Timmermans MCP. Spatiotemporal developmental trajectories in the Arabidopsis root revealed using high-throughput single-cell RNA sequencing. Dev Cell. 2019:**48**(6):840–852.e5. https://doi.org/10.1016/j.devcel.2019.02.022

Doidy J, Li Y, Neymotin B, Edwards MB, Varala K, Gresham D, Coruzzi GM. "Hit-and-Run" transcription: de novo transcription initiated by a transient bZIP1 "hit" persists after the "run." BMC Genom. 2016:**17**(1):1–10. https://doi.org/10.1186/s12864-016-2410-2

Dorrity MW, Alexandre CM, Hamm MO, Vigil AL, Fields S, Queitsch C, Cuperus JT. The regulatory landscape of Arabidopsis thaliana roots at single-cell resolution. Nat Commun. 2021:**12**(1):3334. https://doi.org/10.1038/s41467-021-23675-y

Du H, Ning L, He B, Wang Y, Ge M, Xu J, Zhao H. Cross-species root transcriptional network analysis highlights conserved modules in response to nitrate between maize and sorghum. Int J Mol Sci. 2020:**21**: 1445. https://doi.org/10.3390/ijms21041445

Farmer A, Thibivilliers S, Ryu KH, Schiefelbein J, Libault M. Single-nucleus RNA and ATAC sequencing reveals the impact of chromatin accessibility on gene expression in Arabidopsis roots at the single-cell level. Mol Plant. 2021:**14**:372–383. https://doi.org/10.1016/j.molp.2021.01.001

Fredes I, Moreno S, Díaz FP, Gutiérrez RA. Nitrate signaling and the control of Arabidopsis growth and development. Curr Opin Plant Biol. 2019:**47**:112–118. https://doi.org/10.1016/j.pbi.2018.10.004

Fu LY, Zhu T, Zhou X, Yu R, He Z, Zhang P, Wu Z, Chen M, Kaufmann K, Chen D. ChIP-Hub provides an integrative platform for exploring plant regulome. Nat Commun. 2022:**13**(1):3413. https://doi.org/10.1038/s41467-022-30770-1

Gabaldón T, Koonin EV. Functional and evolutionary implications of gene orthology. Nat Rev Genet. 2013:**14**(5):360–366. https://doi.org/10.1038/nrg3456

Gao T, Fisch A, Chen D. *Making pre-trained language models better few-shot learners*. ACL-IJCNLP 2021 Proceedings of the 59th Annual Meeting of the Association for Computational Linguistics and the 11th International Joint Conference on Natural Language Processing; 2021. p. 3816–3830.

Gao Y, Qi S, Wang Y. Nitrate signaling and use efficiency in crops. Plant Commun. 2022:**3**(5):100353. https://doi.org/10.1016/j.xplc.2022.100353

Gaudinier A, Rodriguez-Medina J, Zhang L, Olson A, Liseron-Monfils C, Bågman A-M, Foret J, Abbitt S, Tang M, Li B, et al. Transcriptional regulation of nitrogen-associated metabolism and growth. Nature. 2018:**563**(7730):259–264. https://doi.org/10.1038/s41586-018-0656-3

Gautrat P, Laffont C, Frugier F, Ruffel S. Nitrogen systemic signaling: from symbiotic nodulation to root acquisition. Trends Plant Sci. 2021:**26**(4):392–406. https://doi.org/10.1016/j.tplants.2020.11.009

Gifford ML, Dean A, Gutierrez RA, Coruzzi GM, Birnbaum KD. Cell-specific nitrogen responses mediate developmental plasticity.

Proc Natl Acad Sci USA. 2008:**105**(2):803–808. https://doi.org/10.1073/pnas.0709559105

Goodstein DM, Shu S, Howson R, Neupane R, Hayes RD, Fazo J, Mitros T, Dirks W, Hellsten U, Putnam N, et al. Phytozome: a comparative platform for green plant genomics. Nucleic Acids Res. 2012:**40**:1178–D1186. https://doi.org/10.1093/nar/gkr944

Graeff M, Rana S, Wendrich JR, Dorier J, Eekhout T, Aliaga Fandino AC, Guex N, Bassel GW, De Rybel B, Hardtke CS. A single-cell morpho-transcriptomic map of brassinosteroid action in the Arabidopsis root. Mol Plant. 2021:**14**(12):1985–1999. https://doi.org/10.1016/j.molp.2021.07.021

Guan P, Ripoll JJ, Wang R, Vuong L, Bailey-Steinitz LJ, Ye D, Crawford NM. Interacting TCP and NLP transcription factors control plant responses to nitrate availability. Proc Natl Acad Sci USA. 2017:**114**(9):2419–2424. https://doi.org/10.1073/pnas.1615676114

Gulko B, Siepel A. An evolutionary framework for measuring epigenomic information and estimating cell type-specific fitness consequences. Nat Genet. 2019:**51**(2):335–342. https://doi.org/10.1038/s41588-018-0300-z

Guo FQ, Young J, Crawford NM. The nitrate transporter AtNRT1.1 (CHL1) functions in stomatal opening and contributes to drought susceptibility in Arabidopsis. Plant Cell. 2003:**15**(1):107–117. https://doi.org/10.1105/tpc.006312

Gutierrez-Triana JA, Mateo JL, Ibberson D, Ryu S, Wittbrodt J. iDamIDseq and iDEAR: an improved method and computational pipeline to profile chromatin-binding proteins. Dev. 2016:**143**(22):4272–4278.

Gutiérrez RA, Lejay LV, Dean A, Chiaromonte F, Shasha DE, Coruzzi GM. Qualitative network models and genome-wide expression data define carbon/nitrogen-responsive molecular machines in Arabidopsis. Genome Biol. 2007:**8**(1):R7. https://doi.org/10.1186/gb-2007-8-1-r7

Hachiya T, Sakakibara H. Interactions between nitrate and ammonium in their uptake, allocation, assimilation, and signaling in plants. J Exp Bot. 2017:**68**:2501–2512. https://doi.org/10.1093/jxb/erw449

Hartman K, Tringe SG. Interactions between plants and soil shaping the root microbiome under abiotic stress. Biochem J. 2019:**476**(19):2705–2724. https://doi.org/10.1042/BCJ20180615

Heerah S, Molinari R, Guerrier S, Marshall-Colon A. Granger-causal testing for irregularly sampled time series with application to nitrogen signalling in Arabidopsis. Bioinformatics. 2021:**37**(16):2450–2460. https://doi.org/10.1093/bioinformatics/btab126

Ho CH, Lin SH, Hu HC, Tsay YF. CHL1 functions as a nitrate sensor in plants. Cell. 2009:**138**(6):1184–1194. https://doi.org/10.1016/j.cell.2009.07.004

Hou M, Yu M, Li Z, Ai Z, Chen J. Molecular regulatory networks for improving nitrogen use efficiency in rice. Int J Mol Sci. 2021:**22**(16):9040. https://doi.org/10.3390/ijms22169040

Hu QQ, Shu JQ, Li WM, Wang GZ. Role of auxin and nitrate signaling in the development of root system architecture. Front Plant Sci. 2021:**12**:690363. https://doi.org/10.3389/fpls.2021.690363

Hu B, Wang W, Chen J, Liu Y, Chu C. Genetic improvement toward nitrogen-use efficiency in rice: lessons and perspectives. Mol Plant. 2023:**16**(1):64–74. https://doi.org/10.1016/j.molp.2022.11.007

Huerta-Cepas J, Szklarczyk D, Heller D, Hernández-Plaza A, Forslund SK, Cook H, Mende DR, Letunic I, Rattei T, Jensen LJ, et al. EggNOG 5.0: a hierarchical, functionally and phylogenetically annotated orthology resource based on 5090 organisms and 2502 viruses. Nucleic Acids Res. 2019:**47**(D1):D309–D314. https://doi.org/10.1093/nar/gky1085

Hummel NFC, Zhou A, Li B, Markel K, Ornelas IJ, Shih PM. The trans-regulatory landscape of gene networks in plants. Cell Syst. 2023:**14**(6):501–511.e4. https://doi.org/10.1016/j.cels.2023.05.002

Huynh-Thu VA, Geurts P. DynGENIE3: dynamical GENIE3 for the inference of gene networks from time series expression data. Sci Rep. 2018:**8**(1):1–12. https://doi.org/10.1038/s41598-018-21715-0

Jean-Baptiste K, McFaline-Figueroa JL, Alexandre CM, Dorrity MW, Saunders L, Bubb KL, Trapnell C, Fields S, Queitsch C, Cuperusa JT. Dynamics of gene expression in single root cells of *Arabidopsis thaliana*. Plant Cell. 2019:**31**(5):993–1011. https://doi.org/10.1105/tpc.18.00785

Jia Z, von Wirén N. Signaling pathways underlying nitrogen-dependent changes in root system architecture: from model to crop species. J Exp Bot. 2020:**71**(15):4393–4404. https://doi.org/10.1093/jxb/eraa033

Jouannet V, Brackmann K, Greb T. (Pro)cambium formation and proliferation: two sides of the same coin? Curr Opin Plant Biol. 2015:**23**:54–60. https://doi.org/10.1016/j.pbi.2014.10.010

Jumper J, Evans R, Pritzel A, Green T, Figurnov M, Ronneberger O, Tunyasuvunakool K, Bates R, Žídek A, Potapenko A, et al. Highly accurate protein structure prediction with AlphaFold. Nature. 2021:**596**(7873):583–589. https://doi.org/10.1038/s41586-021-03819-2

Kasianov AS, Klepikova AV, Mayorov AV, Buzanov GS, Logacheva MD, Penin AA. Interspecific comparison of gene expression profiles using machine learning. PLoS Comput Biol. 2023:**19**(1):e1010743. https://doi.org/10.1371/journal.pcbi.1010743

Katari MS, Nowicki SD, Aceituno FF, Nero D, Kelfer J, Thompson LP, Cabello JM, Davidson RS, Goldberg AP, Shasha DE, et al. VirtualPlant: a software platform to support systems biology research. Plant Physiol. 2010:**152**:500–515. https://doi.org/10.1104/pp.109.147025

Kim JY, Symeonidi E, Pang TY, Denyer T, Weidauer D, Bezrutczyk M, Miras M, Zöllner N, Hartwig T, Wudick MM, et al. Distinct identities of leaf phloem cells revealed by single cell transcriptomics. Plant Cell. 2021:**33**(3):511–530. https://doi.org/10.1093/plcell/koaa060

Klimasmith IM, Kent AD. Micromanaging the nitrogen cycle in agroecosystems. Trends Microbiol. 2022:**30**(11):1045–1055. https://doi.org/10.1016/j.tim.2022.04.006

Krouk G, Mirowski P, LeCun Y, Shasha DE, Coruzzi GM. Predictive network modeling of the high-resolution dynamic plant transcriptome in response to nitrate. Genome Biol. 2010:**11**(12):R123. https://doi.org/10.1186/gb-2010-11-12-r123

Krouk G, Tranchina D, Lejay L, Cruikshank AA, Shasha D, Coruzzi GM, Gutiérrez RA. A systems approach uncovers restrictions for signal interactions regulating genome-wide responses to nutritional cues in Arabidopsis. PLoS Comput Biol. 2009:**5**(3):e1000326. https://doi.org/10.1371/journal.pcbi.1000326

Lamig L, Moreno S, Álvarez JM, Gutiérrez RA. Molecular mechanisms underlying nitrate responses in plants. Curr Biol. 2022:**32**(9):R433–R439. https://doi.org/10.1016/j.cub.2022.03.022

Lana RP, Goes RHTB, Moreira LM, Mâncio AB, Fonseca DM, Tedeschi LO. Application of Lineweaver-Burk data transformation to explain animal and plant performance as a function of nutrient supply. Livest Prod Sci. 2005:**98**(3):219–224. https://doi.org/10.1016/j.livprodsci.2005.03.008

Langfelder P, Horvath S. WGCNA: an R package for weighted correlation network analysis. BMC Bioinformatics. 2008:**9**(1):559. https://doi.org/10.1186/1471-2105-9-559

Lecun Y, Bengio Y, Hinton G. Deep learning. Nature. 2015:**521**:436–444. https://doi.org/10.1038/nature14539

Lee TA, Nobori T, Illouz-Eliaz N, Xu J, Jow B, Nery JR, Ecker JR. A single-nucleus atlas of seed-to-seed development in Arabidopsis. bioRxiv. 2023.03.23.533992. https://doi.org/10.1101/2023.03.23.533992, 2023

Lhamo D, Luan S. Potential networks of nitrogen-phosphorus-potassium channels and transporters in Arabidopsis roots at a single cell resolution. Front Plant Sci. 2021:**12**:1–25. https://doi.org/10.3389/fpls.2021.689545

Li L, Liu KH, Sheen J. Dynamic nutrient signaling networks in plants. Annu Rev Cell Dev Biol. 2021:**37**(1):341–367. https://doi.org/10.1146/annurev-cellbio-010521-015047

Li X, Zhang X, Gao S, Cui F, Chen W, Fan L, Qi Y. Single-cell RNA sequencing reveals the landscape of maize root tips and assists in identification of cell type-specific nitrate-response genes. Crop J. 2022:**10**(6):1589–1600. https://doi.org/10.1016/j.cj.2022.02.004

Lin Z, Akin H, Rao R, Hie B, Zhu Z, Lu W, Smetanin N, Verkuil R, Kabeli O, Shmueli Y, et al. Evolutionary-scale prediction of atomic-level protein structure with a language model. Science. 2023:**379**(6637):1123–1130. https://doi.org/10.1126/science.ade2574

Liu KH, Liu M, Lin Z, Wang ZF, Chen B, Liu C, Guo A, Konishi M, Yanagisawa S, Wagner G, et al. NIN-like protein 7 transcription factor is a plant nitrate sensor. Science 2022:**377**(6613):1419–1425. https://doi.org/10.1126/science.add1104

Liu KH, Niu Y, Konishi M, Wu Y, Du H, Sun Chung H, Li L, Boudsocq M, McCormack M, Maekawa S, et al. Discovery of nitrate-CPK-NLP signalling in central nutrient-growth networks. Nature. 2017:**545**(7654):311–316. https://doi.org/10.1038/nature22077

Liu Y, Von Wirén N. Ammonium as a signal for physiological and morphological responses in plants. J Exp Bot. 2017:**68**(10):2581–2592. https://doi.org/10.1093/jxb/erx086

Liu B, Wu J, Yang S, Schiefelbein J, Gan Y. Nitrate regulation of lateral root and root hair development in plants. J Exp Bot. 2020:**71**(15):4405–4414. https://doi.org/10.1093/jxb/erz536

Ma X, Denyer T, Timmermans MCP. Pscb: a browser to explore plant single cell RNA-Sequencing data sets. Plant Physiol. 2020:**183**(2):464–467. https://doi.org/10.1104/pp.20.00250

Mahmud K, Makaju S, Ibrahim R, Missaoui A. Current progress in nitrogen fixing plants and microbiome research. Plants. 2020:**9**(1):97–17. https://doi.org/10.3390/plants9010097

Marchive C, Roudier F, Castaings L, Bréhaut V, Blondet E, Colot V, Meyer C, Krapp A. Nuclear retention of the transcription factor NLP7 orchestrates the early response to nitrate in plants. Nat Commun. 2013:**4**(1):1–9. https://doi.org/10.1038/ncomms2650

McNickle GG, Brown JS. When Michaelis and Menten met Holling: towards a mechanistic theory of plant nutrient foraging behaviour. AoB Plants. 2014:**6**:1–14. https://doi.org/10.1093/aobpla/plu066

Medici A, Krouk G. The primary nitrate response: a multifaceted signalling pathway. J Exp Bot. 2014:**65**(19):5567–5576. https://doi.org/10.1093/jxb/eru245

Menegat S, Ledo A, Tirado R. Greenhouse gas emissions from global production and use of nitrogen synthetic fertilisers in agriculture. Sci Rep. 2022:**12**:1–13. https://doi.org/10.1038/s41598-021-99269-x

Michaelis L, Menten M. Die kinetik der invertinwirkung. Biochem Zeitung. 1913:**79**:333–369.

Michaelis L, Menten ML, Johnson KA, Goody RS. The original Michaelis constant: translation of the 1913 Michaelis–Menten paper. Biochemistry. 2011:**50**(39):8264–8269. https://doi.org/10.1021/bi201284u

Murashige T, Skoog F. A revised medium for rapid growth and bio assays with tobacco tissue cultures. Physiol Plant. 1962:**15**(3):473–497. https://doi.org/10.1111/j.1399-3054.1962.tb08052.x

Muratore C, Espen L, Prinsi B. Nitrogen uptake in plants: the plasma membrane root transport systems from a physiological and proteomic perspective. Plants. 2021:**10**(4):681. https://doi.org/10.3390/plants10040681

Nguyen P, Braun R. Time-lagged ordered Lasso for network inference. BMC Bioinformatics. 2018:**19**(1):1–16. https://doi.org/10.1186/s12859-018-2558-7

Nguyen TH, Thiers L, Van Moerkercke A, Bai Y, Fernández-Calvo P, Minne M, Depuydt T, Colinas M, Verstaen K, Van Isterdael G, et al. A redundant transcription factor network steers spatiotemporal *Arabidopsis triterpene* synthesis. Nat Plants. 2023:**9**(6):926–937. https://doi.org/10.1038/s41477-023-01419-8

Niemann V, Koch-Singenstreu M, Neu A, Nilkens S, Götz F, Unden G, Stehle T. The NreA protein functions as a nitrate receptor in the staphylococcal nitrate regulation system. J Mol Biol. 2014:**426**(7):1539–1553. https://doi.org/10.1016/j.jmb.2013.12.026

Nobori, T, Monell A, Lee TA, Zhou J, Nery J, Ecker JR. Time-resolved single-cell and spatial gene regulatory atlas of plants under pathogen attack. bioRxiv. 2023.04.10.536170. https://doi.org/10.1101/2023.04.10.536170, 2023.

Nolan TM, Vukašinović N, Hsu C-W, Zhang J, Vanhoutte I, Shahan R, Taylor IW, Greenstreet L, Heitz M, Afanassiev A, et al. Brassinosteroid gene regulatory networks at cellular resolution in the Arabidopsis root. Science 2023:**379**(6639):eadf4721. https://doi.org/10.1126/science.adf4721

Obertello M, Shrivastava S, Katari MS, Coruzzi GM. Cross-species network analysis uncovers conserved nitrogen-regulated network modules in rice. Plant Physiol. 2015:**168**(4):1830–1843. https://doi.org/10.1104/pp.114.255877

O'Brien JAA, Vega A, Bouguyon E, Krouk G, Gojon A, Coruzzi G, Gutiérrez RAA. Nitrate transport, sensing, and responses in plants. Mol Plant. 2016:**9**(6):837–856. https://doi.org/10.1016/j.molp.2016.05.004

Ohkubo Y, Tanaka M, Tabata R, Ogawa-Ohnishi M, Matsubayashi Y. Shoot-to-root mobile polypeptides involved in systemic regulation of nitrogen acquisition. Nat Plants. 2017:**3**(4):1–6. https://doi.org/10.1038/nplants.2017.29

Oldroyd GED, Leyser O. A plant's diet, surviving in a variable nutrient environment. Science 2020:**368**(6486):eaba0196. https://doi.org/10.1126/science.aba0196

O'Malley RC, Huang SSC, Song L, Lewsey MG, Bartlett A, Nery JR, Galli M, Gallavotti A, Ecker JR. Cistrome and epicistrome features shape the regulatory DNA landscape. Cell. 2016:**165**(5):1280–1292. https://doi.org/10.1016/j.cell.2016.04.038

Osugi A, Kojima M, Takebayashi Y, Ueda N, Kiba T, Sakakibara H. Systemic transport of trans-zeatin and its precursor have differing roles in Arabidopsis shoots. Nat Plants. 2017:**3**(8):1–6. https://doi.org/10.1038/nplants.2017.112

Ota R, Ohkubo Y, Yamashita Y, Ogawa-Ohnishi M, Matsubayashi Y. Shoot-to-root mobile CEPD-like 2 integrates shoot nitrogen status to systemically regulate nitrate uptake in Arabidopsis. Nat Commun. 2020:**11**(1):1–9. https://doi.org/10.1038/s41467-020-14440-8

Ötvös K, Marconi M, Vega A, O'Brien J, Johnson A, Abualia R, Antonielli L, Montesinos JC, Zhang Y, Tan S, et al. Modulation of plant root growth by nitrogen source-defined regulation of polar auxin transport. EMBO J. 2021:**40**(3):1–21. https://doi.org/10.15252/embj.2020106862

Palmgren M. Plant epithelia: what is the role of the mortar in the wall? PLoS Biol. 2018:**16**(12):e3000073. https://doi.org/10.1371/journal.pbio.3000073

Para A, Li Y, Marshall-Colón A, Varala K, Francoeur NJ, Moran TM, Edwards MB, Hackley C, Bargmann BOR, Birnbaum KD, et al. Hit-and-run transcriptional control by bZIP1 mediates rapid nutrient signaling in Arabidopsis. Proc Natl Acad Sci U S A. 2014:**111**:10371--10376. https://doi.org/10.1073/pnas.1404657111

Patterson K, Walters LA, Cooper AM, Olvera JG, Rosas MA, Rasmusson AG, Escobar MA. Nitrate-regulated glutaredoxins control arabidopsis primary root growth. Plant Physiol. 2016:**170**(2):989–999. https://doi.org/10.1104/pp.15.01776

Plett DC, Ranathunge K, Melino VJ, Kuya N, Uga Y, Kronzucker HJ. The intersection of nitrogen nutrition and water use in plants: new paths toward improved crop productivity. J Exp Bot. 2020:**71**(15):4452–4468. https://doi.org/10.1093/jxb/eraa049

Poitout A, Crabos A, Petřík I, Novák O, Krouk G, Lacombe B, Ruffel S. Responses to systemic nitrogen signaling in Arabidopsis roots involve trans-zeatin in shoots. Plant Cell. 2018:**30**(6):1243–1257. https://doi.org/10.1105/tpc.18.00011

Ramakrishna P, Ruiz Duarte P, Rance GA, Schubert M, Vordermaier V, Vu LD, Murphy E, Vilches Barro A, Swarup K, Moirangthem K, et al. EXPANSIN A1-mediated radial swelling of pericycle cells positions anticlinal cell divisions during lateral root initiation. Proc Natl Acad Sci USA. 2019:**116**(17):8597–8602. https://doi.org/10.1073/pnas.1820882116

Rich-Griffin C, Stechemesser A, Finch J, Lucas E, Ott S, Schäfer P. Single-Cell transcriptomics: a high-resolution avenue for plant functional genomics. Trends Plant Sci. 2020:**25**(2):186–197. https://doi.org/10.1016/j.tplants.2019.10.008

Ristova D, Carré C, Pervent M, Medici A, Kim GJ, Scalia D, Ruffel S, Birnbaum KD, Lacombe B, Busch W, et al. Combinatorial interaction network of transcriptomic and phenotypic responses to nitrogen and hormones in the Arabidopsis thaliana root. Sci Signal. 2016:**9**(451):rs13. https://doi.org/10.1126/scisignal.aaf2768

Ruffel S, Krouk G, Ristova D, Shasha D, Birnbaum KD, Coruzzi GM. Nitrogen economics of root foraging: transitive closure of the nitrate-cytokinin relay and distinct systemic signaling for N supply vs. demand. Proc Natl Acad Sci USA. 2011:**108**(45):18524–18529. https://doi.org/10.1073/pnas.1108684108

Ryu KH, Huang L, Kang HM, Schiefelbein J. Single-cell RNA sequencing resolves molecular relationships among individual plant cells. Plant Physiol. 2019:**179**(4):1444–1456. https://doi.org/10.1104/pp.18.01482

Sakakibara H. Cytokinin biosynthesis and transport for systemic nitrogen signaling. Plant J. 2021:**105**:421–430. https://doi.org/10.1111/tpj.15011

Sandhu N, Sethi M, Kumar A, Dang D, Singh J, Chhuneja P. Biochemical and genetic approaches improving nitrogen use efficiency in cereal crops: a review. Front Plant Sci. 2021:**12**:675629. https://doi.org/10.3389/fpls.2021.657629

Schaffner W. A hit-and-run mechanism for transcriptional activation? Nature. 1988:**336**(6198):427–428. https://doi.org/10.1038/336427a0

Schrynemackers M, Küffner R, Geurts P. On protocols and measures for the validation of supervised methods for the inference of biological networks. Front Genet. 2013:**4**:1–16. https://doi.org/10.3389/fgene.2013.00262

Shahan R, Hsu CW, Nolan TM, Cole BJ, Taylor IW, Greenstreet L, Zhang S, Afanassiev A, Vlot AHC, Schiebinger G, et al. A single-cell Arabidopsis root atlas reveals developmental trajectories in wild-type and cell identity mutants. Dev Cell. 2022:**57**:543–560.e9. https://doi.org/10.1016/j.devcel.2022.01.008

Shanks CM, Huang J, Cheng CY, Shih HJS, Brooks MD, Alvarez JM, Araus V, Swift J, Henry A, Coruzzi GM. Validation of a high-confidence regulatory network for gene-to-NUE phenotype in field-grown rice. Front. Plant Sci. 2022:**13**:1–18. https://doi.org/10.3389/fpls.2022.1006044

Steensel BV, Henikoff S. Identification of in vivo DNA targets of chromatin proteins using tethered Dam methyltransferase. Nat Biotechnol. 2000:**18**(4):424–428. https://doi.org/10.1038/74487

Stokes DE. Pasteur's quadrant—basic science and technological innovation. Brookings Inst. Press; 1997.

Swift J, Adame M, Tranchina D, Henry A, Coruzzi GM. Water impacts nutrient dose responses genome-wide to affect crop production. Nat Commun. 2019:**10**(1):1374. https://doi.org/10.1038/s41467-019-09287-7

Swift J, Alvarez JM, Araus V, Gutierrez RA, Coruzzi GM, Gutiérrez RA, Coruzzi GM. Nutrient dose-responsive transcriptome changes driven by Michaelis–Menten kinetics underlie plant growth rates. Proc Natl Acad Sci USA. 2020:**117**(23):12531–12540. https://doi.org/10.1073/pnas.1918619117

Swift J, Greenham K, Ecker JR, Coruzzi GM, Robertson McClung C. The biology of time: dynamic responses of cell types to developmental, circadian and environmental cues. Plant J. 2022:**109**(4):764–778. https://doi.org/10.1111/tpj.15589

Tabata R, Sumida K, Yoshii T, Ohyama K, Shinohara H, Matsubayashi Y. Perception of root-derived peptides by shoot LRR-RKs mediates systemic N-demand signaling. Science 2014:**346**(6207):343–346. https://doi.org/10.1126/science.1257800

Tegeder M, Masclaux-Daubresse C. Source and sink mechanisms of nitrogen transport and use. New Phytol. 2018:**217**(1):35–53. https://doi.org/10.1111/nph.14876

The White House Office of Science and Technology Policy (2023). Bold Goals for U.S. Biotechnology and Biomanufacturing Harnessing Research and Development to Further Societal Goals.

Ueda Y, Ohtsuki N, Kadota K, Tezuka A, Nagano AJ, Kadowaki T, Kim Y, Miyao M, Yanagisawa S. Gene regulatory network and its constituent transcription factors that control nitrogen-deficiency responses in rice. New Phytol. 2020:**227**(5):1434–1452. https://doi.org/10.1111/nph.16627

Varala K, Marshall-Colón A, Cirrone J, Brooks MD, Pasquino AV, Léran S, Mittal S, Rock TM, Edwards MB, Kim Grace J, et al. Temporal transcriptional logic of dynamic regulatory networks underlying nitrogen signaling and use in plants. Proc Natl Acad Sci USA. 2018:**115**(25):6494–6499. https://doi.org/10.1073/pnas.1721487115

Vidal EA, Alvarez JM, Araus V, Riveras E, Brooks MD, Krouk G, Ruffel S, Lejay L, Crawford NM, Coruzzi GM, et al. Nitrate in 2020: thirty years from transport to signaling networks. Plant Cell. 2020:**32**(7):2094–2119. https://doi.org/10.1105/tpc.19.00748

Walker L, Boddington C, Jenkins D, Wang Y, Grønlund JT, Hulsmans J, Kumar S, Patel D, Moore JD, Carter A, et al. Changes in gene expression in space and time orchestrate environmentally mediated shaping of root architecture. Plant Cell. 2017:**29**(10):2393–2412. https://doi.org/10.1105/tpc.16.00961

Wang YY, Cheng YH, Chen KE, Tsay YF. Nitrate transport, signaling, and use efficiency. Annu Rev Plant Biol. 2018:**69**(1):85–122. https://doi.org/10.1146/annurev-arplant-042817-040056

Wang R, Guegler K, LaBrie ST, Crawford NM. Genomic analysis of a nutrient response in Arabidopsis reveals diverse expression patterns and novel metabolic and potential regulatory genes induced by nitrate. Plant Cell. 2000:**12**(8):1491–1509. https://doi.org/10.1105/tpc.12.8.1491

Wang Y, Huan Q, Li K, Qian W. Single-cell transcriptome atlas of the leaf and root of rice seedlings. J Genet Genomics. 2021b:**48**(10):881–898. https://doi.org/10.1016/j.jgg.2021.06.001

Wang W, Li A, Zhang Z, Chu C. Posttranslational modifications: regulation of nitrogen utilization and signaling. Plant Cell Physiol. 2021a:**62**(4):543–552. https://doi.org/10.1093/pcp/pcab008

Weber K, Burow M. Nitrogen—essential macronutrient and signal controlling flowering time. Physiol Plant. 2018:**162**(2):251–260. https://doi.org/10.1111/ppl.12664

Wen A, Havens KL, Bloch SE, Shah N, Higgins DA, Davis-Richardson AG, Sharon J, Rezaei F, Mohiti-Asli M, Johnson A, et al. Enabling biological nitrogen fixation for cereal crops in fertilized fields. ACS Synth. Biol. 2021:**10**(12):3264–3277. https://doi.org/10.1021/acssynbio.1c00049

Wendel JF, Jackson SA, Meyers BC, Wing RA. Evolution of plant genome architecture. Genome Biol. 2016:**17**(1):1–14. https://doi.org/10.1186/s13059-016-0908-1

Wendrich JR, et al. Supplementary materials for phosphate conditions. 2020a:2:1–43.

Wendrich JR, Yang B, Vandamme N, Verstaen K, Smet W, Van de Velde C, Minne M, Wybouw B, Mor E, Arents HE, et al. Vascular transcription factors guide plant epidermal responses to limiting phosphate conditions. Science. 2020b:**370**(6518):eaay4970. https://doi.org/10.1126/science.aay4970

Wijewardane NK, Zhang H, Yang J, Schnable JC, Schachtman DP, Ge Y. A leaf-level spectral library to support high-throughput plant phenotyping: predictive accuracy and model transfer. J Exp Bot. 2023:**74**(14):4050–4062. https://doi.org/10.1093/jxb/erad129

Xiao Q, Wu N, Tang W, Zhang C, Feng L, Zhou L, Shen J, Zhang Z, Gao P, He Y. Visible and near-infrared spectroscopy and deep learning application for the qualitative and quantitative investigation of nitrogen status in cotton leaves. Front Plant Sci. 2022:**13**:1–13. https://doi.org/10.3389/fpls.2022.1080745

Xing J, Cao X, Zhang M, Wei X, Zhang J, Wan X. Plant nitrogen availability and crosstalk with phytohormones signallings and their biotechnology breeding application in crops. Plant Biotechnol. J. 2023:**21**(7):1320–1342. https://doi.org/10.1111/pbi.13971

Xu Z, Wang Q, Zhu X, Wang G, Qin Y, Ding F, Tu L, Daniell H, Zhang X, Jin S. Plant single cell transcriptome hub (PsctH): an integrated online tool to explore the plant single-cell transcriptome landscape. Plant Biotechnol J. 2022:**20**(1):10–12. https://doi.org/10.1111/pbi.13725

Yang BJ, Minne M, Brunoni F, Plačková L, Petřík I, Sun Y, Nolf J, Smet W, Verstaen K, Wendrich JR, et al. Non-cell autonomous and spatiotemporal signalling from a tissue organizer orchestrates root vascular development. Nat Plants. 2021a:**7**(11):1485–1494. https://doi.org/10.1038/s41477-021-01017-6

Yang X, Zhang C, Ma X, Liu Q, An J, Xu S, Xie X, Geng J. Combining organic fertilizer with controlled-release urea to reduce nitrogen leaching and promote wheat yields. Front Plant Sci. 2021b:**12**: 802137. https://doi.org/10.3389/fpls.2021.802137

Yang Y, Costa A, Leonhardt N, Siegel RS, Schroeder JI. Isolation of a strong Arabidopsis guard cell promoter and its potential as a research tool. Plant Methods. 2008:**4**(1):1–15. https://doi.org/10.1186/1746-4811-4-6

Zhang TQ, Xu ZG, Shang GD, Wang JW. A single-cell RNA sequencing profiles the developmental landscape of Arabidopsis root. Mol Plant. 2019:**12**(5):648–660. https://doi.org/10.1016/j.molp.2019.04.004

Zhang Z, Hu B, Chu C. Towards understanding the hierarchical nitrogen signalling network in plants. Curr Opin Plant Biol. 2020:**55**:60–65. https://doi.org/10.1016/j.pbi.2020.03.006

https://doi.org/10.1093/plcell/koad326
THE PLANT CELL
THE PLANT CELL 2024: 36: 1504–1523

Review

Milestones in understanding transport, sensing, and signaling of the plant nutrient phosphorus

Shu-Yi Yang [1,†] Wei-Yi Lin [2,†] Yi-Min Hsiao [3,†] and Tzyy-Jen Chiou [3,*]

1 Institute of Plant Biology, National Taiwan University, Taipei 106319, Taiwan
2 Department of Agronomy, National Taiwan University, Taipei 106319, Taiwan
3 Agricultural Biotechnology Research Center, Academia Sinica, Taipei 115201, Taiwan

*Author for correspondence: tjchiou@gate.sinica.edu.tw
†These authors contributed equally.
The author responsible for distribution of materials integral to the findings presented in this article in accordance with the policy described in the Instructions for Authors (https://academic.oup.com/plcell/pages/General-Instructions) is: Tzyy-Jen Chiou (tjchiou@gate.sinica.edu.tw).

Abstract

As an essential nutrient element, phosphorus (P) is primarily acquired and translocated as inorganic phosphate (Pi) by plant roots. Pi is often sequestered in the soil and becomes limited for plant growth. Plants have developed a sophisticated array of adaptive responses, termed P starvation responses, to cope with P deficiency by improving its external acquisition and internal utilization. Over the past 2 to 3 decades, remarkable progress has been made toward understanding how plants sense and respond to changing environmental P. This review provides an overview of the molecular mechanisms that regulate or coordinate P starvation responses, emphasizing P transport, sensing, and signaling. We present the major players and regulators responsible for Pi uptake and translocation. We then introduce how P is perceived at the root tip, how systemic P signaling is operated, and the mechanisms by which the intracellular P status is sensed and conveyed. Additionally, the recent exciting findings about the influence of P on plant-microbe interactions are highlighted. Finally, the challenges and prospects concerning the interplay between P and other nutrients and strategies to enhance P utilization efficiency are discussed. Insights obtained from this knowledge may guide future research endeavors in sustainable agriculture.

Introduction

Phosphorus (P) is a macronutrient that is indispensable for growth, development, and reproduction because it is an essential component of several important biomolecules, such as DNA, RNA, ATP, NADPH, and phospholipids, and is also involved in respiration and photosynthesis (Raghothama 1999; Jiang *et al.* 2019a). Orthophosphate (Pi, $H_2PO_4^-/HPO_4^{2-}$) is the predominant form of P absorbed by plant roots through loading into the xylem and translocation into shoots. Still, it is quickly fixed by Fe^{3+} and Al^{3+} at low pH (<5 or acidic soils) or by tricalcium at high pH (>7 or alkaline soils), resulting in low availability and mobility in the soil (Holford 1997; Prathap *et al.* 2022). Modern agriculture leans heavily on P fertilizer to cope with the low available soil Pi. However, due to the low efficiency of applied P absorbed by plants and the finite nature of P fertilizer, which is resourced from natural P rock reserves, increasing the P use efficiency (PUE) of crops has become an important issue to fulfill the demand for food security and agri-cultural sustainability (Baker *et al.* 2015; van de Wiel *et al.* 2016; Heuer *et al.* 2017). Endeavors to increase PUE will rely on knowledge of how plants sense and respond to environmental P levels.

In response to P deficiency, plants have developed sophisticated mechanisms to enhance the external acquisition and improve the internal utilization of P (Fig. 1), designated as P starvation responses (PSRs). Local PSRs are regulated by the external P concentration in the soil surrounding the root

Received August 02, 2023. Accepted December 19, 2023. Advance access publication January 2, 2024

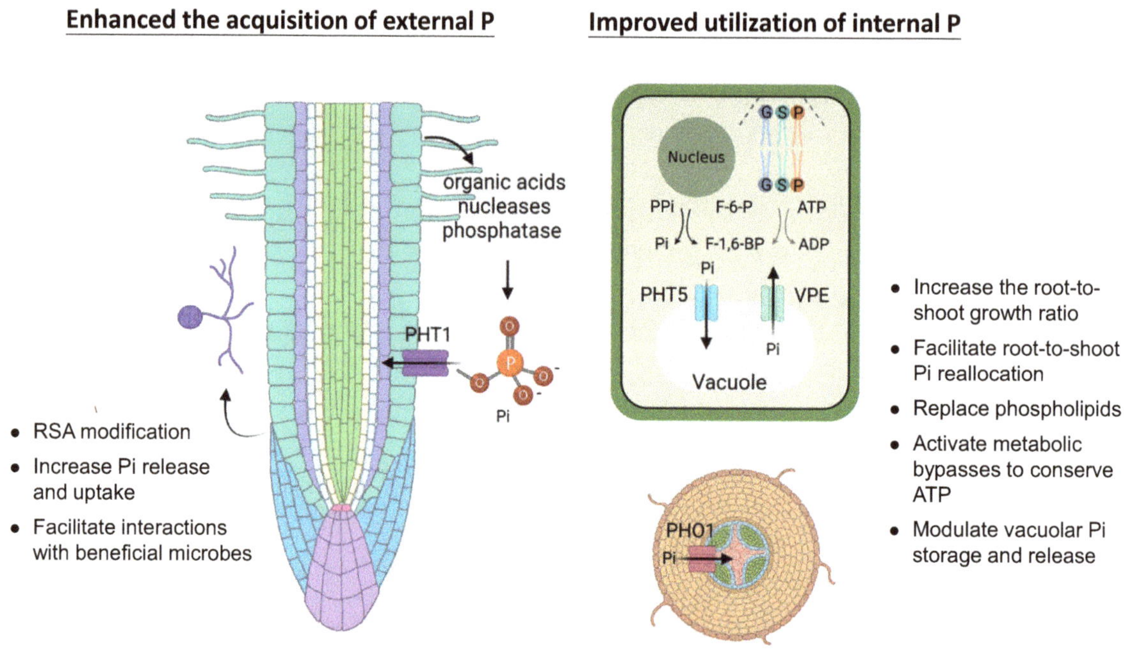

Figure 1. An overview of adaptive P starvation responses in plants. There are 2 major categories of adaptive PSRs. One is to enhance the acquisition of external P, including root system architecture (RSA) modification; increased Pi release via acid/enzyme secretion; elevated Pi uptake through PHT1 transporters; and facilitated interactions with beneficial microbes such as arbuscular mycorrhizal fungi. The other is to improve utilization of internal P, including increased root-to-shoot growth ratio, PHO1-mediated root-to-shoot Pi reallocation, phospholipid replacement with galacto- and sulfolipids, ATP-conserving metabolic bypasses, and regulation of vacuolar Pi storage/release by PHT5 and VPE. Created with BioRender.com.

and are responsible for enhancing the acquisition of external P. Systemic PSRs depend on the adjustment of internal P concentration and improve the utilization of internal P through P recycling and reallocation via coordinating with external P availability (Bates and Lynch 1996; Thibaud *et al.* 2010; Lin *et al.* 2014; Chien *et al.* 2018). Regarding the enhanced acquisition of external P, the modification of the root system architecture, including an increase in the number of lateral roots and root hairs and a reduction of the primary root growth, allows plants to forage for P in the topsoil and enlarge root-soil surface efficiently (Bates and Lynch 2001; López-Bucio *et al.* 2003; Lynch 2011). In addition, increased high-affinity Pi transporter (PHT1) activity in the root surface facilitates Pi uptake. The organic acids, nucleases, and phosphatase secreted from the root enable Pi release from insoluble or organic matter (Taylor *et al.* 1993; Bariola *et al.* 1994; Raghothama 1999; Plaxton and Tran 2011). Interactions with beneficial microorganisms triggered by P starvation could also promote P acquisition (Oldroyd and Leyser 2020). Regarding the improved utilization of internal P, an increase in the root-to-shoot growth ratio, a reduction in shoot branching, and an increase in the erectness of leaves could allow plants to adjust growth and photosynthetic efficiency to adapt to P deficiency (Ruan *et al.* 2018; López-Bucio *et al.* 2003). The increase in the xylem loading activity of Pi facilitates root-to-shoot Pi reallocation (Hamburger *et al.* 2002). The replacement of phospholipids by galacto- and sulfolipids, the activation of metabolic bypasses to conserve ATP, and the maintenance of cytosolic Pi homeostasis by modulating vacuolar Pi storage and release also occur (Härtel *et al.* 2001; Kelly and Dörmann 2002; Gaude *et al.* 2008; Plaxton and Tran 2011; Wang *et al.* 2012; Nakamura 2013; Okazaki *et al.* 2013; Liu *et al.* 2015, 2016; Wang *et al.* 2015; Xu *et al.* 2019).

In this review, we first summarize the historical key discoveries pertaining to the molecular players involved in P sensing, signaling, uptake, and utilization in plants since PHT1 Pi transporters were identified in 1996 (Fig. 2). We then highlight the major players and regulators in Pi uptake and translocation in coordinating PSRs. We illustrate local P sensing at the root tip, systemic P signaling, and intracellular P sensing. Furthermore, the recent exciting findings about P in plant-microbe interactions are discussed. Finally, the remaining challenges and perspectives on crosstalk with other nutrients and how to improve PUE are presented.

Historical overview—the key players in P transport, sensing, and signaling

This section briefly introduces the discoveries of key players in Pi transport, sensing, and signaling chronologically (Fig. 2). The details are elaborated afterward.

Since the identification of the genes encoding Pi transporters, molecular studies on P acquisition and its regulation have been inspired and enthusiastically pursued. Members

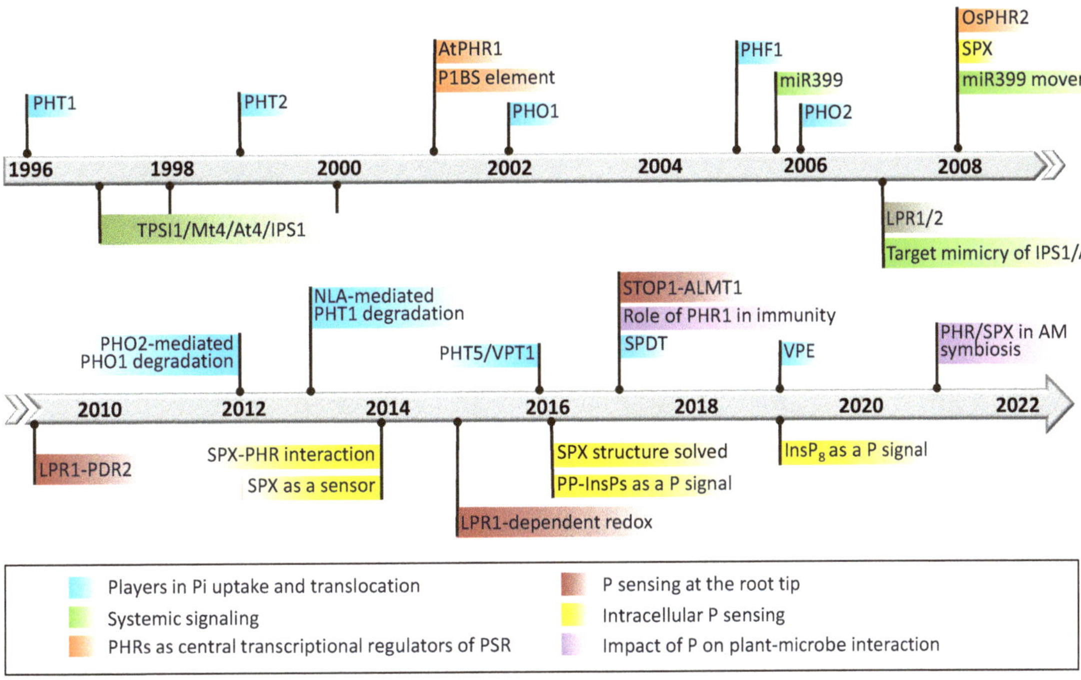

Figure 2. Milestones in understanding the genes involved in P transport, sensing, and signaling pathways. The genes with different roles in Pi uptake and translocation; local (root tip), systemic, and intracellular P sensing and signaling; and P-mediated plant-microbe interactions are depicted along the timeline and chronologically arranged according to the original publication dates.

of the plasma membrane–localized Pi transporter PHT1 family were first identified from Arabidopsis by yeast functional complementation in 1996 (Muchhal *et al.* 1996), followed by the second Pi transporter PHT2 in 1999, which is localized in the chloroplast (Daram *et al.* 1999). Later, in 2002, PHOSPHATE 1 (PHO1) was identified as a Pi efflux transporter mediating root-to-shoot Pi translocation (Hamburger *et al.* 2002). Then, in 2017, an additional group of Pi transpor-ters belonging to the family of sulfate transporters (SULTRs), named SULTR-like phosphorus distribution transporter (SPDT), was uncovered (Yamaji *et al.* 2017). More recently, Pi transporters facilitating Pi translocation between the cytoplasm and vacuole have been identified, including Arabidopsis PHT5;1/VPT1 (VACUOLAR PHOSPHATE TRANSPORTER), a Pi influx transporter; and rice VPE1 and VPE2 (VACUOLAR PHOSPHATE EFFLUX), Pi efflux transporters (Liu *et al.* 2015, 2016; Xu *et al.* 2019). After uncovering the molecular identity of these Pi transporters, their regulation at the transcriptional and posttranslational levels was revealed. For example, upregulation of PHT1 transcription by low P is mediated by PHOSPHATE STARVATION RESPONSE (PHR) proteins binding to PHR1 binding sequence (P1BS, GNATATNC) (Rubio *et al.* 2001; Zhou *et al.* 2008), the proper targeting of PHT1 Pi transporters to plasma mem-branes requires PHOSPHATE TRANSPORTER TRAFFIC FACILITATOR 1 (PHF1) (Gonzalez *et al.* 2005), and the protein abundance of PHT1 and PHO1 in response to Pi availability is controlled by PHOSPHATE 2 (PHO2/UBC24), a ubiquitin E2 conjugase, (Aung *et al.* 2006; Bari *et al.* 2006; Liu *et al.* 2012; Huang *et al.* 2013), and/or NITROGEN LIMITATION ADAPTATION (NLA), a RING-type E3 ligase (Lin *et al.* 2013). It is worth mentioning that PHO1 and PHO2 were identified initially from genetic screenings through having significantly reduced or increased shoot Pi content, respectively (Poirier *et al.* 1991; Delhaize and Randall 1995).

To coordinate the P acquisition in the roots and P utilization at the whole-plant level for optimizing plant growth and development, sensing and signaling P availability are requisite. P sensing and signaling can be classified into local and systemic. The local signaling to sense external P at the root tips was shown to be mediated by Low Phosphate Root1 (LPR1) and LPR2, 2 multicopper oxidases (Svistoonoff *et al.* 2007). Other components, such as PHOSPHATE DEFICIENCY RESPONSES 2 (PDR2) and ALUMINUM ACTIVATED MALATE TRANSPORTER 1 (ALMT1), were later found to work together with LPR1 to regulate root meristem differentiation (Ticconi *et al.* 2009; Balzergue *et al.* 2017). On the other hand, systemic signaling plays an essential role in the improved utilization of internal P. From 1997 to 2000, noncoding RNAs (*TPSI1*, *Mt4*, *At4* and *IPS1*) were discovered in different plant species that are highly induced by P starvation and share a conserved 22-nucleotide sequence (Liu *et al.* 1997; Burleigh and Harrison 1999; Martín *et al.* 2000; Shin *et al.* 2006). Their biological function remained unknown un-til the discovery of microRNA399 (miR399), the first miRNA identified to be involved in P starvation response (Fujii *et al.* 2005). Upon P starvation, miR399 is upregulated, which suppresses *PHO2* expression and leads to the activation of Pi uptake and translocation (Aung *et al.* 2006; Bari *et al.* 2006; Chiou *et al.* 2006). *AT4/IPS1* homologs function to modulate the action of miR399 by sequestering miR399 through the 22 conserved nucleotides via a mechanism of target mimicry (Franco-Zorrilla *et al.* 2007). Interestingly, miR399 was found to serve as a shoot-to-root systemic signal (Lin *et al.* 2008; Pant *et al.* 2008).

With regard to intracellular Pi sensing and signaling, SPX proteins were discovered to be involved in regulating PSR genes in 2008 (Duan *et al.* 2008), and their role as P sensors through SPX-PHR interaction was reported in 2014 (Puga *et al.* 2014; Wang *et al.* 2014b). SPX proteins function as repressors to sequester PHRs from transcriptional activation of PSRs. Notably, a structure-function analysis revealed that inositol polyphosphate (InsP) and inositol pyrophosphate (PP-InsP) rather than Pi could be bound by SPX in vitro, but PP-InsPs could potentially serve as the pertinent signaling molecules in vivo (Wild *et al.* 2016). Later genetic analyses revealed that bis-diphosphoinositol tetrakisphosphate 1,5$(PP)_2$-$InsP_4$ (1,5-$InsP_8$) acts as a bona fide intracellular signaling molecule to regulate PSRs in plants (Dong *et al.* 2019a; Zhu *et al.* 2019).

Plants actively modulate the interactions with various microbes to adapt to the Pi availability. Strikingly, PHR was discovered to serve as a negative and positive regulator to mediate immune-related genes and arbuscular mycorrhizal-related genes, respectively (Castrillo *et al.* 2017; Isidra-Arellano *et al.* 2021; Shi *et al.* 2021; Das *et al.* 2022). The role of SPX in arbuscular mycorrhizal (AM) symbiosis was also revealed (Shi *et al.* 2021; Wang *et al.* 2021; Liao *et al.* 2022), suggesting the central role of SPX-PHR in integrating signals involved in nutritional response and biotic interaction.

Players in Pi uptake and translocation

Pi is initially acquired by the root epidermal and/or cortical cells and subsequently transported radially to the central vascular tissues where Pi is loaded into the xylem for translocation up to the shoot. Pi is then allocated among different tissues and distributed in organelles to sustain growth and development or stored inside vacuoles when excessive. Therefore, Pi transporters in the plasma membrane or organellar membranes are essential to facilitate Pi transport across membranes for uptake, distribution and remobilization (Fig. 3). While vegetative cells generally store Pi inside the vacuole, the primary form of P stored in the vacuole of seeds is inositol hexakisphosphate ($InsP_6$, phytate) mediated by MULTIDRUG RESISTANCE-ASSOCIATED PROTEIN 5 (MRP5), a vacuolar $InsP_6$ transporter (Nagy *et al.* 2009).

Pi transporters comprise different families, such as the phosphate transporter (PHT) family, the SYG1/Pho81/XPR1 (SPX) domain-containing protein family and the SPDT family. Members in the PHT family are classified into 5 types according to sequence identity and subcellular localization (Versaw and Garcia 2017), comprising PHT1 (plasma membrane), PHT2 (chloroplast), PHT3 (mitochondria), PHT4 (the Golgi apparatus, chloroplast and non-photosynthetic plastid) and PHT5 (vacuole). SPX domain-containing proteins are divided into 4 subfamilies: SPX, SPX-EXS (ERD1/XPR1/SYG1), SPX-MFS (Major Facilitator Superfamily), and SPX-RING (Really Interesting New Gene). Among them, SPX-MFS domain- and SPX-EXS domain-proteins have been identified as Pi transporters (e.g. members of the PHT5 family and PHO1, respectively) (Wang *et al.* 2012, 2015; Liu *et al.* 2015, 2016).

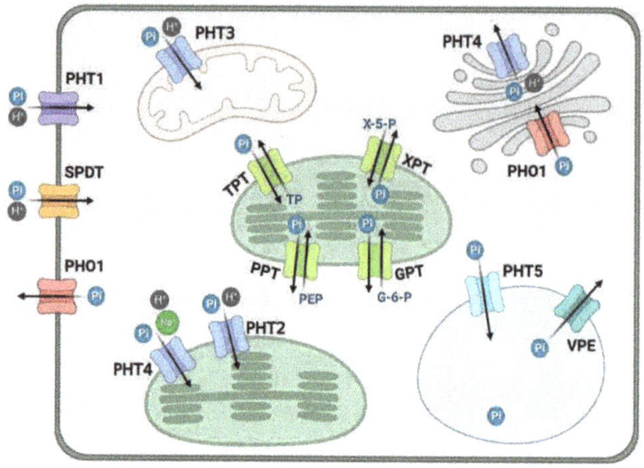

Figure 3. Subcellular localization of players in Pi uptake and translocation. Arrows indicate the direction of Pi transport. Plasma membrane-localized PHT1 and SPDT are responsible for Pi uptake. PHT2 (chloroplast), PHT3 (mitochondria), and PHT4 (the Golgi apparatus, chloroplast, and nonphotosynthetic plastid) mediate Pi import or export from the corresponding organelles. Pi influx and efflux through the tonoplast are operated by PHT5 and VPE, respectively. PHO1 protein localizes to the plasma membrane and Golgi bearing the Pi efflux activity for Pi allocation. Pi translocators (PT) localized in the inner envelope membrane of plastids transport Pi in exchange with different substrates, including triose-phosphate/phosphate translocator (TPT), phosphoenolpyruvate/phosphate translocator (PPT), glucose-6-phosphate/phosphate translocator (GPT), and pentose xylulose-5-phosphate/phosphate translocator (XPT). Coupling with proton (H^+) or sodium (Na^+) transport is indicated. Created with BioRender.com.

PHT1 transporters

PHT1 transporters are proton-Pi symporters and play a critical role in Pi acquisition from the soil and in Pi translocation between tissues (Nussaume *et al.* 2011; Remy *et al.* 2012; Lopez-Arredondo *et al.* 2014; Ayadi *et al.* 2015). There are 13 members in rice (*Oryza sativa*) and 9 in *Arabidopsis thaliana*. Most of them are expressed in roots and upregulated by the PHR transcription factor under P starvation (Karthikeyan *et al.* 2002; Mudge *et al.* 2002). In Arabidopsis, Pi uptake is mainly mediated by PHT1;1 and PHT1;4 (Shin *et al.* 2004; Remy *et al.* 2012; Ayadi *et al.* 2015), and root-to-shoot mobilization is facilitated by PHT1;8/PHT1;9 (Lapis-Gaza *et al.* 2014). In rice, OsPT2, a low-affinity Pi transporter exclusively expressed in the stele, facilitates the movement of Pi from roots to shoots. In contrast, OsPT6 and OsPT8 function as high-affinity Pi transporters responsible for Pi uptake and subsequent translocation (Ai *et al.* 2009; Jia *et al.* 2011). OsPT9 and OsPT10, 2 other high-affinity Pi transporters, are also involved in Pi uptake (Wang *et al.* 2014a). The expres-sion of these 5 rice PHT1 transporters were all induced by Pi starvation in the roots (Ai *et al.* 2009; Jia *et al.* 2011; Wang *et al.* 2014a). The members of *PHT1* are also expressed in the reproductive tissues responsible for anther and embryo development, for example, Arabidopsis *PHT1;6* and *PHT1;7* and rice *OsPht1;7* in the anthers (Mudge *et al.* 2002; Dai *et al.* 2022) and *OsPht1;4* in the embryo (Zhang *et al.* 2015).

Regarding the proper localization of PHT1 proteins, PHOSPHATE TRANSPORTER TRAFFIC FACILITATOR 1 (PHF1), related to yeast SECRETORY 12 (SEC12), facilitates the exit of PHT1 proteins from the endoplasmic reticulum (ER) to plasma membranes in rice and Arabidopsis (Gonzalez *et al.* 2005; Bayle *et al.* 2011; Chen *et al.* 2011b). Additionally, the phosphorylation of rice OsPHT1;8 at Ser-517 by Casein Kinase 2 (OsCK2) interfered with the inter-action between OsPHT1;8 and PHF1 and resulted in ER re-tention of OsPHT1;8 (Chen *et al.* 2015). Intriguingly, PHT1 protein abundance in the ER is controlled by PHO2-mediated degradation according to cellular P status, which determines the PHT1 amount directed to the plasma membrane (Huang *et al.* 2013).

Once at plasma membranes, PHT1 proteins are regulated by NLA-mediated degradation, a RING-type ubiquitin E3 ligase containing the SPX domain that can perceive cellular P levels (as discussed below) (Lin *et al.* 2013; Park *et al.* 2014; Yue *et al.* 2017; Yang *et al.* 2020). It was proposed that the cooperation between NLA and PHO2 controls the final protein level of PHT1 (Lin *et al.* 2013; Park *et al.* 2014). Intriguingly, different from Arabidopsis AtNLA, which is cleaved by P starvation-induced miR827 (Hsieh *et al.* 2009), rice *OsNLA1* is not the target of miR827, but its upstream open reading frame (uORF) is essential for its P-induced transcriptional expression (Lin *et al.* 2018; Yang *et al.* 2020).

PHT2/3/4 transporters

PHT2, PHT3, and PHT4 are localized to organelles, such as chloroplasts/plastids, mitochondria, and the Golgi apparatus. In Arabidopsis, PHT4;1, PHT4;2, and PHT2;1 control the Pi concentration in chloroplasts and regulate ATP synthesis and starch accumulation (Versaw and Harrison 2002; Karlsson *et al.* 2015; Miyaji *et al.* 2015). Similarly, mitochondrial localized PHT3 also plays a role in ATP metabolism (Zhu *et al.* 2012). Unexpectedly, PHT4;4 was shown to trans-port ascorbate into chloroplasts required to tolerate high light stress (Miyaji *et al.* 2015). In rice, OsPHT2;1 is also characterized as a chloroplast Pi influx transporter. Overexpression and loss of OsPHT2;1 result in altered Pi con-tent in the shoot under low P conditions (Liu *et al.* 2020). How different organelles coordinate with each other to mediate cellular Pi homeostasis between the cytosol and organelles awaits more studies.

Vacuolar Pi transporters—PHT5 and VPE

Inside plant vegetative cells, the vacuole is the largest organelle and stores about 70% to 95% of the intracellular Pi (Yang *et al.* 2017), thus serving as an important reservoir in buffering cytoplasm ic Pi concentrations. The Pi influx into the vacuole is operated by Arabidopsis PHT5 or VPT and rice SPX-MFS (Wang *et al.* 2012; Liu *et al.* 2015, 2016). These transporters contain N-term inal SPX and C-term inal MFS dom ains. OsSPX-MFS1 controls Pi hom eostasis in leaves and mediates Pi influx to yeast vacuoles when heterologously expressed (Wang *et al.* 2012; Liu *et al.* 2016). Loss of Arabidopsis PHT5;1/VPT1 reduces total Pi level but results in necrotic leaves during P replenishm ent after starvation (Liu *et al.* 2015, 2016). Patch clamp analysis revealed a reduced Pi influx current in the vacuoles of *vpt1* mutants compared with wild-type ones (Liu *et al.* 2015). In addition, decreased vacuolar Pi and increased cytosolic Pi levels were observed in *pht5* mutants by ^{31}P-NMR analysis or FRET-based Pi sensor (Liu *et al.* 2016; Sahu *et al.* 2020). Notably, the expression of *AtPHT1s* is affected in *pht5* mutants and *PHT5* overexpressor, implying that coordination between the storage capacity and acquisition activity would be achieved to maintain cytosolic Pi concentration (Liu *et al.* 2016). Conversely, rice vacuolar Pi efflux proteins VPE1/2 operate Pi efflux from the vacuole. The *vpe1/2* double m utant and overexpression plants show higher and lower vacuolar Pi levels, respectively (Xu *et al.* 2019). Unlike *VPE1/2*, the expression of *PHT5* is not responsive to P status, suggesting that the activity of PHT5 might be regulated at the protein level (Liu *et al.* 2016; Xu *et al.* 2019). Indeed, a recent study showed that the activity of PHT5 is im paired under Pi-deficient conditions or when the InsP/PP-InsP binding pocket in its SPX dom ain is mutated, thus linking the cellular P status to modulate the vacuolar Pi storage activity (Luan *et al.* 2022).

PHO1

Arabidopsis PHO1 and its close homolog PHO1;H1 with Pi efflux activity were able to facilitate the unloading of Pi into the xylem apoplastic space and allow for the subsequent translocation of Pi to the shoots (Poirier *et al.* 1991; Hamburger *et al.* 2002; Arpat *et al.* 2012; Liu *et al.* 2012). The significance of PHO1 in the root-to-shoot Pi allocation is supported by the higher Pi accumulation in roots and reduced Pi levels in shoots of *pho1* mutants in Arabidopsis and rice (Poirier *et al.* 1991; Secco *et al.* 2010; Che *et al.* 2020). AtPHO1 also facilitates the transfer of Pi from maternal tissues to developing seeds in the chalazal seed coat (Vogiatzaki *et al.* 2017). It is also expressed in leaf guard cells to play a role in the stomatal response to abscisic acid (ABA) (Zimmerli *et al.* 2012). In rice, OsPHO1;2 are involved in the allocation of Pi during grain filling (Che *et al.* 2020; Ma *et al.* 2021). In *Medicago truncatula*, PHO1 mediates the transfer of Pi from nodule-infected cells to bacteroids (Nguyen *et al.* 2021). Earlier results showed that the subcellular localization of Arabidopsis PHO1 is in endomembrane systems, such as the Golgi and trans-Golgi networks, different from the plasma membrane localization of OsPHO1;2 (Arpat *et al.* 2012; Liu *et al.* 2012; Che *et al.* 2020). Nevertheless, a recent report showed that AtPHO1 is constitutively internalized from the plasma membrane (Vetal and Poirier 2023).

Arabidopsis PHO1 is negatively regulated at the translational level by its uORF (Reis *et al.* 2020) or at the transcriptional level by WRKY6 (Chen *et al.* 2009). In rice, the OsPHO1;2 mRNA translation can be enhanced by interacting with its cis-natural antisense transcript (Jabnoune *et al.* 2013; Reis *et al.* 2021). PHO1 has a cytosolic N-terminal SPX domain, 4 transmembrane domains, and a C-terminal EXS domain (Wege *et al.* 2016). The EXS domain is required for PHO1 membrane localization and Pi efflux activity (Wege *et al.* 2016). The SPX-spanning region of PHO1 facilitates its interaction with PHO2, which is responsible for PHO1 degradation (Liu *et al.* 2012). The absence of the SPX domain in PHO1 does not affect its efflux activity in transient expression assay. However, it fails to restore the impaired root-to-shoot Pi translocation activity in *pho1* mutants, indicating that the SPX domain is crucial for PHO1's functionality regulated by InsP/PP-InsP binding (Wege *et al.* 2016; Wild *et al.* 2016).

PHO2-mediated degradation of PHO1 occurs in endomembranes and is involved in vacuole proteolysis mediated by multi-vesicular bodies (Liu *et al.* 2012). Similarly, rice OsPHO2 also interacts with OsPHO1 to direct its degradation via a multi-vesicular body–mediated pathway (Wang *et al.* 2020), suggesting the conservation of the PHO2-PHO1 regula-tory module among plants. The expression of *PHO2* is nega-tively regulated by miR399 (Bari *et al.* 2006) and positively regulated by the HD–ZIP III transcription factor PHB, which is inhibited by SHORT-ROOT (SHR) (Xiao *et al.* 2022). Studies have also revealed that OsCK2, a protein kinase, can enhance the degradation of OsPHO2 through the phosphor-ylation of OsPHO2 by OsCK2α3 (Wang *et al.* 2020).

SULTR-like phosphorus distribution transporter (SPDT)

SPDT was first shown to play a crucial role in regulating P allocation in rice (Yamaji *et al.* 2017). SPDT is a plasma-membrane-localized transporter that facilitates Pi transport and is expressed in the xylem region of both enlarged- and diffuse-vascular bundles of the nodes. Loss-of-function of SPDT in rice decreased P levels in grains but increased P levels in leaves. These findings indicate that SPDT functions as a switch in the rice node to allocate Pi to the grains preferentially. Likewise, barley HvSPDT, mainly expressed in the nodes, also plays an essential role in loading P into grains (Gu *et al.* 2022). In addition, the primary function of vascular cambium-localized Arabidopsis AtSPDT was shown to preferentially distribute Pi to the growing tissues (Ding *et al.* 2020).

Phosphate translocators

Phosphate translocators (PTs) localized in the inner envelope membrane of plastids play a crucial role in establishing the metabolic connection between the plastid stroma and cytosol. Different from the Pi transporters mentioned above, these PT facilitate a precise counter-exchange of Pi with phosphorylated intermediates, including triose-phosphate/phosphate translocator, phosphoenolpyruvate/phosphate translocator, glucose-6-phosphate/phosphate translocator, and pentose xylulose-5-phosphate/phosphate translocator. They were reported to be involved in vegetative tissue and gametophyte development, redox signaling in leaves, and acclimation responses of photosynthesis (Młodzińska and Zboińska 2016; Fabiańska *et al.* 2019).

PHR1 as a central transcriptional regulator of PSRs

Arabidopsis PHR1 or rice PHR2 encoding an MYB coiled-coil transcription factor is a central positive regulator of the PSRs. PHR1 was initially identified from a genetic screen in Arabidopsis harboring a reporter gene driven by the IPS1 promoter highly induced by P starvation (Rubio *et al.* 2001). *phr1* mutants displayed multiple impairments in PSRs, including altered root-to-shoot Pi allocation, reduced accumulation of anthocyanin and carbohydrate, altered lipid composition and lipid remodeling genes, and impaired induction of several P starvation–induced (PSi) genes under P starvation (Rubio *et al.* 2001; Nilsson *et al.* 2007; Pant *et al.* 2015). Arabidopsis PHR1 transcript and protein accumulation are weakly responsive to P starvation; however, its activity is regulated by interacting with SPX proteins coupled with the cellular P status (Rubio *et al.* 2001; Puga *et al.* 2014).

PHR1 forms homodimers or heterodimers with its close homolog PHR1-like 1 (PHL1) and binds to P1BS elements present in the promoter of many PSi genes (Rubio *et al.* 2001; Bustos *et al.* 2010; Jiang *et al.* 2019b). Arabidopsis PHR1 and par-tially redundant PHL1 regulate the majority of transcriptional responses to P starvation, and P1BS elements are enriched in the promoter of PSi genes but not in P starvation-repressed (PSr) genes (Bustos *et al.* 2010). In P-starved *phr1phl1* mutants, nearly 70% of PSi genes and 50% of PSr genes showed decreased and increased expression, respectively, suggesting a dir-ect and indirect regulation of transcriptional activation and repression responses (Bustos *et al.* 2010).

The other 3 Arabidopsis PHR1 homologs, PHL2, PHL3, and PHL4, have also been found to participate in PSRs to varying extents. PHL4 has a redundant function with PHR1/PHL1 but a minimal effect (Wang *et al.* 2018). PHL2 and PHL3 exhibit a specific interaction, and their expression is induced by P starvation, unlike PHR1 and PHL1. The transcriptomic analysis further suggests that PHR1/PHL1 and PHL2/PHL3 regulate various aspects of PSRs as separate regulatory modules (Wang *et al.* 2022). PHR1/PHL1 regulates genes involved in the cellular response to P starvation, Pi transport, and galactolipid biosynthetic process, whereas PHL2/PHL3 regulates genes related to the cellular response to P starvation, translation, and ribosome biogenesis. In rice, Arabidopsis *PHR1* homolog *OsPHR2* plays a crucial role in Pi homeostasis, and overexpressing *OsPHR2* increases PSi gene expression and results in P toxicity (Zhou *et al.* 2008). PHR1 homologs participating in cellular Pi homeostasis have also been identified in various plant species (Wang *et al.* 2022), suggesting that PHR proteins have a universal and predominant role in regulating PSRs.

Despite not being as prevalent as PHR1, other transcription factors have been identified as regulators of PSRs to different extents. MYB62, as a negative regulator of PSRs, is involved in the interplay between PSRs and gibberellic acid responses (Devaiah *et al.* 2009), and WRKY75 is a positive regulator of Pi uptake and a subset of PSi genes (Devaiah *et al.* 2007a). As mentioned, WRKY6 is a repressor of *PHO1*. It in-teracts with 2 W-box motifs in the promoter of *PHO1* and is degraded by 26S proteasome to relieve the suppression upon P starvation (Chen *et al.* 2009). Basic Helix-Loop-Helix 32 and zinc finger of *Arabidopsis thaliana* 6 were also re-ported to have roles in regulating Pi content and root archi-tecture (Chen *et al.* 2007; Devaiah *et al.* 2007b).

Intracellular P sensing and signaling

SPX proteins as intracellular P sensors

Proteins harboring a hydrophilic and poorly conserved SPX domain have been identified in many eukaryotes, such as yeasts, plants, and mammals (Secco *et al.* 2012). Many yeast SPX domain–containing proteins are involved in Pi homeostasis, including Pi transporters Pho87, Pho90, and Pho91; polyphosphate synthase subunits of vacuole transporter chaperone VTC2-5; the cyclin-dependent kinase inhibitor Pho81; and the glycerophosphocholine phosphodiesterase 1 (Secco *et al.* 2012; Desfougères *et al.* 2016). Yeast Pho87 and Pho90 are plasma membrane–localized low-affinity Pi influx transporters, and their SPX domains act as auto-inhibitory domains to regulate Pi import (Hürlimann *et al.* 2009). The SPX domain of vacuolar Pi exporter Pho91 also regulates its activity through binding with PP-InsP (Potapenko *et al.* 2018). Additionally, PP-InsP-dependent binding on the SPX domains of yeast VTC complex is required for its polyphosphate poly-merase activity (Gerasimaite *et al.* 2017).

Arabidopsis SPX1 and SPX2 are nuclear proteins and function to repress PHR1 activity through physical interaction in the nucleus to sequester PHR1 from binding to the P1BS element of PSi genes (Fig. 4A) (Duan *et al.* 2008; Puga *et al.* 2014). Notably, the interaction between PHR1 and SPX1/2 highly depends on the cellular P status, with a strong interaction under P-sufficient conditions. Intriguingly, *SPX1/2* transcription is induced by PHR1 during P starvation, revealing a negative feedback regulatory loop between SPX1/2 and PHR1 (Puga *et al.* 2014). This mechanism is conserved in rice; OsSPX1/2 interacts with OsPHR2 to suppress the binding of OsPHR2 to the P1BS element (Zhou *et al.* 2008; Wang *et al.* 2014b). Other than in the nucleus, Arabidopsis SPX4 and rice OsSPX4 and OsSPX6 interact with PHR in the cytoplasm to restrain the nuclear translocation of PHR when P is replete (Lv *et al.* 2014; Zhong *et al.* 2018). Upon P deficiency, OsSPX4 and OsSPX6 undergo ubiquitin-mediated proteasomal degradation, thus releasing OsPHR2.

$InsP_8$ as a metabolic messenger for intracellular P signaling

Pi was initially considered to be a signaling molecule because of the attenuation of PSRs by phosphite (Phi, a nonmetabolized homolog of Pi) (Ticconi *et al.* 2001; Varadarajan *et al.* 2002). However, this concept was revised when the InsP/ PP-InsP binding site of the SPX domain was revealed after solving the structure (Wild *et al.* 2016). The conserved amino acid residues defined as Pi binding cluster and lysine surface cluster together in the α-helix 2 and 4 of SPX domains form a positively charged surface, with higher affinities for $InsP_6$ and 5-$InsP_7$ than Pi. Moreover, the binding of $InsP_6$ and 5-$InsP_7$ but not Pi to OsSPX4 can facilitate the interaction with OsPHR2, with 5-$InsP_7$ showing a higher affinity than $InsP_6$ (Wild *et al.* 2016), supporting the notion that PP-InsP rather than Pi is the signaling molecule representing intracellular P status.

In Arabidopsis, the synthesis and hydrolysis of 5-$InsP_7$ and 1,5-$InsP_8$ are catalyzed by 2 bifunctional enzymes, inositol 1,3,4-trisphosphate 5-/6-kinase ITPK1/2 and diphosphoinositol pentakisphosphate kinase VIH1/2, respectively (Desai *et al.* 2014; Laha *et al.* 2015, 2019; Zhu *et al.* 2019; Whitfield *et al.* 2020; Riemer *et al.* 2021). The activity of ITPK1/2 and VIH1/2 can respond to changes in cellular ATP and Pi levels: high ATP favors kinase activity to synthesize 1,5-$InsP_8$, as ATP levels decrease in response to P starvation, and ADP phosphotransferase activity of ITPK1 and phosphatase activity of VIH1/2 are stimulated to catalyze the hydrolysis of 5-$InsP_7$ and 1,5-$InsP_8$ (Laha *et al.* 2019; Zhu *et al.* 2019; Whitfield *et al.* 2020; Riemer *et al.* 2021). Inhibition of VIH1/2 phosphatase activity by Phi may explain the early observations of the Phi-mediated suppression of PSRs. Arabidopsis *vih1vih2* mutants displaying severe growth retardation exhibit an undetectable level of $InsP_8$ but overaccumulation of Pi and constitutive expression of PSi genes when grown under P-sufficient conditions, implying that $InsP_8$ is critical for plants to sense cellular P status (Dong *et al.* 2019b; Zhu *et al.* 2019). Furthermore, whereas the levels of $InsP_6$ and $InsP_7$ were relatively unchanged in response to P starvation, the level of $InsP_8$ was significantly decreased (Kuo *et al.* 2018; Dong *et al.* 2019a). 1,5-$InsP_8$ but not 5-$InsP_7$ can restore the interaction between SPX1 and PHR1 in the Pi-depleted cell lysate (Dong *et al.* 2019b). These findings indicate that $InsP_8$ is a signaling molecule communicating the cellular P level to control cellular Pi homeostasis through the SPX domain. A recent yeast study also revealed that 1,5-$InsP_8$ specifically inactivates Pho81 through its SPX domain and that the reduction of 1,5-$InsP_8$ upon P starvation triggers the yeast PHO pathway (Chabert *et al.* 2023). The authors further proposed 1,5-$InsP_8$ as a universal signaling molecule in P sensing and signaling pathways among fungi, plants, and mammals (Chabert *et al.* 2023). It is worth noting that defective

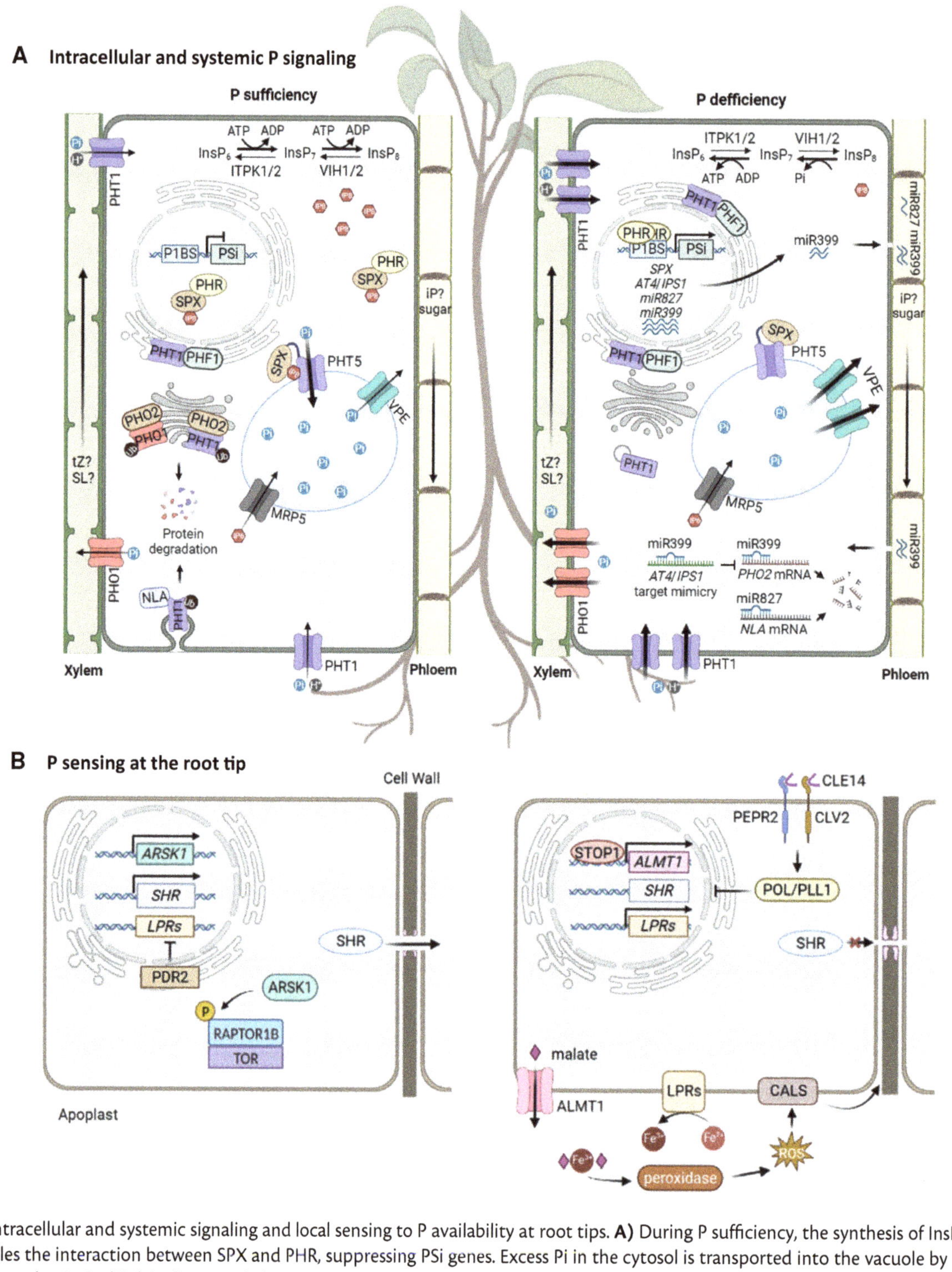

Figure 4. Intracellular and systemic signaling and local sensing to P availability at root tips. **A)** During P sufficiency, the synthesis of $InsP_8$ is favored, which enables the interaction between SPX and PHR, suppressing PSi genes. Excess Pi in the cytosol is transported into the vacuole by PHT5. In the seeds, P is stored as $InsP_6$ (IP6) in the vacuole by MRP5. PHO2 and NLA promote the degradation of Pi transporters PHT1 and PHO1 via ubiquitination (ub). Upon P deficiency, $InsP_8$ (IP8) level is reduced. PHR is released and binds to the P1BS element as a dimer to activate PSi genes. PHF1 facilitates the targeting of PHT1 Pi transporters to plasma membranes. miR399 and miR827 repress *PHO2* and *NLA*, respectively, to increase the abundance of PHT1 and PHO1. *AT4/IPS1* transcripts can sequester miR399 to antagonize the effects of miR399. As indicated, miR399, miR827, sugar, strigolactone (SL), and cytokinin (tZ and iP) are potential systemic signals traveling via vascular tissues. **B)** Under P-sufficient conditions, ARSK1 is induced to phosphorylate RAPTOR1B, stabilizing the TOR1 complex and promoting root growth. Under P-deficient conditions, STOP1-ALMT1 modulates LPR1 activity, which controls Fe distribution in the root apical meristem, blocking SHR cell-to-cell movement by callose deposition. Meanwhile, the reception of CLE14 peptide activates POL/PLL1 phosphatases to inhibit root cell proliferation at transcript levels. Abbreviations: CALS, callose synthase; ROS, reactive oxygen species. Created with BioRender.com.

vacuolar storage of $InsP_6$, as seen in *mrp5* mutants, results in elevated $InsP_7$ and $InsP_8$ levels (Desai *et al.* 2014; Riemer *et al.* 2021), suggesting $InsP_6$ compartmentation may also play a role in P signaling.

Recent studies have investigated the structure-functional relationship of the SPX-PP-InsP-PHR complex. $InsP_8$ promotes the binding between the $InsP_8$-SPX1 complex and the coiled-coil domain of PHR1 (Ried *et al.* 2021). In the presence of $InsP_6$ (used as a substitute for $InsP_8$), OsSPX1 interacts with the MYB and the coiled-coil domains of OsPHR2 to form a complex with 1:1 stoichiometry, which disrupts the OsPHR2 dimer and further inhibits its binding to the P1BS element (Zhou *et al.* 2021). On the other hand, Guan *et al.* (2022) proposed that upon binding with $InsP_6$, OsSPX2 forms a domain-swapped dimer and blocks the Myb and coiled-coil domains of OsPHR2 to abolish OsPHR2 dimerization and DNA binding; the SPX-$InsP_6$-PHR complex was described as a 2:2:2 complex. This discrepancy might be due to the difference in conformational changes be-tween different SPX proteins (Guan *et al.* 2022).

P sensing and signaling at the root tips

Root tips are critical in sensing environmental P status and regulating root meristem activities in response to external stimuli. P starvation causes premature differentiation in the root meristem, inhibiting primary root growth (Sanchez-Calderon *et al.* 2005). The increase of cell wall stiffness in the root transition zone can be observed 2 hours after plants transfer to P-deficient conditions. This results in the restriction of root cell extension and inhibition of root growth (Balzergue *et al.* 2017). LPR1 and LPR2, 2 multicopper oxidases with ferroxidase activity, which function with PDR2, a P5-type ATPase, are involved in the control of iron (Fe) distribution in the root apical meristem (RAM) (Reymond *et al.* 2006; Svistoonoff *et al.* 2007; Ticconi *et al.* 2009; Müller *et al.* 2015; Naumann *et al.* 2022). Intriguingly, Fe distribution in P-depleted roots overlays with the callose deposition pattern in the cell wall, restricting the movement of SHR for stem cell specification (Müller *et al.* 2015). Loss of LPR significantly reduced class III peroxidase activity, which might affect ROS-mediated callose deposition, leading to cell wall softening (Müller *et al.* 2015; Balzergue *et al.* 2017). These studies demonstrate the role of the LPR1-PDR2 module in mediating Fe- and peroxidase-dependent P starvation–induced root meristem differentiation (Fig. 4B) (Balzergue *et al.* 2017; Mora-Macias *et al.* 2017). Later studies revealed the involvement of PDR2-mediated ER stress-dependent autophagy in controlling Pi starvation–induced root growth retardation (Naumann *et al.* 2019).

Organic acids such as malate and citric acid in root exudates are needed for solubilizing metal ion-conjugated Pi in soils (Ryan *et al.* 2001). The activation of ALMT1, a malate efflux channel, by SENSITIVE TO PROTON RHIZOTOXICITY (STOP1) is not only critical for aluminum tolerance (Liu *et al.* 2009; Sawaki *et al.* 2009) but also influences malate export in the apoplast of root tips and Fe distribution. The reduction of *LPR1* mRNA in *stop1* and *almt1* mutants and ineffective malate treatment on the long primary root phenotype in *lpr1* indicate that ALMT1-mediated malate-triggered responses promote LPR1-dependent root meristem differentiation (Fig. 4B) (Balzergue *et al.* 2017).

Arabidopsis CLAVATA3/ENDOSPERM SURROUNDING REGION 14 (CLE14) is a low-P–induced peptide signal found specifically in the cortex, endodermis, and stele of RAM, functioning downstream of LPR1/LPR2. Applying CLE14 peptides can trigger RAM differentiation even under P-sufficient conditions but does not affect callose deposition. Instead, CLE14 perceived by CLAVATA 2 (CLV2) and PEP1 RECEPTOR2 (PEPR2) activates 2 phosphatases, POLTERGEIST (POL) and POL-LIKE1, to repress *SHR* and *SCARECROW* at mRNA levels (Fig. 4B) (Gutierrez-Alanis *et al.* 2017); however, the underlying mechanism is still missing. Moreover, the crosstalk of these signaling pathways with phytohormone-mediated root growth regulation remains to be studied.

Although Fe distribution in root tips determines the root meristem activities, an Fe-independent regulatory pathway is suggested because the reduction of root growth was observed before changes in Fe accumulation in the short-term responses to P deficiency. Arabidopsis-root-specific kinase 1 (ARSK1), a receptor-like kinase, is expressed explicitly in P-sufficient roots and can phosphorylate regulatory- associated protein of TOR 1B (RAPTOR1B), a scaffold protein of the target of rapamycin complex 1 (TOR1) that integrates environmental cues to modulate cellular growth (Wu *et al.* 2019). Overexpressing ARSK1 or phosphorylated RAPTOR1B inhibited P deficiency–triggered primary root retardation, indicating the crucial role of ARSK1 in maintaining primary root growth under P-sufficient conditions via stabilizing TOR1 complex (Fig. 4B) (Cho *et al.* 2023). Therefore, PDR2-LPR1 and STOP1-ALMT1 modules are activated by low P and collectively regulate RAM activities after gradually reducing *ARSK1* levels.

Systemic signaling

In addition to local sensing and signaling, optimizing adaptive responses to P starvation requires communication between roots and shoots via systemic signals traveling long distances in the vasculature. Several molecules are considered long-distance signals that regulate PSRs systemically (Fig. 4A).

miR399 and miR827 are evolutionarily conserved and induced explicitly by P starvation (Fujii *et al.* 2005; Hsieh *et al.* 2009; Lin *et al.* 2018). Their levels are highly increased in phloem sap during P starvation (Pant *et al.* 2008, 2009), suggesting their potential roles as P starvation signals moving from shoot to root to coordinate the shoot Pi demand and root Pi acquisition and translocation activity. miR399 and its target gene *PHO2* are coexpressed in vascular tissues (Fujii *et al.* 2005; Aung *et al.* 2006; Chiou *et al.* 2006). Suppression of *PHO2* by overexpression of miR399 results in overaccumulation of Pi in shoots due to increased uptake and root-to-shoot translocation of Pi (Chiou *et al.* 2006).

The wild-type root grafted with miR399 overexpressing scion shows a high level of mature miR399, suppressing root *PHO2* transcripts and enhancing shoot Pi accumulation (Lin *et al.* 2008; Pant *et al.* 2008). Recently, the molecular basis underlying the miR399-mediated long-distance silencing was uncovered, by which miR399f/miR399f* duplex moves as a mobile entity and is unloaded in phloem pore pericycle in the root through plasmodesmata in a dose-dependent manner independent of its biogenesis, sequence context, and length (Chiang *et al.* 2023). Arabidopsis miR827 has also been reported to move from shoot to root (Huen *et al.* 2017). It is interesting to note that miR827 shifted its target preference during evolution; it targets *NLA* in Brassicaceae and Cleomaceae but targets PHT5 homologs in other species (Lin *et al.* 2018).

Plants accumulate sugars and starch in leaves upon P starvation (Nielsen *et al.* 1998; Nilsson *et al.* 2007), and the induction of PSi genes increased with increasing concentrations of exogenous sugars (Karthikeyan *et al.* 2007). Increased sucrose loading and shoot-to-root translocation enhanced the expression of PSi genes (Lei *et al.* 2011; Dasgupta *et al.* 2014); in contrast, impairment of sucrose translocation via stem girdling of white lupin suppressed the expression of PSi genes in Pi-starved roots (Liu *et al.* 2005). These findings support the systemic role of sugar in P signaling, but the underlying mechanism requires further investigation.

Phytohormones also play a role in P signaling. Strigolactones (SLs), a group of carotenoid-derived terpenoid lactones, inhibit shoot branching in plants (Gomez-Roldan *et al.* 2008; Umehara *et al.* 2008). It has been demonstrated that the reduced shoot branching under P starvation could be attributed to the increased levels of SL in xylem sap (Kohlen *et al.* 2011), indicating a systemic role for SLs in regulating shoot branching during P starvation to coordinate the shoot development with nutrient supply from roots. N^6-(Δ^2-isopentenyl) adenine (iP) and *trans*-zeatin (tZ) are primary forms of natural isoprenoid cytokinins in Arabidopsis with strong physiological activities (Sakakibara 2006). Early studies proposed cytokinins to be negative regulators of PSRs because the exogenous application of cytokinin suppresses the expression of PSi genes in Arabidopsis (Martín *et al.* 2000; Wang *et al.* 2006). iP and tZ are predominantly present in the phloem and xylem, respectively, suggesting cytokinins are able to function as systemic signals (Hirose *et al.* 2007). Silva-Navas *et al.* (2019) reported that even though *tZ* has a greater substantial repression effect on PSR genes, an increase in the cis-zeatin (cZ):tZ ratio in P-deprived plants to improve root architecture and sustain necessary cytokinin responses was observed. Nevertheless, the detailed systemic role of cytokinins in response to P starvation remains to be determined.

Involvement of PSR factors in biotic interaction

Plants live with a diverse composition of microorganisms, and their relationship can be affected by plant internal nutrient levels or immune responses. It has been shown that plant P status significantly affects the interaction with beneficial and pathogenic microbes (Paries and Gutjahr 2023). Moreover, recent evidence reveals the involvement of PSR genes in regulating biotic relationships, providing profound insights into plant-microbe interaction.

Regulation of arbuscular mycorrhizal (AM) symbiosis via SPX-PHR modules

Due to the scarcity of Pi in soil, plants not only modulate Pi uptake efficiency but also form beneficial mutualistic relationships with microbes to enhance Pi acquisition (Fig. 5). Arbuscular mycorrhizal fungi (AMF) are soil-born microbes that can establish endophytic symbiosis with more than 80% of land plant species and provide mineral nutrients via highly branched structures, termed arbuscules, in root cor-tical cells in exchange for carbon source for survival (Chiu and Paszkowski 2019). Interestingly, P1BS elements are pre-sent in the promoter regions of several AM symbiosis- responsive genes, such as symbiosis-induced *PHT1* genes (Chen *et al.* 2011a; Lota *et al.* 2013). Rice *PHR2* is activated in arbuscule-containing cells (Shi *et al.* 2021), and overexpres-sing OsPHR2 induces symbiosis-responsive genes even with-out fungal infection (Shi *et al.* 2021; Das *et al.* 2022). Mutation at 3 rice PHR proteins severely impairs AMF colonization and arbuscular development. It alters the expression pattern of more than 60% symbiosis-responsive genes, including genes involved in pre-contact signaling, fungal entry, arbuscular development, and nutrient exchange (Shi *et al.* 2021; Das *et al.* 2022), demonstrating that PHR also functions in the center of AM symbiotic regulatory network (Fig. 5C).

As mentioned, modulation of PHR activity via the interaction with SPX proteins is widely validated in many plant species. Similarly, *Medicago SPX1* and *SPX3* are induced by low P, and SPX-PHR modules work in a P-dependent manner. In mycorrhizal roots, the activation of these 2 *SPX* genes is more confined to arbuscule-containing cells. Although the colonization efficiency in *spx1spx3* mutants is reduced, the enrichment of large arbuscules and the repression of arbuscular degeneration-related genes indicate that these 2 genes function redundantly in the control of arbuscular degradation (Wang *et al.* 2021). In addition, MtSPX1 and MtSPX3 control SL biosynthesis genes, the critical phytohormone in modulating shoot branching and AMF hyphae branching. The effects of MtSPX1 and MtSPX3 on the regulation of PSR and symbiosis were shown to be partially through the control of SL levels (Wang *et al.* 2021). In contrast, in rice *spx1/2/3/5* quadruple mutants, AMF coloniza-tion efficiency and arbuscular abundance are increased (Shi *et al.* 2021). In tomatoes, *Slspx1* mutants enhance AM colonization under Pi-replete conditions, whereas overexpression of *SlSPX1* inhibits the formation of AM symbiosis (Liao *et al.* 2022). Further characterization is required to determine whether the distinct effects of SPX protein are due to the difference of family members or the interacting partners in different species.

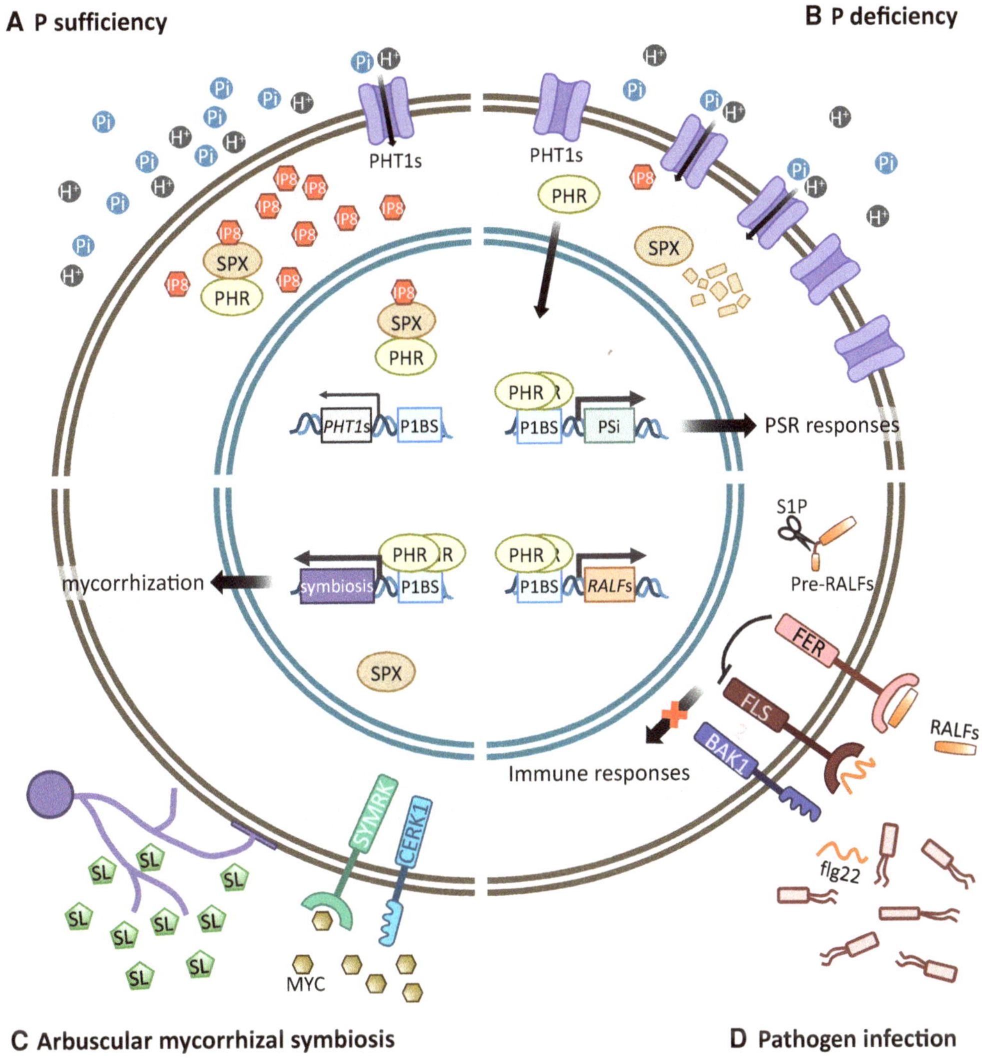

Figure 5. Versatility of SPX-PHR modules in response to P status and plant-microbe interaction. The SPX-PHR module not only regulates adaptive responses to P availability **A)** and **B)**, but it also regulates arbuscular mycorrhizal symbiosis **C)**, and the responses to pathogen infection **D)**, through controlling various sets of genes (see text for details). H in a circle represents a proton. Abbreviations: bbreviations: BAK1, BRI1-associated kinase 1; CERK1, Chitin Elicitor Receptor Kinase 1; flg22, flagellin peptide 22; FLS2, FLAGELLIN SENSING 2; MYC, Myc factor; S1P, site-one protease; SL, strigolactone; SYMRK, Symbiosis Receptor-like Kinase.

Control of rhizobium-mediated nitrogen fixation via P status

Rhizobium-mediated nitrogen (N) fixation is critical for N supply and cycling in agriculture and ecosystems. In nodulating legumes, P supply to rhizobia is essential for maintaining nodule function and mutualistic relationship. Depletion of environmental P impairs nodulation and N fixation (Tang *et al.* 2001; Valverde *et al.* 2002). Soybean (*Glycine max*) PHT1;14, a high-affinity PHT1 family member, is induced by low P treatment, mainly in the junction area between roots and young nodules and in the nodule vascular bundles. It is one of the key players in transporting plant Pi to nodules (Qin *et al.* 2012). In addition to taking up Pi indirectly from host plants, Pi can be acquired directly by nodules from growth environments. Soybean PHT1;1 is upregulated in the plasma membrane of the outer cortex and the fixation zone of nodules in P-depleted roots. Overexpressing *GmPHT1;1* increases the fresh weight of nodules and the nitrogenase activity both under normal and P-deficient conditions, leading to the enhancement of N and P accumulation in plants and the final yield, supporting the significance of Pi transporters in Pi absorption by nodules and N fixation (Chen *et al.* 2019). *GmPHR1* and *GmPHR4* are expressed in the entire nodules and non-N-fixation region, respectively. Overexpressing *GmPHR1* enhances the expression of *GmPHT1;11* in nodules, leading to an increase in nodule Pi content and the size of the nodule, showing the importance of PHR-PHT1 module in nodules (Lu *et al.* 2020).

SPX proteins also regulate Pi homeostasis in nodules but not always via their association with PHR proteins. Soybean *SPX5* and *SPX8* are predominantly expressed in nodules and upregulated by P deficiency. Overexpressing either

GmSPX5 or *GmSPX8* increases the number and the fresh weight of nodules, resulting in higher N and P content in nodules than in the wild type (Zhuang *et al.* 2021; Xing *et al.* 2022). GmNF-YC4, a member of nuclear factor Y transcription factors, physically interacts with GmSPX5, and the association with GmSPX5 enhances the binding capability of GmNF-YC4 to the promoters of downstream target genes, such as asparagine synthetase-related genes. The phenotype of nodulation in overexpressing *GmNF-YC4* is largely overlaid with overexpressing *GmSPX5*, supporting the involvement of GmSPX5-GmNF-YC4 in nodule development and functions (Zhuang *et al.* 2021).

Modulation of plant immunity against pathogens via P status

Excess P supply or Pi overaccumulation reduced the expression of defense-related genes, increasing the susceptibility to fungal pathogen *Magnaporthe oryzae* in rice (Campos-Soriano *et al.* 2020) and indicating the influence of external P in plant responses to pathogens. In contrast, P deficiency also suppresses plant immune response to promote the association with plant growth-promoting bacteria, which might facilitate Pi uptake efficiency or relieve the pressure of P scarcity (Tang *et al.* 2022). Thus, Pi uptake is tightly regulated to optimize the cellular P content for recruiting beneficial microbes. Among Arabidopsis PHT1 members, PHT1;4 is the major player in Pi uptake after recovering from P deficiency. Mutation at PHT1;4 reduces the susceptibility to the pathogen and upregulates pathogen elicitor-induced defense-related genes only under low P conditions (Dindas *et al.* 2022). Coincidently, overexpressing rice PT8 reduces the pathogen resistance (Dong *et al.* 2019b). These studies reveal the direct link between PHT1-dependent Pi uptake and immune responses.

The influences of PSR genes on the structure of root microbe communities also point out the connection between Pi signaling and immune responses. Results of transcriptomic and chromatin immunoprecipitation-sequencing analyses disclose that PHR1 regulates not only PSR genes but also a set of salicylic acid– and jasmonic acid–inducible genes, which are under the transcriptional control of PHR1 via the direct binding on the promoter regions (Castrillo *et al.* 2017). Moreover, the flagellin peptide flg22-responsive defense genes are upregulated in *phr1phl1* mutants. FERONIA (FER), a *Catharanthus roseus* receptor-like kinase 1-like (CrRLK1L) family member, and its peptide ligand, rapid alkalinization factor 23 (RALF23), can inhibit flg22-induced immune signaling (Stegmann *et al.* 2017). Interestingly, *RALF23* and several *RALF* genes are also the direct targets of PHR1 that are upregulated by low P treatment. Studies of overexpressing RALF23 or *fer* mutants support the involvement of the FER-RALF23 module in P starvation-mediated immune inhibition. All these observations together revealed that PHR1 has a negative role in regulating immune responses through the enhancement of the FER-RALK module (Fig. 5D) (Tang *et al.* 2022).

Challenges and perspectives

Facing the limitation of Pi availability in nature, an increase in P fertilizer supply promotes crop yield, but plants acquire less than 30% of it (Vance *et al.* 2003). Tremendous progress has been made in establishing the regulatory network of Pi sensing and signaling over years of study, providing opportunities to accelerate high PUE crop development. Yet, many funda-mental questions and gaps remain if such information is to be usefully translated to agriculture and ecological systems. For example, the underlying mechanisms of apoplastic responses to low external P to adjust the growth in root tips have been established; however, it is still unclear whether other plant cells can also sense extracellular P availability and how. Concerning intracellular sensing, the perception of $InsP_8$ by SPX proteins determines the activation of PHR-mediated PSRs (Wild *et al.* 2016; Ried *et al.* 2021). Multiple enzymes are required to catalyze PP-InsPs synthesis, but understanding of the coordination of the enzyme activities in response to extra- or intracellular P levels remains limited. In addition, the direct link between external and internal P sensing is still missing.

Many studies have focused on increasing Pi acquisition to improve PUE. These include modifying root architecture to enlarge root surface area for Pi scavenging (Jia *et al.* 2018; Sun *et al.* 2018; Wang *et al.* 2019), enhancing the expression of PHTs to increase Pi acquisition and translocation (Song *et al.* 2014; Ouyang *et al.* 2016; Pontigo *et al.* 2023), modulating the secretion of root exudates involved in soil Pi solubilization (Pang *et al.* 2018; Robles-Aguilar *et al.* 2019; Wen *et al.* 2019) and promoting the interaction of plants with beneficial microbes to extend Pi source (Wen *et al.* 2019; Yahya *et al.* 2021).

Natural germplasms are valuable genetic resources for identifying novel alleles to improve the current cultivars' PUEs. Phenotypic and molecular characterization of various germplasm lines in crop species identified many quantitative trait loci associated with low P tolerance. Genome-wide association studies of soybean germplasms pinpointed a variant in the uORF of *GmPHF1* that contributes to P acquisition diversity (Guo *et al.* 2022b). Rice phosphate-starvation tolerance 1 (PSTOL1) is an enhancer of root growth and Pi uptake of aus-type cultivars grown in poor soils (Wissuwa *et al.* 2002; Gamuyao *et al.* 2012). Sequencing *PSTOL1* alleles in wild rice (*Oryza rufipogon*) grown in diverse environments found novel alleles associated with high Pi content, which might be used for breeding high PUE rice cultivars (Neelam *et al.* 2017). Ectopic expression of *OsPSTOL1* in wheat enhanced plant growth and root plasticity and promoted the induction of PHR-mediated P-starvation responses under low P conditions (Kettenburg *et al.* 2023). However, overex-pressing wheat endogenous *TaPSTOL* homolog decreased PUE due to reduced yield but increased P accumulation in grains (Milner *et al.* 2018). Genome-wide association studies of soybean germplasm identified a variant in the uORF of soybean PHF1 as a key determinant for P acquisition efficiency in the field (Guo *et al.* 2022b). Nevertheless, most studies were conducted in greenhouses. Reproducing the traits in field conditions and integrating the knowledge on P sensing/uptake/recycling to improve plant PUE will be the main challenges for further application in breeding programs.

Coordinated acquisition and distribution of P and other nutrients are also required for optimizing plant growth and development. Thus, the interplay between P and other nutri-ents must be considered when modulating PUE via genetic strategies. Perception of nitrate strengthens the interaction of the nitrate sensor, OsNRT1;1B, with OsSPX4, resulting in OsSPX4 degradation, which leads to the activation of OsPHR2-mediated PSRs and NIN-like protein NLP3- mediated nitrate signalings (Hu *et al.* 2019). Meanwhile, Medici *et al.* (2019) identified PHO2 as another integrator of N signals to PSR pathways. These findings give insights into nitrate-regulated PSR responses, but little is known about the effects of ammonium and the role of P signaling in nitrate responses. Additionally, diverse functions of SPX and PHR homologs imply the different SPX-PHR modules present in vivo to exert transcriptional regulation. The under-lying mechanism for controlling SPX-PHR formation and their roles in the interplay between P and other nutrient signaling requires further investigation.

As mentioned, Fe mediates P sensing in the root tips (Fig. 4B). Reciprocally, transcriptomic analysis also revealed the importance of P level in regulating Fe deficiency- or excess-responsive genes (Sanchez-Calderon *et al.* 2005; Thibaud *et al.* 2010). Rice HRZ, a Fe starvation-induced ubiquitin E3 ligase, mediates the protein degradation of OsPHR2. Conversely, OsPHR2 negatively regulates the *HRZ* transcript level to modulate Fe-deficient responses (Guo *et al.* 2022a), pointing out the regulatory loop in P-Fe interaction. Zn-dependent regulation of Pi accumulation has also been observed (Khan *et al.* 2014; Kisko *et al.* 2018), but knowledge about the underlying signaling and the coordination of P and Zn homeostasis is still lacking.

In terms of the role of plant P status in plant-microbe inter-action, it is crucial to integrate PSR and immune responses for recruiting a specified set of microbes for survival. How to identify friends or foes and balance pathogen resistance and symbiotic interaction are fascinating topics about which little is known. Interestingly, PHR can directly activate genes required for AM symbiosis but repress genes involved in de-fense, suggesting that PHR appears to be a hub that coordi-nates immunity and symbiosis. However, the diversity and complexity of soil microbe composition challenge the inte-gration of genetic studies and crop breeding programs. Large-scale studies and field trials are required to elucidate the interaction between P signaling and immune responses.

In nature, plants undergo multiple stresses. The rising tem-perature and CO_2 levels aggravate the pressure and fre-quency of such stresses. However, the coordination between Pi signaling and other stress responses is rarely discussed. Recent research reported that overexpressing a chloroplast-localized Pi importer resulted in the overaccumulation of chloroplast Pi and phytic acids, leading to growth retardation under CO_2-elevated conditions (Bouain *et al.* 2022). This result points out the importance of chloroplast Pi homeostasis in maintaining plant growth when plants face elevated CO_2 levels. Despite being a challenging topic, more studies must be conducted to investigate the interplay between P and other stresses and underlying mechanisms, which need to be scrutinized in the field when applied to crop improvement.

Acknowledgments

We thank the anonymous reviewers for their constructive comments. We apologize to our colleagues whose work could not be cited due to space constraints.

Author contributions

S.-Y.Y., W.-Y.L., Y.-M.H., and T.-J.C. conceptualized the review outline. S.-Y.Y., W.-Y.L. and Y.-M.H. prepared the figures and wrote the original draft, in which S.-Y.Y. contributed to the introduction and P transport, Y.-M.H. to intracellular P sensing and signaling and W.-Y.L. to local sensing, biotic interaction, and perspectives. T.-J.C. compiled, reviewed, and edited the manuscript. All authors contributed to the critical revision and its final approval.

Funding

We thank Academia Sinica (AS-GCS-112-L03) for funding to T.-J.C laboratory and the National Science and Technology Council, Taiwan for funding to T.-J.C. (MOST 110-2311-B-001 -021 -MY3, NSTC 112-2311-B-001 -040 -MY3), S.-Y.Y. (MOST 111-2311-B-002 -005 -MY3) and W.-Y.L. (MOST 111-2313-B-002 -009 -MY3) laboratories. Y.-M.H. was supported by the Taiwan International Graduate Program (TIGP) scholarship.

Conflict of interest statement. None declared.

References

Ai P, Sun S, Zhao J, Fan X, Xin W, Guo Q, Yu L, Shen Q, Wu P, Mille A, et al. Two rice phosphate transporters, OsPht1; 2 and OsPht1; 6 have different functions and kinetic properties in uptake and translocation. Plant J. 2009:**57**(5):798–809. https://doi.org/10.1111/j.1365-313X.2008.03726.x

Arpat AB, Magliano P, Wege S, Rouached H, Stefanovic A, Poirier Y. Functional expression of PHO1 to the Golgi and trans-Golgi network and its role in export of inorganic phosphate. Plant J. 2012:**71**(3): 479–491. https://doi.org/10.1111/j.1365-313X.2012.05004.x

Aung K, Lin SI, Wu CC, Huang YT, Su CL, Chiou TJ. Pho2, a phosphate overaccumulator, is caused by a nonsense mutation in a microRNA399 target gene. Plant Physiol. 2006:**141**(3):1000–1011. https://doi.org/10.1104/pp.106.078063

Ayadi A, David P, Arrighi JF, Chiarenza S, Thibaud MC, Nussaume L, Marin E. Reducing the genetic redundancy of Arabidopsis PHOSPHATE TRANSPORTER1 transporters to study phosphate uptake and signaling. Plant Physiol. 2015:**167**(4):1511–1526. https://doi.org/10.1104/pp.114.252338

Baker A, Ceasar SA, Palmer AJ, Paterson JB, Qi W, Muench SP, Baldwin SA. Replace, reuse, recycle: improving the sustainable use of phosphorus by plants. J Exp Bot. 2015:**66**(12):3523–3540. https://doi.org/10.1093/jxb/erv210

Balzergue C, Dartevelle T, Godon C, Laugier E, Meisrimler C, Teulon JM, Creff A, Bissler M, Brouchoud C, Hagege A, et al. Low phosphate activates STOP1-ALMT1 to rapidly inhibit root cell elongation. Nat Commun. 2017:**8**(1):15300. https://doi.org/10.1038/ncomms15300

Bari R, Datt Pant B, Stitt M, Scheible W-R. PHO2, microRNA399, and PHR1 define a phosphate-signaling pathway in plants. Plant Physiol. 2006:**141**(3):988–999. https://doi.org/10.1104/pp.106.079707

Bariola PA, Howard CJ, Taylor CB, Verburg MT, Jaglan VD, Green PJ. The Arabidopsis ribonuclease gene RNS1 is tightly controlled in response to phosphate limitation. Plant J. 1994:**6**(5):673–685. https://doi.org/10.1046/j.1365-313x.1994.6050673.x

Bates TR, Lynch JP. Stimulation of root hair elongation in Arabidopsis thaliana by low phosphorus availability. Plant, Cell Environ. 1996:**19**(5): 529–538. https://doi.org/10.1111/j.1365-3040.1996.tb00386.x

Bates TR, Lynch JP. Root hairs confer a competitive advantage under low phosphorus availability. Plant Soil. 2001:**236**(2):243–250. https://doi.org/10.1023/A:1012791706800

Bayle V, Arrighi JF, Creff A, Nespoulous C, Vialaret J, Rossignol M, Gonzalez E, Paz-Ares J, Nussaume L. Arabidopsis thaliana high-affinity phosphate transporters exhibit multiple levels of posttranslational regulation. Plant Cell. 2011:**23**(4):1523–1535. https://doi.org/10.1105/tpc.110.081067

Bouain N, Cho H, Sandhu J, Tuiwong P, Prom-u-thai C, Zheng LQ, Shahzad Z, Rouached H. Plant growth stimulation by high CO_2 depends on phosphorus homeostasis in chloroplasts. Curr Biol. 2022:**32**(20):4493–4500. https://doi.org/10.1016/j.cub.2022.08.032

Burleigh SH, Harrison MJ. The down-regulation of Mt4-like genes by phosphate fertilization occurs systemically and involves phosphate translocation to the shoots. Plant Physiol. 1999:**119**(1):241–248. https://doi.org/10.1104/pp.119.1.241

Bustos R, Castrillo G, Linhares F, Puga MI, Rubio V, Pérez-Pérez J, Solano R, Leyva A, Paz-Ares J. A central regulatory system largely controls transcriptional activation and repression responses to phosphate starvation in Arabidopsis. PLoS Genet. 2010:**6**(9):e1001102. https://doi.org/10.1371/journal.pgen.1001102

Campos-Soriano L, Bundo M, Bach-Pages M, Chiang SF, Chiou TJ, San Segundo B. Phosphate excess increases susceptibility to pathogen infection in rice. Mol Plant Pathol. 2020:**21**(4):555–570. https://doi.org/10.1111/mpp.12916

Castrillo G, Teixeira PJPL, Paredes SH, Law TF, de Lorenzo L, Feltcher ME, Finkel OM, Breakfield NW, Mieczkowski P, Jones CD, et al. Root microbiota drive direct integration of phosphate stress and immunity. Nature. 2017:**543**(7646):513–518. https://doi.org/10.1038/nature21417

Chabert V, Kim G, Qiu D, Liu G, Mayer L, Jamsheer K, Jessen H, Mayer A. Inositol pyrophosphate dynamics reveals control of the yeast phosphate starvation program through 1,5-IP8 and the SPX domain of Pho81. Elife. 2023:**12**:RP87956. https://doi.org/10.7554/eLife.87956.2

Che J, Yamaji N, Miyaji T, Mitani-Ueno N, Kato Y, Shen R, Ma J. Node-localized transporters of phosphorus essential for seed development in rice. Plant Cell Physiol. 2020:**61**(8):1387–1398. https://doi.org/10.1093/pcp/pcaa074

Chen AQ, Gu MA, Sun SB, Zhu LL, Hong SA, Xu GH. Identification of two conserved cis-acting elements, MYCS and P1BS, involved in the regulation of mycorrhiza-activated phosphate transporters in eudicot species. New Phytol. 2011a:**189**(4):1157–1169. https://doi.org/10.1111/j.1469-8137.2010.03556.x

Chen J, Liu Y, Ni J, Wang Y, Bai Y, Shi J, Gan J, Wu Z, Wu P. OsPHF1 regulates the plasma membrane localization of low- and high-affinity inorganic phosphate transporters and determines inorganic phosphate uptake and translocation in rice. Plant Physiol. 2011b:**157**(1): 269–278. https://doi.org/10.1104/pp.111.181669

Chen L, Qin L, Zhou L, Li X, Chen Z, Sun L, Wang W, Lin Z, Zhao J, Yamaji N, et al. A nodule-localized phosphate transporter GmPT7 plays an important role in enhancing symbiotic N_2 fixation and yield in soybean. New Phytol. 2019:**221**(4):2013–2025. https://doi.org/10.1111/nph.15541

Chen J, Wang Y, Wang F, Yang J, Gao M, Li C, Liu Y, Liu Y, Yamaji N, Ma JF, et al. The rice CK2 kinase regulates trafficking of phosphate transporters in response to phosphate levels. Plant Cell. 2015:**27**(3): 711–723. https://doi.org/10.1105/tpc.114.135335

Chen YF, Li LQ, Xu Q, Kong YH, Wang H, Wu WH. The WRKY6 transcription factor modulates PHOSPHATE1 expression in response to low Pi stress in Arabidopsis. Plant Cell. 2009:**21**(11):3554–3566. https://doi.org/10.1105/tpc.108.064980

Chen ZH, Nimmo GA, Jenkins GI, Nimmo HG. BHLH32 modulates several biochemical and morphological processes that respond to Pi starvation in Arabidopsis. Biochem J. 2007:**405**(1):191–198. https://doi.org/10.1042/bj20070102

Chiang CP, Li JL, Chiou TJ. Dose-dependent long-distance movement of microRNA399 duplex regulates phosphate homeostasis in Arabidopsis. New Phytol. 2023:**240**(2):802–814. https://doi.org/10.1111/nph.19182

Chien PS, Chiang CP, Leong SJ, Chiou TJ. Sensing and signaling of phosphate starvation: from local to long distance. Plant Cell Physiol. 2018:**59**(9):1714–1722. https://doi.org/10.1093/pcp/pcy148

Chiou TJ, Aung K, Lin SI, Wu CC, Chiang SF, Su CL. Regulation of phosphate homeostasis by microRNA in Arabidopsis. Plant Cell. 2006:**18**(2):412–421. https://doi.org/10.1105/tpc.105.038943

Chiu CH, Paszkowski U. Mechanisms and impact of symbiotic phosphate acquisition. Cold Spring Harb Perspect Biol. 2019:**11**(6): a034603. https://doi.org/10.1101/cshperspect.a038281

Cho H, Banf M, Shahzad Z, Van Leene J, Bossi F, Ruffel S, Bouain N, Cao P, Krouk G, De Jaeger G, et al. ARSK1 activates TORC1 signaling to adjust growth to phosphate availability in *Arabidopsis*. Curr Biol. 2023:**33**(9):1778–1786. https://doi.org/10.1016/j.cub.2023.03.005

Dai C, Dai X, Qu H, Men Q, Liu J, Yu L, Gu M, Xu G. The rice phosphate transporter OsPHT1; 7 plays a dual role in phosphorus redistribution and anther development. Plant Physiol. 2022:**188**(4): 2272–2288. https://doi.org/10.1093/plphys/kiac030

Daram P, Brunner S, Rausch C, Steiner C, Amrhein N, Bucher M. Pht2; 1 encodes a low-affinity phosphate transporter from Arabidopsis. Plant Cell. 1999:**11**(11):2153–2166. https://doi.org/10.1105/tpc.11.11.2153

Das D, Paries M, Hobecker K, Gigl M, Dawid C, Lam H, Zhang J, Chen M, Gutjahr C. PHOSPHATE STARVATION RESPONSE transcription factors enable arbuscular mycorrhiza symbiosis. Nat Commun. 2022:**13**(1):447. https://doi.org/10.1038/s41467-022-27976-8

Dasgupta K, Khadilkar AS, Sulpice R, Pant B, Scheible W-R, Fisahn J, Stitt M, Ayre BG. Expression of sucrose transporter cDNAs specifically in companion cells enhances phloem loading and long-distance transport of sucrose but leads to an inhibition of growth and the perception of a phosphate limitation. Plant Physiol. 2014:**165**(2): 715–731. https://doi.org/10.1104/pp.114.238410

Delhaize ER, Randall PJ. Characterization of a phosphate-accumulator mutant of Arabidopsis thaliana. Plant Physiol. 1995:**107**(1):207–213. https://doi.org/10.1104/pp.107.1.207

Desai M, Rangarajan P, Donahue JL, Williams SP, Land ES, Mandal MK, Phillippy BQ, Perera IY, Raboy V, Gillaspy GE. Two inositol hexakisphosphate kinases drive inositol pyrophosphate synthesis in plants. Plant J. 2014:**80**(4):642–653. https://doi.org/10.1111/tpj.12669

Desfougères Y, Gerasimaitė RU, Jessen HJ, Mayer A. Vtc5, a novel subunit of the vacuolar transporter chaperone complex, regulates polyphosphate synthesis and phosphate homeostasis in yeast. J Biol Chem. 2016:**291**(42):22262–22275. https://doi.org/10.1074/jbc.M116.746784

Devaiah BN, Karthikeyan AS, Raghothama KG. WRKY75 transcription factor is a modulator of phosphate acquisition and root development in Arabidopsis. Plant Physiol. 2007a:**143**(4):1789–1801. https://doi.org/10.1104/pp.106.093971

Devaiah BN, Madhuvanthi R, Karthikeyan AS, Raghothama KG. Phosphate starvation responses and gibberellic acid biosynthesis are regulated by the MYB62 transcription factor in Arabidopsis. Mol Plant. 2009:**2**(1):43–58. https://doi.org/10.1093/mp/ssn081

Devaiah BN, Nagarajan VK, Raghothama KG. Phosphate homeostasis and root development in Arabidopsis are synchronized by the zinc finger transcription factor ZAT6. Plant Physiol. 2007b:**145**(1):147–159. https://doi.org/10.1104/pp.107.101691

Dindas J, DeFalco TA, Yu G, Zhang L, David P, Bjornson M, Thibaud MC, Custodio V, Castrillo G, Nussaume L, et al. Direct inhibition of phosphate transport by immune signaling in Arabidopsis. Curr Biol. 2022:**32**(2):488–495e485. https://doi.org/10.1016/j.cub.2021.11.063

Ding G, Lei GJ, Yamaji N, Yokosho K, Mitani-Ueno N, Huang S, Ma JF. Vascular cambium-localized AtSPDT mediates xylem-to-phloem transfer of phosphorus for its preferential distribution in Arabidopsis. Mol Plant. 2020:**13**(1):99–111. https://doi.org/10.1016/j.molp.2019.10.002

Dong J, Ma G, Sui L, Wei M, Satheesh V, Zhang R, Ge S, Li J, Zhang T-E, Wittwer C, et al. Inositol pyrophosphate InsP8 acts as an intracellular phosphate signal in Arabidopsis. Mol Plant. 2019a:**12**(11):1463–1473. https://doi.org/10.1016/j.molp.2019.08.002

Dong Z, Li W, Liu J, Li L, Pan S, Liu S, Gao J, Liu L, Liu X, Wang GL, et al. The rice phosphate transporter protein OsPT8 regulates disease resistance and plant growth. Sci Rep. 2019b:**9**(1):5408. https://doi.org/10.1038/s41598-019-41718-9

Duan K, Yi K, Dang L, Huang H, Wu W, Wu P. Characterization of a sub-family of Arabidopsis genes with the SPX domain reveals their diverse functions in plant tolerance to phosphorus starvation. Plant J. 2008:**54**(6):965–975. https://doi.org/10.1111/j.1365-313X.2008.03460.x

Fabiańska I, Bucher M, Häusler R. Intracellular phosphate homeostasis—a short way from metabolism to signaling. Plant Sci. 2019:**286**:57–67. https://doi.org/10.1016/j.plantsci.2019.05.018

Franco-Zorrilla JM, Valli A, Todesco M, Mateos I, Puga MI, Rubio-Somoza I, Leyva A, Weigel D, Garcı´a JA, Paz-Ares J. Target mimicry provides a new mechanism for regulation of microRNA activity. Nat Genet. 2007:**39**(8):1033–1037. https://doi.org/10.1038/ng2079

Fujii H, Chiou TJ, Lin SI, Aung K, Zhu JK. A miRNA involved in phosphate-starvation response in Arabidopsis. Curr Biol. 2005:**15**(22):2038–2043. https://doi.org/10.1016/j.cub.2005.10.016

Gamuyao R, Chin JH, Pariasca-Tanaka J, Pesaresi P, Catausan S, Dalid C, Slamet-Loedin I, Tecson-Mendoza EM, Wissuwa M, Heuer S. The protein kinase Pstol1 from traditional rice confers tolerance of phosphorus deficiency. Nature. 2012:**488**(7412):535. https://doi.org/10.1038/nature11346

Gaude N, Nakamura Y, Scheible N, Ohta H, Dörmann P. Phospholipase C5 (NPC5) is involved in galactolipid accumulation during phosphate limitation in leaves of Arabidopsis. Plant J. 2008:**56**(1):28–39. https://doi.org/10.1111/j.1365-313X.2008.03582.x

Gerasimaite R, Pavlovic I, Capolicchio S, Hofer A, Schmidt A, Jessen HJ, Mayer A. Inositol pyrophosphate specificity of the SPX-dependent polyphosphate polymerase VTC. ACS Chem Biol. 2017:**12**(3):648–653. https://doi.org/10.1021/acschembio.7b00026

Gomez-Roldan V, Fermas S, Brewer PB, Puech-Pagès V, Dun EA, Pillot J-P, Letisse F, Matusova R, Danoun S, Portais J-C, et al. Strigolactone inhibition of shoot branching. Nature. 2008:**455**(7210):189–194. https://doi.org/10.1038/nature07271

Gonzalez E, Solano R, Rubio V, Leyva A, Paz-Ares J. PHOSPHATE TRANSPORTER TRAFFIC FACILITATOR1 is a plant-specific SEC12-related protein that enables the endoplasmic reticulum exit of a high-affinity phosphate transporter in Arabidopsis. Plant Cell. 2005:**17**(12):3500–3512. https://doi.org/10.1105/tpc.105.036640

Gu M, Huang H, Hisano H, Ding G, Huang S, Mitani-Ueno N, Yokosho K, Sato K, Yamaji N, Ma J. A crucial role for a node-localized transporter, HvSPDT, in loading phosphorus into barley grains. New Phytol. 2022:**234**(4):1249–1261. https://doi.org/10.1111/nph.18057

Guan Z, Zhang Q, Zhang Z, Zuo J, Chen J, Liu R, Savarin J, Broger L, Cheng P, Wang Q, et al. Mechanistic insights into the regulation of plant phosphate homeostasis by the rice SPX2—PHR2 complex. Nat Commun. 2022:**13**(1):1581. https://doi.org/10.1038/s41467-022-29275-8

Guo MN, Ruan WY, Zhang YB, Zhang YX, Wang XQ, Guo ZH, Wang L, Zhou T, Paz-Ares J, Yi KK. A reciprocal inhibitory module for Pi and iron signaling. Mol Plant. 2022a:**15**(1):138–150. https://doi.org/10.1016/j.molp.2021.09.011

Guo ZL, Cao HR, Zhao J, Bai S, Peng WT, Li J, Sun LL, Chen LY, Lin ZH, Shi C, et al. A natural uORF variant confers phosphorus acquisition diversity in soybean. Nat Commun. 2022b:**13**(1):3796. https://doi.org/10.1038/s41467-022-31555-2

Gutierrez-Alanis D, Yong-Villalobos L, Jimenez-Sandoval P, Alatorre-Cobos F, Oropeza-Aburto A, Mora-Macias J, Sanchez-Rodriguez F, Cruz-Ramirez A, Herrera-Estrella L. Phosphate starvation-dependent iron mobilization induces CLE14 expression to trigger root meristem differentiation through CLV2/PEPR2 signaling. Dev Cell. 2017:**41**(5):555–570.e3. https://doi.org/10.1016/j.devcel.2017.05.009

Hamburger D, Rezzonico E, MacDonald-Comber Petétot J, Somerville C, Poirier Y. Identification and characterization of the Arabidopsis PHO1 gene involved in phosphate loading to the xylem. Plant Cell. 2002:**14**(4):889–902. https://doi.org/10.1105/tpc.000745

Härtel H, Dörmann P, Benning C. Galactolipids not associated with the photosynthetic apparatus in phosphate-deprived plants. J Photochem Photobiol B: Biol. 2001:**61**(1-2):46–51. https://doi.org/10.1016/S1011-1344(01)00144-0

Heuer S, Gaxiola R, Schilling R, Herrera-Estrella L, López-Arredondo D, Wissuwa M, Delhaize E, Rouached H. Improving phosphorus use efficiency: a complex trait with emerging opportunities. Plant J. 2017:**90**(5):868–885. https://doi.org/10.1111/tpj.13423

Hirose N, Takei K, Kuroha T, Kamada-Nobusada T, Hayashi H, Sakakibara H. Regulation of cytokinin biosynthesis, compartmentalization and translocation. J Exp Bot. 2007:**59**(1):75–83. https://doi.org/10.1093/jxb/erm157

Holford ICR. Soil phosphorus: its measurement, and its uptake by plants. Soil Res. 1997:**35**(2):227–240. https://doi.org/10.1071/S96047

Hsieh L-C, Lin S-I, Shih AC-C, Chen J-W, Lin W-Y, Tseng C-Y, Li W-H, Chiou T-J. Uncovering small RNA-mediated responses to phosphate deficiency in Arabidopsis by deep sequencing. Plant Physiol. 2009:**151**(4):2120–2132. https://doi.org/10.1104/pp.109.147280

Hu B, Jiang Z, Wang W, Qiu Y, Zhang Z, Liu Y, Li A, Gao X, Liu L, Qian Y, et al. Nitrate-NRT1.1B-SPX4 cascade integrates nitrogen and phosphorus signalling networks in plants. Nat Plants. 2019:**5**(4):401–413. https://doi.org/10.1038/s41477-019-0384-1

Huang TK, Han CL, Lin SI, Chen YJ, Tsai YC, Chen YR, Chen JW, Lin WY, Chen PM, Liu TY, et al. Identification of downstream components of ubiquitin-conjugating enzyme PHOSPHATE2 by quantitative membrane proteomic in Arabidopsis roots. Plant Cell. 2013:**25**(10):4044–4060. https://doi.org/10.1105/tpc.113.115998

Huen AK, Rodriguez-Medina C, Ho AYY, Atkins CA, Smith PMC. Long-distance movement of phosphate starvation-responsive microRNAs in Arabidopsis. Plant Biology. 2017:**19**(4):643–649. https://doi.org/10.1111/plb.12568

Hürlimann HC, Pinson B, Stadler-Waibel M, Zeeman SC, Freimoser FM. The SPX domain of the yeast low-affinity phosphate transporter Pho90 regulates transport activity. EMBO Rep. 2009:**10**(9):1003–1008. https://doi.org/10.1038/embor.2009.105

Isidra-Arellano M, Delaux P, Valdés-López O. The phosphate starvation response system: its role in the regulation of plant–microbe interactions. Plant Cell Physiol. 2021:**62**(3):392–400. https://doi.org/10.1093/pcp/pcab016

Jabnoune M, Secco D, Lecampion C, Robaglia C, Shu Q, Poirier Y. A rice cis-natural antisense RNA acts as a translational enhancer for its cognate mRNA and contributes to phosphate homeostasis and plant fitness. Plant Cell. 2013:**25**(10):4166–4182. https://doi.org/10.1105/tpc.113.116251

Jia H, Ren H, Gu M, Zhao J, Sun S, Zhang X, Chen J, Wu P, Xu G. The phosphate transporter gene OsPht1; 8 is involved in phosphate homeostasis in rice. Plant Physiol. 2011:**156**(3):1164–1175. https://doi.org/10.1104/pp.111.175240

Jia XC, Liu P, Lynch JP. Greater lateral root branching density in maize improves phosphorus acquisition from low phosphorus soil. J Exp Bot. 2018:**69**(20):4961–4970. https://doi.org/10.1093/jxb/ery252

Jiang M, Caldararu S, Zaehle S, Ellsworth DS, Medlyn BE. Towards a more physiological representation of vegetation phosphorus processes in land surface models. New Phytol. 2019a:**222**(3):1223–1229. https://doi.org/10.1111/nph.15688

Jiang M, Sun L, Isupov MN, Littlechild JA, Wu X, Wang Q, Wang Q, Yang W, Wu Y. Structural basis for the target DNA recognition and binding by the MYB domain of phosphate starvation response 1. FEBS J. 2019b:**286**(14):2809–2821. https://doi.org/10.1111/febs.14846

Karlsson PM, Herdean A, Adolfsson L, Beebo A, Nziengui H, Irigoyen S, Ünnep R, Zsiros O, Nagy G, Garab G, et al. The Arabidopsis thylakoid transporter PHT4; 1 influences phosphate availability for ATP synthesis and plant growth. Plant J. 2015:**84**(1):99–110. https://doi.org/10.1111/tpj.12962

Karthikeyan A, Varadarajan D, Mukatira U, Paino D'Urzo M, Damsz B, Raghothama K. Regulated expression of Arabidopsis phosphate transporters. Plant Physiol. 2002:**130**(1):221–233. https://doi.org/10.1104/pp.020007

Karthikeyan AS, Varadarajan DK, Jain A, Held MA, Carpita NC, Raghothama KG. Phosphate starvation responses are mediated by sugar signaling in Arabidopsis. Planta. 2007:**225**(4):907–918. https://doi.org/10.1007/s00425-006-0408-8

Kelly AA, Dörmann P. DGD2, an arabidopsis gene encoding a UDP-galactose-dependent digalactosyldiacylglycerol synthase is expressed during growth under phosphate-limiting conditions. J Biol Chem. 2002:**266**(2):1166–1173. https://doi.org/10.1074/jbc.M110066200

Kettenburg AT, Lopez MA, Yogendra K, Prior MJ, Rose T, Bimson S, Heuer S, Roy SJ, Bailey-Serres J. PHOSPHORUS-STARVATION TOLERANCE 1 (OsPSTOL1) is prevalent in upland rice and enhances root growth and hastens low phosphate signaling in wheat. Plant Cell Environ. 2023:**46**(7):2187–2205. https://doi.org/10.1111/pce.14588

Khan GA, Bouraine S, Wege S, Li YY, de Carbonnel M, Berthomieu P, Poirier Y, Rouached H. Coordination between zinc and phosphate homeostasis involves the transcription factor PHR1, the phosphate exporter PHO1, and its homologue PHO1; H3 in *Arabidopsis*. J Exp Bot. 2014:**65**(3):871–884. https://doi.org/10.1093/jxb/ert444

Kisko M, Bouain N, Safi A, Medici A, Akkers RC, Secco D, Fouret G, Krouk G, Aarts MGM, Busch W, et al. *LPCAT1* controls phosphate homeostasis in a zinc-dependent manner. Elife. 2018:**7**:e32077. https://doi.org/10.7554/eLife.32077

Kohlen W, Charnikhova T, Liu Q, Bours R, Domagalska MA, Beguerie S, Verstappen F, Leyser O, Bouwmeester H, Ruyter-Spira C. Strigolactones are transported through the xylem and play a key role in shoot architectural response to phosphate deficiency in nonarbuscular mycorrhizal host Arabidopsis. Plant Physiol. 2011:**155**(2):974–987. https://doi.org/10.1104/pp.110.164640

Kuo H, Hsu Y, Lin W, Chen K, Munnik T, Brearley C, Chiou T. Arabidopsis inositol phosphate kinases IPK1 and ITPK1 constitute a metabolic pathway in maintaining phosphate homeostasis. Plant J. 2018:**95**(4):613–630. https://doi.org/10.1111/tpj.13974

Laha D, Johnen P, Azevedo C, Dynowski M, Weiß M, Capolicchio S, Mao H, Iven T, Steenbergen M, Freyer M, et al. VIH2 regulates the synthesis of inositol pyrophosphate InsP8 and jasmonate-dependent defenses in Arabidopsis. Plant Cell. 2015:**27**(4):1082–1097. https://doi.org/10.1105/tpc.114.135160

Laha D, Parvin N, Hofer A, Giehl RFH, Fernandez-Rebollo N, von Wirén N, Saiardi A, Jessen HJ, Schaaf G. Arabidopsis ITPK1 and ITPK2 have an evolutionarily conserved phytic acid kinase activity. ACS Chem Biol. 2019:**14**(10):2127–2133. https://doi.org/10.1021/acschembio.9b00423

Lapis-Gaza H, Jost R, Finnegan P. Arabidopsis PHOSPHATE TRANSPORTER1 genes PHT1; 8 and PHT1; 9 are involved in root-to-shoot translocation of orthophosphate. BMC Plant Biol. 2014:**14**(1):334. https://doi.org/10.1186/s12870-014-0334-z

Lei M, Liu Y, Zhang B, Zhao Y, Wang X, Zhou Y, Raghothama KG, Liu D. Genetic and genomic evidence that sucrose is a global regulator of plant responses to phosphate starvation in Arabidopsis. Plant Physiol. 2011:**156**(3):1116–1130. https://doi.org/10.1104/pp.110.171736

Liao D, Sun C, Liang H, Wang Y, Bian X, Dong C, Niu X, Yang M, Xu G, Chen A, et al. SlSPX1-SlPHR complexes mediate the suppression of arbuscular mycorrhizal symbiosis by phosphate repletion in tomato. Plant Cell. 2022:**34**(10):4045–4065. https://doi.org/10.1093/plcell/koac212

Lin SI, Chiang SF, Lin WY, Chen JW, Tseng CY, Wu PC, Chiou TJ. Regulatory network of MicroRNA399 and PHO2 by systemic signaling. Plant Physiol. 2008:**147**(2):732–746. https://doi.org/10.1104/pp.108.116269

Lin W, Lin Y, Chiang S, Syu C, Hsieh L, Chiou T. Evolution of microRNA827 targeting in the plant kingdom. New Phytol. 2018:**217**(4):1712–1725. https://doi.org/10.1111/nph.14938

Lin WY, Huang TK, Chiou TJ. NITROGEN LIMITATION ADAPTATION, a target of microRNA827, mediates degradation of plasma membrane–localized phosphate transporters to maintain phosphate homeostasis in Arabidopsis. Plant Cell. 2013:**25**(10):4061–4074. https://doi.org/10.1105/tpc.113.116012

Lin WY, Huang TK, Leong SJ, Chiou TJ. Long-distance call from phosphate: systemic regulation of phosphate starvation responses. J Exp Bot. 2014:**65**(7):1817–1827. https://doi.org/10.1093/jxb/ert431

Liu C, Muchhal U, Raghothama K. Differential expression of TPS11, a phosphate starvation induced gene in tomato. Plant Mol Biol. 1997:**33**(5):867–874. https://doi.org/10.1023/a:1005729309569

Liu J, Samac DA, Bucciarelli B, Allan DL, Vance CP. Signaling of phosphorus deficiency-induced gene expression in white lupin requires sugar and phloem transport. Plant J. 2005:**41**(2):257–268. https://doi.org/10.1111/j.1365-313X.2004.02289.x

Liu J, Yang L, Luan M, Wang Y, Zhang C, Zhang B, Shi J, Zhao FG, Lan W, Luan S. A vacuolar phosphate transporter essential for phosphate homeostasis in Arabidopsis. Proc Natl Acad Sci U S A. 2015:**112**(47):E6571–E6578. https://doi.org/10.1073/pnas.1514598112

Liu JP, Magalhaes JV, Shaff J, Kochian LV. Aluminum-activated citrate and malate transporters from the MATE and ALMT families function independently to confer Arabidopsis aluminum tolerance. Plant J. 2009:**57**(3):389–399. https://doi.org/10.1111/j.1365-313X.2008.03696.x

Liu TY, Huang TK, Tseng CY, Lai YS, Lin SI, Lin WY, Chen JW, Chiou TJ. PHO2-dependent degradation of PHO1 modulates phosphate homeostasis in Arabidopsis. Plant Cell. 2012:**24**(5):2168–2183. https://doi.org/10.1105/tpc.112.096636

Liu TY, Huang TK, Yang SY, Hong YT, Huang SM, Wang FN, Chiang SF, Tsai SY, Lu WC, Chiou TJ. Identification of plant vacuolar transporters mediating phosphate storage. Nat Commun. 2016:**7**(1):11095. https://doi.org/10.1038/ncomms11095

Liu XL, Wang L, Wang XW, Yan Y, Yang XL, Xie MY, Hu Z, Shen X, Ai H, Lin H, et al. Mutation of the chloroplast-localized phosphate transporter OsPHT2; 1 reduces flavonoid accumulation and UV tolerance in rice. Plant J. 2020:**102**(1):53–67. https://doi.org/10.1111/tpj.14611

Lopez-Arredondo DL, Leyva-Gonzalez MA, Gonzalez-Morales SI, Lopez-Bucio J, Herrera-Estrella L. Phosphate nutrition: improving low-phosphate tolerance in crops. Annu Rev Plant Biol. 2014:**65**(1):95–123. https://doi.org/10.1146/annurev-arplant-050213-035949

López-Bucio J, Cruz-Ramı´rez A, Herrera-Estrella L. The role of nutrient availability in regulating root architecture. Curr Opin Plant Biol. 2003:**6**(3):280–287. https://doi.org/10.1016/s1369-5266(03)00035-9

Lota F, Wegmuller S, Buer B, Sato S, Brautigam A, Hanf B, Bucher M. The cis-acting CTTC-P1BS module is indicative for gene function of LjVTI12, a Qb-SNARE protein gene that is required for arbuscule formation in *Lotus japonicus*. Plant J. 2013:**74**(2):280–293. https://doi.org/10.1111/tpj.12120

Lu M, Cheng Z, Zhang XM, Huang P, Fan C, Yu G, Chen F, Xu K, Chen Q, Miao Y, et al. Spatial divergence of PHR-PHT1 modules maintains phosphorus homeostasis in soybean nodules. Plant Physiol. 2020:**184**(1):236–250. https://doi.org/10.1104/pp.19.01209

Luan M, Zhao F, Sun G, Xu M, Fu A, Lan W, Luan S. A SPX domain vacuolar transporter links phosphate sensing to homeostasis in Arabidopsis. Mol Plant. 2022:**15**(10):1590–1601. https://doi.org/10.1016/j.molp.2022.09.005

Lv Q, Zhong Y, Wang Y, Wang Z, Zhang L, Shi J, Wu Z, Liu Y, Mao C, Yi K, et al. SPX4 negatively regulates phosphate signaling and homeostasis through its interaction with PHR2 in rice. Plant Cell. 2014:**26**(4):1586–1597. https://doi.org/10.1105/tpc.114.123208

Lynch JP. Root phenes for enhanced soil exploration and phosphorus acquisition: tools for future crops. Plant Physiol. 2011:**156**(3):1041–1049. https://doi.org/10.1104/pp.111.175414

Ma B, Zhang L, Gao Q, Wang J, Li X, Wang H, Liu Y, Lin H, Liu J, Wang X, et al. A plasma membrane transporter coordinates phosphate reallocation and grain filling in cereals. Nat Genet. 2021:**53**(6):906–915. https://doi.org/10.1038/s41588-021-00855-6

Martín AC, Del Pozo JC, Iglesias J, Rubio V, Solano R, De La Peña A, Leyva A, Paz-Ares J. Influence of cytokinins on the expression of phosphate starvation responsive genes in Arabidopsis. Plant J. 2000:**24**(5):559–567. https://doi.org/10.1046/j.1365-313x.2000.00893.x

Medici A, Szponarski W, Dangeville P, Safi A, Dissanayake IM, Saenchai C, Emanuel A, Rubio V, Lacombe B, Ruffel S, et al. Identification of molecular integrators shows that nitrogen actively controls the phosphate starvation response in plants. Plant Cell. 2019:**31**(5):1171–1184. https://doi.org/10.1105/tpc.18.00656

Milner MJ, Howells RM, Craze M, Bowden S, Graham N, Wallington EJ. A PSTOL-like gene, controls a number of agronomically important traits in wheat. BMC Plant Biol. 2018:**18**(1):115. https://doi.org/10.1186/s12870-018-1331-4

Miyaji T, Kuromori T, Takeuchi Y, Yamaji N, Yokosho K, Shimazawa A, Sugimoto E, Omote H, Ma J, Shinozaki K, et al. AtPHT4; 4 is a chloroplast-localized ascorbate transporter in Arabidopsis. Nat Commun. 2015:**6**(1):11. https://doi.org/10.1038/ncomms6928

Młodzińska E, Zboińska M. Phosphate uptake and allocation—a closer look at *Arabidopsis thaliana* L. and *Oryza sativa* L. Front Plant Sci. 2016:**7**:1198. https://doi.org/10.3389/fpls.2016.01198

Mora-Macias J, Ojeda-Rivera JO, Gutierrez-Alanis D, Yong-Villalobos L, Oropeza-Aburto A, Raya-Gonzalez J, Jimenez-Dominguez G, Chavez-Calvillo G, Rellan-Alvarez R, Herrera-Estrella L. Malate-dependent Fe accumulation is a critical checkpoint in the root developmental response to low phosphate. Proc Natl Acad Sci U S A. 2017:**114**(17):E3563–E3572. https://doi.org/10.1073/pnas.1701952114

Muchhal US, Pardo JM, Raghothama KG. Phosphate transporters from the higher plant *Arabidopsis thaliana*. Proc Natl Acad Sci U S A. 1996:**93**(19):10519–10523. https://doi.org/10.1073/pnas.93.19.10519

Mudge S, Rae A, Diatloff E, Smith F. Expression analysis suggests novel roles for members of the Pht1 family of phosphate transporters in Arabidopsis. Plant J. 2002:**31**(3):341–353. https://doi.org/10.1046/j.1365-313x.2002.01356.x

Müller J, Toev T, Heisters M, Teller J, Moore KL, Hause G, Dinesh DC, Bürstenbinder K, Abel S. Iron-dependent callose deposition adjusts root meristem maintenance to phosphate availability. Dev Cell. 2015:**33**(2):216–230. https://doi.org/10.1016/j.devcel.2015.02.007

Nagy R, Grob H, Weder B, Green P, Klein M, Frelet-Barrand A, Schjoerring J, Brearley C, Martinoia E. The Arabidopsis ATP-binding cassette protein AtMRP5/AtABCC5 is a high affinity inositol hexakisphosphate transporter involved in guard cell signaling and phytate storage. J Biol Chem. 2009:**284**(48):33614–33622. https://doi.org/10.1074/jbc.M109.030247

Nakamura Y. Phosphate starvation and membrane lipid remodeling in seed plants. Prog Lipid Res. 2013:**52**(1):43–50. https://doi.org/10.1016/j.plipres.2012.07.002

Naumann C, Heisters M, Brandt W, Janitza P, Alfs C, Tang N, Nienguesso AT, Ziegler J, Imre R, Mechtler K, et al. Bacterial-type ferroxidase tunes iron-dependent phosphate sensing during Arabidopsis root development. Curr Biol. 2022:**32**(10):2189–2205. https://doi.org/10.1016/j.cub.2022.04.005

Naumann C, Mueller J, Sakhonwasee S, Wieghaus A, Hause G, Heisters M, Buerstenbinder K, Abel S. The local phosphate deficiency response activates endoplasmic reticulum stress-dependent autophagy. Plant Physiol. 2019:**179**(2):460–476. https://doi.org/10.1104/pp.18.01379

Neelam K, Thakur S, Neha , Yadav IS, Kumar K, Dhaliwal SS, Singh K. Novel alleles of phosphorus-starvation tolerance 1 gene (*PSTOL1*) from *Oryza rufipogon* confers high phosphorus uptake efficiency. Front Plant Sci. 2017:**8**:509. https://doi.org/10.3389/fpls.2017.00509

Nguyen N, Clua J, Vetal P, Vuarambon D, De Bellis D, Pervent M, Lepetit M, Udvardi M, Valentine A, Poirier Y. PHO1 family members transport phosphate from infected nodule cells to bacteroids in *Medicago truncatula*. Plant Physiol. 2021:**185**(1):196–209. https://doi.org/10.1093/plphys/kiaa016

Nielsen TH, Krapp A, Röper-Schwarz U, Stitt M. The sugar-mediated regulation of genes encoding the small subunit of Rubisco and the regulatory subunit of ADP glucose pyrophosphorylase is modified by phosphate and nitrogen. Plant Cell Environ. 1998:**21**(5):443–454. https://doi.org/10.1046/j.1365-3040.1998.00295.x

Nilsson L, Müller R, Nielsen TH. Increased expression of the MYB-related transcription factor, PHR1, leads to enhanced phosphate uptake in *Arabidopsis thaliana*. Plant Cell Environ. 2007:**30**(12):1499–1512. https://doi.org/10.1111/j.1365-3040.2007.01734.x

Nussaume L, Kanno S, Javot H, Marin E, Pochon N, Ayadi A, Nakanishi T, Thibaud M. Phosphate import in plants: focus on the PHT1 transporters. Front Plant Sci. 2011:**2**:83. https://doi.org/10.3389/fpls.2011.00083

Okazaki Y, Otsuki H, Narisawa T, Kobayashi M, Sawai S, Kamide Y, Kusano M, Aoki T, Hirai MY, Saito K. A new class of plant lipid is essential for protection against phosphorus depletion. Nat Commun. 2013:**4**(1):1510. https://doi.org/10.1038/ncomms2512

Oldroyd GED, Leyser O. A plant's diet, surviving in a variable nutrient environment. Science. 2020:**368**(6486):eaba0196. https://doi.org/10.1126/science.aba0196

Ouyang X, Hong X, Zhao X, Zhang W, He X, Ma W, Teng W, Tong Y. Knock out of the PHOSPHATE 2 gene TaPHO2-A1 improves phosphorus uptake and grain yield under low phosphorus conditions in common wheat. Sci Rep. 2016:**6**(1):29850. https://doi.org/10.1038/srep29850

Pang JY, Bansal R, Zhao HX, Bohuon E, Lambers H, Ryan MH, Ranathunge K, Siddique KHM. The carboxylate-releasing phosphorus-mobilizing strategy can be proxied by foliar manganese concentration in a large set of chickpea germplasm under low phosphorus supply. New Phytol. 2018:**219**(2):518–529. https://doi.org/10.1111/nph.15200

Pant BD, Buhtz A, Kehr J, Scheible W-R. MicroRNA399 is a long-distance signal for the regulation of plant phosphate homeostasis. Plant J. 2008:**53**(5):731–738. https://doi.org/10.1111/j.1365-313X.2007.03363.x

Pant BD, Burgos A, Pant P, Cuadros-Inostroza A, Willmitzer L, Scheible W-R. The transcription factor PHR1 regulates lipid remodeling and triacylglycerol accumulation in *Arabidopsis thaliana* during phosphorus starvation. J Exp Bot. 2015:**66**(7):1907–1918. https://doi.org/10.1093/jxb/eru535

Pant BD, Musialak-Lange M, Nuc P, May P, Buhtz A, Kehr J, Walther D, Scheible W-Rd. Identification of nutrient-responsive Arabidopsis and rapeseed MicroRNAs by comprehensive real-time polymerase chain reaction profiling and small RNA sequencing. Plant Physiol. 2009:**150**(3):1541–1555. https://doi.org/10.1104/pp.109.139139

Paries M, Gutjahr C. The good, the bad, and the phosphate: regulation of beneficial and detrimental plant-microbe interactions by the plant phosphate status. New Phytol. 2023:**239**(1):29–46. https://doi.org/10.1111/nph.18933

Park BS, Seo JS, Chua NH. NITROGEN LIMITATION ADAPTATION recruits PHOSPHATE2 to target the phosphate transporter PT2 for degradation during the regulation of Arabidopsis phosphate homeostasis. Plant Cell. 2014:**26**(1):454–464. https://doi.org/10.1105/tpc.113.120311

Plaxton WC, Tran HT. Metabolic adaptations of phosphate-starved plants. Plant Physiol. 2011:**156**(3):1006–1015. https://doi.org/10.1104/pp.111.175281

Poirier Y, Thoma S, Somerville C, Schiefelbein J. Mutant of Arabidopsis deficient in xylem loading of phosphate. Plant Physiol. 1991:**97**(3):1087–1093. https://doi.org/10.1104/pp.97.3.1087

Pontigo S, Parra-Almuna L, Luengo-Escobar A, Poblete-Grant P, Nunes-Nesi A, Mora M, Cartes P. Biochemical and molecular responses underlying the contrasting phosphorus use efficiency in Ryegrass cultivars. Plants. 2023:**12**(6):1224. https://doi.org/10.3390/plants12061224

Potapenko E, Cordeiro CD, Huang G, Storey M, Wittwer C, Dutta AK, Jessen HJ, Starai VJ, Docampo R. 5-Diphosphoinositol pentakisphosphate (5-IP7) regulates phosphate release from acidocalcisomes and yeast vacuoles. J Biol Chem. 2018:**293**(49):19101–19112. https://doi.org/10.1074/jbc.ra118.005884

Prathap V, Kumar A, Maheshwari C, Tyagi A. Phosphorus homeostasis: acquisition, sensing, and long-distance signaling in plants. Mol Biol Rep. 2022:**49**(8):8071–8086. https://doi.org/10.1007/s11033-022-07354-9

Puga MI, Mateos I, Charukesi R, Wang Z, Franco-Zorrilla JM, Lorenzo L, Irigoyen ML, Masiero S, Bustos R, Rodrı´guez J, et al. SPX1 is a phosphate-dependent inhibitor of PHOSPHATE STARVATION RESPONSE 1 in Arabidopsis. Proc Natl Acad Sci U S A. 2014:**111**(41):14947–14952. https://doi.org/10.1073/pnas.1404654111

Qin L, Zhao J, Tian J, Chen LY, Sun ZA, Guo YX, Lu X, Gu MA, Xu GH, Liao H. The high-affinity phosphate transporter GmPT5 regulates phosphate transport to nodules and nodulation in soybean. Plant Physiol. 2012:**159**(4):1634–1643. https://doi.org/10.1104/pp.112.199786

Raghothama KG. Phosphate acquisition. Annu Rev Plant Physiol Plant Mol Biol. 1999:**50**(1):665–693. https://doi.org/10.1146/annurev.arplant.50.1.665

Reis R, Deforges J, Schmidt R, Schippers J, Poirier Y. An antisense noncoding RNA enhances translation via localised structural rearrangements of its cognate mRNA. Plant Cell. 2021:**33**(4):1381–1397. https://doi.org/10.1093/plcell/koab010

Reis R, Deforges J, Sokoloff T, Poirier Y. Modulation of shoot phosphate level and growth by PHOSPHATE1 upstream open reading frame. Plant Physiol. 2020:**183**(3):1145–1156. https://doi.org/10.1104/pp.19.01549

Remy E, Cabrito TR, Batista RA, Teixeira MC, Sa-Correia I, Duque P. The Pht1; 9 and Pht1; 8 transporters mediate inorganic phosphate acquisition by the *Arabidopsis thaliana* root during phosphorus starvation. New Phytol. 2012:**195**(2):356–371. https://doi.org/10.1111/j.1469-8137.2012.04167.x

Reymond M, Svistoonoff S, Loudet O, Nussaume L, Desnos T. Identification of QTL controlling root growth response to phosphate starvation in *Arabidopsis thaliana*. Plant Cell Environ. 2006:**29**(1):115–125. https://doi.org/10.1111/j.1365-3040.2005.01405.x

Ried MK, Wild R, Zhu JS, Pipercevic J, Sturm K, Broger L, Harmel RK, Abriata LA, Hothorn LA, Fiedler D, et al. Inositol pyrophosphates promote the interaction of SPX domains with the coiled-coil motif of PHR transcription factors to regulate plant phosphate homeostasis. Nat Commun. 2021:**12**(1):384. https://doi.org/10.1038/s41467-020-20681-4

Riemer E, Qiu D, Laha D, Harmel RK, Gaugler P, Gaugler V, Frei M, Hajirezaei M-R, Laha NP, Krusenbaum L, et al. ITPK1 is an InsP6/ADP phosphotransferase that controls phosphate signaling in *Arabidopsis*. Mol Plant. 2021:**14**(11):1864–1880. https://doi.org/10.1016/j.molp.2021.07.011

Robles-Aguilar AA, Pang JY, Postma JA, Schrey SD, Lambers H, Jablonowski ND. The effect of pH on morphological and physiological root traits of *Lupinus angustifolius* treated with struvite as a recycled phosphorus source. Plant Soil. 2019:**434**(1-2):65–78. https://doi.org/10.1007/s11104-018-3787-2

Ruan W, Guo M, Xu L, Wang X, Zhao H, Wang J, Yi K. An SPX-RLI1 module regulates leaf inclination in response to phosphate availability in rice. Plant Cell. 2018:**30**(4):853–870. https://doi.org/10.1105/tpc.17.00738

Rubio V, Linhares F, Solano R, Martín AC, Iglesias J, Leyva A, Paz-Ares J. A conserved MYB transcription factor involved in phosphate starvation signaling both in vascular plants and in unicellular algae. Genes Dev. 2001:**15**(16):2122–2133. https://doi.org/10.1101/gad.204401

Ryan PR, Delhaize E, Jones DL. Function and mechanism of organic anion exudation from plant roots. Annu Rev Plant Physiol Plant Mol Biol. 2001:**52**(1):527–560. https://doi.org/10.1146/annurev.arplant.52.1.527

Sahu A, Banerjee S, Raju A, Chiou T-J, Garcia L, Versaw W. Spatial profiles of phosphate in roots indicate developmental control of uptake, recycling, and sequestration. Plant Physiol. 2020:**184**(4):2064–2077. https://doi.org/10.1104/pp.20.01008

Sakakibara H. Cytokinins: activity, biosynthesis, and translocation. Annu Rev Plant Biol. 2006:**57**(1):431–449. https://doi.org/10.1146/annurev.arplant.57.032905.105231

Sanchez-Calderon L, Lopez-Bucio J, Chacon-Lopez A, Cruz-Ramirez A, Nieto-Jacobo F, Dubrovsky JG, Herrera-Estrella L. Phosphate starvation induces a determinate developmental program in the roots of *Arabidopsis thaliana*. Plant Cell Physiol. 2005:**46**(1):174–184. https://doi.org/10.1093/pcp/pci011

Sawaki Y, Iuchi S, Kobayashi Y, Kobayashi Y, Ikka T, Sakurai N, Fujita M, Shinozaki K, Shibata D, Kobayashi M, et al. STOP1 regulates multiple genes that protect Arabidopsis from proton and aluminum toxicities. Plant Physiol. 2009:**150**(1):281–294. https://doi.org/10.1104/pp.108.134700

Secco D, Baumann A, Poirier Y. Characterization of the rice PHO1 gene family reveals a key role for OsPHO1; 2 in phosphate homeostasis and the evolution of a distinct clade in dicotyledons. Plant Physiol. 2010:**152**(3):1693–1704. https://doi.org/10.1104/pp.109.149872

Secco D, Wang C, Arpat BA, Wang Z, Poirier Y, Tyerman SD, Wu P, Shou H, Whelan J. The emerging importance of the SPX domain-containing proteins in phosphate homeostasis. New Phytol. 2012:**193**(4):842–851. https://doi.org/10.1111/j.1469-8137.2011.04002.x

Shi J, Zhao B, Zheng S, Zhang X, Wang X, Dong W, Xie Q, Wang G, Xiao Y, Chen F, et al. A phosphate starvation response-centered network regulates mycorrhizal symbiosis. Cell. 2021:**184**(22):5527–5540. https://doi.org/10.1016/j.cell.2021.09.030

Shin H, Shin HS, Chen R, Harrison MJ. Loss of At4 function impacts phosphate distribution between the roots and the shoots during phosphate starvation. Plant J. 2006:**45**(5):712–726. https://doi.org/10.1111/j.1365-313X.2005.02629.x

Shin H, Shin HS, Dewbre GR, Harrison MJ. Phosphate transport in Arabidopsis: Pht1; 1 and Pht1; 4 play a major role in phosphate acquisition from both low- and high-phosphate environments. Plant J.

2004:**39**(4):629–642. https://doi.org/10.1111/j.1365-313X.2004.02161.x

Silva-Navas J, Conesa CM, Saez A, Navarro-Neila S, Garcia-Mina JM, Zamarreño AM, Baigorri R, Swarup R, Del Pozo JC. Role of cis-zeatin in root responses to phosphate starvation. New Phytol. 2019:**224**(1):242–257. https://doi.org/10.1111/nph.16020

Song HN, Yin ZT, Chao MN, Ning LH, Zhang D, Yu DY. Functional properties and expression quantitative trait loci for phosphate transporter GmPT1 in soybean. Plant Cell Environ. 2014:**37**(2):462–472. https://doi.org/10.1111/pce.12170

Stegmann M, Monaghan J, Smakowska-Luzan E, Rovenich H, Lehner A, Holton N, Belkhadir Y, Zipfel C. The receptor kinase FER is a RALF-regulated scaffold controlling plant immune signaling. Science. 2017:**355**(6322):287–289. https://doi.org/10.1126/science.aal2541

Sun BR, Gao YZ, Lynch JP. Large crown root number improves topsoil foraging and phosphorus acquisition. Plant Physiol. 2018:**177**(1):90–104. https://doi.org/10.1104/pp.18.00234

Svistoonoff S, Creff A, Reymond M, Sigoillot-Claude C, Ricaud L, Blanchet A, Nussaume L, Desnos T. Root tip contact with low-phosphate media reprograms plant root architecture. Nat Genet. 2007:**39**(6):792–796. https://doi.org/10.1038/ng2041

Tang C, Hinsinger P, Drevon JJ, Jaillard B. Phosphorus deficiency impairs early nodule functioning and enhances proton release in roots of *Medicago truncatula* L. Ann Bot. 2001:**88**(1):131–138. https://doi.org/10.1006/anbo.2001.1440

Tang J, Wu D, Li X, Wang L, Xu L, Zhang Y, Xu F, Liu H, Xie Q, Dai S, et al. Plant immunity suppression via PHR1-RALF-FERONIA shapes the root microbiome to alleviate phosphate starvation. EMBO J. 2022:**41**(6):e109102. https://doi.org/10.15252/embj.2021109102

Taylor CB, Bariola PA, delCardayre´ SB, Raines RT, Green PJ. RNS2: a senescence-associated RNase of Arabidopsis that diverged from the S-RNases before speciation. Proc Natl Acad Sci U S A. 1993:**90**(11):5118–5122. https://doi.org/10.1073/pnas.90.11.5118

Thibaud MC, Arrighi JF, Bayle V, Chiarenza S, Creff A, Bustos R, Paz-Ares J, Poirier Y, Nussaume L. Dissection of local and systemic transcriptional responses to phosphate starvation in Arabidopsis. Plant J. 2010:**64**(5):775–789. https://doi.org/10.1111/j.1365-313X.2010.04375.x

Ticconi CA, Delatorre CA, Abel S. Attenuation of phosphate starvation responses by phosphite in Arabidopsis. Plant Physiol. 2001:**127**(3):963–972. https://doi.org/10.1104/pp.010396

Ticconi CA, Lucero RD, Sakhonwasee S, Adamson AW, Creff A, Nussaume L, Desnos T, Abel S. ER-resident proteins PDR2 and LPR1 mediate the developmental response of root meristems to phosphate availability. Proc Natl Acad Sci U S A. 2009:**106**(33):14174–14179. https://doi.org/10.1073/pnas.0901778106

Umehara M, Hanada A, Yoshida S, Akiyama K, Arite T, Takeda-Kamiya N, Magome H, Kamiya Y, Shirasu K, Yoneyama K, et al. Inhibition of shoot branching by new terpenoid plant hormones. Nature. 2008:**455**(7210):195–200. https://doi.org/10.1038/nature07272

Valverde C, Ferrari A, Wall LG. Phosphorus and the regulation of nodulation in the actinorhizal symbiosis, between *Discaria trinervis* (Rhamnaceae) and *Frankia* BCU110501. New Phytol. 2002:**153**(1):43–51. https://doi.org/10.1046/j.0028-646X.2001.00298.x

van de Wiel CCM, van der Linden CG, Scholten OE. Improving phosphorus use efficiency in agriculture: opportunities for breeding. Euphytica. 2016:**207**(1):1–22. https://doi.org/10.1007/s10681-015-1572-3

Vance CP, Uhde-Stone C, Allan DL. Phosphorus acquisition and use: critical adaptations by plants for securing a nonrenewable resource. New Phytol. 2003:**157**(3):423–447. https://doi.org/10.1046/j.1469-8137.2003.00695.x

Varadarajan DK, Karthikeyan AS, Matilda PD, Raghothama KG. Phosphite, an analog of phosphate, suppresses the coordinated expression of genes under phosphate starvation. Plant Physiol. 2002:**129**(3):1232–1240. https://doi.org/10.1104/pp.010835

Versaw WK, Garcia LR. Intracellular transport and compartmentation of phosphate in plants. Curr Opin Plant Biol. 2017:**39**:25–30. https://doi.org/10.1016/j.pbi.2017.04.015

Versaw WK, Harrison MJ. A chloroplast phosphate transporter, PHT2;1, influences allocation of phosphate within the plant and phosphate-starvation responses. Plant Cell. 2002:**14**(8):1751–1766. https://doi.org/10.1105/tpc.002220

Vetal P, Poirier Y. The Arabidopsis PHOSPHATE 1 exporter undergoes constitutive internalization via clathrin-mediated endocytosis. Plant J. 2023:**116**(5):1477–1491. https://doi.org/10.1111/tpj.16441

Vogiatzaki E, Baroux C, Jung J-Y, Poirier Y. PHO1 exports phosphate from the chalazal seed coat to the embryo in developing Arabidopsis seeds. Curr Biol. 2017:**27**(19):2893–2900. https://doi.org/10.1016/j.cub.2017.08.026

Wang C, Huang W, Ying Y, Li S, Secco D, Tyerman S, Whelan J, Shou H. Functional characterization of the rice SPX-MFS family reveals a key role of OsSPX-MFS1 in controlling phosphate homeostasis in leaves. New Phytol. 2012:**196**(1):139–148. https://doi.org/10.1111/j.1469-8137.2012.04227.x

Wang C, Yue W, Ying Y, Wang S, Secco D, Liu Y, Whelan J, Tyerman SD, Shou H. Rice SPX-major facility superfamily3, a vacuolar phosphate efflux transporter, is involved in maintaining phosphate homeostasis in rice. Plant Physiol. 2015:**169**(4):2822–2831. https://doi.org/10.1104/pp.15.01005

Wang F, Deng M, Chen J, He Q, Jia X, Guo H, Xu J, Liu Y, Zhang S, Shou H, et al. CASEIN KINASE2-dependent phosphorylation of PHOSPHATE2 fine-tunes phosphate homeostasis in rice. Plant Physiol. 2020:**183**(1):250–262. https://doi.org/10.1104/pp.20.00078

Wang P, Snijders R, Kohlen W, Liu J, Bisseling T, Limpens E. Medicago SPX1 and SPX3 regulate phosphate homeostasis, mycorrhizal colonization, and arbuscule degradation. Plant Cell. 2021:**33**(11):3470–3486. https://doi.org/10.1093/plcell/koab206

Wang W, Ding GD, White PJ, Wang XH, Jin KM, Xu FS, Shi L. Mapping and cloning of quantitative trait loci for phosphorus efficiency in crops: opportunities and challenges. Plant Soil. 2019:**439**(1-2):91–112. https://doi.org/10.1007/s11104-018-3706-6

Wang X, Wang Y, Piñeros M, Wang Z, Wang W, Li C, Wu Z, Kochian L, Wu P. Phosphate transporters OsPHT1; 9 and OsPHT1; 10 are involved in phosphate uptake in rice. Plant Cell Environ. 2014a:**37**:1159–1170. https://doi.org/10.1111/pce.12224

Wang X, Yi K, Tao Y, Wang F, Wu Z, Jiang D, Chen X, Zhu L, Wu P. Cytokinin represses phosphate-starvation response through increasing of intracellular phosphate level. Plant Cell Environ. 2006:**29**(10):1924–1935. https://doi.org/10.1111/j.1365-3040.2006.01568.x

Wang Z, Ruan W, Shi J, Zhang L, Xiang D, Yang C, Li C, Wu Z, Liu Y, Yu Y, et al. Rice SPX1 and SPX2 inhibit phosphate starvation responses through interacting with PHR2 in a phosphate-dependent manner. Proc Natl Acad Sci U S A. 2014b:**111**(41):14953–14958. https://doi.org/10.1073/pnas.1404680111

Wang Z, Zheng Z, Song L, Liu D. Functional characterization of Arabidopsis PHL4 in plant response to phosphate starvation. Front Plant Sci. 2018:**9**:1432. https://doi.org/10.3389/fpls.2018.01432

Wang Z, Zheng Z, Zhu Y, Kong S, Liu D. PHOSPHATE RESPONSE 1 family members act distinctly to regulate transcriptional responses to phosphate starvation. Plant Physiol. 2022:**191**(2):1324–1343. https://doi.org/10.1093/plphys/kiac521

Wege S, Khan G, Jung J, Vogiatzaki E, Pradervand S, Aller I, Meyer A, Poirier Y. The EXS domain of PHO1 participates in the response of shoots to phosphate deficiency via a root-to-shoot signal. Plant Physiol. 2016:**170**(1):385–400. https://doi.org/10.1104/pp.15.00975

Wen ZH, Li HB, Shen Q, Tang XM, Xiong CY, Li HG, Pang JY, Ryan MH, Lambers H, Shen JB. Tradeoffs among root morphology, exudation and mycorrhizal symbioses for phosphorus-acquisition strategies of 16 crop species. New Phytol. 2019:**223**(2):882–895. https://doi.org/10.1111/nph.15833

Whitfield H, White G, Sprigg C, Riley AM, Potter BVL, Hemmings AM, Brearley CA. An ATP-responsive metabolic cassette comprised of inositol tris/tetrakisphosphate kinase 1 (ITPK1) and inositol pentakisphosphate 2-kinase (IPK1) buffers diphosphosphoinositol phosphate levels. Biochem J. 2020:**477**(14):2621–2638. https://doi.org/10.1042/BCJ20200423

Wild R, Gerasimaite R, Jung J-Y, Truffault V, Pavlovic I, Schmidt A, Saiardi A, Jessen HJ, Poirier Y, Hothorn M, et al. Control of eukaryotic phosphate homeostasis by inositol polyphosphate sensor domains. Science. 2016:**352**(6288):986–990. https://doi.org/10.1126/science.aad9858

Wissuwa M, Wegner J, Ae N, Yano M. Substitution mapping of Pup1: a major QTL increasing phosphorus uptake of rice from a phosphorus-deficient soil. Theor Appl Genet. 2002:**105**(6):890–897. https://doi.org/10.1007/s00122-002-1051-9

Wu Y, Shi L, Li L, Fu LW, Liu YL, Xiong Y, Sheen J. Integration of nutrient, energy, light, and hormone signalling via TOR in plants. J Exp Bot. 2019:**70**(8):2227–2238. https://doi.org/10.1093/jxb/erz028

Xiao X, Zhang J, Satheesh V, Meng F, Gao W, Dong J, Zheng Z, An G-Y, Nussaume L, Liu D, et al. SHORT-ROOT stabilizes PHOSPHATE1 to regulate phosphate allocation in Arabidopsis. Nat Plants. 2022:**8**(9):1074–1081. https://doi.org/10.1038/s41477-022-01231-w

Xing X, Du H, Yang Z, Li X, Kong Y, Li W, Zhang C. GmSPX8, a nodule-localized regulator confers nodule development and nitrogen fixation under phosphorus starvation in soybean. BMC Plant Biol. 2022:**22**(1):161. https://doi.org/10.1186/s12870-022-03556-2

Xu L, Zhao H, Wan R, Liu Y, Xu Z, Tian W, Ruan W, Wang F, Deng M, Wang J, et al. Identification of vacuolar phosphate efflux transporters in land plants. Nat Plants. 2019:**5**(1):84–94. https://doi.org/10.1038/s41477-018-0334-3

Yahya M, ul Islam E, Rasul M, Farooq I, Mahreen N, Tawab A, Irfan M, Rajput L, Amin I, Yasmin S. Differential root exudation and architecture for improved growth of wheat mediated by phosphate solubilizing bacteria. Front Microbiol. 2021:**12**:744094. https://doi.org/10.3389/fmicb.2021.744094

Yamaji N, Takemoto Y, Miyaji T, Mitani-Ueno N, Yoshida KT, Ma JF. Reducing phosphorus accumulation in rice grains with an impaired transporter in the node. Nature. 2017:**541**(7635):92–95. https://doi.org/10.1038/nature20610

Yang SY, Huang TK, Kuo HF, Chiou TJ. Role of vacuoles in phosphorus storage and remobilization. J Exp Bot. 2017:**68**(12):3045–3055. https://doi.org/10.1093/jxb/erw481

Yang SY, Lu WC, Ko SS, Sun CM, Hung JC, Chiou TJ. Upstream open reading frame and phosphate-regulated expression of rice OsNLA1 controls phosphate transport and reproduction. Plant Physiol. 2020:**182**(1):393–407. https://doi.org/10.1104/pp.19.01101

Yue W, Ying Y, Wang C, Zhao Y, Dong C, Whelan J, Shou H. OsNLA1, a RING-type ubiquitin ligase, maintains phosphate homeostasis in *Oryza sativa* via degradation of phosphate transporters. Plant J. 2017:**90**(6):1040–1051. https://doi.org/10.1111/tpj.13516

Zhang F, Sun Y, Pei W, Jain A, Sun R, Cao Y, Wu X, Jiang T, Zhang L, Fan X, et al. Involvement of OsPht1; 4 in phosphate acquisition and mobilization facilitates embryo development in rice. Plant J. 2015:**82**(4):556–569. https://doi.org/10.1111/tpj.12804

Zhong Y, Wang Y, Guo J, Zhu X, Shi J, He Q, Liu Y, Wu Y, Zhang L, Lv Q, et al. Rice SPX6 negatively regulates the phosphate starvation response through suppression of the transcription factor PHR2. New Phytol. 2018:**219**(1):135–148. https://doi.org/10.1111/nph.15155

Zhou J, Hu Q, Xiao X, Yao D, Ge S, Ye J, Li H, Cai R, Liu R, Meng F. Mechanism of phosphate sensing and signaling revealed by rice SPX1-PHR2 complex structure. Nat Commun. 2021:**12**(1):7040. https://doi.org/10.1038/s41467-021-27391-5

Zhou J, Jiao F, Wu Z, Li Y, Wang X, He X, Zhong W, Wu P. OsPHR2 is involved in phosphate-starvation signaling and excessive phosphate accumulation in shoots of plants. Plant Physiol. 2008:**146**(4):1673–1686. https://doi.org/10.1104/pp.107.111443

Zhu J, Lau K, Puschmann R, Harmel RK, Zhang Y, Pries V, Gaugler P, Broger L, Dutta AK, Jessen HJ, et al. Two bifunctional inositol pyrophosphate kinases/phosphatases control plant phosphate homeostasis. Elife. 2019:**8**:e43582. https://doi.org/10.7554/eLife.43582

Zhu W, Miao Q, Sun D, Yang G, Wu C, Huang J, Zheng C. The mitochondrial phosphate transporters modulate plant responses to salt stress via affecting ATP and gibberellin metabolism in *Arabidopsis thaliana*. PLoS One. 2012:**7**(8):e43530. https://doi.org/10.1371/journal.pone.0043530

Zhuang Q, Xue Y, Yao Z, Zhu S, Liang C, Liao H, Tian J. Phosphate starvation responsive GmSPX5 mediates nodule growth through interaction with GmNF-YC4 in soybean (*Glycine max*). Plant J. 2021:**108**(5):1422–1438. https://doi.org/10.1111/tpj.15520

Zimmerli C, Ribot C, Vavasseur A, Bauer H, Hedrich R, Poirier Y. PHO1 expression in guard cells mediates the stomatal response to abscisic acid in Arabidopsis. Plant J. 2012:**72**(2):199–211. https://doi.org/10.1111/j.1365-313X.2012.05058.x

https://doi.org/10.1093/plcell/koad277
THE PLANT CELL
THE PLANT CELL 2024: 36: 1524–1539
Review

Should I stay or should I go? Trafficking of plant extra-nuclear transcription factors

Sarika Marathe [1,2] Erich Grotewold [3] and Marisa S. Otegui [1,2,*]

1 Department of Botany, University of Wisconsin-Madison, Madison, WI 53706, USA
2 Center for Quantitative Cell Imaging, University of Wisconsin-Madison, Madison, WI 53706, USA
3 Department of Biochemistry and Molecular Biology, Michigan State University, East Lansing, MI 48824-6473, USA

*Author for correspondence: otegui@wisc.edu

Abstract

At the heart of all biological processes lies the control of nuclear gene expression, which is primarily achieved through the action of transcription factors (TFs) that generally contain a nuclear localization signal (NLS) to facilitate their transport into the nucleus. However, some TFs reside in the cytoplasm in a transcriptionally inactive state and only enter the nucleus in response to specific signals, which in plants include biotic or abiotic stresses. These extra-nuclear TFs can be found in the cytosol or associated with various membrane systems, including the endoplasmic reticulum and plasma membrane. They may be integral proteins with transmembrane domains or associate peripherally with the lipid bilayer via acylation or membrane-binding domains. Although over 30 plant TFs, most of them involved in stress responses, have been experimentally shown to reside outside the nucleus, computational predictions suggest that this number is much larger. Understanding how extra-nuclear TFs are trafficked into the nucleus is essential for reconstructing transcriptional regulatory networks that govern major cellular pathways in response to biotic and abiotic signals. Here, we provide a perspective on what is known on plant extranuclear-nuclear TF retention, nuclear trafficking, and the post-translational modifications that ultimately enable them to regulate gene expression upon entering the nucleus.

Introduction

The response of plant cells to developmental cues as well as abiotic and biotic inputs is carefully orchestrated through intricate signal recognition and relay pathways, which ultimately trigger differential gene expression. This dynamic process of gene expression regulation is tightly controlled by transcription factors (TFs), proteins that bind DNA in a sequence-specific manner and either promote or repress transcription of specific genes. TFs are usually classified into 50 to 70 different families according to the structure of their DNA-binding domains, with some TF families being unique to the plant kingdom. TFs represent 5% to 7% of all the plant protein-coding genes, with *Arabidopsis thaliana* predicted to contain about 2,000 TFs (Riechmann et al. 2000; Davuluri et al. 2003; Riaño-Pachón et al. 2007; Yilmaz et al. 2010). However, additional TFs are likely yet to be identified from the large number of genes for which functions remain to be determined. Just as a reference, even in the best described plant genome, *Arabidopsis*, there are still more than 5,000 genes with unknown functions (Shang et al. 2023).

Although TFs regulate gene transcription inside the nucleus, a fraction of them reside in the cytoplasm (extra-nuclear TFs) as transcriptionally inactive forms. Upon activation, they are transported into the nucleus. Some TFs are retained in the cytoplasm because they either form complexes with cytoplasmic proteins or are unable to interact with proteins required for their nuclear translocation. In other instances, TFs associate with particular membrane compartments, and only upon activation they are released for nuclear import (Liu et al. 2018). This phenomenon has been well characterized for a relatively small number of animal and plant TFs, yet computational studies suggest that the number of TFs that are regulated in this way is much larger (Zupicich et al. 2001).

Received June 23, 2023. Accepted September 21, 2023. Advance access publication January 2, 2024

How does cytoplasmic TF retention affect the kinetics of the regulatory cascade in which it is involved? Simple intuition suggests that having a TF preformed in the cytoplasm might allow for a faster response than having to synthesize it de novo (involving transcription, splicing, translation). Consistent with this idea, most extranuclear TFs are activated and translocated into the nucleus upon stress conditions that require rapid transcriptional responses. In mammalian cells, transcription elongation rates vary between 30 to 100 nts/s (Fuchs et al. 2014), RNA splicing takes in average 5 to 10 min (Singh and Padgett 2009), and translation elongation proceeds at 6 amino acid/s (Ingolia et al. 2011). Newly synthesized TFs may need to undergo post-translational modifications before they are ready to be translocated to the nucleus. Therefore, we can estimate that synthesizing TFs de novo can require between a few minutes and a few hours. How long is required to release and activate extranuclear TF from the cytoplasm? Unfortunately, the information on extranuclear TF kinetics is rather limited but based on the few cases that have been studied in plants, it may require even longer than de novo protein synthesis. For example, the release of ANAC013 from membranes has been estimated to occur 1.5 h after the onset of low-oxygen conditions (hypoxia) (Eysholdt-Derzsó et al. 2023). However, ANAC017, another extranuclear TF that also responds to hypoxia, activates the transcription of ANAC013 within minutes of mitochondrial disfunction onset (He et al. 2022), suggesting that activation of preformed TFs and de novo synthesis of the same TFs may represent regulatory circuits to balance the timing and amplitude of complex TF kinetics.

Here, we build on a number of studies that have discussed the extranuclear retention of TFs to provide a perspective of what is known in plants, with particular emphasis given to membrane retention and mechanisms associated with TF activation.

General mechanisms of TF nuclear import and export

Most TFs are synthesized in the cytosol and translocated into the nucleus through the recognition of a nuclear localization signal (NLS), which consists of 1 or more short sequences with basic amino acids (Lu et al. 2021). Whereas proteins smaller than ~40 kDa can enter and exit the nucleus by diffusion, larger proteins require an NLS or a nuclear export signal (NES) to move through nuclear pores. Certain amino acid features are common to many NLSs, but they are not characterized by strict consensus sequences. The general translocation mechanism into the nucleus typically begins with interaction between the NLS sequence in the client protein (e.g. TF) and 2 nuclear import proteins, importin α and importin β, in the cytosol. The complex is then channeled through the nuclear pore complex. Within the nucleus, the GTP-bound Ran (Ras-related nuclear) G protein dissociates the importin-TF complex and releases the TF that can then bind DNA and regulate gene expression (Lu et al. 2021). Proteins that shuttle between the nucleus and cytoplasm also contain an NES. Inside the nucleus, the NES in the client protein is recognized by the GTP-bound Ran-exportin complex, which mediates exit back to the cytoplasm (Merkle 2011) (Fig. 1). As the complex moves through the nuclear pore and into the cytoplasm, the Ran hydrolyzes GTP, causing the disassembly of the complex and the release of the client protein. Occasionally, the NLS and NES sequences can be found in a partner protein of the TF and not in the TF itself. In such cases, partner proteins are essential to control the movement of TFs in and/or out of the nucleus. A good example in plants is provided by the JAZ9 transcriptional repressor, which requires MYC2 to translocate to the nucleus (Withers et al. 2012).

There are several mechanisms by which extra-nuclear TFs are retained outside the nucleus; some of these TFs are soluble in the cytosol (Figs. 2 and 3), and others are associated with membranes. Extra-nuclear TFs have been identified in numer-ous TF families, including the plant-specific TFs NAC (NO APICAL MERISTEM, ATAF1/2, CUP-SHAPED COTYLEDON 2), SBP (SQUAMOSA-PROMOTER BINDING PROTEIN), WRKY, ARF (AUXIN RESPONSE FACTOR) families, and TF families shared with other eukaryotes, such as bHLH (basic HELIX-LOOP-HELIX), bZIP (basic LEUCINE ZIPPER), MADS, MYB, and C2H2. In addition, some TF families, such as the bZIP and bHLH TFs, have evolved multiple mechanisms of re-tention outside the nucleus.

Over 30 extranuclear TFs have been experimentally identified in *Arabidopsis*, including 14 with transmembrane domains. However, it is difficult to estimate how many more there really are due to the limitation in predicting and even detecting extranuclear TFs. In this review, we discuss the main mechanisms that mediate retention of TFs outside the nucleus as an important regulatory component of gene expression and comment on new experimental techniques and computational methods to address challenges in identifying new extra-nuclear TFs. We focus primarily on *Arabidopsis* extra-nuclear TFs because they have been best studied but refer to examples from other plant species when contributing new knowledge.

Cytoplasmic retention of TFs through protein-protein interactions

Eukaryotes use diverse mechanisms to retain TFs in the cytosol (cTFs) as a mechanism to control their transcriptional activity. These mechanisms usually involve interactions with cytosolic proteins and may require specific post-translational modifications such as phosphorylation. Nucleo-cytoplasmic trafficking of cTFs in plants is commonly mediated by hormone signaling and environmental stresses. In *Arabidopsis*, approximately 16 cTFs across 8 TF families have been experimentally characterized (Fig. 2).

The most common mechanism of cytosolic retention of cTFs is binding to 14-3-3 proteins in the cytosol upon phosphorylation. The highly conserved 14-3-3 proteins can interact with phosphorylated forms of their client proteins (Huang et al. 2022). Twelve 14-3-3 genes are expressed in

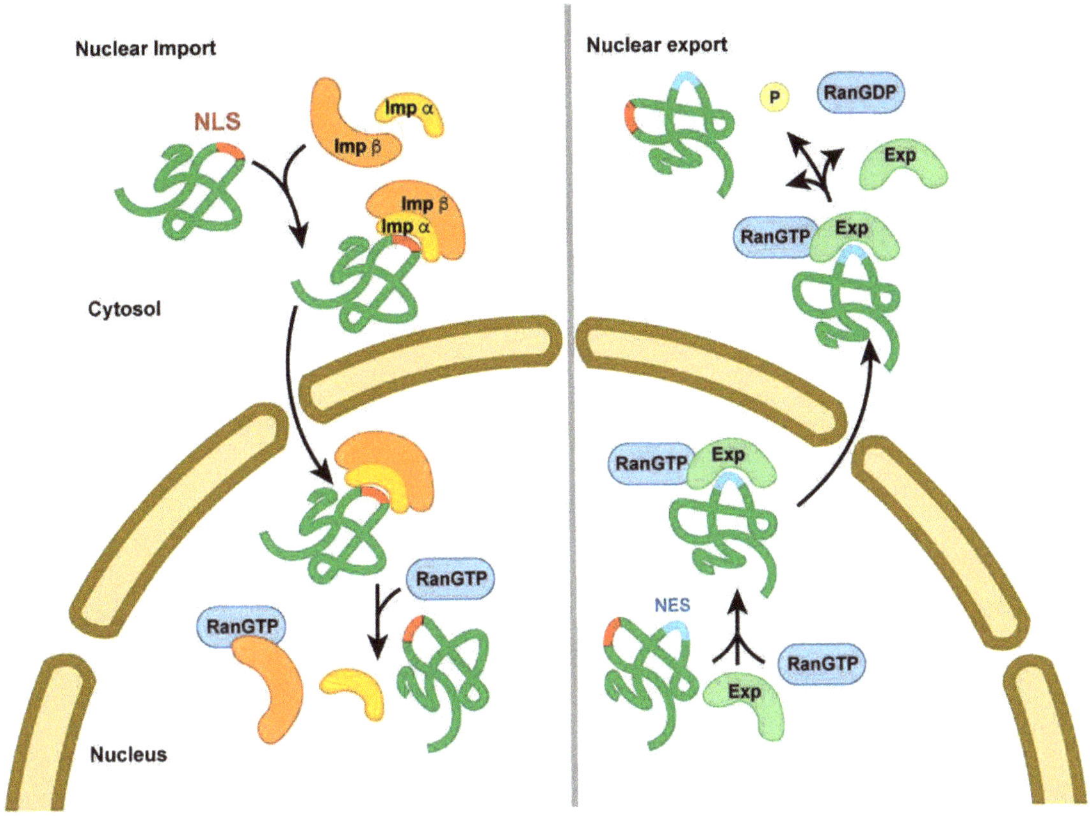

Figure 1. General mechanisms of nuclear protein import and export. NLS sequences are recognized by importins (importins a and b, indicated as Imp a and Imp b, respectively) in the cytosol and mediate the translocation into the nucleus. RanGTP mediates the dissociation of the importin complex and the NLS, facilitating the release of the client protein inside the nucleus. Exportins binds the NES in client proteins and, together with RanGTP, form a complex that moves through the nuclear pore. At the cytosol, GTP hydrolysis in the Ran GTPase facilitates the disassembly of the complex and the release of the client protein.

Arabidopsis, with several involved in gene transcription regulation (Sehnke et al. 2002), including those retaining cTFs outside the nucleus (see Fig. 2).

For example, in *Arabidopsis*, BZR1 (BRASSINAZOLE RESISTANT 1) and BES1 (bri1-EMS-SUPPRESSOR 1)/BRZ2 are 2 important brassinosteroid (BR)-responsive cTFs that directly or indirectly control the expression of thousands of genes during plant development and in response to environ-mental signals (Sun et al. 2010; Nolan et al. 2020). When BR levels are low, BZR1 and BES1 are phosphorylated by the GSK3-like kinase BIN2 (BRASSINOSTEROID INSENSITIVE 2) (Li and Nam 2002; Zhao et al. 2002), which increases their affinity for several 14-3-3 proteins in the cytosol (Gampala et al. 2007) and promotes their degradation, by either the 26S proteasome or autophagy (He et al. 2002; Nolan et al. 2017; Kim et al. 2019). BR perception by the cell surface receptor BRI1 (BRASSINOSTEROID INSENSTIVE 1) and its co-receptor BAK1 (BRI1-ASSOCIATED RECEPTOR KINASE 1) leads to the inhibition and degradation of BIN2 (Li and Nam 2002; Yin et al. 2002) and the enhanced dephosphorylation of BES1 and BRZ1 by PP2A-type phosphatases (Tang et al. 2011). Dephosphorylated BES1 and BRZ1 dissociate from 14-3-3 proteins and are translocated into the nucleus (Ryu et al. 2007) to drive expression of BR-responsive genes.

Interestingly, BIN2 itself partitions between the nucleus and cytosol and can phosphorylate BES1 and BZR1 in both locations (Ryu et al. 2010a). The nuclear phosphorylation of BZR1 triggers their exportin-dependent nuclear export back to the cytosol (Ryu et al. 2007, 2010b) (Fig. 3). Regulation of BZR1's nuclear translocation can also be cell specific. For example, in differentiating phloem cells, the plasma membrane–associated protein OCTOPUS binds and inhibits BIN2 at the cell surface (Anne et al. 2015), promoting the dephosphorylation of BES1 and BZR1 and their nuclear import.

At least 5 isoforms of 14-3-3 proteins were shown to interact with BZR1; however, the analysis of single and double mutants for 14-3-3 genes has failed to confirm the role of 14-3-3 proteins in BR signaling (Gampala et al. 2007). This could be due to functional redundancy within the 14-3-3 protein family or the occurrence of additional mechanism of cytosolic retention. In fact, several studies support the idea that larger molecular scaffolds hold phosphorylated BZR1 in the cytosol. For example, BZR1, together with other BR-signaling proteins, interacts with TTL3 (TETRATRICOPEPTIDE THIOREDOXIN-LIKE 3) in the cytosol. *Arabidopsis* TTL proteins contain 6 tetratricopeptide repeat commonly involved in protein-protein interactions and a C-terminal domain similar to thioredoxins

		cytosolic form						
TF Family	TF	phosphorylated	bound to 14-3-3 proteins	bound to other proteins	in condensates	other features	Signal for nuclear translocation	References
BES1/BZR	BES1	✓	✓				BR	Ryu et al., 2010
	BZR1	✓	✓	✓ BSS1 TLL3		sumolylation causes nuclear accumulation	BR	Ryu et al., 2007; Srivastava et al., 2020
bZIP	bZIP51	✓	✓				mechanical or hypo-osmotic stress	Takeo and Ito, 2017
	bZIP52	✓	✓				heat stress	Wiese et al., 2021
	bZIP18	✓	✓					
bHLH	PIF7	✓	✓				shade	Huang et al., 2018
	bHLH039					Phosphoryl. triggers nuclear localization	low iron	Gratz et al., 2019; Trofimov et al., 2019
	FIT							
MYB	PHR1	✓		✓ SPX4			low Pi	Puga et al., 2014; Osorio et al., 2019
	RVE4; RVE8						cold and heat stress	Kidokoro et al., 2021
HSF	HSFA8			✓ HSP		Cys disulfide brigdes needed for cytosolic retention	oxidative stress	Giesguth et al., 2015
NAC	VND6			✓ XND1			xylem differentiation	Zhong et al., 2020
ARF	ARF7				✓		auxin	Powers et al., 2019; Jing et al., 2022
	ARF19				✓			
E2F	E2Fa					Moves to nucleus upon binding DPa	cell differentiation	Kosugi and Ohashi, 2002

Figure 2. Table of *Arabidopsis* cTFs retained in the cytosol in their transcriptionally inactive forms and known mechanisms of cytosolic retention and activation. BSS1 (BRZ-SENSITIVE-SHORT HYPOCOTYL1); HSF, heat stress factor; HSP, heat shock proteins; SPX4, SYG1/Pho81/XPR1 (SPX) domain-containing protein; RSG, REPRESSION OF SHOOT GROWTH); TTL3 (TETRATRICOPEPTIDE THIOREDOXIN-LIKE 3); XND1 (XYLEM NAC DOMAIN 1).

(Amorim-Silva et al. 2019). When BR signaling is activated, TTL3 relocates and brings several BR signaling proteins to the plasma membrane, where it promotes the dephosphorylation of BZR1 and its nuclear translocation (Fig. 3). BZR1 also binds to the BROAD COMPLEX/TRAMTRACK/BRIC-A-BRAC family protein BSS1 (BRZ-SENSITIVE-SHORT HYPOCOTYL1). BSS1 localizes to both the cytosol and nucleus and forms oligomers that bind BZR1 under low BR levels. These BSS1 oligomers bind BRZ1 and trap it in the cytosol. Increased BR concentration leads to the disassembly of BSS1 oligomers, the release of BRZ1, and its translocation into the nucleus (Shimada et al. 2015). How the 14-3-3 proteins, TLL3, and BSS1 cooperate to retain BZR1 and BES1 in the cytosol is unclear, but they could provide a multilayered mechanism of BR signaling through the dynamic control of BES1 and BZR1 localization (Fig. 3).

BES1 and BZR1 can also be subjected to SUMOylation, that is, the reversible conjugation of SUMO (small ubiquitin-like modifier) through biochemical reactions similar to ubiquitination. However, the effects of SUMOylation in the 2 TFs are very different. SUMOylated BES1 is rapidly degraded (Zhang et al. 2019), whereas SUMOylation of BZR1 prevents its interaction with the BIN2 kinase and causes its accumulation within the nucleus (Srivastava et al. 2020) (Fig. 3). Salt stress stimulates the deSUMOylation of BZR1, which results in the phosphorylation of BZR1 by BIN2 and its nuclear export and degradation (Srivastava et al. 2020).

For other phosphorylated cTFs retained in the cytosol by 14-3-3 proteins, the signal that triggers their dephosphorylation and nuclear translocation is environmental (Fig. 2). For example, activation of photomorphogenesis in response to light depends on the repression of several PIFs (PHYTOCHROME INTERACTING FACTORs) (Pham et al. 2018). These are bHLH TFs that physically interact with phytochromes. Light induces phosphorylation of PIFs, promoting their ubiquitylation and degradation by the 26S proteasome (Al-Sady et al. 2006). However, phosphorylated *Arabidopsis* PIF7 is not degraded, and instead interacts with 14-3-3 proteins in the cytosol. When plants are under shade conditions, PIF7 is dephosphorylated and unable to bind 14-3-3 proteins, resulting in the translocation of PIF7 into the nucleus, where it activates expression of shade-induced genes (Huang et al.

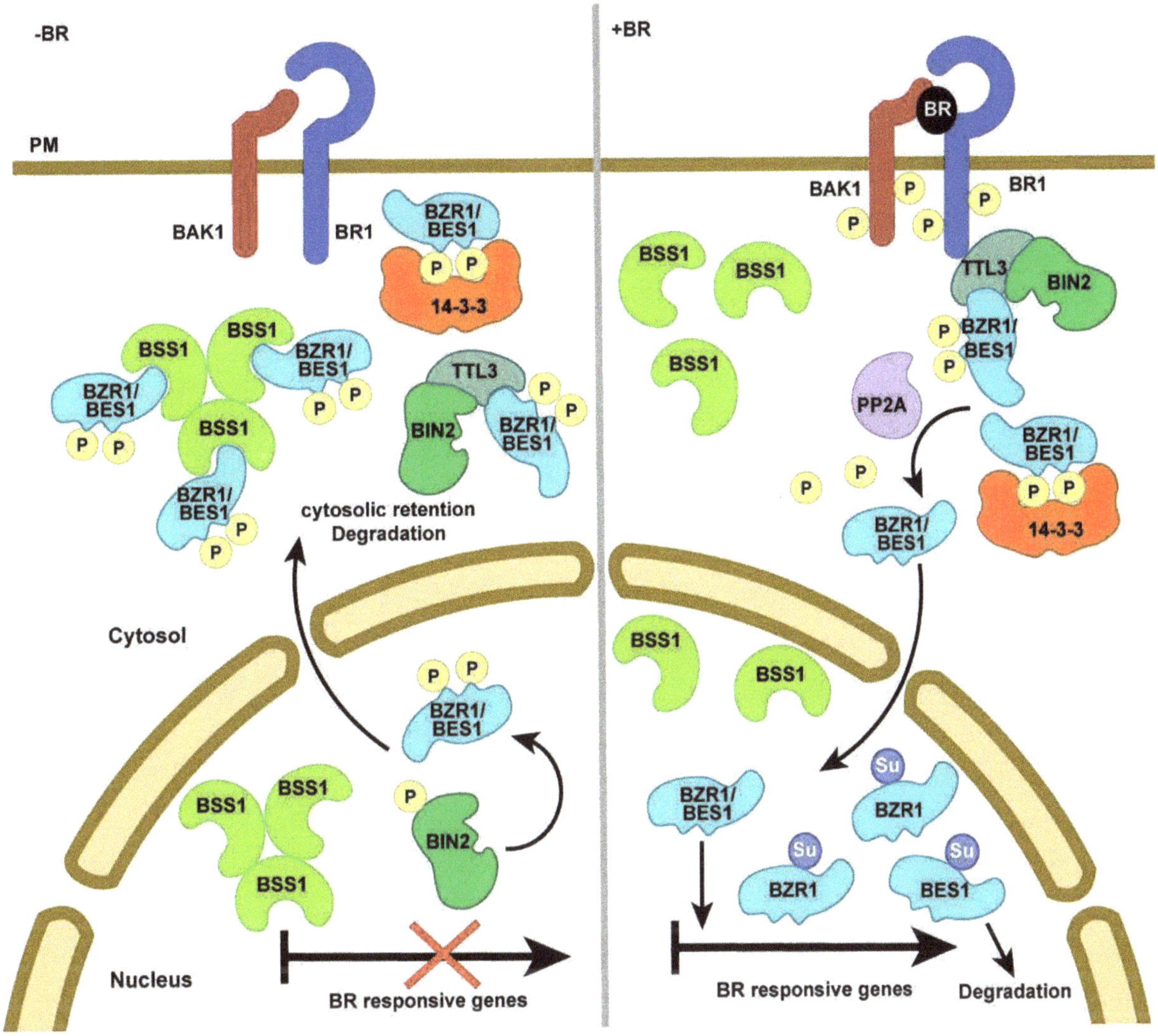

Figure 3. Main cytosolic retention mechanisms for the regulation of BR-responsive TFs BRZ1 and BES1. BZR1/BES1 are retained in the cytosol by various mechanisms involving interaction with 14-3-3 proteins, BSS1, and TTL3. How these 3 mechanisms are interconnected is currently unclear. Some components of the BR signaling pathways are omitted in this figure for clarity. P, phosphorylation; Su, SUMOylation.

2018). Interestingly, PIF7 can bind BES1 and BZR1 to regulate crosstalk between BR and light signaling (Martínez et al. 2018). PIF7 was also recently identified as being important in the in-tegration of growth responses to high temperatures in the shade (Burko et al. 2022).

Some bZIP TFs (e.g. *Arabidopsis* bZIP18, bZIP51/VIP1, bZIP52, and *Nicotiana tabacum* RSG) are also retained in the cytosol in their transcriptional inactive, phosphorylated form through interactions with 14-3-3 proteins (Igarashi et al. 2001; Nakata et al. 2009; Takeo and Ito 2017; Wiese et al. 2021) (Fig. 1). VIP1/bZIP51 (VIRE2-INTERACTING PROTEIN 1/BASIC-LEUCINE ZIPPER 51) is dephosphorylated and translocated into the nucleus upon mechanical or hypo-osmotic stress (Takeo and Ito 2017; Tsugama et al. 2019), whereas dephosphorylation and nuclear translocation of bZIP18 and bZIP52 are triggered by heat stress (Wiese et al. 2021).

RSG (REPRESSION OF SHOOT GROWTH), a *Nicotiana tabacum* bZIP TF closely related to *Arabidopsis* VIP1/bZIP1, represents a special case within plant TFs retained in the cytosol by 14-3-3 proteins. Different from other plant TFs statically retained in the cytosol until activation, RSG undergoes continuous nucleo-cytoplasmic shuttling mediated by importins and exportins (Igarashi et al. 2001; Nakata et al. 2009; Ito et al. 2014). How exactly the dynamic shuttling of RSG is controlled or what functional advantage it provides is unclear. Whereas RSG seems to be the only identified plant cTF that shows continuous nucleo-cytoplasmic shuttling in plants, several examples are well known for animal cTFs. For example, under unstimulated conditions, the animal TF NF-κB (nuclear factor κB) binds IκBα, which promotes the nuclear translocation of both proteins, mediated by the NLS of NF-κB. Because IκBα contains an NES, the NF-κB-IκBα complex quickly exits the nucleus (Huang et al. 2000; Tam et al. 2000). Thus, NF-κB–IκBα complex mostly re-sides in the cytosol because the effect of the NES in IκBα dominates that of the NLS in NF-κB. However, upon activation by perception of pathogens or cytokines, IκBα is phosphorylated and degraded, which enables the rapid localization of NF-κB to the nucleus for transcriptional regulation. The biological relevance of continuous nucleo-cytoplasmic shuttling of TFs in not always evident (Ghosh and Karin 2002) but could allow for very fast responses by keeping readily available pools of TFs ready for nuclear accumulation.

For other plant cTFs retained in the cytosol, such as PHR1 (PHOSPHATE STARVATION RESPONSE1), HSFA8, and VND6 (Puga et al. 2014; Giesguth et al. 2015; Zhong et al. 2020), no 14-3-3 binding partners have been identified. However, these 3 cTFs interact with other types of cytosolic proteins that mediate their cytosolic retention (Fig. 2). For example, PHR1, which together with other MYB TFs control gene expression under low inorganic phosphate (Pi), binds SPX (SYG1/PHO81/XPR1) domain-containing proteins. In both *Arabidopsis* and rice (*Oryza sativa*), in the presence of Pi, SPX4 binds PHR1 and retains it in the cytosol (Puga et al. 2014; Wang et al. 2014), whereas under low Pi, SPX4 is degraded, allowing free PHR1 to enter the nucleus and activate Pi-starvation response gene expression (Lv et al. 2014; Osorio et al. 2019).

Another mechanism of cytosolic retention discovered recently for some ARF (AUXIN RESPONSE FACTOR) cTFs is protein condensation. In root cells with reduced auxin responses, **ARF7** and **ARF19** form condensates in the cytosol that prevent them from entering the nucleus (Powers et al. 2019; Jing et al. 2022). In *Arabidopsis* root tips, ARF19 is predominantly inside the nucleus, but in mature root tissues it forms cytoplasmic condensates through its Phox and Bem1p (PB1) domains and its intrinsically disordered middle region. ARF7 and ARF19 interact with AFF1 (AUXIN RESPONSE FACTOR F-BOX1), an E3 ligase that mediates ubiquitylation of target proteins. In mutants lacking AFF1 function, ARF proteins showed elevated condensation. Although the mechanism by which AFF1 alters ARF dynamics is not entirely clear, it has been postulated that ubiquitination of ARF7 and ARF19 blocks their NLS, affecting their nucleo-cytoplasmic parti-tion (Jing et al. 2022).

The *Arabidopsis* MYB TFs RVE4 (REVEILLE4) and RVE8 move quickly and reversibly from the cytoplasm into the nucleus in response to either low or high temperatures to activate the expression of *DEHYDRATION-RESPONSIVE ELEMENT BINDING PROTEIN 1* (Kidokoro et al. 2021). What molecular mechanism promotes the nuclear import of RVE4 and RVE8 under unfavorable temperature conditions is currently unknown.

Although not considered a bona fide TF because it does not appear to bind DNA directly, the salicylic acid receptor NPR1 (NONEXPRESSOR OF PATHOGENESIS-RELATED GENES1) shows 1 of the most complex mechanisms of dynamic cytosolic-nuclear localization that affects directly gene expression. Inside the nucleus and in the presence of salicylic acid, NPR1 forms a complex with either transcriptional activators (e.g. TGAs within the bZIP TF family) (Zhang et al. 1999) for the expression of *PR* (*PATHOGENESIS-RELATED*) genes (Cao et al. 1997) or transcriptional repressors (e.g. WRKY TFs), reprograming gene expression in response to pathogens and other stresses (Zhou et al. 2023). NPR1 is approximately 66 kDa in size and contains an NLS sequence at its C-terminal domain (Kinkema et al. 2000). In its uninduced state, NPR1 is an oligomer formed by intermolecular disulfide bridges between cysteine (Cys) residues (Mou et al. 2003) (Fig. 4). Oligomerization of NPR1 in the cytosol is further promoted by S-nitrosylation of Cys156 from S-nitrosoglutathione, a natural nitric oxide donor that accumulates during innate immune responses (Tada et al. 2008). Salicylic acid accumulation causes a change in cellular redox potential, leading to the reduction of Cys thiol groups in NPR1 mediated by thioredoxins and the disassembly of NPR1 oligomers (Tada et al. 2008) (Fig. 4). Reduced NPR1 can then enter the nucleus, associates with TFs, and orchestrates plant immune responses. Consistent with the critical role of Cys residues in the retention of NPR1 in the cytosol, mutations of Cys82 or Cys216 result in constitutive nuclear localization and upregulated expression of defense genes (Mou et al. 2003; Tada et al. 2008). Nuclear import of monomeric NPR1 is also promoted by phosphorylation through the action of SnRK2.8 (SNF1-RELATED PROTEIN KINASE 2.8) in the cytosol (Fig. 4). SnrK2.8 phosphorylates Thr 373 and Ser 589, the latter located in the NLs region of NPR1. Phosphorylation of either residue is required for NPR1 nuclear import in distal leaves (systemic signaling) but not in the infected leaves (local immunity). Inside the nucleus, activated NPR1 can be SUMOylated, which switches NPR1 interactions from transcription repressors to activators and promotes its ubiquitination and degradation (Saleh et al. 2015) (Fig. 4).

Interestingly, under high levels of salicylic acid, NPR1 condensates both inside the nucleus and in the cytosol, forming salicylic acid–induced NPR1 condensates that contain stress response proteins. In the nucleus, within these condensates, NPR1 is a major target of ubiquitylation and degradation, whereas, in the cytoplasm, NPR1 condensates associate with CULLIN 3 to assemble a E3 ligase complex that mediates ubiquitination of other proteins, such as the lipase-like protein EDS1 involved in effector-triggered immunity (Zavaliev et al. 2020).

Cytoplasmic retention of TFs through membrane associations

Membrane-associated TFs can be fully integrated into membranes through transmembrane domains or peripherally associated with membranes through various means (lipidation/acylation or hydrophobic membrane-binding domains).

TFs with transmembrane domains (tmTFs)

Although a relatively small number of tmTFs have been experimentally characterized in plants and algae, there are

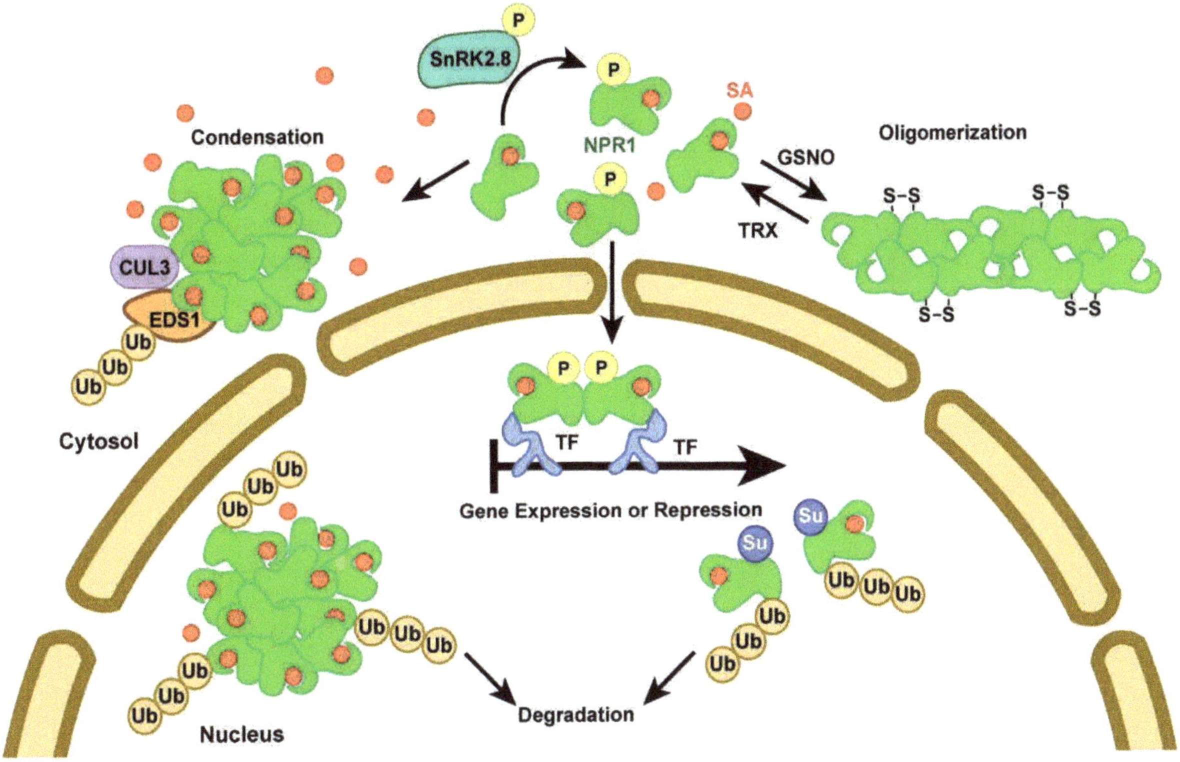

Figure 4. Trafficking of NPR1. Depending on the concentration of salicylic acid (SA) and redox state, NPR1 can be found as an oligomer, monomer, or condensate. At low SA concentration, the oligomeric form is stabilized by S-nitrosylation mediated by GSNO (S-nitrosoglutathione) and disulfide bridges (S-S) between Cys residues. Increased concentration of SA and action of thioredoxins (TRX) facilitate the disassembly of the oligomers and release of monomeric NPR1, which can be phosphorylated by the SNRK2.8 kinase and translocated into the nucleus to mediate either gene activation or repression. SUMOylation or condensation within the nucleus promote ubiquitination and degradation. NPR1 condensates in the cytosol mediate the ubiquitination and degradation of stress-related proteins such as the lipase-like protein EDS1. Su, SUMO; Ub, ubiquitin.

reasons to believe they represent an important fraction of all TFs. In silico studies of 14 species spanning from algae to flowering plants predicted that 16% of the 230 TFs in *Chlamydomonas reinhardtii* and 5% to 10% of the TFs in flowering plants contain 1 to 13 transmembrane domains (Yao et al. 2017) and that from the approximately 2,000 *Arabidopsis* TFs, 64 to 85 TFs corresponding to 24 to 27 families are tmTFs (Kim et al. 2010; Yao et al. 2017) (Fig. 5, A and B). Their representation within these families is variable, being as much as 75% of the SiF2-like (3 of 4), 50% of the NF-X1 (2 of 4), 25% of the YABBY (2 of 8), 20% of the CAMTA (2 of 10), 16% of the SBP (5 of 30), 12% of the NAC (17 of 138), and 5% of the bZIP (5 of 127) TF families. Based on their general topology, tmTFs are classified as type I and type II, both with a single transmembrane domain, and type III, with more than 1 transmembrane domain (Fig. 5C). Type I and II integral proteins differ in their orientations; the domain exposed to the cytosol is the C terminus for type I proteins and the N terminus for type II proteins (Fig. 5C), and this distinction is important because they involve different insertion mechanisms. There is a broad variation in the distribution of transmembrane domains among TF families (Kim et al. 2010; Yao et al. 2017). Most *Arabidopsis* tmTFs have 1 or 2 (a few with 3 or 4) transmembrane domains at either the C- or the N-terminal regions (Fig. 5A). The bZIP and NAC families have the largest number of identified tmTFs (Fig. 5B; Fig. 6). All the NAC tmTFs are type II proteins, whereas other families contain both type I and type II proteins (Kim et al. 2010).

Localization of tmTFs

Similar to other integral proteins, tmTFs are synthesized in the endoplasmic reticulum (ER)-associated polysomes, from where they can be trafficked to other membranes, such as the plasma membrane or be retained in the ER as an inactive form. The topology and orientation of membrane proteins is primarily determined by the direction through which the first transmembrane helix is translocated across ER membranes. However, 50% to 70% (Kim et al. 2010; Yao et al. 2017) of the *Arabidopsis* tmTFs, including many NAC and bZIP tmTFs, are predicted to be inserted post-translationally into membranes as type II proteins. These tmTFs shared features with the so-called tail-anchored proteins, which contain N-terminal cytosolic domains with diverse functions (e.g. membrane trafficking, redox reactions, signal transduction) (Kutay et al. 1993) and a single

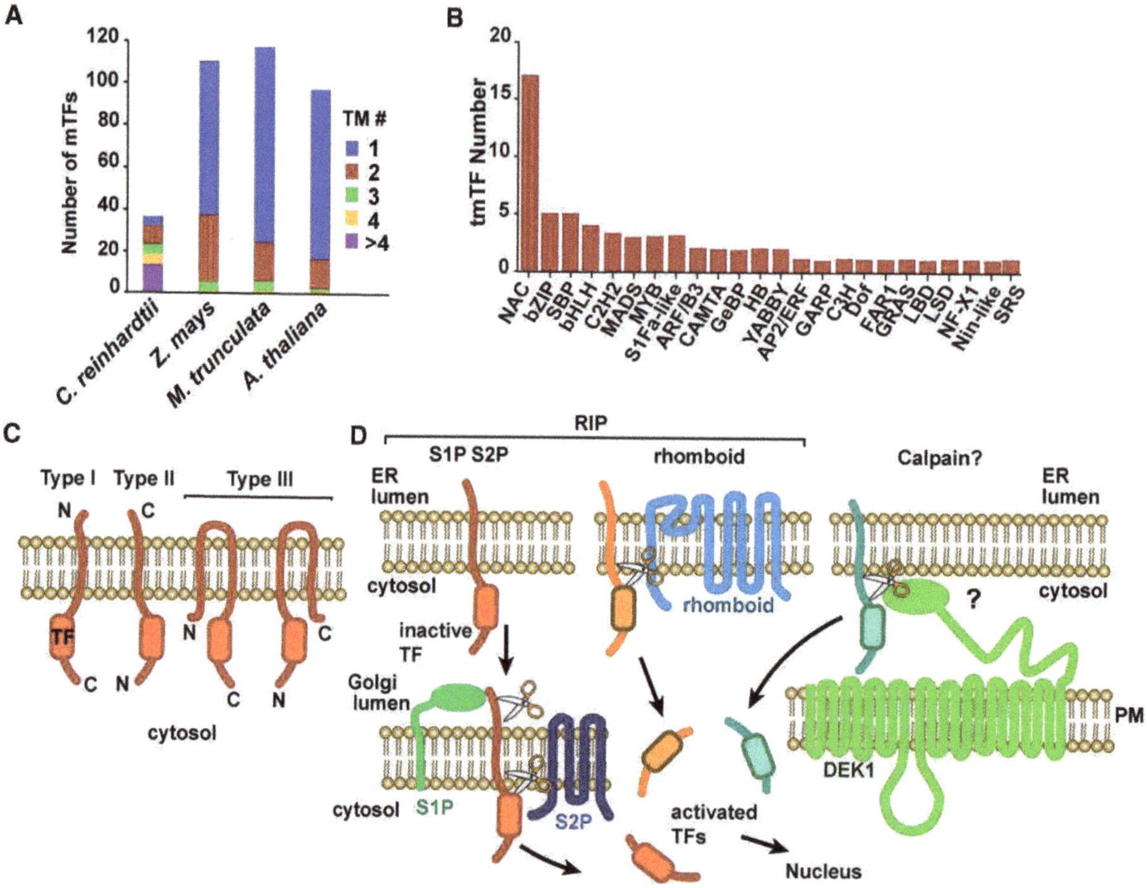

Figure 5. TFs with transmembrane domains (tmTFs). **A)** Number of predicted tmTFs in different plant and algal species, modified from Yao et al. (2017). **B)** Number of predicted tmTFs across different *Arabidopsis* TF families according to Yao et al. (Yao et al. 2017). **C)** Most predicted tmTFs in *Arabidopsis* contain either 1 (type I and type II) or 2 (type III) transmembrane domains at either the N or C terminus. **D)** Mechanism of tmTF ac-tivation by proteolytic cleavage and release from membranes. RIP, regulated intramembrane proteolysis; S1P, Site-1-Protease.

C-terminal transmembrane domain that is inserted post-translationally into many different types of membranes (e.g. ER, outer chloroplast and mitochondrial envelopes, peroxisomal membranes) (Chio et al. 2017). Because of this mechanism of membrane insertion, it is very hard to predict the subcellular localization of tail-anchored tmTFs.

Seventeen of the approximately 20 *Arabidopsis* tmTFs whose localization has been tested by expression of fluorescently tagged proteins seem to localize to the ER (Fig. 2). Four of the 75 *Arabidopsis* bZIP tmTFs (bZIP17, bZIP28, bZIP49, and bZIP60) are predicted to contain a single transmembrane domain, and 3 of them (bZIP17, bZIP28, and bZIP60) have been experimentally shown to localize to the ER in their inactive forms (Fig. 2). NAC tmTFs also show preference for ER localization. Twelve *Arabidopsis* NAC (ANAC) TFs with predicted single transmembrane domains (also called NAC with TMmotif1-like or NTL) (Liang et al. 2015) and 15 NAC TFs from soybean expressed as GFP fusions in *Arabidopsis* protoplasts localize to the ER (Ng et al. 2013; Li et al. 2016). In addition, maMYB, a member of the plant-specific R2R3-MYB TF family that controls root hair elongation (Slabaugh et al. 2011), also localizes to the ER, whereas SPL1 (SQUAMOSA PROMOTER BINDING PROTEIN-LIKE 1), involved in thermotolerance, localizes to the cell periphery (ER or plasma membrane) (Chao et al. 2017). As an example of tmTFs that localize to other organelles besides ER, *Arabidopsis* PTM (PHD type TF with transmembrane domains) localizes to the chloroplast envelope and mediates chloroplast retrograde signaling to the nucleus (Sun et al. 2011), whereas NACMTF3, involved in responses to stress in tomato (*Solanum lycopersicum*) (Bhattacharjee et al. 2017), localizes to the plasma membrane when transiently expressed as a GFP fusion.

Other predicted *Arabidopsis* tmTFs, such as CALMODULIN-BINDING TRANSCRIPTION ACTIVATOR 5, AGAMOUS-LIKE 69, SPL16, among others have been detected outside the nucleus in proteomic studies of isolated subcellular compartments such as plasma membrane, Golgi, mitochondria, and chloroplasts (De Backer et al. 2022), but their extra-nuclear localization has not been confirmed through other methods.

Once the transmembrane domains are removed from the tmTFs, the rest of the protein is free to move throughout the cytosol. According to in silico predictions, 73% of the *Arabidopsis* tmTFs contain NLSs, but it is likely that NLS motifs in other tmTFs have gone undetected (De Backer et al. 2022).

TF Family	TF	Localization	Alternative splicing	Proteolytic processing by	Signal for activation	References
bZIP	bZIP17	ER		S1P and likely S2P	ER stress, salt stress	Liu et al 2007
	bZIP28	ER		S1P and ?	ER stress, heat, BR, SA	Iwata et al 2017
	bZIP60	ER	✓		ER stress, heat, SA	Nagashima et al., 2011
NAC	ANAC013	ER		RBL	hypoxia mitoch. retrograde signaling	Eysholdt-Derzsó et al., 2023 De Clercq et al (2013)
	ANAC016 ANAC017	ER		RBL?	hypoxia hypoxia; ER stress	Eysholdt-Derzsó et al., 2023 Meng et al., 2019
	ANAC040	ER			Salt stress	Kim et al 2007 and 2008
	ANAC053	ER			heat stress anther maturation	Lee et al., 2014 Shih et al., 2014
	ANAC060	no TM in Col-0	✓		ABA signaling	Li et al., 2014
	ANAC062	ER			cold	Seo et al., 2010a Seo et al., 2010b
	ANAC068	ER		Calpain?	cell division	Kim et al (2006)
	ANAC069	ER			Salt stress	Park et al 2011
	ANAC078	ER			flavonoid synthesis	Moshita et al (2009) Yabuta et al (2010)
	ANAC089	ER			ER stress	Klein et al., 2012
	NTL9	ER			oxidative stress	Hye-Kyung et al., 2008
PTM	PHD	Chloropl. envelope			stress retrograde signaling	Sun et al., 2011
maMYB	R2R3-MYB	ER			root hair elongation?	Slabaugh et al., 2011
SBP	SPL1	ER or PM			heat?	Chao et al., 2017

Figure 6. Table of *Arabidopsis* tmTFs with subcellular localization experimentally tested. The table also includes the tmTF-activating protease, when known, and the activation signal.

There are many gaps in our knowledge about tmTFs to fully appreciate their cellular distribution and dynamics. First, in most studies, fluorescently tagged tmTFs are only characterized at low resolution; for tmTFs showing a peripheral signal in vacuolated epidermal cells or protoplasts, it is difficult to distinguish clearly between plasma membrane and ER localization. Second, most localization studies of tmTFs have been restricted to transient overexpression experiments, many times in heterologous systems. It is possible that the final and functional localization of many tmTFs requires additional binding partners expressed in specific cell types and certain developmental stages. More systematic efforts to characterize the subcellular distribution and dynamics of tmTFs will be needed to fully understand the range of cellular localization of plant tmTFs.

Activation of tmTFs

The active form of a tmTF lacks the transmembrane domains and is free to move into the nucleus through recognition of their NLSs and regulate gene expression. This is typically achieved through the proteolytic cleavage of the tmTF protein to release the cytosolic domain. Activation of tmTFs by proteolytic processing is triggered by specific signals and involves either regulated intramembrane proteolysis (Fig. 5D) or regulated ubiquitin/proteasome-dependent processing (Seo et al. 2008). The handful of tmTFs studied in plants are activated by regulated intramembrane proteolysis (no examples of regulated ubiquitin/proteasome-dependent processing–mediated activation have been yet reported in plants), but in most cases, the specific proteases required for their release are unknown. The proteases known to process tmTFs in eukaryotes include Site-1-Proteases (S1P), S2P, rhomboid proteases, and presenilins, all of which are present in plants.

In animals, the bZIP tmTF ATF6 localizes to the ER and activates gene expression as part of the Unfolded Protein Response (UPR) during ER stress. For activation, ATF6 moves from the ER to the Golgi, where it is sequentially processed by S1P and S2P (Ye et al. 2000). S1P is located in the lumen of the *cis* and medial Golgi cisternae and cleaves the lumenal domains of the tmTFs. This first cleavage by S1P facilitates the second cleavage by the intramembrane protease S2P, which releases the cytosolic domain of the tmTFs in the cyto-sol to enter the nucleus (Ye et al. 2000; Shen and Prywes 2004). All the tmTFs that share this proteolytic activation mechanism, including ATF6, have in common the presence of a cytosolic domain able to act as a TF, a S1P recognition motif RXX(R/L) in the lumen, and a helical destabilization motif, such as an NP, NXXP, or PXXP sequence in the trans-membrane helix for S2P cleavage (Ye 2020).

In plants, the UPR signaling response is largely coordinated by ER-localized tmTFs of bZIP family and consists of 2 pathways, one mediated by bZIP17 and bZIP28 and the other by bZIP60. Similar to animal ATF6, *Arabidopsis* bZIP17 and bZIP18 are released from the ER by proteolytic cleavage. The first cleavage step of bZIP17 is mediated by S1P (Liu et al. 2007), and although the identity of the protease that releases bZIP17 from the membrane has not been experimentally determined, bZIP17 contains a S2P cleavage motif close to its transmembrane segment (De Backer et al. 2022). For bZIP28, instead, the first cleavage of the lumenal domain is not mediated by S1P but by an unknown protease, whereas S2P cleaves the cytosolic domain and releases the active form for nuclear translocation (Iwata et al. 2017). Just like ATF6, the C-terminal domain of bZIP28 binds the ER chaperon BiP (Binding Protein), preventing its relocalization to the Golgi under unstressed condition (Shen et al. 2002). In response to ER stress, bZIP28 dissociates from BiP, facilitating its transport out of the ER (Srivastava et al. 2013). A conserved mechanism of bZIP tmTF activation in UPR has been reported for maize (*Zea mays*) (Yang et al. 2013) and other vascular plant species (Howell 2021).

As mentioned above, many NAC tmTFs also localize to the ER. *Arabidopsis* ANAC013 interacts with the RHOMBOID-LIKE 2 (RBL2) intramembrane protease at the ER. Under low-oxygen conditions (hypoxia), dysfunctional mitochondria initiate a retrograde signaling that leads to the cleavage of the cytosolic domain of ANAC013 by RBLs and its translocation into the nucleus (Eysholdt-Derzsó et al. 2023). ANAC017, a closely related NAC tmTF, also interacts with RBL2 (Eysholdt-Derzsó et al. 2023) and has been predicted to be cleaved by rhomboid-type proteases (Ng et al. 2013).

Besides the intramembrane S2P and rhomboid proteases, there is evidence that other proteases may mediate the release of tmTFs in plants. For example, based on pharmacological studies using specific protease inhibitors, it was proposed that the release from ER membranes of *Arabidopsis* NAC068, a tmTF that mediates cytokinin sig-naling during cell division, is catalyzed by a calpain protease (Kim et al. 2006). Interestingly, the only functional calpain in plants is DEK1 (DEFECTIVE KERNEL 1), with 21 trans-membrane domains and localized to the plasma membrane (Liang et al. 2013) (Fig. 5D). However, whether DEK1 direct-ly cleaves ANAC068 has not been experimentally proven. For some tmTFs, phosphorylation after proteolytic cleavage is important for nuclear translocation. For example, ANAC062, which upon cold induction is cleaved from the ER membrane and regulates pathogen-resistance responses, needs to be phosphorylated by SnRK2.8 to enter the nucleus (Kim et al. 2012).

Although most NAC tmTFs are released from the ER in response to stress, there seems to be some specificity for the activation of at least some of them. For example, ANAC013 and ANAC017 are released during hypoxic conditions and ER stress (Meng et al. 2019; Eysholdt-Derzsó et al. 2023), ANAC053 is released in response to heat stress (Lee et al. 2014), ANAC062 in response to low temperatures (Seo et al. 2010a, 2010b) and ER stress (Yang et al. 2014), NTL9 in response to osmotic stress (Hye-Kyung et al. 2008), ANAC089 in response to ER stress (Klein et al. 2012), and ANAC068 to promote cell division (Kim et al. 2006). Considering that only a few proteases are known to cleave tmTFs, it is unclear what provides the specificity to release only the pertinent TF from the ER membrane. Recent evidence suggests that the timing of mTF release from the ER varies among tmTFs. For example, although ANAC013, ANAC016, and ANAC017 interact with the same RBL2 rhomboid protease at the ER membrane, ANAC013 seems to be cleaved by RBL2 before the other 2 TFs, at the onset of hypoxia (Eysholdt-Derzsó et al. 2023). *Arabidopsis* contains 13 RBLs (RBL1-7 and RBL10-15) with different predicted subcellular localizations (ER, mitochondria, chloroplasts) (Kmiec-Wisniewska et al. 2008), which could also provide additional specificity to tmTF activation.

Adding to the complexity of the regulation of these ER-localized TFs, many of them are part of the same gene regulatory networks. For example, ANAC017 binds the promoters and mediates the expression of both ANAC013 and bZIP60 dur-ing mitochondrial retrograde signaling (He et al. 2022).

PTM, the only tmTF known to localize to the chloroplast envelope, contains a DNA-binding domain and an NLS at its N-terminal cytosolic domain and 4 transmembrane domains at its C terminus. Upon chloroplast stress induction by factors such as high light, PTM is cleaved by an unknown protease, released in the cytosol, and translocated into the nucleus to activate *ABI4* transcription and mediate retrograde signaling from plastids (Sun et al. 2011).

As mentioned above, bZIP60 is also an ER-localized tmTF mediating UPR such as bZIP28. However, different from bZIP28, bZIP60 is activated by alternative splicing and not by proteolytic cleavage. Upon perceiving ER stress, the ER-localized ribonuclease IRE1 (INOSITOL REQUIRING ENZYME1) cleaves the unspliced *bZIP60* mRNA in the cytosol, which is different from conventional spliceosome-dependent intranuclear RNA splicing. The spliced *bZIP60* transcript encodes a protein without transmembrane domain that moves freely into the nucleus (Nagashima et al. 2011). This signaling mechanism is conserved across eukaryotes with IRE1 mediating the unconventional splicing of bZIP60 in plants, *HOMOLOGOUS TO ATF/CREB 1* in yeast, and *X-BOX BINDING PROTEIN 1* in metazoans.

The other 18 predicted tmTFs, including LSD ONE LIKE 1, a zinc finger TF regulating oxidative stress-induced cell death, have annotated alternative transcripts that would result in TF variants without a transmembrane domain (Yao et al. 2017; De Backer et al. 2022). However, the lack of detection of alter-natively spliced tmTF transcripts should be taken with cau-tion because they may be produced only under very specific conditions, such as ER stress for the splicing of *bZIP60*. Interestingly, whereas in most *Arabidopsis* accessions ANAC060 is predicted to have a transmembrane domain as many other members of the NAC family, a single-nucleotide polymorphism in the ecotype Columbia alters the splicing pattern of the ANAC060 transcript, resulting in a shorter protein without a transmembrane domain that constitutively localizes to the nucleus and confers sugar insensitivity (Li et al. 2014). Although in principle alternative splicing and proteolytic cleavage are not mutually exclusive, there are no known examples of plant tmTFs activated by both mechanisms.

TFs peripherally associated with membranes *Acylated TFs (acTFs)*

Protein acylation is the covalent attachment of often short fatty acid chains to glycine (Gly), serine (Ser), lysine (Lys), or Cys residues (Shang et al. 2022). The most important protein acylations are myristoylation, prenylation, and, the best studied, palmitoylation (Fig. 7A). S-palmitoylation is a reversible cysteine modification tightly regulated by palmitoyltransferases and acyl protein thioesterases (Spinelli et al. 2018). Very few acTFs have been identified in eukaryotes. The only known plant acTF is *Mf*NACsa from *Medicago falcata* (Duan et al. 2017), which regulates tolerance to drought stress. Under normal conditions, *Mf*NACsa is palmitoylated and associates with the plasma membrane. Under drought stress, MfNACsa is depalmitoylated by *Mt*APT1 and translocated to the nucleus. *Mt*APT1 (Duan et al. 2017) is the only APT characterized in plants (Smokvarska et al. 2021). Interestingly, protein palmitoylation levels are affected by the nutritional stress and metabolic status in mammalian cells (Spinelli et al. 2018), opening the possibility that acTFs could be responsive to metabolic stress (Shang et al. 2022).

Protein domains that mediate membrane association

Some protein domains can mediate peripheral association with membranes (Fig. 7B). *Arabidopsis* XVP/ANAC003 has been experimentally characterized as a plant membrane-bound TF (mbTF). XVP is involved in vasculature differentiation, localizes to the plasma membrane, and does not contain any transmembrane domain (Yang et al. 2020). Instead, a positively charged domain at the C terminus enriched in Lys and arginine (Arg) is required for its plasma membrane local-ization (Lee et al. 2021a). In this case, the release of ANAC003 from the plasma membrane does not depend on proteolytic processing but on the phosphorylation status of a threonine (Thr) residue located close to Lys-Arg-rich domain (Lee et al. 2021a). In silico prediction of mbTFs is challenging and re-quires a more in-depth analysis of putative membrane- binding domains among plant TFs.

Unknown mechanisms

Rice OsMADS18 has been localized to the cell periphery/ plasma membrane and can be translocated to the nucleus in response to abscisic acid signaling (Yin et al. 2019). However, OsMADS18 does not seem to have any predicted transmembrane domain, which suggests that it is retained close to the plasma membrane by membrane tethering (either by acylation or a membrane binding domain) or by interacting with a membrane-associated protein.

Computational identification of plant extra-nuclear TFs—not that easy

There are many challenges associated with the computational prediction of TF subcellular localization, including (1) limitations in determining if a protein is a TF, particularly when no conserved domains are present; (2) limitations in the prediction of NLSs (De Backer et al. 2022); and (3) limitations associated with establishing whether a TF associates with cytoplasmic proteins that prevents translocation into the nucleus. The prediction of tmTFs is probably the most reliable (>80% accuracy) as a number of algorithms are available to detect the presence of trans-membrane domains in proteins (Tsirigos et al. 2018). Predicting whether a TF will be acylated remains challenging, although computational tools are becoming available for at least some of these post-translational modifications [e.g. GPS-Lipid; (Xie et al. 2016)]. The prediction of mbTFs remains very difficult because whether a protein domain is able to bind a membrane is highly dependent on the protein conformation and cellular environment. Thus, we are a long way to accurately predicting whether a particular TF will have extra-nuclear residence; experimental determination, primarily through subcellular localization, biochemistry, and proteomics (Walley et al. 2016; Lee et al. 2021b), remains essential.

Future challenges and perspectives

The retention of TFs outside the nucleus, either by interaction with other proteins or through attachment to membranes, provides one of several mechanisms by which the regulatory activity of a TF is controlled. It is unclear how many TFs are regulated in this fashion, but given what is currently known in plants and other organisms, as many as 10% to 20% of all TFs could be controlled by such mechanisms. This is a significant number, but as discussed here, it is remarkable how much remains to be discovered regarding the mechanisms of the cytoplasmic retention of TFs and their biological implications.

There are several reasons why it is particularly challenging to determine if a TF is regulated through cytoplasmic retention using standard high-throughput methods such as GFP-tagging and ectopic overexpression as well as mass spectrometry detection. For example, low stability outside the nucleus makes it difficult to detect by mass spectrometry and establish TF extranuclear residency. It is also possible that cytoplasmic

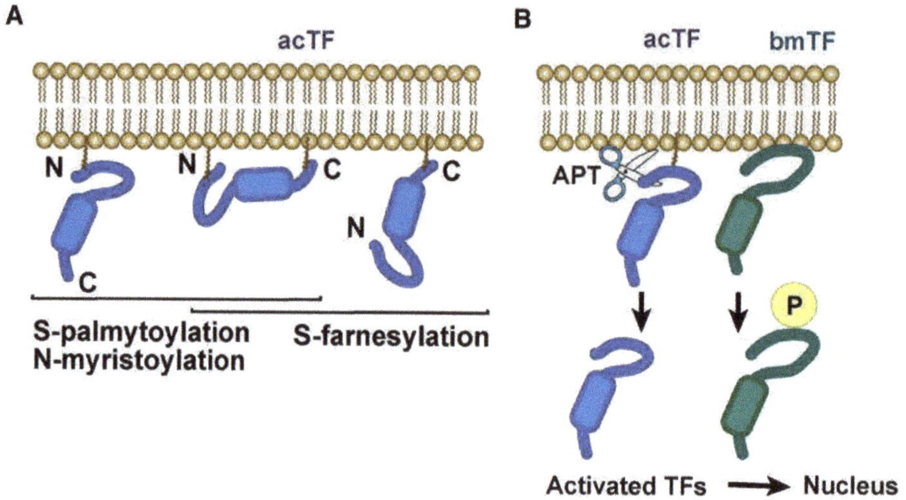

Figure 7. TFs peripherally associated with membranes. **A)** Types of possible acylation in TFs: S-palmitoylation, N-myristoylation, S-farnesylation. **B)** Mechanisms of acTF and mbTF release from membranes. De-acylation of a S-palmitoylated protein and phosphorylation of a mbTF. APT, acyl protein thioesterase.

retention is deployed only in some cell types or under specific developmental stages and therefore is not detectable under ectopic overexpression of fluorescently tagged TFs.

Orthology is often used to infer similar regulatory mechanisms across species, but this might be more challenging for predicting cytoplasmic retention, as shown for *Arabidopsis* ANAC060, which even within the same species has alleles that include or lack the transmembrane domain due to a polymorphism that alters splicing patterns (Li et al. 2014). Indeed, alternative splicing of exons harboring transmembrane domains can result in the formation of isoforms with nuclear and non-nuclear localizations, as described for bZIP60 (Nagashima et al. 2011) but are also potentially found in a number of other TFs (De Backer et al. 2022). While not formally an instance of cytoplasmic retention, as this mechanism involves a different translation product being directed to the nucleus, experimentally it can certainly be confused with nuclear shuttling if experiments are not conducted carefully. Alternate transcription start site selection can potentially also result in distinct protein isoforms, one, for example, with a N-terminal transmembrane domain, and the other without, as has been found for other plant protein domains (Mejía-Guerra et al. 2015).

The cytoplasmic retention of tmTFs offers another opportunity to control TF activity comparable to the use of glucocorticoid receptor fusions (Yamaguchi et al. 2015). TFs could be fused to a transmembrane domain containing a proteolyt-ic site corresponding to a protease whose expression can be tightly controlled.

The impact of TF cytoplasmic retention on the kinetics of the regulatory cascade controlled by the TF is unclear and will likely need to be studied on a case-by-case basis, as the extra time involved in de novo TF synthesis could easily be comparable or even shorter than the activation of a signal transduction cascade necessary to activate and release the TF from the cytoplasm. Moreover, it cannot be ruled out that the inactive extranuclear TF plays other yet to be determined functions that would provide a better reason than just simple kinetic considerations to the question of why so many TFs are controlled in this fashion.

Accession numbers

Gene identifications of TF listed in Figs. 1 and 2: BES1 (AT1G19350), BZR1 (AT1G75080), bZIP51 (AT1G43700), bZIP52 (AT1G06850), bZIP18 (AT2G40620), PIF7 (AT5G61270), bHLH039 (AT3G56980), FIT (AT2G28160), PHR1 (AT4G28610), RVE4 (AT5G02840), RVE8 (AT3G09600), HSFA8 (AT1G67970), VND6 (AT5G62380), ARF7 (AT5G2 0730), ARF19 (AT1G19220), E2Fa (AT2G36010), bZIP17 (AT2G40950), bZIP28 (AT3G10800), bZIP60 (AT1G42990), ANAC053 (AT3G10500), ANAC062 (AT3G49530), ANAC013 (AT1G69850), ANAC017 (AT1G34190), ANAC016 (AT1G 34180), ANAC078 (AT5G04410), ANAC089 (AT5G22290), ANAC040 (AT2G27300), (AT4G35580), ANAC069 (AT4G01 550), ANAC068 (AT4G01540), ANAC060 (AT3G44290), PHD (AT5G35210), R2R3-MYB (AT5G45420), SPL1(AT2G47070).

Acknowledgment

E.G. thanks David Arnosti for perspectives on TF activation.

Author contributions

SM, EG, and MSO wrote this paper.

Funding

The authors acknowledge the support provided by the US National Science Foundation through grants NSF IOS-1733633 and MCB-1822343 to E.G. and IOS-1840687 and MCB 2114603 to M.S.O.

Conflict of interest statement. None declared.

References

Al-Sady B, Ni W, Kircher S, Schäfer E, Quail PH. Photoactivated phytochrome induces rapid PIF3 phosphorylation prior to proteasome-mediated degradation. Mol Cell. 2006:**23**(3):439–446. https://doi.org/10.1016/j.molcel.2006.06.011

Amorim-Silva V, García-Moreno Á, Castillo AG, Lakhssassi N, Esteban Del Valle A, Pérez-Sancho J, Li Y, Posé D, Pérez-Rodriguez J, Lin J, et al. TTL Proteins scaffold brassinosteroid signaling components at the plasma membrane to optmize signal transduction in Arabidopsis. Plant Cell. 2019:**31**(8):1807–1828. https://doi.org/10.1105/tpc.19.00150

Anne P, Azzopardi M, Gissot L, Beaubiat S, Hématy K, Palauqui JC. OCTOPUS Negatively regulates BIN2 to control phloem differentiation in *Arabidopsis thaliana*. Curr Biol. 2015:**25**(19):2584–2590. https://doi.org/10.1016/j.cub.2015.08.033

Bhattacharjee P, Das R, Mandal A, Kundu P. Functional characterization of tomato membrane-bound NAC transcription factors. Plant Mol Biol. 2017:**93**(4–5):511–532. https://doi.org/10.1007/s11103-016-0579-z

Burko Y, Willige BC, Seluzicki A, Novák O, Ljung K, Chory J. PIF7 Is a master regulator of thermomorphogenesis in shade. Nat Commun. 2022:**13**(1):4942. https://doi.org/10.1038/s41467-022-32585-6

Cao H, Glazebrook J, Clarke JD, Volko S, Dong X. The Arabidopsis NPR1 gene that controls systemic acquired resistance encodes a novel protein containing ankyrin repeats. Cell. 1997:**88**(1):57–63. https://doi.org/10.1016/S0092-8674(00)81858-9

Chao L-M, Liu Y-Q, Chen D-Y, Xue X-Y, Mao Y-B, Chen X-Y. Arabidopsis transcription factors SPL1 and SPL12 confer plant thermotolerance at reproductive stage. Mol Plant. 2017:**10**(5):735–748. https://doi.org/10.1016/j.molp.2017.03.010

Chio US, Cho H, Shan SO. Mechanisms of tail-anchored membrane protein targeting and insertion. Annu Rev Cell Dev Biol. 2017:**33**:417–438. https://doi.org/10.1146/annurev-cellbio-100616-060839

Davuluri RV, Sun H, Palaniswamy SK, Matthews N, Molina C, Kurtz M, Grotewold E. AGRIS: arabidopsis gene regulatory information server, an information resource of Arabidopsis cis-regulatory elements and transcription factors. BMC Bioinform. 2003:**4**(1):25. https://doi.org/10.1186/1471-2105-4-25

De Backer J, Van Breusegem F, De Clercq I. Proteolytic activation of plant membrane-bound transcription factors. Front Plant Sci. 2022:**13**:927746. https://doi.org/10.3389/fpls.2022.927746

De Clercq I, Vermeirssen V, Van Aken O, Vandepoele K, Murcha MW, Law SR, Inzé A, Ng S, Ivanova A, Rombaut D, et al. The membrane-bound NAC transcription factor ANAC013 functions in mitochondrial retrograde regulation of the oxidative stress response in Arabidopsis. Plant Cell. 2013:**25**(9):3472–3490. https://doi.org/10.1105/tpc.113.117168

Duan M, Zhang R, Zhu F, Zhang Z, Gou L, Wen J, Dong J, Wang T. A lipid-anchored NAC transcription factor is translocated into the nucleus and activates glyoxalase I expression during drought stress. Plant Cell. 2017:**29**(7):1748–1772. https://doi.org/10.1105/tpc.17.00044

Eysholdt-Derzsó E, Renziehausen T, Frings S, Frohn S, von Bongartz K, Igisch CP, Mann J, Häger L, Macholl J, Leisse D, et al. Endoplasmic reticulum-bound ANAC013 factor is cleaved by RHOMBOID-LIKE 2 during the initial response to hypoxia in Arabidopsis thaliana. Proc Natl Acad Sci U S A. 2023:**120**(11): e2221308120. https://doi.org/10.1073/pnas.2221308120

Fuchs G, Voichek Y, Benjamin S, Gilad S, Amit I, Oren M. 4sUDRB-seq: measuring genomewide transcriptional elongation rates and initiation frequencies within cells. Genome Biol. 2014:**15**(5):R69. https://doi.org/10.1186/gb-2014-15-5-r69

Gampala SS, Kim TW, He JX, Tang W, Deng Z, Bai MY, Guan S, Lalonde S, Sun Y, Gendron JM, et al. An essential role for 14-3-3 proteins in brassinosteroid signal transduction in Arabidopsis. Dev Cell. 2007:**13**(2):177–189. https://doi.org/10.1016/j.devcel.2007.06.009

Ghosh S, Karin M. Missing pieces in the NF-kappaB puzzle. Cell. 2002:**109**(2):S81–S96. https://doi.org/10.1016/S0092-8674(02)00703-1

Giesguth M, Sahm A, Simon S, Dietz K-J. Redox-dependent translocation of the heat shock transcription factor AtHSFA8 from the cytosol to the nucleus in Arabidopsis thaliana. FEBS Lett. 2015:**589**(6): 718–725. https://doi.org/10.1016/j.febslet.2015.01.039

He C, Liew LC, Yin L, Lewsey MG, Whelan J, Berkowitz O. The retrograde signaling regulator ANAC017 recruits the MKK9-MPK3/6, ethylene, and auxin signaling pathways to balance mitochondrial dysfunction with growth. Plant Cell. 2022:**34**(9):3460–3481. https://doi.org/10.1093/plcell/koac177

He JX, Gendron JM, Yang Y, Li J, Wang ZY. The GSK3-like kinase BIN2 phosphorylates and destabilizes BZR1, a positive regulator of the brassinosteroid signaling pathway in Arabidopsis. Proc Natl Acad Sci U S A. 2002:**99**(15):10185–10190. https://doi.org/10.1073/pnas.152342599

Howell SH. Evolution of the unfolded protein response in plants. Plant Cell Environ. 2021:**44**(8):2625–2635. https://doi.org/10.1111/pce.14063

Huang TT, Kudo N, Yoshida M, Miyamoto S. A nuclear export signal in the N-terminal regulatory domain of IkappaBalpha controls cytoplasmic localization of inactive NF-kappaB/IkappaBalpha complexes. Proc Natl Acad Sci U S A. 2000:**97**(3):1014–1019. https://doi.org/10.1073/pnas.97.3.1014

Huang X, Zhang Q, Jiang Y, Yang C, Wang Q, Li L. Shade-induced nuclear localization of PIF7 is regulated by phosphorylation and 14-3-3 proteins in Arabidopsis. eLife. 2018:**7**:e31636. https://doi.org/10.7554/eLife.31636

Huang Y, Wang W, Yu H, Peng J, Hu Z, Chen L. The role of 14-3-3 proteins in plant growth and response to abiotic stress. Plant Cell Rep. 2022:**41**(4):833–852. https://doi.org/10.1007/s00299-021-02803-4

Hye-Kyung Y, Sang-Gyu K, Sun-Young K, Chung-Mo P. Regulation of leaf senescence by NTL9-mediated osmotic stress signaling in Arabidopsis. Mol Cells. 2008:**25**:438–445. PMID: 18443413

Igarashi D, Ishida S, Fukazawa J, Takahashi Y. 14-3-3 Proteins regulate intracellular localization of the bZIP transcriptional activator RSG. Plant Cell. 2001:**13**(11):2483–2497. https://doi.org/10.1105/tpc.010188

Ingolia NT, Lareau LF, Weissman JS. Ribosome profiling of mouse embryonic stem cells reveals the complexity and dynamics of mammalian proteomes. Cell. 2011:**147**(4):789–802. https://doi.org/10.1016/j.cell.2011.10.002

Ito T, Nakata M, Fukazawa J, Ishida S, Takahashi Y. Scaffold function of Ca2+-dependent protein kinase: tobacco Ca2+-DEPENDENT PROTEIN KINASE1 transfers 14-3-3 to the substrate REPRESSION OF SHOOT GROWTH after phosphorylation. Plant Physiol. 2014:**165**(4):1737–1750. https://doi.org/10.1104/pp.114.236448

Iwata Y, Ashida M, Hasegawa C, Tabara K, Mishiba K-I, Koizumi N. Activation of the Arabidopsis membrane-bound transcription factor bZIP28 is mediated by site-2 protease, but not site-1 protease. Plant J. 2017:**91**(3):408–415. https://doi.org/10.1111/tpj.13572

Jing H, Korasick DA, Emenecker RJ, Morffy N, Wilkinson EG, Powers SK, Strader LC. Regulation of AUXIN RESPONSE FACTOR condensation and nucleo-cytoplasmic partitioning. Nat Commun. 2022:**13**(1): 4015. https://doi.org/10.1038/s41467-022-31628-2

Kidokoro S, Hayashi K, Haraguchi H, Ishikawa T, Soma F, Konoura I, Toda S, Mizoi J, Suzuki T, Shinozaki K, et al. Posttranslational regulation of multiple clock-related transcription factors triggers cold-inducible gene expression in Arabidopsis. Proc Natl Acad Sci U S A. 2021:**118**(10):e2021048118. https://doi.org/10.1073/pnas.2021048118

Kim EJ, Lee SH, Park CH, Kim SH, Hsu CC, Xu S, Wang ZY, Kim SK, Kim TW. Plant U-Box40 mediates degradation of the brassinosteroid-responsive transcription factor BZR1 in Arabidopsis roots. Plant Cell. 2019:**31**(4):791–808. https://doi.org/10.1105/tpc.18.00941

Kim M-J, Park M-J, Seo PJ, Song J-S, Kim H-J, Park C-M. Controlled nuclear import of the transcription factor NTL6 reveals a cytoplasmic role of SnRK2.8 in the drought-stress response. Biochem J. 2012:**448**(3):353–363. https://doi.org/10.1042/BJ20120244

Kim S-G, Kim S-Y, Park C-M. A membrane-associated NAC transcription factor regulates salt-responsive flowering via FLOWERING LOCUS T in Arabidopsis. Planta. 2007:**226**(3):647–654. https://doi.org/10.1007/s00425-007-0513-3

Kim S-G, Lee S, Seo PJ, Kim S-K, Kim J-K, Park C-M. Genome-scale screening and molecular characterization of membrane-bound transcription factors in Arabidopsis and rice. Genomics. 2010:**95**(1): 56–65. https://doi.org/10.1016/j.ygeno.2009.09.003

Kim SG, Park CM. Gibberellic acid-mediated salt signaling in seed germination. Plant Signal Behav. 2008:**3**(10):877–879. https://doi.org/10.4161/psb.3.10.6247

Kim YS, Kim SG, Park JE, Park HY, Lim MH, Chua NH, Park CM. A membrane-bound NAC transcription factor regulates cell division in Arabidopsis. Plant Cell. 2006:**18**(11):3132–3144. https://doi.org/10.1105/tpc.106.043018

Kinkema M, Fan W, Dong X. Nuclear localization of NPR1 is required for activation of PR gene expression. Plant Cell. 2000:**12**(12): 2339–2350. https://doi.org/10.1105/tpc.12.12.2339

Klein P, Seidel T, Stöcker B, Dietz KJ. The membrane-tethered transcription factor ANAC089 serves as redox-dependent suppressor of stromal ascorbate peroxidase gene expression. Front Plant Sci. 2012:**3**:247. https://doi.org/10.3389/fpls.2012.00247

Kmiec-Wisniewska B, Krumpe K, Urantowka A, Sakamoto W, Pratje E, Janska H. Plant mitochondrial rhomboid, AtRBL12, has different substrate specificity from its yeast counterpart. Plant Mol Biol. 2008:**68**(1–2):159–171. https://doi.org/10.1007/s11103-008-9359-8

Kosugi S, Ohashi Y. Interaction of the Arabidopsis E2F and DP proteins confers their concomitant nuclear translocation and transactivation. Plant Physiol. 2002:**128**(3):833–843. https://doi.org/10.1104/pp.010642

Kutay U, Hartmann E, Rapoport TA. A class of membrane proteins with a C-terminal anchor. Trends Cell Biol. 1993:**3**(3):72–75. https://doi.org/10.1016/0962-8924(93)90066-A

Lee KH, Wang S, Du Q, Chhetri GT, Qi L, Wang H. The XVP/NAC003 protein associates with the plasma membrane through KR rich regions and translocates to the nucleus by changing phosphorylation status. Plant Signal Behav. 2021a:**16**(11):1970449. https://doi.org/10.1080/15592324.2021.1970449

Lee S, Lee HJ, Huh SU, Paek KH, Ha JH, Park CM. The Arabidopsis NAC transcription factor NTL4 participates in a positive feedback loop that induces programmed cell death under heat stress conditions. Plant Sci. 2014:**227**:76–83. https://doi.org/10.1016/j.plantsci.2014.07.003

Lee Y, Okita TW, Szymanski DB. A co-fractionation mass spectrometry-based prediction of protein complex assemblies in the developing rice aleurone-subaleurone. Plant Cell. 2021b:**33**(9): 2965–2980. https://doi.org/10.1093/plcell/koab182

Li J, Nam KH. Regulation of brassinosteroid signaling by a GSK3/SHAGGY-like kinase. Science. 2002:**295**(5558):1299–1301. https://doi.org/10.1126/science.1065769

Li P, Zhou H, Shi X, Yu B, Zhou Y, Chen S, Wang Y, Peng Y, Meyer RC, Smeekens SC, et al. The ABI4-induced Arabidopsis ANAC060 transcription factor attenuates ABA signaling and renders seedlings sugar insensitive when present in the nucleus. PLoS Genet. 2014:**10**(3): e1004213. https://doi.org/10.1371/journal.pgen.1004213

Li S, Wang N, Ji D, Xue Z, Yu Y, Jiang Y, Liu J, Liu Z, Xiang F. Evolutionary and functional analysis of membrane-bound NAC transcription factor genes in soybean. Plant Physiol. 2016:**172**(3): 1804–1820. https://doi.org/10.1104/pp.16.01132

Liang M, Li H, Zhou F, Li H, Liu J, Hao Y, Wang Y, Zhao H, Han S. Subcellular distribution of NTL transcription factors in *Arabidopsis thaliana*. Traffic. 2015:**16**(10):1062–1074. https://doi.org/10.1111/tra.12311

Liang Z, Demko V, Wilson RC, Johnson KA, Ahmad R, Perroud P-F, Quatrano R, Zhao S, Shalchian-Tabrizi K, Otegui MS, et al. The catalytic domain CysPc of the DEK1 calpain is functionally conserved in land plants. Plant J. 2013:**75**(5):742–754. https://doi.org/10.1111/tpj.12235

Liu JX, Srivastava R, Che P, Howell SH. Salt stress responses in Arabidopsis utilize a signal transduction pathway related to endoplasmic reticulum stress signaling. Plant J. 2007:**51**(5):897–909. https://doi.org/10.1111/j.1365-313X.2007.03195.x

Liu Y, Li P, Fan L, Wu M. The nuclear transportation routes of membrane-bound transcription factors. Cell Commun Signal. 2018:**16**(1):12. https://doi.org/10.1186/s12964-018-0224-3

Lu J, Wu T, Zhang B, Liu S, Song W, Qiao J, Ruan H. Types of nuclear localization signals and mechanisms of protein import into the nucleus. Cell Commun Signal. 2021:**19**(1):60. https://doi.org/10.1186/s12964-021-00741-y

Lv Q, Zhong Y, Wang Y, Wang Z, Zhang L, Shi J, Wu Z, Liu Y, Mao C, Yi K, et al. SPX4 Negatively regulates phosphate signaling and homeostasis through its interaction with PHR2 in rice. Plant Cell. 2014:**26**(4):1586–1597. https://doi.org/10.1105/tpc.114.123208

Martínez C, Espinosa-Ruíz A, de Lucas M, Bernardo-García S, Franco-Zorrilla JM, Prat S. PIF4-induced BR synthesis is critical to diurnal and thermomorphogenic growth. EMBO J. 2018:**37**(23): e99552. https://doi.org/10.15252/embj.201899552

Mejía-Guerra MK, Li W, Galeano NF, Vidal M, Gray J, Doseff AI, Grotewold E. Core promoter plasticity between maize tissues and genotypes contrasts with predominance of sharp transcription initiation sites. Plant Cell. 2015:**27**(12):3309–3320. https://doi.org/10.1105/tpc.15.00630

Meng X, Li L, De Clercq I, Narsai R, Xu Y, Hartmann A, Claros DL, Custovic E, Lewsey MG, Whelan J, et al. ANAC017 Coordinates organellar functions and stress responses by reprogramming retrograde signaling. Plant Physiol. 2019:**180**(1):634–653. https://doi.org/10.1104/pp.18.01603

Merkle T. Nucleo-cytoplasmic transport of proteins and RNA in plants. Plant Cell Rep. 2011:**30**(2):153–176. https://doi.org/10.1007/s00299-010-0928-3

Morishita T, Kojima Y, Maruta T, Nishizawa-Yokoi A, Yabuta Y, Shigeoka S. Arabidopsis NAC transcription factor, ANAC078, regulates flavonoid biosynthesis under high-light. Plant Cell Physiol. 2009:**50**(12):2210–2222. https://doi.org/10.1093/pcp/pcp159

Mou Z, Fan W, Dong X. Inducers of plant systemic acquired resistance regulate NPR1 function through redox changes. Cell. 2003:**113**(7): 935–944. https://doi.org/10.1016/S0092-8674(03)00429-X

Nagashima Y, Mishiba K, Suzuki E, Shimada Y, Iwata Y, Koizumi N. Arabidopsis IRE1 catalyses unconventional splicing of bZIP60 mRNA to produce the active transcription factor. Sci Rep. 2011:**1**(1):29. https://doi.org/10.1038/srep00029

Nakata M, Yuasa T, Takahashi Y, Ishida S. CDPK1, A calcium-dependent protein kinase, regulates transcriptional activator RSG in response to gibberellins. Plant Signal Behav. 2009:**4**(5):372–374. https://doi.org/10.4161/psb.4.5.8229

Ng S, Ivanova A, Duncan O, Law SR, Van Aken O, De Clercq I, Wang Y, Carrie C, Xu L, Kmiec B, et al. A membrane-bound NAC transcription factor, ANAC017, mediates mitochondrial retrograde signaling in Arabidopsis. Plant Cell. 2013:**25**(9):3450–3471. https://doi.org/10.1105/tpc.113.113985

Nolan TM, Brennan B, Yang M, Chen J, Zhang M, Li Z, Wang X, Bassham DC, Walley J, Yin Y. Selective autophagy of BES1 mediated by DSK2 balances plant growth and survival. Dev Cell. 2017:**41**(1): 33–46.e37. https://doi.org/10.1016/j.devcel.2017.03.013

Nolan TM, Vukašinović N, Liu D, Russinova E, Yin Y. Brassinosteroids: multidimensional regulators of plant growth, development, and stress responses. Plant Cell. 2020:**32**(2):295–318. https://doi.org/10.1105/tpc.19.00335

Osorio MB, Ng S, Berkowitz O, De Clercq I, Mao C, Shou H, Whelan J, Jost R. SPX4 Acts on PHR1-dependent and -independent regulation of shoot phosphorus status in Arabidopsis. Plant Physiol. 2019:**181**(1):332–352. https://doi.org/10.1104/pp.18.00594

Park J, Kim YS, Kim SG, Jung JH, Woo JC, Park CM. Integration of auxin and salt signals by the NAC transcription factor NTM2 during seed germination in Arabidopsis. Plant Physiol. 2011:**156**(2):537–549. https://doi.org/10.1104/pp.111.177071

Pham VN, Kathare PK, Huq E. Phytochromes and phytochrome interacting factors. Plant Physiol. 2018:**176**(2):1025–1038. https://doi.org/10.1104/pp.17.01384

Powers SK, Holehouse AS, Korasick DA, Schreiber KH, Clark NM, Jing H, Emenecker R, Han S, Tycksen E, Hwang I, et al. Nucleo-cytoplasmic partitioning of ARF proteins controls auxin responses in *Arabidopsis thaliana*. Mol Cell. 2019:**76**(1):177–190.e175. https://doi.org/10.1016/j.molcel.2019.06.044

Puga MI, Mateos I, Charukesi R, Wang Z, Franco-Zorrilla JM, de Lorenzo L, Irigoyen ML, Masiero S, Bustos R, Rodríguez J, et al. SPX1 Is a phosphate-dependent inhibitor of phosphate starvation response 1 in Arabidopsis. Proc Natl Acad Sci U S A. 2014:**111**(41): 14947–14952. https://doi.org/10.1073/pnas.1404654111

Riaño-Pachón DM, Ruzicic S, Dreyer I, Mueller-Roeber B. PlnTFDB: an integrative plant transcription factor database. BMC Bioinform. 2007:**8**(1):42. https://doi.org/10.1186/1471-2105-8-42

Riechmann JL, Heard J, Martin G, Reuber L, Jiang CZ, Keddie J, Adam L, Pineda O, Ratcliffe OJ, Samaha RR, et al. Arabidopsis transcription factors: genome-wide comparative analysis among eukaryotes. Science. 2000:**290**(5499):2105–2110. https://doi.org/10.1126/science.290.5499.2105

Ryu H, Cho H, Kim K, Hwang I. Phosphorylation dependent nucleocytoplasmic shuttling of BES1 is a key regulatory event in brassinosteroid signaling. Mol Cells. 2010b:**29**(3):283–290. https://doi.org/10.1007/s10059-010-0035-x

Ryu H, Kim K, Cho H, Hwang I. Predominant actions of cytosolic BSU1 and nuclear BIN2 regulate subcellular localization of BES1 in brassinosteroid signaling. Mol Cells. 2010a:**29**(3):291–296. https://doi.org/10.1007/s10059-010-0034-y

Ryu H, Kim K, Cho H, Park J, Choe S, Hwang I. Nucleocytoplasmic shuttling of BZR1 mediated by phosphorylation is essential in Arabidopsis brassinosteroid signaling. Plant Cell. 2007:**19**(9): 2749–2762. https://doi.org/10.1105/tpc.107.053728

Saleh A, Withers J, Mohan R, Marqués J, Gu Y, Yan S, Zavaliev R, Nomoto M, Tada Y, Dong X. Posttranslational modifications of the master transcriptional regulator NPR1 enable dynamic but tight control of plant immune rßesponses. Cell Host Microbe. 2015:**18**(2): 169–182. https://doi.org/10.1016/j.chom.2015.07.005

Sehnke PC, DeLille JM, Ferl RJ. Consummating signal transduction: the role of 14-3-3 proteins in the completion of signal-induced transitions in protein activity. Plant Cell. 2002:**14**(suppl 1):S339–S354. https://doi.org/10.1105/tpc.010430

Seo PJ, Kim MJ, Song JS, Kim YS, Kim HJ, Park CM. Proteolytic processing of an Arabidopsis membrane-bound NAC transcription factor is triggered by cold-induced changes in membrane fluidity. Biochem J. 2010a:**427**(3):359–367. https://doi.org/10.1042/BJ20091762

Seo PJ, Kim MJ, Park J-Y, Kim S-Y, Jeon J, Lee Y-H, Kim J, Park C-M. Cold activation of a plasma membrane-tethered NAC transcription factor induces a pathogen resistance response in Arabidopsis. Plant J. 2010b:**61**(4):661–671. https://doi.org/10.1111/j.1365-313X.2009.04091.x

Seo PJ, Kim SG, Park CM. Membrane-bound transcription factors in plants. Trends Plant Sci. 2008:**13**(10):550–556. https://doi.org/10.1016/j.tplants.2008.06.008

Shang B, Wang L, Yan X, Li Y, Li C, Wu C, Wang T, Guo X, Choi SW, Zhang T, et al. Intrinsically disordered proteins SAID1/2 condensate on SERRATE for dual inhibition of miRNA biogenesis in Arabidopsis. Proc Natal Acad Sci U S A. 2023:**120**(14):e2216006120. https://doi.org/10.1073/pnas.2216006120

Shang S, Liu J, Hua F. Protein acylation: mechanisms, biological functions and therapeutic targets. Sig Transduct Target Ther. 2022:**7**(1): 396. https://doi.org/10.1038/s41392-022-01245-y

Shen J, Chen X, Hendershot L, Prywes R. ER Stress regulation of ATF6 localization by dissociation of BiP/GRP78 binding and unmasking of Golgi localization signals. Dev Cell. 2002:**3**(1):99–111. https://doi.org/10.1016/S1534-5807(02)00203-4

Shen J, Prywes R. Dependence of site-2 protease cleavage of ATF6 on prior site-1 protease digestion is determined by the size of the luminal domain of ATF6. J Biol Chem. 2004:**279**(41):43046–43051. https://doi.org/10.1074/jbc.M408466200

Shimada S, Komatsu T, Yamagami A, Nakazawa M, Matsui M, Kawaide H, Natsume M, Osada H, Asami T, Nakano T. Formation and dissociation of the BSS1 protein complex regulates plant development via brassinosteroid signaling. Plant Cell. 2015:**27**(2):375–390. https://doi.org/10.1105/tpc.114.131508

Singh J, Padgett RA. Rates of in situ transcription and splicing in large human genes. Nat Struct Mol Biol. 2009:**16**(11):1128–1133. https://doi.org/10.1038/nsmb.1666

Slabaugh E, Held M, Brandizzi F. Control of root hair development in *Arabidopsis thaliana* by an endoplasmic reticulum anchored member of the R2R3-MYB transcription factor family. Plant J. 2011:**67**(3): 395–405. https://doi.org/10.1111/j.1365-313X.2011.04602.x

Smokvarska M, Jaillais Y, Martinière A. Function of membrane domains in rho-of-plant signaling. Plant Physiol. 2021:**185**(3):663–681. https://doi.org/10.1093/plphys/kiaa082

Spinelli M, Fusco S, Grassi C. Nutrient-dependent changes of protein palmitoylation: impact on nuclear enzymes and regulation of gene expression. Int J Mol Sci. 2018:**19**(12):3820. https://doi.org/10.3390/ijms19123820

Srivastava M, Srivastava AK, Orosa-Puente B, Campanaro A, Zhang C, Sadanandom A. SUMO Conjugation to BZR1 enables brassinosteroid signaling to integrate environmental cues to shape plant growth. Curr Biol. 2020:**30**(8):1410–1423.e1413. https://doi.org/10.1016/j.cub.2020.01.089

Srivastava R, Deng Y, Shah S, Rao AG, Howell SH. BINDING PROTEIN is a master regulator of the endoplasmic reticulum stress sensor/transducer bZIP28 in Arabidopsis. Plant Cell. 2013:**25**(4): 1416–1429. https://doi.org/10.1105/tpc.113.110684

Sun X, Feng P, Xu X, Guo H, Ma J, Chi W, Lin R, Lu C, Zhang L. A chloroplast envelope-bound PHD transcription factor mediates chloroplast signals to the nucleus. Nat Commun. 2011:**2**(1):477. https://doi.org/10.1038/ncomms1486

Sun Y, Fan XY, Cao DM, Tang W, He K, Zhu JY, He JX, Bai MY, Zhu S, Oh E, et al. Integration of brassinosteroid signal transduction with the transcription network for plant growth regulation in Arabidopsis. Dev Cell. 2010:**19**(5):765–777. https://doi.org/10.1016/j.devcel.2010.10.010

Tada Y, Spoel SH, Pajerowska-Mukhtar K, Mou Z, Song J, Wang C, Zuo J, Dong X. Plant immunity requires conformational changes [corrected] of NPR1 via S-nitrosylation and thioredoxins. Science. 2008:**321**(5891):952–956. https://doi.org/10.1126/science.1156970

Takeo K, Ito T. Subcellular localization of VIP1 is regulated by phosphorylation and 14-3-3 proteins. FEBS Lett. 2017:**591**(13): 1972–1981. https://doi.org/10.1002/1873-3468.12686

Tam WF, Lee LH, Davis L, Sen R. Cytoplasmic sequestration of rel proteins by IkappaBalpha requires CRM1-dependent nuclear export. Mol Cell Biol. 2000:**20**(6):2269–2284. https://doi.org/10.1128/MCB.20.6.2269-2284.2000

Tang W, Yuan M, Wang R, Yang Y, Wang C, Oses-Prieto JA, Kim TW, Zhou HW, Deng Z, Gampala SS, et al. PP2A Activates brassinosteroid-responsive gene expression and plant growth by dephosphorylating BZR1. Nat Cell Biol. 2011:**13**(2):124–131. https://doi.org/10.1038/ncb2151

Tsirigos KD, Govindarajan S, Bassot C, Västermark Å, Lamb J, Shu N, Elofsson A. Topology of membrane proteins-predictions, limitations and variations. Curr Opin Struct Biol. 2018:**50**:9–17. https://doi.org/10.1016/j.sbi.2017.10.003

Tsugama D, Yoon HS, Fujino K, Liu S, Takano T. Protein phosphatase 2A regulates the nuclear accumulation of the Arabidopsis bZIP protein VIP1 under hypo-osmotic stress. J Exp Bot. 2019:**70**(21): 6101–6112. https://doi.org/10.1093/jxb/erz384

Walley JW, Sartor RC, Shen Z, Schmitz RJ, Wu KJ, Urich MA, Nery JR, Smith LG, Schnable JC, Ecker JR, et al. Integration of omic networks in a developmental atlas of maize. Science. 2016:**353**(6301):814–818. https://doi.org/10.1126/science.aag1125

Wang Z, Ruan W, Shi J, Zhang L, Xiang D, Yang C, Li C, Wu Z, Liu Y, Yu Y, et al. Rice SPX1 and SPX2 inhibit phosphate starvation responses through interacting with PHR2 in a phosphate-dependent manner. Proc Natl Acad Sci U S A. 2014:**111**(41):14953–14958. https://doi.org/10.1073/pnas.1404680111

Wiese AJ, Steinbachová L, Timofejeva L, Čermák V, Klodová B, Ganji RS, Limones-Mendez M, Bokvaj P, Hafidh S, Potěšil D, et al. Arabidopsis bZIP18 and bZIP52 accumulate in nuclei following heat stress where they regulate the expression of a similar set of genes. Int J Mol Sci. 2021:**22**(2):530. https://doi.org/10.3390/ijms22020530

Withers J, Yao J, Mecey C, Howe GA, Melotto M, He SY. Transcription factor-dependent nuclear localization of a transcriptional repressor in jasmonate hormone signaling. Proc Natl Acad Sci U S A. 2012:**109**(49):20148–20153. https://doi.org/10.1073/pnas.1210054109

Xie Y, Zheng Y, Li H, Luo X, He Z, Cao S, Shi Y, Zhao Q, Xue Y, Zuo Z, et al. GPS-Lipid: a robust tool for the prediction of multiple lipid modification sites. Sci Rep. 2016:**6**(1):28249. https://doi.org/10.1038/srep28249

Yabuta Y, Morishita T, Kojima Y, Maruta T, Nishizawa-Yokoi A, Shigeoka S. Identification of recognition sequence of ANAC078 protein by the cyclic amplification and selection of targets technique. Plant Signal Behav. 2010:**5**(6):695–697. https://doi.org/10.4161/psb.5.6.11577

Yamaguchi N, Winter CM, Wellmer F, Wagner D. Plant functional genomics: methods and protocols. In: **Alonso JM, Stepanova AN,** editors. Identification of direct targets of plant transcription factors using the GR fusion technique. New York (NY): Springer New York; 2015. p. 123–138.

Yang JH, Lee KH, Du Q, Yang S, Yuan B, Qi L, Wang H. A membrane-associated NAC domain transcription factor XVP interacts with TDIF co-receptor and regulates vascular meristem activity. New Phytol. 2020:**226**(1):59–74. https://doi.org/10.1111/nph.16289

Yang Y-G, Lv W-T, Li M-J, Wang B, Sun D-M, Deng X. Maize membrane-bound transcription factor Zmbzip17 is a key regulator in the cross-talk of ER quality control and ABA signaling. Plant Cell Physiol. 2013:**54**(12):2020–2033. https://doi.org/10.1093/pcp/pct142

Yang Z-T, Lu S-J, Wang M-J, Bi D-L, Sun L, Zhou S-F, Song Z-T, Liu J-X. A plasma membrane-tethered transcription factor, NAC062/ANAC062/NTL6, mediates the unfolded protein response in Arabidopsis. Plant J. 2014:**79**(6):1033–1043. https://doi.org/10.1111/tpj.12604

Yao S, Deng L, Zeng K. Genome-wide in silico identification of membrane-bound transcription factors in plant species. PeerJ. 2017:**5**:e4051. https://doi.org/10.7717/peerj.4051

Ye J. Transcription factors activated through RIP (regulated intramembrane proteolysis) and RAT (regulated alternative translocation). J Biol Chem. 2020:**295**(30):10271–10280. https://doi.org/10.1074/jbc.REV120.012669

Ye J, Rawson RB, Komuro R, Chen X, Davé UP, Prywes R, Brown MS, Goldstein JL. ER Stress induces cleavage of membrane-bound ATF6 by the same proteases that process SREBPs. Mol Cell. 2000:**6**(6): 1355–1364. https://doi.org/10.1016/S1097-2765(00)00133-7

Yilmaz A, Mejia-Guerra MK, Kurz K, Liang X, Welch L, Grotewold E. AGRIS: the Arabidopsis gene regulatory information server, an update. Nucleic Acids Res. 2010:**39**(Database):D1118–D1122. https://doi.org/10.1093/nar/gkq1120

Yin XM, Liu X, Lu B, Domg T, Yang D, Ye T, Feng Y-Q, Wu Y. OsMADS18, a membrane-bound MADS-box transcription factor, modulates plant architecture and the abscisic acid response in rice. J Exp Bot. 2019:**70**(15):3895–3909. https://doi.org/10.1093/jxb/erz198

Yin Y, Wang ZY, Mora-Garcia S, Li J, Yoshida S, Asami T, Chory J. BES1 Accumulates in the nucleus in response to brassinosteroids to regulate gene expression and promote stem elongation. Cell. 2002:**109**(2):181–191. https://doi.org/10.1016/S0092-8674(02)00721-3

Zavaliev R, Mohan R, Chen T, Dong X. Formation of NPR1 condensates promotes cell survival during the plant immune response. Cell. 2020:**182**(5):1093–1108.e1018. https://doi.org/10.1016/j.cell.2020.07.016

Zhang L, Han Q, Xiong J, Zheng T, Han J, Zhou H, Lin H, Yin Y, Zhang D. Sumoylation of BRI1-EMS-SUPPRESSOR 1 (BES1) by the SUMO E3 ligase SIZ1 negatively regulates brassinosteroids signaling in *Arabidopsis thaliana*. Plant Cell Physiol. 2019:**60**(10):2282–2292. https://doi.org/10.1093/pcp/pcz125

Zhang Y, Fan W, Kinkema M, Li X, Dong X. Interaction of NPR1 with basic leucine zipper protein transcription factors that bind sequences required for salicylic acid induction of the PR-1 gene. Proc Natl Acad Sci U S A. 1999:**96**(11):6523–6528. https://doi.org/10.1073/pnas.96.11.6523

Zhao J, Peng P, Schmitz RJ, Decker AD, Tax FE, Li J. Two putative BIN2 substrates are nuclear components of brassinosteroid signaling. Plant Physiol. 2002:**130**(3):1221–1229. https://doi.org/10.1104/pp.102.010918

Zhong R, Kandasamy MK, Ye Z-H. XND1 Regulates secondary wall deposition in xylem vessels through the inhibition of VND functions. Plant Cell Physiol. 2020:**62**(1):53–65. https://doi.org/10.1093/pcp/pcaa140

Zhou P, Zavaliev R, Xiang Y, Dong X. Seeing is believing: understanding functions of NPR1 and its paralogs in plant immunity through cellular and structural analyses. Curr Op Plant Biol. 2023:**73**:102352. https://doi.org/10.1016/j.pbi.2023.102352

Zupicich J, Brenner SE, Skarnes WC. Computational prediction of membrane-tethered transcription factors. Genome Biol. 2001:**2**(12):RESEARCH0050.0051. https://doi.org/10.1186/gb-2001-2-12-research0050

https://doi.org/10.1093/plcell/koad197

THE PLANT CELL

THE PLANT CELL 2024: 36: 1540–1564

What, where, and how: Regulation of translation and the translational landscape in plants

Hsin-Yen Larry Wu [1], **Joey Jen** [1] and **Polly Yingshan Hsu** [1,*]

1 Department of Biochemistry and Molecular Biology, Michigan State University, East Lansing, MI 48824, USA

*Author for correspondence: pollyhsu@msu.edu

The author responsible for distribution of materials integral to the findings presented in this article in accordance with the policy described in the Instructions for Authors (https://academic.oup.com/plcell/pages/General-Instructions) is: Polly Yingshan Hsu (pollyhsu@msu.edu).

Review

Abstract

Translation is a crucial step in gene expression and plays a vital role in regulating various aspects of plant development and environmental responses. It is a dynamic and complex program that involves interactions between mRNAs, transfer RNAs, and the ribosome machinery through both cis- and trans-regulation while integrating internal and external signals. Translational control can act in a global (transcriptome-wide) or mRNA-specific manner. Recent advances in genome-wide techniques, particularly ribosome profiling and proteomics, have led to numerous exciting discoveries in both global and mRNA-specific translation. In this review, we aim to provide a "primer" that introduces readers to this fascinating yet complex cellular process and provide a big picture of how essential components connect within the network. We begin with an overview of mRNA translation, followed by a discussion of the experimental approaches and recent findings in the field, focusing on unannotated translation events and translational control through cis-regulatory elements on mRNAs and trans-acting factors, as well as signaling networks through 3 conserved translational regulators TOR, SnRK1, and GCN2. Finally, we briefly touch on the spatial regulation of mRNAs in translational control. Here, we focus on cytosolic mRNAs; translation in organelles and viruses is not covered in this review.

Introduction

Basics of mRNA translation

Eukaryotic gene expression begins with the transcription of genes in the nucleus. The resulting transcripts are conceptually classified into protein-coding RNAs (or mRNAs) and noncoding RNAs, although translation from some noncoding RNAs has been reported (see section on 'Unannotated translation events'). Following transcription, the primary transcripts or pre-mRNAs undergo processing, including the addition of a 5′ cap and a 3′ poly-A tail and splicing. Once the mature mRNAs are transported out of the nucleus, they are generally ready for translation. However, mRNAs may be degraded or stored in some circumstances, preventing them from being translated.

The major players in translation include mRNAs, ribosomes, transfer RNAs (tRNAs), and translation factors involved in different phases of translation. Below we briefly describe the main steps of translation and core translation factors. For detailed composition of the translation machinery in plants, including ribosomal proteins and translation factors, we refer readers to this comprehensive review (Browning and Bailey-Serres 2015).

Eukaryotic mRNAs typically have a 5′ cap (7-methylguanosine) and a 3′ poly-A tail (Fig. 1A: an overview of translation initiation; Fig. 1B: a summary of core translation factors). The cap serves as an anchor for translational initiation. Cap-dependent translation initiation is governed by 2 large protein complexes: the eIF4F complex (consisting of eukaryotic initiation factor 4E (eIF4E) and eIF4G) and the 43S preinitiation complex (PIC) (consisting of the 40S ribosomal small subunit, ternary complex: eIF2/GTP/Met-$tRNA_i$, eIF1, eIF1A, eIF3, and eIF5). First, the 5′ cap is recognized by eIF4E, which

Received May 04, 2023. Accepted June 15, 2023. Advance access publication July 12, 2023.

A

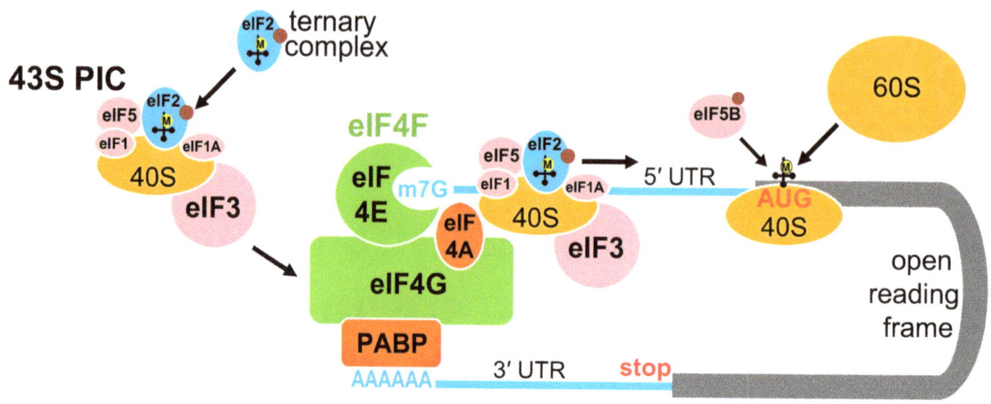

B

Initiation factor (eIF)		Main function
eIF1		PIC formation, scanning, start codon recognition
eIF1A		PIC formation, scanning, start codon recognition
eIF2	α, β, γ	Small GTPase, forms ternary complex with GTP and Met-tRNAi
eIF2B		GDP/GTP exchange factor for eIF2
eIF3	13 subunits	PIC formation, scanning, start codon recognition, reinitiation
eIF4A		Helicase, unwinds mRNA
eIF4B		Cofactor of eIF4A, unwinds mRNA
eIF4F	eIF4E	Binds the 5' Cap
	eIF4G	Scaffold protein
eIF5		Place Met-tRNAi at start codon
eIF5B		GTPase, joining of 60S subunit
PABP		Binds the 3' poly-A tail, interacts with eIF4G

Elongation factor (eEF)		Main function
eEF1A		Small GTPase, binds aminoacyl-tRNA and GTP
eEF1B	α, β, γ	GDP/GTP exchange factor for eEF1A
eEF2		GTPase, translocation

Termination factor	Main function
eRF1	Termination, peptide release
eRF3	Termination, peptide release
ABCE1	Ribosome recycling

Figure 1. Overview of eukaryotic translation initiation complexes and functions of core translation factors. **A)** Eukaryotic mRNAs typically have a 5′ m7G cap and a 3′ poly-A tail. Translation initiation is controlled by 2 large protein complexes, including the eIF4F complex and the 43S PIC. eIF4F consists of eIF4E (cap binding) and eIF4G (a scaffold protein). The 43S PIC consists of the 40S ribosomal subunit, the ternary complex (eIF2, GTP, Met-tRNA$_i$), eIF1, eIF1A, eIF3, and eIF5. The eIF4F complex binds to the cap and recruits the 43S PIC, which then scans the 5′ UTR until a start codon is identified. Subsequently, most initiation factors dissociate, and the 60S ribosomal subunit joins. GTP is shown in dark red circles. The figure was adapted from (Castellano and Merchante 2021; Passmore and Coller 2022). **B)** Summary of core translation factors involved in canonical initiation, elongation, and termination in Arabidopsis. The information was simplified and adapted from Browning and Bailey-Serres (2015) and Raabe et al. (2019).

forms a complex with eIF4G, a large scaffold protein that interacts with the RNA helicase eIF4A. The 43S PIC is then recruited to the cap through the interaction between multimeric eIF3 and eIF4G. eIF4G also interacts with poly-A binding protein (PABP), which binds to the poly-A tail at the 3′ end of the mRNA. Thus, the 5′ and 3′ ends of the mRNAs are connected, forming a circular structure. The circularization of mRNAs enhances their efficient translation, facilitates ribosome recycling, and increases mRNA stability by protecting mRNAs from degradation via exonucleases (reviewed in Browning and Bailey-Serres 2015; Passmore and Coller 2022).

Translation can be divided into 3 phases: initiation, elongation, and termination. Translation initiation is considered the key point of regulation and is the most well studied of these 3 phases. Plants, like other eukaryotes, use the scanning model for translation initiation (Fig. 1A). After being recruited to the cap, the 43S PIC scans along the 5′ leader sequence (i.e. the region upstream of the main ORF, also commonly called 5′ UTR, although extensive translation may occur in this region; see section on 'Unannotated translation events') until an optimal start codon is identified. Then, the 60S ribosomal large subunit is recruited, with the Met-tRNA$_i$ positioned at the peptidyl site (P-site) of the ribosome, and the codon on the mRNA and anti-codon on the tRNA base pair. New charged tRNAs corresponding to the next codons then join at the aminoacyl site (A-site). The amino acids at the P-site and A-site form a peptide bond. Subsequently, the amino acid (or the growing peptide) at the P-site is transferred to the amino acid at the A-site. Next, the ribosome translocates by shifting 3 nucleotides (3 nts; 1 codon). As a result, the uncharged tRNA is now located at the exit site, and the tRNA carrying the peptide is at the P-site, which makes the A-site available for the incoming charged tRNAs. This cycle repeats until a stop codon is encountered at the A-site, where release factors, rather than a charged tRNA, join the A-site and trigger translation

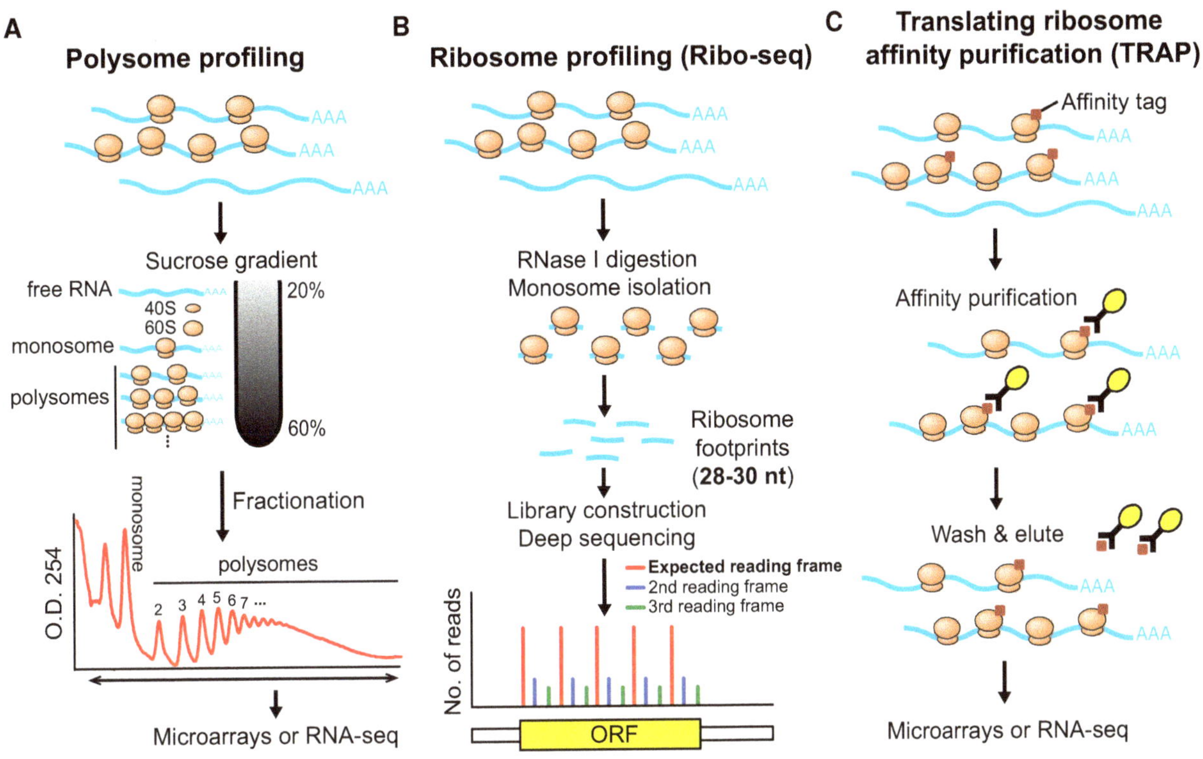

Figure 2. Common genomic techniques for studying translation. **A)** Polysome profiling uses a sucrose gradient to separate mRNAs associated with different numbers of ribosomes. The resulting mRNA populations corresponding to various translational activities can be fractionated and subjected to microarray or RNA-seq for identifying and quantifying the mRNAs. **B)** Ribo-seq uses ribonuclease digestion to obtain ribosome footprints. Sequencing of ribosome footprints reveals the precise position and quantity of ribosomes on mRNAs globally. High-quality datasets display a strong 3-nt periodicity that corresponds to ribosomes translocating by 3 nt per codon within coding regions. **C)** TRAP uses antibodies to isolate a subpopulation of ribosomes by expressing an epitope-tagged ribosomal large subunit protein driven by either a constitutive or a tissue-specific promoter. The mRNAs associated with the purified ribosomes can be subjected to microarray or RNA-seq for identifying and quantifying the mRNAs.

termination. Subsequently, the ribosome subunits dissociate from the mRNA and are recycled for additional rounds of translation (reviewed in Schuller and Green 2018).

Unique features of plant mRNA translation

Although the machinery for plant mRNA translation is largely similar to that of other eukaryotes, several unique characteristics have evolved in plants. In addition to the canonical eIF4F complex, consisting of eIF4E and eIF4G, plants possess eIFiso4F complexes consisting of eIFiso4E and eIFiso4G (encoded by *eIFiso4G1* and *eIFiso4G2* in Arabidopsis [*Arabidopsis thaliana*]). eIFiso4G is deeply conserved throughout plant lineages, whereas eIFiso4E is present only in flowering plants. The eIFiso4F complex is smaller compared with eIF4F complex (~80 kD vs ~180 kD) but more abundant (Patrick and Browning 2012). eIF4G and eIFiso4G have different mRNA sequence preferences and interact differently with various PABP isoforms (Gallie and Browning 2001; Gallie 2018); thus, these complexes may favor specific sets of mRNAs and control distinct physiological responses. For example, eIFiso4G has been shown to prefer mRNAs with unstructured sequences, which are characteristics of mRNAs that are preferentially translated under hypoxia (Gallie and Browning 2001; Branco-Price et al. 2005). More recently, eIFiso4G has been reported to control specific mRNAs involved in chloroplast functions, hypoxia, and abscisic acid (ABA) signaling (Bi et al. 2019; Cho et al. 2019; Lellis et al. 2019).

Plant ribosomal proteins have also expanded due to multiple rounds of genome duplication. In Arabidopsis, there are 80 ribosomal protein (RP) families, and each RP is encoded by 2 to 6 genes (Barakat et al. 2001; Browning and Bailey-Serres 2015; Scarpin et al. 2023). The large number of potential combinations of these RPs in ribosomes is proposed to provide heterogeneity to ribosome properties and enable specific mRNA translation in response to various internal and external cues (Browning and Bailey-Serres 2015).

Genome-wide techniques to study translation

As translation involves mRNAs, which can be amplified and sequenced, much of our knowledge of translational regulation has expanded thanks to the development of next-generation sequencing. Below we discuss 3 common genome-wide techniques for studying translation. For other techniques, we refer readers to a comprehensive review (Mazzoni-Putman and Stepanova 2018).

The first genomic technique used to study translation is polysome profiling (Fig. 2A), which involves separating mRNAs bound by various numbers of ribosomes through a sucrose gradient by ultracentrifugation. The distribution of specific mRNAs in different fractions can be quantified using reverse-transcription PCR, microarrays, or RNA-sequencing (RNA-seq) to estimate their relative translational activity. Although this technique is powerful and has allowed countless discoveries (e.g. Coate et al. 2014; Missra et al. 2015; Bai et al. 2017a; Zhang et al. 2017), polysome profiling cannot reveal the exact ribosome occupancy on mRNAs, and the resolution of polysome separation typically decreases as the number of bound ribosomes increases (Fig. 2A).

The second technique, termed ribosome profiling (a.k.a. Ribo-seq; Fig. 2B), solves the limitation of polysome profiling. Ribo-seq involves ribonuclease digestion of ribosome-bound mRNAs and sequencing of the resulting ribosome-protected mRNA fragments (i.e. ribosome footprints), typically 28 to 30 nts in length (Ingolia et al. 2009; Brar and Weissman 2015). The sequencing of these ribosome footprints reveals the precise quantity and position of ribosome occupancy on mRNAs throughout the transcriptome. Translation efficiency can be inferred by normalizing the Ribo-seq to RNA-seq read density. Moreover, in high-quality Ribo-seq data, ribosome footprints display strong 3-nt periodicity, consistent with translating ribosomes moving every 3 nts along coding regions. This periodic property can be used to identify actively translated regions (Ingolia et al. 2009). The first ribosome profiling studies in Arabidopsis were reported a decade ago; one was applied to study translational control of light responses (Liu et al. 2013), and a variation of the technique was applied to study chloroplast translation (Zoschke et al. 2013). Optimizing the resolution of Ribo-seq led to the discovery of previously unannotated translated ORFs, which dramatically expanded the landscape of the Arabidopsis translatome and improved genome annotation (Hsu et al. 2016). Ribo-seq has been used to study translational responses to many environmental stimuli, such as hypoxia, hormonal responses, pathogen responses, and phosphate deficiency in Arabidopsis (Juntawong et al. 2014; Merchante et al. 2015; Bazin et al. 2017; Xu et al. 2017a) and to characterize the translatomes in crops such as maize (*Zea mays*), soybean (*Glycine max*), tomato (*Solanum lycopersicum*), rice (*Oryza sativa*), sorghum (*Sorghum bicolor*), and wheat (*Triticum aestivum*) (Lei et al. 2015; Chotewutmontri and Barkan 2016; Shamimuzzaman and Vodkin 2018; Wu et al. 2019; Li and Liu 2020; Zhu et al. 2021; Sotta et al. 2022; Guo et al. 2023b). However, the short length of ribosome footprints limits their mapping capability. Therefore, it is challenging to implement Ribo-seq for polyploid genomes where short reads are less likely to be uniquely mapped (Jiang et al. 2022). Nevertheless, Ribo-seq has been recently carried out in hexaploid bread wheat (Guo et al. 2023b).

The third technique, translating ribosome affinity purification (TRAP; Fig. 2C), involves the expression of an epitope-tagged ribosomal protein (usually a large ribosomal subunit protein to ensure the capture of fully assembled ribosomes), followed by affinity purification of the tagged ribosomes and associated mRNAs using antibodies. The mRNAs associated with translating ribosomes can then be identified and quantified by microarrays or RNA-seq (Inada et al. 2002; Mustroph et al. 2009; Juntawong et al. 2014). Using a constitutive promoter to express the tagged ribosomal protein, TRAP-seq has been applied to study translational responses to different treatments in several species (Juntawong et al. 2014; Bazin et al. 2017; Meteignier et al. 2017; Reynoso et al. 2019; Bonnot and Nagel 2021). When combined with tissue-specific promoters, TRAP enables the study of tissue-specific translatomes (Mustroph et al. 2009; Traubenik et al. 2020; Kajala et al. 2021; Reynoso et al. 2022). It should be noted that TRAP requires generating stable transgenic plants, and the accumulation of tagged ribosomal proteins may alter ribosome properties and physiology (Zhao et al. 2019a). Although TRAP-seq can reveal mRNAs engaged in translation within specific tissue types, it is difficult to estimate translation efficiency (average ribosome number per mRNA) because it is technically challenging to quantify mRNA levels from the exact same tissue used to produce the TRAP-seq data. Although, to our knowledge, there are no reports in plants yet, it is possible to combine tissue-specific TRAP with ribosome footprinting (Gonzalez et al. 2014; Chen and Dickman 2017; Doroudgar et al. 2019) to quantify ribosome density and study actively translated ORFs on tissue-specific mRNAs.

Unannotated translation events

The above sequencing techniques, especially Ribo-seq, have transformed our view of translational landscapes. In addition to refining previously annotated ORF models, Ribo-seq has led to 3 major discoveries in the translatomes: (1) the translation of small ORFs (sORFs) in RNAs presumed to be noncoding, (2) the widespread presence of upstream ORFs (uORFs) on protein-coding mRNAs, and (3) the usage of alternative or non-AUG start sites. Unlike canonical protein-coding genes, the act of translating these new ORFs, specifically sORFs and uORFs, may serve a regulatory function, even if the resulting protein products do not play a direct role in regulation (see below).

Newly discovered translated sORFs among presumed noncoding RNAs

To minimize false discoveries, computational genome annotations typically assume that coding sequences are longer than 100 codons. This assumption means that transcripts encoding small proteins can be overlooked and annotated as noncoding RNAs (Basrai et al. 1997). Based on the 3-nt periodicity corresponding to ribosomes translocating 3 nt per codon, dozens or hundreds of translated sORFs within annotated noncoding RNAs or previously unannotated transcripts have been identified in Arabidopsis, tomato, maize, and wheat (Hsu et al. 2016; Wu et al. 2019; Kurihara et al. 2020; Liang et al. 2021; Palos et al. 2022; Guo et al. 2023b). Similarly, an alternative approach based on ribosome dissociation from mRNAs upon translation termination identified over 200 translated sORFs in Arabidopsis (Bazin et al.

2017). The translation efficiency of sORF mRNAs in tomato is indistinguishable from that of canonical protein-coding genes, suggesting that most of these sORF transcripts are likely to be true protein-coding RNAs (Wu et al. 2019). Epitope-tagging followed by western blots and mass spectrometry–based proteomic analysis confirmed that at least some of these sORFs produce stable proteins in planta (Hsu et al. 2016; Bazin et al. 2017; Wu et al. 2019).

Although most of these sORF mRNAs were likely misannotated as noncoding RNAs, surprisingly, some well-characterized noncoding RNAs also contain translated sORFs. The best documented examples are the primary transcripts of trans-acting small interfering RNAs (tasiRNAs). The biogenesis of tasiRNAs involves microRNA (miRNA) cleavage of primary *TAS* transcripts, which are then converted to double-stranded RNAs (dsRNAs) and processed into 21- or 22-nt tasiRNAs. Arabidopsis has 4 *TAS* gene families (*TAS1* to *TAS4*). The primary *TAS3* transcript contains a 50-amino acid (aa) sORF upstream of the first miRNA390 binding site; strong 3-nt periodicity of Ribo-seq reads was unambiguously detected in this sORF, supporting that it is actively translated (Hsu et al. 2016; Hsu and Benfey 2018). The ribosome occupancy on the *TAS3* sORF is also supported by a genome-wide study of 5′ truncated mRNA ends (Hou et al. 2016). By contrast, the primary *TAS2* transcript has a 57-aa sORF that contains a miRNA173 binding site. This sORF was shown to be translated in vitro, and introducing a premature stop codon to this sORF reduces tasiRNA levels. These results suggest that the translation and position of the sORF relative to the miRNA binding site are critical for tasiRNA biogenesis (Yoshikawa et al. 2016). In addition, primary *TAS1A/B/C*, *TAS2*, and *TAS3* transcripts are associated with membrane-bound polysomes, suggesting that they are translated by endoplasmic reticulum (ER)-localized ribosomes (Li et al. 2016). A recent study further demonstrated that the dsRNA-binding protein SUPPRESSOR OF GENE SILENCING 3 (SGS3) regulates ribosome stalling at the sORFs upstream of the miRNA binding sites on *TAS3* and *TAS1C* by forming a complex with ARGONAUTE (AGO) and the relevant miRNA (Iwakawa et al. 2021).

Another surprising example of noncoding RNA translation was found in the primary transcripts of miRNAs. The first study reported that *pri-miR165a* encodes an 18-aa sORF in Arabidopsis and *pri-miR171b* encodes a 9-aa sORF in *Medicago truncatula* (Lauressergues et al. 2015). Application of synthetic peptides corresponding to these sORFs led to an increase in *pri-miRNA* levels (Lauressergues et al. 2015; Sharma et al. 2020). However, the translation levels of these sORFs appear to be low or condition dependent, as no significant ribosome footprints in these sORF regions have been detected in available Ribo-seq datasets (Hsu and Benfey 2018).

Prevalent uORFs on mRNAs

Conventionally, it was assumed that 1 mRNA produces 1 protein from the longest ORF. However, when the 43S PIC scans along the 5′ UTR searching for the optimal start site (usually AUG), it may encounter 1 or multiple smaller ORFs upstream of the annotated ORFs (thus the name upstream ORFs). As a result, the ribosome may translate the uORF before reaching the longest/annotated ORF (main ORF; mORF) (Fig. 3). Ribo-seq analysis has revealed numerous translated uORFs in plants, for example, 2,093 in Arabidopsis, 1,329 in tomato, 2,558 in maize, and 1,254 in wheat (Lei et al. 2015; Wu et al. 2019; Wu and Hsu 2022; Guo et al. 2023b). The identification of these uORFs suggests that the translation of their downstream mORFs is likely repressed by the uORFs (see section 'Translational regulation through mRNA features – uORFs' for details).

Alternative or non-AUG translation start sites

Like uORFs, the 43S PIC may consider codons different from the annotated start site to initiate translation, which can lead to the production of proteins different from the annotated ORFs through the use of an alternative AUG or a near-cognate start site that differs from AUG by 1 nucleotide. Recently, a specialized Ribo-seq termed translation initiation site sequencing (TI-seq) has been applied in plants to study translation initiation and amino-terminal (N-terminal) proteomics (Willems et al. 2017; Li and Liu 2020). TI-seq uses the translation inhibitor lactimidomycin to capture initiating ribosomes, in contrast to regular Ribo-seq that uses cycloheximide, which captures both initiating and elongating ribosomes. Note that even snap freezing in liquid nitrogen during sample harvest can capture many elongating ribosomes

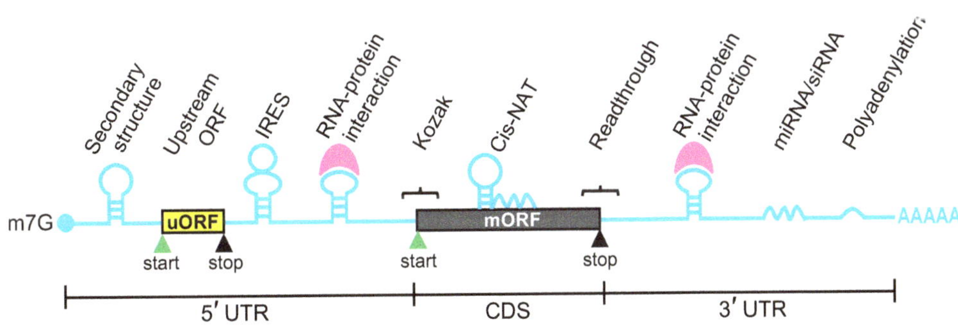

Figure 3. Common cis-regulatory elements or mechanisms contributing to translational control in mRNAs. Cis-regulatory elements or associated mechanisms commonly present in the 5′ UTR, the mORF, or 3′ UTR are presented. The figure was adapted from Liu and Qian (2014).

(Douka et al. 2022). Thus, a technical challenge with TI-seq is to ensure efficient runoff of elongating ribosomes to avoid false positives. Therefore, stringent criteria and careful follow-up validation are necessary (Gao et al. 2015; Li and Liu 2020). Studies in Arabidopsis cultured cells and tomato leaves have revealed that alternative or near-cognate translation start sites are prevalent. The most common near-cognate codon is CUG, whereas AAG and AGG are the least used (Willems et al. 2017; Li and Liu 2020). Many alternative translation start sites identified in plants to date are in frame with the corresponding annotated ORF, resulting in changes in the protein N-terminal sequences (Willems et al. 2017; Wu et al. 2019; Li and Liu 2020). Follow-up experiments have demonstrated that several alternative or non-AUG translation start sites lead to different protein subcellular localizations (Li and Liu 2020).

Translational regulation through mRNA features

Translational control can be classified conceptually into cis- and trans-regulation. An mRNA molecule may contain specific cis-regulatory elements that contribute to translational control. Many cis-regulatory elements have been found within the 5′ UTR, 3′ UTR, or surrounding the start or stop codons of mRNAs (Fig. 3). Some of these cis-regulatory elements are known to be bound or regulated by specific RNA-binding proteins (i.e. trans-acting factors). In this section, we will discuss gene-specific cis-regulatory elements involved in translational control and their corresponding RNA-binding proteins, if known. In the next section 'Translational regulation through modulation of translation factors and master translational regulators', we will focus on translation factors, mechanisms, or signaling that can affect global translation.

Kozak

During its search for the start codon, the 43S PIC may encounter multiple AUG and near-cognate codons present in the 5′ UTR. How does the ribosome decide where to start? An optimal start codon typically possesses a consensus sequence flanking the AUG known as the Kozak consensus sequence (Kozak 1981). In plants, as well as in other eukaryotes, the Kozak consensus sequence is 5′-A/GNNAUGG- 3′, meaning that the upstream third nt (−3) of the AUG is enriched for A or G, and the downstream fourth nt (+4) is enriched for G (Lütcke et al. 1987; Nakagawa et al. 2008). Analysis of actively translated ORFs from Ribo-seq data in Arabidopsis, tomato, maize, and sorghum has confirmed the conserved sequence bias toward −3/+4 positions (Liu et al. 2013; Lei et al. 2015; Wu et al. 2019; Sotta et al. 2022). Thus, mRNA se-quences containing an optimal Kozak consensus sequence are expected to have better translation efficiency.

uORF

uORF properties

As 30% to 70% of plant genes contain potential AUG-initiated uORFs (Zhang et al. 2021), uORFs represent a common mechanism for translational control. If a uORF is translated, it may lead to ribosome stalling or dissociation before the ribosome reaches the mORF, resulting in repression of mORF translation (Fig. 4A). Moreover, when a uORF is translated, the transcript will appear to have a long 3′ UTR, which is expected to activate nonsense-mediated decay (NMD), a conserved mechanism to degrade aberrant mRNAs with a premature termination codon (Kurosaki et al. 2019). Although evidence in animals supports the idea that uORF translation triggers NMD (Hurt et al. 2013; Chew et al. 2016; Johnstone et al. 2016; Baird et al. 2018), whether uORFs trigger NMD in plants is still under debate, particularly with new evidence in Arabidopsis challenging this idea (Wu and Hsu 2022).

The mORF downstream of a uORF may still get translated via leaky scanning or re-initiation (Fig. 4A). Leaky scanning occurs when the 43S PIC passes a uORF without translating it, and, thus, not all predicted uORFs are translated. In the case of re-initiation, after the uORF is translated, the riboso-mal small subunit remains attached to the mRNA and con-tinues scanning the 5′ UTR, ultimately translating the mORF. Factors that determine whether a uORF is translated or the efficiency of re-initiation remain areas of great interest.

uORFs are enriched in the transcripts of genes encoding transcription factors and protein kinases (Kim et al. 2007; Wu and Hsu 2022) and are present in the transcripts of numerous metabolic genes (reviewed in Van Der Horst et al. 2020). uORFs are also associated with allelic variation and agricultural traits (Gage et al. 2022; Guo et al. 2022), highlighting the importance of uORFs in plant physiology. In some uORF-containing transcripts, such as those for the transcription factor genes *ARABIDOPSIS HOMEOBOX 1* (*AtHB1*) and *BASIC LEUCINE-ZIPPER 11* (*bZIP11*), overproduction of their mORFs leads to deleterious growth phenotypes, suggesting that uORF-mediated suppression plays key roles in regulating plant growth (Ribone et al. 2017). Recently, sev-eral studies have manipulated uORFs of genes of interest to improve specific traits. For example, manipulating the uORF in the *bZIP11* transcript increased sugar content in tomato and strawberry (*Fragaria × ananassa*; Sagor et al. 2016; Xing et al. 2020), and manipulating the uORF in the *TL1-BINDING FACTOR* (*TBF1*) transcript enhanced pathogen response in Arabidopsis and rice (Xu et al. 2017b).

Using Ribo-seq with improved coverage, 2,093 actively translated uORFs were identified in Arabidopsis seedlings based on statistically significant 3-nt periodicity (Wu and Hsu 2022). However, this approach may not efficiently identify short uORFs or uORFs that partially overlap with their associated mORFs or with other uORFs. Thus, more uORFs likely remain to be identified. Conversely, the number of potential uORFs may be overestimated due to annotation errors. It has been reported that the current Arabidopsis annotation (Araport11) tends to overestimate the length of 5′ UTRs (Parker et al. 2020). If the 5′ UTR length of a transcript is overestimated, a uORF may be predicted from the extended 5′ UTR sequence, which does not exist in nature. A common experiment to validate uORF functions is to

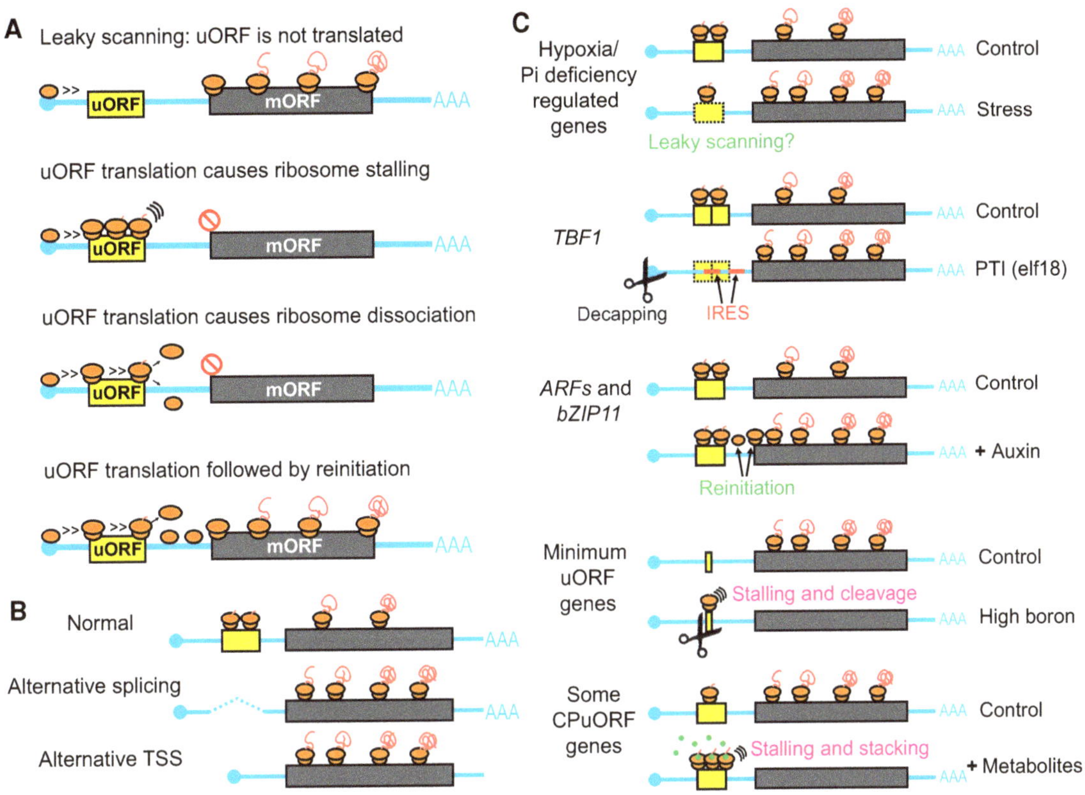

Figure 4. Overview of uORF regulation and expected mORF translation levels. **A)** Effects of uORFs on mORF translation may include leaky scanning, ribosome stalling, ribosome dissociation, and re-initiation. **B)** Mechanisms that regulate the presence of uORFs in mRNA sequences. uORFs may be omitted through alternative splicing or using an alternative transcription start site of mRNAs. uORFs are shown in yellow boxes; mORFs are shown in grey boxes. **C)** Examples of uORF regulation under different environmental conditions and expected mORF translation levels. mRNA degradation is shown by scissors. Metabolites are shown by green dots. PTI: pattern-triggered immunity.

compare the 5′ UTRs possessing a wild-type uORF or a mutated uORF (such as by mutating its start codon) in a reporter assay. However, researchers should be cautious, as those false-positive uORFs may still affect translation when artificially expressing non-existing 5′ UTRs. To avoid this issue, researchers can implement 5′ rapid amplification of cDNA ends experiment or exploit available Cap Analysis of Gene Expression-seq data (Takahashi et al. 2012a) to confirm transcription start sites, thereby increasing the confidence in identifying functional uORFs.

The large number of detected uORFs exhibit diverse properties, such as variation in their length and extent of sequence conservation. The smallest possible uORFs are minimum uORFs (AUG-stop; a start codon immediately followed by a stop codon). As translation initiation and termination are the major rate-limiting steps in translation (Shah et al. 2013), translation of minimum uORFs is likely to cause ribosome stalling. The first minimum uORFs reported were identified in Arabidopsis boric acid channel gene *NODULIN26-LIKE INTRINSIC PROTEIN5;1* and in several boron-responsive mRNAs. In response to high boron levels, these minimum uORFs cause strong translational suppression through ribosome stalling and mRNA degradation of these boron transporter and responsive mRNAs, reducing overall boron uptake (Tanaka et al. 2016). Subsequent Ribo-seq experiments have identified the translation of 153 minimum uORFs in Arabidopsis adult leaves, and 119 of these minimum uORFs have increased Ribo-seq levels under high-boron conditions. Interestingly, high boron also leads to global ribosome stalling at the stop codon of mORFs. Therefore, the increase of ribosome footprints at minimum uORFs under high-boron conditions may be partially due to the stalling effects of boron on stop codons (Sotta et al. 2021).

Most identified uORFs lack sequence conservation at the peptide level across plant species, although the presence of uORFs may be conserved between homologous genes. In Arabidopsis, 120 conserved peptide uORFs (CPuORFs) have been identified using bioinformatics approaches (Hayden and Jorgensen 2007; Jorgensen and Dorantes-Acosta 2012; Takahashi et al. 2012b; Vaughn et al. 2012; Van Der Horst et al. 2019; Takahashi et al. 2020). These CPuORFs were further grouped into 81 homology groups based on the peptide sequence similarity of the CPuORFs. This sequence conservation suggests that the CPuORF peptides play important functions. Seven CPuORFs have been reported to interact with specific metabolites (reviewed in Van Der Horst et al. 2020), and 5 CPuORF-encoding peptides have been reported to cause ribosomal arrest, possibly through the interaction between the nascent peptide and the ribosome exit channel (Uchiyama-Kadokura et al. 2014; Hayashi et al. 2017; Yamashita et al. 2017). Interestingly, the peptide sequence of the CPuORF in homology group 30 is similar to CELL DIVISION CYCLE PROTEIN 26 (CDC26) found in fungi and animal species, which is a component of the anaphase promoting complex involved in cell cycle regulation (Takahashi et al. 2012b; Vaughn et al. 2012). A recent study demonstrated that this CPuORF peptide in Arabidopsis indeed acts like CDC26 as a cell cycle regulator, controlling cell division and embryo development (Lorenzo-Orts et al. 2019).

Regulation of uORF presence by alternative transcription start sites or alternative splicing

The presence or absence of uORFs in mRNA sequences may be regulated at transcriptional and posttranscriptional levels by altering the length and sequence of 5′ UTRs (Fig. 4B) (Von Arnim et al. 2014). At the transcriptional level, different tran-scription start sites (TSSs) can influence whether uORFs are included in the mRNA sequences. Some genes were shown to use different TSSs in response to various light conditions, leading to the gain or loss of uORFs in the mRNA sequences (Kurihara et al. 2018). At the posttranscriptional level, an alternative spli-cing event at the 5′ UTR can determine whether a uORF is pre-sent in *HOLOCARBOXYLASE SYNTHASE 1* mRNA, which encodes a protein-biotin ligase that catalyzes the biotinylation of several key metabolic enzymes in Arabidopsis (Puyaubert et al. 2008). Similarly, an intron retention event in the 5′ UTR of *PHYTOCHROME-INTERACTING FACTOR3* (*PIF3*) in response to phytochrome B signaling results in the inclusion of a uORF that represses translation of the *PIF3* mRNA (Dong et al. 2020). It remains an open question whether uORF presence and regulation are widely controlled by alternative TSSs and/or alternative splicing.

Regulation of uORF repressiveness

The repressiveness of uORFs can be modulated by stress, phytohormones, nutrients, or metabolites (Fig. 4C).

uORF regulation by stress. The translational control of uORFs was first studied in yeast (*Saccharomyces cerevisiae*) and animals, where the translational repression mediated by certain uORFs is alleviated (i.e. derepression), under stress conditions through 2 mechanisms. First, the stringency for uORF initiation can be increased to reduce uORF translation. Second, after the first uORF is translated, a phenomenon called "delayed reinitiation" allows the 40S ribosome to bypass multiple repressive uORFs and reinitiate translation at the mORF (Palam et al. 2011; Young and Wek 2016). Under stress, global translation is typically suppressed to save energy, which is accompanied by selective translation of mRNAs critical for survival (reviewed in Liu and Qian 2014). Therefore, derepression of uORF-containing transcripts provides a way for a rapid and specific increase of certain proteins from the translation of downstream mORFs.

As sessile organisms, fast responses through derepression of uORF-containing genes are likely critical for plants. Two examples of this phenomenon in stress response have been reported. The first example is involved in hypoxia stress (Juntawong et al. 2014). Using Ribo-seq, it was shown that a high fraction of genes with increased translation efficiency under hypoxia contain long uORFs. Additionally, the number of Ribo-seq reads associated with these uORFs decreases under hypoxia, whereas those reads associated with their mORFs increase (Fig. 4C). These results suggest that translation of these long uORFs is reduced, implying that the uORFs are frequently skipped during translation under hypoxia. Alternatively, the reduced translation of uORFs may be due to a global reduction of translation, whereas mORF translation is increased due to enhanced re-initiation.

The second example of derepression involves *TBF1*, which en-codes a heat-shock factor-like transcription factor required for immune response. The *TBF1* transcript contains 2 uORFs (Fig. 4C), and the translation of the *TBF1* mORF is derepressed after pathogen challenge (Pajerowska-Mukhtar et al. 2012). Interestingly, the *TBF1* transcript is decapped during the im-mune response, and 2 R-motifs (guanine/adenine-rich motifs) in the 5′ UTR of *TBF1* can serve as Internal Ribosome Entry Sites (IRESs) (see section on 'IRES' below) to facilitate the translation of *TBF1* mORF. Notably, the 2 R motifs in the *TBF1* mRNA overlap with or are downstream of the uORFs. Therefore, the IRES sequences can allow ribosomes to avoid the uORFs and directly proceed to the *TBF1* mORF (Wang et al. 2022). Whether this IRES-mediated uORF skipping mechanism allows the derepression of other uORF-containing transcripts in different stress conditions in plants is an exciting direction for future research.

uORF regulation by phytohormone(s). The phytohormone auxin can also derepress uORF-containing transcripts by activating the kinase TARGET OF RAPAMYCIN (TOR) (see section 'Translational regulation through modulation of translation factors and master translational regulators – TOR' for details). In Arabidopsis, TOR promotes the translational re-initiation of uORF-containing mRNAs by directly phosphorylating eIF3h, a factor known to promote re-initiation. Activated TOR and eIF3h facilitate the polysome loading of several uORF-containing mRNAs, including those of *AUXIN RESPONSE FACTOR*s (*ARF*s) and *bZIP11* (Fig. 4C) (Schepetilnikov et al. 2013). Notably, TOR derepresses uORF-containing transcripts to facilitate growth, while other examples of derepression respond to stress conditions. Because TOR is activated or repressed by many environmental cues or phytohormones, such as light, glucose, nutrients, auxin, and ABA (reviewed in Schepetilnikov and Ryabova 2018; McCready et al. 2020), it will be interesting to learn which genes with uORF-containing transcripts are controlled by TOR under different growth conditions.

uORF regulation by nutrients and metabolites. In addition to boron control of minimum uORFs mentioned above (Tanaka et al. 2016), phosphate deficiency facilitates the derepression of several uORF-containing transcripts, including those in *ARF4* and *CYTOKININ RESPONSE FACTOR 10*, which might promote root growth for phosphate uptake (Bazin et al. 2017).

It is well documented that the repressiveness of several CPuORFs is modulated by metabolites. For example, sucrose, thermospermine, and galactinol regulate the activities of the CPuORFs on the mRNAs of the transcription factor genes *bZIP11*, *SUPPRESSOR OF ACAULIS 51*, and *TBF1*, respectively (Rook et al. 1998; Wiese et al. 2004; Imai et al. 2006; Rahmani et al. 2009; Cai et al. 2016; Yamashita et al. 2017; Zhu et al. 2018). Both choline and phosphocholine activate the CPuORF of the *XIPOTL1* transcript, encoding a phosphoethanolamine N-methyltransferase, to repress phosphatidylcholine biosynthesis (Alatorre-Cobos et al. 2012). Interestingly, the biosynthesis and degradation of polyamine are modulated by 2 homology groups of CPuORFs in the transcripts of *S-ADENOSYLMETHIONINE DECARBOXYLASE* (*SAMDC*) and *POLYAMINE OXIDASEs* (*PAOs*), respectively (Hanfrey et al. 2005; Guerrero-González et al. 2014). Finally, under high-ascorbate conditions, a conserved uORF with a noncanonical start site on *GDP-L-galactose phosphorylase* (*GGP*) mRNA suppresses downstream mORF translation (Laing et al. 2015; Hsu et al. 2016). *GGP* encodes a key enzyme for ascorbate biosynthesis. Thus, the suppression of GGP translation by the conserved uORF reduces ascorbate biosynthesis (Laing et al. 2015). The mechanism for uORF-dependent metabolite sensing remains unclear but is possibly due to interactions between the ribosome exit channel, the uORF peptide, and the metabolites (Van Der Horst et al. 2020).

Most of the CPuORFs mentioned above respond to high metabolite levels to repress the translation of their mORFs and thus negatively regulate the biosynthesis of the metabolites, except the CPuORF of *PAOs*. In the presence of high polyamine levels, the *PAO* CPuORF is suppressed, promoting mORF translation for production of PAO for polyamine degradation, whereas the *SAMDC* CPuORF suppresses *SAMDC* mORF translation, preventing the biosynthesis of polyamine. Thus, the 2 groups of CPuORFs act together to balance cellular polyamine levels (Hanfrey et al. 2005; Guerrero-González et al. 2014).

Intriguingly, in contrast to derepression where uORFs are repressive by default, CPuORF33 in the *AtHB1* mRNA is not repres-sive until activated. *AtHB1* encodes a transcription factor that regulates hypocotyl and leaf development. Several observations suggest that the repressiveness of CPuORF33 requires photosynthetic activity: (1) the uORF is not repressive in dark-grown seed-ings; (2) the uORF is repressive only in the aerial tissues of plants grown under light-dark cycles; and (3) chemicals inhibiting chloroplast function abolish CPuORF33 repressiveness, as shown by a reporter assay (Ribone et al. 2017). Together, this finding provides an interesting example of uORF regulation linking the environment and plant development.

Why uORF regulation?

Regulating gene expression by uORFs is energy inefficient because these mRNAs are not translated to their full potential. An alternative to producing the same amount of proteins is to generate fewer mRNAs but without uORF repression. Is there an advantage to having uORFs repress translation? Advances on this topic suggest that uORF-mediated regulation offers at least 3 benefits: First, uORFs may provide feedback regulation for maintaining homeostasis of certain nutrients or metabolites (reviewed by Van Der Horst et al. 2020). Second, regulating gene expression at the translational level allows for quick responses to changing environments (Juntawong et al. 2014; Xu et al. 2017b). Finally, translational suppression could be a general mechanism to reduce gene expression noise (see below).

Gene expression is inherently stochastic. Each step of the central dogma, including the synthesis and decay of mRNAs and proteins, fluctuates. This stochasticity results in variance in protein numbers among genetically identical cells, known as gene expression noise (Raj and van Oudenaarden 2008). Recently, using both an experimental approach and mathematical modeling, uORFs of a circadian regulator, *TIMING OF CAB2 EXPRESSION 1* (*TOC1*), were shown to be critical to suppress the gene expression noise of TOC1 and facilitate the homogeneity of the circadian clock in Arabidopsis (Wu et al. 2022). This finding is consistent with prior studies in prokaryotes and other eukaryotes that lower translation can counteract the fluctuation of transcription levels and reduces the variance of protein abundance. Because protein kinases and transcription factors are enriched among uORF-containing genes in plants (Kim et al. 2007; Wu and Hsu 2022), noise reduction for these critical regulators is likely important for stabilizing their regulatory networks despite the associated energy cost.

Internal ribosome entry site (IRES)

An IRES allows the 43S PIC to bind to the internal region of the mRNA instead of binding to the 5′ cap first (Fig. 3). Besides recent examples in defense mRNAs (Wang et al. 2022; see section on 'uORFs' above), another plant gene reported to possess an IRES is *WUSCHEL* (*WUS*; Cui et al. 2015). *WUS* encodes a transcription factor that maintains the stem cell pool in the shoot apical meristem. The IRES-dependent translation of *WUS* transcripts is controlled by La1 RNA-binding protein. La1 is mainly located in the nucleus, but it relocates to the cytoplasm in response to environmental stress, such as UV and the radiomimetic drug zeocin. In the cytoplasm, La1 enhances the translation of *WUS* transcripts, which correlates with the proliferation of stem cells. Plants overexpressing *La1* continue developing true leaves in the presence of zeocin, while wild-type plants cannot. Thus, the regulation of *WUS* by La1 through IRES-mediated translation can control stem cell proliferation in response to environmental stress.

RNA secondary structure

One recent example of translational control through RNA secondary structures is related to development (Cho et al. 2018). Based on the hypothesis that phloem development is regulated by sucrose, a comparative transcriptome analysis was conducted on poplar (*Populus tremula*) phloem/cambium-specific genes, Arabidopsis herbaceous phloem/cambium-specific genes, and Arabidopsis sucrose-upregulated genes. One gene that met all 3 criteria encodes an RNA-binding protein named JULGI ("stream" or "plant shoot" in Korean; JUL). A screen using a random single-stranded RNA library revealed that JUL preferentially binds to G-rich sequences and G-quadruplex. A search for phloem-specific mRNAs with potential G-quadruplex sequences in their 5′ UTRs identified *SUPPRESSOR OF MAX2 1-LIKE4* (*SMXL4*) and *SMXL5* as potential targets of JUL. SMXL4 and SMXL5 are key regulators of phloem differentiation. Follow-up experiments demonstrated that JUL binds to the G-quadruplex region in *SMXL4* and *SMXL5* mRNAs and represses their translation and thus negatively regulates phloem differentiation.

Another example of translational control mediated by RNA secondary structures is related to temperature sensing (Chung et al. 2020). A Ribo-seq experiment comparing Arabidopsis seedlings kept at 17 °C or transferred to 27 °C for 15 minutes identified hundreds of transcripts with enhanced or reduced translation. A subset of transcripts that were translationally upregulated contained a predicted hairpin structure located approximately 30 nt upstream of the start codon. Among these, *PIF7* was a promising candidate, as it is homologous to *PIF4*, a master regulator of the thermosensing pathway. Further experiments revealed that the translation of *PIF7* mRNA is mediated by the hairpin structure responsive to temperature changes in the *PIF7* mRNA and that PIF7 protein levels are positively correlated with hypocotyl growth in response to a warmer temperature, likely through upregulation of auxin biosynthesis genes by PIF7.

Sequences recognized by RNA-binding proteins

Two independent studies have described an elegant example of translational regulation acting on the 3′ UTR of *EIN3-BINDING F-BOX 1* (*EBF1*) and *EBF2* mRNAs in the ethylene signaling pathway (Li et al. 2015; Merchante et al. 2015). EBF1 and EBF2 are key repressors of the ethylene signaling pathway by targeting the master transcription factors of ethylene-regulated genes for proteasome-mediated degradation. It was a mystery that *EBF1* and *EBF2* mRNAs were upregulated in response to ethylene, but no increased protein levels were detected. In 1 of the 2 studies, Ribo-seq revealed that the translation of *EBF1* and *EBF2* mRNAs is downregulated by ethylene (Merchante et al. 2015). In the second study, it was observed that small RNAs recognizing the *EBF1* and *EBF2* 3′ UTRs overly accumulate in a mutant defective in the 5′-to-3′ decay enzyme *EXORIBONUCLEASE4* (*XRN4*) gene (also named *ETHYLENE INSENSITIVE* 5). Additionally, overexpression of the *EBF1* 3′ UTR fragment results in ethylene insensitivity (Li et al. 2015). Both studies revealed that the translational control of *EBF1* and *EBF2* by ethylene depends on the 3′ UTR of their mRNAs, particularly poly-U elements. Furthermore, EIN2, another signaling protein in the ethylene pathway, binds to the 3′ UTR of *EBF1* and *EBF2* mRNAs and brings them to the processing bodies (P-bodies, see section 'Stability and spatial dynamic of mRNAs – P-bodies and stress granules') with the help of NMD factors. Thus, the localization to the P-bodies prevents *EBF1* and *EBF2* mRNAs from being translated (Li et al. 2015; Merchante et al. 2015).

Another example of translational control through 3′ UTR has been discovered in rice panicle development. *FRIZZY PANICLE* (*FZP*) was identified as a quantitative trait locus for grain number per panicle, and it was shown that an increase in *FZP* 3′ UTR length was associated with increased grain number, but the underlying mechanism was unknown (Bai et al. 2017b; Fujishiro et al. 2018). Recently, the presence of a conserved CU-rich element (CURE) was identified in the *FZP* 3′ UTR. The deletion of the CURE in the *FZP* 3′ UTR resulted in an increase in FZP translation levels, without affecting *FZP* mRNA levels. Similar results were observed by overexpressing the *FZP* 3′ UTR fragment, which presumably competes with the endogenous *FZP* 3′ UTR. Furthermore, the CUREs were shown to be bound by POLYPYRIMIDINE TRACT-BINDING PROTEIN 1 (PTB1) and PTB2. Manipulating *PTB1* or *PTB2* expression altered FZP translation and grain number (Chen et al. 2022).

RNA-binding proteins may also act on elements in the 5′ UTR for translational control. LA AND RELATED PROTEINs (LARPs) are conserved RNA-binding proteins in eukaryotes. Plants possess LARP1 and LARP6 families, each encoded by 3 genes (A to C). *LARP6C* is specifically expressed in pollen. Recently, using RNA-immunoprecipitation sequencing followed by motif analysis, LARP6C was discovered to binds to a B-box motif (either U-rich or UC/UCC-rich) within the 5′ UTRs of its target mRNAs. Interestingly, analysis of a reporter construct for its target revealed that LARP6C can switch from a translational activator to a translation repressor when pollen grains transition from dormancy to germination (Billey et al. 2021).

Another example of RNA-binding protein acting on a 5′ UTR to regulate translation is related to phytochrome signaling. Phytochromes are photoreceptors that sense red and far-red light and switch between Pr and Pfr forms. When exposed to red light, the Pr form switches to the Pfr form, which relocates to the nucleus, promoting gene expression for photomorphogenesis (reviewed in Cheng et al. 2021). In the cytosol, the Pfr form also interacts with the RNA-binding protein PENTA1 (PNT1). PNT1 binds to the 5′ UTR of *PROTOCHLOROPHYLLIDE REDUCTASE* (*PORA*) in-volved in chlorophyll biosynthesis and represses its transla-tion in the presence of Pfr. The inhibition of *PORA* translation is critical for adjusting chlorophyll contents when seedlings transition from far-red light to white light conditions (Paik et al. 2012). The specific mRNA features on *PORA* mRNA that contribute to the translational control remain to be identified.

Readthrough

A special mechanism of translational control is readthrough at the stop codon. It occurs when a near-cognate tRNA, instead of release factors, binds to the stop codon at the

A-site of the ribosome and continues translation. The resulting C-terminal extensions of the proteins may alter the protein subcellular localization, function, or stability.

Interestingly, the first known readthrough example in Arabidopsis is *eukaryotic RELEASE FACTOR 1* (*eRF1*). The *eRF1* 3′ UTR possesses sequence features predicted to be targeted by NMD, including a long 3′ UTR and an intron distantly downstream of the stop codon. It was shown that when eRF1 levels are high, readthrough is unfavored, promoting NMD and reducing *eRF1* mRNA levels. By contrast, when eRF1 levels are low, readthrough is favored, preventing the *eRF1* 3′ UTR from being targeted by NMD. Thus, the *eRF1* expression forms an autoregulatory circuit (Nyikó et al. 2017).

More recently, mining available Arabidopsis Ribo-seq data revealed that 144 transcripts exhibit stop codon readthrough (Sahoo et al. 2022). Among the 3 possible stop codons, UGA is the preferred stop codon for translational readthrough. In contrast to the consensus readthrough context (stop-CARYYA) defined by a previous mutational analysis in *Nicotiana benthamiana* (Skuzeski et al. 1991), the nt immediately upstream and downstream of the stop codons was found to be enriched for A or U in the readthrough cases (Sahoo et al. 2022). Remarkably, among the 144 transcripts with readthrough, 34 encode RNA-binding proteins. Although the biological significance of these readthrough events remains to be determined, it was demonstrated that the readthrough of *CURVATURE THYLAKOID 1B* results in the inclusion of a nuclear localization signal peptide in the encoded protein, causing the new protein to localize to the nucleus, in contrast to the original isoform that localizes in the cytoplasm (Sahoo et al. 2022).

G-content in the poly(A) tail

Contrary to the common belief that poly(A) tails are composed exclusively of adenosines, work in human cell lines has revealed the existence of non-A residues in poly(A) tails (Chang et al. 2014). Similarly, ~16% of Arabidopsis poly(A) tails contain at least 1 non-A residue, and the majority of these are guanines. These G residues break the poly(A) tails into multiple A tracts. Genome-wide analysis showed that the G content in poly(A) tails is negatively correlated with the efficiency of PABP binding and translation, without affecting mRNA stability. Consistent with these observations, transcripts with high PABP binding display a significant reduction in translation efficiency in *pabp* mutants (Zhao et al. 2019b).

Small interfering RNAs (miRNAs and siRNAs)

In addition to mRNA cleavage, small interfering RNAs (miRNAs and siRNAs) can target specific mRNA sequences for translational repression (Brodersen et al. 2008). Consistent with the expected roles of miRNAs in translational repression, Ribo-seq analysis showed that predicted miRNA targets are associated with lower translation efficiency in Arabidopsis and tomato (Liu et al. 2013; Wu et al. 2019).

Once miRNAs are produced, they bind to AGO and form the RNA-induced silencing complex (RISC), which can trigger mRNA cleavage or inhibit translation. miRNA-mediated translational inhibition was shown to occur at the ER through interaction between AGO1 and an integral ER membrane protein, ALTERED MERISTEM PROGRAM1 (AMP1; Li et al. 2013).

More recently, HYPONASTIC LEAVES1 (HYL1), a core component of miRNA processing machinery, was shown to also function as a miRNA-mediated translational inhibitor (Yang et al. 2021). HYL1 is a double-stranded RNA (dsRNA)-binding protein that is dually localized to the nucleus and cytoplasm, but the function of cytosolic HYL1 was unknown. Interestingly, examining the role of cytosolic HYL1 by introducing a construct encoding a nucleus-localized form of HYL1 in a null *hyl1* mutant background revealed that cytosolic HYL1 does not regulate miRNA biogenesis or the cleavage of the target mRNAs. Instead, the target mRNAs accumulate more proteins in the absence of cytosolic HYL1. Further experiments showed that cytosolic HYL1 localized on the ER and associated with AGO1, AMP1, and miRNA-targeted mRNAs. Importantly, HYL1 facilitates AGO1 loading onto polysomes without affecting the distribution of the target mRNAs in the polysome profile. This result suggests that miRNA-mediated translational inhibition occurs after translation initiation.

Recent work showed that siRNAs, specifically 22-nt siRNAs, play a role in translational regulation (Wu et al. 2020). Plants produce 21-, 22-, and 24-nt siRNAs. Although the functions of the 21- and 24-nt siRNAs are well studied, the function of 22-nt siRNAs was largely unknown due to their low abundance. In double mutants lacking either the 5′-to-3′ decay enzyme XRN4, or the 3′-to-5′ mRNA decay enzyme SUPER KILLER2 (SKI2), along with DICER-LIKE 4 (DCL4), i.e. *xrn4 dcl4* and *ski2 dcl4*, high levels of 22-nt siRNAs are produced, particularly from 2 genes encoding nitrate reductases NIA1 and NIA2. Surprisingly, overaccumulation of 22-nt siRNAs does not lead to a decrease in *NIA1* and *NIA2* mRNA levels but instead causes a dramatic reduction in their protein levels. Polysome profile revealed global translational repression as well as specific translational repression of *NIA1* and *NIA2* mRNAs. Consistent with a role for 22-nt siRNAs in translational repression, a mutant defective in siRNA loading to RISC (*ago1-27*) rescues the overaccumulation of 22-nt siRNAs and restored NIA1 and NIA2 protein levels.

Cis-natural antisense RNAs

Cis-natural antisense transcripts (Cis-NATs) are noncoding RNAs that overlap with protein-coding genes and are transcribed from the opposite DNA strand. Cis-NATs and their mRNA partners are expected to form dsRNA and lead to RNA degradation; however, some cis-NATs were found to increase the translation of their cognate mRNAs (Deforges et al. 2019; Reis et al. 2021). In the case of rice *PHOSPHATE1.2* (*PHO1.2*), instead of long-range base-pairing between the sense-antisense transcripts, both sense (*PHO1.2*) and antisense transcript (*cis-NAT*$_{PHO1.2}$) alone have extensive intramolecular structures. Upon interaction between the sense-antisense transcripts, local structural changes occur and reduce inhibition of a high GC region located ~350 nt downstream of the start codon, which is associated with translational repression. It was suggested that the interaction between the sense-antisense transcripts leads to increased 80S formation and possibly enhanced translation elongation, resulting in increased *PHO1.2* translation (Reis et al. 2021).

Translational regulation through modulation of translation factors and master translational regulators

Translation initiation is the rate-limiting step of translation and is critical for controlling the overall translation rates (Shah et al. 2013). Two points of regulation for translation initiation have been the research focus: (1) recognition of the 5′ cap, which involves the eIF4F complexes; and (2) formation of the ternary complex, which is a component of the PIC (Fig. 5). To date, studies have revealed that both points of regulation are controlled through the phosphorylation of the relevant translation factors. Mass spectrometry and phosphoproteomics have enabled exciting breakthroughs in identifying new factors and regulatory mechanisms involved in translation initiation. Below, we discuss the approaches and advancements in this area. Several newly discovered players directly phosphorylate or interact with core translation initiation factors, such as eIF(iso)4Es and eIF(iso)4Gs. In addition to defining the interactors, recent studies have also extended the upstream or downstream signaling components of 3 conserved master translational regulators, namely TOR, SNF1 RELATED PROTEIN KINASE 1 (SnRK1), and GENERAL CONTROL NONDEREPRESSIBLE 2 (GCN2; Fig. 5).

Regulation of eIF4F complex assembly through 4E-binding proteins

As eIF4E directly binds to the 5′ cap and recruits eIF4G, which in turn recruits other translation factors, the regulation of eIF4E plays a critical role in translational control, and identifying eIF4E-binding proteins has been of great interest. Most eIF4E-binding proteins in other eukaryotes are translational repressors that compete with eIF4G for binding to eIF4E (Grüner et al. 2018). However, no obvious homologs to yeast or mammalian 4E-binding proteins (4E-BPs) have been identified in plants. Although Arabidopsis LIPOXYGENASE 2 and BASIC TRANSCRIPTION FACTOR 3 have been shown to bind to eIFiso4E in vitro and by yeast 2-hybrid, respectively (Freire et al. 2000; Freire 2005), they do not possess the evolutionarily conserved 4E-binding site (4E-BS: YXXXXLΦ; X being any

A

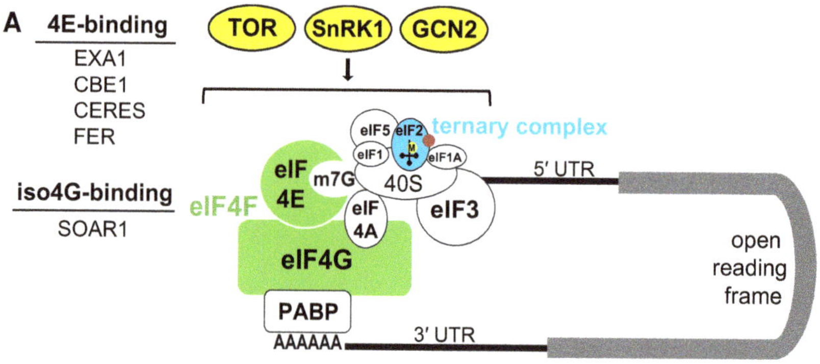

B

	TOR	SnRK1	GCN2
Activated by	High energy (light, glucose, nitrogen, etc.), auxin	Stress/ low energy	Reactive oxygen species, stress, stress hormones (ethylene, SA, JA) Amino acid starvation
Phosphorylation target(s) that links to translation	1. eIF3h 2. LARP1-5′TOP-RP 3. S6K-MRF1-eIF4A 4. (S6K-eS6)	eIF(iso)4E eIFiso4G (TOR)	α subunit of eIF2
Effects on translation	Promote global translation; promote reinitiation of some uORF-containing mRNAs	Repress global and specific mRNA translation	Repress global and specific mRNA translation

Figure 5. Overview of translational regulation through modulation of translation factors and signaling through conserved master translational regulators. **A)** Two key points of translation initiation are under intensive study: assembly of the eIF4F complex (highlighted in green) and formation of the ternary complex (highlighted in blue). Proteins interacting with and/or phosphorylating the eIF4F complex are listed. The conserved master translational regulators, including TOR, SnRK1 and GCN2, are highlighted in yellow. **B)** Summary of activation, target proteins, and effects on translation of the 3 master translational regulators, TOR, SnRK1, and GCN2. Note that TOR–S6K–eS6 signaling may not have observable effects on translation as new evidence suggests (Dasgupta et al. 2023); SnRK1 also connects to translation through TOR, but currently no evidence supports that TOR is directly phosphorylated by SnRK1.

amino acid and Φbeing a hydrophobic amino acid), and their roles in translational control remain unclear.

One eIF4E-binding protein, ESSENTIAL FOR POTEX-VIRUS ACCUMULATION (EXA1), was identified from a mutant screen that showed enhanced autoimmunity (Wu et al. 2017). EXA1 binds to eIF4E as shown by split-luciferase complementation and immunoprecipitation in *N. benthamiana*. EXA1 contains an eIF4E-binding motif as well as a conserved glycine-tyrosine-phenylalanine domain that binds to proline-rich sequences. Consistent with its potential role as a translational repressor, the *exa1* mutant has increased protein levels for several defense-related nucleotide-binding leucine-rich receptors. However, *exa1* does not affect general protein synthesis. How EXA1 specifically regulates a particular set of mRNAs remains to be determined.

A plant-specific eIF4E-binding protein, known as CONSERVED BINDING OF eIF4E1 (CBE1), was identified by searching for the canonical 4E-BS motif among potential cap-binding proteins (Patrick et al. 2018). CBE1 is encoded by a single-copy gene in Arabidopsis and is highly conserved across land plants. CBE1 can form cap-binding complexes in vitro and in vivo. *cbe1* mutant plants are viable but exhibit delayed flowering and upregulated expression of mitosis-related genes. Further investigation is needed to determine the role of CBE1 in translation.

More recently, an eIF4E-binding protein named CERES (named after the Roman goddess of agriculture) was identified through yeast 2-hybrid (Toribio et al. 2019). Like CBE1, CERES is encoded by a single copy gene in Arabidopsis, is specific to plants, and is highly conserved from mosses to angiosperms. CERES interacts with eIF4E via the canonical 4E-binding site, which is also present in eIF4G (Miras et al. 2017). Interestingly, CERES appears to replace the role of eIF4G, forming a translation initiation complex with eIF4E, eIF4A, eIF3, and PABP. Furthermore, in vitro analysis revealed that CERES promotes translation, rather than repressing translation like most 4E-binding proteins in other organisms. CERES also promotes translation of specific mRNAs involved in light response and starch accumulation (Toribio et al. 2019).

In addition to proteins containing the canonical 4E-BS, other proteins may interact with 4E to regulate translation. An interesting regulation of eIF4E controlling local translation is the FERONIA (FER) pathway (Zhu et al. 2020). FER is a receptor-like kinase located at the plasma membrane at the tip of root hair cells. Reanalysis of previous yeast 2-hybrid and immunoprecipitation-mass spectrometry screens revealed that FER interacts with several translation factors involved in initiation and elongation, as well as ribosomal proteins. The interaction with FER sequesters eIF4E to the plasma membrane of root hair tips. Upon binding to its ligand peptide RAPID ALKALINIZATION FACTOR 1 (RALF1), FER undergoes autophosphorylation and triggers eIF4E phosphorylation. The phosphorylated eIF4E moves away from the plasma membrane and binds to the 5′ cap and eIF4G to translate specific mRNAs involved in root hair growth, including those for *ROOT HAIR DEFECTIVE 6-LIKE 4* (*RSL4*). RSL4 is a transcription factor that negatively regulates root hair growth and represses *RALF1* expression. Thus, the RALF1–FER–eIF4E–*RSL4* signaling axis forms a sophisticated negative feedback loop to control local translation and root hair growth. It is worth further investigation to elucidate how the specific translation of *RSL4* mRNA is achieved once eIF4E is released from the plasma membrane. One possibility is that *RSL4* mRNA is also specifically localized at the root hair tip.

Similar to the competition for binding to eIF4E and disrupting eIF4F complex assembly, the RNA-binding protein SUPPRESSOR OF ABAR-OVEREXPRESSOR 1 (SOAR1) interacts with eIFiso4G1 and eIFiso4G2, preventing the assembly of eIFiso4F complexes and repressing translation initiation (Bi et al. 2019). SOAR1 was initially discovered from a suppressor mutant screen in plants overexpressing the putative ABA receptor ABAR (Mei et al. 2014). In addition to eIFiso4G1 and eIFiso4G2, SOAR1 can bind to the mRNA of a key ABA signaling component, *ABA INSENSITIVE5* (*ABI5*). The binding of *ABI5* mRNA releases eIFiso4G1 and eIFiso4G2 from SOAR1 and thus allows the assembly of eIFiso4F complexes and translation initiation (Bi et al. 2019). The mechanisms by which SOAR1 effi-ciently sequesters high-abundance eIFiso4G1 and eIFiso4G2 and how *ABI5* mRNA binding to SOAR1 releases eIFiso4G1 and eIFiso4G2 remain unclear.

TOR acts on multiple aspects of translation

TOR is a master translational regulator. TOR belongs to the phosphatidylinositol 3-kinase–related kinase family and is conserved throughout eukaryotes. TOR is regularly reviewed (Schepetilnikov and Ryabova 2018; Shi et al. 2018; Pacheco et al. 2021; Scarpin et al. 2022). Here, we summarize critical knowledge for understanding TOR regulation in translational control and discuss recent advances in the field.

In mammals, TOR regulates at least 3 aspects of translation: (1) translation initiation is regulated by 4E-BPs, which compete with eIF4G for binding to eIF4E. When cells are in a high energy state, 4E-BPs are phosphorylated by TOR, preventing them from binding to eIF4E, which allows the efficient formation of the eIF4F complex and promotes translation initiation. (2) TOR phosphorylates S6 Kinase 1 (S6K1), which in turn phosphorylates the ribosomal protein of small subunit 6, RPS6, or eS6 ac-cording to the updated ribosomal protein nomenclature (Scarpin et al. 2023). Note that the significance of eS6 phosphor-ylation in translation remains controversial (reviewed in Meyuhas 2015). (3) TOR regulates ribosome biogenesis by controlling the translation of ribosomal protein mRNAs through 5′terminal oligopyrimidine tract (5′ TOP) motifs (reviewed in Burkart and Brandizzi 2021).

In plants, although no sequence homologous 4E-BPs have been identified, the TOR–S6K–eS6 axis is well conserved, and variations of TOR-5′ TOP-ribosomal protein regulation have recently been identified (reviewed in Pacheco et al. 2021; Scarpin et al. 2022). TOR clearly plays an important role in plant translational control, as global polysome levels are reduced in *tor* RNA interference (RNAi) lines or when TOR activity is disturbed (Deprost et al. 2007).

Organization of the TOR complex

Most eukaryotes contain 2 TOR complexes (TORC): TORC1 and TORC2. Only TORC1 is conserved throughout eukaryotes (Brunkard 2020). TORC1 is composed of TOR, REGULATORY-ASSOCIATED PROTEIN OF TOR (RAPTOR), and LETHAL WITH SEC13 PROTEIN 8 (LST8; Scarpin et al. 2020). In Arabidopsis, TOR is encoded by a single gene, whereas RAPTOR and LST8 are both encoded by 2 genes (*RAPTOR1* and *RAPTOR2*; *LST8-1* and *LST8-2*; Menand et al. 2002; Mahfouz et al. 2006; Moreau et al. 2012). RAPTOR assists in the recruitment and phosphorylation of TOR substrates, and LST8 binds to the TOR kinase domain, likely regulating TOR activity. Downstream of TOR, Arabidopsis S6K is also encoded by 2 genes (*S6K1* and *S6K2*). Another conserved TOR substrate that regulates translation is TYPE 2A-PHOSPHATASE-ASSOCIATED PROTEIN 46 (TAP46), which is encoded by a single gene (Ahn et al. 2011).

As *TOR* knockout mutants are embryonic lethal, many studies of TOR have used inducible RNAi lines and treatment of TOR inhibitors, such as rapamycin, Torin-1, and the second-generation inhibitor AZD8055 (Deprost et al. 2007; Xiong and Sheen 2012; Xiong et al. 2013). TOR activity is often monitored through the phosphorylation status of S6Ks or eS6.

Upstream regulators, interactors, and downstream targets of TOR

TOR senses both intracellular and environmental cues, and its activity is regulated by nutrients (nitrogen, nucleotide, and amino acid availability; Cao et al. 2019; Busche et al. 2021; Liu et al. 2021), energy (glucose, light, and the circadian clock; Xiong et al. 2013; Chen et al. 2018; Li et al. 2019), phytohormones (auxin, brassinosteroids, and ABA; Schepetilnikov et al. 2013; Zhang et al. 2016; Wang et al. 2018; Montes et al. 2022), and stress (Meteignier et al. 2017). Under favorable conditions, TOR promotes translation, growth, and anabolism (Pfeiffer et al. 2016; Li et al. 2017) and suppresses catabolism, such as autophagy (Zhang et al. 2016; Pu et al. 2017).

A phosphoproteomic analysis in response to TOR inhibitors (rapamycin and AZD8055) and a targeted interactome study of TOR complex components (TOR, RAPTOR1, LST8-1) and TOR targets (S6K1, TAP46) revealed an intricate TOR signaling network in plants (Van Leene et al. 2019). In addition to several translation initiation factors, the study determined that CBE1, which interacts with eIF(iso)4E, is regulated and phosphorylated by TOR, suggesting that TOR also mediates the formation of cap-binding complexes in plants (Van Leene et al. 2019). Moreover, parallel profiling of the TOR transcriptome, translatome, proteome, and phosphoproteome placed Arabidopsis LARP1 downstream of TOR in the signaling pathway, which is consistent with other eukaryotes (Scarpin et al. 2020). Together these phosphoproteome studies identified nearly 160 TOR-sensitive phosphoproteins; among them, translation factors and ribosomal proteins were enriched, highlighting the critical role of TOR in translational control (Van Leene et al. 2019; Scarpin et al. 2020, 2022).

Additionally, a suppressor screen performed in the *lst8* mutant background identified YET ANOTHER KINASE 1 (YAK1) as a TOR target. YAK1 interacts with RAPTOR and is directly phosphorylated by TOR (Forzani et al. 2019), which is also supported by phosphoproteomic analysis (Van Leene et al. 2019). More recently, screening mutants with altered hypocotyl elongation in response to TOR inhibition in the dark discovered that EIN2, a key component of ethylene signaling, is also targeted by TOR (Fu et al. 2021). EIN2 can move between the cytoplasm and the nucleus depending on its phosphorylation status, which is controlled by ethylene (Ju et al. 2012; Qiao et al. 2012). Glucose-activated TOR directly phosphorylates EIN2 through sites distinct from those targeted by the canonical ethylene signaling pathway, and the phosphorylated EIN2 cannot enter the nucleus to orchestrate transcription of ethylene-responsive genes (Fu et al. 2021). Because EIN2 can bind to the 3′ UTR of *EBF2* and regulates its translation (Li et al. 2015; Merchante et al. 2015), an interesting question to explore is whether the TOR–EIN2 axis is also involved in translational regulation.

Roles of TOR in translational control

TOR and eIF3h in uORF-containing mRNAs. As in mammalian systems, eIF3h in Arabidopsis can serve as a scaffold for either TOR or S6K1, depending on TOR activity. When TOR is inactive, S6K1 binds to eIF3h; when TOR is activated by auxin, TOR binds to and phosphorylates eIF3h. Furthermore, the TOR–eIF3h complex is bound to polysomes, which is important for the reinitiation of uORF-containing mRNAs, including that of *ARFs* and *bZIP11*. By contrast, treatment with Torin-1 leads to dephosphorylation of TOR and S6K1, which promotes the formation of the S6K1–eIF3h complex on polysomes. Consistent with this notion, the polysome loading of uORF-containing mRNAs is significantly reduced in *TOR*-RNAi seedlings. These results demonstrate the role of TOR in the translational control of uORF-containing mRNAs (Schepetilnikov et al. 2013).

A small GTPase, RHO OF PLANTS 2 (ROP2), is also involved in auxin–TOR signaling. In response to auxin, ROP2 binds to and activates TOR. Plants with high levels of active ROP2 exhibit high levels of active TOR and S6K1. By contrast, TOR activation by auxin is abolished in *rop2 rop6 ROP4*-RNAi seedlings. ROP2 also determines the localization of TOR in the endosome. It was proposed that the transient binding of ROP2 to TOR in the endosome is associated with TOR activation (Li et al. 2017; Schepetilnikov et al. 2017).

TOR and S6K. Several lines of evidence suggest that the TOR–S6K–eS6 pathway is involved in light-enhanced translation: (1) the correlation between eS6 phosphorylation and photosynthesis activity during the light period of the diurnal cycle; and (2) the association of phosphorylated eS6 with polysomes (Turkina et al. 2011; Boex-Fontvieille et al. 2013; Enganti et al. 2018). When comparing Arabidopsis seedlings kept in the dark with those exposed to light, it was shown that light increases the fraction of mRNAs bound to polysomes (Liu et al. 2012). This light-enhanced translation is decreased in the photoreceptor mutant *phyA*, while high translation is maintained in the negative regulator of light-signaling mutant *cop1* in the dark. This finding suggests that light-enhanced translation is under the control of light signaling. In agreement, light-induced eS6 phosphorylation is abolished in the red/far-red and blue light photoreceptor mutants *phyA*, *phyA cry1 cry2*, and eS6 phosphorylation remains high in the light-signaling mutant *cop1* in darkness and in the light. It was further shown that light-induced eS6 phosphorylation is decreased in *TOR*-RNAi seedlings. By contrast, in the *cop1* mutant, TOR signaling remains active in both dark and light conditions and is reduced by a TOR inhibitor or in *TOR*-RNAi seedlings. Together, these data reveal that TOR–S6K1–eS6 is involved in light-enhanced translation, which is critical for photomorphogenesis (Chen et al. 2018).

Note that the biochemical and physiological significance of eS6 phosphorylation remains controversial in yeast and mammals (Meyuhas 2015). New evidence in Arabidopsis shows that an eS6 phospho-deficient variant largely does not affect translation, as shown by polysome profiling (Dasgupta et al. 2023). This negative result is consistent with the findings in yeast using Ribo-seq (Yerlikaya et al. 2016). eS6 phosphorylation may still provide physiological advantages, as plants accumulating the eS6 phospho-deficient isoform showed certain growth defects compared with plants with the eS6 phospho-mimic isoform (Dasgupta et al. 2023).

Additional downstream targets of S6K have been discovered (Lee et al. 2017). Arabidopsis MA3-CONTAINING TRANSLATION REGULATORY FACTORs (MRF1 to MRF4) are homologs of the human tumor suppressor PROGRAMMED CELL DEATH 4, which binds to eIF4A through its MA3 domains, inhibiting eIF4A activity as well as the eIF4A–eIF4G interaction (Loh et al. 2009). Consistent with an expected role in regulating translational initiation, Arabidopsis MRF1 interacts with eIF4A and is associated with ribosomal subunits and monosomes, as revealed by polysome profiling. ^{35}S-methionine labeling and polysome profiling analysis showed that *MRF*-RNAi plants exhibit reduced translation under low-energy conditions (dark and starvation) compared with the wild type. Interestingly, *MRF1* expression, MRF1 phosphorylation, and ribosome association are regulated not only by the energy status but also by TOR activity. Furthermore, MRF1 interacts with S6K and is a dir-ect substrate of S6K (Lee et al. 2017).

TOR and TAP46. TAP46 is a regulatory subunit of protein phosphatase 2A, which is a downstream target of TOR in yeast and mammals. Arabidopsis TAP46 can also be phosphorylated by TOR. *TAP46*-RNAi plants display similar phenotypes as TOR silencing, including a reduction in global translation, activation of autophagy, and growth arrest. This observation supports TAP46 as a downstream effector of TOR signaling (Ahn et al. 2011). However, how the TOR–TAP46 axis acts on translational regulation remains unclear.

TOR–LARP1–5′ TOP. Multi-omics profiling of TOR signaling in Arabidopsis revealed that LARP1 acts downstream of TOR (Scarpin et al. 2020). In mammals, TOR–LARP1 signaling regulates ribosomal protein mRNAs through their 5′ TOP motifs. Analyzing Arabidopsis downstream targets of TOR–LARP1 identified both conserved and plant-specific 5′ TOP mRNAs. Although these 5′ TOP mRNAs do not strictly encode ribosomal proteins, many are related to ribosome biogenesis or translation (Scarpin et al. 2020, 2022).

As near 160 TOR-sensitive phosphoproteins have been recently identified (Van Leene et al. 2019; Scarpin et al. 2020), further characterization of these proteins will facilitate our understanding of the detailed mechanisms of mRNA translational regulation by TOR.

SnRK1 integrates environmental cues/phytohormones with translational control

SnRK1 is an evolutionarily conserved protein kinase that functions in energy sensing in plants. It is homologous to mammalian AMP-activated protein kinase (AMPK) and yeast sucrose non-fermenting 1 (SNF1). SnRK1, AMPK, and SNF1 are heterotrimeric enzymes that consist of a catalytic subunit (α) with kinase activity and 2 regulatory subunits (βand γ). In Arabidopsis, the αsubunit is encoded by 2 genes, *SnRK1α1* and *SnRK1α2*, which have been the focus of much SnRK1 research (reviewed in Broeckx et al. 2016; Peixoto and Baena-González 2022).

SnRK1 interacts with translation factors eIF(iso)4E and eIFiso4G1

Searching for the conserved SnRK1/AMPK/SNF1 binding sites among Arabidopsis translation initiation factors revealed that both eIF4E and eIFiso4E contain 2 SnRK1 consensus binding sites (Bruns et al. 2019). In vitro kinase assays and in vivo interaction studies support the idea that SnRK1α2 dir-ectly interacts with and phosphorylates eIF4E and eIFiso4E. Furthermore, SnRK1-mediated phosphorylation of eIF(iso) 4E is associated with global translational repression. Consistent with this idea, the reduced polysome formation caused by overexpression of Arabidopsis *SnRK1α2* in yeast could be remedied by a nonphosphorylatable variant of eIF(iso)4E but not by a phosphomimic variant of eIF(iso)4E.

In addition to interacting with eIF4E, a phosphoproteo-mic analysis comparing wild-type and *snrk1α1* mutant plants under submergence identified eIFiso4G1 as a poten-tial SnRK1 target (Cho et al. 2016). It was further demonstrated that SnRK1 phosphorylates eIFiso4G1 in response to submergence, which, in turn, regulates global translational repression and the specific translation of mRNAs important for hypoxia responses. The *eifiso4g1* knockout mutant and variants lacking phosphorylation are more sensitive to submergence and affect the translation of hypoxia-responsive mRNAs (Cho et al. 2019).

SnRK1 integrates ABA signaling and plant growth/stress responses

Recent studies on SnRK1 provide an elegant example connecting the phytohormone ABA, TOR, and plant growth.

As the *snrk1α1 snrk1α2* double mutant is not viable, insights into the role of SnRK1αwere gained by studying a sesquimutant (*snrk1α1*$^{-/-}$ *snrk1α2*$^{+/-}$) or inducible knockdown lines (Belda-Palazón et al. 2020; Henninger et al. 2022). Because the sesquimutant shows reduced ABA-mediated growth repression, the role of TOR, a growth-promoting kinase, was examined under ABA treatment. It was shown that SnRK1α, but not SnRK2 (a key player in ABA signaling), directly interacts with TOR; this interaction is enhanced in the presence of ABA. Furthermore, SnRK1α is required for TOR repression, as TOR repression in response to ABA is slower in the *snrk1α* mutant. Under non-ABA treatment conditions, SnRK2 forms a complex with SnRK1 and PROTEIN PHOSPHATASE TYPE 2C (PP2C, a repressor of ABA signaling), promoting growth. By contrast, in the presence of ABA, PP2C binds to ABA and the ABA receptor, thus releasing SnRK1, which represses TOR and reduces growth (Belda-Palazón et al. 2020).

An independent study has shown that, in the presence of ABA, SnRK2 is activated and phosphorylates RAPTOR, which also inactivates TOR, repressing growth. In nonstress conditions, TOR phosphorylates the ABA receptors, repressing the ABA/stress response (Wang et al. 2018). Thus, SnRK1 and SnRK2 coordinately regulate TOR activity, as well as plant growth and stress responses.

More recently, it was shown that SnRK1 changes its subcellular localization from the nucleus to the cytoplasm in response to ABA. Chemicals blocking nuclear export or fusing SnRK1α1 to a nucleus localization signal abolished ABA-dependent TOR repression (Belda-Palazón et al. 2022). SnRK1α is also recruited into stress granules in response to heat stress through interaction with the Tudor staphylococcal nuclease (Gutierrez-Beltran et al. 2021). Thus, the subcellular localization of SnRK1α connects environmental input and downstream signaling responses.

GCN2 controls ternary complex formation and global translation initiation

Another important and conserved protein kinase involved in translational control is GCN2, which phosphorylates the α subunit of heterotrimeric eIF2. eIF2 plays a unique role in translation initiation as it is part of the ternary complex [eIF2, GTP, Met-tRNA(i)] that carries the initiator tRNA to the 40S ribosome subunit. In yeast and mammals, the formation of the ternary complex requires efficient GDP/GTP exchange on eIF2, with the help of the guanine nucleotide exchange factor eIF2B. Under stress conditions, a highly conserved serine residue in eIF2α is phosphorylated. The phosphorylated eIF2α tightly binds to eIF2B, inhibiting eIF2B-facilitated GDP-GTP exchange. As a result, the available ternary complex for translation is reduced and leads to global translational repression (reviewed in Browning and Bailey-Serres 2015). Unlike the multiple protein kinases regulating eIF2α phosphorylation in other eukaryotes, only 1 protein kinase, GCN2, is known to phosphorylate eIF2α in plants and yeast (Lageix et al. 2008; Zhang et al. 2008).

In plants, eIF2αphosphorylation has been associated with numerous abiotic/biotic stressors and phytohormones, including reactive oxygen species, UV light, cold, salt, wounding, bacterial infection, ABA, ethylene, salicylic acid, and methyl-jasmonate (Lageix et al. 2008; Liu et al. 2019; Lokdarshi et al. 2020b). Consistent with the role of GCN2 activation and eIF2α phosphorylation in translational repression, amino acid or purine starvation, which activates GCN2, represses global translation; this repression is reduced in *gcn2* mutant plants (Lageix et al. 2008; Wang et al. 2017).

However, other evidence has challenged whether eIF2α phosphorylation has a strong inhibitory effect on translation. Plant eIF2 is known to exhibit a weaker affinity for GDP than mammalian eIF2, leading to the suggestion that plant eIF2 GDP/GTP exchange does not necessarily need the help of eIF2B (Browning and Bailey-Serres 2015). Moreover, the conditions that activate eIF2α phosphorylation do not always lead to translational repression or depend on GCN2 (Izquierdo et al. 2018; Lokdarshi et al. 2020a; Zhigailov et al. 2020). For a com-prehensive discussion on this topic, see (Lokdarshi and von Arnim 2022). Below, we highlight recent work related to the role of GCN2 in global translational control.

First, transcriptome and polysome profiling analyses were performed in wild-type and *gcn2* plants under a condition that activates GCN2. These analyses uncovered a set of mRNAs that escape translational repression dependent on GCN2. Surprisingly, unlike the findings in yeast and mammals, plant mRNAs differentially regulated by GCN2 are not enriched for specific functions or possess distinct mRNA features, such as uORFs (Lokdarshi et al. 2020a). Further investigation is required to understand how GCN2 specifically regulates the translation of these mRNAs.

A recent study underscores the importance of GCN2 in global translational control in response to submergence (Cho et al. 2022). Upon submergence, ethylene is trapped in the submerged tissues, and it is known that ethylene triggers eIF2α phosphorylation (Lageix et al. 2008), suggesting a role for GCN2/eIF2αphosphorylation in submergence. Under submergence, eIF2αphosphorylation is enhanced and global polysome loading of mRNAs is reduced. To characterize the effect of GCN2/eIF2αphosphorylation and ethylene signaling on ribosome loading of mRNAs, transcriptome and polysome profiling analyses were performed in *gcn2* and *ein2* (defective in ethylene signaling) compared with wild-type plants under submergence. The results revealed that global translational repression is less effective in *gcn2* and *ein2*. Importantly, GCN2 and EIN2 enhance translation of distinct sets of mRNAs under submergence. The data suggest that ethylene activates 2 pathways, 1 dependent on GCN2, and the other 1 dependent on EIN2, to regulate the production of hypoxia-responsive proteins for acclimation to submergence. Furthermore, translation of several hypoxia-responsive mRNAs under submergence is affected in an eIF2αnonphosphorylatable variant in which the conserved serine residue was mutated to alanine, highlighting a role for eIF2α phosphorylation in translational control in response to submergence (Cho et al. 2022).

Other previously missed types of regulation

tRNA-derived fragments

In addition to carrying amino acids to translating ribosomes during translation, tRNAs have other regulatory functions. One interesting function is that tRNA-derived fragments (tRFs) can regulate translation. Most tRFs are derived from mature tRNAs and can be grouped into 2 categories based on their size and where the cleavage occurs within the tRNA: long tRFs ranging 30 to 35 nt are cleaved at the anticodon region, and short tRFs ranging 19 to 25 nt are cleaved at the D or T arms. Analysis of small RNA-seq data has revealed that plants contain a diverse population of tRFs, and their composition fluctuates depending on developmental stages or environmental conditions (Cognat et al. 2017). Testing 10 synthesized tRFs has shown that 2 of them, derived from $tRNA^{Ala}$(AGC) and $tRNA^{Asn}$, strongly repress the translation of a reporter mRNA in vitro. Interestingly, this translational repression does not require sequence complementarity between the tRF and the translated mRNA. Furthermore, it was shown that the tRFs are associated with active polysomes but not efficiently with ribosome subunits. Thus, tRFs are unlikely to interfere with the binding of ribosome subunits to the mRNA (Lalande et al. 2020). The mechanism by which tRFs repress translation remains to be resolved.

A $tRNA^{His}$ guanylyltransferase

Another regulation related to tRNA was discovered serendipitously from a genetic screen searching for a thermosensitive mutant in rice, identifying the mutant *adaptation to environmental temperature 1* (*aet1*) (Chen et al. 2019). *AET1* encodes a $tRNA^{HIS}$ guanylyltransferase, which modifies pre-$tRNA^{HIS}$ and is required for tRNA maturation and homeostasis. Interestingly, yeast 2-hybrid screening for AET1-interacting partners revealed that AET1 interacts with the ribosome-associated protein RECEPTOR FOR ACTIVATED C KINASE 1 A and eIF3h. The interaction was confirmed in planta, and it was demonstrated that all 3 proteins localize to the ER. To investigate a potential role for AET1 in mRNA translation, RNA-immunoprecipitation sequencing was performed, and *OsARF* mRNAs were highly enriched. As discussed above, Arabidopsis *ARF* mRNAs harbor uORFs and the re-initiation of their mORFs is regulated by eIF3h (Schepetilnikov et al. 2013). Consistent with the potential role of AET1 modulating translation re-initiation through eIF3h interaction, the fraction of *OsARF19* and *OsARF23* mRNAs in active polysomes is reduced in *aet1*.

eEF2

During translation elongation, eukaryotic Elongation Factor 2 (eEF2) regulates the transfer of peptidyl tRNA from the ribosomal P-site to the A-site. A post-translational modification on a histidine residue of eEF2, termed diphthamide (a target of diphtheria toxin in humans), is conserved in Arabidopsis (Zhang et al. 2022). Arabidopsis mutants of Dph1, which catalyzes the first step of diphthamide biosynthesis, lose the modification on the H700 residue and increase −1 frameshift errors, as shown by a reporter assay. The mutant also displays a decrease in TOR activity, cell proliferation, and plant growth and is accompanied by an increased autophagy. Interestingly, in response to copper and cadmium, wild-type plants accumulate the eEF2 form with unmodified H700 residue, suggesting that an alteration of translational fidelity is involved in acclimating to heavy metal stress.

Stability and spatial dynamic of mRNAs

In the cytosol, mRNAs might be subjected to degradation through various mechanisms or stored in RNA-protein conjugates, including P-bodies and stress granules. Both degradation and storage can prevent mRNAs from being translated.

Decay of mRNAs

Most mRNA degradation is initiated by removing the 3′ poly-A tail, followed by either 5′-3′ or 3′-5′ decay mechanisms. The 5′-3′ decay is governed by the decapping complex and 5′-3′ exonuclease XRN4, and 3′-5′ decay is governed by SUPPRESSOR OF VARICOSE or the RNA exosomes (reviewed in Brunkard and Baker 2018; Sieburth and Vincent 2018). Eukaryotic cells also possess 3 conserved translation-dependent RNA quality control mechanisms to remove aberrant RNAs: NMD, no-go decay, and non-stop decay. NMD recognizes mRNAs containing a premature stop codon, no-go decay targets mRNAs with stalled ribosomes, and non-stop decay recognizes mRNAs lacking a stop codon (reviewed in Powers et al. 2020). In plants, NMD is the best-characterized mechanism among the 3 (reviewed in Ohtani and Wachter 2019), whereas no-go decay and non-stop decay have just started to be characterized (Szádeczky-Kardoss et al. 2018a, 2018b).

It was long believed that mRNAs were either engaged in translation, or subjected to degradation or storage. However, prevalent co-translational decay was reported in yeast (Hu et al. 2009; Pelechano et al. 2015). Consistent with this idea, heat induces the accumulation of several decapped mRNAs in light polysome fractions in the Arabidopsis *xrn4* mutant, supporting the notion that some mRNAs are subjected to degradation while bound by ribosomes (Merret et al. 2015). Furthermore, genome-wide analysis of RNA degradation intermediates in wild-type Arabidopsis plants detected a 3-nt periodicity within the coding regions of mRNAs, a signature of actively translating ribosomes, indicating that the transcripts are also occupied by translating ribosomes when degradation occurs (Hou et al. 2016; Yu et al. 2016). Comparative analysis of 10 angiosperm species confirmed that co-translational decay is widespread and conserved across diverse plant lineages (Guo et al. 2023a).

P-bodies and stress granules

P-bodies and stress granules are cytosolic RNA-protein assemblies that play an important role in balancing mRNA translation, storage, and degradation. Unlike P-bodies, which are constitutively present in cells, stress granules are induced by stress conditions. These topics are reviewed regularly, for example, (Chantarachot and Bailey-Serres 2018; Kearly et al. 2022). Here, we focus on their functions in translational control.

P-bodies were thought primarily to be involved in mRNA degradation, as they are associated with decapping enzymes and the 5′-3′ exonuclease XRN4. However, a recent study has highlighted the storage capacity of P-bodies. As mentioned above, light can enhance the translation of many transcripts (Liu et al. 2012; Chen et al. 2018). Monitoring P-body accumulation in seedlings kept in the dark or transferred to light for 4 h revealed that light reduces P-body accumulation. Furthermore, transcriptome and translatome analysis comparing a mutant defective in P-body formation (*dcp5-1*, a weak mutant allele of *DECAPPING* 5) and wild-type seedlings identified over 2,000 mRNAs that remain actively translated in *dcp5-1* in the dark without changes in their mRNA abundance. These results suggest that in the dark, plants repress the translation of specific mRNAs involved in the translation machinery, photosynthesis, and apical hook formation by transiently storing these mRNAs in P-bodies (Jang et al. 2019). Similarly, in response to hypoxia, OLIGOURIDYLATE BINDING PROTEIN 1 (UBP1) can sequester thousands of mRNAs into stress granules, preventing them from being translated. Importantly, once the stress is removed, these mRNAs reassociate with polysomes to resume active translation (Sorenson and Bailey-Serres 2014).

EBF2 mentioned above is also an elegant example of P-body regulation of mRNA localization. When the 3′ UTR of *EBF2* is bound by EIN2-C (the C-terminal fragment of EIN2, following cleavage by a protease in response to ethylene signaling), the mRNA is targeted to P-bodies with the assistance of NMD factors (Li et al. 2015; Merchante et al. 2015).

A recent exciting study linked P-bodies, NMD, and translational control (Cairo et al. 2022). *THREE-DIVISION MUTANT 1* (*TDM1*) is specifically expressed during the second meiotic division of pollen development and is required for the termination of meiosis. TDM1 was shown to be recruited into P-bodies by interacting with SUPPRESSOR WITH MORPHOGENETIC EFFECTS ON GENITALIA 7 (SMG7), an NMD factor. A suppressor mutant screen in *smg7-6* background and a yeast 2-hybrid screen for TDM1-interacting proteins led to the identification of canonical eIF4G, as well as the plant-specific eIFiso4G1 and eIFiso4G2, as additional factors involved in meiotic exit. Follow-up experiments revealed that TMD1 recruits eIF(iso)4Gs and 4A into P-bodies, suggesting that translation is temporarily repressed at meiotic exit by moving these translation factors into P-bodies. Consistent with this idea, treating *tmd1* mutants with the translation inhibitor cycloheximide rescues the pollen meiotic exit defect. Thus, reduced translation of meiotic mRNAs is proposed to be critical for completing pollen meiosis.

Conclusions

Technological advancements have offered unparalleled opportunities to discover noncanonical translation events and dissect the detailed mechanisms of translational control in various aspects of plant development and physiology. As our understanding of translational control still lags behind that of transcriptional control, ample exciting opportunities for new discoveries are within reach.

Acknowledgments

We thank Isaiah Kaufman for his critical reading of and thoughtful comments on the manuscript. This work was supported by a National Science Foundation grant (2051885) to PYH. We apologize to colleagues whose work could not be cited owing to space limitations.

Author contributions

H.L.W. and P.Y.H. conceived the review, wrote the manuscript, and prepared the figures. J.J. organized the literature. All authors discussed the literature and edited the review.

Conflict of interest statement: None declared.

References

Ahn CS, Han JA, Lee HS, Lee S, Pai HS. The PP2A regulatory subunit Tap46, a component of the TOR signaling pathway, modulates growth and metabolism in plants. Plant Cell. 2011:**23**(1):185–209. https://doi.org/10.1105/tpc.110.074005

Alatorre-Cobos F, Cruz-Ramírez A, Hayden CA, Pérez-Torres CA, Chauvin AL, Ibarra-Laclette E, Alva-Cortés E, Jorgensen RA, Herrera-Estrella L. Translational regulation of Arabidopsis XIPOTL1 is modulated by phosphocholine levels via the phylogenetically conserved upstream open Reading frame 30. J Exp Bot. 2012:**63**(14):5203–5221. https://doi.org/10.1093/jxb/ers180

Bai B, Peviani A, van der Horst S, Gamm M, Snel B, Bentsink L, Hanson J. Extensive translational regulation during seed germination revealed by polysomal profiling. New Phytol. 2017a:**214**(1):233–244. https://doi.org/10.1111/nph.14355

Bai X, Huang Y, Hu Y, Liu H, Zhang B, Smaczniak C, Hu G, Han Z, Xing Y. Duplication of an upstream silencer of FZP increases grain yield in rice. Nat Plants. 2017b:**3**(11):885–893. https://doi.org/10.1038/s41477-017-0042-4

Baird TD, Cheng KCC, Chen YC, Buehler E, Martin SE, Inglese J, Hogg JR. ICE1 promotes the link between splicing and nonsense-mediated mRNA decay. eLife. 2018:**7**:e33178. https://doi.org/10.7554/eLife.33178

Barakat A, Szick-Miranda K, Chang IF, Guyot R, Blanc G, Cooke R, Delseny M, Bailey-Serres J. The organization of cytoplasmic ribosomal protein genes in the Arabidopsis genome. Plant Physiol. 2001:**127**(2):398–415. https://doi.org/10.1104/pp.010265

Basrai MA, Hieter P, Boeke JD. Small open Reading frames: beautiful needles in the haystack. Genome Res. 1997:**7**(8):768–771. https://doi.org/10.1101/gr.7.8.768

Bazin J, Baerenfaller K, Gosai SJ, Gregory BD, Crespi M, Bailey-Serres J. Global analysis of ribosome-associated noncoding RNAs unveils new modes of translational regulation. Proc Natl Acad Sci USA. 2017:**114**:E10018–E10027. https://doi.org/10.1073/pnas.1708433114

Belda-Palazón B, Adamo M, Valerio C, Ferreira LJ, Confraria A, Reis-Barata D, Rodrigues A, Meyer C, Rodriguez PL, Baena-González E. A dual function of SnRK2 kinases in the regulation of SnRK1 and plant growth. Nat Plants. 2020:**6**(11):1345–1353. https://doi.org/10.1038/s41477-020-00778-w

Belda-Palazón B, Costa M, Beeckman T, Rolland F, Baena-González E. ABA represses TOR and root meristem activity through nuclear exit of the SnRK1 kinase. Proc Natl Acad Sci USA. 2022:**119**(28): e2204862119. https://doi.org/10.1073/pnas.2204862119

Bi C, Ma Y, Jiang SC, Mei C, Wang XF, Zhang DP. Arabidopsis translation initiation factors eIFiso4G1/2 link repression of mRNA cap-binding complex eIFiso4F assembly with RNA-binding protein SOAR1-mediated ABA signaling. New Phytol. 2019:**223**(3): 1388–1406. https://doi.org/10.1111/nph.15880

Billey E, Hafidh S, Cruz-Gallardo I, Litholdo Jr CG, Jean V, Carpentier MC, Picart C, Kumar V, Kulichova K, Maréchal E, et al. LARP6C orchestrates posttranscriptional reprogramming of gene expression during hydration to promote pollen tube guidance. Plant Cell. 2021:**33**(8):2637–2661. https://doi.org/10.1093/plcell/koab131

Boex-Fontvieille E, Daventure M, Jossier M, Zivy M, Hodges M, Tcherkez G. Photosynthetic control of Arabidopsis leaf cytoplasmic translation initiation by protein phosphorylation. PLoS ONE. 2013:**8**(7):e70692. https://doi.org/10.1371/journal.pone.0070692

Bonnot T, Nagel DH. Time of the day prioritizes the pool of translating mRNAs in response to heat stress. Plant Cell. 2021:**33**(7):2164–2182. https://doi.org/10.1093/plcell/koab113

Branco-Price C, Kawaguchi R, Ferreira RB, Bailey-Serres J. Genome-wide analysis of transcript abundance and translation in Arabidopsis seedlings subjected to oxygen deprivation. Ann Bot. 2005:**96**(4):647–660. https://doi.org/10.1093/aob/mci217

Brar GA, Weissman JS. Ribosome profiling reveals the what, when, where and how of protein synthesis. Nat Rev Mol Cell Biol. 2015:**16**(11):651–664. https://doi.org/10.1038/nrm4069

Brodersen P, Sakvarelidze-Achard L, Bruun-Rasmussen M, Dunoyer P, Yamamoto YY, Sieburth L, Voinnet O. Widespread translational inhibition by plant miRNAs and siRNAs. Science. 2008:**320**(5880): 1185–1190. https://doi.org/10.1126/science.1159151

Broeckx T, Hulsmans S, Rolland F. The plant energy sensor: evolutionary conservation and divergence of SnRK1 structure, regulation, and function. J Exp Bot. 2016:**67**(22):6215–6252. https://doi.org/10.1093/jxb/erw416

Browning KS, Bailey-Serres J. Mechanism of cytoplasmic mRNA translation. The Arabidopsis Book. 2015:**13**:e0176. https://doi.org/10.1199/tab.0176

Brunkard JO. Exaptive evolution of target of rapamycin signaling in multicellular eukaryotes. Dev Cell. 2020:**54**(2):142–155. https://doi.org/10.1016/j.devcel.2020.06.022

Brunkard JO, Baker B. A two-headed monster to avert disaster: hbs1/ski7 is alternatively spliced to build eukaryotic RNA surveillance complexes. Front Plant Sci. 2018:**9**:1333. https://doi.org/10.3389/fpls.2018.01333

Bruns AN, Li S, Mohannath G, Bisaro DM. Phosphorylation of Arabidopsis eIF4E and eIFiso4E by SnRK1 inhibits translation. FEBS J. 2019:**286**(19):3778–3796. https://doi.org/10.1111/febs.14935

Burkart GM, Brandizzi F. A tour of TOR complex signaling in plants. Trends Biochem Sci. 2021:**46**(5):417–428. https://doi.org/10.1016/j.tibs.2020.11.004

Busche M, Scarpin MR, Hnasko R, Brunkard JO. TOR coordinates nucleotide availability with ribosome biogenesis in plants. Plant Cell. 2021:**33**(5):1615–1632. https://doi.org/10.1093/plcell/koab043

Cai Q, Fukushima H, Yamamoto M, Ishii N, Sakamoto T, Kurata T, Motose H, Takahashi T. The SAC51 family plays a central role in thermospermine responses in Arabidopsis. Plant Cell Physiol. 2016:**57**(8):1583–1592. https://doi.org/10.1093/pcp/pcw113

Cairo A, Vargova A, Shukla N, Capitao C, Mikulkova P, Valuchova S, Pecinkova J, Bulankova P, Riha K. Meiotic exit in Arabidopsis is driven by P-body-mediated inhibition of translation. Science. 2022:**377**(6606):629–634. https://doi.org/10.1126/science.abo0904

Cao P, Kim SJ, Xing A, Schenck CA, Liu L, Jiang N, Wang J, Last RL, Brandizzi F. Homeostasis of branched-chain amino acids is critical for the activity of TOR signaling in Arabidopsis. eLife. 2019:**8**: e50747. https://doi.org/10.7554/eLife.50747

Castellano MM, Merchante C. Peculiarities of the regulation of translation initiation in plants. Curr Opin Plant Biol. 2021:**63**:102073. https://doi.org/10.1016/j.pbi.2021.102073

Chang H, Lim J, Ha M, Kim VN. TAIL-seq: genome-wide determination of poly(A) tail length and 3′ end modifications. Mol Cell. 2014:**53**(6): 1044–1052. https://doi.org/10.1016/j.molcel.2014.02.007

Chantarachot T, Bailey-Serres J. Polysomes, stress granules, and processing bodies: a dynamic triumvirate controlling cytoplasmic mRNA fate and function. Plant Physiol. 2018:**176**(1):254–269. https://doi.org/10.1104/pp.17.01468

Chen GH, Liu MJ, Xiong Y, Sheen J, Wu SH. TOR And RPS6 transmit light signals to enhance protein translation in deetiolating Arabidopsis seedlings. Proc Natl Acad Sci USA. 2018:**115**(50):12823–-12828. https://doi.org/10.1073/pnas.1809526115

Chen K, Guo T, Li XM, Zhang YM, Yang YB, Ye WW, Dong NQ, Shi CL, Kan Y, Xiang YH, et al. Translational regulation of plant response to high temperature by a dual-function tRNAHis guanylyltransferase in rice. Mol Plant. 2019:**12**(8):1123–1142. https://doi.org/10.1016/j.molp.2019.04.012

Chen Q, Tian F, Cheng T, Jiang J, Zhu G, Gao Z, Lin H, Hu J, Qian Q, Fang X, et al. Translational repression of FZP mediated by CU-rich element/OsPTB interactions modulates panicle development in rice. Plant J. 2022:**110**(5):1319–1331. https://doi.org/10.1111/tpj.15737

Chen X, Dickman D. Development of a tissue-specific ribosome profiling approach in Drosophila enables genome-wide evaluation of translational adaptations. PLoS Genet. 2017:**13**(12):e1007117. https://doi.org/10.1371/journal.pgen.1007117

Cheng M-C, Kathare PK, Paik I, Huq E. Phytochrome signaling networks. Annu Rev Plant Biol. 2021:**72**(1):217–244. https://doi.org/10.1146/annurev-arplant-080620-024221

Chew GL, Pauli A, Schier AF. Conservation of uORF repressiveness and sequence features in mouse, human and zebrafish. Nat Commun. 2016:**7**(1):11663. https://doi.org/10.1038/ncomms11663

Cho H, Cho HS, Nam H, Jo H, Yoon J, Park C, Dang TV, Kim E, Jeong J, Park S, et al. Translational control of phloem development by RNA G-quadruplex-JULGI determines plant sink strength. Nat Plants. 2018:**4**(6):376–390. https://doi.org/10.1038/s41477-018-0157-2

Cho HY, Chou MY, Ho HY, Chen WC, Shih MC. Ethylene modulates translation dynamics in Arabidopsis under submergence via GCN2 and EIN2. Sci Adv. 2022:**8**(22):7863. https://doi.org/10.1126/sciadv.abm7863

Cho HY, Lu MYJ, Shih MC. The SnRK1-eIFiso4G1 signaling relay regulates the translation of specific mRNAs in Arabidopsis under submergence. New Phytol. 2019:**222**(1):366–381. https://doi.org/10.1111/nph.15589

Cho HY, Wen TN, Wang YT, Shih MC. Quantitative phosphoproteomics of protein kinase SnRK1 regulated protein phosphorylation in Arabidopsis under submergence. J Exp Bot. 2016:**67**(9):2745–2760. https://doi.org/10.1093/jxb/erw107

Chotewutmontri P, Barkan A. Dynamics of chloroplast translation during chloroplast differentiation in maize. PLoS Genet. 2016:**12**(7): e1006106. https://doi.org/10.1371/journal.pgen.1006106

Chung BYW, Balcerowicz M, Di Antonio M, Jaeger KE, Geng F, Franaszek K, Marriott P, Brierley I, Firth AE, Wigge PA. An RNA thermoswitch regulates daytime growth in Arabidopsis. Nat Plants. 2020:**6**(5):522–532. https://doi.org/10.1038/s41477-020-0633-3

Coate JE, Bar H, Doyle JJ. Extensive translational regulation of gene expression in an allopolyploid (*Glycine dolichocarpa*). Plant Cell. 2014:**26**(1):136–150. https://doi.org/10.1105/tpc.113.119966

Cognat V, Morelle G, Megel C, Lalande S, Molinier J, Vincent T, Small I, Duchene AM, Marechal-Drouard L. The nuclear and organellar tRNA-derived RNA fragment population in *Arabidopsis thaliana* is highly dynamic. Nucleic Acids Res. 2017:**45**(6):3460–3472. https://doi.org/10.1093/nar/gkw1122

Cui Y, Rao S, Chang B, Wang X, Zhang K, Hou X, Zhu X, Wu H, Tian Z, Zhao Z, et al. AtIa1 protein initiates IRES-dependent translation of WUSCHEL mRNA and regulates the stem cell homeostasis of

Arabidopsis in response to environmental hazards. Plant Cell Environ. 2015:**38**(10):2098–2114. https://doi.org/10.1111/pce.12535

Dasgupta A, Camacho RAU, Enganti R, Cho SK, Tucker LL, Torreverde JS, Abraham PE, von Arnim AG. A phosphorylation-deficient ribosomal protein eS6 is largely functional in Arabidopsis thaliana, rescuing mutant defects from global translation and gene expression to photosynthesis and growth. bioRxiv: 2023.05.30.542942; 2023.

Deforges J, Reis RS, Jacquet P, Sheppard S, Gadekar VP, Hart-Smith G, Tanzer A, Hofacker IL, Iseli C, Xenarios I, et al. Control of cognate sense mRNA translation by cis-natural antisense RNAs. Plant Physiol. 2019:**180**(1):305–322. https://doi.org/10.1104/pp.19.00043

Deprost D, Yao L, Sormani R, Moreau M, Leterreux G, Nicolaï M, Bedu M, Robaglia C, Meyer C. The Arabidopsis TOR kinase links plant growth, yield, stress resistance and mRNA translation. EMBO Rep. 2007:**8**(9):864–870. https://doi.org/10.1038/sj.embor.7401043

Dong J, Chen H, Deng XW, Irish VF, Wei N. Phytochrome B induces intron retention and translational inhibition of phytochrome-interacting factor3. Plant Physiol. 2020:**182**(1):159–166. https://doi.org/10.1104/pp.19.00835

Doroudgar S, Hofmann C, Boileau E, Malone B, Riechert E, Gorska AA, Jakobi T, Sandmann C, Jürgensen L, Kmietczyk V, et al. Monitoring cell-type–specific gene expression using ribosome profiling in vivo during cardiac hemodynamic stress. Circ Res. 2019:**125**(4):431–448. https://doi.org/10.1161/CIRCRESAHA.119.314817

Douka K, Agapiou M, Birds I, Aspden JL. Optimization of ribosome footprinting conditions for Ribo-Seq in human and *Drosophila melanogaster* tissue culture cells. Front Mol Biosci. 2022:**8**:791455. https://doi.org/10.3389/fmolb.2021.791455

Enganti R, Cho SK, Toperzer JD, Urquidi-Camacho RA, Cakir OS, Ray AP, Abraham PE, Hettich RL, von Arnim AG. Phosphorylation of ribosomal protein RPS6 integrates light signals and circadian clock signals. Front Plant Sci. 2018:**8**:2210. https://doi.org/10.3389/fpls.2017.02210

Forzani C, Duarte GT, Van Leene J, Clément G, Huguet S, Paysant-Le-Roux C, Mercier R, De Jaeger G, Leprince AS, Meyer C. Mutations of the AtYAK1 kinase suppress TOR deficiency in Arabidopsis. Cell Rep. 2019:**27**(12):3696–3708.e5. https://doi.org/10.1016/j.celrep.2019.05.074

Freire MA. Translation initiation factor (iso) 4E interacts with BTF3, the β subunit of the nascent polypeptide-associated complex. Gene. 2005:**345**(2):271–277. https://doi.org/10.1016/j.gene.2004.11.030

Freire MA, Tourneur C, Granier F, Camonis J, El Amrani A, Browning KS, Robaglia C. Plant lipoxygenase 2 is a translation initiation factor-4E-binding protein. Plant Mol Biol. 2000:**44**(2):129–140. https://doi.org/10.1023/A:1006494628892

Fu L, Liu Y, Qin G, Wu P, Zi H, Xu Z, Zhao X, Wang Y, Li Y, Yang S, et al. The TOR–EIN2 axis mediates nuclear signaling to modulate plant growth. Nature. 2021:**591**(7849):288–292. https://doi.org/10.1038/s41586-021-03310-y

Fujishiro Y, Agata A, Ota S, Ishihara R, Takeda Y, Kunishima T, Ikeda M, Kyozuka J, Hobo T, Kitano H. Comprehensive panicle phenotyping reveals that qSrn7/FZP influences higher-order branching. Sci Rep. 2018:**8**(1):12511. https://doi.org/10.1038/s41598-018-30395-9

Gage JL, Mali S, McLoughlin F, Khaipho-Burch M, Monier B, Bailey-Serres J, Vierstra RD, Buckler ES. Variation in upstream open reading frames contributes to allelic diversity in maize protein abundance. Proc Natl Acad Sci USA. 2022:**119**(14):e2112516119. https://doi.org/10.1073/pnas.2112516119

Gallie DR. Plant growth and fertility requires functional interactions between specific PABP and eIF4G gene family members. PLoS ONE. 2018:**13**(1):e0191474. https://doi.org/10.1371/journal.pone.0191474

Gallie DR, Browning KS. eIF4G functionally differs from eIFiso4G in promoting internal initiation, cap-independent translation, and translation of structured mRNAs. J Biol Chem. 2001:**276**(40):36951–36960. https://doi.org/10.1074/jbc.M103869200

Gao X, Wan J, Liu B, Ma M, Shen B, Qian SB. Quantitative profiling of initiating ribosomes in vivo. Nat Methods. 2015:**12**(2):147–153. https://doi.org/10.1038/nmeth.3208

Gonzalez C, Sims JS, Hornstein N, Mela A, Garcia F, Lei L, Gass DA, Amendolara B, Bruce JN, Canoll P, et al. Ribosome profiling reveals a cell-type-specific translational landscape in brain tumors. J Neurosci. 2014:**34**(33):10924–10936. https://doi.org/10.1523/JNEUROSCI.0084-14.2014

Grüner S, Weber R, Peter D, Chung MY, Igreja C, Valkov E, Izaurralde E. Structural motifs in eIF4G and 4E-BPs modulate their binding to eIF4E to regulate translation initiation in yeast. Nucleic Acids Res. 2018:**46**(13):6893–6908. https://doi.org/10.1093/nar/gky542

Guerrero-González ML, Rodríguez-Kessler M, Jiménez-Bremont JF. uORF, a regulatory mechanism of the Arabidopsis polyamine oxidase 2. Mol Biol Rep. 2014:**41**(4):2427–2443. https://doi.org/10.1007/s11033-014-3098-5

Guo R, Yu X, Gregory BD. The identification of conserved sequence features of co-translationally decayed mRNAs and upstream open reading frames in angiosperm transcriptomes. Plant Direct. 2023a:**7**(1):e479. https://doi.org/10.1002/pld3.479

Guo Y, Chen Y, Wang Y, Wu X, Zhang X, Mao W, Yu H, Guo K, Xu J, Ma L, et al. The translational landscape of bread wheat during grain development. Plant Cell. 2023b:**35**(6):1848–1867. https://doi.org/10.1093/plcell/koad075

Guo Z, Cao H, Zhao J, Bai S, Peng W, Li J, Sun L, Chen L, Lin Z, Shi C, et al. A natural uORF variant confers phosphorus acquisition diversity in soybean. Nat Commun. 2022:**13**:1–14. https://doi.org/10.1038/s41467-022-31555-2

Gutierrez-Beltran E, Elander PH, Dalman K, Dayhoff GW, Moschou PN, Uversky VN, Crespo JL, Bozhkov PV. Tudor staphylococcal nuclease is a docking platform for stress granule components and is essential for SnRK1 activation in Arabidopsis. EMBO J. 2021:**40**(17):e105043. https://doi.org/10.15252/embj.2020105043

Hanfrey C, Elliott KA, Franceschetti M, Mayer MJ, Illingworth C, Michael AJ. A dual upstream open reading frame-based autoregulatory circuit controlling polyamine-responsive translation. J Biol Chem. 2005:**280**(47):39229–39237. https://doi.org/10.1074/jbc.M509340200

Hayashi N, Sasaki S, Takahashi H, Yamashita Y, Naito S, Onouchi H. Identification of Arabidopsis thaliana upstream open reading frames encoding peptide sequences that cause ribosomal arrest. Nucleic Acids Res. 2017:**45**(15):8844–8858. https://doi.org/10.1093/nar/gkx528

Hayden CA, Jorgensen RA. Identification of novel conserved peptide uORF homology groups in Arabidopsis and rice reveals ancient eukaryotic origin of select groups and preferential association with transcription factor-encoding genes. BMC Biol. 2007:**5**(1):32. https://doi.org/10.1186/1741-7007-5-32

Henninger M, Pedrotti L, Krischke M, Draken J, Wildenhain T, Fekete A, Rolland F, Müller MJ, Fröschel C, Weiste C, et al. The evolutionarily conserved kinase SnRK1 orchestrates resource mobilization during Arabidopsis seedling establishment. Plant Cell. 2022:**34**(1):616–632. https://doi.org/10.1093/plcell/koab270

Hou CY, Lee WC, Chou HC, Chen AP, Chou SJ, Chen HM. Global analysis of truncated RNA ends reveals new insights into Ribosome Stalling in plants. Plant Cell. 2016:**28**(10):2398–2416. https://doi.org/10.1105/tpc.16.00295

Hsu PY, Benfey PN. Small but mighty: functional peptides encoded by small ORFs in plants. Proteomics. 2018:**18**(10):1700038. https://doi.org/10.1002/pmic.201700038

Hsu PY, Calviello L, Wu HYL, Li FW, Rothfels CJ, Ohler U, Benfey PN. Super-resolution ribosome profiling reveals unannotated translation events in Arabidopsis. Proc Natl Acad Sci USA. 2016:**113**:E7126–E7135. https://doi.org/10.1073/pnas.1614788113

Hu W, Sweet TJ, Chamnongpol S, Baker KE, Coller J. Co-translational mRNA decay in *Saccharomyces cerevisiae*. Nature. 2009:**461**(7261):225–229. https://doi.org/10.1038/nature08265

Hurt JA, Robertson AD, Burge CB. Global analyses of UPF1 binding and function reveal expanded scope of nonsense-mediated mRNA decay. Genome Res. 2013:**23**(10):1636–1650. https://doi.org/10.1101/gr.157354.113

Imai A, Hanzawa Y, Komura M, Yamamoto KT, Komeda Y, Takahashi T. The dwarf phenotype of the Arabidopsis acl5 mutant is suppressed by a mutation in an upstream ORF of a bHLH gene. Development. 2006:**133**(18):3575–3585. https://doi.org/10.1242/dev.02535

Inada T, Winstall E, Tarun SZ, Yates JR, Schieltz D, Sachs AB. One-step affinity purification of the yeast ribosome and its associated proteins and mRNAs. RNA. 2002:**8**(7):948–958. https://doi.org/10.1017/S1355838202026018

Ingolia NT, Ghaemmaghami S, Newman JRS, Weissman JS. Genome-wide analysis in vivo of translation with nucleotide resolution using ribosome profiling. Science. 2009:**324**(5924):218–223. https://doi.org/10.1126/science.1168978

Iwakawa H, Lam AYW, Mine A, Fujita T, Kiyokawa K, Yoshikawa M, Takeda A, Iwasaki S, Tomari Y. Ribosome stalling caused by the Argonaute-microRNA-SGS3 complex regulates the production of secondary siRNAs in plants. Cell Rep. 2021:**35**(13):109300. https://doi.org/10.1016/j.celrep.2021.109300

Izquierdo Y, Kulasekaran S, Benito P, López B, Marcos R, Cascón T, Hamberg M, Castresana C. Arabidopsis nonresponding to oxylipins locus NOXY7 encodes a yeast GCN1 homolog that mediates noncanonical translation regulation and stress adaptation. Plant Cell Environ. 2018:**41**(6):1438–1452. https://doi.org/10.1111/pce.13182

Jang GJ, Yang JY, Hsieh HL, Wu SH. Processing bodies control the selective translation for optimal development of Arabidopsis young seedlings. Proc Natl Acad Sci USA. 2019:**116**(13):6451–6456. https://doi.org/10.1073/pnas.1900084116

Jiang M, Ning W, Wu S, Wang X, Zhu K, Li A, Li Y, Cheng S, Song B. Three-nucleotide periodicity of nucleotide diversity in a population enables the identification of open reading frames. Brief Bioinformatics. 2022:**23**(4):1–13. https://doi.org/10.1093/bib/bbac210

Johnstone TG, Bazzini AA, Giraldez AJ. Upstream ORFs are prevalent translational repressors in vertebrates. EMBO J. 2016:**35**(7):706–723. https://doi.org/10.15252/embj.201592759

Jorgensen RA, Dorantes-Acosta AE. Conserved peptide upstream open reading frames are associated with regulatory genes in angiosperms. Front Plant Sci. 2012:**3**:191. https://doi.org/10.3389/fpls.2012.00191

Ju C, Yoon GM, Shemansky JM, Lin DY, Ying ZI, Chang J, Garrett WM, Kessenbrock M, Groth G, Tucker ML, et al. CTR1 Phosphorylates the central regulator EIN2 to control ethylene hormone signaling from the ER membrane to the nucleus in Arabidopsis. Proc Natl Acad Sci USA. 2012:**109**(47):19486–19491. https://doi.org/10.1073/pnas.1214848109

Juntawong P, Girke T, Bazin J, Bailey-Serres J. Translational dynamics revealed by genome-wide profiling of ribosome footprints in Arabidopsis. Proc Natl Acad Sci USA. 2014:**111**(1):E203–E212. https://doi.org/10.1073/pnas.1317811111

Kajala K, Gouran M, Shaar-Moshe L, Mason GA, Rodriguez-Medina J, Kawa D, Pauluzzi G, Reynoso M, Canto-Pastor A, Manzano C, et al. Innovation, conservation, and repurposing of gene function in root cell type development. Cell. 2021:**184**(12):3333–3348.e19. https://doi.org/10.1016/j.cell.2021.04.024

Kearly A, Nelson ADL, Skirycz A, Chodasiewicz M. Composition and function of stress granules and P-bodies in plants. Semin Cell Dev Biol. 2022:S1084-9521(22)00350-0. Online ahead of print. https://doi.org/10.1016/j.semcdb.2022.11.008

Kim BH, Cai X, Vaughn JN, Von Arnim AG. On the functions of the h subunit of eukaryotic initiation factor 3 in late stages of translation initiation. Genome Biol. 2007:**8**(4):R60. https://doi.org/10.1186/gb-2007-8-4-r60

Kozak M. Possible role of flanking nucleotides in recognition of the aug initiator codon by eukaryotic ribosomes. Nucleic Acids Res. 1981:**9**(20):5233–5252. https://doi.org/10.1093/nar/9.20.5233

Kurihara Y, Makita Y, Kawashima M, Fujita T, Iwasaki S, Matsui M. Transcripts from downstream alternative transcription start sites evade uORF-mediated inhibition of gene expression in Arabidopsis. Proc Natl Acad Sci USA. 2018:**115**(30):7831–7836. https://doi.org/10.1073/pnas.1804971115

Kurihara Y, Makita Y, Shimohira H, Fujita T, Iwasaki S, Matsui M. Translational landscape of protein-coding and non-protein-coding RNAs upon light exposure in Arabidopsis. Plant Cell Physiol. 2020:**61**(3):536–545. https://doi.org/10.1093/pcp/pcz219

Kurosaki T, Popp MW, Maquat LE. Quality and quantity control of gene expression by nonsense-mediated mRNA decay. Nat Rev Mol Cell Biol. 2019:**20**(7):406–420. https://doi.org/10.1038/s41580-019-0126-2

Lageix S, Lanet E, Pouch-Pélissier MN, Espagnol MC, Robaglia C, Deragon JM, Pélissier T. Arabidopsis eIF2α kinase GCN2 is essential for growth in stress conditions and is activated by wounding. BMC Plant Biol. 2008:**8**(1):134. https://doi.org/10.1186/1471-2229-8-134

Laing WA, Martínez-Sánchez M, Wright MA, Bulley SM, Brewster D, Dare AP, Rassam M, Wang D, Storey R, Macknight RC, et al. An upstream open reading frame is essential for feedback regulation of ascorbate biosynthesis in Arabidopsis. Plant Cell. 2015:**27**(3):772–786. https://doi.org/10.1105/tpc.114.133777

Lalande S, Merret R, Salinas-Giegé T, Drouard L. Arabidopsis tRNA-derived fragments as potential modulators of translation. RNA Biol. 2020:**17**(8):1137–1148. https://doi.org/10.1080/15476286.2020.1722514

Lauressergues D, Couzigou JM, San Clemente H, Martinez Y, Dunand C, Bécard G, Combier JP. Primary transcripts of microRNAs encode regulatory peptides. Nature. 2015:**520**(7545):90–93. https://doi.org/10.1038/nature14346

Lee DH, Park SJ, Ahn CS, Pai HS. MRF Family genes are involved in translation control, especially under energy-deficient conditions, and their expression and functions are modulated by the TOR signaling pathway. Plant Cell. 2017:**29**(11):2895–2920. https://doi.org/10.1105/tpc.17.00563

Lei L, Shi J, Chen J, Zhang M, Sun S, Xie S, Li X, Zeng B, Peng L, Hauck A, et al. Ribosome profiling reveals dynamic translational landscape in maize seedlings under drought stress. Plant J. 2015:**84**(6):1206–1208. https://doi.org/10.1111/tpj.13073

Lellis AD, Patrick RM, Mayberry LK, Lorence A, Campbell ZC, Roose JL, Frankel LK, Bricker TM, Hellmann HA, Mayberry RW, et al. EIFiso4g augments the synthesis of specific plant proteins involved in normal chloroplast function. Plant Physiol. 2019:**181**(1):85–96. https://doi.org/10.1104/pp.19.00557

Li B, Wang Y, Zhang Y, Tian W, Chong K, Jang JC, Wang L. PRR5, 7 and 9 positively modulate TOR signaling-mediated root cell proliferation by repressing TANDEM ZINC FINGER 1 in Arabidopsis. Nucleic Acids Res. 2019:**47**(10):5001–5015. https://doi.org/10.1093/nar/gkz191

Li S, Le B, Ma X, Li S, You C, Yu Y, Zhang B, Liu L, Gao L, Shi T, et al. Biogenesis of phased siRNAs on membrane-bound polysomes in Arabidopsis. eLife. 2016:**5**:e22750. https://doi.org/10.7554/eLife.22750

Li S, Liu L, Zhuang X, Yu Y, Liu X, Cui X, Ji L, Pan Z, Cao X, Mo B, et al. MicroRNAs inhibit the translation of target mRNAs on the endoplasmic reticulum in Arabidopsis. Cell. 2013:**153**(3):562–574. https://doi.org/10.1016/j.cell.2013.04.005

Li W, Ma M, Feng Y, Li H, Wang Y, Ma Y, Li M, An F, Guo H. EIN2-directed translational regulation of ethylene signaling in Arabidopsis. Cell. 2015:**163**(3):670–683. https://doi.org/10.1016/j.cell.2015.09.037

Li X, Cai W, Liu Y, Li H, Fu L, Liu Z, Xu L, Liu H, Xu T, Xiong Y. Differential TOR activation and cell proliferation in Arabidopsis root and shoot apexes. Proc Natl Acad Sci USA. 2017:**114**(10):2765–2770. https://doi.org/10.1073/pnas.1618782114

Li YR, Liu MJ. Prevalence of alternative AUG and non-AUG translation initiators and their regulatory effects across plants. Genome Res. 2020:**30**(10):1418–1433. https://doi.org/10.1101/gr.261834.120

Liang Y, Zhu W, Chen S, Qian J, Li L. Genome-wide identification and characterization of small peptides in maize. Front Plant Sci. 2021:**12**:695439. https://doi.org/10.3389/fpls.2021.695439

Liu B, Qian SB. Translational reprogramming in cellular stress response. Wiley Interdiscipl Rev RNA. 2014:**5**(3):301–305. https://doi.org/10.1002/wrna.1212

Liu M-J, Wu S-H, Chen H-M, Wu S-H. Widespread translational control contributes to the regulation of *Arabidopsis photomorphogenesis*. Mol Syst Biol. 2012:**8**(1):566. https://doi.org/10.1038/msb.2011.97

Liu MJ, Wu SH, Wu JF, Lin WD, Wu YC, Tsai TY, Tsai HL, Wu SH. Translational landscape of photomorphogenic Arabidopsis. Plant Cell. 2013:**25**(10):3699–3710. https://doi.org/10.1105/tpc.113.114769

Liu X, Afrin T, Pajerowska-Mukhtar KM. Arabidopsis GCN2 kinase contributes to ABA homeostasis and stomatal immunity. Commun Biol. 2019:**2**(1):302. https://doi.org/10.1038/s42003-019-0544-x

Liu Y, Duan X, Zhao X, Ding W, Wang Y, Xiong Y. Diverse nitrogen signals activate convergent ROP2-TOR signaling in Arabidopsis. Dev Cell. 2021:**56**(9):1283–1295.e5. https://doi.org/10.1016/j.devcel.2021.03.022

Loh PG, Yang HS, Walsh MA, Wang Q, Wang X, Cheng Z, Liu D, Song H. Structural basis for translational inhibition by the tumour suppressor Pdcd4. EMBO J. 2009:**28**(3):274–285. https://doi.org/10.1038/emboj.2008.278

Lokdarshi A, von Arnim AG. Review: emerging roles of the signaling network of the protein kinase GCN2 in the plant stress response. Plant Sci. 2022:**320**:111280. https://doi.org/10.1016/j.plantsci.2022.111280

Lokdarshi A, Guan J, Urquidi Camacho RA, Cho SK, Morgan PW, Leonard M, Shimono M, Day B, von Arnim AG. Light activates the translational regulatory kinase GCN2 via reactive oxygen species emanating from the chloroplast. Plant Cell. 2020a:**32**(4):1161–1178. https://doi.org/10.1105/tpc.19.00751

Lokdarshi A, Morgan PW, Franks M, Emert Z, Emanuel C, von Arnim AG. Light-dependent activation of the GCN2 kinase under cold and salt stress is mediated by the photosynthetic status of the chloroplast. Front Plant Sci. 2020b:**11**:431. https://doi.org/10.3389/fpls.2020.00431

Lorenzo-Orts L, Witthoeft J, Deforges J, Martinez J, Loubéry S, Placzek A, Poirier Y, Hothorn LA, Jaillais Y, Hothorn M. Concerted expression of a cell cycle regulator and a metabolic enzyme from a bicistronic transcript in plants. Nat Plants. 2019:**5**(2):184–193. https://doi.org/10.1038/s41477-019-0358-3

Lütcke HA, Chow KC, Mickel FS, Moss KA, Kern HF, Scheele GA. Selection of AUG initiation codons differs in plants and animals. EMBO J. 1987:**6**(1):43–48. https://doi.org/10.1002/j.1460-2075.1987.tb04716.x

Mahfouz MM, Kim S, Delauney AJ, Verma DPS. Arabidopsis TARGET of RAPAMYCIN interacts with RAPTOR, which regulates the activity of S6 kinase in response to osmotic stress signals. Plant Cell. 2006:**18**(2):477–490. https://doi.org/10.1105/tpc.105.035931

Mazzoni-Putman SM, Stepanova AN. A plant biologist's Toolbox to study translation. Front Plant Sci. 2018:**9**:873. https://doi.org/10.3389/fpls.2018.00873

McCready K, Spencer V, Kim M. The importance of TOR kinase in plant development. Front Plant Sci. 2020:**11**:16. https://doi.org/10.3389/fpls.2020.00016

Mei C, Jiang SC, Lu YF, Wu FQ, Yu YT, Liang S, Feng XJ, Comeras SP, Lu K, Wu Z, et al. Arabidopsis pentatricopeptide repeat protein SOAR1 plays a critical role in abscisic acid signalling. J Exp Bot. 2014:**65**(18):5317–5330. https://doi.org/10.1093/jxb/eru293

Menand B, Desnos T, Nussaume L, Berger F, Bouchez D, Meyer C, Robaglia C. Expression and disruption of the Arabidopsis TOR (target of rapamycin) gene. Proc Natl Acad Sci USA. 2002:**99**(9):6422–6427. https://doi.org/10.1073/pnas.092141899

Merchante C, Brumos J, Yun J, Hu Q, Spencer KR, Enríquez P, Binder BM, Heber S, Stepanova AN, Alonso JM. Gene-specific translation regulation mediated by the hormone-signaling molecule EIN2. Cell. 2015:**163**(3):684–697. https://doi.org/10.1016/j.cell.2015.09.036

Merret R, Nagarajan VK, Carpentier MC, Park S, Favory JJ, Descombin J, Picart C, Charng YY, Green PJ, Deragon JM, et al. Heat-induced ribosome pausing triggers mRNA co-translational decay in *Arabidopsis thaliana*. Nucleic Acids Res. 2015:**43**(8):4121–4132. https://doi.org/10.1093/nar/gkv234

Meteignier LV, El-Oirdi M, Cohen M, Barff T, Matteau D, Lucier JF, Rodrigue S, Jacques PE, Yoshioka K, Moffett P. Translatome analysis of an NB-LRR immune response identifies important contributors to plant immunity in Arabidopsis. J Exp Bot. 2017:**68**(9):2333–2344. https://doi.org/10.1093/jxb/erx078

Meyuhas O. Ribosomal protein S6 phosphorylation: four decades of research. Int Rev Cell Mol Biol. 2015:**320**:41–73. https://doi.org/10.1016/bs.ircmb.2015.07.006

Miras M, Truniger V, Silva C, Verdaguer N, Aranda MA, Querol-Audí J. Structure of eiF4E in complex with an eiF4G peptide supports a universal bipartite binding mode for protein translation. Plant Physiol. 2017:**174**(3):1476–1491. https://doi.org/10.1104/pp.17.00193

Missra A, Ernest B, Lohoff T, Jia Q, Satterlee J, Ke K, von Arnim AG. The circadian clock modulates global daily cycles of mRNA ribosome loading. Plant Cell. 2015:**27**(9):2582–2599. https://doi.org/10.1105/tpc.15.00546

Montes C, Wang P, Liao CY, Nolan TM, Song G, Clark NM, Elmore JM, Guo H, Bassham DC, Yin Y, et al. Integration of multi-omics data reveals interplay between brassinosteroid and Target of Rapamycin Complex signaling in Arabidopsis. New Phytol. 2022:**236**(3):893–910. https://doi.org/10.1111/nph.18404

Moreau M, Azzopardi M, Clé ment G, Dobrenel T, Marchive C, Renne C, Martin-Magniette ML, Taconnat L, Renou JP, Robaglia C, et al. Mutations in the Arabidopsis homolog of LST8/GβL, a partner of the target of Rapamycin kinase, impair plant growth, flowering, and metabolic adaptation to long days. Plant Cell. 2012:**24**(2):463–481. https://doi.org/10.1105/tpc.111.091306

Mustroph A, Zanetti ME, Jang CJH, Holtan HE, Repetti PP, Galbraith DW, Girke T, Bailey-Serres J. Profiling translatomes of discrete cell populations resolves altered cellular priorities during hypoxia in Arabidopsis. Proc Natl Acad Sci USA. 2009:**106**(44):18843–18848. https://doi.org/10.1073/pnas.0906131106

Nakagawa S, Niimura Y, Gojobori T, Tanaka H, Miura K. Diversity of preferred nucleotide sequences around the translation initiation codon in eukaryote genomes. Nucleic Acids Res. 2008:**36**(3):861–871. https://doi.org/10.1093/nar/gkm1102

Nyikó T, Auber A, Szabadkai L, Benkovics A, Auth M, Mérai Z, Kerényi Z, Dinnyés A, Nagy F, Silhavy D. Expression of the eRF1 translation termination factor is controlled by an autoregulatory circuit involving readthrough and nonsense-mediated decay in plants. Nucleic Acids Res. 2017:**45**:4174–4188. https://doi.org/10.1093/nar/gkw1303

Ohtani M, Wachter A. NMD-based gene regulation—a strategy for fitness enhancement in plants? Plant Cell Physiol. 2019:**60**(9):1953–1960. https://doi.org/10.1093/pcp/pcz090

Pacheco JM, Canal MV, Pereyra CM, Welchen E, Martínez-Noël GMA, Estevez JM. The tip of the iceberg: emerging roles of TORC1, and its regulatory functions in plant cells. J Exp Bot. 2021:**72**(11):4085–4101. https://doi.org/10.1093/jxb/eraa603

Paik I, Yang S, Choi G. Phytochrome regulates translation of mRNA in the cytosol. Proc Natl Acad Sci USA. 2012:**109**(4):1335–1340. https://doi.org/10.1073/pnas.1109683109

Pajerowska-Mukhtar KM, Wang W, Tada Y, Oka N, Tucker CL, Fonseca JP, Dong X. The HSF-like transcription factor TBF1 is a major molecular switch for plant growth-to-defense transition. Curr Biol. 2012:**22**(2):103–112. https://doi.org/10.1016/j.cub.2011.12.015

Palam LR, Baird TD, Wek RC. Phosphorylation of eIF2 facilitates ribosomal bypass of an inhibitory upstream ORF to enhance CHOP translation. J Biol Chem. 2011:**286**(13):10939–10949. https://doi.org/10.1074/jbc.M110.216093

Palos K, Nelson Dittrich AC, Yu L, Brock JR, Railey CE, Wu H-YL, Sokolowska E, Skirycz A, Hsu PY, Gregory BD, et al. Identification and functional annotation of long intergenic noncoding RNAs in Brassicaceae. Plant Cell. 2022:**34**(9):3233–3260. https://doi.org/10.1093/plcell/koac166

Parker MT, Knop K, Sherwood AV, Schurch NJ, Mackinnon K, Gould PD, Hall AJW, Barton GJ, Simpson GG. Nanopore direct RNA sequencing maps the complexity of arabidopsis mRNA processing and m6A modification. eLife. 2020:**9**:e49658. https://doi.org/10.7554/eLife.49658

Passmore LA, Coller J. Roles of mRNA poly(A) tails in regulation of eukaryotic gene expression. Nat Rev Mol Cell Biol. 2022:**23**(2):93–106. https://doi.org/10.1038/s41580-021-00417-y

Patrick RM, Browning KS. The eIF4F and eIFiso4F complexes of plants: an evolutionary perspective. Comp Funct Genomics. 2012:**2012**: 287814. https://doi.org/10.1155/2012/287814

Patrick RM, Lee JCH, Teetsel JRJ, Yang SH, Choy GS, Browning KS. Discovery and characterization of conserved binding of eIF4E 1 (CBE1), a eukaryotic translation initiation factor 4E-binding plant protein. J Biol Chem. 2018:**293**(44):17240–17247. https://doi.org/10.1074/jbc.RA118.003945

Peixoto B, Baena-González E. Management of plant central metabolism by SnRK1 protein kinases. J Exp Bot. 2022:**73**(20):7068–7082. https://doi.org/10.1093/jxb/erac261

Pelechano V, Wei W, Steinmetz LM. Widespread co-translational RNA decay reveals ribosome dynamics. Cell. 2015:**161**(6):1400–1412. https://doi.org/10.1016/j.cell.2015.05.008

Pfeiffer A, Janocha D, Dong Y, Medzihradszky A, Schöne S, Daum G, Suzaki T, Forner J, Langenecker T, Rempel E, et al. Integration of light and metabolic signals for stem cell activation at the shoot apical meristem. eLife. 2016:**5**:e17023. https://doi.org/10.7554/eLife.17023

Powers KT, Szeto JYA, Schaffitzel C. New insights into no-go, non-stop and nonsense-mediated mRNA decay complexes. Curr Opin Struct Biol. 2020:**65**:110–118. https://doi.org/10.1016/j.sbi.2020.06.011

Pu Y, Luo X, Bassham DC. Tor-dependent and -independent pathways regulate autophagy in *Arabidopsis thaliana*. Front Plant Sci. 2017:**8**: 1204. https://doi.org/10.3389/fpls.2017.01204

Puyaubert J, Denis L, Alban C. Dual targeting of Arabidopsis holocarboxylase synthetase1: a small upstream open reading frame regulates translation initiation and protein targeting. Plant Physiol. 2008:**146**(2):323–324. https://doi.org/10.1104/pp.107.111534

Qiao H, Shen Z, Huang SSC, Schmitz RJ, Urich MA, Briggs SP, Ecker JR. Processing and subcellular trafficking of ER-tethered EIN2 control response to ethylene gas. Science. 2012:**338**(6105):390–393. https://doi.org/10.1126/science.1225974

Raabe K, Honys D, Michailidis C. The role of eukaryotic initiation factor 3 in plant translation regulation. Plant Physiol Biochem. 2019:**145**: 75–83. https://doi.org/10.1016/j.plaphy.2019.10.015

Rahmani F, Hummel M, Schuurmans J, Wiese-Klinkenberg A, Smeekens S, Hanson J. Sucrose control of translation mediated by an upstream open reading frame-encoded peptide. Plant Physiol. 2009:**150**(3):1356–1367. https://doi.org/10.1104/pp.109.136036

Raj A, van Oudenaarden A. Nature, nurture, or chance: stochastic gene expression and its consequences. Cell. 2008:**135**(2):216–226. https://doi.org/10.1016/j.cell.2008.09.050

Reis RS, Deforges J, Schmidt RR, Schippers JHM, Poirier Y. An antisense noncoding RNA enhances translation via localized structural rearrangements of its cognate mRNA. Plant Cell. 2021:**33**(4): 1381–1397. https://doi.org/10.1093/plcell/koab010

Reynoso MA, Borowsky AT, Pauluzzi GC, Yeung E, Zhang J, Formentin E, Velasco J, Cabanlit S, Duvenjian C, Prior MJ, et al. Gene regulatory networks shape developmental plasticity of root cell types under water extremes in rice. Dev Cell. 2022:**57**(9): 1177–1192.e6. https://doi.org/10.1016/j.devcel.2022.04.013

Reynoso MA, Kajala K, Bajic M, West DA, Pauluzzi G, Yao AI, Hatch K, Zumstein K, Woodhouse M, Rodriguez-Medina J, et al. Evolutionary flexibility in flooding response circuitry in angiosperms. Science. 2019:**365**(6459):1291–1295. https://doi.org/10.1126/science.aax8862

Ribone PA, Capella M, Arce AL, Chan RL. A uORF represses the transcription factor AtHB1 in aerial tissues to avoid a deleterious phenotype. Plant Physiol. 2017:**175**(3):1238–1253. https://doi.org/10.1104/pp.17.01060

Rook F, Gerrits N, Kortstee A, Van Kampen M, Borrias M, Weisbeek P, Smeekens S. Sucrose-specific signaling represses translation of the Arabidopsis ATB2 bZIP transcription factor gene. Plant J. 1998:**15**(2): 253–263. https://doi.org/10.1046/j.1365-313X.1998.00205.x

Sagor GHM, Berberich T, Tanaka S, Nishiyama M, Kanayama Y, Kojima S, Muramoto K, Kusano T. A novel strategy to produce sweeter tomato fruits with high sugar contents by fruit-specific expression of a single bZIP transcription factor gene. Plant Biotechnol J. 2016:**14**(4):1116–1126. https://doi.org/10.1111/pbi.12480

Sahoo S, Singh D, Singh A, Pandit M, Vasu K, Som S, Pullagurla NJ, Laha D, Eswarappa SM. Identification and functional characterization of mRNAs that exhibit stop codon readthrough in *Arabidopsis thaliana*. J Biol Chem. 2022:**298**(8):102173. https://doi.org/10.1016/j.jbc.2022.102173

Scarpin MR, Busche M, Martinez RE, Harper LC, Reiser L, Szakonyi D, Merchante C, Lan T, Xiong W, Mo B, et al. An updated nomenclature for plant ribosomal protein genes. Plant Cell. 2023:**35**(2): 640–643. https://doi.org/10.1093/plcell/koac333

Scarpin MR, Leiboff S, Brunkard JO. Parallel global profiling of plant tor dynamics reveals a conserved role for LARP1 in translation. eLife. 2020:**9**:1–34. https://doi.org/10.7554/eLife.58795

Scarpin MR, Simmons CH, Brunkard JO. Translating across kingdoms: target of rapamycin promotes protein synthesis through conserved and divergent pathways in plants. J Exp Bot. 2022:**73**(20): 7016–7025. https://doi.org/10.1093/jxb/erac267

Schepetilnikov M, Dimitrova M, Mancera-Martínez E, Geldreich A, Keller M, Ryabova LA. TOR And S6K1 promote translation reinitiation of uORF-containing mRNAs via phosphorylation of eIF3h. EMBO J. 2013:**32**(8):1087–1102. https://doi.org/10.1038/emboj.2013.61

Schepetilnikov M, Makarian J, Srour O, Geldreich A, Yang Z, Chicher J, Hammann P, Ryabova LA. GTP Ase ROP 2 binds and promotes activation of target of rapamycin, TOR, in response to auxin. EMBO J. 2017:**36**(7):886–903. https://doi.org/10.15252/embj.201694816

Schepetilnikov M, Ryabova LA. Recent discoveries on the role of tor (Target of rapamycin) signaling in translation in plants. Plant Physiol. 2018:**176**(2):1095–1105. https://doi.org/10.1104/pp.17.01243

Schuller AP, Green R. Roadblocks and resolutions in eukaryotic translation. Nat Rev Mol Cell Biol. 2018:**19**(8):526–541. https://doi.org/10.1038/s41580-018-0011-4

Shah P, Ding Y, Niemczyk M, Kudla G, Plotkin JB. Rate-limiting steps in yeast protein translation. Cell. 2013:**153**(7):1589–1601. https://doi.org/10.1016/j.cell.2013.05.049

Shamimuzzaman M, Vodkin L. Ribosome profiling reveals changes in translational status of soybean transcripts during immature cotyledon development. PLoS ONE. 2018:**13**(3):e0194596. https://doi.org/10.1371/journal.pone.0194596

Sharma A, Badola PK, Bhatia C, Sharma D, Trivedi PK. Primary transcript of miR858 encodes regulatory peptide and controls flavonoid biosynthesis and development in Arabidopsis. Nat Plants. 2020:**6**(10): 1262–1274. https://doi.org/10.1038/s41477-020-00769-x

Shi L, Wu Y, Sheen J. TOR Signaling in plants: conservation and innovation. Development. 2018:**145**(13):dev160887. https://doi.org/10.1242/dev.160887

Sieburth LE, Vincent JN. Beyond transcription factors: roles of mrna decay in regulating gene expression in plants [version 1; referees: 3 approved]. F1000Res. 2018:**7**:F1000 Faculty Rev- 1940. https://doi.org/10.12688/f1000research.16203.1

Skuzeski JM, Nichols LM, Gesteland RF, Atkins JF. The signal for a leaky UAG stop codon in several plant viruses includes the two downstream codons. J Mol Biol. 1991:**218**(2):365–373. https://doi.org/10.1016/0022-2836(91)90718-L

Sorenson R, Bailey-Serres J. Selective mRNA sequestration by OLIGOURIDYLATEBINDING PROTEIN 1 contributes to translational control during hypoxia in Arabidopsis. Proc Natl Acad Sci USA. 2014:**111**(6):2373–2378. https://doi.org/10.1073/pnas.1314851111

Sotta N, Chiba Y, Aoyama H, Takamatsu S, Suzuki T, Miwa K, Yamashita Y, Naito S, Fujiwara T. Translational landscape of a C4Plant, *Sorghum bicolor*, under normal and sulfur-deficient conditions. Plant Cell Physiol. 2022:**63**(5):592–604. https://doi.org/10.1093/pcp/pcac023

Sotta N, Chiba Y, Miwa K, Takamatsu S, Tanaka M, Yamashita Y, Naito S, Fujiwara T. Global analysis of boron-induced ribosome stalling reveals its effects on translation termination and unique regulation by AUG-stops in Arabidopsis shoots. Plant J. 2021:**106**(5):1455–1467. https://doi.org/10.1111/tpj.15248

Szádeczky-Kardoss I, Csorba T, Auber A, Schamberger A, Nyikó T, Taller J, Orbán TI, Burgyán J, Silhavy D. The nonstop decay and the RNA silencing systems operate cooperatively in plants. Nucleic Acids Res. 2018a:**46**(9):4632–4648. https://doi.org/10.1093/nar/gky279

Szádeczky-Kardoss I, Gál L, Auber A, Taller J, Silhavy D. The No-go decay system degrades plant mRNAs that contain a long A-stretch in the coding region. Plant Sci. 2018b:**275**:19–27. https://doi.org/10.1016/j.plantsci.2018.07.008

Takahashi H, Hayashi N, Hiragori Y, Sasaki S, Motomura T, Yamashita Y, Naito S, Takahashi A, Fuse K, Satou K, et al. Comprehensive genome-wide identification of angiosperm upstream ORFs with peptide sequences conserved in various taxonomic ranges using a novel pipeline, ESUCA. BMC Genomics. 2020:**21**(1):260. https://doi.org/10.1186/s12864-020-6662-5

Takahashi H, Lassmann T, Murata M, Carninci P. 5′ end-centered expression profiling using cap-analysis gene expression and next-generation sequencing. Nat Protoc. 2012a:**7**(3):542–561. https://doi.org/10.1038/nprot.2012.005

Takahashi H, Takahashi A, Naito S, Onouchi H. BAIUCAS: a novel BLAST-based algorithm for the identification of upstream open reading frames with conserved amino acid sequences and its application to the *Arabidopsis thaliana* genome. Bioinformatics. 2012b:**28**(17):2231–2241. https://doi.org/10.1093/bioinformatics/bts303

Tanaka M, Sotta N, Yamazumi Y, Yamashita Y, Miwa K, Murota K, Chiba Y, Hirai MY, Akiyama T, Onouchi H, et al. The minimum open reading frame, AUG-stop, induces boron-dependent ribosome stalling and mRNA degradation. Plant Cell. 2016:**28**(11):2830–2849. https://doi.org/10.1105/tpc.16.00481

Toribio R, Muñoz A, Castro-Sanz AB, Merchante C, Castellano MM. A novel eIF4E-interacting protein that forms non-canonical translation initiation complexes. Nat Plants. 2019:**5**(12):1283–1296. https://doi.org/10.1038/s41477-019-0553-2

Traubenik S, Reynoso MA, Hobecker K, Lancia M, Hummel M, Rosen B, Town C, Bailey-Serres J, Blanco F, Zanetti ME. Reprogramming of root cells during nitrogen-fixing symbiosis involves dynamic polysome association of coding and noncoding RNAs. Plant Cell. 2020:**32**(2):352–373. https://doi.org/10.1105/tpc.19.00647

Turkina MV, Årstrand H, Vener AV. Differential phosphorylation of ribosomal proteins in *Arabidopsis thaliana* plants during day and night. PLoS ONE. 2011:**6**(12):e29307. https://doi.org/10.1371/journal.pone.0029307

Uchiyama-Kadokura N, Murakami K, Takemoto M, Koyanagi N, Murota K, Naito S, Onouchi H. Polyamine-responsive ribosomal arrest at the stop codon of an upstream open reading frame of the AdoMetDC1 gene triggers nonsense-mediated mRNA decay in *Arabidopsis thaliana*. Plant Cell Physiol. 2014:**55**(9):1556–1567. https://doi.org/10.1093/pcp/pcu086

Van Der Horst S, Filipovska T, Hanson J, Smeekens S. Metabolite control of translation by conserved peptide uORFs: the ribosome as a metabolite multisensor. Plant Physiol. 2020:**182**(1):110–122. https://doi.org/10.1104/pp.19.00940

Van Der Horst S, Snel B, Hanson J, Smeekens S. Novel pipeline identifies new upstream ORFs and non-AUG initiating main ORFs with conserved amino acid sequences in the 5′ leader of mRNAs in *Arabidopsis thaliana*. RNA. 2019:**25**(3):292–304. https://doi.org/10.1261/rna.067983.118

Van Leene J, Han C, Gadeyne A, Eeckhout D, Matthijs C, Cannoot B, De Winne N, Persiau G, Van De Slijke E, Van de Cotte B, et al. Capturing the phosphorylation and protein interaction landscape of the plant TOR kinase. Nat Plants. 2019:**5**(3):316–327. https://doi.org/10.1038/s41477-019-0378-z

Vaughn JN, Ellingson SR, Mignone F, Von Arnim A. Known and novel post-transcriptional regulatory sequences are conserved across plant families. RNA. 2012:**18**(3):368–384. https://doi.org/10.1261/rna.031179.111

Von Arnim AG, Jia Q, Vaughn JN. Regulation of plant translation by upstream open reading frames. Plant Sci. 2014:**214**:1–12. https://doi.org/10.1016/j.plantsci.2013.09.006

Wang J, Zhang X, Greene GH, Xu G, Dong X. PABP/purine-rich motif as an initiation module for cap-independent translation in pattern-triggered immunity. Cell. 2022:**185**(17):3186–3200.e17. https://doi.org/10.1016/j.cell.2022.06.037

Wang L, Li H, Zhao C, Li S, Kong L, Wu W, Kong W, Liu Y, Wei Y, Zhu JK, et al. The inhibition of protein translation mediated by AtGCN1 is essential for cold tolerance in *Arabidopsis thaliana*. Plant Cell Environ. 2017:**40**(1):56–68. https://doi.org/10.1111/pce.12826

Wang P, Zhao Y, Li Z, Hsu C-C, Liu X, Fu L, Hou YJ, Du Y, Xie S, Zhang C, et al. Reciprocal regulation of the TOR kinase and ABA receptor balances plant growth and stress response. Mol Cell. 2018:**69**(1):100–112.e6. https://doi.org/10.1016/j.molcel.2017.12.002

Wiese A, Elzinga N, Wobbes B, Smeekens S. A conserved upstream open reading frame mediates sucrose-induced repression of translation. Plant Cell. 2004:**16**(7):1717–1729. https://doi.org/10.1105/tpc.019349

Willems P, Ndah E, Jonckheere V, Stael S, Sticker A, Martens L, Van Breusegem F, Gevaert K, Van Damme P. N-terminal proteomics assisted profiling of the unexplored translation initiation landscape in *Arabidopsis thaliana*. Mol Cell Proteomics. 2017:**16**(6):1064–1080. https://doi.org/10.1074/mcp.M116.066662

Wu H, Li B, Iwakawa H, Pan Y, Tang X, Ling-hu Q, Liu Y, Sheng S, Feng L, Zhang H, et al. Plant 22-nt siRNAs mediate translational repression and stress adaptation. Nature. 2020:**581**(7806):89–93. https://doi.org/10.1038/s41586-020-2231-y

Wu HW, Fajiculay E, Wu JF, Yan CCS, Hsu CP, Wu SH. Noise reduction by upstream open reading frames. Nat Plants. 2022:**8**(5):474–480. https://doi.org/10.1038/s41477-022-01136-8

Wu H-YL, Hsu PY. Translation of upstream open reading frames generally does not trigger nonsense-mediated decay in Arabidopsis. bioRxiv: 2021.09.16.460672; 2022.

Wu HYL, Song G, Walley JW, Hsu PY. The tomato translational landscape revealed by transcriptome assembly and ribosome profiling. Plant Physiol. 2019:**181**(1):367–380. https://doi.org/10.1104/pp.19.00541

Wu Z, Huang S, Zhang X, Wu D, Xia S, Li X. Regulation of plant immune receptor accumulation through translational repression by a glycine–tyrosine–phenylalanine (GYF) domain protein. eLife. 2017:**6**:e23684. https://doi.org/10.7554/eLife.23684

Xing S, Chen K, Zhu H, Zhang R, Zhang H, Li B, Gao C. Fine-tuning sugar content in strawberry. Genome Biol. 2020:**21**(1):230. https://doi.org/10.1186/s13059-020-02146-5

Xiong Y, McCormack M, Li L, Hall Q, Xiang C, Sheen J. Glucose-TOR signalling reprograms the transcriptome and activates meristems. Nature. 2013:**496**(7444):181–186. https://doi.org/10.1038/nature12030

Xiong Y, Sheen J. Rapamycin and glucose-target of rapamycin (TOR) protein signaling in plants. J Biol Chem. 2012:**287**(4):2836–2842. https://doi.org/10.1074/jbc.M111.300749

Xu G, Greene GH, Yoo H, Liu L, Marqués J, Motley J, Dong X. Global translational reprogramming is a fundamental layer of immune regulation in plants. Nature. 2017a:**545**(7655):487–490. https://doi.org/10.1038/nature22371

Xu G, Yuan M, Ai C, Liu L, Zhuang E, Karapetyan S, Wang S, Dong X. UORF-mediated translation allows engineered plant disease resistance without fitness costs. Nature. 2017b:**545**(7655):491–494. https://doi.org/10.1038/nature22372

Yamashita Y, Takamatsu S, Glasbrenner M, Becker T, Naito S, Beckmann R. Sucrose sensing through nascent peptide-meditated ribosome stalling at the stop codon of Arabidopsis bZIP11 uORF2. FEBS Lett. 2017:**591**(9):1266–1277. https://doi.org/10.1002/1873-3468.12634

Yang X, Dong W, Ren W, Zhao Q, Wu F, He Y. Cytoplasmic HYL1 modulates miRNA-mediated translational repression. Plant Cell. 2021:**33**(6):1980–1996. https://doi.org/10.1093/plcell/koab090

Yerlikaya S, Meusburger M, Kumari R, Huber A, Anrather D, Costanzo M, Boone C, Ammerer G, Baranov PV, Loewith R. TORC1 And TORC2 work together to regulate ribosomal protein S6 phosphorylation in *Saccharomyces cerevisiae*. Mol Biol Cell. 2016:**27**(2):397–409. https://doi.org/10.1091/mbc.e15-08-0594

Yoshikawa M, Iki T, Numa H, Miyashita K, Meshi T, Ishikawa M. A short open reading frame encompassing the microRNA173 target site plays a role in trans-acting small interfering RNA biogenesis. Plant Physiol. 2016:**171**(1):359–368. https://doi.org/10.1104/pp.16.00148

Young SK, Wek RC. Upstream open reading frames differentially regulate genespecific translation in the integrated stress response. J Biol Chem. 2016:**291**(33):16927–16935. https://doi.org/10.1074/jbc.R116.733899

Yu X, Willmann MR, Anderson SJ, Gregory BD. Genome-wide mapping of uncapped and cleaved transcripts reveals a role for the nuclear mRNA cap-binding complex in cotranslational RNA decay in Arabidopsis. Plant Cell. 2016:**28**(10):2385–2397. https://doi.org/10.1105/tpc.16.00456

Zhang H, Quintana J, Ütkür K, Adrian L, Hawer H, Mayer K, Gong X, Castanedo L, Schulten A, Janina N, et al. Translational fidelity and growth of Arabidopsis require stress-sensitive diphthamide biosynthesis. Nat Commun. 2022:**13**(1):4009. https://doi.org/10.1038/s41467-022-31712-7

Zhang H, Wang Y, Wu X, Tang X, Wu C, Lu J. Determinants of genome-wide distribution and evolution of uORFs in eukaryotes. Nat Commun. 2021:**12**(1):1076. https://doi.org/10.1038/s41467-021-21394-y

Zhang L, Liu X, Gaikwad K, Kou X, Wang F, Tian X, Xin M, Ni Z, Sun Q, Peng H, et al. Mutations in eIF5b confer thermosensitive and pleiotropic phenotypes via translation defects in *Arabidopsis thaliana*. Plant Cell. 2017:**29**(8):1952–1969. https://doi.org/10.1105/tpc.16.00808

Zhang Y, Wang Y, Kanyuka K, Parry MAJ, Powers SJ, Halford NG. GCN2-dependent Phosphorylation of eukaryotic translation initiation factor-2α in Arabidopsis. J Exp Bot. 2008:**59**(11):3131–3141. https://doi.org/10.1093/jxb/ern169

Zhang Z, Zhu JY, Roh J, Marchive C, Kim SK, Meyer C, Sun Y, Wang W, Wang ZY. TOR signaling promotes accumulation of BZR1 to balance growth with carbon availability in Arabidopsis. Curr Biol. 2016:**26**(14):1854–1860. https://doi.org/10.1016/j.cub.2016.05.005

Zhao J, Qin B, Nikolay R, Spahn CMT, Zhang G. Translatomics: the global view of translation. Int J Mol Sci. 2019a:**20**(1):212. https://doi.org/10.3390/ijms20010212

Zhao T, Huan Q, Sun J, Liu C, Hou X, Yu X, Silverman IM, Zhang Y, Gregory BD, Liu CM, et al. Impact of poly(A)-tail G-content on Arabidopsis PAB binding and their role in enhancing translational efficiency. Genome Biol. 2019b:**20**(1):1–12. https://doi.org/10.1186/s13059-019-1799-8

Zhigailov AV, Alexandrova AM, Nizkorodova AS, Stanbekova GE, Kryldakov RV, Karpova OV, Polimbetova NS, Halford NG, Iskakov BK. Evidence that phosphorylation of the α-subunit of eIF2 does not essentially inhibit mRNA translation in wheat germ cell-free system. Front Plant Sci. 2020:**11**:936. https://doi.org/10.3389/fpls.2020.00936

Zhu S, Estévez JM, Liao H, Zhu Y, Yang T, Li C, Wang Y, Li L, Liu X, Pacheco JM, et al. The RALF1–FERONIA complex phosphorylates eIF4E1 to promote protein synthesis and polar root hair growth. Mol Plant. 2020:**13**(5):698–716. https://doi.org/10.1016/j.molp.2019.12.014

Zhu W, Xu J, Chen S, Chen J, Liang Y, Zhang C, Li Q, Lai J, Li L. Large-scale translatome profiling annotates the functional genome and reveals the key role of genic 3′ untranslated regions in translatomic variation in plants. Plant Commun. 2021:**2**(4):100181. https://doi.org/10.1016/j.xplc.2021.100181

Zhu X, Li Y, Fang W, Kusano T. Galactinol is involved in sequence-conserved upstream open reading frame-mediated repression of Arabidopsis HsfB1 translation. Environ Exp Bot. 2018:**156**:120–129. https://doi.org/10.1016/j.envexpbot.2018.09.006

Zoschke R, Watkins KP, Barkan A. A rapid ribosome profiling method elucidates chloroplast ribosome behavior in vivo. Plant Cell. 2013:**25**(6):2265–2275. https://doi.org/10.1105/tpc.113.111567

www.ingramcontent.com/pod-product-compliance
Lightning Source LLC
LaVergne TN
LVHW071344210926
848486LV00003B/226